AF615332

B-P-H/S

Botanico-Periodicum-Huntianum/Supplementum

Gavin D. R. Bridson
Compiler & Editor

Elizabeth R. Smith
Editorial Assistant

Hunt Institute for Botanical Documentation
Carnegie Mellon University

Pittsburgh, 1991

Camera-ready copy produced from magnetic
file using Apple Macintosh IIcx and LaserWriter II NTX

Printed on acid-free paper and bound by
Allen Press Inc., Lawrence, Kansas

ISBN 0-913196-54-1

Introduction

This work forms both a supplement to and a partial revision of *B-P-H Botanico-Periodicum-Huntianum,* published by the Hunt Institute in 1968. Since B-P-H is not fully superseded, the present work is designed as a "key" to the locations of entries in both volumes. B-P-H included entries for "more than 12,000 titles" published down to the end of 1967. With the new titles now added, the total number of entries has risen to over 25,000 and the period of coverage is extended to the end of 1990, with the addition of a few new titles from 1991.

Subject coverage

The editors of B-P-H interpreted "botanical literature" as including periodicals that dealt with agriculture, biology, ecology, floriculture, forestry, fruit growing, genetics and plant breeding, horticulture, hydrobiology and limnology, microbiology, plant pathology and vegetable crops. They also attempted to cover peripheral disciplines, *i. e.* agronomy, bacteriology, geography, microscopy, palaeontology (especially palaeobotany), pharmacology and pharmacognosy, but for this Supplement *(B-P-H/S)* it has not been possible to give the same emphasis to these interests. This is especially true for palaeobotany, a subject that is scattered through a multitude of earth-science periodicals. Since 1968, when B-P-H appeared, there has been an explosive growth of literature in areas that were then relatively insignificant, e.g. molecular biology, biotechnology, environmental studies and conservation. All the relevant periodicals in these areas have been included in B-P-H/S. In addition to plant-science periodicals, B-P-H included entries for any that published historical, bibliographical and biographical articles on the plant and general life sciences, and B-P-H/S continues to cover these topics. In short, B-P-H and B-P-H/S attempt to record every periodical that might reasonably be consulted in the course of a wide spectrum of plant-science activities.

Presentation

Like the original work, B-P-H/S provides a title catalogue of "all periodical (serial) publications that regularly contain (or, in some period of their history, included) articles dealing with the plant sciences and botanical literature." The entries are arranged alphabetically by full title, disregarding articles, prepositions and conjunctions. Each entry in B-P-H that remains valid is cross-referenced in B-P-H/S with a citation of the relevant B-P-H page and entry number. For the sake of user convenience, these cross-references also include the citation abbreviations.

Special efforts have been made to provide as many cross-references as possible from alternative titles (or forms of the titles). The lack of sufficient cross-references in B-P-H was a weakness that often led users to conclude incorrectly that periodicals had been omitted. The user's ease in finding titles in B-P-H was also hindered by the fact that periodicals were filed not by full title but by the citation abbreviation. Unfortunately, many of these citation abbreviations were specially created for B-P-H and, because they had no previous existence, were unfamiliar to the user. Thus, one was obliged to "translate" the full title into its B-P-H-style citation abbreviation and then to search for that form of the title.

Present status of B-P-H

Although many of the entries in B-P-H have been replaced by revised ones in B-P-H/S, the majority retain full validity. With the present volume serving as the "key" to the whole file, B-P-H now serves as a reservoir for older periodical titles. The user will be directed there from B-P-H/S for only the original entries that remain valid, and adherence to this order of priority will be essential for efficient information retrieval.

Transliteration of Cyrillic titles

B-P-H used the International Organization for Standardization's "ISO-1" system for transliteration of Slavic Cyrillic characters, a system that maximized the number of diacritic marks. The intervening years have seen a growing preference for the "ISO-2" system, which minimizes diacritics in favour of diagraphs. In consequence, we decided to employ the latter system for *B-P-H/S*, and all "ISO-1" entries in *B-P-H* have been rewritten in "ISO-2." See Nicolson for a discussion of the methods and problems of transliteration.

Title abbreviations

Citation abbreviations have been provided for all titles in accordance with the principles fully described in *B-P-H*. Since the total number of titles has grown to more than double the 1968 figure, it is not surprising to find a commensurate growth in the titular vocabulary, especially with regard to novel branches of research. The thesaurus of words listed in Appendix I of *B-P-H* has been corrected and expanded to include this entire vocabulary and is similarly appended in this work.

HI numbers

Users will notice that every entry in *B-P-H/S* concludes with an "HI 00000" number. This domestic control number was created for database handling and has no authority in any other context. If users care to quote this number in any informal communication relating to *B-P-H* it may have some value because it forms a unique identifier in this context, but we cannot guarantee that it will not be changed with the subsequent development of our database.

Sources of information

Many sources were consulted in preparing the new entries for *B-P-H/S*. Notices of new periodicals were regularly sought in the principal botanical bibliographical services, e.g. *Boletín bibliográfico agricola* (later *Boletín para bibliotecas agrícolas*), *Flora Malesiana Bulletin*, *Index to Australian taxonomic literature*, *Index to European taxonomic literature*, *Kew record of taxonomic literature*, and *Taxon*. General bibliographical services have also been surveyed selectively, e.g. *Irregular serials & annuals*, *New serial titles*, *Periodicals news*, *Scientific and technical books and serials in print*, *Serials directory: An international directory*, and the *Standard periodical directory*. In addition, publishers' catalogues, brochures and "fliers" have revealed many new titles.

Catalogues of major libraries, especially in the plant sciences, have been consulted and, in some cases, searched comprehensively for relevant titles, e.g. British Library, Science Reference Library; British Museum (Natural History); Conservatoire et Jardin Botaniques de la Ville de Genève; Instituto Interamericano de Ciencias Agricolas; Linda Hall Library, Science & Technology; Longwood Gardens, Inc.; Missouri Botanical Garden; Naturhistorische Verein der Rheinlande und Westfalen; New York Botanical Garden; Royal Botanic Garden Edinburgh; Royal Botanic Gardens Kew; and the United States Department of Agriculture, National Agricultural Library.

A few of the relevant libraries issue regular accessions lists and these have been consulted methodically for new titles, e.g., *Agricultural library information notes; Current awareness list, library, Royal Botanic Gardens Kew; List of accessions to the museum library, British Museum (Natural History);* and *Nouvelles acquisitions, bibliothèque centrale du Muséum National d'Histoire Naturelle.*

Union lists of periodicals have been consulted for bibliographical information, e.g. *Catalogo colectivo nacional de publicaciones periodicas (Chile); Catalogo colectivo de publicaciones periodicas existentes en bibliotecas cientificas y tecnicas Argentinas; Scientific serials in Australian libraries; Union list of serials in Canada Department of Agriculture libraries; Union list of scientific, technical, agricultural and forestry periodicals and serials in scientific libraries in Finland; Union list of serials in the science area, Oxford;* and the Online Computer Library Center (oclc) database.

Additional information on certain types of periodicals was obtained from special bibliographies, e.g. the Owen, British Library, and Science Reference Library lists of abstracting and indexing periodicals, and the *National directory of newsletters and reporting services, Standard directory of newsletters*, and National Agricultural Library list of these rapidly multiplying periodicals.

Periodicals in particular language groups were researched in special sources, e.g. for Chinese and Japanese titles, the British Library, Gibson & Kunkel, and Science Reference Library lists; for German, Kirchner's bibliographies; for Indian, the D. K. Agencies trade catalogues; for Russian, the Center for Research Libraries, and Coffey's list; for Slavic, the United States Department of Agriculture list; and for South America, Angely's bibliography.

Special subject groups of periodicals were surveyed in some special lists, e.g. Cassie on algae; Hawksworth on lichens; Watling on fungi; Bernstein & Finley, Eggli and Neumann on cacti and succulents; Frodin and Tutin *et al.* on floristics; the American Horticultural Society, Doorenbos and National Agricultural Library on horticulture; *Agricultural library information notes*, Blanchard & Farrell, *Boletín bibliográfico agricola*, British Library, Science Reference Library and *Union list of serials in Canada Department of Agriculture libraries* on agricultural botany; British Library and Science Reference Library on biochemistry; and Davis and Gray on environmental science.

Desiderata

Despite the best efforts of the compiler, there are all too numerous entries for which the exact volume and date-span data could not be ascertained by the time of going to press. The incredible literature explosion of recent decades has included the launching of thousands of new periodicals. Rapid scientific change has caused a seemingly endless accumulation of alterations in periodical titles, and similar changes in the nomenclature of their sponsoring institutions. All the while, libraries have been forced to trim budgets in the face of this growing tide, constantly having to alter their acquisition, documentation and preservation policies as they chase more literature with fewer dollars. The result is a nightmare for the bibliographer, who has to trace the threads of ever more numerous periodicals and details of title changes through the ever-changing tapestry of library resources.

b-p-h, like any large-scale reference work, was not without its share of errors. Most of its regular users have discovered mistakes in numerous individual entries. Indeed, a number of people communicated their findings to the Institute. When the present compiler inherited the project in 1983, there was already a substantial file of corrigenda awaiting inclusion, and through day-to-day use this file has been augmented constantly since. In addition, the preparation of "continuation" entries for periodicals covered in *b-p-h* has highlighted errors in the original data, and consequently a great many old entries were rewritten for *b-p-h/s*. The warrant for preparing *b-p-h/s* could not permit the luxury of comprehensive revision of all *b-p-h* entries. However, in the event, the combined processes of fortuitous discovery and the methodical chronicling of periodicals' marriages, divorces and deaths actually produced a considerable amount of *ad hoc* revision. Many errors undoubtedly remain, and users will perform a great service for the future development of our database, and eventually the preparation of a fully revised and consolidated second edition, if they will notify the compiler of any additional data or corrigenda they may have available.

Acknowledgements

Over the years since 1968, many correspondents have contributed various bits of data to correct or augment the *B-P-H* record, and to them all we extend grateful thanks. Hunt Institute staff have also added substantially to the file of corrigenda, particularly Nancy Teater, Hildegard Pohl, Mary Jo Lilly and Dr. A. F. Günther Buchheim. The Institute's library, though very modest in its periodicals resources, has provided invaluable bibliographical help, and special mention must be made of Bernadette Callery, Elisabeth Mosimann, Charlotte Tancin and Sarah Leroy. During recent years, access to the OCLC database has been extremely helpful, and special thanks are due to Sarah for her skill and patience in making many hundreds of OCLC searches. Work-study students assisted for various periods over the past few years, and I thank Lauren Eames, Marcy Baughman, Michael Greelish and Susan Gondrand for their efforts.

Amongst the 1968 acknowledgements was one to Elizabeth Smith. Throughout the history of this project, Betsy has sustained a continuous record of skilled contribution, and I have special pleasure in acknowledging her experience, her accurate memory of *B-P-H* history, and her cheerful help with the tedious work that this project entailed. To Dr. Robert Kiger I add a special word of thanks for his decision to allow me to perform this project, for providing unerring logic and critical insight to the solution of several problems, and for his patience in waiting so long for this labour to reach a state fit for publication.

Despite so much help from colleagues, errors and omissions surely remain. They are my responsibility, and I ask the user's indulgence. This Supplement is but another step towards an ideal bibliography.

Gavin Bridson

Additional Selected References

Aerschot, P. van — Catalogue de la bibliothèque collective réunie au Jardin Botanique de l'État, à Bruxelles. *Bulletin du jardin botanique de l'état à Bruxelles* 3(1). 1911. 252+xxxiii pp.

Agricultural library information notes — Beltsville, United States National Agricultural Library, 1975+. ("New serials received at NAL.")

American Horticultural Society — *North American horticulture: A reference guide.* New York, Scribner's Sons, 1982. xvi+367 pp.

Angely, J. — Lista das abreviaturas usadas neste trabalho com dados explicativos dos livros e revistas. In: *South America botanical bibliography*, pp. 1107-1401. São Paulo, 1980.

Bernstein, J. R., & D. Finley — Worldwide cactus and succulent serials. *Serials review* 14(4): 27-34. 1988.

Blanchard, J. R., & L. Farrell (eds.) — *Guide to sources for agricultural and biological research.* Berkeley, University of California Press, 1981. xi+735 pp.

Boletín bibliográfico agricola (later *Boletín para bibliotecas agrícolas*). — Vol. 6+. Turrialba, Instituto Interamericano de Ciencias Agrícolas, 1969+.

British Library, Science Reference Library — *Abstracting and indexing periodicals in the Science Reference Library.* Ed. 3. London, 1985. ii+149 pp.

— *Japanese journals in English: Scientific, technical & commercial journals held by the British Library and/or the British Library Lending Division.* [By] B. Smith & S. V. King. London, 1985. iii+138 pp.

— *Periodicals on agriculture held by the SRL.* Part 1: *Agricultural research and industry.* London, 1981. v+143+ix pp.; *Addenda and corrigenda.* London, 1982. 16 pp.

— *Periodicals on agriculture held by the SRL.* Part 2: *Agronomy and other aspects of plant sciences.* London, 1984. x+206 pp.

— *Periodicals on biochemistry held by the SRL.* London, 1977. 27 pp.

— *Periodicals on botany held by the SRL.* London, 1977. 49 pp.

— *Periodicals current in mainland China held by the SRL.* 2nd ed. London, 1984. 91 pp.

British Museum (Natural History) — *[List of serial publications in the British Museum (Natural History) library].* London, 1982 (Nov.). Microfiche, unpaged.; *[List of serial publications in the British Museum (Natural History) library].* London, 1986 (Jan.). Microfiche, unpaged; *[List of serial publications in the British Museum (Natural History) library].* London, 1987 (Jul.). Microfiche, unpaged.

Bruhn, P. — *Gesamtverzeichnis russischer und sowjetischer Periodika und Serienwerke in Bibliotheken der Bundesrepublik Deutschland und West-Berlins.* (=Bibliographische Mitteilungen des Osteuropa-Instituts an der Freien Universität Berlin, 3). 4 vols. Wiesbaden, Harrassowitz, 1962(-76).

Cassie, V. — Bibliography of the freshwater Algae of New Zealand 1849-1980. *New Zealand journal of botany* 18: 433-447. 1980; Additions to the "Bibliography ... 1980." *Ibid.* 19: 389-391. 1981.

Catalogo colectivo nacional de publicaciones periodicas — (= Publicacion tecnica, consejo de rectores, universidades chilenas, 5). Santiago, Chile, Centro Nacional de Informacion y Documentacion, 1968. Various paging.

Catalogo colectivo de publicaciones periodicas existentes en bibliotecas cientificas y tecnicas Argentinas — Ed. 2. 2 vols. (1331 pp.). Buenos Aires, Consejo Nacional de Investigaciones Cientificas y Tecnicas, 1962; Suplemento a la 2a ed. de 1962. 2 vols. (563 pp.). Buenos Aires, Consejo Nacional de Investigaciones Cientificas y Tecnicas, 1972.

Catalogue collectif des périodiques du début du XVIIe siècle à 1939 — 5 vols. Paris, Bibliothèque Nationale, 1967-82.

Center for Research Libraries — *Soviet serials currently received ... A checklist.* Ed. 4., June 1, 1990. Chicago, 1990. vi+305 pp.

Coffey, J. C. — Soviet journals important for taxonomic botany: A translation of Zaikonnikowa's list, with emendations. *Huntia* 5: 85-106. 1984.

Conservatoire et Jardin Botaniques de la Ville de Genève — *Catalogue des périodiques de la Bibliothèque des Conservatoire et Jardin Botaniques de la Ville de Genève.* Genève, 1980. 271 pp.; *Catalogue des périodiques de la Bibliothèque des Conservatoire et Jardin Botaniques de la Ville de Genève. Supplément 1980-1987.* Genève, 1988. 107 pp.

Current awareness list, library, Royal Botanic Gardens, Kew — Kew, 1978+. (Monthly issues. "Additions to periodical publications currently received by the Library.")

D. K. Agencies (P) Ltd. — [Trade catalogues.] New Delhi, various dates. (For information on back sets of modern Indian periodicals.)

Davis, E. — Selected list of environmental science journals for a science collection: An annotated bibliography. *Serials librarian* 17: 149-198. 1989.

Doorenbos, J. — Nederlandse tijdschriften op tuinbouwgebied. Dutch horticultural periodicals. *Mededelingen, directeur van de tuinbouw* 14(6): 372-395. 1951.

Eggli, U. — A bibliography of succulent plant periodicals. *Bradleya* 3: 103-119. 1985; More bibliographical data on succulent plant periodicals. *Ibid.* 5: 101-102. 1987; Newton, L. E.—More bibliographical data on succulent plant periodicals. *Ibid.* 5: 102-104. 1987.

Elsevier Antiquarian Department — *Out of print and rare periodicals and serials and books on botany and agriculture. Catalogue 43, 1983/1984.* Amsterdam, 1984. 179 pp.

Flora Malesiana Bulletin — Vols. 21+. Leiden, Flora Malesiana Foundation, 1966+. ("Research & publications," vols. 21-25; "New journals," vol. 26+.)

Frodin, D. G. — Abbreviations of serials cited. In: *Guide to standard floras of the world*, pp. 545-558. Cambridge, Cambridge University Press, 1984.

Gibson, R. W., Jr., & B. K. Kunkel — *Japanese scientific and technical literature: A subject guide.* Westport, CT, Greenwood Press, 1981. 560 pp.

Gray, D. A. — Voices for wilderness: Conservation society serials. *Serials review* 1989 (Summer): 23-33. 1989

Hawksworth, D. L. — Lichenological journals. In: Seaward, M. R. D. (ed.)—*Lichen ecology*, pp. 497-498. London, etc., Academic Press, 1977.

Index to Australian taxonomic literature for 1968(-70) — (= Regnum Vegetabile, 66, 75 & 83). I. K. Ferguson (ed.) 3 vols. Utrecht, International Association for Plant Taxonomy, 1970-72.

Index to European taxonomic literature for 1965(-69) — (= Regnum Vegetabile, 45, 53, 61, 70 & 80). R. K. Brummitt *et al.* (eds.) 5 vols. Utrecht, International Association for Plant Taxonomy, 1966(-71).

Instituto Interamericano de Ciencias Agricolas — Directorio de las publicaciones periodicas de la Biblioteca Commemorativa Orton. Por O. Arboleda-Sepúlveda. Turrialba, 1966. 486 pp.; Suplemento 1. *Boletín bibliográfico agrícola* 3?: 235-296, 1966; Suplemento 2. *Ibid.* 4: 83-111. 1967; Suplemento 3. *Ibid.* 4: 217-233. 1967; Suplemento 4. *Ibid.* 5: 118-169. 1968; Suplemento 5. *Ibid.* 5: 293-324. 1968.

International Organization for Standardization (iso). — 1968. *ISO Recommendation R 9. International system for the transliteration of Slavic Cyrillic characters.* Geneva. 8 pp.

Irregular Serials & Annuals: An international directory — Ed. 4. 1976-1977[+]. New York, R. R. Bowker, 1976+

Kew record of taxonomic literature relating to vascular plants for 1971 [+]. — London, Her Majesty's Stationery Office, 1974+.

Kirchner, J. — *Bibliographie der Zeitschriften des deutschen Sprachgebietes bis 1900.* Band 1: ... *bis 1830.* Stuttgart, A. Hiersemann, 1966-69. xv+489 pp.; Band 2: ... *von 1831 bis 1870.* Stuttgart, A. Hiersemann, 1971-77. xi+400 pp.; Band 3: ... *von 1871 bis 1900.* Stuttgart, A. Hiersemann, 1971-77. xi+730 pp.

Kurata, S. (ed.) — *A bibliography of forest botany in Japan (1940-1963).* Tokyo, University of Tokyo Press, [1966]. xiii+146 pp.

Linda Hall Library, Science & Technology — *Serials holdings in the Linda Hall Library.* Kansas City, MO, 1983. ix+739 pp.

List of accessions to the museum library, British Museum (Natural History) — 1955-74. 20 vols. London.

Longwood Gardens, Inc. — [*List of periodicals.*] [Kennet Square, PA, 1986.] [21] pp. (Photocopy of "in-house" list.)

Missouri Botanical Garden — [*Lists of "BPH new" periodical titles, with abbreviations, in the TROPICOS data base system.*] [St. Louis, 1987.] 312 pp.; [*Supplement. Monthly lists.*] [St. Louis, 1988-89.] Various paging. (Computer printouts.)

National directory of newsletters and reporting services — Ed. 2. Thomas, R. C. & B. T. Darnay (eds.). Parts 1-8 [continuing]. Detroit, Gale Research Co., 1978+.

Naturhistorische Verein der Rheinlande und Westfalen — Katalog der Zeitschriften und Schriftenreihen der Bibliothek des Naturhistorischen Vereins der Rheinlande und Westfalens. *Decheniana - Beiheft* 14: v+140 pp. 1968

Neumann, D. — Annotated list of succulent serial publications. Parts 1-9. *Succulentarum bibliographia* 1: 1-5, 8-11, 15-20, 22-26. 1971; *Ibid.* 2: 1-5, 8-12, 15-18. 1972; *Ibid.* 3: 11-12, 16-18. 1973.

New serial titles: A union list of serials commencing publication after December 31, 1949. 1950-1970 cumulative — 4 vols. Washington, DC. & New York/ London, Library of Congress & R. R. Bowker Co., 1973. (And continuations.)

New York Botanical Garden, Library — *Current periodicals. January 1987.* New York, 1987 (Jan). 60 pp.

Nicolson, D. H. — Report on the Special Committee on Romanization of Authors' Names from Non-Roman Scripts. *Taxon* 30: 168-183. 1981.

Nouvelles acquisitions, bibliothèque centrale du Muséum National d'Histoire Naturelle — Paris, 1974+.

Online Computer Library Center Inc. (oclc) — [Online bibliographic database system. Dublin, OH.] Current.

Owen, D. B. — *Abstracts and indexes in science and technology: A descriptive guide.* Ed. 2. Metuchen, Scarecrow Press, 1985. xv+235 pp.

Periodicals news — London, British Museum, National Reference Library of Science & Invention (later British Library, Science Reference Library), 1970-81. Monthly issues.

Royal Botanic Garden Edinburgh — *Periodicals list.* Edinburgh, 1969. 108 pp.

Royal Botanic Gardens Kew — *List of periodical publications in the library.* London, Her Majesty's Stationery Office, 1978. 90 pp.

Scientific serials in Australian libraries — J. A. Conochie (ed.). 3 vols. (loose leaf). Melbourne, Commonwealth Scientific, Industrial & Research Organization, 1975.

Scientific and technical books and serials in print 1983[+] — 3 vols. (3,640 pp.). New York, R. R. Bowker, 1982+.

Serials directory: An international directory — Ed. 1[+], 1986[+]. 3 vols. Birmingham, AL, EBSCO Publishing, 1986+ (annually).

Standard directory of newsletters — New York, Oxbridge Publishing Co., 1972. 210 pp.

Standard periodical directory — Ed. 9. 1985-1986. New York, Oxbridge Communications, Inc., 1985. xvii+1452 pp.

Taxon — Vol. 1+. Utrecht, International Association for Plant Taxonomy, 1951+. (Reports of new periodicals.)

Tutin, T. G. *et al.* (eds.) — *Flora Europaea.* 5 vols. Cambridge, Cambridge University Press, 1964-80. (Appx. III in each vol. is a list of periodical sources.)

Union list of scientific, technical, agricultural and forestry periodicals and serials in scientific libraries in Finland — [Vol. 1]. Helsinki, 1950. xii+698 pp.; Vol. 2: *Publications printed in Slavonic characters.* Helsinki, Suomalainen Tiedeakatemia, 1953. viii+183 pp.

Union list of serials in Canada Department of Agriculture libraries — Ed. 2. [Ottawa], Canada Department of Agriculture, 1977. [v]+745 pp.

Union list of serials in the science area, Oxford — Oxford, Bodleian Library, 1968. xvii+398 pp.

United States Department of Agriculture — *Slavic serials in agricultural sciences – A bibliography.* (= Library list, 102). N.p., [1978]. iii+751 pp.

— National Agricultural Library — *Horticultural journals currently received at the National Agricultural Library: October 1988.* By J. P. Gates. Beltsville, 1988. 27 pp.; *Horticultural journals currently received at the National Agricultural Library.* Rev. ed. May 1990. By J. P. Gates. Beltsville, 1990. [ii]+30 pp.

— — *List of journals indexed in AGRICOLA 1990.* By C. L. Dowling, A. M. Funkhouser & T. Lehnert. Beltsville. vi+195 pp.

— — *Newsletters pertaining to agricultural biotechnology. January 1991.* By R. D. Warmbrodt & V. Stone. Beltsville. iii+14 pp.

Watling, R. & A. E. — Journals and periodicals. In: *A literature guide for identifying mushrooms,* pp. 101-110. Eureka, CA, Mad River Press, 1980. vi+121 pp.

Catalogue

A A A S bulletin = American association for the advancement of science bulletin. Lancaster, PA.

A A B G A newsletter = News letter, american association of botanical gardens and arboreta. Philadelphia, PA, Lisle, IL.

A A M bulletin = Bulletin, american association of museums. Washington, DC.

A A S P contributions series = Contributions series, american association of stratigraphic palynologists. Various places.

A A S P newsletter. College Station, TX. Vol. ?-12(4)+, 19??-79+. A. A. S. P. Newslett. Preceded by: Newsletter, american association of stratigraphic palynologists. HI 60839

A B bookman's weekly. Newark, NJ. Vol. 39(23/24)+, 1967+. A. B. Bookman's Weekly. Preceded by: Antiquarian bookman. HI 60840

A B C = Abstracts in biocommerce. Slough.

A B M R = Antiquarian book monthly review. Brayfield, etc.

A B R; Agropecuaria brasileira. Resumos. Brasilia. No. 00, 1982; no. 1+, 1982+. A. B. R. HI 60841

A B T; Abstracts of bioanalytic technology. San Francisco, CA. Vol. 1+, 1953+. A. B. T. HI 60842

A C I A R proceedings series. Canberra, A.C.T. No. 1+, 1984+. A. C. I. A. R. Proc. Ser. HI 62317

A D A S quarterly review. London. Nos. 1-39, 1971-80. A. D. A. S. Quart. Rev. Preceded by: N A A S quarterly review. HI 60843

A E form, agricultural extension station, Purdue university. Lafayette, IN. Nos. 1-19, 1910-11. A. E. Form, Purdue Univ. Superseded by: Leaflet, department of agricultural extension, Purdue university. HI 60844

A E M = Applied and environmental microbiology. Washington, DC.

A E T F A T bulletin = Association pour l'étude taxonomique de la flore d'Afrique tropicale bulletin. Brussels, Geneva.

A E T F A T index; Relève des travaux de phanérogamie systématique et des taxons nouveaux concernant l'Afrique au sud du Sahara et Madagascar. Brussels. 1953-76, 1954-77. A. E. T. F. A. T. Index. HI 60845

A F R I miscellaneous report. Syracuse, NY. Vol. 1+, 1968+. A. F. R. I. Misc. Rep. HI 60846

A F R I research report. Syracuse, NY. Vol. 1+, 1969+. A. F. R. I. Res. Rep. HI 60847

A G bulletin, department of agriculture, New South Wales. Rydalmere, N.S.W. No. ?-2+, 19??-78+. A. G. Bull., N. S. W. HI 60848

A G G S news views. [Magnolia, NJ.] Vol. 1+, 1988+. A. G. G. S. News Views. HI 62914

A G P news = Agp news. Melbourne, Vic.

A G R E P; permanent inventory of agricultural research projects in the European communities. Luxembourg. 1979 (June)+, 1979+. A. G. R. E. P. HI 60849

A H C news; news bulletin of the american horticultural council. Ithaca, NY. Nos. 1-30, 1953-59. A. H. C. News. HI 60850

A H S gardeners' forum. Washington, DC. Vols. 1-2, 1958-59. A. H. S. Gard. Forum. Superseded by: American horticultural society gardeners' forum. HI 60851

A H T A newsletter. Gaithersburg, MD. Vol. 14+, 1987+. A. H. T. A. Newslett. Preceded by: N C T R H newsletter. HI 60852

A I A newsletter. Edmonton. 1947+. A. I. A. Newslett. HI 60853

A I B D A boletín especial. Turrialba, Costa Rica. Vol. 1+, 1966+. A. I. B. D. A. Bol. Espec. HI 60854

A I B S bulletin. Washington, DC. Vols. 1-13, 1951-63. A. I. B. S. Bull. Preceded & superseded by: A I B S newsletter. 1-4-2. HI 60855

A I B S education review. Arlington, VA. Vols. 1+, 1972+. A. I. B. S. Educ. Rev. HI 60856

A I B S forum; an exchange of information on biology in the public interest. Arlington, VA. Vols. 1-2?, 1978-79? A. I. B. S. Forum. HI 60857

A I B S news. Washington, DC. Vol. 1+, 1972+. A. I. B. S. News. Preceded by: C U E B S news. HI 60858

A I B S newsletter. Washington, DC. Vols. 1-3(2), 1949-50; 1967+ [first number lacks volumation and date]. A. I. B. S. Newslett. For 1951-63 see: A I B S bulletin. HI 60859

A I C newsletter. Ottawa. [Dates of publication not ascertained.] A. I. C. Newslett. HI 60861

A I C review. Ottawa. Vol. 24(3)-26, 1969-71. A. I. C. Rev. Preceded by: Agricultural institute review. Ottawa. Superseded by: Agrostologist. HI 60862

A I C E survey of U S S R air pollution literature. Silver Spring, MD. Vols. 1-21, 1969-73. A. I. C. E. Surv. U.S.S.R. Pollut. Lit. HI 60863

A I H P notes. Madison, 1955-57. A. I. H. P. Notes. Superseded by: Pharmacy in history. HI 60864

A I N news = A I N newsletter. East Lansing, MI.

A I N newsletter. East Lansing, MI. 1961+. A. I. N. Newslett. HI 60865

A M A american journal of diseases of children = American journal of diseases of children. Chicago IL. Amer. J. Dis. Children. See B–P–H 81/12.

A M A archives of internal medicine = Archives of internal medicine. Chicago, IL. Arch. Intern. Med. See B–P–H 142/21.

A M A archives of pathology = Archives of pathology. Chicago, IL. Arch. Pathol. See B–P–H 147/9.

A M I news = Mushroom news. Kennett Square, PA.

A P P S newsletter = Newsletter, australian plant pathology society. Sydney, NSW.

A R C research review. London. Vol. 1-5(2), 1975-79. A. R. C. Res. Rev. HI 60866

A R I newsletter. Washington, DC. 1951+. A. R. I. Newslett. HI 60867

A S B bulletin. Chapel Hill, NC. Vol. 1+, 1954+. A. S. B. Bull. HI 60868

A S B P communication. Manila. Vol. 1+, 1984+. A. S. B. P. Commun. HI 60869

A S B S newsletter. No. 1, 1974. A. S. B. S. Newslett. Superseded by: Newsletter, australian systematic botany society. HI 60870

A S C newsletter. Lawrence, KS. Vol. 1+, 1973+. A. S. C. Newslett. HI 60871

A S F A = Aquatic sciences and fisheries abstracts. London.

A S F A aquaculture abstracts = Aquatic sciences and fisheries abstracts. Aquaculture abstracts. Bethesda, MD.

A S F A marine biotechnology abstracts. Bethesda, MD. Vol. 2+, 1990+. A. S. F. A. Mar. Biotechnol. Abstr. Preceded by: Marine biotechnology abstracts. HI 75273

A S F A, 1 = Aquatic sciences and fisheries abstracts. Part 1, biological sciences and living resources. London.

A S F A 3 = Aquatic sciences and fisheries abstracts. Part 3, aquatic pollution and environmental quality. Bethesda, MD.

A S H S newsletter. Alexandria, VA. Vol. 1+, 1985+. A. S. H. S. Newslett. Preceded by: Newsletter, american society for horticultural science. HI 60872

A S L I B book list. London. Nos. 1-50, 1935-85. Aslib Book List. HI 60873

A S L I B proceedings. London. Vol. 1+, 1949+. Aslib Proc. 1-507-1. HI 60874

A S M news. Washington, DC. Vol. 29+, 1963+. A. S. M. News. Preceded by: Bacteriological news. HI 60875

A S P B news. Lexington, KY. [Data for volumation and dates unavailable.] A. S. P. B. News. HI 60876

A S P newsletter. 1972+. A. S. P. Newslett. HI 60877

A S P P newsletter. Rockville, MD. 1924+. A. S. P. P. Newslett. HI 60878

A S P T newsletter. Ithaca, NY. Vol. 1+, 1987+. A. S. P. T. Newslett. HI 60879

A S T C newsletter. Washington, DC. Vol. ?-15(4)+, 19??-87+. A. S. T. C. Newslett. HI 60880

A S T I S bibliography. Calgary. 1979+. A. S. T. I. S. Bibliogr. HI 60881

A S T I S current awareness bulletin. Calgary. 1978+, 1979+. A. S. T. I. S. Curr. Aware. Bull. Preceded by: Library accessions, arctic institute of north America. HI 60882

A S T I S occasional publication. Calgary. No. ?-3+, 19??-80+. A. S. T. I. S. Occas. Publ. HI 60883

A V C. Agricultural & veterinary chemicals. London. Vol. 1+, 1960+. A. V. C. HI 60884

A V R D C highlights. Tainan. 1980+. A. V. R. D. C. Highlights. HI 60885

A V R D C progress report. Tainan. 1975+, 1976+. A. V. R. D. C. Progr. Rep. Preceded by: Report (Annual), asian vegetable research and development center. HI 60886

A W A newsletter = Newsletter, Alberta wilderness association. Calgary.

A W A I R; Arab world agri and agro-industrial review. Vol. ?-3+, 19??-86+. Arab World Agri- Agro-Industr. Rev. HI 60887

Aalstersche pachter. Alost, Belgium. Aalstersche Pachter. See B–P–H 25/12. HI 50002

Aanteekeningen van het verhandelde in de vergaderingen der sectie's van het provinciaal utrechtsch genootschap van kunsten en wetenschappen ter gelegenheid van de algemeene vergadering. Utrecht. Aanteek. Verh. Prov. Utrechtsch Genootsch. Kunsten. See B–P–H 25/13. HI 50003

Aarbog for gartneri; till gartner-tidende. Copenhagen. Aarbog Gartn. See B–P–H 25/15. HI 50004

Aarbok for universitet i Bergen. Matematisk-naturvitenskapelig serie. Bergen. 1960+. Aarbok Univ. Bergen, Mat.-Naturvitensk. Ser. Preceded by: Universitet i Bergen årbok. Naturvidenskabelig rekke. HI 60888

Aardappeldagen; nederlandsche landbouwweek. Wageningen. Aardappeldagen. See B–P–H 25/17.

HI 50005

Aardappelstudiecentrum voor de kleinhandel. The Hague. Aardappelstudiecentrum Kleinhandel. See B–P–H 25/18. HI 50006

Aarsberetning, kongelige norske videnskabers selskabs = Kongelige norske videnskabers selskabs aarsberetning. Trondheim.

Abeille médicale. Montreal. Abeille Méd. See B–P–H 25/23. HI 50009

Abeille ou recueil de philosophie, de litterature et d'histoire. The Hague. Abeille. See B–P–H 25/22. HI 50008

Abeilles et fleurs. Paris. Abeilles & Fl. See B–P–H 25/24. HI 50010

Aberdeen university studies. Aberdeen. Vol. 1+, 1900+. Aberdeen Univ. Stud. HI 60889

Abhandlungen aus der Abeilung für Phytomorphogenese des Timiriaseff-Instituts für Biologie am Zentralen Exekutivkomitee der U.S.S.R. Moscow. Abb. Abt. Phytomorphogenese Timiriaseff-Inst. Biol. See B–P–H 26/1. HI 50011

Abhandlungen über Ägypten, welche während des Feldzuges des Generals Bonaparte an das National-Institut zu Kairo bekannt gemacht worden sind. Aus dem Französischen. Berlin. Abh. Ägypten. See B–P–H 26/3. HI 50012

Abhandlungen der Akademie gemeinnütziger Wissenschaften zu Erfurt. Erfurt. Abh. Akad. Gemeinnütz. Wiss. Erfurt. See B–P–H 26/5. HI 50013

Abhandlungen der Akademie der Wissenschaften der DDR. Abteilung Mathematik, Naturwissenschaften, Technik. Berlin. 1973+. Abh. Akad. Wiss. DDR. Preceded by: Abhandlungen der Deutschen Akademie der Wissenschaften zu Berlin. Klasse für Chemie, Geologie und Biologie. HI 60890

Abhandlungen der Akademie der Wissenschaften, Göttingen. Mathematisch-Physikalische Klasse. Göttingen. Abh. Akad. Wiss. Göttingen, Math.-Phys. Kl. See B–P–H 26/13. HI 50015

Abhandlungen der Arbeitsgemeinschaft für tier- und pflanzengeographische Heimatforschung im Saarland. Saarbrücken. Nos. 1-10, 1969-80. Abh. Arbeitsgem. Tier- Pflanzengeogr. Heimatf. Saarland. Superseded by: Abhandlungen der DELATTINIA. HI 60891

Abhandlungen aus dem Archiv für Geschichte der Naturforschung und Medizin der Deutschen Akademie der Naturforscher Leopoldina = Acta historica leopoldina. Leipzig. Acta Hist. Leop. See B–P–H 43/12.

Abhandlungen der Bayerischen Akademie der Wissenschaften. Mathematisch-naturwissenschaftliche Abteilung. Munich. Abh. Bayer. Akad. Wiss., Math.-Naturwiss. Abt. See B–P–H 26/20. HI 50017

Abhandlungen der Bayerischen Akademie der Wissenschaften. Mathematisch-naturwissenschaftliche Klasse. Munich. Abh. Bayer. Akad. Wiss., Math.-Naturwiss. Kl. See B–P–H 26/21. HI 50018

Abhandlungen der Bayerischen Akademie der Wissenschaften. Mathematisch-physikalische Klasse. = Abhandlungen der Königlich Bayerischen Akademie der Wissenschaften. Mathematisch-physikalische Klasse. Munich. Abh. Königl. Bayer. Akad. Wiss., Math.-Phys. Kl. See B–P–H 30/3.

Abhandlungen und Beobachtungen durch die ökonomische Gesellschaft zu Bern gesammelt. Bern. Abh. Beob. Ökon. Ges. Bern. See B–P–H 27/1. HI 50019

Abhandlungen und Berichte. Magdeburg = Museum für Naturkunde und Heimatkunde zu Magdeburg. Abhandlungen und Berichte. Magdeburg. Mus. Natur-Heimatk. Magdeburg Abh. Ber. See B–P–H 622/2.

Abhandlungen und Berichte des Museums der Natur Gotha. Gotha. 1971+. Abh. Ber. Mus. Natur Gotha. Preceded by: Abhandlungen und Berichte des Naturkundemuseums Gotha. HI 60892

Abhandlungen und Berichte aus dem Museum für Natur- und Heimatkunde und dem Naturwissenschaftlichen Verein in Magdeburg. Magdeburg. Abh. Ber. Mus. Natur- Heimatk. Naturwiss. Verein Magdeburg. See B–P–H 27/3. HI 50021

Abhandlungen und Berichte aus dem Museum für Naturkunde und Vorgeschichte in Magdeburg. Magdeburg. Abh. Ber. Mus. Naturk. Magdeburg. See B–P–H 27/4. HI 50022

Abhandlungen und Berichte aus dem Museum für Naturkunde und Vorgeschichte und dem Naturwissenschaftlichen Verein in Magdeburg. Magdeburg. Abh. Ber. Mus. Naturk. Naturwiss. Verein Magdeburg. See B–P–H 27/5. HI 50023

Abhandlungen und Berichte für Naturkunde und Vorgeschichte = Museum für Kulturgeschichte Magdeburg. Abhandlungen und Berichte für Naturkunde und Vorgeschichte. Leipzig. Mus. Kulturgesch. Magdeburg Abh. Ber. Naturk. Vorgesch. See B–P–H 621/24.

Abhandlungen und Berichte für Naturkunde und Vorgeschichte = Museum für Kulturgeschichte. Abhandlungen und Berichte für Naturkunde und Vorgeschichte. Magdeburg. Mus. Kulturgesch. Abh. Ber. Naturk. Vorgesch. See B–P–H 621/23.

Abhandlungen und Berichte des Naturkundemuseums-Forschungsstelle - Görlitz = Abhandlungen und Berichte des Naturkundemuseums Görlitz. Abh. Ber. Naturkundemus. Görlitz. See B–P–H 27/7.

Abhandlungen und Berichte des Naturkundemuseums Görlitz. Görlitz. Abh. Ber. Naturkundemus. Görlitz. See B–P–H 27/7. HI 50024

Abhandlungen und Berichte des Naturkundemuseums Gotha; Gothaer Museums Hefte. Gotha. 1963-69. Abh. Ber. Naturkundemus. Gotha. Superseded by: Abhandlungen und Berichte des Museums der Natur Gotha. HI 60893

Abhandlungen und Berichte des naturkundlichen Museums "Mauritianum". Altenburg. Vol. 1+, 1958+. Abh. Ber. Naturk. Mus. "Mauritianum" Altenburg. Preceded by: Wissenschaftliche Beiträge aus dem Osterlande. HI 60860

Abhandlungen und Berichte der Naturwissenschaftlichen Abteilung der Grenzmärkischen Gesellschaft zur Erforschung und Pflege der Heimat (e.V.), Schneidemühl. Schneidemühl, Germany [=Pila, Poland]. Abh. Ber. Naturwiss. Abt. Grenzmärk. Ges. Erforsch. Heimat Schneidemühl. See B–P–H 27/8. HI 50025

Abhandlungen und Berichte. Naturwissenschaftliche Gesellschaft Bayreuth. Bayreuth. Abh. Ber. Naturwiss. Ges. Bayreuth. See B–P–H 27/9. HI 50026

Abhandlungen und Berichte der Pommerschen Naturforschenden Gesellschaft. Stettin [Poland]. Abh. Ber. Pommerschen Naturf. Ges. See B–P–H 27/10. HI 50027

Abhandlungen und Berichte des Vereins der Naturfreunde zu Greiz. Greiz. Vols. ?-4-7-?, ?-1902-26-? Abh. Ber. Ver. Naturfr. Greiz. HI 60894

Abhandlungen und Bericht des Vereins für Naturkunde zu Kassel. Kassel. Abh. Ber Vereins Naturk. Kassel. See B–P–H 27/13. HI 50028

Abhandlungen der Böhmischen Gesellschaft der Wissenschaften, nebst der Geschichte derselben. Prague. Abh. Böhm. Ges. Wiss. See B–P–H 27/15. HI 50029

Abhandlungen der Böhmischen Gesellschaft der Wissenschaften in Prag = Abhandlungen der Böhmischen Gesellschaft der Wissenschaften, nebst der Geschichte derselben. Prague. Abh. Böhm. Ges. Wiss. See B–P–H 27/15.

Abhandlungen der Churfürstlich-baierischen Akademie der Wissenschaften. Munich. Abh. Churfürstl.-Baier. Akad.Wiss. See B–P–H 28/3. HI 50030

Abhandlungen der chursächsischen Weinbaugesellachaft. Meissen, Germany. Abh. Chursächs. Weinbauges. See B–P–H 28/4. HI 50031

Abhandlungen der DELATTINIA. Saarbrucken. Vol. 11+, 1982+. Abh. DELATTINIA. Preceded by: Abhandlungen der Arbeitsgemeinschaft für tier- und pflanzengeographische Heimatforschung im Saarland. HI 60895

Abhandlungen der Deutschen Akademie der Naturforscher Leopoldina = Nova acta leopoldina. Halle. Nova Acta Leop. See B–P–H 672/24.

Abhandlungen der Deutschen Akademie der Naturforscher (Leopoldina) = Nova acta leopoldina. Halle. Nova Acta Leop. See B–P–H 672/24.

Abhandlungen der Deutschen Akademie der Naturforscher (Leopoldina) zu Halle/Saale = Nova acta leopoldina. Halle. Nova Acta Leop. See B–P–H 672/24.

Abhandlungen der deutschen Akademie der Wissenschaften zu Berlin. Klasse für Chemie, Geologie und Biologie. Berlin. 1955-66. Abh. Deutsch. Akad. Wiss. Berlin, Kl. Chem. Preceded by: Abhandlungen der deutschen Akademie der Wissenschaften zu Berlin. Klasse für Mathematik und allgemeine Naturwissenschaften. Superseded by: Abhandlungen der Akademie der Wissenschaften der D D R. HI 60896

Abhandlungen der Duetschen Akademie der Wissenschaften zu Berlin. Klasse für Mathematik und allgemeine Naturwissenschaften. Berlin. Abh. Deutsch. Akad. Wiss. Berlin, Kl. Math. See B–P–H 28/10. HI 50032

Abhandlungen der Deutschen Akademie der Wissenschaften zu Berlin. Mathematisch-naturwissenschaftliche Klasse. Berlin. Abh. Deutsch. Akad. Wiss. Berlin, Math.-Naturwiss. Kl. See B–P–H 28/11. HI 50033

Abhandlungen der Deutschen Akademie der Wissenschaften in Prag. Mathematisch-naturwissenschaftliche Klasse. Prague. Abh. Deutsch. Akad. Wiss. Prag, Math.-Naturwiss. Kl. See B–P–H 28/12. HI 50034

Abhandlungen der Deutschen Gesellschaft der Wissenschaften und Künste in Prag. Mathematisch-naturwissenschaftliche Klasse. Prague. Abh. Deutsch. Ges. Wiss. Prag, Math.-Naturwiss. Kl. See B–P–H 28/13. HI 50036

Abhandlungen des Deutschen Naturwissenschaftlich-Medizinischen Vereins für Böhmen "Lotos." Prague. Abh. Deutsch. Naturwiss.-Med. Vereins Böhmen "Lotos". See B–P–H 28/14. HI 50037

Abhandlungen aus dem Forst- und Jagdwesen. Prague. Abh. Forst- Jagdwesen. See B–P–H 28/17. HI 50038

Abhandlungen der freyen Oekonomischen Gesellschaft in St. Petersburg zur Aufmunterung des Ackerbaues und der Hauswirthschaft in Russland. St. Petersburg. Abh. Freyen Oekon. Ges. St. Petersburg. See B–P–H 28/18. HI 50039

Abhandlungen aus dem Gebiet der Auslandskunde. Reihe C: Naturwissenschaften. Hamburg. Abh. Auslandsk., Reihe C, Naturwiss. See B–P–H 26/17. HI 50016

Abhandlungen aus dem Gebiete der Naturwissenschaften herausgegeben von dem naturwissenschaftlichen Verein in Hamburg. Hamburg. Abh. Naturwiss. Naturwiss. Verein Hamburg. See B–P–H 32/18. HI 50088

Abhandlungen aus dem Gebiete der Naturwissenschaften herausgegeben vom naturwissenschaftlichen Verein zu Hamburg-Altona = Abhandlungen aus dem Gebiete der Naturwissenschaften herausgegeben von dem naturwissenschaftlichen Verein in Hamburg. Hamburg. Abh. Naturwiss. Naturwiss. Verein Hamburg. See B–P–H 32/18.

Abhandlungen der geologischen Bundesanstalt, Wien. Vienna. 1925+. Abh. Geol. Bundesanst. Wien. Preceded by: Abhandlungen der kaiserlich-koniglichen geologischen Reichsanstalt, Wien. HI 60897

Abhandlungen des Geologischen Dienstes Berlin. Berlin. Abh. Geol. Dienstes Berlin. See B–P–H 28/20. HI 50040

Abhandlungen der Geologischen Landesanstalt Berlin. Berlin. Abh. Geol. Landesanst. Berlin. See B–P–H 28/21. HI 50041

Abhandlungen zur geologischen Specialkarte von Elsass-Lothringen. Strasbourg. Vols. 1-5, 1875-97; n.s. vols.1-6, 1898-1905. Abh. Geol. Specialkarte Elsass-Lothringen. Superseded by: Mémoires du service de la carte géologique d'Alsace et de Lorraine. 1-151-3. HI 60898

Abhandlungen zur Geologischen Spezialkarte von Preussen und den Thüringischen Staaten. Berlin. Abh. Geol. Spezialkarte Preussen Thüring. Staaten. See B–P–H 29/2. HI 50042

Abhandlungen der Gesellschaft der Künste und Wissenschaften in Batavia. Leipzig. Abh. Ges. Künste Batavia. See B–P–H 29/4. HI 50043

Abhandlungen der Gesellschaft der Naturforscher und Ärzte an der Universität Smolensk = Trudy Smolenskogo obshchestva estestvoispytatelei i vrachei pri Smolenskom gosudarstvennom universitete. Smolensk.

Abhandlungen über die Geschichte und Alterthümer, die Künste, Wissenschaften und Literatur Asiens ... Riga [Latvian S S R]. Abh. Gesch. Asiens. See B–P–H 29/10. HI 50044

Abhandlungen zur Geschichte der Medizin und der Naturwissenschaften. Berlin. Abh. Gesch. Med. Naturwiss. See B–P–H 29/11. HI 50045

Abhandlungen zur Geschichte der Naturwissenschaften und Medizin. Erlangen. Abh. Gesch. Naturwiss. Med. See B–P–H 29/12. HI 50046

Abhandlungen der Hallischen Naturforschenden Gesellschaft. Dessau & Leipzig. Abh. Hallischen Naturf. Ges. See B–P–H 29/13. HI 50047

Abhandlungen des Hamburgischen Kolonialinstituts. Hamburg. Abh. Hamburg. Kolonialinst. See B–P–H 29/14. HI 50048

Abhandlungen der Heidelberger Akademie der Wissenschaften. Stiftung Heinrich Lanz. Mathematisch-naturwissenschaftliche Klasse. Heidelberg. Abh. Heidelberger Akad. Wiss., Math.-Naturwiss. Kl. See B–P–H 29/16. HI 50049

Abhandlungen herausgegeben vom naturwissenschaftlichen Vereine zu Bremen. Bremen. Abh. Naturwiss. Vereine Bremen. See B–P–H 32/21. HI 50089

Abhandlungen herausgegeben von der Senckenbergischen Naturforschenden Gesellschaft. Frankfurt a. M. Abh. Senckenberg. Naturf. Ges. See B–P–H 34/13. HI 50108

Abhandlungen der holländischen Gesellschaft der Wissenschaften zu Harlem. Altenburg. Abh. Holl. Ges. Wiss. Harlem. See B–P–H 29/17. HI 50050

Abhandlungen über interessante Gegenstände beim Forst- und Jagdwesen. Berlin. Abh. Interessante Gegenstände Forst- Jagdwesen. See B–P–H 29/19. HI 50051

Abhandlungen der kaiserlich-koniglichen geologischen Reichsanstalt, Wien. Vienna. 1852-1914. Abh. K. K. Geol. Reichsanst. Wien. Superseded by: Abhandlungen der geologischen Bundesanstalt, Wien. HI 60899

Abhandlungen der kaiserlich-königlichen zoologisch-botanischen Gesellschaft in Wien. Vienna. Vols. 1-16, 1901-56. Abh. K. K. Zool.-Bot. Ges. Wien. 5-4641-3. HI 60900

Abhandlungen der Kaiserlich Leopoldinisch-Carolinisch Deutschen Akademie der Naturforscher = Nova acta leopoldina. Halle. Nova Acta Leop. See B–P–H 672/24.

Abhandlungen der Kaiserlichen Leopoldinisch-Carolinischen Deutschen Akademie der Naturforscher. Dresden = Nova acta academiae caesareae leopoldino-carolinae germanicae naturae curiosorum. Dresden. Nova Acta Acad. Caes. Leop.-Carol. German. Nat. Cur. See B–P–H 672/15.

Abhandlungen der Königlich Bayerischen Akademie der Wissenschaften. Mathematisch-physikalische Klasse. Munich. Abh. Königl. Bayer. Akad. Wiss., Math.-Phys. Kl. See B–P–H 30/3. HI 50054

Abhandlungen der Königlich Preussischen Akademie der Wissenschaften. Physikalisch-mathematische Classe. Berlin. Abh. Königl. Preuss. Akad. Wiss., Phys.-Math. Cl. See B–P–H 30/7. HI 50058

Abhandlungen der Königlich Preussischen geologischen Landesanstalt. Berlin. Abh. Königl. Preuss. Geol. Landesanst. See B–P–H 30/8. HI 50059

Abhandlungen der Königlich Sächsischen Gesellschaft der Wissenschaften. Leipzig = Abhandlungen der Mathematisch-Physischen Classe der Königlich Sächsischen Gesellschaft der Wissenschaften. Leipzig. Abh. Math.-Phys. Cl. Königl. Sächs. Ges. Wiss. See B–P–H 31/11.

Abhandlungen der Königlichen Akademie der Wissenschaften in Berlin. Berlin. Abh. Königl. Akad. Wiss. Berlin. See B–P–H 30/2. HI 50053

Abhandlungen der königlichen Böhmischen Gesellschaft der Wissenschaften. Prague. Abh. Königl. Böhm. Ges. Wiss. See B–P–H 30/4. HI 50055

Abhandlungen der Königlichen Gesellschaft der Wissenschaften zu Göttingen. Göttingen. Abh. Königl. Ges. Wiss. Göttingen. See B–P–H 30/5. HI 50056

Abhandlungen der Königlichen Gesellschaft der Wissenschaften zu Göttingen. Mathematisch-Physikalische Klasse. Berlin. Abh. Königl. Ges. Wiss. Göttingen, Math.-Phys. Kl. See B–P–H 30/6. HI 50057

Abhandlungen der Kurfürstlich-Mainzischen Akademie nützlicher Wissenschaften zu Erfurt. Nova acta academiae electoralis moguntinae scientiarum utilium quae Erfurti est. Erfurt. Abh. Kurfürstl.-Mainz. Akad. Nützl. Wiss. Erfurt. See B–P–H 30/10. HI 50060

Abhandlungen des internationalen Verbandes forstlicher Forschungsanstalten. Rome. Vol. 1+, 1954+. Abh. Int. Verbandes Forstl. Forschungsanst. HI 60901

Abhandlungen zur Landeskunde der Provinz Westpreussen. Danzig [=Gdansk, Poland]. Abh. Landesk. Prov. Westpreussen. See B–P–H 30/12. HI 50061

Abhandlungen aus dem Landesmuseum für Naturkunde zu Münster in Westfalen. Munster. Vols. 12-43, 1949-81. Abh. Landesmus. Naturk. Münster Westfalen. Superseded by: Abhandlungen der Westfälischen Museum für Naturkunde. HI 60902

Abhandlungen aus dem Landesmuseum für Naturkunde der Provinz Westfalen. Münster. Abh. Landesmus. Naturk. Prov. Westfalen. See B–P–H 31/2. HI 50062

Abhandlungen aus dem Landesmuseum der Provinz Westfalen. Museum für Naturkunde. Münster. Abh. Landesmus. Prov. Westfalen Mus. Naturk. See B–P–H 31/3. HI 50064

Abhandlungen der liefländischen gemeinnützigen ökonomischen Societät; hauptsächlich die Landwirthschaft in Liefland betreffend. Riga [Latvian S S R]. Abh. Liefl. Gemeinnütz. Ökon. Soc. See B–P–H 31/4. HI 50066

Abhandlungen der Mathematisch-Naturwissenschaftlichen Klasse der Sächsischen Akademie der Wissenschaften. Leipzig. Abh. Math.-Naturwiss. Kl. Sächs. Akad. Wiss. See B–P–H 31/7. HI 50069

Abhandlungen der Mathematisch-Physikalischen Classe der Königlich Bayerischen Akademie der Wissenschaften. Munich. Abh. Math.-Phys. Cl. Königl. Bayer. Akad. Wiss. See B–P–H 31/10. HI 50070

Abhandlungen der Mathematisch-Physischen Classe der Königlich Sächsischen Gesellschaft der Wissenschaften. Leipzig. Abh. Math.-Phys. Cl. Königl. Sächs. Ges. Wiss. See B–P–H 31/11. HI 50071

Abhandlungen der Mathematisch-Physischen Klasse der Sächsischen Akademie der Wissenschaften. Leipzig. Abh. Math.-Phys. Kl. Sächs. Akad. Wiss. See B–P–H 31/14. HI 50072

Abhandlungen und monatliche Mittheilungen aus dem Gesammtgebiete der Naturwissenschaften. Berlin. Abh. Monatl. Mitth. Gesammtgeb. Naturwiss. See B–P–H 31/15. HI 50073

Abhandlungen der naturforschenden Gesellschaft in Danzig. Danzig. Vol. 1(1), 1924. Abh. Naturf. Ges. Danzig. Incorporated in: Schriften der Naturforschenden Gesellschaft in Danzig. HI 60903

Abhandlungen der naturforschenden Gesellschaft zu Görlitz. Görlitz. Abh. Naturf. Ges. Görlitz. See B–P–H 31/17. HI 50075

Abhandlungen der Naturforschenden Gesellschaft zu Halle. Halle. Abh. Naturf. Ges. Halle. See B–P–H 32/1. HI 50077

Abhandlungen der Naturforschenden Gesellschaft in Zürich. Zurich. Abh. Naturf. Ges. Zürich. See B–P–H 32/2. HI 50078

Abhandlungen zur Naturgeschichte, Chemie, Anatomie, Medicin und Physik, aus den Schriften des Instituts der Künste und Wissenschaften zur Bologna. Brandenburg. Abh. Naturgesch Chem. See B–P–H 32/6. HI 50080

Abhandlungen zur Naturgeschichte, Physik und Oekonomie, aus den Philosophischen Transaktionen und Sammlungen. Leipzig. Abh. Naturgesch. Phys. See B–P–H 32/7. HI 50081

Abhandlungen aus der Naturgeschichte, praktischen Arzneykunst und Chirurgie, aus den Schriften der Haarlemer und anderer holländischen Gesellschaften. Leipzig. Abh. Naturgesch. Prakt. Arzneykunst. See B–P–H 32/8. HI 50082

Abhandlungen zur Naturgeschichte der Thiere und Pflanzen, welche ehemals der Königlich Französischen

Akademie der Wissenschaften vorgetragen wurden. Leipzig. Abh Naturgesch. Thiere Pflanzen. See B–P–H 32/9. HI 50084

Abhandlungen der Naturhistorischen Gesellschaft zu Nürnberg. Nuremburg. Abh. Naturhist. Ges. Nürnberg. See B–P–H 32/12. HI 50085

Abhandlungen des Naturkunde- und Tiergartenvereins für Schwaben in Augsburg. Augsburg. Abh. Naturkunde-Tiergartenvereins Schwaben Augsburg. See B–P–H 32/13. HI 50086

Abhandlungen aus der Naturlehre, Haushaltungskunst und Mechanik, Königl[ich]. Schwedischen Akademie der Wissenschaften. Hamburg & Leipzig = Königl[ich]. Schwedischen Akademie der Wissenschaften Abhandlungen aus der Naturlehre, Haushaltungskunst und Mechanik. Hamburg & Leipzig.

Abhandlungen der Naturwissenschaftlichen Gesellschaft "Saxonia" zu Gross- und Neuschönau. Gross Schönau, Germany. Abh. Naturwiss. Ges. "Saxonia" Gross-Neuschönau. See B–P–H 32/17. HI 50087

Abhandlungen des naturwissenschaftlichen Vereins in Hamburg. Hamburg. N.s. vol. 23+, 1980+. Abh. Naturwiss. Vereins Hamburg. Preceded by: Abhandlungen und Verhandlungen des naturwissenschaftlichen Vereins in Hamburg. HI 60904

Abhandlungen des Naturwissenschaftlichen Vereines für Sachsen und Thüringen in Halle. Berlin. Abh. Naturwiss. Vereines Sachsen Halle. See B–P–H 32/22. HI 50090

Abhandlungen des Naturwissenschaftlichen Vereins zu Magdeburg. Magdeburg. Abh. Naturwiss. Vereins Magdeburg. See B–P–H 33/1. HI 50091

Abhandlungen des Naturwissenschaftlichen Vereins für Schwaben e.V. in Augsburg. Augsburg. Abh. Naturwiss. Vereins Schwaben Augsburg. See B–P–H 33/2. HI 50092

Abhandlungen des Naturwissenschaftlichen Vereins e.V. Würzburg mit Fränkischem Museum für Naturkunde. Würzburg. Abh. Naturwiss. Vereins Würzburg. See B–P–H 33/3. HI 50093

Abhandlungen des Offenbacher Vereins für Naturkunde. Offenbach am Main. No. 1+, 1970+. Abh. Offenbacher Vereins Naturk. HI 60905

Abhandlungen von der ökonomischen Gesellschaft in Basel. Basel. Abh. Ökon. Ges. Basel. See B–P–H 33/6. HI 50094

Abhandlungen ökonomischen, technologischen, naturwissenschaftlichen und vermischten Inhalts. Erfurt. Abh. Ökon. Inhalts. See B–P–H 33/7. HI 50095

Abhandlungen der physikalisch-mathematischen Klasse der Königlichen Akademie der Wissenschaften zu Berlin = Abhandlungen der Königlichen Akademie der Wissenschaften in Berlin. Berlin. Abh. Königl. Akad. Wiss. Berlin. See B–P–H 30/2.

Abhandlungen der Physikalisch-Medicinischen Societät zu Erlangen. Frankfurt a. M. Abh. Phys.-Med. Soc. Erlangen. See B–P–H 33/13. HI 50096

Abhandlungen der physikalischen Klasse der Königlich-Preussischen Akademie der Wissenschaften = Abhandlungen der Königlichen Akademie der Wissenschaften in Berlin. Berlin. Abh. Königl. Akad. Wiss. Berlin. See B–P–H 30/2.

Abhandlungen der physikalischen Klasse der Königlichen Akademie der Wissenschaften zu Berlin = Abhandlungen der Königlichen Akademie der Wissenschaften in Berlin. Berlin. Abh. Königl. Akad. Wiss. Berlin. See B–P–H 30/2.

Abhandlungen der Preussischen Akademie der Wissenschaften. Mathematisch-naturwissenschaftliche Klasse. Berlin. Abh. Preuss. Akad. Wiss., Math.-Naturwiss. Kl. See B–P–H 33/15. HI 50098

Abhandlungen der Preussischen Akademie der Wissenschaften. Physikalisch-mathematische Klasse. Berlin. Abh. Preuss. Akad. Wiss., Phys.-Math. Kl. See B–P–H 33/17. HI 50099

Abhandlungen der Preussischen geologischen Landesanstalt = Abhandlungen der Königlich Preussischen geologischen Landesanstalt. Berlin. Abh. Königl. Preuss. Geol. Landesanst. See B–P–H 30/8.

Abhandlungen einer Privatgesellschaft in Böhmen, zur Aufnahme der Mathematik, der vaterländischen Geschichte, und der Naturgeschichte. Prague. Abh. Privatges. Böhmen. See B–P–H 33/19. HI 50100

Abhandlungen einer Privatgesellschaft von Naturforschern und Oekonomen in Oberdeutschland. Munich. Abh. Privatges. Naturf. Oberdeutschl. See B–P–H 33/20. HI 50101

Abhandlungen des Reichsamts für Bodenforschung Berlin. Berlin. Abh. Reichsamts Bodenf. Berlin. See B–P–H 33/21. HI 50102

Abhandlungen der Reichsstelle für Bodenforschung. Berlin. Abh. Reichsstelle Bodenf. See B–P–H 34/1. HI 50104

Abhandlungen der Sächsischen Akademie der Wissenschaften zu Leipzig. Mathematisch-naturwissenschaftliche Klasse. Berlin. Abh. Sächs. Akad. Wiss. Leipzig, Math.-Naturwiss. Kl. See B–P–H 34/3. HI 50105

Abhandlungen der Schlesischen Gesellschaft für vaterländische Cultur. Abtheilung für Naturwissenschaften un Medicin. Breslau. Abh. Schles. Ges. Vaterl. Cult., Abth. Naturwiss. See

B–P–H 34/6. HI 50106

Abhandlungen der schweizerischen paläontologischen Gesellschaft. Basel, Zurich. Vols. 1-62, 1874-1939. Abh. Schweiz. Paläontol. Ges. Superseded by: Schweizerische paläontologische Abhandlungen. 5-3812-2. HI 60906

Abhandlungen der Seeländischen Gesellschaft der Wissenschaften zu Vlissingen. Giessen, Germany. Abh. Seel. Ges. Wiss. Vlissingen. See B–P–H 34/10. HI 50107

Abhandlungen des Siebenbürgischen Vereins für Naturwissenschaften. Hermannstadt [=Sibiu, Rumania]. Abh. Siebenbürg. Vereins Naturwiss. See B–P–H 34/15. HI 50109

Abhandlungen des Thüringischen Botanischen Vereins Irmischia. Sondershausen, Germany. Abh. Thüring. Bot. Vereins Irmischia. See B–P–H 34/16. HI 50110

Abhandlungen die Verbesserung der Landwirthschaft betreffend. Prague. Abh. Verbess. Landw. See B–P–H 34/18. HI 50111

Abhandlungen des Vereins für forstwissenschaftliche Ausbildung. Tübingen. Abh. Vereins Forstwiss. Ausbild. See B–P–H 34/19. HI 50112

Abhandlungen des Vereins für naturwissenschaftliche Erforschung des Niederrheins. Krefeld, Germany. Abh. Vereins Naturwiss. Erforsch. Niederrheins. See B–P–H 34/20. HI 50113

Abhandlungen und Verhandlungen des naturwissenschaftlichen Vereins in Hamburg. Hamburg. N.s. vol. 1-21/22, 1937-78; Supplement 3-20, 1937-77. Abh. Verh. Naturwiss. Vereins Hamburg. Preceded by: Abhandlungen aus dem Gebiete der Naturwissenschaften herausgegeben von dem naturwissenschaftlichen Verein in Hamburg. Superseded by: Abhandlungen des naturwissenschaftlichen Vereins in Hamburg and Verhandlungen des naturwissenschaftlichen Vereins in Hamburg. 4-2936-3. HI 60907

Abhandlungen und Vorträge aus dem Gesammtgebiete der Naturwissenschaften. Berlin. Vol. 4, 189?-92. Abh. Vorträge Gesamtgeb. Naturwiss. Preceded by: Sammlung naturwissenschaftlicher Vorträge. HI 60908

Abhandlungen der Westfälischen Museum für Naturkunde. Munster. Vol. 44+, 1982+. Abh. Westfälischen Mus. Naturk. Preceded by: Abhandlungen aus dem Landesmuseum für Naturkunde zu Münster in Westfalen. HI 60909

Abhandlungen aus dem westfälischen Provinzial-Museum für Naturkunde. Münster. Abh. Westfäl. Prov.-Mus. Naturk. See B–P–H 35/1. HI 50115

Abhandlungen des zentralen geologischen Instituts. Paläontologische Abhandlungen. Berlin. 1965+. Abh. Zentr. Geol. Inst., Paläontol. Abh. HI 60910

Abhandlungen der zoologisch-botanischen Gesellschaft in Wien = Abhandlungen der kaiserlich-königlichen zoologisch-botanischen Gesellschaft in Wien. Vienna.

Abhandlungen der zoologisch-botanischen Gesellschaft in Österreich. Vienna. Vol. ?-23+, ?-1989+. Abh. Zool.-Bot. Ges. Österreich. Preceded by?: Abhandlungen der zoologisch-botanischen Gesellschaft in Wien. HI 74803

Abiks loodusevaatlejale = V pomoshch′ nablyudatelyam prirody. Tartu.

Abrégé de l'histoire et des mémoires. [Académie royale des sciences, Paris]. Paris. Abr. Hist. Mém. See B–P–H 35/7. HI 50116

Abrégé de physiologie végétale. Paris. Vol. 1+, 1977+. Abr. Physiol. Vég. HI 60911

Abrégé des transactions philosophiques de la société royale de Londres. Botanique. Paris. Abr. Trans. Philos. Soc. Roy. Londres, Bot. See B–P–H 35/9. HI 50117

Abrégé des transactions philosophiques de la société royale de Londres, ... Huitième partie. Matière médicale et pharmacie. Paris. Abr. Trans. Philos. Soc. Roy. Londres, Pt. 8, Matière Méd. See B–P–H 35/10. HI 50118

Abridged report, central sericultural research station = Report (Annual), central sericultural research station. Berhampore.

Abstract, see Abstracts

Abstracta botanica; operum ex instituto taxonomiae-oikologiae plantarum univ. sci. de l'Eötvös = Növenyrendszertani és novényföldrajzi tanszék. Eötvös loránd tudományegyetem, Budapest. Budapest.

ABStracts; forum for members. Arlington, MA, etc. Vol. 1+, 1973+. ABStracts. HI 60912

Abstracts in biocommerce. Slough. Vol. 1+, 1982+. Abstr. Biocommerce. HI 60913

Abstracts, botanical society of America = Abstracts of papers, botanical society of America. Lawrence, KS, etc.

Abstracts of bulgarian scientific literature. Agriculture and forestry, veterinary medicine. Sofia. Vols. 5-16, 1959-71. Abstr. Bulg. Sci. Lit., Agric. Forest. Veterin. Med. Preceded by: Referativnyi byulleten′ bolgarskoi nauchnoi literatury. Superseded by: Abstracts of bulgarian scientific literature. Series A, plant breeding and forest economy. HI 60914

Abstracts of bulgarian scientific literature. Biology. Sofia. Vol. 17+, 1978+. Abstr. Bulg. Sci. Lit., Biol. Preceded by: Abstracts of bulgarian scientific literature. Biology and biochemistry. HI 60915

Abstracts of bulgarian scientific literature. Biology and biochemistry. Sofia. Vols. 1-16, 1963-78. Abstr. Bulg. Sci. Lit., Biol. Biochem. Preceded by: Abstracts of bulgarian scientific literature. Biology and medicine. Superseded by: Abstracts of bulgarian scientific literature. Biology. HI 60916

Abstracts of bulgarian scientific literature. Biology and medicine. Sofia. Vols. 1-5, 1958-62. Abstr. Bulg. Sci. Lit., Biol. Med. Superseded by: Abstracts of bulgarian scientific literature. Biology and biochemistry. HI 60917

Abstracts of bulgarian scientific literature. Series A, plant breeding and forest economy. Sofia. Vol. 17+, 1972+. Abstr. Bulg. Sci. Lit., A. Pl. Breed. Forest. Econ. Preceded by: Abstracts of bulgarian scientific literature. Agriculture and forestry, veterinary medicine. HI 60918

Abstracts on cassava (Manihot esculenta Crantz). Cali. 1975+. Abstr. Cassava. HI 60919

Abstracts, conference on great lakes research. 18th+, 1975+. Abstr., Conf. Great Lakes Res. Preceded by: Proceedings, conference on great lakes research. HI 60920

Abstracts of dissertations approved for the Ph.D., M.Sc., and M. Litt. degrees in the university of Cambridge. Cambridge. 1925-57. Abstr. Diss., Univ. Cambridge. Superseded by: Titles of dissertations approved for the Ph.D., M.Sc., and M. Litt. degrees in the university of Cambridge. HI 60921

Abstracts of dissertations for the degree of doctor of philosophy, Oxford university. Oxford. 1928+. Abstr. Diss., Univ. Oxford. HI 60922

Abstracts on field beans (Phaseolus vulgaris L.). Cali. Vol. 1+, 1976+. Abstr. Field Beans. HI 60923

Abstracts of Japanese literature in forest genetics and related fields. Part A. Maguro, Tokyo. Vol. 1+, 1970+. Abstr. Jap. Lit. Forest Genet., A. HI 60924

Abstracts of Japanese literature in forest genetics and related fields. Part B. Maguro, Tokyo. Vol. 1+, 1972+. Abstr. Jap. Lit. Forest Genet., B. HI 60925

Abstracts of literature of horticultural crops, special crops, and feed crops. [Engei, kogii, shihiryo sakumotsu bunken shorokushu]. Tokyo. Abstr. Lit. Hort. Crops. See B–P–H 35/15. HI 50119

Abstracts of microbiology and hygiene. 1 Abteilung, abstracts = Zentralblatt für Bakteriologie, Mikrobiologie und Hygiene. 1 Abteilung. Referate. Medizinische Mikrobiologie, Parasitologie, Hygiene, präventive Medizin. Stuttgart.

Abstracts of mycology. Philadelphia, PA. Vol. 1+, 1967 +. Abstr. Mycol. HI 60926

Abstracts of new publications, Iowa agricultural experiment station = Iowa agricultural experiment station. Abstracts of new publications. Ames, IA. Iowa Agric. Exp. Sta. Abstr. New Publ. See B–P–H 436/32.

Abstracts of new publications, Oklahoma agricultural experiment station = Oklahoma agricultural experiment station. Abstracts of new publications. Stillwater, OK. Oklahoma Agric. Exp. Sta. Abstr. New Publ. See B–P–H 686/35.

Abstracts of papers, botanical society of America. Lawrence, KS, etc. 1977+. Abstr. Bot. Soc. Amer. Previously contained in: American journal of botany. HI 60927

Abstract of the papers communicated to the royal society of London. London. Abstr. Pap. Roy. Soc. London. See B–P–H 35/22. HI 50121

Abstracts of the papers printed in the philosophical transactions of the royal society of London, ... London. Abstr. Pap. Philos. Trans. Roy. Soc. London. See B–P–H 35/21. HI 50120

Abstracts of papers read before the Brighton and Hove natural history and philosophical society. Hove. 1898-1938. Abstr. Pap. Brighton Hove Nat. Hist. Soc. Preceded by: Abstracts of papers read before the Brighton and Sussex natural history and philosophical society. Superseded by: Report, Brighton and Hove natural history and philosophical society. 1-771-3. HI 60928

Abstracts of papers read before the Brighton and Sussex natural history and philosophical society. Hove. 1888-97. Abstr. Pap. Brighton Sussex Nat. Hist. Soc. Preceded by: Report of the Brighton and Sussex natural history and philosophical society. Superseded by: Abstracts of papers, Brighton and Hove natural history and philosophical society. 1-771-3. HI 60929

Abstracts of papers presented at annual meetings, american association of stratigraphic palynologists. Dallas, TX. 1968+. Abstr. Pap., Amer. Assoc. Stratigr. Palynologists. HI 60930

Abstracts of papers presented at the annual meeting of the Kansas academy of science. Lawrence, Hays, KS. Vol. 1+, 1982+. Abstr. Pap., Kansas Acad. Sci. HI 60931

Abstract of the proceedings of the association of american geologists and naturalists. Boston, MA. Vols. 4-6, 1843-45. Abstr. Proc. Assoc. Amer. Geol. Naturalists. Preceded by: Report of the association of american geologists and naturalists. Superseded by: Proceedings of the american association for the advancement of science. 1-526-1. HI 60932

Abstract of proceedings of the linnean society of New South Wales. Sydney. Abstr. Proc. Linn. Soc. New South Wales. See B–P–H 35/24. HI 50122

Abstract of proceedings of the royal society of New South Wales. Sydney. Abstr. Proc. Roy. Soc. New South Wales. See B–P–H 36/3. HI 50123

Abstracts of the proceedings of the south London entomological and natural history society. London. 1885-96. Abstr. Proc. S. London Entomol. Nat. Hist. Soc. Preceded by: Report of the south London entomological and natural history society. Superseded by: Proceedings of the south London entomological and natural history society. HI 60933

Abstracts of published papers and list of translations, C S I R O, Australia. East Melbourne, Vic. Vols. 1-4, 1952-56. Abstr. Published Pap. List Transl., C. S. I. R. O. Superseded by: C S I R O science index. HI 60934

Abstracts of recent publications, Maine agricultural experiment station = Maine agricultural experiment station. Abstracts of recent publications. Orono, ME. Maine Agric. Exp. Sta. Abstr. Recent Publ. See B–P–H 546/5.

Abstracts of reports of the british pteridological society. Kendal, England. Abstr. Rep. Brit. Pteridol. Soc. See B–P–H 36/5. HI 50124

Abstracts of reports, national weed committee, Canada, eastern section = Proceedings, national weed committee, Canada, eastern section.

Abstracts of romanian scientific and technical literature. Bucharest. Vol. 7+, 1971+. Abstr. Romanian Sci. Techn. Lit. Preceded by: Abstracts of rumanian technical literature. HI 60935

Abstracts of rumanian technical literature. Bucharest. Vols. 1-6, 1965-70. Abstr. Rumanian Techn. Lit. Superseded by: Abstracts of romanian scientific and technical literature. HI 60936

Abstracts on rural development in the tropics. Amsterdam. Vol. 0(0),1+, 1985; vol. 1+, 1986+. Abstr. Rural Developm. Tropics. HI 60937

Abstracts of scientific and technical papers published in Egypt. [Forms part of: Documentation bulletin of the national research centre. Part 2.] Dokki-Cairo. Vols. 1-4, 1955-59. Abstr. Sci. Techn. Pap. Published Egypt. Superseded by: Abstracts of scientific and technical papers published in U A R. HI 60938

Abstracts of scientific and technical papers published in U A R [United Arab Republic]. [Forms part of: Documentation bulletin of the national research centre. Part 2.] Dokki-Cairo. Vol. 5+, 1959+. Abstr. Sci. Techn. Pap. Published UAR. Preceded by: Abstracts of scientific and technical papers published in Egypt. HI 60939

Abstracts of theses ... for the doctor's degree, Cornell university = Cornell university abstracts of theses ... for the doctor's degree. Ithaca, NY. Cornell Univ. Abstr. Theses. See B–P–H 332/4.

Abstracts on tropical agriculture. Amsterdam. Vol. 1+, 1975+. Abstr. Trop. Agric. Preceded by: Tropical abstracts. HI 60940

Abstracts of Uppsala dissertations from the faculty of science. [Forms part of: Acta universitatis upsaliensis.] Stockholm. Vols. 1-?, 1961-? Abstr. Uppsala Diss. Fac. Sci. Superseded by: Comprehensive summaries of Uppsala dissertations from the faculty of science. HI 60941

Abstracts of Uppsala dissertations from the faculty of pharmacy. [Forms part of: Acta universitatis upsalienses.] Stockholm. 19??-75+. Abstr. Uppsala Diss. Fac. Pharm. HI 60942

Abstracts of Uppsala dissertations in science = Abstracts of Uppsala dissertations from the faculty of science. Stockholm.

Abstracts, weed science society of America. Champaign, IL. 1968+. Abstr. Weed Sci. Soc. Amer. Preceded by: Abstracts, weed society of America. HI 60943

Abstracts, weed society of America. Champaign, IL. 1956-67. Abstr. Weed Soc. Amer. Superseded by: Abstracts, weed science society of America. HI 60944

ABT, abstracts of bioanalytic technology. San Francisco, CA. = A B T; abstracts of bioanalytic technology. San Francisco, CA.

Abteilung für Zoologie und Botanik am Landesmuseum Joanneum, Graz. Mitteilungsblatt. Graz. Abt. Zool. Bot. Landesmus. Joanneum Graz Mitteilungsbl. See B–P–H 36/8. HI 50125

Abteilung für Zoologie und Botanik am Landesmuseum Joanneum, Graz. Mitteilungsheft. Graz. Abt. Zool. Bot. Landesmus. Joanneum Graz Mitteilungsh. See B–P–H 36/9. HI 50127

Academia romăna. Memoriile secţiunei ştiinţifice. Bucharest. Acad. Romăna, Mem. Secţ. Şti. See B–P–H 36/25. HI 50133

Academia scientiarum fennica vuosikirja = Vuosikirja, suomalainen tiedeakatemia. Helsinki.

Academiae caesareo-leopoldinae carolinae naturae curiosorum ephemerides sive observationum medico-physicarum ... Nuremberg. Acad. Caes.-Leop. Carol. Nat. Cur. Ephem. See B–P–H 36/12. HI 50128

Academiae caesareo-leopoldinae naturae curiosorum ephemerides sive observationum medico-physicarum ... Frankfurt a. M. & Leipzig. Acad. Caes-Leop. Nat. Cur. Ephem. See B–P–H 36/13. HI 50129

Academic review. [Hsüeh shu chi-k'an]. Taipei, Taiwan. Acad. Rev. See B–P–H 36/23. HI 50132

Académie royale du Gard. Nîmes. Acad. Roy. Gard. See B–P–H 37/2. HI 50134

Académie royale des sciences, belles-letteres et arts de Bordeaux. Séance publique. Bordeaux. Acad. Roy. Sci. Bordeaux Séance Publique. See B–P–H 37/3. HI 50135

Académie des sciences, arts et belles-lettres de Dijon. Séance publique. Dijon. Acad. Sci. Dijon Séance Publique. See B–P–H 37/8. HI 50139

Academie des sciences, belles-lettres et arts, de Besançon. Séance publique. Besançon. Acad Sci. Besançon Séance Publique. See B–P–H 37/5. HI 50136

Academy of natural sciences of Philadelphia monographs. Philadelphia, PA. Acad. Nat. Sci. Philadelphia Monogr. See B–P–H 36/21. HI 50130

Academy news, academy of natural sciences. Philadelphia, PA. 1978+. Acad. News, Acad. Nat. Sci. Philadelphia. HI 60945

Academy news letter of the California academy of sciences. San Francisco, CA. Acad. News Lett. Calif. Acad. Sci. See B–P–H 36/22. HI 50131

Acadian naturalist; bulletin of the natural history society of New Brunswick. Fredericton, New Brunswick. Acadian Naturalist. See B–P–H 37/12. HI 50140

Acadian scientist. Wolfville, Nova Scotia. Acadian Sci. See B–P–H 37/13. HI 50141

Acanthus; an international newsletter to encourage interest in the acanthaceae. San Francisco, CA. No. 1+, 1987+. Acanthus. HI 60946

Accession bulletin, international coffee organization. Leiden. Vol. 1+, 1972+. Accession Bull., Int. Coffee Organ. HI 60947

Acción agraria. Cuzco, Peru. Acción Agrar. (Cuzco). See B–P–H 37/14. HI 50144

Acción agraria. Mexico City. Acción Agrar. (Mexico City). See B–P–H 37/15. HI 50145

Acclimatation; journal des éleveurs. Paris. Acclimatation. See B–P–H 37/18. HI 50147

Acclimatation des animaux et des plantes. Rennes. Acclim. Anim. Pl. See B–P–H 37/17. HI 50146

Acid rain abstracts annual. New York. 1988+, 1989+. Acid Rain Abstr. Annual. HI 60948

Acker- und Gartenbauzeitung. New York, NY. Acker-Gartenbauzeitung. See B–P–H 37/19. HI 50148

Acqua ed aria, ecologia. Milan. Vol. 4, 1974. Acqua Aria, Ecol. Preceded by: Ecologia. Milan. Superseded by: Ecologia, acqua, aria, suolo. HI 60949

Acta academiae agriculturae ac technicae olstenensis. Agricultura. Olsztyn. Vol. 41+, 1985+. Acta Acad. Agric. Techn. Olsten., Agric. Preceded by: Zeszyty naukowe akademii rolniczo-technicznej w Olsztynie. Rolnictwo. HI 60950

Acta academiae electoralis moguntinae scientiarum utilium quae Erfordiae est. Erfurt & Gotha. Acta Acad. Elect. Mogunt. Sci. Util. Erfordiae. See B–P–H 39/3. HI 50149

Acta academiae electoralis moguntinae scientiarum utilium quae Erfurti est. Erfurt. Acta Acad. Elect. Mogunt. Sic. Util. Erfurti. See B–P–H 39/4. HI 50151

Acta academiae paedagogicae agriensis-nova = Egri Ho Si Minh tanárképzö föiskola tudományos közleményei. Eger.

Acta academiae paedagogicae agriensis = Egri pedagógiai föiskola évkönyve. Eger, Hungary. Egri Pedagóg. Föisk. Évk. See B–P–H 355/16.

Acta academiae paedagogicae agriensis = Egri tanárképzö föiskola tudományos közleményei. Eger, Hungary. Egri Tanárképzö Föisk. Tud. Közlem. See B–P–H 355/17.

Acta academiae paedagogicae szegediensis = A Szegedi pedagógiai föiskola évkönyve. Szeged, Hungary. Szegedi Pedagóg. Föisk. Évk. See B–P–H 863/11.

Acta academiae scientiarum čechoslovenicae basis brunensis = Práce brněnske základny československe akademie věd. Brno.

Acta academiae scientiarum imperialis petropolitanae. St. Petersburg. Acta Acad. Sci. Imp. Petrop. See B–P–H 39/11. HI 50152

Acta academiae scientiarum naturalium moravo-silesiacae = Práce moravsko-slezské akademie věd přirodních. Brno, Czechoslovakia. Práce Morav.-Slez. Akad. Věd Přír. See B–P–H 718/6.

Acta academiae szegediensis = A Szegedi tanárképsö föiskola tudományos közleményei. Szeged, Hungary. Szegedi Tanárképzö Föisk. Tud. Közlem. See B–P–H 863/12.

Acta adriatica. Split, Yugoslavia. Acta Adriat. See B–P–H 39/14. HI 50153

Acta agraria. Mexico City. Acta Agrar. See B–P–H 39/15. HI 50154

Acta agraria et sylvestria. Series agraria. Cracow. Vol. 6+, 1966+. Acta Agrar. Silvestria, Ser. Agrar. Preceded by: Acta agraria et sylvestria. Seria rolnicza. HI 60951

Acta agraria et sylvestria. Seria leśna. Cracow. Vols. 1-5, 1961-65. Acta Agrar. Silvestria, Ser. Leśn. Superseded by: Acta agraria et sylvestria. Series sylvestris. HI 60952

Acta agraria et sylvestria. Seria rolnicza. Cracow. Vols. 1-5, 1961-65. Acta Agrar. Silvestria, Ser. Roln. Superseded by: Acta agraria et sylvestria. Series agraria. HI 60953

Acta agraria et sylvestria. Series sylvestris. Cracow. Vol. 6+, 1966+. Acta Agrar. Silvestria, Ser. Sylvest. Preceded by: Acta agraria et sylvestria. Seria leśna. HI 60954

Acta agriculturae scandinavica. Stockholm. Acta Agric. Scand. See B–P–H 39/17. HI 50155

Acta agriculturae sinica. [Nung yeh hsüeh pao]. Peking. Acta Agric. Sin. See B–P–H 39/18. HI 50156

Acta agriculturae suecana. Stockholm. Acta Agric. Suec. See B–P–H 39/19. HI 50157

Acta agriculturae universitatis jilinensis. [Jilin nongye daxue xuebao.] Changchun. 1979+. Acta Agric. Univ. Jilin. HI 60955

Acta agrobotanica. Warsaw. Acta Agrobot. See B–P–H 39/20. HI 50158

Acta agrobotanica hungarica. Budapest. Acta Agrobot. Hung. See B–P–H 39/21. HI 50159

Acta agronómica. Palmira, Colombia. Acta Agron. See B–P–H 39/22. HI 50161

Acta agronomica academiae scientiarum hungaricae. Budapest. Acta Agron. Acad. Sci. Hung. See B–P–H 39/24. HI 50162

Acta agronomica sinica. [Zuowu xuebao.] Beijing. Vol. 1+, 1962?+. Acra Agron. Sin. HI 60956

Acta albertina ratisbonensia. Regensburger Naturwissenschaften. Regensburg. Acta Albertina Ratisb. See B–P–H 39/28. HI 50163

Acta allergologica. Copenhagen. Acta Allergol. See B–P–H 39/30. HI 50164

Acta amazonica. Manaus. Vol. 1+, 1971+. Acta Amazon. Preceded by: Boletim do instituto nacional de pesquisas da Amazônia. Serie avulsa; Boletim do instituto nacional de pesquisas da Amazônia. Série botanica; and Boletim do instituto nacional de pesquisas da Amazônia. Série pesquisas florestais. HI 60957

Acta americana. Washington, DC. Acta Amer. See B–P–H 40/2. HI 50165

Acta anthropologica. Mexico City. Acta Anthropol. (Mexico City). See B–P–H 40/5. HI 50166

Acta archaeologica. Copenhagen. Vol. 1+, 1930+. Acta Archaeol. HI 60958

Acta arctica. Copenhagen. Acta Arctica. See B–P–H 40/7. HI 50167

Acta biochimica et biophysica. Budapest. Acta Biochim. Biophys. See B–P–H 40/9. HI 50169

Acta biochimica et biophysica. Supplementum. Budapest. Vol. 1+, 1967+. Acta Biochim. Biophys. Suppl. HI 75175

Acta biochemica et biophysica sinica. [Sheng wu hua yu sheng wu li hsüeh pao]. Peking. Acta Biochem. Biophys. Sin. See B–P–H 40/8. HI 50168

Acta biochimica polonica. Warsaw. Acta Biochim. Polon. See B–P–H 40/11. HI 50170

Acta biologiae experimentalis sinica. [Shiyan shengwu xuebao.] Peking (Beijing). 1953+ [suspended from 1965-79]. Acta Biol. Exp. Sin. Preceded by: Chinese journal of experimental biology. HI 60959

Acta biologiae et medicinae experimentalis. Pristina. Vol. 1+, 1976+. Acta Biol. Med. Exp. HI 60960

Acta biologica. Istanbul = Biyoloji dergisi. Istanbul.

Acta biologica. Katowice. Vols. 1-?, 1975-84? Acta Biol. (Katowice). Superseded by: Acta biologica silesiana. HI 60961

Acta biologica (1928-37). Szeged, Hungary. Acta Biol. (Szeged 1928-37). See B–P–H 40/13. HI 50171

Acta biologica (1955+). Szeged, Hungary. Acta Biol. (Szeged 1955+). See B–P–H 40/14. HI 50172

Acta biologica academiae scientiarum hungaricae. Budapest. Vol. 1-33, 1950-82; Supplement 1-6, 1957-64. Acta Biol. Acad. Sci. Hung. Preceded by: Hungarica acta biologica. Superseded by: Acta biologica hungarica. HI 60962

Acta biologica Colombiana. Bogota. Vol. 1+, 1982+. Acta Biol. Colomb. HI 60963

Acta biologica cracoviensia. Series botanica. Cracow. Acta Biol. Cracov., Ser. Bot. See B–P–H 40/19. HI 50173

Acta biologica debrecina. Budapest. Acta Biol. Debrecina. See B–P–H 40/21. HI 50174

Acta biologica hungarica. Budapest. Vol. 34+, 1983+. Acta Biol. Hung. Preceded by: Acta biologica academiae scientiarum hungaricae. HI 60964

Acta biologica jugoslavica. Serija A, zemljište i biljka = Zemljište i biljka. Belgrade.

Acta biologica jugoslavica. Serija B, mikrobiologija = Mikrobiologija. Belgrade.

Acta biologica jugoslavica. Serija D, ekologija = Ekologija. Belgrade.

Acta biologica jugoslavica. Serija F, genetika = Genetika. Belgrade.

Acta biologica jugoslavica. Serija G, biosistematika = Biosistematika. Belgrade.

Acta biologica latvica. Riga, Latvia [Latvian S S R]. Acta Biol. Latv. See B–P–H 40/23. HI 50175

Acta biologica leopoldensia. Ca.1989-? Acta Biol. Leopoldensia. HI 63243

Acta biologica et medica. Gdansk. Vols. 1-17, 1957-74. Acta Biol. Med. Superseded by: Acta biologica, societas scientiarum gedanensis. HI 60965

Acta biologica et medica germanica. Berlin, Zurich. Vols. 1-41, 1958-82. Acta Biol. Med. German. Superseded by: Biomedica biochimica acta. HI 60966

Acta biológica paranaense. Curitiba. Vol. 1+, 1972+. Acta Biol. Paran. Preceded by: Boletim da universidade (federal) do Paraná. Botânica. HI 60967

Acta biologica plateau sinica. [Gaoyuan shengwuxue jikan.] Xining. No. 1+, 1982+. Acta Biol. Plateau Sin. HI 60968

Acta biologica silesiana. Katowice. 1985+. Acta Biol. Siles. Preceded by: Acta biologica. Katowice. HI 60969

Acta biologica, societas scientiarum gedanensis. Gdansk. Vols. 4-5, 1979. Acta Biol. Soc. Sci. Gedan. Preceded by: Acta biologica et medica [Societas scientiarum gedansis]. HI 60970

Acta biologica venezuelica. Caracas. Acta Biol. Venez. See B–P–H 41/2. HI 50176

Acta biotechnologica. Berlin. Vol. 0(0), 1980; vol. 1+, 1981+. Acta Biotechnol. HI 65410

Acta biotheoretica. Leiden. Acta Biotheor. See B–P–H 41/3. HI 50178

Acta bolyaiana. Facultas scientiarum naturalium. Cluj, Rumania. Acta Bolyaiana. See B–P–H 41/4. HI 50179

Acta borealia. A. Scientia. Tromso. Vols. 1-30, 1951-75. Acta Boreal., A. Preceded by: Tromso museum aarshefter. HI 60971

Acta botanica. Szeged, Hungary. Acta Bot. (Szeged). See B–P–H 41/18. HI 50180

Acta botanica academiae scientiarum hungaricae. Budapest. Vols. 1-28, 1954-82. Acta Bot. Acad. Sci. Hung. Superseded by: Acta botanica hungarica. HI 60972

Acta botanica austro sinica. [Zhongguo kexueyuan huanan zhiwu yanjiusuo jikan.] Guangzhou. No. 1+, 1983+. Acta Bot. Austro Sin. HI 60973

Acta botanica barcinonensia. Barcelona. Vol. 30+, 1978+. Acta Bot. Barcinon. Preceded by: Acta geobotanica barcinonensia and Acta phytotaxonomica barcinonensia. HI 60974

Acta botanica bohemica. Prague. Acta Bot. Bohem. See B–P–H 41/12. HI 50182

Acta botanica boreali-occidentalis sinica. Yangling. Vol. 1+, 1981+. Acta Bot. Boreal.-Occid. Sin. HI 60975

Acta botanica brasilica. Porto Alegre. Vol. 1+, 1987+. Acta Bot. Brasil. HI 60976

Acta botanica croatica. Zagreb, Yugoslavia. Acta Bot. Croat. See B–P–H 41/13. HI 50183

Acta botanica cubana. Havana. No. 1+, 1980+. Acta Bot. Cub. HI 60977

Acta botanica fennica. Helsinki. Acta Bot. Fenn. See B–P–H 41/14. HI 50185

Acta botanica horti bucurestiensis unacum delectu seminum, quae hortus botanicus pro mutua commutatione offert = Lucrările grădinii botanice din Bucureşti. Bucharest. Lucr. Grăd. Bot. Bucureşti. See B–P–H 536/16.

Acta botanica hungarica. Budapest. Vol. 29+, 1983+. Acta Bot. Hung. Preceded by: Acta botanica academiae scientiarum hungarica. HI 60978

Acta botanica indica. Meerut. Vol. 1+, 1973+. Acta Bot. Indica. HI 60979

Acta botanica instituti botanici universitatis zagrebiensis. Zagreb, Yugoslavia. Acta Bot. Inst. Bot. Univ. Zagreb. See B–P–H 41/17. HI 50186

Acta botanica islandica. Reykjavik. No. 1+, 1972+. Acta Bot. Islandica. Preceded by: Flóra, tímarit um íslenzka grasafraedi. HI 60980

Acta botanica malacitana. Malaga. Vol. 1+, 1975+. Acta Bot. Malac. HI 60981

Acta botanica mexicana. Patzcuaro. ?-1988+. Acta Bot. Mex. HI 60982

Acta botanica neerlandica. Amsterdam. Vol. 1+, 1952+. Acta Bot. Neerl. Preceded by: Recueil des travaux botaniques néerlandais. HI 60983

Acta botanica sinica. [Zhiwu xuebao.] Peking (Beijing). Vol. 1+, 1951+ [publication suspended from vol. 14(2)-15, 1960-1973]. Acta Bot. Sin. HI 60984

Acta botanica sinica; Translated from Chinese. [Translation of: Acta botanica sinica. Zhiwu xuebao.] Vol. 15+, 1973+. Acta Bot. Sin. (Transl.) HI 60985

Acta botanica slovaca. Series A, taxonomica, geobotanica. Bratislava. Vol. 3+, 1978+. Acta Bot. Slov., A. Preceded by: Acta instituti botanici, academiae scientiarum slovacae. HI 60986

Acta botanica slovaca. Series A, taxonomica, geobotanica. Supplementum. Bratislava. Vol. 1+, 1984+. Acta Bot. Slov., A. Suppl. HI 60987

Acta botanica slovaca. Series B, physiologica, pathophysiologica. Bratislava. Vol. 2+, 1978+. Acta Bot. Slov., B. Preceded by: Acta instituti botanici, academiae scientiarum slocavae. Series B, physiologica, phytophysiologica. HI 60988

Acta botanica taiwanica. [Chih wu hsüeh pao]. Taipei, Taiwan. Acta Bot. Taiwan. See B–P–H 41/24. HI 50187

Acta botánica venezuelica. Caracas. Acta Bot. Venez. See B–P–H 41/25. HI 50188

Acta botanica Yunnanica. [Yunnan zhiwu yanjiu.] Kunming. Vol. 1+, 1979+. Acta Bot. Yunnan.

HI 60989

Acta brevia neerlandica de physiologia, pharmacologia, microbiologia. Amsterdam. Acta Brevia Neerl. Physiol. See B–P–H 41/28. HI 50189

Acta brevia sinensia. Chungking, China. Acta Brevia Sin. See B–P–H 41/29. HI 50190

Acta bryologica asiatica. Taipei. 1990+. Acta Bryol. Asiat. HI 74804

Acta carsologica. Ljubljana. Vol. 1+, 1955+. Acta Carsol. HI 60990

Acta chemica fennica. Helsinki. Acta Chem. Fenn. See B–P–H 42/1. HI 50191

Acta chemica scandinavica. Copenhagen. Acta Chem. Scand. See B–P–H 42/2. HI 50192

Acta cientifica potosina. San Luis Potosi. (Mexico.) Vols. 1-?, 1957-66. Acta Ci. Potos. HI 60991

Acta científica venezolana. Caracas. Acta Ci. Venez. See B–P–H 42/4. HI 50193

Acta et commentationes imperialis universitatis jurievensis (olim dorpatensis) = Uchenyya Zapiski Imperatorskago Yur'evskago Universiteta. Yur'ev.

Acta et commentationes universitatis tartuensis (dorpatensis). A. Mathematica, physica, medica. Tartu, Estonia [Estonian S S R]. Acta Commentat. Univ. Tartuensis, A. Math. See B–P–H 42/12. HI 50194

Acta dendrobiologica. Bratislava. Vol. 1+, 1978/79+, 1979+. Acta Dendrol. Preceded by: Zborník prác arboréta mlyňany. HI 60992

Acta dendrologica Čechoslovaca = Dendrologicky sborník. Opava, Czechoslovakia. Dendrol. Sborn. See B–P–H 340/21.

Acta dermatologica. [Hifuka Kiyo]. Kyoto, Japan. Acta Dermatol. See B–P–H 42/16. HI 50195

Acta dermato-venerologica. Stockholm. Acta Dermato-Venereol. See B–P–H 42/17. HI 50196

Acta ecologica. Bratislava. Vol. 6+, no.15+, 1977+. Acta Ecol. Preceded by: Acta geobiologica. HI 60993

Acta ecologica iranica. Tehran. No. 1+, 1976+. Acta Ecol. Iran. HI 60994

Acta ecologica sinica. [Shangtai xuebao.] Vol. ?-2+, 19??-82+. Acta Ecol. Sin. HI 60995

Acta embryologiae et morphologiae experimentalis. Palermo. Acta Embryol. Morphol. Exp. See B–P–H 42/21. HI 50197

Acta eruditorum. Leipzig. Acta Erud. See B–P–H 42/23. HI 50198

Acta eruditorum quae Lipsiae publicantur supplementa. Leipzig. Acta Erud. Suppl. See B–P–H 42/24. HI 50199

Acta facultatis forestalis Zvolen = Sborník vědeckých prác, Zvolen. Bratislava.

Acta facultatis rerum naturalium universitatis comenianae. Botanica. Bratislava. No. 1+, 1956+. Acta Fac. Rerum Nat. Univ. Comenianae, Bot. HI 60996

Acta facultatis rerum naturalium universitatis comenianae. Formatio et protectio naturae. Bratislava. Vol. 1+, 1976+. Acta Fac. Rerum Nat. Univ. Comen., Format. Protect. Nat. HI 60997

Acta facultatis rerum naturalium universitatis comenianae. Genetica. Bratislava. Vol. 1+, 1966+. Acta Fac. Rerum Nat. Univ. Comen., Genet. HI 60998

Acta facultatis rerum naturalium universitatis comenianae. Microbiologica. Bratislava. Vol. 1+, 1971+. Acta Fac. Rerum Nat. Univ. Comen., Microbiol. HI 60999

Acta facultatis rerum naturalium universitatis comenianae. Physiologia plantarum. Bratislava. Vol. 1+, 1970+. Acta Fac. Rerum Nat. Univ. Comen., Physiol. Pl. HI 61000

Acta pro fauna et flora universali. Ser. 2: Botanica. Bucharest. Acta Fauna Fl. Universali, Ser. 2, Bot. See B–P–H 42/28. HI 50200

Acta finska vetenskaps-societeten. Helsinki. Acta Finska Vetensk.-Soc. See B–P–H 42/29. HI 50201

Acta florae rossicae = Věstnik Russkoi flory. Yuryev [= Tartu]

Acta florae sueciae. Stockholm. Vol. 1, 1921. Acta Fl. Sueciae. HI 61001

Acta forestalia fennica. Helsinki. Acta Forest. Fenn. See B–P–H 43/1. HI 50202

Acta fytotechnica. Bratislava. Vol. 13+, 1966+. Acta Phytotechn. Preceded by: Sborník vysokej školy pol'nohospodárskej v Nitre. Seria A. HI 61002

Acta genetica sinica. [Yichuan xuebao.] Peking (Beijing). Vol. 1+, 1974+. Acta Genet. Sin. HI 61003

Acta genetica et statistica medica. Basel. Acta Genet. Statist. Med. See B–P–H 43/2. HI 50203

Acta geobiologica. Bratislava. Vols. 1-6, nos. 1-14, 1972-77. Acta Geobiol. Superseded by: Acta ecologica. Bratislava. HI 61004

Acta geobotanica barcinonensia. Barcelona. Vols. 1-8, 1964-73. Acta Geobot. Barcinon. Superseded by: Acta botanica barcinonensia. HI 61005

Acta geobotanica hungarica. Debrecen, Hungary. Acta Geobot. Hung. See B–P–H 43/3. HI 50204

Acta geographica. Helsinki. Acta Geogr. (Helsinki). See B–P–H 43/4. HI 50206

Acta geographica sinica. [Ti li hsüeh pao]. Peiping [=Peking]. Acta Geogr. Sin. See B–P–H 43/5. HI 50208

Acta geologica academiae scientiarum hungaricae. Budapest. Acta Geol. Acad. Sci. Hung. See B–P–H 43/6. HI 50209

Acta helvetica, physico-mathematico-anatomico-botanico-medica. Basel. Acta Helv. Phys.-Math. See B–P–H 43/10. HI 50210

Acta histochemica et cytochemica. Kyoto. Vol. 1+, 1968+. Acta Histochem. Cytochem. HI 61006

Acta historica leopoldina. Leipzig. Acta Hist. Leop. See B–P–H 43/12. HI 50211

Acta historica rerum naturalium nec non technicarum. Prague = Sborník pro dějinyk přírodních věd a techniky. Prague. Sborn. Dějiny Přír. Věd Techn. See B–P–H 814/2.

Acta historica scientiarum naturalium et medicinalium. Copenhagen. Vol. 1+, 1942+. Acta Hist. Sci. Nat. Med. HI 61007

Acta horti bergiani. Stockholm. Vols. 1-20, [1890]1891-1966. Acta Horti Berg. 1-54-2. HI 61008

Acta horti botanici pragensis. Prague. Acta Horti Bot. Prag. See B–P–H 43/24. HI 50214

Acta horti botanici tadshikistanici. Moscow. Acta Horti Bot. Tadshik. See B–P–H 43/25. HI 50215

Acta horti botanici universitatis imperialis jurjevensis = Trudy botanicheskogo Sada Imperatorskago Yur'evskago Universiteta. Yur'ev.

Acta horti botanici universitatis latviensis. Riga. Vols. 1-13, 1926-40. Acta Horti Bot. Univ. Latv. Superseded by: Acta horti botanici universitatis. Riga. 4-3678-3. HI 61009

Acta horti botanici universitatis. Riga, Latvian S S R. Vol. 14, 1944. Acta Horti Bot. Univ. Preceded by: Acta horti botanici universitatis latviensis. Superseded by: Trudy botanicheskogo sada Latviiskogo gosudarstvennogo universiteta Petra Stuchki. HI 61010

Acta horti gothoburgensis. Göteborg. Vols. 1-28, 1924-67. Acta Horti Gothob. 2-1750-3. HI 61011

Acta horti petropolitani = Trudy glavnago botanicheskago sada. Petrograd, Moscow.

Acta horti pisani. Pisa. Vol. 1+, 1939/40+, 1940+. Acta Horti Pisani. HI 61012

Acta horticulturae. The Hague. Acta Hort. See B–P–H 43/14. HI 50212

Acta horticulturae sinica. [Yuanyi xuebao.] Peking (Beijing). Vol. 1-5(2), 1962-66. Acta Hort. Sin. HI 61013

Acta horticulturalia sinica. Peking. Acta Hort. Sin. See B–P–H 43/18. HI 50213

Acta Humboldtiana. Series geologica et palaeontologica. Wiesbaden. Vol. 1, 1961. Acta Humboldt., Ser. Geol. Palaeontol. Superseded by: Acta Humboldtiana. Series geologica, palaeontologica et biologica. HI 50216

Acta Humboldtiana. Series geologica, palaeontologica et biologica. Wiesbaden. Vol. 2, 1972. Acta Humboldt., Ser. Geol. Palaeontol. Biol. Preceded by: Acta Humboldtiana. Series geologica et palaeontologica Superseded by: Humboldtiana. HI 61014

Acta hydrobiologica. Cracow, Warsaw. Vol. 1+, 1959+; Supplement vol. 7+, 1963+. Acta Hydrobiol. Preceded by: Biuletyn, zakładu biologii stawów, polska akademia nauk. HI 61015

Acta hydrobiologica Lituanica. Vilnius. Vol. 1+, 1980+. Acta Hydrobiol. Lituanica. HI 61016

Acta hydrobiologica sinica. [Shui sheng sheng wu hsüeh chi k'an]. Peking. Acta Hydrobiol. Sin. (Peking). See B–P–H 44/13. HI 50218

Acta hydrobiologica sinica. [Chung-kuo shui shing wu hui pao]. Shanghai. Acta Hydrobiol. Sin. (Shanghai). See B–P–H 44/14. HI 50219

Acta instituti agriculturae nankinensis. [Nan Ching nung hsüeh yüan hsüeh pao]. Nanking. Acta Inst. Agric. Nankin. See B–P–H 44/15. HI 50220

Acta instituti botanici academiae scientiarum slovacae. Bratislava. Nos. 1-2, 1974. Acta Inst. Bot. Acad. Sci. Slov. Superseded by: Acta botanica slovaca. Series A, taxonomica, geobotanica. HI 61017

Acta instituti botanici academiae scientiarum slovacae. Series B, physiologica, phytophysiologica. Bratislava. No. 1, 1976. Acta Inst. Bot. Acad. Sci. Slov., B. Superseded by: Acta botanica slovaca. Series B, physiologica, pathophysiologica. HI 61018

Acta instituti botanici academiae scientiarum URPSS. Botanica experimentalis = Trudy botanicheskogo instituta akademii nauk S S S R. Ser. 4, eksperimental'naya botanika. Moscow & Leningrad.

Acta instituti botanici academiae scientiarum URPSS. Flora et systematica plantae vasculares = Trudy botanicheskogo instituta akademii nauk S S S R. Ser. 1, flora i sistematika vysshikh rastenii. Moscow & Leningrad.

Acta instituti botanici academiae scientiarum URPSS. Geobotanica = Trudy botanicheskogo instituta akademii nauk S S S R. Ser. 3, geobotanika. Moscow & Leningrad.

Acta instituti botanici academiae scientiarum URPSS. Materiae rudes plantarum = Trudy botanicheskogo instituta akademii nauk S S S R. Ser. 5, rastitel'noe syr'e. Moscow & Leningrad.

Acta instituti botanici academiae scientiarum URPSS. Plantae cryptogamae = Trudy botanicheskogo instituta akademii nauk S S S R. Ser. 2, sporovye rasteniya. Moscow & Leningrad.

Acta instituti botanici nomine V. L. Komarovi academiae scientiarum URPSS. Botanica experimentalis = Trudy botanicheskogo instituta akademii nauk S S S R. Ser. 4, eksperimental'naya botanika. Moscow & Leningrad.

Acta instituti botanici nomine V. L. Komarovi academiae scientiarum URPSS. Flora et systematica plantae vasculares = Trudy botanicheskogo instituta akademii nauk S S S R. Ser. 1, flora i sistematika vysshikh rastenii. Moscow & Leningrad.

Acta instituti botanici nomine V. L. Komarovi academiae scientiarum URPSS. Geobotanica = Trudy botanicheskogo instituta akademii nauk S S S R. Ser. 3, geobotanika. Moscow & Leningrad.

Acta instituti botanici nomine V. L. Komarovi academiae scientiarum URPSS. Materiae rudes plantarum = Trudy botanicheskogo instituta akademii nauk S S S R. Ser. 5, rastitel'noe syr'e. Moscow & Leningrad.

Acta instituti botanici nomine V. L. Komarovi academiae scientiarum URPSS. Plantae cryptogamae = Trudy botanicheskogo instituta akademii nauk S S S R. Ser. 2, sporovye rasteniya. Moscow & Leningrad.

Acta instituti defensionnis plantarum latviensis = Latvijas augu aizsardzibas instituta raksti. Riga.

Acta instituti forestalis zvolenensis. Bratislava. Vol. 1+, 1968+. Acta Inst. Forest. Zvolen. HI 61019

Acta instituti et horti botanici tartuensis (dorpatensis). Tartu, Estonia [Estonian S S R]. Acta Inst. Horti Bot. Tartuensis. See B–P–H 45/2. HI 50221

Acta instituti et musei zoologici universitatis atheniensis. Athens. Acta Inst. Mus. Zool. Univ. Athen. See B–P–H 45/3. HI 50223

Acta institutum botanicum nomine V. L. Komarovi academiae scientiarum URSS. Botanica experimentalis = Trudy botanicheskogo instituta akademii nauk S S S R. Ser. 4, eksperimental'naya botanika. Moscow & Leningrad.

Acta institutum botanicum nomine V. L. Komarovi academiae scientiarum URSS. Flora et systematica plantae vasculares = Trudy botanicheskogo instituta akademii nauk S S S R. Ser. 1, flora i sistematika vysshikh rastenii. Moscow & Leningrad.

Acta institutum botanicum nomine V. L. Komarovi academiae scientiarum URSS. Geobotanica = Trudy botanicheskogo instituta akademii nauk S S S R. Ser. 3, geobotanika. Moscow & Leningrad.

Acta institutum botanicum nomine V. L. Komarovi academiae scientiarum URSS. Materiae rudes plantarum = Trudy botanicheskogo instituta akademii nauk S S S R. Ser. 5, rastitel'noe syr'e. Moscow & Leningrad.

Acta institutum botanicum nomine V. L. Komarovi academiae scientiarum URSS. Plantae cryptogamae = Trudy botanicheskogo instituta akademii nauk S S S R. Ser. 2, sporovye rasteniya. Moscow & Leningrad.

Acta der königlich-bestätigten Ostpreussisch-Mohrungschen physikalisch-ökonomischen mit dem Preussisch-ökonomischen Lese-Institut zu Königsberg verbundenen Gesellschaft, zur Nachricht der Mitglieder. Königsberg [=Kaliningrad, Russian S F S R]. Acta Königl.-Bestätigten Ostpreuss.-Mohrungschen Phys.-Ökon. Ges. See B–P–H 45/4. HI 50227

Acta der Königlich-Ostpreussisch-Mohrungschen physicalisch-oeconomischen Gesellschaft. Zur Nachricht der Mitglieder. Königsberg. Acta Königl.-Ostpreuss.-Mohrungschen Phys.-Oecon. Ges. See B–P–H 45/5. HI 50228

Acta leidensia, edita cura et sumptibus medicinae tropicae. Leiden. Acta Leidensia. See B–P–H 45/6. HI 50230

Acta limnologica. Lund. Vols. 1-4, 1948-62. Acta Limnol. Preceded by: Meddelanden från Lunds universitets limnologiska institution. 1-54-2. HI 61021

Acta limnologica indica. Delhi. Vol. 1+, 1981+. Acta Limnol. Indica. HI 61022

Acta literaria et scientiarum Sueciae. Uppsala. Acta Lit. Sci. Sueciae. See B–P–H 45/9. HI 50231

Acta literaria universitatis hafniensis. Copenhagen. Acta Lit. Univ. Hafn. See B–P–H 45/12. HI 50232

Acta litteraria. Altenburg. Acta Litt. (Altenburg). See B–P–H 45/13. HI 50233

Acta litteraria musei nationalis hungarici. Budapest. Acta Litt. Mus. Natl. Hung. See B–P–H 45/14. HI 50234

Acta litteraria societatis rheno-trajectinae. Leiden. Acta Litt. Soc. Rheno-Traject. See B–P–H 45/16. HI 50236

Acta litterarum ac scientiarum regiae universitatis hungaricae francisco-josephinae. Sectio scientiarum naturalium. Szeged, Hungary. Acta Litt. Sci. Regiae Univ. Hung. Francisco-Joseph., Sect. Sci. Nat. See B–P–H 45/15. HI 50235

Acta manilana. Manila. Nos. 1-4, 1965-67. Acta Manilana. Superseded in part by: Acta Manilana. Ser. A, natural and applied sciences HI 61023.

Acta manilana. Series A, natural and applied sciences. Manila. Vol. 1+, 1967+. Acta Manilana, A. Preceded by: Acta Manilana. HI 61024

Acta matritensia. 1806. Acta Matr. HI 61025

Acta medicorum berolinensium in incrementum artis et

scientiarum collecta et digesta. Berlin. Acta Med. Berol. See B–P–H 45/17. HI 50237

Acta medica et biologica. Niigata. Vol. 1+, 1953+. Acta Med. Biol. HI 61026

Acta medica hafniensia. Copenhagen. Acta Med. Hafn. See B–P–H 45/20. HI 50238

Acta medica et philosophica hafniensia. Copenhagen. Acta Med. Philos. Hafn. See B–P–H 46/2. HI 50240

Acta medicorum svecicorum. Stockholm. Acta Med. Svecicorum. See B–P–H 46/4. HI 50241

Acta microbiologia sinica. [Wei shêng wu hsüeh p'ao]. Peking. Acta Microbiol. Sin. See B–P–H 46/10. HI 50243

Acta microbiologica academiae scientiarum hungaricae. Budapest. Vols. 1-29, 1954-82. Acta Microbiol. Acad. Sci. Hung. Superseded by: Acta microbiologica hungarica. HI 61027

Acta microbiologica bulgarica. Sofia. No. 1+, 1978+. Acta Microbiol. Bulg. Preceded by: Acta microbiologica, virologica et immunologica. HI 61028

Acta microbiologica hellenica. Athens. Acta Microbiol. Hellen. See B–P–H 46/6. HI 50242

Acta microbiologica hungarica. Budapest. Vol. 30+, 1983+. Acta Microbiol. Hung. Preceded by: Acta microbiologica academiae scientiarum hungaricae. HI 61029

Acta microbiologica polonica. Warsaw. Vols. 1-17, 1952-68; vol. 25+, 1976+. Acta Microbiol. Polon. For 1969-75 divided into: Acta microbiologica polonica. Series A, microbiologia generalis and Acta microbiologica polonica. Series B, microbiologia applicata. HI 61030

Acta microbiologica polonica. Series A, microbiologia generalis. Warsaw. Vols. 1(18)-2(24), 1969-75. Acta Microbiol. Polon., A. Preceded and superseded by: Acta microbiologica polonica. HI 61031

Acta microbiologica polonica. Series B, microbiologia applicata. Warsaw. Vols. 1(18)-7(24), 1969-75. Acta Microbiol. Polon., B. Preceded and superseded by: Acta microbiologica polonica. HI 61032

Acta microbiologica, virologica et immunologica. Sofia. Vols. 1-6, 1975-77. Acta Microbiol. Virol. Immunol. Preceded by: Izvestiya na mikrobiologičeskija institut. Superseded by: Acta microbiologica bulgarica. HI 61033

Acta micropalaeontologica sinica. [Wei ti gu sheng wu xue-bao.] Beijing. Vol. 1+, 1984+. Acta Micropalaeontol. Sin. HI 61034

Acta musei Devensis = Sargetia. Series scientia naturae. [Deva.]

Acta musei historiae naturalis. Cracow = Prace muzeum przyrodniczego. Cracow. Prace Muz. Przyr. See B–P–H 718/7.

Acta musei et horti botanici Bohemiae borealis. Historia naturalis. Liberec = Sborník severočeskeho musea. Přírodní vědy. Liberec, Czechoslovakia. Sborn. Severočesk. Mus., Přír. Vědy. See B–P–H 815/19.

Acta musei macedonici scientiarum naturalium. Skoplje, Yugoslavia. Acta Mus. Maced. Sci. Nat. See B–P–H 46/13. HI 50244

Acta musei moraviae = Časopis zemského musea v Brně. Brno.

Acta musei Moraviae. Scientiae naturales = Časopis moravského musea v Brně. Vědy přírodni. Brno.

Acta musei moraviensis = Časopis moravského zemského musea v Brně. Brno.

Acta musei nationalis Pragae = Sborník národního muzea v Praze. Řada B: Prírodni vědy (Přírodovědný). Prague.

Acta musei pardubicensis. Pardubice, Czechoslovakia. Acta Mus. Pardubic. See B–P–H 46/18. HI 50245

Acta musei reginaehradecensis. Ser. A, scientiae naturales = Práce krajského musea v Hradci Králové. Serie A, vedy přírodni. Hradec Králové.

Acta musei silesiae. Ser. A, historia naturalis = Časopis slezského musea. Ser. A, historia naturalis. Opava.

Acta musei Silesiae. Ser. A, historiae naturalis = Časopis slezského musea v Opavě. Ser. A. Historia naturalis. Opava, Czechoslovakia. Čas. Slez. Mus. v Opavě, Ser. A, Hist. Nat. See B–P–H 301/8.

Acta musei silesiae. Serie C, dendrologie = Časopis slezského musea v Opava. Série C, dendrologia. Opava.

Acta musei slovaciae regionis Košice = Zbornik východoslovenského muzea v Košiciach. Prírodné vedy. Košice.

Acta musei slovaciae regionis orientalis. A, historia naturalis = Sborník východoslovenského múzea v Košiciach. Séria A, prirodné vedy. Kosice.

Acta musei tyrnaviensis = Sborník krajského múzea v Trnave. Trnava, Czechoslovakia. Sborn. Krajsk. Múz. v Trnave. See B–P–H 814/8.

Acta mycologica. Warsaw. Acta Mycol. See B–P–H 46/23. HI 50247

Acta mycologica hungariae = Magyar gombászati lapok. Budapest.

Acta mycologica sinica. [Zhenjun xuebao] Beijing. Vol. 1+, 1982+. Acta Mycol. Sin. HI 61035

Acta naturalia de "l'ateneo parmense." Parma. Vol. 17+, 1981+. Acta Nat. Ateneo Parmense. Preceded by: Ateneo parmense. Acta naturalia. HI 61036

Acta naturalia islandica. Reykjavik, Iceland. Acta Nat. Islandica. See B–P–H 46/25. HI 50248

Acta oceanographica taiwanica. Taipei. No. 1+, 1971+. Acta Oceanogr. Taiwan. HI 61037

Acta oceanologica sinica. [Haiyang xuebao.] Beijing. Vol. 1+, 1979+. Acta Oceanogr. Sin. HI 61038

Acta oecologia. Oecologia applicata; revue internationale d'écologie fondamentale et appliquée. Paris. Vol. 1+, 1980+. Acta Oecol., Oecol. Appl. Preceded in part by: Oecologia plantarum. HI 61039

Acta oecologia. Oecologia generalis; revue internationale d'écologie fondamentale et appliquée. Paris. Vol. 1+, 1980+. Acta Oecol., Oecol. Gen. Preceded in part by: Oecologia plantarum. HI 61040

Acta oecologia. Oecologia plantarum; revue internationale d'écologie fondamentale et appliquée. Paris. Vol. 1+, 1980+. Acta Oecol., Oecol. Pl. Preceded in part by: Oecologia plantarum. HI 61041

Acta operum facultatis rerum naturalium universitatis slovacae Bratislava = Sborník prác prírodovedeckej fakulty slovenskej university v Bratislave. Bratislava. Sborn. Prác Prír. Fak. Slov. Univ. v Bratislave. See B–P–H 815/8.

Acta palaeobotanica. Cracow. Acta Palaeobot. See B–P–H 46/27. HI 50249

Acta palaeontologica polonica. Warsaw. Acta Palaeontol. Polon. See B–P–H 47/1. HI 50250

Acta palaeontologica sinica. [Ku shêng wu hsüeh pao]. Peking. Acta Palaeontol. Sin. See B–P–H 47/2. HI 50251

Acta palynologica; an international review of palynology. Montpellier. Vol. 1+, 1989+. Acta Palynol. HI 75250

Acta pathologica, microbiologica et immunologica scandinavica. Section B, microbiology. Copenhagen. Vol. 90B+, 1982+. Acta Pathol. Microbiol. Immunol. Scand., B. Preceded by: Acta pathologica et microbiologica scandinavica. Section B, microbiology. HI 61042

Acta pathologica, microbiologica et immunologica scandinavica. Supplementum. Copenhagen. No. 1+, 1988+. Acta Pathol. Microbiol. Immunol. Scand., Suppl. HI 61423

Acta pathologica et microbiologica scandinavica. Copenhagen. Vols. 1-77, 1924-69. Acta Pathol. Microbiol. Scand. Superseded by: Acta pathologica et microbiologica scandinavica. Section B, microbiology. 1-56-3. HI 61043

Acta pathologica et microbiologica scandinavica. Section B, microbiology. Copenhagen. Vols. 78B-89B, 1970-82; Supplementum 19?? Acta Pathol. Microbiol. Scand., B. Preceded by: Acta pathologica et microbiologica scandinavica. Superseded by: Acta pathologica, microbiologica et immunologica scandinavica. Section B, microbiology. HI 61044

Acta pedologica sinica. [T'u jang hsüeh hsüeh p'ao]. Peking. Acad Pedol. Sin. See B–P–H 47/7. HI 50252

Acta phaenologica. The Hague. Acta Phaenol. See B–P–H 47/8. HI 50253

Acta pharmaceutica hungarica = Gyógyszerészetudományi értesitö. Budapest. Gyógyszerésztud. Értes. See B–P–H 409/17.

Acta pharmaceutica jugoslavica. Zagreb. 1951+. Acta Pharm. Jugoslav. HI 61045

Acta pharmaceutica sinica. [Yao hsüeh hsüeh pao]. Peking. Acta Pharm. Sin. See B–P–H 47/11. HI 50255

Acta pharmaciae historica. The Hague. Acta Pharm. Hist. See B–P–H 47/9. HI 50254

Acta pharmacologica sinica. [Zhongguo yaoli xuebao.] Shanghai. Vol. 1+, 1980+. Acta Pharm. Sin. HI 61046

Acta pharmacologica et toxicologica. Copenhagen. Acta Pharmacol. Toxicol. See B–P–H 47/13. HI 50256

Acta physico-medica academiae caesareae leopoldino-carolinae naturae curiosorum exhibentia ephemerides sive observationes historias et experimenta ... Nuremberg. Acta Phys.-Med. Acad. Caes. Leop.-Carol. Nat. Cur. See B–P–H 47/14. HI 50257

Acta physico-medica academiae caesareae leopoldino-franciscanae naturae curiosorum exhibentia ephemerides sive observationes historias et experimenta ... Nuremberg. Acta Phys.-Med. Acad. Caes. Leop.-Francisc. Nat. Cur. See B–P–H 47/15. HI 50258

Acta physiologiae plantarum. Warsaw. Vol. 1+, 1978/79+ 1978+. Acta Physiol. Pl. HI 61047

Acta physiologica latinoamericana. Buenos Aires. Vols. 1-33, 1950-83; Suplemento, no. 1+, 1966+. Acta Physiol. Latinoamer. Superseded by: Acta physiologica et pharmacologica latinoamericana. HI 61048

Acta physiologica et pharmacologica latinoamericana. Buenos Aires. Vol. 34+, 1984+. Acta Physiol. Pharm. Latinoamer. Preceded by: Acta physiologica latinoamericana. HI 61049

Acta physiologica et pharmacologica neerlandica. Amsterdam. Vols. 1-15, 1950-70. Acta Physiol. Pharmacol. Neerl. Preceded by: Archives néerlandaises de physiologie de l'homme et des animaux. HI 61050

Acta physiologica scandinavica. Stockholm. Acta Physiol. Scand. See B–P–H 47/18. HI 50259

Acta physiologica sinica. [Sheng li hsüeh pao]. Peiping

[=Peking]. Acta Physiol. Sin. See B–P–H 47/19. HI 50260

Acta phytochimica. [Shokubutsu kwagaku zasshi]. Tokyo. Acta Phytochim. See B–P–H 47/20. HI 50261

Acta phytoecologia et geobotanica sinica. [Zhiwu shengtaixue yu dizhixue xuebao.] Beijing. Vol. 1+, 1977+. Acta Phytoecol. Geobot. Sin. HI 61051

Acta phytogeographica suecica. Uppsala. Acta Phytogeogr. Suec. See B–P–H 47/22. HI 50262

Acta phytomedica. [Supplement to: Phytopathologische Zeitschrift.] Berlin, Hamburg. Vol. 1+, 1973+. Acta Phytomed. HI 61052

Acta phytopathologica academiae scientiarum hungaricae. Budapest. Acta Phytopathol. Acad. Sci. Hung. See B–P–H 48/2. HI 50263

Acta phytopathologica sinica. [Zhiwu bingli xuebao.] Peking (Beijing). Vol. 1+, 1955+. Acta Phytopathol. Sin. HI 61053

Acta phytophylacica sinica. [Zhiwu baohu xuebao.] Beijing. Vol. 1-5(1), 1962?-66. Acta Phytophyl. Sin. HI 61054

Acta phytophysiologia sinica. [Zhiwu shengli xuebao.] Shanghai. Vol. ?-5+, 19??-79+. Acta Phytophysiol. Sin. HI 61055

Acta phytotaxonomica barcinonensia. Barcelona. Vol. 1-21, 1968-76. Acta Phytotax. Barcinon. Superseded by: Acta botanica barcinonensia. HI 61056

Acta phytotaxonomica et geobotanica. [Shokubutsu bunrui chiri]. Kyoto, Japan. Acta Phytotax. Geobot. See B–P–H 48/5. HI 50264

Acta phytotaxonomica sinica. [Chih wu fen lei hsüeh pao]. Peking. Acta Phytotax. Sin. See B–P–H 48/6. HI 50265

Acta phytotaxonomica sinica. Additamentum. [Zhiwu fenlei xuebao.] Peking (Beijing). Vol. 1+, 1965+. Acta Phytotax. Sin., Addit. HI 61057

Acta phytotherapeutica; scientific journal on botanical medicine. (Périodique scientifique consacré aux drogues d'origine végétale. Wissenschaftliche Zeitschrift für vegetabilische Drogenkunde.) Amsterdam. Vols. 1-19(9), 1954-72. Acta Phytotherap. Superseded by: International journal of crude drug research. HI 61058

Acta poloniae pharmaceutica. Warsaw. 1937+. Acta Polon. Pharm. HI 61059

Acta pruhoniciana. Pruhonice, Czechoslovakia. Acta Pruhon. See B–P–H 48/11. HI 50266

Acta radiobotanika et genetika. [Hoshasen ikushujo kenkyu hokoku.] Ohmiya-Machi, Ibaraki. No. 1+, 1967+. Acta Radiobot. Genet. HI 61060

Acta radiologica. Stockholm. Acta Radiol. See B–P–H 48/12. HI 50267

Acta radiologica. Therapy, physics, biology. Stockholm. Acta Radiol., Therap. See B–P–H 48/13. HI 50268

Acta regiae societatis scientiarum et litterarum gothoburgensis. Botanica. Göteborg. Vol. 1+, 1972+. Acta Regiae Soc. Sci. Litt. Gothob., Bot. Preceded in part by: Göteborgs kungliga vetenskaps- och vitterhets samhälles handlingar. Ser. B, matematiska och naturvetenskapliga skrifter. HI 61061

Acta rerum naturalium districtus ostraviensis. Ostrava= Přírodovědecký sborník ostravského kraje. Ostrava, Czechoslovaia. Přír. Sborn. Ostravsk. Kraje. See B–P–H 722/8.

Acta rerum naturalium districtus Silesiae = Přírodovědecký časopis Slezský. Opava, Czechoslovakia. Přír. Cas. Slezský. See B–P–H 721/27.

Acta rerum naturalium musei nationalis slovaci Bratislava = Sborník slovenského národného múzea. Přírodné vědy. Bratislava. and Zbornik slovenského narodného muzea. Přírodné vědy. Bratislava.

Acta rerum naturalium musei slovenica = Prírodovedny sborník slovenského múzea. Bratislava.

Acta rerum naturalium museorum slovenicorum. Trnava = Prírodovedné práce slovenských múzei. Trnava, Czechoslovakia. Přír. Práce Slov. Múz. See B–P–H 722/5.

Acta ad res naturae estonicae perscrutandas edita a societate rebus naturae investigandis in universitate tartuensi constituta. Ser. 2 = Eesti loodusteaduse arhiiv. 2 Seeria. Tartu, Estonia [Estonian S S R]. Eesti Loodustead. Arh., Seer. 2. See B–P–H 355/5.

Acta saecularia academiae groninganae. Groningen. Acta Saecularia Acad. Groning. See B–P–H 48/19. HI 50269

Acta sanctorum quotquot toto orbe coluntur, vel a catholicis scriptoribus celebrantur ... Rome. Acta Sanctorum. See B–P–H 48/20. HI 50270

Acta scientia sinica. [Chung kuo k'o hsüeh]. Peking. Acta Sci. Sin. See B–P–H 48/26. HI 50275

Acta scientiarum litterarumque, universitas iagellonica; folia botanica = Zeszyty naukowe uniwersitetu jagiellónskiego; prace botaniczne. Cracow.

Acta scientiarum mathematicarum et naturalium universitatis Kolozsvár. Kolozsvar [=Cluj, Rumania]. Acta Sci. Math. Nat. Univ. Kolozsvár. See B–P–H 48/21. HI 50271

Acta scientiarum naturalium academiae scientiarum bohemoslovacae = Přírodovědné práce ústavů československé akademie věd v Brné. Brno.

Acta scientiarum naturalium estonicarum. Series biologica = Eesti loodusteaduste arhiiv. Bioloogiline seeria. Tartu, Estonian S S R. Eesti Loodustead. Arh., Biol. Seer. See B–P–H 355/4.

Acta scientiarum naturalium universitatis amoiensis. [Hsia men ta hsüeh hsüeh p'ao, tsu jan k'o hsüeh pao]. Amoy?, China. Acta Sci. Nat Univ. Amoiensis. See B–P–H 48/23. HI 50272

Acta scientiarum naturalium universitatis intramongolicae. [Neimenggu daxue xuebao ziran kexue.] Beijing. Vol. ?-12+, 19??-81+. Acta Sci. Nat. Univ. Intramongol. HI 61062

Acta scientiarum naturalium universitatis jilinensis. [Chi lin ta hsüeh tzu ran kon hsüeh pao.] Chi-lin. 1979+. Acta Sci. Nat. Univ. Jilin. Preceded by: Jirin university journal. Natural science edition. HI 61063

Acta scientiarum naturalium universitatis Nankaiensis. [Nankai daxue xuebao. Ziran kexue ban.] Nanking. 1981+. Acta Sci. Nat. Univ. Nankai. HI 61064

Acta scientiarum naturalium universitatis pekinensis. [Pei ching ta hsüeh hsüeh p'ao, tzujan k'o hsüeh (chi kan)]. Peking. Acta Sci. Nat. Univ. Pekin. See B–P–H 48/24. HI 50273

Acta scientiarum naturalium universitatis sunyatseni. [Zhongshan daxue xuebao. Ziran kexue ban.] Guangzhou. 1957+. Acta Sci. Nat. Univ. Sunyatseni. HI 61065

Acta scientiarum naturalium universitatis szechuanensis. [Seu ch'uan ta hsüeh hsüeh pao, tzu jân k'o hsüeh]. Chengtu, China. Acta Sci. Nat. Univ. Szechuan. See B–P–H 48/25. HI 50274

Acta silesiaca = Slezsky sborník. Opava, Czechoslovakia. Slez. Sborn. See B–P–H 843/21.

Acta Sluko. Scientiae naturales = Sborník s[tudejniho a] l[idovychovného] ú[stavu] k[raje] o[lomouckého]. Oddil přír. vědy. Olomouc, Czechoslovakia. Sborn. Stud. Lidovychovného Ústavu Kraje Olomouc., Odd. Přír. Vědy. See B–P–H 815/21.

Acta societas medicorum fennica Duodecim. Helsinki. Acta Soc. Med. Fenn. Duodecim. See B–P–H 49/13. HI 50279

Acta societas medicorum fennica Duodecim, series A. Helsinki. Acta Soc. Med. Fenn. Duodecim, Ser. A. See B–P–H 49/14. HI 50280

Acta societatis biologiae Latviae = Latvijas biologijas biedribas raksti. Riga, Latvia [Latvian S S R]. Latv. Biol. Biedribas Raksti. See B–P–H 526/13.

Acta societatis botanicorum Poloniae. Warsaw. Acta Soc. Bot. Poloniae. See B–P–H 49/5. HI 50276

Acta societatis pro fauna et flora fennica. Helsinki. Vols. 1-82, 1875/77-1980. Acta Soc. Fauna Fl. Fenn. 5-3923-1. HI 61066

Acta societatis jablonovianae. Leipzig. Acta Soc. Jablonov. See B–P–H 49/11. HI 50277

Acta societatis jablonovianae nova. Leipzig. Acta Soc. Jablonov. Nova. See B–P–H 49/12. HI 50278

Acta societatis mycologicae čechoslovenicae = Mykologický sborník. Prague.

Acta societatis physiologiae scandinavicae = Skandinavisches Archiv für Physiologie. Leipzig. Skand. Arch. Physiol. See B–P–H 842/12.

Acta societatis regiae scientiarum indo-neerlandicae. Batavia, Dutch E. Indies [=Jakarta, Indonesia]. Acta Soc. Regiae Sci. Indo-Neerl. See B–P–H 49/20. HI 50281

Acta societatis scientiarum naturalium Moraviae = Práce moravské přírodovědecké společnosti. Brno, Czechoslovakia. Práce Morav. Přír. Společn. See B–P–H 718/5.

Acta societatis regiae scientiarum upsaliensis. Stockholm. Acta Soc. Regiae Sci. Upsal. See B–P–H 49/21. HI 50282

Acta societatis scientiarum fennicae. Helsinki. Acad. Soc. Sci. Fenn. See B–P–H 49/22. HI 50283

Acta societatis scientiarum fennica. Series B. Opera biologica. Helsinki. Acta Soc. Sci. Fenn., Ser. B, Opera Biol. See B–P–H 49/23. HI 50284

Acta della societed elvetica per las scienzias naturelas = Verhandlungen der schweizerischen naturforschenden Gesellschaft. Basel.

Acta stratigraphica sinica. [Dicengxue zazhi.] Nanking. Vol. 1+, 1966+. Acta Stratigr. Sin. HI 61067

Acta succulentologica. Nos. 70-87, 1941-43. Acta Succulentol. Preceded by: Shaboten. Superseded by: Cactus. Kushimoto. HI 61068

Acta tropica; review of tropical science and tropical medicine. Basel. Acta Trop. See B–P–H 49/27. HI 50285

Acta universitatis agriculturae. Brno. Ser. C, facultas silviculturae. Brno. 1967+. Acta Univ. Agric., Brno, C. Preceded by: Sborník vysoké školy zemědělské v Brne. Ser. C. HI 61069

Acta universitatis agriculturae Praha = Sborník vysoké školy zemědělské v Praze. Prague. Sborn. Vysoké Školy Zeměd. v Praze. See B–P–H 817/2.

Acta universitatis Asiae mediae = Trudy sredne-Aziatskogo gosudarstvennogo universiteta imeni V. I. Lenina. Tashkent.

Acta universitatis Asiae mediae. Botanica = Trudy Sredne-Aziatskogo gosudarstvennogo universiteta. Seriya 8b. Botanika. Moscow.

Acta universitatis Asiae mediae. Ser. 8c, oekologia = Trudy sredne-Aziatskogo gosudarstvennogo universiteta. Seriya 8c, ekologiya. Moscow.

Acta universitatis bergensis. Ser. mathematica rerumque naturalium = Aarbok for universitet i Bergen. Matematisk-naturvitenskapelig serie. Bergen.

Acta universitatis carolinae. Biologica. Prague. Acta Univ. Carol., Biol. See B–P–H 50/1. HI 50286

Acta universitatis debreceniensis de Ludovico Kossuth nominatae. Series biologica. Budapest. Acta Univ. Debrecen. Ludovico Kossuth, Ser. Biol. See B–P–H 50/3. HI 50287

Acta universitatis latviensis = Latvijas augstskolas raksti. Riga and Latvijas universitates raksti. Riga, and Latvijas universitates raksti. Lauksaimniecibas fakultates serija. Riga.

Acta universitatis leopolitanae in Galicia anno 1784 inauguratae. Lvov, Galicia [Ukrainian S S R]. Acta Univ. Leop. Galicia. See B–P–H 50/4. HI 50288

Acta universitatis Lodziensis. Folia botanica. Łódź. Vol. 1+, 1981+. Acta Univ. Lodz., Folia Bot. Preceded by: Acta universitatis Lodziensis. Seria 2, nauki matematyczno-przyrodnicze. Botanika. HI 61070

Acta universitatis Lodziensis. Seria 2, nauki matematyczno-przyrodnicze. Botanika. Łódź. Nos. 1-34, 1975-80. Acta Univ. Lodz., Ser. 2. Preceded by: Zeszyty naukowe uniwerszytetu łodzkiego. Seria 2, nauki matematyczno przyrodnicze. Superseded by: Acta universitatis Lodziensis. Folia botanica. HI 61071

Acta universitatis lundensis. Nova series. Sectio 2, medica, mathematica, scientiae rerum naturalium. Lund. Vols. [1]-40, 1864-1904; n.s. vols. 1-?, 1905/06-1968. Acta Univ. Lund. 3-2486-2. HI 61072

Acta universitatis nankinensis scientiarum naturalium. [Nan ching ta hsüeh pao.] Nanking. Acta Univ. Nankin. Sic. Nat. See B–P–H 50/6. HI 50290

Acta universitatis Nicolai Copernici. Biologia. Torun. Vol. 32+, 1973+. Acta Univ. Nicolai Copernici, Biol. Preceded by: Zeszyty naukowe universytetu Mikołaja Kopernika w Toruniu. Nauki matematyczno-przyrodnicze. Biologia. HI 61073

Acta universitatis ouluensis. Series A, scientiae rerum naturalium. Oulu. No. 1+, 1974+. Acta Univ. Oulu., A. HI 61074

Acta universitatis palackianae olomucensis. Facultas rerum naturalium. Series 2. Biologica. Prague. Acta Univ. Palack. Olomuc. Fac. Rerum Nat., Ser. 2, Biol. See B–P–H 50/7. HI 50291

Acta universitatis stockholmiensis. Stockholm. Acta Univ. Stockholm. See B–P–H 50/8. HI 50292

Acta universitatis stockholmiensis. Stockholm studies in English = Stockholm studies in English. Stockholm.

Acta universitatis szegediensis. Sectio scientiarum naturalium. Pars botanica. Szeged, Hungary. Acta Univ. Szeged., Sect. Sci. Nat., Pars Bot. See B–P–H 50/10. HI 50293

Acta universitatis voronegiensis (olim jurievensis-dorpatensis) = Trudy Voronezhskogo gosudarstvennogo universiteta. Botanicheskii otdel. Leningrad.

Acta universitatis voronegiensis (olim jurievensis-dorpatensis) = Trudy Voronezhskogo gosudarstvennogo universiteta. Leningrad.

Acta universitatis wratislaviensis. Prace botaniczne. Wroclaw. Vol. 1+, 1964+. Acta Univ. Wratislav., Prace Bot. HI 61075

Acta universitatis wratislaviensis. Studia geograficzne. Wroclaw. Vol. 1+, 1963+. Acta Univ. Wratislav., Stud. Geogr. HI 61076

Acta venezolana. Caracas. Acta Venez. See B–P–H 50/13. HI 50294

Acta virologica. Prague. Acta Virol. See B–P–H 50/14. HI 50295

Acta virologica sinica. [Bingduxue jikan.] Beijing. Vol. 1+, 1982+. Acta Virol. Sin. HI 61077

Acta zoologica mexicana. Mexico City. 1955-71; n.s. 19??+. Acta Zool. Mex. HI 61078

Actas abreviadas de academia general de ciencias, bellas-letras y nobles artes. Córdoba, Spain. Actas Abr. Acad. Gen. Ci. See B–P–H 50/17. HI 50296

Actas de la academia de ciencias exactas, físicas y naturales de Lima. Lima. Actas Acad. Ci. Exact. Lima. See B–P–H 50/18. HI 50297

Actas de la academia nacional de ciencias de Córdoba. Córdoba, Spain. Actas Acad. Nac. Ci. Córdoba. See B–P–H 50/19. HI 50298

Actas del congreso internacional de biologia de Montevideo. [Supplement to: Archivos de la sociedad de biología de Montevideo.] Montevideo. Vols. 1-7, 1930-32. Actas Congr. Int. Biol. Montevideo. HI 61079

Actas del congreso nacional de ciencias naturales. Lisbon. 1942+. Actas Congr. Nac. Ci. Nat. HI 61080

Actas de las juntas generales; real sociedad económica de amigos del país. Havana. Actas Juntas Gen. Real Soc. Econ. Amigos País. See B–P–H 50/21. HI 50299

Actas y memorias: sociedad económica de los amigos del país de la provincia de Segovia. [Spain]. Actas Mem. Soc. Econ. Amigos País Prov. Segovia. See B–P–H 51/1. HI 50300

Actas y relación de los premios que distribuyó en su junita pública; real academia de matemáticas y nobles artes. Valladolid, Spain. Actas Relac. Premios Real

Acad. Mat. See B–P–H 51/2. HI 50301

Actes de l'académie royale des sciences, belles-lettres et arts de Bordeaux. Bordeaux. Actes Acad. Roy. Sci. Bordeaux. See B–P–H 51/6. HI 50302

Actes de l'institut botanique de l'université d'Athènes. Athens. Vol. 1, 1940. Actes Inst. Bot. Univ. Athènes. HI 61081

Actes de l'institut national genevois. Geneva. No. 1+, 1965+. Actes Inst. Natl. Genevois. Preceded by: Bulletin de l'institut national genevois. HI 61082

Actes du muséum d'histoire naturelle. Rouen. Actes Mus. Hist. Nat. See B–P–H 51/8. HI 50303

Actes du muséum de Rouen. Rouen. Vol. 1+, 1978+. Actes Mus. Rouen. HI 61083

Actes de la réserve biologique de la Dombes. Villars-les-Dombes. 1963/64+, 1964?+. Actes Rés. Biol. Dombes. HI 61084

Actes de la réserve zoologique et botanique de Camargue. Annexe du bulletin de la scoiété national d'acclimatation. Paris. Actes Réserve Zool. Bot. Camargue. See B–P–H 51/9. HI 50304

Actes de la société helvétique des sciences naturelles. Lausanne. Actes Soc. Helv. Sci. Nat. See B–P–H 51/10. HI 50305

Actes de la société d'histoire naturelle de Paris. Paris. Actes Soc. Hist. Nat. Paris. See B–P–H 51/12. HI 50306

Actes de la société jurassienne d'émulation. Porrentruy, Délémont. Vols. 9-25, 1857-74; vols. 28-34, 1878-84; 2e sér. vol. 1+, 1885/86+. Actes Soc. Jurass. Émul. Preceded by: Coup-d'oeil sur les travaux de la société jurassienne d'émulation. For 1876-77 see: Émulation jurassienne. HI 61085

Actes de la société linnéenne de Bordeaux. Bordeaux. Vols. 4-101, 1830-1968. Actes Soc. Linn. Bordeaux. Preceded by: Bulletin d'histoire naturelle de la société linnéenne de Bordeaux. Superseded by: Actes de la société linnéenne de Bordeaux. Sér. A, and Actes de la société linnéenne de Bordeaux. Sér. B. 5-3965-1. HI 61086

Actes de la société linnéenne de Bordeaux. Sér. A. Bordeaux. Vols. 102-107, 1965-70. Actes Soc. Linn. Bordeaux, A. Preceded by: Actes de la société linnéenne de Bordeaux. Superseded by: Bulletin de la société linnéenne de Bordeaux. HI 61087

Actes de la société linnéenne de Bordeaux. Sér. B. Bordeaux. Vols. 102-106, 1965-69. Actes Soc. Linn. Bordeaux, B. Preceded by: Actes de la société linnéenne de Bordeaux. Superseded by: Bulletin de la société linnéenne de Bordeaux. HI 61088

Actes de la société médicale des hôpitaux de Paris. Paris. Actes Soc. Méd. Hôp. Paris. See B–P–H 51/14. HI 50307

Actes de la société philologique. Paris. Actes Soc. Philol. See B–P–H 51/15. HI 50308

Actes de la société suisse des sciences naturelles. Lausanne. Actes Soc. Suisse Sci. Nat. See B–P–H 51/16. HI 50309

Actinidia enthusiasts newsletter. Chelan, WA. No. 1+, 1984+. Actinidia Enthus. Newslett. HI 61089

Activité, musées d'histoire naturelle. La Rochelle. 1963-69. Activité Mus. Hist. Nat. Incorporated in: Annales de la société des sciences naturelles de la Charente-Maritime. HI 61090

Activities in the institute of plant sciences, Harvard University. [Reprinted from: Report of the president of Harvard college and reports of departments.] Cambridge, MA. 19??-8?+. Activities Inst. Pl. Sci., Harvard Univ. HI 61091

Activities report, plant products division, department of agriculture, Canada. Ottawa. 1966?+. Activities Rep. Pl. Prod. Div., Canada. Preceded by: Report of activities and special developments during the month, plant products division, department of agriculture, Canada. HI 61092

Activities at Turrialba = C A T I E, actividades en Turrialba. Turrialba.

Actualidades biológicas. Medellin. Vol. 1+, 1972+. Actual. Biol. HI 61093

Actualité scientifique. Paris. Actual. Sci. (Paris 1904-12). See B–P–H 51/20. HI 50312

Actualité scientifique. Revue mensuelle des sciences pures et appliquées. Paris. Actual. Sci. (Paris 1912-19). See B–P–H 51/21. HI 50313

Actualités biochimiques. Liége. Actual Biochim. See B–P–H 51/18. HI 50310

Actualités botaniques. [Forms part of: Bulletin de la société botanique de France.] Paris. 1978+. Actual. Bot. HI 61094

Actualités des eaux douces et de l'aquaculture = Freshwater and aquaculture contents tables. Rome.

Actualités marines. Quebec. Actual. Mar. See B–P–H 51/19. HI 50311

Actualités de phytochimie fondamentale. Paris. Sér. 1-3, 1964-68. Actual. Phytochem. Fondam. HI 61095

Adansonia; recueil périodique d'observations botaniques. Paris. Vols. 1-12, 1860-79; n.s. vol. 1-20, 1961-81. Adansonia. Preceded by: Phanérogamie. Superseded by: Bulletin du muséum national d'histoire naturelle. Section B, Adansonia. 1-64-2 . HI 61096

Adansonia. Mémoire. Paris. 1964+. Adansonia, Mém. HI 61097

Addisonia; colored illustrations and popular descriptions of plants. New York. Vols. 1-24, 1916-64. Addisonia. 1-64-2. HI 61098

Address delivered on the occasion of its anniversary, Massachusetts horticultural society = Massachusetts horticultural society. Address delivered on the occasion of its anniversary. Boston, MA. Mass. Hort. Soc. Address. See B–P–H 549/18.

Administration report (Annual), agricultural department, Mysore = Report (Annual), agricultural department, Mysore. Bangalore.

Administration report of conservator of forests, Trinidad and Tobago = Report, forest department, Trinidad and Tobago. Port of Spain.

Administration report, department of agriculture, Mesopotamia = Report, department of agriculture, Mesopotamia. Baghdad.

Administration report (Annual), department of agriculture, Mysore = Report of the department of agriculture, Mysore. Bangalore.

Administration report of the director of agriculture, British Guiana. Georgetown. 1928-49. Admin. Rep. Director Agric., British Guiana. Preceded by: Report of the department of science and agriculture, British Guiana. Superseded by: Report of the director of agriculture, British Guiana. HI 61099

Administration report of the director of agriculture, Ceylon. [Forms part IV of: Ceylon administrative reports.] Colombo. 1911/12+, 1913+. Admin. Rep. Director Agric., Ceylon. HI 61100

Administration report of the director of agriculture, Trinidad and Tobago. Port of Spain. 1926+. Admin. Rep. Director Agric., Trinidad Tobago. Preceded by: Report of the department of agriculture, Trinidad and Tobago. HI 61101

Administration report (Annual), forest department, Ajmer-Merwara. Abu. 1875-1940, 1946-48 [suspended 1941-45]. Admin. Rep. Forest Dept., Ajmer-Merwara. HI 61102

Administration report (Annual), forest department, Andaman Islands. Port Blair. 1884-1940, 1946-47 [suspended 1941-45]. Admin. Rep. Forest Dept., Andaman Islands. HI 61103

Administration report (Annual), forest department, Assam. Shillong. 1874-1949. Admin. Rep. Forest Dept., Assam. HI 61104

Administration report (Annual), forest department, Baluchistan. Quetta. 1883-1952. Admin. Rep. Forest Dept., Baluchistan. HI 61105

Administration report (Annual), forest department, Bangalore. Bangalore. 1865-1917. Admin. Rep. Forest Dept., Bangalore. HI 61106

Administration report (Annual), forest department, Bengal. Calcutta. 1868-1947. Admin. Rep. Forest Dept., Bengal. HI 61107

Administration report (Annual), forest department, Bihar. Ranchi. 1936-47. Admin. Rep. Forest Dept., Bihar. HI 61108

Administration report (Annual), forest department, Bombay. Bombay. 1849-1960. Admin. Rep. Forest Dept., Bombay. Superseded, in part, by: Administration report (Annual), forest department, Maharashtra. HI 61109

Administration report (Annual), forest department, Ceylon. Colombo. 1890+. Admin. Rep. Forest Dept., Ceylon. HI 61110

Administration report (Annual), forest department, Cochin State. Ernakulam. 1936-48. Admin. Rep. Forest Dept., Cichin State. HI 61111

Administration report (Annual), forest department, Coorg. Merkara. 1870+. Admin. Rep. Forest Dept., Coorg. HI 61112

Administration report (Annual), forest department, Jeypore. 1883-96. Admin. Rep. Forest Dept., Jeypore. HI 61113

Administration report (Annual), forest department, Jodhpur (and Marwar). Jodhpur. 1890-1932. Admin. Rep. Forest Dept., Jodpur. HI 61114

Administration report (Annual), forest department, Kerala. Trivandrum. 1955+. Admin. Rep. Forest Dept., Kerala. HI 61115

Administration report (Annual), forest department, Madhya Pradesh. Nagpur. 1862-1940, 1946-48 [suspended 1941-45]. Admin. Rep. Forest Dept., Madhya Pradesh. HI 61116

Administration report (Annual), forest department, Madras. Madras. 1858+. Admin. Rep. Forest Dept., Madras. HI 61117

Administration report (Annual), forest department, Maharashtra. Poona. 1960+. Admin. Rep. Forest Dept., Maharashtra. Preceded by: Administration report (Annual), forest department, Bombay and Report on forest administration, forest department, Poona. HI 61118

Administration report (Annual), forest department, Oudh. 1863-77. Admin. Rep. Forest Dept., Oudh. HI 61119

Administration report (Annual), forest department, Sind. Karachi. 1858-1947. Admin. Rep. Forest Dept., Sind. Superseded by: Report on administration, forest department, Sind. HI 61120

Administration report (Annual), forest department, Travancore-Cochin. Trivandrum. 1939-55. Admin. Rep. Forest Dept., Travancore-Cochin. HI 61121

Administration report (Annual), forest department, Uttar Pradesh. Naini Tal. 1876+. Admin. Rep. Forest Dept., Uttar Pradesh. HI 61122

Administration report, government cinchona department, Tamil Nadu. 1968/69+, 1969?+. Admin. Rep. Gov. Cinchona Dept., Tamil Nadu. HI 61123

Administration report of the government marine biologist. Peradeniya, Ceylon. Admin. Rep. Gov. Mar. Biol. See B–P–H 52/5. HI 50314

Administration report (Annual), tea section, united planters' association of southern India. Madras, Coimbatore. 1st+, 1933+. Admin. Rep. Tea Sect. Unit. Planters' Assoc. S. India. HI 61124

Administrative report of the agricultural department, Madras = Report, department of agriculture, Madras. Madras.

Administrative report of the forest department, Trinidad and Tobago = Report, forest department, Trinidad and Tobago. Port of Spain.

Admiralty marine science publication. London. Admiralty Mar. Sci. Publ. See B–P–H 52/7. HI 50315

Adumbratio florae Aethiopicae. Florence. No. 1+, 1953+. Adumbr. Fl. Aethiop. HI 61125

Advancement of science; report of the british association for the advancement of science. London. Vols. 1(1)-27(134), 1939-71; nos. 1-2, 1975-76. Advancem. Sci. Preceded by: Reports of the british association for the advancement of science. 1-67-3. HI 61126

Advances in agronomy. New York, NY. Advances Agron. See B–P–H 52/24. HI 50316

Advances in agronomy and crop science = Fortschritte im Acker- und Pflanzenbau. Berlin, etc.

Advances in applied biology. New York, London, etc. Vol. 6+, 1981+. Advances Appl. Biol. Preceded by: Applied biology. HI 61127

Advances in applied microbiology. New York, NY. Advances Appl. Microbiol. See B–P–H 52/25. HI 50317

Advances in aquatic microbiology. New York & London. Vol. 1+, 1977+. Advances Aquatic Microbiol. Preceded by: Advances in microbiology of the sea. HI 61128

Advances in biochemical engineering. New York, NY., Secaucus, NJ. Vols. 1-25, 1971-82. Advances Biochem. Engin. HI 61129

Advances in biochemical engineering / biotechnology. Berlin. Vol. 1+, 1971+. Advances Biochem. Engin./ Biotechnol. HI 75173

Advances in biological and medical physics. New York, NY. Advances Biol. Med. Phys. See B–P–H 52/26. HI 50319

Advances in biological research. New Delhi. Vol. 1+, 1983+. Advances Biol. Res. HI 61130

Advances in biophysics. Tokyo. Vol. 1+, 1970+. Advances Biophys. HI 61131

Advances in the biosciences. Oxford & New York. No. 1+, 1969+. Advances Biosci. HI 61132

Advances in biotechnological processes. New York. Vol. 1+, 1983+. Advances Biotechnol. Processes. HI 61133

Advances in botanical research. New York, NY. Advances Bot. Res. See B–P–H 52/27. HI 50320

Advances in bryology. Vaduz. Vol. 1+, 1981+. Advances Bryol. HI 61134

Advances in carbohydrate chemistry. New York, NY. Advances Carbohyd. Chem. See B–P–H 52/28. HI 50321

Advances in catalysis and related subjects. New York, NY. Advances Catal. Related Subj. See B–P–H 52/29. HI 50322

Advances in cell biology. New York. Vols. 1-2, 1970-71. Advances Cell. Biol. (New York). HI 61135

Advances in cell biology. Greenwich, CT. Vol. 1+, 1987+. Advances Cell. Biol. (Greenwich). HI 61136

Advances in cell culture. Orlando, FL. Vol. 1+, 1981+. Advances Cell. Cult. HI 62616

Advances in cell and molecular biology. New York & London. Vols. 1-3, 1971-74. Advances Cell. Molec. Biol. HI 61137

Advances in cladistics. New York. Vol. 1+, 1980+. Advances Cladist. HI 61138

Advances in colloid science. New York, NY. Advances Colloid Sci. See B–P–H 52/30. HI 50323

Advances in comparative and environmental physiology. New York, Berlin, etc. Vol. 1+, 1988+. Advances Comp. Environm. Physiol. HI 61139

Advances in drug research. New York. Vol. 1+, 1964+. Advances Drug Res. HI 61140

Advances in ecological research. London & New York, NY. Advances Ecol. Res. See B–P–H 52/31. HI 50324

Advances in economic botany. New York. Vol. 1+, 1984+. Advances Econ. Bot. HI 61141

Advances in environmental sciences. New York & London. Vol. 1, 1969. Advances Environm. Sci. Superseded by: Advances in environmental science and technology. HI 61142

Advances in environmental science and technology. New York. 1971+. Advances Environm. Sci. Technol. Preceded by: Advances in environmental sciences.

HI 61143

Advances in enzyme regulation. Oxford, England. Advances Enzyme Regulat. See B–P–H 52/32. HI 50325

Advances in enzymology and related subjects of biochemistry. New York, NY. Advances Enzymol. Related Subj. Biochem. See B–P–H 53/1. HI 50326

Advances in food and nutrition research. San Diego, CA. Vol. 33+, 1989+. Advances Food Nutr. Res. Preceded by: Advances in food research. HI 74805

Advances in food research. New York. Vols. 1-32, 1948-88. Advances Food Res. Superseded by: Advances in food and nutrition research. 1-69-1. HI 61144

Advances in food research. Supplememt New York. Vol. 1+, 1969+. Advances Food Res., Suppl. HI 74889

Advances in genetics. New York, NY. Advances Genet. See B–P–H 53/2. HI 50327

Advances in genetics. Supplement. New York. Vol. 1+, 1966+. Advances Genet., Suppl. HI 61145

Advances in horticultural science. Florence. 1987+. Advances Hort. Sci. Preceded by: Rivista della ortoflorofrutticoltura italiana. HI 63283

Advances in hydroscience. New York, NY. Advances Hydrosci. See B–P–H 53/3. HI 50328

Advances in immunology. New York, NY. Advances Immunol. See B–P–H 53/4. HI 50329

Advances in marine biology. New York, NY. Advances Mar. Biol. See B–P–H 53/5. HI 50330

Advances in microbial ecology. New York & London. Vol. 1+, 1977+. Advances Microbial Ecol. HI 61146

Advances in microbial physiology. New York. Vol. 1+, 1967+. Advances Microbial Physiol. HI 61147

Advances in microbiology of the sea. New York & London. Vol. 1, 1968. Advances Microbiol. Sea. Superseded by: Advances in aquatic microbiology. HI 61148

Advances in modern biology. Moscow = Uspekhi sovremennoi biologii. Moscow.

Advances in morphogenesis. New York. Vols. 1-10, 1961-73. Advances Morphogen. HI 61149

Advances in optical and electron microscopy. London. Vol. 1+, 1966+. Advances Opt. Electron Microscop. HI 61150

Advances in pest control research. New York. Vols. 1-8, 1957-68. Advances Pest Control Res. HI 61151

Advances in pharmacology. New York. Vols. 1-6, 1962-68. Advances Pharmacol. Superseded by: Advances in pharmacology and chemotherapy. HI 61152

Advances in pharmacology and chemotherapy. New York & London. Vol. 7+, 1969+. Advances Pharmacol. Chemotherapy. Preceded by: Advances in chemotherapy [not entered, and] Advances in pharmacology. HI 61153

Advances in plant breeding = Fortschritte der Pflanzenzüchtung. Berlin, Hamburg.

Advances in plant nutrition. New York. Vol. 1+, 1984+. Advances Pl. Nutr. HI 61154

Advances in plant pathology. London & New York. Vol. 1+, 1982+. Advances Pl. Pathol. HI 61155

Advances in pollen-spore research. New Delhi. Vol. 1+, 1974+. Advances Pollen-Spore Res. HI 61156

Advances in protein chemistry. New York, NY. Advances Protein Chem. See B–P–H 53/9. HI 50331

Advances in radiation biology. New York, NY. Advances Radiat. Biol. See B–P–H 53/10. HI 50332

Advances in research and technology of seeds. Wageningen. No. 1+, 1975+. Advances Res. Technol. Seeds. HI 61157

Advances in science of China. Biology. Beijing. Vol. 1+, 1986+. Advances Sci. China, Biol. HI 61158

Advances in steroid biochemistry and pharmacology. New York. Vol. 1+, 1972+. Advances Steroid Biochem. Pharmacol. HI 61159

Advances in strawberry production. Ithaca, NY. Vol. 1+, 1982+. Advances Strawberry Prod. HI 61160

Advances in virus research. New York, NY. Advances Virus Res. See B–P–H 53/11. HI 50333

Advancing frontiers of plant sciences. New Delhi. Vols. 1-30, 1962-75. Advancing Frontiers Pl. Sci. HI 61161

Advisory bulletin, Tocklai experimental station. Jorhat. No. 1+, 1971+. Advis. Bull. Tocklai Exp. Stat. HI 61162

Advisory bulletin, welsh plant breeding station. Aberystwyth. Nos. ?-2-6, 19??-29-44. Advis. Bull. Welsh Pl. Breed. Stat. HI 61163

Advisory circular, rubber research institute of Ceylon. Dartonfield, Agalawatta. Nos. 31-80, 1951-70. Advic. Circ. Rubber Res. Inst. Ceylon. Preceded by: Advisory circular, rubber research scheme, Ceylon. Superseded by: Advisory circular, rubber research institute of Sri Lanka. HI 61164

Advisory circular, rubber research scheme, Ceylon. Peradeniya, etc. Nos. 1-30, 1938-50. Advis. Circ. Rubber Res. Scheme Ceylon. Superseded by: Advisory circular, rubber research institute of Ceylon. HI 61165

Advisory circular, rubber research institute of Sri Lanka. Dartonfield, Agalawatta. No. 81+, 1975+. Advis. Circ. Rubber Res. Inst. Sri Lanka. Preceded by: Advisory circular, rubber reaearch institute of Ceylon.

HI 61166

Advisory leaflet, department of agriculture, Mauritius. Port Louis. 1951+. Advis. Leafl. Dept. Agric., Mauritius. HI 61167

Advisory leaflet, department of agriculture, Scotland. Edinburgh. 1949+. Advis. Leafl. Dept. Agric., Scotland. Preceded by: Leaflet, department of agriculture, Scotland. HI 61168

Advisory leaflet, division of plant industry, department of agriculture and stock, Queensland. Brisbane. No. 1+, 1933+. Advis. Leafl. Div. Pl. Industr., Queensland. HI 61169

Advisory leaflet, forest service, Queensland. Brisbane, Qld. 1938+. Advis. Leafl. Forest Serv., Queensland. HI 61170

Advisory leaflet, Tocklai experimental station. Jorhat. No. 1+, 1970+. Advis. Leafl. Tocklai Exp. Stat. HI 61171

Advisory leaflet, U P A S I scientific department. Madras. No. 1+, 1953+. Advis. Leafl. U.P.A.S.I. Sci. Dept. HI 61172

Advisory leaflet, west of Scotland agricultural college. Glasgow. 1949+. Advis. Leafl. W. Scotland Agric. Coll. HI 61173

Advisory pamphlet, biological branch, department of agriculture, Victoria. Melbourne, Vic. 1942+. Advis. Pam. Biol. Branch Dept. Agric., Victoria. HI 61174

Advisory report, Harper Adams agricultural college = Report of the advisory department, Harper Adams agricultural college. Newport, Shropshire.

Advocate of science; a popular scientific journal. Philadelphia, PA Vol. 1, 1833-34. Advocate Sci. Superseded by: Advocate of science, and annals of natural history. 1-72-2. HI 53012

Advocate of science, and annals of natural history. Philadelphia, PA. Vol. 1(1-9), 1834-35. Advocate Sci. & Ann. Nat. Hist. Preceded by: Advocate of science. 1-72-3. HI 61175

Aesculape. Paris. Aesculape. See B–P–H 53/18. HI 50336

Affiliate reporter, cactus and succulent society of America. South Gate, CA. Vols. 1-?, 1965-82. Affil. Reporter Cact. Succ. Soc. Amer. Superseded by: C S S A newsletter. HI 61176

Affleck's southern rural almanac, and plantation and garden calendar. New Orleans, LA. Affleck's S. Rural Alman. See B–P–H 53/20. HI 50337

Afghanistan journal. Graz. Vol. 1+, 1974+. Afghanistan J. HI 61177

Afhandlingar rörande natur-wetenskaperne. Falun, Sweden. Afh. Natur-Wetensk. See B–P–H 53/21. HI 50338

Afocel = Annales de recherches sylvicoles. Nangis.

African affairs. London. 1944+. African Affairs. Preceded by: Journal of the royal african society. 1-80-3. HI 61178

African environment. Dakar. Vol. 1+, 1975+. African Environm. HI 75206

African journal of agricultural sciences. Addis Ababa. Vol. 5+, 1978+. African J. Agric. Sci. Preceded by: Journal, association for the advancement of agricultural sciences in Africa. HI 61179

African journal of ecology. Oxford. Vol. 17+, 1979+. African J. Ecol. Preceded by: East african wildlife journal. HI 61180

African journal of plant protection. Yaounde. Vol. 1+, 1976+. African J. Pl. Protect. Preceded by: Interafrican phytosanitary bulletin. HI 61181

African journal of tropical hydrobiology and fisheries. Nairobi. Vol. 1+, 1971+. African J. Trop. Hydrobiol. Fish. HI 61182

African research and documentation. Birmingham. Vol. 1+, 1973+. African Res. Doc. Preceded by: Bulletin, african studies association of the United Kingdom. HI 61183

African scientist. Nairobi. Nos. 1-?, 1969-70. African Sci. HI 61184

African soils = Sols africains. Paris. Sols Africains. See B–P–H 846/22.

African sugar and cotton journal. Durban, South Africa. African Sugar Cotton J. See B–P–H 53/30. HI 50339

African sugar and cotton planter. Durban, South Africa. African Sugar Cotton Pl. See B–P–H 54/1. HI 50340

African violet. London. African Violet. See B–P–H 54/2. HI 50341

African violet magazine. Knoxville, TN. African Violet Mag. See B–P–H 54/3. HI 50342

African wild life. Johannesburg. Vol. 1+, 1946+. African Wild Life. 1-82-1. HI 61185

Afrique agriculture. Paris. No. 1- 1975+. Afrique Agric. HI 61186

Ag bioethics forum; an interdisciplinary newsletter in agricultural bioethics. Ames, IA. Vol. 1+, 1988+. Ag Bioethics Forum. HI 74806

Agarica; mykologisk tidsskrift utgitt av Fredrikstad soppforening. Fredrikstad. Vol. 1+, 1980+. Agarica. HI 61187

Agave; quarterly magazine of the desert botanical garden. Phoenix, AZ. Vol. 1+, 1983+. Agave. HI 61188

Agazen. Addis Ababa. Nos. 1-6, 1973-76. Agazen. HI 61189

Agbiotech news and information. Wallingford. Vol. 1+, 1989+. Agbiotech. News Inform. HI 74807

Agchem age. [Noyaku jidai.] Tokyo. Agchem Age. See B–P–H 54/4. HI 50345

Agenda de l'agriculteur et du vigneron. Lausanne. Agenda Agric. Vigneron. See B–P–H 54/5. HI 50346

Agerdyrknings-tidende for Danmark og Slesvig. Copenhagen. 1857. Agerdyrkn.-Tidende Danmark Slesvig. Preceded by: Landoeconomisk tidende for Danmark og Slesvig. HI 61190

Agp news; proceedings of the arabinogalactan protein club. Melbourne, Vic. Vol. 1+, 1977+. Agp News. HI 61191

Agra university journal of research (science). Agra, India. Agra Univ. J. Res., Sci. See B–P–H 54/54. HI 50348

Agrargeschichte; Zeitschrift für Agrargeschichte und Agrarsoziologie. Frankfurt am Main. Vol. 1+, 1953+. Agrargeschichte. HI 61192

Agrarian. Clemson, SC. Agrarian. See B–P–H 55/2. HI 50349

Agrario. Santiago. Agrario. See B–P–H 55/3. HI 50351

Agrárirodalmi szemle. Budapest. Agrárirod. Szemle. See B–P–H 55/4. HI 50352

Agrarista. Asunción. Agrarista (Asunción). See B–P–H 55/5. HI 50354

Agrarista. Mexico City. Agrarista (Mexico City). See B–P–H 55/6. HI 50355

Agrarjahr. Würzburg. Agrarjahr. See B–P–H 55/7. HI 50356

Agrártörténeti i szemle. Budapest. Vol. 1+, 1957+. Agrártört. Szemle. HI 61193

Agrártudományi egyetem agronómiai kar kiadványai. Budapest. Agrártud. Egyet. Agron. Kar Kiadv. See B–P–H 55/8. HI 50357

Agrártudományi egyetem erdömérnöki karának évkönyve. Sopron, Hungary. Agrártud. Egyet. Erdömérn. Karának Évk. See B–P–H 55/10. HI 50358

Agrártudományi egyetem kert- és szölögazdaságtudományi karának évkönyve. Budapest. Agrártud. Egyet. Kert- Szölögazdaságtud. Karának Évk. See B–P–H 55/12. HI 50359

Agrártudományi egyetem kert- és szölögazdaságtudományi karának közleményei. Budapest. Agrártud. Egyet. Kert- Szölögazdaságtud. Karának Közlem. See B–P–H 55/13. HI 50360

Agrártudományi egyetem mezögazdaságtudományi karának évkönyve. Budapest. Agrártud. Egyet. Mezögazdaságtud. Karának Évk. (Budapest). See B–P–H 55/14. HI 50361

Agrártudományi egyetem mezögazdaságtudományi karának évkönyve. Gödöllö, Hungary. Agrártud. Egyet. Mezögazdaságtud. Karának Évk. (Gödöllö). See B–P–H 55/15. HI 50362

Agrártudományi szemle. Budapest. Agrártud. Szemle. See B–P–H 55/16. HI 50363

Agrártudomány. Budapest. Agrártudomány. See B–P–H 56/1. HI 50471

Agressologie; revue internationale de physiobiologie et de pharmacologie appliquées aux effets de l'agression. Paris. Agressologie. See B–P–H 56/2. HI 50365

Agri-Cher. Bourges, France. Agri-Cher. See B–P–H 61/15. HI 50459

Agri-Holland. The Hague. 1977+. Agri-Holland. Preceded by: Agricultural newsletter from the Netherlands. HI 61194

Agri hortique genetica. Landskrona, Sweden. Agri Hort. Genet. See B–P–H 56/3. HI 50366

Agri-science. Ottawa. 1989+. Agri-Sci. HI 61195

Agri 7 jours. Paris. Agri 7 Jours. See B–P–H 56/4. HI 50367

Agriasia; a current bibliography of southeast asian agricultural literature. College, Philippines. Vol. 1+, 1977+. Agriasia. HI 61196

Agricell report. Shrub Oak, NY. Vol. 1+, 1983+. Agricell Rep. HI 61197

Agrichemical west. San Francisco, CA. Agrichem. W. See B–P–H 61/14. HI 50458

Agrícola; revista mensual ilustrada de agricultura, comércio e industria = Vinos, viñas y frutas. Buenos Aires. Vinos Viñas Frutas. See B–P–H 965/22.

Agricola aridus. Fort Collins, CO. Agric. Aridus. See B–P–H 56/11. HI 50374

Agricole. Paris. No. 1+, 1974+. Agricole. Preceded by: Figaro agricole. HI 61198

Agricoltor. Bogotá. Vols. 1-17, 1880-1901. Agricoltor (Bogotá, 1880-1901). Superseded by: Revista nacional de agricultura. 1-85-2. HI 61199

Agricoltor. Bogotá. Vol. 1+, 1930+. Agricoltor (Bogotá, 1930+). 1-85-2. HI 61200

Agricoltor. Quito. Vols. 1+ [vols. from 1949+ not numbered], 1948-54 [Publication suspended 1950-54]. Agricoltor (Quito). Preceded by: Revista de la cámara de agricultura de la primera zona. Superseded by: Agricultura. 1-85-3. HI 61201

Agricoltor. Santiago. Vols. 45-52, 1916-21. Agricoltor (Santiago). Preceded and superseded by: Boletín de la sociedad nacional de agricoltura (Santiago). 2-898-3.

HI 61202

Agricoltura; revista mensile di attualità italiane e straniere. Rome. Vol. 1+, 1952+. Agricoltura (Rome). HI 61203

Agricoltura coloniale. Florence. Vols. 1-38, 1907-44. Agric. Colon. Superseded by: Rivista di agricoltura subtropicale e tropicale. 4-3686-2. HI 61204

Agricoltura e veterinaria. Rome. No. 1+, 1945+. Agric. & Veterin. Preceded by: Bollettino agricolo. HI 61205

Agriculteur belge et étranger. Alost, Belgium. Agric. Belge Étranger. See B–P–H 56/12. HI 50375

Agriculteur haitien. Port au Prince. 1826. Agric. Haitien. HI 61206

Agricultor. Asunción. Agricultor (Asunción). See B–P–H 61/19. HI 50460

Agricultor. Lima. Agricultor (Lima). See B–P–H 61/20. HI 50463

Agricultor. Vitoria, Brazil. Agricultor (Vitoria). See B–P–H 61/2l. HI 50464

Agricultor boliviano. La Paz, Bolivia. Agricultor Boliv. See B–P–H 61/22. HI 50465

Agricultor costarricense. San José, Costa Rica. Agric. Costarricense. See B–P–H 56/20. HI 50382

Agricultor lagunero. Torreón, Mexico. Agric. Lagunero. See B–P–H 59/12. HI 50416

Agricultor mexicano. Ciudad Juárez, Mexico. Agric. Mex. See B–P–H 59/17. HI 50421

Agricultor salvadoreño. Santa Tecla, Salvador. Agric. Salvadoreño. See B–P–H 60/14. HI 50439

Agricultor venezolano. Caracas. Agricultor Venez. See B–P–H 61/24. HI 50466

Agricultura. Bogota. Vols. 8-12, 1937?-40. Agricultura (Bogota). Preceded by: Boletín de agricultura. Superseded by: Agricultura y ganadería. Bogota. 1-86-1. HI 50467

Agricultura. Ciudad Trujillo. Vol. 20+ [also numbered ser. 2, no. 4+], 1946+ [suspended February-June 1922; March 1925-January 1929]. Agricultura (Ciudad Trujillo). Preceded by: Revista agricultura. Santo Domingo. 1-86-1. HI 61207

Agricultura. Guatemala City, Guatemala. Agricultura (Guatemala City). See 62/1. HI 50468

Agricultura; revista de agricultura. Revista mensuel. Havana. Vol. 1, 1917. Agricultura (Havana). Incorporated in: Revista de agricultura, comércio y trabajo. 1-85-3. HI 61208

Agricultura; comptes rendus des recherches patronées par l'institut agronomique de l'université de Louvain. Heyerlee. Vol. 1+, 1953+. Agricultura (Heyerlee). HI 61209

Agricultura. Lima. Agricultura (Lima). See B–P–H 62/3. HI 50469

Agricultura; revista da direcção-geral dos serviços agrícolas. Lisbon. Agricultura (Lisbon). See B–P–H 62/4. HI 50470

Agricultura. Heyerlee. Vol. 32(2)+, 1930+. Agricultura (Louvain). Preceded by: Revue générale agronomique. 1-86-1. HI 61210

Agricultura. [Departamento autónomo de prensa y publicidad.] Mexico City. Agricultura (Mexico City, Dept. Autón. Prensa). See B–P–H 62/5. HI 50472

Agricultura. [Secretaría de agricultura y fomento.] Mexico City. Agricultura (Mexico City, Secr. Agric.). See B–P–H 62/6. HI 50473

Agricultura. [Nung hsiao [hsüeh] tsa chih.] Nanking. 1928. Agricultura (Nanking). Preceded by: Agricultural science. Nanking. HI 61211

Agricultura. Quito. Vol. 1+, 1955+. Agricultura (Quito). Preceded by: Agricoltor. Quito. 1-85-3. HI 61212

Agricultura. San Salvador, Salvador. Agricultura (San Salvador). See B–P–H 62/9. HI 50474

Agricultura. Santiago de las Vegas, Cuba. Agricultura (Santiago de las Vegas). See B–P–H 62/10. HI 50475

Agricultura. Santo Domingo, Dominican Republic. Agricultura (Santo Domingo). See B–P–H 62/11. HI 50476

Agricultura. São Paulo. Agricultura (São Paulo). See B–P–H 62/12. HI 50477

Agricultura; revista de la secretaría de agricultura. Tegucigalpa, Honduras. Agricultura (Tegucigalpa). See B–P–H 62/13. HI 50480

Agricultura de las Américas. Kansas City, MO. Vol. 5+, 1956+. Agric. Amér. Preceded by: Implementos y tractores [not entered]. HI 50370

Agricultura boliviana. La Paz, Bolivia. Agricultura Boliv. See B–P–H 62/15. HI 50481

Agricultura en Chiriquí. Panama City, Panama. Agric. Chiriquí. See B–P–H 56/18. HI 50380

Agricultura al dia. Santurce, Puerto Rico. Agric. al Dia. See B–P–H 56/6. HI 50369

Agricultura en El Salvador. San Salvador, Salvador. Agric. El Salvador. See B–P–H 56/21. HI 50383

Agricultura española. [Supplement to: Anuario estadístico de las producciones agrícolas.] Valencia, Madrid. 1960+. Agric. Esp. HI 61213

Agricultura experimental. Río Piedras, Puerto Rico. Agric. Exp. See B–P–H 57/1. HI 50386

Agricultura y ganaderia; organo de la federación nacional de asociaciones de productores agropecuarios. Caracas.

Vol. 1 -(2)+, 19??-61+. Agric. Ganad. (Caracas). HI 61214

Agricultura y ganadería. Bogota. No. 1+, 1940+. Agric. Ganad. (Bogotá). Preceded by: Agricultura. Bogota. 1-86-2. HI 61215

Agricultura y ganaderia; publición del ministerio de agricultura. Santiago de Chile. Vols. 1+, 1955+. Agric. Ganad. (Santiago). HI 61216

Agricultura y química. Mexico City. Agric. Quím. See B–P–H 60/4. HI 50432

Agricultura em São Paulo. São Paulo. Agric. São Paulo. See B–P–H 60/15. HI 50440

Agricultura sinica. Nanking. Agric. Sin. See B–P–H 61/2. HI 50447

Agricultura téchnica en México. Mexico, DF. Vol. 1+, 1955+. Agric. Tecn. Mexico. HI 61217

Agricultura tropical. Bogotá. Agric. Trop. See B–P–H 61/9. HI 50451

Agricultura venezolana. Caracas. Vol.1+, 1936+. Agricultura Venez. HI 61218

Agricultural bimonthly. [Nung shih shuang yüeh k'an.] Canton. Agric. Bimonthly. See B–P–H 56/13. HI 50376

Agricultural and biological chemistry. Tokyo. Vol. 25+, 1961+. Agric. Biol. Chem. Preceded by: Bulletin of the agricultural chemical society of Japan. HI 61219

Agricultural biotechnology news. Cedar Falls, IA. Vol. 1+, 1984+. Agric. Biotechnol. News. HI 61220

Agricultural bulletin. The Hague. Agric. Bull. (The Hague). See B–P–H 56/14. HI 50377

Agricultural bulletin, Bermuda department of agriculture. Hamilton, Bermuda. Vols. [1]-33, 1922-63 [suspended 1946-54]. Agric. Bull., Bermuda. Superseded by: Monthly bulletin of the Bermuda department of agriculture (and fisheries). HI 61221

Agricultural bulletin, department of agriculture, Bermuda = Agricultural bulletin, Bermuda department of agriculture. Hamilton, Bermuda.

Agricultural bulletin, department of agriculture, Iraq = Bulletin, department of agriculture, Iraq. Basrah, Baghdad.

Agricultural bulletin, department of agriculture, Northern Rhodesia = Bulletin, department of agriculture, Northern Rhodesia. Lusaka.

Agricultural bulletin of the Federated Malay States. Kuala Lumpur. Vols. 1-9(4), 1912-21. Agric. Bull. Fed. Malay States. Preceded by: Agricultural bulletin of the Straits and Federated Malay States. Superseded by: Malayan agricultural journal. 3-2527-1. HI 61222

Agricultural bulletin of the Malay Peninsula. Singapore. Nos. 1-9, 1891-1900. Agric. Bull. Malay Peninsula. Superseded by: Agricultural bulletin of the Straits and Federated Malay States. 2-1667-3. HI 61223

Agricultural bulletin, Palestine. Jerusalem. 1940(July)-1941(Sept.). Agric. Bull., Palestine. Preceded by: Monthly agricultural bulletin, Palestine. HI 61224

Agricultural bulletin of the Saga university. [Saga daigaku nogakubu iho.] Saga. 1952+. Agric. Bull. Saga Univ. HI 61225

Agricultural bulletin of the Straits and Federated Malay States. Singapore. [Ser. 2], vols. 1-10, 1901-11; ser. 3, vol. 1(1-5), 1912. Agric. Bull. Straits Fed. Malay States. Preceded by: Agricultural bulletin of the Malay Peninsula. Superseded by: Agricultural bulletin of the Federated Malay States and Gardens' bulletin, Straits Settlements. 2-1667-3. HI 61226

Agricultural chemicals. [Noyaku hyakuten.] Tokyo. Agric. Chem. See B–P–H 56/17. HI 50379

Agricultural communications research report, college of agriculture, university of Illinois at Urbana-Champaign. Urbana, IL. 1960+. Agric. Commun. Res. Rep., Univ. Illinois. HI 61227

Agricultural conservation, program accomplishments. Washington, DC. 1975+. Agric. Conserv. Program Accomp. Preceded by: Rural environmental conservation. HI 61228

Agricultural economic review. Salonika. Vols. 1-8(1), 1965-72. Agric. Econ. Rev. Superseded by: Hellenic agricultural economic review. HI 61229

Agricultural experiment research report. [Noji shiken chosa shiryo.] Sapporo. 121-130go, 1967-72. Agric. Exp. Res. Rep. HI 61230

Agricultural experiment station of the agricultural college of Colorado. Annual report. Fort Collins, CO. Agric. Exp. Sta. Agric. Coll. Colorado Annual Rep. See B–P–H 57/3. HI 50387

Agricultural experiment station of the agricultural college of Colorado. Bulletin. Fort Collins, CO. Agric. Exp. Sta. Agric. Coll. Colorado Bull. See B–P–H 57/5. HI 50389

Agricultural experiment station of the agricultural college of Colorado. Press bulletin. Fort Collins, CO. Agric. Exp. Sta. Agric. Coll. Colorado Press Bull. See B–P–H 57/4. HI 50388

Agricultural experiment station at Arkansas industrial university. Bulletin. Fayetteville, AR. Agric. Exp. Sta. Arkansas Industr. Univ. Bull. See B–P–H 57/6. HI 50391

Agricultural experiment station bulletin. Fort Collins, CO. Agric. Exp. Sta. Bull. See B–P–H 57/7. HI 50392

Agricultural experiment station progress report, College,

AK = Alaska agricultural college and school of mines. Agricultural experiment station progress report. College, AK. Alaska Agric. Coll. School MInes Agric. Exp. Sta. Progr. Rep. See B–P–H 68/3.

Agricultural extension news. [Nung yeh t'ui kuang t'ung hsin.] Chungking, China. Agric. Extens. News. See B–P–H 57/12. HI 50394

Agricultural and forestry quarterly. [Nung lin chi k'an.] Canton?, China. Agric. Forest. Quart. See B–P–H 57/14. HI 50396

Agricultural gazette of Canada. Ottawa. Agric. Gaz. Canada. See B–P–H 57/17. HI 50398

Agricultural gazette of New South Wales. Sydney. Agric. Gaz. New South Wales. See B–P–H 58/1. HI 50399

Agricultural genetics report. New York. Vol. 1+, 1982+. Agric. Genet. Rep. HI 61231

Agricultural history. Chicago. Vol. 1+, 1927+. Agric. Hist. Preceded by: Papers of the agricultural history society. 1-83-3. HI 61232

Agricultural history review. London. Agric. Hist. Rev. See B–P–H 58/6. HI 50404

Agricultural, horticultural and botanical society of Jefferson college. Natchez, MS. Agric. Soc. Jefferson Coll. See B–P–H 61/3. HI 50448

Agricultural index. New York, NY. Agric. Index See B–P–H 58/13. HI 50408

Agricultural information. [Nung shêng.] Canton. Agric. Inform. See B–P–H 58/14. HI 50409

Agricultural institute review. Ottawa. Vols. 1-24(2), 1945-69. Agric. Inst. Rev. Preceded by: C S T A review. Superseded by: A I C review. 1-89-1. HI 61233

Agricultural journal. [Nung lin kung pao.] [China]. Agric. J. (China). See B–P–H 58/18. HI 50411

Agricultural journal of the Bihar & Orissa department of agriculture. Patna. Vols. 1-5, 1913-17. Agric. J. (Patna). 1-89-2. HI 61234

Agricultural journal of British Columbia. Victoria, British Columbia. Agric. J. British Columbia. See B–P–H 58/21. HI 50412

Agricultural journal of British Guiana. Georgetown. Vols. 1-10, 1928-39 [suspended 1932-33]. Agric. J. British Guiana. Preceded by: Journal of the board of agriculture of British Guiana. 1-89-2. HI 61235

Agricultural journal of the Cape of Good Hope. Cape Town. Vols. 1-37, 1888-1910. Agric. J. Cape of Good Hope. Superseded by: Agricultural journal of the Union of South Africa. 1-89-3. HI 61236

Agricultural journal, department of science and agriculture, Barbados. Bridgetown. Vols. 1-9, 1931-40. Agric. J. (Barbados). Preceded and superseded by: Report of the department of science and agriculture, Barbados. 1-89-2. HI 61237

Agricultural journal of Egypt. Cairo. Agric. J. Egypt. See B–P–H 59/3. HI 50413

Agricultural journal of India. Pusa, Calcutta. Vols. 1-25, 1906-30. Agric. J. India. Superseded by: Agriculture and livestock in India [not entered, and] Indian journal of agricultural science. 1-89-2. HI 61238

Agricultural journal of the Kusunoki society. [Kusunoki noho.] Kochi, Japan. Agric. J. Kusunoki Soc. See B–P–H 59/5. HI 50414

Agricultural journal and mining record. Pietermaritzburg. 1898-1903. Agric. J. Mining Rec. Superseded by: Natal agricultural journal. HI 61239

Agricultural journal. Suva. Vols. 1-31, 1928-61. Agric. J. (Suva). Superseded by: Fiji agricultural journal. 1-89-2. HI 61240

Agricultural journal of the Union of South Africa. Pretoria. Vols. 1-8, 1911-12. Agric. J. Union South Africa. Preceded by: Agricultural journal of the Cape of Good Hope and Natal agricultural journal and mining journal. Superseded by: Journal of the department of agriculture, Union of South Africa. 1-89-3. HI 61241

Agricultural leaflet, department of agriculture, Iraq. Basrah, Baghdad. Nos. 1-25, 1920?-32. Agric. Leafl. Dept. Agric., Iraq. HI 61242

Agricultural ledger. Calcutta. Agric. Ledger. See B–P–H 59/13. HI 50419

Agricultural libraries information notes. Beltsville, MD. Vol. 1+, 1975+. Agric. Libr. Inform. Notes. HI 61243

Agricultural library notes. U.S. department of agriculture library. Washington, DC. Agric. Libr. Notes U.S.D.A. Libr. See B–P–H 59/15. HI 50420

Agricultural literature of Czechoslovakia. Prague. 1960+. Agric. Lit. Czechoslovakia. HI 61244

Agricultural magazine. Copenhagen. 1982+. Agric. Mag. (Copenhagen). HI 61245

Agricultural magazine. [Kuo li pei ching nung yeh chuan mên hsüeh hsiao tsa chih.] Peking. No. 3, 1918 [no further data available]. Agric. Mag. (Peking). HI 61246

Agricultural meteorology; an international journal. Amsterdam. Vol. 1+, 1964+. Agric. Meteorol. HI 61247

Agricultural monographs. U.S. dept. agric. Washington, DC. Agric. Monogr. U.S.D.A. See B–P–H 59/18. HI 50422

Agricultural monthly. Canton = Agricultural bimonthly. [Nung shih shuang yüeh k'an.] Canton. Agric.

Bimonthly. See B–P–H 56/13.

Agricultural monthly. [Noji geppo.] Sendai, Japan. Agric. Monthly. See B–P–H 59/19. HI 50423

Agricultural news. Bridgetown, Barbados. Agric. News (Barbados). See B–P–H 59/21. HI 50424

Agricultural news. [Nung yeh t'ung hsin.] Nanking. Agric. News (Nanking). See B–P–H 59/22. HI 50425

Agricultural newsletter for arid and semiarid areas. Wugong. Vol. 1+, 1981+. Agric. Newslett. Arid Semiarid Areas. HI 61248

Agricultural newsletter from the Netherlands. The Hague. Nos. 1-?, 1951-76. Agric. Newslett. Netherlands. Superseded by: Agri-Holland. HI 61249

Agricultural record. London. Vols. 36-39 [also numbered old ser. vols. 8-11; n.s. vols. 1-4], 1908-11. Agric. Rec. Preceded by: County council and agricultural record. Superseded by: Journal of the central and associated chambers of agriculture, London. 2-956-3. HI 61250

Agricultural record of the central agricultural board. [Trinidad.] Port of Spain, Trinidad. Agric. Rec. Centr. Agric. Board. See B–P–H 60/5. HI 50433

Agricultural record, South Australia department of agriculture. Adelaide, S.A. Vol. 1+, 1974+. Agric. Rec. S. Australia Dept. Agric. Preceded by: Experimental record, South Australia department of agriculture. HI 61251

Agricultural records, agricultural experiment station. Tel-Aviv. 1927+. Agric. Rec., Agric. Exp. Sta., Tel-Aviv. HI 61252

Agricultural research. New Delhi. Vols. 1-5, 1961-65. Agric. Res. (New Delhi). HI 61253

Agricultural research at the agricultural research institute, Peshawar. Peshawar. 1966/67+, 1968+. Agric. Res. Agric. Res. Inst., Peshawar. HI 61254

Agricultural research in the arid areas. Wugong. 1984+. Agric. Res. Arid Areas. HI 61255

Agricultural research bulletin, Jealott's Hill research station. Bracknell. Vols. 1-5(4), 1932?-36. Agric. res. Bull. Jealott's Hill Res. Sta. HI 61256

Agricultural research; college of agriculture, National Central university, Nanking. [Nung hsüeh ts'ung k'an.] Nanking. Agric. Res. Nanking. See B–P–H 60/7. HI 50435

Agricultural research, department of agricultural technical services. South Africa. Pretoria. 1959+, 1960?+. Agric. Res. Dep. Agric. Techn. Serv., S. Africa. HI 61257

Agricultural research, department of agriculture and fisheries, South Africa = Agricultural research, department of agricultural technical services. South Africa. Pretoria.

Agricultural research journal of Kerala. Trivandrum. Vol. 1+, 1963+. Agric. Res. J. Kerala. HI 61258

Agricultural research. Ohara institute for agricultural research. [Nogaku kengyu.] Kurashiki, Japan. Agric. Res. (Kurashiki). See B–P–H 60/6. HI 50434

Agricultural research report, agricultural experiment station, Blacksburg. Blacksburg, VA. 1947/48+, 1948+. Agric. Res. Rep. Agric. Exp. Sta. Blacksburg. Preceded by: Report of the Virginia agricultural experiment station. HI 61259

Agricultural research reports, center for agricultural publishing and documentation. Wageningen. 1907+. Agric. Res. Rep. Center Agric. Publishing Doc. HI 61260

Agricultural research results, agricultural research service, United States department of agriculture. Southern series. (ARR - S). New Orleans, LA. No. 1+, 1979+. Agric. Res. Results, Agric. Res. Serv., U.S.D.A., S. Ser. HI 61261

Agricultural research results, agricultural research service, United States department of agriculture. Western series. (ARR - W). Berkeley, Oakland, CA. No. 1+, 1978+. Agric. Res. Results, Agric. Res. Serv., U.S.D.A., W. Ser. HI 61262

Agricultural research review. Cairo. Agric. Res. Rev. See B–P–H 60/8. HI 50436

Agricultural research service, United States department of agriculture. Peoria, IL. 1973+. Agric. Res. Serv., U.S.D.A. HI 61263

Agricultural research in Texas. College Station, TX. Agric. Res. Texas. See B–P–H 60/9. HI 50437

Agricultural research, United States department of agriculture. Washington, DC. Vols. 1+, 1953+. Agric. Res., U.S.D.A. HI 61264

Agricultural review. London. Agric. Rev. See B–P–H 60/10. HI 50438

Agricultural reviews; a half-yearly publication devoted to all the branches of agricultural & animal science. Karnal. Vol. 1+, 1980+. Agric. Rev. (Karnal). HI 61265

Agricultural reviews and manuals, United States department of agriculture. Oakland, CA. Vol. 1+, 1978+. Agric. Rev. Manuals, U.S.D.A. HI 61266

Agricultural science. Geneva, NY. Agric. Sci. (Geneva). See B–P–H 60/16. HI 50441

Agricultural science. [Lun shu.] Nanking. Vols. 1-3, 1925-27. Agric. Sci. (Nanking). Superseded by: Agricultura. Nanking. 1-91-1. HI 61267

Agricultural science. Saskatoon. 1972+. Agric. Sci. (Saskatchewan). HI 61268

Agricultural science, Hong Kong. Hong Kong. Vol. 1, 1968-72. Agric. Sci., Hong Kong. Superseded by: Agriculture Hong Kong. HI 61269

Agricultural science. National university of Peiping. [Nung hsüeh yüeh k'an.] Peiping [=Peking]. Agric. Sci. (Peiping). See B–P–H 60/18. HI 50442

Agricultural science of the north temperate region. [Kanchi nogaku. Sapporo kasiwa-ba shoin.] Sapporo, Japan. Agric. Sci. N. Temp. Region. See B–P–H 60/20. HI 50444

Agricultural science progress. New Delhi. Vol. 1+, 1983+. Agric. Sci. Progr. HI 61270

Agricultural science review. Washington, DC. Agric. Sci. Rev. See B–P–H 60/21. HI 50445

Agricultural science in South Africa. Agroplantae = Agroplantae. Pretoria.

Agricultural science in South Africa. Phytophylactica = Phytophylactica. Pretoria.

Agricultural science in South Africa. Plant protection sciences and microbiology = Phytophylactica. Pretoria.

Agricultural sciences. [Nogaku.] Tokyo. Agric. Sci. (Tokyo). See B–P–H 60/19. HI 50443

Agricultural series. [Nogyo sosho.] Naha, Okinawa. Agric. Ser. See B–P–H 61/1. HI 50446

Agricultural statistics. Washington, DC. 1936+. Agric. Statist. HI 61271

Agricultural systems. Barking. Vol. 1+, 1976+. Agric. Systems. HI 61272

Agricultural technology. [Nogyo gijutsu kyokai.] Tokyo. Agric. Technol. See B–P–H 61/5. HI 50449

Agricultural university Wageningen papers. Wageningen. Vol. 84+, 1984+. Agric. Univ. Wageningen Pap. Preceded by: Mededelingen van te landbouwhogeschool te Wageningen. HI 61273

Agricultural world. [Nogyo sekai.] Tokyo. Agric. World. See B–P–H 61/12. HI 50455

Agricultural yearbook, United States department of agriculture. Washington, DC. 1923-25. Agric. Yearb. Preceded by: Yearbook of the United States department of agriculture. Superseded by: Yearbook of agriculture, United States department of agriculture. HI 61274

Agriculturchemiske meddelelser. Copenhagen. Vols. 1-2, 1855-56. Agriculturchem. Meddel. HI 61275

Agriculture. Arras, France. Agriculture (Arras). See B–P–H 62/21. HI 50482

Agriculture. London. Agriculture (London). See B–P–H 62/22. HI 50484

Agriculture. Montreal. Agriculture (Montreal). See B–P–H 62/23. HI 50485

Agriculture and agro-industries. Bombay. Vol. 1+, 1968+. Agric. Agro-Industr. HI 61276

Agriculture in Aichi. [Nogyo Aichi] Aichi, Japan. Agric. Aichi. See B–P–H 56/5. HI 50368

Agriculture in the americas. Washington, DC. Agric. Amer. (Washington). See B–P–H 56/8. HI 50371

Agriculture and animal husbandry. Lucknow, India. Agric. Anim. Husb. See B–P–H 56/9. HI 50372

Agriculture in Aomori. [Aomori nogyo.] Aomori, Japan. Agric Aomori. See B–P–H 56/10. HI 50373

Agriculture bulletin, university of Alberta. Edmonton. Nos. 1-32, 1962-77. Agric. Bull. Univ. Alberta. Superseded by: Agriculture and forestry bulletin. HI 61277

Agriculture, eco-systems and environment. Amsterdam. Vol. 9+, 1983+. Agric. Eco-Syst. Environm. Preceded by: Agro-ecosystems. HI 61278

Agriculture et élévage au Congo Belge et dans les colonies tropicales et subtropicales. Supplement to: Bulletin, association des intérêts coloniaux belges. Brussels. Vols. 1-?, 1927-31. Agric. Élevage Congo Belge Colon. Trop. Subtrop. 1-91-3. HI 61279

Agriculture and environment; international journal for scientific research on the relationship of agriculture and food production to the biosphere. Amsterdam. Vol. 1+, 1974+. Agric. Environm. (Amsterdam). HI 61280

Agriculture and the environment. Washington, DC. No. 1+, 1980+. Agric. Environm. (Washington). HI 61281

Agriculture and forestry bulletin, university of Alberta. Edmonton. Nos. 33-35; vol. 1+, 1977+. Agric. Forest. Bull. Univ. Alberta. Preceded by: Agriculture bulletin, university of Alberta. HI 61282

Agriculture and forestry news. [Nung lin t'ung hsin.] Taipei, Taiwan. Agric. Forest. News. See B–P–H 57/13. HI 50395

Agriculture and forestry series. [Pan american union.] Washington, DC. Agric. Forest. Ser. See B–P–H 57/15. HI 50397

Agriculture of Gifu prefecture. [Gifu-ken no nogyo.] Gifu, Japan. Agric. Gifu Prefect. See B–P–H 58/2. HI 50401

Agriculture in Gunma. [Nogyo Gunma] Maebashi, Japan. Agric. Gunma. See B–P–H 58/3. HI 50402

Agriculture handbook. Washington, DC. 1949+. Agric. Handb. HI 61283

Agriculture in Hiroshima. [Hiroshima nogyo.] Hiroshima. Agric. Hiroshima. See B–P–H 58/4. HI 50403

Agriculture in Hokkaido. [Hokuno.] Sapporo, Japan.

Agric. Hokkaido (Hokuno). See B–P–H 58/8. HI 50405

Agriculture in Hokkaido. [Nogyo Hokkaido.] Sapporo, Japan. Agric. Hokkaido (Nogyo Hokkaido). See B–P–H 58/9. HI 50406

Agriculture Hong Kong. Hong Kong. Vol. 1+, 1972+. Agric. Hong Kong. Preceded by: Agricultural science, Hong Kong. HI 61284

Agriculture and horticulture. [Nogyo oyobi engei.] Tokyo. Agric. & Hort. See B–P–H 58/10. HI 50407

Agriculture and horticulture index. ?-1982+. Agric. Hort. Index. HI 61285

Agriculture information bulletin. U.S. department of agriculture. Washington, DC. Agric. Inform. Bull. U.S.D.A. See B–P–H 58/15. HI 50410

Agriculture in Kagawa. [Nogyo Kagawa] Kagawa, Japan. Agric. Kagawa. See B–P–H 59/11. HI 50415

Agriculture leaflet, department of agriculture, Iraq = Agricultural leaflet, department of agriculture, Iraq. Basrah, Baghdad.

Agriculture in Northern Ireland. Belfast. Vol. 40+, 1960+. Agric. N. Ireland. Preceded by: Monthly agricultural report, ministry of agriculture, Northern Ireland. HI 61286

Agriculture Pakistan. Karachi, Pakistan. Agric. Pakistan. See B–P–H 59/25. HI 50427

Agriculture pratique. Paris. Agric. Prat. See B–P–H 60/ 1. HI 50428

Agriculture pratique des pays chauds. Paris. Agric. Prat. Pays Chauds. See B–P–H 60/2. HI 50429

Agriculture propagation. [Gojo noyu] Sendai, Japan. Agric. Propag. See B–P–H 60/3. HI 50431

Agriculture romande; revue mensuelle d'agriculture, de viticulture et d'arboriculture. Lausanne. Vols. 1-7, 1961-68. Agric. Romande. Preceded by: Revue romande d'agriculture, de viticulture et d'arboriculture. Superseded by: Revue suisse d'agriculture and Revue suisse de viticulture et d'arboriculture. HI 61287

Agriculture today. [Nogyo showa] Tokyo. Agric. Today. See B–P–H 61/8. HI 50450

Agriculture of warm regions. [Danchi nogaku.] Japan. Agric. Warm Regions. See B–P–H 61/11. HI 50452

Agriculture in Yamaguchi. [Nogyo Yamaguchi.] Yamaguchi, Japan. Agric. Yamaguchi. See B–P–H 61/13. HI 50456

Agriculturist and canadian journal. Toronto. 1848. Agric. Canad. J. Preceded by: British-american cultivator and Canada Farmer (Toronto 1847). Superseded by: Canadian agriculturist. 1-92-3. HI 61288

Agrikultúra. Nitra, Czechoslovakia. Agrikultúra. See B–P–H 63/5. HI 50486

Agrimensura. Montevideo. Agrimensura. See B–P–H 63/6. HI 50487

Agrindex; international index to agricultural literature [later International information system for the agricultural sciences and technology]. Rome. Experimental issue vol. 1, 1973; vol. 1+, 1975+. Agrindex. HI 61289

Agrinter = Indice agrícola de América latina y el Caribe. San Jose, Costa Rica.

Agrishell; revista de fitopatologia. Madrid. 1973+. Agrishell. HI 61290

Agrisul. Pelotas, Brazil. Agrisul. See B–P–H 63/7. HI 50488

Agritrigo. Rio de Janeiro. Agritrigo. See B–P–H 63/8. HI 50489

Agritrop. Paris. Experimental issue 1976+, 1976+. Agritrop. HI 61291

Agro. El Valle, Venezuela. Vols. 1-15, 1946-61. Agro (El Valle). 1-92-3. HI 50490

Agro. La Plata, Argentina. Agro (La Plata). See B–P–H 63/10. HI 50491

Agro. Maracay, Venezuela. Agro (Maracay). See B–P–H 63/11. HI 50492

Agro. Quito, Ecuador. Agro (Quito). See B–P–H 63/12. HI 50493

Agro. Tegucigalpa, Honduras. Agro (Tegucigalpa). See B–P–H 63/13. HI 50494

Agro, la mejor revista del campo. Mexico City. Agro Mejor Revista Campo. See B–P–H 63/14. HI 50495

Agro en México. Mexico City. Agro México. See B–P–H 63/15. HI 50496

Agro nacional i revista de agricultura. Guayaquil. Vol. 1+, 1935+. Agro Nac. Rev. Agric. HI 61292

Agro-ecosystems. Amsterdam. Vols. 1-8, 1974-83. Agro-Ecosystems. Superseded by: Agriculture, ecosystems and environment. HI 61293

Agro noticias. Lima. Agro Not. See B–P–H 63/16. HI 50497

Agro sur. Valdivia. Vol. 1+, 1973+. Agro Sur. HI 61294

Agrobiologija. Moscow = Agrobiologiya. Moscow.

Agrobiologija. Moscow & Leningrad = Materialy k biobibliografii uchenykh S S S R. Seriya biologicheskikh nauk; agrobiologiya. Moscow & Leningrad.

Agrobiologiya. Dvukhmesyachnyi nauchno-teoreticheskii zhurnal. Moscow. 1946-65. Agrobiologiya. Preceded by: Yarovizatsiya. Superseded by: Sel'skokhozyaistvennaya biologiya. 1-92-3. HI 61295

Agroborealis. Fairbanks, AK. Vol. 1+, 1969+. Agroborealis. HI 61296

Agrobotanika. Tapioszele, Hungary. Agrobotanika. See B–P–H 63/19. HI 50498

Agrochimica. Pisa. Agrochimica. See B–P–H 63/22. HI 50499

Agrociencia. Chapingo. Vols. 1-5, 1966-70; vol. 11+, 1972+. Agrociencia. For vols. 6-10 see: Agrociencia. Serie B. HI 61297

Agrochemie und Bodenkunde = Agrokémia és talajtan. Budapest. Agrokém. & Talajtan. See B–P–H 63/23.

Agrochimie et pédologie. Budapest = Agrokémia és talajtan. Budapest. Agrokém. & Talajtan. See B–P–H 63/23.

Agrociencia. Serie B [plant science and genetics]. Chapingo. Vols. 6-10, 1971-72. Agrociencia, B. Preceded and superseded by: Agrociencia. HI 61298

Agroecologia neotropical; una revista internacional dedicada al estudio ecológico de agroecosistemas en los neotrópicos. Ann Arbor, MI. Vol. 1+, 1990+. Agroecol. Neotrop. HI 61312

Agroforestry abstracts. Wallingford. Vol. 1+, 1988+. Agroforest. Abstr. HI 61424

Agroforestry review; quarterly of the international tree crops institute. Gravel Switch, KY. Vol. 1+, 1979+. Agroforest. Rev. HI 61300

Agroforestry systems. Dordrecht. Vol. 1+, 1982+. Agroforest. Systems. HI 61301

Agroforestry today. Nairobi. Vol. 1+, 1989+. Agroforest. Today. HI 74512

Agrokémia és talajtan. Budapest. Agrokém. & Talajtan. See B–P–H 63/23. HI 50500

Agrologist. Ottawa. Vol. 1+, 1972+. Agrologist. Preceded by: A I C review. 1-89-1. HI 61302

Agronomía. Caracas. Agronomía (Caracas). See B–P–H 64/18. HI 50516

Agronomía. Havana. Agronomía (Havana). See B–P–H 64/19. HI 50517

Agronomía. Lima. Agronomía (Lima). See B–P–H 64/ 20. HI 50518

Agronomía. Manizales, Colombia. Agronomía (Manizales). See B–P–H 64/21. HI 50519

Agronomía. Monterrey, Mexico. Agronomía (Monterrey). See B–P–H 64/22. HI 50520

Agronomía Puerto Bertoni, Paraguay. Agronomía (Puerto Bertoni). See B–P–H 64/23. HI 50521

Agronomia. [Escola nacional de agronomia.] Rio de Janeiro. Agronomia (Rio de Janeiro, Esc. Nac. Agron.). See B–P–H 64/24. HI 50522

Agronomía. [Centro de estudiantes de la facultad de agronomía de la Universidad de Chile.] Santiago. Agronomía (Santiago). See B–P–H 65/2. HI 50524

Agronomía. [Sociedad agronómica de Chile.] Santiago. Agronomía (Santiago, 1915-29). See B–P–H 65/1. HI 50523

Agronomia. [Sociedade brasileira de agronomia.] Rio de Janeiro. Agronomia (Soc. Brasil. Agron.). See B–P–H 65/3. HI 50525

Agronomia angolana. Luanda. Nos. 1-33, 1948-72. Agron. Angol. HI 61303

Agronomía. Buenos Aires. Vol. 25+, 1932+. Agronomía (Buenos Aires). Preceded by: Centro de estudiantes de agronomía. 1-93-1. HI 61304

Agronomia costarricense. San José, Costa Rica. Vol. 1+, 1977+. Agron. Costarric. HI 61305

Agronomia lusitana. Alcobaça, Portugal. Agron. Lusit. See B–P–H 64/5. HI 50505

Agronomia moçambicana. Lourenço Marques. Vol. 1+, 1967+. Agron. Moçamb. HI 61306

Agronomia sul rio grandense. Pôrto Alegre, Brazil. Agron. Sul Rio Grandense. See B–P–H 64/9. HI 50512

Agronomía tropical. Maracay, Venezuela. Agron. Trop. (Maracay). See B–P–H 64/11. HI 50513

Agronómica. Valencia, Venezuela. Agronómica (Valencia). See B–P–H 65/8. HI 50526

Agronómico. Campinas, Brazil. Agronómico (Campinas). See B–P–H 65/9. HI 50527

Agronómico. São Paulo. Agronómico (São Paulo). See B–P–H 65/10. HI 50529

Agronomie. Versailles. 1981+. Agronomie. Preceded by: Annales agronomiques. HI 61307

Agronomie coloniale; bulletin mensuel de l'institut national d'agronomie de la France d'outre-mer. Paris. N.s. vols. 1-28, 1913-39. Agron. Colon. Preceded by: Agronomie tropicale. Uccle. Superseded by: Agronomie tropicale. Nogent-sur-Marne. 1-93-2. HI 61308

Agronomie tropicale. Uccle. Vols. 1-6, 1909-13. Agron. Trop. (Uccle). Superseded by: Agronomie coloniale. 1-93-2. HI 61309

Agronomie tropicale. Nogent-sur-Marne. Vol. 1+, 1946+. Agron. Trop. (Nogent-sur-Marne). Preceded by: Agronomie coloniale. From vol. 17 onwards included the following series: Série, riz et riziculture et cultures vivrières tropicales, Série, agronomie générale, études techniques and Série, agronomie générale, étude scientifiques. 1-93-2. HI 61310

Agronomisch-historische bijdragen. Wageningen. Agron.-Hist. Bijdr. See B–P–H 64/15. HI 50514

Agronomisch-historisch jaarboek. Wageningen. Agron.-Hist. Jaarb. See B–P–H 64/16. HI 50515

Agronomy. New York, NY. Agronomy. See B–P–H 65/11. HI 50530

Agronomy abstracts. Madison, WI. Agron. Abstr. See B–P–H 63/24. HI 50502

Agronomy current literature. Washington, DC. Agron. Curr. Lit. See B–P–H 63/27. HI 50503

Agronomy department pamphlet = South Dakota agricultural experiment station. Agronomy department pamphlet. Brookings, SD. South Dakota Agric. Exp. Sta. Agron. Dept. Pam. See B–P–H 847/24.

Agronomy information circular, North Carolina agricultural experiment station = North Carolina agricultural experiment station agronomy information circular. Raleigh, NC. North Carolina Agric. Exp. Sta. Agron. Inform. Circ. See B–P–H 663/4.

Agronomy journal. Madison, WI. Agron. J. See B–P–H 64/3. HI 50504

Agronomy news. Madison, WI. 1956+. Agron. News. HI 61311

Agronomy notes. Washington, DC. Agron. Notes. See B–P–H 64/7. HI 50507

Agronomy review. Lincoln, New Zealand. Agron. Rev. See B–P–H 64/ 8. HI 50510

Agro-noticias. Paraná, Argentina. Agro-Not. See B–P–H 65/12. HI 50531

Agropecuaria brasileira: Resumos = A B R. Brasilia.

Agroplantae; agricultural science in South Africa. Pretoria. Vol. 1+, 1969+. Agroplantae. Preceded by: South african journal of agricultural science. HI 61313

Agros. Cochabamba, Bolivia. Agros (Cochabamba). See B–P–H 65/13. HI 50532

Agros. Lisbon. Agros (Lisbon). See B–P–H 65/14. HI 50533

Agros. Madrid. Agros (Madrid). See B–P–H 65/15. HI 50535

Agros. Montevideo. Agros (Montevideo). See B–P–H 65/16. HI 50536

Agros. Pelotas, Brazil. Agros (Pelotas). See B–P–H 65/ 17. HI 50537

Agrostologist. London. Agrostologist. See B–P–H 65/ 19. HI 50538

Agrostology bulletin, United States department of agriculture = Bulletin, division of agrostology, United States department of agriculture. Washington, DC.

Agrotecnia de Cuba. Havana. Agrotecn. Cuba. See B–P–H 65/20. HI 50539

Agrotécnica. Havana Agrotécnica. See B–P–H 65/21. HI 50540

Agrotekhnika i biologiya sel'skokhozyaistvennykh kul'tur; sbornik nauchnykh trudov. Ul'yanovsk. Vol. 1+, 1976+. Agrotekhn. Biol. Sel'skokhoz. Kul't. HI 61314

Agrotike zoe. Athens. Agrotike Zoe. See B–P–H 65/22. HI 50541

Agway cooperator. Syracuse, NY. Agway Coop. See B–P–H 65/23. HI 50542

Ahalgazrda meçnier mušakţe šromebis krebuli = Sbornik trudov molodykh nauchnykh rabotnikov, institut botaniki, akademiya nauk gruzinskoi S S R. Tiflis.

A.H.C. news; news bulletin of the american horticultural council. Ithaca, NY = A H C news; news bulletin of the american horticultural council. Ithaca, NY.

Ährenlese des Georgikons. Bécs [=Vienna]. Ährenlese Georgikons. See B–P–H 65/25. HI 50543

AIBDA boletín especial. [Associación interamericana de bibliotecarios y documentalistas agrícolas.] Turrialba, Costa Rica = A I B D A boletín especial. [Associación interamericana de bibliotecarios y documentalistas agrícolas.] Turrialba, Costa Rica.

Aichi gakugei daigaku kenkyu hokoku = Bulletin of the Aichi gakugei university, natural science. Okazaki, Japan. Bull. Aichi Gakugei Univ. Nat. Sci. See B–P–H 237/9.

Aichi-ken engei shikenjo kenkyu hokoku = Bulletin of Aichi horticultural experiment station. Aichi.

Aichi-ken engei shikenjo kenkyu hokoku = Research bulletin of the Aichi-ken agricultural research center. Series A, field crops. Aichi.

Aichi-ken engei shikenjo kenkyu hokoku = Research bulletin of the Aichi-ken agricultural research center. Series B, horticulture. Aichi.

Aichi-ken nogyo shikenjo iho = Bulletin of the Aichi prefecture agricultural experiment station. Aichi, Japan. Bull. Aichi Prefect. Agric. Exp. Sta. See 237/ 11.

Aichi-ken nogyo sogo shikenjo tokubetsu hokoku = Special bulletin of the Aichi-ken agricultural research center. Nagakute.

Aichi-ken ringyo shikenjo hokoku = Bulletin of the Aichi prefecture forest experiment station. Aichi.

Aika. Helsinki. Aika. See B–P–H 66/8. HI 50544

A'in Shams science bulletin. Cairo. No. 1+, 1956+. A'in Shams Sci. Bull. HI 61315

Air repair. Vols. 1-4, 1951-55. Air Repair. Superseded by: Journal of the air pollution control association.

HI 74808

Airone. Milan. 1981+. Airone. HI 61316

Akadémiai értesitö. Budapest. Akad. Értes. See B–P–H 66/10. HI 50545

Akademičeskija Izvěstija. St. Petersburg = Akademicheskiya Izvěstiya. St. Petersburg.

Akademicheskiya Izvěstiya. St. Petersburg. Vols. 1-8, 1779-81. Akad. Izv. 1-111-1. HI 61317

Akademie der Wissenschaften und der Literatur, Mainz. Abhandlungen der mathematisch-naturwissenschaftlichen Klasse. Mainz & Wiesbaden. 1950+. Akad. Wiss. Mainz, Abh. Math.-Naturwiss. Kl. HI 61318

Akademie der Wissenschaften und der Literatur, Mainz. Jahrbuch. Akad. Wiss. Jahrb. See B–P–H 66/23. HI 50546

Akademie der Wissenschaften in Wien. Mathematisch-Naturwissenschaftliche Klasse. Denkschriften. Vienna. Akad. Wiss. Wien, Math.-Naturwiss. Kl., Denkschr. See B–P–H 66/24. HI 50547

Akademie der Wissenschaften in Wien. Sitzungsberichte. Mathematisch-naturwissenschaftliche Klasse. Abteilung 1. Vienna. Akad. Wiss. Wien Sitzungsber., Math.-Naturwiss, Kl., Abt. 1. See B–P–H 66/26. HI 50548

Akita daigaku gakugei gakubu kenkyu kiyo = Memoirs of the faculty of (liberal arts and) education, Akita university. Natural science. Akita.

Akita daigaku gakugei gakubu kenkyu kiyo. Shizen kagaku = Memoirs of gakugei faculty, Akita university. Natural science. Akita, Japan. Mem. Gakugei Fac. Akita Univ., Nat. Sci. See B–P–H 573/3.

Akita-ken kaju shikenjo. Kenkyu hokoku = Bulletin of the Akita fruit-tree experiment station. Akita.

Akita-kenritsu nogyo tanki daigaku kenkyu hokiku = Bulletin of the Akita prefectural college of agriculture. Ogata-Mura.

Akiyoshi-dai kagaku hakubutsukan hokoku = Bulletin of the akiyoshi-dai museum of natural history. Yamaguchi.

Akkerbouw. Ghent. Akkerbouw. See B–P–H 67/3. HI 50549

Akklimatizacija. Ežeměsjačnoe izdanie Komiteta Akklimatizacija = Akklimatizatsiya. Ezhemĕsyachnoe izdanie Komiteta Akklimatizatsiya. Moscow.

Akklimatizatsiya. Ezhemĕsyachnoe izdanie Komiteta Akklimatizatsiya. Moscow. Vols. 1-4, 1860-63. Akklimatizatsiya. Preceded by: Zapiski Komiteta Akklimatizatsiya, uchrezhdennago pri Imperatorskom Moskovskom Obshchestvě Sel'skago Khozyaistva. HI 61319

Akklimatizatsiya roslin, vols. 4-5, 1953-58 = Trudy Botanicheskogo sada. Kiev.

Aklimatizatsiya roslin, vols. 6-7, 1959-60 = Trudy Botanicheskogo sada. Kiev.

Aktuelle Literaturinformation aus dem Ostbau. Berlin. No. 1+, 1972+. Aktuelle LitInform. Ostbau. HI 61321

Akvaristické listy. Prague. Akvar. Listy. See B–P–H 67/8. HI 50552

Akvarium i Komnatnye Rastenija = Akvarium i Komnatnyya Rasteniya. Moscow.

Akvarium i Komnatnye Rasteniya. Moscow. 1908-16. Akvar. Komnatn. Rast. HI 61322

Akvárium és terrárium. Budapest. Akvár. & Terrár. See B–P–H 67/6. HI 50551

Al-biah. Khartoum. Vol. 1+, 1980+. Al-biah. HI 61323

Al dia; boletín informativo de la organización de estudios tropicales en Costa Rica. San José, Costa Rica. Vol. 1+, 1983+. Al Dia. HI 61324

Al-khalij al-Arabi = Journal of the centre for arab gulf studies. Basrah.

Al-khalij al-arabi = Arab gulf. Basrah.

Al-ma' had al-' ilmi = Documents de l'institut scientifique, université Mohamed V. Rabat.

Al-Ma'had al-qaumi li-ăl-buhut al-ğabiyyah = Note de recherche, institut national de recherches forestières. Ariana.

Al-magallah al-misriyyah li-al-basatin = Egyptian journal of horticulture. Dokki, Cairo.

Al-magallah al-misriyyah li-amrad an-nabat = Egyptian journal of phytopathology. Dokki, Cairo.

Al-majallah al-ilmiyah li-kullivat al-ulum bi-jamiat al-Riyad = Bulletin of the faculty of science, university of Riyadh. Riyadh.

Al-reem = Al-rim. Amman.

Al-rim. Amman. No. ?-10+, 19??-82+. Al-rim. HI 61325

Alabama agricultural experiment station of the agricultural and mechanical college. Annual report. Auburn, AL. Alabama Agric. Exp. Sta. Agric. Coll. Annual Rep. See B–P–H 67/10. HI 50553

Alabama agricultural experiment station of the agricultural and mechanical college. Bulletin. Auburn, AL. Nos. 61-319, 1895-1960? Alabama Agric. Exp. Sta. Agric. Coll. Bull. Preceded by: Bulletin, new series. Agricultural experiment station of the agricultural and mechanical college. Superseded by: Bulletin, agricultural experiment station, Auburn university. HI 61326

Alabama agricultural experiment station of the Alabama Polytechnic Institute. Bulletin = Alabama agricultural experiment station of the agricultural and mechanical college. Bulletin. Auburn, AL.

Alabama agricultural experiment station of the Alabama polytechnic institute, circular. Auburn, AL. Nos. 1-136, 1906-59. Alabama Agric. Exp. Sta. Alabama Polytechn. Inst. Circ. Superseded by: Circular, agricultural experiment station, Alabama. HI 61327

Alabama agricultural experiment station of the Alabama polytechnic institute. Leaflet. Auburn, AL. Alabama Agric. Exp. Sta. Alabama Polytechn. Inst. Leafl. See B–P–H 67/13. HI 50554

Alabama agricultural experiment station. Forestry departmental series. Auburn, AL. Alabama Agric. Exp. Sta., Forest. Dept. Ser. See B–P–H 67/17. HI 50555

Alabama agricultural experiment station. Horticulture series. Auburn, AL. Alabama Agric. Exp. Sta., Hort. Ser. See B–P–H 67/18. HI 50556

Alabama agricultural experiment station. Progress report series. Auburn, AL. 1944+. Alabama Agric. Exp. Sta. Progr. Rep. Ser. HI 61328

Alabama agricultural experiment station. Research report series. Auburn, AL. 1983+. Alabama Agric. Exp. Sta. Res. Rep. Ser. HI 61329

Alabama conservation. Montgomery, AL. Vol. 1+, 1940+. Alabama Conserv. Preceded by: Alabama game and fish news [not entered]. HI 61330

Alabama forest products. Montgomery, AL. Vols. ?-20, 19??-77. Alabama Forest Prod. Superseded by: Alabama forests. HI 61331

Alabama forests. Montgomery, AL. Vol. 21+, 1978+. Alabama Forests. Preceded by: Alabama forest products. HI 61332

Alabama marine resources bulletin. Dauphin Island, AL. No. 2+, 1969+. Alabama Mar. Resources Bull. Preceded by: Marine resources bulletin. HI 61333

Alafua agricultural bulletin. Alafua, W. Samoa. Vol. 1+, 1976+. Alafua Agric. Bull. HI 61334

Alaska agricultural college and school of mines. Agricultural experiment station progress report. College, AK. Alaska Agric. Coll. School MInes Agric. Exp. Sta. Progr. Rep. See B–P–H 68/3. HI 50557

Alaska agricultural experiment stations. Annual report. Washington, DC. 1907-32. Alaska Agric. Exp. Sta. Annual Rep. Preceded by: U S department of agriculture. Annual report of the Alaska agricultural experiment station. HI 61335

Alaska agricultural experiment stations. Bulletin. Washington, DC. Alaska Agric. Exp. Sta. Bull. See B–P–H 68/5. HI 50559

Alaska agricultural experiment stations. Circular. Washington, DC. Alaska Agric. Exp. Sta. Circ. See B–P–H 68/6. HI 50560

Alaska agricultural experiment stations. Report. Washington, DC. Alaska Agric. Exp. Sta. Rep. See B–P–H 68/7. HI 50561

Alaska university. Biological papers. College, AK. Alaska Univ. Biol. Pap. See B–P–H 68/8. HI 50562

Alberta horticulturist. Lacombe, Alberta. Alberta Hort. See B–P–H 68/10. HI 50671

Alberta naturalist. Edmonton. Vol. 1+, 1971+; Special issue, no. 1+, 1981+. Alberta Naturalist. HI 61336

Alberta wilderness association newsletter = Newsletter, Alberta wilderness association. Calgary.

Albertina. Belje, Yugoslavia. Albertina. See B–P–H 68/13. HI 50563

Albertoa. Rio de Janeiro. Vol. 1+, 1986+. Albertoa. HI 61337

Albrecht von Graefe's Archiv für Ophthalmologie. Berlin. Albrecht von Graefe's Arch. Ophthalmol. See B–P–H 68/15. HI 50564

Albrecht von Graefe's Archiv für Ophthalmologie, vereinigt mit Archiv für Augenheilkunde. Berlin. Albrecht von Graefe's Arch. Ophthalmol & Arch. Augenheilk. See B–P–H 68/16. HI 50565

Albrecht-Thaer Archiv. Berlin. Vols. 1-14, 1956-70. Albrecht-Thaer-Arch. Superseded by: Archiv für Bodenfruchtbarkeit und Pflanzenproduktion. HI 61338

Album van Eeden. Haarlems flora. Afbeeldingen in kleurendruk van verschillende bol- en knolgewassen. Haarlem. Album Eeden. See B–P–H 68/20. HI 50566

Album der natuur. Haarlem. Album Natuur (Haarlem). See B–P–H 68/22. HI 50568

Album der natuur ter verspreiding van natuurkennis voor iedere stand. Antwerp. Album Natuur (Antwerp). See B–P–H 68/21. HI 50567

Album van natuurmonumenten van Nederlandsch-Indië. Batavia, Dutch E. Indies [=Jakarta, Indonesia]. Album Natuurmonum. Ned.-Indië. See B–P–H 69/1. HI 50569

Album de pomologie. Brussels. Album Pomol. See B–P–H 69/2. HI 50570

Alcance. Facultad de agricultura, universidad central, Venezuela. Caracas. Vol. 1+, 1956+. Alcance. HI 61339

Aldabra newsletter. London. No. ?-12+, 19??-76+. Aldabra Newslett. HI 61340

Alert. Philadelphia, PA. Vol. 1+, 1977+. Alert. HI 61341

Alexandria journal of agricultural research. Alexandria, Egypt [United Arab Republic]. Alexandria J. Agric. Res. See B–P–H 69/5. HI 50571

Alföldi tudományos gyüjtemény. Az alföldi tudományos intézet évkönyve. Budapest. Alföldi Tud. Gyüjt. See B–P–H 69/6. HI 50573

Algae abstracts; a guide to the literature. London, etc. Vol. 1+, 1969+. Algae Abstr. HI 61342

Algal newsletter. Odessa, TX. 1977+. Algal Newslett. HI 61343

Algemeen agrarisch archief. Amsterdam. Alg. Agrar. Arch. See B–P–H 69/7. HI 50574

Algemeen hollandsch landbouwblad. The Hague. Alg. Holl. Landbouwbl. See B–P–H 69/10. HI 50577

Algemeen landbouwweekblad voor Nederlandsch-Indië. Bandoeng, Dutch E. Indies [=Bandung, Indonesia]. Alg. Landbouwweekbl. Ned.-Indië. See B–P–H 69/12. HI 50578

Algemeen magazijn van wetenschap, konst en smaak. Amsterdam. Alg. Mag. Wetensch. See B–P–H 69/13. HI 50580

Algemeene genees- natuur- en huishoudkundige jaarboeken. Dordrecht & Amsterdam. Alg. Genees-Natuur- Huishoudk. Jaarb. See B–P–H 69/8. HI 50576

Algemeene konst- en letter-bode = Algemene konst- en letter-bode. Haarlem.

Algemeene konst- en letterbode = Algemene konst- en letter-bode. Haarlem.

Algemene konst- en letter-bode. Haarlem. Vols. 1-11, 1788-93; 1801-62. Alg. Konst- Lett.-Bode. For 1794-1800 see: Nieuwe algemene konst- en letter-bode. 1-136-2. HI 51496

Algemeene vaderlandsche letteroefeningen. Amsterdam. Alg. Vaderl. Letteroefen. See B–P–H 69/14. HI 50581

Algodón. Lima. Algodón (Lima). See B–P–H 69/15. HI 50583

Algodón mexicano. Mexico City. Algodón Mex. See B–P–H 69/16. HI 50585

Algological studies. [Archiv für Hydrobiologie, Supplementband.] Stuttgart. Vol. 1, 1970. Algol. Stud. Later vols. published as: Archiv für Hydrobiologie, Supplementband. HI 61344

Alicmar. Cumana. No. 1+, 1979+. Alicmar. HI 61345

Aliso. Claremont, CA. Aliso. See B–P–H 69/17. HI 50586

Alkaloids; chemistry and physiology [later and pharmacology]. New York, London. Vol. 1+, 1950+. Alkaloids. HI 61346

Allahabad farmer. Allahabad Farmer. See B–P–H 69/18. HI 50587

Allahabad university studies. Allahabad, India. Allahabad Univ. Stud. See B–P–H 69/19. HI 50588

Államilag minösített növényfajták jegyzéke. Budapest. Államilag Minös. Növényfajt. Jegyzéke. See B–P–H 69/20. HI 50589

Allan Hancock foundation. Publications. Occasional papers. Los Angeles, CA. Allan Hancock Found. Publ. Occas. Pap. See B–P–H 69/22. HI 50590

Allan Hancock monographs in marine biology. Los Angeles, CA. Allan Hancock Monogr. Mar. Biol. See B–P–H 70/2. HI 50591

Állattani közlemények. Budapest. Állatt. Közlem. See B–P–H 70/3. HI 50592

Allergnädigst privilegierte Realzeitung der Wissenschaften, Künste und der Commercien. Vienna. Allergnäd. Privileg. Realzeitung Wiss. See B–P–H 70/4. HI 50593

Allergnädigst-privilegirte Anzeigen aus sämmtlich-kaiserlich-königlichen Erbländern, herausgegeben von einer Gesellschaft. Vienna. Allergnäd.-Privileg. Anz. Sämmtl.-K.-K. Erbländern. See B–P–H 70/ 5. HI 50594

Allerneueste Mannigfaltigkeiten. Eine gemeinnützige Wochenschrift. Berlin. Allerneueste Mannigfaltigk. See B–P–H 70/7. HI 50595

Allertonia; a series of occasional papers. Lawai, HI. Vol. 1+, 1975+. Allertonia. HI 61347

Allegemeine Annalen der Gewerbskunde, oder allgemeines physikalisches. botanisches, mechanisches, ... Magazin. Leipzig. Allg. Ann. Gewerbsk. See B–P–H 70/8. HI 50597

Allgemeine Bibliographie für Deutschland. Leipzig. Allg. Bibliogr. Deutschl. See B–P–H 70/10. HI 50599

Allgemeine Botanische Zeitschrift für Systematik, Floristik, Pflanzengeographie etc. Karlsruhe. Allg. Bot. Z. Syst. See B–P–H 70/13. HI 50601

Allgemeine deutsche Bibliothek. Berlin & Stettin [Poland]. Allg. Deutsche Biblioth. See B–P–H 70/19. HI 50602

Allgemeine deutsche Garten-Zeitung. Passau, Germany. Allg. Deutsche Gart.-Zeitung See B–P–H 70/22. HI 50606

Allgemeine Deutsche Gärtner-Zeitung. Berlin. Allg. Deutsche Gärtn.-Zeitung. See B–P–H 70/20. HI 50604

Allgemeine Deutsche Gärtner-Zeitung & Stellen-Anzeiger für Gärtner. Berlin. Allg. Deutsche Gärtn.-Zeitung & Stellen-Anz. Gärtn. See B–P–H 70/21.

HI 50605

Allgemeine deutsche naturhistorische Zeitung. Dresden. Allg. Deutsche Naturhist. Zeitung. See B–P–H 70/24. HI 50607

Allgemeine forst- und holzwirtschaftliche Zeitung. Vienna. Allg. Forst- Holzw. Zeitung. See B–P–H 71/1. HI 50608

Allgemeine Forst- und Jagd-Zeitung. Frankfurt. Allg. Forst- Jagd-Zeitung. See B–P–H 71/4. HI 50611

Allgemeine Forst- und Jagd-Zeitung. Vienna. Allg. Forst- Jagd-Zeitung (Vienna). See B–P–H 71/5. HI 50612

Allgemeine Forstzeitschrift. Munich. Allg. Forstz. See B–P–H 71/9. HI 50614

Allgemeine Forstzeitung. Vienna. Allg. Forstzeitung. See B–P–H 71/10. HI 50615

Allgemeine Gartenzeitung. Berlin. Allg. Gartenzeitung. See B–P–H 71/14. HI 50616

Allgemeine geographische Ephemeriden. Weimar. Allg. Geogr. Ephem. See B–P–H 71/17. HI 50618

Allgemeine Historie der Reisen zu Wasser und zu Lande oder Sammlung von Reisebeschreibungen ... in Europa, Asia, Africa und America ... durch eine Gesellschaft gelehrter Männer übersetzt. Leipzig. Allg. Hist. Reisen. See B–P–H 71/20. HI 50620

Allgemeine Jahrbücher der Forst- und Jagdkunde = Zeitschrift für das Jagd- und Forstwesenn mit besonderer Rücksicht auf Baiern. Bamberg.

Allgemeine land- und forstwirthschaftliche Zeitung. Vienna. Allg. Land- Forstw. Zeitung. See B–P–H 71/24. HI 50622

Allgemeine landwirthschaftliche Zeitung. Halle. Allg. Landw. Zeitung. See B–P–H 71/25. HI 50624

Allgemeine Literatur-Zeitung. Halle & Leipzig. 1804-49. Allg. Lit.-Zeitung (Halle & Leipzig). Preceded by: Allgemeine Literatur-Zeitung. Jena. 1-144-1. HI 53869

Allgemeine Literatur-Zeitung. Jena. 1785-1803. Allg. Lit.-Zeitung (Jena). Superseded by: Jenaische allgemeine Literatur-Zeitung and Allgemeine Literatur-Zeitung. Halle & Leipzig. 1-144-1. HI 50074

Allgemeine medizinische Annalen. Altenburg. 1800. Allg. Med. Ann. Preceded by: Medizinische National-Zeitung für Deutschland, und die mit selbigem zunächst verbundenen Staaten. Superseded by: Allgemeine medizinische Annalen des Neunzehnten Jahrhunderts. 1-144-1. HI 61348

Allgemeine medizinische Annalen des neunzehnten Jahrhunderts. Altenburg. 1801-30; Suppl. for 1801/10, 1810, and for 1821-25, 1825. [Vols. for 1816-20 also numbered "Zweites Jahrzehend. Zweites Quinquennium" vols. 1-5; vols. for 1826-30 marked as "Neue Folgereihe"]. Allg. Med. Ann. Preceded by: Allgemeine medizinische Annalen. Superseded by: Allgemeine medizinische Zeitung mit Berücksichtigung des Neuesten und Interessantesten aus der allgemeinen Naturkunde. 1-144-1. HI 52015

Allgemeine medizinische Annalen des zweiten Jahrzendes des neunzehnten Jahrhunderts. Altenburg = Allgemeine medizinische Annalen des neunzehnten Jahrhunderts. Altenburg.

Allgemeine medizinische Zeitschrift des chinesischen Reichs = General medical journal. Peking. Gen. Med. J. See B–P–H 396/3.

Allgemeine medizinische Zeitung; mit Berücksichtigung des Neuesten und Interessantesten aus der allgemeinen Naturkunde. Altenburg. 1831-38. Allg. Med. Zeitung. Preceded by: Allgemeine medizinische Annalen des neunzehnten Jahrhunderts. 1-144-1. HI 61349

Allgemeine nordische Annalen der Chemie für die Freunde der Naturkunde und Arzneiwissenschaft, insbesondere der Pharmacie, Arzneimittellehre, Physiologie, Physik, Mineralogie und Technologie im Russischen Reiche. St. Petersburg. Allg. Nord. Ann. Chem. Freunde Naturk. Arzneiwiss. See B–P–H 72/9. HI 50627

Allgemeine Oesterreichische Zeitschrift für den Landwirth, Forstmann und Gärtner. Vienna. Allg. Oesterr. Z. Landwirth. See B–P–H 72/11. HI 50630

Allgemeine oesterreichische Zeitung für Forstkultur. Prague. Allg. Oesterr. Zeitung Forstkult. See B–P–H 71/13. HI 50631

Allgemeine polytechnische Zeitung. Nuremberg. Allg. Polytechn. Zeitung See B–P–H 72/15. HI 50632

Allgemeine teutsche Vaterlandskunde. Wochenschrift der Geschichte, Natur- und Landeskunde, Natur und Kunst, dem Alterthum gewidmet ... Erfurt. Allg. Teutsche Vaterlandsk. See B–P–H 73/3. HI 50638

Allgemeine Thüringische Gartenzeitung. Erfurt. Allg. Thüring. Gartenzeitung. See B–P–H 73/5. HI 50639

Allgemeine Thüringische Vaterlandskunde, Wochenschrift für Geschichte, Natur- und Landeskunde Thüringens. Erfurt. Allg. Thüring. Vaterlandsk. See B–P–H 73/6. HI 50640

Allgemeine Wiener medicinische Zeitung. Vienna. Allg. Wiener Med. Zeitung. See B–P–H 73/8. HI 50641

Allgemeine Zeitschrift für die Heilwissenschaft. St. Petersburg. Vol. 1, 1818. Allg. Z. Heilwiss. HI 61350

Allgemeine Zeitung für Land- und Forstwirthschaft, Gartenbau und Obstbaumzucht ... Leipzig. Allg. Zeitung Land- Forstw. See B–P–H 73/10. HI 50642

Allgemeiner Kameral-, Oekonomie-, Forst- und Technologie-Korrespondent. Erlangen. Vol. [5], 1808. Allg. Kameral- Technol-Korresp. Preceded by: Allgemeiner Kameral-, Oekonomie, Forst- und Technologie-Korrespondent für Deutschland. Superseded by: Allgemeiner Kameral-, Polizei-, Forst-, Technologie- und Handels-Korrespondent. HI 61351

Allgemeiner Kameral-, Oekonomie, Forst- und Technologie-Korrespondent für Deutschland. Erlangen. Vols. 1-[4], 1806-07. Allg. Kameral- Technol.-Korresp. Deutschl. Superseded by: Allgemeiner Kameral-, Oekonomie-, Forst- und Technologie-Korrespondent. HI 61352

Allgemeiner Kameral-, Polizei-, Forst-, Technologie- und Handels-Korrespondent. Erlangen. Vols. 6-12, 1808-12; vols. 14-18, 1812-14. Allg. Kameral- Handels-Korresp. For vol. 13 see: Kameral-Korrespondent oder allgemeiner Kameral-, Polizei, Oekonomie-, Forst-, Technologie- und Handels-Korrespondent für denkende Geschäftsmänner und gebildete Leser. Superseded by: Allgemeiner Kameral-, Polizei-, Oekonomie-, Forst-, Technologie- und Handels-Korrespondent für Teutschland. HI 61353

Allgemeiner Kameral-, Polizei-, Oekonomie-, Forst-, Technologie- und Handels-Korrespondent für Teutschland. Erlangen. Vols. 19-20, 1815. Allg. Kameral- Handels-Korresp. Teutschl. Preceded by: Allgemeiner Kameral-, Polizei, Forst-, Technologie- und handels-Korrespondent. HI 61354

Allgemeiner litterarischer Anzeiger, oder Annalen der gesammten Litteratur für die geschwinde Bekanntmachung verschiedener Nachrichten aus dem Gebiete der Gelehrsamkeit und Kunst. Leipzig. Allg. Litt. Anz. See B–P–H 72/1. HI 50625

Allgemeines Archiv für die Länder-, Völker- und Staatenkunde. Göttingen. Allg. Arch. Länder- Völker-Staatenk. See B–P–H 70/9. HI 50598

Allgemeines Forst- und Jagd-Archiv. Stuttgart. Allg. Forst- Jagd-Arch. See B–P–H 71/2. HI 50609

Allgemeines Forst- und Jagd-Journal. Zeitblatt für Forst- und Landwirthe, Jagdliebhaber, Herrschaftsbesitzer und Freunde der Industrie. Prague. Allg. Forst- Jagd-J. See B–P–H 71/3. HI 50610

Allgemeines Forst- und Seidenbau-Journal. Prague. Allg. Forst- Seidenbau-J. See B–P–H 71/6. HI 50613

Allgemeines Helvetsiches Magazin zur Beförderung der Inländischen Naturkunde. Zurich. Allg. Helv. Mag. Beförd. Inl. Naturk. See B–P–H 71/19. HI 50619

Allgemeines Journal der Chemie. Berlin. Allg. J. Chem. See B–P–H 71/21. HI 50621

Allgemeines Magazin der Natur, Kunst und Wissenschaften. Leipzig. Allg. Mag. Natur. See B–P–H 72/3. HI 50626

Allgemeines oeconomisches Forst-Magazin ... Frankfurt a. M. & Leipzig. Allg. Oecon. Forst-Mag. See B–P–H 72/10. HI 50629

Allgemeines Repertorium der Kritik oder vollständiges, systematisch geordnetes Verzeichniss aller Werke, welche seit dem Jahre 1826 erschienen und in Deutschlands kritischen Blättern beurtheilt wordensind. Berlin. Allg. Repert. Krit. See B–P–H 72/16. HI 50633

Allgemeines Repertorium der Literatur. Jena. Allg. Repert. Lit. See B–P–H 72/17. HI 50634

Allgemeines Repertorium der neuesten in- und ausländischen Literatur. Leipzig & Vienna. Allg. Repert. Neuesten In- Ausl. Lit. See B–P–H 72/19. HI 50635

Allgemeines schwedisches Gelehrsamkeits-Archiv unter Gustafs des Dritten Regierung. Leipzig. Allg. Schwed. Gelehrsamk.-Arch. See B–P–H 72/20 HI 50636

Allgemeines teutsches Garten-Magazin oder gemeinnützige Beiträge für alle Theile des praktischen Gartenwesens. Weimar. Allg. Teutsch. Gart.-Mag. See B–P–H 73/2. HI 50637

Allionia. Bolletino dell'istituto ed orto botanico dell'università di Torino. Turin. Allionia. See B–P–H 73/17. HI 50643

Allmänna journalen. Stockholm. Almänna J. See B–P–H 73/22. HI 50644

Allmänna svenska utsädesföreningens tidskrift. Malmo, Sweden. Allmänna Svenska Utsädesfören. Tidskr. See B–P–H 73/23. HI 50645

Almanach aller um Hamburg liegenden Gärten. Hamburg. Alman. Aller um Hamburg Liegenden Gärt. See B–P–H 74/1. HI 50648

Almanach du bon jardinier. Paris. Alman. Bon Jard. See B–P–H 74/2. HI 50649

Almanach de Carlsbad, ou mélanges médicaux, scientifiques et littéraires, relatifs à ces thermes et au pays. Prague. Alman. Carlsbad. See B–P–H 74/4. HI 50651

Almanach české akademie Císaře Františka Josefa pro vědy, slovesnost a umění. Prague. Alman. Českě Akad. Císaře Františka Josefa Vědy. See B–P–H 74/5. HI 50652

Almanach české akademie věd a uměni. Prague. Alman. České Akad. Věd. See B–P–H 74/6. HI 50653

Almanach de la chimie. Rouen. Alman. Chim. See B–P–H 74/7. HI 50654

Almanach der Georg-Augusts-Universität zu Göttingen. Lüneburg, German. Alman. Georg-Augusta-Univ. Göttingen. See B–P–H 74/10. HI 50657

Almanach für das Jahr ... ed. Heinrich Berghaus.

Stuttgart. Alman. (Berghaus). See B–P–H 73/34. HI 50646

Almanach der Königlich-Bayerischen Akademie der Wissenschaften. Munich. Alman. Königl.-Bayer. Akad. Wiss. See B–P–H 74/1. HI 50658

Almanach der neuesten Forstschritte, Erfindungen und Entdeckungen in den spekulativen und positiven Wissenschaften. Erfurt. Alman. Neuesten Fortschr. Wiss. See B–P–H 74/14. HI 50660

Almanach de la physique instructive et amusante. Rouen. Alman. Phys. Instruct. Amusante. See B–P–H 74/15. HI 50661

Almanach der Reisen, oder unterhaltende Darstellung der Entdeckungen des 18. Jahrhunderts. Leipzig. Alman. Reisen. See B–P–H 74/16. HI 50662

Almanach des sciences. Paris. Alman. Sci. See B–P–H 74/17. HI 50663

Almanak der akademie van Groningen. Groningen. Alman. Akad. Groningen. See B–P–H 73/35. HI 50647

Almanak van het utrechtsche landbouw-genootschap. Utrecht. Alman. Utrechtscher Landb.-Genootsch. See B–P–H 74/18. HI 50665

Almanaque cafetero. Bogotá. Alman. Cafetero. See B–P–H 74/3. HI 50650

Almanaque cientifico. Sierre, Switzerland. Alman. Ci. See B–P–H 74/8. HI 50656

Almanaque del ministerio de agricultura y ganadería de la nación. Buenos Aires. Alman. Minist. Agric. See B–P–H 74/13. HI 50659

Almeennytt samler. Copenhagen. 18??- pre.1824. Almeennytt Samler. Superseded by: Ny almeennyttig samler. HI 61355

Almindelig dansk landbotidende; organ for landoekonomiske og almene landbo-interesser. Aalborg. No. 1[+?], 1846. Almind. Dansk Landbotid. HI 61356

Almond facts. San Francisco, CA. Almond Facts. See B–P–H 74/19. HI 50666

Aloe; mondstuk van die Suid-Afrikaanse Aalwyn- en Vetplant Vereniging. Journal of the South African Aloe and Succulent Society. Pretoria. Vol. 1+, 1963+. Aloe. HI 61357

Aloha. Honolulu, Hi. Aloha. See B–P–H 74/20. HI 50670

Alpe; rivista forestale italiana. Florence, Milan. Vols. 1-11, 1903-13; ser. 2, vols. 1-25, 1914-38. Alpe. Superseded by: Rivista forestale italiana. 1-151-1. HI 61358

Alpengarten. Zeitschrift für Freunde der Alpenwelt, der Alpenpflanzen- und Alpentierwelt, des Alpengartens und des Alpinums. Graz. Alpengarten. See B–P–H 75/1. HI 50673

Alpenrosen, ein Schweizer-Almanach. Bern & Leipzig. Alpenrosen. See B–P–H 75/2. HI 50674

Alpi; rivista mensile del centro alpinistico italiano. Turin, Rome. Vol. 58+, 1938+. Alpi. Preceded by: Rivista mensile del centro alpinistico italiano. 1-151-2. HI 61359

Alpin-biologische Studien. [Subseries of: Veroffentlichungen, Universität Innsbruck.] Innsbruck. Vol. 1+, 1975+. Alpin-Biol. Stud. HI 61360

Alpina. Eine Schrift der genauern Kenntniss der Alpen gewiedmet. Winterthur, Switzerland. Alpina. See B–P–H 75/3. HI 50675

Alpine garden society newsletter = Newsletter, alpine garden society.

Alpine garden society year book = Yearbook, alpine garden society. London.

Alpine gardening. Woking. Vol. 1+, 1986+. Alpine Gard. HI 61361

Alpine journal; a record of mountain adventure and scientific observation. London. Alpine J. See B–P–H 75/6. HI 50676

Alsófehérmegyei történelmi, régészeti és természettudományi egylet évkönyve. Gyulafehervar, Hungary [=Alba-Iulia, Rumania]. Alsófehérm. Tört. Rég. Term. Egyl. Évk. See B–P–H 75/8. HI 50678

Altaiskii Sbornik. Tomsk, Russian S F S R. Vols. 1-12, 1894-1930. Altaisk. Sborn. HI 61362

Altajskij Sbornik = Altaiskii Sbornik. Tomsk.

Altenburger naturwissenschaftliche Forschungen. Altenburg. No. 1+, 1981+. Altenburg. Naturwiss. Forsch. HI 61363

Ältere und neuere wöchentliche Beyträge zur Geschichte der Gebräuche und Sitten, der Kunst und Natur. Zurich. Ältere Neuere Wöchentl. Beytr. Gesch. Gebräuche Sitten. See B–P–H 75/13. HI 50679

Alternatives. Forest Park, GA. Vol. 1+, 1971+. Alternatives. HI 61364

Alumni report of the Taihoku imperial university agronomy graduates. [Takara kai kaiho.] Taihoku [=Taipei, Taiwan]. Alumni Rep. Taihoku Imp. Univ. Agron. Graduates. See B–P–H 75/14. HI 50680

Amakusa rinkai jikkenjo = Publications from the Amakusa marine biological laboratory, Kyushu university. Tomioka.

Amaranth newsletter. Caracas. No. 1+, 1984+. Amaranth Newslett. HI 61365

Amaranth today. Emmaus, PA. Vol. 1+, 1985+. Amaranth Today. HI 61366

Amaryllis bulletin. Westfield, IN. Vol. 1+, 1979+. Amaryllis Bull. HI 61367

Amaryllis year book. [Contained in: Plant life, vol. 16 onwards.] La Jolla, CA, etc. Vols. 12-39, 1960-83. Amaryllis Year Book. Preceded and superseded by: Herbertia. HI 61368

Amateur; a magazine devoted to the interest of the Philadelphia chapter of the american society of natural science. Philadelphia, PA. Amateur. See B–P–H 77/9. HI 50681

Amateur de champignons. Journal consacré à la connaissance populaire de champignons. Paris. Amateur Champignons. See B–P–H 77/10. HI 50682

Amateur gardener and gardeners chronicle. Sydney. Amateur Gard. & Gard. Chron. See B–P–H 77/11. HI 50683

Amateur gardening. London, etc. Vol. 1+, 1884+. Amateur Gard. (London). 1-156-3. HI 61369

Amateur gardening annual; a review of the year's work in garden and greenhouse. London. Amateur Gard. Annual. See B–P–H 77/13. HI 50686

Amateur gardening, for the lovers and cultivators of flowers and fruits. Springfield, MA. Amateur Gard. (Springfield). See B–P–H 77/12. HI 50684

Amateur gardening for town and country = Amateur gardening. London.

Amateur des jardins: revue bi-mensuelle illustrée d'horticulture générale. Paris. Amateur Jard. See B–P–H 77/15. HI 50687

Amateur naturalist. Ashland, ME. Amateur Naturalist (Ashland). See B–P–H 77/16. HI 50688

Amateur naturalist. Germantown, PA. [=Philadelphia, in part]. Amateur Naturalist (Germantown). See B–P–H 77/17. HI 50689

Amatores herbarii. [Shokubutsu shumi.] Kobe, Japan. Amatores Herb. See B–P–H 77/18. HI 50690

Amazon research newsletter. Gainesville, FL. No. 1+, 1979+. Amazon Res. Newslett. HI 61370

Amazônia. Belem. Vol. 1+, 1969+. Amazonia. HI 61371

Amazonia peruana. Lima. 1977+. Amazonia Peruana. HI 61372

Amazoniana. Kiel. Amazoniana. See B–P–H 77/19. HI 50691

Ambas Américas; revista de educación, bibliografía i agricultura. New York, NY. Ambas Amér. See B–P–H 77/20. HI 50692

Ambiente naturale ed urbano. Genoa. No. 1+, 1973+. Amb. Nat. Urb. HI 61373

Ambio; journal of the human environment, research and management. Stockholm. Vol. 1+, 1972+. Ambio. HI 61374

Ambio special report. Stockholm. No. 1+, 1972+. Ambio Special Rep. HI 61375

Ameghiniana; revista de la asociación paleontológica argentina. Buenos Aires. Vol. 1+, 1957+. Ameghiniana. HI 61376

Amenagement et nature. Paris. 1965+. Amenagem. Nat. HI 61377

América agraria. Mexico City. Amér. Agrar. See B–P–H 77/24. HI 50693

American agriculturist. Ithaca, NY. Vol. 1+, 1842+. Amer. Agric. 1-164-2. HI 61378

American anthropologist. Lancaster, PA. Amer. Anthropol. See B–P–H 77/28. HI 50694

American antiquarian society. Bulletin. Worcester, MA. Amer. Antiq. Soc. Bull. See B–P–H 78/2. HI 50696

American antiquarian society. Proceedings. Worcester, MA. Amer. Antiq. Soc. Proc. See B–P–H 78/3. HI 50698

American antiquarian society. Proceedings at the annual meeting. Worcester, MA. Amer. Antiq. Soc. Proc. Annual Meeting. See B–P–H 78/4. HI 50699

American antiquity. Menasha, WI. Amer. Antiquity. See B–P–H 78/6. HI 50700

American aquarist. Brooklyn [=New York, in part], NY. Amer. Aquar. See B–P–H 78/7. HI 50701

American archivist. Washington, DC., Chicago, IL. Vol. 1+, 1938+. Amer. Archivist. 1-170-3. HI 61379

American artist. Stamford, CT, New York. Vol. 4+, 1940+. Amer. Artist. Preceded by: Art instruction in drawing, painting, illustration, advertising art, design [not entered]. 1-172-1. HI 74809

American association for the advancement of science bulletin. Lancaster, PA. Vols. 1-5, 1942-46. A. A. A. S. Bull. 1-1-1. HI 50001

American association of stratigraphic palynologists contributions series = Contributions series, american association of stratigraphic palynologists. Various places.

American bee journal. Philadelphia, PA. Amer. Bee J. See B–P–H 78/14. HI 50702

American biology teacher. Lancaster, PA, etc. Vol. 1+, 1938+. Amer. Biol. Teacher. 1-189-2. HI 61380

American book collector. Chicago, IL. 1950+. Amer. Book Collect. HI 61381

American botanical and horticultural magazine. New York, NY. Amer. Bot. Hort. Mag. See B–P–H 78/18. HI 50705

American botanical register. Washington, DC. Amer.

Bot. Reg. See B–P–H 78/19. HI 50706

American botanist; devoted to economic and ecological botany. Binghamton, NY. Amer. Bot. (Binghamton). See B–P–H 78/16. HI 50703

American botanist. San Diego, CA. Amer. Bot. (San Diego) See B–P–H 78/17. HI 50704

American breeders associaton proceedings. Washington, DC. Amer. Breed. Assoc. Proc. See B–P–H 78/23. HI 50709

American breeders association report. Washington, DC. Amer. Breed. Assoc. Rep. See B–P–H 78/24. HI 50710

American breeders magazine. Washington, DC. Amer. Breed. Mag. See B–P–H 78/25. HI 50711

American brewer. New York, NY. Amer. Brewer. See B–P–H 78/27. HI 50712

American camellia quarterly. Gainesville, FL. Amer Camellia Quart. See B–P–H 78/28. HI 50714

American camellia yearbook. Gainesville, FL. Amer. Camellia Yearb. See B–P–H 79/1. HI 50715

American chemical journal. Baltimore, MD. Amer. Chem. J. See B–P–H 79/3. HI 50716

American Christmas tree growers' journal. Milwaukee, WI. Vol. 1+, 1956+. Amer. Christmas Tree Growers' J. HI 61382

American christmas tree journal = American Christmas tree growers' journal. Milwaukee, WI.

American chrysanthemum annual. Floral Park, NY. Amer. Chrysanthemum Annual. See B–P–H 79/6. HI 50717

American cosmetics and perfumery. Oak Park, IL. Vol. 87, 1972. Amer. Cosmet. Perfumery. Preceded by: American perfumer and cosmetics. Superseded by: Cosmetics and perfumery. HI 61383

American cultivator. Boston, MA. Amer. Cultivator. See B–P–H 79/7. HI 50718

American daffodil year book. Washington, DC. Amer. Daffodil Year Book (Washington). See B–P–H 79/9. HI 50720

American daffodil yearbook. Arlington, VA. Amer. Daffodil Yearb. (Arlington). See B–P–H 79/10. HI 50721

American doctoral dissertations. Ann Arbor, MI. 1964+. Amer. Doct. Diss. Previously contained in: Dissertation abstracts. HI 61384

American druggist. New York, NY. Amer. Druggist. See B–P–H 79/15. HI 50722

American druggist and pharmaceutical record. New York, NY. Amer. Druggist & Pharm. Rec. See B–P–H 79/16. HI 50723

American economic review. Ithaca, NY. Vol.1 +, 1911+. Amer. Econ. Rev. 1-214-3. HI 61385

American entomologist. St. Louis, MO. Amer. Entomol. See B–P–H 79/22. HI 50724

American entomologist and botanist = American entomologist. St. Louis, MO. Amer. Entomol. See B–P–H 79/22.

American environmemt. Hanover, NH. Vol. 16+, 1987+. Amer. Environm. Preceded by: Environmental education report and newsletter. HI 61386

American fern journal; a quarterly devoted to ferns. Port Richmond, NY. Amer. Fern J. See B–P–H 79/24. HI 50725

American florist. Chicago, IL. Amer. Florist (Chicago 1885-1931). See B–P–H 79/25. HI 50726

American florist. Chicago, IL. Amer. Florist (Chicago 1938-53). See B–P–H 79/26. HI 50728

American fruit grower. Charlottesville, VA. Amer. Fruit Grower. See B–P–H 80/1. HI 50729

American forests [title varies]. Washington, DC. Vol. 36(2)+, 1931+. Amer. Forests. Preceded by: American forests and forest life. 1-223-2. HI 61387

American forests and forest life. Washington, DC. Vols. 30-36(1), 1925-31. Amer. Forests Forest Life. Preceded by: American forestry. Superseded by: American forests. 1-223-2. HI 61388

American forestry. Washington, DC. Vols. 16-29, 1911-24. Amer. Forest. Preceded by: Conservation. Superseded by: American forests and forest Life. 1-223-2. HI 61389

American forests and forest life. Washington, DC. Vols. 30-36(1), 1925-31. Amer. Forests Forest Life. Preceded by: American forestry. Superseded by: American forests. 1-223-2. HI 61390

American fruit and nut journal. Petersburg, VA. Amer. Fruit Nut J. See B–P–H 80/2. HI 50730

American fruits. Rochester, NY. Amer. Fruits. See B–P–H 80/3. HI 50731

American fuchsia society. Miscellaneous publications. San Francisco, CA. Amer Fuchsia Soc. Misc. Publ. See B–P–H 80/5. HI 50732

American garden; an illustrated journal of horticulture. New York, NY. Amer. Garden. See B–P–H 80/8. HI 50735

American gardener's magazine and register of useful discoveries and improvements in horticulture and rural affairs. Boston, MA. Amer. Gard. Mag. See B–P–H 80/7. HI 50734

American gardening. New York, NY. Amer. Gardening. See B–P–H 80/9. HI 50736

American geologist. Minneapolis, MN. Vols. 1-36, 1888-1905. Amer. Geol. For vols. 37-77 see: Pan-american geologist. Superseded by: Economic geology (1905-21 only) [not entered]. 1-229-3. HI 61391

American gladiolus society official bulletin. Rochester, NY. Amer. Gladiolus Soc. Off. Bull. See B–P–H 80/10. HI 50737

American gladiolus society official review. Goshen, IN. Amer. Gladiolus Soc. Off. Rev. See B–P–H 80/11. HI 50738

American herb grower; a bi-monthly magazine for gardeners and gourmets. Falls Village, CT. Vol. 1, 1947/48. Amer. Herb Grower. Superseded by: Herb grower magazine. HI 61392

American home. Garden City, NY. Amer. Home. See B–P–H 80/13. HI 50739

American horticultural council news = American horticultural council news letter.

American horticultural council news letter. West Grove, PA. Nos. 1-21-?, 1950-59-? Amer. Hort. Council News Lett. 1-237-2. HI 61393

American horticultural magazine. Washington, DC. Vols. 39-50, 1960-71. Amer. Hort. Mag. Preceded by: National Horticultural Magazine. Superseded by: American horticulturist. HI 61394

American horticultural society gardeners' forum. Washington, DC. Vols. 3-10, 1960-67. Amer. Hort. Soc. Gard. Forum. Preceded by: A H S gardeners' forum. Superseded by: News and views, american horticultural society. HI 61395

American horticulturist. Alexandria, Mount Vernon, Va. Vol. 51+, 1972+. Amer. Hort. (Alexandria). Preceded by: American Horticultural Magazine. HI 61396

American horticulturist; devoted to horticulture, floriculture, agriculture and kindred subjects. Detroit, MI. Amer. Hort. (Detroit). See B–P–H 80/14. HI 50740

American horticulturist; a monthly magazine. Fowler, IN. Amer. Hort. (Fowler). See B–P–H 80/15. HI 50742

American hosta society bulletin = Bulletin of the american hosta society. Baldwin, NY.

American hosta society newsletter. Baldwin, NY. 1968-85? Amer. Hosta Soc. Newslett. Superseded by: Hosta journal. HI 61397

American institute of biological sciences. Publication. Washington, DC. Amer. Inst. Biol. Sci. Publ. See B–P–H 80/21. HI 50743

American iris society bulletin = Bulletin of the american iris society. Wellesley Farms, MA. Bull. Amer. Iris Soc. See B–P–H 238/26.

American journal of agriculture and science. Albany, NY. Amer. J. Agric. Sci. See B–P–H 81/3. HI 50744

American journal of botany. Lancaster, PA. Amer. J. Bot. See B–P–H 81/4. HI 50745

American journal of cancer. New York, NY. Amer. J. Cancer. See B–P–H 81/5. HI 50746

American journal of clinical pathology. Baltimore, MD. Amer. J. Clin. Pathol. See B–P–H 81/6. HI 50747

American journal of dermatology and genito-urinary diseases. St. Louis, MO. Amer. J. Dermatol. Genito-Urin. Dis. See B–P–H 81/7. HI 50748

American journal of digestive diseases. Fort Wayne, IN. Amer. J. Digestive Dis. See B–P–H 81/9. HI 50749

American journal of digestive diseases and nutrition. Fort Wayne, IN. Amer. J. Digestive Dis. Nutr. See B–P–H 81/10. HI 50750

American journal of diseases of children. Chicago IL. Amer. J. Dis. Children. See B–P–H 81/12. HI 50751

American journal of enology and viticulture. St. Helena, CA. Amer. J. Enol. Vitic. See B–P–H 81/13. HI 50752

American journal of horticulture and florists' companion. Boston, MA. 1867-69. Amer. J. Hort. & Florist's Companion. Superseded by: Tilton's journal of horticulture. 1-254-3 & 5-4220-3. HI 61398

American journal of the medical sciences. Philadelphia, PA. Amer. J. Med. Sci. See B–P–H 81/16. HI 50753

American journal of microscopy. Chicago, IL. Amer. J. Microscop. See B–P–H 81/17. HI 50754

American journal of microscopy and popular science. New York, NY. Amer. J. Microscop. Popular Sci. See B–P–H 81/18. HI 50755

American journal of obstetrics and gynecology. St. Louis, MO. Amer. J. Obstet. Gyn. See B–P–H 81/19. HI 50756

American journal of ophthalmology. St. Louis, MO. Amer. J. Ophthalmol. See B–P–H 81/22. HI 50757

American journal of orthopedic surgery. Boston, MA. Amer. J. Orthop. Surg. See B–P–H 81/23. HI 50758

American journal of pathology. Boston, MA. Amer. J. Pathol. See B–P–H 82/1. HI 50759

American journal of pharmacy. Philadelphia, PA. Ser. 2, vols. 1-18, 1836-52; ser. 3, vols. 1-18, 1853-70; ser. 4, vols. 1-24, 1871-95; vol. 68+, 1896+. Amer. J. Pharm. Preceded by: Journal of the Philadelphia college of pharmacy. HI 50760

American journal of physiology. Boston, MA. Amer. J. Physiol. (Boston). See B–P–H 82/4. HI 50762

American journal of physiology. Indianapolis, IN. Amer.

J. Physiol. (Indianapolis). See B–P–H 82/5. HI 50763

American journal of public health and the nation's health. Boston, MA. Amer. J. Public Health Nation's Health. See B–P–H 82/6. HI 50764

American journal of science. New Haven, CT. Amer. J. Sci. See B–P–H 82/7. HI 50766

American journal of science, and arts. New Haven, CT. Amer. J. Sci. Arts. See B–P–H 82/8. HI 50767

American journal of tropical diseases and preventive medicine. New Orleans, LA. Amer. J. Trop. Dis. Prev. Med. See B–P–H 82/10. HI 50768

American journal of tropical medicine. Baltimore, MA. Amer. J. Trop. Med. See B–P–H 82/11. HI 50769

American journal of tropical medicine and hygiene. Baltimore, MA. Amer. J. Trop. Med. Hyg. See B–P–H 82/12. HI 50770

American landscape architect. Chicago, IL. Amer. Landscape Architect. See B–P–H 82/33. HI 50771

American lily year book. Washington, DC. 1939+. Amer. Lily Year Book. HI 61399

American magazine of useful and entertaining knowledge. Boston, MA. Amer. Mag. Useful Entertaining Knowl.. See B–P–H 82/34. HI 50773

American mechanic's magazine. New York, NY. Amer. Mech. Mag. See B–P–H 83/1. HI 50774

American medicine. Philadelphia, PA. & New York, NY. Amer. Med. See B–P–H 83/2. HI 50775

American medical and philosophical register: or, annals of medicine, natural history, agriculture, and the arts. New York, NY. Amer. Med. Philos. Reg. See B–P–H 83/6. HI 50776

American medical recorder of original papers and intelligence in medicine and surgery. Philadelphia, PA. Amer. Med. Rec. See B–P–H 83/8. HI 50779

American midland naturalist; devoted to natural history, primarily that of the prairie states. Notre Dame, IN. Vol. 1(5)+, 1909+. Amer. Midl. Naturalist. Preceded by: Midland naturalist. Notre Dame. 1-285-2. HI 61400

American midland naturalist. Monograph series. Notre Dame, IN. Vol. 1+, 1944+. Amer. Midl. Naturalist, Monogr. 1-285-2. HI 61401

American monthly magazine and critical review. New York, NY. Amer. Monthly Mag. & Crit. Rev. See B–P–H 83/18. HI 50780

American monthly microscopical journal. New York, NY. Amer. Monthly Microscop. J. See B–P–H 83/20. HI 50781

American museum journal. New York, NY. Amer. Mus. J. See B–P–H 83/21. HI 50782

American museum novitates. New York, NY. Amer. Mus. Novit. See B–P–H 83/24. HI 50783

American naturalist. Boston, MA. Amer. Naturalist. See B–P–H 83/30. HI 50784

American nature study society newsletter = Newsletter, american nature study society. Elmsford, NY, etc.

American nurseryman. Rochester, NY. Amer. Nurseryman. See B–P–H 84/1. HI 50786

American nut journal. Petersburg, Va. Amer. Nut J. See B–P–H 84/2. HI 50787

American orchid society awards quarterly. Cambridge, MA. Vol. 1+, 1970+. Amer. Orchid Soc. Awards Quart. HI 61402

American orchid society bulletin. Cambridge, MA. Amer. Orchid Soc. Bull. See B–P–H 84/6. HI 50788

American pecan journal. Waco, TX. Amer. Pecan J. See B–P–H 84/7. HI 50789

American peony society bulletin. Clinton, NY. Amer. Peony Soc. Bull. See B–P–H 84/9. HI 50790

American perfumer. Pontiac, IL. Vols. 75(6)-77(5), 1960-62. Amer. Perfumer. Preceded by: American perfumer and aromatics. Superseded by: American perfumer and cosmetics. 1-297-3. HI 61403

American perfumer and aromatics. Pontiac, IL. Vols. 67-75(5), 1956-60. Amer. Perfumer Aromat. Preceded by: American perfumer and essential oil review. Superseded by: American perfumer. 1-297-3. HI 61404

American perfumer and cosmetics. Oak Park, IL. Vols. 77(6)-86, 1962-71. Amer. Perfumer Cosmet. Preceded by: American perfumer. Superseded by: American cosmetics and perfumery. HI 61405

American perfumer, cosmetics, toilet preparations. New York. Vols. ?-39, 1937?-39? Amer. Perfumer Cosmet. Toilet Prepar. Preceded and superseded by: American perfumer and essential oil review. 1-297-3. HI 61406

American perfumer and essential oil review. Bristol, CT., New York. Vols. 1-?, 1906-36; vols. 40-66, 1940-55. Amer. Perfumer Essential Oil Rev. For vols. ?-39 see: American perfumer, cosmetics, toilet preparations.. Superseded by: American perfumer and aromatics. 1-297-3. HI 61407

American pharmacy. Washington, DC. Vol. n.s. 18+, 1978+. Amer. Pharm. Preceded by: Journal of the american pharmaceutical association (New series). HI 61408

American philosophical society. Year book. Philadelphia = Year book of the american philosophical society. Philadelphia, PA. Yearb. Amer. Philos. Soc. See B–P–H 982/12.

American pioneer. Chillicothe, OH. Amer. Pioneer. See B–P–H 84/15. HI 50792

American plants. San Diego, CA. Amer. Pl. See B–P–H 84/16. HI 50793

American poinsettia society bulletin. Mission, TX. Amer. Poinsettia Soc. Bull. See B–P–H 84/17. HI 50794

American poinsettia society newsletter. Mission, TX. Amer. Poinsettia Soc. Newslett. See B–P–H 84/18. HI 50796

American pomologist; containing finely colored drawings, accompanied by letter-press descriptions of fruits of american origin. Philadelphia, PA. Amer. Pomol. (Philadelphia). See B–P–H 84/20. HI 50672

American pomologist. Bulletin of the american pomological society. Washington, DC. Amer. Pomol. (Washington). See B–P–H 84/21. HI 50798

American pomology. Ames, IA. Amer. Pomol. (Ames). See B–P–H 84/19. HI 50797

American potato journal. Washington, DC. Amer. Potato J. See B–P–H 84/24. HI 50799

American primrose society quarterly = Quarterly of the american primrose society. Portland, OR., Chenalis, WA.

American quarterly journal of agriculture and science. Albany, NY. Amer. Quart. J. Agric. Sci. See B–P–H 84/28. HI 50800

American quarterly microscopical journal. New York, NY. Amer. Quart. Microscop. J. See B–P–H 85/1. HI 50801

American repertory of arts, sciences, and manufactures. New York, NY. Amer. Repert. Arts Sci. Manufactures. See B–P–H 85/2. HI 50802

American repertory of arts, sciences and useful literature. Philadelphia, PA. Amer. Repert. Arts Sci. Useful Lit. See B–P–H 85/3. HI 50803

American review of respiratory diseases. Baltimore, Md. Amer. Rev. Resp. Dis. See B–P–H 85/4. HI 50805

American review of science, art, inventions, etc. New York, NY. Amer. Rev. Sci. See B–P–H 85/5. HI 50806

American review of tropical agriculture. Mexico City. Amer. Rev. Trop. Agric. See B–P–H 85/8. HI 50808

American review of tuberculosis and pulmonary diseases. Baltimore, MA. Amer. Rev. Tuberc. Pumon. Dis. See B–P–H 85/9. HI 50809

American rose, The. Shreveport, LA., etc. Vols. 19(8)-23(6), 1968-75. Amer. Rose. Preceded and superseded by: American rose magazine. HI 61409

American rose annual. Harrisburg, PA. Amer. Rose Annual. See B–P–H 85/14. HI 50810

American rose magazine. Harrisburg, PA., Columbus, OH., etc. Vols. 1-19(7), 1933-68; vol. 23(7)+, 1975+. Amer. Rose Mag. For vols. 19(8)-23(6) see: American rose. 1-315-2. HI 61410

American rose quarterly. Harrisburg, Pa. Amer. Rose Quart. See B–P–H 85/16. HI 50811

American scandinavian review. New York, NY. Vol. 1+, 1913+. Amer. Scand. Rev. 1-316-1. HI 61411

American scientific monthly. Iowa City, IA. Amer. Sci. Monthly. See B–P–H 85/19. HI 50813

American scientist; the Sigma Xi quarterly. Burlington, Vt. Amer. Sci. See B–P–H 85/18. HI 50812

American seedsman. Chicago, IL. Amer. Seedsman. See B–P–H 85/22. HI 50815

American silk and fruit culturist. Philadelphia, PA. Amer. Silk Fruit Cult. See B–P–H 85/23. HI 50816

American silk grower and agriculturist. Keene, NH. Amer. Silk Grower Agric. See B–P–H 85/24. HI 50817

American silk grower and farmer's manual. Philadelphia, PA. Amer. Silk Grower Farmer's Manual. See B–P–H 85/25. HI 50818

American society for horticultural science journal = Journal of the american society for horticultural science. Geneva, NY, & St. Joseph, MI.

American tomato yearbook. Westfield, NJ. 1949-73. Amer. Tomato Yearb. HI 61412

American tung news. Picayune, MS. Amer. Tung News. See B–P–H 86/6. HI 50819

American tung oil. Pensacola, FL. Amer. Tung Oil. See B–P–H 86/7. HI 50820

American vegetable grower. Willoughby, OH. Amer. Veg. Grower. See B–P–H 86/8. HI 50821

American wine and grape grower. Little Silver, NJ. Amer. Wine Grape Grower. See B–P–H 86/9. HI 50822

American woods. United States forest service = United States forest service. America's woods (series). Washington, DC.

American zoologist. Utica, NY, Thousand Oaks, CA. Vol. 1+, 1961+. Amer. Zoologist. HI 61413

Amerigold bulletin. New Britain, PA. Nos. 1-10/11, 1978-81. Amerigold Bull. Superseded by: Amerigold newsletter. HI 61414

Amerigold newsletter. New Britain, PA. Vol. 5(82-1)+, 1982+. Amerigold Newslett. Preceded by: Amerigold bulletin. HI 61415

Amerikanische Bierbrauer = American brewer. New York, NY. Amer. Brewer. See B–P–H 78/27.

Amgueddfa; bulletin of the national museum of Wales. Cardiff. Nos. 1-?, 1969-76. Amgueddfa. HI 61416

Ami des champs. Bordeaux. Ami Champs. See B–P–H 86/12. HI 50823

Ami des jardins. Lyons. Ami Jard. See B–P–H 86/15. HI 50824

Ami des jardins et de la maison. Paris. ?-1982+. Ami Jard. Maison. HI 61417

Ami des sciences. Paris. Ami Sci. See B–P–H 86/16. HI 50825

Amino acids, peptide and protein abstracts. London & Washington. 1972-79. Amino Acids Peptide Protein Abstr. Superseded by: Biochemistry abstracts. Part 3, amino acids, peptides and proteins. HI 61418

Amis des roses; journal de la société française des rosiéristes. Lyons. Sér. 2, vol. 13+, 1928+. Amis Roses. Preceded by: Roses. 1-350-3. HI 61419

Amoeba; mededelingenblad van de nederlandse jeugdbond voor natuurstudie. Amsterdam. Nos. 1+, 1963+. Amoeba (Amsterdam). HI 61420

Amoeba. Tokyo. Amoeba (Tokyo). See B–P–H 86/19. HI 50826

Amoeba. Zwolle, Netherlands. Amoeba (Zwolle). See B–P–H 86/20. HI 50828

Amtliche Pflanzenschutzbestimmungen. Berlin, Brunswick. 1924-43; n.s. 1949+. Amtl. Pflanzenschutzbestimm. HI 61421

Amoy marine biological bulletin. Amoy, China. Amoy Mar. Biol. Bull. See B–P–H 86/21. HI 50829

Amtlicher Bericht über die Versammlung deutscher Naturforscher und Aerzte. Hamburg [Place of publication changing each year]. Amtl. Ber. Versamml. Deutsch. Naturf. Aerzte. See B–P–H 86/25. HI 50830

Amtlicher Bericht über die Versammlung deutscher Naturforscher und Ärzte. Berlin [Place of publication changing each year]. Amtl. Ber. Versamml. Deutsch. Naturf. Ärzte. See B–P–H 86/26. HI 50831

An die Zürcherische Jugend. Zurich. An die Zürcherische Jugend. See B–P–H 87/19. HI 50833

An foras talúntais. Horticulture = Research report, horticulture, agricultural institute, Dublin. Dublin.

An foras talúntais. Plant sciences and crop husbandry = Research report, plant sciences and crop husbandry division, agricultural institute. Dublin.

An hui shih ta hsüeh pao = Anhui shida xuebao. Ziran kexue ban.

An hwei shih fan hsüeh yuan hsüeh pao = Journal of Anhwei pedagogical institute. [China]. J. Anhwei Pedagog. Inst. See B–P–H 457/20.

Anacampseros; Australian national journal. Gawler, S. A. Introductory issue, 1983; vol. 1+, 1985+. Anacampseros. HI 61422

Anais da academia brasileira de ciências. Rio de Janeiro. Anais Acad. Brasil. Ci. See B–P–H 88/34. HI 50834

Anais da academia brasileira de ciências. Suplemento do programma antartico brasileiro. Supplement to: Anais da academia brasileira de ciências. Rio de Janeiro. Vol. 1+, 1986+. Anais Acad. Brasil. Ci. Supl. Programma Antarc. Brasil. HI 61425

Anais botânicos do herbário "Barbosa Rodrigues." Hajaí, Brazil. Anais Bot. Herb. "Barbosa Rodrigues". See B–P–H 88/35. HI 50835

Anais brasileiros de ginecologia. Rio de Janeiro. Anais Brasil. Ginecol. See B–P–H 88/36. HI 50836

Anais de ciências naturais. Oporto, Portugal. Anais Ci. Nat. See B–P–H 88/38. HI 50837

Anais, congresso nacional de botanica, sociedade botânica do Brasil = Anais do [1+] congresso da sociedade de botânica do brasil. [Place varies].

Anais, congresso nacional do cafe. (Brazil.) No. ?-3+, 19??-70+. Anais Congr. Nac. Cafe. HI 61426

Anais do [1+] congresso da sociedade de botânica do brasil. [Forms part of: Anais da sociedade botánica do Brasil.] [Place varies]. No. 1+, 1950?+. Anais Congr. Soc. Bot. Brasil. HI 61427

Anais de escola superior de agricultura "Luiz de Queiroz." Piracicaba, Brazil. Anais Esc. Super. Agric. "Luiz de Queiroz". See B–P–H 88/41. HI 50838

Anais [dos] estudos de defesa fitosanitária. Lisbon. Vol. 11+, 1956+. Anais Estud. Defesa Fitosan. HI 61428

Anais da faculdade de ciências do Porto = Anais da faculdade de sciências do Porto. Oporto.

Anais da faculdade de ciências, universidade do Porto. Oporto. Vol. 47?+, 1964?+. Anais. Fac. Ci. Univ. Porto. Preceded by: Anais da faculdade de sciências do Porto [89/2]. HI 61429

Anais. Faculdade de medicina e cirurgia. São Paulo. Anais Fac. Med. See B–P–H 88/43. HI 50839

Anais. Faculdade de medicina de Pôrto Alegre. Pôrto Alegre, Brazil. Anais Fac. Med. Pôrto Alegre. See B–P–H 89/1. HI 50840

Anais da faculdade de sciências do Porto. Oporto. Vols. 15-46, 1927-63. Anais Fac. Sci. Porto. Preceded by: Annaes scientificos da academia polytecnica do Porto. Superseded by: Anais da faculdade de ciências, universidade do Porto. HI 50841

Anais do instituto de ciências biologicas, universidade federal rural de Pernambuco. Recife. Vols. 1-2, 1971-72. Anais Inst. Ci. Biol., Univ. Fed. Rural Pernambuco. Superseded by: Anais da universidade federal rural de Pernambuco. Ciências biologicas. HI 61430

Anais. Instituto superior de agronomia. Lisbon. Anais Inst. Super. Agron. See B–P–H 89/6. HI 50842

Anais do instituto do vinho do Porto. Oporto, Portugal. Anais Inst. Vinho Porto. See B–P–H 89/8. HI 50843

Anais paulistas de medicina e cirurgia. São Paulo. Anais Paul. Med. Cirurgia. See B–P–H 89/9. HI 50844

Anais. Reunião sul-americana de botânica. Rio de Janeiro. Anais Reunião Sul-Amer. Bot. See B–P–H 89/11. HI 50845

Anais de sociedade de biologia de Pernambuco. Recife, Brazil. Anais Soc. Biol. Pernambuco. See B–P–H 89/14. HI 50846

Anais da sociedade botânica do Brasil. Rio de Janeiro. Vol. 1+, 1950?+. Anais Soc. Bot. Brasil. HI 61431

Anais da universidade federal rural de Pernambuco. Ciências biologicas. Recife. Vol. 3+, 1976+. Anais Univ. Fed. Rural Pernambuco, Ci. Biol. Preceded by: Anais do instituto de ciências biologicas, universidade federal rural de Pernambuco. HI 61432

Analecta hassiaca. Marburg. Analecta Hassiaca. See B–P–H 90/5. HI 50847

Analecta transalpina. Venice. Analecta Transalpina. See B–P–H 90/6. HI 50848

Analecten zur Belustigung, Belehrung und Unterhaltung für Leser aus allen Ständen. Hamburg. Analecten Belust. See B–P–H 90/7. HI 50849

Analecten zur Natur- und Heilkunde. Würzburg. Analecten Natur- Heilk. See B–P–H 90/8. HI 50850

Analekten für Erd- und Himmelskunde. Munich. Analekten Erd- Himmelsk. See B–P–H 90/10. HI 50851

Analekten zur Naturkunde und Oekonomie für Naturforscher, Aerzte und Oekonomen. Zittau, Germany. Analekten Naturk. Oekon. Naturf. See B–P–H 90/13. HI 50852

Analele academiei Republicii Populare Române (Romîne). Bucharest. Analele Acad. Republ. Populare Romȃne. See B–P–H 90/14. HI 50853

Analele academiei române. Bucharest. Analele Acad. Române. See B–P–H 90/15. HI 50854

Analele, facultătii de agronomie din Cluj. Cluj. 1944-47. Analele Fac. Agron. Cluj. Preceded by: Buletinul, facultătii de agronomie din Cluj. Superseded by: Anuar lucrărilor ştiinţifice, institutul agronomic "Dr. Petru Coroza." HI 61433

Analele, institutului central de cercetări agricole. Sectie de protectia plantelor. Bucharest. Vols. 1-3, 1965-66. Analele Inst. Centr. Cercet. Agric., Protectia Pl. Superseded by: Analele, institutului de cercetări pentru protectia plantelor. HI 61434

Analele institutului de cercetări agronomice. Bucharest. Vols. 1-23, 1930-57. Analele Inst. Cercet. Agron. Superseded by: Analele institutului de cercetări agronomice. Seria C, fiziologie, genetica, ameliorare, protectiia plantelor si tecnologie agricola. HI 61435

Analele institutului de cercetări agronomice. Seria C, fiziologie, genetica, ameliorare, protectiia plantelor si tecnologie agricola. Bucharest. Vol. 24-28, 1958-60. Analele Inst. Cercet. Agron., C. Superseded by: Analele institutului de cercetări pentru cereale şi plante tehnice - Fundulea. Seria C, ameliorare, genetică, fiziologie si tehnologie agricolă. HI 61436

Analele institutului de cercetări agronomice al Romaniei = Analele institutului de cercetări agronomice.

Analele institutului de cercetări şi experimentaţ forestiera. Bucharest. Analele Inst. Cercet. Exp. Forest. See B–P–H 90/16. HI 50856

Analele, institutului de cercetări pentru cereale şi plante tehnice - Fundulea. Bucharest. Vol. 41+, 1976+. Analele Inst. Cercet. Pentru Cereale Pl. Tehn. Fundul. Preceded by: Analele institutului de cercetări pentru cereale şi plante tehnice - Fundulea. Seria C, ameliorare, genetică, fiziologie si tehnologie agricolă. HI 61437

Analele institutului de cercetări pentru cereale şi plante tehnice - Fundulea. Seria C, ameliorare, genetică, fiziologie si tehnologie agricolă. Bucharest. Vols. 29-40, 1961-75. Analele Inst. Cercet. Pentru Cereale Pl. Tehn. Fundul., C. Preceded by: Analele institutului de cercetări agronomice. Seria C, fiziologie, genetica, ameliorare, protectiia plantelor si tecnologie agricola. Superseded by: Analele, institutului de cercetări pentru cereale şi plante tehnice - Fundulea. HI 61438

Analele institutului de cercetări pentru legumicultura si floricultura. Bucharest. Vol. 1+, 1971+. Analele Inst. Cercet. Pentru Legumic. Floric. HI 61439

Analele institutului de cercetări pentru pomicultură pitesti. Bucharest. Vol. 1+, 1968+. Analele Inst. Cercet. Pentru Pomic. Pitesti. HI 61440

Analele, institutului de cercetări pentru protectia plantelor. Bucharest. Vols. 4-9, 1966-73. Analele Inst. Cercet. Pentru Protect. Pl. Preceded by: Analele, institutului central de cercetări agricole. Sectie de protectia plantelor. HI 61441

Analele institutului de cercetări pentru viticultură şi vinificaţie. Bucharest. Vol. 1+, 1968+. Analele Inst. Cercet. Pentru Vitic. Vinif. HI 61442

Analele institutului de cercetări silvice. Bucaresti. Ser. 1, vols. 16-19, 1955-1958. Analele Inst. Cercet. Silvice. Preceded by: Studii si cercetări, institutul de cercetări silvice. Superseded by: Studii si cercetări, institutul de cercetări forestiere. HI 61443

Analele ştiinţifice ale universităţii "Al. I. Cuza" din Iaşi. Seria nouă. Secţiunea 2.a, biologie. Iaşi. Vols. 8-21, 23+ [vol.22 not published], 1962+. Analele Şti. Univ. Al. I. Cuza Iaşi, Ser. Nouă, 2.a. Preceded by: Analele ştiinţifice ale universităţii "Al. I. Cuza" din Iaşi, Ser. nouă. Secţiunea 2, stiinţe naturale. HI 61507

Analele ştiinţifice ale universităţii "Al. I Cuza" din Iaşi. Seria nouă. Secţiunea 2.a, biologie monogr. Iaşi. Vol. 1+, 1965+. Analele Şti. Univ. Al. I. Cuza Iaşi, Ser. Nouă, 2.b. Preceded by: Annales scientifiques de l'université de Jassy. Pt.2, sciences naturelles. HI 61444

Analele ştiinţifice ale universităţii "Al. I. Cuza" din Iaşi. Seria nouă. Secţiunea 2, sţiinţe naturale. Iaşi. N.s. vols. 1-7, 1955-61. Analele Şti., Univ. Al. I. Cuza Iaşi, Sect. 2. Preceded by: Annales scientifiques de l'université de Jassy. Pt.2, sciences naturelles. Superseded by: Analele ştiinţifice ale universităţii "al. I. Cuza" din Iaşi. Ser. nouă. Secţiunea 2.a, biologie. HI 61445

Analele universitătii bucureşti. Biologie. Bucharest. Vol. 26+, 1977+. Analele Univ. Bucureşti, Biol. Preceded by: Analele universitătii bucureşti. Seria ştiinţele naturii. HI 61446

Analele universitătii bucureşti. Biologie vegetală. Bucharest. Vols. 18-22, 1969-73. Analele Univ. Bucureşti, Biol. Veg. Preceded by: Analele universitati C.I. Parhon bucareşti. Seria ştiinţelor naturii. Biologie. Superseded by: Analele universitatii bucareşti. Seria ştiinţele naturii. HI 61447

Analele universitătii bucureşti. Geografie. Bucharest. Vol. 18+, 1969+. Analele Univ. Bucureşti, Geogr. Preceded by: Analele universitati C. I. Parhon bucareşti. Seria ştiinţelor naturii. HI 61448

Analele universitătii bucureşti. Seria ştiinţele naturii. Bucharest. Vols. 23-25, 1974-76. Analele Univ. Bucureşti, Şti. Nat. Preceded by: Analele universitătii bucureşti. Biologie vegetală. Superseded by: Analele universitătii bucureşti. Biologie. HI 61449

Analele universitatii C. I. Parhon bucureşti. Seria ştiinţor naturii. Bucharest. Vols. 9-17, 1956-68. Analele Univ. C. I. Parhon Bucureşti, Şti. Nat. Preceded by: Revistă universitatii C. I. Parhon Bucureşti. Ser. ştiinţelor naturi. Superseded by: Analele universitatii bucureşti. Biologie vegetala and Analele universitatii bucureşti. Geografie. HI 61450

Analele universitatii din Craiova. Biologie, agronomie, horticultura. Bucharest. Vol. 7+, 1976 [also numbered vol. 17+]. Analele Univ. Craiova, Biol. Agron. Hort. Preceded by: Analele universitatii din Craiova. Biologie, medicina, stiinte agricole. HI 61451

Analele universitatii din Craiova. Biologie, medicina, stiinte agricole. Bucharest. Vol. 6, 1975 [also numbered vol. 16]. Analele Univ. Craiova, Biol. Med. Şti Agric. Preceded by: Analele universitatii din Craiova. Ser. III-a, biologie, stiinte agricole. Superseded by: Analele universitatii din Craiova. Biologie, agronomie, horticultura. HI 61452

Analele universitatii din Craiova. Ser. III-a, biologie, stiinte agricole. Bucharest. Vols. 1-5, 1969-73 [also numbered vols. 11-15]. Analele Univ. Craiova, III.a, Biol. Şti Agric. Superseded by: Analele universitatii din Craiova. Biologie, medicina, stiinte agricole. HI 61453

Anales de la academia de biología. Santiago. Anales Acad. Biol. See B–P–H 90/18. HI 50857

Anales de la academia chilena de ciencias naturales. Forms part of: Revista universitaria. Santiago. Vols. 1-30, 1936-68. Anales Acad. Chilena Ci. Nat. 1-16-2. HI 61454

Anales de la academia de ciencias médicas, físicas y naturales de la Habana. Havana. Anales Acad. Ci. Méd. Habana. See B–P–H 90/20. HI 50858

Anales, academia de farmacia = Anales, real academia de farmacia. Madrid.

Anales de la academia de geografia e historia de Guatemala. Guatemala City. Vol. 53+, 1980+. Anales Acad. Geogr. Hist. Guatemala. Preceded by: Anales de la sociedad de geografía e historia de Guatemala. HI 61455

Anales, academia nacional de farmacia = Anales, real academia de farmacia. Madrid.

Anales. Academia mexicana de ciencias exactas, físicas y naturales. Mexico City. Anales Acad. Mex. Ci. Exact. See B–P–H 90/21. HI 50859

Anales de la academia nacional de ciencias exactas, físicas y naturales de Buenos Aires. Buenos Aires. Anales Acad. Nac. Ci. Exact. Buenos Aires. See B–P–H 90/22. HI 50860

Anales de agricultura e industria rural. Havana. Anales Agric. Industr. Rural. See B–P–H 90/23. HI 50861

Anales de la asociación española para el progreso de la ciencias. Madrid. Anales Asoc. Esp. Progr. Ci. See B–P–H 90/24. HI 50862

Anales de la asociación de palinologos de lengua español. Cordoba. Vol. 1+, 1984+. Anales Asoc. Palin. Langua Esp. HI 61456

Anales de la asociación química argentina. Buenos Aires. 1955+. Anales Asoc. Quím. Argent. Preceded by: Anales de la sociedad química argentina. HI 61457

Anales de biologia, facultad de biologia, universidad de Murcia. Murcia. Vol. 1+, 1984+ [Subdivided into sections: Especial; biologia ambiental; biologia general; biologia vegetal; etc.]. Anales Biol., Fac. Biol., Univ. Murcia. HI 61458

Anales del centro de ciencias del mar y limnologia, universidad nacional autonomia de Mexico. Mexico City. Vols. 1-7, 1974-80. Anales Centro Ci. Mar Limnol., Univ. Nac. Auton. Mexico. Preceded by: Anales del instituto de biologia. Serie ciencias del mar y limnologia. Superseded by: Anales del instituto de ciencias del mar y limnologia, universidad nacional autonoma de Mexico. HI 61459

Anales de ciencias. Murcia. Vol. 44+, 1985+, 1986+. Anales Ci., Murcia. Preceded by: Anales de la universidad de Murcia. Ciencias. HI 61460

Anales de ciencias, agricultura, comércio y artes. Havana. Anales Ci. Agric. See B–P–H 91/2. HI 50864

Anales de ciencias, literatura y artes. Madrid. Anales Ci. Lit. See B–P–H 91/3. HI 50865

Anales de ciencias naturales. Madrid. Anales Ci. Nat. See B–P–H 91/4. HI 50866

Anales de ciencias naturales. Instituto José de Acosta. Madrid. Anales Ci. Nat. Inst. José de Acosta. See B–P–H 91/5. HI 50867

Anales. Ciencias. Universidad de Madrid. Madrid. Anales Ci. Univ. Madrid See B–P–H 91/7. HI 50869

Anales científicos. Lima. Anales Ci. See B–P–H 91/1. HI 50863

Anales científicos paraguayos. Asunción. Anales Ci. Parag. See B–P–H 91/6. HI 50868

Anales cientificos, universidad agraria, Lima = Anales cientificos, universidad nacional agraria, Lima.

Anales cientificos, universidad nacional agraria, Lima. Lima. 1963+. Anales Ci., Univ. Nac. Agrar., Lima. HI 61461

Anales, o colección de memorias sobre las artes, la artillera, la historia natural de España y Américas, la docimastica de sus minas, etc. Segovia, Spain. Anales Colecc. Mem. Artes España Amér. See B–P–H 91/11. HI 50870

Anales de la comisíon de investigacion cientifica, provincia de Buenos Aires gobernacion. La Plata. Vols. 1-6?, 1960-65?, 1963-67; n.s. vol. 1+, 1969+. Anales Comis. Invest. Ci., Buenos Aires. HI 61462

Anales de edafologia y agrobiologia. Madrid. 1960+. Anales Edafol. Agrobiol. Preceded by: Anales de edafologia y fisiologiya vegetal. HI 61463

Anales de edafologia y fisiologiya vegetal. Madrid. 1949-59. Anales Edafol. Fisiol. Veg. Preceded by: Anales del instituto espanol de edafologia, ecologia y fisiologia vegetal. Superseded by: Anales de edafologia y agrobiologia. HI 61464

Anales de la escuela nacional de ciencias biológicas. Mexico City. Anales Esc. Nac. Ci. Biol. See B–P–H 91/13. HI 50872

Anales de la escuela de peritos agricolas y superior de agricultura. Barcelona. Vols. 1-?, 1941-60. Anales Esc. Peritos Agric. Super. Agric. 1-603-3. HI 61465

Anales de la estación experimental de Aula Dei. Saragossa, Spain. Anales Estac. Exp. Aula Dei. See B–P–H 91/14. HI 50873

Anales. Facultad de ciencias naturales y museo. Universidad nacional. La Plata. Sección botánica. La Plata, Argentina. Anales Fac. Ci. Nat. Mus. Univ. Nac. La Plata, Secc. Bot. See B–P–H 91/15. HI 50874

Anales de farmacia y bioquímica. Buenos Aires. Anales Farm. Bioquím. See B–P–H 91/16. HI 50875

Anales de física y química. Madrid. Vols. 37-43, 1941-47. Anales Fis. Quím. Preceded by: Anales de la real sociedad española de fisica y quimica. Superseded by: Anales de la real sociedad española de física y química. Ser. B, química. HI 61466

Anales hidrograficos. Argentina. Anales Hidrogr. See B–P–H 91/17. HI 50876

Anales de historia natural. Madrid. Anales Hist. Nat. See B–P–H 91/18. HI 50877

Anales del hospital general de Oaxaca. Oaxaca. Vol. 1+, 1938/39+. Anales Hosp. Gen. Oaxaca. 4-3126-2. HI 61467

Anales del I N I A = Anales del instituto nacional de investigaciones agronómicas. Madrid.

Anales del I N I A. Serie agrícola = Anales del instituto nacional de investigaciones agrarias. Serie agrícola. Madrid.

Anales del I N I A. Serie forestal = Anales del instituto nacional de investigaciones agrarias. Serie forestal. Madrid.

Anales del I N I A. Serie general = Anales del instituto nacional de investigaciones agrarias. Serie general. Madrid.

Anales del I N I A. Serie produccion vegetal = Anales del instituto nacional de investigaciones agrarias. Serie produccion vegetal. Madrid.

Anales del I N I A. Serie proteccion vegetal = Anales del instituto nacional de investigaciones agrarias. Serie proteccion vegetal. Madrid.

Anales del I N I A. Serie recursos naturales = Anales del instituto nacional de investigaciones agrarias. Serie recursos naturales. Madrid.

Anales del instituto de biológia de la universidad nacional autónoma de México. Série biologia. Mexico, D.F. Vol. 38+, 1967+. Anales Inst. Biol. Univ. Nac. Auton. Mexico, Biol. Preceded by: Anales del instituto de biología de la universidad nacional de México.

HI 61468

Anales del instituto de biológia de la universidad nacional autónoma de México. Série botánica. Mexico, D.F. Vol. 38+, 1967+. Anales Inst. Biol. Univ. Nac. Auton. Mexico, Bot. Preceded by: Anales del instituto de biología de la universidad nacional de México. HI 61469

Anales del instituto de biología de la universidad nacional autónoma de México. Série ciencias del mar y limnología. Mexico, D.F. Vol. 38-43, 1967-72. Anales Inst. Biol. Univ. Nac. Auton. Mexico, Ci. Mar Limnol. Preceded by: Anales del instituto de biología de la universidad nacional de México. Superseded by: Anales del centro de ciencias del mar y limnologia. HI 61470

Anales del instituto de biológia de la universidad nacional de México. Mexico, D.F. Vol. 1-27, 1930-66. Anales Inst. Biol. Univ. Nac. México. Superseded by: Anales del instituto de biológia, universidad nacional autónomia de México. Série biologia; Anales del instituto de biológia, universidad nacional autónomia de México. Série botánica; and Anales del instituto de biológia, universidad nacional autónomia de México. Série ciencias del mar y limnologica. 3-2629-3. HI 61471

Anales del instituto botánico A. J. Cavanilles. Madrid. Vols. 10-35, 1951-78. Anales Inst. Bot. Cavanilles. Preceded and superseded by: Anales del jardin botánico de Madrid. 5-4037-3. HI 61472

Anales del instituto de ciencias del mar y limnologia, universidad nacional autonoma de Mexico. Mexico City. Vol. 8+, 1981+. Anales Inst. Ci. Mar Limnol., Univ. Nac. Auton. Mexico. Preceded by: Anales del centro de ciencias del mar y limnologia, universidad nacional autonomia de Mexico. HI 61473

Anales del instituto espanol de edafologia, ecologia y fisiologia vegetal. Madrid. Vols. 1-?, 1942-48. Anales Inst. Edafol. Superseded by: Anales de edafologia y fisiologia vegetal. 1-355-3. HI 61474

Anales. Instituto de etnografía americana. Mendoza, Argentina. Anales Inst. Etnogr. Amer. See B–P–H 91/23. HI 50878

Anales, instituto fisico-geográfico y del museo nacional de Costa Rica. San José, Costa Rica. Vols. 1-9, 1887-96. Anales Inst. Fis.-Geogr. Nac. Costa Rica. Preceded by: Anales del museo nacional de Costa Rica. 2-1213-3. HI 61475

Anales del instituto fitotécnico de Santa Catalina. Buenos Aires. Vols. 1-4, 1939-42. Anales Inst. Fitotécn. Santa Catalina. HI 61476

Anales del instituto forestal de investigaciones y experiencias. Madrid. Vols. 1-10(2), 1956-67 [also numbered vols. 28-37?]. Anales Inst. Forest. Invest. HI 61477

Anales de instituto de geología (del) universidad nacional (de) Mexico City. Mexico City. Anales Inst. Geol. Univ. Nac. Mexico City. See B–P–H 92/3. HI 50879

Anales del instituto de investigaciones científicas. Monterrey, Mexico. Anales Inst. Invest. Ci. See B–P–H 92/4. HI 50880

Anales del instituto de investigaciones marinas de Punta de Betin. Santa Marta. (Colombia.) Vol. 9+, 1977+; Suplemento 1+, 1977+. Anales Inst. Invest. Mar. Punta de Betin. Preceded by: Mitteilungen aus dem instituto colombo-aleman de investigaciones cientificas "Punta de Betin". HI 61478

Anales del instituto médico-nacional de México. Mexico City. Anales Inst. Méd.-Nac. México. See B–P–H 92/5. HI 50881

Anales. Instituto nacional de antropología e historia. Mexico City. Anales Inst. Nac. Antropol. See B–P–H 92/6. HI 50882

Anales del instituto nacional de investigaciones agrarias. Serie agrícola. Madrid. Vols. 13-28, 1980-86. Anales Inst. Nac. Invest. Agrar., Agric. Preceded by: Anales del instituto nacional de investigaciones agrarias. Serie produccion vegetal and Anales del instituto nacional de investigaciones agrarias. Serie proteccion vegetal. Superseded by: Investigación agraria. Producción y protección. HI 61479

Anales del instituto nacional de investigaciones agrarias. Serie forestal. Madrid. Vols. 5-9, 1982-85. Anales Inst. Nac. Invest. Agrar., Forest. Preceded by: Anales del instituto nacional de investigaciones agrarias. Serie, recursos naturales. Superseded by: Investigacion agraria. Produccion y proteccion vegetales. HI 61480

Anales del instituto nacional de investigaciones agrarias. Serie general. Madrid. No. 1+, 1971+. Anales Inst. Nac. Invest. Agrar., Gen. Preceded by: Anales del instituto nacional de investigaciones agronómicas. HI 61481

Anales del instituto nacional de investigaciones agrarias. Serie produccion vegetal. Madrid. Nos. 1-12, 1971-80. Anales Inst. Nac. Invest. Agrar., Prod. Veg. Preceded by: Anales del instituto nacional de investigaciones agronómicas. HI 61482

Anales del instituto nacional de investigaciones agrarias. Serie proteccion vegetal. Madrid. Nos. 1-12, 1971-79. Anales Inst. Nac. Invest. Agrar., Protecc. Veg. Preceded by: Boletín de patología vegetal y entomología agrícola. Superseded by: Anales del instituto nacional de investigaciones agrarias. Serie agrícola. HI 61483

Anales del instituto nacional de investigaciones agrarias. Serie recursos naturales. Madrid. Vols. 1-4, 1971-80. Anales Inst. Nac. Invest. Agrar., Recurs. Nat. Preceded by: Anales, instituto nacional de investigaciones agronomicas. Superseded by: Anales del instituto nacional de investigaciones agrarias. Serie forestal. HI 61484

Anales del instituto nacional de investigaciones agronómicas. Madrid. Vols. 1-19, 1952-70. Anales Inst. Nac. Invest. Agron. Superseded by: Anales del instituto nacional de investigaciones agrarias. Serie, general; Anales del instituto nacional de investigaciones agrarias. Serie, produccion vegetal; Anales del instituto nacional de investigaciones agrarias. Serie, proteccion vegetal; and Anales del instituto nacional de investigaciones agrarias. Serie, recursos naturales. 1-378-2. HI 61485

Anales del instituto nacional de investigaciones biologico-pesqueras. Mexico, D.F. Vol. 1+, 1965+. Anales Inst. Nac. Invest. Biol.-Pesq. HI 61486

Anales del instituto nacional de microbiología. Buenos Aires. Anales Inst. Nac. Microbiol. See B–P–H 92/8. HI 50884

Anales del instituto de la Patagonia. Punta Arenas. Vols. 1-14, 1970-83. Anales Inst. Patagonia. Superseded by: Anales del instituto de la Patagonia. Serie ciencias naturales. HI 61487

Anales del instituto de la Patagonia. Serie ciencias naturales. Punta Arenas. Vol. 15+, 1984+. Anales Inst. Patagonia, Ci. Nat. Preceded by: Anales del instituto de la Patagonia. HI 61488

Anales del instituto de segunda enseñanza. Havana. Vols. 1-2, 1894-95. Anales Inst. Segunda Enseñ. 3-1819-3. HI 61489

Anales del jardin botánico de Madrid. Madrid. Vols. 1-9, 1941-50; vol. 36+, 1979+. Anales Jard. Bot. Madrid. For 1950-78 see: Anales del instituto botanico A. J. Cavanilles. 5-4037-3. HI 61490

Anales mexicanos de ciencias, literatura, minería, agricultura, artes, industria y comércio en la republica mexicana. Mexico City. Anales Mex. Ci. See B–P–H 92/10. HI 50885

Anales de minas. Madrid. Anales Minas. See B–P–H 92/11. HI 50886

Anales del museo argentino ce ciencias naturales "Bernardino Rivadavia". Buenos Aires. Vols. 37-42, 1931-47. Anales Mus. Argent. Ci. Nat. "Bernardino Rivadavia". Preceded by: Anales del museo nacional de historia natural de Buenos Aires. Superseded by: Revista del instituto nacional de investigación de las ciencias naturales anexo al museo argentino de ciencias naturales "Bernardino Rivadavia". Ciencias botánicas. 1-818-1. HI 61491

Anales del museo de historia natural de Montevideo. Montevideo. Ser. 2, vol. 2+, 1926+. Anales Mus. Hist. Nat. Montevideo. Preceded by: Anales del museo nacional de Montevideo. 3-2742-1. HI 61492

Anales del museo de historia natural de Valparaiso. Valparaiso. No. 1+, 1968+. Anales Mus. Hist. Nat. Valparaiso. HI 61493

Anales. Museo Michoacano. Morelia, Mexico. Anales Mus. Michoacano. See B–P–H 91/14. HI 50887

Anales del museo nacional de Buenos Aires. Buenos Aires. Vols. 1-14, 1875-1911. Anales Mus. Nac. Buenos Aires. Preceded by: Anales del museo público de Buenos Aires. Superseded by: Anales del museo nacional de historia natural de Buenos Aires. 1-181-3. HI 61494

Anales del museo nacional de Costa Rica. San José, Costa Rica. Vol. 1, 1888. Anales Mus. Nac. Costa Rica. Superseded by: Anales, instituto fisico-geográfico y del museo nacional de Costa Rica. HI 61495

Anales del museo nacional de historia natural Bernardino Rivadavia. Buenos Aires. Vols. 32-36, 1923-31. Anales Mus. Nac. Hist. Nat. Bernardino Rivadavia. Preceded by: Anales del museo nacional de historia natural de Buenos Aires. Superseded by: Anales del museo argentino de ciencias naturales "Bernardino Rivadavia". HI 61496

Anales del museo nacional de historia natural de Buenos Aires. Buenos Aires. Vols. 14-31, 1911-31. Anales Mus. Nac. Hist. Nat. Buenos Aires. Preceded by: Anales del museo nacional de Buenos Aires. Superseded by: Anales del museo nacional de historia natural Bernardino Rivadavia. 1-818-1. HI 61497

Anales de museo nacional de México. Mexico City. Anales Mus. Nac. México. See B–P–H 92/17. HI 50888

Anales del museo nacional de Montevideo. Montevideo. Vols. 1-7, 1894-1909; ser. 2, vol. 1, 1925. Anales Mus. Nac. Montevideo. Superseded by: Anales del museo de historia natural de Montevideo. 3-2742-1. HI 61498

Anales del museo nacional, Republica de El Salvador. San Salvador. 1903-11. Anales Mus. Nac., El Salvador. HI 61499

Anales del museo nacional. Santiago de Chile. Vols. 1-18, 1891-1910. Anales Mus. Nac. Santiago de Chile. 2-1016-2. HI 61500

Anales del museo Nahuel Huapí. Buenos Aires. Vols. 2-3, 1950-53. Anales Mus. Nahuel Huapí. Preceded by: Anales del museo de la Patagonia. Superseded by: Anales de parques nacionales. 5-3771-2. HI 61501

Anales, museo de la Patagonia. Buenos Aires. Vol. 1, 1945. Anales Mus. Patagonia. Superseded by: Anales del museo Nahuel Huapí. HI 61502

Anales del museo de La Plata. Sección B, paleobotanica.

Buenos Aires. Nos. 1-2, 1944-47. Anales Mus. La Plata, B. Preceded by: Anales del museo de La Plata. Segunda serie. HI 61503

Anales del museo de La Plata. Sección botanica. La Plata, Buenos Aires. 1897-1902. Anales Mus. La Plata, Bot. Superseded by: Anales del museo de La Plata. Segunda serie. HI 61504

Anales del museo de La Plata. Segunda serie. La Plata, Buenos Aires. Vols. 1-4(1), 1907-30. Anales Mus. La Plata, 2 Ser. Preceded by: Anales del museo de La Plata. Sección botanica. Superseded by: Anales del museo de La Plata. Sección B, paleobotanica. HI 61505

Anales del museo público de Buenos Aires. Buenos Aires. 1864-74. Anales Mus. Público Buenos Aires. Superseded by: Anales del museo nacional de Buenos Aires. 1-818-1. HI 61506

Anales de parques nacionales. Buenos Aires. Vols. 4-13, 195?-74. Anales Parques Nac. Preceded by: Anales del museo Nahuel Huapí. HI 61653

Anales de química. Madrid. Vols. 64-75, 1968-79. Anales Quím. Preceded by: Anales de la real sociedad española de física y química. Serie B, química. Superseded by: Anales de química. Serie C, quimica organica i bioquimica. HI 61508

Anales de química. Serie C, quimica organica i bioquimica. Madrid. Vol. 76+, 1980+. Anales Quím., C. Preceded by: Anales de química. HI 61509

Anales, real academia de farmacia. Madrid. Vol. 1+, 1932+. Anales Real Acad. Farm. HI 61510

Anales de la real academia nacional de medicina. Madrid. 1879+. Anales Real Acad. Nac. Med. HI 61511

Anales de la real sociedad española de física y química. Madrid. Vols. 1-36, 1903-40. Anales Real Soc. Esp. Fís. Quím. Superseded by: Anales de física y química. HI 61512

Anales de la real sociedad española de física y química. Serie B, química. Madrid. Vols. 44-63, 1948-67. Anales Real Soc. Esp. Fís. Quím., B. Preceded by: Anales de física y química. Superseded by: Anales de química. HI 61513

Anales de la sociedad agronómica. Santiago. Anales Soc. Agron. See B–P–H 93/2. HI 50889

Anales de la sociedad de biología de Bogotá. Bogotá. Anales Soc. Biol. Bogotá. See B–P–H 93/3. HI 50890

Anales de sociedad científica argentina. Buenos Aires. Anales Soc. Ci. Argent. See B–P–H 93/4. HI 50891

Anales de la sociedad cientifica Argentina. Seccion Santa Fe. Vols. 6(3)+, 1934+. Anales Soc. Ci. Argent., Secc. Santa Fe. Preceded by: Anales de la sociedad cientifica de Santa Fe. 5-3904-1. HI 61514

Anales de la sociedad cientifica de Santa Fe. Vols. 1-6(3), 1929-34. Anales Soc. Ci. Santa Fe. Superseded by: Anales de la sociedad cientifica Argentina. Seccion Santa Fe. 5-3904-1. HI 61515

Anales de la sociedad española de historia natural. Madrid. Anales Soc. Esp. Hist. Nat. See B–P–H 93/6. HI 50892

Anales de la sociedad de geografía e historia de Guatemala. Guatemala City. Vols. 1-52, 1924-79. Anales Soc. Geogr. Guatemala. Superseded by: Anales de la academia de geografia e historia de Guatemala. HI 61516

Anales de la sociedad Humboldt. Mexico City. Anales Soc. Humboldt. See B–P–H 93/9. HI 50893

Anales de la sociedad mexicana de historia de la ciencia y de la tecnologia. Mexico, D.F. Vol. 1+, 1969+. Anales Soc. Mex. Hist. Ci. Tecnol. HI 61517

Anales de la sociedad química argentina. Buenos Aires. 1913-54. Anales Soc. Quím. Argent. Superseded by: Anales de la asociación química argentina. HI 61518

Anales de la sociedad rural argentina. Buenos Aires. Anales Soc. Rural Argent. See B–P–H 93/10. HI 50894

Anales de la universidad autonoma de Santo Domingo. Santo Domingo. Vol. 27+, 1961+. Anales Univ. Auton. Santo Domingo. Preceded by: Anales de la universidad de Santo Domingo. HI 61519

Anales de la universidad central del Ecuador. Quito, Ecuador. Anales Univ. Centr. Ecuador. See B–P–H 93/11. HI 50895

Anales de la universidad de Chile. Santiago. Anales Univ. Chile. See B–P–H 93/12. HI 50897

Anales de la universidad de Costa Rica. San José, Costa Rica. Anales Univ. Costa Rica. See B–P–H 93/13. HI 50898

Anales de la universidade de Guayaquil. Guayaquil, Educador. Anales Univ. Guayaquil. See B–P–H 93/14. HI 50899

Anales de la universidad de Murcia. Murcia. Vols. 1-12, 19??-53. Anales Univ. Murcia. Superseded by: Anales de la universidad de Murcia. Ciencias. HI 61520

Anales de la universidad de Murcia. Ciencias. Murcia. Vols. 13-43, 1954/55-84/85, 1955-84. Anales Univ. Murcia, Ci. Preceded by: Anales de la universidad de Murcia. Superseded by: Anales de ciencias. Murcia. HI 61521

Anales de universidad nacional. Córdoba, Argentina. Anales Univ. Nac. See B–P–H 93/15. HI 50900

Anales. Universidad nacional de La Plata. Instituto de

museo. Sección B, paleobotánica. La Plata, Argentina. Anales Univ. Nac. La Plata, Secc. B, Paleobot. See B–P–H 93/16. HI 50901

Anales de de la universidad de la Patagonia "San Juan Bosco". Comodoro Rivadavia. Nos. 1-?, 1964-67. Anales Univ. Patagonia San Juan Bosco. HI 61522

Anales de la universidad de Santo Domingo. Cuidad Trujillo [Santo Domingo]. Nos. 1-26, 1937-60. Anales Univ. Santo Domingo. Superseded by: Anales de la universidad autonoma de Santo Domingo. 2-1063-1. HI 61523

Anali za eksperimentalno šumarstvo. Odjel za prirodne nauke. Zagreb. Vol. 2, 1957. Anali Eksper. Šumarstvo, Odjel Prir. Nauke. Preceded by: Anali instituta za eksperimentalno šumarstvo. Superseded by: Anali za šumarstvo. HI 61524

Anali instituta za eksperimentalno šumarstvo. Jugoslavenske akademije akademije znanosti i umjetnosti. Zagreb. Vol. 1, 1955. Anali Inst. Eksper. Šumarstvo Jugoslav. Akad. Superseded by: Anali za eksperimentalno šumarstvo. Odjel za prirodne nauke. HI 61525

Anali za šumarstvo. Zagreb. Vol. 3+, 1965+. Anali Šumarstvo. Preceded by: Anali za eksperimentalno šumarstvo. Odjel za prirodne nauke. HI 61526

Analilie societâţei academice române. Bucharest. Analilie Soc. Acad. Române. See B–P–H 93/18. HI 50902

Analisis forestal. Mexico City. Analisis Forest. See B–P–H 93/19. HI 50903

Analiticka flora jugoslavije. Zagreb. Vol. 1+, 1967+. Analitika Fl. Jugoslav. HI 61527

Analyse des travaux. Partie physique. Paris. Analyse Trav., Pt. Phys. See B–P–H 93/21. HI 50905

Analyse des travaux de la classe des sciences mathématiques et physiques. Paris. Analyse Trav. Cl. Sci. Math. See B–P–H 93/20. HI 50904

Analyse des travaux de la société d'émulation d'Abbeville. Abbeville. 1828. Analyse Trav. Soc. Émul. Abbeville. Preceded by: Mémoriaux de la société d'émulation d'Abbeville. Superseded by: Mémoires de la société royale d'émulation d'Abbeville. 5-3938-3. HI 61528

Analyst; a monthly journal of science, literature, and the fine arts. London. Analyst. See B–P–H 93/22. HI 50906

Analytical biochemistry. New York, NY. Analytical Biochem. See B–P–H 94/2. HI 50907

Analytical methods for pesticides and plant growth regulators. Orlando, FL. Vol. 6+, 1972+. Analytical Meth. Pestic. Pl. Growth Regulators. Preceded by: Analytical methods for pesticides, plant growth regulators, and food additives. HI 63401

Analytical methods for pesticides, plant growth regulators, and food additives. Orlando, FL. Vols. 1-5, 1963-67. Analytical Meth. Pestic. Pl. Growth Regulators Food Addit. Superseded by: Analytical methods for pesticides and plant growth regulators. HI 63589

Analytical and quantitative cytology. St. Louis, MO. Vol. 1+, 1979+. Analytical Quant. Cytol. HI 61529

Analytical review, or history of literature, domestic and foreign, on an enlarged plan. London. Analytical Rev. See B–P–H 94/3. HI 50908

Analytical review of publications, department of forest research, CNRF = Revue analytique des publications, département des recherches forestières, CNRF. Seichamps.

Anatomische, Chymische und Botanische Abhandlungen, der Königl. Akademie der Wissenschaften in Paris = Der Königl. Akademie der Wissenschaften in Paris Anatomische, Chymische und Botanische Abhandlungen. Breslau [=Wroclaw, Poland]. Königl. Akad. Wiss. Paris Anat. Abh. See B–P–H 517/16.

Ancient science of life; journal of international institute of Ayurveda. Coimbatore. Vol. 1+, 1981+. Ancient Sci. Life. HI 61530

Andhra agricultural journal. Bapatla, India. Andhra Agric. J. See B–P–H 94/7. HI 50909

Andrias; Landessammlungen für Naturkunde. Karlsruhe. Vol. 1+, 1981+. Andrias. HI 61531

Anémone. Geneva. 1978+. Anémone. HI 61532

Anexo, commissão das linhas telegráficas estratégicas de Matto Grosso ao Amazonas = Relatório, commissão das linhas telegráficas estratégicas de Matto Grosso ao Amazonas. Anexo. Rio de Janeiro.

Anexos das memórias do instituto de Butantan. Secção de botânica. [Supplement to: Memórias do instituto Butantan.] São Paulo. Vol. 1(1-6), 1921-22. Anexos Mem. Inst. Butantan, Secç. Bot. Superseded by: Archivos de botânica do São Paulo. 1-855-3. HI 61533

Angewandte Botanik. Berlin. Angew. Bot. See B–P–H 94/14. HI 50910

Angewandte Chemie. Vienna. Angew. Chem. See B–P–H 94/15. HI 50911

Angewandte Pflanzensoziolgie. Stolzenau/Weser, German. Angew. Pflanzensoziol. (Stolzenau). See B–P–H 94/17. HI 50912

Angewandte Pflanzensoziologie. Vienna. Angew. Pflanzensoziol. (Vienna). See B–P–H 94/18. HI 50913

Angus wildlife review. Dundee. 1974+. Angus Wildlife

Rev. HI 61534

Anhaltische Gartenbau-Zeitung. Mit Berücksichtigung der Landwirthschaft. Dessau. Anhalt. Gartenbau-Zeitung. See B–P–H 94/21. HI 50914

Anhui shida xuebao. Ziran kexue ban. No. ?-7+, 19??-81+. Anhui Shida Xuebao, Ziran Kexue Ban. HI 61535

An-hui shih ta hsüeh pao = Anhui shida xuebao. Ziran kexue ban.

Ankara üniversitesi eczacilik fakültesi mecmuasi. Ankara. 1971+. Ankara Univ. Eczac. Fak. Mecm. HI 61536

Ankara universitesi fen fakultesi yayinlari. Botanik. Ankara. 1952. Ankara Univ. Fen Fak. Yayinl., Bot. HI 61537

Anlaegsgartneren. Copenhagen. Anlaegsgartneren. See B–P–H 94/25. HI 50883

Annaes da academia brasileira de sciencias. Rio de Janeiro. Ann. Acad. Brasil. Sci. See B–P–H 95/1. HI 50915

Annaes. Academia nacional de medicina do Rio de Janeiro. Rio de Janeiro. Ann. Acad. Nac. Med. Rio de Janeiro. See B–P–H 95/10. HI 50925

Annaes brasileiros de gynecologia. Rio de Janeiro. Ann. Brasil. Gynecol. See B–P–H 99/1. HI 50983

Annaes. Faculdade de medicina e cirurgia. São Paulo. Ann. Fac. Med. See B–P–H 102/2. HI 51018

Annaes de oculista do Rio de Janeiro. Rio de Janeiro. Ann. Ocul. Rio de Janerio. See B–P–H 109/7. HI 51104

Annaes das sciencias, das artes, e das letras; por huma sociedade de portuguezes residentes em Parîs. Paris. Ann. Sic. Soc. Portug. Parîs. See B–P–H 113/22. HI 51151

Annaes de sciencias naturaes. Oporto, Portugal. Ann. Sci. Nat. (Oporto). See B–P–H 113/15. HI 51147

Annaes scientificos de academia polytecnica do Porto. Coimbra, Portugal. Ann. Sci. Acad. Polytecn. Porto. See B–P–H 113/2. HI 51138

Annale van die uniwersiteit [later universiteit] van Stellenbosch. Reeks A, wis- en natuurkunde. Kaapstad. Vols. 1-50, 1923-75. Ann. Univ. Stellenbosch, Reeks A, Wis- Naturk. Superseded by: Annale van die universiteit van Stellenbosch. Serie A3, landbou and Annale van die universiteit van Stellenbosch. Serie A4, bosbou. 5-4076-1. HI 61538

Annale van die universiteit van Stellenbosch. Serie A3, landbou. Kaapstad. Vol. 1+, 1979+. Ann. Univ. Stellenbosch, A3. Preceded by: Annale van die universiteit van Stellenbosch. Reeks A, wis- en natuurkunde. HI 61539

Annale van die universiteit van Stellenbosch. Serie A4, bosbou. Kaapstad. Vol. 1+, 1979+. Ann. Univ.. Stellenbosch, A4. Preceded by: Annale van die universiteit van Stellenbosch. Reeks A. HI 61540

Annalen der Allgemeinen Schweizerischen Gesellschaft für die gesammten Naturwissenschaften. Bern. Ann. Allg. Schweiz. Ges. Gesammten Naturwiss. See B–P–H 96/14. HI 50945

Annalen der Altenburgischen pomologischen Gesellschaft. Altenburg. Ann. Altenburg. Pomol. Ges. See B–P–H 96/15. HI 50947

Annalen der Baierischen Litteratur. Nuremberg. Ann. Baier. Litt. See B–P–H 97/9 HI 50957.

Annalen der Blumisterei für Gartenbesitzer, Kunstgärtner, Samenhändler und Blumenfreunde. Nuremberg. Ann. Blumisterei Gartenbesitz. See B–P–H 97/12. HI 50970

Annalen der Botanik = Annalen der Botanick. ed. Usteri. Zurich. Ann. Bot. (Usteri). See B–P–H 98/12.

Annalen der Botanick. ed. Usteri. Zurich. Ann. Bot. (Usteri). See B–P–H 98/12. HI 50976

Annalen der Braunschweig-Lüneburgischen Churlande. Hanover. Ann. Braunschweig-Lüneburg. Churl. See B–P–H 99/3. HI 50985

Annalen der Chemie und Pharmacie. Heidelberg. Ann. Chem. Pharm. See B–P–H 99/13. HI 50990

Annalen der deutschen Landwirthschaft in allen Zweigen. Brunswick. Vol. 5, 1836. Ann. Deutsch. Landw. Allen Zweigen. Preceded by: Land- und forstwirthschaftliche Zeitschrift für Norddeutschland. 1-369-2. HI 61541

Annalen der Erd-, Vökler- und Staatenkunde. Berlin. Ann. Erd- Völker- Staatenk. See B–P–H 101/22. HI 51015

Annalen der Forst- und Jagd-Wissenschaft. Darmstadt, Germany. Ann. Forst- Jagd-Wiss. See B–P–H 102/26. HI 51029

Annalen der Fortschritte der Landwirthschaft in Theorie und Praxis. Berlin. Ann. Fortschr. Landw.. See B–P–H 103/2. HI 51030

Annalen der Gärtnerey. Nebst einem allgemeinen Intelligenzblatt für Garten- und Blumen-Freunde. Erfurt. Ann. Gärtnerey. See B–P–H 103/7. HI 51033

Annalen der Geographie und Statistik. Brunswick, Germany. Ann. Geogr. Statist. See B–P–H 103/13. HI 51037

Annalen für die gesammte Heilkunde. Karlsruhe. Ann. Gesammte Heilk. See B–P–H 103/15. HI 51039

Annalen der gesammten Litteratur. Erlanagen. Ann. Gesammten Litt. See B–P–H 103/16. HI 51040

Annalen der gesammten Medicin als Wissenschaft und als Kunst, zur Beurtheilung ihrer neuesten Erfindungen,

Theorien, System und Heilmethoden. Leipzig. Ann. Gesammten Med. See B–P–H 103/17. HI 51041

Annalen der Gewächskunde = Botanische Literatur-Blätter zur periodischen Darstellung der Fortschritte der Pflanzenkunde in steter Beziehung zur gesammten Naturkunde, ... Nuremberg. Bot. Lit.-Blätt. See B–P–H 221/9.

Annalen der Gewerbkunde, oder das Neueste aus dem Gebiete der Manufacturen, des Ackerbaus und Handels. Hamburg. Ann. Gewerbk. See B–P–H 103/19. HI 51043

Annalen des grossherzoglichen Acker- und Gartenbau-Vereins. Luxembourg. Ann. Grossherzogl. Acker-Gartenbau-Vereins. See B–P–H 103/20. HI 51044

Annalen für Hydrographie. Berlin. Ann. Hydrogr. (Berlin). See B–P–H 104/13. HI 51053

Annalen des K. K. Naturhistorischen Hofmuseums. Vienna. Ann. K. K. Naturhist. Hofmus. See B–P–H 106/9. HI 51071

Annalen koninklijk museum voor midden-Afrika = Annales de musée royal de l'Afrique centrale. Sciences economiques. Tervuren.

Annalen für die Landwirthschaft und das Landwirthschafts-Recht. Posen [=Poznan, Poland] & Leipzig. Ann. Landw. Landw.-Recht. See B–P–H 106/10. HI 51072

Annalen der Märkischen Oekonomischen Gesellschaft zu Potsdam. Potsdam. Ann. Märk. Oekon. Ges. Potsdam. See B–P–H 106/25. HI 51078

Annalen der Mecklenburgischen Landwirthschaftsgesellschaft. Rostock. Ann Mecklenburg. Landwirthschaftsges. See B–P–H 106/26. HI 51079

Annalen des National-Museums der Naturgeschichte in Paris. Hamburg & Mainz. Ann. Natl.-Mus. Naturgesch. Paris. See B–P–H 108/18. HI 51096

Annalen der Naturgeschichte. Göttingen. Ann. Naturgesch. See B–P–H 108/20. HI 51098

Annalen des Naturhistorischen Hofmuseums. Vienna. Ann. Naturhist. Hofmus. See B–P–H 108/23. HI 51099

Annalen des naturhistorischen Museums in Wien. Vienna. Vols. 33-83, 1919-79. Ann. Naturhist. Mus. Wien. Preceded by: Annalen des naturhistorischen Hofmuseums. Superseded by: Annalen des naturhistorischen Museums in Wien. Serie B, Bot. Zool. and Annalen des naturhistorischen Museums in Wien. Serie C, Jahresberichte. 5-4395-1. HI 61542

Annalen des naturhistorischen Museums in Wien. Serie B, Botanik und Zoologie. Vienna. Vol. 84+, 1980+. Ann. Naturhist. Mus. Wien, B. Preceded by: Annalen des naturhistorischen Museums in Wien. HI 61543

Annalen des naturhistorischen Museums in Wien. Serie C, Jahresberichte. Vienna. Vol. 84+, 1980+. Ann. Naturhist. Mus. Wien, C. Preceded by: Annalen des naturhistorischen Museums in Wien. HI 61544

Annalen der Niedersächsischen Landwirthschaft. Celle. Ann. Niedersächs. Landw. See B–P–H 109/1. HI 51102

Annalen der Obstkunde, herausgegeben von der Altenburgischen pomologischen Gesellschaft. Leipzig. Ann. Obstk.. See B–P–H 109.4. HI 51103

Annalen der Pharmacie. Lemgo & Heidelberg. Ann. Pharm. (Lemgo & Heidelberg). See B–P–H 109/27. HI 51111

Annalen der Physik. Halle. Ann. Phys. See B–P–H 110/3. HI 51118

Annalen der Physik und Chemie. Halle = Annalen der Physik. Halle. Ann. Phys. See B–P–H 110/3.

Annalen der Physik und Chemie. Leipzig. Ann. Phys. Chem. See B–P–H 110/5. HI 51120

Annalen der Physik und der physikalischen Chemie = Annalen der Physik. Halle. Ann. Phys. See B–P–H 110/3.

Annalen der Rostockschen Academie. Rostock. Ann. Rostockschen Acad. See B–P–H 112/17. HI 51131

Annalen der Samenprüfungsanstalt am Kaiserlichen Botanischen Garten zu St. Petersburg = Zapiski Stantsii dlya Ispytaniya Sĕmyan pri Imperatorskom Botanicheskom Sadĕ. St. Petersburg.

Annalen der Societät der Forst- und Jagdkunde = Annalen der Forst- und Jagd-Wissenschaft. Darmstadt, Germany. Ann. Forst- Jagd-Wiss. See B–P–H 102/26.

Annalen der Staatsarzneykunde. Züllichau, Germany [=Sulechow, Poland]. Ann. Staatsarzneyk. See B–P–H 116/18. HI 51184

Annalen der teutschen Akademien. Stuttgart. Ann. Teutsch. Akad. See B–P–H 117/3. HI 51191

Annalen aller Verhandlungen und Arbeiten der öconomisch-patriotischen Societät des Fürstenthumes Schweidnitz. Jauer, Germany [=Jawor, Poland]. Ann. Verh. Arbeiten Öcon.-Patriot. Soc. Fürstenth. Schweidnitz. See B–P–H 118/7. HI 51208

Annalen der weisruthenischen staatlichen Akademie für Landwirtschaft in Gorky = Zapiski belorusskoi gosudarstvennoi akademii sel'skogo khozyaistva im. oktyabr'skoi revolyutsii. Gorki.

Annalen der Wetterauischen Gesellschaft für die gesammte Naturkunde. Frankfurt a. M. Ann. Wetterauischen Ges. Gesammate Naturk. See B–P–H 118/12. HI 51210

Annalen des Wiener Museums der Naturgeschichte.

Vienna. Ann. Wiener Mus. Naturgesch. See B–P–H 118/18. HI 51211

Annalen der württembergischen Landwirthschaft. Stuttgart. Ann. Württemberg. Landw. See B–P–H 118/19. HI 51212

Annaler, kongl. svenska landtbruks-academiens. Stockholm = Kongl. svenska landtbruks-academiens annaler. Stockholm. Kongl. Svenska Landbr.-Acad. Ann. See B–P–H 517/9.

Annales de l'A C F A S. Montreal. Vol. 1+, 1935+. Ann. A. C. F. A. S. 1-517-2. HI 61545

Annales de l'abeille. Paris. Vols. 4-11, 1961-68. Ann. Abeille. Preceded by: Annales de l'institut national de la recherche agronomique, série C bis, annales de l'abeille. Superseded by: Apidologie. HI 61546

Annales academia médico-quirúrgica espanola. Madrid. Ann. Acad. Méd.-Quir. Esp. See B–P–H 95/9. HI 50924

Annales academiae gandavensis. Ghent. Ann. Acad. Gand. See B–P–H 95/2. HI 50916

Annales academiae groninganae. Groningen. Ann. Acad. Groning. See B–P–H 95/3. HI 50917

Annales academiae horti- et viticulturae = Kertészeti és szölészeti föiskola évkönyve. Budapest.

Annales academiae jenensis. Jena. Ann. Acad. Jenensis. See B–P–H 95/5. HI 50919

Annales academiae leodiensis. Liége. Ann. Acad. Leod. See B–P–H 95/6. HI 50921

Annales academiae lovaniensis. Brussels. Ann. Acad. Lovan. See B–P–H 95/7. HI 50922

Annales academiae lugduno-batavae. Leiden. Ann. Acad. Lugduno-Batavae. See B–P–H 95/8. HI 50923

Annales academiae medicae gedanensis. Gdansk. Vol. 1+, 1971+. Ann. Acad. Med. Gedanensis. HI 61547

Annales academiae rheno-trajectinae. Utrecht. Ann. Acad. Rheno-Traject. See B–P–H 95/14. HI 50928

Annales academiae rheno-trajectinae supplementum. Utrecht. Ann. Acad. Rheno-Traject. Suppl. See B–P–H 95/15. HI 50929

Annales academiae scientiarum fennicae. Ser. A. Helsinki. 1909-46. Ann. Acad. Sci. Fenn., Ser. A. Superseded by: Annales academiae scientiarum fennicae. Ser. A, III, geologica-geographica and Annales academiae scientiarum fennicae. Ser. A, IV, biologica. 5-4116-3. HI 61548

Annales academiae scientiarum fennicae. Ser. A, III, geologica-geographica. Helsinki. 1941+. Ann. Acad. Sci. Fenn., Ser. A, III, Geol.-Geogr. Preceded by: Annales academiae scientiarum fennicae. Ser. A. HI 61549

Annales academiae scientiarum fennicae. Ser. A, IV, biologica. Helsinki. 1945+. Ann. Acad. Sci. Fenn., Ser. A, IV, Biol. Preceded by: Annales academiae scientiarum fennicae. Ser. A. HI 61550

Annales academici. The Hague & Leiden. Ann. Acad. (The Hague & Leiden). See B–P–H 94/27. HI 50896

Annales, académie de La Rochelle. Section des sciences naturelles. La Rochelle. 1854-78. Ann. Acad. La Rochelle, Sect. Sci. Nat. Superseded by: Annales de la société des sciences naturelles de la Charente-inférieure. HI 61551

Annales de l'académie nationale de Reims. Rheims, France. Ann. Acad. Natl. Reims. See B–P–H 95/11. HI 50927

Annales. Académie techéchoslovaque de l'agriculture. Prague = Sborník československé akademie zemědělských věd. Prague. Sborn. Českoslov. Akad. Zeměd. Věd. See B–P–H 813/31.

Annales de l'ACFAS = Annales de l'A C F A S. Montreal.

Annales administratives et scientifiques de l'agriculture française. Paris. Ann. Admin. Sci. Agric. Franç. See B–P–H 95/23. HI 50932

Annales agricoles du département de l'Aisne. St. Quentin, France. Ann. Agric. Dép. Aisne. See B–P–H 96/3. HI 50936

Annales agriculturae Fenniae. Helsinki. Ann. Agric. Fenniae. See B–P–H 96/6. HI 50938

Annals of the agricultural experiment station, government general of Chosen. [Noji shikenjo iho.] Suigen [=Suwon], Korea. Ann. Agric. Exp. Sta. Gov. Gen. Chosen. See B–P–H 96/4. HI 50937

Annales de l'agriculture des colonies et des régions tropicales. Paris. Ann. Agric. Colon. Régions Trop. See B–P–H 96/2. HI 50935

Annales de l'agriculture française = Annales de l'agriculture françoise, ... Paris. Ann. Agric. Franç. See B–P–H 96/7.

Annales de l'agriculture françoise, ... Paris. Ann. Agric. Franç. See B–P–H 96/7. HI 50939

Annales agronomiques. [Institut des recherches agronomiques.] Paris. Ann. Agron. (Inst. Rech. Agron.). See B–P–H 96/10. HI 50942

Annales agronomiques. [Ministére de l'agriculture.] Paris. Ann. Agron. (Minist. Agric.). See B–P–H 96/11. HI 50943

Annales agronomiques. Paris. Vols. 12-31, 1961-80. Ann. Agron. (Paris). Preceded by: Annales de l'institut national de la recherche agronomique. Série A, annales agronomiques. Superseded by: Agronomie. HI 61552

Annales de l'amélioration des plantes. Paris. Ann. Amélior. Pl. See B–P–H 96/16. HI 50948

Annales d'anatomie pathologique. Paris. Ann. Anat. Pathol. See B–P–H 96/18. HI 50949

Annales des arts et manufactures; ou, mémoires technologiques sur les découvertes modernes concernant les arts, les manufactures, l'agriculture et le commerce. Paris. Vols. 1(1)-56(168), 1800-15; 2nd collection vols. 1(1)-6(18) [sér. gén. vols. 57-62], 1815-18. Ann. Arts Manufactures. 1-378-2. HI 52672

Annales de l'association canadienne-française pour l'avancement des sciences = Annales de l'A C F A S. Montreal.

Annales de l'association des naturalistes de Levallois-Perret. Levallois-Perret, France. Ann. Assoc. Naturalistes Levallois-Perret. See B–P–H 97/7. HI 50955

Annales de l'association philomatique vogéso-rhénane, faisant suite à la Flore d'Alsace de F. Kirschleger. Strasbourg. Ann. Assoc. Philom. Vogéso-Rhénane. See B–P–H 97/8. HI 50956

Annales belgiques des sciences, arts et littératures. Ghent. Ann. Belg. Sci. See B–P–H 97/10. HI 50958

Annales de biologie. Paris. Ann. Biol. (Paris). See B–P–H 97/13. HI 50961

Annales de biologie appliquée. Paris. Ann. Biol. Appl. See B–P–H 97/14. HI 50962

Annales de biologie lacustre. Brussels. Ann. Biol. Lacustre. See B–P–H 97/15. HI 50963

Annales biologicae universitatis budapestinensis. Budapest. Ann. Biol. Univ. Budapest. See B–P–H 97/16. HI 50965

Annales biologicae universitatis debreceniensis. Debrecen, Hungary. Ann. Biol. Univ. Debrecen. See B–P–H 97/17. HI 50967

Annales biologicae universitatum Hungariae. Budapest. Ann. Biol. Univ. Hungariae. See B–P–H 97/18. HI 50968

Annales biologicae universitatis szegediensis. Szeged, Hungary. Ann. Biol. Univ. Szeged. See B–P–H 97/19. HI 50969

Annales biologiques. Copenhagen. Ann. Biol. (Copenhagen). See B–P–H 97/11. HI 50959

Annales bogorienses. A journal of tropical general botany being a continuation of the Annals of the botanic gardens Buitenzorg. Bogor, Indonesia. Ann. Bogor. See B–P–H 98/2. HI 50971

Annales botanicae systematicae. [Edited by Walpers.] Leipzig. Ann. Bot. Syst. See B–P–H 98/26. HI 50982

Annales botanici. Genoa = Annali di botanica. Genoa. Ann. Bot. (Genoa). See B–P–H 98/5.

Annales botanici fennici. Helsinki. Vol. 1+, 1964+. Ann. Bot. Fenn. Preceded by: Suomalaisen elain-ja kasvitieteellisen seuran vanamon kasvitieteellisia julkaisuja. HI 61553

Annales botanici societatis zoologicae-botanicae fennicae "Vanamo." Helsinki. Ann. Bot. Soc. Zool.-Bot. Fenn. "Vanamo". See B–P–H 98/25. HI 50981

Annales de botanique et d'horticulture. Liége. Vols. 25-35, 1875-85. Ann. Bot. Hort. Preceded by: Annales d'horticulture Belge et étrangère. HI 61554

Annales de la brasserie et de la distillerie. Paris. Ann. Brass. Distill. See B–P–H 99/2. HI 50984

Annales bryologici; yearbook devoted to the study of mosses and hepatics. The Hague. Vols. 1-12, 1928-39; Supplement, vols. 1-4, 1930-34. Ann. Bryol. Superseded by: Annales cryptogamici et phytopathologici. 1-372-2. HI 61555

Annales bryologici. Supplement. The Hague. Vols. 1-4, 1930-34. Ann. Bryol., Suppl. HI 61556

Annales et bulletin de la société de médecine d'Anvers. Antwerp. Ann. & Bull. Soc. Méd. Anvers. See B–P–H 99/7. HI 50986

Annales et bulletin de la société de médecine de Gand = Annales de la société de médecine de Gand. Ghent.

Annales et bulletin de la société royale de médecine d'Anvers = Annales et bulletin de la société de médecine d'Anvers. Antwerp. Ann. & Bull. Soc. Méd. Anvers. See B–P–H 99/7.

Annales et bulletin de la société royale des sciences médicales et naturelles de Bruxelles. Brussels. Vols. 69(9)-75[+, vol. 75, 1921 is the last numbered volume], 1911-39. Ann. & Bull. Soc. Roy. Sci. Méd. Bruxelles. Preceded by: Bulletin de la société royale des sciences médicales et naturelles de Bruxelles. 5-3972-3. HI 61557

Annales du centre d'enseignement supérieur de Chambéry. Chambéry. Vols. 1-8, 1963-7? Ann. Centre Enseignem. Supér. Chambéry. Superseded by: Annales du centre universitaire de Savoie. Sciences naturelles. HI 61558

Annales du centre régional de documentation pedagogique de Clermont-Ferrand; bulletin régional de liaison et d'information des professeurs de sciences naturelles. Clermont-Ferrand. No. ?-4+, 19??-67+. Ann. Centre Régional Doc. Pedagog. Clermont-Ferrand. HI 61559

Annales du centre universitaire de Savoie. Sciences naturelles. Chambéry. Nos. 1-3, 1973-78. Ann. Centre Univ. Savoie, Sci. Nat. Preceded by: Annales du centre d'emseignement supérieur de Chambéry. Superseded by: Annales de l'université de Savoie.

Sciences naturelles. HI 61560

Annales du cercle agricole et horticole du Grand-Duché de Luxembourg. Luxembourg. Ann. Cercle Agric. Grand-Duché Luxembourg. See B–P–H 99/12. HI 50989

Annales de chimie; ou recueil de mémoires concernant la chimie et les arts qui en dépendent. Paris. Ann. Chim. (Paris). See B–P–H 99/18. HI 50992

Annales cryptogamici et phytopathologici. Waltham, MA. Vols. 1-11, 1944-54. Ann. Cryptog. Phytopathol. Preceded by: Annales bryologici. 1-372-3. HI 61561

Annales de cryptogamie exotique. Paris. Ann. Cryptog. Exot. See B–P–H 100/12. HI 51001

Annales de dermatologie et de syphiligraphie. Paris. Ann. Dermatol. Syphiligr. See B–P–H 100/30. HI 51003

Annales of the Durban museum. Durban, South Africa. Ann. Durban Mus. See B–P–H 101/3. HI 51004

Annales de l'école nationale d'agriculture. Paris. Ann. École Natl. Agric. (Paris). See B–P–H 101/8. HI 51005

Annales, école nationale d'agriculture d'Alger. Alger. Vols. 1-3(3), 1958-62. Ann. Écol. Natl. Agric. Alger. Preceded by: Annales de l'institut agricole et des services de recherches d'experimentation agricole de l'Algérie. Superseded by: Annales, institut national agronomique (El Harrach). HI 61562

Annales de l'école nationale d'agriculture de Montpellier. Montpellier. Ann. École Natl. Agric. Montpellier. See B–P–H 101/9. HI 51006

Annales de l'école nationale d'agriculture de Rennes. Rennes. Ann. École Natl. Agric. Rennes. See B–P–H 101/10. HI 51008

Annales de l'école nationale des eaux et forêts et de la station de recherches et expériences forestières. Nancy. Ann. École Natl. Eaux. See B–P–H 101/11. HI 51009

Annales de l'école nationale supérieure agronomique de Toulouse. Toulouse. Ann. École Natl. Supér. Agron. Toulouse. See B–P–H 101/12. HI 51010

Annales de l'école des sciences, université d'Abidjan. Abidjan. Vols. 1-3, 1965-66. Ann. Écol. Sci. Univ. Abidjan. Superseded by: Annales de l'université d'Abidjan. Série C. Sciences. HI 61563

Annales de l'école des sciences, université du Benin. Lome. Vol. 1, 1972. Ann. Écol. Sci. Univ. Benin. Superseded by: Annales de l'université du Benin. Sciences. HI 61564

Annales de l'école supérieure d'agriculture et amélioration du Don, à Novotcherkassk = Izvestiya donskogo instituta sel'skogo khozyaistva i melioratsii. Novocherkassk.

Annales de l'école supérieure de sciences. Dakar, Senegal [Republic of Senegal]. Ann. École Supér. Sci. See B–P–H 101/13. HI 51011

Annales encyclopédiques. [Edited by A. L. Millin.] Paris. Ann. Encycl. See B–P–H 101/14. HI 51012

Annales de l'enseignement supérieur de Grenoble. Paris & Grenoble. Ann. Enseignem. Supér. Grenoble See B–P–H 101/15. HI 51014

Annales des épiphyties. Paris. Vols. 7-19, 1920-33; n.s. vols. 1-14, 1934/35-48; vols. 12-19, 1961-68. Ann. Épiphyt. Preceded by: Annales du service des épiphyties. For 1950-60 see: Annales de l'institut national de la recherche agronomique, Série C, annales des épiphyties. Superseded by: Annales de phytopathologie. 1-378-2. HI 61565

Annales des épiphyties et de phytogénétique = Annales des épiphyties. Paris.

Annales d'essais de semences = Zapiski po semenovedeniyu. Leningrad.

Annales européennes de physique végétale et d'économie publique. Paris. Ann. Eur. Phys. Vég. Écon. Publique. See B–P–H 101/26. HI 51016

Annales pro experimentis foresticis = Glasnik za šumske pokuse. Zagreb, Yugoslavia. Glasn. Šumske Pokuse. See B–P–H 405/12.

Annales facultatis agronomicae universitatis agriculturae. Budapest = Agrártudományi egyetem mezögazdaságtudományi karának évkönyve. Budapest. Agrártud. Egyet. Mezögazdaságtud. Karának Évk. (Budapest). See B–P–H 55/14.

Annales de la faculté forestière, université de Belgrade = Glasnik šumarskog fakulteta, univerzitet u Beogradu. Belgrade.

Annales de la faculté des sciences de Marseille. Marseille & Paris. Vols. 1-26, 1891-1920; sér. 2. vol. 1-44, 1921-71. Ann. Fac. Sci. Marseille. Superseded by: Annales de l'université de Provence. Sciences. HI 61566

Annales de la faculté des sciences de l'université de Clermont. Clermont-Ferrand. Nos. 1-45, 1959-71. Ann. Fac. Sci. Univ. Clermont. Superseded by: Annales scientifiques de l'université de Clermont. HI 61567

Annales de la faculté des sciences, université de Dakar. Dakar, Senegal [Republic of Senegal]. Ann. Fac. Sci. Univ. Dakar See B–P–H 102/5. HI 51019

Annales, faculté des sciences, université fédérale du Cameroun. Yaoundé. Nos. 1-?, 1968-74; numéro hors série, 1970. Ann. Fac. Sci. Univ. Féd. Cameroun. Superseded by: Annales de la faculté des sciences de Yaoundé. HI 61568

Annales de la faculté des sciences, université nationale du Viêt-Nam. Saigon. Ann. Fac. Sci. Univ. Natl. Viêt-Nam. See B–P–H 102/7. HI 51020

Annales, faculté des sciences, université nationale de Zaire. Section biologie, chemie, sciences de la terre. Kinshasa. Vol. 1+, 1976+. Ann. Fac. Sci. Univ. Natl. Zaire, Sect. Biol. Chemie Sci. Terre. HI 61569

Annales de la faculté des sciences, université de Saigon. [Khao cu'u niên-san khoa-hoc dai-hoc du'ong. Viên dai-hoc Saigon.] Saigon. Ann. Fac. Sci. Univ. Saigon. See B–P–H 102/8. HI 51021

Annales de la faculté des sciences de Yaoundé. Yaoundé. No. 19+, 1975+. Ann. Fac. Sci. Yaoundé. Preceded by: Annales de la faculté des sciences, université fédérale du Cameroun. HI 61570

Annales des fermentations. Paris. Ann. Ferment. See B–P–H 102/15. HI 51024

Annales de flore et de pomone; ou journal des jardins et des champs. Paris. Ann. Fl. Pomone. See B–P–H 102/19. HI 51026

Annales forestières. Paris. Ann. Forest. See B–P–H 102/23. HI 51028

Annales forestières et métallurgiques = Annales forestières. Paris. Ann. Forest. See B–P–H 102/23.

Annales françaises et etrangères d'anatomie et de physiologie. appliquées à la médecine et l'histoire naturelle. Paris. Ann. Franç. Etrangères Anat. Physiol. See B–P–H 103/4. HI 51032

Annales de Gembloux. Brussels. Vol. 15+ 1905+. Ann. Gembloux. Preceded by: Ingénieur agricole de Gembloux [not entered]. HI 61571

Annales générales d'horticulture. Gand (Ghent). Vols. 16-23, 1865-83. Ann. Gén. Hort. Preceded by: Journal général d'horticulture. HI 74810

Annales générales des sciences physiques. Brussels. Ann. Gén. Sci. Phys. See B–P–H 103/10. HI 51034

Annales de génétique. Paris. Ann. Génét. See B–P–H 103/11. HI 51035

Annales de géographie. Paris. Ann. Géogr. See B–P–H 103/12. HI 51036

Annales géologiques des pays helléniques. Athens. Ann Géol. Pays Hellén. See B–P–H 103/14. HI 51038

Annales d'histochemie. Nancy, Paris. Vols. 1-21, 1956-76. Ann. Histochem. Superseded by: Cellular and molecular biology. HI 61572

Annales d'histoire naturelle. Paris. Ann. Hist. Nat. See B–P–H 103/21. HI 51045

Annales historico-naturales musei nationalis hungarici. Budapest. Ann. Hist.-Nat. Mus. Natl. Hung. See B–P–H 103/22. HI 51046

Annales de l'horticulture. Brussels = Annales de l'horticulture en Belgique. Brussels. Ann. Hort. Belgique. See B–P–H 104/3.

Annales d'horticulture Belge et étrangère. Liége. Vols. 15-24, 1865-74. Ann. Hort. Belge Étrangère. Preceded by: Belgique horticole. Superseded by: Annales de botanique et d'horticulture. HI 61573

Annales de l'horticulture en Belgique. Brussels. Ann. Hort. Belgique. See B–P–H 104/3. HI 51049

Annales d'horticulture et de botanique ou flore des jardins du royaume des Pays-Bas. Leiden. Ann. Hort. Bot. See B–P–H 104/4. HI 51050

Annales d'hydrobiologie. Paris. Vols. 1-8, 1970-77. Ann. Hydrobiol. Preceded by: Recherches d'hydrobiologie continentale. HI 61574

Annales hydrographiques. Paris. Ann. Hydrogr. (Paris). See B–P–H 104/14. HI 51054

Annales de l'institut agricole et des services de recherches d'experimentation agricole de l'Algérie. Alger. 1939-57. Ann. Inst. Agric. Serv. Rech. Exp. Agric. Algérie. Superseded by: Annales, école nationale d'agriculture d'Alger. HI 61575

Annales de l'institut agronomique de Moscou = Izvestiya moskovskogo sel'skokhozyaistvennago instituta. Moscow.

Annales de l'institut ampélologique royal hongrois = A magyar királyi központi szölészeti kísérleti állomás és ampelologiai intézet évkönyve. Budapest. Magyar Kir. Közp. Szölész. Kísérl. Állomás Ampelol. Intéz. Évk. See B–P–H 544/2.

Annales de l'institut botanico-géologique colonial de Marseille. Paris. Ann. Inst. Bot.-Géol. Colon. Marseille. See B–P–H 104/22. HI 51056

Annales de l'institut d'essais de semences au jardin botanique de Pierre le Grand = Zapiski Stantsii dlya Ispytaniya Sĕmyan pri Imperatorskom Botanicheskom Sadĕ Petra Velikago. St. Petersburg.

Annales de l'institut d'essais de semences au jardin principal botanique de la république. Petrograd = Zapiski Laboratorii po semenovedeniyu pri Glavnom botanicheskom sade R S F S R. Petrograd.

Annales de l'institut d'études maritimes d'Ostende. Mémoires. Ostend, Belgium. Ann. Inst. Etudes Marit. Ostende, Mém. See B–P–H 104/29. HI 51057

Annales de l'institut fédéral de recherches forestières = Mitteilungen der Schweizerischen Anstalt für das forstliche Versuchswesen. Zurich. Mitt. Schweiz. Anst. Forstl. Versuchswesen. See B–P–H 609/8.

Annales de l'institut géologique de Hongrie. Budapest = A magyar állami földtani intézet évkönyve. Budapest. Magyar Állami Földt. Intéz. Évk. See B–P–H 542/19.

Annales de l'institut horticole de Fromont. Paris = Annales de l'institut royale horticole de Fromont. Paris. Ann. Inst. Roy. Hort. Fromont. See B–P–H 105/17.

Annales de l'institut national agronomique. Paris, Alencon. 1 vol., 1852; vols. 1-16, 1876-1900; sér. 2, vols. 1-47, 1901-30?; sér. 3, vols. 1-6, 19??-?; n.s. vol. 1+, 19??-65+. Ann. Inst. Natl. Agron. 4-3266-3. HI 61576

Annales, institut national agronomique (El Harrach). Alger. Vol. 1+, 1972+. Ann. Inst. Natl. Agroc. (El Harrach). Preceded by: Annales, école nationale d'agriculture d'Alger. HI 61577

Annales de l'institut national de la recherche agronomique. Série A. Annales agronomiques. Paris. Ann. Inst. Natl. Rech. Agron., Sér. A, Ann. Agron. Sewe: 105/6. HI 51058

Annales de l'institut national de la recherche agronomique. Série A bis. Annales de physiologie végétale. Paris. Ann. Inst. Natl. Rech. Agron., Sér. A bis, Ann. Physiol. Veg. See B–P–H 105/7. HI 51059

Annales de l'institut national de la recherche agronomique. Série B. Annales de l'améloration des plantes. Paris. Ann. Inst. Natl. Rech. Agron., Sér. B, Ann. Amélior. Pl. See B–P–H 105/8. HI 51060

Annales de l'institut national de la recherche agronomique. Série C. Annales des épiphyties. Paris. Ann. Inst. Natl. Rech. Agron., Sér. C, Ann. Épiphyt. See B–P–H 105/9. HI 51061

Annales de l'institut national de la recherche agronomique. Série C bis. Annales de l'abeille. Paris. Ann. Inst. Natl. Rech. Agron., Sér. C bis, Ann. Abeille. See B–P–H 105/10. HI 51062

Annales, institut national de la recherche agronomique de Tunisie. Ariana. Vol. 32+, 1959+. Ann. Inst. Natl. Rech. Agron. Tunisie. Preceded by: Annales du service botanique et agronomique de la direction générale de l'agriculture Tunisie. HI 61578

Annales de l'institut océanographique. Monaco. Ann. Inst. Océanogr. See B–P–H 105/11. HI 51063

Annales de l'institut Pasteur. Paris. Vols. 1-123, 1887-1972. Ann. Inst. Pasteur. Superseded by: Annales de microbiologie and Annales d'immunologie [not entered]. 3-2002-2. HI 61579

Annales de l'institut phytopathologique Benaki. Athens. Ann. Inst. Phytopathol. Benaki. See B–P–H 105/13. HI 51064

Annales de l'institut de recherches agronomiques de Roumanie = Analele institutului de cercetări agronomice. Bucharest.

Annales de l'institut royal horticole de Fromont. Paris. Ann. Inst. Roy. Hort. Fromont. See B–P–H 105/17. HI 51065

Annales de l'institut scientifique de l'Alföld = Az alföldi tudományos gyüjtemény. Az alföldi tudományos intézet évkönyve. Budapest. Alföldi Tud. Gyüjt. See B–P–H 69/6.

Annales instituti biologici (Tihany) hungaricae academiae scientiarum = A magyar tudományos akadémia tihanyi biológiai kutatóintézetének évkönyve. Tihany.

Annales instituti biologiae pervestigandae hungarici = Magyar biológiai kutatóintézet évkönyve. Tihany, Hungary. Magyar Biol. Kutatóint. Évk. See B–P–H 542/21.

Annales instituti geologici publici (instituti publici geologiae) hungarici. Budapest = A magyar állami földtani intézet évkönyve. Budapest. Magyar Állami Földt. Intéz. Évk. See B–P–H 542/19.

Annales instituti ad investigandum horticulturae. Budapest = A kertészeti kutató intézet évkönyve. Budapest. Kert. Kutató Intéz. Évk. See B–P–H 512/6.

Annales instituti ad investigandum viticulturae = Szölészeti kutató intézet évkönyve. Budapest. Szölész. Kutató Intéz. Évk. See B–P–H 863/16.

Annales instituti protectionis plantarum hungarici. Budapest = Növényvédelmi kutató intézet évkönyve. Budapest. Növényvéd. Kutató Intéz. Evk. See B–P–H 674/11.

Annales instituti regii hungarici geologici. Budapest = A magyar kir[ályi] állami földtani intézet évkönyve. Budapest. Magyar Kir. Állami Földt. Intéz. Évk. See B–P–H 543/15.

Annales du jardin botanique de Buitenzorg. Batavia, Dutch E. Indies [=Jakarta, Indonesia]. Ann. Jard. Bot. Buitenzorg. See B–P–H 106/3. HI 51070

Annales du jardin botanique de Buitenzorg. Supplement. Batavia. [= Jakarta]. Vols. 1-4, 1897-1918. Ann. Jard. Bot. Buitenzorg, Suppl. 1-723-2. HI 61580

Annales du jardin botanique de Nikita = Zapiski Gosudarstvennogo Nikitskogo botanicheskogo sada. Yalta.

Annales du jardin botanique de Nikita = Zapiski Gosudarstvennogo Nikitskogo opytnogo botanicheskogo sada. Yalta.

Annales des jardiniers amateurs. Paris. Ann. Jard. Amateurs. See B–P–H 106/1. HI 51069

Annales de limnologie. Toulouse. Vol. 1+, 1965+. Ann. Limnol. Preceded by: Recueil, travaux de la station biologique du Lac d'Oredon. HI 61581

Annales literarii. Helmstedt, Germany. Ann. Lit. See B–P–H 106/14. HI 51074

Annales des maladies de l'oreille, du larynx, du nez et du pharynx. Paris. Ann. Malad. Oreille. See

B–P–H 106/23. HI 51076

Annales malgaches. Série sciences. Tananarive. No. 1, 1963. Ann. Malgaches, Sér. Sci. Superseded by: Annales de l'université de Madagascar. Série sciences et techniques. HI 61582

Annales maritimes et coloniales. Paris. Ann. Marit. Colon. See B–P–H 106/24. HI 51077

Annales de la médecine physiologique. Paris. Ann. Méd. Physiol. See B–P–H 107/7. HI 51083

Annales de médecine vétérinaire. Brussels. Ann. Méd. Vétérin. See B–P–H 107/11. HI 51086

Annales medicinae experimentalis et biologiae Fenniae. Helsinki. Vols. 25-51, 1947-73? Ann. Med. Exp. Biol. Fenniae. Preceded by: Acta societas medicorum fennica duodecim. Series A. Superseded by: Medical biology. 1-382-2. HI 61583

Annales de microbiologie. Paris. Vol. 124A+, 1973+. Ann. Microbiol. Preceded in part by: Annales de l'institut Pasteur. HI 61584

Annales de micrographie. Paris. Ann. Microgr. See B–P–H 107/14. HI 51087

Annales du musée du Congo Belge. Botanique. Brussels. Sér. 1, 1 vol., 1898/1902; sér. 2, vol. 1, 1899/1900; sér. 3, vol. 1, 1901; sér. 4, vols. 1-2, 1902/03-13; sér. 5, vols. 1-4, 1903/06-?; sér. 6, vol. 1, 1934. Ann. Mus. Congo Belge, Bot. 5-4184-2. HI 61585

Annales du musée F. Móra, Szeged = A Móra Ferenc múzeum évkönyve. Szeged, Hungary. Móra Ferenc Múz. Évk. See B–P–H 620/7.

Annales du musée d'histoire naturelle de Marseille. Marseilles. 1882-1937. Ann. Mus. Hist. Nat. Marseille. Superseded by: Bulletin du muséum d'histoire naturelle de Marseille. HI 61586

Annales de musée royal de l'Afrique centrale. Sciences economiques. Tervuren. Vol. 3+, 1964+. Ann. Mus. Roy. Afrique Centr., Sci. Econ. Preceded by: Annales du musée royal du Congo Belge. HI 61587

Annales du musée royal du Congo Belgo. Sciences historique et economique. Brussels, Tervuren. Vols. 1-?, 1947-50. Ann. Mus. Roy. Congo Belge, Sci. Hist. Econ. Superseded by: Annales du musée royal de l'Afrique centrale. Sciences economiques. 5-4184-3. HI 61588

Annales musei Goulandris; contributiones ad historiam naturalem graeciae et regionis mediterraneae. Kifisia, Athens. Vol. 1+, 1973+. Ann. Mus. Goulandris. HI 61589

Annales musei botanici lugduno-batavi. Amsterdam. Vols. 1-4, 1863-69. Ann. Mus. Bot. Lugduno-Batavi. 3-2401-2. HI 61590

Annales musei comit[ait] Castriferrei. Sectio historico-naturalis = A vasvármegyei múzeum természetrajzi osztályának évi jelentése. Szombathely, Hungary. Vasvárm. Múz. Term. Oszt. Évi Jel. See B–P–H 947/9.

Annales musei miskolciensis de Herman Ottó nominati = A Herman Ottó múzeum évkönyve. Miskolc, Hungary. Herman Ottó Múz. Évk. See B–P–H 415/13.

Annales. Museum Francisceum. Brünn [Brno, Czechoslovakia]. Ann. Mus. Francisc. See B–P–H 108/4. HI 51090

Annales du muséum du Havre. Le Havre. Vol. 1+, 1974+. Ann. Mus. Havre. HI 61591

Annales du muséum d'histoire naturelle. Paris = Annales du muséum national d'histoire naturelle. Paris. Ann. Mus. Natl. Hist. Nat. See B–P–H 108/11.

Annales du muséum d'histoire naturelle de Nice. Nice. Vol. 1+, 1972+. Ann. Mus. Hist. Nat. Nice. HI 61592

Annales du muséum national d'histoire naturelle ["national" dropped with vol. 5]. Paris. Vols. 1-21, 1802-13; 1827 [vol. 21, 1827, is an index to vols. 1-20]. Ann. Mus. Natl. Hist. Nat. Superseded by: Mémoires du muséum d'histoire naturelle. 4-3267-3. HI 51092

Annales mycologici editi in notitiam scientiae mycologicae universalis. Berlin. Ann. Mycol. See B–P–H 108/14. HI 51093

Annales de paléontologie. Paris. Ann. Paléontol. See B–P–H 109/17. HI 51109

Annales de parasitologie humaine et comparée. Paris. Ann. Parasitol. Humaine Comp. See B–P–H 109/23. HI 51110

Annales du parc national des Cévennes. Florac. Vol. 1+, 1979+. Ann. Parc Nat. Cévennes. HI 61593

Annales pharmaceutiques françaises. Paris. Vol. 1+, 1943+. Ann. Pharm. Franç. Preceded by: Bulletin des sciences pharmacologiques and Journal de pharmacie et de chimie. HI 61594

Annales de physiologie et de physicochimie biologique. Paris. Ann. Physiol. Physicochim. Biol. See B–P–H 110/8. HI 51121

Annales de physiologie végétale. Paris. Vols. 1-11(2), 1959-69. Ann. Physiol. Vég. Preceded by: Annales de l'institut national de la recherche agronomique. Série A bis, annales de physiologie végétale. Incorporated in: Physiologie végétale. HI 61595

Annales de physiologie végetale de l'université de Bruxelles. Brussels. Ann Physiol. Vég. Univ. Bruxelles. See B–P–H 110/10. HI 51123

Annales de physique biologique et médicale. Paris. Ann. Phys. Biol. Méd. See B–P–H 110/4. HI 51119

Annales de phytopathologie. Paris. Vols. 1-12, 1969-80. Ann. Phytopathol. Preceded by: Annales des épiphytiés. HI 61596

Annales de pomologie belge et étrangère. Brussels. Ann. Pomol. Belge Étrangère. See B–P–H 110/12. HI 51125

Annales pro experimentis foresticis = Glasnik za šumske pokuse. Zagreb, Yugoslavia. Glasn. Šumske Pokuse. See B–P–H 405/12.

Annales de protistologie; recueil des travaux originaux concernant la biologie et la systématique des protistes. Paris. Vols. 1-5, 1928-36 [suspended 1931/32-33]. Ann. Protistol. 1-378-1. HI 51126 HI 61597

Annales de la recherche forestière du Maroc. Rabat, Morocco. Ann. Rech. Forest. Maroc. See B–P–H 111/3. HI 51129

Annales de recherches sylvicoles. Nangis. 1976+. Ann. Rech. Sylvic. HI 61598

Annales et resumé de la société nantaise d'horticulture. Nantes. Ann. Res. Soc. Nantaise Hort. See B–P–H 112/9. HI 51130

Annales sabarienses. Szombathely, Hungary = Vasvármegye és Szombathely város kultúregyesülete és a vasvármegyei múzeum évokönyve. Szombathely, Hungary. Vasvárm. Szombathely Város Kultúregyes. Vasvárm. Múz. Évk. See B–P–H 947/10.

Annales sabarienses. Szombathely, Hungary. Ann. Sabar. See B–P–H 112/29. HI 51135

Annales de la science agronomique française et étrangère. Paris. Ann. Sci. Agron. Franç. Étrangére. See B–P–H 113/4. HI 51139

Annales des sciences et arts appliques aux industries textiles; bulletin scientifique. Verviers, Belqium. Ann. Sci. Arts. See B–P–H 113/5. HI 51140

Annales des sciences forestières. Versailles. Ann. Sci. Forest. See B–P–H 113/8. HI 51141

Annales des sciences géologiques. Paris. Ann. Sci. Géol. (Paris 1842-43). See B–P–H 113/10. HI 51143

Annales des sciences géologiques. Paris. Ann. Sci Géol. (Paris 1869-89). See B–P–H 113/11. HI 51144

Annales des sciences naturelles. Paris. Ann. Sci. Nat. (Paris). See B–P–H 113/16. HI 51148

Annales des sciences naturelles. Botanique. Paris. Sér. 2, vol. 1-sér. 11, vol. 20, 1834-1959. Ann. Sci. Nat., Bot. Preceded by: Annales des sciences naturelles, Paris. Superseded by: Annales des sciences naturelles. Botanique et biologie végétale. 1-380-2. HI 51900

Annales des sciences naturelles. Botanique et biologie végétale. Paris. Sér. 12, vol. 1+, 1960+. Ann. Sci. Nat., Bot. Biol. Vég. Preceded by: Annales des sciences naturelles. Botanique. HI 61599

Annales des sciences d'observation, ... Paris. Ann. Sci. Observ. See B–P–H 113/19. HI 51149

Annales des sciences physiques et naturelles, d'agriculture et de l'industrie, publiées par la société royale d'agriculture, etc., de Lyon. Lyons. Ann. Sic. Phys. Nat. Lyon. See B–P–H 113/21. HI 51150

Annales scientifiques, école normale supérieure. Paris. Vols. 1-7, 1864-70; ser. 2, vols. 1-12, 1872-83; ser. 3, vol. 1+, 1884+. Ann. Sci. École. Norm. Supér. 4-3265-2. HI 61600

Annales scientifiques de Franche-Comté. Besaçon. Ann. Sci. Franche-Comté. See B–P–H 113/9. HI 51142

Annales scientifiques du Limousin. Limoges. Vol. 1+, 1985+. Ann. Sci. Limousin. HI 61601

Annales scientifiques, littéraires et industrielles de l'Auvergne. Clermont-Ferrand. Vols. 1-31, 1828-58. Ann. Sci. Auvergne. Superseded by: Mémoires de l'académie des sciences, belles-lettres et arts de Clermont-Ferrand. HI 61602

Annales scientifiques, littéraires et industrielles et statistiques de l'Auvergne = Annales scientifiques, littéraires et industrielles de l'Auvergne. Clermont-Ferrand.

Annales scientifiques de l'université de Besançon. Biologie végétale. Besançon. Sér. 3, vol. 18+, 1977+. Ann. Sci. Univ. Besançon, Biol. Vég. Preceded by: Annales scientifiques de l'université de Besançon. Botanique. HI 61603

Annales scientifiques de l'université de Besançon. Botanique. Besançon. Vols. 5-9, 1946-54; sér. 2, vols. 1-21, 1955-?; sér. 3, vols. 1-17, 19??-76. Ann. Sci. Univ. Besançon, Bot. Preceded by: Annales scientifiques de Franche Comte. Superseded by: Annales scientifiques de l'université de Besançon. Biologie végétale. 1-652-2. HI 61604

Annales scientifiques de l'université de Clermont. Clermont-Ferrand. No. 46-?, 1972-?. Ann. Sci. Univ. Clermont. Preceded by: Annales de la faculté des sciences de l'université de Clermont. Superseded by: Annales scientifiques de l'université de Clermont-Ferrand II. Série biologie végétale. HI 61605

Annales scientifiques de l'université de Clermont-Ferrand II. Série biologie végétale. Clermont-Ferrand. ?-1983+. Ann. Sci. Univ. Clermont-Ferrand II, Sér. Biol. Vég. Preceded by: Annales scientifiques de l'université de Clermont. HI 61606

Annales scientifiques de l'université de Jassy. Jassy, Rumania. Ann. Sci. Univ. Jassy. See B–P–H 114/3. HI 51152

Annales scientifiques de l'université de Jassy. Partie 2, sciences naturelles. Jassy. Vols. 23-31, 1937-48. Ann. Sci. Univ. Jassy, Pt. 2, Sci. Nat. Preceded by: Annales sciientifiques de l'université de Jassy. Superseded by: Analele ştiinţifice ale universitătii "Al. I Cuza" din Iaşi. Sectiunea 2, ştiinţe naturale. 3-2158-

1. HI 61607

Annales scientifiques de l'université de Reims et de l'A R E R S. Reims. Vol. 11+, 1973+. Ann. Sci. Univ. Reims A. R. E. R. S. Preceded by: Annales de l'université et de l'A R E R S. HI 61608

Annales de la section dendrologique de la société botanique de Pologne = Rocznik sekcji dendrologicznej polskiego towarzystwa botanicznego. Warsaw.

Annales sectionis horti- et viticulturae universitatis scientiae agriculturae. Budapest = Agrártudományi egyetem kert- és szölögazdaságtudományi karának évkönyve. Budapest. Agrártud. Egyet. Kert-Szölögazdaságtud. Karának Évk. See B–P–H 55/12.

Annales sectionis horti- et viticulturae universitatis scientiae agriculturae. Budapest = Agrártudományi egyetem kert- és szölögazdaságtudományi karának közleményei. Budapest. Agrártud. Egyet. Kert-Szölögazdaságtud. Karának Közlem. See B–P–H 55/13.

Annales du service botanique et agronomique de la direction générale de l'agriculture. Tunis. Ann. Serv. Bot. Direct. Gén. Agric. See B–P–H 114/17. HI 51154

Annales du service botanique et agronomique de la direction générale de l'agriculture Tunisie. Tunis. Vols. 1-31, 1921-58. Ann. Serv. Bot. Tunisie. Superseded by: Annales, institut national de la recherche agronomique de Tunisie. HI 61609

Annales du service des épiphyties. Paris. Vols. 1-6, 1912-19. Ann. Serv. Épiphyt. Superseded by: Annales des épiphyties. 1-378-2. HI 61610

Annales societatis culturalis comit[ati] Castriferrei et musei comit[ati] Castriferrei = Vasvármegye és Szombathely város kultúregyesülete és a vasvármegyei múzeum évokönyve. Szombathely, Hungary. Vasvárm. Szombathely Város Kultúregyes. Vasvárm. Múz. Évk. See B–P–H 947/10.

Annales societatis rebus naturae investigandis in universitate tartuensis constitutae = Tartu ülikooli juures oleva loodusuurijate seltsi aruanded.

Annales societatis rebus naturae investigandis in universitate tartuensis constitutae = Tartu ülikooli juures oleva loodusuurijate seltsi aruanded.

Annales societatis scientiarum faeroensis. Torshavn = Frodskaparrit. Torshavn.

Annales societatis scientiarum faeroensis. Supplementum. Torshavn = Frodskaparrit. Supplementum. Torshavn.

Annales societatis zoolog[icae] botanicae fennicae Vanamo. Helsinki. Ann. Soc. Zool. Bot. Fenn. Vanamo. See B–P–H 116/14. HI 51182

Annales de la société d'agriculture, science, arts et commerce du Puy. Le Puy, France. Ann. Soc. Agric. Puy. See B–P–H 114/23. HI 51155

Annales de la société agriculture, sciences et industrie de Lyon. Lyons. Sér. 4, vols. 1-10, 1869-78; sér. 5, vols. 1-10, 1879-88; sér. 6, vols. 1-5, 1889-93; sér. 7, vols. 1-10, 1894-1903; sér. 8, vols. 1-2; 1904-05; 1906-21. Ann. Soc. Agric. Lyon. Preceded by: Annales des sciences physiques et naturelles, d'agriculture et de l'industrie, publiées par la société royale d'agriculture, etc. de Lyon. 5-3937-1. HI 61612

Annales de la société belge de microscopie. Brussels. Vols. 1-28, 1874-1907. Ann. Soc. Belge Microscop. HI 61613

Annales de la société botanique de Lyon. Lyons. Ann. Soc. Bot. Lyon. See B–P–H 114/25. HI 51156

Annales de la société d'émulation, agriculture, sciences, lettres et arts de l'Ain. Bourg-en-Bresse, France. Ann. Soc. Émul. Ain. See B–P–H 114/28. HI 51157

Annales de la société d'émulation du département des Vosges. Epinal, France. Ann. Soc. Émul. Dép. Vosges. See B–P–H 114/29. HI 51158

Annales de la société entomologique de France. Paris. Ann. Soc. Entomol. France. See B–P–H 115/2. HI 51159

Annales de la société géologique de Belgique. Liége. Ann. Soc. Géol. Belgique. See B–P–H 115/7. HI 51160

Annales de la société géologique du Nord. Lille. Ann. Soc. Géol. Nord. See B–P–H 115/9. HI 51161

Annales de la société d'histoire naturelle de Toulon. Toulon. 1910-45. Ann. Soc. Hist. Nat. Toulon. Superseded by: Annales de la société des sciences naturelles de Toulon. HI 61614

Annales de la société d'horticulture de la Gironde. Bordeaux. Ann. Soc. Hort. Gironde. See B–P–H 115/11. HI 51162

Annales de la société d'horticulture de la Haute-Garonne. Toulouse. Ann. Soc. Hort. Haute-Garonne. See B–P–H 115/12. HI 51163

Annales de la société d'horticulture et d'histoire naturelle de l'Hérault. Montpellier. Ann. Soc. Hort. Hérault. See B–P–H 115/13. HI 51164

Annales de la société d'horticulture de Paris, et journal spécial de l'état et des progrès du jardinage. Paris. Ann. Soc. Hort. Paris. See B–P–H 115/14. HI 51165

Annales de la société des lettres, sciences et arts des Alpes-Maritimes. Nice. Ann. Soc. Lett. Alpes-Maritimes. See B–P–H 115/16. HI 51166

Annales de la société linnéenne du département de Maine-et-Loire. Angers. Ann. Soc. Linn. Dép. Maine-et-

Loire. See B–P–H 115/19. HI 51167

Annales de la société linnéenne de Lyon. Lyons. Ann. Soc. Linn. Lyon. See B–P–H 115/21. HI 51168

Annales de la société linnéenne de Paris = Mémoires de la société linnéenne de Paris, précédés de son histoire. Paris. Mém. Soc. Linn. Paris. See B–P–H 585/3.

Annales de la société de médecine d'Anvers. Antwerp. Ann. Soc. Méd. Anvers. See B–P–H 115/24. HI 51169

Annales de la société de médecine de Gand. [Published with: Bulletin de la société de médicine de Gand.] Ghent. Vols. 1-2, 1835-36; 1837-43; vols. 14-89, 1844-1909; année 76-85, 1910-19; ed.2, vols. 1-2, 1835-36. [Année 81-83, 1915-17, not published.] Ann. Soc. Méd. Gand. Superseded by: Annales de la société royale de Gand [not entered]. 5-3972-1. HI 61615

Annales de la société médico-chirurgicale de Liège. Liége. Ann. Soc. Méd.-Chir. Liège. See B–P–H 115/26. HI 51170

Annales de la société nationale de l'horticulture de France. Paris. Vols. 1(1)-6(21), 1955-60. Ann. Soc. Natl. Hort. France. Incorporated in: Revue horticole. HI 61616

Annales de la société royale d'agriculture et de botanique de Gand; journal d'horticulture et des sciences accessoires. Ghent. Ann. Soc. Roy. Agric. Gand. See B–P–H 116/3. HI 51171

Annales de la société royale des sciences médicales et naturelles de Bruxelles. Brussels. Vols. 1-20, 1948-67. Ann. Soc. Roy. Sci. Méd. Bruxelles. 5-3972-3. HI 61617

Annales de la société des sciences, belles-lettres et arts d'Orléans. Orléans. Ann. Soc. Sci. Orléans. See B–P–H 116/13. HI 51181

Annales de la société des sciences médicales et naturelles de Bruxelles. Brussels. Ann. Soc. Sci. Méd. Bruxelles. See B–P–H 116/11. HI 51178

Annales de la société des sciences naturelles et d'archéologie de Toulon et du Var. Toulon. 1959+. Ann. Soc. Sci. Nat. Archéol. Toulon Var. Preceded by: Annales de la société des sciences naturelles de Toulon et du Var. HI 61618

Annales de la société de sciences naturelles de Bruges. Bruges, Belgium. Ann. Soc. Sci. Nat. Bruges. See B–P–H 116/12. HI 51179

Annales de la société des sciences naturelle de la Charente-inférieure. La Rochelle. 1879-1938. Ann. Soc. Sci. Nat. Charente-Infér. Preceded by: Annales, académie de La Rochelle. Section des sciences naturelles. Superseded by: Annales de la société des sciences naturelles de la Charente maritime. HI 61619

Annales de la société des sciences naturelles de la Charente-maritime. La Rochelle. 1946+. Ann. Soc. Sci. Nat. Charente-Marit. Preceded by: Annales de la société des sciences naturelles de la Charente-inférieure. HI 61620

Annales de la société des sciences naturelles de Toulon. Toulon. 1946-47. Ann. Soc. Sci. Nat. Toulon. Preceded by: Annales de la société d'histoire naturelle de Toulon. Superseded by: Annales de la société des sciences naturelles de Toulon et du Var. HI 61621

Annales de la société des sciences naturelles de Toulon et du Var. Toulon. 1948-56. Ann. Soc. Sci. Nat. Toulon Var. Preceded by: Annales de la société des sciences naturelles de Toulon. Superseded by: Annales de la société des sciences naturelles et d'archéologie de Toulon et du Var. HI 61622

Annales de la société scientifique de Bruxelles. Brussels. Ann. Soc. Sci. Bruxelles. See B–P–H 116/9. HI 51172

Annales de la société scientifique de Bruxelles. Série B. Brussels. Ann. Soc. Sci. Bruxelles, Sér. B. See B–P–H 116/10. HI 51175

Annales de la société de zymologie pure et appliquée. Ghent. Vols. 1-2, 1929-30. Ann. Soc. Zymol. Pure Appl. Superseded by: Annales de zymologie. 1-378-2. HI 61623

Annales de spéléologie. Toulouse. Vol. 14+, 1959+. Ann. Spéléol. Preceded by: Notes biospéologiques. HI 61624

Annales de la station biologique de Besse-en-Chandesse. Besse-en-Chandesse. No. 1+, 1966+. Ann. Stat. Biol. Besse-en-Chandesse. Preceded by: Annales de la station limnologique de Besse. HI 61625

Annales de la station centrale d'hydrobiologie appliqueé. Paris. Ann. Stat. Centr. Hydrobiol. Appl. See B–P–H 116/20. HI 51185

Annales de la station fédérale de recherches forestières = Mittheilungen der Schweizerischen Centralanstalt für das forstliche Versuchswesen. Zurich. Mitth. Schweiz. Centralanst. Forstl. Versuchswesen. See B–P–H 612/16.

Annales de la station limnologique de Besse. Clermont-Ferrand. Vols. 1-2, 1909-10. Ann. Stat. Limnol. Besse. Superseded by: Annales de la station biologique de Besse-en-Chandesse. 2-1072-3. HI 61626

Annales, station océanographique de Salammbô. Tunis. Ann. Stat. Océanogr. Salammbô. See B–P–H 116/23. HI 51188

Annales des travaux publics de Belgique. Brussels. Ann. Trav. Publics Belgique. See B–P–H 117/7. HI 51194

Annales universitatis fennicae åboensis. Series A, physico-mathematica-biologica. Turku. Vols. 1-4(1),

1922-33. Ann. Univ. Fenn. Åbo., A. Superseded by: Annales universitatis turkuensis. Ser. A, biologica, geographica, geologica. 1-8-3. HI 61627

Annales universitatis Mariae Curie-Skłodowska. Roczniki universytetu Marii Curie-Skłodowskiej. Sect. 3: Biologia. Lublin, Poland. Ann. Univ. Mariae Curie-Skłodowska, Sect. 3, Biol. See B–P–H 117/15. HI 51200

Annales universitatis saraviensis. Naturwissenschaften--Sciences. Saarbrücken. Ann. Univ. Sarav., Naturwiss. Sci. See B–P–H 117/16. HI 51201

Annales universitatis saraviensis. Reihe: Mathematisch-Naturwissenschaftliche Fakultät. Berlin. Ann. Univ. Sarav., Reihe Math.-Naturwiss. Fak. See B–P–H 117/17. HI 51202

Annales universitatis saraviensis. Scientia. Saarbrücken. Ann. Univ. Sarav., Sci. See B–P–H 117/18. HI 51203

Annales universitatis saraviensis. Wissenschaften--Sciences. Saarbrücken. Ann. Univ. Sarav., Wiss. Sci. See B–P–H 117/19. HI 51204

Annales universitatis scientiarum budapestinensis de Rolando Eötvös nominatae. Sectio biologica. Budapest. Ann. Univ. Sci. Budapest. Rolando Eötvös, Sect. Biol. See B–P–H 118/1. HI 51205

Annales universitatis turkuensis. Ser. A, biologica-geographica. Turku. Vols. 4(2)-22, 1935-56; ser. A2, biologica-geographica-geologica, vol. 23+, 1957+. Ann. Univ. Turku., A. Preceded by: Annales universitatis fennicae åboensis. Series A, physico-mathematica-biologica. HI 61628

Annales de l'université et de l'A R E R S. Reims. Vols. ?-10, 1963-72. Ann. Univ. A. R. E. R. S. Preceded by: Annales de l'A R E R S. Superseded by: Annales scientifique de l'université de Reims et de l'A R E R S. HI 61629

Annales de l'université d'Abidjan. Sciences. Abidjan. Vols. 1-?, 1965-67. Ann. Univ. Abidjan, Sci. Preceded by: Annales de l'école des sciences, université d'Abidjan. Superseded by: Annales de l'université d'Abidjan. Série C, sciences; annales de l'université d'Abidjan. Série E, écologie and Annales de l'université d'Abidjan. Série G, géographie. HI 61630

Annales de l'université d'Abidjan. Série C, sciences. Abidjan. Vol. 1+, 1968+. Ann. Univ. Abidjan, C. Preceded by: Annales de l'université d'Abidjan. Sciences. HI 61631

Annales de l'université d'Abidjan. Série E, écologie. Abidjan. Vol. 1+, 1968+. Ann. Univ. Abidjan, E. Preceded by: Annales de l'université d'Abidjan. Sciences. HI 61632

Annales de l'université d'Abidjan. Série G, géographie. Abidjan. Vol. 1+, 1969+. Ann. Univ. Abidjan, G. Preceded by: Annales de l'université d'Abidjan. Sciences. HI 61633

Annales de l'université du Benin. Sciences. Lome. 1975+. Ann. Univ. Benin, Sci. Preceded by: Annales de l'école des sciences, université du Benin. HI 61634

Annales de l'université de Brazzaville. Brazzaville. Vols. ?-8?, 19??-73? Ann. Univ. Brazzaville. Superseded by: Annales de l'université de Brazzaville. Serie C, sciences. HI 61635

Annales de l'université de Brazzaville. Serie C, sciences. Brazzaville. Vols. ?-9+, 19??-74+. Ann. Univ. Brazzaville, C. Preceded by: Annales de l'université de Brazzaville. HI 61636

Annales de l'université de Grenoble. Paris & Grenoble. Ann. Univ. Grenoble. See B–P–H 117/12. HI 51198

Annales de l'université de Lyon. Paris. Ann. Univ. Lyon. See B–P–H 117/13. HI 51199

Annales de l'université de Madagascar. Série sciences et techniques (sciences de la nature et mathématique). Antananarivo. No. 2+, 1965+. Ann. Univ. Madagascar, Sér. Sci. Techn. (Sci. Nat. Math.) Preceded by: Annales malgaches. HI 61637

Annales de l'université de Minsk = Pratsy belaruskaga dzerzhnaunaga universytetu u Mensku. Minsk.

Annales de l'université de Montpellier et du Languedoc Méditerranéen-Roussillon. Supplément scientifique, série botanique. Recueil des travaux de l'institut botanique = Recueil des travaux de l'institut botanique. Annales de l'université de Montpellier et du Languedoc Méditerranéen-Roussillon. Supplément scientifique, série botanique. Montpellier.

Annales de l'université de Provence. Sciences. Marseille. Vols. 45-46, 1971. Ann. Univ. Provence, Sci. Preceded by: Annales de la faculté des sciences de Marseille. Superseded by: Biologie et écologie méditerranéenne. HI 61638

Annales de l'université de Savoie. Sciences naturelles. Chambéry. Vol. 4+, 1979+. Ann. Univ. Savoie, Sci. Nat. Preceded by: Annales du centre universitaire de Savoie. Sciences naturelles. HI 61639

Annales des universités de Belgique. Brussels. Ann. Univ. Belgique. See B–P–H 117/9. HI 51195

Annales des voyages, de la géographie et de l'histoire. Paris. Ann. Voyages. See B–P–H 118/10. HI 51209

Annales de zymologie. Ghent. Ser. 2, vols. 1-3, 1931-36; ser. 3, vols. 4-6, 1937-40. Ann. Zymol. Preceded by: Annales de la société de zymologie pure et appliquée. Superseded by: Revue des fermentations et des industries alimentaires. 1-378-2. HI 61640

Annali dell'accademia degli aspiranti naturalisti. Naples.

Vols. 1-3, 1843-46; ser. 2, vols. 1-2, 1847; ser. 3, vols. 1-6, 1861-66; ser. 4, vol. 1, 1867; 2d. era vols. 1-2, 1868-69; 3d era vol. 1, 1887. Ann. Accad. Aspir. Naturalisti. Preceded by: Bullettino dell'accademia degli aspiranti naturalisti. 1-39-2. HI 61641

Annali dell'accademia italiana di scienze forestali. Florence. Ann. Accad. Ital. Sci. Forest. See B–P–H 95/21. HI 50930

Annali di agricoltura. Rome. Ann. Agric. (Rome). See B–P–H 95/25. HI 50933

Annali dell'agricoltura de regno d'Italia. Milan. Ann. Agric. Regno Italia. See B–P–H 96/8. HI 50940

Annali dell'agricoltura del regno d'Italia. Serie seconda. Milan. Ann. Agric. Regno Italia, Ser. 2. See B–P–H 96/9. HI 50941

Annali di botanica. Genoa. Ann. Bot. (Genoa). See B–P–H 98/5. HI 50973

Annali di botanica. Rome. Vol. 1+, 1903+. Ann. Bot. (Rome). Preceded by: Annuario del reale istituto botanico di Roma. 1-385-1. HI 61643

Annali di chimica applicata alla farmacia ed alla medicina. Milan. Ann. Chim. Appl. Farm. See B–P–H 99/20. HI 50993

Annali di chimica applicata alla medicina cioé alla farmacia. Milan. Ann. Chim. Appl. Med. See B–P–H 99/21. HI 50994

Annali di chimica e di farmacologia. Milan. Ann. Chim. Farmacol. See B–P–H 99/22. HI 50995

Annali di chimica e di medico-farmaceutica. Milan. Ann. Chim. Med.-Farm. See B–P–H 99/23. HI 50996

Annales de chimie et de physique. Paris. Ann. Chim. Phys. See 100/1. HI 50997

Annali di chimica e storia naturale ovvero raccolta di memorie dulle scienze, arti e manufatture ad esse relative. Pavia. Ann. Chim. Storia Nat. See B–P–H 100/2. HI 50998

Annali civili del regno delle due Sicilie. Naples. Ann. Civili Regno Due Sicilie. See B–P–H 100/3. HI 50999

Annali di clinica medica e di medicine sperimentale. Palermo. Ann. Clin. Med. Med. Sperim. See B–P–H 100/4. HI 51000

Annali della facolta di agraria. Rome. Ann. Fac. Agrar. See B–P–H 101/29. HI 51017

Annali della facoltà di agraria di Portici della reale università di Napoli. Portici. Ser. 3, vols. ?-19, 1936-51. Ann. Fac. Agrar. Portici. Preceded by: Annali del reale istituto superiore agrario di Portici. Superseded by: Annali della facoltà di scienze agrarie della università degli studi di Napoli. 4-2820-2. HI 61644

Annali della facoltà di agraria, università Bari. Bari. Vol. 1+, 1939+. Ann. Fac. Agrar. Univ. Bari. 1-605-2. HI 61645

Annali della facoltà di agraria, università cattolica del Sacro Cuore, Milano. Milan. Vols. 1-6, 1955-58. Ann. Fac. Agrar. Univ. Cattolica S. Cuore, Milano. Superseded by: Annali della facoltà di agraria. Università degli studi, Milano. HI 61646

Annali della facoltà agraria della università di Pisa. Pisa. N. s vol.1+, 1938+. Ann. Fac. Agrar. Univ. Pisa. Preceded by: Bollettino, reale istituto superiore agraria. 4-3364-1. HI 61647

Annali della facoltà di agraria, università degli studi Milano. Milan. N.s. vol. 7-9, 1958-60. Ann. Fac. Agrar. Univ. Milano. Preceded by: Annali della facoltà di agraria, università cattolica del S. Cuore, Milano. HI 61648

Annali della facoltà de agraria, università degli studi' Perugia. Perugia. 1942+. Ann. Fac. Agrar. Univ. Perugia. HI 61649

Annali della facoltà di medicina veterinaria di Torino. Turin. 1950+. Ann. Fac. Med. Veterin. Torino. HI 61650

Annali, facoltà di scienze agrarie della università degli studi di Torino. Turin. Vol. 1+, 1961/62+, 1962+. Ann. Fac. Sci. Agrar. Univ. Torino. HI 61651

Annali della facoltà di scienze agrarie della università degli studi di Napoli. Portici. Ser. 3, vols. 20-30, 1951/53-1964/65, 1951-65; ser. 4, vol. 1+, 1966+. Ann. Fac. Sci. Agrar. Univ. Napoli. Preceded by: Annali della facoltà di agraria di Portici della reale università di Napoli. HI 61652

Annali di farmacoterapia e chimica. Milan. Ann. Farmacot. Chim. See B–P–H 102/13. HI 51022

Annali di farmacoterapia e chimica biologica. Milan. Ann. Farmacot. Chim. Biol. See B–P–H 102/14. HI 51023

Annali di fitopatologia. Palermo. Ann. Fitopatol. See B–P–H 102/16. HI 51025

Annali idrografici. Genoa. Ann. Idrogr. See B–P–H 104/15. HI 51055

Annali d'igiene. Turin, Rome. Vols. 1-3, 1889-90; n.s. vols. 1-?, 1891-94; n.s. vols. 26?-59, 1916-1949. Ann. Ig. For 1895-1915 see: Annali d'igiene sperimentale. 1-384-1. HI 61654

Annali d'igiene sperimentale. Turin, Rome. N.s. vols. 5-25, 1895-1915. Ann. Ig. Sperim. Preceded by: Annali d'igiene. Superseded by: Nuovi annali d'igiene e microbiologia. 1-384-1. HI 61655

Annali. Istituto Carlos Forlanini. Rome. Ann. Ist. Carlos Forlanini. See B–P–H 105/24. HI 51067

Annali dell'istituto e museo de storia della scienza di Firenze. Florence. 1976+. Ann. Ist. Mus. Storia Sci. Firenze. HI 61656

Annali, istituto sperimentale agronomico. Bari. Vol. 1+, 1970+. Ann. Ist. Sperim. Agron. HI 61657

Annali dell'istituto sperimentale per l'agrumicoltura. Acireale. Vol. 1+, 1968/69+, 1969+. Ann. Ist. Sperim. Agrumic. HI 61658

Annali dell'istituto sperimentale per la cerealicoltura. Rome. Vol. 1+, 1970+. Ann. Ist. Sperim. Cerealicol. HI 61659

Annali dell'istituto sperimentale per la floricoltura. San Remo. Vol. 1+, 1968+. Ann. Ist. Sperim. Floric. HI 61660

Annali dell'istituto sperimentale per la frutticoltura. Rome. Vol. 1+, 1970+. Ann. Ist. Sperim. Fruttic. Preceded by: Pubblicazioni dell'istituto sperimentale per la frutticoltura. HI 61661

Annali dell'istituto sperimentale per la nutrizione delle piante. Rome. Vol. 1+, 1968/69+, 1970+. Ann. Ist. Sperim. Nutr. Piante. HI 61662

Annali dell'istituto sperimentale per l'olivicoltura. Vol. ?-2+, 19??-74+. Ann. Ist. Sperim. Olivicultura. HI 61663

Annali dell'istituto sperimentale per l'orticoltura. Vols. 1-3, 1970-72. Ann. Ist. Sperim. Ortic. HI 61664

Annali dell'istituto sperimentale per la selvicoltura. Arezzo. Vol. 1+, 1970+. Ann. Ist. Sperim. Selvic. Preceded by: Pubblicazioni dell'istituto sperimentale per la selvicoltura. HI 61665

Annali dell'istituto sperimentale per lo studio e la difesa del suolo. Florence. Vol. 1+, 1954+. Ann. Ist. Sperim. Stud. Difesa Suolo. HI 61666

Annali, istituto sperimentale per il tabacco. Scafati. Vol. 1+, 1974+. Ann. Ist. Sperim. Tabacco. HI 61667

Annali dell'istituto superiore di scienze e lettere "S. Chiara" dell'ordine dei frati minori. Naples. 1948-59. Ann. Ist. Super. Sci. Lett. S. Chiara. Superseded by: Annali del pontificio istituto superiore di scienze e lettere "S. Chiara" dell'ordine dei frati minori. HI 61668

Annali italiani di chirurgia. Naples. Ann. Ital. Chir. See B–P–H 105/28. HI 51068

Annali di laringologia, otologia, rinologia faringologia. Genoa. Ann. Laringol. See B–P–H 106/11. HI 51073

Annali di medicina. Milan. Ann. Med. (Milan). See B–P–H 107/1. HI 51080

Annali medico-chirurgici. Rome. Ann. Med.-Chir. See B–P–H 107/2. HI 51081

Annali di medicina straniera. Milan. Ann. Med. Straniera. See B–P–H 107:8. HI 51084

Annali del museo civico di storia naturale di Genova. Genoa. Vols. 1-56, 1870-1913/15. Ann. Mus. Civico Storia Nat. Genova. Superseded by: Annali del museo civico di storia naturale "Giacomo Doria". HI 61669

Annali del museo civico di storia naturale "Giacomo Doria." Genoa. Vol. 57+, 1916+. Ann. Mus. Civico Storia Nat. Giacomo Doria. Preceded by: Annali del museo civico di storia naturale di Genova. HI 61670

Annali del musèo imperiale di fisica e storia naturale di Firenze. Florence. Ann. Mus. Imp. Fis. Firenze See B–P–H 108/7. HI 51091

Annali del museo libico si storia naturale. Tripoli. 1939-53. Ann. Mus. Libico Storia Nat. HI 61671

Annali di ottalmologia. Pavia. Ann. Ottalmol. See B–P–H 109/15. HI 51107

Annali di ottalmologia e clinica oculistica. Pavia. Ann. Ottalmol. Clin. Oculist. See B–P–H 109/16. HI 51108

Annali del pontificio istituto superiore di scienze e lettere "S. Chiara" dell'ordine dei frati minori. Naples. 1960-66. Ann. Pontif. Ist. Super. Sci. Lett. S. Chiara. Preceded by: Annali dell'istituto superiore di scienze e lettere "S. Chiara" dell'ordine dei frati minori. HI 61672

Annali di radiologia diagnostica. Bologna. Ann. Radiol. Diagn. See B–P–H 110/23. HI 51127

Annali di radiologia e fisica medica. Bologna. Ann. Radiol. Fis. Med. See B–P–H 110/24. HI 51128

Annali del r[eale] istituto superiore agrario e forestale. Ser. 2, vols. 1-?, 1924/25-?, 1925?-? Ann. Reale Ist. Super. Agrar. Forest. HI 61673

Annali del r[eale] istituto superiore agrario di Portici. Naples. Ser. 3, vols. 1-?, 1926-35. Ann. Reale Ist. Super. Agrar. Portici. Preceded by: Annali della reale scuola superiore di agricoltura di Portici. Superseded by: Annali della facoltà di agraria di Portici della università di Napoli. 4-2820-2. HI 61674

Annali r[eale] scuola superiore di agricoltura di Portici. Naples. Ser. 2, vols. 1-20, 1899-1925. Ann. Reale Scuola Super. Agric. Portici. Preceded by: Annuario del r[eale] scuola superiore di agricoltura di Portici. Superseded by: Annuario del r[eale] istituto superiore agrario di Portici. 4-2820-2. HI 61675

Annali della r[eale] scuola di viticoltura e di enologia in Conegliano. Conegliano. 1892. Ann. Reale Scuola Vitic. Enol. Conegliano. Preceded by: Nuova rassegna di viticoltura ed enologia della r[eale] scuola di Conegliano. Superseded by: Rivista periodica della r[eale] scuola di viticoltura e di enologia di Conegliano. HI 61676

Annali di scienze e lettere. Milan. Ann. Sci. Lett. See

B–P–H 113/12. HI 51145

Annali Sclavo; rivista di microbiologia e di immunologia. Siena. Vol. 1+, 1959+. Ann. Sclavo. HI 61677

Annali della sperimentazione agraria. Rome. Ann. Sperim. Agrar. See B–P–H 116/16. HI 51183

Annali di storia naturale. Bologna. Ann. Storia Nat. See B–P–H 116/25. HI 51189

Annali universali di medicina. Milan. Ann. Universali Med. See B–P–H 118/3. HI 51206

Annali universali di medicina e chirurgia. Milan. Vols. 231-286, 1874-89. Ann. Universali Med. Chir. Preceded by: Annali universali di medicina. 1-386-2. HI 61678

Annali universali di tecnologia, di agricoltura, di economia rurale e domestica. Milan. Ann. Universali Tecnol. See B–P–H 118/4. HI 51207

Annali dell'università di Ferrara. N.s., sezione biologia. Ferrara. Vol. ?-1(2)+, 19??-78+. Ann. Univ. Ferrara, N.S., Sez. Biol. Preceded by: Annali dell'università di Ferrara. Sezione 1, ecologia. HI 61679

Annali dell'università di Ferrara. Sezione 1, ecologia. Ferrara. N.s., vols. 1-?, 1972-76. Ann. Univ. Ferrara, Sez. 1, Ecol. Superseded by: Annali dell'università di Ferrara. N.s., Sezione biologia. HI 61680

Annali dell'università di Ferrara. N.s., sezione IV, botanica. Ferrara, Italy. Ann. Univ. Ferrara, Sez. 4, Bot. See B–P–H 117/10. HI 51196

Annali dell'università di Ferrara. N.s., sezione VI, fisiologia e chimica biologica. Ferrara, Italy. Ann. Univ. Ferrara, Sez. 6, Fisiol. Chim. Biol. See B–P–H 117/11. HI 51197

Annals of the agricultural college of Sweden = Lantbrukshögskolans annaler. Uppsala.

Annals of allergy. St. Louis, MO. Ann. Allergy. See B–P–H 96/13. HI 50944

Annals of anatomy and physiology. Edinburgh. Ann. Anat. Physiol. See B–P–H 96/19. HI 50950

Annals of the Andersonian naturalists' society. Glasgow. Ann. Andersonian Naturalists' Soc. See B–P–H 97/1. HI 50951

Annals of applied biology. London. Ann. Appl. Biol. See B–P–H 97/2. HI 50952

Annals of arid zone. Jodhpur, India. Ann. Arid Zone. See B–P–H 97/3. HI 50953

Annals of the association of american geographers. Chicago, IL. Ann. Assoc. Amer. Geogr. See B–P–H 97/5. HI 50954

Annaly biologii. Moscow. Ann. Biol. (Moscow). See B–P–H 97/12. HI 50960

Annals of biology. New Delhi. Vol. 1+, 1985+. Ann. Biol. (New Delhi). HI 61681

Annals of the Bolus herbarium. Cambridge, England. Ann. Bolus Herb. See B–P–H 98/3. HI 50972

Annals of the botanic gardens, Buitenzorg. Buitenzorg, Dutch E. Indies [=Bogor, Indonesia]. Ann. Bot. Gard. Buitenzorg. See B–P–H 98/15. HI 50978

Annals of the botanical society of Canada. Kingston, Ontario. Ann. Bot. Soc. Canada. See B–P–H 98/23. HI 50980

Annals of botany. [Edited by König & Sims.] London. Ann. Bot. (König & Sims). See B–P–H 98/7. HI 50974

Annals of botany. Oxford. Ann. Bot. (Oxford). See B–P–H 98/9. HI 50975

Annals of botany memoirs. Oxford, England. Ann. Bot. Mem. See B–P–H 98/22. HI 50979

Annals of the Cape provincial museums. Natural history. Grahamstown. Vol. 1+, 1961+. Ann. Cape Prov. Mus., Nat. Hist. HI 61682

Annals of the Carnegie museum. Pittsburgh, PA. Vol. 1+, 1901+. Ann. Carnegie Mus. 2-934-1. HI 50988

Annals of chemistry and practical pharmacy. London. Ann. Chem. Pract. Pharm. See B–P–H 99/14. HI 50991

Annals of the Cyprus natural history society. Nicosia, Cyprus. Ann. Cyprus Nat. Hist. Soc. See B–P–H 100/15. HI 51002

Annals of the Fundulea research institute for cereals and technical plants. Series C, breeding, genetics, physiology and agricultural technology = Analele institutului de cercetări pentru cereale şi plante tehnice - Fundulea. Seria C, ameliorare, genetică, fiziologie si tehnologie agricolă. Bucharest.

Annals of the Goulandris museum = Annales musei Goulandris. Kifisia, Athens.

Annals of horticulture in North America. New York, NY. Ann. Hort. N. Amer. See B–P–H 104/8. HI 51051

Annals of horticulture and year-book of information on practical gardening. London. Ann. Hort. See B–P–H 104/1. HI 51047

Annals of the hungarian geological institute. Budapest = A magyar állami földtani intézet évkönyve. Budapest. Magyar Állami Földt. Intéz. Évk. See B–P–H 542/19.

Annals of internal medicine. Ann Arbor, MI. Ann. Intern. Med. See B–P–H 105/21. HI 51066

Annals of the Japan association for philosophy of science. [Kagaku kisoron gakkai.] Tokyo. 1954+. Ann. Japan Assoc. Philos. Sci. HI 61683

Annals of Kentucky natural history. Louisville, KY. Vol. 1, 1941. Ann. Kentucky Nat. Hist. Superseded by:

Annals of the Kentucky society of natural history. 1-389-2. HI 61684

Annals of the Kentucky society of natural history. Louisville, KY. Vol. 2+, 1968+. Ann. Kentucky Soc. Nat. Hist. Preceded by: Annals of Kentucky natural history. HI 61685

Annals of Kirstenbosch botanic gardens. Supplement to: South african journal of botany. Claremont. Vol. 14+, 1986+. Ann. Kirstenbosch Bot. Gard. HI 61686

Annals. Leningrad state university of the name A. S. Boubnoff. Series of biological science = Uchenye zapiski. Leningradskii gosudarstvennyi universitet imeni A. S. Bubnova. Seriya biologicheskikh nauk. Leningrad.

Annals of the lyceum of natural history of New York. New York, NY. Ann. Lyceum Nat. Hist. New York. See B–P–H 106/19. HI 51075

Annals and magazine of natural history, including zoology, botany, and geology. London. Vols. 6-20, 1841-47; ser. 2, vols. 1-20, 1848-57; ser. 3, vols. 1-20, 1858-67; ser. 4, vols. 1-20, 1868-77; ser. 5, vols. 1-20, 1878-87; ser. 6, vols. 1-20, 1888-97; ser. 7, vols. 1-20, 1898-1907; ser. 8, vols. 1-20, 1908-17; ser. 9, vols. 1-20, 1918-27; ser. 10, vols. 1-20, 1928-37; ser. 11, vols. 1-14, 1938-47?; ser. 12, vols. 1-10, 1948?-57; ser. 13, vols. 1-9, 1958-67. Ann. Mag. Nat. Hist. Preceded by: Annals of natural history. In 1942 absorbed Journal of botany. Superseded by: Journal of natural history. 1-387-1. HI 61687

Annals of medical history. New York, NY. Ann Med. Hist. See B–P–H 107/6. HI 51082

Annals of medicine and surgery, or records of the occurring improvements and discoveries in medicine and surgery, and the immediately connected arts and sciences. London. Ann. Med. Surg. See B–P–H 107/9. HI 51085

Annals of the Missouri botanical garden. St. Louis, MO. Ann. Missouri Bot. Gard. See B–P–H 107/17. HI 51088

Annals of the Missouri historical and philosophical society. Jefferson City, MO. Ann. Missouri Hist. Soc. See B–P–H 107/19. HI 51089

Annals of the Natal government museum = Annals of the Natal museum. Pietermaritzburg.

Annals of the Natal museum. Pietermaritzburg. Vol. 1+ [vol. 7(3)+ also numbered no. 21+], 1906+. Ann. Natal Mus. 4-2825-2. HI 61688

Annals of natural history; or, magazine of zoology, botany and geology. London. Ann. Nat. Hist. See B–P–H 108/17. HI 51095

Annals of nature; or, annual synopsis of new genera and species of animals, plants, etc. discovered in North America. Lexington, KY. Ann. Nat. See B–P–H 108/16. HI 51094

Annals of the New York academy of sciences. New York, NY. Ann. New York Acad. Sci. See B–P–H 108/26. HI 51100

Annals of the Oklahoma academy of science. Norman, OK. Nos. 1-6, 1970-76. Ann. Oklahoma Acad. Sci. HI 61689

Annals of ophthalmology and otology. St. Louis, MO. Ann. Ophthalmol. Otol. See B–P–H 109/11. HI 51105

Annals of otology, rhinology and laryngology. St. Louis, MO. Ann. Otol. See B–P–H 109/12. HI 51106

Annals of pharmacy and practical chemistry. London. Ann. Pharm. Pract. Chem. See B–P–H 109/28. HI 51112

Annals of philosophical discovery and monthly reporter of the progress of science and art. Manchester & London. Ann. Philos. Disc. & Monthly Reporter Progr. Sci. Art. See B–P–H 109/33. HI 51114

Annals of philosophy. London. Ann. Philos. See B–P–H 109/32. HI 51113

Annals of philosophy, or magazine of chemistry, mineralogy, mechanics, natural history, agriculture, and the arts. London. Ann. Philos. Mag. Chem. See B–P–H 110/1. HI 51116

Annals of philosophy, natural history, chemistry, literature, agriculture and the mechanical and fine arts. London. Ann. Philos. Nat. Hist. See B–P–H 110/2. HI 51117

Annals of the phytopathological society of Japan. [Nippon shokubutsu byori-gakkai ho.] Tokyo. Ann. Phytopathol. Soc. Japan. See B–P–H 110/11. HI 51124

Annals of the royal botanic garden. Calcutta. Ann. Roy. Bot. Gard. (Calcutta). See B–P–H 112/18. HI 51132

Annals of the royal botanic gardens. Peradeniya, Ceylon. Ann. Roy. Bot. Gard. (Peradeniya). See B–P–H 112/19. HI 51133

Annals of the Saitama horticultural experiment station. [Saitama-ken engei shikenjo nenpo.] Kuki. 1965/66+. Ann. Saitama Hort. Exp. Sta. HI 61020

Annals of science. Cleveland, OH. Ann. Sci. (Cleveland). See B–P–H 112/37. HI 51136

Annals of science. London. Ann. Sci. (London). See B–P–H 112/38. HI 51137

Annals of science, college of liberal arts, Kanazawa University. [Kanazawa daigaku kyoyobu ronshu, shizenkagaku-hen.] Kanazawa. Vol. 1+, 1964+. Ann. Sci., Kanazawa Univ. HI 61690

Annals of scottish natural history. Edinburgh. Ann.

Scott. Nat. Hist. See B–P–H 114/12. HI 51153

Annals of the south african museum. Cape Town. Ann. S. African Mus. See B–P–H 112/28. HI 51134

Annals of surgery. St. Louis,MO & Philadelphia, PA. Ann. Surg. See B–P–H 117/1. HI 51190

Annals of the Transvaal museum. Pretoria. Ann. Transvaal Mus. See B–P–H 117/5. HI 51193

Annals of the Tsukuba botanical garden. [Tsukuba jikken shokubutsuen kenkyu hokoku] Ibaraki. No. 1+, 1983+. Ann. Tsukuba Bot. Gard. HI 61691

Annals of the university of Stellenbosch. Series 3 = Annale van die uniwersiteit [later universiteit] van Stellenbosch. Reeks A, wis- en natuurkunde. Kaapstad.

Annals of the university of Stellenbosch. Series A3, agriculture = Annale van die universiteit van Stellenbosch. Serie A3, landbou. Kaapstad.

Annals of the university of Stellenbosch. Series A3, forestry = Annale van die universiteit van Stellenbosch. Serie A4, bosbou. Kaapstad.

Annals of the V. I. Lenin state university of Azerbaijan. Sekt., natural history and medicine = Izvestiya azerbaidzhanskogo gosudarstvennogo universiteta imeni V. I. Lenina. Otdel., estestvoznanie i meditsina. Baku.

Annals of the white russian agricultural institute = Trudy belaruskaga sel'ska-haspadarchaga instytuta. Gorki.

Annaly sel'skokhozyaistvennych nauk. Seriya E, rastitel'naya produktsiya = Roczniki nauk rolniczych. Seria A, produkcja roslinna. Warsaw.

Annaly sel'skokhozyaistvennych nauk. Seriya E, zashchita rastenii = Roczniki nauk rolniczych. Seria E, ochrona roslin. Warsaw.

Annamalai university agricultural research annual = Auara. Annamalamagar.

Année, association nationale des professeurs de biologie de Belgique. Brussels. Vols. ?-9-23, 19??-63-77. Année Assoc. Natl. Profess. Biol. Belgique. Superseded by: Drimaandelijke tijdskrift, vereniging voor het onderwijs in de biologie. HI 61692

Année biologique. Comptes rendus annuels des travaux de biologie générale. Paris. Année Biol. See B–P–H 119/48. HI 51213

Années de l'académie des sciences, belles-lettres et arts, de Besançon. Besançon. Années Acad. Sci. Besançon. See B–P–H 119/49. HI 51214

Anniversary of the Massachusetts horticultural society. [Title varies.] Boston. MA. Vols. [1-9], 1829-37. Anniv. Mass. Hort. Soc. Superseded by: Transactions of the Massachusetts horticultural society. HI 61693

Annonaceae newsletter. Utrecht. No. 1+, 1985+. Annonaceae Newslett. HI 61694

Annotated bibliography of medical mycology. Kew. 1944-50. Annot. Bibliogr. Med. Mycol. Superseded by: Review of medical and veterinary mycology. HI 61695

Annotated bibliography, weed research organization. Kidlington. No. 1+, 1966+. Annot. Bibliogr. Weed Res. Organ. HI 61696

Annotation of the oceanographical research. [Imperial fisheries institute.] Tokyo. Annot. Oceanogr. Res. See B–P–H 120/49. HI 51215

Annotationes zoologicae et botanicae. Bratislava. Annot. Zool. Bot. See B–P–H 120/50. HI 51216

Annotationes zoologicae japonenses. [Nippon dobutsugaku iho.] Tokyo. Annot. Zool. Jap. See B–P–H 120/51. HI 51217

Annuaire de l'académie royale de Belgique. Brussels. 1933+. Ann. Acad. Roy. Belgique. Preceded by: Annuaire de l'académie des sciences, des lettres, et des beaux-arts de Belgique. HI 61697

Annuaire de l'académie royale des sciences et belles-lettres de Bruxelles. Brussels. Vols. 1-11, 1835-45. Annuaire Acad. Roy. Sci. Bruxelles. Superseded by: Annuaire de l'académie royale des sciences et belles-lettres de Bruxelles. 1-33-3. HI 61698

Annuaire de l'académie royale des sciences, des lettres et des beaux arts de Belgique. Brussels. Vols. 12-?, 1846-1932. Annuaire Acad. Roy. Sci. Belgique. Preceded by: Annuaire de l'académie royale des sciences et belles-lettres de Bruxelles. Superseded by: Annuaire de l'académie royale de Belgique. 1-33-3. HI 61699

Annuaire de l'académie des sciences. Paris. Annuaire Acad. Sci. See B–P–H 121/7. HI 51218

Annuaire agricole de la Suisse. Bern. Annuaire Agric. Suisse. See B–P–H 121/8. HI 51219

Annuaire agricole de la Suisse = Landwirthschaftliches Jahrbuch der Schweiz. Bern. Landw. Jahrb. Schweiz. See B–P–H 525/8.

Annuaire des agrumes. Algiers. Annuaire Agrumes. See B–P–H 121/9. HI 51220

Annuaire de l'arboriculture fruitiére = Annuaire des agrumes. Algiers. Annuaire Agrumes. See B–P–H 121/9.

Annuaire d'art, de sciences et de technologie militaires. Brussels. Annuaire Art. See B–P–H 121/10. HI 51221

Annuaire-bulletin de la société franco-japonaise de Paris = Bulletin de la société franco-japonaise de Paris. Paris. Bull. Soc. Franco-Jap. Paris. See B–P–H 277/9.

Annuaire C N R S. Sciences de la vie = Annuaire:

sciences de la vie. Paris.

Annuaire centre national de la recherche scientifique. Sciences de la vie = Annuaire: sciences de la vie. Paris.

Annuaire de chimie. Paris. Annuaire Chim. See B–P–H 121/12. HI 51222

Annuaire du conservatoire du jardin botaniques de Genève. Geneva. Annuaire Conserv. Jard. Bot. Genève. See B–P–H 121/14. HI 51223

Annuaire des eaux et forêts. Paris. Annuaire Eaux Forêts. See B–P–H 121/15. HI 51224

Annuaire de la faculté d'agriculture et de sylviculture de l'université Skopje. Agronomie = Godišen zbornik na zemjodelsko-šumarski fakultet na univerzitetet Skopje. Zemjodelstvo. Skopje.

Annuaire de la faculté d'agriculture et de sylviculture de l'université Skopje. Sylviculture = Godišen zbornik na zemjodelsko-šumarskiot fakultet na univerzitetot Skopje. Šumarstvo. Skopje.

Annuaire de la faculté d'agronomie et de sylviculture de l'université Skopje. Agronomie = Godišen zbornik na zemjodelsko-šumarski fakultet na univerzitetet Skopje. Zemjodelstvo. Skopje.

Annuaire de la faculté d'agronomie et de sylviculture de l'université Skopje. Sylviculture = Godišen zbornik na zemjodelsko-šumarskiot fakultet na univerzitetot Skopje. Šumarstvo. Skopje.

Annuaire de la faculté agronomique et forestière, université de Belgrade = Godišnjak poljoprivredno-šumarskog fakulteta, universitet u Beogradu. Belgrade.

Annuaire de la faculté de philosophie de l'université Skopje. Section des sciences naturelles = Godišen zbornik, filozofski fakultet na univerzitetot. Prirodno-matematički oddel. Skopje.

Annuaire de la faculté des sciences naturelles de l'université de Skopje. Biologie = Godišen zbornik, prirodno-matematički fakultet na univerzitetot- Skopje. Biologija. Skopje.

Annuaire horticole international. Nice. Annuaire Hort. Int. See B–P–H 121/17. HI 51226

Annuaire de l'horticulture belge. Ghent. Annuaire Hort. Belge. See B–P–H 121/16. HI 51225

Annuaire, institut agronomiques, Skopje = Zbornik, zemjodelskli institut, Skopje. Skopje.

Annuaire de l'institut biologique à Sarajevo = Godišnjak biološkog instituta u Sarajevu. Sarajevo, Yugoslavia. God. Biol. Inst. u Sarajevu. See B–P–H 406/5.

Annuaire: institut de France. Paris. Annuaire Inst. France. See B–P–H 121/19. HI 51227

Annuaire de l'institut géologique royale hongroise. Budapest = A magyar kir[ályi] állami földtani intézet évkönyve. Budapest. Magyar Kir. Állami Földt. Intéz. Évk. See B–P–H 543/15.

Annuaire de l'institut national de la recherche agronomique. Paris. 1979/80+, 1979+. Ann. Inst. Natl. Rech. Agron. HI 61700

Annuaire, institut des recherches agronomiques, Skopje = Zbornik, zemjodelsko-ispitatelen institut, Skopje. Skopje.

Annuaire et liste générale des membres de la société nationale d'horticulture de France = Annuaire, société nationale d'horticulture de France. Paris.

Annuaire du muséum d'histoire naturelle. Paris. Annuaire Mus. Hist. Nat. See B–P–H 121/21. HI 51229

Annuaire scientifique. Paris. Annuaire Sci. See B–P–H 121/22. HI 51230

Annuaire des sciences historiques. Paris. Annuaire Sci. Hist. See B–P–H 121/23. HI 51231

Annuaire: sciences de la vie. Paris. 1976/77+, 1977+. Annuaire Sci. Vie. HI 61701

Annuaire de la société dendrologique de la société botanique de Pologne = Rocznik polskiego towarzystwa dendrologicznego. Lwów.

Annuaire, société française de microbiologie. Paris. 1961+. Ann. Soc. Franç. Microbiol. HI 61702

Annuaire de la société franco-japonaise de Paris = Bulletin de la société franco-japonaise de Paris. Paris. Bull. Soc. Franco-Jap. Paris. See B–P–H 277/9.

Annuaire de la société helvetique des sciences naturelles. Partie scientifique = Jahrbuch der schweizerischen naturforschenden Gesellschaft. Wissenschaftlicher Teil. Bern.

Annuaire, société nationale d'horticulture de France. Paris. 1886+. Ann. Soc. Natl. Hort. France. Preceded by: Journal de la société nationale d'horticulture de France. Annuaire. HI 61704

Annuaire de la société royale belge du dahlia. Linkebeek, Belgium. Annuaire Soc. Roy. Belge Dahlia. See B–P–H 121/25. HI 51232

Annuaire de la société suédoise d'histoire des sciences = Lychnos; lärdomshistoriska samfundets årsbok. Uppsala & Stockholm. Lychnos. See B–P–H 537/11.

Annuaire de la station agronomique de l'état. Brussels. Annuaire Stat. Agron. État. See B–P–H 121/27. HI 51233

Annuaire de l'université de Sofia = Godishnik sofiiskiya universitet. Sofia.

Annuaire de l'université de Sofia, faculté de biologie, géologie et géographie. Livre 1, biologie = Godishnik sofiiskiya universitet, biologi-geologi-geografiski fakultet. Kniga 1, biologija, botanika. Sofia.

Annuaire de l'université de Sofia, faculté de biologie. Livre 2, botanique, microbiologie, physiologie et biochemie des plantes = Godishnik sofiiskiya universitet, biologičeski fakultet. Kniga 2, botanika, mikrobiologiya, fiziologiya i biokhimiya na rastenyata. Sofia.

Annuaire de l'université de Sofia, faculté physico-mathématique = Godišnik sofiiskiya universitet, fiziko-matematicheski fakultet. Sofia.

Annual address before the New Orleans academy of sciences. New Orleans, LA. 1856. Annual Address New Orleans Acad. Sci. HI 61705

Annual administration report, see Administration report (Annual)

Annual announcement, marine biological laboratory of Woods Hole. Woods Hole, MA. Nos. 1-91, 1888-1978. Annual Announc. Mar. Biol. Lab. Woods Hole. Superseded by: Informational bulletin, marine biological laboratory, Woods Hole. HI 61706

Annual bibliography of the history of natural history. London. Vol. 1+, 1982+, 1985+. Annual Bibliogr. Hist. Nat. Hist. HI 61707

Annual bibliography of japanese agriculture = Report (Annual) of the development of agriculture in Japan. Annual bibliography of japanese agriculture. [Nihon nogaku shinpo nenpo.] Tokyo.

Annual biology colloquium, Oregon state college. Corvallis, OR. Nos. 1-21, 1940-60. Annual Biol. Colloq. Oregon State Coll. Superseded by: Proceedings of the annual biology colloquium, Corvallis. HI 61708

Annual biosciences colloquia, Ohio state university. Columbus, OH. 1975+. Annual Biosci. Colloq. Ohio State Univ. HI 61709

Annual of the botanical society of western Pennsylvania. Pittsburgh, Pa. Annual Bot. Soc. W. Pennsylvania. See B–P–H 122/1. HI 51234

Annual bulletin of the agricultural department, Nigeria. Lagos. Vols. 1-15, 1922-36. Annual Bull. Agric. Dept. (Nigeria). HI 61710

Annual bulletin of the Aomori prefecture agricultural experiment station. [Aomori-ken nogyo shikenjo gyomu nenpo.] Aomori, Japan. Annual Bull. Aomori Prefect. Agric. Sta. See B–P–H 122/2. HI 51235

Annual bulletin, Connecticut forestry association. Hartford, CT. 1902-05. Annual Bull. Connecticut Forest. Assoc. Superseded by: Publications of the Connecticut forestry association. HI 61711

Annual bulletin, department of agriculture, Nigeria = Annual bulletin of the agricultural department, Nigeria. Lagos.

Annual bulletin of divisional reports of the department of agriculture, Fiji. Suva. 1931-38. Annual Bull. Div. Rep. Dept. Agric., Fiji. HI 61712

Annual bulletin of horticultural technological society. [Engei gijutsu kondankai nenpo.] Tokyo. Annual Bull. Hort. Technol. Soc. See B–P–H 122/3. HI 51236

Annual bulletin, marine biological laboratory, Woods Hole. Woods Hole, MA. 1980+. Annual Bull. Mar. Biol. Lab., Woods Hole. Preceded by: Informational bulletin, marine biological laboratory, Woods Hole. HI 61713

Annual bulletin, poplar and fast growing forest trees research institute. Izmir. Nos. 1-10?, 1966-75? Annual Bull. Poplar Fast Growing Forest Trees Res. Inst. HI 61714

Annual bulletin, société jersiaise. Jersey, CI. 1957+. Annual Bull. Soc. Jersiaise. Preceded by: Bulletin annuel de la société jersiaise. HI 61715

Annual, canadian gladiolus society. Various places. 1938+. Annual Canad. Gladiolus Soc. Preceded by: Quarterly, canadian gladiolus society. 2-908-1. HI 61716

Annual of the club of natural science and geography of the Young Men's Christian Associations. Harbin = Ezhegodnik Kluba estestvoznaniya i geografii Khristianskogo Soyuza Molodykh Lyudei. Harbin.

Annual conference, Queensland cane growers' association = Report of proceedings, Queensland cane growers' association. Brisbane, Qld.

Annual dahlia register, for 1836; containing particulars of the introduction of the dahlia into this country, etc. London. 1836. Annual Dahlia Reg. HI 61717

Annual departmental report, director of agriculture and fisheries. Hong Kong. 1964+. Annual Dept. Rep. Director Agric. Fish. Preceded by: Annual departmental report, director of agriculture and forestry, Hong Kong. HI 61718

Annual departmental report, director of agriculture, fisheries and forestry. Hong Kong. 1950-60. Annual Dept. Rep. Director Agric. Fish. Forest., Hong Kong. Preceded by: Report of the agricultural department, Hong Kong and Report of the forestry department, Hong Kong. Superseded by: Annual department report, director of agriculture and forestry, Hong Kong. HI 61719

Annual department report, director of agriculture and forestry. Hong Kong. 1960-64. Annual Dept. Rep. Director Agric. Forest., Hong Kong. Preceded by: Annual departmental report, director of agriculture, fisheries and forestry. Superseded by: Annual departmental report, director of agriculture and fisheries. HI 61720

Annual detailed research report, indian coffee board =

Report (Annual), research department, indian coffee board. Bangalore.

Annual forest administration report, Bombay = Administration report (Annual), forest department, Bombay.

Annual general activity report of the California division of forestry = Report, California division of forestry. Sacramento, CA.

Annual index of the reports of plant chemistry. [Hirokawa shoten] Tokyo. 1952-62. Annual Index Rep. Pl. Chem. HI 61721

Annual journal, royal New Zealand institute of horticulture. Wellington, NZ. No. 1+, 1973+. Annual J. Roy. New Zealand Inst. Hort. Preceded by: Journal of the royal New Zealand institute of horticulture. HI 61722

Annual journal of the Taiwan museum. Taipei. ?-1985+. Annual J. Taiwan Mus. HI 61723

Annual letter, institute of tropical forestry. Rio Piedras, PR. 1978+. Annual Lett. Inst. Trop Forest. HI 61724

Annual meeting, Connecticut forest and park association. East Hartford, CT. ?-ca.1979+. Annual Meeting Connecticut Forest Park Assoc. HI 61725

Annual of the national academy of sciences. Cambridge, MA. 1863/64-66, 1864-67? Annual Natl. Acad. Sci. 4-2827-1. HI 61726

Annual, Nepal nature conservation society. 1977+. Annual Nepal Nat. Conservation Soc. HI 61727

Annual newsletter, society of irish plant pathologists. No. ?-7+, 19??-76+. Annual Newslett. Soc. Irish Pl. Pathologists. HI 61728

Annual orchid show, Sandakan orchid society. Sandakan. ?-1983+. Annual Orchid Show, Sandakan Orchid Soc. HI 61729

Annual of the Peking biological science association. [Pe ching shêng wu k'o hsüeh hui nien pao.] Peking. Annual Peking Biol. Sci. Assoc. See B–P–H 122/7. HI 51238

Annual plant resistance to insects newsletter. West Lafayette, IN. 1975+. Annual Pl. Resist. Insects Newslett. HI 61730

Annual proceedings and bulletin of the american rose society. Boston, MA. Annual Proc. & Bull. Amer. Rose Soc. See B–P–H 122/8. HI 51239

Annual proceedings of the phytochemical society of Europe. Oxford, London. No. 1+, 1969+. Annual Proc. Phytochem. Soc. Eur. HI 61731

Annual programme of work, forest products research institute, Ghana. Kumasi. 1971+. Annual Programme Wk. Forest Prod. Res. Inst., Ghana. HI 61732

Annual progress report, division of forest research, department of lands and forests, Ontario. Toronto. 1949-50. Annual Progr. Rep. Div. Forest Res., Ontario. Superseded by: Annual research progress report, division of forest research, department of lands and forests, Ontario. HI 61733

Annual progress report on forest administration in the presidency of Bengal = Administration report (Annual), forest department, Bengal. Calcutta.

Annual progress report on forest administration in the province of Bihar. 1938+. Annual Progr. Rep. Forest Admin. Prov. Bihar. Preceded by: Annual progress report on forest administration in the province of Bihar and Orissa. HI 61734

Annual progress report on forest administration in the province of Bihar and Orissa. 1913-36. Annual Progr. Rep. Forest Admin. Prov. Bihar Orissa. Superseded by: Annual progress report on forest administration in the province of Bihar and Annual progress report on forest administration in the province of Orissa. HI 61735

Annual progress report on forest administration in the province of Orissa. Cuttack. 1938+. Annual Progr. Rep. Forest Admin. Prov. Orissa. Preceded by: Annual progress report on forest administration in the province of Bihar and Orissa. HI 61736

Annual progress report, forest board, South Australia. Adelaide, S.A. 1881/82-? Annual Progr. Rep. Forest Board, S. Australia. Superseded by: Report (Annual), woods and forests department, South Australia. HI 61737

Annual progress report, Pakistan forest college and research institute. Peshawar. 1950/55+, 1955+. Annual Progr. Rep. Pakistan Forest Coll. Res. Inst. HI 61738

Annual progress report, Pakistan forest institute = Annual progress report, Pakistan forest college and research institute. Peshawar.

Annual progress report on research and technical work, department of agriculture, Northern Ireland = Report (Annual) on research and technical work of the department (ministry) of agriculture for Northern Ireland. Belfast.

Annual progress report upon state forest administration in South Australia = Report, woods and forests department, South Australia. Adelaide, S.A.

Annual record of science and industry. New York, NY. Annual Rec. Sci. Industr. See B–P–H 122/9. HI 51240

Annual report ..., see Report (Annual) ..., (N.B. "(Annual)" is disregarded in the filing order).

Annual research programme, forest research institute. Chittagong. 1962/63+, 1963?+. Annual Res.

Programme Forest Res. Inst., Chittagong. HI 61739

Annual research progress report, division of forest research, department of lands and forests, Ontario. Toronto. 1951/52-56/57, 1951-57. Ann. Res. Progr. Rep. Div. Forest Res., Ontario. Preceded by: Annual progress report, division of forest research, department of lands and forests, Ontario. HI 61740

Annual research report, department of agriculture, Fiji. Suva. 1974/75+, 1976+. Annual Res. Rep. Dept. Agric., Fiji. HI 61741

Annual research report, forest research institute. Chittagong. 1960/61+, 196?+. Annual Res. Rep. Forest Res. Inst., Chittagong. HI 61742

Annual research report, division of life sciences research, Louisiana tech university = Research bulletin, division of life sciences research, Louisiana tech university. Ruston, LA.

Annual research report, north central weed control conference = Research report, north central weed control conference. Champaign, IL.

Annual research report, Tatura horticultural research station. Tatura, Vic. 1964/65+, 1965+. Annual Res. Rep. Tatura Hort. Res. Sta. HI 61743

Annual research report, west african rice development association. 1976+. Annual Res. Rep. W. African Rice Developm. Assoc. HI 61744

Annual review of the academy of natural sciences of Philadelphia = Review of the academy of natural sciences of Philadelphia. Philadelphia, PA.

Annual review of biochemical and allied research in India. Bangalore. Vols. 7-35, 1936-65. Annual Rev. Biochem. Allied Res. India. Preceded by: Biochemical and allied research in India. Superseded by: Biochemical reviews. 5-3983-3. HI 61745

Annual review of biochemistry. Palo Alto, CA. Annual Rev. Biochem. See B–P–H 129/12. HI 51242

Annual review of biophysics and bioengineering. Palo Alto, CA. Vols. 1-13, 1972-84. Annual Rev. Biophys. Bioengin. Superseded by: Annual review of biophysics and biophysical chemistry. HI 61746

Annual review of biophysics and biophysical chemistry. Palo Alto, CA. Vol. 14+, 1985+. Annual Rev. Biophysics Biophys. Chem. Preceded by: Annual review of biophysics and bioengineering. HI 75195

Annual review of cell biology. Palo Alto, CA. Vol. 1+, 1985+. Annual Rev. Cell Biol. HI 61747

Annual review, commonwealth mycological institute. Kew. 1976/77+, 1977+. Annual Rev. Commonw. Mycol. Inst. Preceded by: Year, commonwealth mycological institute (The). HI 61748

Annual review of ecology and systematics. Palo Alto, CA. Vol. 1+, 1970+. Annual Rev. Ecol. Syst. HI 61749

Annual review, Efford experimental horticulture station. Lymington. 1979+. Annual Rev. Efford Exp. Hort. Sta. Preceded by: Report (Annual), Efford experimental horticulture station. HI 61750

Annual review of genetics. Palo Alto, CA. Vol. 1+, 1967+. Annual Rev. Genet. HI 61751

Annual review and history of literature. London. Annual Rev. Hist. Lit.. See B–P–H 129/13. HI 51243

Annual review, Kirton experimental horticulture Station. Boston. 1979+. Annual Rev. Kirton Exp.. Hort. Sta. Superseded by: Report (Annual), Kirton experimental horticulture station. HI 61752

Annual review, Lee Valley experimental horticulture station. Hoddesdon. 19??-79+. Annual Rev. Lee Valley Exp. Hort. Sta. Preceded by: Report, Lee Valley experimental horticulture station. HI 61753

Annual review of microbiology. Stanford, CA. Annual Rev. Microbiol. See B–P–H 129/14. HI 51244

Annual review, national fruit trials, Brogdale. Faversham. 1979+, 1980+. Annual Rev. Natl. Fruit Trials, Brogdale. Preceded by: Report (Annual), national fruit trials, Brodgale. HI 61754

Annual review of pharmacology. Palo Alto, CA. Vols. 1-15, 1961-75. Annual Rev. Pharmacol. Superseded by: Annual review of pharmacology and toxicology. HI 61755

Annual review of pharmacology and toxicology. Palo Alto, CA. Vol. 16+, 1976+. Annual Rev. Pharmacol. Toxical. Preceded by: Annual review of pharmacology. HI 61756

Annual review of physiology. Stanford, CA. Annual Rev. Physiol. See B–P–H 129/16. HI 51245

Annual review of phytopathology. Palo Alto, CA. Vol. 1+, 1963+. Annual Rev. Phytopathol. HI 61757

Annual review of plant physiology. Stanford, CA. Vols. 1-38, 1950-87. Annual Rev. Pl. Physiol. Superseded by: Annual review of plant physiology and plant molecular biology. HI 51246

Annual review of plant physiology and plant molecular biology. Palo Alto, CA. Vol. 39+, 1988+. Annual Rev. Pl. Physiol. Pl. Molec. Biol. Preceded by: Annual review of plant physiology. See B–P–H 129/18. HI 61758

Annual reviews of plant sciences. New Delhi. Vol. 1+, 1979+. Annual Rev. Pl. Sci. HI 61759

Annual review of pteridological research. Lawrence, KS. Vols. 1+, 1987+. Annual Rev. Pteridol. Res. Superseded by: Annual review of plant physiology and plant molecular biology. HI 61760

Annual review, Redesdale experimental husbandry farm. Otterburn. No. 1+, 1968+. Annual Rev. Redesdale Exp. Husb. Farm. HI 61761

Annual review, Rosewarne experimental horticulture station. Camborne. 1979+, [1980]+. Annual Rev. Rosewarne Exp. Hort. Sta. Preceded by: Report (Annual), Rosewarne experimental horticulture station. HI 61762

Annual review, rubber research institute of Ceylon. Agalawatta. 1966-71. Annual Rev. Rubber Res. Inst. Ceylon. Preceded by: Report (Annual), rubber research institute of Ceylon. Superseded by: Annual review, rubber research institute of Sri Lanka. HI 61763

Annual review, rubber research institute of Sri Lanka. Agalawatta. 1972+. Annual Rev. Rubber Res. Inst. Sri Lanka. Preceded by: Annual review, rubber research institute of Ceylon. HI 61764

Annual review, Stockbridge House experimental horticulture station. Cawood. 1979+. Annual Rev. Stockbridge House Exp. Hort. Sta. Preceded by: Report, Stockbridge House experimental horticulture station HI 61765.

Annual reviews, see Annual review

Annual ring, The = Louisiana university and agricultural and mechanical college. Forestry department. The annual ring. Baton Rouge, LA. Louisiana Univ. Agric. Coll. Forest. Dept. Annual Ring. See B–P–H 536/9.

Annual scientific report, central potato research institute = Scientific report of the central potato research institute. Simla.

Annual of scientific discovery; or, year-book in science and art, exhibiting the most important discoveries and improvements in mechanics, ... botany ... [etc.]. Boston, MA. Vols. 1-21, 1850-71. Annual Sci. Disc. HI 68655

Annual scientific report, tea research association. Jorhat. 1963+. Annual Sci. Rep. Tea Res. Assoc. HI 61766

Annual survey of research in pharmacy ... Baltimore, MD. Annual Surv. Res. Pharm. See B–P–H 129/20. HI 51247

Annual of the swedish history of science society = Lychnos; lärdomshistoriska samfundets årsbok. Uppsala & Stockholm. Lychnos. See B–P–H 537/11.

Annual systematics symposium, Missouri botanical garden. [Abstracts]. St. Louis, MO. 29th+, 1982+. Annual Syst. Symp. Missouri Bot. Gard. HI 74521

Annual of Taiwan museum. Taipei. Vol. ?-28+, 19??-85+. Annual Taiwan Mus. HI 61767

Annual of the Taiwan provincial museum. [Taiwan shêng li po wu k'uan k'o hsüeh nien k'an.] Taipei, Taiwan. Annual Taiwan Prov. Mus. See B–P–H 129/22. HI 51248

Annual technical report, central rice research institute. New Delhi. 1948/49-67/68, 1949-68; 1972. Annual Techn. Rep. Centr. Rice Res. Inst. Preceded and superseded by: Report (Annual), central rice research institute, Cuttack. HI 61768

Annual technical report, sugar cane agronomy section, department of agriculture, Trinidad and Tobago. Port of Spain. 1960-68. Annual Techn. Rep. Sugar Cane Agron. Sect., Trinidad Tobago. Preceded by: Report (Annual), sugar cane section, department of agriculture, Trinidad and Tobago. HI 61769

Annuario della accademia delle scienze di Torino. Turin. 1971+. Annuario Accad. Sci. Torino. HI 61770

Annuario chimico italiano. Modena, Italy. Annuario Chim. Ital. See B–P–H 129/23. HI 51249

Annuario, giardino botanico alpino dell ordine Mauriziano al Picolo San Bernardo = Chanousia. Rome.

Annuario del reale istituto botanico di Roma. Milan. Vols. 1-10, 1884-1902. Annuario Reale Ist. Bot. Roma. Superseded by: Annali di botanica. HI 61771

Annuario, r[eale] istituto di sperimentazione per la chimica agraria in Torino. Turin. 1941-45. Annuario Reale Ist. Sperim. Cghim. Agrar. Torino. Preceded by: Annuario, reale stazione chimico-agraria di Torino. HI 61772

Annuario r[eale]. scuola superiore di agricoltura di Portici. Naples. Annuario Reale Scuola Super. Agric. Portici. See B–P–H 130/4. HI 51250

Annuario, r[eale] stazione chimico-agraria di Torino. Turin. 1871-? Annuario Reale Staz. Chim.-Agrar. Torino. Superseded by: Annuario r[eale] istituto di sperimentazione per la chimica agraria di Torino. 5-4278-2. HI 61773

Annuario della società dei naturalisti di Modena. Modena, Italy. Annuario Soc. Naturalisti Modena. See B–P–H 13/5. HI 51251

Annuelles et légumes, résultats des cultures d'essai. Montreal. 1963+. Annuelles Légumes Résult. Cult. Essai. HI 61774

Antananarivo annual and Madagascar magazine; a record of information on the topography and natural productions of Madagascar [etc.]. Tananarive. Vols. 1-6(4), nos. 1-24, 1875-78; nos. 1-24, 1881-1900 [nos. 1-4, 1875-78 reprinted 1885, nos. 5-8, 1881-84 reprinted 1896]. Antananarivo Annual Madagascar Mag. HI 61775

Antarctic bibliography. Washington, DC. Vol. 1+, 1965+. Antarc. Bibliogr. Each annual volume preceded by monthly issues of: Current antarctic literature. HI 61776

Antarctic journal of the United States. Washington, DC. Antarc. J. U.S. See B–P–H 130/6. HI 51253

Antarctic record. Christchurch, NZ. Vol. 1+, 1978+. Antarc. Rec. (Christchurch). HI 61777

Antarctic record. [Nankyoku shiryo]; reports of the japanese antarctic research expedition (JARE). Tokyo. 1957+. Antarc. Rec. (Tokyo). HI 61778

Antarctic research series. Washington, DC. 1964+. Antarc. Res. Ser. HI 61779

Antartida. Buenos Aires. No. 1+, 1971+. Antartida. HI 61780

Anthos; Garten- und Landschaftsgestaltung. Zurich. Vol. 1+, 1962+. Anthos. HI 61781

Anthropological papers of the american museum of natural history. New York, NY. Anthropol. Pap. Amer. Mus. Nat. Hist.. See B–P–H 130/8. HI 51254

Anthropological papers, museum of anthropology, university of Michigan. Ann Arbor, MI. 1949+. Anthropol. Pap. Mus. Anthropol. Univ. Michigan. HI 61782

Anthropos. Mexico City. Anthropos. See B–P–H 130/9. HI 51255

Antibiotics annual. New York, NY. Antibiot. Annual. See B–P–H 130/10. HI 51256

Antimicrobic newsletter. New York. Vol. 1+, 1984+. Antimicrob. Newslett. HI 61783

Antioqúia agrícola. Medellín. Vols. 1-4, 1916-22; ser. 2, vol. 1, 1925. Antioquiá Agric. Superseded by: Boletín semanal, sociedad antioqueña de agricultores. 5-3902-2. HI 61784

Antioquía médica. Medellín, Colombia. Antioquía Méd. See B–P–H 130/12. HI 51258

Antiquarian book monthly review. Brayfield, etc. Vol. 1+, 1974+. Antiq. Book Monthly Rev. HI 61785

Antiquarian bookman. New York. Vols. 1-39(22), 1948-67. Antiq. Bookman. Superseded by: A B bookman's weekly. HI 61786

Antiquity; a quarterly review of archaeology. Gloucester, England. Antiquity. See B–P–H 130/13. HI 51259

Antiseptic. A monthly medical journal. Edinburgh & Madras. Antiseptic. See B–P–H 130/14. HI 51260

Antologia; giornale di scienze, lettere e arti. Florence. Antologia (Florence). See B–P–H 130/16. HI 51261

Antologia italiana. Turin. Antologia Ital. See B–P–H 130/17. HI 51262

Antologia straniera; giornale di scienze, lettere ed arti. Turin. Antologia Straniera See B–P–H 130/18. HI 51263

Antonie van Leeuwenhoek journal of microbiology = Antonie van Leeuwenhoek journal of microbiology and serology. Amsterdam, Delft.

Antonie van Leeuwenhoek journal of microbiology and serology. Amsterdam, Delft. 1940?+. Antonie van Leeuwenhoek J. Microbiol. Serol. Preceded by: Antonie van Leeuwenhoek: nederlandsch tijdschrift voor hygiëne, microbiologie en serologie. HI 61787

Antonie van Leeuwenhoek: nederlandsch tijdschrift voor hygiëne, microbiologie en serologie. Amsterdam. Vols. 1-?, 1934-39. Antonie van Leeuwenhoek Ned. Tijdschr. Hyg. Preceded by: Nederlandsch tijdschrift voor hygiëne, microbiologie en serologie. Superseded by: Antonie van Leeuwenhoek journal of microbiology and serology. 1-412-2. HI 61788

Anuar lucrărilor ştiinţifice, institutul agronomic "Dr. Petru Coroza." Cluj. 1957+. Anuar Lucr. Şti. Inst. Agron. Dr. Petru Coroza. Preceded by: Analele, facultătii de agronomie din Cluj. HI 61789

Anuari, junta de ciències naturals de Barcelona. Barcelona. Vols. 1-3, 1916-18. Anuari Junta Ci. Nat. Barcelona. Superseded by: Memórias anuales, junta de ciències naturals. 3-2254-1. HI 61790

Anuario de la academia de ciencias exactas, fisicas y naturales = Anuario de la real academia de ciencias exactas, fisicas y naturales. Madrid.

Anuario de la academia de ciencias de la republica dominicana. Botanica. Santo Domingo. Vol. 1+, 1975+. Anuar. Acad. Ci. Republ. Dominic., Bot. HI 61791

Anuario. Academia mexicana de ciencias exactas, físicas y naturales. Mexico City. Anuario Acad. Mex. Ci. Exact. See B–P–H 131/1. HI 51264

Anuário de actividades, instituto de investigação cientifica tropical. Lisbon. 1986+. Anuário Actividades Inst. Invest. Ci. Trop. HI 61792

Anuário agrícola brasileiro. São João & São Paulo. Anuário Agríc. Brasil. See B–P–H 131/2. HI 51265

Anuário brasileiro de economia florestal. Rio de Janeiro. Vols. 1-?, 1948-? Anuário Brasil. Econ. Florest. Superseded by: Brasil florestal. 1-413-2. HI 61793

Anuario, centro de investigaciones agronómicas. Maracay. 1963+, 1969+. Anuario Centro Invest. Agron. HI 61794

Anuario del colegio nacional de la mineria. Mexico City. Anuario Colegio Nac. Mineria. See B–P–H 131/4. HI 51266

Anuario. Comisión impulsora y coordinadora de la investigación científica. Mexico City. Anuario Comis. Impuls. Invest. Ci. See B–P–H 131/5. HI 51267

Anuario de estudiòs atlanticos. Madrid, Las Palmas. No. 1+, 1955+. Anuario Estud. Atlanticos. HI 61795

Anuario hidrológico. Quito, Ecuador. Anuario Hidrol.

See B–P–H 131/6. HI 51268

Anuario de la provincia de Carácas; sociedad económica de amigos del páis. Caracas. Anuario Prov. Carácas Soc. Econ. Amigos País. See B–P–H 131/7. HI 51269

Anuario de la real academia de ciencias exactas, fisicas y naturales. Madrid. 1857-1936; 1942+. Anuario Real Acad. Ci. Exact. Fis. Nat. Madrid. HI 61796

Anuário de sociedade broteriana. Coimbra, Portugal. Anuário Soc. Brot. See B–P–H 131/8. HI 51270

Anuário técnico do instituto de pesquisas zootécnicas "Francisco Osorio". Porto Alegre. Vol. 1+, 1973+. Auário Técn. Inst. Pesq. Zootécn. Francisco Osorio. HI 61797

Anuario de la universidad de los Andes en los estados unidos de Venezuela. Merida. Vols. 1-4-?, 1891-94-? Anuario Univ. Andes Estados Unidos Venezuela. HI 61798

Anuarul muzeului judetean Suceava. Fascicola ştiintele naturii. Suceava. Vol. 4+, 1977+. Anuario Muz. Judet., Fasc. Şti. Nat. Preceded by: Studii si comunicări, muzeul judetean Suceava. Ştiinţe naturale. HI 61799

Anuarul muzeului de ştiinte naturale. Seria botanica-zoologie. Piatra Neamt. Vol. 3+, 1977+. Anuario Muz. Şti. Nat., Ser. Bot.-Zool. Preceded by: Studii şi cercetări de geologie, geografie, biologie. Seria botanica-zoologie. HI 61800

Anzeige von der Leipziger ökonomischen Societät. Dresden. Anz. Leipziger Ökon. Soc. See B–P–H 132/2. HI 51277

Anzeige der Leipziger ökonomischen Gesellschaft = Anzeige von der Leipziger ökonomischen Societät. Dresden. Anz. Leipziger Ökon. Soc. See B–P–H 132/2.

Anzeige von der neu errichteten Leipziger ökonomischen Societät = Anzeige von der Leipziger ökonomischen Societät. Dresden. Anz. Leipziger Ökon. Soc. See B–P–H 132/2.

Anzeige von der Samlungen einer Privatgesellachaft in der Oberlausiz. Görlitz. Anz. Saml. Privatges. Oberlausiz. See B–P–H 132/6. HI 51280

Anzeigen der Churfürstl. Sächsischen Leipziger ökonomischen Societät. Dresden. Anz. Churfürstl. Sächs. Leipziger Ökon. Soc. See B–P–H 131/14. HI 51272

Anzeigen der Churfürstlich Sächsischen Oberlausitzischen Gesellschaft der Wissenschaften. Görlitz. Anz. Churfürstl. Sächs. Oberlausitz. Ges. Wiss.. See B–P–H 131/15. HI 51273

Anzeigen der Königlich-Sächsischen Leipziger ökonomischen Societät. Dresden. Anz. Königl-Sächs. Leipziger Ökon. Soc. See B–P–H 131/16. HI 51274

Anzeigen der Kurfürstlich Sächsischen Oberlausizischen Gesellschaft der Wissenschaften. Görlitz. Anz. Kurfürstl. Säcks. Oberlausiz. Ges. Wiss. See B–P–H 131/19. HI 51276

Anzeigen der Oberlausizischen Gesellschaft der Wissenschaften. Lauban, Germany [=Luban, Poland]. Anz. Oberlausiz. Ges. Wiss. (Lauban). See B–P–H 132/5. HI 51279

Anzeigen der Oberlausitzischen Gesellschaft der Wissenschaften zu Görlitz. Görlitz. Anz. Oberlausitz. Ges. Wiss. Görlitz. See B–P–H 132/4. HI 51278

Anzeiger der Akademie der Wissenschaften in Krakau. Cracow. Anz. Akad. Wiss. Krakau. See B–P–H 131/11. HI 51271

Anzeiger der Akademie der Wissenschaften in Krakau. Mathematisch-naturwissenschaftliche Classe = Bulletin international de l'académie des sciences de Cracovie. Classe des sciences mathématiques et naturelles. Cracow. Bull. Int. Acad. Sci. Cracovie, Cl. Sci. Math. See B–P–H 257/13.

Anzeiger der Akademie der Wissenschaften in Krakau. Mathematisch-naturwissenschaftliche Classe. Serie B. Naturwissenschaften = Bulletin international de l'académie des sciences de Cracovie. Classe des sciences mathématiques et naturelles. Série B. Sciences naturelles. Cracow. Bull. Int. Acad. Sci. Cracovie, Cl. Sci. Math., Sér. B, Sci. Nat. See B–P–H 257/14.

Anzeiger der Akademie der Wissenschaften in Wien. Mathematische-naturwissenschaftliche Klasse. Wien. Vols. 55-83, 1918-46. Anz. Akad. Wiss. Wien, Math.-Naturwiss. Kl. Preceded by: Anzeiger, kaiserliche Akademie der wissenschaften in Wien. Mathematisch-naturwissenschaftliche Klasse. Superseded by: Anzeiger, Österreichische Akademie der Wissenschaften. Mathematisch- naturwissenschaftliche Klasse. HI 61801

Anzeiger der Kaiserlichen Akademie der Wissenschaften. Mathematisch-Naturwissenschaftliche Klasse. Wien. Vols. 1-51, 1864-1914. Anz. Kaiserl. Akad. Wiss. Wien, Math.-Naturwiss. Kl. Superseded by: Anzeiger, kaiserliche Akademie der wissenschaften in Wien. Mathematisch-naturwissenschaftliche Klasse. HI 61802

Anzeiger, kaiserliche Akademie der wissenschaften in Wien. Mathematisch-naturwissenschaftliche Klasse. Wien. Vols. 52-54, 1915-17. Anz. Kaiserl. Akad. Wiss. Wien, Math.-Naturwiss. Kl. Preceded by: Anzeiger der Kaiserlichen Akademie der Wissenschaften. Mathematisch-Naturwissenschaftliche Klasse. Superseded by: Anzeiger der Akademie der Wissenschaften in Wien. Mathematische-naturwissenschaftliche Klasse. HI 61803

Anzeiger für Kunst- und Gewerbefleiss im Königreiche Baiern. Munich. Anz. Kunst- Gewerbefl. Königr. Baiern. See B–P–H 131/17. HI 51275

Anzeiger, Österreichische Akademie der Wissenschaften. Mathematisch- naturwissenschaftliche Klasse. Wien. Vol. 84+, 1947+. Anz. Österr. Akad. Wiss., Math.-Naturwiss. Kl. Preceded by: Anzeiger der Akademie der Wissenschaften in Wien. Mathematische-naturwissenschaftliche Klasse. HI 61804

Anzeiger für Schädlingskunde. Berlin. Vols. 1-?, 1925-68. Anz. Schädlingsk. Superseded by: Anzeiger für Schädlingskunde und Pflanzenschutz. 1-415-1. HI 61805

Anzeiger für Schädlingskunde, Pflanzen- und Umweltschutz. Berlin. Vol. 46+, 1973+. Anz. Schädlingsk., Pflanzen Umweltschutz. Preceded by: Anzeiger für Schädlingskunde und Pflanzenschutz. HI 61806

Anzeiger für Schädlingskunde und Pflanzenschutz. Berlin. Vols. ?-45. 1968-72. Anz. Schädlingsk. Pflanzenschutz. Preceded by: Anzeiger für Schädlingskunde. Superseded by: Anzeiger für Schädlingskunde, Pflanzen- und Umweltschutz. HI 61807

Anzeiger des Schlesischen Landesmuseums in Troppau. Opava, Czechoslovakia. Anz. Schles. Landesmus. Troppau. See B–P–H 132/7. HI 51281

Aomori nogyo = Agriculture in Aomori. Aomori, Japan. Agric Aomori. See B–P–H 56/10.

Aomori-ken hatasaku engei shikenjo kenkyu hokoku = Bulletin of the Aomori field crops and horticultural experiment station.

Aomori-ken nogyo shikenjo gyomu nenpo = Annual bulletin of the Aomori prefecture agricultural experiment station. Aomori, Japan. Annual Bull. Aomori Prefect. Agric. Sta. See B–P–H 122/2.

Aomori-ken ringo shikenjo gyomu nenpo = Report (Annual) of the Aomori apple experiment station. Aomori.

Aomori-ken ringo shikenjo hokoku = Bulletin of the Aomori apple experiment station. Aomori, Japan. Bull. Aomori Apple Exp. Sta. See B–P–H 239/9.

Aomori-ken tennen kinenbutsu chosa hokokusho. Aomori. Vol. 1+, 1971+. Aomori-ken tennen kinenbutsu chosa hokokusho. HI 66485

Apiacta. Bucharest. Apiacta. See B–P–H 132/11. HI 51282

Apicoltore moderno. Turin. Vol. 1+, 1909+. Apic. Moderno. HI 61808

Apicultural abstracts. Gerrards Cross. Vol. 13+, 1962+. Apic. Abstr. [Previously, 1950-61, contained in: Bee world.] HI 61809

Apidologie. Paris. Vol. 1+, 1970+. Apidologie. Preceded by: Annales de l'abeille. HI 61810

Apollo. Linz. Apollo. See B–P–H 132/12. HI 51283

Apotheker-Zeitung. Berlin. Apotheker-Zeitung. See B–P–H 132/13. HI 51284

Appalachia. Boston, MA. Appalachia. See B–P–H 132/15. HI 51286

Appita. [Australia]. Appita. See B–P–H 132/16. HI 51287

Apple newsletter. Fredericton. Vol. ?-9+, 19??-77+. Apple Newslett. HI 61811

Apple research digest. Yakima, WA. Apple Res. Digest. See B–P–H 132/20. HI 51290

Apple specialist. Quincy, IL. Apple Specialist. See B–P–H 132/21. HI 51291

Appleton's journal; a magazine of general literature. New York, NY. Appleton's J. See B–P–H 133/1. HI 51292

Applied agricultural research. New York. Vol. 1+, 1986+. Appl. Agric. Res. HI 61812

Applied biochemistry and bioengineering. New York, NY. Vol. 1+, 1976+. Appl. Biochem. Bioengin. HI 61813

Applied biochemistry and biotechnology. Clifton, NJ. Vol. 6+, 1981+. Appl. Biochem. Biotechol. Preceded by: Journal of solid-phase biochemistry. HI 61814

Applied biochemistry and microbiology. Translation of: Prikladnoi biokhimiya i mikrobiologiya. New York. Vol. 1+, 1965+. Appl. Biochem. Microbiol. HI 51288

Applied biology. London, New York, NY. Vols. 1-5, 1976-80. Appl. Biol. Superseded by: Advances in applied biology. HI 61815

Applied botany abstracts. Lucknow. Vol. 1+, 1981+. Appl. Bot. Abstr. HI 61816

Applied ecology abstracts. London. Vols. 1-5, 1975-79. Appl. Ecol. Abstr. Superseded by: Ecology abstracts. HI 61817

Applied and environmental microbiology. Washington, DC. Vol. 31+, 1976+. Appl. Environm. Microbiol. Preceded by: Applied microbiology. HI 61818

Applied forestry notes, northern Rocky Mountain experiment station. Missoula, MT. 1934-39. Appl. Forest. Notes, N. Rocky Mountain Exp. Sta. Superseded by: Research notes, northern Rocky Mountain forest experiment station. HI 61819

Applied forestry notes, northern Rocky Mountain and range experiment station = Applied forestry notes, northern Rocky Mountain experiment station. Missoula, MT.

Applied genetics news. Stamford, CT. Vol. 1+, 1980+. Appl. Genet. News. HI 61820

Applied microbiology. Baltimore, MD. Vols. 1-30, 1953-75. Appl. Microbiol. Superseded by: Applied and environmental microbiology. HI 61821

Applied microbiology and biotechnology. New York, etc. Vol. 19+, 1984+. Appl. Microbiol. Biotechnol. Preceded by: European journal of applied microbiology and biotechnology. HI 61822

Applied plant science. Sunnyside, Botswana. Vol. 1+, 1987?+. Appl. Pl. Sci. HI 61823

Applied science and technology index. New York, NY. Vol. 46+, 1958+. Appl. Sci. Technol. Index. Preceded in part by: Industrial arts index. 1-141-3. HI 61824

Approved research programme, federal department of agricultural research, Nigeria. Ibadan. 1973/74+, 1973+. Approv. res. Programme, Fed. Dep. Agric. Res., Nigeria. Preceded by: Index of agricultural research and related subjects in Nigeria. HI 61825

Apteryx; a New England quarterly of natural history. Providence, RI. Apteryx. See B–P–H 133/3. HI 51293

Apuntes para al flora de la pampa. La Pampa. No. ?-93+, ?-1985+. Apuntes Fl. Pampa. HI 61826

Apuntes forestales tropicales. Rio Piedras, Puerto Rico. Apuntes Forest. Trop. See B–P–H 133/7. HI 51294

Apuntes de historia natural. Buenos Aires. Vols. 1-2, 1909-10. Apuntes Hist. Nat. 1-420-1. HI 61827

Aqua planta; Informationsschrift des V D A - Arbeitskreises Wasserpflanzen. Douruth, Dusseldorf, Frankfurt am Main. Vol. 1+, 1976+; Sonderheft no. 1+, 1987+. Aqua Pl. HI 61828

Aqua terra = Aquaterra. Solothurn.

Aqua revue; magazine de l'aquaculture. Bordeaux. Vol. 1+, 1985+. Aqua Rev. HI 61829

Aquabyte; newsletter of the network of tropical aquaculture scientists. Manila. Vol. 1+, 1988+. Aquabyte. HI 61830

Aquaculture. Amsterdam. Vol. 1+, 1972+. Aquaculture. HI 61831

Aquaculture today. Vancouver, B.C. Vol. 1+, 1988+. Aquac. Today. HI 75217

Aquaphyte; newsletter of the I P P C aquatic weed program of the university of Florida. Gainesville, FL. Vol. 1+, 1981+. Aquaphyte. HI 61832

Aquarama. Strasbourg, etc. ?-1968+. Aquarama. HI 61833

Aquariculture and aquatic sciences journal. Kansas City, MO. 1980+. Aquaric. & Aquatic Sci. J. HI 61834

Aquarien Magazine. Stuttgart. Heft 1+, 1968+. Aquar. Mag. HI 61835

Aquarien- und Terrarien-Zeitschrift. Stuttgart. Aquarien-Terrar.-Z. See B–P–H 133/19. HI 51302

Aquarienfreunde. Wilhelmshaven. Vol. 1+, 1972+. Aquarienfreunde. HI 61836

Aquarist and pondkeeper. London. Vol. 1+, 1928+. Aquar. & Pondkeeper. HI 61837

Aquarium. Brooklyn [=New York, in part], NY. Aquarium (Brooklyn). See B–P–H 133/21. HI 51303

Aquarium. Hilversum, Den Haag. Vol. ?-23+, 19??-52+. Aquarium (Netherlands). HI 61838

Aquarium. Paris. Aquarium (Paris). See B–P–H 133/22. HI 51304

Aquarium. Philadelphia, PA. Aquarium (Philadelphia 1912-14). See B–P–H 133/23. HI 51305

Aquarium, Das; Zeitschrift für Aquarien- und Terrarienfreunde. Wuppertal, Minder. Vols. ?-2(7)-7(45), 19??-68-73. Aquarium (Germany). Superseded by: Aquarium mit Aquaterra. HI 61839

Aquarium mit Aquaterra; Zeitschrift für Aquarien- und Terrarienfreunde. Wuppertal. Vol. 7(46)+, 1973+. Aquar. Aquaterra. Preceded by: Aquarium. Wuppertal [and incorporating] Aquaterra. HI 61840

Aquarium bulletin of the Brooklyn aquarium society = Bulletin of the Brooklyn aquarium society. Brooklyn, NY. Bull. Brooklyn Aquar. Soc. See B–P–H 242/33.

Aquarium digest international. Melle, & Hayward, CA. Vol. 1+, 1972+. Aquar. Digest Int. HI 61841

Aquarium journal. San Francisco, CA. Aquar. J. See B–P–H 133/11. HI 51296

Aquarium und Mikroskop. Brunswick, Germany. Aquar. & Mikroskop. See B–P–H 133/9. HI 51295

Aquarium news. Brooklyn [=New York, in part], NY. Aquar. News (Brooklyn). See B–P–H 133/12. HI 51297

Aquarium news. Rochester, NY. Aquar. News (Rochester). See B–P–H 133/13. HI 51298

Aquarium news from far and near = Aquarium news. Rochester, NY. Aquar. News (Rochester). See B–P–H 133/13.

Aquarium newsletter. Tacoma, WA. Aquar. Newslett. See B–P–H 133/14. HI 51299

Aquarium review. London. Aquar. Rev. See B–P–H 133/15. HI 51300

Aquaterra. Monatsschrift für Aquaristik und Terraristik sowie für Pflanzen- und Tierpflege in Heim. Solothurn. Vols. 1-10(3), 1964-73. Aquaterra. Incorporated in: Aquarium mit Aquaterra. HI 61842

Aquatic biology abstracts. London. Vols. 1-3(6), 1969-71. Aquatic Biol. Abstr. Superseded by: Aquatic sciences & fisheries abstracts. HI 61843

Aquatic botany; international scientific journal dealing with applied and fundamental research on submerged floating and emergent plants in marine and freshwater ecosystems. Amsterdam. Vol. 1+, 1975+. Aquatic Bot. HI 61844

Aquatic life. Philadelphia, PA. Aquatic Life. See B–P–H 133/25. HI 51306

Aquatic microbiology newsletter. Baton Rouge, LA. 1962+. Aquatic Microbiol. Newslett. HI 61845

Aquatic plant studies. New York. No. 1+, 1987+. Aquatic Pl. Stud. HI 61846

Aquatic sciences; multidisciplinary journal for theoretical and applied limnology. Basel. Vol. 51?+, 1989?+. Aquatic Sci. Preceded by: Schweizerische Zeitschrift für Hydrologie. HI 61847

Aquatic sciences and fisheries abstracts. London. Vols. 1-7, 1971-77. Aquatic Sci. Fish. Abstr. Preceded by: Aquatic biology abstracts and Current bibliography for aquatic sciences and fisheries. Superseded by: Aquatic sciences and fisheries abstracts. Part 1, biological sciences and living resources and Aquatic sciences and fisheries abstracts. Part 3, aquatic pollution and environmental quality. HI 61848

Aquatic sciences and fisheries abstracts. Aquaculture abstracts. Bethesda, MD. Vol. 1+, 1984+. Aquatic Sci. Fish. Abstr., Aquac Abstr. HI 61849

Aquatic sciences and fisheries abstracts. Part 1, biological sciences and living resources. London. Vol. 8+, 1978+. Aquatic Sci. Fish. Abstr., 1, Biol. Sci. Resources. Preceded by: Aquatic sciences and fisheries abstracts. HI 61850

Aquatic sciences and fisheries abstracts. Part 3, aquatic pollution and environmental quality. Bethesda, MD. Vol. 20+, 1990+. Aquatic Sci. Fish. Abstr., 3, Aquatic Pollut. Environm. Qual. Preceded by: Aquatic sciences and fisheries abstracts. HI 75235

Aquatic world. Baltimore, MD. Aquatic World. See B–P–H 134/1. HI 51307

Aquatica. Port of Stockton, CA. Aquatica. See B–P–H 134/2. HI 51308

Aquila. Budapest. Aquila. See B–P–H 134/3. HI 51309

Aquilegia; newsletter, Colorado native plant society. Fort Collins, CO. Vol. ?-12+, ?-1988+. Aquilegia. HI 68665

Aquilo. Ser. botanica. Oulu. Vol. 1+, 1963+. Aquilo, Ser. Bot. HI 61851

Arab gulf; an academic journal dealing with affairs of the arab gulf and arabian peninsula. Basrah. Vol. 7+, 1977+. Arab Gulf. Preceded by: Journal of the centre for arab gulf studies. HI 61852

Arab gulf journal of scientific research. Riyadh. Vols. 1-4, 1983-86; vol. 7(2)+, 1989+. Arab Gulf J. Sci. Res. For 1987-89 see: Arab gulf journal of scientific research. B, agricultural and biological sciences. HI 61853

Arab gulf journal of scientific research. B, agricultural and biological sciences. Riyadh. Vol. 5-7(1), 1987-89. Arab Gulf J. Sci. Res., B. Preceded & superseded by: Arab gulf journal of scientific research. HI 61854

Arab gulf journal of scientific research. Special publication. Riyadh. No. 1+, 1985+. Arab Gulf J. Sci. Res., Special Publ. HI 61855

Arab journal of plant protection. Beirut. Vol. ?-3+, 19??-85+. Arab J. Pl. Protect. HI 61856

Arabian gulf magazine. Basrah. Vol. 1, 1973. Arab Gulf Mag. Superseded by?: Journal of the centre for arab gulf studies. HI 61857

Arabidopsis information service. Göttingen, Frankfurt am Main. No. 1+, 1964?+. Arabidopsis Inform. Serv. HI 61858

Arabinogalactan protein news = Agp news. Melbourne, Vic.

Arable farmer and vegetable grower. Ipswich. 1972-73. Arable Farming Veg. Grower. Superseded by: Arable farming. HI 61859

Arable farming. Ipswich. Vol. 1+, 1974+. Arable Farming. Preceded by: Arable farmer and vegetable grower. HI 61860

Arago; travaux du laboratoire. [Université de Paris.] Paris. Arago. See B–P–H 134/4. HI 51310

Araneta journal of agriculture. Malabon, Philippines. Araneta J. Agric. See B–P–H 134/5. HI 51313

Araucaria. Santiago. Araucaria. See B–P–H 134/6. HI 51314

Araucariana, botânica. Curitiba. Vols. 1-?, 1967-72. Araucariana, Bot. Preceded by: Boletim, instituto de defesa do patrimônio natural. HI 61861

Araucarilandia. O pinheiro. Curitiba, Brazil. Araucarilandia. See B–P–H 134/7. HI 51315

Arbeider fra den danske arktische station pa Disko = Arbejder fra den danske arktische station pa Disko. Copenhagen.

Arbeidsrapport, norges byggforskningsinstitutt. Oslo. No. 1+, 1976+. Arbeidsrapp. Norg. Byggforskn. Inst. Preceded by: Rapport, norges byggforskningsinstitutt. HI 61862

Arbeiten der Berliner Provinzialstelle für Naturschutz. Berlin. Arbeiten Berliner Provinzialstelle Naturschutz. See B–P–H 134/21. HI 51316

Arbeiten aus der biologischen Bundesanstalt für Land- und Forstwirtschaft = Arbeiten aus der biologischen Reichsanstalt für Land- und Forstwirtschaft. Berlin.

Arbeiten der Biologischen Oka-Station. Murom = Raboty Okskoi biologicheskoi stantsii v Gorode Murome. Murom.

Arbeiten der Biologischen Oka-Station in Niznij-Novgorod. Murom = Raboty Okskoi biologicheskoi stantsii v Nizhnem Novogorode. Murom.

Arbeiten aus der biologischen Reichsanstalt für Land- und Forstwirtschaft. Berlin. Vols. 1-23, 1900-43. Arbeiten Biol. Reichsanst. Land- Forstw. HI 61863

Arbeiten (aus) der Biologischen Wolga-Station = Raboty Volzhskoi Biologicheskoi Stantsii. Saratov.

Arbeiten des Botanischen Instituts der K. K. Deutschen Universität in Prag. Prague. Arbeiten Bot. Inst. K. K. Deutsch. Univ. Prag. See B–P–H 134/29. HI 51317

Arbeiten aus dem Botanischen Institut des Kgl. Lyceums Hosianum in Braunsberg. Braunsberg, Germany [=Braniewo, Poland]. Arbeiten Bot. Inst. Königl. Lyceums Hosianum Braunsberg. See B–P–H 134/30. HI 51318

Arbeiten aus dem Botanischen Institut der Staatlichen Akademie Braunsberg = Arbeiten aus dem Botanischen Institut des Kgl. Lyceums Hosianum in Braunsberg. Braunsberg, Germany [=Braniewo, Poland]. Arbeiten Bot. Inst. Königl. Lyceums Hosianum Braunsberg. See B–P–H 134/30.

Arbeiten des Botanischen Instituts in Würzburg. Leipzig. Arbeiten Bot. Inst. Würzburg. See B–P–H 134/32. HI 51319

Arbeiten des Botanischen Kabinets der Central-Moor-Versuchsstation zu Minsk = Pratsy Botanichnaga gabinetu Menskae tsėntral'nae das'ledchae balotnae stanctsi. Minsk.

Arbeiten aus der Botanischen Station in Hallstatt (Salzkammergut), Oberösterreich. Hallstadt. Vols. 1-297?, 1925-68. Arbeiten Bot. Stat. Hallstatt. HI 61864

Arbeiten der Deutschen Landwirtschafts-Gesellschaft. Prenzlau, Germany. Arbeiten Deutsch. Landw.-Ges. See B–P–H 135/4. HI 51320

Arbeiten der forstwissenschaftlichen Gesellschaft in Finnland = Acta forestalia fennica. Helsinki. Acta Forest. Fenn. See B–P–H 43/1.

Arbeiten aus dem Gebiet der Experimentellen Biologie. Berlin. Arabeiten Exp. Biol. See B–P–H 135/5. HI 51321

Arbeiten der gelehrten Gesellschaft zur Erforschung Weissrusslands bei der Weissrussischen Staatlichen Akademie für Landwirtschaft in Gorky = Pratsa navukovaga tovarystva pa vyvuchen'nyu Belarusi pry Belaruskai dzyarzheaūnai akademii sel'skae khaspadarki ū Gorkakh. Gorki.

Arbeiten der Gory-Goretzkischen Gelehrten Gesellschaft = Pratsy Gory-Goretskaga navukovaga tovarystva. Gory-Gorky.

Arbeiten, hessische Bauernkammer = Phänologische Mitteilungen. Darmstadt.

Arbeiten der Hydrobiologischen Station am See "Glubokoje" = Trudy gidrobiologicheskoi stantsii na glubokom ozere. Moscow.

Arbeiten aus dem Institut für allgemeine Botanik der Universität Zürich. Zurich. Arbeiten Inst. Allg. Bot. Univ. Zürich. See B–P–H 135/11. HI 51322

Arbeiten aus dem Institut für Geschichte der Nasturwissenschaft. Heidelberg. Arbeiten Inst. Gesch. Nasturwiss. See B–P–H 135/12. HI 51323

Arbeiten aus dem Institut für Paläobotanik und Petrographie der Brennsteine. Berlin. Arbeiten Inst. Paläobot. See B–P–H 135/13. HI 51324

Arbeiten aus dem Kieler Physiologischen Institut. Kiel. Arbeiten Kieler Physiol. Inst. See B–P–H 135/14. HI 51326

Arbeiten der Kommission für biologische und wasserhygienische Erforschung des Flusses N.-Donjez und seiner Nebenflüsse (Lopan und Udy) = Trudy komissii po sanitarno-biologicheskomu obsledovaniyu reka S.-Donca i ego pritikov (Lopani i Udy). Kharkov.

Arbeiten der Kommission für biologische und wasserhygienische Erforschung des Flusses N.-Donjez und seiner Nebenflüsse <Lopan und Udy> = Trudy komissii po izucheniyu ozera Baikala. Petrograd.

Arbeiten aus dem Königl. Botansichen Garten zu Breslau. Breslau [=Wroclaw, Poland]. Arbeiten Königl. Bot. Gart. Breslau. See B–P–H 135/16. HI 51327

Arbeiten der Landwirtschaftlichen Hochschule Hohenheim. Stuttgart. Vols. 1-41, 1961-67. Arbeiten Landw. Hochschule Hohenheim. Superseded by: Arbeiten der Universität Hohenheim. HI 61865

Arbeiten der Leningradschen Naturforscher-Gesellschaft = Trudy Leningradskogo obshchestva estestvoispytatelei. Leningrad.

Arbeiten aus dem limnologischen Institut der Universität Freiburg. [Archiv für Hydrobiologie, Supplement.] Stuttgart. Vol. 1+, 1954+. Arbeiten Limnol. Inst. Univ. Freiburg. HI 61866

Arbeiten der Limnologischen Station zu Kossino der Hydrometeorologischen Administration der U S S R = Trudy Limnologicheskoi stantsii v Kosine. Moscow.

Arbeiten der mittelasiatischen wissenschaftlichen Versuchsstation der ätherischen Ölpflanzen O M P K = Trudy sredne-Aziatskoi nauchno-issledovatel'skoi opytnoi stantsii efirno-maslichnykh rastenii O M P K. Tashkent.

Arbeiten des Naturforschenden Vereins zu Riga. Rudolstadt, Germany. Arbeiten Naturf. Vereins Riga. See B–P–H 135/21. HI 51328

Arbeiten des Naturforscher-Vereins zu Riga = Arbeiten des Naturforschenden Vereins zu Riga. Rudolstadt, Germany. Arbeiten Naturf. Vereins Riga. See B–P–H 135/21.

Arbeiten der Nordkaukasischen Assoziation wissenschaftlicher Institute = Trudy Severo-Kavkazskoi assotsiatsii nauchno-issledovatel'skikh institutov. Krasnodar.

Arbeiten der Ost-Sibirischen Staatsuniversität = Trudy Vostochno-sibirskogo gosudarstvennogo universiteta. Moscow.

Arbeiten aus dem pharmazeutischen Institut der Universität Berlin. Berlin. Vols. 1-13, 1903-27. Arbeiten Pharm. Inst. Univ. Berlin. HI 61867

Arbeiten aus der phytopalaeontologischen Laboratoriums der Universitat Graz. Graz. 1924-26. Arbeiten Phytopalaeontol. Lab. Univ. Graz. HI 61868

Arbeiten der Phytopathologischen Versuchsstation der Universität Tartu = Tartu ülikooli taimehaiguste-katsejaama tööd. Tartu, Estonia [Estonian S S R]. Tartu Ülik. Taimeh.-Katsej. Tööd. See B–P–H 866/13.

Arbeiten der Tomsker Staatsuniversität = Trudy Tomskogo gosudarstvennogo universiteta.

Arbeiten aus dem Treub-Laboratorium. Weltevreden. Vol. 1, 1927. Arbeiten Treub-Lab. HI 61869

Arbeiten des Ukrainischen Instituts für Angewandte Botanik = Trudy Ukrayins'kogo instytutu prykladnoi botaniky. Kharkov.

Arbeiten des ungarischen biologischen Forschungsinstituts = A magyar biológiai kutatóintézet munkái. Tihany, Hungary. Magyar Biol. Kutatóint. Munkái. See B–P–H 542/22.

Arbeiten der universität Hohenheim. Stuttgart. Vol. 42+, 1968+. Arbeiten Univ. Hohenheim. Preceded by: Arbeiten der Landwirtschaften Hochschule Hohenheim. HI 61870

Arbeiten einer vereinigten Gesellschaft in der Oberlausitz zu den Geschichten und der Gelahrtheit überhaupt gehörende. Leipzig & Lauaban [=Luban, Poland]. Arbeiten Vereinigten Ges. Oberlausitz. See B–P–H 136/1. HI 51329

Arbeiten der Versuchsstation für Pflanzenzüchtung am Moskauer Landwirtschaftlichen Institut = Trudy Selektsionnoi Stantsii pri Moskovskom Sel'skokhozyaistvennom Institutě. Moscow.

Arbeiten der W. W. Kuibyschews-Staatsuniversität in Tomsk = Trudy Tomskogo gosudarstvennogo universiteta imeni V. V. Kuibysheva. Tomsk.

Arbeiten, Würzburger geographische Gesellschaft. Würzburg, No. 1+, 1953+. Arbeiten Wüzburger Geogr. Ges. HI 61871

Arbeiten aus der Zentralstelle für Vegetationskartierung = Angewandte Pflanzensoziolgie. Stolzenau/Weser, German. Angew. Pflanzensoziol. (Stolzenau). See B–P–H 94/17.

Arbeits- und Forschungsberichte zur Sächsischen Bodendenkmalpflege. Dresden. Arbeits-Forschungsber. Sächs. Bodendenkmalpflege. See B–P–H 136/4. HI 51330

Arbeitsgemeinschaft für Forschung des Landes Nordrhein-Westfalen = Veröffentlichungen der Arbeitsgemeinschaft für Forschung des Landes Nordrhein-Westfalen. Abt. Naturwissenschaften. Cologne & Opladen.

Arbeitsmaterial der zentralen Arbeitsgemeinschaft "Echinopseen". Gothal. 1982+, 1983+. Arbeitsmater. Zentr. Arbeitsgem. Echinopseen. HI 61872

Arbejder fra den danske arktische station pa Disko. Copenhagen. Vol. 1+, 1910+. Arbejder Danske Arktische Stat. Disko. HI 61873

Arbejder. Universitet: botanisk have. Copenhagen. Arbejder Univ. Bot. Have. See B–P–H 136/6. HI 51331

Årbok, kongelige norske videnskabers selskabs museet = Kongelige norske videnskabers selskabs museet. Årbok. Trondheim.

Årbok, Kristiansand museets = Kristiansand museets årbok. Kristansand.

Årbok. Norske videnskaps-akademi i Oslo. Oslo. Årbok Norske Vidensk.-Akad. Oslo. See B–P–H 136/8. HI 51332

Arbol. [Instituto de colonización rural.] San Salvador, Salvador. Arbol (San Salvador). See B–P–H 136/11. HI 51333

Arboles. Medellín, Colombia. Arboles (Medellín). See B–P–H 136/12. HI 51334

Arboles de Costa Rica. San José, Costa Rica. 1975+. Arboles Costa Rica. HI 61874

Arbor. Aberdeen, Scotland. Arbor (Aberdeen). See B–P–H 136/13. HI 51335

Arbor. Madrid. Arbor (Madrid). See B–P–H 136/14. HI 51336

Arbor action. Wantagh, NY. 1941+. Arbor Action. HI 61875

Arbor age. Encino, CA. Vol. 1+, 1981. Arbor Age. HI 61876

Arborea. Worcester, MA. Arborea. See B–P–H 136/24. HI 51341

Arboreta phaenologica. Münden. No. ?-18+, 19??-73+. Arbor. Phaenol. HI 61877

Arboreto carioca; trabalho pelo centro des pesquisas florestais e conservação da natureza. Rio de janeiro. Vol. 1+, 1963+. Arbor. Carioca. HI 61878

Arboretum and botanical garden bulletin. New Providence, NJ., Las Cruces, NM. Vols. 1-7, 1967-73. Arbor. Bot. Gard. Bull. Preceded by: Quarterly news letter, American association of botanical gardens and arboretums. Superseded by: Bulletin of the american association of botanical gardens and arboreta and Newsletter of the american association of botanical gardens and arboreta. HI 61880

Arboretum bulletin, arboretum foundation, university of Washington = Arboretum bulletin, university of Washington. Seattle, WA.

Arboretum bulletin of the associates. [Morris arboretum.] Philadelphia, Pa. Arbor. Bull. Assoc. See B–P–H 136/17. HI 51337

Arboretum bulletin, university of Washington; journal of general horticultural information. Seattle, WA. Vols. 1-49(1), 1936-86. Arbor Bull., Washington. Superseded by: Washington Park arboretum bulletin. 1-427-1. HI 61881

Arboretum kórnickie. Poznan, Poland. Arbor. Kórnickie. See B–P–H 136/19. HI 51338

Arboretum leaves. [The Holden arboretum.] Mentor, OH. Arbor. Leaves. See B–P–H 136/20. HI 51339

Arboretum news and friends newsletter, university of Wisconsin. Madison, WI. Vol. 31+, 1982+. Arbor. News Friends Newslett., Univ. Wisconsin. Preceded by: Arboretum news, university of Wisconsin and Newsletter of the friends of the arboretum, university of Wisconsin [not entered]. HI 61882

Arboretum news, Iowa arboretum. Ames, IA. Vols. 1-4(2), 1981-84. Arbor. News, Iowa Arbor. Superseded by: Iowa arboretum news. Ames, IA. HI 66074

Arboretum news, university of Wisconsin. Madison, WI. Vols. 1-30?, 1952-82? Arbor. News, Univ. Wisconsin. Superseded by: Arboretum news and friends newsletter, university of Wisconsin. HI 61883

Arboretum news, West Virginia university. Morgantown, WV. Vols. 1-3, 1951-53. Arbor. News, West Virginia Univ. Superseded by: Newsletter, West Virginia university arboretum. HI 61884

Arboretum newsletter, West Virginia university. Morgantown, WV. Vols. 5(3)-22, 1955-74. Arbor. Newslett., West Virginia Univ. Preceded by: Newsletter, West Virginia university arboretum. Superseded by: Core arboretum bulletin, West Virginia university. HI 61885

Arboretum notes, botany and plant pathology division, Canada. Ottawa. Vols. 12-?, 1957-59. Arbor. Notes, Bot. Pl. Pathol. Div., Canada. Preceded by: This week in the arboretum, botany and plant pathology division, Canada. Superseded by: Arboretum notes, plant research institute, Ottawa. HI 61886

Arboretum notes, plant research institute, Ottawa. Ottawa. Vols. 13-18(5), 1960?-64. Arbor. Notes Pl. Res. Inst., Ottawa. Preceded by: Arboretum notes, botany and plant pathology division, Canada. HI 61887

Arboretum society bulletin. Oak Ridge, TN. Arbor. Soc. Bull. See B–P–H 136/23. HI 51340

Arboretum Waasland bulletin = Bulletin, arboretum Waasland. Nieuwkerken-Waas.

Arboricultural abstracts. Section 1, plant pathology. Urbana, IL. 1981+. Arboric. Abstr., 1, Pl. Pathol. HI 61888

Arboricultural abstracts. Section 2, physiology and pollution injury. Urbana, IL. 1981+. Arboric. Abstr., 2, Physiol. HI 61889

Arboricultural abstracts. Section 3, horticulture. Urbana, IL. 1981+. Arboric. Abstr., 3, Hort. HI 61890

Arboricultural abstracts. Section 4, entomology. Urbana, IL. 1981+. Arboric. Abstr., 4, Entomol. HI 61891

Arboricultural abstracts. Section 5, genetics, breeding, selection and propogation. Urbana, IL. 1981+. Arboric. Abstr., 5, Genet. HI 61892

Arboricultural association journal. Stansted. Vols. 1-2(6), 1965-74. Arbor. Assoc. J. Superseded by: Arboricultural journal. HI 61893

Arboricultural consultant. Milltown, New Brunswick, NJ. 1967+. Arbor. Consultant. HI 61894

Arboricultural journal. Worplesdon, Ampfield Romsey, Stansted. Vol. 2(7)+, 1974+. Arbor. J. Preceded by: Arboricultural association journal. HI 61895

Arboricultural leaflet. Farnham. No. [1]+, 1977+. Arbor. Leafl. HI 61896

Arboricultural society of South Africa. Johannesburg. Arboric Soc. South Africa. See B–P–H 137/9. HI 51343

Arboriculture. Chicago, IL. Arboriculture. See B–P–H 137/10. HI 51344

Arboriculture fruitière. Paris. Arboric. Fruitière. See B–P–H 137/8. HI 51342

Arboriculture research note. Wrecclesham. No. ?-7+, 19??-79+. Arbor. Res. Note. HI 61897

Arborist's news. Columbus, OH, & Urbana, Ill. Vols. 1-39, 1935-74. Arborist's News. Superseded by: Journal of arboriculture. 1-427-1. HI 61898

Arborvirusy; sbornik trudov. Moscow. Vol. 1+, 1974+. Arborvirusy. HI 61899

Arbre. Paris. Arbre. See B–P–H 137/12. HI 51345

Arbres et fruits. Lille. Arbres & Fruits. See B–P–H 137/13. HI 51346

Arcana; or, the museum of natural history. London. Arcana. See B–P–H 137/17. HI 51347

Arcana naturae. Paris. Arcana Nat. See B–P–H 137/18. HI 51348

Arcana of science and art. London. Arcana Sci. Art. See B–P–H 137/19. HI 51349

Archaeologica lundensia; investigationes de antiquitibus urbis Lundae. Lund. Archaeol. Lund. See B–P–H 151/4. HI 51541

Archaeology Cambridge, MA. Archaeology. See B–P–H 151/5. HI 51542

Archeion Rome. Archeion. See B–P–H 151/6. HI 51543

Archief van de cacao en andere kleine cultures in Nederlandsch-Indië. Salatiga, Dutch E. Indies [Indonesia]. Arch. Cacao Ned.-Indië. See B–P–H 139/18. HI 51388

Archief voor de Java suikerindustrie. Surabaya, Dutch E. Indies [Indonesia]. Arch. Java Suikerindustr. See B–P–H 143/18. HI 51445

Archief voor de koffiecultuur in Nederlandsch-Indië. Surabaya, Dutch E. Indies [Indonesia]. Arch. Koffiecult. Ned.-Indië. See B–P–H 143/26. HI 51446

Archief voor den landbouw der bergstreken in Nederlandsch-Indië. Semarang, Dutch E. Indies [Indonesia]. Arch. Landb. Bergstreken Ned.-Indië. See B–P–H 144/3. HI 51447

Archief voor de rubbercultuur. Batavia, Dutch E. Indies [=Jakarta, Indonesia]. Arch. Rubbercult. See B–P–H 148/17. HI 51507

Archief voor de suikerindustrie in Nederland en Nederlandsch-Indië. Pasuruan, Dutch E. Indies [Indonesia]. Arch. Suikerindustr. Ned. Ned.-Indië. See B–P–H 149/21. HI 51522

Archief voor de suikerindustrie in Nederlandsch-Indië. Surabaya, Dutch E. Indies [Indonesia]. Arch. Suikerindustr. Ned.-Indië. See B–P–H 149/22. HI 51523

Archief voor de theecultuur in Nederlandsch-Indië. Batavia, Dutch E. Indies [=Jakarta, Indonesia]. Arch. Theecult. Ned.-Indië. See B–P–H 150/4. HI 51526

Archiv für Acker- und Pflanzenbau und Bodenkunde. Berlin. Vol. 15(7)+, 1971+. Arch. Acker- Pflanzenbau Bodenk. Preceded by: Archiv für Bodenfruchtbarkeit und Pflanzenproduktion. HI 61900

Archiv der Agriculturchemie für denkende Landwirthe. Berlin. Arch. Agriculturchem. Denkende Landwirthe. See B–P–H 137/21. HI 51351

Archiv für die allgemeinen Fragen der Lebensforschung = Biologia generalis. Vienna & Leipzig. Biol. Gen. See B–P–H 193/18.

Archiv für Anatomie und Physiologie. Leipzig. Arch. Anat. Physiol. See B–P–H 137/26. HI 51356

Archiv für Anatomie, Physiologie und wissenschaftliche Medicin. [Edited by Müller.] Berlin. Arch. Anat. Physiol. Wiss. Med. See B–P–H 137/27. HI 51357

Archiv des Apotheker-Vereins im Nördlichen Teutschland. Schmalkalden, Germany. Arch. Apotheker-Vereins Nördl. Teutschl. See B–P–H 138/2. HI 51362

Archiv des Apothekervereins im nördlichen Teutschland = Pharmaceutische Monatsblätter. Schmalkalden, Germany. Pharm. Monatsbl. See B–P–H 704/12.

Archiv für Augen- und Ohrenheilkunde. Karlsruhe. Arch. Augen- Ohrenheilk. See B–P–H 138/5. HI 51365

Archiv für Augenheilkunde. Karlsruhe. Arch. Augenheilk. See B–P–H 138/6. HI 51366

Archiv für Bergbau und Hüttenwesen. Breslau [=Wroclaw, Poland]. Arch. Bergbau Hüttenwesen. See B–P–H 138/10. HI 51369

Archiv biologicheskikh nauk. Belgrade = Arhiv bioloških nauka. Belgrade. Arh. Biol. Nauka. See B–P–H 152/36.

Archiv für Biontologie. Berlin. Arch. Biontol. See B–P–H 138/23. HI 51378

Archiv für Bodenfruchtbarkeit und Pflanzenproduktion. Berlin. Vol. 15(1-6), 1971. Arch. Bodenfruchtb. Pflanzenprod. Preceded by: Albrecht-Thaer-Archiv. Superseded by: Archiv für Acker- und Pflanzenbau und Bodenkunde. HI 61901

Archiv für die Botanik. Leipzig. Arch. Bot. (Leipzig). See B–P–H 139/5. HI 51381

Archiv aller bürgerlichen Wissenschaften zum Nutzen und Vergnügen wie auch zum Selbstunterricht in reiferen Jahren. Hamburg. Arch. Aller Bürgerl. Wiss. See B–P–H 137/22. HI 51353

Archiv für Chemie und Meteorologie = Archiv für die gesammte Naturlehre. Nuremberg. Arch. Gesemmte Naturl. See B–P–H 141/5.

Archiv für Dermatologie und Syphilis. Vienna, Leipzig, Berlin. Vols. 1-5, 1869-73; vols. 21-200, 1889-1955. Archiv Dermatol. Syph. For vols. 6-20 see: Vierteljahresschrift für Dermatologie und Syphilis. Superseded by: Archiv für klinische und experimentelle Dermatologie. 1-445-2. HI 61902

Archiv für dermatologische Forschung. Berlin, New York. Vols. 240-252, 1971-75. Archiv Dermatol. Forsch. Preceded by: Archiv für klinische und experimentelle Dermatologie. Superseded by: Archives of dermatological research. HI 61903

Archiv für Entwicklungs-Mechanik der Organismen. Leipzig. Vols. 1-52/97, 1894-1923. Arch. Entwicklungs-Mech. Organismen. Superseded by: Archiv für mikroskopische Anatomie und Entwicklungsmechanik. 5-4509-2. HI 61904

Archiv für Forstwesen. Berlin. 1952-71. Arch. Forstwesen. Superseded by: Beiträge für die Forstwirtschaft. HI 61905

Archiv der Freunde der Nasturgeschichte in Mecklenburg. Rostock. Arch. Freunde Naturgesch. Mecklenburg. See B–P–H 140/24. HI 51403

Archiv des Garten- und Blumenbau-Vereins für Hamburg, Altona, und deren Umgegenden. Hamburg. Arch. Garten-Blumenbau-Vereins Hamburg. See B–P–H 140/25. HI 51404

Archiv für Gartenbau. Berlin. Arch. Gartenbau. See B–P–H 140/26. HI 51405

Archiv gemeinnütziger physischer und medizinischer Kenntnisse. Zurich. Arch. Gemeinnütz. Phys. Med. Kenntn. See B–P–H 140/28. HI 51406

Archiv für Genetik. Zurich. Vols. 45-52, 1972-79. Arch. Genet. Preceded by: Archiv, Julius Klaus-Stiftung für Vererbungsforschung, sozialanthropologie und rassenhygiene. HI 61906

Archiv für Geographie, Historie, Staats- und Kriegskunst. Vienna. Arch. Geogr. See B–P–H 141/2. HI 51409

Archiv für Geographie und Statistik, ihre Hilfswissenschaften und Literatur, mit vorzüglicher Rücksicht auf die österreichischen Staaten. Prague. Arch. Geogr. Statist. See B–P–H 141/3. HI 51410

Archiv für die gesammte Naturlehre. Nuremberg. Arch. Gesemmte Naturl. See B–P–H 141/5. HI 51411

Archiv für gesamte Virusforschung. New York. Vols. 1-46, 1939-74? [suspended 1944-47]. Arch. Gesamte Virusforsch. Superseded by: Archives of virology. HI 61907

Archiv für Geschichte und Landeskunde Vorarlbergs. Dornbirn, Austria. Arch. Gesch. Landesk. Vorarlbergs. See B–P–H 141/7. HI 51413

Archiv für die Geschichte der Mathematik, der Naturwissenschaften und der Technik. Leipzig. Arch. Gesch. Math. See B–P–H 141/8. HI 51414

Archiv für Geschichte der Medizin. Leipzig. Arch. Gesch. Med. See B–P–H 141/9. HI 51415

Archiv für Geschichte der Naturwissenschaften. Vienna. No. 1+, 1981+. Arch. Gesch. Naturwiss. In continuation of: Mayerhöfer, J., edit. Lexikon der Geschichte der Naturwissenschaften [A-Edel.], nos. 1-8, Vienna, 1959-79? [not entered]. HI 61908

Archiv für die Geschichte der Naturwissenschaften und der Technik. Leipzig. Arch. Gesch. Naturwiss. Techn. See B–P–H 141/10. HI 51416

Archiv für Geschichte, Statistik, Literatur und Kunst. Vienna. Arch. Gesch. See B–P–H 141/6. HI 51412

Archiv der Gewächskunde, von Leopold Trattinick. Vienna. Vols. 1-2, 1812-14. Arch. Gesächsk. Leopold Trattinick. HI 61909

Archiv für Gynäkologie. Berlin. Arch. Gyn.. See B–P–H 141/13. HI 51418

Archiv for haugevaesen. Copenhagen. 1848. Arch. Haugevaes. Superseded by: Dansk Havetid. (Copenhagen, 1849-87). HI 61910

Archiv für Hydrobiologie. Stuttgart. Arch. Hydrobiol. See B–P–H 141/17. HI 51421

Archiv für Hydrobiologie und Planktonkunde. Stuttgart. Arch. Hydrobiol. Planktonk. See B–P–H 141/18. HI 51422

Archiv für Hygiene = Archiv für Hygiene und Bakteriologie. Munich.

Archiv für Hygiene und Bakteriologie. Munich. Vols. 1-154, 1883-1971. Arch. Hyg. Bakteriol. Superseded by: Zentralblatt für Bakteriologie, Parasitenkunde, Infektionskrankheiten und Hygiene. 1 Abteilung, Originale. Reihe B, Hygiene, präventive Medizin. 1-444-3. HI 61911

Archiv, Julius Klaus-Stiftung für Vererbungsforschung, sozialanthropologie und rassenhygiene. Zurich. Vols. 1-44, 1925-69. Arch. Julius Klaus-Stiftung Vererbungsf. Superseded by: Archiv für Genetik. 3-2251-2. HI 61912

Archiv für klinische und experimentelle Dermatologie. Prague, Vienna, Berlin, New York. Vols. 201-239, 1955-71. Arch. Klin. Exp. Dermatol. Preceded by: Archiv für Dermatologie und Syphilis. Superseded by: Archiv für dermatologische Forschung. 1-445-2. HI 61913

Archiv für Landes- und Volkskunde der Provinz Sachsen (nebst angrenzenden Landesteilen). Halle. Arch. Landes- Volksk. Prov. Sachsen. See B–P–H 144/5. HI 51448

Archiv Mecklenburgischer Naturforscher. Rostock. Arch. Mecklenburg. Naturf. See B–P–H 144/9. HI 51451

Archiv der Medizin, Chirurgie und Pharmacie. Aarau, Switzerland. Arch. Med. See B–P–H 144/10. HI 51452

Archiv für den Menschen und Bürger in allen Verhältnissen. Leipzig. Arch. Menschen Bürger. See B–P–H 144/20. HI 51458

Archiv, Merkwürdigkeiten aus dem Reiche der Natur und dem Gebiete der Künste und Wissenschaften. Brunswick, Germany. Arch. Merkwürdigk. Natur Künste. See B–P–H 144/22. HI 51459

Archiv für Mikrobiologie. Berlin. Vols. 1-94, 1930-73 [publication suspended 1944-47]. Arch. Mikrobiol. Superseded by: Archives of microbiology. 1-446-3. HI 61914

Archiv für mikroskopische Anatomie. Bonn. Vols. 1-43, 1865-94. Arch. Mikroskop. Anat. Superseded by: Archiv für mikroskopische Anatomie und Entwicklungsgeschichte. 1-446-3. HI 61915

Archiv für mikroskopische Anatomie und Entwicklungsgeschichte. Bonn. Vols. 44-76, 1895-1910. Arch. Mikroskop. Anat. Entwicklungsgesch. Preceded by: Archiv für mikroskopische Anatomie. Superseded by: Archiv für mikroskopische Anatomie und Entwicklungsgeschichte, Abteilung für vergleichende und experimentelle Histologie. 1-446-3. HI 61916

Archiv für mikroskopische Anatomie und Entwicklungsgeschichte, Abteilung für vergleichende und experimentelle Histologie. Bonn. Vols. 77-97, 1911-1923. Arch. Mikroskop. Anat. Entwicklungsgesch., Abt. Vergl. Exp. Histol. Preceded by: Archiv für mikroskopische Anatomie und Entwicklungsgeschichte. Superseded by: Archiv für mikroskopische Anatomie Entwicklungsgeschichte, Abteilung Zeugungs- und Vererbungslehre. 1-446-3. HI 61917

Archiv für mikroskopische Anatomie und Entwicklungsgeschichte, Abteilung Zeugungs- und Vererbungslehre. Bonn. Vols. 77-97, 1911-23. Arch. Mikroskop. Anat. Entwicklungsgesch., Abt. Zeugubngs- Vererbungsl. Preceded by: Archiv für mikroskopische Anatomie und Entwicklungsgeschichte. Superseded by: Archiv für mikroskopische Anatomie und Entwicklungsmechanik. 1-446-3. HI 61918

Archiv für mikroskopische Anatomie und Entwicklungsmechanik. Leipzig. Vols. 98-104, 1923-25. Arch. Mikroskop. Anat. Entwicklungsmech. Preceded by: Archiv für mikroskopische Anatomie und Entwicklungsgeschichte, Abteilung vergleichende und experimentelle Histologie; Archiv für mikroskopische Anatomie und Entwicklungsgeschichte, Abteilung Zeugungs- Vererbungslehre and Archiv für Entwicklungsmechanik der Organismen. Superseded by: Wilhelm Roux Archiv für Entwicklungsmechanik der Organismen. 5-4509-2. HI 61919

Archiv für Mineralogie, Geognosie, Bergbau und Hüttenkunde. Berlin. Arch. Mineral. See B–P–H 145/2. HI 51461

Archiv für Natur, Kunst, Wissenschaft und Leben. Brunswick. Arch. Natur. See B–P–H 145/14. HI 51466

Archiv für Naturgeschichte. Berlin. Arch. Naturgesch. See B–P–H 145/15. HI 51467

Archiv für die Naturkunde Estlands <vormals Liv-, Ehst- und Kurlands> herausgegeben von der Naturforscher-Gesellschaft bei der Universität Tartu <Dorpat>. 2 Ser. Biologica = Eesti loodusteaduse arhiiv välja antud loodusuurijate seltsi poolt Tartu ülikooli juures. Tartu, Estonia [Estonian S S R]. Eesti Loodustead. Arh., Ser. 2, Biol. See B–P–H 355/6.

Archiv für die Naturkunde Estlands <vormals Liv-, Ehst- und Kurlands> herausgegeben von der Naturforscher-Gesellschaft bei der Universität Tartu <Dorpat>. 2 Ser. Biologische Naturkunde = Eesti loodusteaduse arhiiv välja antud loodusuurijate seltsi poolt Tartu ülikooli juures. Tartu, Estonia [Estonian S S R]. Eesti Loodustead. Arh., Ser. 2, Biol. Naturk. See B–P–H 355/7.

Archiv für die Naturkunde Liv-, Ehst- und Kurlands. Serie 2, biologische Naturkunde. Dorpat [=Tartu, Estonian S S R]. Vols. 1-13(1), 1859-1905. Arch. Naturk. Liv- Ehst- Kurlands, Ser. 2, Biol. Naturk. Superseded by: Archiv für die Naturkunde des Ostbaltikums. Serie 2, biologische Naturkunde. 2-1417-3. HI 61920

Archiv für die Naturkunde des Ostbaltikums. Serie 2, biologische Naturkunde. Dorpat [=Tartu, Estonian S S R]. Vols. 14(1-3), 1920-22. Arch. Naturk. Ostbaltik., Ser. 2, Biol. Naturk. Preceded by: Archiv für die Naturkunde Liv-, Ehst- und Kurlands. Serie 2, biologische Naturkunde. Superseded by: Eesti loodusteaduse arhiiv välja antud loodusuurijate seltsi poolt Tartu ülikooli juures. Ser. 2. HI 61921

Archiv für Naturschutz und Landschaftsforschung. Berlin. Arch. Naturschutz Landschaftsf. See B–P–H 145/22. HI 51468

Archiv für die naturwissenschaftliche Landesdurchforschung von Böhmen. Prague. Arch. Naturwiss. Landesdurchf. Böhmen. See B–P–H 145/26. HI 51469

Archiv zur neuern Geschichte, Geographie, Natur- und Menschenkenntniss. Leipzig. Arch. Neuern Gesch. See B–P–H 146/15. HI 51473

Archiv für die neuesten Entdeckungen aus der Urwelt. Quedlinburg & Leipzig. Arch. Neuesten Entdeck. Urwelt. See B–P–H 146/17. HI 51474

Archiv po ochrane sredy = Archiwum ochrony srodowiska. Wroclaw, Warsaw.

Archiv für Ophthalmologie. Berlin. Arch. Ophthalmol.

See B–P–H 147/1. HI 51480

Archiv für pathologische Anatomie und Physiologie und klinische Medizin. Berlin. Arch. Pathol. Anat. See B–P–H 147/10. HI 51485

Archiv für Pflanzenbau. Berlin. Arch. Pflanzenbau See B–P–H 147/12. HI 51487

Archiv für Pflanzenschutz. Berlin. Vols. 1-8, 1965-72. Arch. Pflanzenschutz. Superseded by: Archiv für Phytopathologie und Pflanzenschutz. HI 61922

Archiv for pharmaci og chemi. Copenhagen. Vol. 1+, 1894+ [from vol. 80(3)+, 1973+ includes a quarterly section titled Sci. ed, vol. 1+]. Arch. Pharm. Chemi. Preceded by: Archiv for pharmacie (Copenhagen). 1-439-2. HI 61923

Archiv for pharmacie. [Vols. 4-50, 1847-93 = vols. 1-47 of: Archiv for pharmacie og technisk chemie med deres grundvidenskaber.] Copenhagen. Vols. 1-50, 1844-93. Arch. Pharm. (Copenhagen). Superseded by: Archiv for pharmacie og chemi. HI 61924

Archiv der Pharmacie. Lemgo, Germany. Arch. Pharm. (Berlin). See B–P–H 147/15. HI 51488

Archiv der Pharmacie des Apotheker-Vereins im nördlichen Teutschland. Lemgo, Germany. Arch. Pharm. Apotheker-Vereins Nördl. Teutschl. See B–P–H 147/16. HI 51489

Archiv für die Pharmacie und ärztliche Naturkunde. Kassel. Arch. Pharm. Ärztl. Naturk. See B–P–H 147/17. HI 51490

Archiv der Pharmazie. Weinheim. Vol. 305+, 1972+. Arch. Pharm. (Weinheim). Preceded by: Archiv der Pharmazie und Berichte der deutschen pharmazeutischen Gesellschaft. HI 61925

Archiv der Pharmazie und Berichte der deutschen pharmazeutischen Gesellschaft. Berlin. Vols. 262-304, 1924-71. Arch. Pharm. & Ber. Deutsch. Pharm. Ges. Preceded by: Archiv der Pharmacie (Berlin) and Berichte der deutschen pharmaceutischen Gesellschaft. Superseded by: Archiv für Pharmazie. Weinheim. 1-438-3. HI 61926

Archiv für die Physiologie. Halle. Arch. Physiol. See B–P–H 148/1. HI 51495

Archiv für Phytopathologie und Pflanzenschutz. Berlin. Vol. 9+, 1973+. Arch. Phytopathol. Pflanzenschutz. Preceded by: Archiv für Pflanzenschutz. HI 61927

Archiv pro přírodovědecké prozkoumání Čech = Archiv für die naturwissenschaftliche Landesdurchforschung von Böhmen. Prague. Arch. Naturwiss. Landesdurchf. Böhmen. See B–P–H 145/26.

Archiv pro přírodovědecký výzkum Čech. Prague. Arch. Přír. Výzk. Čech. See B–P–H 148/6. HI 51500

Archiv für Protistenkunde. Jena. Arch. Protistenk. See B–P–H 148/8. HI 51501

Archiv für Rassen- und Gesellschafts-Biologie einschliesslich Rassen- und Gesellschafts-Hygiene; Zeitschrift für die Erforschung des Wesens von Rasse und Gesellschaft und ihres gegenseitigen Verhältnisses, etc. Berlin. Vols. 1-37(4), 1904-44. Arch. Rassen-Ges.-Biol. Einschl. Rassen- Ges.-Hyg. HI 61928

Archiv für Schiffs- und Tropenhygiene, unter besonderer Berücksichtigung der Pathologie und Therapie. Leipzig. Arch. Schiffs- Tropenhyg. See B–P–H 148/22. HI 51508

Archiv der Schweizerischen Geschichte und Landeskunde. Zurich. Arch. Schweiz. Gesch. Landesk. See B–P–H 148/24. HI 51509

Archiv skandinavischer Beiträge zur Naturgeschichte. Greifswald, Germany. Arch. Skand. Beitr. Naturgesch. See B–P–H 149/11. HI 51516

Archiv für die systematische Naturgeschichte. Leipzig. Arch. Syst. Naturgesch. See B–P–H 150/2. HI 51524

Archiv der teutschen Landwirthschaft. Leipzig. Arch. Teutsch. Landw. See B–P–H 150/3. HI 51525

Archiv für Toxicologie. Leipzig, Berlin. Vols. 1-31, 1930-73. Archiv Toxicol. Preceded by: Archives of toxicology. 1-450-2. HI 75207

Archiv des Vereins der Freunde der Naturgeschichte in Mecklenburg. Neubrandenburg. Arch. Vereins Freunde Naturgesch. Mecklenburg. See B–P–H 150/12. HI 51533

Archiv des Vereins für die Geschichte des Herzogthums Lauenburg. Möllin = Vaterländisches Archiv für das Herzogthum Lauenburg. Ratzeburg, Germany. Vaterl. Arch. Herzogth. Lauenburg. See B–P–H 947/12.

Archiv des Vereins für siebenbürgische Landeskunde. Hermannstadt & Kronstadt [=Sibiu & Brasov, Rumania]. Arch. Vereins Siebenbürg. Landesk. See B–P–H 150/14. HI 51534

Archiv für Welt-, Erd- und Staatenkunde und ihre Hilfswissenschaften und Litteratur. Vienna. Arch. Welt- Erd- Staatenk. See B–P–H 150/16. HI 51536

Archiv für Wissenschaften, Künste und Gewerbe. Hamburg. Arch. Wiss. See B–P–H 150/17. HI 51537

Archiv für wissenschaftliche Kunde von Russland. Berlin. Arch. Wiss. Kunde Russland. See B–P–H 150/19. HI 51538

Archiv für wissenschaftliche und praktische Thierheilkunde. Berlin. Arch. Wiss. Prakt. Thierheilk. See B–P–H 150/21. HI 51539

Archiva biologica hungarica. Tihany, Hungary. Arch. Biol. Hung. See B–P–H 138/18. HI 51376

Archives d'anatomie microscopique. Paris. Arch. Anat.

Microscop. See B–P–H 137/23. HI 51354

Archives d'anatomie microscopique et de morphologie expérimentale. Paris. Arch. Anat. Microscop. Morphol. Exp. See B–P–H 137/24. HI 51355

Archives belges de dermatologie et de syphiligraphie. Brussels. Arch. Belges Dermatol. Syphiligr. See B–P–H 138/9. HI 51368

Archives of biochemistry. New York, NY. Arch. Biochem. See B–P–H 138/11. HI 51370

Archives of biochemistry and biophysics. New York, NY. Arch. Biochem. Biophys. See B–P–H 138/12. HI 51371

Archives biographiques contemporaines. Paris. Arch. Biogr. Contemp. See B–P–H 138/14. HI 51372

Archives of biological sciences. Belgrade = Arhiv bioloških nauka. Belgrade. Arh. Biol. Nauka. See B–P–H 152/36.

Archives de biologie. Paris. Arch. Biol. (Paris). See B–P–H 138/16. HI 51373

Archives de biologie de la société des sciences et lettres de Varsovie = Archiwum nauk biologhznych towarzystwa natukowego warszawskiego. Warsaw. Arch. Nauk Biol. Towarz. Nauk. Warszawsk. See B–P–H 146/3.

Archives de biologie végétale pure et appliquée. Athens. Vol. 1, 1901. Arch. Biol. Vég. Pure Appl. HI 61929

Archives de botanique. Paris. Arch. Bot. (Paris). See B–P–H 139/6. HI 51382

Archives de botanique, bulletin mensuel. Caen. Arch. Bot. Bull. Mens. See B–P–H 139/8. HI 51384

Archives de botanique, mémoires. Caen. Arch. Bot. Mém. See B–P–H 139/10. HI 51385

Archives botaniques du nord de la France. Revue botanique mensuelle. Lille. Arch. Bot. N. France. See B–P–H 139/11. HI 51386

Archives cliniques de Bordeaux. Bordeaux & Paris. Arch. Clin. Bordeaux. See B–P–H 139/20. HI 51389

Archives of cocoa research. Washington, DC. Vol. 1+, 1981+. Arch. Cocoa Res. HI 61930

Archives of comparative medicine and surgery. New York, NY. Arch. Comp. Med. Surg. See B–P–H 139/21. HI 51390

Archives des découvertes et inventions nouvelles. Paris. Arch. Découv. Invent. Nouv. See B–P–H 140/1. HI 51392

Archives dermato-syphiligraphiques de la clinique de l'hôpital Saint Louis. Paris. Arch. Dermato-Syphiligr. Clin. Hôp. Saint Louis. See B–P–H 140/2. HI 51393

Archives of dermatological research. Berlin, New York. Vol. 253+, 1975+. Arch. Dermatol. Res. Preceded by: Archiv für dermatologische Forschung. HI 61931

Archives entomologiques. Paris. Arch. Entomol. See B–P–H 140/4. HI 51394

Archives of environmental protection = Archiwum ochrony srodowiska. Wroclaw, Warsaw.

Archives de la flore de France et d'Allemagne. Bitche, France. Arch. Fl. France Allemagne. See B–P–H 140/19. HI 51400

Archives de la flore jurassienne. Besançon. Arch. Fl. Jurass. See B–P–H 140/20. HI 51402

Archives franco-belges de chirurgie = Archives provinciales de chirurgie. Brussels. Arch. Prov. Chir. See B–P–H 148/9.

Archives générales de chirurgie. Paris. Arch. Gén. Chir. See B–P–H 140/29. HI 51407

Archives générales de médecine. Paris. Arch. Gén. Méd. See B–P–H 140/30. HI 51408

Archives of gynaecology, obstetrics and paediatrics. New York, NY. Arch. Gyn. Obstet. See B–P–H. 141/11. HI 51417

Archives d'histoire naturelle. Paris. Arch. Hist. Nat. See B–P–H 141/14. HI 51419

Archives de l'institut botanique de l'université de Liège. Liège. Vols. 1-27, 1897-1960/61, 1897-1961. Arch. Inst. Bot. Univ. Liège. Superseded by: Archives de l'institut de botanique de la station scientifique des Hautes-Fagnes et du laboratoire du phytotron (IRSIA). 3-2414-3. HI 61932

Archives de l'institut de botanique de la station scientifique des Hautes-Fagnes et du laboratoire du phytotron (IRSIA) Liège. Vols. 28-34, 1961/62-1968/70, 1962-72. Arch. Inst. Bot. Stat. Sci. Hautes-Fagnes Lab. Phytotron. Preceded by: Archives de l'institut de botanique, université de Liège. HI 61933

Archives de instituto bacteriológico Càmara Pestana. Lisbon. Arch. Inst. Bacteriol. Càmara Pestana. See B–P–H 142/4. HI 51424

Archives des instituts Pasteur de l'Afrique du Nord. Tunis. Arch. Inst. Pasteur Afrique N. See B–P–H 142/10. HI 51425

Archives de l'institut Pasteur d'Algérie. Algiers. Arch. Inst. Pasteur Algérie. See B–P–H 142/11. HI 51426

Archives de l'institut Pasteur de Tunis. Tunis. Arch. Inst. Pasteur Tunis. See B–P–H 142/12. HI 51427

Archives de l'institut des recherches agronomiques de l'Indochine. Saigon. Arch. Inst. Rech. Agron. Indochine. See B–P–H 142/13. HI 51428

Archives internationales de chirurgie. Ghent. Arch. Int. Chir. See B–P–H 142/14. HI 51429

Archives internationales d'histoire des sciences. Paris.

Arch. Int. Hist. Sci. See B–P–H 142/15. HI 51432

Archives internationales de laryngologie, d'otologie, de rhinologie et de bronchooesophagoscopie. Paris. Arch. Int. Laryngol. See B–P–H 142/16. HI 51433

Archives internationales de physiologie = Archives internationales de physiologie et de biochimie. Liége & Brussels. Arch. Int. Physiol. Biochim. See B–P–H 142/19.

Archives internationales de physiologie et de biochimie. Liége & Brussels. Arch. Int. Physiol. Biochim. See B–P–H 142/19. HI 51434

Archives of internal medicine. Chicago, IL. Arch. Intern. Med. See B–P–H 142/21. HI 51435

Archives italiennes de biologie. Revues, resumés, reproductions des travaux scientifiques italiens. Pisa. Arch. Ital. Biol. See B–P–H 143/1. HI 51436

Archives of laryngology. New York, NY. Arch. Laryngol. See B–P–H 144/6. HI 51449

Archives des maladies de l'appareil digestif et de la nutrition. Paris. Arch. Malad. Appar. Digestif Nutr. See B–P–H 144/ 7. HI 51450

Archives de médecine des enfants. Paris. Arch. Méd. Enfants. See B–P–H 144/12. HI 51453

Archives de médecine expérimentale et d'anatomie pathologique. Paris. Arch. Méd. Exp. Anat. Pathol. See B–P–H 144/14. HI 51454

Archives de médecine navale. Paris. Arch. Méd. Navale. See B–P–H 144/15. HI 51455

Archives de médecine navale et coloniale = Archives de médecine navale. Paris. Arch. Méd. Navale. See B–P–H 144/15.

Archives de médecine et de pharmacie navales = Archives de médecine navale. Paris. Arch. Méd. Navale. See B–P–H 144/15.

Archives médico-chirurgicales de l'appareil respiratoire. Paris. Arch. Méd.-Chir. Appareil Resp. See B–P–H 144/18. HI 51456

Archives of microbiology. Berlin, New York, etc. Vol. 95+, 1974+. Arch. Microbiol. Preceded by: Archiv für mikrobiologie. HI 61934

Archives des missions scientifiques et littéraires. Paris. Arch. Missions Sci. Litt. See B–P–H 145/5. HI 51462

Archives de morphologie générale et expérimentale. Paris. Arch. Morphol. Gén. Exp. See B–P–H 145/6. HI 51463

Archives. Musée Teyler. Haarlem. Arch. Mus. Teyler. See B–P–H 145/12. HI 51465

Archives du muséum d'histoire naturelle. Paris. Vols. 1-10, 1839-61. Arch. Mus. Hist. Nat. Preceded by: Nouvelles annales du muséum d'histoire naturelle. Paris. Superseded by: Nouvelles archives du muséum d'histoire naturelle. Paris. 4-3267-3. HI 51464

Archives muséum national d'histoire naturelle. Paris. Vols. 1+, 1926+. Arch. Mus. Natl. Hist. Nat. Preceded by: Nouvelles archives du muséum d'histoire naturelle. Paris. 4-3267-3. HI 74525

Archives of natural history. London. Vol. 10+, 1981+. Arch. Nat. Hist. Preceded by: Journal of the society for the bibliography of natural history. HI 61935

Archives néerlandaises de physiologie de l'homme et des animaux. [Series 3c of: Archives néerlandaises des sciences exactes et naturelles.] The Hague. Vols. 1-28, 1916-47. Arch. Néerl. Physiol. Homme Anim. Superseded by: Acta physiologica et pharmacologica neerlandica. 1-458-2. HI 61936

Archives néerlandaises des sciences exactes et naturelles. The Hague. Arch. Néerl. Sci. Exact. Nat. See B–P–H 146/11. HI 51472

Archives of neurology and psychiatry. London. Arch. Neurol. Psychiat. See B–P–H 146/19. HI 51475

Archives of neurology and psychopathology. Utica, NY. Arch. Eeurol. Psychopathol. See B–P–H 146/20. HI 51476

Archives de oftalmología hispano-americanos. Madrid. Arch. Oftalmol. Hispano-Amer. See B–P–H 146/25. HI 51478

Archives d'ophtalmologie. Paris. Arch. Ophtalmol. See B–P–H 146/26. HI 51479

Archives of oral biology. New York, Elmsford, NY. Vol. 1+, 1959+. Arch. Oral Biol. HI 61937

Archives orientales de médecine et de chirurgie. Paris. Arch. Orient. Méd. Chir. See B–P–H 147/3. HI 51481

Archives of otology. New York, NY. Arch. Otol. See B–P–H 147/5. HI 51482

Archives de parasitologie. Paris. Arch. Parasitol. See B–P–H 147/7. HI 51483

Archives of pathology. Chicago, IL. Arch. Pathol. See B–P–H 147/9. HI 51484

Archives of pathology and laboratory medicine = Archives of pathology. Chicago, IL. Arch. Pathol. See B–P–H 147/9.

Archives of pharmaceutical research. Seoul. Vol. 1+, 1978+. Arch. Pharm. Res. HI 61938

Archives de pharmacie; recueil pratique. Paris. Arch. Pharm. Recueil Prat. See B–P–H 147/19. HI 51491

Archives de physiologie normale et pathologique. Paris. Arch. Physiol. Norm. Pathol. See B–P–H 148/2. HI 51497

Archives de physique biologique. Paris. Arch. Phys. Biol. See B–P–H 147/20. HI 51493

Archives de physique biologique et de chimie-physique de corps organisés. Paris. Arch. Phys. Biol. Chim.-Phys. Corps Organ. See B–P–H 147/21. HI 51494

Archives de plasmologie générale. Paris. Arch. Plasmol. Gén. See B–P–H 148/3. HI 51498

Archives portugaises des sciences biologiques. Lisbon. Arch. Portug. Sci. Biol. See B–P–H 148/4. HI 51499

Archives provinciales de chirurgie. Brussels. Arch. Prov. Chir. See B–P–H 148/9. HI 51502

Archives des recherches agronomiques au Cambodge, au Laos et au Vietnam. Saigon. Arch. Rech. Agron. Cambodge Laos Vietnam. See B–P–H 148/10. HI 51503

Archives roumaines de pathologie expérimentale et de microbiologie. Bucharest. Arch. Roumaines Pathol. Exp. Microbiol. See B–P–H 148/15. HI 51505

Archives of rubber cultivation. Bogor, Indonesia. Arch. Rubber Cultivation. See B–P–H 148/16. HI 51506

Archives russes de protistologie = Russkii arkhiv protistologii. Moscow.

Archives of science and transactions of the Orleans county society of natural sciences. Newport, VT. Arch. Sci. & Trans. Orleans County Soc. Nat. Sci. See B–P–H 149/8. HI 51515

Archives des sciences. Geneva. Vol. 1+ [also numbered vol. 153+], 1948+. Arch. Sci. Preceded by: Bibliothèque universelle: revue suisse (et étrangère); Archives des sciences physiques et naturelles. 1-456-1. HI 51510

Archives des sciences biologiques. Belgrade = Arhiv bioloških nauka. Belgrade. Arh. Biol. Nauka. See B–P–H 152/36.

Archives des sciences biologiques. Moscow. Arch. Sci. Biol. (Moscow). See B–P–H 149/3. HI 51511

Archives des sciences biologiques. Moscow = Arkhiv Biologicheskikh Nauk. Moscow.

Archives des sciences physiques et naturelles. Geneva & Paris = Bibliothèque universelle de Genève; Archives des sciences physiques et naturelles. Geneva & Paris.

Archives des sciences physiques et naturelles. Geneva & Paris = Bibliothèque universelle: revue suisse (et étrangère); Archives des sciences physiques et naturelles. Geneva & Paris.

Archives des sciences physiques et naturelles. Geneva & Paris = Supplément à la bibliothèque universelle de Genève; Archives des sciences physiques et naturelles. Geneva & Paris.

Archives des sciences physiologiques. Paris. Arch. Sci. Physiol. See B–P–H 149/7. HI 51514

Archives slaves de biologie. Paris. Arch. Slaves Biol. See B–P–H 149/12. HI 51518

Archives de la société russe de protistologie = Arkhiv Russkogo protistologicheskogo obshchestva.

Archives statistiques du ministére des travaux publics de l'agriculture et du commerce. Paris. Arch. Statist. Minist. Trav. Publics Agric. Commerce. See B–P–H 149/19. HI 51520

Archives of toxicology. Berlin. Vol. 32+, 1974+. Arch. Toxicol. Preceded by: Archiv für Toxicologie. HI 75208

Archives trimestrielles de l'institut Grand-Ducal de Luxembourg; section des sciences naturelles, physiques et mathématiques. Luxembourg. Arch. Trimestrielles Inst. Grand-Ducal Luxembourg, Sect. Sci. Nat. See B–P–H 150/8. HI 51531

Archives de l'union médicale balkanique. Bucharest. Vol. 1+, 1963+. Arch. Union Méd. Balkan. HI 61939

Archives of virology. New York. Vol. 47+, 1975+. Arch. Virol. Preceded by: Archiv für gesamte Virusforschung. HI 61940

Archives des voyages. Paris. Arch. Voyages. See B–P–H 150/15. HI 51535

Archives de zoologie expérimentale et générale. Paris. Arch. Zool. Exp. Gén. See B–P–H 151/1. HI 51540

Archivio per l'antropologia e l'etnologia. Florence. Arch. Antropol. Etnol. See B–P–H 137/29. HI 51358

Archivio ed atti. Società italiana di chirurgia. Rome. Arch. & Atti Soc. Ital. Chir. See B–P–H 137/20. HI 51350

Archivio botanico. Forlì, Italy. Vols. 11-31, 1935-55. Arch. Bot. (Forlì). Preceded by: Archivio botanico per la sistematica, fitogeografia e genetica [etc.]. Superseded by: Archivio botanico e biogeografico italiano. 1-463-3. HI 51380

Archivio botanico e biogeografico italiano. Forlì, Italy. Vols. 32-63, 1956-87? Arch. Bot. Biogeogr. Ital. Preceded by: Archivio botanico. Forlì. Superseded by: Archivio botanico italiano. 1-463-3. HI 51383

Archivio botanico italiano. Forlì, Italy. Vol. 64+, 1988+. Arch. Bot. Ital. Preceded by: Archivio botanico e biogeografico italiano. HI 74811

Archivio botanico per la sistematica, fitogeografic e genetica (storia e sperimentale) e Bulletino dell'istituto botanico della r. università di Modena. Forlí, Italy. Arch. Bot. Sist. See B–P–H 139/15. HI 51387

Archivio di erboristeria e delle piante officinale utili alimurgiche industriali velenose, fitotecnica, farmacognostica, cure vegetali. Vercelli. No. 1+, 1945+. Arch. Erborist. Piante Off. HI 61941

Archivio di farmacologia sperimentale e scienze affini. Rome. Arch. Farmacol. Sperim. Sci. Affini. See B–P–H 140/17. HI 51397

Archivio italiano di chirurgia. Bologna. Arch. Ital. Chir. See B–P–H 143/2. HI 51437

Archivio italiano di dermatologia, sifilografia i venereologia. Bologna. Arch. Ital. Dermatol. See B–P–H 143.7. HI 51438

Archivio italiano di laringologia. Naples. Arch. Ital. Laringol. See B–P–H 143/8. HI 51439

Archivio italiano per le malattie nervose. Milan. Arch. Ital. Malatt. Nerv. See B–P–H 143/10. HI 51440

Archivio italiano di scienze farmacologiche. Milan. Arch. Ital. Sci. Farmacol. See B–P–H 143/11. HI 51441

Archivio italiano di scienze mediche coloniali e di parassitologia = Archivio italiano di scienze mediche tropicali e di parassitologia. Tripoli [Libya] & Rome. Arch. Ital. Sci. Med. Trop. Parassitol. See B–P–H 143/14.

Archivni zpravy. Prague. Vol. 1+, 1970+. Arch. Zpravy. HI 61942

Archivio italiano di scienze mediche tropicali e di parassitologia. Tripoli [Libya] & Rome. Arch. Ital. Sci. Med. Trop. Parassitol. See B–P–H 143/14. HI 51442

Archivio italiano di urologia. Bologna. Arch. Ital. Urol. See B–P–H 143/15. HI 51443

Archivio di oceanografia e limnologia. Rome. Arch. Oceanogr. Limnol. See B–P–H 146/21. HI 51477

Archivio di scienze biologiche. Naples. Arch. Sci. Biol. (Naples). See B–P–H 149/4. HI 51512

Archivio di storia della scienza. Rome. Arch. Storia Sci. See B–P–H 149/20. HI 51521

Archivo de biología vegetal teórica y aplicada. Buenos Aires. Arch. Biol. Veg. Teór. Aplicada. See B–P–H 138/22. HI 51377

Archivio triennale del laboratorio di botanica crittogamica. Milan. Arch. Triennale Lab. Bot. Crittog. See B–P–H 150/6. HI 51529

Archivo de ciencias biologicas y naturales teoricas y aplicadas. Buenos Aires. Vol. 1+, 1956+. Arch. Ci. Biol. Nat. Teor. Aplicadas. HI 61943

Archivo fitotécnico del Uruguay. Montevideo. Arch. Fitotécn. Uruguay. See B–P–H 140/18. HI 51398

Archivo; revista de ciencias historicas. Denia, Spain. Arch. Revista Ci. Hist. See B–P–H 148/12. HI 51504

Archivos argentinos de neurologia. Buenos Aires. Arch. Argent. Neurol. See B–P–H 138/3. HI 51363

Archivos de la asociación peruana para el progreso de la ciencia. Lima. Arch. Asoc. Peruana Progr. Ci. See B–P–H 138/4. HI 51364

Archivos de biologia. São Paulo. Arch. Biol. (São Paulo). See B–P–H 138/17. HI 51375

Archivos de biología andina. Lima. Vol. 7+, 1977+. Arch. Biol. Andina. Preceded by: Archivos del instituto de biologia andina. HI 61944

Archivos de bioquímica, química y farmacía. Tucumán, Argentina. Arch. Bioquím. See B–P–H 139/1. HI 51379

Archivos de botânica do estado de São Paulo = Archivos de botânica do São Paulo. São Paulo. and Arquivos de botânica do estado de São Paulo. São Paulo.

Archivos de botânica do São Paulo. São Paulo. 1925/27. Arch. Bot. São Paulo. Preceded by: Anexos das memórias do instituto de Butantan. Secção de botânica. Superseded by: Arquivos de botânica do estado de São Paulo. 1-489-1. HI 61945

Archivos de ciências do mar = Arquivos de ciências do mar. Fortaleza.

Archivos de la conferencia de médicos del hospital Ramos-Mejia. Buenos Aires. Arch. Conf. Méd. Hosp. Ramos-Mejia. See B–P–H 139/22. HI 51391

Archivos de la escuela de farmacia de la facultad de ciencias medicas de Cordoba. Cordoba. 1936-39. Arch. Esc. Farm. Fac. Ci. Med. Cordoba. HI 61946

Archivos da escuela superior da agricultura e medicina veterinaria, Niteroi. Rio de Janeiro. Vols. 1-10(1), 1917-33. Arch. Esc. Super. Agric. Med. Veterin. Niteroi. HI 61947

Archivos da estação de biologia marinha da universidade do Ceará = Arquivos da estação de biologia marinha da universidade do Ceará. Fortaleza.

Archivos de farmacía y bioquímica de Tucumán. Tucumán, Argentina. Arch. Farm. Bioquím. Tucumán. See B–P–H 140/15. HI 51395

Archivos. Hospital Rosales. San Salvador, Salvador. Arch. Hosp. Rosales. See B–P–H 141/15. HI 51420

Archivos, instituto de aclimatación, consejo superior de investigaciones cientificas. Almeria. Vols. 1-20, 1953-75. Arch. Inst. Aclim. Cons. Super. Invest. Ci. HI 61948

Archivos do instituto biologico. São Paulo. Vols. 5-8, 1934-37. Arch. Inst. Biol. (São Paulo). Preceded by: Archivos do instituto biologico de defesa agricola e animal. Superseded by: Arquivos do instituto biologico. São Paulo. 5-3779-3. HI 61949

Archivos, instituto de biologia andina. Lima. Vols. 1-6, 1965-73. Arch. Inst. Biol. Andina. Superseded by: Archivos de biología andina. HI 61950

Archivos do instituto biologico de defesa agricola e animal. São Paulo. Vols. 1-4, 1928-33. Arch. Inst. Biol. Defesa Agric. Superseded by: Archivos do instituto biologico. São Paulo. 5-3779-3. HI 61951

Archivos do jardim botânico do Rio de Janeiro. Rio de Janeiro. Arch. Jar. Bot. Rio de Janeiro. See B–P–H 143/16. HI 51444

Archivos mineiros de dermato-syphiligraphia. Belo Horizonte, Brazil. Arch. Mineir. Dermato-Syphiligr. See B–P–H 144/24. HI 51460

Archivos do museo paranaense = Arquivos do museo paranaense. Curitiba.

Archivos de la sociedad de biología de Montevideo. Montevideo. Vols. 1-27, 1929-69. Arch. Soc. Biol. Montevideo. 5-3905-2. HI 61952

Archivos de la sociedad de biología de Montevideo. Suplemento. Montevideo. Arch. Soc. Biol. Montevideo Supl. See B–P–H 149/14. HI 51519

Archivos uruguayos de medicina, cirugía y especialidades. Montevideo. Arch. Uruguayos Med. See B–P–H 150/10. HI 51532

Archivum balatonicum. Tihany, Hungary. Arch. Balaton. See B–P–H 138/7. HI 51367

Archivum melitense. Valletta, Malta. Arch. Melit. See B–P–H 144/19. HI 51457

Archivum societatis zoologicae botanicae fennicae Vanamo = Suomalaisen elain-ja kasvitieteellisen seuran vanamon tiedonannot. Helsinki.

Archiwum hydrobiologji i rybactwa. Suwalki & Warsaw. Arch. Hydrobiol. Rybactwa. See B–P–H 141/19. HI 51423

Archiwum nauk biologhznych towarzystwa natukowego warszawskiego. Warsaw. Arch. Nauk Biol. Towarz. Nauk. Warszawsk. See B–P–H 146/3. HI 51470

Archiwum naukowe. Towarzystwo dla popierania nauki polskiej. Dział 2, Mat.-Przyrod. Lvov, Galicia [Ukrainian S S R]. Arch. Nauk., Dział 2, Mat.-Przyr. See B–P–H 146/4. HI 51471

Archiwum ochrony srodowiska. Wroclaw, Warsaw. 1977+. Arch. Ochr. Srodow. HI 61953

Archiwum towarzystwa naukowego we Lwowie. Dział 3, matematyczno-przyrodniczy. Lvov, Galicia [Ukrainian S S R]. Arch. Towarz. Nauk. we Lwowie, Dział 3. Mat.-Przyr. See B–P–H 150/5. HI 51528

Arctic; journal of the arctic institute of north America. Montreal, Calgary & Washington, DC. Vol. 1+, 1948+. Arctic. HI 61954

Arctic and alpine research. Boulder, CO. Vol. 1+, 1969+. Arctic Alpine Res. HI 61955

Arctic bibliography. Washington, DC. Artic Bibliogr. See B–P–H 152/32. HI 51544

Arctic bulletin. Washington, DC. Vol. 1+, 1973+. Arctic Bull. HI 61956

Ardenne et famenne; art - archéologie - histoire - folklore. Remicourt, Brussels. Vol. 1+, 1958+. Ardenne & Famenne. HI 61957

Ardenne et Gaume. Monographie = Bulletin mensuel d'information, Ardenne et Gaume. Brussels.

Arealy derev'ev i kustarnikov S S S R. Leningrad. 1977+. Arealy Derev. Kustarnik. S.S.S.R. HI 61958

Arealy rastenii flory S S S R. Leningrad. No. 1+, 1965+. Arealy Rast. Fl. S.S.S.R. HI 61959

Arecanut journal. Calicut. Vol. 1+, 1950+. Arecanut J. HI 61960

Argumenta palaeobotanica. Lehre. No. 1+, 1966+. Argum. Palaeobot. HI 61961

Arhiv bioloških nauka. Belgrade. Arh. Biol. Nauka. See B–P–H 152/36. HI 51545

Arhiv Biologičeskih Nauk. Moscow = Arkhiv Biologicheskikh Nauk. Moscow.

Arhiv istorii nauki i tehniki. Moscow & Leningrad = Trudy instituta istorii nauki i tekhniki. Seriya 1, arkhiv istorii nauki i tekhniki. Moscow & Leningrad.

Arhiv ministarstva poljoprivredne i šumarstva. Belgrade. 1934-40. Arh. Minist. Poljopr. Šumarstvo. Superseded by: Arhiv za poljoprivredne nauke i tehniku. HI 61962

Arhiv za poljoprivredne nauke. Belgrade. Vol. 4+, 1948+. Arh. Poljopr. Nauke. Preceded by: Arhiv za poljoprivredne nauke i tehniku. 1-475-3. HI 61963

Arhiv za poljoprivredne nauke i tehniku. Belgrade. Vols. 1-3, 1946-48. Arh. Poljopr. Nauke Tehn. Preceded by: Arhiv ministarstva poljoprivrede i šumarstva. Superseded by: Arhiv za poljoprivredne nauke. 1-475-3. HI 61964

Arhiv Russkogo protistologičeskogo obščestva = Arkhiv Russkogo protistologicheskogo obshchestva.

Arid land plant resources. Lubbock, TX. Vol. 1+, 197?+. Arid. Land Pl. Resources. HI 61965

Arid lands abstracts. Tucson, AZ, Farnham Royal. Vols. 1-2, 1972-81. Arid Lands Abstr. Superseded by: Arid lands development abstracts. HI 61966

Arid lands development abstracts. Farnham Royal. Vol. 3+, 1982+. Arid Lands Developm. Abstr. Preceded by: Arid lands abstracts. HI 61967

Arid lands newsletter. Tucson, AZ. Vol. 1+, 1975+. Arid Lands Newslett. HI 61968

Arid lands research newsletter. Washington, DC. Nos. 1-40, 1959-71. Arid Lands Res. Newslett. HI 61969

Arid lands resource information paper. Tucson, AZ. No. 1+, 1972+. Arid Lands Resource Newslett. HI 61970

Arid zone; news about Unesco's major project on scientific research on arid lands. Paris. Nos. 1-26, 1958-64. Arid Zone. Preceded by: Arid zone research. Superseded by: Nature and resources. HI 61971

Arid zone programme. Paris. 1952-53. Arid Zone Programme. Superseded by: Arid zone research. HI 61972

Arid zone research. Paris. 1955-57. Arid Zone Res. Preceded by: Arid zone programme. Superseded by: Arid zone. HI 61973

Aril society international yearbook. Tujunga, CA. Aril Soc. Int. Yearb. See B–P–H 152/39. HI 51546

Arizona agricultural experiment station. Bulletin. Tucson, AZ. Arizona Agric. Exp. Sta. Bull. See B–P–H 152/43. HI 51547

Arizona agricultural experiment station, university of Arizona, bulletin. Tucson, AZ. Arizona Agric. Exp. Sta. Univ. Arizona Bull. See B–P–H 152/44. HI 51548

Arizona forestry notes. Flagstaff, AZ. Vol. 1+, 1966+. Arizona Forest. Notes. HI 61974

Arizona highways; the friendly journal of life and travel in the old west. Phoenix, AZ. Vols. 1-2, 1921-22; n.s. vol.1+, 1925+. Arizona Highways. 1-478-1. HI 61975

Arkansas academy of science bulletin = Bulletin, Arkansas academy of science. Fayetteville, AR.

Arkansas farm research. Report series. University of Arkansas agricultural experiment station. Fayetteville, AR. Arkansas Farm Res., Rep. Ser. See B–P–H 153/15. HI 51552

Arkansas farm research. University of Arkansas agricultural experiment station. Fayetteville, AR. Arkansas Farm Res. See B–P–H 153/14. HI 51551

Arkansas gardener. Batesville, AR. Arkansas Gard. See B–P–H 153/16. HI 51553

Arkansas industrial university. Agricultural experiment station. Bulletin. Fayetteville, AR. Arkansas Industr. Univ. Agric. Exp. Sta. Bull. See B–P–H 153/17. HI 51554

Arkansas naturalist. Fayetteville, AR. Vol. 1+, 1983+. Arkansas Naturalist. HI 61976

Arkansas rice news. Stuttgart, AR. Arkansas RIce News. See B–P–H 153/18. HI 51555

Arkhiv biologicheskikh nauk. Moscow. Vols. 1-64, 1892-1941. Arkh. Biol. Nauk. 1-482-2. HI 61977

Arkhiv Russkogo protistologicheskogo obshchestva. Petrograd, Moscow. Vols. 1-2, 1922-23. Arkh. Russk. Protistol. Obshch. Superseded by: Russkii arkhiv protistologii. 4-3743-2. HI 61978

Arkiv för botanik utgivet av k. svenska vetenskapsakademien. Stockholm. Ark. Bot. See B–P–H 153/5. HI 51549

Arkif för landtmän och trägårds-odlare. Stockholm. Ark. Landtm. Träg.-Odlare. See B–P–H 153/9. HI 51550

Armchair naturalist. Poynton, Cheshire. Nos. 1-?, 1973-74. Armchair Naturalist. HI 61979

Arnold arboretum newsletter. Jamaica Plain, NY. 1972(Jan.)-1977(Sept.). Arnold Arbor. Newslett. Preceded by: Arnold arboretum staff notes and news [not published?]. HI 61980

Arnoldia. Jamaica Plain, MA. Arnoldia (Jamaica Plain). See B–P–H 153/27. HI 51556

Arnoldia; series of miscellaneous publications, national museum of Southern Rhodesia. Salisbury, Rhodesia. Vols. 1-8(34), 1964-79. Arnoldia (Salisbury). Superseded by: Arnoldia Zimbabwe. HI 61981

Arnoldia Zimbabwe. Series of miscellaneous publications by the national museum and monuments. Harare. Vol. 8(35)+, 1979+. Arnoldia Zimbabwe. Preceded by: Arnoldia. Salisbury. HI 61982

Arnoldia Zimbabwe Rhodesia = Arnoldia Zimbabwe. Harare.

Aroideana; journal of the international aroid society. Sarasota, FL. Vol. 1+, 1978+. Aroideana. HI 61983

Aromatics = Perfumers' journal and essential oil recorder. New York, NY. Perfumers' J. See B–P–H 701/7.

Arquipélago. Série ciências da natureza; revista da universidade dos Açores. Ponta Delgada. Vol. 1+, 1980+. Arquipélago, Sér. Ci. Nat. HI 61984

Arquivos de biologia. São Paulo. Arq. Biol. See B–P–H 153/30. HI 51557

Arquivos de biologia e tecnologia. Curitiba, Brazil. Arq. Biol. Tecnol. See B–P–H 153/31. HI 51558

Arquivos de botânica do estado de São Paulo. São Paulo. Vols. 1-4, 1938-69. Arq. Bot. Estado São Paulo. Preceded by: Archivos de Botânica do São Paulo. Superseded by: Hoehnia. 1-489-1. HI 61985

Arquivos de bromatologia. Rio de Janeiro. Arq. Bromatol. See B–P–H 154/3. HI 51559

Arquivos de ciências do mar. Fortaleza. Vol. 9+, 1969+. Arq. Ci. Mar. Preceded by: Arquivos da estação de biologia marinha da universidade federal do Ceara. HI 61986

Arquivos de cirurgia clinical e experimental. São Paulo. Arq. Cirurgia Clin. Exp. See B–P–H 154/5. HI 51560

Arquivos da estação de biologia marinha da universidade do Ceará. Fortaleza. Vols. 1-8(2), 1961-68. Arq. Estaç. Biol. Mar. Univ. Ceará. Superseded by: Arquivos de ciências do mar. HI 61987

Arquivos de instituto bacteriológico Càmara Pestana. Lisbon. Arq. Inst. Bacteriol. Càmara Pestana. See B–P–H 154/6. HI 51561

Arquivos do instituto de biologia vegetal. Rio de Janeiro. Arq. Inst. Biol. Veg. See B–P–H 154/8. HI 51562

Arquivos do instituto biologico. São Paulo. Vol. 9+, 1938+. Arq. Inst. Biol. (São Paulo). Preceded by: Archivos do instituto biologico. São Paulo. 5-3779-3. HI 61988

Arquivos do instituto de pesquisas agronómicas. [Universidade rural de Pernambuco.] Recife, Brazil. Arq. Inst. Pesq. Agron. See B–P–H 154/9. HI 51563

Arquivos do jardim botânico do Rio de Janeiro = Archivos do jardim botânico do Rio de Janeiro. Rio de Janeiro. Arch. Jar. Bot. Rio de Janeiro. See B–P–H 143/16.

Arquivos do museo paranaense. Curitiba. Vols. 1-10, 1941-54. Arq. Mus. Paranaense. 2-1241-3. HI 61989

Arquivos do museu nacional do Rio de Janeiro. Rio de Janeiro. Arq. Mus. Nac. Rio de Janeiro. See B–P–H 154/11. HI 51564

Arquivos de patologia. Lisbon. Arq. Patol. See B–P–H 154/14. HI 51565

Arquivos rio-grandenses de medicina. Pôrto Alegre, Brazil. Arq. Rio-Grandenses Med. See B–P–H 154/16. HI 51566

Arquivos do servico florestal. Rio de Janeiro. Arq. Serv. Florest. See B–P–H 154/18. HI 51567

Arquivos da universidade federal rural do Rio de Janeiro. Itaguaí. Vol. 1+, 1971+. Arq. Univ. Fed. Rural Rio de Janeiro. HI 61990

Arquivos. Universidade do Recife. Instituto da terra, divisão de ciências geograficas. Recife, Brazil. Arq. Univ. Recife Inst. Terra Div. Ci. Geogr. See B–P–H 154/19. HI 51568

Arrabona. A györi múzeum évkönyve. Györ, Hungary. Arrabona. See B–P–H 154/25.

Arran naturalist. Brodick. 1978+. Arran Naturalist. HI 61991

Arroz. Bogota. Vol. 1+, 1952+. Arroz (Bogota). HI 51569

Arroz. Lima. Vol. 1+, 1966+. Arroz (Lima). HI 61992

Arroz. Santo Domingo. Vol. 1+, 1971+. Arroz (Santo Domingo). HI 61993

Ars pharmaceutica; revista de la facultad de farmacia, universidad de Granada. Granada. Vol. 1+, 1960+. Ars Pharm. HI 61994

Årsberättelse om botaniska arbeten och upptäckter. Stockholm. 1828-52. Årsberätt. Bot. Arbeten Upptäckter. Preceded by: Årsberättelse om framstegen uti botanik. Superseded by: Berättelse om botaniska arbeten och upptäckter. 5-4131-1. HI 50244

Årsberättelse om framstegen uti botanik. Stockholm. 1826-27. Årsberätt. Framsteg. Bot. Preceded by: Oversigt af botaniska arbeten och upptäckter. Superseded by: Årsberättelse om botaniska arbeten och upptäckter. HI 61995

Årsberättelse, svenska läkare-sällskapet = Svenska läkare-sällskapet. Årsberättelse. Stockholm. Svenska Läkare-Sällsk. Årsberätt. See B–P–H 861/13.

Årsberättelser om vetenskapernas framsteg, afgifne af kongl. vetenskaps-academiens embetsmän. Stockholm. Årsberätt. Vetensk. Framsteg Kongl. Vetensk.-Acad. Embetsmän. See B–P–H 155/2. HI 51570

Årsberetning fra höiere landbrugsskolw. Oslo. 1862/63-1896/97. Årsberetn. Höiere Landbrugsskole. Superseded by: Årsberetning fra norges landbrukshøgskole. 4-3075-2. HI 61996

Årsberetning, kongelige norske videnskabers selskabs museet = Kongelige norske videnskabers selskabs museet. Årsberetning. Trondheim.

Årsberetning fra Norges landbrukshøgskole; [For some vols. "landbrukshøgskole" as "landbrukshøiskole"] Oslo. 1879/98-1929/30. Årsberetn. Norg. Landbrukshøgskole. Preceded by: Åsberetning fra hoïere landbrugsskole. Superseded by: Åksmelding fra norges landbrukshøgskole. 4-3075-2. HI 61997

Årsberetning, norsk institutt for skogforskning. Ås. 1972+, 1973+. Årsberetn. Norsk Inst. Skogsforskning. Preceded by: Norske skogsforsøksvesens virksomhet. HI 61998

Årsberetning fra statens plantetilsyn vedrørende fropatologisk kontrol. Copenhagen. Vols. 1-3-?, 1948/51-52/53-?, 1951-53-? Årsberetn. Statens Plantetilsyn Vedrør. Kontrol. Superseded by: Beretning, statens plantetilsyn. HI 61999

Årsberetning, universitet i Bergen, botanisk museum. Bergen. 1977+. Årsberetn. Univ. Bergen, Bot. Mus. HI 62000

Årsberetning vedrørende fropatologisk kontrol = Årsberetning fra statens plantetilsyn vedrørende fropatologisk kontrol. Copenhagen.

Årsbok, finska mosskulturföreningen = Finska mosskulturföreningen, årsbok. Helsinki. Finska Mosskulturfören. Årsbok. See B–P–H 371/13.

Årsbok, kungliga vetenskaps- ock vitterhets-samhället. Gothenburg. 1967+, 1967+. Årsbok Kungl. Vetensk. Vitterh.-Samhället. HI 62001

Årsbok, Malmö museum. Malmö. Vol. 1+, 1970+. Årsbok Malmö Mus. HI 62002

Årsbok, statens naturvårdsverk. Solna. 1969+, 1970+. Årsbok Statens Naturvårdsverk. HI 62003

Årsbok-vuosikirja. Societas scientiarum fennica. Helsinki. Vol. 1+, 1922/23+. Årsbok-Vuosik. Soc. Sci. Fenn. 2-1567-2. HI 62004

Årsmelding, botanisk museum og botanisk hage. Bergen. 1971+. Årsmeld. Bot. Mus. Bot. Hage. HI 62005

Årsmelding ... fra det akademiske kollegium. Bergen. 1949/50+, 1951+. Årsmeld. Akad. Koll. Previously formed part of: Universitet i Bergen årbok. HI 62006

Årsmelding fra Norges landbrukshøgskole. Oslo. 1931. Årsmeld. Norges Landbrukshøgskole. Preceded by: Årsberetning fra norges landbrukshøgskole. 4-3075-2. HI 62007

Årsmelding, skogavdelingen, landbruksdepartementet. Norway. Oslo. 1974+. Årsmeld. Skogavd. Landbr. Dep. Norway. Preceded by: Årsmelding, skogdirektoralet. Norway. HI 62008

Årsmelding, skogdirektoratet. Norway. Oslo. 19??-73, 19??-75. Årsmeld. Skogdirektorat. Norway. Superseded by: Årsmelding, skogavdelingen, landbruksdepartementet. Norway. HI 62009

Årsskrift, fredericksborgmuseet. Ny Carlsbergfondet. Copenhagen. 1977+. Årsskr. Fredericksborgmus. HI 62010

Årsskrift, Göteborgs svampklubb. Gothenberg. 1971-81. Årsskr. Göteborgs Svampklubb. Superseded by: Windhalia. HI 62011

Årsskrift. K. vetenskaps-societeten i Upsala. Uppsala. Årsskr. Kongl. Vetensk.-Soc. Upsala. See B–P–H 155/5. HI 51571

Årsskrift universitet. Uppsala. Årsskr. Univ. See B–P–H 155/6. HI 51572

Art canadien de la nature = Canadian nature art. Ottawa.

Artedi. Amsterdam. ?-1975+. Artedi. HI 62012

Arte y la ciencia. Mexico City. Arte & Ci. See B–P–H 155/7. HI 51573

Artist. London. Vol. 1+, 1931+. Artist. 1-499-2. HI 74812

Arve-Léman nature. Bonneville. Vols. 11-15, 197?-??. Arve-Léman Nat. HI 62013

Arve-Léman-Savoie nature. Bonneville. Vol. 16+, 198?+. Arve-Léman-Savoie Nat. HI 62014

Arve nature. Bonneville. Vols. 1-10, 1972-? Arve Nat. HI 62015

Arvernia biologica. Clermont-Ferrand. Nos. 1-22?, 1930-52. Arvernaria Biol. Superseded by: Arvernaria biologica. Botanique. HI 62016

Arvernia biologica. Botanique; recueil des travaux des laboratoire de botanique de la faculté des sciences de Clermont-Ferrand et de la station de Besse. Clermont-Ferrand. N.s. no. 1+, 1954+. Arvernaria Biol., Bot. Preceded by: Arvernaria biologica. HI 62017

Arxius de l'institut de ciéncias. Barcelona. Vols. 1-12, 1911-24 [suspended 1924-47]. Arxius Inst. Ci. Superseded by: Arxius de la secció de ciéncias; institut d'estudis catalans. 3-1995-3. HI 62018

Arxius de la secció de ciéncias; institut d'estudis catalans. Barcelona. Vols. 13-?, 1947-67. Arxius Secc. Ci. Inst. Estud. Catalans. Preceded by: Arxius de l'institut de ciéncias. HI 62019

Arya Swapatra. Bombay. Arya Swapatra. See B–P–H 155/13. HI 51574

Arzneykundige Annalen. Copenhagen. Arzneyk. Ann. See B–P–H 155/14. HI 51575

Asa Gray bulletin; a botanical quarterly published in the interests of the Gray memorial botanical association, the botanical gardeners association of the university of Michigan botanical club. Ann Arabor, MI. Asa Gray Bull. See B–P–H 155/17. HI 51576

Asarho, akademijai fanhoi R S S toğikiston = Trudy, akademiya nauk tadzhikskoi S S R. Stalinabad [= Dushanbe].

ASB bulletin. [Association of the southeastern biologists.] Chapel Hill, NC = A S B bulletin. Chapel Hill, NC.

Asclepiadaceae; official organ of the asclepiadaceae society. Epsom, etc. Nos. 1-22, 1974-81. Asclepiadaceae. Superseded by: Asklepios. HI 62020

Asean orchid newsletter. Kuala Lumpur. ?-1986+. Asean Orchid Newslett. HI 74513

Asher's guide to botanical periodicals. Amsterdam. Vols. 1-2, 1973-75. Asher's Guide Bot. Period. HI 62021

Ashingtonia. Ashington. Vol. 1+, 1973+. Ashingtonia. HI 62022

Asia. New York, NY. Asia. See B–P–H 155/23. HI 51577

Asia major; journal devoted to the study of the languages, arts and civilization of the far east and central Asia. London. Asia Major. See 155/25. HI 51579

Asian aquaculture. Tisbauan, Iloilo, Philippines. Vol. 1+, 1978+. Asian Aquac. HI 62023

Asian ecological society newsletter = Newsletter, asian ecological society. Taichung.

Asian environmental review. Taichung. Vol. 1(1), 1977. Asian Environm. Rev. Superseded by: Journal of asian ecology. HI 62024

Asian forestry science and technology. [Yalin Keji.] Vol. 1+, 1985+. Asian Forest. Sci. Technol. HI 75287

Asian marine biology. Hong Kong. Vol. 1+, 1984+. Asian Mar. Biol. HI 62025

Asian orchids and ornamentals. Taytay, Rizal. Vol. 1+,

1984+. Asian Orchids Ornam. HI 62026

Asian-Pacific weed science society conference, proceedings = Proceedings of the conference, asian-Pacific weed science society.

Asian perspectives; bulletin of the far eastern prehistory association [later Journal of archaeology and prehistory of Asia and the Pacific]. Tucson, AZ., Honolulu, HI. Vol. 1+, 1957+. Asian Perspect. HI 62027

Asiatic researches, or transactions of the society. Calcutta. Asiat. Res. See B–P–H 155/27. HI 51580

Askläpieion. Berlin = Asklepieion. Berlin.

Asklepieion. Allgemeines medicinisch-chirurgisches Wochenblatt für alle Thiele der Heilkunde und ihre Hülfswissenschaften. Berlin. Vols. 1- 1811-12; n.s. vols. 1-4, 1812. Asklepieion. 1-507-1. HI 52868

Asklepios. Burgess Hill. Vol. 23+, 1981+. Asklepios. Preceded by: Asclepiadiaceae. HI 62028

Aslib book list. London = A S L I B book list. London.

Aslib proceedings. London = A S L I B proceedings. London.

Asoprole. Caracas. Asoprole. See B–P–H 156/4. HI 51581

Asparagus research newsletter. Palmerston North, N.Z. Vol. 1+, 1983+. Asparagus Res. Newslett. HI 62029

Aspas de la patata. Madrid. No. 1+, 1951+. Aspas Patata. HI 62030

ASPB news. Lexington, KY = A S P B news. Lexington, KY.

Aspects of plant sciences. New Delhi. Vol. 1+, 1976+. Aspects Pl. Sci. HI 62031

Aspergillus news letter. Sheffield. No. 1+, 1960+. Aspergillus News Lett. HI 62032

Asplundh tree. Jenkintown, PA. ?-1957+. Asplund Tree. HI 62033

Assam department of agriculture. Fruit series. Shillong, India. Assam Dept. Agric., Fruit Ser. See B–P–H 156/8. HI 51582

Assemblée générale, société des pharmaciens de la Côte d'Or = Bulletin des pharmaciens de la Côte d'Or. Dijon.

Associates N A L today. Beltsville, MD. N.s. vols. 1-3, 1976-78. Assoc. N. A. L. Today. Superseded by: Journal of N A L associates. HI 62034

Association bulletin, american economic association = American economic review. Ithaca, NY.

Association pour l'étude taxonomique de la flore d'Afrique tropicale bulletin. Brussels, Geneva. No. 1+, 1953+. Assoc. Étude Taxon. Fl. Afrique Trop. Bull. 1-252-1. HI 62035

Association française pour l'avancement des sciences; conférences faites en ... Paris. Assoc. Franç. Avancem. Sci. Conf. See B–P–H 156/14. HI 51583

Association nationale des professeurs de biologie de Belgique, année = Année, association nationale des professeurs de biologie de Belgique. Brussels.

Association of Pacific systematists newsletter = Newsletter, association of Pacific systematists. Honolulu, HI.

Association Strasbourgeoise des amis de l'histoire naturelle. Compte rendu de la séance générale. Strasbourg. Assoc. Strasbourg. Amis Hist. Nat. Compt. Rend. Séance Gén. See B–P–H 156/22. HI 51584

Association for tropical biology bulletin. Port of Spain. Nos. 1-8, 1962-67. Assoc. Trop. Biol. Bull. Superseded by: Biotropica. HI 62036

Association for tropical biology newsletter. Washington, DC., etc. No. 1+, 1966+. Assoc. Trop. Biol. Newslett. HI 62037

Astarte; short papers published by the zoological department, Tromsø museum. Tromsø. Vols. 1-12(1), 1951-79. Astarte. HI 62038

Astrologer; composed of twelve monthly parts treating on the science of astrology, medical botany, etc. London. Astrologer. See B–P–H 156/24. HI 51585

Asturnatura; revista asturiana de ciencias naturales. Oviedo. Vol. 1+, 1973+. Asturnatura. HI 62039

Asuntos agrarias. La Plata. Vol. 1+, 1953/54+, 1954+. Asuntos Agrar. HI 62040

Atas, instituto de micologia, universidade federal de Pernambuco. Recife. Ano. 2-5, 1965-67. Atas Inst. Micol. Univ. Pernambuco. Preceded by: Atas, instituto de micologia, universidade de Recife. HI 74813

Atas, instituto de micologia, universidade de Recife. Recife. 1960. Atas Inst. Micol. Univ. Recife. Superseded by: Atas, instituto de micologia, universidade federal de Pernambuco. HI 51586

Atas da sociedade de biologia do Rio de Janeiro. Rio de Janeiro. 1957+. Atas Soc. Biol. Rio de Janeiro. HI 62041

Atas da sociedad botânica do Brasil. Secção Rio de Janeiro. Rio de Janeiro. Vol. 1+, 1981+. Atas. Soc. Bot. Brasil, Secç. Rio de Janeiro. HI 62042

Ateneo científico; memoria leída en la junta general. Madrid. Ateneo Ci. See B–P–H 156/28. HI 51587

Ateneo parmense. Sezione 2, acta naturalia. Parma. Vols. 1-8, 1929-72? Ateneo Parmense, Sez. 2. Preceded by: Bollettino della società medica di Parma [not entered]. Superseded by: Ateneo parmense. Acta naturalia. 1-545-3. HI 62043

Ateneo parmense. Acta naturalia. Parma. Vols. 9-16, 1973-80. Ateneo Parmense, Acta Nat. Preceded by: Ateneo parmense. Superseded by: Acta naturalia de "l'ateneo parmense." HI 62044

Athena. Berlin. Athena (Berlin). See B–P–H 157/2. HI 51588

Athena. Paris. Athena (Paris). See B–P–H 157/3. HI 51589

Athenaeum. The Hague. Athenaeum (The Hague). See B–P–H 157/5. HI 51591

Athenaeum. London. Athenaeum (London). See B–P–H 157/4. HI 51590

Athenaeum für Wissenschaft, Kunst und Leben. Nuremberg. Athenaeum Wiss. See B–P–H 157/6. HI 51594

Athene; a journal of natural history and microscopy. Oldham. 1961-66. Athene. HI 62045

Atkinson's casket. Philadelphia, PA. Atkinson's Casket. See B–P–H 157/9. HI 51595

Atkinson's Saturday evening post. Philadelphia, PA. Atkinson's Saturday Eve. Post. See B–P–H 157/10. HI 51596

Atlanta botanical garden clippings. Atlanta, GA. Vol. ?-9(3)+, 19??-86+. Atlanta Bot. Gard. Clippings. Preceded by: Atlanta botanical garden [newsletter]. HI 62046

Atlanta botanical garden [newsletter]. Atlanta, GA. Vols. ?-2(2)-?, 19??-79-? Atlanta Bot. Gard. Newslett. Superseded by: Atlanta botanical garden clippings. HI 62047

Atlantic journal, and friend of knowledge. [Edited by C. S. Rafinesque.] Philadelphia, PA. Atlantic J. See B–P–H 157/14. HI 51597

Atlantic naturalist. Washington, DC. Atlantic Naturalist. See B–P–H 157/17. HI 51598

Atlantic slope naturalist. Narberth, PA. Atlantic Slope Naturalist. See B–P–H 157/18. HI 51599

Atlantica. Rio Grande. 1976+. Atlantica. HI 62048

Atlantide reports; scientific results of the danish expedition to the coasts of tropical west Africa, 1945-46. Copenhagen. No. 1+, 1950+. Atlantide Rep. HI 62049

Atlantis; a register of literature and science. Dublin & London. Atlantis. See B–P–H 157/19. HI 51600

Atoll research bulletin. Washington, DC. Atoll Res. Bull. See B–P–H 157/20. HI 51601

Atompraxis. Karlsruhe. Atompraxis. See B–P–H 157/21. HI 51602

Atti dell'accademia de' georgofili. Florence. Atti Accad. Georgof. See B–P–H 158/5. HI 51603

Atti dell'accademia Gioenia di scienze naturali di Catania. Catania, Italy. Atti Accad. Gioenia Sci. Nat. Catania. See B–P–H 158/8. HI 51604

Atti della accademia italiana. Florence. Atti Accad. Ital. See B–P–H 158/9. HI 51605

Atti della accademia ligure di scienze e lettere. Genoa. 1942+. Atti Accad. Ligure Sci. Lett. Preceded by: Atti della società di scienze e lettere di Genova. HI 62050

Atti dell'accademia di medicina de Torino. Turin. Atti Accad. Med. Torino. See B–P–H 158/11. HI 51609

Atti della accademia nazionale dei lincei. Memorie di classe di scienze fisiche, matematiche e naturale. Sezione IIIa: botanica, zoologia, fisiologia, patologia. Rome. Ser. 8, vol. 1+, 1948+. Atti Accad. Naz. Lincei, Mem. Cl. Sci. Fis., Sez. 3a, Bot. Zool. Preceded by: Atti della reale accademia d'Italia. Memorie delle classe di scienze fisiche, matematiche e naturali. 1-43-2. HI 62051

Atti della accademia nazionale dei lincei. Rendiconti delle adunanze solenni. Rome. Vol. 5+, 1946/57+, 1957?+. Atti Accad. Naz. Lincei, Rendiconti Adunanze Solenni. Preceded by: Atti della reale accademia dei lincei. Rendiconti delle sedute solenni. HI 62052

Atti della accademia nazionale dei lincei. Rendiconti, classe di scienze fisiche, matematiche e naturali. Rome. Ser. 8, vol. 1+, 1948+. Atti Accad. Naz. Lincei, Rendiconti Cl. Sci. Fis. Mat. Nat. Preceded by: Atti della reale accademia d'Italia. Rendiconti. 1-43-3. HI 62053

Atti dell'accademia olimpica d'agricoltura, scienze, lettere ed arti. Vicenza, Italy. Atti Accad. Olimpica Agric. See 158/12. HI 51610

Atti dell'accademia pontaniana. Naples. Atti Accad. Pontan. See B–P–H 158/13. HI 51611

Atti dell'accademia pontificia delle scienze de' (dei) nuovi lincei. Rome. Atti Accad. Pontif. Sci. Nuovi Lincei. See B–P–H 158/14. HI 51612

Atti dell'accademia scientifica veneto-trentino-istriana. Padua. Ser. 3, vols. 1-25, 1908-35. Atti Accad. Sci. Veneto-Trentino-Istriana. Preceded by: Atti della società veneto-trentino di scienze naturali in Padova. Incorporated in: Atti e memorie della (reale) accademia di scienze, lettere ed arti, Padova. 1-44-3. HI 62054

Atti dell'accademia delle scienze fisiche e matematiche. Naples. Atti Accad. Sci. Fis. See B–P–H 158/16. HI 51613

Atti dell'accademia delle scienze dell'istituto di Bologna. Classe di scienze fisiche. Memorie. Bologna. Ser. 1, nos. 1-10, 1954-63; ser. 2, vol. 1+, 1964+. Atti Accad. Sci. Ist. Bologna, Cl. Sci. Fis., Mem. Preceded by:

Memorie della reale accademia delle scienze dell'istituto di Bologna. Classe di scienze fisiche. HI 62055

Atti dell'accademia delle scienze dell'istituto di Bologna. Classe di scienze fisiche. Rendiconti. Bologna. Ser. 11, vols. 1-10, 1953/54-63; ser. 12, vol. 1+, 1963+. Atti Accad. Sci. Ist. Bologna, Cl. Sci. Fis., Rendiconti. Preceded by: Rendiconto delle sessioni della reale accademia delle scienze dell'istituto di Bologna. Classe di scienze fisiche. HI 62056

Atti dell'accademia delle scienze di Siena detta de' fisiocritici. Siena. Atti Accad. Sci. Siena. See B–P–H 158/19. HI 51614

Atti della accademia delle scienze di Torino. Classe di scienze fisiche, matematiche e naturali. Turin. 1945+. Atti Accad. Sci. Torino, Cl. Sci. Fis. Mat. Nat. Preceded by: Atti della reale accademia delle scienze di Torino. HI 62057

Atti delle adunanze dell'Imperiale reale istituto veneto di scienze, lettere ed arti = Atti del reale istituto veneto di scienze, lettere ed arti. Venice.

Atti, associazione genetica italiana. Pisa. 1955+. Atti Assoc. Genet. Ital. HI 62058

Atti della clinica otorinolaringologica della R. università di Napoli. Naples. Atti Clin. Otorinolaringol. Reale Univ. Napoli. See B–P–H 158/21. HI 51615

Atti della congresso degli scienziati italiani. Turin. Atti Congr. Sci. Ital. See B–P–H 158/22. HI 51616

Atti dei convegni lincei, accademia nazionale dei lincei. Rome. No. 1+, 1974+. Atti Conv. Lincei, Accad. Naz. Lincei. HI 62059

Atti, giornate fitopatologiciche. 1969-71. Atti Giorn. Fitopatol. HI 62060

Atti dell'imperiale accademia pistojese di scienze e lettere. Pistoia, Italy. Atti Imp. Accad. Pistojese Sci. See B–P–H 159/1. HI 51617

Atti dell'imperiale reale società agraria di Gorizia. Udine, Austria [Italy]. Atti Imp. Reale Soc. Agrar. Gorizia. See B–P–H 159/2. HI 51618

Atti dell'imperiale regia accademia di scienze, lettere ed arti degli agiati di Rovereto. Rovereto, Austria [Italy]. Atti Imp. Regia Accad. Rovereto. See B–P–H 159/3. HI 51619

Atti della imp. società economica di Firenze ossia de' georgofili. Florence. Atti Imp. Soc. Econ Firenze. See B–P–H 159/4. HI 51620

Atti dell'istituto botanico "Giovanni Briosi" e laboratorio crittogamica italiano della reale università di Pavia Siena. Ser. 4, vols. 1-13, 1929-41. Atti Ist. Bot. "Giovani Briosi". Preceded by: Atti dell'istituto botanico dell'università di Pavia. Superseded by: Atti dell'istituto botanico e del laboratorio crittogamico dell'universita di Pavia. 4-3280-2. HI 62061

Atti dell'istituto botanico e del laboratorio crittogamico dell'università di Pavia. Forlì. Ser. 5, vols. 1-21, 1943-64; suppl. vols. A-J, 1944-55; ser. 6, vols. 1-?, 1965-?; ser. 7+, 19??+. Atti Ist. Bot. Lab. Crittog. Univ. Pavia. Preceded by: Atti dell'istituto botanico "Giovanni Briosi" e laboratorio crittogamica italiano della reale università di Pavia. 4-3280-2. HI 62062

Atti dell'istituto botanico dell'università di Pavia. Milan. Atti Ist. Bot. Univ. Pavia. See B–P–H 159/15. HI 51621

Atti istituto veneto di scienze, lettere ed arti. Venice. 1942-53. Atti Ist. Veneto Sci. Lett. Arti. Preceded by: Atti del reale istituto veneto di scienze, lettere ed arti. Superseded by: Atti istituto veneto di scienze, lettere ed arti. Classe di scienze matematiche e naturali. HI 62063

Atti istituto veneto di scienze, lettere ed arti. Classe di scienze matematiche e naturali. Venice. 1953+. Atti Ist. Veneto Sci. Lett. Arti, Cl. Sci. Mat. Nat. Preceded by: Atti istituto veneto di scienze, lettere ed arti. HI 62064

Atti e memorie, accademia patavina di scienze, lettere ed arti. Padua. 1947+. Atti Mem. Patavina Sci. Lett. Arti. Preceded by: Atti e memorie della (reale) accademia di scienze, lettere ed arti, Padova. HI 62065

Atti e memorie della (reale) accademia di scienze, lettere ed arti, Padova. Padua. 1885-1943. Atti Mem. Accad. Sci. Lett. Arti, Padova. Superseded by: Atti e memorie, accademia patavina di scienze, lettere ed arti. HI 62066

Atti e memorie inedite dell'accademia del cimento e notizie aneddoti dei progressi delle scienze in Toscana. Atti & Mem. Ined. Accad. Cimento. See B–P–H 159/23. HI 51623

Atti e memorie della reale accademia de scienze, lettere ed arti, Modena. Modena. 1926+. Atti Mem. Reale Accad. Sci. Lett. Arti, Modena. Preceded by: Memorie della reale accademia di scienze, lettere e d'arti di Modena. HI 62067

Atti e memorie della reale accademia di scienze, lettere ed arti, Padova = Atti e memorie della accademia di scienze, lettere ed arti, Padova. Padua.

Atti e memorie della reale accademia virgiliana di Mantova. Mantua. 1863, 1868-1907; n.s. vol. 1+, 1908+. Atti Mem. Reale Accad. Virgil. Mantova. HI 62068

Atti del museo civico di storia naturale di Trieste. Trieste, Austria [Italy]. Atti Mus. Civico Storia Nat. Trieste. See B–P–H 159/24. HI 51624

Atti della pontificia accademia romana dei nuovi lincei. Rome. Atti Pontif. Accad. Romana Nuovi Lincei. See

B–P–H 159/25. HI 51625

Atti della reale accademia economico-agraria dei georgofili di Firenze. Florence. Atti Reale Accad. Econ.-Agrar. Georgof. Firenze. See B–P–H 160/9. HI 51628

Atti della reale accademia dei fisiocritici Siena. Siena. Atti Reale Accad. Fisiocrit. Siena. See B–P–H 160/10. HI 51629

Atti della reale accademia dei georgofili di Firenze. Florence. Atti Reale Accad. Georgof. Firenze. See B–P–H 160/11. HI 51630

Atti della reale accademia d'Italia. Memorie delle classe di scienze fisiche, matematiche e naturali. Rome. Ser. 7, 1940-44. Atti Reale Accad. Italia, Mem. Cl. Sci. Fis. Preceded by: Atti della reale accademia nazionale dei lincei, memorie di classe di scienze fisiche, matematiche e naturale. Superseded by: Atti della accademia nazionale dei lincei. Memorie di classe di scienze fisiche, matematiche e naturale. Sezione IIIa: botanica, zoologia, fisiologia, patologia. 1-43-2. HI 62069

Atti della reale accademia d'Italia. Rendiconti. Rome. Ser. 7, 1940-43. Atti Reale Accad. Italia, Rendiconti. Preceded by: Atti della reale accademia nazionale dei lincei. Rendiconti di classe di scienze fisiche, matematische e naturale. Superseded by: Atti della accademia nazionale dei lincei. Rendiconti, classe di scienze fisiche, matematiche e naturali. 1-43-3. HI 62070

Atti della reale accademia ligure di scienze e lettere = Atti della accademia ligure di scienze e lettere. Genoa.

Atti della reale accademia dei lincei. Rome. Atti Reale Sccad. Lincei. See B–P–H 160/13. HI 51631

Atti della reale accademia dei lincei. Memorie di classe di scienze fisiche, matematiche e naturale. Rome. Atti Reale Accad. Lincei, Mem. Cl. Sci. Fis. See B–P–H 160/14. HI 51632

Atti della reale accademia dei lincei. Rendiconti di classe di scienze fisiche, matematiche e naturale. Rome. Atti Reale Accad. Lincei, Rendiconti Cl. Sci. Fis. See B–P–H 160/15. HI 51633

Atti della reale accademia dei lincei. Rendiconti delle sedute solenni. Rome. Vols. 1-11, 1892/1901-1929/39, 1901-39. Atti Reale Accad. Lincei, Rendiconti Sedute Solenni. Superseded by: Atti della accademia nazionale dei lincei. Rendiconti delle adunanze solenni. HI 62071

Atti della reale accademia lucchese di scienze, lettere ed arti. Lucca, Italy. Atti Reale Accad. Lucchese Sci. See B–P–H 160/16. HI 51634

Atti della reale accademia medico-chirurgico de Torino = Atti dell'accademia di medicina de Torino. Turin. Atti Accad. Med. Torino. See B–P–H 158/11.

Atti della reale accademia nazionale dei lincei. Memorie di classe di scienze fisiche, matematiche e naturale. Rome. Atti Reale Accad. Naz. Lincei, Mem. Cl. Sci. Fis. See B–P–H 160/18. HI 51635

Atti della reale accademia nazionale dei lincei. Rendiconti delle adunanze solenni = Atti della reale accademia dei lincei. Rendiconti delle sedute solenni. Rome.

Atti della reale accademia nazionale dei lincei. Rendiconti di classe di scienze fisiche, matematiche e naturale. Rome. Ser. 6, vols. 1-29, 1925-39. Atti Reale Accad. Naz. Lincei, Rendiconti Cl. Sci. Fis. Preceded by: Atti della reale accademia dei lincei, rendiconti di classe fisiche, matematiche e naturale. Superseded by: Atti della reale accademia d'Italia. Rendiconti. 1-43-3. HI 62072

Atti della r. accademia pistojese di scienze, lettere, ed arti. Memorie di matematica e fisica de' socj corrispondenti. Pistoia, Italy. Atti Reale Accad. Pistojese Sci. Mem. Mat. See B–P–H 161/3. HI 51636

Atti della reale accademia delle scienze e belle-lettere di Napoli. Naples. Atti Reale Accad. Sci. Napoli. See B–P–H 161/4. HI 51637

Atti della reale accademia delle scienze. sezione della società reale borbonica. Naples. Atti Reale Accad. Sci. Sez. Soc. Reale Borbon. See B–P–H 161/5. HI 51638

Atti della reale accademia delle scienze di Torino. Turin. 1865-1907. Atti Reale Accad. Sci. Torino. Superseded by: Atti della reale accademia delle scienze di Torino. Classe di scienze, fisiche, matematiche e naturali. HI 62073

Atti della reale accademia delle scienze di Torino. Classe di scienze, fisiche, matematiche e naturali. Turin. 1908-45. Atti Reale Accad. Sci. Torino, Cl. Sci. Fis. Mat. Nat. Preceded by: Atti della reale accademia delle scienze di Torino. Superseded by: Atti della accademia delle scienze di Torino. Classe di scienze fisiche, matematiche, e naturali. HI 62074

Atti del real istituto d'incoraggimento alle scienze naturali di Napoli. Naples. Atti Real Ist. Incoragg. Sci. Nat. Napoli. See 160/7. HI 51626

Atti dell reale istituto lombardo di scienze, lettere ed arti. Milan. Vols. 1-3, 1858-62. Atti Reale Ist. Lombardo Sci. Lett. Arti. Preceded by: Giornale dell'imperiale e reale istituto lombardo di scienze ed arti e biblioteca italiana. Superseded by: Rendiconti delle reale istituto lombardo di scienze e lettere. Classe di scienze, matematiche e naturali. 3-2127-3. HI 62076

Atti del reale istituto veneto di scienze, lettere ed arti. Venice. Vols. 1-97, 1841-1937/38, 1841-1942. Atti Reale Ist. Veneto Sci. Lett. Arti. Superseded by: Atti istituto veneto di scienze, lettere ed arti. 3-2130-1.

HI 62077

Atti della real società economica di Firenze ossia de' georgofili. Florence. Atti Real Soc. Econ. Firenze. See B–P–H 160/8. HI 51627

Atti della r[eale] università di Genova. Genoa. Vols. 1-24?, 1869-1923. Atti Reale Univ. Genova. HI 62078

Atti della riunione degli scienziati italiani. Pisa. Atti Riunione Sci. Ital. See B–P–H 161/6. HI 51639

Atti della società crittogamologica italiana. Milan. Vols. 1-3, 1878-85. Atti Soc. Crittog. Ital. Preceded by: Commentario della società crittogamalogica italiana. 5-3916-3. HI 62079

Atti della società elvetica delle scienze naturali. Lugano, Switzerland. Atti Soc. Elvet. Sci. Nat. See B–P–H 161/9. HI 51640

Atti della società geologica residente in Milano = Atti della società italiana di scienze naturali e del museo civico di storia naturale di Milano. Milan.

Atti della società geologica residente in Milano = Atti della società italiana di scienze naturali. Milan.

Atti della società italiana di scienze naturali. Milan. Vols. 1-34, 1859-96? Atti Soc. Ital. Sci. Nat. Superseded by: Atti della società italiana di scienze naturali e del museo civico di storia naturale di Milano. 5-3919-3. HI 62080

Atti della società italiana di scienze naturali e del museo civico di storia naturale di Milano. Milan. Vol. 35+, 1896+. Atti Soc. Ital. Sci. Nat. Mus. Civico Storia Nat. Milano. Preceded by: Atti della società italiana di scienze naturali. HI 62081

Atti della società ligustica di scienze naturali e geografiche. Genoa. 1890-1935. Atti Soc. Ligustica Sci. Nat. Geogr. Superseded by: Atti della società di scienze e lettere di Genova. HI 62082

Atti, società medico-chirurgica. Turin = Atti dell'accademia di medicina de Torino. Turin. Atti Accad. Med. Torino. See B–P–H 158/11.

Atti della società dei naturalisti e matematici di Modena. Modena, Italy. Atti Soc. Naturalisti Mat. Modena. See B–P–H 161/15. HI 51641

Atti della società dei naturalisti di Modena. Modena. Vols. 23-31 [also numbered ser. 3, vols. 8-16], 1888-99. Atti Soc. Naturalisti Modena. Preceded by: Memorie e rendiconti della società di naturalisti di Modena. Superseded by: Atti della società dei naturalisti e matematici di Modena. 5-3917-1. HI 62083

Atti della società patriotica di Milano diretta all'aranzamento dell'agricoltura, delle arti, e delle manifatture. Milan. Atti Soc. Patriot. Milano. See B–P–H 161/17. HI 51642

Atti della società di scienze e lettere di Genova. Genoa. 1936-40. Atti Soc. Sci. Lett. Genova. Preceded by: Atti della società ligustica di scienze naturali e geografiche. Superseded by: Atti della (reale) accademia ligure di scienze e lettere. HI 62084

Atti della società toscana di orticultura = Bullettino della società toscana di orticultura. Florence. Bull. Soc. Tosc. Ortic. See B–P–H 282/16.

Atti della società toscana di scienze naturali residente in Pisa. Memorie. [Vols. 1-4(1) without title Memorie]. Pisa. 1875-77?; vol. 55+, 1978+. Atti Soc. Tosc. Sci. Nat. Pisa Mem. From 1878-1948 appeared as: Atti della società toscana di scienze naturali residente in Pisa, processi verbali. 5-3922-1. HI 62085

Atti della società toscana di scienze naturali residente in Pisa, processi verbali. Pisa. Vols. ?-54, 1878-1948. Atti Soc. Tosc. Sci. Nat. Pisa Processi Verbali. Preceded and superseded by: Atti della società toscana di scienze naturali residente in Pisa. Memorie. 5-3922-1. HI 62086

Atti della società veneto-trentino di scienze naturali in Padova. Padua. Vols.1-12, 1872-92; ser. 2, vols. 1-4, 1893-1902; nuova ser. vols. 1-5(1), 1904-08. Atti Soc. Veneto-Trentino Sci. Nat. Padova. Superseded by: Atti dell'accademia scientifica veneto-trentino-istriana. 1-44-3. HI 62087

Atti della solenne distribuzione di premi d'agricoltura e d'industria. Milan. Atti Solenne Distrib. Premi Agric. Industr. See B–P–H 162/3. HI 51643

Atti della università di Genova = Atti della r[eale] università di Genova. Genoa.

Auara. Annamalamagar. Vol. 1+, 1969+. Auara. HI 62089

Auckland botanical society journal = Journal, Auckland botanical society. Auckland.

Audubon. New York. Vols. 68(5)+, 1966+. Audubon. Preceded by: Audubon magazine. HI 62090

Audubon magazine. Harrisburg, PA & New York. vols. 43-68(4), 1941-66. Audubon Mag. Preceded by: Bird lore [not entered]. Superseded by: Audubon. HI 62091

Audubon nature bulletin. New York. Vol. ?+, 193?+. Audubon Nat. Bull. Preceded by: School nature league bulletin. 1-555-3. HI 62092

Audubon nature yearbook. New York. 1987+. Audubon Nat. Yearb. HI 62093

Audubon wildlife report. New York. 1985+. Audubon Wildlife Rep. HI 62094

Aufsätze und Reden, Senckenbergische Naturforschenden Gesellschaft. Frankfurt am Main. 1947+. Aufsätze Reden, Senckenberg. Naturf. Ges. HI 62095

Augustana library publications. Rock Island, IL. Nos. 1-

7,1898-1910. Augustana Libr. Publ. HI 62096

Auk. Boston, MA. Vol. 1+, 1884+. Auk. HI 62097

Aurora. Mexico City. Aurora. See B–P–H 162/6. HI 51644

Aurora; periódico de ciencias, artes y literatura (later Periódico semanal de ciencias, literatura y artes). Zaragoza. Vols. 1-2, 1839-40. Aurora (Zaragoza). 1-558-3. HI 52856

Aus der Heimat. Stuttgart. Aus der Heimat (Stuttgart). See B–P–H 162/10. HI 51646

Aus der Heimat - für die Heimat. Bremerhaven. Aus der Heimat - Für die Heimat. See B–P–H 162/9. HI 51645

Aus der Natur. Berlin. Aus der Natur (Berlin). See B–P–H 162/11. HI 51647

Aus der Natur. Leipzig. Aus der Natur (Leipzig). See B–P–H 162/12. HI 51648

Aus Natur und Heimat von Guben und Umgebung. Guben, Germany. Aus Natur & Heimat Guben. See B–P–H 162/13. HI 51649

Aus Natur und Museum. Frankfurt a. M. Aus Natur & Mus. See B–P–H 162/14. HI 51650

Aus dem Walde. Hanover, Frankfurt am Main, & Tübingen. Vols. 1-17, 1865-1900. Aus dem Walde (Hanover). 1-559-2. HI 62098

Aus dem Walde; Mitteilungen aus der niedersächischen Landesforstverwaltung. Wiesbaden. Vol. 1+, 1953+. Aus dem Walde (Wiesbaden). HI 62099

Auserlesene Abhandlungen für Aerzte, Naturforscher und Psychologen aus den Schriften der Literarisch-Philosophischen Gesellschaft zu Manchester. Leipzig. Auserlesene Abh. Aerzte. See B–P–H 162/16. HI 51651

Auserlesene Abhandlungen, welche an die kgl. Akademie der Wissenschaften zu Paris von einigen Gelehrten eingesendet, in ihren Versammlungen abgelesen und von ihr herausgegeben wurden. Leipzig. Auserlesene Abh. Königl. Akad. Wiss. Paris. See B–P–H 162/17. HI 51652

Auserlesene Bibliothek der neuesten deutschen Literatur. Lemgo, Germany. Auserlesene Biblioth. Neuesten Deutsch. Lit. See B–P–H 162/18. HI 51653

Ausgrabungen und Funde; Nachrichtenblatt für Vor- und Frühgeschichte. Berlin. Vol. 1+, 1956+. Ausgrabungen & Funde. HI 62100

Auskunftsblatt des Biologen = Spravočnyj Listok Biologa. Yur'ev [=Tartu, Estonian S S R]. Sprav. Listok Biol. See B–P–H 853/8.

Ausland. Stuttgart. Ausland. See B–P–H 162/20. HI 51654

Australasian beekeeper. Maitland, N.S.W. Vol. 1+, 1899+. Australas. Beekeeper. HI 62101

Australasian bryological newsletter. Kensington, N.S.W. No. 2+, 1980+. Australas. Bryol. Newslett. Preceded by: Bryological newsletter. Kensington, N.S.W. HI 62102

Australasian chemist and druggist. Melbourne. Australas. Chem. Druggist. See B–P–H 165/5. HI 51691

Australasian herbarium news. Adelaide. Australas. Herb. News. See B–P–H 165/6. HI 51692

Australasian international nurseryman, seedsman, and florist. Melbourne. Vols. 1-25(3), 1899-1927. Australas. Int. Nurseryman, Seedsman Florist. Superseded by: Seed and nursery trader of Australia and New Zealand. 5-3840-1. HI 62103

Australasian journal of pharmacy. Melbourne. Australas. J. Pharm. See B–P–H 165/7. HI 51693

Australasian plant pathology. South Perth, W.A. Vol. 1+, 1972+. Australas. Pl. Pathol. Preceded by: Newsletter, australian plant pathology society. HI 62104

Australian acacias. Canberra, A.C.T. 1975+. Austral. Acacias. HI 62105

Australian beekeeper. Sydney. Austral. Beekeeper. See B–P–H 163/20. HI 51655

Australian canegrower. Brisbane, Qld. Vol. 1+, 1979+. Austral. Canegrower. HI 62106

Australian conservation foundation, newsletter. Eastwood, N.S.W. [Vol. 1], no. 1+, 1967+. Austral. Conservation Found. Newslett. HI 62107

Australian fern journal. Essendon, Vic. Vol. 1+, 1984+. Austral. Fern J. HI 62108

Australian flora and fauna series. Canberra, A.C.T. No. 1+, 1984+. Austral. Fl. Fauna Ser. HI 62109

Australian forest research. East Melbourne, Vic. Vol. 1+, 1964+. Austral. Forest. Res. Preceded by: Forestry research notes, forestry and timber bureau, Australia. HI 51657

Australian forest resources. Canberra, A.C.T. 1976+, 1977+. Austral. Forest Resources. Preceded by: Forest resources. HI 62110

Australian forestry. Melbourne, Vic., etc. Vol. 1+, 1936+. Austral. Forest. HI 62111

Australian forestry journal. Sydney. Austral. Forest. J. See B–P–H 163/22. HI 51656

Australian fruit grower and smallholder. Brisbane. Austral. Fruit Grower Smallholder. See B–P–H 163/25. HI 51658

Australian fruitgrower, fertiliser and poultry farmer.

Brisbane. Austral. Fruitgrower. See B–P–H 163/26. HI 51659

Australian garden and field. Adelaide. Austral. Gard. Field. See B–P–H 163/28. HI 51661

Australian garden journal. Bowral, N.S.W. Vol. 2(7)+, 1983+. Austral. Gard. J. Preceded by?: Journal of the australian garden history society. HI 62112

Australian garden lover. Melbourne. Austral. Gard. Lover. See B–P–H 163/29. HI 51662

Australian gardener. Adelaide. Austral. Gard. See B–P–H 163/27. HI 51660

Australian geographer. Sydney, N.S.W. Vol. 1+, 1928/32+, 1932?+. Austral. Geogr. HI 62113

Australian horticultural magazine and garden guide. Melbourne. Austral. Hor. Mag. & Gard. Guide. See B–P–H 163/30. HI 51663

Australian horticulture. North Melbourne, Vic. Vol. 79(6)+, 1981+. Austral. Hort. Preceded by: Seed and nursery trader of Australia and New Zealand. HI 62114

Australian journal of agricultural research. Melbourne. Austral. J. Agric. Res. See B–P–H 163/32. HI 51664

Australian journal of applied science. Melbourne. Austral. J. Appl. Sci. See B–P–H 163/33. HI 51665

Australian journal of biological sciences. Melbourne. Austral. J. Biol. Sci. See B–P–H 163/34. HI 51666

Australian journal of biotechnology. Adelaide, S.A. Vol. 1+, 1987+. Austral. J. Biotechnol. HI 75236

Australian journal of botany. Melbourne. Austral. J. Bot. See B–P–H 163/35. HI 51667

Australian journal of botany. Supplementary series. [Supplement to: Australian journal of botany. See B–P–H 163/35]. East Melbourne, Vic. No. 1+, 1971+. Austral. J. Bot., Suppl. Ser. HI 62115

Australian journal of chemistry. Melbourne, Vic. Vol. 6+, 1953+. Austral. J. Chem. Preceded by: Australian journal of scientific research. Series A, physical sciences [not entered]. HI 62116

Australian journal of dermatology. Sydney. Austral. J. Dermatol. See B–P–H 163/36. HI 51668

Australian journal of ecology. Melbourne, Vic. & Oxford. Vol. 1+, 1976+. Austral. J. Ecol. HI 62117

Australian journal of experimental agriculture and animal husbandry. Melbourne? Austral. J. Exp. Agric. Anim. Husb. See B–P–H 164/1. HI 51669

Australian journal of experimental biology and medical science. Adelaide. Austral. J. Exp. Biol. Med. Sci. See B–P–H 164/2. HI 51670

Australian journal of marine and freshwater research. Melbourne. Austral. J. Mar. Freshwater Res. See B–P–H 164/3. HI 51671

Australian journal of plant physiology; journal for the publication of original research in all aspects of plant physiology. Melbourne, Vic. Vol. 1+, 1974+. Austral. J. Pl. Physiol. HI 62118

Australian journal of science. Sydney, N.S.W. Vols. 1-32, 1938-70. Austral. J. Sci. Superseded by: Search. Sydney. 1-566-2. HI 62119

Australian journal of scientific research. Series B. Biological sciences. Melbourne. Austral. J. Sci. Res., Ser. B, Biol. Sci. See B–P–H 164/5. HI 51672

Australian journal of soil research. Melbourne. Austral. J. Soil Res. See B–P–H 164/6. HI 51675

Australian lilium society bulletin. Melbourne. Austral. Lilium Soc. Bull. See B–P–H 164/9. HI 51676

Australian lilium society quarterly. Glenhuntly, Australia. Austral. Lilium soc. Quart. See B–P–H 164/10. HI 51677

Australian marine science bulletin. Sydney, N.S.W. 1970+. Austral. Mar. Sci. Bull. Preceded by: Australian marine science newsletter. HI 62120

Australian marine science newsletter. Sydney, N.S.W. Nos. 1-?, 1963-69. Austral. Mar. Sci. Newslett. Superseded by: Australian marine science bulletin. HI 62121

Australian museum magazine; [Vol. 1 titled Australian museum.] Sydney, N.S.W. Vols. 1-13. 1921-61. Austral. Mus. Mag. Superseded by: Australian natural history. HI 62122

Australian museum memoirs. Sydney. Austral. Mus. Mem. See B–P–H 164/13. HI 51678

Australian national botanic gardens occasional publication. Canberra, A.C.T. = Occasional publications, australian national botanic gardens. Canberra, A.C.T.

Australian natural history. Sydney, N.S.W. Vol. 14+, 1962+. Austral. Nat. Hist. Preceded by: Australian museum magazine. HI 62123

Australian naturalist. Sydney. Austral. Naturalist. See B–P–H 164/16. HI 51679

Australian and New Zealand rose annual. Melbourne, Vic. Vols. 16-36, 1943-63. Austral. New Zealand Rose Annual. Preceded by: Australian rose annual. Superseded by: Australian rose annual and New Zealand rose annual. HI 62124

Australian oil seed grower. Sydney. Austral. Oil Seed Grower. See B–P–H 164/17. HI 51680

Australian orchid review. Sydney. Austral. Orchid Rev. See B–P–H 164/18. HI 51681

Australian parks. Melbourne, Vic. Vols. 1-11(1), 1964-74. Austral. Parks. Superseded by: Australian parks and recreation. HI 62125

Australian parks and recreation. Braddon, A.C.T. 1974+. Austral. Parks & Recreation. Preceded by: Australian parks HI 62126.

Australian plant disease recorder. Sydney. Austral. Pl. Dis. Rec. See B–P–H 164/20. HI 51684

Australian plant introduction review. Canberra City, A.C.T. Vol. 10(2)+, 1974+. Austral. Pl. Introd. Rev. Preceded by: Plant introduction review, division of plant industry, CSIRO. HI 62127

Australian plant pathology. Vol. 1+, 1972+. Austral. Pl. Pathol. HI 62128

Australian plant society newsletter. Pasadena, CA. Vol. 1+, 1989+. Austral. Pl. Soc. Newslett. HI 62129

Australian plants. Picnic Point, Australia. Austral. Pl. See B–P–H 164/19. HI 51683

Australian rangeland journal. Cottesloe, W.A. Vol. 1+, 1976+. Austral. Rangeland J. HI 62130

Australian rose annual. Melbourne, Vic. Vols. 1-15, 1928-41; vol. 37+, 1964+. Austral. Rose Annual. For vols. 16-36 see: Australian and New Zealand Rose Annual. HI 62131

Australian science abstracts. [From 1938-57 formed a supplement to: Australian journal of science.] Sydney, N.S.W. Vols. 1-35(5), 1922-57. Austral. Sci. Abstr. Superseded by: Australian science index. HI 62132

Australian science index. [Vol. 1 contained in: C S I R O science index.] Melbourne, Vic. Vol. [1]-21, 1957-77. Austral. Sci. Index. Preceded by: Australian science abstracts. Superseded by: C S I R O index. HI 62133

Australian seed producers' review. Melbourne, Vic., Hamilton. Vol. 1(4)+, 1969+. Austral. Seed Producers' Rev. Preceded by: Seed producers' review. HI 62134

Australian sugar journal. Brisbane. Austral. Sugar J. See B–P–H 164/24. HI 51685

Australian sugar year book. Brisbane. Austral. Sugar Year Book. See B–P–H 164/26. HI 51686

Australian systematic botany. East Melbourne, Vic. Vol. 1+, 1988+. Austral. Syst. Bot. Preceded by: Brunonia. HI 62135

Australian systematic botany society newsletter = Newsletter, Australian systematic botany society. South Perth, W.A.

Australian tobacco grower's bulletin. Canberra. Austral. Tobacco Grower's Bull. See B–P–H 165/1. HI 51687

Australian veterinary journal. Sydney. Austral. Veterin. J. See B–P–H 165/3. HI 51688

Australian weeds. North Clayton, Vic. Vols. 1-?, 1981-84? Austral. Weeds. Superseded by: Plant protection quarterly. HI 62136

Australian wild life. Sydney. Austral. Wild Life. See B–P–H 165/4. HI 51689

Austrobaileya. Brisbane, Qld. Vol. 1+, 1977+. Austrobaileya. Preceded by: Contributions from the Queensland herbarium. HI 62138

Auswahl der besten ausländischen geographischen und statistischen Nachrichten zur Aufklärung der Völker- und Landeskunde. Halle. Auswahl Besten Ausl. Geogr. Statist. Nachr. Völker- Landesk. See B–P–H 165/12. HI 51694

Auswahl aller eigenthümlichen Abhandlungen und Beobachtungen aus den neuesten Entdeckungen in der Chemie. Leipzig. Auswahl Eigenthüml. Abh. Beob. Neuesten Entdeck. Chem. See B–P–H 165/13. HI 51695

Auswahl der kleinen Schriften, welche in der Arzneiwissenschaft, Chemie, Chirurgie und Botanik herausgekommen. Göttingen. Auswahl Kleinen Schriften Arzneiwiss. See B–P–H 165/14. HI 51696

Auswahl kleiner Reisebeschreibungen und anderer statistischen und geographischen Nachrichten. Leipzig. Auswahl Kleiner Reisebeschreib. See B–P–H 165/15. HI 51697

Auswahl der medicinischen Aufsätze und Beobachtungen aus den Nürnbergischen gelehrten Unterhandlungen. Halle. Auswahl Med. Aufsätze Beob. Nürnberg. Gel. Unterhandl. See B–P–H 165/16. HI 51698

Auswahl ökonomischer Abhandlungen, welche die Freye ökonomische Gesellschaft zu St. Petersburg in teutscher Sprache erhalten hat. St. Petersburg. Auswahl Ökon. Abh. Freye Ökon. Ges. St. Petersburg. See B–P–H 165/17. HI 51699

Auszug aus den Protocollen über die Versammlungen der ersten Classe. [City not stated]. Auszug Protoc. Versamml. Ersten Cl. See B–P–H 165/18. HI 51700

Auszug aus den Protokollen über die ... Versammlungen von der Landwirtschaft. Leipzig. Auszug Protok. Versamml. Landw. See B–P–H 165/20. HI 51703

Auszug aus den Protocollen über die Versammlungen der zwoten Classe. [City not stated]. Auszug Protoc. Versamml. Zwoten Cl. See B–P–H 165/19. HI 51702

Auszug aus den Sitzungs-Protokollen des Naturwissenschaftlichen Vereins in Halle. Halle. Auszug Sitzungs-Protok. Naturwiss. Vereins Halle. See B–P–H 166/1. HI 51704

Auszüge aus den Protokollen der Gesellschaft für Natur- und Heilkunde in Dresden. Dresden. Auszüge Protok. Ges. Natur- Heilk. Dresden. See B–P–H 166/2. HI 51705

Auszüge aus den Verhandlungen der Mecklenburgischen Naturforschenden Gesellschaft. Rostock. Auszüge

Verh. Mecklenburg. Naturf. Ges. See B–P–H 166/3. HI 51706

Avance conservacionista. [Venezuela]. Vol. 1+, 1966+. Avance Concervacionista. HI 62139

Avant gardener. New York. Vol. 1+, 1968+. Avant Gard. HI 62140

AVC. Agricultural & veterinary chemicals. London = A V C. Agricultural & veterinary chemicals. London.

Avhandlinger utgitt av det norske videnskaps-akademi i Oslo. [Ser.] 1, matematisk-naturvidenskapelig klasse = Norske videnskaps-akademi, matematisk-naturvidenskapelig klasse. Avhandlinger. Christiania.

Aviso; monthly dispatch from the american association of museums. Washington, DC. No. 1+, 1975+. Aviso. Preceded by: Bulletin, american association of museums. HI 62141

Avocado grower. Fallbrook, CA, Vista, CA. Vols. 1-9(8), 1977-85. Avocado Grower. Superseded by: California grower: Avocados, citrus, subtropicals. HI 62142

Avvenire agricolo. Palma, Majorca. Avvenire Agric. See B–P–H 166/7. HI 51709

Awair = A W A I R; arab world agri and agro-industrial review.

Awamia (El); revue de la recherche agronomique marocaine. Rabat, Morocco. Awamia. See B–P–H 166/9. HI 51710

Awards quarterly, american orchid society = American orchid society awards quarterly. Cambridge, MA.

Az alföldi tudományos gyüjtemény. Az alföldi tudományos intézet évkönyve = Alföldi tudományos gyüjtemény. Az alföldi tudományos intézet évkönyve. Budapest. Alföldi Tud. Gyüjt. See B–P–H 69/6.

Az egri pedagógiai föiskola évkönyve. Eger = Egri pedagógiai föiskola évkönyve. Eger, Hungary. Egri Pedagóg. Föisk. Évk. See B–P–H 355/16.

Az Egri tanárképzö föiskola tudományos közleményei. Eger = Egri tanárképzö föiskola tudományos közleményei. Eger, Hungary. Egri Tanárképzö Föisk. Tud. Közlem. See B–P–H 355/17.

Az erdo = Erdo, Az.

Azalean. Silver Spring, MD. 1979+. Azalean. HI 62143

Azärbajğan filialynyn häbärläri = Izvestiya Azerbaidzhanskogo filiala [Akademii nauk S S S R.]. Baku.

Azärbajğan S S R èlmär akademijasynyn häbärläri = Izvestiya Akademii nauk Azerbaidzhanskoi S S R. Baku.

Azärbajğan S S R èlmlär akademijasynyn häbärläri = Izvestiya Akademii nauk Azerbaidzhanskoi S S R. Seriya biologicheskikh i meditsinskikh nauk. Baku.

Azärbajğan S S R èlmlär akademijasynyn häbärläri = Izvestiya Akademii nauk Azerbaidzhanskoi S S R. Seriya biologicheskikh i sel'skokhozyaistvennykh nauk. Baku.

Azerbaidzhanskii filial Akademii nauk S S S R. Baku, Azerbaijan S S R. Vol. 19, 1935. Azerbaidzhansk. Fil. Akad. Nauk S.S.S.R. Preceded & superseded by: Trudy Azerbaidzhanskogo filiala, akademiya nauk S S S R. 1-109-3. HI 62144

Azerbajdžanskij filial Akademii nauk S S S R = Azerbaidzhanskii filial Akademii nauk S S S R. Baku.

Aztekia. Prague. No. 1+, 1978+. Aztekia. HI 62145

B A S I C; key to the world's biological research. Philadelphia, PA. Vol. 1+, 1962+ [appendix to Biological abstracts vol. 38+.] B. A. S. I. C. HI 62146

B C forest history magazine = Whistle punk. Victoria, BC.

B C G report = Report, biology curators' group.

B C naturalist. Vancouver, B.C. Vol. 18+, 1980+. B. C. Naturalist. Preceded by: Newsletter, federation of British Columbia naturalists. HI 62147

B C G newsletter. Newcastle-upon-Tyne. Vols. [1](5)-2(10), 1977-81. B. C. G. Newslett. Preceded by: Biological curators' group newsletter. Superseded by: Newsletter B C G. HI 62148

B I O news. Plymouth. No. ?-17+, 19??-86+. B. I. O. News. HI 62149

B I O review. Dartmouth, N.S. 1981+. B. I. O. Rev. Preceded by: Biennial review, Bedford Institute of Oceanography. HI 62150

B I O S I S, see BIOSIS

B I O T R O P special publications = Biotrop special publications. Bogor.

B L L announcement bulletin. Boston Spa. 1973-74. B. L. L. Announc. Bull. Preceded by: N L L announcement bulletin. Superseded by: B L L D announcement bulletin. HI 62168

B L L D announcement bulletin. Boston Spa. 1975-80. B. L. L. D. Announc. Bull. Preceded by: B L L announcement bulletin. Superseded by: British reports, translations and theses. HI 62169

B M. Tokyo = Bifidobacteria and microflora. Tokyo.

B O B S I = Bulletin of the botanical survey of India. Calcutta. Bull. Bot. Surv. India. See B–P–H 242/20.

B P P. Jena = Biochemie und Physiologie der Pflanzen. Jena.

B S B I abstracts: Abstracts of literature relating to the

vascular plants of the British Isles. London. Vol. 1+, 1971+. B. S. B. I. Abstr. HI 62170

B S B I conference reports = Conference reports of the botanical society of the British Isles. London. and Report of the conference of the botanical society of the British Isles. London.

B S B I news. London. Vol. 1+, 1972+. B. S. B. I. News. Preceded by: Newsletter, botanical society of the British Isles. HI 62171

B S B I scottish newsletter. Glasgow. No. 1+, 1979+. B. S. B. I. Scott. Newslett. HI 62172

B S C P communique. Washington, D.C. Nos. 1-28, 1961-68. B. S. C. P. Communique. HI 62173

B S C S journal. Boulder, CO. 1978+. B. S. C. S. Journal. Preceded by: B S C S newsletter. HI 62174

B S C S newsletter. Boulder, CO. Nos. 1-69, 1959-77. B. S. C. S. Newslett. Superseded by: B S C S journal. HI 62175

B S E news. Edinburgh. Vol. 1+, 1970+. B. S. E. News. HI 62176

B S I newsletter = Newsletter, B S I. Howrah.

B T catalyst; North Carolina biotechnology center. Research Triangle Park, NC. Vol. 1+, 1987+. B. T. Catalyst. HI 62177

B Z G Berichte = Bericht, botanisch-zoologische Gesellschaft Liechtenstein-Sargens-Werdenberg. Vaduz.

Babylonian & oriental record. London. Babylonian Orient. Rec. See B–P–H 166/18. HI 51711

Bacillaria; international journal for diatom research. Lehre. Vols. 1-7, 1978-84. Bacillaria. Superseded by: Diatom research. HI 62178

Back to Eden; monthly flower and garden magazine. De Queen, AR. Back to Eden. See B–P–H 166/19. HI 51712

Bacteriological news. Baltimore, MD. Vols. 12-28, 1948-62. Bacteriol. News. Preceded by: Newsletter of the society of american bacteriologists. Superseded by: A S M news. 1-583-1. HI 62179

Bacteriological proceedings. [American society for microbiology.] Baltimore, MD. Bacteriol. Proc. See B–P–H 166/24. HI 51713

Bacteriological reviews. Baltimore, MD. Vols. 1-41, 1937-77. Bacteriol. Rev. Superseded by: Microbiological reviews. 1-583-1. HI 62180

Badania fizjograficzne nad Polska zachodnia. Ser. B, botanika. Poznan. 1948+. Badan. Fizjogr. Polsk. Zachodn., B. HI 62181

Badger common 'tater. Antigo, Wi. Badger Common 'Tater. See B–P–H 167/2. HI 51716

Badische geologische Abhandlungen. Karlsruhe. Bad. Geol. Abh. See B–P–H 166/26. HI 51714

Badische Obst- und Gartenbauer; Monatsschrift für Obst-, Gemüse, Wein- und Gartenbau: Organ des Badischen Landesobstbauverbandes. Karlsruhe. Vols. 1-20, 1948-67. Bad. Obst- Gartenbauer. Superseded by: Obst und Garten (Karlsruhe). 1-584-2. HI 51715

Baer's garden newsletter. Lancaster, PA. Vol. 1+, 1980+. Baer's Gard. Newslett. HI 62182

Baessler-Archiv. Berlin. 1959+. Baessler-Arch. HI 62183

Bahce. Istanbul. Vol. 8+, 1977+. Bahce. Preceded by: Yalova bahce kulturleri arastirma ve egitim merkezi dergis. HI 62184.

Bahia rural. Salvador, Brazil. Bahia Rural. See B–P–H 167/5. HI 51717

Baierische Beyträge zur schönen und nützlichen Litteratur. Munich. Baier. Beytr. Schönen Nützl. Litt. See B–P–H 167/7. HI 51719

Baierischen Akademie der Wissenschaften in München, meteorologische Ephemeriden. Munich. Baier. Akad. Wiss. München Meteorol. Ephem. See B–P–H 167/6. HI 51718

Baileya; a quarterly journal of horticultural taxonomy. Ithaca, NY. Baileya. See B–P–H 167/8. HI 51720

Baileygram. Ithaca, NY. Vols. ?-3(2-4)-?, 19??-60-62-? Baileygram. HI 62185

Baiosaiensu to indasutori. Tokyo. Vol. 46(4)+, 1988+. HI 62186

Baiotoronikusu = Biotronics. Kukuoka.

Bakawan. Quezon City. Vol. ?-3(2)+, 19??-85?+. Bakawan. HI 62187

Bakony természettudományi kutatásának eredménye kutatásának eredményei. Veszprem, Hungary. Bakony Term. Kutatás. Eredm. See B–P–H 167/9. HI 51721

Bakovsko. Bakov, Czechoslovakia. Bakovsko. See B–P–H 167/10. HI 51722

Balai. Pergumuman istimewa. Balai penjeli dihan hehutanan. Bogor, Indonesia. Balai. See B–P–H 167/ 11. HI 51724

Balatoni múseum egyesület évkönyve. Keszthely, Hungary. Balatoni Múz. Egyes. Évk. See B–P–H 167/12. HI 51725

Bălgarsko ovoshtarstvo i gradinarstvo. Sofia, Bulgaria. Vols. 1-18, 1920-37. Bălg. Ovoshtarstvo Gradinarstvo. HI 62188

Bălgarsko ovoštarstvo i gradinarstvo = Bălgarsko ovoshtarstvo i gradinarstvo. Sofia.

Balkán-kutatásainak tudományos eredményei. Budapest. Balkán-Kutat. Tud. Eredm. See B–P–H 167/14.

HI 51726

Ball state teachers' college bulletin = Bulletin, Ball state teachers' college. Muncie, IN.

Baltimore cactus journal. Baltimore, MA. Baltimore Cact. J. See B–P–H 167/19. HI 51728

Baltimore medical and physical recorder. Baltimore, MA. Baltimore Med. Phys. Rec. See B–P–H 167/20. HI 51729

Baltische Studien. Stettin [Poland]. Balt. Stud. See B–P–H 167/18. HI 51727

Balwant vidyapeeth journal of agricultural and scientific research. Bichiri, Agra. Vols. 1-19, 1959-77. Balwant Vidyapeeth J. Agric. Sci. Res. Superseded by: Journal of agricultural and scientific research. HI 62189

Bamboo journal. Kyoto. No. ?-4+, ?-1987+. Bamboo J. HI 62190

Bamboo leaves = Bambusblätter. Marktheidenfeld.

Bamboo research. Nanjing. Vol. 1+, 1981+. Bamboo Res. HI 62191

Bamboos of the world. Odenthal. Pt. 1+, 1983+. Bamboos World. HI 62192

Bambou. Mons, Belgium. Bambou (Mons). See B–P–H 167/22. HI 51730

Bambou. Paris. Bambou (Paris). See B–P–H 167/23. HI 51731

Bambusblätter. Marktheidenfeld. Vol. 1+, 1984+. Bambusblätter. HI 62193

Banana bulletin. Murnillumbah, Australia. Banana Bull. See B–P–H 167/25. HI 51732

Banano; agronomia, tecnologia, comercializacion. Guayaquil. Vol. 1+, 1964+. Banano. HI 62194

Banater Zeitschrift für Landwirthschaft, Handel, Künste und Gewerbe. Temeschburg [=Timisoara, Rumania]. Banater Z. Landw. See B–P–H 168/1. HI 51733

Bangladesh agricultural abstracts. Dacca. Vol. 1+, 1983+. Bangladesh Agric. Abstr. HI 62195

Bangladesh agricultural sciences abstracts. Mymensingh. Vol. 1+, 1974+. Bangladesh Agric. Sci. Abstr. HI 62196

Bangladesh horticulture. Mymensingh. Vol. 1+, 1973+. Bangladesh Hort. HI 62197

Bangladesh journal of agricultural sciences. Dacca. Vol. 1+, 1974+. Bangladesh J. Agric. Sci. Preceded by: Bangladesh journal of biological and agricultural sciences. HI 62198

Bangladesh journal of biological sciences. Dacca. Vol. 3+, 1974?+. Bangladesh J. Biol. Sci. Preceded by: Bangladesh journal of biological and agricultural sciences. HI 62199

Bangladesh journal of biological and agricultural sciences. Dacca. Vols. 1-2, 1972-73? Bangladesh J. Biol. Agric. Sci. Preceded by: Pakistan journal of biological and agricultural sciences. Superseded by: Bangladesh journal of biological sciences and Bangladesh journal of agricultural sciences. HI 62200

Bangladesh journal of botany. Dacca. Vol. 1+, 1972+. Bangladesh J. Bot. HI 62201

Bangladesh journal of scientific and industrial research. Dacca. Vol. 8+, 1974+. Bangladesh J. Sci. Industr. Res. Preceded by: Scientific researches. Bangladesh council of scientific and industrial research, laboratories. HI 62202

Bangladesh journal of scientific research. Dacca. Vol. 1+, 1978+. Bangladesh J. Sci. Res. HI 62203

Bangladesh science and technology index. Dacca. Vol. 1+, 1977+. Bangladesh Sci. Technol. Index. HI 62204

Banko janakari; a journal of forestry information for Nepal. Katmandu. Vol. 1+, 1987+. Banko Janak. HI 63828

Banksia. Uppsala. Vol. 1+, 1979+. Banksia. HI 62205

Banksia study. n.p. No. 1+, 1972+. Banksia Study. HI 62206

Banksia study report = Banksia study.

Baracoa; revista forestal. Havana. Vol. 1+, 1971+. Baracoa. HI 62207

Barbados agricultural reporter and planters' scientific journal. Bridgetown, Barbados. Barbados Agric. Rep. See B–P–H 168/2. HI 51734

Barcena. Villa Nueva, Guatemala. Barcena. See B–P–H 168/3. HI 51735

Barley genetics newsletter; an international communication medium. Fort Collins, CO. Vol. 1+, 1971+. Barley Genet. Newslett. HI 62208

Bartlett arboretum newsletter = Newsletter, Bartlett arboretum.

Bartlettia; notes from the Matthaei botanical gardens of the university of Michigan. Ann Arbor, MI. No. 1+, 1975+. Bartlettia. HI 62209

Bartonia; a botanical annual. Philadelphia, PA. Bartonia. See B–P–H 168/4. HI 51736

B.A.S.I.C. Philadelphia, PA = B A S I C. Philadelphia, PA.

Basic life sciences. New York, NY. Vol. 1+, 1973+. Basic Life Sci. HI 62210

Basutoland notes and records = Lesotho. Morija.

Batatinha; resumos informativos. Brasilia. ?-1978+. Batatinha. HI 62211

Batumskii Sel'skii Khozyain. Batum, Georgian S S R. 1908-11. Batumsk. Sel'sk. Khozyain. Superseded by:

Russkie Subtropiki. HI 62212

Batumskij Sel'skij Hozjain = Batumskii Sel'skii Khozyain. Batum.

Bauhinia. Basel. Bauhinia. See B–P–H 168/9. HI 51737

Baum-Zeitung. Schriftenreihe für Baumfreude und Naturschützer. Frankfurt a. M. Baum-Zeitung. See B–P–H 168/10. HI 51738

Baumschulpraxis. Aachen. No. 1+, 1971+. Baumschulpraxis. HI 62213

Bauwelt. Berlin. Vols. 1-36, 1910-45; vol. 43(18)+, 1952+. Bauwelt. From 1946-52 titled: Neue Bauwelt. 1-611-3. HI 62214

Bay cities garden monthly. Berkeley, CA. Bay Cities Gard. Monthly. See B–P–H 168/11. HI 51739

Bay leaf. Berkeley, CA. 1979+. Bay Leaf. HI 62215

Bayer agrochem courier. Bury St. Edmunds. No. 1+, 1976+. Bayer Agrochem Courier. HI 62216

Bayer Pflanzenschutz-Kurier. Leverkusen. Vols. 1-12, 1956-67. Bayer Pflanzenschutz-Kurier. Superseded by: Pflanzenschutz-Kurier. HI 62217

Bayerische Akademie der Wissenschaften. Jahrbuch. Munich. Bayer. Akad. Wiss. Jahrb. See B–P–H 168/12. HI 51741

Bayerische Akademie der Wissenschaften. Mathematisch-naturwissenschaftliche Klasse. Abhandlungen. Munich. Bayer. Akad. Wiss., Math.-Naturwiss. Kl., Abh. See B–P–H 168/13. HI 51742

Bayerische Akademie der Wissenschaften. Mathematisch-naturwissenschaftliche Klasse. Sitzungsberichte = Sitzungsberichte der bayerischen Akademie der Wissenschaften. Mathematisch-naturwissenschaftliche Klasse. Munich.

Bayerische Kleingärtner. Ausgabe A. Munich. Bayer. Kleingärtn., Ausgabe A. See B–P–H 168/15. HI 51744

Bayerische landwirtschaftliches Jahrbuch. Munich. Vol. 32(5)+, 1955+. Bayer. Landw. Jahrb. Preceded by: Landwirtschaftliches Jahrbuch für Bayern. 1-613-3. HI 62218

Bayerischer literärischer und merkantilischer Anzeiger für Literatur- und Kunstfreunde ... im In- und Auslande. Munich. Bayer. Lit. Merkantil. Anz. Literatur-Kunstfr. See B–P–H 168/16. HI 51745

Bean bag; current research on legumes [sub-title changes]. Beltsville, MD. No. 1+, 1975+. Bean Bag. HI 62219

Bedford institute of oceanography review = B I O review. Dartmouth, N.S.

Bedfordshire naturalist. Bedford, England. Bedfordshire Naturalist. See B–P–H 168/17. HI 51746

Bedrijfsontwikkeling. The Hague. Vol. 3+, 1972+. Bedrijfsontwikkeling. Preceded by: Bedrijfsontwikkeling: Editie tuinbouw. HI 62220

Bedrijfsontwikkeling: Editie tuinbouw. The Hague. Vols. 1-2, 1970-71. Bedrijfsontwikkeling, Ed. Tuinb. Preceded by: Tuinbouw mededelingen. Superseded by: Bedrijfsontwikkeling. HI 62221

Bee, or literary weekly intelligencer. Edinburgh. Bee. See B–P–H 168/19. HI 51747

Bee world. Benson, England. Bee World. See B–P–H 168/21. HI 51748

Beekeeper. Brantford, Ontario. Beekeeper. See B–P–H 168/22. HI 51749

Beekeepers' item. New Braunfels, TX. Beekeepers' Item. See B–P–H 169/1. HI 51750

Beekeepers' magazine. Lansing, MI. Beekeepers' Mag. See B–P–H 169/2. HI 51751

Bees; devoted to all phases of honey production and honeybee culture. Hapeville, GA. Bees. See B–P–H 169/3. HI 51752

Begonian. Long Beach, CA. Begonian. See B–P–H 169/4. HI 51753

Beihefte zu den Berichten den naturhistorische Gesellschaft zu Hannover. Hannover. Vol. 5+, 1968+. Beih. Ber. Naturhist. Ges. Hannover. Preceded by: Beihefte zu den Jahresberichten der naturhistorischen Gesellschaft zu Hannover. HI 62222

Beihefte zu den Berichten den naturwissenschaftlichen Gesellschaft (zu) Bayreuth. Bayreuth. Vol. 1+, 1974+. Beih. Ber. Naturwiss. Ges. Bayreuth. HI 62223

Beihefte zum Botanischen Centralblatt. Kassel. Beih. Bot. Centralbl. See B–P–H 169/9. HI 51754

Beihefte zum geologischen Jahrbuch. Hanover. 1951-72. Beih. Geol. Jahrb. HI 62224

Beihefte zu den Jahresberichten der naturhistorischen Gesellschaft zu Hannover. [Also formed vols. 1,2,4 & 5 of: Mitteilungen der floristisch- soziologischen Arbeitsgemeinschaft in Niedersachsen.] Hannover. Nos. 1-4, 1928-38. Beih. Jahresber. Naturhist. Ges. Hannover. Superseded by: Beihefte zu den Berichten den naturhistorische Gesellschaft zu Hannover. HI 62225

Beihefte, Naturschutzarbeit in Berlin und Brandenburg = Naturschutzarbeit in Berlin und Brandenburg. Beiheft. Potsdam & Cottbus. Naturschutzarbeit Berlin Brandenburg Beih. See B–P–H 636/3.

Beihefte zur Nova Hedwigia. Weinheim, Germany. Beih. Nova Hedwigia. See B–P–H 169/13. HI 51755

Beihefte der "Pharmazie." Berlin. Beih. "Pharmazie". See B–P–H 169/15. HI 51756

Beihefte, die Pharmazie. Ergänzungsband = Die Pharmazie. Beiheft. Ergänzungsband. Berlin. Pharmazie Beih. Ergänzungsband. See B–P–H 704/23.

Beihefte, Schriften der Thüringischen Landesarbeitsgemeinschaft für Heilpflanzenkunde und Heilpflanzenbeschaffung in Weimar und Mitteilungen der Thüringischen Botanischen Gesellschaft <ehemals Thüringischer Botanischer Verein> = Schriften der Thüringischen Landesarbeitsgemeinschaft für Heilpflanzenkunde und Heilpflanzenbeschaffung in Weimar und Mitteilungen der Thüringischen Botanischen Gesellschaft <ehemals Thüringischer Botanischer Verein>. Beiheft. Weimar. Schriften Thüring. Landesarbeitsgem. Heilpflanzenk. Weimar & Mitt. Thüring. Bot. Ges. Beih. See B–P–H 820/15.

Beihefte zur Sydowia. Horn. Ser. 2, vol. 1+, 1957+. Beih. Sydowia. HI 62226

Beihefte zum Tübinger Atlas des vorderen Orients. Reihe A, Naturwissenschaften. Wiesbaden. No. 1+, 1977+. Beih. Tübinger Atlas Vorderen Orients, A. HI 62227

Beihefte zu den Veröffentlichungen der Landesstelle für Naturschutz und Landespflege Baden-Württemberg. Ludwigsburg. Vols. 1-6, 1973-74. Beih. Veröff. Landesstelle Naturschutz Landespflege Baden-Württemburg. Superseded by: Beihefte zu den Veröffentlichungen für Naturschutz und Landschaftspflege in Baden-Württemberg. HI 62228

Beihefte zu den Veröffentlichungen für Naturschutz und Landschaftspflege in Baden-Württemberg. Ludwigsburg. Vol. 7+, 1975+. Beih. Veröff. Naturschutz Landschaftspflege Baden-Württemberg. Preceded by: Beihefte zu den Veröffentlichungen der Landesstelle für Naturschutz und Landespflege Baden-Württemberg. HI 62229

Beihefte zur Willdenowia = Willdenowia. Beihefte. Berlin-Dahlem.

Beihefte zur Zeitschrift Geologie. Berlin. 1952+. Beih. Z. Geol. HI 62230

Beihefte zur Zeitschrift für Mykologie. Durlangen. Vol. 2+, 1980+. Beih. Z. Mykol. Preceded by: Beihefte zur Zeitschrift für Pilzkunde. HI 62231

Beihefte zur Zeitschrift für Pilzkunde. Vol. 1, 1976. Beih. Z. Pilzk. Superseded by: Beihefte zur Zeitschrift für Mykologie. HI 62232

Beihefte zu den Zeitschriften des schweizerischen Forstversins. Bern. Vol. 1+, 1925+. Beih. Z. Schweiz. Forstvereins. HI 62233

Beijing linxueyuan xuebao= Journal of Beijing forestry college. Beijing.

Beijing shifan daxue xuebao. Ziran kexue ban = Journal of natural science of Beijing normal university. Peking (Beijing).

Beijing ziran bowuguan yanjiu baogao = Memoirs of Beijing natural history museum. [Beijing.]

Beilage zu: Jahreshefte des Vereins für Vaterländische Naturkunde in Württemberg = Ergebnisse der pflanzengeographischen Durchforschung von Württemberg, Baden und Hohenzollern. Stuttgart. Ergebn. Pflanzengeogr. Durchforsch. Württemberg. See B–P–H 361/21.

Beilage zu: Mitteilungen des Badischen Botanischen Vereins = Ergebnisse der pflanzengeographischen Durchforschung von Württemberg, Baden und Hohenzollern. Stuttgart. Ergebn. Pflanzengeogr. Durchforsch. Württemberg. See B–P–H 361/21.

Beinn bhreagh recorder. Baddeck, Vic. N.s. vols. 1-25, 1909-1921/22. Beinn Bhreagh Rec. HI 62234

Beiträge zur allgemeinen Botanik. Berlin. Beitr. Allg. Bot. See B–P–H 169/21. HI 51757

Beiträge des Bezirks-Naturkundemuseums Stralsund. Stralsund. Beitr. Bez.-Naturkundemus. Stralsund. See B–P–H 169/22. HI 51758

Beiträge zur Biologie der Pflanzen. Breslau [=Wroclaw, Poland]. Beitr. Biol. Pflanzen. See B–P–H 169/25. HI 51759

Beiträge zur Erforschung der Flora und Fauna Weissrusslands (Weissrutheniens) = Matar′yaly da vyvuchen′iya flëry i faŭny Belarusi. Minsk.

Beiträge zur Erforschung Mecklenburgischer Naturschutzgebiete. Greifswald, Germany. Beitr. Erforsch. Mecklenburg. Naturschutzgeb. See B–P–H 169/28. HI 51760

Beiträge zur Erforschung der Natur der Lausitz <Sachsen> = Natura lusatica. Bautzen. Nat. Lusatica. See B–P–H 629/13.

Beiträge zur Flora und Fauna Aschaffenburgs und seiner Umgebung. Aschaffenburg. Beitr. Fl. Fauna Aschaffenburgs. See B–P–H 169/29. HI 51761

Beiträge für die Forstwirtschaft. Berlin. 1971+. Beitr. Forstw. Preceded by: Archiv für Forstwesen. HI 62235

Beiträge zur geobotanischen Landesaufnahme. Zurich & Leipzig. Beitr. Geobot. Landesaufn. See B–P–H 170/1. HI 51762

Beiträge zur geobotanischen Landesaufnahme der Schweiz. Bern & Berlin. Beitr. Geobot. Landesaufn. Schweiz. See B–P–H 170/3. HI 51764

Beiträge zur Geographie, Geschichte und Staatenkunde. Nuremberg. Beitr. Geogr. See B–P–H 170/4. HI 51765

Beiträge zur gesammten Natur- und Heilwissenschaft. Prague. Beitr. Gesammten Natur- Heilwiss. See B–P–H 170/6. HI 51766

Beiträge zur Geschichte, Statistik, Naturkunde und Kunst von Tyrol und Vorarlberg. Innsbruck. Beitr. Gesch. Tyrol Vorarlberg. See B–P–H 170/8. HI 51767

Beiträge zur Heimatkunde. Nordhausen. Vols. ?-48, 19??-78? Beitr. Heimatk. Nordhausen. Superseded by: Beitrage zur Heimatkunde des Sensebezirks. HI 62236

Beiträge zur Heimatkunde des Sensebezirks. Basel. Vols. 49-52, 1979-84. Beitr. Heimatk. Sensebez. Preceded by: Beiträge zur Heimatkunde. Superseded by: Deutschfreiburger Beiträge zur Heimatkunde. HI 62237

Beiträge zur Kartierung der schweizer Flora. Bern. No. 1+, 1969+. Beitr. Kart. Schweiz. Fl. HI 62238

Beiträge zur Kenntnis des Russischen Reiches und der angränzenden Länder Asiens ... St. Petersburg. Beitr. Kenntn. Russ. Reiches. See B–P–H 170/10. HI 51768

Beiträge zur Klinik der Tuberkulose und spezifischen Tuberkulose-Forschung. Würzburg. Beitr. Klin. Tuberkulose Spezif. Tuberkulose-Forsch. See B–P–H 170/14. HI 51770

Beiträge zur klinischen Chirurgie. Tübingen. Beitr. Klin. Chir. See B–P–H 170/11. HI 51769

Beiträge zur Kolonialforschung. Berlin. Beitr. Kolonialf. See B–P–H 170/15. HI 51771

Beiträge zur Kolonialforschung. Ergänzungsband. Berlin. Beitr. Kolonialf. Ergänzungsband. See B–P–H 170/16. HI 51772

Beiträge zur Kryptogamenflora der Schweiz. Bern. Vols. 1-15(1), 1898-1977. Beitr. Kryptogamenfl. Schweiz. Superseded by: Cryptogamica helvetica. 1-625-2. HI 51773

Beiträge zur Kunde Preussens. Königsberg [=Kaliningrad, Russian S F S R]. Beitr. Kunde Preussens. See B–P–H 171/1. HI 51774

Beiträge zur Landeskunde von Oberösterreich. Naturwissenschaftliche Reihe. Linz. Vol. ?-2+, 19??-75+. Beitr. Landesk. Oberösterr., Naturw. Reihe. HI 62239

Beiträge zur Landeskunde Oesterreichs unter der Enns. Vienna. Beitr. Landesk. Oesterreichs unter der Enns. See B–P–H 171/2. HI 51775

Beiträge zur Landespflege. Stuttgart. Vols. 1-3, 1963-67. Beitr. Landespflege. Superseded by: Landschaft und Stadt. HI 62240

Beiträge zur Morphologie und Physiologie der Pflanzenzelle. Tübingen. Beitr. Morphol. Physiol.. Pflanzenzelle. See B–P–H 171/3. HI 51777

Beiträge zur näheren Kenntnis und Verbreitung der Algen. [Edited by L. Rabenhorst.] Leipzig. Beitr. Näheren Kenntn. Verbreit. Algen. See B–P–H 171/4. HI 51778

Beiträge zur Natur- und Heilkunde. Heilbronn. Baitr. Natur- Heilk. (Heilbronn). See B–P–H 171/6. HI 51780

Beiträge zur Naturdenkmalpflege. Berlin. Beitr. Naturdenkmalpflege. See B–P–H 171/8. HI 51781

Beiträge zur Naturgeschichte [ed. by Link]. Rostock & Leipzig. Beitr. Naturgesch. See B–P–H 171/10. HI 51782

Beiträge zur Naturgeschichte, Landwirthschaft und Topographie des Herzogthums Krain. Laibach [=Ljubljana, Yugoslavia]. Beitr. Naturgesch. Herzogth. Krain. See B–P–H 171/11. HI 51783

Beiträge zu der Naturgeschichte des Schweizerlandes. Bern. Beitr. Naturgesch. Schweizerl. See B–P–H 171/12. HI 51784

Beiträge zur Naturkunde. Kiel. Beitr. Naturk. See B–P–H 171/13. HI 51785

Beiträge zur Naturkunde Niedersachsens. Hanover. Vols. 1-17, 1948-64; vol. 27+, 1974+. Beitr. Naturk. Niedersachsens. From 1965-73 title changed to: Natur, Kultur und Jagd. 1-626-3. HI 62241

Beiträge zur Naturkunde in Osthessen. Fulda, BDR. 1969+. Beitr. Naturk. Osthessen. HI 62242

Beiträge zur Naturkunde aus den Ostseeprovinzen Russlands. Dorpat [=Tartu, Estonian S S R]. Beitr. Naturk. Ostseeprov. Russlands. See B–P–H 171/20. HI 51787

Beiträge zur Naturkunde Preussens, herausgegeben von der (Königlichen) Physikalisch-ökonomischen Gesellschaft zu Königsberg. Königsberg [=Kaliningrad, Russian S F S R]. Beitr. Naturk. Preussens. See B–P–H 171/22. HI 51788

Beiträge zur naturkundlichen Forschung im Oberrheingebiet. Karlsruhe. Beitr. Naturk. Forsch. Oberrheingeb. See B–P–H 171/14. HI 51786

Beiträge zur naturkundlichen Forschung in Südwestdeutschland. Karlsruhe. Vols. 1-4, 1936-39; vols. 8-39, 1949-80. Beitr. Naturk. Forsch. Südwestdeutschl. For vols. 5-7 see: Beiträge zur naturkundlichen Forschung im Oberrheingebiet. Superseded by: Carolinea. 1-626-3. HI 62243

Beiträge zum Naturschutz in der Schweiz. Basel. 1982+. Beitr. Naturschutz Schweiz. HI 62244

Beiträge zur naturwissenschaftlichen Erforschung Badens. Freiburg im Breisgau. Beitr. Naturwiss. Erforsch. Badens. See B–P–H 172/1. HI 51790

Beiträge zur Ökologie und Landeskultur. [Forms part of: Wissenschaftliche Zeitschrift der Friedrich-Schiller-Universität Jena. Mathematisch-Naturwissenschaftliche Reihe, vol. 22(3/4)+.] Jena. Vol. 1+, 1973+. Beitr.

Ökol. Landeskult. HI 62245

Beiträge zur Petrefacten-Kunde. Bayreuth. Beitr. Petrefacten-Kunde. See B–P–H 172/3. HI 51791

Beiträge zur Pflanzenkunde des Russischen Reiches. St. Petersburg. Beitr. Pflanzenk. Russ. Reiches. See B–P–H 172/6. HI 51792

Beiträge zur Pflanzenzucht. Berlin. Beitr. Pflanzenzucht. See B–P–H 172/7. HI 51793

Beiträge zur Phytotaxonomie. [Forms part of: Wissenschaftliche Zeitschrift der Friedrich-Schiller-Universität Jena. Mathematisch-Naturwissenschaftliche Reihe, vol. 17(3)+.] Jena. Vol. 1+, 1969+. Beitr. Phytotax. HI 62246

Beiträge zur rheinischen Naturgeschichte, herausgegeben von der Gesellschaft für Beförderung der Naturwissenschaften zu Freiburg im Breisgau. Greiburg im Breisgau. Beitr. Rhein. Naturgesch. See B–P–H 172/10. HI 51794

Beiträge zur Sittenlehre, Oekonomie, Arzneywissenschaft, Naturlehre und Geschichte in ihrem allgemeinen Umfange ... Mannheim. Beitr. Sittenl. See B–P–H 172/11. HI 51795

Beiträge zur Sukkulentenkunde und -pflege. Berlin. Beitr. Sukkulentenk. Sukkulentenpflege. See B–P–H 172/14. HI 51796

Beiträge zur Systematik und Pflanzengeographie. Berlin = Repertorium specierum novarum regni vegetabilis. Berlin. Repert. Spec. Nov. Regni Veg. Beih. See B–P–H 772/21.

Beiträge zur tropischen Landwirtschaft und Veterinärmedizin. Leipzig. Vol. 1+, 1963+. Beitr. Trop. Landw. Veterinärmed. HI 62247

Beiträge zur tropischen und subtropischen Landwirtschaft und Tropenveterinärmedizin = Beiträge zur tropischen Landwirtschaft und Veterinärmedizin. Leipzig.

Beiträge zur Vaterlandskunde für Inner-Oesterreichs Einwohner. Graz. Beitr. Vaterlandsk. Inner-Oesterreichs Einwohner. See B–P–H 172/16. HI 51797

Beiträge zur Verbesserung der Landwirthschaft durch alle ihre Theile. Prague. Beitr. Verbess. Landw. See B–P–H 172/17. HI 51798

Beiträge zu verschiedenen Wissenschaften von einigen oesterreichischen Gelehrten. Vienna. Beitr. Verschiedenen Wiss. See B–P–H 172/18. HI 51799

Beiträge zur Völker- und Länderkunde. Leipzig. Beitr. Völker-Länderk. See B–P–H 172/19. HI 51800

Beiträge der Wetterauischen Gesellschaft für die gesammte Naturkunde. Zur Botanik. Hanau. Beitr. Wetterauischen Ges. Gesammte Naturk. Bot. See B–P–H 172/20. HI 51801

Beiträge zur wissenschaftlichen Botanik. Leipzig. Beitr. Wiss. Bot. (Leipzig). See B–P–H 172/21. HI 51802

Beiträge zur wissenschaftlichen Botanik. Stuttgart. Beitr. Wiss. Bot. (Stuttgart). See B–P–H 172/22. HI 51803

Békés megyei természetvédelmi evkönyv. Békéscsaba. Vol. 1+, 1976+. Békés Megyei Természetvédel. Evk. HI 62248

Belehrende Herbarsbeilage. Prague. Belehrende Herbarsbeil. See B–P–H 173/1. HI 51804

Belgique coloniale. Brussels. Belgique Colon. See B–P–H 173/6. HI 51806

Belgique horticole; annales de botanique et d'horticulture. Liège, The Hague, & Ghent. Vols. 1-14, 1850-64. Superseded by: Annales d'horticulture belge et étrangère. 1-631-3. HI 62249

Belgique maritime et coloniale. Brussels. Belgique Marit. Colon. See B–P–H 173/8. HI 51807

Belgische fruit-revue: organ der Vlaamsche pomologische vereniginge. Sint-Amandsberg, Belgium. Belg. Fruit-Rev. See B–P–H 173/4. HI 51805

Belgische tuinbouw. Internationale editie. Ghent. Vol. ?-58+, 19??-??+. Belg. Tuinb., Int. Ed. HI 62250

Belmontia. Miscellaneous publications in botany. I. Taxonomy. Wageningen. Vols. 1-14, 1957-73. Belmontia, 1, Taxon. Superseded by: Belmontia. New series. HI 62251

Belmontia. Miscellaneous publications in botany. II. Ecology. Wageningen. Vols. 1-17, 1957-72. Belmontia, 2, Ecol. Superseded by: Belmontia. New series. HI 62252

Belmontia. Miscellaneous publications in botany. III. Horticulture. Wageningen. Vols. 1-10, 1957-1970. Belmontia, 3, Hort. Superseded by: Belmontia. New series. HI 62253

Belmontia. Miscellaneous publications in botany. IV. Incidental. Wageningen. Vols. 1-13, 1957-1970. Belmontia, 4, Incidental. Superseded by: Belmontia. New series. HI 62254

Belmontia. Miscellaneous publications in botany. New series. Wageningen. Vol. 1+, [1974]+. Belmontia, N.S. Preceded by: Belmontia, 1, Taxonomy; Belmontia, 2, Ecology; Belmontia, 3, Horticulture; and Belmontia, 4, Incidental. HI 62255

Bemerkungen der Kuhrpfälzischen physikalisch-ökonomischen Gesellschaft. Mannheim. Bemerk. Kuhrpfälz. Phys.-Ökon. Ges. See B–P–H 173/13. HI 51808

Bendigo naturalist. Bendigo, Vic. 1967+. Bendigo Naturalist. HI 62256

Bengal agricultural journal. Dacca, [Pakistan]. Bengal Agric. J. See B–P–H 173/15. HI 51809

Beobachtungen und Entdeckungen aus der Naturkunde von der Gesellschaft naturforschender Freunde zu Berlin = Schriften der Gesellschaft naturforschender Freunde zu Berlin. Berlin. Schriften Ges. Naturf. Freunde Berlin. See B–P–H 819/11.

Beobachtungen über das Phytoplankton der Wolga = Nablyudeniya nad Fitoplanktonum Volgi. Saratov.

Berättelse om botaniska arbeten och upptäckter. Stockholm. 1853-54. Berätt. Bot. Arbeten Upptäckter. Preceded by: Årsberättelse om botaniska arbeten och upptäckter. 5-4131-1. HI 62257

Beretning, bioteknisk institut. Kolding? No. 72+, 1975+. Beretn. Biotekn. Inst. Preceded by: Beretning, forskningsinstituttet for handels-og industrieplanter. HI 62258

Beretning om botanisk haves virksomhed. Copenhagen. Beretn. Bot. Haves Virks. See B–P–H 182/12. HI 51923

Beretning fra faellesudvalget for statens planteavls- og husdyrbrugsforsoeg. Copenhagen. No. 1+, 1971+. Beretn. Faellesudv. Statens Planteavl- Husdyrbrugsfors. HI 62259

Beretning, forskningsinstituttet for handels-og industrieplanter. Nos. ?-31-71, 19??-59-74. Beretn. Forskningsinst. Handels- Industriepl. Superseded by: Beretning, bioteknisk institut. HI 62260

Beretning om det skandinaviske naturforskermøde. Copenhagen. Vols. 18-19, 1925-36. Beretn. Skand. Naturforskermøde. Preceded by: Förhandlinger ved de skandinaviske naturforskeres möte. HI 62261

Beretning fra statens forsøgsvirksomhed i plantekultur. Kristiania, Lyngby. Nos. 1-513, 1895-1955. Beretn. Statens Forsøgsvirksomh. Plantekult. HI 62262

Beretning, statens plantetilsyn. Hellerup. Vol. 17+, 1967/68+, 1968+. Beretn. Statens Plantetilsyn. Preceded by: Årsberetning fra statens plantetilsyn vedrørende fropatologisk kontrol. HI 62263

Beretning fra statens planteavlsudvalg. Copenhagen. 1898/99+, 1899?+. Beretn. Statens Planteavlsudv. HI 62264

Bergcultures, organ van hat algemeen landbouw syndicaat en het Zuid en West Sumatra syndicaat. Batavia, Dutch E. Indies [=Jakarta, Indonesia]. Bergcultures. See B–P–H 182/14. HI 51924

Bergens museums årbok; afhandlimnger og aarsberetning. Bergen. 1883-1914/15; naturvidenskabelig rekke, 1915/16-47. Bergens Mus. Årbok. Superseded by: Universitet i Bergen årbok. Naturvidenskabelig rekke. 1-638-3. HI 62265

Bergens museums skrifter. Bergen. Nos. 1-7, 1878-1905; n.s. vols. 1-22, 1909-43. Bergens Mus. Skr. Superseded by: Universitet i Bergen skrifter. 1-639-1. HI 62266

Bergmännisches Journal. Freiberg, Germany. Bergmänn. J. See B–P–H 182/20. HI 51925

Bericht und Abhandlungen des Clubs für Naturkunde (Section des Brünner Lehrervereins). Brünn [=Brno, Czechoslovakia]. Ber. Abh. Clubs. Naturk. See B–P–H 173/19. HI 51810

Bericht van de afdeeling handelsmuseum van het koloniaal instituut = Bericht, Koninklijk instituut voor de tropen, afdeling agrarisch onderzoek. Amsterdam.

Berichte der Akademiker Borodin Biologischen Süsswasser Station = Trudy Borodinskoi presnovodnoi biologicheskoi stantsii v Karelii. Leningrad.

Bericht van het algemeen proefstation der A V R O S. Medan. Vols. 1-?, 1956-57. Ber. Alg. Proefstat. A. V. R. O. S. Superseded by: Bulletin of the research institute of the S P A. HI 62267

Bericht über den Annaberg-Buchholzer-Verein für Naturkunde. Annaberg, Germany. Ber. Annaberg-Buchholzer-Verein Naturk. See B–P–H 173/20. HI 51811

Berichte über den Arbeiten der Königlich Baierischen Akademie der Wissenschaften in München. Munich. Ber. Arbeiten Königl. Baier. Akad. WIss. München. See B–P–H 173/21. HI 51812

Berichte über die Arbeiten an der mathematisch-physicalischen Klasse der Königlich Bayerischen Akademie der Wissenschaften. Munich. Ber. Arbeiten Math.-Phys. Kl. Königl. Bayer. Akad. Wiss. See B–P–H 173/22. HI 51813

Berichte der Arbeitsgemeinschaft sächsischer Botaniker. Dresden. Ber. Arbeitsgem. Sächs. Bot. See B–P–H 173/23. HI 51814

Berichte aus den Arbeitskreisen heimische Orchideen; Beiträge zur Erhaltung, Erforschung und Verbreitung heimischer Orchideen. Friedberg. Vol. 1+, 1984+; Beiheft 1+, 1986+. Ber. Arbeitskreis. Heimische Orchid. HI 62268

Berichte der Bayerischen Botanischen Gesellschaft zur Erforschung der heimischen Flora. Munich. Ber. Bayer. Bot. Ges. See B–P–H 174/1. HI 51816

Bericht über die zur Bekanntmachung geeigneten Verhandlungen der Königlich Preussischen Akademie der Wissenschaften zu Berlin. Berlin. Ber. Bekanntm. Verh. Königl. Preuss. Akad. Wiss. Berlin. See B–P–H 174/5. HI 51817

Berichte Biochemie und Biologie. Berlin. Vols. 306-521, 1969-80. Ber. Biochem. Biol. Preceded by: Bericht über die gesamte Biologie. Abteilung A, Berichte über die wissenschaftliche Biologie. HI 62269

Berichte über die biologisch-geographischen Untersuchungen in den Kaukasusländern. Tiflis

[Georgian S S R]. Ber. Biol.-Geogr. Untersuch. Kaukasusländern. See B–P–H 174/7. HI 51818

Berichte der Biologischen Borodin Station = Trudy Borodinskoi biologicheskoi stantsii v Karelii. Leningrad.

Bericht, botanisch-zoologische Gesellschaft Liechtenstein-Sargens-Werdenberg. Vaduz. Vol. ?-14+, 19??-85+. Ber. Bot.-Zool. Ges. Liechtenstein-Sargens-Werdenberg. HI 62270

Bericht über den botanischen Garten in Bern. Bern. Ber. Bot. Gart. Bern. See B–P–H 174/9. HI 51819

Bericht über den botanischen Garten und das botanische Institut in Bern. Bern. Ber. Bot. Gart. Bot. Inst. Bern. See B–P–H 174/10. HI 51820

Bericht über den botanischen Garten und das botanische Institut der Universität Bern. Bern. Ber. Bot. Gart. Bot. Inst. Univ. Bern. See B–P–H 174/11. HI 51821

Bericht des botanischen Vereines in Landshut (Bayern) (anerkannter Verein). Landshut, Germany. Ber. Bot. Vereines Landshut. See B–P–H 174/12. HI 51822

Bericht der Bundesanstalt für Pflanzenbau und Samenprüfung in Wien. Vienna. 1924-31. Ber. Bundesanst. Pflanzenbau Samenprüfung Wien. Superseded by: Jahresbericht der Bundesanstalt für Pflanzenbau und Samenprüfung in Wien. HI 62271

Berichte der deutschen botanischen Gesellschaft. Berlin. Vols. 1-100, 1883-1986. Ber. Deutsch. Bot. Ges. Superseded by: Botanica acta. 2-1296-3. HI 51823

Berichte der deutschen chemischen Gesellschaft. Berlin. 1868-1944 [suspended 1945-46]. Ber. Deutsch. Chem. Ges. Superseded by: Chemische Berichte. HI 62272

Bericht, deutsche wissenschaftliche Kommission für Meeresforschung. Berlin. N.s. 1-24, 1919-76. Ber. Deutsch. Wiss. Kommiss. Meeresf. Superseded by: Meeresforschung: Reports on marine research. HI 62273

Bericht über den Deutschen Naturschutztag. Bad Godesberg. Ber. Deutsch. Naturschutztag. See B–P–H 174/18. HI 51824

Bericht der Deutschen Orchideen-Gesellschaft. Hamburg. Ber. Deutsch. Orchideen-Ges. See B–P–H 174/19. HI 51825

Berichte der Deutschen Pharmaceutischen Gesellschaft. Berlin. Ber. Deutsch. Pharm. Ges. See B–P–H 174/20. HI 51826

Berichte der Deutschen Pharmaceutischen Gesellschaft. Bericht über die pharmacognostische Literatur aller Länder. Berlin. Ber. Deutsch. Pharm. Ges. Ber. Pharmacogn. Lit. Aller Länder. See B–P–H 174/21. HI 51827

Bericht, eidgenössische Anstalt für das forstliche Versuchswesen. Birmensdorf. Vol. 1+, 1968+. Ber. Eidgenöss. Anst. Forstl. Versuchswesen. HI 62274

Berichte aus der Forschungsstelle nedri As = Skýrska, rannssóknastofnunin nedri As. Hveragerdi.

Berichte der Forstwirtschaftlichen Arbeitsgemeinschaft an der Hochschule für Bodenkultur in Wien. Vienna. Ber. Forstw. Arbeitsgem. Hochschule Bodenkult. Wien. See B–P–H 175/4. HI 51828

Bericht der Freien Vereinigung für Pflanzengeographie und systematische Botanik. Leipzig. Ber. Freien Vereinigung Pflanzengeogr. See B–P–H 175/7. HI 51829

Bericht über das Geobotanische Forschungsinstitut Rübel in Zürich. Zurich. Ber. Geobot. Forschungsinst. Rübel Zürich. See B–P–H 175/9. HI 51830

Berichte des Geobotanischen Instituts der Eidg. Techn. Hochschule Stiftung Rübel. Zurich. Ber. Geobot. Inst. E. T. H. Stiftung Rübel. See B–P–H 175/10. HI 51831

Bericht der geologischen Gesellschaft in der Deutschen Demokratischen Republik für das Gesamtgebiet der geologischen Wissenschaften. Berlin. Vol. 1-10, 1955/56-65; Sonderheft 1-3, 1963-65. Ber. Geol. Ges. D.D.R. Superseded by: Bericht der deutschen Gesellschaft für Geologische Wissenschaften. HI 62275

Bericht über die gesamte Biologie. Abteilung A, Berichte über die wissenschaftliche Biologie. Berlin. Vols. 1-305, 1926-69 [suspended 1944-48]. Ber. Gesamte Biol., Abt. A, Ber. Wiss. Biol. Superseded by: Berichte Biochemie und Biologie. 1-640-2. HI 62276

Berichte über die gesamte Biologie. Abteilung B, Berichte über die gesamte Physiologie und experimentelle Pharmakologie. Berlin. Vols. 1-?, 1920-80. Ber. Gesamte Biol., Abt. B, Ber. Gesamte Physiol. Superseded by: Berichte Physiologie, physiologische Chemie und Pharmakologie. 1-640-1. HI 62277

Bericht über die Gesamtsitzung des Preussischen Botanischen Vereins. Königsberg. Vol. 27, 1889? Ber. Gesamtsitzung Preuss. Bot. Ver. Preceded by: Bericht über die Versammlung des Preussischen Botanischen Vereins. Superseded by: Bericht über die wissenschaftlichen Verhandlungen der Jahresversammlung des Preussischen Botanischen Vereins. HI 62278

Bericht, Hamburgischer Lehrer-Verein für Naturkunde = Hamburgischer Lehrer-Verein für Naturkunde. Bericht. Hamburg. Hamburg. Lehrer-Verein Naturk. Ber. See B–P–H 410/28.

Berichte, Hauptstelle für Pflanzenschutz in Baden. Stuttgart. 1911-18?, 1912-19? Ber. Hauptstelle Pflanzenschutz Baden. HI 62279

Bericht über die Hauptversammlung, deutscher Forstverein. Berlin. Vols. 1-21, 1901-34. Ber. Hauptversamml. Deutsch. Forstverein. Superseded by: Jahresberichte, deutscher Forstverein. 2-1323-1. HI 62280

Bericht aus dem Haus der Natur in Salzburg. Abteilung A, zoologische und botanische Sammlungen sowie Allgemeines. Salzburg. Heft 1+, 1970+. Ber. Haus der Natur Salzburg, A. HI 62281

Bericht, höhere staatliche Lahranstalt för Obst- und Gartenbau. Merseburg, etc. 1918+. Ber. Höhere Staatl. Lahranst. Obst- Gartenbau. Preceded by: Bericht der Lehranstalt für Obst- und Gartenbau zu Proskau. HI 62282

Bericht über die internationalen Symposia der internationalen Vereinigung für Vegetationskunde. The Hague. 1959+, 1960+. Ber. Int. Symp. Int. Vereinigung Vegetationsk. HI 62283

Berichte der japanischen Gesellschaft für Balneologie. [Onsen kagaku]. Ber. Jap. Ges. Balneol. See B–P–H 175/15. HI 51833

Berichte der Kaiserlich Deutschen Akademie der Naturforscher zu Halle = Leopoldina. Amtliches Organ der Kaiserlichen Leopoldinisch-Carolinischen Detuschen Akademie der Naturforscher. Jena. Leopoldina. See B–P–H 529/1.

Bericht der Kaiserlichen Akademie der Wissenschaften zu St. Petersburg über die Zuerkennung der von dem Kammerherrn Paul von Demidov gestifteten Preise. St. Petersburg. Ber. Kaiserl. Akad. Wiss. St. Petersburg Zuerkenn. Paul von Demidov Gestifiteten Preise. See B–P–H 175/17. HI 51834

Bericht der Kommission zur wissenschaften Erforschung des Nationalparks = Jahresbericht der Eidgenössischen Nationalparkkommission. Grosshöchstetten.

Berichte aus der Königlich Bayerischen Biologischen Versuchsstation in München. Munich. Ber. König. Bayer. Biol. Versuchsstat. München. See B–P–H 175/18. HI 51835

Bericht der königliche Gärtnerlehranstalt zu Dahlem. Berlin. 1906-50. Ber. Königl. Gärtnerlehranst. Dahlem. Preceded by: Jahresbericht der königliche Gärtnerlehreanstalt zu Dahlem. HI 62284

Bericht der königliche Lehranstalt für Obst- und Gartenbau zu Proskau. Merseburg. 18??-1900-11? Ber. Königl. Lehranst. Obst- Gartenbau Proskau. Preceded by: Jahresbericht der königliche Lehranstalt für Obst- und Gartenbau zu Proskau. Superseded by: Bericht der Lehranstalt für Obst- und Gartenbau zu Proskau. HI 62285

Bericht der königliche Lehranstalt für Wein, Obst- und Gartenbau zu Geisenheim. Berlin. 1897+. Ber. Königl. Lahranst. Wein Obst- Gartenbau Geisenheim. HI 62286

Bericht, Koninklijk instituut voor de tropen, afdeling agrarisch onderzoek. Amsterdam. Nos. 1-275, 1920-61. Ber. Kon. Inst. Tropen, Afd. Agrar. Onderz. Superseded by: Bulletin, department of agricultural research of the royal tropical institute. HI 62287

Bericht, koninklijke vereeniging indisch instituut = Bericht, Koninklijk instituut voor de tropen, afdeling agrarisch onderzoek. Amsterdam.

Berichte über Land- und Forstwirtschaft in Deutsch-Ostafrika. Heidelberg. Ber. Land- Forstw. Deutsch-Ostafrika. See B–P–H 175/20. HI 51836

Bericht aus der Land- und Forstwissenschaftlichen Forschung. Bad Godesberg. Vol. 1+, 1956?+. Ber. Land- Forstw. Forsch. HI 62288

Berichte über Landwirtschaft; Zeitschrift für Agrarpolitik und Landwirtschaft. Berlin, Hamburg. Nos. 1-41, 1907-19; n.s. vol. 1+, 1923+ [publication suspended 1944-51]; Sonderhefte n.s. vol. 1+, 1924+. Ber. Landw. (Berlin). HI 62289

Berichte der landwirtschaftlichen Fakultät der Universität für Agrarwissenschaften. Gödöllö, Hungary = Agrártudományi egyetem mezögazdaságtudományi karának évkönyve. Gödöllö, Hungary. Agrártud. Egyet. Mezögazdaságtud. Karának Évk. (Gödöllö). See B–P–H 55/15.

Bericht der Lehranstalt für Obst- und Gartenbau zu Proskau. Oppeln. 1916-17. Ber. Lehranst. Obst-Gartenbau Proskau. Preceded by: Bericht der königliche Lehranstalt für Obst- und Gartenbau zu Proskau. Superseded by: Bericht, höhere staatliche Lahranstalt för Obst- und Gartenbau. HI 62290

Bericht des Lehrerklubs für Naturkunde. Brünn [=Brno, Czechoslovakia]. Ber. Lehrerklubs Naturk. See B–P–H 176/1. HI 51837

Bericht über die Leistungen des vaterländischen Vereines zur Bildung eines Museums für das Erzherzogthum Oesterreich ob der Enns und das Herzogthum Salzburg. Linz. Ber. Leist. Vaterl. Vereines Erzherzogth. Oesterreich ob der Enns. See B–P–H 176/3. HI 51838

Berichte der Limnologischen Flussstation Freudenthal. Aussenstelle der Hydrobiologischen Anstalt der Max-Planch-Gesellschaft. Münden, Germany. Ber. Limnol. Flussstat. Freudenthal. See B–P–H 176/4. HI 51839

Berichten en mededeelingen van het genootschap voor landbouw en kruidkunde te Utrecht. Utrecht. Ber Meded. Genootsch. Landb. Utrecht. See B–P–H 176/5. HI 51840

Berichte über die Mittheilungen von Freunden der Naturwissenschaften in Wien. Vienna. Ber. Mitth. Freunden Naturwiss. Wien. See B–P–H 176/7. HI 51841

Bericht über das Museum Francisco-Carolinum. Linz. Ber. Mus. Francisco-Carol. See B–P–H 176/8. HI 51842

Bericht des Museumsvereins oder Vereins für Kunde der Natur und Kunst im Fürstenthum Hildesheim und in der Stadt Goslar. Hildesheim. Ber Museumsvereins Fürstenth. Hildesheim See B–P–H 176/9. HI 51843

Bericht, naturforschende Gesellschaft in Bamberg = Naturforschende Gesellschaft in Bamberg. Bericht. Bamberg, Germany. Naturf. Ges. Bamberg Ber. See B–P–H 634/16.

Bericht der naturforschenden Gesellschaft Augsburg. Augsburg. Nos. 1-31, 1948-76. Ber. Naturf. Ges. Augsburg. 4-2931-3. HI 62291

Bericht der Naturforschenden Gesellschaft zu Bamberg. Bamberg, Germany. Ber. Naturf. Ges. Bamberg. See B–P–H 176/11. HI 51844

Berichte der Naturforschenden Gesellschaft zu Freiburg i. B. Freiburg im Breisgau. Ber. Naturf. Ges. Freiburg. See B–P–H 176/12. HI 51845

Berichte der Naturforschenden Gesellschaft Uri. Altdorf, Switzerland. See B–P–H 176/13. HI 51846

Bericht der Naturhistorischen Gesellschaft zu Hannover. Hanover. Ber. Naturhist. Ges. Hannover. See B–P–H 176/15. HI 51847

Bericht der Naturhistorischen Kantonal-Gesellschaft in Solothurn. Solothurn, Switzerland. Ber. Naturhist. Kantonal-Ges. Solothurn. See B–P–H 176/18. HI 51848

Bericht des naturhistorischen Vereins in Augsburg. Augsburg. Ber. Naturhist. Vereins Augsburg. See B–P–H 176/20. HI 51849

Bericht des Naturhistorischen Verein(e)s in Passau. Passau, Germany. Ber. Naturhist. Vereins Passau. See B–P–H 177/1. HI 51850

Bericht des Naturhistorischen Vereins von Wisconsin. Milwaukee, Wisconsin. Ber. Naturhist. Vereins Wisconsin. See B–P–H 177/2. HI 51851

Bericht des Naturwissenschaftlich-Medizinischen Vereins Innsbruck. Innsbruck. Ber. Naturwiss.-Med. Vereins Innsbruck. See B–P–H 177/7. HI 51984

Bericht über das naturwissenschaftliche Seminar der Universität zu Königsberg. Königsberg [=Kaliningrad, Russian S F S R]. Ber Naturwiss. Seminar Univ. Königsberg. See B–P–H 177/9. HI 51855

Bericht(e) der Naturwissenschaftlichen Gesellschaft (zu) Bayreuth. Bayreuth. Ber. Naturwiss. Ges. Bayreuth. See B–P–H 177/5. HI 51852

Bericht der Naturwissenschaftlichen Gesellschaft zu Chemnitz. Chemnitz [=Karl-Marx-Stadt]. Ber. Naturwiss. Ges. Chemnitz. See B–P–H 177/6. HI 51853

Bericht der Naturwissenschaftlichen Sektion des Vereins "Botanischer Garten" in Olmütz. Olmütz. Ber. Naturwiss. Sekt. Vereins Bot. Gart. Olmütz. See B–P–H 177/8. HI 51854

Bericht über den Naturwissenschaftlichen Verein zu Zerbst. Zerbst, Germany. Ber. Naturwiss. Verein Zerbst. See B–P–H 177/10. HI 51856

Bericht des naturwissenschaftlichen Vereines in Aussig. Aussig, Bohemia [=Usti nad Labem, Czechoslovakia]. Ber. Naturwiss. Vereines Aussig. See B–P–H 177/11. HI 51857

Berichte des Naturwissenschaftlichen Vereines zu Regensburg. Regensburg. Ber. Naturwiss. Vereines Regensburg. See B–P–H 177/13. HI 51859

Bericht des naturwissenschaftlichen Vereins in Bayreuth. Bayreuth. ?-1982+. Ber. Naturwiss. Ver. Bayreuth. HI 62292

Bericht des Naturwissenschaftlichen Vereins für Bielefeld und Umgebung. Bielefeld. Ber. Naturwiss. Vereins Bielefeld. See B–P–H 177/14. HI 51860

Berichte des Naturwissenschaftlichen Vereins in Dessau. Dessau. Ber. Naturwiss. Vereins Dessau. See B–P–H 177/15. HI 51861

Bericht des Naturwissenschaftlichen Vereins des Harzes. Blankenburg, Germany. Ber. Naturwiss. Vereines Harzes. See B–P–H 177/12. HI 51858

Berichte des Naturwissenschaftlichen Vereins des Harzes zu Blankenburg. Wernigerode, Germany. Ber. Naturwiss. Vereins Harzes Blankenburg. See B–P–H 177/17. HI 51862

Bericht des naturwissenschaftlichen Vereins Landshut (Bayern). Landshut, Germany. Ber. Naturwiss. Vereins Landshut. See B–P–H 177/18. HI 51863

Berichte des Naturwissenschaftlichen Vereins und des Museums für Naturkunde und Vorgeschichte in Dessau. Dessau. Ber. Naturwiss. Vereins Mus. Naturk. Dessau. See B–P–H 178/1. HI 51864

Bericht des Naturwissenschaftlichen Vereins für Schwaben e. V. Augsburg. Ber. Naturwiss. Vereins Schwaben. See B–P–H 178/2. HI 51865

Bericht des Naturwissenschaftlichen Vereins für Schwaben und Neuburg (a. V.) in Augsburg, früher Naturhistorischen Vereins in Augsburg. Augsburg. Ber. Naturwiss. Vereins Schwaben Augsburg. See B–P–H 178/3. HI 51866

Bericht des Naturwissenschaftlichen Vereins für Schwaben und Neuburg (a. V.), früher Naturhistorischen Vereins in Augsburg. Augsburg. Ber. Naturwiss. Vereins Schwaben Neuburg. See B–P–H 178/4. HI 51867

Bericht des naturwissenschaftlicher Vereins, Darmstadt. Darmstadt. 1957/58+, 1958+. Ber. Naturwiss. Ver. Darmstadt. HI 62293

Bericht des Nordoberfränkischen Vereins für Natur-, Geschichts- und Landeskunde in Hof. Hof, Germany. Ber. Nordoberfränk. Vereins Nartur- Landesk. Hof. See B–P–H 178/7. HI 51868

Bericht der Oberhessischen Gesellschaft für Natur- und Heilkunde. Giessen, Germany. Ber. Oberhess. Ges. Natur- Heilk. See B–P–H 178/9. HI 51869

Bericht der Oberhessischen Gesellschaft für Natur- und Heilkunde zu Giessen. Naturwissenschaftliche Abteilung. Giessen. N.s. vols. 1-37, 1904/06-70, 1907-1970. Ber. Oberhess. Ges. Natur- Heilk. Giessen, Naturwiss. Abt. Preceded by: Bericht der Oberhessischen Gesellschaft für Natur- und Heilkunde. Superseded by: Oberhessische naturwissenschaftliche Zeitschrift. 4-3126-3. HI 62294

Bericht über die oesterreichische Literatur der Zoologie, Botanik und Paleontologie. Vienna. Ber. Oesterr. Lit. Zool. Bot. See B–P–H 178/12. HI 51870

Bericht des Offenbacher Vereins für Naturkunde über seine Thätigkeit. Offenbach. Ber. Offenbacher Vereins Naturk. Thätigk. See B–P–H 178/13. HI 51871

Berichte des Ohara Instituts für landwirtschaftliche Biologie. Okayama Universität. Kurashiki, Japan. Ber. Ohara Inst. Landw. Biol. Okayama Univ. See B–P–H 178/14. HI 51873

Berichte des Ohara Instituts für landwirthschaftliche Forschungen in Kurashiki, Provinz Okayama, Japan. Kurashiki, Japan. Ber. Ohara Inst. Landw. Forsch. Kurashiki. See B–P–H 178/15. HI 51874

Bericht über Pflanzenschutz. Dielsdorf. No. 1+, 1942+. Ber. Pflanzenschutz. 3-2495-3. HI 62295

Berichte der Pharmaceutischen Gesellschaft. Berlin. Ber. Pharm. Ges. See B–P–H 178/17. HI 51875

Bericht über die pharmacognostische Literatur aller Länder = Berichte der Deutschen Pharmaceutischen Gesellschaft. Berlin. Ber. Deutsch. Pharm. Ges. Ber. Pharmacogn. Lit. Aller Länder. See B–P–H 174/21.

Berichte über die pharmakognostische Litteratur aller Länder, herausgegeben von der Deutschen Pharmaceutischen Gesellschaft. Berlin. Ber. Pharmakogn. Litt. Aller Länder. See B–P–H 179/1. HI 51876

Berichte der Physikalisch-Medizinischen Gesellschaft zu Würzburg. Würzburg. Ber. Phys.-Med. Ges. Würzburg. See B–P–H 179/3. HI 51878

Berichte Physiologie, physiologische Chemie und Pharmakologie. Berlin. Vols. 302-373, 1969-80. Ber. Physiol. Physiologischen Chem. Pharmakol. Preceded by: Berichte über die gesamte Biologie. Abteilung B, Berichte über die gesamte Physiologie und experimentelle Pharmakologie. HI 62296

Berichte aus dem Physiologischen Laboratorium und der Versuchsanstalt des Landwirthschaftlichen Instituts. Halle. Ber. Physiol. Lab. Versuchsanst. Landw. Inst. See B–P–H 179/2. HI 51877

Berichte zur Polarforschung. Bremerhaven. Vol. 1+, 1982+. Ber. Polarforsch. HI 62297

Bericht des Preussischen Botanischen Vereins E. V. Königsberg [=Kaliningrad, Russian S F S R]. Ber. Preuss. Bot. Vereins. See B–P–H 179/4. HI 51879

Bericht over rassenkreuze. Wageningen. Nos. 1-438, 1948-70. Ber. Rassenkreuze. Superseded by: Rassenbericht. HI 62298

Berichte der Saratower Naturforschergesellschaft = Izvestiya Saratovskogo obshchestva estestvoispytatelei. Saratov.

Bericht der schweizerischen botanischen Gesellschaft. Basel, Geneva, Zurich. Vols. 1-90, 1891-1981. Ber. Schweiz. Bot. Ges. Superseded by: Botanica helvetica. 5-3809-3. HI 62299

Berichte der Schwyz. Naturforschenden Gesellschaft. Einsiedeln, Switzerland. Ber. Schwyz. Naturf. Ges. See B–P–H 179/8. HI 51881

Bericht über die Senckenbergische Naturforschende Gesellschaft. Frankfurt a. M. Ber. Senckenberg. Naturf. Ges. See B–P–H 179/9. HI 51882

Bericht über die Senckenbergische Naturforschende Gesellschaft in Frankfurt am Main. Frankfurt a. M. Ber. Senckenberg. Naturf. Ges. Frankfurt. See B–P–H 179/10. HI 51883

Berichte über die Sitzungen der Gesellschaft für Botanik zu Hamburg. Kassel. Ber. Sitzungen Ges. Bot. Hamburg. See B–P–H 179/11. HI 51884

Bericht über die Sitzungen der Naturforschenden Gesellschaft zu Halle. Halle. Ber. Sitzungen Naturf. Ges. Halle. See B–P–H 179/12. HI 51885

Bericht, staatliches Museum für Naturkunde in Stuttgart. [Sonderdrucke aus: Jahreshefte des Vereins für vaterländische Naturkunde in Württemberg (1950-66), and Jahreshefte Gesellschaft für Naturkunde in Württemberg (1967/68+).] Stuttgart. 1950+, 1951+. Ber. Staatl. Mus. Naturk. Stuttgart. Preceded by: Berichte der Württembergischen Naturaliensammlung in Stuttgart. HI 62300

Bericht über die Tätigkeit der Biologischen Wolga-Station (in Saratow) = Otchet o Dĕyatel′nosti Volzhskoi Biologicheskoi Stantsii. Saratov.

Bericht über die Tätigkeit, Bundesanstalt für Pflanzenbau und Samenprüfung in Wien = Bericht der Bundesanstalt für Pflanzenbau und Samenprüfung in Wien. Vienna.

Bericht über die Tätigkeit des Fränkischen Gartenbauvereins. Würzburg. Ber. Tätigk. Fränk. Gartenbauvereins. See B–P–H 179/19. HI 51886

Bericht über die Tätigkeit (Jahrbuch) der St. Gallischen Naturwissenschaftlichen Gesellschaft. St. Gallen, Switzerland. Ber. Tätigk. (Jahrb.) St. Gallischen Naturwiss. Ges.. See B–P–H 179/20. HI 51887

Bericht über die Tätigkeit der Naturwissenschaftlichen Gesellschaft Isis in Bautzen. Bautzen. Ber. Tätigk. Naturwiss. Ges. Isis Bautzen. See B–P–H 179/21. HI 51888

Bericht über die Tätigkeit des Offenbacher Vereins für Naturkunde = Bericht über die Thätigkeit des Offenbacher Vereins für Naturkunde. Offenbach. Ber. Thätigk. Offenbacher Vereins Naturk. See B–P–H 180/4.

Bericht über die Tätigkeit der St. Gallischen Naturwissenschaftlichen Gesellschaft. St. Gallen, Switzerland. Ber. Tätigk. St. Gallischen Naturwiss. Ges. See B–P–H 180/1. HI 51889

Bericht über die Thätigkeit der Biologischen Wolga-Station = Otchet o Dĕyatel'nosti Volzhskoi Biologicheskoi Stantsii. Saratov.

Bericht über die Thätigkeit des Naturwissenschaftlichen Vereins in Lüneburg. Lüneburg, Germany. Ber. Thätigk. Naturwiss. Vereins Lüneburg. See B–P–H 180/3. HI 51890

Bericht über die Thätigkeit des Offenbacher Vereins für Naturkunde. Offenbach. Ber. Thätigk. Offenbacher Vereins Naturk. See B–P–H 180/4. HI 51891

Bericht über die Thätigkeit der St. Gallischen Naturwissenschaftlichen Gesellschaft. St. Gallen, Switzerland. Ber. Thätigk. St. Gallischen Naturwiss. Ges. See B–P–H 180/5. HI 51892

Bericht über die Thätigkeit des Thier- und Pflanzenschutz-Vereins für das Herzoghtum Coburg. Coburg, Germany. Ber. Thätigk. Their-Pflanzenschutz-Vereins Herzogth. Coburg. See B–P–H 180/6. HI 51893

Bericht über die Thätigkeit des Vereins für Naturkunde in (zu) Cassel. Cassel [=Kassel]. Ber. Thätigk. Vereins Naturk. Cassel. See B–P–H 180/7. HI 51894

Berichte der Tomsker Staats-Universität = Izvestiya Tomskogo gosudarstvennogo universiteta.

Berichte der Tomsker Staatsuniversität = Izvestiya Imperatorskago Tomskago Universiteta. Tomsk.

Berichte der ungarischen pharmazeutischen Gesellschaft = Magyar gyógyszerésztudományi társaság értesitöje, vol. 21, 1947. Magyar Gyógyszerésztud. Társ. Értes. See B–P–H 543/7.

Bericht des Vereines für Naturkunde zu Cassel. Kassel. Ber. Vereins Naturk. Cassel. See B–P–H 180/19.

Bericht des Vereines für Naturkunde zu Kassel. Kassel. Ber. Vereines Naturk. Kassel. See B–P–H 180/14. HI 51895

Bericht des Vereines zum Schutze und zur Pflege der Alpenpflanzen (e. V.). Bamberg, Germany. Ber. Vereines Schutze Pflege Alpenpfl. See B–P–H 180/15. HI 51896

Bericht des Vereins zur Erforschung der heimischen Pflanzenwelt, Halle. Halle. Ber. Vereins Erforsch. Heimischen Pflanzenwelt Halle. See B–P–H 180/16. HI 51897

Bericht des Vereins für Kunde der Natur und der Kunst im Fürstenthum Hildesheim und in der Stadt Goslar. Hildesheim. Ber. Vereins Kunde Natur Fürstenth. Hildesheim. See B–P–H 180/17. HI 51898

Berichte des Vereins "Natur und Heimat" und des Naturhistorischen Museums zu Lübeck. Lübeck. No. 1+, 1959+. Ber. Vereins "Natur und Heimat" Lübeck. Preceded by: Mitteilungen der geographischen Gesellschaft Lübeck. HI 51899

Bericht des Vereins für Naturkunde zu Cassel. Kassel. Ber. Vereins Naturk. Cassel. See B–P–H 180/19. HI 51902

Bericht des Vereins für Naturkunde zu Fulda. Fulda. Ber. Vereins Naturk. Fulda. See B–P–H 180/20. HI 51904

Bericht des Vereins für Naturkunde zu Kassel. Kassel. Ber. Vereins Naturk. Kassel. See B–P–H 181/1. HI 51905

Bericht des Vereins zum Schutze der Alpenpflanzen (e. V.). Bamberg, Germany. Ber. Vereins Schutze Alpenpfl. See B–P–H 181/2. HI 51906

Bericht über die Verhandlungen des academischen naturwissenschaftlichen Vereins zu Breslau. Breslau [=Wroclaw, Poland]. Ber. Verh. Acad. Naturwiss. Vereins Breslau. See B–P–H 181/3. HI 51907

Berichte über die Verhandlungen der königlich Sächsischen Gesellschaft der Wissenschaften zu Leipzig. Mathematisch-physische Classe. Leipzig. Ber. Verh. König. Sächs. Ges. Wiss. Leipzig, Math.-Phys. Cl. See B–P–H 181/4. HI 51908

Bericht über die Verhandlungen der Naturforschenden Gesellschaft in Basel. Basel. Ber. Verh. Naturf. Ges. Basel. See B–P–H 181/5. HI 51909

Berichte über die Verhandlungen der Naturforschenden Gesellschaft zu Freiburg i. Br. Freiburg im Breisgau. Ber. Verh. Naturf. Ges. Freiburg. See B–P–H 181/6. HI 51910

Bericht über die Verhandlungen der Naturforschenden Gesellschaft in Zürich. Zurich. Ber. Verh. Naturf. Ges. Zürich. See B–P–H 181/7. HI 51911

Berichte über Verhandlungen der Sächsischen Akademie

der Wissenschaften zu Leipzig. Mathematisch-naturwissenschaftliche Klasse. Leipzig. Ber. Verh. Sächs. Akad. Wiss. Leipzig, Math.-Naturwiss. Kl. See B–P–H 181/8. HI 51912

Berichte über Verhandlungen der Sächsischen Akademie der Wissenschaften zu Leipzig. Mathematisch-physische Klasse. Leipzig. Ber. Verh. Sächs. Akad. Wiss. Leipzig, Math.-Phys. Kl. See B–P–H 181/9. HI 51913

Berichte über die Verhandlungen der Sächsischen Gesellschaft der Wissenschaften zu Leipzig. Mathematisch-physische Klasse. Leipzig. Ber. Verh. Sächs. Ges. Wiss. Leipzig, Math.-Phys. Kl. See B–P–H 181/10. HI 51914

Berichte über die Versammlung Deutscher Forstmänner. Berlin. Ber. Versamml. Deutsch. Forstmänner. See B–P–H 181/11. HI 51915

Bericht über die Versammlung deutscher Naturforscher und Aerzte. Prague. Ber. Versamml. Deutsch. Naturf. Aerzte See B–P–H 181/12. HI 51916

Bericht über die Versammlung deutscher Naturforscher und Ärzte. Vienna. Ber. Versamml. Deutsch. Naturf. Ärzte. See B–P–H 181/13. HI 51917

Bericht über die Versammlung von Freunden der Flora Preussens und Stiftung des Preussischen Botanischen Vereins. Königsberg. 1862. Ber. Versamml. Freunden Fl. Preuss. Superseded by: Bericht über die Versammlung des Preussischen Botanischen Vereins. HI 62301

Bericht über die Versammlung des naturwissenschaftlichen Vereins für thUuringen. Erfurt. Ber. Versamml. Naturwiss. Vereins Thüringen. See B–P–H 181/14. HI 51918

Bericht über die Versammlung des Preussischen Botanischen Vereins. Königsberg. Vols. [2]-26, 1863-88? Ber. Versamml. Preuss. Bot. Ver. Preceded by: Bericht über die Versammlung von Freunden der Flora Preussens und Stiftung des Preussischen Botanischen Vereins. Superseded by: Bericht über die Gesamtsitzung des Preussischen Botanischen Vereins. HI 62302

Bericht(e über die Versammlung) des Westpreussischen Botanisch-Zoologischen Vereins (zu) Danzig. Danzig [=Gdansk, Poland]. Ber. Versamml. Westpreuss. Bot.-Zool. Vereins Danzig. See B–P–H 182/2. HI 51920

Bericht über die Versammlung, württembergischer Forstverein. Stuttgart. Nos. ?-9-36, 18??-86-1929. Ber. Versamml. Württemberg. Forstverein. HI 62303

Bericht des Westpreussischen Botanisch-Zoologischen Vereins, Danzig = Bericht(e über die Versammlung) des Westpreussischen Botanisch-Zoologischen Vereins (zu) Danzig. Danzig [=Gdansk, Poland]. Ber. Versamml. Westpreuss. Bot.-Zool. Vereins Danzig. See B–P–H 182/2.

Bericht der Wetterauischen Gesellschaft für die gesammte Naturkunde zu Hanau. Hanau. Ber. Wetterauischen Ges. Gesammte Naturk. Hanau. See B–P–H 182/4. HI 51921

Berichte des wissenschaftlichen Meeresinstituts = Trudy morskogo nauchnogo instituta. Moscow, Leningrad. and Trudy plovuchego Morskogo nauchnogo instituta. Moscow.

Bericht über die wissenschaftlichen Verhandlungen der Jahresversammlung des Preussischen Botanischen Vereins. Königsberg. Vol. 29, 1890. Ber. Wiss. Verh. Jahresversamml. Preuss. Bot. Ver. Preceded by: Bericht über die Gesamtsitzung des Preussischen Botanischen Vereins. Superseded by: Jahresbericht des Preussischen Botanischen Vereins. HI 62304

Berichte der wissenschaftsgeschichte. Weinheim. Vol. 11+, 19??-88+. Ber. Wiss HI 62305

Berichte der Württembergischen Naturaliensammlung in Stuttgart. [Sonderdruck aus: Jahreshefte des Vereins für vaterländische Naturkunde in Württemberg.] Stuttgart. 1932-46/49, 1934-50. Ber. Württemberg. Naturaliensamml. Stuttgart. Superseded by: Bericht, staatliches Museum für Naturkunde in Stuttgart. HI 62306

Bericht der Zürcherischen botanischen Gesellschaft. Zurich. Vols. 6-20, 1895/96-1941/43, 1896-1943? Ber. Zürcherischen Bot. Ges. Preceded by: Jahresbericht der Zürcherischen botanischen Gesellschaft. 5-4644-3. HI 62307

Bericht über die Zusammenkunft der freien Vereinigung der systematischen Botaniker und Pflanzengeographen zu ... Leipzig. Ber. Zusammenk. Freien Vereinigung Syst. Bot. See B–P–H 182/8. HI 51922

Berichte, Berichten, see Bericht.

Berigt nopens de gouvernments kina-onderneming op Java. Batavia, Dutch E. Indies [=Jakarta, Indonesia]. Ber. Gouv. Kina-Ondern. Java. See B–P–H 175/13. HI 51832

Berita, balai penjelidikan perusahaan gula. Pasuruan. 1958+. Berita Balai Penjel. Perusah. Gula. Preceded by: Mededelingen uit de Java-suikerindustrie. HI 62308

Berita biologi. Bogor. Vol. 1+, 1968+. Berita Biol. HI 62309

Berkeley newsletter and quarterly, university of California botanical garden. Berkeley, CA. 19??-77+. Berkeley Newslett. Quart., Univ. Calif. Bot. Gard. HI 62310

Berliner allgemeine Gartenzeitung. Berlin. Berliner Allg. Gartenzeitung See B–P–H 183/19. HI 51934

Berliner Beyträge zur Landwirthschaftswissenschaft.

Berlin. Berliner Beytr. Landwirthschaftswiss. See B–P–H 183/20. HI 51936

Berliner entomologische Zeitschrift. Berlin. Berliner Entomol. Z. See B–P–H 183/23. HI 51937

Berliner geographische Abhandlungen. Berlin. Vol. 1+, 1962?+. Berliner Geogr. Abh. HI 62311

Berliner klinische Wochenschrift; Organ für praktische Aerzte. Berlin. Berliner Klin. Wochenschr. See B–P–H 183/24. HI 51938

Berliner Naturschutzblätter. Berlin. Berliner Naturschutzbl. See B–P–H 183/25. HI 51939

Berlinische Blätter. Berlin. Berlin. Blätt. See B–P–H 183/6. HI 51927

Berlinische Monatsschrift. Berlin. Berlin. Monatsschr. See B–P–H 183/17. HI 51932

Berlinische Sammlungen zur Beförderung der Arzneywissenschaft, der Naturgeschichte, der Haushaltungskunst, Cameralwissenschaft und der dahin einschlagenden Litteratur. Berlin. Berlin. Samml. Beförd. Arzneywiss. See B–P–H 183/18. HI 51933

Berlinisches Archiv der Zeit und ihres Geschmacks. Berlin. Berlin. Arch. Zeit Geschmacks. See B–P–H 183/5. HI 51926

Berlinisches Iahrbuch der Pharmacie = Berlinisches Jahrbuch der Pharmacie. Berlin. Berlin. Jahrb. Pharm. See B–P–H 183/10.

Berlinisches Iahrbuch für die Pharmacie und für die damit verbundenen Wissenschaften = Berlinisches Jahrbuch für die Pharmacie und für die damit verbundenen Wissenschaften. Berlin. Berlin. Jahrb. Pharm. Verbundenen Wiss. See B–P–H 183/11.

Berlinisches Jahrbuch der Pharmacie. Berlin. Berlin. Jahrb. Pharm. See B–P–H 183/10. HI 51928

Berlinisches Jahrbuch für die Pharmacie und für die damit verbundenen Wissenschaften. Berlin. Berlin. Jahrb. Pharm. Verbundenen Wiss. See B–P–H 183/11. HI 51929

Berlinisches Magazin, oder gesammlete Schriften und Nachrichten für die Liebhaber der Arzneywissenschaft, Naturgeschichte und der angenehmen Wissenschaften überhaupt. Berlin. Berlin. Mag. See B–P–H 183/15. HI 51930

Berlinisches Magazin der Wissenschaaften und Künste. Berlin. Berlin. Mag. Wiss. Künste. See B–P–H 183/16. HI 51931

Bermuda biological station for research. Contributions. Cambridge, MA. Bermuda Biol. Sta. Res. Contr. See B–P–H 183/26. HI 51940

Berner Beiträge zur Geschichte der Medizin und Naturwissenschaften. Bern. Berner Beitr. Gesch. Med. Naturwiss. See B–P–H 184/4. HI 51943

Bernerisches Magazin der Natur, Kunst und Wissenschaften. Bern. Berner. Mag. Natur. See B–P–H 184/5. HI 51944

Bernice P. Bishop museum bulletin. Honolulu, HI. 1922-87. Bernice P. Bishop Mus. Bull. Superseded by: Bishop museum bulletins in botany. HI 62312

Bernische Blätter für Landwirthschaft. Bern. Bern. Blätt. Landw. See B–P–H 184/1. HI 51942

Berry botanic garden [series]. Portland, OR. Vols. ?-2(2)-9, 19??-79-86. Berry Bot. Gard. Superseded by: Bulletin of the Berry botanic garden. HI 62313

Beschäftigungen der Berlinischen Gesellschaft Naturforschender Freunde. Berlin. Beschäft. Berlin. Ges. Naturf. Freunde. See B–P–H 184/8. HI 51945

Beschrijvende rassenlijst voor fruit. Wageningen. Beschr. Rassenlijst Fruit. See B–P–H 184/9. HI 51946

Beschrijvende rassenlijst voor fruitgewassen. Wageningen. Beschr. Rassenlijst Fruitgew. See B–P–H 184/10. HI 51947

Beschrijvende rassenlijst voor groentegewassen. Wageningen. Beschr. Rassenlijst Groentegew. See B–P–H 184/11. HI 51948

Beschrijvende rassenlijst voor landbouwgewassen van de rijkscommissie voor de samenstelling van de rassenlijst. Wageningen. Beschr. Rassenlijst Landbouwgew. See B–P–H 184/12. HI 51949

Besorgte Forstmann. Weimar. Besorgte Forstmann. See B–P–H 184/13. HI 51950

Besseres Obst. Vienna. Besseres Obst. See B–P–H 184/14. HI 51951

Better crops. New York, NY. Better Crops. See B–P–H 184/18. HI 51952

Better crops with plant food. New York, NY. Better Crops with Pl. Food. See B–P–H 184/20. HI 51953

Better delphiniums. San Rafael, CA. Better Delphiniums. See B–P–H 184/21. HI 51954

Better farming. Chicago, IL. Better Farming. See B–P–H 184/22. HI 51955

Better flowers. Portland, OR. Better Fl. See B–P–H 184/23. HI 51956

Better fruit. Portland, OR. Better Fruit. See B–P–H 184/24. HI 51957

Better homes and gardens. Des Moines, IA. Better Homes Gard. See B–P–H 185/1. HI 51958

Better New York hops. Geneva, NY. Better New York Hops. See B–P–H 185/2. HI 51959

Betteravier français. Paris. N.s. no. 1+, 1952+. Betterav. Franç. Preceded by: Planteurs de betteraves. HI 62314

Betula; devoted to trees and shrubs. Westminster [=London, in part]. Betula. See B–P–H 185/3. HI 51960

Betuwsche rassenlijst voor fruit van de nederlandsche fruittelers organisatie. The Hague. Betuwsche Rassenlijst Fruit. See B–P–H 185/4. HI 51961

Beyträge zur Beförderung der Haushaltungskunde und anderer damit verwandter Wissenschaften. Münster. Beytr. Beförd. Haushaltungsk. Verwandter Wiss. See B–P–H 185/7. HI 51962

Beyträge zur Beförderung der Naturkunde. Halle. Beytr. Beford. Naturk. See B–P–H 185/8. HI 51963

Beyträge zur Botanik. Bremen. Beytr. Bot. See B–P–H 185/9. HI 51964

Beyträge zur Bürgerlichen Naturgeschichte, Naturlehre und dem Feldbau, aus dem Schreiben der Brüssler Akademie. Leipzig. Vols. 1-?, 1783-? Beytr. Bürgerl. Natur-Gesch. Naturl. Feldbau. HI 62315

Beyträge zu den Chemischen Annalen. Helmstedt & Leipzig. Beytr. Chem. Ann. See B–P–H 185/10. HI 51965

Beyträge zur Geschichte der Erfindungen. Leipzig. Beytr. Gesch. Erfind See B–P–H 185/11. HI 51966

Beyträge zur Kultur der Oekonomie in Preussen = Acta der Königlich-Ostpreussisch-Mohrungschen physicalisch-oeconomischen Gesellschaft. Königsberg. Acta Königl.-Ostpreuss.-Mohrungschen Phys.-Oecon. Ges. See B–P–H 45/5.

Beyträge zur Landeskunde von Oesterreich ob der Enns. Linz. Beytr. Landesk. Oesterreich ob der Enns. See B–P–H 185/13. HI 51967

Beyträge zur Literärgeschichte und Bibliographie. Munich. Beytr. Literärgesch. Bibliogr. See B–P–H 185/14. HI 51968

Beyträge zur Naturkunde des Herzogthums Zelle. Celle. Beytr. Naturk. Herzogth. Zelle. See B–P–H 185/15. HI 51969

Beyträge zur natürlichen ökonomischen und politischen Geschichte der Ober- und Niederlausiz. Zittau, Germany. Beytr. Natürl. Ökon. Polit. Gesch. OBer-Niederlausiz. See B–P–H 185/16. HI 51970

Beyträge zur Naturwissenschaft. Leipzig. Beytr. Naturwiss. See B–P–H 185/17. HI 51971

Beyträge zur neuesten Litteraturgeschichte oder Rothenburgische gelehrte Intelligenzblätter. Rothenburg ob der Tauber, Germany. Beytr. Neuesten Litteraturgesch. See B–P–H 185/18. HI 51972

Beyträge zur Oekonomie, Technologie, Polizey und Cameralwissenschaft. Göttingen. Vols. 1-12, 1779-91. Beytr. Oekon. Technol. Polizey Cameralwiss. HI 62316

Bi-monthly research notes, department of forestry, Canada = Bi-monthly research notes, forestry service, Canada. Ottawa.

Biblio-mer; bibliographie de la mer, des marines, des eaux de mer et interieures: sciences, enseignement, littérature: livres et périodiques. Ostende. 1967/68+, 1969+. Biblio-Mer. HI 62318

Bibliografía agrícola chilena. Santiago de Chile. 1960/77+, 1979+. Bibliogr. Agríc. Chilena. HI 62319

Bibliografía agrícola latinoamericana. San José, Costa Rica. Vols. 1-9, 1966-74. Bibliogr. Agríc. Latinoamer. Superseded by: Indice agrícola de Amerіca latina y el Caribe. HI 62320

Bibliografia agricolă curentă română. Bucharest. Vol. 1+, 1967+. Bibliogr. Agric. Curentă Română. HI 62321

Bibliografía agrícola del Uruguay. Montevideo. 1965/66+, [1966]+. Bibliogr. Agríc. Uruguay. HI 62322

Bibliografía argentina de agronomía y veterinaria. Buenos Aires. Vols. 1-3, 1967-70. Bibliogr. Argent. Agron. Veterin. HI 62323

Bibliografía do arroz. Brasilia. Vol. 1+, 1869/1975+, 1975+. Bibliogr. Arroz. HI 62324

Bibliografia bibliotecologica, bibliografica y de obras de refencia colombianas. Medellín, Colombia. Bibliogr. Bibliotecol. See B–P–H 186/20. HI 51976

Bibliografia brasileira de agricultura. Brasilia. Vol. 1+, 1975/77+, 1978+. Bibliogr. Brasil. Agric. Preceded by: Bibliografia brasileira de ciências agrícolas. HI 62325

Bibliografia brasileira de botânica. Rio de Janeiro. Vol. 1+, 1950/55+, [1956]+. Bibliogr. Brasil. Bot. HI 62326

Bibliografia brasileira de ciências agrícolas. Rio de Janeiro. Vols. 1-8, 1967-74/75, 1969-77. Bibliogr. Brasil. Ci. Agríc. Superseded by: Bibliografia brasileira de agricultura. HI 62327

Bibliografía do cacau. Brasilia. Vol. 1+, 1977+. Bibliogr. Cacau. HI 62328

Bibliografía do café. Brasilia. Vol. 1+, 1975+. Bibliogr. Café. HI 62329

Bibliografia de ciencias históricas. Madrid. Bibliogr. Ci. Hist. See B–P–H 186/25. HI 51980

Bibliografia florestal de interese africano. Lourenço Marques. No. 1+, 1965+. Bibliogr. Florest. Interese Africano. HI 62330

Bibliografia forestal latinoamericana. Merida. No. 01+, 1982+. Bibliogr. Florest. Latinoamer. HI 62331

Bibliografia forestal Venezuela. [Forms part of: Bibliografia forestal latinoamericana.] Merida. Vol. 1+, 1982+. Bibliogr. Forest. Venez. HI 62332

Bibliografia forestiera romana. Bucharest. 1965+. Bibliogr. Forest. Romana. HI 62333

Bibliografia lucrărilor publicati de cadrele didactice din institutul politechnic. Serie A, silvicultură şi industria lemnului. Braşov. Vol. 1+, 1968+. Bibliogr. Lucr. Publ. Cadr. Didact. Inst. Politechn., A. HI 62334

Bibliografía de la silvicultura y productos forestales = Bibliography of forestry and forest products. Washington, DC. Bibliogr. Forest. Forest Prod. See B–P–H 187/3.

Bibliografičeskie materialy, [vsesojuznyj institut rastenievodstva, biblioteka] = Bibliograficheskie materialy, [vsesoyuznyi institut rastenievodstva, biblioteka]. Leningrad.

Bibliograficheskie materialy, [vsesoyuznyi institut rastenievodstva, biblioteka]. Leningrad. Vols. 1-4, 1932-34. Bibliogr. Mater. 3-2391-1. HI 62335

Bibliografie botaniczne. Cracow. Vol. 1+, 1983+. Bibliogr. Bot., Cracow. HI 62336

Bibliografija jugoslavije. Serija B, prirodne i primenjene nauke. Belgrade. 1952+. Bibliogr. Jugoslav., B. HI 62337

Bibliografija Vostoka = Bibliografiya Vostoka. Moscow & Leningrad.

Bibliografiya Vostoka. Moscow & Leningrad. Vols. 1-10, 1932-37. Bibliogr. Vostoka (Moscow & Leningrad). 1-117-3. HI 62338

Bibliographia biotheoretica. [Acta biotheoretica. Series C.] Leiden. Vols. 1-8 (1925/29-1960/64), 1929-74. Bibliogr. Biotheor. 1-644-3. HI 62339

Bibliographia botanica čechoslovaca. Prague. 1952+. Bibliogr. Bot. Čechoslov. HI 62340

Bibliographia evolutionis. Paris. Bibliogr. Evol. See B–P–H 187/1. HI 51985

Bibliographia forestalis. Berlin. Berliogr. Forest. See B–P–H 187/2. HI 51986

Bibliography of forestry and forest products. Washington, DC. Bibliogr. Forest. Forest Prod. See B–P–H 187/3. HI 51987

Bibliographia genetica. [Supplement to: Genetica.] The Hague. Vols. 1-20, 1925-70. Bibliogr. Genet. 1-665-1. HI 62341

Bibliographia historiae rerum rusticarum internationalis. Budapest. 1960/61+, 1964+. Bibliogr. Hist. Rerum Rustic. Int. HI 62342

Bibliographia Humboldtiana. Bonn. 1986+. Bibliogr. Humboldtuana. Preceded by: Humboldtiana. HI 62343

Bibliographia oceanographica. Venice & Rome. Bibliogr. Oceanogr. See B–P–H 187/8. HI 51989

Bibliographia phytosociologica syntaxonomica. Lehre. Vol. 1+, 1971+; supplementum vol. 1+, 1976+. Bibliogr. Phytosoc. Syntax. HI 62344

Bibliographia scientiae naturalis helveticae. Bern. Bibliogr. Sci. Nat. Helv. See B–P–H 187/14. HI 51993

Bibliographia syntaxonomica cechoslovaca. Pruhonice. Vol. 1+, 1983+. Bibliogr. Syntax. Cechoslov. HI 62345

Bibliographical bulletin, instituto forestal Latino-Americano de investigación y capacitación. Information on forestry in latin America. [English edition of: Boletin bibliográfico, instituto forestal Latino-Americano de investigación y capitación.] Merida. Nos. 1(34)-11(52), 1976?-80. Bibliogr. Bull. Ist. Forest. Latinoamer. Invest. Capac. Superseded by: Latin american forest bibliography. HI 62346

Bibliographical bulletin, latin american forestry research and training institute = Bibliographical bulletin, instituto forestal Latino-Americano de investigación y capacitación. Merida.

Bibliographical bulletin. United States department of agriculture. Washington, DC. Bibliogr. Bull. U.S.D.A. See B–P–H 185/24. HI 51979

Bibliographical contributions, institute of plant industry, Leningrad, library = Bibliograficheskie materialy, [vsesoyuznyi institut rastenievodstva, biblioteka]. Leningrad.

Bibliographical contributions. Lloyd library and museum. Cincinnati, OH. Bibliogr. Contr. Lloyd Libr. Mus. See B–P–H 186/27. HI 51981

Bibliographical series on coconut. Series 1. 1967/68+, 1969+. Bibliogr. Ser. Coconut, Ser. 1. HI 62347

Bibliographical series, science library = Science library. Bibliographical series. London. Sci. Libr., Bibliogr. Ser. See B–P–H 824/23.

Bibliographie agricole courante roumaine = Bibliografia agricolă curentă română. Bucharest.

Bibliographie agricole yougoslave contemporaive = Savrema jugoslovenska poljoprivredny bibliografija. Belgrade.

Bibliographie des ausländischen forst- und holzwirtschaftlichen Schrifttums. Berlin. Vols. 1-8, 1950-54. Bibliogr. Ausl. Forst- Holzw. Schrifttums. Superseded by: Bibliographie des forst-holzwirtschaftlichen Schrifttums. HI 62348

Bibliographie de la Belgique. Brussels. Bibliogr. Belgique. See B–P–H 186/19. HI 51975

Bibliographie der Biochemie und Biophysik. Berlin. Bibliogr. Biochem. Biophys. See B–P–H 186/21. HI 51977

Bibliographie der Biologie. Berlin. Bibliogr. Biol. See B–P–H 186/22. HI 51978

Bibliographie courante, association pour le développement de la riziculture en Afrique de l'ouest = Current bibliography, west Africa rice development association. [Monrovia.]

Bibliographie der deutschen naturwissenschaftlichen Literatur. Jena. Bibliogr. Deutsch. Naturwiss. Lit. See B–P–H 186/28. HI 51982

Bibliographie von Deutschland. Leipzig. Bibliogr. Deutschl. See B–P–H 186/29. HI 51983

Bibliographie de l'empire français. Paris. Bibliogr. Empire Franç. See B–P–H 186/30. HI 51984

Bibliographie de l'empire français. Paris = Journal général de l'imprimerie et de la librairie. Paris. J. Gén. Imprimerie Libr. See B–P–H 467/5.

Bibliographie de l'empire français, ou journal de l'imprimerie et de la libriarie = Bibliographie de la France ou journal général de l'imprimerie de la librairie. Paris. Bibliogr. France. See B–P–H 187/5.

Bibliographie des forêts et products forestières = Bibliography of forestry and forest products. Washington, DC. Bibliogr. Forest. Forest Prod. See B–P–H 187/3.

Bibliographie des forst- und holzwirtschaftlichen Schrifttums. Hamburg. Vol. 1+, 1955+. Bibliogr. Forst- Holzw. Schrifftums. Preceded by: Bibliographie des ausländischen forst- und holzwirtschaftlichen Schrifttums. HI 62349

Bibliographie des forstlichen Schrifttums Deutschlands. Freiburg im Breisgau. Vol. 19+, 1955+. Bibliogr. Forstl. Schrifttums Deutschlands. Preceded by: Internationale Bibliographie für Forstwirtschaft. Deutschland. HI 62350

Bibliographie de la France ou journal général de l'imprimerie de la librairie. Paris. Bibliogr. France. See B–P–H 187/5. HI 51988

Bibliographie. Palynologie. Paris. 1974/75+. Bibliogr. Palynol. Preceded by: Pollen et spores. Supplement. Bibliographie. HI 62351

Bibliographie der Pflanzenschutzliteratur. Berlin. 1914-58; n.s. vols. 1-21, 1970-85. Bibliogr. Pflanzenschutzlit. Preceded by: Jahresbericht über das Gebiet der Pflanzenkrankheiten. Superseded by: Bibliography of plant protection. 1-669-3. HI 62352

Bibliographie de la protection des plantes = Bibliographie der Pflanzenschutzliteratur. Berlin. and Bibliography of plant protection. Berlin.

Bibliographie der schweizerischen naturwissenschaftlichen und geographischen Literatur. Bern. Bibliogr. Schweiz. Naturwiss. Geogr. Lit. See B–P–H 187/11. HI 51990

Bibliographie der schweizerischen naturwissenschaftlichen Literatur. Bern. Bibliogr. Schweiz. Naturwiss. Lit. See B–P–H 187/12. HI 51991

Bibliographie scientifique française. Paris. Bibliogr. Sci. Franç. See B–P–H 187/13. HI 51992

Bibliographie scientifique suisse = Bibliographie der schweizerischen naturwissenschaftlichen Literatur. Bern. Bibliogr. Schweiz. Naturwiss. Lit. See B–P–H 187/12.

Bibliographie scientifique suisse = Bibliographie der schweizerischen naturwissenschaftlichen und geographischen Literatur. Bern. Bibliogr. Schweiz. Naturwiss. Geogr. Lit. See B–P–H 187/11.

Bibliographie des travaux publiées par le corps enseignant de l'institut politechnique. Série A, sylviculture et industrie du bois = Bibliografia lucrărilor publicati de cadrele didactice din institutul politechnic. Serie A, silvicultură şi industria lemnului. Braşov.

Bibliographies and literature of agriculture. Beltsville, MD. BLA 1+, 1978+. Bibliogr. Lit. Agric. HI 62353

Bibliographische Mitteilungen der Universitäts-Bibliothek Jena. Jena. No. 1+, 1962+. Bibliogr. Mitt. Univ. Biblioth. Jena. HI 62354

Bibliographische Reihe der Technischen Universität Berlin. Berlin. ?-1978+. Bibliogr. Reihe Techn. Univ. Berlin. HI 62355

Bibliographischer Anzeiger. Leipzig. Bibliogr. Anz. See B–P–H 186/18. HI 51974

Bibliography of agricultural bibliographies; a categorized listing of bibliographies indexed in AGRICOLA. 1977+, 1978+. Bibliogr. Agric. Bibliogr. HI 62356

Bibliography of agricultural sciences in Japan. Tokyo. Nos. 1-9, 1971-76. Bibliogr. Agric. Sci. Japan. HI 62357

Bibliography of agriculture. Washington, DC. Bibliogr. Agric. (Washington). See B–P–H 186/17. HI 51973

Bibliography of american paleobotany. Iowa City, IA. 1952/57+, 1959+. Bibliogr. Amer. Paleobot. HI 62358

Bibliography of american pteridology. Durham, NC. Vol. 1+, 1974/76+, 1976?+. Bibliogr. Amer. Pteridol. HI 62359

Bibliography of australian references to eucalypts. Canberra. 1956/66-1969/70, 1966?-73. Bibliogr. Austral. Refer. Eucalypts. Superseded by: Bibliography of references to eucalypts. HI 62360

Bibliography documentation terminology. Paris. Vol. 1+, 1961+. Bibliogr. Doc. Terminol. HI 62361

Bibliography of the history of medicine. Bethesda, MD. No. 1+, 1964+, 1965+. Bibliogr. Hist. Med.

HI 62362

Bibliography and index of micropaleontology. New York. Vol. 1+, 1972+. Bibliogr. Index Micropaleontol. HI 62363

Bibliography of north american paleobotany. n.p. 1985+, 1986+. Bibliogr. N. Amer. Paleobot. HI 62364

Bibliography: photosynthesis and related subjects. Tokyo. 1955-77. Bibliogr. Photosyn. HI 62365

Bibliography of plant protection. Berlin. N.s. vol. 22+, 1986+. Bibliogr. Pl. Protect. Preceded by: Bibliographie der Pflanzenschutzliteratur. HI 62366

Bibliography of references to eucalypts. Canberra. 1971/72+, 1975+. Bibliogr. Refer. Eucalypts. Preceded by: Bibliography of australian references to eucalypts. HI 62367

Bibliography of scientific publications of south and south east Asia. New Delhi. Bibliogr. Sci. Publ. S. & S.E. Asia. See B–P–H 187/15. HI 51994

Bibliography of scientific publications of south Asia. New Delhi. Bibliogr. Sci. Publ. S. Asia. See B–P–H 187/16. HI 51995

Bibliography series: Nature conservancy council. Banbury. No. 1+, 1979+. Bibliogr. Ser. Nat. Conservancy Council. HI 62368

Bibliography of soil science, fertilizers and general agronomy. Farnham Royal, England. Bibliogr. Soil Sci. See B–P–H 188/3. HI 51997

Bibliography of systematic mycology. Kew, England. Bibliogr. Syst. Mycol. See B–P–H 188/4. HI 51998

Bibliography of translations from russian scientific and technical literature. Washington, DC. Bibliogr. Transl. Russ. Sci. Techn. Lit. See B–P–H 188/5. HI 51999

Bibliography of tropical agriculture. Rome. Bibliogr. Trop. Agric. See B–P–H 188/6. HI 52000

Bibliography, utilization of seaweed. Halifax, Nova Scotia. No. 2, 1953. Bibliogr. Utiliz. Seaweed. Preceded & superseded by: Selected bibliography on algae. HI 62369

Biblioteca analitica de scienze, letteratura e belle-arti. Naples. Ser. 1-4, pts. 1 & 2, 1810-23. Bibliot. Analitica Sci. HI 52984

Biblioteca argentina de ciencias naturales del museo argentino de ciencias naturales "Bernardino Rivadavia." Buenos Aires. Bibliot. Argent. Ci. Nat. Mus. Argent. Ci. Nat. "Bernardino Rivadavia". See B–P–H 188/8. HI 52001

Biblioteca española de divulgación científica. Madrid. Bibliot. Esp. Divulg. Ci. See B–P–H 188/10. HI 52002

Bibliotheca de farmacia - chimica - fisica - medicina - chirurgia - terapeutica - storia naturale, ecc. Milan. Bibliot. Farm. See B–P–H 188/11. HI 52003

Biblioteca fisica d'Europa. Pavia. Bibliot. Fis. Eur. See B–P–H 188/12. HI 52004

Biblioteca germanica di lettere, arti e scienze. Padua. Bibliot. German. Lett. See B–P–H 188/13. HI 52005

Biblioteca italiana. Davos, Switzerland. Bibliot. Ital. (Davos). See B–P–H 188/14. HI 52006

Biblioteca italiana. Florence. Bibliot. Ital. (Florence). See B–P–H 188/15. HI 52007

Biblioteca italiana ossia giornale di letteratura, scienze ed arti Milan. Vols. 1-100 [Also numbered "Anno" 1-25, each "Anno" composed of 4 vols.] 1816-1840. Bibliot. Ital. Giorn. Lett. Superseded by: Giornale dell'imperiale reale istituto lombardo di scienze, lettere ed arti e biblioteca italiana. 1-681-1. HI 50335

Biblioteca José Jerónimo Triana. Bogota. No. 1+, 1983+. Bibliot. José Jerónimo Triana. HI 62370

Bibliotheca biotheoretica. [Acta biotheoretica. Series D.] Leiden. Vols. 1-12, 1941-54. Biblioth. Biotheor. 1-685-1. HI 62371

Bibliotheca botanica. Kassel. Biblioth. Bot. See B–P–H 189/1. HI 52009

Bibliotheca critica. Amsterdam. Biblioth. Crit. See B–P–H 189/5 HI 52013.

Bibliotheca critica nova. Leiden. Biblioth. Crit. Nova. See B–P–H 189/6. HI 52014

Bibliotheca diatomologica. Vaduz, Stuttgart. Vol. 1+, 1983+. Biblioth. Diatomol. HI 62372

Bibliotheca genetica. Leipzig. Biblioth. Genet. See B–P–H 189/8. HI 52017

Bibliotheca lichenologica. Lehre, Vaduz, Stuttgart. Vol. 1+, 1973+. Biblioth. Lichenol. HI 62373

Bibliotheca mycologica. Lehre, Vaduz, Stuttgart. Vol. 1+, 1967+. Biblioth. Mycol. HI 62374

Bibliotheca nicotiana. Stockholm. Biblioth. Nicotiana. See B–P–H 189/19. HI 52028

Bibliotheca orientalis. London & Leipzig. Biblioth. Orient. See B–P–H 190/1 HI 52029.

Bibliotheca phycologica. Vaduz, Stuttgart. Vol. 1+, 1967+. Biblioth. Phycol. HI 62375

Bibliotheca pteridologica. Vaduz, Stuttgart. Vol. 1+, 1979+. Biblioth. Pteridol. HI 62376

Bibliotheca zoologica. Cassel, Stuttgart. Vols. 1-8, 1888-97. Biblioth. Zool. Superseded by: Zoologica. Stuttgart [not entered]. HI 62377

Bibliothek von Anzeigen und Auszügen kleiner, meist akademischer, Schriften ... Jena. Biblioth. Anz. Auszügen Kleiner Schriften. See B–P–H 188/20. HI 52008

Bibliothek der gesammten Naturgeschichte. Frankfurt a. M. & Mainz. Biblioth. Gesammten Naturgesch. See B–P–H 189/9. HI 52018

Bibliothek for laeger. Copenhagen. Biblioth. Laeger. See B–P–H 189/11. HI 52020

Bibliothek der neuesten und interessantesten Reisebeschreibungen. Berlin. Biblioth. Neuesten Interessantesten Reisebeschreib. See B–P–H 189/13. HI 52022

Bibliothek der neuesten Länder- und Völkerkunde. Tübingen. Biblioth. Neuesten Länder- Völkerk. See B–P–H 189/14. HI 52023

Bibliothek der neuesten physisch-chemischen, metallurgischen, technologischen und pharmaceutischen Literatur. Berlin. Biblioth. Neuesten Phys.-Chem. Lit. See B–P–H 189/15. HI 52024

Bibliothek der neuesten Reisebeschreibungen. Frankfurt a. M. & Leipzig. Biblioth. Neuesten Reisebeschreib. See B–P–H 189/16. HI 52025

Bibliothek der neuesten Reisen in die classischen Länder der Vorwelt. Meiningen, Germany. Biblioth. Neuesten Reisen Class. Länder Vorwelt. See B–P–H 189/17. HI 52026

Biblikothek der neuesten und wichtigsten Reisebeschreibungen zur Erweiterung der Erdkunde. Weimar. Biblioth. Neuesten Wichtigsten Reisebeschreib. Erweit. Erdk. See B–P–H 189/18. HI 52027

Bibliothek for physik, medicin og oekonomie. Copenhagen. Biblioth. Phys. Med. See B–P–H 190/5. HI 52033

Bibliothek für Physiker. Königsberg [=Kaliningrad, Russian S F S R]. Biblioth. Phys. (Königsberg). See B–P–H 190/2. HI 52030

Bibliothèque britannique, ou histoire des ouvrages des savans de la Grande-Bretagne. The Hague. Biblioth. Brit. Hist. Savans Grande-Bretagne. See B–P–H 189/3. HI 52011

Bibliothèque britannique; ou recueil extrait des ouvrages ... Agriculture anglaise. Geneva. Vols. 1-20, 1796-1815. Biblioth. Brit., Agric. Anglaise. Superseded by: Bibliothèque universelle des sciences, belles-lettres, et arts, ... Agriculture. 1-698-3. HI 52010

Bibliothèque britannique; ou recueil extrait des ouvrages ... Sciences et arts. Geneva. Vols. 1-60, 1796-1815 [Also numbered Années [1]-20]. Biblioth. Brit., Sci. Arts. Superseded by: Bibliothèque universelle des sciences, belles-lettres, et arts, ... Sciences et arts. 1-698-3. HI 52012

Bibliothèque de l'école des hautes études. Section des sciences naturelles. Paris. Biblioth. École Hautes Études, Sect. Sci. Nat. See B–P–H 189/7. HI 52016

Bibliothèque italienne, ou tableau des progrès des sciences et des arts en Italie. Turin. Biblioth. Ital. (Turin). See B–P–H 189/10. HI 52019

Bibliothèque médico-physique du nord. Lausanne. Biblioth. Méd.-Phys. N. See B–P–H 189/12. HI 52021

Bibliothèque physico-économique, instructive et amusante. Paris. Biblioth. Phys.-Écon. See B–P–H 190/3. HI 52031

Bibliothèque de physique et d'histoire naturelle. Paris. Biblioth. Phys. Hist. Nat. See B–P–H 190/4. HI 52032

Bibliothèque scientifique internationale. Paris. Biblioth. Sci. Int. See B–P–H 190/6. HI 52034

Bibliothèque universelle. Agriculture. Geneva & Paris = Bibliothèque universelle des sciences, belles-lettres, et arts, ... Agriculture. Geneva & Paris.

Bibliothèque universelle de Genève. Geneva & Paris. N.s. vols. 1-60, 1836-45. Biblioth. Universelle Genève. Preceded by: Bibliothèque universelle des sciences, belles-lettres, et arts, ... Sciences et arts. Superseded by: Supplément à la bibliothèque universelle de Genève; Archives des sciences physiques et naturelles. 1-698-1. HI 52036

Bibliothèque universelle de Genève. Archives des sciences physiques et naturelles. Geneva & Paris. Vols. 7-36, 1848-57. Biblioth. Universelle Genève, Arch. Sci. Phys. Nat. Preceded by: Supplément à la bibliothèque universelle de Genève. Archives des sciences physiques et naturelles. Superseded by: Bibliothèque universelle: revue suisse (et étrangère). 1-698-1. HI 74542

Bibliothèque universelle: revue suisse (et étrangère); Archives des sciences physiques et naturelles [Revue suisse omitted from vols. 61-64]. Geneva & Paris. N.s. vols. 1-64, 1858-78; 3e période, vols. 1-34, 1879-95; 4e période, vols, 1-46, 1896-1918; 5e période, vols. 1-29, 1919-47. Biblioth. Universelle Rev. Suisse. Preceded by: Bibliothèque universelle de Genève. Archives des sciences physiques et naturelles. Superseded by: Archives des sciences. 1-698-1. HI 74543

Bibliothèque universelle des sciences, belles-lettres, et arts, ... Agriculture. Geneva & Paris. Vols. 1-14, 1816-29. Biblioth. Universelle, Agric. Preceded by: Bibliothèque britannique; ou recueil extrait des ouvrages ... Agriculture anglaise. Incorporated in: Bibliothèque universelle des sciences, belles-lettres, et arts, ... Sciences et arts vols. 43-60. 1-698-2. HI 52035

Bibliothèque universelle des sciences, belles-lettres, et arts, ... Sciences et arts. Geneva & Paris. Vols. 1-60, 1816-35 [Also numbered Année [1]-20, vols. 43-60 with identical second t.p. but different numbering, 1830.I-1835.III]. Biblioth. Universelle Sci. Belles-Lettres Arts, Sci. Arts. Preceded by: Bibliothèque britannique; ou recueil extrait des ouvrages ... Sciences et arts. Superseded by: Bibliothèque universelle de Genève. 1-698-2. HI 52037

Bidrag till kännedom of Finlands natur och folk. Helsinki. Bidrag Kännedom Finlands Natur Folk. See B–P–H 190/31. HI 52038

Bidrag till kundskab over naturvidensckaberne. Christiania [=Oslo, Norway]. Bidrag Kundskab Naturvidensk. See B–P–H 190/32. HI 52040

Bidrag till kungl. svenska vetenskapsakademiens historia. Stockholm. 1963+. Bidrag Kungl. Svenska Vetensk.-Akad. Hist. HI 62378

Bielerseebuch. Biel, Switzerland. Bielerseebuch. See B–P–H 191/1. HI 52041

Biene. Giessen. Vols. 1-74(7), 1862-1936; vol. 91+, 1955+. Biene (Giessen). For vols. 74(8)-90, 1936-54 see: Hessische Biene. 1-700-1. HI 62379

Biene. Holyoke, MA. Vols. 1-27, 1893-1920. Biene (Holyoke). 1-700-1. HI 52042

Bienenvater. Vienna. Vol. 1+, 1869+. Bienenvater. HI 62380

Bienenwelt. Graz. Vol. 1+, 1959+. Bienenwelt. HI 62381

Biennial report, agricultural experiment stations, university of Georgia = Biennial report, Georgia agricultural experiment stations. Athens, GA.

Biennial report, american type culture collection = Report (Annual), american type culture collection. Rockville, MD.

Biennial report, C S I R O division of forest research = Biennial report, division of forest research, C S I R O. Melbourne, Vic.

Biennial report, California division of forestry = Biennial report of the California state board of forestry. Sacramento, CA.

Biennial report of the California state board of forestry. Sacramento, CA. 1885/86-1923-?, 1886-19?? [suspended 1893-1905]. Bienn. Rep. Calif. State Board Forest. Superseded by: Report, California division of forestry. Not in ULS. HI 62382

Biennial report, Cawthron institute = Report (Annual), Cawthron institute. Nelson, N.Z.

Biennial report, department of agriculture, California. Sacramento. 1967/68+, 1969+. Bienn. Rep. Dept. Agric. Calif. HI 62383

Biennial report, department of arboreta and botanic gardens, Los Angeles county = Biennial report, Los Angeles state and county arboretum. Los Angeles, CA.

Biennial report, division of forest research, C S I R O. Melbourne, Vic. 1979/81+, 1981+. Bienn. Rep. Div. Forest Res. C.S.I.R.O. Preceded by: Report, division of forest research, C S I R O. HI 62384

Biennial report, division of plant industry, department of agriculture, Florida. Gainesville, FL. Vol. 25+, (1962/64+), 1965+. Bienn. Rep. Div. Pl. Industr. Dept. Agric. Florida. Previously contained in: Bulletin, department of plant industry, Florida. HI 62385

Biennial report, Field museum of natural history. Chicago, IL. 1979/80, 1981. Bienn. Rep. Field Mus. Nat. Hist. Preceded by: Report (Annual) of the Field museum of natural history. Incorporated in: Bulletin, Field museum of natural history. HI 62386

Biennial report, Florida department of natural resources. Tallahassee, FL. 1969/70+, 1970+. Bienn. Rep. Florida Dept. Nat. Resources. Preceded by: Biennial report, Florida state board of conservation. HI 62387

Biennial report, Florida state board of conservation. Tallahassee, FL. Vols. 1-[18], 1933-68. Bienn. Rep. Florida State Board Conserv. Superseded by: Biennial report, Florida department of natural resources. HI 62388

Biennial report, Georgia agricultural experiment stations. Athens, GA. 1969/70-73/74, 1970-74. Bienn. Rep. Georgia Agric. Exp. Sta. Preceded by: Report (Annual) of the experiment stations, college of agriculture, Georgia university Superseded by: Report (Annual) of the Georgia agricultural experiment stations. HI 62389

Biennial report of the Illinois state museum of natural history. Springfield, IL. 1908. Bienn. Rep. Illinois State Mus. Nat. Hist. Superseded by: Report (Annual) of the Illinois state museum of natural history. HI 62390

Biennial report, Kansas agricultural experiment station = Kansas agricultural experiment station. Biennial report. Manhattan, KS.

Biennial report, Los Angeles county museum of natural history. Los Angeles, CA. 1967/68+, 1968+. Bienn. Rep. Los Angeles County Mus. Nat. Hist. HI 62391

Biennial report, Los Angeles state and county arboretum. Los Angeles, CA. 1959/61+, 1961+. Bienn. Rep. Los Angeles State County Arbor. Preceded by: Report (Annual), Los Angeles state and county arboretum. HI 62392

Biennial report, Mildura horticultural research station. Mildura. 1975/77+, 1977+. Bienn. Rep. Mildura Hort. Res. Sta. Preceded by: Report (Annual), plant research laboratory, victorian plant research institute. HI 62393

Biennial report of the New York state science service. Albany, NY. 1977/78+, 1978?+. Bienn. Rep. New

York State Sci. Serv. HI 62394

Biennial report, Santa Barbara botanic garden. Santa Barbara, CA. 19??+. Bienn. Rep. Santa Barbara Bot. Gard. Preceded by: Report (Annual) of the Santa Barbara botanic garden. HI 62395

Biennial report, Sunraysia horticultural research institute. 1979/81+, 1981?+. Bienn. Rep. Sunraysia Hort. Res. Inst. HI 62396

Biennial report of the Waite agricultural research institute, South Australia. Adelaide, S.A. 1966/67+, 1967?+. Bienn. Rep. Waite Agric. Res. Inst., S. Australia. Preceded by: Report of the Waite agricultural research institute, South Australia. HI 62397

Biennial research report, Sunraysia horticultural research institute = Biennial report, Sunraysia horticultural research institute.

Biennial review, Bedford institute of oceanography. Dartmouth, N.S. 1967/68-1977/78. 1968-78. Bienn. Rev. Bedford Inst. Oceanogr. Preceded by: Report (Annual), Bedford institute of oceanography. Superseded by: B I O review. HI 62398

Bifidobacteria and microflora. Tokyo. Vol. 1+, 1982+. Bifidobact. & Microfl. HI 62399

Bifidus, flores et fructus. Tokyo. Vol. 1+, 1981+. Bifidus Fl. Fructus. HI 62400

Bihang til kongliga svenska vetenskaps-akademiens handlingar. Stockholm. Bih. Kongl. Svenska Vetensk.-Akad. Handl. See B–P–H 191/14. HI 52044

Bijblad. Landbouw courant. Zwolle, Netherlands. Bijbl. Landb. Courant. See B–P–H 191/17. HI 52045

Bijdragen tot de geschiedenis der geneeskunde. Haarlem. Bijdr. Gesch. Geneesk. See B–P–H 191/20. HI 52046

Bijdragen tot de kennis der boomsoorten van Java. Batavia, Dutch E. Indies [=Jakarta, Indonesia]. Bijdr. Kennis Boomsoorten Java. See B–P–H 191/21. HI 52047

Bijdragen van de natuurkundig genootschapp te Groningen. Groningen. Bijdr. Natuurk. Genootsch. Groningen. See B–P–H 191/22. HI 52048

Bijdragen tot de natuurkundige wetenschappen. Amsterdam. Bijdr. Natuurk. Wetensch. See B–P–H 191/23. HI 52049

Bijzondere publicaties van het bosbouwproefstation. Buitenzorg, Dutch E. Indies [=Bogor, Indonesia]. Bijzondere Publ. Bosbouwproefstat. See B–P–H 191/24. HI 52050

Biken journal. Osaka, Japan. Biken J. See B–P–H 191/25. HI 52051

Biljeske, institut za oceanografiju i ribarstvo Split. Split. 1951+. Biljeske Inst. Oceanogr. Ribarst. Split. HI 62401

Biljna zaštita. Zagreb. Vols. 1-21, 1957-77. Biljna Zaštita. Superseded by: Glasnik zaštite bilja. HI 62402

Biljni lekar. Vols. ?-15-21, 19??-70-76. Biljni Lekar. Superseded by: Glasnik zaštite bilja. HI 62403

Billotia; ou, notes de botanique. Besançon. Billotia. See B–P–H 191/27. HI 52052

Bilten dokumentacija za poljoprivredu, sumarstvo drvnu i duvansku industriju = Bilten; dokumentacija stručne literature. Ser. A, bilten dokumentacija za poljoprivredu, sumarstvo drvnu i duvansku industriju. Belgrade.

Bilten; dokumentacija stručne literature. Ser. A, bilten dokumentacija za poljoprivredu, šumarstvo drvnu i duvansku industriju. Belgrade. Vols. 1-7, 1950-56. Bilten Dokument. Stručne Lit., Ser.A, Bilten Dokument. Poljopr. Superseded by: Bilten dokumentacije; šumarstvo i drvna industrija. HI 62404

Bilten dokumentacije; šumarstvo i drvna industrija. Belgrade. Vol. 8+, 1957+. Bilten Dokument. Šumarstvo Drvna Industr. Preceded by: Bilten; dokumentacija stručne literature. Ser. A, bilten dokumentacija za poljoprivredu, šumarstvo drvnu i duvansku industriju. HI 62405

Biltmore botanical studies; a journal of botany embracing papers by the director and associates of the Biltmore Herbarium. Biltmore, NC. Biltmore Bot. Stud. See B–P–H 191/28. HI 52053

Bimarihaye guiahi = Iranian journal of plant pathology. Teheran.

Bi-monthly progress report, department of forestry, Canada. Ottawa. 1965-66. Bi-Monthly Progr. Dept. Forest. Canada. Superseded by: Bi-monthly research notes, forestry service, Canada. HI 62406

Bi-monthly report of the agricultural department. Washington, DC. Bi-Monthly Rep. Agric. Dept. See B–P–H 192/4. HI 52054

Bi-monthly research notes, department of forestry, Canada = Bi-monthly research notes, forestry service, Canada. Ottawa.

Bi-monthly research notes, forestry service, Canada. Ottawa. Vols. ?-25-36, 19??-66-80. Bi-Monthly Res. Notes Forest. Serv. Canada. Preceded by: Bi-monthly progress report, department of forestry, Canada. Superseded by: Research notes, canadian forestry service. HI 62407

Bingduxue jikan = Acta virologica sinica. Beijing.

Binnengewässer. Stuttgart. Binnengewässer. See B–P–H 192/5. HI 52055

Bio. Montreal. Vols. 1-2(1), 1981-84. Bio. Superseded

by: Bulletin d'information, association des biologistes du Québec. HI 62408

Bioactive molecules. Amsterdam, New York. Vol. 1+, 1986+. Bioactive Molec. HI 62409

Bioactive plants. Lawrence, MA. Vol. 1+, 1982+. Bioactive Pl. HI 62410

Biocatalysis. New York. Vol. 1+, 1987+. Biocatalysis. HI 62411

Biocenoses; bulletin d'écologie terrestre. Alger. No. 1+, 1982+. Biocenoses. HI 62412

Biocenosis; revista de educación ambiental. San Jose, Costa Rica. Vol. ?-4(2)+, 19??-83+. Biocenosis. HI 62413

Biochemical and allied research in India. Bangalore. Vols. 1-6, 1930-35. Biochem. Allied Res. India. Superseded by: Annual review of biochemical and allied research in India. HI 62414

Biochemical and biophysical perspectives in marine biology. Orlando, FL. Vol. 1+, 1974+. Biochem. Biophys. Perspect. Mar. Biol. HI 62415

Biochemical and biophysical research communications. New York, NY. Biochem. Biophys. Res. Commun. See B–P–H 192/8. HI 52056

Biochemical bulletin. New York, NY. Biochem. Bull. See B–P–H 192/10. HI 52057

Biochemical genetics. New York & London. Vol. 1+, 1967+. Biochem. Genet. HI 62416

Biochemical journal. Liverpool, London. Vols. 1-130, 1906-72. Biochem. J. Superseded by: Biochemical journal, molecular aspects and Biochemical journal, cellular aspects. 1-703-3. HI 62417

Biochemical journal, cellular aspects. London. Vols. 132-216, 1973-83. Biochem. J., Cell. Aspects. Preceded by: Biochemical journal. HI 62418

Biochemical journal, molecular aspects. London. Vol. 131+, 1973+. Biochem. J., Molec. Aspects. Preceded by: Biochemical journal. HI 62419

Biochemical pharmacology, London & N New York, NY. Biochem. Pharmacol. See B–P–H 192/15. HI 52059

Biochemical reviews. Bangalore. Vol. 36+, 1966+. Biochem. Rev. Preceded by: Annual review of biochemical and allied research in India. HI 62420

Biochemical systematics. Oxford, etc. Vol. 1, 1973. Biochem. Syst. Superseded by: Biochemical systematics and ecology. HI 62421

Biochemical systematics and ecology. Oxford, etc. Vol. 2+, 1973+. Biochem. Syst. & Ecol. Preceded by: Biochemical systematics. HI 62422

Biochemisches Centralblatt. Berlin. Biochem. Centralbl. See B–P–H 192/11. HI 52058

Biochemie und Physiologie der Pflanzen. Jena. Vol. 161+, 1970+. Biochem. Physiol. Pflanzen. Preceded by: Flora. Abteilung A, Physiologie und Biochemie. HI 62423

Biochemische Zeitschrift. Berlin. Vols. 1-346, 1906-66. Biochem. Z. Superseded by: European journal of biochemistry. 1-704-2. HI 62424

Biochemistry. [Translation of: Biokhimiya.] New York. Vol. 21+, 1956+. Biochemistry (Akad. Nauk S.S.S.R.) HI 75209

Biochemistry. Washington, DC. Vol. 1+, 1962+. Biochemistry (Amer. Chem. Soc.). HI 52060

Biochemistry abstracts. Part 1, biological membranes. Bethesda, MD. Vols. 8-12, 1980-84. Biochem. Abstr., 1. Biol. Membr. Preceded by: Biological membrane abstracts. Superseded by: Cambridge scientific biochemistry abstracts. Part 1, biological membranes. HI 62425

Biochemistry abstracts. Part 2, nucleic acids. Bethesda, MD. Vols. ?-14, 1980-84. Biochem. Abstr., 2. Nucl. Acids. Preceded by: Nucleic acids abstracts. Superseded by: Cambridge scientific biochemistry abstracts. Part 2, nucleic acids. HI 62426

Biochemistry abstracts. Part 3, amino acids, peptides and proteins. Bethesda, MD. Vols. ?-13, 1980-84. Biochem. Abstr., 3. Amino Acids. Preceded by: Amino acids, peptide and protein abstracts. Superseded by: Cambridge scientific biochemistry abstracts. Part 3, amino acids, peptides and proteins. HI 62427

Biochemistry and cell biology. Ottawa. Vol. 64+, 1986+. Biochem. & Cell Biol. Preceded by: Canadian journal of biochemistry and cell biology. HI 62428

Biochemistry international. Sydney, London, etc. Vol. 1+, 1980+. Biochem. Int. HI 62429

Biochimica et biophysica acta. New York, NY. & Amsterdam. Biochim. Biophys. Acta. See B–P–H 192/23. HI 52061

Biochimie. Paris. Vol. 53+, 1971+. Biochimie. Preceded by: Bulletin de la société de chimie biologique. HI 62430

Biochimie et biologie cellulaire = Biochemistry and cell biology. Ottawa.

Biocontrol news and information. Farnham Royal. Vol. 1+, 1979+. Biocontrol News Inform. HI 62431

Biocosme mésogéen; revue d'histoire naturelle. Nice. Vol. 1+, 1984+. Biocosme Mésogéen. HI 62432

BioCycle. Emmaus, PA. Vol. 22+, 1981+. BioCycle. Preceded by: Compost science/land utilization. HI 62433

Biodeterioration abstracts. Kew. Vol. 1+, 1987+. Biodeterior. Abstr. Previously contained in:

International biodeterioration. HI 62434

Biodeterioration research titles. Birmingham. 1972+. Biodeterior. Res. Titles. Preceded by: International biodeterioration bulletin reference index supplement. Superseded by: International biodeterioration. HI 62435

Biodynamica. A scientific journal for the elaboration and the experimental study of working hypotheses on the nature of life. Normandy, MS. Biodynamica. See B–P–H 192/24. HI 52062

Biodynamica monographs. Normandy, MS. Biodynamica Monogr. See B–P–H 192/25. HI 52063

Bioengineering abstracts. Vols. 1+, 1974+. Bioengin. Abstr. HI 62436

Bioengineering news; biotechnology news source. San Francisco, CA. 1980+. Bioengin. News. HI 62437

Bioessays. Cambridge, New Rochelle, NY. 1984+. Bioessays. HI 62438

Biofeedback and self-regulation. New York, NY. Vol. 1+, 1976+. Biofeedb. Self-Regulat. HI 62439

Biofizika. Moscow. Biofizika. See B–P–H 192/26. HI 52064

Biofouling. Chur, London, New York, etc. Vol. 1+, 1988+. Biofouling. HI 62440

Biofutur. Puteaux, Paris. No. ?-67+, ?-1988+. Biofutur. HI 75251

Biogeochemistry; an international journal. Dordrecht & Hingham, MA. Vol. ?-7+, 19??-89+. Biogeochemistry. HI 62441

Biogeografia. São Paulo. Vol. 1+, 1969+. Biogeographia. HI 62442

Biogeographica. The Hague. Vol. 1+, 1972+. Biogeographica (The Hague). HI 62443

Biogeographica. Tokyo. Vols. 1-4, 1935-41. Biogeographica (Tokyo). 1-705-2. HI 52065

Biogeografiya i kraevedenie. Perm'. Vol. 1+, 1971+. Biogeogr. & Kraev. HI 62444

Biogeotsenologiya. Novosibirsk. Vol. 1+, 1973+. Biogeotsenologiya. HI 62445

Biograms. Uniform reports of biological activity. Princeton, NJ. Biograms. See B–P–H 193/2. HI 52067

Biographical memoirs of fellows of the royal society. London. 1955+. Biogr. Mem. Fellows Roy. Soc. Preceded by: Obituary notices of fellows of the royal society. HI 62446

Biographical memoirs of the national academy of sciences of the United States of America. Washington, DC. Biogr. Mem. Natl. Acad. Sci. U.S.A. See B–P–H 192/30. HI 52066

Biographien bedeutender Biologen. Jena. 1974+. Biogr. Bedeut. Biol. HI 62447

Biohimija = Biokhimiya. Moscow.

BioIndonesia. Bogor, Jakarta. No. 1+, 1975+. BioIndonesia. HI 62448

Bio-information. Lille. 1969+. Bio-information. HI 62449

Biokhimicheskaya ekologiya i meditsina. Sverdlovsk. Vol. 1+, 1978+. Biokhim. Ekol. Med. HI 64653

Biokhimii kul'turnykh rastenii Moldavii. Kishinev. Vol. 1, 1962. Biokhim. Kul't. Rast. Moldav. Superseded by: Voprosy fiziologii i biokhimii kul'turnykh rastenii. HI 62450

Biokhimichnyi zhurnal. Kiev. 1937-41 [publication suspended 1942-45]. Biokim. Zhurn. Preceded and superseded by: Ukrayins'kyi biokhimichnyi zhurnal. HI 62451

Biokhimiya. Moscow. 1936+. Biokhimiya. 1-706-1. HI 64916

Bioklimatische Beiblätter der Meteorologischen Zeitschrift. Brunswick, Germany. Bioklimat. Beibl. Meteorol. Z. See B–P–H 193/3. HI 52068

BioLlania. Guanare. No. 1+, 1984+. BioLlania. HI 62452

Biologe. Monatsschrift zur Wahrung der Belange der deutschen Biologen. Munich. Biologe. See B–P–H 196/4. HI 52106

Biologi; Turk biologi dernegi'nin yayin organi. Istanbul. Vols. 1-7, 1950-57. Biologi. Superseded by: Türk biologi dergesi. HI 62453

Biologi gunaan Malaysia = Malaysian applied biology. Kuala Lumpur.

Biologia. Bratislava. Biologia (Bratislava). See B–P–H 196/5. HI 52107

Biológia. Budapest. Vol. 22+, 1974+. Biológia (Budapest). Preceded by: Biologiai közlemények. HI 62454

Biologia. Lahore, Pakistan. Biologia (Lahore). See B–P–H 196/6. HI 52108

Biologia. Waltham, MA. Biologia (Waltham). See B–P–H 196/7. HI 52109

Biologia acuatica. La Plata. No. 1+, 1981+. Biol. Acuatica. HI 62455

Biologia gabonica. Libreville, Gabon. Biol. Gabon. See B–P–H 193/17. HI 52076

Biologia gallo-hellenica; travaux des groupes franco-helléniques et de la station de Kéramou. Ergasia tôn gall-ellênikôon etaireiôn kai tou stathmou Keramous. Toulouse, Kéramou. Vol. 1+, 1967+. Biol. Gallo-Hellen. HI 62456

Biologia generalis. Vienna & Leipzig. Biol. Gen. See B–P–H 193/18. HI 52078

Biologia plantarum. Prague. Biol. Pl. See B–P–H 194/20. HI 52088

Biologia w szkole. Warsaw. Biol. Szkole. See B–P–H 195/22. HI 52102

Biológia tanítása. Budapest. Biol. Tanít. See B–P–H 195/23. HI 52103

Biológiai kozlemények. Pars biologica. Budapest. Vols. 1-21, 1954-73. Biol. Közlem., Pars Biol. Superseded by: Biológia. HI 62457

Biológiai szemle. Kecskemet, Hungary. Biol. Szemle. See B–P–H 195/21. HI 52101

Biologica. Mexico, DF. Vol. 3+, 1973+. Biologica, (Mexico). HI 62458

Biologica. Navsari. Vol. 8+, 1975+. Biologica (Navsari). Preceded by: Biology news bulletin. HI 62459

Biologica. Oulu = Acta universitatis ouluensis. Series A, scientiae rerum naturalium. Oulu.

Biológica. Santiago. Nos. 1-40, 1944-67. Biológica (Santiago). 1-706-2. HI 62460

Biologica didactica. Bad Salzdetfurth. 1978+. Biol. Didact. HI 62461

Biologica hungarica. Budapest. Biol. Hung. See B–P–H 193/20. HI 52079

Biologica latina. Milan. Biol. Latina. See B–P–H 194/4. HI 52082

Biological abstracts. Philadelphia, PA. Biol. Abstr. See B–P–H 193/6. HI 52069

Biological abstracts/R R M [reports, reviews, meetings]; references and indexes to the world's life science research literature. Philadelphia, PA. Vol. 18+, 1980+. Biol. Abstr. R. R. M. Preceded by: Bioresearch index. HI 62462

Biological and agricultural index. New York, NY. Biol. Agric. Index. See B–P–H 193/7. HI 52070

Biological agriculture and horticulture; an international journal. Berkhamsted. Vol. 1+, 1982+. Biol. Agric. & Hort. HI 62463

Biological bulletin; department of biology, college of science, Tunghai university [Tung-hai ta hsüeh.] Taichung, Taiwan. Biol. Bull. Dept. Biol. Coll. Sci. Tunghai Univ. See B–P–H 193/9. HI 52071

Biological bulletin of Fukien christian university. [Hsieh ta shêng wu hsüeh pao.] Foochow, China. Biol. Bull. Fukien Christian Univ. See B–P–H 193/11. HI 52072

Biological bulletin of India. Bhagalpur. Vol. 1+, 1979+. Biol. Bull. India. HI 62464

Biological bulletin of the Marine biological laboratory, Woods Hole, Mass. Boston, MA. Biol. Bull. Mar. Biol. Lab. Woods Hole. See B–P–H 193/12. HI 52073

Biological bulletin of national Taiwan normal university. [Shih ta sheng wu hsüeh pao.] Taipei. Nos. 1-2, 1966-67. Biol. Bull. Natl. Taiwan Norm. Univ. HI 62465

Biological bulletin of Taiwan normal university = Biological bulletin of national Taiwan normal university. Taipei.

Biological conservation; international quarterly journal devoted to scientific protection of plant and animal wildlife and all nature throughout the world, [etc.]. Barking. Vol. 1+, 1968+. Biol. Conservation. HI 62466

Biological conservation newsletter. Washington, DC. No. 1+, 1981+. Biol. Conservation Newslett. HI 62467

Biological curators' group newsletter. Nos. 1-4, 1975-76. Biol. Curators' Group Newslett. Superseded by: B C G newsletter. HI 62468

Biological fraternity. Oklahoma City, OK. Biol. Fraternity. See B–P–H 193/16. HI 52075

Biological inventory news. Asuncion. Nos. 1-3, 1981-82. Biol. Invent. News. Superseded by: Boletín del inventario biológico. HI 62469

Biological journal; publication of the natural history society of the university of the West Indies. St. Augustine, Trinidad. Vols. 1-?, 1966-67. Biol. J. (St. Augustine). HI 62470

Biological journal of the linnean society. London. Vol. 1+, 1969+. Biol. J. Linn. Soc. Preceded by: Proceedings of the linnean society of London. HI 62471

Biological journal of Nara women's university. [Nara joshi daigaku seibutsu gakkai-shi.] Nara. No. 1+, 1951+. Biol. J. Nara Women's Univ. HI 62472

Biological journal of Okayama university. [Okayama daigaku rigakubu seibutsu gaku kiyo.] Okayama. Vols. 1-17(2), 1952-76. Biol. J. Okayama Univ. HI 62473

Biological laboratory of Owens college (Manchester university), studies. Manchester, England. Biol. Lab. Owens Coll. Stud. See B–P–H 194/3. HI 52081

Biological leaflet. Washington, DC. Biol. Leafl. See B–P–H 194/5. HI 52083

Biological magazine. [Okinawa seibutsu gekhai.] Naha, Okinawa. Biol. Mag. See B–P–H 194/7. HI 52085

Biological membrane abstracts. London, Bethesda, MD. Vols. 1-7, 1973-79? Biol. Membr. Abstr. Superseded by: Biochemistry abstracts. Part 1, biological membranes. HI 62474

Biological membranes. [Translation of: Biologicheskie membrany.] New York. Vol. 1+, 1985+. Biol. Membr. (New York). HI 62475

Biological memoirs; international journal of biological disciplines. Lucknow. Vol. 1+, 1976+. Biol. Mem. HI 62476

Biological memoirs. Angiosperm taxonomy series. Lucknow. Vol. 1+, 1977+. Biol. Mem., Angiosp. Taxon. Ser. HI 62477

Biological monographs of the Canary Islands = Monographiae biologicae canarienses. Las Palmas.

Biological news. [Shêng wu hsüeh t'ung pao.] Peking. Biol. News. See B–P–H 194/16. HI 52086

Biological notes. Urbana, IL. Biol. Notes. See B–P–H 194/18. HI 52087

Biological notes, Ohio biological survey. Columbus, OH. No. 1+, 1964+. Biol. Notes Ohio Biol. Surv. HI 62478

Biological oceanography. New York, NY. Vol. 1+, 1981+. Biol. Oceanogr. HI 62479

Biological papers of the university of Alaska. College, Fairbanks, AK. 1957+. Biol. Pap. Univ. Alaska. HI 62480

Biological records. Berkhamsted. Pts. 6-8, 1954-55. Biol. Rec. Preceded by: Biological supplement, cave research group of Great Britain. Incorporated in: British hypogean fauna and Biological records of the cave research group of Great Britain. HI 62481

Biological regulation and development. New York. 1979+. Biol. Regulat. & Developm. HI 62482

Biological research reports from the university of Jyväskylä. Jyväskylä. No. 1+, 1975+. Biol. Res. Rep. Univ. Jyväskylä. HI 62483

Biological resources and natural conditions of the Mongolian People's Republic = Biologicheskie resursy i prirodnye usloviya Mongol'skoi Narodnoi Respubliki. Leningrad.

Biological results of the Japan antarctic research expedition. [Forms part of: Special publications from the Seto marine biological laboratory.] Nos. 1-17, 1959-62. Biol. Res. Japan Antarc. Res. Exped. Superseded by: Scientific reports, Japan antarctic research expedition. Series E, biology. HI 62484

Biological review. London. Biol. Rev. (London). See B–P–H 194/26. HI 52090

Biological review of the city college. New York, NY. Biol. Rev. City Coll. See B–P–H 195/1. HI 52091

Biological reviews and biological proceedings of the Cambridge philosophical society. Cambridge. Vols. 3-10, 1926-35. Biol. Rev. Biol. Proc. Cambridge Philos. Soc. Preceded by: Proceedings of the Cambridge philosophical society. Biological sciences. Superseded by: Biological reviews of the Cambridge philosophical society. 2-895-1. HI 62485

Biological reviews of the Cambridge philosophical society. Cambridge. Vol. 11+, 1935+. Biol. Rev. Cambridge Philos. Soc. Preceded by: Biological reviews and biological proceedings of the Cambridge philosophical society. 2-895-1. HI 62486

Biological rhythms. Sheffield. Vol. 3+, 1972+. Biol. Rhythms. Preceded by: Circadian rhythms. HI 62487

Biological science. [Seibutsu kagaku.] Tokyo. Biol. Sci. (Tokyo). See B–P–H 195/7. HI 52094

Biological science bulletin of the university of Arizona. Tucson, Arizona. Biol. Sci. Bull. Univ. Arizona. See B–P–H 195/9. HI 52095

Biological sciences curriculum study bulletin. Washington, DC. No. 1, 1961. HI 63706

Biological sciences news. Japan. No. ?-10+, 19??-72+. Biol. Sci. News, Japan. HI 62488

Biological sciences. Stanford university. University series. Palo Alto, CA. Biol. Sci. Stanford Univ., Univ. Ser. See B–P–H 195/10. HI 52096

Biological series of the bulletin of the state university of Montana. Missoula, MT. Biol. Ser. Bull. State Univ. Montana. See B–P–H 195/12. HI 52097

Biological series of the bulletin of the university of New Mexico. Albuquerque, NM. Biol. Ser. Bull. Univ. New Mexico. See B–P–H 195/13. HI 52098

Biological studies. Catholic university of America. Washington, DC. Biol. Stud. Catholic Univ. Amer. See B–P–H 195/20. HI 52100

Biological supplement. Cave research group of Great Britain. 1938-53, 1955-60. Biol. Suppl. Cave Res. Group Great Britain. Superseded by: Biological records. HI 62489

Biological trace element research. Clinton, Passaic, NJ. Vol. 1+, 1979+. Biol. Trace Elem. Res. HI 75176

Biological weed control newsletter. No. 1+, 1969+. Biol. Weed Control Newslett. HI 62490

Biologicals; journal of the international association of biological standardization. London, San Diego, CA. Vol. 18+, 1990+. Biologicals. Preceded by: Journal of biological standardization HI 75237

Biologicheskie membrany. Moscow. 1984+. Biol. Membrany. HI 62491

Biologicheskie resursy dagestana. Makhachkala. Vol. 1+, 1975+. Biol. Resursy Dagestana. HI 62492

Biologicheskie resursy dagestanskogo pribrezhya Kaspiiskogo morya. Makhachkala. Vol. 1+, 1982+. Biol. Resursy Dagestansk. Pribr. Kaspiisk. Morya.

HI 65341

Biologicheskie resursy i prirodnye usloviya Mongol'skoi Narodnoi Respubliki. Leningrad. Vol. 1+, 1972+. Biol. Resursy Prir. Uslov. Molgol'sk. Nar. HI 62494

Biologicheskie resursy vodoemov Moldavii. Kishinev. 1962-75. Biol. Resursy Vod. Moldavii. HI 62493

Biologicheskii nauki. Alma-Ata. 1971+. Biol. Nauki (Alma-Ata). HI 62495

Biologicheskii zhurnal. Bucharest = Revue de biologie. Bucharest.

Biologicheskii zhurnal Armenii. Erevan. Vol. 19+, 1966+. Biol. Zhurn. Armenii. Preceded by: Izvestiya, akademiya nauk armyanskoi S S R. HI 62496

Biologicheskoe deistvie bystrykh neitronov. Kiev. Vol. 1+, 1969+. Biol. Deistvie Bystrykh Neitron. HI 62497

Biologichnii zbirnyk. Kiev, Ukrainian S S R. Vols. 1-4, 1935-49. Biol. Zbirn. Superseded by: Pratsy Biologo-gruntovogo fakul'tetu. HI 62498

Biologické listy. Prague. Biol. Listy. See B–P–H 194/6. HI 52084

Biologické práce, Slovenskej akademie vied. Bratislava. Biol. Práce Slov. Akad. Vied. See B–P–H 194/24. HI 52089

Biologický sborník, Slovenskej akadémie vied a umení. Bratislava. Biol. Sborn. Slov. Akad. Vied. See B–P–H 195/6. HI 52093

Biológico. São Paulo. Biológico (São Paulo). See B–P–H 196/14. HI 52110

Biologie cellulaire; an international journal of cellular biology. Ivry sur Seine. Vols. 28-39, 1977-80. Biol. Cellulaire. Preceded by: Journal de microscopie et de biologie cellulaire. Superseded by: Biology of the cell. HI 62499

Biologie et écologie Méditerranéenne; revue de biologie et d'écologie Méditerranéenne. Marseille St. Jerome, Aix-en-Provence. Vol. 1+, 1974+. Biol. & Écol. Médit. Preceded by: Annales de l'université de Provence. Sciences. HI 62500

Biologie médicale. Paris. Vols. 1-60, 1903-71; n.s. vol. 1+, 1972+. Biol. Méd. HI 62501

Biologie in der Schule. Berlin. Vol. 1+, 1952+. Biol. Schule. HI 62502

Biologie du sol. Paris. N.s. vols. 1-14, 1964-71. Biol. Sol. Superseded by: Biologie du sol. Partie microbiologie. HI 62503

Biologie du sol. Partie microbiologie. Paris. N.s. vols. 15-22, 1972-75/76. Biol. Sol, Microbiol. Preceded by: Biologie du sol. Superseded by: Pedofauna [not entered]. HI 62504

Biologie in unserer Zeit. Weinheim. Vol. 1+, 1971+. Biol. Uns. Zeit. HI 62505

Biologija. Vilnius = Lietuvos T S R aukstuju mokyklu mokslo darbai. Biologija. Vilnius.

Biologijos instituto darbai. Lietuvos T S R mokslu akademija. Vilnius = Lietuvos T S R mokslu akademijos biologijos instituto darbai. Vilnius.

Biologisch jaarboek; uitgegeven door het k. natuurwetenschappelijk genootschap Dodonaea te Gent. Antwerp. Biol. Jaarb. See B–P–H 193/23. HI 52080

Biologische Rundschau. Zeitschrift für die gesamte Biologie und ihre Grenzgebiete. Jena. Biol. Rundschau. See B–P–H 195/4. HI 52092

Biologisches Centralblatt. Erlangen. Biol. Centralbl. See B–P–H 193/15. HI 52074

Biologisches Zentralblatt. Leipzig. Biol. Zentralbl. See B–P–H 196/1. HI 52105

Biologiske meddelelser. Kongelige danske videnskabernes selskab. Copenhagen. Vols. 1-24, 1917-71. Biol. Meddel. Kongel. Danske Vidensk. Selsk. 2-1261-3. HI 62506

Biologiske skrifter. Copenhagen. Vol. 1+, 1939+. Biol. Skr. Preceded by: Kongelige danske videnskabernes selskabs naturvidenskabelige og mathematiske afhandlinger. 2-1261-3. HI 62507

Biologist. Denver, CO. Vol. 1+, 1916+. Biologist (Denver). HI 52111

Biologist. London. Vol. 16+, 1969+. Biologist (London). Preceded by: Journal of the institute of biology. HI 62508

Biologiya; nauchnye doklady. Leningrad. Vol. 1+, 1976+. Biologiya. HI 62509

Biologiya baltiiskogo morya. Riga. Vol. 1+, 1974+. Biol. Balt. Morya. HI 62510

Biologiya morya. Kiev. Vol. 1+, 1965+. Biol. Morya (Kiev). Preceded by: Trudy, sevastopol'skoi biologicheskoi stantsii. HI 62511

Biologiya morya. Vladivostok. 1975+. Biol. Morya (Vladivostok). HI 62512

Biologiya vnutrennykh vod, informatsionnii byulleten'. Leningrad. 1967+. Biol. Vnutrenn. Inform. Byull. HI 62513

Biologiya zhivotnykh i rastenii turkmenistana. Ashkhabad. Vol. 3+, 1976+. Biol. Zhivot. Rast. Turkmen. Preceded by: Voprosy biologii, turkmenskii gosudarstvennyi universytet. HI 62514

Biology '83 [etc.]; current titles in the biological sciences. Lawrence, KS. 1983+. Biology '83 [etc.] HI 62515

Biology bulletin of the academy of sciences of the U.S.S.R. [Translation of: Izvestiya akademii nauk S S

S R. Seriya biologicheskaya. See B–P–H 441/13.] New York. Vol. 1+, 1974+. Biol. Bull. Acad. Sci. U.S.S.R. HI 62516

Biology of carbohydrates. New York, NY. Vol. 1+, 1981+. Biol. Carbohyd. HI 62517

Biology of the cell. Ivry sur Seine, Paris. Vol. 40+, 1981+. Biol. Cell. Preceded by: Biologie cellulaire. HI 62518

Biology digest. Medford, NJ. Vol. 1+, 1974+. Biol. Digest. HI 62520

Biology and fertility of soils. Secaucus, NJ. Vol. 1+, 1985+. Biol. Fertil. Soils. HI 62521

Biology international; I U B S newsmagazine. Paris. No. 1+, 1980+; special issue no. 1+, 1983+. Biol. Int. Preceded by: I U B S newsletter. HI 62522

Biology news bulletin. Navsari. Vols. 1-2, 1968-69. Biol. News Bull. Superseded by: Biologica. Navsari. HI 62523

Biology newsletter, Utah state university. Logan, UT. No. 1+, 1978+. Biol. Newslett., Utah State Univ. HI 62524

Biology and philosophy. Dordrecht, Boston, etc. Vol. 1+, 1986+. Biol. & Philos. HI 62525

Biološki glasnik. Zagreb. Vols. 8-21, 1955-68. Biol. Glasn. Preceded by: Glasnik biološke sekcije. Superseded by: Periodicum biologorum. HI 62526

Biološki vestnik; glasilo slovenskih biologov. Ljubljana. Vol. 1+, 1952+. Biol. Vestn. HI 52104

Biomass; an international journal. New York. 1981+. Biomass. HI 62527

Biomass digest. Englewood Cliffs, NJ. Vols. 1-6, 1979-84? Biomass Digest. Superseded by: Bioprocessing technology. HI 75252

Biome. Ottawa. No. 1+, 1981+. Biome. HI 62528

Biomedica biochimica acta. Berlin. Vol. 42+, 1983?+. Biomed. Biochim. Acta. Preceded by: Acta biologica et medica germanica. HI 62529

Bio-medical preview. Rochester, MN. Bio-Med. Preview. See B–P–H 196/16. HI 52112

Biometrical journal. Berlin. Vol. 19+, 1977+. Biometr. J. Preceded by: Biometrische Zeitschrift. HI 62530

Biometrics. New Haven, CT. Biometrics. See B–P–H 196/17. HI 52113

Biometrics and ecology = Biométrie et écologie. Versailles.

Biométrie et écologie. Versailles. No. 1+, 1978+. Biométr. & Ecol. HI 62531

Biométrie - praximetrie. Gembloux. Vol. 1+, 1960+. Biométr. Praximetr. HI 62532

Biometrika: A journal for the statistical study of biological problems. Cambridge, England. Biometrika. See B–P–H 196/18. HI 52114

Biometrische Zeitschrift. Berlin. Vols. 1-18, 1959-76. Biometr. Z. Superseded by: Biometrical journal. HI 62533

Bionature. Bhopal, New Delhi. Vol. 1+, 1981+. Bionature. HI 62534

Bionika. Kiev. Vol. [1-2], 3+, 1965-68, 1969+. Bionika. HI 65442

Bionomics briefs. Davis, CA. Vol. 1+, 1967+. Bionom. Briefs. HI 62535

Bionomie. Stuttgart. 1913. Bionomie. HI 62536

Bio-nouvelles. Montreal. Vol. 1+, 1973+. Bio-nouv. HI 62537

Bioorganicheskaya khimiya. Moscow. Vol. 1+, 1975+. Bioorg. Khim. HI 62538

Biopedagos. Caen. 1986+. Biopedagos. HI 75274

Biophysical journal. New York, NY. Biophys. J. See B–P–H 196/19. HI 52115

Biophysical society abstracts, annual meetings. New York, NY. Biophys. Soc. Abstr. Ann. Meetings. See B–P–H 196/21. HI 52116

Biophysics. New York, NY. Biophysics New York). See B–P–H 196/22. HI 52117

Biophysics. [Seibutsu butsuri.] Kyoto. Vol. 1+, 1961+. Biophysics (Kyoto). HI 61299

Biophysics of structure and mechanism. Berlin, etc. Vol. 1+, 1974+. Biophys. Struct. Mechanism. HI 62539

Biophysik. Berlin, etc. Vols. 1-10, 1963-73. Biophysik. Superseded by: Radiation and environmental biophysics. HI 62540

Bioprocess engineering. Berlin, New York. Vol. 1+, 1986+. Bioprocess Engin. HI 68115

Bioprocessing technology. Englewood, NJ. Vol. 7+, 1985+. Bioprocess. Technol. Preceded by: Biomass digest. HI 75235

Bioresearch. Ujjain. 1977+. Bioresearch. HI 62541

Bioresearch index; reporting the worlds research literature in the life sciences. Philadelphia, PA. Vol. 1-17, 1967-79. Biores. Index. Preceded by: Bioresearch titles. Superseded by: Biological abstracts/R R M. HI 62542

Bioresearch titles; reporting the world's research literature in the life sciences. Philadelphia, PA. Vols. [1-2], 1965-66. Biores. Titles. Superseded by: Bioresearch index. HI 62543

Bioresearch today: Food microbiology. Philadelphia, PA. 1972+. Biores. Today, Food Microbiol. HI 62544

Biorheology. Oxford, England & New York, NY.

Biorheology. See B–P–H 196/25. HI 52118

Bios; revista bimensual de ciencias médicas. Barcelona. Bios (Barcelona). See B–P–H 196/26. HI 52119

Bios. Rivista di biologia sperimentale e generale. Genoa. Bios (Genoa). See B–P–H 196/27. HI 52120

Bios; Abhandlungen zur theoretischen Biologie und ihrer Geschichte ... Leipzig. Bios (Leipzig). See B–P–H 197/1. HI 52121

Bios; revista del seminario de estudios biologicos. Mexico City. Vol. 1(1-3), 1968-69. Bios (Mexico City). HI 62546

Bios. Mount Vernon, IA. Bios (Mount Vernon). See B–P–H 197/2. HI 52123

Bios; revista de ciencias naturales. San José, Costa Rica. Bios (San José). See B–P–H 197/3. HI 52124

Bios. Swansea. Vol. 1+, 1967+. Bios (Swansea). HI 62547

Bioscans; plant tissue culture. Nottingham. 1979+. Bioscans, Pl. Tissue Cult. HI 62548

Bioscans; physiology and biochemistry of seeds. Nottingham. 1979+. Bioscans, Physiol. Biochem. Seeds. HI 62549

BioScene; the BIOSIS newsletter. Philadelphia, PA. Vol. 1+, 1971+. BioScene. HI 62550

BioScience. Washington, DC. BioScience. See B–P–H 197/4. HI 52125

Bioscience and industry. Tokyo = Baiosaiensu to indasutori. Tokyo.

Bioscience reports. Colchester. Vol. 1+, 1981+. Biosci. Rep. HI 62551

Biosciences communications. Basel, London, etc. Vol. 1+, 1975+. Biosci. Commun. HI 62552

BioSearch 83 [84 etc.]. Philadelphia, PA. Vol. 1+, 1983+. BioSearch 83 [84 etc.] HI 62553

Biosensors. Barking. Vols. 1-4, 1985-89. Biosensors. Superseded by: Biosensors and bioelectronics. HI 70290

Biosensors and bioelectronics. Barking. Vol. 5+, 1990+. Biosensors & Bioelectronics. Preceded by: Biosensors. HI 75174

Bioseparation; international journal of separation science in biotechnology. Dordrecht & Hingham, MA. Vol. 1+, 1989+. Bioseparation. HI 62554

BIOSIS / C A S selects: antibacterial agents. Philadelphia, PA. 1990+. BIOSIS C. A. S. Selects, Antibacterial Agents. HI 62151

BIOSIS / C A S selects: antifungal agents. Philadelphia, PA. 1988+. BIOSIS C. A. S. Selects, Antifungal Agents. HI 62152

BIOSIS / C A S selects: antiviral agents. Philadelphia, PA. 1986+. BIOSIS C. A. S. Selects, Antiviral Agents. HI 62153

BIOSIS / C A S selects: Bacterial and viral genetics. Philadelphia, PA. 1982+. BIOSIS C. A. S. Selects, Bact. Viral Genet. HI 62154

BIOSIS / C A S selects: Biological clocks. Philadelphia, PA. 1981+. BIOSIS C. A. S. Selects, Biol. Clocks. HI 62155

BIOSIS / C A S selects: Biological information transfer. Philadelphia, PA. 198?+. BIOSIS C. A. S. Selects, Biol. Inform. Transfer. HI 62156

BIOSIS / C A S selects: DNA replication. Philadelphia, PA. 1982+. BIOSIS C. A. S. Selects, DNA Replic. HI 62157

BIOSIS / C A S selects: Environmental pollution. Philadelphia, PA. 198?+. BIOSIS C. A. S. Selects, Environm. Pollut. HI 62158

BIOSIS / C A S selects: Fungicides. Philadelphia, PA. 198?+. BIOSIS C. A. S. Selects, Fungicides. HI 62159

BIOSIS / C A S selects: Genetic manipulation in plants. Philadelphia, PA. 1985+. BIOSIS C. A. S. Selects, Genet. Manipul. Pl. HI 62160

BIOSIS / C A S selects: Herbicides. Philadelphia, PA. 198?+. BIOSIS C. A. S. Selects, Herbicides. HI 62161

BIOSIS / C A S selects: Histochemistry and cytochemistry. Philadelphia, PA. 1981+. BIOSIS C. A. S. Selects, Histochem. Cytochem. HI 62162

BIOSIS / C A S selects: Membrane separation. Philadelphia, PA. 198?+. BIOSIS C. A. S. Selects, Membr. Separation. HI 62163

BIOSIS / C A S selects: Photobiochemistry. Philadelphia, PA. 198?+. BIOSIS C. A. S. Selects, Photobiochemistry. HI 62164

BIOSIS / C A S selects: Plant genetics. Philadelphia, PA. 1981+. BIOSIS C. A. S. Selects, Pl. Genet. HI 62165

BIOSIS / C A S selects: Transportation. Philadelphia, PA. 1981+. BIOSIS C. A. S. Selects, Transport. HI 62166

BIOSIS technical topics. Philadelphia, PA. Vol. 1+, 1976+. BIOSIS Techn. Topics. HI 62167

Biosistematika. Belgrade. Vol. 1+, 1975+. Biosistematika. HI 62555

Biosources digest. Santa Monica, CA. Vol. 1+, 1979+. Biosources Digest. HI 62556

Biosphaera. [Seibutsukai.] Tokyo. Biosphaera. See B–P–H 197/6. HI 52127

Biosphere; bulletin of the international biological programme. London. Nos. 1-10, 1967-71. Biosphere (London). HI 62557

Biosphere. Montreal. Vol. 1+, 1980+. Biosphere (Montreal). HI 62558

Biosphere bulletin. Singapore. No. 1+, 1972+. Biosphere Bull. HI 62559

Biosphere reserve research report. St. Thomas, VI. No. ?-2+, ?-1986+. Biosphere Reserve Res. Rep. HI 62560

Biosynthesis; a review of the literature. London. Vols. 1-7, 1970/71-79/81, 1972-82? Biosynthesis. HI 62561

BioSystems. Amsterdam. Vol. 6+, 1974+. BioSystems. Preceded by: Currents in modern biology. HI 62562

Biota. Magdalena del Mar [= Lima, in part]. Vol. 1(1)+, 1954+. Biota. HI 62563

Biota newsletter. Ferndale, MI. 1980+. Biota Newslett. HI 62564

Biotec. Stuttgart, New York. Vol. 1+, 1987+. Biotec. HI 62565

Biotech patent news. Metuchen, NJ. Vol. ?-2+, ?-1988+. Biotech Patent News. HI 75236

Biotech-update; biotechnology as it relates to the clinical laboratory - now and in the future. Anaheim, CA. Vol. 1+, 1982+. Biotech-Update. HI 62566

Biotechniques. Natick, MA. Vol. 1+, 1983+. Biotechniques. HI 62567

Bio/technology. New York. Vol. 1+, 1983+. Bio/Technology. HI 62568

Biotechnology advances; research reviews and patent abstracts. Oxford, Elmsford, NY., etc. Vol. 1+, 1983+. Biotechnol. Advances. HI 62569

Biotechnology in agriculture and forestry. New York. No. 1+, 1986+. Biotechnol. Agric. Forest. HI 62570

Biotechnology and applied biochemistry. San Diego, CA, Duluth, MN. Vol. 8+, 1986+. Biotechnol. & Appl. Biochem. Preceded by: Journal of applied biochemistry. HI 75189

Biotechnology and bioengineering. New York. Vol. 4+, 1962+. Biotechnol. & Bioengin. Preceded by: Journal of biochemical and microbiological technology. HI 62571

Biotechnology business. New York. 19??-85+. Biotechnol. Business. HI 62572

Biotechnology directory. New York. 1985+. Biotechnol. Directory. HI 62573

Biotechnology education. Oxford, Elmsford, NY. Vol. 1+, 1989+. Biotechnol. Educ. HI 62574

Biotechnology education report. New York. 1983+. Biotechnol. Educ. Rep. HI 62575

Biotechnology and genetic engineering reviews. Newcastle upon Tyne. Vol. 1+, 1984+. Biotechnol. Genet. Engin. Rev. HI 62576

Biotechnology information news. London. No. 16+, 1988+. Biotechnol. Inform. News. Preceded by: News, european biotechnology information project. HI 75237

Biotechnology in Japan newsservice. Palo Alto, CA. Vol. ?-4(12)+, ?-1986+. Biotechnol. Japan Newsserv. HI 75238

Biotechnology letters. London. Vol. 1+, 1979+. Biotechnol. Lett. HI 62577

Biotechnology news. Maplewood, NJ. Vol. 1+, 1981+. Biotechnol. News. HI 62578

Biotechnology newswatch. New York. Vol. 7(23)+, 1987+. Biotechnol. Newswatch. Preceded by: McGraw-Hill's biotechnology newswatch. HI 74814

Biotechnology notes. Washington, DC. Vol. 1+, 1988+. Biotechnol. Notes. HI 75239

Biotechnology patent digest. Washington, DC. Vol. 1+, 1982+. Biotechnol. Patent Digest. HI 62579

Biotechnology progress. New York. Vol. 1+, 1985+. Biotechnol. Progr. HI 62580

Biotechnology research abstracts. Bethesda, MD. Vol. 1+, 1984+. Biotechnol. Res. Abstr. HI 62581

Biotechnology software. New York. Vol. 1+, 1984+. Biotechnol. Software. HI 62582

Biotechnology techniques. Kew. 1987+. Biotechnol. Techn. HI 62583

Biotekhnologiya; teoreticheskii nauchno-prakticheskii zhurnal. Moscow. 1985+. Biotekhnologiya. HI 62584

Biótica; publicación del instituto nacional de investigaciones sobre recursos bióticos. Xalapa. Vol. 2+, 1977+. Biótica. Preceded by: Publicaciones, Instituto de Investigatores sobre Recursos Bióticos. HI 62585

Biotico. San José, (Costa Rica). Vol. 1+, 1966+. Biotico. HI 62586

BioTrack. Pittsburgh, PA. Vol. 1+, 1986+. BioTrack. HI 62587

Biotronics. Fukuoka. Vol. 1+, 1970+. Biotronics. HI 62588

Biotrop bulletin. Bogor. Nos. 1-21, 1970-82. Biotrop Bull. Superseded by: Biotrop bulletin in tropical biology. HI 62589

Biotrop bulletin in tropical biology. Bogor. No. 22+, 1983+. Biotrop. Bull. Trop. Biol. Preceded by: Biotrop bulletin. HI 62590

Biotrop newsletter. Bogor. Vol. ?-11+, 19??-75+.

Biotrop Newslett. HI 62591

Biotrop special publications. Bogor. No. 1+, 1976+. Biotrop. Special Publ. HI 62592

Biotropia. Bogor. Vol. 1+, 1987+. Biotropia. HI 62593

Biotropica. Pullman, WA. Vol. 1+, 1969+. Biotropica. Preceded by: Association for tropical biology bulletin. HI 62594

Biotypologie. Paris. Biotypologie. See B–P–H 197/8. HI 52129

Biovigyanan; journal of life sciences. Poona. Vol. 1+, 1975+. Biovigyanan. HI 62595

Birmingham botanic garden, or Midland floral magazine; ... London. Birmingham Bot. Gard. See B–P–H 197/9. HI 52130

Birmingham botanical gardens newsletter = Newsletter, Birmingham botanic gardens. Birmingham.

Birmingham iris, and Midland counties monthly magazine. Birmingham, England. Birmingham Iris & Midl. Counties Monthly Mag. See B–P–H 197/10. HI 52131

Birmingham and Midland gardeners' magazine. London. Birmingham Midl. Gard. Mag. See B–P–H 197/11. HI 52132

Birøkteren. Christiania [=Oslo, Norway]. Birøkteren. See B–P–H 197/12. HI 52133

Biruniya, Al; revue marocaine de pharmacognosie, d'études ethnomédicales et de botanique appliquée. Rabat. Vol. 1+, 1985+. Biruniya. HI 62596

Bishop museum bulletins in botany. Honolulu, HI. No. 1+, 1988+. Bishop Mus. Bull. Bot. Preceded by: Bernice P. Bishop museum bulletin. HI 62597

Bishop museum news = Ka 'elele. Honolulu, HI.

Bishop museum occasional papers. Honolulu, HI. Vol. 26+, 1986+. Bishop Mus. Occas. Pap. Preceded by: Occasional papers of the Bernice Pauahi Bishop museum of polynesian ethology and natural history. HI 62598

Bitidningen. Huddinge, Mantorp. 1902+. Bitidningen. HI 62599

Bitki; Turkish journal of plant science. Izmir. Vol. 1+, 1974+. Bitki. HI 62600

Bitki koruma bulteni. Ek yayin. Ankara. Bitki Koruma Bult. See B–P–H 197/18. HI 52134

Biuletyn informacyjny. Cracow. Biul. Inform. See B–P–H 197/19. HI 52135

Biuletyn, instytut przemyslu zielarskiego. Poznan. Vols. 6(2)-10, 1960-64. Biul. Inst. Przemyslu Zielarsk. Preceded by: Biuletyn, panstwowy instytut naukowy leczniczych surowcow roslinnych. Superseded by: Herba polonica. HI 62601

Biuletyn instytutu hodowii i aklimatyzacji roślin. Warsaw. Biul. Inst. Hodowii Rośl. See B–P–H 197/20. HI 52136

Biuletyn instytutu ochrony roślin. Poznan, Poland. Biul. Inst. Ochr. Rośl. See B–P–H 197/21. HI 52137

Biuletyn naukowy, panstwowy instytut naukowy leczniczych surowcow roslinnych. Poznan. Vols. 1-2, 1955-56. Biul. Nauk. Panstw. Inst. Nauk. Leczn. Surow. Rosl. Superseded by: Biuletyn, panstwowy instytut naukowy leczniczych surowcow roslinnych. HI 62602

Biuletyn, panstwowy instytut naukowy leczniczych surowcow roslinnych. Poznan. Vols. 3-6(1), 1957-60. Biul. Panstw. Inst. Nauk. Leczn. Surow. Rosl. Preceded by: Biuletyn naukowy, panstwowy instytut naukowy leczniczych surowcow roslinnych. Superseded by: Biuletyn, instytut przemyslu zielarskiego. HI 62603

Biuletyn Warzywniczy. Skierniewice. Vol. 1+, 1953+. Biul. Warzywn. HI 62604

Biuletyn, zakładu biologii stawów, polska akademia nauk. Cracow. Nos. 1-7, 1954-58. Biul. Zakładu Biol. Stawów Polska Akad Nauk. Superseded by: Acta hydrobiologica. Cracow. HI 62605

Biyoloji dergisi; bioloji derneği'nin yayin organi. Istanbul. Vols. 24-28, 1974-78. Biyol. Derg. Preceded by: Türk biyoloji dergisi. HI 62606

Black Hills engineer. Rapid City, SD. Vols. 11-28, 1923-42. Black Hills Engin. Preceded by: Pahasapa quarterly [not entered]. HI 62607

Black Rock forest bulletin. Cornwall-on-the-Hudson, NY. Black Rock Forest Bull. See B–P–H 199/10. HI 52138

Black Rock forest papers. Cornwall, NY. Vol. 1+, 1935+. Black Rock Forest Pap. HI 62608

Blackwood's Edinburgh magazine. Edinburgh & London. Vols. 1-178, 1817-37. Blackwood's Edinburgh Mag. Superseded by: Blackwood's magazine. HI 50353

Blackwood's magazine. London. Vol. 179+, 1906+. Blackwood's Mag. (London). Preceded by: Blackwood's Edinburgh Magazine. HI 62609

Blancoana. Jaen. Vol. 1+, 1983+. Blancoana. HI 62610

Blätter und Blüten. St. Louis, MO. Blätt. & Blüten. See B–P–H 199/13. HI 52139

Blätter für Kakteenforschung. Volksdorf [=Hamburg, in part]. Blätt. Kakteenf. See B–P–H 199/14. HI 52140

Blätter für Literatur, Kunst und Kritik. Zur oesterr. Zeitschrift für Geschichts- und Staatskunde. Vienna. Blätt. Lit. Kunst. See B–P–H 199/15. HI 52141

Blätter für literarische Unterhaltung. Leipzig. Blätt. Lit.

Unterhalt. See B–P–H 199/16. HI 52142

Blatter für Natur und Umwelt Schutz. Munich. Jahrg. 51-55, 1971-75. Blatt. Natur Umwelt Schutz. Preceded by: Blatter für Naturschutz. Superseded by: Natur und Umwelt. HI 62611

Blätter für Naturkunde und Naturschutz. Vienna. Blätt. Naturk. Naturschutz. See B–P–H 199/17. HI 52143

Blätter für Naturkunde und Naturschutz Niederösterreichs. Vienna. Blätt. Naturk. Naturschutz Niederösterreichs. See B–P–H 199/18. HI 52144

Blätter für Naturschutz. München. Vol. 1+, 1910+. Blatt. Naturschutz. Superseded by: Blatter für Natur und Umwelt Schutz. HI 62612

Blätter für Naturschutz und Naturpflege. Munich. Blätt. Naturschutz Naturpflege. See B–P–H 199/19. HI 52145

Blätter für Obst-, Wein- und Gartenbau. Brünn [=Brno, Czechoslovakia]. Blätt. Obst- Wein- Gartenbau. See B–P–H 199/20. HI 52146

Blätter für Obst-, Wein-, Gartenbau und Kleintierzucht. Brünn [=Brno, Czechoslovakia]. Blätt. Obst- Wein- Gartenbau Kleintierzucht. See B–P–H 200/1. HI 52147

Blätter für Obst-, Wein- und Gemüsebau. Brünn [=Brno, Czechoslovakia]. Blätt. Obst- Wein- Gemüsebau. See B–P–H 200/2. HI 52148

Blätter für Pflanzenbau und Pflanzenzüchtung. Decin, Czechoslovakia. Blätt. Pflanzenbau Pflanzenzücht. See B–P–H 200/3. HI 52149

Blätter für Staudenkunde. Berlin. Blätt. Staudenk. See B–P–H 200/4. HI 52150

Blätter für Sukkulentenkunde. Volksdorf. Vol. 1(1), 1949. Blatt. Sukkulentk. HI 62613

Blätter für Untersuchungs- und Forschungs-Instrumente. Rathenow, Germany. Blätt. Untersuchungs- Forsch.-Instrum. See B–P–H 200/5. HI 52151

Blätter des Vereins für Landeskunde in Niederösterreich. Vienna. Blätt. Vereins Landesk. Niederösterreich. See B–P–H 200/7. HI 52152

Blätter vermischten Inhalts. Oldenburg in Oldenburg, Germany. Blätt. Vermischten Inhalts. See B–P–H 200/8. HI 52153

Blekinge läns hushållstidningen. Karlskrona, Sweden. Blekinge Läns Hushållstidn. See B–P–H 200/9. HI 52154

Blister rust news. U.S. department of agriculture. Washington, DC. Blister Rust News See B–P–H 200/10. HI 52156

Blithewold gardens and arboretum newsletter = Newsletter, Blithewold gardens and arboretum. Bristol, RI.

Bloem en blad; vakblad voor de bloemendetailhandel. Doetinchem. Vol. 1+, 1979+. Bloem & Blad. HI 62614

Bloembollencultuur. Hillegom. Vol. ?-81+, ?-1970+. Bloembollencultuur. HI 62615

Bloemisterij; weeklad gewijd aan de teelt van en den hadel in bloemen, planten en heesters. Doetinchem, Netherlands. Bloemisterij. See B–P–H 200/11. HI 52157

Bloemisterij onderzoek in Nederland. Aalsmeer. 1975+. Bloemist. Onderz. Ned. Preceded by: Jaarverslag, proefstation voor de bloemisterij in Nederland. HI 62617

Blomster. Nordisk tidskrift for binderi og blomster. Copenhagen. Blomster. See B–P–H 200/12. HI 52158

Blue jay; bulletin of the Yorktown natural history society (later A journal of natural history and conservation for Saskatchewan and adjacent region). Regina, Saskatchewan. Vol. 1+, 1942-48, 1961+. Blue Jay. HI 62618

Blue jay news. Saskatoon. No. 63+, 1983+. Blue Jay News. Preceded by: Newsletter, Saskatchewan natural history society. HI 62619

Blühende Kakteen. Nuedamm [=Debno, Poland]. Blüh. Kakteen. See B–P–H 200/14. HI 52159

Blühende Kakteen und andere sukkulenten Pflanzen. Leipzig & Neudamm. Nos. 7-42, 1932-39. Blüh. Kakteen And. Sukk. Pflanzen. Preceded by: Blühende Sukkulenten. HI 62620

Blühende Sukkulenten. Leipzig & Neudamm. Nos. 1-6, 1930-31. Blüh. Sukk. Superseded by: Blühende Kakteen und andere sukkulenten Pflanzen. HI 62621

Blumea; tijdschrift voor de systematiek en de geografie der planten (A journal of plant taxonomy and plant geography.) Leiden. Blumea. See B–P–H 200/15. HI 52160

Blumen-Kalender. Zurich. Blumen-Kalender. See B–P–H 201/1. HI 52167

Blumen und Palmen. Frankfurt a. M. Blumen & Palmen. See B–P–H 200/16. HI 52161

Blumen und Pflanzen im Gewerbe. Vienna. Blumen Pflanzen im Gewerbe. See B–P–H 200/17. HI 52162

Blumen- und Pflanzenbau. Berlin. Blumen- Pflanzenbau. See B–P–H 200/18. HI 52163

Blumen- und Pflanzenbau vereinigt mit Die Gartenwelt. Berlin. Blumen- Pflanzenbau & Gartenwelt. See B–P–H 200/19. HI 52164

Blumen-Zeitung. Weissensee, Germany. Blumen-Zeitung. See B–P–H 201/2. HI 52168

Blumenfreund. Bautzen. Blumenfreund. See

B–P–H 200/21. HI 52165

Blumengärtner. Stuttgart. Blumengärtner. See B–P–H 200/22. HI 52166

Blumenzwiebel. Hillegom, Netherlands. Blumenzwiebel. See B–P–H 201/3. HI 52169

Blyttia. Oslo. Blyttia. See B–P–H 201/4. HI 52170

Board of greenkeeping research, british golf unions, journal. Bingley, England. Board Greenkeeping Res. Brit. Golf Unions J. See B–P–H 201/5. HI 52171

Bocagiana. Funchal, Madeira. Bocagiana. See B–P–H 201/7. HI 52172

Bochumer geographische Arbeiten. Bochum. No. 1+, 1965+. Bochum.Geogr. Arbeiten. HI 62622

Bodenbiologie = Biologie du sol. Paris.

Bodenkultur. Österreichisches Zentralorgan der Landwirtschaftswissenschaften und Ernährungsforschung. Vienna. Bodenkultur. See B–P–H 201/12. HI 52174

Bodenkunde und Pflanzenernährung. Neue Folge der Zeitschrift für Pflanzenernährung, Düngung und Bodenkunde. Leipzig & Berlin. Bodenk. & Pflanzenernähr. See B–P–H 201/9. HI 52173

Boerderij. Weekblad gewijd aan den landbouw en tuinbouw, veeteelt, pluimveehouderij ens. Doetinchem, Netherlands. Boerderij. See B–P–H 210/13. HI 52175

Boekzaal der geleerde wereld. Amsterdam. 1812-63. Boekzaal Gel. Wereld. Preceded by: Maendelyke uittreksels, of boekzael der geleerde werelt. 3-2496-1. HI 62623

Bohemia. Jahrbuch des collegium carolinum. Munich. Vol. 1+, 1960+. Bohemia. HI 62624

Bohemia centralis. Prague. Bohemia Centr. See B–P–H 201/15. HI 52176

Bois, Le. Paris. Vol. 1+, 1883+. Bois (Paris). HI 62625

Bois et forêts des tropiques. Paris. Bois Forêts Trop. See B–P–H 210/18. HI 52177

Bois national. Colmar, France. Bois Natl. See B–P–H 201/19. HI 52178

Boissiera. Geneva. Boissiera. See B–P–H 201/20. HI 52179

Bokin bobai = Journal of research society for antibacterial and antifungal agents, Japan.

Boletim da academia das ciências de Lisboa. Lisbon. 1929-42, 1951+. Bol. Acad. Ci. Lisboa. Preceded by: Jornal de sciencias mathematica, physicas e naturaes. HI 62626

Boletim de agricultura. São Paulo. Bol. Agric (São Paulo). See B–P–H 202/11. HI 52189

Boletim de agricultura do estado de São Paulo. São Paulo. Bol. Agric. Estado São Paulo. See B–P–H 202/13. HI 52190

Boletim de agricultura Minas Gerais. Belo Horizonte, Brazil. Bol. Agric. Minas Gerais. See B–P–H 202/14. HI 52191

Boletim agrostológico Brasileiro. São Paulo = Correio agrostologico. São Paulo.

Boletim anual do serviço de fitopatologia. Pôrto Alegre, Brazil. Bol. Anual Serv. Fitopatol. See B–P–H 202/18. HI 52195

Boletim de biologia e pesquisas technologicas. Parana. No. ?-43+, 19??-66+. Bol. Biol. Pesq. Technol. HI 62627

Boletim biologico. São Paulo. Bol. Biol. (São Paulo). See B–P–H 203/5. HI 52207

Boletim de botânica, departamento de botânica, instituto de biociencias, universidade de São Paulo. São Paulo. Vols. 1-4, 1973-76. Bol. Bot. Univ. São Paulo. Preceded by: Boletim de faculdade de filosofia, ciencias e letras, universidade de São Paulo. Botanica. HI 62628

Boletim cearense de agronomia. Fortaleza, Ceara. 1969+. Bol. Cear. Agron. Preceded by: Boletim da sociedade cearense de agronomia. HI 62629

Boletim de ciências do mar. Fortaleza. No. 21+, 1969+. Bol. Ci. Mar. Preceded by: Boletím da estação de biologia marinha da universidade do Ceará. HI 62630

Boletim. Circulo gaucho de orquidofilos. Pôrto Alegre, Brazil. Bol. Circ. Gaucho Orquidofilos. See B–P–H 204/4. HI 52223

Boletim da commissão geographica e geologica do estado de São Paulo. São Paulo. Nos. 1-21, 1889-1906. Bol. Commiss. Geogr. Estado São Paulo. HI 62631

Boletim commissão nacional ambiente. ?-1981+. Bol. Commiss. Nac. Amb. HI 62632

Boletim do departamento de agricultura. Rio de Janeiro. Bol. Dept. Agric. See B–P–H 204/10. HI 52226

Boletim do departamento de oceanografia e limnologica de centro de biociencias da universidade federal do Rio Grande do Norte. Natal. Vol. 6+, 1978+. Bol. Dep. Oceanogr. Limnol. Univ. Rio Grande do Norte. Preceded by: Boletim do instituto de biologia marinha da universidade (federal) do Rio Grande do Norte. HI 62633

Boletim didatico, escola superior de agricultura "Luiz de Queiroz", universidade de São Paulo. Piracicaba. No. 1+, 1962+. Bol. Didat. Esc. Super. Agric. Luiz de Queiroz. HI 62634

Boletim de divulgaçao, escola superior de agricultura "Luiz de Queiroz", universidad de São Paulo.

Piracicaba. No. 1+, 1962+. Bol. Divulg. Esc. Super. Agric. Luiz de Queiroz. HI 62635

Boletim, escola de agronomia da Amazônia. Belem. Nos. 1-4, 1971. Bol. Esc. Agron. Amazôn. Superseded by: Boletim da faculdade de ciências agrárias do Pará. HI 62636

Boletim. Escola superior de agricultura "Luiz de Queiroz." Universidade. São Paulo. Bol. Esc. Super. Agric. Luiz de Queiroz. See B–P–H 204/17. HI 52231

Boletím da estação de biologia marinha da universidade do Ceará. Fortaleza. Vols. 1-20, 1961-68. Bol. Estac. Biol. Mar. Univ. Ceará. Superseded by: Boletím de ciências do mar. HI 62637

Boletim F B C N. Rio de Janeiro. Vol. 13+, 1978+. Bol. F. B. C. N. Preceded by: Boletim informativo, fundação brasileira para a conservação da natureza. HI 62638

Boletim da faculdade de ciências agrárias do Pará. Belem. No. 5+, 1972+. Bol. Fac. Ci. Agrár. Pará. Preceded by: Boletim, escola de agronomia da Amazônia. HI 62639

Boletim da facultade de farmacia e odontologia de Pelotas. Pelotas. 1940-49. Bol. Fac. Farm. Odont. Pelotas. Superseded by: Boletim sobre plantas medicinais e ornamentais. HI 62640

Boletim da facultade de filosofia, ciências e letras; universidade de São Paulo; biologia geral. São Paulo. 1937-68. Bol. Fac. Filos. Univ. São Paulo, Biol. Geral. HI 62641

Boletim de faculdade de filosofia, ciencias e letras, universidade de São Paulo; botânica. São Paulo. Nos. 1-27, 1937-69. Bol. Fac. Filos. Univ. São Paulo, Bot. Superseded by: Boletim de botánica, departamento de botanica, instituto de biociencias, universidade de São Paulo. HI 62642

Boletim fitossanitário. Rio de Janeiro. Bol. Fitossan. See B–P–H 205/7. HI 52241

Boletim I C B. Séria botânica = Boletim, instituto central de biociências, universidade federal do Rio Grande do Sul. Séria botânica. Porto Alegre.

Boletim I G A. São Paulo. Vols. 1-2, 1970-71. Bol. Inst. Geoci. Astron. Univ. São Paulo. Superseded by: Boletim, instituto geociencias, universidade de São Paulo. HI 62643

Boletim I G - U S P = Boletim, instituto geociencias, universidade de Sao Paulo. São Paulo.

Boletim do I N P A. Serie botânica = Boletim do instituto nacional de pesquisas da Amazônia. Serie botânica. Manaus.

Boletim do I N P A. Serie florestais = Boletim do instituto nacional de pesquisas da Amazônia. Serie pesquisas florestais. Manaus.

Boletim informativo, biblioteca I F = Boletim informativo, biblioteca instituto florestal. São Paulo.

Boletim informativo, biblioteca instituto florestal. São Paulo. Vol. 1+, 1982+. Bol. Inform. Biblot. Inst. Florest. HI 62645

Boletim informativo, fundação brasileira para a conservação de natureza. Rio de Janeiro. 1966-77. Bol. Inform. Fund. Brasil. Conserv. Nat. Superseded by: Boletim F B C N. HI 62646

Boletim informativo do instituto de cacau de Bahia. Salvador, Brazil. Bol. Inform. ILnst. Cacau Bahia See B–P–H 205/24. HI 52253

Boletim informativo do instituto de biologia marítima. Lisboa. No. 1+, 1971+. Bol. Inform. Inst. Biol. Marít. HI 62647

Boletim do InPaBo. Curitiba. Nos. 1-15, 1955-61. Bol. InPaBo. HI 62644

Boletim de inspectoria regional de fomento agrícola do estado do Pará. Belém, Brazil. Bol. Inspectoria Regional Fomento Agríc. Estado Pará. See B–P–H 206/8. HI 52260

Boletim; instituto agronomico do estado de São Paulo. São Paulo. Bol. Inst. Agron. Estado São Paulo. See B–P–H 206/10. HI 52262

Boletim do instituto de biociências, universidade federal do Rio Grande do Sul. Porto Alegre. No. 37+, 1977+. Bol. Inst. Bioci. Univ. Fed. Rio Grande do Sul. Preceded by: Boletim, instituto central de biociências, universidade federal do Rio Grande do Sul. HI 62648

Boletim do instituto biológico da Bahia. Sao Salvador. Vols. 1-15, 1954-76. Bol. Inst. Biol. Bahia. HI 62649

Boletim. Instituto biologico de defesa agrícola. Rio de Janeiro. Bol. Inst. Biol. Defesa Agríc.. See B–P–H 206/13. HI 52264

Boletim do instituto de biologia marinha da universidade (federal) do Rio Grande do Norte. Natal. Vols. 1-5, 1964-71. Bol. Inst. Biol. Mar. Univ. Rio Grande do Norte. Superseded by: Boletim do departamento de oceanografia e limnologia do centro de biociencias da universidade federal do Rio Grande do Norte. HI 62650

Boletim, instituto de biologia e pesquisas tecnologicas. [Curitiba.] Nos. 40-47, 1959-70. Bol. Inst. Biol. Pesq. Tecnol. Curitiba. HI 62651

Boletim do instituto de botânica. São Paulo. Nos. 1-7, 1963-70. Bol. Inst. Bot. (São Paulo). HI 62652

Boletim. Instituto brasileiro de sciencias. Rio de Janeiro. Bol. Inst. Brasil. Sci. See B–P–H 206/20. HI 52268

Boletim do instituto de café do estado de São Paulo. São

Paulo. Vols. 1-5, 1926-30. Bol. Inst. Café Estado São Paulo. Superseded by: Revista do instituto de café de estado de São Paulo. HI 62653

Boletim, instituto central de biociências, universidade federal do Rio Grande do Sul. Porto Alegre. Nos. 29-36, 1971-77. Bol. Inst. Centr. Bioci. Univ. Fed. Rio Grande do Sul. Preceded by: Boletim, instituto de ciencias naturais. Superseded by: Boletim do instituto de biociências, universidade federal do Rio Grande do Sul. HI 62654

Boletim, instituto central de biociências, universidade federal do Rio Grande do Sul. Séria botânica. Porto Alegre. No. 29+, 197?+. Bol. Inst. Centr. Bioci. Univ. Fed. Rio Grande do Sul, Sér. Bot. Preceded by: Boletim, instituto de ciencias naturais. HI 62655

Boletim do instituto central de fomento económica da Bahia. Salvador, Brazil. Bol. Inst. Centr. Fomento Econ. Bahia. See B–P–H 207/1. HI 52269

Boletim, instituto de ciencias naturais, universidade federal do Rio Grande do Sul = Boletim, instituto de ciencias naturais, universidade do Rio Grande do Sul. Porto Alegre.

Boletim, instituto de ciencias naturais, universidade do Rio Grande do Sul. Porto Alegre. Nos. 1-28, 1954-70. Bol. Inst. Ci. Nat. Univ. Rio Grande do Sul. Superseded by: Boletim, instituto central de biociências, universidade federal do Rio Grande do Sul. HI 62656

Boletim do instituto de ciências naturais da universidade de Santa Maria. Rio Grande do Sul. Vols. 1-3, 1962-68. Bol. Inst. Ci. Nat. Univ. Santa Maria. HI 62657

Boletim, instituto de defesa do patrimônio natural. Curitiba. Vols. 4-5, 1963-64. Bol. Inst. Defesa Patrim. Nat. Superseded by: Araucariana, botànica. HI 62658

Boletim do instituto de ecologia e experimentação agrícolas. Rio de Janeiro. Bol. Inst. Ecol. Exp. Agríc. See B–P–H 207/7. HI 52272

Boletim, instituto geociencias, universidade de São Paulo. São Paulo. Vols. 3-14, 1972-84. Bol. Inst. Geoci. Univ. São Paulo. Preceded by: Boletim I G A. HI 62659

Boletim, instituto de geociencias e astronomia, universidade de São Paulo = Boletim I G A. São Paulo.

Boletim do instituto de história natural. Botânica. Curitiba, Brazil. Bol. Inst. Hist. Nat., Bot. See B–P–H 207/17. HI 52274

Boletim do instituto de investigação cientifica de Angola. Luanda. Vol. 1+, 1962+. Bol. Inst. Invest. Ci. Angola. HI 62660

Boletim. Instituto Kuribara de ciência natural brasileira. São Paulo. Bol. Inst. Kuribara Ci. Nat. Brasil. See B–P–H 207/21. HI 52277

Boletim de instituto nacional do mate. Rio de Janeiro. Bol. Inst. Nac. Mate. See B–P–H 208/1. HI 52281

Boletim do instituto nacional de pesquisas da Amazônia. Serie avulsa. Manaus. Vol. 1, 1968. Bol. Inst. Nac. Pesq. Amazôn. Ser. Avulsa. Superseded by: Acta amazonica. HI 62661

Boletim do instituto nacional de pesquisas da Amazônia. Serie, botânica. Manaus. No. 29+, 1969+. Bol. Inst. Nac. Pesq. Amazôn. Ser. Bot. Preceded by: Publicação, instituto nacional de pesquisas da Amazônia. Botânica. HI 62662

Boletim do instituto nacional de pesquisas da Amazônia. Serie, pesquisas florestais. Manaus. Nos. 1-17, 1970. Bol. Inst. Nac. Pesq. Amazôn. Ser. Pesq. Florest. Preceded by: Publicação, instituto nacional de pesquisas da Amazônia. Botânica. Superseded by: Acta amazônica. HI 62663

Boletim do instituto de oceanografia. São Paulo. Bol. Inst. Oceanogr. (São Paulo). See B–P–H 208/4. HI 52283

Boletim do instituto paulista de oceanografia. São Paulo. Bol. Inst. Paul. Oceanogr. See B–P–H 208/5. HI 52284

Boletim do instituto dos produtos florestais. Lisbon. Vol. 35(411)+, 1973+. Bol. Inst. Prod. Florest. Preceded by: Boletim da junta nacional da cortica. HI 62664

Boletim do instituto de química agrícola. Rio de Janeiro. Nos. 1-62, 1938-61. Bol. Inst. Quím. Agríc. HI 62665

Boletim do jardim botânico. Rio de Janeiro. Bol. Jard. Bot. (Rio de Janeiro). See B–P–H 208/11. HI 52287

Boletim da junta nacional da cortica. Lisbon. Vols. ?-34(410), 19??-73. Bol. Junta Nac. Cort. Superseded by: Boletim do instituto dos produtos florestais. HI 62666

Boletim. Junta nacional das frutas. Lisbon. Bol. Junta Nac. Frutas. See B–P–H 208/12. HI 52288

Boletim mensal. Commisão do comércio de cacau da Bahia. Salvador, Brazil. Bol. Mensal Comiss. Comércio Cacau Bahia. See B–P–H 208/18. HI 52294

Boletim do ministerio de agricultura. Bol. Minist. Agric. (Rio de Janeiro). See B–P–H 209/1. HI 52297

Boletim do museu de biologia Prof. Mello-Leitão. Sér. Bot. Santa Teresa, Estado E. Santo. 1949+. Bol. Mus. Biol. Prof. Mello-Leitão. Sér. Bot. HI 62667

Boletim do museu botânico Kuhlmann. Rio de Janeiro. Vol. 1+, 1978+. Bol. Mus. Bot. Kuhlmann. HI 62668

Boletim do museu botânico municipal. Curitiba. Vols. 1-26, [1971]-[76]. Bol. Mus. Bot. Munic. HI 62669

Boletim do museo Goeldi de historia natural e ethnographia. Belém 1904-14. Bol. Mus. Goeldi Paraense Hist. Nat. Ethnogr. Preceded by: Boletim do museu paraense de historia natural e ethnographia. HI 62670 Superseded by: Boletim do museu paraense "Emilio Goeldi".

Boletim do museu de historia natural da universidade federal de Minas Gerais. Botanica. Belo Horizonte. No. 1+, 1968+. Bol. Mus. Hist. Nat. Univ. Fed. Minas Gerais. HI 62671

Boletim do museu municipal do Funchal. Funchal, Madeira. Bol. Mus. Munic. Funchal. See B–P–H 209/14. HI 52302

Boletim do museu nacional de Rio de Janeiro. Rio de Janeiro. Bol. Mus. Nac. Rio de Janeiro. See B–P–H 209/18. HI 52304

Boletim do museu nacional de Rio de Janeiro. Botânica. Rio de Janeiro. Bol. Mus. Nac. Rio de Janeiro, Bot. See B–P–H 209/19. HI 52305

Boletim do museu paraense "Emilio Goeldi". Belém. 1933-56. Bol. Mus. Paraense Emilio Goeldi. Preceded by: Boletim do museu Goeldi de historia natural e ethnographia. Superseded by: Boletim do museu paraense "Emilio Goeldi". Nova serie, botanica. HI 62672

Boletim do museu paraense "Emilio Goeldi". Nova série, botânica. Belém. Vol. 1+, 1958+. Bol. Mus. Paraense Emilio Goeldi, N. S., Bot. Preceded by: Boletim do museu paraense de historia natural e ethnographia. HI 62673

Boletim do museu paraense de historia natural e ethnographia. Belém. Vols. 1-3, 1894-1902. Bol. Mus. Paraense Hist. Nat. Ethnogr. Superseded by: Boletim do museo Goeldi de historia natural e ethnographia. HI 62674

Boletim do muzeu Rocha. Ceara. Vol. 1+, 1908+. Bol. Muz. Rocha. HI 62675

Boletim de la oficina de agricultura y ganaderia, provincia de Buenos Aires. Buenos Aires. Vols. 1-2, 1901-02. Bol. Of. Agric. Ganad. Prov. Buenos Aires. HI 62676

Boletim de la oficina quim. municipal de Tucuman. Tucuman. Vols. 1-?, 1888-1916? Bol. Of. Quim. Munic. Tucuman. HI 62677

Boletim paranense de geografia. Curitiba, Brazil. Bol. Paran. Geogr. See B–P–H 210/1. HI 52307

Boletim paulista de geografia. São Paulo. No. 1+, 1949+. Bol. Paul. Geogr. HI 62678

Boletim de pesquisa, centro de pesquisa agropecuária do trópico úmido. Belém. No. 1+, 1980+. Bol. Pesq. Centro Pesq. Agropecu. Tróp. Umido. HI 62679

Boletim de pesquisa florestal. Curitiba. No. 1+, 1980+. Bol. Pesq. Florest. HI 62680

Boletim da secretaria de agricultura, industria e comércio. Recife, Brazil. Bol. Secr. Agric. (Recife). See B–P–H 210/9. HI 52310

Boletim da secretaria da agricultura, industria e comércio. Salvador, Brazil. Bol. Secr. Agric. (Salvador). See B–P–H 210/10. HI 52311

Boletim do serviço nacional de pesquisas agronómicas. Rio de Janeiro. Bol. Serv. Nac. Pesq. Agron. See B–P–H 210/11. HI 52312

Boletim sobre plantas medicinais e ornamentais. Pelotas. No. 5, 1952. Bol. Pl. Med. Ornam. Preceded by: Boletim da faculdade de farmacia e odontologia de Pelotas. HI 62681

Boletim de recursos naturais, SUDENE. Recife. 1963+. Bol. Recurs. Nat. SUDENE. HI 62682

Boletim da secretaria de agricultura, indústria e comércio, estado de Pernambuco. Vol. ?-2-24+, 19??-37-56+. Bol. Secr. Agric. (Pernambuco). HI 62683

Boletim, servicio florestal do estado de São Paulo. São Paulo. No. 1+, 1954+. Bol. Serv. Florest. São Paulo. HI 62684

Boletim, setor de inventarios florestais. Rio de Janeiro. No. 1+, 1959+. Bol. Setor Invent. Florest. HI 62685

Boletim da sociedade broteriana. Coimbra, Portugal. Bol. Soc. Brot. See B–P–H 211/1. HI 52319

Boletim da sociedade cearense de agronomia. Fortaleza, Ceara. 1960-68. Bol. Soc. Cearense Agron. Supeseded by: Boletim cearense de agronomia. HI 62686

Boletim da sociedade de estudios da Cólonia de Moçambique. Lourenço Marques. Vols. 1-40, 1931-71. Bol. Soc. Estud. Cólon. Moçambique. HI 62687

Boletim de sociedade de medicina e cirurgia de São Paulo. São Paulo. Bol. Soc. Med. São Paulo. See B–P–H 211/26. HI 52329

Boletim da sociedade portuguesa de ciencias naturais. Lisbon. 1948+. Bol. Soc. Portug. Ci. Nat. Preceded by: Bulletin de la société portugaise des sciences naturelles. HI 62688

Boletim da superintendencia dos serviços do café. São Paulo. Vol. 17+, 1942+. Bol. Superintend. Serv. Café. Preceded by: Revista do instituto de café de estado de São Paulo. HI 62689

Boletim técnico da C O D E C A P. Recife. No. ?-3+, 19??-74+. Bol. Técn. C. O. D. E. C. A. P. HI 62690

Boletim técnico, centro de pesquisas agropecuárias do trópico úmido. Belém. Vol. 1+, 1976+. Bol. Técn. Centro Pesq. Agropecu. Tróp. Umido. Preceded by: Boletim técnico do instituto de pesquisas agropecuarias do norte. HI 62691

Boletim técnico, centro pesquisas do cacau. Itabuna.

Vols. 1-59, 1970-77. Bol. Técn. Centro Pesq. Cacau. Incorporated in: Revista theobroma. HI 62692

Boletim técnico commissão executiva do plano da lavoura cacaueira = Boletim técnico, centro pesquisas do cacau. Itabuna.

Boletim tecnico do departamento de produçao vegetal. Curitiba. No. 1+, 1959+. Bol. Tecn. Dep. Prod. Veg. HI 62693

Boletim técnico-cientifico, escola superior de agricultura "Luiz de Queiroz", universidade de São Paulo. Piracicaba. No. 1+, 1962+. Bol. Técn.-Ci. Esc. Super. Agric. Luiz de Queiroz. HI 62694

Boletim técnico. Instituto agronómico do Leste. Cruz das Almas, Brazil. Bol. Técn. Inst. Agron. Leste. See B–P–H 212/21. HI 52336

Boletim técnico do instituto agronômico do norte. Belém. Vols. 1-43, 1943-61. Bol. Técn. Inst. Agron. N. Superseded by: Boletim técnico do instituto de pesquisas e experimentação agropecuarias do norte. HI 62695

Boletim técnico, instituto agronômico, São Paulo. São Paulo. Nos. 1-85, 1933-40. Bol. Técn. Inst. Agron. Estado São Paulo. Superseded by: Bragantia. 5-3779-3. HI 62696

Boletim técnico de instituto agronômico do sul. Pelotas. Vols. 1-70, 1947-70. Bol. Técn. Inst. Agron. S. Superseded by: Boletim técnico, instituto de pesquisas agropecuarias do sul. HI 62697

Boletim técnico, instituto brasileiro do café. São Paulo. No. 1+, 1970+. Bol. Técn. Inst. Brasil. Café. HI 62698

Boletim técnico do instituto de cacau da Bahia. São Paulo. Bol. Técn. Inst. Cacau Bahia. See B–P–H 212/25. HI 52337

Boletim técnico, instituto florestal. São Paulo. 1972+. Bol. Técn. Inst. Florest. HI 62699

Boletim técnico, instituto de pesquisas agropecuarias do sul. Pelotas. Vol. 71+, 1971+. Bol. Técn. Inst. Pesq. Agropecu. S. Preceded by: Boletim técnico de instituto agronômico do sul. HI 62700

Boletim técnico do instituto de pesquisas e experimentação agropecuárias do norte. Belém. Vols. 44-55, 1964-72. Bol. Técn. Inst. Pesq. Exp. Agropecu. N. Preceded by: Boletim técnico do instituto agronomico do norte. Superseded by: Boletim técnico do instituto de pesquisas agropecuárias do norte (IPEAN). HI 62701

Boletim técnico do instituto de pesquisas agropecuarias do norte. Belém. Vols. 56-64, 1973-74. Bol. Técn. Inst. Pesq. Agropecu. N. Preceded by: Boletim técnico do instituto de pesquisas e experimentação agropecuárias do norte; Série quimica de solos, instituto de pesquisas e expérimentaçao agropecuárias do norte and Série técnologia do instituto de pesquisas e expérimentaçao agropecuárias do norte. Superseded by: Boletim técnico, centro de pesquisas agropecuárias do trópico úmido. HI 62702

Boletim técnico do instituto de pesquisas e experimentação do sul = Boletim técnico, instituto de pesquisas agropecuarias do sul. Pelotas.

Boletim técnico, projeto Radambrasil. Serie vegetação. Salvador. No. 1+, 1982+. Bol. Técn. Proj. Radambrasil, Ser. Veg. HI 62703

Boletim da universidade (federal) do Paraná. Botânica. Curitiba. 1960-72. Bol. Univ. Parará, Bot. Superseded by: Acta biologica paranaense. HI 62715

Boletim da universidade rural de Pernambuco. Dois Irmãos, Brazil. Bol. Univ. Rural Pernambuco. See B–P–H 213/6. HI 52344

Boletim da universidade de São Paulo. Botânica. São Paulo. Bol. Univ. São Paulo, Bot. See B–P–H 213/7. HI 52345

Boletim de zoologia e biologia marinha. Nova serie. São Paulo. 1969-73. Bol. Zool. Biol. Mar., N. S. Preceded by: Boletim da facultade de filosofia, ciências e letras, universidade do São Paulo; zoologia [not entered]. Superseded by: Boletim de zoologia etc. HI 62704

Boletín de la academia de ciencias y artes de Barcelona. Madrid. Epoca 3, vol. 6(1-4), 1932-33. Bol. Acad. Ci. Nat. Barcelona. Preceded by: Boletín de la real academia de ciencias y artes de Barcelona. Superseded by: Butlletí de l'acadèmia de ciencias i arts de Barcelona. 1-19-1. HI 62705

Boletín de academia de ciencias, bellas lettras y nobles artes. Córdoba, Spain. Bol. Acad. Ci. (Córdoba). See B–P–H 201/22. HI 52180

Boletín de academia de ciencias exactas, físicas y naturales de Madrid. Madrid. Bol. Acad. Ci. Exact. Madrid. See B–P–H 201/23. HI 52181

Boletín de la academia de ciencias físicas, matemáticas y naturales. Caracas. Bol. Acad. Ci. Fís. See B–P–H 201/24. HI 52182

Boletín de academia de ciencias naturales y artes de Barcelona. Barcelona. Bol. Acad. Ci. Nat. Barcelona. See B–P–H 201/25. HI 52184

Boletín de la academia de la historia. Madrid. Bol. Acad. Hist. See B–P–H 202/3. HI 52185

Boletín de academia nacional de ciencias. Córdoba, Argentina. Bol. Acad. Nac. Ci. See B–P–H 202/5. HI 52186

Boletín, academia nacional de ciencias (exactas existente en la universidad de Córdoba) = Boletín de academia nacional de ciencias. Córdoba, Argentina.

Boletín aereo. Mexico City. Bol. Aereo. See B–P–H 202/8. HI 52187

Boletín agrícola, sociedad antioqueña de agricultores. Medellín. Ser. 3, nos. 66-221, 1927-36; no. 222+, 1937+. Bol. Agric. (Medellín). Preceded by: Boletín semanal, sociedad antioqueña de agricultores. HI 52192

Boletín de agricultura. Bogota. Vols. 1-7, 1927-36. Bol. Agric. (Bogota). Superseded by: Agricultura. Bogota. HI 62706

Boletín de agricultura. Suplemento. Bogota. 19??+. Bol. Agric (Bogota), Supl. HI 62707

Boletín de agricultura. Guatemala City. Bol. Agric. (Guatemala City). See B–P–H 202/10. HI 52188

Boletín de agricultura, mineria e industria. Mexico City. Vols. 1-10, 1891-1901. Bol. Agric. Mineria Industr. Superseded by: Boletín, secretaría de agricultura y ganadería. Mexico. HI 62708

Boletín de agricultura tropical. San José, Costa Rica. Bol. Agric. Trop. See B–P–H 202/16. HI 52193

Boletín de algódon. Buenos Aires. Bol. Algódon. See B–P–H 202/17. HI 52194

Boletín antártico chileno. Santiago. Vol. 1+, 1981+. Bol. Antárt. Chileno. Preceded by?: Revista de difusión, instituto antártico chileno. HI 62709

Boletín anual, sociedad de orquideologia de Carabodo. (Venezuela). 19??-66-70. Bol. Anual, Soc. Orquideol. Carabodo. HI 62710

Boletín de la asociación cafetalera de El Salvador. Santa Ana, Salvador. Bol. Asoc. Cafetalera El Salvador. See B–P–H 202/19. HI 52196

Boletín de la asociación general de agricoltures, Guatemala. Guatemala City, Guatemala. Bol. Aoc. Gen. Agric. Guatemala. See B–P–H 202/20. HI 52197

Boletín de la asociacion latinoamericana de paleobotanica y palinologica. Buenos Aires. 1973+. Bol. Asoc. Latinoamer. Paleobot. Palin. HI 62711

Boletín, asociacion mexicana de orquideologica. 1973+. Bol. Asoc. Mex. Orquideol. HI 62712

Boletín, asociación nacional del café. Guatemala. No. 1+, 1962+. Bol. Asoc. Nac. Café. HI 62713

Boletín de la asociación nacional de ingenieros agrónomos. Madrid. Bol. Asoc. Nac. Ing. Agrón. See B–P–H 202/21. HI 52198

Boletín . Asociación de peritos forestales. [Venezuela]. Bol. Asoc. Peritos Forest. See B–P–H 202/22. HI 52200

Boletín azúcarero mexicano. Lambayeque, Peru. No. 1+, 1949+. Bol. Azúcarero Mex. HI 62714

Boletín de bibliografía botánica. San Isidro, Argentina. Bol. Bibliogr. Bot. See B–P–H 203/2. HI 52204

Boletín bibliográfico. [Banco central de Venezuela.] Caracas. Bol. Bibliogr. See B–P–H 202/23. HI 52201

Boletín bibliografico de la academia colombiana de ciencias exactas, fisicas y naturales. Bogota. No. 1+, 1968+. Bol. Bibliogr. Acad. Colomb. Ci. Exact. Fis. Nat. HI 62527

Boletín bibliográfico agrícola. Madrid. Bol. Bibliogr. Agríc. (Madrid). See B–P–H 202/24. HI 52202

Boletín bibliográfico agricola. Turrialba. Vols. 1-5, 1964-68. Bol. Bibliogr. Agric. (Turrialba). Superseded by: Boletín para bibliotecas agrícolas. HI 62528

Boletín bibliográfico agropecuario. Pasto, Colombia. Bol. Bibliogr. Agropecu. See B–P–H 203/1. HI 52203

Boletín bibliográfico, biblioteca, universidad nacional mayor de San Marcos de Lima. Lima. Vol. 1+, 1923+. Bol. Bibliogr. Bibliot. Univ. Nac. Mayor San Marcos de Lima. HI 62529

Boletín bibliografico, centro nacional de investigaciones agropecuarias, Venezuela. Maracay. Vol. 1+, 1973+. Bol. Bibliogr. Centro Nac. Invest. Agropecu. Venezuela. HI 62530

Boletín bibliografico, escuela nacional de ciencias biologicas. Mexico, DF. No. 1+, 1965+. Bol. Bibliogr. Esc. Nac. Ci. Biol. HI 62531

Boletín bibliografico informativo. Bogotá. No. 1+, 1971+. Bol. Bibliogr. Inform. Bogota. HI 62532

Boletín bibliográfico. Instituto forestal latinoamericano de investigación y capacitación. Mérida, Venezuela. Bol. Bibliogr. Inst. Forest. Lationamer. Invest. See B–P–H 203/3. HI 52205

Boletín bibliografico, instituto del mar del Peru. Callao. 1974+. Bol. Bibliogr. Inst. Mar Peru. HI 62533

Boletín bibliográfico, instituto nacional de tecnología agropecuaria. Buenos Aires. No. ?-95+, 19??-84+. Bol. Bibliogr. Inst. Nac. Tecnol. Agropecu. HI 62534

Boletín bibliográfico mexicano. Mexico City. Bol. Bibliogr. Mex. See B–P–H 203/4. HI 52206

Boletín para bibliotecas agrícolas. Turrialba. Vol. 6+, 1969+. Bol. Bibliot. Agríc. (Turrialba) Preceded by: Boletín bibliográfico agricola. HI 62535

Boletín de biologia vegetal. Santiago de Chile. Vol. 1+, 1978+. Bol. Biol. Veg. HI 62536

Boletín biológico. Cátedra de microbiología y micología. Facultad de ciencias médicas. La Plata, Argentina. Bol. Biol. Cátedra Microbiol. Fac. Ci. Méd. See B–P–H 203/6. HI 52208

Boletín biológico. Universidad de Puebla. Puebla, Mexico. Bol. Biol. Univ. Puebla. See B–P–H 203/8. HI 52209

Boletín de bosques, pesca i caza. Santiago. Bol. Bosques. See B–P–H 203/10. HI 52210

Boletín botánico latinoamericano; publicacion oficial de la asociación Latinoamericano de botánica. Bogotá. No. 1+, 1978+. Bol. Bot. Latinoamer. HI 62537

Boletín cafetalero. San Pedro Sula, Honduras. Bol. Cafetalero. See B–P–H 203/11. HI 52211

Boletín de la cámara de agricultura de la primera zona. Quito. Vol. 1(1-?), 1937. Bol. Cámara Agric. Primera Zona. Superseded by: Revista de la cámara de agricultura de la primera zona. 2-892-2. HI 62538

Boletín. Cátedra de fitopatología. Universidad nacional La Plata. La Plata, Argentina. Bol. Cátedra Fitopatol. Univ. Nac. La Plata. See B–P–H 203/12. HI 52212

Boletín. Centro de documentación científica y técnica. Mexico City. Bol. Centro Doc. Ci. Técn. See B–P–H 203/13. HI 52213

Boletín. Centro de documentación científica y técnica. Sección 4. Medicina. Mexico City. Bol. Centro Doc. Ci. Técn., Secc. 4, Med. See B–P–H 203/14. HI 52214

Boletín, centro de investigacion en fruticultura, horticultura y vitivinicultura. Montevideo. No. 1+, 1917+. Bol. Centro Invest. Frutic. Hort. Vitivinic. HI 62539

Boletín del centro de investigaciones antropologicas de México. Mexico City. Bol. Centro Invest. Antropol. México. See B–P–H 203/15. HI 52215

Boletín del centro de investigaciones biologicas, universidad de Zulia. Maracaibo. 1967+. Bol. Centro Invest. Biol. Univ. Zulia. HI 62540

Boletín, centro nacional de experimentacion y extension agricola. Cuba. Vol. ?-64+, 19??-66+. Bol. Centro Nac. Exp. Extens. Agric., Cuba. HI 62541

Boletín de ciencia y technologica. Venezuela. No. 6+, 19??+. Bol. Ci. Technol. HI 62542

Boletín de ciencias médicas. Guadalajara, Mexico. Bol. Ci. Méd. (Guadalajara). See B–P–H 203/18. HI 52218

Boletín de ciencias médicas. Mexico City. Bol. Ci. Méd. (Mexico City). See B–P–H 203/19. HI 52220

Boletín de ciencias de la naturaleza I D E A. ?-1980+. Bol. Ci. Naturaleza I. D. E. A. HI 62543

Boletín científico. Havana. Bol. Ci. See B–P–H 203/16. HI 52216

Boletín científico, centro de investigaciones oceanográficas e hidrográficas. Cartagena. Vol. 1+, 1977+. Bol. Ci. Centro Invest. Oceanogr. Hidrogr. HI 62544

Boletín científico, instituto de fomento pesquero. Santiago. Bol. Ci. Inst. Fomento Pesq. See B–P–H 203/17. HI 52217

Boletín científico de la sociedad Sánchez Oropeza. Orizaba, Mexico. Bol. Ci. Soc. Sánchez Oropeza. See B–P–H 203/20. HI 52221

Boletín científico y técnico. [Instituto nacional de pesca del Ecuador.] Guayaquil, Ecuador. Bol. Ci. Técn. See B–P–H 203/21. HI 52222

Boletín, colegio de ingenieros de Venezuela. Venezuela. No. 37+, 1927+. Bol. Colegio Ing. Venezuela. HI 62545

Boletín comisión forestal del estado de Michoacán. Morelia. Nos 1-14, 1960-63. Bol. Com. Forest. Estado Michoacán. HI 62546

Boletín del comité nacional del café. Tegucigalpa, Honduras. Bol. Comité Nac. Café. See B–P–H 204/5. HI 52224

Boletín, comite de orquideologica, sociedad venezolana de ciencias naturales. Caracas. No. ?-20+, 19??-81+. Bol. Comité Orquideol. Soc. Venez. Ci. Nat. HI 62547

Boletín de la corporación venezolana de fomento. [Venezuela]. Bol. Corp. Venez. Fomento. See B–P–H 204/9. HI 52225

Boletín, departamento de biologia, universidad nacional de Colombia. Bogota. Vol. ?-1(5)+, 19??-83+. Bol. Dep. Biol. Univ. Nac. Colombia. HI 62548

Boletín del departamento de biologia, universidad del Valle. Cali. Vols. 1-4(1), 1968-1971. Bol. Dep. Biol. Univ. Valle. HI 62549

Boletín, departamento forestal, ministerio de agricultura. No. 1+, 1964?+. Bol. Dep. Forest. Minist. Agric., Chile. HI 62550

Boletín del departamento forestal. [Universidad de la república. Facultad de agronomía.] Montevideo, Uruguay. Bol. Dept. Forest. See B–P–H 204/11. HI 52227

Boletín de difusión, instituto antártico chileno. Santiago. Nos. 7-8, 1972-75. Bol. Difusión Inst. Antárt. Chileno. Preceded by: Boletín, instituto antartico chileno. Superseded by: Revista de difusión, instituto antártico chileno. HI 62551

Boletín de la dirección de estudios biológicos. Mexico City. Bol. Dirección Estud. Biol. See B–P–H 204/13. HI 52228

Boletín. Dirección de frutas, hortalizas y flores. Buenos Aires. Bol. Dirección Frutas. See B–P–H 204/14. HI 52229

Boletín, dirección general de agricultura. Mexico. Vols.

1-5(2), 1911-15; n.s. Vols. 1-3(5), 1915-17. Bol. Dirección Gen. Agric., Mexico. HI 62552

Boletín de divulgacion, centro de investigaciones agricolas "Alberto Boerger". La Estanzuela. No. 1+, 1970+. Bol. Divulg. Centro Invest. Agric. Alberto Boerger. HI 62553

Boletín de divulgacion, estaçión experimental agrícola de Saavedra. Santa Cruz. No. 1+, 1970+. Bol. Divulg. Estaç. Exp. Agríc. Saavedra. HI 62554

Boletín de divulgacion, instituto nacional de investigaciones forestales. Mexico, D.F. No. 1+, 1961+. Bol. Divulg. Inst. Nac. Invest. Forest., Mexico. HI 62555

Boletín divulgativo e informativo del instituto venezolano de investigaciones cientificas. Caracas. Vol. 1+, 1966+. Bol. Divulg. Inform. Inst. Venez. Invest. Ci. HI 62556

Boletín divulgativo, instituto nacional de investigaciones forestales = Boletín de divulgacion, instituto nacional de investigaciones forestales. Mexico, D.F.

Boletín ecotrópica; ecosistemas del tropico. Bogota. No. 12+, 1985+. Bol. Ecotróp. Preceded by: Boletín del museo del mar, universidad de Bogotá Jorge Tadeo Lozano. HI 62557

Boletín. Escuela nacional de ciencias biológicas. Mexico City. Bol. Esc. Nac. Ci. Biol. See B–P–H 204/15. HI 52230

Boletín. Estación agrícola experimental. Chihuahua, Mexico. Bol. Estac. Agríc. Exp. (Chihuahua). See B–P–H 204/18. HI 52232

Boletín. Estación agrícola experimental. Oaxaca, Mexico. Bol. Estac. Agríc. Exp. (Oaxaca). See B–P–H 204/19. HI 52233

Boletín. Estación agrícola experimental. San Juan Bautista [=Villahermosa], Mexico. Bol. Estac. Agríc. Exp. (San Juan Bautista). See B–P–H 204/20. HI 52234

Boletín, estaçion agricola experimental de Palmira. Palmira (Colombia). No. 1+, 196?+. Bol. Estaçion Agric. Exp. Palmira. HI 62558

Boletín, estaçión central agronómica de Cuba. Havana No. 1+, 1905+. Bol. Estaçión Agron. Cuba. HI 62559

Boletín de la estación central de ecología. Madrid. Vol. 1+, 1972+. Bol. Estaçión Centr. Ecol. Madrid. HI 62560

Boletín. Estación experimental agrícola "La Molina." Lima. Bol. Estac. Exp. Agríc. "La Molina". See B–P–H 204/22. HI 52235

Boletín, estaçión experimental agricola de Tingo Maria. Lima. No. 1+, 1954+. Bol. Estaçión Agric. Tingo Maria. HI 62561

Boletín de la estaçión experimental agrícola de Tucumán. San Miguel de Tucumán. No. 1+, 1924+. Bol. Estaçion Exp. Agric. Tucumán. HI 62562

Boletín, estaçión experimental agronomica de Santiago de las Vegas. Havana. No. 1+, 1905+. Bol. Estaçión Exp. Agron Santiago de las Vegas. HI 62563

Boletín, estaçión experimental agropecuaria de Lambayeque. Lima. No. 1+, 1961+. Bol. Estaçión Exp. Agropecu. Lambayeque. HI 62564

Boletín, estaçión experimental agropecuaria presidencia Roque Saenz Peña. Chaco. No. 1+, 1959+. Bol. Estaçión Exp. Agropecu. Presid. Roque Saenz Peña. HI 62565

Boletín, estaçión experimental de Aula Dei. Zaragoza. Nos. [1-4] 5+, 1949+ [Nos. 1-4 were not titled "Boletín"]. Bol. Estaçión Exp. Aula Dei. HI 62566

Boletín. Estacíon experimental de occidente. [Venezuela]. Bol. Estac. Exp. Occid. See B–P–H 204/23. HI 52236

Boletín, estaçión experimental de Paysandú. Paysandú. No. 1+, 1963?+. Bol. Estaçión Exp. Paysandú. HI 62567

Boletín de estadistica de la federación nacional de cafeteros. Bogotá. Bol. Estadist. Fed. Nac. Cafeteros. See B–P–H 204/24. HI 52237

Boletín de estadistica meteorologica e hidrologica. Perú. Vols. 5+, 1966+. Bol. Estadfíst. Meteorol. Hidrol., Peru. HI 62568

Boletín de estudios médicos y biológicos. Mexico City. Vol. 25+, 1967+. Bol. Est. Méd. Biol. Preceded by: Boletín del instituto de estudios médicos y biológicos. HI 62569

Boletín experimental, ministerio de agricultura. Bolivia. No. 1+, 1952?+. Bol. Exp. Minist. Agric., Bolivia. HI 62570

Boletín de extension, instituto nacional de agricultura. Maracay. No. 1+, 1952+. Bol. Extens. Inst. Nac. Agric. Maracay. HI 62571

Boletín de la facultad de agronomía. [Universidad de San Carlos.] Guatemala City. Bol. Fac. Agron. See B–P–H 204/25. HI 52238

Boletín de la facultad de agronomia de universidad de la republica, Montevideo. Montevideo. No. 1+, 1957+. Bol. Fac. Agron. Univ. Montevideo. HI 62572

Boletín, facultad de ciencias agrarias, universidad austral de Chile. Valdivia. No. 1+, 1966+. Bol. Fac. Ci. Agrar. Univ. Austral Chile. HI 62573

Boletín, facultad de ciencias agropecuarias, universidad central de las villas. Santa Clara. No. 1+, 1970+. Bol. Fac. Ci. Agropecu. Univ. Centr. Villas Santa Clara. HI 62574

Boletín de la facultad de ciencias exactas, fisicas y naturales. Cordoba. Vols. 1-7, 1938-47. Bol. Fac. Ci. Exact. Fis. Nat. Córdoba. Superseded by: Revista de la facultad de ciencias exactas, fisicas y naturales. 2-1200-1. HI 62575

Boletín. Facultad nacional de agronomía. Medellín, Colombia. Bol. Fac. Nac. Agron. See B–P–H 205/5. HI 52239

Boletín. Federación nacional de cultivadores de cereales. Bogotá. Bol. Fed. Nac. Cultivadores Cereales. See B–P–H 205/6. HI 52240

Boletín, field science office for Latin America, Unesco. Montevideo. Nos. 1-8, 19??-74? Bol. Field Off. Latin Amer., Unesco. Superseded by: Boletín, regional office for science and technology for Latin America and the Caribbean, Unesco. HI 62576

Boletín de fomento, ciencias, agricultura, artes y comercio. Madrid. Bol. Fomento (Madrid). See B–P–H 205/8. HI 52242

Boletín de fomento. [Secretaría de fomento y agricultura.] San José, Costa Rica. Bol. Fomento (San José). See B–P–H 205/9. HI 52243

Boletín forestal. Caracas. Bol. Forest. (Caracas). See B–P–H 205/10. HI 52244

Boletín forestal. Chihuahua, Mexico. Bol. Forest. (Chihuahua). See B–P–H 205/11. HI 52245

Boletín forestal. La Paz, Bolivia. Bol. Forest. (La Paz). See B–P–H 205/12. HI 52246

Boletín forestal. Quito, Ecuador. Bol. Forest. (Quito). See B–P–H 205/13. HI 52247

Boletín forestal y de industrias forestales para América latina. [FAO] Santiago. Bol. Forest. Industr. Forest. Amér. Latina. See B–P–H 205/16. HI 52248

Boletín forestal, universidad de Tolima. Tolima. Vol. 1+, 1967+. Bol. Forest. Univ. Tolima. HI 62577

Boletín genético. Castelar, Argentina. Bol. Genet. See B–P–H 205/17. HI 52249

Boletín de geografía y estadística de la república méxicana. Mexico City. Vol. 1, 1849. Bol. Geogr. Estadist. Repübl. Méx. Superseded by: Boletín de la sociedad de geografía y estadística de la república méxicana. 5-3911-1. HI 62578

Boletín de geologico. Bogota. Vol. 1+, 1953+. Bol. Geol. (Bogota). HI 62579

Boletín de geologico. Bucaramanga. Vol. 1+, 1958+. Bol. Geol. (Bucaramanga). HI 62580

Boletín de historia natural de la sociedad "Felipe Poey," universidad de la Habana. Havana. Bol. Hist. Nat. Soc. "Felipe Poey" Univ. Habana. See B–P–H 205/20. HI 52250

Boletín de los hospitales. Caracas. Bol. Hosp. See B–P–H 205/21. HI 52251

Boletín de información científica cubana. Havana. Vols. 1-3, 1969-72. Bol. Inform. Ci. Cubana. Superseded by: Boletín de informacion cientifica y tecnica cubana. HI 62581

Boletín de informacion cientifica y tecnica cubana. Havana. Vols. 4+, 1973+. Bol. Inform. Ci. Tecn. Cubana. Preceded by: Boletín de información científica cubana. HI 62582

Boletín de informacíon del ministerio de agricultura. Madrid. Vol. 1+, 1948+. Bol. Inform. Minist. (Industr.) Agric., Madrid. HI 62583

Boletín de informacíon del ministerio de industria y agricultura = Boletín de informacíon del ministerio de agricultura. Madrid.

Boletín informativo de la biblioteca del mac. Bol. Inform. Bibliot. Mac. See B–P–H 205/22. HI 52252

Boletín informativo, centro nacional de investigaciones de café. Chinchiná. Vols. 1-7, 1949-56. Bol. Inform. Centro Nac. Invest. Café. Superseded by: Cenicafé. 2-955-2. HI 62584

Boletín informativo, circulo de coleccionistas de cactus de la republica Argentina. Buenos Aires. Vol. 1+, 1966+. Bol. Inform. Curculo Colecç. Cact. Republ. Argent. HI 62585

Boletín informativo divulgativo, instituto forestal latino-americano de investigación y capacitación. Merida. Nos. 1-[5], 1957-59. Bol. Inform. Inst. Forest. Latinoamer. Invest. Capac. HI 62586

Boletín informativo, estaçión experimental agricola de los Llanos. Santa Cruz, Bolivia. No. 1+, 1961+. Bol. Inform. Estaçión Exp. Agric. Llanos. HI 62587

Boletín informativo, estaçión experimental agrícola de Tucumán. San Miguel de Tucumán. No. 1+, 1963+. Bol. Inform. Estaçión Exp. Agríc. Tucumán. HI 62588

Boletín informativo, instituto agropecuario nacional, Guatemala. Guatemala. 1964+. Bol. Inform. Agropecu. Nac. Guatemala. HI 62589

Boletín informativo del instituto de botanica, universidad de Guadalajara. Guadalajara. 1974+. Bol. Inform. Inst. Bot. Univ. Guadalajara. HI 62590

Boletín informativo del instituto de fitotécnia. Castelar, Argentina. Bol. Inform. Inst. Fitotécn. See B–P–H 206/1. HI 52254

Boletín informativo. Instituto forestal. Santiago. Bol. Inform. Inst. Forest. See B–P–H 206/2. HI 52255

Boletín informativo, instituto geobiológico "La Salle." Canoas, Brazil. Bol. Infomr. Inst. Geobiol. "La Salle". See B–P–H 206/3. HI 52256

Boletín informativo. Instituto interamericano de ciencias

agrícolas. Turrialba, Costa Rica. Bol. Inform. Inst. Interamer. Ci. Agríc. See B–P–H 206/4. HI 52257

Boletín informativo, instituto nacional del trigo. Bolivia. No. 1+, 1972+. Bol. Inform. Inst. Nac. Trigo. HI 62591

Boletín informativo de plagas. Madrid. Nos. 89-112, 1972-73. Bol. Inform. Plagas. Preceded by: Boletín del servicio de plagas forestales. Superseded by: Boletín, servicio de defensa contra plagas e inspecccion fitopatologica. HI 62592

Boletín informativo, sociedad botánica de México. Mexico, DF. Nos. 1-22, 1968-70. Bol. Inform. Soc. Bot. México. Superseded by: Macpalxochitl. HI 62593

Boletín informativo. Sociedad dasonomica de la América tropical. San José, Costa Rica. Bol. Inform. Soc. Dasonomica Amér. Trop. See B–P–H 206/5. HI 52258

Boletín informativo, sociedad ecuatoriana Francisco Campos de amigos de la naturaleza. Quito. No. 1+, 1974+. Bol. Inform. Soc. Ecuat. Francisco Campos Amigos Naturaleza. HI 62594

Boletín informativo sociedad latinoamericana de historia de las ciencias y la tecnología. Mexico City. No. 1+, 1983+. Bol. Inform. Soc. Latinoamer. Hist. Ci. Tecnol. HI 62595

Boletín informativo de la sociedad mexicana de micologia. Mexico, D.F. Vols. 1-3, 1968-69. Bol. Inform. Soc. Mex. Micol. Superseded by: Boletín de la sociedad mexicana de micologia. HI 62596

Boletín informativo, sociedad venezolana de ciencias naturales biblioteca "Alfredo Jahn". Caracas. ?-1977+. Bol. Inform. Soc. Venez. Ci. Nat. Bibliot. Alfredo Jahn. HI 62597

Boletín informativo. Universidad de Los Andes, facultad de ingeniería. Los Andes, Venezuela. Bol. Inform. Univ. Los Andes Fac. Ing. See B–P–H 206/6. HI 52259

Boletín, institut française d'études andines = Bulletin, institut française d'études andines. Lima.

Boletín del instituto agrario nacional. Caracas. Bol. Inst. Agrar. Nac. See B–P–H 206/9. HI 52261

Boletín, instituto antártico chileno. Santiago de Chile. Nos. 1-3, 1965-71. Bol. Inst. Antárt. Chileno. Superseded by: Boletín de difusión, instituto antártico chileno. HI 62598

Boletín. Instituto bacteriológico de Chile. Santiago. Bol. Inst. Bacteriol. Chile. See B–P–H 206/11. HI 52263

Boletín del instituto de biología marina. Mar del Plata, Argentina. Bol. Inst. Biol. Mar. See B–P–H 206/14. HI 52265

Boletín del instituto botánico de ciencias naturales. Quito, Ecuador. Bol. Inst. Bot. Ci. Nat. See B–P–H 206/17. HI 52266

Boletín del instituto botánico de la universidad central. Quito, Ecuador. Bol. Inst. Bot. Univ. Centr. See B–P–H 206/19. HI 52267

Boletín del instituto de ciencias naturales, universidad central del Ecuador. Quito. Vol. 1(1-2), 1952-55. Bol. Inst. Ci. Nat. Univ. Central Ecuador. HI 62599

Boletín de instituto científico y literario "Porfirio Diaz." Toluca, Mexico. Bol. Inst. Ci. "Porfirio Diaz". See B–P–H 207/2. HI 52270

Boletín de instituto de clínica quirúrgica de facultad de ciencias médicas de universidad nacional. Buenos Aires. Bol. Inst. Clín. Quir. Fac. Ci. Méd. Univ. Nac. See B–P–H 207/5. HI 52271

Boletín del instituto ecuatoriano de ciencias naturales = Boletín del instituto de ciencias naturales, universidad central del Ecuador. Quito.

Boletín del instituto español de oceanografia. Madrid. Vol. 1+, 1948+. Bol. Inst. Esp. Oceanogr. Preceded by: Notas y resumes, instituto español de oceanografia. HI 62600

Boletín del instituto de estudios asturianos. Oviedo. Vol. 1+, 1947+. Bol. Inst. Estud. Asturianos. Superseded in part by: Boletín del instituto de estudios asturianos. Suplemento de ciencias. 3-2023-1. HI 62601

Boletín del instituto de estudios asturianos. Suplemento de ciencias. Oviedo. Nos. 1-17, 1960-73. Bol. Inst. Estud. Adsturianos, Supl. Ci. Preceded by: Boletín del instituto de estudios asturianos. Superseded by: Suplemento de ciencias del Boletín del instituto de estudios asturianos. 3-2023-1. HI 62602

Boletín del instituto de estudios Giennénses. Jaen. ?-1983+. Bol. Inst. Estud. Giennén. HI 62603

Boletín del instituto de estudios médicos y biológicos. Mexico City. Vols. 3-24, 1944-66? Bol. Inst. Estud. Méd. Preceded by: Boletín del laboratorio de estudios médicos y biológicos. Superseded by: Boletín de estudios médicos y biológicos. 3-2630-1. HI 62604

Boletín del instituto fisico-geografico de Costa Rica. San José. Vols. 1-3, 1901-04. Bol. Inst. Fis.-Geogr. Costa Rica. HI 62605

Boletín del instituto fisico-geografico y organo de la sociedad nacional de agricultura de Costa Rica = Boletín del instituto fisico-geografico de Costa Rica. San José.

Boletín. Instituto de fomento algodonero. Bogotá. Bol. Inst. Fomento Algodonero. See B–P–H 207/13. HI 52273

Boletín del instituto forestal de investigaciónes y experiencias. Madrid. Vol. 1+, 1928+. Bol. Inst.

Forest. Invest. Exp. 3-2506-1. HI 62606

Boletín, instituto forestal latino-americano de investigación y capitación. Merida. Nos. 1-53, 1957-78. Bol. Inst. Forest. Latinoamer. Invest. Superseded by: Revista forestal latinoamericana. HI 62607

Boletín, instituto de investigación de los recursos marinos. Lima? Bol. Inst. Invest. Recurs. Mar. See B–P–H 207/20. HI 52276

Boletín, instituto de investigaciones agricolas. Mexico. No. 1+, 1953+. Bol. Inst. Invest. Agric., Mexico City. HI 62608

Boletín del instituto de investigaciones científicas. Monterrey, Mexico. Bol. Inst. Invest. Ci. See B–P–H 207/19. HI 52275

Boletín, instituto de investigaciones forestales La Molina. Lima. No. 1+, 1965+. Bol. Inst. Invest. Forest. La Molina. HI 62609

Boletín, instituto del mar del Peru. Callao, Peru. Bol. Inst. Mar Peru. See B–P–H 207/22. HI 52278

Boletín. Instituto mexicana del café. Mexico City. Bol. Inst. Mex. Café. See B–P–H 207/23. HI 52279

Boletín del instituto (nacional) de investigaciones agronómicas. Madrid. Bol. Inst. Nac. Invest. Agron. See B–P–H 207/24. HI 52280

Boletín del instituto oceanográfico. Cumaná, Venezuela. Bol. Inst. Oceanogr. (Cumaná). See B–P–H 208/2. HI 52282

Boletín del instituto de química de la universidad nacional autónoma de México. Mexico City. Bol. Inst. Quím. Univ. Nac. Autón. México. See B–P–H 208/7. HI 52285

Boletín, instituto superior de agricultura. Santiago de los Caballeros. No. 1+, 1966+. Bol. Inst. Super. Agric. HI 62610

Boletín del inventario biológico. Asuncion. Nos. 4-5, 1983-84. Bol. Invent. Biol. Preceded by: Biological inventory news. Superseded by: Boletín del inventario biológico nacional. HI 62611

Boletín del inventario biológico nacional. Asuncion. No. 6+, 1984+. Bol. Invent. Biol. Nac. Preceded by: Boletín del inventario biológico. HI 62612

Boletín. Jardín botánico. Mexico City. Bol. Jard. Bot. (Mexico City). See B–P–H 208/10. HI 52286

Boletín, jardín botánico La Laguna. San Salvador. Vols. 1-6, 1982-87. Superseded by: Pankia. Bol. Jard. Bot. La Laguna. HI 62613

Boletín del jardin botanico nacional "Dr. Rafael M. Moscoso". Santo Domingo. Vol. 1+, 1974+. Bol. Jard. Bot. Nac. Dr. Raphael M. Moscoso. HI 62614

Boletín del jardin botanico nacional, Santo Domingo = Boletín del jardin botanico nacional "Dr. Rafael M. Moscoso". Santo Domingo.

Boletín. Laboratorio de bacteriología Tucumán. Tucumán, Argentina. Bol. Lab. Bacteriol. Tucumán. See B–P–H 208/13. HI 52289

Boletín. Laboratorio de biología marina. Universidad católica de Santo Tomás de Villanueva. Marianao, Cuba. Bol. Lab. Biol. Mar. Univ. Católica Santo Tomás de Villanueva. See B–P–H 208/14. HI 52290

Boletín del laboratorio de estudios médicos y biológicos. Universidad nacional. Mexico City. Bol. Lab. Estud. Méd. See B–P–H 208/15. HI 52291

Boletín. Laboratorio químico nacional. Bogotá. Bol. Lab. Quím. Nac. See B–P–H 208/16. HI 52292

Boletín de Lima; revista cultural cientifica. Lima. No. 1+, 1979+. Bol. Lima. HI 62615

Boletín de medicina. Guanajuato, Mexico. Bol. Med. See B–P–H 208/17. HI 52293

Boletín mensual, museo de productos argentinos. Buenos Aires. Nos. 1-31, 1888-90? Bol. Mens. Prod. Argent. HI 62616

Boletín micológico. Valparaiso. Vol. 1+, 1982+. Bol. Micol. HI 62617

Boletín de microbiología. [Venezuela]. Bol. Microbiol. See B–P–H 208/20. HI 52295

Boletín del ministerio de agricultura. Buenos Aires. Bol. Minist. Agric. (Buenos Aires). See B–P–H 208/21. HI 52296

Boletín del ministerio de fomento. Caracas? Bol. Minist. Fomento. See B–P–H 209/2. HI 52298

Boletín del museo argentino de ciencias naturales "Bernardino Rivadavia" e instituto nacional de investigacíon de ciencias naturales. Buenos Aires. Nos. 1-19, 1956-58. Bol. Mus. Argent. Cienc. Nat. Bernardino Rivadavia. HI 62618

Boletín del museo de ciencias naturales. Caracas. Bol. Mus. Ci. Nat. See B–P–H 209/3. HI 52299

Boletín, museo de ciencias naturales, universidad nacional de Tucuman = Boletín, museo de historia natural, universidad nacional de Tucuman. Tucuman.

Boletín del museo de historia natural "Javier Prado." Lima. Bol. Mus. Hist. Nat. "Javier Prado". See B–P–H 209/9. HI 52300

Boletín. Museo de historia natural "Javier Prado." Publicationes series B, botánica. Lima. Bol. Mus. Hist. Nat. "Javier Prado," Publ. Ser. B, Bot. See B–P–H 209/11. HI 52301

Boletín, museo de historia natural, universidad nacional de Tucuman. Tucuman. Vols. 1-2(11), 1924-36. Bol. Mus. Hist. Nat., Tucuman. 5-4274-3. HI 62619

Boletín del museo del hombre dominicano. Santo

Domingo. Vol. 1+, 1972+. Bol. Mus. Hombre Dominic. HI 62620

Boletín, museo del mar, universidad de Bogotá Jorge Tadeo Lozano. Bogota. Nos. 1-11, 1970-83. Bol. Mus. Mar. Univ. Bogotá Jorge Tadeo Lozano. Superseded by: Boletín ecotrópica. HI 62621

Boletín del museo nacional, Costa Rica. San José. Vol. 1(1-2), 1945; ser. 2, no. 1, 1948; ser. 3, no. 1+, 1950+. Bol. Mus. Nac. Costa Rica. 5-3773-1. HI 62622

Boletín del museo nacional de historia natural. Santiago. Bol. Mus. Nac. Hist. Nat. See B–P–H 209/16. HI 52303

Boletín del museo nacional de Mexico. Mexico City. Vol. 1(1-3), 1903; ser. 2, vol. 1(1-10/12), 1903-04. Bol. Mus. Nac. Mexico. 3-2629-1. HI 62623

Boletín de noticias de la asociacion de palinologos de lengua espanola. Madrid. No. 1+, 1978+. Bol. Not. Asoc. Palin. Lenga Esp. HI 62624

Boletín oficial, secretaría de agricultura y fomento. Mexico = Boletín oficial, secretaría de agricultura y ganadería. Mexico. Mexico City.

Boletín oficial, secretaría de agricultura y ganadería. Mexico. Mexico City. 3. época, vols. 1-7, 1908-11; 4. época. vols. 1-9, 1916-25. Bol. Ofic. Secr. Agric. Ganad. Mexico. Preceded by: Boletín, secretaría de agricultura y ganadería. Mexico. HI 62625

Boletín, oficina regional de ciencia y tecnología de la Unesco para América latina y el Caribe = Boletín, regional office for science and technology for Latin America and the Caribbean, Unesco. Montevideo.

Boletín de oleicultura internacional. Madrid. Bol. Oleic. Int. See B–P–H 209/28. HI 52306

Boletín de patología vegetal y entomología agrícola. Madrid. Vols. 1-31, 1926-69. Bol. Patol. Veg. Entomol. Agric. Superseded by: Anales del instituto nacional de investigaciones agrarias. Serie, produccion vegetal and Serie, proteccion vegetal. 1-729-2. HI 62626

Boletín popular, estaçión experimental agrícola de Tingo María. Lima. No. 1+, 1946+, 1947+. Bol. Popular Estaçión Exp. Agric. Tingo María. HI 62627

Boletín de producción y fomento agrícola. Buenos Aires. Vol. 1+, 1949+. Bol. Prod. Fomento Agríc. 1-729-3. HI 62628

Boletín, proyecto de evaluacion forestal. Guatemala. No. 1+, 1967+. Bol. Proy. Eval. Forest. HI 62629

Boletín R O S T L A C. Montevideo. No. 21+, 1981+. Bol. R. O. S. T. L. A. C. Preceded by: Boletín, regional office for science and technology for Latin America and the Caribbean, Unesco. HI 62630

Boletín de la real academia de ciencias y artes de Barcelona. Madrid. Vol. 1, 1840; epoca 3, vols. 1-5, 1892-1931. Bol. Real Acad. Ci. Nat. Barcelona. Superseded by: Boletín de la academia de ciencias y artes de Barcelona. HI 62631

Boletín de la real sociedad española de historia natural. Madrid. Vols. 4-47, 1904-49 [suspended 1937-40]. Bol. Real Soc. Esp. Hist. Nat. Preceded by: Boletín de la sociedad española de historia natural. Superseded by: Boletín de la real sociedad española de historia natural. Seccion biologica. HI 62632

Boletín de la real sociedad española de historia natural. Seccion biologica; organo de los institutos "José de Acosta" de zoologia y "Antonio J. de Cavanilles", de botânica. Madrid. Vol. 48+, 1950+. Bol. Real Soc. Esp. Hist. Nat., Secc. Biol. Preceded by: Boletín de la real sociedad española de historia natural. HI 62633

Boletín; real sociedad geográfica. Madrid. Bol. Real Soc. Geogr. See B–P–H 210/7. HI 52308

Boletín, regional office for science and technology for Latin America and the Caribbean, Unesco. Montevideo. Nos. 9-20, 1974-80. Bol. Regional Off. Sci. Technol. Latin America & Caribbean, Unesco. Preceded by: Boletín, field science office for Latin America, Unesco. Superseded by: Boletín R O S T L A C. HI 62634

Boletín de reseñas. Plantas medicinales. Havana. No. ?-5+, 19??-83+. Bol. Reseñas, Pl. Med. HI 62635

Boletín rural I N T A. San Pedro, Argentina. Vol. 1+, 1966+. Bol. Rural I. N. T. A. HI 62636

Boletín de sanidad vegetal. Santiago. Bol. Sanid. Veg. See B–P–H 210/8. HI 52309

Boletín, secretaría de agricultura y ganadería. Mexico. Mexico City. 2. época, vols. 1-7, 1901-08. Bol. Secr. Agric. Ganad., Mexico. Preceded by: Boletín de agricultura, mineria e industria. Mexico City. Superseded by: Boletín oficial, secretaría de agricultura y ganadería. Mexico. HI 62637

Boletín semanal de precios agrarios, ministerio de agricultura. Madrid. 1964+. Bol. Semanal Precios Agrar. HI 62638

Boletín semanal, sociedad antioqueña de agricultores. Medellin. Ser. 2, nos. 1-65, 1926-27. Bol. Semanal. Soc. Antioqueña Agric. Preceded by: Antioqúia agrícola. Superseded by: Boletín agrícola, sociedad antioqueña de agricultores. 5-3902-2. HI 62639

Boletín, servicio de defensa contra plagas e inspección fitopatológica. Madrid. Vol. 1+, 1975+. Bol. Serv. Defensa Plagas Inspecc. Fitopalól. Preceded by: Boletín, servicio de plagas forestales and Boletín informativo de plagas. HI 62640

Boletín del servicio de plagas forestales. Madrid. Nos. 1-27, 1958-71. Bol. Serv. Plagas Forest. Superseded by:

Boletín, servicio de defensa contra plagas e inspeccion fitopatologica. HI 62641

Boletín de la sociedad agrícola mexicana. Mexico City. Bol. Soc. Agríc. Mex. See B–P–H 210/15. HI 52313

Boletín de la sociedad aragonesa de ciencias naturales. Saragossa, Spain. Bol. Soc. Aragonesa Ci. Nat. See B–P–H 210/18. HI 52314

Boletín de la sociedad argentina de botánica. La Plata, Argentina. Bol. Soc. Argent. Bot. See B–P–H 210/20. HI 52315

Boletín de la sociedad argentina de horticultura. Buenos Aires. Vols. 1(1)-34(184), 1943-76. Bol. Soc. Argent. Hort. Superseded by: Jardin y sus plantas. 5-3902-3. HI 62642

Boletín de la sociedad de biologia de Concepción. Concepción. 1927+. Bol. Soc. Biol. Concepcion. HI 62643

Boletín de la sociedad de biología de Santiago de Chile. Santiago. Bol. Soc. Biol. Santiago de Chile. See B–P–H 210/22. HI 52316

Boletín de la sociedad botánica del estado de Jalisco. Guadalajara, Mexico. Bol. Soc. Bot. Estado Jalisco. See B–P–H 210/24. HI 52317

Boletín de la sociedad botanica de La Libertad. Trujillo, Peru. Vol. 1+, 1969+. Bol. Soc. Bot. La Libertad. HI 62644

Boletín de la sociedad botánica de México. Mexico City. Bol. Soc. Bot. México. See B–P–H 210/26. HI 52318

Boletín de la sociedad de ciencias naturales del instituto de La Salle. Bogota. Nos. 1-?, 1913-18. Bol. Soc. Ci. Nat. Inst. La Salle. Superseded by: Boletín de la sociedad colombiana de ciencias naturales. HI 62645

Boletín de la sociedad de ciencias naturales de Jalisco. Guadalajara. No. 1+, 1969?+. Bol. Soc. Ci. Nat. Jalisco. HI 62646

Boletín de la sociedad científica, literaria y artistica. Ciudad de Las Casas, Mexico. Bol. Soc. Ci. Lit. Artist. See B–P–H 211/4. HI 52320

Boletín de la sociedad científica y literaria "Manuel R. Gutierrez." Bol. Soc. Ci. Lit. "Manuel R. Gutierrez". See B–P–H 211/5. HI 52321

Boletín de la sociedad colombiana de ciencias naturales. Bogota. Nos. ?-99, 1918-29. Bol. Soc. Colomb. Ci. Nat. Preceded by: Boletín de la sociedad de ciencias naturales del instituto de La Salle. Superseded by: Revista de la sociedad colombiana de ciencias naturales. HI 62647

Boletín de la sociedad cubana de dermatología y sifiligrafía. Havana. Bol. Soc. Cub. Dermatol. See B–P–H 211/8. HI 52322

Boletín de la sociedad cubana des orquideas. Havana. Vols. 1-5, 1943-59. Bol. Soc. Cub. Orquideas. HI 62648

Boletín de la sociedad dominicana de orquideología. Santo Domingo. Vol. 1(2)+, 1981+. Bol. Soc. Dominic. Orquideol. Preceded by: Orquidiota. HI 62649

Boletín de la sociedad española de historia de la farmacia. Madrid. 1950+. Bol. Soc. Esp. Hist. Farm. HI 62650

Boletín de la sociedad española de historia natural. Madrid. Vols. 1-3, 1901-03. Bol. Soc. Esp. Hist. Nat. Superseded by: Boletín de la real sociedad española de historia natural Seccion biologica. 5-3908-3. HI 62651

Boletín de la sociedad de estudios biológicos. Mexico City. Vol. 1(1-3), 1923. Bol. Soc. Estud. Biol. HI 62652

Boletín de la sociedad de geografía y estadística de la república méxicana. Mexico City. Vol. 2, 1850-51. Bol. Soc. Geogr. Estadist. Repúbl. Mex. Preceded by: Boletín de geografía y estadística de la república méxicana. Superseded by: Boletín de la sociedad méxicana de geografía y estadística. 5-3911-1. HI 62653

Boletín de la sociedad geográfica de Colombia. Bogota. 1907; vol. 1+, 1934+. Bol. Soc. Geogr. Colomb. 5-3909-2. HI 62654

Boletín de la sociedad geográfica de Lima. Lima. Bol. Soc. Geogr. Lima. See B–P–H 211/16. HI 52323

Boletín de la sociedad geográfica nacional. Madrid. Bol. Soc. Geogr. Nac. See B–P–H 211/17. HI 52324

Boletín de la sociedad geológica del Perú. Lima. Bol. Soc. Geol. Perú. See B–P–H 211/18. HI 52325

Boletín de la sociedad de historia, geografía y estadística de Aguascalientes. Aguascalientes, Mexico. Bol. Soc. Hist. Aquascalientes. See B–P–H 211/19. HI 52326

Boletín de la sociedad de historia natural de Baleares. Palma de Mallorca. 1955+. Bol. Soc. Hist. Nat. Baleares. HI 62655

Boletín de la sociedad hortícola para América tropical. San José, Costa Rica. Bol. Soc. Hort. Amér. Trop. See B–P–H 211/21. HI 52327

Boletín de la sociedad ibérica de ciencias naturales. Saragossa, Spain. Bol. Soc. Ibér. Ci. Nat. See B–P–H 211/23. HI 52328

Boletín (de la) sociedad médico-quirúrgica del centro de la república. Florida, Uruguay. Bol. Soc. Méd.-Quir. Centro Repúbl. See B–P–H 212/1. HI 52330

Boletín de la sociedad méxicana de geografía y estadística. Mexico City. Vols 3-12, 1852-68; ser. 2, vols. 1-4, 1869-72; ser. 3, vols. 1-6, 1873-87; ser. 4, vols. 1-4, 1888-1901; ser. 5, vols. 1-18, 1902-34. Bol. Soc. Méx. Geogr. Estadist. Preceded by: Boletín de la sociedad de geografía y estadística de la república méxicana. 5-3911-1. HI 62656

Boletín de la sociedad mexicana de historia natural. Mexico City. Bol. Soc. Mex. Hist. Nat. See B–P–H 212/2. HI 52331

Boletín de la sociedad méxicana de micologia. Mexico, D.F. Vols. 4-?, 1970-8? Bol. Soc. Méx. Micol. Preceded by: Boletín informativo de la sociedad méxicana de micologia. Superseded by: Revista mexicana de micologia. HI 62657

Boletín de la sociedad micologia castellana. Madrid. Vol. ?-5+, 19??-80+. Bol. Soc. Micol. Castellana. HI 62658

Boletín de la sociedad nacional agraria. Lima. Bol. Soc. Nac. Agrar. See B–P–H 212/3. HI 52332

Boletín de la sociedad nacional de agricultura. San José, Costa Rica. Vols. 1-4, 1906-10. Bol. Soc. Nac. Agric. (San José). 5-3911-3. HI 62659

Boletín de la sociedad nacional de agricoltura. Santiago. Vols. 1-44, 1869-1915; vols. 53-65, 1922-33 [for vols. 45-52 see Agricultor (Santiago).] Bol. Soc. Nac. Agric. (Santiago). Superseded by: Campesino. 2-898-3. HI 62660

Boletín de la sociedad nuevo leonesa de historia natural "Dr. J. Eleuterio Gonzalez". Monterrey, Mexico. Vols. 1-?, 1966-67. Bol. Soc. Nuevo Leonesa Hist. Nat. HI 62661

Boletín de la sociedad peruana de botanica. Lima. Bol. Soc. Peruana Bot. See B–P–H 212/7. HI 52333

Boletín de la sociedad physis para el cultivo y difusion de las ciencias naturales en la Argentina. Buenos Aires. 1912-15. Bol. Soc. Physis. Superseded by: Physis. Buenos Aires. HI 62662

Boletín de la sociedad venezolana de ciencias naturales. Caracas. Bol. Soc. Venez. Ci. Nat. See B–P–H 212/11. HI 52334

Boletín, sociedad venezolana de ciencias naturales, comite de orquideologica = Boletín, comite de orquideologica, sociedad venezolana de ciencias naturales. Caracas.

Boletín técnico, centro de investigaciones agricolas "Alberto Boerger". La Estanzuela. No. 1+, 1965?+. Bol. Técn. Centro Invest. Agric. Alberto Boerger. HI 62663

Boletín técnico, centro nacional de investigaciones agropecuarias, Venezuela. Maracay. No. ?-4+, 19??-77+. Bol. Técn. Centro Nac. Invest. Agropecu., Venezuela. HI 62664

Boletín técnico, comite de defensa técnica del algodon, sociedad nacional agraria. Lima. No. 1+, 1957?+. Bol. Técn. Comite Defensa Técn. Algodon. HI 62665

Boletín técnico, departamento de fitotecnia, facultad de ciencias agronomicas, universidad de El Salvador = Boletín técnico de la facultad de ciencias agronomicas, departamento de fitotecnia, universidad de El Salvador. Salvador.

Boletín técnico, dirección general de agraria, Peru. Lima. Nos. 78-80, 1973-74. Bol. Técn. Direccion Gen. Agrar., Peru. Preceded by: Boletín técnico, dirección general de investigaciones agropecuarias, Peru. HI 62666

Boletín técnico, dirección general de investigaciones agropecuarias, Peru. Nos. ?-72-77, 19??-69-72. Bol. Técn. Direccion Gen. Invest. Agropecu., Peru. Superseded by: Boletín técnico, dirección general de agraria, Peru. HI 62667

Boletín técnico, dirección general de investigación y capacitación forestales. Mexico, D.F. Nos. 55-59, 1978. Bol. Técn. Dirección Gen. Invest. Capac. Forest., Mexico. Preceded by: Boletín técnico, secretaria de agricultura y ganaderia, subsecretaria forestal y de la fauna. Superseded by: Boletín técnico, instituto nacional de investigaciones forestales. HI 62668

Boletín técnico, escuela de ciencias forestales, universidad de Chile. Santiago, Chile. No. 63+, 1981+. Bol. Técn. Esc. Ci. Forest. Univ. Chile. Preceded by: Boletín técnico, facultad de ciencias forestales, universidad de Chile. HI 62669

Boletín técnico, escuela de ingeniería forestal, universidad de Chile. Santiago, Chile. Nos. 1-7-?, 1963-64-? Bol. Técn. Esc. Ing. Forest. Univ. Chile. Superseded by: Boletín técnico, facultad de ciencias forestales, universidad de Chile. HI 62670

Boletín técnico, estaçión agrostologica. Palmira, Colombia. No. 1+, 1949+. Bol. Técn. Estaç. Agrostol., Palmira. HI 62671

Boletín técnico, estaçión experimental agrícola de Saavedra. Santa Cruz. No. 1+, 1971+. Bol. Técn. Estaç. Exp. Agríc. Saavedra. HI 62672

Boletín técnico, estaçión experimental agronomica, facultad de agronomia, universidad de Chile. Maipu. Nos. 1-35, 1958-72. Bol. Técn. Estaç. Exp. Agron. Fac. Agron. Univ. Chile. HI 62673

Boletin técnico, estaçión experimental agropecuaria Balcarce. Balcarce. No. 1+, 1962+. Bol. Técn. Estaç. Exp. Agropecu. Balcarce. HI 62674

Boletín técnico, estaçión experimental agropecuaria Trelew. Trelew. No. 1+, 1964+. Bol. Técn. Estaç. Exp. Agropecu. Trelew. HI 62675

Boletín técnico, facultad de agronomia, universidad de

Costa Rica. San José. No. 1+, 1968+. Bol. Técn. Fac. Agron. Univ. Costa Rica. HI 62676

Boletín técnico de la facultad de ciencias agronomicas, departamento de fitotecnia, universidad de El Salvador. Salvador. No. 1+, 196?+. Bol. Técn. Fac. Ci. Agron. Dep. Fitotecn. Univ. El Salvador. HI 62677

Boletín técnico; facultad de ciencias biológicas. Universidad de Nuevo Leon. Monterrey, Mexico. Bol. Técn. Fac. Ci. Biol. Univ. Nuevo Leon. See B–P–H 212/18. HI 52335

Boletín técnico, facultad de ciencias forestales, universidad de Chile. Santiago, Chile. Nos. ?-62, 19??-80. Bol. Técn. Fac. Ci. Forest. Univ. Chile. Preceded by: Boletín técnico, escuela de ingeniería forestal, universidad de Chile. Superseded by: Boletín técnico, escuela de ciencias forestales, universidad de Chile. HI 62678

Boletín técnico. Instituto de fitotécnia. Buenos Aires. Bol. Técn. Inst. Fitotécn. See B–P–H 212/26. HI 52338

Boletín técnico, instituto de fomento algodonero, departamento de experimentacion. Bogotá. No. 1+, 1962?+. Bol. Técn. Inst. Fomento Algodonero Dep. Exp. HI 62679

Boletín técnico, instituto de fomento algodonero, division de algodon. Bogotá. No. 1+, 1964+. Bol. Técn. Inst. Fomento Agrodonero Div. Algodon. HI 62680

Boletín técnico de la instituto interamericano de ciencias agrícolas. Turrialba, Costa Rica. Bol. Técn. Inst. Interamer. Ci. Agríc. See B–P–H 212/27. HI 52339

Boletín técnico, instituto de investigaciones agropecuarias. Osorno. No. 1+, 1965+. Bol. Técn. Inst. Invest. Agropecu., Osorno. HI 62681

Boletín técnico, instituto de investigaciones agropecuarias universidad de oriente, nucleo de Monagas. Monagas. No. 1+, 1969+. Bol. Técn. Inst. Invest. Agropecu. Univ. Oriente Nucleo Monagas. HI 62682

Boletín técnico, instituto nacional de investigaciones forestales. Mexico, D.F. No. 60+, 1979+. Bol. Técn. Inst. Nac. Invest. Forest., Mexico. Preceded by: Boletín técnico, dirección general de investigación y capacitación forestales. HI 62683

Boletín técnico, instituto de pesquisas e experimentação agropecuarias do nordeste. Recife. No. 1+, 1966+. Bol. Técn. Inst. Pesq. Exp. Agropecu. Nordeste. HI 62684

Boletín técnico. Ministerio de agricultura y cria. Caracas. Bol. Técn. Minist. Agric. See B–P–H 213/1. HI 52340

Boletín técnico. Ministerio de agricultura. Departamento de genética fitotécnica. Santiago. Bol. Técn. Minist. Agric. Dept. Genét. Fitotécn. See B–P–H 213/2. HI 52341

Boletín técnico, oficina del café. San José, Costa Rica. Bol. Técn. Oficina Café. See B–P–H 213/3. HI 52342

Boletín técnico, proyecto regional de desarrollo pesquero en centroamerica. San Salvador. Vols. ?-2-4, 19??-68-71. Bol. Técn. Proy. Regional Desarr. Pesq. Centroamer. HI 62685

Boletín técnico, seccion de divulgacion cientifica, centro nacional de investigaciones de cafe. Chinchiná. No. 1+, 1972+. Bol. Técn. Seccion Divulg. Ci. Centro Nac. Invest. Cafe. HI 62686

Boletín técnico, secretaria de agricultura y ganaderia, subsecretaria forestal y de la fauna. Mexico City. Nos. 1-54, 1961-76. Bol. Técn. Secr. Agric. Ganad. Subsecr. Forest. Fauna. Superseded by: Boletín técnico, dirección general de investigación y capacitación forestales. HI 62687

Boletín de la U I C N. Madrid. No. 1+, 1968+. Bol. U. I. C. N. HI 62688

Boletín de la universidad de Granada. Granada. Vol. 1+, 1928+. Bol. Univ. Granada. HI 62689

Boletín de la zona andina. Bol. Zona Andina. See B–P–H 213/8. HI 52346

Boletinos técnicos. Sociedad agronómica mexicana. Mexico City. Bol. Técn. Soc. Agron. Mex. See B–P–H 213/4. HI 52343

Boletíns, see Boletim [i.e. singular].

Boletus; Kulturband der deutschen demokratischen Republik. Erfurt, Halle. Vol. 1+, 1977+. Boletus. HI 62690

Bolezni rastenii. Leningrad. Vols. 12-29, 1823-30. Bolezni Rast. Preceded by: Zhurnal "Bolĕzni Rastenii". 1-731-2. HI 52347

Bolivar; revista colombiana de cultura. Bogotá. Bolivar. See B–P–H 213/13. HI 52348

Bollettino, see also Bullettino

Bollettino agricolo. Rome. Nos. 1-10, 1945-? Boll. Agric. (Rome). Superseded by: Agricoltura e veterinaria. 1-85-1. HI 62691

Bollettino agricolo e commerciale della colonia Eritrea. Asmara. Vols. 1(1)-3(9), 1903-05. Boll. Agric. Commerciale Colon. Eritrea. HI 62692

Bollettino dell'arboricoltura italiana. Acireale, Italy. Boll. Arboric. Ital. See B–P–H 213/15. HI 52350

Bollettino di associazione italiana pro piante medicinali, aromatiche ed altre piante utili. Milan. Boll. Assoc. Ital. Piante Med. See B–P–H 213/17. HI 52351

Bollettino di bibliografia e di storia delle scienze matematiche e fisiche. Rome. Boll. Bibliogr. Storia

Sci. Mat. See B–P–H 213/20. HI 52352

Bollettino della biblioteca e dei musei civici e delle biennali d'arte antica. Udine. Vol. 1+, 1907+. Boll. Bibliot. Mus. Civici Bienn Arte Antica, Udine. 5-4289-3. HI 62693

Bollettino bibliografico della botanica italiana. Florence. Vols. 1-11, 1904-39/51, 1904-54. Boll. Bibliogr. Bot. Ital. 1-732-1. HI 62694

Bollettino chimico-farmaceutico. Milan. Boll. Chim.-Farm. See B–P–H 213/21. HI 52353

Bollettino, club alpino italiano. Turin. Vol. 1+, 1865+. Boll. Club Alpino Ital. 2-962-2. HI 62695

Bollettino del gruppo micologico G. Bresadola. Trento. Vols. 1-29, 1957-86. Boll. Gruppo Micol. G. Bresadola. Superseded by: Rivista di micologia. HI 62696

Bollettino del gruppo ricerche di biologia marina "Cerianthus". Genoa. 1976+. Boll. Gruppo Ric. Biol. Mar. Cerianthus. HI 62697

Bollettino d'informazioni, consiglio nazionale delle richerche. Rome. Anno 1-2, 1930-31. Boll. Inform. Cons. Naz. Ric. Superseded by: Ricerca scientifica ed il progresso tecnico nell'economia nazionale. 4-3676-3. HI 62698

Bollettino dell'istituto botanico della università di Catania. Catania. Ser. 1, vols. 1-2, 1955-59; ser. 2, vols. 1-3, 195?-59; ser. 3, vols. 1-4, 195?-63. Boll. Ist. Bot. Univ. Catania. HI 62699

Bollettino dell'istituto botanico della reale università di Sassari. Sassari. Vol. 1, 1909/23. Boll. Ist. Bot. Reale Univ. Sassari. HI 62700

Bollettino dell'istituto sieroterapico milanese. Milan. Boll. Ist. Sieroterap. Milan. See B–P–H 214/1. HI 52354

Bollettino del laboratorio sperimentale e osservatorio di fitopatologia. Turin. 1935-42; 1956+. Boll. Lab. Sperim. Osserv. Fitopatol. Preceded by: Difesa delle piante contra le malattie ed i parassiti. HI 62701

Bollettino e memorie della società piemontese di chirurgia. Turin. Boll. & Mem. Soc. Piemont. Chir. See B–P–H 214/4. HI 52355

Bollettino mensile della r[eale] stazione di patologia vegetale. Rome. 1920-25. Boll. Mens. Reale Staz. Patol. Veg. Superseded by: Bollettino della stazione di patologia vegetale di Roma. HI 62702

Bollettino del museo civico di storia naturale di Venezia. Venice. Vol. 7+, 1954+. Boll. Mus. Civico Storia Nat. Venezia. Preceded by: Bollettino della società veneziana di storia naturale e del museo civico di storia naturale. HI 62703

Bollettino del museo civico di storia naturale di Verona. Verona. Vol. 1+, 1974+. Boll. Mus. Civico Storia Nat. Verona. Preceded by: Memorie del museo civico di storia naturale di Verona. HI 62704

Bollettino del museo regionale di scienze naturali, Torino. Turin. Vol. 1+, 1983+. Boll. Mus. Regionale Sci. Nat. Torino. HI 62705

Bollettino del naturalista, collettore, allevatore, coltivatore. Siena. Boll. Naturalista. See B–P–H 214/7. HI 52356

Bollettino di pesca, di piscicoltura et di idrobiologia. Rome. Boll. Pesca. See B–P–H 214/14. HI 52357

Bollettino delle r[eale] istituto botanico dell'università parmense. Parma. Boll. Reale Ist. Bot. Univ. Parmense. See B–P–H 214/18. HI 52358

Bollettino, r[eale] istituto superiore agraria. Pisa. Vols. 1-13, 1925-37. Boll. Reale Ist. Super. Agrar. Superseded by: Annali della facoltà agraria della università di Pisa. 4-3364-1. HI 62706

Bollettino del r[eale] orto botanico di Palermo. Palermo. Vols. 1-11, 1897-1912; n.s. vols. 1-2, 1914-21. Boll. Reale Orto Bot. Palermo. Superseded in part by: Lavori del r[eale] istituto botanico di Palermo. 4-3245-1. HI 62707

Bollettino della r[eale] società toscana d'orticultura. Florence. Vols. 1-10, 1876-85; ser. 2, vols. 1-10, 1886-95; ser. 3, vols. 1-10, 1896-1915; ser. 4, vols. 1-?, 1916-38. Boll. Reale Soc. Tosc. Ortic. Superseded by: Rivista della r[eale] società toscana di orticultura. HI 62708

Bollettino della r[eale] stazione sperimentale di agrumicoltura e frutticoltura in Acireale. Acireale. Nos. 1-59, 1912-32; n.s. nos. 1-71?, 1932-38. Boll. Reale Staz. Sperim. Agrumic. Fruttic. Acireale. HI 62709

Bollettino scientifico della facoltà di chimica industriale, università di Bologna. Bologna. 1940+. Boll. Sci. Fac. Chim. Industr. Univ. Bologna. Preceded by: Giornale di biologia industriale, agraria ed alimentare. HI 62710

Bollettino scientifico della facoltà di zootecnica e veterinaria, universita nazionale Somala. Pisa. 1980+. Boll. Sci. Fac. Zootecn. Veterin. Univ. Naz. Somala. HI 62711

Bollettino delle sedute accademia Gioenia di scienze naturali, Catania. Catania, Italy. Boll. Accad. Gioenia Sci. Nat. Catania. See B–P–H 213/14. HI 52349

Bollettino della sezione italiana; società internazionale di microbiologia. Milan. Boll. Sez. Ital. Soc. Int. Microbiol. See B–P–H 214/21. HI 52359

Bollettino delle sezioni regionale di società italiana di dermatologia e sifilografia. Milan. Boll. Sez. Regionale Soc. Ital. Dermatol. See B–P–H 214/25.

HI 52360

Bollettino della società adriatica di scienze naturali di Trieste. Trieste, Austria [Italy]. Boll. Soc. Adriat. Sci. Nat. Trieste. See B–P–H 214/30. HI 52361

Bollettino della società botanica italiana. Florence. Boll. Soc. Bot. Ital. See B–P–H 214/31. HI 52362

Bollettino de società geologica italiana. Rome. Boll. Soc. Geol. Ital. See B–P–H 215/2. HI 52363

Bollettino, società internazionale di microbiologia. Sezione italiana. Milan. Vols. 1-12, 1929-40. Boll. Soc. Int. Microbiol., Sez. Ital. Superseded by: Bollettino, società italiana di microbiologia. 5-3919-2. HI 62712

Bollettino della società italiana di biologia sperimentale. Naples. Boll. Soc. Ital. Biol. Sperim. See B–P–H 215/4. HI 52364

Bollettino società italiana dell'iris. Florence. Vol. ?-3+, 19??-63+. Boll. Soc. Ital. Iris. HI 62713

Bollettino, società italiana di microbiologia. Milan. Vol. 13+, 1941+. Boll. Soc. Ital. Microbiol. Preceded by: Bollettino, società internazionale di microbiologia, sezione italiana. 5-3919-2. HI 62714

Bollettino di società lancisiana degli ospedali di Roma = Bullettino di società lancisiana degli ospedali di Roma. Rome. Bull. Soc. Lancisiana Osped. Roma. See B–P–H 278/16.

Bollettino de società medico-chirurgica di Pavia. Pavia. Boll. Soc. Med.-Chir. Pavia. See B–P–H 215/9. HI 52365

Bollettino della società di naturalisti di Napoli. Naples. Boll. Soc. Naturalisti Napoli. See B–P–H 215/11. HI 52366

Bollettino della società paleontologica italiana. Modena, Italy. Boll. Soc. Paleontol. Ital. See B–P–H 215/13. HI 52367

Bollettino della società sarda di scienze naturali. Sassari. Vol. 1+, 1967+. Boll. Soc. Sarda Sci. Nat. HI 62715

Bollettino della società ticinese di scienze naturali. Lugano, Switzerland. Boll. Soc. Ticinese Sci. Mat. See B–P–H 215/15. HI 52368

Bollettino della società veneziana di storia naturale (e del museo civico di storia naturale). Venice. 1932-52. Boll. Soc. Veneziana Storia Nat. (Mus. Civico Storia Nat.) Superseded by: Bollettino del museo civico di storia naturale di Venezia. HI 62716

Bollettino della stazione di patologia vegetale di Roma. Rome. Ser. 2, vol. 6+, 1926+. Boll. Staz. Patol. Veg. Roma. Preceded by: Bollettino mensile della r[eale] stazione di patologia vegetale. HI 62717

Bollettino. Stazione sperimentale di agrumicoltura e frutte-coltura. Acireale, Italy. Boll. Staz. Sperim. Agrumic. See B–P–H 215/18. HI 52369

Bollettino di studi ed informazioni. R. giardino coloniale. Palermo. Boll. Stud. Inform. Reale Giardino Colon. See B–P–H 215/20. HI 52370

Bollettino tecnico della coltivazione del tabacchi. Scafati, Italy. Boll. Tecn. Coltiv. Tabacchi. See B–P–H 215/22. HI 52371

Bolotovedenie; vestnik minskoi bolotnoi opytnoi stantsii. (Zeitschrift fer Moorversuchsstation zu Minsk.) Minsk. Vol. 1(1), 1912. Bolotovedenie. HI 62718

Bolton naturalist. Bolton. 1963+. Bolton Naturalist. HI 62719

Bombay university journal. Bombay. Bombay Univ. J. See B–P–H 216/53. HI 52372

Le bon jardinier, almanach. Paris. Bon Jard. See B–P–H 216/55. HI 52373

Bondens magasin; et maanedsblad till nytte og fornøielse. Odense. Vols. 1-9, 1855-64. Bondens Mag. HI 62720

Bonner Geographische Abhandlungen. Bonn. Bonner Geogr. Abh. See B–P–H 216/56. HI 52374

De bononiensi scientiarum et artium instituto atque academia commentarii. Bologna. Bononiensi Sci. Inst. Acad. Comment. See B–P–H 216/57 HI 52375.

Bonplandia. Corrientes. Vol. 1+, 1960+. Bonplandia. HI 62721

Bonplandia. Hanover. Bonplandia. See B–P–H 217/2. HI 52376

Bonsai; journal of the american bonsai society. Bedford, NY. Vols. 1-7, 1967-73. Bonsai (Bedford). Superseded by: Bonsai journal. HI 62722

Bonsai; magazine of bonsai, japanese gardens, saikei and suiseki. Mountain View, Sunnyvale, CA. Vols. 5(10)-?, 1966-79? Bonsai (Mountain View). Preceded by: Bonsai newsletter. Superseded by: Bonsai international. HI 52377

Bonsai. Philadelphia = Bonsai bulletin. Philadelphia, PA. Bonsai Bull. (Philadelphia). See B–P–H 217/6.

Bonsai. Prescott, AZ. Vol. 1+, 1976+. Bonsai (Prescott). HI 62723

Bonsai in Australia. Sydney. Vol. 1+, 1970+. Bonsai Australia. HI 62724

Bonsai bulletin. New York. Vol. 1+, 1963+. Bonsai Bull. (New York). HI 62725

Bonsai bulletin. Philadelphia, PA. Bonsai Bull. (Philadelphia). See B–P–H 217/6. HI 52378

Bonsai bulletin. Sydney. Vol. ?-4(1)+, 19??-69+. Bonsai Bull. (Sydney). HI 62726

Bonsai clubs international. Tallahassee, FL. No. 20+, 1981+. Bonsai Clubs Int. Preceded by: Bonsai international. HI 62727

Bonsai forum. [Supplement to: Bonsai, journal of the american bonsai society. See B–P–H 217/3.] Norfolk, VA. Vol. 1+, 1972+. Bonsai Forum. HI 62728

Bonsai international; magazine of bonsai, japanese gardens, saikei and suiseki. Palo Alto, CA. Nos. ?-19, 1979-80. Bonsai Int. Preceded by: Bonsai. Mountain View, CA. Superseded by: Bonsai clubs international. HI 62729

Bonsai journal. Keene, NH. Vol. 8+, 1974+. Bonsai J. Preceded by: Bonsai, journal of the american bonsai society. HI 62730

Bonsai newsletter. [Bonsai clubs association, inc.] Mountain View, CA. Bonsai Newslett. See B–P–H 217/7. HI 52379

Bonsai sekai. Vol. ?-1(9)+, 19??-71+. Bonsai Sekai. HI 62731

Bonsai today. Sudbury, MA. No. 1+, 1989+. Bonsai Today. HI 62732

Bontebok. Cape Town. Vol. 1, 1977; n.s. vol. 1+, 1981. Bontebok. Preceded by: Investigational report, department of nature conservation, Cape of Good Hope. HI 62733

Booklet, forestry commission. London. No. 1+, 1947+. Booklet Forest. Commiss. HI 62734

Books and libraries at the university of Kansas. Lawrence, KS. Vol. 1+, 1952+. Books Libr. Univ. Kansas. HI 62735

Boomkweekerij: tijdschrift voor boomkweekerij en vasteplantencultuur. The Hague. Boomkweekerij. See B–P–H 217/8. HI 52380

De boomkweker. Haarlem. Boomkweker. See B–P–H 217/9. HI 52381

De boomkweker voor Nederland en Belgie. Haarlem. Boomkweker New. Belgie. See B–P–H 217/10. HI 52382

Boor en spade. Verspreide bijdragen tot de kennis van de bodem van Nederland. Utrecht. Boor & Spade See B–P–H 217/11. HI 52383

Borászati füzetek. Pest [=Budapest, in part]. Borász. Füz. See B–P–H 217/12. HI 52384

Borászati lapok. Budapest. Borász. Lapok (Budapest). See B–P–H 217/14. HI 52385

Borászati lapok. Pest [=Budapest, in part]. Borász. Lapok (Pest). See B–P–H 217/15. HI 52386

Borbásia. Budapest. Borbásia. See B–P–H 217/16. HI 52387

Borbásia. nova. Budapest. Borbásia Nova. See B–P–H 217/17. HI 52388

Borneo research bulletin. Phillips, ME. Vol. 1+, 1969+. Borneo Res. Bull. HI 62736

Boron and plant life. London. Boron Pl. Life. See B–P–H 217/24. HI 52389

Borsenblatt für den deutschen Buchhandel. Frankfurter Ausgabe. Wiesbaden. N.S. vol. [1]+ [vols. 1-2, 1945-46 as vols. "112-113"; this volumation refers to Börsenbl. Deutsch. Buchhandel published in Leipzig from which the "Frankfurter Ausgabe" split off in October 1945.] 1945+. Bösenbl. Deutsch. Buchhandel Frankfurter Ausg. 1-722-3. HI 62737

Borsodi szemle. A társodalom és természettudomámyi társulat Borsod-Abauj-Zemplén megyei szevezetének tudományos folyóirata. Miskolc, Hungary. Borsodi Szemle. See B–P–H 217/25. HI 52390

Bosbou in Suid-Afrika = Forestry in South Africa. Pretoria.

Boschbouwkundig tijdschrift. Buitenzorg, Dutch E. Indies [=Bogor, Indonesia]. Boschbouwk. Tijdschr. See B–P–H 218/1. HI 52391

Bosque. Amatitlán, Guatemala. Bosque. See B–P–H 218/2. HI 52392

Bosques. Mexico City. Bosques. See B–P–H 218/3. HI 52393

Bosques y fauna. Nueva epoca. Mexico, D.F. No. 1+, 1978+. Bosques Fauna, N. E. HI 62738

Bosques y maderas. Bogotá. Bosques & Maderas. See B–P–H 218/4. HI 52394

Boston flower market and New England florist. Boston, MA. Boston Fl. Market. See B–P–H 218/9. HI 52395

Boston journal of natural history. Boston, MA. Boston J. Nat. Hist. See B–P–H 218/10. HI 52396

Boston journal of philosophy and the arts. Boston, MA. Boston J. Philos. Arts See B–P–H 218/11. HI 52397

Boston medical and surgical journal. Boston, MA. Boston Med. Surg. J. See B–P–H 218/17. HI 52398

Botaanilised uurimused. Tartu. Vols. 1-4, 1961-67. Bot. Uurim. HI 62739

Botanic advertiser and Rhode Island record of modern medical reform. Providence, RI. Bot. Advertiser & Rhode Island Rec. Med. Reform. See B–P–H 218/26. HI 52402

Botanic advocate and journal of health. Montpeller, VT. Bot.Advocate & J. Health. See B–P–H 218/27. HI 52403

Botanic advocate and thomsonian family physician. New Haven, CT. Bot. Advocate & Thomsonian Family Physician. See B–P–H 218/28. HI 52404

Botanic annual. London. Bot. Annual. See B–P–H 218/31. HI 52405

Botanic garden; consisting of highly finished representations of hardy ornamental flowering plants, cultivated in Great Britain. [Edited by Maund.] London. Bot. Gard. See B–P–H 220/6. HI 52419

Botanic gardens conservation news; magazine of the I U C N botanic gardens conservation secretariat. Kew. Vol. 1+, 1987+. Bot. Gard. Conservation News. HI 62740

Botanic investigator. Vicksburg, MS. Bot. Investigator. See B–P–H 220/12. HI 52422

Botanic journal. Boston, MA. Bot. J. (Boston). See B–P–H 220/13. HI 52423

Botanic ledger and family journal of health. Oxford, OH. Bot. Ledger & Family J. Health. See B–P–H 221/7. HI 52432

Botanic luminary. Saline & Adrian, MI. Bot. Luminary. See B–P–H 221/12. HI 52436

Botanic magazine. Melbourne, Vic. Ca.1989+. Bot. Mag. (Melbourne). HI 62741

Botanic medical reformer and home physician. Philadelphia, PA. Bot. Med. Reformer & Home Physician. See B–P–H 223/7. HI 52451

Botanic practitioner. London. Bot. Practitioner. See B–P–H 224/10. HI 52466

Botanic sentinel and literary gazette. Philadelphia, PA. Bot. Sentinel & Lit. Gaz. See B–P–H 224/19. HI 52470

Botanica, The; magazine of Delhi university botanical society. New Delhi. Vol. 1+, 1950+. Botanica. HI 62742

Botânica. São Paulo. Botânica. See B–P–H 226/26. HI 52492

Botanica acta; Berichte der deutschen botanischen Gesellschaft (Journal of the German botanical society). Stuttgart, New York. Vol. 101+, 1987+. Bot. Acta. Preceded by: Berichte der deutschen botanischen Gesellschaft. HI 62743

Botanica complutensis. Madrid. Ca.1989+. Bot. Complut. HI 63275

Botanica experimentalis = Trudy botanicheskogo instituta akademii nauk S S S R. Ser. 4, eksperimental′naya botanika. Moscow & Leningrad.

Botanica gothoburgensia. Acta universitatis gothoburgensis. Goteborg. Nos. 1-6, 1963-68. Bot. Gothob. 2-1750-3. HI 62744

Botanica helvetica. Zurich. Vol. 91+, 1981+. Bot. Helv. Preceded by: Bericht der schweizerischen botanischen Gesellschaft. HI 62745

Botanica macaronésica. IV, ciencias. Las Palmas. Vol. 1+, 1976+. Bot. Macaronés., IV Ci. HI 62746

Botanica marina. Hamburg. Bot. Mar. See B–P–H 221/24. HI 52439

Botanica oeconomica. Hamburg. Bot. Oecon. See B–P–H 224/7. HI 52464

Botanica rhedonica. Série A, bulletin des laboratoires de botanique de la faculté des sciences de Rennes. Rennes. Vol. 1+, 1966+. Bot. Rhedon., A. HI 62747

Botanica rhedonica. Série B, recueil de tirés à part. Rennes. Vol. 1+, 1966+. Bot. Rhedon, B. HI 62748

Botanica texana. Austin, TX. 1981+. Bot. Texana. HI 62749

Botanical abstracts; a monthly serial furnishing abstracts and citations of publications in the international field of botany in its broadest sense. Baltimore, MD. Bot. Abstr. See B–P–H 218/25. HI 52401

Botanical bulletin. Hanover, IN. Vol. 1, 1875-76. Bot. Bull. (Hanover). Superseded by: Botanical gazette. 1-754-3. HI 62750

Botanical bulletin of academia sinica. A quarterly journal containing scientific contributions from the institut of botany, academia sinica. [Kuo li chung yang yen chiu yuan chih wu hsüeh hui pao.] Shanghai. Bot. Bull. Acad. Sin. See B–P–H 219/9. HI 52409

Botanical bulletin, Northern Territory. Darwin, N.T. No. 1+, 1975+. Bot. Bull. Northern Territory. HI 62751

Botanical bulletin of the herbarium, forest department, Sabah. Sandakan. Nos. 10-11, 1968. Bot. Bull. Herb. Forest Dept., Sabah. Preceded by: Botanical new bulletin, forestry department. HI 62752

Botanical cabinet; consisting of coloured delineations of plants from all countries. [Edited by Loddiges.] London. Bot. Cab. See B–P–H 219/14. HI 52410

Botanical center newsletter, Des Moines. Des Moines, IA. Vol. 1+, 1980+. Bot. Center Newslett. Des Moines. HI 62753

Botanical exchange club of the British Isles. Report. Manchester, England. Bot. Exch. Club Brit. Isles Rep. See B–P–H 220/1. HI 52416

Botanical exchange club and society of the British Isles. Manchester, England. Bot. Exch. Club Soc. Brit. Isles See B–P–H 220/2. HI 52417

Botanical garden newsletter, university of south Florida. Tampa, FL. No. ?-2+, 19??-79+. Bot. Gard. Newslett. Univ. S. Florida. HI 62754

Botanical garden quarterly, university of California. Berkeley, CA. Vol. 6+, 1981+. Bot. Gard. Quart. Univ. Calif. Preceded by: Newsletter, friends of the botanical garden, university of California. HI 62755

Botanical gazette; paper of botanical notes. Crawfordsville, IN, Chicago, IL. Vol. 2+, 1878+. Bot. Gaz. Preceded by: Botanical bulletin. Hanover. 1-

754-3 HI 62756

Botanical gazette. London. Bot. Gaz. (London). See B–P–H 220/9. HI 52420

Botanical journal. London. Bot. J. (London) See B–P–H 220/14. HI 52424

Botanical journal of the linnean society. London. Vol. 62+, 1969+. Bot. J. Linn. Soc. Preceded by: Journal of the linnean society. Botany. HI 62757

Botanical leaflets, a series of studies in the systematic botany of miscellaneous dicotyledonous plants. Chicago, IL. Bot. Leafl. See B–P–H 221/5. HI 52431

Botanical magazine; or, flower-garden displayed ... [Edited by Wm. Curtis]; [with vol. 15, 1801, title became Curtis's botanical magazine; or, ...] London. Vols. 1-184, 1787-1983 [vols. 54-70, 1827-44, also as n.s. vols. 1-17; vols. 71-130, 1845-1904, also as ser. 3, vols. 1-60; vols. 131-146, 1905-20, also as ser. 4, vols. 1-16; continuous volumation only, without series identity, vols. 147-184.] [Plates in vols. 1-164 numbered consecutively 1-9688; plates in vols. 165-184, 1948-83, numbered 1-882]. Bot. Mag. Superseded by: Kew magazine. 2-1247-2. HI 62758

Botanical magazine, Peking = Plant magazine. Peking (Beijing).

Botanical magazine. [Shokubutsu-gaku zasshi.] Tokyo. Bot. Mag. (Tokyo). See B–P–H 221/17. HI 52438

Botanical magazine. Special issue. [Supplement to: Botanical magazine, Tokyo. See B–P–H 221/17.] Tokyo. No. 1+, 1978+. Bot. Mag. (Tokyo), Special Issue. 1-755-1. HI 62759

Botanical memoirs. London. Bot. Mem. See B–P–H 223/8. HI 52452

Botanical memoirs, Maharaja Sayajirao university of Baroda, faculty of science. Baroda, India. Bot. Mem. Maharaja Sayajirao Univ. Baroda Fac. Sci. See B–P–H 223/9. HI 52453

Botanical memoirs of the university of Bombay. Bombay. Bot. Mem. Univ. Bombay See 223/10. HI 52454

Botanical miscellany. [Editor W. J. Hooker.] London. Bot. Misc. See B–P–H 223/11. HI 52455

Botanical monographs, council of scientific and industrial research, India. Oxford, New Delhi. Vol. 1+, 1961+. Bot. Monogr. Council Sci. Industr. Res., India. HI 62760

Botanical museum leaflets. [Harvard university.] Cambridge, MA. Vols. 1-30(4), 1932-86 Bot. Mus. Leafl. Superseded by: Harvard papers in botany. 3-1811-2. HI 52457

Botanical news bulletin, forest department, Sabah. Sandakan. Nos. 1-9, 1964-67. Bot. News Bull. Forest Dept., Sabah. Superseded by: Botanical bulletin, herbarium, forest department. HI 62761

Botanical notes. Nigerian college of technology, Ibadan branch. Ibadan, Nigeria. Bot. Notes. See B–P–H 224/4. HI 52462

Botanical papers, Fairchild tropical gardens. Coconut Grove, FL. Nos. 1-9, 1950-52. Bot. Pap. Fairchild Trop. Gard. HI 62762

Botanical records and monographs; a subsidiary of the New botanist. New Delhi. No. 1+, 1976+. Bot. Rec. Monogr. HI 62763

Botanical register; consisting of coloured figures of exotic plants cultivated in british gardens; with their history and mode of treatment. London. Bot. Reg. See B–P–H 224/11. HI 52467

Botanical research: contributions from the institute of botany, academia sinica. [Chih wu hsüeh chi k'an.] Beijing. No. 1+, 1983+. Bot. Res. Academia Sinica. HI 62764

Botanical review, interpreting botanical progress. Lancaster, PA. Bot. Rev. (Lancaster). See B–P–H 224/17. HI 52468

Botanical review, or the beauties of flora; ... London. Bot. Rev. (London). See B–P–H 224/18. HI 52469

Botanical series, Field museum of natural history = Field museum of natural history. Chicago, IL. Field Mus. Nat. Hist., Bot. Ser. See B–P–H 370/21.

Botanical society of the British Isles. Proceedings. Arbroath. Vols. 1-7, 1954/55-69. Bot. Soc. Brit. Isles Proc. Preceded by: Year book of the botanical society of the British Isles. Incorporated in: Watsonia. HI 62765

Botanical society of the British Isles welsh regional bulletin = Welsh regional bulletin, botanical society of the British Isles. Aberystwyth.

Botanical society and exchange club of the British Isles. Arbroath, Scotland. Bot. Soc. Exch. Club Brit. Isles See B–P–H 224/24. HI 52472

Botanical studies, Butler university = Butler university botanical studies. Indianapolis, IN.

Botanical studies, Oklahoma agricultural and mechanical college = Oklahoma agricultural and mechanical college. Botanical studies. Stillwater, OK. Oklahoma Agric. Coll. Bot. Stud. See B–P–H 686/34.

Botanical survey of Nebraska. Lincoln, NE. Nos. 1-7, 1892-04; n.s. vol. 1+, 1917+. Bot. Surv. Nebraska. 4-2945-3. HI 62766

Botanical systematics; an occasional series of monographs. London. Vol. 1+, 1976+. Bot. Syst. HI 62767

Botanical transactions of the Yorkshire naturalists union. Leeds. Bot. Trans. Yorkshire Naturalists Union. See B–P–H 225/12. HI 52475

Botanical world. [Shokubutsu-kai.] [Japan]. Bot. World. See B–P–H 225/16. HI 52479

Botaničeskie materialy Gerbarija Botaničeskogo instituta Uzbekistanskogo filiala Akademii nauk S S S R = Botanicheskiye materialy Gerbariya Botanicheskogo instituta Uzbekistanskogo filiala Akademii nauk S S S R. Tashkent.

Botaničeskie materialy gerbarija botaničeskogo instituti imeni V. L. Komarova, akademii nauk S S S R. Leningrad = Botanicheskie materialy gerbariya botanicheskogo instituti imeni V. L. Komarova, akademii nauk S S S R. Leningrad.

Botaničeskie materialy Gerbarija Glavnogo botaničeskogo sada RSFSR = Botanicheskiye materialy Gerbariya Glavnogo botanicheskogo sada R S F S R. Petrograd.

Botaničeskie materialy Gerbarija Glavnogo botaničeskogo sada S S S R = Botanicheskiye materialy Gerbariya Glavnogo botanicheskogo sada S S S R. Leningrad.

Botaničeskie materialy Gerbarija Instituta botaniki Akademii nauk Kazahskoj S S R = Botanicheskiye materialy Gerbariya Instituta botaniki Akademii nauk Kazakhskoi S S R. Alma-Ata.

Botaničeskie materialy Gerbarija Instituta botaniki Akademii nauk Uzbekskoj S S R = Botanicheskiye materialy Gerbariya Instituta botaniki Akademii nauk Uzbekskoi S S R. Tashkent.

Botaničeskie materialy Gerbarija Instituta botaniki i zoologii Akademii nauk Uzbekskoj S S R = Botanicheskiye materialy Gerbariya Instituta botaniki i zoologii Akademii nauk Uzbekskoi S S R. Tashkent.

Botaničeskie materialy Instituta sporovyh rastenij Glavnogo botaničeskogo sada RSFSR = Botanicheskiye materialy Instituta sporovykh rastenii Glavnogo botanicheskogo sada R S F S R. Petrograd.

Botaničeskie materialy Otdela sporovyh rastenij = Botanicheskiye materialy Otdela sporovykh rastenii. Moscow & Leningrad.

Botaničeskie materialy Otdela sporvyh rastenij Botaničeskogo instituta Akademii nauk S S S R = Botanicheskiye materialy Otdela sporvykh rastenii Botanicheskogo instituta Akademii nauk S S S R. Moscow & Leningrad.

Botaničeskie materialy otdela sporovyh rastenij botaničeskogo instituta imeni V. L. Komarova akademii nauk S S S R. Moscow & Leningrad = Botanicheskie materialy otdela sporovyh rastenii botanicheskogo instituta imeni V. L. Komarova akademii nauk S S S R. Moscow & Leningrad.

Botaničeskij žurnal S S S R = Botanicheskii zhurnal S S S R. Moscow & Leningrad.

Botaničeskij žurnal. Moscow & Leningrad = Botanicheskii zhurnal. Moscow & Leningrad.

Botaničeskija zapiski. St. Petersburg = Botanicheskiya zapiski. St. Petersburg.

Botaničeskoe obozrenie = Botanicheskoye obozrenie. Petrograd.

Botaničnyj žurnal = Botanichnyi zhurnal. Kiev.

Botanicheskie issledovaniya. Tartu = Botaanilised uurimused. Tartu.

Botanicheskie issledovaniya za polyarnym krugom. Apatity. Vol. 1+, 1969+. Bot. Issl. Polyarn. Krugom. HI 62768

Botanicheskie materialy gerbariya botanicheskogo instituti imeni V. L. Komarova, akademii nauk S S S R. Leningrad. Vols. 7-23, 194?-63. Bot. Mater. Gerb. Bot. Inst. Komarova Akad. nauk S.S.S.R. Preceded by: Botanicheskie materialy gerbariya glavnogo botanicheskkogo sada S S S R. Superseded by: Novosti sistematiki vysshikh rastenii. 1-114-2. HI 62769

Botanicheskie materialy Gerbariya Botanicheskogo instituta Uzbekistanskogo filiala Akademii nauk S S S R. Tashkent, Uzbek S S R. Vols. 1-7, 1940-41. Bot. Mater. Gerb. Bot. Inst. Uzbekistansk. Fil. Akad. Nauk S.S.S.R. Superseded by: Botanicheskie materialy Gerbariya Instituta botaniki i zoologii Akademii nauk Uzbekskoi S S R. HI 52440

Botanicheskie materialy Gerbariya Glavnogo botanicheskogo sada R S F S R. Petrograd. Vols. 1-5, [1919]-24. Bot. Mater. Gerb. Glavn. Bot. Sada RSFSR. Preceded by: Botanicheskie materialy gerbariya botanicheskogo instituti imeni V. L. Komarova, akademii nauk S S S R. Superseded by: Botanicheskie materialy Gerbariya Glavnogo botanicheskogo sada S S S R. 1-114-2. HI 52441

Botanicheskie materialy Gerbariya Glavnogo botanicheskogo sada S S S R. Leningrad. Vol. 6, 1926. Bot. Mater. Gerb. Glavn. Bot. Sada S.S.S.R. Preceded by: Botanicheskie materialy Gerbariya Glavnogo botanicheskogo sada R S F S R. Superseded by: Botanicheskie materialy gerbariya botanicheskogo instituti imeni V. L. Komarova, akademii nauk S S S R. HI 52442

Botanicheskie materialy Gerbariya Instituta botaniki Akademii nauk Kazakhskoi S S R. Alma-Ata, Kazakh S S R. Vol. 1+, 1963. Bot. Mater. Gerb. Inst. Bot. Akad. Nauk Kazakhsk. S.S.R. HI 52443

Botanicheskie materialy Gerbariya Instituta botaniki Akademii nauk Uzbekskoi S S R. Tashkent, Uzbek S S

R. Vol. 13+, 1952+. Bot. Mater. Gerb. Inst. Bot. Akad. Nauk Uzbeksk. S.S.R. Preceded by: Botanicheskie materialy Gerbariya Instituta botaniki i zoologii Akademii nauk Uzbekskoi S S R. HI 52444

Botanicheskie materialy Gerbariya Instituta botaniki i zoologii Akademii nauk Uzbekskoi S S R. Tashkent, Uzbek S S R. Vols. 8-12, 1947-48. Bot. Mater. Gerb. Inst. Bot. Zool. Akad. Nauk Uzbeksk. S.S.R. Preceded by: Botanicheskie materialy Gerbariya Botanicheskogo instituta Uzbekistanskogo filiala Akademii nauk S S S R. Superseded by: Botanicheskie materialy Gerbariya Instituta botaniki Akademii nauk Uzbekskoi S S R. HI 52445

Botanicheskie materialy Instituta sporovykh rastenii Glavnogo botanicheskogo sada R S F S R. Petrograd. Vols. 1-4(7), 1922-26. Bot. Mater. Inst. Sporov. Rast. Glavn. Bot. Sada RSFSR. Superseded by: Botanicheskie materialy Otdela sporvykh rastenii Botanicheskogo instituta Akademii nauk S S S R. HI 52446

Botanicheskie materialy Otdela sporovykh rastenii. Moscow & Leningrad. Vol. 6(1/6), 1949. Bot. Mater. Otd. Sporov. Rast. Preceded & superseded by: Botanicheskie materialy otdela sporovyh rastenii botanicheskogo instituta imeni V. L. Komarova akademii nauk S S S R. HI 52447

Botanicheskie materialy Otdela sporvykh rastenii Botanicheskogo instituta Akademii nauk S S S R. Moscow & Leningrad. Vols. 4(8/9)-5(1/3), 1937-40. Bot. Mater. Otd. Sporov. Rast. Bot. Inst. Akad. Nauk S.S.S.R. Preceded by: Botanicheskie materialy Instituta sporovykh rastenii Glavnogo botanicheskogo sada R S F S R. Superseded by: Botanicheskie materialy otdela sporovyh rastenii botanicheskogo instituta imeni V. L. Komarova akademii nauk S S S R. HI 52448

Botanicheskie materialy otdela sporovyh rastenii botanicheskogo instituta imeni V. L. Komarova akademii nauk S S S R. Moscow & Leningrad. Vols. 5(4/6)-5(10/12), 1940-45; vols. 6(7/12)-16, 1950-63. Bot. Mater. Otd. Sporov. Rast. Bot. Inst. Komarova Akad. Nauk S.S.S.R. Preceded by: Botanicheskie materialy Otdela sporvykh rastenii Botanicheskogo instituta Akademii nauk S S S R. For vol. 6(1/6), 1949 see: Botanicheskie materialy Otdela sporovykh rastenii. Superseded by: Novosti sistematiki nizshikh rastenii. HI 62770

Botanicheskie i zoologicheskie issledovaniya na dal'nem vostoke. Vladivostok. Vol. 1+, 1968+. Bot. Zool. Issl. Dal'nem Vostoke. HI 62771

Botanicheskii zhurnal. Moscow & Leningrad. Vol. 33+, 1948+. Bot. Zhurn. (Moscow & Leningrad). Preceded by: Botanicheskii zhurnal S S S R. 1-756-1. HI 52490

Botanicheskii zhurnal. St. Petersburg. Vols. 1-7, 1906-12. Bot. Zhurn. (St. Petersburg). Replaces: Trudy imperatorskago S.-Peterburgskago obshchestva estestvoispytatelei. Vypusk 3, otdělenie botaniki, vols. 35-37. Superseded by: Trudy imperatorskago S.-Peterburgskago obshchestva estestvoispytatelei. Vypusk 4, otdělenie botaniki. 1-756-1. HI 62772

Botanicheskii zhurnal S S S R. Moscow & Leningrad. Vols. 17-32, 1932-47. Bot. Zhurn. S.S.S.R. Preceded by: Zhurnal Russkogo botanicheskogo obshchestva. Superseded by: Botanicheskii zhurnal. Moscow & Leningrad. 1-756-1. HI 52491

Botanicheskiya zapiski. St. Petersburg. Vols. 1-30, 1886-1916. Bot. Zap. 3-2393-2. HI 52482

Botanicheskoe obozrenie; referiruyushchii organ Glavnogo botanicheskogo sada v Petrograde. Petrograd. Nos. 1-3, 1919/20-23. Bot. Obozr. 3-2389-2. HI 52463

Botanichnyi zhurnal. Kiev, Ukrainian S S R. Vols. 1-12, 1940-55. Bot. Zhurn. (Kiev). Preceded by: Zhurnal Instytuto botaniky Vseukraïns'koi akademii nauk. Superseded by: Ukrayins'kyi botanichnyi zhurnal. 5-4293-2. HI 52489

Botanico-medical recorder, ... Columbus, OH. Bot.-Med. Rec. See B–P–H 223/5. HI 52449

Botanico-medical reformer, or a course of lectures introductory to a knowledge of true medical science. Mount Vernon, OH. Bot.-Med. Reformer (Mount Vernon). See B–P–H 223/6. HI 52450

Botanicus brief. Frankfurt am Main. 1977+. Bot. Brief. HI 62773

Botanika. Lvov, Galicia [Ukrainian S S R]. Vol. 1+, 1925+. Botanika (Lvov). 1-757-3. HI 52493

Botanika; issledovaniya belorusskoe otdelenie, vsesoyuznogo botanicheskogo obshchestvo. Minsk. Vol. 5+, 1963+. Botanika (Minsk). Preceded by: Sbornik botanicheskikh rabot, belorusskoe otdelenie, vsesoyuznogo botanicheskogo obshchestvo. HI 62774

Botanika. Moscow & Leningrad = Materialy k biobibliografii uchenykh S S S R. Seriya biologicheskikh nauk; botanika. Moscow & Leningrad.

Botanika. Tiflis = Botanikis institutis šromebi. Tiflis.

Botanika chronika. Patras. Vol. 1+, 1981+. Bot. Chron. HI 62775

Botanika institutynyn Ishleri = Trudy instituta botaniki, akademiya nauk Turkmenskoi S S R. Ashkhabad.

Botanikai közlemények. Budapest. Bot. Közlem. See B–P–H 221/4. HI 52430

Botanikai múzeumi füzetek. Kolozsvar [=Cluj, Rumania]. Bot. Múz. Füz. See B–P–H 223/21. HI 52458

Botanikis institutis šromebi. Tiflis. Vols. 24-29, 1965-79. Bot. Inst. Šromebi. Superseded by: N. Kec′xovelis sax. botanikis institutis šromebi. HI 65506

Botanikos klausimai. Vilna, Lithuanian S S R. Bot. Klaus. See B–P–H 221/1. HI 52429

Botanique. Nagpur. Vol. 1+, 1970+. Botanique. HI 62777

Botanisch jaarboek, uitgegeven door het Kruidkundig genootschap Dodonaea te Gent. Ghent. Bot. Jaarb. See B–P–H 220/15. HI 52425

Botanisch-phaenologische Beobachtungen in Böhmen. Prague. Bot.-Phaenol. Beob. Bohmen. See B–P–H 224/9. HI 52465

Botanische Abhandlungen. [Edited by K. Goebel.] Jena. Bot. Abh. See 218/21. HI 52399

Botanische Abhandlungen aus dem Gebiet der Morphologie und Physiologie. Bonn. Bot. Abh. Morphol. Physiol. See B–P–H 218/23. HI 52400

Botanische Blätter zur Beförderung des Selbststudiums der Pflanzenkunde auch besonders für Frauenzimmer. Hamburg. Bot. Blätt. Beförd. Selbststud. Pflanzenk. See B–P–H 219/6. HI 52408

Botanische Exkursionen. Stuttgart. Vols. 1-2, 1979-81. Bot. Exkurs. HI 62778

Botanische Exkursionen und pflanzengeographische Studien in der Schweiz. Zurich. Bot. Exkurs. Pflanzengeogr. Stud. Schweiz. See B–P–H 220/4. HI 52418

Botanische Hefte. Forschungen aus dem Botanischen Garten zu Marburg. Marburg. Bot. Hefte. See B–P–H 220/11. HI 52421

Botanische Jahrbücher für Jedermann. Lüneburg, Germany. Bot. Jahrb. Jedermann. See B–P–H 220/19. HI 52426

Botanische Jahrbücher für Systematik, Pflanzengeshichte und Pflanzengeographie. Leipzig. Bot. Jahrb. Syst. See B–P–H 220/20. HI 52427

Botanische Literatur-Blätter zur periodischen Darstellung der Fortschritte der Pflanzenkunde in steter Beziehung zur gesammten Naturkunde, ... Nuremberg. Bot. Lit.-Blätt. See B–P–H 221/9. HI 52433

Botanische Mitteilungen, Süd-west Afrika wissenschaftliche Gesellschaft. Windhoek. Vol. 1+, 1973+. Bot. Mitt. Süd-West Afrika Wiss. Ges. HI 62779

Botanische Mitteilungen aus den Tropen. Jena. Bot. Mitt. Tropen. See B–P–H 223/16 HI 52456.

Botanische Museumshefte. Kolozsvar = Botanikai múzeumi füzetek. Kolozsvar [=Cluj, Rumania]. Bot. Múz. Füz. See B–P–H 223/21.

Botanische Nachrichten. Österreichischer Lehrerverein für Naturkunde. Vienna. Bot. Nachr. See B–P–H 223/22. HI 52459

Botanische Sitzungsberichte, societas pro fauna et flora fennica. Helsinki. Bot. Sitzungsber. Soc. Fauna Fl. Fenn. See B–P–H 224/20. HI 52471

Botanische studien. Jena. Nos. 1-19, 1954-72. Bot. Stud. HI 62780

Botanische Untersuchungen. Berlin. Bot. Untersuch. (Berlin). See B–P–H 225/13. HI 52476

Botanische Untersuchungen. Heidelberg. Bot. Untersuch. (Heidelberg). See B–P–H 225/14. HI 52477

Botanische Zeitung. Berlin. Bot. Zeitung (Berlin). See B–P–H 225/20. HI 52483

Botanische Zeitung. Leipzig = Botanische Zeitung. Berlin. Bot. Zeitung (Berlin). See B–P–H 225/20.

Botanische Zeitung welche Recensionen, Abhandlungen, Aufsaetze, Neuigkeiten und Nachrichten, die Botanik betreffend, enthaelt. Regensburg. Bot. Zeitung (Regensburg). See B–P–H 225/21. HI 52485

Botanische Zeitung. 2. Abteilung. Leipzig. Bot. Zeitung, 2. Abt. See B–P–H 225/22. HI 52486

Botanischer Jahresbericht. Systematisch geordnetes Repertorium der Botanischen Literatur aller Länder. [Edited by L. Just.] Berlin. Bot. Jahresber. (Just). See B–P–H 220/22. HI 52428

Botanisches Archiv. Königsberg [=Kaliningrad, Russian SFSR] & Dahlem bei Berlin [=Berlin-Dahlem]. Bot. Arch. See B–P–H 219/1. HI 52406

Botanisches Archiv der Gartenbaugesellschaft des österreichischen Kaiserstaates. Vienna. Bot. Arch. Gartenbauges. Österr. Kaiserstaates. See B–P–H 219/4. HI 52407

Botanisches Centralblatt. Kassel Bot. Centralbl. See B–P–H 219/19. HI 52411

Botanisches Centralblatt für Deutschland. Leipzig. Bot. Centralbl. Deutschl. See B–P–H 219/21. HI 52412

Botanisches Echo. Literatur-Beilage zum Botanischen Archiv. Berlin-Dahlem. Bot. Echo. See B–P–H 219/29. HI 52415

Botanisches Literaturblatt. Organ für Autor- und Instituts-Referate aus dem Gesamtgebiet der botanischen Literatur. Innsbruck. Bot. Literaturbl. (Innsbruck). See B–P–H 221/10. HI 52434

Botanisches Magazin. [Edited by Römer & Usteri.] Zurich. Bot. Mag. (Römer & Usteri). See B–P–H 221/16. HI 52437

Botanisches Taschenbuch. [Edited by Trattinick.] Vienna. Bot. Taschenb. (Trattinick). See B–P–H 225/5. HI 52473

Botanisches Taschenbuch für die Anfänger dieser Wissenschaft und der Apothekerkunst. Regensburg. Bot. Taschenb. Anfänger Wiss. Apothekerkunst. See B–P–H 225/6. HI 52474

Botanisches Zentralblatt. Jena. Bot. Zentralbl. See B–P–H 225/23. HI 52487

Botanisk tidsskrift. Copenhagen. Vols. 1-4, 1866-70/71 [for vols. "5-10" see ser. 2 & 3]; ser. 2, vols. 1-4, 1872-74/76; ser. 3, vols. 1-3, 1876/77-79/80 [vol.3, 1879/80 also numbered vol.11]; vol.12-75, 1880/81-1981. Bot. Tidsskr. Superseded by: Nordic journal of botany. 1-757-3. HI 62781

Botaniska notiser. Lund. Vols. 1-133, [1839]1841-1980 [suspended 1847-48; 1859-62; 1864; 1869-70]. Bot. Not. Superseded by: Nordic journal of botany. 1-757-3. HI 50385

Botaniska notiser. Supplement. Lund. Bot. Not. Suppl. See B–P–H 224/3. HI 52461

Botaniska utflygter. Uppsala. Bot. Utflygter. See B–P–H 225/15. HI 52478

Botaniske litteraturblade. Copenhagen. Bot. Litteraturbl. (Copenhagen). See B–P–H 221/11. HI 52435

Botanist; containing accurately coloured figures, of tender and hardy ornamental plants; ... London. Botanist. See B–P–H 226/29. HI 52494

Botaniste, Le. Caen. Vols. 1-61, 1889-1975 [suspended 1913-20; 1943-47]. Botaniste. Superseded by: Revue de cytologie et de biologie végétales — le botaniste. 1-757-3. HI 62782

Botanists' chronicle. Chelsea [=London, in part]. Bot. Chron. See B–P–H 219/22. HI 52413

Botanists' repository, for new, and rare plants. London. Vols. 1-10, 1797-1814. Bot. Repos. 1-758-1. HI 62783

Botany bulletin, department of agriculture, Queensland. Brisbane, Qld. Nos. 1-22, 1890-1920. Bot. Bull. Dept. Agric., Queensland. HI 62784

Botany bulletin, department of forests, Papua New Guinea. Lae. Vol. 1+, 1969+. Bot. Bull. Dep. Forests Papua New Guinea. HI 62785

Botany circular, division of plant industry, Florida. Gainesville, FL. No. ?-20+, 19??-83+. Bot. Circ. Div. Pl. Industr. Florida. HI 62786

Botany, current literature. Library U S Department of agriculture. Washington, DC. Bot. Curr. Lit. Libr. U.S.D.A. See B–P–H 219/23. HI 52414

Botany department report, New Jersey agricultural college experiment station = New Jersey agricultural college experiment station. Botany department report. New Brunswick, NJ. New Jersey Agric. Coll. Exp. Sta. Bot. Dept. Rep. See B–P–H 653/14.

Botany focus. 1987/88+, 1988+. Bot. Focus. Preceded by: Report (Annual) of the botany branch and Queensland herbarium. HI 62787

Botany, Iowa state university. Ames, IA. 19??-1981+. Bot. Iowa State Univ. HI 62788

Botany leaflet. London. 1974+. Bot. Leafl. Brit. Mus. Nat. Hist. HI 62789

Botany leaflet, Field museum of natural history = Leaflet, department of botany, Field museum of natural history. Chicago, IL.

Botany news. Amherst, MA. Vol. 1+, 1981+. Bot. News. HI 62790

Botany newsletter of the Iowa univerisity. Iow City, IA. Bot. Newslett. Iowa Univ. See 224/1. HI 52460

Botany pamphlet, Carnegie museum = Carnegie museum botany pamphlet. Pittsburgh, PA. Carnegie Mus. Bot. Pam. See B–P–H 299/18.

Botoşani; studii şi communicări. Dorohoi. Vol. 1+, 1972+. Botoşani. HI 62791

Botany and zoology; theoretical and applied. [Syokubutsu oyobi dobutsu.] Tokyo. Bot. & Zool. See B–P–H 226/6. HI 52488

Bothalia; a record of contributions from the national herbarium, Union of South Africa. Pretoria. Bothalia. See B–P–H 226/34. HI 52495

Botrychium newsletter. Ann Arbor, MI. No. 1+, 1981+. Botrychium Newslett. HI 62792

Botswana notes and records. Gaberones. Vol. 1+, 1968+. Botswana Notes Rec. HI 62793

Botswana notes and records. Special issue. Gaberones. Vol. 1+, 1971+. Botswana Notes Rec., Special Issue. HI 62794

Bouquet. Litchfield, IL. Bouquet. See B–P–H 227/1. HI 52496

Bowdoin scientific review. A fortnightly journal. Brunswick, ME. Bowdoin Sci. Rev. See B–P–H 227/3. HI 52497

Bowyseri rhadiozgaynowt yown eew mowtaownakowt yowne = Radiochuvstvitel′nost′ i mutabil′nost′ rastenii. Erevan.

Boxwood bulletin; a quarterly devoted to man's oldest garden ornamental. Boyce, VA. Boxwood Bull. See B–P–H 227/4. HI 52498

Brackishwater aquaculture abstracts. Tigbauan, Iloilo, Philippines. Vol. ?-2+, 19??-85+. Brackiswater Aquac. Abstr. HI 62795

Bradea; boletim do herbarium bradeanum. Rio de Janeiro. Vol. 1+, 1969+. Bradea. HI 62796

Bradford scientific journal. Bradford. Vols. 1-3, 1904-12. Bradford Sci. J. HI 62797

Bradleya; yearbook of the british cactus and succulent society. Botley. Vol. 1+, 1983+. Bradleya. HI 62798

Braga, vaterländische Blätter für Kunst und Wissenschaft. Heidelberg. Braga. See B–P–H 227/20. HI 52499

Bragantia. São Paulo. Vol. 1+, 1941+. Bragantia. Preceded by: Boletim tecnico, instituto agronômico, São Paulo. 1-761-3. HI 62799

Brasil açucareio. Rio de Janeiro. Brasil Açucareio:. See B–P–H 227/22. HI 52500

Brasil florestal, instituto brasileiro de desenvolvimento florestal. Rio de Janeiro. Vol. 1+, 1970+. Brasil Florest. Preceded by: Anuário brasileiro de economia florestal. HI 62800

Brasil florestal, instituto brasileiro de desenvolvimento florestal. Boletim tecnico. Rio de Janeiro. Vol. 1+, 1971+. Brasil Florest. Bol. Tecn. HI 62801

Bratislavai <Pozsonyi> orvos- és természettudományi egyesület közleményei = Verhandlungen des heil- und naturwissehschaftlichen Vereins zu Bratislava <Pressburg>. Pressburg [=Bratislava, Czechoslovakia]. Verh. Heil- Naturwiss. Vereins Bratislava. See B–P–H 951/11.

Braun-Blanquetia; recueil de travaux de géobotanique. Bailleul. Vol. 1+, 1984+. Braun-Blanquetia. HI 62802

Braunkohle; Zeitschrift für Gewinnung und Verwertung der Braunkohle. Halle. Braunkohle. See B–P–H 227/25. HI 52501

Braunschweiger naturkundliche Schriften. Brunswick. Vol. 1+, 1980+. Braunschweig. Naturk. Schriften. HI 62803

Braunschweigisches Magazin. Brunswick, Germany. Braunschweig. Mag. See B–P–H 227/27. HI 52502

Brazil agrícola, industrial, commercial, scientifico, literario e noticioso. Recife, Brazil. Brazil Agríc. See B–P–H 227/29. HI 52503

Brazil-médico. Rio de Janeiro. Brazil-Méd. See B–P–H 227/30. HI 52504

Brazilian journal of genetics = Revista brasileira de genética. Ribeirão Preto.

Brazilian journal of medical and biological research. São Bernardo do Campo. Vol. 14+, 1981+. Brazil. J. Med. Biol. Res. Preceded by: Revista brasileira de pesquisas médicas e biológicas. HI 75210

Brazilian orchids newsletter = Brazilian orchids. Edinburgh.

Brazilian orchids. Edinburgh. No. 1+, 1984+. Brazil. Orchids. HI 62804

Brébissonia. Paris. Brébissonia. See B–P–H 227/32. HI 52505

Breconshire naturalist. Brecon. No. 1+, 1971+. Breconshire Naturalist. Preceded by: Newsletter, Brecknock county naturalists' Trust. HI 62805

Brem- und Verdische Bibliothek, worin zur Aufnahme der Wissenschaften, ... allerley brauchbare Abhandlungen und Anmerkungen mitgetheilet werden. Hamburg. Brem- Verdische Biblioth. See B–P–H 227/33. HI 52506

Bremer Beiträge zur Naturwissenschaft. Bremen. Bremer Beitr. Naturwiss. See B–P–H 227/34. HI 52507

Bremisches Magazin zur Ausbreitung der Wissenschaften, Künste und Tugend ... Hanover. Bremisches Mag. Ausbreit. Wiss. See B–P–H 228/1. HI 52508

Brenesia. San José, Costa Rica. Vol. 1+, [1972]+. Brenesia. HI 62806

Breviora. Cambridge, MA. No. 1+, 1952+. Breviora. HI 62807

Breviora geológica astúrica. Oviedo, Spain. Breviora Geol. Astúrica. See B–P–H 228/8. HI 52509

Brief record of work of the Connecticut pomological society = Proceedings of the Connecticut pomological society. New Haven, CT. Proc. Connecticut Pomol. Soc. See B–P–H 728/10.

Briefe aus dem botanischen Garten, Zürich. Zurich. No. 1+, 1967+. Briefe Bot. Gart. Zürich. HI 62808

Brigham Young university. Science bulletin. Biological series. Provo, UT. Brigham Young Univ. Sci. Bull., Biol. Ser. See B–P–H 228/10. HI 52511

Brigham Young university. Science bulletin. Geological series. Provo, UT. Brigham Young Univ. Sci. Bull., Geol. Ser. See B–P–H 228/11. HI 52512

Brimleyana. Raleigh, NC. 1979+. Brimleyana. HI 62809

Briolatina. Mexico, D.F. No. 1+, 1983+. Briolatina. HI 62810

Bristol cactus society newsletter. Bristol. 1971+. Bristol Cact. Soc. Newslett. HI 62811

British agricultural bulletin. London. Brit. Agric. Bull. See B–P–H 228/13. HI 52513

British-american cultivator. Toronto. Vols. 1-2, 1842-44; n.s. vols. 1-3, 1845-47. Brit.-Amer. Cultivator. Superseded by: Agriculturist and canadian journal. 1-774-3. HI 62812

British antarctic survey bulletin = Bulletin, british antarctic survey. London.

British antarctic survey scientific reports = Scientific reports, british antarctic survey. London.

British beet grower and empire producer. London. Brit. Beet Grower. See 228/16. HI 52514

British cactus and succulent journal. Botley. Vol. 1+, 1983+. Brit. Cactus Succ. J. Preceded by: Cactus and succulent journal of Great Britain and National cactus and succulent journal. HI 62813

British Columbia naturalist = B C naturalist. Vancouver, B.C.

British Columbia orchardist. West Summerland, B.C. Vols. 1-16, 1959-76. British Columbia Orchardist. HI 62814

British delphinium society's yearbook. London. Brit. Delphinium Soc. Yearb. See B–P–H 228/20. HI 52515

British dental journal. London. Brit. Dental J. See B–P–H 228/21. HI 52516

British farmer and journal of agriculture. [Vol. 4, 1923 includes as a supplement: Journal of the central and associated chambers of agriculture, London.] London. Vols. 1-4, 1920-23. Brit. Farmer & J. Agric. 1-781-3. HI 62815

British farmer's magazine. London. Brit. Farmer's Mag. See B–P–H 228/24. HI 52517

British fern gazette. Kendal. Vols. 1-10, 1909-73. Brit. Fern Gaz. Superseded by: Fern gazette. 1-782-1. HI 62816

British florist; or lady's journal of horticulture. London. Brit. Florist. See B–P–H 228/26. HI 52518

British flower. London. 1973+. Brit. Fl. Preceded by: Journal, british flower association. HI 62817

British food journal and hygienic review. London. Brit. Food J. See B–P–H 228/27. HI 52519

British and foreign medical review. London. Brit. Foreign Med. Rev. See B–P–H 229/4. HI 52520

British and foreign medico-chirurgical review. London. Brit. Foreign Med.-Chir. Rev. (London). See B–P–H 229/5. HI 52521

British and foreign medico-chirurgical review or quarterly journal of practical medicine and surgery. Philadelphia, PA. Brit. Foreign Med.-Chir. Rev. (Philadelphia). See B–P–H 229/6. HI 52522

British and foreign review; or European quarterly journal. London. Brit. Foreign Rev. See B–P–H 229/7. HI 52523

British and foreign scientific magazine and journal of scientific inventions. London. Brit. Foreign Sci. Mag. See B–P–H 229/8. HI 52524

British Guiana timbers: leaflet = Leaflet, forest department, British Guiana. Georgetown.

British hosta and hemerocallis society bulletin = Bulletin, british hosta and hemerocallis society. Boldre, Eltham.

British iris society newsletter. London. No. 1+, 1946+. Brit. Iris Soc. Newslett. HI 62818

British journal of dermatology (and syphilis). London. Brit. J. Dermatol. See B–P–H 229/14. HI 52525

British journal of entomology and natural history. London. Vol. 1+, 1988+. Brit. J. Entomol. Nat Hist. HI 74514

British journal of experimental biology. Edinburgh. Brit. J. Exp. Biol. See B–P–H 229/15. HI 52526

British journal for the history of science. Ravenswood, Oxford & London. Vol. 1+, 1962+. Brit. J. Hist. Sci. Preceded by: Bulletin of the british society for the history of science. HI 62819

British journal of ophthalmology. London. Brit. J. Ophthalmol. See B–P–H 229/17. HI 52527

British journal of photography. London. Vol. 6+, 1860+. Brit. J. Photogr. Preceded by: Liverpool and Manchester photographic journal [not entered]. 1-785-3. HI 74815

British lichen society bulletin = Bulletin, british lichen society. London, etc.

British medical journal. London. Brit. Med. J. See B–P–H 229/20. HI 52528

British micropalaeontologist. Leeds. 1976+. Brit. Micropalaeontol. HI 62820

British naturalist. Baldock. No. 1+, 1980+. Brit. Naturalist. HI 62821

British naturalist. London. Brit. Naturalist. See B–P–H 229/24. HI 52529

British New Guinea = Report (Annual) of British New Guinea. Melbourne.

British newsletter, society for the bibliography of natural history. London. Nos. 1-7, 1977-78. Brit. Newslett. Soc. Bibliogr. Nat. Hist. Superseded by: Newsletter, society for the bibliography of natural history. HI 62822

British phycological bulletin. Glasgow. Vols. 1(7)-3, 1959-68. Brit. Phycol. Bull. Preceded by: Phycological bulletin. Superseded by: British phycological journal. HI 62823

British phycological journal. Leeds, London. Vol. 4+, 1969+. Brit. Phycol. J. Preceded by: British phycological bulletin. HI 62824

British pteridological society newsletter. [Loughton], etc. Nos. 1-10, 1963-72. Brit. Pteridol. Soc. Newslett. Superseded by: Bulletin, british pteridological society. HI 62825

British reports, translations and theses. Boston Spa. Nos. BRTT 81/1-85/12, 1981-85. Brit. Rep. Transl. Theses. Preceded by: B L L D announcement bulletin. Superseded by microfiche cumulations. HI 62826

British science news. London. Brit. Sci. News See B–P–H 228/28. HI 52530

British sugar beet review. London. Brit. Sugar Beet Rev. See B–P–H 229/29. HI 52531

British wildlife; the magazine for the modern naturalist. Basingstoke. Vol. 1+, 1989?+. Brit. Wildlife. HI 74816

Brittonia; a series of botanical papers. New York, NY. Vol. 1+, 1931/35+. Brittonia. 1-795-2. HI 62827

Bromel-ana of the greater New York chapter of the bromeliad society. Long Island City, NY. Vol. 1(1), 1963. Bromel-ana. Superseded by: Bromeliana of the greater New York chapter, bromeliad society. HI 62828

Bromeliads; journal of the british bromeliad society. [Bristol.] Vol. 1+, 1968+. Bromeliads. HI 62829

Bromeliana of the greater New York chapter of the bromeliad society. Long Island City, N.Y. Vol. 1(2)+, 1963+. Bromeliana. Preceded by: Bromel-ana of the greater New York chapter of the bromeliad society. HI 62830

Brookgreen bulletin. Murrels Inlet, SC. Vols. 1-15(1), 1971-84. Brookgreen Bull. Superseded by: Brookgreen journal. HI 62831

Brookgreen gardens newsletter. Murrels Inlet, SC. 1987+. Brookgreen Gard. Newslett. HI 62832

Brookgreen journal. Murrels Inlet, SC. Vol. 15(2)+, 1985+. Brookgreen J. Preceded by: Brookgreen bulletin. HI 62833

Brooklyn botanic garden leaflets. Brooklyn, NY. Brooklyn Bot. Gard. Leaft. See B–P–H 230/5. HI 52532

Brooklyn botanic garden memoirs. Brooklyn, NY. Brooklyn Bot. Gard. Mem. See B–P–H 230/6. HI 52533

Brooklyn botanic garden newsletter. New York. 1972-85. Brooklyn Bot. Gard. Newslett. Superseded by: Plants and gardens news. HI 62834

Brooklyn botanic garden record. Brooklyn, NY. Brooklyn Bot. Gard. Rec. See B–P–H 230/7. HI 52534

Broom and broom corn news. Arcola, Il. Broom Broom Corn News. See B–P–H 230/10. HI 52535

Broom corn review. Wichita, KS. Broom Corn Rev. See B–P–H 230/11. HI 52536

Brotéria. Lisbon. Brotéria. See B–P–H 230/12. HI 52537

Brotéria: genética; Orgão da sociedade Portuguesa der genética. Lisbon. Vol. 1(76)+, 1980+. Brotéria Genét. Preceded by: Broteria: ciências naturais. HI 62835

Brotéria. Série botânica; revista de sciencias naturaes do collegio de S. Fiel. Lisbon. Vols. 6-25, 1907-31. Brotéria, Sér. Bot. Preceded by: Brotéria. Superseded by: Brotéria: ciências naturais. 1-802-3. HI 62836

Brotéria: ciências naturais; revista de sciencias naturaes do collegio de S. Fiel. Lisbon. Vols. 3(34)-48(75), 1938-75. Brotéria Ci. Nat. Preceded by: Brotéria. Série botânica. Superseded by: Broteria genetica. HI 62837

Brunei museum journal. Brunei. Vol. 1+, 1969+. Brunei Mus. J. HI 62838

Brunonia. East Melbourne, Vic. Vols. 1-10(2), 1978-87. Brunonia. Preceded by: Contributions from herbarium australiense. Superseded by: Australian systematic botany. HI 62839

Bruns Beiträge zur klinischen Chirurgie = Beiträge zur klinischen Chirurgie. Tübingen. Beitr. Klin. Chir. See B–P–H 170/11.

Bruxelles-médical. Brussels. Bruxelles-Méd. See B–P–H 230/27. HI 52538

Brycheiniog. Brecknock, Wales. Brycheiniog. See B–P–H 230/28. HI 52539

Bryological news letter, indian bryological society. Lucknow. No. 1+, 1984+. Bryol. News Lett. Indian Bryol. Soc. HI 62840

Bryological newsletter. Kensington, N.S.W. No. 1+, 1979. Bryol. Newslett. Superseded by: Australasian bryological newsletter. HI 62841

Bryological times; newsletter of the international association of bryologists. Utrecht, Penicuik. No. 1+, 1980+. Bryol. Times. HI 62842

Bryologische Beiträge. Duisburg. Vol. 1+, 1982+. Bryol. Beitr. HI 62843

Bryologische Zeitschrift. Berlin-Schöneberg. Bryol. Z. See B–P–H 230/29. HI 52540

Bryologist. [Vols. 1-2 formed part of vols. 6-7 of: Fern bulletin. See B–P–H 369/15.] Binghampton, NY, Brooklyn, NY, etc. Vol. 1+, 1898+. Bryologist. 1-809-3. HI 62844

Bryomania; an irregular journal for the irrelevant. St. John's, Nfld. No. 9+ [actually the first no. published], 1980+ [irregular numbering and dating]. Bryomania. HI 62845

Bryophytorum bibliotheca. Lahre, Vaduz. Vol. 1+, 1973+. Bryophyt. Biblioth. HI 62846

Buckskin totem. Covington, KY. Buckskin Totem. See B–P–H 230/35. HI 52541

A Budapesti magyar királyi állami vetömagvizsgáló állomás évi müködése. Budapest. Budapesti Magyar Kir. Állami Vetöm. Állomás Évi Mük. See B–P–H 231/1. HI 52543

Budapesti szemle. Budapest. Budapesti Szemle. See B–P–H 231/2 HI 52544.

Budapesti tudományegyetem biológiai intézeteinek évkönyve = Annales biologicae universitatis budapestinensis. Budapest. Ann. Biol. Univ. Budapest. See B–P–H 97/16.

Budapestikirályi magyar tudományegyetemi természettudományi szövetseg évkönyve. Budapest. Budapesti Kir. Magyar Tudományegyet. Term. Szöv. Évk. See B–P–H 230/37. HI 52542

Buffalo medical journal. Buffalo, NY. Buffalo Med. J. See 231/6. HI 52545

Buffalo medical & surgical journal = Buffalo medical journal. Buffalo, NY. Buffalo Med. J. See 231/6.

Buffalo medical and surgical journal and reporter = Buffalo medical journal. Buffalo, NY. Buffalo Med. J. See 231/6.

Bugd nairamdakh Mongol Ard Ulsyn baigaliin bayalag = Biologicheskie resursy i prirodnye usloviya Mongol'skoi Narodnoi Respubliki. Leningrad.

Buiten; geillustreerd weekblad. Amsterdam. Buiten. See B–P–H 231/9. HI 52546

Buitenleven. The Hague. Buitenleven. See B–P–H 231/10. HI 52547

Buletin i institutit të shkencave. Tirana, Albania. Bul. Inst. Shkencave. See B–P–H 232/18. HI 52555

Buletin i institutit të studimeve. Tirana, Albania. Bul. Inst. Stud. See B–P–H 232/19. HI 52556

Buletin kebun raya. Bogor. Vol. 1+, 1973+. Bul. Kebun Raya. HI 62847

Buletin penelitian hortikultura. Jakarta. ?-ca.1988+. Bul. Penelitian Hort. HI 62848

Buletin penelitian hutan. Bogor. No. 462+, 1984+. Bul. Penelitian Hutan. Preceded by: Laporan pusat penelitian hutan. HI 74818

Buletin për shkencat biologijke. Tirana, Albania. Bul. Shkencat Biol. See B–P–H 233/8. HI 52557

Buletin për shkencat natyrore. Tirana, Albania. Bul. Shkencat Nat. See B–P–H 233/9. HI 52558

Buletin i shkencavet te natyres. Tirana. 1972+. Bul. Shkencacet Nat. Preceded by: Buletin i universitetit shtetëror te Tiranës. HI 62849

Buletin ştiinţific. Academia republicii populare române [or romîne]. Bucharest. Bul. Sti. Acad. Republ. Populare Române. See B–P–H 234/1. HI 52564

Buletin ştiinţific, institutul pedagogic, Baiă-Mare. Seria B, biologie, fizică-chemie, matematică. Baiă-Mare. No. 1+, 1969+. Bul. Şti. Inst. Pedagog. Baiă-Mare, B. HI 62850

Buletin i universitetit shtetëror të Tiranës. Seria shkencat natyrore. Tirana. Vols. 11(3)-25, 1957-71. Bul. Univ. Shtetëror Tiranës, Ser. Shkencat Nat. Preceded by: Buletin për shkencat natyrore. Superseded by: Buletin i shkencavet te natyres. HI 62851

Buletinul academiei de inalte studii agronomice din Cluj. Cluj, Rumania. Bul. Acad. Stud. Agron. Cluj. See B–P–H 231/14. HI 52548

Buletinul akademiej de štiince a R S S moldovenšt'. Serija biologičeskich i chimičeskich nauk = Izvestiya akademiya nauk moldavskoi S S R. Seriya biologicheskikh i khimicheskikh nauk. Kishinev.

Buletinul comisíeri pentru ocrotirea monumentelor naturii = Ocrotirea naturii. Bucharest.

Buletinul cultivări şi fermentări tutunului. Bucharest. Bul. Cult. Ferment. Tutunului. See B–P–H 231/33. HI 52549

Buletinul erbarului institutului botanic din Bucureşti. Bucharest. Bul. Erb. Inst. Bot. Bucureşti. See B–P–H 231/39. HI 52550

Buletinul, facultătii de agronomie din Cluj. Cluj. 1938-43. Bul. Fac. Agron. Cluj. Preceded by: Buletinul academiei de înalte studii agronomice din Cluj. Superseded by: Analele, facultătii de agronomie din Cluj. HI 62852

Buletinul facultaţii de ştiinţe din Cernăuti. Cernăuti, Rumania [=Chernovtsy, Ukrainian S S R]. Bul. Fac Şti. Cernăuti. See B–P–H 232/1. HI 52551

Buletinul grădinii botanice şi al muzeului botanic dela universitatea din Cluj. Cluj, Rumania. Bul. Grăd. Bot. Univ. Cluj. See B–P–H 232/6. HI 52552

Buletinul grădinii botanice şi al muzeului botanic dela universitatea din Cluj la Timişoara. Cluj, Rumania. Bul. Grăd. Bot. Univ. Cluj. See B–P–H 232/6.

Buletinul de informaţii al grădinii botanice şi al muzeului botanic dela universitatea din Cluj. Cluj, Rumania. Bul. Inform. Grăd. Bot. Univ. Cluj. See B–P–H 232/15. HI 52553

Buletinul de informaţii al societăţii naturalliştilor din România. Bucharest. Bul. Inform. Soc. Nat. România. See B–P–H 232/16. HI 52554

Buletinul institutului agronomic Cluj-Napoca. Cluj. Vol. 29+, 1975+. Bul. Inst. Agron. Cluj-Napoca. Preceded by: Lucrari stiintifice, institutul agronomic "Dr. Petru Groza". HI 62853

Buletinul institutului politehnic din Braşov. Ser. B, economie forestiera. Braşov. Vols. 1-13, 1959-71. Bul. Inst. Politehn. Braşov, B. Superseded by: Buletinul universitatii din Braşov. Ser. B, economie forestiera. HI 62854

Buletinul muzeului national de istorie naturala din Chisinau. Chisinau. 1926-38. Bul. Muz. Natl. Istorie Nat. Chisinau. HI 62855

Buletinul muzeului regional al Basarabiei din Chisinau. Chisinau. 1938-42. Bul. Muz. Regional Basarabiei Chisinau. HI 62856

Buletinul politecnicii "GH. Asachi" din Jasi = Bulletin de l'école polytechnique de Jassy. Jassy, Rumania. Bull. École Polytechn. Jassy. See B–P–H 248/20.

Buletinul societăţii naturaliştilor din Romania. Bucharest. Bul. Soc. Naturaliştilor Romania. See B–P–H 233/32. HI 52559

Buletinul societăţii române de ştiinţe. Bucharest. Bul. Soc. Române Şti. See B–P–H 233/35. HI 52560

Buletinul societăţii de ştiinţe din Bucureşti. Bucharest. Bul. Soc. Şti. Bucureşti. See B–P–H 233/47. HI 52561

Buletinul societăţii de ştiinţe din Cluj. Cluj. Vols. 1-?, 1921-48. Bul. Soc. Şti. Cluj. HI 62857

Buletinul societăţii studenţilor in ştiinţe naturale din Bucureşti. Bucharest. Bul. Soc. Stud. Şti. Nat. Bucureşti. See B–P–H 233/49. HI 52562

Buletinul tutunului. Bucharest. Bul. Tutunului. See B–P–H 234/7. HI 52565

Buletinul universitatii din Braşov. Ser. B, economie forestiera. Braşov. Vols. 14-16, 1972-74; vol. 19+, 1977+. Bul. Univ. Braşov, B. Preceded by: Buletinul institutului politehnic din Braşov. Ser. B, economie forestiera. HI 62858

Buletinul universitătii din Braşov. Seria C, matematica, fizica, chimia, stiinte naturale. Braşov. Vol. 14+, 1972+. Bul. Univ. Braşov, C. Preceded by: Buletinul institutului politehnic din Brasov. HI 62859

Buletinul universitătilor "V. Babes" si "Bolyai". Ser. stiintele naturii. Cluj. Vol. 2(1-2), 1957. Bul. Univ. V. Babes Bolyai, Ser. Şti. Nat. Superseded by: Studia universitatis Babes-Bolyai. Ser. 2, biologia. HI 62860

Bulletin A F A. Paris. No. ?-4+, 19??-76+. Bull. A. F. A. HI 62861

Bulletin de l'A R E R S = Bulletin, association régionale pour l'étude et la recherche scientifiques. Reims.

Bulletin de l'académie impériale des sciences de Saint-Pétersbourg. St. Petersburg. Bull. Acad. Imp. Sci. Saint-Pétersbourg. See B–P–H 234/30. HI 52566

Bulletin de l'académie impériale des sciences de Saint Pétersbourg. St. Petersburg. [Sér. 3], vols. 1-32, 1860-88; n.s. [= sér. 4] vols. 1-4 [also numbered vols. 33-36], 1890-94. Bull. Acad. Imp. Sci. Saint-Pétersbourg. Preceded by: Bulletin de la classe physico-mathématique de l'académie impériale des sciences de Saint-Pétersbourg. Superseded by: Izvestiya Imperatorskoi Akademii Nauk. 1-111-1. HI 62862

Bulletin de l'académie internationale de géographie botanique. Le Mans. Bull. Acad. Int. Géogr. Bot. See B–P–H 234/32. HI 52567

Bulletin de l'académie malgache = Bulletin trimestriel de l'académie malgache. Tananarive, Madagascar. Bull. Trimestriel Acad. Malgache. See B–P–H 285/5.

Bulletin de l'académie royale hongroise d'horticulture. Budapest = A m[agyar] kir[ályi] kertészeti akadémia közleményei. Budapest. Magyar Kir. Kert. Akad. Közlem. See B–P–H 543/17.

Bulletin de l'académie royale de médecine. Paris. Bull. Acad. Roy. Méd. See B–P–H 235/6. HI 52568

Bulletins de l'académie royale des sciences et belles-lettres de Bruxelles. Brussels. Bull. Acad. Roy. Sci. Burxelles. See B–P–H 235/9. HI 52571

Bulletins de l'académie royale des sciences, des lettres et des beaux arts de Belgique. Brussels. Bull. Acad. Roy. Sci. Belgique. See B–P–H 235/7. HI 52569

Bulletins de l'académie royale des sciences, des lettres et des beaux arts de Belgique. Classe des sciences. Brussels. Bull. Acad. Roy. Sci. Belgique, Cl. Sci. See B–P–H 235/8. HI 52570

Bulletin de l'académie des sciences. Petrograd = Izvestiya Akademii Nauk. Petrograd.

Bulletin de l'académie des sciences, inscriptions et belles-lettres de Toulouse. Toulouse. Bull. Acad. Sci. Toulouse. See 235/21. HI 52574

Bulletin de l'académie des sciences mathématiques et naturelles. B. Sciences naturelles. Belgrade. Bull. Acad. Sci. Math., B, Sci. Nat. See B–P–H 235/15. HI 52572

Bulletin de l'académie des sciences de Russie. = Izvestiya Rossiiskoi akademii nauk. Petrograd.

Bulletin de l'académie des sciences de l'U R S S = Izvestiya Akademii nauk S S S R. Moscow & Leningrad.

Bulletin de l'académie des sciences de l'U R S S. Classe des sciences mathématiques et naturelles = Izvestiya Akademii nauk S S S R. Ser. 7, Otdeleniye matematicheskikh i estestvennykh nauk. Moscow & Leningrad.

Bulletin de l'académie des sciences de l'U R S S; classe des sciences mathématiques et naturelles. Série biologique = Izvestiya Akademii nauk S S S R; otdeleniye matematicheskikh i estestvennykh nauk. Seriya biologicheskaya. Moscow & Leningrad.

Bulletin de l'académie des sciences de l'URRS; classe des sciences mathématiques et naturelles. Série géologique = Izvestiya Akademii nauk S S S R; otdeleniye matematicheskikh i estestvennykh nauk. Seriya geologicheskaya. Moscow & Leningrad.

Bulletin de l'académie des sciences de l'U R S S. Classe des sciences physico-mathématiques = Izvestiya

Akademii nauk S S S R. Ser. 7, Otdeleniye fiziko-matematicheskikh nauk. Moscow & Leningrad.

Bulletin de l'académie des sciences de l'U R S S. Série biologique = Izvestiya Akademii nauk S S S R. Seriya biologicheskaya. Moscow & Leningrad.

Bulletin de l'académie des sciences de l'U R S S. Série géologique = Izvestiya Akademii nauk S S S R. Seriya geologicheskaya. Moscow & Leningrad.

Bulletin of the academy of science of St. Louis. St. Louis, MO. Bull. Acad. Sci. St. Louis. See B–P–H 235/19. HI 52573

Bulletin de l'académie polonaise des sciences. Classe III, mathématique, astronomie, physique, chimie, géologie et géographie. Warsaw. Vols. 1-5, 1953-57. Bull. Acad. Polon. Sci., Cl. 3. Superseded by: Bulletin de l'académie polonaise des sciences. Serie des sciences chimiques, géologiques et géographiques. HI 62863

Bulletin de l'académie polonaise des sciences. Série des sciences biologiques. Warsaw. Vols. 1-30, 1953-82. Bull. Acad. Polon. Sci., Sér. Sci. Biol. Preceded by: Bulletin international de l'académie polonaise des sciences et des lettres. Classe des sciences mathématiques et naturelles. Série B 1. Botanique. Superseded by: Bulletin of the polish academy of sciences. Biology. HI 62864

Bulletin de l'académie polonaise des sciences. Serie des sciences chimiques, géologiques et géographiques. Warsaw. Vols. 6-7(11), 1958-59. Bull. Acad. Polon. Sci., Sér. Sci. Chim. Géol. Géogr. Preceded by: Bulletin de l'académie polonaise des sciences. Classe III, mathématique, astronomie, physique, chimie, géologie et géographie. Superseded by: Bulletin de l'académie polonaise des sciences. Série des sciences géologiques et géographiques. HI 62865

Bulletin de l'académie polonaise des sciences. Série des sciences géologiques et géographiques. Warsaw. Vols. 8(3)-18, 1960-70. Bull. Acad. Polon. Sci., Sér. Sci. Géol. Géogr. Preceded by: Bulletin de l'académie polonaise des sciences. Serie des sciences chimiques, géologiques et géographiques. Superseded by: Bulletin de l'académie polonaise des sciences. Série des sciences de la terre. HI 62866

Bulletin de l'académie polonaise des sciences. Série des sciences de la terre. Warsaw. Vols. 19-30, 1971-82. Bull. Acad. Polon. Sci., Sér. Sci. Terre. Preceded by: Bulletin de l'académie polonaise des sciences. Série des sciences géologiques et géographiques. Superseded by: Bulletin of the polish academy of sciences. Earth sciences. HI 62867

Bulletin de l'académie des sciences agricoles et forestières. Bucharest. No. 1+, 1972+. Bull. Acad. Sci. Agric. Forest. HI 62868

Bulletin de l'académie serbe des sciences. B. Sciences naturelles. Belgrade. Bull. Acad. Serbe Sci., B, Sci. Nat. See B–P–H 236/3. HI 52575

Bulletin de l'académie et de la société lorraine des sciences. Nancy. Bull. Acad. Soc. Lorraine Sci. See B–P–H 236/4. HI 52576

Bulletin de l'académie vétérinaire de France. Paris. Bull. Acad. Vérérin. France. See B–P–H 236/8. HI 52577

Bulletin of the academy of agricultural and forestry sciences = Bulletin de l'académie des sciences agricoles et forestières. Bucharest.

Bulletin of the academy of sciences of the Armenian S S R = Izvestiya akademiya nauk armyanskoi S S R. Biologicheskie i sel'skokhozyaistvennie nauki. Erevan.

Bulletin of the academy of sciences of the Armenian S S R = Izvestiya: Estestvennye nauki. Erivan.

Bulletin of the academy of sciences of the united provinces of Agra and Oudh. Allahabad. Vols. 1-3, 1931-34. Bull. Acad. Sci. Unit. Prov. Agra Oudh. Superseded by: Proceedings of the academy of sciences of the united provinces of Agra and Oudh. HI 62869

Bulletin of the academy of sciences of the Georgian S S R = Soobshcheniya Akademii nauk Gruzinskoi S S R. Tiflis.

Bulletin, african studies association of the United Kingdom. Birmingham. Vols. 1-22, 1964-72. Bull. African Stud. United Kingdom. Superseded by: African research and documentation. HI 62870

Bulletin of the african succulent plant society. London. Vols. 1-12(1), 1966-77. Bull. African Succ. Pl. Soc. HI 62871

Bulletin agricole. Antananarivo. 1948-51. Bull. Agric. (Antananarivo). HI 62872

Bulletin agricole. Port-au-Prince, Haiti. Bull. Agric. (Port-au-Prince). See B–P–H 236/15. HI 52579

Bulletin agricole du Congo Belge. Brussels. Bull. Agric. Congo Belge. See B–P–H 236/18. HI 52580

Bulletin agricole de l'institut scientifique de l'Indochine. Saigon. Bull. Agric. Inst. Sci. Indochin. See B–P–H 236/27. HI 52586

Bulletin agricole de la Martinique. Fort-de-France. Vols. 1-3-?, 1898-99-?; [n.s.] nos. 1-9, 1921-27?; n.s. vols. 1-10, 1930-41. Bull. Agric. Martinique HI 62873

Bulletin agricole de l'Algérie et de la Tunisie. Algiers. Vols. 1-19, 1895-1913. Bull. Agric. Algérie Tunisie Superseded by: Bulletin agricole de l'Algérie-Tunisie-Maroc. HI 62874

Bulletin agricole de l'Algérie-Tunisie-Maroc. Algiers. Vols. 20-35, 1914-29. Bull. Agric. Algérie Tunisie Maroc. Preceded by: Bulletin agricole de l'Algérie et de la Tunisie. HI 62875

Bulletin agricole de la Martinique. Fort-de-France. Vols. 1-3-?, 1898-99-?; [n.s.] nos. 1-9, 1921-27?; n.s. vols. 1-10, 1930-41. Bull. Agric. Martinique HI 62876

Bulletin agricole de la société d'agriculture et des quatre comices du Bas-Rhin. Neue Ackerbau-Zeitung der Ackerbau-Gesellschaft und der vier Comitien des Niederrheins. Strasbourg. Bull. Agric. Soc. Agric. Bas-Rhin. See B–P–H 237/3. HI 52588

Bulletin agricole de la société d'agriculture et des quatre comices du département du Bas-Rhin. Strasbourg. Bull. Agric. Soc. Agric. Dép. Bas-Rhin. See B–P–H 237/4. HI 52589

Bulletin agricole de la société des sciences, agriculture et arts du département du Bas-Rhin. Strasbourg. Bull. Agric. Soc. Sci. Dép. Bas-Rhin. See B–P–H 237/5. HI 52590

Bulletin des agriculteurs. Montreal. Bull. Agric. (Montreal). See B–P–H 236/14. HI 52578

Bulletin of the agricultural chemical society of Japan. [Nihon nogei kagakkai.] Tokyo. Vols. 2(5)-24, 1926-60. Bull. Agric. Chem. Soc. Japan. Preceded by: Journal of the agricultural chemical society of Japan. Superseded by: Agricultural and biological chemistry. HI 62877

Bulletin of the agricultural department, Assam. Shillong. Nos. 1-23, 1894-1912; n.s. vol. 1+, 1915+. Bull. Agric. Dept., Assam. HI 62878

Bulletin, agricultural department, Bahamas. Vols. 1-6(1)?, 1906-11. Bull. Agric. Dept., Bahamas. HI 62879

Bulletin, agricultural department, Bengal = Bulletin, department of land records and agriculture, Bengal. Calcutta.

Bulletin, agricultural department, Bengal. Agricultural series = Bulletin, department of land records and agriculture, Bengal. Agricultural series. Calcutta.

Bulletin of the agricultural department, British North Borneo. Jesselton. Ca.1950s. Bull. Agric. Dept., Brit. N. Borneo. HI 62880

Bulletin, agricultural directorate, Mesopotamia = Bulletin, department of agriculture, Iraq. Basrah, Baghdad.

Bulletin of the agricultural department, Nigeria. Lagos. Nos. 1-11, 1922-36. Bull. Agric. Dept., Nigeria. HI 62881

Bulletin of the agricultural department, Tasmania. Launceston, Hobart, Tas. Nos. 1-123, 1904-27; n.s. vol. 1+, 1928+. Bull. Agric. Dept., Tasmania. HI 62882

Bulletin of the agricultural department of the university college of Wales. Aberystwyth. Nos. 1-4, 1911-22. Bull. Agric. Dept. Univ. Coll. Wales. HI 62883

Bulletin, agricultural division, ministry of food and agriculture, Ghana. Accra. No. 1, 1959. Bull. Agric. Div. Minist. Food Agric., Ghana. Superseded by: Bulletin, ministry of agriculture, Ghana. HI 62884

Bulletin, agricultural experiment station of the agricultural college of Colorado = Agricultural experiment station of the agricultural college of Colorado. Bulletin. Fort Collins, CO. Agric. Exp. Sta. Agric. Coll. Colorado Bull. See B–P–H 57/5.

Bulletin, agricultural experiment station at Arkansas industrial university = Agricultural experiment station at Arkansas industrial university. Bulletin. Fayetteville, AR. Agric. Exp. Sta. Arkansas Industr. Univ. Bull. See B–P–H 57/6.

Bulletin, agricultural experiment station, Auburn university. Auburn, AL. No. 320+, 1960+. Bull. Agric. Exp. Sta., Auburn Univ. Preceded by: Alabama agricultural experiment station of the agricultural and mechanical college. Bulletin. HI 62885

Bulletin of the agricultural experiment station, Baton Rouge. Baton Rouge, LA. Bull. Agric. Exp. Sta. (Baton Rouge). See B–P–H 236/19. HI 52581

Bulletin, agricultural experiment station. Fayetteville, AR. = Agricultural experiment station at Arkansas industrial university. Bulletin. Fayetteville, AR. Agric. Exp. Sta. Arkansas Industr. Univ. Bull. See B–P–H 57/6.

Bulletin, agricultural experiment stations, Georgia. Athens, GA. N.s. nos. 1-183, 1954-66. Bull. Agric. Exp. Sta., Georgia. Preceded by: Bulletin, college experiment stations, university of Georgia and Bulletin, Georgia coastal plain experiment station. HI 62886

Bulletin of the agricultural experiment station, government general of Chosen. Suwon. Nos. 1-4, 1922-28. Bull. Agric. Exp. Sta., Chosen. HI 62887

Bulletin of the agricultural experiment station, Kungchulinjg. [Manchukuo kung chu ling noji shikenjo hokoku.] Kungchuling, Manchoukuo [China]. Bull. Agric. Exp. Sta. Kungchuling. See B–P–H 236/21. HI 52582

Bulletin, agricultural experiment station, Laramie. Laramie. No. 1+, 1891+. Bull. Agric. Exp. Sta., Laramie. From 1966+, incorporated in: Bulletin of the agricultural extension service, university of Wyoming. HI 62888

Bulletin, agricultural experiment station, Rehovoth. [Nos. 63-65 form a supplement to: Ktavim. See B–P–H 520/4.] Rehovot. Nos. 59-65, 1953-64. Bull. Agric. Exp. Sta., Rehovoth. Preceded by: Bulletin, institute of agriculture and natural history, Tel-Aviv. HI 62889

Bulletin, agricultural experiment station, university of

Tennessee. Knoxville, TN. Vol. 1+, 1888+. Bull. Agric. Exp. Sta., Univ. Tennessee. HI 62890

Bulletin, agricultural experiment station, West Virginia university. Morgantown, WV. Nos. 1-646, 1888-1976. Bull. Agric. Exp. Sta., West Virginia Univ. Superseded by: Bulletin, agricultural and forestry experiment station, West Virginia university. HI 62891

Bulletin of the agricultural extension service, university of Wyoming. Laramie. No. 1+, 1928+. Bull. Agric. Extens. Serv., Univ. Wyoming. HI 62892

Bulletin of the agricultural and forestry college, Suwon. Suwon. 1925-31. Bull. Agric. Coll. Suwon. HI 62893

Bulletin, agricultural and forestry experiment station, West Virginia university. Morgantown, WV. No. 647+, 1976+. Bull. Agric. Forest. Exp. Sta., West Virginia Univ. Preceded by: Bulletin, agricultural experiment station, West Virginia university. HI 62894

Bulletin of agricultural information, botanical department, Trinidad = Bulletin of miscellaneous information, department of agriculture, Trinidad. Saint Augustine, Port-of-Spain.

Bulletin. Agricultural research and experiment station. Tuskegee institute, Alabama. Tuskegee, AL. Bull. Agric. Res. Sta. Tuskegee Inst. See B–P–H 237/2. HI 52587

Bulletin of the agricultural research institute of Kanagawa prefecture. [Kanagawa-ken nogyo sogo kenkyujo kenkyu hokoku.] Kamakura. No. 108+, 1970+. Bull. Agric. Res. Inst. Kanagawa Pref. Preceded by: Bulletin of Kanagawa agricultural experiment station. HI 62895

Bulletin, agricultural research institute, Pusa. Calcutta. Nos. 1-207, 1906-31. Bull. Agric. Res. Inst., Pusa. Superseded by: Bulletin of the imperial institute of agricultural research, Pusa. 3-1954-2. HI 62896

Bulletin of the agricultural section of the department of natural resources, Newfoundland. St. John's. Nos. 1-15, 1934-44. Bull. Agric. Sect. Dep. Nat. Resources, Newfoundland. HI 62897

Bulletin, agricultural series, department of land records and agriculture, Bengal = Bulletin, department of land records and agriculture, Bengal. Agricultural series. Calcutta.

Bulletin of the agricultural society, Cairo. Technical section. Cairo. 1920+. Bull. Agric. Soc., Cairo, Techn. Sect. HI 62898

Bulletin, agriculture and fisheries department, Hong Kong. Hong Kong. No. 1+, 1974+. Bull. Agric. Fish. Dep., Hong Kong. HI 62899

Bulletin of agriculture and forestry. [Nung lin hui pao.] Bull. Agric. Forest. See B–P–H 236/24. HI 52583

Bulletin de l'agriculture et de l'horticulture. Brussels. Bull. Agric. Hort. See B–P–H 236/25. HI 52584

Bulletin of the agri-horticultural society of Madras. Madras. Bull. Agri-Hort. Soc. Madras. See B–P–H 237/6. HI 52591

Bulletin of the agri-horticultural society of western India. Poona. 1904-08. Bull. Agri-Hort. Soc. W. India. HI 62900

Bulletin of agri-horticulture. Singapore. Vol. 1+, 1965+. Bull. Agri-Hort. HI 62901

Bulletin agronomique. Section technique d'agriculture tropicale; Annales du centre de recherches agronomiques de Bambey au Sénégal. Nogent-sur-Marne. No. 1+, 1946+. Bull. Agron. (Nogent-sur-Marne). HI 62902

Bulletin agronomique. Paris. Nos. 1-23, 1946-72. Bull. Agron. (Paris). HI 62903

Bulletin of the Aichi gakugei university, natural science. [Aichi gakugei daigaku kenkyu hokoku.] Okazaki, Japan. Bull. Aichi Gakugei Univ. Nat. Sci. See B–P–H 237/9. HI 52592

Bulletin of Aichi horticultural experiment station. [Aichi-ken engei shikenjo kenkyu hokoku.] Aichi. Vols. 1-6, 1960-67. Bull. Aichi Hort. Exp. Sta. Superseded by: Research bulletin of the Aichi-ken agricultural research center. Series A, field crops and Series B, horticulture. HI 62904

Bulletin of the Aichi prefecture agricultural experiment station. [Aichi-ken nogyo shikenjo iho.] Aichi, Japan. Bull. Aichi Prefect. Agric. Exp. Sta. See 237/11. HI 52593

Bulletin of the Aichi prefecture forest experiment station. [Aichi-ken ringyo shikenjo hokoku.] Aichi. 1950+. Bull. Aichi Pref. Forest Exp. Sta. HI 62776

Bulletin of the Akita fruit-tree experiment station. [Akita-ken kaju shikenjo. Kenkyu hokoku.] Akita. No. 1+, 1969+. Bull. Akita Fruit-Tree Exp. Sta. HI 62905

Bulletin of the Akita prefectural college of agriculture. [Akita-kenritsu nogyo tanki daigaku kenkyu hokiku.] Ogata-Mura. No. 1+, 1975+. Bull. Akita Pref. Coll. Agric. HI 62906

Bulletin of the akiyoshi-dai museum of natural history. [Akiyoshi-dai kagaku hakubutsukan hokoku.] Yamaguchi. No. 1+, 1961+. Bull. Akiyoshi-dai Mus. Nat. Hist. HI 62907

Bulletin, Alabama agricultural experiment station of the agricultural and mechanical college = Alabama agricultural experiment station of the agricultural and mechanical college. Bulletin. Auburn, AL.

Bulletin, Alabama agricultural experiment station of the Alabama Polytechnic Institute = Alabama agricultural experiment station of the agricultural and mechanical college. Bulletin. Auburn, AL.

Bulletin of the Alabama museum of natural history. University City, AL. Vol. 1+, 1975+. Bull. Alabama Mus. Nat. Hist. HI 62908

Bulletin, Alaska agricultural experiment stations = Alaska agricultural experiment stations; bulletin. Washington, DC. Alaska Agric. Exp. Sta. Bull. See B–P–H 68/5.

Bulletin of the Alexandria horticultural society. Alexandria, Egypt [United Arab Republic]. Bull. Alexandria Hort. Soc. See B–P–H 237/18. HI 52594

Bulletin van het algemeen proefstation voor de landbouw. Buitenzorg, Dutch E. Indies [=Bogor, Indonesia]. Bull. Alg. Proefstat. Landb. See B–P–H 237/19. HI 52595

Bulletin van het algemeen proefstation te Salatiga. Semarang, Dutch E. Indies [Indonesia]. Bull. Alg. Proefstat. Salatiga. See B–P–H 237/20. HI 52596

Bulletin, alpine garden club of British Columbia. Vancouver, B.C. 1986?+. Bull. Alpine Garden Club British Columbia. Preceded by: Monthly bulletin, alpine garden club of British Columbia. HI 74515

Bulletin of the Alpine garden society of Great Britain. London. Bull. Alpine Gard. Soc. Gr. Brit. See B–P–H 238/1. HI 52597

Bulletin of the alumni association, Utsunomiya agricultural college. Utsunomiya, Japan. Bull. Alumni Assoc. Utsunomiya Agric. Coll. See B–P–H 238/2. HI 52598

Bulletin of the amateur orchid growers' society. 1952-57. Bull. Amateur Orchid Growers' Soc. Superseded by: Journal of the amateur orchid growers' society. HI 62909

Bulletin of the american academy of arts and sciences. Boston, MA. Bull. Amer. Acad. Arts See B–P–H 238/12. HI 52599

Bulletin, american antiquarian society = American antiquarian society. Bulletin. Worcester, MA. Amer. Antiq. Soc. Bull. See B–P–H 78/2.

Bulletin of the american association of botanical gardens and arboreta. Las Cruces, NM, Lima, PA. Vols. 8-19, 1974-85. Bull. Amer. Assoc. Bot. Gard. Arbor. Preceded by: Arboretum and botanical garden bulletin. Superseded by: Public garden. HI 62910

Bulletin, american association of museums. Washington, DC. 1970-75. Bull. Amer. Assoc. Mus. Superseded by: Aviso. HI 62911

Bulletin of the american association of petroleum geologists. Norman, Ok. Bull. AmK.r. Assoc. Petrol. Geol. See B–P–H 238/15. HI 52600

Bulletin of the american begonia society. Long Beach, CA. Bull. Amer. Begonia Soc. See B–P–H 238/16. HI 52601

Bulletin of the american conifer society. Philadelphia, PA. Vol.1+, 1983+. Bull. Amer. Conifer Soc. HI 62912

Bulletin of the american dahlia society. Newark, NJ. Bull. Amer. Dahlia Soc. See B–P–H 238/18. HI 52602

Bulletin of the american delphinium society. Morgantown, WV. Bull. Amer. Delphinium Soc. (Morgantown). See B–P–H 238/20. HI 52604

Bulletin of the american delphinium society. San Rafal, CA. Bull. Amer. Delphinium Soc. (San Rafael). See 238/21. HI 52605

Bulletin of the american forestry association. Washington, DC. Bull. Amer. Forest. Assoc. See B–P–H 238/23. HI 52606

Bulletin american fuchsia society. San Francisco, CA. Bull. Amer. Fuchsia Soc. See B–P–H 238/24. HI 52607

Bulletin of american garden history. New York. Vol. 1+, 1986+. Bull. Amer. Gard. Hist. HI 62913

Bulletin of the american horticultural society. Washington, DC. Bull. Amer. Hort. Soc. See B–P–H 238/25. HI 52608

Bulletin of the american hosta society. Baldwin, NY. Nos. 1-16, 1969-85. Bull. Amer. Hosta Soc. Superseded by: Hosta journal. HI 62915

Bulletin, american institute of biological sciences = A I B S bulletin. Washington, DC.

Bulletin of the american iris society. Wellesley Farms, MA. Bull. Amer. Iris Soc. See B–P–H 238/26. HI 52609

Bulletin of the american meteorological society. Easton, PA. Bull. Amer. Meteorol. Soc. See B–P–H 238/27. HI 52610

Bulletin of the american museum of natural history. New York, NY. Bull. Amer. Mus. Nat. Hist. See B–P–H 238/28. HI 52611

Bulletin, american ivy society = Ivy bulletin. Mt. Vernon, VA, Dayton, OH.

Bulletin, American orchid society = American orchid society bulletin. Cambridge, MA. Amer. Orchid Soc. Bull. See B–P–H 84/6.

Bulletin; american oriental poppy society. Bull. Amer. Orient. Poppy Soc. See B–P–H 239/1. HI 52612

Bulletin of the american penstemon society. Arlington, VA. Bull. Amer. Penstemon Soc. See B–P–H 239/2. HI 52613

Bulletin, American peony society = American peony society bulletin. Clinton, NY. Amer. Peony Soc. Bull. See B–P–H 84/9.

Bulletin, american pharmaceutical association. Columbus, OH. Vols. 1-6, 1906-11. Bull. Amer. Pharm. Assoc. Superseded by: Journal of the american pharmaceutical association. 3-2232-2. HI 62916

Bulletin, American poinsettia society = American poinsettia society bulletin. Mission, TX. Amer. Poinsettia Soc. Bull. See B–P–H 84/17.

Bulletin of the american pomological society. Madison, WI. Bull. Amer. Pomol. Soc. See B–P–H 239/4. HI 52615

Bulletin of the american rhododendron society. Portland, OR. Bull. Amer. Rhododendron Soc. See B–P–H 239/5. HI 52616

Bulletin of the american rock garden society. Plainfield, NJ. Bull. Amer. Rock Gard. Soc. See B–P–H 239/6. HI 52617

Bulletin of the american soil survey association. Ames, Iowa. Vols. 1-17, 1921-36 [vol. 7, no. 2 not published.] Bull. Amer. Soil Surv. Assoc. Preceded by: Bulletin of the american association of soil survey workers (not botanical). Superseded by: Proceedings of the soil science society of America. 1-332-3. HI 52614

Bulletin analytique. Paris. Bull. Analytique. See B–P–H 239/7. HI 52618

Bulletin für Angewandte Botanik. St. Petersburg = Trudy byuro po prikladnoi botanike. St. Petersburg.

Bulletin annuel de la fédération centre-est d'histoire naturelle et de mycologie. Lons Le Saunier. No. 1+, 1979+. Bull. Annuel Féd. Centre-Est Hist. Nat. Mycol. Preceded by: Bulletin, société des naturalistes d'Oyonnax. HI 62917

Bulletin annuel de la société jersiaise. 1875-1956. Bull. Annuel Soc. Jersiaise. Superseded by: Annual bulletin, société jersiaise. HI 62918

Bulletin of the Aomori apple experiment station. [Aomori-ken ringo shikenjo hokoku.] Aomori, Japan. Bull. Aomori Apple Exp. Sta. See B–P–H 239/9. HI 52619

Bulletin of the Aomori field crops and horticultural experiment station. [Aomori-ken hatasaku engei shikenjo kenkyu hokoku.] Gonohe. No. 1+, 1976+. Bull. Aomori Field Crops Hort. Exp. Sta. HI 63283

Bulletin of applied botany. St. Petersburg = Trudy byuro po prikladnoi botanike. St. Petersburg.

Bulletin of applied botany, of genetics, and plant-breeding. Moscow & Leningrad = Trudy po prikladnoi botanike, genetike i selektsii. Moscow & Leningrad.

Bulletin of applied botany, of genetics, and plant-breeding. Moscow & Leningrad = Trudy po prikladnoi botanike, genetike i selektsii. Seriya 14. Osvoenie pustyn'. Moscow & Leningrad.

Bulletin of applied botany, of genetics, and plant-breeding. Abstracts and bibliography. Moscow & Leningrad = Trudy po prikladnoi botanike, genetike i selektsii. Seriya 13. Referaty i bibliografiya. Moscow & Leningrad.

Bulletin of applied botany, of genetics, and plant-breeding. Dendrology and ornamental horticulture. Moscow & Leningrad = Trudy po prikladnoi botanike, genetike i selektsii. Seriya 10. Dendrologiya i dekorativnoe sadovodstvo. Moscow & Leningrad.

Bulletin of applied botany, of genetics, and plant-breeding. Forage crops. Moscow & Leningrad = Trudy po prikladnoi botanike, genetike i selektsii. Seriya 7. Kormovye kul'tury. Moscow & Leningrad.

Bulletin of applied botany, of genetics, and plant-breeding. Fruit and berry growing. Moscow & Leningrad = Trudy po prikladnoi botanike, genetike i selektsii. Seriya 8. Plodovye i yagodnye kul'tury. Moscow & Leningrad.

Bulletin of applied botany, of genetics, and plant-breeding. Fruit and small fruits. Moscow & Leningrad = Trudy po prikladnoi botanike, genetike i selektsii. Seriya 8. Plodovye i yagodnye kul'tury. Moscow & Leningrad.

Bulletin of applied botany, of genetics, and plant-breeding. Fruit growing and berry growing. Moscow & Leningrad = Trudy po prikladnoi botanike, genetike i selektsii. Seriya 8. Plodovye i yagodnye kul'tury. Moscow & Leningrad.

Bulletin of applied botany, of genetics, and plant-breeding. Genetics, plant-breeding, and cytology. Moscow & Leningrad = Trudy po prikladnoi botanike, genetike i selektsii. Seriya 2. Genetika selektsiya i tsitologiya rastenii. Moscow & Leningrad.

Bulletin of applied botany, of genetics, and plant-breeding. Grain crops. Moscow & Leningrad = Trudy po prikladnoi botanike, genetike i selektsii. Seriya 5. Zernovye kul'tury. Moscow & Leningrad.

Bulletin of applied botany, of genetics, and plant-breeding. New cultures and questions of introduction. Moscow & Leningrad = Trudy po prikladnoi botanike, genetike i selektsii. Seriya 11. Novye kul'tury i voprosy introduktsii. Moscow & Leningrad.

Bulletin of applied botany, of genetics, and plant-breeding. Northern subarctic agriculture. Moscow & Leningrad = Trudy po prikladnoi botanike, genetike i selektsii. Seriya 15. Severnoe (pripolyarnoe) zemledelie. Moscow & Leningrad.

Bulletin of applied botany, of genetics, and plant-breeding. Papers on the taxonomy, ecology and geography of plants. Moscow & Leningrad = Trudy po prikladnoi botanike, genetike i selektsii. Seriya 1. Sistematika, ekologiya i geografii rastenii. Moscow & Leningrad.

Bulletin of applied botany, of genetics, and plant-breeding. Physiology, biochemistry, and anatomy of plants. Moscow & Leningrad = Trudy po prikladnoi botanike, genetike i selektsii. Seriya 3. Fiziologiya, biokhimiya i anatomiya rastenii. Moscow & Leningrad.

Bulletin of applied botany, of genetics, and plant-breeding. Problems of plant industry in the extreme North. Moscow & Leningrad = Trudy po prikladnoi botanike, genetike i selektsii. Seriya 15. Problemy severnogo rastenievodstva. Moscow & Leningrad.

Bulletin of applied botany, of genetics, and plant-breeding. Review [later: Abstracts] and bibliography. Moscow & Leningrad = Trudy po prikladnoi botanike, genetike i selektsii. Seriya 13. Referaty i bibliografiya. Moscow & Leningrad.

Bulletin of applied botany, of genetics, and plant-breeding. Seed science and seed testing. Moscow & Leningrad = Trudy po prikladnoi botanike, genetike i selektsii. Seriya 4. Semenovedenie i semennoi kontrol′. Moscow & Leningrad.

Bulletin of applied botany, of genetics, and plant-breeding. Systematics, geography, ecology of plants, and plant industry in general. Moscow & Leningrad = Trudy po prikladnoi botanike, genetike i selektsii. Seriya 1. Sistematika, ekologiya i geografii rastenii i obshchie voprosy rastenievodstva. Moscow & Leningrad.

Bulletin of applied botany, of genetics, and plant-breeding. Technical plants. Moscow & Leningrad = Trudy po prikladnoi botanike, genetike i selektsii. Seriya 9. Tekhnicheskie kul′tury. Moscow & Leningrad.

Bulletin of applied botany, of genetics, and plant-breeding. Vegetables. Moscow & Leningrad = Trudy po prikladnoi botanike, genetike i selektsii. Seriya 6. Ovoshchnye kul′tury. Moscow & Leningrad.

Bulletin of applied botany, of genetics, and plant-breeding. Wheat. Moscow & Leningrad = Trudy po prikladnoi botanike, genetike i selektsii. Seriya 5a. Pshenica. Moscow & Leningrad.

Bulletin of applied botany and plant-breeding. Petrograd = Trudy po prikladnoi botanike i selektsii. Petrograd.

Bulletin of applied botany and selection. Petrograd = Trudy po prikladnoi botanike i selektsii. Petrograd.

Bulletin of aquatic biology. Amsterdam. Vols. 1(1)-3(35), 1957-67. Bull. Aquatic Biol. HI 62919

Bulletin, arboretum Waasland. Nieuwkerken-Waas. Vol. 1+, 1981+. Bull. Arbor. Waasland. HI 62920

Bulletin d'arboriculture, de culture potagère et de floriculture. Ghent. Sér. 8, vols. 2-4, 1903-05. Bull. Arboric. Culture Potag. Floric. Preceded by: Bulletin de l'arboriculture, de floriculture et de culture potagère. HI 62921

Bulletin d'arboriculture, de floriculture et de culture potagère. Ghent. Vols. 1-5, 1872-76; sér. 3, vols. 1-5, 1877-81; sér. 4, vols. 1-5, 1882-86; sér. 5, vols. 1-5, 1887-91; sér. 6, vol. 1-5, 1892-96; sér. 7, vols. 1-5, 1897-1901; sér. 8, vol. 1, 1902. Bull. Arboric. Floric. Culture Potag. Preceded by: Bulletin du cercle d'arboriculture de Belgique. Superseded by: Bulletin d'arboriculture de culture-potagère et de floriculture. HI 62922

Bulletin of the Arctic institute of North America. Montreal. Bull. Arctic Inst. N. Amer. See B–P–H 240/10. HI 52620

Bulletin, Arizona agricultural experiment station = University of Arizona. Arizona agricultural experiment station. Bulletin. Tucson, AZ. Univ. Arizona Agric. Exp. Sta. Bull. See B–P–H 933/14.

Bulletin, Arizona agricultural experiment station, university of Arizona = Arizona agricultural experiment station, university of Arizona, bulletin. Tucson, AZ. Arizona Agric. Exp. Sta. Univ. Arizona Bull. See B–P–H 152/44.

Bulletin, Arizona native plant society. Tucson, AZ. Vol. 5(1), 1981. Bull. Arizona Native Pl. Soc. Preceded by: Newsletter of the Arizona native plant society. Superseded by: Plant press. Tucson, AZ. HI 70075

Bulletin, Arkansas academy of science. Fayetteville, AR. ?-1976+ Bull. Arkansas Acad. Sci. HI 62923

Bulletin, Arkansas agricultural experiment station = University of Arkansas. The college of agriculture. Arkansas agricultural experiment station. Bulletin. Fayetteville, AR. Univ. Arkansas Coll. Agric. Arkansas Agric. Exp. Sta. Bull. See B–P–H 934/1.

Bulletin, Arkansas industrial university. Agricultural experiment station = Arkansas industrial university. Agricultural experiment station. Bulletin. Fayetteville, AR. Arkansas Industr. Univ. Agric. Exp. Sta. Bull. See B–P–H 153/17.

Bulletin of arts & science division, Ryukyu university. [Ryukyu daigaku bunrui gakubu kiyo.] Naha, Okinawa. Bull. Arts Sci. Div. Ryukyu Univ. See B–P–H 240/14. HI 52621

Bulletin, asiatic society of Bengal. Calcutta. No. 1+, 1947+. Bull. Asiat. Soc. Bengal. HI 62924

Bulletin de l'association des amis du jardin botanique de Fribourg. Fribourg. Vol. 1+, 1967+. Bull. Assoc. Amis Jard. Bot. Fribourg. HI 62925

Bulletin de l'association de l'arboretum du vallon de l'Aubonne. Lausanne. Vol. 1+, 1971+. Bull. Assoc. Arbor. Vallon Aubonne. HI 62926

Bulletin de l'association pour l'avancement des sciences naturelles au Sénégal. Dakar. Nos. ?-28-63, 19??-69-78. Bull. Assoc. Avancem. Sci. Nat. Sénégal. HI 62927

Bulletin, l'association botanique du Canada = Bulletin, canadian botanical association. Vancouver, B.C., Guelph, Ont.

Bulletin, association des diplômés de microbiologie de la faculté de pharmacie de Nancy. Nancy. Nos. 1-136, 1931-74. Bull. Assoc. Diplômés Microbiol. Fac. Pharm. Univ. Nancy. Superseded by: Microbia. 4-2818-2. HI 62928

Bulletin de l'association florimontane d'Annecy et revue savoisienne. Annecy. Vols. 1-3, 1855-59. Bull. Assoc. Florimont. Annecy. HI 62929

Bulletin de l'association française pour l'étude du quarternaire. Paris. Bull. Assoc. Franç. Étude Quartern. See B–P–H 240/19. HI 52622

Bulletin de l'association française de botanique. Le Mans. Bull. Assoc. Franç. Bot. See B–P–H 240/21. HI 52623

Bulletin, association française pour l'étude du sol. Paris, Versailles. Nos. 1-93, 1935-57; [n.s.] 1958+. Bull. Assoc. Franç. Étude Sol. HI 62930

Bulletin de l'association française pomologique de l'Ouest = Bulletin de l'association pomologique de l'Ouest. Le Mans. Bull. Assoc. Pomol. Ouest. See B–P–H 240/28.

Bulletin de l'association du musée Slovénie = Glasnik muzejskega društva za Slovenijo. Ljubljana, Yugoslavia. Glasn. Muz. Društva Slovenijo. See B–P–H 405/5.

Bulletin de l'association des musées de Nyon. Nyon. No. 1+, 1983+. Bull. Assoc. Mus. Nyon. HI 62931

Bulletin of the association of natural science, Senshu university. Kanagawa. No. ?-6+, 19??-73+. Bull. Assoc. Nat. Sci. Senshu Univ. HI 62932

Bulletin de l'association des naturalistes du Mali. Bamaka. No. ?-3+, 19??-72+. Bull. Assoc. Naturalistes Mali. HI 62933

Bulletin de l'association des naturalistes orléanais. Orléans. 1958-65. Bull. Assoc. Naturalistes Orléanais. Superseded by: Bulletin de l'association des naturalistes orléanais et de la Loire Moyenne. HI 62934

Bulletin de l'association des naturalistes orléanais et de la Loire Moyenne. Orléans. 1966-71. Bull. Assoc. Naturalistes Orléanais Loire Moyenne. Preceded by: Bulletin de l'association des naturalistes orléanais. Superseded by: Naturalistes orleannais. HI 62935

Bulletin de l'association des naturalistes de la Vallée du Loing. Moret-sur-Loing. Vols. 1-?, 1913-39. Bull. Assoc. Naturalistes Vallée du Loing. Superseded by?: Bulletin mensuel de l'association des naturalistes de la Vallée du Loing (et de la Fôret de Fontainebleau). 1-519-3. HI 62936

Bulletin de l'association des parcs botaniques de France. Paris. No. 1+, 1979+. Bull. Assoc. Parcs Bot. France. HI 62937

Bulletin de l'association philomatique d'Alsace et de Lorraine. Strasbourg. Vol. 6+, 1920+. Bull. Assoc. Philom. Alsace Lorraine. Preceded by: Mittheilungen der philomatischen Gesellschaft in Elsass-Lothringen. 1-536-3. HI 62938

Bulletin de l'association des planteurs de caoutchouc et autres produits coloniaux. Antwerp. Bull. Assoc. Pl. Caoutchouc. See B–P–H 240/27. HI 52625

Bulletin de l'association pomologique de l'Ouest. Le Mans. Bull. Assoc. Pomol. Ouest. See B–P–H 240/28. HI 52626

Bulletin de l'association pour la protection des plantes. Geneva. Bull. Assoc. Protect. Pl. See B–P–H 240/29. HI 52627

Bulletin de l'association pyrénéenne pour l'échange des plantes. Poitiers, France. Bull. Assoc. Pyrén. Échange Pl. See B–P–H 240/30. HI 52628

Bulletin, association régionale des amis de l'université et de l'enseignement supérieur pour la promotion de l'étude et la recherche scientifiques, en liaison avec le monde de l'economie. Reims. Vol. 14(55)+, 197?+. Bull. Assoc. Régionale Amis Univ., Reims. Preceded by: Bulletin, association régionale pour l'étude et la recherche scientifiques. HI 62939

Bulletin, association régionale pour l'étude et la recherche scientifiques. Reims. Vols. 1(1)-13(54), 1960-72. Bull. Assoc. Régionale Étude Rech. Sci., Reims. Superseded by: Bulletin, association régionale des amis de l'université et de l'enseignement supérieur pour la promotion de l'étude et la recherche scientifiques, en liaison avec le monde de l'economie. HI 62940

Bulletin de l'association de recherches scientifiques à la faculté des sciences de la Première université de Moscou = Izvestiya Assotsiatsii nauchno-issledovatel'skikh institutov pri Fiziko-Matematicheskom fakul'tete Pervogo Moskovskogo gosudarstvennogo universiteta. Ser. A. Otdel biologicheskikh i geologicheskikh nauk. Moscow.

Bulletin de l'association de recherches scientifiques. Faculté physico-mathématique de la première université d'état de Moscou = Izvestiya Assotsiatsii nauchno-issledovatel'skikh institutov pri Fiziko-Matematicheskom fakul'tete Pervogo Moskovskogo gosudarstvennogo universiteta. Moscow.

Bulletin of association of scientific institutes on North Caucasus = Trudy Severo-Kavkazskoi assotsiatsii nauchno-issledovatel'skikh institutov. Krasnodar.

Bulletin de l'association scientifique algérienne. Algiers. Bull. Assoc. Sci. Algér. See B–P–H 241/3. HI 52629

Bulletin, Auckland botanical society. Auckland. No. ?-15+, 19??-85+. Bull. Auckland Bot. Soc. HI 62940

Bulletin of the Auckland institute and museum. Auckland. Bull. Auckland Inst. Mus. See B–P–H 241/6. HI 52630

Bulletin, Australian lilium society = Australian lilium society bulletin. Melbourne. Austral. Lilium Soc. Bull. See B–P–H 164/9.

Bulletin, australian littoral society. Toowong, Qld. 1978+. Bull. Austral. Littoral Soc. HI 62941

Bulletin, australian society for limnology. Clayton, Vic. Nos. 1-6, 1969-75. Bull. Austral. Soc. Limnol. HI 62942

Bulletin, Badlands natural history society. Interior, SD. 1968+. Bull. Badlands Nat. Hist. Soc. HI 62943

Bulletin, balai penelitian perkebunan Medan. Medan. Vol. 1+, 1970+. Bull. Balai Penelitian Perkebunan Medan. HI 62944

Bulletin, Ball state teachers' college. Muncie, IN. Vol. ?-15+, 19??-39+. Bull. Ball State Teachers' Coll. HI 62945

Bulletin of the Bartlett tree research laboratories. Satamford, CT. Bull. Bartlett Tree Res. Lab. See B–P–H 241/8. HI 52631

Bulletin of basic science research. Cincinnati, OH. Bull. Basic Sci. Res. See B–P–H 241/9. HI 52632

Bulletin, Basildon natural history society. Basildon. No. 1+, 1968+. Bull. Basildon Nat. Hist. Soc. HI 62946

Bulletin of the Basrah natural history museum. Basrah. 1974+. Bull. Basrah Nat. Hist. Mus. HI 62947

Bulletin, Berkshire, Buckinghamshire & Oxfordshire naturalists' trust. [Chesham.] No. ?-25+, 19??-68+. Bull. Berkshire, Buckinghamshire & Oxfordshire Naturalists' Trust. HI 62948

Bulletin, Bernice P. Bishop museum = Bernice P. Bishop museum bulletin. Honolulu, Hawaii.

Bulletin of the Berry botanical garden. Portland, OR. Vol. 10+, 1987+. Bull. Berry Bot. Gard. Preceded by: Berry botanic garden [series]. HI 62949

Bulletin of the bibliographical society of America. Chicago, IL. Vols. 1-4, 1907-12. Bull. Bibliogr. Soc. Amer. Incorporated in: Papers, bibliographical society of America. 1-667-1. HI 62950

Bulletin of the bibliographical society of Australia and New Zealand. Clayton, Adelaide. Vol. 1+, 1970+. Bull. Bibliogr. Soc. Australia New Zealand. HI 62951

Bulletin bibliographique, centre de documentation, institut national du bois. Paris. Vol. 1+, 1950+. Bull. Bibliogr. Centre Doc. Inst. Natl. Bois. HI 62952

Bulletin bibliographique des publications périodiques ... académie des sciences. Paris. Bull. Bibliogr. Publ. Périod. Acad. Sci. See B–P–H 241/13. HI 52633

Bulletin bibliographique de la société d'ethnozoologie et d'ethnobotanique. Paris. Vol. ?-13/14+, 19??-78+. Bull. Bibliogr. Soc. Ethnozool. Ethnobot. Preceded by?: Bulletin d'information de la société d'ethnozoologie et d'ethnobotanique. Supplement. HI 62953

Bulletin bimestriel, association des naturalistes de la Vallée du Loing et du massif de Fontainebleau. Fontainebleau. 1960+. Bull. Bimestr. Assoc. Naturalistes Vallée du Loing & Fontainebleau. Preceded by: Bulletin mensuel de l'association des naturalistes de la Vallée du Loing (et de la fôret de Fontainebleau). HI 62954

Bulletin bimestriel de la société d'histoire naturelle du Doubs = Bulletin de la société d'histoire naturelle du Doubs. Besançon.

Bulletin bimestriel de la société d'histoire naturelle de Haute-Savoie. Annecy. Sér. 4, vol.5, 19?? Bull. Bimestr. Soc. Hist. Nat. Haute-Savoie. Preceded by: Bulletin mensuel, société d'histoire naturelle de Haute-Savoie. Superseded by: Bulletin trimestriel de la société d'histoire naturelle de Haute-Savoie. HI 62955

Bulletin of the Bingham oceanographic collection. New Haven, CT. Vols. 1-19, 1927-67. Bull. Bingham Oceanogr. Collect. Incorporated in: Peabody museum natural history bulletin. 5-4567-2. HI 62956

Bulletin of the biogeographical society of Japan. Tokyo. Bull. Biogeogr. Soc. Japan. See B–P–H 241/19. HI 52634

Bulletin of the biological association; university of Amoy. [Hsia-ta shêngqwu-hsüeh-hui ch'i k'an.] Amoy, China. Bull. Biol. Assoc. Univ. Amoy. See B–P–H 241/21. HI 52635

Bulletin of the biological institute, department of biology, college of liberal arts and sciences; Seoul (national) university. [Murhak yon'gu saengmurkak yonguhoe. Mulligwataehak Sôul Taehakkyo.] Seoul, Korea. Bull. Biol. Inst. Seoul Natl. Univ. See 241/27. HI 52637

Bulletin of the biological research centre; the official annual publication of the biological research centre, university of Baghdad. Baghdad. 1965+. Bull. Biol. Res. Centre, Baghdad. HI 62957

Bulletin of the biological society of Hiroshima university.

[Hiroshima daigaku seibutsu gakkai shi.] Hiroshima. Vol. 1+, 1949+. Bull. Biol. Soc. Hiroshima Univ. HI 62958

Bulletin of the biological society of Washington. Washington, DC. Vol. 1+, 1918+ [publication suspended between no. 1, 1918 & no. 2, 1972]. Bull. Biol. Soc. Wash. 1-707-3. HI 62959

Bulletin biologique. Yur'ev = Spravočnyj Listok Biologa. Yur'ev [=Tartu, Estonian S S R]. Sprav. Listok Biol. See B–P–H 853/8.

Bulletin biologique de l'association scientifique de Côte d'Ivoire. Abidjan. No. 1, 1967. Bull. Biol. Assoc. Sci. Côte d'Ivoire. HI 62960

Bulletin biologique de la France et de la Belgique. Paris. Bull. Biol. France Belgique. See B–P–H 241/25. HI 52636

Bulletin of the Boston mycological club. Boston, MA. Bull. Boston Mycol. Club. See B–P–H 242/9 HI 52639

Bulletin of the Boston society of natural history. Boston, MA. Bull. Boston Soc. Nat. Hist. See B–P–H 242/10. HI 52640

Bulletin, botanical club of Wisconsin. Stevens Point, WI. Vol. 11+, 1979+. Bull. Bot. Club Wisconsin. Preceded by: Newsletter, botanical club of Wisconsin. HI 62961

Bulletin of the botanical department. Kingston, Jamaica. Nos. 1-50, 1887-94; n.s. vols. 1-9, 1894-1902. Bull. Bot. Dept., Jamaica. Superseded by: Bulletin of the department of agriculture, Jamaica. HI 62962

Bulletin from the botanical department of the state agricultural college, Ames, Iowa. Cedar Rapids, IA. Nos. 1-3, 1884-88. Bull. Bot. Dept. State Agric. Coll., Ames. HI 62963

Bulletin, botanical division, U S department of agriculture = Department of agriculture. Botanical division. Bulletin. Washington, DC. Dept. Agric. Bot. Div. Bull. See B–P–H 342/17.

Bulletin of botanical laboratory of north-eastern forestry institute. [Chih wu yen chiu shih hui kan.] Harbin. Nos. 1-9, 1959-80. Bull. Bot. Lab. N. E. Forest. Inst., Harbin. Superseded by: Bulletin of botanical research. Harbin. HI 62964

Bulletin of botanical research. [Zhiwi yanjiu.] Harbin. Vol. 1+, 1981+. Bull. Bot. Res., Harbin. Preceded by: Bulletin of botanical laboratory of north-eastern forestry institute. HI 62965

Bulletin of the botanical society of Bengal. Calcutta. Bull. Bot. Soc. Bengal. See B–P–H 242/19. HI 52641

Bulletin, botanical society, college of science, Nagpur. Nagpur. Vols. 1-4(2), 1960-63. Bull. Bot. Soc. Coll. Sci., Nagpur. HI 62966

Bulletin of the botanical society of Nagano. [Nagano-ken shokubutsu kenkyukai-shi.] Nagano, Honshu. No. 1+, 1968+. Bull. Bot. Soc. Nagano. HI 62967

Bulletin of the botanical society of the university of Saugar. Saugar. Vol. 1+, 1948+. Bull. Bot. Soc. Univ. Saugar. HI 62968

Bulletin of the botanical survey of India. Calcutta. Bull. Bot. Surv. India. See B–P–H 242/20. HI 52642

Bulletin botanique ou collection de notices originales et d'extraits des ouvrages botaniques. Geneva. Nos. 1-12, 1830-32. Bull. Bot., Geneva. HI 62969

Bulletin van de botanische tuinen Wageningen. Wageningen. No. 1+, 1978+. Bull. Bot. Tuinen Wageningen. HI 62970

Bulletin des botanistes liègeois. Mémoire = Lejeunia; revue de botanique. Mémoire. Liège.

Bulletin, botany division, New Zealand department of scientific and industrial research. [Forms part of, and also numbered as: Bulletin, New Zealand department of scientific and industrial research.] Wellington, N.Z. Nos. 1-15, 1940-52. Bull. Bot. Div., New Zealand Dep. Sci. Industr. Res. HI 62971

Bulletin, boxwood society of the midwest. St. Louis, MO. Vol. 1+, 1976+. Bull. Boxwood Soc. Midw. HI 62972

Bulletin of the Bradford natural history and microscopical society. Bradford. Nos. ?-4, 19??-51. Bull. Bradford Nat. Hist. Microsc. Soc. Superseded by: Bulletin of the Bradford naturalists' society. HI 62973

Bulletin of the Bradford naturalists' society. Bradford. Nos. 5-57, 1951-55. Bull. Bradford Naturalist's Soc. Preceded by: Bulletin of the Bradford natural history and microscopical society. HI 62974

Bulletin of brewing science. Tokyo. Bull. Brew. Sci. See B–P–H 242/23. HI 52643

Bulletin, british antarctic survey. London. Vol. 1+, 1963+. Bull. Brit. Antarc. Surv. HI 62975

Bulletin, british bryological society. Cardiff. No. 1+, 1963+. Bull. Brit. Bryol. Soc. HI 62976

Bulletin of the british carnation society. London. Bull. Brit. Carnation Soc. See B–P–H 242/24. HI 52644

Bulletin of the British Columbia fruit-growers' association. Vancouver. Bull. Brit. Columbia Fruit-Growers' Assoc. See B–P–H 242/25. HI 52645

Bulletin of the british ecological society. Reading. No. 1+, 1970+. Bull. Brit. Ecol. Soc. HI 62977

Bulletin, british hosta and hemerocallis society. Boldre, Eltham. Vol. 1+, 1983+. Bull. Brit. Hosta Hemerocallis Soc. HI 62978

Bulletin, british lichen society. London, etc. No. 1+, 1958+. Bull. Brit. Lichen Soc. HI 62979

Bulletin of the british museum (natural history). Botany. Bull. Brit. Mus. (Nat. Hist.), Bot. See B–P–H 242/28. HI 52646

Bulletin of the british museum (natural history). Geology. London. Bull. Brit. Mus. (Nat. Hist.), Geol. See B–P–H 242/30. HI 52647

Bulletin of the british museum (natural history). Historical. London. Vol. 1+, 1953+. Bull. Brit. Mus. (Nat. Hist.), Hist. HI 62980

Bulletin of the british mycological society. Bury St. Edmunds. Vols. 1-20, 1967-86? Bull. Brit. Mycol. Soc. Preceded by: News bulletin, british mycological society. Superseded by: Mycologist. HI 62981

Bulletin, british naturalists' association. Alton. 1975+. Bull. Brit. Naturalists' Assoc. HI 62982

Bulletin, british pelargonium and geranium society. Vols. 14-16(1), 1965-67. Bull. Brit. Pelargonium Geranium Soc. Preceded by: Bulletin, geranium society. Superseded by: Pelargonium news. HI 62983

Bulletin of the british pteridological society. [Loughton], etc. Vol. 1+, 1973+. Bull. Brit. Pteridol. Soc. Preceded by: British pteridological society newsletter. HI 62984

Bulletin of the bromeliad society. Los Angeles, CA, Orlando, FL. Vol. 1-20, 1951-70. Bull. Bromeliad Soc. Superseded by: Journal of the bromeliad society. HI 62985

Bulletin of the Brookline society of natural history. Brookline, MA. Bull. Brookline Soc. Nat. Hist. See B–P–H 242/32. HI 52648

Bulletin of the Brooklyn aquarium society. Brooklyn, NY. Bull. Brooklyn Aquar. Soc. See B–P–H 242/33. HI 52649

Bulletin of the Brookville society of natural history. Richmond, IN. Bull. Brookville Soc. Nat. Hist. See B–P–H 243/1. HI 52650

Bulletin of the Buffalo naturalists' field club. Buffalo, NY. Bull. Buffalo Naturalists' Field Club. See B–P–H 243/2. HI 52651

Bulletin of the Buffalo society of natural sciences. Buffalos, NY. Bull. Buffalo Soc. Nat. Sci. See B–P–H 243/3. HI 52652

Bulletin of the bureau of agricultural intelligence and of plant diseases. Rome. Vols. 1-3, 1910-12. Bull. Bur. Agric. Intelligence Pl. Dis. Superseded by: Monthly bulletin of agricultural intelligence and plant diseases. HI 62986

Bulletin of the bureau of agricultural microbiology = Trudy otdela sel′sko-khozyaistvennoi mikrobiologii, gosudarstvennyi institut opytnoi agronomii. Leningrad.

Bulletin des Bureau für Angewandte Botanik. St. Petersburg = Trudy byuro po prikladnoi botanike. St. Petersburg.

Bulletin of the bureau of forestry, Philippine Islands. Manila. Bull. Bur. Forest. Philipp. Islands. See B–P–H 243/ 8. HI 52653

Bulletin, bureau of plant industry, U S department of agriculture = Bulletin, U S department of agriculture, Bureau of plant industry. Washington, DC. Bull. U.S.D.A. Bur. Pl. Industr. See B–P–H 286/30.

Bulletin, bureau of plant industry, United States department of agriculture = U S department of agriculture. Bureau of plant industry. Bulletin. Washington, DC. U.S.D.A. Bur. Pl. Industr. Bull. See B–P–H 940/24.

Bulletin du bureau des renseignements agricoles et des maladies des plantes. [French language edition of: Bulletin of the bureau of agricultural intelligence and of plant diseases.] Rome. Vols. 1-3, 1910-12. Bull. Bur. Renseign. Agric. Malad. Pl. Superseded by: Bulletin mensuel des renseignements agricoles et des maladies des plantes. HI 62987

Bulletin of the Bussey institution. Jamaica Plain, MA. Bull. Bussey Inst. See B–P–H 243/13. HI 52654

Bulletin of cactus research = Blätter für Kakteenforschung. Volksdorf [=Hamburg, in part]. Blätt. Kakteenf. See B–P–H 199/14.

Bulletin of the cactus and succulent society of Australia. Melbourne. Vols. 1-6, 1949-52. Bull. Cactus Succ. Soc. Australia. Preceded and superseded by: Spine. HI 62988

Bulletin of the Caithness field club. Thurso. Vol. 1+, 1973+. Bull. Caithness Field Club. HI 62989

Bulletin of the California academy of sciences. San Francisco, CA. Bull. Calif. Acad. Sci. See B–P–H 243/16. HI 52655

Bulletin of the California dahlia society. San Francisco, CA. Bull. Calif. Dahlia Soc. See B–P–H 243/17. HI 52656

Bulletin of the California department of agriculture. Sacramento, CA. Vols. 24(3)-56, 1935-67. Bull. Calif. Dep. Agric. Preceded by: Monthly bulletin, California department of agriculture. HI 62990

Bulletin, California native plant society. Berkeley, CA., Sacramento, CA Vol. ?-11(6)+, 19??-81+. Bull. Calif. Native Pl. Soc. HI 62991

Bulletin, canadian botanical association. Vancouver, B.C., Guelph, Ont. Vol. 1+, 1968+. Bull. Canad. Bot. Assoc. HI 62992

Bulletin, canadian gladiolus society. [Canada]. Vols. 1-

7, 1923-30 [none published 1927]. Bull. Canad. Gladiolus Soc. Superseded by: Canadian gladiolus annual. 2-908-1. HI 62993

Bulletin of the canadian peat society. Ottawa. Bull. Canad. Peat Soc. See B–P–H 243/20. HI 52657

Bulletin of canadian petroleum geology. Calgary, Alberta. Bull. Canad. Petrol. Geol. See B–P–H 243/21. HI 52658

Bulletin, canadian society for cell biology. Toronto. Vol. 1+, 1974+. Bull. Canad. Soc. Cell Biol. HI 62994

Bulletin canadien des orchidées = Canadian orchid journal. St. Johns.

Bulletin, Canebrake agricultural experiment station = Canebrake agricultural experiment station bulletin. Montgomery, AL. Canebrake Agric. Exp. Sta. Bull. See B–P–H 298/9.

Bulletin, Canebrake experiment station = Canebrake agricultural experiment station bulletin. Montgomery, AL. Canebrake Agric. Exp. Sta. Bull. See B–P–H 298/9.

Bulletin. Canton Christian college. Canton. Bull. Canton Christian Coll. See B–P–H 243/23. HI 52659

Bulletin of the Carnegie museum of natural history. Pittsburgh. No. 1+, 1976+. Bull. Carnegie Mus. Nat. Hist. HI 62995

Bulletin of the central cotton improvement institute. [Shaling-wei.] Nanking. Bull. Centr. Cotton Improv. Inst. See B–P–H 243/26. HI 52660

Bulletin of the central national museum of Manchoukuo. [Manshu teikoku kokuritsu chuo hakubutsu kan runso.] Hsinginjg, Manchoukuo [=Changchun, China]. Bull. Centr. Natl. Mus. Manchoukuo. See B–P–H 244/2. HI 52662

Bulletin of the central research institute, university of Kerala. Series C, natural science. Trivandrum. 1957-61. Bull. Centr. Res. Inst. Univ. Kerala, C. Preceded by: Bulletin of the central research institute, university of Travancore. Series C, natural science. Superseded by: Bulletin of the department of marine biology and oceanography, university of Kerala. HI 62996

Bulletin of the central research institute, university of Travancore. Series C, natural science. Trivandrum. 1950-57. Bull. Centr. Res. Inst. Univ. Travancore, C. Superseded by: Bulletin of the central research institute, university of Kerala. Series C, natural science. HI 62997

Bulletin of the central scientific research institute of plant protection = Izvestiya na tsentralniya nauchno-izsledovatelski institut za zashtita na rasteniyata. Sofia.

Bulletin of the central sugar cane breeding station, British West Indies. Barbados. No. 1+, 1933+. Bull. Centr. Sugar Cane Breed. Stat., Brit. W. Indies. HI 62998

Bulletin de centre d'études et de recherches scientifiques. Biarritz, France. Bull. Centr. Études Rech. Sci. See B–P–H 243/29. HI 52661

Bulletin. Centre de recherches agronomiques de Bingerville. Abidjan, Ivory Coast [=Republic of the Ivory Coast]. Bull. Centr. Rech. Agron. Bingerville. See B–P–H 244/3. HI 52663

Bulletin du cercle d'arboriculture de Belgique. Ghent. 1871. Bull. Cercle Arboric. Belgique. Preceded by: Bulletin du cercle professoral pour le progrès de l'arboriculture en Belgique. Superseded by: Bulletin d'arboriculture de floriculture et de culture potagère. HI 62999

Bulletin du cercle floral d'Anvers. Antwerp. Bull. Cercle Fl. Anvers. See B–P–H 244/5. HI 52664

Bulletin. Cercle général d'horticulture. Paris. Bull. Cercle Gén. Hort. See B–P–H 244/6. HI 52665

Bulletin. Cercle horticole et viticole de la Brie. Lagny, France. Bull. Cercle Hort. Brie. See B–P–H 244/7. HI 52666

Bulletins. Cercle pratique d'horticulture et de l'arrondissement du Havre. Le Havre. Bull. Cercle Prat. Hort. Arrondissement Havre. See 244/8. HI 52667

Bulletin du cercle professoral pour le progrès de l'arboriculture en Belgique. Ghent. Ann. 1865-70. Bull. Cercle Profess. Progr. Arboric. Belgique. Superseded by: Bulletin du cercle d'arboriculture de Belgique. HI 63000

Bulletin de cercle vaudois de botanique. Lausanne. Bull. Cercle Vaud. Bots. See B–P–H 244/9. HI 52669

Bulletin CETIOM. [Centre technique interprofessionnel des oléagineux metropolitains.] Paris. Bull. CETIOM. See B–P–H 244/10. HI 52670

Bulletin, Ceylon rubber research scheme. Colombo. Nos. 1-?, 1913-20. Bull. Ceylon Rubber Res. Scheme. Superseded by: Bulletin of the rubber research scheme, Ceylon. HI 63001

Bulletin de la chaire d'histoire naturelle de la faculté mixte de médecine générale et coloniale, et de pharmacie de l'université d'Aix-Marseille. Bulletin d'échanges du jardin botanique. Marseilles. 1941. Bull. Chaire Hist. Nat., Univ. Aix-Marseille. HI 63002

Bulletin of the chamber of horticulture. London. Bull. Chamber Hort. See B–P–H 244/11. HI 52671

Bulletin of the Charleston museum. Charleston, SC. Vols. 117, 1907-22. Bull. Charleston Mus. Preceded by: Bulletin of the college of Charleston museum. HI 52673

Bulletin of Chekiang provincial fisheries experiment station. [Chekiang, China]. Bull. Chekiang Prov.

Fish. Exp. Sta. See B–P–H 244/13. HI 52674

Bulletin of the Chiba college of horticulture. [Chiba koto engeigakko gakujutsu hokoku.] Matsudo, Japan. Bull. Chiba Coll. Hort. See B–P–H 244/14. HI 52675

Bulletin of the Chiba prefecture agricultural experiment station. [Chiba-ken nogyo shikenjo kenkyu hokoku.] Chiba, Japan. Bull. Chiba Prefect. Agric. Exp. Sta. See B–P–H 244/15. HI 52676

Bulletin of the Chicago academy of sciences. Chicago, IL. Bull. Chicago Acad. Sci. See B–P–H 244/17. HI 52677

Bulletin of the Chicago natural history museum. Chicago. Vols. 15-37(2), 1944-66. Bull. Chicago Nat. Hist. Mus. Preceded by: Field museum news. Superseded by: Bulletin, Field museum of natural history. 2-998-1. HI 63003

Bulletin of the Chichibu museum of natural history. [Chichibu shizenkagaku hakubutsukan kenkyu hokoku.] Chichibu. Nos. 1-18, 1950-78. Bull. Chichibu Mus. Nat. Hist. HI 63004

Bulletin of the chinese association for the advancement of science. Taipei, Taiwan. Bull. Chin. Assoc. Advancem. Sci. See B–P–H 244/22. HI 52678

Bulletin of the chinese botanical society. [Chung kuo chih-wu hsüeh hui hui-pao.] Peiping [=Peking]. Bull. Chin. Bot. Soc. See B–P–H 245/1. HI 52679

Bulletin of chinese materia medica. [Zhongyao tongbao.] Peking (Beijing). Vol. 6+, 1981+. Bull. Chin. Mater. Med. HI 63005

Bulletin of chinese studies. [Chung kuo wên hua yen chini hui kan.] Chengtu, China. Bull. Chin. Stud. See B–P–H 245/2. HI 52680

Bulletin of the Choshi marine laboratory, Chiba university. [Chiba daigaku bunrigakubu Choshi rinkai kenkyu bunshitsu kenkyu hokoku.] Chiba. 1959+. Bull. Choshi Mar. Lab. Chiba Univ. HI 63006

Bulletin of the chrysanthemem society of America. New York, NY. Bull. Chrysanthemum Soc. Amer. See B–P–H 245/4. HI 52681

Bulletin of the Chugoku agricultural experiment station. Ser. A, crop division and environment division. [Chugoku nogyo shikenjo hokoku. A, sakumotsubu, kankyobu.] Fukuyama. Vols. 1-13, 1952-66. Bull. Chugoku Agric. Exp. Sta., A. Superseded by: Bulletin of the Chugoku national agricultural experiment station. Ser. A, crop division and Ser. E, environment division. HI 63007

Bulletin of the Chugoku national agricultural experiment station. Ser. A, crop division. [Chugoku nogyo shikenjo hokoku. A, sakumotsubu.] Fukuyama. No. 14+, 1967+. Bull. Chugoku Natl. Agric. Exp. Sta., A. Preceded by: Bulletin of the Chugoku agricultural experiment station. Ser. A, crop division and environment division. HI 63008

Bulletin of the Chugoku national agricultural experiment station. Ser. E, environment division. [Chugoku nogyo shikenjo hokoku. E, kankyobu.] Fukuyama. No. 1+, 1967+. Bull. Chugoku Natl. Agric. Exp. Sta., E. Preceded by: Bulletin of the Chugoku agricultural experiment station. Ser. A, crop division and environment division. HI 63009

Bulletin of the Claremont pomological club. Claremont, CA. Bull. Claremont Pomol. Club. See B–P–H 245/10. HI 52684

Bulletin de la classe d'agriculture de la société des arts de Genève. Geneva. Bull. Cl. Agric. Soc. Arts Genève. See B–P–H 245/7. HI 52682

Bulletin de la classe physico-mathématique de l'académie impériale des sciences de Saint-Pétersbourg. St. Petersburg. Bull. Cl. Phys.-Math. Acad. Imp. Sci. Saint-Pétersbourg. See B–P–H 245/8. HI 52683

Bulletin de la classe des sciences naturelles et techniques. Kiev = Zapysky Pryrodnychco-tekhnichnogo viddulu. Kiev.

Bulletin of the classification society. Cambridge, England. Bull. Classific. Soc. See B–P–H 245/11. HI 52685

Bulletin, Clemson agricultural college of South Carolina. Agricultural experiment station = Clemson agricultural college of South Carolina. Agricultural experiment station. Bulletin. Clemson, SC. Clemson Agric. Exp. Sta. Bull. See B–P–H 314/34.

Bulletin of the Cleveland medical library. Cleveland, OH. Bull. Cleveland Med. Libr. See B–P–H 245/14. HI 52686

Bulletin of the Cleveland museum of natural history. Cleveland, OH. Bull. Cleveland Mus. Nat. Hist. See B–P–H 245/15. HI 52687

Bulletin. Clinique national ophtalmologique. Paris. Bull. Clin. Natl. Ophtalmol. See B–P–H 245/16. HI 52688

Bulletin, club des amis de la nature en Mauritanie. Nouakchott. No. ?-4, 19??-76+. Bull. Club Amis Nat. Mauritanie. HI 63010

Bulletin, college of agriculture and forestry, Miyazaki university. Miyazaki. Vol. 1+, 1955+. Bull. Coll. Agric. Forest. Miyazaki Univ. HI 63011

Bulletin of the college of agriculture and forestry, university of Nanking. Nanking. Bull. Coll. Agric. Univ. Nanking. See 245/22. HI 52689

Bulletin, college of agriculture research center, Washington state university. Pullman, WA. No. 793+, 1974+. Bull. Coll. Agric. Res. Center Washington State Univ. Preceded by: Washington (state) agricultural experiment station. Bulletin. HI 63012

Bulletin of the college of agriculture, Tokyo imperial university. [Noka daigaku gakujutsu hokoku.] Tokyo. Vols. 1-8, 1887-1909. Bull. Coll. Agric. Tokyo Imp. Univ. Superseded by: Journal of the college of agriculture, imperial university of Tokyo. 5-4231-1. HI 63013

Bulletin, college of agriculture, university of Tehran. Teheran. No. 93+. 1967+. Bull. Coll. Agric. Univ. Tehran. Preceded by: Bulletin, département de protection des plantes, université de Tehran. HI 63014

Bulletin of the college of agriculture, Utsunomiya university. [Utsunomiya daigaku nogakubu gakujutsu hokoku.] Utsunomiya. Vol. 1/2+, 1951+. Bull. Coll. Agric. Utsunomiya Univ. Preceded by: Bulletin of the Utsunomiya agricultural college. Series A and Series B. HI 63015

Bulletin of the college of arts and sciences, Baghdad. Baghdad. Bull. Coll. Arts Baghdad. See B–P–H 246/1. HI 52690

Bulletin of the college of Charleston museum. Charleston, SC. Vols. 1-2, 1905-06. Bull. Coll. Charleston Mus. Superseded by: Bulletin of the Charleston museum. HI 63016

Bulletin, college experiment stations, university of Georgia. Athens, GA. Nos. 1-6, 1952-54. Bull. Coll. Exp. Sta. Univ. Georgia. Superseded by: Bulletin, agricultural experiment stations, Georgia. HI 63017

Bulletin, college of foreign studies. Natural science. Yokohama. No. 1+, 1969+. Bull. Coll. Foreign Stud., Yokohama, Nat. Sci. HI 63018

Bulletin of college of forestry, university of Belgrade = Glasnik šumarskog fakulteta, univerzitet u Beogradu. Belgrade.

Bulletin, college of forestry, wildlife and range sciences, university of Idaho. Moscow, ID. No. 1+, 1955+. Bull. Coll. Forest. Wildlife Range Sci. Univ. Idaho. HI 63019

Bulletin, college of forestry, Washington state university. Seattle, WA. 1946+. Bull. Coll. Forest. Washington State Univ. HI 63020

Bulletin, college of Hawaii = College of Hawaii publications. Bulletins. Honolulu, HI. Coll. Hawaii Publ. Bull. See B–P–H 316/23.

Bulletin of the college of science, university of Baghdad. Baghdad. Vols. 4-6, 1959-61; part 2, botany, zoology & geology vols. 7?-18(1), 1962?-77. Bull. Coll. Sci. Univ. Baghdad. Preceded by: Bulletin of the college of arts and sciences, Baghdad. Superseded by: Iraqi journal of science. HI 63021

Bulletin of the college of science, university of the Ryukyus. Okinawa. No. 30+, 1980+. Bull. Coll. Sci. Univ. Ryukyus. Preceded by: Bulletin of the science and engineering division, university of the Ryukyus. HI 63022

Bulletin Colorado agricultural college, Colorado experiment station. Fort Collins, CO. Bull. Colorado Agric. Coll. Colorado Exp. Sta. See B–P–H 246/4. HI 52692

Bulletin of the Colorado biological association. Washington, DC.? Bull. Colorado Biol. Assoc. See B–P–H 246/6. HI 52694

Bulletin, Colorado experiment station = Colorado experiment station bulletin. Fort Collins, CO. Colorado Exp. Sta. Bull. See B–P–H 318/11.

Bulletin of the Colorado scientific society. Denver, CO. Bull. Colorado Sci. Soc. See 246/8. HI 52696

Bulletin, Colorado flower growers' association. Denver, CO. Nos. 1-354, 1949-79. Bull. Colorado Fl. Growers' Assoc. Superseded by: Research bulletin, Colorado greenhouse growers' association. HI 75135

Bulletin, Colorado seed laboratory = Colorado seed laboratory bulletin; Colorado agricultural experiment station. Fort Collins, CO. Colorado Seed Lab. Bull. See B–P–H 318/18.

Bulletin, Colorado state flower growers' association = Bulletin, Colorado flower growers' association. Denver, CO.

Bulletin Colorado state university experiment station. Fort Collins, CO. Bull. Colorado State Univ. Exp. Sta. See B–P–H 246/10. HI 52697

Bulletin du comité pour l'étude des sapropélites = Izvestiya Sapropelevogo komiteta. Petrograd.

Bulletin du comité d'études historiques et scientifiques de l'Afrique occidentale française. Paris. Bull. Comité Études Hist. Sci. Afrique Occidentale Franç. See B–P–H 246/13. HI 52698

Bulletin du comité de Madagascar = Revue de Madagascar. Paris.

Bulletin de la commission pour l'étude du quaternaire. Moscow & Leningrad = Byulleten' Komissii po izucheniyu chetvertichnogo perioda. Moscow & Leningrad.

Bulletin, commission d'études des ennemis des arbres, des bois abattus et des bois mis en oeuvre. Nancy. Nos. 1-31, 1929-46. Bull. Commiss. Études Ennemis Arbres. HI 63023

Bulletin de la commission géologique de la Finlande. Helsinki. Bull. Commiss. Géol. Finlande. See B–P–H 246/17. HI 52699

Bulletin de la commission internationale pour l'exploration de la mer Méditerranée. Monaco. Nos. 1-10, 1920-24. Bull. Commiss. Int. Explor. Mer. Médit.

Superseded by: Rapport et procès-verbaux des réunions de la commission internationale pour l'exploration scientifique de la mer. HI 63024

Bulletin de la commission internationale pour l'exploration scientifique de la mer méditerranée. Monaco. Bull. Commiss. Int. Explor. Sci. Mer Médit. See B–P–H 246/18. HI 52700

Bulletin of the commonwealth bureau of pastures and crops. Aberystwyth, Wales. Bull. Commonw. Bur. Pastures. See 246/21. HI 52701

Bulletin, commonwealth forestry and timber bureau, Australia. Canberra, A.C.T. Nos. 1-28, 1931-45? Bull. Commonw. Forest. Timber Bur., Australia. Superseded by: Bulletin, forestry and timber bureau, Australia. HI 63025

Bulletin of the Connecticut arboretum. New London, CT. Bull. Connecticut Arbor. See B–P–H 246/25. HI 52702

Bulletin, Connecticut state geological and natural history survey = Connecticut state geological and natural history survey bulletin. Hartford, CT. Connecticut State Geol. Surv. Bull. See B–P–H 325/4.

Bulletin, conservation and survey division of university of Nebraska. Lincoln, NE. Nos. 1-17, 1916-25. Bull. Conservation Surv. Div. Univ. Nebraska. Superseded by: Nebraska conservation bulletin. 4-2946-1. HI 61320

Bulletin, Cornell university agricultural experiment station = Cornell university agricultural experiment station bulletin. Ithaca, NY. Cornell Univ. Agric. Exp. Sta. Bull. See B–P–H 332/6.

Bulletin, Cornwall naturalists' trust. St. Ives, Penzance. No. 1+, [1965]+. Bull. Cornwall Naturalists' Trust. HI 63026

Bulletin of the Cranbrook institute of science. Bloomfield Hills, MI. Bull. Cranbrook Inst. Sci. See B–P–H 246/28. HI 52703

Bulletin of the Croydon microscopical and natural history club. Croydon. No. ?-28+, 19??-67-72+. Bull. Croydon Microscop. Nat. Hist. Club. Superseded by?: Bulletin of the Croydon natural history and scientific society. HI 63027

Bulletin of the Croydon natural history and scientific society. Croydon. 19??-7?+. Bull. Croydon Nat. Hist. Sci. Soc. Preceded by?: Bulletin of the Croydon microscopical and natural history club. HI 63028

Bulletin of the Cyprus natural history society. Nicosia, Cyprus. Bull. Cyprus Nat. Hist. Soc. See B–P–H 247/1. HI 52704

Bulletin of the dahlia society of California. San Rafael, CA. Bull. Dahlia Soc. Calif. See B–P–H 247/2. HI 52705

Bulletin of the dahlia society of Michigan. East Lansing, MI. Bull. Dahlia Soc. Michigan. See B–P–H 247/3. HI 52706

Bulletin of the dahlia society of San Francisco = California dahlia news. San Francisco, CA. Calif. Dahlia News. See B–P–H 293/28.

Bulletin, Delaware agricultural experiment station = Delaware college, agricultural experiment station. Newark, DE. Delaware Coll. Agric. Exp. Sta. Bull. See B–P–H 340/11.

Bulletin, Delaware college, agricultural experiment station = Delaware college, agricultural experiment station. Newark, DE. Delaware Coll. Agric. Exp. Sta. Bull. See B–P–H 340/11.

Bulletin van het deli proefstation Medan. Medan, Dutch E. Indies [Indonesia]. Bull. Deli Proefstat. Medan. See B–P–H 247/10. HI 52707

Bulletin de département de l'agriculture aux Indes Néerlandaises. Buitenzorg, Dutch E. Indies [=Bogor, Indonesia]. Bull. Dép. Agric. Indes Néerl. See B–P–H 247/14. HI 52708

Bulletin van het departement van landbouw in Suriname. Paramaribo, Dutch Guiana [Surinam]. Bull. Dept. Landb. Suriname. See B–P–H 248/7. HI 52719

Bulletin, departement de protection des plantes, université de Tehran. Karadj. No. ?-65+, 19??-65+. Bull. Dép. Protect. Pl. Univ. Tehran. Superseded by: Bulletin, college of agriculture, university of Tehran. HI 63029

Bulletin, department of agricultural research, Nigeria = Bulletin of the agricultural department, Nigeria. Lagos.

Bulletin, department of agricultural research of the royal tropical institute. Amsterdam. No. 276+, 1962+. Bull. Dept. Agric. Res. Roy. Trip. Inst. Preceded by: Bericht, Koninklijk instituut voor de tropen, afdeling agrarisch onderzoek. HI 63030

Bulletin of the department of agricultural technical services, South Africa = Bulletin of the department of agriculture and forestry, Union of South Africa. Pretoria.

Bulletin, department of agriculture, Bengal = Bulletin, department of land records and agriculture, Bengal. Calcutta.

Bulletin, department of agriculture, Bengal. Agricultural series = Bulletin, department of land records and agriculture, Bengal. Agricultural series. Calcutta.

Bulletin, department of agriculture, Bihar and Orissa. Patna. 1929+. Bull. Dept. Agric. Bihar & Orissa. HI 63031

Bulletin, department of agriculture, Bombay = Bulletin of the department of land records and agriculture, Bombay. Bombay.

Bulletin, department of agriculture, British Columbia. Victoria, B.C. No. 1+, 1892+. Bull. Dept. Agric. British Columbia. HI 63032

Bulletin, department of agriculture, British East Africa. Nairobi. Nos. 1-4, 1914-19. Bull. Dept. Agric. Brit. E. Africa. Superseded by: Bulletin, department of agriculture, Kenya colony. HI 63033

Bulletin of the department of agriculture, Burma. Rangoon. Nos. 1-35, 1909-37. Bull. Dept. Agric. Burma. HI 63034

Bulletin of the department of agriculture, Cape of Good Hope. Cape Town. 18??-98+. Bull. Dept. Agric. Cape of Good Hope. HI 63035

Bulletin of the department of agriculture, Central Provinces. Nagpur. Nos. 1-9, 1895-1902; [ser. 2], nos. 1-5, 1908-16; [ser. 3], no. 1+, 1919+. Bull. Dept. Agric. Central Provinces. HI 63036

Bulletin of the department of agriculture, Ceylon. Colombo. Nos. 1-94, 1912-38. Bull. Dept. Agric. Ceylon. Preceded by: Circular and agricultural journal of the royal botanic gardens, Peradeniya. HI 63037

Bulletin, department of agriculture, colony of the Gambia. Bathurst. Nos. 1-4, 1929-30. Bull. Dept. Agric. Gambia. HI 63038

Bulletin, department of agriculture and commerce, Virginia = Bulletin, department of agriculture and immigration, Virginia. Richmond, VA.

Bulletin: Department of agriculture, Cyprus. Horticultural series. Nicosia, Cyprus. Bull. Dept. Agric. Cyprus, Hort. Ser. See B–P–H 247/26. HI 52710

Bulletin. Department of agriculture, Cyprus. Mycological series. Nicosia, Cyprus. Bull. Dept. Agric. Cyprus, Mycol. Ser. See B–P–H 247/27. HI 52711

Bulletin of the department of agriculture of the dominion of Canada. New series. Ottawa. Nos. 1-181, 1922-35. Bull. Dept. Agric. Domin. Canada. Superseded by: Farmer's bulletin, department of agriculture, Canada. HI 63039

Bulletin of the department of agriculture; Federated Malay States. Kuala Lumpur. Nos. 1-?, 1907-27. Bull. Dept. Agric. Fed. Malay States. Superseded by: Bulletin, department of agriculture, Straits Settlements and Federated Malay States. HI 63040

Bulletin, department of agriculture, Federation of Malaya. Kuala Lumpur. Nos. 101-114, 1956-63. Bull. Dept. Agric. Fed. Malaya. Preceded by: Bulletin, department of agriculture, Straits Settlements and Federated Malay States. HI 63041

Bulletin of the department of agriculture, Fiji. Suva. Nos. 1+, 1912+. Bull. Dept. Agric. Fiji. HI 63042

Bulletin, department of agriculture and forests, Sudan. Wad Medani. Nos. 1-12-?, 19??-41-55. Bull. Dept. Agric. Forests Sudan. Superseded by: Bulletin, ministry of agriculture, Sudan. HI 63043

Bulletin of the department of agriculture and forestry, Union of South Africa. Pretoria. 1910-22; no. 1+, 1925+. Bull. Dept. Agric. Forest. Union S. Africa. HI 63044

Bulletin, department of agriculture, Gold Coast colony. Accra. Nos. 1-36, 1925-38. Bull. Dept. Agric. Gold Coast Colony. HI 63045

Bulletin of the department of agriculture, government research institute, Formosa. [Taiwan sotokufu chuo kendyu jo, nogyo-bu iho.] Taihoku [=Taipei, Taiwan]. Bull. Dept. Agric. Gov. Res. Inst. Formosa. See B–P–H 247/29. HI 52712

Bulletin, department of agriculture and immigration, Virginia. Richmond, VA. No. 1+, 1900+. Bull. Dept. Agric. Immigr. Virginia. HI 63046

Bulletin, department of agriculture, Iraq. Basrah, Baghdad. Nos. 1-27, 1918?-33. Bull. Dept. Agric. Iraq. HI 63047

Bulletin of the department of agriculture, Jamaica. Kingston, Jamaica. Nos. 1-6, 1903-06; n.s. vols. 1-2, 1909-15; 1941-56. Bull. Dept. Agric. Jamaica. Preceded by: Bulletin botanical department, Jamaica. For 1934-43 see: Bulletin, department of science and agriculture, Jamaica. Superseded by: Bulletin of the ministry of agriculture and lands, Jamaica. HI 52709

Bulletin of the department of agriculture, Kenya colony. Nairobi. Nos. 5-13, 1921-23; n.s. nos. 1-28, 1925-29; [ser. 3], 1929-35. Bull. Dept. Agric. Kenya Colony. Preceded by: Bulletin, department of agriculture, British East Africa. HI 63048

Bulletin, department of agriculture and lands, Southern Rhodesia. [Forms part of: Rhodesia agricultural journal. See B–P–H 799/3.] Salisbury, Rhodesia. Nos. 1-2518?, 1909?-? Bull. Dept. Agric. Lands Southern Rhodesia. HI 63049

Bulletin, department of agriculture, Mauritius. General series. Port Louis, Reduit. Nos. 1-97, 1914-68. Bull. Dept. Agric. Mauritius, Gen. Ser. HI 63050

Bulletin, department of agriculture, Mauritius. Scientific series. Reduit. Nos. 1-30, 1915-48. Bull. Dept. Agric. Mauritius, Sci. Ser. Incorporated in: Bulletin, department of agriculture, Mauritius. General series. HI 63051

Bulletin. Department of agriculture, Mysore state. Botanical series. Bangalore, India. Bull. Dept. Agric. Mysore State, Bot. Ser. See B–P–H 247/31. HI 52714

Bulletin, department of agriculture, Natal. Pietermaritzburg. 1902-10. Bull. Dept. Agric. Natal. HI 63052

Bulletin, department of agriculture, New Guinea = Leaflet, department of agriculture, New Guinea. Rabaul.

Bulletin, department of agriculture, New Zealand. Wellington, N.Z. No. 1+, 1910+. Bull. Dept. Agric. New Zealand. HI 63053

Bulletin, department of agriculture, New Zealand. Division of biology and horticulture. Wellington, N.Z. Nos. 1-23, 1904-09. Bull. Dept. Agric. New Zealand, Div. Biol. Hort. Incorporated in: Bulletin, department of agriculture, New Zealand. HI 63054

Bulletin, department of agriculture, Northern Rhodesia. Lusaka. Nos. 1-18, 1950-61. Bull. Dept. Agric. Northern Rhodesia. HI 63055

Bulletin of the department of agriculture, North-West Territories. Regina. Nos. 1-16, 1898-05. Bull. Dept. Agric. N. W. Territories. Superseded by: Bulletin, department of agriculture of the province of Saskatchewan. HI 63056

Bulletin, department of agriculture, Nova Scotia. Halifax, N.S. No. 1+, 1908+. Bull. Dept. Agric. Nova Scotia. HI 63057

Bulletin, department of agriculture, Nyasaland. Zomba. N.s. nos. 1-14, 1932-36. Bull. Dept. Agric. Nyasaland. HI 63058

Bulletin, department of agriculture, Nyasaland Protectorate = Bulletin, department of agriculture, Nyasaland. Zomba.

Bulletin, department of agriculture, Orange River Colony. Blomfontein. 1???-1905-? Bull. Dept. Agric. Orange River Colony. HI 63059

Bulletin, department of agriculture of the province of Saskatchewan. Regina. Nos. 1-97 1906-44. Bull. Dept. Agric. Saskatchewan. Preceded by: Bulletin of the department of agriculture, North-West Territories. HI 63060

Bulletin of the department of agriculture, Punjab. Lahore. 1896-1910. Bull. Dept. Agric. Saskatchewan. HI 63061

Bulletin, department of agriculture, Queensland. Brisbane. No. 1+, 1887+. Bull. Dept. Agric. Queensland. HI 63062

Bulletin, department of agriculture, Saskatchewan = Bulletin, department of agriculture of the province of Saskatchewan. Regina.

Bulletin of the department of agriculture of South Australia. Adelaide. Nos. 1-473, 1905-67. Bull. Dept. Agric. S. Australia. HI 63063

Bulletin, department of agriculture, state of California = Bulletin of the California department of agriculture. Sacramento, CA.

Bulletin, department of agriculture, Straits Settlements and Federated Malay States. Kuala Lumpur. Nos. ?-39, 1923-27. Bull. Dept. Agric. Straits Settlements & Fed. Malay States. Preceded by: Bulletin of the department of agriculture, Federated Malay States. Superseded by: Bulletin, department of agriculture, Federation of Malaya. HI 63064

Bulletin, department of agriculture, Tasmania = Bulletin of the agricultural department, Tasmania. Launceston, Hobart, Tas.

Bulletin, department of agriculture, Thailand = Technical bulletin, department of agriculture, Thailand. Bangkok.

Bulletin, department of agriculture, Transvaal. A, horticulture. Pretoria. No. 1+, 1905+. Bull. Dept. Agric. Transvaal, A. HI 63065

Bulletin of the department of agriculture, Trinidad and Tobago. Port-of-Spain. Vol. 13-21(2) [vol. 13 also numbered 77-84], 1914-27; n.s. no. 1+, 1947+. Bull. Dept. Agric. Trinidad & Tobago. Preceded by: Bulletin of miscellaneous information, department of agriculture, imperial college of agriculture, Trinidad. HI 63066

Bulletin, department of agriculture, United Provinces of Agra and Oudh. Allahabad. Nos. 1-104?, 1895-1916. Bull. Dept. Agric. Unit. Prov. Agra & Oudh. HI 63067

Bulletin, department of agriculture, United Provinces of Agra and Oudh. Fruit series. Allahabad. No. 1+, 1933+. Bull. Dept. Agric. Unit. Prov. Agra & Oudh, Fruit Ser. HI 63068

Bulletin of the department of agriculture, university college of Reading = Bulletin of the department of agriculture, university of Reading.

Bulletin of the department of agriculture, university of Reading. Reading. 1907-23. Bull. Dept. Agric. Univ. Reading. Superseded by: Bulletin of the faculty of agriculture and horticulture, university of Reading. HI 63069

Bulletin, department of agriculture, Uttar Pradesh. Allahabad. Nos. 1-39, 1895-1916; [n.s.] no. 40+, 1921+. Bull. Dept. Agric. Uttar Pradesh. HI 63070

Bulletin, department of agriculture, Uttar Pradesh. Fruit series. Allahabad. No. 1+, 1933+. Bull. Dept. Agric. Uttar Pradesh, Fruit Ser. HI 63071

Bulletin of the department of agriculture, Victoria. Melbourne, Vic. Nos. 1-14, 1888-91; [2 ser.], nos. 1-30, 1902-11; n.s. no. 1+, 1912+ [n.s. nos. 28 & 37 not published]. Bull. Dept. Agric. Victoria. HI 63072

Bulletin, department of agriculture, Virginia = Bulletin, department of agriculture and immigration, Virginia. Richmond, VA.

Bulletin, department of agriculture, Western Australia.

[Forms part of: Journal of agriculture of Western Australia.] Perth, South Perth, W.A. Nos. 1-42, 1903?-11; n.s. nos. 1-?, 1911-22? Bull. Dept. Agric. W. Australia. Superseded by: Leaflet, department of agriculture, Western Australia. HI 63073

Bulletin, department of agriculture, Zanzibar. Zanzibar. 1952?+. Bull. Dept. Agric. Zanzibar. HI 63074

Bulletin, department of archaeology and palaeontology, free museum of science and art of the university of Pennsylvania = Bulletin of the free museum of science and art of the university of Pennsylvania. Philadelphia, PA.

Bulletin of department of biology, college of science, Sun Yatsen university. [Kuo li chung shan ta hsüeh li k'o shêng hse ts'ung k'an.] Canton. Bull. Dept. Biol. Sun Yatsen Univ. See B–P–H 248/1. HI 52715

Bulletin of the department of biology, Yenching university. [Yen ta shêng wu pu ts'ung k'an.] Peiping [=Peking]. Bull. Dept. Biol. Yenching Univ. See B–P–H 248/2. HI 52716

Bulletin, department of botany, ministry of agriculture and agrarian reform, republic of Iraq. Baghdad. Ca.1980. Bull. Dept. Bot. Minist. Agric. Agrar. Reform Iraq. HI 63075

Bulletin, department of forestry, British North Borneo. Sandakan. Nos. 1-3, 1916-17. Bull. Dept. Forest. Brit. N. Borneo. HI 63076

Bulletin of the department of forestry, Canada. Ottawa. Nos. ?-128?, 1960-62. Bull. Dept. Forest. Canada. Preceded by: Bulletin of the forestry branch, department of northern affairs and natural resources, Canada. Superseded by: Publications, department of forestry and rural development, Canada. HI 63077

Bulletin. Department of forestry, government research institute, Taihoku, Taiwan [Taiwan sotokufu chuo kendyu-sho, ringyobu iho.] Taihoku [=Taipei, Taiwan]. Bull. Dept. Forest. Gov. Res. Inst. Taihoku. See B–P–H 248/4. HI 52717

Bulletin of the department of forestry, Pennsylvania. Harrisburg, PA. Nos. 1-55, 1902-36. Bull. Dept. Forest. Pennsylvania. HI 63078

Bulletin, department of forestry, Queensland = Bulletin, forest service, Queensland. Brisbane, Qld.

Bulletin, department of forestry, Stephen F. Austen state college, Texas. Nacogdoches, TX. 1957+. Bull. Dept. Forest. Stephen F. Austen State Coll. Texas. HI 63079

Bulletin of the department of forestry, Union of South Africa = Bulletin of the department of forests, Union of South Africa.

Bulletin, department of forestry, university of Adelaide. Adelaide, S.A. Nos. 1-9, 1918-22. Bull. Dept. Forest. Univ. Adelaide. Superseded by: Bulletin, woods and forests department, South Australia. HI 63080

Bulletin of the department of forestry, university of Ibadan. Ibadan. No. ?-2+, 19??-70+. Bull. Dept. Forest. Univ. Ibadan. HI 63081

Bulletin of the department of forests, Union of South Africa. Cape Town. No. 1+, 1920+. Bull. Forest. Union S. Africa. Preceded by?: Bulletin of the department of agriculture and forestry, Union of South Africa. HI 63082

Bulletin of the department of forests and waters, Pennsylvania = Bulletin of the department of forestry, Pennsylvania. Harrisburg, PA.

Bulletin, department of general education, Nagoya city university. Natural science section. [Nagoya-shiritsu daigaku kyoyobu kiyo, shizenkagaku-hen.] Nagoya. Vol. 1+, 1955+. Bull. Dept. Gen. Educ. Nagoya City Univ., Nat. Sci. Sect. HI 63083

Bulletin, department of horticulture, Lincoln college. Lincoln, N.Z. Nos. 1-28, 1967-79. Bull. Dept. Hort. Lincoln Coll. HI 63084

Bulletin, department of land records and agriculture, Bengal. Calcutta. 1895-1937? Bull. Dept. Land. Rec. Bengal. HI 63085

Bulletin, department of land records and agriculture, Bengal. Agricultural series. Calcutta. 1895-1905. Bull. Dept. Land. Rec. Bengal., Agric. Ser. HI 63086

Bulletin of the department of land records and agriculture, Bombay. Bombay. No. 1+, 1884+. Bull. Dept. Land. Rec. Agric. Bombay. HI 63087

Bulletin of the department of land records and agriculture, Central Provinces = Bulletin of the department of agriculture, Central Provinces. Nagpur.

Bulletin of the department of land records and agriculture. Madras. Bull. Dept. Land Rec. (Madras). See B–P–H 248/5. HI 52718

Bulletin, department of lands and forests, Nova Scotia. Halifax, N.S. No. 1+, 1951+. Bull. Dept. Lands Forests Nova Scotia. HI 63088

Bulletin, department of marine biology and oceanography, university of Cochin. Ernakulam. Vol. 5, 1971. Bull. Dept. Mar. Biol. Oceanogr. Univ. Cochin. Preceded by: Bulletin, department of marine biology and oceanography, university of Kerala. Superseded by: Bulletin of the department of marine sciences, university of Cochin. HI 63089

Bulletin, department of marine biology and oceanography, university of Kerala. Ernakulam. Vols. 1-4, 1963-68. Bull. Dept. Mar. Biol. Oceanogr. Univ. Kerala. Preceded by: Bulletin of the central research institute, university of Kerala. Series C, natural science. Superseded by: Bulletin, department of marine biology and oceanography, university of Cochin. HI 63090

Bulletin of the department of marine sciences, university of Cochin. Ernakulam. Vol. 6+, 1973+. Bull. Dept. Mar. Sci. Univ. Cochin. Preceded by: Bulletin of the department of marine biology and oceanography, university of Cochin. HI 63091

Bulletin, department of medicinal plants. Thapathali, Kathmandu. No. 1+, 1967+. Bull. Dept. Med. Pl., Thapathali. HI 63092

Bulletin of the department of science and agriculture, Barbados. Bridgetown. Nos. 1-32, 1945-63. Bull. Dept. Sci. Agric. Barbados. Superseded by: Bulletin, ministry of agriculture, lands and fisheries, Barbados. HI 63093

Bulletin of the department of science and agriculture, Jamaica. Kingston, Jamaica. Nos. 1-55?, 1934-40. Bull. Dept. Sci. Agric. Jamaica. Preceded and superseded by: Bulletin of the department of agriculture, Jamaica. HI 63094

Bulletin, department of woods and forests, South Australia = Bulletin, South Australia woods and forests department. Adelaide.

Bulletin of the departments of agriculture and horticulture of the university. Bristol, England. Bull. Dept. Agric. Hort. Univ. See B–P–H 247/30. HI 52713

Bulletin of the Devon trust for nature conservation. Exeter. 1964+. Bull. Devon Trust Nat. Conservation. HI 63095

Bulletin, direction générale des bois et forêts, Québec = Bulletin, forest service, department of lands and forests, Quebec.

Bulletin, division of agricultural sciences, university of California. Berkeley, CA. No. 871+, 1975+. Bull. Div. Agric. Sci. Univ. Calif. Preceded by: University of California agricultural experiment station, bulletin. HI 63096

Bulletin, division of agrostology, United States department of agriculture. Washington, DC. Nos. 1-25, 1895-1901. Bull. Div. Agrostol. U.S.D.A. HI 63097

Bulletin, division of biology and horticulture, department of agriculture, New Zealand = Bulletin, department of agriculture, New Zealand. Division of biology and horticulture. Wellington, N.Z.

Bulletin, division of botany, United States department of agriculture = United States department of agriculture. Division of botany. Bulletin. Washington, DC. U.S.D.A. Div. Bot. Bull. See B–P–H 941/21.

Bulletin, division of forestry, United States department of agriculture = U S department of agriculture, division of forestry bulletin. Washington, DC. U.S.D.A. Div. Forest. Bull. See B–P–H 941/24.

Bulletin, division of pathology and physiology, Hawaiian sugar planters' association experiment station. Honolulu, HI. Nos. 1-12, 1905-12. Bull. Div. Pathol. Physiol. Hawaiian Sugar Planters Assoc. Exp. Sta. Superseded by: Bulletin, hawaiian sugar planters' association experiment station. Botanical series. HI 63098

Bulletin of the division of plant industry, Florida. Gainesville, FL. No. 1+, 1961+. Bull. Div. Pl. Industr. Florida. Preceded by: Bulletin, state plant board of Florida. HI 63099

Bulletin. Division of plant pathology. Agricultural experiment station. Formosa. [Taiwan nosakubutsu byogai mokuroku.] Taihoku [=Taipei, Taiwan]. Bull. Div. Pl. Pathol. Agric. Exp. Sta. Formosa. See B–P–H 248/11. HI 52720

Bulletin, division of pomology, United States department of agriculture = U S department of agriculture. Division of pomology. Bulletin. Washington, DC. U.S.D.A. Div. Pomol. Bull. See B–P–H 941/27.

Bulletin, division of research, college of life sciences, Louisiana tech university. Ruston, LA. Nos. 6-8, 1972-73. Bull. Div. Res. Coll. Life Sci. Louisiana Tech Univ. Preceded by: Bulletin, division of research, school of agriculture and forestry, Louisiana polytechnic institute. Superseded by: Bulletin, school of forestry, Louisiana tech university. HI 75137

Bulletin, division of research, school of agriculture and forestry, Louisiana polytechnic institute. Ruston, LA. Nos. 1-5, 1967-70. Bull. Div. Res. School Agric. Forest. Louisiana Polytechn. Inst. Superseded by: Bulletin, division of research, college of life sciences, Louisiana tech university. HI 75136

Bulletin, division of vegetable pathology, United States department of agriculture = U S department of agriculture. Division of vegetable pathology. Bulletin. Washington, DC. U.S.D.A. Div. Veg. Pathol. Bull. See B–P–H 942/2.

Bulletin, division of vegetable physiology and pathology, United States department of agriculture = U S department of agriculture. Division of vegetable pathology. Bulletin. Washington, DC. U.S.D.A. Div. Veg. Pathol. Bull. See B–P–H 942/2.

Bulletin of documentation for agriculture, forestry, wood and tobacco industry = Bilten; dokumentacija stručne literature. Ser. A, bilten dokumentacija za poljoprivredu, sumarstvo drvnu i duvansku industriju. Belgrade.

Bulletin of documentation: Forestry and wood industry = Bilten dokumentacije: Šumarstvo i drvna industrija. Belgrade.

Bulletin, duna-tisza közi mezögazdasági kisérleti intézet. Kecskemét. Vols. 1-5, 1967-70. Bull. Duna-Tisza

Közi Mezögazd. Kisérl. Intéz. Superseded by: Bulletin, zöldségtermesztési kutató intézet. HI 63100

Bulletin, Dunedin rhododendron group. Dunedin. ?-1987+. Bull. Dunedin Rhododendron Group. HI 63101

Bulletin of the Durham county conservation trust. Sunderland. 1971+. Bull. Durham County Conservation Trust. HI 63102

Bulletin E U C A R P I A. No. 4+, 1971+. Bull. E. U. C. A. R. P. I. A. Preceded by: Bulletin d'information E U C A R P I A. HI 63103

Bulletin of the East Africa natural history society = E A N H S bulletin. Nairobi.

Bulletin d'échanges du jardin botanique, Marseilles = Bulletin de la chaire d'histoire naturelle de la faculté mixte de médecine générale et coloniale, et de pharmacie de l'université d'Aix-Marseille.

Bulletin de l'école française d'Extrême-Orient. Hanoi, Saigon, Paris. Vol. 1+, 1901+. Bull. École Franç. Extrême-Orient. HI 63104

Bulletin, école nationale supériore agronomique de Nancy. Nancy. Vols. 1-13, 1959-71. Bull. Écol. Natl. Supér. Agron. Nancy. Superseded by: Bulletin, école nationale supériore d'agronomie et des industries alimentaires. HI 63105

Bulletin, école nationale supériore d'agronomie et des industries alimentaires. Nancy. Vol. 14+, 1972+. Bull. Écol. Natl. Supér. Agron. Industr. Aliment. Preceded by: Bulletin, école nationale supériore d'agronomie et des industries alimentaires. HI 63106

Bulletin de l'école polytechnique de Jassy. Jassy, Rumania. Bull. École Polytechn. Jassy. See B–P–H 248/20. HI 52722

Bulletin, école pratique des hautes études, museum national d'histoire naturelle (Paris) = Bulletin muséum national d'histoire naturelle, école pratique des hautes études. Papeete.

Bulletin de l'école royale hongroise d'horticulture. Budapest = A m[agyar] kir[ályi] kertészeti tanintézet közleményei. Budapest. Magyar Kir. Kert. Tanintéz. Közlem. See B–P–H 544/1.

Bulletin de l'école supérieure d'agronomie. Brno = Sborník vysoké školy zemědělské v Brně. Brno, Czechoslovakia. Sborn. Vysoké Školy Zeměd. v Brně. See B–P–H 817/1.

Bulletin de l'école supérieure royale hongroise d'horticulture et viticulture. Budapest = M[agyar] kir[ályi] kertészeti és szölészeti föiskola közleményei. Budapest. Magyar Kir. Kert. Szölész. Föisk. Közlem. See B–P–H 543/18

Bulletin, ecological research committee, swedish natural science research committee = Bulletin, ekologikommitten, statens naturvetenskapliga forskningsråd. Stockholm.

Bulletin of the ecological society of America. Tucson, Arizona. Bull. Ecol. Soc. Amer. See B–P–H 248/18. HI 52721

Bulletin d'écologie. Brunoy. Vol. 4+, 1973+. Bull. Écol., Brunoy. Preceded by: Bulletin de la société d'écologie. HI 63107

Bulletin économique (de l'Indochine). Saigon, Hanoi. Vols. 1-52, 1898-1949 [from 1930-32 divided into series; sér. B, agriculture, élevage, forêts]. Bull. Écon. Indochine. HI 63108

Bulletin, Edgbaston botanic gardens. Birmingham. 1963-66. Bull. Edgbaston Bot. Gard. HI 63109

Bulletin, education faculty, Shizuoka university. Natural science series. [Shizuoka daigaku kyoikugakubu kenkyu hokoku. Shizenkagaku-hen.] Shizuoka. No. ?-16+, 19??-65+. Bull. Educ. Fac. Shizuoka Univ., Nat. Sci. Ser. HI 63110

Bulletin of the educational research institute, faculty of education, university of Kagoshima. Kagoshima. Vols. 1-12, 1949-60. Bull. Educ. Res. Inst. Fac. Educ. Univ. Kagoshima. Superseded by: Bulletin, faculty of education, university of Kagoshima. Natural science. HI 63111

Bulletin of the egyptian society of cactus amateurs. Cairo. Nos. 1-3, 1941. Bull. Egypt. Soc. Cactus Amateurs. HI 63112

Bulletin of the Ehime university forest. [Ehime daigaku nogakubu enshurin hokoku.] Matsayama. 1963+. Bull. Ehime Univ. Forest. HI 63113

Bulletin, ekologikommitten, statens naturvetenskapliga forskningsråd. Stockholm. Nos. 1-18, 1968-73. Bull. Ekol. Kommitten. Statens Naturvetensk. Forskningsråd. Superseded by: Ecological bulletins. Stockholm. HI 63114

Bulletin, Emirates natural history group (Abu Dhabi). Abu Dhabi. No. 1+, 1977+. Bull. Emirates Nat. Hist. Group. HI 63115

Bulletin of entomological research. London. Vol. 1+, 1910+. Bull. Entomol. Res. 1-840-2. HI 63116

Bulletin of the entomological society of America. Washington, DC. 1955+. Bull. Entomol. Soc. Amer. HI 63117

Bulletin of the epiphyllum society of America. Los Angeles, Monrovia, CA. Vol. 1+, 1940+. Bull. Epiphyllum Soc. Amer. HI 63118

Bulletin, Essex naturalist's trust. 1967-81. Bull. Essex Naturalist's Trust. Superseded by: Watch over Essex. HI 63119

Bulletin, estuarine and brackish-water sciences

association. Norwich, etc. No. 1+, 1972+. Bull. Estuarine Brackish-Water Sci. Assoc. HI 63120

Bulletin of the experiment forests, Tokyo university of agriculture and technology. [Tokyo noko daigaku nogakubu enshurin hokoku.] Tokyo. 1958+. Bull. Exp. Forests Tokyo Univ. Agric. Technol. Preceded by?: Bulletin of the Tokyo university forest. HI 63121

Bulletin of the experiment station of Florida at the state agricultural college. Lake City, FL. Bull. Exp. Sta. Florida. See B–P–H 249/1. HI 52723

Bulletin, experiment station, south african sugar association. [Supplement to: South african sugar journal.] Mount Edgcombe. No. 1+, 1956+. Bull. Exp. Sta. S. African Sugar Assoc. HI 63122

Bulletin experiment station Tuskegee normal & industrial institute. [Tuskegee institute, Alabama.] Tuskegee, Alabama. Bull. Exp. Sta. Tuskegee Normal Industr. Inst. See B–P–H 249/2 HI 52724.

Bulletin, experimental farms branch, Canada. Ottawa. Nos. 1-98 [no. 46 withdrawn from circulation], 1887-1921; 2nd ser. nos. 1-49, 1898-21. Bull. Exp. Farms Branch Canada. HI 63123

Bulletin of the experimental forest of national Taiwan university. [Yen chiu pao kao, kuo li Taiwan ta hsüeh shih yen lin.] Taipei. No. ?-113+, 19??-74+. Bull. Exp. Forest Natl. Taiwan Univ. HI 63124

Bulletin, extension series, Michigan state university = Extension bulletin, cooperative extension service, Michigan state university. East Lansing, MI.

Bulletin de la faculté d'agriculture. Brno = Sborník vysoké školy zemědělské v Brně. Brno, Czechoslovakia. Sborn. Vysoké Školy Zeměd. v Brně. See B–P–H 817/1.

Bulletin, faculté forestière, université de Teheran. Karadj. Nos. 17-25, 1969/70-71. Bull. Fac. Forest. Univ. Teheran. Preceded by: Bulletin faculté des forêts et paturages, université de Teheran. Superseded by: Bulletin de la faculté des ressources naturelles, université de Teheran. HI 63125

Bulletin, faculté des forêts et paturages, université de Téhéran. Karadj. Nos. ?-13-14, 19??-69-70. Bull. Fac. Forêts Patur. Univ. Téhéran. Superseded by: Bulletin, faculté forestière, université de Téhéran. HI 63126

Bulletin de la faculté horticole et viticole de l'université des sciences agricoles. Budapest = Agrártudományi egyetem kert- és szölögazdaságtudományi karának közleményei. Budapest. Agrártud. Egyet. Kert-Szölögazdaságtud. Karának Közlem. See B–P–H 55/13.

Bulletin de la faculté des ressources naturelles, université de Téhéran. Karadj. No. 26+, 1971+. Bull. Fac. Ressources Nat. Univ. Téhéran. Preceded by: Bulletin, faculté forestière, université de Téhéran. HI 63127

Bulletin de la faculté des sciences de l'université. Cernăuti = Buletinul facultaţii de ştiinţe din Cernăuti. Cernăuti, Rumania [=Chernovtsy, Ukrainian S S R]. Bul. Fac Şti. Cernăuti. See B–P–H 232/1.

Bulletin of the faculty for agricultural sciences of agrarian university. Gödöllö, Hungary = Agrártudományi egyetem mezögazdaságtudományi karának évkönyve. Gödöllö, Hungary. Agrártud. Egyet. Mezögazdaságtud. Karának Évk. (Gödöllö). See B–P–H 55/15.

Bulletin, faculty of agriculture, Hirosaki university. [Hirosaki daigaku nogakubu gakujutsu hokoku.] Hirosaki. No. 1+, 1955+. Bull. Fac. Agric. Hirosaki Univ. HI 63128

Bulletin, faculty of agriculture and horticulture, university of Reading. Reading. 1925+. Bull. Fac. Agric. Hort. Univ. Reading. Preceded by: Bulletin of the department of agriculture, university of Reading. HI 63129

Bulletin of the faculty of agriculture, Kagoshima university. Kagoshima, Japan. Bull. Fac. Agric. Kagoshima Univ. See B–P–H 249/9. HI 52725

Bulletin of the faculty of agriculture; Mie university. [Mie kenritsu daigaku nogakubu, gakujutsu hokoku.] Tsu, Japan. Bull. Fac. Agric. Mie Univ. See B–P–H 249/10. HI 52726

Bulletin of the faculty of agriculture, Niigata university. [Niigata daigaku nogakubu gakujutsu hokoku.] Niigata, Japan. Bull. Fac. Agric. Niigata Univ. See B–P–H 249/11. HI 52727

Bulletin of the faculty of agriculture, Shinshu university. [Shinshu daigaku nogakubu gakujitsu hokoku.] Ina, Japan. Bull. Fac. Agric. Shinshu Univ. See B–P–H 249/13. HI 52728

Bulletin of the faculty of agriculture, Shizuoka university. Iwata, Japan. Bull. Fac. Agric. Shizuoka Univ. See B–P–H 249/14. HI 52729

Bulletin of the faculty of agriculture; university of Miyazaki. [Miyazaki daigaku nogaku-bu kenkyo jiho.] Miyazaki, Japan. Bull. Fac. Agric. Univ. Miyazaki. See B–P–H 249/15. HI 52730

Bulletin, faculty of agriculture, university of Shimane. Matsue. ?-1976+. Bull. Fac. Agric. Univ. Shimane. Preceded by?: Bulletin of the Shimane agricultural college. HI 63130

Bulletin of faculty of the agriculture, Tamagawa university. [Tamagawa daigaku nogakabu kenkyu hokoku.] Machida. Vol. 1+, 1960+. Bull. Fac. Agric. Tamagawa Univ. HI 63131

Bulletin of the faculty of agriculture, Tokyo university of agriculture and technology. [Tokyo noko daigaku nogakubu gakujutsu hokoku.] Tokyo. 1950+. Bull. Fac. Agric. Tokyo Univ. Agric. Technol. HI 63132

Bulletin of the faculty of agriculture, Tottori university. [Tottori daigaku nogakubu kenkyu hokoku.] Tottori. Vol. 22+, 1970+. Bull. Fac. Agric. Tottori Univ. Preceded by: Transactions of the Tottori society of agricultural science. HI 63133

Bulletin of the faculty of agriculture; Yamagata university. [Yamagata daigaku kiyo nogaku.] Shimonoseki, Japan. Bull. Fac. Agric. Yamagata Univ. See B–P–H 249/16. HI 52731

Bulletin of the faculty of agriculture, Yamaguchi university. [Yamaguchi daigaku nogakubu gakujutsu hokoku.] Shimonoseki. No. 1+, 1950+. Bull. Fac. Agric. Yamaguchi Univ. HI 63134

Bulletin of the faculty of arts and sciences, Ibaraki university, natural science. [Ibaraki daigaku bunrigakubu kiyo, shizendagaku.] Mito, Japan. Bull. Fac. Arts Ibaraki Univ., Nat. Sic. See B–P–H 249/17. HI 52732

Bulletin, faculty of education, Kanazawa university. Series, natural science. [Kanazawa daigaku kyoikugakubu kiyo. Shizenkagaku-hen.] Kanazawa. No. 1+, 1953+. Bull. Fac. Educ. Kanazawa Univ. HI 63135

Bulletin of the faculty of education; Kobe university. [Kobe daigaku kyoiku gakubu kenkyu shuroku.] Kobe, Japan. Bull. Fac. Educ. Kobe Univ. See B–P–H 249/18. HI 52733

Bulletin of the faculty of education; Kochi university. [Kochi daigaku kyoikugakubu kenkyu hokoku.] Kochi, Japan. Bull. Fac. Educ. Kochi Univ. See B–P–H 249/19. HI 52734

Bulletin, faculty of education, university of Kagoshima. Natural science. [Kagoshima daigaku kyoikugakubu kenkyu kiyo, shizenkagaku-hen.] Kagoshima. Vol. 1+, 1949+. Bull. Fac. Educ. Univ. Kagoshima, Nat. Sci. Preceded by: Bulletin of the educational research institute, faculty of education, university of Kagoshima. HI 63136

Bulletin of the faculty of fisheries, Nagasaki university. [Nagasaki daigaku suisangakubu kenkyu hokoku.] Nagasaki, Japan. Bull. Fac. Fish. Nagasaki Univ. See B–P–H 249/20. HI 52735

Bulletin, faculty of forestry, Teheran university = Bulletin, faculté forestière, université de Teheran. Karadj.

Bulletin, faculty of forestry, university of British Columbia. Vancouver, B.C. No. 5+, 1968+. Bull. Fac. Forest. Univ. British Columbia. Preceded by: Forestry bulletin, faculty of forestry, university of British Columbia. HI 63137

Bulletin of the faculty of horticulture and viticulture of the university of agriculture. Budapest = Agrártudományi egyetem kert- és szölögazdaságtudományi karának közleményei. Budapest. Agrártud. Egyet. Kert- Szölögazdaságtud. Karának Közlem. See B–P–H 55/13.

Bulletin of the faculty of liberal arts and education, Shiga university. Part 2, natural science. [Shiga daigaku gakugeibu kenkyu ranshu.] Hikone, Japan. Bull. Fac. Liberal Arts Shiga Univ., Pt. 2, Nat. Sci. See B–P–H 250/3. HI 52736

Bulletin of the faculty of liberal arts, Nagasaki university. Natural science. [Nagasaki daigaku kyoyobu kiyo, shizenkagaku.] Nagasaki. Vol. 1+, 1960+. Bull. Fac. Liberal Arts Nagasaki Univ., Nat. Sci. HI 63138

Bulletin of the faculty of pharmacy. Cairo. Bull. Fac. Pharm. See B–P–H 250/4. HI 52737

Bulletin of the faculty of science, Cairo university. Cairo. Vol. 31+, 1956+. Bull. Fac. Sci. Cairo Univ. Preceded by: Bulletin of the faculty of science, egyptian (Fouad I) university. HI 63139

Bulletin of the faculty of science, egyptian university. Cairo. Vols. 1-?, 1934-59. Bull. Fac. Sci. Egypt. Univ. Superseded by: Bulletin of the faculty of science, Fouad I university. 2-869-2. HI 63140

Bulletin of the faculty of science, Fouad I university. Cairo. Vols. ?-30, 1938-51? Bull. Fac. Sci. Fouad I Univ. Preceded by: Bulletin of the faculty of science, egyptian university. Superseded by: Bulletin of the faculty of science, Cairo university. 2-869-2. HI 63141

Bulletin of the faculty of science, King Abdul Aziz university. Jeddah. Vol. 1+, 1977+. Bull. Fac. Sci. King Abdul Aziz Univ. HI 63142

Bulletin of the faculty of science, Riyad university = Bulletin of the faculty of science, university of Riyadh. Riyadh.

Bulletin of the faculty of science, university of Riyadh. Riyadh. Vols. 1-8, 1969-77. Bull. Fac. Sci. Univ. Riyadh. Superseded by: Journal of the faculty of science, university of Riyadh. HI 63143

Bulletin of the faculty of textile fibers, Kyoto university of industrial arts and textile fibers. [Kyoto kogei sen'i daigaku sen'igakubu gakujutsu hokoku.] Kyoto. 1954+. Bull. Fac. Textile Fibers Kyoto Univ. Industr. Arts Textile Fibers. HI 63144

Bulletin, Fairchild tropical garden = Fairchild tropical garden bulletin. Coconut Grove, FL. Fairchild Trop. Gard. Bull. See B–P–H 366/22.

Bulletin of the Fan memorial institute of biology. [Ts'ing cheng cheng wou tiao tch'a so houei pao.] Peiping [=Peking]. Vols. 1-4, 1929-33; Botany vols. 5-9, 1934-

39; n.s. vols. 1-?, 1943-48. Bull. Fan Mem. Inst. Biol. HI 63145

Bulletin of the Far Eastern branch of the academy of sciences of the U S S R = Vestnik Dal'nevostochnogo filiala Akademii nauk S S S R. Vladivostok.

Bulletin of the Far Eastern branch of the academy of sciences of the U S S R = Vestnik Dal'nevostochnogo otdeleniya Akademii nauk S S S R. Vladivostok.

Bulletin de la fédération canadienne. London, Ont. Vols. 14-23, 1972-81. Bull. Féd. Canad. Preceded by: Canadian federation news. Superseded by: Newsletter, canadian federation of biological societies. HI 63146

Bulletin de la fédération canadienne des sociétés de biologie = Canadian federation news. Montreal, London, Ont.

Bulletin de la féderation centre-est d'histoire naturelle et de mycologie = Bulletin annuel de la fédération centre-est d'histoire naturelle et de mycologie. Lons Le Saunier.

Bulletin of the federation of Ontario naturalists. Toronto. Nos. 51-98, 19??-62. Bull. Fed. Ontario Naturalists. Preceded by: Circular, federation of Ontario naturalists. Superseded by: Ontario naturalist. 2-1549-3. HI 63147

Bulletin de la fédération des sociétés d'histoire naturelle de Franche-Comté. Besançon. Vols. 70-78, 1968-77. Bull. Féd. Soc. Hist. Nat. Franche-Comté. Preceded and superseded by: Bulletin de la société d'histoire naturelle du Doubs. HI 63148

Bulletin de la fédération des sociétés d'horticulture de Belgique. Ghent. Bull. Féd. Soc. Hort. Belgique. See 250/18. HI 52738

Bulletin de la fédération des sociétés d'horticulture et d'écologie du Québec. Montreal. Vols. 1-4(2), 1980-84. Bull. Féd. Soc. Hort. Écol. Québec. Superseded by: Loisir horticole. HI 63149

Bulletin, Field museum of natural history. Chicago, IL. Vol. 37(3)+, 1966+. Bull. Field Mus. Nat. Hist. Preceded by: Bulletin of the Chicago natural history museum. HI 63150

Bulletin of the fiftieth anniversary of the faculty of agriculture, Miyazaki university. [Forms part of: Bulletin of the faculty of agriculture, Miyazaki university. See B–P–H 249/15.] Miyazaki. 1976. Bull. Fiftieth Anniv. Fac. Agric. Miyazaki Univ. HI 63151

Bulletin, Florida agricultural experiment station = Florida agricultural experiment station bulletin. Lake City, FL. Florida Agric. Exp. Sta. Bull. See B–P–H 375/5.

Bulletin of the Florida Audubon society. Winter Park, FL. Bull. Florida Audubon Soc. See B–P–H 250/21. HI 52739

Bulletin Florida nurserymen and growers association. Leesburg, FL. Bull. Florida Nurserymen Growers Assoc. See B–P–H 250/22. HI 52740

Bulletin of the Florida rose seocity. Deland, FL. Bull. Florida Rose Soc. See B–P–H 250/23. HI 52741

Bulletin of the Florida state museum. Biological sciences. Tallahassee, FL. Bull. Florida State Mus., Biol. Sci. See B–P–H 250/24. HI 52743

Bulletin, fonds de recherches et de developpement forestier. Quebec. No. 14 [i.e. 16]+, 1983+. Bull. Fonds Rech. Developpem. Forest. Preceded by: Bulletin, fonds de recherches forestières de l'université Laval. HI 63152

Bulletin, fonds de recherches forestières de l'université Laval. Québec. Nos. 1-15, 1956-77. Bull. Fonds Rech. Forest. Univ. Laval. Superseded by: Bulletin, fonds de recherches et de developpement forestier. Quebec. HI 63153

Bulletin of foreign plant introductions, United States department of agriculture = U S department of agriculture. Bulletin of foreign plant introductions. Washington, DC. U.S.D.A. Bull. Foreign Pl. Introd. See B–P–H 940/19.

Bulletin of the forest biological research station, Cawthron institute. Nelson, N.Z. Vol. 1+, 1932+. Bull. Forest Biol. Res. Sta. Cawthron Inst. HI 63154

Bulletin, forest and conservation experiment station, university of Montana. Missoula, MT. Nos. 2-39, 1949-70 (nos. 1, 3 & 4 not published]. Bull. Forest Conservation Exp. Sta. Univ. Montana. HI 63155

Bulletin of the forest department, British Honduras. Belize. No. 1+, 1946+. Bull. Forest Dept. British Honduras. HI 63156

Bulletin, forest department, British North Borneo = Bulletin, department of forestry, British North Borneo. Sandakan.

Bulletin, forest department, Sarawak. Kuching. 1925+. Bull. Forest Sarawak. HI 63157

Bulletin of the forest department, Uganda. Entebbe. 1934+. Bull. Forest Dept. Uganda. HI 63158

Bulletin, forest department, Uttar Pradesh. Allahabad. 1928+. Bull. Forest Dept. Uttar Pradesh. HI 63159

Bulletin of forest experiment station. Government of Taiwan. [Taiwan ringyo shikenjo hokoku.] Taihoku [=Taipei, Taiwan]. Bull. Forest Exp. Sta. Gov. Taiwan. See B–P–H 251/1. HI 52744

Bulletin, forest experiment station of the imperial household. Tokyo = Bulletin of the imperial forest experiment station. [Ringyo shiken hokoku.] Tokyo. Bull. Imp. Forest Exp. Sta. See B–P–H 254/19.

Bulletin of the forest experiment station, national Taiwan

university = Bulletin of the experimental forest of national Taiwan university. Taipei.

Bulletin of the forest experimental station. Seoul. Nos. 5-9, 1956-60. Bull. Forest Exp. Sta. (Seoul). Superseded by: Research reports of the forest experiment station, Seoul. HI 63161

Bulletin of forest information; Sun Yat-sen university. Canton. Bull. Forest. Inform. Sun Yat-sen Univ. See B–P–H 251/2. HI 52747

Bulletin, forest and park service, Florida. Tallahassee, FL. Vol. 1+, 1929+. Bull. Forest Park. Serv. Florida. HI 63162

Bulletin, forest research foundation, Laval university = Bulletin, fonds de recherches forestières de l'université Laval. Québec.

Bulletin of the forest research institute, Sudan. Khartoum. No. 1+, 1973+. Bull. Forest Res. Inst. Sudan. HI 63163

Bulletin of forest research institute. Taipei, Taiwan. Bull. Forest Res. Inst. See B–P–H 251/3. HI 52748

Bulletin, forest service, department of lands and forests, Quebec. Quebec. N.s. no. 1+, 1942+. Bull. Forest Dept. Lands Forests Quebec. HI 63164

Bulletin, forest service, Queensland. Brisbane, Qld. No. 8+, 192?+. Bull. Forest Serv. Queensland. Preceded by: Forestry bulletin, forest service, Queensland. HI 63165

Bulletin of the forest tree breeding institute. [Rinkobu ikushujo kenkyu hokoku.] Ibaraki. No. 1+, 1983+. Bull. Forest Tree Breed. Inst. HI 66267

Bulletin, forest, wildlife and range experiment station, university of Idaho = Bulletin, college of forestry, wildlife and range sciences, university of Idaho. Moscow, ID.

Bulletin forestier polonais. Warsaw. Nos. 1-3?, 1931-33. Bull. Forest. Polon. 1-837-3. HI 63166

Bulletin of the forestry branch, department of northern affairs and natural resources, Canada. Ottawa. Nos. 1-127, 1904-60. Bull. Forest. Branch Dept. N. Affairs Nat. Resources Canada. Superseded by: Bulletin of the department of forestry, Canada. HI 63167

Bulletin of the forestry commission. London. No. 1+, 1919+. Bull. Forest. Commiss. HI 63168

Bulletin, forestry commission, Tasmania = Bulletin, forestry department, Tasmania. Hobart, Tas.

Bulletin of the forestry department, Gold Coast. Accra. Nos. 1-4, 1935-52. Bull. Forest. Dept. Gold Coast. Superseded by: Bulletin of the forestry division, Ghana. HI 63169

Bulletin, forestry department, Nigeria. Lagos, Ibadan. 1928+. Bull. Forest. Dept. Nigeria. HI 63170

Bulletin, forestry department, Tasmania. Hobart, Tas. No. 1+, 1933 [suspended between no. 1 and no. 2, 1962]. Bull. Forest. Dept. Tasmania HI 63171

Bulletin of the forestry department of the university of Edinburgh. Edinburgh. Nos. 1-8, 1955-61. Bull. Forest. Dept. Univ. Edinburgh. HI 63172

Bulletin of the forestry division, Ghana. Accra. No. 5+, 1957+. Bull. Forest. Div. Ghana. Preceded by: Bulletin of the forestry department, Gold Coast. HI 63173

Bulletin, forestry division, United States department of agriculture = U S department of agriculture, forestry division bulletin. Washington, DC. U.S.D.A. Forest. Div. Bull. See B–P–H 942/16.

Bulletin of the forestry and forest products research institute. Ushiku. No. 299+, 1978+. Bull. Forestry Forest Prod.. Res. Inst., Ushiku. Preceded by: Bulletin of the government forest experiment station. Tokyo. HI 63174

Bulletin, forestry service, Kentucky = Bulletin, state forest service, Kentucky. Frankfort, KY.

Bulletin of the forestry society of Hokkaido. [Hokkaido ringyo kai kaiho.] Sapporo?, Japan. Bull. Forest. Soc. Hokkaido. See B–P–H 251/4. HI 52749

Bulletin of the forestry society of Korea. [Chosen sanrin-kaiho.] Bull. Forest. Soc. Korea. See B–P–H 251/5. HI 52750

Bulletin, forestry and timber bureau, Australia. Canberra, A.C.T. No. 29-47, 1945?-75. Bull. Forest. Timber Bur. Australia. Preceded by: Bulletin, commonwealth forestry and timber bureau. HI 63175

Bulletin, forestry, wildlife and range experiment station, university of Idaho = Bulletin, college of forestry, wildlife and range sciences, university of Idaho. Moscow, ID.

Bulletin, forests commission, Victoria. Melbourne, Vic. Nos. 1-3, 1922-25; [n.s.] no. 1+, 1937+. Bull. Forest. Commiss. Victoria. HI 63176

Bulletin, forests department, Sudan. No. 8+, 1964+. Bull. Forests Dept. Sudan. Preceded by: Forests bulletin, forestry division, Sudan. HI 63177

Bulletin of the forests department, Western Australia = Bulletin, woods and forests department, Western Australia. Perth, W.A.

Bulletin of the free museum of science and art of the university of Pennsylvania. Philadelphia, PA. Vols. 1-3, 1897-1902. Bull. Free Mus. Sci. Art Univ. Pennsylvania. HI 63178

Bulletin, fruit branch, department of agriculture, Canada. Ottawa. Nos. 1-12, 1915-17. Bull. Fruit Branch Dept. Agric. Canada. Superseded by: Bulletin, department of agriculture, Canada. New series. HI 63179

Bulletin of the fruit tree research station. Series A, (Hiratsuka). [Kaju shikenjo. Hokoku. A (Hiratsuka).] Hiratsuka. No. 1+, 1974+. Bull. Fruit Tree Res. Sta., A (Hiratsuka). Preceded by: Bulletin, horticultural research station, Hiratsuka. HI 63180

Bulletin of the fruit tree research station. Series B, (Okitsu). [Kaju shikenjo hokoku. B (Okitsu).] Shimizu. No. 1+, 1974+. Bull. Fruit Tree Res. Sta., B (Okitsu). Preceded by: Bulletin of the horticultural research station. Series B, Okitsu. HI 63181

Bulletin of the fruit tree research station. Series C, (Morioka). [Kaju shikenjo hokoku. C (Morioka).] Morioka. No. 1+, 1974+. Bull. Fruit Tree Res. Sta., C (Morioka). Preceded by: Bulletin of the horticultural research station. Series C, Morioka. HI 63182

Bulletin of the fruit tree research station. Series D, (Kuchinotsu). [Kaju shikenjo hokoku. D (Kuchinotsu).] Kuchinotsu. No. 1+, 1977+. Bull. Fruit Tree Res. Sta., D (Kuchinotsu). HI 63183

Bulletin of the fruit tree research station. Series E, (Akitsu). [Kaju shikenjo hokoku. E (Akitsu).] Akitsu. No. 1+, 1976+. Bull. Fruit Tree Res. Sta., E (Akitsu). HI 68282

Bulletin, Fuji women's college. Series 2. Sapporo. No. ?-5+, 19??-67+. Bull. Fuji Women's Coll., Ser. 2. HI 63184

Bulletin of Fukui prefecture agricultural experiment station. [Fukui-ken nogyo shikenjo hokoku.] Fukui, Japan. Bull. Fukui Prefect. Agric. Exp. Sta. See B–P–H 251/6. HI 52751

Bulletin Fukuoka gakugei university. Part III. Natural science. [Fukuoka gakugei daigaku kiyo III. Rikakeito.] Fukuoka, Japan. Bull. Fukuoka Gakugei Univ., Pt. 3, Nat. Sci. See B–P–H 251/8. HI 52752

Bulletin of the Fukuoka horticultural experiment station. [Fukuoka kenritsu engei shikenjo, kenkyu hokoku.] Fukuoka. No. ?-16+, 19??-78+. Bull. Fukuoka Hort. Exp. Sta. HI 63185

Bulletin of Fukuoka prefectural forest experiment station. [Fukuoka-ken ringyo shikenjo jiho.] Kuroki. 1946+. Bull. Fukuoka Pref. Forest Exp. Sta. HI 63186

Bulletin of the Fukushima horticultural experiment station. [Fukushima-ken engei shikenjo, kenkyu hokoku.] Fukushima. No. 1+, 1968+. Bull. Fukushima Hort. Exp. Sta. HI 63187

Bulletin of the garden center of Greater Cincinnati. Cincinnati, OH. 19??-53+. Bull. Gard. Center Gr. Cincinnati. HI 63188

Bulletin of the garden center of Rochester. Rochester, NY. Vols. 1-7, 1945-52. Bull. Gard. Center Rochester. Superseded by: Garden center bulletin. Rochester, NY. HI 63189

Bulletin of the garden club of America. Chicago, IL. Bull. Gard. Club Amer. See B–P–H 251/10. HI 52753

Bulletin général de thérapeutique médicale et chirurgicale. Paris. Bull. Gén. Thérap. Méd. Chir. See B–P–H 251/12. HI 52754

Bulletin général et universel des annonces et des nouvelles scientifiques, ... [Edited by Férussac.] Paris. Bull. Gén. Universel Annonces Nouv. Sci. See B–P–H 251/13. HI 52755

Bulletin, genetics society of Canada. Ottawa. Vol. 1+,Bull. Genet. Soc. Canada. HI 63190

Bulletin of the geological institute = Izvestiya na geologicheskiya institut. Sofia.

Bulletin of the geological institute. Series paleontology = Izvestiya na geologicheskiya institut. Seriya paleontologiya. Sofia.

Bulletin of the geological and natural history survey. St. Paul, MN. Bull. Geol. Nat. Hist. Surv. See B–P–H 251/16. HI 52756

Bulletin of the geological society of America. Washignton, DC. Bull. Geol. Soc. Amer. See B–P–H 251/18. HI 52757

Bulletin of the gological society of China. [Chung kuo ti chih hsüeh hui chih.] Nanking. Bull. Geol. Soc. China See B–P–H 251/19. HI 52758

Bulletin of the geological survey of Canada. Ottawa. Bull. Geol. Surv. Canada. See B–P–H 251/21. HI 52759

Bulletin of the geological survey of Chosen. Keijo. Vols. 1-11, 1919-39. Bull. Geol. Surv. Chosen. HI 63191

Bulletin of the Georgia academy of science. Atlanta. Vols. 1-34, 1943-76. Bull. Georgia Acad. Sci. Superseded by: Georgia journal of science. 2-1705-1. HI 63192

Bulletin, Georgia coastal plain experiment station. Tifton, GA. Nos. 1-54, 1921-54. Bull. Georgia Coastal Plain Exp. Sta. Superseded by: Bulletin, agricultural experiment stations, Georgia. HI 63193

Bulletin of the Georgia society of naturalists. Atlanta, GA. Bull. Georgia Soc. Naturalists. See B–P–H 252/1. HI 52760

Bulletin of the georgian experiment station of plant protection. Ser. A, phytopathology = Izvestiya gruzinskoi opytnoi stantsii zashchity rastenii. Seriya A, fitopatologia. Tiflis.

Bulletin geranium society. Vols. ?-1(2)-13(1), 19??-52-64. Bull. Geranium Soc. Superseded by: Bulletin, british pelargonium and geranium society. HI 63194

Bulletin, Glamorgan county naturalists' trust. [Swansea.] No. 1+, 1962+. Bull. Glamorgan County Naturalists'

Trust. HI 63195

Bulletin of the government agricultural research institute of Formosa. [Taiwan sotukufu nogyobu, shikenjo iho.] Taihoku. Nos. 1-224, 1922-44. Bull. Gov. Agric. Res. Inst. Formosa. Preceded by: Bulletin of the Taihoku agricultural experiment station. Superseded by: Bulletin of the Taiwan agricultural research institute. 5-4150-2. HI 63196

Bulletin, government botanical garden Nikita = Byulleten′, gosudarstvennyi nikitskii opytnii botanicheskii sad. Yalta.

Bulletin of the government forest experiment station. [Ringyo shikenjo kenkyu hokoku.] Meguro, Tokyo, etc. Nos. 40-298, 1948-77. Bull. Gov. Forest Exp. Sta. Preceded by: Bulletin of the imperial forest experiment station. Tokyo. Superseded by: Bulletin of the forestry and forest products research institute. HI 52761

Bulletin of the government museum. New series. Natural history section. Madras. Bull. Gov. Mus., New Ser., Nat. Hist. Sect. See B–P–H 252/6. HI 52762

Bulletin of the Gray memorial botanical association. Fayetteville, NC. Vols. 1-7, 1933-39. Bull. Gray Memorial Bot. Assoc. 2-1759-1. HI 63197

Bulletin of the Gray memorial chapter of the Agassiz association. [Place of publication varies.] Vol. [1](1), 1893. Bull. Gray Mem. Chapt. Agassiz Assoc. Superseded by: Asa Gray bulletin. HI 63198

Bulletin, groupe d'études des rythmes biologiques. Clermond-Ferrand. No. 1+, 1969+. Bull. Groupe Études Rhythmes Biol. HI 63199

Bulletin, groupe internationale pour l'étude des mimosoideae. Toulouse-Cédex. Vol. 1+, 1973+. Bull. Groupe Int. Étude Mimosoideae. HI 63200

Bulletin, Guam agricultural experiment station, office of experiment stations, United States department of agriculture = U S department of agriculture, office of experiment stations; the Guam agricultural experiment station, bulletin. Washington, DC. U.S.D.A. Off. Exp. Sta. Guam Agric. Exp. Sta. Bull. See B–P–H 943/25.

Bulletin of the Gulf biologic station. Cameron, LA. Bull. Gulf Biol. Sta. See B–P–H 252/10. HI 52763

Bulletin, gulf fauna and flora = Gulf fauna and flora bulletin. Ruston, LA. Gulf Fauna Fl. Bull. See B–P–H 409/8.

Bulletin of the Hadley climatological laboratory of the university of New Mexico. Albuquerque, NM. Bull. Hadley Climatol. Lab. See B–P–H 252/14. HI 52764

Bulletin of the Hakodate marine observatory. Hakodate, Japan. Bull. Hakodate Mar. Observ. See B–P–H 252/15. HI 52765

Bulletin of the Hancock museum, Newcastle-upon-Tyne. Newcastle-upon-Tyne. Bull. Hancock Mus. Newcastle-upon-Tyne. See B–P–H 252/16. HI 52766

Bulletin of the hardy plant society. London, Reading. Vols. 1-6?, 1957-84? Bull. Hardy Pl. Soc. Superseded by: Hardy plant. HI 52767

Bulletin of the Harpswell laboratory for biological research & study. Salisbury Cove, ME. Bull. Harpswell Lab. Biol. Res. See B–P–H 252/18. HI 52768

Bulletin of the Harvard forest. Petersham, MA. Bull. Harvard Forest. See B–P–H 252/20. HI 52769

Bulletin of the Harvard forestry club. Cambridge, MA. Bull. Harvard Forest. Club. See B–P–H 252/21. HI 52770

Bulletin of the Hatano tobacco experiment station. [Nihon senbai kosha hatano tabako shikenjo hokoku.] Kanagawa, Japan. Bull. Hatano Tobacco Exp. Sta. See B–P–H 252/22. HI 52771

Bulletin, Hatch experiment station of Massachusetts agricultural college = Hatch experiment station of Massachusetts agricultural college. Bulletin. Amherst, MA. Hatch Exp. Sta. Mass. Agric. Coll. Bull. See B–P–H 412/18.

Bulletin, Hawaii agricultural experiment station. Honolulu, HI. Nos. 1-142, 1901-67. Bull. Hawaii Agric. Exp. Sta. Superseded by: Research bulletin, Hawaii agricultural experiment station. HI 63201

Bulletin of the Hawaii orchid society. Hilo, HI. Bull. Hawaii Orchid Soc. See B–P–H 253/1. HI 52772

Bulletin, hawaiian sugar planters' association experiment station. Botanical series. Honolulu, HI. Vol. 3-55, 1921-45. Bull. Hawaiian Sugar Planters' Assoc. Exp. Sta., Bot. Ser. Preceded by: Bulletin, division of pathology and physiology, Hawaiian sugar planters' association experiment station. 3-1824-1. HI 63202

Bulletin, the heather society. Horley. No. 1-2(15), 1967-78. Bull. Heather Soc. Superseded by: Heather society bulletin. HI 63203

Bulletin hebdomadaire de l'association scientifique de France. Paris. Bull. Hebd. Assoc. Sci. France See B–P–H 253/2. HI 52773

Bulletin, heliconia society international. Sarasota, FL. Vol. 1+, 1985+. Bull. Heliconia Soc. Int. HI 63204

Bulletin of the Hemlock arboretum at "Far Country." Philadelphia, PA. Bull. Hemlock Arbor. See B–P–H 253/3. HI 52774

Bulletin of the herb society of America. Boston, MA. Bull. Herb Soc. Amer. See B–P–H 253/8. HI 52776

Bulletin of the herbarium of the north-eastern forestry academy. [Tung pei lin hsüeh yuan chih wu piao pen shih hui ken.] Harbin. Nos. 1-3, 1959-61. Bull.

Herbarium N. E. Forest. Acad., Harbin. HI 63205

Bulletin de l'herbier Boissier. Geneva. Vols. 1-7, 1893-99; sér. 2, vols., 1-8, 1901-02. Bull. Herb. Boissier. For 1900 see: Mémoires de l'herbier Boissier suite au bulletin de l'herbier Boissier. 2-975-3. HI 63206

Bulletin de l'herbier de l'institut botanique de Bucarest. Bucharest. Bull. Herb. Inst. Bot. Bucarest. See B–P–H 253/7. HI 52775

Bulletin of the Hiroshima agricultural college. [Hiroshima nogyo tanki daigaku kenkyu.] Saijo. Vol. 1+, 1958+. Bull. Hiroshima Agric. Coll. HI 63207

Bulletin of the Hiroshima botanical garden. Kurashige. Vol. 1+, 1977+. Bull. Hiroshima Bot. Gard. HI 63208

Bulletin of Hiroshima prefectural experiment station. [Hiroshima-kenritsu nogyo shikenjo hokoku.] Hiroshima. Bull. Hiroshima Prefect. Exp. Sta. See B–P–H 253/9. HI 52777

Bulletin of the Hiroshima prefectural forest experiment station. [Hiroshima-kenritsu ringyo shikenjo kenkyu hokoku.] Miyoshi. 1955+. Bull. Hiroshima Pref. Forest Exp. Sta. HI 63411

Bulletin of the Hiroshima women's university. Series 2, natural science. [Hiroshima joshi daigaku kiyo. Dai-2-bu, shizenkagaku.] Hiroshima. Nos. 1-3, 1966-68. Bull. Hiroshima Women's Univ., Ser. 2. Superseded by: Bulletin of the faculty of home economics, Hiroshima women's university [not entered]. HI 63209

Bulletin d'histoire naturelle de France. Paris. Bull. Hist. Nat. France See B–P–H 253/11. HI 52779

Bulletin de l'histoire naturelle de la société linnéenne de Bordeaux. Bordeaux. Bull. Hist. Nat. Soc. Linn. Bordeaux. See B–P–H 253/12. HI 52781

Bulletin of the history of medicine. Baltimore, MD. Bull. Hist. Med. See B–P–H 253/10. HI 52778

Bulletin, Hoblitzelle agricultural laboratory. Texas research foundation. Renner, TX. Nos. 1-27, 1950-69. Bull. Hoblitzelle Agric. Lab. Texas Res. Found. HI 63210

Bulletin of the Hokkaido agricultural experiment station. Sapporo. Nos. 1-61, 1905-37. Bull. Hokkaido Agric. Exp. Sta. Superseded by: Research bulletin of the Hokkaido national agricultural research station. HI 63211

Bulletin of the Hokkaido forest experiment station. [Hokkaido ringyo shikenjo hokoku.] Bibai. 1962+. Bull. Hokkaido Forest Exp. Sta. HI 64058

Bulletin of the Hokkaido prefectural agricultural experiment station. [Hokkaidoritsu nogyo shikenjo shuho.] Sapporo, Japan. Bull. Hokkaido Prefect. Agric. Exp. Sta. See B–P–H 253/15. HI 52782

Bulletin of the Hokkaido university forest. [Hokkaido daigaku nogaku-bu enshurin kenkyu hokoku.] Sapporo, Japan. Bull. Hokkaido Univ. Forest. See B–P–H 253/17. HI 52783

Bulletin of the Hokuriku agricultural experiment station. [Hokuriku nogyo shikenjo hokoku.] Niigata, Japan. Bull. Hokuriku Agric. Exp. Sta. See B–P–H 253/18. HI 52784

Bulletin of the holly society of America. Baltimore, MD. Bull. Holly Soc. Amer. See B–P–H 253/19. HI 52785

Bulletin horticole, agricole et apicole. Liége. Bull. Hort. See B–P–H 253/20. HI 52786

Bulletin of the horticultural research station. Series A, Hiratsuka. [Engei shikenjo hokoku. A, Hiratsuka.] Hiratsuka. Vols. 1-12, 1962-73. Bull. Hort. Res. Sta., Ser. A, Hiratsuka. Superseded by: Bulletin of the fruit tree research station. Series A, (Hiratsuka). HI 63212

Bulletin of the horticultural research station. Series B, Okitsu. [Engei shikenjo hokoku. B, Okitsu.] Shimizu. Vols. 1-12, 1962-72. Bull. Hort. Res. Sta., Ser. B, Okitsu. Superseded by: Bulletin of the fruit tree research station. Series B, (Okitsu). HI 63213

Bulletin of the horticultural research station. Series C, Morioka. [Engei shikenjo hokoku. C, Morioka.] Morioka. Vols. 1-8, 1963-73. Bull. Hort. Res. Sta., Ser. C., Morioka. Preceded by: Bulletin of the fruit tree research station. Series C, (Morioka) and Bulletin of the horticultural research station. Series D, Kurume. HI 63214

Bulletin of the horticultural research station. Series D, Kurume. [Engei shikenjo hokoku. D, Kurume.] Fukuoka. Vols. 1-7, 1965-72. Bull. Hort. Res. Sta., Ser. D, Kurume. Superseded by: Bulletin of the vegetable and ornamental crops research station. Series C, (Kurume). HI 63215

Bulletin of the horticultural research station, Series D, Kurume, miscellaneous publication. [Norin-sho engei shikenjo kurume shijo rinji hokoku.] Fukuoka, Japan. Bull. Hort. Res. Sta., Ser. D, Kurume, Misc. Publ. See B–P–H 254/4. HI 52787

Bulletin, horticultural society of New York. New York, NY. Vols. 7(2)-19, 1957-69; n.s. vol. 1+, 1970+. Bull. Hort. Soc. New York. Preceded by: News, horticultural society of New York. HI 63216

Bulletin, Hull natural history society. Hull. Vol. 3(2)+, 1971+. Bull. Hull Nat. Hist. Soc. Preceded by: Bulletin, Hull scientific and field naturalists' club. HI 63217

Bulletin, Hull scientific and field naturalists' club. Hull. Vols. 1-3(1), 1947-69; Supplement no. 1, 1970. Bull. Hull. Sci. Field Naturalists' Club. Superseded by: Bulletin, Hull natural history society. HI 63218

Bulletin of the Hunt institute for botanical documentation. Pittsburgh, PA. Vol. 1+, 1979+. Bull. Hunt Inst. Bot. Doc. HI 63219

Bulletin hydrographique. Conseil international pour l'exploration de la mer. Copenhagen. 1908-56. Bull. Hydrogr. Preceded by: Bulletin trimestriel des résultats acquis pendant les crosières périodique. Conseil permenent international pour l'exploration de la mer. Superseded by: I C E S oceanographic data lists. 3-2056-2. HI 63220

Bulletin of the Hyogo prefectural agricultural experiment station. [Hyogo kenritsu nogyo shikenjo. Kenkyu hokoku.] Akashi. Vols. 1-24, 1954-76. Bull. Hyogo Prefect. Agric. Exp. Sta. Superseded by: Hyogo-ken nogyo sogo senta. Kenkyu hokoku. 5-4397-3. HI 63221

Bulletin, I F C C = I F C C bulletin. Paris.

Bulletin of the Ibaraki agricultural experiment station. [Ibaraki-ken nogyo shikenjo kenkyui hokoku.] Mito, Japan. Bull. Ibaraki Agric. Exp. Sta. See B–P–H 254/9. HI 52788

Bulletin of the Ibaraki horticultural experiment station. [Ibaraki-ken engei shikenjo kenkyu hokoku.] Ibaraki. No. 1+, 1964+. Bull. Ibaraki Hort. Exp. Sta. HI 64984

Bulletin of the Ibaraki prefectural forest experiment station. [Ibaraki-ken ringyo shikenjo kenkyu hokoku.] Mito. No. 1+, 1966+. Bull. Ibaraki Pref. Forest Exp. Sta. HI 63222

Bulletin of the Illinois natural history survey. Urbana, IL. Bull. Illinois Nat. Hist. Surv. See B–P–H 254/12. HI 52789

Bulletin of the Illinois state laboratory of natural history. Urbana, IL. Bull. Illinois State Lab. Nat. Hist. See B–P–H 254/14. HI 52790

Bulletin of the Illinois state microscopical society. Bull. Illinois State Microscop. Soc. See B–P–H 254/15. HI 52791

Bulletin of immediate information, Connecticut agricultural experiment station = Connecticut agricultural experiment station. Bulletin of immediate information. New Haven, CT. Connecticut Agric. Exp. Sta. Bull. Immed. Inform. See B–P–H 324/22.

Bulletin of the imperial college of agriculture and forestry. [Morioka koto norin gakko gajukutsu hokoku.] Morioka, Japan. Bull. Imp. Coll. Agric. See B–P–H 254/17. HI 52792

Bulletin of the imperial forest experiment station. [Ringyo shiken hokoku.] Tokyo. Bull. Imp. Forest Exp. Sta. See B–P–H 254/19. HI 52793

Bulletin of the imperial institute of agricultural research, Pusa. Calcutta. Nos. ?-207, 1928?-31. Bull. Imp. Inst. Agric. Res. Pusa. Preceded by: Bulletin of the agricultural research institute, Pusa. 3-1954-2. HI 63223

Bulletin of the imperial institute of Great Britain. London. Nos. 3-46, 1905-48. Bull. Imp. Inst. Gr. Brit. Preceded by: Quarterly supplement of the board of trade journal. Superseded by: Colonial plant and animal products. 2-1760-2. HI 63224

Bulletin of the imperial sericultural experiment station, Japan. [Suginami-machi.] Tokyo. Bull. Imp. Seric. Exp. Sta. Japan See B–P–H 255/3. HI 52794

Bulletin of the independent biological laboratories, Palestine. Kefar-Malal, Ramatayim. Vols. 1-14(1), 1932-60. Bull. Indep. Biol. Lab. Palestine. HI 63225

Bulletin of the India (republic) geological survey. Series A. Economic geology. Calcutta. Bull. India Republ. Geol. Surv., Ser. A, Econ. Geol. See B–P–H 255/4. HI 52795

Bulletin of the India (republic) geological survey. Series B. Engineering geology and ground water. Calcutta. Bull. India Republ. Geol. Surv., Ser. B, Engin. Geol. See B–P–H 255/5. HI 52796

Bulletin, indian central coconut committee. Ernakulam. 1949-57. Bull. Indian Centr. Coconut Committee. Superseded by: Coconut bulletin. HI 63226

Bulletin of the indian national science academy. New Delhi. No. 41+, 1970+. Bull. Indian Natl. Sci. Acad. Preceded by: Bulletin, national institute of sciences of India. HI 63227

Bulletin of the indian phytopathological society. New Delhi. Nos. 1-6, 1963-70. Bull. Indian Phytopathol. Soc. HI 63228

Bulletin of indian raw materials and their utilization; a quarterly list. New Delhi. Vols. 1-2, 1974-76. Bull. Indian Raw Mater. Utiliz. Superseded by: Bulletin of indian raw materials and their utilization. Series A, current literature on medicinal and aromatic plants. HI 63229

Bulletin of indian raw materials and their utilization. Series A, current literature on medicinal and aromatic plants. New Delhi. Vols. 3A-4A, 1977-78. Bull. Indian Raw Mater. Utiliz., A. Preceded by: Bulletin of indian raw materials and their utilization. A quarterly list. Superseded by: Medicinal and aromatic plants abstracts. HI 63230

Bulletin, indigenous bulb growers' association of South Africa = I B S A bulletin. Durbanville.

Bulletin d'information de l'académie suisse des sciences humaines et de la société helvétique des sciences naturelles = Mitteilungsblatt der schweizerischen Akademie der Geisteswissenschaften und der schweizerischen naturforschenden Gesellschaft. Bern.

Bulletin d'information, association des biologistes du Québec. Montreal. 1984?+. Bull. Inform. Assoc. Biol. Québec. Preceded by: Bio. HI 63231

Bulletin d'information de l'association française de lichénologie. Paris. No. 1+, 1976+. Bull. Inform. Assoc. Franç. Lichénol. HI 63232

Bulletin d'information, comité central d'océanographie et d'étude des côtes. Paris. Bull. Inform. Comité Centr. Océanogr. See B–P–H 255/8. HI 52798

Bulletin d'information C O R E S T A. Centre de coopération pour les recherches scientifiques rélatives au tabac. Paris. Bull. Inform. CORESTA. See B–P–H 255/9. HI 52799

Bulletin d'information E U C A R P I A. Nos. 1-3, 19??-71? Bull. Inform. E. U. C. A. R. P. I. A. Superseded by: Bulletin E U C A R P I A. HI 63233

Bulletin d'information sur l'environnement, commission du Pacifique sud. Noumea. Vol. 1+, 1975+. Bull. Inform. Environm. Commiss. Pacifique Sud. HI 63234

Bulletin d'information, institut belge pour l'amélioration de la betterave. Tirlemont. Vols. 1-?, 1966?-83. Bull. Inform. Inst. Belge Amélior. Betterave. Superseded by: Bulletin d'information, institut royal belge pour l'amélioration de la betterave. HI 63235

Bulletin d'information de l'institut pour l'étude agronomique du Congo. Prussels. Bull. Inform. Inst. Étude Agron. Congo. See B–P–H 255/10. HI 52800

Bulletin d'information, institut royal belge pour l'amélioration de la betterave. Tirlemont. 1983+. Bull. Inform. Inst. Roy. Belge Amélior. Betterave. Preceded by: Bulletin d'information, institut belge pour l'amélioration de la betterave. HI 63236

Bulletin d'information, muséum national d'histoire naturelle. Paris. No. 1+, 1974+. Bull. Inform. Mus. Natl. Hist. Nat. HI 63237

Bulletin d'information, office national des forêts. Paris. Vol. ?-25+, 19??-73+. Bull. Inform. Off. Natl. Forêts. HI 63238

Bulletin of information, Pennsylvania state college agricultural experiment station = Pennsylvania state college agricultural experiment station. Bulletin of information. State College, PA. Pennsylvania State Coll. Agric. Exp. Sta. Bull. Inform. See B–P–H 700/28.

Bulletin d'information phytosanitaires interafricain = Interafrican phytosanitary bulletin. Yaounde.

Bulletin d'information des planteurs de betteraves = Planteurs de betteraves. Paris.

Bulletin d'information, société d'écologie. Brunoy. 1974+. Bull. Inform. Soc. Écol. HI 63239

Bulletin d'information de la société d'ethnozoologie et d'ethnobotanique. Supplément. Paris. Nos. 1-12, 1969-77. Bull. Inform. Soc. Ethnozool. Ethnobot. Superseded by?: Bulletin bibliographique de la société d'ethnozoologie et d'ethnobotanique. HI 63240

Bulletin d'information, société française d'écologie. Brunoy. Vol. 1(1), 1969. Bull. Inform. Soc. Franç. Écol. Superseded by: Bulletin, société d'écologie. HI 63241

Bulletin d'information, société mycologie du Béarn. Pau. No. ?-71+, 19??-80+. Bull. Inform. Soc. Mycol. Béarn. HI 63242

Bulletin d'information de la société suisse des sciences humaines et de la société helvétique des sciences naturelles = Mitteilungsblatt der schweizerischen Geisteswissenschaftlichen Gesellschaft und der schweizerischen naturforschenden Gesellschaft. Bern.

Bulletin inspectie van den landbouw. Paramaribo, Dutch Guiana [Surinam]. Bull. Inspect. Landb. See B–P–H 255/14. HI 52801

Bulletin de l'institut agronomique de l'état et des stations de recherches de Gembloux. Bull. Inst. Agron. État Gembloux. See B–P–H 255/18. HI 52803

Bulletin de l'institut botanique de Buitenzorg. Buitenzorg, Dutch E. Indies [=Bogor, Indonesia]. Bull. Inst. Bot. Buitenzorg. See B–P–H 255/24. HI 52804

Bulletin de l'institut botanique de Caen. Caen. Nos. 1-9, 1936-56. Bull. Inst. Bot. Caen. HI 63244

Bulletin de l'institut botanique. Sofia = Izvestiya na botanicheskaya institut. Sofia.

Bulletin de l'institut coloniale de Marseille. Section des matières grasses. Marseilles. Vols. 1-30, 1917-46. Bull. Inst. Colon. Marseille, Sect. Matières Grasses. HI 63245

Bulletin de l'institut du désert d'Egypte. Heliopolis. Vols. 3-15, 1953-65. Bull. Inst. Désert Egypte. Preceded by: Bulletin de l'institut Fouad I du désert. Superseded by: Desert institute bulletin, U. A. R. HI 63246

Bulletin de l'institut d'écologie appliquée. Orléans. 1974+. Bull. Inst. Écol. Appl. HI 63247

Bulletin de l'institut d'études centrafricaines. Brazzaville, French Equatorial Africa [Congo Republic]. Bull. Inst. Études Centrafr. See B–P–H 255/26. HI 52805

Bulletin de l'institut expérimental par la culture et la fermentation du tabac. Bucharest = Buletinul cultivări şi fermentări tutunului. Bucharest. Bul. Cult. Ferment. Tutunului. See B–P–H 231/33.

Bulletin de l'institut fondamental d'Afrique Noire. Sér. A. Sciences naturelles. Paris & Dakar, Republic of Senegal. Bull. Inst. Fondam. Afrique Noire, Sér. A., Sci. Nat. See B–P–H 256/1. HI 52806

Bulletin de l'institut Fouad I du désert. Heliopolis. Vols. 1-2, 1951-52. Bull. Inst. Fouad I Désert. Superseded by: Bulletin de l'institut du désert d'Egypte. HI 63248

Bulletin, institut français du café et du cacao = I F C C bulletin. Paris.

Bulletin de l'institut française d'Afrique Noire. Paris & Dakar, Senegal. Bull. Inst. Franç. Afrique Noire. See B–P–H 256/3. HI 52807

Bulletin, institut française d'études andines. Lima. Vol. 1+, 1972+. Bull. Inst. Franç. Études Andines. HI 63249

Bulletin de l'institut et du jardin botaniques de l'université de Belgrade. Belgrade. Bull. Inst. Jard. Bot. Univ. Belgrade. See B–P–H 256/6. HI 52809

Bulletin de l'institut microbiologique de l'académie bulgare des sciences = Izvestiya na mikrobiologicheskiya institut. Sofia.

Bulletin de l'institut national agronomique. Brno = Sborník vysoké školy zemědělské v Brně. Brno, Czechoslovakia. Sborn. Vysoké Školy Zeměd. v Brně. See B–P–H 817/1.

Bulletin de l'institut national genevois. Geneva. Vols. 1-62, 1853-1964. Bull. Inst. Natl. Genevois. Superseded by: Actes de l'institut national genevois. HI 63250

Bulletin de l'institut national scientifique et technique d'océanographie et de pêche de Salammbô. Tunis. N.s. 1966+. Bull. Inst. Natl. Sci. Techn. Océanogr. Pêche Salammbô. Preceded by: Bulletin, station océanographique de Salammbô. HI 63251

Bulletin de l'institut oceanographique, Monaco. Paris. Bull. Inst. Océanogr. Monaco. See B–P–H 256/9. HI 52810

Bulletin de l'institut Oinoue de recherches agronomiques et biologiques. [Oinoue rinogaku kenkyusho. Kenkyu hokoku.] Simo-Omi, Japan. Bull. Inst. Oinoue Rech. Agron. See B–P–H 256/10. HI 52811

Bulletin de l'institut Pasteur. Paris. Bull. Inst. Pasteur (Paris). See B–P–H 256/13. HI 52812

Bulletin de l'institut de pédologie et de géobotanique de l'université de l'Asie centrale = Izvestiya Instituta pochvovedeniya i geobotaniki Sredne-Aziatskogo gosudarstvennogo universiteta. Tashkent.

Bulletin de l'institut des recherches biologiques de Perm = Izvestiya Biologicheskogo nauchno-issledovatel'skogo instituta pri Permskom gosudarstvennom universitete (imeni A. M. Gor'kogo.) Perm.

Bulletin de l'institut des recherches biologiques de Perm = Izvestiya Permskogo biologicheskogo nauchno-issledovatel'skogo instituta. Perm.

Bulletin de l'institut des recherches biologiques et de la station biologique à l'université de Perm = Izvestiya Biologicheskogo nauchno-issledovatel'skogo instituta i Biologicheskoi stantsii pri Permskom gosudarstvennom universitete. Perm.

Bulletin de l'institut de recherches scientifiques au Congo. Brazzaville, Congo Republic. Bull. Inst. Rech. Sci. Congo. See B–P–H 256/20. HI 52814

Bulletin de l'institut royal des sciences naturelles de Belgique. Brussels. Vols. 25-47, 1949-71. Bull. Inst. Roy. Sci. Nat. Belgique. Preceded by: Bulletin du Musée Royal d'Histoire Naturelle de Belgique. Superseded by: Bulletin de l'institut royal des sciences naturelles de Belgique. Biologie. 1-807-1. HI 63252

Bulletin de l'institut royal des sciences naturelles de Belgique. Biologie. Brussels. Vol. 48+, 1972+. Bull. Inst. Roy. Sci. Nat. Belgique, Biol. Preceded by: Bulletin de l'institut royal des sciences naturelles de Belgique. HI 63253

Bulletin de l'institut scientifique de biologie et de géographie à l'université d'Irkoutsk = Izvestiya Biologo-geograficheskogo nauchno-issledovatel'skogo instituta pri Gosudarstvennom Irkutskom universitete. Irkutsk.

Bulletin de l'institut scientifique de biologie et de géographie à l'université d'Irkoutsk = Izvestiya Biologo-geograficheskogo nauchno-issledovatel'skogo instituta pri Vostochno-Sibirskom gosudarstvennom universitete. Moscow.

Bulletin de l'institut scientifique Lesshaft = Izvestiya Leningradskogo nauchnogo instituta imeni P. F. Lesgafta. Moscow & Leningrad.

Bulletin de l'institut scientifique Lesshaft = Izvestiya Nauchnogo instituta imeni P. F. Lesgafta. Leningrad.

Bulletin de l'institut scientifique de Saint-Pétersbourg (de l'institut Leshaft) = Izvestiya Petrogradskogo nauchnogo instituta imeni P. F. Lesgafta. Petrograd.

Bulletin de l'institut scientifique, université Mohammed V. Rabat. No. 1+, 1976+. Bull. Inst. Sci. Univ. Mohammed V. HI 63254

Bulletin of the institute for advanced research of Fukuoka university. Natural sciences. Fukuoka. 1977+. Bull. Inst. Advanced Res. Fukuoka Univ., Nat. Sci. HI 63255

Bulletin of the institute of agricultural microbiology = Trudy instituta sel'skokhozyaistvennoi mikrobiologii. Moscow, Leningrad.

Bulletin of the institute of agricultural research; Tohoku university. [Tohoku daigaku nogaku kenkyujo iho.] Sendai, Japan. Bull. Inst. Agric. Res. Tohoku Univ. See B–P–H 255/17. HI 52802

Bulletin, institute of agriculture and natural history, Tel-Aviv. Tel-Aviv. Nos. 1-58, 1924-50. Bull. Inst.

Agric. Nat. Hist. Tel-Aviv. Superseded by: Bulletin, agricultural experiment station, Rehovoth. HI 63256

Bulletin of the institute of biology in Shiga Heights. [Shiga kogen seibutsu kenkyusho kenkyu gyoseki.] Nagano. Nos. 1-4, 1962-65. Bull. Inst. Biol. Shiga Heights. Superseded by: Bulletin, institute of natural [nature?] education in Shiga Heights. HI 63257

Bulletin of the institute of biology. Sofia = Izvestiya na biologicheskiya institut. Bulgarska akademiya na naukite, otdelenie za biologicheskiye i meditsinski nauk. Sofia.

Bulletin, institute of environmental science and technology, Yokohama national university. [Yokohama kokuritsu daigaku. Kankyo kagaku kenkyu senta.] Kiyo. Vol. 1+, 1974+. Bull. Inst. Environm. Sci. Technol. Yokohama Natl. Univ. HI 63258

Bulletin of the institute of experimental agriculture of Georgia = Vestnik instituta experimental′noi agronomii Gruzii. Tiflis.

Bulletin of the institute of forestry and pedology, academia sinica. [Zhongguo kexueyuan linye turang yanjiusuo jikan.] Shenyang. No. 1+, 1964+. Bull. Inst. Forest. Pedol. Acad. Sin. HI 63259

Bulletin of the institute of genetics, Moscow = Trudy instituta genetiki, akademiya nauk S S S R. Moscow, Leningrad.

Bulletin of the institute of the history of medicine. Baltimore, MD. Bull. Inst. Hist. Med. See B–P–H 256/4. HI 52808

Bulletin of the institute of Jamaica. Science series. Kingston, Jamaica. Nos. 1-21, 1940-72. Bull. Inst. Jamaica, Sci. Ser. 3-2010-1. HI 63260

Bulletin of the institute of marine biology and oceanography, university of Sierra Leone. Fourah Bay. Vol. 1+, 1976+. Bull. Inst. Mar. Biol. Oceanogr. Univ. Sierra Leone. HI 63261

Bulletin of institute of natural education in Shiga Heights. [Shiga shizen kyoiku kenkyu shisetsu kenkyu gyoshi.] Nagano. No. 5+, 1966+. Bull. Inst. Nat. Educ. Shiga Heights. Preceded by: Bulletin of the institute of biology in Shiga Heights. HI 63262

Bulletin, institute of oceanography and fisheries, Cairo. Cairo. Vol. 1+, 1970+. Bull. Inst. Oceanogr. Fish. Cairo. Preceded by: Publications of the marine biological station, Ghardaqa, Red Sea and Notes and memoirs, hydrobiological department, ministry of scientific research. Cairo. HI 63263

Bulletin of the institute of radiation breeding = Acta radiobotanika et genetika. Ohmiya-Machi, Ibaraki.

Bulletin of the institute of scientific research, Manchoukuo. [Ta lu k'o hsüeh yuan hui pao.] Hsinking, Manchoukuo [=Changchun, China]. Bull. Inst. Sci. Res. Manchoukuo. See B–P–H 257/4. HI 52815

Bulletin, institute of south-east asian biology, university of Aberdeen. Aberdeen. No. ?-21+, 19??-76+. Bull. Inst. S. E. Asian Biol. Univ. Aberdeen. HI 63264

Bulletin, institute for the study of science in human affairs. New York. No. 1+, 1968+. Bull. Inst. Study Sci. Human Affairs. HI 63265

Bulletin of the institute of tropical agriculture, Kyushu university. [Kyushu daigaku. Nettai nogaku kenkyu senta.] Fukuoka. Vol. 1+, 1975+. Bull. Inst. Trop. Agric. Kyushu Univ. HI 63266

Bulletin des institutions r[oyales] d'histoire naturelle à Sophia = Izvestiya na tsarskite prirodonauchni instituti v Sofiya. Sofia.

Bulletin van het instituut voor plantenziekten, uitgegeven door het departement voor landbouw, nijverheid en handel in Nederlandsch-Indië. Buitenzorg, Dutch E. Indies [=Bogor, Indonesia]. Bull. Inst. Plantenziekten. See B–P–H 256/14. HI 52813

Bulletin of the inter-american institute of agricultural science of the O A S = C A T I E, actividades en Turrialba.

Bulletin international de l'académie croate des sciences et des beaux-arts. Classe des sciences mathématiques et naturelles. Zagreb, Yugoslavia. Bull. Inst. Acad. Croate Sci., Cl. Sci. Math. See B–P–H 257/8. HI 52816

Bulletin international de l'académie polonaise des sciences et des lettres. Classe des sciences mathématiques et naturelles. Série B 1. Botanique. Cracow. Bull. Int. Acad. Polon. Sci., Cl. Sci. Math., Sér. B 1, Bot. See B–P–H 257/10. HI 52817

Bulletin international de l'académie polonaise des sciences et des lettres. Classe des sciences mathématiques et naturelles. Série B. Sciences naturelles. Cracow. Bull. Int. Acad. Polon. Sci., Cl. Sci. Math., Sér. B, Sci. Nat. See 257/11. HI 52818

Bulletin international de l'académie des sciences de Cracovie. Cracow. Bull. Int. Acad. Sci. Cracovie. See B–P–H 257/12. HI 52819

Bulletin international de l'académie des sciences de Cracovie. Classe des sciences mathématiques et naturelles. Cracow. Bull. Int. Acad. Sci. Cracovie, Cl. Sci. Math. See B–P–H 257/13. HI 52820

Bulletin international de l'académie des sciences de Cracovie. Classe des sciences mathématiques et naturelles. Série B. Sciences naturelles. Cracow. Bull. Int. Acad. Sci. Cracovie, Cl. Sci. Math., Sér. B, Sci. Nat. See B–P–H 257/14. HI 52821

Bulletin international, académie des sciences de l'empereur François Joseph. Classe des sciences mathématiques, naturelles et de la médecine. Prague. Vols. 1-22, 1894-1920. Bull. Int. Acad. Sci. Emp. François Joseph, Cl. Sci. Math. Nat. Méd. Superseded by: Bulletin international, académie tchéque des sciences. Classe des sciences mathématiques, naturelles et de la médecine. HI 63267

Bulletin international, académie tchéque des sciences. Classe des sciences mathématiques, naturelles et de la médecine. Prague. Vols. 23-53, 1923-53. Bull. Int. Acad. Tchéque Sci., Cl. Sci. Math. Nat. Méd. Preceded by: Bulletin international, académie des sciences de l'empereur François Joseph. HI 63268

Bulletin international de l'académie Yougoslave des sciences et des beaux-arts. Classe des sciences mathématiques et naturelles Zagreb. Vols. 24-33, 1930-39; n.s. vols. 1+, 1948+. Bull. Int. Acad. Yougoslave Sci., Cl. Sci. Math. Preceded by: Izvješća o raspravama matematičko-prirodoslovnoga razreda, jugoslavenska akademija znanosti i umjetnosti o Zagrebu. For vols. 34-35 see: Bulletin international de l'académie Croate des sciences et des beaux-arts. Classe des sciences mathématiques et naturelles. HI 63269

Bulletin of the international association of scientific hydrology. London. Bull. Int. Assoc. Sci. Hydrol. See B–P–H 258/2. HI 52822

Bulletin, international association of wood anatomists = I A W A bulletin. Leiden, Zurich.

Bulletin, international association of wood anatomists = International association of wood anatomists, news bulletin. Zurich.

Bulletin of the international group for the study of mimosoideae = Bulletin, groupe internationale pour l'étude des mimosoideae. Toulouse-Cédex.

Bulletin, international institute for land reclamation and improvements. Wageningen No. 1+, 1958+. Bull. Int. Inst. Land Reclam. Improv. HI 63270

Bulletin, international peat society. Helsinki. No. 1+, 1970+. Bull. Int. Peat Soc. HI 63271

Bulletin of the international society of arboriculture. Connersville, IN. Bull. Int. Soc. Arboric. See B–P–H 258/3. HI 52823

Bulletin of the international society for tropical ecology. Poona. Vol. 1, 1960. Bull. Int. Soc. Trop. Ecol. Superseded by: Tropical ecology. HI 63272

Bulletin, international union for conservation of nature and natural resources. Brussels, Morges. Vols. 5(4)-9, 1956-60. Bull. Int. Union Conservation Nature Nat. Resources. Preceded by: Bulletin, international union for the protection of nature. Superseded by: I U C N bulletin. HI 63273

Bulletin, international union for the protection of nature. Brussels. Vols. 1-5(2/3), 1952-56. Bull. Int. Union Prot. Nat. Superseded by: Bulletin, international union for conservation of nature and natural resources. HI 63274

Bulletin, Iowa agricultural experiment station = Iowa agricultural experiment station. Bulletin. Ames, IA. Iowa Agric. Exp. Sta. Bull. See B–P–H 436/33.

Bulletin of Iowa university's laboratories of natural history. Iowa City, IA. Bull. Iowa Univ. Lab. Nat. Hist. See B–P–H 258/7. HI 52824

Bulletin of the Iraq natural history museum. Baghdad. Vols. 1-5, 1961-72. Bull. Iraq Nat. Hist. Mus. Superseded by: Bulletin of the natural history research centre. HI 63276

Bulletin, irish biogeographical society. Bray. 1976/77+. Bull. Irish Biogeogr. Soc. HI 63277

Bulletin of the Ishikawa prefectural marine culture station. [Ishikawa-ken zoshoku shikenjo kenkyu hokoku.] Notojima. No. 1+, 1970+. Bull. Ishikawa Pref. Mar. Cult. Sta. HI 63278

Bulletin, Ishikawa prefecture college of agriculture. [Ishikawa-ken nogyo tanki daigaku. Kenkyu hokoku.] Ishikawa. No. 1+, 1972+. Bull. Ishikawa Pref. Coll. Agric. HI 63279

Bulletin of the Iwate prefecture agricultural experiment station. [Iwate-ken nogyo shikenjo kenkyu hokoku.] Morioka, Japan. Bull. Iwate Prefect. Agric. Exp. Sta. See B–P–H 258/8. HI 52825

Bulletin of the Iwate university forests. [Iwate daigaku nogakubu enshurin hokoku.] Morioka. ?-1983+. Bull. Iwate Univ. Forests. HI 63280

Bulletin of the Japan alpine rock garden society. Hyogo-ken. Vol. 1+, 1979+. Bull. Japan Alpine Rock Gard. Soc. HI 63281

Bulletin, Japan association of botanical gardens. Tokyo. 1946+. Bull. Japan Assoc. Bot. Gard. HI 63282

Bulletin, Japan orchid society = Japan orchid society bulletin. Osaka.

Bulletin of japanese association of benthology. [Nihon bentosu kenkyukai shi.] No. 21/22+, 1981+. Bull. Jap. Assoc. Benthol. HI 63284

Bulletin of the japanese society of phycology. [Sorui.] Sapporo. Vols. 1-25, 1953-77. Bull. Jap. Soc. Phycol. Superseded by: Japanese journal of phycology. HI 63285

Bulletin of the japanese society of scientific fisheries. [Nippon suisangakkaishi.] Tokyo. Bull. Jap. Soc. Sci. Fish. See B–P–H 258/10. HI 52826

Bulletin de jardin botanique de l'académie des sciences de l'URSS = Izvestiya Botanicheskogo sada Akademii nauk S S S R. Leningrad.

Bulletin du jardin botanique de Bruxelles = Bulletin du jardin botanique de l'état. Brussels.

Bulletin du jardin botanique de Buitenzorg. Buitenzorg, Dutch E. Indies [=Bogor, Indonesia]. Bull. Jard. Bot. Buitenzorg. See B–P–H 258/19. HI 52827

Bulletin du jardin botanique de l'état. Brussels. Vols. 1-36, 1902/05-66. Bull. Jard. Bot. État. Superseded by: Bulletin du jardin botanique national de Belgique. 1-807-2. HI 63286

Bulletin de jardin botanique de Kieff = Izvestiya Kievskogo botanicheskogo sada. Kiev.

Bulletin du jardin botanique national de Belgique. Brussels. Vol. 37+, 1967+. Bull. Jard. Bot. Belg. Preceded by: Bulletin du jardin botanique de l'état Bruxelles. HI 63287

Bulletin du jardin botanique principal de l'URSS = Izvestiya Glavnogo botanicheskogo sada S S S R. Leningrad.

Bulletin du jardin botanique de la république russe = Izvestiya Glavnogo botanicheskogo sada R S F S R. Petrograd.

Bulletin du jardin impériale botanique de Saint-Péterbourg = Izvestiya Imperatorskogo S.-Peterburgskogo Botanicheskogo Sada. St. Petersburg.

Bulletin du jardin et du musée botaniques de l'université de Cluj = Buletinul de informaţii al grădinii botanice şi al muzeului botanic dela universitatea din Cluj. Cluj, Rumania. Bul. Inform. Grăd. Bot. Univ. Cluj. See B–P–H 232/15.

Bulletin du jardin et du musée botaniques de l'université de Cluj = Buletinul grădinii botanice şi al muzeului botanic dela universitatea din Cluj. Cluj, Rumania. Bul. Grăd. Bot. Univ. Cluj. See B–P–H 232/6.

Bulletin du jardin et du musée botaniques de l'université de Cluj à Timisoara = Buletinul grădinii botanice şi al muzeului botanic dela universitatea din Cluj. Cluj, Rumania. Bul. Grăd. Bot. Univ. Cluj. See B–P–H 232/6.

Bulletin du jardin principal de l'URSS = Izvestiya Glavnogo botanicheskogo sada S S S R. Leningrad.

Bulletin des jardins botaniques, A R T J B. Nancy, Antibes. No. 1+, 1971+. Bull. Jard. Bot. A. R. T. J. B. HI 63288

Bulletin, Jealott's Hill research station. Bracknell. Nos. 1-10, 1940-60. Bull. Jealott's Hill Res. Sta. HI 63289

Bulletin, John J. Tyler arboretum. Lima, PA. No. 1+, 1958+. Bull. John J. Tyler Arbor. HI 63290

Bulletin of the Johns Hopkins hospital. Baltimore, MD. Bull. Johns Hopkins Hosp. See B–P–H 259/5. HI 52828

Bulletin of the Josselyn botanical society of Maine. Portland ME. Bull. Josselyn Bot. Soc. Maine. See B–P–H 259/6. HI 52829

Bulletin of Kagawa agricultural experiment station. [Kagawa-ken nogyo shikenjo kenkyu hokoku.] Kagawa, Japan. Bull. Kagawa Agric. Exp. Sta. See B–P–H 259/8. HI 52830

Bulletin of the Kagoshima agricultural college. [Kagoshima novin-senmon gakko.] Kagoshima, Japan. Bull. Kagoshima Agric. Coll. See B–P–H 259/9. HI 52831

Bulletin of the Kagoshima agricultural experiment station. [Kagoshima-ken nogyo shikenjo kenkyu hokoku.] Kamifukumoto. No. 1+, 1973+. Bull. Kagoshima Agric. Exp. Sta. HI 63291

Bulletin of the Kagoshima imperial college of agriculture and forestry. Kagoshima, Japan. Bull. Kagoshima Imp. Coll. Agric. See B–P–H 259/10. HI 52832

Bulletin of the kagoshima tobacco experiment station. [Kagoshima tabako shikenjo hokoku.] Kagoshima, Japan. Bull. Kagoshima Tobacco Exp. Sta. See B–P–H 259/11. HI 52833

Bulletin of Kanagawa agricultural experiment station. [Kanagawa-ken nogyo shikenjo kenkyu hokoku.] Kamakura. Nos. 1-107, 1962-68. Bull. Kanagawa Agric. Exp. Sta. Superseded by: Bulletin of the agricultural research institute of Kanagawa prefecture. HI 63292

Bulletin of Kanagawa agricultural experiment station. Horticultural branch. Kamakura. Nos. 1-9, 1953-61. Bull. Kanagawa Agric. Exp. Sta., Hort. Branch. Superseded by: Bulletin of Kanagawa horticultural experiment station. HI 63293

Bulletin of Kanagawa horticultural experiment station. [Kanagawa-ken engei shikenjo kenkyu hokoku.] Kamakura. No. 10+, 1962+. Bull. Kanagawa Hort. Exp. Sta. Preceded by: Bulletin of Kanagawa agricultural experiment station. Horticultural branch. HI 63294

Bulletin of the Kanagawa prefectural agricultural corporated experimental research organization. [Kanagawa-ken nogyo shiken kenkyu kikan kyodo kenkyu hokoku.] Hiratsuki. Nos. 1-5, 1971-72? Bull. Kanagawa Pref. Agric. Corp. Exp. Res. Organ. HI 63295

Bulletin of the Kanagawa prefectural museum. Natural science. [Kanagawa kenritsu hakubutsukan kenkyu hokoku, shizen kagaku.] Naka-ku, Yokohama. Vol. 1+, 1968+. Bull. Kanagawa Pref. Mus., Nat. Sci. HI 63296

Bulletin, Kanagawa-ken agricultural experiment station. [Kanagawa ken noji shiken seiseki.] Yokohama?, Japan. Bull. Kanagawa-ken Agric. Exp. Sta. See

B–P–H 259/14. HI 52834

Bulletin of the Kansas academy of science. Topeka, KS. Bull. Kansas Acad. Sci. See B–P–H 259/16. HI 52835

Bulletin of the Kansas geological survey. Topeka, KS. No. 1+, 1913+. Bull. Kansas Geol. Surv. HI 63670

Bulletin of the Kansas state geological survey = Bulletin of the Kansas geological survey. Topeka, KS.

Bulletin of the Kansas university geological survey = Bulletin of the Kansas geological survey. Topeka, KS.

Bulletin, Kent trust for nature conservation. Maidstone. 1975+. Bull. Kent Trust Nat. Conservation. HI 63297

Bulletin of the Kentucky agricultural experiment station. [Nos. 1-410 also issued in: Report (Annual) of the Kentucky agricultural experiment station.] Lexington, KY. No. 1+, 1885+. Bull. Kentucky Agric. Exp. Sta. HI 63298

Bulletin, Kenya orchid society. Nairobi. No. ?-20+, 19??-73+. Bull. Kenya Orchid Soc. HI 63299

Bulletin of the Kitakyushi museum of natural history. Kitakyushi. 1979+. Bull. Kitakyushi Mus. Nat. Hist. HI 63300

Bulletin of the Kochi women's college. Kochi. Vols. 1-3, 1952-55. Bull. Kochi Women's Coll. Superseded by: Bulletin of the Kochi women's university. HI 63301

Bulletin of the Kochi women's university. Series of natural sciences. [Kochi joshi daigaku kiyo. Shizenkagaku-hen.] Kochi. Vol. 5(1)+, 1957+. Bull. Kochi Women's Univ., Ser. Nat. Sci. Preceded by: Bulletin of the Kochi women's college. HI 63302

Bulletin, koloniaal museum. Haarlem. Nos. 1-52, 1892-1913. Bull. Kolon. Mus. Preceded by: Verslagen, koloniaal museum. Superseded by: Mededeeling van de afdeeling handelsmuseum van het koloniaal instituut. 3-1781-3. HI 63303

Bulletin, koninklijk belgisch instituut voor natuurwetenschappen. Biologie = Bulletin de l'institut royal des sciences naturelles de Belgique. Biologie. Brussels.

Bulletin of Korea ocean research and development institute. [Hainyan nyenguso sobo.] Seoul. Vol. 3+, 1981+. Bull. Korea Ocean Res. Developm. Inst. Preceded by: Haeyang kaebal yonguso sobo. HI 63304

Bulletin of the Kwanak arboretum. Suwon. No. 1+, 1976+. Bull. Kwanak Arbor. HI 63305

Bulletin of the Kyoto agricultural experiment station. [Kyoto-furitsu nogyo shikenjo kenkyu hokoku.] Kyoto. Vols. 1-6, 1966-73. Bull. Kyoto Agric. Exp. Sta. Superseded by: Bulletin of the Kyoto prefectural institute of agriculture. HI 63306

Bulletin of the Kyoto educational university. Series B, mathematics and natural science. [Kyoyo gakugei daigaku hokoku.] Kyoto. 1951+. Bull. Kyoto Educ. Univ., B. HI 63307

Bulletin of the Kyoto prefectural institute of agriculture. [Kyoto-furitsu nogyo kenkujo kenkyu hokoku.] No. 7+, 1975+. Bull. Kyoto Pref. Inst. Agric. Preceded by: Bulletin of the Kyoto agricultural experiment station. HI 63308

Bulletin of the Kyoto prefectural university forests. [Kyoto-furitsu daigaku nogakubu enshurin hokoku.] Kyoto. 1957+. Bull. Kyoto Pref. Univ. Forests. HI 65312

Bulletin of the Kyoto university forests. [Kyoto daigaku nogakubu fuzoku enshurin hokoku.] Kyoto. No. 1+, 1930+. Bull. Kyoto Univ. Forests. HI 63309

Bulletin of the Kyushu agricultural experiment station. [Kyushu nogyo shikenjo iho.] Fukuoka, Japan. Bull. Kyushu Agric. Exp. Sta. See B–P–H 260/3. HI 52836

Bulletin of the Kyushu university forests. [Kyushu daigaku nogaku-bu enshurin hokoku.] Fukuoka, Japan. Bull. Syushu Univ. Forests. See B–P–H 260/5. HI 52837

Bulletin du laboratoire biologique de Petrograd = Izvestiya Petrogradskoi Biologicheskoi Laboratorii. Petrograd.

Bulletin du laboratoire biologique de Saint-Pétersbourg = Izvestiya S.-Peterburgskoi Biologicheskoi Laboratorii. St. Petersburg.

Bulletin du laboratoire de botanique générale de l'universite de Genève. Geneva. Bull. Lab. Bot. Gén. Univ. Genève. See B–P–H 260/9. HI 52838

Bulletin du laboratoire maritime du Dinard. Dinard, France. Bull. Lab. Marit. Dinard. See B–P–H 260/14. HI 52840

Bulletin du laboratoire maritime du muséum d'histoire naturelle à Saint Servan. St.-Servan, France. Bull. Lab. Marit. Mus. Hist. Nat. Saint Servan. See B–P–H 260/15. HI 52841

Bulletin du laboratoire maritime de Saint-Servan. St.-Servan, France. Bull. Lab. Marit. Saint-Servan. See B–P–H 260/16. HI 52842

Bulletin des laboratoires de gêologie, gêographie, physique, minéralogie et paléontologie de l'université de Lausanne. Lausanne. Bull. Lab. Géol. Univ. Lausanne. See B–P–H 260/13. HI 52839

Bulletin of the laboratories of natural history, Iowa state university. Iowa City, IA. Bull. Lab. Nat. Hist. Iowa State Univ. See B–P–H 260/17. HI 52843

Bulletin of the laboratory of genetics = Trudy laboratorii po genetike. Leningrad, Moscow.

Bulletin of the laboratory of systematic botany and plant ecology, Taihoku imperial university. Taihoku [=Taipei, Taiwan]. Bull. Lab. Syst. Bot. taihoku Imp. Univ. See B–P–H 260/22. HI 52845

Bulletin van het landbouwproefstation Suriname. Paramaribo, Dutch Guiana [Surinam]. Bull. Landbouwproefstat. Suriname See B–P–H 261/1. HI 52846

Bulletin, Leicester museum and art gallery (and library). Leicester. 1924-52. Bull. Leicester Mus. Art Gall. HI 63310

Bulletin de liaison des amis des plantes de serres et d'acclimatation. (France). No. 1+, 1973+. Bull. Liais. Amis Pl. Serres Acclim. HI 63311

Bulletin de liaison des chercheurs de L A M T O. N'Douci. 1970+. Bull. Liais. Cherch. L. A. M. T. O. HI 63312

Bulletin de liaison des musées d'histoire naturelle. Paris. No. ?-16+, 19??-73. Bull. Liais. Mus. Hist. Nat. HI 63313

Bulletin de liaison des professeurs de biologie et de géologie, centre regional de recherche et de documentation pedagogique de Besançon. Besançon. 1970+. Bull. Liais. Profess. Biol. Géol. Besançon. HI 63314

Bulletin de liaison, société d'ethnozoologie et d'ethnobotanique. Paris. No. 1+, 1976+. Bull. Liais. Soc. Ethnozool. Ethnobot. HI 63315

Bulletin, liberal arts college, Wakayama university. Natural science. Wakayama. No. 1+, 1950+. Bull. Liberal Arts Coll. Wakayama Univ., Nat. Sci. HI 63316

Bulletin de la ligue pour la conservation de la Suisse pittoresque = Heimatschutz. Bümpliz, Olten.

Bulletin, lily group, royal horticultural society = Bulletin, R H S lily group. Pangbourne.

Bulletin, Liverpool botanical society. Liverpool. No. 1+, 1951+. Bull. Liverpool Bot. Soc. HI 63317

Bulletin of the Lloyd library of botany, pharmacy and materia medica. Cincinnati, OH. Bull. Lloyd Libr. Bot. See B–P–H 261/5. HI 52848

Bulletin, lodzkie towarzystwo naukowe. Łodz. Vol. 11+, 1960+. Bull. Lodz. Towarz. Nauk. Preceded by: Bulletin de la société des sciences et des lettres de Lødz. Classe III, de sciences mathématiques et naturelles. HI 63318

Bulletin of the Long Island horticultural society. Farmingdale, NY. Bull. Long Island Hort. Soc. See B–P–H 261/6. HI 52849

Bulletin of the Los Angeles county museum (of natural history). Los Angeles, CA. Nos. 1-12, 1965-71. Bull. Los Angeles County Mus. Superseded by: Science bulletin, natural history museum of Los Angeles county. HI 63319

Bulletin of the Los Angeles international fern society = L A I F S. Los Angeles, La Mirada, CA.

Bulletin, Louisiana agricultural experiment station = Louisiana agricultural experiment station bulletin. Baton Rouge, LA. Louisiana Agric. Exp. Sta. Bull. See B–P–H 535/23.

Bulletin of the Louisiana society for horticultural research. Ferriday, LA. Vols. 1-?, 1956-76. Bull. Louisiana Soc. Hort. Res. HI 63320

Bulletin of the Lucknow national botanic gardens. Lucknow, India. Bull. Lucknow Natl. Bot. Gard. See B–P–H 261/8. HI 52850

Bulletin of the Luther Burbank society. Santa Rosa, CA. Bull. Luther Burbank Soc. See B–P–H 261/9. HI 52851

Bulletin of the Madras government museum. Madras. Bull. Madras Gov. Mus. See B–P–H 261/10. HI 52853

Bulletin, Maine agricultural experiment station = Maine agricultural experiment station. Bulletin. Orono, ME.

Bulletin, Maine life sciences and agricultural experiment station. Orono, ME. No. 685+, 1970+. Bull. Maine Life Sci. Agric. Exp. Sta. Preceded by: Maine agricultural experiment station. Bulletin. HI 63321

Bulletin of the Maine state college laboratory of natural history. Augusta, ME. Vol. 1(1-3), 1888-97. Bull. Maine Sta. Coll. Lab. Nat. Hist. HI 63322

Bulletin de la maison franco-japonaise. Tokyo. Bull. Maison Franco-Jap. See B–P–H 261/13. HI 52854

Bulletin, maize section, European association for research on plant breeding = European association for research on plant breeding. Maize section. Bulletin. Paris. Eur. Assoc. Res. Pl. Breed., Maize Sect. Bull. See B–P–H 364/15.

Bulletin of the Maizuru marine observatory. Maizuru, Japan. Bull. Maizuru Mar. Observatory. See B–P–H 261/15. HI 52855

Bulletin, Manitoba naturalists' society. Winnipeg. 1974+. Bull. Manitoba Naturalists' Soc. HI 63323

Bulletin, Marie Selby botanical gardens. Sarasota, FL. Vol. 1+, 1974+. Bull. Marie Selby Bot. Gard. HI 63324

Bulletin of the marine biological station of Asamushi. Nonai, Japan. Bull. Mar. Biol. Sta. Asamushi. See B–P–H 261/16. HI 52857

Bulletins of marine ecology. Edinburgh. Bull. Mar.

Ecol. See B–P–H 261/17. HI 52858

Bulletin of the marine park research stations. [Kaichu koen kenkyujo kenkyu hokoku.] Kushimoto. Vol. 1+, 1975+. Bull. Mar. Park Res. Sta. HI 63325

Bulletin of the marine research centre, Saudi Arabia. Jeddah & Bangor. 1972-75. Bull. Mar. Res. Centre Saudi Arabia. HI 63326

Bulletin of marine science. Coral Gables, Miami, FL. Vol. 15+, 1965+. Bull. Mar. Sci. Preceded by: Bulletin of marine science of the Gulf and Caribbean. HI 63327

Bulletin of marine science of the Gulf and Caribbean. Coral Gables, FL. Vols. 1-14, 1951-64. Bull. Mar. Sci. Gulf Caribbean. Superseded by: Bulletin of marine science. HI 63328

Bulletin, Maryland agricultural experiment station = Maryland agricultural experiment station. Bulletin. College Park, MD. Maryland Agric. Exp. Sta. Bull. See B–P–H 549/1.

Bulletin, Maryland state board of agriculture = Maryland state board of agriculture bulletin. College Park, MD. Maryland State Board Agric. Bull. See B–P–H 549/7.

Bulletin of the Massachusetts agricultural experiment station = Massachusetts agricultural experiment station bulletin. Amherst, MA.

Bulletin, Massachusetts Audubon society = Massachusetts Audubon bulletin. Boston, MA.

Bulletin of the Massachusetts horticultural society. Boston, MA. 1919-23. Bull. Mass. Hort. Soc. Preceded by part 1 of: Transactions of the Massachusetts horticultural society. HI 63329 Superseded by: Yearbook of the Massachusetts horticultural society. 3-2552-2.

Bulletin of Massachusetts natural history. Amherst, MA. Bull. Mass. Nat. Hist. See B–P–H 261/25. HI 52859

Bulletin. Massey agricultural college (University of New Zealand). Palmerston North, New Zealand. Bull. Massey Agric. Coll. See B–P–H 261/26. HI 52860

Bulletin of mathematical biology. Oxford, Elmsford, NY. Vol. 35+. 1973+. Bull. Math. Biol. Preceded by: Bulletin of mathematical biophysics. HI 63330

Bulletin of mathematical biophysics. Chicago, IL. Vols. 1-34, 1939-72. Bull. Math. Biophys. Superseded by: Bulletin of mathematical biology. HI 63331

Bulletin der mathematisch-physikalischen Klasse der königlichen Akademie der Wissenschaften. Munich. Bull. Math.-Phys. Kl. Königl. Akad. Wiss. See B–P–H 261/27. HI 52861

Bulletin des matières grasses, institut colonial de Marseille = Bulletin de l'institut coloniale de Marseille. Section des matières grasses. Marseilles.

Bulletin, Max C. Fleischmann college of agriculture, Nevada University = Publications, Max C. Fleischmann college of agriculture. Series B [bulletin series]. Reno, NV.

Bulletin de Mayenne - sciences. Laval, France. Bull. Mayenne Sci. See B–P–H 262/1. HI 52862

Bulletin médical. Paris. Bull. Méd. See B–P–H 262/2. HI 52863

Bulletin médical du Katanga. Elizabethville, Belgian Congo [=Lubumbaski, Republic of the Congo]. Bull. Méd. Katanga See B–P–H 262/3. HI 52864

Bulletin of medico-ethno-botanical research. New Delhi. Vol. 1+, 1980+. Bull. Med.-Ethno-Bot. Res. HI 63332

Bulletins et mémoires de la société anatomique de Paris. Paris. Bull. & Mém. Soc. Anat. Paris. See B–P–H 262/7. HI 52865

Bulletins et mémoires de la société française d'ophtalmologie. Paris. Bull. & Mém. Soc. Franç. Ophtalmol. See B–P–H 262/10. HI 52866

Bulletins et mémoires de la société médicale des hôpitaux de Paris. Paris. Bull. & Mém. Soc. Méd. Hôp. Paris. See B–P–H 262/12. HI 52867

Bulletin mensuel de l'association des naturalistes de la Vallée du Loing (et de la Forêt de Fontainbleau.) Moret-sur-Loing. 1926-39, 1947-59. Bull. Mens. Assoc. Naturalistes Vallée du Loing. Preceded by?: Bulletin de l'association des naturalistes de la Vallée du Loing. Superseded by: Bulletin bimestriel, association des naturalistes de la Vallée du Loing et du massif de Fontainebleau. HI 63333

Bulletin mensuel d'information, Ardenne et Gaume. Brussels. Vols. 1-11, 1968-75. Bull. Mens. Inform. Ardenne Gaume. HI 63334

Bulletin mensuel météorologique de l'association scientifique de France. Paris. Bull. Mens, Météorol. Assoc. Sci. France. See B–P–H 262/14. HI 52869

Bulletin mensuel, les naturalistes de Mons et du Borinage. Mons. Vols. 27(3)-39(4), 1944-56. Bull. Mens. Naturalistes Mons & Borinage. Preceded by: Bulletin des naturalistes de Mons et du Borinage. Superseded by: Bulletin mensuel, société royale les naturalistes de Mons at du Borinage. HI 63335

Bulletin mensuel des renseignements agricoles et des maladies des plantes. [French language edition of: Monthly bulletin of agricultural intelligence and of plant diseases.] Rome. Vols. 4-6, 1913-15. Bull. Mens. Renseign. Agric. Malad. Pl. Preceded by: Bulletin du bureau des renseignements agricoles et des maladies des plantes. Superseded by: Revue internationale de renseignements agricoles. HI 63336

Bulletin mensuel de la société d'agriculture,

d'horticulture et d'acclimatation du Var à Toulon. Toulon. Vols. 1-4, 1880-83. Bull. Mens. Soc. Agric. Hort. Acclim. Var à Toulon. Preceded by: Bulletin trimestriel du comice Agricole de l'arrondissement de Toulon and Bulletin société d'horticulture et d'acclimatation du département du Var à Toulon. Superseded by: Provence agricole et horticole illustrée. HI 63337

Bulletin mensuel, société d'histoire naturelle de Haute-Savoie. Annecy. N.s. vols. 11-14, 19??-?? Bull. Mens. Soc. Hist. Nat. Haute-Savoie. Preceded by: Bulletin, société d'histoire naturelle de Haute-Savoie. Superseded by: Bulletin bimestriel de la société d'histoire naturelle de Haute-Savoie. HI 63338

Bulletin mensuel de la société linnéenne de Lyon. Lyons. Bull. Mens Soc. Linn. Lyon. See B–P–H 262/16. HI 52870

Bulletin mensuel de la société linnéenne de Paris. Paris. Bull. Mens Soc. Linn. Paris. See B–P–H 262/18. HI 52871

Bulletin mensuel de la société linnéene de la Seine-Maritime. Le Havre. Ser. 2, vols. 1-?, 1915-39. Bull. Mens. Soc. Linn. Seine-Marit. Preceded by: Bulletin de la société linnéenne de la Seine-Maritime. 5-3965-1. HI 63339

Bulletin mensuel de la société linnéenne et des sociétés botanique de Lyon, d'anthropologie et de biologie de Lyon reunies. Lyons. Bull. Mens. Soc. Linn. Soc. Bot. Lyon. See B–P–H 262/19. HI 52872

Bulletin mensuel de la société des naturalistes luxembourgeois. Luxemburg. Nos. 17-53 [nos. 17-53 = n.s. nos. 1-42,] 1907-48. Bull. Mens. Soc. Naturalistes Luxemb. Preceded by: Comptes-rendus des séances, société des naturalistes luxembourgeois. Superseded by: Bulletin de la société des naturalistes luxembourgeois. HI 63340

Bulletin mensuel, société royale les naturalistes de Mons et du Borinage. Mons. Vols. 39(5)-45(4), 1956-62. Bull. Mens. Soc. Roy. Naturalistes Mons & Borinage. Preceded by: Bulletin mensuel, les naturalistes de Mons et du Borinage. Superseded by: Bulletin de la société royale les naturalistes de Mons et du Borinage. HI 63341

Bulletin mensuel, société des sciences, agriculture et arts de la Basse-Alsace <Gesellschaft zur Beförderung der Wissenschaften, des Ackerbaues und der Künste im Unter-Elsass> = Société des sciences, agriculture et arts de la Basse-Alsace <Gesellschaft zur Beförderung der Wissenschaften, des Ackerbaues und der Künste im Unter-Elsass>. Bulletin mensuel. Strasbourg. Soc. Sci. Basse-Alsace Bull. Mens. See B–P–H 845/30.

Bulletin mensuel de la société des sciences de Nancy. Nancy. Bull. Mens. Soc. Sci. Nancy. See B–P–H 262/22. HI 52873

Bulletin mensuel de la société des sciences naturelles de Toulon. Toulon. 1946-47. Bull. Mens. Soc. Sci. Nat. Toulon. Superseded by: Bulletin mensuel, société des sciences naturelles de Toulon et du Var. HI 63342

Bulletin mensuel, société des sciences naturelles de Toulon et du Var. Toulon. 1948-50. Bull. Mens. Soc. Sci. Nat. Toulon & Var. Preceded by: Bulletin mensuel de la société naturelles de Toulon et du Var. Superseded by: Bulletin, société des sciences naturelles de Toulon et du Var. HI 63343

Bulletin mensuel de l'union coloniale française = Bulletin de l'union coloniale française. Paris.

Bulletin, mesemb study group = Mesemb study group bulletin. Belen, NM.

Bulletin of the Methodi Popoff institute of biology = Izvestiya na instituta po biologi "Metodii Popov". Sofia.

Bulletin, microbiology research group, south african council for scientific and industrial research. Pretoria. Vol. 1+, 1968+. Bull. Microbiol. Res. Group S. African Council Sci. Industr. Res. HI 63344

Bulletin, microscopical society of Canada. Mississauga. No. 1+, 1973+. Bull. Microscop. Soc. Canada. HI 63345

Bulletin de microscopie appliquée. Paris. Bull. Microscop. Appl. See B–P–H 262/27. HI 52874

Bulletin, Michigan agricultural experiment station = Michigan agricultural experiment station. Bulletin. East Lansing, MI. Michigan Agric. Exp. Sta. Bull. See B–P–H 592/8.

Bulletin of the middle siberian section of the state russian geographical society = Izvestiya Sredne-Sibirskogo otdela Gosudarstvennogo russkogo geograficheskogo obshchestva. Krasnoyarsk.

Bulletin of the middle siberian state geographical society = Izvestiya Sredne-Sibirskogo gosudarstvennogo geograficheskogo obshchestva. Krasnoyarsk.

Bulletin of the Mie university forests. [Mie daigaku nogaku-bu enshurin hokuku.] Tsu, Japan. Bull. Mie Univ. Forests. See B–P–H 263/4. HI 52875

Bulletin; mineralogisk-geologiska institut. Upsala universitet. Uppsala. Bull. Mineral.-Geol. Inst. Upsala Univ. See B–P–H 263/5. HI 52876

Bulletin, ministère de l'agriculture et de la colonisation, Québec = Publication, ministère de l'agriculture et de la colonisation, Québec. Quebec.

Bulletin of the ministère de l'agriculture, Québec = Publication, ministère de l'agriculture et de la colonisation, Québec. Quebec.

Bulletin, ministry of agriculture, Egypt. Technical and

scientific service. Cairo. Nos. 1-236?, 1916-39. Bull. Minist. Agric. Egypt, Techn. Sci. Serv. HI 63346

Bulletin, ministry of agriculture, Federation of Malaya = Bulletin, department of agriculture, Federation of Malaya. Kuala Lumpur.

Bulletin, ministry of agriculture and fisheries, Barbados = Bulletin, ministry of agriculture, lands and fisheries, Barbados. Bridgetown.

Bulletin of the ministry of agriculture, fisheries and food. London = Bulletin of the ministry of agriculture and fisheries. London.

Bulletin of the ministry of agriculture and fisheries. London. Nos. 1-209, 1930-65 [some numbers were published in revised editions]. Bull. Minist. Agric. Fish., London. Preceded by: Miscellaneous publications, ministry of agriculture and fisheries. London. HI 63347

Bulletin, ministry of agriculture, lands and fisheries, Barbados. Bridgetown. Nos. 33-38, 1963; n.s. no. 1+, 1970+. Bull. Minist. Agric. Lands Fish. Barbados. Preceded by: Bulletin of the department of science and agriculture, Barbados. HI 63348

Bulletin, ministry of agriculture and fisheries, Malaysia. Kuala Lumpur. No. 126+, 1972+. Bull. Minist. Agric. Fish. Malaysia. Preceded by: Bulletin, ministry of agriculture and lands, Malaysia. HI 63349

Bulletin, ministry of agriculture, Ghana. Accra. No. 2+, 1960?+. Bull. Minist. Agric. Ghana. Preceded by: Bulletin, agricultural division, ministry of food and agriculture, Ghana. HI 63350

Bulletin, ministry of agriculture, Iraq. Baghdad. No. 110+, 1965+. Bull. Minist. Agric. Iraq. Preceded by: Technical bulletin, ministry of agriculture, Iraq. HI 63351

Bulletin, ministry of agriculture, Israel. Tel-Aviv. 1949+. Bull. Minist. Agric. Israel. HI 63352

Bulletin, ministry of agriculture, Jamaica. Kingston, Jamaica. N.s. no. 63+, 1973+. Bull. Minist. Agric. Jamaica. Preceded by: Bulletin of the ministry of agriculture and lands, Jamaica. HI 63353

Bulletin of the ministry of agriculture and lands, Jamaica. Kingston, Jamaica. N.s. nos. ?-62, 1959-63. Bull. Minist. Agric. Lands Jamaica. Preceded by: Bulletin of the department of agriculture, Jamaica. Superseded by: Bulletin, ministry of agriculture, Jamaica. HI 63354

Bulletin, ministry of agriculture and lands, Malaysia. Kuala Lumpur. Nos. 115-125, 1964-71. Bull. Minist. Agric. Lands Malaysia. Preceded by: Bulletin, department of agriculture, Federation of Malaya. Superseded by: Bulletin, ministry of agriculture and fisheries. HI 63355

Bulletin, ministry of agriculture, Sudan. Khartoum. No. 13+, 1956+. Bull. Minist. Agric. Sudan. Preceded by: Bulletin, department of agriculture and forests, Sudan. HI 63356

Bulletin of the ministry of education. [Chiao-yü kung pao.] Nanking. Bull. Minist. Educ. See B–P–H 263/6. HI 52877

Bulletins of the Minnesota academy of natural sciences. Minneapolis, MN. Bull. Minnesota Acad. Nat. Sci. See B–P–H 263/7. HI 52878

Bulletin, Minnesota agricultural experiment station = Minnesota agricultural experiment station. Bulletin. St. Paul, MN. Minnesota Agric. Exp. Sta. Bull. See B–P–H 596/7.

b. Bulletin of the Misaki marine biological institute, Kyoto university. Kyoto. Vols. 1-?, 1962-68. Bull. Misaki Mar. Biol. Inst. HI 63357

Bulletin of miscellaneous information, botanic garden, Grenada. St. George's. Nos. 1-36, 1891-93. Bull. Misc. Inform. Bot. Gard. Grenada. Preceded by: Grenada agricultural botanical bulletin. HI 63358

Bulletin of miscellaneous information, botanical department, imperial college of agriculture. St. Augustine, Trinidad = Bulletin of miscellaneous information, botanical department, Trinidad.

Bulletin of miscellaneous information, botanical department, Trinidad. Trinidad. Vols. 4-8 [also numbered as 18-60], 1899-1908. Bull. Misc. Inform. Bot. Dept., Trinidad. Preceded by: Bulletin of miscellaneous information, royal botanic gardens, Trinidad. Superseded by: Bulletin of miscellaneous information, department of agriculture, imperial college of agriculture, Trinidad. HI 52879

Bulletin of miscellaneous information, department of agriculture, imperial college of agriculture, Trinidad = Bulletin of miscellaneous information, department of agriculture, Trinidad. Saint Augustine, Port-of-Spain.

Bulletin of miscellaneous information, department of agriculture, Trinidad. Saint Augustine, Port-of-Spain. Vols. 9-12 [also numbered as 61-76], 1909-13. Bull. Misc. Inform. Dept. Agric. Imp. Coll. Agric. Preceded by: Bulletin of miscellaneous information, botanical department, imperial college of agriculture, Trinidad. Superseded by: Bulletin of the department of agriculture, Trinidad and Tobago. HI 63359

Bulletin of miscellaneous information, royal botanic gardens, Kew = Bulletin of miscellaneous information, royal gardens, Kew.

Bulletin of miscellaneous information, royal botanic gardens, Trinidad. Nos. 1-17 [also numbered as vols. 1-3(9)], 1888-98. Bull. Misc. Inform. Roy. Bot. Gard. Trinidad. Superseded by: Bulletin of miscellaneous information, botanical department, Trinidad. HI 63360

Bulletin of miscellaneous information, royal gardens, Kew. Kew. 1887-1942. Bull. Misc. Inform. Kew. Superseded by: Kew bulletin. 3-2284-1. HI 63361

Bulletin of miscellaneous information, Trinidad royal botanic gardens = Bulletin of miscellaneous information, royal botanic gardens, Trinidad.

Bulletin, Mississippi agricultural experiment station. Mississippi agricultural and mechanical college = Mississippi agricultural experiment station. Mississippi agricultural and mechanical college. Bulletin. Jackson, MS. Mississippi Agric. Exp. Sta. Bull. See B–P–H 599/24.

Bulletin, Mississippi agricultural experiment station. Mississippi state college = Mississippi agricultural experiment station. Mississippi agricultural and mechanical college. Bulletin. Jackson, MS. Mississippi Agric. Exp. Sta. Bull. See B–P–H 599/24.

Bulletin, Mississippi department of agriculture = Mississippi department of agriculture, bulletin. Jackson, MS. Mississippi Dept. Agric. Bull. See B–P–H 599/27.

Bulletin, Missouri academy of science. Columbia, MO. Vols. 1-2, 1973. Bull. Missouri Acad. Sci. Superseded by: Occasional papers, Missouri academy of science. HI 63362

Bulletin, Missouri agricultural experiment station = Missouri agricultural experiment station. Bulletin. Columbia, MO.

Bulletin, Missouri botanical garden = Missouri botanical garden bulletin. St. Louis, MO. Missouri Bot. Gard. Bull. See B–P–H 601/2.

Bulletin, Missouri state agricultural college = Missouri state agricultural college, bulletin. Columbia, MO. Missouri State Agric. Coll. Bull. See B–P–H 601/4.

Bulletin, Missouri state museum. Jefferson City, MO. Nos. 1-12 [no. 6(6-8) never published], 1932-35. Bull. Missouri State Mus. Superseded by: Bulletin, Missouri state resources museum. HI 63363

Bulletin, Missouri state resources museum. Jefferson City, MO. No. 14+, 1935+. Bull. Missouri State Resource Mus. Preceded by: Bulletin, Missouri state museum. HI 63364

Bulletin of the Miyazaki agricultural experiment station. [Miyazaki-ken sogo nogyo shikenjo kenkyu hokoku.] Miyazaki, Japan. Bull. Miyazaki Agric. Exp. Sta. See B–P–H 263/20. HI 52880

Bulletin of the Miyazaki college of agriculture and forestry. [Miyazaki koto norin gakko gakujutsu hokoku.] Miyazaki, Japan. Bull. Miyazaki Coll. Agric. See B–P–H 263/22. HI 52881

Bulletin of the Miyazaki university. Natural science. [Miyazaki daigaku ziho. Shizen kagaku.] Miyazaki, Japan. Bull. Miyazaki Univ., Nat. Sci. See B–P–H 263/24. HI 52882

Bulletin, Montana college of agriculture and mechanic arts. Agricultural experiment station = Montana college of agriculture and mechanic arts. Agricultural experiment station. Bulletin. Bozeman, MT. Montana Agric. Exp. Sta. Bull. See B–P–H 618/12.

Bulletin. Montana state university. Biological series. Missoula, Montana. Bull. Montana State Univ., Biol. Ser. See B–P–H 263/28. HI 52885

Bulletin of the Montreal botanical gardens. Montreal. Bull. Montreal Bot. Gard. See B–P–H 264/1. HI 52886

Bulletin of the Morioka college of agriculture and forestry [Iwate daigaku Morioka norin senmon gakko gakujutsu hokoku] = Bulletin of the imperial college of agriculture and forestry. Morioka, Japan. Bull. Imp. Coll. Agric. See B–P–H 254/17.

Bulletin of the Morioka tobacco experiment station. [Morioka tabako shikenjo hokoku.] Morioka, Japan. Bull. Morioka Tobacco Exp. Sta. See B–P–H 264/4. HI 52887

Bulletin of moss flora of arctic north America = Bulletin of moss flora of arctic north America and Greenland. Copenhagen.

Bulletin of moss flora of arctic north America and Greenland. Copenhagen. No. 1+, 1981+. Bull. Moss Fl. Arctic N. Amer. Greenland. HI 63365

Bulletin of the Mount Desert Island biological laboratory. Salisbury Cove, ME. Vol. 1+, 1924+. Bull. Mount Desert Island Biol. Lab. 3-2775-3. HI 63366

Bulletin of the municipal museum of natural history. Nishiku, Osaka. 1954-56. Bull. Munic. Mus. Nat. Hist., Osaka. Superseded by: Bulletin of the Osaka museum of natural history. HI 63367

Bulletin de la Murithienne; société valaisanne des sciences naturelles. Sion, Switzerland. Bull. Murith. Soc. Valais. Sci. Nat. See B–P–H 264/10. HI 52889

Bulletin du musée du Caucase = Izvestiya kavkazskago muzeya. Tiflis.

Bulletin du musée de Géorgie = Byulleten′ gosudarstvennogo muzeya Gruzii. Tiflis.

Bulletin du musée de Géorgie = Vestnik, gosudarstvennogo muzeya Gruzii. Tiflis.

Bulletin du musée d'histoire naturelle de Marseille. Marseilles. Vol. 36+, 1976+. Bull. Mus. Hist. Nat. Marseille. Preceded by: Bulletin du muséum d'histoire naturelle. HI 63368

Bulletin de musée oceanographique, Monaco = Bulletin de l'institut oceanographique, Monaco. Paris. Bull.

Inst. Océanogr. Monaco. See B–P–H 256/9.

Bulletin du musée de la république populaire de Bosnie et Hercégovine à Sarajevo = Glasnik zemaljskog muzeja u Sarajevu. Sarajevo, Yugoslavia. Glasn. Zemaljsk. Muz. u Sarajevu. See B–P–H 405/16.

Bulletin du musée de la république socialiste de Bosnie-Hercégovine à Sarajevo. Sciences naturelles. Sarajevo = Glasnik zemalskog muzeja Bosne i Hercegovine u Sarajevu. Prirodne nauke. Sarajevo.

Bulletin du musée royal d'histoire naturelle de Belgique. Brussels. Bull. Mus. Roy. Hist. Nat. Belgique. See B–P–H 264/24. HI 52890

Bulletin du muséum d'histoire naturelle. Paris. Vols. 1-12, 1895-1906. Bull. Mus. Hist. Nat. (Paris). Superseded by: Bulletin du muséum national d'histoire naturelle. HI 63369

Bulletin du muséum d'histoire naturelle. Marseilles. Nos. 1-35, 1941-75. Bull. Mus. Hist. Nat. (Marseille). Preceded by: Annales du musée d'histoire naturelle de Marseille. Superseded by: Bulletin du musée d'histoire naturelle de Marseille. 3-2543-2. HI 63370

Bulletin du muséum d'histoire naturelle Belgrade. Série B. Sciences biologiques = Glasnik prirodnaučkog museja u Beogradu. Serija B. Biološke nauke. Belgrade. Glasn. Prir. Mus. u Beogradu, Ser. B, Biol. Nauke. See B–P–H 405/6.

Bulletin du muséum d'histoire naturelle du pays serbe. Serija B. Biološke nauke = Glasnik prirodnjačkog muzeja srpske zemlje. Serija B. Biološke nauke. Belgrade. Glasn. Prir. Muz. Srpske Zemlje, Ser. B, Biol. Nauke. See B–P–H 405/7.

Bulletin of the museum of the life sciences, Louisiana state university. Shreveport, LA. No. 1+, 1979+. Bull. Mus. Life Sci. Louisiana State Univ. HI 63371

Bulletin du muséum national d'histoire naturelle. Paris. Vols. 13-34, 1907-28; sér. 2, vol. 1-42, 1929-70. Bull. Mus. Natl. Hist. Nat. Preceded by: Bulletin du muséum d'histoire naturelle. Superseded by: Bulletin du muséum national d'histoire naturelle. Paris. Sér. 3, botanique, Bulletin du muséum national d'histoire naturelle. Paris. Sér. 3, ecologie générale and Bulletin du muséum national d'histoire naturelle. Paris. Sér. 3, supplément. HI 63372

Bulletin muséum national d'histoire naturelle, école pratique des hautes études. Papeete. ?-1980+. Bull. Mus. Natl. Hist. Nat. École Prat. Hautes Études. HI 63373

Bulletin du muséum national d'histoire naturelle. Section B, Adansonia: Botanique phytochimie. Paris. Vol. 3+, 1981+. Bull. Mus. Natl. Hist. Nat., B, Adansonia. Preceded by: Bulletin du muséum national d'histoire naturelle. Paris. Sér. B, botanique [etc.] and Adansonia. HI 63374

Bulletin du muséum national d'histoire naturelle. Section B, botanique, biologie et écologie végétales, phytochimie. Paris. Vols. 1-2, 1979-80. Bull. Mus. Natl. Hist. Nat., B. Preceded by: Bulletin du muséum national d'histoire naturelle. Paris. Sér. 3, botanique. Superseded by: Bulletin du muséum national d'histoire naturelle. Paris. Sér. B, Adansonia HI 63375

Bulletin du muséum national d'histoire naturelle. Sér. 3, botanique. Paris. Nos. 1-35, 1971-78. Bull. Mus. Natl. Hist. Nat., Sér. 3, Bot. Preceded by: Bulletin du muséum national d'histoire naturelle. Paris. Superseded by: Bulletin de muséum national d'histoire naturelle. Paris. Section B, botanique [etc.]. HI 63376

Bulletin du muséum nationale d'histoire naturelle. Sér. 3, écologie générale. Paris. Nos. 1-42, 1973-78. Bull. Mus. Natl. Hist. Nat., Sér. 3, Écol. Preceded by: Bulletin du muséum national d'histoire naturelle. Paris. HI 63377

Bulletin du muséum national d'histoire naturelle. Sér. 3, supplément, travaux faits dans les laboratoires et accroissement des collections. Paris. 1972-77. Bull. Mus. Natl. Hist. Nat., Sér. 3, Suppl. Preceded by: Bulletin du muséum national d'histoire. Paris. HI 63378

Bulletin du muséum national d'histoire naturelle. Sér. 4, miscellanea. Paris. Vol. 1+, 1979+. Bull. Mus. Natl. Hist. Nat., Sér. 4, Misc. Preceded by: Bulletin du muséum national d'histoire naturelle. HI 63379

Bulletin, museum of natural history, Los Angeles county. Science. Los Angeles, CA. Vols. 1-12, 1965-71. Bull. Mus. Nat. Hist. Los Angeles County, Sci. Superseded by: Science bulletin, natural history museum, Los Angeles county. HI 63380

Bulletin, museum of natural history, university of Oregon. Eugene, OR. 1965+. Bull. Mus. Nat. Hist. Univ. Oregon. HI 63381

Bulletin, museum of northern Arizona. Flagstaff, AZ. 1932+. Bull. Mus. N. Arizona. HI 63382

Bulletin, museum of technology and applied science. Sydney, N.S.W. 1948+. Bull. Mus. Technol. Appl. Sci. Preceded by: Bulletin, technological museum. Sydney. HI 63383

Bulletin of the Mysore coffee experiment station. Bangalore. Nos. 1-?, 1930-45. Bull. Mysore Coffee Exp. Sta. HI 63384

Bulletin of the Nagano college of agriculture and forestry. [Nagano kenritsu norin semmon gakko gakujutsu hokoku.] Ina. 1947+. Bull. Nagano Coll. Agric. Forest. HI 63385

Bulletin of the Nagasaki agricultural and forestry experiment station. [Nagasaki-ken sogo norin shikenjo. Nogyo bumon. Kenkyu hokoku.] Isahaya. No. 1+, 1973+. Bull. Nagasaki Agric. Forest. Exp. Sta.

HI 63386

Bulletin of the Nagasaki agricultural and forestry experiment station. Section of agriculture. [Nagasaki-ken sogo norin shikenjo. Kenkyu hokoku: nogyo bumon.] Ishihaya. No. 1+, 1973+. Bull. Nagasaki Agric. Forest. Exp. Sta., Agric. HI 63387

Bulletin of the Nagasaki municipal science museum. Nagasaki. ?-1979+. Bull. Nagasaki Munic. Sci. Mus. HI 63388

Bulletin of the Nagoya university forests. [Nagoya daigaku nogakubu enshurin hokoku.] Anjo. No. 1+, 1958+. Bull. Nagoya Univ. Forests. HI 63389

Bulletin of the Naniwa university. Series B. Agricultural and natural sciences. Sakai, Japan. Bull. Naniwa Univ., Ser. B, Agric. Sci. See B–P–H 265/10. HI 52892

Bulletin of the Nanjing botanical garden, Mem. Sun Yat Sen [Sun Yat-Sen tomb and memorial]. [Nanjing zhongshan zhiwuyuan yanjiu lunwenji.] Nanjing. 19??-80+. Bull. Nanjing Bot. Gard. HI 63390

Bulletin of Nara university of education. Series B, natural sciences. [Nara kyoiku daigaku kiyo shizen kagaku.] Nara. Vol. 15+, 1967+. Bull. Nara Univ. Educ., B. Preceded by: Journal of Nara gakugei university. HI 63391

Bulletin of the Natal cactus and succulent club. Durban. Vols. 1-6, 1957-65. Bull. Natal Cactus Succ. Club. HI 63392

Bulletin of the national academy of Peiping. [Kuo li Peip'ing yen chiu yüan yüan yu hui pao.] Peiping [=Peking]. Bull. Natl. Acad. Peiping. See B–P–H 265/20. HI 52897

Bulletin, National botanic gardens = National botanic gardens. Bulletin. Lucknow, India. Natl. Bot. Gard. Bull. See B–P–H 630/20.

Bulletin, national cactus and succulent society, London branch. London. Vol. 1+, 1968+. Bull. Natl. Cactus Succ. Soc., London Branch. HI 63393

Bulletin of the national chrysanthemum society. South orange, NJ [etc.]. Vols. 1-28, 1945-72. Bull. Natl. Chrysanthemum Soc. Superseded by: Chrysanthemum. 4-2852-1. HI 63394

Bulletin of the national council of state garden clubs. Pleasantville, N.Y. Vols. 1-19(4), 1930-48. Bull. Natl. Council State Gard. Clubs. Superseded by: National gardener. 4-2877-2. HI 63395

Bulletin, national dahlia society. Leamington Spa, London. 1974+. Bull. Natl. Dahlia Soc. HI 63396

Bulletin, national grassland research institute. [Sochi shikenjo kenkyu hokoku.] Nishinasuno. No. 1+, 1972+. Bull. Natl. Grassand Res. Inst. HI 63397

Bulletin of the national hygienic laboratory. Tokyo. Nos. 1-77, 1952-59. Bull. Natl. Hyg. Lab., Tokyo. Superseded by: Bulletin of the national institute of hygienic sciences. HI 63398

Bulletin of the national institute of agricultural sciences, Kanagawa. Series E, Horticulture. [Nogyo gijutsu kenkyusho hokoku, E. Engei.] Kanagawa, Japan. Bull. Natl. Inst. Agric. Sci., Ser. E, Hort. See B–P–H 265/25. HI 52898

Bulletin of the national institute of agricultural sciences, Nishigahara. Series C, phytopathology and entomology. [Nogyo gijutsu kenkyujo hokoku, C.] Tokyo. Vols. 1-?, 1952-83. Bull. Natl. Inst. Agric. Sci. Nishigahara, Ser. C, Pl. Pathol. HI 63399

Bulletin of the national institute of agricultural sciences, Nishigahara. Series C, plant pathology and entomology = Bulletin of the national institute of agricultural sciences, Nishigahara. Series C, phytopathology and entomology. Tokyo.

Bulletin of the national institute of agricultural sciences, Nishigahara. Series D, plant physiology, genetics and crops in general. [Nogyo gijutsu kenkyujo hokoku, D.] Tokyo. 1951-83. Bull. Natl. Inst. Agric. Sci. Nishigahara, Ser. D, Pl. Physiol. HI 63400

Bulletin of the national institute of agrobiological resources. [Nogyo seibutsu shigen kenkyujo kenkyu hokoku.] Yatabe. 1985+. Bull. Natl. Inst. Agrobiol. Resources. HI 63402

Bulletin of the national institute of hygienic sciences. [Eisei shikenjo hokoku.] Tokyo. No. 78+, 1960+. Bull. Natl. Inst. Hyg. Sci., Tokyo. Preceded by: Bulletin of the national hygienic laboratory. Tokyo. HI 63403

Bulletin of the national institute of sciences of India. New Delhi. Vols. 1-40, 1952-69. Bull. Natl. Inst. Sci. India. Superseded by: Bulletin of the indian national science academy. HI 63404

Bulletin of the national museum of Canada. Ottawa. Nos. 1-237, 1913-68. Bull. Natl. Mus. Canada. Superseded by: Publications in biological oceanography, national museum of Canada and Publications in botany, national museum of Canada. 2-900-1. HI 63405

Bulletin of the national museum, Singapore. Singapore. Vols. 30-35(1), 1961-70. Bull. Natl. Mus. Singapore. Preceded by: Bulletin of the Raffles museum. HI 63406

Bulletin of the national museum, state of Singapore = Bulletin of the national museum, Singapore.

Bulletin of the national museum of Wales = Amgueddfa. Cardiff.

Bulletin of the national parks association. Washington,

DC. Vols. [1]-[4?] (= nos. 1-36), 1919-23. Bull. Natl. Parks Assoc. Superseded by: National parks bulletin. 4-2896-1. HI 63407

Bulletin of national Pingtung institute of agriculture. [Nung chuan hsüeh pao.] Pingtung. Vol. 1+, 1958+. Bull. Natl. Pingtung Inst. Agric. HI 63408

Bulletin of the national research council of the Philippines. Quezon City, Rizal. Vols. 1-35, 1934-53. Bull. Natl. Res. Council Philippines. Superseded by: Research bulletin, national research council of the Philippines. HI 63409

Bulletin of national research institute of aquaculture. [Yoshoku kenkyujo kenkyu hokoku.] Mie. 1980+. Bull. Natl. Res. Inst. Aquac., Mie. Preceded by: Tansuiku suisan kenkyujo kenkyu hokoku. HI 63410

Bulletin, national research institute of tea. [Chagyo shikenjo kenkyu hokoku.] Shizuoka. 1962+. Bull. Natl. Res. Inst. Tea. HI 53085

Bulletin, national science foundation. Washington, DC. No. 1+, 1973+. Bull. Natl. Sci. Found., Wash. HI 63412

Bulletin of the national science museum, Tokyo. [Kokuritsu kagaku hakubutsukan kenkyu hokoku.] Tokyo. Nos. 25-33, 1949-53; n.s. nos. 34-95 [also numbered vols. 1-17], 1954-74. Bull. Natl. Sci. Mus. Tokyo. Preceded by: Bulletin of the Tokyo science museum. Superseded by: Bulletin of the national science museum, Tokyo. Ser. B, botany and Ser. C, geology. HI 63413

Bulletin of the national science museum, Tokyo. Ser. B, botany. [Kokuritsu kagaku hakubutsukan kankyu hokoku. B-riu, shokubutsugaku.] Tokyo. Vol. 1+, 1975+. Bull. Natl. Sci. Mus., Tokyo, B. Preceded by: Bulletin of the national science museum, Tokyo. HI 63414

Bulletin of the national science museum, Tokyo. Ser. C, geology. Tokyo. Vol. 1, 1975. Bull. Natl. Sci. Mus. Tokyo, C, Geol. Preceded by: Bulletin of the national science museum, Tokyo. HI 63415 Superseded by: Bulletin of the national science museum, Tokyo. Ser. C, geology and paleontology.

Bulletin of the national science museum, Tokyo. Ser. C, geology and paleontology. Tokyo. Vol. 2+, 1976+. Bull. Natl. Sci. Mus. Tokyo, C, Geol. Paleontol. Preceded by: Bulletin of the national science museum, Tokyo. Ser. C, geology. HI 63416

Bulletin of the national snapdragon society. Smethport, PA. Bull. Natl. Snapdragon Soc. See B–P–H 266/3. HI 52899

Bulletin of the national sweet pea society. Didcot. 1977+. Bull. Natl. Sweet Pea Soc. HI 63417

Bulletin, national tropical botanical garden. Lawai, Kauai, HI. Vol. 19(2)+, 1989+. Bull. Natl. Trop. Bot. Gard. Preceded by: Bulletin of the Pacific tropical botanical garden. HI 74817

Bulletin, national and university institute of agriculture, Rehovot = Bulletin, agricultural experiment station, Rehovoth. Rehovot.

Bulletin van de nationale plantentuin van Belgie = Bulletin du jardin botanique national de Belgique. Brussels.

Bulletin of the native plant society of Oregon. Eugene, OR. Vol. 14+, 1980+, 1981+. Bull. Native Pl. Soc. Oregon. Preceded by: Native plant society of Oregon [series]. HI 63418

Bulletin of the natural history museum, Balboa Park. San Diego, CA. 1922-67. Bull. Nat. Hist. Mus. Balboa Park. HI 63419

Bulletin of the natural history museum of Belgrade. Series B, biological sciences. [Translation of: Glasnik prirodnaučkog muzeja u Beogradu. Seriya B, biološke nauke.] Belgrade. Vol. 23+, 1968(1973)+. Bull. Nat. Hist. Mus. Belgrade, B. HI 63420

Bulletin of the natural history research centre. Baghdad. Vol. 6+, 1975+. Bull. Nat. Hist. Res. Centre, Baghdad. Preceded by: Bulletin of the Iraq natural history museum. HI 63421

Bulletin of the natural history society of British Columbia. Vancouver. Bull. Nat. Hist. Soc. British Columbia See B–P–H 265/12. HI 52893

Bulletin, natural history society of Maryland. Baltimore, MD. 1930-43. Bull. Nat. Hist. Soc. Maryland. Superseded by: Maryland. 3-2547-1. HI 63422

Bulletin of the natural history society of New Brunswick. St. John, New Brunswick. Bull. Nat. Hist. Soc. New Brunswick. See B–P–H 265/13. HI 52894

Bulletin of the natural history survey; Chicago academy of sciences. Chicago, IL. Bull. Nat. Hist. Surv. Chicago Acad. Sci. See B–P–H 265/14. HI 52895

Bulletin, natural history survey, Louisiana state museum. New Orleans, LA. No. 1+, 1910+. Bull. Nat. Hist. Surv. Louisiana State Mus. HI 63423

Bulletin of the natural history survey, state museum, New Orleans. New Orleans, LA. Bull. Nat. Hist. Surv. State Mus. New Orleans. See B–P–H 265/15. HI 52896

Bulletin des naturalistes des Alpes-Maritimes. Nice. 1911-13. Bull. Naturalistes Alpes-Maritimes. Superseded by: Rivièra scientifique. HI 63424

Bulletin des naturalistes de Mons et du Borinage. Mons. Vols. 1-?, 1918-34. Bull. Naturalistes Mons & Borinage. Superseded by: Bulletin mensuel, les naturalistes de Mons et du Borinage. HI 63425

Bulletin, nature conservation branch, Transvaal provincial administration. Pretoria. Nos. 1-2, 196?-68. Bull. Nat. Conservation Branch Transvaal. HI 63426

Bulletin, nature conservation division, Transvaal provincial administration = Bulletin, nature conservation branch, Transvaal provincial administration. Pretoria.

Bulletin, la nature à l'école. Paris. No. ?-17+, 19??-72+. Bull. Nat. École. HI 63427

Bulletin der Naturwissenschaftlichen Section der Schlesischen Gesellschaft für vaterländische Cultur. Breslau [=Wroclaw, Poland]. Bull. Naturwiss. Sect. Schles. Ges. Vaterl. Cult. See B–P–H 266/8. HI 52900

Bulletin, Nebraska agricultural experiment station = Nebraska agricultural experiment station bulletin. Lincoln, NB. Nebraska Agric. Exp. Sta. Bull. See B–P–H 639/25.

Bulletin of the nerine society. Uxbridge, Welland. No. 1+, 1966+. Bull. Narine Soc. HI 63429

Bulletin des Neuesten und Wissenswürdigsten aus der Naturwissenschaft, so wie den Künste, Manufakturen, technischen Gewerben, der Landwirthschaft und der bürgerlichen Haushaltung; ... Berlin. Bull. Neuesten Wissenswürd. Naturwiss. See B–P–H 266/13. HI 52903

Bulletin of the New England museum of natural history. Boston, MA. Bull. New England Mus. Nat. Hist. See B–P–H 266/16. HI 52904

Bulletin of the New Hampshire academy of science. Durham, NH. Bull. New Hampshire Acad. Sci. See B–P–H 266/18. HI 52905

Bulletin, New Hampshire agricultural experiment station = New Hampshire agricultural experiment station. Bulletin. Durham, NH. New Hampshire Agric. Exp. Sta. Bull. See B–P–H 653/1.

Bulletin of the New Jersey academy of science. Newark, NJ. Bull. New Jersey Acad. Sci. See B–P–H 266/21. HI 52906

Bulletin, New Jersey agricultural college experiment station = New Jersey agricultural college experiment station. Bulletin. New Brunswick, NJ. New Jersey Agric. Coll. Exp. Sta. Bull. See B–P–H 653/15.

Bulletin, New Jersey plant and flower growers. 19??-62-69+. Bull. New Jersey Pl. Fl. Growers. HI 63430

Bulletin, New Mexico agricultural experiment station = New Mexico agricultural experiment station, bulletin. Las Cruces & Santa Fe, NM. New Mexico Agric. Exp. Sta. Bull. See B–P–H 654/7.

Bulletin, New Mills natural history society. New Mills. No. 1+, 1971+. Bull. New Mills Nat. Hist. Soc. HI 63431

Bulletin, new series. Agricultural experiment station of the agricultural and mechanical college. Auburn, AL. Bull. New Ser. Agric. Exp. Sta. Agric. Coll. See B–P–H 266/24. HI 52907

Bulletin, new series. Report of the agricultural experiment station, agricultural and mechanical college. Auburn AL. Bull. New Ser. Rep. Agric. Exp. Sta. Agric. Coll. See B–P–H 266/25. HI 52908

Bulletin of the New York botanical garden. Lancaster, PA. Bull. New York Bot. Gard. See B–P–H 267/3. HI 52909

Bulletin of the New York christmas tree growers' association. Freeville, NY. Bull. New York Christmas Tree Growers' Assoc. See B–P–H 267/4. HI 52912

Bulletin of New York department of agriculture and markets = Proceedings of the New York state agricultural society. Albany, NY. Proc. New York State Agric. Soc. See B–P–H 733/14.

Bulletin of the New York public library. New York. Vols. 1-80, 1897-77. Bull. New York Public Libr. Superseded by: Bulletin of research in the humanities. 4-3024-3. HI 63432

Bulletin of the New York state college of forestry. Syracuse, NY. Bull. New York State Coll. Forest. See B–P–H 267/6. HI 52913

Bulletin of New York state department of agriculture and markets = Proceedings of the New York state agricultural society. Albany, NY. Proc. New York State Agric. Soc. See B–P–H 733/14.

Bulletin, New York state flower growers incorporated. New York. Nos. 1-295, 1945-70. Bull. New York State Fl. Growers. Superseded by: N Y S F I bulletin. HI 63433

Bulletin of the New York state museum of natural history. Albany, NY. Nos. 1-350?, 1887-1955. Bull. New York State Mus. Nat. Hist. Superseded by: Bulletin of the New York state museum and science service. 4-2046-2. HI 63434

Bulletin of the New York state museum and science service. Albany, NY. No. 351?+, 1956+. Bull. New York State Mus. Sci. Serv. Preceded by: Bulletin of the New York state museum of natural history. HI 63435

Bulletin, New Zealand department of scientific and industrial research. Wellington, NZ. No. 1+, 1927+. Bull. New Zealand Dept. Sci. Industr. Res. HI 63436

Bulletin of the New Zealand forest service. Wellington, N.Z. Nos. 1+, 1923+. Bull. New Zealand Forest Serv. HI 63437

Bulletin, New Zealand institute. Wellington, N.Z. Nos. 1-3, 1910-29 [suspended 1930-53]. Bull. New Zealand Inst. Superseded by: Bulletin, royal society of New

Zealand. 4-3730-2. HI 63438

Bulletin, New Zealand iris society = New Zealand iris society bulletin. Christchurch, N.Z., Upper Hutt.

Bulletin of the New Zealand orchid society. [New Zealand]. Bull. New Zealand Orchid Soc. See B–P–H 267/9. HI 52914

Bulletin, Niagara Falls nature club. Niagara Falls, Ont. Nos. 1-172, 1965?-85. Bull. Niagara Falls Nat. Club. Superseded by: Nature Niagara news. HI 63439

Bulletin of the nigerian forestry departments = Bulletin, forestry department, Nigeria. Lagos, Ibadan.

Bulletin of the Niigata horticultural experiment station. [Niigata-ken engei shikenjo kenkyu hokoku.] Niitsu. No. 1+, 1965+. Bull. Niigata Hort. Exp. Sta. HI 65460

Bulletin of the Niigata prefectural forest experiment station. [Niigata-ken ringyo shikenjo kenkyu hokoku.] Murakami. [Dates of publication not ascertained.] Bull. Niigata Pref. Forest Exp. Sta. HI 63440

Bulletin of the Niigata university forests. [Niigata daigaku nogakubu enshurin hokoku.] Niigata. 1962+. Bull. Niigata Univ. Forests. HI 65897

Bulletin of the Nippon agricultural research institute. [Nihon nogyo kenkyusho hokoku.] Tokyo. Bull. Nippon Agric. Res. Inst. See B–P–H 267/11. HI 52915

Bulletin of nippon dental college, general education. Tokyo. Nos. 1-4, 1972-75. Bull. Nippon Dental Coll., Gen. Educ. Superseded by: Bulletin of nippon dental university. HI 63441

Bulletin of nippon dental university. Tokyo. No. 5+, 1976+. Bull. Nippon Dental Univ. Preceded by: Bulletin of nippon dental college general education. HI 63442

[Bulletin, nippon fernist club.] Tokyo. No. ?-66+, 19??-64+. Bull. Nippon Fernist Club. HI 63443

Bulletin, Norfolk botanical garden society = Norfolk botanical garden society bulletin. Norfolk, VA.

Bulletin, north american gladiolus council. Bel Air, MD, etc. No. 1+, 1945+. Bull. N. Amer. Gladiolus Council. 4-3080-3. HI 52891

Bulletin, North Carolina agricultural experiment station = North Carolina agricultural experiment station bulletin. Raleigh, NC. North Carolina Agric. Exp. Sta. Bull. See B–P–H 663/6.

Bulletin of the north caucasian institute for plant protection = Trudy Severo-Kavkazskogo instituta zashchity rastenii. Rostov.

Bulletin of the north caucasian plant protection station = Izvestiya Severo-Kavkazskoi kraevoi stantsii zashchity rastenii. Rostov.

Bulletin, North Dakota agricultural experiment station = North Dakota agricultural experiment station bulletin. Fargo, ND. North Dakota Agric. Exp. Sta. Bull. See B–P–H 664/10.

Bulletin, North Dakota agricultural experiment station, Dickinson sub-station = North Dakota agricultural experiment station, Dickinson sub-station annual report. Dickinson, ND. North Dakota Agric. Exp. Sta. Dickson Sub-Sta. Annual Rep. See B–P–H 664/12.

Bulletin, north Florida botanical society. Vol. ?-3+, 19??-80+. Bull. N. Florida Bot. Soc. HI 63444

Bulletin, north of Scotland college of agriculture. Aberdeen. No. 1-11, 1970-76. Bull. N. Scotland Coll. Agric. Superseded by: College bulletin, north of Scotland college of agriculture. HI 63445

Bulletin, Northamptonshire naturalists' trust ltd. [Northampton.] Nos. 1-24, 1967-80. Bull. Northamptonshire Naturalist's Trust. Superseded by: Newsletter, Northamptonshire trust for nature conservation. HI 63446

Bulletin de nouvelles, biological council of Canada = Newsletter, biological council of Canada. Ottawa.

Bulletin de nouvelles, société canadienne des microbiologistes = Newsletter, canadian society of microbiologists. London, Ont.

Bulletin de nouvelles, société canadienne de physiologie végétale = Newsletter, canadian society of plant physiologists. Edmonton.

Bulletin O E P P. Paris. N.s. vol. 1+, 1971+. Bull. O. E. P. P. Preceded by: Publications, european and mediterranean plant protection organisation. Series A, reports of technical working parties and conferences and Publications, european and mediterranean plant protection organisation. Series C, reports of working meetings. HI 63447

Bulletin de l'observatoire de la mer, Ile des Embiez. Vol. 1+, 1974+. Bull. Observatiore Mer Ile Embiez. HI 63448

Bulletin of the ocean research institute, university of Tokyo. Tokyo. No. 1+, 1967+. Bull. Ocean Res. Inst. Univ. Tokyo. HI 63449

Bulletin of the oceanographical institute of Taiwan. Taipei, Taiwan. Bull. Oceanogr. Inst. Taiwan. See B–P–H 267/16. HI 52916

Bulletin, office of experiment stations, United States department of agriculture. Washington, DC. Nos. 1-256, 1889-1913. Bull. Off. Exp. Sta. U. S. D. A. HI 63450

Bulletin de l'office international (de la vigne et) du vin. Paris. Bull. Off. Int. Vin. See B–P–H 267/17. HI 52917

Bulletin. Office national du café. Port-au-Prince, Haiti.

Bull. Off. Natl. Café. See B–P–H 267/18. HI 52918

Bulletin, Ohio agricultural experiment station = Ohio agricultural experiment station bulletin. Wooster, OH.

Bulletin, Ohio biological survey = Ohio biological survey bulletin. [Ohio state university.] Columbus, OH. Ohio Biol. Surv. Bull. See B–P–H 685/11.

Bulletin, Ohio botanic garden society = Ohio botanical garden society. [Bulletin.] Cincinnati, OH.

Bulletin, Ohio florists association = Ohio florists association, bulletin. Columbus, OH. Ohio Florists Assoc. Bull. See B–P–H 685/14.

Bulletin of the Ohio state university. Columbus, OH. Bull. Ohio State Univ. See B–P–H 267/22. HI 52919

Bulletin of the Oita prefectural agricultural research center. [Oita-ken nogyo gijutsu senta kenkyu hokoku]. Usashi. 1970+. Bull. Oita Prefect. Agric. Exp. Sta. HI 66266

Bulletin of the Oita prefectural forest experiment station. [Oita-ken ringyo shikenjo kenkyu hokoku]. Oita. Vol. 1+, 1974+. Bull. Oita Pref. Forest Exp. Sta. HI 63671

Bulletin of the Oji institute for forest tree development. [Oji seishi k. k. rinkobu ikushu kenkyujo, kenkyu hokoku.] No. 1+, 1966+. Bull. Oji Inst. Forest Tree Developm. HI 63451

Bulletin of the Oji institute for forest tree improvement. Kuriyama, Japan. Bull. Oji Inst. Forest Tree Improv. See B–P–H 267/23. HI 52920

Bulletin of Okayama prefectural agricultural experiment station. [Okayama-kenritsu nogyo shikenjo gyomu kotel.] Okayama, Japan. Bull. Okayama Prefect. Agric. Exp. Sta. See B–P–H 267/24. HI 52921

Bulletin of the Okayama prefecture forest experiment station. [Okayama-ken ringyo hokoku.] Okayama. 1959+. Bull. Okayama Pref. Forest Exp. Sta. HI 66036

Bulletin of the Okayama tobacco experiment station. [Okayama tabako shikenjo hokoku.] Okayama, Japan. Bull. Okayama Tobacco Exp. Sta. See B–P–H 267/25. HI 52922

Bulletin. Okinawa prefectural agricultural experiment station. [Okinawa kenritsu noji shikenjo.] Naha?, Okinawa. Bull. Okinawa Prefect. Agric. Exp. Sta. See B–P–H 267/27. HI 52923

Bulletin, Oklahoma agricultural experiment station = Oklahoma agricultural experiment station bulletin. Stillwater, OK. Oklahoma Agric. Exp. Sta. Bull. See B–P–H 687/2.

Bulletin of the Old Dominion horticultural society. Norfolk, VA. Bull. Old Dominion Hort. Soc. See B–P–H 268/1. HI 52924

Bulletin, Ontario agricultural college = Ontario agricultural college bulletin. Guelph, Ontario. Ontario Agric. Coll. Bull. See B–P–H 688/9.

Bulletin of the orchid society of south east Asia. Singapore. No. 11/67+, 1967+. Bull. Orchid. Soc. S. E. Asia. Preceded by: Monthly bulletin, malayan orchid society. HI 63452

Bulletin, orchid society of Thailand. Bangkok. Vol. 1+, 1967+. Bull. Orchid Soc. Thailand. HI 63453

Bulletin, Oregon agricultural experiment station = Oregon agricultural experiment station bulletin. Corvallis, OR. Oregon Agric. Exp. Sta. Bull. See B–P–H 690/11.

Bulletin, Oregon orchid society = Oregon orchid society bulletin. Portland, OR. Oregon Orchid Soc. Bull. See B–P–H 690/19.

Bulletin. Organisatie voor natuurwetenschappelijk onderzoek in Indonesië. Jakarta, Indonesia. Bull. Organ. Natuurw. Onderz. Indonesië. See B–P–H 268/5. HI 52925

Bulletin of the organization for scientific research in Indonesia = Bulletin. Organisatie voor natuurwetenschappelijk onderzoek in Indonesië. Jakarta, Indonesia. Bull. Organ. Natuurw. Onderz. Indonesië. See B–P–H 268/5.

Bulletin of the Orkney field club. Kirkwall. No. 1+, 1968+. Bull. Orkney Field Club. HI 63454

Bulletino dell'orto botanico della regia università di Napoli. Naples. Bull. Orto Bot. Regia Univ. Napoli. See B–P–H 268/10. HI 52926

Bulletin of the Osaka agricultural center. [Osaka-fu norin gijutsu senta kenkyuh okoku.] Kibikino. No. 1+, 1964+. Bull. Osaka Agric. Center. HI 63455

Bulletin of the Osaka museum of natural history. [Osaka-shiritsu shizenkagaku hakubutsukan kenkyu hokoku.] Osaka. No. 11+, 1959+. Bull. Osaka Mus. Nat. Hist. Preceded by: Bulletin of the municipal museum of natural history. HI 63456

Bulletin of the Ottawa horticultural society. Ottawa. Bull. Ottawa Hort. Soc. See B–P–H 268/12. HI 52927

Bulletin, Oxford university exploration club. Oxford. No. 1-?, 1948-74?; n.s. vol. 1+, 1976+. Bull. Oxford Univ. Explor Club. Preceded by: Report (Annual), Oxford university exploration club. HI 63457

Bulletin of the Pacific committee of the academy of sciences of the U S S R = Byulleten' Tikhookeanskogo komiteta Akademii nauk S S S R. Leningrad.

Bulletin of the Pacific orchid society of Hawaii. Honolulu, HI Vol. 1- 29(3), 1941-71. Bull. Pacific

Orchid Soc. Hawaii. Superseded by: Hawaii orchid journal. 4-3235-2. HI 63458

Bulletin of the Pacific scientific fishery institute = Izvestiya Tikhookeanskogo nauchnogo instituta rybnogo khozyaistva. Voronezh.

Bulletin of the Pacific scientific institute of fisheries and oceanography = Izvestiya Tikhookeanskogo nauchno-issledovatel'skogo instituta rybnogo khozyaistva i okeanografii. Vladivostok.

Bulletin of the Pacific scientific research station = Izvestiya Tikhookeanskoi nauchno-promyslovoi stantsii. Vladivostok.

Bulletin of the Pacific tropical botanical garden. Lawai, Kauai, HI. Vols. 1-19, 1971-89. Bull. Pacific Trop. Bot. Gard. Superseded by: Bulletin, national tropical botanical garden. HI 63459

Bulletin of the Pacific tropical botanical garden. Lawai, Kauai, HI. Vol. 1+, 1971+. Bull. Pacific Trop. Bot. Gard. HI 63459

Bulletin du Pacifique sud; revue officielle de la commission du Pacifique sud. [French edition of: South Pacific bulletin.] Nouméa, Sydney, N.S.W. Vol. ?-20(2)+, 19??-70+. Bull. Pacifique Sud. HI 63460

Bulletin, palm society. Daytona Beach, FL. Nos. 1-6, 1956. Bull. Palm Soc. Superseded by: Principes. HI 63461

Bulletin, Papua and New Guinea department of forests = Papua and New Guinea department of forests. Bulletin. Port Moresby, New Guinea. Papua New Guinea Dept. Forests Bull. See B–P–H 698/22.

Bulletin, Peabody museum of natural history = Peabody museum of natural history bulletin. New Haven, CT. Peabody Mus. Nat. Hist. Bull. See B–P–H 699/27.

Bulletin of the Peking society of natural history. Peking. Bull. Peking Soc. Nat. Hist. See B–P–H 268/23. HI 52928

Bulletin penelitian hortikultura. Jakarta. Vol. 1+, 1973+. Bull. Penelitian Hort. HI 63462

Bulletin, Pennsylvania flower growers. Kennett Square, Bloomsburg, PA. No. 1+, 1950+. Bull. Pennsylvania Fl. Growers. HI 63463

Bulletin, Pennsylvania state college agricultural experiment station = Pennsylvania state college agricultural experiment station. Bulletin. State College, PA. Pennsylvania State Coll. Agric. Exp. Sta. Bull. See B–P–H 700/27.

Bulletin of peony news. Clinton, NY. Bull. Peony News. See B–P–H 268/25. HI 52929

Bulletin périodique. Chambre de commerce, d'agriculture et d'industrie. Lomé, Republic of Togo. Bull. Périod. Chambre Commerce. See B–P–H 268/26. HI 52930

Bulletin périodique des sociétés d'agriculture et d'horticulture du Doubs. Besançon. 1857?-6?; n.s. vols. 1-2, 1867-68. Bull. Périod. Soc. Agric. Hort. Doubs. HI 63464

Bulletin of the pharmaceutical research institute. Takatsuki, Osaka. Nos. 1-100, 1950-73. Bull. Pharm. Res. Inst. Superseded by: Euphoria et cacophoria. HI 63465

Bulletin de pharmacie. Paris. Bull. Pharm. (Paris). See B–P–H 268/28. HI 52931

Bulletin de pharmacie et des sciences accessoires = Bulletin de pharmacie. Paris. Bull. Pharm. (Paris). See B–P–H 268/28.

Bulletin des pharmaciens de la Côte d'Or. Dijon. Nos. 1-?, 1882-? Bull. Pharmaciens Côte d'Or. Superseded by: Bulletin, société syndicale des pharmaciens de la Côte d'Or. HI 63466

Bulletin of the philosophical society of Washington. Washington, DC. Bull. Philos. Soc. Wash. See B–P–H 268/29. HI 52932

Bulletin of the physiographical science research institute. [Tokyo daigaku ritchi shizen kagaku kenkyujo hokoku.] Tokyo. 1950+. Bull. Physiogr. Sci. Res. Inst. HI 63467

Bulletin, phytopathological experiment station of the university of Tartu = Tartu ülikooli taimehaiguste-katsejaama teated. Tartu, Estonia [Estonian S S R]. Tartu Ülik Taimeh.-Katsej. Teated. See B–P–H 866/12.

Bulletin of the Pictou academy scientific association. Pictou, N.S. Vol. 1(1-4), 1906-09. Bull. Pictou Acad. Sci. Assoc. HI 63468

Bulletin of the plankton society of Japan. [Nihon purankuton gakkaiho.] Hakodate. 1968+. Bull. Plankt. Soc. Japan. Preceded by: Information bulletin on planktology in Japan. HI 63469

Bulletin planktonique. Copenhagen. 1908-12. Bull. Planktonique. Preceded by: Bulletin trimestriel des résultats acquis pendant les croisières périodique. Partie D. 3-2056-2. HI 63470

Bulletin on plant ecology and taxonomy. Ife. No. 1+, 1969+. Bull. Pl. Ecol. Taxon. HI 63471

Bulletin of plant galls. Leyburn. No. 1+, 1984+. Bull. Pl. Galls. HI 63472

Bulletin, plant growth regulator = Plant growth regulator bulletin. St. Paul, MN, etc.

Bulletin of plant protection. Moscow & Leningrad = Vestnik zashchity rastenii. Moscow & Leningrad.

Bulletin of plant protection. Phytopathology. Leningrad = Trudy po zashchite rastenii. Seriya 2. Fitopatologiya. Leningrad.

Bulletin, plant protection station. Osaka. No. ?-37+, 19??-58+. Bull. Pl. Protect. Sta., Osaka. Preceded by?: Osaka plant protection. HI 63473

Bulletin of the polish academy of sciences. Biological sciences = Bulletin of the polish academy of sciences. Biology. Warsaw.

Bulletin of the polish academy of sciences. Biology. Warsaw. Vol. 31+, 1983+. Bull. Polish Acad. Sci., Biol. Preceded by: Bulletin de l'académie polonaise des sciences. Série des sciences biologiques. HI 63474

Bulletin of the polish academy of sciences. Earth sciences. Warsaw. Vol. 31+, 1983+. Bull. Polish Acad. Sci., Earth Sci. Preceded by: Bulletin de l'académie polonaise des sciences. Série des sciences de la terre. HI 63475

Bulletin of popular information: Arnold Arboretum Harvard University. Jamaica Plain, MA. Bull. Popular Inform. Arnold Arbor. See B–P–H 269/13. HI 52934

Bulletin of popular information, Morton arboretum. Lisle, IL. Bull. Popular Inform. (Morton Arbor.). See B–P–H 269/12. HI 52933

Bulletin, Porto Rico agricultural experiment station = Porto Rico agricultural experiment station bulletin. Mayagüez, Puerto Rico. Porto Rico Agric. Exp. Sta. Bull. See B–P–H 715/24.

Bulletin van het proefstation voor cacao te Salatiga. Semarang, Dutch E. Indies [Indonesia]. Bull. Proefstat. Cacao Salatiga. See B–P–H 269/16. HI 52935

Bulletin van het proefstation voor indigo te Klaten. Semarang, Dutch E. Indies [Indonesia]. Bull. Proefstat. Indigo Klaten. See B–P–H 269/17. HI 52936

Bulletin van het proefstation voor suikerriet in West-Java "Kagok" te Pekalongan. Tegal, Dutch E. Indies [Indonesia]. Bull. Proefstat. Suikerriet W.-Java "Kagok" Pekalongan. See B–P–H 269/18. HI 52937

Bulletin du programme canadien pour la conservation des plantes = Canadian plant conservation programme newsletter. Edmonton, Alb.

Bulletin de la protection des végétaux. Dakar, Senegal [Republic of Senegal]. See B–P–H 269/21. HI 52938

Bulletin of the public museum of Milwaukee. Milwaukee, WI. Vols. 1-10, 1910-53. Bull. Public. Mus. Milwaukee. Superseded by?: Milwaukee public museum publications in botany. 3-2663-2. HI 63476

Bulletin of pure and applied sciences. Modinagar. Vol. 1(1-2), 1982. Bull. Pure Appl. Sci., Modinagar. Superseded by: Bulletin of pure and applied sciences, Modinagar. Sect. B. plant science. HI 63477

Bulletin of pure and applied sciences, Modinagar. Sect. B, plant science. An international research journal of science. Modinagar. Vol. 2+, 1983+. Bull. Pure Appl. Sci., Modinagar, B. Preceded by: Bulletin of pure and applied sciences. Modinagar. HI 63478

Bulletin of Pusan fisheries college. [Pusan susan taehak yongu poko]. Pusan, Korea. Bull. Pusan Fish. Coll. See B–P–H 269/23. HI 52939

Bulletin, R H S lily group. Pangbourne. 1981+. Bull. R. H. S. Lily Group. Preceded by: Lilies and other liliaceae. HI 63479

Bulletin R P C = P G R C newsletter. Ottawa.

Bulletin of the Raffles museum. Singapore. Vols. 1-29, 1928-60. Bull. Raffles Mus. Superseded by: Bulletin of the national museum, state of Singapore. HI 63480

Bulletin, rattan information centre = R I C bulletin. Kepong.

Bulletin. Regional research laboratory. Jammu, Kashmir. Bull. Regional Res. Lab. See B–P–H 269/29. HI 52940

Bulletin of the republic institution for the protection of nature and the museum of natural history in Titograd = Glasnik, republičkog zavoda za zaštitu prirode i prirodnjăcke zbirke u Titogradu. Titograd.

Bulletin of the research council of Israel. Jerusalem. Bull. Res. Council Israel. See 270/1. HI 52941

Bulletin of the research council of Israel. Section B. Biology and geology. Jerusalem. Bull. Res. Council Israel, Sect. B, Biol. Geol. See B–P–H 270/2. HI 52942

Bulletin of the research council of Israel. Section D, botany. Jerusalem. Vols. 5D-11D, 1955-63. Bull. Res. Council Israel. Sect. D, Bot. Preceded by: Bulletin of the research council of Israel. Palestine journal of botany, Jerusalem series and Palestine journal of botany, Rehovot series. Superseded by: Israel journal of botany. HI 63481

Bulletin, research department, indian coffee board = Report (Annual), research department, indian coffee board. Bangalore.

Bulletin, research division, Virginia polytechnic institute = Research division bulletin, Virginia polytechnic institute.

Bulletin of research in the humanities. New York. Vol. 81+, 1978+. Bull. Res. Humanit. Preceded by: Bulletin of the New York public library. HI 63482

Bulletin of the research institut for natural resources. [Sigen kagaku kenkyusyo hokoku.] Tokyo. Bull. Res. Inst. Nat. Resources. See B–P–H 270/4. HI 52943

Bulletin of the research institut; university of Kerala. Series C, natural sciences. Trivandrum, India. Bull. Res. Isnt. Univ. Kerala, Ser. C, Nat. Sci. See B–P–H 270/5. HI 52944

Bulletin of the research institute nedri As = Skýrska, rannssóknastofnunin nedri As. Hveragerdi.

Bulletin of the research institute of the S P A. Medan. 1958+. Bull. Res. Inst. S. P. A. Preceded by: Bericht van de algemeen proefstation der A V R O S. HI 63483

Bulletin, ressources phytogénétiques du Canada = P G R C newsletter. Ottawa.

Bulletin des résultats acquis pendant les courses périodiques. Copenhagen. 1902-05. Bull. Résult. Acquis Pendant Courses Périod. Superseded by: Bulletin trimestriel des résultats acquis pendant les croisières périodiques. HI 63484

Bulletin, rhododendron society of Canada. Burlington, Ont. Vol. 1+, 1972+. Bull. Rhododendron Soc. Canada. HI 63485

Bulletin van den rijksplantentuin = Bulletin du jardin botanique de l'état. Brussels.

Bulletin romand de mycologie. Lausanne. Bull. Romand Mycol. See B–P–H 270/7. HI 52945

Bulletin of the royal hungarian college for horticulture and viniculture. Budapest = M[agyar] kir[ályi] kertészeti és szölészeti föiskola közleményei. Budapest. Magyar Kir. Kert. Szölész. Föisk. Közlem. See B–P–H 543/18

Bulletin of the royal hungarian horticultural college. Budapest = A m[agyar] kir[ályi] kertészeti tanintézet közleményei. Budapest. Magyar Kir. Kert. Tanintéz. Közlem. See B–P–H 544/1.

Bulletin of the royal hungarian horticultural college. (New series.) Budapest = A m[agyar] kir[ályi] kertészeti akadémia közleményei. Budapest. Magyar Kir. Kert. Akad. Közlem. See B–P–H 543/17.

Bulletin, royal society of New Zealand. Wellington, N.Z. Vol. 4+, 1954+. Bull. Roy. Soc. New Zealand. Preceded by: Bulletin, New Zealand institute. 4-3730-2. HI 63486

Bulletin, royal tropical institute, department of agricultural research = Bulletin, department of agricultural research of the royal tropical institute. Amsterdam.

Bulletin, rubber growers' association. London. Vols. 1-23, 1919-49. Bull. Rubber Growers' Assoc. HI 63487

Bulletin, rubber growers' association. Rubber and agriculture series. London. Vols. 1-23, 1919-41; 1945?-49. Bull. Rubber Growers' Assoc., Rubber Agric. Ser. HI 63488

Bulletin of the rubber research institute of Ceylon. Dartonfield, Agalawatta. Nos. ?-57, 1955-63. Bull. Rubber Res. Inst. Ceylon. Preceded by: Bulletin of the rubber research scheme, Ceylon. Superseded by: R R I C bulletin. HI 63489

Bulletin of the rubber research institute of Malaya. Kuala Lumpur. Bull. Rubber Res. Inst. Malaya . See B–P–H 270/13. HI 52946

Bulletin of the rubber research scheme, Ceylon. Dartonfield, Agalawatta. 1921-37. Bull. Rubber Res. Scheme, Ceylon. Preceded by: Bulletin, Ceylon rubber research scheme. Superseded by: Bulletin of the rubber research institute of Ceylon. HI 63490

Bulletin, Rutgers University agricultural experiment station = New Jersey agricultural college experiment station. Bulletin. New Brunswick, NJ. New Jersey Agric. Coll. Exp. Sta. Bull. See B–P–H 653/15.

Bulletin de la S A J I B = Bulletin de la société d'animation du jardin et de l'institut botaniques. Montreal.

Bulletin, S R O P. Wageningen. 1973+. Bull. S. R. O. P. HI 63491

Bulletin S, seed branch, department of agriculture, Canada = Bulletin, seed branch, department of agriculture, Canada. Ottawa.

Bulletin, Sabah orchid society. Kota Kinabalu. Ca.1980? Bull. Sabah Orchid Soc. HI 63492

Bulletin of the Sacramento cactus and succulent society. Sacramento, CA. Vol. 1+, 1960+. Bull. Sacramento Cact. Succ. Soc. HI 63493

Bulletin of the Saga agricultural experiment station. [Saga-ken nogyo shikenjo kenkyu hokoku.] Saga, Japan. Bull. Saga Agric. Exp. Sta. See B–P–H 270/27. HI 52951

Bulletin of the Saghalien central experiment station. Series 2, forestry. [Karafutocho chuo shikenjo iho.] Konuma, Saghalien. [Dates of publication not ascertained.] Bull. Saghalien Centr. Exp. Sta., Ser. 2 Forest. HI 63494

Bulletin of the Saitama horticultural experiment station. [Saitama-ken engei shikenjo kenkyu hokoku.] Kuki. No. 1+, 1967+. Bull. Saitama Hort. Exp. Sta. HI 66746

Bulletin of the Saitama museum of natural history. [Saitama kenritsu shizenshi hakubutsukan kenkyu hokoku.] Saitama. Vol. 1+, 1983+. Bull. Saitama Mus. Nat. Hist. HI 63495

Bulletin of the Sapporo branch, forestry experiment station. [Hokkaido ringyo shiken hokoku.] Sapporo, Japan. Bull. Sapporo Branch Forest. Exp. Sta. See B–P–H 271/2. HI 52952

Bulletin of the school of agriculture and forestry, Taihoku imperial university. [Taihoku teikoku daigaku fuzoku norin senmon-bu Gajutsu hokoku.] Taihoku [=Taipei, Taiwan]. Bull. School Agric. Taihoku Imp. Univ. See B–P–H 271/13. HI 52953

Bulletin of the school of forestry and conservation, university of Michigan. Ann Arbor, MI. Nos. 1-11, 1932-44. Bull. School Forest. Conservation Univ.

Michigan. HI 63496

Bulletin of the school of forestry of Duke university. Durham, NC. Bull. School Forest. Duke Univ. See B–P–H 271/14. HI 52954

Bulletin, school of forestry, Louisiana tech university. Ruston, LA. No. 9+, 1976+. Bull. School Forest. Louisiana Tech Univ. Preceded by: Bulletin, division of research, college of life sciences, Louisiana tech university. HI 75138

Bulletin, school of forestry, Montana state university = Bulletin, forest and conservation experiment station, university of Montana. Missoula, MT.

Bulletin, school of forestry, Stephen F. Austen state university = Bulletin, department of forestry, Stephen F. Austen state college, Texas. Nacogdoches, TX.

Bulletin, school of forestry, university of Idaho. Moscow, ID. Nos. 1-9, 1921-39. Bull. School Forest. Univ. Idaho. HI 63497

Bulletin, school of forestry, university of Melbourne. Melbourne, Vic. No. 1+, 1960+. Bull. School Forest. Univ. Melbourne. HI 63498

Bulletin, school nature league. New York = School nature league bulletin.

Bulletin, school of science, Purdue university. Lafayette, IN. No. 1, 1886. Bull. School Sci. Purdue Univ. HI 63499

Bulletin of the science and engineering division, university of the Ryukyus. Mathematics and natural sciences. Naha, Okinawa. Nos. 1-26, 19??-78. Bull. Sci. Engin. Div. Univ. Ryukyus, Math. Nat. Sci. Superseded by: Bulletin of the college of science, university of the Ryukyus. HI 63500

Bulletin of the science and engineering research laboratory, Waseda university. [Waseda daigaku rikogaku kenkyujo hokoku.] Tokyo. 1955+. Bull. Sci. Engin. Res. Lab. Waseda Univ. HI 63501

Bulletin de science et technique de la polytechnique de Timisoara. Timisoara, Rumania. Bull. Sci. Techn. Polytechn. Timisoara. See B–P–H 272/15. HI 52970

Bulletin des séances de l'académie royale des sciences d'Outre Mer. Brussels. Bull. Séances Acad. Roy. Sci. Outre Mer. See B–P–H 272/25. HI 52972

Bulletin des sciences agricoles et économiques. Paris. Bull. Sci. Agric. Écon. See B–P–H 271/20. HI 52956

Bulletin des sciences géographiques, économie publique, voyages. Paris. Bull. Sci. Géogr. See B–P–H 272/3. HI 52962

Bulletin, sciences géologiques, université Louis Pasteur de Strasbourg. Strasbourg. Vol. 25+, 1972+. Bull. Sci. Géol. Univ. Louis Pasteur Strasbourg. Preceded by: Bulletin du service de la carte géologique d'Alsace et de Lorraine. HI 63502

Bulletin des sciences médicales. Paris. Bull. Sci. Méd. See B–P–H 272/5. HI 52964

Bulletin des sciences naturelles et de géologie. Deuxième section du bulletin universel des sciences et de l'industrie. Paris. Bull. Sci. Nat. Géol. See B–P–H 272/6. HI 52965

Bulletin des sciences pharmacologiques. Paris. Vols 1-42, 1899-1942. Bull. Sci. Pharmacol. Superseded by: Annales pharmaceutiques françaises. 1-836-1. HI 63503

Bulletin des sciences physiques, médicales et d'agriculture d'Orléans. Orléans. Bull. Sci. Phys. Orléans. See B–P–H 272/11. HI 52967

Bulletin des sciences physiques et naturelles en Néerlande. Rotterdam. Bull. Sci. Phys. Nat. Néerl. See B–P–H 272/10. HI 52966

Bulletin des sciences, par la société philomatique. Paris. Bull. Sci. Soc. Philom. Paris. See B–P–H 272/14. HI 52969

Bulletin des sciences, par la société philomatique de Paris = Bulletin des sciences, par la société philomatique. Paris. Bull. Sci. Soc. Philom. Paris. See B–P–H 272/14.

Bulletin des sciences technologiques. Paris. Bull. Sci. Technol. See B–P–H 272/16. HI 52971

Bulletin of the scientific laboratory of Denison university. Granville, OH. Bull. Sci. Lab. Denison Univ. See B–P–H 272/4. HI 52963

Bulletin of the scientific researches of the alumni association of the Morioka college of agriculture and forestry. [Morioka kono dosokai gakujutsu iho.] Morioka, Japan. Bull. Sci. Res. Alumni Assoc. Morioka Coll. Agric. See B–P–H 272/12. HI 52968

Bulletin scientifique. Académie de la république populaire roumaine = Buletin ştiintific. Academia republicii populare române [from 1954 ... romîne] Bucharest. Bul. Sti. Acad. Republ. Populare Române. See B–P–H 234/1.

Bulletin scientifique de Bourgogne. Toulouse. Bull. Sci. Bourgogne. See B–P–H 271/21. HI 52957

Bulletin scientifique, conseil des académies de la R P F de Yougoslavie. Sect. A, sciences naturelles, techniques et médicales. Ljubljana & Zagreb. Vols. 1-10, 1953-65. Bull. Sci. Cons. Acad. RPF. Yougoslavie, A. Superseded by: Bulletin scientifique, conseil des académies des sciences at des arts de la R S F de Yougoslavie. HI 63504

Bulletin scientifique, conseil des académies des sciences et des arts de la R S F de Yougoslavie. Sect. A. Zagreb. Vol. 11?+, 1966+. Bull. Sci. Cons. Acad. Sci. Arts

RSF. Yougoslavie, A. Preceded by: Bulletin scientifique, conseil des académies de la R P F de Yougoslavie. HI 63505

Bulletin scientifique du départment du nord et des pays voisins = Bulletin scientifique, historique et litéraire du départment du nord et des pays voisins. Paris. Bull. Sci. Dép. Nord Pays Voisins. See B–P–H 271/23.

Bulletin scientifique de l'école polytechnique de Timisoara. Timisoara, Rumania. Bull. Sci. École Polytechn. Timisoara. See B–P–H 271/25. HI 52960

Bulletin scientifique de la France et de la Belgique. Paris. Bull. Sci. France Belgique. See B–P–H 272/2. HI 52961

Bulletin scientifique, historique et litéraire du départment du nord et des pays voisins. Paris. Bull. Sci. Dép. Nord Pays Voisins. See B–P–H 271/23. HI 52958

Bulletin scientifique publié par l'académie imperiale des sciences de Saint-Pétersbourg. St. Petersburg & Leipzig. Bull. Sci. Acad. Imp. Sci. Saint-Pétersbourg. See B–P–H 271/15. HI 52955

Bulletin scientifique. Université d'état de Kiev = Naukovi zapysky. Kyiivs'kyi derzhavnyi universytet imeny T. G. Shevchenka. Kiev.

Bulletin, scottish horticultural research institute. Invergowrie. No. 1+, 1968+. Bull. Scott. Hort. Res. Inst. HI 63506

Bulletin des séances de l'institut royal colonial belge. Brussels. Bull. Séances Inst. Roy. Colon. Belge. See B–P–H 273/1. HI 52973

Bulletin des séances de la société d'agriculture, sciences, arts et commerce du Puy. Le Puy. Vol. 5, 1847. Bull. Séances Soc. Agric. Puy. Preceded by: Bulletin agronomique et Industriel de la société d'agriculture, sciences, arts et commerce du Puy. 5-3936-3. HI 63507

Bulletin des séances de la société centrale d'agriculture de France, compte rendu mensuel. Paris. Sér. 3, vols. 6-9, 1871-74?; [sér. 4] vols. 35-38 [continues volumation of 1st sér.], 1875-78. Bull Séances Soc. Centr. Agric. France. Preceded by: Bulletin des séances de la société impériale et centrale d'agriculture de France, compte rendu mensuel. Superseded by: Bulletin des séances de la société nationale d'agriculture de France; compte rendu mensuel. 1-25-3. HI 63508

Bulletin des séances de la société impériale et centrale d'agriculture, compte rendu mensuel. Paris. Sér. 2, vols. 9-20, 1853-65?; sér. 3, vols. 1-5, 1866?-69. Bull. Séances Soc. Imp. Centr. Agric. Preceded by: Bulletin des séances de la société royale et centrale d'agriculture, compte rendu mensuel. Superseded by: Bulletin des séances de la société centrale d'agriculture de France, compte rendu mensuel. 1-25-3. HI 63509

Bulletin des séances de la société nationale d'agriculture de France, compte rendu mensuel. Paris. Vols. 39-74, 1879-1914. Bull. Séances Soc. Natl. Agric. France. Preceded by: Bulletin des séances de la société centrale d'agriculture de France, compte rendu mensuel. Superseded by: Comptes rendus des séances de l'académie d'agriculture de France. 1-25-3. HI 63510

Bulletin des séances de la société royale et centrale d'agriculture, compte rendu mensuel. Paris. Vols. 1-8?, 1841-52? Bull. Séances Soc. Roy. Centr. Agric. Paris. Superseded by: Bulletin des séances de la société impériale et centrale d'agriculture, compte rendu mensuel. 1-25-3. HI 63511

Bulletin de la section d'agriculture coloniale. Paris. 1902-08. Bull. Sect. Agric. Colon. HI 63512

Bulletin, section régionale ouest paléarctique. Wageningen = Bulletin, S R O P. Wageningen.

Bulletin de la section scientifique de l'académie roumaine. Bucharest. Bull. Sect. Sci. Acad. Roumaine. See B–P–H 273/4. HI 52974

Bulletin de la section du Turkestan de la société russe de géographie = Izvestiya Turkestanskogo otdela Russkogo geograficheskogo obshchestva. Tashkent.

Bulletin, section of vegetable pathology, U S department of agriculture = Department of agriculture. Botanical division. Bulletin. Washington, DC. Dept. Agric. Bot. Div. Bull. See B–P–H 342/17.

Bulletin, seed branch, department of agriculture, Canada. Ottawa. Nos. 1-9, 1905-15. Bull. Seed Branch Dept. Agric. Canada. Superseded by: Pamphlet, seed commissioner's branch, department of agriculture, Canada. HI 63513

Bulletin, Seoul national university forests. [Soul taehakkyo nongkwa taehak yonsumnim yongu pogo.] Seoul. Vol. 1+, 1962+. Bull. Seoul Natl. Univ. Forests. HI 63514

Bulletin of the sericultural experiment station. [Sanshi shikenjo hokoku.] Tokyo. Bull. Seric. Exp. Sta. See B–P–H 273/7. HI 52975

Bulletin of the serological museum, Rutgers university. New Brunswick, NJ. Bull. Serol. Mus. See B–P–H 273/9. HI 52976

Bulletin du service de biogéographie; université de Montréal. Montreal. Bull. Serv. Biolgéogr. Univ. Montréal. See B–P–H 273/13. HI 52977

Bulletin du service botanique et agronomique de Tunisie. Tunis. Nos. 1-21, 1946-51. Bull. Serv. Bot. Agron. Tunisie. Superseded by: Rapport sur les travaux de recherche effectues, service botanique et agronomique de Tunisie. HI 63515

Bulletin du service de la carte géologique d'Alsace et de Lorraine. Strasbourg. Vols. 1-24, 1920-71. Bull.

Serv. Carte Géol. Alsace Lorraine. Preceded by: Mittheilungen der geologischen Landesanstalt von Elsass-Lothringen. Superseded by: Bulletin, sciences géologiques, université Louis Pasteur de Strasbourg. HI 63516

Bulletin du service de la carte phytogéographique. Sér. A, carte de la végétation. Paris. Vols. 1-7, 1956-62. Bull. Serv. Carte Phytogéogr., Sér. A, Carte Vég. HI 52978

Bulletin du service de la carte phytogéographique. Sér. B, carte des groupements végétaux. Paris. Vols. 1-7, 1956-62. Bull. Serv. Carte Phytogéogr., Sér. B, Carte Groupements Vég. HI 63517

Bulletin, service forestier, Québec = Bulletin, forest service, department of lands and forests, Quebec.

Bulletin of the Shanghai science institute. [Shanghai shizen kwagaku kenkyu-sho iho.] Shanghai. Bull. Shanghai Sci. Inst. See B–P–H 273/16. HI 52979

Bulletin, Shenandoah natural history association. Luray, VA. 1950+. Bull. Shenandoah Nat. Hist. Assoc. HI 63518

Bulletin of the Shikoku agricultural experiment station. [Shikoku nogyo shikenjo hokoku.] Kagawa, Japan. Bull. Shikoku Agric. Exp. Sta. See B–P–H 273/17. HI 52980

Bulletin of the Shimane agricultural college. [Shimane noka daigaku kenkyu hokoku.] Matsue, Japan. Bull. Shimane Agric. Coll. See B–P–H 273/19. HI 52981

Bulletin of the Shimane agricultural experiment station. [Shimane-kenritsu noji shikenjo kenkyu hokoku.] Shimane, Japan. Bull. Shimane Agric. Exp. Sta. See B–P–H 273/20. HI 52982

Bulletin of the Shimane university. Natural science. [Shimane daigaku ronshu. Shizen kagaku.] Matsue. Nos. 1-15, 1951-66. Bull. Shimant Univ., Nat Sci. Superseded by: Memoirs of the faculty of literature and science, Shimane university. HI 63519

Bulletin of the Shimane university forests. [Shimane daigaku nogakubu enshurin hokoku.] Matsue. No. 1+, 1972+. Bull. Shimane Univ. Forests. HI 63520

Bulletin of the Shinshu university forest. [Shinshu daigaku nogaku-bu enshurin hokoku.] Uyeda, Japan. Bull. Shinshu Univ. Forest. See B–P–H 273/22. HI 52983

Bulletin of the Shizuoka agricultural college. Morioka, Japan. Bull. Shizuoka Agric. Coll. See B–P–H 273/ 23. HI 52985

Bulletin of Shizuoka prefectural citrus experiment station. [Shizuoka-ken kankitsu shikenjo kenkyu hokoku.] Shizuoka, Japan. Bull. Shizuoka Prefect. Citrus Exp. Sta. See B–P–H 274/2. HI 52987

Bulletin of Shizuoka prefecture agricultural experiment station. [Shizuoka-ken nogyo shikenjo kenkyu hokoku.] Shizuoka, Japan. Bull. Shizuoka Prefect Agric. Exp. Sta. See B–P–H 274/1. HI 52986

Bulletin of the Shizuoka prefecture forestry experiment station = Research report of the Shizuoka prefecture forestry experiment station. Hamakita.

Bulletin of the Shizuoka prefecture tea experiment station. [Shizuoka-ken chagyo shikenjo kenkyu hokoku.] No. 1+, 1966+. Bull. Shizuoka Pref. Tea Exp. Sta. HI 66937

Bulletin of the Shropshire conservation trust. Shrewsbury. No. 1+, 1963+. Bull. Shropshire Conservation Trust. HI 63521

Bulletin signalétique. 380. Agronomie, zootechnie, phytiatrie et phytopharmacie, aliments et industries alimentaires. Paris. Vol. 32+, 1971+. Bull. Signal., 380, Agron. Preceded by: Bulletin signalétique. 380. Sciences agricoles, etc. HI 63522

Bulletin signalétique. 320. Biochimie, biophysique (later Biochimie, biophysique moléculaire, biologie moléculaire et cellulaire). Paris. Vol. 33-44, 1972-83. Bull. Signal., 320, Biochim. Biophys. Preceded by: Bulletin signalétique. 320 Biochimie, biophysique, génie biologique et medical. Superseded by: P A S C A L folio. F52, biochimie; biophysique moléculaire; biologie moléculaire et cellulaire. HI 63523

Bulletin signalétique. 320. Biochimie, biophysique, genie biologique et médical (later Biochimie, biophysique, genie bio-médical). Paris. Vols. 30-32, 1969-71. Bull. Signal., 320, Biochim. Biophys. Genie Biol. Méd. Preceded by: Bulletin signalétique. 12. Biophysique, biochimie, chimie analytique biologique. Superseded by: Bulletin signalétique. 320 Biochimie, biophysique. HI 63524

Bulletin signalétique. 17. Biologie et physiologie végétales. Paris. Vols. 17-29, 1956-68. Bull. Signal., 17, Biol. Physiol. Vég. Preceded by: Bulletin analytique. Superseded by: Bulletin signalétique. 370. Biologie et physiologie végétales. HI 63525

Bulletin signalétique. 370. Biologie et physiologie végétales (later et Sylviculture). Paris. Vol. 30-41-?, 1969-83? Bull. Signal., 370, Biol. Physiol. Vég. Preceded by: Bulletin signalétique. 17. Biologie et physiologie végétales. Superseded, in part, by: P A S C A L folio. F55, biologie végétale. HI 63526

Bulletin signalétique. 12. Biophysique, biochimie, chimie analytique biologique. Paris. Vols. 22-29, 1961-68. Bull. Signal., 12, Biophys. Biochim. Chim. Analytique Biol. Superseded by: Bulletin signalétique. 320. Biochimie, biophysique, genie biologique et medical. HI 63527

Bulletin signalétique. 215. Biotechnologies. Paris. Vol. 43+, 1982+. Bull. Signal., 215, Biotechnol. Preceded by: Bulletin signalétique. Informascience. HI 63528

Bulletin signalétique. 401. Congrès, rapports, thèses. Paris. Vol. 42+, 1981+. Bull. Signal., 401, Congr. Rapp. Thèses. HI 63529

Bulletin signalétique. 363. Génétique. Paris. Vols. 33-37, 1972-76. Bull. Signal., 363, Génét. Preceded in part by: Bulletin signalétique. 361. Endocrinologie et reproduction, génétique [not entered]. Superseded by: P A S C A L explore. E60, génétique. HI 63530

Bulletin signalétique. 22. Histoire des sciences et des techniques. Paris. Vols. ?-22, 196??-68. Bull. Signal., 22, Hist. Sci. Techn. Superseded by: Bulletin signalétique. 522. Histoire des sciences et des techniques. HI 63531

Bulletin signalétique. 522. Histoire des sciences et des techniques. Paris. Vol. 23+, 1969+. Bull. Signal., 522, Hist. Sci. Techn. Preceded by: Bulletin signalétique. 22. Histoire des sciences et des techniques. HI 63532

Bulletin signalétique. 14. Microbiologie, virus, bactériophages, immunologie. Paris. Vol. 22-29, 1961-68. Bull. Signal., 14, Microbiol. Virus Bactérioph. Immunol. Superseded by: Bulletin signalétique. 340. Microbiologie, virologie, immunologie. HI 63533

Bulletin signalétique. 340. Microbiologie, virologie, immunologie. Paris. Vols. 30-44, 1969-83. Bull. Signal., 340, Microbiol. Virol. Immunol. Preceded by: Bulletin signalétique. 14. Microbiologie, virus, bactériophages, immunologie. Superseded by: P A S C A L explore. E61, microbiologie, [etc.]. HI 63534

Bulletin signalétique. 18. Sciences agricoles, zootechnie, phytiatrie et phytopharmacie, aliments et industries alimentaires. Paris. Vols. 17-29, 1956-68. Bull. Signal., 18, Agric. Zootechn. Phytiat. Phytopharm. Ailments Industr. Aliment. Preceded by: Bulletin analytique. Superseded by: Bulletin signalétique. 380. Sciences agricoles, etc. HI 63535

Bulletin signalétique. 380. Sciences agricoles, etc. Paris. Vols. 30-31, 1969-70. Bull. Signal., 380, Sci. Agric., etc. Preceded by: Bulletin signalétique. 18. Sciences agricoles, zootechnie, phytiatrie et phytopharmacie, aliments et industries alimentaires. Superseded by: Bulletin signalétique. 380. Agronomie, zootechnie, phytiatrie et phytopharmacie, aliments et industries alimentaires. HI 63536

Bulletin signalétique. 381. Sciences agronomiques. Productions (produits) végétales. Paris. Vol. 41+, 1980+. Bull. Signal., 381, Sci. Agron. Prod. Vég. Preceded, in part, by: Bulletin signalétique. 380. Sciences agricoles, etc. HI 63537

Bulletin signalétique. 13. Sciences pharmacologiques, toxicologie. Paris. Vols. 22-29, 1961-68. Bull. Signal., 13, Sci. Pharmacol. Toxicol. Superseded by: Bulletin signalétique. 330. Sciences pharmacologiques, toxicologie. HI 63538

Bulletin signalétique. 330. Sciences pharmacologiques, toxicologie. Paris. Vols. 30+, 1969+. Bull. Signal., 330, Sci. Pharmacol. Toxicol. Preceded by: Bulletin signalétique. 13. Sciences pharmacologiques, toxicologie. HI 63539

Bulletin signalétique. [Section] 11. Sciences de la terre II. Physique du globe. Géologie. Paléontologie. Paris. Bull. Signal., 11, Sci. Terre. See B–P–H 274/3. HI 52988

Bulletin de la société académique de l'arrondissement de Boulogne-sur-Mer. Boulogne-sur-Mer, France. Bull. Soc. Acad. Arrondissement Boulogne-sur-Mer. See B–P–H 274/7. HI 52989

Bulletin de la société académique de Brest. Bull. Soc. Acad. Brest See B–P–H 274/8. HI 52990

Bulletin de la société d'acclimatation et de protection de la nature = Bulletin de la société nationale d'acclimatation de France. Paris.

Bulletin, société d'agriculture de l'arrondissement de Mayenne. Mayenne. Vol. 1, 1859/62. Bull. Soc. Agric. Arrondissement Mayenne HI 63540

Bulletin de la société d'agriculture du départment de l'Ardèche. Paris. Bull. Soc. agric. Dép. Ardèche. See B–P–H 274/13. HI 52991

Bulletin, société d'agriculture et d'horticulture de l'arrondissement de Poitoise (Seine et Oise). Pontoise. Vols. 1(1)-8(148), 1851-98. Bull. Soc. Agric. Hort. Arrondissement Poitoise. HI 63541

Bulletin de la société d'agriculture, industrie, sciences, arts et lettres du départment de l'Ardèche = Bulletin de la société d'agriculture du départment de l'Ardèche. Paris. Bull. Soc. agric. Dép. Ardèche. See B–P–H 274/13.

Bulletin de la société d'agriculture, sciences et arts de la Sarthe. Le Mans. Vols. 5-8, 1844-49; ser. 2, vols. 1-41, 1850-1924; ser. 3, vol. 1+, 1925+. Bull. Soc. Agric. Sarthe. 5-3936-3. HI 63542

Bulletin, société des amis du jardin Van Den Hende Inc. Quebec. Vol. 1+, 1981+. Bull. Soc. Amis Jard. Van Den Hende. HI 63543

Bulletin, société des amis du muséum de Chartres et des naturalistes d'Eure-et-Loire. Chartres. Vol. 1+, 1983+. Bull. Soc. Amis. Mus. Chartres Naturalistes Eure-Et-Loire. HI 63545

Bulletin de la société des amis des sciences et des lettres de Poznań. Série B: Sciences mathématiques et naturelles. Poznan, Poland. Bull. Soc. Amis Sci. Lett. Poznań, Sér. B, Sci. Math. See B–P–H 274/16. HI 52992

Bulletin de la société des amis des sciences et des lettres de Poznań. Série D. Sciences biologiques. Poznan,

Poland. Bull. Soc. Amis Sci. Lett. Poznań, Sér. D, Sci. Biol. See B–P–H 274/17. HI 52993

Bulletin de la société des amis des sciences naturelles de Rouen. Rouen. Bull. Soc. Amis Sci. Nat. Rouen. See B–P–H 274/18. HI 52994

Bulletin de la société des amis des sciences de Poznań. Série B: Sciences mathématiques et naturelles. Poznan, Poland. Bull. Soc. Amis Sci. Poznań, Sér. B, Sci. Math. See B–P–H 274/20. HI 52995

Bulletins de la société anatomique de Nantes. Nantes. Bull. Soc. Anat. Nantes. See B–P–H 275/3. HI 52996

Bulletin de la société d'anatomie pathologique. Brussels. Bull. Soc. Anat. Pathol. See B–P–H 275/4. HI 52997

Bulletin de la société d'animation du jardin et de l'institut botaniques. Montreal. Vols. 1-10, 1975-86. Bull. Soc. Animat. Jard. Inst. Bot., Montreal. Superseded by: Quatre-temps. HI 63546

Bulletin de la société archéologique du Morbihan. Mémoires et procès-verbaux. Vannes, Château-Gaillard. 1857+. Bull. Soc. Archéol. Morbihan, Mém. Procès-Verbaux. HI 63547

Bulletin, société autunoise d'horticulture. Autun. Vols. 1+, 1859+. Bull. Soc. Autunoise Hort. HI 63548

Bulletin de la société belge d'études coloniales. Brussels. Bull. Soc. Belge Études Colon. See B–P–H 275/9. HI 52998

Bulletin de la société belge de géologie, de paléontologie et d'hydrologie. Brussels. Bull. Soc. Belge Géol. See B–P–H 275/11. HI 52999

Bulletin de la société belge de microscopie. Brussels. Bull. Soc. Belge Microscop. See B–P–H 275/12. HI 53000

Bulletin de la société de biologie de Lettonie = Latvijas biologijas biedribas raksti. Riga, Latvia [Latvian S S R]. Latv. Biol. Biedribas Raksti. See B–P–H 526/13.

Bulletin de la société botanique de Bulgarie = Izvestiya na bulgarskoto botanichesko druzhestvo. Sofia.

Bulletin de la société botanique du Centre-Ouest. Niort. 193?-46; n.s. vol. 1+, 1970+; Num. spec. 2+, 1978+. Bull. Soc. Bot. Centre-Ouest. Preceded by: Bulletin de la société botanique des Deux-Sèvres. HI 63549

Bulletin de la société botanique de Copenhague = Meddelelser fra den botaniske forening i Kjøbenhavn. Copenhagen.

Bulletin de la société botanique des Deux-Sèvres, [etc.]. Niort. Vols. 1-18, 20-26, 1889-1907, 1910-31, 1936. Bull. Soc. Bot. Deux-Sèvres. For vol. 19 see: Bulletin de la société régionale de botanique. Superseded by: Bulletin de la société botanique du Centre-Ouest. HI 63550

Bulletin de la société botanique, fondée dans les Deux-Sèvres [etc.] = Bulletin de la société botanique des Deux-Sèvres, [etc.]. Niort.

Bulletin de la société botanique de France. [Mémoires, vols.1-7 (1905-21), were issued as part of Bulletin series with separate pagination.] Paris. Vol. 1-?, 1854-1977. Bull. Soc. Bot. France. Re-organised into two sections in 1978: Bulletin de la société botanique de France. Lettres botaniques and Bulletin de la société botanique de France. Actualitiés botaniques. 5-3933-2. HI 63551

Bulletin de la société botanique de France. Actualitiés botaniques. Paris. Vol. 125+, 1978+. Bull. Soc. Bot. France, Actual. Bot. Preceded by: Bulletin de la société botanique de France. HI 63552

Bulletin de la société botanique de France. Lettres botaniques. Paris. Vol. 126+, 1978+. Bull. Soc. Bot. France, Lett. Bot. Preceded by: Bulletin de la société botanique de France. HI 63553

Bulletin de la société botanique de Genève. Geneva. Bull. Soc. Bot. Genève. See B–P–H 275/26. HI 53001

Bulletin de la société de botanique, de géologie et d'entomologie du Var et de la Corse. Toulon. Vols. 41-54, 19??-43. Bull. Soc. Bot. Géol. Entomol. Var. & Corse. Preceded by: Bulletin trimestriel de la société de botanique, de d'entomologie du Var et de la Corse. HI 63554

Bulletin, société botanique et horticole de Provence. Marseille. Vols. 1-5, 1879-83. Bull. Soc. Bot. Hort. Provence. 5-3934-1. HI 63555

Bulletin de la société botanique de Lyon. Comptes rendus des séances. Lyons. Bull. Soc. Bot. Lyon Compt. Rend. Séances. See B–P–H 275/29. HI 53002

Bulletin de la société de botanique du nord de la France. Lille. Bull. Soc. Bot. N. France. See B–P–H 275/32. HI 53003

Bulletin de la société botanique de la Provence centrale et de la Corse. Toulon. Vol. 15, 19?? Bull. Soc. Bot. Provence Centr. Corse. Preceded by: Bulletin de la société botanique du Var. Superseded by: Bulletin de la société botanique du Var et de la Corse. HI 63556

Bulletin de la société botanique du Québec. Quebec. No. 1+, 1981+. Bull. Soc. Bot. Québec. HI 63557

Bulletin, société botanique de rochelaise. La Rochelle. Vols. 11-25, 1889-1904. Bull. Soc. Bot. Rochelaise. Preceded by: Comptes rendus des excursions botaniques, société botanique de rochelaise. HI 63558

Bulletin de la société botanique suisse = Bericht der schweizerischen botanischen Gesellschaft. Basel, Geneva, Zurich.

Bulletin de la société botanique suisse = Botanica helvetica. Zurich.

Bulletin de la société botanique du Var. Toulon. Vols. 1-14, 1916-? Bull. Soc. Bot. Var. Superseded by: Bulletin de la société botanique de la Provence centrale et de la Corse. HI 63559

Bulletin de la société botanique tchécoslovaque à Prague = Preslia. Věstník (Časopis) československé botanické společnosti. Prague. Preslia. See B–P–H 720/16.

Bulletin de la société botanique du Var et de la Corse. Toulon. Vols. 16-17, 19??-? Bull. Soc. Bot. Var & Corse. Preceded by: Bulletin de la société botanique de la Provence centrale et de la Corse. Superseded by: Bulletin trimestriel de la société botanique et géologique du Var et de la Corse. HI 63560

Bulletin de la société de botanique et de zoologie congolaise. Leopoldville. Vols. 3(2)-5(1/2), 1940-42. Bull. Soc. Bot. Zool. Congol. Preceded and superseded by: Zooleo. 5-4640-1. HI 63561

Bulletin de la société centrale d'ágriculture du département de l'Hérault. Montpellier. Bull. Soc. Centr. Agric. Dép. Hérault. See B–P–H 276/3. HI 53004

Bulletin de la société centrale d'aquiculture et de pêche. Paris. Vols. 1-2, 1889-45. Bull. Soc. Centr. Acquicult. Pêche. 5-3934-2. HI 63562

Bulletin de la société centrale d'arboriculture en Belgique. Brussels. Bull. Soc. Centr. Arboric. Belgique. See B–P–H 276/4. HI 53005

Bulletin de la société centrale forestière de Belgique. Brussels. Vols. 1-57, 1893-1950. Bull. Soc. Centr. Forest. Belgique. Superseded by: Bulletin de la société royale forestière de Belgique. 5-3973-1. HI 53006

Bulletin de la société de chimie biologique. Paris. Vol. 1-52, 1914-70 [suspended 1914-19]. Bull. Soc. Chim. Biol. Superseded by: Biochemie. 5-3944-2. HI 63563

Bulletin de la société climatologique algérienne = Bulletin de la société des sciences physiques, naturelles et climatologiques de l'Algérie. Algiers. Bull. Soc. Sci. Phys. Algérie. See B–P–H 282/7.

Bulletin de la société dauphinoise pour l'échange des plantes. Grenoble. Bull. Soc. Dauphin. Échange Pl. See B–P–H 276/12. HI 53007

Bulletin de la société dauphinoise d'études biologiques, (bio-club). Grenoble. Vols. 1-4, 1906-12; n.s. vol. 1-?, 19??-76? Bull. Soc. Dauphin. Etudes Biol. HI 63564

Bulletin de la société dauphinoise d'études biologiques et de protection de la nature (Bio-club.) = Bulletin de la société dauphinoise d'études biologiques, (Bio-club). Grenoble.

Bulletin de la société dendrologique de France. Paris. Bull. Soc. Dendrol. France. See B–P–H 276/14. HI 53009

Bulletin, société pour l'échange des plantes vasculaires de l'Europe occidentale et du bassin méditerranéen. Liège. Fasc. 14(2)+, 1972+. Bull. Soc. Échange Pl. Vasc. Eur. Occid. Bassin Médit. Preceded by: Bulletin, société française pour l'échange des plantes vasculaires. HI 63565

Bulletin de la société d'écologie. Brunoy. Vol. 1(2)-3, 1970-72. Bull. Soc. Écol. Preceded by: Bulletin d'information, société française d'écologie. Superseded by: Bulletin d'écologie. HI 63566

Bulletin de la société d'écophysiologie. Montpellier. Vol. 1+, 1976+. Bull. Soc. Écophys. HI 63567

Bulletin de la société d'émulation d'Abbeville. Abbeville. 1797-1814. Bull. Soc. Émul. Abbeville. Superseded by: Mémoriaux de la société d'émulation d'Abbeville. 5-3938-3. HI 63568

Bulletin de la société d'encouragement pour l'industrie nationale. Paris. Bull. Soc. Encour. Industr. Natl. See B–P–H 276/15. HI 53010

Bulletin de la société d'enseignement mutuel des sciences naturelles d'Elbeuf. Elbeuf. Vols. 1-?, 1881-1947. Bull. Soc. Enseignem. Mutuel Sci. Nat. 5-3939-3. HI 63569

Bulletin de la société d'étude des sciences naturelles de Béziers. Beziers, France. Bull. Soc. Étude Sci. Nat. Béziers. See B–P–H 276/18. HI 53011

Bulletin de la société d'étude des sciences naturelles d'Elbeuf = Bulletin de la société d'enseignement mutuel des sciences naturelles d'Elbeuf. Elbeuf.

Bulletin de la société d'étude des sciences naturelles de Nîmes. Nîmes. Vol. 1+, 1874+. Bull. Soc. Étude Sci. Nat. Nîmes. 5-3939-3. HI 63570

Bulletin de la société d'étude des sciences naturelles de Reims. Rheims. Vols. 1-20, 1892-1914; n.s., vol. 1+, 1922-37, 1946-47, 1956+. Bull. Soc. Étude Sci. Nat. Reims. 5-3939-3. HI 63571

Bulletin de la société d'étude des sciences naturelles de Vaucluse. Avignon. Vol. 1+, 1930+. Bull. Soc. Étude Sci. Nat. Vaucluse. HI 63572

Bulletin de la société des études océaniennes. Papeete, Tahiti. Bull. Soc. Études Océanien. See B–P–H 276/19. HI 53013

Bulletin de la société d'études scientifiques d'Angers. Angers. Vols. 1-14, 1871-84 [t-p imprint dates, 1872, 1885]; n.s. vols. 15-?, 1885 [t-p imprint, 1886]-1959 [?suspended 1946-54]. Bull. Soc. Études Sci. Angers. Superseded by: Bulletin de la société d'études scientifiques de l'Anjou. 5-3940-3. HI 63573

Bulletin de la Société d'études scientifiques de l'Anjou. Anjou. N.s. vol. 3+, 1960+. Bull. Soc. Études Sci.

Anjou. Preceded by: Bulletin de la société d'études scientifiques d'Angers. Superseded by?: Bulletin trimestriel de la société d'études scientifiques de l'Anjou. HI 63574

Bulletin de la société d'études scientifiques de l'Aude. Carcassonne. Vol. 1+, 1890+. Bull. Soc. Études Sci. Aude. 5-3940-3. HI 63575

Bulletin de la société d'études scientifiques de Lyon. Lyons. Bull. Soc. Études Sci. Lyon. See B–P–H 276/22. HI 53014

Bulletin, société de la flore Valdôtaine. Aoste, Turin. Nos. 1-24?, 1902-71? Bull. Soc. Fl. Valdôt. Superseded by: Revue Valdôtaine d'histoire naturelle. 5-3947-1. HI 63576

Bulletin, société forestière de Franche-Comté et Belfort (et des provinces de l'est). Besançon. 1891+. Bull. Soc. Forest. Franche-Comté & Belfort. HI 63577

Bulletin de la société française de dermatolgie et de syphiligraphie. Paris. Bull. Soc. Franç. Dermatol. See B–P–H 277/1. HI 53016

Bulletin, société française pour l'échange des plantes. Agen. Vols. 1-26, 1911-36. Bull. Soc. Franç. Échange Pl. Superseded by?: Bulletin de la société française pour l'échange des plantes vasculaires. HI 63578

Bulletin de la société française pour l'échange des plantes vasculaires. Versailles. Vols. 1-14(1), 1947-71. Bull. Soc. Franç. Échange Pl. Vasc. Preceded by?: Bulletin, société française pour l'échange des plantes. Superseded by: Bulletin, société pour l'échange des plantes vasculaires de l'Europe occidentale et du bassin méditerranéen. 5-3960-3. HI 63579

Bulletin de la société française d'économie rurale. Paris. Vols. 1-4, 1949-52. Bull. Soc. Franç. Écon. Rurale. Superseded by: Économie rurale. 2-1399-1. HI 63580

Bulletin, société française d'histoire des sciences et des techniques. Paris. Vol. 1, 1980+. Bull. Soc. Franç. Hist. Sci. Techn. HI 63581

Bulletin de la société française de mycologie médicale. Paris. Vol. 1+, 1960+. Bull. Soc. Franç. Mycol. Méd. HI 63582

Bulletin de la société française de physiologie végétale. Paris. Vols. 1-15(2), 1955-69. Bull. Soc. Franc. Physiol. Vég. Incorporated in: Physiologie végétale. HI 63583

Bulletin de la société française de systématique. Paris. No. 0+, 1985+. Bull. Soc. Franç. Syst. HI 63584

Bulletin de la société franco-japonaise de Paris. Paris. Bull. Soc. Franco-Jap. Paris. See B–P–H 277/9. HI 53017

Bulletin de la société fribourgeoise des sciences naturelles. Fribourg, Switzerland. Bull. Soc. Fribourg. Sci. Nat. See B–P–H 277/10. HI 53018

Bulletin de la société de géographie. Paris. Bull. Soc. Géogr. See B–P–H 277/11. HI 53020

Bulletin de la société de géographie de Québec. Quebec. Vols. 1-28(2), 1880-34 [publication suspended 1898-1907]. Bull. Soc. Géogr. Québec. Superseded by: Bulletin des sociétés de géographie de Québec et Montréal. HI 63585

Bulletin de la société de géographie et d'archéologie de la province d'Oran. Oran. Vols. 1-5, 1878/81-85. Bull. Soc. Géogr. Archéol. Oran. Superseded by: Bulletin trimestriel de géographie et d'archéologie. HI 63586

Bulletin de la société de géographie de la province d'Oran = Bulletin de la société de géographie et d'archéologie de la province d'Oran. Oran.

Bulletin de la société géologique de France. Paris. Vols. 1-14, 1830-43; sér. 2, vols. 1-29, 1843-72; sér. 3, vols. 1-28, 1872-1900; sér. 4, vols. 1-30, 1901-30; sér. 5, vols. 1-20, 1931-50; sér. 6, vols. 1-8, 1951-58; sér. 7, vol. 1+, 1959+. Bull. Soc. Géol. France. 5-3961-2. HI 63587

Bulletin de la société géologique de Normandie. Le Havre. 1873-1973. Bull. Soc. Géol. Normandie. Superseded by: Bulletin trimestriel de la société géologique de Normandie et des amis du muséum du Havre. HI 63588

Bulletin de la société d'histoire naturelle de l'Afrique du nord. Algiers. Bull. Soc. Hist. Nat. Afrique N. See B–P–H 277/17. HI 53022

Bulletin de la société d'histoire naturelle des Ardennes. Charleveille, France. Bull. Soc. Hist. Nat. Ardennes. See B–P–H 277/19. HI 53023

Bulletin de la société d'histoire naturelle d'Autun. Autun. Vols. 1-30, 1888-1949. Bull. Soc. Hist. Nat. Autun. Superseded by: Eduen. 5-3942-2. HI 63590

Bulletin de la société d'histoire naturelle d'Auvergne. Clermont Ferrand. 1921-34. Bull. Soc. Hist. Nat. Auvergne. Superseded by: Revue des sciences naturelles d'Auvergne. HI 63591

Bulletin de la société d'histoire naturelle de Colmar. Colmar, France. Bull. Soc. Hist. Nat. Colmar. See B–P–H 277/21. HI 53024

Bulletin de la société d'histoire naturelle du Creusot. Le Cruesot. Vols. 1-7, 1932-49; vol. 32+, 1974+. Bull. Soc. Hist. Nat. Creusot. From 1950-73 incorporated in: Revue de la féderation française des sociétés des sciences naturelles. HI 63592

Bulletin de la société d'histoire naturelle du département de la Moselle. Metz. Bull. Soc. Hist. Nat. Dép. Moselle. See B–P–H 277/22. HI 53025

Bulletin de la société d'histoire naturelle du Doubs. Besançon. Nos. ?-69. 1914-67; vol. 79+, 1981+. Bull. Soc. Hist. Nat. Doubs. Preceded by: Mémoires de la société d'histoire naturelle du Doubs. For 1968-80 see: Bulletin de la féderation des sociétés d'histoire naturelle de Franche-Comte. HI 63593

Bulletin, société d'histoire naturelle de Haute-Savoie. Annecy. Vols. ?-7-35, 19??-60-? Bull. Soc. Hist. Nat. Haute-Savoie. Superseded by: Bulletin mensuel, société d'histoire naturelle de Haute-Savoie. HI 63594

Bulletin de la société d'histoire naturelle du Jura. Lons-le-Saunier. 1919-36. Bull. Soc. Hist. Nat. Jura. HI 63595

Bulletin de la société d'histoire naturelle de Metz. Metz. Bull. Soc. Hist. Nat. Metz. See B–P–H 277/24. HI 53026

Bulletin de la société d'histoire naturelle de la Moselle. Metz. Bull. Soc. Hist. Nat. Moselle. See B–P–H 278/1. HI 53027

Bulletin, société d'histoire naturelle du pays de Montbéliard. Montbéliard. 1966/67+, 1969+. Bull. Soc. Hist. Nat. Pays Montbéliard. HI 63596

Bulletin de la société d'histoire naturelle, Reims. Rheims. 1877-82. Bull. Soc. Hist. Nat. Reims. HI 63597

Bulletin de la société d'histoire naturelle de Savoie. Chambéry. 1850-54; 1882+. Bull. Soc. Hist. Nat. Savoie. HI 63598

Bulletin de la société d'histoire naturelle de Toulouse. Toulouse. Bull. Soc. Hist. Nat. Toulouse. See B–P–H 278/2. HI 53028

Bulletin de la société d'histoire de la pharmacie. Paris. Bull. Soc. Hist. Pharm. See B–P–H 278/3. HI 53029

Bulletin de la société hongroise de géographie = Földrajzi közlemények. Budapest. Földr. Közlem. See B–P–H 377/10.

Bulletin, société d'horticulture et d'acclimatation du département du Var à Toulon. Toulon? 18??-7? Bull. Soc. Hort. Acclim. Dép. Var Toulon. Superseded by: Bulletin mensuel de la société d'agriculture, d'horticulture et d'acclimatation du Var à Toulon. HI 63599

Bulletin de la société d'horticulture et d'acclimation du Maroc. Casablanca, Morocco. Bull. Soc. Hort. Maroc. See B–P–H 278/5. HI 53030

Bulletin, société d'horticulture de Cherbourg. Cherbourg. 1846-48; n.s. no. 1+, 1869+. Bull. Soc. Hort. Cherbourg. HI 63600

Bulletin de la société d'horticulture de la côte. Nyon. Vols. ?-5-12, 18??-92-99. Bull. Soc. Hort. Côte. HI 63601

Bulletin de la société d'horticulture de la Côte-d'Or agrégée à la société d'acclimatation. Dijon. 1862-73; 2. ser., vols. 1-6, 1873-74. Bull. Soc. Hort. Côte-d'Or Agrég. Soc. Acclim. HI 63602

Bulletin de la société d'horticulture de Genève. Geneva. Vols. 1-91?, 1868-1946. Bull. Soc. Hort. Genève. 5-3943-1. HI 63603

Bulletin de la société d'horticulture du Maroc = Bulletin de la société d'horticulture et d'acclimation du Maroc. Casablanca, Morocco. Bull. Soc. Hort. Maroc. See B–P–H 278/5.

Bulletin de la société d'horticulture d'Orléans et du Loiret. Orléans. Bull. Soc. Hort. Orléans. See B–P–H 278/6. HI 53031

Bulletin, société d'horticulture pratique du département du Rhône. Lyons. Vol. 1+, 1844+. Bull. Soc. Hort. Prat. Dép. Rhône. 5-3943-3. HI 63604

Bulletin de la société d'horticulture de la Sarthe. Le Mans. Bull. Soc. Hort. Sarthe. See B–P–H 278/7. HI 53032

Bulletin, société d'horticulture et de viticulture d'Eure-et-Loire. Chartres. Vols. 1-16?, 1853-89? Bull. Soc. Hort. Vitic. Eure-et-Loire. HI 63605

Bulletin de la société impériale des naturalistes de Moscou. Moscow. Vols. 1-62, 1829-86 [-1887]; n.s. vols. 1-30, 1887-1917. Bull. Soc. Imp. Naturalistes Moscou. Superseded by: Byulleten′ Moskovskogo obshchestva ispytatelei prirody. Otdel biologicheskii and Byulleten′ Moskovskogo obshchestva ispytatelei prirody. Otdel geologicheskii. 3-2770-1. HI 51593

Bulletin de la société impériale russe de géographie = Izvestiya Imperatorskogo Russkogo Geograficheskogo Obshchestva. St. Petersburg.

Bulletin de la société industrielle et agricole d'Angers et du département de Maine et Loire = Bulletin de la société industrielle d'Angers et du département de Maine et Loire. Angers. Bull. Soc. Industr. Angers. See B–P–H 278/14.

Bulletin de la société industrielle d'Angers et du département de Maine et Loire. Angers. Bull. Soc. Industr. Angers. See B–P–H 278/14. HI 53034

Bulletin, société internationale des amis des arbres de Tunisie. Tunis. Vols. ?-24-32, 19??-30-38. Bull. Soc. Int. Amis Arbres Tunisie. Superseded by: Sylvicole tunisienne. HI 53033

Bulletin de la société khédiviale de géographie. Cairo. Bull. Soc. Khédiv. Géogr. See B–P–H 278/15. HI 53035

Bulletin de la société libre d'émulation de Rouen. Rouen. Bull. Soc. Libre Emul. Rouen. See B–P–H 278/17. HI 53037

Bulletin de la société linnéenne de Bordeaux. Bordeaux. Vol. 1+, 1971+. Bull. Soc. Linn. Bordeaux. Preceded by: Actes de la société linnéenne de Bordeaux. Sér. A, and Actes de la société linnéenne de Bordeaux. Sér. B. HI 63606

Bulletin de la société linnéenne de la Charente-Inférieure. Saint-Jean-d'Angely. Vols. 1-2, 1877-82. Bull. Soc. Linn. Charente-Inférieure. 5-3965-1. HI 63607

Bulletin de la société linnéenne du nord de la France. Abbeville, France. Bull. Soc. Linn. N. France. See B–P–H 278/19. HI 53038

Bulletin de la société linnéenne de Normandie. Caen. Bull. Soc. Linn. Normandie. See B–P–H 278/22. HI 53039

Bulletin de la société linnéenne de Provence. Marseilles. Bull. Soc. Linn. Provence. See B–P–H 279/3. HI 53040

Bulletin de la société linnéene de la Seine-Maritime. Le Havre. Vol. 1(1-2), 1913-14. Bull. Soc. Linn. Seine-Maritime. Superseded by: Bulletin mensuel de la société linnéene de la Seine-maritime. 5-3965-1. HI 63608

Bulletin de la société littéraire et scientifique de Castres. Castres. [Dates of publication not ascertained.] Bull. Soc. Litt. Sci. Castres. HI 63609

Bulletin de la société lorraine des sciences. Nancy. Bull. Soc. Lorraine Sci. See B–P–H 279/4. HI 53041

Bulletin de la société de médecine d'Anvers. Antwerp. Bull. Soc. Méd. Anvers. See B–P–H 279/5. HI 53042

Bulletin de la société médicale des hôpitaux universitaires de Quebec. Quebec. Bull. Soc. Méd. Hôp. Univ. Quebec. See B–P–H 279/7. HI 53043

Bulletin de la société mycologique de France Paris. 1885-1903. Bull. Soc. Mycol. France. Superseded by: Bulletin trimestriel de la société mycologique de France. 5-3967-3. HI 63610

Bulletin de la société mycologique de Genève. Geneva. Vols. 1-13, 1914-36. Bull. Soc. Mycol. Genève. 5-3967-3. HI 63611

Bulletin, société mycologique du Poitou. Poitiers. 1978+. Bull. Soc. Mycol. Poitou. HI 63612

Bulletin de la société nationale d'acclimatation de France. Paris. Vols. 43-97(3), 1896-1950. Bull. Soc. Natl. Acclim. France. Preceded by: Revue des sciences naturelles appliquées. Incorporated in: Terre et vie. HI 63613

Bulletin de la société nationale d'horticulture de France. Paris. Bull. Soc. Natl. Hort. France. See B–P–H 279/20. HI 53044

Bulletin de la société des naturalistes de l'Ain. Bourg. Vol. [1]-37, 1896-1921? Bull. Soc. Naturalistes Ain. Superseded by: Bulletin de la société des naturalistes et des archéologiques de l'Ain. 5-3954-3. HI 63614

Bulletin de la société des naturalistes et des amis de la nature en Crimée = Zapiski Krymskago Obshchestva Estestvoispytatelei i Lyubitelei Prirody. Simferopol.

Bulletin de la société des naturalistes et des archéologiques de l'Ain. Bourg. Vol. 38+, 1925?+. Bull. Soc. Naturalistes Archéol. Ain. Preceded by: Bulletin de la société des naturalistes de l'Ain. 5-3954-3. HI 63615

Bulletin de la société des naturalistes et archéologiques du nord de la Meuse. Montmédy? Vols. 18-24, 1906-11. Bull. Soc. Naturalistes N. Meuse. Preceded by: Mémoires de la société des amateurs-naturalistes du nord de la Meuse. Superseded by: Recueil de la société des naturalistes et archéologiques du nord de la Meuse. 5-3954-3. HI 63616

Bulletin de la société des naturalistes dinantais. Dinant, Belgium. Bull. Soc. Naturalistes Dinantais. See B–P–H 279/23. HI 53046

Bulletin de la société des naturalistes luxembourgeois. Luxemburg. No. 54+ [nos. 54-58 = n.s. nos. 43-47,] 1949+. Bull. Soc. Naturalistes Luxemb. Preceded by: Bulletins mensuels de la société des naturalistes luxembourgeois. HI 63617

Bulletin de la société des naturalistes de Moscou. Section biologique = Byulleten′ Moskovskogo obshchestva ispytatelei prirody. Otdel biologicheskii. Moscow.

Bulletin de la société des naturalistes de Moscou. Section géologique = Byulleten′ Moskovskogo obshchestva ispytatelei prirody. Otdel geologicheskii. Moscow & Leningrad.

Bulletin de la société des naturalistes d'Oyonnax pour l'étude et la diffusion des sciences naturelles dans la region. Bourg. Nos. 1-21, 1947-72. Bull. Soc. Naturalistes Oyonnax. Superseded by: Bulletin annuel de la fédération centre-est d'histoire naturelle et de mycologie. HI 63618

Bulletin de la société les naturalistes parisiens. Paris. Bull. Soc. Naturalistes Parisiens. See B–P–H 279/26 HI 53047

Bulletin de la société des naturalistes de Voronèje = Byulleten′ Obshchestva estestvoispytatelei pri Voronezhskom gosudarstvennom universitete. Voronezh.

Bulletin de la société des naturalistes de Voronèje = Byulleten′ Voronezhshkogo obshchestva estestvoispytatelei pri Gosudarstvennom universitete. Voronezh.

Bulletin de la société neuchâteloise de géographie. Neuchatel. Vols. 1-49, 1885-1943; n.s. vol. 1+ [also numbered vol. 50+], 1944+. Bull. Soc. Neuchâtel. Géogr. 5-3968-2. HI 63619

Bulletin de la société neuchâteloise de sciences naturelles. Neuchâtel. Bull. Soc. Neuchâteloise Sci. Nat. See B–P–H 280/1. HI 53048

Bulletin de la société des océanistes. Paris. Vol. 1(1-2), 1937. Bull. Soc. Océanistes. Superseded by: Journal de la société des océanistes. HI 74819

Bulletin de la société d'océanographie de France. Paris. Bull. Soc. Océanogr. France. See B–P–H 280/3. HI 53049

Bulletin de la société ouralienne d'amateurs des sciences naturelles = Zapiski Ural'skago Obshchestva Lyubitelei Estestvozaniya. Ekaterinburg.

Bulletin de la société ouralienne d'amis des sciences naturelles = Zapiski Ural'skago obshchestva lyubitelei estestvozaniya v gorode Sverdlovske. Sverdlovsk.

Bulletin de la société de pathologie exotique. Paris. Bull. Soc. Pathol. Exot. See B–P–H 280/9. HI 53050

Bulletin, société de pathologie végétale de France = Revue de pathologie végétale et d'entomologie agricole de France. Paris. Rev. Pathol. Vég. Entomol. Agric. France. See B–P–H 784/5.

Bulletin de la société de pharmacie de Bordeaux = Bulletin des travaux de la société de pharmacie de Bordeaux. Bordeaux. Bull. Trav. Soc. Pharm. Bordeaux. See B–P–H 285/1.

Bulletin de la société de pharmacie de Marseille. Vol. 1+, 1952+. Bull. Soc. Pharm. Marseille. HI 63620

Bulletin, société de pharmacie de Nancy. Nancy. Nos. 1-95, 1949-72. Bull. Soc. Pharm. Nancy. Superseded by: Sciences pharmaceutiques et biologiques de Lorraine. 5-3949-1. HI 63621

Bulletin de la société philomathique de Bordeaux. Bordeaux. Bulll. Soc. Philom. Bordeaux. See B–P–H 280/11. HI 53051

Bulletin de la société philomatique de Paris. Paris. Bull. Soc. Philom. Paris. See B–P–H 280/12. HI 53052

Bulletin de la société philomatique de Perpignan. Perpignan, France. Bull. Soc. Philom. Perpignan. See B–P–H 280/13. HI 53055

Bulletin de la société des Phyrénées-orientales; sciences, belles-lettres, arts industriels et agricoles. Perpignan, France. Bull. Soc. Pyrénées-Orientales. See B–P–H 280/20. HI 53057

Bulletin de la société phycologique de France. Paris. Nos. 1-23, 1963-78. Bull. Soc. Phycol. France. Superseded by: Cryptogamie - algologie. HI 63624

Bulletin de la société "plantes et rocailles"; par monts, vaux et jardins. Lausanne. 1976+. Bull. Soc. Pl. Rocailles. HI 63625

Bulletin de la société polymathique du Morbihan. Mémoires et procès-verbaux. = Bulletin de la société archéologique du Morbihan. Mémoires et procès-verbaux. Vannes, Château-Gaillard.

Bulletin de la société portugaise des sciences naturelles. Lisbon. Vols. 1-14, 1907-43. Bull. Soc. Portug. Sci. Nat. Superseded by: Boletim da sociedade portuguesa de ciências naturais. HI 63622

Bulletin de la société préhistorique française. Paris. Bull. Soc. Préhist. Franç. See B–P–H 280/19. HI 53056

Bulletin de la société régionale de botaniques. Niort. Vol. 19, 19?? Bull. Soc. Régionale Bot. Preceded and superseded by: Bulletin de la société botanique des Deux-Sèvres. HI 63626

Bulletin de la société royale d'agriculture, sciences et arts du Mans. Le Mans. Vols. 1-4, 1833-4? Bull. Soc. Roy. Agric. Mans. Superseded by: Bulletin de la sociéte d'agriculture, sciences et arts de la Sarthe. 5-3936-3. HI 63627

Bulletin de la société royale belge de géographie. Brussels. Vols. 1-85, 1876-1961. Bull. Soc. Roy. Belge Géogr. Superseded by: Revue belge de géographie. 5-3970-3. HI 63628

Bulletin de la société royale de botanique de Belgique. Brussels. Bull. Soc. Roy. Bot. Belgique. See B–P–H 280/26. HI 53058

Bulletin de la société royale forestière de Belgique. Brussels. Vol. 58+, 1951+. Bull. Soc. Roy. Forest. Belgique. Preceded by: Bulletin de la société centrale forestière de Belgique. 5-3973-1. HI 63629

Bulletin de la société royale d'horticulture de Liège. Liege. 1860-97. Bull. Soc. Roy. Hort. Liège. HI 63630

Bulletin de la société royale linnéenne de Bruxelles. Brussels. Vols. 1-31, 1872-1906. Bull. Soc. Roy. Linn. Bruxelles. 5-3973-1. HI 63631

Bulletin, société royale linnéenne et de flore = Bulletin de la société royale linnéenne de Bruxelles. Brussels.

Bulletin de la société royale des naturalistes de Mons et du Borinage. Mons. Vol. 45(5)+, 1962+. Bull. Soc. Roy. Naturalistes Mons & Borinage. Preceded by: Bulletin mensuel, société royale des naturalistes de Mons et du Borinage. HI 63632

Bulletin. Société royale de pharmacie de Bruxelles. Brussels. Bull. Soc. Roy. Pharm. Bruxelles. See B–P–H 281/2. HI 53059

Bulletin de la société royale des sciences de Liège. Liége. Bull. Soc. Roy. Sci. Liège. See B–P–H 281/3. HI 53060

Bulletin de la société royale des sciences médicales et naturelles de Bruxelles. Brussels. Vols. 27-69(8), 1866-1911. Bull. Soc. Roy. Sci. Méd. Nat. Bruxelles. Preceded by: Bulletin de la société des sciences médicales et naturelles de Bruxelles. Superseded by: Annales et bulletin de la société royale des sciences médicales et naturelles de Bruxelles. 5-3972-3. HI 63633

Bulletin de la société royale des sciences naturelles du Laos. Vientiane. 1961-68. Bull. Soc. Roy. Sci. Nat. Laos. HI 63634

Bulletin des société savantes det littéraires de Belgique. Tournai, Belgium. Bull. Soc. Savantes Litt. Belgique. See B–P–H 281/5. HI 53061

Bulletin de la société des sciences, agriculture et arts du département du Bas-Rhin. Strasbourg. Bull. Soc. Sci. Dép. Bas-Rhin. See B–P–H 281/10. HI 53063

Bulletin de la société des sciences et arts de l'Île Réunion. St.-Denis, Réunion. Bull. Soc. Sci. Île Réunion See B–P–H 281/12. HI 53064

Bulletin de la société des sciences de Chatellerault. Chatellerault. No. 1+, 1965+. Bull. Soc. Sci. Chatellerault. HI 63635

Bulletin de la société des sciences historiques et naturelles de la Corse. Bastia. Vols. 1(1)-56(530), 1881-1938. Bull. Soc. Sci. Hist. Nat. Corse. 5-3955-3. HI 63636

Bulletin de la société des sciences historiques et naturelles de Semur-en-Auxois. Semur-en-Auxois. Vols. 1-38, 1865-1927 [suspended 1923-27]. Bull. Soc. Hist. Nat. Semur-en-Auxois. 5-3955-3. HI 63637

Bulletin de la société des sciences historiques et naturelles de l'Yvonne. Auxerre. Vol. 1+, 1847+. Bull. Soc. Sci. Hist. Nat. Yvonne. 5-3955-3. HI 63638

Bulletin de la société des sciences et des lettres de Lødz. Classe III, de sciences mathématiques et naturelles. Lødz. Vols. 1-10, 1946-59. Bull. Soc. Sci. Lett. Lødz, Cl. 3. Superseded by: Bulletin, lodzkie towarzystwo naukowe. HI 63639

Bulletin de la société des sciences médicales et naturelles de Bruxelles. Brussels. Vols. 1-26, 1840-65. Bull. Soc. Sci. Méd. Nat. Bruxelles. Superseded by: Bulletin de la société royale des sciences médicales et naturelles de Bruxelles. HI 63640

Bulletin de la société des sciences de Nancy. Nancy. Bull. Soc. Sci. Nancy. See B–P–H 281/15. HI 53067

Bulletin de la société des sciences naturelles de l'Ain. Bourg. No. 1+, 1894+. Bull. Soc. Sci. Nat. Ain. Superseded by: Bulletin de la société des sciences naturelles de l'Ain. HI 63641

Bulletin de la société des sciences naturelles et d'archéologie de l'Ain = Bulletin de la société des sciences naturelles de l'Ain. Bourg.

Bulletin, société des sciences naturelles et d'archéologie de Toulon et du Var. Toulon. 1960+. Bull. Soc. Sci. Nat. Archéol. Toulon & Var. Preceded by: Bulletin, société des sciences naturelles de Toulon et du Var. HI 63642

Bulletin de la société des sciences naturelles de France. Paris. 1835. Bull. Soc. Sci. Nat. France. 5-3956-2. HI 63643

Bulletin de la société de sciences naturelles de la Haute-Marne. Langres. Vols. 1-9, 1904-12. Bull. Soc. Sci. Nat. Haute-Marne. HI 63644

Bulletin de la société des sciences naturelles et historiques de l'Ardèche. Privas, France. Bull. Soc. Sci. Nat. Ardèche. See B–P–H 281/16. HI 53068

Bulletin de la société des sciences naturelles du Maroc. Rabat. Morocco. Bull. Soc. Sci. Nat. Maroc. See B–P–H 281/17. HI 53069

Bulletin de la société des sciences naturelles et médicales de Seine-et-Oise. Versailles. Ser. 2, vols. 1-13, 1919-32; ser. 3, vols. 1-8, 1933-46; ser. 4, vol. 1(1-4), 1948. Bull. Soc. Sci. Nat. Méd. Seine-et-Oise. Preceded by: Mémoires de la société des sciences naturelles de Seine-et-Oise. Incorporated in: Bulletin trimestriel, union des sociétés françaises d'histoire naturelle. 5-3957-1. HI 63645

Bulletin de la société des sciences naturelles de Neuchâtel = Bulletin de la société neuchâteloise de sciences naturelles. Neuchâtel. Bull. Soc. Neuchâteloise Sci. Nat. See B–P–H 280/1.

Bulletin de la société des sciences naturelles de l'ouest de la France. Nantes. Bull. Soc. Sci. Nat. Ouest France. See B–P–H 281/20. HI 53070

Bulletin de la société des sciences naturelles et physiques du Maroc = Bulletin de la société des sciences naturelles du Maroc. Rabat. Morocco. Bull. Soc. Sci. Nat. Maroc. See B–P–H 281/17.

Bulletin de la société des sciences naturelles de Saône-et Loire. Chalon-sur Saône, France. Bull. Soc. Sci. Nat. Saône-et Loire. See B–P–H 282/1. HI 53071

Bulletin de la société des sciences naturelles de Strasbourg. Strasbourg. Bull. Soc. Sci. Nat. Strasbourg. See B–P–H 282/3. HI 53072

Bulletin de la société des sciences naturelles du sud-est. Grenoble. Vols. 1-4 1882-85. Bull. Soc. Sci. Nat. Sud-Est. 5-3956-3. HI 63646

Bulletin de la société de sciences naturelles du Tarn-et-Garonne. Montauban. No. ?-8+, 19??-69+. Bull. Soc. Sci. Nat. Tarn-et-Garonne HI 63647

Bulletin de la société des sciences naturelles de Toulon et du Var. Toulon. 1950-51. Bull. Soc. Sci. Nat. Toulon & Var. Preceded by: Bulletin mensuel, société des sciences naturelles de Toulon et du Var. Superseded by: Bulletin, société des sciences naturelles et d'archéologie de Toulon et du Var. HI 63648

Bulletin de la société des sciences naturelles de Tunisie. Tunis. Bull. Soc. Sci. Nat. Tunisie. See B–P–H 282/5. HI 53073

Bulletin de la société des sciences physiques, naturelles et climatologiques de l'Algérie. Algiers. Bull. Soc. Sci. Phys. Algérie. See B–P–H 282/7. HI 53074

Bulletin de la société des sciences physiques & naturelles de Toulouse. Toulouse & Paris. Bull. Soc. Sci. Phys. Nat. Toulouse. See B–P–H 282/8. HI 53075

Bulletin de la société scientifique de Bretagne. Rennes. Bull. Soc. Sci. Bretagne. See B–P–H 281/8. HI 53062

Bulletin de la société scientifique du Dauphiné = Bulletin de la société scientifique de l'Isère. Grenoble. Bull. Soc. Sci. Isère. See 281/13.

Bulletin de la société scientifique de l'Isère. Grenoble. Bull. Soc. Sci. Isère. See 281/13. HI 53065

Bulletin de la société scientifique et médicale de l'ouest. Rennes. Bull. Soc. Sci. Méd Ouest. See B–P–H 281/14. HI 53066

Bulletin de la société scientifique de Skopje = Glasnik skopskog naučnog društva. Skoplje, Yugoslavia. Glasn. Skopsk. Naučn. Društva. See B–P–H 405/10.

Bulletin de la société et de la station agronomique, société des sciences, agriculture et arts de la Basse-Alsace = Société des sciences, agriculture et arts de la Basse-Alsace. Bulletin de la société et de la station agronomique. Strasbourg. Soc. Sci. Basse-Alsace Bull. Soc. Stat. Agron. See B–P–H 845/31.

Bulletin de la société de statistique, des sciences naturelles, et des arts industrielles du département de l'Isère. Grenoble. Bull. Soc. Statist. Dép. Isère. See B–P–H 282/12. HI 53076

Bulletin de la société suisse de physiologie végétale. Lausanne. Bull. Soc. Suisse Physiol. Vég. See B–P–H 282/14. HI 53077

Bulletin, société syndicale des pharmaciens de la Côte d'Or. Dijon. No. ?-25, 18??-1906/07. Bull. Soc. Syndic. Pharmaciens Côte d'Or. Preceded by: Bulletin des pharmaciens de la Côte d'Or. HI 63649

Bulletin de la société vaudoise d'agriculture et de viticulture. Lausanne. Nos. 1-226, 1870-1909. Bull. Soc. Vaud. Agric. Vitic. 5-3975-1. HI 63650

Bulletin de la société vaudoise des sciences naturelles . Lausanne. Bull. Soc. Vaud. Sci. Nat. See B–P–H 282/19. HI 53079

Bulletin, société versaillaise de sciences naturelles. Versailles. Sér. 4, vol. 1+, 1974+. Bull. Soc. Versaill. Sci. Nat. Preceded by: Revue de la fédération française des sociétés des sciences naturelles. HI 63651

Bulletin des sociétés de géographie de Québec et Montréal. Quebec. N.s. vols. 1-3, 1942-44. Bull. Soc. Géogr. Québec & Montréal. Preceded by: Bulletin de la société de géographie de Québec. Superseded by: Revue canadienne de géographie. HI 63652

Bulletin of the society of forestry; Kyoto imperial university. [Kyoto teikoku daigaku enshu-rin hokoku.] Kyoto, Japan. Bull. Soc. Forest. Kyoto Imp. Univ. See B–P–H 276/24. HI 53015

Bulletin of the society of plant ecology. [Shokubutsu seitai-gaku kaiho.] Sendai. Vols. 1-3, 1951-54. Bull. Soc. Pl. Ecol. Superseded by: Japanese journal of ecology. HI 63653

Bulletin, society of wetland scientists. Wilmington, NC. Vol. ?-1(4)+, 19??-82+. Bull. Soc. Wetland Sci. HI 636454

Bulletin des soies Kinugasa. [Sinugasa san-ho.] Bull. Soies Kinugasa. See B–P–H 282/23. HI 53080

Bulletin, soil survey of Great Britain. England and Wales. Harpenden. No. 1+, 1964+. Bull. Soil Surv. Gr. Brit., England Wales HI 63655

Bulletin; south african association for marine research. Durban, South Africa. Bull. S. Agrican Assoc. Mar. Res. See B–P–H 270/16. HI 52947

Bulletin of the south african biological society. Pretoria. Bull. S. African Biol. Soc. See B–P–H 270/17. HI 52948

Bulletin, South Australia woods and forests department. Adelaide. No. 1+, 1928+ [no. 2 published in 1950]. Bull. S. Australia Woods Forests Dept. HI 63656

Bulletin, South Dakota agricultural experiment station = South Dakota agricultural experiment station. Bulletin. Brookings, SD. South Dakota Agric. Exp. Sta. Bull. See B–P–H 848/1.

Bulletin of the south-eastern union of scientific societies. 1912-46. Bull. S. E. Union Sci. Soc. HI 63657

Bulletin, south Essex natural history society. Southend-on-Sea. 1938-39. Bull. S. Essex Nat. Hist. Soc. Preceded by: News sheet, south Essex natural history society. HI 63658

Bulletin of the southern California academy of sciences. Los Angeles, CA. Bull. S. Calif. Acad. Sci. See B–P–H 270/22. HI 52949

Bulletin of the southern California cactus exchange. Los Angeles, CA. Nos. 1-10, 1936-42. Bull. S. Calif. Cact. Exchange. HI 63659

Bulletin of the southern California camellia society. Pasadena, CA. Vols. 1-11 [some vols. misnumbered], 1940-50. Bull. S. Calif. Camellia Soc. Superseded by: Camellia review. 2-896-2. HI 63660

Bulletin of the southern Pacific general hospital. San Francisco, CA. Bull. S. Pacific Gen. Hosp. See B–P–H 270/25. HI 52950

Bulletin of the Southern Ussuri branch of the russian geographical society = Izvestiya Yuzhno-Ussuriiskogo otdela Gosudarstvennogo russkogo geograficheskogo obshchestva. Nikolsk-Ussuriski.

Bulletin of the Southern Ussuri branch of the russian

geographical society = Izvestiya Yuzhno-Ussuriiskogo otdeleniya Russkogo geograficheskogo obshchestva. Nikolsk-Ussuriski.

Bulletin of special products in Taiwan. [Taiwan t'ê ch'an ts'ung k'an.] See B–P–H 283/1. HI 53081

Bulletin spécial, société mycologique du Poitou. Poitiers. Vol. 1+, 1984+. Bull. Soc. Mycol. Poitou. HI 63661

Bulletin, Springfield museum of natural history = Springfield museum of natural history bulletin. Springfield, MA. Springfield Mus. Nat. Hist. Bull. See B–P–H 853/20.

Bulletin, state agricultural college agricultural experiment station, Colorado = State agricultural college agricultural experiment station bulletin. Fort Collins, CO. State Agric. Coll. Agric. Exp. Sta. Bull. See B–P–H 855/3.

Bulletin, state agricultural college of Colorado = State agricultural college of Colorado. Bulletin. Fort Collins, CO. State Agric. Coll. Colorado Bull. See B–P–H 855/4.

Bulletin, state agricultural college experiment station, Colorado = State agricultural college experiment station bulletin. Fort Collins, CO. State Agric. Coll. Exp. Sta. Bull. See B–P–H 855/5.

Bulletin of the state agricultural experiment station. Auburn, AL. Bull. State Agric. Exp. Sta. (Alabama). See B–P–H 283/7. HI 53082

Bulletin, state forest service, Kentucky. Frankfort, KY. No. 1+, 1926+. Bull. State Forest Serv., Kentucky. HI 63662

Bulletin of the state institute of agricultural microbiology = Trudy vsesoyuznogo instituta sel'skokhozyaistvennoi mikrobiologii. Leningrad, Moscow.

Bulletin, state plant board of Florida. Gainesville, FL. Nos. 1-14, 1953-60. Bull. State Pl. Board Florida. Superseded by: Bulletin of the division of plant industry, Florida. HI 63663

Bulletin de la station biologique de la société des amis des sciences naturelles, d'anthropologie et d'éthnographie à Bolchowo du Gouvernement de Moscou = Zapiski Biologičeskoj stancii Obščestva ljubitelej estestvoznanija, antropologii i etnografii v Bolševe Moskovskoj gubernii. Moscow. Zap. Biol. Stancii Obšč. Ljubit. Estestv. Bolševe Moskovsk. Gub. See B–P–H 989/16.

Bulletin, station océanographique de Salammbô. Tunis. Nos. 2-54, 1924-59. Bull. Stat. Océanogr. Salammbô. Preceded by: Notes et mémoires, station océanographique de Salammbô. Superseded by: Bulletin de l'institut national scientifique et technique d'océanographie et de pêche de Salammbô. HI 63664

Bulletin de la station régionale protectrice des plantes à Pétrograd = Izvestiya Petrogradskoi oblastnoi stantsii zashchity rastenii ot vreditelei. Petrograd.

Bulletin des stations d'expérimentation agricole hongroises. Budapest = Kísérletügyi kölemények. Budapest. Kísérl. Közlem. See B–P–H 514/5.

Bulletin, Storrs agricultural experiment station = Storrs agricultural experiment station bulletin. Mansfield, CT. Storrs Agric. Exp. Sta. Bull. See B–P–H 856/6.

Bulletin, Storrs agricultural experiment station at Connecticut agricultural college = Storrs agricultural experiment station bulletin. Mansfield, CT. Storrs Agric. Exp. Sta. Bull. See B–P–H 856/6.

Bulletin, Storrs agricultural experiment station at state agricultural college = Storrs agricultural experiment station bulletin. Mansfield, CT. Storrs Agric. Exp. Sta. Bull. See B–P–H 856/6.

Bulletin, Storrs agricultural experiment station at university of Connecticut = Storrs agricultural experiment station bulletin. Mansfield, CT. Storrs Agric. Exp. Sta. Bull. See B–P–H 856/6.

Bulletin, Storrs school agricultural experiment station = Storrs school agricultural experiment station bulletin. Mansfield, CT. Storrs School Agric. Exp. Sta. Bull. See B–P–H 856/8.

Bulletin, Sudan natural history museum Khartoum. No. 1+, 1960+. Bull. Sudan. Nat. Hist. Mus. HI 63665

Bulletin of the Sugadaira biological laboratory. Nagano-ken. Nos. 1-7, 1967-75. Bull. Sugadaira Biol. Lab. HI 63666

Bulletin of sugar beet research. [Tensai kenkyu hokoku.] Tokyo. Vols. 1-17, 1964-73; Supplement vols. 1-14, 19??-73. Bull. Sugar Beet Res. HI 63667

Bulletin of sugar beet research. Supplement. [Tensai kenkyu hokoku.] Tokyo. Vols. 1-14, 1963-72. Bull. Sugar Beet Res., Suppl. Superseded by: Proceedings of the sugar beet research association. [Tensai kenkyu kaiho.] HI 75085

Bulletin, sugarcane research station. Mauritius. Nos. 1-19, 1933-47. Bull. Sugarcane Res. Sta., Mauritius. HI 63668

Bulletin suisse de mycologie = Schweizerische Zeitschrift für Pilzkunde. Bern-Bümpliz. Schweiz. Z. Pilzk. See B–P–H 822/13.

Bulletin of the Suzugamine women's college. Natural science. Hiroshima. Vols. 1-12, 1957-66. Bull. Suzugamine Women's Coll., Nat. Sci. HI 63669

Bulletin of the Taihoku agricultural experiment station. Taihoku. 19??-16-20-? Bull. Taihoku Agric. Exp. Sta. Superseded by: Bulletin of the government agricultural research institute of Formosa. 5-4150-2. HI 63672

Bulletin of the Taiwan agricultural research institute. Taipei. Nos. 1-6, 1946-47. Bull. Taiwan Agric. Res. Inst. Preceded by: Bulletin of the government agricultural research institute of Formosa. 5-4150-2. HI 63673

Bulletin of the Taiwan forest research institute. [Taiwan shêng lin yeh shih yen so pao kao.] Taipei, Taiwan. Bull. Taiwan Forest Res. Inst. See B–P–H 283/15. HI 53083

Bulletin of Taiwan normal university. [Shih ta hsüeh pao.] Taipei, Taiwan. Bull. Taiwan Normal Univ. See B–P–H 283/16. HI 53084

Bulletin of tall timbers research station. Tallahassee, FL. No. 1+, 1962+. Bull. Tall Timbers Res. Sta. HI 63674

Bulletin, tasmanian field naturalists' club. Hobart. No. ?-90+, 19??-65+. Bull. Tasmanian Field Naturalists' Club. HI 63675

Bulletin of the tea research station, ministry of agriculture and forestry = Bulletin, national research institute of tea. Shizuoka.

Bulletin, technical and scientific service, ministry of agriculture, Egypt = Bulletin, ministry of agriculture, Egypt. Technical and scientific service. Cairo.

Bulletin, technical series. Maine agricultural experiment station = Maine agricultural experiment station. Bulletin, technical series. Orono, ME.

Bulletin of technical services, Ohio agricultural experiment station = Ohio agricultural experiment station, bulletin of technical services. Wooster, OH. Ohio Agric. Exp. Sta. Bull. Techn. Serv. See B–P–H 684/23.

Bulletin technique, office nationale des forêts. Paris. No. 1+, 1971+. Bull. Techn., Off. Natl. Forêts. HI 63676

Bulletin technique, Pyrénées-orientales, chambre d'agriculture. Montpellier. Bull. Techn. (Montpellier). See B–P–H 283/18. HI 53086

Bulletin, technological museum. Sydney, N.S.W. 1922-46 [some numbers were re-issued in revised editions]. Bull. Technol. Mus., Sydney. Superseded by: Bulletin, museum of technology and applied science. Sydney. HI 63677

Bulletin of the Tennessee agricultural experiment station. Knoxville, TN. No. 1+, 1888+. Bull. Tennessee Agric. Exp. Sta. HI 63678

Bulletin, Texas agricultural experiment station. College Station, TX. No. 1+, 1888+. Bull. Texas Agric. Exp. Sta. HI 63679

Bulletin, Texas agricultural extension service. College Station, TX. Nos. 1-259, 1914-55. Bull. Texas Agric. Extens. Serv. Incorporated with: Bulletin, Texas agricultural experiment station. HI 63680

Bulletin, Texas technological college. Scientific series. Lubbock, TX. Nos. 1-2, 1931-36. Bull. Texas Technol. Coll., Sci. Ser. Superseded by: Research publication, Texas technological college. HI 63681

Bulletin, Thorne ecological foundation. Boulder, CO. No. 9, 1967. Bull. Thorne Ecol. Found. Preceded by: Bulletin Thorne ecological research station. Superseded by: Technical bulletin, Thorne ecological institute. HI 63682

Bulletin, Thorne ecological research station. Boulder, CO. 1956-60. Bull. Thorne Ecol. Res. Sta. Superseded by: Bulletin, Thorne ecological foundation. HI 63683

Bulletin of the tobacco sub-station, Connecticut agricultural experiment station = Connecticut agricultural experiment station. Bulletin of the tobacco sub-station. New Haven, CT. Connecticut Agric. Exp. Sta. Bull. Tobacco Sub-Sta. See B–P–H 324/23.

Bulletin of the Tochigi agricultural experiement station. [Tochigi-ken nogyo shikenjo kenkyu hokoku.] Tochigi, Japan. Bull. Tochigi Agric. Exp. Sta. See B–P–H 283/23. HI 53087

Bulletin of the Tohoku national agricultural experiment station. [Tohoku nogyo shikenjo kenkyu hokoku.] Morioka, Japan. Bull. Tohoku Natl. Agric. Exp. Sta. See B–P–H 283/24. HI 53088

Bulletin of the Tokai-kinki national agricultural experiment station. [Tokai kinki nogyo shikenjo kenkyu hokoku.] Tsu. Vols. 1-27, 1953-74. Bull. Tokai-Kinki Natl. Agric. Exp. Sta. Superseded by: Bulletin of the vegetable and ornamental crops research station. Series C (Kurume). HI 63684

Bulletin, Tokyo gakugei university. Series, chemistry and biology. Tokyo. Vol. 1+, 1949+. Bull. Tokyo Gakugei Univ., Ser. Chem. Biol. HI 63685

Bulletin of the Tokyo imperial university forests. [Nogaku-bu enshurin hokoku.] Tokyo. Bull. Tokyo Imp. Univ. Forests. See B–P–H 284/2. HI 53089

Bulletin of the Tokyo science museum. Tokyo. Bull. Tokyo Sci. Mus. See B–P–H 284/4. HI 53090

Bulletin of the Tokyo university forest. [Tokyo daigaku nogakubu enshurin hokoku.] Tokyo. Bull. Tokyo Univ. Forest See B–P–H 284/6. HI 53091

Bulletin of the Tokyo-to agricultural experiment station. [Tokyo-to nogyo shikenjo kenkyu hokoku.] Tokyo. Bull. Tokyo-To Agric. Exp. Sta. See B–P–H 284/8. HI 53092

Bulletin of Tomsk state university = Trudy Tomskogo gosudarstvennogo universiteta.

Bulletin of the Torrey botanical club. Lancaster, Pa. Bull. Torrey Bot. Club. See B–P–H 284/15. HI 53093

Bulletin of the Tottori agricultural experiment station. [Tottori-ken nogyo shikenjo kenkyu hokoku.] Tottori, Japan. Bull. Tottori Agric. Exp. Sta. See B–P–H 284/18. HI 53094

Bulletin of the Tottori fruit tree experiment station. [Tottori-ken kaju shikenjo kenkyu hokoku.] Tottori, Japan. Bull. Tottori Fruit Tree Exp. Sta. See B–P–H 284/19. HI 53095

Bulletin of the Tottori university forests. [Tottori daigaku nogaku-bu Fusoku enshurin hokoku.] Tottori, Japan. Bull. Tottori Univ. Forests. See B–P–H 284/21. HI 53096

Bulletin of the Toyama science museum. Toyama. No. 1+, 1979+. Bull. Toyama Sci. Mus. HI 63686

Bulletin of the Transvaal museum. Pretoria. Nos. 1-9, 1955-62. Bull. Transvaal Mus. Superseded by: Transvaal museum bulletin. HI 63687

Bulletin des travaux de la classe des sciences mathématiques et naturelles, académie des sciences et des arts des slaves du sud de Zagreb (Croatie) = Izvješća o raspravama matematičko-prirodoslovnoga razreda, jugoslavenska akademija znanosti i umjetnosti o Zagrebu. Zagreb.

Bulletin des travaux de la société d'agriculture, sciences et belles-lettres d'Agen. Agen. Vol. 1, 1804. Bull. Travaux Soc. Agric. Sci. Belles-Lettres Agen. HI 63688

Bulletin des travaux de la société botanique de Genève. Geneva. Bull. Trav. Soc. Bot. Genève. See B–P–H 284/24. HI 53097

Bulletin des travaux de la société départementale d'agriculture de la Drôme. Valence, France. Bull. Trav. Soc. Dép. Agric. Drôme. See B–P–H 284/26. HI 53098

Bulletin des travaux de la société Murithienne = Bulletin de la Murithienne; société valaisanne des sciences naturelles. Sion, Switzerland. Bull. Murith. Soc. Valais. Sci. Nat. See B–P–H 264/10.

Bulletin des travaux de la société de pharmacie de Bordeaux. Bordeaux. Bull. Trav. Soc. Pharm. Bordeaux. See B–P–H 285/1. HI 53099

Bulletin trimestriel de l'académie malgache. Tananarive, Madagascar. Bull. Trimestriel Acad. Malgache. See B–P–H 285/5. HI 53100

Bulletin trimestriel; les amis du musée océanographique de Monaco. Monaco. Bull. Trimestriel Amis Mus. Océanogr. Monaco. See B–P–H 285/6. HI 53101

Bulletin trimestriel de l'association pour l'avancement de la recherche scientifique concernant Madagascar et la region malgache. Paris. No. 1+, 197?+. Bull. Trimestriel Assoc. Avancem. Rech. Sci. Madagascar. HI 63689

Bulletin trimestriel de l'association internationale des bibliothecaires et documentalistes agricoles = Quarterly bulletin of the international association of agricultural librarians and documentalists. Harpenden, etc.

Bulletin trimestriel, cercle des naturalistes hutois. Huy. Vols. 1-25, 1883-1908. Bull. Trimestriel Cercle Naturalistes Hutois. HI 63690

Bulletin trimestriel du Comice agricole de l'arrondissement de Toulon; [title varies.] Toulon. Vols. 1-31, 1850-80. Bull. Trimestriel Comice Agric. Arrondissement Toulon. Superseded by: Bulletin mensuel de la société d'agriculture, d'horticulture et d'acclimatation du Var à Toulon. HI 63691

Bulletin trimestriel, fédération française des sociétés des sciences naturelles. Versailles. 1956-61. Bull. Trimestriel. Féd. Franç Soc. Sci. Nat. Preceded by: Bulletin trimestriel, union des sociétés françaises d'histoire naturelle. Superseded by: Revue de la fédération française des sociétés des sciences naturelles. HI 63692

Bulletin trimestriel de la fédération mycologique Dauphiné-Savoie. Reaumont. No. 1+, 1961+. Bull. Trimestriel Féd. Mycol. Dauphiné-Savoie. HI 63693

Bulletin trimestriel de géographie et d'archéologie. Oran. Vol. 6+, 1886+. Bull. Trimestriel Géogr. Archéol. Preceded by: Bulletin de la société de géographie et d'archéologie de la province d'Oran. HI 63694

Bulletin trimestriel des résultats acquis pendant les croisières périodiques. Copenhagen. 1905-08. Bull. Trimestriel Résult. Crois. Périod. Preceded by: Bulletin des résultats acquis pendant les courses périodiques. Parties A, B. & C superseded by: Bulletin hydrographique. Partie D superseded by: Bulletin planktonique. HI 63695

Bulletin trimestriel de la section mycologique de Bordeaux. Bordeaux. Vol. 1+, 1978+. Bull. Trimestriel Sect. Mycol. Bordeaux. HI 63696

Bulletin trimestriel, société des amis des arbres et du reboisement et société d'histoire naturelle de Haute-Provence. Nice. No. 1-2, 1965-66. Bull. Trimestriel Soc. Amis Arbres Rebois. Soc. Hist. Nat. Haute-Provence. Superseded by: Revue trimestrielle des société des amis des arbres et du reboisement à Nice et d'histoire naturelle de Haute-Provence à Digne. HI 63697

Bulletin trimestriel de la société botanique et géologique du Var et de la Corse. Toulon. Vols. 18-36, 19??-? Bull. Trimestriel Soc. Bot. Géol. Var. & Corse. Preceded by: Bulletin de la société botanique du Var et de la Corse. Superseded by: Bulletin trimestriel de la société de botanique, de géologie et d'entomologie du Var et de la Corse. HI 63698

Bulletin trimestriel de la société de botanique, de géologie et d'entomologie du Var et de la Corse. Toulon. Vols. 37-40, 19??-? Bull. Trimestriel Soc. Bot. Géol. Entomol. Var & Corse. Preceded by: Bulletin trimestriel de la société botanique et géologique du Var et de la Corse. Superseded by: Bulletin de la société de botanique, de géologie et d'entomologie du Var et de la Corse. HI 63699

Bulletin trimestriel de la société d'études scientifiques de l'Anjou. Angers. 1966+. Bull. Trimestriel Soc. Etudes Sci. Anjou. Preceded by?: Bulletin de la Société d'études scientifiques de l'Anjou. HI 63700

Bulletin trimestriel de la société forestière française des amis des arbres. Paris. Bull. Trimestriel Soc. Forest. Franç. Amis Arbres. See B–P–H 285/7. HI 53102

Bulletin trimestriel de la société de géographie et d'archéologie de la province d'Oran = Bulletin trimestriel de gégraphie et d'archéologie. Oran.

Bulletin trimestriel de la société géologique de Normandie et des amis du muséum du Havre. Le Havre. Vol. 62+, 1973/75+. Bull. Trimestriel. Soc. Géol. Normandie Amis Mus. Havre. Preceded by: Bulletin de la société géologique de Normandie. HI 63701

Bulletin trimestriel de la société d'histoire naturelle et des amis du muséum d'Autun. Autun. Vol. 46+, 1968+. Bull. Trimestriel Soc. Hist. Nat. Amis Mus. Autun. Preceded by: Eduen, feuille de liaison périodique de la société eduenne des lettres, sciences et arts et de la société d'histoire naturelle de l'Autun. HI 63702

Bulletin trimestriel de la société d'histoire naturelle de Haute-Savoie. Annecy. 1972+. Bull. Trimestriel Soc. Hist. Nat. Haute-Savoie. Preceded by: Bulletin bimestriel de la société d'histoire naturelle de Haute-Savoie. HI 63703

Bulletin trimestriel de la société mycologique de France. Paris. 1903+. Bull. Trimestriel Soc. Mycol. France. Preceded by: Bulletin de la société mycologique de France. HI 63704

Bulletin trimestriel de la société Ramond. Bagnères-de-Bigorre, France. Bull. Trimestriel Soc. Ramond. See B–P–H 285/8. HI 53103

Bulletin trimestriel, société des sciences, agriculture et arts de la Basse-Alsace = Société des sciences, agriculture et arts de la Basse-Alsace. Bulletin trimestriel. Strasbourg. Soc. Sci. Basse-Alsace Bull. Trimestriel. See B–P–H 845/32.

Bulletin trimestriel de la société des sciences, arts et belles-lettres du département du Var, séant à Toulon. Toulon. Bull. Trimestriel Soc. Sci. Dép. Var Toulon. See B–P–H 285/9. HI 53104

Bulletin trimestriel de la société et de la station agronomique. Strasbourg = Société des sciences, agriculture et arts du Bas-Rhin. Bulletin trimestriel de la société et de la station agronomique. Strasbourg. Soc. Sci. Dép. Bas-Rhin Bull. Trimestriel Soc. Stat. Agron. See B–P–H 846/1.

Bulletin trimestriel de la société et de la station agronomique. Strasbourg = Société des sciences, agriculture et arts de la Basse-Alsace. Bulletin trimestriel de la société et de la station agronomique. Strasbourg. Soc. Sci. Basse-Alsace Bull. Trimestriel Soc. Stat. Agron. See B–P–H 845/33.

Bulletin trimestriel, union des sociétés françaises d'histoire naturelle. Versailles. Vols. 1-?, 1950-55. Bull. Union Soc. Franç. Hist. Nat. Superseded by: Bulletin trimestriel, féderation française des sociétés des sciences naturelles. HI 63705

Bulletin Tuskegee normal and industrial institute experiment station. Tuskegee, Alabama. Bull. Tuskegee Normal Industr. Inst. Exp. Sta. See B–P–H 285/12. HI 53105

Bulletin, U S department of agriculture, bureau of forestry = U S department of agriculture, bureau of forestry. Bulletin. Washington, DC.

Bulletin U I C N. [Union international pour la conservation de la nature et de ses resources.] Morges, Switzerland. Bull. U. I. C. N. See B–P–H 285/18. HI 53107

Bulletin of the U S department of agriculture. Washington, DC. Bull. U.S.D.A. See B–P–H 286/27. HI 53119

Bulletin, U S department of agriculture, Bureau of plant industry. Washington, DC. Bull. U.S.D.A. Bur. Pl. Industr. See B–P–H 286/30. HI 53120

Bulletin of the U S S R geographical society = Izvestiya Vsesoyuznogo geograficheskogo obshchestva. Moscow & Leningrad.

Bulletin of the Uganda society. Kampala. 1943-45. Bull. Uganda Soc. Preceded and superseded by: Uganda journal. HI 63707

Bulletin de l'union des agriculteurs d'Égypte. Cairo. Bull. Union Agric. Égypte. See B–P–H 285/20. HI 53108

Bulletin de l'union coloniale française. Paris. N.s. vols. 22(451)-26(510), 1923-27. Bull. Union. Colon. Franç. Preceded and superseded by: Quinzaine coloniale. 5-4298-3. HI 63708

Bulletin de l'union horticole professionelle international. The Hague. Bull. Union Hort. Profess. Int. See B–P–H 285/21. HI 53109

Bulletin de l'union medicale balkanique. Bucharest = Archives de l'union médicale balkanique. Bucharest.

Bulletin, union des océanographes de France. ?-1975? Bull. Union Océanographes France. Superseded by:

Journal de recherche océanographique. HI 63709

Bulletin, united avocado growers bulletin. La Habra, CA = United avocado growers bulletin. La Habra, CA. United Avocado Growers Bull. See B–P–H 933/6.

Bulletin, United States department of agriculture = United States department of agriculture. Bulletin. Washington, DC. U.S.D.A. Bull. (1895-1901). See B–P–H 940/15.

Bulletin, United States forest service. Washington, DC. Nos. 64-127, 1905-13 [nos. 120 & 124 never published]. Bull. U.S. Forest Serv. Preceded by: U S department of agriculture, bureau of forestry. Bulletin. Incorporated in: Bulletin of the United States department of agriculture. HI 63710

Bulletin of the United States geological survey. Washington, D.C. Vol. 1+, 1883+. Bull. U.S. Geol. Surv. HI 63711

Bulletin of the United States national museum. Washington, DC. Bull. U.S. Natl. Mus. See B–P–H 286/26. HI 53118

Bulletin universel des sciences et de l'industrie, deuxième section = Bulletin des sciences naturelles et de géologie. Deuxième section du bulletin universel des sciences et de l'industrie. Paris. Bull. Sci. Nat. Géol. See B–P–H 272/6.

Bulletin universel des sciences, des lettres et des arts. Paris. Bull. Universel Sci. See B–P–H 286/22. HI 53117

Bulletin de l'université de l'Asie centrale = Byulleten' Sredne-Aziatskogo gosudarstvennogo universiteta. Tashkent.

Bulletin de l'université l'Aurore. [Si-ka-wei, Shanghai.] Shanghai. Bull. Univ. Aurore. See B–P–H 285/27. HI 53110

Bulletin de l'universitet de Tirana. Série sciences naturelles = Buletin i universitetit shtetëror të Tiranës. Seria shkencat natyrore. Tirana.

Bulletin, university of Delaware agricultural experiment station = University of Delaware agricultural experiment station bulletin. Newark, DE. Univ. Delaware Agric. Exp. Sta. Bull. See B–P–H 935/12.

Bulletin, university of Florida agricultural experiment station = University of Florida agricultural experiment station bulletin. Gainesville, FL. Univ. Florida Agric. Exp. Sta. Bull. See B–P–H 935/14.

Bulletin of the university forest; Kyushu imperial university. Fukuoka, Japan. Bull. Univ. Forest Kyushu Imp. Univ. See B–P–H 286/5. HI 53111

Bulletin, university of Hawaii = University of Hawaii bulletin. Honolulu, HI. Univ. Hawaii Bull. (Honolulu, 1917-19). See B–P–H 935/17.

Bulletin, university of Hawaii = University of Hawaii bulletin. Honolulu, HI. Univ. Hawaii Bull. (Honolulu, 1927+). See B–P–H 935/18.

Bulletin, university of Illinois agricultural experiment station = University of Illinois agricultural experiment station bulletin. Champaign, IL. Univ. Illinois Agric. Exp. Sta. Bull. See B–P–H 936/7.

Bulletin, university museum, university of Tokyo. [Sogo kenkyu shiryokan.] Tokyo. No. 1+, 1970+. Bull. Univ. Mus. Univ. Tokyo. HI 63712

Bulletin of the university of New Mexico, anthropology series. Albuquerque, NM. Vols. 1-4, 1930-43. Bull. Univ. New Mexico, Anthropol. Ser. Superseded by: Publications in anthropology of the university of New Mexico. 4-3004-3. HI 63713

Bulletin of the university of New Mexico, archaeology series = Bulletin of the university of New Mexico, anthropology series. Albuquerque, NM.

Bulletin, university of New Mexico. Biological series = University of New Mexico. Bulletin. Biological series. Albuquerque, NM.

Bulletin, university of New Mexico. Bulletin. Monograph series = University of New Mexico. Bulletin. Monograph series. Albuquerque, NM. Univ. New Mexico Bull., Monogr. Ser. See B–P–H 937/18.

Bulletin of the university of Osaka prefecture. Series B, agriculture and biology. Osaka, Japan. Bull. Univ. Osaka Prefect., Ser. B, Agric. See B–P–H 286/13. HI 53112

Bulletin, university of Pretoria, faculty of agriculture = University of Pretoria, faculty of agriculture. Bulletin. Pretoria. Univ. Pretoria Fac. Agric. Bull. See B–P–H 937/24.

Bulletin of the university of the state of New York = Bulletin of the New York state museum of natural history. Albany, NY.

Bulletin, university of Tennessee arboretum society = University of Tennessee arboretum society bulletin. Oak Ridge, TN.

Bulletin of the university of Texas. Austin, TX. Bull. Univ. Texas. See B–P–H 286/15. HI 53113

Bulletin of the university of Texas. Scientific series. Forms part of: Bulletin of the university of Texas. Austin, TX. Nos. 1-29 [nos. 7, 22, 25 & 26 are incorrectly numbered 6, 21, 35 & 36 respectively], 1904-14. Bull. Univ. Texas, Sci. Ser. HI 63714

Bulletin of the university of Utah. Biological series. Salt Lake City, UT. Bull. Univ. Utah, Biol. Ser. See B–P–H 286/18. HI 53114

Bulletin of the university of Wisconsin. Science series. Madison, WI. Bull. Univ. Wisconsin, Sci. Ser. See B–P–H 286/20. HI 53115

Bulletin of the university of Wisconsin. Studies in science. Madison, WI. Bull. Univ. Wisconsin Stud. Sci. See B–P–H 286/21. HI 53116

Bulletin, Utah agricultural college, agricultural experiment station = Utah agricultural college, agricultural experiment station. Bulletin. Logan, UT. Utah Agric. Exp. Sta. Bull. See B–P–H 945/29.

Bulletin of the Utsunomiya agricultural college. [Utsunomiya koto norin gakko gakujutsu hokoku.] Utsunomiya, Japan. Bull. Utsunomiya Agric. Coll. See B–P–H 287/8. HI 53121

Bulletin of the Utsunomiya agricultural college. Series A, agricultural sciences, forestry, veterinary science, agricultural engineering, agricultural chemistry [Name of sectional title varies]. Utsunomiya. Vols. 2-4, 1934-48. Bull. Utsunomiya Agric. Coll., Ser. A. Preceded by: Bulletin of the Utsunomiya agricultural college. Superseded by: Bulletin of the college of agriculture, Utsunomiya university. 5-4348-1. HI 63715

Bulletin of the Utsunomiya agricultural college. Series B, agricultural and forest economics. Utsunomiya. Vol. 2-[4], 1937-48. Bull. Utsunomiya Agric. Coll., Ser. B. Preceded by: Bulletin of the Utsunomiya agricultural college. Superseded by: Bulletin of the college of agriculture, Utsunomiya university. 5-4348-1. HI 63716

Bulletin of Utsunomiya tobacco experiment station. [Utsunomiya tabako shikenjo hokoku.] Tochigi, Japan. Bull. Utsunomiya Tobacco Exp. Sta. See B–P–H 287/11. HI 53122

Bulletin of the Utsunomiya university forests. [Utsunomiya daigaku nogakubu enshurin hokoku.] Tochigi. 1961+. Bull. Utsunomiya Univ. Forests. HI 67063

Bulletin, Vancouver botanical gardens association. Vancouver, B.C. 1972+. Vancouver Bot. Gard. Assoc. HI 63717

Bulletin of the Vanderbilt marine museum. Huntington, NY. Bull. Vanderbilt Mar. Mus. See B–P–H 287/12. HI 53123

Bulletin des variétés. Céréals et fiches descriptives provisiores des variétés. Guyancourt. 1979?-83? Bull. Var., Céréals Fiche Descr. Provis. Var. Preceded by: Fiches descriptives provisoires des nouvelles variétés de cereals proposées à l'inscription au catalogue (Liste de commercialisation en France). HI 63718

Bulletin des variétés. Plantes fourragères. Guyancourt. 1982?-83? Bull. Var., Pl. Fourrag. Preceded by: Fiches descriptives provisoires des nouvelles variétés de plantes fourragères proposées à l'inscription au catalogue (Liste de commercialisation en France). Superseded by: Bulletin des variétés. Plantes fourragères annuelles and Bulletin des variétés. Plantes fourragères pérennes. HI 63719

Bulletin des variétés. Plantes fourragères annuelles. Guyancourt. 1984+. Bull. Var., Pl. Fourrag. Annuelles. Preceded by: Bulletin des variétés. Plantes fourragères. HI 63720

Bulletin des variétés. Plantes fourragères pérennes. Guyancourt. 1984+. Bull. Var., Pl. Fourrag. Pérennes. Preceded by: Bulletin des variétés. Plantes fourragères. HI 63721

Bulletin des varietés. Pommes de terre. 1980+. Bull. Var., Pommes de Terre. HI 63722

Bulletin of varieties. Annual forage plants = Bulletin des variétés. Plantes fourragères annuelles. Guyancourt.

Bulletin of varieties. Cereals and descriptive entries of varieties = Bulletin des variétés. Céréals et fiches descriptives provisiores des variétés. Guyancourt.

Bulletin of varieties. Forage plants = Bulletin des variétés. Plantes fourragères. Guyancourt.

Bulletin of varieties. Perennial forage plants = Bulletin des variétés. Plantes fourragères pérennes. Guyancourt.

Bulletin of the vegetable and ornamental crops research station, Tsu. Series A. [Yasai shikenjo hokoku. A.] Tsu. No. 1+, 1974+. Bull. Veg. Ornam. Crops Res. Sta. Tsu., A. HI 63723

Bulletin of the vegetable and ornamental crops research station. Series B. [Yasai shikenjo hokoku. B, Morioka.] Morioka. No. 1+, 1977+. Bull. Bull. Veg. Ornam. Crops Res. Sta. Tsu., B. HI 63724

Bulletin of the vegetable and ornamental crops research station. Series C, (Kurume). [Yasai shikenjo hokoku. C, Kurume.] Kurume. No. 1+, 1974+. Bull. Bull. Veg. Ornam. Crops Res. Sta. Tsu., C. Preceded by: Bulletin of the horticultural research station. Series D, Kurume and Bulletin of the Tokai-kinki national agricultural experiment station. HI 63725

Bulletin of the Vermont botanical club. Burlington, VT. Nos. 1-9, 1906-14. Bull. Vermont Bot. Club. Superseded by: Joint Bulletin, Vermont botanical and bird club. 5-3478-1. HI 63726

Bulletin, Vermont state agricultural college, agricultural experiment station = Vermont state agricultural college, agricultural experiment station. Bulletin. Burlington, VT. Vermont Agric. Exp. Sta. Bull. See B–P–H 956/22.

Bulletin de vulgarisation des sciences naturelles; organe de la société botanique et entomologique de Gers. Auch. Vols. 1-6, 1901-23. Bull. Vulg. Sci. Nat. HI 63727

Bulletin of the Washburn college laboratory of natural history. Topeka, KS. Nos. 1-11, 1884-90. Bull. Washburn Coll. Lab. Nat. Hist. HI 63728

Bulletin of the Washburn laboratory of natural history. Topeka, KS. Bull. Washburn Lab. Nat. Hist. See B–P–H 287/18. HI 53124

Bulletin, Washington agricultural experiment station, special series = Washington (state) agricultural experiment station. Bulletin, special series. Pullman, WA. Wash. State Agric. Exp. Sta. Bull., Special Ser. See B–P–H 971/13.

Bulletin of Washington college extension service = Washington agricultural experiment station (later Stations). Bulletin. Pullman, WA.

Bulletin, Washington state agricultural experiment station, special series = Washington (state) agricultural experiment station. Bulletin, special series. Pullman, WA. Wash. State Agric. Exp. Sta. Bull., Special Ser. See B–P–H 971/13.

Bulletin of Washington state college extension service = Washington state agricultural experiment station (later Stations). Bulletin. Pullman, WA.

Bulletin, Washington state university college of forestry = Bulletin, college of forestry, Washington state university. Seattle, WA.

Bulletin of Washington university, Henry Shaw school of botany. [Forms part of: Publications of Washington university. Series 2.] St. Louis, MO. 1910. Bull. Wash. Univ., Henry Shaw School Bot. HI 63729

Bulletin, Wellington botanical society. Wellington, NZ. Vol. 1+, 1941+. Bull. Wellington Bot. Soc. HI 63730

Bulletin, welsh plant breeding station. Series C. Aberystwyth. Nos. 1-3, 1920-22, 1921-23. Bull. Welsh Pl. Breed. Sta., C. HI 63731

Bulletin, welsh plant breeding station. Series H. Aberystwyth. Nos. 1-18, 1922-56. Bull. Welsh Pl. Breed. Sta., H. HI 63732

Bulletin, West Virginia agricultural experiment station. Morgantown, WV. Nos. 1-646, 1888-1976. Bull. West Virginia Agric. Exp. Sta. Superseded by: Bulletin, West Virginia agricultural and forestry experimental station. HI 63733

Bulletin, West Virginia agricultural and forestry experimental station. Morgantown, WV. No. 647+, 1976?+. Bull. West Virginia Agric. Forest. Exp. Sta. Preceded by: Bulletin, West Virginia agricultural experiment station. HI 63734

Bulletin of the west Wales naturalists' trust. Haverfordwest. 1972+. Bull. W. Wales Naturalists' Trust. HI 63735

Bulletin of the white-ruthenian Lenin's institute of scientific research of agriculture and forestry at the soviet people's commissaries of W-R S S R = Pratsy belaruskaga navukova-das'ledchaga instytutu liasnoi gaspadarki i liasnoi pramyslovas'chi. Trudy po lesnomu opytnomu delu B S S R. Minsk.

Bulletin, Wiltshire archaeological and natural history society. Natural history section. Swindon. Nos. 1-16, 1973-81. Bull. Wiltshire Archaeol. Nat. Hist. Soc., Nat. Hist. Sect. HI 63736

Bulletin, Wisconsin agricultural experiment station = Wisconsin agricultural experiment station. Bulletin. Madison, WI. Wisconsin Agric. Exp. Sta. Bull. See B–P–H 976/1.

Bulletin of the Wisconsin natural history society. Milwaukee, WI. Bull. Wisconsin Nat. Hist. Soc. See B–P–H 287/22. HI 53125

Bulletin, woods and forests department, South Australia. Adelaide, S.A. No. 1+, 1928+. Bull. Woods Forests Dept. South Australia. Preceded by: Bulletin, department of forestry, university of Adelaide. HI 63737

Bulletin, woods and forests department, Western Australia. Perth, W.A. No. 1+, 1919+ [publication suspended between no. 53, 1939 & no. 55, 1947]. Bull. Woods Forests Dept., Western Australia. HI 63738

Bulletin of the woody plant society. Minneapolis, MS. Vol. 1+, 1984+. Bull. Woody Pl. Soc. HI 63739

Bulletin, Wyoming agricultural college agricultural experiment station = Wyoming agricultural college agricultural experiment station. Bulletin. Laramie, WY. Wyoming Agric. Exp. Sta. Bull. See B–P–H 980/25.

Bulletin of the Yale university school of forestry. New Haven CT. Bull. Yale Univ. School Forest. See B–P–H 285/27. HI 53126

Bulletin of Yamagata prefectural museum. [Yamagata kenritsu hakabutsukan. Kenkyu hokoku.] Yamagata. No. 1+, 1973+. Bull. Yamagata Pref. Mus. HI 63740

Bulletin of the Yamagata university. [Yamagata daigaku kiyo.] Tsuruoka, Japan. Bull. Yamagata Univ. See B–P–H 287/30. HI 53127

Bulletin of the Yamagata university. Natural science. Yamagata. Vol. 1+, 1950+. Bull. Yamagata Univ., Nat. Sci. HI 63741

Bulletin of the Yamaguchi agricultural experiment station. [Yamaguchi-ken nogyo shikenjo kenkyu hokoku.] Yamaguchi, Japan. Bull. Yuamaguchi Agric. Exp. Sta. See B–P–H 287/31. HI 53128

Bulletin of the Yamanashi fruit tree experiment station. [Yamanashi-ken kaju shikenjo kenkyu hokoku.] Yamanashi. 1970+. Bull. Yamanashi Fruit Tree Exp. Sta. HI 72484

Bulletin of the Yamanashi prefectural agricultural experiment station. [Yamanashi-ken nogyo shikenjo hokoku.] Kofu, Japan. Bull. Yamanashi Prefect. Agric. Exp. Sta. See B–P–H 288/1. HI 53129

Bulletin of the Yamanashi prefectural forest experiment station. [Yamanashi-ken ringyo shikenjo hokoku.] Fujiyoshida. 1952+. Bull. Yamanashi Pref. Forest Exp. Sta. HI 63742

Bulletin, Yokohama municipal (later city) university society. Natural science. [Yokohama-shiritsu daigaku ronso. Shizenkagaku keiritsu.] Yokohama. Vol. 1+, 1949+. Bull. Yokohama Munic. Univ. Soc., Nat. Sci. HI 63743

Bulletin, Yorkshire naturalists' union. York. No. 1+, 1984+. Bull. Yorkshire Naturalists' Union. HI 63744

Bulletin, young explorers' trust. No. 1-?, 1972-77. Bull. Young Explorers' Trust. Incorporated in: Yetmag. HI 63745

Bulletin of the Yunnan botanical research institute. [Yunnan nung lin chih wu yen chiu so t'sung k'an.] Kunming, China. Bull. Yunnan Bot. Res. Inst. See B–P–H 288/2. HI 53130

Bulletin der zittingen. Koninklijk belgisch koloniaalinstitut = Bulletin des séances de l'institut royal colonial belge. Brussels. Bull. Séances Inst. Roy. Colon. Belge. See B–P–H 273/1.

Bulletin, zöldségtermesztési kutató intézet. Kecskemét. Vol. 6+, 1971. Bull. Zöldségterm. Kutató Intéz. Preceded by: Bulletin, duna-tisza közi mezögazdasági kisérleti intézet. HI 63746

Bullettino see also Bollettino

Bullettino dell'accademia degli aspiranti naturalisti. Naples. Vol. 1, 1842. Bull. Accad. Aspir. Naturalisti. Preceded by: Esercitazioni accademiche degli aspiranti naturalisti. Superseded by: Annali dell'accademia degli aspiranti naturalisti. 1-32-2 HI 63747

Bullettino di associazione napolitana di medici e naturalisti. Naples. Bull. Assoc. Napol. Med. See B–P–H 240/24. HI 52624

Bulletino dell'istituto botanico della r. università di Modena = Archivio botanico per la sistematica, fitogeografic e genetica (storia e sperimentale). Modena. Forlí, Italy. Arch. Bot. Sist. See B–P–H 139/15.

Bullettino dell'istituto botanico della reale università di Sassari. Sassari. Vol. 1+, 1909+. Bull. Ist. Bot. Reale. Univ. Sassari. HI 63748

Bullettino del laboratorio ed orto botanico della r. università di Siena. Siena. Bull. Lab. Orto Bot. Reale Univ. Siena. See B–P–H 260/20. HI 52844

Bullettino mensile delle sedute accademia Gioenia di scienze naturali, Catania = Bollettino delle sedute accademia Gioenia di scienze naturali, Catania. Catania, Italy. Boll. Accad. Gioenia Sci. Nat. Catania. See B–P–H 213/14.

Bulletino nautico e geografico di Roma. Rome. Bull. Naut. Geogr. Roma. See B–P–H 266/9. HI 52901

Bullettino di società lancisiana degli ospedali di Roma. Rome. Bull. Soc. Lancisiana Osped. Roma. See B–P–H 278/16. HI 53036

Bullettino della società toscana di orticultura. Florence. Bull. Soc. Tosc. Ortic. See B–P–H 282/16. HI 53078

Bulletino ufficiale della associazione orticola professionale italiana. San Remo, Italy. Bull. Uff. Assoc. Ortic. Profess. Ital. See B–P–H 285/17. HI 53106

Bulletins of association of scientific institutes on North Caucasus = Trudy Severo-Kavkazskoi assotsiatsii nauchno-issledovatel'skikh institutov. Krasnodar.

Bulletins, college of Hawaii = College of Hawaii publications. Bulletins. Honolulu, HI. Coll. Hawaii Publ. Bull. See B–P–H 316/23.

Bulletins of the laboratory of genetics = Trudy laboratorii po genetike. Leningrad, Moscow.

Bulletins of New York (state) department of agriculture and markets = Proceedings of the New York state agricultural society. Albany, NY. Proc. New York State Agric. Soc. See B–P–H 733/14.

Bulletins of the Pacific scientific fisheries institute = Izvestiya Tikhookeanskogo nauchnogo instituta rybnogo khozyaistva. Voronezh.

Bulletins de la société botanique de Copenhague = Meddelelser fra den botaniske forening i Kjøbenhavn. Copenhagen.

Bulletins de la station biologique de la société des amis des sciences naturelles, d'anthropologie et d'éthnographie à Bolchowo du Gouvernement de Moscou = Zapiski Biologičeskoj stancii Obščestva ljubitelej estestvoznanija, antropologii i etnografii v Bolševe Moskovskoj gubernii. Moscow. Zap. Biol. Stancii Obšč. Ljubit. Estestv. Bolševe Moskovsk. Gub. See B–P–H 989/16.

Bulletins des travaux de la société Murithienne = Bulletin de la Murithienne; société valaisanne des sciences naturelles. Sion, Switzerland. Bull. Murith. Soc. Valais. Sci. Nat. See B–P–H 264/10.

Bulteno scienca de la fakultato terkultura. Kjusu imperia universitato. [Gakugei zasshi.] Fukuoka, Japan. Bult. Sci. Fak. Terk. Kjusu Imp. Univ. See B–P–H 288/7. HI 53131

Bündnerwald. Chur, Switzerland. Bündnerwald. See B–P–H 288/9. HI 53132

Bureau of plant industry library. Current author entries. Washington, DC. Bur. Pl. Industr. Libr. Curr. Author Entries. See B–P–H 288/17. HI 53133

Bureau of plant industry library. Current botanical

literature. Washington, DC. Bur. Pl. Industr. Libr. Curr. Bot. Lit. See B–P–H 288/18. HI 53134

Burgenländische Forschungen. Vienna. ?-1975+. Burgenl. Forsch. HI 63749

Burgenländische Heimatblätter. Eisenstadt, Austria. Burgenl. Heimatbl. See B–P–H 288/29. HI 53135

Burma forest bulletin. Rangoon. 1921-40 [in various series, see below]. Burma Forest Bull. HI 63750

Burma forest bulletin. Botanical series. [Forms: Burma forest bulletin, nos. 19, 20, 23 & 27.] Rangoon. Nos. 1-4 1927-32. Burma Forest Bull., Bot. Ser. HI 63751

Burma forest bulletin. Ecology series. [Forms: Burma forest bulletin, no. 25.] Rangoon. No. 1, 1931. Burma Forest Bull., Ecol. Ser. HI 63752

Burma forest bulletin. Economic series. [Forms: Burma forest bulletin, nos. 12, 16, 21, 22, 28-30 & 33.] Rangoon. Nos. 1-8, 1925-40. Burma Forest Bull., Econ. Ser. HI 63753

Burma forest bulletin. Silvicultural series. [Forms: Burma forest bulletin, nos. 1-6, 8-10, 14-15, 17-18, 24 & 31.] Rangoon. Nos. 1-15, 1921-34. Burma Forest Bull., Silvic. Ser. HI 63754

Burmese forester. Rangoon. 1951+. Burmese Forester. HI 63755

Burrendong arboretum brigge. Wellington, N.S.W. No. 1+, 1968+. Burrendong Arbor. Brigge. HI 63756

Bu$iness of herbs. Madison, VA. Vol. 1+, 1983+. Business Herbs. HI 63757

Butler university botanical studies. Indianapolis, IN. Vols. 1-14, 1929-64. Butler Univ. Bot. Stud. 1-856-1. HI 63758

Butlletí de l'acadèmia de ciencias i arts de Barcelona. Madrid. Epoca 3, vol. 6(5-7), 1934-36. Butl. Acad. Ci. Arts Barcelona. Preceded by: Boletín de la academia de ciencias y artes de Barcelona. HI 63759

Butlletí d'informació, institut botànic de Barcelona. No. 1+, 1937+. Butl. Inform. Bot. Barcelona. HI 63760

Butlletí de la institució catalana d'historia natural. Barcelona. 1901-37/49, 1901-49? Butl. Inst. Catalana Hist. Nat. Superseded by: Butlletí de la institució catalana d'historia natural. Seccio de botanica. 3-1994-2. HI 63761

Butlletí de la institució catalana d'historia natural. Seccio de botanica. Barcelona. 1974+. Butl. Inst. Catalana Hist. Nat., Secc. Bot. Preceded by: Butlletí de la institució catalana d'historia natural. HI 63762

Búvár. [Edited by Lambrecht.] Budapest. Búvár (1935-44). See B–P–H 289/4. HI 53136

Búvár. [Tudományos ismeretterjeszdö társulat.] Budapest. Búvár (1960+). See B–P–H 289/5. HI 53137

Buxbaumia; mededelingen van de bryologische werkgroep van de [kon.] nederlandsche natuurhistorische vereniging. Amsterdam. Vols. 1-23, 1947-69. Buxbaumia. Incorporated in: Lindbergia. HI 63763

Buxbaumiella. Various. No. 1+, 1972+. Buxbaumiella. HI 63764

Buxtonian. Boston, MA. Vol. 1+, 1972+; Supplement 1972+. Buxtonian. HI 63765

Buyseri pordzararakan mutageneze = Eksperimental'nyi mutagenez rastenii. Erevan.

Byggforskningen smaskrift. Oslo. No. ?-2+, 19??-75+. Byggforskn. Smaskr. HI 63766

Byochu-gai zasshi = Journal of plant protection. Tokyo. J. Pl. Protect. See B–P–H 477/4.

Byulleten' Akademii nauk Uzbekskoi S S R. Tashkent, Uzbek S S R. 1944-47. Byull. Akad. Nauk Uzbeksk. S.S.R. Superseded by: Doklady Akademii nauk Uzbekskoi S S R. 1-123-3. HI 63767

Byulleten' Botanicheskogo sada. Erivan [=Erevan], Armenian S S R. Vol. 1+, 1939+. Byull. Bot. Sada, Akad. Nauk Armyanskoi S.S.R. 2-1473-1. HI 63768

Byulleten' po fiziologii rastenii. Kiev. Nos. 1-4, 1957-58. Byull. Fiziol. Rast. HI 63769

Byulleten' Glavnogo botanicheskogo sada. Moscow & Leningrad. Vol. 1, 1948+. Byull. Glavn. Bot. Sada. 3-2763-1. HI 63770

Byulleten' Glavnogo botanicheskogo sada, Akademiya nauk S S S R = Byulleten' Glavnogo botanicheskogo sada. Moscow & Leningrad.

Byulleten' gosudarstvennogo muzeya Gruzii. Tiflis. 1920-35; ser. A & B, 1937-47. Byull. Gosud. Muz. Gruzii. Preceded by: Izvestiya kavkazskogo muzeya, Tiflis. Superseded by: Vestnik gosudarstvennogo muzeya Gruzii. Tiflis. HI 63771

Byulleten' gosudarstvennogo Tiranskogo universiteta. Seriya estestvennykh nauk = Buletin i universitetit shtetëror të Tiranës. Seria shkencat natyrore. Tirana.

Byulleten', gosudarstvennyi nikitskii opytnii botanicheskii sad. Yalta. Vols. 1-19, 1929-38. Byull. Gosud. Nikitsk. Opytn. Bot. Sad. Superseded by: Byulleten' nauchnoi informatsii, gosudarstvennyi nikitskii botanicheskii sad. 3-1912-2. HI 63772

Byulleten', gosudarstvennyi nikitskii opytnii botanicheskii sad imeni V. M. Molotova = Byulleten', gosudarstvennyi nikitskii opytnii botanicheskii sad.

Byulleten' gosudarstvennogo nikitskogo botanicheskogo sada. Yalta. [N.s.] vol. 1(7)+, 1968+. Byull. Gosud. Nikitsk. Bot. Sada. Preceded by: Byulleten' nauchno-tekhnicheskoi informatsii, gosudarstvennogo nikitskogo botanicheskogo sada. HI 63773

Byulleten' instituta biologii, belarusskaya akademiya navuk. Minsk. Vols. ?-2+, 19??-56+. Byull. Inst. Biol. Belaruss. Akad. Navuk. HI 63774

Byulleten' instituta biologii vodokhranilishch. Moscow. 1958-62. Byull. Inst. Biol. Vodokhr. HI 63775

Byulleten' Khabarovskogo lesnogo pitomnika. Vladivostok, Russian S F S R. 1925. Byull. Khabarovsk. Lesn. Pitomn. HI 63776

Byulleteni Khar'kovskago obshchestva lyubitelei prirody. Kharkov, Ukrainian S S R. 1912-16. Byull. Khar'kovsk. Obshch. Lyubit. Prir. HI 63777

Byulleten' Komissii po izucheniyu chetvertichnogo perioda. Moscow & Leningrad. Vol. 1+, 1929+ [suspended 1932-37]. Byull. Komiss. Izuch. Chetvert. Perioda. 1-118-1. HI 63778

Byulleten', mironovskii nauchno-issledovatel'skii institut selektsii i semenovodstva pshenitsy. Moscow. Vol. ?-5+, 19??-74+. Byull. Mironovskii Nauchno-Issl. Inst. Selektsii Semenov. Pshenitsy. HI 63779

Byulleten' Moskovskogo obshchestva ispytatelei prirody. Otdel biologicheskii. Moscow. N.s. vol. 31+, 1922/23+. Byull. Moskovsk. Obshch. Isp. Prir., Otd. Biol. Preceded by: Bulletin de la société impériale des naturalistes de Moscou. 3-2770-2. HI 63780

Byulleten' Moskovskogo obshchestva ispytatelei prirody. Otdel geologicheskii. Moscow & Leningrad. N.s. vol. 31+ [vol. 33, 1925 also numbered vol. 3], 1922+. Byull. Moskovsk. Obshch. Isp. Prir., Otd. Geol. Preceded by: Bulletin de la société impériale des naturalistes de Moscou. HI 63781

Byulleten' nauchno-tekhnicheskoi informatsii, gosudarstvennyi nikitskii botanicheskii sad = Byulleten' nauchnoi informatsii, gosudarstvennyi nikitskii botanicheskii sad. Yalta.

Byulleten' nauchnoi informatsii, gosudarstvennyi nikitskii botanicheskii sad. Yalta. Nos. 1-6, 1956-57. Byull. Nauchnoi Inform. Gosud. Nikitsk. Bot. Sad. Preceded by: Byulleten', gosudarstvennii nikitskii opytnii botanicheskii sad. Superseded by: Byulleten' gosudarstvennyi nikitskii botanicheskii sad. HI 63782

Byulleten' nauchnoi informatsii, tsentral'naya geneticheskaya laboratoriya imeni I. V. Michurina. Vols. ?-13-19, 19??-67-72. Byull. Nauchnoi Inform. Tsentr. Genet. Lab. I. V. Michurina. HI 63783

Byulleten' nauchno-tekhnicheskoi informatsii po maslichnym kul'turam. Vol. 1+, 1980+. Byull. Nauchno-Tekhn. Inform. Maslich. Kul't. HI 63784

Byulleten' nauchno-tekhnicheskoi informatsii po zashchite rastenii. Leningrad. Vols. 1-2, 1956-58. Byull. Nauchno-Tekhn. Inform. Zashch. Rast. Superseded by: Byulleten' Vsesoyuznogo nauchno-issledovatel'skogo instituta zashchity rastenii. HI 63785

Byulleten' nauchno-tekhnicheskoi informatsii, vsesoyuznyi nauchno-issledovatel'skii institut zernobobovykh kul'tur. Nos. 1-7, 1971-73. Byull. Nauchno-Tekhn. Inform., Vsesoyuzn. Nauchno-Issl. Inst. Zernob. Kul't. Superseded by: Byulleten' nauchno-tekhnicheskoi informatsii, vsesoyuznyi nauchno-issledovatel'skii institut zernobobovykh y krupyanykh kul'tur. HI 63786

Byulleten' nauchno-tekhnicheskoi informatsii, vsesoyuznyi nauchno-issledovatel'skii institut zernobobovykh i krupyanykh kul'tur. Orel. No. 8+, 1974+. Byull. Nauchno-Tekhn. Inform. Vsesoyuzn. Nauchno-Issl. Inst. Zernob. Krupyanykh Kul't. Preceded by: Byulleten' nauchno-tekhnicheskoi informatsii, vsesoyuznyi nauchno-issledovatel'skii instittut zernobobovykh kul'tur. HI 63787

Byulleten' Obshchestva estestvoispytatelei pri Voronezhskom gosudarstvennom universitete. Voronezh, Russian S F S R. Vols. 1-2(2), 1925-28; vols. 3-9,1939-55. Byull. Obshch. Estestvoisp. Voronezhsk. Gosud. Univ. For vol. 2(3/4) see: Byulleten' Voronezhskogo obshchestva estestvoispytatelei pri Gosudarstvennom universitete. 5-4430-1. HI 63788

Byulleten' Sibirskogo botanicheskogo sada. Tomsk, Russian S F S R. Vol. 1+. 1947+. Byull. Sibirsk. Bot. Sada. HI 63789

Byulleten' Sibirskogo botanicheskogo sada, Tomskii gosudarstvennyi universitet im. V. V. Kuibysheva = Byulleten' Sibirskogo botanicheskogo sada. Tomsk.

Byulleten' Sredne-Aziatskogo gosudarstvennogo universiteta. Tashkent, Uzbek S S R. Vols. 1/6-30, 1923/24-49. Byull. Sredne-Aziatsk. Gosud. Univ. 5-4155-3. HI 63790

Byulleten', sredne-volzhskaya kraevaya stantsiya zashchity rastenii ot vreditelei pri kraizemupravlenii. Samara. 1926/28. Byull. Sredne-Volzhsk. Kraev. Zashch. Rast., Samara. HI 63791

Byulleten' Tikhookeanskogo komiteta Akademii nauk S S S R. Leningrad. Vols. 1-3, 1929-34. Byull. Tikhookeansk. Komiteta Akad. Nauk S.S.S.R. 1-120-3. HI 63792

Byulleten' Voronezhskogo obshchestva estestvoispytatelei pri Gosudarstvennom universitete. Voronezh, Russian S F S R. Vol. 2(3/4), 1929. Byull. Voronezhsk. Obshch. Estestvoisp. Gosud. Univ. Preceded & superseded by: Byulleten' Obshchestva estestvoispytatelei pri Voronezhskom gosudarstvennom universitete. 5-4430-1. HI 63793

Byulleteni Voronezhskoi stantsii zashchity rastenii. Voronezh, Russian S F S R. Vols. 6-7, 1926-2?; vols.

9-11, 192?-28. Byull. Voronezhsk. Stantsii Zashch. Rast. Preceded by: Stantsiya zashchity sel'skokhozyaistvennykh rastenii ot vreditelei Voronezhskogo gubernskogo zemel'nogo upravleniya. For vol. 8 see: Voronezhskaya stantsiya zashchity rastenii. Superseded by: Trudy Voronezhskoi stantsii zashchity rastenii. HI 63794

Byulleten', vsesoyuznii nauchno-issledovatel'skii institut zernobovykh kul'tur. Moscow. Nos. 1-3, 1932-33; [n.s.] nos. 1-6, 1971-73. Byull. Vsesoyuzn. Nauchno-Issl. Inst. Zernobov. Kul't. HI 63795

Byulleten', vsesoyuznii selektsionno-geneticheskii institut. Vol. ?-9+, 19??-68+. Byull. Vsesoyuzn. Selekts.-Genet. Inst. HI 63796

Byulleten' vsesoyuznogo instituta rastenievodstva. Leningrad. Vols. 1-15, 1956-70. Byull. Vsesoyuzn. Inst. Rasteniev. Superseded by: Nauchno-tekhnicheskii byulleten' vsesoyuznogo ordena Lenina i ordena Druzhby Narodov instituta rastenievodstva im N. I. Vavilova. HI 63797

Byulleten' Vsesoyuznogo nauchno-issledovatel'skogo instituta zashchity rastenii. Leningrad. Vol. 3/4, 1961+. Byull. Vsesoyuzn. Nauchno-Issl. Inst. Zashch.Rast. Preceded by: Byulleten' nauchno-tekhnicheskoi informatsii po zashchite rastenii. HI 63798

Byulleten' vsesoyuznogo ordena Lenina instituta rastenievodstva imeni N.I. Vavilova = Byulleten' vsesoyuznogo instituta rastenievodstva. Leningrad.

Byulleten' yarovizatsii. Moscow. 1932. Byull. Yarov. Superseded by: Yarovizatsiya. 4-3136-3. HI 63799

Byvšego Zapadno-Sibirskogo otdela Gosudarstvennogo russkogo geografičeskogo obščestva = Izvestiya Zapadno-Sibirskogo geograficheskogo obshchestva. Omsk.

C A B abstracts online. Farnham Royal. No. 1+, 1981+. C. A. B. Abstr. Online. HI 63800

C A B international news. Farnham Royal. 1986+. C. A. B. Int. News. Preceded by: C A B news. HI 68540

C A B news. Farnham Royal. 1976-86. C. A. B. News. Superseded by: C A B international news. HI 63801

C A B S = Chemical abstracts - biochemistry sections. Columbus, OH.

C A M C O R E bulletin on tropical forestry. Raleigh?, NC. No. 1+, 1984+. C. A. M. C. O. R. E. Bull. Trop. Forest. HI 63813

C A N = Conservation administration news. Laramie, WY, Tulsa, OK.

C A N O P; Canadian Association of Nature and Outdoor Photographers. Toronto. 1988+. C. A. N. O. P. HI 74821

C A review titles. 2, chemical-biological interactions. Nottingham. Vol. 75+, [1975]+. C. A. Rev. Titles, 2, Chem.-Biol. Interact. HI 63802

C A review titles. 3, general biochemistry. Nottingham. Vol. 75+, [1975]+. C. A. Rev. Titles, 3, Gen. Biochem. HI 63803

C A review titles. 4, biochemical processes in organisms. Nottingham. Vol. 75+, [1975]+. C. A. Rev. Titles, 4, Biochem. Processes Organisms. HI 63804

C A review titles index. 1, pharmacology and pharmaceutical chemistry. Nottingham. Vol. 82+, 1975+. C. A. Rev. Titles Index, 1, Pharmacol. Pharm. Chem. HI 63805

C A review titles index. 2, chemical biological interactions. Nottingham. Vol. 82+, 1975+. C. A. Rev. Titles Index, 2, Chem. Biol. Interactions. HI 63806

C A review titles index. 3, general biochemistry. Nottingham. Vol. 82+, 1975+. C. A. Rev. Titles Index, 3, Gen. Biochem. HI 63807

C A review titles index. 4, biochemical processes in organisms. Nottingham. Vol. 82+, 1975+. C. A. Rev. Titles Index, 4, Biochem. Processes Organisms. HI 63808

C A selects. Biological information transfer. Columbus, OH. 1979+. C. A. Selects, Biol. Inform. Transfer. HI 63809

C A selects. Environmental pollution. Columbus, OH. 1978+. C. A. Selects, Environm. Pollut. HI 74522

C A selects. Fungicides. Columbus, OH. 1978+. C. A. Selects, Fungic. HI 63810

C A selects. Herbicides. Columbus, OH. 1978+. C. A. Selects, Herbic. HI 63811

C A selects. Membrane separation. Columbus, OH. 1989+. C. A. Selects, Membr. Separation. HI 74820

C A selects. Novel pesticides and herbicides. Columbus, OH. 1985+. C. A. Selects, Novel Pestic. Herbic. HI 63812

C A selects. Photobiochemistry. Columbus, OH. 1978+. C. A. Selects, Photobiochemistry. HI 74523

C A T I E, actividades en Turrialba. Turrialba. Vol. 1+, 1972+. C. A. T. I. E. Actividades Turrialba. HI 63814

C B A = Current biotechnology abstracts. London, Letchworth.

C B A - A B C bulletin = Bulletin, canadian botanical association. Vancouver, B.C., Guelph, Ont.

C B E views; official publication of the council of biology editors. Washington, DC. Vol. 1+, 1978+. C. B. E. Views. Preceded by: Newsletter, council of biology editors. HI 63815

C B H L newsletter = Newsletter, council on botanical and horticultural libraries. Beltsville MD., etc.

C B S newsletter. Baarn. No. 1+, 1983+. C. B. S. Newslett. (Baarn). HI 63816

C B S newsletter. Storrs, CT. = Connecticut botanical society newsletter. Storrs, CT.

C C - A F V = Current contents: agriculture, food and veterinary sciences. Philadelphia, PA.

C C H mykologický sbornik v Praze = Mykologický sbornik. Prague.

C C N B newsletter. Fredericton, N. B. Vols. 1-10(4), 1969?-79. C. C. N. B. Newslett. Superseded by: Conservation. Fredericton, N. B. HI 63817

C F B S newsletter = Newsletter, canadian federation of biological societies. Saskatoon.

C F I occasional papers. Oxford. No. 1+, 1978+. C. F. I. Occas. Pap. HI 63818

C F R U progress report = Progress report, cooperative forestry research unit, university of Maine at Orono. Orono, ME.

C F R U research bulletin = Research bulletin, cooperative forestry research unit, university of Maine at Orono. Orono, ME.

C F R U research note = Research notes, cooperative forestry research unit, university of Maine at Orono. Orono, ME.

C I K A R D news. Ames, IA. Vol. 1+, 1989+. C. I. K. A. R. D. News. HI 63819

C I M M Y T annual report = Report, international maize and wheat improvement center. Mexico City.

C I M M Y T maize improvement = C I M M Y T report on maize improvement. Mexico City.

C I M M Y T report on maize improvement. Mexico City. 1973+. C. I. M. M. Y. T. Rep. Maize Improv. HI 63820

C I M M Y T report on wheat improvement. Mexico City. 1973+. C. I. M. M. Y. T. Rep. Wheat Improv. HI 63821

C I M M Y T research bulletin = Research bulletin international maize and wheat improvement center. Chapingo.

C I M M Y T review. Mexico City. 1974+. C. I. M. M. Y. T. Rev. HI 63822

C I M M Y T world wheat facts and trends. Mexico City. No. ?-3+, 19??-85+. C. I. M. M. Y. T. World Wheat Facts Trends. HI 63823

C I P circular. Lima. Vol. 5+, 1976+. C. I. P. Circ. Preceded by?: International potato center newsletter. HI 63824

C I S. Tokyo = Chromosome information service. Tokyo.

C I T R E newsletter. [Washington, DC.] No. 1, 1971. C. I. T. R. E. Newslett. HI 63825

C M I/A A B descriptions of plant viruses. Kew. No. 1+, 1970+. C. M. I./A. A. B. Descr. Pl. Viruses. HI 63826

C M I descriptions of pathogenic fungi and bacteria. Kew, England. Set. 1+, 1964+. C. M. I. Descript. Pathog. Fungi Bact. HI 63827

C N F R A, publications = Publications de la comité national français des recherches antartiques. Biologie. Paris.

C O B I newsletter. London. No. 1+, 1973+. C. O. B. I. Newslett. HI 63829

C P notes. Ithaca, NY. No. 11+, 1983+. C. P. Notes. Preceded by: Cornell plantations notes. HI 63830

C P S news. Ottawa, Saskatoon. Vol. 1+, 1949+. C. P. S. News. HI 63831

C P T newsletter = Central plains newsletter. Manhattan, KS.

C R C critical reviews in biochemistry. Boca Raton, FL. Vol. 1+, 1972+. C. R. C. Crit. Rev. Biochem. HI 63832

C R C critical reviews in biotechnology. Boca Raton, FL. Vol. 1+, 1983+. C. R. C. Crit. Rev. Biotechnol. HI 63833

C R C critical reviews in microbiology. Boca Raton, FL. Vol. 1+, 1971+. C. R. C. Crit. Rev. Microbiol. HI 63834

C R C critical reviews in plant sciences. Boca Raton, FL. Vol. 1+, 1983+. C. R. C. Crit. Rev. Pl. Sci. HI 63835

C R F G yearbook = California rare fruit growers' yearbook. Bonsall, Fullerton, CA.

C R U S news. Sheffield. No. 1+, 1977+. C. R. U. S. News. HI 63836

C S C technical publication series. London. No. ?-202+, 19??-86+. C. S. C. Techn. Publ. Ser. HI 63837

C S I E = Cactus and other succulent information exchange. Burnaby, B. C.

C S I R publications. Pretoria. No. 1+, 1973+. C. S. I. R. Publ. Preceded by: C S I R research review. HI 63838

C S I R research review. Pretoria. Vols. 3-23 [issues for 1963-73 also numbered 49-69], 1953-73. C. S. I. R. Res. Rev. Preceded by: Research review of the south african council for scientific and industrial research. Superseded by: C S I R publications. HI 63839

C S I R O abstracts Melbourne, Vic. Vol. 6+, 1958+. C. S. I. R. O. Abstr. Preceded by: C S I R O science index. HI 63840

C S I R O index. East Melbourne, Vic. Vol. 5+, 1979+. C. S. I. R. O. Index. Preceded by: Australian science index. HI 63841

C S I R O science index. Melbourne, Vic. Vol. 5, 1957 [Includes vol. 1 of Australian science index, q.v.]. C. S. I. R. O. Sci. Index. Preceded by: Abstracts of published papers and list of translations, C S I R O, Australia. Superseded by: C S I R O abstracts. HI 63842

C S S A newsletter. Laundale. No. 19+, 1983+. C. S. S. A. Newslett. Preceded by: Affiliate reporter, cactus and succulent society of America. HI 63843

C S T A review. Ottawa. Nos. 1-45, 1935-45. C. S. T. A. Rev. Superseded by: Agricultural institute review. Ottawa. HI 63844

C U E B S news. Washington, DC. Vol. 1-8, 1964-72. C. U. E. B. S. News. Superseded by: A I B S news. HI 63845

C U E B S publication. Washington, DC. No. 1+, 1964+. C. U. E. B. S. Publ. HI 63846

Cabinet biologique. Rostov = Issledovatel'skaya kafedra biologii. Rostov.

Cabinet botanique. Rostov = Issledovatel'skaya kafedra botaniki. Rostov.

Cabinet of natural history and american rural sports, with illustrations. Philadelphia, PA. Cab. Nat. Hist. & Amer. Rural Sports. See B–P–H 290/3. HI 53138

Cabios = Computer applications in the biosciences. Washington, DC.

Cacao. Turrialba, Costa Rica. Cacao. See B–P–H 290/4. HI 53139

Cacao en Colombia. Palmira, Colombia. Cacao en Colombia. See B–P–H 290/6. HI 53140

Cacaotero. San José, Costa Rica. Cacaotero. See B–P–H 290/8. HI 53141

Cacau atualidades. Itabuna. Vols. 1-14, 1964-77. Cacau Atual. HI 63847

Cactaceae; Jahrbücher der Deutschen Kakteen-Gesellschaft e. V. Berlin. Berlin. 1937-43/44, 1936-44. Cactaceae (Berlin). 2-863-1. HI 53149

Cactaceae. Long Beach, CA. No. [1], 1935. Cactaceae (Long Beach). HI 63848

Cactaceae brugensis. No. ?-4+, 19??-82+. Cactaceae Brugensis. HI 63849

Cactaceas y suculentes mexicanas. Azcapotzalco. Vol. 1+, 1955/56+, 1955+. Cact. Suc. Mex. HI 63850

Cactastrophes; bulletin de club genevois des amateurs de cactus. Geneva. No. 0, 1982; no. 1+, 1982+. Cactastrophes. HI 63851

Cactea; revista mensile per coltivatori, commercianti, amatori e collezionisti. Ventimiglia. Vol. 1(1-9), 1934. Cactea. HI 63852

Cacti. [Saboten.] Hyogo. Vol. 1+, 1957+. Cacti (Hyogo). HI 53150

Cacti. Rotorua. Vols. 1-2, 1964-65. Cacti (Rotorua). HI 63853

Cacticulture series. London. Cacticult. Ser. See B–P–H 291/10. HI 53151

Cactochat; bulletin of the Christchurch cactus and succulent society. Christchurch, N.Z. Vol. 1+, 1959+. Cactochat. HI 63854

Cactofilo; revista de cactus-club de España. No. 0, 1973; no. 1, 1974. Cactofilo. HI 63855

Cactography. National City & San Diego, CA. No. 1-7, 1906-26. Cactography. HI 63856

Cactominga. Wanganui, N. Z. Nos. 1-6, 1957-62. Cactominga. HI 63857

Cactophile, The. Bristol. Vol. 1(1-4), 1960-61. Cactophile. HI 63858

Cactophiles, Les. Rouen. [Nos. 1-6], 1941-42. Cactophiles. HI 63859

Cactophiles cactivities; monthly bulletin of the Colorado cactophiles. Denver, CO. Vols. 1-6, 1962-66. Cactophiles Cactivities. Superseded by: Colorado cactophiles cactivities. HI 63860

Cactos y suculentos. San Diego, CA. Vols. 1-2, 1965-66. Cactos Suc. HI 63861

Cactulent. London. Vols. 1-18, 1949-66. Cactulent. HI 63862

Cactus. Belgian edition subtitled: Revue pour amateurs de cactacées et autres succulentes. Dutch edition subtitled: Tijdschrift voor liefhebbers van cactussen en andere sukkulenten. Antwerp, etc. Vol. 1-7(5), 1969-75. Cactus (Antwerp). Preceded, in part, by: Tijdschrift voor internationaal succulentenliefhebbers. Superseded by: Cactus. (Brussels) and Cactus. (Wijnegem etc.) HI 63863

Cactus; revue bimensuelle pour amateurs de cactacées et autres succulentes. Brussels. Vol. 1+, 1976+. Cactus (Brussels). Preceded by: Cactus. (Antwerp). HI 63864

Cactus. (Gent) = Cactus. Sint-Amandsberg.

Cactus; journal of the cactus and succulent society of Japan. Kushimoto. No. 88+, 1954+. Cactus (Kushimoto). Preceded by: Acta succulentologica. HI 63865

Cactus; organe de l'association française de cactus et plantes grasses. Paris. Vols. 1(1)-22(88), 1946-67. Cactus (Paris). 2-863-1. HI 63866

Cactus. Regensberg. 19??-82+. Cactus (Regensberg). HI 63867

Cactus; bulletin bimestriel du cercle des cactéophiles belges. (Tweemaandelijksch tijdschrift van den kring der belgische cactusliefhebbers). Sint-Amandsberg, Ghent. Vols. 1-11(1), 1931-41. Cactus (Sint-Amandsberg). HI 63868

Cactus. Wijnegem, Gent. Vol. 8+, 1976+. Cactus (Wijnegem). Preceded by: Cactus. (Antwerp). HI 63869

Cactus Anjou; bulletin d'information de la section de l'Anjou de l'association française des amateurs de cactées et plantes grasses. Angers. Nos. 1-9, 1949-?; n.s. nos. 1-4, 19??-58. Cactus Anjou. HI 63870

Cactus bulletin and hobbyist. Holme, England. Cact. Bull. Hobbyist. See B–P–H 290/10. HI 53142

Cactus capital chatter; newsletter of Tucson cactus and botanical society. Tucson, AZ. Vol. 1+, 1965+. Cact. Capital Chatter. HI 63871

Cactus chatter. Long Beach, CA. Vol. 1, 1940-48. Cact. Chatter (Long Beach). HI 63872

Cactus chatter; Oregon cactus and succulent society newsletter. Portland, OR. 19??-83+. Cact. Chatter (Portland) HI 63873

Cactus Chiba. Chiba. No. ?-7+, 19??-70+. Cactus Chiba. HI 63874

Cactus club bulletin. Delta, AL. 1955-57. Cact. Club. Bull. Superseded by: Cactus points. HI 63875

Cactus chronicle. Hamilton, N. Z. 19??-62+. Cact. Chron. HI 63876

Cactus comments. Brooklyn [=New York, In part], NY. Cact. Comments. See B–P–H 290/11. HI 53143

Cactus contact; a journal for connecting cactophiles (Une circulaire pour les relations entre cactéophiles. Ein Rundschreiben zur Verbindung von Kakteenfreunden). Pieterlen. Nos. 1-5, 1954-58. Cact. Contact. HI 63877

Cactus courier. San Jose, CA. Vol. 1+, 1968+. Cact. Courier (San Jose). HI 63878

Cactus courier; official newsletter of the Adelaide and country cactus club. South Plympton, S.A. Vol. 1+, 1969+. Cact. Courier (South Plympton). HI 63879

Cactus digest. St. Louis, MO. Cact. Digest. See B–P–H 290/13. HI 53144

Cactus-flora. [Netherlands.] No. ?-7+, 19??-82+. Cact.-Flora. HI 63880

Cactus journal. Croydon. Vols. 1-8(2), 1932-39. Cact. J. (Croydon). Superseded by: Cactus and succulent journal of Great Britain. 2-863-1. HI 63881

Cactus journal; devoted exclusively to cacti and other succulent plants. London. Cact. J. (London). See B–P–H 290/15. HI 53145

Cactus Kenkyu. Tokyo. 1954-56. Cact. Kenkyu. HI 63882

Cactus Kyushu. Kyushu. No. 1+, 1956+. Cactus Kyushu. HI 63883

Cactus needle. Tucson, AZ. Cact. Needle. See B–P–H 290/18. HI 53146

Cactus news reel. Te Awamutu. No. 1+, 1959+. Cact. News Reel. HI 63884

Cactus et plantes grasses. Rouen. Nos. 1-9, 1951-54. Cactus Pl. Grasses. HI 63885

Cactus points. Delta, AL. Vols. ?-11, 1957-65. Cact. Points. Preceded by: Cactus club bulletin. HI 63886

Cactus; revue pour amateurs de cactacées et autres succulentes = Cactus. Antwerp, etc.

Cactus romand. Prilly. 1967-68. Cact. Romand. HI 63887

Cactus sticker. Las Vegas, NV. 1972-76. Cact. Sticker. HI 63888

Cactus and succulent bulletin. Adelaide, S.A. Nos. 1-2?, 1968-70? Cact. Succ. Bull. HI 63889

[Cactus and succulent garden news.] [Japan.] 19??-55+. Cact. Succ. Gard. News. HI 63890

Cactus and other succulent information exchange. Burnaby, B. C. Vol. 1+, 1967+. Cact. Succ. Inform. Exch. HI 63891

Cactus and succulent information exchange = Cactus and other succulent information exchange. Burnaby, B. C.

Cactus and succulent journal. Los Angeles, CA. Cac. Succ. J. (Los Angeles). See B–P–H 290/23. HI 53147

Cactus and succulent journal. Woollahra, Australia. Cact. Succ. J. (Woollahra). See B–P–H 290/24. HI 53148

Cactus and succulent journal. Amateur bulletin section. Pasadena, CA. Vol. 1(1-12), 1942. Cact. Succ. J., Amateur Bull. Sect. HI 63892

Cactus and succulent journal. Auckland. Vols. 3(8)-6(5), 1950-53. Cact. Succ. J. (Auckland). Preceded by: Journal of the cactus and succulent society of New Zealand. Superseded by: New Zealand cactus and succulent journal. 2-863-1. HI 63893

Cactus and succulent journal of Great Britain. London. Vol. 8(3)-44, 1946-82. Cact. Succ. J. Gr. Brit. Preceded by: Cactus journal. Croydon. Superseded by: British cactus and succulent journal. 2-863-1. HI 63894

Cactus and succulent journal of the gumma cactus club. [Shaboten gumma.] Gumma. Vol. 1+, 1958+. Cact.

Succ. J. Gumma Cact. Club. HI 63895

Cactus and succulent notes. Auckland. Vol. 1(1-3), 1947. Cact. Succ. Notes. Superseded by: Journal of the cactus and succulent society of New Zealand. HI 63896

Cactus and succulent yearbook. Southampton. Vol. 1+, 1983+. Cact. Succ. Yearb. HI 63897

Cactus; tijdschrift voor liefhebbers van cactussen en andere sukkulenten = Cactus. Antwerp, etc.

Cactus Tokyo. [Saboten.] Tokyo. No. 1+, 1958+. Cactus Tokyo. Preceded by?: Shaboten sha. HI 63898

Cactus Toyama. Toyama. No. ?-10+, 19??-58+. Cactus Toyama. HI 63899

Cactus wereld in 't klein; kringblas voor liefhebbers van cactussen en andere succulenten. Zaandam. Vol. 1-?, 1973-75? Cact. Wereld Klein. HI 63900

Cactussen en vetplanten; maanblad van de nederl. vereeniging van liefhebbers van cactussen en vetplanten. Westzaan. Vols. 1-9(1), 1935-43. Cact. & Vetpl. Superseded by: Succulenta. HI 63901

Cactusvrienden. Antwerp. Vol. 1+, 1961+. Cactusvrienden. HI 63902

Cactusweelde; maanblad voor en door liefhebbers ter bevoerdering van de cactus- en vetplantenkweek. Antwerp. Vols. 1-12, 1955-66. Cactusweelde. Superseded by: Tijdschrift voor internationaal succulentenliefhebbbers. HI 63903

Cactusweelde nieuwsblad. Antwerp. 1967+. Cactusweelde Nieuwsblad. HI 63904

Cadernos da Amazonia. Manaus. Vol. 1-11, 1962-68. Cad. Amazonia. HI 63905

Cadernos F E E M A. Série tecnica: Espécies raras ou ameaçadas de extinção. Rio de Janeiro. 1982?+. Cad. F. E. E. M. A., Sér. Técn. HI 63906

Cadernos F E E M A. Série trabalhos ténicos: Flora, alguns estudos. Rio de Janeiro. No. ?-8/79+, 19??-79+. Cad. F. E. E. M. A., Sér. Trab. Técn. HI 63907

Cadernos da faculdade de filosofia, universidade federal de Pernambuco. Ser. 5, historia natural. Recife. No. ?-11+, 19??-65+. Cad. Fac. Filos. Univ. Fed. Pernambuco, Ser. 5. HI 63908

Cadernos do conselho de desenvolvimento de Pernambuco. Serie 1, agricultura. Recife. No. 1+, 1970+. Cad. Cons. Desenvolv. Pernambuco, Ser. 1. HI 63909

Cadernos de pesquisas. 2, série botânica. Teresina. No. 1+, 1982+. Cad. Pesq., 2, Sér. Bot. HI 63910

Cadia. [Centro argentino de ingenieros agrônomos.] Buenos Aires. Cadia. See B–P–H 291/20. HI 53152

Café. Lima. Café (Lima). See B–P–H 291/21. HI 53153

Cafe. Turrialba. Vols. 6-9, 1965-68. Cafe (Turrialba). Preceded by: Coffee. HI 63911

Café, cacao, thé. Paris. Café, Cacao, Thé. See B–P–H 291/22. HI 53154

Café de El Salvador. San Salvador, Salvador. Café El Salvador. See B–P–H 291/23. HI 53155

Café de Nicaragua. Managua, Nicaragua. Café Nicaragua. See B–P–H 291/24. HI 53156

Café peruano. Lima. Café Peruano. See B–P–H 291/25. HI 53157

Café vert. Paris. Café Vert. See B–P–H 291/26. HI 53158

Cafeicultor. São Paulo. Cafeicultor. See B–P–H 291/27. HI 53159

Cafeicultura. Rio de Janeiro. Cafeicultura. See B–P–H 291/28. HI 53160

Cafetal. Havana. Cafetal. See B–P–H 291/29. HI 53161

Cafetero. Medellin, Colombia. Cafetero. See B–P–H 292/1. HI 53162

Caffer. Florence. Caffer. See B–P–H 292/2. HI 53163

Cahiers d'agriculture pratique des pays chauds. Paris. Cah. Agric. Prat. Pays Chauds. See B–P–H 292/3. HI 53165

Cahier bimestriel, centre d'études internationales pour la préservation de la vie et la protection de la nature. St.-Cloud. No. 1+, 1973+. Cah. Bimestr. Centre Int. Préserv. Vie Protect. Nat. HI 63912

Cahiers de biologie marine. Roscoff, France. Cah. Biol. Mar. See B–P–H 292/4. HI 53166

Cahiers du centre d'études et de recherches de biologie et d'océanographie médicale. Nice. Cah. Centr. Études Rech. Biol. See B–P–H 292/5. HI 53167

Cahiers de l'environnement. Berne. No. 1+, 1982+. Cah. Environm. (Berne). HI 63913

Cahiers de l'environnement. Besançon. 1985+. Cah. Environm. (Besançon). HI 75253

Cahiers de la faculté des sciences, université de Genève. Geneva. No. 1+, 1980+. Cah. Fac. Sci. Univ. Genève. HI 63914

Cahiers de la faculté des sciences de l'université Mohammed V. Série biologie végétale. Rabat. Nos. 1-5, 1963-64. Cah. Fac. Sci. Univ. Mohammed V, Sér. Biol. Vég. HI 63915

Cahiers de l'Indo-Pacifique. Paris. No. 1+, 1979+. Cah. Indo-Pacifique. Preceded by: Cahiers du Pacifique. HI 63916

Cahier, informations annuelles de caryosystématique et

cytogénétique; travaux, laboratoires de phytogénétique, Strasbourg et Lille. Strasbourg. No. 1+, 1967+. Cah. Inform. Annuelles Caryosyst. Cytogén. HI 63917

Cahier d'information, département d'océanographique, université du Québec à Rimouski. Rimouski. No. 4+, 1979+. Cah. Inform. Dép. Océanogr. Univ. Québec Romouski. Preceded by: Cahier d'information, section d'océanographie, université du Québec à Rimouski. HI 63918

Cahier d'information, section d'océanographie, université du Québec à Rimouski. Rimouski. Nos. ?-3, 19??-78. Cah. Inform. Sect. Océanogr. Univ. Québec Rimouski. Superseded by: Cahier d'information, département d'océanographie, université du Québec à Rimouski. HI 63919

Cahier du journal d'agriculture tropicale et de botanique appliquée. Paris. Vol. 1+, 1972+. Cah. J. Agric. Trop. Bot. Appl. HI 63920

Cahiers du laboratoire d'hydrobiologie de Montereau. Grenoble. Vol. 1+, 1974+. Cah. Lab. Hydrobiol. Montereau. HI 63921

Cahiers lyonnais d'histoire de la médecine. Lyons. Cah. Lyon. Hist. Méd. See B–P–H 292/6. HI 53168

Cahiers de la Maboké. Paris. Cah. Maboké. See B–P–H 292/7. HI 53169

Cahier de micropaléontologie. Paris. 1974+. Cah. Micropaléontol. HI 63922

Cahiers des naturalistes; bulletin des naturalistes Parisiens. Paris. N.s. vol. 8+, 1953+. Cah. Naturalistes. Preceded by: Feuille des naturalistes. 2-867-2. HI 63923

Cahier népalais, documents. Paris. Vol. 1+, 1969+. Cah. Népal. Doc. HI 63924

Cahiers océanographiques. Paris. Cah. Océanogr. See B–P–H 292/10. HI 53170

Cahiers office de la recherche scientifique et technique d'outre-mer = Cahiers O R S T O M. Paris.

Cahiers O P T I M A = O P T I M A leaflets. Geneva, Chambésy.

Cahiers O R S T O M. Série géologie. Paris. Vols. 1-14, 1969-84. Cah. O. R. S. T. O. M., Sér. Géol. Superseded by: Géodynamique [not entered]. HI 63925

Cahiers O R S T O M. Série océanographique. Bondy, Paris. Vols. 1-16, 1963-78. Cah. O. R. S. T. O. M., Sér. Océanogr. Superseded by: Océanographie tropicale. HI 63926

Cahiers O R S T O M. Physiologie des plantes tropicales cultivées. Paris. Vols. 1-2, 1964-65. Cah. O. R. S. T. O. M. Physiol. Pl. Trop. Cult. Superseded by: Cahiers O R S T O M. Série biologie. HI 63927

Cahiers O R S T O M. Série biologie. Paris. Vol. 1+, 1966+. Cah. O. R. S. T. O. M., Sér. Biol. Preceded by: Cahiers O R S T O M. Physiologie des plantes tropicales cultivées. HI 63928

Cahiers O R S T O M. Série hydrobiologie. Paris. Vol. 1-13, 1967-79/80. Cah. O. R. S. T. O. M., Sér. Hydrobiol. Superseded by: Revue d'hydrobiologie tropicale. HI 63929

Cahiers du Pacifique. Paris. Nos. 1-21, 1958-78. Cah. Pacifique. Superseded by: Cahiers de l'Indo-Pacifique. HI 63930

Cahiers de pathologie végétale et entomologie agricole. Paris. Cah. Pathol. Vég. Entomol. Agric. See B–P–H 292/16. HI 53172

Cahiers de la recherche agronomique, direction de la recherche agronomique et de l'enseignement agricole. Rabat. Vol. 1+, 1948+. Cah. Rech. Agron. 2-866-2. HI 63931

Cahier de la société de chimie organique et biologique. Paris. Cah. Soc. Chim. Org. Biol. See B–P–H 292/17. HI 53173

Cahiers, see Cahier

Cairo scientific journal. Cairo. Cairo Sci. J. See B–P–H 292/20. HI 53174

Calafia; revista de la universidad autonomia de Baia California. Mexicali. 19??-6?+. Calafia. HI 63932

Calanus. Aitsu. No. 1+, [1968]+. Calanus. HI 63933

Calavo newsletter. Los Angeles, CA. 1959+. Calavo Newslett. HI 63934

Calcutta journal of natural history, and miscellany of the arts and sciences in India. Calcutta. Calcutta J. Nat. Hist. See B–P–H 292/27. HI 53175

Caldasia. Boletín del instituto de ciencias naturales de la universidad nacional de Colombia. Bogotá. Caldasia. See B–P–H 293/3. HI 53176

Calendar of the exhibitions, Huntington library, art gallery and botanical gardens. San Marino, CA. 1936-? Calend. Exhib. Huntington Lib. Art Gall. Bot. Gard. Superseded by: Huntington calendar. HI 63936

Calendrier; institut de France. Paris. Calend. Inst. France. See B–P–H 293/4. HI 53177

Calendrina. Port Elizabeth, S. A. Vol. 1+, 1980+. Calendrina. HI 63937

Calgary field naturalist. Calgary. 1969-79. Calgary Field Naturalist. Preceded by: Calgary bird club bulletin [not entered]. Superseded by: Pica. HI 63938

Calicut university research journal. Kerala. Vol. 1+, 1980+. Calicut Univ. Res. J. HI 63939

California agricultural extension service circular = Circular, California agricultural extension service.

Berkeley, CA.

California agriculture. Progress reports of agricultural research. University of California, college of agriculture. Agricultural experiment station. Berkeley, CA. Calif. Agric. See B–P–H 293/12. HI 53178

California-Arizona cotton. Fresno, CA. 1965+. Calif.-Arizona Cotton. HI 63940

California art and nature. San Diego, CA. Calif. Art Nat. See B–P–H 293/17. HI 53179

California calavo growers exchange. Los Angeles, CA. Vols. 1-3, 1924-26. Calif. Calavo Growers Exch. Superseded by: Report (Annual), calavo growers of California. HI 63941

California citrograph. Riverside, CA. Vols. 1-54(4), 1915-69. Calif. Citrogr. Superseded by: Citrograph. 2-883-3. HI 63942

California cultivator. Los Angeles, CA. Calif. Cultivator. See B–P–H 293/26. HI 53182

California culturist; a journal of agriculture, horticulture, mechanism, and mining. San Francisco, CA. Calif. Cult. See B–P–H 293/24. HI 53181

California dahlia grower. Los Angeles, CA. Calif. Dahlia Grower. See B–P–H 293/27. HI 53183

California dahlia news. San Francisco, CA. Calif. Dahlia News. See B–P–H 293/28. HI 53184

California farmer and journal of useful sciences. San Francisco, CA. Calif. Farmer & J. Useful Sci. See B–P–H 293/29. HI 53185

California florist. Santa Barbara, CA. Calif. Florist. See B–P–H 293/31. HI 53186

California forestry. Berkeley, CA. Calif. Forest. See B–P–H 293/34. HI 53187

California forestry and forest products. Berkeley, CA. No. 1+, 1957+. Calif. Forest. Forest Prod. HI 63943

California fruit and grape grower. San Francisco, CA. 1948-50. Calif. Fruit Grape Grower. Preceded by: Grape grower. Superseded by: Western fruit grower. 5-4486-3. HI 63944

California fruit grower. San Francisco, CA. Calif. Fruit Grower. See B–P–H 294/1. HI 53188

California fruit news. San Francisco, CA. Calif. Fruit News. See B–P–H 294/2. HI 53189

California garden. San Diego, CA. Calif. Gard. See B–P–H 294/3. HI 53190

California horticulturist and floral magazine. San Francisco, CA. Calif. Hort. & Fl. Mag. See B–P–H 294/4. HI 53191

California grower: Avocados, citrus, subtropicals. Vista, CA. Vol. 9(9)+, 1985+. Calif. Grower: Avocados Citrus Subtrop. Preceded by: Avocado grower. HI 63945

California horticultural journal. San Francisco, CA. Vols. 29-36, 1968-75. Calif. Hort. J. Preceded by: Journal, California horticultural society and Notes from Strybing arboretum. Superseded by: Pacific horticulture. HI 63946

California macdamia society. Yearbook. Carlsbad, CA. Calif. Macadamia Soc. Yearb. See B–P–H 294/6. HI 53192

California magazine and mountaineer. San Francisco, CA. Vols. 1-7, 1861-63? Calif. Mag. Mountaineer. Preceded by: Hutching's illustrated California magazine and California mountaineer [not entered]. 2-887-3. HI 63947

California medicine. San Francisco, CA. Calif. Med. See B–P–H 294/7. HI 53193

California native plant society bulletin = Bulletin, California native plant society. Berkeley, CA., Sacramento, CA

California newsletter, palm society. Walnut Creek, CA. Vol. 1+, 1982+. Calif. Newslett. Palm Soc. HI 63948

California olive industry news. San Francisco, CA. Calif. Olive Industr. News. See B–P–H 294/8. HI 53194

California pear grower. San Francisco, CA. Calif. Pear Grower. See B–P–H 294/10. HI 53195

California plant pathology. Berkeley, CA. No. 1+, 1971+. Calif. Pl. Pathol. HI 63949

California rare fruit growers' yearbook. Bonsall, Fullerton, CA. Vols. 1-17, 1969-85; 1987+. Calif. Rare Fruit Growers' Handb. Superseded, in part, by: Journal, California rare fruit growers. HI 63950

California rosarian. Point Loma, CA. Calif. Rosarian. See B–P–H 294/11. HI 53196

California state journal of medicine. San Francisco. CA. Calif. State J. Med. See B–P–H 294/15. HI 53197

California tomato grower. Stockton, CA. 1958+. Calif. Tomato Grower. HI 63951

California turf grass culture. Experimental program in turf culture. California. University. University at Los Angeles. Los Angeles, CA. Calif. Turf Grass Cult. See B–P–H 294/17. HI 53198

California and western medicine. San Francisco, CA. Calif. W. Med. See B–P–H 294/18. HI 53199

California and western states grape grower. Fresno, CA. Vol. 1+, 1969+. Calif. W. States Grape Grower. HI 63952

Callaway gardens newsletter. Pine Mountain, GA. 1982?+. Callaway Gard. Newslett. HI 63953

Calochortus; newsletter of the consortium of California

herbaria and reference collections. Berkeley, CA. No. 1+, 1980+. Calochortus. HI 63954

Cambient newsletter. No. 43+, 1978+. Cambient Newslett. Preceded by: Newsletter, Cambridge and Isle of Ely naturalists' trust (CAMBIENT). HI 63955

Cambridge expeditions journal. Cambridge. 1965+. Cambridge Exped. J. HI 63956

Cambridge scientific biochemistry abstracts. Part 1, biological membranes. Bethesda, MD. Vol. 13+, 1985+. Cambridge Sci. Biochem. Abstr., Part 1, Biol. Menbr. Preceded by: Biochemistry abstracts. Part 1, biological membranes. HI 63957

Cambridge scientific biochemistry abstracts. Part 2, nucleic acids. Bethesda, MD. Vol. 15+, 1985+. Cambridge Sci. Biochem. Abstr., Part 2, Nucleic Acids. Preceded by: Biochemistry abstracts. Part 2, nucleic acids. HI 63958

Cambridge scientific biochemistry abstracts. Part 3, amino acids, peptides and proteins. Bethesda, MD. Vol. 14+, 1985+. Cambridge Sci. Biochem. Abstr., Part 3, Amino Acids. Preceded by: Biochemistry abstracts. Part 3, amino acids, peptides and proteins. HI 63959

Camellia annual. Gordon, N.S.W. Vol. 1(1-7), 1954-60. Camellia Annual. Incorporated in: Camellia news. HI 63960

Camellia bulletin, South Auckland camellia society. Hamilton, N.Z. Nos. 1-3, 1957-58. Camellia Bull. S. Auckland Camellia Soc. Superseded by: New Zealand camellia bulletin. HI 63961

Camellia journal; official publication of the american camellia society. Tifton, Fort Valley, GA. Vol. 1+, 1945+. Camellia J. HI 63962

Camellia news. Naremburn, N.S.W. N.s. no. 1+, 1961+. Camellia News. HI 63963

Camellia nomenclature. Arcadia, CA. 1947+. Camellia Nomencl. HI 63964

Camellia review; a publication of the southern California camellia society. Pasadena, CA. Vol. 12+, 1950+. Camellia Rev. Preceded by: Bulletin of the southern California camellia society. 2-896-2. HI 63965

Camellian. Columbia, SC. Camellian. See B–P–H 294/29. HI 53200

Camellias. Gainesville, FL. Camellias. See B–P–H 295/1. HI 53201

Campanula. Ostrava. Nos. 1-3, 1971-72. Campanula. HI 63966

Campesino. Santiago. Vol. 66+, 1934+. Campesino. Preceded by: Boletín de la sociedad nacional de agricoltura (Santiago). 2-898-3. HI 63967

Campo; revista agraria nacional. Havana. Campo (Havana). See B–P–H 295/4. HI 53204

Campo. La Paz, Bolivia. Campo (La Paz). See B–P–H 295/5. HI 53205

Campo. Pôrto Alegre, Brazil. Campo (Pôrto Alegre). See B–P–H 295/6. HI 53206

Campo; revista mensuel illustrada de lavoura, criação, industria, comercio. Rio de Janeiro. Vol. 1+, 1930+. Campo (Rio de Janeiro). HI 63968

Campo y arado. Buenos Aires. Campo & Arado. See B–P–H 295/2. HI 53202

Campo y suelo argentino. Buenos Aires. Campo Suelo Argent. See B–P–H 295/7. HI 53207

Campo y vida. Caracas. Campo & Vida. See B–P–H 295/3. HI 53203

Canada agriculture. Ottawa. Vol. 11(2)+, 1966+. Canada Agric. Preceded by: Research for farmers. HI 63969

Canada department of agriculture monograph = Monographs, research branch, Canada department of agriculture. Ottawa.

Canada department of agriculture publications = Publications, department of agriculture, Canada. Ottawa.

Canada farmer. Toronto. Vol. 1, 1847. Canada Farmer (Toronto 1847). Superseded by: Agriculturist and canadian journal. 2-900-3. HI 63970

Canada farmer. Toronto. Vols. 1-5, 1864-68; n.s. vols. 1-4, 1869-72; ser. 3, vols. 1-4, 1873-76. Canada Farmer (Toronto 1864-76). 2-900-3. HI 63971

Canada green. Rexdale. Vol. 1+, 1976+. Canada Green. HI 63972

Canada medical and surgical journal. Montreal. Canada Med. Surg. J. See B–P–H 297/31. HI 53229

Canadian agriculturist. Toronto. Vols. 1-15, 1849-63. Canad. Agriculturist. Preceded by: Agriculturist and canadian journal. 2-902-2. HI 63973

Canadian Audubon. Toronto. Vols. 21-33, 1958-71. Canadian Audubon. Preceded by: Canadian nature. Superseded by: Nature Canada. 2-903-1. HI 63974

Canadian bee journal. Brantford, Ont. Canad. Bee J. See B–P–H 295/25. HI 53208

Canadian botanic garden newsletter. Edmonton, Alb. 19??-86. Canad. Bot. Gard. Newslett. Superseded by: Canadian plant conservation programme newsletter. HI 63975

Canadian dairy and ice cream journal. Toronto. Canad. Dairy Ice Cream J. See B–P–H 295/26. HI 53209

Canadian entomologist. London, Ont. Canad. Entomol. See B–P–H 295/29. HI 53210

Canadian federation news. Montreal, London, Ont. Vols.

1-13, 1958-71. Canad. Fed. News. Superseded by: Bulletin de la fédération canadienne. HI 63976

Canadian field-naturalist. Ottawa. Vols. 33+, 1919+. Canad. Field-Naturalist. Preceded by: Ottawa naturalist. 2-907-1 HI 63977

Canadian florist. Peterborough, Ont. Canad. Florist. See B–P–H 295/32. HI 53211

Canadian forestry journal. Ottawa. Vols. 1-16(11), 1905-20. Canad. Forest. J. Superseded by: Illustrated canadian forestry magazine. HI 63978

Canadian forestry magazine = Canadian forestry journal. Ottawa.

Canadian forestry service research notes = Research notes, canadian forestry service. Ottawa.

Canadian fruitgrower. Niagara. Vols. 1-37, 1944-81. Canad. Fruitgrower. 2-907-3. HI 63979

Canadian geographer. Toronto. No. 1+, 1951+. Canad. Geographer. HI 63980

Canadian geographical journal. Montreal. Canad. Geogr. J. See B–P–H 295/34. HI 53212

Canadian gladiolus annual. Guelph. Nos. 8-10, 1931-33. Canad. Gladiolus Annual. Preceded by: Bulletin of the canadian gladiolus society. Superseded by: Quarterly of the canadian gladiolus society. HI 63981

Canadian grower. Toronto. Vols. 70(2)-74(8), 1947-51. Canad. Grower. Preceded by: Canadian horticulture and home magazine. Growers' edition. Superseded by: Grower. Toronto. HI 63982

Canadian horticultural history; an interdisciplinary journal. Hamilton, Ont. Vol. 1+, 1986+. Canad. Hort. Hist. HI 63983

Canadian horticultural magazine. Montreal. Canad. Hort. Mag. See B–P–H 296/7. HI 53214

Canadian horticulture and home magazine. Fruit edition. Toronto. Vols. 38-61, 1915-38. Canad. Hort. Home Mag., Fruit Ed. Preceded by: Canadian horticulturist. Superseded by: Canadian horticulture and home magazine. Growers' edition. 2-909-1. HI 63984

Canadian horticulture and home magazine. Growers' edition. Toronto. Vols. 62-71(1), 1939-47. Canad. Hort. Home Mag., Growers' Ed. Preceded by: Canadian horticulture and home magazine. Fruit edition. Superseded by: Canadian grower. 2-909-1. HI 63985

Canadian horticulturist. Oshawa, Ont. Vols. 1-56, 1877-1933. Canad. Horticulturist. Superseded by: Canadian horticulture and home magazine. Fruit edition. 2-909-1. HI 63986

Canadian horticulturist and beekeeper. Brantford, Ont. Canad. Hort. Beekeeper. See B–P–H 296/6. HI 53213

Canadian iris society newsletter. Hannon, Willowdale, Ont. 1958+. Canad. Iris. Soc. Newslett. HI 63987

Canadian journal; a repertory of industry, science and art. Toronto. Canad. J. See B–P–H 296/8. HI 53215

Canadian journal = Proceedings of the royal canadian institute. Toronto. Proc. Roy. Canad. Inst. See B–P–H 735/11.

Canadian journal of agricultural science. Ottawa. Vols. 33-36, 1953-56. Canad. J. Agric. Sci. Preceded by: Scientific agriculture. Ottawa. Superseded by: Canadian journal of plant science and Canadian journal of soil science. 2-910-3. HI 63988

Canadian journal of biochemistry. Ottawa. Vols. 42-60, 1964-82? Canad. J. Biochem. Preceded by: Canadian journal of biochemistry and physiology. Superseded by: Canadian journal of biochemistry and cell biology. HI 63989

Canadian journal of biochemistry and cell biology. Ottawa. Vol. 61+, 1983+. Canad. J. Biochem Cell Biol. Preceded by: Canadian journal of biochemistry. HI 63990

Canadian journal of biochemistry and physiology. Ottawa. Vols. 32-41, 1954-63. Canad. J. Biochem. Physiol. Preceded by: Canadian journal of medical sciences [not entered]. Superseded by: Canadian journal of biochemistry and Canadian journal of physiology and pharmacology. HI 63991

Canadian journal of botany. Ottawa. Vol. 29+, 1951+. Canad. J. Bot. Preceded by: Canadian journal of research. Section C, botanical sciences. HI 63992

Canadian journal of chemistry. Ottawa Canad. J. Chem. See B–P–H 296/14. HI 53216

Canadian journal of earth sciences. Ottawa. Vol. 1+, 1964+. Canad. J. Earth Sci. HI 63993

Canadian journal of fisheries and aquatic sciences. Ottawa. Vol. 37+, 1980+. Canad. J. Fish. Aquatic Sci. Preceded by: Journal of the fisheries research board of Canada. HI 63994

Canadian journal of forest research. Ottawa. Vol. 1+, 1971+. Canad. J. Forest Res. HI 63995

Canadian journal of genetics and cytology. Ottawa. Vols. 1-28, 1959-86. Canad. J. Genet. Cytol. Preceded by: Proceedings, genetics society of Canada. Superseded by: Genome. HI 63996

Canadian journal of industry, sciences and art. Toronto. Canad. J. Industr. See B–P–H 296/16. HI 53217

Canadian journal of microbiology. Ottawa. Canad. J. Microbiol. See B–P–H 296/17. HI 53218

Canadian journal of physiology and pharmacology. Ottawa. Vol. 42+, 1964+. Canad. J. Physiol. Pharmacol. Preceded by: Canadian journal of biochemistry and physiology. HI 63997

Canadian journal of plant pathology. Ottawa. Vol. 1+, 1979+. Canad. J. Pl. Pathol. HI 63998

Canadian journal of plant science. Ottawa. Canad. J. Pl. Sci. See B–P–H 296/19. HI 53219

Canadian journal of public health. Toronto. Canad. J. Public Health. See B–P–H 296/21. HI 53220

Canadian journal of research. Ottawa. Canad. J. Res. See B–P–H 296/22. HI 53221

Canadian journal of research, section B. Chemical sciences. Ottawa. Canad. J. Res., Sect. B, Chem. Sci. See B–P–H 296/23. HI 53222

Canadian journal of research. Section C. Botanical sciences. Ottawa. Canad. J. Res., Sect. C, Bot. Sci. See B–P–H 296/24. HI 53223

Canadian journal of science, literature and history. Toronto. Vols. 12-15, 1870-78. Canad. J. Sci. Lit. Hist. Preceded by: Canadian journal of industry, science and art. Superseded by: Proceedings of the canadian institute HI 63999

Canadian journal of soil science. Ottawa. Vols. 37-62, 1957-62. Canad. J. Soil Sci. Preceded by: Canadian journal of agricultural science. 2-910-3. HI 64000

Canadian natural science news. Baden, Ont. Canad. Nat. Sci. News. See B–P–H 297/15. HI 53225

Canadian naturalist and geologist. Montreal. Vols. 1-8, n.s. vols. 1-3, 1856-66? Canad. Naturalist Geol. Superseded by: Canadian naturalist and quarterly journal of science. 2-914-1. HI 64001

Canadian naturalist and quarterly journal of science; [Subtitle varies]. [Issued with: Proceedings of the natural history society of Montreal.] Montreal. N.s. vols. 4-10, 1867?-83. Canad. Naturalist & Quart. J. Sci. Preceded by: Canadian naturalist and geologist. Superseded by: Canadian record of natural history and geology. 2-915-1. HI 64002

Canadian nature. Toronto. Canad. Nat. See B–P–H 297/12. HI 53224

Canadian nature art. Ottawa. 1976+. Canad. Nature Art. HI 64003

Canadian nurseryman. Burlington, Ont. Vols. 2-10(5), 1965?-73. Canad. Nurseryman. Preceded by: Nurseryman. Waterdown. Superseded by: Landscape Canada. HI 64004

Canadian orchid journal. St. Johns. Vol. 1+, 1981+. Canad. Orchid J. HI 64005

Canadian phytopathological society news = C P S news. Ottawa, Saskatoon.

Canadian plains proceedings. Regina. Vol. ?-2+, ?-1975+. Canad. Plains Proc. HI 75205

Canadian plant conservation programme newsletter. Edmonton, Alb. Vol. 1+, 1987+. Canad. Pl. Conservation Programme Newslett. Preceded by: Canadian botanic garden newsletter. HI 64006

Canadian plant disease survey. Ottawa. Vol. 40+, 1960+. Canad. Pl. Dis. Surv. Preceded by: Report (Annual) of the canadian plant disease survey. HI 64007

Canadian public health journal. Toronto. Canad. Public Health J. See B–P–H 297/21. HI 53226

Canadian record of natural history and geology. [Issued with: Proceedings of the natural history society of Montreal.] Montreal. 1 vol., 1884. Canad. Rec. Nat. Hist. Geol. Preceded by: Canadian naturalist and quarterly journal of science. Superseded by: Canadian record of science. 2-917-2. HI 64008

Canadian record of science. [Including: Proceedings of the natural history society of Montreal.] Montreal. Vols. 1-9, 1885-1916 [suspended 1905-13]. Canad. Rec. Sci. Preceded by: Canadian record of natural history and geology. 2-917-2. HI 64009

Canadian rosarian. Don Mills. Vol. 20+, 1975+. Canad. Rosarian. Preceded by: Rose bulletin of the canadian rose society. HI 64010

Canadian rose annual. Toronto. Canad. Rose Annual. See B–P–H 297/25. HI 53227

Canadian science digest. London, Ont. Canad. Sci. Digest. See B–P–H 297/27. HI 53228

Canadian science monthly = Acadian scientist, vol. 3. Wolfville, Nova Scotia. Acadian Sci. See B–P–H 37/13.

Cancer research. Chicago, IL. Cancer Res. See B–P–H 298/7. HI 53230

Candollea; organe du conservatoire et du jardin botaniques de la ville de Genève. Geneva. Candollea. See B–P–H 298/8. HI 53231

Cane grower's bulletin. New Delhi. Vols. 1-4(1), 1972-77. Cane Growers' Bull. Superseded by: Indian sugar crops journal. HI 64011

Cane growers quarterly bulletin. Indooroopilly (Brisbane), Qld. Vol. 1+, 1933+. Cane Growers' Quart. Bull. HI 64012

Canebrake agricultural experiment station bulletin. Montgomery, AL. Canebrake Agric. Exp. Sta. Bull. See B–P–H 298/9. HI 53232

Canopy. Laguna. Vols. 1-5(5), 1975-79. Canopy (Laguna). Superseded by: Canopy international. HI 64014

Canopy. London. Vol. 1+, 1969+. Canopy (London). HI 64013

Canopy international. Laguna. Vol. 5(6)+, 1979+. Canopy Int. Preceded by: Canopy. HI 64015

Canstatt's Jahresbericht über die Fortschritte in der Biologie. Erlangen. Canstatt's Jahresber. Fortschr.

Biol. See B–P–H 298/14. HI 53233

Canstatt's Jahresbericht über die Fortschritte in der Pharmacie und verwandten Wissenschaften. Erlangen & Würzburg. Canstatt's Jahresber. Fortschr. Pharm. Verwandten Wiss. See B–P–H 298/15. HI 53234

Canstatt's Jahresbericht über die Leistungen in den physiologischen Wissenschaften. Erlangen. Canstatt's Jahresber. Leist. Physiol. Wiss. See B–P–H 298/16. HI 53235

Canterbury botanical society journal = Journal, Canterbury botanical society. Christchurch, N.Z.

Caoutchouc et la gutta-percha. Paris. Caoutchouc & Gutta-Percha. See B–P–H 298/19. HI 53236

Cape Haze marine laboratory; collected papers. Sarasota, FL. Cape Haze Mar. Lab. Collect. Pap. See B–P–H 298/22. HI 53237

Cape monthly magazine. Cape Town. Vols. 1-9, 1857-61; n.s. vols. 1-18, 1870-79; n.s. [2] vols. 1-4, 1879-81. Cape Monthly Mag. Superseded by: Cape quarterly review [not entered]. HI 64016

Cape naturalist. Cape Town. Cape Naturalist. See B–P–H 298/24. HI 53238

Cape naturalist, Cape Cod museum of natural history. Brewster, MA. 1972+. Cape Naturalist, Cape Cod Mus. Nat. Hist. HI 64017

Capitula; publicación destinada a difundir trabajos breves y comunicaciones que signifiquen nuevos avances en el conocimiento de la familia Asteraceae. La Plata. No. 1+, 1984+. Capitula. HI 64018

Capsicum newsletter. Turin. No. 1+, 1982+. Capsicum Newslett. HI 64019

Caradoc record of bare facts. Shrewsbury. 1892-94. Caradoc Rec. Bare Facts. Superseded by: Record of bare facts. Caradoc and Severn Valley field club. HI 64020

Carbohydrate research. Amsterdam. Carbohyd. Res. See B–P–H 298/25. HI 53239

Caribbean; the newsletter of the caribbean commission. Hato Rey, PR. Caribbean (Puerto Rico). See B–P–H 298/31. HI 53241

Caribbean. Port-of-Spain, Trinidad. Caribbean (Trinidad). See B–P–H 299/1. HI 53242

Caribbean agriculture. Hato Rey, PR. Vols. 1-3(1), 1962-64. Caribbean Agric. HI 64021

Caribbean agriculture and science. St. Croix, Virgin Islands. Vol. 1+, 1966+. Caribbean Agric. & Sci. HI 64022

Caribbean conservation news. St. Michael, Barbados. Vol. 1+, 1975+. Caribbean Conservation News. HI 64023

Caribbean farming. Kingston, Jamaica. No. 1+, 1969+. Caribbean Farming. HI 64024

Caribbean forester. Rio Piedras, PR. Vols. 1-24(2), 1939-65. Caribbean Forester. 2-927-2. HI 64025

Caribbean journal of science. San Juan, PR. Caribbean J. Sci. See B–P–H 299/5. HI 53243

Caribbean journal of science and mathematics. St. Thomas, Cullowhee, NC. Vol. 1+, 1968+. Caribbean J. Sci. Math. HI 64026

Caribbean studies. Río Piedras, PR. Caribbean Stud. See B–P–H 299/6. HI 53244

Carinthia. Klagenfurt, Austria. Carinthia. See B–P–H 299/7. HI 53245

Carinthia II. Klagenfurt, Austria. Carinthia II. See B–P–H 299/8. HI 53246

Carlsberg research communications. Copenhagen. Vol. 41+, 1976+. Carlsberg. Res. Commun. Preceded by: Comptes rendus des travaux du Carlsberg laboratoriet. HI 64027

Carnation craft. Chicago, IL. Carnation Craft. See B–P–H 299/9. HI 53247

Carnation year book. London. Carnation Year Book. See B–P–H 299/11. HI 53248

Carnegie institution of Washington year book. Washington, DC. Carnegie Inst. Wash. Year Book. See B–P–H 299/14. HI 53249

Carnegie magazine. Pittsburgh, PA. Vol. 1+, 1927+. Carnegie Mag. HI 64029

Carnegie museum botany pamphlet. Pittsburgh, PA. Carnegie Mus. Bot. Pam. See B–P–H 299/18. HI 53250

Carnet del vinicultor. Montevideo. Carnet Vinic. See B–P–H 299/19. HI 53251

Carniola. Laibach [=Ljubljana, Yugoslavia]. Carniola. See B–P–H 299/20. HI 53252

Carnivorous plant newsletter. Statesville, Fullerton, CA. Vol. 1+, 1972+. Carniv. Pl. Newslett. HI 64030

Carnivorous plant society journal. Rusthall. Vol. 1+, 1979+. Carniv. Pl Soc. J. HI 64031

Carnivorous plants digest. Kelly Corners, NY. Vol. 1-2, 1978-80. Carniv. Pl. Digest. HI 64032

Carolina camellias. Columbia, SC. Carolina Camellias. See B–P–H 299/22. HI 53254

Carolina journal of medicine, science and agriculture. Charleston, SC. Carolina J. Med. See B–P–H 299/23. HI 53255

Carolina tips. Elon College, NC. Vol. 1+, 1938+. Carolina Tips. 2-937-3. HI 64033

Carolinea; Beiträge zur naturkundlichen Forschung in

Südwestdeutschland. Karlsruhe. Vol. 40+, 1982+. Carolinea. Preceded by: Beiträge zur naturkundlichen Forschung in Südwestdeutschland. HI 64034

Carpaţii. Cluj, Rumania. Carpaţii. See B–P–H 299/24. HI 53256

Carreta. San José, Costa Rica. Carreta. See B–P–H 299/25. HI 53257

Carta informativa, asociacion latinoamericana de educacion agricola superior. Mexico D.F. No. 1+, 1968+. Carta Inform. Asoc. Latinoamer. Educ. Agric. Super. HI 64035

Carta informativa, estación científica Charles Darwin. Santa Cruz, Galapagos. No. ?-11+, 19??-83+. Carta Inform. Estación. Ci. Charles Darwin. HI 64036

Carta informativa, instituto salvadoreño de investigaciones del café. Nueva, San Salvador. Vol. 1+, 1978+. Carta. Inform. Inst. Salvadoreño Invest. Café. HI 64037

Cartilla, ministeria de fomento, dirección de agricultura, ganadería y colonización. Lima. Vols. ?-26-33?, 19??-36-40? Cartilla Minist. Fomento Direccion Agric. Ganad. Coloniz. HI 64038

Cary arboretum of the New York botanical garden; newsletter published for friends of the arboretum by the public affairs department. Millbrook, NY. Vol. 1+, 1975+. Cary Aboretum, Newslett. HI 64039

Cary arboretum of the New York botanical garden [report]. Millbrook, NY. "The first 4 years, 1971/75"; year 5+, 1975/76+. Cary Aboretum, Rep. HI 64040

Caryologia; giornale di citologia, citosistematica e citogenetica. Pisa. Caryologia. See B–P–H 300/1. HI 53258

Cascadian = Bulletin of the Berry botanical garden. Portland, OR.

Case's botanical index. Richmond, IN = L B Case's botanical index; an illustrated quarterly botanical magazine. Richmond, IN. L. B. Case's Bot. Index. See B–P–H 523/1.

Casket. [Vol. [4], (1829) with the subtitle: or, flowers of literature, wit & sentiment.] Philadelphia, PA. Vols. 1-5, 1825-30; vols. 14(5)-17, 1839-40. Casket (Philadelphia). For 1831-1839(4) see: Atkinson's casket. Superseded by: Graham's lady's and gentleman's magazine. 2-1753-2. HI 64041

Časopis českého lékárnictva. Prague. Čas. Českého Lékárn. See B–P–H 300/5. HI 53259

Časopis českého museum. Prague. Čas. Českého Mus. See B–P–H 300/6. HI 53260

Časopis českého ovocnictví. Prague-Troja. Čas. Českého Ovocn. See B–P–H 300/7. HI 53261

Časopis československých houbařů. Prague. 1919-34. Čas. Českoslov. Houb. Superseded by: Mykologický sbornik. 3-2808-2. HI 64042

Časopis československého lékárnictva. Prague. Čas. Českoslov. Lékárn. See B–P–H 300/9. HI 53262

Časopis českých zahradníku. Prague. Čas. Českých Zahradn. See B–P–H 300/10. HI 53263

Časopis pre farmaceuticku vedu a prax = Farmaceutický obzor. Bratislava.

Časopis lékařuv českých. Prague. Čas. Lékařuv Českých. See B–P–H 300/17. HI 53264

Časopis matice moravské. Brünn [=Brno, Czechoslovakia]. Čas. Matice Morav. See B–P–H 300/19. HI 53265

Časopis moravského musea v Brně. Vědy přírodni. Brno. Vols. 34+, 1949+. Čas. Morav. Mus. Zemsk. Preceded by: Časopis zemského musea v Brně. 1-795-3. HI 64043

Časopis moravského zemského musea v Brně. Brno. 1901-32; Čast. 2, prirodovedna 1933-46. Čas. Morav. Zemsk. Brně. Superseded by: Časopis zemského musea v Brně. 1-795-3. HI 64044

Časopis musea království českého. Oddíl přírodovědný. Prague. Čas. Mus. Král. Českého, Odd. Přír. See B–P–H 300/29. HI 53266

Časopis musejního (muzehjního) spolku olomouckého. Olmütz, Moravia [=Olomouc, Czechoslovakia]. Čas. Mus. Spolku Olomouck. See B–P–H 300/31. HI 53267

Časopis muzeálnej (museálnej) slovenskej společnosti. Turocszentmarton, Hungary [=Turciansky Svaty Martin, Czechoslovakia]. Čas. Mus. Slov. Společn. See B–P–H 301/2. HI 53268

Časopis národního musea. Oddíl přírodovědný. Prague. Čas. Nár. Mus., Odd. Přír. See B–P–H 301/3. HI 53269

Časopis ovocnického spolku pro království České. Prague. Čas. Ovocn. Spolku Král. České. See B–P–H 301/6. HI 53270

Časopis slezského musea. Ser. A, historia naturalis. Opava. Vol. 5+, 1956+. Čas. Slez. Mus., Ser. A, Hist. Nat. Preceded by: Časopis slezského musea v Opavě. Ser. A, historia naturalis. HI 64045

Časopis slezského musea. Ser. A, vedy prirodni = Časopis slezského musea. Ser. A, historia naturalis. Opava.

Časopis slezského musea v Opavě. Ser. A. Historia naturalis. Opava, Czechoslovakia. Čas. Slez. Mus. v Opavě, Ser. A, Hist. Nat. See B–P–H 301/8. HI 53271

Časopis slezského musea v Opavă. Série C, dendrologia. Opava. Nos. 1-28, 1962-79. Čas. Slez. Mus. Opava,

C, Dendrol. Continued in: Časopis slezského musea. Ser. A. Historia naturalis. HI 64046

Časopis společnosti vlasteneckého museum w (v) Čechách. Prague. Čas. Společn. Vlasten. Mus. w Čechách. See B–P–H 301/9. HI 53273

Časopis spolku pěstitelú kaktusú v R Č S. Prague. Vol. 1, 1925. Čas. Spolku Pěstitelú Kaktusú R Č S. Superseded by: Kaktusarske listy. HI 64047

Časopis vědecko-zahradnický = Flora. Prague. Flora (Prague). See B–P–H 374/9.

Časopis vlasteneckého musejního (muzejního) spolku olomouckého. Olmütz, Moravia [=Olomouc, Czechoslovakia]. Čas. Vlasten. Mus. Spolku Olomouc. See B–P–H 301/11. HI 53274

Časopis vlasteneckého spolku musejního v Olomouci. Olmütz, Moravia [=Olomouc. Czechoslovakia]. See B–P–H 301/13. HI 53275

Časopis zemského musea v Brně. Brno. Vols. 31-33, 1947-48. Čas. Zemsk. Mus. Brně. Preceded by: Časopis moravskeho zemskeho musea v Brně. Superseded by: Časopis moravského musea v Brně. Vědy přírodni. HI 64048

Časopis zemského spolku štěpařkého pro království České. Prague. Čas. Zemsk. Spolku Štěp. Král. Řeské. See B–P–H 301/14. HI 53276

Casas y jardines. Buenos Aires. Casas & Jard. See B–P–H 301/15. HI 53277

Cassava newsletter. Cali. No. 1+, 1977+. Cassava Newslett. HI 64049

Cassava program annual report = Report (Annual), cassava program.

Castalia; rivista di storia della medicina. Milan. Vols. 1-21, 1945-65. Castalia. Superseded by: Episteme. HI 64050

Castanea; journal of the southern appalachian botanical club. Morgantown, WV. Castanea. See B–P–H 301/21. HI 53278

Cat-tales; environmental education center news. Somerville, NJ. Vol. 1+, 1973?+. Cat-Tales. HI 64051

Catalog of landscape records in the United States newsletter. Bronx, NY. Vol. ?-1(2)+, 19??-87+. Cat. Landscape Rec. U.S. Newslett. HI 64052

Catalog de seminţe. Note de botanică = Note botanice. Cluj.

Catalogo ilustrado de las plantas de cundinamarca Bogota. Bogota. Vol. 1+, 1966+. Cat. Ill. Pl. Cundinamarca Bogota. HI 64053

Catalogue of fossil spores and pollen. [Palynological laboratory, Pennsylvania state university.] University Park, PA. Cat. Fossil Spores Pollen. See B–P–H 301/22. HI 53279

Catalogue [des] plantes tropicales. Villers-lès-Nancy. 1979+. Cat. Pl. Trop. HI 64054

Catalyst; quarterly publication of the Belle W. Baruch foundation. New York. Vol. 1(1), 1966. Catalyst. Superseded by: Conservation catalyst. HI 64055

Catalyst for environment/energy. New York. Vol. 6(4)+, 1979+. Catalyst Environm./Energy. Preceded by: Catalyst for environmental quality. HI 64056

Catalyst for environmental quality. New York. Vols. 1-6(3), 1970-78. Catalyst Environm. Qual. Preceded by: Conservation catalyst. Superseded by: Catalyst for environment/energy. HI 64057

Catholic taehac uihak-bu nonmunjip = Holy Ghost medical college theses. [Korea]. Holy Ghost Med. Coll. Theses. See B–P–H 419/27.

Catholic university of America. Biological series. Washington, DC. Catholic Univ. Amer., Biol. Ser. See B–P–H 302/3. HI 53280

Causeries scientifiques. Paris. Causeries Sci. See B–P–H 302/7. HI 53281

Cavanillesia, rerum botanicarum acta. Barcelona. Cavanillesia. See B–P–H 302/8. HI 53282

Cawthron institute annual report = Report (Annual), Cawthron institute. Nelson, N.Z.

Caza fotografica; revista del instituto de la caza fotografica y ciencias de la naturaleza. Madrid. Nos. 0 [introductory number], 1-?, 1973-? Caza Fotogr. Superseded by: Periplo. HI 64059

Cecidologia indica; bulletin of the cecidological society of India. Allahabad. Vols. 1-14, 1966-79. Cecidol. Indica. Superseded by: Cecidologia internationale. HI 64060

Cecidologia internationale. Allahabad. Vol. 1+, 1980+. Cecidol. Int. Preceded by: Cecidologia indica. HI 64061

Cedro; revista del instituto de estudios de jardineria y arte paisajista. Madrid. Vols. 1-3-?, 1954-56-? Cedro. HI 64062

Ceiba; a scientific journal issued by the Escuela agricola panamericana. Tegucigalpa, Honduras. Ceiba. See B–P–H 302/13. HI 53284

Celas bulletins. Paramaribo. No. 1+, 1966+. Celas Bull. HI 64063

Cell. Cambridge, MA. Vol. 1+, 1974+. Cell (Cambridge). HI 64064

Cell biochemistry and function. Chichester. Vol. 1+, 1983+. Cell Biochem. Function. HI 64065

Cell biology international reports. London, etc. Vol. 1+, 1977+. Cell Biol. Int. Rep. HI 64066

Cell biophysics; an international journal. Clifton, NJ. Vol. 1+, 1979+. Cell Biophys. HI 64067

Cell chromosome newsletter. Calcutta. Vols. 1-4, 1978-81. Cell Chromosome Newslett. Superseded by: Cell and chromosome research. HI 64068

Cell and chromosome research. Calcutta. Vol. 5+, 1982+. Cell Chromosome Res. Preceded by: Cell and chromosome newsletter. HI 64069

Cell differentiation; international journal for the rapid publication of original research papers on the aspects of cellular differentiation in eukaryotes. Amsterdam. Vols. 1-4?, 1972-87? Cell Different. Superseded by?: Cell differentiation and development. HI 64070

Cell differentiation and development; official journal of the international society of developmental biologists. Shannon. Vol. 5+, 1988+. Cell Different. Developm. ?Preceded by: Cell differentiation. HI 64071

Cell membranes, methods and reviews. New York. Vol. 1+, 1983+. Cell. Membr. Meth. Rev. Preceded by: Methods in membrane biology. HI 64072

Cell monograph series. Cambridge, MA. Vol. 1+, 1977+. Cell Monogr. Ser. HI 64073

Cell motility. New York. Vol. 1+, 1980+. Cell Motility. HI 64074

Cell regulation. Bethesda, MD. Vol. 1+, 1989+. Cell Regulat. HI 74822

Cell structure and function. Okayama. Vol. 1+, 1975+. Cell Struct. Function. HI 64075

Cell and tissue kinetics. Oxford. Vols. 1-7, 1968-74. Cell Tissue Kinet. HI 64076

Cell and tissue research. Berlin, etc. Vol. 148+, 1974+. Cell Tissue Res. Preceded by: Zeitschrift für Zellforschung und mikroskopische Anatomie. HI 64077

Cellular and molecular biology. Elmsford, NY. Vol. 22+ 1977+. Cell. Molec. Biol. Preceded by: Annales d'histochemie. HI 64078

Cellular signalling. Oxford, Elmsford, NY. Vol. 1+, 1989+. Cell. Signalling. HI 64079

Cellule; recueil de cytologie et d'histologie générale. Louvain. Cellule. See B–P–H 302/15. HI 53285

Cellulosa e carta. Rome. Cellulosa & Carta. See B–P–H 302/16. HI 53287

Cellulose-Chemie. Berlin. Cellulose-Chem. See B–P–H 302/18. HI 53288

Cellulose chemistry and technology. Bucharest. Vol. 1+, 1967+. Cellulose Chem. Technol. HI 75177

Čelovek i priroda. Moscow & Leningrad = Chelovek i priroda. Moscow & Leningrad.

Cenicafé. Chinchiná. Vol. 8+, 1957?+. Cenicafé. Preceded by: Boletín informativo, centro nacional de investigaciones de café. 2-955-2. HI 64080

Cenni storici del museo civico di storia naturale di Trieste. Trieste, Austria [Italy]. Cenni Storici Mus. Civico Storia Nat. Trieste. See B–P–H 301/21. HI 53289

Census report, forestry commission. London. Nos. 1-5, 1952-53. Census Rep. Forest. Commiss. HI 64081

Centaurus; internatonal magazine of the history of science and medicine. Copenhagen. Centaurus. See B–P–H 302/22. HI 53290

Center for plant conservation [newsletter]. Jamaica Plain, NY. Vol. 1+, 1986+. Center Pl. Conservation. HI 64083

Central african planter. Songani, Zomba, Blantyre. Vols. 1-2, 1895-97. Centr. African Planter. HI 64084

Central mediterranean naturalist. Sliema. Vol. 1+, 1979+. Centr. Medit. Naturalist. Preceded by: Maltese naturalist. HI 64085

Central plains newsletter. Manhattan, KS. 1951+. Centr. Plains Newslett. HI 64086

Centralblatt für allgemeine Pathologie und pathologische Anatomie. Jena. Centralbl. Allg. Pathol. Pathol. Anat. See B–P–H 302/27. HI 53291

Centralblatt für allgemeine Pathologie und pathologische Anatomie. Ergänzungsheft = Verhandlungen des Deutschen Pathologischen Gesellschaft. Berlin. Verh. Deutsch. Pathol. Ges. See B–P–H 950/8.

Centralblatt für Bakteriologie und Parasitenkunde. Jena. Centralbl. Bakteriol. Parasitenk. See B–P–H 303/3. HI 53292

Centralblatt für Bakteriologie, Parasitenkunde und Infektionskrankheiten. 1 Abteilung, medizinisch-hygienische Bakteriologie, Virusforschung und tierische Parasitologie. Originale. [From 1932: Centralblatt as Zentralblatt.] Jena. Vols. 31-151, 1902-45. Centralbl. Bakteriol., 1. Abt., Originale. Preceded by: Centralblatt für Bakteriologie, Parasitenkunde und Infektionskrankheiten. Abtheilung 1, medizinisch-hygienische Bakteriologie, Virusforschung und tierische Parasitologie. Superseded by: Zentralblatt für Bakteriologie, Parasitenkunde, Infektionskrankheiten und Hygiene. 1 Abteilung, medizinisch-hygienische Bakteriologie, Virusforschung und tierische Parasitologie. Originale. 5-4626-3. HI 64087

Centralblatt für Bakteriologie, Parasitenkunde und Infektionskrankheiten. 1 Abteilung. Referate. Medizinisch-hygienische Bakteriologie, Virusforschung und Parasitologie. [From 1932: Centralblatt as Zentralblatt.] Jena, Stuttgart. Vols. 31-270, 1902-81 [suspended 1945-47]; Supplementheft vols. 1-?, 1965-? Centralbl. Bakteriol., 1. Abt., Ref. Preceded by:

Centralblatt für Bakteriologie, Parasitenkunde und Infektionskrankheiten. 1 Abtheilung. Medizinisch-hygienische Bakteriologie, Virusforschung und tierische Parasitologie. Superseded by: Zentralblatt für Bakteriologie, Mikrobiologie und Hygiene. 1 Abteilung. Referate. Medizinische Mikrobiologie, Parasitologie, Hygiene, präventive Medizin. 5-4627-1. HI 64088

Centralblatt für Bakteriologie und Parasitenkunde. Erste Abheilung. Medizinisch-hygienische Bakteriologie, Virusforschung und tierische Parasitologie. Jena. Centralbl. Bakteriol. Parasitenk., 1. Abth. See B–P–H 303/4. HI 53293

Centralblatt für Bakteriologie und Parasitenkunde. Zweite Abheilung. Jena. Centralbl. Bakteriol. Parasitenk., 2. Abth. See B–P–H 303/5. HI 53294

Centralblatt für Bakteriologie, Parasitenkunde und Infektionskrankheiten. Erste Abtheilung. Medizinisch-hygienische Bakteriologie, Virusforschung und tierische Parasitologie. Jena. Centralbl. Bakteriol., 1. Abth. See B–P–H 303/8. HI 53295

Centralblatt für Bakteriologie, Parasitenkunde und Infektionskrankheiten. Zweite Abtheilung. Jena. Centralbl. Bakteriol., 2. Abth. See B–P–H 303/9. HI 53296

Centralblatt für Chirurgie. Leipzig. Centralbl. Chir. See B–P–H 303/10. HI 53297

Centralblatt für das gesammte Forstwesen. Vienna & Munich. Centralbl. Gesammte Forstwesen. See B–P–H 303/13. HI 53298

Centralblatt für die gesammte Landescultur des In- und Auslandes. Prague. Centralbl. Gesammte Landescult. In- Ausl. See B–P–H 303/14. HI 53300

Centralblatt des landwirthschaftlichen Vereins in Bayern. Munich. Centralbl. Landw. Vereins Bayern. See B–P–H 303/16. HI 53301

Centralblatt für Landwirthschaft und verwandte Gewerbe. Leipzig. Centralbl. Landw. Verwandte Gewerbe. See B–P–H 304/1. HI 53302

Centralblatt für die mährischen Landwirthe. Brünn [=Brno, Czechoslovakia]. Centralbl. Mähr. Landwirthe. See B–P–H 304/2. HI 53303

Centralblatt für Mineralogie, Geologie und Paläontologie in Verbindung mit dem Neuen Jahrbuch für Mineralogie, Geologie und Paläontologie. Stuttgart. Centralbl. Mineral. See B–P–H 304/3. HI 53304

Centralblatt für Mineralogie, Geologie und Paläontologie in Verbindung mit dem Neuen Jahrbuch für Mineralogie, Geologie und Paläntologie. Abteilung B. Geologie und Paläntologie. Stuttgart. Centralbl. Mineral., Abt. B, Geol. Paläontol. See B–P–H 304/4. HI 53305

Centralblatt für Sammlung und Veröffentlichung von Einzeldiagnosen neuer Pflanzen = Repertorium specierum novarum regni vegetabilis. Berlin. Repert. Spec. Nov. Regni Veg. See B–P–H 772/20.

Centralblatt für Sammlung und Veröffentlichung von Einzeldiagnosen neuer Pflanzen. Beihefte. = Repertorium specierum novarum regni vegetabilis. Vol. 1, 1914. Berlin. Repert. Spec. Nov. Regni Veg. Beih. See B–P–H 772/21.

Centre national de géologie houillère. Publication. Brussels. Centr. Natl. Géol. Houillère Publ. See B–P–H 304/5. HI 53306

Centro de estudiantes de agronomía. Buenos Aires. Vol. 24, 1931. Centro Estud. Agron. Preceded by: Revista del centro de estudiantes de agronomía y veterinaría. Superseded by: Agronomía (Buenos Aires). 1-93-1. HI 64089

Cercetari marine. Constanta. No. 1+, 1971+. Cercet. Mar. HI 64090

Cereal chemistry. St. Paul, MN. Cereal Chem. See B–P–H 304/7. HI 53307

Cereal foods world. Minneapolis, MN. Vol. 20+, 1975+. Cereal Foods World. Preceded by: Cereal science today. HI 64091

Cereal news. Christchurch, N.Z. = D S I R cereal news. Christchurch, N.Z.

Cereal news, research branch, department of agriculture, Canada. Ottawa. Vols. 1-15(1), 1952-71. Cereal News., Dept. Agric., Canada. HI 64092

Cereal rust bulletin. St. Paul, MN. 1963+. Cereal Rust Bull. (St. Paul). HI 64093

Cereal rusts bulletin. Wageningen. No. 1+, 1973+. Cereal Rusts Bull. (Wageningen). HI 64094

Cereal rusts news from everybody to everybody = Robigo. Castelar.

Cereal science today. Minneapolis, MN. Vols. 1-19, 1956-74. Cereal Sci. Today. Preceded by: Transactions, american association of cereal chemists. Superseded by: Cereal foods world. HI 64095

Cereals; annual report on field experiments. Bracknell. 1979+, [1980?]+. Cereals. HI 64096

Ceres; F A O review on agriculture and development. Rome. Vol. 1(2)+, 1968+. Ceres (Rome). Preceded by: F A O review. HI 64097

Ceres; revista de agricultura. São Paulo. Vol. ?-2(9 &10)+, 19??-27+. Ceres (São Paulo). HI 64098

Cerrado. Abstracts = Cerrado. Resumos informativos. Brasilia.

Cerrado. Bibliografia analítica. Brasilia. Vol. 1, 1976. Cerrado, Bibliograf. Analítica. Superseded by: Cerrado. Resumos informativos. HI 64099

Cerrado. Resumos informativos. Brasilia. Vol. 2+, 1979+. Cerrado, Resumos Inform. Preceded by: Cerrado. Bibliografia analítica. HI 64100

Česká dermatologie. Prague. Česká Dermatol. See B–P–H 304/16. HI 53308

Česká flora. Prague. Česká Fl. See B–P–H 304/17. HI 53310

Česká mykologie. Prague. Česká Mykol. See B–P–H 304/19. HI 53311

Česká věda. Prague. Česká Věda. See B–P–H 304/20. HI 53313

Česká zemědělska bibliografie. Prague. Vols. 1-9, 1941-49. Česká Zeměd. Bibliogr. 2-969-3. HI 64101

České lesnické rozhledy. Pisek, Bohemia [Czechoslovakia]. České Lesn. Rozhl. See B–P–H 304/21. HI 53314

České listy hospodářské. Prague. České Listy Hospod. See B–P–H 304/22. HI 53315

Ceské listy zahradnické. Prague. 1904-09. Ceské Listy Zahradn. Preceded by: Listy zahradnické. Superseded by: Ovocnické rozhledy. HI 64102

České zahradnické listy. Prague. České Zahradn. Listy. See B–P–H 304/24. HI 53316

Československá biologie. Prague. Českoslov. Biol. See B–P–H 304/26. HI 53317

Československá mikrobiologie. Prague. Českoslov. Mikrobiol. See B–P–H 3304/28. HI 53320

Československá ochrana prírody. Bratislava. Českoslov. Ochr. Prír. See B–P–H 305/1. HI 53321

Československé botanické listy. Prague. Českoslov. Bot. Listy. See B–P–H 304/27. HI 53319

Československé zahradnické listy. Prague. Českoslov. Zahradn. Listy. See B–P–H 305/3. HI 53323

Československý zahrandník. Prague. Českoslov. Zahradn. See B–P–H 305/2. HI 53322

Český včelař. Prague. Český Včelař. See B–P–H 305/5. HI 53324

Český vinař. Dolni Berkovice, Bohemia [Czechoslovakia]. Český Vinař. See B–P–H 305/6. HI 53325

Český vinař a věstník vinařského spolku okolí Mělníka. Dolni Berkovice, Bohemia [Czechoslovakia]. Český Vinař Věstn. Vinařsk. Spolku Okolí Mělníka. See B–P–H 305/7. HI 53326

Cespedesia; boletin cientifico del departamento del valle de Cauca. Cali. Vol. 1+, 1972+. Cespedesia. HI 64103

Cévennes; revue du parc national des Cévennes. Florac. Vol. 1+, [1973]+. Cévennes. HI 64104

Ceylon administration reports. Agriculture = Administration report of the director of agriculture, Ceylon. Colombo.

Ceylon coconut journal. Colombo, Ceylon. Ceylon Coconut J. See B–P–H 305/9. HI 53327

Ceylon coconut quarterly. Lunuwila. Vols. 1-?, 1950-82? Ceylon Coconut Quart. Superseded by: Cocos. HI 64105

Ceylon forester. Colombo. 1895-98; Vols. 1-10(1/2), 1955-71. Ceylon Forest. Superseded by: Sri Lanka forester. HI 64106

Ceylon journal of science. Biological sciences. Colombo, Ceylon. Ceylon J. Sci., Biol. Sci. See B–P–H 305/11. HI 53328

Ceylon journal of science. Section A. botany. Colombo, Ceylon. Ceylon J. Sci., Sect. A, Bot. See B–P–H 305/12. HI 53329

Ceylon tea review. Colombo, Ceylon. Ceylon Tea Rev. See B–P–H 305/16. HI 53330

Chácaras e quintais. São Paulo. Chácaras & Quintais. See B–P–H 305/17. HI 53331

Chagyo gijutsu kenkyu = Study of tea. Shizuoka, Japan. Study Tea. See B–P–H 857/15.

Chagyo kenkyu hokoku = Research report of tea.

Chagyo shikenjo kenkyu hokoku = Bulletin, national research institute of tea. Shizuoka.

Chagyo shikenjo. Nyusu. Kanaya. No. 1+, 1969+. Chagyo Shikinjo, Nyusu. HI 64107

Chairman's report, Gwent trust for nature conservation. Newport, Gwent. 1973/74+, 1974+. Chairm. Rep., Gwent Trust Nat. Conservation. HI 64108

Champignon. Erlangen. Champignon (Erlangen). See B–P–H 305/22. HI 53332

Champignon. [Netherlands]. Champignon (Netherlands). See B–P–H 305/23. HI 53333

Champignoncultuur. Contactorgaan van de nederlandse champignonkwekers. Wageningen. Champignoncultuur. See B–P–H 305/24. HI 53334

Champignonteelt. Wageningen. Champignonteelt. See B–P–H 305/25. HI 53335

Changing scene; a review of natural history in the north west of England. The joint transactions of the Eden field club, Penrith and district natural history society, Kendal natural history society, the Grange natural history society, and the Ambleside field club. Penrith, Kirby Lonsdale. 1957-66. Changing Scene. HI 64109

Changing seasons; an occasional journal devoted to botanical research, exploration, conservation and horticulture in California. Orinda, CA. Vol. 1+, 1979+. Changing Seasons. HI 64110

Chanousia. Rome. Vols. 1-3, 1928-37. Chanousia. 2-977-3. HI 64111

Chanoyu = Chanoyu quarterly. Kyoto.

Chanoyu quarterly. Kyoto. No. 1+, 1970+. Chanoyu Quart. HI 64112

Chaparra agrícola. San Manuel, Cuba. Chaparra Agríc. See B–P–H 306/1. HI 53337

Chapingo, Mexico City. 1945-64; n.s. nos. 1-31/32, 1976-81. Chapingo. Superseded by: Revista Chapingo. HI 64113

Chapters in the history of science. London. Chapt. Hist. Sci. See B–P–H 306/4. HI 53339

Charité-Annalen. Berlin. Charité-Ann. See B–P–H 306/6. HI 53340

Chatter; journal of the african violet society of Canada. Various places. 1955?+. Chatter. HI 64114

Chayon = Science. Seoul.

Chayon taehak. Soul taehakkyo = Proceedings of the college of natural sciences. Section 4, life (biological) sciences. Seoul.

Chayon poho = Conservation of nature. Seoul.

Chayon pojon = Nature conservation. Seoul.

Che-chiang lin hsüeh yüan hsüeh pao = Journal of Zhejiang forestry college. Che-chiang Lin-an.

Che chiang t'u shu kuan pao = Journal of the Chekiang provincial library. Hangchow, China. J. Chekiang Prov. Libr. See B–P–H 461/17.

Cheekwood mirror. Nashville, TN. 1961-81. Cheekwood Mirror. Superseded by: Cheekwood quarterly. HI 64115

Cheekwood quarterly. Nashville, TN. 1981+. Cheekwood Quart. Preceded by: Cheekwood mirror. HI 64116

Chekiang agricultural quarterly. [Che-tai nung hsüeh chi k'an.] Peiping [=Peking]. Chekiang Agric. Quart. See B–P–H 306/12. HI 53341

Chekiang sheng k'un ch'ung chü nien k'an = Year book of the bureau of entomology, Hangchow. Hangchow, China. Yearb. Bur. Entomol. Hangchow. See B–P–H 982/16.

Chelovek i biosfera; materialy, zasedaniya rabuchei gruppy po proektu No.18, vid i ego produktivnost' v areale. Vilnius. Vol. 1+, 1975+. Chelovek Biosfera. HI 64117

Chelovek i priroda. Moscow & Leningrad. 1920-26. Chelovek & Prir. 2-983-2. HI 64118

Cheltenham and district naturalists' society journal. Cheltenham. Vols. 1-7, 1950-56. Cheltenham Distr. Naturalists' Soc. J. Superseded by: North Gloucestershire naturalists' society journal. HI 64119

Cheltenham magazine. London. Cheltenham Mag. See B–P–H 306/16. HI 53342

Chemical abstracts. Easton, PA, Washington, DC. Vols. 1-51, 1907-62. Chem. Abstr. Superseded by: Chemical abstracts - biochemistry sections. 2-983-2. HI 64120

Chemical abstracts - biochemistry sections. Columbus, OH. Vol. 52+, 1963+. Chem. Abstr., Biochem. Sect. Preceded by: Chemical abstracts. HI 64121

Chemical abstracts review titles = C A review titles.

Chemical abstracts selects = C A selects. Columbus, OH.

Chemical-biological activities; index to current literature on the biological activity of organic compounds. Easton, PA. Sample issue, 1962; Vol. 1+, 1965+. Chem.-Biol. Activities. HI 64122

Chemical and pharmaceutical bulletin. Tokyo. Vol. 6+, 1958+. Chem. Pharm. Bull. Preceded by: Pharmaceutical bulletin. Tokyo. HI 64123

Chemical plant taxonomy newsletter. Cambridge, Liverpool, & Reading, PA. No. 1+, 1964+. Chem. Pl. Taxon. Newslett. HI 64124

Chemical regulation in plants. [Shokubutsu no kagaku chosetsu.] Tokyo. Chem. Regulat. Pl. See B–P–H 307/2. HI 53349

Chemical reviews. Baltimore, MD. Chem. Rev. See B–P–H 307/3. HI 53350

Chemico-biological interactions; international journal devoted to studies of the mechanisms by which exogenous chemicals produce changes in biological systems. Amsterdam. Vol. 1+, 1969+. Chem.-Biol. Interact. HI 64125

Chemie der Pflanzenschutz- und Schädlingsbekämpfungsmittel. Berlin, etc. Vol. 1+, 1970+. Chem. Pflanzenschutz Schädlingsbekämpf. HI 64126

Chemie der Zelle und Gewebe. Berlin. Chem. Zelle Gewebe. See B–P–H 307/5. HI 53351

Chemisch-pharmaceutisch archief. Schoonhoven, Netherlands. Chem.-Pharm. Arch. See B–P–H 306/27. HI 53348

Chemische Annalen für die Freunde der Naturlehre, Arzneygelahrtheit, Haushaltungskunst und Manufacturen. Helmstedt & Leipzig. Chem. Ann. Freunde Naturl. See B–P–H 306/21. HI 53344

Chemische Berichte. Heidelberg & Berlin, etc. Vol. 1+, 1947+. Chem. Ber. Preceded by: Berichte der deutschen chemischen Gesellschaft. HI 64127

Chemisches Archiv. Leipzig. Chem. Arch. See B–P–H 306/22. HI 53345

Chemisches Journal für die Freunde der Naturlehre, Arzneygelahrtheit, Haushaltungskunst und

Manufacturen. Lemgo, Germany Chem. J. Freunde Naturl. See B–P–H 306/24. HI 53346

Chemist and druggist. London. Chem. & Druggist. See B–P–H 306/17. HI 53343

Chemistry and biology. [Kagaku to seibutsu.] Tokyo. Vols. 1+, 1962+. Chem. & Biol. HI 64128

Chemistry of natural compounds. New York, NY. Chem. Nat. Compounds. See B–P–H 306/26. HI 53347

Chemosphere. Oxford, Elmsford, NJ. 1972+. Chemosphere.

Chen chun hsüeh pao = Acta mycologica sinica. Beijing.

Chêne, Le. Marseille. Vols. 1+, 1909+. Chêne. HI 64129

Chesapeake science; a regional journal of research. Solomons Island, MD. Vols. 1-18, 1960-77. Chesapeake Sci. Superseded by: Estuaries. HI 64130

Chestnutworks. Alachue, FL. Vol. 1+, 1985+. Chestnutworks. HI 64131

Che-tai nung hsüeh chi k'an = Chekiang agricultural quarterly. Peiping [=Peking]. Chekiang Agric. Quart. See B–P–H 306/12.

Chi lin ta hsüeh tzu ran kon hsüeh pao = Acta scientiarum naturalium universitatis jilinensis. Chi-lin.

Chi nan lii hsueh pao = Journal of science and medicine of Jinan university. Kuang-chou.

Chi-ta yuan-i = Horticultural journal. Hangchow, China. Hort. J. (Hangchow). See B–P–H 421/10.

Chiao wu tsa chih = Chinese recorder; journal of the christian movement in China. Foochow, China. Chin. Rec. See B–P–H 308/26.

Chiao-yü kung pao = Bulletin of the ministry of education. Nanking. Bull. Minist. Educ. See B–P–H 263/6.

Chiba daigaku bunrigakubu Choshi rinkai kenkyu bunshitsu kenkyu hokoku = Bulletin of the Choshi marine laboratory, Chiba university. Chiba.

Chiba daigaku bunrigakubu kiyo. Shizenkagaku = Journal of the college of arts and sciences, Chiba university. Natural science. Chiba.

Chiba daigaku engei gakubu gakujutsu hokoku = Technical bulletin of the faculty of horticulture; Chiba university. Matsudo, Japan. Techn. Bull. Fac. Hort. Chiba Univ. See B–P–H 868/25.

Chiba daigaku engei gakabu tokubetsu hokoku = Transactions of faculty of horticulture, Chiba university. Matsudo.

Chiba daigaku fuhai kenkyusho hokoku = Report (Annual) of the institute of food microbiology, Chiba university. Chiba.

Chiba-ken danchi engei shikenjo kenkyu hokoku = Report of Chiba horticultural experiment station. Tateyama.

Chicago journal of nervous and mental diseases. Chicago, IL. Chicago J. Nerv. Mental Dis. See B–P–H 307/24. HI 53352

Chicago natural history museum bulletin = Bulletin of the Chicago natural history museum. Chicago.

Chicago naturalist. Chicago, IL. Chicago Naturalist. See B–P–H 307/26. HI 53353

Chichibu shizenkagaku hakubutsukan kenkyu hokoku = Bulletin of the Chichibu museum of natural history. Chichibu.

Chief scientist's team notes, nature conservancy council. London. ?-1980+. Chief Sci. Team Notes, Nat. Conservancy Council. HI 64132

Chigaku zasshi = Journal of geography. Tokyo. J. Geogr. (Tokyo). See B–P–H 467/19.

Chih p'ing chih shih = Plant disease knowledge. [China]. Pl. Dis. Knowl. See B–P–H 712/1.

Chih wu fên lei hsüeh pao = Acta phytotaxonomica sinica. Peking. Acta Phytotax. Sin. See B–P–H 48/6.

Chih-wu fen-liu hsüeh-pao = Journal of botanical analysis. Peking. J. Bot. Analysis. See B–P–H 460/11.

Chih wu hsüeh chi k'an = Botanical research: contributions from the institute of botany, academia sinica. Beijing.

Chih wu hsüeh pao = Acta botanica sinica. Peking (Beijing).

Chih wu hsüeh pao = Acta botanica taiwanica. Taipei, Taiwan. Acta Bot. Taiwan. See B–P–H 41/24.

Chih-wu hsüeh-pao = Journal of botany. Peking. J. Bot. (Peking). See B–P–H 460/5.

Chih wu hsüeh tsa chih = Botanical magazine. Peking (Beijing).

Chih wu hsueh tsa chih = Plant magazine. Peking (Beijing).

Chih wu pao hu hsueh hui k'an = Plant protection bulletin. Taipei.

Chih wu pao hu hsueh pao = Acta phytophylacica sinica. Beijing.

Chih wupao hu Zhiwu baohu = Plant protection. Peking. Pl. Protect. (Peking). See B–P–H 712/20.

Chih wu ping li hsüeh p'ao = Acta phytopathologica sinica. Peking (Beijing).

Chih wu sheng li hsüeh pao = Acta phytophysiologia sinica. Shanghai.

Chih wu sheng li hsueh tung hsun = Plant physiology

communications. Shanghai.

Chih wu sheng t'ai hsüeh yu ti chih wu hsüeh ts'ung k'an = Acta phytoecologia et geobotanica sinica. Beijing.

Chih wu shêng t'ai hsüeh yü ti chih wu hsüeh tzu liao t'sung k'an = Plant ecology and phytogeography research series. Peking. Pl. Ecol. Phytogeogr. Res. Ser. See B–P–H 712/5.

Chih wu tsa chih = Plant magazine. Peking (Beijing). and Plants. Peking (Beijing).

Chih wu yen chiu = Bulletin of botanical research. Harbin.

Chih wu yen chiu shih hui kan = Bulletin of botanical laboratory of north-eastern forestry institute. Harbin.

Chih wupao hu Zhiwu baohu = Plant protection. Peking. Pl. Protect. (Peking). See B–P–H 712/20.

Chihuahuan desert discovery. Alpine, TX. No. 1+, 1975+. Chihuahuan Desert Disc. HI 64133

Chihuahuan desert newsbriefs; bulletin of the Chihuahuan desert research institute. Alpine, TX. No. ?-2+, 19??-83+. Chihuahuan Desert Newsbr. HI 64134

Chiigaku zappo = Lichenological miscellanies. Mito, Japan. Lichenol. Misc. See B–P–H 530/15.

Chile forestal. Santiago. Vol. 1+, 1975+. Chile Forest. HI 64135

Chile triguero. Santiago de Chile. 1956-57. Chile Trig. Preceded by: Trigo. Superseded by: Trigo y cereales. HI 64136

Chilean forestry news. Santiago. 1978+. Chilean Forest. News. HI 64137

Chilean university life. Washington, DC. No. 19+, 1985+. Chilean Univ. Life. Preceded by: Report on chilean university life. HI 64138

Chileans, The. Hetton le Hole, Bristol, Guildford, etc. Vol. 1+, 1966+. Chileans. HI 64139

Chileans - yearbook. 1967+, 1968+. Chileans-Yearb. HI 64140

Chilin nungyeh tahsueh hsüeh pao = Acta agriculturae universitatis jilinensis. Changchun.

Chimia. Zurich. Chimia. See B–P–H 307/39. HI 53354

Chin ling hsüeh pao = Nanking journal. [Chin ling hsüeh pao.] Nanking. Nanking J. See B–P–H 627/17.

Chin ling kuang = University of Nanking magazine. Nanking. Univ. Nanking Mag. See B–P–H 937/11.

China cotton journal. [Shang Ch'ang ho chi hua sha lien hui k'an.] Shanghai. China Cotton J. See B–P–H 309/4. HI 53371

China cottons. [Zhongguo mianhua.] Honan. Vol. ?-10+, 19??-83+. China Cottons. HI 64141

China exchange news. Washington, DC. Vol. 8+, 1980+. China Exch. News. Preceded by: China exchange newsletter. HI 64142

China exchange newsletter. Washington, DC. Vols. 1-7, 1973-79. China Exch. Newslett. Preceded by: China science notes. Superseded by: China exchange news. HI 64143

China historical materials of science and technology. [Zhongguo kekue shiliao.] Peking (Beijing). Vol. 1+, 1963+. China Hist. Mater. Sci. Technol. HI 64144

China institute bulletin; a monthly publication on chinese thought and cluture. New York, NY. China Inst. Bull. See B–P–H 309/8. HI 53372

China journal. [Chung kuo k'o hsüeh mei-shu tsa chi.] Shanghai. China J. See B–P–H 309/9. HI 53373

China journal of science and arts. Shanghai. China J. Sci. Arts. See B–P–H 309/10. HI 53374

China medical journal. Shanghai. China Med. J. See B–P–H 309/14. HI 53375

China medical missionary journal. Peking. China Med. Missionary J. See B–P–H 309/16. HI 53376

China reconstructs. Peking. China Reconstructs. See B–P–H 309/17. HI 53377

China review, or notes and queries on the Far East ... Hong Kong. China Rev. See B–P–H 309/18. HI 53378

China science notes. [Chung-kuo k'o hsüeh t'ung hsin.] Washington, DC. Vols. 1-3(1), 1969-72. China Sci. Notes. Superseded by: China exchange newsletter. HI 64145

Chinchilla; anales históricos de la medicina en general y biográfico-bibliográficos de la Española en particular. Chinchilla. See B–P–H 309/20. HI 53379

Chinese agricultural monthly. [Chung nung yüeh k'an.] Chin. Agric. Monthly. See B–P–H 308/1. HI 53355

Chinese agricultural science. [Zhongguo nongye kexue.] Peking (Beijing). 1962-66. Chin. Agric. Sci. Superseded by: Scientia agricultura sinica. HI 64146

Chinese economic bulletin. Peking. Chin. Econ. Bull. See B–P–H 308/2. HI 53356

Chinese economic journal. Shanghai. Chin. Econ. J. See B–P–H 308/3. HI 53357

Chinese economic journal and bulletin. [Chung kuo ching chi hüeh k'an.] Peiping [=Peking]. Chin. Econ. J. Bull. See B–P–H 308/4. HI 53358

Chinese economic monthly. Peking. Chin. Econ. Monthly. See B–P–H 308/8. HI 53359

Chinese flowers. [Chung-kuo hua hui.] Taipei. ?-1975+. Chin. Fl. HI 64147

Chinese forestry. [Zhongguo linye.] Peking (Beijing).

1950+. Chin. Forest. HI 64148

Chinese forestry science. [Zhongguo linye kexue.] Peking (Beijing). 1975-78. Chin. Forest. Sci. Preceded and superseded by: Scientia silvae sinicae. HI 64149

Chinese and Japanese repository of facts and events in science, history, and art, relating to eastern Asia. London. Chin. Jap. Repos. Facts Events Sci. See B–P–H 308/16. HI 53364

Chinese journal of botany. [Chung kuo chih wu hsüeh tsa chih.] Peiping [=Peking]. Chin. J. Bot. See B–P–H 308/11. HI 53360

Chinese journal of experimental biology. [Shih yen sheng wu hsueh pao.] Peking (Beijing). Vols. 1-3(1-3), 1936-51 [suspended 1945-50]. Chin. J. Exp. Biol. Superseded by: Acta biologiae experimentalis sinica. HI 64150

Chinese journal of genetics. [Translation of: Acta genetica sinica. [Yichuan xuebao].] New York. Vol. 15+, 1988+. Chin. J. Genet. HI 74823

Chinese journal of microbiology. [Chung-hua min kuo wei sheng wu hsueh tsa chih.] Taipei. Vols. 1-12, 1968-79. Chin. J. Microbiol. Superseded by: Chinese journal of microbiology and immunology. HI 64151

Chinese journal of microbiology and immunology. [Zhonghua minguo wei shengwu ji mianyixue zazhi.] Taipei. Vol. 1+, 1980+. Chin. J. Microbiol. Immunol. Preceded by: Chinese journal of microbiology. HI 64152

Chinese journal of physiology. [Chung kuo sheng li hsüeh ts'a chih.] Peking. Chin. J. Physiol. See B–P–H 308/13. HI 53362

Chinese journal of tropical crops. [Redai zuowu xuebao.] Tan-hsien. Vol. 1+, 1980+. Chin. J. Trop. Crops. HI 64153

Chinese journal of zoology. [Chung kuo tung wu hsüeh tsa chih.] Nanking. Chin. J. Zool. See B–P–H 308/14. HI 53363

Chinese marine biological bulletin, Amoy, China = Amoy marine biological bulletin. Amoy, China. Amoy Mar. Biol. Bull. See B–P–H 86/21.

Chinese medical journal = Chinese medicine. Taipei.

Chinese medical journal. Shanghai. Chin. Med. J. See B–P–H 308/25. HI 53365

Chinese medicine. [Chung-hua i hsueh tsa chih.] Taipei. Vol. 1+, 1954?+. Chin. Med. HI 64154

Chinese medicine magazine = Chinese medicine. Taipei.

Chinese pharmaceutical bulletin. [Yaoxue tongbao.] Peking (Beijing). Vol. 6+, 1959?+ [publication suspended 1966-79]. Chin. Pharm. Bull. Preceded by: Chung yao t'ung pao. HI 64155

Chinese recorder; journal of the christian movement in China. [Chiao wu tsa chih.] Foochow, China. Chin. Rec. See B–P–H 308/26. HI 53366

Chinese recorder and missionary journal. Foochow, China. Chin. Rec. & Missionary J. See B–P–H 308/28. HI 53367

Chinese repository. Canton. Chin. Repos. See B–P–H 308/29. HI 53368

Chinese republic. Shanghai. Chin. Republ. See B–P–H 308/30. HI 53369

Chinese science abstracts. Part B. Peking (Beijing). Vol. 1+, 1982+. Chin. Sci. Abstr., B. HI 64156

Chinese students' monthly. Baltimore, MD. Chin. Stud. Monthly. See B–P–H 309/2. HI 53370

Chinese traditional and herbal drugs. [Zhong cao yao.] Shao-yang. Vol. ?-12+, 19??-81+. Chin. Tradit. Herbal Drugs. Preceded by?: Chung tsao yao t'ung hsun. HI 64157

Chinesische Zeitschrift für die gesamte Medizin = General medical journal. Peking. Gen. Med. J. See B–P–H 396/3.

Ching chi pu chung yang kung yeh shih yen so mu to'ai shih yen kuan chuan pao = Technical bulletin of the forests products laboratory; national bureau of industrial research. Kaiting, China. Techn. Bull. Forests Prod. Lab. See B–P–H 868/28.

Ching-hai nung ling. Quingbai nonglin = Tsinghai agriculture and forestry. Sining, China. Tsinghai Agric. Forest. See B–P–H 925/13.

Ch'ing-hua chou k'an = Tsing Hua weekly. Peking. Tsing Hua Weekly. See B–P–H 925/12.

Ch'ing-hua hsüeh pao = Tsing Hua journal. Peking. Tsing Hua J. See B–P–H 925/9.

Chirigaku hyoron = Geographical review of Japan. Tokyo. Geogr. Rev. Japan. See B–P–H 398/1.

Chirurgia degli organi di movimento. Bologna. Chir. Organi Movim. See B–P–H 309/35. HI 53380

Chisan rindo koho = Journal of Japan association of forestry conservation. Tokyo.

Chmelařské listy. Rakonitz, Bohemia [=Rakovnik, Czechoslovakia]. Chmel. Listy. See B–P–H 310/4. HI 53381

Chmelařský věstník. Laun, Bohemia [=Louny, Czechoslovakia]. Chmel. Věstn. See B–P–H 310/5. HI 53382

Choix de mémoires sur divers objets d'histoire naturelle. Paris. Choix Mém. Divers Objets Hist. Nat. See B–P–H 310/6. HI 53383

Chon nong = Journal of agriculture. Suwon, Korea. J. Agric. (Suwon). See B–P–H 455/1.

Choroby roślin. Warsaw. 1931. Choroby Rośl. Preceded by: Choroby i szkodniki roślin. HI 64158

Choroby i szkodniki roślin. Warsaw. Vols. 1-2(1), 1925-26. Choroby Szkodn. Rośl. Superseded by: Choroby roślin. HI 64159

Chosen bulletin. [Chosen iho.] Seoul?, Korea. Chosen Bull. See B–P–H 310/8. HI 53384

Chosen hakubutsu gakkai kaiho = Transactions of the Chosen natural history society. [Korea]. Trans. Chose Nat. Hist. Soc. See B–P–H 882/17.

Chosen hakubutsu gakkai zasshi = Journal of the Chosen natural history society. Seoul, Korea. J. Chosen Nat. Hist. Soc. See B–P–H 462/4.

Chosen iho = Chosen bulletin. Seoul?, Korea. Chosen Bull. See B–P–H 310/8.

Chosen nokai ho. Keijo = Journal of the agricultural society of Korea. Seoul?, Korea. J. Agric. Soc. Korea. See B–P–H 456/13.

Chosen nokai ho. Keijo = Journal of the Chosen agricultural society. Seoul?, Korea. J. Chosen Agric. Soc. See B–P–H 462/3.

Chosen sanrin-kaiho = Bulletin of the forestry society of Korea. Bull. Forest. Soc. Korea. See B–P–H 251/5.

Chosen yakugaku kai zasshi = Journal of the Chosen pharmaceutical society. Seoul?, Korea. J. Chosen Pharm. Soc. See B–P–H 462/5.

Christianias physicalske aarbog. Christiana [=Oslo, Norway]. Christianias Phys. Aarbog. See B–P–H 310/19. HI 53386

Christmas tree growers' journal = American Christmas tree growers' journal.

Chromatographia; international journal for rapid communication in chromatography and related techniques. Brunswick. Vol. 1/2+, 1968+. Chromatographia. HI 64160

Chromatographic reviews; progress in chromatography, electrophoresis and related methods. Amsterdam, London. Vol. 1-15, 1959-71. Chromatogr. Rev. Incorporated in: Journal of chromatography. HI 64161

Chromosoma. Vienna. Chromosoma. See B–P–H 310/22. HI 53387

Chromosome information service. Tokyo. No. 1+, 1960+. Chromosome Inform. Serv. HI 64162

Chromosomes today; proceedings of the Oxford [etc.] (later international) chromosome conference. New York, etc. Vol. 1+, 1964+. Chromosomes Today. HI 64163

Chronica botanica; [Subtitle varies]. Leiden, Waltham, MA, New York. Vols. 1-33, 1935-62 [From vol. 17 onwards became a Ronald Press book series]. Chron. Bot. HI 64164

Chronica horticulturae; bulletin of the international society for horticultural science. Wageningen, The Hague. Vol. 1+, 1961+. Chron. Hort. HI 64165

Chronica naturae. Batavia, Dutch E. Indies [=Jakarta, Indonesia]. Chron. Nat. See B–P–H 310/25. HI 53389

Chronica nicotiana. Bremen. Vols. 1-?, 1940-50. Chron. Nocotiana. Preceded by: Tabak. Berlin. Superseded by: Tabacologia. 5-4146-3. HI 64166

Chronicle; newsletter of Los Angeles cactus and succulent society. Los Angeles, CA. 19??-83+. Chronicle. HI 64167

Chronicle, Northamptonshire naturalists' trust. Northampton. 1977+. Chron. Northamptonshire Naturalists' Trust. Preceded by: Report, Northamptonshire naturalists' trust. HI 64168

Chronicle, university of Maryland graduate school = University of Maryland graduate school chronicle. College Park, MD.

Chronique agricole. Brussels. Chron. Agric. See B–P–H 310/23. HI 53388

Chronique agricole du canton de Vaud. Lausanne. Vols. 1-21, 1888-1908. Chron. Agric. Vaud. Superseded by: Terre vaudoise. HI 64169

Chrońmy przyrode ojczysta. Cracow. Chrońmy Przyr. Ojczysta. See B–P–H 310/27. HI 53390

Chronobiologia; organ of the international society for chronobiology. Milan. Vol. 1+, 1974+. Chronobiologia. HI 64170

Chronobiology international. Oxford, Elmsford, NY. Vol. 1+, 1984+. Chronobiol. Int. HI 64171

Chrudimsky kaktusar. Chrudim. 1979?-82+. Chrudim. Kakt. HI 64173

Chrysanthème; journal de la société française des chrysanthémistes. Lyons. Chrysanthème. See B–P–H 310/28. HI 53391

Chrysanthemum. Las Vegas, NV., [etc.]. Vol. 29+, 1972+. Chrysanthemum. Preceded by: Bulletin of the national chrysanthemum society. HI 64174

Chrysanthemum annual. London. Chrysanthemum Annual See B–P–H 310/30. HI 53394

Chrysanthemum and dahlia. London. Chrysanthemum & Dahlia. See B–P–H 310/29. HI 53392

Chrysanthemum year book. London. 1967+. Chrysanthemum Year Book. Preceded by: Yearbook of the national chrysanthemum society. HI 64175

Chu tzu yen chiu hui k'an = Journal of bamboo research. Hangzhou.

Chugoku agricultural research. [Chugoku nogyo kenkyu.]

Hiroshima. Vols. 1-47, 1965-73. Chugoku Agric. Res. Superseded by: Kauki chugoku nogyo kenkyu. HI 64176

Chugoku nogyo kenkyu = Chugoku agricultural research. Hiroshima.

Chugoku nogyo shikenjo hokoku. A, sakumotsubu, kankyobu = Bulletin of the Chugoku agricultural experiment station. Ser. A, crop division and environment division. Fukuyama.

Chugoku nogyo shikenjo hokoku. A, sakumotsubu = Bulletin of the Chugoku national agricultural experiment station. Ser. A, crop division. Fukuyama.

Chung Chi hsueh pao = Chung Chi journal; Shatin, Hong Kong.

Chung Chi journal [Chung Chi hsueh pao]; primarily devoted to far eastern studies. Shatin, Hong Kong. Vol. 1+, 1961+. Chung Chi J. HI 64177

Chung-hua chiao yü chieh = Shanghai educational review. Shanghai. Shanghai Educ. Rev. See B–P–H 834/25.

Chung-hua i hsueh tsa chih = Chinese medicine. Taipei.

Chung-hua lin hsüeh chi k'an = Quarterly journal of chinese forestry. Taipei.

Chung-hua min kuo wei sheng wu chi mien i hsüeh tsa chih = Chinese journal of microbiology and immunology. Taipei.

Chung-hua min kuo wei sheng wu hsueh tsa chih = Chinese journal of microbiology. Taipei.

Chung kuo chih-wu hsüeh hui hui-pao = Bulletin of the chinese botanical society. Peiping [=Peking]. Bull. Chin. Bot. Soc. See B–P–H 245/1.

Chung kuo chih wu hsüeh tsa chih = Chinese journal of botany. Peiping [=Peking]. Chin. J. Bot. See B–P–H 308/11.

Chung kuo chih wu hsüeh tsa chih = Journal of the botanical society of China. Peiping [=Peking]. J. Bot. Soc. China. See B–P–H 460/16.

Chung kuo ching chi hüeh k'an = Chinese economic journal and bulletin. Peiping [=Peking]. Chin. Econ. J. Bull. See B–P–H 308/4.

Chung kuo hsi pei chih wu tiao cha so ts'ung k'an = Contributions from the botanical survey of northwestern China. Wukung, China. Contr. Bot. Surv. NorthW. China. See B–P–H 326/22.

Chung-kuo hua hui = Chinese flowers. Taipei.

Chung-kuo hua hui p'en ching. Beijing. 1984+. Chung-kuo Hua Hui P'en Ching. HI 64178

Chung-kuo je-tai nung hsüeh hui-pao = Journal of the society of chinese tropical agriculture. Taipei, Taiwan. J. Soc. Chin. Trop. Agric. See B–P–H 481/24.

Chung kuo k'o hsüeh = Acta scientia sinica. Peking. Acta Sci. Sin. See B–P–H 48/26.

Chung kuo k'o hsüeh. B = Scientia sinica. Series B, Chem. biol. ag. med. earth sci. [European edition; also published in a Chinese edition]. Beijing.

Chung kuo k'o hsüeh mei-shu tsa chi = China journal. Shanghai. China J. See B–P–H 309/9.

Chung kuo k'o hsüeh shê yen chin ts'ung k'an = Memoirs of the science society of China. Shanghai. Mem. Sci. Soc. China. See B–P–H 581/5.

Chung-kuo k'o hsüeh t'ung hsin = China science notes. Washington, DC.

Chung kuo k'o hsüeh yü chien she = Science and technology in China. Nanking. Sci. Technol. China. See B–P–H 829/15.

Chung-kuo k'o hsüeh yüan. Chih wu yüan kung tso wei yüan hui = Plant breeding. Peking.

Chung kuo k'o hsüeh yuan. Hua nan chih we yen chiu so, ping chung chuan k'an, ti i hao = Special bulletin. South China botanical institute. Academia sinica. Series C. [China]. Special Bull. S. China Bot. Inst., Ser. C. See B–P–H 850/19.

Chung-kuo k'o hsüeh yuan hua nan chih wu yen chiu so chi k'an = Acta botanica austro sinica. Guangzhou.

Chung-kuo k'o hsüeh yuan lin yeh t'u jang yen chiu so chi k'an = Bulletin of the institute of forestry and pedology, academia sinica. Shenyang.

Chung-kuo lin yeh = Chinese forestry. Peking (Beijing).

Chung-kuo lin yeh ko hsueh = Chinese forestry science. Peking (Beijing).

Chung-kuo mien hua = China cottons. Honan.

Chung-kuo nung yeh k'o hsüeh = Chinese agricultural science. Peking (Beijing).

Chung-kuo nung yeh k'o hsüeh yüan = Grassland of China. Beijing.

Chung kuo sheng li hsüeh ts'a chih = Chinese journal of physiology. Peking. Chin. J. Physiol. See B–P–H 308/13.

Chung kuo sheng wu hua hsueh hui hui chih = Journal of the chinese biochemical society. Taipei.

Chung kuo shih yen sheng wu shüeh tsa chih = Chinese journal of experimental biology. [Shih yen sheng wu hsueh pao.] Peking (Beijing).

Chung-kuo shui shing wu hui pao = Acta hydrobiologica sinica. Shanghai. Acta Hydrobiol. Sin. (Shanghai). See B–P–H 44/14.

Chung kuo ti chih hsüeh hui chih = Bulletin of the geological society of China. Nanking. Bull. Geol. Soc. China See B–P–H 251/19.

Chung kuo tung wu hsüeh tsa chih = Chinese journal of

zoology. Nanking. Chin. J. Zool. See B–P–H 308/14.

Chung kuo wên hua yen chini hui kan = Bulletin of chinese studies. Chengtu, China. Bull. Chin. Stud. See B–P–H 245/2.

Chung nung yüeh k'an = Chinese agricultural monthly. Chin. Agric. Monthly. See B–P–H 308/1.

Chung-shan ta hsueh hsueh pao = Journal of Sun Yatsen university. Canton.

Chung-shan ta hsüeh hsüeh pao. Tzu jan k'o hsüeh pan = Acta scientiarum naturalium universitatis sunyatseni. Guangzhou.

Chung tsao yao = Chinese traditional and herbal drugs. Shao-yang.

Chung tsao yao t'ung hsün. Shao-yang? 19??-75. Chung Tsao Yao T'ung Hsün. HI 64179

Chung-yang wei-shong shih-yen so nein pao. Yen tiao ch'a pao kao = Report (Annual) of the central hygiene experiment station. Report of investigations. Nanking?

Chung yao t'ung pao. Peking (Beijing). Vols. 1-5(3), 1955-59. Chung Yao T'ung Pao. Superseded by: Chinese pharmaceutical bulletin. HI 64180

Chung yao tung pao = Bulletin of chinese materia medica. Peking (Beijing).

Chung yo t'ung pao = Information on chinese medicine. Peking. Inform. Chin. Med. (Peking). See B–P–H 432/15.

Chungang imop sihomjang = Bulletin of the forest experimental station. Seoul.

Chuo engei = Horticultural journal. [Japan]. Hort. J. (Japan). See B–P–H 421/11.

Chuo shiken-sho hokoku = Report of the central research institute of the South Manchuria railway co[mpany]. Rep. Centr. Res. Inst. S. Manchuria Railway Co. See B–P–H 763/19.

Ciba foundation bulletin. Basle, London. No. ?-19+, ?-1988+. Ciba Found. Bull. HI 74824

Ciba-Geigy weed tables = Geigy weed tables. Basle.

Ciba review. Basle. Vol. 1+, 1937+. Ciba Rev. HI 64181

Ciba Rundschau. Basle. Vol. 1+, 1937+. Ciba Rundschau. HI 64182

Ciba symposia. Summit, NJ. Ciba Symp. See B–P–H 311/32. HI 53400

Ciba-tijdschrift. Basel. Ciba-Tijdschr. See B–P–H 311/34. HI 53402

Ciba Zeitschrift. Basel. Ciba Z. See B–P–H 311/33. HI 53401

Ciel et terre. Brussels. Ciel & Terre. See B–P–H 311/35. HI 53403

Ciencia; revista hispano americana de ciencias puras y aplicadas. Mexico City. Ciencia (Mexico). See B–P–H 311/40. HI 53404

Ciência biológica. Biologia molecular e celular. Coimbra. Vol. 2+ [vol. 1 not published], 1972+ [vol. numbering alternates with Ciência biólogica. Ecologia e sistemática q.v.]. Ci. Biol. Biol. Molec. Cel. HI 64183

Ciência biológica. Ecologia e sistemática. Coimbra. Vol. 1+, 1973+ [vol. numbering alternates with Ciência biólogica. Biologia molecular e celular q.v.]. Ci. Biol. Ecol. Sist. HI 64184

Ciência biológica. Ecology and systematics = Ciência biológica. Ecologia e sistemática. Coimbra.

Ciência e cultura; orgao da sociedade brasileira para o progresso da ciencia. São Paulo. Vol. 1+, 1949+. Ci. & Cult. 2-1053-1. HI 64185

Ciencia al día; notociero de divulgación cientifica y ténica. Caracas. Vol. 1+, 1961+. Ci. Dia. HI 64186

Ciencia forestal. Coyoacan. Vol. ?-2(7)+, 19??-77+. Ci. Forest. (Coyoacan). HI 64187

Ciencia forestal. Mexico, D.F. Vol. 1+, 1976+. Ci. Forest. (Mexico). HI 64188

Ciência hoje; revista de divulgação cientifica da sociedade brasileira para o progresso da ciência. Rio de Janeiro. Vol. 1+, 1982+. Ci. Hoje. HI 64189

Ciencia interamericana. Washington, DC. Ci. Interamer. See B–P–H 311/31. HI 53399

Ciencia e investigación. Buenos Aires. Ci. & Invest. See B–P–H 311/28. HI 53396

Ciencia y natura. Santa Maria. No. 1+, 1979+. Ci. & Nat. HI 64191

Ciencia y naturaleza. Quito, Ecuador. Vol. 1+, 1957+. Ci. & Naturaleza. HI 64192

Ciencia y naturaleza de ciencias naturales. Quito. Vol. 1+, 1957+. Ci. Naturaleza Ci. Nat. HI 64193

Ciencia nueva; revista mensual de ciencia i tecnología. Buenos Aires. Vol. 1+, 1970+. Ci. Nueva (Buenos Aires). HI 64194

Ciência e práctica. Lavras. Vol. 1+, 1977+. Ci. & Pract. HI 64195

Ciencia y tecnica en la agricultura. Havana. Vol. 1+, 1978+. Ci. Tecn. Agric. HI 64196

Ciencia y tecnologia. San José, Costa Rica. Vol. 1+, 1977+. Ci. & Tecnol. HI 64197

Ciencia y tecnologia del mar. Valparaiso. No. 1+, 1975+. Ci. & Tecnol. Mar. HI 64198

Ciencia y tecnologia de Venezuela. Caracas. Vol. 1+, 1977+. Ci. & Tecnol. Venezuela. HI 64199

Ciencias. Serie 2, sanidad vegetal. Havana. No. 1+, 1974+. Ciencias (Havana), Ser. 2. 64172

Ciencias. Serie 4, ciencias biologicas. Havana. No. 1+, 1968+. Ciencias (Havana), Ser. 4. HI 64200

Ciencias. Serie 8, investigaciones marinas. Havana. No. 1-47, 1972-79. Ciencias (Havana) Ser. 8. Superseded by: Revista de investigaciones marinas. HI 64201

Ciencias. Serie 10, botánica. Havana. No. 1+, 1975+. Ciencias (Havana), Ser. 10. HI 64202

Ciencias; asociación española por el progreso de las ciencias. Madrid. Ciencias (Madrid). See B–P–H 312/1. HI 53406

Ciencias de la agricultura. Havana. No. 1+, 1977+. Ci. Agric. Preceded by: Revista de agricultura. HI 64203

Ciencias y artes. Mexico City. Ci. & Artes. See B–P–H 311/27. HI 53395

Ciencias biologicas, academia de ciencias de Cuba. Havana. No. 1+, 1977+. Ci. Biol. Acad. Ci. Cuba. HI 64204

Ciências biológicas. Luanda. Vol. 1+, 1970+. Ci. Biol. (Luanda). HI 64205

Ciencias forestales. Santiago. Vol. 1+, 1978+. Ci. Forest. HI 64206

Ciencias y letras. San Luis Potosí, Mexico. Ci. & Letras. See B–P–H 311/29. HI 53397

Ciencias marinas. Ensenada. Vol. 1+, 1974+. Ci. Mar. HI 64207

Cientifica. Jaboticabal. Vol. 1+, 1974+. Cientifica. HI 64208

CIKARD news = C I K A R D news. Ames, IA.

Cimbebasia. Windhoek. Nos. 1-22, 1962-67; ser. A, natuurwetenskappe (natural history) vol. 1+, 1967+. Cimbebasia. HI 64209

Cimbebasia memoirs. Windhoek. Vol. 1+, 1967+. Cimbebasia Mem. HI 64210

Cinchona. Archief voor de kina-cultuur. Bandoeng, Dutch E. Indies [=Bandung, Indonesia]. Cinchona. See B–P–H 312/5. HI 53407

Cincinnati observer. Cincinnati, OH. Cincinnati Observ. See B–P–H 312/6. HI 53408

Cincinnati quarterly journal of science. Cincinnati, OH. Cincinnati Quart. J. Sci. See B–P–H 312/7. HI 53409

Cincinnatus; devoted to scientific agriculture, horticulture, education and improvement of rural taste. Cincinnati, OH. Cincinnatus. See B–P–H 312/9. HI 53410

Circadian rhythms. Vols. ?-2, 19??-71. Circad. Rhythms. Superseded by: Biological rhythms. HI 64211

Circolare, osservatore per le malattie delle piante. Sez. entomologia. Pisa. 1958+. Circ. Osserv. Malatt. Piante, Sez. Entomol. HI 64212

Circolare osservatorio regionale de fitopatologia per la Calabria. Catanzaro, Italy. Circ. Osserv. Regionale Fitopatol. Calabria. See B–P–H 313/27. HI 53414

Circoli; rivista di poesia. Genoa. Circoli. See B–P–H 314/18. HI 53419

Circulaire, biohistorisch instituut der rijksuniversiteit te Utrecht. Utrecht. Nos. ?-86-?, 19??-72-? Circ. Biohist. Inst. Rijksuniv. Utrecht. Superseded by: Incidentele mededelingen van het biohistorisch instituut der rijksuniversiteit te Utrecht. HI 64213

Circulaire de la station botanique. Brignoles, France. Circ. Stat. Bot. See B–P–H 313/37. HI 53415

Circular, agricultural experiment station, Alabama. Auburn, AL. No. 137+, 1960+. Circ. Agric. Exp. Sta., Alabama. Preceded by: Alabama agricultural experiment station of the Alabama polytechnic institute, circular. HI 64214

Circular, agricultural experiment station of the Alabama polytechnic institute = Alabama agricultural experiment station of the Alabama polytechnic institute, circular. Auburn, AL.

Circular, agricultural experiment station, university of California = University of California, college of agriculture, agricultural experiment station. Circular. Berkeley, CA. Univ. Calif. Coll. Agric. Exp. Sta. Circ. See B–P–H 934/14.

Circular, agricultural experiment station, university of Tennessee. Knoxville, TN. No. 1+, 1926+. Circ. Exp. Sta. Univ. Tennessee. HI 64215

Circular and agricultural journal of the royal botanic gardens, Peradeniya. Colombo. Vols. 2-6, 1902-12. Circ. Agric. J. Roy. Bot. Gard. Peradeniya. Preceded by: Circular, royal botanic gardens, Ceylon. Superseded by: Bulletin of the department of agriculture, Ceylon. HI 64216

Circular, Alaska agricultural experiment stations = Alaska agricultural experiment stations. Circular. Washington, DC. Alaska Agric. Exp. Sta. Circ. See B–P–H 68/6.

Circular, Arizona agricultural experiment station = University of Arizona. Arizona agricultural experiment station. Circular. Tucson, AZ. Univ. Arizona Agric. Exp. Sta. Circ. See B–P–H 933/15.

Circular, Arkansas agricultural experiment station = University of Arkansas. The college of agriculture. Arkansas agricultural experiment station. Circular. Fayetteville, AR. Univ. Arkansas Coll. Agric. Arkansas Agric. Exp. Sta. Circ. See B–P–H 934/2.

Circular, Auburn university agricultural experiment

station = Alabama agricultural experiment station of the Alabama polytechnic institute, circular. Auburn, AL.

Circular bulletin, Maryland agricultural experiment station = Maryland agricultural experiment station. Circular bulletin. College Park, MD. Maryland Agric. Exp. Sta. Circ. Bull. See B–P–H 549/2.

Circular bulletin, Michigan agricultural experiment station = Michigan agricultural experiment station. Circular bulletin. East Lansing, MI. Michigan Agric. Exp. Sta. Circ. Bull. See B–P–H 592/10.

Circular bulletin, Oregon agricultural experiment station = Oregon agricultural experiment station circular. Corvallis, OR. Oregon Agric. Exp. Sta. Circ. See B–P–H 690/12.

Circular, bureau of forestry, Philippine Islands. Manila. Nos. 1-237?, 1908-29. Circ. Bur. Forest. Philippine Islands. HI 64217

Circular, bureau of forestry, U S department of agriculture = Department of agriculture. Forestry division. Circular. Washington, DC. Dept. Agric. Forest. Div. Circ. See B–P–H 343/1.

Circular, bureau of plant industry, United States department of agriculture = U S department of agriculture. Bureau of plant industry. Circular. Washington, DC. U.S.D.A. Bur. Pl. Industr. Circ. See B–P–H 940/25.

Circular, California agricultural extension service. Berkeley, CA. Nos. 1-183, 1926-52. Circ. Calif. Agric. Extens. Serv. HI 64218

Circular, Clemson agricultural college of South Carolina. Agricultural experiment station = Clemson agricultural college of South Carolina. Agricultural experiment station. Circular. Clemson, SC. Clemson Agric. Exp. Sta. Circ. See B–P–H 315/1.

Circular of the commissioner of agriculture of the state of Alabama; embracing bulletin no. 3 of the state agricultural experiment station at the agricultural and mechanical college. Auburn, AL. Circ. Commiss. Agric. State Alabama. See B–P–H 312/21. HI 53411

Circular, Connecticut agricultural experiment station = Connecticut agricultural experiment station. Circular. Storrs, CT. Connecticut Agric. Exp. Sta. Circ. See B–P–H 325/1.

Circular, Cornell university agricultural experiment station = Cornell university agricultural experiment station circular. Ithaca, NY. Cornell Univ. Agric. Exp. Sta. Circ. See B–P–H 332/7.

Circular, Croydon natural history and scientific society. Croydon. Nos. ?-349-366, 19??-72-80. Circ. Croydon Nat. Hist. Sci. Soc. Superseded by: Programme, Croydon natural history and scientific society. HI 64219

Circular, department of agriculture, Canada. Ottawa. Nos. 1-189, 1922-50. Circ. Dept. Agric. Canada. Preceded by: Circular, experimental farms service, Canada and Circular, fruit branch, department of agriculture, Canada. Superseded by: Publication, department of agriculture, Canada. HI 64220

Circular, department of agriculture, Nyasaland. Zomba. 1913-23. Circ. Dept. Agric. Nyasaland. Superseded by: Circular, department of agriculture, Nyasaland. Agronomic series. HI 64221

Circular, department of agriculture, Nyasaland. Agronomic series. Zomba. Nos. 1-4, 1927-29. Circ. Dept. Agric. Nyasaland, Agron Ser. Preceded by: Circular, department of agriculture, Nyasaland. HI 64222

Circular, department of agriculture, Uganda protectorate. Entebbe. Nos. 1-28, 1914-39. Circ. Dept. Agric. Uganda. HI 64223

Circular, division of agrostology, United States department of agriculture. Washington, DC. Nos. 1-36, 1895-1901. Circ. Div. Agrostol. U.S.D.A. HI 64224

Circular, division of botany, United States department of agriculture = United States department of agriculture. Division of botany. Circular. Washington, DC. U.S.D.A. Div. Bot. Circ. See B–P–H 941/22.

Circular, division of forestry, U S department of agriculture = Department of agriculture. Forestry division. Circular. Washington, DC. Dept. Agric. Forest. Div. Circ. See B–P–H 343/1.

Circular, division of pomology, United States department of agriculture = U S department of agriculture. Division of pomology. Circular. Washington, DC. U.S.D.A. Div. Pomol. Circ. See B–P–H 942/1.

Circular, division of systematic biology, Stanford university. Stanford, CA. 1962-64. Circ. Div. Syst. Biol., Stanford Univ. Preceded by: Circular, natural history museum, Stanford university. HI 64225

Circular, division of vegetable pathology, United States department of agriculture = U S department of agriculture. Division of vegetable pathology. Circular. Washington, DC. U.S.D.A. Div. Veg. Pathol. Circ. See B–P–H 942/3.

Circular, division of vegetable physiology and pathology, United States department of agriculture = U S department of agriculture. Division of vegetable pathology. Circular. Washington, DC. U.S.D.A. Div. Veg. Pathol. Circ. See B–P–H 942/3.

Circular. Estación experimental agrícola "La Molina." Lima. Circ. Estac. Exp. Agríc. "La Molina". See B–P–H 312/28. HI 53412

Circular, estaçión experimental agronómica de Cuba. Santiago de las Vegas. Nos. 1-96, 1904-66. Circ. Estaç. Exp. Agron. Cuba. HI 64226

Circular, experimental farms of Canada. Ottawa. No. 6+, 1914+. Circ. Exp. Farms Canada. Preceded by: Farmers' circular, experimental farms service, Canada. HI 64227

Circular farmacéutica. Barcelona. No. 1+, 1943+. Circ. Farm. HI 64228

Circular. Federal experiment station in Puerto Rico; United States department of agriculture. Mayagüez, Puerto Rico. Circ. Fed. Exp. Sta. Puerto Rico U.S.D.A. See B–P–H 312/29. HI 53413

Circular, federation of Ontario naturalists. Toronto. Nos. 1-50, 1933-? Circ. Fed. Ontario Naturalists. Superseded by: Bulletin of the federation of Ontario naturalists. 2-1549-3. HI 64229

Circular of the fish and wildlife service, U S department of the interior. Washington, DC. 1941-70. Circ. Fish Wildlife Serv. U.S. Dept. Interior. Superseded by: Circular, United States national marine fisheries service [not entered]. HI 64230

Circular, forestry branch, Canada. Ottawa. Nos. 1-65, 1901-48. Circ. Forest. Branch Canada. HI 64231

Circular, forestry division, U S department of agriculture = Department of agriculture. Forestry division. Circular. Washington, DC. Dept. Agric. Forest. Div. Circ. See B–P–H 343/1.

Circular, fruit branch, department of agriculture, Canada. Ottawa. Nos. 1-3, 1915-21. Circ. Fruit Branch Dep. Agric. Canada. Superseded by: Circular, department of agriculture, Canada. HI 64232

Circular, Guam agricultural experiment station, office of experiment stations, United States department of agriculture = U S department of agriculture, office of experiment stations; the Guam agricultural experiment station, circular. Washington, DC. U.S.D.A. Off. Exp. Sta. Guam Agric. Exp. Sta. Circ. See B–P–H 943/26.

Circular, Hawaii agricultural experiment station = Hawaii agricultural experiment station. Circular. Honolulu, HI. Hawaii Agric. Exp. Sta. Circ.. See B–P–H 412/27.

Circular, hawaiian sugar planters' association experiment station. Honolulu, HI. Nos. 1-91, 1907-48. Circ. Hawaiian Sugar Planters' Assoc. Exp. Sta. HI 64233

Circular I P E A N = Circular do instituto de pesquisas e experimentaçao agropecuárias do norte. Belém.

Circular, Illinois natural history survey. Urbana, IL. No. 1+, 1934+. Circ. Illinois Nat. Hist. Surv. Preceded by: Forestry circular, Illinois natural history survey. HI 64234

Circular of information, Missouri agricultural experiment station = Missouri agricultural experiment station. Circular of information. Columbia, MO.

Circular of information, Oklahoma agricultural experiment station = Oklahoma agricultural experiment station circular of information. Stillwater, OK. Oklahoma Agric. Exp. Sta. Circ. Inform. See B–P–H 687/3.

Circular information, Oregon agricultural experiment station = Oregon agricultural experiment station circular information. Corvallis, OR. Oregon Agric. Exp. Sta. Circ. Inform. See B–P–H 690/14.

Circular of information, Wisconsin (state) agricultural experiment station = Wisconsin (state) agricultural experiment station. Circular of information. Madison, WI. Wisconsin Agric. Exp. Sta. Circ. Inform. See B–P–H 976/2.

Circular informativo, sociedad argentina de botánica. Cordoba. No. 1+, 1983+. Circ. Inform. Soc. Argent. Bot. HI 64235

Circular do instituto agronómico do norte. Belém. Nos. 1-6, [1943]-[62]. Circ. Inst. Agron. N. Superseded by: Circular do instituto de pesquisas e experimentaçao agropecuárias do norte. HI 64236

Circular, instituto agronómico de nordeste. Recife, Pernambuco. 1954+. Circ. Inst. Agron. Nordeste. HI 64237

Circular do instituto de pesquisas e experimentaçao agropecuárias do norte. Belém. No. 7-18?, [1963]-[73]? Circ. Inst. Pesq. Exp. Agropecu. N. Preceded by: Circular do instituto agronomico do norte. HI 64238

Circular, international organization of palaeobotany. London. 1982+. Circ. Int. Organ. Paleobot. HI 64239

Circular, international organization for the study of succulent plants = I O S circular.

Circular, Iowa agricultural experiment station = Iowa agricultural experiment station. Circular. Ames, IA. Iowa Agric. Exp. Sta. Circ. See B–P–H 436/34.

Circular, malayan orchid society. Nos. 1-11, 1956-57. Circ. Malayan Orchid Soc. Superseded by: Monthly bulletin, malayan orchid society. HI 64240

Circular, Massachusetts agricultural experiment station bulletin = Massachusetts agricultural experiment station bulletin. Circular. Amherst, MA. Mass. Agric. Exp. Sta. Bull. Circ. See B–P–H 549/14.

Circular, Max C. Fleischmann college of agriculture = Publications, Max C. Fleischmann college of agriculture. Series C [circular series]. Reno, NV.

Circular, Michigan agricultural experiment station. = Michigan agricultural experiment station. Circular. East Lansing, MI. Michigan Agric. Exp. Sta. Circ.

See B–P–H 592/9.

Circular, Mississippi agricultural experiment station. Mississippi agricultural and mechanical college = Mississippi agricultural experiment station. Mississippi agricultural and mechanical college. Circular. Jackson, MS. Mississippi Agric. Exp. Sta. Circ. See B–P–H 599/25.

Circular, Mississippi agricultural experiment station. Mississippi state college = Mississippi agricultural experiment station. Mississippi agricultural and mechanical college. Circular. Jackson, MS. Mississippi Agric. Exp. Sta. Circ. See B–P–H 599/25.

Circular, Missouri agricultural experiment station = Missouri agricultural experiment station. Circular. Columbia, MO.

Circular, Montana college of agriculture and mechanic arts. Agricultural experiment station = Montana college of agriculture and mechanic arts. Agricultural experiment station. Circular. Bozeman, MT. Montana Agric. Exp. Sta. Circ. See B–P–H 618/13.

Circular, Montana state college. Agricultural experiment station = Montana college of agriculture and mechanic arts. Agricultural experiment station. Circular. Bozeman, MT. Montana Agric. Exp. Sta. Circ. See B–P–H 618/13.

Circular, natural history museum, Stanford university. Stanford, CA. 1955-61. Circ. Nat. Hist. Mus. Stanford Univ. Superseded by: Circular, division of systematic biology, Stanford university. HI 64241

Circular, Nebraska agricultural experiment station = Nebraska agricultural experiment station circular. Lincoln, NB. Nebraska Agric. Exp. Sta. Circ. See B–P–H 639/26.

Circular, New Hampshire agricultural experiment station = New Hampshire agricultural experiment station. Circular. Durham, NH. New Hampshire Agric. Exp. Sta. Circ. See B–P–H 653/2.

Circular, New Jersey agricultural college experiment station = New Jersey agricultural college experiment station. Circular. New Brunswick, NJ. New Jersey Agric. Coll. Exp. Sta. Circ. See B–P–H 653/16.

Circular, New York (state) agricultural experiment station = New York (state) agricultural experiment station. Circular. Geneva, NY. New York Agric. Exp. Sta. Circ. See B–P–H 655/12.

Circular of the New York state museum. Albany, NY. Vols. 1-45 [vols. 3 and 4 not issued], 1928-56. Circ. New York State Mus. Superseded by: Circular, New York state museum and science service. 4-3045-3. HI 64242

Circular, New York state museum and science service. Albany, NY. No. 46+, 1973+. Circ. New York State Mus. Sci. Serv. Preceded by: Circular of the New York state museum. HI 64243

Circular of the New Zealand state forest service. Wellington, N.Z. Nos. 1-35, 1922-35. Circ. New Zealand State Forest Serv. HI 64244

Circular, North Carolina agricultural experiment station = North Carolina agricultural experiment station circular. Raleigh, NC. North Carolina Agric. Exp. Sta. Circ. See B–P–H 663/7.

Circular, North Dakota agricultural experiment station = North Dakota agricultural experiment station circular. Fargo, ND. North Dakota Agric. Exp. Sta. Circ. See B–P–H 664/11.

Circular note, botanical department of Trinidad. Trinidad. No. ?-6+, 18??-94+. Circ. Note Bot. Dep. Trinidad. HI 64245

Circular, office of experiment stations, United States department of agriculture. Washington, DC. Nos. 1-118, 1889-1913. Circ. Off. Exp. Sta. U.S.D.A. HI 64246

Circular, office of the secretary, United States department of agriculture = United States department of agriculture. Office of the secretary. Circular. Washington, DC. U.S.D.A. Off. Secr. Circ. See B–P–H 944/6.

Circular, Oregon agricultural experiment station = Oregon agricultural experiment station circular. Corvallis, OR. Oregon Agric. Exp. Sta. Circ. See B–P–H 690/12.

Circular, Pennsylvania state college agricultural experiment station = Pennsylvania state college agricultural experiment station circular. State College, PA. Pennsylvania State Coll. Agric. Exp. Sta. Circ. See B–P–H 701/1.

Circular, Porto Rico agricultural experiment station = Porto Rico agricultural experiment station circular. Mayagüez, Puerto Rico. Porto Rico Agric. Exp. Sta. Circ. See B–P–H 715/25.

Circular, Puerto Rico experiment station. United States department of agriculture = Puerto Rico experiment station. United States department of agriculture. Circular. Mayagüez, Puerto Rico. Puerto Rico Agric. Exp. Sta. U.S.D.A. Circ. See B–P–H 749/16.

Circular, Rutgers University agricultural experiment station = Rutgers University agricultural college experiment station. Circular. New Brunswick, NJ. New Jersey Agric. Coll. Exp. Sta. Circ. See B–P–H 653/16.

Circular, royal botanic gardens, Ceylon. Colombo. Vol. 1(1-25), 1897-1901. Circ. Roy. Bot. Gard. Ceylon. Superseded by: Circular and agricultural journal of the royal botanic gardens, Peradeniya. HI 64247

Circular, school of forestry, Oregon state college. Corvallis, OR. No. 1+, 1949+. Circ. School Forest. Oregon State Coll. HI 64248

Circular, section of seed and plant introduction, United States department of agriculture = U S department of agriculture, section of seed and plant introduction. Circular. Washington, DC.

Circular, South Dakota agricultural experiment station = South Dakota agricultural experiment station. Circular. Brookings, SD. South Dakota Agric. Exp. Sta. Circ. See B–P–H 848/2.

Circular técnica, centro de pesquisa agropecuária do trópica semi-arido. Petrolina. No. 1+, 1980+. Circ. Técn. Centro Pesquisa Agropecu. Tróp. Semi-Arido. HI 64249

Circular técnica, centro de pesquisa agropecuária do trópica úmido. Belem. No. 1+, 1980+. Circ. Técn. Centro Pesquisa Agropecu. Trop. Úmido. HI 64250

Circular technica, centro nacional de pesquisa de Gado de Corte. Campo Grande. No. 1+, 1980+. Circ. Techn. Centro Nac. Pesquisa Gado de Corte. HI 64251

Circular técnica, unidade regional de pesquisa florestal centro-sul. Curitiba. No. ?-4+, ?-1981+ Circ. Técn. Unid. Regional Pesquisa Florest. Centro-Sul. HI 64252

Circular, Texas forest service. College Station, TX. No. 1+, 1953+. Circ. Texas Forest Serv. HI 75240

Circular. United States department of agriculture. Washington, DC. Circ. U.S.D.A. See B–P–H 313/44. HI 53417 See also: United States department of agriculture. Circular. Washingon, DC. U.S.D.A. Circ. See B–P–H 941/6.

Circular, university of Hawaii agricultural experiment station = University of Hawaii agricultural experiment station. Circular. Honolulu, HI. Univ. Hawaii Agric. Exp. Sta. Circ. See B–P–H 935/16.

Circular. University of Idaho agricultural experiment station. Moscow, ID. Circ. Univ. Idaho Agric. Exp. Sta. See B–P–H 313/41. HI 53416

Circular, university of Michigan school of forestry and conservation. Ann Arbor, MI. Nos. 1-6, 1937-42. Circ. Univ. Michigan School Forest. Conservation. HI 64253

Circular, Utah agricultural college, agricultural experiment station = Utah agricultural college, agricultural experiment station. Circular. Logan, UT. Utah Agric. Exp. Sta. Circ. See B–P–H 945/30.

Circular, Vermont state agricultural college, agricultural experiment station = Vermont state agricultural college, agricultural experiment station. Circular. Burlington, VT. Vermont Agric. Exp. Sta. Circ. See B–P–H 956/23.

Circular, Washington (state) agricultural experiment station = Washington (state) agricultural experiment station. Circular. Pullman, WA. Wash. State Agric. Exp. Sta. Circ. See B–P–H 971/14.

Circular of the wild flower preservation society. Cincinnati, OH. Circ. Wild Fl. Preserv. Soc. See B–P–H 314/14. HI 53418

Circular, Wyoming agricultural college agricultural experiment station = Wyoming agricultural college agricultural experiment station. Circular. Laramie, WY. Wyoming Agric. Exp. Sta. Circ. See B–P–H 980/26.

Cistula entomologica. London. Cistula Entomol. See B–P–H 314/19. HI 53420

Citologija = Tsitologiya. Moscow & Leningrad.

Citrograph. Riverside, CA. Vol. 54(5)+, 1969+. Citrograph. Preceded by: California citrograph. HI 64254

Citrus. [Tachibana.] Hiroshima. Citrus (Hiroshima). See B–P–H 314/22. HI 53421

Citrus. [Kankitsu.] Shizuoka, Japan. Citrus (Shizuoka). See B–P–H 314/23. HI 53422

Citrus grower and sub-tropical fruit journal. Johannesburg. Nos. 453-468, 1971-72. Citrus Grower Sub-Trop. Fruit J. Preceded & superseded by: Citrus and sub-tropical fruit journal. HI 64255

Citrus industry. Tampa, FL. Citrus Industr. See B–P–H 314/26. HI 53423

Citrus leaves. Los Angeles, CA. Citrus Leaves. See B–P–H 314/27. HI 53424

Citrus news. Melbourne. Citrus News. See B–P–H 314/28. HI 53425

Citrus and sub-tropical fruit journal. Westcliffe. No. 451, 1971; no. 469+, 1973+. Citrus Sub-Trop. Fruit J. Preceded by: South african citrus journal. For nos. 453-468, 1971-72 see: Citrus grower and sub-tropical fruit journal. HI 64256

Citrus and vegetable magazine. Tampa, FL. 1938+. Citrus Veg. Mag. HI 64257

City trees. St. Paul, MN. 1972+. City Trees. HI 64258

Cladistics; the international journal of the Willi Hennig society. Westport, CT., London. Vol. 1+, 1985+. Cladistics. HI 64259

Clamer informa. Milan. Vol. ?-6(4)+, 19??-81+. Clamer Informa. HI 64260

Classification of dahlias. Wayne, NJ. 19??-85? Classific. Dahlias. Superseded by: Classification and handbook of dahlias. HI 64261

Classification and handbook of dahlias. [Forms a

supplement to: Bulletin of the american dahlia society.] n.p. 1986+. Classific. Handb. Dahlias. Preceded by: Classification of dahlias. HI 64262

Classification programs newsletter. Leicester. No. 1+, 1966+. Classific. Programs Newslett. HI 64263

Classification society bulletin. Cambridge, etc. No. 1+, 1965+. Classific. Soc. Bull. HI 64264

Claytonia. Lynchburg, VA. Claytonia. See B–P–H 314/30. HI 53426

Clemson agricultural college, agronomy department. Mimeographed series. Clemson, SC. Clemson Agric. Coll. Agron. Dept., Mimeogr. Ser. See B–P–H 314/31. HI 53427

Clemson agricultural college. Department of forestry, forest research series. Clemson, SC. Clemson Agric. Coll. Dept. Forest., Forest Res. Ser. See B–P–H 314/32. HI 53428

Clemson agricultural college of South Carolina. Agricultural experiment station. Annual report. Clemson, SC. Clemson Agric. Exp. Sta. Annual Rep. See B–P–H 314/33. HI 53429

Clemson agricultural college of South Carolina. Agricultural experiment station. Bulletin. Clemson, SC. Clemson Agric. Exp. Sta. Bull. See B–P–H 314/34. HI 53430

Clemson agricultural college of South Carolina. Agricultural experiment station. Circular. Clemson, SC. Clemson Agric. Exp. Sta. Circ. See B–P–H 315/1. HI 53431

Clemson agricultural college of South Carolina. Agricultural experiment station. Miscellaneous series. Clemson, SC. Clemson Agric. Exp. Sta., Misc. Ser. See B–P–H 315/2. HI 53432

Clemson agricultural college of South Carolina. Agricultural experiment station. Special circular. Clemson, SC. Clemson Agric. Exp. Sta. Special Circ. See B–P–H 315/3. HI 53433

Cleveland clinic quarterly. Cleveland, OH. Cleveland Clin. Quart. See B–P–H 315/5. HI 53434

Climatological bulletin, department of geography, McGill university. Montreal. No. 1+, 1967+. Climatol. Bull. Dep. Geogr. McGill Univ. HI 64265

Cling peach quarterly. San Francisco, CA. Cling Peach Quart. See B–P–H 315/24. HI 53447

Clinica, higiene e hidrologia. Lisbon. Clin. Hig. Hidrol. See B–P–H 315/11. HI 53437

Clinica oculistica. Palermo. Clin. Oculist. See B–P–H 315/16. HI 53440

Clinica otorinolaringoiatrica. Rome. Clin. Otorinolaringoiatr. See B–P–H 315/18. HI 53443

Clinica veterinaria. Milan. Clin. Veterin. (Milan). See B–P–H 315/22. HI 53445

Clinical microbiology newsletter. New York. Vol. 1+, 1979+. Clin. Microbiol. Newslett. HI 64266

Clinics in dermatology. Philadelphia, PA. Vol. 1+, 1983+. Clin. Dermatol. HI 64267

Clinique. Brussels. Clinique (Brussels). See B–P–H 316/1. HI 53448

Clinique. Chicago, IL. Clinique (Chicago). See B–P–H 316/2. HI 53449

Clinique. Montreal. Clinique (Montreal). See B–P–H 316/3. HI 53450

Clinique européenne. Paris. Clin. Eur. See B–P–H 315/9. HI 53435

Clinique française. Paris. Clin. Franç. See B–P–H 315/10. HI 53436

Clinique des hôpitaux des enfants et revue rétrospective médico-chirugicale et hygiénique. Paris. Clin. Hôp. Enfants. See B–P–H 315/12. HI 53438

Clinique des hôpitaux et de la ville. Paris. Clin. Hôp. Ville. See B–P–H 315/14. HI 53439

Clinique ophtalmologique. Paris. Clin. Ophtalmol. See B–P–H 315/17. HI 53442

Clinique pratique médico-chirurgicale et spéciale. Paris. Clin. Prat. Méd.-Chir. Spéciale. See B–P–H 315/19. HI 53444

Clinique vétérinaire. Paris. Clin. Vétérin. (Paris). See B–P–H 315/23. HI 53446

Clintonia; magazine of the Niagara frontier botanical society. Buffalo, NY. Vol. 1+, 1986+. Clintonia. Preceded by: Newsletter, Niagara frontier botanical society. HI 64268

Clio medica; acta academiae internationalis historiae medicinae. Oxford, Liège. Vol. 1+, 1966+. Clio Med. HI 64269

Clippings, Atlanta botanical garden = Atlanta botanical garden clippings. Atlanta, GA.

Clover information index. 1975+. Clover Inform. Index. HI 64270

Clover and special purpose legumes research newsletter = Newsletter, clover and special purpose legumes research. Madison, WI.

Clujul medical. Bucharest. Clujul Med. See B–P–H 316/6. HI 53451

C.M.I descriptions of pathogenic fungi and bacteria. Kew, England = C M I descriptions of pathogenic fungi and bacteria. Kew, England.

Coco y palma. Caracas. No. 1+, 1973+. Coco & Palma. HI 64271

Cocoa growers' bulletin. Bournville, England. Cocoa

Growers' Bull. See B–P–H 316/9. HI 53452

Coconut bulletin. Ernakulam. Vols. ?-20, 1957-66; n.s. 1-7, 1970-77. Coconut Bull. Preceded by: Bulletin, indian central coconut committee. Superseded by: Indian coconut journal. HI 64272

Cocos; journal of the coconut research institute of Sri Lanka. Lunuwila. Vol. 1+, 1983+. Cocos. Preceded by: Ceylon coconut quarterly. HI 64273

Coedwigwr: the forester; magazine of the forestry society of the university college of North Wales. Bangor, Wales. Coedwigwr. See B–P–H 316/10. HI 53453

CoEvolution quarterly. Sausalito, CA. Nos. 1-43, 1974-85? CoEvol. Quart. Superseded by: Whole earth review. HI 73350

Coffee. Turrialba. Vols. 1(1)-5(19), 1959-65. Coffee. Superseded by: Cafe. Turrialba. HI 64274

Coffee board of Kenya monthly bulletin = Monthly bulletin, coffee board of Kenya. Nairobi.

Coffee and cacao journal. Manila. Coffee Cacao J. See B–P–H 316/12. HI 53454

Coffin's botanical journal and medical reformer. Manchester, England. Coffin's Bot. J. Med. Reformer. See B–P–H 316/13. HI 53455

Cold Spring Harbor monographs. Cold Spring Harbor, NY. Cold Spring Harbor Monogr. See B–P–H 316/17. HI 53456

Cold Spring Harbor symposia on quantitative biology. Cold Spring Harbor, NY. Cold Spring Harbor Symp. Quant. Biol. See B–P–H 316/18. HI 53457

Colecção natura. Lisbon. 1922-28; 1964+. Colecç. Nat. HI 72381

Colecçion botanica Canaria. Las Palmas. Vol. 1+, 1981+. Colecç. Bot. Canaria. HI 64275

Colecçión ciencias biologicas. Caracas. Vol. 1+, 1963+. Colecç. Ci. Biol. (Caracas). HI 64276

Colecçion ciencias biologicas. Pamplona. Vol. ?-10+, 19??-80+. Colecç. Ci. Biol. (Pamplona). HI 64277

Colecçión científica, instituto nacional de antropología e historia. No. ?-79+, 19??-79+. Colecç. Ci. Inst. Nac. Antropol. Hist. HI 64278

Colecçion cientifica del instituto nacional de tecnologia agropecuaria. Buenos Aires. No. 1+, 1959+. Colecç. Ci. Inst. Nac. Tecnol. Nac. Tecnol. Agropecu. HI 64279

Colecçión docencia. Tegucigalpa. No. ?-5+, ?-ca.1984+. Colecç. Docen. HI 64280

Colecçion naturaleza espanola. Madrid. Vol. ?-2+, 19??-76+. Colecç. Naturaleza Esp. HI 64281

Colectânea; missão de biolgia marítima; junta de investigações do Ultramar. Lisbon. Colectânea. See B–P–H 316/19. HI 53458

Coletanea, museu de historia natural. Belo Horizonte. 1975+. Colet. Mus. Hist. Nat. HI 64282

Coleopterists' bulletin. Dryden, NY. Vol. 1+, 1947+. Coleopterists' Bull. HI 64283

Collana biologica, stazione sperimentale del sughero. Tempio Pausania. No. 1+, 1975+. Collana Biol. Staz. Sperim. Sughero. HI 64284

Collana culturale scientifica dell'istituto tecnico agrario di Vercelli. Vercelli. Vols. 1-9, 1948-59. Collana Cult. Sci. Ist. Tecn. Agrar. Vercelli. HI 64285

Collecção de livros ineditos de historia portugueza. Lisbon. Collecç. Livros Ined. Hist. Portug. See B–P–H 316/25. HI 53461

Collecção parques naturais. Lisbon. No. ?-7+, 19??-80+. Collecç. Parques Nat. HI 64286

Collectanea botanica; a barcionensi botanico instituto edita. Barcelona. Collect. Bot. (Barcelona). See B–P–H 317/4. HI 53464

Collectanea ad botanicam, chemiam, et historiam naturalem spectantia. Vienna. Collect. Bot. Spectantia (Vienna). See B–P–H 317/6. HI 53465

Collectanea medico-chirurgica (caesareae academiae medico-chirurgicae Vilnae). Vilna [Lithuanian S S R]. Collect. Med.-Chir. See B–P–H 317/11. HI 53470

Collectanea ad omnem rem botanicam spectantia. Zurich. Collect. Rem Bot. Spectantia (Zurich). See B–P–H 317/12. HI 53471

Collectaneorum supplementum = Collectanea ad botanicam, chemiam, et historiam naturalem spectantia. Vienna. Collect. Bot. Spectantia (Vienna). See B–P–H 317/6.

Collected papers, caraïbisch marien-biologisch instituut. Curaçao. Vol. 1+, 1956+. Collect. Pap. Caraïb. Marien-Biol. Inst. HI 64287

Collected papers, caribbean marine biological institute = Collected papers, caraïbisch marien-biologisch instituut. Curaçao.

Collected papers, freshwater biological association of the british empire = Freshwater biological association of the british empire; collected papers. Ambleside, England. Freshwater Biol. Assoc. Brit. Empire Collect. Pap. See B–P–H 385/4.

Collected papers, New Zealand oceanographic Institute = Collected reprints, New Zealand oceanographic Institute. Wellington, N.Z.

Collected reprints, Aitsu marine biological station. Aitsu. Vol. 1+, 1954/62+, 1968+. Collect. Repr. Aitsu Mar. Biol. Sta. HI 64288

Collected reprints, department of oceanography, Oregon state university = Oregon state university, school of science, department of oceanography, collected reprints. Corvallis, OR. Oregon State Univ. School Sci. Dept. Oceanogr. Collect. Repr. See B–P–H 690/25.

Collected reprints, Institut für Meereskunde an der Universität Kiel. Kiel. 1974+. Collec. Repr. Inst. Meeresk. Univ. Kiel. HI 64289

Collected reprints, institut za oceanografiju i ribarstvo = Zbornik radova, institut za oceanografiju i ribarstvo. Split.

Collected reprints, institute for oceanographic sciences. Wormley. 1974+. Collec. Repr. Inst. Oceanogr. Sci. Preceded by: National institute of oceanography, collected reprints. HI 64290

Collected reprints, Israel oceanographic and limnological research ltd. Haifa. Vol. 1+, 1971/72+, 1972+. Collec. Repr. Israel Oceanogr. Limnol. Res. HI 64291

Collected reprints, Lerner marine laboratory = Lerner marine laboratory; collected reprints. [American museum of natural history.] Bailey's Town, Bahamas. Lerner Mar. Lab. Collect. Repr. See B–P–H 529/2.

Collected reprints, Narragansett marine laboratory = Narragansett marine laboratory; collected reprints. Kingston, RI. Narragansett Mar. Lab. Collect. Repr. See B–P–H 627/25.

Collected reprints, New Zealand oceanographic institute = New Zealand oceanographic institute; collected reprints. Wellington. New Zealand Oceanogr. Inst. Collect. Repr. See B–P–H 657/21.

Collected reprints, ocean research institute, university of Tokyo. Tokyo. Nos. 1-10, 1962-71. Collec. Repr. Ocean Res. Inst. Univ. Tokyo. HI 64292

Collected reprints, scottish marine biological association = Scottish marine biological association, collected reprints. Edinburgh, Millport.

Collected reprints, university of Georgia marine institute. Athens, GA. Vol. 1+, 1955/58+, 1958?+. Collec. Repr. Univ. Georgia Mar. Inst. HI 64293

Collected reprints, Woods Hole oceanographic institution = Woods Hole oceanographic institution; collected reprints. Lancaster, PA. Woods Hole Oceanogr. Inst. Collect. Repr. See B–P–H 979/16.

Collected studies of the Mlyňany arboretum = Zbornik prác arboréta Mlyňany. Slepčany.

Collecting and breeding. [Saishu to shiiku.] Tokyo. Collect. & Breed. See B–P–H 317/7. HI 53466

Collection académique, composée des mémoires, actes ou journaux des plus célébres académies & sociétés littéraires, ... Partie étrangère. Dijon. Collect. Acad., Pt. Étrangère. See B–P–H 317/1. HI 53462

Collection académique, composée des mémoires, actes ou journaux des plus célébres académies & sociétés littéraires, ... Partie française. Dijon. Collect. Acad., Pt. Franç. See B–P–H 317/2. HI 53463

Collection forum; bulletin for the society for scientific collections (later society for the preservation of natural history collections). Ottawa, Trenton, NJ. Vol. [1]+, 1985+. Collect. Forum. HI 75264

Collection for improvement of husbandry and trade. London. Collect. Improv. Husb. Trade. See B–P–H 317/8. HI 53467

Collection of letters for the improvement of husbandry and trade. Collect. Lett. Improv. Husb. Trade. See B–P–H 317/9. HI 53468

Collection sauvegarde de la nature. Strasbourg. Vol. ?-6+. 19??-74+. Collec. Sauveg. Nat. HI 64294

Collections. Buffalo, NY. Vol. 56+, 1976+. Collections. Preceded by: Science on the march. HI 64295

Collections of the Massachusetts historical society. Boston, MA. Collect. Mass. Hist. Soc. See B–P–H 317/10. HI 53469

Collections of the New York Historical Society. New York. Vols. 1-5, 1809-30; ser. 2 vols. 1-4, 1841-59 [ser. 2 vol. 3(2) not published]. Collect. New-York Hist. Soc. 4-3033-2. HI 52480

Collections of the Ryukyu forestry experiment station. [Ryukyu ringyo shikenjo shuho.] Naha?, Okinawa. Collect. Ryukyu Forest. Exp. Sta. See B–P–H 317/14. HI 53472

College bulletin, north of Scotland college of agriculture. [Aberdeen.] No. 12+, 1976+. Coll. Bull. N. Scotland Coll. Agric. Preceded by: Bulletin, north of Scotland college of agriculture. HI 64296

College of Hawaii publications. Bulletins. Honolulu, HI. Coll. Hawaii Publ. Bull. See B–P–H 316/23. HI 53459

Collegit et communicavit; occasional newsletter of the australian orchid foundation. Essendon, Vic. Vol. 1+, 1985+. Colleg. Commun. HI 64297

Collegium. Frankfurt a. M. Collegium. See B–P–H 317/18. HI 53473

Collezione d'opuscoli scientifici e letterarj. Florence. Collez. Opusc. Sci. Lett. See B–P–H 317/20. HI 53474

Colloid journal = Kolloidnyi zhurnal. Voronezh, Moscow.

Colloid journal of the U S S R. [Translation of: Kolloidnyi zhurnal.] New York. Vol. 15+, 1953+. Colloid J. U.S.S.R. HI 64298

Colloques de l'I N R A. Paris. 1981+. Colloq. I. N. R. A. HI 64299

Colloques de l'institut national de la recherche

agronomique = Colloques de l'I N R A. Paris.

Colloques phytosociologiques. Vaduz, Stuttgart. Vol. 1+, 1971+. Colloq. Phytosoc. HI 64300

Colloquis, societat catalana de biologia. Barcelona. No. 1+, 1964+. Colloq. Soc. Catalana Biol. HI 64301

Colloquium geographicum; Vorträge des Bonner geographischen Kolloquiums zum Gedächtnis an Ferdinand von Richtofen. Bonn. Vol. 1+, 1951+. Colloq. Geogr. HI 64302

Colombia amazónica. Bogota. Vol. 1+, 1982+. Columbia Amazon. HI 64303

Colombia geográfica; revista del instituto geográfico "Agustin Codazzi." Bogota. Vol. 1+, 1970+. Colombia Geogr. HI 64304

Colonial plant and animal products. London. Vols. 1-9, 1950-56. Colon. Pl. Anim. Prod. Preceded by: Bulletin of the imperial institute. Superseded by: Tropical science. 2-1760-2. HI 64305

Colonial research publications. London. Nos. 1-25, 1944-61. Colon. Res. Publ. Superseded by: Overseas research publications. HI 64306

Colonial Williamsburg; the journal of the Colonial Williamsburg foundation. Williamsburg, VA. Vol. 6(3)+, 1984+. Colon. Williamsburg. Preceded by: Colonial Williamsburg today. HI 64307

Colonial Williamsburg today. Williamsburg, VA. Vols. 1-6(2), 1978-84. Colon. Williamsburg Today. Superseded by: Colonial Williamsburg. HI 64308

Colonias y foresta. Lima. Colon. & Foresta. See B–P–H 318/1. HI 53475

Colorado cactophiles cactivities. Denver, CO. 1967-71; nos. 1-7, 1972-78. Colorado Cactivities. Preceded by: Cactophiles cactivities. Superseded by: Colorado cactus and succulent society (newsletter). HI 64309

Colorado cactus and succulent society (newsletter). Denver, CO. No. 1+, 1981+. Colorado Cact. Succ. Soc. (Newslett.) HI 64310

Colorado college publication. Science series = Colorado college studies. Science series. Colorado Springs, CO.

Colorado college studies. Colorado Springs, CO. Vols. 1-10, 1890-1903. Colorado Coll. Stud. Superseded by: Colorado college studies. Science series. 2-1110-3. HI 64311

Colorado college studies. Science series. Colorado Springs, CO. Vols. 11-13(3), 1904-26. Colorado Coll. Stud. Sci. Ser. Preceded by: Colorado college studies. Superseded by: Colorado college studies. Studies series. 2-1111-1. HI 64312

Colorado college studies. Studies series. Colorado Springs, CO. Nos. 1-33, 1929-42. Colorado Coll. Stud. Stud. Ser. Preceded by: Colorado college studies. Science series. 2-1111-2. HI 64313

Colorado experiment station, annual report = Report (Annual) of the Colorado agricultural experiment station. Fort Collins, CO.

Colorado experiment station bulletin. Fort Collins, CO. Colorado Exp. Sta. Bull. See B–P–H 318/11. HI 53476

Colorado experiment station press bulletin = Agricultural experiment station of the agricultural college of Colorado. Press bulletin. Fort Collins, CO. Agric. Exp. Sta. Agric. Coll. Colorado Press Bull. See B–P–H 57/4.

Colorado farm and home research. Colorado agricultural experiment station. Fort Collins, CO. Colorado Farm Home Res. See B–P–H 318/12. HI 53477

Colorado forester. Fort Collins, CO. Colorado Forester. See B–P–H 318/13. HI 53478

Colorado fruit grower = Irrigation fruit grower and agriculturist. Paonia, CO. Irrig. Fruit Grower Agric. See B–P–H 438/8.

Colorado magazine. Denver, CO. Colorado Mag. See B–P–H 318/15. HI 53479

Colorado native plant society newsletter = Aquilegia; newsletter, Colorado native plant society. Fort Collins, CO.

Colorado school of mines quarterly. Golden, CO. 1906+. Colorado School Mines Quart. HI 64314

Colorado scientific society proceedings. Denver, CO. Colorado Sci. Soc. Proc. See B–P–H 318/16. HI 53480

Colorado scientific society yearbook = Colorado scientific society proceedings. Denver, CO. Colorado Sci. Soc. Proc. See B–P–H 318/16.

Colorado seed laboratory bulletin; Colorado agricultural experiment station. Fort Collins, CO. Colorado Seed Lab. Bull. See B–P–H 318/18. HI 53481

Colorado wheat grower. Denver, CO. Colorado Wheat Grower. See B–P–H 318/19. HI 53482

Coloured icones of rare and interesting fungi. [Supplement to: Nova hedwigia.] Lehre. Nos. 1-4, 1966-69. Coloured Icon. Rare Interesting Fungi. Superseded by: Fungorum rariorum icones coloratae. HI 64315

Coltivatore. Casale Monferrato, Italy. Coltivatore (Casale Monferrato). See B–P–H 318/23. HI 53484

Coltivatore. Rome. Coltivatore (Rome). See B–P–H 318/24. HI 53485

Coltivatore e giornale vinicolo italiano. Casale Monferrato, Italy. Coltiv. Giorn. Vinicolo Ital. See B–P–H 318/22. HI 53483

Colton's journal of geography and collateral sciences; a record of discovery, exploration, and survey. New York, NY. Colton's J. Geogr. Collat. Sci. See B–P–H 318/25. HI 53486

Coltsfoot. Shipman, VA. 1980+. Coltsfoot. HI 64316

Columbia college natural science series. New York, NY. Columbia Coll. Nat. Sci. Ser. See B–P–H 319/2. HI 53487

Columbia university studies in the history of american agriculture. New York, NY. Columbia Univ. Stud. Hist. Amer. Agric. See B–P–H 319/4. HI 53488

Combined annual report on the Partabgarh and Benares agricultural stations= Report on the agricultural stations in the eastern circle, United Provinces of Agra and Oudh. Allahabad.

Combined proceedings, international plant propogators' society. Vol. 13+, 1964+. Comb. Proc. Int. Pl. Propag. Soc. Preceded by: Combined proceedings, plant propogators' society. HI 64317

Combined proceedings, plant propogators society. Cleveland, OH. Vol. 10-12, 1960-62. Comb. Proc. Pl. Propogag. Soc. Preceded by: Proceedings of the plant propagators' society. Superseded by: Combined proceedings, international plant propogators' society. HI 64318

Combined rose list. Irvington, NY. No. 1+, 1984+. Comb. Rose List. HI 64319

Combined rose list. Cumulative supplement. Irvington, NY. No. 1+, 1984+. Comb. Rose List. Cumulative Suppl. HI 64320

Commentaries in plant science. Oxford, Elmsford, NY. Vol. ?-2+, 197?+. Comment. Pl. Sci. HI 64321

Commentarii academiae scientiarum imperialis petropolitanae. St. Petersburg. Comment. Acad. Sci. Imp. Petrop. See B–P–H 319/22. HI 53489

Commentarii de rebus in scientia naturali et medicina gestis. Leipzig. Comment. Rebus Sci. Nat. Med. Gestis. See B–P–H 319/31. HI 53493

Commentarii societatis regiae scientiarum gottingensis. Göttingen. Comment. Soc. Regiae Sci. Gott. See B–P–H 320/1. HI 53409

Commentario della fauna, flora e gea del Veneto e del Trentino. Venice. Comment. Fauna Veneto Trentino. See B–P–H 319/28. HI 53492

Commentario della società crittogamologica italiana. Genoa. Vols. 1-2, 1861-67. Comment. Soc. Crittog. Ital. Superseded by: Atti della società crittogamologica italiana. HI 64322

Commentarj della accademia di scienze, lettere, agricultura, ed arti del dipartimento del Mella. Brescia, Italy. Comment. Accad. Sci. Dip. Mella. See B–P–H 319/23. HI 53490

Commentarj dell'ateneo di Brescia. Brescia, Italy. Comment. Ateneo Brescia. See B–P–H 319/24. HI 53491

Commentationes biologicae. Helsinki. Nos. 1-90, 1922-77. Commentat. Biol. 2-1567-2. HI 64323

Commentationes forestales. Helsinki. Nos. 1-6, 1928-33. Commentat. Forest. HI 64324

Commentationes pontificiae academiae scientiarum. Rome. Commentat. Pontif. Acad. Sci. See B–P–H 320/5. HI 53410

Commentationes societatis physico-medicae apud universitatem literarum caesaream mosquensem institutae. Moscow. Commentat. Soc. Phys.-Med. Univ. Lit. Caes. Mosq. See B–P–H 320/6. HI 53411

Commentationes societatis regiae scientiarum gottingensis. Göttingen. Commentat. Soc. Regiae Sci. Gott. See B–P–H 320/7. HI 53412

Commentationes societatis regiae secientiarum gottingensis recentiores. Göttingen. Commentat. Soc. Regiae Sci. Gott. Recent. See B–P–H 320/8. HI 53413

Commentationes societatis scientiarum fennicae = Acta societatis scientiarum fennicae. Helsinki. Acad. Soc. Sci. Fenn. See B–P–H 49/22.

Comments on agricultural and food chemistry. London. Vol. 1+, 1987+. Comments Agric. Food Chem. HI 64325

Comments on modern biology. Part A, comments on molecular and cellular biophysics = Comments on molecular and cellular biophysics. New York, London.

Comments on molecular and cellular biophysics. New York, London. Vol. 1+, 1980+. Comments Molec. Cell. Biophys. HI 64326

Comments on theoretical biology. London. Vol. 1+, 198?+. Comments Theor. Biol. HI 64327

Commercial floriculture notes. Vol. 1+, 1967+. Commercial Floric. Notes. HI 64328

Commercial grower. London. No. 2971-4176, 1952-76. Commercial Grower. Preceded by: Fruit grower. Superseded by: Horticulture industry. HI 64329

Commercium litterarium ad rei medicae et scientiae naturalis incrementum institutum. Nuremberg. Commercium Litt. Rei Med. Sci. Nat. Incrementum. See B–P–H 320/11. HI 53414

Commissão das linhas telegráficas estratégicas de Matto Grosso ao Amazonas. Historia natural: anexo 5, botanica = Relatório, commissão das linhas telegráficas estratégicas de Matto Grosso ao Amazonas. Anexo. Rio de Janeiro.

Common property resource digest. St. Paul, MN. No. 1+,

1986+. Common Prop. Resource Digest. HI 64330

Common sense pest control quarterly. Berkeley, CA. Vol. 1+, 1984+. Common Sense Pest Control Quart. HI 64331

Commonwealth forestry handbook. London, Oxford. 1962+. Commonw. Forest. Handb. Preceded by: Empire forestry handbook. HI 64332

Commonwealth forestry review. London. Vol. 41(4)+, 1962+. Commonw. Forest Rev. Preceded by: Empire forestry review. HI 64333

Commonwealth journal of the society for growing australian plants. Padslois, Australia. Commonw. J. Soc. Growing Austral. Pl. See B–P–H 320/16. HI 53415

Commonwealth mycological institute. Miscellaneous publications. Kew, England. Commonw. Mycol. Inst. Misc. Publ. See B–P–H 320/18. HI 53416

Commonwealth phytopathological news. Kew. Vol. 1-10, 1955-71. Commonw. Phytopath. News. HI 64334

Communicationes instituti forestalis Čechosloveniae. Zbraslav, Czechoslovakia. Commun. Inst. Forest. Čechosloveniae. See B–P–H 321/3. HI 53421

Communicationes instituti forestalis fenniae. Helsinki. Vol. 101+, 1982+. Commun. Inst. Forest. Fenniae. Preceded by: Metsätutkimuslaitoksen julkausija. HI 64335

Communications, afdeling agrarisch onderzoek, koninklijk instituut voor de tropen = Communications of the department of agricultural research of the royal tropical institute. Amsterdam.

Communications, agricultural university, Wageningen = Mededeelingen van de landbouwhoogeschool te Wageningen. Wageningen.

Communicatons to the board of agriculture; on subjects relative to the husbandry, and internal improvement of the country. London. Commun. Board Agric. See B–P–H 320/21. HI 53417

Communications central research institute of forestry = Pengumuman lembaga pusat penjelidikan kehutanan. Bogor.

Communications, centre de cartographie phytosociologique et centre de recherche écologiques et phytosociologique de Gembloux. Brussels. Nos. 1-?, 1942-? Commun. Centre Cartogr. Phytosoc. Centre Rech. Ecol. Phytosoc. Gembloux. Superseded by: Communications du centre de recherches écologiques et phytosociologiques de Gembloux. HI 64336

Communications du centre d'écologie forestière et rurale. Brussels. N.s. vol. ?-24+, 19??-79+. Commun. Centre Ecol. Forest. Rurale. Preceded by?: Communications du centre de recherches écologiques et phytosociologique de Gembloux. HI 64337

Communications du centre de recherches écologiques et phytosociologiques de Gembloux. Brussels. 1948+. Commun. Centre Rech. Écol. Phytosoc. Gembloux. Preceded by: Communications, centre de cartographie phytosociologique et centre de recherches écologiques. Superseded by?: Communications du centre d'écologie forestière et rurale. HI 64338

Communications of the department of agricultural research of the royal tropical institute. Amsterdam. Vol. 53+, 1961+. Commun. Dept. Agric. Res. Roy. Trop. Inst. Preceded by: Mededelingen van de afdeling tropische producten van het koninklijk instituut voor de tropen. HI 64339

Communications, department of plant physiology, rijksuniversiteit, Groningen. Groningen. No. ?-10+, 19??-70+. Commun. Dep. Pl. Physiol. Rijks-Univ. Groningen. HI 64340

Communications, department of systematic botany, rijksuniversiteit, Groningen [Nos. 1-5 published without a serial title]. Groningen. No. 1+, 1961+. Commun. Dep. Syst. Bot. Rijks-Univ. Groningen. HI 64341

Communications de la faculté des sciences de l'université d'Ankara. Istanbul. Commun. Fac. Sci. Univ. Ankara. See B–P–H 320/22. HI 53418

Communications de la faculté des sciences de l'université d'Ankara. Série C. Sciences naturelles. Ankara. Commun. Fac. Sci. Univ. Ankara, Sér. C, Sci. Nat. See B–P–H 320/23. HI 53419

Communications, forest products research institute = Pengumuman, lembaga penelitian hasil hutan. Bogor.

Communications of the forest research institute (Indonesia). Bogor. Nos. 33-38, 1951-53. Commun. Forest Res. Inst., Bogor. Preceded by: Mededeelingen van het bosbouwproefstation. Superseded by: Pengumuman balai besar penjelidikan kehutanan, Indonesia. HI 64342

Communication from the horticultural institute, Taihoku imperial university. Taihoku [=Taipei, Taiwan]. Commun. Hort. Inst. Taihoku Imp. Univ. See B–P–H 321/2. HI 53420

Communications of the institute of phytopathological research = Mededelingen van het instituut voor plantenziektenkundig onderzoek. Wageningen.

Communications from the national botanic garden. Siniloan, Laguna. No. 1+, 1971+. Commun. Natl. Bot. Gard., Siniloan. HI 64343

Communications of the research institute for estate crops = Laporan tahunan, balai penelitian perkebunan. Bogor.

Communications on the science and practice of brewing. New York, NY. Commun. Sci. Pract. Brew. See B–P–H 321/4. HI 53422

Communications in soil science and plant analysis. New York. Vol. 1+, 1970+. Commun. Soil Sci. Pl. Analysis. HI 64344

Communication de la station internationale géobotanique méditeranéenne et alpine, Montpellier. Montpellier. Commun. Stat. Int. Gébot. Médit. Montpellier. See B–P–H 321/8. HI 53423

Communications, Wallerstein laboratories = Wallerstein laboratories communications. New York, NY. Wallerstein Lab. Commun. See B–P–H 971/1.

Comoara satelor. Blaj, Rumania. Comoara Satelor. See B–P–H 321/15. HI 53424

Compact fruit tree; newsletter, dwarf fruit tree association. East Lansing, MI. Vol. 1+, 1958+. Compact Fruit Tree. HI 64345

Companion to the botanical magazine. London. Companion Bot. Mag. See B–P–H 321/20. HI 53425

Comparative biochemistry and physiology. Oxford, New York. Vols. 1-37, 1961-70. Comp. Biochem. Physiol. Superseded by: Comparative biochemistry and physiology. B, comparative biochemistry and Comparative biochemistry and physiology. C, comparative pharmacology and toxicology. HI 53361

Comparative biochemistry and physiology. B, comparative biochemistry. New York. Vol. 38+, 1971+. Comp. Biochem. Physiol., B, Comp. Biochem. Preceded by: Comparative biochemistry and physiology. HI 53398

Comparative biochemistry and physiology. C, comparative pharmacology and toxicology. New York. Vol. 50+, 1975+. Comp. Biochem. Physiol., C, Comp. Pharmacol. Preceded by: Comparative biochemistry and physiology. HI 64190

Comparative and general pharmacology. Bristol, London, etc. Vols. 1-5, 1970-74. Comp. Gen. Pharmacol. HI 64525

Comparative immunology, microbiology and infectious diseases. Elmsford, NY. Vol. 1+, 1978+. Comp. Immunol. Microbiol. Infect. Dis. HI 64526

Comparative pathobiology. New York. Vol. 1+, 1969+. Comp. Pathobiol. HI 75212

Comparative physiology and ecology. Jodhpur. Vol. 1+, 1976+. Comp. Physiol. Ecol. HI 64527

Compendium historiae litterariae novissimae; oder Erlangische Gelehrte Anmerkungen und Nachrichten. Erlangen. Compend. Hist. Litt. Noviss. See B–P–H 321/21. HI 53426

Compositae newsletter. Columbus, OH, [etc.]. Vol. 1+, [1975]+. Compositae Newslett. HI 64528

Compost science. Emmaus, PA. Vols. 1-18, 1960-77. Compost Sci. Superseded by: Compost science/land utilization. HI 64529

Compost science/land utilization. Emmaus, PA. Vols. 19-21, 1978-80. Compost Sci. Land Utiliz. Preceded by: Compost science. Superseded by: BioCycle. HI 64530

Comprehensive summaries of Uppsala dissertations from the faculty of science. [Forms part of: Acta universitatis upsaliensis.] Stockholm. Vol. 1+, 1985+. Compreh. Summ. Uppsala Diss. Fac. Sci. Preceded by: Abstracts of Uppsala dissertations from the faculty of science. HI 64531

Compte rendu, see Comptes rendus

Comptes rendus, A E T F A T = Comptes rendus, réunion plénière, association pour l'étude taxonomique de la flore d'Afrique tropicale.

Comptes rendu, académie agricole "Georgi Dimitrov" = Doklady, selskostopanska akademiya "George Dimitrov." Sofia.

Comptes rendus de l'académie bulgare des sciences. Sciences mathématiques et naturelles = Doklady bulgarskoi akademiia nauk. Sofia.

Compte-rendu de l'académie impériale des sciences (de Saint Pétersbourg). St. Petersburg. 1824-58. Compt.-Rend. Acad. Imp. Sci. Superseded by: Otchet o Prisuzhdenii Akademieyu Nauk. 1-111-2. HI 64532

Compte rendu de l'académie des sciences agricoles en Bulgarie = Doklady, akademiya na selskostopanskite nauki, Sofia. Sofia.

Compte rendu de l'académie des sciences, Paris. Série générale, vie des sciences. Paris. Vol. 1+, 1984+. Compt. Rend. Acad. Sci. Paris, Sér. Gén. Vie Sci. HI 64533

Compte rendu de l'académie des sciences, Paris. Série 3, Sciences de la vie. Paris. Vol. 298+, 1984+. Compt. Rend. Acad. Sci. Paris, Sér. 3, Sci. Vie. Preceded by: Compte rendu des séances de l'académie des sciences, Paris. Série 3, sciences de la vie. HI 64534

Comptes-rendus de l'académie des sciences de Russie = Doklady Rossiiskoi akademii nauk. Petrograd.

Comptes-rendus (Doklady) de l'académie des sciences de l'URSS. Moscow & Leningrad. Compt.-Rend. (Dokl.) Acad. Sci. URSS. See B–P–H 322/3. HI 53428

Comptes-rendus de l'académie des sciences de l'URSS = Doklady akademii nauk S S S R. Moscow & Leningrad.

Compte-rendu annuel de la société royale des lettres et des sciences de Bohême = Výrochni zpráva královské České společnosti nauk. Prague.

Compte rendu de l'association pour la création et l'entretien de réserves naturelles dans le canton de Genève. Geneva. Nos. 1-21, 1930-51. Compt. Rend.

Assoc. Pour Création Entret. Réserves Nat. Cant. Genève. HI 64535

Comptes rendus, association pour l'étude taxonomique de la flore d'Afrique tropicale = Comptes rendus, réunion plénière, association pour l'étude taxonomique de la flore d'Afrique tropicale.

Compte rendu de l'association française pour l'avancement des sciences. Paris. 1872+. Compt. Rend. Assoc. Franç. Avancem. Sci. HI 64536

Compte-rendu de l'association strasbourgeoise des amis de l'histoire naturelle. Strasbourg. Compt.-Rend. Assoc. Strasbourg. Amis Hist. Nat. See B–P–H 322/2. HI 53427

Compte rendu, congrès d'hiver, institut international de recherches betteravières. Brussels. 21st-27th, 1958-64; 39th+, 1976+ [reports of 1st-20th congresses not published]. Comp. Rend. Congr. Hiver Inst. Int. Rech. Betterav. For 28th-38th, 1965-75 see: I I R B. Journal of the international institute for sugar beet research. HI 64537

Compte rendu du congrès national de la culture des plantes médicinales. Lons-le-Saulnier, Paris. Vols. 3-6, 192?-26. Comp. Rend. Cong. Natl. Cult. Pl. Méd. Preceded by: Mémoires et comptes rendus du congrès national de la culture des plantes médicinales. HI 64538

Comptes rendus du congrès national des sociétés savantes. Section des sciences. Limoges, Paris. 1961+. Compt. Rend. Congr. Natl. Soc. Savantes, Sec. Sci. Preceded by: Comptes rendus du congrès des sociétés savantes de Paris et des départements. Section des sciences. HI 64539

Comptes rendus du congrès des sociétés savantes de Paris et des départements. Section des sciences. Limoges, Paris. 1896-1960 [suspended 1915-19, 1931-32, 1940-49]. Compt. Rend. Congr. Soc. Savantes Paris Dép., Sec. Sci. Superseded by: Comptes rendus du congrès national des sociétés savantes. Section des sciences. HI 64540

Comptes rendus, descriptions, notes et communications, société botanique rochelaise = Comptes rendus des excursions botaniques, société botanique rochelaise. La Rochelle.

Comptes rendus des excursions botaniques, société botanique rochelaise. La Rochelle. Vols. 1-10, 1879-88. Compt. Rend. Excurs. Bot. Soc. Bot. Rochel. Superseded by: Bulletin de la société botanique rochelaise. HI 64541

Comptes rendus hebdomadaires des séances de l'académie d'agriculture de France. Paris. Vol. 1+, 1932+. Compt. Rend. Hebd. Séances Acad. Agric. France. Preceded by: Comptes rendus des séance de l'académie d'agriculture de France. HI 64542

Comptes rendus hebdomadaires des séances de l'académie des sciences. Paris. Compt. Rend. Hebd. Séances Acad. Sci. See B–P–H 322/4. HI 53429

Compte rendu hebdomadaire des séances de l'académie des sciences. Série D, sciences naturelles. Paris. Vols. 262-291, 1966-80. Compt. Rend. Hebd. Séances Acad. Sci., Sér. D. Preceded by: Compte rendu hebdomadaire des séances de l'académie des sciences. Superseded by: Compte rendu des séances de l'académie des sciences. Paris. HI 64543

Comptes-rendus hebdomadaires des séances et mémoires de la société de biologie et de ses filiales. Paris. Vols. 84+, 1921+. Compt.-Rend. Hebd. Séances Mém. Soc. Biol. Preceded by: Comptes-rendus des séances et mémoires de la société de biologie [title varies]. 5-3944-1. HI 64544

Comptes rendus mensuels des séances; académie polonaise des sciences et des lettres. Classe des sciences mathématiques et naturelles. Cracow. Compt. Rend. Mens. Séances Acad. Polon. Sci., Cl. Sci. Math. See B–P–H 322/7. HI 53430

Comptes rendus mensuels des séances de l'académie des sciences coloniales. Paris. Vols. 1-17, 1941-57. Compt. Rend. Mens. Séances Acad. Sci. Colon. Preceded by: Comptes rendus des séances de l'académie des sciences coloniales. Communications. Superseded by: Comptes rendus mensuels des séances de l'académie des sciences d'Outre-Mer. HI 64545

Comptes rendus mensuels des séances de l'académie des sciences d'Outre-Mer. Paris. Vols. 18-30, 1958-70. Compt. Rend. Mens. Séances Acad. Sci. Outre-Mer. Preceded by: Comptes rendus mensuels des séances de l'académie des sciences coloniales. Superseded by: Comptes rendus trimestriels des séances de l'académie des sciences d'Outre-Mer. HI 64546

Comptes-rendus mensuels des séances de la société des sciences naturelles et physiques de Maroc = Comptes-rendus des séances de la société des sciences naturelles et physiques de Maroc. Rabat.

Compte rendu de recherches, institut pour l'encouragement de la recherche scientifique dans l'industrie et l'agriculture. Brussels. No. 1+, 1950+. Compt. Rend. Rech. Inst. Pour Encour. Rech. Sci. Industr. Agric. HI 64547

Compte rendu des recherches, station d'amélioration des plantes, Gembloux. Gembloux. 1980+. Compt. Rend Rech. Stat. Amélior. Pl. Gembloux. HI 64548

Compte rendu des recherches. Travaux du comité pour l'établissement de la carte des sols et de la végétation de la Belgique. Brussels. Vol. 1+, 194?+. Compt. Rend. Rech. Trav. Comité établiss. Carte Sols Vég. Belgique. HI 64549

Comptes rendus, réunion plénière, association pour

l'étude taxonomique de la flore d'Afrique tropicale. Lisbon. 1st+, [1957?]+. Compt. Rend. Réun. Plén. Assoc. Pour Étude Taxon. Fl. Afrique Trop. HI 64550

Comptes rendus des séances de l'académie d'agriculture de France. Paris. Vol. 1-17, 1915-31. Compt. Rend. Séances Acad. Agric. France. Preceded by: Bulletin des séances de la société nationale d'agriculture. Superseded by: Comptes rendus hebdomadaires des séances de l'académie d'agriculture de France. 1-25-3. HI 64551

Compte rendu des séances de l'académie des sciences. Série 3, sciences de la vie. Paris. Vol. 292-297, 1981-83. Comp. Rend. Séances Acad. Sci., Sér. 3, Sci. Vie. Preceded by: Compte rendu hebdomadaire des séances de l'académie des sciences. Série D, sciences naturelles. Superseded by: Compte rendu de l'académie des sciences. Paris. HI 64552

Comptes rendus des séances de l'académie des sciences coloniales. Communications. Paris. Vols. 1-29, 1922-38 [vols. 20, 22, 24, 26 & 28 not published]. Compt. Rend. Séances Acad. Sci. Colon., Commun. Superseded by: Comptes rendus mensuels des séances de l'académie des sciences coloniales. 1-30-2. HI 64553

Comptes-rendus des séances. Communications de l'académie des sciences coloniales. Paris. Compt.-Rend. Séances Commun. Acad. Sci. Colon. See B–P–H 323/10. HI 53439

Compte rendu de la séance générale, association Strasbourgeoise des amis de l'histoire naturelle = Association Strasbourgeoise des amis de l'histoire naturelle. Compte rendu de la séance générale. Strasbourg. Assoc. Strasbourg. Amis Hist. Nat. Compt. Rend. Séance Gén. See B–P–H 156/22.

Comptes-rendus des séances et mémoires de la société de biologie [title varies]. Paris. Vols. 1-83, 1849-1920 [also numbered in series of 5 vols. each]. Compt.-Rend. Séances Mém. Soc. Biol. Superseded by: Comptes-rendus hebdomadaires des séances et mémoires de la société de biologie et de ses filiales. 5-3944-1. HI 64554

Comptes-rendus des séances mensuelles de la société des sciences naturelles et physiques de Maroc = Comptes-rendus des séances de la société des sciences naturelles et physiques de Maroc. Rabat.

Comptes rendus des séances publiques de l'institut royal des Pays-Bas. Amsterdam. Compt. Rend. Séances Publiques Inst. Roy. Pays-Bas. See B–P–H 322/11. HI 53431

Compte rendu des séances de la société de biogéographie. Paris. Vol. 45+, 1969+. Compt. Rend. Séances Soc. Biogéogr. Preceded by: Compte rendu sommaire des séances de la société de biogéographie. HI 64555

Compte-rendu des séances de la société de Biologie = Comptes-rendus des séances et mémoires de la société de biologie [title varies]. Paris.

Compte rendu des séances de la société de biologie et ses filiales = Comptes-rendus hebdomadaires des séances et mémoires de la société de biologie et de ses filiales. Paris.

Comptes rendus des séances, société botanique de Genève = Recueil des comptes-rendus de la société botanique de Genève. Geneva.

Compte rendu des séances, société des naturalistes luxembourgeois. Luxembourg. 1891-1906. Comp. Rend. Séances Soc. Naturalistes Luxemb. Superseded by: Bulletin mensuel de la société des naturalistes luxembourgeois. HI 64556

Compte rendu des séances de la société de physique et d'histoire naturelle de Genève. Geneva. Vol. 1+, 1885+. Compt. Rend. Séances Soc. Phys. Genève. From 1932-65 contained in: Archives des sciences physiques et naturelles and Archives des sciences. 5-3949-1. HI 64557

Comptes-rendus des séances de la société des sciences naturelles et physiques de Maroc. Rabat. Vols. 1-38, 1935-72. Compt.-Rend. Séances Soc. Sci. Nat. Maroc. 5-3957-1. HI 53440

Comptes rendus des séances de la société scientifique de Timisoara = Bulletin scientifique de l'école polytechnique de Timisoara. Timisoara, Rumania. Bull. Sci. École Polytechn. Timisoara. See B–P–H 271/25.

Compte-rendu de la session, société linnéenne du nord de la France = Société linnéenne du nord de la France. Compte-rendu de la session. Amiens, France. Soc. Linn. N. France Compt.-Rend. Session. See B–P–H 845/14.

Comptes-rendus de la société botanique de Genève = Recueil des comptes-rendus de la société botanique de Genève. Geneva.

Comptes rendus, société botanique rochelaise = Comptes rendus des excursions botaniques, société botanique rochelaise. La Rochelle.

Comptes-rendus de la société helvétique des sciences naturelles = Verhandlungen der schweizerischen naturforschenden Gesellschaft. Basel.

Compte rendu de la société linnéenne du nord de la France. Abbeville, France. Compt. Rend. Soc. Linn. N. France. See B–P–H 322/18. HI 53432

Compte rendu à la société royale d'agriculture de Paris. Paris. Compt. Rend. Soc. Roy. Agric. Paris. See B–P–H 322/21. HI 53433

Comptes rendus de la société des sciences et des lettres de Wrocław. Wroclaw, Poland. Compt. Rend. Soc. Sci.

Wrocław. See B–P–H 322/22. HI 53434

Compte rendu sommaire des séances de la société de biogéographie. Paris. Vols. 1-44, 1924-68. Compt. Rend. Sommaire Séances Soc. Biogéogr. Superseded by: Compte rendu des séances de la société de biogéographie. 5-3944-1. HI 53435

Comptes-rendus des travaux du Carlsberg laboratoriet. Copenhagen. Vols. 1-20, 1878-1935; vols. 31-40, 1959-76. Compt.-Rend. Trav. Carlsberg Lab. Preceded by: Meddelelser fra Carlsberg laboratoriet. From 1935-58 published as: Comptes-rendus des travaux du Carlsberg laboratiet; série physiologique. Superseded by: Carlsberg research communications. 2-928-3. HI 64558

Comptes-rendus des travaux du Carlsberg laboratoriet; série physiologique. Copenhagen. Vols. 21-30 1935-58. Compt.-Rend. Trav. Carlsberg Lab., Sér. Physiol. Preceded and superseded by: Comptes-rendus des travaux du Carlsberg laboratoriet. 2-929-1. HI 64559

Comptes-rendus des travaux de la faculté des sciences de Marseille. Marseille. Compt.-Rend. Trav. Fac. Sci. Marseille. See B–P–H 323/16. HI 53442

Comptes-rendus des travaux du laboratoire Carlsberg = Comptes-rendus des travaux du Carlsberg laboratoriet. Copenhagen.

Comptes-rendus des travaux du laboratoire Carlsberg; série physiologique = Comptes-rendus des travaux du Carlsberg laboratoriet; série physiologique. Copenhagen.

Compte rendu des travaux de la société d'agriculture, histoire naturelle et arts utiles de Lyon. Lyons. Compt. Rend. Trav. Soc. Agric. Lyon See B–P–H 323/4. HI 53436

Compte-rendu des travaux de la société hallérienne. Geneva. Compt.-Rend. Trav. Soc. Hallér. See B–P–H 323/17. HI 53443

Comptes-rendus des travaux de la société helvétique des sciences naturelles = Verhandlungen der schweizerischen naturforschenden Gesellschaft. Basel.

Comptes rendus des travaux de la société linnéenne de Lyon. Lyons. Compt. Rend. Trav. Soc. Linn. Lyon. See B–P–H 323/5. HI 53437

Compte rendu des travaux de la société linnéenne de Paris. Paris. Vols. ?-2, 18??-23. Comp. Rend. Trav. Soc. Linn. Paris. HI 64560

Compte rendu des travaux de la société nationale havraise d'études diverses. Le Havre. Compt. Rend. Trav. Soc. Natl. Havraise Études Diverses. See B–P–H 323/6. HI 53438

Compte-rendu des travaux des vacances de la station biologique du Volga, organisée par la société des naturalistes à Saratow = Otchet Volzhskoi Biologicheskoi Stantsii Saratovskogo Obshchestva Estestvoispytatelei i Lyubitelei Estestvoznaniya. Saratov.

Compte rendu trimestriels des séances de l'académie des sciences d'Outre-Mer. Paris. Vols. 31-37, 1971-77. Comp. Rend. Trimestriels Séances Acad. Sci. Outre-Mer. Preceded by: Compte rendu mensuels des séances de l'académie des sciences d'Outre-Mer. Superseded by: Mondes et culture. HI 64561

Compte rendu, union internationale des sciences biologiques = Proceedings, international union of biological sciences, general assemblies.

Comptes rendus see Compte rendu

Computer applications in the biosciences. Washington, DC. Vol. 1+, 1985+. Computer Applic. Biosci. HI 64562

Computers in biology and medicine. New York, Oxford, etc. Vol. 1+, 1970+. Computers Biol. Med. HI 64563

Computers in the environmental sciences. Norwich. Vols. 1+, 1971+. Computers Environm. Sci. HI 64564

Computers in forestry series. Florence, AL. Vol. ?-2+, 19??-84+. Computers Forest. Ser. HI 64565

Comunicaciones botánicas del museo de historia natural de Montevideo. Montevideo. Comun. Bot. Mus. Hist. Nat. Montevideo. See B–P–H 324/1. HI 53446

Comunicaciones científicas del museo de La Plata. La Plata, Argentina. Comun. Ci. Mus. La Plata. See B–P–H 324/2. HI 53447

Comunicaciones, herbarium Cornelius Osten. Montevideo. Vol. 1-2, 1925-32. Comun. Herb. Cornelius Osten. 3-1842-3. HI 64566

Comunicaciones I N I A. Série general. Madrid. No. 1+, 1976+. Commun. I. N. I. A., Sér. Gen. HI 64567

Comunicaciones I N I A. Série produccion vegetal. Madrid. No. 1+, 1973+. Comun. I. N. I. A., Sér. Prod. Veg. HI 64567

Comunicaciones I N I A. Série protección vegètal. Madrid. Vol. 3+, 1975+. Comun. I. N. I. A., Sér. Protecc. Veg. Preceded by: Comunicaciones instituto nacional de investigaciones agrarias. HI 64568

Comunicaciones I N I A. Série recursos naturales. Madrid. No. 1+, 1974+. Comun. I. N. I. A., Sér. Recurs. Nat. HI 64569

Comunicaciones instituto nacional de investigaciones agrarias. Madrid. Vols. 1-2, 1972-7? Comun. Inst. Nac. Invest. Agrar. Superseded by: Comunicaciones I N I A. Série protección Vegetal. HI 64570

Comunicaciones, instituto nacional de investigaciones agrarias. Série general = Comunicaciones I N I A.

Série general. Madrid.

Comunicaciones, instituto nacional de investigaciones agrarias. Série produccion vegetal = Comunicaciones I N I A. Série produccion vegetal. Madrid.

Comunicaciones del instituto nacional de investigaciones de las ciencias naturales anexo al museo argentino de ciencias naturales "Bernardino Rivadavia". Serie ciencias botánicas. Buenos Aires. Vols. 1(1-6), 1948-54. Comun. Inst. Nac. Invest. Ci. Nat., Cer. Ci. Bot. Superseded by: Comunicaciones del museo argentino de ciencias naturales "Bernardino Rivadavia" e instituto nacional de investigaciones de las ciencias naturales. Ciencias botanicas. HI 64571

Comunicaciones. Instituto tropical de investigaciones científicas, universidad de El Salvador. San Salvador, Salvador. Comun. Inst. Trop. Invest. Ci. Univ. El Salvador. See B–P–H 324/6. HI 53448

Comunicaciones del museo argentino de ciencias naturales "Bernardino Rivadavia" e instituto nacional de investigaciones de las ciencias naturales. Ciencias botánicas. Buenos Aires. Vol. 1+, 1960+. Comun. Mus. Argent. Ci. Nat. "Bernardino Rivadavia", Ci. Bot. Preceded by: Comunicaciones del instituto nacional de investigaciones de la ciencias naturales anexo al museo argentino de ciencias naturales "Bernardino Rivadavia". Serie ciencias botánicas. 1-818-2. HI 64572

Comunicaciones del museo argentino de ciencias naturales "Bernardino Rivadavia" e instituto nacional de investigaciones de las ciencias naturales. Ecologia. Buenos Aires. Vol. 1+, 1968+. Comun. Mus. Argent. Ci. Nat. "Bernardino Rivadavia", Ecol. HI 64573

Comunicaciones del museo argentino de ciencias naturales "Bernardino Rivadavia" e instituto nacional de investigaciones de las ciencias naturales. Hidrobiología. Buenos Aires. Comun. Mus. Argent. Ci. Nat. "Bernardino Rivadavia," Hidrobiol. See B–P–H 324/9. HI 53449

Comunicaciones del museo argentino de ciencias naturales "Bernardino Rivadavia" e instituto nacional de investigación de las ciencias naturales. Parasitología. Buenos Aires. Vol. 1+, 1965+. Comun. Mus. Argent. Cienc. Nat. "Bernardino Rivadavia", Parasitol. HI 64574

Comunicaciones del museo nacional de Buenos Aires. Buenos Aires. Vols. 1-2, 1898-1901. Comun. Mus. Nac. B. Aires. Superseded by: Comunicaciones del museo nacional de historia natural Bernardino Rivadavia. 1-818-2. HI 64575

Comunicaciones del museo nacional de historia natural Bernardino Rivadavia. Buenos Aires. 1923-25. Comun. Mus. Nac. Hist. Nat. Bernardino Rivadavia. Preceded by: Comunicaciones del museo nacional de Buenos Aires. HI 64576

Comunicaciones del museo provincial de ciencias naturales "Florentine Ameghino". Santa Fe. No. 1+, 1967+. Comun. Mus. Prov. Ci. Nat. "Florentine Ameghino". HI 64577

Communicaciones; revista del instituto tropical de investigaciones científicas. San Salvador, Salvador. Comunicaciones. See B–P–H 324/11. HI 53451

Comunicações, instituto de investigaçao agronomica de Moçambique. Lourenço Marques. Vol. ?-81+, 19??-73+. Comun. Inst. Invest. Agron. Moçambique. HI 64578

Comunicações do museu de ciências da P U C R G S. Série botânica. Porto Alegre. No. 30+, 1985+. Comun. Mus. Ci. P. U. C. R. G. S., Sér. Bot. HI 64579

Comunicações dos serviços geológicos de Portugal. Lisbon. Comun. Serv. Geol. Portugal. See B–P–H 324/10. HI 53450

Comunicações técnicas, projecto de desenvolvimento e pesquisa florestal. Brasilia. Nos. 1-19, 1976-77. Comun. Técn. Proj. Desenvolv. Pesq. Florest. HI 64580

Comunicado cientifico do instituto de micologia. Recife. Nos. 1-3, 1957. Comunicado Ci. Inst. Micol. HI 64581

Comunicado técnico, empresa de pesquisa agropecuária da Bahia. Salvador. 19??-79+. Comunicado Técn. Empresa Pesq. Agropecu. Bahia. HI 64582

Comunicări de botanică. Soc. de ştiinţe naturale şi geografie din Republica Populară Romînă. Bucharest. Comun. Bot. See B–P–H 323/22. HI 53445

Comunicările academiei Republicii Populare Romîne. Bucharest. Comun. Acad. Republ. Populare Romîne. See B–P–H 323/21. HI 53444

ConCiencia. Panama. 1974+. ConCiencia. HI 64583

Concise report, central sericultural research station = Report (Annual), central sericultural research station. Berhampore.

Conference on maple products. Report of proceedings. Philadelphia, Pa. Conf. Maple Prod. Rep. Proc. See B–P–H 324/14. HI 53452

Conference papers index: life sciences, physical sciences, engineering. Chestnut Hills, MA., Bethesda, MD. Vol. 6+, 1978+. Conf. Pap. Index, Life Sci. Phys. Sci. Engin. Preceded by: Current programe: materials science and technology [not entered]. HI 64584

Conference reports, agriculture, food, and biology. London. Vol. 1+, 1977+. Conf. Rep. Agric Food Biol. HI 64585

Conference reports of the botanical society of the British

Isles. London. [4th]+, 1954+. Conf. Rep. Bot. Soc. Brit. Isles. Preceded by: Report of the conference of the botanical society of the British Isles. 1-756-1. HI 64586

Conférences et documents, laboratoire d'océanographie biologique, université de Rennes. Rennes. 1971-72. Conf. Doc. Lab. Océanogr. Biol. Univ. Rennes. Preceded by: Conferences, laboratoire d'océanographie biologique, université de Rennes. HI 64587

Conférences, laboratoire d'océanographie biologique, université de Rennes. Rennes. Nos. 1-?, 1969-70. Conf. Lab. Océanogr. Biol. Univ. Rennes. Superseded by: Conferences et documents, laboratoire d'océanographie biologique, université de Rennes. HI 64588

Conferencias, instituto nacional de investigaciones agronomicas. Madrid. 1957/58-66/67, 1957-67. Conf. Inst. Nac. Invest. Agron. HI 64589

Conferencias y reseñas cientificas de la real sociedad española de historia natural. Madrid. Vols. 1-5, 1926-30. Conf. Reseñas Ci. Real Soc. Esp. Hist. Nat. Superseded by: Reseñas cientificas de la sociedad española de historia natural. HI 64590

Congrès scientifiques de France. Rouen [Place of publication changes each year]. Congr. Sci. France. See B–P–H 324/17. HI 53453

Congress gazette, international horticultural congress. Nos. 1-8, 1962. Congr. Gaz. Int. Hort. Congr. HI 64591

Congress reports, international plant breeders association for the protection of new varieties. Paris. Vol. 1+, 1975+. Congr. Rep. Int. Pl. Breed. Assoc. Protect. New Var. HI 64592

Connecticut agricultural experiment station. Bulletin of immediate information. New Haven, CT. Connecticut Agric. Exp. Sta. Bull. Immed. Inform. See B–P–H 324/22. HI 53454

Connecticut agricultural experiment station. Bulletin of the tobacco sub-station. New Haven, CT. Connecticut Agric. Exp. Sta. Bull. Tobacco Sub-Sta. See B–P–H 324/23. HI 53455

Connecticut agricultural experiment station. Circular. Storrs, CT. Connecticut Agric. Exp. Sta. Circ. See B–P–H 325/1. HI 53456

Connecticut agricultural experiment station. Forestry publication. New Haven, CT. Connecticut Agric. Exp. Sta. Forest. Publ. See B–P–H 325/2. HI 53457

Connecticut botanical society newsletter. Storrs, CT. Vols. 1-7(2), 1973-79. Connecticut Bot. Soc. Newslett. Superseded by: Newsletter, Connecticut botanical society. HI 64593

Connecticut greenhouse newsletter. Storrs, CT. No. 30+, 1969?+. Connecticut Greenh. Newslett. Preceded by: Florists news letter. HI 64594

Connecticut state geological and natural history survey bulletin. Hartford, CT. Connecticut State Geol. Surv. Bull. See B–P–H 325/4. HI 53458

Connecticut woodlands. East Hartford, CT. Vol. 1+, 1936+. Connecticut Woodlands. HI 64595

Conophytum journal. Tokyo. No. 1+, 1966+. Conophytum J. HI 64596

Conselho nacional de pesquisas, instituto nacional de pesquisas da Amazônia. Serie avulsa = Boletim do instituto nacional de pesquisas da Amazônia. Serie avulsa. Manaus.

Conselho nacional de pesquisas, instituto nacional de pesquisas da Amazônia. Serie pesquisas florestais = Boletim do instituto nacional de pesquisas da Amazônia. Serie, pesquisas florestais. Manaus.

Conservation. Fredericton, N. B. Vol. 10(5)+, 1979+. Conservation (Fredericton). Preceded by: C C N B newsletter. HI 64597

Conservation. Washington, DC. Vols. 14(9)-15, 1909-10. Conservation (Washington). Preceded by: Forestry and irrigation. Superseded by: American forestry. 1-223-2. HI 64598

Conservation administration news. Laramie, WY., Tulsa, OK. No. 1+, 1979+. Conservation Admin. News. HI 64599

Conservation biology; journal of the society for conservation biology. Boston, MA. Vol. 1+, 1987+. Conservation Biol. HI 64600

Conservation Canada. Ottawa. Vol. 1-5(4), 1974-80. Conservation Canada. HI 64601

Conservation catalyst. New York. Vols. 1(2)-2, 1966-68. Conservation Catalyst. Superseded by: Catalyst for environmental quality. HI 64602

Conservation current awareness list. Kew. 1979+. Conservation Curr. Aware. List. HI 64603

Conservation directory. Parkville. 1970+. Conservation Directory (Parkville). HI 64604

Conservation directory. Washington, DC. 1956+. Conservation Directory (Washington). HI 64605

Conservation foundation letter. Washington, DC. 1966+. Conservation Found. Lett. HI 64606

Conservation Indonesia. Bogor. Vol. 2(4)+, 1978+. Conservation Indonesia. Preceded by: Newsletter, world wildlife fund Indonesia programme. HI 64607

Conservation Malaysia. Kuala Lumpur. Vol. 1+, 1980+. Conservation Malaysia. HI 64608

Conservation of nature. [Chayon poho.] Seoul. 19??-83+. Conservation Nat. HI 64609

Conservation news. London. No. 1+, 1967+. Conservation News (London). HI 64610

Conservation news. Toronto. Vols. 1-3(3), 1974-75. Conservation News (Toronto). Preceded by: Newsletter, conservation council of Ontario. Superseded by: Ontario conservation news. HI 64612

Conservation news. Washington, DC. 1938-79. Conservation News (Washington). Incorporated in: National wildlife. 2-1179-3. HI 64611

Conservation news, south east Asia. Bangkok. 1963+. Conservation News S. E. Asia. Preceded by: Conservation news from Thailand. HI 64613

Conservation news from Thailand. Bangkok. 1961. Conservation News Thailand. Superseded by: Conservation news, south east Asia. HI 64614

Conservation report; national wildlife federation's weekly update of conservation legislation. Washington, DC. 1938-82. Conservation Rep. Nat. Wildlife Fed. Superseded by: National wildlife federation's conservation. HI 64615

Conservation research report, United States department of agriculture, forest service. Washington, DC. No. 1+, 1965+. Conservation Res. Rep. U.S.D.A. Forest. Serv. HI 64616

Conservation review. Alford. No. 1-21, 1970-80. Conservation Rev. Superseded by: Natural world. HI 64617

Conservation and scientific report, Charles Darwin foundation for the Galapagos Islands. Brussels. 1964+. Conservation Sci. Rep. Charles Darwin Found. Galapagos Islands. HI 64618

Conservation studies, Dorset naturalists' trust. Bournemouth. No. 1+, 1974+. Conservation Stud. Dorset Naturalists' Trust. HI 64619

Conservation yearbook. Washington, DC. No. [1]-6, 1952-62. Conservation Yearb. HI 64620

Conservationist. Albany, NY. Vol. 14(5)+, 1960+. Conservationist. Preceded by: New York state conservationist. HI 64621

Conserve; publication of western Pennsylvania conservancy. Pittsburgh, PA. Vol. 13(3)+, 1971+. Conserve. Preceded by: Water, land and life. HI 64622

Conspectus literaturae botanicae = Botanicheskoye obozrenie. Petrograd.

Consultatio de universali commercio litterario. Nuremberg. Consultatio Universali Commercio Litt. See B–P–H 325/6. HI 53459

Consulting arborist. Milltown, NJ. 19??-77+. Consulting Arborist. HI 64623

Contemporary japanese agriculture. [Atarashi nihon no nogoyo.] Tokyo. Contemp. Jap. Agric. See B–P–H 325/9. HI 53463

Contemporary review. London. 1866+. Contemp. Rev. 2-1187-1. HI 64624

Contemporary yugoslave bibliography of agriculture = Savrema jugoslovenska poljoprivredny bibliografija. Belgrade.

Continental shelf research. Elmsford, NY. Vol. 1+, 1982+. Continental Shelf Res. HI 64625

Continuazione degli atti dell'imp. e reale accademia economico-agraria dei georgofili di Firenze. Florence. Cont. Atti Imp. Reale Accad. Econ.-Agrar. Georgof. Firenze. See B–P–H 325/7. HI 53461

Continuazione del Nuovo giornale de' letterati d' Italia. Modena, Italy. Cont. Nuovo Giorn. Lett. Italia. See B–P–H 325/8. HI 53462

Contribuciones C O N A = Ciencia y tecnologia del mar. Valparaiso.

Contribuciones cientificas, centro de investigacíon de biologia marina. Buenos Aires. No. 1+, 1962+. Contr. Ci. Centro Invest. Biol. Mar. HI 64626

Contribuciones científicas; universidad de Buenos Aires; facultad de ciencias exactes y naturales. Serie botánica. Buenos Aires. Contr. Ci. Univ. Buenos Aires Fac. Ci. Exact., Ser. Bot. See B–P–H 327/9. HI 53484

Contribuciones de Colombia a las ciencias i a las artes; boletin de la sociedad de naturalistas neo-granadinos. Bogota. Vols. 1-4, 1860-63. Contr. Colombia Ci. Artes. 2-1189-2. HI 64627

Contribuciones, estación de investigaciones y extensión agropecuaria, fundación La Salle de ciencias naturales. Caracas, Sucre. No. 1+, 1978+. Contrib. Estac. Invest. Extens. Agropecu. Fund. La Salle Ci. Nat. HI 64628

Contribuciones, flora de Veracruz. Veracruz, Mexico. [1978+. Contrib. Fl. Veracruz. HI 64629

Contribuciones à la historia natural colombiana. Barranquilla, Colombia. Contr. Hist. Nat. Colomb. See B–P–H 328/8. HI 53494

Contribuciones del instituto antarctico argentino. Buenos Aires. Contr. Inst. Antarc. Argent. See B–P–H 328/13. HI 53496

Contribuciones; instituto de biología marina; universidades nacionales de Buenos Aires, La Plata y del Sur. Buenos Aires. Contr. Inst. Biol. Mar. Univ. Nac. Buenos Aires. See B–P–H 328/17. HI 53499

Contribuciones ocasionales del museo de historia natural del colegio "de la Salle." Vedado, Havana. Nos. 1-18, 1944-60. Contr. Ocas. Mus. Hist. Nat. Colegio "De La Salle". HI 53519

Contribuciones ocasionales, museo nacional de historia natural. Santo Domingo. No. 1+, 1981+. Contr. Ocas.

Mus. Nac. Hist. Nat. HI 64630

Contribuciones, instituto ecuatoriano de ciencias naturales. Quito. No. 1+, 1942+. Contr. Inst. Ecuat. Ci. Nat. 4-3513-1. HI 64631

Contribuciones técnicas, centro de investigacíon de biologia marina. Buenos Aires. No. 1+, 1965+. Contr. Técn Centro Invest. Biol. Mar. HI 64632

Contribuições avulsas do instituto avulsas. Oceanografia biológica. São Paulo. Nos. 1-25, 1955-71. Contr. Avulsas Inst. Avulsas, Oceanogr. Biol. HI 64633

Contribuições do instituto geobiologica La Salle de Canoas. Canoas, Brazil. 1951-62. Contr. Inst. Geobiol. La Salle Canoas. HI 53502

Contributi, istituto di ricerche agrarie. Milan. Vol. 1+, 1956+. Contr. Ist. Ric. Agrar. HI 64634

Contributi scientifico-pratici per una migliore conoscenza ed utilizzazione del legno. Florence. Vol. 1+, 1957+. Contr. Sci-Pract. Migl. Conosc. Util. Legno. HI 64635

Contributi per lo studio della flora crittogama svizzera = Beiträge zur Kryptogamenflora der Schweiz. Bern.

Contributi allo studio geobotanico della Svizzera. Zurich. Contr. Stud. Geobot. Svizzera. See B–P–H 330/21. HI 53526

Contributii botanice universitatea "Babes-Bolyai" din Cluj-Napoca. Cluj-Napoca. 1958+. Contr. Bot. Univ. "Babes-Bolyai" Cluj-Napoca. Preceded by: Contributiuni botanice din Cluj la Timişoara. HI 53475

Contribution, see Contributions

Contributiones pro fauna et flora U R P S S, a societate naturae curiosorum mosquensi editae. Sectio botanica = Materialy k poznaniyu fauny i flory S S S R, izdavaemye Moskovskim obshchestvom ispytatelei prirody. Otdel botanicheskii. Moscow.

Contributions of the Allan Hancock foundation for scientific research. Los Angeles, CA. Contr. Allan Hancock Found. Sci. Res. See B–P–H 325/10. HI 53464

Contributions from the Amakusa marine biological laboratory; Kyushu university. Fukuoka, Japan. Contr. Amakusa Mar. Biol. Lab. See B–P–H 325/11. HI 53465

Contributions of the Ames botanical laboratory. Easton, MA. Contr. Ames Bot. Lab. See B–P–H 325/12. HI 53466

Contributions from the anatomical laboratory of Brown University. Providence, RI. Contr. Anat. Lab. Brown Univ. See B–P–H 325/13. HI 53467

Contributions to the anthropology, botany, geology and zoology of the Papuan region. Botany. Leiden. Contr. Anthropol. Papuan Region, Bot. See B–P–H 325/14. HI 53468

Contribution of the arctic institute. Washington, DC. Contr. Arctic Inst. See B–P–H 325/17. HI 53469

Contributions from the Arnold arboretum of Harvard university. Jamaica Plain, Ma. Contr. Arnold Arbor. See B–P–H 325/19. HI 53470

Contributions, Bermuda biological station for research = Bermuda biological station for research. Contributions. Cambrideg, MA. Bermuda Biol. Sta. Res. Contr. See B–P–H 183/26.

Contributions from the biological laboratory of Brown university. Providence, RI. Contr. Biol. Lab. Brown Univ. See B–P–H 325/20. HI 53471

Contributions from the biological laboratory of the chinese association for the advancement of science. Nanking. Contr. Biol. Lab. Chin. Assoc. Advancem. Sci. See B–P–H 326/3. HI 53472

Contributions from the biological laboratory of the chinese association for the advancement of science. Section botany. Nanking. Vols. 6-12, 1930-39. Contr. Biol. Lab. Chin. Assoc. Advancem. Sci., Sect. Bot. Preceded by: Contributions from the biological laboratory of the chinese association for the advancement of science. 2-1023-3. HI 64636

Contributions from the biological laboratory; Kyoto university. Kyoto, Japan. Contr. Biol. Lab. Kyoto Univ. See B–P–H 326/4. HI 53473

Contributions in biology and geology, Milwaukee public museum. Milwaukee, WI. No. 1+, 1974+. Contr. Biol. Geol Milwaukee Public Mus. HI 64637

Contributions from the Bolus herbarium. Rondebosch. Vol. 1+, 1969+. Contr. Bolus Herb. HI 64638

Contributions, botanical department, Iowa state college of agriculture. Ames, IA. Nos. 1-125, 1896-1925. Contr. Bot. Dep. Iowa State Coll. Agric. HI 64639

Contributions from the botanical department, Nebraska university. Lincoln, NE. N.s. nos. 1-18, 1892-1902. Contr. Bot. Dep. Nebraska Univ. HI 64640

Contributions from the botanical institute of the university of Amsterdam. Amsterdam. 1944/47-53?, 1948-53. Contr. Bot. Inst. Univ. Amsterdam. HI 64641

Contributions of the botanical laboratory; Johns Hopkins university. Baltimore, MD. Contr. Bot. Lab. Johns Hopkins Univ. See B–P–H 326/14. HI 53477

Contributions from the botanical laboratory, Miami university. Oxford, OH. No. 1+, [1908]+. Contr. Bot. Lab. Miami Univ. HI 64642

Contributions from the botanical laboratory and the Morris arboretum of the university of Pennsylvania. Philadelphia, PA. Contr. Bot. Lab. Morris Abor.

Univ. Pennsylvania. See B–P–H 326/15. HI 53478

Contributions from the botanical laboratory of Ohio state university. Columbus, OH. 1951+. Contr. Bot. Lab. Ohio State Univ. HI 64643

Contributions from the botanical laboratory, university of Missouri. Columbia, MO. No. ?-4+, ?-1906+. Contr. Bot. Lab. Univ. Missouri. HI 64644

Contributions of the botanical laboratory. University of Oklahoma. Norman, OK. Contr. Bot. Lab. Univ. Oklahoma See B–P–H 326/18. HI 53479

Contributions from the botanical laboratory, university of Tennessee. Knoxville, TN. 1935+. Contr. Bot. Lab. Univ. Tennessee. HI 64645

Contributions from the botanical survey of northwestern China. [Chung kuo hsi pei chih wu tiao cha so ts'ung k'an.] Wukung, China. Contr. Bot. Surv. NorthW. China. See B–P–H 326/22. HI 53480

Contributions botaniques de Cluj, Roumanie = Contribuţiuni botanice din Cluj. Cluj, Rumania. Contr. Bot. Cluj. See B–P–H 326/10.

Contributions to the botany of Vermont. Burlington, VT. Contr. Bot. Vermont. See B–P–H 326/23. HI 53481

Contributions of the Boyce Thompson institute for plant research. Yonkers, NY. Vols. 1-24, 1925/27-67/71, 1925-71. Contr. Boyce Thompson Inst. Pl. Res. 1-759-3. HI 64646

Contributions of the Brooklyn botanical garden. Brooklyn, NY. Vols. 1-208?, 1911-77. Contrib. Brooklyn Bot. Gard. 1-800-2. HI 64647

Contributions to canadian biology; being studies from the biological stations of Canada. Toronto. 1901-21; vols. 1-2, 1922-25. Contr. Canad. Biol. Superseded by: Contributions to canadian biology and fisheries. HI 64648

Contributions to canadian biology and fisheries. Toronto. Vols. 3-8, 1926-34. Contr. Canad. Biol. Fish. Preceded by: Contributions to canadian biology. Superseded by: Journal of the biological board of Canada. HI 64649

Contributions to canadian botany. Nos. 13-18, 1899-1906. Contr. Canad. Bot. Preceded by: Contributions from the herbarium of the geological survey of Canada. HI 64650

Contributions, Catholic university of America, biological laboratory. Biological series = Catholic university of America. Biological series. Washington, DC. Catholic Univ. Amer., Biol. Ser. See B–P–H 302/3.

Contributions, Catholic university of America. Biological series = Catholic university of America. Biological series. Washington, DC. Catholic Univ. Amer., Biol. Ser. See B–P–H 302/3.

Contribution, central research institute for agriculture = Lembaga pusat penelitian pertanian. Bogor.

Contributions du centre de recherches et d'études océanographiques. Paris. Contr. Centr. Rech. Études Océanogr. See B–P–H 327/6. HI 53482

Contribution, cereal division, Canada. Ottawa. Nos. 1-241?, 19??-59. Contr. Cereal Div. Canada. HI 64651

Contributions, Chesapeake biological laboratory. Solomons Island, MD. Contr. Chesapeake Biol. Lab. See B–P–H 327/7. HI 53483

Contributions à la connaissance de la faune et flore de l'URSS, publiés par la société des naturalistes de Moscou. Section botanique = Materialy k poznaniyu fauny i flory S S S R, izdavaemye Moskovskim obshchestvom ispytatelei prirody. Otdel botanicheskii. Moscow.

Contributions from the cryptogamic laboratory of Harvard university. Cambridge, MA. Nos. 1-125, 1883-1935. Contr. Cryptog. Lab. Harvard Univ. Superseded by: Contributions from the laboratories of cryptogamic botany and the Farlow herbarium of Harvard university. HI 53485

Contributions from the danish government institute of seed pathology for developing countries. Copenhagen. No. 1+, 1969+. Contr. Danish Gov. Inst. Seed Pathol. Developing Countries. HI 64652

Contributions; department of botany, Columbia university. New York, NY. Contr. Dept. Bot. Columbia Univ. See B–P–H 327/15. HI 53486

Contributions, department of botany, school of agriculture, Pennsylvania state college = Pennsylvania state college, school of agriculture, department of botany. Contributions. State College, PA. Pennsylvania State Coll. School Agric. Dept. Bot. Contr. See B–P–H 701/2.

Contributions, department of pathology and bacteriology; university of Minnesota. Minneapolis, MN. Contr. Dept. Pathol. Univ. Minnesota. See B–P–H 327/16. HI 53487

Contributions from the Dudley herbarium of Stanford university. Stanford, CA. Contr. Dudley Herb. See B–P–H 327/17. HI 53488

Contribution, fonds de recherches forestières de l'université Laval. Québec. No. 1+, 1958+. Contr. Fonds Rech. Forest. Univ. Laval. HI 64654

Contributions, forest research foundation, Laval university = Contribution, fonds de recherches forestières de l'université Laval. Québec.

Contributions of the general agricultural research station. Bogor. 1950-61. Contr. Gen. Agric. Res. Sta., Bogor. Preceded by: Mededelingen van het algemeen proefstation voor den landbouw. Superseded by:

Pemberitaan, balai besar penjelidikan pertanian. HI 64655

Contributions of the Gray herbarium of Harvard university. Cambridge, MA. Nos. 1-214, 1891-1984. Contr. Gray Herb. Superseded by: Harvard papers in botany. 3-1813-3. HI 53489

Contributions, Hawaii marine laboratory = Hawaii marine laboratory; contributions. [University of Hawaii.] Honolulu, HI. Hawaii Mar. Lab. Contr. See B–P–H 413/9.

Contributions of the Henry Shaw school of botany, Washington university. St. Louis, MO. Contr. Henry Shaw School Bot. See B–P–H 327/22. HI 53490

Contributions from herbarium australiense. East Melbourne, Vic. Vols. 1-18/26, 1972-76. Contr. Herb. Austral. Superseded by: Brunonia. HI 64656

Contributions from the herbarium, college of agriculture and forestry, university of Nanking. Nanking. Contr. Herb. Coll. Agric. Univ. Nanking. See B–P–H 327/23. HI 53491

Contributions; herbarium of Columbia college = Contributions; department of botany, Columbia university. New York, NY. Contr. Dept. Bot. Columbia Univ. See B–P–H 327/15.

Contributions from the herbarium of Franklin and Marshall college. Lancaster, PA. No. 1, 1895. Contr. Herb. Franklin Marshall Coll. Superseded by: Muhlenbergia. 2-1629-1. HI 64657

Contributions from the herbarium of the geological survey of Canada. Nos. 1-12, 1894-98. Contr. Herb. Geol. Surv. Canada. Superseded by: Contributions to canadian botany. HI 64658

Contributions from the herbarium of northeast Louisiana university. Monroe, LA. No. 1+, 1980+. Contr. Herb. NorthE. Louisiana Univ. HI 64660

Contributions from the herbarium of Taihoku imperial university. Taihoku [=Taipei, Taiwan]. Contr. Herb. Taihoku Imp. Univ. See B–P–H 328/3. HI 53492

Contributions from the herbarium of the university of West Virginia. Morgantown, WV. Contr. Herb. Univ. West Virginia. See B–P–H 328/7. HI 53493

Contributions from the herbarium, university of Wisconsin = Contributions, university of Wisconsin herbarium. La Crosse, Madison, WI.

Contributions from the horticultural institute of Taihoku imperial university. [Taihoku teikoku daigaku rinogakubu engeigaku kyoshitsu kiyo.] Taihoku [=Taipei, Taiwan]. Contr. Hort. Inst. Taihoku Imp. Univ. See B–P–H 328/9. HI 53495

Contributions from the Hull botanical laboratory. Chicago, IL. Vol. 1+, 1895+. Contr. Hull Bot. Lab. 2-1005-1. HI 64661

Contributions de l'institut botanique de l'université de Montréal. Montreal. Vol. 31+, 1938+. Contr. Inst. Bot. Univ. Montréal. Preceded by: Contributions du laboratoire de botanique de l'université de Montréal. 3-2751-3. HI 64662

Contributions, institut océanographique. Nhatrang = Nhatrang institut océanographique, contributions. Hue?, Indochina [South Vietnam]. Nhatrang Inst. Océanogr. Contr. See B–P–H 659/7.

Contributions de l'institut d'Oka. Université de Montréal. La Trappe, Quebec. Contr. Inst. Oka. See B–P–H 329/2. HI 53505

Contributions, institute of algological research. Muroran. No. 1+, 1965+. Contr. Inst. Algol. Res. HI 64663

Contributions de l'institute de biologie générale et de zoologie. Montreal. Contr. Inst. Biol. Gén. See B–P–H 328/16. HI 53498

Contributions from the institute of biology, national central university. [Kuo li chung yang ta hsüeh, yen chiu yuan li k'o yen chiu sho sheng wu hsüeh pu chuam k'an.] Chungking, China. Contr. Inst. Biol. Natl. Centr. Univ. See B–P–H 328/18. HI 53500

Contributions from the institute of botany, national academy of Peiping. Peiping [=Peking]. Contr. Inst. Bot. Natl. Acad. Peiping. See B–P–H 328/22. HI 53501

Contributions from the institute of geology and paleontology, Tohoku university. Sendai, Japan. Contr. Inst. Geol. Tohoku Univ. See B–P–H 328/26. HI 53503

Contributions from the institute of low-temperature science. Series B, biological science. [Teion kagaku kenkyujo obun hokoku.] Sapporo. No. 40?+, 1982?+. Contr. Inst. Low-Temp. Sci., Ser. B. Preceded by: Low temperature science. Series B, biological sciences. HI 64664

Contributions; institute of marine biology; university of Puerto Rico. Mayagüez, Puerto Rico. Contr. Inst. Mar. Biol. Univ. Puerto Rico. See B–P–H 329/1. HI 53504

Contributions of the Iowa corn research institute. Ames, Iowa. Contr. Iowa Corn Res. Inst. See B–P–H 329/3. HI 53506

Contributions du jardin botanique de Rio de Janeiro. Rio de Janeiro. Vols. 1-4, 1901-07. Contr. Jard. Bot. Rio de Janeiro. Preceded by: Plantas novas cultivadas no jardim botanico do Rio de Janeiro. 4-3682-1. HI 64665

Contributions from the Kansu science education institute. [Kan-su k'o hsüeh chiao yü kuan chuan k'an.] Lanchow, China. Contr. Kansu Sci. Educ. Inst. See B–P–H 329/6. HI 53507

Contributions du laboratoire de botanique de l'université de Montréal. Montreal. Nos. 1-30, 1922-38. Contr. Lab. Bot. Univ. Montréal. Superseded by: Contributions de l'instutut de botanique de l'université de Montréal. 3-2751-3. HI 64666

Contributions from the laboratories of cryptogamic botany and the Farlow herbarium of Harvard university. Cambridge, MA. Nos. 126-225, 1934-44. Contr. Lab. Cryptog. Bot. Farlow Herb. Harvard Univ. Preceded by: Contributions from the cryptogamic laboratory of Harvard university. HI 64667

Contributions from the laboratory of botany, national academy of Peiping. Peiping [=Peking]. Contr. Lab. Bot. Natl. Acad. Peiping. See B–P–H 329/8. HI 53508

Contributions from the laboratory for general botany, plant physiology and pharmacognosie of the university of Amsterdam. Amsterdam. 1954/58+, 1958?+. Contr. Lab. Gen. Bot. Pl. Physiol. Pharmacogn. Univ. Amsterdam. HI 64668

Contributions from the laboratory of the marine biological association. San Diego, CA. Contr. Lab. Mar. Biol. Assoc. See B–P–H 329/10. HI 53510

Contributions from the laboratory of plant genetics, Bussey institution of applied biology = Contributions on plant genetics. Jamaica Plain, Boston, MA.

Contributions from the laboratory of silviculture, Yamagata university. Yamagata, Japan. Contr. Lab. Silvic. Yamagata Univ. See B–P–H 329/11. HI 53511

Contributions from the laboratory of systematic botany of the faculty of science, Kanazawa university. [Kanazawa daigaku rigaku-bu shokubutsu bunruigaku-bu kenkyu shitsu.] Kanazawa, Japan. Contr. Lab. Syst. Bot. Fac. Sci. Kanazawa Univ. See B–P–H 329/13. HI 53512

Contributions from the laboratory of systematic botany and plant oecology; Taihoku imperial university. Taihoku [=Taipei, Taiwan]. Contr. Lab. Syst. Bot. Taihoku Imp. Univ. See B–P–H 329/15. HI 53513

Contributions, life sciences division, royal Ontario museum = Contributions, life sciences, royal Ontario museum. Toronto.

Contributions, life sciences, royal Ontario museum. Toronto. No. 1+, 1960+. Contr. Life Sci. Roy. Ontario Mus. 5-4237-2. HI 64669

Contributions to marine biology from the university of Wales. Menai Bridge, Wales. Contr. Mar. Biol. Univ. Wales. See B–P–H 329/18. HI 53514

Contributions of the marine botanical institute. Goteborg, Sweden. Contr. Mar. Bot. Inst. See B–P–H 328/19. HI 53515

Contributions in marine science. Port Aransas, TX. Vol. 12+, 1967+. Contr. Mar. Sci. Preceded by: Publications of the institute of marine science, university of Texas. HI 64670

Contributions in marine science, royal university of Malta. Msida. 1974+. Contr. Mar. Sci. Roy. Univ. Malta. HI 64671

Contributions to microbiology and immunology. Basel, London, etc. Vol. 1+, 1973+. Contr. Microbiol. Immunol. Preceded by: Bibliotheca microbiologica. HI 64672

Contributions, Misaki marine biological station = Misaki marine biological station; contributions. Tokyo. Misaki Mar. Biol. Sta. Contr. See B–P–H 596/28.

Contributions from the Mt. Hakkoda botanical laboratory of the Tohoku imperial university. Sendai. Nos. 1-?, 1930-33. Contr. Mt. Hakkoda Bot. Lab. Tohoku Imp. Univ. HI 64673

Contributions from the museum of geology. University of Michigan. Ann Arbor, MI. Contr. Mus. Geol. Univ. Michigan. See B–P–H 328/20. HI 53516

Contributions of the museum of natural history; university of Illinois. Urbana, IL. Contr. Mus. Nat. Hist. Univ. Illinois. See B–P–H 329/21. HI 53517

Contributions from the museum of paleontology. University of Michigan. Ann Arbor, MI. Contr. Mus. Paleontol. Univ. Michigan. See B–P–H 330/1. HI 53518

Contributions, national botanic gardens, Glasnevin. Dublin. Vol. 1, 1976. Contr. Nat. Bot. Gard. Glasnevin. Superseded by: Glasra. HI 64674

Contributions to natural science. Victoria, B.C. No. 1+, 1985+. Contr. Nat. Sci. HI 64675

Contributions from the nature conservancy. Arlington, VA. No. 1+, 1970+. Contr. Nat. Conservancy. HI 64676

Contributions from the New South Wales national herbarium. Sydney, N.S.W. Vol. 1-4, 1939-73. Contr. New South Wales Natl. Herb. Superseded by: Telopea. 5-4137-1. HI 64677

Contributions from the New South Wales national herbarium. Flora series. Sydney, N.S.W. Vol. 1-211?, 1961-74. Contr. New South Wales Nat. Herb., Fl. Ser. Superseded by: Flora of New South Wales (separate floristic monographs). HI 64678

Contributions of the New York botanical garden. New York. No. 1+, 1899+ [publication suspended 1933-83]. Contr. New York Bot. Gard. 4-3029-3. HI 64679

Contributions in oceanography and meteorology, Texas A. & M. university. College Station, TX. Contr. Oceanogr. Meteorol. See B–P–H 330/12. HI 53520

Contributions from the Osborn botanical laboratory, Yale

university. New Haven, CT. 1916-57. Contr. Osborn Bot. Lab. Yale Univ. HI 64680

Contributions to palaeontology (from the Carnegie Institution of Washington). [Forms part of: Publications of the Carnegie institution of Washington. Washington, DC. 1925+. Contr. Palaeontol. Carnegie Inst. Wash. HI 64682

Contributions on the paleolimnology of Lake Biwa and the japanese pleistocene = Paleolimnology of Lake Biwa and the japanese pleistocene. Otsu.

Contributions from the phanerogamic laboratories of Harvard university. Boston, MA. No. 1+, [1904]+. Contr. Phanerog. Lab. Harvard Univ. HI 64683

Contributions of the phytopathological laboratory; Taihoku imperial university. Taihoku [=Taipei, Taiwan]. Contr. Phytopathol. Lab. Taihoku Imp. Univ. See B–P–H 330/15. HI 53522

Contributions on plant genetics. Jamaica Plain, Boston, MA. Nos. 1-3, 1909/13-1920/28, 1909-28. Contr. Pl. Genet. 3-1811-3. HI 53523

Contribution of the plant pathology laboratory, botany department, university of Nanking. Nanking. Contr. Pl. Pathol. Lab. Bot. Dept. Univ. Nanking. See B–P–H 330/18. HI 53524

Contributions à la protection de la nature en Suisse = Beiträge zum Naturschutz in der Schweiz. Basel.

Contributions to the protection of nature in Switzerland = Beiträge zum Naturschutz in der Schweiz. Basel.

Contributions from the Queensland herbarium. Brisbane, Qld. Nos. 1-10, 1968-77. Contr. Queensland Herb. Superseded by: Austrobaileya. HI 64684

Contributions from the Reed herbarium. Baltimore, MD. No. ?-19+, 19??-71+. Contr. Reed Herb. HI 64685

Contributions to research, Smithsonian tropical research institute. [Panama.] 19??-79+, 19??-80+. Contr. Res. Smithsonian Trop. Res. Inst. HI 64686

Contributions of the royal Ontario museum of life sciences = Contributions, life sciences, royal Ontario museum. Toronto.

Contributions in science, museum of natural history, Los Angeles. Los Angeles, CA. Nos. 92-218, 1966-71. Contr. Sci. Mus. Nat. Hist. Los Angeles. Preceded by: Los Angeles county museum contributions in science. Superseded by: Contributions in science, natural history museum of Los Angeles county. HI 64687

Contributions in science, natural history museum of Los Angeles county. Los Angeles, CA. No. 219+, 1972+. Contr. Sci. Nat. Hist. Mus. Los Angeles County. Preceded by: Contributions in science, museum of natural history, Los Angeles. HI 64688

Contributions in science, Santa Barbara museum of natural history. Santa Barbara, CA. Nos. 1-3, 1970-71. Contr. Sci. Santa Barbara Mus. Nat. Hist. HI 64689

Contributions, Scripps institute of oceanography = Scripps insitutue of oceanography, contributions. La Jolla, CA. Scripps Inst. Oceanogr. Contr. See B–P–H 831/13.

Contributions series, american association of stratigraphic palynologists. Various places. No. 1+, 1970+. Contr. Ser. Amer. Assoc. Stratigr. Palynologists. HI 64690

Contributions of the Seto marine biological laboratory, Kyoto university. Kyoto, Japan. Contr. Seto Mar. Biol. Lab. See B–P–H 330/19. HI 53525

Contributions from the Shaw school of botany. St. Louis, MO. Nos. 1-9, 1882-92. Contr. Shaw School Bot. HI 64691

Contributions suisses à la dendrologie = Schweizerische Beiträge zur Dendrologie. Horgen, Switzerland. Schweiz. Beitr. Dendrol. See B–P–H 822/3.

Contributions from the Tamano marine laboratory, Okoyama university. [Okoyama daigaku Tamano rinkai jikkensho gyoseki.] Tamano. Vols. 1?-17(219), 1956-77. Contr. Tamano Mar. Lab. HI 64692

Contributions of the Tennessee university, botanical laboratory. Knoxville, TN. Contr. Tennessee Univ. Bot. Lab. See B–P–H 330/24. HI 53527

Contributions from the Texas research foundation, botanical studies. Renner, TX. Contr. Texas Res. Found., Bot. Stud. See B–P–H 331/1. HI 53528

Contributions from the United States national herbarium. Smithsonian institution. Washington, DC. Vols. 1-38, 1890-1974. Contr. U. S. Natl. Herb. Superseded by: Smithsonian contributions to botany. 5-4318-1. HI 64693

Contributions from the university of Kansas herbarium. Lawrence, KS. No. 1+, 1982+. Contr. Univ. Kansas Herb. HI 64694

Contributions from the university of Michigan herbarium. Ann Arbor, MI. Contr. Univ. Michigan Herb. See B–P–H 331/4. HI 53529

Contributions, university of Wisconsin herbarium. La Crosse, Madison, WI. Vol. 1+, 1980+. Contr. Univ. Wisconsin Herb. HI 64695

Contributions, Virgin Islands ecological research station, caribbean research institute. St. Thomas, VI. Nos. 1-?, 1969-71. Contr. Virgin Islands Ecol. Res. Sta. Caribbean Res. Inst. HI 64696

Contributions to western botany. [By Marcus E. Jones.] San Francisco, CA. Contr. W. Bot. See B–P–H 331/7. HI 53530

Contribuţiuni botanice din Cluj. Cluj, Rumania. Contr. Bot. Cluj. See B–P–H 326/10. HI 53476

Contributiuni botanice din Cluj la Timişoara. Cluj-Napoca. Vol. 4, 1941-43. Contr. Bot. Cluj Timişoara. Preceded by: Contributiuni botanice din Cluj. Superseded by: Contributii botanice universitatea "Babes-Bolyai" din Cluj-Napoca. HI 64697

Contributors' communiqué, vascular flora of the southeastern United States. Chapel Hill, NC. Vol. 1+, 1981+. Contributors' Communiqué Vasc. Fl. SouthW. United States. HI 64698

Contribuzioni alla biologia vegetale. Istituto botanico de Palermo. Palermo. Contr. Biol. Veg. See B–P–H 326/8. HI 53474

Contribuzioni dell'istituto botanico della università di Catania = Bollettino dell'istituto botanico della università di Catania. Catania.

Coolia; contactblad van de nederlandse mycologische vereniging. Leiden. Coolia. See B–P–H 331/37. HI 53531

Cooperative extension publication, Louisiana state university and agricultural and mechanical college = Extension publication, Louisiana state university and agricultural and mechanical college. Baton Rouge, LA.

Cooperative plant pest control programs, U S department of agriculture. Hyattsville, MD. 1967/68. Coop. Pl. Pest Control Programs. U.S.D.A. Superseded by: Progress report, plant protection, U S department of agriculture. HI 64699

Cooperative plant pest report, United States department of agriculture. Hyattsville, MD. Vols. 1+, 1976+. Coop. Pl. Pest Rep. U.S.D.A. HI 64700

Cooperative research report, international council for the exploration of the sea. Charlottenlund Slot. Nos. 1-6, 1962-65; ser. A, no.7+, 1966+. Coop. Res. Rep. Int. Council Explor. Sea. HI 64701

Coral reef newsletter. Sydney, N.S.W, Mangilas, GU. No. 1+, 1972+. Coral Reef Newslett. HI 64702

Core arboretum bulletin, West Virginia university. Morgantown, WV. Vol. 1+, 1975+. Core Arbor. Bull. West Virginia Univ. Preceded by: Arboretum newsletter, West Virginia university. HI 64703

Corn-cutter's journal. London. Corn-Cutter's J. See B–P–H 331/41. HI 53532

Cornell countryman. Ithaca, NY. Vol. 1+, 1903+. Cornell Countryman. 2-1201-2. HI 64704

Cornell extension bulletin, New York state college of agriculture. Ithaca, NY. Nos. 1-1227, 1916-71. Cornell Extens. Bull. New York State Coll. Agric. Superseded by: Information bulletin, New York state college of agriculture and life sciences. HI 64705

Cornell international agricultural bulletin. Ithaca, NY. Vols. 24+, 1973+. Cornell Int. Agric. Bull. Preceded by: Cornell international agricultural development bulletin. HI 64706

Cornell international agricultural development bulletin. Ithaca, NY. Vols. 1-23, 1963-72. Cornell Int. Agric. Developm. Bull. Superseded by: Cornell international agricultural bulletin. HI 64707

Cornell nature study quarterly = Teacher's leaflets on nature-study. [Cornell university. New York state college of agriculture.] Ithaca, NY. Teacher's Leafl. Nat.-Study. See B–P–H 868/3.

Cornell plantations. Ithaca, NY. Cornell Plantations. See B–P–H 332/3. HI 53533

Cornell plantations notes. Ithaca, NY. Nos. Minus 1, issue 0, 1-10, 1981-83. Cornell Plantations Notes. Superseded by: C P notes. HI 64708

Cornell recommendations for commercial floriculture crops. Ithaca, NY. 1954+. Cornell Recommend. Commercial Floric. Crops. HI 64709

Cornell recommendations for commercial production and maintenance of trees and shrubs. Ithaca, NY. 1968-70. Cornell Recommend. Commercial Prod. Maint. Trees Shrubs. Preceded by: Cornell recommendations for trees and shrubs. Superseded by: Cornell recommendations for pest control for commercial production and maintenance of trees and shrubs. HI 64710

Cornell recommendations for florist crops = Cornell recommendations for commercial floriculture crops. Ithaca, NY.

Cornell recommendations for pest control for commercial production and maintenance of trees and shrubs. Ithaca, NY. 1972+. Cornell Recommend. Pest Control Commercial Prod. Maint. Trees Shrubs. Preceded by: Cornell recommendations for commercial production and maintenance of trees and shrubs. HI 64711

Cornell recommendations for trees and shrubs. Ithaca, NY. 1962-69? Cornell Recommend. Trees Shrubs. Superseded by: Cornell recommendations for commercial production and maintenance of trees and shrubs. HI 64712

Cornell university abstracts of theses ... for the doctor's degree. Ithaca, NY. Cornell Univ. Abstr. Theses. See B–P–H 332/4. HI 53534

Cornell university agricultural experiment station bulletin. Ithaca, NY. Cornell Univ. Agric. Exp. Sta. Bull. See B–P–H 332/6. HI 53535

Cornell university agricultural experiment station circular. Ithaca, NY. Cornell Univ. Agric. Exp. Sta. Circ. See B–P–H 332/7. HI 53537

Cornell university agricultural experiment station memoir. Ithaca, NY. Cornell Univ. Agric. Exp. Sta. Mem. See B–P–H 332/8. HI 53538

Cornell university agricultural experiment station press

bulletin. Ithaca, NY. Cornell Univ. Agric. Exp. Sta. Press Bull. See B–P–H 332/9. HI 53539

Cornell university nature-study leaflets. Ithaca, NY. Cornell Univ. Nat.-Study Leafl. See B–P–H 332/12. HI 53540

Cornell university science bulletin. Ithaca, NY. Cornell Univ. Sci. Bull. See B–P–H 332/13. HI 53542

Cornell veterinarian. Ithaca, NY. Cornell Veterin. See B–P–H 332/14. HI 53543

Cornhill magazine. London. 1860+. Cornhill Mag. 2-1204-3. HI 64713

Cornish biological records. Redruth. 1977+. Cornish Biol. Rec. HI 64714

Cornish garden. St. Mawes. Vol. ?-28+, 19??-85+. Cornish Gard. HI 64715

Cornucopia project newsletter. Emmaus, PA. Vols. 1-?, 1981-84? Cornucopia Proj. Newslett. Superseded by: Regeneration. HI 64716

Corny bags information. [Komebukuro.] Tokyo. Corny Bags Inform. See B–P–H 332/16. HI 53544

Correio agro-pecuario. Rio de Janeiro. Correio Agro-Pecuario. See B–P–H 332/17. HI 53545

Correio agrostologico. São Paulo. No. 1+, 1984+. Correio Agrostol. HI 64717

Correspondance botanique. Liége. Corresp. Bot. See B–P–H 332/18. HI 53546

Correspondentieblad ten dienste van de floristiek en het vegetatie-onderzoek van Nederland. Leiden. Correspondentiebl. Dienste Florist. Veg.-Onderz. Ned. See B–P–H 332/21. HI 53548

Correspondentieblad Zuiderzeeonderzoek. [Netherlands]. Correspondentiebl. Zuiderzeeonderz. See B–P–H 332/22. HI 53549

Correspondenz der Schlesischen Gesellschaft für vaterländische Cultur. Breslau [=Wroclaw, Poland]. Corresp. Schles. Ges. Vaterl. Cult. See B–P–H 332/20. HI 53547

Correspondenzblatt der Gesellschaft zur Beförderung der Naturkunde und Industrie in Schlesien. Breslau [=Wroclaw, Poland]. Correspondenzbl. Ges. Beförd. Naturk. Schlesien. See B–P–H 332/23. HI 53550

Correspondenzblatt des Naturforschenden Vereins zu Riga. Riga [Latvian S S R]. Correspondenzbl. Naturf. Vereins Riga. See B–P–H 332/24. HI 53551

Correspondenzblatt des Naturforscher-Vereins zu Riga. Riga [Latvian S S R]. Correspondenzbl. Naturf.-Vereins Riga. See B–P–H 333/1. HI 53552

Correspondenzblatt des Naturhistorischen Vereines für die preussischen Rheinlande. Bonn. Correspondenzbl. Naturhist. Vereines Preuss. Rheinl. See B–P–H 333/2. HI 53553

Correspondenzblatt der Schlesischen Gesellschaft für vaterländische Cultur. Breslau [=Wroclaw, Poland]. Correspondenzbl. Schles. Ges. Vaterl. Cult. See B–P–H 333/3. HI 53554

Correspondenzblatt des Vereins für Naturkunde zu Presburg. Pressburg [=Bratislava, Czechoslovakia]. Correspondenzbl. Vereins Naturk. Presburg. See B–P–H 333/4. HI 53555

Correspondenzblatt des Württembergischen landwirthschaftlichen Vereins. Stuttgart. Correspondenzbl. Württemberg. Landw. Vereins. See B–P–H 333/5. HI 53556

Corse historique, archéologique, littéraire, scientifique. Ajaccio. Vol. 2(5/6)-6(36), 1962-69. Corse Hist. Archéol. Litt. Sci. Preceded by: Revue d'études historiques, littéraires et scientifiques corses. HI 64718

Cortica = Boletim do instituto dos produtos florestais. Lisbon.

Cosmetics and perfumery. Oak Park, IL. Vols. 88-90, 1973-75. Cosmet. & Perfumery. Preceded by: American cosmetics and perfumery. Superseded by: Perfumer and flavorist international. HI 64719

Costa azzurra agricol-floreale; rivista mensile di floricoltura ed orticoltura. San Remo, Italy. Costa Azzurra Agric.-Fl. See B–P–H 333/6. HI 53557

Côte d'azur agricole et horticole et la revue oléicole. Nice. Côte d'Azur Agric. Hort. See B–P–H 333/8. HI 53558

Coton et culture cotonnière; périodique consacré questions cotonnières. Paris. Vols. 1-11-?, 1926-37-? Coron Cult. Cotonn. HI 64720

Coton et fibres tropicales. Paris. Coton Fibres Trop. See B–P–H 333/10. HI 53559

Coton et fibres tropicales. Bulletin analytique. Paris. Coton Fibres Trop. Bull. Analytique. See B–P–H 333/11. HI 53560

Coton et fibres tropicales. Bulletin bibliographique. Paris. Coton Fibres Trop. Bull. Bibliogr. See B–P–H 333/12. HI 53561

Cottage gardener; or, amateur and cottager's guide to outdoor gardening. London. Cottage Gardener. See B–P–H 333/14. HI 53562

Cottage gardening. London. Cottage Gardening. See B–P–H 333/15. HI 53563

Cotton growing review. London. Vols. 44-52, 1967-75. Cotton Growing Rev. Preceded by: Empire cotton growing review. HI 64721

Cotton research report, Kenya. London. 1970/71-73/74, 1971-74. Cotton Res. Rep. Kenya. Preceded by: Progress report from experiment stations, cotton research corporation. HI 64722

Cotton research report, Malawi. London. 1970/71-?, 1971-? Cotton Res. Rep. Kenya. Preceded by: Progress report from experiment stations, cotton research corporation. HI 64723

Cotton research report, Northern States, Nigeria. London. 1970/71-73/74, 1971-74. Cotton Res. Rep. N. States Nigeria. Preceded by: Progress report from experiment stations, cotton research corporation. HI 64724

Cotton research report, Republic of the Sudan. London. 1970/71-73/74, 1971-74. Cotton Res. Rep. Sudan. Preceded by: Progress report from experiment stations, cotton research corporation. HI 64725

Cotton research report, Swaziland. London. 1970/71-73/74, 1971-74. Cotton Res. Rep. Swaziland. Preceded by: Progress report from experiment stations, cotton research corporation. HI 64726

Cotton research report, Tanzania. London, Dar es Salaam. 1970/71-?, 1971-? Cotton Res. Rep. Tanzania. Preceded by: Progress report from experiment stations, cotton research corporation. HI 64727

Cotton research report, Uganda. London. 1971-72, 1972. Cotton Res. Rep. Uganda. Preceded by: Progress report from experiment stations, cotton research corporation. HI 64728

Cotton research report, Zambia. London. 1970/71-73/74, 1971-74. Cotton Res. Rep. Zambia. Preceded by: Progress report from experiment stations, cotton research corporation. HI 64729

Cotton and tropical fibres abstracts. Farnham Royal. Vol. 1+, 1976+. Cotton Trop. Fibres Abstr. HI 64730

Country gentleman. Dublin. Country Gent. (Dublin). See B–P–H 333/19. HI 53564

Country gentleman. Philadelphia, PA. Country Gent. (Philadelphia). See B–P–H 333/20. HI 53565

Country journal; or the craftsman. London. Country J. See B–P–H 333/21. HI 53566

Country life. Garden City, NY. Country Life (Garden City). See B–P–H 334/2. HI 53567

Country life. London. Country Life (London). See B–P–H 334/3. HI 53568

Country life in America. Garden City, NY. Country Life Amer. See B–P–H 334/4. HI 53569

Country life and the sportsman. Garden City, NY. Country Life & Sportsman. See B–P–H 334/5. HI 53570

Country life in war = Country life. Garden City, NY. Country LIfe (Garden City). See B–P–H 334/2.

Countryman. Idbury, London. Vol. 1+, 1927+. Countryman (London). HI 64731

Countryman. Nicosia. Vol. 1-32, 1947-63. Countryman (Nicosia). HI 64732

Country-side; journal of the british naturalists' association. London. Country-Side. See B–P–H 334/7. HI 53571

Country-side monthly. London. Country-Side Monthly. See B–P–H 334/8. HI 53572

County council and agricultural record. London. 1906-07. County Council & Agric. Rec. Preceded by: County council times. Superseded by: Agricultural record. 2-956-3. HI 64733

County council times. London. Vols. 1-22? [vols. 12-22 also numbered n.s. vols. 1-11], 1889-1906. County Council Times. Superseded by: County council and agricultural record. 2-956-3. HI 64734

Coup-d'oeil sur les travaux de la société jurassienne d'émulation. Porrentruy. Vols. 1-8, 1849-56. Coup-d'Oeil Trav. Soc. Jurass. Emul. Superseded by: Actes de la société jurassienne d'émulation. HI 64735

Courier Forschungsinstitut Senckenberg. Frankfurt am Main. No. 1+, 1973+. Courier Forschungsinst. Senckenberg. HI 64736

Courrier des chercheurs. Paris. Vols. 1-?, 1949-56? Courrier Cherch. HI 64737

Courrier horticole. Brussels. Vols. 1-22, 1932-60. Courrier Hort. Superseded by: Jardins et logis. HI 64738

Courrier de la nature. Paris. N.s. vol. 1/2+, 1967+. Courrier Nat. HI 64739

Courrier phytochimique. Paris. Courrier Phytochim. See B–P–H 334/11. HI 53573

Cours et documents de biologie. New York. 1969+. Cours Doc. Biol. HI 64740

Cours pratique de mycologie. [Supplement to: Revue de mycologie.] Paris. Nos. 1-7, 1937-39. Cours Prat. Mycol. HI 64741

Covent garden gazette. London. Covent Gard. Gaz. See B–P–H 334/12. HI 53575

Cox arboretum foundation newsletter = Cox arboretum newsletter. Dayton, OH.

Cox arboretum newsletter. Dayton, OH. 19??-76+. Cox Arbor. Newslett. HI 64742

Crab gab. Lisle, IL. Vol. 1(1), 1985. Crab Gab. Superseded by: Malus. HI 64743

Craftsman. London. Nos. 1-44, 1726-27. Craftsman. Superseded by: Country journal. 2-1218-3. HI 64744

Cranberries; the national cranberry magazine. Wareham, MA. Cranberries. See B–P–H 334/13. HI 53576

Cranberry grower. Cranmoor, WI. Cranberry Grower. See B–P–H 334/14. HI 53577

Cranberry world. New York, NY. Cranberry World. See B–P–H 334/15. HI 53578

Creation/evolution. Buffalo, NY. No. 1+, 1980+. Creation & Evol. Superseded in part by: Creation/evolution newsletter. HI 64745

Creation/evolution newsletter. Athens, WV. Vol. 4+, 1984+. Creation & Evol. Newslett. Previously part of: Creation evolution. HI 64746

Crell's chemical journal. London. Crell's Chem. J. See B–P–H 334/18. HI 53579

Criador paulista. Publicação official da secretaria da agricultura, commercio e obras publicas, do estado de São Paulo. São Paulo. Criador Paul. See B–P–H 334/19. HI 53580

Critica e scienza positiva; rivista scientifica. Naples. Crit. Sci. Positiva. See B–P–H 335/1. HI 53583

Critical review; or, annals of literature. London. Crit. Rev. See B–P–H 334/23. HI 53582

Critical reviews in biochemistry = C R C critical reviews in biochemistry. Boca raton, FL.

Critical reviews in biotechnology = C R C critical reviews in biotechnology. Boca Raton, FL.

Critical reviews in microbiology = C R C critical reviews in microbiology. Boca Raton, FL.

Critical reviews in plant sciences = C R C critical reviews in plant sciences. Boca Raton, FL.

Critische Bibliothek. Leipzig. Crit. Biblioth. See B–P–H 334/21. HI 53581

Crónica agrícola. Buenos Aires. Crón. Agríc. See B–P–H 335/2. HI 53585

Crónica científica y literaria. Madrid. Crón. Ci. Lit. See B–P–H 335/3. HI 53586

Crónica médica. Lima. Crón. Méd. See B–P–H 335/5. HI 53587

Crónica médico-quirúrgica. Havana. Crón. Méd.-Quir. See B–P–H 335/7. HI 53588

Crop bulletin, grain research laboratory. Winnipeg. No. 1+, 1941+. Crop Bull. Grain Res. Lab., Winnipeg. HI 64747

Crop improvement. Ludhiana. Vol. 1+, 1974+. Crop Improv. HI 64748

Crop physiology abstracts. Farnham Royal. Vol. 1+, 1975+. Crop Physiol. Abstr. HI 64749

Crop production. Budapest = Növénytermelés. Budapest. Növénytermelés. See B–P–H 674/7.

Crop production. Pretoria. Vol. 1+, 1972+. Crop Prod. (Pretoria). HI 64750

Crop protection. Guildford. Vol. 1+, 1982+. Crop Protect. HI 64751

Crop protection newsletter. Brooks, Alta. Vol. 1+, 1970+. Crop Protect. Newslett. (Brooks). Preceded by: Newsletter, provincial horticultural station, Brooks. HI 64752

Crop protection newsletter. Laguna. Vol. ?-2+, ?-1980+. Crop Protect. Newslett. (Laguna). HI 64753

Crop protection research. Washington, DC. 1980+. Crop Protect. Res. HI 64754

Crop research. Edinburgh. Vol. 22+, 1982+. Crop Res. (Edinburgh). Preceded by: Horticultural research. HI 64755

Crop research. Hissar. Vol. 1+, 1988+. Crop Res. (Hissar). HI 74825

Crop research news. Wellington, Lincoln, N.Z. Vol. 1+, 1969+. Crop Res. News. HI 64756

Crop science. Madison, WI. Crop Sci. (Madison). See B–P–H 335/10. HI 53589

Crop science. [Tso wu hsüeh pao. Zuown zuebao.] Peking. Crop Sci. (Peking). See B–P–H 335/11. HI 53590

Crops in India. Delhi. Vols. 1-2, 1968-69. Crops India. HI 64757

Crops and soils. Madison, WI. Vols. 11-18, 1958-66. Crops & Soils. Preceded by: What's new in crops and soils. Superseded by: Crops and soils magazine. 2-1233-2. HI 64758

Crops and soils magazine. Madison, WI. Vols. 19-?, 1966-? Crops & Soils Mag. Preceded by: Crops and soils. HI 64759

Croquis du naturaliste; publication du cercle des naturalistes corbeillois (later et de la vallée de l'Essonne). Corbeil. Vols. [1]-13(8), 1926-39. Croq. Naturaliste. HI 64760

Crosby gardens [newsletter]. Toledo, OH. ?-1985+. Crosby Gard. Newslett. HI 64761

Crossosoma. Claremont, CA. 1978+. Crossosoma. HI 64762

CrossWords; the gesneriad hybridizers' association newsletter. Hyde Park, MA., Warwick, RI. Vol. 1+, 1977+. CrossWords. HI 64763

Cruciferae newsletter. E U C A R P I A. Pentlandfield, etc. No. 1+, 1976+. Cruciferae Newslett. E U C A R P I A. HI 64764

Cryo letters. Cambridge. Vol. 1+, 1979+. Cryo Lett. HI 75204

Cryobiology. Kiev = Kriobiologiya. Kiev.

Cryobiology. Rockville, MD. Cryobiology. See B–P–H 335/16. HI 53592

Cryptanthus society journal. Marietta, GA. Vol. 1+, 1986+. Cryptanthus Soc. J. HI 64765

Cryptogamic botany. Stuttgart, New York. Vol. 1+, 1989+. Cryptog. Bot. HI 74826

Cryptogamic studies. Stuttgart. Vol. 1+, 1987+. Cryptog. Stud. HI 64082

Cryptogamica helvetica. Geneva, Teufen. Vol. 16+, 1985+, 1986+. Cryptog. Helv. Preceded by: Beiträge zur Kryptogamenflora der Schweiz. HI 64766

Cryptogamie - algologie. Paris. Vol. 1+, 1980+. Cryptog. Algol. Preceded by: Revue algologique and Bulletin, société phycologique de France. HI 64767

Cryptogamie - bryologie et lichénologie. Paris. Vol. 1+, 1980+. Cryptog. Bryol. Lichénol. Preceded by: Revue bryologique et lichénologique. HI 64768

Cryptogamie - mycologie. Paris. Vol. 1+, 1980+. Cryptog. Mycol. Preceded by: Revue de mycologie (Paris). HI 64769

Csili. Kaktuszgyujto szakkor. Budapest. 19??-64+. Csili. Kaktuszg. Szakkor. HI 64770

Cuaderno, see Cuadernos

Cuadernos academia nacional de ciencias de Bolivia. La Paz. No. ?-65+, ?-ca.1987+. Cuad. Acad. Nac. Ci. Bolivia. HI 64771

Cuadernos de botánica; anexos de la revista el museo canario. Las Palmas. Vols. 1-4, 1967-68. Cuad. Bot. Superseded by: Cuadernos de botánica canaria. HI 64772

Cuadernos de botánica canaria; comunicaciones sobre flora y vegetacion del Archipielago Canario. Las Palmas. Vols. 5-28, 1969-77. Cuad. Bot. Canaria. Preceded by: Cuadernos de botánica. HI 64773

Cuadernos de botánica canaria. Suplemento. Las Palmas. Vols. 1-4, 1971-75. Cuad. Bot. Canaria. Supl. HI 64774

Cuadernos C V F. Caracas. Cuad. C.V.F. See B–P–H 335/26. HI 53593

Cuadernos de la C V F. Caracas. Cuad. de la C.V.F. See B–P–H 335/27. HI 53594

Cuadernos de ciencias biológicas; anejo del boletin de la universidad de Granada. Granada. Vol. 1+, 1971+. Cuad. Ci. Biol. HI 64775

Cuadernos fitotécnicos. Cochabamba. No. 1+, 1947+. Cuad. Fitotécnicos. HI 64777

Cuaderno del instituto nacional de investigaciones agronómicas. Madrid. Cuad. Inst. Nac. Invest. Agron. See B–P–H 335/28. HI 53595

Cuadernos de investigación biologica. Bilbao. Vol. ?-2+, 19??-81+. Cuad. Invest. Biol. HI 64778

Cuadernos oceanográficos. Cumaná, Venezuela. Cuad. Oceanogr. See B–P–H 335/29. HI 53596

Cuadernos de la revista de la facultad de ciencias naturales. Salta. Vol. 1+, 1961+. Cuad. Revista Fac. Ci. Nat. HI 64779

Cuadernos téchnicas, I A D I Z A = Cuadernos téchnicas, instituto argentino de investigacion de las zonas áridas. Mendoza.

Cuadernos téchnicas, instituto argentino de investigacion de las zonas áridas. Mendoza. No. 1+, 1977+. Cuad. Técn. Inst. Argent. Invest. Zonas Aridas. HI 64780

Cuba agrícola. Havana. Cuba Agríc. See B–P–H 335/30. HI 53597

Cuba bibliotecologica. Havana. Cuba Bibliotecol. See B–P–H 335/31. HI 53598

Cuban journal of agricultural science. Havana. Vol. 7+, 1973+. Cuban J. Agric. Sci. Preceded by: Revista cubana de ciencia agricola. HI 64781

Cukorrépa. Budapest. Cukorrépa (Budapest). See B–P–H 336/1. HI 53599

Cukorrépa. Magyarovar [=Mosonmagyarovar, in part.] Cukorrépa (Magyarovar). See B–P–H 336/2. HI 53600

Cultivador moderno, El. Barcelona. Vol. 1+, 1910+. Cultivador Moderno. 2-1238-2. HI 64782

Cultivateur. Journal belge d'économie rurale. Brussels. Cultivateur. See B–P–H 336/3. HI 53601

Cultivation of neglected tropical fruits with promise. New Orleans, LA. 1976+. Cultivation Negl. Trop. Fruits Promise. HI 64783

Cultivator. Albany, NY. Vols. 1-10, 1834-43; n.s. vols. 1-9, 1844-52; ser. 3, vols. 1-13, 1853-65. Cultivator (Albany). 2-1238-2. HI 53602

Cultivator. Salt Lake City, UT. 1979+. Cultivator (Salt Lake City). HI 64784

Cultivator and country gentleman. Philadelphia, PA. Cultivator Country Gent. See B–P–H 336/5. HI 53603

Cultivator's leaflet, department of agriculture, Burma. Rangoon. Nos. 1-82, 1907-37. Cultivator's Leafl. Dept. Agric. Burma. HI 64785

Cultura. Orgaan van het nederlandsch instituut van landbouwkundigen. Wageningen. Cultura. See B–P–H 336/6. HI 53604

Cultuur en handel. Brussels. No. ?-5+, 19??-37+. Cultuur & Handel. 2-1240-1. HI 64786

Cultuurgids. Malang, Dutch E. Indies [Indonesia]. Cultuurgids. See B–P–H 336/7. HI 53605

Cunninghamia; ecological contributions from the national herbarium of New South Wales. Sydney, N.S.W. Vol. 1+, 1981+. Cunninghamia. HI 64787

Curator. New York, Westport, CT. Vol. 1+, 1958+. Curator. HI 64788

Curator's report; Thirsk natural history society. Thirsk, England. Curator's Rep. Thirsk Nat. Hist. Soc. See B–P–H 336/11. HI 53606

Current accessions list, royal botanic gardens, Kew. Kew. 1968-70. Curr. Accessions List. Roy. Bot. Gard. Kew. Superseded by: Current awareness list, royal botanic gardens, Kew. HI 64789

Current advances in biochemistry; current awareness in biological sciences (C A B S) database. Oxford, Elmsford, NY, etc. Vol. 1+, 1984+. Curr. Advances Biochem. HI 64790

Current advances in cell and development biology; current awareness in biological sciences (C A B S) database. Oxford, Elmsford, NY, etc. Vol. 1+, 1984+. Curr. Advances Cell Developm. Biol. HI 64791

Current advances in ecological and environmental sciences. Oxford, Elmsford, NY. Vol. 15+, 1989+. Curr. Advances Ecol. Environm. Sci. Preceded by: Current advances in ecological sciences? HI 74827

Current advances in ecological sciences; current awareness in biological sciences (C A B S) database. Oxford, Elmsford, NY. Vol. 1+, 1975+. Curr. Advances Ecol. Sci. HI 64792

Current advances in genetics. Oxford, Elmsford, NY, etc. Vol. 1+, 1976+. Curr. Advances Genet. HI 64793

Current advances in genetics and molecular biology; current awareness in biological sciences (C A B S) database. Oxford, Elmsford, NY, etc. Vol. 1+, 1984+. Curr. Advances Genet. Molec. Biol. HI 64794

Current advances in microbiology; current awareness in biological sciences (C A B S) database. Oxford, Elmsford, NY. Vol. 1+, 1984+. Curr. Advances. Microbiol. HI 64795

Current advances in plant reproductive biology. New Delhi. Vol. 1+, 1979+. Curr. Advances Pl. Reprod. Biol. HI 64796

Current advances in plant science; a monthly subject categorised listing of titles in plant science compiled from the current literature. Current awareness in biological sciences (C A B S) database. Oxford, Elmsford, NY, etc. Vol. 1+, 1972+. Curr. Advances Pl. Sci. HI 64797

Current agricultural literature. London. Vol. 1+, 1953+. Curr. Agric. Lit. HI 64798

Current agriculture; quarterly journal of soil and plant sciences. Karnal. Vol. 1+, 1977+. Curr. Agric. HI 64799

Current antarctic literature. Washington, DC. 1966+. Curr. Antarc. Lit. Cumulated annually as: Antarctic bibliography. HI 64800

Current author entries, bureau of plant industry library = . Bureau of plant industry library. Current author entries. Washington, DC. Bur. Pl. Industr. Libr. Curr. Author Entries. See B–P–H 288/17.

Current awareness in biological sciences. Elmsford, NY. Vol. 100+, 1983+. Curr. Aware. Biol. Sci. Preceded by: International abstracts of biological sciences. HI 64801

Current awareness in biotechnology. Chicago, IL. Vol. 1+, 1981+. Curr. Aware. Biotechnol. HI 64802

Current awareness bulletin, arctic science and technology information system = A S T I S current awareness bulletin. Calgary.

Current awareness list, royal botanic gardens, Kew. Kew. 1971+. Curr. Aware. List Roy. Bot. Gard. Kew. Preceded by: Current accessions list, royal botanic gardens, Kew. HI 64803

Current bibliography of agriculture in China. Wageningen. 1979-81. Curr. Bibliogr. Agric. China. HI 64804

Current bibliography for aquatic sciences and fisheries. Rome. Vols. 1-17, 1958-71. Curr. Bibliogr. Aquatic Sci. Fish. Superseded by: Aquatic sciences and fisheries abstracts. HI 64805

Current bibliography for fisheries science = Current bibliography for aquatic sciences and fisheries. Rome.

Current bibliography, west Africa rice development association. [Monrovia.] No. 1+, 1974+. Curr. Bibliogr. W. Africa Rice Developm. Assoc. HI 64806

Current biotechnology abstracts. London, Letchworth. No. 1+, 1983+. Curr. Biotechnol. Abstr. HI 64807

Current botanical literature, bureau of plant industry library = Bureau of plant industry library. Current botanical literature. Washington, DC. Bur. Pl. Industr. Libr. Curr. Bot. Lit. See B–P–H 288/18.

Current contents: agriculture, food and veterinary sciences. Philadelphia, PA. Vols. 1-3, 1970-72. Curr. Contents, Agric. Food. Veterin. Sci. Superseded by: Current contents: agriculture, biology and environmental sciences. HI 64808

Current contents: agriculture, biology and environmental sciences. Philadelphia, PA. Vol. 4+, 1973+. Curr. Contents, Agric. Biol. Environm. Sci. Preceded by: Current contents: agriculture, food and veterinary sciences. HI 64809

Current contents: life sciences; including weekly subject index. Philadelphia, PA. Vol. 1+, 1958+. Curr. Contents, Life Sci. HI 64810

Current contents of journals received by nature conservancy council and institute of terrestrial ecology. Banbury. ?-1976+. Curr. Contents J. Receiv. Nat. Conservancy Council. HI 64811

Current contents in marine science. Rome. 1966-67.

Curr. Contents Mar. Sci. Superseded by: Marine sciences contents tables. HI 64812

Current genetics. Berlin, New York, etc. Vol. 1+, 1979+. Curr. Genet. HI 64813

Current geographical publications. New York, NY. Curr. Geogr. Publ. See B–P–H 336/19. HI 53607

Current index to conference papers in life sciences. New York. Vol. 1+, 1969+. Curr. Index Conf. Pap. Life Sci. HI 64814

Current information report, forest service, United States department of agriculture. Washington, DC. No. ?-14+, 19??-75+. Curr. Inform. Rep. Forest Serv., U.S. Dept. Agric. HI 64815

Current microbiology; an international journal. New York. Vol. 1+, 1978+. Curr. Microbiol. HI 64816

Current opinion in cell biology; review articles, recommended reading, bibliography of world literature. Philadelphia, PA. Vol. 1+, 1989+. Curr. Opin. Cell Biol. HI 64817

Current plant taxonomic research in Australia. Indooroopilly, Qld. 1975/76+. Curr. Pl. Taxon. Res. Australia. HI 64818

Current practices in dryland resources and technology. Jodphur. Vol. 2+, 1985+. Curr. Pract. Dryland Resources Technol. Preceded by: Desert resources and technology. HI 64819

Current projects on plant ecology in the Pacific area. Canberra, A.C.T. Nos. 1-3?, 1963-? Curr. Proj. Pl. Ecol. Pacific Area. HI 64820

Current report, agricultural experiment station, West Virginia university. Morgantown, WV. Nos. 1-67, 1952-76. Curr. Rep. Agric. Exp. Sta. West Virginia Univ. Superseded by: Current report, agricultural and forestry experiment station, West Virginia university. HI 64821

Current report, agricultural and forestry experiment station, West Virginia university. Morgantown, WV. No. 68+, 1976+. Curr. Rep. Agric. Forest. Exp. Sta. West Virginia Univ. Preceded by: Current report, agricultural experiment station, West Virginia university. HI 64822

Current research in Britain. Biological sciences. Wetherby. 1985+. Curr. Res. Brit., Biol. Sci. Preceded by: Research in british universities, polytechnics and colleges. HI 64823

Current research on medicinal and aromatic plants. Lucknow. Vol. 1+, 1979+. Curr. Res. Med. Aromat. Pl. HI 64879

Current science. Bangalore, India. Curr. Sci. See B–P–H 336/21. HI 53608

Current tissue culture literature. New York, NY. Curr. Tissue Cult. Lit. See B–P–H 336/23. HI 53609

Current titles in the biological sciences = Biology '83 [etc.]. Lawrence, KS.

Current titles in turkish science. Ankara. Vol. 3+, 1976+. Curr. Titles Turkish Sci. Preceded by: Monthly current awareness service on scientific publications. Series B, foreign edition. HI 64824

Current topics in biochemistry. New York. Vol. 1+, 1971+. Curr. Topics Biochem. HI 64825

Current topics in bioenergetics. New York, NY. Curr. Topics Bioenerget. See B–P–H 336/24. HI 53610

Current topics in chinese science. Section D, biology. New York. Vol. 1+, 1982+. Curr. Topics Chin. Sci., D. HI 64826

Current topics in developmental biology. New York, NY. Curr. Topics Developm. Biol. See B–P–H 336/25. HI 53611

Current topics in microbiology and immunology. Berlin, New York, etc. Vol. 40+, 1967+. Curr. Topics Microbiol. Immunol. Preceded by: Ergebnisse der Mikrobiologie und Immunitätsforschung. HI 64827

Current topics in plant biochemistry and physiology. Columbia, MO. Vol. 1+, 1983+. Curr. Topics Biochem. Physiol. HI 64828

Current topics in vector research. New York. Vol. 1+, 1983+. Curr. Topics Vector Res. HI 64829

Current trends in geology. New Delhi. Vol. 1+, 1978+. Curr. Trends Geol. HI 64830

Current trends in life sciences. New Delhi. Vol. ?-2+, 19??-76+. Curr. Trends Life Sci. HI 64831

Current work in the history of medicine. London. No. 1+, 1954+. Curr. Work Hist. Med. HI 64832

Currents in modern biology. Amsterdam. Vol. 1-5, 1967-74. Currents Modern Biol. Superseded by: BioSystems. HI 64833

Curtis's botanical magazine = Botanical magazine; or, flower-garden displayed ... [Edited by Wm. Curtis]. London.

Cuscatlania. San Salvador, El Salvador. No. 1+, 1989+. Cuscatlania. HI 74828

Custos. Pretoria. Vol. 1+, 1971+. Custos. HI 64834

Cuzco agronómico. Cuzco, Peru. Cuzco Agron. See B–P–H 337/4. HI 53612

Cybele columbiana, a series of studies in botany, chiefly north american. Washington, DC. Cybele Columb. See B–P–H 337/5. HI 53613

Cycad newsletter. Baton Rouge, Lafayette, LA. Vol. 1+, 1977+. Cycad Newslett. HI 64835

Cycad society newsletter = Cycad newsletter. Baton Rouge, Lafayette, LA.

Cyclamen society's journal. Haslemere. Vol. 1+, 1977+. Cyclamen Soc. J. HI 64836

Cycles; a monthly report. Riverside, CT, Pittsburgh, PA. Vol. 1+, 1950+. Cycles. HI 64837

Cymbidium society news. Pasadena, Arcadia, CA. Vols. 1-29, 1946-74. Cymbidium Soc. News. Continued as: Orchid advocate. HI 64838

Cymmrodor. London. Cymmrodor. See B–P–H 337/7. HI 53615

Cyperaceae newsletter. Gent. No. 1+, 1987+. Cyperaceae Newslett. HI 64839

Cyprus agricultural journal; a quarterly journal, a quarterly review of the agriculture and industry of Cyprus. Nicosia. Nos. 40-47, 1916-17; vols. 13-35(1), 1918-40 [vol. nos. 4-12 not used]. Cyprus Agric. J. Superseded by: Cyprus agricultural journal. 2-1249-3. HI 64840

Cyprus agricultural research institute. Miscellaneous publications. Nicosia, Cyprus. Cyprus Agric. Res. Inst. Misc. Publ. See B–P–H 337/9. HI 53616

Cyprus journal. Nicosia. Vols. 1-3(1), 1904-06; nos. 1-39, 1906-15. Cyprus J. Superseded by: Cyprus agricultural journal. HI 64841

Cytobiologie; Zeitschrift für experimentelle Zellforschung. Organ der deutschen Gesellschaft für Elektronmikroskopie e. V. Stuttgart. Vols. 1-18, 1969-79. Cytobiologie. Superseded by: European journal of cell biology. HI 64842

Cytobios; a prestige international biomedical journal of cell biology. Cambridge. Vol. 1+, 1969+. Cytobios. HI 64843

Cytogenetics. Basel, London, etc. Vols. 1-11, 1962-72. Cytogenetics. Superseded by: Cytogenetics and cell genetics. HI 64844

Cytogenetics and cell genetics. Basel, London, etc. Vol. 12+, 1973+. Cytogenet. & Cell Genet. Preceded by: Cytogenetics. HI 64845

Cytokines. Basle, New York. Vol. 1+, 1989+. Cytokines. HI 74829

Cytologia; international journal of cytology. Tokyo. Cytologia. See B–P–H 337/11. HI 53617

Cytology and genetics = Tsitologiya i genetika. Kiev.

Cytology and genetics. [Translation of: Tsitologiya i genetika.] New York. 1974+. Cytol. & Genet. HI 64846

Cytometry. [Supplement to: Journal of the society for analytical cytologists.] New York. Vol. 1+, 1980+. Cytometry. HI 64847

Cytotaxonomical atlases. Lehre. 1974+. Cytotax. Atlases. HI 64848

Cytotechnology. Dordrecht & Hingham, MA. Vol. 1+, 1988+. Cytotechnology. HI 64849

Czechoslovak research work. Second section = Revue des travaux scientifiques tchécoslovaques. Section deuxième. Prague. Rev. Trav. Sci. Tchécoslov., Sect. 2. See B–P–H 785/28.

Czechoslovak scientific and technical periodicals contents. Prague. Vol. 1+, 1971+. Czech. Sci. Techn. Period. Contents. HI 64850

D I A boletín técnico. Bogotá. No. 1+, 1957+. D. I. A. Bol. Técn. HI 64851

D L G Mitteilungen. Frankfurt on Main. 1975(9)+, 1975+. D. L. G. Mitt. Preceded by: Mitteilungen der deutschen Landwirtschaftsgesellschaft. HI 64852

D N A; a journal of molecular and cellular biology. New York. Vols. 1-8, 1981-89. D. N. A. Superseded by: D N A and cell biology. HI 64853

D N A and cell biology. New York. Vol. 9+, 1990+. D. N. A. Cell Biol. Preceded by: D N A. HI 75276

D N T journal. Plymouth. 1964. D. N. T. J. Preceded by: Journal of the Devon trust for nature conservation. Superseded by: Journal of the Devon trust for nature conservation. HI 64854

D S I R cereal news. Christchurch, N.Z. No. ?-4+, 19??-77+. D. S. I. R. Cereal news. HI 64855

D S I R information series, New Zealand. Wellington, N.Z. Nos. 127+, 1978+. D. S. I. R. Inform. Ser., New Zealand. Preceded by: Information series, New Zealand department of scientific and industrial research. HI 64505

D V T = Dějiny věd a techniky. Prague.

Dabas un vestures kalendars. Riga. 1970+. Dabas Vestures Kalend. HI 64856

Dacca university studies. Arts and sciences. Dacca. Vols. 1-?, 1935-? Dacca Univ. Stud., Arts & Sci. Superseded by: Dacca university studies. Part B, science. 2-1252-3. HI 64857

Dacca university studies. Part B, science. Dacca. Vol. ?-11+, 19??-63+. Dacca Univ. Stud., B. Preceded by: Dacca university studies. HI 64858

Daedalus. Cambridge, MA. Vol. 86+, 1955+. Daedalus. Preceded by: Proceedings of the american academy of arts and sciences. 2-1252-3. HI 64859

Daffodil bulletin. Lutherville, MD. Daffodil Bull. See B–P–H 337/16. HI 53618

Daffodil journal. West Hartford, CT. Daffodil J. See B–P–H 337/17. HI 53619

Daffodil and tulip year book. London. Nos. 12-26, 1946-71. Daffodil Tulip Year Book. Preceded by: Daffodil Year-Book. Superseded by: Daffodils. 2-1253-1. HI 64860

Daffodil year-book. London. Daffodil Year-Book. See B–P–H 337/20. HI 53620

Daffodils; an annual for amateurs and specialists. London. 1972+. Daffodils. Preceded by: Daffodil and tulip year book. HI 64861

Dahlia. Chicago, IL. Dahlia. See B–P–H 337/21. HI 53621

Dahlia "chatta". Chattanooga, TN. Dahlia "Chatta". See B–P–H 337/22. HI 53622

Dahlia digest. Madison, WI. Dahlia Digest. See B–P–H 336/23. HI 53623

Dahlia news. Boston, MA. Dahlia News (Boston). See B–P–H 336/24. HI 53624

Dahlia news. Madison, WI. Dahlia News (Madison). See B–P–H 336/25. HI 53625

Dahlia year book. London. Dahlia Year Book. See B–P–H 338/1. HI 53626

Dahlie. Zeitschrift für Land- und Forstwirtschaft. Königsberg [=Kaliningrad, Russian S F S R]. Dahlie. See B–P–H 338/2. HI 53627

Dahlien und Gladiolen, Jahrbuch. Berlin. Dahlien & Gladiolen. See B–P–H 338/3. HI 53628

Dai Nippon nokwai = Journal of the agricultural society of Japan. Tokyo. J. Agric. Soc. Japan. See B–P–H 456/12.

Daihan misainmur haghoi ji = Journal of the korean society for microbiology. Seoul.

Daildarzniecība. Augu introdukcija un zala celtniecība Latvijas P S R. Riga. Nos. ?-2-10, 19??-50-75. Daildarzniecība. HI 64862

Dana reports. Copenhagen. Vol. 1+, 1932+. Dana Rep. HI 64863

Danchi nogaku = Agriculture of warm regions. Japan. Agric. Warm Regions. See B–P–H 61/11.

Dänische Bibliothek oder Sammlung von alten und neuen gelehrten Sachen aus Dänemark. Copenhagen & Leipzig. Dän Biblioth. See B–P–H 338/6. HI 53629

Dänisches Journal. Copenhagen, Odensee & Leipzig. Vols. 1-2, 1767-70. Dän. J. HI 53180

Dänisches Journal für Politik, Natur- und Menschenkunde. [Denmark]. Dän. J. Polit. See B–P–H 338/7. HI 53630

Danish arctic research. Charlottenlund, Denmark. Danish Arctic Res. See B–P–H 338/10. HI 53631

Danish ecological abstracts. Risskov. 1980+, 1981+. Danish Ecol. Abstr. HI 64864

Danish journal of plant and soil science = Tidsskrift for planteavl. Copenhagen, Lyngby.

Danmarks geologiske undersøgelse. Række 1-5. Copenhagen. 1890-1983; serie A-D, 1976+. Danmarks Geol. Undersøl., Række 1-5. HI 64865

Dansk botanisk arkiv udgivet af dansk botanisk forening. Copenhagen. Vols. 1-34, 1913-81. Dansk Bot. Ark. Incorporated in: Opera botanica. 2-1259-2. HI 64866

Dansk dendrologisk årsskrift. Copenhagen. Vol. 1+, 1950+. Dansk Dendrol. Årsskr. HI 64867

Dansk frøanl. [Samvirkende dansk frøanlerforeninger.] Copenhagen. Dansk Frøanl. See B–P–H 338/14. HI 53632

Dansk frugtavleren. Copenhagen. Vols. 1-43, 1935-71. Dansk Frugtavl. Superseded by: Frugtavleren. HI 64868

Dansk literatur-tidende. Copenhagen. 1811-37. Dansk Lit.-Tidende. Preceded by: Kjøbenhavnske Laerde Efterretninger. 2-1260-1. HI 53790

Dansk havebrugs-tidende; Tidsskrift for havebrug og biavl. Copenhagen. 1890-98. Dansk Havebr.-Tidende. HI 64869

Dansk havetidende. Copenhagen. Vols. 1-22, 1918-39. Dansk Havetid. (Copenhagen, 1918-39). 2-1259-3. HI 64870

Dansk havetidende; [Edited by Dalskov.] Copenhagen. Vols. 1-2, 1904-06. Dansk Havetid. (Dalskov). HI 64871

Dansk landbotidende. Copenhagen. Vols. 1-9, 1866-74. Dansk Landbotid. HI 64872

Dansk landbrug. Copenhagen. Vol. 56+, 1937+. Dansk Landbr. Preceded by: Vort landbrug. 2-1260-1. HI 64873

Dansk landmands-bog; et ugeskrift for større og mindre agerdyrkere. Copenhagen. Vols. 1-25, 1871-72. Dansk Landmands-Bog. HI 64874

Dansk litteratur-tidende. Copenhagen. 1811-37. Dansk Litt.-Tidende. Preceded by: Kjøbenhavnske laende efterretninger. 2-1260-1. HI 64875

Dansk skovforenings tidsskrift. Copenhagen. 1916+. Dansk Skovforen. Tidsskr. HI 64876

Dansk tidsskrift. Copenhagen. Dansk Tidsskr. See B–P–H 338/15. HI 53633

Dansk tidsskrift for farmaci. Copenhagen. Vols. 1-46(10), 1926-72. Dansk Tidsskr. Farm. Incorporated in: Archiv for pharmaci og chemi. Sci. ed. HI 64877

Dansk ugeskrift. Copenhagen. Dansk Ugeskr. See B–P–H 338/16. HI 53634

Danske haugetidende = Danske havetidende. Copenhagen.

Danske havetidende. Copenhagen. Vols. 1-39, 1849-87. Danske Havetid. (Copenhagen, 1849-87). Preceded by: Archiv for haugevaesen. In 1887 merged with:

Gartner-tidende. HI 64878

Darling Downs naturalist. Toowoomba, Qld. No. 1+, 1952+. Darling Downs Naturalist. HI 64880

Darstellung der Verichtungen und des Zustandes der naturforschenden Gesellschaft zu Emden. Emden. Darstellung Verrichtungen Zustandes Naturf. Ges. Emden. See B–P–H 338/25. HI 53635

Darwin. Budapest. Darwin (Budapest). See B–P–H 338/27. HI 53636

Darwiniana; carpeta del "Darwinion." Buenos Aires. Darwiniana. See B–P–H 338/28. HI 53637

Dasht-e-kavir, dasht-e-lut, jaz murian = Contribution a l'étude de la flore et de la végétation des deserts d'Iran.

Dasonomia interamericana. Turrialba, Costa Rica. Dasonomia Interamer. See B–P–H 339/1. HI 53638

Data reports, Japan Antarctic research expedition = J A R E data reports. Marine biology. Tokyo.

Data reports; Virginia institute of marine science. Gloucester Point, VA. Data Rep. Virginia Inst. Mar. Sci. See B–P–H 339/2. HI 53639

Date palm journal. Baghdad. Vol. 1+, 1981+. Date Palm J. HI 64882

Davidsonia. Vancouver. Vol. 1+, 1970+. Davidsonia. HI 64883

Dawes arboretum newsletter = Newsletter, the Dawes arboretum. Newark, OH.

Daylily journal. Waterboro, SC. Vol. 35(3)+, 1981+. Daylily J. Preceded by: Hemerocallis journal. HI 64884

Daziran = Nature. Beijing.

De re rustica; or, the repository for select papers on agriculture, arts and manufactures. London. De Re Rustica. See B–P–H 339/7. HI 53641

Debreceni agrártudományi föiskola tudományos közleményei. Debrecen, Hungary. Debreceni Agrártud. Föisk. Tud. Közlem. See B–P–H 339/9. HI 53642

Debreceni mezögazdasági akadémia évkönyve. Debrecen, Hungary. Debreceni Mezögazd. Akad. Évk. See B–P–H 339/10. HI 53643

Debreceni mezögazdasági kísérleti intézet évkönyve. Debrecen, Hungary. Debreceni Mezögazd. Kísérl. Intéz. Évk. See B–P–H 339/11. HI 53644

Debreceni szemle. Debrecen, Hungary. Debreceni Szemle. See B–P–H 339/12. HI 53645

Debreceni Tisza István tudományegyetem évkönyve és almanachja. Debrecen, Hungary. Debreceni Tisza István Tudományegyet. Évk. Alman. See B–P–H 339/15. HI 53648

Debreceni Tisza István tudományos társaság II. (orvos-természettudományi) osztályának munkái. Debrecen, Hungary. Debreceni Tisza István Tud. Társ. 2 (Orv.-Term.) Oszt. Munkái. See B–P–H 339/14. HI 53647

Debreceni Tisza István tudományos társaság honismertetö bizottságának kiadványai. Debrecen, Karcag & Budapest. Debreceni Tisza István Tud. Társ. Honism. Bizott. Kiadv. See B–P–H 339/13. HI 53646

Debreceni tudományegyetem biológiai intézeteinek évkönyve = Annales biologicae universitatis debreceniensis. Debrecen, Hungary. Ann. Biol. Univ. Debrecen. See B–P–H 97/17.

Décade egyptienne. Cairo. Décade Egypt. See B–P–H 339/18. HI 53649

Décade philosophique, littéraire et politique. Paris. Décade Philos. See B–P–H 339/19. HI 53650

Decheniana. Bonn. Decheniana. See B–P–H 339/20. HI 53651

Decheniana. B. Biologische Abteilung. Bonn. Decheniana, B, Biol. Abt. See B–P–H 339/21. HI 53652

Decheniana. Beihefte. Bonn. Decheniana Beih. See B–P–H 340/1. HI 53653

Deciduous fruit grower. Bellville. Vol. 26(3)+, 1976+. Decid. Fruit Grower. HI 64885

Deep-sea research. London. Vols. 1-8, 1953-61; vols. 24-25, 1977-78. Deep-Sea Res. For vols. 9-23 see: Deep-sea research and oceanographic abstracts. Superseded by: Deep sea research. Part A, oceanographic research papers. HI 64886

Deep-sea research. Part A, oceanographic research papers. Oxford, etc. Vol. 26A+, 1979+. Deep Sea Res., A. Preceded by: Deep-sea research. HI 64887

Deep-sea research. Part B, oceanographic literature review. Oxford, etc. Vol. 26B+, 1979+. Deep Sea Res., B. Preceded by: Oceanographic abstracts and bibliography. HI 64888

Deep-sea research and oceanographic abstracts. Oxford, etc. Vols. 9-23, 1962-76. Deep Sea Res. Oceanogr. Abstr. Preceded by: Deep-sea research. Superseded by: Deep-sea research and Oceanographic abstracts and bibliography. HI 64889

Deeside field. Aberdeen. 1922-35. Deeside Field. HI 64890

Defensa de la naturaleza. Caracas. Vol. 1+, 1970+. Defensa Naturaleza. HI 64891

Défense des plantes. Leningrad = Zashchita rastenii. Leningrad.

Défense des végétaux. Paris. Défense Vég. See B–P–H 340/4. HI 53654

Deforestation and development. Brussels. 1982+. Deforest. Developm. HI 64892

Dějiny věd a techniky. Prague. Vol. 1+, 1968+. Dějiny Věd Techn. HI 64893

Dela. Akademija znanosti in umetnosti v Ljubljani. Matematično-prirodoslovni razred. Ljubljana, Yugoslavia. Dela Akad. Znan. v Ljubljani, Mat.-Prir. Razred. See B–P–H 340/7. HI 53655

Dela institut za biologijo. Slovenska akad. znanosti in umietnosti. Ljubljana, Yugoslavia. Dela Inst. Biol. Slov. Akad. Znan. See B–P–H 340/8. HI 53656

Delaware college, agricultural experiment station. Bulletin. Newark, DE. Delaware Coll. Agric. Exp. Sta. Bull. See B–P–H 340/11. HI 53657

Déliberation et mémoires de la société royale d'agriculture de la généralité de Rouen. Rouen. Vols. 1-3, 1763-87. Délib. Mém. Soc. Roy. Agric. Généralité Rouen. HI 64894

Delphinium; the book of the American delphinium society. Morgantown, WV. Delphinium. See B–P–H 340/13. HI 53658

Delphinium news. Willowdale, Ont. Delphinium News. See B–P–H 340/14. HI 53659

Delphinium society yearbook. London. Delphinium Soc. Yearb. See B–P–H 340/16. HI 53660

Delpinoa; nuova serie del bulletino dell'orto botanico della università di Napoli. Naples. Delpinoa. See B–P–H 340/17. HI 53661

Delta newsletter. Beltsville, MD. No. 1+, 1988+. Delta Newslett. HI 64895

Deltion' ellenikes mikrobiologikes etaireias. Vol. ?-29(5)+, 19??-84+. Deltion' Ellen. Mikrobiol. Etair. HI 64896

Deltion geórgikon. Athens. Deltion Geórgikon. See B–P–H 340/18. HI 53662

Dendroflora. Boskoop, Netherlands. Dendroflora. See B–P–H 340/19. HI 53663

Dendrological notes. Oxford. Nos. 1-16, 1936-39. Dendrol. Notes. HI 64897

Dendrologian seuran tiedotuksia. Helsinki. Vols. 1-12, 19??-81? Dendrol. Seuran Tiedot. Superseded by: Sorbifolia. HI 64898

Dendrologicky sborník. Opava, Czechoslovakia. Dendrol. Sborn. See B–P–H 340/21. HI 53664

Dendrologiese tydskrif. Amptelike tydskrif van die dendrologiese vereniging van suider-Afrika = Journal of dendrology. Pretoria.

Dendrologiska sällskapets notiser. ?-1980+. Dendrol. Sällsk. Notiser. HI 64899

Dendrologists' newsletter. Logan, UT. No. 1+, 1980+. Dendrologists' Newslett. HI 64900

Dendrologiya uzbekistana. Tashkent. Vols. 1+, 1965+. Dendrol. Uzbekist. HI 64901

Dendron. News bulletin of the international dendrology union. Oisterwijk, Netherlands. Dendron. See B–P–H 340/22. HI 53665

Denkschriften, Akademie der Wissenschaften in Wien. Mathematisch-Naturwissenschaftliche Klasse = Akademie der Wissenschaften in Wien. Mathematisch-Naturwissenschaftliche Klasse. Denkschriften. Vienna. Akad. Wiss. Wien, Math.-Naturwiss. Kl., Denkschr. See B–P–H 66/24.

Denkschriften der allgemeinen Schweizerischen Gesellschaft für die gesammten Naturwissenschaften. Zurich. Denkschr. Allg. Schweiz. Ges. Gesammten Naturwiss. See B–P–H 341/2. HI 53666

Denkschriften der Bayer. Botanischen Gesellschaft in Regensburg. Regensburg. Denkschr. Bayer. Bot. Ges. Regensburg. See B–P–H 341/5. HI 53667

Denkschriften der holländischen Societät der Wissenschaften. n.p. Vols. ?-2, 18??-26. Denkschr. Holl. Soc. Wiss. HI 64902

Denkschriften der Kaiserlichen Akademie der Wissenschaften, Wien. Mathematisch-naturwissenschaftliche Klasse. Vienna. Vols. 1-92, 1850-1916. Denkschr. Kaiserl. Akad. Wiss., Wien. Math.-Naturwiss. Kl. Superseded by: Kaiserliche Akademie der Wissenschaften in Wien. Mathematisch-naturwissenschaftliche Klasse. Denkschriften. 1-106-3. HI 64903

Denkschriften der Königlichen Akademie der Wissenschaften zu München. Munich. Denkschr. Königl. Akad. Wiss. München. See B–P–H 341/10. HI 53668

Denkschriften der Königlich-Baierischen Botanischen Gesellschaft in Regensburg. Regensburg. Denkschr. Königl.-Baier. Bot. Ges. Regensburg. See B–P–H 341/11 HI 53669

Denkschriften der Kgl. Bayr. botanischen Gesellschaft in Regensburg. Regensburg. Denkschr. Königl. Bayr. Bot. Ges. Regensburg. See B–P–H 341/13. HI 53670

Denkschriften der Königlich-Bayerischen Botanischen Gesellschaft zu Regensburg = Denkschriften der Königlich-Baierischen Botanischen Gesellschaft in Regensburg. Regensburg. Denkschr. Königl.-Baier. Bot. Ges. Regensburg. See B–P–H 341/11

Denkschriften der Kgl. botanischen Gesellschaft in Regensburg. Regensburg. Denkschr. Königl. Bot. Ges. Regensburg See B–P–H 341/14. HI 53671

Denkschriften, österreichische Akademie der Wissenschaften. Mathematisch-naturwissenschaftliche Klasse = Österreichische Akademie der Wissenschaften. Mathematisch-naturwissenschaftliche Klasse. Denkschriften. Vienna. Österr. Akad. Wiss.,

Math.-Naturwiss. Kl., Denkschr. See B–P–H 692/12.

Denkschriften der regensburgischen botanischen Gesellschaft. Regensburg. Vol. 21, 1940; vols. 23-28, 1953-71. Denkschr. Regensburg. Bot. Ges. Preceded by: Denkschriften der bayerischen botanischen Gesellschaft in Regensburg. For vol. 22, 1946, see: Denkschriften der bayerischen botanischen Gesellschaft in Regensburg. Superseded by: Hoppea. 4-3555-3. HI 64904

Denkschriften der Russischen Geographischen Gesellschaft zu St. Petersburg. Weimar. Denkschr. Russ. Geogr. Ges. St. Petersburg See B–P–H 341/19. HI 53672

Denkschriften der Schweizerischen Naturforschenden Gesellschaft. Zurich. Denkschr. Schweiz. Naturf. Ges. See B–P–H 342/2. HI 53673

Denkschriften der Vaterländischen Gesellschaft der Aerzte und Naturforscher Schwabens. Tübingen. Denkschr. Vaterl. Ges. Aerzte Schwabens See B–P–H 342/4. HI 53674

Denkwürdigkeiten aus dem Felde der Geschichte und dem Gebiete der Kunst und Natur. Leipzig. Denkwürdigk. Gesch. Kunst Natur. See B–P–H 342/6. HI 53675

Denshi kenbikyo gakkai-shi = Journal of electron microscopy. Tokyo.

Dental items of interest. Philadelphia, PA. Dental Items of Interest. See B–P–H 342/7. HI 53676

Department of agriculture. Annual report of the directors. Washington, DC. Dept. Agric. Annual Rep. Directors See B–P–H 342/12. HI 53677

Department of agriculture. Botanical division. Bulletin. Washington, DC. Dept. Agric. Bot. Div. Bull. See B–P–H 342/17. HI 53679

Department of agriculture. Botany division. Annual report. Washington, DC. Dept. Agric. Bot. Div. Annual Rep. See B–P–H 342/16. HI 53678

Department of agriculture. Department reports. Washington, DC. Dept. Agric. Dept. Rep. See B–P–H 342/21. HI 53680

Department of agriculture, division of forestry, report of forester. Washington, DC. 1899-1911? Dept. Agric. Div. Forest. Dep. Forester. Preceded by: Department of agriculture, forestry division, report of the chief of the forestry division. Superseded by: Report of the forest service, United States department of agriculture. HI 64905

Department of agriculture, forestry divison bulletin. Washington, DC. Dept. Agric. Forest. Div. Bull. See B–P–H 342/27. HI 53681

Department of agriculture. Forestry division. Circular. Washington, DC. Dept. Agric. Forest. Div. Circ. See B–P–H 343/1. HI 53682

Department of agriculture, forestry division. Report of the chief of the forestry division. Washington, DC. Dept. Agric. Forest. Div. Rep. Chief Forest. Div. See B–P–H 343/2. HI 53683

Department of agriculture. Microscopy division. Food products. Washington, DC. Dept. Agric. Microscop. Div. Food Prod. See B–P–H 343/4. HI 53684

Department of agriculture. Microscopy division. Report of the microscopist. Washington, DC. Dept. Agric. Microscop. Div. Rep. Microscop. See B–P–H 343/5. HI 53685

Department of agriculture. Miscellaneous special report. Washington, DC. Dept. Agric. Misc. Special Rep. See B–P–H 343/6. HI 53686

Department of agriculture. Report of the botanist. Washington, DC. Dept. Agric. Rep. Bot. See B–P–H 343/8. HI 53687

Department of agriculture. Report on forestry. Washington, DC. Dept. Agric. Rep. Forest. See B–P–H 343/11. HI 53688

Department of agriculture. Seed division. Annual report. Washington, DC. Dept. Agric. Seed Div. Annual Rep. See B–P–H 343/15. HI 53689

Department of agriculture. Special report. Washington, DC. Dept. Agric. Special Rep. See B–P–H 343/16. HI 53690

Department bulletin, United States department of agriculture = United States department of agriculture. Department bulletin. Washington, DC. U.S.D.A. Dept. Bull. See B–P–H 941/17.

Department circular, United States department of agriculture = United States department of agriculture. Department circular. Washington, DC. U.S.D.A. Dept. Circ. See B–P–H 941/18.

Department of forestry publications = Publications, department of forestry and rural development, Canada. Ottawa.

Department information circular, Montana state college. Agricultural experiment station = Montana state college. Agricultural experiment station. Department information circular. Bozeman, MT. Montana Agric. Exp. Sta. Dept. Inform. Circ. See B–P–H 618/14.

Department reports, U S department of agriculture = Department of agriculture. Department reports. Washington, DC. Dept. Agric. Dept. Rep. See B–P–H 342/21.

Departmental circular, United States department of agriculture = United States department of agriculture. Departmental circular. Washington, DC. U.S.D.A. Departmental Circ. See B–P–H 941/16.

Departmental herbarium publications, university of Dar-es-Salaam. Dar-es-Salaam. Vol. 1+, 1971+. Dept. Herb. Publ. Univ. Dar-es-Salaam. HI 64906

Departmental publications, department of forestry and rural development, Canada = Publications, department of forestry and rural development, Canada. Ottawa.

Departmental report series, Bernice Pauahi Bishop museum. Honolulu, HI. 1980+. Dept. Rep. Ser., Bernice Pauahi Bishop Mus. Preceded by: Report, department of anthropology, Bernice Pauahi Bishop museum [not entered]. HI 64907

Dergisi, yalova bahce kulturleri arastirma ve egitim merkezi = Yalova bahce kulturleri arastirma ve egitim merkezi dergisi. Istanbul.

Dermatologische Studien. Hamburg & Leipzig. Dermatol. Stud. See B–P–H 343/27. HI 53691

Dermatologische Wochenschrift. Leipzig. Dermatol. Wochenschr. See B–P–H 343/28. HI 53692

Dermatologische Zeitschrift. Berlin. Dermatol. Z. See B–P–H 344/1. HI 53693

Dermatologica. Berlin. Dermatologica. See B–P–H 344/2. HI 53694

Derwent biotechnology abstracts. London. Vol. 1+, 1982+. Derwent Biotechnol. Abstr. HI 64908

Descriptions of pathogenic fungi and bacteria = C M I descriptions of pathogenic fungi and bacteria. Kew.

Descriptions of plant viruses = C M I/A A B descriptions of plant viruses. Kew.

Desert. San Diego, Palm Desert, CA. Vols. 15-31(11), 1951-68. Desert. Preceded and superseded by: Desert magazine. HI 64909

Desert magazine. El Centro, Palm Desert, CA. Vols. 1-14, 1937-51; vol. 31(12)+, 1968+. Desert Mag. For 1952-68 see: Desert. HI 53695

Desert plant life. Pasadena, Ca. Desert Pl. Life. See 344/6. HI 53696

Desert plants. Superior, AZ. Vol. 1+, 1979+. Desert Pl. HI 64910

Desert resources and technology. Jodphur. Vol. 1, 1983. Desert Resources Technol. Superseded by: Current practices in dryland resources and technology. HI 64911

Deserta; anales del instituto de investigaciones de las zonas aridas y semiaridas. Mendoza. No. 1+, 1971+. Deserta. HI 64912

Desinfektion. Monatsschrift für Desinfektion, Sterilisation und Konservierung. Dresden & Berlin. Desinfektion. See B–P–H 344/8. HI 53697

Deutsche Acta eruditorum oder Geschichte der Gelehrten, welche den gegenwärtigen Zustand der Litteratur in Europa begreifen. Leipzig. Deutsche Acta Erud. See B–P–H 345/7. HI 53704

Deutsche Akademie der Landwirtschaftswissenschaften zu Berlin. Wissenschaftliche Abhandlungen. Berlin. Deutsche Akad. Landwirtschaftswiss. Berlin Wiss. Abh. See B–P–H 345/9. HI 53705

Deutsche Akademie der Wissenschaften zu Berlin. Vorträge und Schriften. Berlin. Deutsche Akad. Wiss. Berlin Vorträge Schriften. See B–P–H 345/11. HI 53706

Deutsche Apotheker-Zeitung. Berlin. Deutsche Apotheker-Zeitung. See B–P–H 345/12. HI 53707

Deutsche Aquarien- und Terrarien-Zeitschrift. Stuttgart. Deutsche Aquarien- Terrar.-Z. See B–P–H 345/13. HI 53708

Deutsche Baumschule. Aachen. Deutsche Baumschule. See B–P–H 345/14. HI 53709

Deutsche Blätter für Pilzkunde. Vienna. Deutsche Blätt. Pilzk. See B–P–H 345/15. HI 53710

Deutsche botanische Monatsschrift. Sondershausen, Germany. Deutsche Bot. Monatsschr. See B–P–H 345/16. HI 53711

Deutsche Chirurgie. Stuttgart. Deutsche Chir. See B–P–H 345/17. HI 53712

Deutsche Dahlien-Gesellschaft. Jahrbuch. Altona-Bahrenfeld [=Hamburg, in part]. Deutsche Dahlien-Ges. Jahrb. See B–P–H 345/18. HI 53713

Deutsche Entomologische Zeitschrift. Berlin. Deutsche Entomol. Z. See B–P–H 345/19. HI 53714

Deutsche Erwerbsgartenbau. Berlin. Deutsche Erwerbsgartenbau. See B–P–H 345/20. HI 53715

Deutsche Forschung. Aus der Arbeit der Notgemeinschaft der Deutschen Wissenschaft. Berlin. Deutsche Forsch. See B–P–H 345/21. HI 53717

Deutsche Garten. Monatsschrift für Gärtner und Gartenfreunde. Berlin. Deutsche Gart. See B–P–H 345/22. HI 53718

Deutsche Garten-Zeitung. Wochenschrift für Gärtner und Gartenfreunde. Berlin. Deutsche Gart.-Zeitung. See B–P–H 346/1. HI 53720

Deutsche Gartenbau. Berlin. Vols. 1-18(6), 1954-71. Deutsche Gartenbau. Superseded by: Gartenbau. 6-593-1. HI 64913

Deutsche Gartenbauwirtschaft. Stuttgart. Vols. 1-15, 1960-67. Deutsche Gartenbauwirt. Preceded by: Zentralblatt für den deutschen Erwerbsgartenbau. Incorporated in: Erwerbsgartner. HI 64914

Deutsche Gartenkunst. Vienna. Deutsche Gartenkunst. See B–P–H 346/3. HI 53721

Deutsche Gartenzeitung. Organ der vereinigten

Gartenbau-Gesellschaften von ... Leipzig. Deutsche Gartenzeitung. See B–P–H 346/4. HI 53722

Deutsche Gärtner-Verbands-Zeitung. Berlin. Deutsche Gärtn.-Verbands-Zeitung See B–P–H 346/5. HI 53723

Deutsche Gärtner-Zeitung. Kassel. Deutsche Gärtn.Zeitung. See B–P–H 346/6. HI 53724

Deutsche Gärtnerbörse. Aachen. Vols. 1-77(25), 1948-77. Deutsche Gärtnerb. Superseded by: Gb + Gw: Gärtnerbörse und Gartenwelt. 2-1300-1. HI 64915

Deutsche Gesellschaft für Kunst und Wissenschaft in Posen. Naturwissenschaftliche Abteilung (Naturwiss. Verein). Zeitschrift der Sektion für Botanik. Posen [=Poznan, Poland]. Deutsche Ges. Kunst Posen Naturwiss. Abt. Z. Sekt. Bot. See B–P–H 346/7. HI 53725

Deutsche Gesellschaft für Kunst und Wissenschaft in Posen. Zeitschrift der naturwissenschaftlichen Abteilung (des naturwissenschaftlichen Vereins). Posen [=Poznan, Poland]. Deutsche Ges. Kunst Posen Z. Naturwiss. Abt. See B–P–H 346/8. HI 53726

Deutsche Gesellschaft für Natur- und Völkerkunde Ostasiens, Tokio. Nachrichten. Tokyo. Deutsche Ges. Natur- Völkerk. Ostasiens Tokio Nachr. See B–P–H 346/9. HI 53727

Deutsche Heilpflanze. Fach- und Nachrichtenblatt der deutschen Arbeitsgemeinschaft zur Förderung der Beschaffung heimischer Hiel-, Gewürz- und Duftpflanzen. Stollberg, Germany. Deutsche Heilpflanze See B–P–H 346/10. HI 53729

Deutsche hydrographische Zeitschrift. Hamburg. Deutsche Hydrogr. Z. See B–P–H 346/11. HI 53730

Deutsche Iris Gesellschaft. Jahrbuch. Leonberg, Germany. Deutsche Iris Ges. Jahrb. See B–P–H 346/12. HI 53731

Deutsche Iris-Gesellschaft e. V. Berlin-Dahlem. Nachrichtenblätter. Berlin-Dahlem. Deutsche Iris-Ges. Berlin-Dahlem Nachrichtenbl. See B–P–H 346/15. HI 53734

Deutsche Iris-Gesellschaft e. V. Nachrichtenblatt. Leonberg, Germany. Deutsche Iris-Ges. Nachrichtenbl. See B–P–H 346/16. HI 53735

Deutsche Iris- und Liliengesellschaft e. V. Jahrbuch. Leonberg, Germany. Deutsche Iris- Lilienges. Jahrb. See B–P–H 346/13. HI 53732

Deutsche Iris- und Lilien-Gesellschaft e. V. Nachrichtenblatt. Leonberg, Germany. Deutsche Iris-Lilien-Ges. Nachrichtenbl. See B–P–H 346/14. HI 53733

Deutsche Jahrbücher für Wissenschaft und Kunst. Halle. Deutsche Jahrb. Wiss. Kunst. See B–P–H 346/17. HI 53736

Deutsche Landwirtschaft. Berlin. Deutsche Landw. See B–P–H 347/2. HI 53737

Deutsche landwirtschaftliche Presse. Berlin & Hamburg. Vol. 1+, 1874+. Deutsche Landw. Presse. 2-1307-1. HI 64917

Deutsche landwirtschaftliche Rundschau. Neudamm [=Debno, Poland]. Vols. 1-11, 1928-34. Deutsche Landw. Rundschau. Superseded by: Forschungsdienst. HI 64918

Deutsche Literaturzeitung. Berlin. Vols. 1-44, 1880-1923. Deutsche Literaturzeitung. Superseded by: Deutsche Literaturzeitung, für Kritik der internationalen Wissenschaft. 2-1310-2. HI 64919

Deutsche Literaturzeitung. Für Kritik der internationalen Wissenschaft. Berlin. Deutsche Literatureitung Krit. Int. Wiss. See B–P–H 347/3. HI 53738

Deutsche medizinische Wochenschrift. Leipzig & Stuttgart. Deutsche Med. Wochenschr. See B–P–H 347/4. HI 53739

Deutsche Obst- und Gemüsebauzeitung. Stuttgart. Deutsche Obst- Gemüsebauzeitung. See B–P–H 347/5. HI 53740

Deutsche Obstbauzeitung. Stuttgart. Deutsche Obstbauzeitung. See B–P–H 347/6. HI 53741

Deutsche Rhododendron Gesellschaft. Jahresbericht. Deutsche Rhododendron Ges. Jahresber. See B–P–H 347/7. HI 53742

Deutsche Rundschau für Geographie. Vienna & Leipzig. Deutsche Rundschau Geogr. See B–P–H 347/8. HI 53743

Deutsche Rundschau für Geographie und Statistik. Vienna & Leipzig. Deutsche Rundschau Geogr. Statist. See B–P–H 347/9. HI 53744

Deutsche tierärztliche Wochenschrift. Karlsruhe & Hanover. Deutsche Tierärztl. Wochenschr. See B–P–H 347/10. HI 53745

Deutsche Tropenmedizinische Zeitschrift; Archiv für Schiffs- und Tropenhygiene. Leipzig. Deutsche Tropenmed. Z. See B–P–H 347/11. HI 53746

Deutsche Wasserwirtschaft. Berlin. Deutsche Wasserw. See B–P–H 347/12. HI 53747

Deutsche Zeitschrift für Thiermedizin und vergleichende Pathologie. Leipzig. Deutsche Z. Thiermed. Vergleichende Pathol. See B–P–H 347/13. HI 53748

Deutschen Kolonialblatt; Amtsblatt für die Schutzgebiete in Afrika und in der Südsee. Berlin. Vols. 1-32(6), 1890-1921. Deutsch. Kolonialbl. 2-1330-3. HI 64920

Deutscher Garten. Holzminden, Germany. Deutsch. Gart. See B–P–H 344/25. HI 53698

Deutscher Gartenbau. Stuttgart. Vol. 29+, 1975+. Deutsch. Gartenbau. Preceded by: Erwerbsgärtner,

Der. HI 64921

Deutscher Geographentag. Tagungsbericht und wissenschaftliche Abhandlungen. Landshut. Deutsch Geographentag. See B–P–H 344/31. HI 53700

Deutsches gemeinnütziges Magazine. Leipzig. Deutsch. Gemeinnütz. Mag. See B–P–H 344/30. HI 53699

Deutsches Jahrbuch für die Pharmacie = Berlinisches Jahrbuch für die Pharmacie und für die damit verbundenen Wissenschaften. Berlin. Berlin. Jahrb. Pharm. Verbundenen Wiss. See B–P–H 183/11.

Deutsches Kolonialblatt. Wissenschaftliche Beihefte = Mittheilungen von Forschungsreisenden und Gelehrten aus den Deutschen Schutzgebieten. Berlin. Mitth. Forschungsreisenden Gel. Deutsch. Schutzgeb. See B–P–H 611/12.

Deutsches Magazin. Hamburg. Deutsch. Mag. See B–P–H 344/35. HI 53701

Deutsches Magazin für Garten- und Blumenkunde. Stuttgart. Deutsch. Mag. Garten- Blumenk. See B–P–H 344/37. HI 53702

Deutsches Museum. Leipzig. Deutsch. Mus. (Leipzig). See B–P–H 345/2. HI 53703

Deutschfreiburger Beiträge zur Heimatkunde. Freiburg. Vol. 53+, 1985+. Deutschfreiburger Beitr. Heimatk. Preceded by: Beiträge zur Heimatkunde des Sensebezirks. HI 64922

Deutschlands Obstsorten. Stuttgart. Vols. 1-26, 1905-33? Deutschl. Obstsorten. 2-1332-3. HI 64923

Deutschösterreichische Monatsschrift für natuwissenschaftliche Fortbildung. Vienna. Deutschösterr. Monatsschr. Naturwiss. Fortbild. See B–P–H 347/14. HI 53749

Developmental biology. New York, NY. Developm. Biol. See B–P–H 347/15. HI 53750

Developmental genetics. New York. Vol. 1+, 1979+. Developmental Genet. HI 64924

Developments in biological standardization. Basel. Vol. 23+, 1974+. Developm. Biol. Standardiz. Preceded by: Symposia series in immunobiological standardization. HI 64925

Developments in food microbiology. Englewood, NJ. Vol. 1+, 1982+. Developm. Food Microbiol. HI 64926

Developments in genetics. Amsterdam, New York. Vol. 1+, 1979+. Developm. Genet. HI 64927

Developments in hydrobiology. The Hague, Boston, MA. Vol. 1+, 1980+. Developm. Hydrobiol. HI 64928

Developments in industrial microbiology. New York, NY. Developm. Industr. Microbiol. See B–P–H 347/16. HI 53751

Developments in plant biology. Amsterdam. Vol. 1+, 1978+. Developm. Pl. Biol. HI 64929

Developments in plant genetics and breeding. New York. No. 1+, 1983+. Developm. Pl. Genet. Breed. HI 64930

Developments in plant and soil sciences. The Hague, Boston, MA. Vol. 1+, 1981+. Developm. Pl. Soil Sci. HI 64931

Dhaka university studies. Part E, biological sciences. Dhaka. Vol. 1+, 1986+. Dhaka Univ. Stud., E. HI 64932

Dhauner Echo. Mitteilungsblatt der Heimvolkshochschule Schloss Dhaun. [Germany]. Dhauner Echo. See B–P–H 347/19. HI 53752

D.I.A. boletín técnico. Bogotá = D I A boletín técnico. Bogotá.

Diagnostic microbiology and infectious disease. New York. Vol. 1+, 1983+. Diagn. Microbiol. Infect. Dis. HI 64933

Diagnosi. Naples. Diagnosi (Naples). See B–P–H 348/3. HI 53754

Diagnosi. Pisa & Bologna. Diagnosi (Pisa) See B–P–H 348/4. HI 53755

Diagnostica e tecnica de laboratorio. Naples. Diagn. Tecn. Lab. See B–P–H 348/2. HI 53753

Diamond brand news = Diamond walnut news. Los Angeles, CA. Diamond Walnut News. See B–P–H 348/7.

Diamond light. Vol. ?-2(5)+, 19??-75+. Diamond Light. HI 64934

Diamond walnut news. Los Angeles, CA. Diamond Walnut News. See B–P–H 348/7. HI 53756

Diana; revue mensuelle d'information cynégétiques: Faune, flore, nature, chasse, pêche, mycologie, ornithologie, entomologie, cynologie. Bern, Geneva. Vols. 1-104, 1883-1987. Diana (Bern). HI 64935

Diana; oder Gesellschaftsschrift zur Erweiterung und Berichtigung der Natur-, Forst- und Jagdkunde. Waltershausen, Germany. Vols. 1-4, 1797-1816. Diana (Waltershausen). 2-1336-2. HI 53757

Diatom research; journal of the international society for diatom research. Stuttgart. Vol. 1+, 1986+. Diatom Res. Preceded by: Bacillaria HI 64936

Diatomiste; journal spécial s'occupant exclusivement des diatomées et de tout ce qui s'y rattache. Paris. Diatomiste. See B–P–H 348/9. HI 53758

Dicengxue zazhi = Acta stratigraphica sinica. Nanking.

Difesa delle piante. Bologna. Vol. 1+, 1978+. Difesa Piante. HI 64937

Difesa delle piante contra le malattie ed i parassiti. Turin.

1924-35. Difesa Piante Malatt. Parassiti. Preceded by: Pubblicazioni mensili del reale osservatorio di fitopatologia. Superseded by: Bollettino del laboratorio sperimentale e osservatorio di fitopatologia. HI 64938

Differentiation. London, New York. Vol. 1+, 1973+. Differentiation. HI 64939

Digest des rapports d'information = Information reports digest, canadian forestry service. Hull, Quebec.

Dili zhishi. Ti li chih shih = Geographical knowledge. Peking. Geogr. Knowl. (Peking). See B–P–H 397/22.

Dinteria; contributions to the flora of South West Africa. (Beiträge zur Flora von Südwestafrika. Bydraes tot die flora van Suidwes-Afrika). Windhoek. Vol. 1+, 1968+. Dinteria. HI 64940

Dirasat. Amman. Vols. 1-10(2), 1974-83. Dirasat. Preceded by: Dirasat: natural sciences. HI 64941

Dirasat: al-ulum al-tabiyah = Dirasat: natural sciences. Amman.

Dirasat an al-ahya fi al-Iraq = Prospects of Iraq biology. [Baghdad.]

Dirasat: natural sciences. Amman. Vols. 1-10(2), 1974-83. Dirasat Nat. Sci. Superseded by: Dirasat. HI 64942

Direccion de investigaciones, instituto de sanidad vegetal. Ser. A. Buenos Aires. Nos. 1-58, 1945-54. Direccion Invest. Inst. Sanid. Veg., A. Superseded by: Publicaciones tecnicas, instituto de patologia vegetal. HI 64943

Director's report for the Arnold arboretum = Report, Arnold arboretum. Jamaica Plain, MA.

Director's report, Kansas agricultural experiment station = Kansas agricultural experiment station. Director's report. Manhattan, KS.

Directory of organizations and officials concerned with the protection of wildlife and other natural resources = Conservation directory. Washington, DC.

Discours d'ouverture de la sessions de la société helvétique des sciences naturelles = Discours ..., en ouvrant la première séance de la réunion périodique de la société helvétique des sciences naturelles. Lausanne. Discours Ouvrant Prem. Séance Soc. Helv. Sci. Nat. See B–P–H 348/15.

Discours ..., en ouvrant la première séance de la réunion périodique de la société helvétique des sciences naturelles. Lausanne. Discours Ouvrant Prem. Séance Soc. Helv. Sci. Nat. See B–P–H 348/15. HI 53759

Discover; magazine of science and technology: their wonders, their uses, their impact on our lives. New York. Vol. 1+, 1980+. Discover. HI 64944

Discovery; the popular journal of knowledge [subtitle varies.] London. Vols. 1-19, 1920-38; n.s. vols. 1-27(9), 1938-66. Discovery (London). Superseded by: Science journal. 2-1344-1. HI 64945

Discovery; magazine of the Peabody museum of natural history. New Haven, CT. Vol. 1+, 1965+. Discovery (New Haven). HI 64946

Discovery. Vancouver, B.C. 1943+. Discovery (Vancouver). HI 64947

Disease notice, plant pathology laboratory. Harpenden. No. ?-13+, 19??-76+. Dis. Not. Pl. Pathol. Lab. HI 64948

Dissertation abstracts; abstracts of dissertations and monographs in microfilm [From vol. 16+ includes: Index to american doctoral dissertations]. Ann Arbor, MI. Vols. 12-26, 1952-66. Diss. Abstr. Preceded by: Microfilm abstracts. Index to american doctoral dissertations superseded by: American doctoral dissertations. Superseded by: Dissertation abstracts. Section B, sciences and engineering. 2-1345-2. HI 64949

Dissertation abstracts. Section B, sciences and engineering. Ann Arbor, MI. Vols. 27-29, 1966-69. Diss. Abstr., B. Preceded by: Dissertation abstracts. Superseded by: Dissertation abstracts international. Section B, sciences and engineering. HI 64950

Dissertation abstracts international. Section B, sciences and engineering. Ann Arbor, MI. Vol. 30+, 1969+. Diss. Abstr. Int., B. Preceded by: Dissertation abstracts. Section B, sciences and engineering. HI 64951

Dissertation abstracts international. C, european abstracts. Ann Arbor, MI. Vol. 37+, 1976+. Diss. Abstr. Int., C. Preceded by: Dissertation abstracts. HI 64952

Dissertationes, academia scientiarum et artium slovenica. Classis 4, historia naturalis et medicina = Razprave, slovenska akademija znanosti in umetnosti. Razrad za prirodoslovne in medicinske vede. Ljubljana.

Dissertationes, academia scientiarum et artium slovenica. Classis 4, historia naturalis et medicina. Pars historico naturalis = Razprave, slovenska akademija znanosti in umetnosti. Razrad za prirodoslovne in medicinske vede. Oddelek za prirodoslovne vede. Ljubljana.

Dissertationes botanicae. Budapest = Borbásia. Budapest. Borbásia. See B–P–H 217/16.

Dissertationes botanicae. Lehre. Vol. 1+, 1968+. Diss. Bot. HI 64953

Distribution maps of plant diseases. Kew, England. Distrib. Maps Pl. Dis. See B–P–H 348/18. HI 53760

Distributiones plantarum africanarum. Brussels. Vol. 1+, 1969+. Distrib. Pl. African. HI 64954

Dittany. Invercargill. ?-1987+. Dittany. HI 64955

Diversidata; centros de datos para la conservación de America Latina y el Caribe. Washington, DC. 1984+. Diversidata. HI 64956

Diversity. Fort Collins, CO. Vol. 1+, 1982+. Diversity. HI 64957

Divisional report, division of forest research, C S I R O = Report (Annual), division of forest research, commonwealth scientific and industrial research organisation. Canberra, A.C.T.

Divisional report, division of tropical crops and pastures, C S I R O. Brisbane, Qld. 1975/76+, [1976?]+. Div. Rep. Div. Trop. Crops Pastures, C. S. I. R. O. Preceded by: Report (Annual), division of tropical agronomy, C S I R O. HI 64958

Divulgação agronómica. Rio de Janeiro. Divulg. Agron. See B–P–H 349/2. HI 53761

Divulgacion agropecuaria, universidad nacional agraria de la selva. Tingo Maria. No. 1+, 1970+. Divulg. Agropecu. Univ. Nac. Agrar. Selva. HI 64959

Djela jugoslavenske akademije znanosti i umjetnosti. Zagreb, Croatia [Yugoslavia]. Djela Jugoslav. Akad. Znan. See B–P–H 349/4. HI 53762

Dnevnik ėkskursii, [Volzhkaya biologicheskaya stantsiya]. Saratov, Russian S F S R. 1906-17. Dnevn. Ėkskurs. HI 64960

Dnevnik S″ezda Russkih Estestvoispytatelej i Vračej = Dnevnik S″ezda Russkikh Yestyestvoispytatelei i Vrachei.

Dnevnik S″ezda Russkikh Yestyestvoispytatelei i Vrachei. Moscow, etc. [Place of publication changes each year]. Vols. 1-12, 18??-1910. Dnevn. S″ezda Russk. Yestyestvoisp. Vrachei. HI 64961

Dnevnik Vserossijskogo s″ezda russkih botanikov = Dnevnik Vserossiiskogo s″ezda russkikh botanikov. Petrograd.

Dnevnik Vserossiiskogo s″ezda russkikh botanikov. Petrograd. Vol. 1, 1921. Dnevn. Vserossiisk. S″ezda Russk. Bot. Superseded by: Dnevnik Vsesoyuznogo s″ezda botanikov. HI 64962

Dnevnik Vsesojuznogo s″ezda botanikov = Dnevnik Vsesoyuznogo s″ezda botanikov. Leningrad.

Dnevnik Vsesoyuznogo s″ezda botanikov. Leningrad. 1926-28 Dnevn. Vsesoyuzn. S″ezda Bot. Preceded by: Dnevnik Vserossiiskogo s″ezda russkikh botanikov. HI 64963

Docama; bulletin de l'association de micro-aquaculture. Trouville. Vol. 1+, 1984+. Docama. HI 64964

Docencia = Collección docencia. Tegucigalpa.

Doctoral dissertations accepted by american universities. New York, NY. Doct. Diss. Amer. Univ. See B–P–H 349/13. HI 53763

Document, see Documents

Documenta kungl. vetenskaps-akademien. Stockholm. Vol. 9+, 1974+. Doc. Kungl. Vetensk.-Akad., Stockholm. Preceded by: Svenska vetenskapsakademiens årsbok. HI 64965

Documenta physiographica polonica = Materiały do fizjografii kraju. Cracow. Mater. Fizjogr. Kraju. See B–P–H 550/19.

Documentatie van de afdeling tropische producten, indisch instituut. Amsterdam. Vols. [1](12)-7(?), 1949-52 [vol. 1(1-11) not published]. Doc. Afd. Trop. Prod. Indisch Inst. Superseded by: Documentatieblad van de afdeling tropische produktion, koninklijk instituut voor de tropen. HI 64966

Documentatie van de afdeling tropische producten, koninklijk instituut voor de tropen = Documentatie van de afdeling tropische producten, indisch instituut. Amsterdam.

Documentatie van de afdeling tropische producten, koninklijke vereeniging indisch instituut = Documentatie van de afdeling tropische producten, indisch instituut. Amsterdam.

Documentatie oosteuropese landbouwkundige literatuur. Wageningen. Vols. 1-11, 1960-71. Doc. Oosteur. Landbouwk. Lit. HI 64967

Documentatieblad van de afdeling tropische production, koninklijk instituut voor de tropen. Amsterdam. Vol. 7(?), 1952. Documentatieblad Afd. Trop. Prod. Kon. Inst. Tropen. Preceded by: Documentatie van de afdeling tropische producten, indische instituut. Superseded by: Tropical abstracts. HI 64968

Documentation analytique, institut français de recherches fruitières outremer. Paris. 1966+. Doc. Analytique Inst. Franç. Rech. Fruitières Outremer. Preceded by: Documentations analytique et signaletique, institut français de recherches fruitieres outremer. HI 64969

Documentation east-european agricultural literature = Documentatie oosteuropese landbouwkundige literatuur. Wageningen.

Documentations analytique et signaletique, institut français de recherches fruitières outremer. [Supplement to: Fruits. See B–P–H 386/17.] Paris. 1948-66. Doc. Analytique Signal. Inst. Franç. Rech. Fruitières Outremer. Superseded by: Documentation analytique, institut français de recherches fruitieres outremer. HI 64970

Documentos, centro de pesquisa agropecuaria do tropico umido. Petrolina. No. ?-3+, 19??-80+. Doc. Centro Pesq. Agropecu. Trop. Umido. HI 64971

Documentos tecnicos, base oceanografica Atlantica. Rio Grande. No. 1+, 1980+. Doc. Tecn. Base Oceanogr. Atlantica. HI 64972

Documentos de trabajos, investigacion del desarrollo de los recursos forestales del noreste de Nicaragua. Managua. No. ?-2+, 19??-72+. Doc. Trab. Invest. Desarr. Recurs. Forest. Noreste Nicaragua. HI 64973

Documents pour les cartes des productions végétales. Toulouse. ?-1966+. Doc. Cartes Prod. Vég. HI 64974

Documents pour la carte de la végétation des Alpes. Grenoble. Nos. 1-10, 1963-72. Doc. Carte Vég. Alpes. Superseded by: Documents de cartographie écologique. HI 64975

Documents de cartographie écologique. Grenoble. Vol. 11+, 1973+. Doc. Cartogr. Ecol. Preceded by: Documents pour la Carte de la végétation des Alpes. HI 64976

Documents, centre national de recherches océanographiques, ministère de l'enseignement supérieur et de la recherche scientifique, république democratique malgache. Nosy-Be. No. 1+, 1978+. Doc. Centre Natl. Rech. Océanogr. Minist. Enseignem. Supér. Rech. Sci. Malgache. Preceded by?: Documents, centre O R S T O M de Nosy-Be. HI 64977

Documents, centre O R S T O M de Nosy-Be. -Nosy-Be. 1968-? Doc. Centre O. R. S. T. O. M. Nosy-Be. Superseded by?: Documents, centre national de recherche océanographiques, ministère de l'enseignement supérieur et de la recherche scientifique, république democratique malgache. HI 64978

Documents, centre de recherches sur les ressources biologiques terrestres. Alger. No. 1+, 1980+. Doc. Centre Rech. Ressources Biol. Terrestr. HI 64979

Documents floristiques. Saint Valery sur Somme. 1977+. Doc. Florist. HI 64980

Documents pour l'histoire du vocabulaire scientifique. Besançon-Nancy. Vol. 1+, 1980+. Doc. Hist. Vocab. Sci. HI 74830

Documents de l'institut scientifique, université Mohamed V. Rabat. No. 1+, 1977+. Doc. Inst. Sci. Univ. Mohamed V. HI 64981

Documents mycologiques. Lille. Vol. 1+, 1971+. Doc. Mycol. HI 64982

Documents phytosociologiques. Lille, Vaduz, Stuttgart. Vols. 1-20, 1972-7?; n.s. vol. 1+, 1978+. Doc. Phytosoc. HI 64983

Documents sur les sciences de la nature. Saint-Etienne. Vol. ?-6+, 19??-80+. Doc. Sci. Nat. HI 64985

Documents scientifiques, centre de recherche océanographiques Abidjan. Abidjan. No. ?-009+, 19??-66+. Doc. Sci. Centre Rech. Océanogr. Abidjan. HI 64986

Documents scientifiques du parc national des Pyrénées. Tarbes. Vol. 1+, 1984+. Doc. Sci. Parc Natl. Pyrénées. HI 75277

Documents scientifiques provisoires, centre de recherche océanographiques Abidjan = Documents scientifiques, centre de recherche océanographiques Abidjan. Abidjan.

Dodatek do biblioteki umietnosci przyrodniczych. Cracow. Dodatek Bibliot. Umietn. Przyr. See B–P–H 349/14. HI 53764

Dodonaea. Biologisch jaarboek uitgegeven door het koninklijk natuurwetenschappelijk genootschap "Dodanaea" te Gent. Brussels. Dodonaea. See B–P–H 349/15. HI 53765

Dodonaeus; drijmaandelijkes tijdschrift van afdeeling duffel en omgeving van de belgische vereniging cactusweelde. (Périodiques bimensuel de cactéophiles). Brussels. Vols. 1-6, 1963-68. Dodonaeus. Incorporated in: Cactus (Antwerp). HI 64987

Doga. Ankara. Vols. 1-3, 1976-79?. Doga. Superseded by: Doga bilim dergisi. Seri A, temel bilimler. HI 63935

Doga bilim dergisi. Seri A, temel bilimler. Ankara. Vol. 4+, 1980+. Doga Bilim Derg. A. Preceded by: Doga. HI 64988

Dogwood's bark; a bulletin of wild flower notes and outings. Washington, DC. Dogwood's Bark. See B–P–H 349/17. HI 53766

Dohrniana. Abhandlungen und Berichte der Pommerschen Naturforschenden Gesellschaft und des Naturkunde-Museums der Stadt Stettin. Stettin [Poland]. Dohrniana. See B–P–H 349/17. HI 53767

Dokladchoi akademijai fachnoi RSS Tochikiston = Doklady akademii nauk Tadzhikskoi S S R. Dushanbe.

Doklady Akademii nauk Tadžikskoj S S R = Doklady Akademii nauk Tadzhikskoi S S R. Stalinabad.

Doklady Akademii nauk Ukrainskoj S S R = Dopovidi Akademiï nauk Ukrayins'koi R S R. Kiev.

Doklady Akademii nauk Ŭzbekskoj S S R = Doklady Akademii nauk Uzbekskoi S S R. Tashkent.

Doklady akademiya nauk armyanskoi S S R. Erevan. Vol. 1+, 1944+. Dokl. Akad. Nauk Armyanskoi S.S.R. HI 64989

Doklady akademiya nauk azerbaidzhanskoi S S R. Baku. Vol. 1+, 1945+. Dokl. Akad. Nauk Azerbaidzhansk. S.S.R. HI 64990

Doklady akademiya nauk belorusskoi S S R. Minsk. Vol. 1+, 1957+. Dokl. Akad. Nauk Belorussk. S.S.R. HI 64991

Doklady akademii nauk S S S R. Moscow & Leningrad. Ser. A, 1925[4]-33; n.s. 1933+ [numbering of n.s.

commences with vol. 8+, 1935+]. Dokl. Akad. Nauk S.S.S.R. Preceded by: Doklady Rossiiskoi akademii nauk. For English editions see: Doklady, biochemistry section; Doklady, biological sciences; Doklady, biophysics; and Doklady, botanical sciences. 1-111-2. HI 64992

Doklady Akademii nauk Tadzhikskoi S S R. Stalinabad [=Dushanbe], Tadzhik S S R. Vols. 1-22, 1951-57; vol. 1+, 1958+. Dokl. Akad. Nauk Tadzhiksk. S.S.R. HI 64993

Doklady akademii nauk ukrainskoi S S R. Seriya B, geologicheskie, khimicheskie i biologicheskie nauki. Kiev. 1979+. Dokl. Akad. Nauk Ukrainsk. S.S.R., B. Preceded by: Dopovidi akademiji nauk Ukrajins'koji R S R. HI 64994

Doklady Akademii nauk Uzbekskoi S S R. Tashkent, Uzbek S S R. 1948+. Dokl. Akad. Nauk Uzbeksk. S.S.R. Preceded by: Byulleten' Akademii nauk Uzbekskoi S S R. 1-123-3. HI 64995

Doklady, akademiya na selskostopanskite nauki, Sofia. Sofia. Vols. 1-4, 1968-71. Dokl. Akad. Selskost. Nauki Sofia. Superseded by: Doklady, selskostopanska akademiya "George Dimitrov." HI 64996

Doklady, biochemistry section. Proceedings of the academy of sciences of the U S S R. Biochemistry section. [Translation of: Doklady akademii nauk S S S R.] New York, Washington, DC. Vol. 118-141, 1958-62; vol. 154/56+, 1964+ [for vols. 142-153 see Doklady, biological sciences]. Dokl. Biochem. Sect. Preceded by: Proceedings of the academy of sciences of the U S S R. Biochemistry section. HI 64997

Doklady, biological sciences. Proceedings of the academy of sciences of the U S S R. Biological sciences. [Translation of: Doklady akademii nauk S S S R.] New York, Washington, DC. Vols. 112+, 1957+. Dokl. Biol. Sci. Vols. 142-153 incorporate: Doklady, biochemistry section and Doklady, botanical sciences. Superseded in part, from 1964+, by: Doklady, biophysics. HI 64998

Doklady, biophysics. Proceedings of the academy of sciences of the U S S R. Biophysics section. [Translation of: Doklady akademii nauk S S S R.] New York. Vol. 154/56+, 1964+. Dokl. Biophys. Preceded, in part, by: Doklady, biological sciences. HI 64999

Doklady bolgarskoi akademiia nauk. Sofia = Doklady bulgarskoi akademiia nauk. Sofia.

Doklady, botanical sciences. Proceedings of the academy of sciences of the U S S R. Botanical sciences. [Translation of: Doklady akademii nauk S S S R.] New York, Washington, DC. Vols. 112-141, 1957-62; vol. 154/56+, 1964+ [for vols. 142-153 see Doklady, biological sciences]. Dokl. Bot. Sci. HI 65000

Doklady bulgarskoi akademiia nauk. Sofia. 1948-51, 1962+. Dokl. Bulg. Akad. Nauk. HI 65001

Doklady M O I P = Doklady, moskovskoe obshchestvo ispytatelei prirody. Moscow.

Doklady, moskovskoe obshchestvo ispytatelei prirody. Obschaya biologiya. Moscow. 1967+, 1969+. Moskovsk. Obshch. Isp. Prir., Obschaya Biol. HI 65002

Doklady, moskovskoe obshchestvo ispytatelei prirody. Zoologiya i botanika. Moscow. 1967+, 1969+. Dokl. Moskovsk. Obshch. Isp. Prir., Zool. Bot. HI 65003

Doklady, moskovskoi sel'sko-khozyaistvennoi akademii imeni K.A. Timiryazeva. Moscow. Vol. 1+, 1943+. Dokl. Moskovsk. Sel'skokhoz. Akad. Im. K. A. Timiryazeva. HI 65004

Doklady Rossiiskoi akademii nauk. Petrograd. Ser. A, 1923-25 [(3)]. Dokl. Rossiisk. Akad. Nauk. Superseded by: Doklady akademii nauk S S S R. 1-111-3. HI 65005

Doklady Rossijskoj akademii nauk = Doklady Rossiiskoi akademii nauk. Petrograd.

Doklady sel'skokhozyaistvennoi akademii imeni Georgiya Dimitrova = Doklady, selskostopanska akademiya "George Dimitrov." Sofia.

Doklady, selskostopanska akademiya "George Dimitrov." Sofia. Vol. 5+, 1972+. Dokl. Selskost. Akad. "George Dimitrov." Preceded by: Doklady, akademiya na selskostopanskite nauki, Sofia. HI 65006

Doklady Vsesojuznogo soveščanija po fiziologii rastenij = Doklady Vsesoyuznogo soveshchaniya po fiziologii rastenii. Moscow & Leningrad.

Doklady Vsesojuznoj akademii sel'skohozjajstvennyh nauk imeni V. I. Lenina = Doklady Vsesoyuznoi akademii sel'skokhozyaistvennykh nauk imeni V. I. Lenina. Moscow.

Doklady Vsesojuznoj ordena Lenina akademii sel'skohozjajstvennyh nauk imeni V. I. Lenina = Doklady Vsesoyuznoi ordena Lenina akademii sel'skokhozyaistvennykh nauk imeni V. I. Lenina. Moscow.

Doklady Vsesoyuznoi akademii sel'skokhozyaistvennykh nauk imeni V. I. Lenina. Moscow. Vols. 1-14(10), 1936-49. Dokl. Vsesoyuzn. Akad. Sel'skokhoz. Nauk Lenina. Superseded by: Doklady Vsesoyuznoi ordena Lenina akademii sel'skokhozyaistvennykh nauk imeni V. I. Lenina. 5-4433-3. HI 65007

Doklady Vsesoyuznoi ordena Lenina akademii sel'skokhozyaistvennykh nauk imeni V. I. Lenina. Moscow. Vol. 14(11)+, 1949+. Dokl. Vsesoyuzn. Ordena Lenina Akad. Sel'skokhoz. Nauk Lenina. Preceded by: Doklady Vsesoyuznoi akademii sel′skokhozyaistvennykh nauk imeni V. I. Lenina. 5-4433-3. HI 65008

Doklady Vsesoyuznogo soveshchaniya po fiziologii rastenii. Moscow & Leningrad. Vol. 2, 1945. Dokl. Vsesoyuzn. Soveshch. Fiziol. Rast. HI 65009

Dokumentation für Umweltschutz und Landespflege. Bad Godesberg. Vol. 11+, 1971+. Dokument. Umweltschutz Landespflege. Preceded by: Mitteilungen zur Landespflege. Dokumentation-Information. HI 65010

Dominguezia. Buenos Aires. No. 1+, 1978+. Dominguezia. HI 65011

Dongbei linxueyuan xuebao = Journal of north-eastern forestry institute. Harbin.

Dopovidi Akademiï nauk Ukrayins′koi R S R. Kiev, Ukrainian S S R. Vol. 1+, 1939+. Dopov. Akad. Nauk Ukraïns′k. R.S.R. 1-121-2. HI 65012

Dopovidi Akademiji nauk Ukrajins′koji R S R = Dopovidi Akademiï nauk Ukrayins′koi R S R. Kiev.

Dopovidi Akademiyi nauk Ukrayins′koyi R S R = Dopovidi Akademiï nauk Ukrayins′koi R S R. Kiev.

Dörfleria. Internationale Zeitschrift für Förderung praktischer Interessen der Botaniker und der Botanik. Vienna. Dörfleria. See B–P–H 350/13. HI 53768

Doriana. Genoa. Doriana. See B–P–H 350/14. HI 53769

Dorpater Jahrbücher für Litteratur, Statistik, und Kunst, besonders Russlands. Riga [Latvian S S R] & Dorpat [=Tartu, Estonian S S R]. Dorpater Jahrb. Litt. See B–P–H 350/18. HI 53770

Dortmundisches Magazin. Dortmund. Dortmund. Mag. See B–P–H 350/19. HI 53771

Dosyahnennya botanichnoyi nauky na Ukrayini. Kiev. 1965/66+, 1968+. Dosyahn. Bot. Nauky Ukrayni. Preceded by: Shchorichnyk ukrayins′kogo botanichnogo tovarystva. HI 65013

Dossier. Vol. 1+, 1977+. Dossier. HI 65014

Douglasia; newsletter of the Washington native plant society. Seattle, WA. Vol. ?-8(2)+, 19??-84+. Douglasia. HI 65015

Down to earth; a review of agricultural chemical progress. Midland, MI. Down to Earth. See B–P–H 350/21. HI 53772

Doyo kai. Tokyo. Ca.1966?+. Doyo Kai. HI 65016

Dr. A. Petermann's Mittheilungen aus Justus Perthes' Geographischer Anstalt = Petermann's Mittheilungen aus Justus Perthes' Geographischer Anstalt. Gotha. Petermann's Mitth. Justus Perthes' Geogr. Anst. See B–P–H 702/13.

Dr. A. Petermann's Mittheilungen aus Justus Perthes' Geographischer Anstalt. Ergänzungsband = Petermann's Mittheilungen aus Justus Perthes' Geographischer Anstalt. Ergänzungsband. Gotha. Petermann's Mitth. Justus Perthes' Geogr. Anst. (Ergänzungsband). See B–P–H 702/14.

Dr. Neubert's Deutsches Garten-Magazin = Neubert's Deutsches Garten-Magazin. Stuttgart. Neubert's Deutsch. Gart.-Mag. See B–P–H 641/24.

Dr. Neubert's Garten-Magazin = Neubert's Garten-Magazin. Munich. Neubert's Gart.-Mag. See B–P–H 641/25.

Dragoco report. Totowa, NJ. Vol. 1+, 1954+. Dragoco Rep. HI 65017

Dresdner naturwissenschaftliches Jahrbuch. Leipzig. Dresdner Naturwiss. Jahrb. See B–P–H 351/6. HI 53773

Dresdnisches Magazin, oder Ausarbeitungen und Nachrichten zum Behuf der Naturlehre, der Arzneykunst, der Sitten und der schönen Wissenschaften. Dresden. Dresdnisches Mag. See B–P–H 351/7. HI 53774

Driemaandelijke publikatie, belgisch instituut tot verbetering van de biet = Publications trimestrielle, institut belge pour l'amélioration de la betterave. Tirlemont.

Driemaandelijke tijdskrift, vereeniging voor het onderwijs in de biologie. Mol. Vols. 1-2, 1978-79. Driemaand. Tijdskr. Ver. Onderwijs Biol. Preceded by: Année, association nationale des professeurs de biologie de Belgique. Superseded by: Jaarboek, vereniging voor het onderwijs in de biologie. HI 65018

Drontheimischen Gesellschaft Schriften. Copenhagen. Drontheim. Ges. Schriften. See B–P–H 351/8. HI 53775

Drosera; naturkundliche Mitteilungen aus Norwestdeutschland. Oldenburg. 1976+. Drosera. HI 65019

Drudea; Mitteilungen des geobotanischen Arbeitskreises Sachsen-Thüringen. Jena. Vols. 1-5, 1961-65. Drudea. HI 65020

Drugs and medicines of North America. A publication devoted to the historical and scientific discussion of the botany, pharmacy, chemistry and therapeutics of the medicinal plants of North America. Cincinnati, OH. Drugs Med. N. Amer. See B–P–H 351/10. HI 53776

Društva za nauka i umjetnost Crne Gore. Otdjeljenje prirodnih nauka. Titograd = Glasnik, odjeljenje prirodnikh nauka, društvo za nauku i umjetnost Crne Gore. Titograd.

Dublin journal of medical and chemical science. Dublin. Dublin J. Med. Chem. Sci. See B–P–H 351/27. HI 53777

Dublin journal of medical science. Dublin. Dublin J. Med. Sci. See B–P–H 351/28. HI 53778

Dublin medical and physical essays. Dublin. Dublin Med. Phys. Essays. See B–P–H 351/32. HI 53779

Dublin philosophical journal and scientific review. Dublin. Dublin Philos. J. Sci. Rev. See B–P–H 351/33. HI 53781

Dublin quarterly journal of medical science. Dublin. Dublin Quart. J. Med. Sci. See B–P–H 351/34. HI 53782

Dublin review. London. Dublin Rev. See B–P–H 351/36. HI 53783

Duisburgischen gelehrten Gesellschaft Deutsche Schriften. Duisburg, Germany. Duisburg. Gel. Ges. Deutsche Schriften. See B–P–H 352/5. HI 53784

Duke university marine station. Bulletin. Durham, NC. Duke Univ. Mar. Sta. Bull. See B–P–H 352/10. HI 53785

Dumortiera; revue consacré à la flore et à la végétation de la Belgique et des territoires voisins. Meise. Vol. 1+, 1975+. Dumortiera. HI 65021

Dump heap; the journal of diverse unsung miracle plants for healthy evolution among people. St. Raphael, CA. 1979+. Dump Heap. HI 65022

Dunántúli szemle. Szombathely, Hungary. Dunántúli Szemle. See B–P–H 352/12. HI 53786

Düngungsversuche in dem Deutschen Kolonien. Berlin. Nos. 1-4, 1913-14. Düngungsversuche Deutsch. Kolon. HI 65023

Durban museum novitates. Natal, South Africa. Durban Mus. Novit. See B–P–H 352/13. HI 53787

Durham colleges natural history society journal. Durham. 1953-56. Durham Coll. Nat. Hist. Soc. J. Superseded by: Journal of the Durham colleges natural history society. HI 65024

Dusenia; publicatio periodica de scientia naturali. Curitibá. Vols. 1-8, 1950-68. Dusenia. HI 53788

Dwarf conifer notes. Little Compton, RI. Vol. 1+, 1980+. Dwarf Conifer Notes. HI 65025

Dziennik rolniczy. Cracow. Dziennik Roln. See B–P–H 352/15. HI 53789

Dziennik wileński. Vilna [Lithuanian S S R]. Dziennik Wileński. See B–P–H 352/16. HI 53791

E A N H S bulletin. Nairobi. 1971+. E. A. N. H. S. Bull. Preceded by: E A N H S newsletter. HI 74520

E A N H S newsletter. Nairobi. 1970 (Aug.-Dec.). E. A. N. H. S. Newslett. Superseded by: E A N H S bulletin. HI 74890

E A P R abstracts of conference papers. Wageningen. 1975+. E. A. P. R. Abstr. Conf. Pap. Preceded by: Proceedings of the triennial conference, european association for potato research. HI 74891

E E report. Washington, DC = Environmental education report. Washington, DC.

E I S cumulative = [annual cumulation of:] E I S, key to environmental impact statements and later E I S, digests of environmental impact statements. Bethesda, MD.

E I S, digests of environmental impact statements. Bethesda, MD. Vol. 3(7)+, 1979+. E. I. S. Digests Environm. Impact Statem. Superseded by: E I S, key to environmental impact statements. HI 74892

E I S, key to environmental impact statements. Bethesda, MD. Vols. 1-3(6), 1977-79. E. I. S. Key Environm. Impact Statem. Superseded by: E I S, digests of environmental impact statements. HI 74893

E M B O journal. Arlington, VA. Vol. 1+, 1982+. E. M. B. O. J. HI 74894

E P A journal. Washington, DC. Vol. 1+, 1975. E. P. A. J. HI 74895

E P P O bulletin = Bulletin O E P P. Paris.

E P P O publications. Series A = Publications, european and mediterranean plant protection organisation. Series A, reports of technical working parties and conferences. Paris.

E P P O publications. Series A = Publications, european plant protection organisation. Series A. Paris.

E P P O publications. Series B = Publications, european and mediterranean plant protection organisation. Series B. Paris.

E P P O publications. Series B = Publications, european plant protection organisation. Series B. Paris.

E P P O publications. Series C = Publications, european and mediterranean plant protection organisation. Series C, reports of working parties. Paris.

E P P O publications. Series C = Publications, european plant protection organisation. Series C. Paris.

E P P O publications. Series D = Publications, european and mediterranean plant protection organisation. Series D. Paris.

E phusis. Athens. 1975+. E phusis. HI 74896

E S A bulletin = Bulletin of the epiphyllum society of America. Los Angeles, Monrovia, CA.

E S N information bulletin. London. 1987+. E. S. N. Inform. J. Preceded by: European science notes, office of naval research, branch office, London. HI 74897

E S S A. Rockville, MD. Vol. 5+, 1970+. E. S. S. A. Preceded by: E S S A world. HI 74898

E S S A professional papers. Washington, DC. No. 1+, 1967+. E. S. S. A. Profess. Pap. HI 74899

E S S A world. Rockville, MD. Vols. 1-4, 1966-69. E.

S. S. A. World. Superseded by: E S S A. HI 74900

E S & T = Environmental science and technology. Washington, DC.

E T P. Chicago, IL = Evolutionary trends in plants. Chicago, IL.

E U C A R P I A cruciferae newsletter = Cruciferae newsletter. E U C A R P I A. Pentlandfield, etc.

Earth research. Moscow = Zemlevedenie. Moscow. Zemlevedenie. See B–P–H 998/4.

East african agricultural and forestry journal. Nairobi. Vol. 1+, 1960+. E. African Agric. Forest. J. Preceded by: East african agricultural journal of Kenya, Tanganyika, Uganda, and Zanzibar. HI 74901

East african agricultural journal of Kenya, Tanganyika, Uganda, and Zanzibar. Nairobi. Vols. 1-25, 1935-60. E. Afric. Agric. J. Kenya. Superseded by: East african agricultural and forestry journal. 2-1379-3. HI 74902

East african wildlife journal. Nairobi. Vols. 1-16, 1963-78. E. African Wildlife J. Superseded by: African journal of ecology. HI 74903

East anglian magazine. Ipswich. 1935+. E. Anglian Mag. HI 74904

East asian tertiary/quaternary newsletter. Hong Kong. Vol. 1+, 1984+. E. Asian Tert. Quatern. Newslett. HI 74905

East Yorkshire field studies; joint publication of the Hull geological society, Hull scientific and field naturalists' club and field studies association (Hull and East Yorkshire). Hull. 1968-73. E. Yorkshire Field Stud. Superseded by: Humberside geologist [not entered]. HI 74906

Eastern Cape naturalist; journal of the eastern cape section of the wild life protection society. Port Elizabeth. No. 1+, 1957+. E. Cape Naturalist. HI 74907

Eastern fruit grower. Charles Town, WV. Vol. 1-34(5), 1938-71. E. Fruit Grower. Incorporated in: American fruitgrower. 2-1383-1. HI 74908

Eastern New York horticulturist. Chatham, NY. E. New York Hort. See B–P–H 352/20. HI 53792

Eastern region begonia news. Southampton, NY. Vol. 1+, 1985+. E. Region Begonia News. HI 74909

Eastern shore farmer and fruit culturist. Salisbury, MD. E. Shore Farmer Fruit Cult. See B–P–H 352/21. HI 53793

Eau pure; protection de la nature. Paris. Nos. ?-24-61, 19??-73-81?. Eau Pure. Superseded by: Lettre de l'eau pure. HI 74910

Échange; journal mensuel paraissant. Moulins, France. Échange. See B–P–H 352/28. HI 53794

Echinocereanae; official bulletin of the echinocereanae society. Mitcham. Vol. 1+, 1968+. Echinocereanae. HI 74911

Écho des Alpes. Geneva. Écho Alpes. See B–P–H 352/29. HI 53795

Écho médical du Nord. Lille. Écho Méd. Nord. See B–P–H 352/31. HI 53796

Écho du monde savant. Paris. Écho Monde Savant. See B–P–H 353/2. HI 53797

Écho des sociétés et associatons vétérinaires de France. Lyons. Écho Soc. Assoc. Vétérin. France. See B–P–H 353/3. HI 53798

Éclaireur agricole et horticule; revue pratique des cultures mériodionales et de l'Afrique du nord et des industries annexes. Nice. Éclaireur Agric. Hort. See B–P–H 353/7. HI 53802

Eclectic journal of science. Columbus, OH. Ecl. J. Sci. See B–P–H 353/4. HI 53799

Eclectic and medical botanist. Columbus, OH. Ecl. Med. Bot. See B–P–H 353/5. HI 53800

Eclectic review. London. Ecl. Rev. See B–P–H 353/6. HI 53801

Eclogae geologicae Helvetiae. Lausanne. Eclog. Geol. Helv. See B–P–H 353/8. HI 53803

Ecoandes; studies on tropical andean ecosystems. (Estudios de ecosistemas tropandinos.) Brunswick. Vol. 1+, 1983+. Ecoandes. HI 74912

Ecoforum; revue du centre de liaison pour l'environnement. Nairobi. Vol. ?-5+, 19??-80+. Ecoforum. HI 74913

Ecologia; rivista di studi e analisi sull'inquinamento la pianificazione e la conservazione ambientale. Milan. Vols. 1-3, 1971-73. Ecologia. Superseded by: Acqua ed aria, ecologia. HI 74914

Ecologia agraria. Perugia, Italy. Ecol. Agrar. See B–P–H 353/10. HI 53804

Ecologia, aque, aria, suolo. Milan. 1975+. Ecol. Aque Aria Suolo. Preceded by: Acqua ed aria, ecologia. HI 74915

Ecologia en Bolivia; revista del instituto de ecología. La Paz. No. 1+, 1982+. Ecol. Bolivia. HI 74916

Ecologia mediterranea; revue d'écologie terrestre et limnique. Aix-Marseille. Vol. 1+, 1975+. Ecol. Medit. HI 74917

Ecological abstracts. Norwich. 1974+. Ecol. Abstr. Preceded by: Geo abstracts. B, biogeography and climatology. HI 74918

Ecological bulletins. Stockholm. No. 19+, 1975+. Ecol. Bull. Preceded by: Bulletin, ekologikommittén, statens naturvetenskapliga forskningsråd. HI 74919

Ecological economics; the journal of the international society for ecological economics. New York. Vol. 1+, 1989+. Ecol. Econ. HI 74920

Ecological modelling. Amsterdam. Vol. 1+, 1975+. Ecol. Modelling. HI 74921

Ecological monographs. Durham, NC. Ecol. Monogr. See B–P–H 353/12. HI 53805

Ecological research. Tokyo. Vol. 1+, 1986+. Ecol. Res (Tokyo). HI 74922

Ecological research. [Research reporting series.] Washington, DC. 197?+. Ecol. Res. (Washington). HI 74923

Ecological review. [Seitaigaku kenkyu.] Sendai, Japan. Ecol. Rev. See B–P–H 353/14. HI 53806

Ecological studies: Analysis and synthesis. Heidelberg, New York. Vol. 1+, 1970+. Ecol. Stud. Analysis Synth. HI 74924

Ecological studies leaflet. Washington, DC. Ecol. Stud. Leafl. See B–P–H 353/18. HI 53807

Ecological studies of nature conservation of the Ryukyus Islands. Naha, Okinawa. Vol. 1+, 1974+. Ecol. Stud. Nat. Conservation Ryukyus Islands. HI 74925

Ecologie et conservation = Ecology and conservation. Paris.

Ecologie marina. Bucharest. Vol. 1+, 1965+. Ecol. Mar. HI 74926

Ecologist. London, Camelford. Vols. 1-7, 1970-77; vol. 9(3)+, 1979+. Ecologist. For 1978-79 see: Ecologist quarterly and New ecologist. HI 74927

Ecologist quarterly. Woodbridge. Nos. 1-4, 1978. Ecologist Quart. Preceded and superseded by: Ecologist. HI 74928

Ecology; quarterly journal devoted to all phases of ecological biology. Brooklyn, NY. Vol. 1+, 1920+. Ecology. Preceded by: Plant world. 2-1391-3. HI 74929

Ecology. [Translation of: Ekologiya, Sverdlovsk.] New York, London. No. 1+, 1970+. Ecology (U.S.S.R.). Superseded by: Soviet journal of ecology. HI 74930

Ecology abstracts. London, Bethesda, MD. Vol. 6+, 1980+. Ecol. Abstr. Preceded by: Applied ecology abstracts. HI 74931

Ecology and conservation. Paris. Vol. 1+, 1970+. Ecol. Conservation. HI 74932

Ecology and conservation studies. Sanderstead. Vol. 1+, [1980]+. Ecol. Conservation Stud. HI 74933

Ecology of food and nutrition. New York. Vol. 1+, 1971+. Ecol. Food Nutr. HI 74934

Ecology handbook. London. No. 1+, 1984+. Ecol. Handb. HI 74935

Ecology U S A. Silver Spring, MD. 1972+. Ecol. U.S.A. HI 74936

Ecology of western north America. Vancouver, B.C. Vols. 1-2(2), 1965-70. Ecol. W. N. Amer. Incorporated in: Syesis. HI 74937

Economia e agricultura. Rio de Janeiro = Brasil açucareio. Rio de Janeiro. Brasil Açucareio:. See B–P–H 227/22.

Economia colombiana. Bogotá. Vols. 1-5, 1954-58. Econ. Colomb. Superseded by: Economia grancolombiana. HI 74938

Economia grancolombiana. Bogotá. Vols. 1-7, 1959-63. Econ. Grancolomb. Preceded by: Economia colombiana. HI 74939

Economía y técnica agrícola. Madrid. Econ. Técn. Agríc. See B–P–H 354/1. HI 53811

Economic botany; devoted to applied botany and plant utilization. Lancaster, PA. Econ. Bot. See B–P–H 353/22. HI 53809

Economic proceedings of the royal Dublin society. Dublin. Econ. Proc. Roy. Dublin Soc. See B–P–H 353/26. HI 53810

Économie rurale. Paris. No. 15+, 1953+. Écon. Rurale. Preceded by: Bulletin de la société française d'économie rurale. 2-1399-1. HI 74940

Economiska annaler; med kongl. maj:ts nådigste tillstånd utgifna af kongl. vetenskaps-academien. Stockholm. Econ. Ann. See B–P–H 353/21. HI 53808

Econoticiero. Santa Fe. 1981+. Econoticiero. HI 74941

Ecos. Caracas. ?-1977+. Ecos (Caracas). HI 74942

Ecos; review of conservation. London. Vol. 1+, 1980+. Ecos (London). HI 74943

Ecos; C S I R O environmental research. Melbourne, Vic. Vol. 1+, 1974+. Ecos (Melbourne). HI 74944

Ecosur. Corrientes. No. 1+, 1974+. Ecosur. HI 74945

Ecosystems of the world. Amsterdam. Vol. 1+, 1977+. Ecosyst. World. HI 74946

Ecotoxicology and environmental safety; official journal of the international society of ecotoxicology and environmental safety. New York, London. Vol. 1+, 1977+. Ecotoxicol. Environm. Safety. HI 74947

Edinburgh journal of botany. Edinburgh. Vol. 47+, 1990+. Edinburgh J. Bot. Preceded by: Notes from the Royal Botanic Garden of Edinburgh. HI 74831

Edinburgh journal of medical science. Edinburgh. Edinburgh J. Med. Sci. See B–P–H 354/29. HI 53812

Edinburgh journal of natural and geographical science. Edinburgh. Edinburgh J. Nat. Geogr. Sci. See B–P–H 354/30. HI 53813

Edinburgh journal of natural history and physical sciences. Edinburgh J. Nat. Hist. Phys. Sci. See B–P–H 354/31. HI 53814

Edinburgh journal of science. Edinburgh. Edinburgh J. Sci. See B–P–H 354/32. HI 53815

Edinburgh medical journal. Edinburgh. Edinburgh Med. J. See B–P–H 354/34. HI 53816

Edinburgh medical and surgical journal: exhibiting a concise view of the latest and most important discoveries in medicine, surgery, and pharmacy. Edinburgh Edinburgh Med. Surg. J. See B–P–H 354/35. HI 53817

Edinburgh new philosophical journal. Edinburgh. Edinburgh New Philos. J. See B–P–H 354/37. HI 53818

Edinburgh philosophical journal. Edinburgh. Edinburgh Philos. J. See B–P–H 354/40. HI 53819

Edinburgh review, or critical journal. Edinburgh. Edinburgh Rev. See B–P–H 354/41. HI 53820

Editions of the faculty of agronomy of the university of agricultural sciences, Budapest = Agrártudományi egyetem agronómiai kar kiadványai. Budapest. Agrártud. Egyet. Agron. Kar Kiadv. See B–P–H 55/8.

Éditions hor-série, muséum d'histoire naturelle, Belgrade = Posebna izdanja, prirodnjacki musej u Beogradu. Belgrade.

Éditions speciales. Station hydrobiologique, Ochrida = Posebna izdanja. Filoz. fak. na univerzitetot Skopje. Hidrobiološki zavod Ohrid. Ochrida.

Education review, american institute of biological sciences = A I B S education review. Arlington, VA.

Educational leaflet, New York state museum and science service = Educational leaflet series, New York state museum. New York.

Educational leaflet series, New York state museum. New York. Nos. 1-20, 1948-66. Educ. Leafl. Ser. New York State Mus. Superseded by: New York state museum leaflet. 4-3045-3. HI 74948

Educational series, Harold L. Lyon arboretum. Honolulu, HI. No. 1+, 1980+. Educ. Ser. Harold L. Lyon Arbor. HI 74949

Educational series, West Virginia geological survey = West Virginia geological survey. Educational series. Morgantown, WV. West Virginia Geol. Surv., Educ. Ser. See B–P–H 973/13.

Eduen; feuille de liaison périodique de la société eduenne des lettres, sciences et arts et de la société d'histoire naturelle d'Autun. Autun. Vols. 1-36, 1957-?; n.s. vol. 1-45, 19??-68. Eduen. Preceded by: Bulletin de la société d'histoire naturelle d'Autun. Superseded by: Bulletin trimestriel de la société d'histoire naturelle et des amis du muséum d'Autun. HI 74951

Edwards's botanical register; or, flower garden and shrubbery. London. Edwards's Bot. Reg. See B–P–H 355/1. HI 53821

Eesti loodus. Eesti NSV teaduste akadeemia populaarteaduslik ajakiri. Tallinn, Estonian S S R. Esti Loodus (Tallinn). See B–P–H 355/2. HI 53822

Eesti loodus. Tartu ülikooli juures oleva loodusuurijate seltsi teataja. Eesti loodus. Tartu, Estonia [Estonian S S R]. Eesti Loodus (Tartu). See B–P–H 355/3. HI 53823

Eesti looduskaitse; Informationsjournal der Bestätigung des Naturschutzes. Tallinn. Vol. 1, 1938. Eesti Loodusk. HI 74952

Eesti loodusteaduse arhiiv. 2 Seeria. Tartu, Estonia [Estonian S S R]. Eesti Loodustead. Arh., Seer. 2. See B–P–H 355/5. HI 53826

Eesti loodusteaduse arhiiv välja antud loodusuurijate seltsi poolt Tartu ülikooli juures. Tartu, Estonia [Estonian S S R]. Eesti Loodustead. Arh., Ser. 2, Biol. See B–P–H 355/6. HI 53827

Eesti loodusteaduse arhiiv välja antud loodusuurijate seltsi poolt Tartu ülikooli juures. Ser. 2, biologische Naturkunde. Dorpat [=Tartu, Estonian S S R]. Vols. 13(2-3), 15-16(2), 1931-34. Eesti Loodustead. Arh., Ser. 2, Biol. Naturk. Preceded by: Archiv für die Naturkunde des Ostbaltikums. Serie 2, biologische Naturkunde. For vol. 14(4) see: Eesti loodusteaduse arhiiv välja antud loodusuurijate seltsi poolt Tartu ülikooli juures. Ser. 2, Biologica. Superseded by: Eesti loodusteaduse arhiiv. 2 seeria. HI 53828

Eesti loodusteaduste arhiiv. Bioloogiline seeria. Tartu, Estonian S S R. Eesti Loodustead. Arh., Biol. Seer. See B–P–H 355/4. HI 53824

Eesti loodusuurijate seltsi aastaraamat. Tallinn. No. 64+, 1976+. Eesti Loodusuur. Seltsi Aastar. Preceded by: Loodusuurijate seltsi aastaraamat. HI 74953

Eesti metsanduse aastaraamat. Tartu. Vols. 1-9-?, 1926-29-? Eesti Metsand. Aastar. HI 74954

Eesti N S V teaduste akadeemia toimetised = Izvestiya Akademii nauk Ėstonskoi S S R.

Eesti N S V teaduste akadeemia toimetised. Bioloogiline seeria. = Izvestiya Akademii nauk Ėstonskoi S S R. Seriya biologii, sel'skokhozyaistvennykh nauk i meditsinskikh nauk.

Eesti vabariigi Tartu ülikooli toimetused = Acta et commentationes universitatis tartuensis <dorpatensis>. A. Mathematica, physica, medica. Tartu, Estonia [Estonian S S R]. Acta Commentat. Univ. Tartuensis, A. Math. See B–P–H 42/12.

Effemeridi letterarie di Roma. Rome. Effem. Lett. Roma. See B–P–H 355/14. HI 53829

Effemeridi scientifiche e letterarie per la Sicilia. Palermo. Effem. Sci. Lett. Sicilia. See B–P–H 355/15. HI 53830

Ege üniversitesi fen fakültesi dergisi. Ser. B. Izmir. Vol. 1+, 1977+. Ege Üniv. Fen Fak. Derg., B. HI 74955

Ege üniversitesi fen fakultesi ilmi raporlar serisi. Bornova, Izmir. 1962+. Ege Üniv. Fen Fak. Ilmi Raporlar Ser. HI 74956

Egerton college agricultural bulletin. Njoro. Vol. 1+, 1977+. Egerton Coll. Agric. Bull. HI 74957

Egri Ho Si Minh tanárképzö föiskola tudományos közleményei. Eger. Vol. 10+, 19??+. Egri Ho Si Minh Tanárképzö Föisk. Tud. Közlem. Preceded by: Egri tanárképzö föiskola tudományos közleményei. HI 74959

Egri pedagógiai föiskola évkönyve. Eger, Hungary. Egri Pedagóg. Föisk. Évk. See B–P–H 355/16. HI 53831

Egri tanárképzö föiskola tudományos közleményei. Eger. N.s. vols. 1-?, 1963-? Egri Tanárképzö Föisk. Tud. Közlem. Preceded by: Egri pedagógiai föiskola évkönyve. Superseded by: Egri Ho Si Minh tanárképzö föiskola tudományos közleményei. HI 53834

Égypte contemporaine. Cairo. Égypte Contemp. See B–P–H 356/3. HI 53836

Egyptian botanical society yearbook. Cairo. No. 1+, 1969/70+, 1971+. Egypt. Bot. Soc. Yearb. HI 74960

Egyptian journal of agronomy. Cairo. Vol. 1+, 1976+. Egypt. J. Agron. HI 74961

Egyptian journal of botany. Dokki, Cairo. Vols. 1-2, 1958-59; vol. 15+, 1972+. Egypt. J. Bot. From 1960-71 superseded by: United Arab Republic journal of botany. HI 74962

Egyptian journal of chemistry. Cairo. Egypt. J. Chem. See B–P–H 356/1. HI 53835

Egyptian journal of genetics and cytology. Alexandria. Vol. 1+, 1972+. Egypt. J. Genet. Cytol. HI 74963

Egyptian journal of horticulture. Dokki, Cairo. Vol. 1+, 1974+. Egypt. J. Hort. HI 74964

Egyptian journal of microbiology. Cairo. Vol. 7+, 1972+. Egypt. J. Microbiol. Preceded by: United Arab Republic journal of microbiology. HI 74965

Egyptian journal of pharmaceutical sciences. Dokki, Cairo. Vol. 13+, 1972+. Egypt. J. Pharm. Sci. Preceded by: United Arab Republic journal of pharmaceutical sciences. HI 74966

Egyptian journal of phytopathology. Dokki, Cairo. Vol. 1+, 1969+. Egypt. J. Phytopathol. HI 74967

Egyptian reviews of science. Cairo. 1957+. Egypt. Rev. Sci. HI 74968

Ehime daigaku kiyo, dai-2-bu shizenkagaku, B-shirizu, seibutsugaku = Memoirs of Ehime university; section 2, natural science, series B. [Biology]. Matsuyama, Japan. Mem. Ehime Univ., Sect. 2, Nat. Sci., Ser. B. See B–P–H 571/17.

Ehime daigaku nogakubu = Memoirs of the college of agriculture, Ehime university. Matsuyama.

Ehime daigaku nogakubu enshurin hokoku = Bulletin of the Ehime university forest. Matsayama.

Ehime kenritsu norin senmon gakko gakujutsu hokoku = Scientific reports of the Ehime agricultural college. Matsuyama, Japan. Sci. Rep. Ehime Agric. Coll. See B–P–H 826/19.

Eidike ekdose, institouton okeanographikon kai halieutikon ereunon. Athens. No. 1+, 1979+. Eidike Ekd. Inst. Okeanogr. Halieut. Ereunon. HI 74969

Eipeldauer's Zimmerpflanzen-Zeitschrift. Vienna. Eipeldauer's Zimmerpflanzen-Z. See B–P–H 356/6. HI 53837

Eisei shikenjo hokoku = Bulletin of the national institute of hygienic sciences. Tokyo.

Eivissa; revista de l'institut d'estudis Eivissencs. Eivissa, Ibiza. ?-1973+. Eivissa. HI 74970

Eiszeitalter und Gegenwart. Ohringen, Germany. Eiszeitalter & Gegenwart. See B–P–H 356/7. HI 53838

Ekológia Č S S R. Bratislava. Vol. 1+, 1982+. Ekól. Č.S.S.R. HI 74971

Ekologia polska. Warsaw. Vol. 18+, 1970+. Ekol. Polska. Preceded by: Ekologia polska. Seria A. HI 74972

Ekologia polska. Seria A. Warsaw. Vols. 3-17, 1955-69. Ekol. Polska, Ser. A. Superseded by: Ekologia polska. HI 74973

Ekologia polska. Seria B. Warsaw. Vols. 1-15, 1955-69. Ekol. Polska, Ser. B. Superseded by: Wiadomości ekologiczne. HI 74974

Ekologija. Belgrade. Vol. 1+, 1966+. Ekologija. HI 74975

Ekologiya. Sofia. Vol. 1+, 1975+. Ekologiya (Sofia). HI 74976

Ekologiya. Sverdlovsk. Vol. 1+, 1970+. Ekologiya (Sverdlovsk). HI 74977

Ekologiya i fiziologiya rastenii. Kalinin. Vol. 1+, 1974+. Ekol. Fiziol. Rast. HI 74978

Ekologiya morya; respublikanskii mezhvedomstvennyi sbornik. Kiev. Vol. 1+, 1980+. Ekol. Morya. HI 74979

Ekologiya opyleniya. Perm'. Vol. 1+, 1975+. Ekol. Opyl. HI 74980

Ekologiya virusov sbornik trudov. Moscow. Vol. 1, 1973. Ekol. Virusov Sborn. Trudov. HI 74981

Eksperimental'naya botanika = Trudy botanicheskogo instituta akademii nauk S S S R. Ser. 4, eksperimental'naya botanika. Moscow & Leningrad.

Eksperimental'nyi mutagenez. Erevan. No. 4+, 1978+. Eksper. Mutagen. Preceded by: Eksperimental'nyi mutagenez rastenii. HI 74982

Eksperimental'nyi mutagenez rastenii. Erevan. Nos. 1-3, 19??-7? Eksper. Mutagen. Rast. Superseded by: Eksperimental'nyi mutaganez. HI 74983

El palacio, see Palacio

El Paso cactus and rock club bulletin. El Paso, TX. Vols. 1-3, 1959-61. El Paso Cact Rock Club Bull. HI 74984

Elaeis; journal of Igalaland. Kaduna. Vol. 1+, 1979+. Elaeis. HI 74985

Elepaio; journal of the Hawaii Audubon society. Honolulu, HI. Elepaio. See B–P–H 356/18. HI 53839

Elementary science bulletin, Michigan agricultural experiment station = Michigan agricultural experiment station. Elementary science bulletin. East Lansing, MI. Michigan Agric. Exp. Sta. Element. Sci. Bull. See B–P–H 592/11.

Élet és tudomány. Budapest. Élet & Tud. See B–P–H 356/19. HI 53840

Eleutheria oder Freyburger literarische Blätter. Freiburg im Breisgau. Eleutheria. See B–P–H 356/20. HI 53841

Elliotta. Notes from the herbarium of Georgia southern college. Statesboro, GA. Elliotta. See B–P–H 356/23. HI 53842

Elm bulletin. Waldwick, NY. 19??-67+. Elm Bull. HI 74986

Élövilag. Budapest. Élövilag. See B–P–H 356/25. HI 53843

Embo journal = E M B O journal. Arlington, VA.

Emerald necklace. Cleveland, OH. Vols. 1-24(5), 1952-75; vol. 31+, 1982. Emerald Necklace. From 1976-82 see: Metroparks emerald necklace. HI 74987

Emlékbeszédek a m[agyar] tud[ományos] akadémia tagjai felett. Budapest. Emlékbeszédek Magyar Tud. Akad. Tagjai Felett. See B–P–H 356/27. HI 53844

Emlékbeszédek m[agyar] t[udományos] akadémia tagjai fölött. Budapest. Emlékbeszédek Magyar Tud. Akad. Tagjai Fölött. See B–P–H 356/28. HI 53845

Emlékbeszédek a m[agyar] t[udományos] akadémia tagjairól. Budapest. Emlékbeszédek Magyar Tud. Akad. Tagjairól. See B–P–H 356/29. HI 53846

Empire cotton growing corporation, research memoirs. Nos. 1-63, 1946-66. Empire Cotton Growing Corp. Res. Mem. Superseded by: Research memoirs, cotton research corporation. 2-1440-3. HI 74988

Empire cotton review. London. Vols. 1-43, 1924-66 [suspended 1939-40]. Empire Cotton Growing Rev. Superseded by: Cotton growing review. HI 74989

Empire forestry. London. Empire Forest. See B–P–H 357/11. HI 53847

Empire forestry handbook. London. Vols. 1-7, 1930-57. Empire Forest. Handb. Superseded by: Commonwealth forestry handbook. HI 74990

Empire forestry journal. London. Empire Forest. J. See B–P–H 357/12. HI 53848

Empire forestry review. London. Vols. 25-41(3), 1946-62. Empire Forest. Rev. Preceded by: Empire forestry journal. 2-1441-1. HI 74991

Empire journal of experimental agriculture. Oxford, England. Empire J. Exp. Agric. See B–P–H 357/16. HI 53849

Empire violet magazine. Flushing, NY. Vol. ?-32+, 19??-86+. Empire Violet Mag. HI 74992

Emporium of arts & sciences. Philadelphia, PA. Empor. Arts Sci. See B–P–H 357/19. HI 53850

Emu. Melbourne, Vic. Vol. 1+, 1901+. Emu. HI 74993

Émulation jurassienne. Porrentruy. Vols. 1-2, 1876-77. Émul. Jurass. Preceded and superseded by: Actes de la société jurassienne d'émulation. HI 74994

Encephalartos; journal of the cycad society of southern Africa. Phalaborwa. ?-1987+. Encephalartos. HI 74995

Encyclia; journal of the Utah academy of sciences, arts and letters. Provo, UT. Vol. 54+, 1977+. Encycla. Preceded by: Proceedings of the Utah academy of sciences, arts and letters. HI 74996

Encyclopädisches Journal. Cleves & Düsseldorf. Encycl. J. See B–P–H 357/20. HI 53851

Encyclopédie biologique. Paris. Vol. 1+, 1927+. Encycl. Biol. HI 74997

Encyklopädischer Almanach für die Geschichte, Geographie, Naturforschung und Alterthumskunde. Leipzig & Vienna. Encykl. Alman. Gesch. See B–P–H 357/21. HI 53852

Endangered species act reauthorization bulletin. Washington, DC. No. 1+, 1981+. Endang. Spec. Act Reauthoriz. Bull. HI 74998

Endangered species technical bulletin. Washington, DC. Vol. 1+, 1976+. Endang. Spec. Techn. Bull. HI 74999

Endangered species update. Ann Arbor, MI. 1987+.

Endang. Spec. Update. HI 75240

Endangered and threatened wildlife and plants. Washington, DC. 19??-86+. Endang. Threat. Wildlife Pl. HI 75000

Endeavour; a quarterly review designed to record the progress of the sciences in the service of mankind. London. Vols. 1-35, 1942-76. Endeavour. Superseded by: Endeavour. New series. 2-1444-1. HI 75001

Endeavour. New series. Oxford. Vol. 1+, 1977+. Endeavour, N.S. Preceded by: Endeavour. HI 75002

Engei gakkai zasshi = Journal of the horticultural association of Japan. Tokyo. J. Hort. Assoc. Japan. See B–P–H 469/5.

Engei gakkai zasshi = Journal of the japanese society for horticultural science. Tokyo. J. Jap. Soc. Hort. Sci. See B–P–H 471/2.

Engei gijutsu kondankai nenpo = Annual bulletin of horticultural technological society. Tokyo. Annual Bull. Hort. Technol. Soc. See B–P–H 122/3.

Engei, kogii, shihiryo sakumotsu bunken shorokushu = Abstracts of literature of horticultural crops, special crops, and feed crops. Tokyo. Abstr. Lit. Hort. Crops. See B–P–H 35/15.

Engei nippon = Japanese gardening. Tokyo. Jap. Gard. See B–P–H 498/12.

Engei shikenjo hokoku. A, Hiratsuka = Bulletin of the horticultural research station. Series A, Hiratsuka. Hiratsuka.

Engei shikenjo hokoku. B, Okitsu = Bulletin of the horticultural research station. Series B, Okitsu. Shimizu.

Engei shikenjo hokoku. C, Morioka = Bulletin of the horticultural research station. Series C, Morioka. Morioka.

Engei shikenjo hokoku. D, Kurume = Bulletin of the horticultural research station. Series D, Kurume, miscellaneous publication. Fukuoka.

Engei shinchishiki = New information on horticulture. Kyoto.

Engei shinchishiki. Hana no go = [New information on horticulture. Flowers.] Kyoto.

Engei shinchishiki. Yasai go = [New information on horticulture. Vegetables.] Kyoto.

Engei tsushin = Horticultural times. Kanagawa, Japan. Hort. Times (Kanagawa). See B–P–H 422/21.

Engelhardtia. Central Islip, NY. 1968+. Engelhardtia. HI 75003

Engineering index bioengineering abstracts = Bioengineering abstracts.

Englera; Veröffentlichungen aus dem botanischen Garten und botanischen Museum, Berlin-Dahlem. Berlin. Vol. 1+, 1979+. Englera. Preceded by: Willdenowia. Beiheft. HI 75004

Englische Blätter. Erlangen. Engl. Blätt. See B–P–H 357/33. HI 53853

English hops. Paddock Wood. Vol. 1+, 1981+. Engl. Hops. HI 75005

Ensayo de la sociedad bascongada de los amigos del país. Vitoria, Spain. Esayo Soc. Bascongada Amigos País. See B–P–H 358/5. HI 53854

Enshurin. Tokyo daigaku nogakubu enshurin = Miscellaneous information. Tokyo university forests. [Japan]. Misc. Inform. Tokyo Univ. Forests. See B–P–H 597/15.

Ensino agrícola; boletim tecnico. São Paulo. Ensino Agríc. See B–P–H 358/8. HI 53855

Ente nazionale per la cellulosa e per la carta, pubblicazioni = Pubblicazioni del centro di sperimentazione agricola e forestale Rome.

Entomologia experimentalis et applicata. Amsterdam. Vol. 1+, 1958+. Entoml. Exp. Appl. HI 75006

Entomological magazine. Kyoto, Japan. Entomol. Mag. (Kyoto). See B–P–H 358/25. HI 53858

Entomological magazine. London. Entomol. Mag. (London). See B–P–H 358/26. HI 53859

Entomological news. Philadelphia, PA. See B–P–H 358/29. HI 53862

Entomological world; organ of the insect lovers association. Tokyo. Entomol. World. See B–P–H 358/33. HI 53866

Entomologie et phytopathologie appliquées; publication trimestrielle du laboratoire du département général de la protection des plantes. Teheran. Vol. 1+, 1946+. Entoml. Phytopathol. Appl. HI 75007

Entomologische Hefte. Frankfurt a. M. Entomol. Hefte. See B–P–H 358/24. HI 53857

Entomologische Monatsblätter. Berlin. Entomol. Monatsbl. See B–P–H 358/27. HI 53860

Entomologische Zeitung. Stettin [Poland]. Entomol. Zeitung. See B–P–H 358/35. HI 53867

Entomologisk tidskrift. Uppsala & Stockholm. Entomol. Tidskr. See B–P–H 358/31 HI 53864

Entomologist. London. Entomologist. See B–P–H 359/1. HI 53868

Entomologist's annual. London. Entomol. Annual. See B–P–H 358/23. HI 53856

Entomologist's gazette. London. Vol. 1+, 1950+. Entomol. Gaz. HI 75008

Entomologists' monthly magazine. London. Entomol. Monthly Mag. See B–P–H 358/28. HI 53861

Entomologist's record and journal of variation. London. Vol. 1+, 1890+. Entomol. Rec. J. Variat. HI 75009

Entomologist's weekly intelligencer. London. Entomol. Weekly Intelligencer. See B–P–H 358/32. HI 53865

Entomology and phytopathology. [K'un Ch'ung yü chih ping.] Hangchow, China. Entomol. & Phytopathol. See B–P–H 358/30. HI 53863

Entomophaga; mémoire hors série. Paris. Entomophaga. See B–P–H 359/4. HI 53870

Entopath news. Wrecclesham, Farnham. No. 1+, 1953+. Entopath News. HI 75010

Entwicklungsgeschichte und Systematik der Pflanzen = Plant systematics and evolution. Vienna, New York.

Environment. St. Louis, MO, Washington, DC. Vol. 1+, 1959+. Environment. HI 75011

Environment abstracts. New York. Vol. 4+, 1974+. Environm. Abstr. Preceded by: Environment information access. HI 75012

Environment abstracts annual. New York. Vol. 10+, 1980+. Environm. Abstr. Annual. HI 75013

Environment and change. London. Vols. 2(1-6), 1973-74. Environm. Change. Preceded by: Environment this month. HI 75014

Environment control in biology. [Seibutsu kankyo chosetsu.] Tokyo. Vol. 1+, 1963+. Environm. Control Biol. HI 75015

Environment and ecology. Kalyani. Vol. 1+, 1983+. Environm. Ecol. HI 75016

Environment index. New York. Vol. 1+, 1971+. Environm. Index. HI 75017

Environment India. [Special issue of: India international centre quarterly.] New Delhi. 1982. Environm. India. HI 75018

Environment information access. New York. Vol. 1-3, 1971-73. Environm. Inform. Access. Superseded by: Environment abstracts. HI 75019

Environment international. A journal of science technology, health monitoring and policy. Oxford, Elmsford, NJ. Vol. 1+, 1978+. Environm. Int. HI 75020

Environment newsletter, south Pacific commission. Noumea. Vol. 1+, 1975+. Environm. Newslett. S. Pacific Commiss. HI 75021

Environment resource. Des Plaines, IL. Vol. 1+, 1973+. Environm. Resource. HI 75022

Environment southwest. San Diego, CA. No. 393+, 1967+. Environm. SouthW. Preceded by: Natural history museum bulletin. San Diego. HI 75023

Environment this month. Lancaster. Vol. 1(1-2), 1972. Environm. This Month. Superseded by: Environment and change. HI 75024

Environment tomorrow. Victoria, B.C. 1970 (Autumn)-71, 1970-71. Environm. Tomorrow. HI 75025

Environment update. Hull, Quebec. Vol. 1+, 1980+. Environm. Update. HI 75026

Environmental action. Washington, DC. Vol. 1+, 1970+. nvironm. Action. HI 75027

Environmental awareness. Baroda. Vol. 1+, 1977+. Environm. Aware. HI 75028

Environmental biology and medicine. New York, London, etc. Vols. 1-2(3), 1971-76. Environm. Biol. Med. HI 75029

Environmental bulletin. = Environmental science bulletin. Washington, DC. & London.

Environmental conservation. Lausanne. Vol. 1+, 1974+. Environm. Conservation. HI 75030

Environmental education. London. Vol. ?-1972/73+, 19??-73+. Environm. Educ. HI 75031

Environmental education. Madison, WI. Vols. 1-2(2), 1969-70. Environm. Educ. Superseded by: Journal of environmental education. HI 75032

Environmental education and information. London. Vol. 1+, 1981+. Environm. Educ. Inform. HI 75033

Environmental education report. Washington, DC. Vols. 1-11, 1973-83. Environm. Educ. Rep. Superseded by: Environmental education report and newsletter. HI 75034

Environmental education report and newsletter. Hanover, NH. Vols. 12-15, 1983-87. Environm. Educ. Rep. Newslett. Preceded by: Environmental education report. Superseded by: American environment. HI 75035

Environmental enhancement. Ithaca, NY. Vol. 17(4)+, 1973+. Environm. Enchancem. Preceded by: Ornamentals newsletter. HI 75036

Environmental and experimental botany; an international journal. Oxford, Elmsford, NY, etc. Vol. 16+, 1976+. Environm. Exp. Bot. Preceded by: Radiation botany. HI 75037

Environmental focus. Bronx, NY. 197?+. Environm. Focus (Bronx). HI 67149

Environmental focus. St. Paul, MN. ?-ca.1975+. Environm. Focus (St. Paul). HI 75038

Environmental health perspectives. Washington, DC., Durham, NC. No. 1+, 1972+. Environm. Health Perspect. HI 75039

Environmental impact assessment review. New York. Vol. 1+, 1980+. Environm. Impact Assessm. Rev. HI 75040

Environmental letters. Part A, environmental science and

engineering. New York. Vols. 1-10, 1971-75. Environm. Lett., A. Superseded by: Journal of environmental science and health. Part A. HI 75041

Environmental management; an international journal for decision makers and scientists. New York, Heidelberg. Vol. 1+, 1976+. Environm. Managem. HI 75042

Environmental monitoring and assessment; an international journal. Dordrecht & Hingham, MA. Vol. 1+, 1981+. Environm. Monit. Assessm. HI 75043

Environmental mutagenesis. New York. Vol. 1+, 1979+. Environm. Mutagen. HI 75044

Environmental periodicals bibliography: indexed article titles. Santa Barbara, CA. Vol. 1+, 1972+. Environm. Period. Bibliogr. Indexed Article Titles. HI 75045

Environmental physiology; nutrition, pollution, toxicology. Copenhagen. Vol. 1, 1971. Environm. Physiol. Superseded by: Environmental physiology and biochemistry. HI 75046

Environmental physiology and biochemistry. Copenhagen. Vols. 2-5, 1972-75. Environm. Physiol. Biochem. Preceded by: Environmental physiology. HI 75047

Environmental plant life services newsletter. Versailles, Lexington, KY. Vol. 1+, 1982+. Environm. Pl. Life Serv. Newslett. HI 75048

Environmental pollution. London. Vols 1-20, 1970-79. Environm. Pollut. Superseded by: Environmental pollution. Series A. and Series B. HI 75049

Environmental pollution. Series A, ecological and biological. London. Vol. 21+, 1980+. Environm. Pollut., A. Preceded by: Environmental pollution. HI 75050

Environmental pollution. Series B, chemical and physical. London. Vol. 1+, 1980+. Environm. Pollut., B. Preceded by: Environmental pollution. HI 75051

Environmental protection agency journal = E P A journal. Washington, DC.

Environmental protection bulletin. Ottawa. Nos. 1-2, 1977-78. Environm. Protect. Bull. HI 75052

Environmental quality; annual report of the council on environmental quality. Washington, DC. No. 1+, 1970+. Environm. Qual. HI 75053

Environmental research. New York & London. Vol. 1+, 1967+. Environm. Res. (New York). HI 75054

Environmental research. Solna. 1969/70+, [1971]+. Environm. Res. (Solna). HI 75055

Environmental research center papers. Ibaraki. No. 1+, 1982+. Environm. Res. Center Pap. HI 75056

Environmental research perspectives in South Africa; annual report to the national committee for environmental sciences. Pretoria. 1981/82+, 1982?+. Environm. Res. Perspect. S. Africa. HI 75057

Environmental resources abstracts. Bangalore. Vol. 2+, 1984+. Environm. Resources Abstr. HI 75058

Environmental review. Denver, CO. Vol. 1+, 1976+. Environm. Rev. HI 74832

Environmental science bulletin. Washington, DC & London. 1971+. Environm. Sci. Bull. HI 75059

Environmental science and forestry newsletter. Syracuse, NY. 1973+. Environm. Sci. Forest. Newslett. HI 75060

Environmental science and technology. Washington, DC. Vol. 1+, 1967+. Environm. Sci. Technol. HI 75061

Environmental systems = Journal of environmental systems. Farmingdale, NY.

Environmental toxicology and chemistry. Oxford, Elmsford, NY. Vol. 1+, 1982+. Environm. Toxicol. Chem. HI 75062

Environmental toxicology newsletter. Berkeley, CA. Vol. 1+, 1980+. Environm. Toxicol. Newslett. HI 74833

Environmentalist. London. Vol. 1+, 1981+. Environmentalist. HI 75063

Environnement. Brussels. 1970(3)-71, 1970-71. Environnement. Preceded by: Maison (Bruxelles). HI 75064

Environnement à la une = Environment update. Hull, Quebec.

Enzymologia; acta biocatalytica. The Hague. Vols. 1-43, 1936-72. Enzymologia. Superseded by: Molecular and cellular biochemistry. HI 75065

Eötvös Loránd tudományegyetem évkönyve. Budapest. Eötvös Loránd Tudományegyet. Évk. See B–P–H 359/7. HI 53871

Epetêris mouseion goulandrê = Annales musei Goulandris. Kifisia, Athens.

Ephemeriden der Menschheit; oder, Bibliothek der Sittenlehre, der Politik und der Gesetzgebung. Basel. Ephem. Menschh. See B–P–H 359/8. HI 53872

Ephemeriden van de natuurkundige wetenschappen. The Hague. Ephem. Natuurk. Wetensch. See B–P–H 359/10. HI 53873

Epiphyllum bulletin = Bulletin of the epiphyllum society of America. Los Angeles, Monrovia, CA.

Epiphytes; newsletter of the epiphytic plant study group. Bristol, Cannock, etc. Vol. 1+, 1968+. Epiphytes. HI 75066

Episteme; rivista critica di storia delle scienze, medicine e biologiche. Milan. Vols. 1-10, 1967-76. Episteme. Preceded by: Castalia. Superseded by: History and philosophy of the life sciences. HI 75067

Epistemonike epeteris tes scholes ton physikon kai mathematikon epistemon = Scientific annals of the faculty of physics and mathematics, aristotelian university of Thessaloniki. Salonika.

Erboristeria domani; periodico sulle piante officinali, l'agricoltura, l'alimentazione e la medicina naturali. Milan. Vol. 1+, 1978+. Eborist. Domani. HI 75069

Erde. Zeitschrift der Gesellschaft für Erdkunde zu Berlin. Berlin. Erde. See B–P–H 359/15. HI 53874

Erdo, Az. Budapest. Vol. 1+, 1952+. Erdo. HI 75070

Erdély. Kolozsvar [=Cluj, Rumania]. Erdély. See B–P–H 359/17. HI 53875

Erdélyi gazda. Kolozsvar [=Cluj, Rumania]. Erdélyi Gazda. See B–P–H 359/18. HI 53876

Erdélyi múzeum. Kolozsvar [=Cluj, Rumania]. Erdélyi Múz. (Kolozsvar). See B–P–H 359/19. HI 53877

Erdélyi múzeum. Pest [=Budapest, in part] Erdélyi Múz. (Pest). See B–P–H 359/20. HI 53878

Erdélyi múzeum-egyesület évkönyve. Kolozsvar [=Cluj, Rumania]. Erdélyi Múz.-Egyes. Évk. See B–P–H 359/21. HI 53879

Erdélyi múzeum-egylet évkönyvei. Kolozsvar [=Cluj, Rumania]. Erdélyi Múz.-Egyl. Évk. See B–P–H 359/22. HI 53880

Erdészeti értesitö. Budapest. Erdész. Értes. See B–P–H 359/23. HI 53881

Erdészeti és faipari egyetem tudományos közleményei. Sopron, Hungary. Erdész. Faip. Egyet. Tud. Közlem. See B–P–H 359/24. HI 53882

Erdészeti és gazdászati lapok. Pest [=Budapest, in part]. Erdész. Gazd. Lapok. See B–P–H 359/25. HI 53883

Erdészeti kísérletek. Selmec, Hungary [=Banska Stiavnica, Czechoslovakia] & Sopron, Hungary. Erdész. Kísérl. See B–P–H 360/1. HI 53884

Erdészeti kutatások. Budapest. Erdész. Kutatás. See B–P–H 360/2. HI 53885

Erdészeti lapok. Selmec, Hungary [=Banska Stiavnica, Czechoslovakia], Pozsony [=Bratislava, Czechoslovakia], & Budapest. Erdész. Lapok. See B–P–H 360/4. HI 53886

Erdészeti tudományos intézet évkönyve. Budapest. Erdész. Tud. Intéz. Évk. See B–P–H 360/5. HI 53887

Erdészeti újság. Szaszsebes, Hungary [=Sebes, Rumania]. Erdész. Újs. See B–P–H 360/7. HI 53888

Erdészettudományi közlemények. Sopron, Hungary. Erdészettud. Közlem. See B–P–H 360/11. HI 53889

Erdkunde. Archiv für wissenschaftliche Geographie. Bonn. Erdkunde. See B–P–H 360/14. HI 53890

Erdö. Budapest. Erdö (Budapest). See B–P–H 360/19. HI 53891

Erdö. Miskolc, Budapest, Esztergom & Tata. Erdö (Miskolc). See B–P–H 360/20. HI 53892

Erdö. Temesvar [=Timisoara, Rumania]. Erdö (Temesvar). See B–P–H 360/21. HI 53893

Erdögazdaság. Budapest. Erdögazdaság. See B–P–H 360/24. HI 53895

Erdögazdaság és faipar. Budapest. Erdögazd. & Faip. See B–P–H 360/22. HI 53894

Erdömérnöki föiskola évkönyve. Sopron, Hungary. Erdömérn. Föisk. Évk. See B–P–H 360/25. HI 53896

Erdömérnöki föiskola közleményei. Sopron, Hungary. Erdömérn. Föisk. Közlem. See B–P–H 361/2. HI 53897

Erdöör. Sebes, Rumania. Erdöör. See B–P–H 361/3. HI 53898

Erdwissenschaftliche Forschung. Wiesbaden. Vol. 1+, 1968+. Erdwiss. Forschung. HI 75071

Erfelijkheid in praktijk. Orgaan van de nederlandsche genetische vereeniging. Soesterberg, Netherlands. Erfelijkheid Prakt. See B–P–H 361/4. HI 53899

Erfolg- und Tätigkeitsbericht des Hauses der Natur in Salzburg. Salzburg. Erfolg- Tätigkeitsber. Hauses Natur Salzburg. See B–P–H 361/5. HI 53900

Erfolgreicher Obstbau. Vienna. Erfolgr. Obstbau. See B–P–H 361/6. HI 53901

Erfurter Führer im Gartenbau. Erfurt. Vols. 1-5, 1900/01-1904/05, 1900-05. Erfurter Führer Gartenbau. Superseded by: Erfurter Führer im Obst- und Gartenbau. HI 75072

Erfurter Führer im Obst- und Gartenbau. Erfurt. Vols. 6-5, 1905/06-1944, 1905-44. Erfurter Führer Obst-Gartenbau. Preceded by: Erfurter Führer im Gartenbau. HI 75073

Erfurtische Gelehrte Nachrichten, die unter der Aufsicht der Churf.-Mayntzischen Academie nützlicher Wissenschaften ... herausgegeben worden. Erfurt. Erfurt. Gel. Nachr. See B–P–H 361/7. HI 53902

Erfurtische gelehrte Zeitung. Erfurt. Erfurt. Gel. Zeitung. See B–P–H 361/9. HI 53903

Erfurtische gelehrte Zeitungen. Erfurt. Erfurt. Gel. Zeitungen. See B–P–H 361/10. HI 53904

Erfurtische Nachrichten von gelehrten Sachen, herausgegeben unter Aufsicht der Akademie. Erfurt. Erfurt. Nachr. Gel. Sachen. See B–P–H 361/11. HI 53905

Ergänzungsblätter zur jenaischen allgemeinen Literatur-Zeitung. Leipzig. Vols. 1-29, 1813-41. Ergänzungsbl.

Jenaischen Allg. Lit.-Zeitung. 3-2160-1. HI 52746

Ergänzungshefte, Mitteilungen aus den deutschen Schutzgebieten = Mitteilungen aus den deutschen Schutzgebieten. Ergänzungshefte. Berlin. Mitt. Deutsch. Schutzgeb. Ergänzungsh. See B–P–H 603/8.

Ergebnisse der Biologie. Berlin. Ergebn. Biol. See B–P–H 361/18. HI 53906

Ergebnisse der Chirurgie und Orthopädie. Berlin. Ergebn. Chir. Orthopäd. See B–P–H 361/19. HI 53907

Ergebnisse der Hygiene, Bakteriologie, Immunitätsforschung und experimentellen Therapie. Berlin. Vols. 2-29, 1917-55. Ergebn. Hyg. Bakteriol. Immunitätsforsch. Exp. Therapie. Preceded by: Ergebnisse der Immunitätsforschung, experimentellen Therapie, Bakteriologie und Hygiene. Superseded by: Ergebnisse der Mikrobiologie, Immunitätsforschung und experimentellen Therapie. HI 75074

Ergebnisse der Immunitätsforschung, experimentellen Therapie, Bakteriologie und Hygiene. Berlin. Vol. 1, 1914. Ergebn. Immunitätsforsch. Exp. Therapie Bakteriol. Hyg. Preceded by: Jahresbericht über die Ergebnisse der Immunitätsforschung. Superseded by: Ergebnisse der Hygiene, Bakteriologie, Immunitätsforschung und experimentellen Therapie. HI 75075

Ergebnisse der Limnologie. Stuttgart. Ergebn. Limnol. See B–P–H 361/20. HI 53908

Ergebnisse der Mikrobiologie, Immunitätsforschung und experimentellen Therapie. Berlin, New York, etc. Vols. 30-39, 1956-65. Ergebn. Mikrobiol. Immunitätsforsch. Exp. Therapie. Preceded by: Ergebnisse der Hygiene, Bakteriologie, Immunitätsforschung und experimentellen Therapie. Superseded by: Current topics in microbiology and immunology. HI 75076

Ergebnisse der pflanzengeographischen Durchforschung von Württemberg, Baden und Hohenzollern. Stuttgart. Ergebn. Pflanzengeogr. Durchforsch. Württemberg. See B–P–H 361/21. HI 53909

Ergebnisse der wissenschaftlichen Untersuchung des schweizerischen Nationalparkes. Aarau, Switzerland. Ergebn. Wiss. Untersuch. Schweiz. Nationalparkes. See B–P–H 361/22. HI 53910

Erhvervsfrugtavleren. Odense, etc. Vols. 1-38, 1934-71. Erhvervsfrugtavleren. Superseded by: Frugtavleren. HI 75077

Erigenia; journal of the southern Illinois native plant society. Carbondale, IL. No. 1+, 1982+. Erigenia. HI 75078

Erlanger Forschungen. Reihe B, naturwissenschaften. Erlangen. Vol. 1+, 1955+. Erlanger Forsch., Reihe B. Naturwiss. HI 75079

Erlanger gelehrte Zeitungen. Erlangen. Erlanger Gel. Zeitungen. See B–P–H 362/1. HI 53911

Erlangische Abhandlungen zur Beförderung der Wissenschaften. Erlangen & Nuremberg. Erlangische Abh. Beförd. Wiss. See B–P–H 362/2. HI 53912

Erlangische gelehrte Anmerkungen und Nachrichten. Erlangen. Erlangische Gel. Anmerk. Nachr. See B–P–H 362/3. HI 53913

Erlangische Gelehrte Anzeigen, darinnen kurtze und zur Verbesserung derer Wissenschaften ausgearbeitete Materien befindlich. Erlangen. Erlangische Gel. Anz. See B–P–H 362/4. HI 53914

Ernährung der Pflanze. Berlin. Ernähr. Pflanze See B–P–H 362/5. HI 53915

Erneuerte Vaterländische Blätter für den Oesterreichischen Kaiserstaat. Vienna. Erneuerte Vaterl. Blätt. Oesterr. Kaiserstaat. See B–P–H 362/6. HI 53916

Ernstia. Maracay. No. 1+, 1981+. Ernstia. HI 75080

Eröffnungs-Rede der Jahhres-Versammlung der allgem. schweizerischen Gesellschaft für gesammte Naturwissenschaften = Eröffnungsrede der Jahresversammlung der allgemeinen schweizerischen Gesellschaft für die gesammten Naturwissenschaften. Zurich. Eröffnungsrede Jahresversamml. Allg. Schweiz. Ges. Gesammten Naturwiss. See B–P–H 362/8.

Eröffnungs-Rede der Jahres-Versammlung der schweizerischen Gesellschaft für die gesammten Naturwissenschaften = Eröffnungsrede der Jahresversammlung der allgemeinen schweizerischen Gesellschaft für die gesammten Naturwissenschaften. Zurich. Eröffnungsrede Jahresversamml. Allg. Schweiz. Ges. Gesammten Naturwiss. See B–P–H 362/8.

Eröffnungsrede der Jahresversammlung der allgemeinen schweizerischen Gesellschaft für die gesammten Naturwissenschaften. Zurich. Eröffnungsrede Jahresversamml. Allg. Schweiz. Ges. Gesammten Naturwiss. See B–P–H 362/8. HI 53917

Erste Verhandlungen der freyen Landwirthschafts-Gesellschaft des Niederrheinischen Departements zu Strasburg. Strasbourg. Erste Verh. Freyen Landw.-Ges. Niederrhein. Dept. Strasburg. See B–P–H 362/9. HI 53918

Értekezések a természettudományi köréböl magyar tudományos akadémia. Budapest. Értek. Term. Köréb. Magyar Tud. Akad. See B–P–H 362/13. HI 53919

Ertesitö az erdélyi muzeum-egylet orvos-természet tudomanyi szakosztályából. Kolozsvar [= Cluj, Rumania]. 1890-1905. Ertes. Erdélyi Muz.-Egyl.

Orv.-Term. Tud. Szakosztályából. Preceded by: Orvos-természettudományi értesitö. Superseded by: Múzeumi füzetek. HI 75081

Ertesitö a kolozsvári orvos-természettudományi társulat. Lolozsvart. 1876-78. Ertes. Kolozsvári Orv.-Term. Társ. Superseded by: Orvos-természettudományi ertesitó. HI 75082

Erwerbsgärtner. Stuttgart. Vol. 22-28, 1968-74. Erwerbsgärtner. Preceded by: Süddeutscher Erwerbsgärtner. Superseded by: Deutscher Gartenbau. HI 75083

Erwerbsobstbau. Berichte aus Wissenschaft und Praxis. Berlin. Erwerbsobstbau. See B–P–H 362/16. HI 53920

Erythea; a journal of botany, west american and general. Berkeley, CA. Erythea. See B–P–H 362/17. HI 53922

Erzgebirgs-Zeitung. Teplitz, Bohemia [=Teplice, Czechoslovakia]. Erzgebirgs-Zeitung. See B–P–H 362/18. HI 53923

Erzgebirgszeitung = Erzgebirgs-Zeitung. Teplitz, Bohemia [=Teplice, Czechoslovakia]. Erzgebirgs-Zeitung. See B–P–H 362/18.

Escuela de agricultura; revista mensual. San José, Costa Rica. Esc. Agric. See B–P–H 362/19. HI 53924

Esercitazioni accademiche degli aspiranti naturalisti. Naples. Vols. 1-2(1), 1839-40. Esercit. Accad. Aspir. Naturalisti. Superseded by: Bullettino dell'accademia degli aspiranti naturalisti. 1-39-2. HI 65026

Esercitazioni scientifiche e letterarie dell'ateneo di Venezia. Venice. Esercit. Sci. Lett. Ateneo Venezia. See B–P–H 363/1. HI 53925

Esitelmiä ja pöytäkirjoja. Helsinki. Esitelmiä Pöytäkirj. See B–P–H 363/2. HI 53926

Esperienzi e ricerche, istituto di agronomia e coltivazioni erbacee = Istituto di agronomia e coltivazioni erbacee. Esperienzi e ricerche. Università di Pisa. Pisa. Ist. Agron. Esper. Ric. See B–P–H 439/5.

Espiga. [Venezuela]. Espiga. See B–P–H 363/3. HI 53927

Espinas y flores. San Diego, CA. [11 unnumbered issues], 1966-67; vol. 3+, 1968+. Espinas & Fl. Preceded by: Cactos y suculentos. HI 65027

Esprit des journaux, français et étrangers. Paris. Esprit J. Franç. Étrangers. See B–P–H 363/4. HI 53928

Essais de botanique. Moscow & Leningrad = Voprosy botaniki. Moscow & Leningrad. Vopr. Bot. See B–P–H 968/1.

Essays in biochemistry. London. Essays Biochem. See B–P–H 363/10. HI 53929

Essays and observations, physical and literary, read before a society in Edinburgh, and published by them. Edinburgh. Essays Observ. Phys. Lit. Soc. Edinburgh. See B–P–H 363/11. HI 53930

Essex institute. Annual report. Salem, MA. Essex Inst. Annual Rep. See B–P–H 363/12. HI 53931

Essex institute historical collections. Salem, MA. Vol. 1+, 1859+. Essex Inst. Hist. Collect. 2-1480-3. HI 65028

Essex naturalist. Buckhurst Hill. Vol. 1+, 1887+. Essex Naturalist. Preceded by: Transactions of the Essex field club. 2-1481-1. HI 65029

Esta semana; que ver, que hacer, donde y como en México. Mexico City. Esta Semana. See B–P–H 363/17. HI 53932

Estanzuela; investigacion agricola. Colonia. Vol. 1+, 1966+. Estanzuela. HI 65030

Estestvenno-istoričeskaja Biblioteka: Priroda = Estestvenno-istoricheskaya Biblioteka: Priroda. Moscow.

Estestvenno-istoričeskij sbornik. Kostroma = Estestvenno-istoricheskii sbornik. Kostroma.

Estestvennye proizvoditel'nye sily Rossii = Estestvennye proizvoditel'nye sily Rossii. Petrograd.

Estländisches Forstwirtschaftliches Jahrbuch = Eesti metsanduse aastaraamat. Tartu.

Estnischer Naturschutz = Eesti looduskaitse. Tallinn.

Estonian nature = Eesti loodus. Tartu, Estonia [Estonian S S R]. Eesti Loodus (Tartu). See B–P–H 355/3.

Estuaries; journal of research on any aspect of natural science applied to estuaries. Solomons, MD. Vol. 1+, 1978+. Estuaries. Preceded by: Chesapeake science. HI 65031

Estuarine bulletin. Newark, DE. Estuarine Bull. See B–P–H 363/25. HI 53933

Estuarine and coastal marine science. London, New York, Orlando, FL. Vols. 1-11, 1973-80. Estuarine Coastal Mar. Sci. Superseded by: Estuarine, coastal and shelf science. HI 65032

Estuarine, coastal and shelf science. London, Orlando, FL. Vol. 12+, 1981+. Estuarine Coastal Shelf Sci. Preceded by: Estuarine and coastal marine science. HI 65033

Estuarine research. New York. Vol. 1+, 1975. Estuarine Res. HI 65034

Estudio, see Estudios

Estudio. Bucaramanga, Colombia. Estudio (Bucaramanga). See B–P–H 364/1. HI 53937

Estudio. Mexico City. Estudio (Mexico City). See B–P–H 364/2. HI 53938

Estudio. Puebla, Mexico. Estudio (Puebla). See

B–P–H 364/3. HI 53939

Estudios agrarios. Mexico City. Estud. Agrar. See B–P–H 363/26. HI 53934

Estudios botanicos en el sur del Ecuador. Loja. Vol. 1+, 1948+. Estud. Bot. Sur Ecuador. HI 65035

Estudios geológicos. Madrid. Estud. Geol. See B–P–H 363/28. HI 53936

Estudos agronómicos. Lisbon. Estud. Agron. See B–P–H 363/27. HI 53935

Estudos de biologia, universidade católica do Paraná. Curitiba. Vol. 1+, 1978+. Estud. Biol. Univ. Católica Paraná. HI 65036

Estudos de botanica = Trabalhos do centro de botanica da junta de investigaçoes do ultramar. Lisbon.

Estudos e divulgação tecnica, serviços florestais e aquícolas. Grupo A, secção aquicultura. Lisbon. 1970?+. Estud. Divulg. Tecn. Serv. Florest. Aquíc., A, Secç. Aquic. HI 65037

Estudos e divulgação tecnica, serviços florestais e aquícolas. Grupo A, secção protecção da natureza. Lisbon. 1970?+. Estud. Divulg. Tecn. Serv. Florest. Aquíc., A, Secç. Protecç. Nat. HI 65038

Estudos e divulgação tecnica, serviços florestais e agricolas. Grupo D, secção botânica. Lisbon. ?-1976+. Estud. Divulg. Tecn. Serv. Florest. Aquíc., D, Secç. Bot. HI 65039

Estudos, ensaios e documentos, junta de investigações cientificas do ultramar. Lisbon. Vol. 1+, 1950+. Estud. Ensaios Doc. Junta Invest. Ci. Ultramar. HI 65040

Estudos leopoldenses. São Leopoldo. No. 1+, 1966+. Estud. Leop. HI 65041

Estudos e pesquisas, departamento de botânica, universidade federal de Pernambuco. Recife. Vol. ?-1(3)+, 19??-70+. Estud. Pesq. Dep. Bot. Univ. Fed. Pernambuco. HI 65042

Ethnographisches Archiv. Jena. Ethnogr. Arch. See B–P–H 364/4. HI 53940

Ethnology; an international journal of cultural and social anthropology. Pittsburgh, PA. Vol. 1+, 1962+. Ethnology. HI 65043

Etnobiologica. Corrientes. Vols. 1-14, 1967-70. Etnobiologica. HI 65044

ETP. Chicago, IL = Evolutionary trends in plants. Chicago, IL.

Études de biologie agricole. London, Brussels. Vols. 1-4, 1915-18. Etudes Biol. Agric. 2-1488-3. HI 65045

Étude botanique de l'institut d'élevage et de médecine vétérinaire des pays tropicaux (I E M V P T). Maisons-Alfort. 1972+. Étude Bot. Inst. Élevage Méd. Vétérin. Pays Trop. Superseded by?: Études et synthéses de l'institut d'élevage et de médecine vétérinaire des pays tropicaux (I E M V P T). HI 65046

Études camerounaises. Yaoundé, French Cameroons [Cameroon Republic]. Études Cameroun. See B–P–H 364/7. HI 53941

Études du continent africain. Brussels. No. 1+, 1973+. Études Continent Africain. HI 65047

Études corses; revue trimestrielle. Bastia. Vols. 74 (1954)-80 (1960) [also numbered as n.s. vols. 1-27/28, 1954-1960. Études Corses. Superseded by: Revue d'études historiques, litteraires et scientifiques corses. HI 65048

Études éburnéennes. Abidjan. Nos. 1-9, 1950-68. Études Eburn. HI 65049

Études écologiques, laboratoire d'écologie forestière, universitaire Laval. Quebec. No. 1+, 1979+. Études Écol. Lab. Ecol. Forest. Univ. Laval. HI 65050

Études entomologiques. Helsinki. Études Entomol. See B–P–H 364/8. HI 53942

Études de morphologie et de physiologie cellulaires. Geneva. Nos. 1-7, 1898-1905. Études Morphol. Physiol. Cell. HI 65051

Études et recherches, association forêt-cellulose. Nangis. Nos. ?-12+, 19??-79+. Études Rech. Assoc. Forêt-Cellulose. HI 65052

Études et recherches de biologie = Studii şi cercetări de biologie. Bucharest.

Études et recherches de biologie = Studii şi cercetări de biologie. Seria biologie vegetală. Bucharest.

Études et recherches de biologie = Studii şi cercetări de biologie. Seria botanică. Bucharest.

Études sénégalaises. Dakar. Vol. 1+, 1949+. Études Sénégal. 2-1491-1. HI 65053

Études et synthéses de l'institut d'élevage et de médecine vétérinaire des pays tropicaux (I E M V P T). Maisons-Alfort. [Dates of publication not ascertained.] Etudes Synthéses Inst. Elevage Méd. Vétérin. Pays Trop. Preceded by?: Étude botanique de l'institut d'élevage et de médecine vétérinaire des pays tropicaux (I E M V P T). HI 65054

Études togolaises; revue togolaise des sciences. Lomé. Vols. 1-?, 1965-?; n.s. vol. 1+, 1971+. Études Togolaises. HI 65055

Etwelche meistens bayrische Denck- und Lesswürdigkeiten zur Fortführung des sog. Parnassus Boicus aufgesetzet ... Ingolstadt, Germany. Etwelche Bayr. Denck- Lesswürdigk. See B–P–H 364/10. HI 53943

Eucarpia. Bulletin. Wageningen. Vol. 1+, 1959+. Eucarpia Bull. HI 65056

Eugeniana; boletim científico do colégio anchieta Nova Friburgo - R.J. Nova Friburgo, R.J. No. 1+, 1981+. Eugeniana. HI 65057

Euphorbia journal. Mill Valley, CA. Vol. 1+, 1983+. Euphorbia J. HI 65058

Euphorbia review. Los Angeles, CA. Euphorbia Rev. See B–P–H 364/11. HI 53944

Euphoria et cacophoria. Gifu. No. 1+, 1974+. Euphoria & Cacophoria. Preceded by: Bulletin of the pharmaceutical research institute. HI 65059

Euphytica. Netherlands journal of plant breeding. Wageningen. Euphytica. See B–P–H 364/12. HI 53945

Eupopäische Annalen. Tübingen. Eur. Ann. See B–P–H 364/14. HI 53946

Eurogardeners' news. Kew. ?-1983+. Eurogardeners' News. HI 65060

Europäische Zeitschrift für Forstpathologie = European journal of forest pathology. Hamburg, etc.

European association for research on plant breeding. Maize section. Bulletin. Paris. Eur. Assoc. Res. Pl. Breed., Maize Sect. Bull. See B–P–H 364/15. HI 53947

European bamboo network newsletter. Morton Bourne. Vol. 1+, 1985+. Eur. Bamboo Network Newslett. HI 65061

European biotechnology newsletter. Paris. 1986+. Eur. Biotechnol. Newslett. HI 75241

European journal of applied microbiology. Berlin, New York, etc. Vols. 1-4, 1975-77. Eur. J. Appl. Microbiol. Superseded by: European journal of applied microbiology and biotechnology. HI 65062

European journal of applied microbiology and biotechnology. Berlin, New York, etc. Vols. 5-18, 1978-83. Eur. J. Appl. Microbiol. Biotechnol. Preceded by: European journal of applied microbiology. Superseded by: Applied microbiology and biotechnology. HI 65063

European journal of biochemistry. Berlin, Heidelberg, & New York. Vol. 1+, 1967+. Eur. J. Biochem. Preceded by: Biochemische Zeitschrift. HI 65064

European journal of cell biology. Stuttgart. Vol. 19+, 1979+. Eur. J. Cell Biol. Preceded by: Cytobiologie. HI 65065

European journal of cell biology. Supplement: Abstracts. Stuttgart. Vol. 1+, 1983+. Eur. J. Cell Biol., Suppl. Abstr. HI 65066

European journal of clinical microbiology. Philadelphia, PA. Vol. 1+, 1982+. Eur. J. Clin. Microbiol. HI 65067

European journal of forest pathology. Hamburg, etc. Vol. 1+, 1971+. Eur. J. Forest Pathol. HI 65068

European magazine and London review. London. Eur. Mag. & London Rev. See B–P–H 364/16. HI 53948

European potato journal. Wageningen. Vols. 1-2, 1958-69. Eur. Potato J. Superseded by: Potato research. HI 65069

European science notes information bulletin = E S N information bulletin. London.

European science notes, office of naval research, branch office, London. London. Vols. 39(3)-41(10), 1985-87. Eur. Sci. Notes Off. Naval Res. Preceded by: European scientific notes, office of naval research, branch office, London. Superseded by: E S N information bulletin. HI 65070

European scientific notes, office of naval research, branch office, London. London. Vols. 3?-39(2), 1948-85. Eur. Sci. Notes Off. Naval Res. Preceded by: London newsletter, office of naval research, London branch. Superseded by: European science notes, office of naval research, branch office, London. HI 65071

Evansia. New York. Vol. 1+, 1984+. Evansia. HI 65072

Evansville medical journal = Indiana medical journal; a quarterly record of the medical sciences of the south and the west. Evansville, IN. Indiana Med. J. (Evansville). See B–P–H 431/16.

Everglades natural history; a magazine of natural history of south Florida. Homestead, FL. Everglades Nat. Hist. See B–P–H 364/22. HI 53949

Evergreen. Peechi. No. 5+, 1979+. Evergreen. Preceded by: Newsletter, Kerala forest research institute. HI 65073

Evolución. Bogotá. Evolución. See B–P–H 364/25. HI 53951

Evolucion biologica. Revista internacional de la asociación iberoamericana de biología evolutiva. Bogota. Vol. 1+, 1987+. Evol. Biol. (Bogota). HI 65074

Evolución de las ciencias en la república Argentina. Buenos Aires. Vols. 2-7, 11 & 13, 1922-36 [vols. 1, 8-10 & 12 never published]. Evol. Ci. Republ. Argentina. HI 65075

Evolution; international journal of organic evolution. Lancaster, PA. Evolution. See B–P–H 364/26. HI 53952

Evolutionary biology. New York, NY. Evol. Biol. See B–P–H 364/24. HI 53950

Excerpta botanica. Sectio A: Taxonomica et Chorologica. Stuttgart. Excerpta Bot., Sect. A, Taxon. See B–P–H 364/27 HI 53953

Excerpta botanica. Sectio B: Sociologica. Stuttgart.

Excerpta Bot., Sect. B, Sociol. See B–P–H 365/1. HI 53954

Evolutionary ecology. London. Vol. 1+, 1987+. Evol. Ecol. HI 65076

Evolutionary genetics research reports. Stony Brook, N.Y. Nos. 1-20?, 1968-76. Evol. Genet. Res. Rep. HI 65077

Evolutionary monographs. Chicago, IL. Vol. 1+, 1979+. Evol. Monogr. HI 65078

Evolutionary theory; international journal of fact and interpretation. Chicago, IL. Vol. 1+, 1973+. Evol. Theory. HI 65079

Evolutionary trends in plants. Zurich. Vol. 1+, 1987+. Evol. Trends Pl. HI 65080

Excellence in ecology. Oldendorf/Luhe. No. 1+, [1987?]+. Excellence Ecol. HI 65081

Excelsa; journal of the aloe, cactus and succulent society of Rhodesia (later Zimbabwe). Salisbury (later Harare), Rhodesia (later Zimbabwe). No. 1+, 1971+. Excelsa. HI 65082

Excelsa. Taxonomic series. Salisbury (Harare), Rhodesia (Zimbabwe). No. 1+, 1978+. Excelsa, Taxon. Ser. HI 65083

Excerpta medica. Section II-C, pharmacology and toxicology. Amsterdam. Excerpta Med., Sect. 2-C, Pharmacol. Toxicol. See B–P–H 365/2. HI 53955

Excerpta medica. Section 26, medical microbiology, immunobiology and serology. Amsterdam. Vols. 1-20(6), 1948-67. Excerpta Med., Sect. 26, Microbiol. Superseded in part by: Excerpta medica. Section 4, microbiology, bacteriology, virology, mycology. HI 65084

Excerpta medica. Section 4, microbiology, bacteriology, virology, mycology. Amsterdam. Vol. 20(7)+, 1967+. Excerpta Med., 4, Microbiol. Previously part of: Excepta medica. Section 26, medical microbiology, immunology and serology. HI 65085

Excerpta medica. Section 47, virology. Amsterdam. 1971+. Excerpta Med., Sect. 47, Virol. HI 65086

Excursionen Tagebuch, [biologische Wolga-Station] = Dnevnik Ėkskursii, [Volzhkaya biologicheskaya stantsiya]. Saratov.

Exkursionstagebuch, [biologische Wolga-Station] = Dnevnik Ėkskursii, [Volzhkaya biologicheskaya stantsiya]. Saratov.

Experientia. Basel. Experientia (Basel). See B–P–H 365/20. HI 53964

Experientia supplementum. [Forms a supplement to: Experientia.] Basel. Vol. 1+, 1953+. Experientia Suppl. HI 65087

Experientiae. Viçosa, Brazil. Experientiae (Viçosa). See B–P–H 365/21. HI 53965

Experiment station bulletin. Auburn, Alabama. Exp. Sta. Bull. (Auburn, Alabama). See B–P–H 365/12. HI 53960

Experiment station bulletin. Montgomery, Alabama. Exp. Sta. Bull. (Montgomery, Alabama). See B–P–H 365/13. HI 53961

Experiment station record. Washington, DC. Exp. Sta. Rec. See B–P–H 365/14. HI 53962

Experiment station work, office of experiment stations, United States department of agriculture = U S department of agriculture. Office of experiment stations. Experiment station work. Washington, DC. U.S.D.A. Off. Exp. Sta. Exp. Sta. Work. See B–P–H 943/24.

Experimenta. Universidad nacional de Cuyo. Mendoza, Argentina. Experimenta. See B–P–H 365/22. HI 53966

Experimental agriculture. London. Exp. Agric. See B–P–H 365/6. HI 53957

Experimental biology. Berlin, New York. Vol. 43+, 1984+. Exp. Biol. Preceded by: Revue canadienne de biologie experimentelle. HI 75196

Experimental cell research. New York & London. Vol. 1+, 1950+. Exp. Cell. Res. 2-1509-3. HI 65088

Experimental control in biology. 1963+. Exp. Control Biol. HI 65089

Experimental horticulture. London. Nos. 1-32, 1957-81. Exp. Hort. HI 65090

Experimental husbandry. London. Nos. 1-38, 1956-82. Exp. Husb. HI 65091

Experimental and molecular pathology. New York, NY. Exp. Molec. Pathol. See B–P–H 365/9. HI 53958

Experimental mycology; international journal. New York & London. Vol. 1+, 1977+. Exp. Mycol. HI 65092

Experimental parasitology. New York, London. Vol. 1+, 1951+. Exp. Parasitol. HI 65093

Experimental programme, Efford experimental horticulture station. Lymington. 195?-59. Exp. Programme Efford Exp. Hort. Sta. Preceded by: Station guide, Efford experimental horticulture station. Superseded by: Station guide and programme of experiments, Efford experimental horticulture station. HI 65094

Experimental record. Grassland husbandry department. [West of Scotland agricultural college.] Auchincruive, Scotland. Exp. Rec. Grassland Husb. Dept. See B–P–H 365/10. HI 53959

Experimental record, South Australia department of agriculture. Adelaide, S.A. Nos. 1-7, 1963-72. Exp. Rec. S. Australia Dept. Agric. Superseded by:

Agricultural record, South Australia department of agriculture. HI 65095

Experimental report, Rokupr rice research station. Rokupr. 1969-71. Exp. Rep. Rokupr Rice Res. Sta. Superseded by: Report, rice research station, Rokupr. HI 65096

Experiments in progress, grassland improvement (research) station. Stratford-on-Avon, Hurley. Nos. 1-?, 1948-55. Exp. Progr. Grassland Improv. Res. Sta. Superseded by: Experiments in progress, grassland research institute and Report (Annual), grassland research institute. HI 65097

Experiments in progress, grassland research institute. Hurley. Nos. ?-17, 1955-64. Exp. Progr. Grassland Res. Inst. Preceded by: Experiments in progress, grassland improvement station. Superseded by: Report, grassland research institute. HI 65098

Experiments with seedlings and other canes, Barbados = Report of sugar-cane experiments, department of agriculture, Barbados. Bridgetown.

Explants. Bronx, NY. Vols. 1-7(1), 1963-71. Explants. HI 65099

Exploration and scientific research. [Pan-American society of tropical research.] New Orleans, LA. Explor. Sci. Res. See B–P–H 365/24. HI 53968

Explorer. Cleveland, OH. Explorer. See B–P–H 366/2. HI 53969

Explorer naturalist. Los Altos, CA. Explor. Naturalist. See B–P–H 365/23. HI 53967

Explorers' journal. New York. Vol. 1+, 1921+. Explor. J. HI 65100

Exsiccatorum genavensium e conservatorio botanico distributorum fasciculus. Geneva. Vol. 1+, 1970+. Exsicc. Genav. Conserv. Bot. Distrib. Fasc. HI 65101

Extension bulletin, agricultural extension service, university of Wyoming = Bulletin of the agricultural extension service, university of Wyoming. Laramie.

Extension bulletin, cooperative extension service, Michigan state university. East Lansing, MI. No. 1+, 1917+. Extens. Bull. Coop. Extens. Serv. Michigan State Univ. HI 65102

Extension bulletin, economic botany information service, national botanical research institute. Lucknow. No. 1+, 1979+. Extens. Bull. Econ. Bot. Inform. Serv. Natl. Bot. Res. Inst. HI 65103

Extension bulletin, New York state college of agriculture = Cornell extension bulletin, New York state college of agriculture. Ithaca, NY.

Extension bulletin of Rutgers university. New Brunswick, NJ. No. 1+, 1915+. Extens. Bull. Rutgers Univ. HI 65104

Extension circular, Guam agricultural experiment station, office of experiment stations, United States department of agriculture = U S department of agriculture, office of experiment stations; the Guam agricultural experiment station, extension circular. Washington, DC. U.S.D.A. Off. Exp. Sta. Guam Agric. Exp. Sta. Extens. Circ. See B–P–H 943/27.

Extension leaflet, department of agricultural extension, Purdue university. Lafayette, IN. Nos. ?-385, 1947?-56. Extens. Leafl. Agric. Extens. Purdue Univ. Preceded by: Leaflet, department of agricultural extension, Purdue university. HI 65105

Extension publication, Louisiana state university and agricultural and mechanical college. Baton Rouge, LA. No. 1001+, 1949+. Extens. Publ. Louisiana St. Univ. Agric. Mech. Coll. HI 65106

Extract af det kongel. svenske videnskaps academiets afhandlinger; (fra 1739 aars begyndelse til og med 1750). Kristiania [= Oslo]. Vol. 1, 1762. Extr. Kongel. Svenske Vidensk. Acad. Afh. HI 65107

Extracts from proceedings of the royal horticultural society and Wisley trial reports. London. Vol. 78+, 1953+. Extr. Proc. Roy. Hort. Soc. Wisley Trial Rep. Previously contained in: Journal of the royal horticultural society. HI 65108

Extrait des mémoires du musée royal d'histoire naturelle de Belgique = Mémoires du musée royal d'histoire naturelle de Belgique. Brussels. Mém. Mus. Roy. Hist. Nat. Belgique. See B–P–H 577/24.

Extraits des mémoires de l'académie royale de Metz. Agriculture. Metz. Extr. Mém. Acad. Roy. Metz, Agric. See B–P–H 366/5. HI 53970

Extraits des procès-verbaux des séances ..., academie des sciences et lettres de Montpellier. Paris. Extr. Procès-Verbaux Séances Acad. Sci. Montpellier. See B–P–H 366/7. HI 53971

Extraits des procès-verbaux des séances de la société philomatique de Paris. Paris. Extr. Procés-Verbaux Séances Soc. Philom. Paris. See B–P–H 366/8. HI 53972

Ežegodnik Imperatorskago Russkago Geografičeskago Obščestva = Ezhegodnik Imperatorskago Russkago Geograficheskago Obshchestva. St. Petersburg.

Ežegodnik Kluba estestvoznanija i geografii Hristianskogo Sojuza Molodyh Ljudej. Harbin = Ezhegodnik Kluba estestvoznaniya i geografii Khristianskogo Soyuza Molodykh Lyudei. Harbin.

Ežegodnik počvovedenija. Warsaw = Roczniki gleboznawcze. Warsaw.

Ežegodnik Svědĕnij o Bolĕznjah i Povreždenijah Kul'turnyh i Dikorastuščih Poleznyh Rastenij = Ezhegodnik Svĕdĕnii o Bolĕznyakh i Povrezhdeniyakh

Kul'turnykh i Dikorastushchikh Poleznykh Rastenii. St. Petersburg.

Ežegodnik Volžskoj Biologičeskoj Stancii Saratovskago Obščestva Estestvoispytatelej i Ljubitelej Estestvoznanija = Ezhegodnik Volzhskoi Biologicheskoi Stantsii Saratovskago Obshchestva Estestvoispytatelei i Lyubitelei Estestvoznaniya. Saratov.

Ezhegodnik baltiiskago obshchestva dlya pooshchreniya kul'tury bolot. Tartu. Vols. 1-5 1911-16. Ezheg. Balt. Obshch Pooshchr. Kul't. Bolot. 1-590-3. HI 65109

Ezhegodnik Imperatorskago Russkago Geograficheskago Obshchestva. St. Petersburg. Vols. 1-8, 1890-99. Ezheg. Imp. Russk. Geogr. Obshch. HI 65110

Ezhegodnik izdanii akademii nauk tadzhikskoi S S R. Dushanbe. 1970+. Ezheg. Izd. Akad. Nauk Tadzhikskoi S.S.R. HI 65111

Ezhegodnik Kluba estestvoznaniya i geografii Khristianskogo Soyuza Molodykh Lyudei. Harbin, Manchoukuo [China]. 1 vol., 1934. Ezheg. Kluba Estestv. Khristiansk. Soyuza Molod. Lyudei. HI 65112

Ezhegodnik obshchestva estestvoispytatelei = Loodusuurijate seltsi aastaraamat. Tallinn.

Ezhegodnik pochvovedeniya. Warsaw = Roczniki gleboznawcze. Warsaw.

Ezhegodnik russkago paleontologicheskogo obshchestva. Leningrad, Moscow. 1917-30. Ezheg. Russk. Paleontol. Obshch. Superseded by: Ezhegodnik vsesoyuznogo paleontologicheskogo obshchestva. HI 65113

Ezhegodnik Svěděnii o Bolěznyakh i Povrezhdeniyakh Kul'turnykh i Dikorastushchikh Poleznykh Rastenii. St. Petersburg. Vols. 1-6, 19??-12. Ezheg. Svěd. Bolězny. Povrezhden. Kul't. Dikorast. Polezn. Rast. HI 65114

Ezhegodnik Volzhskoi biologicheskoi stantsii Saratovskago Obshchestva Estestvoispytatelei i Lyubitelei Estestvoznaniya. Saratov, Russian S F S R. Vol. 2(1), 1903. Ezheg. Volzhsk. Biol. Stantsii Saratovsk. Obshch. Estyestvoisp. Preceded by: Raboty Volzhskoi Biologicheskoi Stantsii. Superseded by: Lětniya raboty. HI 65115

Ezhegodnik vsesoyuznogo paleontologicheskogo obshchestva. Leningrad, Moscow. 1931+. Ezheg. Vsesoyuzn. Paleontol. Obshch. Preceded by: Ezhegodnik russkago paleontologicheskogo obshchestva. HI 65116

F A B I S newsletter. Aleppo. No. 1+, 1979+. F. A. B. I. S. Newslett. HI 65117

F A C E N A. Corrientes. No. 4+, 1980/81+, 1981+. F. A. C. E. N. A. Preceded by: F A C E N A. Serie ciencias naturales. HI 65118

F A C E N A. Serie ciencias naturales. Corrientes. Nos. 1-3, 1977-7? F. A. C. E. N. A., Ser. Ci. Nat. Preceded by: F A C E N A. HI 65119

F A O forestry development paper. Rome. Nos. 1-18?, 1954-74? F. A. O. Forest. Developm. Pap. Superseded by: F A O forestry series. HI 65120

F A O forestry occasional paper. Rome. No. 1+, 1955+. F. A. O. Forest. Occas. Pap. HI 65121

F A O forestry paper. Rome. No. 1+, 1977+. F. A. O. Forest. Pap. HI 65122

F A O forestry series. Rome. 19??-7?+. F. A. O. Forest. Ser. Preceded by: F A O forestry development papers. HI 65123

F A O plant production and protection papers. Rome. No. 1+, 1976+. F. A. O. Pl. Prod. Protect. Pap. HI 65124

F A O plant protection bulletin. Rome. Vol. 1+, 1952+. F.A.O. Pl. Protect. Bull. HI 65125

F A O review. Rome. No. 1, 1968. F. A. O. Rev. Superseded by: Ceres. HI 65126

F A O statistics series. Rome. No. 1+, 1976+. F. A. O. Statist. Ser. HI 65127

F A O training series. Rome. No. 1+, 1980+. F. A. O. Training Ser. HI 65128

F A S. Grafenhausen-Balzhausen. 19??-77+. F. A. S. HI 65129

F A S E B federation proceedings. Bethesda, MD. Vols. 1-46, 1942-87. F. A. S. E. B. Fed. Proc. Superseded by: F A S E B journal. HI 65130

F A S E B journal; official publication of the federation of american societies for experimental biology. Bethesda, MD. Vol. 1+, 1987+. F. A. S. E. B. J. Preceded by: F A S E B federation proceedings. HI 75213

F B C N informativo. Rio de Janeiro. Vol. ?-2(3)+, 19??-78+. F. B. C. N. Inform. HI 65131

F C A P informe didatico = Informe didatico, F C A P.

F C A P informe tecnico = Informe tecnico, F C A P.

F E B S letters = Federation of european biochemical societies letters. Amsterdam.

F E L actualités; marches européens des fruits et legumes. Paris. No. 0+, 1984+. F. E. L. Actual. HI 65132

F E M S microbiology letters. Amsterdam. Vol. 1+, 1977+. F. E. M. S. Microbiol. Lett. HI 65133

F E M S symposium. Amsterdam. No. ?-2+, ?-1977+. F. E. M. S. Symp. HI 75214

F E S P P newsletter. Copenhagen. No. 1+, 1982+. F. E. S. P. P. Newslett. HI 65134

F L K Berichte. Hamburg-Altona. Vol. 1+, 1958+. F. L.

K. Ber. Preceded by: Seminarium transmarinum. HI 65135

F N G A bulletin. Leesburg, FL. Vol. 1+, 1965+. F. N. G. A. Bull. HI 65136

F O N A news. Washington, DC. 1987?+. F. O. N. A. News. HI 65137

F P I: Flowering plant index. Chanhassen, MN. Vol. 1+, 1990?+. Fl. Pl. Index. HI 74834

F P R D I journal. Laguna. Vol. 12+, 1983+. F. P. R. D. I. J. Preceded by: Forpride digest. HI 65138

F P R L papers. Princes Risborough. Nos. 1, 2 & 7, 1969. F. P. R. L. Pap. Nos. 3-6 entitled: Timberlab papers. Superseded by: Timberlab papers. HI 65139

F R I bulletin. Rotorua. No. 1+, 1982+. F. R. I. Bull., Rotorua. HI 64681

F R I messenger. Lae. No. 1+, 1989+. F. R. I. Messenger. HI 74836

Faba. Paris. Faba. See B–P–H 366/19. HI 53973

Faba bean abstracts. Farnham Royal. Vol. 1+, 1981+. Faba Bean Abstr. HI 65140

Faculty papers, Seoul national university. (E), biology and agriculture series = Kungrip soul taehakkyo. Seoul.

Fairchild tropical garden bulletin. Coconut Grove, FL. Fairchild Trop. Gard. Bull. See B–P–H 366/22. HI 53974

Fairchild tropical garden newsletter. Coconut Grove, Fl. Fairchild Trop. Gard. Newslett. See B–P–H 366/23. HI 53975

Fairfield newsletter. Kirkham. Nos. ?-7-13, 19??-73-76. Fairfield Newslett. Superseded by: Newsletter, Fairfield experimental horticulture station. HI 65141

Fajtakiserletezes-fajtaminosites. Budapest. Vol. 23+, 1972+. Fajtakiserletezes-Fajtaminosites. Preceded by: Országos fajtakiserletek. HI 65142

Falusi gazda. Budapest. Falusi Gazda (Budapest). See B–P–H 366/27. HI 53976

Falusi gazda. Pest [=Budapest, in part]. Falusi Gazda (Pest). See B–P–H 367/1. HI 53977

Fang-chih yüeh-k'an = Geographical review. Nanking. Geogr. Rev. (Nanking). See B–P–H 397/25.

F.A.O. = F A O

Far eastern medical journal. [Tung fong hsüeh tsa chih.] Peking. Far E. Med. J. See B–P–H 367/4. HI 53978

Far eastern review; engineering, commerce, finance. Manila. Far E. Rev. See B–P–H 367/5. HI 53979

Farbenfabriken Bayer. Pflanzenschutz Abteilung, Bayer Pflanzenschutz-Kurier = Bayer Pflanzenschutz-Kurier. Leverkusen.

Farbenfabriken Bayer. Pflanzenschutz-Abteilung, Pflanzenschutz Kurier = Pflanzenschutz-Kurier. Leverkusen.

Farlowia; a journal of cryptogamic botany. Cambridge, MA. Farlowia. See B–P–H 367/7. HI 53980

Farm bulletin. Colorado agricultural college. Colorado experiment station. Fort Collins, CO. Farm Bull. Colorado Agric. Coll. Colorado Exp. Sta. See B–P–H 367/8. HI 53981

Farm bulletin, Missouri agricultural college = Missouri agricultural college. Farm bulletin. Columbia, MO. Missouri Agric. Coll. Farm Bull. See B–P–H 600/13.

Farm and forest; land use and rural planning in West Africa. Ibadan. Vols. 1-11, 1941-52. Farm & Forest. Preceded by: Nigerian Forester. 2-1529-1. HI 65143

Farm forestry. Wellington, N.Z. Vol. 1+, 1958+. Farm Forestry HI 65144

Farm and garden. Philadelphia, PA. Vols. 1-7, 1881-88. Farm & Gard. 2-1529-1. HI 65145

Farm and garden index. Mankato, MN. 1979+. Farm Gard. Index. HI 65146

Farm and home science. Logan, UT. Vols. 1-22, 1940-61. Farm Home Sci. Superseded by: Utah farm and home science. HI 65147

Farm and home science. Utah resources series. Utah state university, college of agricculture, agricultural experiment station. Logan, UT. Farm Home Sci., Utah Resources Ser. See B–P–H 367/13. HI 53985

Farm journal. Philadelphia, PA. Farm J. See B–P–H 367/14. HI 53986

Farm journal of British Guiana. Georgetown, British Guiana [Guiana]. Farm J. Brit. Guiana. See B–P–H 367/15. HI 53987

Farm pond harvest; dedicated to successful farm pond planning, construction, management, fishing and harvesting. Kankakee, IL. Vol. 1+, 1967+. Farm Pond Harvest. HI 65148

Farm, ranch and home quarterly. Lincoln, NE. Vol. 20(2)+, 1975+. Farm Ranch Home Quart. Preceded by: Quarterly, college of agriculture and home economics, university of Nebraska. HI 65149

Farm research. Geneva, NY. Vols. 1-33, 1934-67. Farm Res. Superseded by: New York's food and life sciences quarterly. 2-1531-3. HI 65150

Farm research news. Dublin. Farm Res. News. See B–P–H 367/18. HI 53988

Farm science reporter. Ames, IA. Farm Sci. Reporter. See B–P–H 367/20. HI 53989

Farmaceutický obzor. Bratislava. Vol. 1+, 1961+. Farm. Obzor. HI 65151

Farmaceutisk tidende. Copenhagen. Vol. 1+, 1890+.

Farm. Tidende (Copenhagen). 2-1532-2. HI 65152

Farmaceutisk tidende. Oslo. Vols. 1-23, 1893-1915. Farm. Tidende (Oslo). Superseded by: Norsk farmaceutisk tidsskrift. 4-3076-2. HI 65153

Farmaceutisk tidskrift. Stockholm. Farm. Tidskr. See B–P–H 367/21. HI 53990

Farmacia; revista a societatii de farmacie. Bucharest. Vol. 1+, 1953+. Farmacia (Bucharest). HI 65154

Farmacia chilena. Santiago de Chile. Vols. 1-29, 1927-55. Farm. Chilena. 2-1532-2. HI 65155

Farmacia espanola. Madrid. Farm. Esp. See B–P–H 367/10. HI 53983

Farmatsevticheskii zhurnal. Kharkiv. Vols. 1-2, 1928-30. Farm. Zhurn. (Kharkiv). Superseded by: Farmatsevtychnii zhurnal. HI 65156

Farmatsevticheskii zhurnal. St. Petersburg. Vols. 1+, 1864/65-66, 1879+, 1865+. Farm. Zhurn. (St. Petersburg). 2-1533-1. HI 65157

Farmatsevtychnii zhurnal. Kiev. Vol. 3+, 1930+ [publication suspended from June 1941 to Dec. 58]. Farm. Zhurn. (Kiev). Preceded by: Farmatsevticheskii zhurnal. Kharkov. HI 65158

Farmatsiya. Moscow. Vol. 1+, 1938+. Farmatsiya. 2-1533-1. HI 65159

Farmer; journal of the Jamaica agricultural society. Kingston, Jamaica. Vol. 52+, 1948+. Farmer. Preceded by: Journal of the Jamaica agricultural society. HI 65160

Farmer and gardener. Philadelphia, PA. Farmer & Gard. See B–P–H 367/22. HI 53991

Farmers' bulletin, department of agriculture, Canada. Ottawa. Nos. 1-164, 1936-53. Farmers' Bull. Dept. Agric. Canada. Preceded by: Bulletin, department of agriculture, dominion of Canada. HI 65161

Farmers bulletin, United States department of agriculture = United States department of agriculture. Farmers bulletin. Washington, DC. U.S.D.A. Farmers Bull. See B–P–H 942/13.

Farmers' circular, experimental farms service, Canada. Ottawa. Nos. 1-5, 1912-14. Farmers' Circ. Exp. Farms Serv. Canada. Superseded by: Circular, experimental farms service, Canada. HI 65162

Farmers' journal. Carthage, NY. Farmers' J. See B–P–H 367/25. HI 53992

Farmer's leaflet of the national institute of agricultural botany. Cambridge, England. Farmer's Leafl. Natl. Inst. Agric. Bot. See B–P–H 367/26. HI 53993

Farmers' magazine. London. Farmers' Mag. (London 1832-33). See B–P–H 368/2. HI 53995

Farmers' magazine. London. Farmers' Mag. (London 1834-81). See B–P–H 368/3. HI 53996

Farmers' magazine. London. Farmers' Mag. (London 1889-90). See B–P–H 368/4. HI 53997

Farmer's magazine, and useful family companion. London. Farmer's Mag. & Useful Family Companion. See B–P–H 368/1. HI 53994

Farmer's own bulletin. Bombay = Arya Swapatra Bombay. Arya Swapatra. See B–P–H 155/13.

Farming. Hastings, New Zealand. Farming. See B–P–H 368/5. HI 53998

Farming in South Africa. Pretoria. Vol. 1+, 1926+. Farming South Africa. Preceded by: Journal of the department of agriculture, Union of South Africa. 2-1537-2. HI 65163

Farnblätter. Zurich. 1978+. Farnblätter. HI 65164

Farol. [Venezuela]. Farol. See B–P–H 368/10. HI 53999

FASEB federation proceedings. Bethesda, MD = F A S E B federation proceedings. Bethesda, MD.

Faserforschung. Zeitschrift für Wissenschaft und Technik der Faserpflanzen und der Bastfaserindustrie. Leipzig. Faserforschung. See B–P–H 368/13. HI 54000

Fast-growing timber trees of the lowland tropics. Oxford. Nos. 1-6, 1968-73. Fast-Growing Timber Trees Lowl. Trop. Superseded by: Tropical forest papers. HI 65165

Fathom. Gainesville, FL. Vol. 1+, 1989+. Fathom. HI 75241

Fauna and flora; an official publication of the Transvaal provincial administration. Pretoria. No. 1+, 1950+. Fauna & Fl. (Pretoria). HI 65166

Fauna och flora; populär tidskrift för biologi. Uppsala & Stockholm. Vol. 1+, 1906+. Fauna & Fl. (Uppsala). 2-1539-2. HI 65167

Fauna et flora laurentianae. Quebec. Fauna Fl. Laurent. See B–P–H 368/16. HI 54001

Faunistisch-floristische Notizen aus dem Saarland. Saarbrücken. Vol. 1+, 1968+. Faunist. Florist. Not. Saarland. HI 65168

Fava bean abstracts. Farnham Royal. 1981+. Fava Bean Abstr. HI 65169

Fed news. Cheltenham. No. 1+, 1980+. Fed News. HI 65170

Feddes Repertorium. Berlin. Feddes Repert. See B–P–H 368/30. HI 54002

Feddes Repertorium. Beiheft. Berlin. Feddes Repert. Beih. See B–P–H 368/32. HI 54003

Feddes Repertorium specierum novarum regni vegetabilis. Berlin. Feddes Repert. Spec. Nov. Regni Veg. See B–P–H 369/2. HI 54004

Feddes Repertorium specierum novarum regni vegetabilis. Beiheft. Berlin. Feddes Repert. Spec. Nov. Regni Veg. Beih. See B–P–H 369/3. HI 54005

Fedecame; boletín semanel sobre café. San Salvador, Salvador. Fedecame. See B–P–H 369/5. HI 54006

Fedemoa. [Federación mexicana de organizaciónes agrícolas.] Mexico City. Fedemoa. See B–P–H 369/6. HI 54007

Federal museums journal, Malaya = Federation museums journal. Kuala Lumpur.

Federal register. Washington, DC. Vol. 1+, 1936+. Fed. Reg. 2-1542-3. HI 75242

Federation of european biochemical societies letters. Amsterdam. Vol. 1+, 1968+. Fed. Eur. Biochem. Soc. Lett. HI 65171

Federation of european microbiological societies microbiology letters = F E M S microbiology letters. Amsterdam.

Federation museums journal. Kuala Lumpur. N.s. vol. 1/2+, 1954/55+, 1955+. Fed. Mus. J. Preceded by: Journal of the Federated Malay States museums. 3-2242-3. HI 65172

Federation proceedings, federation of american societies for experimental biology = F A S E B federation proceedings. Bethesda, MD.

Feld und Garten. Vienna. Feld & Gart. See B–P–H 369/8. HI 54008

Feld und Wald. Essen. 1882+ [suspended 1945-48]. Feld & Wald. 2-1551-3. HI 65173

Feld, Wald und Wasser. Bern. Feld Wald Wasser. See B–P–H 369/9. HI 54009

Felsö-magyarországi szölészeti, borászati és gazdasági lap. Kassa, Hungary [=Kosice, Czechoslovakia]. Felsö-Magyarorsz. Szölész. Borász. Gazd. Lap. See B–P–H 369/10. HI 54010

Felsöoktatási szakirodalmi tajékoztató sorozat müszake és természet-tudományok. Budapest. Felsöokt. Szakirod. Tájékozt. Soroz. Müsz. Természet-Tud. See B–P–H 369/11. HI 54011

Fen fakültesi dergisi, Ege universitesi. Ser. B = Ege universitesi fen fakültesi dergisi. Ser. B. Izmir.

Fennia. Helsinki. Fennia. See B–P–H 369/12. HI 54012

Fenogenetičeskaja izmenčivost′ = Fenogeneticheskaya izmenchivost′. Moscow.

Fenogeneticheskaya izmenchivost′. Moscow. Vols. 1-2, 1933-35. Fenogenet. Izmenchiv. HI 65174

Ferme extérieure de Grignon. Paris. 1924+. Ferme Extér. Grignon. Preceded by: Rapport, technique, centre national d'experimentation agricole de Grignon. HI 65175

Ferme suisse; revue générale d'agriculture, d'industrie laitière, d'apiculture, et mercuriale des produits agricoles. Lausanne. Vols. ?-2+, 18??-75+. Ferme Suisse. HI 65176

Ferment. London. Vol. 1+, 1988+. Ferment. HI 74516

Fermentation and industry = Hakko to kogyo. Tokyo.

Fermentation research institute news. [Hakken nyusu.] Chiba, Japan. Ferment. Res. Inst. News. See B–P–H 369/14. HI 54013

Fern bulletin; a quarterly devoted to ferns. Binghamton, NY. Fern Bull. See B–P–H 369/15. HI 54014

Fern gazette; journal of the british pteridological society. London. Vol. 11+, 1974+. Fern Gaz. Preceded by: British fern gazette. HI 65177

Fernbank quarterly. Atlanta, GA. 1975+. Fernbank Quart. HI 65178

Fernhurst bulletin. Fernhurst. Nos. 1-3, 1953-56. Fernhurst Bull. HI 65179

Fernwood annual report. Niles, MI. 19??-77+. Fernwood Annual Rep. HI 65180

Fernwood news. Niles, MI. 19??-83+. Fernwood News. HI 65181

Fernwood notes. Niles, MI. No. 1+, 1966+. Fernwood Notes. HI 65182

Fertödi növénynemesítési és növénytermesztési kutató intézet közleményei. Budapest. Fertödi Növénynemes. Növényterm. Kutató Intéz. Közlem. See B–P–H 369/16. HI 54015

Feuilles d'agriculture et d'économie générale, publiées par la société d'agriculture et d'économie du canton de Vaud. Lausanne. Feuilles Agric. Écon. Gén. See B–P–H 370/8. HI 54019

Feuille de bambou = Bambusblätter. Marktheidenfeld.

Feuille du canton de Vaud, ou journal de Vaud, ou journal d'agriculture pratique, des sciences naturelles et d'économie publique. Lausanne. Feuille Canton Vaud. See B–P–H 369/19. HI 54016

Feuille de correspondance de la seconde classe de la société d'agriculture, sciences et arts du département du Bas-rhin, séante à Strasbourg — Korrespondenz-Blatt der zweyten Klasse der Gesellschaft des Ackerbaues, der Wissenschaften und Künste, zu Strassburg. Strasbourg. Feuille Corresp. Seconde Cl. Soc. Agric. Dép. Bas-Rhin Strasbourg. See B–P–H 370/1. HI 54018

Feuille de correspondence du libraire. Paris. Feuille Corresp. Libr. See B–P–H 369/20. HI 54017

Feuille pour l'étude des cactées = Blätter für Kakteenforschung. Volksdorf [=Hamburg, in part].

Blätt. Kakteenf. See B–P–H 199/14.

Feuille des jeunes naturalistes; revue mensuelles d'histoire naturelle. Rennes & Paris. Vols. 1-44, 1870-1914. Feuille Jeunes Naturalistes. Superseded by: Feuille des naturalistes. HI 65183

Feuille des naturalistes; bulletin de la société des naturalistes parisiens. Paris. 1924-27; n.s. vols. 1-7, 1946-52. Feuille Naturalistes. Preceded by: Feuille des jeunes naturalistes. Superseded by: Cahiers des naturalistes. 2-1555-1 HI 65184

Feuille pour la protection de la nature = Schweizerische Blätter für Naturschutz. Basel.

Feuille verte. Geneva. No. 1+, 1982+. Feuille Verte. HI 65185

Feuilles, see Feuille

Fiber crops in China. [Zhongguo mazuo.] No. ?-13+, 19??-82+. Fiber Crops China. HI 65186

Fibra; international journal on growing and processing of flax and hemp. Wageningen. Vol. 1+, 1956+. Fibra. HI 65187

Fibre crops = Rostnövények. Kompolt.

Fichero bibliografico hispanoamericano. New York, NY. Fichero Bibliogr. Hispanoamer. See B–P–H 370/10. HI 54020

Fiches descriptives provisoires des nouvelles variétés de cereales et de lin proposées à l'inscription au catalogue (Liste de commercialisation en France) = Fiches descriptives provisoires des nouvelles variétés de cereales proposées à l'inscription au catalogue (Liste de commercialisation en France). Versailles.

Fiches descriptives provisoires des nouvelles variétés de cereales proposées à l'inscription au catalogue (Liste de commercialisation en France). Versailles. 1973-78? Fiches Descr. Provis. Nouv. Var. Cereales Prop. Inscript. Cat.

Superseded by: Bulletin des variétés. Céréales et fiches descriptives provisiores des variétés. HI 65189

Fiches descriptives provisoires des nouvelles variétés de plantes fourragères proposées à l'inscription au catalogue (Liste de commercialisation en France). Versailles. 1975?-81? Fiches Descr. Provis. Nouv. Var. Pl. Fourrag. Prop. Inscript. Cat. Superseded by: Bulletin des variétés. Plantes fourragères. HI 65190

Fichier biologique. Trois-Rivières. Vol. 1+, 1977+. Fichier Biol. HI 65191

Fiddlehead forum; bulletin of the american fern society. Bronx, NY. No. 1+, 1974+. Fiddlehead Forum. Preceded by: News and views, american fern society. HI 65192

Field crop abstracts. Aberystwyth, Wales. Filed Crop Abstr. See B–P–H 370/14. HI 54021

Field crop news. Christchurch, N.Z. No. ?-2+, 19??-78+. Field Crop News. HI 65193

Field crops research. Amsterdam. Vol. 1+, 1978+. Field Crop Res. HI 65194

Field facts: Soils, insects, diseases, weeds, crops. Brookings, SD. Vol. 1+, 1986+. Field Facts. Preceded by: Pest-o-gram: Insects, diseases, weeds. HI 65195

Field, the farm, the garden. London. Field Farm Garden. See B–P–H 370/15. HI 54022

Field and forest. Washington, DC. Field & Forest. See B–P–H 370/16. HI 54023

Field & laboratory. Dallas, TX. Field & Lab. See B–P–H 370/18. HI 54024

Field manual, agricultural experiment station, university of Delaware. Newark, DE. Vol. 1+, 1956+. Field Manual Agric. Exp. Sta. Univ Delaware. HI 65196

Field museum of natural history. Botanical series. Chicago, IL. Field Mus. Nat. Hist., Bot. Ser. See B–P–H 370/21. HI 54025

Field museum of natural history bulletin = Bulletin, Field museum of natural history. Chicago, IL.

Field museum news. Chicago, IL. Field Mus. News. See B–P–H 370/24. HI 54026

Field naturalist. London. Field Naturalist. See B–P–H 370/27. HI 54027

Field naturalist; quarterly review of natural history observations in the north west. Penrith. Vols. 1-13, 1954-69. Field Naturalist (Penrith). Preceded by: Monthly bulletin, Penrith and district natural history society. HI 65197

Field naturalist and scientific student. Manchester, London. Nos. 1-9, 1882-83. Field Naturalist Sci. Student. 2-1557-2. HI 65198

Field naturalists' quarterly. Edinburgh. Field Naturalists' Quart. See B–P–H 370/28. HI 54028

Field station record, division of plant industry, C S I R O. Canberra, A.C.T. Vol. 1+, 1962+. Field Sta. Rec. Div. Pl. Industry C. S. I. R. O. HI 65199

Field studies. London. Field Stud. See B–P–H 371/1. HI 54029

Field trip guidebook, american association of stratigraphic palynologists. Dallas, TX. 1971+. Field Trip Amer. Assoc. Stratigr. Palynologists. HI 65200

Fieldiana. Anthropological series. Natural history museum. Chicago, IL. Fieldiana, Anthropol. Ser. See B–P–H 371/2. HI 54030

Fieldiana: botany. Chicago, IL. Vols. 24-41, 1946-78; n.s. vol. 1+, 1979+. Fieldiana, Bot. Preceded by: Publications, Field museum of natural history. Botanical series. HI 65201

Fierdingsaarskriftet ny minerva. Copenhagen. 1808. Fierdingaarsskr. Ny Minerva. Preceded by: Maanedskriftet ny Minerva. 4-3123-3. HI 65202

Fig leaf. Fresno, CA. Fig Leaf. See B–P–H 371/5. HI 54031

Fig and olive journal. Los Angeles, CA. Fig Olive J. See B–P–H 371/6. HI 54032

Fig research. Fresno, CA. Fig Res. See B–P–H 371/7. HI 54033

Figaro agricole. Paris. Nos. 1-267, 1952-74. Figaro Agric. Superseded by: Agricole. HI 65203

Figaro documents agricole = Figaro agricole. Paris.

Figyelmezö. Pest [=Budapest, in part]. Figyelmezö. See B–P–H 371/8. HI 54034

Fiji agricultural journal. Suva. Vol. 32+ 1970+. Fiji Agric. J. Preceded by: Agricultural journal, Suva. HI 65204

Fiji farmer. Suva. Vols. 1-3, 1965-67. Fiji Farmer. HI 65205

Filipinas journal of science and culture. Metro Manila. Vol. 1+, 1981+. Filipinas J. Sci Cult. HI 65206

Filipino forester. Manila. Filip. Forester See B–P–H 371/10. HI 54035

Finnish marine research. Helsinki. No. 244+, 1978+. Finn. Mar. Res. Preceded by: Merentutkimuslaitos, menentutkimuslaitoksen julkaisu. HI 65207

Finnländische hydrographisch-biologische Untersuchungen. Helsinki. Nos. 1-14 [no. 11 not published], 1907-17. Finnl. Hydrogr.-Biol. Untersuch. 2-1566-3. HI 65208

Finska läkaresällskapet, handlingar. Helsinki. Finska Läkaresällsk. Handl. See B–P–H 371/11. HI 54036

Finska mosskulturföreningen, årsbok. Helsinki. Finska Mosskulturfören. Årsbok. See B–P–H 371/13. HI 54037

Fiori. Rome. Vols. [1]-12, 1958-69. Fiori. HI 65209

Fiori e piante, in giardino in terrazza in casa = Fiori. Rome.

Fish and wildlife reference service newsletter. Denver, CO. No. 28+, 1974+. Newslett. Fish Wildlife Ref. Serv. Preceded by: Newsletter, library reference service, federal aid in fish and wildlife restoration [not entered]. HI 68705

Fish and wildlife research. Washington, DC. Vol. 1+, 1986+. Fish Wildlife Res. Preceded by: Wildlife research report, U S fish and wildlife service [not entered]. HI 75243

Fisheries research bulletin, fisheries research division, New Zealand. Wellington, N.Z. N.s. no. 1+, 1968+. Fish. Res. Bull., New Zealand. Preceded by: Fisheries bulletin of the New Zealand marine department [not entered]. HI 65210

Fison's arable and grassland research. Felixstowe. 1979+. Fison's Arable Grassland Res. Preceded by: Fison's grassland research. HI 65211

Fison's grassland research. Felixstowe. 1976. Fison's Grassland Res. Superseded by: Fison's arable and grassland research. HI 65212

Fitiatria e fitofarmacologia. Laboratorio de fitofarmacologia, direcção geral dos servicos agrícolas. Lisbon. Fitiatria & Fitofarmacol. See B–P–H 371/16. HI 54038

Fitófilo. Dirección general de agricultura. San Jacinto, Mexico. Fitófilo. See B–P–H 371/17. HI 54039

Fitologiya. Sofia. Vol. 1+, 1975+. Fitologiya. Preceded by: Izvestiya botanicheskaya institut, Sofia. HI 65213

Fitopatologia; organo oficial de la asociación latinoamericana de fitopatologia. Santiago. Vol. 1+, 1966+. Fitopatologia. HI 65214

Fitopatologia brasileira. Brasilia. Vol. 1+, 1976+. Fitopatol. Brasil. HI 65215

Fitopatología venezolana; revista de la sociedad Venezolana de fitopatología. Maracay. Vol. 1+, 1988+. Fitopatol. Venez. HI 74863

Fitosanitarias; organo del departamento de sanidad vegetal de la facultad de agronomia de la universidad de La Plata. La Plata. Vol. 1+, 1961+. Fitosanitarias. HI 65217

Fitossanidade; orgão dos fitossanitaristas do estado do Ceará. Fortaleza. Vol. 1+, 1974+. Fitossanidade. HI 65218

Fitotecnia latinoamericana. San José, Costa Rica. Fitotecn. Latinoamer. See B–P–H 371/19. HI 54040

Fitoterapia. Milan. Fitoterapia. See B–P–H 371/20. HI 54041

Fiziologija rastenij. Moscow = Fiziologiya rastenii. Moscow.

Fiziologija rastenij. Moscow & Leningrad = Materialy k biobibliografii uchenykh S S S R. Seriya biologicheskikh nauk; fiziologiya rastenii. Moscow & Leningrad.

Fiziologiya i biokhimiya kul′turnykh rastenii. Kiev. Vol. 1+, 1969+. Fiziol. Biokhim. Kul′t. Rast. HI 65220

Fiziologiya rastenii. Moscow. Vol. 1+, 1954+. Fiziol. Rast. HI 65221

Fiziologiya na rasteniyata. Sofia. Vol. 1+, 1970+. Fiziol. Rast. Preceded by: Izvestiya, institut po fiziologiya na rasteniya "Metodii Popov". HI 65222

Fiziologiya sel′skokhozyaistvennykh rastenii. Moscow. Vols. 1-12, 1967-71. Fiziol. Sel′skokhoz. Rast.

HI 65223

Fiziologiya vodoobmena i ustoichivost′ rastenii. Kazan. Vol. 1+, 1968+. Fiziol. Vodoobmena Ustoichivost′ Rast. HI 65224

Fiziologo-biokhimicheskie osnovy pitaniya rastenii. Kiev. Vols. [1]-4, 1966-68. Fiziol.-Biokhim. Osnovy Pitan. Rast. HI 65225

Fiziologo-biokhimicheskie osnovy vzaimodeistviya rastenii v fitosenozakh. Kiev. Vol. 1+, 1970+. Fiziol.-Biokhim. Osnovy Vzaimod. Rast. Fitosenozakh. HI 65226

Flacourtiaceae. Série científica devotado ao estudo geral das plantas desta família. Curitiba, Brazil. Flacourtiaceae. See B–P–H 373/19. HI 54067

Flavour and fragrance journal. Chichester & New York. Vol. 1+, 1985+. Flav. Fragr. J. HI 65227

Flavour industry. London. Vols. 1-5, 1970-74; supplements, 1969-74. Flav. Industr. Preceded by: Perfumery and essential oil record. Superseded by: International flavours and food additives. HI 65228

Flavours. London. Vols. 1-11, 1938-48. Flavours. Incorporated in: Perfumery and essential oil record. HI 65229

Flavours. London = International flavours and food additives. London.

Flavours, fruit juices and spices review = Flavours. London.

Fleuriste. Zurich = Florist. Zurich. Florist (Zurich). See B–P–H 376/8.

Fleuriste du Québec. Grand'mère, Quebec. Vols. 1-?, 1971-77? Fleur. Québec. Superseded by: Québec vert. HI 75086

Fleurs de France. Paris. No. ?-76+, ?-1985+. Fleurs France. HI 65230

Flintshire historical society publications; or journal of the Flintshire historical society. Prestatyn, Wales. Flintshire Hist. Soc. Publ. See B–P–H 373/25. HI 54068

Flora; oder (allgemeine) botanische Zeitung. Regensburg, Jena. Vols. 1-155, 1818-1965 [Vols. 1-16, 1818-33, include "Beilage" and "Ergänzungsblätter"; vols. 17-25, 1834-42, include "Beiblatt" and "Intelligenzblatt"]. Flora. Superseded by: Flora. Abteilung A, Physiologie und Biochemie and Flora. Abteilung B, Morphologie und Geobotanik. 2-1574-3. HI 50534

Flora. Abteilung A, Physiologie und Biochemie. Jena. Vols. 156-160, 1965-69. Flora, A. Preceded by: Flora. Superseded by: Biochemie und Physiologie der Pflanzen. 2-1574-3. HI 65231

Flora. Abteilung B, Morphologie und Geobotanik. Jena. Vols. 156-158, 1965-69. Flora, B. Preceded by: Flora. Superseded by: Flora. Morphologie, Geobotanik, Oekophysiologie. 2-1574-3. HI 65232

Flóra; tímarit um íslenzka grasafraedi. (A journal of icelandic botany). Akureyri. Vols. 1-6, 1963-68. Flóra (Akureyri). Superseded by: Acta botanica islandica. HI 65233

Flora. Amsterdam. Flora (Amsterdam). See B–P–H 374/2. HI 54069

Flora; illustreret havetidende for damer. Copenhagen. 1877-78. Flora (Copenhagen). HI 65234

Flora. Hillegom, Netherlands. Flora (Hillegom). See B–P–H 374/3. HI 54070

Flora. Instituto ecuatoriano de ciencias naturales. Quito, Ecuador. Flora (Inst. Ecuat.). See B–P–H 374/5. HI 54071

Flora. Lvov, Galicia [Ukrainian S S R]. Flora (Lvov). See B–P–H 374/7. HI 54073

Flora. Lund. Flora (Lund). See B–P–H 374/6. HI 54072

Flora. Milan. Flora (Milan). See B–P–H 374/8. HI 54074

Flora. Prague. Flora (Prague). See B–P–H 374/9. HI 54075

Flora; revista de botánica y farmacognosia. Quito. Vols. 1-7, 1937-50. Flora (Quito). 2-1575-1. HI 65235

Flora; revista tecnica da madeire e do reflorestamento. Rio de Janeiro. Vol. 1+, 1970+. Flora (Rio de Janeiro). HI 65236

Flora. Stuttgart. Flora (Stuttgart). See B–P–H 374/11. HI 54076

Flora. Vienna. Flora (Vienna). See B–P–H 374/12. HI 54077

Flora buttensis. Chico, CA. Vol. 1+, 1980+. Fl. Buttensis. HI 65237

Flora colonia; Mitteilungen des Freundeskreises botanischer Garten Köln Cologne. Vol. 1+, 1984+, 1985+. Fl. Colon. HI 65238

Flora og fauna; Aarbog for naturvenner og naturhistoriske samiere. Esbjerg, Silkeborg, etc. Vol. 1+, 1899+. Fl. & Fauna (Esbjerg). Preceded by: Meddelelser fra "Flora og fauna". HI 54042

Flora and fauna handbook. Gainesville, FL. No. 1+, 1985+. Fl. Fauna Handb. HI 72382

Flora y fauna peruanas. Lima. Fl. Fauna Peruanas. See B–P–H 372/6. HI 54045

Flora in fauna severnega jadrana. Ljubljana. 1975. Fl. Fauna. Severn. Jadrana. HI 65240

Flora fennica. Helsinki. Vols. 1-6, 1923-77 [suspended 1924-58]. Fl. Fenn. 1-1575-1. HI 65241

Flora, giardini, orti, frutteti. Milan. 1949-51. Fl. Giard. Orti & Frutt. Preceded by: Giardini. HI 65242

Flora, Königlich Sächsische Gesellschaft für Botanik und Gartenbau zu Dresden. - Sitzungsberichte und Abhandlungen. Dresden. "Flora" Königl. Sächs. Ges. Bot. Dresden Sitzungsber. Abh. See B–P–H 374/19. HI 54078

Flora malesiana bulletin. Leiden. Fl. Males. Bull. See B–P–H 372/19. HI 54054

Flora. Morphologie, Geobotanik, Oekophysiologie. Jena. Vol. 159+, 1970+. Flora, Morphol. Geobot. Oekophysiol. Preceded by: Flora. Abteilung B, Morphologie und Geobotanik. HI 65243

Flora murmanskoi oblasti. Moscow. Vol. 1+, 1953+. Fl. Mermansk. Obl. HI 65244

Flora neotropica, monograph. New York. Vol. 1+, 1967+. Fl. Neotrop. Monogr. HI 65245

Flora of New South Wales = Contributions from the New South Wales national herbarium. Flora series. Sydney, N.S.W.

Flora of north America newsletter. St. Louis, MO. No. 1+, 1987+. Fl. N. Amer. Newslett. HI 65246

Flora north america reports. Washington, DC. Nos. 1-84, 1966-78. Fl. N. Amer. Rep. HI 65247

Flora online; bulletin board system for systematic botany Buffalo, NY. No. 1+, 1987+ [published online and on computer diskette]. Fl. Online. HI 65248

Flora do Paraná. Curitiba, Brazil. Fl. Paraná. See B–P–H 372/21. HI 54055

Flora polska = Flora Polski. Warsaw, Cracow.

Flora Polski. Rosliny zarodnikowe Polski i ziem osciennych. Warsaw, Cracow. 1957+. Fl. Polski. HI 65249

Flora en pomona. Arnhem, Netherlands. Fl. & Pomona (Arnhem). See B–P–H 372/26. HI 54058

Flora en pomona. Utrecht. Fl. & Pomona (Utrecht). See B–P–H 373/1. HI 54059

Flora i rastitel′nost Belorussii. Ukazetel′ literatury. Minsk. 1971+. Fl. Rastitel′n. Belorussi, Ukaz. Lit. HI 65250

Flora i rastitel′nost dal′nego vostoka. Vladivostok. 1970+. Fl. Rastitel′n. Dal′nego Vostoka. HI 65251

Flora i rastitel′nost′ yugo-vostoka. Saratov. Vol. 1+, 1972+. Fl. Rastitel′n. Yugo-Vostoka. HI 65252

Flora, rastitel′nost′ i rastitel′nye resursy armyanskoi S S R. Erevan. Vol. ?-5+, 19??-70+. Fl. Rastitel′nost′ Rastitel′nye Resursy Armyansk. S.S.R. HI 65253

Flora, rastitel′nost′ i rastitel′nye resursy Zabaikal′ya. Vol. ?-2+, 19??-71+. Fl. Rastitel′nost′ Rastitel′nye Resursy Zabaikal′ya. HI 65254

Flora, Sarah P. Duke gardens. Durham, NC. No. ?-5+, 19??-83+. Fl. Sarah P. Duke Gard. HI 65255

Flora i sistematika vysshikh rastenii = Trudy botanicheskogo instituta akademii nauk S S S R. Ser. 1, flora i sistematika vysshikh rastenii. Moscow & Leningrad.

Flora and sylva. A monthly review for lovers of garden, woodland, tree or flower; new and rare plants, trees, shrubs, and fruits; the garden beautiful, home woods, and home landscape. London. Fl. & Sylva. See B–P–H 373/8. HI 54063

Flora et systematica plantae vasculares = Trudy botanicheskogo instituta akademii nauk S S S R. Ser. 1, flora i sistematika vysshikh rastenii. Moscow & Leningrad.

Flora et vegetatio italica. Memoria; monografie sulla flora e vegetazione d'Italia. Milan. Vol. 1+, 1959+. Fl. Veg. Ital. Mem. HI 65256

Flora et vegetatio mundi. Brunswick. 1960+. Fl. Veg. Mundi. HI 65257

Florae regionum Hungariae criticae = Magyar flóramüvek. Debrecen, Kolozsvar [=Cluj, Rumania], Pecs, & Veszprem, Hungary. Magyar Flóramüvek. See B–P–H 543/4.

Floral cabinet, and magazine of exotic botany. London. Fl. Cab. See B–P–H 371/26 HI 54044.

Floral Magazine. Greenwood, SC. Vol. 67(9), 1968. Fl. Mag. (Greenwood). HI 65258

Floral magazine; comprising figures and descriptions of popular garden flowers. London. Fl. Mag. (London). See B–P–H 372/16. HI 54052

Floral magazine and botanical repository. Philadelphia, PA. Fl. Mag. & Bot. Repos. See B–P–H 372/17. HI 54053

Floral world. Springfield, OH. Fl. World. See B–P–H 373/10. HI 54065

Floral world and garden guide. London. Fl. World Gard. Guide. See B–P–H 373/11. HI 54066

Floralia. Geïllustreerd weekblad voor tuinbouw en fruithandel. Assen & Groningen. Floralia. See B–P–H 374/29. HI 54079

Florascope. Glen Ellyn, IL. Vol. 1+, 1981+. Florascope. HI 65259

Flore de la Guadeloupe et des dépendences. Basse-Terre, Guadeloupe. Fl. Guadeloupe Dépend. See B–P–H 372/13. HI 54050

Flore des serres et jardins de l'Angleterre. Brussels. Fl. Serres Jard. Angleterre. See B–P–H 373/4. HI 54060

Flore des serres et des jardins de l'Europe. Ghent (Gand). Vols. 1-10, 1845-55. Fl. Serres Jard. Eur. Superseded by: Journal général d'horticulture. Gand. 2-1575-3.

HI 54061

Flore des serres et des jardins de Paris. Paris. Fl. Serres Jard. Paris. See B–P–H 373/6. HI 54062

Florentia; annuario generale della orticultura in Italia. Florence. Vols. 1-2, 1880-81. Florentia. HI 65260

Flôres do Brasil. São Paulo. Fl. Brasil. See B–P–H 371/25. HI 54043

Flores y jardines. Santiago de Chile. 1955+. Fl. & Jard. HI 65261

Floresta. Curitiba. Vol. 1+, 1969+. Floresta (Curitiba). HI 65262

Floresta. Guatemala City. Vol. 1+, 1965+. Floresta (Guatemala). HI 54080

Floresta; revista de divulgacion. La Paz. Vol. 1+, 1962+. Floresta (La Paz). HI 65263

Floribunda. Bogor. Vol. 1+, 1987+. Floribunda. HI 65264

Floricoltura pesciatina. Pescia. Vol. 1+, 1958?+. Floric. Pesciat. HI 65265

Floricultural cabinet, and florist's magazine. London. Floric. Cab. & Florist's Mag. See B–P–H 375/2. HI 54081

Floricultural magazine and miscellany of gardening. London. Floric. Mag. & Misc. Gard. See B–P–H 375/3. HI 54082

Floricultural review and florist's and gardener's register. London. Floric. Rev. & Florist's Gard. Reg. See B–P–H 375/4. HI 54083

Floriculture Indiana. West Lafayette, IN. Vol. 1+, 1986+. Floric. Indiana. HI 64776

Florida agricultural experiment station bulletin. Lake City, FL. Florida Agric. Exp. Sta. Bull. See B–P–H 375/5. HI 54084

Florida agricultural research. Gainesville, FL. 1982+. Florida Agric. Res. Preceded by: Sunshine state agricultural research report. HI 65266

Florida buggist = Florida entomologist. Gainesville, FL. Florida Entomol. See B–P–H 375/10.

Florida citrus nurserymens association news. Leesburg, FL. Florida Citrus Nurserymens Assoc. News. See B–P–H 375/9. HI 54085

Florida entomologist. Gainesville, FL. Florida Entomol. See B–P–H 375/10. HI 54086

Florida experiment station press bulletin. Lake City, FL. Florida Exp. Sta. Press Bull. See B–P–H 375/12. HI 54087

Florida farmer. Jacksonville, FL. Florida Farmer. See B–P–H 375/15. HI 54088

Florida farmer and fruit grower. Jacksonville, FL. Florida Farmer Fruit Grower. See B–P–H 375/16. HI 54089

Florida flora newsletter. Gainesville, FL. Florida Fl. Newslett. See B–P–H 375/17 HI 54090.

Florida flower grower. Gainesville. 1964+. Florida Fl. Grower. HI 65267

Florida foliage. Apopka, FL. Vol. 1+, 1978+. Florida Foliage. HI 65268

Florida foliage grower. Gainesville, FL. Vol. 1+, 1964+. Florida Foliage Grower. HI 65269

Florida foliage locator. Aopoka, FL. 19??-85/86+, 19??-86?+. Florida Foliage Locator. HI 65270

Florida foliage news. Vol. 1+, 1980+. Florida Foliage News. HI 65271

Florida fruit digest. Jacksonville, FL. Florida Fruit Digest. See B–P–H 375/18. HI 54091

Florida fruit and produce news. Jacksonville, FL. Florida Fruit Produce News. See B–P–H 375/20. HI 54092

Florida fruit world. Tampa, FL. Florida Fruit World. See B–P–H 375/21. HI 54093

Florida fruits and flowers. Bartow, FL. Florida Fruits Fl. See B–P–H 375/22. HI 54094

Florida gardener. Gainesville, FL. Vol. 1+, 1951+. Florida Gard. HI 65272

Florida grower. Jacksonville, FL. Florida Grower. See B–P–H 375/23. HI 54095

Florida homemaker and gardener. Miami, FL. Florida Homemaker Gard. See B–P–H 375/24. HI 54096

Florida marine research publications. St. Petersburg, FL. Vol. 1+, 1973+. Florida Mar. Res. Publ. Preceded by: Leaflet series, Florida department of natural resources, marine laboratory; Professional papers series, Florida department of natural resources, marine laboratory; Special scientific report, Florida department of natural resources, marine laboratory; and Technical series, Florida state board of conservation, marine laboratory. HI 65273

Florida naturalist. Daytona Beach, FL. Florida Naturalist. See B–P–H 376/1. HI 54097

Florida orchidist. Miami Beach, FL. Vol. 1+, 1958+. Fl. Orchidist. HI 65274

Florida plant finder. Fort Lauderdale, FL. Vols. 1-?, 1980-84. Florida Pl. Finder. Superseded by: Plantfinder. HI 65275

Florida scientist. Gainesville, FL. Vol. 36+. 1973+. Florida Sci. Preceded by: Quarterly journal of the Florida academy of sciences. HI 65276

Florida truck news. Jacksonville, FL. Florida Truck News. See B–P–H 376/5. HI 54098

Florida woods and waters. Tallahassee, FL. Florida

Woods Waters. See B–P–H 376/6. HI 54099

Florist. London. Florist (London). See B–P–H 376/7. HI 54100

Florist. Zurich. Florist (Zurich). See B–P–H 376/8. HI 54101

Florist bulletin, Pennsylvania flower growers = Bulletin, Pennsylvania flower growers. Kennett Square, Bloomsburg, PA.

Florist, fruitist and garden miscellany. London. Florist Fruitist Gard. Misc. See B–P–H 376/9. HI 54102

Florist and garden miscellany. London. Florist Gard. Misc. See B–P–H 376/10. HI 54103

Florist and horticultural journal. Philadelphia, PA. Florist Hort. J. See B–P–H 376/11. HI 54104

Florist and nurseryman. Miami, FL. Florist & Nurseryman. See B–P–H 376/19. HI 54109

Florist and pomologist. London. Florist & Pomol. See B–P–H 376/20. HI 54110

Floristic notes. Tartu = Floristilised mjärkmed. Tartu.

Floristicheskie zametki. Tartu = Floristilised mjärkmed. Tartu.

Floristilised markmed. Tartu = Floristilised mjärkmed. Tartu.

Floristilised mjärkmed. Tartu, Estonian S S R. Vol. 1+, 1958+. Florist. Mjärkmed. HI 54105

Floristische Mitteilungen des botanischen Instituts und des botanischen Gartens der Universität Rostock. [Sonderdruck aus der Wissenschaftliche Zeitschrift der Universität Rostock. Mathematische, naturwissenschaftliche Reihe.] Rostock. Vol. 1-?, 1965-68. Florist. Mitt. Bot. Inst. Bot. Gart. Univ. Rostock. HI 65277

Floristische Mitteilungen aus Salzburg. Salzburg. No. 1+, 1975+. Florist. Mitt. Salzburg. HI 65278

Floristische Rundbriefe. Zeitschrift für floristische Geobotanik, Populationsökologie und Systematik. Göttingen. Vol. 21+, 1987/88+, 1987+. Florist. Rundbr. Preceded by: Göttinger floristische Rundbriefe. HI 65279

Florists' exchange and horticultural trade world. New York, NY. Florists' Exch. & Hort. Trade World. See B–P–H 376/15. HI 54106

Florist's guide and cultivator's directory. London. Vols. 1-2, 1827-32. Florist's Guide Cultivator's Directory. HI 65280

Florists' guide and gardeners' and naturalists' calendar. London. 1850. Florists' Guide Gard. Naturalists' Calend. HI 65281

Florist's journal. London. 1840-47. Florist's J. The 6 vols for 1840-45 were re-issued in 1846 as The british florist. 2-1582-2. HI 65282

Florist's journal and gardener's record = Florist's journal. London.

Florist's magazine. A register of the newest and most beautiful varieties of florist's flowers. London. Florist's Mag. See B–P–H 376/17. HI 54107

Florists' news digest. New York, NY. Florists' News Digest. See B–P–H 376/18. HI 54108

Florists news letter. Storrs, CT. Vols. [1-2] 3(1-6) [forming nos. 1-19 of the whole series], 1964-67; nos. 20-29, 1968-69. Florists News Lett. Superseded by: Connecticut greenhouse newsletter. HI 65283

Florists' review; a weekly journal for florists, seedsmen and nurserymen. Chicago, IL. Florists' Rev. See B–P–H 376/21. HI 54111

Flower essence journal. Nevada City, CA. 1982+. Fl. Essence J. Preceded by: Flower essence quarterly. HI 65284

Flower essence quarterly. Nevada City, CA. 1980-81? Fl. Essence Quart. Superseded by: Flower essence journal. HI 65285

Flower and feather. Chattanooga, TN. Fl. & Feather. See B–P–H 372/7. HI 54046

Flower garden. New York, NY. Fl. Gard. See B–P–H 372/10. HI 54047

Flower and garden. Eastern edition. Kansas City, MO. Vol. 1+, 1957+. Fl. Gard. Mag., E. Ed. HI 65287

Flower and garden. Mid-America edition. Kansas City, MO. Vol. 16+, 1962+. Fl. Gard. Mag., Mid-Amer. Ed. Preceded by: Flower and garden magazine for mid-America. HI 65288

Flower and garden. Western edition. Kansas City, MO. Vol. 1+, 1956+. Fl. Gard., W. Ed. HI 65289

Flower and garden magazine for mid-America. Kansas City, MO. Vols. 1-15, 1957-61. Fl. Gard. Mag. Mid-Amer. Superseded by: Flower and garden. Mid-America edition. HI 65290

Flower grower. Albany, NY. Fl. Grower. See B–P–H 372/12. HI 54049

Flower lore. New York, NY. Fl. Lore. See B–P–H 372/15. HI 54051

Flower and nursery notes. ?-1952-69-? Fl. Nursery Notes. HI 65291

Flower and nursery report for commercial growers. Berkeley, CA. 1970+. Fl. Nursery Rep. Commercial Growers. HI 65292

Flower talks. New York, NY. Fl. Talks. See B–P–H 373/9. HI 54064

Flowering plants of Africa. Pretoria & Ashford. Fl. Pl. Africa. See B–P–H 372/22. HI 54056

Flowering plants of South Africa. London, Johannesburg & Cape town. Fl. Pl. South Africa. See B–P–H 372/25. HI 54057

Flowers. Warsaw = Kwiaty. Warsaw.

Flowers and gardens. Calcium, NY. Fl. & Gard. See B–P–H 372/11. HI 54048

Flugblatt, Abteilung für Pflanzenkrankheiten d. Kaiser Wilhelm-Instituts für Landwirtschaft in Bromberg. Posen. 1909-12. Flugbl. Abt. Pflanzenkrankh. Kaiser Wilhelm-Inst. Landw. Bromberg. HI 65293

Flugblatt, biologische Bundesanstalt für Land- und Forstwirtschaft. Brunswick. 1949+. Flugbl. Biol. Bundesanst. Land- Forstw. HI 65294

Flugblatt, biologisch-landwirtschaftliches Institut in Amani. Tanga, Dar-es-Salaam. 1909-14. Flugbl. Biol.-Landw. Inst. Amani. HI 65295

Flugblatt der biologische Reichsanstalt für Land- und Forstwirtschaft, Berlin. Berlin. 1919-45. Flugbl. Biol. Reichsanst. Land- Forstw. Berlin. Preceded by: Flugblatt, kaiserliche biologische Anstalt für Land- und Forstwirtschaft. Superseded by: Flugblatt der biologischen Zentralanstalt für Land- und Forstwirtschaft, Berlin. HI 65296

Flugblatt der biologischen Zentralanstalt Braunschweig. Stuttgart. 1948-54. Flugbl. Biol. Zentralanst. Braunschweig. Superseded by: Flugblätter der Bundesanstalt für Land- und Forstwirtschaft Braunschweig. HI 65297

Flugblatt der biologischen Zentralanstalt für Land- und Forstwirtschaft, Berlin. Berlin. Vol. 1+, 1951+. Flugbl. Biol. Zentralanst. Land-Forstw. Berlin. Preceded by: Flugblatt der biologische Reichsanstalt für Land- und Forstwirtschaft, Berlin. HI 65298

Flugblatt der Bundesanstalt für Land- und Forstwirtschaft Braunschweig. Stuttgart. 1955+. Flugbl. Bundesanst. Land- Forstw. Braunschweig. Preceded by: Flugblätter der biologischen Zentralanstalt Braunschweig. HI 65299

Flugblatt, Bundesanstalt für Pflanzenschutz. Vienna. No. 1+, 1945+. Flugbl. Bundesanst. Pflanzenschutz. HI 65300

Flugblatt, kaiserliche biologische Anstalt für Land- und Forstwirtschaft. Berlin. 1905-19. Flugbl. Kaiserl. Biol. Anst. Land- Forstw. Superseded by: Flugblatt der biologische Reichsanstalt für Land- und Forstwirtschaft, Berlin. HI 65301

Flugblatt der Pflanzenschutzstelle a. d. K. Landwirtschaftliches Akademie zu Bonn-Poppelsdorf. Bonn. 1915-21. Flugbl. Pflanzenschutzstelle Landwirtschaftliches Akad. Bonn-Poppelsdorf. HI 65302

Flycatcher. Hereford. No. 24+, 1975+. Flycatcher. Preceded by: Newsletter, Herefordshire and Radnorshire nature trust ltd. HI 65303

Flygblad från centralanstalten för försöksväsendet på jordbruksområdet. Stockholm. 1910-33. Flygblad Centralanst. Försöksväs. Jordbruksomr. Superseded by: Flygblad, statens växtskyddsanstalt. HI 65304

Flygblad, statens växtskyddsanstalt. Stockholm. 1933-57. Flygblad Statens Växtskyddsanst. Preceded by: Flygblad från centralanstalten för försöksväsendet på jordbruksområdet. HI 65305

Flygeskrift, statens plantepatologiske institut. Oslo. Vols. 1-38, 1943-45. Flygeskr. Statens Plantepatol. Inst. Superseded by: Statens plantevern flygeskrift. HI 65306

Flytrap news. Redfern, N.S.W. No. 1+, 1985+. Flytrap News. HI 65307

FNGA bulletin. Leesburg, FL = F N G A bulletin. Leesburg, FL.

Focus; on natural renewable resources. Moscow, ID. Vol. 1+, [1975]+. Focus (Moscow). HI 65308

Focus. Rochester, NY? Vol. 33+, 1962?+. Preceded by: Educational focus [not entered]. Focus (Rochester). HI 65309

Focus. Washington, DC. 1979+. Focus (Washington). HI 65310

Focus: biology. Guilford, CT. 1975+. Focus, Biol. HI 65311

Focus on phytochemical pesticides. Boca Raton, FL. Vol. 1+, 1988+. Focus Phytochem. Pestic. HI 74511

Föld és ember. Budapest. Föld & Ember. See B–P–H 377/3. HI 54112

Folder, Michigan agricultural experiment station. = Michigan agricultural experiment station. Folder. East Lansing, MI. Michigan Agric. Exp. Sta. Folder. See B–P–H 592/12.

Földmüvelési érdekeink. Budapest. Földmüv. Érdek. See B–P–H 377/7. HI 54113

Földrajzi értesitö. Budapest. Földr. Értes. See B–P–H 377/8. HI 54114

Földrajzi közlemények. Budapest. Földr. Közlem. See B–P–H 377/10. HI 54115

Földtani ertesitö. Budapest. Földt. Értes. See B–P–H 377/12. HI 54116

Földtani közlöny. Budapest. Földt. Közl. See B–P–H 377/13. HI 54117

Földtani szemle. Budapest. Földt. Szemle. See B–P–H 377/16. HI 54118

Folhas de gramíneas. São Paulo = Correio agrostologico. São Paulo.

Folia Baeriana. Tallinn. Vol. 1+, 1975+. Folia Baeriana.

HI 65313

Folia balcanica. Skoplje, Yugoslavia. Folia Balcan. See B–P–H 377/20. HI 54119

Folia biochemica et biologica graeca. Salonika, Greece. Folia Biochem. Biol. Graeca. See B–P–H 377/21. HI 54120

Folia biologica. Buenos Aires. Folia Biol. (Buenos Aires). See B–P–H 377/22. HI 54121

Folia biologica. Cracow. Folia Biol. (Cracow). See B–P–H 377/23. HI 54122

Folia biologica. Prague. Folia Biol. (Prague). See B–P–H 377/25. HI 54123

Folia biotheoretica. [Forms: Acta biotheoretica. Series B.] Leiden. Nos. 1-6, 1936-66. Folia Biotheor. 2-1586-2. HI 65314

Folia botanica, Cracow = Zeszyty naukowe uniwersitetu jagiellónskiego; prace botaniczne. Cracow.

Folia botanica miscellanea. Barcelona. Vol. 1+, 1979+. Folia Bot. Misc. HI 65315

Folia botanica, universitatis Lodziensis = Acta universitatis Lodziensis. Folia botanica. Łódź.

Folia clinica et biologica. São Paulo. Folia Clin. Biol. See B–P–H 377/30. HI 54124

Folia cryptogamica. Szeged, Hungary & Kolosvar [=Cluj, Rumania]. Folia Cryptog. See B–P–H 377/35. HI 54125

Folia cryptogamica estonica. Tartu. Vol. 1+, 1972+. Folia Cryptog. Estonica. HI 65316

Folia dendrologica. Bratislava. Vol. 1+, 1974+. Folia Dendrol. HI 65317

Folia facultatis scientiarum naturalium universitatis purkynianae brunensis. Brno, Czechoslovakia. Folia Fac. Sci. Nat. Univ. Purkynianae Brun. See B–P–H 378/1. HI 54126

Folia forestalia. Helsinki. Folia Forest. See B–P–H 378/2. HI 54127

Folia forestalia polonica. Ser. A. Warsaw. Folia Forest. Polon., Ser. A. See B–P–H 378/3. HI 54128

Folia forestalia polonica. Ser. B. Warsaw. Folia Forest. Polon., Ser. B. See B–P–H 378/4. HI 54129

Folia geobotanica et phytotaxonomica. Prague. Vol. 2+, 1967+. Folia Geobot. Phytotax. Preceded by: Folia geobotanica et phytotaxonomica bohemoslovaca. HI 65318

Folia geobotanica et phytotaxonomica bohemoslovaca. Prague. Vol. 1, 1966. Folia Geobot. Phytotax. Bohemoslov. Superseded by: Folia geobotanica et phytotaxonomica. HI 65319

Folia geographica. Series geographica physica. Cracow. Vol. 1+, 1967+. Folia Geogr., Ser. Geogr. Phys.

HI 65320

Folia histochemica et cytobiologica. Cracow. Vol. 22+, 1984+. Folia Histochem. Cytobiol. Preceded by: Folia histochemica et cytochemica. HI 65321

Folia histochemica et cytochemica. Cracow. Vols. 1-21, 1963-83; Suppl. 1, 1963. Folia Histochem. Cytochem. Superseded by: Folia histochemica et cytobiologica. HI 65322

Folia japonica pharmacologica = Folia pharmacologica japonica. Kyoto, Japan. Folia Pharmacol. Jap. See B–P–H 378/13.

Folia limnologica scandinavica. Copenhagen. Folia Limnol. Scand. See B–P–H 378/7. HI 54130

Folia medica orientalia. Sectio 1: Folia medicinae internae orientalia. Jerusalem. Folia Med. Orient., Sect. 1. See B–P–H 378/9. HI 54131

Folia mendeliana; papers relating to Mendel and to the early development of genetics. Brno. No. 1+, 1966+. Folia Mendeliana. HI 65323

Folia microbiologica. Delft. Folia Microbiol. (Delft). See B–P–H 378/10. HI 54132

Folia microbiologica. Prague. Folia Microbiol. (Prague). See B–P–H 378/11. HI 54133

Folia musei rerum naturalium Bohemiae occidentalis. Botanica. Pilsen. Vol. 1+, 1972+. Folia Mus. Rerum Nat. Bohemiae Occid., Bot. HI 65324

Folia pharmacologica japonica. Kyoto, Japan. Folia Pharmacol. Jap. See B–P–H 378/13. HI 54134

Folia prirodovedecke fakulta universita J. E. Purkyne. Brno. Vols. ?-2(2)-?, 19??-61-70? Folia Prir. Fak. Univ. J. E. Purkyne. HI 65325

Folia quaternaria. Cracow. Folia Quaternaria. See B–P–H 378/15. HI 54135

Folia savariensia = Vasi szemle. Szombathely, Hungary. Vasi Szemle. See B–P–H 947/7.

Folia universitaria. Cochabamba. Vols. 1-7, 1947-54. Folia Univ., Cochabamba. HI 65326

Foliage digest. Apopka, FL. Vol. 1+, 1978+. Foliage Digest. HI 65327

Folju ta taghrif u ahbarijiet. Zejtun. Vol. 1, 1954-59. Folju Taghrif Ahbarijiet. HI 65328

Folleto de divulgación. Santiago. Folleto Divulg. See B–P–H 378/20. HI 54136

Folletos de divulgación científica del instituto de biología. Mexico City. Folletos Divulg. Ci. Inst. Biol. See B–P–H 378/21. HI 54137

Folleto divulgativo, instituto nacional de investigaciones forestales = Boletín de divulgacion, instituto nacional de investigaciones forestales. Mexico, D.F.

Folleto de investigacion, centro internacional de

mejoramiento de maiz y trigo = Research bulletin international maize and wheat improvement center. Chapingo.

Folletos técnicos forestales. Buenos Aires. Folletos Técn. Forest. See B–P–H 378/22. HI 54138

Folletos, see Folleto

Fomento agrícola. São Paulo. Fomento Agríc. See B–P–H 378/23. HI 54139

Fondo nacional del café y del cacao. Boletín informativo. Caracas? Fondo Nac. Café Bol. Inform. See B–P–H 378/24. HI 54140

Fontes florae hungaricae. Kolozsvar [=Cluj, Rumania]. Fontes Fl. Hung. See B–P–H 378/25. HI 54141

Fontes musei reginaehradecensis. Hradac Kralové. Vol. 1+, 1965+. Fontes Mus. Reginaehrad. HI 65329

Fontes. Towarzystwa pryjaciól nauk na Ślasku. Katowice, Poland. Fontes Towarz. Przyjac. Nauk Ślasku. See B–P–H 378/26. HI 54142

Fontqueria. Madrid. Vol. 1+, 1982+. Fontqueria. HI 65330

Food biotechnology. London, New York. Vol. 1+, 198?+. Food Biotechnol. HI 65331

Food and cosmetics toxicology. London. Food Cosmet. Toxicol. See B–P–H 379/3. HI 54144

Food & flowers. Hong Kong. Food & Fl. See B–P–H 379/1. HI 54143

Food microbiology. Orlando, FL. Vol. 1+, 1984+. Food Microbiol. HI 65332

Food products, microscopy division, U S department of agriculture = Department of agriculture. Microscopy division. Food products. Washington, DC. Dept. Agric. Microscop. Div. Food Prod. See B–P–H 343/4.

Food research. Champaign, IL. Vols. 1-25, 1936-60. Food Res. Superseded by: Journal of food science. 2-1594-3. HI 65333

Food research institute studies. Stanford, CA. Vols. 1-7, 1960-67; vol. 14+, 1975+. Food Res. Inst. Stud. Preceded by: Food research institute studies in agricultural economics, trade, and development. HI 65334

Food research institute studies in agricultural economics, trade, and development. Stanford, CA. Vols. 8-13(3), 1968-74. Food Res. Inst. Stud. Agric. Econ. Trade Developm. Preceded and superseded by: Food research institute studies. HI 65335

For better delphiniums = Better delphiniums. San Rafael, CA. Better Delphiniums. See B–P–H 184/21.

For tre skilling. Ugeblad. Copenhagen. 1855-56. For Tre Skilling. Superseded by: I hjemmet. HI 65336

Forage crops leaflet, Oklahoma agricultural experiment station = Oklahoma agricultural experiment station forage crops leaflet. Stillwater, OK. Oklahoma Agric. Exp. Sta. Forage Crops Leafl.. See B–P–H 687/4.

Forage grasses circular, Nevada agricultural experiment station = Nevada agricultural experiment station. Forage grasses circular. Reno, NV. Nevada Agric. Exp. Sta. Forage Grasses Circ. See B–P–H 651/17.

Forage notes, forage crops division, Canada. Ottawa. Vols. 1-5(1), 1955-59. Forage Notes Forage Crops Div. Canada. Superseded by: Forage notes, research branch, department of agriculture, Canada. HI 54145

Forage notes, research branch, department of agriculture, Canada. Ottawa. Vol. 5(2)+, 1959+. Forage Notes Res. Branch Dep. Agric. Canada. Preceded by: Forage notes, forage crops division, Canada. HI 65337

Forage research. Hissar. Vol. 1+, 1975+. Forage Res. HI 65338

Förderungsdienst. Vienna. Förderungsdienst. See B–P–H 379/18. HI 54146

Foreign agriculture. Washington, DC. Foreign Agric. See B–P–H 379/19. HI 54147

Foreign agriculture circular. Grains. Washington, DC. Vol. 1+, 1980+. Foreign Agric. Circ., Grains. HI 65339

Foreign agriculture circular. Horticultural products. Washington, DC. No. FHORT 1-82+ 1982+. Foreign Agric. Circ., Hort. Prod. HI 65340

Foreign monthly review and continental literary journal. London. Foreign Monthly Rev. & Continental Lit. J. See B–P–H 379/20. HI 54148

Foreign quarterly review. London. Foreign Quart. Rev. See B–P–H 379/22. HI 54149

Foreign review, and continental miscellany. London. Foreign Rev. & Continental Misc. See B–P–H 379/23. HI 54150

Forest. [Hayashi.] Sapporo. ?-1988+. Forest (Sapporo). HI 73785

Forest aerial survey. [Shinrin kosoku.] Tokyo. 1956+. Forest Aerial Surv. HI 67098

Forest and bird. Wellington, N.Z. No. 31+, 1934+. Forest & Bird. Preceded by: Birds [not entered]. HI 65342

Forest bulletin – botany. Bangkok. Nos. 1-4, 1954-58. Forest Bull., Bot. HI 65343

Forest bulletin, Dehra Dun. Calcutta. Nos. 1-11, 1905-07; n.s. nos. 1-91, 1911-35. Forest Bull. Dehra Dun. Superseded by: Indian forest bulletin. HI 65344

Forest and conservation history. Durham, NC. Vol. 34+, 1990+. Forest Conservation Hist. HI 74837

Forest department bulletin, Zambia. Lusaka. No. 1+,

1965+. Forest Dep. Bull., Zambia. HI 65345

Forest department record. Karachi. 1958+. Forest Dept. Rec., Pakistan. HI 65346

Forest ecology and management. Amsterdam. Vol. 1+, 1976+. Forest Ecol. Managem. HI 65347

Forest and environment. [Shinrin ritchi.] Tokyo. 1959+. Forest & Environm. HI 67371

Forest focus. Perth, W.A. No. 1+, 1970+. Forest Focus. HI 65348

Forest history. St. Paul, MN, New Haven, CT, Santa Cruz, CA. Vols. 1-17, 1958-74. Forest Hist. Superseded by: Journal of forest history. HI 65349

Forest history cruiser. Santa Cruz, CA. Vol. 1(3)+, 1978+. Forest Hist. Cruiser. Preceded by: Forest history newsletter. HI 65350

Forest history newsletter. Santa Cruz, CA. Vol. 1(1-2), 1977. Forest Hist. Newslett. Superseded by: Forest history cruiser. HI 65351

Forest leaflets, forest department, United Provinces of Agra and Oudh. Allahabad. 1934-? Forest Leafl. Forest Dep., Unit. Prov. Agra Oudh. HI 65352

Forest leaflets, forest research institute, Dehra Dun. Calcutta. 1908-10, 1941. Forest Leafl. Forest Res. Inst. Dehra Dun. Superseded by: Leaflet, forest research institute, Dehra Dun. HI 65353

Forest leaves. Mechanicsburg, PA. Vols. 1(1)-35(345), 1886-1951. Forest Leaves Superseded by: Pennsylvania forests. 4-3299-1. HI 65354

Forest management bulletin, United States forest service. Atlanta, GA. 1971+. Forest Managem. Bull., U.S. Forest Serv. HI 65188

Forest memoir, Pakistan forest department. Karachi. 1953. Forest Mem. Pakistan Forest Dept. HI 65355

Forest newsletter. Ndola. ?-1971+. Forest Newslett., Zambia. HI 65356

Forest and outdoors. Kingston, Ont. Vols. 33-55, 1937-59. Forest & Outdoors. Preceded by: Illustrated canadian forest and outdoors. HI 65357

Forest pamphlets, Dehra Dun. Botany series. Dehra Dun. 1910. Forest Pam. Dehra Dun, Bot. Ser. HI 65358

Forest pamphlets, Dehra Dun. Sylvicultural series. Dehra Dun. 1910. Forest Pam Dehra Dun, Sylvic. Ser. HI 65359

Forest planning; journal of the nationwide forest planning clearinghouse. Eugene, OR. Vol. 1+, 1980+. Forest Planning. HI 65360

Forest products abstracts. Farnham Royal. Vol. 1+, 1978+. Forest Prod. Abstr. HI 65361

Forest products journal. Madison, WI. Vol. 5+, 1955+. Forest Prod. J. Preceded by: Journal of the forest products research society. HI 65362

Forest products research. London. 1939+. Forest Prod. Res. Preceded by: Report of the forest products research board. HI 65363

Forest products research bulletin. London. Nos. 1-25, 1928-52. Forest Prod. Res. Bull. HI 65364

Forest products research notes. Wellington, N.Z. Vol. 1, nos. 1-12, 1950-54. Forest Prod. Res. Notes. Superseded by: New Zealand forestry research notes. HI 65365

Forest products research records. London. Nos. 1-30, 1935-40. Forest Prod. Res. Rec. HI 65366

Forest products research reports. Lagos. No. 1+, 1966+. Forest Prod. Res. Rep. HI 65367

Forest record. London. No. 1+, 1949+. HI 72383

Forest recreation research. Tokyo. No. 1+, 1977+. Forest Recreation Res. HI 72384

Forest report, Bombay = Administration report (Annual), forest department, Bombay.

Forest research bulletin. Bogor = Buletin penelitian hutan. Bogor.

Forest research bulletin, division of forest research, Zambia. Lusaka. Vol. 1+, 1967+. Forest Res. Bull. Div Forest Res., Zambia. HI 72385

Forest research in India. Delhi. 1907+ [in 1939 divided into two parts: Part 1, forest research institute; Part 2, Reports for Burma and indian states]. Forest Res. India. HI 72386

Forest research information paper, ministry of natural resources, Ontario. Toronto. No. ?-101+, 19??-77+. Forest Res. Inform. Pap. Minist. Nat. Resources, Ontario. HI 72387

Forest research notes, California forest and range experiment station. Berkeley, CA. Nos. 1-147, 1930-59? Forest Res. Notes Calif Forest Range Exp. Sta. Superseded by: Research notes, Pacific southwest forest range experiment station. HI 72388

Forest research notes, northeastern forest experiment station. Upper Darby, PA. Nos. 22-135?, 1950-62. Forest Res. Notes NorthE. Forest Exp. Sta. Preceded by: Northeastern research notes, northeastern forest experiment station. Superseded by: Research notes N E, United States forest service. HI 72389

Forest research notes, Pacific northwest forest and range experiment station. Portland, OR. Nos. 1-45, 1928-48. Forest Res. Notes Pacific NorthW. Forest Range Exp. Sta. Superseded by: Research notes, Pacific northwest forest and range experiment station. HI 72390

Forest research notes, Pacific southwest forest and range experiment station = Research notes, Pacific southwest forest and range experiment station. Berkeley, CA.

Forest research notes. Wellington, N.Z. Vols. 1(1-12), 1950-54. Forest Res. Notes. Superseded by: New Zealand forestry research notes. HI 72391

Forest research review, forest service, British Columbia. Victoria, B.C. 1956+. Forest Res. Rev. Forest Serv. Brit. Columbia. HI 72392

Forest research series, Clemson agricultural college. Department of forestry = Clemson agricultural college. Department of forestry, forest research series. Clemson, SC. Clemson Agric. Coll. Dept. Forest., Forest Res. Ser. See B–P–H 314/32.

Forest research in the southeast. Asheville, NC. 1976+. Forest Res. S. E. Preceded by: Research information digest. HI 72393

Forest researches. Budapest = Erdészeti kutatások. Budapest. Erdész. Kutatás. See B–P–H 360/2.

Forest resource report, United States forest service. Washington, DC. No. 1+, 1950+. Forest Resource Rep. U. S. Forest Serv. HI 72394

Forest resources. Canberra, A.C.T. 1975, 1977. Forest Resources. Preceded by: Report (Annual), forestry and timber bureau. Australia. Superseded by: Australian forest resources. HI 72395

Forest revue. Zagreb = Šumarski list. Zagreb.

Forest scene. Toronto. Vol. 1+, 1970+. Forest Scene. HI 72396

Forest science. Sofia = Gorskostopanska nauka. Sofia.

Forest science. Washington, DC. Forest Sci. See B–P–H 380/22. HI 54165

Forest science. Monograph. Washington, DC. Forest Sci. Monogr. See B–P–H 380/24. HI 54166

Forest science research notes. College Station, TX. No. 1+, 1972+. Forest Sci. Res. Notes. HI 72397

Forest science and technology. [Linye keji tongxun.] Forestry technical newsletter. Beijing. 1980+. Forest Sci Technol. HI 72398

Forest service bulletin, U S department of agriculture = Bulletin, United States forest service. Washington, DC.

Forest service circular, Canada = Circular, forestry branch, Canada. Ottawa.

Forest service research paper I T F. Rio Piedras, PR = Research papers I T F, United States forest service. Rio Piedras, PR.

Forest service research paper, institute of tropical forestry. Rio Pedras. (Puerto Rico) = Research papers I T F, United States forest service. Rio Piedras, PR.

Forest society journal = Journal, Oxford university forest society. Oxford.

Forest soils of Japan. Tokyo. Forest Soils Japan. See B–P–H 380/25. HI 54167

Forest survey notes, British Columbia forest service. Victoria, B.C. No. 1+, 1957+. Forest Surv. Notes Brit. Columbia Forest Serv. HI 72399

Forest survey release, California forest and range experiment station. Berkeley, CA. Nos. 1-37, 1939-60. Forest Surv. Release Calif. Forest Range Exp. St. HI 72400

Forest survey release, central states forest experiment station. Columbus, OH. No. 1+, 1948+. Forest Surv. Release Centr. States Forest Exp. Sta. HI 72401

Forest survey release, intermountain forest and range experiment station. Ogden, UT. Nos. 1-5, 1958-62. Forest Surv. Release Intermount. Forest Range Exp. Sta. Preceded by: Forest survey release, northern Rocky Mountain forest and range experiment station. Superseded by: Resource bulletin I N T, United States forest service. HI 72402

Forest survey release, northern Rocky Mountain forest and range experiment station. Missoula, MT. Nos. 1-21, 1936-43. Forest Surv. Release N. Rocky Mountain Forest Range Exp. Sta. Superseded by: Forest survey release, intermountain forest and range experiment station. HI 72403

Forest survey release, Rocky Mountain forest and range experiment station. Fort Collins, CO. Nos. 1-4, 1959-61. Forest Surv. Release Rocky Mountain Forest Range Exp. Sta. HI 72404

Forest survey release, southeastern forest experiment station. Asheville, NC. No. 1+, 1939+. Forest Surv. Release S. E. Forest Exp. Sta. HI 72405

Forest survey release, southern forest experiment station. New Orleans, LA. Nos. 1-87, 193?-62 Forest Surv. Release S. Forest Exp. Sta. Superseded by: Resource bulletin S O, United States forest service. HI 72406

Forest survey report, Pacific northwest forest and range experiment station. Portland, OR. Nos. 1-146, 1934-62. Forest Surv. Rep. Pacific NorthW. Forest Range Exp. Sta. Superseded by: Resource bulletin P N W, United States forest service. HI 72407

Forest technique. [Ringyo gijitsu.] Tokyo. Forest Techn. See B–P–H 381/1. HI 54169

Forest and timber. Sydney, N.S.W. Vol. 1+, 1963+. Forest Timber. HI 72408

Forest times. Truro, N.S. Vol. 1+, 1979+. Forest Times. HI 72409

Forest treasures of Cyprus. Nicosia. Vols. 1-13, 1932-60. Forest Treas. Cyprus. HI 72410

Forest tree breeding. [Rinboku no ikushu.] Tokyo. No. ?-43+, ?-1967+. Forest Tree Breed. HI 72411

Forest tree improvement. Copenhagen. No. 1+, 1970+. Forest Tree Improv. HI 72412

Forest tree series, australian forestry and timber bureau. Canberra, A.C.T. No. 1+, 1970+. Forest Tree Ser. Austral. Forest. Timber Bur. HI 72413

Forest trees and timbers of the british empire. Oxford, England. Forest Trees Timbers Brit. Empire. See B–P–H 381/3. HI 54170

Forestal. Caracas. Forestal. See B–P–H 381/4. HI 54171

Forester. Belfast. Forester (Belfast). See B–P–H 381/5. HI 54172

Forester. Port Huron, MI. Forester (Port Huron). See B–P–H 381/6. HI 54173

Forester. Washington, DC. Vols. 1(4)-7, 1895-1903. Forester (Washington). Preceded by: New Jersey forester. Superseded by: Forestry and irrigation. HI 72414

Forestry. London. Forestry (London). See B–P–H 381/7. HI 54174

Forestry. Oxford, England. Forestry (Oxford). See B–P–H 381/8. HI 54175

Forestry. [Hsüeh Lin.] Shanghai. Forestry (Shanghai). See B–P–H 381/9. HI 54176

Forestry. [Sanrin.] Tokyo. 1882+. Forestry (Tokyo). HI 67979

Forestry abstracts. Farnham Royal, England. Forest. Abstr. See B–P–H 379/24. HI 54151

Forestry and british timber. Bearsden. Vol. 5(3)+, 1976+. Forest. Brit. Timber. Preceded by: Forestry and home grown timber. HI 72415

Forestry bulletin, department of agriculture and forestry, central forestry bureau, China. [Lin-yeh t'ung-hsun.] Nanking. Forest. Bull. (Nanking). See B–P–H 379/26. HI 54152

Forestry bulletin, department of agriculture and forestry, Palestine. Jerusalem. No. 1, 1934. Forest. Bull. Dep. Agric. Forest., Palestine. HI 72416

Forestry bulletin, department of public lands, Queensland. Brisbane, Qld. Nos. 1-3, 1917-2? Forest. Bull. Dep. Public Lands, Queensland. Superseded by: Forestry bulletin, forest service, Queensland. HI 72417

Forestry bulletin, faculty of forestry, university of British Columbia. Vancouver, B.C. Nos. 1-4, 1951-62. Forest. Bull. Fac. Forest. Univ. Brit. Columbia. Superseded by: Bulletin, faculty of forestry, university of British Columbia. HI 72418

Forestry bulletin, faculty of forestry, university of Toronto. Toronto. Nos. 1-6, 1951-60. Forest. Bull. Fac. Forest. Univ. Toronto. HI 72419

Forestry bulletin, forest department, British Guiana. Georgetown. N.s. no. 1+, 1948+. Forest. Bull. Forest Dep. Brit. Guiana. HI 72420

Forestry bulletin, forest department, Jamaica. Kingston, Jamaica. No. 1+, 1945+. Forest. Bull. Forest Dep., Jamaica. HI 72421

Forestry bulletin, forest service, Queensland. Brisbane, Qld. Nos. 4-7, 192?-2? Forest. Bull. Forest Serv., Queensland. Preceded by: Forestry bulletin, department of public lands, Queensland. Superseded by: Bulletin, forest service, Queensland. HI 72422

Forestry bulletin, forestry experiment station, college of agriculture, national Taiwan university. [Lin yeh chon k'an.] Taipei, Taiwan. Forest. Bull. (Taipei). See B–P–H 379/27. HI 54153

Forestry bulletin R 8 - F B / U, United States forest service, southern region. Atlanta, GA. No. 1+, 1983+. Forest. Bull. R 8 - F B / U. Preceded by: Forestry bulletin S A - F B / U. HI 65216

Forestry bulletin S A - F B / U. Atlanta, GA. Nos. 1-?, 1981-83. Forest. Bull. S. A. - F B / U. Superseded by: Forestry bulletin R 8 - F B / U, United States forest service, southern region. HI 75215

Forestry bulletin of the school of forestry of Duke university. Durham, NC. Forest. Bull. School Forest. Duke Univ. See B–P–H 379/29. HI 54154

Forestry chronicle. Toronto. Forest. Chron. See B–P–H 380/1. HI 54155

Forestry circular, Illinois natural history survey. Urbana, IL. Nos. 1-5, 1920-26. Forest. Circ. Illinois Nat. Hist. Surv. Superseded by: Circular, Illinois natural history survey. HI 72423

Forestry circular, Ohio agricultural experiment station = Ohio agricultural experiment station forestry circular. Wooster, OH. Ohio Agric. Exp. Sta. Forest. Circ. See B–P–H 684/24.

Forestry commission booklet = Booklet, forestry commission. London.

Forestry commission bulletin = Bulletin of the forestry commission. London.

Forestry commission forest record = Forest record. London.

Forestry commission, New South Wales. Technical paper. Sydney, N.S.W. Forest. Commiss. New South Wales Techn. Pap. See B–P–H 380/4. HI 54156

Forestry commission research and development papers = Research and development papers, forestry commission. London, Edinburgh.

Forestry commission research information note = Research information note, forestry commission research and development division. Farnham.

Forestry current literature. Washington, DC. 1930-40. Forest. Curr. Lit. Incorporated in: Bibliography of agriculture. HI 72424

Forestry departmental series, Alabama agricultural experiment station = Alabama agricultural experiment station. Forestry departmental series. Auburn, AL. Alabama Agric. Exp. Sta., Forest. Dept. Ser. See B–P–H 67/17.

Forestry development paper, F A O = F A O forestry development paper. Rome.

Forestry digest. Washington, DC. Forest. Digest. See B–P–H 380/6. HI 54157

Forestry division bulletin, United States department of agriculture = U S department of agriculture, forestry division bulletin. Washington, DC. U.S.D.A. Forest. Div. Bull. See B–P–H 942/16.

Forestry-geological review. Atlanta, GA. Forest.-Geol. Rev. See B–P–H 380/8. HI 54158

Forestry and home grown timber. London. Vols. 1-5(2), 1972-76. Forest. Home Grown Timber. Preceded by: Forestry and home-grown timber (1964-71). Superseded by: Forestry and british timber. HI 72425

Forestry in Indonesia. Bogor. Vol. 1+, 1974+. Forest. Indonesia. HI 72426

Forestry industry monthly. [Lin ch'an yüeh k'an.] Taipei, Taiwan. Forest. Industr. Monthly. See B–P–H 380/11. HI 54159

Forestry industry news. [Lin ch'an tung hsin.] Taipei, Taiwan. Forest. Industr. News. See B–P–H 380/12. HI 54160

Forestry information, forest research institute. Nanking. 1944-45. Forest. Inform. Forest Res. Inst., Nanking. HI 72427

Forestry investigation, department of agriculture and forestry, department of agriculture and forestry, China. [Lin-hsun.] Chungking, China. Forest. Invest. See B–P–H 380/13. HI 54161

Forestry and irrigation. Washington, DC. Vols. 8-14(8), 1903-09. Forest. Irrig. Preceded by: Forester (Washington). Superseded by: Conservation. HI 72428

Forestry leaves. Santa Cruz?, Philippines. Forest. Leaves. See B–P–H 380/14. HI 54162

Forestry memoirs, forests department, Sudan. Khartoum. Nos. 1-10, 1950-57. Forest. Mem. Forests Dep. Sudan. HI 72429

Forestry news. Washington, DC. Forest. News. See B–P–H 380/15. HI 54163

Forestry newsletter, Great Lakes forest research centre. Sault Ste Marie. 1984+. Forest. Newslett. Great Lakes Forest Res. Centre. Preceded by: Forestry research newsletter, Great Lakes forest research centre. HI 72430

Forestry note, Illinois university agricultural experiment station. Urbana, IL. Nos. 1-126, 1948-69. Forest. Note Illinois Univ. Agric. Exp. Sta. Superseded by: Forestry research report, Illinois university agricultural experiment station. HI 72431

Forestry occasional paper, F A O = F A O forestry occasional paper. Rome.

Forestry pamphlet, Trinidad and Tobago. Port-of-Spain. Nos. 1-5, 1934-37. Forest. Pam. Trinidad Tobago. HI 72432

Forestry publication, Connecticut agricultural experiment station = Connecticut agricultural experiment station. Forestry publication. New Haven, CT. Connecticut Agric. Exp. Sta. Forest. Publ. See B–P–H 325/2.

Forestry publication, Ohio agricultural experiment station = Ohio agricultural experiment station forestry publication. Wooster, OH. Ohio Agric. Exp. Sta. Forest. Publ. See B–P–H 684/25.

Forestry quarterly. Ithaca, NY. Forest. Quart. See B–P–H 380/17. HI 54164

Forestry report, forestry service, Canada. Ottawa. No. 1+, 1971+. Forest. Rep. Forest. Serv. Canada. HI 72433

Forestry report, northern forest research centre. Edmonton, Alta. Vol. 1+, 1972+. Forest. Rep. N. Forest Res. Centre. HI 72434

Forestry report R 8 - F R, United States forest service, southern region. Atlanta, GA. No. 1+, 1983+. Forest. Rep. R 8 - F R. Preceded by: Forestry report S A - F R. HI 65450

Forestry report S A - F R. Atlanta, GA. Nos. 1-?, 1978-83? Forest. Rep. S A - F R. Superseded by: Forestry report R 8 - F R, United States forest service, southern region. HI 75216

Forestry research highlights, Rocky Mountains forest and range experiment station. Fort Collins, CO. 1968+. Forest. Res. Highlights Rocky Mountains Forest Range Exp. Sta. Preceded by: Report (Annual), Rocky Mountains forest and range experiment station. HI 72435

Forestry research newsletter. Perth, W.A. No. ?-3+, 19??-76+. Forest. Res. Newslett. HI 72436

Forestry research newsletter, Great Lakes forest research centre. Sault Ste Marie. Vol. 2(4)-5(4) [issues for Spring 1976 onwards were not numbered], 1972-83. Forest. Res. Newslett. Great Lakes Forest Res. Centre. Preceded by: Research newsletter, Great Lakes forest research centre. Superseded by: Forestry newsletter, Great Lakes forest research centre. HI 72437

Forestry research notes, forestry department, university of Wisconsin = Forestry research notes, Wisconsin university college of agriculture and conservation department. Madison, WI.

Forestry research notes, forestry and timber bureau, Australia. Canberra, A.C.T. Vols. 1-2, 1960-63. Forest. Res. Notes Forest. Timber Bur. Australia. Preceded by: Australian forest research. HI 65798

Forestry research notes, Wisconsin university college of agriculture and conservation department. Madison, WI. No. 1+, 1953+. Forest. Res. Notes. Wisconsin Univ. Coll. Agric. Conservation Dep. HI 72438

Forestry research report, Illinois university agricultural experiment station. Urbana, IL. No. 70/1-72/4, 1970-72. Forest. Res. Rep. Illinois Univ. Agric. Exp. Sta. Preceded by: Forestry note, Illinois university agricultural experiment station. HI 72439

Forestry research west. Fort Collins, CO. 1979+. Forest. Res. W. HI 72440

Forestry science. Peking (Beijing) = Scientia silvae. Peking. Sci. Silvae. See B–P–H 829/11.

Forestry in South Africa. Pretoria. Nos. 1-17, 1961-75. Forest. South Africa. HI 54168

Forestry technical notes, east african agriculture and forestry research organization. Miguyu. No. 1+, 1953+. Forest. Techn. Notes E. African Agric. Forest. Res. Organ. HI 72442

Forestry technical papers, forests commission, Victoria. Melbourne, Vic. No. 1, 1959+. Forest. Techn. Papers Forests Commiss. Victoria. Preceded by: Plantation technical papers, Victoria forests commission. HI 72443

Forestry technical publications, canadian forestry service. Ottawa. No. 27+, 1978+. Forest. Techn. Pub. Canad. Forest. Serv. Preceded by: Forestry technical report, canadian forestry service. HI 72444

Forestry technical report, canadian forestry service. Ottawa. Nos. 1-26, 1974-78. Forest. Techn. Rep. Canad. Forest. Serv. Superseded by: Forestry technical publication, canadian forestry service. HI 72445

Forestry topic series, forestry branch, Canada. Ottawa. Nos. 1-6, 1924-29. Forest. Topic Ser. Forest. Branch Canada. HI 72446

Forests bulletin, forestry division, Sudan. Nos. 1-7, 1958-64. Forests Bull. Forest Div., Sudan. Continued as: Bulletin, forests department, Sudan. HI 72447

Forests of Manitoba. Winnipeg. 1974, 1975. Forests Manitoba. HI 72448

Forests and people. Alexandria, LA. 1951+. Forests People. HI 72449

Forêt. Neuchâtel. Forêt. See B–P–H 381/12. HI 54177

Forêt conservation. Quebec. Vol. 18+, 1952+. Forêt Conservation. Preceded by: Forêt et conservation. HI 72450

Forêt et conservation. Quebec. Vols. n.s. 1-3, 1949-51. Forêt & Conservation. Preceded by: Forêt québecoise. Superseded by: Forêt conservation. HI 72451

Forêt méditerranéenne. Aix-en-Provence. Vol. 1+, 1979+. Forêt Médit. HI 72452

Forêt privée, La. Paris. No. 114+, 1977+. Forêt Privée. Preceded by: Forêt privée française et revue forestière européenne. HI 72453

Forêt privée française et revue forestière européenne. Paris. Nos. ?-27-113, 19??-62-76. Forêt Privée Franç. Rev. Forest. Eur. Superseded by: Forêt privée. HI 72454

Forêt québecoise. Quebec. Vols. 1-14(5), 1939-49. Forêt Québec. Superseded by: Forêt et conservation. HI 72455

Förhandlingar; biologiska föreningen i Stockholm. Stockholm. Förh. Biol. Fören. Stockholm. See B–P–H 381/14. HI 54178

Förhandlingar. Botaniska sällskapet i Stockholm. Stockholm. Förh. Bot. Sällsk. Stockholm. See B–P–H 381/15. HI 54179

Förhandlingar; geologiska föreningen i Stockholm. Stockholm. Förh. Geol. Fören. Stockholm. See B–P–H 381/16. HI 54180

Förhandlingar; k. fysiografiska sällskapet. Lund. Förh. Kungl. Fysiogr. Sällsk. See B–P–H 381/17. HI 54181

Förhandlingar, skandinaviske naturforskeres möte. Goteborg, etc. Vols. 1-18, 1839-1929. Förh. Skand. Naturf. Möte. Superseded by: Beretning om det skandinaviske naturforskermøde. HI 72456

Forhandlinger i videnskabs-selskabet i Christiana = Forhandlinger i videnskabs-selskabet i Kristiania. Christiania.

Forhandlinger, det k. norske videnskabers selskabs. Trondheim = Det k. norske videnskabers selskabs forhandlinger. Trondheim. Kongel. Norske Vidensk. Selsk. Forh. See B–P–H 516/17.

Forhandlinger i videnskabs-selskabet i Kristiania. Christiania. 1858-1924. Forh. Vidensk.-Selsk. Kristiania. Superseded by: Norske videnskaps-akademi, matematisk-naturvidenskapelig klasse. Avhandlinger. HI 72457

Forma et functio; an international journal of functional biology. Brunswick, Oxford. Vols. 1-8, 1969-75. Forma & Functio. HI 72458

Formosa journal of aquatic products. [Taiwan suisan zasshi.] Taihoku [=Taipei, Taiwan]. Formosa J. Aquatic Prod. See B–P–H 381/21. HI 54182

Formosan agricultural review. [Taiwan nojiho.] Taihoku [=Taipei, Taiwan]. Formosan Agric. Rev. See

B–P–H 381/24. HI 54183

Formosan science. [Taiwan k'o hsüeh.] Taipei, Taiwan. Formosan Sci. See B–P–H 381/25. HI 54184

Forpride digest; journal for forest products research and industries development. Laguna. Vols. 1-11, 1972-82? Forpride Digest. Superseded by: F P R D I journal. HI 72459

Forscher. Cuxhaven, Germany. Forscher (Cuxhaven). See B–P–H 382/6. HI 54188

Forscher. Hanover. Forscher (Hanover). See B–P–H 382/7. HI 54189

Forschung Wissenschaftsberichte aus der Universität Hamburg. Hamburg. ?-1977+. Forsch. Wissenschaftsber. Univ. Hamburg. HI 72460

Forschungen und Berichte zu Naturschutz und Landschaftspflege. Beiheft zur Schriftenreihe Naturschutz und Landschaftspflege in Niedersachsen. Hannover. Vol. 1+, 1981+. Forschung. Ber. Naturschutz Landschaftspflege, Beih. HI 72461

Forschungen und Fortschritte; Korrespondenzblatt der deutschen Wissenschaft und Technik (later Nachrichtenblatt der deutschen Wissenschaft und Technik). Berlin. Vol. 1-5, 1925-39, 1963-67. Forsch. & Fortschr. HI 72462

Forschungen auf dem Gebiet der Pflanzenkrankheiten. Kyoto = Plant disease research. [Shokubutsu byogai kenkyu.] Kyoto.

Forschungen auf dem Gebiet der Pflanzenkrankheiten und der Immunität im Pflanzenreich. Jena. Forsch. Pflanzenkrankh. Immunität Pflanzenr. See B–P–H 382/5. HI 54187

Forschungen der Geographischen Gesellschaft und des Naturhistorischen Museums in Lübeck. Lübeck. Forsch. Geogr. Ges. Naturhist. Mus. Lübeck. See B–P–H 382/3. HI 54185

Forschungen auf dem Gesamtgebiet der Angewandten Botanik = Botanica oeconomica. Hamburg. Bot. Oecon. See B–P–H 224/7.

Forschungsberichte aus der biologischen Station zu Plön. Berlin. Forschungsber. Biol. Stat. Plön. See B–P–H 382/8. HI 54190

Forschungsberichte, forstliche Forschungsanstalt München. Munich. No. 1+, 1971+. Forschungsber. Forstl. Forschungsanst. München. Superseded by: Forstliche Forschungsberichte München. HI 72463

Forschungsdienst. Neudamm [=Debno, Poland]. Vols. 1-18, 1936-45. Forschungsdienst. Preceded by: Deutsche landwirtschaftliche Rundschau. HI 72464

Forst-Archiv zur Erweiterung der Forst- und Jagd-Wissenschaft und der Forst- und Jagd-Literatur. Ulm. Forst-Arch. Erweit. Forst- Jagd-Wiss. See B–P–H 382/13. HI 54194

Forst-Calender. Leipzig. Forst-Calender. See B–P–H 382/15. HI 54196

Forst und Holz. Hannover. Vols. 1-8, 1946-53; vol. 43+, 1988+. Forst & Holz. Superseded by: Forst und Holzwirt. HI 72465

Forst- und Holzwirt. Hannover. Vols. 1-42, 1954-87. Forst- Holzwirt. Preceded and superseded by: Forst und Holz. HI 72466

Forst und Jagd. Berlin. Vols. 4-11, 1954-61. Forst & Jagd. Preceded by: Wald. Superseded by: Sozialistische Forstwirtschaft. HI 72467

Forst- und Jagd-Archiv von und für Preussen. Berlin. Forst- Jagd-Arch. Preussen. See B–P–H 382/10. HI 54191

Forst- und Jagdbibliothek. Stuttgart. Forst- Jagdbiblioth. See B–P–H 382/11. HI 54192

Forst- und Jagdzeitung. Saaz, Bohemia [=Zatec, Czechoslovakia]. Forst- Jagdzeitung. See B–P–H 382/12. HI 54193

Forst-Journal. [Edited by C. F. Medikus.] Leipzig. Forst-J. See B–P–H 382/16. HI 54197

Forst-tidende; Tidsskrift for skovbrug, mosekultur og jagt [Subtitle varies.] Aakirkeby, Copenhagen. Vols. [1]-14, 1888-1902. Forst-tidende. HI 72468

Forstarchiv. Zeitschrift für wissenschaftlichen und technischen Fortschritt in der Forstwirtschaft. Hanover. Forstarchiv. See B–P–H 382/14. HI 54195

Forstlich-naturwissenschaftliche Zeitschrift. Munich. Forstl.-Naturwiss. Z. See B–P–H 383/1. HI 54203

Forstliche Berichte und Miscellen. Tübingen. Forstl. Ber. Misc. See B–P–H 382/17. HI 54198

Forstliche Bibliographie der Internationalen Forstzentrale = Bibliographia forestalis. Berlin. Berliogr. Forest. See B–P–H 187/2.

Forstliche Blätter für Württemberg. Tübingen. Forstl. Blätt. Württemberg. See B–P–H 382/18. HI 54199

Forstliche Forschungsberichte München. Munich. No. 46+, 1980+. Forstl. Forschungsber. München. Preceded by: Forschungsberichte, forstliche Forschungsanstalt München. HI 72469

Forstliche Mitteilungen. Stuttgart. Forstl. Mitt. See B–P–H 382/19. HI 54200

Forstliche Mitteilungen. Wiesbaden. Vol. 1+, 1948+. Forstl. Mitt. (Wiesbaden). HI 72470

Forstliche Mittheilungen. Munich. Vols. 2(1-3), 1854-56. Forstl. Mitth. (Munich). Preceded by: Forstwirtschaftliche Mittheilungen. Superseded by: Mittheilungen über das Forst- und Jagdwesen in Bayern. HI 65368

Forstliche Mittheilungen; eine Zeitschrift in zwanglosen Folgen. Pesth [= Budapest]. Vols. 1-?, 1835-? Forstl. Mitth. (Pesth). HI 65369

Forstliche Rundschau der Zeitschrift für Weltforstwirtschaft. Naudamm. Nos. 1-16, 1928-44. Forstl. Rundschau Z. Weltforstw. Preceded by: Forstlicher Jahresbericht. Superseded by: Forstliche Umschau. HI 65370

Forstliche Umschau; Referate über das forst- und holzwirtschaftliche Schrifttum. Hamburg, New York. Vol. 1+, 1958+. Forstl. Umschau. Preceded by: Forstliche Rundschau der Zeitschrift für Weltforstwirtschaft. HI 65371

Forstliche Versuche. Budapest = Erdészeti kutatások. Budapest. Erdész. Kutatás. See B–P–H 360/2.

Forstliche Wochenschrift "Silva." Tübingen. Forstl. Wochenschr. "Silva". See B–P–H 382/21. HI 54201

Forstliche Zeitschrift für das Grossherzogthum Baden. Karlsruhe. Forstl. Z. Grossherzogth. Baden. See B–P–H 382/22. HI 54202

Forstlicher Jahresbericht. [Supplement to: Allgemeine Forst- und Jagd-Zeitung.] Frankfurt am Main. 1924-26. Forstl. Jahresber. Superseded by: Forstliche Rundschau der Zeitschrift für Weltforstwirtschaft. HI 65372

Forstlige forsøgvaesen i Danmark. Beretninger utgivne ved den forstlige forsgskommission. Copenhagen. Vol. 1+, 1905+. Forstl. Forsøgvaesen Danmark. HI 65373

Forstpflanzen, Forstsamen; Zentralblatt für Forstpflanzen und Forstkulturen. Strassenhaus über Neuwied. Vol. 1+, 1961+. Forstpflanzen Forstsamen. HI 65374

Forstrevue. Zagreb = Šumarski list. Zagreb.

Forstwirtschaft, Holzwirtschaft. Berlin. Vols. 1-5, 1947-51. Forstw. Holzw. Superseded by: Wald. HI 65375

Forstwissenschaftliche Forschungen. [Beihefte zum Forstwissenschaftlichen Centralblatt.] Berlin, Hamburg. Vol. 1+, 1952+. Forstwiss. Forsch. HI 65376

Forstwirtschaftliche Mittheilungen. Munich. Vol. 1(1-4), 1852. Forstw. Mitth. Superseded by: Forstliche Mittheilungen. HI 65377

Forstwirtschaftliches Jahrbuch. Dresden. Forstw. Jahrb. See B–P–H 383/3. HI 54204

Forstwissenschaftliche Mitteilungen. Sopron = Erdészettudományi közlemények. Sopron, Hungary. Erdészettud. Közlem. See B–P–H 360/11.

Forstwissenschaftliches Centralablatt. Berlin. Forstwiss. Centralbl. See B–P–H 383/5. HI 54205

Forstwissenschaftliches Zentralblatt. Berlin. Forstwiss. Zentralbl. See B–P–H 383/6. HI 54206

Fort Hare papers. Alice, Cape Province. Vol. 1+, 1945+. Fort Hare Pap. HI 65378

Fortgesezte Beiträge zur Naturkunde. Berlin. Fortgesezte Beitr. Naturk. See B–P–H 383/10. HI 54208

Fortgesetzte Nachrichten von dem Zustande der Wissenschaften und Künste in den Königlich Dänischen Reichen und Ländern. Copenhagen & Leipzig. Fortgesetzte Nachr. Wiss. Künste Königl. Dän. Reichen Ländern. See B–P–H 383/9. HI 54207

Forth naturalist and historian. Stirling. 1977+. Forth Naturalist & Historian. HI 65379

Fortschritte im Acker- und Pflanzenbau. [Supplement to: Zeitschrift für Acker- und Pflanzenbau.] Berlin, etc. No. 1+, 1973+. Fortsch. Acker- Pflanzenbau. HI 65380

Fortschritte der Botanik; Anatomie (later Morphologie), Physiologie, Genetik, Systematik, Geobotanik.] Berlin. Vols. 1-35, 1932-73. Fortschr. Bot. (Berlin). Superseded by: Progress in botany. HI 65381

Fortschritte der Botanik. Cologne & Leipzig. Fortschr. Bot. (Cologne & Leipzig). See B–P–H 383/13. HI 54209

Fortschritte der Botanik. Jena = Progressus rei botanicae. Jena. Progr. Rei Bot. See B–P–H 740/1.

Fortschritte der Chemie organischer Naturstoffe. Vienna, New York. Vol. 1+, 1938-39, 1945+. Fortschr. Chem. Organischer Naturst. HI 65382

Fortschritte der Evolutionsforschung. Stuttgart. 1965-76. Fortschr. Evolutionsforsch. HI 65383

Fortschritte in der Geologie von Rheinland und Westfalen. Krefeld, Germany. Fortschr. Geol. Rheinl. Westfalen. See B–P–H 383/17. HI 54210

Fortschritte im intergrierten Pflanzenschutz. Darmstadt. Vol. 1+, 1975+. Fortschr. Intergr. Pflanzenschutz. HI 65384

Fortschritte der naturwissenschaftlichen Forschung. Berlin Vols. 1-12, 1910-27; n.s. vols. 1-11, 1927-32. Fortschr. Naturwiss. Forsch. 2-1614-1. HI 74838

Fortschritte der Pflanzenzüchtung. [Beiheft zur Zeitschrift für Pflanzenzüchtung.] Berlin, Hamburg. Vol. 1+, 1971+. Fortschr. Pflanzenzücht. HI 65385

Fortsetzung des Allgemeinen Teutschen Garten-Magazins ... Weimar. Fortsetz. Allg. Teutsch. Gart.-Mag. See B–P–H 383/18. HI 54211

Forum botanicum; newsletter of the south african association of botanists. Nuusbrief van die suid-afrikaanse genootskap van plantkundiges. Grahamstown. Vol. 8+, 1970+. Forum Bot. Preceded by: South african forum botanicum. HI 65386

Forum mikrobiologie. Darmstadt. Vol. 1+, 1978+. Forum Mikrobiol. HI 65387

Fossilia. Barcelona. Fossilia. See B–P–H 383/19. HI 54212

Fossilium catalogus. Pars plantae. Berlin, The Hague. No. 1+, 1914+. Foss. Cat., Pars Pl. HI 65388

Four C's. Wagga Wagga, N.S.W. No. 1+, 1979+. Four C's HI 65389

Four seasons. Berkeley, CA. Four Seasons. See B–P–H 383/21. HI 54213

Fox forest notes. Hillsboro, NH. Vol. 1+, 1938+. Fox Forest Notes. HI 65390

Fra alle lande; et Maanedsskrift for nyere Reisebeskrivelser, for Skildringer af Lande of Folkeslag, af Dyre -og Planteliv. Copenhagen. Vols. 1-38, 1865-83. Fra Alle Lande. HI 65391

Fragmenta balcanica. Skoplje. Vol. 1+, 1954+. Fragm. Balcan. HI 65392

Fragmenta botanica musei historico-naturalis hungarici. Budapest. Vols. 1-7, 1962-69. Fragm. Bot. Mus. Hist.-Nat. Hung. Superseded by: Studia botanica hungarica. HI 65393

Fragmenta botanica musei macedonici scientiarum naturalium. Skoplje, Yugoslavia. Fragm. Bot. Mus. Maced. Sci. Nat. See B–P–H 383/25. HI 54214

Fragmenta florae philippinae; contributions to the flora of the Philippine Islands. Leipzig. Nos. 1-3, 1904-05. Fragm. Fl. Philipp. HI 65394

Fragmenta floristica et geobotanica. Cracow. Fragm. Florist. Geobot. See B–P–H 384/3 HI 54215

Fragmenta herbologica jugoslavica. Zagreb. Vol. ?-6+, ?-1978+. Fragm. Herbol. Jugoslav. HI 65395

Fragmente, Nachrichten und Abhandlungen zur Beförderung der Finanz-, Polizey-, Oekonomie- und Naturkunde. Berlin. Fragm. Nachr. Abh. Beförd. Finanz- Naturk. See B–P–H 384/4. HI 54216

Frailea. Prague. 1970+. Frailea. HI 65396

France horticole. Nos. ?-237, 19??-70. France Hort. Superseded by: Horticulture française. HI 65397

France viticole. Montpellier. Vol. 1+, 1969+. France Vitic. HI 65398

Franckfurtische Gelehrte Zeitungen, darinnen die merckwürdigsten Neuigkeiten der gelehrten Welt. Frankfurt a. M. Franckfurt. Gel. Zeitungen. See B–P–H 384/7. HI 54217

Frankfurter Kakteenfreund. Frankfurt am Main. No. 1+, 1974+. Frankfurter Kakteenf. HI 65399

Fränkische acta erudita et curiosa. Die Geschichte der Gelehrten in Francken, auch andere in diesem Creyss vorgefallene Curiosa und Merckwürdigkeiten in sich haltend. Nuremberg. Fränk. Acta Erud. Cur. See 384/8. HI 54218

Fränkische Merkur. Schweinfurt. Fränk. Merkur (Schweinfurt). See B–P–H 384/11. HI 54220

Fränkische Sammlungen von Anmerkungen aus der Naturlehre, Arzneygelahrtheit, Oekonomie und den damit verwandten Wissenschaften. Nuremberg. Fränk. Samml. Anmerk. Naturl. See B–P–H 384/13. HI 54221

Fränkisches Magazin für Statistik, Naturkunde und Geschichte. Sonnenberg, Germany. Fränk. Mag. Statist. See B–P–H 384/10. HI 54219

Frankfurter gelehrte Anzeigen. Frankfurt a. M. Frankfurter Gel. Anz. See B–P–H 384/18. HI 54222

Frankfurter geographische Hefte. Frankfurt a. M. Frankfurter Geogr. Hefte. See B–P–H 384/19. HI 54223

Frankfurter medizinische Annalen für Aerzte, Wundärzte, Apotheker und denkende Leser aus allen Ständen. Frankfurt a. M. Frankfurter Med. Ann. Aerzte. See B–P–H 384/20. HI 54224

Frankfurter Zeitschrift für Pathologie. Wiesbaden. Frankfurter Z. Pathol. See B–P–H 384/21. HI 54225

Franklin journal and american mechanic's magazine. Philadelphia, PA. Franklin J. & Amer. Mech. Mag. See B–P–H 384/22. HI 54226

Französische Annalen für die allgemeine Naturgeschichte, Physik, Chemie, Physiologie und ihre gemeinnützigen Anwendungen. Hamburg. Franz. Ann. Allg. Naturgesch. See B–P–H 384/24. HI 54227

Französische Blätter. Basel. Franz. Blätt. See B–P–H 385/1. HI 54228

Fraterna; official bulletin for international hoya association. Central Point, OR. ?-1990+. Fraterna. HI 75267

Freethinker. London. 1881+. Freethinker. HI 65400

Freiberger Forschungshefte. C: Angewandte Naturwissenschaften: Geophysik. Nos. 1-8 issued as a supplement to: Bergakademie. Leipzig. No. 1+, 1951+. Freiberger Forschungsh., C HI 65401

Fremad!; et ugeblad for taenkende og straebsomme jordbrugere. Viborg. Vol. 1, nos. 1-22, 1869-70. Fremad. HI 65402

Fremontia; journal of the California native plant society. Berkeley, CA. Vol. 1+, 1973+. Fremontia. Preceded by?: Newsletter, California native plant society. HI 65403

Freshwater aquatic life research highlights. Duluth, MN. 1973+. Freshwater Aquatic Life Res. Highlights. HI 65404

Freshwater and aquaculture contents tables. Rome. Vol. 1+, 1978+. Freshwater Aquac. Contents Tables. HI 65405

Freshwater biological association of the british empire; collected papers. Ambleside. Freshwater Biol. Assoc. Brit. Empire Collect. Pap. See B–P–H 385/4. HI 54229

Freshwater biology. Oxford, Edinburgh. Vol. 1+, 1971+. Freshwater Biol. HI 65406

Freshwater and marine aquarium Sierra Madre, CA. Vol. 1(4)+, 1978+. Freshwater Mar. Aquar. HI 65407

Freude am Leben. Berlin. Freude am Leben. See B–P–H 385/7. HI 54230

Freye Nachrichten aus dem Reiche der Wissenschaften und der schönen Künste. Hamburg. Freye Nachr. Wiss. Schönen Künste. See B–P–H 385/8. HI 54231

Freywillige Beyträge zu den Hamburgischen Nachrichten aus dem Reiche der Gelehrsamkeit. Hamburg. Freywillige Beytr. Hamburg. Nachr. Gelehrsamk. See B–P–H 385/9. HI 54232

Friciana. Olumouc. No. 1+, 1962+. Friciana. HI 65408

Friends of the arboretum, university of Wisconsin. Madison, WI. Vol. 1+, 1975?+. Friends Arbor. Univ Wisconsin. HI 65409

Friends of the botanical garden newsletter, university of California = Newsletter, friends of the botanical garden, university of California. Berkeley, CA.

Friends of the forest. [Rinyu.] Tokyo. Friends Forest. See B–P–H 385/10. HI 54233

Friends newsletter, friends of planting fields. Oyster Bay, NY. Vol. ?-11+, 19??-83+. Friends Newslett. Friends Pl. Fields. HI 65411

Friends of planting fields newsletter = Friends newsletter, friends of planting fields. Oyster Bay, NY.

Friends of the U N C herbarium newsletter = Newsletter, friends of the U N C herbarium. Chapel Hill, NC.

Friends of the U S national arboretum news = F O N A news. Washington, DC.

Friesia; Nordisk mykologisk tidsskrift. Copenhagen. Vols. 1-11, 1932-78. Friesia. Preceded by: Meddelelser fra foreningen til svampekundskabens fremme. Superseded by: Nordic journal of botany. HI 65412

Frivaldszkya. Budapest. 1974+. Frivaldszkya. HI 65413

Frø-tidende; Organ for frøavl og frøhandel. Copenhagen. Vols. 1-31(1), 1889-1919. Frø-Tidende. HI 65414

Frodskaparrit; annales societatis scientiarum faeroensis. Torshavn. Vol. 1+, 1952+. Frodskaparrit.

Frodskaparrit; annales societatis scientiarum faeroensis. Supplementum. Torshavn. No. ?-2+, ?-1959+. Frodskaparrit, Suppl. HI 61611

From woods and fields. South Minneapolis, MN. 198?+. From Woods & Fields. HI 65415

Frontier nursing service quarterly bulletin. Lexington, KY. Vol. 1+, 1925+. Frontier Nursing Serv. Quart. Bull. 2-1649-2. HI 65416

Frontiers; a magazine of natural history (later Annual of the academy of natural sciences of Philadelphia). Philadelphia, PA. Vols. 1-43(1), 1936-78; [n.s.] vol. 1+, 1979+. Frontiers. HI 65417

Frontiers of biology. Philadelphia, PA. Frontiers Biol. See B–P–H 385/14. HI 54234

Frontiers of plant science. New Haven, CT. See B–P–H 385/15. HI 54235

Froriep's Notizen aus dem Gebiet der Natur- und Heilkunde. Jena. Froriep's Not. Natur- Heilk. See B–P–H 385/17. HI 54236

Fruchtgarten. Vienna. Fruchtgarten. See B–P–H 385/20. HI 54237

Fruechte und Gemuese. Zug. 1933+. Fruechte & Gemuese. HI 65418

Frugtavleren. Odense. Vol. 1+, 1972+. Frugtavleren. Preceded by: Dansk frugtavleren and Erhvervsfrugtavleren. HI 65419

Fruit belge. Liège. Fruit Belge. See B–P–H 385/21. HI 54238

Fruit belt. Washington, DC. Fruit Belt. See B–P–H 385/22. HI 54239

Fruit, garden and home = Better homes and gardens. Des Moines, IA. Better Homes Gard. See B–P–H 185/1.

Fruit gardener. Fullerton, CA. Vol. 18(3)+, 1986+. Fruit Gard. Superseded by: Newsletter, California rare fruit growers. HI 75087

Fruit and grape grower. Charlottesville, VA. Fruit Grape Grower. See B–P–H 385/24. HI 54240

Fruit grower. London. Nos. 1-2970?, 1895-1952. Fruit Grower. Superseded by: Commercial grower. HI 65420

Fruit grower and horticulturist; a monthly magazine of practical horticulture. Lacon, IL. Fruit Grower Hort. See B–P–H 385/25. HI 54241

Fruit recorder and cottage gardener. Rochester, NY. Fruit Rec. Cottage Gard. See B–P–H 386/3. HI 54242

Fruit science reports. Skierniewice. Vol. 1+, 1974+. Fruit Sci. Rep. HI 65421

Fruit tree. [Kaju.] Okayama, Japan. Fruit Tree. See B–P–H 386/4. HI 54243

Fruit trees. [Kwaju.] Hangchow, China. Fruit Trees. See B–P–H 386/5. HI 54244

Fruit and tropical products. London. 1978+. Fruit Trop Prod. HI 65422

Fruit and truck experiment station report, Louisiana agricultural experiment station = Louisiana agricultural experiment station. Fruit and truck experiment station report. [Title varies.] Hammond, LA. Louisiana Agric. Exp. Sta. Fruit Truck Exp. Sta. Rep. See B–P–H 535/24.

Fruit varieties and horticultural digest. Wooster, OH. Vols. 1-26, 1946-72. Fruit Var. Hort. Digest. Superseded by: Fruit varieties journal. HI 65423

Fruit varieties journal. University Park, PA. Vol. 27+, 1973+. Fruit Var. J. Preceded by: Fruit varieties and horticultural digest. HI 65424

Fruit and vegetable digest. Charles Town, WV. Fruit Veg. Digest. See B–P–H 386/8. HI 54245

Fruit and vegetable review. Orange, CA. Fruit Veg. Rev. See B–P–H 386/9. HI 54246

Fruit and vegetable situation. Washington, DC. Vols. TVS1-TVS6, 1937. Fruit Veg. Situation. Superseded by: Vegetable situation. HI 65425

Fruit world. Los Angeles, CA. Fruit World (Los Angeles). See B–P–H 386/12. HI 54247

Fruit world. South Bend, IN. Fruit World (South Bend). See B–P–H 386/13. HI 54248

Fruit world of Australia. Melbourne. Fruit World Australia. See B–P–H 386/14. HI 54249

Fruit world and market grower. Melbourne. Fruit World Market Grower. See B–P–H 386/15. HI 54250

Fruit year book. London. Nos. 1-12, 1947-58. Fruit Year Book. HI 65426

Fruition. Santa Cruz, CA. No. 1+, 1979+. Fruition. HI 65427

Fruitman. Fresno, CA. Fruitman. See B–P–H 386/16. HI 54251

Fruits. Paris. Vol. 6+, 1951+. Fruits. Preceded by: Fruits d'outre mer. HI 65428

Fruits and gardens. Chicago, IL. Fruits & Gard. See B–P–H 386/18. HI 54252

Fruits et légumes. Basel. Fruits & Légumes. See B–P–H 386/19. HI 54253

Fruits et legumes. Zug = Fruechte und Gemuese. Zug.

Fruits d'outre-mer. Paris. Vols. 1-5, 1945-50. Fruits Outre-Mer. Superseded by: Fruits. HI 65429

Fruits et primeurs de l'Afrique du nord et la revue française de l'oranger. Casablanca, Morocco. Fruits Primerus Afrique N. See B–P–H 386/22. HI 54254

Fruitteelt. Organ der nederlandsche pomologische vereeniging. Utrecht. Fruitteelt. See B–P–H 386/23. HI 54255

Frukt og baer. Oslo. Frukt & Baer. See B–P–H 386/24. HI 54256

Fruktodlaren. Stockholm. Vol. 1+, 1900+. Fruktodlaren. HI 65430

Frutos. Madrid. Frutos. See B–P–H 387/1. HI 54257

Frutticoltura. Bologna. Vols. 17-44, 1955-82. Frutticoltura. Preceded by: Rivista di frutticoltura. Superseded by: Rivista de frutticoltura e di ortofloricultura. HI 65431

Fu-Chien nung yeh = Fukien agriculture. Foochow, China. Fukien Agric. See B–P–H 387/10.

Fu tan nsüch pao = Fu-tan journal. Pehpei, China. Fu-tan J. See B–P–H 387/27.

Fu-tan ta hsüeh hsüeh pao. Tzu jan k'o hsüeh = Fu-tan university journal. Natural science. Shanghai. Fu-tan Univ. J., Nat. Sci. See B–P–H 387/31.

Fuaux herbarium bulletin. Rosanna, Vic. Nos. 1-5, 1949-53. Fuaux Herb. Bull. HI 65432

Fuchsia annual. Reading, England. Fuchsia Annual. See B–P–H 387/6. HI 54258

Fuchsia fan. Oxnard, Palos Verdes, Santa Paula, CA. 19??-77+. Fuchsia Fan. HI 65433

Fühlings landwirtschaftliche Zeitung. Stuttgart. Vols. 23-71, 1874-1922. Fühlings Landw. Zeitung. Preceded by: Neue landwirtschaftliche Zeitung. HI 65434

Fuji take-rui shokubutsuen hokoku = Reports of the Fuji bamboo gardens. Shizuoka, Japan. Rep. Fuji Bamboo Gard. See B–P–H 765/15.

Fukien agricultural journal. [Hsieh ta nung pao.] Foochow, China. Fukien Agric. J. See B–P–H 387/11. HI 54260

Fukien agriculture. [Fu-Chien nung yeh.] Foochow, China. Fukien Agric. See B–P–H 387/10. HI 54259

Fukien christian university biological bulletin. Foochow, China. Fukien Christian Univ. Biol. Bull. See B–P–H 387/12. HI 54261

Fukien culture. [Fukien wen hua.] Foochow, China. Fukien Cult. See B–P–H 387/13. HI 54262

Fukien wen hua = Fukien culture. Foochow, China. Fukien Cult. See B–P–H 387/13.

Fukui daigaku, kyoiku gakubu. 5bu, oyo kogaku = Memoirs, faculty of education, Fukui university. Series 5, applied science and agricultural science. Fukui.

Fukui-ken nogyo shikenjo hokoku = Bulletin of Fukui prefecture agricultural experiment station. Fukui, Japan. Bull. Fukui Prefect. Agric. Exp. Sta. See B–P–H 251/6.

Fukuoka acta medica. = Fukuoka ilkwadaigaky zasshi. Fukuoka Ilkwadaigaky Zasshi. See B–P–H 387/21.

Fukuoka gakugei daigaku kiyo III. Rikakeito = Bulletin Fukuoka gakugei university. Part III. Natural science.

Fukuoka, Japan. Bull. Fukuoka Gakugei Univ., Pt. 3, Nat. Sci. See B–P–H 251/8.

Fukuoka hakubutsu-gaku zasshi = Journal of natural history society in Fukuoka. Fukuoka, Japan. J. Nat. HIst. Soc. Fukuoka. See B–P–H 474/9.

Fukuoka ilkwadaigaky zasshi. Fukuoka Ilkwadaigaky Zasshi. See B–P–H 387/21. HI 54263

Fukuoka-ken ringyo shikenjo jiho = Bulletin of Fukuoka prefectural forest experiment station. Kuroki.

Fukuoka kenritsu engei shikenjo, kenkyu hokoku = Bulletin of the Fukuoka horticultural experiment station.

Fukushima cactus; journal of the Fukushima cactus club. Fukushima. No. 1-2-?, 1958-? Fukushima Cact. HI 65435

Fukushima daigaku gakugei gakubu rika hokoku = Science report of the faculty of arts and science; Fukushima university. Fukushima, Japan. Sci. Rep. Fac. Arts Fukushima Univ. See B–P–H 826/25.

Fukushima-ken engei shikenjo, kenkyu hokoku = Bulletin of the Fukushima horticultural experiment station. Fukushima.

Functional ecology. Oxford. Vol. 1+, 1987+. Funct. Ecol. HI 65436

Fundamenta scientiae; international journal for critical analysis of science and the responsibility of scientists. Paris, Sao Paulo, Strasbourg. Vol. 1+, 1980+. Fundam. Sci. HI 74839

Fungi. [Kin-rui.] Tokyo. Fungi. See B–P–H 387/23. HI 54264

Fungi canadenses. Ottawa. Vol. 1+, 1973+. Fungi Canadenses. HI 65440

Fungorum rariorum icones coloratae. Lehre, Vaduz, Berlin & Stuttgart. No. 5+, 1971+. Fungorum Rar. Icon. Color. Preceded by: Coloured icones of rare and interesting fungi. HI 65437

Fungus; populair orgaan voor de leden. Wageningen. Fungus. See B–P–H 387/25. HI 54265

Für Freunde des Obstbaues. Eine Zeitschrift zur Beförderung des Obstbaues in Deutschland. Leipzig. Für Freunde Obstbaues. See B–P–H 386/26. HI 54266

Für die Litteratur und kenntniss der Naturgeschichte; sonderlich der Conchylien und der Steine. Weimar. Vols. 1-2, 1782. Für Litt. Kenntn. Naturgesch. Superseded by: Neue Litteratur und Beyträge zur Kenntniss der Naturgeschichte. HI 65438

Further contributions to canadian biology; being studies from the biological stations of Canada = Contributions to canadian biology; being studies from the biological stations of Canada. Toronto.

Fu-tan journal. [Fu tan nsüch pao.] Pehpei, China. Fu-tan J. See B–P–H 387/27. HI 54267

Fu-tan university journal. Natural science. [Fu-tan ta hsüeh hsüeh pao. Tzu jan k'o hsüeh.] Shanghai. Fu-tan Univ. J., Nat. Sci. See B–P–H 387/31. HI 54268

Futures. East Lansing, MI. No. 1+, 1982+. Futures. Preceded by: Michigan science in action. HI 65443

G B + G W = Gb + Gw. Aachen, Berlin.

G C A newsletter. Washington, DC. 19??-82+. G. C. A. Newslett. HI 65445

G C & H T J; gardeners' chronicle and horticultural trade journal. London. Vols. 180(15)-199(1), 1976-86. G. C. & H. T. J. Preceded by: Gardener's chronicle. Superseded by: Horticulture week. HI 65446

G C V A journal. Richmond, VA. Vol. 1+, 1955+. G. C. V. A. J. HI 65447

G F G report, forestry and forest products. Samarinda/ Kaltim, Indonesia. Vol. 1+, 1985+. G. F. G. Rep. Forestry Forest Prod. HI 65448

G N S I newsletter. Washington, DC. 1969+. G. N. S. I. Newslett. HI 65449

G S N; gesneriad Saintpaulia news. Knoxville, TN, etc. Vol. 1+, 1964+. G. S. N. HI 65451

Gaben der Flora. Conversationsblatt für Blumenfreunde. Berlin. Gaben Fl. See B–P–H 388/9. HI 54269

Gabonaipar. Budapest. Vol. 22+, 1975+. Gabonaipar. Preceded by: Malomipar és terményforgalom. HI 65452

Gaceta algodonera. Publicación defensora de plantadores e industriales del algodón. Buenos Aires. Gac. Algodonera. See B–P–H 388/10. HI 54270

Gaceta de granja. Buenos Aires. Gac. Granja. See B–P–H 388/11. HI 54271

Gaceta médica de México. Mexico City. Gac. Méd. México. See B–P–H 388/13. HI 54272

Gaden raifu = Garden life. Tokyo. Gard. LIfe (Tokyo). See B–P–H 390/11.

Gaea. Berlin. Gaea (Berlin). See B–P–H 388/16. HI 54273

Gaea. Buenos Aires. Gaea (Buenos Aires). See B–P–H 388/17. HI 54274

Gaea. Cologne. Gaea (Cologne). See B–P–H 388/18. HI 54275

Gaea norvegica. Christiania [=Oslo, Norway]. Gaea Norveg. See B–P–H 388/20. HI 54276

Gakugei. Section B. Hokkaido gakugei university. Sapporo. Vols. 1-4(1), 1942-52. Gakugei, Sect. B, Hokkaido Gakugei Univ. Superseded by: Journal of the Hokkaido gakugei university. Section B, natural sciences. HI 65453

Gakugei zasshi = Bulteno scienca de la fakultato terkultura. Kjusu imperia universitato. Fukuoka, Japan. Bult. Sci. Fak. Terk. Kjusu Imp. Univ. See B–P–H 288/7.

Gallica biologica acta. Paris. Gallica Biol. Acta. See B–P–H 388/23. HI 54277

Gamete research. New York. Vol. 1+, 1978+. Gamete Res. HI 65454

Ğami'at al-Basrah. Mathaf at-Tarih at-Tabi'i = Publications, Basrah natural history museum. Basrah.

Ğami'at al-Kuwayt. Kulliyat al'utum = Journal of the university of Kuwait (science). Kuwait.

Gamma field symposia. Ibaraki. No. 1+, 1962+. Gamma Field Symp. HI 65455

Gamophyllous; a monthly magazine devoted to plant life in field, forest and garden. Plainfield, NJ. Gamophyllous. See B–P–H 388/24. HI 54278

Gamta; Lietuvos gamtininku zurnalas. Kaunas. Vols. 1-5, 1936-40. Gamta. HI 65456

Gan Vanof. Tel Aviv, Israel. Gan Vanof. See B–P–H 388/25. HI 54279

Gaoyuan shengwuxue jikan = Acta biologica plateau sinica. Xining.

Garcia de Orta. Lisbon. Garcia de Orta. See B–P–H 388/26. HI 54280

Garcia de orta. Série de botânica. Revista da junta de investigações do ultramar (later ... investigações científica do ultramar). Lisbon. Vol. 1+, 1973+. Garcia de Orta, Sér. Bot. Preceded by: Trabalhos do centro de botânica da junta de investigações do ultramar. HI 65457

Garcia de orta. Série de estudos agronomicos; revista da junta de investigações do ultramar (later ... investigações científica do ultramar). Lisbon. Vol. 1+, 1973+. Garcia de Orta, Sér. Estudos Agron. HI 65458

Garcia de orta. Série de farmacognosia; revista da junta de investigações do ultramar (later ... investigações científica do ultramar). Lisbon. Vol. 1+, 1972+. Garcia de Orta, Sér. Farmacogn. HI 65459

Garden. Bairdstown, KY. Garden (Bairdstown). See B–P–H 391/9. HI 54316

Garden; an illustrated weekly journal of gardening in all its branches. London. Vols. 1-91, 1871-1927. Garden (London 1871-1927). 2-1665-2. HI 54317

Garden; journal of the royal horticultural society. London. Vol. 100(6)+, 1975+. Garden (London 1975+). Preceded by: Journal of the royal horticultural society. HI 65461

Garden. New Orleans, LA. Garden (New Orleans). See B–P–H 391/11. HI 54318

Garden. New York. Vols. 1-3, 1948-59. Garden (New York 1948-50). 2-1665-3. HI 54319

Garden. New York. Vol. 55(2-5), 1951. Garden (New York 1951). Preceded by: Garden and the gardener's chronicle. HI 65462

Garden. New York. Vol. 1+, 1977+ [Also published in a "Regional edition"]. Garden (New York 1977+). Preceded by: Garden journal of the New York botanical garden and Journal of the Chicago horticultural society. HI 65463

Garden. Seattle, WA. Garden (Seattle). See B–P–H 391/13. HI 54320

Garden album and review. London. Gard. Album Rev. See B–P–H 388/27. HI 54281

Garden appeal; newsletter of the Memphis botanic garden. Memphis, TN. 1985+. Preceded by: Newsletter, Memphis botanic garden. Gard. Appeal. HI 65464

Garden beautiful. Vancouver. Gard. Beautiful. See B–P–H 388/28. HI 54282

Garden center bulletin. Cleveland, OH. 1937+. Gard. Center Bull. (Cleveland). HI 65465

Garden center bulletin. Rochester, NY. Vol. 8+, 1952+. Gard. Center Bull. (Rochester). Preceded by: Bulletin of the garden center of Rochester. HI 65466

Garden cities and town planning. London. Gard. Cities Town Planning. See B–P–H 389/11. HI 54285

Garden and city. Boston, MA. Gard. & City. See B–P–H 389/8. HI 54284

Garden city and town planning = Garden cities and town planning. London. Gard. Cities Town Planning. See B–P–H 389/11.

Garden companion and florists' guide. London. Gard. Companion Florists' Guide. See B–P–H 389/14. HI 54286

Garden design. Washington, DC. Vol. 1+, 1982+. Gard. Design. HI 65467

Garden design magazine = Garden design. Washington, DC.

Garden digest. Pleasantville, NY. Gard. Digest. See B–P–H 389/15. HI 54287

Garden and farm. Reed City, MI. Gard. & Farm. See B–P–H 389/16. HI 54288

Garden and field; a monthly journal of agronomy. Adelaide. Gard. & Field. See B–P–H 389/19. HI 54289

Garden, field and forest. Demarara. 1887-89. Gard. Field Forest. HI 65468

Garden flowers. Rochester, NY. Gard. Fl. See B–P–H 389/20. HI 54290

Garden and forest; a journal of horticulture, landscape art and forestry. New York, NY. Gard. & Forest. See B–P–H 389/22. HI 54291

Garden forum. Pattonville, MO. Garden Forum. See B–P–H 391/15. HI 54321

Garden forum = American horticultural society gardeners' forum. Washington, DC.

Garden and the gardener's chronicle. New York. Vols. 54(7)-55(1), 1950-51. Gard. Gard. Chron. Preceded by: Gardener's chronicle of America. Superseded by: Garden. New York. HI 65469

Garden gossip. Richmond, VA. Gard. Gossip. See B–P–H 389/27. HI 54295

Garden guides. 19??-71+. Gard. Guides. HI 65470

Garden history; journal of the garden history society. Oxford. Vol. 1+, 1972+. Preceded by: Newsletter, garden history society. [1st series.] HI 65471

Garden in April (May ...). Edinburgh. 1971-75 [published for April/September only each year]. Gard April [May etc.]. HI 65472

Garden journal, the magazine about plants = Garden journal of the New York botanical garden. New York.

Garden journal of the New York botanical garden. New York. Vols. 1-26, 1951-76. Gard. J. New York Bot. Gard. Preceded by: Journal of the New York botanical garden. Superseded by: Garden. New York. 4-3030-1. HI 65473

Garden and landscape. Tel Aviv = Gan Vanof. Tel Aviv, Israel. Gan Vanof. See B–P–H 388/25.

Garden life. London. Gard. Life (London). See B–P–H 390/10. HI 54296

Garden vine. St. Louis, MO. Vol. 1+, 1984+. Gard. Vine. HI 65482

Garden life. [Gaden raifu.] Tokyo. Gard. LIfe (Tokyo). See B–P–H 390/11. HI 54297

Garden lover. Melbourne. Gard. Lover. See B–P–H 390/12. HI 54298

Garden magazine. New York, NY. Gard. Mag. (New York). See B–P–H 390/16. HI 54301

Garden magazine and home builder. New York, NY. Gard. Mag. & Home Builder. See B–P–H 390/20. HI 54303

Garden news. Peterborough. 1958+. Gard. News. HI 65475

Garden notes. Edinburgh. 1965+. Gard. Notes (Edinburgh). Preceded by: Monthly notes, royal botanic garden, Edinburgh. HI 65476

Garden notes. Van Wert, OH. Nos. 1-22, 1920-26. Gard. Notes (Van Wert). 1-1666-3. HI 54306

Garden notes, university of Georgia botanical garden. Atlanta, GA. Vol. 1+, 1979+. Gard. Notes Univ. Georgia Bot. Gard. HI 65477

Garden, orchard and farm. Vancouver. Gard. Orchard Farm. See B–P–H 390/28. HI 54307

Garden path. Columbus, OH. Vols. 1-8, 1930-38? Gard. Path. 2-1666-3. HI 65478

Garden spotlight. Staten Island, NY. Vol. 1+, 1984?+. Gard. Spotlight. HI 65479

Garden talk. Chicago, IL. Vols. 1-12, 1953-74. Superseded by: Journal of the Chicago horticultural society. Gard. Talk. HI 65480

Garden today. St. Louis, MO. 19??-69-72-? Gard. Today. HI 65481

Garden vine. [Dates of publication not ascertained.] Gard. Vine. HI 65482

Garden work for amateurs. London. Gard. Work Amateurs. See B–P–H 391/6. HI 54313

Garden-work for villa, suburban, town, and cottage gardens. London. Gard.-Work Villa Gard. See B–P–H 391/7. HI 54314

Garden writers' bulletin. New York. 1962-75-? Gard. Writers' Bull. HI 65483

Gardener; magazine of horticulture and floriculture. Edinburgh, London. Vols. 1-16, 1868-82. Gardener (Edinburgh). Preceded by: Scottish gardener. 2-1666-3. HI 65484

Gardener. London. Vols. 1-21, 1899-1920. Gardener (London). Superseded by: Popular gardening. 4-3399-2. HI 65485

Gardener. Poona, India. Gardener (Poona). See B–P–H 391/17. HI 54322

Gardener. Rockford, IL. Gardener (Rockford). See B–P–H 391/18. HI 54323

Gardener and practical florist. London. Gard. Pract. Florist. See B–P–H 391/2. HI 54309

Gardener's abstracts. Goleta, CA. Vols. 1-2(2), 1969-? Gard. Abstr. HI 65486

Gardeners' chronicle. London. [Ser. 1] vols. 1-30, 1841-73; n.s. vols. 1-26, 1874-86; ser. 3, vols. 1-180(14), 1887-1976. Gard. Chron. Incorporated: Gardening illustrated 1956+, Greenhouse 1964+, and Horticultural trade journal 1969+. Superseded by: G C & H T J. 2-1666-3. HI 65487

Gardeners' chronicle and agricultural gazette = Gardeners' chronicle. London.

Gardener's chronicle of America. New York. Vols. 1-54, 1905-50. Gard. Chron. Amer. Superseded by: Garden and the gardener's chronicle. 2-1665-2. HI 65488

Gardeners' chronicle and gardening illustrated = Gardeners' chronicle. London.

Gardeners chronicle and horticultural trade journal = G C & H T J. London.

Gardeners' chronicle and new horticulturist = Gardeners' chronicle. London.

Gardeners' digest; official publication of the gardeners' guild. Toronto. Vol. 1+, 1982+. Gard. Digest. HI 65489

Gardeners' and farmers' journal. London. 1847-80. Gard. Farmers' J. Preceded by: United gardeners' and land-stewards' journal. 2-1666-3. HI 65490

Gardener's and forester's record. London. Gard. Forester's Rec. See B–P–H 389/23. HI 54292

Gardener's gazette. London. Gard. Gaz. See B–P–H 389/26. HI 54294

Gardeners' magazine. London. Gard. Mag. (London). See B–P–H 390/14. HI 54300

Gardeners' magazine of botany, horticulture, floriculture, and natural science. London. Gard. Mag. Bot. See B–P–H 390/17. HI 54302

Gardener's magazine and register of rural and domestic improvement. London. Gard. Mag. & Reg. Rural Domest. Improv. See B–P–H 390/21. HI 54304

Gardener's monthly and horticultural advertiser. Philadelphia, PA. Gard. Monthly & Hort. Advertiser. See B–P–H 390/25. HI 54305

Gardener's monthly and horticulturist. Philadelphia, PA. Vols. 18-30, 1876-88. Gard. Monthly & Hort. Preceded by: Gardener's monthly and horticultural advertiser and Horticulturist and journal of rural art and rural taste. 2-1667-2. HI 65491

Gardener's monthly volume. London. Vols. ?-2-11-?, 1847-? Gard. Monthly Vol. HI 65492

Gardeners' record. Dublin. Gard. Rec. (Dublin). See B–P–H 391/3. HI 54310

Gardener's record and amateur florist's companion. London. Gard. Rec. Amateur Florist's Companion. See B–P–H 391/4. HI 54311

Gardener's weekly magazine and floricultural cabinet. London. Gard. Weekly Mag. & Floric. Cab. See B–P–H 391/5. HI 54312

Gardeners' year book. London. 18??-1862-1930-? Gard. Year Book. HI 65493

Gardenia notes. Atwater, CA. 19??-82+. Gardenia Notes. HI 65494

Gardening. Chicago, IL. Gardening (Chicago). See B–P–H 391/20. HI 54324

Gardening. Morgantown, WV. Gardening (Morgantown). See B–P–H 391/21. HI 54325

Gardening in America. Burlington, VT. 1973-78, 1974-79. Gard. America. Superseded by: National gardening survey. HI 68357

Gardening in British Columbia. Vancouver. Gard. British Columbia. See B–P–H 388/29. HI 54283

Gardening illustrated. London. Vols. 1-73 [vols. 11-16 also numbered ser. 2, vols. 1-6; vols. 17-19 also numbered ser. 3, vols. 1-3], 1879-1956. Gard. Ill. Incorporated in: Gardeners' chronicle. 2-1667-2. HI 65495

Gardening library news. New York. Vol. 1(1-8), 1951. Gard. Libr. News. Preceded by: Monthly bulletin of the horticultural society of New York. Superseded by: News, horticultural society of New York. HI 65496

Gardening magazine. Calcium, NY. Gard. Mag. (Calcium). See B–P–H 390/13. HI 54299

Gardening news; journal of the royal horticultural society of Victoria. Box Hill South, Vic. Vol. ?-4(4)+, ?-1981+. Gard. News. HI 65497

Gardening world. London. Gard. World. See B–P–H 391/8. HI 54315

Gardens' bulletin; royal botanical gardens. Hamilton, Ont. Vol. 1+, 1948+. Gard. Bull. Roy. Bot. Gard. HI 65498

Gardens' bulletin. Singapore. Vol. 11(4)+, 1947[1949]+. Gard. Bull. Singapore. Preceded by: Gardens' bulletin, Straits Settlements. HI 65499

Gardens' bulletin. Straits Settlements. Singapore. Ser. 3, vols. 1(6)-11(3), 1913-47. Gard. Bull. Straits Settlem. Preceded by: Agricultural bulletin of the Straits and Federated Malay States. Superseded by: Gardens' bulletin. Singapore. 2-1666-3. HI 65500

Gardens and countrysides; journal of picturesque travels. San Antonio, TX. Vol. 1+, 1990+. Gard. & Countrysides. HI 74840

Gardens and gardening; the studio gardening annual. London. Gard. & Gardening. See B–P–H 389/25. HI 54293

Gardens for all news. Shelburne, VT. Vol. 1(3)- 8(8)?, 1978-85? Gard. For All News. Preceded by: Spotlight on community gardening. Superseded by: National gardening. HI 68741

Gardens of the pacific. Berkeley, CA. Gard. Pacific. See B–P–H 391/1. HI 54308

Gardenwise; the NParks newsletter. Singapore. No. ?-2+, ?-1990+. Gardenwise. HI 68445

Gardyrjuritid; arsrit gardyrkjufélags íslands. Reykjavík. Vol. 48+, 1968+. Gardyrjuritid. Preceded by: Arsrit gardyrkjufélags islands. HI 65501

Garten und Landschaft. Munich. Gart. & Landschaft. See B–P–H 391/25. HI 54326

Garten-Schönheit. Illustrierte Schrift für den Garten- und Blumenfreund, Liebhaber und Fachmann. Aachen.

Gart.-Schönheit. See B–P–H 391/26. HI 54327

Garten-Schönheit vereinigt mit "Deutscher Garten". Aachen. Gart.-Schönheit & Deutsch. Gart. See B–P–H 392/1. HI 54328

Garten-Zeitschrift für Gärtner und Gartenfreunde, Siedler und Kleingärtner. Illustrierte Flora. Vienna. Gart.-Z. Gärtn. Ill. Fl.. See B–P–H 392/2. HI 54329

Garten-Zeitung. Berlin. Gart.-Zeitung (Berlin). See B–P–H 392/5. HI 54330

Garten-Zeitung. Halle. Gart.-Zeitung (Halle). See B–P–H 392/6. HI 54331

Gartenamt. Hanover, Berlin, & Sarstedt. Gartenamt. See B–P–H 392/8. HI 54332

Gartenbau. Berlin, Neustadt. Vol. 18(7)+, 1971+. Gartenbau (Berlin). Preceded by: Obstbau and Deutsche Gartenbau. HI 65502

Gartenbau. Solothurn. Vol. ?-92+, 19??-71+. Gartenbau (Solothurn). HI 65503

Gartenbau-Forschung. Wiesbaden. Gartenbau-Forschung. See B–P–H 392/11. HI 54334

Gartenbau und Gartenwelt. ?-1988+. Gartenbau & Gartenwelt. HI 65504

Gartenbau im Reich. Berlin. Gartenbau Reich. See B–P–H 392/10. HI 54333

Gartenbau-Taschenkalender. Vienna. Gartenbau-Taschenkalender. See B–P–H 392/12. HI 54335

Gartenbau Wirtschaft; Mitteilungsblatt der landwirtschaftlichen Gemüse- und Obstverwertungs-Genossenschaft für Wien und Umgebung. Vienna. Vol. 1+, 1946+. Gartenbau Wirtsch. HI 65505

Gartenbauwissenschaft. Berlin, Vienna & Munich. Vols. 1-18, 1928-44; n.s. vols. 1(19)-21(39), 1954-74; vol. 40+, 1975+ [issued in two series, Originalien and Referate und Bibliographie]. Gartenbauwissenschaft. 2-1668-3. HI 54336

Gartenbeobachter. Eine Zweitschrift des Neuesten und Interessantesten im Gebiete der Blumistik und Horticultur. Nuremberg. Gartenbeobachter. See B–P–H 392/15. HI 54337

Gartenflora. Erlangen. Gartenflora. See B–P–H 392/17. HI 54338

Gartenfreund. Basel. Gartenfreund (Basel). See B–P–H 392/18. HI 54339

Gartenfreund. Munich. Gartenfreund. (Munich). See B–P–H 392/19. HI 54340

Gartenfreund. Vienna. Gartenfreund (Vienna). See B–P–H 392/20. HI 54341

Gartenkalender. Kiel & Dessau. Gartenkalender. See B–P–H 392/21. HI 54342

Gartenkunst. Berlin. Gartenkunst. See B–P–H 392/22. HI 54343

Gartenpraxis. Stuttgart. Vol. 1+, 1975+. Gartenpraxis. HI 65508

Gartenschönheit. Eine Zeitschrift mit Bildern für Garten- und Blumenfreund, für Liebhaber und Fachmann. Berlin. Gartenschönheit. See B–P–H 392/24. HI 54344

Gartenwelt, Die. Berlin. Vols. 2-37, 1897-1933; vols. 49-77, 1948-77. Gartenwelt. Preceded by: Hesdörffers Monatshefte für Blumen- und Gartenfreunde. For vols. 38-48, 1934-44, see: Blumen-Pflanzenbau & Gartenwelt. Incorporated with: Gb + Gw: Gärtnerbörse und Gartenwelt. 2-1669-1. HI 65509

Gartenzeitschrift Illustrierte Flora. Fachblatt für Gärtner, Gartenfreunde, Kleingärtner und Siedler. Vereinigt mit "Nützliche Blätter". Vienna. Gartenz. Ill. Fl. See B–P–H 393/2. HI 54345

Gartenzeitung. Organ der Oesterreichischen Gartenbau-Gesellschaft in Wien. Vienna. Gartenzeitung (Vienna). See B–P–H 393/5. HI 54346

Gartenzeitung der Österreichischen Gartenbau-Gesellschaft. Vienna. Gartenzeitung Österr. Gartenbau-Ges. See B–P–H 393/8. HI 54348

Gartenzeitung der Oesterreichischen Gartenbaugesellschaft in Wien. Vienna. Gartenzeitung Oesterr. Gartenbauges. Wien. See B–P–H 393/6. HI 54347

Gärtner-Meister. Laufenburg. Vol. 79+, 1976+. Gärtn.-Meister. Preceded by: Schweizerische Gärtnerzeitung. HI 65510

Gartner-tidende. Copenhagen. Vols. 1-65, 1885-1949; [n.s.] vol. 1+, 1952+. Gartn.-Tidende. In 1887 incorporated: Danske havetidende (Copenhagen, 1849-87). 2-1669-2. HI 65511

Gärtnerbörse und Gartenwelt = Gb + Gw. Aachen, Berlin.

Gärtnerisch-botanische Briefe. [Germany.] Gärtn.-Bot. Briefe. See B–P–H 393/9. HI 54349

Gartneryket. Oslo. Gartneryket. See B–P–H 393/14. HI 54350

Gaussenia; travaux du laboratoire forestier de Toulouse, université Paul Sabatier. Toulouse. Vol. 1+, 1984+. Gaussenia. Preceded by: Travaux du laboratoire forestier de Toulouse. HI 65512

Gavnlig og undeholdende laesning i naturvidenskaben. Copenhagen. Gavnlig Undeh. Laesn. Naturvidensk. See B–P–H 393/17. HI 54351

Gayana; instituto central de biologia. Botánica. Santiago, etc. Gayana, Bot. See B–P–H 393/18. HI 54352

Gayana. Miscelania. Concepción. Vol. 1+, 1971+. Gayana, Misc. HI 65513

Gazdasági gyütemény. Pest [=Budapest, in part]. Gazd. Gyüt. See B–P–H 394/12. HI 54362

Gazdasági lapok. Pest [=Budapest, in part]. Gazd. Lapok. See B–P–H 394/14. HI 54364

Gazdasagi lapok és magyar föld = Gazdasági lapok. Pest [=Budapest, in part]. Gazd. Lapok. See B–P–H 394/ 14.

Gazdasági literatura. Pest [=Budapest, in part]. Gazd. Lit. See B–P–H 394/16. HI 54365

Gazdasági tudósitások. Pest [=Budapest, in part]. Gazd. Tud. See B–P–H 395/17. HI 54366

Gazdasági tudósitások és Rohonczi közlemények. Pest [=Budapest, in part]. Gazd. Tud. Rohonczi Közlem. See B–P–H 394/18. HI 54367

Gazdaságot tzélozó ujság. Bécs [=Vienna]. Gazd. Tzél. Ujs. See B–P–H 394/19. HI 54368

Gazdászati közlöny. Pest [=Budapest, in part]. Gazd. Közl. See B–P–H 394/13. HI 54363

Gazeta agrícola de Angola. Luanda, Angola. Gaz. Agríc. Angola. See B–P–H 393/20. HI 54354

Gazeta do agricultor. Lourenco Marques, Mozambique. Gaz. Agric. See B–P–H 393/19. HI 54353

Gazeta rolnicza. Lvov, Galicia [Ukrainian S S R]. Gaz. Roln. See B–P–H 394/11. HI 54361

Gazette apicole. Montfavet, France. Gaz. Apicole. See B–P–H 393/21. HI 54355

Gazette hebdomadaire de médecine et de chirurgie. Paris. Gaz. Hebd. Méd. Chir. See B–P–H 394/2. HI 54356

Gazette hebdomadaire des sciences médicales de Bordeaux. Bordeaux. Gaz. Hebd. Sci. Méd. Bordeaux. See B–P–H 394/3. HI 54357

Gazette médicale de Nantes. Nantes. Gaz. Méd. Nantes. See B–P–H 394/7. HI 54358

Gazette médicale de Paris. Paris. Gaz. Méd. Paris. See B–P–H 394/8. HI 54359

Gazette nationale; ou le moniteur universel. Paris. Gaz. Natl. See B–P–H 394/10. HI 54360

Gazette, peperomia society = Peperomia gazette. St. Louis, MO.

Gazzetta medica italiana. Turin. Gazz. Med. Ital. See B–P–H 394/23. HI 54369

Gazzetta medica italo-argentina. Naples. Gazz. Med. Italo-Argent. See B–P–H 394/25. HI 54370

Gazzetta medica napolitana. Naples. Gazz. Med. Napol. See B–P–H 394/26. HI 54371

Gazzetta degli ospedali e delle cliniche. Milan. Gazz. Osped. Clin. See B–P–H 395/1. HI 54372

Gb + Gw; Zeitschrift für Produktion und Dienstleistung im Erwerbsgartenbau. Aachen, Berlin. Vol. 77(26)+, 1977+. Gb. Gw. Preceded by: Deutsche Gärtnerbörse. Incorporated: Gartenwelt. HI 65514

Geigy weed tables. Basle. No. 1+, 1968+. Geigy Weed Tables. HI 65515

Geillustreerd tuinbouwblad. Leeuwarden, Netherlands. Geill. Tuinbouwbl. See B–P–H 395/8. HI 54373

Gele- roestbericht van de stichting Nederlands graancentrum. Wageningen. Gele- Roestber. Stichting Ned. Graancentrum. See B–P–H 395/18. HI 54381

Gelehrte Anzeigen. Herausgegeben von Mitgliedern der bayer. Akademie der Wissenschaften. Munich. Gel. Anz. See B–P–H 395/9. HI 54374

Gelehrte Beyträge zu den Braunschweigischen Anzeigen. Brunswick, Germany. Gel. Beytr. Braunschweig. Anz. See B–P–H 395/11. HI 54375

Gelehrte Beyträge zu den Rigischen Anzeigen. Riga [Latvian S S R]. Gel. Beytr. Rigischen Anz. See B–P–H 395/12. HI 54376

Gelehrte Ergötzlichkeiten und Nachrichten. Stuttgart. Gel. Ergötzlichk. Nachr. See B–P–H 395/13. HI 54377

Gelehrte Fama, welche den gegenwärtigen Zustand der gelehrten Welt und sonderlich deren deutschen Universitäten entdeckt. Leipzig. Gel. Fama. See B–P–H 395/14. HI 54378

Gelehrte Neuigkeiten Schlesiens. Schweidnitz, Germany [Poland]. Gel. Neuigk. Schlesiens. See B–P–H 395/ 15. HI 54379

Gelehrte Notizen der Saratover Staatsuniversität namens N. G. Tschernischewsky = Uchenye zapiski. Saratovskogo gosudarstvennyogo imeni N. G. Chernyshevskogo universiteta. Saratov.

Gelehrte Zeitung, herausgegeben zu Kiel. Kiel. Gel. Zeitung Kiel. See B–P–H 395/17. HI 54380

Gemeinnützige Blätter. [Edited by Rösler.] Gemeinnütz. Blätt. (Rösler). See B–P–H 395/23. HI 54383

Gemeinnützige Nachrichten aus dem Reiche der Wissenschaften und Künste. Hamburg. Gemeinnütz. Nachr. Wiss. Künste. See B–P–H 395/25. HI 54385

Gemeinnütziger Almanach. Hamburg. Gemeinnütz. Alman. See B–P–H 395/22. HI 54382

Gemeinnütziges Fränkisches Magazin oder Sammlung merkwürdiger nützlicher Grundsätze und Erfahrungen aus der Naturlehre, Naturgeschichte, ... Nuremberg. Gemeinnütz. Fränk. Mag. See B–P–H 395/24. HI 54384

Gemeinnütziges Natur und Kunstmagazin oder Abhandlungen zur Beförderung der Naturkunde, der Künste, Manufacturen und Fabriken. Berlin. Gemeinnütz. Natur. Kunstmag. See B–P–H 395/26. HI 54386

Gemeinnütziges Organ für Botanik = Oesterreichische botanische Zeitschrift. Vienna.

Gemeinnütziges Pesther Journal; Zeitung für Landwirthschaft, Gartenbau, Handel, Industrie und Gewerbe. Pest [=Budapest, in part]. Gemeinnütz. Pesther J. See B–P–H 395/27. HI 54387

Gemeinsame Deutsche Zeitschrift für Geburtskunde. Weimar. Gemeinsame Deutsche Z. Geburtsk. See B–P–H 396/1. HI 54388

Gemüse; Spezialblatt für den Feld- und Intensivgemüsebau. Stuttgart. Vol. 3(7)+, 1967+. Gemüse. Preceded by: Gemusebau. HI 65516

Gemüsebau. Stuttgart. Vols. 1-3(6), 1965-67. Gemüsebau. Superseded by: Gemüse. HI 65517

Gene; international journal devoted to gene cloning and recombinent nucleic acids. Amsterdam. Vol. 1+, 1977+. Gene. HI 65518

Gene analysis techniques. New York. Vols. 1-6, 1984-89. Gene Analysis Techn. Superseded by: Genetic analysis, techniques and applications. HI 65519

Genees-, natuur- en huishoudkundige jaarboeken. Dordrecht, Netherlands. Genees- Natuur- Huishoudk. Jaarb. See B–P–H 396/6. HI 54391

Genees-, natuur- en huishoudkundig kabinet. Leiden. Genees- Natuur- Huishoudk. Kab. See B–P–H 396/7. HI 54392

Geneeskundig tijdschrift voor Nederlandsch-Indië. Batavia, Dutch E. Indies [=Jakarta, Indonesia]. Geneesk. Tijdschr. Ned.-Indië. See B–P–H 396/11. HI 54393

Geneeskundige verhandelingen, aan de kon. sweedsche akademie medegedeelt en door dezelve van het jaar 1739 tot op deezen tijd bekent gamaakt. Leiden. Geneesk. Verh. Kon. Sweedsche Akad. See B–P–H 396/13. HI 54394

General administrative report of the agricultural department, Grenada = Report on the agricultural department, Grenada.

General magazine of arts and sciences, philosophical, philological, mathematical, and mechanical. London. Gen. Mag. Arts Sci. See B–P–H 396/2. HI 54389

General medical journal. [Min-kuo i hsüeh tsa chih.] Peking. Gen. Med. J. See B–P–H 396/3. HI 54390

General pharmacology. London, etc. Vol. 6+, 1975+. Gen. Pharm. Preceded by: Comparative and general pharmacology. HI 65520

General report R 8 - G R, United States forest service, southern region. Atlanta, GA. No. ?-4+, ?-1984+. Gen. Rep. R 8 - G R, U.S. Forest Serv. HI 65899

General report S A, United States forest service. Atlanta, GA. No. GR1-?, 1978-? Gen. Rep. S A, U.S. Forest Serv. Superseded by: General report R 8 - G R, United States forest service, southern region. HI 75218

General repository and review. Cambridge, MA. Vols. 1-4, 1812-13. Gen. Repos. Rev. Preceded by: Monthly anthology and Boston review. 2-1685-2. HI 65521

General science index. New York. Vol. 1+, 1978+. Gen. Sci. Index. HI 65522

General series, department of agriculture, Straits Settlements and Federated Malay States. Kuala Lumpur. Nos. 1-32, 1930-49 [suspended 1941-48]. Gen. Ser. Dep. Agric. Malay States. HI 70245

General technical report FPL. United States forest service, forest products laboratory. Madison, WI. No. FPL 1+, 1973+. Gen. Techn. Rep. F. P. L., U.S. Forest Serv. Forest Prod. Lab. HI 65523

General technical report I N T, United States forest service. Ogden, UT. No. INT 1+, 1972+. Gen. Techn. Rep. I. N. T., U. S Forest Serv. HI 65524

General technical report N C, United States forest service. St. Paul, MN. No. NC 1+, 1972+. Gen. Techn. Rep. N. C., U.S. Forest Serv. HI 65525

General technical report N E, United States forest service. Upper Darby, PA. No. NE 1+, 1973+. Gen. Techn. Rep. N. E., U.S. Forest Serv. HI 65526

General technical report P N W, United States forest service. Portland, OR. No. 1+, 1972?+. Gen. Techn. Rep. P. N. W., U.S. Forest Serv. HI 65527

General technical report P S W, United States forest service. Berkeley, CA. No. 1+, 1972+. Gen. Techn. Rep. P. S. W., U.S. Forest Serv. HI 65528

General technical report R M, United States forest service. Fort Collins, CO. No. RM 1+, 1972+. Gen. Techn. Rep. R. M., U.S. Forest Serv. HI 65529

General technical report S E, United States forest service. Washington DC. No. SE 1+, 19??+. Gen. Rechn. Rep. S. E., U.,S. Forest Serv. HI 65530

General technical report S O, United States forest service. Washington DC. No. SO 1+, 1973+. Gen. Techn. Rep. S. O., U.,S. Forest Serv. HI 65531

General technical report W O, United States forest service. Washington, DC. No. WO 1+, 1975+. Gen. Techn. Rep. W. O., U.S. Forest Serv. HI 65532

Genessee farmer. Rochester, NY. Genessee Farmer. See B–P–H 396/16. HI 54395

Genessee farmer and gardener's journal. Rochester, NY. Genessee Farmer Gard. J. See B–P–H 396/17. HI 54396

Genetic analysis, techniques and applications. New York. Vol. 7+, 1990+. Genet. Analysis Techn. Applic. Preceded by: Gene analysis techniques. HI 74841

Genetic engineering. London, New York, Orlando, FL. Vol. 1+, 1981+. Genet. Engin. (Academic Press). HI 65533

Genetic engineering; principles and methods. New York. Vol. 1+, 1979+. Genet. Engin. (Plenum Press). HI 65534

Genetic engineering and bio-technology monitor. Vienna. Vol. 1+, 1982+. Genet. Engin. Bio-Technnol. Monit. HI 65535

Genetic engineering letter. Washington, DC. 1981+. Gen. Engin. Letter. HI 65536

Genetic engineering news; the information source of the biotechnology industry. New York. 1981+. Genet. Engin. News. HI 65537

Genetic manipulation in crops newsletter. Beijing. Vol. 1+, 1985+. Genet. Manipul. Crops Newslett. HI 65538

Genetic technology news. Fort Lee, NJ. 1981+. Genet. Technol. News. HI 65539

Genetica. The Hague. Genetica. See B–P–H 396/22. HI 54399

Genetica agraria. Rome. Genet. Agrar. See B–P–H 396/19. HI 54397

Genética ibérica. Madrid. Genét. Ibér. See B–P–H 396/20. HI 54398

Genetica polonica; [sub-title varies]. Poznan. Vol. 1+, 1960+. Genet. Polon. HI 65540

Genetical research. Cambridge. Vol. 1+, 1960+. Genet. Res. HI 65541

Genetics. [Translation of: Genetika.] New York. Vol. 1, 1965. Genetics. Superseded by: Soviet genetics. HI 65542

Genetics. Princeton, NJ. Genetics. See B–P–H 396/23. HI 54400

Genetics. Supplement. [Supplement to Genetics.] Vancouver, B.C. 1969+. Genetics, Suppl. HI 65543

Genetics abstracts. London, Bethesda, MD. Vol. 1+, 1966+. Genet. Abstr. HI 65544

Genetics newsletter. Stellenbosch. 1968+. Genet. Newslett. HI 65545

Genetics and plant breeding = Genetika i selektsiya. Sofia.

Genetics, selection, evolution. Paris. Vol. 21+. 1989+. Genet. Select. Evol. Preceded by: Génétique, sélection, évolution. HI 75242

Genetika. Belgrade. Vol. 1+, 1969+. Genetika (Belgrade). HI 65546

Genetika. Moscow. No. 1+, 1965+. Genetika (Moscow). HI 65547

Genetika i evolyutsiya prirodnykh populyatsii rastenii; voprosy obshchiev teorii i kolichestvennoi fenetiki. Mokhochkala. Vol. 1+, 1975+. Genet. Evol. Prir. Populyat. Rast. HI 65548

Genetika nuusbrief = Genetics newsletter. Stellenbosch.

Genetika i selektsiya. Sofia. Vol. 1+, 1968+. Genet. Selekts. HI 65549

Genetika i selektsiya v Azerbaidzhane. Baku. 1971+. Genet. Selekts. Azerbaidzhane. HI 65550

Génétique, sélection, évolution. Paris. Vols. 15-20, 1983-88. Génét. Sélect. Évol. Preceded by: Annales de génétique et sélection animale [not entered]. Superseded by: Genetics, selection, evolution. HI 75278

Geneva special report = Special report, New York agricultural experiment station. Geneva, NY.

Genewatch; newsletter of the committee for responsible genetics. Cambridge, MA. 1983+. Genewatch. HI 65551

Genius des Neunzehnten Jahrhunderts. Altona [=Hamburg, in part]. Genius Neunzehnten Jahrh. See B–P–H 396/26. HI 54401

Genius der Zeit; ein Journal. Altona [=Hamburg, in part]. Genius Zeit. See B–P–H 397/1. HI 54402

Genome. Ottawa. Vol. 29+, 1987?+. Genome. Preceded by: Canadian journal of genetics and cytology. HI 65552

Genossenschaft "Flora", Gesellschaft für Botanik und Gartenbau zu Dresden. Sitzungsberichte und Abhandlungen. Dresden. Genossensch. "Flora" Ges. Bot. Dresden Sitzungsber. Abh. See B–P–H 397/2. HI 54403

Gentes herbarum; occasional papers on the kinds of plants. Ithaca, NY. Gentes Herb. See B–P–H 397/7. HI 54405

Gentlemen's magazine: or, monthly intelligencer. London. Gent. Mag. See B–P–H 397/6. HI 54404

Geo abstracts. B, biogeography and climatology. Norwich. 1972-73. Geo Abstr., B. Preceded by: Geographical abstracts. B, biogeography, climatology. Superseded by: Geo abstracts. B, climatology and hydrology and Ecological abstracts. HI 65553

Geo abstracts. G, remote sensing, photogrammetry and cartography. Norwich. 1978-85. Geo Abstr., G. Superseded by: Geographical abstracts, G, remote sensing, photogrammetry, and cartography. HI 65554

Geobios; an international bimonthly journal of life sciences. Jodhpur. Vol. 1+, 1974+. Geobios (Jodhpur). HI 65555

Geobios. Lyon. 1968+. Geobios (Lyon). HI 65556

Geobios. Mémoires spéciaux; Paléontologie,

stratigraphie, paléoécologie. Lyon. No. 1+, 1977+. Geobios, Mém. Spéc. HI 65557

Geobios. New reports. Jodhpur. Vol. 1+, 1982+. Geobios, New Rep. HI 65558

Geobotanica. Moscow & Leningrad = Trudy botanicheskogo instituta akademii nauk S S S R. Ser. 3, geobotanika. Moscow & Leningrad.

Geobotanica selecta. Stuttgart. Vol. 1+, 1961+. Geobot. Selecta. HI 65559

Geobotanical mapping = Geobotanicheskoe kartografirovanie. Moscow, Leningrad.

Geobotanicheskoe kartografirovanie. Moscow, Leningrad. Vol. 1+, 1963+. Geobot Kartogr. HI 65560

Geobotanichnyi zbirnyk. Kiev, Ukrainian S S R. Vols. [1]-2, 1937-38. Geobot. Zbirn. 1-121-3. HI 65561

Geobotaničnyj zbirnyk. Kiev = Geobotanichnyi zbirnyk. Kiev.

Geobotanika. Moscow & Leningrad = Trudy botanicheskogo instituta akademii nauk S S S R. Ser. 3, geobotanika. Moscow & Leningrad.

Geobotany. The Hague, etc. Vol. 1+, 1981+. Geobotany. HI 65562

Geochimica et cosmochimica acta. London. Geochim. Cosmochim. Acta. See B–P–H 397/10. HI 54406

Geochemistry. [Translation of: Geokhimiya.] Ann Arbor, MI. Vol. 1-?, 1956-63. Geochemistry (Ann Arbor). Superseded by: Geochemistry international. HI 65563

Geochemistry international. Ann Arbor, MI. 1964+. Geochem. Int. Preceded by: Geochemistry (Ann Arbor). HI 65564

Geo-eco-trop; revue internationale d'ecologie et de géographie tropicales. Lubumbashi. Vol. 1+, 1977+. Geo-Eco-Trop. HI 65565

Geoforum; journal of physical, human and regional geosciences. New York, Brunswick, etc. Vols. 1-12, 1970-81. Geoforum. HI 65566

Geografičeskij sbornik. Moscow & Leningrad = Geograficheskii sbornik. Moscow & Leningrad.

Geografičeskija Izvěstija, vydavaemyja ot Russkago Geografičeskago Obščestva = Geograficheskiya Izvěstiya, vydavaemyya ot Russkago Geograficheskago Obshchestva. St. Petersburg.

Geograficheskiya Izvěstiya, vydavaemyya ot Russkago Geograficheskago Obshchestva. St. Petersburg. 1848-50. Geogr. Izv. Superseded by: Věstnik Imperatorskago Russskago Geograficheskago Obshchestva. 2-1692-1. HI 65567

Geograficheskii sbornik. Moscow & Leningrad. Vol. 1+, 1952+. Geogr. Sborn. HI 65568

Geografinis metraštis. Vilnius. Vol. 1+, 1958+. Geogr. Metraštis. HI 65569

Geografisk tidsskrift. Copenhagen. Vol. 1+, 1877+. Geogr. Tidsskr. 2-1693-1. HI 65570

Geografiska annaler utgivne av svenska sällskapet för antropologi och geografi. Stockholm. Geogr. Ann. Svenska Sällsk. Antropol. See B–P–H 397/15. HI 54407

Géographe canadien = Canadian geographer. Toronto.

Géographica helvetica. Berne. 1946+. Géogr. Helv. Preceded by: Mitteilungen der geographisch-ethnographischen Gesellschaft in Zürich. HI 65571

Geographical abstracts. B, biogeography, climatology. London. 1966-71. Geogr. Abstr., B. Superseded by: Geo abstracts. B, biogeography and climatology. HI 65572

Geographical abstracts. G, remote sensing, photogrammetry, and cartography. Norwich. 1986-88. Geogr. Abstr., G. Preceded by: Geo abstracts. G, remote sensing, photogrammetry and cartography. Superseded by: Geographical abstracts. Physical geography. HI 65444

Geographical abstracts. Physical geography. Norwich. 1989+. Geogr. Abstr., Phys. Geogr. Preceded by: Geographical abstracts. A, Landforms & quaternary [not entered], Geographical abstracts. B, Climatology & hydrology [not entered], Geographical abstracts. E, Sedimentology [not entered] and Geographical abstracts. G, remote sensing, photogrammetry, and cartography. HI 69395

Geographical annual = Geografinis metraštis. Vilnius.

Geographical journal. London. Vol. 1+, 1893+. Geogr. J. (London). Preceded by: Proceedings of the royal geographical society of London. 2-1694-1. HI 65573

Geographical journal. [Ti hsüeh tsa-chih.] Peking. Geogr. J. (Peking). See B–P–H 397/20. HI 54408

Geographical knowledge. [Dili zhishi. Ti li chih shih.] Peking. Geogr. Knowl. (Peking). See B–P–H 397/22. HI 54410

Geographical magazine. London. Vol. 1+, 1935+. Geogr. Mag. HI 65574

Geographical review. [Ti-li tsa-chih; later Fang-chih yüeh-k'an.] Nanking. Geogr. Rev. (Nanking). See B–P–H 397/25. HI 54412

Geographical review. New York, NY. Geogr. Rev. (New York). See B–P–H 397/26. HI 54413

Geographical review of Japan. [Chirigaku hyoron.] Tokyo. Geogr. Rev. Japan. See B–P–H 398/1. HI 54414

Géographie. Paris. Géographie. See B–P–H 398/9.

HI 54417

Géographie physique et quaternaire. Montreal. Vol. 31+, 1977+. Géogr. Phys. Quatern. Preceded by: Revue de géographie de Montréal. HI 65575

Geographische Schriften. Weissenburg, Germany. Geogr. Schriften. See B–P–H 398/4. HI 54415

Geographische Zeitschrift. Leipzig & Berlin. Vol. 1+, 1895+ [suspended 1945-82]. Geogr. Z. 2-1696-3. HI 65576

Geographische Zeitung der Hertha. Stuttgart & Tübingen. Geogr. Zeitung Hertha. See B–P–H 398/8. HI 54416

Geographisches Jahrbuch. Gotha. Geogr. Jahrb. See B–P–H 397/21. HI 54409

Geographisches Magazin. Dessau & Leipzig. Geogr. Mag. See B–P–H 397/23. HI 54411

Geography. London. Geography (London). See B–P–H 398/10. HI 54418

Geography. [Ti-li.] Nanking. Geography (Nanking). See B–P–H 398/11. HI 54419

Geohimija. Moscow = Geokhimiya. Moscow.

Geokhimiya. Moscow. Vol. 1+, 1956+. Geokhimiya. HI 65577

Geología colombiana. Bogotá. Geol. Colomb. See B–P–H 398/20. HI 54422

Geologica bavarica. Munich. Geol. Bavar. See B–P–H 398/17. HI 54420

Geologica hungarica. Series geologica. Budapest. Geol. Hung., Ser. Geol. See B–P–H 398/28. HI 54423

Geologica hungarica. Series palaeontologica. Budapest. Geol. Hung., Ser. Palaeontol. See B–P–H 398/29. HI 54424

Geological magazine; or, monthly journal of geology. London. Geol. Mag. See B–P–H 399/2. HI 54425

Geological notes = Geoloogilised märkmed. Tallinn.

Geological review = Geologisches Zentralblatt. Anzeiger für Geologie, Petrographie, Palaeontologie und verwandte Wissenschaften. Leipzig. Geol. Zentralbl. See B–P–H 399/17.

Geological survey paper. Mines and geology branch, Canada. Ottawa. Geol. Surv. Pap., Mines Geol. Branch Canada. See B–P–H 399/15. HI 54428

Geologičeskie zametki = Geoloogilised märkmed. Tallinn.

Geologičnyj žurnal. Kiev = Geologichnyi zhurnal. Kiev.

Geologie. Zeitschrift für das Gesamtgebiet der Geologie und Mineralogie sowie der angewandten Geophysik. Berlin. Geologie. See B–P–H 399/21. HI 54431

Geologie en mijnbouw. The Hague. Geol. & Mijnb. See B–P–H 399/5. HI 54426

Geologie und Naturgeschichte des Ihnern der Erde. Weimar = Zeitung für Geognosie, Geologie und innere Naturgeschichte der Erden. Weimar.

Geologija i geofizika. Novosibirsk = Geologiya i geofizika. Novosibirsk.

Geologische Blätter für Nordost-Bayern und angrenzende Gebiete. Erlangen. Geol. Blätt. Nordost-Bayern. See B–P–H 398/19. HI 54421

Geologische Rundschau. Leipzig. Geol. Rundschau. See B–P–H 399/8. HI 54427

Geologisches Jahrbuch. Hanover. Vols. 64-90, 1950-72. Geol. Jahrb. Preceded by: Jahrbuch der Reichsstelle für Bodenforschung. Superseded by: Geologisches Jahrbuch. Reihe A-F. 2-1700-3. HI 65578

Geologisches Zentralblatt. Anzeiger für Geologie, Petrographie, Palaeontologie und verwandte Wissenschaften. Geol. Zentralbl. See B–P–H 399/17. HI 54429

Geologisches Zentralblatt. B. Palaeontologie. Palaeontologisches Zentralblatt. Berlin. Geol. Zentralbl., B, Palaeontol. See B–P–H 399/18. HI 54430

Geologiya i geofizika. Novosibirsk. 1960+. Geol. & Geofiz. HI 65579

Geologichnyi zhurnal. Kiev, Ukrainian S S R. Vol. 1+, 1934+. Geol. Zhurn. 2-1699-3. HI 65580

Geoloogilised märkmed. Tallinn. Vol. 1+, 1961+. Geol. Märkmed. HI 65581

Geomicrobiology journal; an international journal of geomicrobiology and microbial biochemistry. New York. Vol. 1+, 1978+. Geomicrobiol. J. HI 65582

Geophytology; an international journal of palaeobotany, palynology and allied sciences. Lucknow. Vol. 1+, 1971+. Geophytology. HI 65583

Geoponika. Athens. Geoponika. See B–P–H 399/22. HI 54432

George Landis arboretum newsletter = Newsletter, George Landis arboretum.

Georgia agricultural research. Athens, GA. Georgia Agric. Res. See B–P–H 399/24. HI 54433

Georgia Augusta. Göttingen. No. 1+, 196?+. Georgia Augusta. HI 65584

Georgia botanic journal and college sentinel. Macon, GA. Georgia Bot. J. & Coll. Sentinel. See B–P–H 399/25. HI 54434

Georgia forest lookout = Forestry-geological review. Atlanta, GA. Forest.-Geol. Rev. See B–P–H 380/8.

Georgia forest research paper. Macon, GA. No. 1+, [1961]+. Georgia Forest Res. Pap. HI 65585

Georgia forestry. Dry Branch, GA. Vol. 1+, 1948+. Georgia Forestry. 2-1705-3. HI 65991

Georgia journal of science. Atlanta, GA. Vol. 35+, 1977+. Georgia J. Sci. Preceded by: Bulletin of the Georgia academy of science. HI 65586

Georgikon. Athens. Georgikon. See B–P–H 399/27. HI 54435

Georgikon deltion. Athens. Georgikon Deltion. See B–P–H 399/28. HI 54436

Georgofili. Florence. Georgofili. See B–P–H 399/29. HI 54437

Georgooikonomik e epitheoresis = Agricultural economic review. Salonika.

Geoscience and man. Baton Rouge, LA. Vol. 1+, 1970+. Geosci. & Man. HI 65587

Geotimes. Washington, DC. Geotimes. See B–P–H 399/30. HI 54438

Geraniaceae group news, british pelargonium and geranium society. Uxbridge. No. 1+, 1981+. Geraniaceae Group News. HI 65588

Geranium year book. Hitchin. Vols. 1-9, 1954-64. Geranium Year Book. Superseded by: Year book, british pelargonium and geranium society. HI 65589

Geraniums around the world. Santa Paula, Buena Park, CA. Vol. 1+, 1953+. Geraniums Around World. HI 65590

Gerard's garden. Arlington, VA. ?-1983+. Gerard's Gard. HI 65591

Germ plasm newsletter. Addis Ababa. No. ?-2+, 19??-83+. Germ Plasm Newslett. HI 65592

Germany journal of horticultural science = Gartenbauwissenschaft. Berlin, Vienna & Munich.

Germinacion. Ambato. No. 1+, 1970+. Germinacion. HI 65593

Germinal. Mexico City. Germinal. See B–P–H 400/2. HI 54439

Geschäftliche Mitteilungen, Nachrichten von der Gesellschaft der Wissenschaften zu Göttingen = Nachrichten von der Gesellschaft der Wissenschaften zu Göttingen. Geschäftliche Mitteilungen. Berlin. Nachr. Ges. Wiss. Göttingen Geschäftl. Mitt. See B–P–H 626/3.

Geschäftsbericht der naturwissenschaftlichen Gesellschaft in Dresden. Dresden. Geschäftsber. Naturwiss. Ges. Dresden. See B–P–H 400/14. HI 54448

Geschichte der merkwürdigen Reisen seit dem 12. Jahrhundert. Weimar. Gesch. Merkwürd. Reisen. See B–P–H 400/12. HI 54446

Geschichte der Regensburgischen Botanischen Gesellschaft nebst einigen Aufsätzen, Reden und Abhandlungen. Regensburg. Gesch. Regensburg. Bot. Ges. See B–P–H 400/13. HI 54447

Geschichtliche Darstellung der Naturforschenden Gesellschaft in Emden. Emden. Gesch. Darstellung Naturf. Ges. Emden. See B–P–H 400/10. HI 54444

Geschichtliche Darstellung der Verhandlungen der Naturforschenden Gesellschaft in Emden. Emden. Gesch. Darstellung Verh. Naturf. Ges. Emden. See B–P–H 400/11. HI 54445

Gesellschaft für Natur- und Völkerkunde Ostasiens, e. V., Hamburg. Nachrichten. Hamburg. Ges. Natur-Völkerk. Ostasiens Hamburg Nachr. See B–P–H 400/3. HI 54440

Gesellschaft naturforschender Freunde zu Berlin Magazin für die neuesten Entdeckungen in der gesammten Naturkunde. Berlin. Ges. Naturf. Freunde Berlin Mag. Neuesten Entdeck. Gesammten Naturk. See B–P–H 400/6. HI 54441

Gesellschaft naturforschender Freunde zu Berlin, Neue Schriften. Berlin. Ges. Naturf. Freunde Berlin Neue Schriften. See B–P–H 400/7. HI 54442

Gesellschaft naturforschender Freunde Westphalens neue Schriften. Düsseldorf. Ges. Naturf. Freunde Westphalens Neue Schriften. See B–P–H 400/9. HI 54443

Gesneriad journal. San Leandro, CA. Gesneriad J. See B–P–H 400/15. HI 54449

Gesneriad journal news bulletin. Nashville, TN. No. 1+, 1962+. Gesneriad J. News Bull. Preceded by?: Gesneriad journal news letter. HI 65594

Gesneriad journal news letter. Irving, CA. Vols. 1-9, 1953-61. Gesneriad J. News Lett. Superseded by?: Gesneriad journal news bulletin. HI 65595

Gesnerus. Vierteljahrsschrift für Geschichte der Medizin und der Naturwissenschaften. Aarau, Switzerland. Gesnerus. See B–P–H 400/17. HI 54450

Gesunde Pflanzen. Monatsblatt für Pflanzenschutz und Schädlingsbekämpfung. Frankfurt a. M. Gesunde Pflanzen. See B–P–H 400/18. HI 54451

Getreide und Mehl. Detmold, Germany. Getreide & Mehl. See B–P–H 400/19. HI 54452

Gewasproduksie = Crop production. Pretoria.

Gewässer und Abwässer. Dusseldorf. Nos. 1-68/69, 1953-82. Gewässer & Abwässer. HI 65596

Gewerbefleiss. Berlin. Gewerbefleiss. See B–P–H 401/1. HI 54453

Ghana farmer. Accra, Ghana. Ghana Farmer. See B–P–H 401/2. HI 54454

Ghana journal of agricultural science. Accra. Vol. 1+, 1968+. Ghana J. Agric. Sci. HI 65597

Ghana journal of science. Accra, Ghana. Ghana J. Sci. See B–P–H 401/4. HI 54455

Ghanaian bulletin of agricultural economics. Accra. Vols. 1-2(2), 1961-62. Ghanaian Bull. Agric. Econ. HI 65598

Giardini; rivista mensile. Milan. 1854-1948. Giardini. Superseded by: Flora giardini orti & frutteti. 2-1724-2. HI 65599

Giardino fiorito. Florence & Rome. Giardino Fior. See B–P–H 401/6. HI 54456

Gidrobiologičeskij žurnal. Kiev = Gidrobiologicheskii zhurnal. Kiev.

Gidrobiologičeskij žurnal S S S R = Gidrobiologicheskii zhurnal S S S R. Saratov.

Gidrobiologicheskie issledovaniya = Hüdrobioloogilised uurimused. Tartu.

Gidrobiologicheskii zhurnal. Kiev, Ukrainian S S R. Vol. 1+, 1965+. Gidrobiol. Zhurn., (Kiev). HI 65600

Gidrobiologicheskii zhurnal S S S R. Saratov, Russian S F S R. Vol. 9, 1930. Gidrobiol. Zhurn. S.S.S.R. Preceded by: Russkii gidrobiologicheskii zhurnal. 2-1724-3. HI 65601

Gids. Amsterdam. Gids. See B–P–H 401/11. HI 54457

Gids in het plantenrijk. Tijdschrift voor tuinbouw en plantkunde. Amsterdam. Gids Plantenrijk. See B–P–H 401/12. HI 54458

Giessener naturwissenschaftliche Vorträge. Giessen, Germany. Giessener Naturwiss. Vorträge. See B–P–H 401/13. HI 54459

Gifu daigaku gakugei by kenkyu hokoku. Shizen kagaku = Science report of the faculty of liberal arts and education; Gifu university. Natural science. Gifu, Japan. Sci. Rep. Fac. LIberal Arts Gifu Univ., Nat. Sci. See B–P–H 827/2.

Gifu-ken kanreichi ringyo shikenjo. Shiken hokoku. Takayama. No. 1+, 1972+. Gifu-ken Kanreichi Ringyo Shikenjo, Shiken Hokoku. HI 65602

Gifu-ken no nogyo = Agriculture of Gifu prefecture. Gifu, Japan. Agric. Gifu Prefect. See B–P–H 58/2.

Gifu-ken nogyo shikenjo hokoku = Research bulletin of the Gifu agricultural experiment station. Gifu, Japan. Res. Bull. Gifu Agric. Exp. Sta. See B–P–H 774/9.

Gimtasai kraštas. Siauliai. Vols. ?-3(19)-5(30)-?, 19??-38-42-? Gimt. Kršt. HI 65603

Ginecologia. Recife, Brazil. Ginecologia (Recife). See B–P–H 401/23. HI 54465

Ginecologia. Turin. Ginecologia (Turin). See B–P–H 401/24. HI 54466

Ginecologia moderna. Genoa. Ginecol. Moderna. See B–P–H 401/18. HI 54460

Ginecología y obstetricia de México. Mexico City. Ginecol. Obstet. México. See B–P–H 401/19. HI 54461

Ginecologia e l'ostetricia pratica. Naples. Ginecol. Ostet. Prat. See B–P–H 401/20. HI 54462

Ginecologia pratica. Genoa. Ginecol. Prat. See B–P–H 401/21. HI 54463

Ginecologia: revista pratica. Florence. Ginecologia (Florence). See B–P–H 401/22. HI 54464

Ginkgoana; contributions to the flora of Asia and the Pacific region. Tokyo. Vol. 1+, 1973+. Ginkgoana. HI 65604

Gin'yo = Silvery leaves. Hakodate.

Giornale agrario italiano, industriale e commerciale. Forti, Italy. Giorn. Agrar. Ital. See B–P–H 401/31. HI 54467

Giornale agrario toscano. Florence. Giorn. Agrar. Tosc. See B–P–H 402/1. HI 54468

Giornale di agricoltura. Jesi [=Iesi], Italy. Giorn. Agric. (Jesi). See B–P–H 402/2. HI 54469

Giornale di agricoltura industriale e commercio del regno d'Italia = Giornale di agricoltura del regno d'Italia. Piacenza & Bologna. Giorn. Agric. Regno Italia. See B–P–H 402/3.

Giornale di agricoltura del regno d'Italia. Piacenza & Bologna. Giorn. Agric. Regno Italia. See B–P–H 402/3. HI 54470

Giornale di anatomia e fisiologia patologica. Pavia. Giorn. Anat. Fisiol. Patol. See B–P–H 402/5. HI 54471

Giornale di anatomia, fisiologia e patologia degli animali. Pisa. Giorn. Anat. Fisiol. Patol. Anim. See B–P–H 402/6. HI 54472

Giornale arcadico di scienze, lettere, ed arti. Rome. Giorn. Arcadico Sci. See B–P–H 402/11. HI 54473

Giornale della associazione napolitana di medici e naturalisti. Naples. Giorn. Assoc. Napol. Med. See B–P–H 402/12. HI 54474

Giornale di batteriologia e immunologia. Turin. Giorn. Batteriol. Immunol. See B–P–H 402/14. HI 54475

Giornale di biologia applicata alla industria chimica. Bologna. Ser. 2, vols. 1-4, 1931-34. Giorn. Biol. Appl. Industr. Chim. Preceded by: Zymologica e chemica dei colloidi e degli zuccheri. Superseded by: Giornale di biologia industriale, agraria ed alimentare. 5-4648-3. HI 65605

Giornale di biologia industriale, agraria ed alimentare. Bologna. Ser. 2, vols. 5-7, 1935-37. Giorn. Biol. Industr. Agrar. Aliment. Preceded by: Giornale di biologia applicata alla industria chimica. Superseded by: Bollettino scientifico della facoltà di chimica industriale, università di Bologna. 5-4648-3. HI 65606

Giornale botanico italiano. Florence. Vols. 1-2, 1844-47; [n.s.] vol. 69+, 1962+. Giorn. Bot. Ital. For 1869-1961 see: Nuovo giornale botanico italiano. 2-1727-1. HI 65607

Giornale enciclopedico di Firenze. Florence. Giorn. Encicl. Firenze. See B–P–H 402/18. HI 54476

Giornale enciclopedico di Napoli. Naples. Giorn. Encicl. Napoli. See B–P–H 402/19. HI 54477

Giornale di farmacia-chimica e scienze accessorie ... Milan. Giorn. Farm.-Chim. Sci. Accessorie. See B–P–H 402/21. HI 54478

Giornale di fisica, chimica, storia naturale, medicina ed arti = Giornale di fisica, chimica e storia naturale ossia raccolta di memorie sulle scienze, arti e manifatture ad esse relative. Pavia. Giorn. Fis. See B–P–H 402/22.

Giornale di fisica, chimica, storia naturale, medicina, ed arti del regno italico = Giornale di fisica, chimica e storia naturale ossia raccolta di memorie sulle scienze, arti e manifatture ad esse relative. Pavia. Giorn. Fis. See B–P–H 402/22.

Giornale di fisica, chimica e storia naturale ossia raccolta di memorie sulle scienze, arti e manifatture ad esse relative. Pavia. Giorn. Fis. See B–P–H 402/22. HI 54479

Giornale fisico-chimico italiano. Rome. Giorn. Fis.-Chim. Ital. See B–P–H 402/24. HI 54480

Giornale fisico-medico ossia raccolta di osservazioni sopra la fisica, matematica, chimica, storia naturale, medicina, chirurgia, arti e agricoltura. Pavia. Giorn. Fis.-Med. See B–P–H 403/4. HI 54481

Giornale dell'imperiale reale istituto lombardo di scienze, lettere ed arti e biblioteca italiana. Milan. Vols. 1-16 (= nos. 1-48), 1841-47; n.s. vols. 1-9, 1847-56. Giorn. Imp. Reale Ist. Lombardo Sci. & Bibliot. Ital. Preceded by: Biblioteca italiana ossia giornale di letteratura, scienze ed arti. Superseded by: Atti dell reale istituto lombardo di scienze, lettere ed arti. 3-2127-3. HI 65608

Giornale dell'italiana letteratura; compilato da una società di letterati italiani. Padua. Vols. 1-31, 1802-11; ser. 2, vols. 1-30 [also numbered vols. 32-61], 1812-24; ser. 3, vols. 1-3 [also numbered vols. 62-64], 1825-26; ser. 4, vols. 1-2 [also numbered vols. 65-66], 1828, [publication suspended 1826-27]. Giorn. Ital. Lett. 2-1727-3. HI 52959

Giornale d'Italia agricola. Rome. Giorn. Italia Agric. See B–P–H 403/7. HI 54483

Giornale d'Italia, spettante alla scienze naturale e principalmente all'agricoltura, alle arti ed al commercio. Venice. Giorn. Italia Sci. Nat. See B–P–H 403/8. HI 54484

Giornale italiano di dermatologia e sifilologia. Milan. Giorn. Ital. Dermatol. Sifilol. See B–P–H 403/6. HI 54482

Giornale letterario-scientifico modenese. Modena, Italy. Giorn. Lett.-Sci. Modenese. See B–P–H 403/13. HI 54487

Giornale de' letterati. Pisa. Giorn. Letterati (Pisa). See B–P–H 403/10. HI 54485

Giornale de' letterati pubblicato in Firenze. Florence. Giorn. Letterati Firenze. See B–P–H 403/11. HI 54486

Giornale della letteratura straniera; scientiis et bonis artibus. Mantua. Vols. 1-2, 1793. Giorn. Lett. Straniera. 2-1727-3. HI 52780

Giornale ligustico di scienze, lettere ed arti. Genoa. Giorn. Ligustico Sci. See B–P–H 403/14. HI 54488

Giornale di medicina pratica. Padua. Giorn. Med. Prat. See B–P–H 403/15. HI 54489

Giornale medico-chirurgico di Alessandro Flajani. Rome. Vols. 1-8, 1808-13. Giorn. Med.-Chir. (Roma). HI 52852

Giornale di microbiologia. Milan. Giorn. Microbiol. See B–P–H 403/16. HI 54490

Giornale pisano di letteratura, scienze ed arti. Pisa. Giorn. Pisano Lett. See B–P–H 403/17. HI 54491

Giornale del reale istituto d'incoraggiamento di agricoltura, arti e manifatture in Sicilia. Palermo. Ser. 3, vol. 1, 1863. Giorn. Reale Ist. Incoragg. Agric. Sicilia. HI 65609

Giornale di risicoltura. Vercelli, Italy. Giorn. Risic. See B–P–H 403/18. HI 54492

Giornale scientifico e letterario dell'accademia italiana di scienze, lettere ed arti. Pisa. Giorn. Sci. Lett. Accad. Ital. Sci. See B–P–H 403/20. HI 54494

Giornale scientifico, letterario e delle arti di una società filosofica di Torino. Turin. Giorn. Sci. Soc. Filos. Torino. See B–P–H 403/24. HI 54497

Giornale di scienze ed arti. Florence. Giorn. Sci. Arti. See B–P–H 403/19. HI 54493

Giornale di scienze, letteratura ed arti per la Sicilia. Palermo. Giorn. Sci. Sicilia. See B–P–H 403/23. HI 54496

Giornale di scienze, lettere ed arti per la Sicilia = Giornale di scienze, letteratura ed arti per la Sicilia. Palermo. Giorn. Sci. Sicilia. See B–P–H 403/23.

Giornale de scienze naturali ed economiche di Palermo. Palermo. Giorn. Sci. Nat. Econ. Palermo. See

B–P–H 403/22. HI 54495

Giornale della società d'incoraggiamento delle scienze e delle arti stabilata in Milano. Milan. Giorn. Soc. Incoragg. Sci. Milano See B–P–H 404/2. HI 54498

Giornale della società medico-chirurgica di Parma. Parma. Giorn. Soc. Med.-Chir. Parma. See B–P–H 404/3. HI 54499

Giornale toscano di scienze mediche, fisiche e naturali. Pisa. Giorn. Tosc. Sci. Med. See B–P–H 404/4. HI 54500

Giornale di viticoltura e di enologia, agricoltura e industrie agrarie. Avellino, Italy. Giorn. Vitic. Enol. See B–P–H 404/5. HI 54501

Glad messenger. Belle Vernon, PA. Glad Messenger. See B–P–H 404/8. HI 54502

Gladio grams. Brandenton, FL. No. 1+, 1971+. Gladio Grams. HI 65610

Gladiolus. Boston, MA. Gladiolus. See B–P–H 404/9. HI 54503

Gladiolus annual of the british gladiolus society. Colchester, England. Gladiolus Annual Brit. Gladiolus Soc. See B–P–H 404/11. HI 54504

Gladiolus bulletin. Medway, MA. Gladiolus Bull. See B–P–H 404/12. HI 54505

Gladiolus magazine. Boston, MA. Gladiolus Mag. See B–P–H 404/13. HI 54506

Gladiolus review = American gladiolus society official review. Goshen, IN. Amer. Gladiolus Soc. Off. Rev. See B–P–H 80/11.

Glas srpske kralevske akademije. Belgrade. Glas Srpske Kral. Akad. See B–P–H 404/16. HI 54507

Glasgow naturalist. Glasgow. Glasgow Naturalist. See B–P–H 404/18. HI 54508

Glasnik biološke sekcije. Zagreb, Yugoslavia. Glasn. Biol. Sekc. See B–P–H 404/20. HI 54509

Glasnik botaničkog zavoda i bašte universiteta u Beogradu = Bulletin de l'institut et du jardin botaniques de l'université de Belgrade. Belgrade. Bull. Inst. Jard. Bot. Univ. Belgrade. See B–P–H 256/6.

Glasnik državnog muzeja u Sarajevu. Prirodne nauke. Sarajevo, Yugoslavia. Glasn. Državn. Muz. u Sarajevu, Prir. Nauke. See B–P–H 404/22. HI 54510

Glasnik hrvatskih zemaljskog muzeja u Sarajevu. Sarajevo, Yugoslavia. Glasn. Hrvatsk. Zemaljsk. Muz. u Sarajevu. See B–P–H 405/2. HI 54513

Glasnik hrvatskog prirodoslovnoga društva. Zagreb, Croatia [Yugoslavia]. Glasn. Hrvatsk. Prir. Društva. See B–P–H 405/1. HI 54512

Glasnik hrvatskoga naravoslovnoga društva (družtva). Zagreb, Croatia [Yugoslavia]. Glasn. Hrvatsk. Nar. Društva. See B–P–H 404/23. HI 54511

Glasnik muzeja šumarstva i lova. Belgrade. Nos. 1-6, 1961-66. Glasn. Muz. Šumarstva Lova. Superseded by: Glasnik prirodnjackog muzeja u Beogradu. Seriya C, šumarstvo i lov. HI 65611

Glasnik muzejskega društva za Slovenijo. Ljubljana, Yugoslavia. Glasn. Muz. Društva Slovenijo. See B–P–H 405/5. HI 54514

Glasnik odjeljenja prirodnih nauku, Crnogorska akademija nauka i umjetnosti. Titograd. Vol. 2+, 1977+. Glasn. Odjeljenje Prir. Nauka Crnogorska Akad. Nauka Umjetn. Preceded by: Glasnik, odjeljenje prirodnikh nauka, društvo za nauku i umjetnost Crne Gore. HI 65613

Glasnik, odjeljenje prirodnikh nauka, društvo za nauku i umjetnost Crne Gore. Titograd. Vol. 1, 1974. Glasn. Odjeljenje Prir. Nauka Društvo Nauku Umjetn. Crne Gore. Superseded by: Glasnik odjeljenja prirodnih nauku, Crnogorska akademija nauka i umjetnosti. HI 65612

Glasnik otdelenija estestvennych nauk, Černogorskaja akademija nauk i iskusstv. Titograd = Glasnik odjeljenja prirodnih nauku, Crnogorska akademija nauka i umjetnosti. Titograd.

Glasnik otdeleniya estestvennykh nauk, Chernogorskaya akademiya nauk i iskusstv. Titograd = Glasnik odjeljenja prirodnih nauku, Crnogorska akademija nauka i umjetnosti. Titograd.

Glasnik prirodnaučkog museja u Beogradu. Serija B. Biološke nauke. Belgrade. Glasn. Prir. Mus. u Beogradu, Ser. B, Biol. Nauke. See B–P–H 405/6. HI 54515

Glasnik prirodnjačkog muzeja u Beogradu. Seriya C, šumarstvo i lov. Belgrade. Vol. 7+, 1973+. Glasn. Prir. Muz. Beogradu, C. Preceded by: Glasnik muzeja šumarstva i lova. HI 65615

Glasnik prirodnjačkog muzeja srpske zemlje. Serija B. Biološke nauke. Belgrade. Glasn. Prir. Muz. Srpske Zemlje, Ser. B, Biol. Nauke. See B–P–H 405/7. HI 54516

Glasnik, republičkog zavoda za zaštitu prirode i prirodnjačke zbirke u Titogradu. Titograd. No. 1+, 1968+. Glasn. Republ. Zavoda Zaštitu Prir. Prirodnjačke Zbirke Titogradu. HI 65616

Glasnik, republičkog zavoda za zaštitu prirode i prirodnjäckog muzeja u Titogradu = Glasnik, republičkog zavoda za zaštitu prirode i prirodnjäcke zbirke u Titogradu. Titograd.

Glasnik of the section of natural sciences of the Montenegrin academy of sciences and arts. Titograd = Glasnik odjeljenja prirodnih nauku, Crnogorska akademija nauka i umjetnosti. Titograd.

Glasnik skopskog naučnog društva. Skoplje, Yugoslavia. Glasn. Skopsk. Naučn. Društva. See B–P–H 405/10. HI 54517

Glasnik, société des sciences et des arts de Monténégro, classe des sciences naturelles. Titograd = Glasnik, odjeljenje prirodnikh nauka, društvo za nauku i umjetnost Crne Gore. Titograd.

Glasnik šumarskog fakulteta, univerzitet u Beogradu. Belgrade. Vol. 1+, 1950+. Glasn. Šumarsk. Fak. Univ. u Beogradu. HI 65617

Glasnik za šumske pokuse. Zagreb, Yugoslavia. Glasn. Šumske Pokuse. See B–P–H 405/12. HI 54518

Glasnik zaštite bilja. Zagreb. Vol. 1+, 1978+. Glasn. Zaštite Bilja. Preceded by: Biljni lekar and Biljna zaštita. HI 65618

Glasnik zemaljskog muzeja u Bosni i Hercegovini. Sarajevo, Bosnia and Herzegovina [Yugoslavia]. Glasn. Zemaljsk. Muz. Bosni Hercegovini. See B–P–H 405/14. HI 54519

Glasnik zemaljskog muzeja Bosne i Hercegovine u Sarajevu. Prirodne nauke. Sarajevo. N.s. vol. ?+, 19??+. Glasn. Zemaljsk. Muz. Bosne Hercegovine Sarajevu. Prir. Nauke. Preceded by?: Glasnik zemaljskog muzeja u Sarajevu. HI 65619

Glasnik zemaljskog muzeja kraljevine Jugoslavije. Sarajevo, Yugoslavia. Glasn. Zemaljsk. Muz. Kral. Jugoslavije. See B–P–H 405/15. HI 54520

Glasnik zemaljskog muzeja u Sarajevu. Sarajevo, Yugoslavia. Glasn. Zemaljsk. Muz. u Sarajevu. See B–P–H 405/16. HI 54521

Glasra; contributions from the national botanic gardens, Glasnevin. Dublin. Vol. 2+, 1978+. Glasra. Preceded by: Contributions, national botanic gardens, Glasnevin. HI 65620

Glastechnische Berichte. Frankfurt am Main. 1923+. Glastechn. Ber. HI 65621

Gleanings in bee culture. Medina, OH. Gleanings Bee Cult. See B–P–H 405/24. HI 54522

Gleanings from the garden, Santa Barbara botanical garden. Santa Barbara, CA. Vol. 1+, 1976+. Gleanings Gard. Santa Barbara Bot. Gard. HI 65622

Gleanings in science. Calcutta. Gleanings Sci. See B–P–H 406/2. HI 54523

Gleditschia; Beiträge zur botanischen Taxonomie und deren Grenzgebiete. Berlin. Vol. 1+, 1973+. Gleditschia. HI 65623

Glimpses in plant research. Lucknow. Vol. 1+, 1973+. Glimpses Pl. Res. HI 65624

Global biogeochemical cycles. Washington, DC. Vol. 1+, 1987+. Global Biogeochem. Cycles. HI 75219

Global ecology and biogeography letters. Oxford. Vol. 1+, 1991+. Global Ecol. Biogeogr. Lett. HI 75262

Global and planetary change. [Forms a supplement to: Palaeogeography, palaeoclimatology, palaeoecology.] Amsterdam. Vol. 1+, 1989+. Global Planet. Change. HI 75279

Globus. Illustrierte Zeitschrift für Länder- und Völkerkunde. Hildburghausen, Germany. Globus. See B–P–H 406/3. HI 54524

Gloucestershire naturalists' society journal = Journal, Gloucestershire naturalists' society. Cheltenham.

Gloxinian. Gray, OK. Gloxinian. See B–P–H 406/4. HI 54525

Gnat. Cardiff. No. 1+, 1980+. Gnat. Preceded by: Newsletter, Glamorgan county naturalists' trust. HI 65625

Godičnyj žurnal (ežegodnik) Vengerskogo geologičeskogo instituta. Budapest = A magyar állami földtani intézet évkönyve. Budapest. Magyar Állami Földt. Intéz. Évk. See B–P–H 542/19.

Godišen zbornik, bioloski fakultet na univerzitetot "Kiril i Metodij". Skopje. Vol. 29+, 1976+. Godišen Zborn. Biol. Fak. Univ. Kiril i Metodij. Preceded by: Godišen zbornik, prirodno-matematički fakultet na univerzitetot- Skopje. Biologija. HI 65626

Godišen zbornik, filozofski fakultet na univerzitetot. Prirodno-matematički oddel. Skopje. Vols. 1-12, 1948-59. Godišen Zborn. Filoz. Fak. Univer., Prir.-Mat. Odd. Superseded by: Godišen zbornik, prirodno-matematički fakultet na univerzitetot- Skopje. Biologija. HI 65627

Godišen zbornik, prirodno-matematički fakultet na univerzitetot- Skopje. Biologija. Skopje. Vols. 13-27/28, 1962-75. Godišen Zborn. Prir.-Mat. Fak. Univ. Skopje, Biol. Preceded by: Godišen zbornik, filozofski fakultet na univerzitetot. Prirodno-matematički oddel. Superseded by: Godišen zbornik, bioloski fakultet na univerzitetot "Kiril i Metodij". HI 65628

Godišen zbornik na zemjodelsko-šumarskiot fakultet na univerzitetot Skopje. Šumarstvo. Skopje. Vol. 1+, 1947/48+. God. Zborn. Zemjod.-Šumarsk. Fak. Univ. Skopje, Šumarstvo. 5-3880-1. HI 65629

Godišen zbornik na zemjodelsko-šumarski fakultet na univerzitetet Skopje. Zemjodelstvo. Skopje. Vol. 1+, 1947/48+. God. Zborn. Zemjod.-Šumarsk Fak. Univ. Skopje, Zemjod. 5-3880-1. HI 65630

Godišnik sofijskija universitet. Sofia = Godishnik sofiiskiya universitet. Sofia.

Godišnjak biološkog instituta u Sarajevu. Sarajevo, Yugoslavia. God. Biol. Inst. u Sarajevu. See B–P–H 406/5. HI 54526

Godishnik sofiiskiya universitet. Sofia. Vols. 1-4, 1904/05-06/07, 1906-08. God. Sofiisk. Univ. Superseded by: Godishnik sofiiskiya universitet, fiziko-matematicheski fakultet. 5-3991-1. HI 65631

Godishnik sofiiskiya universitet, biologi-geologi-geografiski fakultet. Kniga 1, biologija, botanika. Sofia. Vols. 50-57, 1955/56-62/63, 1956-63. God. Sofiisk. Univ. Biol.-Geol.-Geogr. Fak., 1. Biol. Bot. Preceded by: Godišnik sofiiskiya universitet, prirodo-matematicheski fakultet. HI 65632

Godishnik sofiiskiya universitet, biologičeski fakultet. Kniga 2, botanika. Sofia. Vol. 68+, 1973/75+ 1975+. God. Sofiisk. Univ. Biol. Fak., 2 Bot. Preceded by: Godishnik sofiiskiya universitet, biologičeski fakultet. Kniga 2, botanika, mikrobiologiya, fiziologiya i biokhimiya na rastenyata. HI 65633

Godishnik sofiiskiya universitet, biologičeski fakultet. Kniga 2, botanika, mikrobiologiya, fiziologiya i biokhimiya na rastenyata. Sofia. Vols. 58-67, 1963/64-72/73. 1964-73. God. Sofiisk. Univ. Biol. Fak., 2. Bot. Mikrobiol. Fiziol. Biokhim. Rast. Preceded by: Godišnik sofiiskiya universitet, biologi-geologi-geografiski fakultet. Kniga 1, biologija, botanika. Superseded by: Godishnik sofiiskiya universitet, biologičeski fakultet. Kniga 2, botanika. HI 65634

Godišnik sofiiskiya universitet, fiziko-matematicheski fakultet. Sofia. Vols. 5-43(1), 1910-47. God. Sofiisk. Univ. Fiz.-Mat. Fak. Preceded by: Godishnik sofiiskiya universitet. Superseded by: Godišnik sofiiskiya universitet, prirodo-matematicheski fakultet. 5-3991-1. HI 65635

Godišnik sofiiskiya universitet, prirodo-matematicheski fakultet. Sofia. Vols. 44-49, 1948?-55? God. Sofiisk. Univ. Prir.-Mat. Fak. Preceded by: Godišnik sofiiskiya universitet, fiziko-matematicheski fakultet. Superseded by: Godišnik sofiiskiya universitet, biologi-geologi-geografiski fakultet. Biologija. 5-3991-1. HI 65636

Godišnjak poljoprivredno-šumarskog fakulteta, universitet u Beogradu. Belgrade. Vols. 1-4, 1948-52. God. Poljopr.-Šumarsk. Fak. Univ. u Beogradu. Superseded by: Zbornik radova poljoprivrednog fakulteta, universitet u Beogradu. 1-633-2. HI 65637

Godišnjak. Srpska kraljevska akademija. Belgrade. God. Srpska Kral. Akad. See B–P–H 406/8. HI 54528

Godišnjak zemaliskog zavoda zaštitu spomenika kulture i prirodnik rijetkosti N. R. Bosne i Hercegovine. Sarajevo, Yugoslavia. God. Zemaljsk. Zavoda Zašt. Spomen. Kult. Bosne Hercegovine See B–P–H 406/9. HI 54529

Godišnji izveštaj državnoga fitopatološka zavoda. Sarajevo, Yugoslavia. God. Izv. Državn. Fitopatol. Zavoda. See B–P–H 406/6. HI 54527

Godovoj otčet Vengerskogo geologičeskogo instituta. Budapest = A magyar állami földtani intézet évi jelentese(i). Budapest. Magyar Állami Földt. Intéz. Évi Jel. See B–P–H 542/18.

Goethe; Vierteljahrs-Schrift der Goethe-Gesellschaft. Neue Folge des Jahrbuchs. Weimar. Vol. 1+, 1936+ [suspended; 1945-46]. Goethe. Preceded by: Jahrbuch der Goethe-Gesellschaft. 2-1739-2. HI 65638

Goethe-Jahrbuch. Frankfurt am Main. Vols. 1-34, 1880-1913. Goethe-Jahrb. Superseded by: Jahrbuch der Goethe-Gesellschaft. 2-1739-3. HI 65639

Gojo noyu = Agriculture propagation. [Miyagi prefecture propagation assoc.] Sendai, Japan. Agric. Propag. See B–P–H 60/3.

Golden bough; newsletter to foster biosystematics of loranthaceae and viscaceae. Kew. No. 1+, 1982+. Golden Bough. HI 65640

Gomu; F N R D monthly = Journal of the foundation for natural rubber research and development. Tokyo.

Good gardening; Australian house and garden magazine. Sydney, N.S.W. 1970-81. Good Gard. HI 65641

Gornoe delo. Moscow. Vols. 96-97, 1920-21. Gornoe Delo. Preceded & superseded by: Gornyi zhurnal. 2-1747-2. HI 65642

Gornyi zhurnal; ili sobranie svěděnii o gornom i solyanom dělě, s prisovokupleniem novykh otkrytii po naukam, k semu predmetu otnosyashchimsya. St. Petersburg. Vols. [1]-93, 1825-1917; vol. 98+, 1922+ [Volumation only in some later vols.; vol. 96, 1920 not published]. Gornyi Zhurn. For 1918 see: Izvestiya gornogo otdela. For 1920-21 see: Gornoe Delo. 2-1747-3. HI 53240

Gornij žurnal. St. Petersburg = Gornyi zhurnal. St. Petersburg.

Gornyj žurnal ili Sobranie Svěděnij o Gornom i Soljanom Dělě, s Prisovokupleniem Novyh Otkrytij po Naukam, k Semu Predmetu Otnosjaščimsja. St. Petersburg = Gornyi zhurnal. St. Petersburg.

Gorosti. ?-1987+. Gorosti. HI 65643

Gorskostopanska nauka. Sofia. Vol. 1+, 1964+. Gorskost. Nauka. HI 65644

Gortania; atti del museo Friulano di storia naturale. Udine. Vol. 1+, 1979+. Gortania. HI 65645

Gorteria; mededelingenblad ten dienste van de floristiek en het vegetatie-onderzoek van Nederland. Leiden. Gorteria. See B–P–H 406/15. HI 54530

Gosamvardhana. New Delhi. Gosamvardhana. See B–P–H 406/16. HI 54531

Göteborgs k. vetenskaps- och vitterhets samhälles handlingar. Ny tidsfoljd. Goteborg. N.s. [= ser. 3], vols. 1-32, 1850-97; ser. 4, vols. 1-33, 1898-1928.

Göteborgs Kungl. Vetensk. Samhälles Handl. Preceded by: Götheborgs kungl. wettenskaps och witterhets samhällets nya handlingar [not entered]. Superseded by: Göteborgs k. vetenskaps- och vitterhets samhälles handlingar. Series B, matematiska och naturvetenskapliga skrifter. 2-1750-1. HI 65646

Göteborgs k. vetenskaps- och vitterhets samhälles handlingar. Series B, matematiska och naturvetenskapliga skrifter. Goteborg. Ser. 5, vols. 1-6, 1928/30-40; ser. 6, vol. 1-9, 1941-66. Göteborgs Kungl. Vetensk. Samhälles handl., Ser. B, Mat. Naturvetensk. Skr. Preceded by: Göteborgs k. vetenskaps- och vitterhets samhälles handlingar. Superseded by: Acta regiae societatis scientiarum et litterarum gothoburgensis. Botanica. 2-1750-2. HI 65647

Gothaische gelehrte Zeitungen. Gotha. 1774-1804. Gothaische Gel. Zeitungen. 2-1750-2. HI 50596

Gothaische gelehrte Zeitungen. Ausländische Literatur. Gotha. 1787-94. Gothaische Gel. Zeitungen Ausl. Lit. 2-1750-2. HI 52603

Gothaisches Magazin der Künste und Wissenschaften. Gotha. Gothaisches Mag. Künste Wiss. See B–P–H 406/21. HI 54532

Götheborgs och Bohusläns hushall-sällskapets handlingar. Goteborg, Sweden. Götheborgs Hush.-Sällsk. Handl. See B–P–H 406/23. HI 54533

Götheborgs kongliga wettenskaps och witterhets samhällets nya handlingar = Nya handlingar af kongliga wettenskaps och witterhets samhället i Götheborg. Goteborg, Sweden. Nya Handl. Kongl. Wettensk. Samhället Götheborg. See B–P–H 677/14.

Götheborgska wetenskaps och witterhets samhällets handlingar. Wetenskaps afdelningen. Goteborg, Sweden. Götheborgska Wetensk. Samhällets Handl., Wetensk. Afd. See B–P–H 406/25. HI 54534

Göttinger Beiträge zur Land- und Forstwirtschaft in den Tropen und Subtropen. Göttingen. No. 1+, 1983+. Göttinger Beitr. Land- Forstw. Tropen Subtrop. HI 65648

Göttinger floristische Rundbriefe. Göttingen. Vols. 1-20?, 1967-87?, 1969-87?. Göttinger Florist. Rundbr. Superseded by: Floristische Rundbriefe. HI 65649

Göttinger Jahrbuch. Göttingen. Göttinger Jahrb. See B–P–H 407/7. HI 54540

Göttingische Anzeigen von gelehrten Sachen unter der Aufsicht der Königl. Gesellschaft der Wissenschaften. Göttingen. Gött. Anz. Gel. Sachen. See B–P–H 406/26. HI 54535

Göttingische gelehrte Anzeigen (unter der Aufsicht der königl. Gesellschaft der Wissenschaften.) Göttingen. Gött. Gel. Anz. See B–P–H 407/1. HI 54536

Göttingische Zeitungen von Gelehrten Sachen. Göttingen. Gött. Zeitungen Gel. Sachen. See B–P–H 407/4. HI 54539

Göttingisches Journal der Naturwissenschaften. Göttingen. Gött. J. Nasturwiss. See B–P–H 407/2. HI 54537

Göttingisches Magazin der Wissenschaften und Literatur. Göttingen. Gött. Mag. Wiss. Lit. See B–P–H 407/3. HI 54538

Gourd. Mt. Gilead, OH. 1971+. Gourd. Preceded by: Gourd seed. HI 65650

Gourd seed. Boston, MA. 1939-70. Gourd Seed. Superseded by: Gourd. 2-1751-2. HI 65651

Gower; journal of the Gower society. 1948+. Gower. HI 65652

Gozdarski vestnik. Ljubljana. 1942+. Gozd. Vestn. 2-1752-2. HI 65653

Grădina şi livada. Bucharest. Grăd. & Livada. See B–P–H 407/9. HI 54541

Grădina, via şi livada. Bucharest. Vols. 3-15, 1954-66. Grăd. Via Livada. Preceded by: Grădina, via şi livada. Superseded by: Revista de horticultura şi viticultură. HI 65655

Gradinarska i lozarska nauka. Sofia, Bulgaria. Grad. Loz. Nauka. See B–P–H 407/10. HI 54542

Gradinarstvo. Sofia, Bulgaria. Gradinarstvo. See B–P–H 407/13. HI 54543

Graduate school chronicle, university of Maryland. College Park, MD. Vols. 1-7(1), 1967-73 [vol. 6(4) not published]. Graduate School Chron. Univ. Maryland. Superseded by: University of Maryland graduate school chronicle. HI 65656

Graham's american monthly magazine of literature, art, and fashion = Graham's lady's and gentlemen's magazine. Philadelphia, PA.

Graham's lady's and gentlemen's magazine. Philadelphia, PA. Vols. 18-22, [22]-[53], 1841-58. Graham's Lady's & Gent. Mag. Preceded by: Casket. 2-1753-2. HI 65657

Graham's magazine of literature and art = Graham's lady's and gentlemen's magazine. Philadelphia, PA.

Gran Sasso d'Italia. Aquila, Italy. Gran Sasso Italia. See B–P–H 407/17. HI 54544

Grana; an international journal of palynology including World pollen and spore flora. Stockholm. Vol. 10+, 1970+. Grana. Preceded by: Grana palynologica. HI 65658

Grana palynologica; an international journal of palynology (later including world pollen and spore flora). Stockholm. 1948-53; n.s., vols. 1-9, 1954-69. Grana Palynol. Superseded by: Grana. HI 65659

Grand Canyon nature notes = Nature notes from the Grand Canyon. Grand Canyon, AZ. Nat. Notes Grand Canyon. See B–P–H 629/21.

Grande; the magazine for bromeliophiles. Vol. 1, 1979. Grande. HI 65660

Granos; semilla selecta. Buenos Aires. Granos. See B–P–H 407/21. HI 54545

Grape belt and Chautauqua farmer. Dunkirk, NY. Grape Belt Chautauqua Farmer. See B–P–H 407/22. HI 54546

Grape culturist. St. Louis, MO. Grape Cult. See B–P–H 407/23. HI 54547

Grape grower. San Francisco, CA. 1949 [no other data available]. Grape Grower. Superseded by: California fruit and grape grower. 5-4486-3. HI 65661

Grape and fruit growers advocate = Fruit and grape grower. Charlottesville, VA. Fruit Grape Grower. See B–P–H 385/24.

Grape vine. Greensburg, PA. Vol. ?-13(8)+, 19??-64+. Grape Vine. HI 65662

Graphia bulletin. Kapellen. ?-1967+. Graphia Bull. HI 65663

Grasas y aceites. Seville. 1950+. Grasas & Aceites. HI 65664

Grass and forage science. Oxford. Vol. 34+, 1979+. Grass Forage Sci. Preceded by: Journal, british grassland society. HI 65665

Grassland of China. [Zhongguo caoyuan.] Beijing. No. ?-16+, 19??-83+. Grassland China. HI 65666

Great basin naturalist. Provo, UT. Great Basin Naturalist. See B–P–H 408/4. HI 54548

Great basin naturalist memoirs. Provo, UT. Vol. 1+, 1976+. Great Basin Naturalist Mem. HI 65667

Great lakes research highlights. Grosse Ile, MI. 1973+. Great Lakes Res. Highlights. HI 65668

Grebe. Northwich. 1974?+. Grebe (Cheshire). Preceded by: News bulletin, Cheshire conservation trust ltd. HI 65669

Grebe; journal of the Hertfordshire & Middlesex trust for nature conservation. [Hemel Hempstead] & St. Albans. 1970-73. Grebe (Herts.). HI 65670

Green America. Washington, DC. Vol. 1+, 1972+. Green Amer. HI 65671

Green guerilla report; a quarterly devoted to the greening and cleaning of our environment. New York. 1979+. Green Guerilla Rep. HI 65672

Green scene. Philadelphia, PA. Vol. 1+, 1972+. Green Scene. HI 65673

Green thumb. Denver, CO. Vol. 1+, 1944+. Green Thumb (Denver). 2-1763-3. HI 54549

Green thumb. Louisville? 19??-76+. Green Thumb (Louisville). HI 65674

Green thumb news. Denver, CO. ?-1983+. Green Thumb News. HI 65675

Greenhouse. Hounslow. Vols. 1-8, 1956-63. Greenhouse. From 1964+ incorporated in: Gardener's chronicle. HI 65676

Greenhouse, garden, grass. Ottawa. Greenh. Gard. Grass. See B–P–H 408/8. HI 54550

Greenhouse grower. Vol. 1+, 19??+. Greenh. Grower. HI 65677

Greenhouse manager. 19??-82+. Greenh. Manager. HI 65678

Greenhouse notes. 19??-68+. Greenh. Notes. HI 65679

Greenland bioscience = Meddelelser om Grønland. Bioscience. Copenhagen.

Greenland geoscience = Meddelelser om Grønland. Geoscience. Copenhagen.

Greifswaldisches academemisches Archiv. Greifswald, Germany. Greifswald. Acad. Arch. See B–P–H 408/10. HI 54551

Greifswaldische academemische Zeitschrift. Greifswald, Germany. Greifswald. Acad. Z. See B–P–H 408/11. HI 54552

Greinar. Reykjavik. 1935+. Greinar. 5-4410-1. HI 65680

Grenada agricultural botanical bulletin. Nos. 1-2, 1890. Grenada Agric. Bot. Bull. Superseded by: Bulletin of miscellaneous information, botanic gardens, Grenada. HI 65681

Grevillea; a monthly record of cryptogamic botany and its literature. London. Grevillea. See B–P–H 408/12. HI 54553

Gribnye Vrediteli Kul'turnyh i Dikorastuščih Poleznyh Rastenij Kavkaza = Gribnye Vrediteli Kul'turnykh i Dikorastushchikh Poleznykh Rastenii Kavkaza. Tiflis.

Gribnye Vrediteli Kul'turnykh i Dikorastushchikh Poleznykh Rastenii Kavkaza. [Supplement to: Trudy Tiflisskogo Botanicheskogo Sada.] Tiflis, Georgian S S R. 1 vol., 1912. Gribn. Vredit. Kul't. Dikorast. Polezn. Rast. Kavkaza. HI 65682

Groei & bloei; maanblad van de koninklijke maatschappij tuinbouw en plantenkunde. The Hague. ?-1982(7)+. Groei & Bloei. HI 65683

Groenten en fruit. Orgaan van den nederlandschen tuinbouw. The Hague. Groenten & Fruit. See B–P–H 408/15. HI 54554

Grower. Adelaide, S.A. Vol. 1+, 1946+. Grower (Adelaide). 2-1770-1. HI 54555

Grower. Toronto, Ont. Vol. 1+, 1952+. Grower

(Toronto). Preceded by: Canadian grower. HI 65684

Grower's digest of current research. London. Grower's Digest Curr. Res. See B–P–H 408/18. HI 54556

Growing native plants. Canberra, A.C.T. Vol. 1+, 1971+. Growing Native Pl. HI 65685

Growth. Lakeland, FL. Growth (Lakeland). See B–P–H 408/19. HI 54557

Growth. Montreal. Growth (Montreal). See B–P–H 408/20. HI 54558

Growth point; the horticultural therapy magazine. Frome. No. 1+, 1979+. Growth Point. HI 65686

Grundlagen und Fortschritte im Garten- und Weinbau. Stuttgart. No. 1+, 1935+. Grundl. Fortschr. Gart.-Weinbau. 2-1770-3. HI 65687

Grusonia. (Netherlands). No. ?-22+, 19??-82+. Grusonia. HI 65688

Guam agricultural experiment station and its work, office of experiment stations, United States department of agriculture = U S department of agriculture, office of experiment stations; the Guam agricultural experiment station and its work. Washington, DC. U.S.D.A. Off. Exp. Sta. Guam Agric. Exp. Sta. Work. See B–P–H 944/1.

Guanabara; revista mensal artistica, scientifica e litteraria. Rio de Janeiro. 1849-56. Guanabara. Superseded by: Revista brazileira (1857-61). 2-1771-3. HI 65689

Guar newsletter. Hissar. Vol. 1+, 1980+. Guar Newslett. HI 65690

Guatemala agrícola. Guatemala City, Guatemala. Guatemala Agríc. See B–P–H 409/1. HI 54559

Guernsey society quarterly review. Guernsey, C.I. ?-1972+. Guernsey Soc. Quart. Rev. HI 65691

Guide du botaniste sur le Grand St. Bernard = Bulletin de la Murithienne; société valaisanne des sciences naturelles, no. 1. Sion, Switzerland. Bull. Murith. Soc. Valais. Sci. Nat. See B–P–H 264/10.

Guide to bulb experiments, Kirton experimental husbandry farm. Boston. 1963-64. Guide Bulb Exp. Kirton Exp. Husb. Farm. Preceded by: Guide to horticultural experiments, Kirton experimental husbandry farm. HI 65692

Guide to experiments, Efford experimental horticulture station. Lymington. 1961-62. Guide Exp. Efford Exp. Hort. Sta. Preceded by: Station guide and programme of experiments, Efford experimental horticulture station. Incorporated in: Report (Annual), Efford experimental horticulture station. HI 65693

Guide to experiments, Jealott's Hill research station. Bracknell. 1930+. Guide Exp. Jealott's Hill Res. Sta. HI 65694

Guide to experiments, Stockbridge house experimental horticulture station. Cawood. 1967-68. Guide Exp. Stockbridge House Exp. Hort. Sta. Preceded by: Station guide, Stockbridge house experimental horticulture station. HI 65695

Guide to field experiments and farm projects, Jealott's Hill research station = Guide to experiments, Jealott's Hill research station. Bracknell.

Guide to graduate study in botany for the United States and Canada. Pullman, WA. 1966+. Guide Graduate Study Bot. U.S. Canada. HI 65696

Guide to horticultural experiments, Kirton experimental husbandry farm Boston. 1954-62. Guide Hort. Exp. Kirton Exp. Husb. Farm. Superseded by: Guide to bulb experiments, Kirton experimental husbandry farm. HI 65697

Guides for science teaching of the Boston natural history society. Boston, MA. Guides Sci. Teaching Boston Nat. Hist. Soc. See B–P–H 409/3. HI 54560

Guihaia. [Kuang-hsi chih wu.] Yanshan, Guilin. No. 1+, 1981+. Guihaia. HI 65698

Guild gardener. Edgware, England. Guild Gard. See B–P–H 409/4. HI 54561

Gulf coast grower. Gulf Coast Grower. See B–P–H 409/6. HI 54562

Gulf coast lumberman. Houston, TX. Gulf Coast Lumberman. See B–P–H 409/7. HI 54563

Gulf fauna and flora bulletin. Ruston, LA. Gulf Fauna Fl. Bull. See B–P–H 409/8. HI 54564

Gulf research reports. Ocean Springs, MS. 1961+. Gulf Res. Rep. Ocean Springs - Gulf Coast Research Laboratory. HI 65699

Gulf states grower = Gulf coast grower. Gulf Coast Grower. See B–P–H 409/6.

Gunma daigaku kiyo, shizen kagaku hen = Science reports of the Gunma university, natural science series. Maebashi.

Gunneria. Trondheim. Vol. 26+, 1976+. Gunneria. Preceded by: Miscellanea, det kongelige norske videnskabernes selskabs museet. HI 65700

Gurukula kangri vishwavidyalaya journal of scientific research. Hardwar. Vol. 1+, 1969+. Gurukula Kangri Vishwavidyalaya J. Sci. Res. HI 65701

Gwyddonydd. Cardiff, Wales. Gwyddonydd. See B–P–H 409/11. HI 54565

Gymnocalcium. Kufstein. Ca.1989+. Gymnocalcium. HI 65702

Gyógyszeres értekezések. Gyógy. Értek. See B–P–H 409/12. HI 54566

Gyógyszerész. Budapest. Gyógyszerész. See B–P–H 409/13. HI 54567

Gyógyszerészet. Budapest. Gyógyszerészet (Budapest). See B–P–H 409/15. HI 54568

Gyógyszerészet. Misztotfalu, Hungary. Gyógyszerészet (Misztotfalu). See B–P–H 409/16. HI 54569

Gyógyszerészetudományi értesitö. Budapest. Gyógyszerésztud. Értes. See B–P–H 409/17. HI 54570

Gyümölcsészeti, borászati és kertészeti újság. Budapest. Gyümölcsész. Borász. Kert. Újs. See B–P–H 409/19. HI 54571

Gyümölcsészeti és konyhakertészeti füzetek. Budapest. Gyümölcsész. Konyhakert. Füz. See B–P–H 409/20. HI 54572

Gyümölcsészeti, szölöszéti és kertészeti újság. Budapest. Gyümölcsész. Szölösz. Kert. Újs. See B–P–H 409/21. HI 54573

Gyümölcskertész. Budapest. Gyümölcskertész. See B–P–H 409/22. HI 54574

Gyümölcskultura. Szeged, Hungary. Gyümölcskultura. See B–P–H 409/23. HI 54575

Gyümölcstermesztés. Budapest. Vol. 1-3, 1974-76. Gyümölcstermesztés. Preceded by: Szölö- es gyümölcstermesztés. HI 65703

H B L newsletter. Pittsburgh, PA. No. 1, 1967. H. B. L. Newslett. HI 65704

H-Briefe für Wissenschaft und Praxis = Höfchen-Briefe für Wissenschaft und Praxis. Leverkusen.

H E A year book. Wye. Vols. 1-2, 1932-33. H. E. A. Year Book. Superseded by: Scientific horticulture. HI 65705

H S T C bulletin; quarterly newsletter for the history of science and technology of Canada (later journal of the history of Canadian sciences, technology, and medicine). Downsview, Thornhill, Ont. Vols. 1-7(3) [nos. 1-25], 1976?-83. H. S. T. C. Bull. Superseded by: Scientia canadensis. HI 75268

Habitat. Carlton, Vic. No. 1+, 1973+. Habitat (Carlton). HI 65708

Habitat. London. Vol. 1+, 1965+. Habitat (London). HI 65706

Habitat. Oxford. Vol. 1+, 1976+. Habitat (Oxford). HI 65707

Hacienda. Revista mensual ilustrada sobre agricultura, ganaderia e industrias rurales. Buffalo, NY. Vol. 1+, 1905+. Hacienda. 3-1782-2. HI 65709

Haenkeana; revista del círculo botánico boliviano. Cochabamba. No. 1(+?), 1973(+?). Haenkeana. HI 65710

Haeyang kaebal yonguso sobo. Seoul. Vols. 1-2, 1979-81. Haeyang Kaebal Yonguso Sobo. Superseded by: Bulletin of Korea ocean research and development institute. HI 65711

Hai yang hsueh pao = Acta oceanologica sinica. Beijing.

Hai yang k'o hsüeh chi K'an = Journal of oceanographic works in China. Studia marina sinica. Peking. J. Oceanogr. Works China. See B–P–H 475/7.

Hai yang yü hu chao = Oceanologia et limnologia sinica. Hong Kong. Oceanol. Limnol. Sin. See B–P–H 682/3.

Hai yang yu hu zhao = Oceanologia et limnologia sinica. Peking (Beijing).

Hainyan nyenguso sobo = Bulletin of Korea ocean research and development institute. Seoul.

Haiyang xuebao = Acta oceanologica sinica. Beijing.

Háj. Prague. Háj. See B–P–H 410/5. HI 54576

Hakken nyusu = Fermentation research institute news. Chiba, Japan. Ferment. Res. Inst. News. See B–P–H 369/14.

Hakko kenkyusho hokoku = Research communications, institute for fermentation, Osaka. Osaka.

Hakko kogaku kaishi. Osaka. Vol. 55+, 1977+. Hakko Kogaku Kaishi. Preceded by: Journal of fermentation technology. HI 65712

Hakko kogaku zasshi = Journal of fermentation technology. Osaka.

Hakko to kogyo. Tokyo. Vols. 34-46(3), 1976-88. Hakko To Kogyo. Preceded by: Journal of the fermentation association, Japan. Superseded by: Baiosaiensu to indasutori. HI 65713

Hakkogogaku = Hakko kogaku kaishi. Osaka.

Hakubutsu gaku zasshi = Journal of natural history. Tokyo.

Haladó gazda. Budapest. Haladó gazda. See B–P–H 410/11. HI 54577

Half-yearly journal of Mysore university. Bangalore. Vols. 1-8, 1927-36. Half-Yearly J. Mysore Univ. Superseded by: Half-yearly journal of Mysore university. New series, section B. 3-2808-3. HI 65714

Half-yearly journal of Mysore university. New series, section B, science, including medicine and engineering. Bangalore. Vols. 1-?, 1940-64. Half-Yearly J. Mysore Univ., N.S., B. Preceded by: Half-yearly journal of Mysore university. Superseded by: Journal of the Mysore university. HI 65715

Half-yearly progress report, cocoa research institute of Ghana. Tafo. Nos. 2-7, 1963-67. Half-Yearly Progr. Rep. Cocoa Res. Inst. Ghana. Preceded by: Half-yearly progress report, west african cocoa research institute. HI 65716

Half-yearly progress report, cocoa research institute of

Nigeria. Ibadan. Nos. 4-8, 1965-69. Half-Yearly Progr. Rep. Cocoa Res. Inst. Nigeria. Preceded by: Half-yearly progress report, west african cocoa research institute. HI 65717

Half-yearly progress report, west african cocoa research institute. Tafo. No. 1, 1962. Half-Yearly Progr. Rep. W. African Cocoa Res. Inst. (Tafo). Superseded by: Half-yearly progress report, cocoa research institute of Ghana. HI 65718

Half-yearly progress report, west african cocoa research institute (Nigeria). Ibadan. Nos. 1-3, 1963-64. Half-Yearly Progr. Rep. W. African Cocoa Res. Inst. (Nigeria). Superseded by: Half-yearly progress report, cocoa research institute of Nigeria. HI 65719

Halifax field naturalists' newsletter. Halifax, Nova Scotia. No. 1+, 1975+. Halifax Field Naturalists' Newslett. HI 65720

Halifax naturalist and record of the scientific society. Halifax, Nova Scotia. Halifax Naturalist & Rec. Sci. Soc. See B–P–H 410/14. HI 54578

Hallische Jahrbücher für deutsche Wissenschaft und Kunst. Halle. Hallische Jahrb. Deutsche Wiss. Kunst. See B–P–H 410/16. HI 54579

Hamburger Garten- und Blumenzeitung. Eine Zeitschrift für Garten- und Blumenfreunde, für Kunst- und Handelsgärtner. Hamburg. Hamburger Garten-Blumenzeitung. See B–P–H 411/3. HI 54589

Hamburgische Berichte von neuen gelehrten Sachen. Hamburg. Hamburg. Ber. Neuen Gel. Sachen. See B–P–H 410/19. HI 54580

Hamburgische Berichte von den neuesten gelehrten Sachen = Hamburgische Berichte von neuen gelehrten Sachen. Hamburg. Hamburg. Ber. Neuen Gel. Sachen. See B–P–H 410/19.

Hamburgische Beyträge zur Aufnahme der gelehrten Historie und der Wissenschaften. Hamburg. Vols. 1-4, 1740-43. Hamburg Beytr. Aufnahme Gel. Hist. Wiss. HI 65721

Hamburgische Gartenbibliothek, ... Hamburg. Hamburg. Gartenbiblioth. See B–P–H 410/24. HI 54582

Hamburgischer Lehrerverein für Naturkunde = Hamburgischer Lehrer-Verein für Naturkunde. Bericht. Hamburg. Hamburg. Lehrer-Verein Naturk. Ber. See B–P–H 410/28.

Hamburgische literarische und kritische Blätter. Hamburg. Hamburg. Lit. Krit. Blätt. See B–P–H 410/29. HI 54586

Hamburgische Nachrichten aus dem Reiche der Gelehrsamkeit. Hamburg. Hamburg. Nachr. Gelehrsamk. See B–P–H 411/2. HI 54588

Hamburgischer Garten-Almanach. Hamburg. Hamburg. Gart.-Alman. See B–P–H 410/23. HI 54581

Hamburgischer gemeinnütziger Allmanach. Hamburg. Hamburg. Gemeinnütz. Allman. See B–P–H 410/25. HI 54583

Hamburgischer Lehrer-Verein für Naturkunde. Bericht. Hamburg. Hamburg. Lehrer-Verein Naturk. Ber. See B–P–H 410/28. HI 54585

Hamburgisches Institut für angewandte Botanik. Jahresbericht. Hamburg. Hamburg. Inst. Angew. Bot. Jahresber. See B–P–H 410/26. HI 54584

Hamburgisches Magazin, oder gesammlete Schriften, aus der Naturforschung und den angenehmen Wissenschaften überhaupt = Hamburgisches Magazin, oder gesammlete Schriften, zum Unterricht und Vergnügen aus der Naturforschung und den angenehmen Wissenschaften überhaupt. Hamburg & Leipzig. Hamburg. Mag. See B–P–H 411/1.

Hamburgisches Magazin, oder gesammlete Schriften, zum Unterricht und Vergnügen aus der Naturforschung und den angenehmen Wissenschaften überhaupt. Hamburg & Leipzig. Hamburg. Mag. See B–P–H 411/1. HI 54587

Hamdard. Karachi. Vols. 1-19, 1957-76; vol. 23+, 1980+. Hamdard. For 1976-79 see: Hamdard medicus. HI 65722

Hamdard medicus. Karachi. Vol. 20-22, 1976-79. Hamdard Med. Preceded and superseded by: Hamdard. HI 65723

Hana no go = [New information on horticulture. Flowers.] Kyoto.

Hanauisches Magazin. Hanau. Hanauisches Mag. See B–P–H 411/4. HI 54590

Handbook and annual report society for the promotion of nature reserves. London = Handbook, society for the promotion of nature reserves. London.

Handbook of the flora and fauna of South Australia. Adelaide, S.A. 1923+. Handb. Fl. Fauna South Australia. HI 65724

Handbook of the flora of Papua New Guinea. Carlton, Vic. Vol. 1+, 1978+. Handb. Fl. Papua New Guinea. HI 65725

Handbook, national cactus and succulent society. Bradford. No. 1+, 1968+. Handb. Natl. Cact. Succ. Soc. HI 65726

Handbook, nature conservancy. London. 1968+, 1968+. Handb. Nat. Conservancy. HI 65727

Handbook, New York state museum = New York state museum handbook. Albany, NY. New York State Mus. Handb. See B–P–H 656/25.

Handbook of the north western naturalists' union. Manchester. 1961-68. Handb. N. W. Naturalists' Union. HI 65728

Handbook, society of natural history. Peking. Nos. 1-6, 1926-38. Handb. Soc. Nat. Hist., Peking. HI 65729

Handbook society for the promotion of nature reserves. London. 1922-69. Handb. Soc. Promot. Nat. Reserves. HI 65730

Handbook, western australian naturalists' club. Nedlands, W.A. 1950+. Handb. W. Austral. Naturalists' Club. HI 65731

Handbuch der Österreichischen Wissenschaft = Österreichische Akademie der Wissenschaften. Handbuch der Österreichischen Wissenschaft. Vienna. Österr. Akad. Wiss. Handb. Öesterr. Wiss. See B–P–H 692/9.

Handbuch der Pflanzenanatomie. Berlin, Stuttgart. Vol. 1+, 1921+. Handb. Pflanzenanat. HI 65732

Handbuch der Pflanzenkrankheiten. Berlin, Hamburg. Vol. 1+, 1905+ [several editions]. Handb. Pflanzenkrankh. HI 65733

Handelinge van die weidingsgenootskap van suidelike Afrika = Proceedings (of the annual congresses), grassland society of southern Africa. Pietermaritzburg, etc.

Handelingen van de hydrobiologische club. Amsterdam. Vols. 1-?, 1938-47. Handel. Hydrobiol. Club. Superseded by: Handelingen van de hydrobiologische vereeniging. HI 65734

Handelingen van de hydrobiologische vereeniging. Amsterdam. 1950-58. Handel. Hydrobiol. Ver. Preceded by: Handelingen van de hydrobiologisches club. HI 65735

Handelingen van de nederlands-belgische vereeniging van graanonderzoekers. Wageningen. Handel. Ned.-Belg. Ver. Graanonderz. See B–P–H 411/10. HI 54592

Handelingen van het nederlandsch natuuren geneeskundig congres. [Netherlands]. Handel. Ned. Natuur-Geneesk. Congr. See B–P–H 411/9. HI 54591

Handelingen van het provinciaal genootschap van kunsten en wetenschappen in Noord Brabant. 's Hertogenbosch, Netherlands. Handel. Prov. Genootsch. Kunsten N. Brabant. See B–P–H 411/11. HI 54593

Handelingen van het vlaamsch natuuren geneeskundig congres. Antwerp. Handel. Vlaamsch Natuur-Geneesk. Congr. See B–P–H 411/12. HI 54594

Handelsblad voor den tuinbouw Sempervirens. Amsterdam. Handelsbl. Tuinb. Sempervirens. See B–P–H 411/15. HI 54596

Handelsblatt für den deutschen Gartenbau und die mit ihm verwandten Zweige. Leipzig. Handelsbl. Deutsch. Gartenbau Verw. Zweige. See B–P–H 411/14. HI 54595

Handlingar, finska läkaresällskapet = Finska läkaresällskapet, handlingar. Helsinki. Finska Läkaresällsk. Handl. See B–P–H 371/11.

Handlingar, kongl. götheborgska wetenskaps och witterhets samhällets. Wetenskaps afdelningen. Goteborg = Kongl. götheborgska wetenskaps och witterhets samhällets handlingar. Wetenskaps afdelningen. Goteborg, Sweden. Kongl. Götheborgska Wetensk. Samhällets Handl., Wetensk. Afd. See B–P–H 517/4.

Handlingar, kongl. svenska vetenskaps academiens. Stockholm = Kongl. svenska vetenskaps academiens handlingar. Stockholm. Kongl. Svenska Vetensk. Acad. Handl. See B–P–H 517/10.

Handlingar, kongl. svenska vetenskapsakademiens. Stockholm = Kongl. svenska vetenskapsakademiens handlingar. Stockholm. Kongl. Svenska Vetenskapsakad. Handl. See B–P–H 517/11.

Handlingar, kongl. swenska wetenskaps academiens. Stockholm = Kongl. swenska wetenskaps academiens handlingar. Stockholm. Kongl. Swenska Wetensk. Acad. Handl. See B–P–H 517/12.

Handlingar, kongl. vetenskaps academiens. Stockholm = Kongl. vetenskaps academiens handlingar. Stockholm. Kongl. Vetensk. Acad. Handl. See B–P–H 517/13.

Handlingar, kungl. svenska vetenskapsakademiens. Stockholm = Kongl. svenska vetenskapsakademiens handlingar. Stockholm. Kongl. Svenska Vetenskapsakad. Handl. See B–P–H 517/11.

Handlingar, svenska läkare-sällskapet = Svenska läkare-sällskapet. Handlingar. Stockholm. Svenska Läkare-Sällsk. Handl. See B–P–H 861/14.

Hang-chou ta hsüeh hsüeh pao. Tzu jan k'o hsüeh pan = Journal of Hangzhou university. Natural science edition.

Han'gok ui suhyongmok = Plus trees in Korea. Suwon.

Hangug nyujen haghoi ji = Korean journal of genetics. Seoul.

Hanguk choji yongu hoe-ji. Seoul. [Vol.] ?-1(2)+, 19??-79+. Hanguk Choji Yongu Hoe-ji. HI 65736

Han'guk chongwon hakhoe chi = Journal of the korean traditional garden research association. Seoul.

Han'guk imhakhoe chi = Journal of the korean forestry society. Suwon.

Hanguk journal of genetic engineering. [Hanguk yujon konghakchi.] Seoul. Vol. 1+, 1984+. Hanguk J. Genet. Engin. HI 65737

Han'guk sikmul poho hak hoe = Korean journal of plant protection. Suwon.

Han'guk singmul poho hakhoe chi = Korean journal of plant protection. Suwon.

Han'guk wonye hakhoe chi = Journal, korean society for horticultural science. Seoul.

Hanguk yujon hakhoe chi = Korean journal of genetics. Seoul.

Hanguk yujon konghakchi = Hanguk journal of genetic engineering. Seoul.

Han'guk yukchong hakhoe = Korean journal of breeding.

Hangzhou daxue xuebao. Ziran kexue ban = Journal of Hangzhou university. Natural science edition.

Hannoverische Beyträge zum Nutzen und Vergnügen. Hanover. Hannover. Beytr. Nutzen Vergnügen. See B–P–H 411/17. HI 54597

Hannoverische Land- und forstwirtschaftliche Zeitung. Hanover. 1847+. Hannover. Land- Forstw. Zeitung. HI 65738

Hannoverisches Magazin worin kleine Abhandlungen, ... gesamlet (gesammelt) und aufbewahret sind. Hanover. Hannover. Mag. See B–P–H 411/18. HI 54598

Hanseatisches Magazin. Bremen. Hamseat. Mag. See B–P–H 411/19. HI 54599

Harbinger. Carbondale, IL. Vols. 1-4, 1982-85. Harbinger. HI 65739

Hardwicke's science gossip. London. Hardwicke's Sci. Gossip. See B–P–H 411/21. HI 54600

Hardy and half-hardy plants. Hampton Wick, England. Hardy Half-Hardy Pl. See B–P–H 412/1. HI 54601

Hardy plant; bulletin of the hardy plant society. Reading. Vol. ?-7+, 19??-85+. Hardy Pl. Preceded by: Bulletin of the hardy plant society. HI 65740

Hardy plant directory. [Nottingham.] No. 1+, 1975+. Hardy Pl. Directory. HI 65741

Hardy plant year book. London. 1912-13. Hardy Pl. Year Book. HI 65742

Harrowsmith. Camden East. Vol. 1+, 1976+. Harrowsmith. HI 65743

Harvard botanical memoirs. Boston, MA. Harvard Bot. Mem. See B–P–H 412/4. HI 54602

Harvard forest papers. Petersham, MA. Harvard Forest Pap. See B–P–H 412/6. HI 54603

Harvard journal of asiatic studies. Cambridge, MA. Harvard J. Asiat. Stud. See B–P–H 412/7. HI 54604

Harvard landscape architecture monographs. Cambridge, MA. Harvard Landscape Architect. Monogr. See B–P–H 412/9. HI 54605

Harvard magazine. Cambridge, MA. Vol. 76+, 1973+. Harvard Mag. Preceded by: Harvard bulletin [not included]. HI 65744

Harvard papers in botany. Cambridge, MA. No. 1+, 1989+. Harvard Pap. Bot. Preceded by: Botanical museum leaflets. [Harvard university]; Contributions of the Gray herbarium of Harvard university and Occasional papers of the Farlow herbarium of cryptogamic botany. HI 72815

Harvest; newsletter of the associates of the national agricultural library. Washington, DC. Vol. 1+, 1989+. Harvest. HI 74842

Harvester. Warner Robins, GA. Vol. 1+, 1965?+. Harvester. HI 65745

Harvests. Marysville, OH, Pleasant Hill, TN. 1957+. Harvests. HI 65746

Haryana journal of horticultural sciences. Hissar. Vol. 1+, 1972+. Haryana J. Hort. Sci. HI 65747

Hasad, Al. Libya. ?-1977+. Hasad (Libya). HI 65748

Hasad, Al. Beirut. Vols. ?-8+, 19??-??+. Hasad (Beirut). HI 65749

Haslemere museum gazette; journal of objective education and field study. Haslemere. Vol. 1, 1906-07. Haslemere Mus. Gaz. HI 65750

Haslo ogrodnicze. Warsaw. Vol. 28+, 1971+. Haslo Ogrodn. Preceded by: Haslo ogrodniczo-rolnicze. HI 65751

Haslo ogrodniczo-rolnicze. Vols. ?-21-27, 19??-64-70. Haslo Ogrodn.-Roln. Superseded by: Haslo ogrodnicze. HI 65752

Hastings and East Sussex naturalist. St. Leonards, England. Hastings E. Sussex Naturalist. See B–P–H 412/13. HI 54606

Hasznos mulatságok. Pest [=Budapest, in part]. Hasznos Mul. See B–P–H 412/15. HI 54607

Hatabako kenkyu = Research of tobacco leaf. Tokyo. Res. Tobacco Leaf. See B–P–H 776/23.

Hatachi nogyo = New development on agriculture. Tokyo. New Developm. Agric. See B–P–H 652/2.

Hatch experiment station of Massachusetts agricultural college. Bulletin. Amherst, MA. Hatch Exp. Sta. Mass. Agric. Coll. Bull. See B–P–H 412/18. HI 54608

Hattori shokubutsu kenkyusho hokoku = Journal of the Hattori botanical laboratory. Obi.

Häussknechtia; Mitteilungen der Thüringischen botanischen Gesellschaft. Jena. Vol. 1+, 1984+. Häussknechtia. Preceded by: Mitteilungen der Thüringischen botanischen Gesellschaft. HI 65754

Haustorium; parasitic plant newsletter. Norfolk, VA. Vol. 1+, 1978+. Haustorium. HI 65755

Hautes fagnes; revue trimestrielle publiée avec l'appui du ministère de l'éducation nationale et de la culture. Verviers. Vol. ?-36+, 19??-70+. Hautes Fagnes. HI 65756

Have-avis; udgivet af det haveselskab for Jylland, Fyen og hertugdømmet Slesvig. Odense. [52 pts.], 1844-47. Have-Avis. HI 65757

Havekust. Copenhagen. Havekunst. See B–P–H 412/20. HI 54609

Haven; medlemsblad før de samvirkende danske haveselskaber. Copenhagen. Haven. See B–P–H 412/21. HI 54610

Havetidende. Copenhagen. Havetidende. See B–P–H 412/22. HI 54611

Hawaii agricultural experiment station. Annual report. Honolulu, HI. Hawaii Agric. Exp. Sta. Annual Rep. See B–P–H 412/26. HI 54612

Hawaii agricultural experiment station. Circular. Honolulu, HI. Hawaii Agric. Exp. Sta. Circ.. See B–P–H 412/27. HI 54613 HI 54614

Hawaii agricultural experiment station. Press bulletin. Honolulu, HI. Hawaii Agric. Exp. Sta. Press Bull. See B–P–H 413/1. HI 54615

Hawaii agricultural experiment station, university of Hawaii, technical bulletin. Honolulu, HI. Hawaii Agric. Exp. Sta. Univ. Hawaii Techn. Bull. See B–P–H 413/5. HI 54616

Hawaii botanical science paper. Honolulu, HI. No. 1+, 1966+. Hawaii Bot. Sci. Pap. HI 65758

Hawaii farm science. Honolulu, HI. Hawaii Farm Sci. See B–P–H 413/7. HI 54617

Hawaii marine laboratory; contributions. Honolulu, HI. Hawaii Mar. Lab. Contr. See B–P–H 413/9. HI 54618

Hawaii nature notes. Vol. ?-4+, ?-1978?+. Hawaii Nat. Notes. HI 65759

Hawaii orchid journal. Honolulu, HI. 1972+. Hawaii Orchid J. Preceded by: Bulletin of the Pacific orchid society of Hawaii. HI 65760

Hawaiian digest = Aloha. Honolulu, HI. Aloha. See B–P–H 74/20.

Hawaiian forester and agriculturist. Honolulu, HI. Hawaiian Forester Agric. See B–P–H 413/16. HI 54619

Hawaiian naturalist. Honolulu, HI. Hawaiian Naturalist. See B–P–H 413/17. HI 54620

Hawaiian planters' record. Honolulu, HI. Hawaiian Pl. Rec. See B–P–H 413/18. HI 54621

Haworthia review. San Carlos. Vols. 1-2(3), 1946-48. Haworthia Rev. HI 65761

Haworthia society newsletter = Newsletter, the haworthia society. Buckden, N. Yorks.

Hayashi = Forest. Sapporo.

Hazai-hiradó. Kolozsvar [=Cluj, Rumania]. Hazai-Hiradó. See B–P–H 413/21. HI 54622

Heath hen; newsletter of the Long Island pine barrens society. Vol. 1+, 1980+. Heath Hen. HI 65762

Heather society bulletin. Vol. 2(16)+, 1979+. Heather Soc. Bull. Preceded by: Bulletin, the heather society. HI 65763

Hebe news; newsletter of the Hebe society. Richmond, Sy. ?-1988+. Hebe News. HI 65764

Hebei daxue xuebao. Ziran kexue ban = Journal of Hebei university. Natural science edition.

Hebridean naturalist. Stornoway. 1978+. Hebridean Naturalist. HI 65765

Hedendaagsche letteroefeningen. Amsterdam. Hedend. Letteroefen. See B–P–H 413/23. HI 54623

Hedeselskab tidsskrift. Viborg, Denmark. Hedeselsk. Tidsskr. See B–P–H 413/24. HI 54624

Hedwigia. Dresden. Hedwigia. See B–P–H 413/25. HI 54625

Heidelberger klinische Annalen. Heidelberg. Heidelberger Klin. Ann. See B–P–H 413/27. HI 54627

Heidelbergische Jahrbücher der Literatur. Heidelberg. Heidelberg. Jahrb. Lit. See B–P–H 413/26. HI 54626

Heil- und Gewürz-Pflanzen. Munich. Heil- Gewürz-Pflanzen. See B–P–H 414/1. HI 54628

Heilpflanze. Beilage zu: Natur und Nahrung. Berlin. Heilpflanze. See B–P–H 414/3. HI 54629

Heimat. Kiel Heimat. See B–P–H 414/4. HI 54630

Heimatschutz. Munich. Vols. 1-12, 1905-17. Heimatschutz (Munich). 3-1834-2. HI 54631

Heimatschutz. Zeitschrift der schweizer Vereinigung für Heimatschutz. Bümpliz, Olten. Vol. 1+, 1906+. Heimatschutz (Olten). HI 65766

Helgoländer Meeresunterscuhungen. Hamburg. Vol. 33+, 1980+. Helgoländer Meeresuntersuch. Preceded by: Helgoländer wissenschaftliche Meeres-untersuchungen. HI 65767

Helgoländer wissenschaftliche Meeresuntersuchungen. Stuttgart. Vols. 1-32, 1937-79. Helgoländer Wiss. Meeresuntersuch. Preceded by: Wissenschaftliche Meeresuntersuchungen. Abteilung Helgoland. Superseded by: Helgoländer Meeresuntersuchungen. 3-1835-1. HI 65768

Helios. Berlin. Helios. See B–P–H 414/7. HI 54632

Hellenic agricultural economic review. Salonika. Vol. 8(2)+, 1972+. Hellen. Agric. Econ. Rev. Preceded by: Agricultural economic review, aristotelian university of Thessaloniki. HI 65769

Hellenic oceanology and limnology = Hellenikè okeanologia kai limnologia tou institoutou okeanographikon kai halieutikon ereunon. Athens.

Hellenike georgia. Athens. Hellen. Georgia. See B–P–H 414/8. HI 54633

Hellenike mikrobiologike kai hygieinologike hetoireia Deltion = Acta microbiologica hellenica. Athens. Acta Microbiol. Hellen. See B–P–H 46/6.

Hellenikė okeanologia kai limnologia tou institoutou okeanographikon kai halieutikon ereunon. Athens. Vols. ?-11, 19??-73. Hellen. Okeanol. Limnol. Inst. Okeanogr. Halieut. Ereunon. Preceded by: Praktika tou hellenikou hydroviologikou institoutou. Superseded by: Thalassographika. HI 65771

Helminthological abstracts. [Vols. 1-3(4) formed a supplement to: Journal of helminthology.] St. Albans, etc. Vols. 1-38, 1932-69. Helminthol. Abstr. Superseded by: Helminthological abstracts. Series B, plant nematology. HI 65772

Helminthological abstracts. Series B, plant nematology. Farnham Royal. Vol. 39+, 1970+. Helminthol. Abst., B. Preceded by: Helminthological abstracts. HI 65773

Helobiae newsletter. Melbourne, Vic. No. 1+, 1976+. Helobiae Newslett. HI 65774

Helsingin yliopiston kasvimuseon monisteita. Helsinki. No. 1+, 1968+. Helsingin Yliop. Kasvimus. Monist. Partly superseded, from 1983 onwards, by: Norrlinia. HI 65775

Helsingfors universitets botaniska institutions publikationer = Helsingin yliopiston kasvitieteen laitoksen julkeisuja. Helsinki.

Helsingin yliopiston kasvitieteen laitoksen julkeisuja. Helsinki. No. 1+, 1973+. Helsingen Yliop. Kasvit. Laitok. Julk. HI 65776

Helvetica chimica acta. Basel & Geneva. Helv. Chim. Acta. See B–P–H 414/9. HI 54634

Helvetischer-Verein für den Austausch von Pflanzen = Société helvétique pour l'echange des plantes. Neuchatel.

Hemerocallis journal; daylily gardener's magazine. Minneapolis, MN., etc. Vols. 1-35(2), 1947-81 [vols. 5(2), 8(2) & 9(2) were issued as yearbooks of the society]. Hemerocallis J. Superseded by: Daylily journal. 3-2838-3 HI 65777

Hemträdgården. Föreningen för dendrologi och parkvård. Stockholm. Hemträdgården. See B–P–H 414/13. HI 54635

Henequén. Mérida, Mexico. Henequén. See B–P–H 414/14. HI 54636

Herb basket. Brattleboro, VT. No. 1+, 1984+. Herb Basket. HI 65778

Herb grower magazine. Falls Village, CT. Vol. 2+, 1948+. Herb Grower Mag. Preceded by: American herb grower. 3-1842-3 HI 65779

Herb magazine. Boulder, CO. 1986+. Herb Mag. HI 65780

Herb quarterly. Wilmington, New Fare, VT. Vol. 1+, 1979+. Herb Quart. HI 65781

Herb, spice and medicinal plant digest. Amherst, MA. Vol. 1+, 1983+. Herb Spice Med. Pl. Digest. HI 65783

Herb walk. Provo, UT. Vol. 1+, 1979+. Herb Walk. HI 65784

Herba. [Edited by F. Darvas.] Budapest. Herba (Darvas). See B–P–H 414/19. HI 54638

Herba. [Edited by R. Giovannini.] Budapest. Herba (Giovannini). See B–P–H 414/20. HI 54639

Herba. The Hague. Herba (The Hague). See B–P–H 415/1. HI 54640

Herba hungarica. Budapest. Herba Hung. See B–P–H 415/2. HI 54641

Herba polonica. Poznan. Vol. 11(41/42)+, 1965+. Herba Polon. Preceded by: Biuletyn, instytut przemyslu zielarskiego. HI 65785

Herba topiara. Orgaan van de Nederlandsche Vereeniging tot bevordering der wetenschappelijke veredeling van siergewassen. Lisse, Netherlands. Herba Topiara. See B–P–H 415/3. HI 54642

Herbage abstracts. Aberystwyth, Wales. Herbage Abstr. See B–P–H 415/4. HI 54643

Herbage reviews. Aberystwyth. Vols. 1-8, 1933-40. Herbage Rev. HI 65786

Herbage seed growers leaflet. Cambridge, England. Herbage Seed Growers Leafl. See B–P–H 415/5. HI 54644

Herbal cure. Hyderabad. Vol. ?-1(9)+, 19??-77+. Herbal Cure. HI 65787

Herbal review. London. Vol. 1+, 1976+. Herbal Rev. HI 65788

Herbalgram; herb news. Austin, TX. 1979+. Herbalgram. HI 65789

Herbalist. Cambridge. Vol. 1+, 1982+. Herbalist. HI 65790

Herbarist; publication of the herb society of America. Boston, MA. No. 1+, 1935+. Herbarist. 3-1842-3. HI 65791

Herbarium. Leipzig. Herbarium. See B–P–H 415/7. HI 54645

Herbarium australiense technical paper. Vol. 1+, 19??+. Herb. Austral. Techn. Pap. HI 65792

Herbarium and library news. Kew. No. 885+, 1974+.

Herb. Libr. News, Kew. Preceded by: Herbarium news, Kew. HI 65794

Herbarium news. Kew. Nos. 1-884, 1957-74. Herb. News (Kew). Superseded by: Herbarium and library news, Kew. HI 65795

Herbarium news. St. Louis, MO. Vol. 1+, 1981+. Herb. News (St. Louis). HI 65796

Herbarium news, plant science laboratories, university of Reading = Plant science laboratories herbarium news, university of Reading. Reading.

Herbarium pacificum news. Honolulu, HI. No. 1+, 1984+. Herb. Pacificum News. HI 65797

Herbarium publications, Youngstown state university. Youngstown, OH. No. 1, 1976. Herb. Publ. Youngstown State Univ. Superseded by: Laitsch, herbarium publications, Youngstown state university. HI 65799

Herbarium sheet. New York. 1980+. Herb. Sheet. HI 65800

Herbertia. Orlando, FL. Vols. 3-15, 1936-48; ser. 2, vols. 16-26, 1949-59; vol. 40+, 1984+. Herbertia. Preceded by: Yearbook of the american amaryllis society. For 1949-59 included in: Plant life. For 1960-83 see: Amaryllis year book. 3-1843-1. HI 65801

Herbier général de l'amateur, contenant la description, l'histoire, propriétés et la culture des végétaux utiles et agréables. Paris. Herb. Gén. Amateur. See B–P–H 414/17. HI 54637

Herbs, spices, and medicinal plants; recent advances in botany, horticulture, and pharmacology. Phoenix, AZ. Vol. 1+, 1986+. Herbs Spices Med. Pl. HI 65802

Hercynia; Abhandlungen der botanischen Vereinigung Mitteldeutschlands (New series subtitled: Zeitschrift für Naturwissenachaften). Halle, Berlin & Leipzig. Vols. 1-4, 1937-44; n.s. vol. 1+, 1963+. Hercynia. 3-1843-1. HI 65803

Hereditas. [Yichuan.] Beijing. Vol. 1+, 1979+. Hereditas (Beijing). Preceded by: Yichuan yu yuzhong. HI 65804

Hereditas. Lund. Vol. 1+, 1920+. Hereditas (Lund). 3-1843-3. HI 54646

Heredity; an international journal of genetics. London. Heredity. See B–P–H 415/11. HI 54647

Heritage; quarterly bulletin of the Loughborough naturalists club. Loughborough. No. ?-21+, 19??-66+. Heritage. HI 65805

Herman Ottó múzeum évkönyve. Miskolc, Hungary. Herman Ottó Múz. Évk. See B–P–H 415/13. HI 54648

Hermann; Zeitschrift von und für Westfalen. Hermann. See B–P–H 415/14. HI 54649

Hermathena; a series of papers on literature, science and philosophy by members of Trinity College, Dublin. Dublin. Vol. 1+, 1873+. Hermathena. 3-1844-2. HI 65806

Hertha, Zeitschrift für Erd-, Völker- und Staatenkunde. Stuttgart & Tübingen. Hertha. See B–P–H 415/16. HI 54650

Herzogia; Zeitschrift der bryologisch-lichenologischen Arbeitsgemeinschaft für Mitteleuropa. Lehre. Vol. 1+, 1968+. Herzogia. HI 65807

Hesdörffers Monatshefte für Blumen- und Gartenfreunde. Berlin. Hesdörffers Monatsh. Blumen- Gartenfr. See B–P–H 415/18. HI 54651

Hesperian; or, western monthly magazine. Columbus, OH. Vols. 1-3, 1838-39. Hesperian (Columbus). 3-1846-2. HI 54652

Hesperian. San Francisco, CA. Vols. 1-10(1), 1858-63. Hesperian (San Francisco). Superseded by: Pacific monthly. 3-1846-1. HI 65808

Hesperus. Brünn [=Brno, Czechoslovakia]. Hesperus. See B–P–H 415/20. HI 54653

Hessische Beyträge zur Gelehrsamkeit und Kunst. Frankfurt a. M. Hess. Beytr. Gelehrsamk. Kunst. See B–P–H 415/21. HI 54654

Hessische Biene. Giessen. Vols. 74(8)-90, 1936-54. Biene. From 1943-44 merged with several other journals to form: Sudwestdeutsche Bienenzeitung. Preceded and superseded by: Biene. Giessen. HI 65809

Hessische floristische Briefe. Offenbach-Bürgel. Hess. Florist. Briefe. See B–P–H 416/1. HI 54655

Hessische landwirtschaftliche Zeitschrift. Darmstadt, Germany. Hess. Landw. Z. See B–P–H 416/2. HI 54656

Hessische landwirtschaftliche Zeitschrift, Beilage = Phänologische Mitteilungen. Darmstadt.

Hickenia; boletín del Darwinion. Lavardén, San Isidro. Vol. 1+, 1976+. Hickenia. HI 65810

Hidrobiologia. Bucharest. Hidrobiologia. See B–P–H 416/3. HI 54657

Hidrológiai közlöny. Budapest. Hidrol. Közl. See B–P–H 416/4. HI 54658

Hifuka Kiyo = Acta dermatologica. Kyoto, Japan. Acta Dermatol. See B–P–H 42/16.

High mountain research = Hochgebirgsforschung. Innsbruck & Munich.

Highlights, research branch, department of agriculture, Canada. Ottawa. Nos. 1-119, 1961-66. Highlights Res. Branch Dep. Agric. Canada. HI 65811

Hikobia; journal of the Hiroshima botanical club.

Hiroshima. Hikobia. See B–P–H 416/6. HI 54659

Hikobia. Supplement. Hiroshima. Vol. 1+, 1981+. Hikobia, Suppl. HI 65812

Hilgardia; a journal of agricultural science. Berkeley, CA. Hilgardia. See B–P–H 416/7. HI 54660

Himalayan journal. Calcutta. Himalayan J. See B–P–H 416/9. HI 54661

Himalayan journal of sciences. Pauri. Vol. 1+, 1981+. Himalayan J. Sci. HI 65813

Himalayan plant journal. Kalimpong. Vol. 1+, 1982+. Himalayan Pl. J. HI 65814

Himmel und Erde. Illustri(e)rte naturwissenschaftliche Monatsschrift. Berlin. Himmel & Erde. See B–P–H 416/10. HI 54662

Himochal journal of agricultural research. Palampur. Vol. 1+, 1974+. Himochal J. Agric. Res. HI 65815

Hindustan antibiotics bulletin. Pune. Vol. 1+, 1958+. Hindustan Antibiot. Bull. HI 65816

Hinweise an die Mitarbeiter. Graz. Nos. 1-4, 1962-64. Hinweise Mitarbeiter. Superseded by: Mitteilungsblatt, floristische Arbeitsgemeinschaft, naturwissenschaftlichen Verein für Steiermark. HI 65817

Hirokawa shoten = Annual index of the reports of plant chemistry. Tokyo.

Hirosaki daigaku nogakubu gakujutsu hokoku = Bulletin, faculty of agriculture, Hirosaki university. Hirosaki.

Hiroshima daigaku seibutsu gakkai shi = Bulletin of the biological society of Hiroshima university. Hiroshima.

Hiroshima daigaku seibutsu seisangakubu kiyo = Journal of the faculty of applied biological science, Hiroshima university. Hiroshima.

Hiroshima hakubutsu gakkai shi = Hiroshima journal of natural history. Hiroshima. Hiroshima J. Nat. Hist. See B–P–H 416/12.

Hiroshima joshi daigaku kiyo. Dai-2-bu, shizenkagaku = Bulletin of the Hiroshima women's university. Series 2, natural science. Hiroshima.

Hiroshima journal of natural history. [Hiroshima hakubutsu gakkai shi.] Hiroshima. Hiroshima J. Nat. Hist. See B–P–H 416/12. HI 54663

Hiroshima jozo gakkai-shi = Journal of Hiroshima society of brewing science. Hiroshima. J. Hiroshima Soc. Brew. Sci. See B–P–H 468/15.

Hiroshima nogyo = Agriculture in Hiroshima. Hiroshima. Agric. Hiroshima. See B–P–H 58/4.

Hiroshima nogyo tanki daigaku kenkyu = Bulletin of the Hiroshima agricultural college. Saijo.

Hiroshima-kenritsu nogyo shikenjo hokoku = Bulletin of Hiroshima prefectural experiment station. Hiroshima. Bull. Hiroshima Prefect. Exp. Sta. See B–P–H 253/9.

Hiroshima-kenritsu ringyo shikenjo kenkyu hokoku = Bulletin of the Hiroshima prefectural forest experiment station. Miyoshi.

Hiryo jiho = Manurial times. Tokyo. Manurial Times. See B–P–H 548/7.

Histoire de l'académie royale des sciences. Amsterdam = Histoire de l'académie royale des sciences [Paris]. Avec les mémoires de mathématique & de physique. Amsterdam. Hist. Acad. Roy. Sci. Mém. Math. Phys. (Amsterdam). See B–P–H 417/1.

Histoire de l'académie royale des sciences. Paris. Hist. Acad. Roy. Sci. (Paris). See B–P–H 416/22. HI 54666

Histoire de l'académie royale des sciences [Paris]. Avec les mémoires de mathématique & de physique. Amsterdam. Hist. Acad. Roy. Sci. Mém. Math. Phys. (Amsterdam). See B–P–H 417/1. HI 54668

Histoire de l'académie royale des sciences. Avec les mémoires de mathématique & de physique. [In duodecimo.] Paris. Hist. Acad. Roy. Sci. Mém. Math. Phys. (Paris, 12mo). See B–P–H 417/2. HI 54669

Histoire de l'académie royale des sciences. Avec les mémoires de mathématique & de physique. [In quarto.] Paris. Hist. Acad. Roy. Sci. Mém. Math. Phys. (Paris, 4to). See B–P–H 417/3. HI 54670

Histoire de l'académie royale des sciences. Avec les mémoires de physique tirés des régistres de cette académie. Paris. Hist. Acad. Roy. Sci. Mém. Phys. See B–P–H 417/4. HI 54671

Histoire de l'académie royale des sciences et belles lettres. Berlin. Hist. Acad. Roy. Sci. (Berlin). See B–P–H 416/21. HI 54665

Histoire de l'académie royale des sciences et belles lettres de Berlin. Avec les mémoires, tirez des registres de cette académie. Berlin. Hist. Acad. Roy. Sci. Berlin Mém. See B–P–H 416/23. HI 54667

Histoire de l'academie des sciences, avec les mémoires de mathématique & de physique. Paris = Histoire de l'académie royale des sciences. Avec les mémoires de mathématique & de physique. [In quarto.] Paris. Hist Acad. Roy. Sci. Mém. Math. Phys. (Paris, 4to). See B–P–H 417/3.

Histoire et biologie; cahiers du cercle d'étude historique des sciences de la vie. Paris. Fasc. 1-2, 1968-69. Hist. & Biol. Superseded by: Histoire et nature. HI 65818

Histoire de l'horticulture au Canada; revue interdisciplinaire = Canadian horticultural history. Hamilton, Ont.

Histoire et mémoires de l'académie royale des sciences, inscriptions et belles lettres de Toulouse. Toulouse. Hist. & Mém.Acad. Roy. Sci. Toulouse. See B–P–H 418/1. HI 54679

Histoire et mémoires de la société royale de médicine = Histoire de la société royale de médicine. Paris.

Histoire et mémoires de la société des sciences physiques de Lausanne. Lausanne. Hist. & Mém.Soc. Sci. Phys. Lausanne. See B–P–H 418/4. HI 54680

Histoire et nature; cahiers du cercle d'étude historique des sciences de la vie [sub-title varies]. Paris. No. 3+ [= n.s. no. 1+], 1973+. Hist. & Nat. Preceded by: Histoire et biologie. HI 65819

Histoire des plantes. Petites nouvelles du jardin botanique Genève. Geneva. Hist. Pl. See B–P–H 418/6. HI 54682

Histoire de la société royale de médicine. Avec les mémoires de médicine et de physique médical. Paris. Vols. 1-10, 1776-89. Hist. Soc. Roy. Méd. HI 65820

Historia agriculturae. Groningen. Hist. Agric. See B–P–H 417/7. HI 54672

Historia et commentationes academiae electoralis scientiarum et elegantiorum literarum theodoro-palatinae. Mannheim. Hist. & Commentat. Acad. Elect. Sci. Theod.-Palat. See B–P–H 417/12. HI 54676

Historia e memórias da academia r(eal) das sciencias de Lisboa. Lisbon. Vols. 4-11(1), 1815-31; ser. 2, vols. 1-3, 1843-56. Hist. & Mem. Acad. Real Sci. Lisboa. Preceded by: Memórias de mathematica e phisica da academia r. das sciencias de Lisboa. For vols. 11(2)-12, 1835-37 see: Memórias da academia r(eal) das sciencias de Lisboa. Superseded by: Memórias da academia real das sciencias de Lisboa. Primeira classe de sciencias mathematicas, physicas e naturaes. 1-17-2. HI 65821

Historia e memórias da academia real das sciencias de Lisboa. Segunda classe de sciencias moraes, politicos e bellas lettras. Lisbon. 1899-1934. Hist. & Mem. Acad. Real Sci. Lisboa, Segunda Classe Sci. Moraes. Preceded by: Memórias da academia real das sciencias de Lisboa. Segunda classe de sciencias moraes, politicos e bellas lettras. Superseded by: Memórias da academia das sciencias de Lisboa. Classe de letras [Not entered]. 1-17-3. HI 65822

Historia natural. Mendoza. No. 1+, 1979/81+. Hist. Nat. HI 65823

Historia natural de Costa Rica. San José, Costa Rica. Vol. 1+, 1977+. Hist. Nat Costa Rica. HI 65824

Historia natural y pro-natura. Guatemala. No. 1+, 1964+. Hist. Nat. Pro-Nat. HI 65825

Historia rerum rusticarum = Agrártörténeti i szemle. Budapest.

Historia scientiarum. Tokyo. No. 19+, 1980+. Historia Sci. Preceded by: Japanese studies in the history of science. HI 65826

Historiallisia tutkimuksia. Helsinki. Hist. Tutkimuksia. See B–P–H 418/9. HI 54684

Historical biology; an international journal of paleobiology. Chur, London. 1987+. Historical Biol. HI 65827

Historical records of australian science. Canberra, A.C.T. Vol. 5+, 1980+. Hist. Rec. Austral. Sci. Preceded by: Records of the australian academy of science. HI 65828

Historical studies in irish science and technology. Dublin. No. 1+, 1980+. Hist. Stud. Irish Sci. Technol. HI 65829

Historisch-geographisches Journal. Halle, Leipzig, & Jena. Hist.-Geogr. J. See B–P–H 417/13. HI 54677

Historische Abhandlungen der königlich-baierischen Akademie der Wissenschaften. Munich. Hist. Abh. Königl.-Baier. Akad. Wiss. See B–P–H 416/18. HI 54664

Historische und geographische Monatsschrift. Halle, Leipzig, Jena, Gotha, Hamburg, Nuremberg & Vienna. Hist. Geogr. Monatsschr. See B–P–H 417/14. HI 54678

Historischer Bericht von den sämmtlichen, durch Engländer geschehenen Reisen um die Welt. Leipzig. Hist. Ber. Sämmtl. Reisen Welt. See B–P–H 417/8. HI 54673

Historiska bibliotheket. Stockholm. Hist. Biblioth. See B–P–H 417/10. HI 54675

History of Australian science newsletter. Parkville, Vic. No. 1+, 1983+. Hist. Austral. Sci. Newslett. HI 65830

History of the Berwickshire naturalists' club Berwick-upon-Tweed, England. Hist. Berwickshire Naturalists' Club. See B–P–H 417/9. HI 54674

History of the Minnesota state horticultural society. St. Paul & Minneapolis, MN. Hist. Minnesota State Hort. Soc. See B–P–H 418/5. HI 54681

History and philosophy of the life sciences. Naples, etc. Vol. 1+, 1979+. Hist. Philos. Life Sci. Preceded by: Episteme. HI 65831

History of photography. London. Vol. 1+, 1977+. Hist. Photogr. HI 65832

History of the royal society of London, in which the most considerable of those papers communicated to the society which have not been published are inserted in their proper order as a supplement to the philosophical transactions. [Edited by Birch.] London. Hist. Roy. Soc. London. See B–P–H 418/8. HI 54683

History of science; an annual review of literature, research and teaching. Cambridge. Vol. 1+, 1962+. Hist. Sci. HI 65833

History of science in America news and views; newsletter to the forum for the history of science in America. Washington, DC. Vol. 1+, 1980+. Hist. Sci. Amer. News Views. HI 74843

Hlasy okresního musea ve Frenštátě p. Radhoštěm. Frenstat pod Radhostem, Czechoslovakia. Hlasy Okresn. Mus. Ve Frenštátě Pod Radhoštěm. See B–P–H 418/12. HI 54685

Ho nan ta hsüeh hsüeh pao = Journal of the national Honan university. Kaifeng, China. J. Natl. Honan Univ. See B–P–H 474/14.

Hoanrin = Journal of the japanese institute of protection forests. Tokyo.

Hobbies. Buffalo, NY. Vols. 1-38(4), 1920-58. Hobbies. Superseded by: Science on the march. HI 65834

Hochgebirgsforschung. Innsbruck & Munich. No. 1+, 1968+. Hochgebirgsforschung. HI 65835

Hochland; monatliches Mitteilungsblatt für die Mitglieder des deutschen Wirtschaftsverbands im Tanganyika Territory. Mufindi. Vols. 1-5, 1930-35; 2nd edn. of vol. 1, 1935. Hochland. HI 65836

Hodowla roślin, aklimatyzacja i nasiennictwo. Warsaw. Hodowla Rośl. See B–P–H 418/15. HI 54686

Hoehnea. São Paulo. Vol. 1+, 1971+. Hoehnea. Preceded by: Arquivos de botânica do estado de São Paulo. HI 65837

Höfchen-Briefe, Bayer Pflanzenschutz-Nachrichten. Leverkusen. Vols. 5-14, 1952-61. Höfchenbriefe Bayer Pflanzenschutznachr. Preceded by: Höfchen-Briefe für Wissenschaft und Praxis. Superseded by: Pflanzenschutz-Nachrichten "Bayer". 3-1869-2 HI 65838

Höfchen-Briefe für Wissenschaft und Praxis. Veröffentlichungen der "Bayer" Pflanzenschutz-Abteilung Leverkusen. Leverkusen. Vols. 1-4, 1948-51. Höfchenbriefe Wiss. Praxis. Superseded by: Höfchen-Briefe, Bayer Pflanzenschutz-Nachrichten. 3-1869-2 HI 65839

Hoffy's orchardist's companion; or, fruits of the United States. Philadelphia, PA. Hoffy's Orchardist's Companion. See B–P–H 418/18. HI 54687

Hoja informativa. Buenos Aires. Hoja Inform. See B–P–H 418/19. HI 54688

Hoja téchnica I N I A. Madrid. Vol. 1+, 1974+. Hoja Techn. I. N. I. A. HI 65840

Hokkaido daigaku nogaku-bu enshurin kenkyu hokoku = Bulletin of the Hokkaido university forest. Sapporo, Japan. Bull. Hokkaido Univ. Forest. See B–P–H 253/17.

Hokkaido daigaku nogakubu kiyo = Journal of the faculty of agriculture, Hokkaido imperial university. Sapporo.

Hokkaido daigaku nogakubu kiyo = Memoirs of the faculty of agriculture, Hokkaido university. Sapporo, Japan. Mem. Fac. Agric. Hokkaido Univ. See B–P–H 572/6.

Hokkaido daigaku rigakubu kaiso kenkyusho obun hokoku = Scientific papers of the institute of algological research; faculty of science of the Hokkaido (imperial) university. Sapporo, Japan. Sci. Pap. Inst. Algol. Res. Fac. Sci. Hokkaido Imp. Univ. See B–P–H 825/21.

Hokkaido daigaku rigakubu kiyo = Journal of the faculty of science, Hokkaido university. Series 5, botany. Sapporo.

Hokkaido gakugei daigaku kiyo 2 = Journal of the Hokkaido gakugei university. Section B, natural science. Sapporo.

Hokkaido kyoiku daigaku kiyo. Dai-2-bu, B, seibutsugaku, chigaku, nogaku-hen = Journal of Hokkaido university of education. Section 2B, biology, geology, agricultural science. Sapporo.

Hokkaido kyoiku daigaku Taisetsuzan kagaku kenkyusho hokoku = Report of the Taisetsuzan institute of science. Asahigawa.

Hokkaido nogyo shikenjo hokoku = Report, Hokkaido national agricultural experiment station. Sapporo, Japan. Rep. Hokkaido Natl. Agric. Exp. Sta. See B–P–H 766/6.

Hokkaido nogyo shikenjo iho = Research bulletin of the Hokkaido national agricultural research station. Sapporo.

Hokkaido nogyo shikenjo kenkyu hokoku = Research bulletin of the Hokkaido national agricultural research station. Sapporo.

Hokkaido ringyo kai kaiho = Bulletin of the forestry society of Hokkaido. Sapporo?, Japan. Bull. Forest. Soc. Hokkaido. See B–P–H 251/4.

Hokkaido ringyo shiken hokoku = Bulletin of the Sapporo branch, forestry experiment station. Sapporo, Japan. Bull. Sapporo Branch Forest. Exp. Sta. See B–P–H 271/2.

Hokkaido ringyo shikenjo hokoku = Bulletin of the Hokkaido forest experiment station. Bibai.

Hokkaido-ritsu kitamu nogyo shikenjo. Hakka, yakuyo sakumotsu. [Kitami.] 1977-78. Hokkaido-Ritsu Kitamu Nogyo Shikenjo Hakka Yakuyo Sakumotsu Superseded by: Hokkaido-ritsu kitamu nogyo shikenjo. Tensai yakuyo sakumutsu shiken seisekisho. HI 65841

Hokkaido-ritsu nogyo shikenjo hokoku = Report of the Hokkaido prefectural agricultural experiment station. Sapporo, Japan. Rep. Hokkaido Prefect. Agric. Exp. Sta. See B–P–H 766/7.

Hokkaido-ritsu nogyo shikenjo shuho = Bulletin of the Hokkaido prefectural agricultural experiment station. Sapporo, Japan. Bull. Hokkaido Prefect. Agric. Exp. Sta. See B–P–H 253/15.

Hokkaido sui san shikenjo hokoku = Scientific papers of the Hokkaido fisheries scientific institution. Yoichi, Japan. Sci. Pap. Hokkaido Fish. Sci. Inst. See B–P–H 825/19.

Hokkaido teikoku daigaku nogakubu enshurin kenkyu hokoku = Research bulletins of the college experiment forests, college of agriculture, Hokkaido imperial university. Sapporo, Japan. Res. Bull. Coll. Exp. Forests Coll. Agric. Hokkaido Imp. Univ. See B–P–H 773/24.

Hokkaido teikoku daigaku nogakubu. Enshurin kenkyu hokoku = Research bulletins of the college experiment forests, Hokkaido university. Sapporo, Japan. Res. Bull. Coll. Exp. Forests Hokkaido Univ. See B–P–H 774/1.

Hokokuriku no shokubutsu no kai = Journal of geobotany. Kanazawa.

Hokuno = Agriculture in Hokkaido. Sapporo, Japan. Agric. Hokkaido (Hokuno). See B–P–H 58/8.

Hokuriku bykogaichu kenkyukai-ho = Proceedings of the association for plant protection of Hokuriku. Niijata, Japan. Proc. Assoc. Pl. Protect. Hokuriku. See B–P–H 726/18.

Hokuriku journal of botany. [Hokuriku no shokubutsu.] Kanazawa, Japan. Hokuriku J. Bot. See B–P–H 419/12. HI 54689

Hokuriku nogyo shikenjo hokoku = Bulletin of the Hokuriku agricultural experiment station. Niigata, Japan. Bull. Hokuriku Agric. Exp. Sta. See B–P–H 253/18.

Hokuriku no shokubutsu = Hokuriku journal of botany. Kanazawa, Japan. Hokuriku J. Bot. See B–P–H 419/12.

Hokusuisi geppo = Journal of the Hokkaido fisheries scientific institution. Sapporo, Japan. J. Hokkaido Fish. Sci. Inst. See B–P–H 468/22.

Holarctic ecology. Copenhagen. Vol. 1+, 1978+. Holarc. Ecol. HI 65842

Holland bulb news. Vol. ?-7(3), 19??-71-80. Holland Bulb News. HI 65843

Holländisches Magazin der Naturkunde. Frankfurt a. M. Holl. Mag. Naturk. See B–P–H 419/19. HI 54691

Hollandsch landbouwweekblad. Officieel orgaan van de hollandsche maatschappij van landbouw. The Hague. Holl. Landbouwweekbl. See B–P–H 419/18. HI 54690

Holly agricultural news. Colorado Springs, CO. Holly Agric. News. See B–P–H 419/21. HI 54692

Holly letter. Baltimore, MD. Nos. 6-75, 1957-83. Holly Lett. Preceded by: Holly newsletter. Superseded by: Holly society journal. HI 65844

Holly newsletter. Baltimore, MD. Nos. 4-5, 1954-57. Holly Newslett. Preceded by: Newsletter, holly society of America. Superseded by: Holly letter. HI 65845

Holly society journal. Baltimore, MD. Vol. 1+, 1983+. Holly Soc. J. Preceded by: Holly letter. HI 65846

Holmbergia. Buenos Aires. Holmbergia. See B–P–H 419/25. HI 54693

Holy Ghost medical college theses. [Catholic taehac uihak-bu nonmunjip.] [Korea]. Holy Ghost Med. Coll. Theses. See B–P–H 419/27. HI 54694

Holz; Schweizerische Holz-Zeitung. Zurich. Vol. 1+, 1887+. Holz. 3-1873-3. HI 65847

Holz-Zentralblatt. Stuttgart. Vol. 1+, 1874+ [suspended 1944-46]. Holz-Zentralbl. 3-1874-1. HI 65848

Holzforschung. Berlin. Holzforschung. See B–P–H 419/29. HI 54696

Holzforschung und Holzverwertung. Vienna. Holzf. & Holzverwert. See B–P–H 419/28. HI 54695

Holzzucht. Reinbek, Germany. Holzzucht. See B–P–H 419/30. HI 54697

Home and foreign review. London. Vols. 1-4, 1862-64. Home Foreign Rev. 3-1875-1. HI 65849

Home garden. Concord, NH. Home Garden. See B–P–H 419/32. HI 54699

Home gardening. New Orleans, LA. Vols. 1-3, 1940-42. Home Gard. Superseded by: Home gardening for the south. 3-1876-2. HI 65850

Home gardening for the south. New Orleans, LA. Home Gard. S. See B–P–H 419/31. HI 54698

Homme d'outre-mer; comptes rendus de missions et études du conseil supérieur des recherches sociologiques outre-mer et de l'office de la recherche scientifique et technique outre-mer (section des sciences humaines). Paris. Nos. 1-4, 1955-58; n.s. no. 1+, 1959+. Homme Outre-Mer. HI 65851

Honan ta hsüeh nung hsüeh yüan yan k'an = Publications of the college of agriculture, Honan university. Kaifeng, China. Publ. Coll. Agric. Honan Univ. See B–P–H 744/18.

Honduras cafetalera. Tegucigalpa, Honduras. Honduras Cafetalera. See B–P–H 420/3. HI 54700

Honey bee monthly = Japanese bee journal. Takikawa?, Japan. Jap. Bee J. See B–P–H 498/10.

Hong Kong naturalist. Hong Kong. Hong Kong Naturalist. See B–P–H 420/7. HI 54701

Hong Kong naturalist, supplement. Hong Kong. Hong Kong Naturalist, Suppl. See B–P–H 420/8. HI 54702

Hong Kong university fisheries journal. Hong Kong. Hong Kong Univ. Fish. J. See B–P–H 420/9. HI 54703

Honi- 's külföldi gazda és kereskedö. Buda [=Budapest, in part]. Honi- Külf. Gazda Keresk. See B–P–H 420/10. HI 54704

Hooker's icones plantarum; or figures, with brief descriptive characters and remarks of new or rare plants. London. Hooker's Icon. Pl. See B–P–H 420/26. HI 54705

Hooker's journal of botany and Kew garden miscellany. London. Hooker's J. Bot. Kew Gard. Misc. See B–P–H 420/27. HI 54706

Hoosier horticulture. Lafayette, IN. Hoosier Hort. See B–P–H 420/32. HI 54707

Hop review, Rosemaund experimental husbandry farm. Preston Wynne. 1974+. Hop Rev. Rosemaund Exp. Husb. Farm. Preceded by: Report, experimental work on hops, Rosemaund experimental husbandry farm. HI 65852

Hopei agricultural and forestry journal. [Hopei nung-lin hsüeh-k'an.] Paoting China. Hopei Agric. Forest. J. See B–P–H 420/33. HI 54708

Hopei nung-lin hsüeh-k'an = Hopei agricultural and forestry journal. Paoting China. Hopei Agric. Forest. J. See B–P–H 420/33.

Hopfenrundschau. [Verband der Hopfenpflanzer.] Wolnzach-Markt. Vol. 1+, 1950+. Hopfenrundschau. HI 65853

Hoppea; Denkschriften der regensburgischen botanischen Gesellschaft. Regensburg. Vol. 29+, 1971+. Hoppea. Preceded by: Denkschriften der regensburgischen botanischen Gesellschaft. HI 65854

Hoppo nogyo = Northern agriculture. Sapporo, Japan. N. Agric. See B–P–H 624/11.

Hoppo ringyo = Journal of the northern forestry society. Sapporo.

Hoppo ringyo kenkyukai koenshu = Journal of the northern forestry society. Sapporo.

Horaijisan chiiki shokubutsu seitai chosa. Nagoya. Showa ?-46+, 19??-72+. Horaijisan Chiiki Shok. Seitai Chosa. HI 65855

Horizons. Washington, DC. 19??-79+. Horizons. HI 65856

Hornbill. Bombay. 1967+. Hornbill. HI 65857

Horticulteur belge, journal des jardiniers et amateurs. Brussels, Paris, & Leipzig. Hort. Belge. See B–P–H 421/2. HI 54711

Horticulteur châlonnais. Chalon-sur-Saône, France. Hort. Châlonnais. See B–P–H 421/4. HI 54713

Horticulteur praticien. Paris & Brussels. Hort. Praticien. See B–P–H 422/2. HI 54727

Horticulteur provençal. Marseilles. Hort. Provençal. See B–P–H 422/4. HI 54728

Horticulteur universel, journal général des jardiniers et amateurs, ... Paris. Hort. Universel. See B–P–H 423/2. HI 54741

Horticulteurs français; journal des amateurs et des intérêts horticoles. Paris. Vols. 1-8, 1851-58; sér. 2, vols. 1-6, 1859-64; sér. 3, vols. 1-7, 1865-72 [suspended 1870-71]. Hort. Franc. 3-1887-1. HI 54742

Horticulteurs et maraîchers romands. Lausanne. Vol. 1+, 1978+. Hort. Maraîch. Romands. HI 65858

Horticultura. Copenhagen. Vols. 1-24, 1947-70. Horticultura (Copenhagen). Incorporated in: Ugeskrift for agronomer og hortonomer. 3-1887-1. HI 65859

Horticultura. Jakarta, Indonesia. Horticultura (Jakarta). See B–P–H 423/10. HI 54743

Horticultural abstracts; compiled from world literature on temperate and tropical fruits; vegetables; ornamentals; plantation crops. East Malling. Vol. 1+, 1931+. Hort. Abstr. 2-1144-2. HI 65860

Horticultural advance. Saharanpur. Vol. 1+, 1957+. Hort. Advance. HI 65861

Horticultural art journal. Rochester, NY. Hort. Art J. See B–P–H 421/1. HI 54710

Horticultural bulletin. College of agriculture, Sun Yatsen university. [Yuan-i chuan k'an.] Canton. Hort. Bull. Coll. Agric. Sun Yatsen Univ. See B–P–H 421/3. HI 54712

Horticultural crops; summary of experiments, studies and surveys. Reading. 1977+, [1978?]+. Hort. Crops. HI 65862

Horticultural digest. Ames, IA. Hort. Digest. See B–P–H 421/5. HI 54714

Horticultural directory for [year 1881?+]. London. 1881+. Hort. Directory 1881 [etc.]. HI 65863

Horticultural gleaner. Austin, TX. Hort. Gleaner. See B–P–H 421/7. HI 54716

Horticultural institute, Taihoku imperial university. Essay. [Taihoku teikoku daigaku rinogakubu engeigaku kyoshitsu ronhyo.] Taihoku [=Taipei, Taiwan]. Hort. Inst. Taihoku Imp. Univ. Essay. See B–P–H 421/8. HI 54717

Horticultural journal. [Chuo engei.] [Japan]. Hort. J. (Japan). See B–P–H 421/11. HI 54719

Horticultural journal, florists' register, and royal lady's magazine. London. Hort. J. Florists' Reg. & Roy.

Lady's Mag. See B–P–H 421/13. HI 54721

Horticultural journal, and florists' register, of useful information connected with floriculture, &c. London. Hort. J. & Florists' Reg. See B–P–H 421/12. HI 54720

Horticultural journal, national university of Chekiang. [Chi-ta yuan-i.] Hangchow, China. Hort. J. (Hangchow). See B–P–H 421/10. HI 54718

Horticultural journal and royal lady's magazine. London. Hort. J. & Roy. Lady's Mag. See B–P–H 421/14. HI 54722

Horticultural leaflet, royal botanical gardens, Ontario. Hamilton, Ont. ?-1988+. Hort. Leafl. Roy. Bot. Gard. Ontario. HI 65864

Horticultural magazine = Annals of horticulture and year-book of information on practical gardening. London. Ann. Hort. See B–P–H 104/1.

Horticultural news. Columbia, MO. Hort. News (Columbia). See B–P–H 421/19. HI 54724

Horticultural news. Dublin. Vol. 1+, 1969+. Hort. News (Dublin). HI 65865

Horticultural news. New Brunswick, NJ. Hort. News (New Brunswick). See B–P–H 421/20. HI 54725

Horticultural newsletter. Ardmore, PA. Nos. 1+, 1953-66+. Hort. Newslett. HI 65866

Horticultural products review. Washington, DC. Vol. 1+, 1987+. Hort. Prod. Rev. HI 65867

Horticultural progress report, agricultural experiment station, Louisiana. Calhoun, LA. No. 1+, 1969+. Hort. Progr. Rep. Agric. Exp. Sta. Louisiana. HI 65868

Horticultural progress report, Nebraska agricultural experiment station = Nebraska agricultural experiment station horticultural progress report. Lincoln, N Nebraska Agric. Exp. Sta. Hort. Progr. Rep. See B–P–H 639/27.

Horticultural record. London. Hort. Rec. See B–P–H 422/5. HI 54729

Horticultural register. London. Hort. Reg. See B–P–H 422/6. HI 54730

Horticultural register, and gardener's magazine. Boston, MA. Hort. Reg. & Gard. Mag. See B–P–H 422/8. HI 54731

Horticultural register, and general magazine. London. Hort. Reg. & Gen. Mag. See B–P–H 422/9. HI 54732

Horticultural register, and general magazine of all useful and interesting discoveries connected with natural history and rural subjects = Horticultural register, and general magazine. London. Hort. Reg. & Gen. Mag. See B–P–H 422/9.

Horticultural report, department of horticultural science, university of Natal. Pietermaritzburg. No. 1+, 1976+. Hort. Rep. Dep. Hort. Sci. Univ Natal. HI 65869

Horticultural research. Athens. Hort. Res. (Athens). See B–P–H 422/10. HI 54733

Horticultural research; an international journal for the publication of scientific work on horticulture. Edinburgh. Vols. 1-21(2), 1961-81. Hort. Res. (Edinburgh). Superseded by: Crop research. HI 65870

Horticultural research circular, Louisiana agricultural experiment station = Louisiana agricultural experiment station. Horticultural research circular. Baton Rouge, LA. Louisiana Agric. Exp. Sta. Hort. Res. Circ. See B–P–H 535/25.

Horticultural review. Cairo. Hort. Rev. (Cairo). See B–P–H 422/14. HI 54735

Horticultural review. Farnham Royal. Nos. 1-5, 1969-77. Hort. Rev. (Farnham Royal). HI 65871

Horticultural review and botanical magazine. Cincinnati, OH. Hort. Rev. & Bot. Mag. See B–P–H 422/13. HI 54734

Horticultural reviews. Westport, CT, New York. Vol. 1+, [1979]+. Hort. Rev. HI 65872

Horticultural science. Calcutta. Vols. 1-2(2), 1969-70. Hort. Sci. HI 65873

Horticultural station, Tokai Kinki agricultural experiment station, bulletin. [Tokai kinki nogyo shikenjo kenkyu hokoku, engei-bu.] Shizuoka, Japan. Hort. Sta. Tokai Kinki Agric. Exp. Sta. Bull. See B–P–H 422/20. HI 54736

Horticultural times. [Engei tsushin.] Kanagawa, Japan. Hort. Times (Kanagawa). See B–P–H 422/21. HI 54737

Horticultural times. London. Hort. Times (London). See B–P–H 422/22. HI 54738

Horticultural topics. New Brunswick, NJ. Hort. Topics. See B–P–H 422/23. HI 54739

Horticultural trade journal. Burnley. Vol. 1+, 1898+. Hort. Trade J. 3-1887-3. HI 65874

Horticultural trends. Brisbane, Qld. 1981+. Hort. Trends. HI 65875

Horticultural and viticultural sciences = Gradinarska i lozarska nauka. Sofia.

Horticulture. Boston, MA. Horticulture (Boston). See B–P–H 423/11. HI 54744

Horticulture. Cuyahoga Falls, OH. Horticulture (Cuyahoga Falls). See B–P–H 423/12. HI 54745

Horticulture alsacienne. Strasbourg. Hort. Alsac. See B–P–H 420/37. HI 54709

Horticulture belge. Editional internationale = Belgische tuinbouw. Internationale editie. Gent.

Horticulture digest. Honolulu, HI. No. 1+, 1970+. Hort. Digest. HI 65876

Horticulture française. Paris. Vol. 1+, 1907+. Horticulture Franç. (1907+). 3-1888-1. HI 54746

Horticulture française. Paris. No. 1+, 1970+. Hort. Franç. Preceded by: France horticole. HI 65877

Horticulture genèvoise. Geneva. Hort. Genèv. See B–P–H 421/6. HI 54715

Horticulture industry. Tunbridge. 1976+. Hort. Industr. Preceded by: Commercial grower. HI 65878

Horticulture moderne. Bourg-la-Reine, France. Hort. Moderne. See B–P–H 421/18. HI 54723

Horticulture in New Zealand. Wellington, N.Z. 1976+. Hort. New Zealand. Preceded by?: Journal of the royal New Zealand institute of horticulture. HI 65879

Horticulture northwest; journal of the northwest ornamental horticultural society. Seattle, WA. Vol. 4+, 1977+. Hort. NorthW. Preceded by: Newsletter, northwest ornamental horticultural society. HI 65880

Horticulture nouvelle. [Société d'horticulture pratique du département du Rhône.] Lyons. Hort. Nouv. See B–P–H 421/22. HI 54726

Horticulture now. London. No. 1+, 1982+. Hort. Now. HI 65881

Horticulture suisse = Gartenbau. Solothurn.

Horticulture tropicale. [Taiwan horticultural society.] Taihoku [=Taipei, Taiwan]. Hort. Trop. See B–P–H 422/25. HI 54740

Horticulture and viticulture. Sofia = Gradinarska i lozarska nauka. Sofia, Bulgaria. Grad. Loz. Nauka. See B–P–H 407/10.

Horticulture week; gardener's chronicle and horticultural trade journal. Teddington. Vol. 199, no. 2+, 1986+. Hort. Week. Preceded by: G C & H T J. HI 65882

Horticulturist. Pilot Point, TX. Horticulturist. See B–P–H 423/16. HI 54747

Horticulturist and journal of rural art and rural taste; devoted to horticulture, landscape gardening, rural architecture, botany, pomology, entomology, rural economy, etc. Albany, NY. Vols. 1(1)-30(354), 1846-75. Hort. & J. Rural Art Rural Taste. Superseded by: Gardener's monthly and horticulturist. 3-1888-1. HI 65883

HortIdeas. Gravel Switch, KY. Vol. 1+, 1984+. HortIdeas. HI 65884

Hortikultura. Zagreb, Yugoslavia. Hortikultura. See B–P–H 423/17. HI 54748

HortScience. St. Joseph, MI., Alexandria, Mount Vernon, VA. Vol. 1+, 1966+. HortScience. HI 65885

Hortulus aliquando. Pasadena, CA. Vols. 1-3, 1975-78. Hortulus Aliquando. HI 65886

Hortus. Beaverton, OR. 1979+. Hortus (Beaverton). HI 65887

Hortus; a gardening journal. Farnham. Vol. 1+, 1987+. Hortus (Farnham). HI 65888

Hortus. [Yüan-i.] Nanking. Vol. 1+, 1935+. Hortus (Nanking). 3-1888-1. HI 54749

Hortus sanitatis. Prague. Hortus Sanit. See B–P–H 423/20. HI 54750

Hortus Van Houtteanus; bulletin périodique faisant suite à la flore des serres et des jardins. Gand. No. 1, 1845-46. Hortus Van Houtteanus. HI 65889

Hoshasen ikushujo kenkyu hokoku = Acta radiobotanika et genetika. Ohmiya-Machi, Ibaraki.

Hospitals-tidende. Copenhagen. Vols. 1-81 [vols. 1-60 also numbered in series as vols. 1-16, 1858-73; ser. 2, vols. 1-9, 1874-83; ser. 3, vols. 1-10, 1883-92; ser. 4, vols. 1-15, 1893-1907; ser. 5, vols. 1-10, 1908-17], 1858-1938. Hosp.-Tidende. Superseded by: Nordisk medicin, formed by the merger of this and other titles. 3-1891-1. HI 65890

Hosta journal. Des Moines, IA. Vol. 17+, 1986+. Hosta J. Preceded by: Bulletin, american hosta society and American hosta society newsletter. HI 65891

House and garden. New York, NY, Philadelphia, PA. House & Gard. See B–P–H 423/22. HI 54751

Houselekes; journal of the sempervivum society. Burgess Hill. No. 10+, 1979+. Houselekes. Preceded by: Journal of the sempervivum society. HI 65892

Hout. The Hague. Hout. See B–P–H 423/23. HI 54752

Hoyan, The; bulletin of the Hoya society international. Atlanta, GA, Kaufman, TX. Vol. 1+, 1979+. Hoyan. HI 65893

Hradecký kraj. Hradec Kralove, Czechoslovakia. Hradecký Kraj. See B–P–H 423/24. HI 54753

Hrvatski šumarski list. Zagre Vols. 65-68, 1941-44. Hrvatski Šumarski List. Preceded and superseded by: Šumarski list. HI 65894

Hsi nan nung yeh k'o hsüeh = Southwest agricultural science. Chungking, China. SouthW. Agric. Sci. See B–P–H 848/15.

Hsi nan pien chiang = Journal of the southwestern frontier. Kunming, China. J. SouthW. Frontier. See B–P–H 482/23.

Hsi pei nung hsüeh yuan hsüeh pao = Northwest agricultural college journal. Sian, China. NorthW. Agric. Coll. J. See B–P–H 665/3.

Hsi pei nung lin. Kuo li hsi pei nung hsüeh yuan pi'en chi

ch'u pan wei yuan hui Pien Hsing = Northwestern journal of agriculture and forestry. The northwestern college of agriculture. Wukung, China. NorthW. J. Agric. Forest. See B–P–H 665/4.

Hsi pei nung pao = Northwest agricultural bulletin; national northwest agricultural college. Wukung, China. NorthW. Agric. Bull. See B–P–H 665/2.

Hsia men ta hsüeh chih tzu jan k'o hsüeh tsung kan = Natural science bulletin of the university of Amoy. Amoy, China. Nat. Sci. Bull. Univ. Amoy. See B–P–H 629/27.

Hsia men ta hsüeh hsüeh p'ao, tsu jan k'o hsüeh pao = Acta scientiarum naturalium universitatis amoiensis. Amoy?, China. Acta Sci. Nat Univ. Amoiensis. See B–P–H 48/23.

Hsia-ta shêngqwu-hsüeh-hui ch'i k'an = Bulletin of the biological association; university of Amoy. Amoy, China. Bull. Biol. Assoc. Univ. Amoy. See B–P–H 241/21.

Hsiang-kang chung wen ta hsueh hsueh pao = Journal of the chinese university of Hong Kong. Shatin, Hong Kong.

Hsieh ta nung pao = Fukien agricultural journal. Foochow, China. Fukien Agric. J. See B–P–H 387/11.

Hsieh ta shêng wu hsüeh pao = Biological bulletin of Fukien christian university. Foochow, China. Biol. Bull. Fukien Christian Univ. See B–P–H 193/11.

Hsin hsi pei = New northwest. [China]. New NorthW. (China). See B–P–H 654/15.

Hsin nung t'ung hsün = New agricultural news. Nanking. New Agric. News. See B–P–H 651/20.

Hsüeh Lin = Forestry. Shanghai. Forestry (Shanghai). See B–P–H 381/9.

Hsüeh shu chi-k'an = Academic review. Taipei, Taiwan. Acad. Rev. See B–P–H 36/23.

Hsüeh yi tsa chih = Wissen und Wissenschaft. Shanghai. Wissen & Wiss. See B–P–H 978/9.

Hu-nan chiao-yü tsa-chih = Special number of the educational miscellany of Hunan. Special Number Educ. Misc. Hunan. See B–P–H 851/3.

Hua-nan nung hsueh yuan hsueh pao = Journal of south China agricultural college. Kuangchou.

Hua nan nung yüeh k'o hsüeh = South China agricultural science. Canton. S. China Agric. Sci. See B–P–H 808/16.

Hua pei nung yeh = North China agriculture. [China]. N. China Agric. See B–P–H 624/16.

Hua pei nung yeh k'o hsüeh yen chiu so. Yen chu tsuan k'an = Special research bulletin. North China agricultural science research institute. Special Res. Bull. N. China Agric. Sci. Res. Inst. See B–P–H 852/5.

Hua-t'ung normal university journal. [Hua-t'ung shih ta hsüeh pao.] [China]. Hua-t'ung Norm. Univ. J. See B–P–H 424/16. HI 54754

Hua-t'ung normal university journal. Natural science series. [Hua-t'ung shih ta hsüeh pao. Tzu jan k'o hsüeh.] [China]. Hua-t'ung Norm. Univ. J., Nat. Sci. Ser. See B–P–H 424/17. HI 54755

Hua tung shih fan ta hsüeh hsüeh pao, tzu jan k'o hsüeh pan = Journal of east China normal university. Natural science edition. Shanghai.

Hua-t'ung shih ta hsüeh pao = Hua-t'ung normal university journal. [China]. Hua-t'ung Norm. Univ. J. See B–P–H 424/16.

Hua-t'ung shih ta hsüeh pao. Tzu jan k'o hsüeh = Hua-t'ung normal university journal. Natural science series. [China]. Hua-t'ung Norm. Univ. J., Nat. Sci. Ser. See B–P–H 424/17.

Huadong shifan daxue xuebao. Ziran kexue ban = Journal of east China normal university. Natural science edition. Shanghai.

Huanan nongxueyuan. Xuebao = Journal of south China agricultural college. Kuangchou.

Huazhong shiyuan xuebao. Ziran kexue ban = Journal of central China teachers' college. Natural science edition. Wuchang.

Hüdrobioloogilised uurimused. Tartu. Vol. 1+, 1958+. Hüdrobiol. Uurim. HI 65895

Huis en Hof. Maandblad voor liefhebbers van bloem en plant. Amsterdam. Huis & Hof. See B–P–H 424/21. HI 54756

Hukui daigaku gakugei gakubu kiyo. Shizen kagaku = Memoirs of the faculty of liberal arts, Fukui university. Ser. 2: Natural science. Fukui, Japan. Mem. Fac. Liberal Arts Fukui Univ., Ser. 2, Nat. Sci. See B–P–H 572/18.

Hull bulletin of marine ecology. Edinburgh. Hull Bull. Mar. Ecol. See B–P–H 424/23. HI 54757

Human ecology; an interdisciplinary journal. New York & London. Vol. 1+, 1972+. Human Ecol. HI 65896

Humboldt. Monatsschrift für die gesammten Naturwissenschaften. Stuttgart. Humboldt. See B–P–H 424/24. HI 54758

Humboldtiana. Bonn. 1978-85, 1979-86. Humboldtiana. Preceded by: Acta Humboldtiana. Superseded by: Bibliographia Humboldtiana. HI 65898

Hungarica acta biologica. Budapest. Hung. Acta Biol. See B–P–H 424/26. HI 54759

Hungarian agricultural review. Budapest. Hung. Agric. Rev. See B–P–H 424/27. HI 54760

Hunter natural history. New Lambton, Waratah, N.S.W. Vol. ?-8+, 19??-6?+. Hunter Nat. Hist. HI 65900

Huntia; a yearbook of botanical and horticultural bibliography (later a journal of botanical history). Pittsburgh, PA. Vol. 1+, 1964-65, 1979+ [publication suspended 1966-78]. Huntia. HI 65901

Hunting group review. London. 1951+. Hunting Group Rev. HI 65902

Huntington calendar = Huntington library, art gallery and botanical gardens calendar. San Marino, CA.

Huntington library, art gallery and botanical gardens calendar. An informal account of happenings in the library, art gallery, botanical gardens. San Marino, CA. 19??-73?+. Huntington Libr. Art Gall. Bot. Gard. Calend. Preceded by: Calendar of the exhibitions, Huntington library, art gallery and botanical gardens. HI 65903

Hunyadmegyei történelmi és régészeti társulat évkönyve. Vajdahunyad, Hungary [=Hunedoara, Rumania]. Hunyadmegyei Tört. Régész. Társ. Évk. See B–P–H 424/29. HI 54761

Hushållnings journal, utgifwen genom kgl. patriotiska sällskapets föranstaltende. Stockholm. Hush. J. See B–P–H 424/30. HI 54762

Husvennen; menigmands Billedblad for Morskabslaesning, Oplysning og Husflid. Copenhagen. 1874-1889/90. Husvennen. HI 65904

Hutching's illustrated California magazine. San Francisco, CA. Vols. 1-5, 1856-61. Hutching's Ill. Calif. Mag. Superseded by: California magazine and mountaineer. 3-1907-1. HI 65905

Hvalrådets skrifter. Oslo. Hvalrådets Skr. See B–P–H 424/31. HI 54763

Hyacinth control journal. Fort Lauderdale, Fort Myers, FL. Vols. 1-13, 1962-75. Hyacinth Control J. Superseded by: Journal of aquatic plant management. HI 65906

Hybrid. Vol. 1+, 1972+. Hybrid. HI 65907

Hybrid grain sorghum yield results. Sydney?, N.S.W. 1983/84+, 1984+. Hybrid Grain Sorg. Yield Results. HI 65908

Hydrobiologia. Acta hydrobiologica, limnologica et protistologica. The Hague. Hydrobiologia. Se: 425/1. HI 54764

Hydrobiological bulletin. Amsterdam. Vol. 7+, 1973+. Hydrobiol. Bull. Preceded by: Mededelingen van de hydrobiologische vereniging. HI 65909

Hydrobiological journal. [Translation of: Gidrobiologicheskii zhurnal.] Washington, DC. Vol. 5+, 1969+. Hydrobiol. J. HI 65910

Hydrobiological studies. Prague. Vol. 1+, 1966+. Hydrobiol. Stud. HI 65911

Hydrobiology newsletter, George Washington University. Washington, DC. Nos. [1]-20, 1965-68. Hydrobiol. Newslett. George Washington Univ. HI 65912

Hydrological journal. Budapest = Hidrológiai közlöny. Budapest. Hidrol. Közl. See B–P–H 416/4.

Hydroponic society of America newsletter. Concord, CA. 19??-82+. Hydrop. Soc. America Newslett. HI 65913

Hydroponics. Detroit, MI. Hydroponics. See B–P–H 425/3. HI 54765

Hydrothérapie = Médicine contemporaine. Paris. Med. Contemp. (Paris). See B–P–H 554/12.

Hygaea. Copenhagen. Hygaea (Copenhagen). See B–P–H 425/5. HI 54766

Hygiea. Stockholm. Hygiea (Stockholm). See B–P–H 425/6. HI 54767

Hyogo biology. [Hyogo seibutsu.] Sasayama, Japan. Hyogo Biol. See B–P–H 425/7. HI 54768

Hyogo hakubutsugakkai kaiho = Hyogo natural history. Sasayama, Japan. Hyogo Nat. Hist. See B–P–H 425/ 9.

Hyogo-ken chuto kyoiku hakubutsugaku zasshi = Hyogo-ken journal of natural history. Hyogo-ken J. Nat. Hist. See B–P–H 425/14.

Hyogo-ken journal of natural history. [Hyogo-ken chuto kyoiku hakubutsugaku zasshi.] Hyogo-ken J. Nat. Hist. See B–P–H 425/14. HI 54770

Hyogo-ken nogyo sogo senta. Kenkyu hokoku. Akashi. No. 25+, 1976+. Hoyogo-Ken Nogyo Sogo Senta Kenkyu Hokoku. Preceded by: Bulletin of the Hyogo prefectural agricultural experiment station. HI 65914

Hyogo kenritsu nogyo shikenjo. Kenkyu hokoku = Bulletin of the Hyogo prefectural agricultural experiment station. Akashi.

Hyogo natural history. [Hyogo hakubutsugakkai kaiho.] Sasayama, Japan. Hyogo Nat. Hist. See B–P–H 425/ 9. HI 54769

Hyogo nogyo tanki daigaku kenkyu shuroku = Research report; Hyogo agricultural college. Kobe, Japan. Res. Rep. Hyogo Agric. Coll. See B–P–H 776/5.

Hyogo seibutsu = Hyogo biology. Sasayama, Japan. Hyogo Biol. See B–P–H 425/7.

I A A P miscellaneous publications. No. 1+, 1979+. I. A. A. P. Misc. Publ. HI 65915

I A B newsletter. Penicuik, etc. 1980+. I. A. B. Newslett. HI 65916

I A P news; the newsletter of the international association of pteridologists. Utrecht. No. 1+, 1986+. I. A. P. News. HI 65917

I A P patrika. [Indian association of

palynostratigraphers.] Bangalore, Lucknow. Vol. ?-3+, 19??-81+. I: A. P. Patrika. HI 65918

I A T E M. Boletín informativo. Posadas, Argentina. No. 1+, 1964?+. I. A. T. E. M. Bol. Inform. HI 65919

I A W A bulletin. Leiden, Zurich. Vols. 1-?, 1977-79. I. A. W. A. Bull. Preceded by: International association of wood anatomists, news bulletin. Superseded by: I A W A bulletin. New series. HI 65920

I A W A bulletin. New series. Quarterly periodical of the international association of wood anatomists. Leiden. Vol. 1+, 1980+. I. A. W. A. Bull., N.S. Preceded by: I A W A bulletin. HI 65921

I B C = Indian botanical contactor.

I B P boletin interamericano = I B P inter-american news. Washington, DC.

I B P G R annual report = Report (Annual), international board for plant genetic resources. Rome.

I B P G R regional committee for southeast Asia newsletter = Newsletter, regional committee for the south east asian programme, international board for plant genetic resources.

I B P handbook. London. No. 1+, 1965+. I. B. P. Handb. HI 65922

I B P inter-american news. Washington, DC. Vol. 1-4(2), 1968-71. I. B. P. Inter-Amer. News. HI 65923

I B P news. Rome. Nos. 1-25, 1964-75. I. B. P. News. HI 65924

I B R = Indian botanical reporter.

I B S A bulletin. Durbanville. 1975?+. I. B. S. A. Bull. Preceded by: I B S A newsletter. HI 65925

I B S A newsletter. Durbanville. 19??-75? I. B. S. A. Newslett. Superseded by: I B S A bulletin. HI 65926

I C A R. Cereal crop series. New Delhi. No. 1+, 1961+. I. C. A. R. Cereal Crop Ser. HI 65927

I C A R D A annual report. Aleppo. 1980/81+, 1981+. I. C. A. R. D. A. Annual Rep. HI 65928

I C A R D A research highlights. Aleppo. 1984+. I. C. A. R. D. A. Res. Highlights. HI 65929

I C A S A L S newsletter. Lubbock, TX. 1967+. I. C. A. S. A. L. S. Newslett. HI 65930

I C E S oceanographic data lists. Charlottenlund-Slot. 1957-66. I. C. E. S. Oceanogr. Data Lists. Preceded by: Bulletin hydrographique. Conseil international pour l'exploration de la mer. Superseded by: I C E S oceanographic data lists and inventories. HI 65931

I C E S oceanographic data lists and inventories. Charlottenlund-Slot. 1971+. I. C. E. S. Oceanogr. Data Lists Invent. Preceded by: I C E S oceanographic data lists. HI 65932

I C N - U C L A symposia on molecular and cellular biology. New York. Vols. 1-23, 1974-81. I. C. N. - U. C. L. A. Symp. Molec. Cell. Biol. Superseded by: U C L A symposia on molecular and cellular biology. HI 75220

I C O M news. Vol. 1+, 1948+. I. C. O. M. News. HI 65933

I C P newsletter = Newsletter, international commission for palynology. Tempe, AZ.

I C S U yearbook = Year book, international council of scientific unions.

I chuan hsüeh pao = Acta genetica sinica. Peking (Beijing).

I D I A. Buenos Aires. No. 1+, 1948+. I. D. I. A. HI 65934

I E S newsletter. Millbrook, NY. Vol. 1+, 1984+. I. E. S. Newslett. HI 65935

I F C C bulletin. Paris. No. 1+, 1961+. I. F. C. C. Bull. HI 65936

I Hjemmet; illustreret Folkeblad. Copenhagen. Vols. 3-10?, 1857-64. I Hjemmet. Preceded by: For Tre Skilling. HI 65937

I I R B. Brussels. Vols. 1-7, 1965-77. I. I. R. B. Preceded and superseded by: Compte rendu, congrès d'hiver, institut international de recherches betteravières. HI 65938

I I T A research briefs. Ibadan. Vol. 1+, 1980+. I. I. T. A. Res. Briefs. HI 65939

I J S B = International journal of systematic bacteriology. Ames, IA, Washington, DC.

I M A news. Kew. Vol. 1+, 1987+. I. M. A. News. HI 65940

I M S newsletter; international marine science. Paris. No. 1+, 1973+. I. M. S. Newslett. Preceded by: International marine science. HI 65941

I N A newsletter. Utrecht. Vol. 1+, 1979+. I. N. A. Newslett. HI 65942

I N I R E B informa: comunicado. Xalapa. No. 1+, 1976+. I. N. I. R. E. B. Informa Comunicado. HI 65943

I O B C newsletter. Zurich. No. 1+, 1972+. I. O. B. C. Newslett. HI 65944

I O P newsletter. London. 1976+. I. O. P. Newslett. HI 65945

I O P B newsletter = International organization of plant biosystematists newsletter. Utrecht, Reading, Copenhagen.

I O P B newsletter = Newsletter, international organization of plant biosystematists. Reading, etc.

I O S bulletin; journal of the international organization for succulent plant study. Monaco, etc. [place of publication varies]. Nos. 1-4, 1958-60; ser. 2, no. 1+, 1961+. I. O. S. Bull. Preceded by: I O S circular. HI 65946

I O S circular. Nos. 1-6, 1955-57. I. O. S. Circ. Superseded by: I O S bulletin. HI 65947

I P E F. Piracicaba. No. 1+, 1971+. I. P. E. F. HI 65948

I P E F international. Piracicaba. No. 01+, 1990+. I. P. E. F. Int. HI 75244

I P M practitioner; newsletter of integrated pest management. Winters, CA. Vol. 1+, 979?+. I. P. M. Practitioner. HI 65949

I P P C infoletter. Corvallis, OR. No. 1+, 1970+. I. P. P. C. Infolett. HI 65950

I P R news. Singapore. ?-1981+. I. P. R. News. HI 65951

I R A T informations. Paris. 1974+. I. R. A. T. Inform. HI 65952

I R R I reporter. Manila. No. 1+, 1965+. I. R. R. I. Reporter. HI 65953

I R R I research paper series. Manila. No. 1+, 1976+. I. R. R. I. Res. Pap. Ser. HI 65954

I S B C = Index to scientific book contents. Philadelphia, PA.

I S H A bulletin = Bulletin, institute for the study of science in human affairs. New York.

I S I atlas of science: Animal and plant sciences. Philadelphia, PA. Vol. 1+, 1988+. I. S. I. Atlas Sci., Anim. Pl. Sci. HI 65955

I S I atlas of science: biochemistry. Philadelphia, PA. Vol. 1+, 1988+. I. S. I. Atlas Sci., Biochem. HI 74844

I S N A newsletter. New Delhi. Vols. 1-3, 1972-74. I. S. N. A. Newslett. Superseded by: Journal of nuclear agriculture and biology. HI 65956

I S N A R newsletter; international service for national agricultural research. The Hague. Vol. 1+, 1984+. I. S. N. A. R. Newslett. HI 65957

I S R = Index to scientific reviews. Philadelphia, PA.

I S R = Interdisciplinary science reviews. London.

I S S N key title register. Washington, DC. 1977+. I. S. S. N. Key Title Reg. HI 65958

I S T A news bulletin. Wageningen, Zürich. No. 1+, 1959+. I. S. T. A. News Bull. HI 65959

I S T F news. Bethesda, MD. 1980+. I. S. T. F. News. HI 65960

I T E symposia = Symposia, institute for terrestrial ecology.

I T S L. Antwerp. Vols. 1-2, 1967-68. I. T. S. L. HI 65961

I U B S monograph series. Eynsham. 1986+. I. U. B. S. Monogr. Ser. HI 64346

I U B S news. Paris. Nos. 1-3, 1971-73. I. U. B. S. News. Preceded by: News-letter from the international union of biological sciences. HI 64347 Superseded by: I U B S newsletter.

I U B S newsletter. Paris. Nos. 4-16, 1973-79. I. U. B. S. Newslett. Preceded by: I U B S news. Superseded by: Biology international. HI 64348

I U C N bulletin. Morges. N.s. nos. 1-20, 1961-66; n.s. vol. 2+, 1966+. I. U. C. N. Bull. Preceded by: Bulletin, international union for conservation of nature and natural resources. HI 64349

I U C N conservation achievements. Gland. 1980+. I. U. C. N. Conservation Achievem. Preceded by: I U C N director general's report. HI 64350

I U C N director general's report. Gland. 1979. I. U. C. N. Director Gen. Rep. Superseded by: I U C N conservation achievements. HI 64351

I U C N environmental law papers. Morges. Nos. 1-3, 1972. I. U. C. N. Environm. Law Pap. Superseded by: I U C N environmental policy and law papers. HI 64352

I U C N environmental policy and law occasional paper. Morges. No. 1+, 1985+. I. U. C. N. Environm. Policy Law Occ. Pap. HI 64353

I U C N environmental policy and law paper. Morges. No. 4+, 1973+. I. U. C. N. Environm. Policy Law Pap. Preceded by: I U C N environmental law papers. HI 64354

I U C N monograph. Morges. No. 1+, 1971+. I. U. C. N. Monogr. HI 64355

I U C N occasional paper. Morges. No. 1+, 1971+. I. U. C. N. Occas. Pap. HI 64356

I U C N publications. New series. Morges. No. 1+, 1963+. I. U. C. N. Publ., N.S. HI 64357

I U C N publications. New series, supplementary. Morges. No. 1+, 1964+. I. U. C. N. Publ., N.S., Suppl. HI 64358

I U C N yearbook; annual report of the international union for conservation of nature and natural resources. Morges. 1970+, 1971+. [Issue for 1978/79 is contained in I U C N bulletin, vol. 10, no. 7; yearbooks for 1976/77 and 1977/78 were not published.] I. U. C. N. Yearb. Preceded by: International union for conservation of nature and natural resources, annual report. HI 64359

Ibaraki-ken engei shikenjo kenkyu hokoku = Bulletin of the Ibaraki horticultural experiment station. Ibaraki.

Icones bogorienses. Leiden. Icon. Bogor. See

B–P–H 425/23. HI 54771

Icones filicum sinicarum. Nanking. Vols. 1-4, 1930-37. Icon. Filicum Sin. HI 64360

Icones florae alpinae plantarum. Paris. Icon. Fl. Alpinae Pl. See B–P–H 425/26. HI 54772

Icones florae japonicae. Tokyo. Icon. Fl. Jap. See B–P–H 425/27. HI 54773

Icones fungorum malayensium. Vienna. Icon. Fungorum Malayensium. See B–P–H 424/30. HI 54774

Icones of japanese algae. Tokyo. Icon. Jap. Algae. See B–P–H 425/31. HI 54775

Icones mycologicae. Meise. No. 1+, 1982+. Icon. Mycol. HI 64361

Icones plantarum. London. Icon. Pl. See B–P–H 425/32. HI 54776

Icones plantarum africanarum. Dakar. No. 1+, 1953+. Icon. Pl. African. HI 64362

Icones plantarum formosanarum nec non et contributiones ad floram formosanam. Taihoku [=Taipei, Taiwan]. Icon. Pl. Formosan. See B–P–H 426/3. HI 54777

Icones plantarum koisikavenses. Tokyo. Icon. Pl. Koisikav. See B–P–H 426/5. HI 54778

Icones plantarum omeiesium. Chengtu, China. Icon Pl. Omeiesium. See B–P–H 426/7. HI 54779

Icones plantarum sinicarum. Shanghai. Icon. Pl. Sin. See B–P–H 426/8. HI 54780

Icones plantarum tropicarum. Sarasota, FL. No. 1+, 1980+. Icon. Pl. Trop. HI 64363

Icones selectae. Tirlemont, Belgium. Icon. Select. See B–P–H 426/9. HI 54781

Iconographia dermatologica, syphilidologica et urologica. Kyoto. Iconogr. Dermatol. See B–P–H 426/10. HI 54782

Iconographia florae sinicae. Kunming, China. Iconogr. Fl. Sin. See B–P–H 426/11. HI 54783

Iconographie des maladies cutanées et syphilitiques. Paris. Iconogr. Malad. Cutan. Syph. See B–P–H 426/12. HI 54784

Iconographia mycologica. Milan. Vols. 1-26, 1927-33. Iconogr. Mycol. (MIlan). 3-1914-2. HI 54785

Iconographia mycologica. [Supplement to: Mycopathologia et mycologia applicata.] The Hague. Vol. 1+, 1959+. Iconogr. Mycol. (The Hague). HI 64364

Iconographia plantarum Asiae orientalis. [Toa shokubutsu zusetsu.] Tokyo. Vols. 1-5(2), 1935-52. Iconogr. Pl. Asiae Orient. HI 64365

Iconographia selecta florae azoricae. Coimbra. No. 1+, 1980+. Iconogr. Selecta Fl. Azor. HI 64366

Iden. The Heredity. Tokyo. Iden. See B–P–H 426/20. HI 54786

Idengaku zasshi = Japanese journal of genetics. Tokyo Jap. J. Genet. See B–P–H 498/22.

Idesia. Arica. No. 1+, 1970+. Idesia. HI 64367

Idia = I D I A. Buenos Aires.

Idöjárás. Budapest. Idöjárás. See B–P–H 426/22. HI 54787

I.F.C.C. bulletin. Paris = I F C C bulletin. Paris.

Ife university herbarium bulletins on plant ecology and taxonomy. Ife-Ife. Nos. 1-10, 1969-75. Ife Univ. Herbar. Bull. HI 64368

Ifjuság és élet. Budapest. Ifjuság & Élet. See B–P–H 426/24. HI 54788

Igaku to seibutsugaku = Medicine and biology. [Japan]. Med. & Biol. See B–P–H 553/26.

Iheringia: Botanica. Porto Alegre, Brazil. Iheringia, Bot. See B–P–H 426/26. HI 54789

Iheringia: Divulgação. Porto Alegre. No. 1+, 1971+. Iheringia Divulg. HI 64369

Iheringia: Miscelânea. Porto Alegre. No. 1+, 1985+. Iheringia Misc. HI 64370

Ikebana International magazine. Tokyo. Ikebana Int. Mag. See B–P–H 427/1. HI 54790

Ikonographia dermatologica, syphilidologica et urologica = Iconographia dermatologica, syphilidologica et urologica. Kyoto. Iconogr. Dermatol. See B–P–H 426/10.

Ilgenfritz orchardist. Monroe, MI. Ilgenfritz Orchardist. See B–P–H 427/7. HI 54791

Illini horticulture. Carbondale, IL. Illini Hort. See B–P–H 428/5. HI 54805

Illinois biological monographs. Urbana, IL. Illinois Biol. Monogr. See B–P–H 428/7. HI 54806

Illinois gladiolus bulletin. Champaign, IL. Illinois Gladiolus Bull. See B–P–H 428/8. HI 54807

Illinois monthly magazine. Vandalia, IL. Illinois Monthly Mag. See B–P–H 428/10. HI 54808

Illinois research. University of Illinois agricultural experiment station. Urbana, IL. Illinois Res. See B–P–H 428/11. HI 54809

1Illinois science news. Normal, IL. Vol. 1(2)+, 1977+. Illinois Sci. News. Preceded by: Newsletter, Illinois state academy of science. HI 64371

Illinois vegetable growers' bulletin. Illinois Veg. Growers' Bull. See B–P–H 428/12. HI 54810

Illustrated botany. [Edited by Comstock.] New York, NY. Ill. Bot. (Comstock). See B–P–H 427/11. HI 54792

Illustrated botany. [Edited by Newman.] New York, NY. Ill. Bot. (Newman). See B–P–H 427/12. HI 54793

Illustrated canadian forest and outdoors. Ottawa. Vols. 19(8)-32(8), 1923-36. Ill. Canad. Forest Outdoors. Preceded by: Illustrated canadian forestry magazine. Superseded by: Forest and outdoors. HI 64372

Illustrated canadian forestry magazine. Ottawa. Vols. 16(12)-19(7), 1920-23. Ill. Canad. Forest. Mag. Preceded by: Canadian forestry journal. Superseded by: Illustrated canadian forest and outdoors. HI 64373

Illustrated farmer and garden almanack. Ill. Farmer Gard. Alman. See B–P–H 427/13. HI 54794

Illustrated horticultural times. London. Ill. Hort. Times. See B–P–H 427/17. HI 54798

Illustrated journal of agriculture. Montreal. Ill. J. Agric. See B–P–H 427/19. HI 54799

Illustrated travels. London. Ill. Travels. See B–P–H 428/3. HI 54804

Illustration horticole. Ghent & Brussels. Ill. Hort. See B–P–H 427/16. HI 54797

Illustrierte Flora. Vienna. Ill. Fl. See B–P–H 427/14. HI 54795

Illustrierte Monatshefte für Obst- und Weinbau. Stuttgart. Ill. Monatsh. Obst- Weinbau. See B–P–H 427/23. HI 54801

Illustrierte nützliche Blätter. Vienna. Ill. Nützl. Blätt. See B–P–H 427/26. HI 54802

Illustrierte Zeitschrift für das Gesamtgebiet des Naturschutzes = Schweizerische Blätter für Naturschutz. Basel.

Illustriertes Jahrbuch der Naturkunde. Teschen. Ill. Jahrb. Naturk. See B–P–H 427/20. HI 54800

Illustrirte Garten-Zeitung. Stuttgart. Ill. Gart.-Zeitung. See B–P–H 427/15. HI 54796

Illustrirter Rosengarten. Stuttgart. Ill. Rosengart. See B–P–H 427/28. HI 54803

Imkerfreund. Munich. Vol. 1+, 1946+. Imkerfreund. HI 64374

Immergrüne Blätter. Bremen. Immergrüne Blätt. See B–P–H 428/21. HI 54811

Imop sihom yon'gu pogo = Bulletin of the forest experimental station. Seoul.

Imop sihom yon'gu pogo = Research report of the forest experiment station. Seoul.

Imop sihomjang yon'gu pogi = Research report of the institute of forest genetics. Suwon.

Imop sihomjang yongu pogo = Research report of the forest research institute, Seoul. Seoul.

Imperial institute journal. London. Imp. Inst. J. See B–P–H 428/24. HI 54812

Imperial mycological institute. Mimeographed publications. Kew, England. Imp. Mycol. Inst. Mimeogr. Publ. See B–P–H 428/25. HI 54813

In natuur. Tijdschrift voor het lager onderwijs ... Amsterdam. In Natuur (Amsterdam). See B–P–H 428/27. HI 54814

In vitro. Lake Placid, NY, Gaithersburg, Rockville, MD. Vols. 1-20, 1965-84. In Vitro. Superseded by: In vitro cellular and developmental biology. HI 64375

In vitro cellular and developmental biology. Gaithersburg, MD. Vol. 21+, 1985+. In Vitro Cell. Developmental Biol. Preceded by: In vitro. HI 64376

Inagro. Revista de investigaciones agronómicas. Buenos Aires. Inagro. See B–P–H 428/28. HI 54815

Incidentele mededelingen van het biohistorisch instituut der rijksuniversiteit te Utrecht. Utrecht. N.s. no. 1+, 1974+. Incidentele Meded. Biohist. Inst. Rijksuniv. Utrecht. Preceded by: Circulaire, biohistorisch instituut der rijksuniversiteit te Utrecht. HI 64377

Incompatibility newsletter. Wageningen. No. 1+, 1972+. Incompat. Newslett. HI 64378

Indeks biologi dan pertanian di Indonesia. Bogor. Vol. 1+, 1969+. Indeks Biol. Pert. Indonesia. HI 64379

Index on agricultural research in the Netherlands. Projects = Index op het landbouwkundig onderzoek in Nederland. Projecten. The Hague.

Index of agricultural research and related subjects in Nigeria. Ibadan. 1966-74. Index Agric. Res. Nigeria. Preceded by: Index of research, department of agricultural research, federal republic of Nigeria. Superseded by: Approved research programme, federal department of agricultural research, Nigeria. HI 64380

Index to american doctoral dissertations. [Forms part 13: Dissertation abstracts, vols. 16-23.] Ann Arbor, MI. 1955-64. Index Amer. Doct. Diss. Superseded by: American doctoral dissertations. HI 64381

Index to australasian taxonomic literature for 1968 (-1970). [Forms part of: Regnum vegetabile. See B–P–H 760/25.] Utrecht. 1968-70, 1970-72. Index Australas. Taxon. Lit. Superseded in part by: Kew record of taxonomic literature. HI 64382

Index bibliographique de botanique tropicale. Paris. Nos. 1-7, 1964-70, 1966-71. Index Bibliogr. Bot. Trop. HI 64383

Index of biochemical reviews. [Supplement to: F E B S letters.] Amsterdam. 1971/72+, 1973+. Index Biochem. Rev. HI 64384

Index to book reviews in the sciences. Philadelphia, PA. No. 1+, 1980+. Index Book Rev. Sci. HI 64385

Index of conference proceedings received by the B L L.

Boston Spa. No. 69+, 1973+. Index Conf. Proc. Receiv. B. L. L. Preceded by: Index of conference proceedings received by the national lending library for science and technology. HI 64386

Index of conference proceedings received by the national lending library for science and technology. Boston Spa. Nos. 1-68, 1964-73. Index Conf. Proc. Receiv. N. L. L. Superseded by: Index of conference proceedings received by the B L L. HI 64387

Index of current horticultural investigations at the experimental horticulture stations and in the regions, ministry of agriculture. 1967-70. Index Curr. Hort. Invest. Minist. Agric. Superseded by: Index of current investigations at the experimental centres and in the regions. Part 2, horticulture. HI 64388

Index of current investigations at the experimental centres and in the regions. Part 2. horticulture. 1971-73. Index Curr. Invest. Exp. Centres Regions, 2. Hort. Preceded by: Index of current horticultural investigations at the experimental horticulture stations and in the regions, ministry of agriculture. HI 64389

Index of current research in arid zone ecology = Indice de proyectos en desarrollo en ecologia de zonas aridas. Xalapa.

Index of current tropical ecological research = Indice de proyectos en desarrollo en ecologia tropical. Xalapa.

Index to european taxonomic literature for 1965 (-1969). [Forms part of: Regnum vegetabile.] Utrecht. 1965-69, 1966-71. Index Eur. Taxon. Lit. Superseded by: Kew record of taxonomic literature. HI 64390

Index of forest research, Nigeria. Ibadan. 1972/73+, 1973?+. Index Forest Res., Nigeria. HI 64391

Index of fungi. Kew. Vol. 1+, 1920+ [earlier parts, also known as "Petrak's lists", are reprinted from: Just's botanischer Jahresbericht and Review of applied mycology, supplement; Supplement, 1969; Supplement: lichens, 1972+. Index Fungi. HI 64392

Index holmensis; world phytogeographic index. Zurich. Vol. 1+, 1969+. Index Holm. HI 64393

Index horti botanici universitatis budapestinensis. Budapest. Index Horti Bot. Univ. Budapest. See B–P–H 429/10. HI 54816

Index kewensis plantarum phanerogamarum. Oxford. Pts. 1-4, 1893-95; Supplementum, vols. 1-16, 1901-81. Index Kew. Superseded by: Kew index for 1986 [etc.]. HI 64394

Index op het landbouwkundig onderzoek in Nederland. Projecten. The Hague. Nos. 1-?, 1968-73? Index Landbouwk. Onderz. Nederland. Proj. Superseded by: Index op het onderzoek voor de landbouw en landinrichting in Nederland. Projecten. HI 64395

Index of the national park system and affiliated areas = Index of the national park system and related areas. Washington, DC.

Index of the national park system and related areas. Washington, DC. 1979+. Index Natl. Park System Related Areas. HI 64396

Index op het onderzoek voor de landbouw en landinrichting in Nederland. Projecten. 1974+. Index Onderz. Landb. Landinricht. Nederland. Proj. Preceded by: Index op het landbouwkundig onderzoek in Nederland. Projecten. HI 64397

Index to plant chromosome numbers. Chapel Hill, NC, Utrecht, St. Louis, MO. 1956+, 1958+. Index Pl. Chromosome Numbers. For the years 1965-73/74 contained in: Regnum vegetabile. HI 64398

Index des projects en développement sur l'écologie des zones aridas = Indice de proyectos en desarrollo en ecologia de zonas aridas. Xalapa.

Index of research: Agriculture - animal husbandry - animal disease - forestry = Index of research, department of agricultural research, Nigeria.

Index of research, department of agricultural research, Nigeria. 1959-66. Index Res. Dept. Agric. Res., Nigeria. Superseded by: Index of agricultural research and related subjects in Nigeria. HI 64399

Index to science and technical articles in chinese periodicals and newspapers. 1962/67+, 1970+. Index Sci. Techn. Articles Period. Newspapers. HI 64400

Index to scientific book contents. Philadelphia, PA. 1985+. Index Sci. Book Contents. HI 64401

Index to scientific reviews; an interdisciplinary index to the review literature of science, medicine, agriculture, technology, and the behavioral sciences. Philadelphia, PA. 1974+. Index Sci. Rev. HI 64402

Index to scientific and technical proceedings. Philadelphia, PA. 1978+. Index Sci. Techn. Proc. HI 64403

Index seminum. [The numerous serially issued "Index seminum" titles published around the world are not detailed here. They can be cited under this "generic" title coupled with the place of the issuing botanical garden as cited in the place headings in: Heywood, C. A. & V. H., & P. W. Jackson. 1990. International directory of botanical gardens V. 5th ed. Koenigstein, Koeltz Scientific Books. xi+1021 pp., e.g. Index seminum (Barcelona).] Index Seminum (...). HI 64404

Index translationum; international bibliography of translations. Repertoire international des traductions Paris. Nos. 1-31, 1932-40; n.s. vol. 1+, 1948+. Index Transl. HI 64405

India journal of medical and physical science. Calcutta. India J. Med. Phys. Sci. See B–P–H 429/16

HI 54817

India review and journal of foreign science and the arts. Calcutta. Vols. 1-3, 1837-39. India Rev. J. Sci. Arts. HI 64406

India weekly. London. 1964+. India Weekly. HI 64407

Indian agriculturist. Calcutta. Indian Agric. See B–P–H 429/18. HI 54818

Indian annals of medical science. Calcutta. Indian Ann. Med. Sci. See B–P–H 429/19. HI 54819

Indian biological journal. Bhagalpur. Vol. 2+, 1980+. Indian Biol. J. Preceded by: Indian biological review. HI 64408

Indian biological review. Bhagalpur. Vol. 1, 1979? Indian Biol. Rev. Superseded by: Indian biological journal. HI 64409

Indian biologist. Madras, Calcutta. Vol. 1+, 1969+. Indian Biologist. HI 64410

Indian botanical contactor. Vol. 1+, 1983+. Indian Bot. Contactor. HI 64411

Indian botanical reporter. Aurangabad. Vol. 1+, 1982+. Indian Bot. Reporter. HI 64412

Indian cashew journal. Cochin, India. Indian Cashew J. See B–P–H 429/20. HI 54820

Indian coconut journal. Cochin. Vol. 8+, 1977+. Indian Coconut J. Preceded by: Coconut bulletin. HI 64413

Indian coffee. Bangalore, India. Indian Coffee. See B–P–H 429/21. HI 54821

Indian cotton growing review. Bombay. Vols. 1-18, 1947-64. Indian Cotton Growing Rev. Superseded by: Indian cotton journal. HI 64414

Indian cotton journal. Bombay. Vols. 19-20(2), 1965-66. Indian Cotton J. Preceded by: Indian cotton growing review. HI 64415

Indian drugs. Dadar, Bombay. 1963+. Indian Drugs. HI 64416

Indian ecologist. Bombay. Vols. 1-[2], 1946-47. Indian Ecol. HI 64417

Indian farmers' digest. Pantnagar. Vol. 1+, 1968+. Indian Farmers' Digest. HI 64418

Indian farming. Delhi. Indian Farming. See B–P–H 429/25. HI 54822

Indian fern journal. Patiala. Vol. 1+, 1984+. Indian Fern J. HI 64419

Indian forest bulletin. Calcutta. N.s. no. 92+, 1941+. Indian Forest Bull. Preceded by: Forest bulletin, Dehra Dun. HI 64420

Indian forest college magazine. Dehra Dun. Vol. ?-3+, 19??-58+. Indian Forest Coll. Mag. HI 64421

Indian forest leaflets. Dehra Dun. No. 7+, 1941+. Indian Forest Leafl. Preceded by: Leaflet, forest research institute. Dehra Dun. HI 64422

Indian forest memoirs. Forest botany series. Calcutta. Vol. 1(1), 1911. Indian Forest Mem., Forest Bot. Ser. HI 64423

Indian forest records. Calcutta. Vols. 1-20, 1907-34. Indian Forest Rec. Superseded by: Indian forest records. New series, botany; Indian forest records. New series, forest influences; Indian forest records. New series, forest management and mensuration; Indian forest records. New series, mycology; Indian forest records. New series, sylvics; Indian forest records. New series, sylviculture; Indian forest records. New series, utilization; and Indian forest records. New series, wood technology. HI 64424

Indian forest records. Botany. Calcutta. Vol. 1+, 1937+. Indian Forest Rec., Bot. Preceded by: Indian forest records. HI 64425

Indian forest records. New series, forest influences. Calcutta. No. 1+, 1982+. Indian Forest Rec., Forest Influences. Preceded by: Indian forest records. HI 64426

Indian forest records. New series, forest management and mensuration. Calcutta. 1978+. Indian Forest Rec., Forest Managem. Preceded by: Indian forest records. HI 64427

Indian forest records. New series, forest pathology. Delhi. Vol. 1+, 1950+. Indian Forest Rec., Forest Pathol. Preceded by: Indian forest records. Mycology. HI 64428

Indian forest records. New series, mycology. Delhi. Vols. 1(5)-?, 1950-58. Indian Forest Rec., Mycol. Preceded by: Indian forest records. HI 64429 Superseded by: Indian forest records. Forest pathology.

Indian forest records. New series, sylvics. Calcutta. 1974+. Indian Forest Rec., Sylvics. Preceded by: Indian forest records. HI 64430

Indian forest records. New series, sylviculture. Calcutta. Vol. 1+, 1936+. Indian Forest Rec., Sylvic. Preceded by: Indian forest records. HI 64431

Indian forest records. New series, utilization. Calcutta. Vol. 1+, 1935+. Indian Forest Rec., Utiliz. Preceded by: Indian forest records. HI 64432

Indian forest records. New series, wood technology (later wood anatomy). Calcutta. Vol. 1+, 1949+. Indian Forest Rec., Wood Technol. (Anat.) Preceded by: Indian forest records. HI 64433

Indian forester. Allahabad, India. Indian Forester. See B–P–H 429/30. HI 54823

Indian horticultural abstracts. New Delhi. Vols. 1-3?, 1951-53? Indian Hort. Abstr. HI 64434

Indian horticulture. New Delhi. Indian Hort. See

B–P–H 430/1. HI 54824

Indian journal of agricultural research. Sadar, Karnal. Vol. 5+, 1971+. Indian J. Agric. Res. Preceded by: Indian journal of science and industry. Section A, agricultural sciences. HI 64435

Indian journal of agricultural science. Calcutta. Vol. 1+, 1931+. Indian J. Agric. Sci. Preceded by: Agricultural journal of India. HI 64436

Indian journal of agronomy. New Delhi. Indian J. Agron. See B–P–H 430/5. HI 54825

Indian journal of applied and pure biology. Bhopal. Vol. 1+, 1986+. Indian J. Appl. Pure Biol. HI 64437

Indian journal of biochemistry. New Delhi. Vols. 1-7, 1964-70. Indian J. Biochem. Superseded by: Indian journal of biochemistry and biophysics. HI 64438

Indian journal of biochemistry and biophysics. New Delhi. Vol. 8+, 1971+. Indian J. Biochem. Biophys. Preceded by: Indian journal of biochemistry. HI 64439

Indian journal of botany; half-yearly journal of research. Hyderabad. Vol. 1+, 1978+. Indian J. Bot. HI 64440

Indian journal of chemistry. Section B, organic and medicinal chemistry. New Delhi. Vol. 1+, 1963+. Indian J. Chem., B. HI 64441

Indian journal of ecology. Ludhiana. Vol. 1+, 1974+. Indian J. Ecol. HI 64442

Indian journal of environmental protection. Varanasi. Vol. 1+, 1981+. Indian J. Environm. Protect. HI 64443

Indian journal of experimental biology. New Delhi. Indian J. Exp. Biol. See B–P–H 430/7. HI 54826

Indian journal of farm sciences. Kanpur. Vol. 1+, 1973+. Indian J. Farm Sci. HI 64444

Indian journal of forestry; quarterly journal of forestry, agriculture, horticulture, natural history, wild life, field botany, and allied subjects. Dehra Dun. Vol. 1+, 1976+. Indian J. Forest. HI 64445

Indian journal of genetics and plant breeding. Indian J. Genet. Pl. Breed. See B–P–H 430/8. HI 54827

Indian journal of heredity. Bareilly. Vol. 1+, 1969+. Indian J. Heredity. HI 64446

Indian journal of the history of medicine. Madras. Vol. 1+, 1956+. Indian J. Hist. Med. HI 64447

Indian journal of history of science. New Delhi. Vol. 1+, 1966+. Indian J. Hist. Sci. HI 64448

Indian journal of horticulture. New Delhi. Indian J. Hort. See B–P–H 430/9. HI 54828

Indian journal of marine sciences. New Delhi. Vol. 1+, 1972+. Indian J. Mar. Sci. HI 64449

Indian journal of medical research. Calcutta. Indian J. Med. Res. See B–P–H 430/10. HI 54829

Indian journal of medical science. Edinburgh. Indian J. Med. Sci. See B–P–H 430/11. HI 54830

Indian journal of microbiology. Bombay. Indian J. Microbiol. See B–P–H 430/12. HI 54831

Indian journal of mushrooms. Solan. Vol. 1+, 1975+. Indian J. Mushr. HI 64450

Indian journal of mycological research. Calcutta. Indian J. Mycol. Res. See B–P–H 430/13. HI 54832

Indian journal of mycology and plant pathology. Udaipur. Vol. 1+, 1971+. Indian J. Mycol. Pl. Pathol. HI 64451

Indian journal of natural products. Sagar. Vol. 1+, [ca. 1985]+. Indian J. Nat. Prod. HI 74510

Indian journal of pharmaceutical sciences. Bombay. Vol. 40(2)+, 1978+. Indian J. Pharm. Sci. Preceded by: Indian journal of pharmacy. HI 64452

Indian journal of pharmacology. Pondicherry. Vol. ?-3+, 19??-71+. Indian J. Pharmacol. HI 64453

Indian journal of pharmacy. Banaras, Bombay. Vol. 1-40(1), 1939-78. Indian J. Pharmacy. Superseded by: Indian journal of pharmaceutical sciences. HI 64454

Indian journal of physiology and pharmacology. New Delhi. Vol. 1+, 1957+. Indian J. Physiol. Pharmacol. HI 64455

Indian journal of plant pathology. Lucknow. Vol. 1+, 1983+. HI 65974

Indian journal of plant physiology. New Delhi. Indian J. Pl. Physiol. See B–P–H 430/14. HI 54833

Indian journal of plant protection. Hyderabad. Vol. 1+, 1973+. Indian J. Pl. Protect. HI 64456

Indian journal of plant sciences. Sagar. Vol. 1+, 1983+. Indian J. Pl. Sci. HI 64457

Indian journal of science and industry. Karnal. Vol. 1, 1967. Indian J. Sci. Industry. Superseded by: Indian journal of science and industry. Section A, agricultural and animal sciences. HI 64458

Indian journal of science and industry. Section A, agricultural and animal sciences. Karnal. Vols. 2-3, 1968-69. Indian J. Sci. Industry, A. Preceded by: Indian journal of science and industry. Superseded by: Indian journal of science and industry. Section A, agricultural sciences. HI 64459

Indian journal of science and industry. Section A, agricultural sciences. Karnal. Vol. 4, 1970. Indian J. Sci. Industry, B. Preceded by: Indian journal of science and industry. Section A, agricultural and animal sciences. Superseded by: Indian journal of agricultural research. HI 64460

Indian journal of sericulture. Bombay. Indian J. Seric.

See B–P–H 430/15. HI 54834

Indian journal of sugarcane research and development. New Delhi. Indian J. Sugarcane Res. Developm. See B–P–H 430/16. HI 54835

Indian journal of technology. New Delhi. Indian J. Technol. See B–P–H 430/18. HI 54836

Indian journal of veterinary science and animal husbandry. Calcutta. Indian J. Veterin. Sci. Anim. Husb. See B–P–H 430/21. HI 54837

Indian journal of weed science. Hissar. Vol. 1+, 1969+. Indian J. Weed Sci. HI 64461

Indian medical gazette. Calcutta. Indian Med. Gaz. See B–P–H 430/31. HI 54838

Indian mushroom journal = Indian journal of mushrooms. Solan.

Indian orchid journal. Kalimpong. Vol. 1+, 1985+. Indian Orchid J. HI 64462

Indian phytopathology. New Delhi. Indian Phytopathol. See B–P–H 431/3. HI 54839

Indian planting and gardening. Calcutta. Indian Pl. Gard. See B–P–H 431/4. HI 54840

Indian potato journal. New Delhi. Indian Potato J. See B–P–H 431/5. HI 54841

Indian review. Madras. Indian Rev. See B–P–H 431/6. HI 54842

Indian review of life sciences. Jodpur. Vol. 1+, 1981+. Indian Rev. Life Sci. HI 64463

Indian science abstracts. Calcutta. Indian Sci. Abst. (Calcutta). See B–P–H 431/7. HI 54843

Indian science abstracts. New Delhi. Indian Sci. Abstr. (New Delhi). See B–P–H 431/8. HI 54844

Indian science index. Haryana. 1975+. Indian Sci. Index. HI 64464

Indian spices. Ernakulam, India. Indian Spices. See B–P–H 431/9. HI 54845

Indian sugar. Cawnpore [=Kanpur], India. Indian Sugar. See B–P–H 431/11. HI 54846

Indian sugar crops journal. Ghaziabad. Vol. 4(2)+, 1977+. Indian Sugar Crops J. Preceded by: Cane grower's bulletin. HI 64465

Indian sugar year book. Calcutta. 1970/71+, 1971+. Indian Sugar Year Book. HI 64466

Indian sugarcane journal. New Delhi. Indian Sugarcane J. See B–P–H 431/12. HI 54847

Indiana medical journal. Evansville, IN. Indiana Med. J. (Evansville). See B–P–H 431/16. HI 54848

Indiana medical journal. Indianapolis, Indiana. Indiana Med. J. (Indianapolis). See B–P–H 431/17. HI 54849

Indiana nursery news. Indianapolis, IN. 19??-83+. Indiana Nursery News. HI 64467

Indiana university publications. Science series. Bloomington, IN. 1942+. Indiana Univ. Publ., Sci. Ser. Preceded by: Indiana university studies. HI 64468

Indiana university studies. Bloomington, IN. Vols. 1-26, 1910-42. Indiana Univ. Stud. Superseded by: Indiana university publications. Science series. HI 64469

Indice agrícola de América latina y el Caribe. San Jose, Costa Rica. Vol. 10+, 1975+. Indice Agric. América Latina & Caribe. Preceded by: Bibliografia agricola latinoamericana. HI 64470

Indice agrícola colombiano. Medellín, Colombia. Indice Agríc. Colomb. See B–P–H 431/18. HI 54850

Indice bibliografico de centro de investigación científica y técnica. Mexico City. Indice Bibliogr. Centro Invest. Ci. Técn. See B–P–H 431/19. HI 54851

Indice de proyectos en desarrollo en ecologia tropical. Xalapa. Vol. 1+, 1975+. Indice Proy. Desarr. Ecol. Trop. HI 64471

Indice de proyectos en desarrollo en ecologia de zonas aridas. Xalapa. Vol. 1+, 1978+. Indice Proy. Desarr. Ecol. Trop. Zonas Aridas. HI 64472

Indice taxonomico. Tucumán, Argentina. Indice Taxon. See B–P–H 431/20. HI 54852

Indices naturwissenschaftlich-medizinischer Periodica bis 1850. Stuttgart. Vol. 1+, 1971+. Indices Naturwiss.-Med. Period. Bis 1850. HI 64473

Indices de revista sobre agua dulce y acuicultura = Freshwater and aquaculture contents tables. Rome.

Indische culturen. Batavia, Dutch E. Indies [=Jakarta, Indonesia]. Indische Cult. See B–P–H 431/21. HI 54853

Indische letterbode. Orgaan gewijd aan de nederlandsch-indische bibliographie. Amsterdam. Indische Letterbode. See B–P–H 431/22. HI 54854

Indische mercuur. Haarlem & Amsterdam. Indische Mercuur. See B–P–H 430/23. HI 54855

Indonesian abstracts. Jakarta, Indonesia. Indones. Abstr. See B–P–H 431/24. HI 54856

Indonesian agricultural research and development journal. Jakarta Selatam. Vol. 1+, 1979+. Indones. Agric. Res. Developm. J.

Indonesian biological and agricultural index = Indeks biologi dan pertanian di Indonesia. Bogor.

Indonesian journal of crop science. Jakarta. Vol. 1+, 1985+. Indones. J. Crop Sci. HI 66008

Indonesian journal for natural science = Madjallah ilmu alam untuk Indonesia. Jakarta, Indonesia. Madj. Ilmu Alam Untuk Indonesia. See B–P–H 538/6.

Indoor citrus and rare fruit society. 19??-83+. Indoor Citrus Rare Fruit Soc. HI 64474

Indoor garden. New York. Vol. 21(4)+, 1984+. Indoor Gard. Preceded by: Light garden. HI 64475

Indoor light gardening news. Reading, PA. Vol. 1+, 1966+. Indoor Light Gard. News. HI 64476

Industria azucarera. Buenos Aires. Industr. Azucarera. See B–P–H 432/3. HI 54857

Industria cafetera. Cali, Colombia. Industr. Cafetera. See B–P–H 432/4. HI 54858

Industrial arts index. New York. Vols. 1-45, 1913-57. Industr. Arts Index. Superseded in part by: Applied science and technology index. HI 64477

Industrial contributions. [Sangyo shiryo.] [China]. Industr. Contr. See B–P–H 432/5. HI 54859

Industrial microbiology abstracts. London. Industr. Microbiol. Abstr. See B–P–H 432/7. HI 54860

Industrial research. [Kung yeh chung hsin.] Nanking. Industr. Res. See B–P–H 432/8. HI 54861

Ine to mugi = Rice and wheat. Mitsubishi agricultural journal. Tokyo. Rice & Wheat. See B–P–H 799/24.

Infection and immunity. Washington, DC. Vol. 1+, 1970+. Infect. Immun. HI 64478

Infoletter, international plant protection center = I P P C infoletter. Corvallis, OR.

Info-nature; bulletin de liaison de la société réunionnaise pour l'étude de la protection de la nature. St. Denis, Réunion. Vol. 1+, 1971+. Info-Nat. HI 64479

Informação, instituto nacional de investigação agraria. Lisbon. No. 6+, 1986+. Inform. Inst. Nac. Investi. Agrar. Preceded by: Informação, instituto nacional de investigação agraria e de extensão rural. HI 74845

Informação, instituto nacional de investigação agraria e de extensão rural. Lisbon. Nos. 1-5, 1985-86. Inform. Inst. Nac. Investi. Agrar. Superseded by: Informação, instituto nacional de investigação agraria. HI 74846

Informacion al dia alerta. I I C A - tropicos, agronomia. Turrialba. No. 1+, 1975+. Inform. Alerta I. I. C. A., Trop. Agron. HI 64480

Informacion tecnica, estacion experimental agropecuaria. Manfredi. No. 1+, 1964+. Inform. Tecn. Estac. Exp. Agropecu. HI 64481

Informacion tematica, instituto de documentacion e informacion cientifica y tecnica. Havana. No. 1+, 1967+. Inform. Temat. Inst. Doc. Inform. HI 64482

Informaciones científicas francesas. [Association pour la diffusion de la pensée française.] Paris. Inform. Ci. Franc. See B–P–H 432/16. HI 54866

Informador agrícola. La Aurora, Guatemala. Inform. Agríc. (La Aurora). See B–P–H 432/13. HI 54863

Informasi, lembaga penelitian hortikultura. Jakarta. No. 1+, 19??+. Inform. Lemb. Penelitian Hort. HI 64483

Informasjon, norsk botanikk forening, nord-norsk avdeling. Tromso? Nos. 1-2, 1977. Inform. Norsk Bot. Foren. Nord-Norsk Avd. Superseded by: Polarflokken. HI 64484

Informateur O P T I M A = O P T I M A newsletter. Chambésy, Berlin, etc.

Information bulletin, aquatic plant control program. Vicksburg, MS. 1977+. Inform. Bull. Aquatic Pl. Control Progam. HI 64486

Information bulletin, aquatic plant management programme: Okanagan Lakes. Victoria, B.C. Vol. 2+, 1977+. Inform. Bull. Aquatic Pl. Control Progamme, Okanagan Lakes. Preceded by: News bulletin, aquatic plant management programme: Okanagan lakes. HI 64487

Information bulletin of the California fig institute. Fresno, CA. Inform. Bull. Calif. Fig Inst. See B–P–H 432/14. HI 54864

Information on chinese medicine. [Chung yo t'ung pao.] Peking. Inform. Chin. Med. (Peking). See B–P–H 432/15. HI 54865

Information bulletin, citrus and subtropical fruit research institute. Nelspruit, Pretoria. No. 1+, 1972+. Inform. Bull. Citrus Subtrop. Fruit Res. Inst. HI 64488

Information bulletin, forest products research institute, Ghana. Kumasi. ?-1976+. Inform. Bull. Forest Prod. Res. Inst., Ghana. HI 64489

Information bulletin, international crops research institute for the semi-arid tropics. Hyderabad. No. 1+, 1978+. Inform. Bull. Int. Crops Res. Inst. Semi-Arid Trop. HI 64490

Information bulletin, international maize and wheat improvement center. Londres. No. 1+, 1972+. Inform. Bull. Int. Maize Wheat Improv. Center. HI 64491

Information bulletin, New York state college of agriculture and life sciences. Ithaca, NY. No. 1+, 1971+. Inform. Bull. New York State Coll. Agric. Life Sci. Preceded by: Cornell extension bulletin, New York state college of agriculture. HI 64492

Information bulletin, Pacific science association = Pacific science association. Information bulletin. Honolulu, HI. Pacific Sci. Assoc. Inform Bull. See B–P–H 695/3.

Information bulletin on planktology in Japan. Hakodate. Nos. 1-14, 1953-67. Inform. Bull. Planktol. Japan. Superseded by: Bulletin of plankton society of Japan. HI 64494

Information bulletin, research institute, Belleville. Belleville, Ont. No. 1+, 1963+. Inform. Bull. Res. Inst. Belleville. HI 64495

Information circular, Ohio biological survey = Informative circular, Ohio biological survey. Columbus, OH.

Information géographique. Paris. Vol. 1+, 1936+. Inform. Géogr. HI 64496

Information letter, plant protection committee for the south east Asia and Pacific region. Bangkok. No. ?-122+, 19??-80+. Inform. Lett. Pl. Protect. Committee S. E. Asia & Pacific. HI 64497

Information notes, institute of terrestrial ecology. Grange-over-Sands. Nos. ?-209-567?, 19??-76-80? Inform. Notes Inst. Terrestr. Ecol. HI 64498

Information pamphlet, west Texas museum = Research publication, Texas technological college. Lubbock, TX.

Information report B C - X. Victoria, B.C. Nos. 1-57, 1965-71. Inform. Rep. B C - X. Superseded by: Information report, Pacific forest research centre. HI 64499

Information report D P C - X. Ottawa. ?-1979+. Inform. Rep. D P C - X. HI 66748

Information report, Great Lakes forest research centre. Saulte Ste. Marie, Ont. 1973-84? Inform. Rep. Great Lakes Forest Res. Centre. Superseded by: Information report, Great Lakes forestry centre HI 75179

Information report, Great Lakes forestry centre. Saulte Ste. Marie, Ont. 1985+. Inform. Rep. Great Lakes Forest. Centre. Preceded by: Information report, Great Lakes forest research centre. HI 75221

Information report M - X. Fredericton, N.B. = Information report, maritimes forest research laboratory. Fredericton, N.B.

Information report, maritimes forest research centre. Fredericton, N.B. 1971+. Inform. Rep. Maritimes Forest Res. Centre. Preceded by: Information report, maritimes forest research laboratory HI 67638

Information report, maritimes forest research laboratory. Fredericton, N.B. Nos. 1-11, 1966-69. Inform. Rep. Maritimes Forest Res. Lab. Superseded by: Information report, maritimes forest research centre. HI 66986

Information report N - X. St. John's. Ca.1977?-84? Inform. Rep. N - X. Preceded by: Information report, Newfoundland forest research centre. Superseded by: Information report, Newfoundland forestry centre. HI 67066

Information report, Newfoundland forest research centre. St. John's. 1971-7? Inform. Rep. Newfoundland Forest Res. Centre. Superseded by: Information report N - X. HI 64500

Information report, Newfoundland forestry centre. St. John's. 1985+. Inform. Rep. Newfoundland Forest. Centre. Preceded by: Information report N - X. HI 61879

Information report O - X. Saulte Ste. Marie, Ont. = Information report, Great Lakes forest research centre.

Information report, Pacific forest research centre. Victoria, B.C. 1971+. Inform. Rep. Pacific Forest Res. Centre. Preceded by: Information report B C - X. HI 64501

Information report, Pacific forestry centre. Victoria, B.C. 1971-8? Inform. Rep. Pacific Forest. Centre. Preceded by: Information report, Pacific forest research centre. HI 75180

Information reports digest, canadian forestry service. Hull, Quebec. Vol. 1+, 1980+. Inform. Rep. Digest Canad. Forest. Serv. HI 64502

Information series, college of forestry, wildlife and range science, university of Idaho. Moscow, ID. No. 1+, 1972+. Inform. Ser. Coll. Forest. Wildlife Range Sci. HI 64503

Information series, forest, wildlife and range experiment station, university of Idaho. = Information series, college of forestry, wildlife and range science, university of Idaho. Moscow, ID.

Information series, natural history, royal scottish museum. Edinburgh. Vol. 1+, 1973+. Inform. Ser. Nat. Hist. Roy. Scott. Mus. HI 64504

Information series, New Zealand department of scientific and industrial research. Wellington, N.Z. Nos. 1-126, 1948-77. Inform. Ser. New Zealand Dept. Sci. Industr. Res. Superseded by: D S I R information series, New Zealand. HI 64505

Information series, New Zealand forest service. Wellington, N.Z. Nos. 1-22, 1953-55. Inform. Ser. New Zealand Forest Serv. Preceded by: Information series, state forest service, New Zealand. HI 64506

Information series, state forest service, New Zealand. Wellington, N.Z. 1949-53. Inform. Ser. State Forest Serv. New Zealand. Superseded by: Information series, New Zealanmd forest service. HI 64507

Information series, western australian museum. Perth, W.A. Vol. 1+, [1971]+. Inform. Ser. W. Austral. Mus. HI 64508

Information service, forest, wildlife and range experiment station, university of Idaho. Moscow, ID. ?-1977+. Inform. Serv. Forest Wildlife Range Exp. Sta. Univ. Idaho. HI 64509

Information sheet of the herb society of America. Boston, MA. Inform. Sheet Herb Soc. Amer. See B–P–H 432/19. HI 54868

Information sheet, Mississippi agricultural experiment station. Mississippi state college = Mississippi agricultural experiment station. Mississippi state college. Information sheet. Jackson, MS. Mississippi Agric. Exp. Sta. Inform. Sheet. See B–P–H 599/26.

Information sheet, nature conservancy council. Banbury. 1978+. Inform. Sheet Nat. Conservancy Council. HI 64510

Informational bulletin, marine biological laboratory. Woods Hole, MA. 1979. Inform. Bull. Mar. Biol. Lab. Preceded by: Annual announcement, marine biological laboratory. Woods Hole. Superseded by: Annual bulletin, marine biological laboratory. Woods Hole. HI 64511

Informationen aus der Berliner Landschaft. Berlin. ?-1982+. Inform. Berliner Landschaft. HI 64512

Informationen Z A G Wasserpflanzen. Waltershausen, Bad Langensalza. Nos. 1-12, 1971-82. Inform. Z. A. G. Wasserpflanzen. Superseded by: Z A G Wasserpflanzen A M. HI 64513

Informationes. Zagreb. Nos. 1-2, 1962-63. Informationes. Preceded by: Informationes botanicae. HI 64514

Informationes annuales hortorum botanicorum facultatis scientiarum universitatis Tokyoensis. Tokyo. 1967+. Inform. Annuales Hort. Bot. Fac. Sci. Univ. Tokyo. HI 64515

Informationes botanicae. Zagreb. No. 3+, 1964+. Inform. Bot. Preceded by: Informationes. Zagreb. HI 64516

Informations agricoles. Algiers. Inform. Agric. (Algiers). See B–P–H 432/11. HI 54862

Informations annuelles de caryosystématique et cytogénétique; travaux des laboratoires de phytogénétique de Strasbourg et de Lille. Strasbourg and Lille. Vols 1-6, 1967-72. Inform. Annuelles Caryosyst. Cytogén. HI 64517

Informations - Bulletin, Gesellschaft für interdisziplinäre Sahara-Forschung. Hallein. Vol. 1+, 1982+. Inform. Bull. Ges. Interdiszipl. Sahara-Forsch. HI 64518

Informations - forêt. Nangis. No. 1+, 1980+. Inform. Forêt. HI 64519

Informations, institut de recherches agronomiques tropicales = I R A T informations. Paris.

Informationsbrief der zentralen Arbeitsgemeinschaft "Echinopseen". Gotha. 1981+. Informationsbrief Zentr. Arbeitsgem. Echinopseen. HI 64520

Informationsbrief der zentralen Arbeitsgemeinschaft "Mammillaria". Potsdam. 1975+. Informationsbrief Zentr. Arbeitsgem. Mammillaria. HI 64521

Informationsbrief der zentralen Arbeitsgemeinschaft "Andere Sukkulenten". Leipzig. 1983+. Informationsbrief Zentr. Arbeitsgem. Andere Sukkulenten. HI 64522

Informative circular, Ohio biological survey. Columbus, OH. No. 1+, 1973+. Inform. Circ. Ohio Biol. Surv. HI 64523

Informativo de investigaciones agricolas = I D I A. Buenos Aires.

Informativo, fundacao brasileira para a conservacao da natureza = F B C N informativo. Rio de Janeiro.

Informatore botanico italiano; bollettino della società botanica italiana. Florence. Vol. 1+, 1969+. Inform. Bot. Ital. HI 65962

Informatore fitopatologico. Bologna. Vol. 1+, 1951+. Inform. Fitopatol. HI 64881

Informatore del giovane entomologo. Genoa. Vol. 1+, 1960+. Inform. Giova. Entomol. HI 65963

Informatsionnyi byulleten', biologiya vnutrennikh vod, akademiya nauk SSSR. Leningrad. 1967+. Inform. Byull. Biol. Vnutrenn. Vod Akad. Nauk S.S.S.R. HI 75128

Informatsionnyi byulleten', nauchnyi tsentr biologicheskikh issledovanii, akademiya nauk SSSR. Pushchino. Vol. ?-19+, ?-1985+. Inform. Byull. Nauchn. Tsentr Biol. Issl. Akad. Nauk S.S.S.R. HI 71424

Informatsionnyi byulleten, sibirskii institut fiziologii i biokhimii rastenii. Irkutsk. Vol. ?-4+, 19??-69+. Inform. Byull. Sibirsk. Inst. FIziol. Biokhim. Rast. HI 65964

Informatsionnyi materialy, institut ekologii rastenii i zhivotnykh. Sverdlovsk. [No.] 1+, 1974+. Inform. Mater. Inst. Ekol. Rast. Zhivot. HI 65965

Informazioni seriche; rivista dell'industria bacologica e serica. Rome. Inform. Seriche See B–P–H 432/18. HI 54867

Informazioni techniche per i floricoltori. San Remo. Vols. ?-9-18, 19??-60-69. Inform. Techn. Floric. HI 65966

Informe agropecuario. Belo Horizonte. 1977+. Informe Agropecu. HI 65967

Informe anual, asociacion latinoamericana de fitotecnistas de frijol. Bogota. Vol. 1+, 1965+. Informe Anual Asoc. Latinoamer. Fitotecn. Frijol. HI 65968

Informe anual de la estación central agronomica de Cuba. Havana. Vol. 1+, 1906+. Informe Anual Estaç. Centr. Agron. Cuba. HI 65969

Informe anual estación científica Charles Darwin. Santa Cruz, Galapagos. 1982+. Informe Anual Estaç. Ci. Charles Darwin. Preceded by: Report (Annual) of the Charles Darwin research station. HI 65970

Informe anual, institute for tropical agriculture, university of Puerto Rico = Report, institute for tropical agriculture, university of Puerto Rico. Mayagüez.

Informe anual, instituto mexicano de recursos naturales renovables. Mexico, D.F. Vol. 1+, 1953+. Informe Anual Inst. Mex. Recurs. Nat. Renov. HI 65971

Informe anual, inter-american institute of agricultural sciences = Report, inter-american institute of agricultural sciences. Washington, DC, Turrialba.

Informe anual de investigacion, facultad de agronomia, universidad agraria. La Molina. No. 1+, 1967+. Informe Anual Invest. Fac. Agron. Univ. Agrar. HI 65972

Informe, centro internacional de mejoramiento de maiz y trigo. Mexico, D.F. 1966/67+, 1967+. Informe Centro Int. Mejoram. Maiz. Trigo. HI 65973

Informe, centro nacional de investigacion y experimentacion agricola "La Molina" = Informe, estación experimental agrícola, "La Molina". Lima.

Informe cientifico-tecnico, academia de ciencias de Cuba. Havana. No. 1+, 1977+. Informe Ci.-Tecn. Acad. Ci. Cuba. HI 65975

Informe cientifico-tecnico, instituto de oceanologia, academia de ciencias de Cuba. Havana. No. ?-11+, 19??-77+. Informe Ci-Tecn. Inst. Oceanol. Acad. Ci. Cuba. HI 65976

Informe didatico, F C A P. Belém. No. 1+, 1979+. Informe Didat. F. C. A. P HI 65977

Informe especial, servicio de investigación y promoción agraria, Peru. Lima. 1962+. Informe Espec. Serv. Invest. Promoc. Agrar. Peru. HI 65978

Informe, estación experimental agrícola, "La Molina". Lima. Vols. 1-?, 1927-? Inform. Estac. Exp. Agríc. La Molina. HI 65979

Informe, estación experimental agrícola, sociedad nacional agraria = Informe, estación experimental agrícola, "La Molina". Lima.

Informe; instituto dell mar del Perú. Callao, Peru. Informe Inst. Mar Perú. See B–P–H 432/22. HI 54869

Informe, instituto mexicano de recursos naturales renovables = Informe anual, instituto mexicano de recursos naturales renovables. Mexico, D.F.

Informe mensual, estación experimental agrícola, "La Molina" = Informe, estación experimental agrícola, "La Molina". Lima.

Informe da pesquisa, fundação instituto agronómico do Paraná. No. ?-21+, 19??-79+. Informe Pesq. Fund. Inst. Agron. Paraná. HI 65980

Informe ... del programa nacional de fisiología vegetal. Bogotá. 1975+. Informe Programa Nac. Fisiol. Veg. HI 65981

Informe técnico de banano; convenio para el programa de investigación en Banano. Quito. 19??-75+. Informe Técn. Banano. HI 65982

Informe técnico, centro de pesquisas do cacau. Itabuna. 1968-71. Informe Técn. Centro Pesq. Cacau. HI 65983

Informe técnico, estacion experimental agropecuaria Pergamino. Pergamino. No. 1+, 1960+. Informe Técn. Estac. Exp. Agropecu. Pergamino. HI 65984

Informe tecnico, estacion experimental agropecuaria San Pedro. San Pedro. No. 1+, 1966+. Informe Tecn. Estac. Exp. Agropecu. San Pedro. HI 65985

Informe tecnico, estacion experimental agropecuaria Trelew. Trelew. No. ?-10+, 19??-66+. Informe Tecn. Estac. Exp. Agropecu. Trelew. HI 65986

Informe tecnico, F C A P. Belém. No. ?-2+, 197?+. Informe Tecn. F. C. A. P. HI 65987

Informe tecnico, instituto interamericano de ciencias agricolas. Turrialba. Vol. 1+, 1942/43+, 1943+. Informe Tecn. Inst. Interamer. Ci. Agric. HI 65988

Infuse. Loughborough. Vol. 1+, 1977+. Infuse. HI 65989

Ingazat al-baht az-zira'i = Cahiers de la recherche agronomique, direction de la recherche agronomique et de l'enseignement agricole. Rabat.

Ingeniería agronómica. Caracas. Ing. Agron. See B–P–H 432/23. HI 54870

Inheemsche nijverheid op Java, Madoera, Bali en Lombok. Jogjakarta, Dutch E. Indies [Indonesia]. Inheemsche Nijverh. Java. See B–P–H 432/25. HI 54871

Initiations-documentations techniques: O R S T O M. Paris. No. 1+, 1963+. Init. Doc. Techn. O. R. S. T. O. M. HI 65990

Inländisches Museum. Dorpat [=Tartu, Estonian S S R]. Inl. Mus. See B–P–H 432/26. HI 54872

Inligtingsbulletin, navorsingsinstituut vir sitrus en subtropiese vrugte = Information bulletin, citrus and subtropical fruit research institute. Nelspruit, Pretoria.

Insect life. Washington, DC. Insect Life. See B–P–H 433/1. HI 54873

Institut. Paris. Institut. See B–P–H 434/12. HI 54879

Institut. Section 1. Paris. Institut, Sect. 1. See B–P–H 434/13. HI 54880

Institut für angewandte Botanik. Jahresbericht. Hamburg. 1912-28. Inst. Angew. Bot. Jahresber. Preceded by: Jahresberichte, botanische Staatsinstitut zu Hamburg. Superseded by: Institut für angewandte Botanik Hamburg. Jahresbericht. HI 65992

Institut de botanique, université de Genève. Geneva. Inst. Bot. Univ. Genève. See B–P–H 433/11. HI 54875

Institut matematiki i estestvoznanija pri Severo-Kavkazskom gosudarstvennom universitete = Institut matematiki i estestvoznaniya pri Severo-Kavkazskom gosudarstvennom universitete. Rostov.

Institut matematiki i estestvoznaniya pri Severo-Kavkazskom gosudarstvennom universitete. Rostov, Russian S F S R. Vols. 1-18 [2 vols. with vol. no. "2" each published in 1926; 2 vols. with vol. no. "7" published, one in 1928, the other in 1929], 1926-30. Inst. Mat. Estestv. Severo-Kavkazsk. Gosud. Univ. HI 65993

Institut für Naturschutz Darmstadt. Schriftenreihe. Darmstadt, Germany. Inst. Naturschutz Darmstadt Schriftenreihe. See B–P–H 433/24. HI 54876

Institut océanographique de l'Indochine. Nhatrang, Indochina [South Vietnam]. Inst. Océanogr. Indochine. See B–P–H 433/25. HI 54877

Institut Pasteur au Viêt-Nam. Rapport. Saigon. Inst. Pasteur Viêt-Nam Rapp. See B–P–H 434/6. HI 54878

Institut rastenievodstva. Leningrad. 1929-34. Inst. Rasteniev. Superseded by: Sbornik rabot po biokhimii kul'turnykh rastenii. 3-2390-3. HI 65994

Institute of mathematics and natural sciences to the North Caucasus state university. Rostov = Institut matematiki i estestvoznaniya pri Severo-Kavkazskom gosudarstvennom universitete. Rostov, Russian S F S R.

Institute paper, commonwealth forestry institute. Oxford. No. 36+, 1963+. Inst. Pap. Commonw. Forest. Inst. Preceded by: Institute paper, imperial forestry institute. HI 65995

Institute paper, imperial forestry institute. Oxford. Nos. 1-35, 1936-60. Inst. Pap. Imp. Forest. Inst. Superseded by: Institute paper, commonwealth forestry institute. HI 65996

Instituto de pesquisas e expérimentaçao agropecuárias do norte. Série, botânica e fisiologia vegetal = Série botanica e fisiologia vegetal. Instituto de pesquisas e expérimentaçao agropecuárias do norte. Belém.

Instituto de pesquisas e expérimentaçao agropecuárias do norte. Série, estudos sôbre forrageiras na Amazônia = Série estudos sôbre forrageiras na Amazônia. Instituto de pesquisas e expérimentaçao agropecuárias do norte. Belém.

Instituto de pesquisas e expérimentaçao agropecuárias do norte. Série, fitotécnica = Série fitotecnia. Instituto de pesquisas e expérimentaçao agropecuárias do norte. Belém.

Instituto de pesquisas e expérimentaçao agropecuárias do norte. Série, solos da Amazônia = Série solos da Amazônia, instituto de pesquisas e expérimentaçao agropecuárias do norte. Belém.

Instituto de pesquisas e expérimentaçao agropecuárias do norte. Série técnico = Série técnico do instituto de pesquisas e expérimentaçao agropecuárias do norte. Belém.

Instituto de pesquisas e expérimentaçao agropecuárias do norte. Série, técnologia = Série técnologia do instituto de pesquisas e expérimentaçao agropecuárias do norte. Belém.

Instituto de sanidad vegetal. [Publicaciones.] Ser. A = Direccion de investigaciones, instituto de sanidad vegetal. Ser. A. Buenos Aires.

Insula; boletim do centro de pesquisas e estudos botânicos (later Boletim do horto botânico). Florianopolis, Santa Catarina. Vol. 1+, 1969+. Insula. HI 65997

Intecol bulletin. London. No. 1+, 1969+. Intecol Bull. HI 65998

Intecol newsletter. Athens, GA. No. ?-14(4)+, 19??-84+. Intecol Newslett. HI 65999

Integrated pest conbtrol management. 19??-81+. Integr. Pest Control Managem. HI 66000

Intellectual observer: Review of natural history, microscopic research, and recreative science. London. Intellectual Observ. See B–P–H 435/23. HI 54901

Intelligenzblatt, Jenaische allgemeine Literatur-Zeitung = Jenaische allgemeine Literatur-Zeitung. Jena.

Intelligenzblatt der jenaischen allgemeinen Literatur-Zeitung. Jena & Leipzig. Vols. 1-38, 1804-41. Intelligenzbl. Jenaischen Allg. Lit.-Zeitung. 3-2160-1. HI 52745

Intensivobstbau. Berlin. Vols. 1-2, 1961-62. Intensivobstbau. Superseded by: Obstbau. Berlin. HI 66001

Interafrican phytosanitary bulletin. Yaounde. Nos. 1-6, 1971-74. Interafr. Phytosan. Bull. Superseded by: African journal of plant protection. HI 66002

Interciencia. Caracas. Vol. 1+, 1976+. Interciencia (Caracas). HI 66003

Interciencia. Elmsford, NY. 1982+. Interciencia (Pergamon Press). HI 66004

Intercom. Kew. No. ?-21+, 19??-86+. Intercom (Kew). HI 66005

INTercom. Ogden, UT. 19??-79+. INTercom (Ogden). HI 66006

Interdisciplinary science reviews. London. Vol. 1+, 1976+. Interdiscipl. Sci. Rev. HI 66007

Interim report of the japanese national committee for I B P. Tokyo. 1967. Interim Rep. Jap. Natl. Committee I. B. P. Superseded by: Report of the japanese national committee for I B P. HI 66009

International abstracts of biological sciences. London. Vols. 4-?, 1954-82. Int. Abstr. Biol. Sci. Preceded by: British abstracts of medical sciences [not entered]. Superseded by: Current awareness in biological sciences. HI 66010

International agricultural development. Crawborough. 1980+. Int. Agric. Developm. HI 66011

International arachis newsletter. Patancheru. No. 1+, 1987+. Int. Arachis Newslett. HI 74847

International archives of normal and pathological biology = Biologica latina. Milan. Biol. Latina. See B–P–H 194/4.

International association for ecology bulletin = Intecol bulletin. London.

International association for ecology newsletter = Intecol newsletter. Athens, GA.

International association of wood anatomists, news bulletin. Zurich. 1957-69. Int. Assoc. Wood Anat. News Bull. Superseded by: I A W A bulletin. HI 66012

International atomic energy agency. Technical report series. Vienna. Int. Atomic Energy Agency, Techn. Rep. Ser. See B–P–H 434/19. HI 54881

International bibliography of rice research. New York. 1951/60+, 1961+. Int. Bibliogr. Rice Res. HI 66013

International biodeterioration. Farnham Royal. Vol. 20+, 1984+. Int. Biodeterior. Preceded by: International biodeterioration bulletin; Biodeterioration research titles and Waste materials biodeterioration research titles. HI 66014

International biodeterioration bulletin. Birmingham. Vols. 1-19, 1965-83? Int. Biodeterior. Bull. Superseded by: International biodeterioration. HI 66015

International biodeterioration bulletin reference index supplement. Birmingham. 1967-71. Int. Biodeterior. Bull. Ref. Index Suppl. Superseded by: Biodeterioration research titles. HI 66016

International bio-energy directory and handbook. Washington, DC. 1984+. Int. Bio-Energy Directory Handb. HI 66017

International bio-energy handbook = International bio-energy directory and handbook. Washington, DC.

International bioscience monographs. Hissar, New Delhi. Vols. 1+, 1974+. Int. Biosci. Monogr. HI 66018

International bioscience series. New Delhi. Vols. 1-9-?, 1975-81-? Int. Biosci. Ser. HI 66019

International biotechnology directory. London, New York. 1984+. Int. Biotechnol. Directory. HI 66020

International bonsai. Rochester, NY. Vol. 1+, 1979+. Int. Bonsai. HI 66021

International bulletin on bacteriological nomenclature and taxonomy. Ames, IA. Vols. 1-15, 1951-65. Int. Bull. Bacteriol. Nomencl. Taxon. Superseded by: International journal of systematic bacteriology. HI 66022

International bulletin of plant protection. Rome. Int. Bull. Pl. Protect. See B–P–H 434/23. HI 54882

International camellia journal. Denbigh, Wales. Int. Camellia J. See B–P–H 434/24. HI 54883

International chickpea newsletter. Patancheru. No. 1+, 1979+. Int. Chickpea Newslett. HI 66023

International cotton bulletin. Manchester, England. Int. Cotton Bull. See B–P–H 434/25. HI 54884

International dendrology society newsletter = Newsletter, international dendrology society.

International dendrology society year book. London. 1966+. Int. Dendrol. Soc. Year Book. HI 66024

International dental journal. New York, NY. Int. Dental J. See B–P–H 434/26. HI 54885

International flavours and food additives. London. Vols. 6-10, 1975-79. Int. Flav. Food Addit. Preceded by: Flavour industry. Superseded by: Food flavourings, ingredients and processing. HI 66025

International genetic resources programme report. Pittsboro, NC. Vol. 1+, 1984+. Int. Genet. Resources Programme Rep. HI 66026

International geology review. Washington, DC. Int. Geol. Rev. See B–P–H 434/27. HI 54886

International journal of applied radiation and isotopes. New York, NY. Int. J. Appl. Radiat. Isotopes. See B–P–H 434/28. HI 54887

International journal of biochemistry. Oxford, Elmsford, NY. Vol. 1+, 1970+. Int. J. Biochem. HI 66027

International journal of biochemistry and biophysics = Biochimica et biophysica acta. New York, NY. & Amsterdam. Biochim. Biophys. Acta. See B–P–H 192/23.

International journal of bioclimatology and biometeorology. The Hague. Vols. 1-4, 1957-60. Int. J. Bioclimatol. Biometeorl. Superseded by: International journal of biometeorology. HI 66028

International journal of biological macromolecules. Guildford. Vol. 1+, 1979+. Int. J. Biol. Macromolec. HI 66029

International journal of biometeorology. The Hague. Vols. 5+, 1961+. Int. J. Biometeorol. Preceded by: International journal of bioclimatology and biometeorology. HI 66030

International journal of cell cloning. Dayton, OH. Vol. 1+, 1983+. Int. J. Cell Cloning. HI 66031

International journal of chronobiology. London. Vols. 1-8, 1973-83. Int. J. Chronobiol. Superseded by: Life chemistry reports. HI 66032

International journal of crude drug research. Lisse. Vols. 20-21(2), 1982-83. Int. J. Crude Drug Res. Preceded by: Quarterly journal of crude drug research and Acta phytotherapeutica. HI 66033

International journal of ecology and environmental sciences. Jaipur. Vol. 1+, 1974+. Int. J. Ecol. Environm. Sci. HI 66034

International journal of environmental studies. New York & London. Vol. 1+, 1970+. Int. J. Environm. Stud. HI 66035

International journal of food microbiology. Amsterdam. Vol. 1+, 1984+. Int. J. Food Microbiol. HI 66037

International journal of microbiology. New Delhi. Vol. 1+, 1983+. Int. J. Microbiol. HI 66038

International journal of microbiology and hygiene. Abstracts, medical microbiology, virology, parasitology, hygiene, preventive medicine. Stuttgart & New York. Vol. 287+, 1984+. Int. J. Microbiol. Hyg., Abstr. Med. Microbiol. Preceded by: Zentralblatt für Bakteriologie, Mikrobiologie und Hygiene. 1 Abteilung. Referate. HI 66039

International journal of microbiology and hygiene. Series B, environmental hygiene, hospital hygiene, industrial hygiene, preventive medicine. Stuttgart & New York. Vol. 182+, 1985+. Int. J. MIcrobiol. Hyg., B. Preceded by: Zentralblatt für Bakteriologie, Mikrobiologie und Hygiene. Abteilung Originale. B, Hygiene. HI 66040

International journal of mycology and lichenology. Brunswick. Vol. 1+, 1982+. Int. J. Mycol. Lichenol. HI 66041

International journal of oceanology and limnology. Haddonfield, NJ. Vol. 1, 1967. Int. J. Oceanol. Limnol. HI 66042

International journal of pharmaceutics. Amsterdam. Vol. 1+, 1978+. Int. J. Pharm. HI 66043

International journal of radiation biology and related studies in physics, chemistry and medicine. London. Int. J. Radiat. Biol. Related Stud. Phys. See B–P–H 435/2. HI 54888

International journal of remote sensing. London. Vol. 1+, 1980+. Int. J. Remote Sensing. HI 75243

International journal of speleology. Weinheim. Vol. 1+, 1964+. Int. J. Speleol. HI 66044

International journal of systematic bacteriology. Ames, IA., Washington, DC. Vols. 16-23, 1966-73. Int. J. Syst. Bacteriol. Preceded by: International bulletin on bacteriological nomenclature and taxonomy. HI 66045

International journal of tropical agriculture. Hissar. Vol. 1+, 1983+. Int. J. Trop Agric. HI 66046

International journal of tropical plant diseases. New Delhi. Vol. 1+, 1983+. Int. J. Trop. Pl. Dis. HI 66047

International laboratory. Green Farms, Fairfield, CT. Vol. 1+, 1971+. Int. Lab. HI 66048

International marine science. Paris. Vols. 1-7, 1963-70. Int. Mar. Sci. Preceded by: Marine sciences newsletter. Superseded by: I M S newsletter. HI 66049

International marine study. London. Vols. ?-2(9/10)-4(8)-?, 19??-67-69?-? Int. Mar. Study. HI 66050

International newsletter on plant pathology. Vol. 1+, 1970+. Int. Newslett. Pl. Pathol. HI 66051

International newsletter on regulated stream limnology. Fort Collins, CO. Vol. 1+, 1980+. Int. Newslett. Regulat. Stream Limnol. HI 66052

International newsletter, sempervivum society. Burgess Hill. No. 1+, 1975+. Int. Newslett. Sempervivum Soc. HI 66053

International nickel magazine. London. 1967+. Int. Nickel Mag. HI 66054

International organization of plant biosystematists newsletter. Utrecht, Reading, Copenhagen. Vols. 1-9, 1969-74. Int. Organ. Pl. Biosyst. Newslett. HI 66055

International permaculture seed yearbook = International permaculture species yearbook. Orange, ME.

International permaculture solutions journal. Wichita, KS. 1990+. Int. Permacult. Solut. J. Preceded by: International permaculture species yearbook. HI 69456

International permaculture species yearbook. Orange, MA. 1986-? Int. Permacult. Spec. Yearb. Superseded by: International permaculture solutions journal. HI 66056

International pest control. London. Int. Pest Control. See B–P–H 435/4. HI 54890

International plant index. New Haven, CT. Nos. 1-2, 1962-65. Int. Pl. Index. HI 66057

International potato center newsletter. Lima. No. 1+, 1971+. Int. Potato Center Newslett. Superseded by?: C I P circular. HI 66058

International review of agricultural economics. Rome. Vols. 1-13, 1910-22; n.s. vols. 1-4, 1923-26. Int. Rev. Agric. Econ. Superseded by: International review of agriculture. HI 66059

International review of agriculture. Rome. N.s. vols. 18-19, 1927-28, vols. 22-37, 1931-46. Int. Rev. Agric. For vols. 20-21 see: Monthly bulletin of agricultural science and practice. Preceded by: International review of the science and practice of agriculture and International review of agricultural economics. HI 66060

International review of cotton and allied industries. Manchester, England. Int. Rev. Cotton Allied Industr. See B–P–H 435/6. HI 54891

International review of cytology. New York, NY. Int. Rev. Cytol. See B–P–H 435/7. HI 54892

International review of forestry research. New York, NY. Int. Rev. Forest. Res. See B–P–H 435/8. HI 54893

International review of plant sociology, ecology and plantgeography. The Hague = Vegetatio; acta geobotanica. The Hague. Vegetatio. See B–P–H 948/18.

International review of the science and practice of agriculture. Rome. Vols. 7-13, 1916-22; n.s. vols. 1-4, 1923-26. Int. Rev. Sci. Pract. Agric. Preceded by: Monthly bulletin of agricultural intelligence and of plant diseases. Superseded by: International review of agriculture. HI 66061

International review of tropical medicine. New York, NY. Int. Rev. Trop. Med. See B–P–H 435/13. HI 54896

International rice research institute; technical bulletin. Los Baños, Philippines. Int. Rice Res. Inst. Techn. Bull. See B–P–H 435/16. HI 54897

International rice research institute. Technical paper. Los Baños, Philippines. Int. Rice Res. Inst. Techn. Pap. See B–P–H 435/17. HI 54898

International rice research newsletter. Manila. No. 1/76+, 1976+. Int. Rice Res. Newslett. Preceded by: Rice entomology newsletter. HI 66062

International scientists' directory = Scientists' international directory. Boston, MA.

International seaweed symposium = Proceedings, international seaweed symposium. London, etc.

International society for tropical root crops symposium, proceedings = Proceedings of the symposium, international society for tropical root crops.

International studio. New York. Vols. 1-99, 1897-1931. Int. Stud. 3-2084-1. HI 74848

International sugar journal; technical and commercial periodical devoted entirely to the sugar industry [subtitle varies]. London. Vol. 1+, 1899+. Int. Sugar J. Preceded by: Sugar cane. HI 66063

International tijdschrift voor plantensociologie, oecologie en algemene plantengeographie. The Hague = Vegetatio; acta geobotanica. The Hague. Vegetatio. See B–P–H 948/18.

International tijdschrift voor succulentenliefhebbers = I T S L. Antwerp.

International traditional medicine newsletter. Chicago, IL. Vol. 1+, 1985+. Int. Tradit. Med. Newslett. HI 66064

International tree crops journal; journal of agroforestry. Berkhamsted. Vol. 1+, 1980+. Int. Tree Crops J. HI 66065

International wildlife. Washington, DC. Vol. 1+, 1971+. Int. Wildlife. HI 66066

International Zeitschrift für Pflanzensoziologie, Oekologie und Pflanzengeographie. The Hague = Vegetatio; acta geobotanica. The Hague. Vegetatio. See B–P–H 948/18.

Internationale Bibliographie für Forstwirtschaft Deutschland. Freiburg im Breisgau. Vols. 1-18, 1934-54. Int. Bibliogr. Forstw. Deutschland. Superseded by: Bibliographie des forstlichen Schrifttums Deutschlands. HI 66067

Internationale phaenologische Zeitschrift = Acta phaenologica. The Hague. Acta Phaenol. See B–P–H 47/8.

Internationale Revue der gesamten Hydrobiologie. Berlin. Int. Rev. Gesamten Hydrobiol. See B–P–H 435/10. HI 54894

Internationale Revue der gesamten Hydrobiologie und Hydrographie. Leipzig. Int Rev. Gesamten Hydrobiol. Hydrogr. See B–P–H 435/11. HI 54895

Internationale Vereinigung für theoretische und angewandte Limnologie. Verhandlungen. Stuttgart. Int. Vereinigung Theor. Limnol. Verh. See B–P–H 435/21. HI 54899

Internationale Zeitschrift für die Erforschung und Auswertung von Meeresalgen = Botanica marina. Hamburg. Bot. Mar. See B–P–H 221/24.

Internationale Zeitschrift für physikalisch-chemische Biologie. Leipzig. Int. Z. Phys.-Chem. Biol. See B–P–H 435/22. HI 54900

Internoto; Zeitschrift für Notokakteenliebhaber. Internationales Mitteilungsblatt für Notokakteenliebhaber. Vol. 1+, 1980+. Internoto. HI 66068

Intersylva. Zeitschrift der internationalen Forstzentrale. Munich. Intersylva. See B–P–H 436/14. HI 54902

Intertropiques agricultures. Paris. No. 00, 1+, 1982+. Intertrop. Agric. HI 66069

Intervirology. Basel. Vol. 1+, 1973+. Intervirology. HI 66070

Introduction aux observations sur la physique, sur l'histoire naturelle et sur les arts. Paris. 1771-72; ed. 2, 1777. Introd. Observ. Phys. Superseded by: Observations et mémoires sur la physique, sur l'histoire naturelle et sur les arts. HI 66071

Introductions à la biologie africaine. Abidjan. Vol. 1+, 1968+. Introd. Biol. Africaine. HI 66072

Introduktsiya i akklimatyzatsiya rastenii. Kiev. Vol. 1+, 1984+. Introd. Akklim. Rast. (Kiev). Preceded by: Introduktsiya ta akklimatyzatsiya roslin na Ukrayini. HI 66073

Introduktsiya i akklimatizatsiya rastenii. Tashkent. Vol. 1+, 1968+. Introd. Akklim. Rast. (Tashkent). HI 66075

Introduktsiya ta aklimatyzatsiya roslyn. Kiev. Vols. 1-2, 1966. Introd. Aklim. Rosl. Superseded by: Introduktsiya ta aklimatyzatsiya roslyn na Ukrayini. HI 66076

Introduktsiya ta aklimatyzatsiya roslin na Ukrayini. Kiev. Vols. 3-23, 1968-83. Introd. Aklim. Rosl. Ukrayini. Preceded by: Introduktsiya ta aklimatyzatsiya roslyn. Superseded by: Introduktsiya i akklimatyzatsiya rastenii. HI 66077

Introduktsiya i ekologiya rastenii. Ashkhabad. Vols. 1-5, 1968-74. Introd. Ekol. Rast. HI 66078

Introduktsiya ta eksperymental'na ecologiya roslyn. Kiev. Vol. 1+, 1972+. Introd. Eksper. Ecol. Rosl. HI 66079

Introduktsiya rastenii i zelenoe stroitel'stvo. Moscow & Leningrad = Trudy botanicheskogo instituta imeni V. L. Komarova akademii nauk S S S R. Ser. 6, introduktsiya rastenii i zelenoe stroitel'stvo. Moscow & Leningrad.

Introduktsiya rastenii i zelenoe stroitel'stvo. Tiflis. Vol. 14(83)+, 1982+. Introd. Rast. Zelen. Stroit. Preceded by: Voprosy introduktsii rastenii i zelenogo stroitel'stva. HI 66080

Introduktsiya rastenii i zelonoe stroitel'stvo v Latviiskoi S S R = Daildarznieciba. Augu introdukcija un zala celtnieciba Latvijas P S R. Riga.

Inventario tecnologico do arroz. Goiânia. 1975+. Invent. Tecn. Arroz. HI 66081

Inventory, division of botany, United States department of agriculture = United States department of agriculture. Division of botany. Inventory. Washington, DC. U.S.D.A. Div. Bot. Invent. See B–P–H 941/23.

Inventory, section of seed and plant introduction, United States department of agriculture = U S department of agriculture. Section of seed and plant introduction. Inventory. Washington, DC. U.S.D.A. Sect. Seed Introd. Invent. See B–P–H 945/5.

Inventory of seeds and plants imported by the office of foreign seed and plant introduction, United States department of agriculture = U S department of agriculture. Bureau of plant industry. Inventory of seeds and plants imported by the office of foreign seed and plant introduction. Washington, DC. U.S.D.A. Bur. Pl. Industr. Invent. Seeds. See B–P–H 940/26.

Inventory, United States department of agriculture = United States department of agriculture. Inventory. Washington, DC. U.S.D.A. Invent. See B–P–H 943/1.

Investigacion agraria. Produccion y proteccion vegetales. Madrid. Vol. 1+, 1986+. Invest. Agrar., Prod. Protecc. Veg. Preceded by: Anales del instituto nacional de investigaciones agrarias. Serie, agrícola and Anales del instituto nacional de investigaciones agrarias. Serie, forestal. HI 66082

Investigación agrícola. Caracas. Invest. Agríc. See B–P–H 436/23. HI 54903

Investigación y progreso agrícola. Santiago de Chile. Vol. 1+, 1967+. Invest. Progr. Agric. HI 66083

Investigaciones agropecuarias. La Aurora. Vol. 1+, 1960+. Invest. Agropecu. (La Aurora). HI 54904

Investigaciones agropecuarias. Lima. Vol. 2+, 1971+. Invest. Agropecu. (Lima). Preceded by: Investigaciones agropecuarias del Perú. HI 66084

Investigaciones agropecuarias del Perú. Lima. Vol. 1(1-2), 1970. Invest. Agropecu. Perú. Superseded by: Investigaciones agropecuarias. HI 66085

Investigaciones del laboratorio de química biológica, universidad nacional de Cordoba. Cordoba. Vol. 1, 1931-33. Invest. Lab. Quím. Biol. Univ. Nac. Cordoba. HI 66086

Investigaciones marinas. Valparaiso. Vol. 1+, 1970+. Invest. Mar. HI 66087

Investigatio et studium naturae. Shanghai. Vol. 1+, 1983+. Invest. Stud. Nat. HI 66088

Investigational report, department of nature conservation, Cape of Good Hope. Cape Town. Vols. 1-19, 1962-74. Invest. Rep. Dep. Nat. Conservation, Cape Good Hope. Superseded by: Bontebok. HI 66089

Investigations, north of Scotland college of agriculture = Investigations, research and field trials, north of Scotland college of agriculture. Aberdeen.

Investigations and reports on nature conservation and landscape protection = Forschungen und Berichte zu Naturschutz und Landschaftspflege. Beiheft zur Schriftenreihe Naturschutz und Landschaftspflege in Niedersachsen. Hannover.

Investigations, research and field trials, north of Scotland college of agriculture. Aberdeen. 1953-64. Invest. Res. Field Trials N. Scotland Coll. Agric. Superseded by: Report on investigations and research, north of Scotland college of agriculture and Report on field trials and observations, north of Scotland college of agriculture. HI 66090

Investigative techniques in medicine and biology. New

York. Vol. 1+, 1984+. Invest. Techn. Med. Biol. HI 66091

Iowa agricultural experiment station. Abstracts of new publications. Ames, IA. Iowa Agric. Exp. Sta. Abstr. New Publ. See B–P–H 436/32. HI 54906

Iowa agricultural experiment station. Bulletin. Ames, IA. Iowa Agric. Exp. Sta. Bull. See B–P–H 436/33. HI 54907

Iowa agricultural experiment station. Circular. Ames, IA. Iowa Agric. Exp. Sta. Circ. See B–P–H 436/34. HI 54908

Iowa agricultural experiment station. Research bulletin. Ames, IA. Iowa Agric. Exp. Sta. Res. Bull. See B–P–H 436/35. HI 54909

Iowa agricultural experiment station. Special report. Ames, IA. Iowa Agric. Exp. Sta. Special Rep. See B–P–H 436/36. HI 54910

Iowa agriculturist. Ames, IA. Iowa Agric. See B–P–H 436/30. HI 54905

Iowa arboretum news; newsletter of the Iowa arboretum. Ames, IA. Vol. 4(3)+, 1984+. Iowa Arbor. News. Preceded by: Arboretum news. Ames, IA. HI 66092

Iowa conservationist. Des Moines, IA. Vol. 1+, 1942+. Iowa Conservationist. HI 66093

Iowa farm science. Ames, IA. Iowa Farm Sci. See B–P–H 436/38. HI 54911

Iowa horticulture. Des Moines, IA. Iowa Hort. See B–P–H 437/2. HI 54912

Iowa naturalist. Iowa City, IA. Iowa Naturalist. See B–P–H 437/4. HI 54913

Iowa state college journal of science. Ames, IA. Vols. 1-33, 1926-59. Iowa State Coll. J. Sci. Superseded by: Iowa state journal of science. HI 66094

Iowa state journal of research. Ames, IA. Vol. 47+, 1972+. Iowa State J. Res. Preceded by: Iowa state journal of science. HI 66095

Iowa state journal of science. Ames, IA. Vols. 34-46, 1959-72. Iowa State J. Sci. Preceded by: Iowa state college journal of science. Superseded by: Iowa state journal of research. HI 66096

Ipar és természetbarát. Kolozsvar [=Cluj, Rumania]. Ipar & Természetbarát. See B–P–H 437/17. HI 54914

Iranian journal of agricultural research. Oxford. Vol. 1+, 1971+. Iranian J. Agric. Res. HI 66097

Iranian journal of botany. Tehran. Vol. 1+, 1976+ [suspended from 1978-82]. Iranian J. Bot. HI 66098

Iranian journal of plant pathology; official organ of the iranian phytopathological society. Teheran. Vols. ?-4-14?, 19??-67-78? Iranian J. Pl. Pathol. HI 66099

Iranian journal of science and technology. Shiraz, Oxford. Vols. 1-7, 1971-78. Iranian J. Sci. Techn. HI 66100

Iraq natural history museum's publications. Baghdad. Nos. 1-29, 1950-72. Iraq Nat. Hist. Mus. Publ. Superseded by: Publications, natural history research center, Baghdad. HI 66101

Iraqi journal of agricultural science. Baghdad. Vol. 1+, 1966+. Iraqi J. Agric. Sci. HI 66102

Iraqi journal of marine science. Basrah. Vol. 1+, 1982+. Iraqi J. Mar. Sci. HI 66103

Iraqi journal of science. Baghdad. Vol. 18(2)+, 1977+. Iraqi J. Sci. Preceded by: Bulletin of the college of science, university of Baghdad. HI 66104

Iregszemcse bulletin. Iregszemcse, Hungary. Iregszemcse Bull. See B–P–H 437/22. HI 54915

Iris. Eldagsen, Germany. Iris. See B–P–H 437/23. HI 54916

Iris og Hebe. Copenhagen. 1796-1806. Iris & Hebe. Preceded by: Maanedskriftet Iris. HI 53272

Iris und Lilien; vierteljährlich erscheinendes Organ der deutschen Iris- und Liliengesellschaft e. V. Leonberg. 19??-68-73. Iris Lilien. Preceded by: Deutsche Iris- und Liliengesellschaft e. V. Nachrichtenblatt. Superseded by: Staudengarten. HI 66105

Iris society year book. Liverpool. Iris Soc. Year Book. See B–P–H 437/24. HI 54917

Iris year book. Tunbridge Wells, England. Iris Year Book. See B–P–H 437/26. HI 54918

Irish agricultural magazine. Dublin. Irish Agric. Mag. See B–P–H 437/28. HI 54919

Irish booklore. Belfast. Vol. 1+, 1971+. Irish Booklore. HI 66106

Irish farmer's and gardener's magazine and register of rural affairs. Dublin. Irish Farmer's Gard. Mag. & Reg. Rural Affairs. See B–P–H 437/29. HI 54920

Irish farmers' journal and weekly intelligencer. Dublin. Irish Farmers' J. & Weekly Intelligencer. See B–P–H 437/30. HI 54921

Irish forestry. Wexford, Ireland. Irish Forest. See B–P–H 437/31. HI 54922

Irish geography. Dublin. Irish Geogr. See B–P–H 437/32. HI 54923

Irish journal of agricultural research. An foras taluntais. Dublin. Irish J. Agric. Res. See B–P–H 438/1. HI 54924

Irish journal of environmental science. Dublin. Vol. 1+, 1980+. Irish J. Environm. Sci. HI 66107

Irish naturalist; a monthly journal of general irish natural history. Dublin. Irish Naturalist. See B–P–H 438/4. HI 54925

Irish naturalists' journal; a magazine of natural history, antiquities and ethnology. Belfast. Irish Naturalists' J. See B–P–H 438/5. HI 54926

Irmischia. Correspondenzblatt des botanischen Vereins für das nördliche Thüringen. Sondershausen, Germany. Irmischia. See B–P–H 438/6. HI 54927

Irregular newsletter of the international golden fossil tree society. Lombard, IL. Vol. ?-2+, ?-1980+. Irreg. Newslett. Int. Golden Fossil Tree Soc. HI 66108

Irrigación en México. Mexico City. Irrig. México. See B–P–H 438/9. HI 54929

Irrigation bulletin, Nevada agricultural experiment station = Nevada agricultural experiment station. Irrigation bulletin. Reno, NV. Nevada Agric. Exp. Sta. Irrig. Bull. See B–P–H 651/18.

Irrigation fruit grower and agriculturist. Paonia, Colorado. Irrig. Fruit Grower Agric. See B–P–H 438/8. HI 54928

Iselya; botanical journal of the X club. Madison, WI. Vol. 1+, 1979+. Iselya. HI 66109

Ishikawa-ken nogyo shikenjo, tokubetsu kenkyu hokoku = Special bulletin of the Ishikawa-ken agricultural experiment station. Nonoichi.

Ishikawa-ken nogyo tanki daigaku. Kenkyu hokoku = Bulletin, Ishikawa prefecture college of agriculture. Ishikawa.

Ishikawa-ken zoshoku shikenjo kenkyu hokoku = Bulletin of the Ishikawa prefectural marine culture station. Notojima.

Isis. Berlin. Isis (Berlin). See B–P–H 438/10. HI 54930

Isis. Haarlem. Isis (Haarlem). See B–P–H 438/11. HI 54931

Isis. Munich. Isis (Munich). See B–P–H 438/12. HI 54932

Isis. [Edited by L. Oken.] Jena. Isis (Oken). See B–P–H 438/13. HI 54933

Isis. Prague. Isis (Prague). See B–P–H 438/14. HI 54934

Isis. Brussels, etc. Isis (Sarton). See B–P–H 438/15. HI 54935

Isis. Zurich. Isis (Zurich). See B–P–H 438/16. HI 54936

Isis budissina. Bautzen. Isis Budissina. See B–P–H 438/18. HI 54937

Isis Moderne. Paris. Isis Moderne. See B–P–H 438/20. HI 54938

Iskolakert. Arad, Hungary [Rumania]. Iskolakert. See B–P–H 438/21. HI 54939

Islamabad journal of sciences. Islamabad. Vol. 1(3)+, 1975+. Islamabad J. Sci. Preceded by: Journal of mathematics and sciences. HI 66110

Island grower. Cooke, B.C. Vol. 1+, 1984+. Island Grower. HI 66111

Island naturalist; newsletter of the Prince Edward Island natural history society. Charlottetown, P.E.I. No. 79+, 1984+. Island Naturalist. Preceded by: Newsletter, Prince Edward Island natural history society. HI 66112

Islenzkar landbunada ramsoknir. Reykjavik. Vol. 1+, 1969+. Islen. Landbun. Ramsok. HI 66113

Ismertetö honi 's külgazdaságban és kereskedésben. Pest [=Budapest, in part]. Ismert. Honi Külgazd. Keresk. See B–P–H 438/22. HI 54940

Ismertetö az összmüvészet és polgári szorgalomban. Pest [=Budapest, in part]. Ismert. Összmüv. Polg. Szorg. See B–P–H 439/1. HI 54942

Ismertetö összmüvészetben, gazdaságban és kereskedésben. Pest [=Budapest, in part]. Ismert. Összmüv. Gazd. Keresk. See B–P–H 438/23. HI 54941

Ismertetö vagyis összmüvészet (polytechnia) polgári szörgalom, 's magasb islésü müveszetek tára. Pest [=Budapest, in part]. Ismert. Vagyis Összmüv. (Polytechn.) Polg. Szorg. Magasb Izlézü Müv. Tára. See B–P–H 439/2. HI 54943

Isozyme bulletin. Chapel Hill, NC, East Lansing, MI. No. 1+, 1967+. Isozyme Bull. HI 66114

Israel journal of agricultural research; quarterly of the national and university institute of agriculture, the Volcani institute of agricultural research. Rehovoth. Vols. 11-23, 1961-74. Israel J. Agric. Res. Preceded by: Ktavim. Superseded in part by: Phytoparasitica. HI 66115

Israel journal of botany. Jerusalem. Vol. 12+, 1963+. Israel J. Bot. Preceded by: Bulletin of the research council of Israel. Section D, botany. HI 66116

Israel journal of chemistry. Jerusalem. Vol. 1+, 1963+. Israel J. Chem. HI 66117

Israel — land and nature. Tel Aviv. Vol. 1+, 1975+. Israel Land Nat. HI 66118

Issledovanie flory severo-zapadnogo prichernomov'ya. Odessa. Vol. 1+, 1975+. Issl. Fl. Severo-Zapadn. Prichern. HI 66119

Issledovaniya po bionike, Kiev, 1965 = Bionika. Kiev.

Issledovaniya po mikrobiologii = Mikrobioloogilised uurimused. Tallinn.

Issledovatel'skaja kafedra botaniki = Issledovatel'skaya kafedra botaniki. Rostov.

Issledovatel'skaya kafedra biologii. Rostov, Russian S F S R. Vol. 3, 1930. Issl. Kafedra Biol. Preceded by:

Issledovatel'skaya kafedra botaniki. HI 66120

Issledovatel'skaya kafedra botaniki. Rostov, Russian S F S R. Vols. 1-2, 1926. Issl. Kafedra Bot. Superseded by: Issledovatel'skaya kafedra biologii. HI 66121

Issues in science and technology. Washington, DC. Vol. 1+, 1984+. Issues Sci. Techn. HI 66122

Istanbul üniversitesi eczacilik fakültesi mecmuasi. Istanbul. Vol. 1+, 1965+. Istanbul Üniv. Eczac. Fak. Mecm. HI 66123

Istanbul üniversitesi eczacilik fakültesi mecmuasi, ayri barki. Istanbul. Vol. 1+, 1965+. Istanbul Üniv. Eczac. Fak. Mecm. HI 66124

Istanbul üniversitesi eczacilik fakültesi yayinlari. Istanbul. ?-1983+. Istanbul Üniv. Eczac. Fak. Yayinl. HI 66125

Istanbul üniversitesi fen fakültesi hidrobiologi arastirma enstitusü yayinlarindan. Ser. B. Istanbul. Vols. 1-?, 1952-70; continued without series 1970+. Istanbul Üniv. Fen Fak. Hidrobiol. Arast. Enst. Yayinl., B. HI 66126

Istanbul üniversitesi fen fakültesi mecmuasi. Istanbul. Vols. 1-8, 1923-31; n.s. vols 1-4, 1935-39. Istanbul Üniv. Fak. Mecm. Superseded by: Istanbul üniversitesi fen fakültesi mecmuasi. Seri B, tabii ilimler. HI 66127

Istanbul üniversitesi fen fakültesi mecmuasi. Seri B, tabii ilimler. Istanbul. N.s. vol. 5+, 1940+. Istanbul Üniv. Fak. Mecm., B. Preceded by: Istanbul üniversitesi fen fakültesi mecmuasi. HI 66128

Istanbul üniversitesi orman fakültesi dergisi. Ser. A. Istanbul. Vol. 1+, 1951+. Istanbul Üniv. Orman Fak. Derg., A. HI 66129

Istanbul üniversitesi orman fakültesi yayinlari. Istanbul. ?-1952+. Istanbul Üniv. Orman Fak. Yayinl. HI 66130

Istituto di agronomia e coltivazioni erbacee. Esperienzi e ricerche. Università di Pisa. Pisa. Ist. Agron. Esper. Ric. See B–P–H 439/5. HI 54944

Istoriko-biologicheskie issledovaniya. Moscow. Vol. 6+, 1978+. Istoriko-Biol. Issl. Preceded by: Iz istorii biologii. HI 66131

Istria agricola. Parenzo [= Porec, Yugoslavia]. Istria Agric. See B–P–H 439/10. HI 54945

Isvesstija der Ost-Sibirischen Abtheilung der Kaiserlich-Russischen Geographischen Gesellschaft = Izvestiya Vostochno-Sibirskago Otděla Imperatorskago Russkago Geograficheskago Obshchestva. Irkutsk.

Italia agricola, giornale di agricoltura. Milan. Italia Agric. See B–P–H 439/13. HI 54946

Italia forestale e montana; rivista di politica economica e tecnica. Florence. Vol. 1+, 1946+. Italia Forest. Mont. HI 66132

Italia orticola. Naples. Italia Ortic. See B–P–H 439/14. HI 54947

Italian journal of biochemistry. Rome. Vol. 1+, 1957+. Ital. J. Biochem. HI 66133

Items of interest. Philadelphia, PA. Items of Interest. See B–P–H 439/15. HI 54948

Itogi naučno-issledovatel'skih rabot Vsesojuznogo instituta zaščity rastenij = Itogi nauchno-issledovatel'skikh rabot Vsesoyuznogo instituta zashchity rastenii. Leningrad.

Itogi nauchno-issledovatel'skikh rabot Vsesoyuznogo instituta zashchity rastenii. Leningrad. 1935-36. Itogi Nauchno-Issl. Rabot Vsesoyuzn. Inst. Zashch. Rast. HI 66137

Itogi nauki. Okeanologiya. Moscow. Vol. 1, 1970. Itogi Nauki, Okeanol. Preceded by: Itogi nauki i tekhniki. Seriya okeanologiya. HI 66141

Itogi nauki. Seriya biologicheskaya khimiya. Moscow. Vols. [1]-[4]-5, 1966-71. Itogi Nauki, Ser. Biol. Khim. Superseded by: Itogi nauki i tekhniki. Seriya biologicheskaya khimiya. HI 66138

Itogi nauki. Serija biologičeskie nauki. Moscow = Itogi nauki. Seriya biologicheskie nauki. Moscow.

Itogi nauki. Seriya biologicheskie nauki. Moscow. Vols. 1-3, 1957-60. Itogi Nauki Biol. Nauki. Superseded by: Itogi nauki, institut nauchnoi informatsii. Seriya biologiya. HI 66139

Itogi nauki. Seriya obschchaya genetika (mutagenez, mutatsii). [Subseries of: Itogi nauki. Seriya biologicheskie nauki.] Moscow. Vol. [1], 1967. Itogi Nauki, Ser. Obschchaya Genet. Superseded by: Itogi nauki i tekhniki. Seriya obshchaya genetika. HI 66140

Itogi nauki. Seriya okhrana prirody i vosproizvodstvo prirodnykh resursov. Moscow. Vol. 1, 1968. Itogi Nauki, Ser. Okhr. Prir. Vosproizv. Prir. Resursov. Superseded by: Itogi nauki i tekhniki. Seriya okhrana prirody i vosproizvodstvo prirodnykh resursov. HI 66142

Itogi nauki. Seriya tsitologiya, obshchaya genetika, genetika cheloveka. [Subseries of: Itogi nauki, institut nauchnoi informatsii. Seriya biologicheskie nauki.] Moscow. Vol. 1, 1970. Itogi Nauki, Ser. Tsitol. Obshchaya Genet. Cheloveka. Superseded by: Itogi nauki i tekhniki. Seriya tsitologiya. HI 66143

Itogi nauki. Seriya virusologiya i mikrobiologiya. Moscow. Vols. [1]-[2]-3, 1967-72. togi Nauki, Ser. Virusol. Mikrobiol. Superseded by: Itogi nauki i tekhniki. Seriya mikrobiologiya and Itogi nauki i tekhniki. Seriya virusologiya. HI 66144

Itogo nauki, institut nauchnoi informatsii. Seriya biologiya. Moscow. Vols. 4-?, 1962-72. Itogo Nauki,

Inst. Nauchn. Inform., Ser. Biol. Preceded by: Itogi nauki. Seriya biologicheskie nauki. Superseded by: Itogi nauki i tekhniki, institut nauchnoi informatsii. Seriya biologicheskie nauki. HI 66145

Itogi nauki i tekhniki. Biofizika = Itogi nauki i tekhniki. Seriya biofizika. Moscow.

Itogi nauki i tekhniki. Biogeografiya = Itogi nauki i tekhniki. Seriya biogeografiya. Moscow.

Itogi nauki i tekhniki. Biologicheskaya khimiya = Itogi nauki i tekhniki. Seriya biologicheskaya khimiya. Moscow.

Itogi nauki i tekhniki. Botanika = Itogi nauki i tekhniki. Seriya botanika. Moscow.

Itogi nauki i tekhniki. Fiziologiya rastenii = Itogi nauki i tekhniki. Seriya fiziologiya rastenii. Moscow.

Itogi nauki i tekhniki. Lesovedenie i lesovodstvo = Itogi nauki i tekhniki. Seriya lesovedenie i lesovodstvo. Moscow.

Itogi nauki i tekhniki. Mikrobiologiya = Itogi nauki i tekhniki. Seriya mikrobiologiya. Moscow.

Itogi nauki i tekhniki. Molekularnaya biologiya = Itogi nauki i tekhniki. Seriya molekularnaya biologiya. Moscow.

Itogi nauki i tekhniki. Obschaya ekologiya, biotsenologiya, gidrobiologiya = Itogi nauki i tekhniki. Seriya obschaya ekologiya, biotsenologiya, gidrobiologiya. Moscow.

Itogi nauki i tekhniki. Obshchaya genetika = Itogi nauki i tekhniki. Seriya obshchaya genetika. Moscow.

Itogi nauki i tekhniki. Okeanologiya = Itogi nauki i tekhniki. Seriya okeanologiya. Moscow.

Itogi nauki i tekhniki. Radiatsionnaya biologiya = Itogi nauki i tekhniki. Seriya radiatsionnaya biologiya. Moscow.

Itogi nauki i tekhniki. Rastenievodstvo = Itogi nauki i tekhniki. Seriya rastenievodstvo. Moscow.

Itogi nauki i tekhniki. Seriya biofizika. Moscow. Vol. 1+, 1971?+. Itogi Nauki Tekhn., Ser. Biofiz. HI 66146

Itogi nauki i tekhniki. Seriya biogeografiya. Moscow. Vol. 1+, 1976+. Itogi Nauki Tekhn., Ser. Biogeogr. HI 66147

Itogi nauki i tekhniki. Seriya biologicheskaya khimiya. Moscow. Vol. 6+, 1973+. Itogi Nauki Tekhn., Ser. Biol. Khim. Preceded by: Itogi nauki. Seriya biologicheskaya khimiya. HI 66148

Itogi nauki i tekhniki. Seriya Biologiya. Moscow. Vol. 1+, 1972+. Itogi Nauki Tekhn., Ser. Biol. Preceded by: Itogi nauki, institut nauchnoi informatsii. Seriya biologiya. HI 66149

Itogi nauki i tekhniki. Seriya bioorganicheskaya khimiya. Moscow. Vol. 1+, 1984+. Itogi Nauki Tekhn., Ser. Bioorg. Khim. HI 72597

Itogi nauki i tekhniki. Seriya biotekhnologiya. Moscow. Vol. 1+, 1983+. Itogi Nauki Tekhn., Ser. Biotekhnol. HI 73387

Itogi nauki i tekhniki. Seriya botanika. Moscow. Vol. 1+, 1972+. Itogi Nauki Techn., Ser. Bot. HI 66150

Itogi nauki i tekhniki. Seriya fiziologiya rastenii. Moscow. Vol. 1+, 1973+. Itogi Nauki Tekhn., Ser. Fiziol. Rast. HI 66151

Itogi nauki i tekhniki. Seriya lesovedenie i lesovodstvo. Moscow. Vol. 1+, 1975+. Itogi Nauki Tekhn., Ser. Lesoved. Lesov. HI 66152

Itogi nauki i tekhniki. Seriya mikrobiologiya. Moscow. Vol. 1+, 1972+. Itogi Nauki Tekhn., Ser. Mikrobiol. Preceded, in part, by: Itogi nauki. Seriya virusologiya i mikrobiologiya. HI 66153

Itogi nauki i tekhniki. Seriya molekularnaya biologiya. Moscow. Vol. 2+, 1973+. Itogi Nauki Tekhn., Ser. Molek. Biol. Preceded, in part, by: Itogi nauki. Seriya virusologiya i mikrobiologiya. HI 66154

Itogi nauki i tekhniki. Seriya obschaya ekologiya, biotsenologiya, gidrobiologiya. Moscow. Vol. 1+, 1973+. Itogi Nauki Tekhn., Ser. Obshchaya Ekol. Biotsenol. Gidrobiol. HI 75107

Itogi nauki i tekhniki. Seriya obshchaya genetika. Moscow. Vol. 2+, 1977+. Itogi Nauki Tekhn., Ser. Obshchaya Genet. Preceded by: Itogi nauki. Seriya obshchaya genetika (mutagenez, mutatsii). HI 66155

Itogi nauki i tekhniki. Seriya obshchie problemy biologii. Moscow. Vol. 1+, 1982+. Itogi Nauki Tekhn., Ser. Obshchie Probl. Biol. HI 75108

Itogi nauki i tekhniki. Seriya obshchie problemy fiziko-khimicheskoi biologii. Moscow. Vol. 1+, 1984+. Itogi Nauki Tekhn., Ser. Obshchie Probl. Fiz.-Khim. Biol. HI 75109

Itogi nauki i tekhniki. Seriya okeanologiya. Moscow. Vol. 2+, 1973+. Itogi Nauki Tekhn., Ser. Okeanol. Preceded by: Itogi nauki. Okeanologiya. HI 66156

Itogi nauki i tekhniki. Seriya okhrana prirody i vosproizvodstvo prirodnykh resursov. Moscow. Vol. 2+, 1978+. Itogi Nauki Tekhn., Ser. Okhrana Prir. Vosproizv. Prir. Resursov. Preceded by: Itogi nauki. Seriya okhrana prirody i vosproizvodstvo prirodnykh resursov. HI 66157

Itogi nauki i tekhniki. Seriya radiatsionnaya biologiya. Moscow. Vol. ?-2+, ?-1974+. Itogi Nauki Tekhn., Ser. Radiats. Biol. HI 66158

Itogi nauki i tekhniki. Seriya rastenievodstvo. Moscow. Vol. ?-2+, ?-1973+. Itogi Nauki Tekhn., Ser. Rasteniev. HI 66159

Itogi nauki i tekhniki. Seriya tsitologiya, obshchaya genetika, genetika cheloveka. Moscow. Vol. 2+, 1975+. Itogi Nauki Tekhn., Ser. Tsitol. Genet. Genet Cheloveka. Preceded by: Itogi nauki. Seriya tsitologiya. HI 66160

Itogi nauki i tekhniki. Seriya virusologiya. Moscow. Vol. 2+, 1973+. Itogi Nauki Tekh., Ser. Virusol. Preceded, in part, by: Itogi nauki. Seriya virusologiya i mikrobiologiya. HI 66161

Itogi nauki i tekhniki. Seriya zashchita rastenii. Moscow. Vol. ?-2+, ?-1980+. Itogi Nauki Tekhn., Ser. Zashch. Rast. HI 66162

Itogi. Summaries of scientific progress, microbiology. [Translation of: Itogi nauki i tekhniki. Seriya mikrobiologiya.] Boston, MA. Vol. 1+, 1974+. Itogi, Summ. Sci. Progr. Microbiol. HI 66134

Itogi. Summaries of scientific progress, oceanology. [Translation of: Itogi nauki. Seriya okeanologiya (later Itogi nauki i tekhniki. Seriya okeanologiya).] Boston, MA. Vol. 1+, 1974+. Itogi, Summ. Sci. Progr. Oceanol. HI 66135

Itogi. Summaries of scientific progress, virology. [Translation of: Itogi nauki i tekhniki. Seriya virusologiya.] Boston, MA. Vol. 1+, 197?+. Itogi, Summ. Sci. Progr. Virol. HI 66136

Itogi Voronežskaja stancii zaščity rastenij = Itogi Voronezhskaya stantsii zashchity rastenii. Voronezh.

Itogi Voronezhskaya stantsii zashchity rastenii. Voronezh, Russian S F S R. Vol. 14, 1948. Itogi Voronezhsk. Stantsii Zashch. Rast. Preceded & superseded by: Trudy Voronezhskoi stantsii zashchity rastenii. HI 66163

Ivy bulletin. Mt. Vernon, VA, Dayton, OH. Vols. 1-7, 1975?-81? Ivy Bull. Superseded by: Ivy journal. HI 66164

Ivy journal. Mount Vernon, VA. Vol. 8+, 1982+. Ivy J. Preceded by: Ivy bulletin. HI 66165

Iwate daigaku, gakugeigakubu kenkyu nenpo = Report (Annual) of the gakugei faculty of the Iwate university. Morioka.

Iwate daigaku kyoikugakubu kenkyu nenpo = Report (Annual) of the college of education, university of Iwate. Morioka.

Iwate daigaku Morioka norin senmon gakko gakujutsu hokoku = Bulletin of the imperial college of agriculture and forestry. Morioka, Japan. Bull. Imp. Coll. Agric. See B–P–H 254/17.

Iwate daigaku nogakubu enshurin hokoku = Bulletin of the Iwate university forests. Morioka.

Iwate daigaku nogakubu hokoku = Journal of the faculty of agriculture, Iwate university. Morioka, Japan. J. Fac. Agric. Iwate Univ. See B–P–H 465/5.

Iwate-ken nogyo shikenjo kenkyu hokoku = Bulletin of the Iwate prefecture agricultural experiment station. Morioka, Japan. Bull. Iwate Prefect. Agric. Exp. Sta. See B–P–H 258/8.

Iwate-ken ringyo shikenjo. Kenkyu hokoku. Takizawa. No. 1+, 1977+. Iwate-ken Ringyo Shikenjo, Kenkyu Hokoku. HI 66166

Iz istorii biologii. Moscow. Vols. 3-5, 1971-75. Iz Istorii Biol. Preceded by: Iz istorii biologicheskii nauk. Superseded by: Istoriko-biologicheskie issledovaniya. HI 66167

Iz istorii biologicheskii nauk. Moscow. Vols. 1-2, 1966-70. Iz Istorii Biol. Nauk. Superseded by: Iz istorii biologii. HI 66168

Izdanija na prirodo-naučniot muzej Skoplje = Acta musei macedonici scientiarum naturalium. Skoplje, Yugoslavia. Acta Mus. Maced. Sci. Nat. See B–P–H 46/13.

Izdanija Severo-Kavkazskoj kraevoj stancii zaščity rastenij. Ser. A. Naučnye i organizacionnye raboty = Izdaniya Severo-Kavkazskoi kraevoi stantsii zashchity rastenii. Ser. A. Nauchnye i organizatsionnye raboty. Rostov.

Izdaniya Severo-Kavkazskoi kraevoi stantsii zashchity rastenii. Ser. A. Nauchnye i organizatsionnye raboty. Rostov, Russian S F S R. Vols. 1-12, 19??-29. Izd. Severo-Kavkazsk. Kraev. Stantsii Zashch. Rast., A. HI 66169

Izvešča botaničkog zavoda kr. sveučilišta u Zagrebu = Acta botanica instituti botanici universitatis zagrebiensis. Zagreb, Yugoslavia. Acta Bot. Inst. Bot. Univ. Zagreb. See B–P–H 41/17.

Izvješča o raspravama matematičko-prirodoslovnoga razreda, jugoslavenska akademija znanosti i umjetnosti o Zagrebu. Zagreb. Vols. 1-23, 1914-29. Izvj. Razpr. Mat.-Prir. Razr. Jogoslav. Akad. Znan. Superseded by: Bulletin international de l'académie Yougoslave des sciences et des beaux-arts. Classe des sciences mathématiques et naturelles. HI 66361

Izvestia de la société de géographie de l'URSS = Izvestiya Vsesoyuznogo geograficheskogo obshchestva. Moscow & Leningrad.

Izvestia de la société russe de géographie = Izvestiya Gosudarstvennogo russkogo geograficheskogo obshchestva. Moscow & Leningrad.

Izvestija Akademii Nauk. Petrograd = Izvestiya Akademii Nauk. Petrograd.

Izvestija akademija nauk armjanskoj S S R. Biologicheskie nauki. Erevan = Izvestiya akademiya nauk armyanskoi S S R. Biologicheskie nauki. Erevan.

Izvestija Akademii nauk Azerbajdžanskoj S S R = Izvestiya Akademii nauk Azerbaidzhanskoi S S R.

Baku.

Izvestija Akademii nauk Azerbajdžanskoj S S R. Serija biologičeskih i medicinskih nauk = Izvestiya Akademii nauk Azerbaidzhanskoi S S R. Seriya biologicheskikh i meditsinskikh nauk. Baku.

Izvestija Akademii nauk Azerbajdžanskoj S S R. Serija biologičeskih i sel'skohozjajstvennyh nauk = Izvestiya Akademii nauk Azerbaidzhanskoi S S R. Seriya biologicheskikh i sel'skokhozyaistvennykh nauk. Baku.

Izvestija Akademii nauk Belorusskoj S S R = Izvestiya Akademii nauk Belorusskoi S S R. Minsk.

Izvestija Akademii nauk Belorusskoj S S R = Vestsi Akademii navuk Belaruskai S S R. Minsk.

Izvestija Akademii nauk Belorusskoj S S R. Serija biologičeskih nauk = Vestsi Akademii navuk Belaruskai S S R. Seryya biyalagichnykh navuk. Minsk.

Izvestija Akademii nauk Belorusskoj S S R. Serija biologičeskih i sel'skohozjajstvennyh nauk = Vestsi Akademii navuk Belaruskai S S R. Seryya biyalagichnykh i sel'skagaspadarchykh navuk. Minsk.

Izvestija Akademii nauk Ėstonskoj S S R = Izvestiya Akademii nauk Ėstonskoi S S R. Tallinn.

Izvestija Akademii nauk Ėstonskoj S S R. Biologičeskaja serija = Izvestiya Akademii nauk Ėstonskoi S S R. Seriya biologii, sel'skokhozyaistvennykh nauk i meditsinskikh nauk. Tallinn.

Izvestija Akademii nauk Ėstonskoj S S R. Serija biologii, sel'skohozjajstvennyh nauk i medicinskih nauk = Izvestiya Akademii nauk Ėstonskoi S S R. Seriya biologii, sel'skokhozyaistvennykh nauk i meditsinskikh nauk. Tallinn.

Izvestija akademii nauk kazahskoj S S R. Alma Ata = Izvestiya akademii nauk kazakhskoi S S R. Alma Ata.

Izvestija Akademii nauk Kazahskoj S S R. Serija botaničeskaja = Izvestiya Akademii nauk Kazakhskoi S S R. Seriya botanicheskaya. Alma-Ata.

Izvestija Akademii nauk Kazahskoj S S R. Serija botaniki i počvovedenija = Izvestiya Akademii nauk Kazakhskoi S S R. Seriya botaniki i pochvovedeniya. Alma-Ata.

Izvestija Akademii nauk Kirgizskoj S S R = Izvestiya Akademii nauk Kirgizskoi S S R. Frunze.

Izvestija Akademii nauk Kirgizskoj S S R. Serija biologičeskih nauk = Izvestiya Akademii nauk Kirgizskoi S S R. Seriya biologicheskikh nauk. Frunze.

Izvestija Akademii nauk Latvijskoj S S R = Izvestiya Akademii nauk Latviiskoi S S R. Riga.

Izvestija akademii nauk moldavskoj S S R. Serija biologičeskich nauk = Izvestiya akademiya nauk moldavskoi S S R. Seriya biologicheskikh i khimicheskikh nauk. Kishinev.

Izvestija Akademii nauk S S S R = Izvestiya Akademii nauk S S S R. Moscow & Leningrad.

Izvestija Akademii nauk S S S R. Serija biologičeskaja = Izvestiya Akademii nauk S S S R. Seriya biologicheskaya. Moscow & Leningrad.

Izvestija Akademii nauk S S S R. Serija geologičeskaja = Izvestiya Akademii nauk S S S R. Seriya geologicheskaya. Moscow & Leningrad.

Izvestija Akademii nauk S S S R. Ser. 7, Otdelenie fiziko-matematičeskih nauk = Izvestiya Akademii nauk S S S R. Ser. 7, Otdeleniye fiziko-matematicheskikh nauk. Moscow & Leningrad.

Izvestija Akademii nauk S S S R. Ser. 7, Otdelenie matematičeskih i estestvennyh nauk = Izvestiya Akademii nauk S S S R. Ser. 7, Otdeleniye matematicheskikh i estestvennykh nauk. Moscow & Leningrad.

Izvestija Akademii nauk S S S R; otdelenie matematičeskih i estestvennyh nauk. Serija biologičeskaja = Izvestiya Akademii nauk S S S R; otdeleniye matematicheskikh i estestvennykh nauk. Seriya biologicheskaya. Moscow & Leningrad.

Izvestija Akademii nauk S S S R; otdelenie matematičeskih i estestvennyh nauk. Serija geologičeskaja = Izvestiya Akademii nauk S S S R; otdeleniye matematicheskikh i estestvennykh nauk. Seriya geologicheskaya. Moscow & Leningrad.

Izvestija Akademii nauk Tadžikskoj S S R. Otdelenie estestvennyh nauk = Izvestiya Akademii nauk Tadzhikskoi S S R. Otdelenie estestvennykh nauk. Stalinabad.

Izvestija akademiya nauk Turkmenskoj S S R. Ashkhabad = Izvestiya akademiya nauk Turkmenskoi S S R. Ashkhabad.

Izvestija Akademii nauk Uzbekskoj S S R = Izvestiya Akademii nauk Uzbekskoi S S R. Tashkent.

Izvestija Akademii nauk Uzbekskoj S S R. Serija biologičeskaja = Izvestiya Akademii nauk Uzbekskoi S S R. Seriya biologicheskaya. Tashkent.

Izvestija. Armjanskij filial Akademii nauk S S S R = Izvestiya. Armyanskii filial Akademii nauk S S S R. Erivan.

Izvestija Armjanskogo filiala Akademii nauk S S S R: Estestvennye nauki = Izvestiya Armyanskogo filiala Akademii nauk S S S R: Estestvennye nauki. Erivan.

Izvestija Associacii naučno-issledovatel'skih institutov pri Fiziko-Matematičeskom fakul'tete Pervogo Moskovskogo gosudarstvennogo universiteta = Izvestiya Assotsiatsii nauchno-issledovatel'skikh institutov pri Fiziko-Matematicheskom fakul'tete Pervogo Moskovskogo gosudarstvennogo universiteta.

Moscow.

Izvestija Associacii naučno-issledovatel'skih institutov pri Fiziko-Matematičeskom fakul'tete Pervogo Moskovskogo gosudarstvennogo universiteta. Ser. A. Otdel biologičeskih i geologičeskih nauk = Izvestiya Assotsiatsii nauchno-issledovatel'skikh institutov pri Fiziko-Matematicheskom fakul'tete Pervogo Moskovskogo gosudarstvennogo universiteta. Ser. A. Otdel biologicheskikh i geologicheskikh nauk. Moscow.

Izvestija Azerbajdžanskogo filiala [Akademii nauk S S S R] = Izvestiya Azerbaidzhanskogo filiala [Akademii nauk S S S R.]. Baku.

Izvestija Batumskogo botaničeskogo sada = Izvestiya Batumskogo botanicheskogo sada. Batum.

Izvestija Batumskogo subtropičeskogo botaničeskogo sada = Izvestiya Batumskogo subtropicheskogo botanicheskogo sada. Batum.

Izvestija na biologičeskija institut. Bulgarska akademija na naukite, otdelenie za biologicne i medicinski nauki = Izvestiya na biologicheskiya institut. Bulgarska akademiya na naukite, otdelenie za biologicheskiye i meditsinski nauk. Sofia.

Izvestija Biologičeskogo naučno-issledovatel'skogo instituta i Biologičeskoj stancii pri Permskom gosudarstvennom universitete = Izvestiya Biologicheskogo nauchno-issledovatel'skogo instituta i Biologicheskoi stantsii pri Permskom gosudarstvennom universitete. Perm.

Izvestija Biologičeskogo naučno-issledovatel'skogo instituta pri Molotovskom gosudarstvennom universitete imeni A. M. Gor'kogo = Izvestiya Biologicheskogo nauchno-issledovatel'skogo instituta pri Molotovskom gosudarstvennom universitete imeni A. M. Gor'kogo. Molotov.

Izvestija Biologičeskogo naučno-issledovatel'skogo instituta pri Permskom gosudarstvennom universitete = Izvestiya Biologicheskogo nauchno-issledovatel'skogo instituta pri Permskom gosudarstvennom universitete. Perm.

Izvestija Biologičeskogo naučno-issledovatel'skogo instituta pri Permskom gosudarstvennom universitete imeni A. M. Gor'kogo = Izvestiya Biologicheskogo nauchno-issledovatel'skogo instituta pri Permskom gosudarstvennom universitete imeni A. M. Gor'kogo. Perm.

Izvestija Biologo-geografičeskogo naučno-issledovatel'skogo instituta pri Gosudarstvennom Irkutskom universitete = Izvestiya Biologo-geograficheskogo nauchno-issledovatel'skogo instituta pri Gosudarstvennom Irkutskom universitete. Irkutsk.

Izvestija Biologo-geografičeskogo naučno-issledovatel'skogo instituta pri Irkutskom gosudarstvennom universitete imeni A. A. Zdanova = Izvestiya Biologo-geograficheskogo nauchno-issledovatel'skogo instituta pri Irkutskom gosudarstvennom universitete imeni A. A. Zdanova. Irkutsk.

Izvestija Biologo-geografičeskogo naučno-issledovatel'skogo instituta pri Vostočno-Sibirskom gosudarstvennom universitete = Izvestiya Biologo-geograficheskogo nauchno-issledovatel'skogo instituta pri Vostochno-Sibirskom gosudarstvennom universitete. Moscow.

Izvestija Biologo-geografičeskogo naučno-issledovatel'skogo instituta pri Vostočno-Sibirskom gosudarstvennom universitete imeni tovarišča Zdanova = Izvestiya Biologo-geograficheskogo nauchno-issledovatel'skogo instituta pri Vostochno-Sibirskom gosudarstvennom universitete imeni tovarishcha Zdanova. Irkutsk.

Izvestija Botaničeskago Sada Petra Velikago = Izvestiya Botanicheskogo Sada Petra Velikogo. Petrograd.

Izvestija na botaničeskaja institut. Sofia = Izvestiya na botanicheskaya institut. Sofia.

Izvestija Botaničeskogo sada Akademii nauk S S S R = Izvestiya Botanicheskogo sada Akademii nauk S S S R. Leningrad.

Izvestija na bulgarskoto botaničesko druzestvo = Izvestiya na bulgarskoto botanichesko druzhestvo. Sofia.

Izvestija na carskite prirodonaučni instituti v Sofija = Izvestiya na tsarskite prirodonauchni instituti v Sofiya. Sofia.

Izvestija na centralnija naučno-izsledovatelski institut po rastenievudstvo = Izvestiya na tsentralniya nauchno-izsledovatelski institut po rastenievudstvo. Sofia.

Izvestija Estestvenno-naučnogo instituta pri Molotovskom universitete imeni A. A. Gor'kogo = Izvestiya Estestvenno-nauchnogo instituta pri Molotovskom universitete imeni A. A. Gor'kogo. Molotov.

Izvestija Estestvenno-naučnogo instituta imeni P. F. Lesgafta = Izvestiya Estestvenno-nauchnogo instituta imeni P. F. Lesgafta. Leningrad.

Izvestija Estestvenno-naučnogo instituta pri Permskom gosudarstvennom universitete imeni A. M. Gor'kogo = Izvestiya Estestvenno-nauchnogo instituta pri Permskom gosudarstvennom universitete imeni A. M. Gor'kogo. Perm.

Izvestija: Estestvennye nauki = Izvestiya: Estestvennye nauki. Erivan.

Izvestija Glavnogo botaničeskogo sada R S F S R = Izvestiya Glavnogo botanicheskogo sada R S F S R. Petrograd.

Izvestija Glavnogo botaničeskogo sada S S S R =

Izvestiya Glavnogo botanicheskogo sada S S S R. Leningrad.

Izvestija Gosudarstvennogo geografičeskogo obščestva = Izvestiya Gosudarstvennogo geograficheskogo obshchestva. Moscow & Leningrad.

Izvestija Gosudarstvennogo russkogo geografičeskogo obščestva = Izvestiya Gosudarstvennogo russkogo geograficheskogo obshchestva. Moscow & Leningrad.

Izvestija Gosudarstvennogo Tomskogo universiteta = Izvestiya Gosudarstvennogo Tomskogo universiteta. Tomsk.

Izvestija Imperatorskago Botaničeskago Sada Petra Velikago = Izvestiya Imperatorskogo Botanicheskogo Sada Petra Velikogo. St. Petersburg.

Izvestija Imperatorskago Kazanskago Universiteta = Izvestiya Imperatorskago Kazanskago Universiteta. Kazan.

Izvestija imperatorskago Obščestva ljubitelej Estestvoznanija, Antropologii i Ėtnografii = Izvestiya imperatorskogo Obshchestva lyubitelei estestvoznaniya, antropologii i etnografii, sostoyashchago pri (Imperatorskom) Moskovskom Universitetě. Moscow.

Izvestija imperatorskago Obščestva ljubitelej Estestvoznanija, Antropologii i Ėtnografii, sostojaščago pri (Imperatorskom) Moskovskom Universitetě = Izvestiya imperatorskogo Obshchestva lyubitelei estestvoznaniya, antropologii i etnografii, sostoyashchago pri (Imperatorskom) Moskovskom Universitetě. Moscow.

Izvestija Imperatorskago Russkago Geografičeskago Obščestva = Izvestiya Imperatorskogo Russkogo Geograficheskogo Obshchestva. St. Petersburg.

Izvestija Imperatorskago S.-Peterburgskago Botaničeskago Sada = Izvestiya Imperatorskogo S.-Peterburgskogo Botanicheskogo Sada. St. Petersburg.

Izvestija Imperatorskago Tomskago Universiteta. = Izvestiya Imperatorskago Tomskago Universiteta. Berichte der Tomsker Staatsuniversität. Tomsk.

Izvestija Imperatorskoj Akademii Nauk. St. Petersburg = Izvestiya Imperatorskoi Akademii Nauk. St. Petersburg.

Izvestija na instituta po biologi "Metodij Popov." Sofia = Izvestiya na instituta po biologi "Metodii Popov." Sofia.

Izvestija Instituta počvovedenija i geobotaniki Sredne-Aziatskogo gosudarstvennogo universiteta. = Izvestiya Instituta pochvovedeniya i geobotaniki Sredne-Aziatskogo gosudarstvennogo universiteta. Tashkent.

Izvestija na instituta po rastenievudstvo = Izvestiya na instituta po rastenievudstvo. Sofia.

Izvestija Irkutskoj stancii zaščity rastenij = Izvestiya Irkutskoi stantsii zashchity rastenii. Irkutsk.

Izvestija Jakutskogo otdela Gosudarstvennogo russkogo geografičeskogo obščestva = Izvestiya Yakutskogo otdela Gosudarstvennogo russkogo geograficheskogo obshchestva. Yakutsk.

Izvestija Jakutskago Otděla Imperatorskago Russkago Geografičeskago Obščestva = Izvestiya Yakutskago Otděla Imperatorskago Russkago Geograficheskago Obshchestva. Yakutsk.

Izvestija Južno-Russkago Obščestva Akklimatizacii = Izvestiya Yuzhno-Russkago Obshchestva Akklimatizatsii. Kharkov.

Izvestija Južno-Ussurijskago Otdělenija Priamurskago Otděla Russkago Geografičeskago Obščestva = Izvestiya Yuzhno-Ussuriiskago Otděleniya Priamurskago Otděla Russkago Geograficheskago Obshchestva. Nikolsk-Ussuriski.

Izvestija Južno-Ussurijskogo otdela Gosudarstvennogo russkogo geografičeskogo obščestva = Izvestiya Yuzhno-Ussuriiskogo otdela Gosudarstvennogo russkogo geograficheskogo obshchestva. Nikolsk-Ussuriski.

Izvestija Južno-Ussurijskogo otdelenija Russkogo geografičeskogo obščestva = Izvestiya Yuzhno-Ussuriiskogo otdeleniya Russkogo geograficheskogo obshchestva. Nikolsk-Ussuriski.

Izvestija Karel'skogo i Kol'skogo filialov Akademii nauk S S R = Izvestiya Karel'skogo i Kol'skogo filialov Akademii nauk S S R. Petrozavodsk.

Izvestija Karelo-Finskogo filiala Akademii nauk S S S R = Izvestiya Karelo-Finskogo filiala Akademii nauk S S S R. Petrozavodsk.

Izvestija Karelo-Finskoj naučno-issledovatel'skoj bazy Akademii nauk S S S R = Izvestiya Karelo-Finskoi nauchno-issledovatel'skoi bazy Akademii nauk S S S R. Petrozavodsk.

Izvestija Kavkazskago Otděla Imperatorskago Russkago Geografičeskago Obščestva = Izvestiya Kavkazskago Otděla Imperatorskago Russkago Geograficheskago Obshchestva. Tiflis.

Izvestija Kazahskogo filiala Akademii nauk S S S R = Izvestiya Kazakhskogo filiala Akademii nauk S S S R. Alma-Ata.

Izvestija Kazahskogo filiala Akademii nauk S S S R. Serija botaničeskaja = Izvestiya Kazakhskogo filiala Akademii nauk S S S R. Seriya botanicheskaya. Alma-Ata.

Izvestija Kievskogo botaničeskogo sada = Izvestiya Kievskogo botanicheskogo sada. Kiev.

Izvestija Kievskogo botaničeskogo sada imeni akad. A. V. Fomina = Izvestiya Kievskogo botanicheskogo sada imeni akad. A. V. Fomina. Kiev.

Izvestija Kirgizskogo filiala Akademii nauk S S S R = Izvestiya Kirgizskogo filiala Akademii nauk S S S R. Frunze.

Izvestija Komiteta Akklimatizacii Imperatorskago Moskovskago Obščestva Sel'skago Hozjajstva. Pribavlenie k Zapiskam Komiteta Akklimatizacija = Izvestiya Komiteta Akklimatizatsii Imperatorskago Moskovskago Obshchestva Sel'skago Khozyaistva. Pribavlenie k Zapiskam Komiteta Akklimatizatsiya. Moscow.

Izvestija Krasnojarskago Podotděla Vostočno-Sibirskago Otděla Imperatorskago Russkago Geografičeskago Obščestva = Izvestiya Krasnoyarskago Podotděla Vostochno-Sibirskago Otděla Imperatorskago Russkago Geograficheskago Obshchestva. Krasnoyarsk.

Izvestija Krasnojarskogo otdela Russkogo geografičeskogo obščestva = Izvestiya Krasnoyarskogo otdela Russkogo geograficheskogo obshchestva. Krasnoyarsk

Izvestija Leningradskogo naučnogo instituta imeni P. F. Lesgafta = Izvestiya Leningradskogo nauchnogo instituta imeni P. F. Lesgafta. Moscow & Leningrad.

Izvestija Leningradskoj oblastnoj stancii zaščity rastenij ot vreditelej = Izvestiya Leningradskoi oblastnoi stantsii zashchity rastenii ot vreditelei. Leningrad.

Izvestija na mikrobiologičeskija institut. Sofia = Izvestiya na mikrobiologicheskiya institut. Sofia.

Izvestija muzejskega društva za Kranjsko. Laibach [=Ljubljana, Yugoslavia]. Izv. Muz. Društva Kranjsko. See B–P–H 448/9. HI 54949

Izvestija Naučnogo instituta imeni P. F. Lesgafta = Izvestiya Nauchnogo instituta imeni P. F. Lesgafta. Leningrad.

Izvestija Obščestva izučenija Vostočno-Sibirskogo kraja = Izvestiya Obshchestva izucheniya Vostochno-Sibirskogo kraya. Irkutsk.

Izvestija Obščestva izučenija Vostočno-Sibirskoj oblasti = Izvestiya Obshchestva izucheniya Vostochno-Sibirskoi oblasti. Irkutsk.

Izvestija Obščestva Ljubitelej Estestvoznanija = Izvestiya Obshchestva Lyubitelei Estestvoznaniya. Moscow.

Izvestija Orenburgskago Otděla Imperatorskago Russkago Geografičeskago Obščestva = Izvestiya Orenburgskago Otděla Imperatorskago Russkago Geograficheskago Obshchestva. Orenburg.

Izvestija Otdela zaščity rastenij = Izvestiya Otdela zashchity rastenii. Tiflis.

Izvestija Otdelenija estestvennyh nauk = Izvestiya Otdeleniya estestvennykh nauk. Stalinabad.

Izvestija Permskogo biologičeskogo naučno-issledovatel'skogo instituta = Izvestiya Permskogo biologicheskogo nauchno-issledovatel'skogo instituta. Perm.

Izvestija Petrogradskogo naučnogo instituta imeni P. F. Lesgafta = Izvestiya Petrogradskogo nauchnogo instituta imeni P. F. Lesgafta. Petrograd.

Izvestija Petrogradskoj Biologičeskoj Laboratorii = Izvestiya Petrogradskoi Biologicheskoi Laboratorii. Petrograd.

Izvestija Petrogradskoj oblastnoj stancii zaščity rastenij ot vreditelej = Izvestiya Petrogradskoi oblastnoi stantsii zashchity rastenii ot vreditelei. Petrograd.

Izvestija Podotdela bor'by s vrediteljami pri Petrogradskom komitete po sel'skomu hozjajstvu = Izvestiya Podotdela bor'by s vreditelyami pri Petrogradskom komitete po sel'skomu khozyaistvu. Petrograd.

Izvestija Rossijskoj akademii nauk. Petrograd = Izvestiya Rossiiskoi akademii nauk. Petrograd.

Izvestija Rostovskoj stancii zaščity rastenij = Izvestiya Rostovskoi stantsii zashchity rastenii. Rostov.

Izvestija Russkago Geograficeskago Obščestva = Izvestiya Russkago Geograficeskago Obshchestva. Petrograd.

Izvestija Russkogo geografičeskogo obščestva = Izvestiya Russkago Geograficeskago Obshchestva. Petrograd.

Izvestija Rybinskago Otdělenija Jaroslavskago Estestvenno-Istoričeskago Obščestva = Izvestiya Rybinskago Otděleniya Yaroslavskago Estestvenno-Istoricheskago Obshchestva. Yaroslavl.

Izvestija Rybinskogo naučnogo obščestva = Izvestiya Rybinskogo nauchnogo obshchestva. Rybinsk.

Izvestija S.-Peterburgskoj Biologičeskoj Laboratorii = Izvestiya S.-Peterburgskoi Biologicheskoi Laboratorii. St. Petersburg.

Izvestija Sapropelevogo komiteta = Izvestiya Sapropelevogo komiteta. Petrograd.

Izvestija Saratovskogo obščestva estestvoispytatelej = Izvestiya Saratovskogo obshchestva estestvoispytatelei. Saratov.

Izvestija Severnoj oblastnoj stancii zaščity rastenij ot vreditelej = Izvestiya Severnoi oblastnoi stantsii zashchity rastenii ot vreditelei. Petrograd.

Izvestija Severo-Kavkazskoj kraevoj stancii zaščity rastenij = Izvestiya Severo-Kavkazskoi kraevoi stantsii zashchity rastenii. Rostov.

Izvestija Sibirskago Otděla Imperatorskago Russkago Geografičeskago Obščestva = Izvestiya Sibirskago Otděla Imperatorskago Russkago Geograficheskago Obshchestva. Irkutsk.

Izvestija Sibirskogo eńtomologičeskogo bjuro = Izvestiya Sibirskogo eńtomologicheskogo byuro. Novosibirsk.

Izvestija sibirskogo otdelenija akademii nauk S S S R. Novosibirsk = Izvestiya sibirskogo otdeleniya akademii nauk S S S R. Novosibirsk.

Izvestija Sibirskoj kraevoj stancii zaščity rastenij ot vreditelej = Izvestiya Sibirskoi kraevoi stantsii zashchity rastenii ot vreditelei. Novosibirsk.

Izvestija Sibirskoj stancii zaščity rastenij ot vreditelej = Izvestiya Sibirskoi stantsii zashchity rastenii ot vreditelei. Novosibirsk.

Izvestija Sredne-Aziatskogo geografičeskogo obščestva = Izvestiya Sredne-Aziatskogo geograficheskogo obshchestva. Tashkent.

Izvestija Sredne-Aziatskogo otdela Gosudarstvennogo russkogo geografičeskogo obščestva = Izvestiya Sredne-Aziatskogo otdela Gosudarstvennogo russkogo geograficheskogo obshchestva. Tashkent.

Izvestija Sredne-Sibirskogo gosudarstvennogo geografičeskogo obščestva = Izvestiya Sredne-Sibirskogo gosudarstvennogo geograficheskogo obshchestva. Krasnoyarsk.

Izvestija Sredne-Sibirskogo otdela Gosudarstvennogo russkogo geografičeskogo obščestva = Izvestiya Sredne-Sibirskogo otdela Gosudarstvennogo russkogo geograficheskogo obshchestva. Krasnoyarsk.

Izvestija Tihookeanskogo naučno-issledovatel'skogo instituta rybnogo hozjajstva i okeanografii = Izvestiya Tikhookeanskogo nauchno-issledovatel'skogo instituta rybnogo khozyaistva i okeanografii. Vladivostok.

Izvestija Tihookeanskogo naučnogo instituta rybnogo hozjajstva = Izvestiya Tikhookeanskogo nauchnogo instituta rybnogo khozyaistva. Voronezh.

Izvestija Tihookeanskoj naučno-promyslovoj stancii = Izvestiya Tikhookeanskoi nauchno-promyslovoi stantsii. Vladivostok.

Izvestija Tomskago Universiteta = Izvestiya Tomskogo Universiteta. Tomsk.

Izvestija Tomskogo gosudarstvennogo universiteta. = Izvestiya Tomskogo gosudarstvennogo universiteta. Tomsk.

Izvestija Tomskogo otdelenija Gosudarstvennogo russkogo botaničeskogo obščestva = Izvestiya Tomskogo otdeleniya Gosudarstvennogo russkogo botanicheskogo obshchestva. Tomsk.

Izvestija Tomskogo otdelenija Gosudarstvennogo russkogo botaničeskogo obščestva = Izvestiya Tomskogo otdeleniya Gosudarstvennogo russkogo botanicheskogo obshchestva. Tomsk.

Izvestija Tomskogo otdelenija Vsesojuznogo botaničeskogo obščestva = Izvestiya Tomskogo otdeleniya Vsesoyuznogo botanicheskogo obshchestva. Novosibirsk.

Izvestija Turkestanskago Otděla Imperatorskago Geografičeskago Obščestva = Izvestiya Turkestanskago Otděla Imperatorskago Geograficheskago Obshchestva. Tashkent.

Izvestija Turkestanskogo otdela Russkogo geografičeskogo obščestva = Izvestiya Turkestanskogo otdela Russkogo geograficheskogo obshchestva. Tashkent.

Izvestija Turkmenskogo filiala Akademii nauk S S S R = Izvestiya Turkmenskogo filiala Akademii nauk S S S R. Ashkhabad.

Izvestija i Učenija Zapiski Imperatorskago Kazanskago Universiteta = Izvestiya i Ucheniya Zapiski Imperatorskago Kazanskago Universiteta. Kazan.

Izvestija Uzbekistanskogo filiala Akademii nauk S S S R = Izvestiya Uzbekistanskogo filiala Akademii nauk S S S R. Tashkent.

Izvestija Voronežskoj stancii po bor'be s vrediteljami rastenij = Izvestiya Voronezhskoi stantsii po bor'be s vreditelyami rastenii. Voronezh.

Izvestija Vostočno-Sibirskago Otděla Imperatorskago Russkago Geografičeskago Obščestva = Izvestiya Vostochno-Sibirskago Otděla Imperatorskago Russkago Geograficheskago Obshchestva. Irkutsk.

Izvestija Vostočno-Sibirskogo otdela Geografičeskogo obščestva S S S R = Izvestiya Vostochno-Sibirskogo otdela Geograficheskogo obshchestva S S S R. Irkutsk.

Izvestija Vostočno-Sibirskogo otdela Gosudarstvennogo russkogo geografičeskogo obščestva = Izvestiya Vostochno-Sibirskogo otdela Gosudarstvennogo russkogo geograficheskogo obshchestva. Irkutsk.

Izvestija Vostočno-Sibirskogo otdela Russkogo geografičeskogo obščestva = Izvestiya Vostochno-Sibirskogo otdela Russkogo geograficheskogo obshchestva. Irkutsk.

Izvestija vostočnyh filialov Akademii nauk S S S R = Izvestiya vostochnykh filialov Akademii nauk S S S R. Novosibirsk.

Izvestija Vsesojuznogo geografičeskogo obščestva = Izvestiya Vsesoyuznogo geograficheskogo obshchestva. Moscow & Leningrad.

Izvestija Zapadno-Sibirskago Otděla Imperatorskago Russkago Geografičeskago Obščestva = Izvestiya Zapadno-Sibirskago Otděla Imperatorskago Russkago Geograficheskago Obshchestva. Omsk.

Izvestija Zapadno-Sibirskogo geografičeskogo obščestva = Izvestiya Zapadno-Sibirskogo geograficheskogo obshchestva. Omsk.

Izvestija Zapadno-Sibirskogo otdela Russkogo

geografičeskogo obščestva = Izvestiya Zapadno-Sibirskogo otdela Russkogo geograficheskogo obshchestva. Omsk.

Izvestija Zapadno-Sibirskoj kraevoj stancii zaščity rastenij = Izvestiya Zapadno-Sibirskoi kraevoi stantsii zashchity rastenii. Tomsk.

Izvestiya abkhazskoi sel'sko-khozyaistvennoi opytnoi stantsii. Sukhum. Vols. ?-42, 1926-30. Izv. Abkhaz. Sel'sko-Khoz. Optyn. Stantsii. Preceded by: Izvestiya sukhumskoi sadovoi i sel'skokhyaistvennoi opytnoi stantsii. HI 66170

Izvestiya akademii krupnogo socialisticheskogo sel'skogokhozyaistva imeni K. A. Timiryazeva = Izvestiya sel'skokhozyaistvennoi akademii imeni K.A. Timiryazeva. Moscow.

Izvestiya Akademii Nauk. Petrograd. Ser. 6, vols. 11(5)-11(11), 1917. Izv. Akad. Nauk. Preceded by: Izvestiya Imperatorskoi Akademii Nauk. Superseded by: Izvestiya Rossiiskoi akademii nauk. HI 66171

Izvestiya akademiya nauk armyanskoi S S R. Biologicheskie nauki. Erevan. Vols. 12(2)-18, 1959-65. Izv. Akad. Nauk Armyansk. S.S.R., Biol. Nauki. Preceded by: Izvestiya akademiya nauk armyanskoi S S R. Biologicheskie i sel'skokhozyaistvennie nauki. Superseded by: Biologicheskii zhurnal Armenii. HI 66172

Izvestiya akademiya nauk armyanskoi S S R. Biologicheskie i sel'skokhozyaistvennie nauki. Erevan. Vols. 1-12(1), 1948-59. Izv. Akad. Nauk Armyansk. S.S.R., Biol. Sel'skokhoz. Nauki. Preceded by: Izvestiya: Estestvennye nauki. Superseded by: Izvestiya akademiya nauk armyanskoi S S R. Biologicheskie nauki. 1-109-2. HI 66173

Izvestiya Akademii nauk Azerbaidzhanskoi S S R. Baku, Azerbaijan S S R. 1945(4)-1957. Izv. Akad. Nauk Azerbaidzhnsk. S.S.R. Preceded by: Izvestiya Azerbaidzhanskogo filiala [Akademii nauk S S S R.] Superseded by: Izvestiya Akademii nauk Azerbaidzhanskoi S S R. Seriya biologicheskikh i sel'skokhozyaistvennykh nauk. 1-109-3. HI 66174

Izvestiya Akademii nauk Azerbaidzhanskoi S S R. Seriya biologicheskikh i meditsinskikh nauk. Baku, Azerbaijan S S R. 1959+. Izv. Akad. Nauk Azerbaidzhnsk. S.S.R., Ser. Biol. Med. Nauk. Preceded by: Izvestiya Akademii nauk Azerbaidzhanskoi S S R. Seriya biologicheskikh i sel'skokhozyaistvennykh nauk. HI 66175

Izvestiya Akademii nauk Azerbaidzhanskoi S S R. Seriya biologicheskikh i sel'skokhozyaistvennykh nauk. Baku, Azerbaijan S S R. 1958-59. Izv. Akad. Nauk Azerbaidzhansk. S.S.R., Ser. Biol. Sel'skokhoz. Nauk. Preceded by: Izvestiya Akademii nauk Azerbaidzhanskoi S S R. Superseded by: Izvestiya Akademii nauk Azerbaidzhanskoi S S R. Seriya biologicheskikh i meditsinskikh nauk. HI 66176

Izvestiya Akademii nauk Belorusskoi S S R. Minsk, Belorussian S S R. 1940-55. Izv. Akad. Nauk Belorussk. S.S.R. HI 66177

Izvestiya Akademii nauk Ėstonskoi S S R. Tallinn, Estonian S S R. Vols. 1-4, 1952-55. Izv. Akad. Nauk Ėstonsk. S.S.R. Superseded by: Izvestiya Akademii nauk Ėstonskoi S S R. Seriya biologii, sel'skokhozyaistvennykh nauk i meditsinskikh nauk. HI 66178

Izvestiya Akademii nauk Ėstonskoi S S R. Seriya biologii, sel'skokhozyaistvennykh nauk i meditsinskikh nauk. Tallinn, Estonian S S R. Vol. 5+, 1956+. Izv. Akad. Nauk Ėstonsk. S.S.R., Ser. Biol. Preceded by: Izvestiya Akademii nauk Ėstonskoi S S R. HI 66179

Izvestiya akademii nauk gruzinskoi S S R. Seriya biologicheskaya. Tiflis. Vol. 1+, 1975+. Izv. Akad. Nauk Gruzinsk. S.S.R., Ser. Biol. HI 66180

Izvestiya akademii nauk kazakhskoi S S R. Alma Ata. Vols. 28-138, 1946-54. Izv. Akad. nauk Kazakh. S.S.R. Preceded by: Izvestiya Kazakhskogo filiala Akademii nauk S S S R. Superseded by: Izvestiya akademiya nauk kazakhskoi S S R. Seriya biologicheskaya and Izvestiya akademiya nauk kazakhskoi S S R. Seriya botanicheskaya. HI 66181

Izvestiya akademiya nauk kazakhskoi S S R. Seriya biologicheskaya. Alma-Ata. Nos. 1-14, 1946-57; 1966+. Izv. Akad. Nauk Kazakh. S.S.R., Ser. Biol. Preceded by: Izvestiya kazahskogo filiala akademii nauk S S S R. For 1963-65? see: Izvestiya akademiya nauk kazakhskoi S S R. Seriya biologicheskikh nauk. HI 66182

Izvestiya akademiya nauk kazakhskoi S S R. Seriya biologicheskikh nauk. Alma-Ata. 1963-65? Izv. Akad Nauk Kazakh. S.S.R., Ser. Biol. Nauk. Preceded & superseded by: Izvestiya akademiya nauk kazakhskoi S S R. Seriya biologicheskaya. HI 66183

Izvestiya Akademii nauk Kazakhskoi S S R. Seriya botanicheskaya. Alma-Ata, Kazakh S S R. Vols. 3-5, 1948-50. Izv. Akad. Nauk Kazakhsk. S.S.R., Ser. Bot. Preceded by: Izvestiya Kazakhskogo filiala Akademii nauk S S S R. Seriya botanicheskaya. Superseded by: Izvestiya Akademii nauk Kazakhskoi S S R. Seriya botaniki i pochvovedeniya. HI 66184

Izvestiya Akademii nauk Kazakhskoi S S R. Seriya botaniki i pochvovedeniya. Alma-Ata, Kazakh S S R. Vol. 1+, 1958+. Izv. Akad. Nauk Kazakhsk. S.S.R., Ser. Bot. Pochvov. Preceded by: Izvestiya Akademii nauk Kazakhskoi S S R. Seriya botanicheskaya. HI 66185

Izvestiya Akademii nauk Kirgizskoi S S R. Frunze, Kirghiz S S R. Vols. 1-6, 1954-58. Izv. Akad. Nauk

Kirgizsk. S.S.R. Preceded by: Izvestiya Kirgizskogo filiala Akademii nauk S S S R. Superseded by: Izvestiya Akademii nauk Kirgizskoi S S R. Seriya biologicheskikh nauk. HI 66186

Izvestiya Akademii nauk Kirgizskoi S S R. Seriya biologicheskikh nauk. Frunze, Kirghiz S S R. Vol. 1+, 1959+. Izv. Akad. Nauk Kirgizsk. S.S.R., Ser. Biol. Nauk. Preceded by: Izvestiya Akademii nauk Kirgizskoi S S R. HI 66187

Izvestiya Akademii nauk Latviiskoi S S R. Riga, Latvian S S R. Vol. 1+, 1947+. Izv. Akad. Nauk Latviisk. S.S.R. 3-2363-3. HI 66188

Izvestiya akademiya nauk moldavskoi S S R = Izvestiya moldavskaya filiala, akademiya nauk S S S R. Kishinev.

Izvestiya akademiya nauk moldavskoi S S R. Seriya biologicheskikh i khimicheskikh nauk. Kishinev. No. 1+, 1968+. Izv. Akad. Nauk Moldavsk. S.S.R., Ser. Biol. Khim. Nauk. Preceded by: Izvestiya moldavskaya filiala, akademiya nauk S S S R. HI 66189

Izvestiya Akademii nauk S S S R. Moscow & Leningrad. Ser. 6, vols. 19[(12)]-21, 1925-27. Izv. Akad. Nauk S.S.S.R., Ser. 6. Preceded by: Izvestiya Rossiiskoi akademii nauk. Superseded by: Izvestiya Akademii nauk S S S R. Ser. 7, Otdeleniye fiziko-matematicheskikh nauk. 1-111-1. HI 66190

Izvestiya Akademii nauk S S S R; otdeleniye matematicheskikh i estestvennykh nauk. Seriya biologicheskaya. Moscow & Leningrad. 1936-38. Izv. Akad. Nauk S.S.S.R., Otd. Mat. Nauk, Ser. Biol. Preceded by: Izvestiya Akademii nauk S S S R. Ser. 7, Otdeleniye matematicheskikh i estestvennykh nauk. Superseded by: Izvestiya Akademii nauk S S S R. Seriya biologicheskaya. 1-112-2. HI 66191

Izvestiya Akademii nauk S S S R; otdeleniye matematicheskikh i estestvennykh nauk. Seriya geologicheskaya. Moscow & Leningrad. 1936-38. Izv. Akad. Nauk S.S.S.R., Ser. 7. Preceded by: Izvestiya Akademii nauk S S S R. Ser. 7, Otdeleniye matematicheskikh i estestvennykh nauk. Superseded by: Izvestiya Akademii nauk S S S R. Seriya geologicheskaya. 1-112-2. HI 66192

Izvestiya Akademii nauk S S S R. Seriya biologicheskaya. Moscow & Leningrad. 1939-56; vol. 22+, 1957+. Izv. Akad. Nauk S.S.S.R., Ser. Biol. Preceded by: Izvestiya Akademii nauk S S S R; otdeleniye matematicheskikh i estestvennykh nauk. Seriya biologicheskaya. 1-112-2. HI 66193

Izvestiya Akademii nauk S S S R. Seriya geologicheskaya. Moscow & Leningrad. 1939+. Izv. Akad. Nauk S.S.S.R., Ser. Geol. Preceded by: Izvestiya Akademii nauk S S S R; otdeleniye matematicheskikh i estestvennykh nauk. Seriya geologicheskaya. 1-112-2. HI 66194

Izvestiya Akademii nauk S S S R. Ser. 7, Otdeleniye fiziko-matematicheskikh nauk. Moscow & Leningrad. 1928-30. Izv. Akad. Nauk S.S.S.R., Ser. 7. Preceded by: Izvestiya Akademii nauk S S S R. Superseded by: Izvestiya Akademii nauk S S S R. Ser. 7, Otdeleniye matematicheskikh i estestvennykh nauk. 1-112-1. HI 66195

Izvestiya Akademii nauk S S S R. Ser. 7, Otdeleniye matematicheskikh i estestvennykh nauk. Moscow & Leningrad. 1931-35. Izv. Akad. Nauk S.S.S.R., Ser. 7. Preceded by: Izvestiya Akademii nauk S S S R. Ser. 7, Otdeleniye fiziko-matematicheskikh nauk. Superseded by: Izvestiya Akademii nauk S S S R; otdeleniye matematicheskikh i estestvennykh nauk. Seriya biologicheskaya. 1-112-1. HI 66196

Izvestiya akademii nauk tadzhikskoi S S R. Otdelenie biologicheskikh nauk. Dushanbe. No. 1+, 1959+. Izv. Akad. Nauk Tadzhiksk. S.S.R., Otd. Biol. Nauk. Preceded by: Izvestiya tadzhikskogo filiala akademii nauk S S S R. HI 66197

Izvestiya Akademii nauk Tadzhikskoi S S R. Otdelenie estestvennykh nauk. Stalinabad [=Dushanbe], Tadzhik S S R. Vol. 1, 1952. Izv. Akad. Nauk Tadzhiksk. S.S.R., Otd. Estestv. Nauk. Superseded by: Izvestiya Otdeleniya estestvennykh nauk. HI 66198

Izvestiya akademiya nauk Turkmenskoi S S R. Ashkhabad. 1951(4)-59. Izv. Akad. Nauk Turkmensk. S.S.R. Preceded by: Izvestiya Turkmenskogo filiala Akademii nauk S S S R. Superseded by: Izvestiya akademiya nauk turkmenskoi S S R. Seriya biologicheskikh nauk. HI 66199

Izvestiya akademiya nauk turkmenskoi S S R. Seriya biologicheskikh nauk. Ashkhabad. No. 1+, 1960+. Izv. Akad. Nauk Turkmensk S.S.R., Ser. Biol. Nauk. HI 66200

Izvestiya Akademii nauk Uzbekskoi S S R. Tashkent, Uzbek S S R. 1947-56. Izv. Akad. Nauk Uzbeksk. S.S.R. Preceded by: Izvestiya Uzbekistanskogo filiala Akademii nauk S S S R. Superseded by: Izvestiya Akademii nauk Uzbekskoi S S R. Seriya biologicheskaya. HI 66201

Izvestiya Akademii nauk Uzbekskoi S S R. Seriya biologicheskaya. Tashkent, Uzbek S S R. 1957+. Izv. Akad. Nauk Uzbeksk. S.S.R., Ser. Biol. Preceded by: Izvestiya Akademii nauk Uzbekskoi S S R. HI 66202

Izvestiya. Armyanskii filial Akademii nauk S S S R. Erivan [=Erevan], Armenian S S R. 1940-42. Izv. Armyansk. Fil. Akad. Nauk S.S.S.R. Preceded by: Izvestiya Armyanskogo filiala Akademii nauk S S S R: Estestvennye nauki. HI 66203

Izvestiya Armyanskogo filiala Akademii nauk S S S R:

Estestvennye nauki. Erivan [=Erevan], Armenian S S R. Vol. 4, 1943. Izv. Armyansk. Fil. Akad. Nauk S.S.S.R., Estestv. Nauki. Preceded by: Izvestiya. Armyanskii filial Akademii nauk S S S R. Superseded by: Izvestiya: Estestvennye nauki. HI 66204

Izvestiya Assotsiatsii nauchno-issledovatel'skikh institutov pri Fiziko-Matematicheskom fakul'tete Pervogo Moskovskogo gosudarstvennogo universiteta. Moscow. Vols. 1-2, 1922-29. Izv. Assots. Nauchno-Issl. Inst. Fiz.-Mat. Fak. Perv. Moskovsk. Gosud. Univ. Superseded by: Izvestiya Assotsiatsii nauchno-issledovatel'skikh institutov pri Fiziko-Matematicheskom fakul'tete Pervogo Moskovskogo gosudarstvennogo universiteta. Ser. A. Otdel biologicheskikh i geologicheskikh nauk. 3-2767-1. HI 66205

Izvestiya Assotsiatsii nauchno-issledovatel'skikh institutov pri Fiziko-Matematicheskom fakul'tete Pervogo Moskovskogo gosudarstvennogo universiteta. Ser. A. Otdel biologicheskikh i geologicheskikh nauk. Moscow. Vol. 3, 1930. Izv. Assots. Nauchno-Issl. Inst. Fiz.-Mat. Fak. Perv. Moskovsk. Gosud. Univ., Ser. A. Preceded by: Izvestiya Assotsiatsii nauchno-issledovatel'skikh institutov pri Fiziko-Matematicheskom fakul'tete Pervogo Moskovskogo gosudarstvennogo universiteta. HI 66206

Izvestiya Azerbaidzhanskogo filiala [Akademii nauk S S S R]. Baku, Azerbaijan S S R. 1936-45(3). Izv. Azerbaidzhansk. Fil. Superseded by: Izvestiya Akademii nauk Azerbaidzhanskoi S S R. 1-109-3. HI 66207

Izvestiya azerbaidzhanskogo gosudarstvennogo universiteta imeni V. I. Lenina. Otdel., estestvoznanie i meditsina. Baku. 1922-30. Izv. Azerbaidzhansk. Gosud. Univ. V. I. Lenina, Otd., Estestv. Med. Preceded by: Izvestiya bakinskogo universiteta. HI 66208

Izvestiya bakinskogo universiteta. Otdel., estestvoznanie i meditsina. Baku. 1921-26. Izv. Bakinsk. Univ., Otd. Estestv. Med. Superseded by: Izvestiya azerbaidzhanskogo gosudarstvennogo universiteta imeni V. I. Lenina. Otdel., estestvoznanie i meditsina. HI 66209

Izvestiya Batumskogo botanicheskogo sada. Batum, Georgian S S R. Vol. 6+, 1956+. Izv. Batumsk. Bot. Sada. Preceded by: Izvestiya Batumskogo subtropicheskogo botanicheskogo sada. HI 66210

Izvestiya Batumskogo subtropicheskogo botanicheskogo sada. Batum, Georgian S S R. Vols. 1-5, 1936-40. Izv. Batumsk. Subtrop. Bot. Sada. Superseded by: Izvestiya Batumskogo botanicheskogo sada. 1-609-2. HI 66211

Izvestiya na biologicheskiya institut. Bulgarska akademiya na naukite, otdelenie za biologicheskiye i meditsinski nauk. Sofia, Bulgaria. Vols. 1-5, 1950-54. Izv. Biol. Inst. Superseded by: Izvestiya na instituta po biologi “Metodii Popov”. HI 66212

Izvestiya Biologicheskogo nauchno-issledovatel'skogo instituta i Biologicheskoi stantsii pri Molotovskom gosudarstvennom universitete = Izvestiya Biologicheskogo nauchno-issledovatel'skogo instituta pri Molotovskom gosudarstvennom universitete imeni A. M. Gor'kogo. Molotov.

Izvestiya Biologicheskogo nauchno-issledovatel'skogo instituta i Biologicheskoi stantsii pri Permskom gosudarstvennom universitete. Perm. Russian S F S R. Vols. 1-7(6), 1922-31. Izv. Biol. Nauchno-Issl. Inst. Biol. Stantsii Permsk. Gosud. Univ. Superseded by: Izvestiya Permskogo biologicheskogo nauchno-issledovatel'skogo instituta. 4-3310-2. HI 66213

Izvestiya Biologicheskogo nauchno-issledovatel'skogo instituta pri Molotovskom gosudarstvennom universitete imeni A. M. Gor'kogo. Molotov [=Perm], Russian S F S R. Vols. 12(1)-12(2), 1941. Izv. Biol. Nauchno-Issl. Inst. Molotovsk. Gosud. Univ. Gor'kogo. Preceded by: Izvestiya Biologicheskogo nauchno-issledovatel'skogo instituta pri Permskom gosudarstvennom universitete imeni A. M. Gor'kogo. Superseded by: Izvestiya Estestvenno-nauchnogo instituta pri Molotovskom universitete imeni A. A. Gor'kogo. 4-3310-2. HI 66214

Izvestiya Biologicheskogo nauchno-issledovatel'skogo instituta pri Permskom gosudarstvennom universitete. Perm, Russian S F S R. Vols. 9-10(7), 1934-36. Izv. Biol. Nauchno-Issl. Inst. Permsk. Gosud. Univ. Preceded by: Izvestiya Permskogo biologicheskogo nauchno-issledovatel'skogo instituta. Superseded by: Izvestiya Biologicheskogo nauchno-issledovatel'skogo instituta pri Permskom gosudarstvennom universitete imeni A. M. Gor'kogo. 4-3310-2. HI 66215

Izvestiya Biologicheskogo nauchno-issledovatel'skogo instituta pri Permskom gosudarstvennom universitete imeni A. M. Gor'kogo. Perm, Russian S F S R. Vols. 10(8)-11, 1936-39. Izv. Biol. Nauchno-Issl. Inst. Permsk. Gosud. Univ. Gor'kogo. Preceded by: Izvestiya Biologicheskogo nauchno-issledovatel'skogo instituta pri Permskom gosudarstvennom universitete. Superseded by: Izvestiya Biologicheskogo nauchno-issledovatel'skogo instituta pri Molotovskom gosudarstvennom universitete imeni A. M. Gor'kogo. 4-3310-2. HI 66216

Izvestiya Biologo-geograficheskogo nauchno-issledovatel'skogo instituta pri Gosudarstvennom Irkutskom universitete. Irkutsk, Russian S F S R. Vols. 1-6(1/2), 1924-34. Izv. Biol.-Geogr. Nauchno-Issl. Inst. Gosud. Irkutsk. Univ. Superseded by: Izvestiya Biologo-geograficheskogo nauchno-issledovatel'skogo instituta pri Vostochno-Sibirskom gosudarstvennom universitete. 3-2116-3. HI 66217

Izvestiya Biologo-geograficheskogo nauchno-issledovatel'skogo instituta pri Irkutskom gosudarstvennom universitete imeni A. A. Zdanova. Irkutsk, Russian S F S R. Vol. 10+, 1948+. Izv. Biol.-Geogr. Nauchno-Issl. Inst. Irkutsk. Gosud. Univ. Zhdanova. Preceded by: Izvestiya Biologo-geograficheskogo nauchno-issledovatel'skogo instituta pri Vostochno-Sibirskom gosudarstvennom universitete imeni tovarishcha Zdanova. 3-2116-3. HI 66218

Izvestiya Biologo-geograficheskogo nauchno-issledovatel'skogo instituta pri Vostochno-Sibirskom gosudarstvennom universitete. Moscow. Vols. 6(3/4)-8, 1935-39. Izv. Biol.-Geogr. Nauchno-Issl. Inst. Vost.-Sibirsk. Gosud. Univ. Preceded by: Izvestiya Biologo-geograficheskogo nauchno-issledovatel'skogo instituta pri Gosudarstvennom Irkutskom universitete. Superseded by: Izvestiya Biologo-geograficheskogo nauchno-issledovatel'skogo instituta pri Vostochno-Sibirskom gosudarstvennom universitete imeni tovarishcha Zdanova. 3-2116-3. HI 66219

Izvestiya Biologo-geograficheskogo nauchno-issledovatel'skogo instituta pri Vostochno-Sibirskom gosudarstvennom universitete imeni tovarishcha Zdanova. Irkutsk, Russian S F S R. Vol. 9, 1942. Izv. Biol.-Geogr. Nauchno-Issl. Inst. Vost.-Sibirsk. Gosud. Univ. Zhdanova. Preceded by: Izvestiya Biologo-geograficheskogo nauchno-issledovatel'skogo instituta pri Vostochno-Sibirskom gosudarstvennom universitete. Superseded by: Izvestiya Biologo-geograficheskogo nauchno-issledovatel'skogo instituta pri Irkutskom gosudarstvennom universitete imeni A. A. Zdanova. 3-2116-3. HI 66220

Izvestiya na botanicheskaya institut. Sofia. Vols. 1-25, 1950-74. Izv. Bot. Inst. (Sofia). Preceded by: Izvestiya na bulgarskoto botanichesko druzhestvo. Superseded by: Fitologiya. Sofia. HI 66221

Izvestiya botanicheskogo sada, akademii nauk kirgizskoi S S R. Frunze. 1964+. Izv. Bot. Sada Akad. Nauk Kirgizsk. S.S.R. HI 66222

Izvestiya Botanicheskogo sada Akademii nauk S S S R. Leningrad. Vol. 30, 1932. Izv. Bot. Sada Akad. Nauk S.S.S.R. Preceded by: Izvestiya Glavnogo botanicheskogo sada S S S R. Superseded by: Sovetskaya botanika. 3-2389-2. HI 66223

Izvestiya Botanicheskogo sada Petra Velikogo. Petrograd. Vol. 17, 1917. Izv. Bot. Sada Petra Velikago. Preceded by: Izvestiya Imperatorskogo Botanicheskogo Sada Petra Velikago. Superseded by: Izvestiya Glavnogo botanicheskogo sada R S F S R. 3-2389-2. HI 66224

Izvestiya na bulgarskoto botanichesko druzhestvo. Sofia, Bulgaria. Vols. 1-9, 1926-43. Izv. Bulg. Bot. Druzh. Superseded by: Izvestiya na botanicheskaya institut. Sofia. HI 66225

Izvestiya donskogo instituta sel'skogo khozyaistva i melioratsii. Novocherkassk. Vol. ?-5 suppl.-?, ?-1925-? Izv. Donsk. Inst. Sel'sk. Khoz. Melior. HI 66226

Izvestiya Estestvenno-nauchnogo instituta imeni P. F. Lesgafta. Leningrad. Vol. 24+, 1951+. Izv. Estestv.-Nauchn. Inst. Lesgafta. Preceded by: Izvestiya Nauchnogo instituta imeni P. F. Lesgafta. HI 66347

Izvestiya Estestvenno-nauchnogo instituta pri Molotovskom universitete imeni A. A. Gor'kogo. Molotov [=Perm], Russian S F S R. Vols. 12(3)-13, 1941-57. Izv. Estestv.-Nauchn. Inst. Molotovsk. Univ. Gor'kogo. Preceded by: Izvestiya Biologicheskogo nauchno-issledovatel'skogo instituta pri Molotovskom gosudarstvennom universitete imeni A. M. Gor'kogo. Superseded by: Izvestiya Estestvenno-nauchnogo instituta pri Permskom gosudarstvennom universitete imeni A. M. Gor'kogo. 4-3310-2. HI 66348

Izvestiya Estestvenno-nauchnogo instituta pri Permskom gosudarstvennom universitete imeni A. M. Gor'kogo. Perm, Russian S F S R. Vol. 14+, 1957+. Izv. Estestv.-Nauchn. Inst. Permsk. Gosud. Univ. Gor'kogo. Preceded by: Izvestiya Estestvenno-nauchnogo instituta pri Molotovskom universitete imeni A. A. Gor'kogo. HI 66349

Izvestiya: Estestvennye nauki. Erivan [=Erevan], Armenian S S R. Vols. 5-8, 1944-47. Izv. Estestv. Nauki. Preceded by: Izvestiya Armyanskogo filiala Akademii nauk S S S R: Estestvennye nauki. Superseded by: Izvestiya akademiya nauk armyanskoi S S R. Biologicheskie i sel'skokhozyaistvennie nauki. HI 66350

Izvestiya Fakulteta sadovodstvennykh i vinogradstvennykh nauk Vengerskogo universiteta agrarnykh nauk. Budapest = Agrártudományi egyetem kert- és szölögazdaságtudományi karának közleményei. Budapest. Agrártud. Egyet. Kert- Szölögazdaságtud. Karának Közlem. See B–P–H 55/13.

Izvestiya na geologicheskiya institut. Sofia. Vols. 1-7, 1951-59. Izv. Geolog. Inst. Superseded by: Izvestiya na geologicheskiya institut "Strashimir Dimitrov", bulgarska akademiya na naukite. HI 66227

Izvestiya na geologicheskiya institut. Seriya paleontologiya. Sofia. Vols. 16-23, 1967-74. Izv. Geolog. Inst., Ser. Paleontol. Preceded by: Izvestiya na geologicheskiya institut "Strashimir Dimitrov", bulgarska akademiya na naukite. Superseded by: Paleontologiya, stratigrafiya, i litologiya. HI 66228

Izvestiya na geologicheskiya institut "Strashimir Dimitrov", bulgarska akademiya na naukite. Sofia. Vols. 8-15, 1960-66. Izv. Geol. Inst. Strashimir Dimitrov Bulg. Akad. Nauk. Preceded by: Izvestiya na geologicheskiya institut. Superseded by: Izvestiya na geologicheskiya institut. Seriya paleontologiya. HI 66229

Izvestiya Glavnogo botanicheskogo sada R S F S R. Petrograd. Vols. 18-24, 1918-25. Izv. Glavn. Bot. Sada R.S.F.S.R. Preceded by: Izvestiya Botanicheskogo Sada Petra Velikogo. Superseded by: Izvestiya Glavnogo botanicheskogo sada S S S R. 3-2389-2. HI 66230

Izvestiya Glavnogo botanicheskogo sada S S S R. Leningrad. Vols. 25-29, 1926-30. Izv. Glavn. Bot. Sada S.S.S.R. Preceded by: Izvestiya Glavnogo botanicheskogo sada R S F S R. Superseded by: Izvestiya Botanicheskogo sada Akademii nauk S S S R. 3-2389-2. HI 66231

Izvestiya gornogo otdela. Moscow. Vol. 94(1-4), 1918. Izv. Gornogo Otd. Preceded & superseded by: Gornyi zhurnal. 3-2136-2. HI 66232

Izvestiya Gosudarstvennogo geograficheskogo obshchestva. Moscow & Leningrad. Vols. 63-71, 1931-39. Izv. Gosud. Geogr. Obshch. Preceded by: Izvestiya Gosudarstvennogo russkogo geograficheskogo obshchestva. Superseded by: Izvestiya Vsesoyuznogo geograficheskogo obshchestva. 2-1692-1. HI 66233

Izvestiya gosudarstvennogo politekhnicheskogo instituta. Tiflis. Vols. 1-3, 1924-28. Izv. Gosud. Politekhn. Inst. Superseded by: Izvestiya politekhnicheskogo instituta gruzii imeni V. I. Lenina. 5-4217-3. HI 66234

Izvestiya Gosudarstvennogo russkogo geograficheskogo obshchestva. Moscow & Leningrad. Vols. 57-62, 1925-30. Izv. Gosud. Russk. Geogr. Obshch. Preceded by: Izvestiya Russkago Geograficeskago Obshchestva. 2-1692-1. Superseded by: Izvestiya Gosudarstvennogo geograficheskogo obshchestva. HI 66235

Izvestiya Gosudarstvennogo Tomskogo universiteta. Tomsk, Russian S F S R. Vols. 71-72, 1921-23. Izv. Gosud. Tomsk. Univ. Preceded by: Izvestiya Tomskogo Universiteta. Superseded by: Izvestiya Tomskogo gosudarstvennogo universiteta. 5-4334-3. HI 66236

Izvestiya gruzinskoi opytnoi stantsii zashchity rastenii. Seriya A, fitopatologia. Tiflis. Vol. 1, 1938. Izv. Gruzinsk. Opytyn. Stantsii Rast., A. HI 66237

Izvestiya imperatorskogo botanicheskogo sada Petra Velikago. St. Petersburg. Vols. 13-16, 1913-16. Izv. Imp. Bot. Sada Petra Velikago. Preceded by: Izvestiya imperatorskogo S.-Peterburgskogo botanicheskogo sada. Superseded by: Izvestiya botanicheskogo sada Petra Velikogo. 3-2389-2. HI 66238

Izvestiya Imperatorskago Kazanskago Universiteta. Kazan, Russian S F S R. Vol. 20, 1884. Izv. Imp. Kazansk. Univ. Preceded by: Izvestiya i Ucheniya Zapiski Imperatorskago Kazanskago Universiteta. 3-2276-1. HI 66239

Izvestiya imperatorskago lesnogo instituta. St. Petersburg, Petrograd. Vols. 10-30, 1903-16. Izv. Imp. Lesn. Inst. Preceded by: Izvestiya Sankt-Peterburgskago lesnogo instituta. Superseded by: Izvestiya, petrogradskago lesnogo instituta. HI 66240

Izvestiya imperatorskogo Obshchestva lyubitelei estestvoznaniya, antropologii i etnografii, sostoyashchago pri (Imperatorskom) Moskovskom Universitetě. Moscow. Vols. 3-130, 1868-1916. Izv. Imp. Obshch. Lyubit. Estestv. Moskovsk. Univ. Preceded by: Izvestiya Obshchestva Lyubitelei Estestvoznaniya. 4-3131-1. HI 66241

Izvestiya imperatorskogo Russkogo geograficheskogo obshchestva. St. Petersburg. Vols. 1-52, 1865-1916. Izv. Imp. Russk. Geogr. Obshch. Preceded by: Zapiski imperatorskago Russkago geograficheskago obshchestva. Superseded by: Izvestiya Russkago geograficeskago obshchestva. 2-1692-1. HI 66242

Izvestiya imperatorskogo S.-Peterburgskogo botanicheskogo sada. St. Petersburg. Vols. 1-12, 1901-12. Izv. Imp. S.-Peterburgsk. Bot. Sada. Superseded by: Izvestiya imperatorskogo botanicheskogo sada Petra Velikago. HI 66243

Izvestiya Imperatorskago Tomskago universiteta. Tomsk, Russian S F S R. Vols. 1-65, 1889-1916. Izv. Imp. Tomsk. Univ. Superseded by: Izvestiya Tomskogo Universiteta. 5-4234-3. HI 66244

Izvestiya imperatorskoi Akademii nauk. St. Petersburg. Ser. 5, vols. 1-15, 1894-1906/07; ser. 6, vols. 1-11(4), 1907-17. Izv. Imp. Akad. Nauk. Preceded by: Bulletin de l'académie impériale des sciences de Saint Pétersbourg. Superseded by: Izvestiya Akademii nauk. 1-111-1. HI 66245

Izvestiya na instituta po biologi "Metodii Popov". Sofia. Vols. 5-13, 1955-63. Izv. Inst. Biol. Metodii Popov. Preceded by: Izvestiya na biologicheskiya institut. Bulgarska akademiya na naukite, otdelenie za biologicheskiye i meditsinski nauk. Superseded by: Izvestiya na institut po fiziologiya na rasteniyata "Metodii Popov". HI 66246

Izvestiya na institut po fiziologiya na rasteniyata "Metodii Popov". Sofia. Vols. 14-18, 1964-73. Izv. Inst. Fiziolog. Rast. Metodii Popov. Preceded by: Izvestiya na instituta po biologi "Metodii Popov." Superseded by: Fiziologiya na rasteniyata. HI 66247

Izvestiya, institut za gorata. Sofia. Vols. ?-3-14, ?-1958-63. Izv. Inst. Gorata. HI 66248

Izvestiya na institut po pochvoznanie i agrotekhnika "Nikola Pushkarov". Sofia. Vol. 6+, 1964+. Izv. Inst. Pochvozn. Nikola Pushkarov. Preceded by: Izvestiya na tsentralniya nauchno-izsledovatelski institut po pochvoznanie i agrotekhnika "Nikola Pushkarov". HI 66249

Izvestiya Instituta pochvovedeniya i geobotaniki Sredne-Aziatskogo gosudarstvennogo universiteta. Tashkent, Uzbek S S R. Vols. 1-4, 1925-28. Izv. Inst. Pochvov. Geobot. Sredne-Aziatsk. Gosud. Univ. 5-4156-3. HI 66250

Izvestiya na instituta po rastenievudstvo. Sofia, Bulgaria. Vols. 1-8, 1953-59. Izv. Inst. Rasteniev. Superseded by: Izvestiya na tsentralniya nauchno-izsledovatelski institut po rastenievudstvo. HI 66251

Izvestiya Irkutskoi stantsii zashchity rastenii ot vreditelei. Irkutsk, Russian S F S R. Vols. 1-2, 1927-30. Izv. Irkutsk. Stantsii Zashch. Rast. Vredit. Superseded by: Trudy po zashchite rastenii Vostochnoi Sibiri. HI 66252

Izvestiya Karelo-Finskogo filiala Akademii nauk S S S R. Petrozavodsk, Russian S F S R. [1948](2)-1951. Izv. Karelo-Finsk. Fil. Akad. Nauk S.S.R. Preceded & superseded by: Izvestiya Karelo-Finskoi nauchno-issledovatel'skoi bazy Akademii nauk S S S R. HI 66253

Izvestiya Karelo-Finskoi nauchno-issledovatel'skoi bazy Akademii nauk S S S R. Petrozavodsk, Russian S F S R. 1948(1); 1957(1)-1957(2) Izv. Karelo-Finsk. Nauchno-Issl. Bazy Akad. Nauk S.S.S.R. For 1948(2)-1951 see: Izvestiya Karelo-Finskogo filiala Akademii nauk S S S R. Superseded by: Izvestiya Karel'skogo i Kol'skogo filialov Akademii nauk S S R. HI 66254

Izvestiya Karel'skogo i Kol'skogo filialov Akademii nauk S S R. Petrozavodsk, Russian S F S R. 1957-59. Izv. Karel'sk. Fil. Akad. Nauk S.S.R. Preceded by: Izvestiya Karelo-Finskoi nauchno-issledovatel'skoi bazy Akademii nauk S S S R. HI 66255

Izvestiya kavkazskago muzeya. Tiflis. Vols. 1-12, 1897-1919. Izv. Kavkazsk. Muz. Superseded by: Byulleten' gosudarstvennogo muzeya Gruzii. 5-4217-3. HI 66256

Izvestiya Kavkazskago Otděla Imperatorskago Russkago Geograficheskago Obshchestva. Tiflis, Georgian S S R. 1-25, 1872-1917. Izv. Kavkazsk. Otd. Imp. Russk. Geogr. Obshch. 2-1692-2. HI 66257

Izvestiya Kazakhskogo filiala Akademii nauk S S S R. Alma-Ata, Kazakh S S R. 1944-45; vols. 1-27, 1946. Izv. Kazakhsk. Fil. Akad. Nauk S.S.S.R. Superseded by: Izvestiya akademii nauk kazakhskoi S S R. HI 66258

Izvestiya Kazakhskogo filiala Akademii nauk S S S R. Seriya botanicheskaya. Alma-Ata, Kazakh S S R. Vol. 2 1945. Izv. Kazakhsk. Fil. Akad. Nauk S.S.S.R., Ser. Bot. Superseded by: Izvestiya Akademii nauk Kazakhskoi S S R. Seriya botanicheskaya. HI 66259

Izvestiya Kievskogo botanicheskogo sada. Kiev, Ukrainian S S R. Vols. 1-17, 1924-34. Izv. Kievsk. Bot. Sada. Superseded by: Izvestiya Kievskogo botanicheskogo sada imeni akad. A. V. Fomina. 3-2289-3. HI 66260

Izvestiya Kievskogo botanicheskogo sada imeni akad. A. V. Fomina. Kiev, Ukrainian S S R. Vol. 18, 1949. Izv. Kievsk. Bot. Sada Fomina. Preceded by: Izvestiya Kievskogo botanicheskogo sada. Superseded by: Trudy botanichnogo sadu imeny akad. O. V. Fomina. HI 66261

Izvestiya Kirgizskogo filiala Akademii nauk S S S R. Frunze, Kirghiz S S R. Vols. 1-11, 1945-54. Izv. Kirgizsk. Fil. Akad. Nauk S.S.S.R. Superseded by: Izvestiya Akademii nauk Kirgizskoi S S R. HI 66262

Izvestiya Komiteta Akklimatizatsii Imperatorskago Moskovskago Obshchestva Sel'skago Khozyaistva. Pribavlenie k Zapiskam Komiteta Akklimatizatsiya. Moscow. Pts. 1-12, 1859. Izv. Komiteta Akklim. Imp. Moskovsk. Obshch. Sel'sk. Khoz. HI 66263

Izvestiya Krasnoyarskogo otdela Russkogo geograficheskogo obshchestva. Krasnoyarsk, Russian S F S R. Vol. 3(1-2), 1923-24. Izv. Krasnoyarsk. Otd. Russk. Geogr. Obshch. Preceded by: Izvestiya Krasnoyarskago Podotděla Vostochno-Sibirskago Otděla Imperatorskago Russkago Geograficheskago Obshchestva. Superseded by: Izvestiya Sredne-Sibirskogo otdela Gosudarstvennogo russkogo geograficheskogo obshchestva. 2-1693-1. HI 66264

Izvestiya Krasnoyarskago Podotděla Vostochno-Sibirskago Otděla Imperatorskago Russkago Geograficheskago Obshchestva. Krasnoyarsk, Russian S F S R. Vols. 1(3)-2, 1902-14. Izv. Krasnoyarsk. Podotd. Vost.-Sibirsk. Otd. Imp. Russk. Geogr. Obshch. Preceded by: Protokoly Zasědanii Rasporyaditel'nago Komiteta Krasnoyarskago Podotděla. Superseded by: Izvestiya Krasnoyarskogo otdela Russkogo geograficheskogo obshchestva. HI 66265

Izvestiya Leningradskago lesnogo instituta. Leningrad. Vols. 32-37, 1925-29. Izv. Leningradsk. Lesn. Inst. Preceded by: Izvestiya Petrogradskago lesnogo instituta. Superseded by: Izvestiya lesotekhnicheskoi akademii. HI 66268

Izvestiya Leningradskogo nauchnogo instituta imeni P. F. Lesgafta. Moscow & Leningrad. Vols. 8-10, 1924. Izv. Leningradsk. Nauchn. Inst. Lesgafta. Preceded by: Izvestiya Petrogradskogo nauchnogo instituta imeni P. F. Lesgafta. Superseded by: Izvestiya Nauchnogo instituta imeni P. F. Lesgafta. 3-2391-3. HI 66269

Izvestiya Leningradskoi oblastnoi stantsii zashchity rastenii ot vreditelei. Leningrad. Vols. 6-7, 1929-36. Izv. Leningradsk. Obl. Stantsii Zashch. Rast. Vredit. Preceded by: Izvestiya Severnoi oblastnoi stantsii zashchity rastenii ot vreditelei. HI 66270

Izvestiya lesotekhnicheskoi akademii. Moscow, Leningrad. Vol. 38, 1931; vols. 42-44, 1934-35. Izv. Lesotekhn. Akad. Preceded by: Izvestiya Leningradskogo lesnogo instituta. For vols. 39-41, 1932-33 see: Trudy lesotekhnicheskoi akademii. Superseded by: Trudy lesotekhnicheskoi akademii imeni S. M. Kirova. HI 66271

Izvestiya na mikrobiologicheskiya institut. Sofia. Vols. 1-25, 1950-74. Izv. Mikrobiol. Inst. Superseded by: Acta microbiologica, virologica et immunologica. HI 66272

Izvestiya moldavskaya filiala, akademiya nauk S S S R. Kishinev. 1951-67. Izv. Moldavsk. Fil. Akad. Nauk S.S.S.R. Superseded by: Izvestiya akademiya nauk moldavskoi S S R. Seriya biologicheskikh i khimicheskikh nauk. HI 66273

Izvestiya moskovskogo sel'skokhozyaistvennago instituta. Moscow. Vols. 1-22, 1895-1916 [1916/17]. Izv. Moskovsk. Sel'skokhoz. Inst. Preceded by: Zhurnaly zasedanii soveta petrovskoi sel'skokhozyaistvennoi akademii. Superseded by: Izvestiya petrovskoi sel'skokhozyaistvennoi akademii. 3-2766-1. HI 66274

Izvestiya Nauchnogo instituta imeni P. F. Lesgafta. Leningrad. Vols. 11-23, 1925-40. Izv. Nauchn. Inst. Lesgafta. Preceded by: Izvestiya Leningradskogo nauchnogo instituta imeni P. F. Lesgafta. Superseded by: Izvestiya Estestvenno-nauchnogo instituta imeni P. F. Lesgafta. 3-2391-3. HI 63160

Izvestiya obshchestva izucheniya olonetskoi gubernii. Petrozavodsk. 1913-16. Izv. Obshch. Izuch. Olonetsk. Gub. HI 66275

Izvestiya obshchestva dlya izsledovaniya prirody orlovskoi gubernii. Kiev. Vols. 1-3, 1907-13. Izv. Obshch. Dlya Izsl. Prir. Orlovsk. Gub. HI 66276

Izvestiya Obshchestva izucheniya Vostochno-Sibirskogo kraya. Irkutsk, Russian S F S R. Vol. 56 [also numbered vol. 1], 1936. Izv. Obshch. Izuch. Vost.-Sibirsk. Kraya. Preceded by: Izvestiya Vostochno-Sibirskogo otdela Gosudarstvennogo russkogo geograficheskogo obshchestva. Superseded by: Izvestiya Obshchestva izucheniya Vostochno-Sibirskoi oblasti. 3-2116-3; 2-1693-1. HI 66277

Izvestiya Obshchestva izucheniya Vostochno-Sibirskoi oblasti. Irkutsk, Russian S F S R. Vol. 57 [also numbered vol. 2], 1937. Izv. Obshch. Izuch. Vost.-Sibirsk. Obl. Preceded by: Izvestiya Obshchestva izucheniya Vostochno-Sibirskogo kraya. Superseded by: Izvestiya Vostochno-Sibirskogo otdela Geograficheskogo obshchestva S S S R. 3-2116-3; 2-1693-1. HI 66278

Izvestiya Obshchestva lyubitelei estestvoznaniya. Moscow. Vols. 1-2, 1866-67. Izv. Obshch Lyubit. Estestv. Superseded by: Izvestiya imperatorskogo Obshchestva lyubitelei estestvoznaniya, antropologii i etnografii, sostoyashchago pri (Imperatorskom) Moskovskom Universitetě. 4-3131-1. HI 66279

Izvestiya Orenburgskago Otděla Imperatorskago Russkago Geograficheskago Obshchestva. Orenburg [=Chkalov], Russian S F S R. Vols. 1-25, 1893-1916. Izv. Orenburgsk. Otd. Imp. Russk. Geogr. Obshch. Superseded by: Trudy Orenburgskogo otdela Gosudarstvennogo geograficheskogo obshchestva. 2-1692-2. HI 66280

Izvestiya Otdeleniya estestvennykh nauk. Stalinabad [=Dushambe], Tadzhik S S R. Vol. 1+, 1952+. Izv. Otd. Estestv. Nauk. Preceded by: Izvestiya Akademii nauk Tadzhikskoi S S R. Otdelenie estestvennykh nauk. HI 66281

Izvestiya Otdela zashchity rastenii. Tiflis, Georgian S S R. Vol. 1, 1930. Izv. Otd. Zashch. Rast. Superseded by: Trudy instituta zashchity rastenii, akademiya nauk Gruzinskoi S S R. HI 66282

Izvestiya Permskogo biologicheskogo nauchno-issledovatel'skogo instituta. Perm, Russian S F S R. Vols. 7(7/8)-8, 1931. Izv. Permsk. Biol. Nauchno-Issl. Inst. Preceded by: Izvestiya Biologicheskogo nauchno-issledovatel'skogo instituta i Biologicheskoi stantsii pri Permskom gosudarstvennom universitete. Superseded by: Izvestiya Biologicheskogo nauchno-issledovatel'skogo instituta pri Permskom gosudarstvennom universitete. HI 66283

Izvestiya Petrogradskago lesnogo instituta. Petrograd. Vol. 31, 1917. Izv. Petrogradsk. Lesn. Inst. Preceded by: Izvestiya imperatorskago lesnogo instituta. Superseded by: Izvestiya Leningradskogo lesnogo instituta. HI 66284

Izvestiya Petrogradskoi Biologicheskoi Laboratorii. Petrograd. Vols. 15-16, 1915-17. Izv. Petrogradsk. Biol. Lab. Preceded by: Izvestiya Sankt-Peterburgskoi Biologicheskoi Laboratorii. Superseded by: Izvestiya Petrogradskogo nauchnogo instituta imeni P. F. Lesgafta. 3-2391-3. HI 66285

Izvestiya Petrogradskogo nauchnogo instituta imeni P. F. Lesgafta. Petrograd. Vols. 1-7, 1919-23. Izv. Petrogradsk. Nauchn. Inst. Lesgafta. Preceded by: Izvestiya Petrogradskoi Biologicheskoi Laboratorii. Superseded by: Izvestiya Leningradskogo nauchnogo instituta imeni P. F. Lesgafta. 3-2391-3. HI 66286

Izvestiya Petrogradskoi oblastnoi stantsii zashchity rastenii ot vreditelei. Petrograd. Vols. 2-3, 1921. Izv. Petrogradsk. Obl. Stantsii Zashch. Rast. Vredit. Preceded by: Izvestiya Podotdela bor'by s vreditelyami pri Petrogradskom komitete po sel'skomu khozyaistvu. Superseded by: Izvestiya Severnoi oblastnoi stantsii zashchity rastenii ot vreditelei. HI 66287

Isvestiya petrovskoi sel'skokhozyaistvennoi akademii. Moscow. Vols. 12(3)-16, 1889-93. Izv. Petrovsk. Sel'skokhoz. Akad. (1889-93). Preceded by: Izvestiya petrovskoi zemledel'cheskoi i lesnoi akademii. Superseded by: Zhurnaly zasedanii soveta petrovskoi sel'skokhozyaistvennoi akademii. HI 66288

Izvestiya petrovskoi sel'skokhozyaistvennoi akademii. Moscow. 1917-20, [1919]-[22]. Izv. Petrovsk. Sel'skokhoz. Akad. (1917-20). Preceded by: Izvestiya petrovskoi zemledel'cheskoi i lesnoi akademii. Superseded by: Izvestiya sel'skokhozyaistvennoi akademii imeni K. A. Timiryazeva. 3-2766-1. HI 66289

Izvestiya petrovskoi zemledel'cheskoi i lesnoi akademii. Moscow. Vols. 1-12(2), 1878-89; 1917. Izv. Petrovsk. Zemled. Akad. Preceded by: Zhurnaly zasedanii soveta petrovskoi zemledel'cheskoi i lesnoi akademii. For vols 12(3)-16, 1889-93 see: Isvestiya petrovskoi sel'skokhozyaistvennoi akademii (1889-93). For 1894 see: Zhurnaly zasedanii soveta petrovskoi sel'skokhozyaistvennoi akademii. For 1895-1916 see: Izvestiya moskovskogo sel'skokhozyaistvennago instituta. Superseded by: Izvestiya petrovskoi sel'skokhozyaistvennoi akademii (1917-20). HI 66290

Izvestiya Podotdela bor'by s vreditelyami pri Petrogradskom komitete po sel'skomu khozyaistvu. Petrograd. Vol. 1, 1919. Izv. Podotd. Bor'by Vredit. Petrogradsk. Komitete Sel'sk. Khoz. Superseded by: Izvestiya Petrogradskoi oblastnoi stantsii zashchity rastenii ot vreditelei. HI 66291

Izvestiya politekhnicheskogo instituta gruzii imeni V. I. Lenina. Tiflis. Ser. 2, vols. 1-3, 1929. Izv. Politekhn. Inst. Gruzii Im. V. I. Lenina. Preceded by: Izvestiya gosudarstvennogo politekhnicheskogo instituta. Tiflis. 5-4217-3. HI 66292

Izvestiya Rossiiskoi akademii nauk. Petrograd, Leningrad. Ser. 6, vols. 11(12)-19(9/11), 1917-25. Izv. Rossiisk. Akad. Nauk. Preceded by: Izvestiya Akademii Nauk. Superseded by: Izvestiya Akademii nauk S S S R. 1-111-1. HI 66293

Izvestiya Rostovskoi stantsii zashchity rastenii. Rostov, Russian S F S R. Vol. 9, 1938. Izv. Rostovsk. Stantsii Zashch. Rast. Preceded by: Trudy Severo-Kavkazskogo instituta zashchity rastenii. 4-3711-2. HI 66294

Izvestiya Russkago Geograficeskago Obshchestva. Petrograd. Vols. 53-56, 1917-24. Izv. Russk. Geogr. Obshch. Preceded by: Izvestiya Imperatorskogo Russkogo Geograficheskogo Obshchestva. Superseded by: Izvestiya Gosudarstvennogo russkogo geograficheskogo obshchestva. 2-1692-1. HI 66295

Izvestiya Rybinskogo nauchnogo obshchestva. Rybinsk [=Shcherbakov], Russian S F S R. Vols. 3-4, 1923-27. Izv. Rybinsk. Nauchn. Obshch. Preceded by: Izvestiya Rybinskago Otdĕleniya Yaroslavskago Estestvenno-Istoricheskago Obshchestva. HI 66296

Izvestiya Rybinskago Otdĕleniya Yaroslavskago Estestvenno-Istoricheskago Obshchestva. Yaroslavl, Russian S F S R. Vols. 1-2, 1915-18. Izv. Rybinsk. Otd. Yaroslavsk. Estestv.-Istorich. Obshch. Superseded by: Izvestiya Rybinskogo nauchnogo obshchestva. HI 66297

Izvestiya Sankt-Peterburgskago lesnogo instituta. St. Petersburg. Vols. 1-9, 1898-1903. Izv. Sank-Peterburgsk. Lesn. Inst. Superseded by: Izvestiya imperatorskago lesnogo instituta. HI 66298

Izvestiya Sankt-Peterburgskoi Biologicheskoi Laboratorii. St. Petersburg. Vols. 1-14, 1896-1914. Izv. S.-Peterburgsk. Biol. Lab. Superseded by: Izvestiya Petrogradskoi Biologicheskoi Laboratorii. HI 66299

Izvestiya Sapropelevogo komiteta. Petrograd. Vols. 1-6, 1923-32. Izv. Sapropel. Komiteta. 1-118-2. HI 66300

Izvestiya Saratovskogo obshchestva estestvoispytatelei. Saratov, Russian S F S R. Vols. 1-3, 1924-29. Izv. Saratovsk. Obshch. Estestvoisp. HI 66301

Izvestiya sel'skokhozyaistvennoi akademii imeni K.A. Timiryazeva. Moscow. Vols. 1-?, 1923-52. Izv. Sel'skokhoz. Akad. Timiryazeva. Preceded by: Izvestiya petrovskoi sel'skokhozyaistvennoi akademii (1917-20). Superseded by: Izvestiya timiryazevskoi sel'skokhozyaistvennoi akademii. 3-2766-1. HI 66302

Izvestiya po sel'sko-khozyaistvennomu opytnomu delu leningradskoi oblasti. Leningrad. Vols. ?-10-?, ?-1929-? Izv. Sel'skokhoz. Opytn. Leningradsk. Obl. HI 66303

Izvestiya Severnoi oblastnoi stantsii zashchity rastenii ot vreditelei. Petrograd. Vols. 4-5, 1922-25. Izv. Severn. Obl. Stantsii Zashch. Rast. Vredit. Preceded by: Izvestiya Petrogradskoi oblastnoi stantsii zashchity rastenii ot vreditelei. Superseded by: Izvestiya Leningradskoi oblastnoi stantsii zashchity rastenii ot vreditelei. HI 66304

Izvestiya Severo-Kavkazskoi kraevoi stantsii zashchity rastenii. Rostov, Russian S F S R. Vols. 1-6/7, 1926-30. Izv. Severo-Kavkazsk. Kraev. Stantsii Zashch. Rast. Superseded by: Trudy Severo-Kavkazskogo instituta zashchity rastenii. 4-3711-2. HI 66305

Izvestiya Sibirskogo ėntomologicheskogo byuro. Novosibirsk, Russian S F S R. Vols. 1-3, 1922-24. Izv. Sibirsk. Ėntomol. Byuro. Superseded by: Izvestiya Sibirskoi stantsii zashchity rastenii ot vreditelei. 5-3868-2. HI 66306

Izvestiya Sibirskago Otdĕla Imperatorskago Russkago Geograficheskago Obshchestva. Irkutsk, Russian S F S R. Vols. 1-8, 1870-77. Izv. Sibirsk. Otd. Imp. Russk. Geogr. Obshch. Superseded by: Izvestiya Vostochno-Sibirskago Otdĕla Imperatorskago Russkago

Geograficheskago Obshchestva. 2-1693-1. HI 66307

Izvestiya sibirskogo otdeleniya akademii nauk S S S R. Novosibirsk. 1958(3)-62. Izv. Sibirsk. Otd. Akad. Nauk S.S.S.R. Preceded by: Izvestiya vostochnykh filialov Akademii nauk S S S R. Superseded by: Izvestiya sibirskogo otdeleniya akademii nauk S S S R. Seriya biologicheskikh nauk. HI 66308

Izvestiya sibirskogo otdeleniya akademii nauk S S S R. Seriya biologicheskikh nauk. Novosibirsk. No. 15+, 1969+. Izv. Sibirsk. Otd. Akad. Nauk S.S.S.R., Ser. Biol. Nauk. Preceded by: Izvestiya sibirskogo otdeleniya akademii nauk S S S R. HI 66309

Izvestiya Sibirskoi kraevoi stantsii zashchity rastenii ot vreditelei. Novosibirsk, Russian S F S R. Vols. 5-7 [also numbered vols. 2-4], 1927-30. Izv. Sibirsk. Kraev. Stantsii Zashch. Rast. Vredit. Preceded by: Izvestiya Sibirskoi stantsii zashchity rastenii ot vreditelei. Superseded by: Trudy po zashchite rastenii Sibiri. 5-3868-2. HI 66310

Izvestiya Sibirskoi stantsii zashchity rastenii ot vreditelei. Novosibirsk, Russian S F S R. Vol. 4. [also numbered Vol. 1], 1924. Izv. Sibirsk. Stantsii Zashch. Rast. Vredit. Preceded by: Izvestiya Sibirskogo entomologicheskogo byuro. Superseded by: Izvestiya Sibirskoi kraevoi stantsii zashchity rastenii ot vreditelei. 5-3868-2. HI 66311

Izvestiya Sredne-Aziatskogo geograficheskogo obshchestva. Tashkent, Uzbek S S R. Vols. 19-20, 19??-32. Izv. Sredne-Aziatsk. Geogr. Obshch. Preceded by: Izvestiya Sredne-Aziatskogo otdela Gosudarstvennogo russkogo geograficheskogo obshchestva. Superseded by: Trudy Uzbekistanskogo geograficheskogo obshchestva. 5-4054-2. HI 66312

Izvestiya Sredne-Aziatskogo otdela Gosudarstvennogo russkogo geograficheskogo obshchestva. Tashkent, Uzbek S S R. Vol. 18, 1929. Izv. Sredne-Aziatsk. Otd. Gosud. Russk. Geogr. Obshch. Preceded by: Izvestiya Turkestanskogo otdela Russkogo geograficheskogo obshchestva. Superseded by: Izvestiya Sredne-Aziatskogo geograficheskogo obshchestva. 5-4054-2. HI 66313

Izvestiya Sredne-Sibirskogo gosudarstvennogo geograficheskogo obshchestva. Krasnoyarsk, Russian S F S R. Vol. 3(4), 1929. Izv. Sredne-Sibirsk. Gosud. Geogr. Obshch. Preceded by: Izvestiya Sredne-Sibirskogo otdela Gosudarstvennogo russkogo geograficheskogo obshchestva. 2-1693-1. HI 66314

Izvestiya Sredne-Sibirskogo otdela Gosudarstvennogo russkogo geograficheskogo obshchestva. Krasnoyarsk, Russian S F S R. Vol. 3(3), 1928. Izv. Sredne-Sibirsk. Otd. Gosud. Russk. Geogr. Obshch. Preceded by: Izvestiya Krasnoyarskogo otdela Russkogo geograficheskogo obshchestva. Superseded by: Izvestiya Sredne-Sibirskogo gosudarstvennogo geograficheskogo obshchestva. 2-1693-1. HI 66315

Izvestiya, subtropicheskii botanicheskii sad = Izvestiya Batumskogo subtropicheskogo botanicheskogo sada. Batum.

Izvestiya sukhumskoi sadovoi i sel'skokhozyaistvennoi opytnoi stantsii. Tiflis. Vols. 1-?, 1911-25. Izv. Sukhumsk. Sadovoi Sel'skokhoz. Opyt. Stantsii. Superseded by: Izvestiya abkhazkoi sel'sko-khozyaistvennoi opytnoi stantsii. HI 66316

Izvestiya tadzhikistansgoi bazy botaniki = Acta horti botanici tadshikistanici. Moscow. Acta Horti Bot. Tadshik. See B–P–H 43/25.

Izvestiya tadzhikskogo filiala akademii nauk S S S R. Stalinabad [= Dushanbe]. 1945-? Izv. Tadzhiksk. Fil. Akad. Nauk S.S.S.R. Superseded by: Izvestiya akademii nauk tadzhikskoi S S R. Otdelenie biologicheskikh nauk and Izvestiya akademii nauk tadzhikskoi S S R. Otdelenie estestvennykh nauk. HI 66317

Izvestiya Tikhookeanskogo nauchnogo instituta rybnogo khozyaistva. Voronezh, Russian S F S R. Vol. 3(6), 1929/30; vols. 4(2)-8, 1929/30-34. Izv. Tikhookeansk. Nauchn. Inst. Rybn. Khoz. For vols. 1-3(5), 1928-29/30, and vol. 4(1), 1929/30 see: Izvestiya Tikhookeanskoi nauchno-promyslovoi stantsii. Superseded by: Izvestiya Tikhookeanskogo nauchno-issledovatel'skogo instituta rybnogo khozyaistva i okeanografii. HI 66318

Izvestiya Tikhookeanskogo nauchno-issledovatel'skogo instituta rybnogo khozyaistva i okeanografii. Vladivostok, Russian S F S R. Vol. 9+, 1937+. Izv. Tikhookeansk. Nauchno-Issl. Inst. Rybn. Khoz. Preceded by: Izvestiya Tikhookeanskogo nauchnogo instituta rybnogo khozyaistva. 5-4415-1. HI 66319

Izvestiya Tikhookeanskoi nauchno-promyslovoi stantsii. Vladivostok, Russian S F S R. Vols. 1-3(6), 1928-29/30; vol. 4(1), 1929/30. Izv. Tikhookeansk. Nauchno-Promysl. Stantsii. For vol. 3(6) see: Izvestiya Tikhookeanskogo nauchnogo instituta rybnogo khozyaistva. Superseded by: Izvestiya Tikhookeanskogo nauchnogo instituta rybnogo khozyaistva. 5-4415-1. HI 66320

Izvestiya. Tikhookeanskii nauchnyi institut rybnogo khoziaistva = Izvestiya Tikhookeanskogo nauchnogo instituta rybnogo khozyaistva. Voronezh.

Izvestiya timiryazevskoi sel'skokhozyaistvennoi akademii. Moscow. Vol. 1+, 1952+. Izv. Timiryazevsk. Sel'skokhoz. Akad. Preceded by: Izvestiya sel'skokhozyaistvennoi akademii imeni K. A. Timiryazeva. HI 66321

Izvestiya Tomskogo gosudarstvennogo universiteta. Tomsk, Russian S F S R. Vols. 73-82 1924-28; vol. 84, 1929. Izv. Tomsk. Gosud. Univ. For vol. [83] see:

Trudy biologicheskogo fakul'teta Tomskogo gosudarstvennogo universiteta. Superseded by: Trudy Tomskogo gosudarstvennogo universiteta. 5-4234-3. HI 66322

Izvestiya Tomskogo otdeleniya Gosudarstvennogo russkogo botanicheskogo obshchestva. Tomsk, Russian S F S R. Vol. 3, 1931. Izv. Tomsk. Otd. Gosud. Russk. Bot. Obshch. Preceded by: Izvestiya Tomskogo otdeleniya Russkogo botanicheskogo obshchestva. Superseded by: Izvestiya Tomskogo otdeleniya Vsesoyuznogo botanicheskogo obshchestva. 4-3744-3. HI 66323

Izvestiya Tomskogo otdeleniya Russkogo botanicheskogo obshchestva. Tomsk, Russian S F S R. Vols. 1-2, 1921-27. Izv. Tomsk. Otd. Russk. Bot. Obshch. Superseded by: Izvestiya Tomskogo otdeleniya Gosudarstvennogo russkogo botanicheskogo obshchestva. 4-3744-3. HI 66324

Izvestiya Tomskogo otdeleniya Vsesoyuznogo botanicheskogo obshchestva. Novosibirsk, Russian S F S R. Vol. 4+, 1959+. Izv. Tomsk. Otd. Vsesoyuzn. Bot. Obshch. Preceded by: Izvestiya Tomskogo otdeleniya Gosudarstvennogo russkogo botanicheskogo obshchestva. HI 66325

Izvestiya Tomskogo Universiteta. Tomsk, Russian S F S R. Vols. 66-70, 1917-19. Izv. Tomsk. Univ. Preceded by: Izvestiya Imperatorskago Tomskago Universiteta. Berichte der Tomsker Staatsuniversität. Superseded by: Izvestiya Gosudarstvennogo Tomskogo universiteta. 5-4234-3. HI 66326

Izvestiya na tsarskite prirodonauchni instituti v Sofiya. Sofia, Bulgaria. Vols. 1-16, 1928-43. Izv. Tsarsk. Prir. Inst. Sofiya. 5-3991-1. HI 66327

Izvestiya na tsentralniya nauchno-izsledovatelski institut po rastenievudstvo. Sofia, Bulgaria. Vol. 9+, 1960+. Izv. Tsentr. Nauchno-Izsl. Inst. Rasteniev. Preceded by: Izvestiya na instituta po rastenievudstvo. HI 66328

Izvestiya na tsentralniya nauchno-izsledovatelski institut za zashtita na rasteniyata. Sofia. Vols. 1-5, 1961-63. Izv. Tsentr. Nauchno-Izsl. Inst. Zasht. Rast. Preceded by: Nauchni trudove, nauchno-izsledovatelski institut za zashtita na rasteniyata. HI 66329

Izvestiya na tsentralniya nauchno-izsledovatelski institut po pochvoznanie i agrotekhnika "Nikola Pushkarov" [title varies.]. Sofia. Vols. 1-5, 1961-63. Izv. Centr. Nauchno-Izsl. Inst. Pochvozn. Nikola Pushkarov. Superseded by: Izvestiya na institut po pochvoznanie i agrotekhnika "Nikola Pushkarov". HI 66330

Izvestiya Turkestanskago Otděla Imperatorskago Geograficheskago Obshchestva. Tashkent, Uzbek S S R. Vols. 1-12, 1898-1916. Izv. Turkestansk. Otd. Imp. Russk. Goegr. Obshch. Superseded by: Izvestiya Turkestanskogo otdela Russkogo geograficheskogo obshchestva. 5-4054-2. HI 66331

Izvestiya Turkestanskogo otdela Russkogo geograficheskogo obshchestva. Tashkent, Uzbek S S R. Vols. 13-17, 1917-24. Izv. Turkestansk. Otd. Russk. Geogr. Obshch. Preceded by: Izvestiya Turkestanskago Otděla Imperatorskago Geograficheskago Obshchestva. Superseded by: Izvestiya Sredne-Aziatskogo otdela Gosudarstvennogo russkogo geograficheskogo obshchestva. 5-4054-2. HI 66332

Izvestiya Turkmenskogo filiala Akademii nauk S S S R. Ashkhabad, Turkmen S S R. 1944-46; 1948-1951(3). Izv. Turkmensk. Fil. Akad. Nauk S.S.S.R. Superseded by: Izvestiya akademiya nauk turkmenskoi S S R. 1-120-3. HI 66333

Izvestiya i Ucheniya Zapiski Imperatorskago Kazanskago Universiteta. Kazan, Russian S F S R. Vols. 32-50 [also numbered vols. 1-19], 1865-1883. Izv. Uchen. Zap. Imp. Kazansk. Univ. Preceded by: Uchenya Zapiski Imperatorskago Kazanskago Universiteta po Otděleniyu Fiziko-matematicheskikh i Meditsinskikh Nauk. Superseded by: Uchenya Zapiski, izdavaemya Imperatorskim Kazanskim Universitetom and Izvestiya Imperatorskago Kazanskago Universiteta. 3-2276-2. HI 66334

Izvestiya Uzbekistanskogo filiala Akademii nauk S S S R. Tashkent, Uzbek S S R. 1940-41. Izv. Uzbekistansk. Fil. Akad. Nauk S.S.S.R. Superseded by: Izvestiya Akademii nauk Uzbekskoi S S R. HI 66335

Izvestiya vengerskikh sel'skokhoziaistvennykh nauchno-issledovatel'skikh institutov. Budapest = Kísérletügyi kölemények. Budapest. Kísérl. Közlem. See B–P–H 514/5.

Izvestiya, voronezhskoe otdelenie, vsesoyuznogo botanicheskogo obshchestva. Voronezh. 1963. Izv. Voronezhsk. Otd. Vsesoyuzn. Bot. Obshch. HI 66336

Izvestiya Voronezhskoi stantsii po bor'be s vreditelyami rastenii. Voronezh, Russian S F S R. Vols. [2]-3, 1918-? Izv. Voronezhsk. Stantsii Bor'be Vredit. Rast. Preceded by: Trudy Voronezhskoi Stantsii po Bor'bě s Vreditelyami Rastenii. Superseded by: Stantsiya zashchity sel'skokhozyaistvennykh rastenii ot vreditelei Voronezhskogo gubernskogo zemel'nogo upravleniya. HI 66337

Izvestiya vostochnykh filialov Akademii nauk S S S R. Novosibirsk, Russian S F S R. 1957-58. Izv. Vost. Fil. Akad. Nauk S.S.S.R. Superseded by: Izvestiya sibirskogo otdeleniya akademii nauk S S S R. HI 66338

Izvestiya Vostochno-Sibirskogo otdela Geograficheskogo obshchestva S S S R. Irkutsk, Russian S F S R. Vol. 58, 1954. Izv. Vost.-Sibirsk. Otd. Geogr. Obshch.

S.S.S.R. Preceded by: Izvestiya Obshchestva izucheniya Vostochno-Sibirskoi oblasti. HI 66339

Izvestiya Vostochno-Sibirskogo otdela Gosudarstvennogo russkogo geograficheskogo obshchestva. Irkutsk, Russian S F S R. Vols. 50-55, 1926-29. Izv. Vost.-Sibirsk. Otd. Gosud. Russk. Geogr. Obshch. Preceded by: Izvestiya Vostochno-Sibirskogo otdela Russkogo geograficheskogo obshchestva. Superseded by: Izvestiya Obshchestva izucheniya Vostochno-Sibirskogo kraya. 2-1693-1. HI 66340

Izvestiya Vostochno-Sibirskago Otděla Imperatorskago Russkago Geograficheskago Obshchestva. Irkutsk, Russian S F S R. Vols. 9-45, 1878-1917. Izv. Vost.-Sibirsk. Otd. Imp. Russk. Geogr. Obshch. Preceded by: Izvestiya Sibirskago Otděla Imperatorskago Russkago Geograficheskago Obshchestva. Superseded by: Izvestiya Vostochno-Sibirskogo otdela Russkogo geograficheskogo obshchestva. 2-1693-1. HI 66341

Izvestiya Vostochno-Sibirskogo otdela Russkogo geograficheskogo obshchestva. Irkutsk, Russian S F S R. Vols. 46-49, 1921-26. Izv. Vost.-Sibirsk. Otd. Russk. Geogr. Obshch. Preceded by: Izvestiya Vostochno-Sibirskago Otděla Imperatorskago Russkago Geograficheskago Obshchestva. Superseded by: Izvestiya Vostochno-Sibirskogo otdela Gosudarstvennogo russkogo geograficheskogo obshchestva. 2-1693-1. HI 66342

Izvestiya Vsesoyuznogo geograficheskogo obshchestva. Moscow & Leningrad. Vol. 72+, 1940+. Izv. Vsesoyuzn. Geogr. Obshch. Preceded by: Izvestiya Gosudarstvennogo geograficheskogo obshchestva. 2-1692-1. HI 66343

Izvestiya vysshikh uchebnykh zavadenii, lesnoi zhurnal. Archangel. 1958+. Izv. Vyssh. Uchebn. Zaved., Lesn. Zhurn. Preceded by: Nauchnye doklady vysshei shkoly. Lesoinzhenernoe delo [not entered]. HI 66344

Izvestiya Yakutskogo otdela Gosudarstvennogo russkogo geograficheskogo obshchestva. Yakutsk, Russian S F S R. Vol. 2, 1928. Izv. Yakutsk. Otd. Gosud. Russk. Geogr. Obshch. Preceded by: Izvestiya Yakutskago Otděla Imperatorskago Russkago Geograficheskago Obshchestva. 2-1692-2. HI 66345

Izvestiya Yakutskago Otděla Imperatorskago Russkago Geograficheskago Obshchestva. Yakutsk, Russian S F S R. Vol. 1, 1915. Izv. Yakutsk. Otd. Imp. Russk. Geogr. Obshch. Superseded by: Izvestiya Yakutskogo otdela Gosudarstvennogo russkogo geograficheskogo obshchestva. 2-1692-2. HI 66346

Izvestiya Yuzhno-Russkago Obshchestva Akklimatizatsii. Kharkov, Ukrainian S S R. 1897-1911. Izv. Yuzhno-Russk. Obshch. Akklim. HI 66351

Izvestiya Yuzhno-Ussuriiskogo otdela Gosudarstvennogo russkogo geograficheskogo obshchestva. Nikolsk-Ussuriski [=Ussuriysk], Russian S F S R. Vols. 12-16, 1926-28. Izv. Yuzhno-Ussuriisk. Otd. Gosud. Russk. Geogr. Obshch. Preceded by: Izvestiya Yuzhno-Ussuriiskogo otdeleniya Russkogo geograficheskogo obshchestva. 2-1692-2. HI 66352

Izvestiya Yuzhno-Ussuriiskago Otděleniya Priamurskago Otděla Russkago Geograficheskago Obshchestva. Nikolsk-Ussuriski [=Ussuriysk], Russian S F S R. Vols. 1-4, 1922. Izv. Yuzhno-Ussuriisk. Otd. Priamursk. Otd. Russk. Geogr. Obshch. Superseded by: Izvestiya Yuzhno-Ussuriiskogo otdeleniya Russkogo geograficheskogo obshchestva. 2-1692-2. HI 66353

Izvestiya Yuzhno-Ussuriiskogo otdeleniya Russkogo geograficheskogo obshchestva. Nikolsk-Ussuriski [=Ussuriysk], Russian S F S R. Vols. 5-11, 1924-25. Izv. Yuzhno-Ussuriisk. Otd. Russk. Geogr. Obshch. Preceded by: Izvestiya Yuzhno-Ussuriiskago Otděleniya Priamurskago Otděla Russkago Geograficheskago Obshchestva. Superseded by: Izvestiya Yuzhno-Ussuriiskogo otdela Gosudarstvennogo russkogo geograficheskogo obshchestva. 2-1692-2. HI 66354

Izvestiya Zapadno-Sibirskogo geograficheskogo obshchestva. Omsk, Russian S F S R. Vol. 7, 1930. Izv. Zapadno-Sibirsk. Obshch. Preceded by: Izvestiya Zapadno-Sibirskogo otdela Russkogo geograficheskogo obshchestva. 2-1693-1. HI 66355

Izvestiya Zapadno-Sibirskoi kraevoi stantsii zashchity rastenii. Tomsk, Russian S F S R. Vol. 9 [also numbered vol. 1], 1935. Izv. Zapadno-Sibirsk. Kraev. Stantsii Zashch. Rast. Preceded by: Trudy po zashchite rastenii Sibiri. Superseded by: Trudy Zapadno-Sibirskoi kraevoi stantsii zashchity rastenii. HI 66356

Izvestiya Zapadno-Sibirskago Otděla Imperatorskago Russkago Geograficheskago Obshchestva. Omsk, Russian S F S R. Vols. 1-3, 1913-15. Izv. Zapadno-Sibirsk. Otd. Imp. Russk. Geogr. Obshch. Superseded by: Izvestiya Zapadno-Sibirskogo otdela Russkogo geograficheskogo obshchestva. 2-1693-1. HI 66357

Izvestiya Zapadno-Sibirskogo otdela Russkogo geograficheskogo obshchestva. Omsk, Russian S F S R. Vols. 4-6, 1924-29. Izv. Zapadno-Sibirsk. Otd. Russk. Geogr. Obshch. Preceded by: Izvestiya Zapadno-Sibirskago Otděla Imperatorskago Russkago Geograficheskago Obshchestva. Superseded by: Izvestiya Zapadno-Sibirskogo geograficheskogo obshchestva. 2-1693-1. HI 66358

Izvestja, gozdarski institut slovenije. Ljubljana. Vols. 1+, 1947+. Izv. Gozd. Inst. Slovenije. Superseded by: Zbornik instituta za gozdo in lesno gospodarstvo Slovanije. 3-2451-1. HI 66359

Izvještaj o radu državnoga filopatološka zavoda.

Sarajevo. 1925-28. Izv. Radu Državn. Fitopatol. Zavoda. Superseded by: Rad fitopatoloskog zavoda u Sarajevu. HI 66360

J A O C S = Journal of the american oil chemists' society. Chicago, IL.

J A R E data reports. Tokyo. No. 1+, 1968+. J. A. R. E. Data Rep. HI 66362

J A R E data reports. Marine biology. Tokyo. No. 1+, 1981+. J. A. R. E. Data Rep, Mar. Biol. HI 66363

J A R E scientific reports. Series D, oceanography. Tokyo. 1964-73. J. A. R. E. Sci. Rep., D. HI 66364

J A R E scientific reports. Series E, biology. Tokyo. Nos. 18-31, 1963-70. J. A. R. E. Sci. Rep., E. Preceded by: Biological results of the Japan antarctic research expedition (contained in: Special publications from the Seto marine biological laboratory). Superseded by: Memoirs of the national institute of polar research. HI 66365

J A R E scientific reports. Special issue. Tokyo. Nos. 1-2, 1967-71. J. A. R. E. Sci. Rep., Special Issue. HI 66366

J A R Q = Japan agricultural research quarterly. Yatabe.

J A T B A. Travaux d'ethnobotanique et d'ethnozoologie, publ. avec le concours du C N R S. Paris. Vol. 24+, 1977+. J. A. T. B. A. Preceded by: Journal d'agriculture tropicale et de botanique appliquée. HI 66367

J C M = Journal of clinical microbiology. Washington, DC.

J E O R. Wheaton, IL = Journal of essential oil research. Wheaton, IL.

J I A S = Journal of the Iowa academy of science. Cedar Falls, IA.

J I B P synthesis. Tokyo. Vol. 1+, 1975+. J. I. B. P. Synth. HI 66368

J I P A. Simla. Vol. 1+, 1974+. J. I. P. A. HI 66369

J N K V V research journal. Jabalpur. Vol. 1+, 1967+. J. N. K. V. V. Res. J. HI 66370

J S P P newsletter = Newsletter, japanese society of plant physiologists.

Jaarblad van die botaniese vereniging van Suid-Afrika = Journal of the botanical society of South Africa. Kirstenbosch.

Jaarboek. Bataviaasch genootschap van kunsten en wetenschappen. Bandoeng, Dutch E. Indies [=Bandung, Indonesia]. Jaarb. Batav. Genootsch. Kunsten. See B–P–H 485/12. HI 55382

Jaarboek van het departement van landbouw in Nederlandsch-Indië. Batavia. 1906-10, 1907-11. Jaarb. Dep. Landb. Ned.-Indië. Preceded by: Verslagen omtrent de te Buitenzorg gevestigde technische afdeelingen van het departement van landbouw. Superseded by: Jaarboek van het departement van landbouw, nijverheid en handel in Nederlandsch-Indië. HI 66371

Jaarboek van het departement van landbouw, nijverheid en handel in Nederlandsch-Indië. Batavia. 1911-19, 1912-20. Jaarb. Dep. Landb. Nijverh. Handel Ned.-Indië. Preceded by: Jaarboek van het departement van landbouw in Nederlandsch-Indië. HI 66372

Jaarboek, institut voor biologisch en scheikundig onderzoek van landbouwgewassen. [Forms part of: Mededelingen, institut voor biologisch en scheikundig onderzoek van landbouwgewassen.]. Wageningen. 1963-69, 1964-70. Jaarb. Inst. Biol. Scheik. Onderz. Landbouwgew. Preceded by: Jaarverslag, institut voor biologisch en scheikundig onderzoek van landbouwgewassen. HI 66373

Jaarboek van de k. vlaamse academie voor wetenschappen, letteren en schone kunsten van België. Ghent. Jaarb. Kon. Vlaamse Acad. Wetensch. België. See B–P–H 485/20. HI 55390

Jaarboek van de koninklijke akademie van wetenschappen, gevestigd te Amsterdam. Amsterdam. Jaarb. Kon. Akad. Wetensch. Amsterdam. See B–P–H 485/15. HI 55385

Jaarboek van de koninklijke nederlandsche maatschappij tot aanmoediging van den tuinbouw. Leiden. Jaarb. Kon. Ned. Maatsch. Aanm. Tuinb. See B–P–H 485/18. HI 55388

Jaarboek van het mijnwezen in Nederlandsch-Indië. Amsterdam. 1872-1939. Jaarb. Mijnw. Ned.-Indië. HI 66374

Jaarboek van het natuurhistorisch genootschap in Limburg. Assen, Netherlands. Jaarb. Natuurhist. Genootsch. Limburg. See B–P–H 485/21. HI 55391

Jaarboek van de natuurwetenschappelijke studiekring voor Suriname en Curaçao. Utrecht. Jaarb. Natuurw. Studiekring Suriname Curaçao. See B–P–H 486/1. HI 55392

Jaarboek. Nederlandsche dendrologische vereeniging. Wageningen. Jaarb. Ned. Dendrol. Ver. See B–P–H 486/2. HI 55393

Jaarboek. Proefstation voor de boomkwekerij. Boskoop, Netherlands. Jaarb. Proefstat. Boomkwekerij. See B–P–H 486/4. HI 55394

Jaarboek van het rijksinstituut voor het onderzoek der zee. Helder, Netherlands. Jaarb. Rijksinst. Onderz. Zee. See B–P–H 486/5. HI 55395

Jaarboek uitgegeven door de koninklijke nederlandse maatschappij voor tuinbouw en plantkunde. Assen, Netherlands. Jaarb. Kon. Ned. Maatsch. Tuinb. See

B–P–H 485/19. HI 55389

Jaarboek uitgegeven door de vereniging de proeftuin. Boskoop, Netherlands. Jaarb. Ver. Proeftuin. See B–P–H 486/6. HI 55396

Jaarboek, vereniging voor het onderwijs in de biologie. Mol. Vol. 3+, 1980+. Jaarb. Ver. Onderwijs Biol. Preceded by: Driemaandelijks tijdschrift, vereniging voor het onderwijs in de biologie. HI 66375

Jaarboek, verslagen en mededelingen van de koninklijke nederlandse botanische vereniging. Amsterdam. Jaarb. Kon. Ned. Bot. Ver. See B–P–H 485/17. HI 55387

Jaarboeken der genees- heel- en natuurkunde. Amsterdam. Jaarb. Genees- Heel- Natuurk. See B–P–H 485/13. HI 55383

Jaarboekje van de hollandsche maatschappij van landbouw. Rotterdam. Jaarb. Holl. Maatsch. Landb. See B–P–H 485/14. HI 55384

Jaarboekje van de koninklijke algemeene vereeniging von bloembollencultuur. Haarlem. Jaarb. Konl. Alg. Ver. Bloembollencult. See B–P–H 485/16. HI 55386

Jaarverslag, die botaniese vereniging van Suid-Afrika = Report (Annual), botanical society of South Africa. Kirstenbosch.

Jaarverslag, caraïbïsch marien-biologisch instituut. Curaçâo. 1970+. Jaarversl. Caraïb. Mar.-Biol. Inst. Preceded by: Verslagen, caraïbïsch marien-biologisch instituut. HI 66376

Jaarverslag van de departement van bosbou, South Africa = Report (Annual), department of forestry, South Africa. Cape Town.

Jaarverslag, geologisch bureau voor het Nederlandsche mijngebied te Heerlen. 1925-37. Jaarversl. Geol. Bur. Ned. Mijngeb. Heerlen. Superseded by: Mededeelingen, geologisch bureau voor het mijngebied te Heerlen. HI 66378

Jaarverslag, institut voor biologisch en scheikundig onderzoek van landbouwgewassen. [Forms part of: Mededelingen, institut voor biologisch en scheikundig onderzoek van landbouwgewassen.] Wageningen. 1957-62. Jaarversl. Inst. Biol. Scheik. Onderz. Landbouwgew. Superseded by: Jaarboek, institut voor biologisch en scheikundig onderzoek van landbouwgewassen. HI 66379

Jaarverslag, instituut voor plantenziektenkundig onderzoek. Wageningen. 1950-71. Jaarversl. Inst. Plantenziektenk. Onderz. Superseded by: Report (Annual), instituut voor plantenziektenkundig onderzoek. HI 66380

Jaarverslag, instituut voor rassenonderzoek van landbouwgewassen. 1958-76, 1959-77. Jaarversl. Inst. Rassenonderz. Landbouwgew. Superseded by: Jaarverslag, rijksinstituut voor het rassenonderzoek van cultuurgewassen. HI 66381

Jaarverslag der nederlandsch-indische vereeniging tot natuurbescherming = Verslagen, nederlandsch-indische vereeniging tot natuurbescherming. Buitenzorg.

Jaarverslag, proefstation voor de bloemisterij in Nederland. Aalsmeer. 1956?-74. Jaarversl. Proefstat. Bloemist. Ned. Superseded by: Bloemisterij onderzoek in Nederland. HI 66382

Jaarverslag, projecten van onderzoek, instituut voor de veredeling van tuinbouwgewassen. Wageningen. 1948+. Jaarversl. Proj. Onderz. Inst. Veredel. Tuinbouwgew. HI 66383

Jaarverslag, koninklijk instituut voor de tropen. Amsterdam. 1951+. Jaarversl. Kon. Inst. Tropen. Preceded by: Jaarverslag, koninklijke vereeniging indisch instituut. 3-2311-1. HI 66384

Jaarverslag, koninklijke vereeniging indisch instituut. Amsterdam. 1945-49. Jaarversl. Kon. Ver. Indisch Inst. Preceded by: Jaarverslag, vereeniging kolonial instituut. Superseded by: Jaarverslag, koninklijke instituut voor de tropen. 3-2311-1. HI 66385

Jaarverslag, koninklijke vereeniging kolonial instituut. Amsterdam. 1925-29; 1932-40. Jaarversl. Kon. Ver. Kolon. Inst. Preceded by: Jaarverslag, vereeniging koloniaal instituut. For vols. for 1930-32 see: Jaarverslag, vereeniging koloniaal instituut. Superseded by: Jaarverslag, vereeniging koloniaal instituut. 3-2311-1. HI 66386

Jaarverslag van het proefstation voor de Java-suikerindustrie. Soerabaia. Vols. 1-24, 1907-34. Jaarversl. Proefstat. Java Suikerindstr. HI 66387

Jaarverslag, rijksinstituut voor het rassenonderzoek van cultuurgewassen. Wageningen. 1977+, 1978?+. Jaarversl. Rijksinst. Rassenonderz. Cultuurgew. Preceded by: Jaarverslag, instituut voor rassenonderzoek van landbouwgewassen. HI 66388

Jaarverslag, rijks instituut voor veldbiologisch onderzoek ten behouve van het natuurbehoud. Bilthoven. 1962-69. Jaarversl. Rijks. Inst. Veldbiol. Onderz. Behouve Natuurbehoud. Preceded by: Verslagen van de werkzaamheden rijksinstituut voor veldbiologische onderzoek ten beohoeven van het natuurbehoud. HI 66389

Jaarverslag, rijksinstituut voor natuurbeheer. Arnhem. 1970/71+, 1971?+. Jaarversl. Rijksinst. Natuurbeheer. HI 66390

Jaarverslag, rijkstuinbouw consultentschap voor champignonteelt = Jaarverslag, stichting proefstation voor de champignoncultur. Horst.

Jaarverslag van die S W A wetenskaplike vereniging = Report (Annual) of the south west african scientific society. Windhoek.

Jaarverslag, stichting proefstation voor de champignoncultur. Horst. 1968+. Jaarversl. Stichting Proefstat. Champignoncultur. HI 66391

Jaarverslag, vereeniging koloniaal instituut. Amsterdam. 1910/11-23; 1930-32; 1940-44. Jaarversl. Ver. Kolon. Inst. For vols. for 1925-29 and for 1932-40 see: Jaarverslag, koninklijke vereeniging koloniaal instituut. Superseded by: Jaarverslag, koninklijke vereeniging indisch instituut. 3-2311-1. HI 66392

Jagd- und Forst-Neuigkeiten. Prague. Jagd- Forst-Neuigk. See B–P–H 486/7. HI 55397

Jahrbuch. [Edited by H. C. Schumacher.] Stuttgart & Tübingen. Jahrb. (Schumacher). See B–P–H 486/8. HI 55398

Jahrbücher der Akademie gemeinnütziger Wissenschaften zu Erfurt = Jahrbücher der Königlichen Akademie gemeinnütziger Wissenschaften zu Erfurt. Erfurt. Jahrb. Königl. Akad. Gemeinnütz. Wiss. Erfurt. See B–P–H 487/20.

Jahrbuch, Akademie der Wissenschaften der D D R. Berlin. 1978+. Jahrb. Akad Wiss. D.D.R. Preceded by: Jahrbuch der deutschen Akademie der Wissenschaften zu Berlin. HI 66393

Jahrbuch der Akademie der Wissenschaften in Göttingen (Societät der Reichsakademie). Göttingen. Jahrb. Akad. Wiss. Göttingen. See B–P–H 486/10. HI 55399

Jahrbuch, Akademie der Wissenschaften und der Literatur. Mainz. 1957+. Jahrb. Akad. Wiss. Lit. HI 66394

Jahrbuch des amtlichen Pflanzengesundheitsdienstes. Budapest = Növényegészégyügyi évkönyv. Budapest. Növényegész. Évk. See B–P–H 674/3.

Jahrbuch der bayerischen Akademie der Wissenschaften. Munich. 1918+. Jahrb. Bayer. Akad. Wiss. 1-104-1. HI 66395

Jahrbuch des biologischen Institute in Sarajevo = Godišnjak biološkog instituta u Sarajevu. Sarajevo, Yugoslavia. God. Biol. Inst. u Sarajevu. See B–P–H 406/5.

Jahrbuch der Biologischen Wolga-Station = Ezhegodnik Volzhskoi Biologicheskoi Stantsii Saratovskago Obshchestva Estestvoispytatelei i Lyubitelei Estestvoznaniya. Saratov.

Jahrbücher des böhmischen Museums für Natur- und Länderkunde, Geschichte, Kunst und Literatur. Prague. Jahrb. Böhm. Mus. Natur- Länderk. See B–P–H 486/15. HI 55400

Jahrbuch der Bundesanstalt für Pflanzenbau und Samenprüfung. [Supplement to: Bodenkultur.] Vienna. 1949+. Jahrb. Bundesanst. Pflanzenbau Samenprüfung. Preceded by: Jahresbericht der Bundesanstalt für Pflanzenbau und Samenprüfung in Wien. HI 66396

Jahrbuch der Chemie und Physik. Als eine Zeitschrift des wissenschaftlichen Vereins zur Verbreitung von Naturkenntniss und höherer Wahrheit, 1825.I-1828.III = Journal für Chemie und Physik, vols. 43-45, 1825-28. Nuremberg. J. Chem. Phys. (Nuremberg). See B–P–H 461/19.

Jahrbuch, Deutsche Dahlien-Gesellschaft = Deutsche Dahlien-Gesellschaft. Jahrbuch. Altona-Bahrenfeld [=Hamburg, in part]. Deutsche Dahlien-Ges. Jahrb. See B–P–H 345/18.

Jahrbuch, Deutsche Iris Gesellschaft = Deutsche Iris Gesellschaft. Jahrbuch. Leonberg, Germany. Deutsche Iris Ges. Jahrb. See B–P–H 346/12.

Jahrbuch, Deutsche Iris- und Liliengesellschaft e. V. = Deutsche Iris- und Liliengesellschaft e. V. Jahrbuch. Leonberg, Germany. Deutsche Iris- Lilienges. Jahrb. See B–P–H 346/13.

Jahrbuch der deutschen Akademie der Wissenschaften zu Berlin. Berlin. 1946/49-68/69, 1950-69. Jahrb. Deutsch. Akad. Wiss. Berlin. Superseded by: Jahrbuch, Akademie der Wissenschaften der D D R. 1-101-2. HI 66397

Jahrbuch der Deutschen Dahlien-Gesellschaft. Altona-Bahrenfeld [=Hamburg, in part]. Jahrb. Deutsch. Dahlien-Ges. See B–P–H 486/24. HI 55402

Jahrbuch der Deutschen Dahlien- und Gladiolen-Gesellschaft. Düsseldorf. Jahrb. Deutsch. Dahlien-Gladiolen-Ges. See B–P–H 486/23. HI 55401

Jahrbuch der Deutschen Iris-Gesellschaft. Berlin-Dahlem. Jahrb. Deutsch. Iris-Ges. See B–P–H 486/25. HI 55403

Jahrbuch der Deutschen Kakteen-Gesellschaft in der Deutschen Gesellschaft für Gartenkultur. Neudamm [=Debno, Poland]. Jahrb. Deutsch. Kakteen-Ges. See B–P–H 487/1. HI 55404

Jahrbuch der deutschen landwirtschaftsgesellschaft. Berlin. Vols. 1-37?, 1886-1922. Jarhb. Deutsch. Landw.-Ges. HI 66398

Jahrbuch der Deutschen Mikrologischen Gesellschaft. Munich. Jahrb. Deutsch. Mikrol. Ges. See B–P–H 487/2. HI 55405

Jahrbuch der deutschen Quartärvereinigung = Eiszeitalter und Gegenwart. Ohringen, Germany. Eiszeitalter & Gegenwart. See B–P–H 356/7.

Jahrbuch des F. Móra Museums, Szeged = A Móra Ferenc múzeum évkönyve. Szeged, Hungary. Móra Ferenc Múz. Évk. See B–P–H 620/7.

Jahrbuch des Forschungsinstitutes für Ampelologie = Szölészeti kutató intézet évkönyve. Budapest. Szölész. Kutató Intéz. Évk. See B–P–H 863/16.

Jahrbuch der Forst- und Jagdkunde. Darmstadt. Vols. ?-10-12, ?-1834-36. Jahrb. Forst- Jagdk. HI 66399

Jahrbuch für Gartenkunde und Botanik. Bonn. Jahrb. Gartenk. Bot. See B–P–H 487/5. HI 55406

Jahrbuch der Geographischen Gesellschaft zu Hannover. Hanover. Jahrb. Geogr. Ges. Hannover. See B–P–H 487/6. HI 55407

Jahrbuch der Geologischen Bundesanstalt. Vienna. Jahrb. Geol. Bundesanst. See B–P–H 487/7. HI 55408

Jahrbuch der geologischen Bundesanstalt. Sonderband. Vienna. 1951+. Jahrb. Geol. Bundesanst. Sonderb. HI 66400

Jahrbuch der Geologischen Landesanstalt. Vienna. Jahrb. Geol. Landesanst. See B–P–H 487/8. HI 55409

Jahrbuch der Geologischen Reichsanstalt. Vienna. Jahrb. Geol. Reichsanst. See B–P–H 487/9. HI 55410

Jahrbuch der Geologischen Staatsanstalt. Vienna. Jahrb. Geol. Staatsanst. See B–P–H 487/10. HI 55411

Jahrbücher der Gewächskunde. Berlin & Leipzig. Jahrb. Gewächsk. See B–P–H 487/11. HI 55412

Jahrbuch der Goethe-Gesellschaft. Weimar. Vols. 1-21, 1914-35. Jahrb. Goethe-Ges. Preceded by: Goethe-Jahrbuch. Superseded by: Goethe. 2-1739-2. HI 66401

Jahrbuch der Hamburgischen Wissenschaftlichen Anstalten. Hamburg. Jahrb. Hamburg. Wiss. Anst. See B–P–H 487/15. HI 55413

Jahrbuch der Hamburgischen Wissenschaftlichen Anstalten. Beihefte. Hamburg. Jahrb. Hamburg. Wiss. Anst. Beih. See B–P–H 487/16. HI 55414

Jahrbuch der k. k. Geologischen Reichsanstalt. Vienna. Jahrb. K. K. Geol. Reichsanst. See B–P–H 487/18. HI 55416

Jahrbücher des Kaiserlich Königlichen polytechnischen Instituts in Wien. Vienna. Jahrb. K. K. Polytechn. Inst. Wien. See B–P–H 487/19. HI 55417

Jahrbuch der Kaiser-Wilhelm-Gesellschaft zur Förderung der Wissenschaften. Berlin. Jahrb. Kaiser-Wilhelm-Ges. Förd. Wiss. See B–P–H 487/17. HI 55415

Jahrbuch der Kaiserlich-Königlichen geologischen Reichsanstalt. Vienna. 1850-1919. Jahrb. Kaiserl.-Konigl. Geol. Reichsanst. Superseded by: Jahrbuch der geologischen Staatsanstalt. HI 66402

Jahrbuch der Königlich Bayerischen Akademie der Wissenschaften. Munich. Jahrb. Königl. Bayer. Akad. Wiss. See B–P–H 488/1. HI 55419

Jahrbuch der königlichen ungarischen geologischen Anstalt. Budapest = A magyar kir[ályi] állami földtani intézet évkönyve. Budapest. Magyar Kir. Állami Földt. Intéz. Évk. See B–P–H 543/15.

Jahrbuch der Königlich Sächsischen Akademie für Forst- und Landwirte zu Tharandt. Leipzig. Jahrb. Königl. Sächs. Akad. Forst- Landwirte Tharandt. See B–P–H 488/3. HI 55421

Jahrbücher der Königlichen Akademie gemeinnütziger Wissenschaften zu Erfurt. Erfurt. Jahrb. Königl. Akad. Gemeinnütz. Wiss. Erfurt. See B–P–H 487/20. HI 55418

Jahrbuch des Königlichen Botanischen Gartens und des Botanischen Museums zu Berlin. Berlin. Jahrb. Königl. Bot. Gart. Berlin. See B–P–H 488/2. HI 55420

Jahrbuch für Landeskunde von Niederdonau = Jahrbuch für Landeskunde von Niederösterreich. Vienna. Jahrb. Landesk. Niederösterreich. See B–P–H 488/6.

Jahrbuch für Landeskunde von Niederösterreich. Vienna. Jahrb. Landesk. Niederösterreich. See B–P–H 488/6. HI 55422

Jahrbuch der Landwirthschaft. Breslau [=Wroclaw, Poland]. Jahrb. Landw. (Breslau). See B–P–H 488/7. HI 55423

Jahrbüch der Literatur. Vienna. Vols. 1-128, 1818-49. Jahrb. Lit. 3-2143-3. HI 52585

Jahrbuch der Max-Planck-Gesellschaft zur Förderung der Wissenschaften e.V. Göttingen. Jahrb. Max-Planck-Ges. Förd. Wiss. See B–P–H 488/8. HI 55424

Jahrbuch für Mikroskopiker. Munich. Jahrb. Mikroskop. See B–P–H 488/9. HI 55425

Jahrbuch für Mineralogie, Geognosie, Geologie und Petrefaktenkunde. Heidelberg. Jahrb. Mineral. See B–P–H 488/10. HI 55426

Jahrbuch der Moorkunde; Bericht über die Fortschritte auf allen Gebieten der Moorkultur und Torfverweertung. Hannover. 1912-40. Jahrb. Moork. HI 66403

Jahrbücher des Nassauischen Vereins für Naturkunde. Wiesbaden. Jahrb. Nassauischen Vereins Naturk. See B–P–H 488/11. HI 55427

Jahrbuch der Naturgeschichte zur Anzeige und Prüfung neuer Entdeckungen und Beobachtungen und zur Aufnahme socher Beyträge, ... Leipzig. Jahrb. Naturgesch. See B–P–H 488/12. HI 55428

Jahrbuch des naturhistorischen Landesmuseums von Kärnten. Klagenfurt, Austria. Jahrb. Naturhist. Landesmus. Kärnten. See B–P–H 488/14. HI 55429

Jahrbuch der naturwissenschaftlichen Abteilungen am Joanneum. Graz. Jahrb. Naturwiss. Abt. Joanneum. See B–P–H 488/15. HI 55430

Jahrbuch der neuesten und wichtigsten Erfindungen und Entdeckungen. Ilmenau, Germany. Jahrb. Neuesten

Wichtigsten Erfind. Entdeck. See B–P–H 488/17. HI 55431

Jahrbuch des Oberösterreichischen Musealvereins. Linz. Jahrb. Oberösterr. Musealvereins. See B–P–H 488/18. HI 55432

Jahrbuch der Österreichischen Wissenschaft = Österreichische Akademie der Wissenschaften. Jahrbuch der Österreichischen Wissenschaft. Vienna. Österr. Akad. Wiss. Jahrb. Österr. Wiss. See B–P–H 692/10.

Jahrbücher für Pflanzenkrankheiten. St. Petersburg = Zhurnal "Bolĕzni Rastenii". St. Petersburg.

Jahrbuch der Pharmacie. Berlin. Vol. 1, 1811. Jahrb. Pharm. HI 66404

Jahrbuch für Pflanzenkrankheiten. Leningrad = Bolezni rastenii. Leningrad.

Jahrbuch für Pomologen, Gärtner und Gartenfreunde. Stuttgart. Vols. 11-13 [also numbered n.s. vols. 1-13], 1871-73. Jahrb. Pomol. Preceded by: Taschenbuch für Pomologen, Gärtner und Gartenfreunde. 3-2142-2. HI 66405

Jahrbuch für practische Pharmacie und verwandte Fächer. Kaiserslautern. Jahrb. Pract. Pharm. Verwandte Fächer. See B–P–H 488/20. HI 55433

Jahrbuch für praktische Pharmacie und verwandte Fächer = Jahrbuch für practische Pharmacie und verwandte Fächer. Kaiserslautern. Jahrb. Pract. Pharm. Verwandte Fächer. See B–P–H 488/20.

Jahrbuch der Preussischen Akademie der Wissenschaften. Berlin. Jahrb. Preuss. Akad. Wiss. See B–P–H 488/21. HI 55434

Jahrbuch der Preussischen Geologischen Landesanstalt. Berlin. Jahrb. Preuss. Geol. Landesanst. See B–P–H 489/1. HI 55435

Jahrbuch der Preussischen Rhein-Universität. Bonn. Jahrb. Preuss. Rhein-Univ. See B–P–H 489/3. HI 55436

Jahrbuch des Reichsamts für Bodenforschung. Berlin. Jahrb. Reichsamts Bodenf. See B–P–H 489/4. HI 55437

Jahrbuch der Reichsstelle für Bodenforschung. Berlin. Jahrb. Reichsstelle Bodenf. See B–P–H 489/5. HI 55438

Jahrbuch der Reisen und neuesten Statistik. Stuttgart. Jahrb. Reisen Neuesten Statist. See B–P–H 489/7. HI 55439

Jahrbuch der Rhododendron-Gesellschaft. Bremen. 1952-54. Jahrb. Rhododendron-Ges. Superseded by: Rhododendrom und Immergrune Laubgeholze Jahrbuch. HI 66406

Jahrbuch des Römisch-Germanischen Zentralmuseums Mainz. Mainz. Jahrb. Röm.-German. Zentralmus. Mainz. See B–P–H 489/8. HI 55440

Jahrbuch des schlesischen Forstvereins. Breslau. 1872-1907, 1873-1908. Jahrb. Schles. Forstvereins. HI 66407

Jahrbuch der Schwedischen Gesellschaft für Geschichte der Wissenschaften = Lychnos; lärdomshistoriska samfundets årsbok. Uppsala & Stockholm. Lychnos. See B–P–H 537/11.

Jahrbuch der schweizerischen naturforschenden Gesellschaft. Wissenschaftlicher Teil. Bern. No. 158+, 1978+. Jahrb. Schweiz. Natruf. Ges. Wiss. Teil. Preceded by: Verhandlungen der schweizerischen naturforschenden Gesellschaft. HI 66408

Jahrbuch des siebenbürgischen Karpathen-Vereins. Hermannstadt [=Sibiu, Rumania]. Jahrb. Siebenbürg. Karpathen-Vereins. See B–P–H 489/11. HI 55441

Jahrbuch der St. Gallischen Naturwissenschaftlichen Gesellschaft. St. Gallen, Switzerland. Jahrb. St. Gallischen Naturwiss. Ges. See B–P–H 489/12. HI 55442

Jahrbuch des staatlichen Museums für Mineralogie und Geologie zu Dresden. Dresden & Leipzig. 1955-64. Jahrb. Staatl. Mus. Mineral. Dresden. Superseded by: Abhandlungen des staatlichen Museums für Mineralogie und Geologie zu Dresden [not entered]. HI 66409

Jahrbuch für Staudenkunde. Langensalza, Germany. Jahrb. Staudenk. See B–P–H 489/14. HI 55443

Jahrbuch vom Thuner- und Brienzersee. Interlaken, Switzerland. Jahrb. Thuner- Brienzersee. See B–P–H 489/15. HI 55444

Jahrbuch des ungarischen Forschungsinstitutes für Pflanzenschutz. Budapest = Növényvédelmi kutató intézet évkönyve. Budapest. Növényvéd. Kutató Intéz. Evk. See B–P–H 674/11.

Jahrbuch des ungarischen forstwissenschaftlichen Instituts = As erdészeti tudományos intézet évkönyve. Budapest. Erdész. Tud. Intéz. Évk. See B–P–H 360/5.

Jahrbuch der ungarischen geologischen Anstalt. Budapest = A magyar állami földtani intézet évkönyve. Budapest. Magyar Állami Földt. Intéz. Évk. See B–P–H 542/19.

Jahrbuch des ungarischen Karpathen-Vereines. Kassa, Kesmark, & Iglo, Hungary [=Kosice, Kezmarok, & Spisska Nova Ves, Czechoslovakia]. Jahrb. Ung. Karpathen-Vereines. See B–P–H 489/19. HI 55445

Jahrbuch der universität Düsseldorf. Triltsch, Düsseldorf. 1968/69+, 1969+. Jahrb. Univ. Düsseldorf. HI 66410

Jahrbuch der Universitäten Deutschlands. Neustrelitz, Germany. Vols. 1-2, 1910-11. Jahrb. Univ. Deutschl. HI 53318

Jahrbuch zur Verbreitung naturwissenschaftlicher Kenntnisse, veranstaltet vom Physikalischen Vereine zu Frankfurt a. M. Frankfurt a. M. Jahrb. Verbreit. Naturwiss. Kenntn. Phys. Vereine Frankfurt. See B–P–H 489/21. HI 55446

Jahrbuch des Vereins für Landeskunde und Heimatpflege im Gau Oberdonau. Linz. Jahrb. Vereins Landesk. Gau Oberdonau. See B–P–H 489/22. HI 55447

Jahrbücher der Vereins für Naturkunde im Herzogthum Nassau. Wiesbaden. Jahrb. Vereins Naturk. Herzogth. Nassau. See B–P–H 490/1. HI 55448

Jahrbuch des Vereins zum Schutz der Bergwelt. Munich. Vol. 42+, 1977+. Jahrb. Vereins Schutz Bergwelt. Preceded by: Jahrbuch des Vereins zum Schutze der Alpenpflanzen und -Tiere. HI 66411

Jahrbuch des Vereins zum Schutze der Alpenpflanzen. Freising & Munich. Jahrb. Vereins Schutze Alpenpfl. See B–P–H 490/2. HI 55449

Jahrbuch des vereins zum Schutze der Alpenpflanzen und -Tiere. Freising & Munich. Vols. 7-41, 1935-76. Jahrb. Vereins Schutze Alpenpfl. Alpentiere. Preceded by: Jahrbuch des vereins zum Schutze der Alpenpflanzen. Superseded by: Jahrbuch des Vereins zum Schutz der Bergwelt. 5-4373-2. HI 66412

Jahrbuch des Vorarlberger Landesmuseumsvereins. Bregenz, Austria. Jahrb. Vorarlberger Landesmuseumsvereins. See B–P–H 490/5. HI 55451

Jahrbuch des Vorarlberger Landesmuseumsvereins Bregenz. Bregenz, Austria. Jahrb. Vorarlberger Landesmuseumsvereins Bregenz. See B–P–H 490/6. HI 55452

Jahrbuch des Vorarlberger Landesmuseums in Bregenz. Dornbirn, Austria. Jahrb. Vorarlberger Landesmus. Bregenz. See B–P–H 490/4. HI 55450

Jahrbuch des Westpreussischen Lehrervereins für Naturkunde. Danzig [=Gdansk. Poland]. Jahrb. Westpreuss. Lehrervereins Naturk. See B–P–H 490/7. HI 55453

Jahrbücher für Wissenschaftliche Botanik. Berlin. Jahrb. Wiss. Bot. See B–P–H 490/8. HI 55454

Jahrbücher für wissenschaftliche Kritik. Herausgegeben von der Societät für wissenschaftliche Kritik zu Berlin. Stuttgart & Tübingen. Jahrb. Wiss. Krit. See B–P–H 490/10. HI 55455

Jahrbuch vom Zürichsee. Zurich. Jahrb. Zürichsee. See B–P–H 490/13. HI 55456

Jahrbuch der Zweigstelle Wien der Reichsstelle für Bodenforschung. Vienna. Jahrb. Zweigstelle Wien Reichsstelle Bodenf. See B–P–H 490/14. HI 55457

Jahrbücher, see Jahrbuch

Jahres-Bericht, see Jahresbericht

Jahresbericht und Abhandlungen des Naturwissenschaftlichen Vereins in Magdeburg. Magdeburg. Jahresber. Abh. Naturwiss. Vereins Magdeburg. See B–P–H 490/22. HI 55458

Jahresbericht des akademischen naturwissenschaftlichen Vereines zu Breslau. Breslau [=Wroclaw, Poland]. Jahresber. Akad. Naturwiss. Vereines Breslau. See B–P–H 490/23. HI 55459

Jahresbericht des Annaberg-Buchholzer Vereins für Naturkunde. Annaberg & Buchholz, Germany. Jahresber. Annaberg-Buchholzer Vereins Naturk. See B–P–H 491/1. HI 55460

Jahresbericht der Arbeitsgemeinschaft sächsischer Botaniker. Dresden. Jahresber. Arbeitsgem. Sächs. Bot. See B–P–H 491/2. HI 55461

Jahresbericht der basler botanischen Gesellschaft. [Basle.] Vol. ?-5+, ?-1956+. Jahresber. Basler Bot. Ges. HI 66413

Jahresbericht der biologischen Bundesanstalt für Land- und Forstwirtschaft in Berlin und Braunschweig. Brunswick. 1945+. Jahresber. Biol. Bundesanst. Land Forstw. HI 66414

Jahresberichte, botanische Staatsinstitut zu Hamburg. Hamburg. 1901-11. Jahresber. Bot. Staatsinst. Hamburg. Superseded by: Institut für angewandte Botanik, Jahresbericht. 3-1787-2. HI 66415

Jahresbericht über den botanischen Garten in Bern. Bern. Jahresber. Bot. Gart. Bern. See B–P–H 491/3. HI 55462

Jahres-Bericht der botanischen Section des Westfälischen Provinzial-Verein für Wissenschaft und Kunst. Münster. 1875+, 1876+. Jahres-Ber. Bot. Sec. Westfäl. Prov.-Verein Wiss. Kunst. HI 66416

Jahresbericht des botanischen Vereines am Mittel- und Niederrheine. Bonn. Jahresber. Bot. Vereines Mittel-Niederrheine. See B–P–H 491/4. HI 55463

Jahresbericht, botanischer Garten der Universität Hamburg. Hamburg. 1974+. Jahresber. Bot. Gart. Univ. Hamburg. HI 66417

Jahresbericht der Bundesanstalt für Pflanzenbau und Samenprüfung in Wien. [From 1949, supplement to: Bodenkultur]. Vienna. 1932-49. Jahresber. Bundesanst. Pflanzenbau Samenprüfung Wien. Preceded by: Bericht der Bundesanstalt für Pflanzenbau und Samenprüfung in Wien. Superseded by: Jahrbuch der Bundesanstalt für Pflanzenbau und Samenprüfung. HI 66418

Jahresbericht, Deutsche Rhododendron Gesellschaft = Deutsche Rhododendron Gesellschaft. Jahresbericht. Deutsche Rhododendron Ges. Jahresber. See B–P–H 347/7.

Jahresberichte, deutscher Forstverein. Berlin. Vols. 1-38,

1925-62. Jahresber. Deutsch. Forstvereins. Preceded by: Bericht über die Hauptversammlung, deutscher Forstverein. 2-1323-1. HI 66419

Jahresbericht der Eidgenössischen Nationalparkkommission. Grosshöchstetten. 1915-23. Jahresber Eidgenöss. Nationalparkkommiss. HI 66420

Jahresbericht der forstlich-phaenologischen Stationen Deutschlands. Berlin. 1885-94. Jahresber. Forstl.-Phaenol. Stat. Deutschl. HI 66421

Jahresberichte über die Fortschritte der Forstwissenschaft und forstlichen Naturkunde. Berlin. Jahresber. Fortschr. Forstwiss. Forstl. Naturk. See B–P–H 491/6. HI 55464

Jahresbericht über die Fortschritte der gesammten Pharmacie und Pharmacologie im In- und Auslande. Erlangen & Würzburg. Jahresber. Fortschr. Gesammten Pharm. Pharmacol. In- Ausl. See B–P–H 491/7. HI 55465

Jahresbericht über die Fortschritte in der Lehre von den Gährungsorganismen und Enzymen. Leipzig. 1890-1911. Jahresber. Fortschr. Lehre Gährungsorganismen Enzymen. HI 66422

Jahresbericht über die Fortschritte in der Lehre von den pathogenen Mikroorganismen, umfassend Bacterien, Pilze und Protozoen. Leipzig. Jahresber. Fortschr. Lehre Pathogenen Mikroorgan. See B–P–H 491/8. HI 55466

Jahresbericht über die Fortschritte in der Pharmacie in allen Ländern. Erlangen & Würzburg. Jahresber. Fortschr. Pharm. in Allen Ländern. See B–P–H 491/9. HI 55467

Jahresbericht über die Fortschritte der Pharmakognosie, Pharmacie und Toxicologie. Göttingen. Jahresber. Fortschr. Pharmakogn. See B–P–H 491/10. HI 55468

Jahres-Bericht über die Fortschritte der physischen Wissenschaften. Tübingen. Jahres-Ber. Fortschr. Phys. Wiss. See B–P–H 491/11. HI 55469

Jahresbericht des Frankfurter Vereins für Geographie und Statistik. Frankfurt a. M. Jahresber. Frankfurter Vereins Geogr. See B–P–H 491/12. HI 55470

Jahresbericht der Fürstlich Jablonowskischen Gesellschaft. Leipzig. Jahresber. Fürstl. Jablonowskischen Ges. See B–P–H 491/13. HI 55471

Jahresbericht des Gartenbau-Vereins für Anhalt. Dessau. Jahresber. Gartenbau-Vereins Anhalt. See B–P–H 491/14. HI 55472

Jahresbericht des Gartenbauvereins in Mainz. Mainz. Jahresber. Gartenbauvereins Mainz. See B–P–H 491/15. HI 55473

Jahresbericht des Gartenbauvereins für die Ober-Lausitz. Görlitz. Jahresber. Gartenbauvereins Ober-Lausitz. See B–P–H 491/16. HI 55474

Jahresbericht über das Gebiet der Pflanzenkrankheiten. Berlin. Vols. 8-16, 1905-13. Jahresber. Geb. Pflanzenkrankh. Preceded by: Jahresbericht über die Neuerungen und Leistungen auf dem Gebiete der Pflanzenkrankheiten. Superseded by: Bibliographie der Pflanzenschutzliteratur. 3-2146-1. HI 66423

Jahrbesbericht der Generaldirektion der staatlichen naturwissenschaftlichen Sammlungen Bayerns. Munich. 1971+, 1972+. Jahresber. Generaldirekt. Staatl. Naturwiss. Samml. Bayerns. HI 66424

Jahresbericht der geographisch-ethnographischen Gesellschaft in Zürich. Zurich. 1899-1917. Jahresber. Geogr.-Ethnogr. Ges. Zürich. Superseded by: Mitteilungen der geographisch-ethnographischen Gesellschaft in Zürich. HI 66425

Jahresbericht der Geographischen Gesellschaft in Hamburg. Hamburg. Jahresber. Geogr. Ges. Hamburg. See B–P–H 491/17. HI 55475

Jahresbericht der Geographischen Gesellschaft zu Hannover. Hanover. Jahresber. Geogr. Ges. Hannover. See B–P–H 492/1. HI 55476

Jahresbericht der geographischen Gesellschaft in München. Munich. Jahresber. Geogr. Ges. München. See B–P–H 492/2. HI 55477

Jahresbericht des Geographischen Vereins zu Frankfurt a./M. Frankfurt am Main. Jahresber. Geogr. Vereins Frankfurt. See B–P–H 492/3. HI 55478

Jahresbericht der Gesellschaft zur Beförderung der Naturwissenschaft zu Freiburg. Freiburg im Breisgau. Jahresber. Ges. Beförd. Naturwiss. Freiburg. See B–P–H 492/4. HI 55479

Jahresbericht der Gesellschaft zur Förderung der naturhistorischen Erforschung des Orients in Wien. Vienna. Jahresber. Ges. Förd. Naturhist. Erforsch. Orients Wien. See B–P–H 492/5. HI 55480

Jahresbericht der Gesellschaft von Freunden der Naturwissenschaften in Gera. Gera, Germany. Jahresber. Ges. Freunden Naturwiss. Gera. See B–P–H 492/6. HI 55481

Jahresbericht (von) der Gesellschaft für Natur- und Heil-Kunde zu Dresden. Dresden. Jahresber. Ges. Natur-Heil-Kunde Dresden. See B–P–H 492/7. HI 55482

Jahresbericht, Gesellschaft für Strahlen- und Umweltforschung MBH. Munich. 1970+. Jahresber. Ges. Strahlen- Umweltforsch. HI 66426

Jahresbericht, Hamburgisches Institut für angewandte Botanik = Hamburgisches Institut für angewandte Botanik. Jahresbericht. Hamburg. Hamburg. Inst. Angew. Bot. Jahresber. See B–P–H 410/26.

Jahresbericht, institut für angewandte Botanik Hamburg = Institut für angewandte Botanik Hamburg.

Jahresbericht. Hamburg. Inst. Angew. Bot. Hamburg. Jahresber. See B–P–H 433/4.

Jahresbericht, Institut für Botanik und botanischer garten der Universität Wien. Vienna. 1972/73+, 1973?+. Jahresber. Inst. Bot. Bot. Gart. Univ. Wien. HI 66427

Jahresbericht des Instituts für angewandte Botanik = Institut für angewandte Botanik. Jahresbericht. Hamburg.

Jahresbericht der Instituts für Geschichte der Naturwissenschaft. Heidelberg. Jahresber. Inst. Gesch. Naturwiss. See B–P–H 492/8. HI 55483

Jahresbericht des Instituts für systematische Botanik und Pflanzengeographie der freien Universität Berlin. Berlin. 1981+, 1982+. Jahresber. Inst Syst. Bot. Pflanzengeogr. Freien Univ. Berlin. HI 66428

Jahres-Bericht der Kgl. Baier. Academie der Wissenschaften in München = Jahresberichte der Kgl. Bayerischen Academie der Wissenschaften in München. Munich. Jahresber. Königl. Bayer. Akad. Wiss. München. See B–P–H 492/11.

Jahresbericht(e) der kgl. ung. geologischen Anstalt. Budapest = A magyar királyi állami földtani intézet évi jelentése(i). Budapest. Magyar Kir. Állami Földt. Intéz. Évi Jel. See B–P–H 543/16.

Jahresbericht der königliche Gärtnerlehranstalt zu Dahlem. Berlin. 190?-05? Jahresber. Königl. Gärtnerlehranst. Dahlem. Superseded by: Bericht der königliche Gärtnerlehranstalt zu Dahlem. HI 66429

Jahresbericht der k. k. Staats-ober-Realschule zu Steyr. Steyr. [Dates of publication not ascertained.] Jahresber. K. K. Staats-Ober-Realschule Steyr HI 66430

Jahresberichte der Kgl. Bayerischen Academie der Wissenschaften in München. Munich. Jahresber. Königl. Bayer. Akad. Wiss. München. See B–P–H 492/11. HI 55484

Jahresberichte der Königl. Schwedischen Akademie der Wissenschaften über die Fortschritte der Botanik. Breslau [=Wroclaw, Poland]. Jahresber. Königl. Schwed. Akad. Wiss. Fortschr. Bot. See B–P–H 492/12. HI 55485

Jahresbericht der königliche Lehranstalt für Obst- und Gartenbau zu Proskau. Oppeln. 19?? [Dates of publication not ascertained.] Jahresber. Königl. Lehranst. Obst- Gartenbau Proskau. Superseded by: Bericht der königliche Lehranstalt für Obst- und Gartenbau zu Proskau. HI 66431

Jahresbericht des Kreismuseums Haldensleben. Haldensleben. Vol. 1+, 1960+. Jahresber. Kreismus. Haldensleben. HI 66432

Jahresbericht, Landesmuseum Joanneum Graz. Graz. N.s. vol. 1+, 1972+. Jahresber. Landesmus. Joanneum Graz. HI 66433

Jahresbericht über die Leistungen und Fortschritte in der Forstwirtschaft. Frankfurt am Main. Vols. 1-9, 1879-87. Jahresber. Leist. Fortschr. Forstw. HI 66434

Jahresbericht über die Leistungen im Gebiete der physiologischen Botanik. 1839-46. Jahresber. Leist. Geb. Physiol. Bot. HI 66435

Jahresbericht der Limnologischen Flussstation Freudenthal. Aussenstelle der Hydrobiologischen Anstalt der Max-Planck-Gesellschaft. Münden, Germany. Jahresber. Limnol. Flussstat. Freudenthal. See B–P–H 492/14. HI 55486

Jahresbericht des Mannheimer Vereins (later Vereines) für Naturkunde. Mannheim. Jahresber. Mannheimer Vereins Naturk. See B–P–H 492/15. HI 55487

Jahresbericht und Mitgliederverzeichnis; internationaler Verband forstlicher Lehranstalten. Stockholm & Zurich. 1929+. Jahresber. Mitgliederverz. Int. Verband Forstl. Forschungsanst. 3-2086-3. HI 66436

Jahresbericht und Mittheilungen des Gartenbau-Vereins im Grossherzogthum Hessen. Darmstadt, Germany. Jahresber. Mitth. Gartenbau-Vereins Grossherzogth. Hessen. See B–P–H 492/16. HI 55488

Jahresbericht und Mittheilungen des Gartenbau-Vereins für Neuvorpommern und Rügen. Greifswald, Germany. Jahresber. Mitth. Gartenbau-Vereins Neuvorpommern. See B–P–H 492/17. HI 55489

Jahresbericht und Mitteilungen des Oberrheinischen geologischen Vereins. Stuttgart. N.s. vol. 1+, 1911+. Jahresber. Mitt. Oberrhein. Geol. Vereins. Preceded by: Berichte über die Versammlung des oberreinischen geologischen Vereins [not entered]. HI 66437

Jahresbericht der Moorkulturstation in Sebastiansberg. Staab, Bohemia [=Stod, Czechoslovakia]. Jahresber. Moorkulturstat. Sebastiansberg. See B–P–H 493/1. HI 55490

Jahresbericht des Museums Francisco-Carolinum. Linz. Jahresber. Mus. Francisco-Carol. See B–P–H 493/2. HI 55491

Jahresbericht, Nachrichten von der Gesellschaft der Wissenschaften zu Göttingen = Nachrichten von der Gesellschaft der Wissenschaften zu Göttingen. Jahresbericht. Berlin. Nachr. Ges. Wiss. Göttingen Jahresber. See B–P–H 626/4.

Jahresbericht der Naturforschenden Gesellschaft in Emden. Emden. Jahresber. Naturf. Ges. Emden. See B–P–H 493/3. HI 55492

Jahresbericht der Naturforschenden Gesellschaft zu Freiburg. Freiburg im Breisgau. Jahresber. Naturf. Ges. Freiburg. See B–P–H 493/4. HI 55493

Jahresbericht der Naturforschenden Gesellschaft Graubündens. Chur, Switzerland. Jahresber. Naturf.

Ges. Graubündens. See B–P–H 493/5. HI 55494

Jahresbericht der naturforschenden Gesellschaft zu Halle. Halle. Jahresber. Naturf. Ges. Halle. See B–P–H 493/6. HI 55495

Jahresbericht des Naturforschenden Vereins zu Riga. Riga [Latvian S S R]. Jahresber. Naturf. Vereins Riga. See B–P–H 493/8. HI 55496

Jahresbericht der Naturhistorischen Gesellschaft zu Hannover. Hanover. Jahresber. Naturhist. Ges. Hannover. See B–P–H 493/9. HI 55497

Jahresberichte der Naturhistorischen Gesellschaft zu Nürnberg. Nuremberg. Jahresber. Naturhist. Ges. Nürnberg. See B–P–H 493/10. HI 55498

Jahresbericht der Naturhistorischen Kantonal-Gesellschaft in Solothurn. Solothurn, Switzerland. Jahresber. Naturhist. Kantonal-Ges. Solothurn. See B–P–H 493/11. HI 55499

Jahres-Bericht des Natur-Historischen Vereines "Lotos". Prague. See B–P–H 493/12. HI 55500

Jahres-Bericht des Naturhistorischen Vereins in Passau. Passau, Germany. Jahres-Ber. Naturhist. Vereins Passau. See B–P–H 493/14. HI 55501

Jahres-Bericht des Naturhistorischen Vereins von Wisconsin. Milwaukee, Wisconsin. Jahres-Ber. Naturhist. Vereins Wisconsin. See B–P–H 493/15. HI 55502

Jahresbericht des Naturhistorischen Verein(e)s in Zweibrücken. Zweibrücken. Jahresber. Naturhist. Vereins Zweibrücken. See B–P–H 493/16. HI 55503

Jahresbericht der naturwissenschaftlichen Gesellschaft in Dresden. Dresden. Jahresber. Naturwiss. Ges. Dresden. See B–P–H 493/17. HI 55504

Jahresbericht der naturwissenschaftlichen Gesellschaft zu Elberfeld. Elberfeld, Germany [=Wuppertal, in part.] Jahresber. Naturwiss. Ges. Elberfeld. See B–P–H 494/1. HI 55505

Jahresbericht der naturwissenschaftlichen Orientvereins in Wien. Vienna. Jahresber. Naturwiss. Orientvereins Wien. See B–P–H 494/2. HI 55506

Jahresbericht über den Naturwissenschaftlichen Verein des Fürstenthums Lüneberg. Lüneburg, Germany. Jahresber. Naturwiss. Verein Fürstenth. Lüneburg. See B–P–H 494/3. HI 55507

Jahresbericht des Naturwissenschaftlichen Vereines zu Bremen. Bremen. Jahresber. Naturwiss. Vereines Bremen. See B–P–H 494/4. HI 55508

Jahresbericht des Naturwissenschaftlichen Vereins Altona. Altona [=Hamburg, in part]. Jahresber. Naturwiss. Vereins Altona. See B–P–H 494/5. HI 55509

Jahres-Berichte des Naturwissenschaftlichen Vereins von Elberfeld und Barmen. Elberfeld, Germany [=Wuppertal, in part]. Jahres-Ber. Naturwiss. Vereins Elberfeld Barmen. See B–P–H 494/7. HI 55510

Jahres-Berichte des Naturwissenschaftlichen Vereins in Elberfeld. Elberfeld, Germany [=Wuppertal, in part]. Jahres-Ber. Naturwiss. Vereins Elberfeld. See B–P–H 494/8. HI 55511

Jahresbericht des Naturwissenschaftlichen Vereins von Elsass-Lothringen und Annales de la société botanique vogéso-rhénane. Strasbourg. Jahresber. Naturwiss. Vereins Elsass-Lothringen & Ann. Soc. Bot. Vogéso-Rhénane. See B–P–H 494/9. HI 55512

Jahresbericht des Naturwissenschaftlichen Vereins in Halle. Berlin. Jahresber. Naturwiss. Vereins Halle. See B–P–H 494/10. HI 55513

Jahresbericht des Naturwissenschaftlichen Vereins zu Magdeburg. Nebst den Sitzungsberichten. Magdeburg. Jahresber. Naturwiss. Vereins Magdeburg Sitzungsber. See B–P–H 494/11. HI 55514

Jahresbericht des Naturwissenschaftlichen Vereins zu Osnabrück. Osnabrück. Jahresber. Naturwiss. Vereins Osnabrück. See B–P–H 494/12. HI 55515

Jahresbericht des Naturwissenschaftlichen Vereins in Wuppertal. Wuppertal, Germany. Jahresber. Naturwiss. Vereins Wuppertal. See B–P–H 494/13. HI 55516

Jahresbericht über die Neuerungen und Leistungen auf dem Gebiete der Pflanzenkrankheiten. Berlin. Vols. 4-7, 1901-04 [1903-05]. Jahresber. Neuerungen Pflanzenkrankh. Preceded by: Jahresbericht über die Neuerungen und Leistungen auf dem Gebiete des Pflanzenschutzes. Superseded by: Jahresbericht über das Gebiet der Pflanzenkrankheiten. 3-2146-1. HI 66438

Jahresbericht über die Neuerungen und Leistungen auf dem Gebiete des Pflanzenschutzes. Berlin. Vols. 1-3, 1898-1900. Jahresber. Neuerungen Leist. Pflanzenschutzes. Superseded by: Jahresbericht über die Neuerungen und Leistungen auf dem Gebiete der Pflanzenkrankheiten. 3-2146-1. HI 66439

Jahresbericht des nieder-österreichischen Landes-Lehrerseminars in Wiener-Neustadt. [Dates of publication not ascertained.] Jahresber. Nieder-Österr. Landes-Lehrersem. Wiener-Neustadt. HI 66440

Jahresbericht des Niedersächsischen botanischen Vereins. (Botanische Abteilung der Naturhistorischen Gesllschaft zu Hannover). Hanover. Jahresber. Niedersächs. Bot. Vereins. See B–P–H 495/1. HI 55517

Jahresbericht des niedersächsischen geologischen Vereins. Hannover. 1911+. Jahresber. Niedersächs. Geol. Vereins. HI 66441

Jahresbericht des Nordoberfränkischen Vereins für Natur-, Geschichts-, Landes- und Familienkunde in Hof a.d.S. Hof, Germany. Jahresber. Nordoberfränk. Vereins Natur- Familienk. Hof. See B–P–H 495/2. HI 55518

Jahresbericht des Oberösterreichischen Musealvereins. Linz. Jahresber. Oberösterr. Musealvereins. See B–P–H 495/3. HI 55519

Jahresberichte der Pflanzenschutzämter. Brunswick. Vol. 1+, 1952+. Jahresber. Pflanzenschutzämter. HI 66442

Jahresbericht der Pharmazie. Göttingen. Jahresber. Pharm. See B–P–H 495/5. HI 55520

Jahresbericht des Physikalischen Vereins zu Frankfurt am Main. Frankfurt a. M. Jahresber. Phys. Vereins Frankfurt. See B–P–H 495/6. HI 55521

Jahresbericht der Pollichia, eines naturwissenschaftlichen Vereins der bayerischen Pfalz. Landau, Germany. Jahresber. Pollichia. See B–P–H 495/7. HI 55522

Jahresbericht der Pollichia, eines naturwissenschaftlichen Vereins der Rheinpfalz = Jahresbericht der Pollichia, eines naturwissenschaftlichen Vereins der bayerischen Pfalz. Landau, Germany. Jahresber. Pollichia. See B–P–H 495/7.

Jahresberichte des pomologischen Instituts zu Proskau. Oppeln. 1908. Jahresber. Pomol. Inst. Proskau. HI 66443

Jahresbericht des Preussischen Botanischen Vereins. Königsberg [= Kaliningrad, R S F S R]. 1891/92-1915/16, 1892-1916; 1930-37. Jahresber. Preuss. Bot. Vereins. Preceded by: Bericht über die wissenschaftlichen Verhandlungen der Jahresversammlung des Preussischen Botanischen Vereins. For 1917-27 see: Bericht des Preussischen Botanischen Vereins. 4-3432-3. HI 66444

Jahresbericht über der Resultate der Arbeiten im Felde der physiologischen Botanik. Berlin. 1837-43. Jahresber. Result. Arbeiten Felde Physiol. Bot. HI 66445

Jahresbericht der S W A wissenschaftlichen Gesellschaft = Report (Annual) of the south west african scientific society. Windhoek.

Jahresbericht über die Schicksale des Vereins für Kunde der Natur und der Kunst im Fürstenthum Hildesheim und der Stadt Goslar. Hildesheim. Jahresber. Schicksale Vereins Kunde Natur Fürstenth. Hildesheim. See B–P–H 495/9. HI 55523

Jahresbericht der Schlesischen Gesellschaft für vaterländische Cultur. Breslau [=Wroclaw, Poland]. Jahresber. Schles. Ges. Vaterl. Cult. See B–P–H 495/10. HI 55524

Jahres-Bericht, Schlesischer Lehrer-Verein für Naturkunde = Schlesischer Lehrer-Verein für Naturkunde. Jahres-Bericht. Görlitz. Schles. Lehrer-Verein Naturk. Jahres-Ber. See B–P–H 818/1.

Jahresbericht der Schwedischen Academie der Wissenschaften über die Fortschritte der Naturgeschichte, Anatomie und Physiologie der Thiere und Pflanzen. Bonn. Jahresber. Schwed. Acad. Wiss. Fortschr. Naturgesch. Thiere Pflanzen. See B–P–H 495/14. HI 55525

Jahresbericht, schweizerische Naturschutzkommission. [Freiburg.] Vols. 1-7, 1906/07-1913/14. Jahresber. Schweiz. Naturschutzkommiss. HI 66447

Jahresbericht der schweizerischen Gesellschaft für Vererbungsforschung. [Largely reprinted from: Archiv der Julius-Klaus-Stiftung für Vererbungsforschung, Sozialanthropologie und Rassenhygeine.] Zurich. Nos. 1-30, 1941-70. Jahresber. Schweiz. Ges. Vererbungsf. HI 66446

Jahresbericht der schweizerischen Nationalpark-kommission = Jahresbericht der Eidgenössischen Nationalparkkommission. Grosshöchstetten.

Jahresbericht des Sonnblick-Vereins. Vienna. Jahresber. Sonnblick-Vereins. See B–P–H 495/16. HI 55526

Jahresbericht, Staatsinstitut für angewandte Botanik, Hamburg = Staatsinstitut für angewandte Botanik, Hamburg. Jahresbericht. Hamburg. Staatsinst. Angew. Bot. Hamburg Jahresber. See B–P–H 854/16.

Jahresbericht, Station für Pflanzenschutz zu Hamburg. Hamburg. Vols. 1-19, 1900?-18. Jahresber. Stat. Pflanzenschutz Hamburg. HI 66448

Jahresbericht der Stiftung "Herbarium Paul Aellen". Basel. 1962-70. Jahresber. Stiftung Herb. Paul Aellen. HI 66449

Jahresbericht, systematisch-geobotanisches Institut der Universität Bern. Berne. [Dates of publication not ascertained.] Jahresber. Syst.-Geobot. Inst. Univ. Bern. HI 66450

Jahresbericht über die Tätigkeit, Bundesanstalt für Pflanzenbau und Samenprüfung in Wien = Jahresbericht der Bundesanstalt für Pflanzenbau und Samenprüfung in Wien. Vienna.

Jahresbericht über die Thätigkeit des Naturwissenschaftlichen Vereins für das Fürstenthum Lüneburg. Lüneburg, Germany. Jahresber. Thätigk. Naturwiss. Vereins Fürstenth. Lüneburg. See B–P–H 495/17. HI 55527

Jahresbericht über die Thätigkeit des Vereins für Naturkunde in Cassel. Cassel [=Kassel]. Jahresber. Thätigk. Vereins Naturk. Cassel. See B–P–H 496/1. HI 55528

Jahresbericht des Thüringer Gartenbau-Vereins zu Gotha. Gotha. Jahresber. Thüringer Gartenbau-Vereins Gotha. See B–P–H 496/2. HI 55529

Jahresbericht(e) der ungarischen geologischen Anstalt. Budapest = A magyar állami földtani intézet évi jelentese(i). Budapest. Magyar Állami Földt. Intéz. Évi Jel. See B–P–H 542/18.

Jahres-Bericht über den Verein für Kunde der Natur und der Kunst im Fürstenthum Hildesheim und in der Stadt Goslar. Hildesheim. Jahres-Ber. Verein Kunde Natur Fürstenth. Hildesheim. See B–P–H 496/4. HI 55530

Jahresbericht der Vereinigung für angewandte Botanik. Berlin. Jahresber. Vereinigung Angew. Bot. See B–P–H 496/5. HI 55531

Jahresbericht der Vereinigung der Vertreter der angewandten Botanik. Berlin. Jahresber. Vereinigung Vertreter Angew. Bot. See B–P–H 496/6. HI 55532

Jahresbericht des Vereins zur Gründung eines naturhistorischen Museums zu Hannover. Hanover. Jahresber. Vereins Gründung Naturhist. Mus. Hannover. See B–P–H 496/7. HI 55533

Jahresbericht des Vereins für Naturkunde zu Mannheim. Mannheim. Jahresber. Vereins Naturk. Mannheim. See B–P–H 496/8. HI 55534

Jahresbericht des Vereins für Naturkunde in Österreich ob der Enns. Linz. Jahresber. Vereins Naturk. Österreich ob der Enns. See B–P–H 496/9. HI 55535

Jahresbericht des Vereins für Naturkunde an der Unterweser. Geestemünde [=Bremerhaven, in part]. Jahresber Vereins Naturk. Unterweser. See B–P–H 496/10. HI 55536

Jahresbericht des Vereins für Naturkunde zu Zwickau (in Sachsen). Zwickau. Jahresber. Vereins Naturk. Zwickau. See B–P–H 496/11. HI 55537

Jahres-Bericht, Verein für Naturkunde Mannheim = Verein für Naturkunde Mannheim. Jahres-Bericht. Mannheim. Verein Naturk. Mannheim Jahres-Ber. See B–P–H 949/1.

Jahresbericht des Vereins für Naturwissenschaft zu Braunschweig. Brunswick, Germany. Jahresber. Vereins Naturwiss. Braunschweig. See B–P–H 496/12. HI 55538

Jahresberichte des Vereins für Pomologie und Gartenbau in Meiningen = Verhandlungen des Vereins für Pomologie und Gartenbau in Meiningen. Meiningen, Germany. Verh. Vereins Pomol. Meiningen. See B–P–H 955/17.

Jahres-Bericht über die Verhandlungen des Naturwissenschaftlichen Vereins in Hamburg. Jahres-Ber. Verh. Naturwiss. Vereins Hamburg. See B–P–H 496/14. HI 55539

Jahresbericht über die Verrichtungen und den Zustand der Naturforschenden Gesellschaft in Emden. Emden. Jahresber. Verrichtungen Zustand Naturf. Ges. Emden. See B–P–H 496/15. HI 55540

Jahres-Bericht des Westfälischen Provinzial-Vereins für Wissenschaft und Kunst. Münster. Jahres-Ber. Westfäl. Prov.-Vereins Wiss. See B–P–H 497/1. HI 55541

Jahres-Bericht der Westphälischen Gesellschaft für vaterländische Cultur. Minden. Jahres-Ber. Westphäl. Ges. Vaterl. Cult. See B–P–H 497/2. HI 55542

Jahresbericht der Wetterauer Gesellschaft für die gesammte Naturkunde zu Hanau. Hanau. Jahresber. Wetterauer Ges. Gesammte Naturk. Hanau. See B–P–H 497/3. HI 55543

Jahresbericht der Wetterauischen Gesellschaft für die gesammte Naturkunde. Hanau. Jahresber. Wetterauischen Ges. Gesammte Naturk. See B–P–H 497/5. HI 55544

Jahresbericht über die Wirksamkeit der beiden Komitees für die naturwissenschaftliche Durchforschung Böhmens. Prague. Jahresber. Wirksamk. Beiden Komitees Naturwiss. Durchforsch. Böhmens. See B–P–H 497/6. HI 55545

Jahresbericht über die Wirksamkeit und den Zustand der naturforschenden Gesellschaft in Emden. Emden. Jahresber. Wirksamk. Zustand Naturf. Ges. Emden. See B–P–H 497/7. HI 55546

Jahresbericht der Zürcherischen Botanischen Gesellschaft. Zurich. Jahresber. Zücherischen Bot. Ges. See B–P–H 497/8. HI 55547

Jahresberichte, see Jahresbericht

Jahresgabe, altmärkisches Museum Stendal. Tangermünde. Vols. 1-20, [1947]-68. Jahresg. Altmärk. Mus. Stendal. HI 66451

Jahresheft des Naturwissenschaftlichen Vereines des Trencsiner Komitates = A trencsénvármegyei természettudományi egylet évkönyve. Trencsén, Hungary [=Trencin, Czechoslovakia]. Trencsénvárm. Term. Egyl. Évk. See B–P–H 896/7.

Jahresheft, Sitzungsberichte der Heidelberger Akademie der Wissenschaften = Sitzungsberichte der Heidelberger Akademie der Wissenschaften. Jahresheft. Heidelberg. Sitzungsber. Heidelberger Akad. Wiss. Jahresh. See B–P–H 840/3.

Jahresheft, Sitzungsberichte der Heidelberger Akademie der Wissenschaften = Sitzungsberichte der Heidelberger Akademie der Wissenschaften. Stiftung Heinrich Lanz. Jahresheft. Heidelberg. Sitzungsber. Heidelberger Akad. Wiss. Stiftung Heinrich Lanz Jahresh. See B–P–H 840/6.

Jahreshefte, Gesellschaft für Naturkunde in Württemberg. Stuttgart. Vol. 124+, 1969+. Jahresh. Ges. Naturk. Württemberg. Preceded by: Jahreshefte des Vereins für vaterländische Naturkunde in Württemberg. HI 66452

Jahreshefte der naturwissenschaftlichen Section der K. K.

Mährisch-Schlesischen Gesellschaft für Ackerbau, Natur- und Landeskunde in Brünn. Brünn [=Brno, Czechoslovakia]. Jahresh. Naturwiss. Sect. K. K. Mähr.-Schles. Ges. Ackerbau Brünn. See B–P–H 497/10. HI 55548

Jahreshefte des Naturwissenschaftlichen Vereins für das Fürstenthum Lüneburg. Lüneburg & Dannenberg, Germany. Jahresh. Naturwiss. Vereins Fürstenth. Lüneburg. See B–P–H 497/11. HI 55549

Jahreshefte des Naturwissenschaftlichen Vereins zu Lüneburg. Lüneburg, Germany. Jahresh. Naturwiss. Vereins Lüneburg. See B–P–H 497/12. HI 55550

Jahreshefte des Vereines des Krainisches Landes-Museums. Laibach. Vols. 1-3, 1856, 1858 & 1862. Jahresh. Vereines Krain. Landes-Mus. 5-4366-2. HI 74849

Jahreshefte des Vereins für Mathematik und Naturwissenschaften in Ulm a/D. Ulm. Jahresh. Vereins Math. Ulm. See B–P–H 497/16. HI 55551

Jahreshefte des Vereins für vaterländische Naturkunde in Württemberg. Stuttgart. Vols. 1-123, 1845-1968. Jahresh. Vereins Vaterl. Naturk. Württemberg. Superseded by: Jahreshefte, Gesellschaft für Naturkunde in Württemberg. 5-4372-1. HI 66453

Jahres-Katalog pro 189 ... des Wiener botanischen Tauschanstalt. Vienna. 18??-1902. Jahreskat. Wiener Bot. Tauschanst. Preceded by: Jahres-Katalog der Wiener botanischen Tauschvereins. HI 66454

Jahres-Katalog pro 189 ... des Wiener botanischen Tauschvereins. Vienna. 1894-95-? Jahreskat. Wiener Bot. Tauschvereins. Superseded by: Jahres-Katalog der Wiener botanischen Tauschanstalt. HI 66455

Jahresschrift für mitteldeutsche Vorgeschichte. Halle. 1954+. Jahresschr. Mitteldeutsche Vorgesch. HI 66456

Jahresverhandlungen: Kurländische Gesellschaft für Literatur und Kunst. Mitau, [=Jelgava, Latvian S S R]. Jahresverh. Kurl. Ges. Lit. See B–P–H 498/2. HI 55552

Jährliche Uebersicht der Thätigkeit der Gesellschaft für Erdkunde in Berlin. Berlin. Jährl. Uebers. Thätigk. Ges. Erdk. Berlin. See B–P–H 498/3. HI 55553

Jahrs-Bericht des Naturwissenschaftlichen Vereins in Hamburg. Hamburg. Jahrs-Ber. Naturwiss. Vereins Hamburg. See B–P–H 498/4. HI 55554

Jamaica journal; quarterly journal of the institute of Jamaica. Kingston, Jamaica. 1967+. Jamaica J. HI 66457

Jamaica naturalist. Kingston, Jamaica. Jamaica Naturalist. See B–P–H 498/6. HI 55555

Jamaica and west indian review. Kingston, Jamaica. Vols. 1-9(1), 1963-72. Jamaica W. Indian Rev. Preceded and superseded by: West indian review. HI 66458

Janus. Breslau [=Wroclaw, Poland]. Janus (Breslau). See B–P–H 498/7. HI 55556

Janus Leiden. Janus (Leiden). See B–P–H 498/8. HI 55557

Janus pannonius múzeum évkönyve. Pecs, Hungary. Janus Pannon. Múz. Évk. See B–P–H 498/9. HI 55558

Janus pannonius múzeum füzetei. Pecs. Vol. 1+, 1962+. Janus Pannon. Múz. Füz. HI 66459

Japan agricultural research quarterly. Yatabe. 1966+. Japan Agric. Res. Quart. HI 66460

Japan fertilizer news. [Nihon hiryo shinbun.] Tokyo. Japan Fertilizer News. See B–P–H 499/22. HI 55574

Japan magazine, a representative monthly of things japanese. Tokyo. Japan Mag. See B–P–H 499/25. HI 55575

Japan orchid society bulletin. [Ran.] Osaka. Vol. ?-3+, ?-1963+. Japan Orchid Soc. Bull. HI 75156

Japan science review. Biological sciences. Tokyo. 1949-64. Japan Sci. Rev., Biol. Sci. HI 66461

Japan society for the promotion of scientific research. [Kagaku nanyo.] Tokyo. Japan Soc. Promot. Sci. Res. See B–P–H 499/26. HI 55576

Japanese agricultural sciences index. [Nihon nogaku bunken kiji sakuin.] Tokyo. 1978+. Jap. Agric. Sci. Index. HI 66462

Japanese antarctic research. Tokyo. 1973+. Jap Antarc. Res. HI 66463

Japanese bee journal. Takikawa?, Japan. Jap. Bee J. See B–P–H 498/10. HI 55559

Japanese forestry. [Nihon ringyo.] Tokyo. 1949+. Jap. Forest. HI 68902

Japanese fruits. [Kajitsu nihon.] Tokyo. Jap. Fruits. See B–P–H 498/11. HI 55560

Japanese gardening. [Engei nippon.] Tokyo. Jap. Gard. See B–P–H 498/12. HI 55561

Japanese journal of bacteriology. [Nihon saikingaku zasshi.] Tokyo. Vol. 1+, 1944+. Jap. J. Bacteriol. HI 66464

Japanese journal of biometeorology. 1966+. Jap. J. Biometeorol. HI 66465

Japanese journal of botany. [Nippon shokubutsugaku shuho.] Tokyo. Vols. 1-20, 1922-75. Jap. J. Bot. Superseded by: Recent progress of natural sciences in Japan. 3-2154-2. HI 66466

Japanese journal of breeding. Tokyo. Jap. J. Breed. See B–P–H 498/14. HI 55562

Japanese journal of crop science. [Nippon sakumotsu gakkai kiji.] Tokyo. Vol. 46+, 1977+. Jap J. Crop Sci. Preceded by: Proceedings, crop science society of Japan. HI 66467

Japanese journal of dermatology. Tokyo. Jap. J. Dermatol. See B–P–H 498/18. HI 55563

Japanese journal of dermatology and urology. Tokyo. Jap. J. Dermatol. Urol. See B–P–H 498/19. HI 55564

Japanese journal of dermatology and venereology. Tokyo. Jap. J. Dermatol. Venereol. See B–P–H 498/20. HI 55565

Japanese journal of ecology. [Nippon seitaigakkai-shi.] Sendai. Vol. 1+. 1954+. Jap. J. Ecol. Preceded by: Bulletin of the society of plant ecology. HI 66468

Japanese journal of genetics. [Idengaku zasshi.] Tokyo Jap. J. Genet. See B–P–H 498/22. HI 55566

Japanese journal of geology and geography. Tokyo. 1922+. Jap J. Geol. Geogr. HI 66469

Japanese journal of the history of biology. [Seibutsugakushi kenkyu.] Tokyo. No. ?-43+, ?-1984+. Jap J. Hist. Biol. HI 66470

Japanese journal of limnology. [Rikusui gakkai shi.] Tokyo. Jap. J. Limnol. See B–P–H 498/23. HI 55567

Japanese journal of medical mycology. [Shinkin to shinkinsho.] Tokyo. 1960+. Jap J. Med. Mycol. HI 66471

Japanese journal of medical science and biology. Tokyo. Jap. J. Med. Sci. Biol. See B–P–H 499/1. HI 55568

Japanese journal of medical sciences. Part 4, pharmacology. Kyoto, Japan. Jap. J. Med. Sci., Pt. 4, Pharmacol. See B–P–H 499/2. HI 55569

Japanese journal of Michurin biology. [Michurin seibutsugaku kankyu.] Tokyo. Vol. ?-11+, ?-1975+. Jap. J. Michurin Biol. HI 66472

Japanese journal of microbiology. Tokyo. Vols. 1-20, 1957-76. Jap. J. Microbiol. Superseded by: Microbiology and immunology. HI 66473

Japanese journal of palynology. [Nihon kafun gakkai kaishi.] Shizuoka. 1965+. Jap. J. Palynol. HI 66474

Japanese journal of pharmacognosy. [Syoyakugaku zasshi.] Chiyoda. 1947+. Jap. J. Pharmacogn. HI 66475

Japanese journal of pharmacology. Kyoto, Japan. Jap. J. Pharmacol. See B–P–H 499/4. HI 55570

Japanese journal of phycology. [Sorui.] Tokyo. Vol. 26+, 1978+. Jap. J. Phycol. Preceded by: Bulletin, japanese society of phycology. HI 66476

Japanese journal of tropical agriculture. Tokyo. Jap. J. Trop. Agric. See B–P–H 499/5. HI 55571

Japanese journal of water treatment biology. 1966+. Jap. J. Water Treatm. Biol. HI 66477

Japanese periodicals index, science and technology. [Zasshi kifi sakuin, kagaku gijutsu-hen.] Tokyo. Vol. 1+, 1950+. Jap. Period. Index Sci. Technol. HI 66478

Japanese scientific contents. Tokyo. Vol. 0(1) sample copy, 1985; vol. 1+, 198?+. Jap. Sci. Contents. HI 66479

Japanese society of grassland science journal. [Nihon sochi gakkai shi.] Jap. Soc. Grassland Sci. J. See B–P–H 499/17. HI 55572

Japanese studies in the history of science. Tokyo. Vols. 1-18, 1962-79; Supplement, vols. 1-2, 1971. Jap. Stud. Hist. Sci. Superseded by: Historia scientiarum. HI 66480

Japanische Zeitschrift für Dermatologie und Urologie. Tokyo. Jap. Z. Dermatol. Urol. See B–P–H 499/20. HI 55573

Japanscan; bio-industry bulletin. 1983+. Japanscan. HI 66481

Jardin. Paris. Jardin. See B–P–H 500/3. HI 55581

Jardín botánico. Mexico City. Jard. Bot. See B–P–H 499/27. HI 55577

Jardin fleuriste; journal général des progrès et des intériêts horticoles et botaniques. Ghent. Jard. Fleur. See B–P–H 499/30. HI 55578

Jardin y paisaje. Madrid. Vol. 1+, 1972+. Jard. Paisaje. HI 66482

Jardin y sus plantas; boletín de la sociedad Argentina de horticultura. Buenos Aires. Vol. 34(185/186)+, 1976+. Jard. & Pl. Preceded by: Boletín de la sociedad argentina de horticultura. HI 66483

Jardinage; comment on soigne son jardin. Versailles. Vol. 1+, 1911+. Jardinage. HI 66484

Jardinier portatif. Paris. Jard. Portatif. See B–P–H 500/2. HI 55580

Jardinier suisse; journal de la société helvétique d'horticulture. Geneva. Vols. 1-55-?, 1873-1927-?. Jard. Suisse. HI 66486

Jardins. Paris. Jardins. See B–P–H 500/6. HI 55582

Jardins d'aujourd'hui; cahiers de l'amateur. Paris. Vols. 1-14, 1952-55. Jard. Aujourd'hui. HI 66487

Jardins de France. Paris. Jard. France. See B–P–H 500/1. HI 55579

Jardins et logis. Brussels. Vol. 23+, 1961+. Jard. & Logis. Preceded by: Courrier horticole. HI 66488

Jare scientific reports = Scientific reports, Japan artarctic research expedition. Series E, biology. Tokyo.

Jarovizacija. Žurnal po biologii i razvitija rastenij = Yarovizatsiya. Zhurnal po biologii i razvitiya rastenii. Moscow & Odessa.

Jaruda. ?-1987+. Jaruda. HI 66489

Je tai hai yang = Tropic oceanology. Beijing.

Je tai tso wu hsüeh pao = Chinese journal of tropical crops. Tan-hsien.

Je tai tso wu yan chiu t'ung hsun = Tropical crops research bulletin. Canton. Trop. Crops Res. Bull. See B–P–H 897/10.

Jeddah journal of marine research. Jeddah. Vol. 1, 1981. Jeddah J. Mar. Res. Superseded by: Journal of the faculty of marine science, King Abdulaziz university. HI 66490

Jeffersonia; newsletter of Virginia botany. Bridgewater, VA. No. 1+, 1986?+. Jeffersonia. HI 66491

Jelentés a magyar nemzeti múzeum ... évi állapotáról. Budapest. Jel. Magyar Nemz. Múz. Évi Állapot. See B–P–H 500/33. HI 55583

Jenaische allgemeine Literatur-Zeitung. [Includes: Intelligenzblatt (1804-41), and Ergänzungsblätter (1813-41)]. Jena. Vols. 1-37, 1804-41. Jenaische Allg. Lit.-Zeitung. Superseded by: Neue jenaische allgemeine Literatur-Zeitung. 3-2160-1. HI 51682

Jenaische Zeitschrift für Medizin herausgegeben von der medicinisch-naturwissenschaftlichen Gesellschaft zu Jena. Leipzig. Jenaische Z Med. See B–P–H 500/37. HI 55584

Jenaische Zeitschrift für Medizin und Naturwissenschaft herausgegeben von der medicinisch-naturwissenschaftlichen Gesellschaft zu Jena = Jenaische Zeitschrift für Medizin herausgegeben von der medicinisch-naturwissenschaftlichen Gesellschaft zu Jena. Leipzig. Jenaische Z Med. See B–P–H 500/37.

Jenaische Zeitschrift für Naturwissenschaft herausgegeben von der medicinisch-naturwissenschaftlichen Gesellschaft zu Jena. Jena. Jenaische Z. Naturwiss. See B–P–H 500/39. HI 55585

Jenaische Zeitungen von gelehrten Sachen. Jena. 1765-71, 1766-72. Jenaische Zeitungen Gel. Sachen. Superseded by: Jenaische gelehrte Zeitungen [not entered]. 3-2160-1. HI 52888

Jepson globe; newsletter from the friends of the Jepson herbarium. Berkeley, CA. No. 1+, 1987+. Jepson Globe. HI 66492

Jernel sains Malaysia = Malaysian journal of science. Kuala Lumpur.

Jerusalem naturalists' club bulletin. Jerusalem. 1945-46. Jerusalem Naturalists' Club Bull. HI 66493

Jeschken- Iserland. Beiträge zur Heimatkunde. Reichenberg [=Liberec, Czechoslovakia]. Jeschken-Iserland. See B–P–H 501/2. HI 55586

Jeune scientifique. Joliette. Vols. 1-8(2), 1962-69. Jeune Sci. Superseded by: Quebec science. HI 66494

Jilin nongye daxue xuebao = Acta agriculturae universitatis jilinensis. Changchun.

Jilindaxue xuebao. Chueh-moon = Kirin university journal. Natural science edition. Chi-lin.

Jinan liyi xuebao = Journal of science and medicine of Jinan university. Kuang-chou.

Jissai engei = Practical horticulture. [Japan]. Pract. Hort. See B–P–H 719/3.

Jodrell laboratory news. Kew = Jodrell newsletter from the laboratories at Kew and Wakehurst Place. Kew.

Jodrell newsletter from the laboratories at Kew and Wakehurst Place. Kew. No. ?-4+, ?-1989+. Jodrell Newslett. Lab. Kew Wakehurst Place. HI 66495

John Innes horticultural institution leaflet. London, Bayfordbury. 1940+. John Innes Hort. Inst. Leafl. HI 66496

Johns Hopkins university studies in geology. [Dates of publication not ascertained.] John Hopkins Univ. Stud. Geol. HI 66497

Joint agricultural research and development project publication. University college of north Wales. Aberystwyth. No. 1+, 1972+. Joint Agric. Res. Developm Proj. Pub. Univ. Coll. N. Wales. HI 66498

Joint bulletin, Vermont botanical and bird club. [Joint bulletin no. 1 is also Bulletin no. 10 of the Vermont botanical club and Bulletin no. 9 of the Vermont bird club]. Burlington, VT. No. 1+, 1915+. Joint Bull. Vermont Bot. Bird Club. Preceded by: Bulletin of the Vermont botanical club. 5-4378-1. HI 66499

Joint contributions from the horticultural institute, Kyushu, and phytotechnical institute, Miyazaki college of agriculture. Fukuoka, Japan. Joint Contr. Hort. Inst. Kyushu Phytotechn. Inst. Miyazaki Coll. Agric. See B–P–H 501/23. HI 55587

Joint publication of the commonwealth agricultural bureaux = Joint publication of the imperial agricultural bureaux. Aberystwyth, Wales. Joint Publ. Imp. Agric. Bur. See B–P–H 501/25.

Joint publication of the imperial agricultural bureaux. Aberystwyth, Wales. Joint Publ. Imp. Agric. Bur. See B–P–H 501/25. HI 55588

Joint report, British Columbia forest service / canadian forestry service. Victoria, B.C. No. 1+, 1974+. Joint Rep. Brit. Columbia Forest Serv., Canad. Forest. Serv. HI 66500

Jojoba happenings. Tucson, Phoenix, AZ. 1972+. Jojoba

Happen. HI 66501

Jolly green gardener = Newsletter, Denver botanical garden. Denver, CO.

Jord och skog. Stockholm. Jord & Skog. See B–P–H 501/27. HI 55589

Jornal de agronomia; orgao do sindicato agronômico do estado do São Paulo. São Paulo. Vols. 1-3, 1938-40. J. Agron. 3-2178-2. HI 66502

Jornal hortícolo-agrícola. Oporto, Portugal. Jorn. Hort.-Agríc. See B–P–H 501/30. HI 55591

Jornal de horticultura pratica. Oporto, Portugal. Jorn. Hort. Prat. See B–P–H 501/29. HI 55590

Jornal dos reflorestadores. São Paulo. Vol. ?-6(38)+, ?-1983+. J. Reflorestad. HI 66503

Jornal de sciencias mathematica, physicas e naturaes. Lisbon. 1866-1927. J. Sci. Math. Phys. Nat. Superseded by: Boletim da academia das ciências de Lisboa. HI 66504

Jornal de sciencias naturais. Lisbon. Jorn. Sci. Nat. See B–P–H 501/31. HI 55592

Journal of A V C. London = A V C. Agricultural & veterinary chemicals. London.

Journal of aberrant botany. Cambridge, MA. Vol. 1(1), 1972. J. Aberr. Bot. HI 66505

Journal of abstracts from research; french literature on tropical and subtropical agriculture. Paris. Experimental issue, 1976. J. Abstr. Res. HI 66506

Journal of the academy of natural sciences of Philadelphia. Philadelphia, PA. J. Acad. Nat. Sci. Philadelphia. See B–P–H 454/14. HI 54950

Journal of the Accrington naturalists' and antiquarians' society. Accrington. No. 1+, 1965+. J. Accrington Naturalists' Antiq. Soc. HI 66507

Journal of the Adelaide botanic gardens. Adelaide, S.A. Vol. 1+, 1975+. J. Adelaide Bot. Gard. HI 66508

Journal of the african society. London. Vols. 1(1)-34(136), 1901-35. J. African Soc. Superseded by: Journal of the royal african society. 1-80-2. HI 66509

Journal of the agricultural association of China. Nanking. J. Agric. Assoc. China (Nanking). See B–P–H 455/2. HI 54956

Journal of the agricultural association of China. Taipei, Taiwan. J. Agric. Assoc. China (Taipei). See B–P–H 455/3. HI 54957

Journal of agricultural botany. Kharkov. Ukrainian S S R = Trudy sil's'ko-gospodars'koi botaniky. Kharkov.

Journal of the agricultural chemical society of Japan. [Nihon nogei kagaku kaisi.] Tokyo. Vols. 1-2(4), 1924-26. J. Agric. Chem. Soc. Japan. Superseded by: Bulletin of the agricultural chemical society of Japan. 1-87-2. HI 66510

Journal of the agricultural experiment station, government general Chosen. [Noji shikenjo kenkyu hokoku.] Suigen [=Suwon], Korea. J. Agric. Exp. Sta. Gov. Gen. Chosen. See B–P–H 455/8. HI 54960

Journal of agricultural and food chemistry. Easton, PA. J. Agric. Food Chem. See B–P–H 455/9. HI 54961

Journal of the agricultural and horticultural society of India. Calcutta. J. Agric. Soc. India. See B–P–H 456/11. HI 54975

Journal of the agricultural and horticultural society of Western Australia. Perth, Australia. J. Agric. Soc. Western Australia. See B–P–H 456/16. HI 54978

Journal of the agricultural research centre. Helsinki = Annales agriculturae Fenniae. Helsinki. Ann. Agric. Fenniae. See B–P–H 96/6.

Journal of agricultural research. Department of agriculture. Washington, DC. J. Agric. Res. See B–P–H 456/3. HI 54970

Journal of the agricultural researches in Tokai-kinki region. [Tokai kinki nogy kenkyu.] Mie, Japan. J. Agric. Res. Tokai-Kinki Region. See B–P–H 456/6. HI 54971

Journal of agricultural science. Cambridge, England. J. Agric. Sci. (Cambridge). See B–P–H 456/7. HI 54972

Journal of agricultural science. Nanking. J. Agric. Sci. (Nanking). See B–P–H 456/8. HI 54973

Journal of agricultural science. [Tokyo nogo daigaku nogyo shuho.] Tokyo. J. Agric. Sci. (Tokyo). See B–P–H 456/9. HI 54974

Journal of agricultural sciences. Yangon. Vol. 1+, 1989+. J. Agric. Sci. (Yangon). HI 74850

Journal of the agricultural society of Japan. [Dai Nippon nokwai.] Tokyo. J. Agric. Soc. Japan. See B–P–H 456/12. HI 54976

Journal of the agricultural society of Korea. [Chosen nokai ho. Keijo.] Seoul?, Korea. J. Agric. Soc. Korea. See B–P–H 456/13. HI 54977

Journal of the agricultural society of Trinidad and Tobago. Port of Spain. Vol. 51+, 1951+. J. Agric. Soc. Trinidad Tobago. Preceded by: Proceedings of the agricultural society of Trinidad and Tobago. 1-91-2. HI 66511

Journal of the agricultural department of the university college of Wales [title varies]. Aberystwyth. Vols. 1-18, 1908-29. J. Agric. Dept. Univ. Coll. Preceded by: Transactions of the agricultural society, university college of Wales. Superseded by: Journal of the agricultural society, university college of Wales. 5-4439-1. HI 66512

Journal of agricultural research in Iceland = Islenzkar landbunada ramsoknir. Reykjavik.

Journal of agricultural science, Tokyo nogyo daigaku. Supplement. [Tokyo nogyo daigaku. Nogaku shuho, tokubetsugo.] Tokyo. Vol. 1+, 1978+. J. Agric. Sci. Tokyo Nogyo Daigaku, Suppl. HI 66513

Journal of agricultural and scientific research. Bichpuri. Vol. 20+, 1978+. J. Agric. Sci. Res. Preceded by: Balvant Vidyapeeth journal of agricultural and scientific research. HI 66514

Journal of the agricultural society, university college of Wales. Aberystwyth. Vol. 19+, 1930+. J. Agric. Soc. Univ. Coll. Wales. Preceded by: Journal of the agricultural department of the university college of Wales. 5-4439-1. HI 66515

Journal of agriculture. Boston, MA. J. Agric. (Boston). See B–P–H 454/19. HI 54951

Journal of agriculture. Edinburgh. J. Agric. (Edinburgh). See B–P–H 454/20. HI 54952

Journal of agriculture. Melbourne, Vic. Vols. 53(6)-76, 1955-79. J. Agric. (Melbourne). Preceded by: Journal of the department of agriculture, Victoria. HI 66516

Journal d'agriculture. Montreal. J. Agric. (Montreal). See B–P–H 454/22. HI 54953

Journal of agriculture. New York, NY. J. Agric. (New York). See B–P–H 454/23. HI 54954

Journal of agriculture of British Guiana. Georgetown, British Guiana [Guiana]. J. Agric. British Guiana. See B–P–H 455/6. HI 54959

Journal d'agriculture, d'économie rurale et des manufactures du royaume des Pays-Bas. Brussels. J. Agric. Pays-Bas. See B–P–H 455/17. HI 54967

Journal of agriculture, faculty of science, Seoul agricultural college. [Chon nong.] Suwon, Korea. J. Agric. (Suwon). See B–P–H 455/1. HI 54955

Journal of agriculture and forestry. Nanking. J. Agric. Forest. (Nanking). See B–P–H 455/10. HI 54962

Journal of agriculture and forestry. [Nung lin hsüeh pao.] Taichung, Taiwan. J. Agric. Forest. (Taichung). See B–P–H 455/11. HI 54963

Journal d'agriculture et horticulture. Montreal. J. Agric. Hort. See B–P–H 455/12. HI 54964

Journal d'agriculture, d'horticulture, d'économie rurale, de botanique et des manufactures de la Belgique, ... Namur, Belgium. J. Agric. Belgique. See B–P–H 455/4. HI 54958

Journal d'agriculture, d'horticulture, ... manufactures des Pays-Bas = Journal d'agriculture, d'économie rurale et des manufactures du royaume des Pays-Bas. Brussels. J. Agric. Pays-Bas. See B–P–H 455/17.

Journal d'agriculture illustré. Montreal. J. Agric. Ill. See B–P–H 455/14. HI 54965

Journal of agriculture and industry of South Australia. Adelaide. Vols. 1-8(6), 1897-1905. J. Agric. Industr. South Australia. Superseded by: Journal of the department of agriculture, South Australia. 3-2196-2. HI 66517

Journal d'agriculture, de médecine et des sciences accessoires. Évreux, France. J. Agric. Méd. See B–P–H 455/16. HI 54966

Journal of agriculture, New Zealand department of agriculture. Wellington, N.Z. Vols. 1-6, 1913-18. J. Agric. New Zealand Dep. Agric. Preceded by: Journal of the department of agriculture, New Zealand. Superseded by: New Zealand journal of agriculture. 4-3054-2 HI 66518

Journal d'agriculture pratique, d'économie rurale ... de Belgique. Liège & Brussels. J. Agric. Prat. Belgique. See B–P–H 456/1. HI 54969

Journal d'agriculture pratique, de jardinage et d'économie domestique. Paris. J. Agric. Prat. See B–P–H 455/18. HI 54968

Journal of agriculture and research. Nanking = Journal of the agricultural association of China. Nanking. J. Agric. Assoc. China (Nanking). See B–P–H 455/2.

Journal of agriculture of South Australia. Adelaide, S.A. Vol. 67(4)+, 1963+. J. Agric. S. Australia. Preceded by: Journal of the department of agriculture, South Australia. HI 66519

Journal d'agriculture du sud-ouest. Toulouse. J. Agric. Sud-Ouest. See B–P–H 456/17. HI 54979

Journal d'agriculture suisse. Geneva. J. Agric. Suisse. See B–P–H 456/18. HI 54980

Journal of agriculture (Tasmania). Hobart. Vols. 49(4)-52, 1978-81. J. Agric. (Tasmania). Preceded by: Tasmanian journal of agriculture. HI 66520

Journal d'agriculture traditionelle et de botanique appliquée = J A T B A. Paris.

Journal d'agriculture tropicale. Paris. Vols. 1-19, 1901-19. J. Agric. Trop. Superseded by: Revue de botanique appliquée e d'agriculture coloniale. 3-2180-2. HI 66521

Journal d'agriculture tropicale et de botanique appliquée. Paris. Vols. 1-23, 1954-76. J. Agric. Trop. Bot. Appl. Preceded by: Revue international de botanique appliquée et d'agriculture tropicale. Superseded by: J A T B A. 3-2180-2. HI 66522

Journal of agriculture of the university of Puerto Rico. Río Piedras, PR. J. Agric. Univ. Puerto Rico. See B–P–H 456/23. HI 54981

Journal of agriculture of Western Australia. South Perth, W.A. Ser. 4, vol. 12+, 1952+. J. Agric. W. Australia.

Preceded by: Journal of the department of agriculture, Western Australia. HI 66523

Journal of the agri-horticultural society of western India. Poona. 1904-07. J. Agri-Hort. Soc. W. India. HI 66524

Journal of agronomy and crop science. Berlin, New York. Vol. 156+, 1986+. J. Agron. Crop Sci. Preceded by: Zeitschrift für Acker- und Pflanzenbau. HI 66525

Journal of the air pollution control association. Vol. 5+, 1955+. J. Air Pollut. Control Assoc. Preceded by: Air repair. HI 74851

Journal of the Akita mining college. Akita, Japan. J. Akita Mining Coll. See B–P–H 457/1. HI 54983

Journal of the Alabama academy of science. Birmingham, AL. J. Alabama Acad. Sci. See B–P–H 457/3. HI 54984

Journal, Alberta society of petroleum geologists = Bulletin of canadian petroleum geology. Calgary, Alta. Bull. Canad. Petrol. Geol. See B–P–H 243/21.

Journal of allergy. St. Louis, MO. J. Allergy. See B–P–H 457/5. HI 54985

Journal of the amateur orchid growers' society. Vols. ?-9(2), 1958-60. J. Amateur Orchid Growers' Soc. Preceded by: Bulletin of the amateur orchid growers' society. Superseded by: Journal of the orchid society of Great Britain. HI 66526

Journal of the american association for the promotion of science, literature, and the arts. New York, NY. J. Amer. Assoc. Promot. Sci. See B–P–H 457/10. HI 54986

Journal of the american bamboo society. Solana Beach, CA. Vol. 1+, 1980+. J. Amer. Bamboo Soc. HI 66527

Journal of the american chemical society. J. Amer. Chem. Soc. See B–P–H 457/11. HI 54987

Journal of the american chestnut foundation. St. Paul, MN. Vol. 1+, 1985+. J. Amer. Chestnut Found. HI 66528

Journal of the american college of toxicology. New York. 1981+. J. Amer. Coll. Toxicol. HI 66529

Journal, american magnolia society. Vol. ?-16+, ?-1980+. J. Amer. Magnolia Soc. HI 66530

Journal of the american medical association. Chicago, IL. J. Amer. Med. Assoc. See B–P–H 457/13. HI 54988

Journal of the american oil chemists' society. Chicago, IL. Vol. 24+, 1947+. J. Amer. Oil Chem. Soc. Preceded by: Oil and soap. 1-294-2. HI 66531

Journal of the american pharmaceutical association. Columbus, OH. Vols. 1-28, 1912-39; n.s. vols. 1-17, 1961-77. J. Amer. Pharm. Assoc. Preceded by: Bulletin, american pharmaceutical association. From 1940 divided into: Journal of the american pharmaceutical association. Practical pharmacy edition and Journal of the american pharmaceutical association. Scientific edition. Superseded by: American pharmacy. 3-2232-2. HI 66532

Journal of the american pharmaceutical association. Practical pharmacy edition. Columbus, OH. Vols. 1-21, 1940-60. J. Amer. Pharm. Assoc., Pract. Pharm. Ed. Preceded and superseded by: Journal of the american pharmaceutical association. 3-2232-2. HI 66533

Journal of the american pharmaceutical association. Scientific edition. Columbus, OH. Vols. 29-49, 1940-60. J. Amer. Pharm. Assoc., Sci. Ed. Preceded by: Journal of the american pharmaceutical association. Superseded by: Journal of pharmaceutical sciences. 3-2232-2. HI 66534

Journal of the american rhododendron society. Tigard, OR. Vol. 36+, 1982+. J. Amer. Rhododendron Soc. Preceded by: Quarterly bulletin of the american rhododendron society. HI 66535

Journal of the american silk society and rural economist. Baltimore, MD. J. Amer. Silk Soc. See B–P–H 457/16. HI 54989

Journal of the american society of agronomy. Geneva, NY. See B–P–H 457/17. HI 54990

Journal of the american society for horticultural science. Geneva, NY, & St. Joseph, MI. Vol. 94+, 1969+. J. Amer. Soc. Hort. Sci. Preceded by: Proceedings of the american society for horticultural science. HI 66536

Journal of the american society of sugar beet technologists. Fort Collins, CO. J. Amer. Soc. Sugar Beet Technol. See B–P–H 457/18. HI 54991

Journal de l'anatomie et de la physiologie normales et pathologiques de l'homme et des animaux. Paris. Vols. 1-50, 1864-1919. J. Anat. Physiol. Norm. Pathol. Homme Anim. HI 66537

Journal, Andra Pradesh academy of sciences. Hyderabad. Vol. ?-6+, ?-1971+. J. Andra Pradesh Acad. Sci. HI 66538

Journal anglais. Paris. J. Anglais. See B–P–H 457/19. HI 54992

Journal of Anhwei pedagogical institute. [An hwei shih fan hsüeh yuan hsüeh pao.] [China]. J. Anhwei Pedagog. Inst. See B–P–H 457/20. HI 54993

Journal of the anthropological institute of New York. New York, NY. J. Anthropol. Inst. New York. See B–P–H 457/22. HI 54994

Journal of the anthropological society of Nippon. [Jinruigaku zasshi.] Tokyo. J. Anthropol. Soc.

Nippon. See B–P–H 457/23. HI 54995

Journal of antibacterial and antifungal agents = Journal of research society for antibacterial and antifungal agents, Japan.

Journal of antibiotics. Tokyo. Vols. 1-5, 1948-52; vol. 21+, 1968+. J. Antibiot. Preceded by: Journal of penicillin. For 1953-67 see: Journal of antibiotics, Series A and Journal of antibiotics. Series B. 3-2198-3. HI 66539

Journal of antibiotics. Series A. Tokyo. Vols. 6-20, 1953-67. J. Antibiot., A. Preceded and superseded by: Journal of antibiotics. 3-2198-3. HI 66540

Journal of antibiotics. Series B. Tokyo. Vols. 6-20, 1953-67. J. Antibiot., B. Preceded and superseded by: Journal of antibiotics. 3-2198-3. HI 66541

Journal of the Aomori-ken biological society. [Aomori-ken seibutsu gakkaishi.] Aomori. ?-1980+. J. Aomori-Ken Biol. Soc. HI 66542

Journal of apicultural research. London. J. Apic. Res. See B–P–H 457/25. HI 54996

Journal of applied bacteriology. Reading, England. J. Appl. Bacteriol. See B–P–H 458/2. HI 54997

Journal of applied biochemistry. New York, Orlando, FL. Vols. 1-7, 1979-85. J. Appl. Biochem. Superseded by: Biotechnology and applied biochemistry. HI 66543

Journal of applied chemistry. Oxford, London, Barking. Vols. 1-20, 1951-70. J. Appl. Chem. Superseded by: Journal of applied chemistry and biotechnology. 3-2199-1. HI 75181

Journal of applied chemistry and biotechnology. Oxford, London, Barking. Vols. 21-28, 1971-78. J. Appl. Chem. Biotechol. Preceded by: Journal of applied chemistry. Superseded by: Journal of chemical technology and biotechnology. HI 75182

Journal of applied ecology, Oxford, England. J. Appl. Ecol. See B–P–H 458/3. HI 54998

Journal of applied microscopy and laboratory methods. Rochester, NY. J. Appl. Microscop. Lab. Meth. See B–P–H 458/4. HI 54999

Journal of applied mycology; Hokkaido university. [Oyo kinkagu.] Sapporo, Japan. J. Appl. Mycol. Hokkaido Univ. See B–P–H 458/5. HI 55000

Journal of applied phycology. Dordrecht & Boston, MA. Vol. 1+, 1989+. J. Appl. Phycol. HI 66544

Journal of applied plant sociology. Worcester, MA. J. Appl. Pl. Sociol. See B–P–H 458/6. HI 55001

Journal of applied science and record of progress in the industrial arts. London. Vols. 1-13(2), 1870-82. J. Appl. Sci. Rec. Progr. Industr. Arts. 3-2200-1. HI 66545

Journal of applied seed production. Corvallis, OR. Vol. 1+, 1983+. J. Appl. Seed Prod. HI 66546

Journal of applied toxicology. Chichester. Vol. 1+, 1981+. J. Appl. Toxicol. HI 75203

Journal of aquaculture in the tropics. Calcutta. Vol. 1+, 1986+. J. Aquac. Trop. HI 66547

Journal of aquatic plant management. Fort Myers, FL, Vicksburg, MS. Vol. 14+, 1976+. J. Aquatic Pl. Managem. Preceded by: Hyacinth control journal. HI 66548

Journal of arboriculture. Urbana, IL. Vol. 1+, 1975+. J. Arboric. Preceded by: Arborist's news and Proceedings of the annual meeting of the international shade tree conference. HI 66549

Journal of arid environments. London, New York, etc. Vol. 1+, 1978+. J. Arid Environm HI 66550

Journal of the Arizona academy of science. Tucson, AZ. Vols. 1-12, 1959-77. J. Arizona Acad. Sci. Superseded by: Journal of the Arizona-Nevada Academy of Science. HI 66551

Journal of the Arizona-Nevada academy of science. Tucson, Tempe, AZ. Vol. 13+, 1978+. J. Arizona-Nevada Acad. Sci. Preceded by: Journal of the Arizona academy of science. HI 66552

Journal of the Arnold arboretum. Cambridge, MA. J. Arnold Arbor. See B–P–H 458/10. HI 55002

Journal of asian ecology. Taichung. Vol. 1+, 1979+. J. Asian Ecol. Preceded by: Asian environmental review. HI 66553

Journal of the asiatic society. Calcutta. 3rd ser. vols. 17-24, 1951-58; 4th ser. vol. 1+, 1959+. J. Asiat. Soc. Preceded by: Journal of the royal asiatic society of Bengal. HI 66554

Journal of the asiatic society of Bengal. Calcutta. J. Asiat. Soc. Bengal. See B–P–H 458/15. HI 55004

Journal of the asiatic society of Bengal. Part 2, natural history. Calcutta. Vols. 34-?, 1865-1904. J. Asiat. Soc. Bengal, Pt. 2, Nat. Hist. Preceded by: Journal of the asiatic society of Bengal. Superseded by: Journal and proceedings of the asiatic society of Bengal. 1-505-3. HI 66555

Journal asiatique, ou recueil de mémoires, d'extraits et de notices relatifs à l'histoire, ... des peuples orientaux. Paris. J. Asiat. See B–P–H 458/13. HI 55003

Journal of the Assam science society. Gauhati. Vol. 3+, 1959+. J. Assam Sci. Soc. Preceded by: Journal of the Gauhati science society. HI 66556

Journal, association for the advancement of agricultural sciences in Africa. Addis Ababa. Vols. 1-4, 1971-77; Supplement, 1973. J. Assoc. Advancem. Agric. Sci. Africa. Superseded by: African journal of agricultural sciences. HI 66556

Journal de l'association des biologistes du Québec. Montreal. Vols. 1-5(2), 1976-80. J. Assoc. Biol. Québec. HI 66557

Journal of the association of military surgeons of the United States. Carlisle, PA. J. Assoc. Military Surgeons U.S. See B–P–H 458/17. HI 55005

Journal of the association of official agricultural chemists. Baltimore, MD. Vols. 1-48, 1915-65. J. Assoc. Off. Agric. Chem. Superseded by: Journal of the association of official analytical chemists. HI 66558

Journal of the association of official analytical chemists. Baltimore, MD. Vol. 49+, 1966+. J. Assoc. Off. Analytical Chem. Preceded by: Journal of the association of official agricultural chemists. HI 66559

Journal, Auckland botanical society. Auckland, N.Z. 198?+. J. Auckland Bot. Soc. HI 66560

Journal of audiovisual media in medicine. London. Vol. 1+, 1978+. J. Audiovis. Media Med. Preceded by: Medical and biological illustration. HI 74532

Journal of the australian garden history society. Launceston, N.S.W. Vols. 1-?, 1980-? J. Austral. Gard. Hist. Soc. Superseded by?: Australian garden journal. HI 66561

Journal of the australian institute of agricultural science. Sydney. J. Austral. Inst. Agric. Sci. See B–P–H 458/19. HI 55006

Journal of the australian rhododendron society. Olinda. Vol. 1(1-3), 1959-61. J. Austral. Rhododendron Soc. Preceded by: Rhododendron. Melbourne. HI 66562

Journal of bacteriology. Baltimore, MA. J. Bacteriol. See B–P–H 458/22. HI 55007

Journal of the Bahama society for the diffusion of knowledge. Nassau, Bahamas. J. Bahama Soc. Diffusion Knowl. See B–P–H 458/23. HI 55008

Journal of the balneological society of Japan. [Onsen kagaku.] J. Balneol. Soc. Japan. See B–P–H 459/1. HI 55009

Journal of bamboo research. [Zhuzi yanjiu huikan.] Hangzhou. Vol. 1+, 1982+. J. Bamboo Res. HI 66563

Journal of the Bangladesh academy of sciences. Dacca. Vol. 1+, 1977+. J. Bangladesh Acad. Sci. HI 66564

Journal of the Barbados museum and historical society. Bridgetown. 1933+. J. Barbados Mus. Hist. Soc. HI 66565

Journal of basic microbiology. Berlin. Vol. 25+, 1985+. J. Basic Microbiol. Preceded by: Zeitschrift für allgemeine Mikrobiologie. HI 66566

Journal des beaux-arts et des sciences. Paris. Vols. 1-6, 1768-78 [4 issues, each called a "tome" were published each year; in 1774 and 1775 an additional 4 issues, each called a "Supplément," were published]. J. Beaux-Arts Sci. Preceded by: Mémoires pour l'histoire des sciences et des beaux arts. 3-2188-3. HI 66567

Journal of the Bedfordshire natural history society and field club = Bedfordshire naturalist. Bedford, England. Bedfordshire Naturalist. See B–P–H 168/17.

Journal of Beijing forestry college. [Beijing linxueyuan xuebao.] Beijing. 1980+. J. Beijing Forest. Coll. HI 66568

Journal of the Bengal natural history society. Darjeeling. 1939-69. J. Bengal Nat. Hist. Soc. Preceded by: Journal of the Darjeeling natural history society. HI 66569

Journal de bibliographie médicale par une société de médecine. Paris. J. Bibliogr. Méd. See B–P–H 459/5. HI 55010

Journal of the Bihar botanical society. Gaya. Vol. 1+, 1972+. J. Bihar Bot. Soc. HI 66570

Journal of biochemical and biophysical methods. Amsterdam, etc. Vol. 1+, 1979+. J. Biochem. Biophys. Meth. HI 66571

Journal of biochemical and microbiological technology. New York. Vols. 1-3, 1959-61. J. Biochem Microbiol. Techn. Superseded by: Biotechnology and bioengineering. HI 66572

Journal of biocommunication. Durham, NC. Vol. 1+, 1974+. J. Biocommun. HI 74852

Journal of bioenergetics. New York. Vols. 1-8(2), 1970-76. J. Bioenerget. Superseded by: Journal of bioenergetics and biomembranes. HI 66573

Journal of bioenergetics and biomembranes. New York. Vol. 8(3)+, 1968+. J. Bioenerget. Biomembr. Preceded by: Journal of bioenergetics. HI 66574

Journal of biogeography. Hull, Oxford, etc. Vol. 1+, 1974+. J. Biogeogr. HI 66575

Journal of the biological board of Canada. Toronto. Vols. 1-3(5), 1934-37. J. Biol. Board Canada. Preceded by: Contributions to canadian biology and fisheries. Superseded by: Journal of the fisheries research board of Canada. HI 66576

Journal of biological chemistry. Baltimore, MD. J. Biol. Chem. See B–P–H 459/8. HI 55012

Journal of biological education. London. Vol. 1+, 1967+. J. Biol. Educ. (London). HI 66577

Journal of the biological institute. Seoul, Korea. J. Biol. Inst. See B–P–H 459/9. HI 55013

Journal, biological photographic association. Baltimore, MD. Vols. 1-47, 1932-79. J. Biol. Photogr. Assoc. Superseded by: Journal of biological photography. 1-707-3. HI 66578

Journal of biological photography. Durham, NC. Vol.

48+, 1980+. J. Biol. Photogr. Preceded by: Journal, biological photographic association. HI 66579

Journal of biological physics. Stillwater, OK. Vol. 1+, 1973+. J. Biol. Phys. HI 66580

Journal of biological rhythms. New York, London. Vol. 1+, 1986+. J. Biol. Rhythms. HI 66581

Journal of biological sciences. Baghdad. Vols. 13-14(1), 1982-83. J. Biol. Sci. Preceded by: Bulletin of the biological research center, university of Baghdad. Superseded by: Journal of biological sciences research. Baghdad. HI 66582

Journal of biological sciences. Bombay. Vol. 1+, 1958+. J. Biol. Sci. (Bombay). HI 55014

Journal, biological sciences curriculum study = B S C S journal. Boulder, CO.

Journal of biological sciences research. Baghdad. Vol. 14(2)+, 1983+. J. Biol. Sci. Res. Preceded by: Journal of biological sciences. Baghdad. HI 66583

Journal of biological standardization. New York, London. Vols. 1-17, 1973-89. J. Biol. Standardiz. Superseded by: Biologicals. HI 66584

Journal of biology. Bucharest = Revue de biologie. Bucharest.

Journal of biology (Vietnam). ?-1987+. J. Biol. (Vietnam). HI 66585

Journal of biology. Osaka city university. Osaka. Vols. 12-17, 1961-66. J. Biol. Osaka City Univ. Preceded by: Journal of the institute of polytechnics, Osaka city university. Series D. Biology. HI 66586

Journal of the biometeorological society of Japan = Japanese journal of biometeorology.

Journal of biomolecular structure and dynamics. Guilderland, NY. Vol. 1+, 1983+. J. Biomolec. Struct. Dynam. HI 66587

Journal of biophysical and biochemical cytology. New York, NY. J. Biophys. Biochem. Cytol. See B–P–H 459/15. HI 55015

Journal of biosciences. Bangalore. Vol. 1+, 1979+. J. Biosci. Preceded by: Proceedings of the indian academy of sciences. Experimental biology section. HI 66588

Journal of the Birmingham natural history and philosophical society. Birmingham, England. J. Birmingham Nat. Hist. Soc. See B–P–H 459/16. HI 55016

Journal of the board of agriculture. London. J. Board Agric. See B–P–H 459/17. HI 55017

Journal of the board of agriculture of British Guiana. Demarara, Georgetown. Vols. 1-20, 1907-27. J. Board Agric. British Guiana. Superseded by: Agricultural journal of British Guiana. 1-89-2. HI 66589

Journal, board of greenkeeping research, british golf unions = Board of greenkeeping research, british golf unions, journal. Bingley, England. Board Greenkeeping Res. Brit. Golf Unions J. See B–P–H 201/5.

Journal of the Bombay branch of the royal asiatic society. Bombay. Vols. 1(1)-26(75), 1841-1923; n.s. vol. 1+, 1925+. J. Bombay Branch Roy. Asiat. Soc. HI 66590

Journal of the Bombay natural history society. Bombay. J. Bombay Nat. Hist. Soc. See B–P–H 459/20. HI 55018

Journal of bone and joint surgery. Boston, MA. J. Bone Joint Surg. See B–P–H 459/21. HI 55019

Journal of the Boston society of medical sciences. Boston, MA. Vols. 1-5, 1896-1901. J. Boston Soc. Med. Sci. Superseded by: Journal of medical research. Boston. 3-2226-1. HI 66591

Journal of botanical analysis. [Chih-wu fen-liu hsüeh-pao.] Peking. J. Bot. Analysis. See B–P–H 460/11. HI 55027

Journal of the botanical society of South Africa. Kirstenbosch. Vols. 1-60, 1915-74. J. Bot. Soc. South Africa. Superseded by: Veld and flora. 1-756-1. HI 66592

Journal of the botanical society of China. [Chung kuo chih wu hsüeh tsa chih.] Peiping [=Peking]. J. Bot. Soc. China. See B–P–H 460/16. HI 55029

Journal für die Botanik. [Edited by H. A. Schrader.] Göttingen. J. Bot. (Schrader). See B–P–H 460/6. HI 55024

Journal de botanique. Copenhagen = Botanisk tidsskrift. Copenhagen.

Journal de botanique. [Edited by Desvaux.] Paris. J. Bot. (Desvaux). See B–P–H 460/1. HI 55020

Journal de botanique. [Edited by L. Morot.] Paris. J. Bot. (Morot). See B–P–H 460/4. HI 55022

Journal botanique de l'académie des sciences de la R S S d'Ukraine = Botanichnyi zhurnal. Kiev.

Journal de botanique, appliquée à l'agriculture, à la pharmacie, à la médecine et aux arts. Paris. Bot. Agric. See B–P–H 460/10. HI 55026

Journal de botanique historique = Tidsskrift för historisk botanik. Copenhagen.

Journal de botanique néerlandaise. Amsterdam. J. Bot. Néerl. See B–P–H 460/14. HI 55028

Journal botanique de l'U R S S = Botanicheskii zhurnal S S S R. Moscow & Leningrad.

Journal of botany. [Chih-wu hsüeh-pao.] Peking. J. Bot. (Peking). See B–P–H 460/5. HI 55023

Journal of botany. Tokyo. J. Bot. (Tokyo). See

B–P–H 460/8. HI 55025

Journal of botany, (being a second series of the Botanical miscellany), containing figures and descriptions ... [Edited by W. J. Hooker.] London. J. Bot. (Hooker). See B–P–H 460/2. HI 55021

Journal of botany, british and foreign. London. Vols. 1-80, 1863-1942 [vols. 10-20, 1872-82, also numbered n.s. vols. 1-11]. J. Bot. In 1942 incorporated in: Annals and magazine of natural history. 3-2202-2. HI 66593

Journal of botany of the United Arab Republic = United Arab Republic journal of botany. Cairo.

Journal of brewing. [Jozogaku zasshi.] Osaka, Japan. J. Brew. See B–P–H 460/22. HI 55030

Journal britannique. The Hague. J. Brit. See B–P–H 460/23. HI 55031

Journal of the british dental association. London. J. Brit. Dental Assoc. See B–P–H 460/24. HI 55032

Journal, british flower association. ?-1972. J. Brit. Fl. Assoc. Superseded by: British flower. HI 66594

Journal of the british gardeners' association. London. J. Brit. Gard. Assoc. See B–P–H 460/25. HI 55033

Journal of the british grassland society. Penglais. Vols. 1-33, 1946-78. J. Brit. Grassland Soc. Superseded by: Grass and forage science. 1-782-2. HI 66595

Journal of the British Honduras agricultural society. Belize. Vols. 1-[6], 1935-42. J. Brit. Honduras Agric. Soc. HI 66595

Journal of the British North Borneo branch of the royal asiatic society. Sandakan. 1895-97. J. Brit. N. Borneo Branch Roy. Asiat. Soc. HI 66596

Journal of the british science guild. London. Nos. 1-23, 1915-27. J. Brit. Sci. Guild. 1-793-1. HI 66597

Journal of the british wood preserving association. London. Vols. 1-5, 1929-35. J. Brit. Wood Preserving Assoc. 1-795-1. HI 66598

Journal of the bromeliad society. Los Angeles, CA, etc. Vol. 21+, 1971+. J. Bromeliad Soc. Preceded by: Bulletin of the bromeliad society [242/31]. HI 66599

Journal of bryology. Oxford. Vol. 7+, 1972+. J. Bryol. Preceded by: Transactions of the british bryological society. HI 66600

Journal of the Burma research society. Rangoon. Vols. 1-?, 1911-41. J. Burma Res. Soc. 1-850-2. HI 66601

Journal of the cactus and succulent club of Tokyo. [Shaboten sha.] Tokyo. No. 1+, 1953+. J. Cact. Succ. Club Tokyo. HI 66603

Journal of the cactus and succulent society of America. Los Angeles, CA. J. Cact. Succ. Soc. Amer. See B–P–H 461/1. HI 55034

Journal of the cactus and succulent society of Japan. [Shaboten.] Vols. 1-? [vol. 5(1) also numbered no. 37], 1932?-?? J. Cact. Succ. Soc. Japan. HI 66604

Journal of the cactus and succulent society of New Zealand. Auckland. Vols. 1(4)-3(7), 1947-50. J. Cact. Succ. Soc. New Zealand. Preceded by: Cactus and succulent notes. Superseded by: Cactus and succulent journal. Auckland. HI 66602

Journal of the cactus and succulent society of South Australia. Norwood, S.A. Vol. 1+, 1967+. J. Cact. Succ. Soc. S. Australia. HI 66605

Journal of California anthropology. Banning, CA. Vols. 1-5, 1974-78. J. Calif. Anthropol. Superseded by: Journal of California and Great Basin anthropology. HI 66606

Journal of California and Great Basin anthropology. Banning, CA. Vol. 1+, 1979+. J. Calif. Great Basin Anthropol. Preceded by: Journal of California anthropology. HI 66607

Journal of the California horticultural society. San Francisco, CA. Vols. 1-23, 1940-67. J. Calif. Hort. Soc. Superseded by: California horticultural journal. 1-886-2. HI 66608

Journal, California rare fruit growers. Fullerton, CA. Vol. 18, 1986. J. Calif. Rare Fruit Growers. Preceded, in part, by: California rare fruit growers' yearbook. HI 66609

Journal of the Camborne-Redruth natural history society. Camborne. Vols. [1]-?, 1963-74. J. Camborne-Redruth Nat. Hist. Soc. HI 66610

Journal of the canadian medical association. Toronto. J. Canad. Med. Assoc. See B–P–H 461/8. HI 55035

Journal of the canadian peat society. Ottawa. J. Canad. Peat Soc. See B–P–H 461/9. HI 55036

Journal canadien de biochimie = Canadian journal of biochemistry. Ottawa.

Journal canadien de botanique = Canadian journal of botany. Ottawa.

Journal canadien de génétique et de cytologie = Canadian journal of genetics and cytology. Ottawa.

Journal canadien de physiologie et de pharmacologie = Canadian journal of physiology and pharmacology. Ottawa.

Journal canadien de la recherche forestière = Canadian journal of forest research. Ottawa.

Journal canadien des sciences halieutiques et aquatiques = Canadian journal of fisheries and aquatic sciences. Ottawa.

Journal of cancer research. Baltimore, MD. J. Cancer Res. See B–P–H 461/10. HI 55037

Journal, Canterbury botanical society. Christchurch, N.Z.

Vol. 1+, 1968+. J. Canterbury Bot. Soc. HI 66611

Journal of cell biology. New York, NY. J. Cell Biol. See B–P–H 461/11. HI 55038

Journal of cell science. Cambridge. 1966+. J. Cell. Sci. Preceded by: Quarterly journal of microscopical science. HI 66612

Journal of cell science. Supplement. Colchester. Vol. 1+, 1984+. J. Cell Sci., Suppl. HI 66613

Journal of cellular biochemistry. New York. Vol. 18+, 1972+. J. Cell. Biochem. Preceded by: Journal of supramolecular structure and cellular biochemistry. HI 66614

Journal of cellular and comparative physiology. Philadelphia, PA. J. Cell. Comp. Physiol. See B–P–H 461/12. HI 55039

Journal of cellular physiology. Supplement. Philadelphia, PA. Vol. 6+, 1982+. J. Cell Physiol., Suppl. Preceded by: Journal of supramolecular structure and cellular biochemistry. Supplement. HI 66615

Journal of the central agricultural experiment station. [Noji shikenjo kenkyu hokoku.] Urawa, Japan. J. Centr. Agric. Exp. Sta. See B–P–H 461/15. HI 55040

Journal of the central and associated chambers of agriculture, London. London. Vols. 40-?, 1912-36 [Jan.-Sept. 1923 published as supplement to British farmer and journal of agriculture.] J. Centr. Assoc. Chambers Agric. London. Preceded by: Agricultural record. 2-956-3. HI 66616

Journal of central China teachers' college. Natural science edition. [Huazhong shiyuan xuebao. Ziran kexue ban.] Wuchang. No. ?-14+, ?-1980+. J. Centr. China Teachers Coll., Nat. Sci. Ed. HI 66617

Journal of the centre for arab gulf studies. Basrah. Vols. 2-6, 1975-76. J. Centre Arab Gulf Stud. Preceded by?: Arabian gulf magazine. Superseded by: Arab gulf. HI 66618

Journal of cereal science. London, New York. Vol. 1+, 1983+. J. Cereal Sci. HI 66619

Journal of the Ceylon branch of the british medical association. Colombo, Ceylon. J. Ceylon Branch Brit. Med. Assoc. See B–P–H 461/16. HI 55041

Journal of the Ceylon branch of the royal asiatic society. Colombo. 1845-1910; 1937-71. J. Ceylon Branch Roy. Asiat. Soc. Superseded by: Journal of the Sri Lanka branch of the royal asiatic society. HI 66620

Journal of the Chekiang provincial library. [Che chiang t'u shu kuan pao.] Hangchow, China. J. Chekiang Prov. Libr. See B–P–H 461/17. HI 55042

Journal, Cheltenham and district naturalists' society = Cheltenham and district naturalists' society journal. Cheltenham.

Journal of chemical ecology. New York, London. Vol. 1+, 1975+. J. Chem Ecol. HI 66622

Journal of chemical technology and biotechnology. Oxford, London, Barking. Vols. 29-32, 1979-82?. J. Chem. Technol. Biotechnol. Preceded by: Journal of applied chemistry and biotechnology. Superseded by: Journal of chemical technology and biotechnology. Biotechnology. HI 75183

Journal of chemical technology and biotechnology. Biotechnology. Oxford, London, Barking. Vols. 33B+, 1983+. J. Chem. Technol. Biotechnol., Biotechnol. Preceded by: Journal of chemical technology and biotechnology. HI 75197

Journal für die Chemie und Physik. Berlin. J. Chem. Phys. (Berlin). See B–P–H 461/18. HI 55043

Journal für Chemie und Physik. Nuremberg. J. Chem. Phys. (Nuremberg). See B–P–H 461/19. HI 55044

Journal für die Chemie, Physik, und Mineralogie. Berlin. J. Chem. Phys. Mineral. See B–P–H 461/20. HI 55045

Journal of chemistry of the United Arab Republic. Cairo. J. Chem. U.A.R. See B–P–H 461/21. HI 55046

Journal of the Chicago horticultural society. Chicago, IL. Vols. 1-3, 1974-76. J. Chicago Hort. Soc. Preceded by: Garden talk. Superseded by: Garden. New York. ("Regional edition"). HI 75088

Journal de chimie médicale, de pharmacie et de toxicologie. Paris. J. Chim. Méd. See B–P–H 461/23. HI 55047

Journal de chimie médicale, de pharmacie, de toxicologie, et revue des nouvelles scientifiques nationales et étrangères = Journal de chimie médicale, de pharmacie et de toxicologie. Paris. J. Chim. Méd. See B–P–H 461/23.

Journal of China university of science and technology. [Zhongguo kexue jishu daxue xuebao.] Hofei. Vol. 1+, 1965+. J. China Univ. Sci. Technol. HI 66623

Journal of the chinese agricultural chemical society. [Zongguo nongye huaxue hui.] Taipei. Vol. 1+, 1963+. J. Chin. Agric. Chem. Soc. HI 66624

Journal of chinese and american engineers. Peking. J. Chin. Amer. Engin. See B–P–H 462/2. HI 55048

Journal of the chinese biochemical society. [Zhongguo shengwu huaxue hui.] Taipei. Vol. 1+, 1972+. J. Chin. Biochem Soc. HI 66625

Journal of the chinese chemical society. Beijing. Vol. 1+, 1933+. J. Chin. Chem. Soc. HI 66626

Journal of the chinese university of Hong Kong. [Hsiang-kang chung wen ta hsueh hsueh pao.] Shatin, Hong Kong. Vol. 1+, 1973+. J. Chin. Univ. Hong Kong.

HI 66627

Journal of the Chosen agricultural society. [Chosen nokai ho. Keijo.] Seoul?, Korea. J. Chosen Agric. Soc. See B–P–H 462/3. HI 55049

Journal of the Chosen natural history society. [Chosen hakubutsu gakkai zasshi.] Seoul, Korea. J. Chosen Nat. Hist. Soc. See B–P–H 462/4. HI 55050

Journal of the Chosen pharmaceutical society. [Chosen yakugaku kai zasshi.] Seoul?, Korea. J. Chosen Pharm. Soc. See B–P–H 462/5. HI 55051

Journal of chromatography. Amsterdam. J. Chromatogr. See B–P–H 462/7. HI 55052

Journal of chronic diseases. St. Louis, MO. J. Chron. Dis. See B–P–H 462/8. HI 55053

Journal of the Cincinnati society of natural history. Cincinnati, OH. J. Cincinnati Soc. Nat. Hist. See B–P–H 462/10. HI 55054

Journal of the city of London college science society. London. 1888-1906. J. City London Coll. Sci. Soc. Preceded by: Report of the city of London college science society. HI 66628

Journal of classification. New York. Vol. 1+, 1984+. J. Classific. HI 66629

Journal of clinical microbiology. Washington, DC. 1975+. J. Clin. Microbiol. HI 66630

Journal of coastal research. Fort Lauderdale, FL. Vol. 1+, 1985+. J. Coastal Res. HI 66631

Journal of coconut industries. Colombo, Ceylon. J. Coconut Industr. See B–P–H 462/11. HI 55055

Journal of coffee research. Balehonnur. Vol. 1+, 1971+. J. Coffee Res. HI 66632

Journal of the college of agriculture, Hokkaido imperial university. Sapporo. Vols. 8-19, 1918-28. J. Coll. Agric. Hokkaido Imp. Univ. Preceded by: Journal of the college of agriculture, Tohoku imperial university. Superseded by: Journal of the faculty of agriculture, Hokkaido imperial university. HI 66633

Journal of the college of agriculture, imperial university of Tokyo. [Nogakubu kiyo.] Tokyo. Vols. 1-15, 1909-43. J. Coll. Agric. Imp. Univ. Tokyo. Preceded by: Bulletin of the college of agriculture, Tokyo imperial university. Superseded by: Record of researches in the faculty of agriculture, university of Tokyo. 5-4231-2. HI 66634

Journal of the college of agriculture, Tohoku imperial university. Sapporo. Vols. 3-7, 1908/09-17. J. Coll. Agric. Tohoku Imp. Univ. Preceded by: Journal of the Sapporo agricultural college. Superseded by: Journal of the college of agriculture, Hokkaido imperial university. HI 66635

Journal of the college of agriculture, university of Riyadh. Riyadh. Vol. 1+, 1979+. J. Coll. Agric. Univ. Riyadh. HI 66636

Journal of the college of arts and sciences, Chiba university. Natural science. [Chiba daigaku bunrigakubu kiyo. Shizenkagaku.] Chiba. Vols. 1-5(2), 1952-69. J. Coll. Arts Chiba Univ. Superseded by: Journal of the college of arts and sciences, Chiba university. HI 66637

Journal of the college of fisheries. Shimonoseki, Japan. J. Coll. Fish. See B–P–H 462/15. HI 55056

Journal of the college of liberal arts, Toyama university. [Toyama daigaku kyoy obu kiyo.] Toyama Vol. 1+, 1968+. J. Coll. Liberal Arts Toyama Univ. HI 66638

Journal of the college of science, imperial university of Tokyo. Tokyo. J. Coll. Sci. Imp. Univ. Tokyo. See B–P–H 462/17. HI 55057

Journal of the college of science, King Saud university. Riyadh. Vol. 12(3)+, 1981+. J. Coll. Sci. King Saud Univ. Preceded by: Journal of the college of science, university of Riyadh. HI 66639

Journal of the college of science, university of Riyadh. Riyadh. Vols. 11-12(2), 1980-81. J. Coll. Sci. Univ. Riyadh. Preceded by: Journal of the faculty of science, university of Riyadh. Superseded by: Journal of the college of science, King Saud university. HI 55064

Journal of colloid and interface science. New York, NY. J. Colloid Interface Sci. See B–P–H 462/18. HI 55058

Journal of colloid science. New York, NY. J. Colloid Sci. See B–P–H 462/19. HI 55059

Journal of the Colorado-Wyoming academy of science. Boulder, CO. J. Colorado-Wyoming Acad. Sci. See B–P–H 462/20. HI 55060

Journal of the Columbus horticultural society = Report (Annual) of the Columbus horticultural society. Columbus, OH.

Journal of community gardening. Ann Arbor, MI. Vol. ?-2+, ?-1983+. J. Community Gard. HI 66640

Journal of comparative medicine and surgery. New York, NY. J. Comp. Med. Surg. See B–P–H 463/3. HI 55061

Journal of comparative medicine and veterinary archives. New York, NY. J. Comp. Med. Veterin. Arch. See B–P–H 463/4. HI 55062

Journal of comparative pathology and therapeutics. Edinburgh & London. J. Comp. Pathol. Therap. See B–P–H 463/5. HI 55063

Journal complémentaire du dictionnaire des sciences médicales. Paris. Vols. 1-36, 1818-29. J. Complém. Dict. Sci. Méd. Superseded by: Journal complémentaire des sciences médicales. 3-2179-3.

HI 52043

Journal complémentaire des sciences médicales. Paris. Vols. 36-44, 1830-32. J. Complém. Sci. Méd. Preceded by: Journal complémentaire du dictionnaire des sciences médicales. 3-2179-3. HI 53281

Journal of the Concord society of natural history. Concord, NH. J. Concord Soc. Nat. Hist. See B–P–H 463/9. HI 55065

Journal des connaissances usuelles et pratiques. Paris. J. Connaissances Usuelles Prat. See B–P–H 463/10. HI 55066

Journal du conseil, conseil permanent international pour l'éxploration de la mer. Copenhagen. Vol. 1+, 1926+. J. Counseil Perman. Int. Explor. Mer. HI 66641

Journal de la Corse agricole. Ajaccio. 1965+. J. Corse Agric. HI 66642

Journal of the council for scientific and industrial research, commonwealth of Australia. Melbourne. J. Council Sci. Res. Australia. See B–P–H 463/12. HI 55067

Journal of cutaneous diseases including syphilis. Chicago, IL. J. Cutan. Dis. Syph. See B–P–H 463/14. HI 55068

Journal of cutaneous and genito-urinary diseases. Chicago, IL. J. Cutan. Genito-Urin. Dis. See B–P–H 463/15. HI 55069

Journal of cutaneous and veneral diseases. Chicago, IL. J. Cutan. Veneral Dis. See B–P–H 463/16. HI 55070

Journal, cyclamen society. Haslemere. Vol. 1+, 1977+. J. Cyclamen Soc. HI 66643

Journal du cycle bio-botanique de l'académie des sciences d'Ukraine = Zhurnal Bio-botanichnogo tsyklu Vseukraïns'koi akademii nauk. Kiev.

Journal of cycle research. New York, East Brady, NY. Vols. 1-10, 1951-61. J. Cycle Research. HI 66644

Journal of cytology and genetics. Delhi. Vol. 1+, 1966+. J. Cytol. Genet. HI 66645

Journal of dairy science. Baltimore, MD. J. Dairy Sci. See B–P–H 463/18. HI 55071

Journal of the Darjeeling natural history society. Darjeeling. 1926-39. J. Darjeeling Nat. Hist. Soc. Superseded by: Journal of the Bengal natural history society. HI 66646

Journal of dendrology; official organ of the dendrological society of southern Africa. Pretoria. Vol. 1+, 1981+. J. Dendrol. HI 66647

Journal of the department of agriculture, Eire. Dublin. 1965+. J. Dept. Agric. Republ. Ireland. Preceded by: Journal of the department of lands and agriculture, Irish Free State. 3-2113-3. HI 66648

Journal of the department of agriculture and fisheries, Northern Ireland. Belfast. 19??+. J. Dept. Agric. Fish. Northern Ireland. Preceded by?: Journal of the ministry of agriculture for Northern Ireland HI 66649.

Journal of the department of agriculture of the Kyushu imperial university. Fukuoka, Japan. J. Dept. Agric. Kyushu Imp. Univ. See B–P–H 463/25. HI 55072

Journal of the department of agriculture, New Zealand. Wellington, N.Z. Vols. 1-5, 1910-12. J. Dept. Agric. New Zealand. Superseded by: Journal of agriculture, New Zealand department of agriculture. HI 66650

Journal of the department of agriculture of Porto Rico. J. Dept. Agric. Porto Rico. See B–P–H 463/26. HI 55073

Journal of the department of agriculture and technical instruction for Ireland. Dublin. Vols. 1-24, 1900-24. J. Dept. Agric. Ireland. Superseded by: Journal of the department of lands and agriculture, Irish Free State. 3-2113-3. HI 66651

Journal of the department of agriculture, South Australia. Adelaide, S.A. Vols. 8(7)-67(3), 1905-63. J. Dept. Agric. South Australia. Preceded by: Journal of agriculture and industry of South Australia. Superseded by: Journal of agriculture of South Australia. 3-2196-2. HI 55074

Journal of the department of agriculture and technical instruction, Irish Free State = Journal of the department of agriculture and technical instruction for Ireland. Dublin.

Journal of the department of agriculture, Union of South Africa. Pretoria. 1920-26. J. Dept. Agric. Union of South Africa. Preceded by: Agricultural journal of the Union of South Africa. Superseded by: Farming in South Africa. HI 66652

Journal of the department of agriculture, Victoria. Melbourne. J. Dept. Agric. Victoria. See B–P–H 464/1. HI 55075

Journal of the department of agriculture, Western Australia. South Perth, W.A. Vols. 1-18, 1899-1909; ser. 2, vols. 1-28, 1924-51; ser. 3, vols. 1-8, 1952-59, ser. 4, vols. 1-11, 1960-70. J. Dept. Agric. Western Australia. Superseded by: Journal of agriculture of Western Australia. HI 66653

Journal of the department of lands and agriculture, Irish Free State. Dublin. 1924-37. J. Dept. Lands (Dublin). Preceded by: Journal of the department of agriculture and technical instruction for Ireland. Superseded by: Journal of the department of agriculture, Eire. 3-2113-3. HI 66654

Journal of the department of science of Calcutta university. Calcutta. Vol. 1+, 1919+. J. Dept. Sci. Calcutta Univ. HI 66655

Journal of the Devon trust for nature conservation. Exeter. Nos. ?-2-23, ?-1964-69; vol. 9(1), 1978. J. Devon Trust Nat. Conservation. Preceded by: D N T journal. For 1970-77 see: Quarterly journal, Devon trust for nature conservation. Superseded by: Nature in Devon. HI 66656

Journal of the Durham colleges natural history society. Durham. Vols. ?-9, 1958-63? J. Durham Coll. Nat. Hist. Soc. Preceded by: Durham colleges natural history society journal. Superseded by: Journal of Durham university natural history society. HI 66657

Journal of Durham university biological society. Durham. Vols. 13-20, 1967-76. J. Durham Univ. Biol. Soc. Preceded by: Journal of the Durham university natural history society. HI 66658

Journal of the Durham university natural history society. Durham. Vols. 10-12, 1964-66. J. Durham Univ. Nat. Hist. Soc. Preceded by: Journal of the Durham colleges natural history society. Superseded by: Journal of the Durham university biological society. HI 66659

Journal of earth sciences, Nagoya university. Nagoya. Vol. 1+, 1953+. J. Earth Sci. HI 66660

Journal of the East Africa natural history society. Nairobi. 1942-61. J. E. Africa Nat. Hist. Soc. Preceded by: Journal of the East Africa and Uganda natural history society. Superseded by: Journal of the East Africa natural history society and Coryndon museum. HI 66661

Journal of the East Africa natural history society and Coryndon museum. Nairobi. Vols. 21-24, 1962-64. J. E. Africa Nat. Hist. Soc. Coryndon Mus. Preceded by: Journal of the East Africa natural history society. Superseded by: Journal of the East Africa natural history society and national museum. HI 66662

Journal of the East Africa natural history society and national museum. Nairobi. Vol. 25+, 1965+. J. E. Africa Nat. Hist. Soc. Natl. Mus. Preceded by: Journal of the East Africa natural history society and Coryndon museum. HI 66663

Journal of the East Africa and Uganda natural history society. Nairobi. Vols. 1-? [vols. 1-6 also numbered nos. 1-12; nos. 13-50 have no vol. numbering; vol. numbering resumed with vol. 12], 1910-42. J. E. Africa Nat. Hist. Soc. Superseded by: Journal of the East Africa natural history society. 2-1379-2. HI 66664

Journal of east China normal university. Natural science edition. [Huadong shifan daxue xuebao. Ziran kexue ban.] Shanghai. 1980+. J. E. China Norm. Univ., Nat. Sci. Ed. HI 66665

Journal of the east India association. London. J. E. India Assoc. See B–P–H 464/6. HI 55077

Journal of eastern Asia. Singapore. J. E. Asia. See B–P–H 464/5. HI 55076

Journal of the echeveria society. Coxcatlan, Puebla. No. 1+, 1985+. J. Echeveria Soc. HI 66666

Journal de l'école polytechnique. Paris. J. Ecole Polytechn. See B–P–H 464/9. HI 55079

Journal of ecology. London. J. Ecol. See B–P–H 464/7. HI 55078

Journal of ecology. [Shengtaixue zazhi.] Beijing. No. 1+, 1982+. J. Ecol. (Beijing). HI 66667

Journal of economic biology. London. J. Econ. Biol. See B–P–H 464/10. HI 55080

Journal of economic entomology. Concord, NH. J. Econ. Entomol. See B–P–H 464/12. HI 55081

Journal of economic and taxonomic botany. Jodpur. Vol. 1+, 1980+. J. Econ. Taxon. Bot. HI 66668

Journal of economic and taxonomic botany. Additional series. Jodpur. Vol. 1+, 1985+. J. Econ. Taxon Bot., Addit. Ser. HI 66669

Journal of the Edinburgh natural history society. Edinburgh. Vol. 1+, 1973+. J. Edinburgh Nat. Hist Soc. Preceded by: Newsletter, Edinburgh natural history society. HI 66670

Journal of egyptian archaeology. London. Vol. 1+, 1914+. J. Egypt. Archaeol. HI 66671

Journal of electron microscopy. [Denshi kenbikyo gakkai-shi.] Tokyo. Vol. 1+, 1953+. J. Electron Microscopy. HI 66672

Journal of electronmicroscopy = Journal of electron microscopy. Tokyo.

Journal of the Elisha Mitchell scientific society. Raleigh, NC. J. Elisha Mitchell Sci. Soc. See B–P–H 464/14. HI 55082

Journal of the Elliott society of natural history. Charleston, SC. Vol. 1(1-3), 1859-60. J. Elliott Soc. Nat. Hist. HI 66673

Journal of the Elliott society of science and art. Charleston, SC = Journal of the Elliott society of natural history. Charleston, SC.

Journal of embryology and experimental morphology. London. J. Embryol. Exp. Morphol. See B–P–H 464/15. HI 55083

Journal encyclopédique. Liège. J. Encycl. See B–P–H 464/16. HI 55084

Journal of the entomological society. New York, NY. J. Entomol. Soc. See B–P–H 464/17. HI 55085

Journal of the entomological society of British Columbia. Victoria, B.C. Vol. 63+, 1966+. J. Entomol. Soc. Brit. Columbia. Preceded by: Proceedings of the entomological society of British Columbia. HI 66674

Journal of environmental biology; an international journal concerned with toxicology and the interrelations of organisms and their environment. Muzaffarnagar. Vol. 1+, 1980+. J. Environm. Biol. HI 66675

Journal of environmental economics and management. San Diego, CA. 1975+. J. Environm. Econ. Managem. HI 66676

Journal of environmental education. Madison, WI. Vol. 2(3)+, 1971+. J. Environm. Educ. Preceded by: Environmental education. HI 66677

Journal of environmental horticulture. Blacksburg, VA, Washington, DC. Vol. 1+, 1983+. J. Environm. Hort. HI 66678

Journal of environmental management. London, New York. Vol. 1+, 1973+. J. Environm. Managem. HI 66679

Journal of environmental quality. Madison, WI. Vol. 1+, 1972+. J. Environm. Qual. HI 66680

Journal of environmental science. Orissa. Vol. ?+, 198?+. J. Environm. Sci. (Orissa). HI 74531

Journal of environmental science and health. Part A, environmental science and engineering. New York. Vol. 11+, 1976+. J. Environm. Sci. Health, A. Preceded by: Environmental letters. Part A, environmental science and engineering. HI 66681

Journal of environmental science and health. Part B, pesticides, food contaminants and agricultural wastes. New York. Vol. 1+, 1975+. J. Environm. Sci Health, B. HI 66682

Journal of environmental sciences. Mt. Prospect, IL. Vol. 2(5)+, 1959+. J. Environm. Sci. (Mt. Prospect). Preceded by: Journal of environmental engineering [not entered]. HI 66683

Journal of environmental systems. Farmingdale, NY. 1971+. J. Environm. Systems. HI 66684

Journal der Erfindungen, Theorien und Widersprüche in der Natur- und Arzneiwissenschaft. Gotha. J. Erfind. Natur- Arzneiwiss. See B–P–H 464/18. HI 55086

Journal of essential oil research. Wheaton, IL. Vol. 1+, 1989+. J. Essential Oil Res. HI 66685

Journal of the Essex county natural history society. Salem, MA. J. Essex County Nat. Hist. Soc. See B–P–H 464/19. HI 55087

Journal of ethnobiology. Flagstaff, AZ. 1981+. J. Ethnobiol. HI 66686

Journal of ethno-pharmacology; interdisciplinary journal devoted to bioscientific research on indigenous drugs. Lausanne, Limerick. Vol. 1+, 1979+. J. Ethno-Pharmacol. HI 66687

Journal européen de pathologie forestière = European journal of forest pathology. Hamburg, etc.

Journal of evolutionary biochemistry and physiology. [Translation of: Zhurnal evolutsionno biokhimii i fiziologii.] New York. Vol. 1+, 1965+. J. Evol. Biochem. Physiol. HI 66688

Journal of evolutionary biology. Basle. Vol. 1+, 1988+. J. Evol. Biol. HI 75260

Journal of experimental biology. London. J. Exp. Biol. See B–P–H 464/20. HI 55088

Journal of experimental botany. London. J. Exp. Bot. See B–P–H 464/21. HI 55089

Journal of experimental marine biology and ecology. Amsterdam. Vol. 1+, 1967+. J. Exp. Mar. Biol. Ecol. HI 66689

Journal of experimental medicine. New York, NY. J. Exp. Med. See B–P–H 464/22. HI 55090

Journal für Fabrik, Manufactur, Handlung und Mode = Journal für Fabrik, Manufactur und Handlungen. Leipzig. J. Fabrik. See B–P–H 465/1.

Journal für Fabrik, Manufactur und Handlungen. Leipzig. J. Fabrik. See B–P–H 465/1. HI 55091

Journal of the faculty of agriculture, Hokkaido imperial university. [Hokkaido daigaku nogakubu kiyo.] Sapporo. Vol. 20+, 1927+. J. Fac. Agric. Hokkaido Univ. Preceded by: Journal of the college of agriculture, Hokkaido imperial university. 5-3781-3. HI 66690

Journal of the faculty of agriculture, Hokkaido university = Journal of the faculty of agriculture, Hokkaido imperial university. Sapporo.

Journal of the faculty of agriculture, Iwate university. [Iwate Daigaku nogakubu hokoku.] Morioka, Japan. J. Fac. Agric. Iwate Univ. See B–P–H 465/5. HI 55092

Journal of the faculty of agriculture; Kyushu university. Fukuoka, Japan. J. Fac. Agric. Kyushu Univ. See B–P–H 465/6. HI 55093

Journal of the faculty of agriculture, Shinshu university. [Shinsu daigaku nogakubu kiyo.] Ina. 1952+. J. Fac. Agric. Shinshu Univ. HI 66691

Journal of the faculty of agriculture, Tottori university. [Tottori daigaku nogakubu kiyo.] Tottori. 1951+. J. Fac. Agric. Tottori Univ. HI 66692

Journal of the faculty of applied biological science, Hiroshima university. [Hiroshima daigaku seibutsu seisangakubu kiyo.] Hiroshima. 1979+. J. Fac. Appl. Biol. Sci. Hiroshima Univ. HI 66693

Journal of the faculty of education, Tottori university. Natural science. [Tottori daigaku kyoikugakubu kenkyu hokoku, shizenkagaku.] Tottori. Vol. 17+, 1966+. J. Fac. Educ. Tottori Univ., Nat. Sci. Preceded by: Liberal arts journal, Tottori university. HI 66694

Journal of the faculty of fisheries and animal husbandry;

Hiroshima university. Fukuyama, Japan. J. Fac. Fish. Hiroshima Univ. See B–P–H 465/9. HI 55094

Journal of the faculty of liberal arts and science, Shinsu university. Matsumoto. 1951-57. J. Fac. Liberal Arts Sci. Shinsu Univ. Superseded by: Journal of the faculty of liberal arts, Shinsu university. Part 2, natural science. HI 66695

Journal of the faculty of liberal arts, Shinsu university = Journal of the faculty of liberal arts and science, Shinsu university. Matsumoto.

Journal of the faculty of liberal arts, Shinsu university. Part 2, natural science. Matsumoto. 1958-65. J. Fac. Liberal Arts Shinsu Univ., Part 2, Nat. Sci. Preceded by: Journal of the faculty of liberal arts and science, Shinsu university. Superseded by: Journal of the faculty of science, Shinsu university. HI 66696

Journal of the faculty of marine science, King Abdulaziz university. Jeddah. Vol. 2+, 1982+. J. Fac. Mar. Sci. King Abdulaziz Univ. Preceded by: Jeddah journal of marine science. HI 66697

Journal of faculty of pharmacy of Ankara university = Ankara üniversitesi eczacilik fakültesi mecmuasi. Ankara.

Journal of faculty of pharmacy of Istanbul university = Istanbul üniversitesi eczacilik fakültesi mecmuasi. Istanbul.

Journal faculty of pharmacy, Istanbul university = Istanbul üniversitesi eczacilik fakültesi mecmuasi, ayri barki. Istanbul.

Journal of the faculty of science, Ege university. Series B = Ege universitesi fen fakültesi dergisi. Ser. B. Izmir.

Journal of the faculty of science of the Hokkaido imperial university. Ser. 4. Geology and mineralogy. Sapporo, Japan. J. Fac. Sci. Hokkaido Imp. Univ., Ser. 4, Geol. See B–P–H 465/11. HI 55095

Journal of the faculty of science of the Hokkaido imperial university. Series 5, botany. [Rigaku-bu kiyo.] Sapporo. Vols. 1-5, 1930-44. J. Fac. Sci. Hokkaido Imp. Univ., Ser. 5, Bot. Superseded by: Journal of the faculty of science, Hokkaido university. Series 5, botany. 5-3782-1. HI 66698

Journal of the faculty of science, Hokkaido university. Series 4, geology and mineralogy. Sapporo. Vol. 6+, 194?+. J. Fac. Sci. Hokkaido Univ., Ser. 4, Geol. Mineral. HI 66699

Journal of the faculty of science, Hokkaido university. Series 5, botany. [Hokkaido daigaku rigakubu kiyo.] Sapporo. Vol. 6+, 194?+. J. Fac. Sci. Hokkaido Univ., Ser. 5, Bot. Preceded by: Journal of the faculty of science of the Hokkaido imperial university. Series 5, botany. HI 66700

Journal of the faculty of science, Niigata university. Series II, biology, geology and mineralogy. [Niigata daigaku rigakubu kenkyu hokoku. Dai-2-rui, seibutsugaku, chishitsugaku kobutsugaku.] Niigata. 1952-63. J. Fac Sci. Niigata Univ., Ser. 11, Biol. Geol. Mineral. Superseded by: Science reports of Niigata university. Series D, (biology). HI 66701

Journal of the faculty of science, Shinsu university. Matsumoto. Vol. 1+, 1966+. J. Fac. Sci. Shinsu Univ. Preceded by: Journal of the faculty of liberal arts, Shinsu university. Part 2, natural science. HI 66702

Journal of the faculty of science, university of Riyadh. Riyadh. Vols. 9-10, 1978-79. J. Fac. Sci. Univ. Riyadh. Preceded by: Bulletin of the faculty of science, university of Riyadh. Superseded by: Journal of the college of science, university of Riyadh. HI 66703

Journal of the faculty of science, university of Tokyo. Section III. Botany. Tokyo. J. Fac. Sci. Univ. Tokyo, Sect. 3, Bot. See B–P–H 465/18. HI 55096

Journal of the faculty of textile science and technology, Shinsu university. Series A, biology. Ueda. 1961+. J. Fac. Textile Sci. Technol. Shinsu Univ., A. Preceded by: Journal of the faculty of textiles and sericulture, Shinsu university. Series A, biology. HI 66704

Journal of the faculty of textile science and technology, Shinsu university. Series E, agriculture and sericulture. Ueda. 1961+. J. Fac. Textile Sci. Technol. Shinsu Univ., E. Preceded by: Journal of the faculty of textiles and sericulture, Shinsu university. Series E, sericulture. HI 66705

Journal of the faculty of textiles and sericulture, Shinsu university. Series A, biology. Ueda. Vols. 1-?, 1951-60. J. Fac. Textiles Seric. Shinsu Univ., A. Superseded by: Journal of the faculty of textile science and technology, Shinsu university. Series A, biology. HI 66706

Journal of the faculty of textiles and sericulture, Shinsu university. Series E, sericulture. Uedo. 1955-60. J. Fac. Textiles Seric. Shinsu Univ., E. Superseded by: Journal of the faculty of textile science and technology, Shinsu university. Series E, agriculture and sericulture. HI 66707

Journal of the Federated Malay States museums. Kuala Lumpur. Vols. 1-59, 1906-36, 41, [vol. 18 published in 1941]. J. Fed. Malay States Mus. Preceded by: Notes, Perak museum. Superseded by: Federation museums journal. 3-2242-3. HI 66708

Journal of the fermentation association of Japan. [Hakko kyokai-shi.] Tokyo. Vols. 1-33(10/11), 1943-75. J. Ferment. Assoc. Japan. Superseded by: Hakko to kogyo. HI 66709

Journal of fermentation and bioengineering. Osaka, Amsterdam. Vol. 67+, 1989+. J. Ferment. Bioengin.

Preceded by: Journal of fermentation technology. HI 75245

Journal of fermentation technology. [Hakko kogaku zasshi.] Osaka. Vols. 1-54, 1924-76; [English language edition] vols. 55-66, 1977-88. J. Ferment. Technol. Preceded by: Journal of brewing. Osaka. In 1977 divided into English language and Japanese language journals, the latter entitled: Hakko kogaku kaishi. English language title superseded by: Journal of fermentation and bioengineering. 3-1784-2. HI 55097

Journal of the field naturalists' club, Trinidad. Port of Spain. Vols. 1(1)-2(12), 1892-96. J. Field Naturalists' Club, Trinidad. HI 66710

Journal of fish biology. London, New York. Vol. 1+, 1969+. J. Fish Biol. HI 66711

Journal of fisheries. [Nippon suisangaku zasshi.] Hakodate, Japan. J. Fish. See B–P–H 465/26. HI 55098

Journal of the fisheries research board of Canada. Toronto. Vols. 4-36, 1938-79. J. Fish. Res. Board Canada. Preceded by: Journal of the biological board of Canada. Superseded by: Canadian journal of fisheries and aquatic sciences. 2-899-3. HI 55099

Journal et flore des jardins. Paris. J. Fl. Jard. See B–P–H 466/3. HI 55100

Journal of food science. Champaign, Chicago, IL. Vol. 26+, 1961+. J. Food Sci. Preceded by: Food research. HI 66712

Journal of food science and technology. Mysore, India. J. Food Sci. Technol. (Mysore). See B–P–H 466/5. HI 55101

Journal of food science and technology. [Nihon shokuhin kogyo gakkai shi.] Tokyo. J. Food Sci. Technol. (Tokyo). See B–P–H 466/6. HI 55102

Journal of food technology. Oxford. Vols. 1-26, 1966-86. J. Food Techn. HI 66713

Journal of forensic sciences; official publication of the american academy of forensic sciences. Mundelein, IL, Philadelphia, PA. Vol. 1+, 1955+. J. Forensic Sci. HI 66714

Journal of forest history. Santa Cruz, CA. Vols. 18-33, 1974-89?. J. Forest Hist. Preceded by: Forest history. Superseded by: Forest and conservation history. HI 66715

Journal of the forest products research society. Madison, WI. Vols. 1-4, 1951-54. J. Forest Prod. Res. Soc. Superseded by: Forest products journal. HI 66716

Journal forestier suisse. Bern. J. Forest. Suisse. See B–P–H 466/10. HI 55105

Journal of forestry. Skopje = Sumarski pregled. Skopje.

Journal of forestry. [Sen lin.] Tsinan, China. J. Forest. (Tsinan). See B–P–H 466/7. HI 55103

Journal of forestry. Washington, DC. Vol. 15+, 1917+. J. Forest. (Washington). Preceded by: Forestry quarterly. For 1917-46 includes: Proceedings of the society of american foresters. 3-2215-2. HI 66717

Journal of the forestry commission. London. Vol. 1+, 1922+. J. Forest. Commiss. HI 66718

Journal of forestry and estates management. London. J. Forest. Estates Managem. See B–P–H 466/9. HI 55104

Journal of the formosan forest research insitute. Taipei, Taiwan. J. Formosan Forest Res. Inst. See B–P–H 466/13. HI 55107

Journal of formosan forestry. [Taiwan sanrinkai kaiho.] Taihoku [=Taipei, Taiwan]. J. Formosan Forest. See B–P–H 466/12. HI 55106

Journal für das Forst-, Jagd-, und Fischereywesen. Marburg. Vols. 1-3, 1806-08. J. Forst- Jagd-Fischereywesen. 3-2191-1. HI 66719

Journal für das Forst- und Jagdwesen. Leipzig. J. Forst-Jagdwesen. See B–P–H 466/14. HI 55108

Journal of the foundation for natural rubber research and development. Tokyo. Vol. 1+, 1954+. J. Found. Nat. Rubber Res. Developm. HI 66720

Journal von und für Franken. Nuremberg. J. Franken. See B–P–H 466/15. HI 55109

Journal of the Franklin institute of the state of Pennsylvania. Philadelphia, PA. J. Franklin Inst. See B–P–H 466/16. HI 55110

Journal of freshwater. Navarre, MN. 1977+. J. Freshwater. HI 66721

Journal of freshwater ecology. Holmen, WI. Vol. 1+, 1981+. J. Freshwater Ecol. HI 66722

Journal of gakugei, Tokushima university. Natural science. [Tokusima daigaku gakubeibu kiyo. Shizen kagaku.] Tokushima, Japan. J. Gakugei Tokushima Univ., Nat. Sci. See B–P–H 466/19. HI 55111

Journal, garden club of Virginia = G C V A journal. Richmond, VA.

Journal of garden history; an international quarterly. London. Vol. 1+, 1981+. J. Gard. Hist. HI 66723

Journal für die Gartenkunst, ... Stuttgart. J. Gartenkunst. See B–P–H 466/20. HI 55112

Journal für die Gärtnerey. Stuttgart. J. Gärtnerey. See B–P–H 466/21. HI 55113

Journal of Gauhati India university. Gauhati, India. J. Gauhati India Univ. See B–P–H 467/1. HI 55114

Journal of the Gauhati science society. Gauhati. 1955. J. Gauhati Sci. Soc. Superseded by: Journal of the

Assam science society. HI 66724

Journal of general and applied microbiology. Tokyo. J. Gen. Appl. Microbiol. See B–P–H 467/2. HI 55115

Journal of general biology. Moscow = Zhurnal obshchei biologii. Moscow.

Journal général d'horticulture. Gand (Ghent). Vols. 11-15 [also numbered as 2e sér. vols. 1-5], 1856-65. J. Gén. Hort. Preceded by: Flore des serres et des jardins de l'Europe. Superseded by: Annales générales d'horticulture. Gand. HI 74853

Journal général de l'imprimerie et de la librairie. J. Gén. Imprimerie Libr. See B–P–H 467/5. HI 55116

Journal général de la littératur étrangère, ou indicateur bibliographique et raisonné des livres nouveaux ... Paris & Strasbourg. J. Gén. Litt. Étrangère. See B–P–H 467/6. HI 55117

Journal général de la littérature de France. Paris & Strasbourg. J. Gén. Litt. France. See B–P–H 467/7. HI 55118

Journal général de médecine, de chirurgie et de pharmacie. Paris. J. Gén. Méd. See B–P–H 467/8. HI 55119

Journal général de médecine, de chirurgie et de pharmacie, françaises et étrangères. Paris. J. Gén. Méd. Franç. Étrangères. See B–P–H 467/9. HI 55120

Journal of general microbiology. Cambridge, England. J. Gen. Microbiol. See B–P–H 456/10. HI 55121

Journal of general physiology. New York, NY. J. Gen. Physiol. See B–P–H 467/11. HI 55122

Journal général des sociétés et travaux scientifiques de la France et de l'étranger. Paris. J. Gén. Soc. Trav. Sci. Franc Étranger. See B–P–H 467/12. HI 55123

Journal général des sociétés et travaux scientifiques de la France et de l'étranger. Section 1. Sciences mathématiques, physiques et naturelles. Paris. J. Gén. Soc Trav. Sci. France Étranger, Sect. 1, Sci. Math. See B–P–H 467/13. HI 55124

Journal of general virology. Cambridge, England. J. Gen. Virol. See B–P–H 467/14. HI 55125

Journal of genetics. Cambridge, England. J. Genet. See B–P–H 467/15. HI 55126

Journal of genetics and breeding. Rome. Vol. ?-43+, ?-1989+. J. Genet. Breed. HI 66725

Journal of geobotany. [Hokokuriku no shokubutsu no kai.] Kanazawa. Vols. 7-26, 1958-79? J. Geobot. Preceded by: Hokuriku journal of botany. Superseded by: Journal of phytogeography and taxonomy. HI 66726

Journal of geobotany and plant systematics. Forlì = Archivio botanico italiano. Forlì.

Journal of geography. Lancaster, PA. J. Geogr. (Lancaster). See B–P–H 467/18. HI 55127

Journal of geography. [Chigaku zasshi.] Tokyo. J. Geogr. (Tokyo). See B–P–H 467/19. HI 55128

Journal of the geographical society of China. [Ti-li hsüeh-pao.] Nanking. J. Geogr. Soc. China. See B–P–H 467/20. HI 55129

Journal of geological education. Columbus, OH. J. Geol. Educ. See B–P–H 468/3. HI 55131

Journal of the geological society of Australia. Adelaide, S.A. Vol. 1+, 1953+. J. Geol. Soc. Australia. HI 66727

Journal of the geological society of Dublin. Dublin. Vols. 1-10, 1832-64. J. Geol. Soc. Dublin. Superseded by: Journal of the royal geological society of Ireland. HI 66728

Journal of the geological society of India. Bangalore, India. J. Geol. Soc. India. See B–P–H 468/4. HI 55132

Journal of the geological society of Japan. Tokyo. Vol. 42+, 1935+. J. Geol. Soc. Japan. Preceded by: Journal of the geological society of Tokyo. HI 66729

Journal of the geological society of Tokyo. Tokyo. Vols. 1-41, 1893-34. J. Geol. Soc. Tokyo. Superseded by: Journal of the geological society of Japan. HI 66730

Journal de géologie. Paris. Vols. 1-3, 1830-31. J. Géol. (Paris). 3-2181-2. HI 53286

Journal of geology. Chicago, IL. J. Geol. See B–P–H 468/1. HI 55130

Journal of geology. Kiev = Geologichnyi zhurnal. Kiev.

Journal, Gloucestershire naturalists' society. Cheltenham. Vol. 25(2)+, 1974+. J. Gloucestershire Naturalists' Soc. Preceded by: North Gloucestershire naturalists' society journal. HI 66731

Journal of the Gold Coast agricultural and commercial society. Accra. Vols. 1-6, 1921-28. J. Gold Coast Agric. Commercial Soc. HI 66732

Journal of the Gold Coast welfare and agricultural society. Accra. Vols. 1-6, 1921-27. J. Gold Coast Welfare Agric. Soc. HI 66733

Journal of the government botanical garden (Nikita, Yalta) = Zapiski Gosudarstvennogo Nikitskogo opytnogo botanicheskogo sada. Yalta.

Journal of the grassland society of southern Africa. Pretoria. Vol. 1+, 1966+. J. Grassland Soc. S. Africa. HI 66734

Journal of the Gray memorial botanical association. Marietta, OH. Vols. 7-10, 1935-39. J. Gray Mem. Bot. Assoc. HI 66735

Journal of Great Lakes research. Buffalo, NY, Toronto.

Vol. 1+, 1975+. J. Great Lakes Res. HI 66736

Journal of the Hampshire and Isle of Wight naturalists' trust. 1967-68? J. Hampshire Isle Wight Naturalists' Trust. HI 66737

Journal of Hangzhou university. Natural science edition. [Hangzhou daxue xuebao. Ziran kexue ban.] Vol. ?-8+, ?-1981+. J. Hangzhou Univ., Nat. Sci. Ed. HI 66738

Journal of the Harrow natural history society. Harrow. 1973+. J. Harrow Nat. Hist. Soc. HI 66739

Journal of the Hattori botanical laboratory. [Hattori shokubutsu kenkyusho hokoku.] Obi. Vol. 3+, 1947+. J. Hattori Bot. Lab. Preceded by: Journal of the Hattori syokubutsu kenkyusyo. 3-1818-2. HI 66740

Journal of the Hattori shokubutsu kenkyusyo = Journal of the Hattori syokobutsu kenkyusyo. Miyazaki.

Journal of the Hattori syokobutsu kenkyusyo. Miyazaki. Vols. 1-2, 1946-47? J. Hattori Syok. Kenkyusyo. Superseded by: Journal of the Hattori botanical laboratory. HI 66741.

Journal for Hawaiian and Pacific agriculture. Hilo, HI. Vol. 1+, 1988?+. J. Hawaiian Pacific Agric. HI 66742

Journal hebdomadaire de médecine. Paris. Vols. 1(1)-8(104), 1828-30. J. Hebd. Méd. Superseded by: Journal universel et hebdomadaire de médecine et de chirurgie pratiques et des institutions médicales. 3-2193-1. HI 66743

Journal hebdomadaire des progrès des sciences et institutions médicales. Paris. J. Hebd. Progr. Sci. Inst. Méd. See B–P–H 468/9. HI 55133

Journal hebdomadaire des progrès des sciences médicales. Paris. J. Hebd. Progr. Sci. Méd. See B–P–H 468/10. HI 55134

Journal of Hebei university. Natural science edition. [Hebei daxue xuebao. Ziran kexue ban.] Vol. 1+, 1981+. J. Hebei Univ., Nat. Sci. Ed. HI 66744

Journal of helminthology. London. J. Helminthol. See B–P–H 468/11. HI 55135

Journal helvétique. Neuchâtel. J. Helv. See B–P–H 468/12. HI 55136

Journal helvétique ou annales littéraires et politiques de l'Europe, & principalement de la Suisse = Journal helvétique. Neuchâtel. J. Helv. See B–P–H 468/12.

Journal of heredity. Washington, DC. J. Heredity. See B–P–H 468/14. HI 55137

Journal of the Hillingdon natural history society. [West Drayton]. 1970, 1971. J. Hillingdon Nat. Hist. Soc. Superseded by: Report, Hillingdon natural history society. HI 66745

Journal of Hiroshima society of brewing science. [Hiroshima jozo gakkai-shi.] Hiroshima. J. Hiroshima Soc. Brew. Sci. See B–P–H 468/15. HI 55138

Journal of histochemistry and cytochemistry. Baltimore, MD. J. Histochem. Cytochem. See B–P–H 468/21. HI 55142

Journal d'histoire naturelle. Paris. J. Hist. Nat. See B–P–H 468/18. HI 55140

Journal of the history of biology. Cambridge, MA., Dordrecht. Vol. 1+, 1968+. J. Hist. Biol. HI 66747

Journal of the history of medicine and allied sciences. New York, NY. J. Hist. Med. Allied Sci. See B–P–H 468/17. HI 55139

Journal of history of science. [Kagaku-shi kenkyu. Nippon Kagaku-shi gakkai.] Tokyo. J. Hist. Sci. See B–P–H 468/19. HI 55141

Journal of the Hokkaido fisheries scientific institution. [Hokusuisi geppo.] Sapporo, Japan. J. Hokkaido Fish. Sci. Inst. See B–P–H 468/22. HI 55143

Journal of the Hokkaido gakugei university. Section B, natural science. [Hokkaido gakugei daigaku kiyo 2.] Sapporo. Vols. 1-16, 1954-65. J. Hokkaido Gakugei Univ., B. Preceded by: Gakugei. Section 2. Hokkaido gakugei university. Superseded by: Journal of the Hokkaido university of education. Section 2B, biology, geology, agricultural science. HI 66749

Journal of Hokkaido university of education. Section 2B, biology, geology, agricultural science. [Hokkaido kyoiku daigaku kiyo. Dai-2-bu, B, seibutsugaku, chigaku, nogaku-hen.] Sapporo. Vol. 17+ 1966+. J. Hokkaido Univ. Educ., 2B. Preceded by: Journal of the Hokkaido gakugei university. Section B, natural science. HI 66750

Journal of the Hongkong fisheries research station. Hong Kong. J. Hongkong Fish. Res. Sta. See B–P–H 469/1. HI 55144

Journal des horticulteurs et maraîchers, pépiniéristes et fleuristes romands. Lausanne. J. Hort. Romands. See B–P–H 469/11. HI 55152

Journal of the horticultural association of Japan. [Engei gakkai zasshi.] Tokyo. J. Hort. Assoc. Japan. See B–P–H 469/5. HI 55147

Journal of horticultural science. London. J. Hort. Sci. See B–P–H 469/12. HI 55153

Journal of the horticultural society of London. London. J. Hort. Soc. London. See B–P–H 469/13. HI 55154

Journal of the horticultural society of New York. New York. Vols. 1-4, 1906-24. J. Hort. Soc. New York. 3-1887-2. HI 66751

Journal of horticulture. Hangchow, China. J. Hort. (Hangchow). See B–P–H 469/3. HI 55145

Journal of horticulture. London = Journal of horticulture and practical gardening. London. J. Hort. Pract. Gard. See B–P–H 469/7.

Journal of horticulture. [Yuan i hsüeh pao.] Peking. J. Hort. (Peking). See B–P–H 469/4. HI 55146

Journal of horticulture, cottage gardener and country gentleman. London. J. Hort. Cottage Gard. See B–P–H 469/6. HI 55148

Journal of horticulture and practical gardening. London. J. Hort. Pract. Gard. See B–P–H 469/7. HI 55149

Journal d'horticulture pratique = Journal d'horticulture pratique de la Belgique. Brussels. J. Hort. Prat. Belgique. See B–P–H 469/9.

Journal d'horticulture pratique de la Belgique. Brussels. J. Hort. Prat. Belgique. See B–P–H 469/9. HI 55150

Journal d'horticulture pratique et de jardinage. Paris. J. Hort. Prat. Jard. See B–P–H 469/10. HI 55151

Journal d'horticulture suisse. Geneva. J. Hort. Suisse. See B–P–H 469/14. HI 55155

Journal d'horticulture et de viticulture suisse = Journal d'horticulture suisse. Geneva. J. Hort. Suisse. See B–P–H 469/14.

Journal of hydrology. Amsterdam. J. Hydrol. (Amsterdam). See B–P–H 469/18. HI 55156

Journal of hydrology. Dunedin, New Zealand. J. Hydrol. (Dunedin). See B–P–H 469/19. HI 55157

Journal of hygiene. Cambridge, England. J. Hyg. See B–P–H 469/20. HI 55158

Journal of the Idaho academy of science. Rexburg, ID. Vol. 1+, 1960+. J. Idaho Acad. Sci. HI 66752

Journal of the Idaho state university museum = Tebiwa. Pocatello, ID.

Journal of the Illinois state agricultural society. Springfield, IL. J. Illinois State Agric. Soc.. See B–P–H 468/22. HI 55159

Journal of immunology. Baltimore, MD. J. Immunol. See B–P–H 469/23. HI 55160

Journal of immunology, virus research and experimental chemotherapy = Journal of immunology. Baltimore, MD. J. Immunol. See B–P–H 469/23.

Journal of the imperial agricultural experiment station, Nishigahara, Tokyo. [Noji shikenjo iho.] Tokyo. J. Imp. Agric. Exp. Sta. Nishigahara Tokyo. See B–P–H 469/24. HI 55161

Journal of the imperial fisheries institute. [Suisan koshujo shiken hokoku.] Tokyo. J. Imp. Fish. Inst. See B–P–H 470/1. HI 55162

Journal, indian academy of wood science. Bangalore. Vol. 1+, 1970+. J. Indian Acad. Wood Sci. HI 66753

Journal of the indian archipelago and eastern Asia. Singapore. Vols. 1-9, 1847-55; n.s. vols. 1-?, 1856-62. J. Indian Archipel. E. Asia. HI 66754

Journal of the indian botanical society. Madras, Wattair. Vols. 3-54 & jubilee vol., 1924-75. J. Indian Bot. Soc. Preceded by: Journal of indian botany. HI 66755

Journal of indian botany. Madras. J. Indian Bot. See B–P–H 470/3. HI 55163

Journal of the indian institute of science. Bangalore, India. J. Indian Inst. Sci. See B–P–H 470/5. HI 55164

Journal of the indian potato association = J I P A. Simla.

Journal, indian society of soil science. New Delhi. Vol. 1+, 1953+. J. Indian Soc. Soil Sci. HI 66756

Journal of industrial microbiology. Amsterdam. Vol. 1+, 1986+. J. Industr. Microbiol. HI 66757

Journal of infectious diseases. Chicago, IL. J. Infect. Dis. See B–P–H 470/6. HI 55165

Journal of inferential and deductive biology, Worcester, MA. Vol. 1+, 1985+. J. Inferential Deduct. Biol. HI 66758

Journal d'information de la société d'histoire naturelle d'Autun. Autun. No. 1+, 1983+. J. Inform. Soc. Hist. Nat. Autun. HI 66759

Journal of insect pathology. New York. Vols. 1-6, 1959-64. J. Insect Pathol. Superseded by: Journal of invertebrate pathology. HI 66760

Journal de l'institut botanique de l'académie des sciences de la R S S d'Ukraine = Zhurnal Instytuto botaniky Vseukraïns'koi akademii nauk. Kiev.

Journal de l'institut botanique de l'académie des sciences d'Ukraine = Zhurnal Instytuto botaniky Vseukraïns'koi akademii nauk. Kiev.

Journal of the institute of biology. London. Vols. 1-15(4), 1953-68. J. Inst. Biol. Superseded by: Biologist. HI 66761

Journal of the institute of brewing. London. J. Inst. Brew. See B–P–H 470/10. HI 55166

Journal of the institute of Jamaica. Kingston, Jamaica. J. Inst. Jamaica. See B–P–H 470/11. HI 55167

Journal, institute of landscape architects. London. Nos. 1-92, 1941-71. J. Inst. Landscape Architects. Superseded by: Landscape design. 3-2010-2. HI 66762

Journal of the institute of polytechnics, Osaka City university. Series D. Biology. [Osaka shiritsu daigaku. Rikogakubu.] Osaka, Japan. J. Inst. Polytechn. Osaka City Univ., Ser. D, Biol. See B–P–H 470/13. HI 55168

Journal of the institute of wood science. London. No. 1+, 1958+. J. Inst. Wood Sci. HI 66763

Journal of interdisciplinary cycle research. Amsterdam, Lisse. Vol. 1+, 1970+. J. Interdiscipl. Cycle Res. HI 66764

Journal of the international garden club. New York. Vols. 1-3, 1917-19. J. Int. Gard Club. HI 66765

Journal for the international institute for sugar beet research = I I R B. Brussels.

Journal of the international society of leather trades' chemists. [England?]. J. Int. Soc. Leather Trades' Chem. See B–P–H 470/14. HI 55169

Journal of interpretation. Derwood, MD. 1976+. J. Interpret. HI 66766

Journal of invertebrate pathology. New York. Vol. 7+, 1965+. J. Invert. Pathol. Preceded by: Journal of insect pathology. HI 66767

Journal of investigative dermatology. Baltimore, MD. J .Invest. Dermatol. See B–P–H 470/17. HI 55170

Journal of the Iowa academy of science. Cedar Falls, IA. Vol. 95+, 1988+. J. Iowa Acad. Sci. Preceded by: Proceedings of the Iowa academy of science. HI 75222

Journal of the Ipswich and district field club. Ipswich. 1908-21. J. Ipswich Distr. Field Club. Superseded by: Journal of the Ipswich and district natural history society. HI 66768

Journal of the Ipswich and district natural history society. Ipswich. 1925-35. J. Ipswich Distr. Nat. Hist. Soc. Preceded by: Journal of the Ipswich and district field club. HI 66769

Journal of the Isle of Man natural history and antiquarian society = Lioar manninagh. Douglas.

Journal of the Jamaica agricultural society. Kingston, Jamaica. Vols. 1-51, 1897-1947. J. Jamaica Agric. Soc. Superseded by: Farmer. Journal of the Jamaica agricultural society. HI 66770

Journal of Japan association of forestry conservation. [Chisan rindo koho.] Tokyo. 1949+. J. Japan Assoc. Forest. Conservation. HI 69137

Journal of the Japan turf grass research association. [Shibakusa kenkyu.] Tokyo. Vol. 1+, 1972+. J. Japan Turf Grass Res. Assoc. HI 70263

Journal of the japanese alpine club. [Sangaku.] J. Jap. Alpine Club. See B–P–H 470/18. HI 55171

Journal of japanese biochemical society. [Seikagaku.] Tokyo. Vol. 1+, 1925+. J. Jap. Biochem. Soc. HI 66771

Journal of japanese botany. [Shokubutsu kenkyu zasshi.] Tokyo. J. Jap. Bot. See B–P–H 470/19. HI 55172

Journal of the japanese forestry society. [Nippon ringaku kaishi.] Tokyo. J. Jap. Forest. Soc. See B–P–H 470/20. HI 55173

Journal of the japanese horticultural society. [Nippon engeikai zasshi.] Tokyo. J. Jap. Hort. Soc. See B–P–H 470/21. HI 55174

Journal of the japanese institute of protection forests. [Hoanrin.] Tokyo. 1955+. J. Jap. Inst. Protect. Forests. HI 70288

Journal of the japanese society of food and nutrition. Tokyo. J. Jap. Soc. Food Nutr. See B–P–H 470/22. HI 55175

Journal of japanese society of grassland science. [Nihon sochi gakkai shi.] = Japanese society of grassland science journal. Jap. Soc. Grassland Sci. J. See B–P–H 499/17.

Journal of the japanese society for horticultural science. [Engei gakkai zasshi.] Tokyo. J. Jap. Soc. Hort. Sci. See B–P–H 471/2. HI 55176

Journal of the japanese wood research society. [Mokuzai gakkaishi.] Tokyo. J. Jap. Wood Res. Soc. See B–P–H 471/3. HI 55177

Journal of the John Innes association. Bayfordbury. Nos. 1-20, 1934-53; ser. 2, nos. 1-4, 1954-57. J. John Innes Assoc. Superseded by: Journal of the John Innes society. HI 66772

Journal of the John Innes society. Bayfordbury. No. 5+, 1958+. J. John Innes Soc. Preceded by: Journal of the John Innes Association. HI 66773

Journal of the Kagoshima fisheries college. [Kagoshima suisan senmon gakko hokoku.] Kagoshima, Japan. J. Kagoshima Fish. Coll. See B–P–H 471/4. HI 55178

Journal of the Kansas entomological society. Manhattan, KS. J. Kansas Entomol. Soc. See B–P–H 471/5. HI 55179

Journal of the Kansas medical society. Columbus, KS. J. Kansas Med. Soc. See B–P–H 471/6. HI 55180

Journal of the Kansu institute of scientific education. Lanchow, China. J. Kansu Inst. Sci. Educ. See B–P–H 471/7. HI 55181

Journal of the Kanto Tosan agricultural experiment station. [Kanto Tosan nogyo shikenjo kenkyu hokoku.] Urawa, Japan. J. Kanto Tosan Agric. Exp. Sta. See B–P–H 471/8. HI 55182

Journal of the Karnatak university. Science. Dharwar. Vol. 1+, 1956+. J. Karnatak Univ. Sci. HI 66774

Journal, Kerala academy of biology. [Trivandrum.] Vol. 1+, 1969+. J. Kerala Acad. Biol. HI 66775

Journal of the Kew guild; an association of Kew gardeners. London. J. Kew Guild. See B–P–H 471/9. HI 55183

Journal of the khedivial agricultural society. Cairo. J. Khediv. Agric. Soc. See B–P–H 471/10. HI 55184

Journal of the Kiangsi agricultural college. [Kiangsi nung hsüeh yüan hsüeh.] Nancheng?, China. J. Kiangsi Agric. Coll. See B–P–H 471/11. HI 55185

Journal of the King's college natural history society.

Newcastle-upon-Tyne. No. 1, 1959. J. King's Coll. Nat. Hist. Soc. HI 66776

Journal of the korean forestry society. [Han'guk imhakhoe chi.] Suwon. ?-1978+. J. Korean Forest. Soc. HI 66777

Journal of korean plant taxonomy. Seoul. Ca.1989. J. Korean Pl. Taxon. Superseded by: Korean journal of plant taxonomy. HI 66778

Journal of korean research institute for better living. Seoul. Vol. 1+, 1968+. J. Korean Res. Inst. Better Living. HI 66779

Journal, korean society for horticultural science. [Han'guk wonye hakhoe chi.] Seoul. Vol. 1+, 1965+. J. Korean Soc. Hort. Sci. HI 66780

Journal of the korean society for microbiology. [Taehan misaengmul hakhoe chi.] Seoul. 1966+. J. Korean Soc. Microbiol. HI 66781

Journal of the korean traditional garden research association. [Han'guk chongwon hakhoe chi.] Seoul. Vol. 1+, 1982+. J. Korean Trad. Gard. Res. Assoc. HI 66782

Journal of laboratory and clinical medicine. St. Louis, MO. J. Lab. Clin. Med. See B–P–H 471/12. HI 55186

Journal of Lanchow university. Natural sciences. [Lanzhou daxue xuebao. Ziran kexue ban.] Lanchow. No. 1+, 1957+. J. Lanchow Univ., Nat Sci. HI 66783

Journal für Landwirtschaft. Berlin. J. Landw. See B–P–H 471/13. HI 55187

Journal für Landwirtschaft und Gartenbau. Erfurt. J. Landw. Gartenbau. See B–P–H 471/14. HI 55188

Journal of the Letchworth and district natural history and antiquarian society. Letchworth. 1941-43. J. Letchworth Distr. Nat. Hist. Antiq. Soc. Superseded by: Journal of the Letchworth naturalists' society. HI 66784

Journal of the Letchworth naturalists' society. Letchworth. 1944-57. J. Letchworth Naturalist's Soc. Journal of the Letchworth and district natural history and antiquarian society. HI 66785

Journal für die Liebhaber des Steinreiches und der Konchyliologie. Weimar. J. Liebhaber Steinreiches Konchyliol. See B–P–H 471/15. HI 55189

Journal of life sciences, royal Dublin society. Dublin. Vol. 1+, 1978+. J. Life Sci. Roy. Dublin Soc. Preceded by: Scientific proceedings of the royal Dublin society. HI 66786

Journal of the limnological society of southern Africa. Bloemfontein. Vol. 1+, 1975+. J. Limnol. Soc. S. Africa. Preceded by: Newsletter, limnological society of southern Africa. HI 66787

Journal of the linnean society. Botany. London. Vols. 8-61, 1865-1968. J. Linn. Soc., Bot. Preceded by: Journal of the proceedings of the linnean society. Botany. Superseded by: Botanical journal of the linnean society. 3-2429-3. HI 66788

Journal of lipid research. Memphis, TN. J. Lipid Res. See B–P–H 471/20. HI 55190

Journal für Literatur, Kunst und geselliges Leben. Weimar & Gotha. J. Lit. Kunst Geselliges Leben. See B–P–H 471/21. HI 55191

Journal für Literatur, Kunst, Luxus und Mode. Weimar & Gotha. J. Lit. Kunst. Luxus. See B–P–H 471/22. HI 55192

Journal för litteraturen og theatern. Stockholm. J. Litt. Theatern. See B–P–H 472/1. HI 55193

Journal of the Louisiana society for horticultural research. Lafayette, LA. Vol. 11-15(1), 1970-74. J. Louisiana Soc. Hort. Res. Preceded by: L S H R news letter. HI 66789

Journal, Louisiana state medical society = Louisiana state medical society journal. New Orleans, LA. Louisiana State Med. Soc. J. See B–P–H 536/6.

Journal für Luxus, Mode und Gegenstände der Kunst. Weimar & Gotha. J. Luxus Mode Gegenstände Kunst. See B–P–H 472/2. HI 55194

Journal des Luxus und der Moden. Weimar & Gotha. J. Luxus Moden. See B–P–H 472/3. HI 55195

Journal, Madras agricultural student's union. Madras. Vols. 1-16, 1912-28. J. Madras Agric. Stud. Union. Superseded by: Madras agricultural journal. 3-2505-1. HI 66790

Journal of the Madras university. Madras. J. Madras Univ. See B–P–H 472/4. HI 55196

Journal of the Maharaja Sayajirao university of Baroda. Baroda, India. J. Maharaja Sayajirao Univ. Baroda. See B–P–H 472/6. HI 55197

Journal of the Maharashtra agricultural universities, college of agriculture. Pune. Vol. 1+, 1976+. J. Maharashtra Agric. Univ. Coll Agric. Preceded by: P K V research journal and Research journal of Mahatma Phule agricultural university. HI 66791

Journal für makromoleculare Chemie = Journal für praktische Chemie. Leipzig. J. Prakt. Chem. See B–P–H 477/8.

Journal of the malayan branch of the royal asiatic society. Singapore. Vol. 1+, 1923+. J. Malayan Branch Roy. Asiat. Soc. 4-3715-3. HI 66792

Journal of the Mammillaria society. Weybridge, England. J. Mammillaria Soc. See B–P–H 472/10. HI 55198

Journal of the Manchester geographical society. Manchester, England. J. Manchester Geogr. Soc. See

B–P–H 472/11. HI 55199

Journal of the Manx museum. Douglas, Isle of Man. J. Manx Mus. See B–P–H 472/13. HI 55200

Journal of the marine biological association of India. Mandapam Camp, India. J. Mar. Biol. Assoc. India. See B–P–H 472/15. HI 55201

Journal of the marine biological association of the United Kingdom. Plymouth, England. J. Mar. Biol. Assoc. U.K. See B–P–H 472/16. HI 55202

Journal of marine research. New Haven, CT. J. Mar. Res. See B–P–H 472/17.

Journal of marine science. Bayou la Batre, AL. Vols. 1-2, 1969-72. J. Mar. Sci. Superseded by: Northeast gulf science. HI 66793

Journal of mathematical biology. Berlin, New York, etc. Vol. 1+, 1974+. J. Math. Biol. HI 66794

Journal of mathematics and sciences. Islamabad. Vol. 1(1-2), 1974. J. Math. Sci. Superseded by: Islamabad journal of sciences. HI 66795

Journal de médecine et de chirurgie. Montreal. J. Méd. Chir. See B–P–H 472/21. HI 55205

Journal de médecine, chirurgie, pharmacie, etc. [Edited by Corvisart, Leroux et Boyer.] Paris. J. Méd. Chir. Pharm. (Corvisart). See B–P–H 472/23. HI 55206

Journal de médecine, de chirurgie et de pharmacie militaires. Paris. J. Méd. Chir. Pharm. Militaires. See B–P–H 472/24. HI 55207

Journal de médecine pratique; ou, recueil des travaux de la société royale de médecine de Bordeaux. Bordeaux. J. Méd. Prat. See B–P–H 473/2. HI 55209

Journal of the medical association of Georgia. Atlanta, GA. J. Med. Assoc. Georgia. See B–P–H 472/18. HI 55203

Journal médical de Bruxelles. Brussels. J. Méd. Bruxelles. See B–P–H 472/20. HI 55204

Journal médical de la Gironde. Bordeaux. J. Méd. Gironde. See B–P–H 473/1. HI 55208

Journal of medical microbiology. London. Vol. 1+, 1968+. J. Med. Microbiol. Preceded by: Journal of pathology and bacteriology. HI 66796

Journal of medical research. Boston, MA. Vols. 6-44, 1901-1924 [Also numbered n.s. Vols. 1-39]. J. Med. Res. Preceded by: Journal of the Boston society of medical sciences. 3-2226-1. HI 66797

Journal of the medical society of New Jersey. J. Med. Soc. New Jersey. See B–P–H 473/3. HI 55210

Journal of medical and veterinary mycology. Abingdon. Vol. 24+, 1986+. J. Med. Veterin. Mycol. Preceded by: Sabouraudia. HI 66798

Journal of medicinal plant research. Stuttgart. Vol. 32A+, 1977+. J. Med. Pl. Res. Preceded by: Planta medica. HI 66799

Journal of membrane biology; international journal for studies on the structure, formation and genesis of biomembranes. New York. Vol. 1+, 1969+. J. Membr. Biol. HI 66800

Journal of the Merioneth historical and record society. Dolgelley, Wales. J. Merioneth Hist. Soc. See B–P–H 473/4. HI 55211

Journal of meteorological observations made in the garden of the horticultural society at Chiswick. [England]. J. Meteorol. Observ. Chiswick. See B–P–H 473/5. HI 55212

Journal of microbiological methods. Amsterdam. Vol. 1+, 1983+. J. Microbiol. Meth. HI 66801

Journal of microbiology and serology. Amsterdam. J. Microbiol. Serol. See B–P–H 473/8. HI 55213

Journal of microbiology of the United Arab Republic. Cairo. Vol. 1-4, 1966-69. J. Microbiol. United Arab Republic. Superseded by: United Arab Republic journal of microbiology. HI 66802

Journal de micrographie. Paris. J. Microgr. See B–P–H 473/9. HI 55214

Journal of micrology and natural history mirror. Reading, England. J. Microl. Nat. Hist. Mirror. See B–P–H 473/10. HI 55215

Journal of micropalaeontology. London. Vol. 1+, 1982+. J. Micropalaeontol. HI 66803

Journal of the microscopical society of Victoria. Melbourne. J. Microscop. Soc. Victoria. See B–P–H 473/13. HI 55216

Journal de microscopie. Paris. Vols. 1-21, 1962-74. J. Microscopie. Superseded by: Journal de microscopie et de biologie cellulaire. HI 66804

Journal de microscopie et de biologie cellulaire. Paris. Vols. 22-27, 1975-76. J. Microscopie Biol. Cell. Preceded by: Journal de microscopie. Superseded by: Biologie cellulaire. HI 66805

Journal of microscopy. Oxford. Vol. 89+, 1969+. J. Microscopy. Preceded by: Journal of the royal microscopical society of London. HI 66806

Journal des mines de Russie. St. Petersburg = Gornyi zhurnal. St. Petersburg.

Journal of the ministry of agriculture. London. J. Minist. Agric. See B–P–H 473/14. HI 55217

Journal of the ministry of agriculture for Northern Ireland. Belfast. Vols. 1-6, 1927-38. J. Minist. Agric. Northern Ireland. Superseded by?: Journal of the department of agriculture of fisheries, Northern Ireland. HI 66807

Journal of the Minnesota academy of science. Minneapolis, MN. Vol. 32+, 1964+. J. Minnesota

Acad. Sci. Preceded by: Proceedings of the Minnesota academy of sciences. 3-2679-2. HI 66808

Journal of the Mississipi academy of sciences. State College & Jackson, MS. Vol. 1+, 1939+. J. Mississipi Acad. Sci. HI 66809

Journal der Moden. Weimar & Gotha. J. Moden. See B–P–H 473/18. HI 55218

Journal of molecular and applied genetics. New York. Vol. 1+, 1981+. J. Molec. Appl. Genet. HI 66810

Journal of molecular biology. London. J. Molec. Biol. See B–P–H 473/20. HI 55219

Journal of molecular evolution. Berlin, New York, etc. 1971+. J. Molec. Evol. HI 66811

Journal of morbid anatomy, ophthalmic medicine, and pharmaceutical analysis; with medico-botanical transactions, communicated by the medico-botanical society of London. London. J. Morbid Anat. See B–P–H 473/21. HI 55220

Journal of morphology. Boston, MA. J. Morphol. See B–P–H 473/22. HI 55221

Journal des Museums Godeffroy. Hamburg. J. Mus. Godeffroy. See B–P–H 473/24. HI 55222

Journal of mycology. Columbus, OH. J. Mycol. See B–P–H 474/1. HI 55223

Journal of the Mysore university. Section B, science. Mysore. 1965+. J. Mysore Univ., B. Preceded by: Half-yearly journal of Mysore university. HI 66812

Journal of N A L associates. Bethesda, MD. N.s. vol. 4+, 1979+. J. N. A. L. Assoc. Preceded by: Associates N A L today. HI 66813

Journal of Nanjing agricultural university. [Nan-ching nung yeh ta hsueh.] Nanching (Nanjing). No. 1+, 1986+. J. Nanjing Agric. Univ. HI 66814

Journal of Nanjing college of pharmacy. [Nanjing yaoxueyuan xuebao.] Nanching (Nanjing). No. ?-12+, ?-1979+. J. Nanjing Coll. Pharm. HI 66815

Journal of Nanjing institute of forestry. [Nan-ching lin hsüeh yuan hsüeh pao.] Nanching (Nanjing). Vol. 1(19)+, 1984+. J. Nanjing Inst. Forest. Preceded by: Journal of Nanjing technological college of forest products. HI 66816

Journal of Nanjing technological college of forest products. [Nanjing linchan gongye xueyuan xuebao.] Nanching (Nanjing). Nos. 1-18, 1979-83. J. Nanjing Technol. Coll. Forest Prod. Superseded by: Journal of Nanjing institute of forestry. HI 66817

Journal of Nanjing university. Natural sciences edition. [Nanjing daxue xuebao. Ziran kexue.] Nanching (Nanjing). 1979+. J. Nanjing Univ., Nat. Sci. Ed. HI 66818

Journal of the Nanki biological society. [Nanki seibutsu.] Wakayama. ?-1988+. J. Nanki Biol. Soc. HI 74188

Journal of Nara gakugei university. Natural science. [Nara gakugei daigaku kiyo, shizenkagaku.] Nara. 1951-66. J. Nara Gakugei Univ., Nat. Sci. Superseded by: Bulletin of Nara university of education. Series B, natural sciences. HI 66819

Journal of the national agricultural experiment station, Nishigahara, Tokyo. [Noji shikenjo iho.] Tokyo. J. Natl. Agric. Exp. Sta. Nishigahara Tokyo. See B–P–H 474/13. HI 55228

Journal, national agricultural society of Ceylon. Peradeniya. Vol. 1+, 1964+. J. Natl. Agric. Soc. Ceylon. HI 66820

Journal of the national cactus and succulent society of India. Chandigarh. Vol. 1+, 1981+. J. Natl. Cact. Succ. Soc. India. HI 66821

Journal of the national Honan university. [Ho nan ta hsüeh hsüeh pao.] Kaifeng, China. J. Natl. Honan Univ. See B–P–H 474/14. HI 55229

Journal of the national institute of agricultural botany. Cambridge, England. J. Natl. Inst. Agric. Bot. See B–P–H 474/15. HI 55230

Journal, national library of Wales = National library of Wales journal. Aberystwyth, Wales. Natl. Libr. Wales J. See B–P–H 631/3.

Journal of the national research council of Thailand. Bangkok. J. Natl. Res. Council Thailand. See B–P–H 474/16. HI 55231

Journal of the national science club. Washington, DC. 1898-99. J. Natl. Sci. Club. HI 66822

Journal of national science council of Sri Lanka. Colombo. Vol. 1+, 1973+. J. Natl. Sci. Council Sri Lanka. HI 66823

Journal, native orchid society of South Australia. Everard Park, S.A. Vol. 2(3)+, 1978+. J. Native Orchid Soc. S. Australia. Preceded by: Newsletter, native orchid society of South Australia. HI 66824

Journal of the natural areas association. Rockford, IL. Vol. 1(1), 1981. J. Nat. Areas Assoc. Superseded by: Natural areas journal. HI 66825

Journal of natural history. London. Vol. 1+, 1967+. J. Nat. Hist. (London). Preceded by: Annals and magazine of natural history. HI 66826

Journal of natural history. [Hakubutsu gaku zasshi.] Tokyo. Nos. 1-72, 1903-41. J. Nat. Hist. (Tokyo). HI 66827

Journal of the natural history and science society of Western Australia. Perth, W.A. Vols. 3-5, 1910-14. J. Nat. Hist. Sci. Soc. Western Australia. Preceded by: Journal of the Western Australia natural history society. Superseded by: Journal and proceedings of the royal society of Western Australia 4-2926-3. HI 66828

Journal of natural history society in Fukuoka. [Fukuoka hakubutsu-gaku zasshi.] Fukuoka, Japan. J. Nat. HIst. Soc. Fukuoka. See B–P–H 474/9. HI 55225

Journal of the natural history society of Siam. Bangkok. Vols. 1-6, 1914-25. J. Nat. Hist. Soc. Siam. Superseded by: Journal of the Siam society. Natural history supplement. 5-3868-1. HI 66829

Journal of the natural history society of Taiwan. Taihoku [=Taipei, Taiwan]. J. Nat. Hist. Soc. Taiwan. See B–P–H 474/11. HI 55226

Journal of natural philosophy, chemistry, and the arts. London. J. Nat. Philos. See B–P–H 474/12. HI 55227

Journal of natural products (Lloydia). Columbus, OH. Vol. 42+, 1979+. J. Nat. Prod. (Lloydia). Preceded by: Lloydia. HI 66830

Journal of natural resource management and interdisciplinary studies. Winnipeg. Vol. 1+, 1976+. J. Nat. Resource Managem. Interdiscipl. Stud. HI 66831

Journal of natural science of Beijing normal university. [Beijing shifan daxue xuebao. Ziran kexue ban.] Peking (Beijing). No. ?-14+, ?-1979+. J. Nat. Sci. Beijing Norm. Univ. HI 66832

Journal of natural science illustration. Washington, DC. Vol. 1+, 1988+. J. Nat. Sci. Ill. HI 74854

Journal, naturalist society of central Africa. Kitwe. Vols. 1-4(2)-?, 1960-63-? J. Naturalist Soc. Centr. Africa. HI 66833

Journal für Naturwissenschaft und Medizin. Frankfurt a. M. J. Naturwiss. Med. See B–P–H 474/17. HI 55232

Journal of nervous and mental disease. Chicago, IL. J. Nerv. Mental Dis. See B–P–H 474/18. HI 55233

Journal für die neueste holländische medizinische und naturhistorische Literatur. Herborn & Hadamar, Germany. J. Neueste Holl. Med. Naturhist. Lit. See B–P–H 474/19. HI 55234

Journal für die neuesten Land- und Seereisen und das Interessanteste aus der Völker- und Länderkunde ... Berlin. J. Neuesten Land- Seereisen. See B–P–H 474/20. 4-3037-2. HI 55235

Journal of the New Brunswick museum. Saint John, N.B. 1977+. J. New Brunswick Mus. HI 66834

Journal of the New York botanical garden. Lancaster, PA. J. New York Bot. Gard. See B–P–H 475/1. HI 55236

Journal of the New York microscopical society. New York. Vols. 1-16, 1885-1903. J. New York Microscop. Soc. HI 66835

Journal of the New Zealand institute of horticulture. Wellington, N.Z. Vols. 1-8(3), 1929-38. J. New Zealand Inst. Hort. Preceded by: Report, New Zealand institute of horticulture. Superseded by: Journal of the royal New Zealand institute of horticulture. HI 66836

Journal of the Newell entomological society. Gainesville, FL. J. Newell Entomol. Soc. See B–P–H 475/2. HI 55237

Journal of the nigerian institute for oil palm research. Benin City. Vol. 4(15)+, 1965+. J. Nigerian Inst. Oil Palm Res. Preceded by: Journal of the west african institute for oil palm research. HI 66837

Journal of the Niigata agricultural experiment station. [Niigata-ken nogyo shikenjo kenkyu hokoku.] Niigata, Japan. J. Niigata Agric. Exp. Sta. See B–P–H 475/3. HI 55238

Journal, nippon fernist club. [Nihon shida no kai kaiho.] Tokyo. No. ?-66+, ?-1964+. J. Nippon Fernist Club. HI 66838

Journal, nippon institute for biological science. [Nisseiken tayori.] Tokyo. 1955+. J. Nippon Inst. Biol. Sci. HI 66839

Journal of the north China branch of the royal asiatic society. Shanghai. J. N. China Branch Roy. Asiat. Soc. See B–P–H 474/3. HI 55224

Journal of north-eastern forestry institute. [Dongbei linxueyuan xuebao.] Harbin. Vol. 1+, 1979+. J. NorthE. Forest. Inst., Harbin. HI 66840

Journal, north Gloucestershire naturalists' society. Cheltenham = North Gloucestershire naturalists' society journal. Cheltenham.

Journal of the Northamptonshire natural history society and field club. Northampton. 1880+. J. Northamptonshire Nat. Hist. Soc. Field Club. HI 66842

Journal of the northern forestry society. [Hoppo ringyo kenkyukai koenshu.] Sapporo. Vol. 1+, 1949+. J. N. Forest. Soc. HI 66843

Journal, northwest agricultural college. Sian, China = Northwest agricultural college journal. Sian, China. NorthW. Agric. Coll. J. See B–P–H 665/3.

Journal of nuclear agriculture and biology. New Delhi. Vol. 4+, 1975+. J. Nucl. Agric. Biol. Preceded by: I S N A newsletter. HI 66844

Journal of nutrition. Springfield, IL. J. Nutr. See B–P–H 475/4. HI 55239

Journal of obstetrics and gynaecology of the British Empire. London. J. Obstet. Gynaecol. Brit. Empire. See B–P–H 475/6. HI 55240

Journal of oceanographic works in China. Studia marina sinica. [Hai Yang k'o hsüeh chi K'an.] Peking. J.

Oceanogr. Works China. See B–P–H 475/7. HI 55241

Journal of the oceanographical society of Japan. [Nihon kaiyo gakkai shi.] Tokyo. Vol. 1+, 1942+. J. Oceanogr. Soc. Japan. HI 66845

Journal, oceanological society of Korea. Seoul. Vol. 1+, 1966+. J. Oceanol. Soc. Korea. HI 66846

Journal oeconomique; ou mémoires, notes et avis sur l'agriculture, les arts, le commerce et tout ce qui peut avoir rapport à la santé, ... Paris. Ser. 1, 1751-57; ser. 2, 1758-72. J. Oecon. 3-2194-1. HI 74524

Journal of oil and fat industries. Chicago, IL. Vols. 1-3, 1924-26. J. Oil Fat Indust. Superseded by: Oil and fat industries. 1-294-2. HI 66847

Journal of Okayama prefectural agricultural experiment station. [Okayama-kenritsu nogyo shikenjo jiho.] Okayama, Japan. J. Okayama Prefect. Agric. Exp. Sta. See B–P–H 475/8. HI 55242

Journal, Oklahoma state medical association = Oklahoma state medical association. Journal. Muskogee, OK. Oklahoma State Med. Assoc. J. See B–P–H 687/11.

Journal of Oman studies. Muscat. Vol. 1+, 1975+. J. Oman Stud. HI 66848

Journal of the Ontario rock garden society. ?-1986+. J. Ontario Rock Gard. Soc. HI 74517

Journal of the orchid society of Great Britain. St. Albans. Vol. 9(3)+, 1960+. J. Orchid Soc. Gr. Brit. Preceded by: Journal of the amateur orchid growers' society. HI 66849

Journal of the orchid society of India. Chandigarh. Vol. 1+, 1987+. J. Orchid Soc. India. HI 74855

Journal des orchidées. Ghent. J. Orchidées. See B–P–H 475/9. HI 55243

Journal of orchidology. N.S.W. (Australia). Vol. ?-1(2)+, ?-1960+. J. Orchidol. HI 66850

Journal of the oriental arts and sciences. [Toyo gakugei zasshi.] Tokyo. J. Orient. Arts Sci. See B–P–H 475/10. HI 55244

Journal of oriental medicine. [Manshu igaku zasshi.] Dairen, China. J. Orient. Med. See B–P–H 475/11. HI 55245

Journal of oriental studies. [Toho gakuho.] Kyoto, Japan. J. Orient. Stud. See B–P–H 475/12. HI 55246

Journal of the Orissa botanical society. Bhubaneswar. Vol. 1+, 1979+. J. Orissa Bot. Soc. HI 66851

Journal of orthopedic surgery. Boston, MA. J. Orthop. Surg. See B–P–H 475/13. HI 55247

Journal of the Osaka cactus club. Senyukai-kaiho. Osaka. No. 1+, 1956+. J. Osaka Cact. Club. HI 66852

Journal of the Osmania university college (science). Haiderabad. Vol. 1+, 1933+. J. Osmania Univ. Coll. (Sci.). 4-3211-2. HI 66853

Journal, Oxford university forest society. Oxford. Ser. 1, nos. 1-10, 1920-29; ser. 2, nos. 11-19, 1930-38; ser. 3, nos. 1-6, 1946-52; ser. 4, nos. 1-6, 1953-58; ser. 5, nos. 7-13, 1959-65; ser. 6, nos. 1-?, 1966; ser. 7, nos.1-3, 196?-?? J. Oxford Univ. Forest Soc. Superseded by: Land. HI 66854

Journal of the palaeontological society of India. Lucknow, India. J. Palaeontol. Soc. India. See B–P–H 475/15. HI 55248

Journal of paleolimnology. Dordrecht & Hingham, MA. Vol. 1+, 1988+. J. Paleolimnol. HI 66855

Journal of paleontology. Chicago, IL. J. Paleontol. See B–P–H 475/18. HI 55249

Journal of palynology. Lucknow. Vol. 1+, 1965+. J. Palynol. From 1972 onwards incorporated: Palynological bulletin. HI 66856

Journal of parasitology. Lancaster, PA, Urbana, IL. 1914+. J. Parasitol. HI 66857

Journal of park administration, horticulture and recreation. London. Vols. 1-21(3), 1936-56. J. Park Admin. Hort. Recreation. Superseded by: Park administration, horticulture and recreation. HI 66858

Journal of pathology and bacteriology Cambridge, London. Vols. 1-96, 1892-1968. J. Pathol. Bacteriol. Superseded by: Journal of medical microbiology. 3-2231-2 HI 66859

Journal of pediatrics. St. Louis, MO. J. Pediatrics. See B–P–H 475/21. HI 55250

Journal of the Peking national university. [Kuo li Pei-p'ing ta hsüeh hsüeh pao.] Peiping [=Peking]. J. Peking Natl. Univ. See B–P–H 475/22. HI 55251

Journal of penicillin. Tokyo. J. Penicillin. See B–P–H 475/23. HI 55252

Journal of pharmaceutical sciences. Easton, PA. Vol. 50+, 1961+. J. Pharm. Sci. Preceded by: Journal of the american pharmaceutical association. Scientific edition. 3-2232-2. HI 66860

Journal of pharmaceutical sciences of the United Arab Republic. Cairo. Vols. 1-10, 1960-69. J. Pharm. Sci. United Arab Republic. Superseded by: United Arab Republic journal of pharmaceutical sciences. HI 55257

Journal of the pharmaceutical society of Japan. [Yakagaku zasshi.] Tokyo. J. Pharm. Soc. Japan. See B–P–H 476/11. HI 55258

Journal de pharmacie. Paris. J. Pharm. See B–P–H 476/1. HI 55253

Journal der Pharmacie für Aerzte, Apotheker und

Chemisten. Leipzig. J. Pharm. Aerzte. See B–P–H 476/2. HI 55254

Journal der Pharmacie für Aerzte und Apotheker. Leipzig. J. Pharm. Aerzte Apotheker. See B–P–H 476/3. HI 55255

Journal de pharmacie et de chimie. Paris. Vols. 1-6, 1809-14; sér. 2, vols. 1-27, 1815-41; sér. 3, vols. 1-46, 1842-64; sér. 4, vols. 1-30, 1865-79; sér. 5, vols. 1-30, 1880-94; sér. 6, vols. 1-30, 1895-1909; sér. 7, vols. 1-30, 1910-24; sér. 8, vols. 1-30, 1925-39; sér. 9, vols. 1-2, 1940-42. J. Pharm. Chim. Superseded by: Annales pharmaceutiques françaises. 3-2184-2. HI 66861

Journal de pharmacie et des sciences accessoires. Paris. J. Pharm. Sci. Accessoires. See B–P–H 476/9. HI 55256

Journal de pharmacie et des sciences accessoires et bulletin de la société de pharmacie de Paris = Journal de pharmacie et des sciences accessoires. Paris. J. Pharm. Sci. Accessoires. See B–P–H 476/9.

Journal de pharmacie et des sciences accessoires contenant le bulletin de la société de pharmacie de Paris = Journal de pharmacie et des sciences accessoires. Paris. J. Pharm. Sci. Accessoires. See B–P–H 476/9.

Journal de pharmacie et des sciences accessoires contenant le bulletin des travaux de la société de pharmacie de Paris = Journal de pharmacie et des sciences accessoires. Paris. J. Pharm. Sci. Accessoires. See B–P–H 476/9.

Journal of pharmacobiodynamics. Tokyo. Vol. 1+, 1978+. J. Pharmacobiodyn. HI 66862

Journal de pharmacologie. Paris. Vol. 1+, 1970+. J. Pharmacol. HI 66863

Journal of pharmacology and experimental therapeutics. Baltimore, MD. J. Pharmacol. Exp. Therap. See B–P–H 476/13. HI 55259

Journal of pharmacy and pharmacognosy. 1966+. J. Pharm. Pharmacogn. HI 66864

Journal of pharmacy and pharmacology. London. Vols. 1+, 1949+. J. Pharm. Pharmacol. Preceded by: Quarterly journal of pharmacy and pharmacology. 3-2232-3. HI 66865

Journal of the Philadelphia college of pharmacy. Philadelphia, PA. Vol. 1, 1825-27; vols. 1-6, 1829-35. J. Philadelphia Coll. Pharm. Superseded by: American journal of pharmacy. 1-258-2. HI 66866

Journal of the philippine pharmaceutical association. Manila. J. Philipp. Pharm. Assoc. See B–P–H 476/15. HI 55260

Journal of photochemistry and photobiology. Part B: Biology. Amsterdam. Vol. ?-3+, ?-1989+. J. Photochem. Photobiol., B. HI 66867

Journal of phycology. Baltimore, MD. Vol. 1+, 1965+. J. Phycol. Preceded by: News bulletin, phycological society of America. HI 66868

Journal der Physik. Halle. J. Phys. (Halle). See B–P–H 476/18. HI 55261

Journal de physique, de chimie, d'histoire naturelle et des arts. Paris. J. Phys. Chim. Hist. Nat. Arts. See B–P–H 476/19. HI 55262

Journal de physique, chimie et histoire naturelle élémentaires. Paris. J. Phys. Chim. Hist. Nat. Élément. See B–P–H 476/20. HI 55263

Journal de physiologie. Paris. J. Physiol. See B–P–H 476/21. HI 55264

Journal de physiologie expérimentale et pathologique. Paris. J. Physiol. Exp. Pathol. See B–P–H 476/22. HI 55265

Journal de physiologie et pathologie générale. Paris. J. Physiol. Pathol. Gén. See B–P–H 477/2. HI 55266

Journal of phytogeography and taxonomy. [Shokubutsu chiri bunrui kenkyu.] Ishikawa. Vol. 27+, 1979+. J. Phytogeogr. Taxon. Preceded by: Journal of geobotany. HI 66869

Journal of phytopathology. Berlin, New York. Vol. 115+, 1986+. J. Phytopathol. Preceded by: Phytopathologische Zeitschrift. HI 66870

Journal of phytopathology U A R. Vol. 1+, 1969+. J. Phytopathol. U.A.R HI 66871

Journal of phytopathology United Arab Republic = Journal of phytopathology U A R.

Journal of plankton research. Oxford. Vol. 1+, 1979+. J. Plankt. Res. HI 66872

Journal of plant anatomy and morphology. Jodhpur. Vol. 1+, 1984+. J. Pl. Anat. Morphol. HI 66873

Journal of plant breeding. Berlin, etc. = Plant breeding. Berlin, etc.

Journal of plant foods. London. Vol. 3+, 1978+. J. Pl. Foods. Preceded by: Plant foods for man. HI 66874

Journal of plant growth regulation. New York, Heidelberg, Berlin. Vol. 1+, 1982+. J. Pl. Growth Regulat. HI 66875

Journal of plant nutrition. New York. Vol. 1+, 1979+. J. Pl. Nutr. HI 66876

Journal of plant physiology. Stuttgart, Deerfield Beach, FL. Vol. 115+, 1984+. J. Pl. Physiol. Preceded by: Zeitschrift für Pflanzenphysiologie. HI 66877

Journal of plant protection. [Sikmul poho.] Suwon. Vols. 1-9, 1962-70. J. Pl. Protect. Superseded by: Korean journal of plant protection. HI 66878

Journal of plant protection. [Byochu-gai zasshi.] Tokyo. J. Pl. Protect. See B–P–H 477/4. HI 55267

Journal of plant protection in the tropics. Kuala Lumpur. Vol. 1+, 1984+. J. Pl. Protect. Tropics. HI 68114

Journal of plant sciences research. New Delhi. Vol. 1+, 1985+. J. Pl. Sci. Res. HI 66879

Journal of plantation crops. Kasaragad. Vol. 1+, 1973+. J. Plantation Crops. HI 66880

Journal of the polynesian society. Wellington. J. Polynes. Soc. See B–P–H 477/5. HI 55268

Journal of pomology. London. J. Pomol. See B–P–H 477/6. HI 55269

Journal of pomology and horticultural science. London. J. Pomol. Hort. Sci. See B–P–H 477/7. HI 55270

Journal of the Portland society of natural history. Portland, ME. Vol. 1(1) 1864. J. Portland Soc. Nat. Hist. HI 66881

Journal of the Portsmouth and district natural history society. Portsmouth. ?-1982+. J. Portsmouth Distr. Nat. Hist. Soc. HI 66882

Journal, post-graduate school, indian agricultural research institute. New Delhi. Vol. 1+, 1963+. J. Post-Graduate School Indian Agric. Res. Inst. HI 66883

Journal für praktische Chemie. Leipzig. J. Prakt. Chem. See B–P–H 477/8. HI 55271

Journal and proceedings of the asiatic society of Bengal. Calcutta. Vols. 1-30, 1905-34. J. Proc. Asiat. Soc. Bengal. Preceded by: Journal of the asiatic society of Bengal. Part 2, natural history. From 1935 onwards the Proceedings were incorporated in: Yearbook of the royal asiatic society of Bengal. Superseded by: Journal of the royal asiatic society of Bengal. 1-505-3. HI 66884

Journal of proceedings of the Essex field club. Buckhurst Hill. Vols. 1-4, 1880-87. J. Proc. Essex Field Club. Superseded by: Essex naturalist. 2-1480-2. HI 66885

Journal and proceedings of the Hamilton association for the cultivation of science = Journal and proceedings of the Hamilton scientific association. Hamilton, Ont.

Journal and proceedings of the Hamilton scientific association. Hamilton, Ont. Vols. 1-30, 1882-1922. J. Proc. Hamilton Sci. Assoc. HI 66886

Journal of the proceedings of the linnean society. Botany. London. J. Proc. Linn Soc., Bot. See B–P–H 477/13. HI 55272

Journal of proceedings of the Mueller botanic society of Western Australia. Perth, W.A. J. Proc. Mueller Bot. Soc. Western Australia. See B–P–H 477/14. HI 55273

Journal of the proceedings of the royal society of New South Wales. Sydney, N.S.W. Vol. 10+, 1876+. J. Proc. Roy. Soc. New S. Wales. Preceded by: Transactions of the royal society of New South Wales. 4-3730-1. HI 66887

Journal and proceedings of the royal society of Western Australia. Perth, W.A. Vols. 1-10, 1914-24. J. Proc. Roy. Soc. Western Australia. Preceded by: Journal of the natural history and science society of Western Australia. Superseded by: Journal of the royal society of Western Australia. 4-3731-2. HI 66888

Journal of the proceedings of the Winchester and Hampshire scientific and literary society. Winchester, England. J. Proc. Winchester Hampshire Sci. Soc. See B–P–H 477/19. HI 55274

Journal des progrès des sciences. Paris. J. Progr. Sci. See B–P–H 477/20. HI 55275

Journal of protozoology. Utica, NY. J. Protozool. See B–P–H 477/21. HI 55276

Journal of psychedelic drugs. San Francisco, CA, Madison, WI. Vols. 1-12, 1967-80 [suspended 1973]. J. Psychedelic Drugs. Superseded by: Journal of psychoactive drugs. HI 66889

Journal of psychoactive drugs. San Francisco, CA. Vol. 13+, 1981+. J. Psychoact. Drugs. Preceded by: Journal of psychedelic drugs. HI 66890

Journal of the Quekett microscopical club. London. Vols. 1-6, 1868/69-79/81; ser. 2, vols. 1-16, 1882/84-1928/33; ser. 3, vol. 1 (nos. 1-8), 1934-37; ser. 4, vols. 1-, 1938-66. J. Quekett Microsc. Club. Superseded by: Microscopy. 4-3507-2. HI 66891

Journal of range management. Baltimore, MD. J. Range Managem. See B–P–H 478/5. HI 55277

Journal de recherche océanographique. Bulletin. Paris. Vol. 1+, 1976+. J. Rech. Océanogr., Bull. Preceded by: Bulletin, union des océanographes de France. HI 66892

Journal of research in ayurveda and sidda. New Delhi. Vol. 1+, 1980+. J. Res. Ayurveda Sidda. Preceded by: Journal of research in indian medicine, yoga and homeopathy HI 66893.

Journal of research, Haryana agricultural university. Hissar. Vol. 1+, 1971+. J. Res. Haryana Agric. Univ. HI 66894

Journal of research in indian medicine. Varanasi. Vols. 1-10, 1966-75. J. Res. Indian Med. Superseded by: Journal of research in indian medicine, yoga and homeopathy. HI 66895

Journal of research in indian medicine, yoga and homeopathy. New Delhi. Vols. 11-14, 1976-79. J. Res. Indian Med. Yoga Homeopathy. Preceded by: Journal of research in indian medicine. Superseded by: Journal of research in ayurveda and siddha. HI 66896

Journal of research in lepidoptera. Arcadia, CA. 1962+. J. Res. Lepid. HI 66897

Journal of research, Punjab agricultural university. Ludhiana. Vol. 1+, 1964+. J. Res. Punjab Agric. Univ. HI 66898

Journal of research science; Agra university. Agra, India. J. Res. Sci. Agra Univ. See B–P–H 478/6. HI 55278

Journal of research society for antibacterial and antifungal agents, Japan. [Bokin bobai.] Vol. 1+, 1973+. J. Res. Soc. Antibact. Antifungal Agents., Japan. HI 66899

Journal of research of the U S geological survey. Washington, DC. 1973+. J. Res. U.S. Geol. Surv. HI 66900

Journal of the Rio Grande valley horticultural society. Weslaco, TX. J. Rio Grande Valley Hort. Soc. See B–P–H 478/7. HI 55279

Journal of root crops. Trivandrum. Vol. 1+, 1975+. J. Root Crops. HI 66901

Journal des roses; revue des jardins. Lyon. Vol. 1, 1855. J. Roses (Lyon). Superseded by: Journal des roses et des vergers. HI 66902

Journal des roses. Paris. Vols. 1-38, 1877-1914. J. Roses (Paris). 3-2188-2. HI 55280

Journal des roses et des vergers; revue des jardins. Lyon. Vols. 2-6, 1856?-59. J. Roses Vergers. Preceded by: Journal des roses. Lyon. HI 66903

Journal of the royal african society. London. Vols. 34(136)-43(171), 1935-44. J. Roy. African Soc. Preceded by: Journal of the african society. Superseded by: African affairs. 1-80-3. HI 66904

Journal of the royal agricultural and commercial society of British Guiana = Timehri. Georgetown, Demarara.

Journal of the royal agricultural society of England. London. J. Roy. Agric. Soc. England. See B–P–H 478/9. HI 55281

Journal of the royal anthropological institute of Great Britain and Ireland. London. J. Roy. Anthropol. Inst. Gr. Brit. See B–P–H 478/10 HI 55282

Journal of the royal army medical corps. London. J. Roy. Army Med. Corps. See B–P–H 478/11. HI 55283

Journal of the royal asiatic society of Bengal. Calcutta. 3rd ser. vols. 1-16, 1935-50. J. Roy. Asiat. Soc. Bengal. Preceded by: Journal and proceedings of the asiatic society of Bengal. Superseded by: Journal of the asiatic society. HI 66905

Journal of the royal asiatic society, Bombay branch = Journal of the Bombay branch of the royal asiatic society. Bombay.

Journal of the royal asiatic society, Ceylon branch = Journal of the Ceylon branch of the royal asiatic society. Colombo.

Journal of the royal asiatic society of Great Britain and Ireland. London. J. Roy. Asiat. Soc. Gr. Brit. See B–P–H 478/12. HI 55284

Journal of the royal asiatic society, Malayan branch = Journal of the Malayan branch of the royal asiatic society. Singapore.

Journal of the royal asiatic society, North China branch = Journal of the North China branch of the royal asiatic society. Shanghai.

Journal of the royal asiatic society, Sri Lanka branch = Journal of the Sri Lanka branch of the royal asiatic society. Colombo.

Journal of the royal asiatic society, Straits branch = Journal of the Straits branch of the royal asiatic society. Singapore.

Journal of the royal botanic garden guild. Edinburgh. J. Roy. Bot. Gard. Guild. See B–P–H 478/13. HI 55285

Journal of the royal caledonian horticultural society. Edinburgh. 1950-67. J. Roy. Caledonian Hort. Soc. Preceded by: Transactions, royal caledonian horticultural society. Superseded by: Year book and journal, royal caledonian horticultural society HI 66906.

Journal of the royal geographical society. London. J. Roy. Geogr. Soc. See B–P–H 478/14. HI 55286

Journal of the royal geological society of Ireland. Dublin. Vols. 11-18 (n.s. vols. 1-8) 1864-89. J. Roy. Geol. Soc. Ireland. Preceded by: Journal of the geological society of Dublin. HI 66907

Journal of the royal horticultural society. London. Vols. 1-100(5), 1866-1975. J. Roy. Hort. Soc. Superseded by: Garden. London. 4-3722-1. HI 66908

Journal of the royal horticultural society gardens club. London. Nos. 1-?, 1908-52. J. Roy. Hort. Soc. Gard. Club. HI 66909

Journal of the royal institution of Cornwall. Truro, England. J. Roy. Inst. Cornwall. See B–P–H 478/17. HI 55287

Journal of the royal institution of Great Britain. London. J. Roy. Inst. Gr. Brit. See B–P–H 478/18. HI 55288

Journal of the royal microscopical society of London containing its transactions and proceedings, with other microscopical information. London. Vols. 1-3, 1878-80; ser. 2, vols. 1-6, 1881-86; 1887-1926; ser. 3, vols. 47-88, 1927-68. J. Roy. Microscop. Soc. London. Superseded by: Journal of microscopy. 4-3725-1. HI 66910

Journal of the royal naval medical service. London. J. Roy. Naval Med. Serv. See B–P–H 478/20. HI 55289

Journal of the royal New Zealand institute of horticulture.

[From 1947-55 published as a section of: New Zealand gardener. From 1955-68 formed a section of: New Zealand plants and gardens.] Auckland, N.Z. Vols. 8(4)-17(2) [vol. 17(1) not issued], 1939-47; n.s. vols.1-2(5), 1968-71. J. Roy. New Zealand Inst. Hort. Preceded by: Journal of the New Zealand institute of horticulture. Superseded by?: Horticulture in New Zealand. 4-3725-3. HI 66911

Journal of the royal society of antiquaries of Ireland. Dublin. J. Roy. Soc. Antiq. Ireland. See B–P–H 478/22. HI 55290

Journal of the royal society of arts. London. Vol. 56+, 1908+. J. Roy. Soc. Arts. Preceded by: Journal of the society of arts. 4-3726-3. HI 66912

Journal of the royal society of New Zealand. Wellington, N.Z. Vol. 1+, 1971+. J. Roy. Soc. New Zealand. Preceded by: Transactions of the royal society of New Zealand. Biological sciences and Transactions of the royal society of New Zealand. New series, general. HI 66913

Journal of the royal society of Western Australia. Perth, W.A. Vol. 11+, 1924+. J. Roy. Soc. W. Australia. Preceded by: Journal and proceedings of the royal society of Western Australia. HI 66914

Journal, royal Swaziland society of science and technology. Kwaluseni. Vol. 1+, 1978+. J. Roy. Swaziland Soc. Sci. Techn. HI 66915

Journal of the rubber research institute of Malaya (later Malaysia). Kuala Lumpur. Vol. 3+, 1932+. J. Rubber Res. Inst. Malaya (Malaysia) Preceded by: Quarterly journal of the rubber research institute of Malaya. 3-2321-2. HI 66918

Journal of the rubber research institute of Malaysia = Journal of the rubber research institute of Malaya. Kuala Lumpur.

Journal, rubber research institute of Sri Lanka. Dartonfield, Agalawatta. Vol. 53+, 1976+. J. Rubber Res. Inst Sri Lanka. Preceded by: Quarterly journal, rubber research institute of Sri Lanka. HI 66919

Journal of the Ruislip and district natural history society. Ruislip. No. 1+, 1951+ [publication suspended, 1972-78]. J. Ruislip Distr. Nat. Hist. Soc. HI 66920

Journal of rural studies. New York. Vol. 1+, 1985+. J. Rural Stud. HI 66921

Journal russe de botanique. St. Petersburg = Russkii Botanicheskii Zhurnal. St. Petersburg.

Journal, S W A scientific society = Journal of the South West Africa scientific society. Windhoek, South-West Africa. J. South West Africa Sci. Soc. See B–P–H 482/22.

Journal, Sabah society. Jesselton. Vol. 1+, 1961+. J. Sabah Soc. HI 66922

Jaurnal of Saitama university, Natural science. [Saitama daigaku kiyo. Shizenkagaku-hen.] Urawa. Vol. 1+, 1965+. J. Saitama Univ., Nat. Sci. HI 55293

Journal of the Sapporo society of agriculture and forestry. [Sapporo noringakkai-ho.] Sapporo, Japan. J. Sapporo Soc. Agric. See B–P–H 479/13. HI 55294

Journal, Saudi arabian natural history society. Jeddah. Vol. 1(2)+, 1971+. J. Saudi Arabian Nat. Hist. Soc. Preceded by: Report, Saudi arabian natural history society. HI 66923

Journal des savans. Paris. J. Savans. See B–P–H 479/14. HI 55295

Journal des savans d'Italie. Amsterdam. J. Savans Italie. See B–P–H 479/15. HI 55296

Journal des savants. Paris. J. Savants. See B–P–H 479/16. HI 55297

Journal des savants de Normandie. Caen. J. Savants Normandie. See B–P–H 479/17. HI 55298

Journal des sçavans. Amsterdam. J. Sçavans (Amsterdam). See B–P–H 479/21. HI 55299

Journal des sçavans. Leipzig. J. Sçavans (Leipzig). See B–P–H 479/22. HI 55300

Journal des sçavans. [In quarto.] Paris. J. Sçavans (Paris, 4to). See B–P–H 479/23. HI 55301

Journal des sçavans. [In duodecimo.] Paris. J. Sçavans (Paris, 12mo). See B–P–H 480/1. HI 55302

Journal of school of education. [Sze ta hsüeh k'an.] Peiping [=Peking]. J. School Educ. See B–P–H 480/2. HI 55303

Journal of science. Louisville, KY. J. Sci. (Louisville). See B–P–H 480/3. HI 55304

Journal of science, and annals of astronomy, biology, geology, industrial arts, manufactures, and technology. London. J. Sci. & Ann. Astron. See B–P–H 480/5. HI 55306

Journal of science and the arts. London. J. Sci. Arts (London). See B–P–H 480/6. HI 55307

Journal of science and the arts. New York, NY. J. Sci. Arts (New York). See B–P–H 480/7. HI 55308

Journal des sciences, arts et métiers; par une société de gens de lettres et d'artistes. Paris. J. Sci. Arts Métiers. See B–P–H 480/8. HI 55309

Journal des sciences et des beaux-arts. Paris. J. Sci. Beaux-Arts. See B–P–H 480/9. HI 55310

Journal of the science faculty of Chiangmai university. [Chiangmai.] Vol. 1+, 1974+. J. Sci. Fac. Chiangmai Univ. HI 66924

Journal of the science of food and agriculture. London. J. Sci. Food Agric. See B–P–H 480/11. HI 55311

Journal of science of Hiroshima university: series B,

division 2 (botany). Hiroshima. J. Sci. Hiroshima Univ., Ser. B, Div. 2, Bot. See B–P–H 480/13. HI 55312

Journal of science of the Kanazawa normal college. [Kanazawa koto shihan gakko rika kioy.] Kanazawa. [Dates of publication not ascertained.] J. Sci. Kanazawa Norm. Coll. HI 66925

Journal of science and medicine of Jinan university. [Jinan liyi xuebao.] Kuang-chou. Vol. 1+, 1980+. J. Sci. Med. Jinan Univ. HI 66926

Journal of science; national university of Shantung. [Kuo-li Shantung ta hsüeh k'o hsüeh ts'ung k'an.] Tsingtao, China. J. Sci. Natl. Univ. Shantung. See B–P–H 480/17. HI 55315

Journal of science of the north-eastern normal university. Biology. [K'o hsüeh chi kan tung pei shih fan ta hsüeh. Sheng wu.] Changchun, China. J. Sci. N.-E. Norm. Univ., Biol. See B–P–H 480/18. HI 55316

Journal of the science of soil and manure. [Nihon dojo-hiryo-gaku zasshi.] Tokyo. J. Sci. Soil Manure. See B–P–H 481/1. HI 55320

Journal of science, university of Kashmir. Srinagar. Vol. 1+, 1973+. J. Sci. Univ. Kashmir. Preceded by: Kashmir science. HI 66927

Journal of the scientific agricultural society. Tokyo. J. Sci. Agric. Soc. See B–P–H 480/4. HI 55305

Journal of the scientific agricultural society of Finland = Maataloustieteellinen aikakauskirja. Helsinki.

Journal of scientific exploration. New York, Oxford, etc. Vol. 1+, 1987+. J. Sci. Explor. HI 66928

Journal of scientific and industrial research. C. Biological sciences. New Delhi. J. Sci. Industr. Res., C, Biol. Sci. See B–P–H 480/14. HI 55313

Journal of the scientific laboratories of Denison university. Granville, OH. J. Sci. Lab. Denison Univ. See B–P–H 480/16. HI 55314

Journal of scientific and practical mycology = Karstenia. Sienitieteellinen ja sienitalondellinen aikakauskirja. Helsinki. Karstenia. See B–P–H 511/18.

Journal of scientific research. Bhopal. Vol. 1+, 1978+. J. Sci. Res. HI 66929

Journal of scientific research. Jakarta, Indonesia. J. Sci. Res. (Jakarta). See B–P–H 480/19. HI 55317

Journal of scientific research. Jwalapur = Gurukula kangri vishwavidyalaya journal of scientific research. Hardwar.

Journal of scientific research of the Benares Hindu university. Banaras, India. J. Sci. Res. Benares Hindu Univ. See B–P–H 480/21. HI 55318

Journal of scientific research, Gurukula Kangri vishwavidyalaya = Gurukula kangri vishwavidyalaya journal of scientific research. Hardwar.

Journal of the scientific research institute. Tokyo. J. Sci. Res. Inst. See B–P–H 480/22. HI 55319

Journal of scientific research in plants and medicines. Hardwar. Vol. 1+, 1979+. J. Sci. Res. Pl. Med. HI 66930

Journal of the scottish rock garden club. Edinburgh, Glenfarg. Vols. 1-18(2), 1937-83. J. Scott. Rock Gard. Club. Superseded by: Rock garden. HI 66931

Journal de la section botanique de l'académie des sciences d'Ukraine = Zhurnal Bio-botanichnogo tsyklu Vseukraïns'koi akademii nauk. Kiev.

Journal of sedimentary petrology. Tulsa, OK. J. Sediment. Petrol. See B–P–H 481/9. HI 55322

Journal of seed technology. Lansing, MI, Reynoldsburg, OH. Vol. 1+, 1976+. J. Seed Technol. Preceded by: Proceedings of the association of official seed analyists of north America. HI 66932

Journal of the sempervivum society. Bishops Stortford, Burgess Hill. Vols. 1-10, 1970-74. J. Sempervivum Soc. Superseded by: Houseleeks. HI 66933

Journal of sericultural science of Japan. [Nihon sanshigaku zasshi.] Tokyo, Tsukubu. Vol. 1+, 1930+. J. Sericult. Sci. Japan. HI 66934

Journal of the Seychelles society. Victoria, Seychelles. No. 1+, 1961+. J. Seychelles Soc. HI 66935

Journal of Shandong college of oceanology. [Shandong haiyangxue yuan xuebao.] Beijing. 1971+. J. Shandong Coll. Oceanol. HI 66936

Journal of the Shanghai literary and scientific society = Journal of the north China branch of the royal asiatic society. Shanghai. J. N. China Branch Roy. Asiat. Soc. See B–P–H 474/3.

Journal of the Shanghai science institute. Shanghai. J. Shanghai Sci. Inst. See B–P–H 481/11. HI 55323

Journal of Shansi normal college. [Shan hsi sze fan hsüeh yuan hsüeh pao.] Taiyüan, China. J. Shansi Norm. Coll. See B–P–H 481/12. HI 55324

Journal of Shanxi university. Natural science edition = Shanxi university journal. Natural science edition.

Journal of the Shimonoseki college of fisheries. [Suisan koshujo kenkyu-hokoku.] Shimonoseki, Japan. J. Shimonoseki Coll. Fish. See B–P–H 481/13. HI 55325

Journal of the Shinano natural science society. [Shinano hakubutsu-gaku zasshi.] Shinano, Japan. J. Shinano Nat. Sci. Soc. See B–P–H 481/14. HI 55326

Journal of the Siam society. Natural history supplement. Bangkok. Vols. 7-22, 1926-38. J. Siam Soc., Nat. Hist. Suppl. Preceded by: Journal of the natural history society of Siam. Superseded by: Journal of the

Thailand research society. Natural history supplement. 5-3868-1. HI 66938

Journal of Sichuan university. Natural science edition. [Sichuan daxue xuebao.] Chengtu. 1955+. J. Sichuan Univ., Nat. Sci. Ed. HI 66939

Journal of the Sigenkagaku Kenkyusyo. [Shigen kagaku kenkyusho obun hokoku.] Tokyo. J. Sigenkagaku Kenkyusyo. See B–P–H 481/16. HI 55327

Journal of the Simla naturalists' society. Vol. 1, nos. 1-2, 1885-86. J. Simla Naturalists' Soc. HI 66940

Journal of the Sind natural history society. Karachi. 1929-41. J. Sind Nat. Hist. Soc. HI 66941

Journal of the Singapore national academy of science. Tecco, Singapore. Vol. 1+, 1969+ [suspended 1971-72]. J. Singapore Natl. Acad. Sci. HI 66942

Journal de la société d'agriculture de Suisse Romande. Lausanne. Vol. ?-31-48-?, ?-1890-1907-? J. Soc. Agric. Suisse Romande. HI 66943

Journal de la société d'agronomie pratique. Paris. J. Soc. Agron. Prat. See B–P–H 481/17. HI 55328

Journal de la société botanique de Russie = Zhurnal Russkago Botanicheskago Obshchestva pri Akademii Nauk. Petrograd.

Journal de société centrale d'horticulture du Nord. Lille. J. Soc. Centr. Hort. Nord. See B–P–H 481/22. HI 55330

Journal. Société française du Dahlia. Biarritz, France. J. Soc. Franç. Dahlia. See B–P–H 482/2. HI 55335

Journal, société d'horticulture et d'arboriculture du Doubs. Besançon. Vols. ?-6-?, ?-1862-? J. Soc. Hort. Arboric. Doubs. HI 66944

Journal de la société d'horticulture du Japon = Journal of the japanese horticultural society. Tokyo. J. Jap. Hort. Soc. See B–P–H 470/21.

Journal de la société impériale et centrale d'horticulture. Paris. J. Soc. Imp. Centr. Hort. See B–P–H 482/4. HI 55336

Journal de la société nationale d'horticulture de France. Paris. J. Soc. Natl. Hort. France. See B–P–H 482/6. HI 55337

Journal de la société des océanistes. Paris. Vol. 1+, 1945+. J. Soc. Océanistes. Preceded by: Bulletin de la société des océanistes. 5-3954-3. HI 74518

Journal de la société physico-chemique russe = Zhurnal Russkago Fiziko-khimicheskago Obshchestva pri Imperatorskom S.-Peterburgskom Universitetě. Chast′ khimicheskaya. St. Petersburg.

Journal de la société régionale d'horticulture du nord de la France. Lille. J. Soc. Régionale Hort. N. France. See B–P–H 482/10. HI 55338

Journal de la société des sciences, agriculture et arts, du département du Bas-Rhin, séant à Strasbourg. Strasbourg. J. Soc. Sci.Dép. Bas-Rhin. See B–P–H 482/11. HI 55339

Journal de la société vaudoise d'utilité publique, faisant suite à la Feuille du canton de Vaud. Lausanne. J. Soc. Vaud. Util. Publique. See B–P–H 482/15. HI 55341

Journal of the society of arts. London. Vols. 1-53, 1852-1907. J. Soc. Arts. Preceded by: Transactions of the society, instituted at London, for the encouragement of arts, manufactures and commerce. Superseded by: Journal of the royal society of arts. HI 66945

Journal of the society for the bibliography of natural history. London. Vols. 1-9, 1936-80. J. Soc. Bibliogr. Nat. Hist. Superseded by: Archives of natural history. 5-3979-2. HI 66946

Journal of the society of brewing, Japan. [Nihon jozo kyokai zasshi.] Tokyo. J. Soc. Brew. Japan. See B–P–H 481/21. HI 55329

Journal of the society of chemical industry. London. J. Soc. Chem. Industr. See B–P–H 481/23. HI 55331

Journal of the society of chinese tropical agriculture. [Chung-kuo je-tai nung hsüeh hui-pao.] Taipei, Taiwan. J. Soc. Chin. Trop. Agric. See B–P–H 481/24. HI 55332

Journal of the society of dyers and colourists. Bradford, England. J. Soc. Dyers. See B–P–H 481/25. HI 55333

Journal of the society of forestry. [Ringakukai zasshi.] [Japan]. J. Soc. Forest. See B–P–H 482/1. HI 55334

Journal, society of Malawi. Blantyre. ?-1976+. J. Soc. Malawi. HI 66947

Journal of the society of tropical agriculture. [Nettai nogaku kwaishi.] Taihoku [=Taipei, Taiwan]. J. Soc. Trop. Agric. See B–P–H 482/13. HI 55340

Journal of soil science. London. J. Soil Sci. See B–P–H 482/16. HI 55342

Journal, soil science society of America. Madison, WI. Vol. 40+, 1976+. J. Soil Sci. Soc. Amer. Preceded by: Proceedings, soil science society of America. HI 66948

Journal of the soil science society of the Philippines. Manila. J. Soil Sci. Soc. Philipp. See B–P–H 482/17. HI 55343

Journal of soil science of the United Arab Republic. J. Soil Sci. U.A.R. See B–P–H 482/18. HI 55344

Journal of soil and water conservation. Baltimore, MD. J. Soil Water Conservation. See B–P–H 482/20. HI 55345

Journal of solid-phase biochemistry. New York. Vols. 1-5, 1976-80. J. Solid-Phase Biochem. Superseded by:

Applied biochemistry and biotechnology. HI 66949

Journal of Soochow university. College of arts and sciences. [Tung-wu hsüeh-pao.] Soochow, China. J. Soochow Univ. Coll. Arts. See B–P–H 482/21. HI 55346

Journal of the south african biological society. Pretoria. J. S. African Biol. Soc. See B–P–H 479/5. HI 55291

Journal of south african botany. Kirstenbosch, Cape Town. Vols. 1-50, 1935-84. J. S. African Bot. Incorporated in: South african journal of botany. 3-2241-2. HI 66950

Journal of the south african chemical institute. Johannesburg. J. S. African Chem. Inst. See B–P–H 479/7. HI 55292

Journal of the south african forestry association. Pretoria. Nos. 1-40, 1938-62. J. S. African Forest. Assoc. Superseded by: South african forestry journal. 5-4000-3. HI 66951

Journal of south China agricultural college. [Huanan nongxueyuan. Xuebao.] Kuangchou. Vol. 1+, 1980+. J. S. China Agric. Coll. HI 66952

Journal of the south-eastern agricultural college. Wye, England. J. S.-E. Agric. Coll. See B–P–H 481/6. HI 55321

Journal of the South West Africa scientific society. Windhoek, South-West Africa. J. South West Africa Sci. Soc. See B–P–H 482/22. HI 55347

Journal of the southwest Scotland grassland society. Auchincruive, Scotland. J. SouthW. Scotland Grassland Soc. See B–P–H 483/1. HI 55349

Journal of the southwestern frontier. [Hsi nan pien chiang.] Kunming, China. J. SouthW. Frontier. See B–P–H 482/23. HI 55348

Journal of the sports turf research institute = Sports turf research institute journal. Bingley, England. Sports Turf Res. Inst. J. See B–P–H 853/7.

Journal of the Sri Lanka branch of the royal asiatic society. Colombo. N.s. vol. 16+, 1972+. J. Sri Lanka Branch Roy. Asiat. Soc. Preceded by: Journal of the Ceylon branch of the royal asiatic society. HI 66953

Journal de la station agronomique de la Guadeloupe. Pointe-à-Pitre, Guadeloupe. J. Stat. Agron. Guadeloupe. See B–P–H 483/2. HI 55350

Journal of stored products research. Oxford. Vol. 1+, 1965/66+. J. Stored Prod. Res. HI 66954

Journal of the Straits branch of the royal asiatic society. Singapore. Vols. 1-86, 1878-1922. J. Straits Branch Roy. Asiat. Soc. Superseded by: Journal of the malayan branch of the royal asiatic society. 4-3717-1. HI 66955

Journal of the Stranahan arboretum. ?-1978+. J. Stranahan Arbor. HI 66956

Journal of structural biology. San Diego, CA. Vol. 103+, 1990+. J. Struct. Biol. Preceded by: Journal of ultrastructure and molecular structure research. HI 75246

Journal of submicroscopic cytology. Bologna. Vol. 1+, 1969+. J. Submicrosc. Cytol. HI 66957

Journal of sugar cane research. [Kan chê yen Chiu.] Pingtung, Taiwan. J. Sugar Cane Res. See B–P–H 483/8. HI 55351

Journal of Sun Yatsen university. [Chung-shan ta hsueh hsueh pao.] Canton. ?-1981+. J. Sun Yatsen Univ. HI 66958

Journal of supramolecular structure. New York. Vols. 1-14, 1972-80. J. Supramolec. Struct. Superseded by: Journal of supramolecular structure and cellular biochemistry. HI 66959

Journal of supramolecular structure. Supplement. New York. Vols. 1-4, 1977-80. J. Supramolec. Struct., Suppl. Superseded by: Journal of supramolecular structure and cellular biochemistry. Supplement. HI 66960

Journal of supramolecular structure and cellular biochemistry. New York. Vol. 15-17, 1981. J. Supramolec. Struct. Cell. Biochem. Preceded by: Journal of supramolecular structure. Superseded by: Journal of cellular biochemistry. Supplement. HI 66961

Journal of supramolecular structure and cellular biochemistry. Supplement. New York. Vol. 5, 1981. J. Supramolec. Struct. Cell. Biochem., Suppl. Preceded by: Journal of supramolecular structure. Supplement. Superseded by: Journal of cellular biochemistry. Supplement. HI 66962

Journal of the Swamy botanical club. Madurai. Vol. 1+, 1984+. J. Swamy Bot. Club. HI 66963

Journal of the swedish seed association = Sveriges utsadesforenings tidskrift. Malmö, Svaloev.

Journal of the Taihoku society of agriculture and forestry. [Taihoku noringakkaiho.] Taihoku [=Taipei]. Vol. 1+, 1936+. J. Taihoku Soc. Agric. Preceded by: Sylvia. 5-4149-1. HI 66964

Journal of the Taiwan museum. [Ta'i-wan sheng li po wu kuan pan nien k'an.] Taipei. Vol. 36+, 1983+. J. Taiwan Mus. Preceded by: Quarterly journal of the Taiwan museum. HI 66965

Journal of the Taiwan museum association. [Kagaku no Taiwan.] Taihoku [=Taipei, Taiwan]. J. Taiwan Mus. Assoc. See B–P–H 483/14. HI 55352

Journal of the Tennessee academy of science. Nashville, TN. J. Tennessee Acad. Sci. See B–P–H 483/16. HI 55353

Journal of the textile institute. Manchester, England. J. Textile Inst. See B–P–H 483/17. HI 55354

Journal of the Thailand research society. Natural history supplement. Bangkok. 1939-40. J. Thailand Res. Soc., Nat. Hist. Suppl. Preceded by: Journal of the Siam society. Natural history supplement. Superseded by: Natural history bulletin of the Thailand research society. HI 66966

Journal of theoretical biology. London. J. Theor. Biol. See B–P–H 483/18. HI 55355

Journal of thermal biology. Oxford & Elmsford, NY. Vol. 1+, 1975+. J. Thermal Biol. HI 66967

Journal of thoracic and cardiovascular surgery. St. Louis, MO. J. Thorac. Cardiovasc. Surg. See B–P–H 483/19. HI 55356

Journal of thoracic surgery. St. Louis, MO. J. Thorac. Surg. See B–P–H 483/20. HI 55357

Journal of the tissue culture association = In vitro. Lake Placid, NY, Gaithersburg, Rockville, MD.

Journal of tissue culture methods. Gaithersburg, MD. 1975+. J. Tissue Cult. Meth. Preceded by: T C A manual. HI 66968

Journal of the Tokyo agricultural college. [Tokyo nogyo daigaku kiyo.] Tokyo. J. Tokyo Agric. Coll. See B–P–H 483/22. HI 55358

Journal of the Tokyo university faculty of science. Tokyo. J. Tokyo Univ. Fac. Sci. See B–P–H 483/23. HI 55359

Journal of the Torquay natural history society. Torquay. 1909-22. J. Torquay Nat. Hist. Soc. Superseded by: Transactions and proceedings, Torquay natural history society. HI 66969

Journal of toxicological sciences. Sapporo. Vol. 1+, 1976+. J. Toxicol. Sci. HI 66970

Journal and transactions of the Eastbourne natural history and archaeological society. Eastbourne. Vol. 13(1-3), 1949-53. J. Trans. Eastbourne Nat. Hist Archaeol. Soc. Preceded by: Transactions and journal of the Eastbourne natural history and archaeological society. HI 66971

Journal of the Transvaal horticultural society. Johannesburg. Vols. 1-31?, 1952-82? J. Transvaal Hort. Soc. Superseded by: Transvaal gardener. HI 55360

Journal of travel and natural history. London & Edinburgh. J. Travel Nat. Hist. See B–P–H 483/26. HI 55361

Journal of tree sciences. New Delhi. Vol. 1+, 1982+. J. Tree Sci. HI 66972

Journal of the Trenton natural history society. Trenton, NJ. J. Trenton Nat. Hist. Soc. See B–P–H 483/27. HI 55362

Journal de Trévoux = Mémoires pour l'histoire des sciences et des beaux arts. Trévoux, Lyons, Paris.

Journal of the Trinidad field naturalists' club. Port of Spain. 1956-57, 1965+. J. Trinidad Field Naturalists' Club. HI 66973

Journal of tropical ecology. Cambridge, New York. Vol. 1+, 1985+. J. Trop. Ecol. HI 66974

Journal of tropical geography. Singapore. Nos. 11-49?, 1958-79? J. Trop. Geogr. Preceded by: Malayan journal of tropical geography. Superseded by: Malaysian journal of tropical geography and Singapore journal of tropical geography. HI 66975

Journal of tropical medicine. London. J. Trop. Med. See B–P–H 484/2. HI 55363

Journal of tropical medicine and hygiene. London. J. Trop. Med. Hyg. See B–P–H 484/3. HI 55599

Journal of the Turkestan branch of the russian geographical society = Izvestiya Turkestanskogo otdela Russkogo geograficheskogo obshchestva. Tashkent.

Journal of the turkish forest research institute = Ormancilik arastirma enstitusu dergisi. Ankara.

Journal of turkish phytopathology. Bornova. Vol. 1+, 1971+. J. Turkish Phytopathol. HI 66976

Journal typographique et bibliographique, ou ... Paris. J. Typogr. Bibliogr. See B–P–H 484/5. HI 55364

Journal for udenlandsk literatur. Copenhagen. Vols. 1-7, 1810-16. J. Udenl. Lit. 3-2190-3. HI 53309

Journal of ultrastructure and molecular structure research. New York. Vol. 102, 1989. J. Ultrastruct. Molec. Struct. Res. Preceded by: Journal of ultrastructure research. Superseded by: Journal of structural biology. HI 66977

Journal of ultrastructure research. New York. Vols. 1-101?, 1957-88?; Supplement, vols. 1-12, 1959-73. J. Ultrastruct. Res. Superseded by: Journal of ultrastructure and molecular structure research. 6-1105-1. HI 66978

Journal universel et hebdomadaire de médecine et de chirurgie pratiques et des institutions médicales. Paris. Vols. 1-13, 1830-33. J. Universel Hebd. Méd. Preceded by: Journal universel des sciences médicales and Journal hebdomadaire de médecine. 3-2246-2. HI 66979

Journal universel des sciences médicales. Paris. Vols. 1-59, 1816-30. J. Universel Sci. Med. Superseded by: Journal universel et hebdomadaire de médecine et de chirurgie pratiques et des institutions médicales. Paris. 3-2246-2. HI 50719

Journal of the university of Bombay. Bombay. J. Univ. Bombay. See B–P–H 484/7. HI 55365

Journal of university of fisheries. Tokyo. J. Univ. Fish. See B–P–H 484/8. HI 55366

Journal of the university of Kuwait (science). Kuwait. Vol. 1+, 1974+. J. Univ. Kuwait (Sci.). HI 66980

Journal of the university of Poona. Science and technology section. Poona. 1952+. J. Univ. Poona., Sci. Technol. Sect. HI 66981

Journal of the university of Saugar. Part 2, section B, natural sciences. Sagar. 1957-59. J. Univ. Saugar, Pt. 2, B. Preceded by: Saugar university journal. Superseded by: Madhya bharati. Journal of the university of Saugar. HI 66982

Journal for utilization of agricultural products. [Nosan Kako Gijutsu kenkyukai Shi.] Tokyo. J. Utiliz. Agric. Prod. See B–P–H 484/12. HI 55367

Journal of vegetation science; official organ of the international association for vegetation science. Uppsala. Vol. 1+, 1990+. J. Veg. Sci. HI 74856

Journal of the Victoria university of Manchester. Manchester, England. J. Victoria Univ. Manchester. See B–P–H 484/13. HI 55368

Journal de Vienne pour l'agriculture et horticulture. Bunzlau, Germany [=Boleslawiec, Poland]. J. Vienne Agric. Hort. See B–P–H 484/14. HI 55369

Journal of virological methods. Amsterdam. 1980+. J. Virol. Meth. HI 66983

Journal of virology. Baltimore, MD. J. Virol. See B–P–H 484/15. HI 55370

Journal de vulgarisation de l'horticulture. Paris. J. Vulg. Hort. See B–P–H 484/16. HI 55371

Journal of the Washington academy of sciences. Baltimore, MD. J. Wash. Acad. Sci. See B–P–H 484/23. HI 55374

Journal of the water pollution control federation. Lancaster, PA. J. Water Pollut. Control Fed. See B–P–H 485/3. HI 55375

Journal of the Wellington field naturalists' club. = Ontario natural science bulletin; journal of the Wellington field naturalists' club. Guelph, Ont. Ontario Nat. Sci. Bull. See B–P–H 688/11.

Journal of the Wellington orchid society. Wellington, N.Z. Vol. ?-2+, ?-1978+. J. Wellington Orchid Soc. HI 66984

Journal of the west african institute for oil palm research. Benin City, London. Vols. 1-14, 1953-63. J. W. African Inst. Oil Palm Res. Superseded by: Journal of the nigerian institute for oil palm research. HI 66985

Journal of the west african science association. London. J. W. African Sci. Assoc. See B–P–H 484/19. HI 55372

Journal of the west China border research society. Chengtu, China. J. W. China Border Res. Soc. See B–P–H 484/21. HI 55373

Journal of the Western Australia natural history society. Perth, W.A. J. Western Australia Nat. Hist. Soc. See B–P–H 485/4. HI 55376

Journal of wildlife management. Menasha, WI. Vol. 1+, 1937+. J. Wildlife Managem. 3-2245-2. HI 66987

Journal, Wimbledon natural history society. London. 1933-39. J. Wimbledon Nat. Hist. Soc. HI 66988

Journal für Wissenschaft und Kunst. Leipzig. J. Wiss. Kunst. See B–P–H 485/5. HI 55377

Journal of the Worcestershire nature conservation trust. Worcester. 1975+. J. Worcestershire Nat. Conservation Trust. Preceded by: Yearbook, Worcestershire nature conservation trust. HI 66989

Journal of the world aquaculture society. Baton Rouge, LA. Vol. 17+, 1986+. J. World Aquac. Soc. HI 66990

Journal of world forest resource management. Berkhamsted. Vol. 1+, 1984+. J. World Forest Resource Managem. HI 66991

Journal of world forestry. Neudamm & Berlin = Zeitschrift für Weltforstwirtschaft. Neudamm & Berlin. Z. Weltforstw. See B–P–H 988/15.

Journal of world forestry and soil science with colonial forestry digest = Zeitschrift für Weltforstwirtschaft waldwirtschaftliche und bodenkundliche Grossraumforschung zugleich Kolonialforstliche Mitteilungen. Neudamm [=Debno, Poland] & Berlin. Z. Weltforstw. Waldw. Bodenk. Grossraumf. & Kolonialforstl. Mitt. See B–P–H 988/16.

Journal of world mariculture society. Baton Rouge, LA. Vol. 12+, 1981+. J. World Maric. Soc. Preceded by: Proceedings of world mariculture society. HI 66992

Journal of Wuhan botanical research. [Wu-han chih wu hsüeh yen chiu.] Wuhan, Hubei. Vol. 3+, 1985+. J. Wuhan Bot. Res. Preceded by: Wuhan botanical research. HI 66993

Journal of Wuhan university. Natural science edition. [Wuhan daxue xuebao. Ziran kexue ban.] Wuhan. No. ?-29+, ?-1981+. J. Wuhan Univ., Nat. Sci. Ed. HI 66994

Journal of the Yalova horticultural research and training center = Yalova bahce kulturleri arastirma ve egitim merkezi dergisi. Istanbul.

Journal of the Yamagata agriculture and forestry society. [Yamagata norin gakkaiho.] Tsuruoka, Japan. J. Yamagata Agric. Soc. See B–P–H 485/8. HI 55378

Journal of the Yokohama municipal university. Series C, natural sciences. [Yokohama shiritsu daigaku kiyo.] Yokohama, Japan. J. Yokohama Munic. Univ., Ser.

C, Nat. Sci. See B–P–H 485/9. HI 55379

Journal of the Yunnan university. Series A. [Yun nan ta hsüeh hsüeh pao. Lui.] Kunming, China. J. Yunnan Univ., Ser. A. See B–P–H 485/10. HI 55380

Journal of the Yunnan university. Series B. Kunming, China. J. Yunnan Univ., Ser. B. See B–P–H 485/11. HI 55381

Journal of Zhejiang forestry college [Zhejiang linxueyuan xuebao]. Che-chiang Lin-an. 1984+. J. Zhejiang Forest. Coll. HI 66995

Journée d'agronomie coloniale. Gembloux. Vol. 1+, 1933+. J. Agron. Colon. HI 66996

Jozo shikenjo hokoku = Report of the research institute of brewing. Tokyo. Rep. Res. Inst. Brew. See B–P–H 770/3.

Jozogaku zasshi = Journal of brewing. Osaka, Japan. J. Brew. See B–P–H 460/22.

Junge Forscher. Cuxhaven, Germany. Junge Forscher. See B–P–H 508/35. HI 55593

Junge Forscher für alle. Cuxhaven, Germany. Junge Forscher für Alle. See B–P–H 508/36. HI 55594

Junta pública: sociedad económica de amigos del país. Valencia, Spain. Junta Pública Soc. Econ. Amigos País. See B–P–H 508/38. HI 55595

Jura nature; bulletin de la fédération de protection de la nature du Jura. Lons-le-Saunier. No. ?-13+, ?-198?+. Jura Nat. HI 66997

Jurnal sains farmosi malaysia. Petaling Jaya. Vol. 1+, 1978+. J. Sains Farm. Malaysia. HI 66998

Jurnal sains, pusat penyelidikan getah Malaysia. Kuala Lumpur. 1977+. J. Pusat Penyel. Getah Malaysia. HI 66999

Just's botanischer Jahresbericht. Systematisch geordnetes Repertorium der botanischen Literatur aller Länder. Berlin. Just's Bot. Jahresber. See B–P–H 508/43. HI 55596

Justus Liebigs Annalen der Chemie. Weinheim, Germany. Justus Liebigs Ann. Chem. See B–P–H 508/45. HI 55597

Justus Liebigs Annalen der Chemie und Pharmacie. Leipzig. Justus Liebigs Ann. Chem. Pharm. See B–P–H 508/46. HI 55598

Jute bulletin. Calcutta. Vols. ?-39, 1955?-77. Jute Bull. Superseded by: Jute development journal. HI 75089

Jute development journal. Calcutta. Vol. 1+, 1981+. Jute Developm. J. Jute bulletin. HI 67000

Jyväskyalän yliopisto = Biological research reports from the university of Jyväskylä. Jyväskylä.

K F R I research report. Peechi. No. ?-4+, ?-1980+. K. F. R. I. Res. Rep. HI 67001

K O R = Kakteen - und Orchideen - Rundschau. Lübeck.

K S R O gylym akademijasynyn Kazak filialynyn habarlary = Izvestiya Kazakhskogo filiala Akademii nauk S S S R. Alma-Ata.

K S R O gylym akademijasynyn Kazak filialynyn habarlary = Izvestiya Kazakhskogo filiala Akademii nauk S S S R. Seriya botanicheskaya. Alma-Ata.

Ka 'elele. Honolulu, HI. Vol. 1+, 1976+. Ka 'elele. HI 67002

Kabinet der natuurlijke historien, wetenschappen, konsten en handwerken. Amsterdam. Kab. Natuurl. Hist. See B–P–H 509/11. HI 55600

Kaffee- und Tee-Markt. Hamburg. Kaffee- Tee-Markt. See B–P–H 509/12. HI 55601

Kagaku = Science. Tokyo. Science (Tokyo/Kagaku). See B–P–H 829/23.

Kagaku chishiki = Scientific knowledge. Tokyo. Sci. Knowl. See B–P–H 824/21.

Kagaku kisoron gakkai = Annals of the Japan association for philosophy of science. Tokyo.

Kagaku nanyo = Japan society for the promotion of scientific research. Tokyo. Japan Soc. Promot. Sci. Res. See B–P–H 499/26.

Kagaku to seibutsu = Chemistry and biology. Tokyo.

Kagaku no Taiwan = Journal of the Taiwan museum association. Taihoku [=Taipei, Taiwan]. J. Taiwan Mus. Assoc. See B–P–H 483/14.

Kagaku-shi kenkyu. Nippon Kagaku-shi gakkai = Journal of history of science. Tokyo. J. Hist. Sci. See B–P–H 468/19.

Kagawa daigaku nogakubu kiyo = Memoirs of faculty of agriculture, Kagawa university. Miki, Japan. Mem Fac. Agric. Kagawa Univ. See B–P–H 572/7.

Kagoshima daigaku kyoikugakubu kenkyu kiyo, shizenkagaku-hen = Bulletin, faculty of education, university of Kagoshima. Natural science. Kagoshima.

Kagoshima daigaku rigakubu kiyo. Chigaku, seibutsugaku = Reports of the faculty of science, Kagoshima university. Earth sciences and biology. Kagoshima.

Kagoshima daigaku rika hokoku = Science reports of the Kagoshima university. Kagoshima.

Kagoshima daigaku. Suisdan gakubu kiyo = Memoirs of the faculty of fisheries, Kagoshima university. Kagoshima, Japan. Mem. Fac. Fish. Kagoshima Univ. See B–P–H 572/15.

Kagoshima-ken bunkazai chosa hokoku = Research report of cultural properties of Kagoshima prefecture. Kagoshima, Japan. Res. Rep. Cult. Propert. Kagoshima Prefect. See B–P–H 776/4.

Kagoshima-ken nogyo shikenjo kenkyu hokoku = Bulletin of the Kagoshima agricultural experiment station. Kamifukumoto.

Kagoshima-ken nogyo shikenjo. Nyusu. No. 1+, 1972+. Kagoshima-ken Nogyo Shikenjo, Nyusu. HI 67003

Kagoshima-ken ringyo gyomu shikenjo hokoku = Report of the Kagoshima prefectural forest experiment station. Gamo.

Kagoshima koto norin-gakko hakubutsu gakkai kaiho = Transactions of the natural history society, Kagoshima imperial college of agriculture and forestry. Kagoshima, Japan. Trans. Nat. Hist. Soc. Kagoshima Imp. Coll. Agric. See B–P–H 886/21.

Kagoshima novin-senmon gakko = Bulletin of the Kagoshima agricultural college. Kagoshima, Japan. Bull. Kagoshima Agric. Coll. See B–P–H 259/9.

Kagoshima suisan senmon gakko hokoku = Journal of the Kagoshima fisheries college. Kagoshima, Japan. See B–P–H 471/4.

Kagoshima tabako shikenjo gyotei hokoku = Report of the Kagoshima tobacco experiment station. Kagoshima, Japan. Rep. Kagoshima Tobacco Exp. Sta. See B–P–H 767/7.

Kagoshima tabako shikenjo hokoku = Bulletin of the kagoshima tobacco experiment station. Kagoshima, Japan. Bull. Kagoshima Tobacco Exp. Sta. See B–P–H 259/11.

Kagwa-ken nogyo shikenjo kenkyu hokoku = Bulletin of Kagawa agricultural experiment station. Kagawa, Japan. Bull. Kagawa Agric. Exp. Sta. See B–P–H 259/8.

Kai fêng shih fan hsüeh yüan hsüeh pao = Kaifeng teachers college journal. Kaifeng, China. Kaifeng Teachers Coll. J. See B–P–H 509/30.

Kaichu koen kenkyujo kenkyu hokoku = Bulletin of the marine park research stations. Kushimoto.

Kaifeng teachers college journal. [Kai fêng shih fan hsüeh yüan hsüeh pao.] Kaifeng, China. Kaifeng Teachers Coll. J. See B–P–H 509/30. HI 55602

Kaiho. [Wakayama.] No. 1+, 1972+. Kaiho. HI 67004

Kaiserliche Akademie der Wissenschaften in Wien. Mathematisch-naturwissenschaftliche Klasse. Anzeiger. = Anzeiger. Kaiserliche Akademie der Wissenschaften in Wien. Mathematisch-naturwissenschaftliche Klasse. Anzeiger. Vienna.

Kaiserliche Akademie der Wissenschaften in Wien. Mathematisch-naturwissenschaftliche Klasse. Denkschriften. Vienna. Vols. 93-95, 1917-18. Kaiserl. Akad. Wiss. Wien, Math.-Naturwiss. Kl., Denkschr. Preceded by: Denkschriften der Kaiserlichen Akademie der Wissenschaften, Wien. Mathematisch-naturwissenschaftliche Klasse. Superseded by: Akademie der Wissenschaften in Wien. Mathematisch-naturwissenschaftliche Klasse. Denkschriften. 1-106-3. HI 67005

Kaiyo no kagaku = Science of the sea. Tokyo. Sci. Sea. See B–P–H 829/9.

Kaiyo seibutsu shinyo kenkyukai. Hokoku. Tosa. No. 1+, 1977+. Kaiyo Seibutsu Shinyo Kenkyukai, Hokoku. HI 67006

Kajitsu nihon = Japanese fruits. Tokyo. Jap. Fruits. See B–P–H 498/11.

Kaju = Fruit tree. [Fruits cultivation association, Okayama prefecture.] Okayama, Japan. Fruit Tree. See B–P–H 386/4.

Kaju byogaichu bojo, handobukku. Otsu. Vol. ?-48+, ?-1973+. Kaju Byogaichu Bojo Handobukku. HI 67007

Kaju shikenjo hokoku. A (Hiratsuka) = Bulletin of the fruit tree research station. Series A, (Hiratsuka). Hiratsuka.

Kaju shikenjo hokoku. B (Okitsu) = Bulletin of the fruit tree research station. Series B, (Okitsu). Shimizu.

Kaju shikenjo hokoku. C (Morioka) = Bulletin of the fruit tree research station. Series C, (Morioka). Morioka.

Kaju shikenjo hokoku. D (Kuchinotsu) = Bulletin of the fruit tree research station. Series D, (Kuchinotsu). Kuchinotsu.

Kaju shikenjo hokoku. E (Akitsu) = Bulletin of the fruit tree research station. Series E, (Akitsu).

Kakteen; Gesamtdarstellung (Monographie) der eingefürten Arten nebst anzucht- und Pflege. Stuttgart. Lief. 1-63, 1956-75. Kakteen. HI 67008

Kakteen und andere Sukkulenten. Berlin. Kakteen Sukk. See B–P–H 509/40. HI 55603

Kakteen und andere Sukkulenten. Dresden = Kakteen / Sukkulenten. Dresden.

Kakteen-Freund. Illustrierte Monatsschrift für Kakteenliebhaber. Mannheim. Kakteen-Freund. See B–P–H 510/1. HI 55604

Kakteen - und Orchideen - Rundschau. Lübeck. No. 1+, 1975+. Kakteen Orchid. Rundschau. Preceded by: Kakteen - Rundschau. HI 67009

Kakteen - Rundschau. Lübeck. Vol. 2-3, 1971-73. Kakteen Rundschau. Preceded by: Lübecker Kakteen-Zeitschrift. Superseded by: Kakteen - und Orchideen - Rundschau. HI 67010

Kakteen / Sukkulenten. Dresden. 1964+. Kakteen Sukk. HI 67011

Kakteenfreund; Mitteilungsblatt der Kakteenfreund. Beil. Nos. 1-12, 1953-55. Kakteenfreund (Beil). HI 67012

Kakteenkunde. Neudamm [=Debno, Poland] & Berlin.

Kakteenkunde. See B–P–H 510/3. HI 55606

Kakteenkunde vereinigt mit dem "Kakteenfreund". Organ der Deutschen Kakteen-Gesellschaft in der Deutschen Gesellschaft für Gartenkultur. Neudamm [=Debno, Poland] & Berlin. Kakteenk. & Kakteenfr. See B–P–H 510/2. HI 55605

Kakti u sukkulenti ohra. Zejtun. No. 2+, 196?+. Kakti Sukk. Ohra. Preceded by: Sukkulenti. HI 67013

Kaktos komments. Houston, TX. Vol. 1+, 1963+. Kaktos Komments. HI 67014

Kaktus; medlemsblad for nordisk selskab. Odense, Hilferød. Vol. 1+, 1965+. Kaktus (Odense). HI 67015

Kaktus. Hanau. 1984+. Kaktus (Hanau). HI 67016

Kaktusář; odborný měsičnik. Astrophytum spolek pestitelu kaktusu a jinych sukkulentu. Brno. Vols. 1-7, 1930-36. Kaktusář. 3-2260-3. HI 67017

Kaktusářské listy. Prague. Vols. 2-16, 1926-51 [suspended 1948-50]. Kaktusář. Listy. Preceded by: Časopis spolku pěstitelu kaktusu v Č S R. For 1946-47 see: Zpravy, ceskoslovenske kaktusarske spolecnosti. HI 67018

Kaktusářský obzor. Prague. Vols. 1-2, 1930-31. Kaktusář. Obzor. HI 67019

Kaktusedvelö szakkör tajekoztatoja. Budapest. Vols. 1-8, 1964-71. Kaktusář. Szakkör Tajekozt. Superseded by: Kaktusz vilag. HI 67020

Kaktusy. Alma Ata. 1972- 79? Kaktusy (Alma Ata). HI 67021

Kaktusy; zpravodaj svazu českých kaktusářu. Brno. No. 1+, 1965+. Kaktusy (Brno). Preceded by: Vestnik, kroyzku kaktusářu. HI 67022

Kaktusy; sukkulenty a jejich pesteni. Prague. Nos. 1-5, 1923-24. Kaktusy (Praha). HI 67023

Kaktusy sukkulenty. Bratislava. Vol. 1+, 1980+. Kaktusy Sukk. HI 67024

Kaktusz vilag; Magyar kaktusz magazin. (Hazai kaktusz tükör. Magyar kaktusz hirado.) Budapest. Vols. 1-?, 1971- 82? Kaktusz Vilag. Preceded by: Kaktusedvelö szakkör tajekoztatoja. HI 67025

Kalikasan; Philippine journal of biology. Quezon City. Vol. 1+, 1972+. Kalikasan. HI 67026

Kalmia; botanic journal. Levittown, Philadelphia, PA. Vol. 1+, 1969+. Kalmia. HI 67027

Kambroo; newsletter of the south african aloe and succulent society. (Nuusbrief van de suid-afrikaanse aalwyn- en vetplantvereniging.) Pretoria. Vol. 1+, 1979+. Kambroo. Preceded by: Newsletter of the south african aloe and succulent society. HI 67028

Kampong notes. Coconut Grove, FL. Vol. 1+, 1964+. Kampong Notes. HI 67029

Kan chê yen Chiu = Journal of sugar cane research. Pingtung, Taiwan. J. Sugar Cane Res. See B–P–H 483/8.

Kan-su k'o hsüeh chiao yü kuan chuan k'an = Contributions from the Kansu science education institute. Lanchow, China. Contr. Kansu Sci. Educ. Inst. See B–P–H 329/6.

Kanagawa-ken engei shikenjo = Special bulletin of the Kanagawa prefectural horticultural experiment station. Ninomiya-machi.

Kanagawa-ken engei shikenjo kenkyu hokoku = Bulletin of Kanagawa horticultural experiment station. Kamakura.

Kanagawa-ken nogyo shiken kenkyu kikan kyodo kenkyu hokoku = Bulletin of the Kanagawa prefectural agricultural corporated experimental research organization. Hiratsuki.

Kanagawa-ken nogyo shikenjo kenkyu hokoku = Bulletin of Kanagawa agricultural experiment station. Kamakura.

Kanagawa-ken nogyo sogo kenkyujo kenkyu hokoku = Bulletin of the agricultural research institute of Kanagawa prefecture. Kamakura.

Kanagawa ken noji shiken seiseki = Bulletin, Kanagawa-ken agricultural experiment station. Yokohama?, Japan. Bull. Kanagawa-ken Agric. Exp. Sta. See B–P–H 259/14.

Kanagawa kenritsu hakubutsukan kenkyu hokoku, shizen kagaku = Bulletin of the Kanagawa prefectural museum. Natural science. Naka-ku, Yokohama.

Kanagawa-kenritsu noji shiken-jo moji shiken seiseki = Report of the Kanagawa-ken agricultural experiment station. Yokohama, Japan. Rep. Kanagawa-Ken Agric. Exp. Sta. See B–P–H 767/9.

Kanagawa shizenshi shiryo = Natural history report of Kanagawa. Yokohama.

Kanazawa daigaku kyoikugakubu kiyo. Shizenkagaku-hen = Bulletin, faculty of education, Kanazawa university. Series, natural science. Kanazawa.

Kanazawa daigaku kyoyobu ronshu, shizenkagaku-hen = Annals of science, college of liberal arts, Kanazawa University. Kanazawa.

Kanazawa daigaku rika hokoku. Seibutsugaku = Science reports of the Kanazawa university; biology. Kanazawa, Japan. Sci. Rep. Kanazawa Univ., Biol. See B–P–H 827/9.

Kanazawa Daigaku Rigakubu Fuzoku Shokubutsuen nenpo = Report (Annual) of botanic garden, faculty of science, Kanazawa university. Kanazawa.

Kanazawa daigaku rigaku-bu shokubutsu bunruigaku-bu kenkyu shitsu = Contributions from the laboratory of systematic botany of the faculty of science, Kanazawa university. Kanazawa, Japan. Contr. Lab. Syst. Bot. Fac. Sci. Kanazawa Univ. See B–P–H 329/13.

Kanazawa koto shihan gakko rika kioy = Journal of science of the Kanazawa normal college. Kanazawa.

Kanchi nogaku. Sapporo kasiwa-ba shoin = Agricultural science of the north temperate region. Sapporo, Japan. Agric. Sci. N. Temp. Region. See B–P–H 60/20.

Kankitsu = Citrus. [Shizuoka prefecture citrus agricultural corporation.] Shizuoka, Japan. Citrus (Shizuoka). See B–P–H 314/23.

Kankitsu kenkyu = Studia citrologica. Fukuóka, Japan. Stud. Citrol. See B–P–H 857/4.

Kankyo ryokka suishin jigyo chosa kenkyu hokoku. [Himeji?] No. 1+, 1970+. Kankyo Ryokka Suishin Jigyo Chosa Kenkyu Hokoku. HI 67030

Kansai byochugai kenkyukai-ho = Proceedings of the Kansai plant protection society. Mie, Japan. Proc. Kansai Pl. Protect. Soc. See B–P–H 731/9.

Kansai rinboku Ikushujo nenpo = Report (Annual) of the Kansai forest tree breeding station, ministry of agriculture and forestry. Okayama.

Kansas agricultural experiment station. Abstracts of new publications. Manhattan, KS. Kansas Agric. Exp. Sta. Abstr. New Publ. See B–P–H 510/24. HI 55607

Kansas agricultural experiment station. Annual report. Manhattan, KS. 1888-1907/08. Kansas Agric. Exp. Sta. Annual Rep. Superseded by: Kansas agricultural experiment station. Director's report. HI 67031

Kansas agricultural experiment station. Biennial report. Manhattan, KS. 1920/22-39/40. Kansas Agric. Exp. Sta. Bienn. Rep. Preceded by: Kansas agricultural experiment station. Director's report. HI 67032

Kansas agricultural experiment station. Director's report. Manhattan, KS. 1908/09-19/20. Kansas Agric. Exp. Sta. Director's Rep. Preceded by: Kansas agricultural experiment station. Annual report. Superseded by: Kansas agricultural experiment station. Biennial report. HI 67033

Kansas school naturalist. Emporia, KS. Vol. 1+, 1954+. Kansas School Naturalist. HI 67034

Kansas state agricultural college, agricultural experiment station, bulletin. Manhattan, KS. Kansas Agaric. Exp. Sta. Bull. See B–P–H 510/27. HI 55608

Kansas state agricultural college, agricultural experiment station circular. Manhattan, KS. Kansas Agric. Exp. Sta. Circ. See B–P–H 510/28. HI 55609

Kansas state agricultural college, agricultural experiment station leaflet. Manhattan, KS. Kansas Agric. Exp. Sta. Leafl. See B–P–H 510/30. HI 55610

Kansas state agricultural college, agricultural experiment station press bulletin. Manhattan, KS. Kansas Agric. Exp. Sta. Press Bull. See B–P–H 510/31. HI 55611

Kansas state agricultural college, agricultural experiment station progress report. Manhattan, KS. Kansas Agaric. Exp. Sta. Progr. Rep. See B–P–H 511/1. HI 55612

Kansas state agricultural college, agricultural experiment station report. Manhattan, KS. Kansas Agric. Exp. Sta. Rep. See B–P–H 511/2. HI 55613

Kansas state agricultural college, agricultural experiment station technical bulletin. Manhattan, KS. Nos. 1-161, 1916-69. Kansas Agric. Exp. Sta. Techn. Bull. Superseded by: Research publication, Kansas agricultural experiment station. HI 67035

Kansas university quarterly. Lawrence, KS. Kansas Univ. Quart. See B–P–H 511/6. HI 55614

Kansas university quarterly. Series A. Science and mathematics. Lawrence, KS. Kansas Univ. Quart., Ser. A, Sci. See B–P–H 511/7. HI 55615

Kansas university science bulletin. Lawrence, KS. Kansas Univ. Sci. Bull. See B–P–H 511/8. HI 55616

Kansas wildflower society newsletter = Newsletter, Kansas wildflower society newsletter.

Kansu agricultural news. [Kan nung lung hsin.] [China]. Kansu Agric. News. See B–P–H 511/10. HI 55617

Kanto rinboku Ikushujo nenpo = Report (Annual) of Kanto forest tree breeding station. Mito.

Kanto Tosan byogaichu kenkyukai nenpo = Proceedings of the Kanto Tosan plant protection society. Urawa, Japan. Proc. Kanto Tosan Pl. Protect. Soc. See B–P–H 731/10.

Kanto Tosan nogyo shikenjo kenkyu hokoku = Journal of the Kanto Tosan agricultural experiment station. Urawa, Japan. J. Kanto Tosan Agric. Exp. Sta. See B–P–H 471/8.

Kao cha yu yen chiu = Investigatio et studium naturae. Shanghai.

Kao yuan sheng wu hsüeh chi k'an = Acta biologica plateau sinica. Xining.

Karafutocho chuo shikenjo iho = Bulletin of the Saghalien central experiment station. Series 2, forestry. Konuma, Saghalien.

Kariba studies. Salisbury (Harare), Rhodesia (Zimbabwe). Nos. 1-10, 1961-82. Kariba Stud. HI 67036

Karlovarská vlastivěda. Cheb, Czechoslovakia. Karlovarská Vlastiv. See B–P–H 511/14. HI 55618

Kärntner Museum Schriften. Klagenfurt. Vol. 1+, 1954+. Kärntner Mus. Schriften. HI 67037

Kärntner Naturschutzblätter. Klagenfurt, Austria. Kärntner Naturschutzbl. See B–P–H 511/15.

HI 55619

Kartofel'; nauchno proizvodstvennyi zhurnal. Moscow. Vols. 1-4, 1957-59. Kartofel'. Superseded by: Kartofel' i ovoshchi. HI 67038

Karstenia. Sienitieteellinen ja sienitalondellinen aikakauskirja. Helsinki. Karstenia. See B–P–H 511/18. HI 55620

Kartofel' i ovoshchi; nauchno proizvodstvennyi zhurnal. Moscow. Vol. 5+, 1960+. Kartofel' & Ovoshchi. Preceded by: Kartofel' and Obmen opytom v sel'skom khozyaistve. Ser. ovoshchevod. HI 67039

Kartoffelbau; Zeitschrft zur Förderung der Kartoffelerzeugung. Düsseldorf. Kartoffelbau. See B–P–H 511/19. HI 55621

Kasetsart journal. Bangkok. 1961+. Kasetsart J. HI 67040

Kasetsart university research activities. Bangkok. 1967. Kasetsart Univ. Res. Activities. Superseded by: Research reports, Kasetsart university. HI 67041

Kashmir science. Srinagar. Vols. 1-8, 1964-71. Kashmir Sci. Superseded by: Journal of science, university of Kashmir. HI 67042

Kaskelot; biologforbundets blad. Gedved, Öland. No. 1+, 1971+. Kaskelot. HI 67043

Kasvitietellisen puutarhan eliömaailmaa. Helsinki. Vol. 1+, 1983+. Kasvit. Puut. Eliöm. HI 67044

Kauchuk i rezina. Moscow. Vol. 11+, 1937+. Kauchuk & Rezina. Preceded by: Zhurnal rezinovoi promyshlennosti and Sinteticheskii kauchuk [not entered]. 3-2275-1. HI 67045

Kautschuk und Gummi. Berlin & Frankfurt am Main. Vol. 1+, 1948+. Kautschuk Gummi. 3-2275-3. HI 67046

Kavaka; transactions of the mycological society of India. Madras. Vol. 1+, 1973+. Kavaka. HI 67047

Kazak S S R ğylym akademijasynyn habarlary. Alma-Ata = Izvestiya Akademii nauk Kazakhskoi S S R. Seriya botanicheskaya. Alma-Ata.

Kazak S S R ğylym akademijasynyn habarlary. Alma-Ata = Izvestiya Akademii nauk Kazakhskoi S S R. Seriya botanicheskaya. Alma-Ata.

Kazak S S R gylym akademiyasynyn habarlary = Izvestiya akademii nauk kazakhskoi S S R. Alma Ata.

Kazak S S R gylym akademijasynyn habaršysy = Vestnik Akademii nauk Kazakhskoi S S R. Alma-Ata.

Kehutanan Indonesia = Forestry in Indonesia. Bogor.

Kemiska växtskyddsmedel. Stockholm. Kem. Växtskyddsmedel. See B–P–H 511/23. HI 55622

Kémlö a gazdaság, ipár és kereskedésben. Pest [=Budapest, in part]. Kémlö Gazd. See B–P–H 511/24. HI 55623

Kendyr-Rami. Moscow. ?-1930. Kendyr-Rami. HI 67048

Kenkyu shiryo. Hokkaido nogyo shikenjo. Sapporo. No. 1+, 1973+. Kenkyu Shiryo, Hokkaido Nogyo Shikenjo. Preceded by: Agricultural experiment research report. HI 67049

Kenya agricultural abstracts. Current series. Nairobi. No. 1+, 19??+. Kenya Agric. Abstr., Curr. Ser. HI 67050

Kenya journal of science and technology. Series B, biological sciences. Nairobi. Vol. 1+, 1980+. Kenya J. Sci. Techn., B. HI 67051

Kentucky farm and home science. Lexington, KY. Kentucky Farm Home Sci. See B–P–H 511/25. HI 55624

Kentucky horticulture. Frankfort, KY. Kentucky Hort. See B–P–H 512/1. HI 55625

Kentucky naturalist. Louisville, KY. Kentucky Naturalist. See B–P–H 512/2. HI 55626

Kern kaktiletter. Bakersfield, CA. 1964-65. Kern Kaktilett. HI 67052

Kert. Budapest. Kert. See B–P–H 512/3. HI 55627

Kert és szölö. Budapest. Kert & Szölö. See B–P–H 512/13. HI 55636

Kertészet. Budapest. Kertészet (Budapest 1913-24). See B–P–H 512/21. HI 55637

Kertészet. A növényvédelem melléklete. Budapest. Kertészet (Budapest 1927-44). See B–P–H 512/22. HI 55638

Kertészet és szölészet. Budapest. Kert. & Szölész. See B–P–H 512/10. HI 55635

Kertészeti füzetek. Budapest. Kert. Füz. See B–P–H 512/4. HI 55628

Kertészeti irodalmi tajékoztató. Budapest. Kert. Irod. Tajékozt. See B–P–H 512/5. HI 55629

Kertészeti kutató intézet évkönyve. Budapest. Kert. Kutató Intéz. Évk. See B–P–H 512/6. HI 55631

Kertészeti kutató intézet közleményei. Budapest. Kert. Kutató Intéz. Közlem. See B–P–H 512/7. HI 55632

Kertészeti lapok. Budapest. Kert. Lapok. See B–P–H 512/8. HI 55633

Kertészeti szemle. Budapest. Kert. Szemle. See B–P–H 512/9. HI 55634

Kertészeti és szölészeti föiskola évkönyve. Budapest. Vols. 17-27, 1953-63; [n.s.] vol. 1(33)+,1969+. Kert. Szölész. Föisk. Évk. Preceded by: Agrártudományi egyetem kert- és szölögazdaságtudományi karának évkönyve. For 1964-68 see: Kertészeti egyetem közleményei. 1-812-3. HI 67053

Kertészeti egyetem közleményei. Budapest. Vols. 1(28)-5(32), 1964-68. Kert. Szölész. Föisk. Közlem. Preceded and superseded by: Kertészeti és szölészeti föiskola évkönyve. HI 67054

Kertészgazda. Pest [=Budapest, in part]. Kertészgazda. See B–P–H 512/25. HI 55640

Kertészgazda nép kertésze. Pest [=Budapest, in part]. Kertészgazda Nép Kert. See B–P–H 512/26. HI 55641

Kertészgazdasági lapok. Kolozsvar [=Cluj, Rumania]. Kertészgazd. Lapok. See B–P–H 512/24. HI 55639

Keszthelyi agrártudományi föiskola kiadványai. Budapest. Keszthelyi Agrártud. Föisk. Kiadv. See B–P–H 513/2. HI 55642

Keszthelyi mezögazdasági akadémia kiadványai. Budapest. Keszthelyi Mezögazd. Akad. Kiadv. See B–P–H 513/3. HI 55643

Keszthelyi mezögazdasági akadémia közleményei. Budapest. Keszthelyi Mezögazd. Akad. Közlem. See B–P–H 513/4. HI 55644

Kevo notes. Turku. No. 1+, 1975+. Kevo Notes. HI 67055

Kew bulletin. Kew, England. Kew Bull. See B–P–H 513/5. HI 55645

Kew bulletin. Additional series. Kew. Vol. 1+, 1958+. Kew Bull., Addit. Ser. HI 67056

Kew index for 1986 [etc.]; names of seed-bearing plants, ferns, and fern allies at the rank of family and below published during 1986 [etc.] Oxford. 1987+. Kew Index 1986 [Etc.]. Preceded by: Index kewensis plantarum phanerogamarum. HI 67057

Kew magazine; incorporating Curtis's botanical magazine. London. Vol. 1+, 1984+. Kew Mag. Preceded by: Botanical magazine. HI 67058

Kew record of taxonomic literature relating to vascular plants for ... [year]. London. 1971+, 1974+. Kew Rec. Taxon. Lit. Vasc. Pl. Preceded in part by: Index to australasian taxonomic literature and Index to european taxonomic literature. HI 67059

Kewmunication. Kew. No. 1+, 1979+. Kewmunication. HI 67060

Kexue shiyan. Peking (Beijing). 1974+. Kexue Shiyan. HI 67061

Kexue Tongbao. Peking (Beijing). Vol. 18+, 1973+. Kexue Tongbao. Preceded by: Scientia. Peking. HI 67062

Key to turkish science. Agriculture. Ankara. Vol. 1+, 1968+. Key Turk. Sci., Agric. HI 67064

Key to turkish science. Biological science. Ankara. Vol. 1+, 1969+. Key Turk. Sci., Biol. Sci. HI 67065

Khao cu'u niên-san khoa-hoc dai-hoc du'ong. Viên dai-hoc Saigon = Annales de la faculté des sciences, université de Saigon. Saigon. Ann. Fac. Sci. Univ. Saigon. See B–P–H 102/8.

Khimiko-farmatsevticheskii zhurnal. Moscow. Vol. 1+, 1967+. Khimiko-Farm. Zhurn. Preceded by: Meditsinkaya promyshlennost S S S R. HI 67067

Khimiya drevesiny. Riga. 1974+. Khim. Drevesiny. HI 67068

Khimiya prirodnykh soedinenii. Tashkent. 1965+. Khim. Prir. Soedin. HI 67069

Khlopkovodstvo. Moscow. 1951+. Khlopkovodstvo. HI 67070

Kiangsi nung hsüeh yüan hsüeh = Journal of the Kiangsi agricultural college. Nancheng?, China. J. Kiangsi Agric. Coll. See B–P–H 471/11.

Kieler Beiträge. Schleswig. Kieler Beitr. See B–P–H 513/24. HI 55649

Kieler Blätter. Kiel. Kieler Blätt. See B–P–H 513/25. HI 55650

Kieler Meeresforschungen. Kiel. Kieler Meeresf. See B–P–H 513/26. HI 55651

Kieler Notizen zur Pflanzenkunde in Schleswig-Holsten und Hamburg. Kiel. Vol. 1+, 1969+. Kieler Not. Pflanzenk. Schelswig-Holsten Hamburg. HI 67072

Kielische Gelehrte Zeitung. Kiel. Kiel. Gel. Zeitung. See B–P–H 513/20. HI 55646

Kielische Gelehrte Zeitungen. Kiel. Kiel. Gel. Zeitungen. See B–P–H 513/21. HI 55647

Kielisches Litteratur-Journal. Altona [=Hamburg, in part]. Kiel. Litt.-J. See B–P–H 513/22. HI 55648

Kingfisher; news and comment about wildlife and conservation at home and abroad. St. Albans, London. Vols. 1-?, 1965-70. Kingfisher. HI 67073

Kingia. South Perth, W.A. Vol. 1+, 1988+. Kingia. Preceded by: Research notes, western australian herbarium. HI 67074

King's Park native plant annual. West Perth, W.A. Vol. 1+, 1987+. King's Park Native Pl. Annual. HI 67075

King's Park research notes. West Perth, W.A. No. 1+, 1973+. King's Park Res. Notes. HI 67076

Kingwood center news. Mansfield, OH. Vol. 1+, 1976+. Preceded by: Kingwood center notes. Kingwood Center News. HI 67077

Kingwood center notes. Mansfield, OH. Vols. 1-?, 1954-75? Kingwood Center Notes. Superseded by: Kingwood center news. HI 67078

Kinjin kenkyusho kenkyu hokoku = Reports of the Tottori mycological institute. Tottori, Japan. Rep. Tottori Mycol. Inst. See B–P–H 771/10.

Kinki chugoku nogyo kenkyu. Fukuyama. No. 48+, 1974+. Kinki Chugoku Nogyo Kenkyu. Preceded by: Chugoku agricultural research. HI 67079

Kinnikinnick; newsletter of the friends of the garden. Edmonton, Alta. Vol. ?-3+, ?-1980+. Kinnikinnick. HI 67080

Kiøbenhavns patriotische Samlinger. Copenhagen. Vol. 1, 1771. Kiøbenhavns Patriot. Saml. HI 52691

Kiøbenhavnske efterretninger om laerde sager. Copenhagen. 1767-77. Kiøbenhavnske Efterretn. Laerde Sager. Preceded by: Kiøbenhavnske nye tidender om laerde sager. Superseded by: Kiøbenhavnske nye efterretninger. Laerde sager. 2-1260-1. HI 67081

Kiøbenhavnske nye efterretninger. Laerde sager. Copenhagen. 1778-82. Kiøbenhavnske Nye Efterretn. Laerde Sager. Preceded by: Kiøbenhavnske efterretninger. Laerde sager. Superseded by: Nyeste kiøbenhavnske efterretninger. Laerde Sager. 2-1260-1. HI 67082

Kiøbenhavnske nye tidender om laerde og curieuse sager. Copenhagen. 1749-59. Kiøbenhavnske Nye Tidender Laerde Cur. Sager. Preceded by: Nye tidender om laerde og curieuse sager. Superseded by: Kiøbenhavnske nye tidender om laerde sager. 2-1260-1. HI 67083

Kiøbenhavnske nye tidender om laerde sager. Copenhagen. 1760-66. Kiøbenhavnske Nye Tidender Laerde Sager. Preceded by: Kiøbenhavnske nye tidender om laerde og curieuse sager. Superseded by: Kiøbenhavnske efterretninger om laerde sager. 2-1260-1. HI 67084

Királyi magyar természettudományi társulat evkönyve. Budapest. Kir. Magyar Term. Társ. Évk. (Budapest). See B–P–H 513/34. HI 55652

Királyi magyar természettudományi társulat evkönyve. Pest [=Budapest, in part]. Kir. Magyar Term. Társ. Évk. (Pest). See B–P–H 513/35. HI 55653

Királyi magyar természettudományi társulat közlönye. Pest [=Budapest, in part]. Kir. Magyar Term. Társ. Közl. See B–P–H 513/36. HI 55654

Kirin university journal. Natural science edition. [Jilindaxue xuebao. Chueh-moon.] Chi-lin. 1977-78. Kirin Univ. J., Nat. Sci. Ed. Superseded by: Acta scientiarum naturalium universitatis jilinensis. HI 67085

Kirkia; journal of the federal herbarium. Salisbury, Rhodesia. Kirkia. See B–P–H 514/1. HI 55655

Kirtlandia. Cleveland, OH. No. 1+, 1967+. Kirtlandia. HI 67086

Kirton agricultural journal. Worcester. Nos. 1-11, 1938-46. Kirton Agric. J. HI 67087

Kísérletügyi kölemények. Budapest. Kísérl. Közlem. See B–P–H 514/5. HI 55656

Kísértletügyi közlemények. Kertészet. Gartenbau. Budapest. Kísérl. Közlem., Kert. See B–P–H 514/6. HI 55657

Kísértletügyi közlemények. Növénytermesztes. Pflanzenbau. Budapest. Kísérl. Közlem., Növényterm. See B–P–H 514/7. HI 55658

Kisetsuhu chosakai shi = Report of the monsoon committee. [Japan]. Rep. Monsoon Committee. See B–P–H 768/15.

Kiso kankyo kagaku kenkyu = Memoirs of the faculty of integrated arts and sciences, Hiroshima university. Studies in environmental science. Hiroshima.

Kist. [Kilberry.] No. 1+, 1971+. Kist. HI 67088

Kitanihon byogaichu kenkyukai kaiho = Report (Annual), society of plant protection of north Japan. Omagari.

Kjøbenhavns k. privilegeride nye kritiske journal. Copenhagen. Kjøbenhavns Kongel. Privileg. Nye Krit. J. See B–P–H 514/12. HI 55659

Kjøbenhavns universitets journal. Copenhagen. Kjøbenhavns Univ. J. See B–P–H 514/13. HI 55660

Kjøbenhavnske laerde efterretninger. Copenhagen. 1791-1810. Kjøbenhavnske Laerde Efterretn. Preceded by: Nyeste kiøbenhavnske efterretninger. Laerde Sager. Superseded by: Dansk literatur-tidende. 2-1260-1. HI 67089

Kleine Gartenbibliothek. Kiel. Kleine Gartenbiblioth. See B–P–H 514/15. HI 55661

Kleine Mitteilungen für die Mitglieder des Vereins für Wasser-, Boden- und Lufthygiene. Berlin. Kleine Mitt. Mitgl. Vereins Wasser- Lufthyg. See B–P–H 514/16. HI 55662

Kleine Schriften der Naturforschenden Gesellschaft in Emden. Emden. Kleine Schriften Naturf. Ges. Emden. See B–P–H 514/17. HI 55663

Kleine Schriftenreihe des Holz-Zentralblatts. Stuttgart. Vol. 1+, 1954+. Kleine Schriftenreihe Holz-Zentralbl. HI 67090

Kleine Senckenberg-Reihe. Frankfurt am Main. 1971+. Kleine Senckenberg Reihe. HI 67091

Kleiner Landwirthschaftskalender. Brünn [=Brno, Czechoslovakia]. Kleiner Landwirthschaftskalend. See B–P–H 514/18. HI 55664

Kleines Magazin von Reisen zur angenehmen und belehrenden Unterhaltung nach den neuesten deutschen und ausländischen Originalwerken. Berlin. Kleines Mag. Reisen. See B–P–H 514/19. HI 55665

Klinkii; journal of the forestry society of Papua New Guinea. Lae. Vol. 1+, 1977+. Klinkii. HI 67092

Kluborgan des Pilzklub Länggasse Bern. Bern. Kluborgan Pilzklub Länggasse Bern. See B–P–H 514/20. HI 55666

Knight's penny magazine. London. Knight's Penny Mag. See B–P–H 514/21. HI 55667

Knizni a dokumentacni zpravodaj, vysoka skola zemedelska v Brne. Brno. 1967+. Knizni Dokument. Zprav. Vysoka Skola Zemed. Brne. HI 67093

Know your hibiscus. Fort Myers, FL. Know Your Hibiscus. See B–P–H 514/22. HI 55668

K'o hsüeh = Science. Science society of China. Shanghai. Science (Shanghai). See B–P–H 829/22.

K'o hsüeh chi k'an = Science quarterly. [National central university.] Chungking, China. Sci. Quart. See B–P–H 826/11.

K'o hsüeh chi kan tung pei shih fan ta hsüeh. Sheng wu = Journal of science of the north-eastern normal university. Biology. Changchun, China. J. Sci. N.-E. Norm. Univ., Biol. See B–P–H 480/18.

K'o hsüeh chi lu = Science record. [Academia sinica.] Chungking, China. Sci. Rec. (Chungking). See B–P–H 826/15.

K'o hsüeh chiao hsüeh chi k'an = Quarterly bulletin of science education. [China]. Quart. Bull. Sci. Educ. See B–P–H 751/9.

K'o hsüeh chiao yü = Science education. [Chinese association for the advancement of natural science.] [Taiwan]. Sci. Educ. (Taiwan). See B–P–H 824/7.

K'o hsüeh fa chan = National science council monthly. Taipei.

K'o hsüeh hui pao = Science bulletin. Taipei, Taiwan. Sci. Bull. (Taipei). See B–P–H 823/10.

K'o hsüeh nung yeh = Scientific agriculture. [Agricultural science club; national Taiwan university.] Taipei, Taiwan. Sci. Agric. (Taiwan). See B–P–H 823/2.

K'o hsüeh shih chieh. Chung hua tsu jao k'o hsüeh shê = Scientific world. Natural science society of China. Chungking & Nanking. Sci. World (Chunking & Nanking). See B–P–H 829/17.

K'o hsüeh shih pao = Science monthly bulletin. [World science association.] Peiping [=Peking]. Sci. Monthly Bull. See B–P–H 825/7.

K'o hsüeh t'ung pao = Science news. [Academia sinica.] Peking. Sci. News (Peking). See B–P–H 825/12.

K'o hsüeh t'ung pao = Scientia. Peking (Beijing).

Kobe daigaku kyoiku gakubu kenkyu shuroku = Bulletin of the faculty of education; Kobe university. Kobe, Japan. Bull. Fac. Educ. Kobe Univ. See B–P–H 249/18.

Kobe daigaku nogakubu kenkyu hokoku = Science reports of the faculty of agriculture, Kobe university. Kobe.

Kobe plant protection and plant quarantine information. [Kobe shobubutsu boeki joho.] Kobe. 1954+. Kobe Pl. Protect. Pl. Quarant. Inform. HI 67094

Kobe shobubutsu boeki joho = Kobe plant protection and plant quarantine informatic n. Kobe.

Kobie; revista de ciencias. Bilbao. Vol. 1+, 1969+. Kobie. HI 67095

Kochi daigaku gakujutsu kenkyu hokoku = Research reports of the Kochi university. Kochi.

Kochi daigaku gakujutsu kenkyu hokoku. Shizen kagaku = Research reports of the Kochi university. 2, natural science. Kochi.

Kochi daigaku. Kaiyo seibutsu kyoiku kenkyu senta = Reports of the Usa marine biological institute, Kochi university.

Kochi daigaku kenkyu hokokui. Shizen kagaku = Reports of the Kochi university; natural science (biology, geology, agriculture). Kochi, Japan. Rep. Kochi Univ., Nat. Sci. See B–P–H 767/12.

Kochi daigaku kyoikugakubu kenkyu hokoku = Bulletin of the faculty of education; Kochi university. Kochi, Japan. Bull. Fac. Educ. Kochi Univ. See B–P–H 249/19.

Kochi joshi daigaku kiyo. Shizenkagaku-hen = Bulletin of the Kochi women's university. Series of natural sciences. Kochi.

Koedoe. Journal for scientific research in the national parks of the Union of South Africa. Pretoria. Koedoe. See B–P–H 515/15. HI 55669

Koffiebessenboeboek-fonds: mededelingen. Surabaya, Dutch E. Indies [Indonesia]. Koffiebessenboeboek-Fonds Meded. See B–P–H 515/16. HI 55670

Koffiegids. Malang, Dutch E. Indies [Indonesia]. Koffiegids. See B–P–H 515/17. HI 55671

Kogoshima daigaku nogakubu kiyo = Memoirs of the faculty of agriculture; Kagoshima university. Kagoshima, Japan. Mem Fac. Agric. Kagoshima Univ. See B–P–H 572/8.

Kogyo gijutsu-in hakko kenkyusho kenkyu hokoku = Report of the fermentation research institute. Chiba.

Kokuritsu idengaku kenkyusho, nenpo = Report (Annual) of the national institute of genetics. Mishima, Sizuoka-ken.

Kokuritsu kagaku hakubutsukan kenkyu hokoku = Bulletin of the national science museum, Tokyo. Tokyo.

Kokuritsu kagaku hakubutsukan kankyu hokoku. B-riu, shokubutsugaku = Bulletin of the national science museum, Tokyo. Ser. B, botany. Tokyo.

Kokuritsu kagaku hakubutsukan senho = Memoirs of the national science museum, Tokyo. Tokyo.

Kokuritsu kagaku hakubutsukan shuho = Miscellaneous reports of the national science museum. Tokyo. Misc. Rep. Natl. Sci. Mus. See B–P–H 599/7.

Kokusai kankyo hozen kagaku kaigi. Kokunai shinpojiumu. Koen yoshi. Tokyo. ?-2 kai+, ?-1974+. Kokusai Kankyo Hozen Kagaku Kaigi, Kokunai Shinpojiumu Koen Yoshi. HI 67096

Kolloid-Zeitschrift. Zeitschrift für reine und angewandte Kolloidwissenschaft. Darmstadt, Germany. Kolloid-Z. See B–P–H 515/19. HI 55672

Kolloidnyi zhurnal. Voronezh, Moscow. Vol. 1+, 1935+. Kolloid. Zhurn. 3-2306-2. HI 67097

Kölnisches encyclopedisches Journal. Cologne. Köln. Encycl. J. See B–P–H 515/20. HI 55673

Kolonial jaarboeken. Maandschrift ter verspreiding van kennis der nederlandsche en buitenlandsche overzeesche bezittingen. Zutphen, Netherlands. Kolon. Jaarb. See B–P–H 515/22. HI 55674

Kolonialdeutsche. Wissenschaftliche Beihefte. Berlin. Kolonialdeutsche Wiss. Beih. See B–P–H 515/24. HI 55676

Koloniale Rundschau. Berlin. Kolon. Rundschau. See B–P–H 515/23. HI 55675

Kolonialforstliche Mitteilungen. Neudamm [=Debno, Poland] & Berlin. Kolonialforstl. Mitt. See B–P–H 515/25. HI 55677

Kolozsvári Babeş és Bolyai egyetemek közleményei. Természettudományi sorozat. Cluj, Rumania. Kolozsvári Babeş Bolyai Egyet. Közlem., Term. Soroz. See B–P–H 515/26. HI 55678

Kolozsvári Bolyai tudományegyetem matematikai és természettudományi karának kiadványai. = Acta bolyaiana. Cluj, Rumania. Acta Bolyaiana. See B–P–H 41/4.

Kolozsvári Ferenc József tudományegyetem évkönyve. Kolozsvar [=Cluj, Rumania]. Kolozsvári Ferenc József Tudományegyet. Évk. See B–P–H 515/29. HI 55680

Kolozsvári szemle. Kolozsvar [=Cluj, Rumania]. Kolozsvári Szemle. See B–P–H 515/30. HI 55681

Kolozsvári V. Babeş és Bolyai egyetemek közleményei. Cluj, Rumania. Kolozsvári Babeş Bolyai Egyet. Közlem. See B–P–H 515/27. HI 55679

Komarovskie čtenija. Moscow & Leningrad = Komarovskie chteniya. Moscow & Leningrad.

Komarovskie čtenija. Vladivostok, Russian S F S R = Komarovskie chteniya. Vladivostok.

Komarovskie chteniya. Moscow & Leningrad. Vol. 1+, 1949+. Komarovskie Chteniya (Moscow & Leningrad). 3-2307-3. HI 67099

Komarovskie chteniya. Vladivostok, Russian S F S R. Vol. 1+, 1947+. Komarovskie Chteniya (Vladivostok). HI 67100

Komebukuro = Corny bags information. Tokyo. Corny Bags Inform. See B–P–H 332/16.

Kompleksnye issledovaniya kaspiiskogo morya. Moscow. Vol. 1+, 1970+. Kompl. Issl. Kaspiisk. Morya. HI 67101

Kongelige danske landhuusholdings-selskabs skrifter. Copenhagen. Kongel. Danske Landhuushold.-Selsk. Skr. See B–P–H 516/12. HI 55682

Kongelige danske videnskabernes selskabs bekiendtgiorelse. Copenhagen. Kongel. Danske Vidensk. Selsk. Bekiendtg. See B–P–H 516/13. HI 55683

Kongelige danske videnskabernes selskabs naturvidenskabelige og mathematiske afhandlinger. Copenhagen. Vols. 1-12, 1824-46. Kongel. Danske Vidensk. Selsk. Naturvidensk. Math. Afh. Preceded by: Kongelige danske videnskabernes-selskabs skrifter. Superseded by: Kongelige danske videnskabernes selskabs skrifter. Naturvidenskabelige og mathematisk afdeling. 2-1262-2. HI 50478

Kongelige danske videnskabernes-selskabs skrifter. Copenhagen. Vols. 1-6, 1802-18. Kongel. Danske Vidensk.-Selsk. Skr. Preceded by: Nye samling af det kongelige danske videnskabers selskabs skrifter. Superseded by: Kongelige danske videnskabernes selskabs naturvidenskabelige og mathematiske afhandlinger. 2-1262-2. HI 67102

Kongelige danske videnskabernes selskabs skrifter. Naturvidenskabelig og mathematisk afdeling. Copenhagen. Ser. 5, vols. 1-12, 1849-80; ser. 6, vols. 1-12, 1880-1904; ser. 7, vols. 1-12, 1904-14/15; ser. 8, vols. 1-12, 1915/17-27; ser. 9, vols. 1-8, 1928-38. Kongel. Danske Vidensk. Selsk. Skr., Naturvidensk. Math. Afd. Preceded by: Kongelige danske videnskabernes selskabs naturvidenskabelige og mathematiske afhandlinger. Superseded by: Biologiske skrifter. 2-1262-2. HI 67103

Kongelige norske videnskabers selskabs aarsberetning. Trondheim. 1874-1925. Kongel. Norske Vidensk. Selsk. Aarsberetn. Superseded by: Kongelige norske videnskabers selskabs. Årsberetning. 4-3079-1. HI 67104

Kongelige norske videnskabers selskabs forhandlinger. Trondheim. Kongel. Norske Vidensk. Selsk. Forh. See B–P–H 516/17. HI 55684

Kongelige norske videnskabers selskabs museet. Årsberetning. Trondheim. 1926-50. Kongel. Norske Vidensk. Selsk. Årsberetn. Preceded by: Kongelige norske videnskabers selskabs museet. Aarsberetning. Superseded by: Kongelige norske videnskabers selskabs museet. Årbok. 4-3079-2. HI 67105

Kongelige norske videnskabers selskabs museet. Årbok. Trondheim. 1951-67. Kongel. Norske Vidensk. Selsk. Mus. Årbok. Preceded by: Kongelige norske videnskabernes selskabs museet. Årsberetning. 4-3079-2. HI 67106

Kongelige norske videnskabers selskabs skrifter. Copenhagen. Kongel. Norske Vidensk. Selsk. Skr. (Copenhagen). See B–P–H 516/20. HI 55685

Kongelige norske videnskabers selskabs skrifter. Trondheim. Kongel. Norske Vidensk. Selsk. Skr. (Trondheim). See B–P–H 516/21. HI 55686

Kongelige norske videnskabersselskabs skrifter i det 19de aarhundrede. Copenhagen. Kongel. Norske Videnskabersselsk. Skr. 19de Aarhundr. See B–P–H 516/22. HI 55687

Kongelige svenske videnskabers academiens oeconomiske, physiske og mechaniske afhandlinger. Copenhagen. Kongel. Svenske Vidensk. Acad. Oecon. Afh. See B–P–H 516/23. HI 55688

Kongl[iga]. götheborgska wetenskaps och witterhets samhällets handlingar. Wetenskaps afdelningen. Goteborg, Sweden. Kongl. Götheborgska Wetensk. Samhällets Handl., Wetensk. Afd. See B–P–H 517/4. HI 55689

Kongl[iga]. svenska landtbruks-academiens annaler. Stockholm. Kongl. Svenska Landbr.-Acad. Ann. See B–P–H 517/9. HI 55690

Kongl[iga]. svenska vetenskaps academiens handlingar. Stockholm. Kongl. Svenska Vetensk. Acad. Handl. See B–P–H 517/10. HI 55691

Kongl[iga]. svenska vetenskapsakademiens handlingar. Stockholm. Kongl. Svenska Vetenskapsakad. Handl. See B–P–H 517/11. HI 55692

Kongl[iga]. swenska wetenskaps academiens handlingar. Stockholm. Kongl. Swenska Wetensk. Acad. Handl. See B–P–H 517/12. HI 55693

Kongl[iga]. vetenskaps academiens handlingar. Stockholm. Kongl. Vetensk. Acad. Handl. See B–P–H 517/13. HI 55694

Kongl[iga]. vetenskaps academiens nya handlingar. Stockholm. Kongl. Vetensk. Acad. Nya Handl. See B–P–H 517/14. HI 55695

Königlich Norwegische Gesellschaft der Wissenschaften Schriften. Leipzig. Königl. Norweg. Ges. Wiss. Schriften. See B–P–H 517/19. HI 55700

Königlich Sächsische Gesellschaft für Botanik und Gartenbau "Flora" zu Dresden. Sitzungsberichte und Abhandlungen. Dresden. Königl. Sächs. Ges. Bot. "Flora" Dresden Sitzungsber. Abh. See B–P–H 518/1. HI 55701

Königliche Akademie der Wissenschaften und das königliche General-Conservatorium der wissenschaftlichen Sammlungen des Staates zu München. Munich. Königl. Akad. Wiss. Königl. Gen.-Conserv. Wiss. Samml. München. See B–P–H 517/15. HI 55696

Königl[iche] Akademie der Wissenschaften in Paris Anatomische, Chymische und Botanische Abhandlungen. Breslau [=Wroclaw, Poland]. Königl. Akad. Wiss. Paris Anat. Abh. See B–P–H 517/16. HI 55697

Königliche Akademie der Wissenschaften in Paris Physische Abhandlungen. Breslau [=Wroclaw, Poland]. Königl. Akad. Wiss. Paris Phys. Abh. See B–P–H 517/17. HI 55698

Königl[iche]. Grossbritt. Churfürstl. Braunschw. Lüneb. Landwirthschaftsgesellschaft Nachrichten von Verbesserung der Landwirthschaft und des Gewerbes. Celle. Königl. Grossbritt. Churfürstl. Braunschweig-Lüneburg. Landwirthschaftsges. Nachr. Verbess. Landw. Gewerbes. See B–P–H 517/18. HI 55699

Königl[iche]. schwedische Akademie der Wissenschaften Abhandlungen aus der Naturlehre, Haushaltungskunst und Mechanik. Hamburg & Leipzig. Vols. 1-41, 1749-83; ed. 2, vols. 1-6, 1768-78. Königl. Schwed. Akad. Wiss. Abh. Naturl. Superseded by: Königl[iche]. schwedische Akademie der Wissenschaften. Neue Abhandlungen aus der Naturlehre, Haushaltungskunst und Mechanik. 1-12-1. HI 50020

Königl[iche]. schwedische Akademie der Wissenschaften. Neue Abhandlungen aus der Naturlehre, Haushaltungskunst und Mechanik. Leipzig. Vols. 1-12, 1784-92. Königl. Schwed. Akad. Wiss. Neue Abh. Naturl. Preceded by: Königl[iche]. schwedische Akademie der Wissenschaften Abhandlungen, aus der Naturlehre, Haushaltungskunst und Mechanik. 4-2965-1. HI 55702

Königsberger Archiv für Naturwissenschaft und Mathematik. Königsberg [=Kaliningrad, Russian S F S R]. Königsberger Arch. Naturwiss. Math. See B–P–H 518/6. HI 55704

Königsberger naturwissenschaftliche Unterhaltungen. Königsberg [=Kaliningrad, Russian S F S R]. Königsberger Naturwiss. Unterhalt. See B–P–H 518/8. HI 55705

Koralpe; Beiträge zur Botanik, Geologie, Klimatologie und Volkskunde. Graz. 1982. Koralpe. HI 55703

Korean journal of biochemistry. [Taehan saenghwahakhoe.] Seoul. Vol. 1+, 1964+. Korean J. Biochem. HI 67107

Korean journal of biology. [Saengmul hak'kae-po.] Seoul, Korea. Korean J. Biol. See B–P–H 518/9.

HI 55706

Korean journal of botany. [Sikmul hak'kae-chi.] Seoul, Korea. Korean J. Bot. See B–P–H 518/10. HI 55707

Korean journal of breeding. [Han'guk yukchong hakhoe.] Vol. 1+, 1969+. Korean J. Breed. HI 67108

Korean journal of ecology. Seoul. ?-1981+. Korean J. Ecol. HI 67109

Korean journal of genetics. [Hanguk yujon hakhoe chi.] Seoul. Vol. 1+, 1979+. Korean J. Genet. HI 67110

Korean journal of limnology. Seoul. Vol. 1+, 1968+. Korean J. Limnol. HI 67111

Korean journal of microbiology. [Misaengmul hakhoe chi.] Seoul. Vol. 1+, 1963+. Korean J. Microbiol. HI 67112

Korean journal of mycology. Seoul. Vol. 1+, 1973+. Korean J. Mycol. HI 67113

Korean journal of plant protection. [Han'guk singmul poho hakhoe chi.] Suwon. Vol. 10+, 1971+. Korean J. Pl. Protect. Preceded by: Journal of plant protection. Suwon. HI 67114

Korean journal of plant taxonomy. Seoul. Ca.1989+. Korean J. Pl. Taxon. Preceded by: Journal of korean plant taxonomy. HI 67115

Korean journal of plant tissue culture. [Singmul chojik paeyang hakhoe chi.] Seoul. ?-1977+. Korean J. Pl. Tissue Cult. HI 67116

Korean journal of virology. [Taehan pairosu hakhoe chi.] Seoul. ?-1979+. Korean J. Virol. HI 67117

Korean nature. Pyongyang. Vol. 1+, 1965+. Korean Nat. HI 67118

Korean scientific abstracts. Pyongyang, Seoul. Vol. 1+, 1969+. Korean Sci. Abstr. HI 67119

Kormo i kormovyrobnytstvo; respublikans'kyi mizhvidomchyi tematychnyi naukovyi zbirnyk. Vol. 1+, 1976+. Kormo Kormovyrobn. HI 67120

Kormovyrobnytstvo. Kiev. Vols. 1-3, 1973-75. Kormovyrobnytstvo. HI 67121

Korrespondenzblatt des Naturforscher-Vereins zu Riga. Riga [Latvian S S R]. Korrespondenzbl. Naturf.-Vereins Riga. See B–P–H 518/14. HI 55708

Korte berichten voor landbouw, nijverheid en handel. Buitenzorg, Dutch E. Indies [=Bogor, Indonesia]. Korte Ber. Landb. See B–P–H 518/15. HI 55709

Korte mededeling. Algemeen proefstation voor de landbouw. Buitenzorg, Dutch E. Indies [=Bogor, Indonesia]. Korte Meded. Alg. Proefstat. Landb. See B–P–H 518/16. HI 55710

Korte mededeling van het instituut voor veredeling van tuinbouwgewassen, Wageningen. Wageningen. Korte Meded. Inst. Veredel. Tuinbouwgew. Wageningen. See B–P–H 518/17. HI 55711

Korte mededeeling van het landbouwproefstation in Suriname. Paramaribo, Dutch Guiana [Surinam]. Korte Meded. Landbouwproefstat. Suriname. See B–P–H 518/18. HI 55712

Korte mededeeling van het proefstation voor cacao. Salatiga, Dutch E. Indies [Indonesia]. Korte Meded. Proefstat. Cacao See B–P–H 518/19. HI 55713

Korte mededeling van het rijksboschbouwproefstation. Wageningen. Vol. 1+, 1949+. Korte Meded. Rijksboschbouwproefstat. HI 67122

Kos; rivista di cultura e storia delle scienze mediche, naturali e umane. Milan. No. 1+, 1984+. Kos. HI 67123

Kosmos. Lvov, Galicia [Ukrainian S S R]. Kosmos (Lvov). See B–P–H 518/22. HI 55714

Kosmos. Stuttgart. Kosmos (Stuttgart). See B–P–H 519/2. HI 55715

Kosmos. Ser. A: Biologia. Warsaw. Kosmos, Ser. A, Biol. (Warsaw). See B–P–H 519/3. HI 55716

Koso no sekai = World of enzyme. Otsu, Japan. World of Enzyme. See B–P–H 980/4.

Kött kertészet közlönye. Pecs, Hungary. Kött Kert. Közl. See B–P–H 519/5. HI 55717

Közlemények az állat- és növényhonosítés köréböl. Budapest. Közlem. Allat- Növényhon. Köréb. See B–P–H 519/6. HI 55718

Köztelek. Budapest. Köztelek. See B–P–H 519/7. HI 55719

Kralupský vlastivědný sborník. Prague. Kralupský Vlastiv. Sborn. See B–P–H 519/9. HI 55720

Kranke Pflanze. Monatsblatt der Sächsischen Pflanzenschutzgesellschaft. Mitteilungsblatt des Verbandes Deutscher Pflanzenärzte. Dresden. Kranke Pflanze. See B–P–H 519/10. HI 55721

Kras v Ceskoslovensku. Prague. Kras v Ceskoslov. See B–P–H 519/11. HI 55722

Krása našeho domova. Prague. Krása Našeho Domova. See B–P–H 519/12. HI 55723

Kratkii obzor nauchno-opytnykh rabot Gosudarstvennogo Nikitskogo opytnogo botanicheskogo sada. Yalta, Ukrainian S S R. 1929. Kratk. Obzor Nauchno-Opytn. Rabot Gosud. Nikitsk. Opytn. Bot. Sada. Preceded by: Kratkii obzor nauchno-opytnykh rabot Gosudarstvennogo Nikitskogo opytnogo sada. 3-1912-2. HI 67124

Kratkii obzor nauchno-opytnykh rabot Gosudarstvennogo Nikitskogo opytnogo sada. Yalta, Ukrainian S S R. 1927. Kratk. Obzor Nauchno-Opytn. Rabot Gosud. Nikitsk. Opytn. Sada. Superseded by: Kratkii obzor nauchno-opytnykh rabot Gosudarstvennogo Nikitskogo opytnogo botanicheskogo sada. 3-1912-2. HI 67125

Kratkij obzor naučno-opytnyh rabot Gosudarstvennogo Nikitskogo opytnogo botaničeskogo sada = Kratkii obzor nauchno-opytnykh rabot Gosudarstvennogo Nikitskogo opytnogo botanicheskogo sada. Yalta.

Kratkij obzor naučno-opytnyh rabot Gosudarstvennogo Nikitskogo opytnogo sada = Kratkii obzor nauchno-opytnykh rabot Gosudarstvennogo Nikitskogo opytnogo sada. Yalta.

Kratos; Zeitschrift für Gymnasien. Prague. Kratos. See B–P–H 519/15. HI 55724

Krelage's bloemhof. Maandblad voor bloemenliefhebbers. Haarlem. Krelage's Bloemhof. See B–P–H 519/16. HI 55725

Kristiansand museets årbok. Kristansand. ?-1975+. Kristiansand Mus. Årbok. HI 67126

Kritische Blätter der Börsenhalle. Hamburg. Krit. Blätt. Börsenhalle. See B–P–H 519/17. HI 55726

Kritische Blätter der Forst- und Jagdwissenschaft. Berlin. Krit. Blätt. Forst- Jagdwiss. See B–P–H 519/18. HI 55727

Kritisches Repertorium für die gesammte Heilkunde. Berlin. Krit. Repert. Gesammte Heilk. See B–P–H 519/19. HI 55728

Krok. Weřegný spis wšenančný pro wzdělance národu českoslowanského. Prague. Krok. See B–P–H 519/20. HI 55729

Kromosomo. [Senshokutai.] Tokyo. Kromosomo. See B–P–H 519/21. HI 55730

Kruipnieuws. Orgaan van de N J N sociologengroep. Helder, Netherlands. Kruipnieuws. See B–P–H 520/1. HI 55731

Kriobiologiya. Kiev. 1985+. Kriobiologiya. HI 75110

Kryptogamische Forschungen. Munich. Kryptog. Forsch. See B–P–H 520/2. HI 55732

Ktavim; quarterly journal of the national university institute of agriculture [of Israel]. Jerusalem. Vols. 10-19, 1960-69. Ktavim. Preceded by: Records of the agricultural research station. Superseded by: Israel journal of agricultural science. HI 67127

Ku shêng wu hsüeh pao = Acta palaeontologica sinica. Peking. Acta Palaeontol. Sin. See B–P–H 47/2.

Kuan ch'a = Observation. Shanghai. Observation. See B–P–H 679/9.

Kuan ch'a = Observer. Shanghai. Observer. See B–P–H 679/10.

Kuang-hsi chih wu = Guihaia. Yanshan, Guilin.

Kuang hsi nung yeh = Kwangsi agriculture. Liuchow, China. Kwangsi Agric. See B–P–H 522/12.

Kuang tung nung yeh tung hsün = Kwangtung agricultural news. [Kwangtung, China]. Kwangtung Agric. News. See B–P–H 522/13.

Kudoa. Taihoku [=Taipei, Taiwan]. Kudoa. See B–P–H 520/9. HI 55733

Kühn-Archiv. Arabeiten aus der Landwirtschaftlichen Fakultät der Universität Halle-Wittenberg. Berlin. Kühn-Arch. See B–P–H 520/10. HI 55734

Kukui leaf. Honolulu, HI. Vol. ?-4(4)+, ?-1978+. Kukui Leaf. HI 67128

Külföldi agrárirodalmi rövid kivonatai. Budapest. Külf. Agrárirod. Rövid Kivonatai. See B–P–H 520/11. HI 55735

Külföldi agrárirodalmi rövid kivonatai. Kertészet, szölészet, borászat. Budapest. Külf. Agrárirod. Rövid Kivonatai, Kert. See B–P–H 520/12. HI 55736

Külföldi agrárirodalmi rövid kivonatai. Növényvédelem. Budapest. Külf. Agrárirod. Rövid Kivonatai, Növényvéd. See B–P–H 520/13. HI 55737

Külföldi agrárirodalmi szemle. Erdészet. Budapest. Külf. Agrárirod. Szemle, Erdész. See B–P–H 520/14. HI 55738

Külföldi agrárirodalmi szemle. Kertészet, szölészet, borászat. Budapest. Külf. Agrárirod. Szemle, Kert. See B–P–H 520/15. HI 55739

Külföldi agrárirodalmi szemle. Növénytermesztés. Budapest. Külf. Agrárirod. Szemle, Növényterm. See B–P–H 520/16. HI 55740

Külföldi agrárirodalmi szemle. Növényvédelem. Budapest. Külf. Agrárirod. Szemle, Növényvéd. See B–P–H 520/17. HI 55741

Külföldi agrárirodalom szemléje. Erdészet. Budapest. Külf. Agrárirod. Szemléje, Erdész. See B–P–H 520/18. HI 55742

Külföldi agrárirodalom szemléje. Kertészet és szölészet. Budapest. Külf. Agrárirod. Szemléje, Kert. Szölész. See B–P–H 520/19. HI 55743

Külföldi agrárirodalom szemléje. Novénytermesztés. Budapest. Külf. Agrárirod. Szemléje, Növényterm. See B–P–H 520/20. HI 55744

Külföldi agrárirodalom szemléje. Novényvédelem. Budapest. Külf. Agrárirod. Szemléje, Növényvéd. See B–P–H 520/21. HI 55745

Külföldi és hazai agrárirodalmi szemle. Budapest. Külf. Hazai Agrárirod. Szemle, Növényterm. See B–P–H 521/3. HI 55748

Külföldi és hazai agrárirodalmi szemle. Erdészet. Budapest. Külf. Hazai Agrárirod. Szemle, Erdész. See B–P–H 521/1. HI 55746

Külföldi és hazai agrárirodalmi szemle. Kertészet és szölészet. Budapest. Külf. Hazai Agrárirod. Szemle, Kert. Szölész. See B–P–H 521/2. HI 55747

Külföldi és hazai agrárirodalmi szemle. Növényvédelem. Budapest. Külf. Hazai Agrárirod. Szemle, Növényvéd. See B–P–H 521/4. HI 55749

Kullabergs natur. Lund. No. 1+, 1960+. Kullabergs Natur. HI 67129

Kulturpflanze. Berichte und Mitteilungen aus dem Institut für Kulturpflanzenforschung der Deutschen Akademie der Wissenschaften zu Berlin in Gatersleben Krs. Aschersleben. Berlin. Kulturpflanze. See B–P–H 521/5. HI 55750

Kumamoto daigaku kyoikugakubu kiyo, shizenkagaku = Memoirs of the faculty of education, Kumamoto university. Section 1, natural science. Kumamoto.

Kumamoto daigaku rigakubu kiyo = Kumamoto journal of science. Series B. Biology and geology. Kumamoto, Japan. Kumamoto J. Sci., Ser. B, Biol. Geol. See B–P–H 521/9.

Kumamoto journal of science. Sect. 1. Geology. Kumamoto, Japan. Kumamoto J. Sci., Sect. 1, Geol. See B–P–H 521/7. HI 55751

Kumamoto journal of science. Sect. 2. Biology. Kumamoto, Japan. Kumamoto J. Sci., Sect. 2, Biol. See B–P–H 521/8. HI 55752

Kumamoto journal of science. Series B. Biology and geology. [Kumamoto daigaku rigakubu kiyo.] Kumamoto, Japan. Kumamoto J. Sci., Ser. B, Biol. Geol. See B–P–H 521/9. HI 55753

Kumamoto-ken nori kenkyujo jigyo hokokusho. 1go+, [1978?]+. Kumamoto-ken Nori Kenkyujo Jigyo Hokokusho HI 67130

K'un Ch'ung yü chih ping = Entomology and phytopathology. Hangchow, China. Entomol. & Phytopathol. See B–P–H 358/30.

Kung yeh chung hsin = Industrial research. Nanking. Industr. Res. See B–P–H 432/8.

Kungl[iga]. fysiografiska sällskapets i Lund förhandlingar. Lund. Kungl. Fysiogr. Sällsk. Lund Förh. See B–P–H 521/14. HI 55754

Kungliga lantbrukshögskolans annaler. Uppsala. Vols. 13-228, 1946-62. Kungl. Lantbrukshögskolans Ann. Preceded and superseded by: Lantbrukshögskolans annaler. HI 67131

Kungliga svenska vetenskapsakademiens avhandlingar i naturskyddsärenden. Stockholm. 1938-66. Kung. Svenska Vetenskapsakad. Avh. Naturskyddsärenden. HI 67132

Kungrip soul taehakkyo. Seoul. Vol. 1+, 1971+. Kungrip Soul Taehakkyo. HI 67133

Kunst- und Gewerbe-Blatt. Herausgegeben von dem polytechnischen Verein für das Königreich Baiern. Munich. Kunst- Gewerbe-Blatt. See B–P–H 521/18. HI 55755

Kuo chia k'o hsüeh wei yüan hui = Proceedings of the national science council, republic of China. Taipei.

Kuo li ch'ing-hua ta hsüeh li k'o pao kao = Science reports of national Tsing Hua university. Series B. Biological and psychological sciences. Peiping [=Peking]. Sci. Rep. Natl. Tsing Hua Univ., Ser. B, Biol. Sci. See B–P–H 827/17.

Kuo li chung shan ta hsüeh li k'o shêng hse ts'ung k'an = Bulletin of department of biology, college of science, Sun Yatsen university. Canton. Bull. Dept. Biol. Sun Yatsen Univ. See B–P–H 248/1.

Kuo li chung shan ta hsüeh tzu jan k'o hsüeh k'o = Monographic series of the natural sciences bulletin, college of science; Sun Yatsen university. Canton. Monogr. Ser. Nat. Sci. Bull. Coll. Sci. Sun Yatsen Univ. See B–P–H 617/20.

Kuo li chung yang ta hsüeh k'o hsüeh chi k'an = Quarterly science bulletin of the national central university. Nanking. Quart. Sci. Bull. Natl. Centr. Univ. See B–P–H 753/5.

Kuo li chung yang ta hsüeh shen lin hsüeh yen chiu so yen chiu pao-kao. Shu mu hsüeh = Research notes, forestry institute; national central university, Nanking. Dendrological series. Nanking. Res. Notes Forest. Inst. Natl. Centr. Univ. Nanking, Dendrol. Ser. See B–P–H 775/14.

Kuo li chung yang yen chiu yuan chih wu hsüeh hui pao = Botanical bulletin of academia sinica. Shanghai. Bot. Bull. Acad. Sin. See B–P–H 219/9.

Kuo li pei ching nung yeh chuan mên hsüeh hsiao tsa chih = Agricultural magazine. Peking.

Kuo li Pei-p'ing ta hsüeh hsüeh pao = Journal of the Peking national university. Peiping [=Peking]. J. Peking Natl. Univ. See B–P–H 475/22.

Kuo li Peip'ing yen chiu yüan yüan yu hui pao = Bulletin of the national academy of Peiping. Peiping [=Peking]. Bull. Natl. Acad. Peiping. See B–P–H 265/20.

Kuo li Tai wan ta hsüeh nung hsüeh yüan yen chiu pao kao = Memoirs of the faculty of agriculture; national Taiwan university. Taipei, Taiwan. Mem. Fac. Agric. Natl. Taiwan Univ. See B–P–H 572/10.

Kuo li Taiwan ta hsüeh nung hsüeh yüan yen chiu pao k'an = Memoirs of the college of agriculture; national Taiwan university. Taipei, Taiwan. Mem Coll. Agric. Natl. Taiwan Univ. See B–P–H 570/12.

Kuo-li Shantung ta hsüeh k'o hsüeh ts'ung k'an = Journal of science; national university of Shantung. Tsingtao, China. J. Sci. Natl. Univ. Shantung. See B–P–H 480/17.

Kuo li Wu-han ta hsüeh li k'o chi k'an = Quarterly journal of science, Wu-han university. Wuchang,

China. Quart. J. Sci. Wu-han Univ. See B–P–H 752/10.

Kuopion luonnon ystavain yhdistyksen julkaisuja. Sarja B. Kuopio. 1927-67. Kuopion Luonnon Ystavain Yhdistyksen Julkaisuja, B. Preceded by: Luonnonystavain yhdistyksen julkaisuja. Superseded by: Savonia. HI 67134

Kurama yama no shokubutsu chosa hokokusho. Kyoto. No. 1+, 1977+. Kurama Yama Shokubutsu Chosa Hokokusho. HI 67135

Kurtziana. Córdoba, Argentina. Kurtziana. See B–P–H 522/7. HI 55756

Kurz und bündig; auslese aus den neuesten landwirtschaftlichen Veröffentlichungen der landwirtschaftlichen Abteilung der Badischen Anilin- und Soda-Fabrik, [Ser. 1]. Ludwigshafen. Vol. 1+, 1948+. Kurz & Bündig (Ser. 1). HI 67136

Kurz und bündig; auslese aus den neuesten weinbaulichen Veröffentlichungen der landwirtschaftlichen Abteilung der Badischen Anilin- und Soda-Fabrik, [Ser. 2]. Ludwigshafen. Vol. 1+, 1959+. Kurz & Bündig (Ser. 2). HI 67137

Kurze Uebersicht der Verhandlungen der allgemeinen Schweizerischen Gesellschaft für die gesammten Naturwissenschaften. Aarau, Switzerland. Kurze Uebers. Verh. Allg. Schweiz. Ges. Gesammten Naturwiss. See B–P–H 522/8. HI 55757

Kurzmitteilungen der deutschen dendrologischen Gesellschaft. [Darmstadt.] Vol. 1+, 1968+. Kurzmitt. Deutsch. Dendrol. Ges. HI 67138

Kusunoki noho = Agricultural journal of the Kusunoki society. Kochi, Japan. Agric. J. Kusunoki Soc. See B–P–H 59/5.

Kuwait bulletin of marine sciences. Salmiya. No. 1+, 1979+. Kuwait Bull. Mar. Sci. HI 67139

Kwaju = Fruit trees. Hangchow, China. Fruit Trees. See B–P–H 386/5.

Kwangsi agriculture. [Kuang hsi nung yeh.] Liuchow, China. Kwangsi Agric. See B–P–H 522/12. HI 55758

Kwangtung agricultural news. [Kuang tung nung yeh tung hsün.] [Kwangtung, China]. Kwangtung Agric. News. See B–P–H 522/13. HI 55759

Kwartalnik historii nauki i techniki. Warsaw. Vol. 1+, 1956+. Kwart. Hist. Nauki Techn. HI 67140

Kwartalnik opolski. Opole. Vols. 1-2, 1961-62. Kwart. Opolski. Superseded by: Zeszyty przyrodnicze. HI 67141

Kwartalnik poswiecony ochronie roślin w Polsce = Choroby i szkodniki roślin. Warsaw.

Kwartalny biuletyn informacyjny o ochronie przyrody. Cracow. Kwart. Biul. Inform. Ochr. Przyr. See B–P–H 522/15. HI 55760

Kweekersblad. Officieel orgaan van het hollandsch bloembollenkwekers genootschap. Heemstede, Netherlands. Kweekersblad. See B–P–H 522/16. HI 55761

Kwiaty. Warsaw. No. 1+, 1976+. Kwiaty. HI 67142

Kyokuchi kenkyu nyusu. Tokyo. No. 1+, 1974+. Kyokuchi Kenkyu Nyusu. HI 67143

Kyoto daigaku nogakubu enshurin iho = Report of the Kyoto university forest. Kyoto, Japan. Rep. Kyoto Univ. Forest. See B–P–H 767/14.

Kyoto daigaku nogakubu fuzoku enshurin hokoku = Bulletin of the Kyoto university forests. Kyoto.

Kyoto daigaku nogakubu kiyo = Memoirs of the college of agriculture; Kyoto university. Kyoto, Japan. Mem. Coll. Agric. Kyoto Univ. See B–P–H 570/11.

Kyoto daigaku uiurusu kenkyusho nenkan liyo = Report (Annual), institute for virus research, Kyoto university. Kyoto.

Kyoto-fu chagyo kenkyusho gyomu hokokusho = Tea research journal, Kyoto. Kyoto, Japan. Tea Res. J. Kyoto. See B–P–H 868/2.

Kyoto-fu ringyo shikenjo. Kenkyu shiryo. No. 1+, 1978+. Kyoto-fu Ringyo Shikenjo Kenkyu Shiryo. HI 67147

Kyoto furitsu daigaku. Gakujutsu hokoku. Nogaku = Scientific reports of the Kyoto prefectural university. Agriculture. Kyoto.

Kyoto furitsu daigaku. Gakujutsu hokoku. Nogaku = Scientific reports of the Kyoto prefectural university. Natural science and living science. Kyoto.

Kyoto-furitsu daigaku nogakubu enshurin hokoku = Bulletin of the Kyoto prefectural university forests. Kyoto.

Kyoto-furitsu nogyo kenkujo kenkyu hokoku = Bulletin of the Kyoto prefectural institute of agriculture.

Kyoto furitsu nogyo kenkyujo. Sokuho. Vols. 1-5 (48-51), 1973-76. Kyoto Furitsu Nogyo Kenkyujo Sokuho. HI 67144

Kyoto furitsu nogyo kenkyujo. Tango bunjo. Kyotan. Takeno-gun. 74/1-74/5, 1974. Kyoto Furitsu Nogyo Kenkyujo, Tango Bunjo, Kyotan. Superseded by: Kyoto furitsu nogyo kenkyujo. Tango bunjo. Shiken seisekisho. HI 67145

Kyoto furitsu nogyo kenkyujo. Tango bunjo. Shiken seisekisho. Takeno-gun. 1975+. Kyoto Furitsu Nogyo Kenkyujo, Tango Bunjo Shiken Seisekisho. Preceded by: Kyoto furitsu nogyo kenkyujo. Tango bunjo. Kyotan. HI 67146

Kyoto-furitsu nogyo shikenjo kenkyu hokoku = Bulletin

of the Kyoto agricultural experiment station. Kyoto.

Kyoto gakugei daigaku hokoku = Bulletin of the Kyoto educational university. Series B, mathematics and natural science. Kyoto.

Kyoto kogei sen'i daigaku sen'igakubu gakujutsu hokoku = Bulletin of the faculty of textile fibers, Kyoto university of industrial arts and textile fibers. Kyoto.

Kyoto teikoku daigaku enshu-rin hokoku = Bulletin of the society of forestry; Kyoto imperial university. Kyoto, Japan. Bull. Soc. Forest. Kyoto Imp. Univ. See B–P–H 276/24.

Kyungpook university theses collection. Taegu, Korea. Kyungpook Univ. Theses Collect. See B–P–H 522/27. HI 55762

Kyushu agricultural research. [Kyushu nogyo kenkyu.] Fukuoka, Japan. Kyushu Agric. Res. See B–P–H 522/28. HI 55763

Kyushu byogaichu kenkyukai kaiho = Proceedings, association for plant protection of Kyushu. Chikugo City.

Kyushu daigaku. Nettai nogaku kenkyu senta = Bulletin of the institute of tropical agriculture, Kyushu university. Fukuoka.

Kyushu daigaku nogakubu enshurin hokoku = Bulletin of the Kyushu university forests. Fukuoka, Japan. Bull. Syushu Univ. Forests. See B–P–H 260/5.

Kyushu daigaku nogakubu enshurin shuho = Report of the Kyushu university forests. Fukuoka.

Kyusu daigaku nogakubu. Gakugei zasshi = Science bulletin of the faculty of agriculture, Kyushu university. Fukuoka, Japan. Sci. Bull. Fac. Agric. Kyushu Univ. See B–P–H 823/17.

Kyushu nogyo kenkyu = Kyushu agricultural research. Fukuoka, Japan. Kyushu Agric. Res. See B–P–H 522/28.

Kyushu nogyo shikenjo iho = Bulletin of the Kyushu agricultural experiment station. Fukuoka, Japan. Bull. Kyushu Agric. Exp. Sta. See B–P–H 260/3.

Kyushu noji shiken kenkyu happyokai koen yoshi = Proceedings of the meeting of agricultural research workers in Kyushu. Tokyo. Proc. Meeting Agric. Res. Workers Kyushu. See B–P–H 732/4.

L A I F S. Los Angeles, La Mirada, CA. Vol. ?-4(11)+, ?-1977+. L. A. I. F. S. HI 67148

L A S C A leaves = Lasca leaves. Los Angeles, Arcadia, CA.

L A S C A miscellanea = Lasca miscellanea. Arcadia, CA.

L B Case's botanical index; an illustrated quarterly botanical magazine. Richmond, IN. L. B. Case's Bot. Index. See B–P–H 523/1. HI 55764

L I S T monthly; national source guide for interior and exterior plants. Columbia, MD. Vol. 1+, 1982+. L. I. S. T. Monthly. HI 67150

L S H R journal = Journal of the Louisiana society for horticultural research. Lafayette, LA.

L S H R news letter. Vols. 8(3)-10, 1967-69. L. S. H. R. News Lett. Preceded by: Monthly news letter, Louisiana society for horticultural research. Superseded by: Journal, Louisiana society for horticultural research. HI 67151

L S S A / I W E newsletter. Pretoria. No. ?-2+, ?-1979+. L. S. S. A./ I. W. E. Newslett. HI 67152

L S U forestry notes. Louisiana agricultural experiment station. Baton Rouge, LA. No. 1+, 1953+. L. S. U. Forest. Notes. HI 67153

La-Ya'aran. Natanya, Israel. La-Ya'aran. See B–P–H 527/8. HI 55828

Laboratoire maritime du muséum national d'histoire naturelles à l'arsenal de Saint Servan. St.-Servan, France. Lab. Marit. Mus. Natl. Hist. Nat. Arsenal Saint Servan. See B–P–H 523/13. HI 55765

Labores musei in Benatky nad Jizerou. Benatky nad Jizerou, Czechoslovakia. 1965-75. Lab. Mus. Benátky nad Jizerou. HI 55766

Ladies' wreath; a magazine devoted to literature, industry and religion. New York. Nos. 1-183, 1846-61?; n.s. nos. 184-189, 1861-62. Ladies' Wreath. HI 67154

Lagascalia. Seville. Vol. 1+, 1971+. Lagascalia. HI 67155

Lagena. Cumana. No. 1+, 1964+. Lagena. HI 67156

Laitsch herbarium publications. Youngstown, OH. No. 2+, 1980+. Laitsch Herb. Publ. Preceded by: Herbarium publications, Youngstown state university. HI 67157

Läkaren och naturforskaren. Stockholm. Läkaren & Naturf. See B–P–H 523/15. HI 55767

Lake and reservoir management; an international journal. Washington, DC. Vol. 1+, 1985+. Lake Reservoir Managem. HI 74857

Lakehead forester. Thunder Bay. Vol. 1+, 1976+. Lakehead Forester. HI 67158

Lakes letter, international association for Great Lakes research. Toronto. Vol. 1+, 1970?+. Lakes Lett. Int. Assoc. Great Lakes Res. HI 67159

Lal-baugh journal; journal of the Mysore horticultural society. Murshidabad [Lal-Bagh]. Vol. 1+, 1956+. Lal-baugh J. HI 67160

Lamatepec. Santa Ana, Salvador. Lamatepec. See B–P–H 523/16. HI 55768

Lammergeyer; journal of the Natal parks game and fish

preservation board. Pietermaritzburg. Vol. 1+, 1960+. Lammergeyer. HI 67161

Lancashire and Cheshire naturalist. Darwen. Vols. 8-17, 1914-25. Lancashire Cheshire Naturalist. Preceded by: Lancashire naturalist. Superseded by: Northwestern naturalist. 3-2346-1. HI 67162

Lancashire naturalist. Darwen. Vols. 1-7, 1907-14. Lancash. Naturalist. Superseded by: Lancashire and Cheshire naturalist. HI 67163

Lancet. Chicago, IL. Lancet (Chicago). See B–P–H 523/19. HI 55769

Lancet. London. Lancet (London). See B–P–H 523/20. HI 55770

Lancet. Newark, NJ. Lancet (Newark). See B–P–H 523/22. HI 55771

Lanchow ta hsüeh hsüeh pao. Tsu jên k'o hsüeh = Lanchow university journal, natural sciences. Lanchow, China. Lanchow Univ. J., Nat. Sci. See B–P–H 523/24.

Lanchow university journal, natural sciences. [Lanchow ta hsüeh hsüeh pao. Tsu jen k'o hsüeh.] Lanchow, China. Lanchow Univ. J., Nat. Sci. See B–P–H 523/24. HI 55772

Land. Oxford. Vol. 1+, 1974+. Land. Preceded by: Journal, Oxford university forest society. HI 67164

Land- und forstwirthschaftliche Zeitschrift für Norddeutschland; Zunächst für Braunschweig, Hannover und die angrenzenden Länder. Brunswick. Vols. 1-4, 1834-36. Land- Forstw. Z. Norddeutschl. Superseded by: Annalen der deutschen Landwirthschaft in allen Zweigen. 1-369-2. HI 67165

Land- und hauswirthschaftliche Zeitung. Halle. Land-Hausw. Zeitung. See B–P–H 523/27. HI 55773

Land research series. Melbourne, Vic. Nos. 1-6, 1953-77. Land Res. Ser. HI 67166

Land resource study. Tolworth. Vol. 1+, 1966+. Land Resource Study. HI 67167

Land- en tuinbouwjaarboek. Vols. ?-16-25, ?-1961-71. Land- Tuinbouwjaarb. HI 67168

Landbouw. Buitenzorg, Dutch E. Indies [=Bogor, Indonesia]. Landbouw (Buitenzorg). See B–P–H 524/2. HI 55778

Landbouw. The Hague. Landbouw (The Hague). See B–P–H 524/4. HI 55779

Landbouw courant. Zwolle, Netherlands. Landb. Courant. See B–P–H 523/29. HI 55775

Landbouw-documentatie. The Hague, Wageningen. Vol. 4+, 1948+. Landb.-Doc. Preceded by: Mededelingen van de afdeeling voor de landbouwvoorlichtingsdienst. HI 67169

Landbouw en plantenziekten. Wageningen. Landb. & Plantenziekten. See B–P–H 523/28. HI 55774

Landbouw-weekblad. Bloemfontein, South Africa. Landb.-Weekbl. See B–P–H 523/30. HI 55776

Landbouw wereldnieuws. The Hague. Landb. Wereldnieuws. See B–P–H 524/1. HI 55777

Landbouwberichten. Groningen. Landbouwberichten. See B–P–H 524/5. HI 55780

Landbouwetenskap in Suid-Afrika. Plantbeskermingswetenskappe en mikrobiologie = Phytophylactica. Pretoria.

Landbouwjournaal van de Kaap de Goede Hoop. Cape Town. Landbouwj. Kaap Goede Hoop. See B–P–H 524/6. HI 55781

Landbouwkundig tijdschrift. The Hague. Lanbouwk. Tijdschr. (The Hague). See B–P–H 524/7. HI 55782

Landbouwkundig tijdschrift. Wageningen & Groningen. Landbouwk. Tijdschr. (Wageningen & Groningen). See B–P–H 524/8. HI 55783

Landbouwkundig tijdschrift voor Nederlandsch-Indië. Buitenzorg, Dutch E. Indies [=Bogor, Indonesia]. Landbouwk. Tijdschr. Ned.-Indië. See B–P–H 524/9. HI 55784

Landbouwkundig tijdschrift voor Suriname en Curaçao. Paramaribo, Dutch Guiana [Surinam]. Landbouwk. Tijdschr. Suriname Curaçao See B–P–H 524/10. HI 55785

Landbouwnieuws. [Department van landbouw, veeteelt en oiserij.] Paramaribo, Dutch Guiana [Surinam]. Landbouwnieuws. See B–P–H 524/11. HI 55786

Landbouwtijdschrift. Brussels. Landbouwtijdschrift. See B–P–H 524/12. HI 55787

Landbrug og naturvidenskab; et maanedsskrift for danske landmaend. Odense. 1867-1871. Landbr. & Naturvidensk. HI 67170

Landes-Museum im Herzogthume Krain Jahresbericht. Laibach [=Ljublijana, Yugoslavia]. Landes-Mus. Herzogth. Krain Jahresber. See B–P–H 524/13. HI 55788

Landis, geo. arboretum newsletter. Vol. 1+, 1982+. Landis Geo. Arbor. Newslett. HI 67171

Landmands-blade. Copenhagen. Vols. 1-46, 1868-1913. Landm.-Blade. 3-2351-1. HI 67172

Landmarks. Toronto. Vol. 1+, 1982+. Landmarks. HI 67173

Landoeconomisk maanedsskrift isaer for mindre jordbrugere. Aarhus & Randers. Vols. 1-4, 1844-47. Landoecon. Maanedsskr. HI 67174

Landoekonomisk tidende [Edited by J. C. Luno]. Copenhagen. Vols. 1-15, 1839-53. Landoekon.

Tidende (Luno). Superseded by: Landoeconomisk tidende for Denmark og Slesvig. HI 67175

Landoeconomisk tidende for Danmark og Slesvig. Copenhagen. Vols. 1-3, 1854-56. Landoecon. Tidende Danmark Slesvig. Preceded by: Landoekonomisk tidende [Edited by J. C. Luno]. Superseded by: Agerdysknings-tidende for Danmark og Slesvig. HI 67176

Landoeconomiske tidender, udgivne af et selskab i Sjaelland. Copenhagen. Landoecon. Tidender. See B–P–H 524/14. HI 55789

Landscape architecture. Louisville, KY. ?-1938+. Landscape Architect. HI 67177

Landscape Canada; serving the landscape/nursery industry coast to coast. Ottawa. Vol. 10+, 1973+. Landsape Canada. Preceded by: Canadian nurseryman. HI 67179

Landscape design. London. No. 93+, 1971+. Landsape Design. Preceded by: Journal, institute of landscape architects. HI 67178

Landscape and garden. London. Landscape & Gard. See B–P–H 524/16. HI 55790

Landscape journal. Madison, WI. Vol. 1+, 1982+. Landscape J. HI 68282

Landscape planning; an international journal on landscape ecology, reclamation and conservation, outdoor recreation and land-use management. Amsterdam. Vol. 1+, 1974+. Landscape Planning. HI 67180

Landscape research. Manchester. Vol. 2+, 1976+. Landscape Res. Preceded by: Landscape research news. HI 67181

Landscape research news. Manchester. Vol. 1(1-12), 1968-76. Landscape Res. News. Superseded by: Landscape research. HI 67182

Landscape trades. Mississauga. Vol. 1-3, 1979-81. Landscape Trades. HI 67183

Landscape and turf industry. Elm Grove, WI. Vol. 25+, 1980+. Landscape Turf Industr. Preceded by: Turfgrass times and Landscape industry [not entered]. HI 67184

Landschaft und Stadt. Stuttgart. 1969+. Landschaft & Stadt. Preceded by: Beiträge zur Landespflege. HI 67185

Landschaftspflege und Naturschutz in Thüringen. Eisenach. Vol. 1+, 1964+. Landschaftspflege Naturschutz Thüringen. HI 67186

Landscope. Perth, W.A. Vol. 1+, 1985+. Landscope. HI 67187

Landtmannen. Linkoping, Sweden. Landtmannen. See B–P–H 524/20. HI 55792

Landtmannen - svensk land. Stockholm. Landtm. - Svensk Land. See B–P–H 524/19. HI 55791

Landwirt; Fachblatt der Südtiroler Bauern und Genossenschaften. Bozen [Bolzano]. Vol. 1+, 1947+. Landwirt (Bozen). HI 67188

Landwirtschaftlich-Historische Blätter. Weimar. Vol. 1+, 1902+. Landw.-Hist. Blätt. HI 67189

Landwirthschaftliche(s), see Landwirtschaftliche(s)

Landwirthschaftliche Annalen des Mecklenburgischen Patriotischen Vereins. Rostock. Landw. Ann. Mecklenburg. Patriot. Vereins. See B–P–H 524/21. HI 55793

Landwirthschaftliche Berichte aus Mitteldeutschland. Ilmenau, Germany. Landw. Ber. Mitteldeutschl. See B–P–H 525/1. HI 55794

Landwirthschaftliche Blätter. Mitau. No. 1, 1827. Landw. Blätt. HI 67190

Landwirthschaftliche Blätter von Hofwyl. Aarau, Switzerland. Landw. Blätt. Hofwyl. See B–P–H 525/2. HI 55795

Landwirthschaftliche Blätter und Obst- und Weinbau-Zeitung für Siebenbürgen. Hermannstadt [=Sibiu, Rumania]. Landw. Blätt. Obst- Weinbau-Zeitung Siebenbürgen. See B–P–H 525/3. HI 55796

Landwirthschaftliche Blätter für Siebenbürgen. Hermannstadt [=Sibiu, Rumania]. Landw. Blätt. Siebenbürgen. See B–P–H 525/4. HI 55797

Landwirthschaftliche Blätter, den Zöglingen der Bildungs-Anstalt in Ungarisch-Altenburg gewidmet. Bécs [=Vienna]. Landw. Blätt. Zögl. Bildungs-Anst. Ungarisch-Altenburg. See B–P–H 525/5. HI 55798

Landwirtschaftliche Forschungen; Zeitschrift des Verbandes deutscher landwirtschaftlicher Untersuchungs- und Forschungsanstalten. Frankfurt am Main & Darmstadt. Vol. 1+, 1949+. Landw. Forsch. 3-2352-3. HI 67191

Landwirthschaftliche Hefte. Bécs [=Vienna]. Landw. Hefte. See B–P–H 525/6. HI 55799

Landwirthschaftliche Jahrbücher. Berlin. Landw. Jahrb. See B–P–H 525/7. HI 55800

Landwirthschaftliche Mitteilungen; Zeitschrift der Königlichen höheren landwirthschaftlichen Lehranstalt und der damit vereinigten Landwirthschaftlichen Versuchsstation zu Poppelsdorf. Bonn. Vols. 1-3, 1858-60. Landw. Mitth. (Bonn). 3-2353-1. HI 67192

Landwirthschaftliche Mitteilungen, besonders für das Fürstenthum Lüneburg, und Verhandlungen des landwirthschaftlichen Provinzial-Vereins zu Uelzen. Lüneburg, Germany. Landw. Mitt. Fürstenth. Lüneburg & Verh. Landw. Prov.-Vereins Uelzen. See B–P–H 525/9. HI 55802

Landwirthschaftliche Zeitschrift, herausgegeben von der

Gesellschaft zur Beförderung des Ackerbaus, der Künste und Wissenschaften im Niederrheinischen Departement. Strasbourg. Landw. Z. Ges. Beförd. Ackerbaus Niederrhein. Dept. See B–P–H 525/12. HI 55805

Landwirtschaftliche Zeitschrift für das oesterreichische Schlesien. Troppau [=Opava, Czechoslovakia]. Landw. Z. Oesterr. Schlesien. See B–P–H 525/13. HI 55806

Landwirthschaftliche Zeitschrift für den Unter-Mainkreis des Königreichs Bayern. Würzburg. Landw. Z. Unter-Mainkreis Königr. Bayern. See B–P–H 525/14. HI 55807

Landwirthschaftliche Zeitung. Augsburg. Landw. Zeitung (Augsburg). See B–P–H 525/15. HI 55808

Landwirthschaftliche Zeitung. Halle. Landw. Zeitung (Halle). See B–P–H 525/16. HI 55809

Landwirthschaftliche Zeitung für Kurhessen. Kassel. Landw. Zeitung Kurhessen. See B–P–H 525/17. HI 55810

Landwirtschaftlichen Versuchsstationen. Berlin. Landw. Versuchsstat. See B–P–H 525/10. HI 55803

Landwirtschaftliches Jahrbuch für Bayern. Munich. Vols. 1-32(4), 1911-55 [Suspended 1935-48]. Landw. Jahrb. Bayern. Superseded by: Bayerisches landwirtschaftliches Jahrbuch. 1-613-3. HI 67193

Landwirthschaftliches Jahrbuch der Schweiz. Bern. Landw. Jahrb. Schweiz. See B–P–H 525/8. HI 55801

Landwirthschaftliches Magazin. Leipzig. Vols. 1-2, 1788-91. Landwirthschaftliches Mag. HI 67194

Landwirthschaftliches Wochenblatt für (das Grossherzogthum) Baden. Karlsruhe. Landw. Wochenbl. Baden. See B–P–H 525/11. HI 55804

Landwirtschaftliches Zentralblatt. Abteilung 2, pflanzliche Produktion. Berlin. Vol. 1+, 1955+. Landw. Zentralbl. Abt. 2 Pflanzl. Prod. HI 67195

Landwirthschaftskalender. Brünn [=Brno, Czechoslovakia]. Landwirthschaftskalender. See B–P–H 525/18. HI 55811

Lantbrukshögskolans annaler. Uppsala. Vols. 1-12, 1933-46; vols. 29-36, 1962-70. Lantbrukshögskolans Ann. For 1946-62 see: Kungliga lantbrukshögskolans annaler. Superseded by: Swedish journal of agricultural research. HI 67196

Lantbrukshögskolans meddelanden. Serie A. Uppsala. Nos. 1-284, 1963-77. Lantbrukshögskolans Meddel., A. HI 67197

Lantbrukshögskolans meddelanden. Serie B. Uppsala. Nos. 1-26, 1963-77. Lantbrukshögskolans Meddel., B. HI 67198

Lantbrukstidskrift för Dalarne. Falun, Sweden. Lantbrukstidskr. Dalarne. See B–P–H 526/2. HI 55812

Lantbrukstidskrift för Jämtland och Härjedalen. Ostersund, Sweden. Lantbrukstidskr. Jämtland Härjedalen. See B–P–H 526/3. HI 55813

Lantbrukstidskrift för Stockholms län och stad. Stockholm. Lantbrukstidskr. Stockholms Län Stad. See B–P–H 526/4. HI 55814

Lantern; journal of knowledge and culture. (Tydskrif vir kennis en kultuur) [subtitle varies.] Pretoria. Vol. 1+, 1949+. Lantern (Pretoria). 3-2356-2. HI 67199

Lapin tutkimusseura, vuosikirja = Vuosikirja, lapin tutkimusseura. Rovaniemi?

Laporan balai penelitian hutan. Bogor. Nos. 366-414, 1981-83. Laporan Balai Penelitian Hutan. Preceded by: Laporan, lembaga penelitian hutan. Superseded by: Laporan, pusat penelitian dan pengembangan hutan. HI 67200

Laporan dari balai penjelidikan perusahaan gula. Pasuruan. 1957+. Laporan Dari Balai Penjel. Perus. Gula. Preceded by: Verslagen van het proefstation voor de Java-Suikerindustrie. HI 67201

Laporan dari lembaga pusat penjelidikan kehutanan. Bogor. Nos. ?-74-?, ?-1959-? Laporan Dari Lemb. Pusat Penjel. Kehut. Preceded by: Rapport van het bosbouwproefstation. Superseded by: Laporan lembaga-lembaga penelitian kehutanan. HI 64493

Laporan, lembaga-lembaga penelitian kehutanan. Bogor. Nos. 100-149, 1969-72. Laporan Lemb.-Lemb. Penelitian Kehut. Preceded by: Laporan dari lembaga pusat penjelidikan kehutanan. Superseded by: Laporan, lembaga penelitian hutan. HI 67202

Laporan, lembaga penelitian hutan. Bogor. Nos. 155-365, 1972-81. Laporan Lemb. Penelitian Hutan. Preceded by: Laporan, lembaga-lembaga penelitian kehutanan. Superseded by: Laporan, balai penelitian hutan. HI 67203

Laporan, pusat penelitian dan pengembangan hutan. Bogor. Nos. 415-461, 1983-84. Laporan Pusat Penelitian Pengemb. Hutan. Preceded by: Laporan, balai penelitian hutan. Superseded by: Buletin penelitian hutan. Bogor. HI 67204

Laporan tahunan, balai penelitian perkebunan. Bogor. Vol. 1+, 1971+. Laporan Tahunan Balai Penelitian Perkebunan. HI 67205

Laporan tahunan, dari balai penjelidikan perusahaan gula = Laporan dari balai penjelidikan perusahaan gula. Pasuruan.

Lapwing. Samlesbury. 1974+. Lapwing. HI 67206

Lärda tidningar. Stockholm. Lärda Tidn. See B–P–H 526/5. HI 55815

Laryngoscope. St. Louis, MO. Laryngoscope. See B–P–H 526/6. HI 55816

Lasca leaves; quarterly publication of the southern California horticultural institute and the California arboretum foundation inc. [subtitle varies.] Los Angeles, Arcadia, CA. Vols. 1-26, 1950-76. Lasca Leaves. From 1977 contained in: Garden [California edition]. 7-1172-3. HI 67207

Lasca miscellanea. Arcadia, CA. Vols. 1-6, 1953-57. Lasca Misc. Superseded by: Report (Annual), Los Angeles state and county arboretum. HI 67208

Läsning för allmogen i Kronobergs län. Hushållssällskapet. Lund. Läsn. Allmogen Kronobergs Län. See B–P–H 526/11. HI 55817

Latin american forest bibliography. Merida. 1981?+. Latin Amer. For. Bibliogr. Preceded by: Bibliographical bulletin, instituto forestal Latino-Americano de investigación y capacitación. HI 67209

Latin american physiology and pharmacology reports = Acta physiologica et pharmacologica latinoamericana. Buenos Aires.

Latin american research review. Austin, TX. Latin Amer. Res. Rev. See B–P–H 526/12. HI 55818

Latvijas augstskolas raksti. Riga. Vols. 1-5, 1921-23. Latv. Augstsk. Raksti. Superseded by: Latvijas universitates raksti. 4-3678-3. HI 67210

Latvijas augu aizsardzibas instituta raksti. Riga. Vols. 1-2, 1930-32. Latv. Augu Aizsard. Inst. Raksti. 4-3678-3. HI 67211

Latvijas biologijas biedribas raksti. Riga = Acta biologica latvica. Riga, Latvia [Latvian S S R]. Acta Biol. Latv. See B–P–H 40/23.

Latvijas biologijas biedribas raksti. Riga, Latvia [Latvian S S R]. Latv. Biol. Biedribas Raksti. See B–P–H 526/13. HI 55819

Latvijas mežu petišanas stacijas raksti. Riga. Vols. 1-10, 1934-38. Latv. Mežu Petiš. Sta. Raksti. HI 67212

Latvijas P S R zinätnu akademijas vëstis = Izvestiya Akademii nauk Latviiskoi S S R. Riga.

Latvijas universitates botaniska darza darbi. Riga. Vol. 1, 1926. Latv. Univ. Bot. Darza Darbi. 4-3678-3. HI 67213

Latvijas universitates botaniska darza raksti = Acta horti botanici universitatis latviensis. Riga.

Latvijas universitates raksti. Riga. Vols. 6-20, 1923-29. Latv. Univ. Raksti. Preceded by: Latvijas augstkolas raksti. Superseded by: Latvijas universitatis raksti. Lauksaimniecibas fakultates serija. 4-3678-3. HI 67214

Latvijas universitates raksti. Lauksaimniecibas fakultates serija. Riga. Vols. 1-3(3), 1929-36. Latv. Univ. Raksti, Lauksaimn. Fak. Ser. Preceded by: Latvijas universitatis raksti. 4-3678-3. HI 67215

Launceston field naturalist. Launceston, Tas. Vols. 1-?, 1967-? Launceston Field Naturalist. Superseded by: Launceston naturalist. HI 67216

Launceston naturalist. Launceston, Tas. Vol. ?-2(8)+, ?-1969+. Launceston Naturalist. Preceded by: Launceston field naturalist. HI 67217

Lausitzisches Magazin oder Sammlung verschiedener Abhandlungen und Nachrichten zum Behuf der Natur-, Kunst-, Welt- und Vaterlands-Geschichte, der Sitten und der schönen Wissenschaften. Görlitz. Lausitz. Mag. See B–P–H 526/20. HI 55820

Lausizische Monatsschrift. Görlitz. Lausiz. Monatsschr. See B–P–H 526/23. HI 55821

Lavori di botanica, istituto di botanica e di fisiologia vegetale dell'università di Padova. Padua. Vol. 13+, 1946/48+. Lav. Bot. Ist. Bot. Univ. Padova. Preceded by: Pubblicazioni, istituto botanico della r[eale]. università di Padova. HI 67218

Lavori dell'istituto botanico e del giardino coloniale di Palermo. Palermo. Vols. 13-?, 1952-? Lav. Ist. Bot. Giard. Colon. Palermo. Preceded by: Lavori del r[eale]. istituto botanico e del r[eale]. giardino coloniale di Palermo. Superseded by?: Lavori dell'istituto ed orto botanico dell'universita di Palermo. HI 67219

Lavori dell'istituto botanico della r[eale]. università di Cagliari. Cagliari. Lav. Ist. Bot. Reale Univ. Cagliari. See B–P–H 526/26. HI 55822

Lavori dell'istituto botanico, università di Modena. Modena. Vol. 1+, 1973+. Lav. Ist. Bot. Univ. Modena. HI 67220

Lavori dell'instituto botanico dell'università di Milano. Milan. Vol. 1+, 1947+. Lav. Inst. Bot. Univ. Milano. HI 67221

Lavori dell'istituto ed orto botanico dell'universita di Palermo. Palermo. ?-1980+. Lav. Ist. Orto Bot. Univ. Palermo. Preceded by: Lavori dell'istituto botanico e del giardino coloniale di Palermo. HI 67222

Lavori del r[eale]. istituto botanico di Modena. Forli & Modena. 1931-34. Lav. Reale Ist. Bot. Modena. HI 67223

Lavori del r[eale]. istituto botanico di Palermo. Palermo. Vols. 1-10, 1930-39. Lav. Reale Ist. Bot. Palermo. Preceded, in part, by: Bollettino delle reale orto botanico di Palermo. Superseded by: Lavori del r[eale]. istituto botanico e del r[eale]. giardino coloniale di Palermo. 4-3245-1. HI 67224

Lavori del r[eale]. istituto botanico e del r[eale]. giardino coloniale di Palermo. Palermo. Vols. 11-12, 1940-44. Lav. Reale Ist. Bot. Reale Giard. Colon. Palermo. Preceded by: Lavori del r[eale]. istituto botanico di

Palermo. Superseded by: Lavori dell'istituto botanico e del giardino coloniale di Palermo. HI 67225

Lavori di r[eale]. orto botanico. R[eale]. museo di fisica e storia naturale. Florence. Lav. Reale Orto Bot. See B–P–H 527/2. HI 55823

Lavori della società italiana di biogeografia. Forli. N.s. vol. 1+, 1970+. Lav. Soc. Ital. Biogeogr. HI 67226

Lavori, società veneziana di scienze naturali. Venice. Vol. ?-2+, ?-1977+. Lav. Soc. Veneziana Sci. Nat. HI 67227

Lavoura. Rio de Janeiro. Lavoura. See B–P–H 527/3. HI 55824

Lavoura arrozeira. Porto Alegre, Brazil. Lavoura Arroz. See B–P–H 527/5. HI 55826

Lavoura e criação. Rio de Janeiro. Lavoura & Criação. See B–P–H 527/4. HI 55825

Lawn care. Marysville, OH. Lawn Care. See B–P–H 527/7. HI 55827

Lawrence review of natural products. Collegeville, PA. Vol. ?-3+, ?-1982+. Lawrence Rev. Nat. Prod. HI 67228

Lazaroa. Madrid. Vol. 1+, 1979+. Lazaroa. HI 67229

Leaflet of acadian biology. Grand Falls. No. 1+, 1953+. Leafl. Acadian Biol. HI 67230

Leaflet, agricultural department, St. Lucia. St. Lucia. [Dates of publication not ascertained.] Leafl. Agric. Dep. St. Lucia. HI 67231

Leaflet, agricultural experiment station, zionist organisation institute. Tel-Aviv. No. 1+, 1922+. Leafl. Agric. Exp. Sta. Zionist Organ. Inst. HI 67232

Leaflet, agricultural and horticultural experiment station, university of Bristol. Long Ashton. 1913+. Leafl. Agric. Hort. Exp. Sta. Univ. Bristol. Preceded by: Leaflet, national fruit and cider institute. HI 67233

Leaflet, Alabama agricultural experiment station of the Alabama polytechnic institute = Alabama agricultural experiment station of the Alabama polytechnic institute. Leaflet. Auburn, AL. Alabama Agric. Exp. Sta. Alabama Polytechn. Inst. Leafl. See B–P–H 67/13.

Leaflet, board of agriculture for Scotland. Edinburgh. 1912-28. Leafl. Board Agric. Scotland. Superseded by: Leaflet, department of agriculture, Scotland. HI 67234

Leaflets of botanical observation and criticism. Washington, DC. Leafl. Bot. Observ. Crit. See B–P–H 527/14. HI 55829

Leaflets, Brooklyn botanic garden = Brooklyn botanic garden leaflets. Brooklyn, NY. Brooklyn Bot. Gard. Leaft. See B–P–H 230/5.

Leaflet, California agricultural experiment station. Berkeley, CA. Nos. 1-217, 1925-75. Leafl. Calif. Agric. Exp. Sta. Superseded by: Leaflet, division of agricultural sciences, university of California. HI 67235

Leaflet, coconut research institute, Ceylon. Lunawila. 1951+. Leafl. Coconut Res. Inst. Ceylon. Preceded by: Leaflet, coconut research scheme, Ceylon. HI 67236

Leaflet, coconut research scheme, Ceylon. Lunawila. 1936-50. Leafl. Coconut Res. Scheme Ceylon. Preceded by: Leaflet, coconut research institute, Ceylon. HI 67237

Leaflet, colonial forest resources development department. London. 1936-39. Leafl. Colon. For. Resource Developm. Dep. HI 67238

Leaflet, commonwealth forestry and timber bureau. Canberra, A.C.T. 1932+. Leafl. Commonw. Forest. Timber Bur. HI 67239

Leaflet, department of agricultural extension, Purdue university. Lafayette, IN. Nos. 20-?, 1912-47. Leafl. Dep. Agric. Extens. Purdue Univ. Preceded by: A E form. Superseded by: Extension leaflet, department of agricultural extension, Purdue university. HI 67240

Leaflets, department of agriculture. Washington, DC. Leafl. Dept. Agric. See B–P–H 527/16. HI 55830

Leaflet, department of agriculture, Bermuda. 1916-27. Leafl. Dep. Agric. Bermuda. HI 67241

Leaflet, department of agriculture, Bombay. Poona. 1908-48. Leafl. Dep. Agric. Bombay. HI 67242

Leaflet, department of agriculture, British East Africa. Nairobi. Nos. 1-27, 1905-07. Leafl. Dep. Agric. Brit. E. Africa. HI 67243

Leaflet, department of agriculture, British Honduras. Belize. Nos. 1-3?, 1933-? Leafl. Dep. Agric. Brit. Honduras. HI 67244

Leaflet, department of agriculture, Burma = Cultivator's leaflet, department of agriculture, Burma. Rangoon.

Leaflet, department of agriculture, Ceylon. Colombo. Nos. 1-151, 1917-38. Leafl. Dep. Agric. Ceylon. HI 67245

Leaflet, department of agriculture, Cyprus. Nicosia. Nos. 1-24, 1920-38. Leafl. Dep. Agric. Cyprus. HI 67246

Leaflet, department of agriculture, Cyprus. Educational series. Nicosia. 1935-48. Leafl. Dep. Agric. Cyprus, Educ. Ser. HI 67247

Leaflet, department of agriculture, Iraq = Agricultural leaflet, department of agriculture, Iraq. Basrah, Baghdad.

Leaflet, department of agriculture, New Guinea. Rabaul. Nos. 1-70, 1924-34. Leafl. Dep. Agric. New Guinea. Incorporated in: New Guinea agricultural gazette.

HI 67248

Leaflet, department of agriculture, Madras. Madras. 1908-28. Leafl. Dep. Agric. Madras. HI 67249

Leaflet, department of agriculture, Mauritius. Reduit. Nos. 1-43, 1917-38. Leafl. Dep. Agric. Mauritius. HI 67250

Leaflet, department of agriculture, New Guinea = Leaflet, department of agriculture, territory of New Guinea. Rabaul.

Leaflet, department of agriculture, Nova Scotia. Halifax, N.S. 1931-41. Leafl. Dep. Agric. Nova Scotia. HI 67251

Leaflet, department of agriculture, Nyasaland Protectorate. Zomba. 1926+. Leafl. Dep. Agric. Nyasaland. HI 67252

Leaflet, department of agriculture, Orange River Colony. Bloemfontein. 1906-07. Leafl. Dep. Agric. Orange River Colony. HI 67253

Leaflet, department of agriculture, Quebec. Quebec. 1935+. Leafl. Dep. Agric. Quebec. HI 67254

Leaflet, department of agriculture, Scotland. Edinburgh. 1929-49. Leafl. Dep. Agric. Scotland. Preceded by: Leaflet, board of agriculture for Scotland. Superseded by: Advisory leaflet, department of agriculture, Scotland. HI 67255

Leaflet, department of agriculture, Tanganyika Territory. Dar-es-Salaam. 1925-28. Leafl. Dep. Agric. Tanganyika. HI 67256

Leaflet, department of agriculture, territory of New Guinea. Rabaul. 1924-34. Leafl. Dep. Agric. New Guinea. HI 67257

Leaflet, department of agriculture, Transvaal. Botany. Pretoria. 1904-07. Leafl. Dep. Agric. Transvaal, Bot. HI 67258

Leaflet, department of agriculture, Transvaal. Horticulture. Pretoria. 1904-06. Leafl. Dep. Agric. Transvaal, Hort. HI 67259

Leaflet, department of agriculture, Western Australia. Perth, W.A. 1922+. [Many publications are unnumbered, some were reprinted with later imprint dates, some are duplicated, e.g. nos. 30, 38, 48, & 122. Nos. 42? & 1100-1999 not issued.] Leafl. Dep. Agric. W. Australia. Preceded by: Bulletin, department of agriculture, Western Australia. HI 67260

Leaflet, department of botany, Field museum of natural history. Chicago, IL. Nos. 1-25, 1922-42. Leafl. Dep. Bot. Field Mus. Nat. Hist. Superseded by: Popular series, Chicago natural history museum. Botany. HI 67261

Leaflet, department of forestry, Israel. Ilanoth. Nos. 1-?, 1956-61. Leafl. Dep. Forest. Israel. Superseded by: Leaflet, division of forestry, ministry of agriculture, Israel. HI 67262

Leaflet, department of forestry, Oxford. Oxford. 1939. Leafl. Dep. Forest. Oxford. HI 67263

Leaflet, division of agricultural sciences, university of California. Berkeley, CA. No. 2200+, 1975+. Leafl. Div. Agric. Sci. Univ. Calif. Preceded by: Leaflet, California agricultural experiment station. HI 67264

Leaflet, division of forestry, ministry of agriculture, Israel. Ilanoth. 1962+. Leafl. Div. Forest. Minist. Agric. Israel. Preceded by: Leaflet, department of forestry, Israel. HI 67265

Leaflet, division of plant industry, Florida department of agriculture and consumer services. Gainesville, FL. No. 1+, 1966+. Leafl. Div. Pl. Industr. Florida Dep. Agric. HI 67266

Leaflet, Edinburgh and east of Scotland college of agriculture. Edinburgh. 1916+. Leafl. Edinburgh E. Scotland Coll. Agric. HI 67267

Leaflet, forest department, British Guiana. Georgetown. Nos. 1-13, 1951. Leafl. Forest Dep. Brit. Guiana. HI 67268

Leaflet, forest department, Sarawak. Kuching. Nos. 1-45, 1953-54. Leafl. Forest Dep. Sarawak. HI 67269

Leaflet. Forest department, Trinidad and Tobago. Port of Spain, Trinidad. Leafl. Forest Dept. Trinidad Tobago. See B–P–H 527/18. HI 55831

Leaflet, forest products research laboratory, department of scientific and industrial research. London. 1927+. Leafl. Forest Prod. Res. Lab. Dep. Sci. Industr. Res. HI 67270

Leaflet, forest research institute, Dehra Dun. Calcutta. Nos. 1-6, 1941. Leafl. Forest Res. Inst. Dehra Dun. Preceded by: Forest leaflets, forest research institute, Dehra Dun. Superseded by: Indian forest leaflets. HI 67271

Leaflet, forest research institute, Ilanoth = Leaflet, department of forestry, Israel. Ilanoth.

Leaflet, forest research institute, Israel = Leaflet, department of forestry, Israel. Ilanoth.

Leaflet, forest research station, Ilanoth = Leaflet, department of forestry, Israel. Ilanoth.

Leaflet, forestry commission. London. Vol. 1+, 1920+. Leafl. Forest. Commiss. HI 67272

Leaflet, forestry division, national and university institute of agriculture. Rehovot. No. ?-24+, ?-1964+. Leafl. Forest. Div. Natl. Univ. Inst. Agric., Rehovot. HI 67273

Leaflet, forests commission of Victoria. Melbourne, Vic. 1941+. Leafl. Forests Commiss. Victoria. HI 67274

Leaflet, George Foster Peabody school of forestry.

Athens, GA. 1943+. Leafl. George Foster Peabody School Forest. HI 67275

Leaflet, horticultural section, ministry of agriculture, Egypt. Cairo. 1916-22. Leafl. Hort. Sect. Minist. Agric. Egypt. HI 67276

Leaflet, imperial forestry institute = Leaflet, department of forestry, Oxford. Oxford.

Leaflet of the indian forest department. Calcutta. 1908-10. Leafl. Indian Forest Dep. HI 67277

Leaflet, institute of plant industry. Indore. 1932+. Leafl. Inst. Pl. Industr., Indore. HI 67278

Leaflet of the Mauritius sugar industry research institute. Reduit, Port Louis. Nos. 1-19, 1959-79. Leafl. Mauritius Sugar Industr. Res. Inst. HI 67279

Leaflet of the Mengel natural history society. Reading, PA. 1943-51. Leafl. Mengel Nat. Hist. Soc. HI 67280

Leaflet, ministry of agriculture, fisheries and food. London. 1894. Leafl. Minist. Agric. Fish. Food. HI 67281

Leaflet, ministry of agriculture, Northern Ireland. Belfast. No. 1+, 1922+. Leafl. Minist. Agric. Northern Ireland. HI 67282

Leaflet, ministry of agriculture, Sudan. Khartoum. 1953-? Leafl. Minist. Agric. Sudan. HI 67283

Leaflet of the Montreal botanical garden. Montreal. No. 1+, 1940+. Leafl. Montreal Bot. Gard. HI 67284

Leaflet, national arboretum, United States forest service = National arboretum leaflet, forest service, U S department of agriculture. Washington, DC.

Leaflet, national fruit and cider institute. Long Ashton, Bristol. 1905-12. Leafl. Natl. Fruit Cider Inst. Superseded by: Leaflet, agricultural and horticultural experiment station, university of Bristol. HI 67285

Leaflet, natural history department, Papua and New Guinea museum and art gallery. Port Moresby. 1976+. Leafl. Nat. Hist. Dep. Papua New Guinea Mus Art Gall. HI 67286

Leaflet, New York state college of forestry. Syracuse, NY. Nos. 1-3, 19??-? Leafl. New York State Coll. Forest. HI 67287

Leaflet, New York state museum = New York state museum leaflet. New York.

Leaflet, New Zealand forest service. Wellington, N.Z. Nos. 1-28, 1926-36. Leafl. New Zealand Forest Serv. HI 67288

Leaflet, Oregon state college school of forestry. Corvallis, OR. 1941-42. Leafl. Oregon State Coll. School Forest. HI 67289

Leaflets of philippine botany. Manila. Leafl. Philipp. Bot. See B–P–H 527/22. HI 55832

Leaflet, plant breeding section, ministry of agriculture, Egypt. Cairo. 1928-29-? Leafl. Pl. Breed. Sect. Minist. Agric. Egypt. HI 67291

Leaflet, research department, indian coffee board. Balehonnur. Nos. 1-19?, 1958-? Leafl. Res. Dep. Indian Coffee Board. HI 67292

Leaflets of the Santa Barbara botanic garden. Santa Barbara, CA. Leafl. Santa Barbara Bot. Gard. See B–P–H 527/24. HI 55833

Leaflet, school of forestry, oregon state college = Leaflet, Oregon state college school of forestry. Corvallis, OR.

Leaflet series, department of agriculture, Mauritius = Leaflet, department of agriculture, Mauritius. Reduit.

Leaflet series, Florida state board of conservation, marine laboratory. St. Petersburg, FL. Vols. 1-7(2), 1964-72. Leafl. Ser., Florida State Board Conservation Mar. Lab. Superseded by: Florida marine research publications. HI 67293

Leaflet series: Plankton. Florida state board of conservation, marine laboratory. St. Petersburg, FL. Vol. 1, 1964. Leafl. Ser. Plankt. Florida State Board Conservation Mar. Lab. HI 67294

Leaflet series: Phytoplankton. Florida state board of conservation, marine laboratory. St. Petersburg, FL. Vol. 1(1-3), 1965-66. Leafl. Ser, Phytoplankt. Florida State Board Conservation Mar. Lab. HI 67295

Leaflet, society for the preservation of native New England plants. Boston, MA. Vols. 1-33, 1901-21? Leafl. Soc. Preserv. Native New England Pl. 5-3979-3. HI 67296

Leaflets, U S department of agriculture. Washington, DC = Leaflets, department of agriculture. Washington, DC.

Leaflet, university of Georgia agricultural experiment station. Athens, GA. Nos. 1-47, 1954-65. Leafl. Univ. Georgia Agric. Exp. Sta. HI 67297

Leaflet, university of Ife herbarium. Ife. ?-ca.1971+. Leafl. Univ. Ife Herb. HI 67298

Leaflet, Utah agricultural college, agricultural experiment station = Utah agricultural college, agricultural experiment station. Leaflet. Logan, UT. Utah Agric. Exp. Sta. Leafl. See B–P–H 945/31.

Leaflet, Volcani center, agricultural research organisation. Bet-Dagani. ?-ca.1981+. Leafl. Volcani Center Agric. Res. Organ. HI 67299

Leaflet, Wau ecology institute. Wau. Nos. 1-2, 1975?-78. Leafl. Wau Ecol. Inst. HI 67300

Leaflet, welsh plant breeding station. Aberystwyth. No. 1+, 1928+. Leafl. Welsh Pl. Breed. Sta. HI 67301

Leaflets of western botany. San Francisco, CA. Leafl. W. Bot. See B–P–H 527/27. HI 55834

Leaflets, see Leaflet

Leandra; revista de informação científica do departamento de botânica, universidade federal do Rio de Janeiro. Rio de Janeiro. Vol. 1+, 1971+. Leandra. HI 67302

Leaves from the garden, Glen Oak botanic garden. Vol. ?-3(7)+, ?-1979+. Leaves Gard., Glen Oak. HI 67303

Leben, Das; Zeitschrift für Biologie und Lebenschutz. Hamburg-Sasel. Vols. 1-11(3), 1964-74. Leben. Superseded by: Leben und Umwelt. HI 67304

Leben und Umwelt. Aarau, Switzerland. Leben & Umwelt. See B–P–H 527/31. HI 55835

Leben und Umwelt. Wiesbaden. Vol. 11(4)+, 1974+. Leben Umwelt. Preceded by: Leben, das. HI 67305

Lebendige Erde. Stuttgart, Darmstadt. 1950+. Lebendige Erde. 3-3430-1. HI 67306

Lebendiges Wissen. Monatsschrift für Freunde der Natur, der wissenschaftlichen Erkenntnis und des technischen Fortschritts. Stuttgart. Lebendiges Wissen. See B–P–H 528/1. HI 55836

Lecointea; communications de dendrolobiologie et de dendrogénétique de l'estuaire de l'Amazone. Belem. Nos. 1-?, 1963-64? Lecointea. HI 67307

Lecture series, Harold Lloyd Lyon arboretum. Honolulu, HI. Vol. 1+, 1970+. Lecture Ser. Harold Lloyd Lyon Arbor. HI 67308

Legume research; an international journal. Karnal. Vol. 1+, 1977+. Legume Res. HI 67310

Lehigh Valley medical magazine. Easton, PA. Lehigh Valley Med. Mag. See B–P–H 528/6. HI 55837

Lehrmeister im Garten und Kleintierhof; illustrierte Wochenschrift für Obst- und gartenbau. Leipzig. Vols. ?-31-33-?, ?-1933-34-? Lehrmeister Gart. Kleintierhof. HI 67311

Leiden botanical series. Leiden. Vol. 1+, 1975+. Leiden Bot. Ser. HI 67312

Leidse geologische mededeelingen. Leiden. Leidse Geol. Meded. See B–P–H 528/8. HI 55949

Leipziger gelehrte Zeitungen. Leipzig. Leipziger Gel. Zeitungen. See B–P–H 528/15. HI 55838

Leipziger Jahrbuch der neuesten Literatur. Leipzig. Leipziger Jahrb. Neuesten Lit. See B–P–H 528/16. HI 55839

Leipziger Literaturzeitung. Leipzig. Leipziger Literaturzeitung. See B–P–H 528/17. HI 55840

Leipziger Magazin zur Naturkunde, Mathematik und Oekonomie. Leipzig & Dessau. Leipziger Mag. Naturk. Math. See B–P–H 528/18. HI 55841

Leipziger Magazin zur Naturkunde und Oekonomie. Leipzig & Dessau. Leipziger Mag. Naturk. Oekon. See B–P–H 528/19. HI 55842

Leipziger Repertorium der deutschen und ausländischen Literatur. Leipzig. Leipziger Repert. Deutsch. Ausl. Lit. See B–P–H 528/20. HI 55843

Leipziger Sammlungen von Wirthschaftlichen, Polizey-, Cammer- und Finanzsachen. Leipzig. Leipziger Samml. Wirthschaftl. Finanzsachen. See B–P–H 528/21. HI 55844

Lejeunia; revue de botanique. [Subtitle varies.] Liège. Vols. 1-23, 1937-59; n.s. vol. 1+, 1961+. Lejeunia. 3-2388-1. HI 67313

Lejeunia; revue de botanique. Mémoire. Liège. Vols. 1-15, 1939-61. Lejeunia Mém. Incorporated in: Lejeunia; revue de botanique. Nouvelle série. 3-2388-1. HI 67314

Lékárnické listy. Prague. Lékárn. Listy. See B–P–H 528/25. HI 55845

Lekovite sirovine zbornik radova. Belgrade. Vols. 1-5, 1951-60. Lekov. Sirovine Zborn. Rad. HI 67315

Lembaga pusat penelitian pertanian. Bogor. No. 1+, 1972+. Lemb. Pusat Penelitian Pert. HI 67316

Lens; a quarterly journal of microscopy and the allied natural sciences. Chicago, IL. Lens. See B–P–H 528/27. HI 55846

Lentil abstracts. Farnham Royal. Vol. 1+, 1981+. Lentil Abstr. HI 67317

Leopard; revue de l'institut national pour la conservation de la nature. Kinshasa. No. 1+, 1973+. Leopard. HI 67318

Leopoldina. Amtliches Organ der Kaiserlichen Leopoldinisch-Carolinischen Deutschen Akademie der Naturforscher. Jena. Leopoldina. See B–P–H 529/1. HI 55847

Lerner marine laboratory; collected reprints. Bailey's Town, Bahamas. Lerner Mar. Lab. Collect. Repr. See B–P–H 529/2. HI 55848

Les. Bratislava. Vol. 10+, 1954+. Les (Bratislava). Preceded by: Pol'ana. HI 55849

Les. Ljubljana. Vol. 1+, 1949+. Les (Ljubljana). 3-2396-2. HI 67319

Les a lov. Pisek. Bohemia [Czechoslovakia]. Les & Lov. See B–P–H 529/5. HI 55850

Les i step'. Moscow. Les & Step'. See B–P–H 529/6. HI 55851

Lesnická práce. Pisek, Czechoslovakia. Lesn. Práce. See B–P–H 529/13. HI 55854

Lesnícky časopis. Prague & Bratislava. Vol. 11+, 1955+. Lesn. Čas. HI 67320

Lesnictví. Prague. Lesnictví. See B–P–H 529/17. HI 55855

Lesnictwo; zaszyty naukowe szkoly glownej

gospodarstwa wiejskiego w Warszawie. Warsaw. Vol. 1+, 1958+. Lesnictwo. HI 67321

Lesnoe khozyaistvo; ezhemesyachnii proizvodstvennyii nauchno-tekhnicheskii zhurnal. Organ ministerstva lesnogo khozyaistva S S S R. Moscow. 1939+. Lesn. Khoz. Preceded by: Zashchita lesa. HI 67322

Lesnoi zhurnal. [Forms part of: Izvestiya vysshikh uchebnykh zavednii.] Archangel. Vol. 1+, 1958+. Lesn. Zhurn. HI 67323

Lĕsnoi zhurnal. St. Petersburg. 1833-51; vols. 1-48, 1871-1918. Lĕsn. Zhurn. 3-2397-1. HI 67324

Lĕsnoi zhurnal, izdaniye Lesnogo obshchestva, S.-Peterburg = Lĕsnoi Zhurnal. St. Petersburg.

Lĕsnoj Žurnal = Lĕsnoi Zhurnal. St. Petersburg.

Lesotho. Morija. Vol. 1+, 1959+. Lesotho. HI 67325

Lesovedenie. Moscow. Vol. 1+, 1967+. Lesovedenie. HI 67326

Lesovodstvennye issledovanija. Budapest = Erdészeti kutatások. Budapest. Erdész. Kutatás. See B–P–H 360/2.

Lesovodstvo, lesnye kul′tury i pochvovedenie. Leningrad. Vol. 1+, 1973+. Lesov. Lesn. Kul′t. Pochvov. HI 67327

Lethaia; international journal of palaeontology and stratigraphy. Oslo, New York. Vol. 1+, 1968+. Lethaia. HI 67328

Lĕtnija raboty = Lĕtniya raboty. Saratov.

Lĕtniya raboty. Saratov, Russian S F S R. Vol. 2(2), 1905. Lĕtn. Raboty. Preceded by: Ezhegodnik Volzhskoi Biologicheskoi Stantsii Saratovskago Obshchestva Estestvoispytatelei i Lyubitelei Estestvoznaniya. Superseded by: Raboty Volzhskoi Biologicheskoi Stantsii. HI 67329

Letopis akademije znanosti in umetnosti v Ljubljani. Ljubljana. 1943-47. Let. Akad. Znan. Umetn. Ljubljani. Superseded by: Letopis slovenska akademija znanosti in umetnosti. HI 67330

Letopis slovenska akademija znanosti in umetnosti. Ljubljana. 1948+. Let. Slov. Akad. Znan. Umetn. Preceded by: Letopis akademije znanosti in umetnosti v Ljubljani. HI 67331

Letras médicas. São Paulo. Letras Méd. See B–P–H 529/20. HI 55856

Letters in applied microbiology. London. Vol. 1+, 1985+. Lett. Appl. Microbiol. HI 67332

Letter, conservation foundation = Conservation foundation letter. Washington, DC.

Letterkundig magazijn van wetenschap, kunst en smaak. Amsterdam. Letterk. Mag. Wetensch. See B–P–H 529/22. HI 55858

Letters and papers on agriculture, planting etc. selected from the correspondence-book of the Bath and west and southern counties society. Bath, England. Lett. Pap. Agric. Bath Soc. See B–P–H 529/21. HI 55857

Lettre de l'eau pure. Versailles. No. 62+, 1981+. Lett. Eau Pure. Preceded by: Eau pure. HI 67333

Lettres botaniques. [Part of: Bulletin de la société botanique de France.] 1978+. Lett. Bot. Preceded by: Bulletin de la société botanique de France. HI 67334

Lettura oftalmologica. Pistoia, Italy. Lettura Oftalmol. See B–P–H 529/23. HI 55859

Letzebueger cacteefren. Luxemburg. Vol. 1+, 1980+. Letzeb. Cacteefren. HI 67335

Letzeburger Kaktussefrenn = Letzebueger cacteefren. Luxemburg.

Leucaena newsletter. Taipei. Vol. 1, 1980. Leucaena Newslett. Superseded by: Leucaena research reports. HI 67336

Leucaena research reports. Taipei. Vol. 2+, 1981+. Leucaena Res. Rep. Preceded by: Leucaena newsletter. HI 67337

Leuchtenbergia. No. ?-3+, ?-1982+. Leuchtenbergia. HI 67338

Levende natuur. Amsterdam. Levende Natuur. See B–P–H 530/2. HI 55860

Lexington leaflets. [Lexington botanic garden.] Lexington, MA. Lexington Leafl. See B–P–H 530/3. HI 55861

Liaison - S A J I B. Montreal. No. 1+, 1976+. Liais. S. A. J. I. B HI 67339

Liaoning daxue xuebao. Ziran kexue ban. Shen-yang. No. 1+. 1974?-81+. Liaoning Daxue Xuebao, Ziran Kexue Ban. HI 67340

Liana. The organization for tropical studies newsletter. Durham, NC. ?-1989+. Liana. HI 67341

Liao-ning ta hsueh hsueh pao. Che hsueh she hui ko hsuek pan = Liaoning daxue xuebao. Ziran kexue ban. Shen-yang.

Liberal arts journal, Tottori university. Natural science. [Tottori daigaku gakugeibu kenkyu hokoku.] Tottori. Vols. 1-16(2), 1950-65. Liberal Arts J. Tottori Univ. Nat. Sci. Superseded by: Journal of the faculty of education, Tottori university. HI 67342

Librarium; Zeitschrift der schweizerischen Bibliophilen-Gesellschaft. (Revue de la société suisse des bibliophiles.) Zurich. Vol. 1+, 1958+. Librarium. Preceded by: Stultifera navis. HI 67343

Library; quarterly review of bibliography and library lore. (From 4th ser. 1920+ subtitled: Transactions of the bibliographical society.) London. Vols. 1-10, 1889-1899; n.s. vols. 1-10, 1899-1909; 3rd ser. vols. 1-10,

1910-1919; 4th ser. vols. 1-26, 1920-46; 5th ser. vols. 1-33, 1947-78; 6th ser. vol. 1+, 1979+. Library. 3-2407-2. HI 67344

Library accessions, arctic institute of north America. Calgary. 1970 (Aug./Dec.)-1978 (Mar./Apr.). Library Accessions, Arctic Inst. N. Amer. Superseded by: A S T I S current awareness bulletin. HI 67345

Library accessions list, BIOTROP. Bogor. No. 1+, 1973+. Library Accessions List, B. I. O. T. R. O. P. HI 67346

Library bulletin, british museum (natural history). London. Vols. 1-6(2), 1976-81. Library Bull. Brit. Mus. Nat. Hist. Superseded by: Newsletter, department of library services, british museum (natural history). HI 67347

Library bulletin, N C C. London, Banbury, Peterborough. No. 1+, 1977+. Library Bull. N. C. C. HI 67348

Library list, United States department of agriculture = United States department of agriculture. Library list. Washington, DC. U.S.D.A. Libr. List. See B–P–H 943/11.

Library quarterly, forest research station, Alice Holt Lodge = Library record, forest research station, Alice Holt Lodge. Wrecclesham, Farnham.

Library record, forest research station, Alice Holt Lodge. Wrecclesham, Farnham. 1949-? Library Rec. Forest Res. Sta. Alice Holt Lodge. Superseded by: Library review, forestry commission. HI 67349

Library review, commonwealth forestry institute. Oxford. Vol. ?-27+, ?-1980+. Library Rev. Commonw. Forest. Inst. HI 67350

Library review, forestry commission. Wrecclesham, Farnham. No. 1+, 1965+. Library Rev. Forest. Commiss. Preceded by: Library record, forest research station, Alice Holt Lodge. HI 67351

Libyan journal of science; an international journal. Tripoli. Vol. 1+, 1971+. Libyan J. Sci. HI 67352

Lichen; [News bulletin of the lichenological society of Japan]. Vol. 1+, 1972+. Lichen. HI 67353

Lichenological miscellanies. [Chiigaku zappo.] Mito, Japan. Lichenol. Misc. See B–P–H 530/15. HI 55862

Lichenologist. Cardiff, Wales. Lichenologist. See B–P–H 530/17. HI 55863

Lidia; a norwegian journal of botany. Aas. Vol. 1+, 1986+. Lidia. HI 67354

Lietuvos T S R aukstuju mokyklu mokslo darbai. Biologija. Vilnius. 1961+. Lietuvos T.S.R. Aukst. Mokslo Darb., Biol. HI 75111

Lietuvos T S R mokslu akademijos biologijos instituto darbai. Vilnius. Vols. 1-3, 1951-58. Lietuvos T.S.R. Mokslu Akad. Biol. Inst. Darb. Superseded by: Trudy Laboratorii fiziologii životnyh Instituta biologii Akademii nauk Litovskoj S S R, vol. 4, 1954 [not entered]. HI 67355

Lietuvos T S R mokslu akademijos darbai. B serija. Vilnius. Vol. 1+. 1955+. Lietuvos T.S.R. Mokslu Akad. Biol. Inst. Darb., B. Superseded, in part, by: Lietuvos T S R mokslu akademijos darbai. C serija. HI 67356

Lietuvos T S R mokslu akademijos darbai. C serija. Vilnius. Vol. 21+. 1960+. Lietuvos T.S.R. Mokslu Akad. Biol. Inst. Darb., C. Preceded by: Lietuvos T S R mokslu akademijos darbai. B serija. HI 67357

Lietuvos T S R mokslu akademijos darbai. C serija, biologijos darbai = Lietuvos T S R mokslu akademijos darbai. C serija. Vilnius.

Lietuvos universiteto matematikos gamtos fakulteto darbai. Kaunas. Vols. [1]-4, 1923-28. Lietuvos Univ. Mat. Gamtos Fak. Darb. Superseded by: Vytauto didžiojo universiteto matematikos-gamtos fakulteto darbai, biologijos skyrius. 3-2275-2. HI 67358

Life chemistry reports. New York, Chur. Vol. 1+, 1982+. Life Chem. Rep. Preceded by: International journal of chronobiology. HI 67359

Life sciences; an international medium for the rapid publication of preliminary communications in the life sciences. Oxford, New York. Vols. [1]-5, 1962-66; vol. 13+, 1973+. Life Sci. From vols. 6-12, 1967-73, divided into two parts. Life sciences. Part 1, physiology and pharmacology and Life sciences. Part 2, biochemistry, general and molecular biology. [Parts number alternately in one series till vol. 8, then separate series for vols. 9-12.] HI 67360

Life sciences. Part 1, physiology and pharmacology. Oxford, New York. Vols. 6-12, 1967-73. Life Sci., 1, Physiol. Pharmacol. Preceded & superseded by: Life sciences. HI 67361

Life sciences. Part 2, biochemistry, general and molecular biology. Oxford, New York. Vols. 6-12, 1967-73. Life Sci., 2, Biochem. Gen. Mol. Biol. Preceded & superseded by: Life sciences. HI 67362

Life sciences contributions, royal Ontario museum = Contributions, life sciences, royal Ontario museum. Toronto.

Life sciences report, national aeronautics and space administration. Washington, DC. 1987+, 1988+. Life Sci. Rep. Nat. Aeron. Space Admin. HI 67363

Life study. [Seibutsu kenkyu.] Fukui. Vols. 1-17, 1957-73. Life Study. HI 67364

Light flash. Vol. 1+, 1972+. Light Flash. HI 67365

Light garden; the magazine for artificial light gardening. New York. Vols. ?-21(3), 1980-84. Light Gard.

Superseded by: Indoor garden. HI 67366

Lilac letter. Durham, NH. Lilac Lett. See B–P–H 530/26. HI 55864

Lilac newsletter; international lilac society newsletter. Rumford, ME. Vols. 1-4(1)?, 1978-81? Lilac Newslett. HI 67367

Lilacs; publication of the international lilac society. Hamilton, Ont. Vol. ?-8+, ?-1979+. Lilacs. HI 67368

Lilies, 1973 [etc.], and other liliaceae. London. 1972[sic]-80. Lilies. Preceded by: Lily year-book. Superseded by: Bulletin, R H S lily group. HI 67369

Lilloa; revista de botánica. Tucumán, Argentina. Lilloa. See B–P–H 530/30. HI 55865

Lilly scientific bulletin. Indianapolis, IN. Lilly Sci. Bull. See B–P–H 530/31. HI 55866

Lily bulletin. Washington, DC. 1941+. Lily Bull. 1-237-3. HI 67370

Lily year-book. London. Nos. [1]-34, 1932-71, 1932-70[sic]. Lily Year-Book. Superseded by: Lilies, 1973 [etc.], and other liliaceae. 4-3722-2. HI 67372

Lily yearbook of the north american lily society. Geneva, NY. Lily Yearb. N. Amer. Lily Soc. See B–P–H 530/34. HI 55867

Lima bean bulletin. Oxnard, CA. Lima Bean Bull. See B–P–H 531/1. HI 55868

Limestone. Washington, DC. Limestone. See B–P–H 531/3. HI 55869

Limnobios; contribuciones cientificas del instituto de limnologia. La Plata. Vol. 1+, 1976+. Limnobios. HI 67373

Limnologica. Berlin. Limnologica. See B–P–H 531/6. HI 55871

Limnological bibliography for Africa south of the Sahara. Grahamstown. Vol. 1+, 1975+. Limnol. Bibliogr. Africa S. Sahara. Until 1974 published in: Newsletter, limnnological society of southern Africa. HI 67374

Limnological data report, Lake Erie. Ottawa. No. 1+, 1970+. Limnol. Data Rep. Lake Erie. HI 67375

Limnological data report, Lake Huron, Lake Superior. Ottawa. No. 1+, 1970+. Limnol. Data Rep. Lake Huron Lake Superior. HI 67376

Limnology and oceanography. Baltimore, MD. Limnol. & Oceanogr. See B–P–H 531/4. HI 55870

Lin ch'an tung hsin = Forestry industry news. Taipei, Taiwan. Forest. Industr. News. See B–P–H 380/12.

Lin ch'an yüeh k'an = Forestry industry monthly. Taipei, Taiwan. Forest. Industr. Monthly. See B–P–H 380/11.

Lin-hsun = Forestry investigation. [China. Department of agriculture and forestry. Central forestry bureau.] Chungking, China. Forest. Invest. See B–P–H 380/13.

Lin yeh chon k'an = Forestry bulletin. [Forestry experiment station, college of agriculture, national Taiwan university.] Taipei, Taiwan. Forest. Bull. (Taipei). See B–P–H 379/27.

Lin yeh k'o hsüeh = Scientia silvae sinicae. Peking (Beijing).

Lin yeh t'ui kuang tsuan k'an = Special issue. Sylviculture and education. [Taiwan forest experiment station.] Taipei, Taiwan. Special Issue Sylvic. Educ. See B–P–H 851/2.

Lin-yeh t'ung-hsun = Forestry bulletin. [China. Department of agriculture and forestry. Central forestry bureau.] Nanking. Forest. Bull. (Nanking). See B–P–H 379/26.

Lindbergia; a journal published by the nordic bryological society and the dutch bryological society. Aarhus, Leiden. Vol. 1+, 1971+. Lindbergia. HI 67377

Lindenia; iconographie des orchidées. Ghent & Brussels. Lindenia. See B–P–H 531/13. HI 55872

Lindleyana; scientific journal of the american orchid society. West Palm Beach, FL. Vol. 1+, 1986+. Lindleyana. HI 67378

Lingnaam agricultural review. Canton. Lingnaam Agric. Rev. See B–P–H 531/17. HI 55873

Lingnan agricultural journal. [Lingnan nung-k'an.] Canton. Lingnan Agric. J. See B–P–H 531/19. HI 55874

Lingnan hsüeh pao = Lingnan journal. Canton. Lingnan J. See B–P–H 531/21.

Lingnan journal. [Lingnan hsüeh pao.] Canton. Lingnan J. See B–P–H 531/21. HI 55875

Lingnan nung-k'an = Lingnan agricultural journal. Canton. Lingnan Agric. J. See B–P–H 531/19.

Lingnan science journal. Canton. Lingnan Sci. J. See B–P–H 531/22. HI 55876

Lingnan university science bulletin. Canton. Lingnan Univ. Sci. Bull. See B–P–H 531/24. HI 55877

Links; bulletin of the international society of biometeorology. Cambridge? Vol. 1+, 1976+. Links. HI 67379

Linnaea. Berlin. Linnaea. See B–P–H 532/4. HI 55880

Linnaea. Geneva. Linnaea (Geneva). See B–P–H 532/5. HI 55881

Linnaean fern bulletin. Bingham, NY. Linn. Fern Bull. See B–P–H 531/27. HI 55878

Linnaean newsletter = Newsletter, Linnaean society of New York. New York.

Linnean; newsletter and proceeedings of the linnean

society of London. London. Vol. 1+, 1984+. Linnean. HI 67380

Linneana belgica; revue belge d'entomologie. Brussels. 1958+. Linneana Belg. HI 67381

Linnéen, Le. Sainte-Foy, Québec. No. 1+, 1976+. Linnéen. HI 67382

Linnéska samfundets handlingar. Stockholm. Linn. Samfund. Handl. See B–P–H 531/28. HI 55879

Linye keji tongxun = Forest science and technology. Beijing.

Linye kexue = Scientia silvae sinicae. Peking (Beijing).

Linzer biologische Beiträge. Linz. Vol. 7+, 1975+. Linzer Biol. Beitr. Preceded by: Mitteilungen der botanischen Arbeitsgemeinschaft am O. Ö. [oberösterreichisches] Landesmuseum Linz. HI 67383

Lioar manninagh. Douglas. 1889-1905. Lioar Manninagh. Superseded by: Proceedings of the Isle of Man natural history and antiquarian society. HI 67384

Lipids. Champaign, IL. Vol. 1+, 1966+. Lipids. HI 67385

List of accessions to the museum library, british museum (natural history). London. 1955-74. List. Accessions Mus. Library Brit. Mus. Nat. Hist. HI 67386

List of cultivars of agricultural crops = Sorter af landbrugsplanter. [Denmark.]

List of indexed articles, BIOTROP. Bogor. 1974?+. List Indexed Articles B. I. O. T. R. O. P. HI 67387

List of literatures related to japanese plant taxonomy and flora. Tokyo. 1974+. List Lit. Relat. Jap. Pl. Taxon. Fl. HI 67388

List of phytopathological publications of Turkey = Turkiye'deki fitopatolojik yayinlar listesi.

List of scientific reports relating to Thailand. Bangkok. 1964-65. List. Sci. Rep. Relat. Thailand. Superseded by: List of scientific and technical literature relating to Thailand. HI 67389

List of scientific and technical literature relating to Thailand. Bangkok. 1964+. List Sci. Techn. Lit. Relat. Thailand. HI 67390

List of varieties of agricultural crops = Sorter af landbrugsplanter. [Denmark.]

Liste des acquisitions, bibliothèque centrale du muséum national d'histoire naturelle. Paris. 1957-70. Liste Acquis. Biblioth. Centr. Mus. Natl. Hist. Nat. Superseded by: Nouvelles acquisitions, bibliothèque centrale du muséum national d'histoire naturelle. HI 67391

Liste bibliographique analytique des publications reçues en collection de reference, O R S T O M. III, domaines biologiques. Bondy. 1976+. Liste Bibliogr. Analytique Publ. Reçues Collect. Refer. O. R. S. T. O. M. 3, Dom. Biol. Preceded by: Liste bibliographique des travaux. Chercheurs en service à l'O R S T O M. HI 67392

Liste bibliographique des travaux, chercheurs en service à l'O R S T O M. Bondy. 1958/62-1969/70. Liste Bibliogr. Trav. Cherch Serv. O. R. S. T. O. M. Superseded by: Liste bibliographique analytique des publications recues en collection de reference, O R S T O M. HI 67393

Listok, astrakhanskaya stantsiya zashchity rastenii ot vreditelei. Astrakhan. Nos. ?-21-43-?, ?-1922-? Listok Astrakh. Zashch. Rast. Vredit. HI 67394

Listok dlja Bor'by s Bolěznjamy i Povreždenijamy Kul'turnyh i Dikorastuščih Poleznyh Rastenij = Listok dlya Bor'by s Bolěznyamy i Povrezhdeniyamy Kul'turnykh i Dikorastushchikh Poleznykh Rastenii. St. Petersburg.

Listok dlya Bor'by s Bolěznyamy i Povrezhdeniyamy Kul'turnykh i Dikorastushchikh Poleznykh Rastenii. St. Petersburg. 1902-06. Listok Bor'by Bolězny. Povrezhden. Kul't. Dikorast. Polezn. Rast. Superseded by: Zhurnal "Bolězni Rastenii". 3-2433-2. HI 67395

Listy cukrovarnické. Prague. Listy Cukrovarn. See B–P–H 532/16. HI 55882

Listy lesnické. Pisek, Bohemia [Czechoslovakia]. Listy Lesn. See B–P–H 532/17. HI 55883

Listy pomologické. Prague-Troja 1878+. Listy Pomol. HI 55884

Listy zahradnické. Prague. 1871-1903. Listy Zahradn. Superseded by: Ceské listy zahradnické. HI 67396

Literarische Beilage von und für Schlesien. Breslau [=Wroclaw, Poland]. Lit. Beil. Schlesien. See B–P–H 532/22. HI 55888

Literarische Berichte für Forstmänner. Tübingen. Vol. 1+, 1832+. Lit. Ber. Forstmänner HI 67397

Literarische Blätter der Börsenhalle. Hamburg. Lit. Blätt. Börsenhalle. See B–P–H 532/23. HI 55889

Literarische und kritische Blätter der Börsenhalle. Hamburg. Lit. Krit. Blätt. Börsenhalle. See B–P–H 532/26. HI 55892

Literarische Zeitung. Berlin. Lit. Zeitung (Berlin). See B–P–H 533/4. HI 55897

Literarischer Anzeiger. Vienna. Lit. Anz. See B–P–H 532/20. HI 55886

Literärischer Anzeiger für Ungarn. Lit. Anz. Ungarn. See B–P–H 532/21. HI 55887

Literarisches Conversations-Blatt. Leipzig. Lit. Conversations-Blatt. See B–P–H 532/24. HI 55890

Literarisches Wochenblatt. Weimar. Lit. Wochenbl. (Weimar). See B–P–H 533/2. HI 55895

Literarisches Wochenblatt der Börsenhalle. Hamburg. Lit. Wochenbl. Börsenhalle. See B–P–H 533/3. HI 55896

Literarisches Wochenblatt oder gelehrte Anzeigen und Abhandlungen. Nuremberg. Lit. Wochenbl. (Nuremberg). See B–P–H 533/1. HI 55894

Literary gazette, and american athenaeum. New York. Vol. 3, 1826-27. Lit. Gaz. (New York). Preceded by: New York literary gazette. 3-2438-7. HI 67398

Literary journal; a review of literature, sciences, manners, politics. London. Lit. J. (London). See B–P–H 532/25. HI 55891

Literary magazine and British review. London. Vols. 1-12, 1788-94. Lit. Mag. & Brit. Rev. 3-2438-3. HI 52128

Literary record and journal of the linnean association of Pennsylvania college. Gettysburg, PA. Lit. Rec. & J. Linn. Assoc. Pennsylvania Coll. See 532/28. HI 55893

Literaturberichte zur Flora oder allgemeinen botanischen Zeitung. Regensburg. Literaturber. Flora. See B–P–H 533/6. HI 55898

Literaturberichte über Wasser, Abwasser, Luft und Boden. Stuttgart. Vol. 1+, 1950+. Literaturber. Wasser. HI 67400

Literaturberichte über Wasser, Abwasser, Luft und Feste Abfallstoffe = Literaturberichte über Wasser, Abwasser, Luft und Boden. Stuttgart.

Literaturblatt von und für Schlesien. Breslau [=Wroclaw, Poland]. Literaturbl. Schlesien. See B–P–H 533/9. HI 55900

Literaturblätter für reine und angewandte Botanik. Nuremberg. Literaturbl. Reine Angew. Bot. See B–P–H 533/8. HI 55899

Literaturschau Kakteen. Berlin. Vols. 1-6, 1977-83. Lit.-Schau Kakteen. HI 67401

Lithoclastia; revue semestrielle éditée par le centre de recherches et d'études océanographiques. Boulogne. 1975+. Lithoclastia. HI 67402

Litoralia; international journal of coastal research. New York. 1984+. Litoralia. HI 67403

Litterarische Denkwürdigkeiten oder Nachrichten von neuen Büchern und kleinen Schriften besonders der Chursächsischen Universitäten, Schulen und Lande. Leipzig. Litt. Denkwürdigk. See B–P–H 533/12. HI 55902

Litterarisches Archiv der Akademie zu Bern. Bern. Litt. Arch. Akad. Bern. See B–P–H 533/11. HI 55901

Litteratur-Bericht zur Linnaea. Berlin. Vols. 1-16, 1826-42. Litt.-Ber. Linnaea. 3-2429-2. HI 52638

Litteratur-Journal. Altona [=Hamburg, in part]. Litt.-J. (Altona). See B–P–H 533/13. HI 55903

Litteratur-Zeitung. Erlangen. Litt.-Zeitung (Erlangen). See B–P–H 533/14. HI 55904

Little cactus corner. Milwaukee, WI. Vols. 1-10, 1965-66. Little Cact. Corner. HI 67404

Liverpool naturalists' journal. Liverpool. Liverpool Naturalists' J. See B–P–H 533/18. HI 55905

Living countryside. London. Vol. 1+, 1981+. Liv. Countryside. HI 67405

Living museum. Springfield, IL. Vol. 1+, 1939+. Liv. Mus. HI 67406

Living wilderness. Washington, DC. Vols. 1(1)-46(157), 1935-82. Liv. Wildern. Superseded by: Wilderness. 3-2450-1. HI 67407

Livländische Jahrbücher der Landwirthschaft. Dorpat [=Tartu, Estonian S S R]. Livl. Jahrb. Landw. See B–P–H 533/19. HI 55906

Livres du mycologue. Paris. Livres Mycol. See B–P–H 533/20. HI 55907

Lizard; a magazine of field studies. Mazarion. N.s. vol. 1+, 1957+. Lizard. HI 67408

Lloydia; a quarterly journal of biological science. Cincinnati, OH. Vols. 1-41, 1938-78. Lloydia. Superseded by: Journal of natural products. 3-2452-1. HI 67409

Lloyd's log. London. 1944+. Lloyd's Log. HI 67410

Lobivia. Vienna. 1975+. Lobivia. HI 67411

Loefgrenia; comunicaçoes avulsas de botânica. São Paulo. 1961+. Loefgrenia. HI 67412

Loisir horticole; revue de la fédération des sociétés d'horticulture et d'écologie du Québec. Montreal. Vol. 4(3)+, 1984+. Loisir Hort. Preceded by: Bulletin de la fédération des sociétés d'horticulture et d'écologie du Québec. HI 67413

London, Edinburgh and Dublin philosophical magazine and journal of science. London. London Edinburgh Dublin Philos. Mag. & J. Sci. See 534/8. HI 55908

London and Edinburgh philosophical magazine and journal of science. London. London Edinburgh Philos. Mag. & J. Sci. See B–P–H 534/9. HI 55909

London environmental bulletin. London? Ca.1989+. London Environm. Bull. HI 67414

London geological journal, and record of discoveries in british and foreign palaeontology. London. London Geol. J. See B–P–H 534/10. HI 55910

London journal. London. Vol. 1+, 1975+. London J. HI 67415

London journal of botany. London. London J. Bot. See B–P–H 534/12. HI 55911

London journal and weekly record of literature, science and art. London, Paris. 1845-1906. London J. Weekly Rec. Lit. Sci. Art. Superseded by: New London journal [not entered]. HI 67416

London medical gazette; being a weekly journal of medicine and the collateral sciences. London. London Med. Gaz. See B–P–H 534/16. HI 55912

London medical journal. London. London Med. J. See B–P–H 534/17. HI 55913

London medical and physical journal. London. London Med. Phys. J. See B–P–H 534/18. HI 55914

London medical press and circular = Medical press and circular. London and Dublin. Med. Press Circ. See B–P–H 555/6.

London medical repository. London. London Med. Repos. See B–P–H 534/20. HI 55915

London medical repository and review. London. London Med. Repos. Rev. See B–P–H 534/21. HI 55916

London medical, surgical and pharmaceutical repository. London. London Med. Surg. Repos. See B–P–H 534/23. HI 55917

London naturalist. London. London Naturalist. See B–P–H 534/26. HI 55918

London newsletter, office of naval research, London branch. London. Vols. 1-2(2), 1947-48. London Newslett. Off. Naval Res., London Branch. Superseded by: European scientific notes, office of naval research, branch office, London. HI 67417

London physiological journal. London. London Physiol J. See B–P–H 534/27. HI 55919

London popular gardening. London. London Popular Gard. See B–P–H 535/2. HI 55920

London review. London. London Rev. See B–P–H 535/5. HI 55921

London review of english and foreign literature. London. London Rev. Engl. Foreign Lit. See B–P–H 535/6. HI 55922

Long Island agronomist. Medford, NY. Long Island Agron. See B–P–H 535/9. HI 55923

Long Island gardening. Riverhead, LI. 1986+. Long Island Gard. HI 67418

Long Island horticulture news. Long Island, NY. ?-1977+. Long Island Hort. News. HI 67419

Long Island journal of philosophy, and cabinet of variety. Huntington, NY. Long Island J. Philos. See B–P–H 535/10. HI 55924

Long room; bulletin of the friends of the library of Trinity College. Dublin. No. 1+, 1970+. Long Room. HI 67420

Looduskaitse. Tallinn. Vol. 1+, 1937+. Looduskaitse. HI 67421

Loodusuurijate seltsi aastaraamat. Tallinn. Vols. 48-63, 1955-75. Loodusuur. Seltsi Aastar. Preceded by: Tartu ülikooli juures oleva loodusuurijate seltsi aruanded. Superseded by: Eesti loodusuurijate seltsi aastaraamat. HI 67422

Lore. Milwaukee, WI. 1952+. Lore. HI 67423

Lorentzia. Córdoba. Vols. 1+, 1970+. Lorentzia. HI 67424

Loris; a journal of Ceylon wild life. Colombo. Vol. 1+, 1936+. Loris. HI 67425

Lorquinia. Los Angeles, CA. Lorquinia. See B–P–H 535/12. HI 55925

Los Angeles cactus chronicle. Los Angeles, CA. Vol. 1+, 1941+. Los Angeles Cact. Chron. HI 67426

Los Angeles county museum contributions in science. Los Angeles, CA. Vols. 1-91, 1945-65. Los Angeles County Mus. Contr. Sci. Superseded by: Contributions in science, museum of natural history, Los Angeles county. HI 67427

Los Angeles county museum of natural history quarterly = Quarterly, Los Angeles county museum of natural history. Los Angeles, CA.

Lotos. Prague. Lotos. See B–P–H 535/16. HI 55926

Lotsya. Waltham, MA. Lotsya. See B–P–H 535/19. HI 55927

Lotus international. Milan. 1974+. Preceded by: Lotus [not entered]. Lotus Int. HI 67428

Lotus newsletter. Quebec, etc. No. 1+, 1970+. Lotus Newslett. HI 67429

Louisiana agricultural experiment station. Baton Rouge, LA. Louisiana Agric. Exp. Sta. See B–P–H 535/21. HI 55929

Louisiana agricultural experiment station bulletin. Baton Rouge, LA. Louisiana Agric. Exp. Sta. Bull. See B–P–H 535/23. HI 55930

Louisiana agricultural experiment station. Fruit and truck experiment station report. Hammond, LA. Louisiana Agric. Exp. Sta. Fruit Truck Exp. Sta. Rep. See B–P–H 535/24. HI 55931

Louisiana agricultural experiment station. Horticultural research circular. Baton Rouge, LA. Louisiana Agric. Exp. Sta. Hort. Res. Circ. See B–P–H 535/25 HI 55932.

Louisiana agricultural experiment station. North Louisiana experiment station, annual report. Calhoun, LA. Louisiana Agric. Exp. Sta. N. Louisiana Exp. Sta. Annual Rep. See B–P–H 536/2. HI 55933

Louisiana agricultural experiment station. Rice experiment station. Annual report. Crowley, LA.

Louisiana Agric. Exp. Sta. Rice Exp. Sta. Annual Rep. See B–P–H 536/4. HI 55934

Louisiana agriculture. Baton Rouge, LA. Louisiana Agric. See B–P–H 535/20. HI 55928

Louisiana state medical society journal. New Orleans, LA. Louisiana State Med. Soc. J. See B–P–H 536/6. HI 55935

Louisiana state univeristy studies. Biological science series. Baton Rouge, LA. Louisiana State Univ. Stud., Biol. Sci. Ser. See B–P–H 536/8. HI 55936

Louisiana university and agricultural and mechanical college. Forestry department. The annual ring. Baton Rouge, LA. Louisiana Univ. Agric. Coll. Forest. Dept. Annual Ring. See B–P–H 536/9. HI 55937

Louisiana university and agricultural and mechanical college. Forestry department. Progress report. Baton Rouge, LA. Louisiana Univ. Agric. Coll. Forest. Dept. Progr. Rep. See B–P–H 536/10. HI 55938

Low temperature science. [Teion kagaku.] Sapporo. Vols. 1-10, 1944-53. Low Temp. Sci. Superseded by: Low temperature science. Series B, biological sciences. HI 67430

Low temperature science. Series B, biological sciences. [Teion kagaku.] Sapporo. Nos. 11-39, 1954-81. Low Temp. Sci., B, Biol. Sci. Preceded by: Low temperature science. Superseded by: Contributions from the institute of low temperature science. Series B, biological science. HI 67431

LSU forestry notes. Louisiana agricultural experiment station. Baton Rouge, LA = L S U forestry notes. Louisiana agricultural experiment station. Baton Rouge, LA.

Lübecker Kakteen-Zeitschrift. Lübeck. Vols. ?-191-198, ?-1969-70. Lübecker Kakteen Z. Superseded by: Kakteen - Rundschau. HI 67432

Lucknow university studies. Allahabad. No. 1+, 935+. Lucknow Univ. Stud. HI 67433

Lucrări, staţiunea de cercetări biologica, geologice şi geografice "Stejarul". Pingarati. Vol. 1+, 1968+. Lucr. Stat. Cercet. Biol. Geol. Geogr. "Stejarul". HI 67434

Lucrări ştiinţifice din cursul anului. Institutul de cercetări horti-viticole. Bucharest. Lucr. Şti. Curs. Anului Inst. Cercet. Horti-Vitic. See B–P–H 536/17. HI 55940

Lucrări ştiinţifice, facultatea de silvicultura, institutul politehnic Braşov = Lucrări ştiinţifice, institutul politehnic Braşov (facultatea de silvicultura). Braşov.

Lucrări ştiinţifice, institutul agronomic "Dr. Petru Groza". Seria agricultura. Cluj. Vols. 1-28, 1958-72/74. Lucr. Şti. Inst. Agron. "Dr. Petru Groza", Ser. Agric. Superseded by: Buletinul institutului agronomic Cluj-Napoca. HI 67435

Lucrări ştiinţifice, institutul agronomic "N. Bălcescu". Bucharest. Vols. 1-3, 1955-60. Lucr. Şti. Inst. Agron. "N. Bălcescu". Superseded by: Lucrări ştiinţifice, institutul agronomic "N. Bălcescu". Seria A, agronomie and Seria B, Agrotecnica şi selecţia plantelor de cimp şi horti-viticole. HI 67436

Lucrări ştiinţifice, institutul agronomic "N. Bălcescu". Imbunatatiri funciare. Bucharest. Vol. 16+, 1973+. Lucr. Şti. Inst. Agron. "N. Bălcescu". Imbunat. Func. Preceded by: Lucrări ştiinţifice, institutul agronomic "N. Bălcescu". Seria A, agronomie. HI 67437

Lucrări ştiinţifice, institutul agronomic "N. Bălcescu". Seria A, agronomie. Bucharest. Vols. 4-15, 1960-72. Lucr. Şti. Inst. Agron. "N. Bălcescu," A. Preceded by: Lucrări ştiinţifice, institutul agronomic "N. Bălcescu". Superseded by: Lucrări ştiinţifice, institutul agronomic "N. Bălcescu". Imbunatatiri funciare. HI 67438

Lucrări ştiinţifice, institutul agronomic "N. Bălcescu". Seria B, Agrotecnica, fitotehnica şi selecţia plantelor de cimp şi horti-viticole (later Horticultura). Bucharest. Vol. 4+, 1960+. Lucr. Şti. Inst. Agron. "N. Bălcescu," B. Preceded by: Lucrări ştiinţifice, institutul agronomic "N. Bălcescu". HI 67439

Lucrări ştiinţifice, institutul agronomic "Prof. Ion Ionescu de la Brad." Iaşi. Vol. 1+, 1957+. Lucr. Sti. Inst. Agron. "Prof. Ion Ionescu de la Brad". HI 67440

Lucrări ştiinţifice, institutul agronomic "Tudor Vladimirescu" Craiova. Bucharest. 1958-60. Lucr. Şti. Inst. Agron "Tudor Vladimirescu" Craiova. HI 67441

Lucrări ştiinţifice, institutul politehnic Braşov (facultatea de silvicultura. Braşov. Vol. 1+, 1959?+. Lucr. Şti. Inst. Politehn. Braşov (Fac. Silvic.) HI 67442

Lucrări ştiinţifice, statiunea centrala de cercetari pentru ameliorarea nisipurilor. Bechet. Vol. ?-3+, ?-1979+. Lucr. Şti. Centr. Cercet. Pentru Amelior. Nisipur. HI 67443

Lucrările grădinii botanice din Bucureşti. Bucharest. Lucr. Grăd. Bot. Bucureşti. See B–P–H 536/16. HI 55939

Ludoviciana. Contributions de l'herbier Louis-Marie, faculté d'agriculture de l'université Laval. Quebec. Ludoviciana. See B–P–H 536/22. HI 55941

Luminescence. Vol. 1+, 1975+. Luminescence. HI 67444

Lun shu = Agricultural science. Nanking.

Lunds universitets årsskrift. Avd. 2, medicin samt matematiska och naturvetenskapliga ämnen = Acta universitatis lundensis. Nova series. Sectio 2, medica, mathematica, scientiae rerum naturalium. Lund.

Luonnon tutkija. Helsinki. Luonnon Tutkija. See B–P–H 537/6. HI 55942

Luonnon ystävä = Luonnon tutkija. Helsinki. Luonnon Tutkija. See B–P–H 537/6.

Luonnon ystavain yhdistyksen julkaisuja. Kuopio. Nos. 1-2, 1901-11. Luonnon Ystavain Yhdistyksen Julkaisuja. Superseded by: Kuopion luonnon ystavain yhdistksen. HI 67445

Lupine newsletter. Lima. Vol. 1+, 1980. Lupine Newslett. HI 67446

Luso; journal of science and technology, university of Malawi. Zomba. Vol. 1+, 1980+. Luso. HI 67447

Lustgården. Årsskrift. Föreningen för dendrologi och parkvård. Stockholm. Lustgården. See B–P–H 537/9. HI 55943

Lutukka. Helsinki. Vol. 1+, 1985+. Lutukka. HI 67448

Lycée armoricain. Nantes. Lycée Armoricain. See B–P–H 537/10. HI 55944

Lychnos; lärdomshistoriska samfundets årsbok. Uppsala & Stockholm. Lychnos. See B–P–H 537/11. HI 55945

Lyon horticole. Lyons. Lyon Hort. See B–P–H 537/12. HI 55946

Lyon-horticole et horticulture nouvelle. Lyons. Lyon-Hort. & Hort. Nouv. See B–P–H 537/13. HI 55947

Lyon médical. Lyons. Lyon Méd. See B–P–H 537/15. HI 55948

Lyonia; occasional papers of the Harold Lloyd Lyon arboretum. Honolulu, HI. Vol. 1+, 1974+. Lyonia. HI 67449

Lyubitel′ prirody. St. Petersburg. 1906-16. Lyubit. Prir. HI 67450

M A F E S research highlights. State College, MS. Vol. 36+, 1973+. M. A. F. E. S. Res. Highlights. Preceded by: Mississipi farm research. HI 67451

M A F F bulletin. London = Bulletin of the ministry of agriculture and fisheries (and food). London.

M A H A magazine. Kuala Lumpur. 1931+. M. A. H. A. Mag. 3-2494-1. HI 67452

M A P A = Medicinal and aromatic plants abstracts. New Delhi.

M A S C A journal. Philadelphia, PA. Vol. 1+, 1978+. M. A. S. C. A. J. HI 67453

M A V I S newsletter. Wooster, OH. 1973+. M. A. V. I. S. Newslett. HI 67454

M C B = Molecular and cellular biology. Washington, DC.

M C C P newsletter = Newsletter, midwest cooperative conservation program. Carbondale, IL.

M E P S = Marine ecology: progress series. Halstenbek, Amelinghausen.

M G A bulletin. Yaxley, Highfield. Nos. 1-276, 1947-72. M. G. A. Bull. Superseded by: Mushroom journal. 3-2494-3. HI 67455

M G C A newsletter. Des Moines, IA. 1981+. M. G. C. A. Newslett. HI 67456

M G G = Molecular and general genetics. Berlin.

M N H N. Paris. No. 1+, 1985+. M.N.H.N. HI 67457

M N H N noticiario mensuel = Noticiário mensual, museo nacional de história natural. Santiago de Chile.

M N R = Michigan natural resources. Lansing, MI.

M P G news. Presque Isle, ME. Vol. 1+, 1946+. M. P. G. News. 3-2495-1. HI 67459

M P M I. St. Paul, MN = Molecular plant-microbe interactions. St. Paul, MN.

Maandblad voor naturwetenschappen. Amsterdam. Maandbl. Natuurw. See B–P–H 537/19. HI 55951

Maandblad van de nederlandsche natuurhistorische vereeniging. Deventer, Netherlands. Maandbl. Ned. Natuurhist. Ver. See B–P–H 537/20. HI 55952

Maandblad voor nederlandsche pomologische vereeniging. Utrecht. Maandbl. Ned. Pomol. Ver. See B–P–H 537/21. HI 55953

Maandblad, uitgegeven door het natuurhistorisch genootschap in Limburg. Maastricht. Maandbl. Natuurhist. Genootsch. Limburg. See B–P–H 537/18. HI 55950

Maandblad van de vereeniging ter bevordering van tuin- en landbouw in Limburg. Maastricht. Maandbl. Ver. Beford. Tuin- Landb. Limburg. See B–P–H 537/22. HI 55954

Maandelijksche mededeelingen van den bond van bloembollenhandelaren. Haarlem. Maandel. Meded. Bond Bloembollenhandelaren. See B–P–H 537/23. HI 55955

Maandschrift van de nederlandsche m[aatschapp]ij voor tuinbouw en plantkunde. The Hague. Maandschr. Ned. Maatsch. Tuinb. See B–P–H 537/24. HI 55956

Maandschrift voor tuinbouw. Dordrecht, Netherlands. Maandschr. Tuinb. See B–P–H 537/25. HI 55957

Maanedskriftet Minerva. Copenhagen. 1799-1805. Maanedskr. Minerva. Preceded by: Minerva et maanedskrivt. Superseded by: Maanedskriftet ny Minerva. 4-3123-3. HI 53956

Maanedskriftet Iris. Copenhagen. 1791-95. Maanedskr. Iris. Superseded by: Iris og Hebe. HI 67460

Maanedsoversigt over plantesygdomme. Copenhagen. Maanedsovers. Plantesygd. See B–P–H 538/1. HI 55958

Maanedsskrift for huusmaend. Randers. Vols. 1-4, 1872-75. Maanedsskr. Huusm. HI 67461

Maanedsskrift for litteratur. Copenhagen. Maanedsskr. Litt. See B–P–H 538/3. HI 55959

Maatalouden tutkimuskeskuksen aikakauskirja = Annales agriculturae Fenniae. Helsinki. Ann. Agric. Fenniae. See B–P–H 96/6.

Maataloustieteellinen aikakauskirja. Helsinki. Vol. 1+, 1929+. Maataloust. Aikak. 3-2496-3. HI 67462

McGraw-Hill's biotechnology newswatch. New York. 1981+. McGraw-Hill's Biotechnol. Newswatch. HI 67463

McIlvainea; official organ of the north american mycological association. Portsmouth, OH. Vol. 1+, 1972+. McIlvanea. Preceded by: Mycophile. HI 67464

Macmillan's magazine. London. Vols. 1-92, 1859-1905; n.s. vols. 1-2, 1905-07. Macmillan's Mag. 3-2503-2. HI 67465

Macpalxochitl; boletin bimestrial de la sociedad botanica de Mexico. Mexico City. No. 23+, 1970+. Macpalxochitl. Preceded by: Boletín informativo, sociedad botánica de México. HI 67466

Madagascar: recherches scientifiques; bulletin trimestriel de l'association pour l'avancement de la recherche scientifique concernant Madagascar et la région malgache. Paris. No. 1+, 1977+. Madagascar Rech. Sci. HI 67467

Madagascar revue de géographie. Antananarivo. ?-1980+. Madagascar Rev. Géogr. HI 67468

Maderero. Santiago. Maderero. See B–P–H 538/5. HI 55960

Madhya bharati. Sagar. Vols. 8-19/21, 1959-71/73. Madhya Bharati. Preceded by: Journal of the university of Saugar. Part 2, section B, natural sciences. HI 67469

Madjalah perusahaan gula. Pasuruan. Vol. 1+, 1965+. Madj. Perus. Gula. HI 67470

Madjallah ilmu alam untuk Indonesia. Jakarta, Indonesia. Madj. Ilmu Alam Untuk Indonesia. See B–P–H 538/6. HI 55961

Madoqua; tydskrif vor natuurbewaringsnavorsing Suidwest-Afrika. (Journal of nature conservation research, South West Africa.) Windhoek. No. 1+, 1969+. Madoqua. HI 67471

Madoqua. Series 2, scientific papers of the Namib desert research station. Windhoek. Vol. 1+, 1972+. Madoqua, 2, Sci. Pap. Namib Desert Res. Sta. Preceded by: Scientific papers of the Namib desert research station. HI 67472

Madras agricultural journal. Coimbatore. Vol. 17+, 1929+. Madras Agric. J. Preceded by: Journal, Madras agricultural student's union. 3-2505-1. HI 67473

Madras journal of literature and science. Madras. Madras J. Lit. Sci. See B–P–H 538/10. HI 55962

Madroño; journal of the California botanical society. Berkeley, CA. Madroño. See B–P–H 538/18. HI 55963

Maendelyke uittreksels, of boekzael der geleerde werelt. Amsterdam. Vols. 1-194, 1715-1811. Maendel. Uittr. Superseded by: Boekzaal der geleerde wereld. 3-2496-1. HI 52693

Magasin encyclopédique, ou journal des sciences, des lettres et des arts = Magazin encyclopédique, ou journal des sciences, des lettres et des arts. Paris. Mag. Encycl. See B–P–H 539/4.

Magasin d'horticulture. Liége. Mag. Hort. (Liége). See B–P–H 539/21. HI 55981

Magazijn voor landbouw en kruidkunde. Utrecht. Mag. Landb. Kruidk. See B–P–H 540/3. HI 55986

Magazijn voor wetenschappen, kunsten en letteren. Amsterdam. Mag. Wetensch. See B–P–H 541/16. HI 56014

Magazin vor Aerzte. Leipzig. Mag. Aerzte. See B–P–H 538/19. HI 55964

Magazin für allgemeine Natur- und Thier-Geschichte. Gottingen & Leipzig. 1788-96. Mag. Allg. Natur Thier-Gesch. HI 67474

Magazin über Asien oder Nachrichten von den Sitten und Gebräuchen, den Wissenschaften und Künsten, den Handwerken und Gewerben, der Denkart und Der Religion der Asiaten, von den Thieren, den Pflanzen, den Mineralien, dem Boden und dem Clima von Asien. Leipzig. Mag. Asien. See B–P–H 538/20. HI 55965

Magazin der ausländischen Literatur der gesammten Heilkunde, und Arbeiten des Aerztlichen Vereins zu Hamburg. Hamburg. Mag. Ausl. Lit. Gesammten Heilk. See B–P–H 538/21. HI 55966

Magazin für die Botanik. Zurich = Botanisches Magazin. [Edited by Römer & Usteri.] Zurich. Bot. Mag. (Römer & Usteri). See B–P–H 221/16.

Magazin des Buch- und Kunsthandels, ... Leipzig. Mag. Buch- Kunsthandels. See B–P–H 538/29. HI 55968

Magazin von und für Deutschland. Mag. Deutschl. See B–P–H 539/1. HI 55970

Magazin von und für Dortmund. Dortmund. Mag. Dortmund. See B–P–H 539/3. HI 55971

Magazin encyclopédique, ou journal des sciences, des lettres et des arts. Paris. Mag. Encycl. See B–P–H 539/4. HI 55972

Magazin för föreldrar och lärare. Mag. Föreldrar Lärare. See B–P–H 539/8. HI 55973

Magazin für Freunde guten Geschmacks, der bildenden und mechanischen Künste, Manufakturen und Gewerbe. Leipzig. Mag. Freunde Guten Geschmacks. See B–P–H 539/10. HI 55974

Magazin für Freunde der Naturlehre und Naturgeschichte, Scheidekunst, Land- und Stadtwirthschaft, Volks- und Staatsarznei. Berlin. Mag. Freunde Naturl. See B–P–H 539/11. HI 55975

Magazin für die Geographie, Staatenkunde und Geschichte. Nuremberg. Mag. Geogr. See B–P–H 539/15. HI 55976

Magazin für die gerichtliche Arzneikunde und medicinsche Polizei. Stendal, Germany. Mag. Gerichtl. Arzneik. Med. Polizei. See B–P–H 539/16. HI 55977

Magazin für die gesammte Heilkunde. Berlin. Mag. Gesammte Heilk. See B–P–H 539/17. HI 55978

Magazin für Heilkunde und Naturwissenschaft in Polen. Mag. Heilk. Naturwiss. Polen. See B–P–H 539/20. HI 55980

Magazin der italienischen LItteratur und Künste. Weimar. Mag. Ital. Litt. Künste. See B–P–H 540/1. HI 55984

Magazin für das Jagd- und Forstwesen. Leipzig. Mag. Jagd- Forstwesen. See B–P–H 540/2. HI 55985

Magazin für die Literatur des Auslandes. Berlin. Mag. Lit. Ausl. See B–P–H 540/4. HI 55987

Magazin für die Naturkunde Helvetiens. Zurich. Mag. Naturk. Helv. See B–P–H 540/13. HI 55992

Magazin für die Naturkunde und Oekonomie Meklenburgs. Schwerin & Leipzig. Mag. Naturk. Oekon. Meklenburgs. See B–P–H 540/14. HI 55993

Magazin for naturkundskab. Copenhagen. Mag. Naturkundsk. See B–P–H 540/16. HI 55994

Magazin for naturvidenskaberne. Christiania [=Oslo, Norway]. Mag. Naturvidensk. See B–P–H 540/18. HI 55995

Magazin aller neuen Erfindungen, Entdeckungen und Verbesserungen, ... Leipzig. Mag. Neuen Erfind. See B–P–H 540/19. HI 55996

Magazin für die neueste Geschichte der evangelischen Missions- und Bibelgesellschaften. Basel. Mag. Neueste Gesch. Evang. Missions- Bibelges. See B–P–H 540/24. HI 55997

Magazin für Neueste aus der Physik und Naturgeschichte. Gotha. Mag. Neueste Phys. Naturgesch. See B–P–H 540/25. HI 55998

Magazin der neuesten und besten ausländischen Reisebeschreibungen. Hamburg & Mainz. Mag. Neuesten Besten Ausl. Reisebeschreib. See B–P–H 541/1. HI 55999

Magazin für die neuesten Entdeckungen in der gesammten Naturkunde, Gesellschaft naturforschender Freunde zu Berlin = Der Gesellschaft naturforschender Freunde zu Berlin Magazin für die neuesten Entdeckungen in der gesammten Naturkunde. Berlin. Ges. Naturf. Freunde Berlin Mag. Neuesten Entdeck. Gesammten Naturk. See B–P–H 400/6.

Magazin für die neuesten Erfahrungen, Entdeckungen und Berichtigungen im Gebiete der Pharmacie. Karlsruhe. Mag. Neuesten Erfahr. Pharm. See B–P–H 541/3. HI 56001

Magazin der neuesten Erfahrungen in der gesammten Naturkunde. Berlin. Mag. Neuesten Erfahr. Gesammten Naturk. See B–P–H 541/2. HI 56000

Magazin der neuesten Erfindungen, Entdeckungen und Verbesserungen, ... Leipzig. Mag. Neuesten Erfind. See B–P–H 541/4. HI 56002

Magazin der neuesten und interessantesten Reisebeschreibungen. Vienna. Mag. Neuesten Interessantesten Reisebeschreib. See B–P–H 541/5. HI 56003

Magazin für den neuesten Zustland der Naturkunde mit Rücksicht auf die dazugehörigen Hülfswissenschaften. Jena. Mag. Neuesten Zustand Naturk. See B–P–H 541/6. HI 56004

Magazin für Pflanzenliebhaber und Maler. Zwickau. Mag. Pflanzenliebh. Maler. See B–P–H 541/7. HI 56005

Magazin des Pflanzenreiches. Erlangen. Mag. Pflanzenr. See B–P–H 541/8. HI 56006

Magazin für Pharmacie und die dahin einschlagenden Wissenschaften. Karlsruhe. Mag. Pharm. See B–P–H 541/9. HI 56007

Magazin für Pharmacie in Verbindung mit einer Experimental-Kritik = Magazin für Pharmacie und die dahin einschlagenden Wissenschaften, vols. 34-36, 1831. Karlsruhe. Mag. Pharm. See B–P–H 541/9.

Magazin für Pharmazie, Botanik und Materia Medica. Halle. Mag. Pharm. Bot. See B–P–H 541/10. HI 56008

Magazin für Westphalen. Dortmund. Mag. Westphalen. See B–P–H 541/15. HI 56013

Magazine of botany and gardening, british and foreign. London. Mag. Bot. Gard. See B–P–H 538/25. HI 55967

Magazine of the cactus and succulent society of Malta = Kakti u sukkulenti ohra. Zejtun.

Magazine; college of agriculture. Dhulia, India. Vol. 1+, 1962+. Mag. Coll. Agric., Dhulia. HI 55969

Magazine, college of agriculture, Nagpur = Nagpur agricultural college magazine. Nagpur.

Magazine for greenhouse culture. [Onshitsu kenkyu.] Mag. Greenh. Cult. See B–P–H 539/18. HI 55979

Magazine of horticulture, botany and all useful discoveries and improvements in rural affairs. Boston, MA, & New York, NY. Mag. Hort. Bot. See B–P–H 539/22. HI 55982

Magazine of natural history. London. Mag. Nat. Hist. See B–P–H 540/6. HI 55988

Magazine of natural history, and journal of zoology, botany, mineralogy, geology, and meteorology. London. Mag. Nat. Hist. & J. Zool. See B–P–H 540/8. HI 55989

Magazine of natural history and naturalist. London. Mag. Nat. Hist. & Naturalist. See B–P–H 540/9. HI 55990

Magazine of natural philosophy. London. Mag. Nat. Philos. See B–P–H 540/11. HI 55991

Magazine of popular science, and journal of the useful arts. London. Mag. Popular Sci. & J. Useful Arts. See B–P–H 541/11. HI 56009

Magazine of science and artists, architects, and builders' (later miners') journal; intended to illustrate the most useful, novel, and interesting parts of natural history and experimental philosophy, artistical processes, ornamental manufactures, and the arts of life. London. Vols. 12-15, 1850-52. Mag. Sci. Artists Architects Builders' J. Preceded by: Magazine of science and school of arts. 3-2514-1. HI 67475

Magazine of science and school of arts; intended to illustrate the most useful, novel, and interesting parts of natural history and experimental philosophy, artistical processes, ornamental manufactures, and the arts of life. London. Vols. 1-11, 1840-49. Mag. Sci. School Arts. Superseded by: Magazine of science and artists, architects and builders' journal. 3-2514-1. HI 67476

Magazine, Suffolk trust for nature conservation. Saxmundham. 1985+. Mag. Suffolk Trust Nat. Conservation. Preceded by: Newsletter, Suffolk trust for nature conservation. HI 67477

Magazine of zoology and botany. Edinburgh. Mag. Zool. Bot. See B–P–H 541/18. HI 56015

Magazzino italiano che contiene storia ... Livorno. Mag. Ital. See B–P–H 539/23. HI 55983

Magazzino toscano. Florence. Mag. Tosc. See B–P–H 541/12. HI 56010

Magazzino toscano d'instruzione e di piacere. Livorno. Mag. Tosc. Instruz. Piacere. See B–P–H 541/13. HI 56011

Magnolia. Nanuet, NY. Vol. 16+, 1980+. Magnolia. Preceded by: Newsletter of the american magnolia society. HI 67478

Magon. Série scientifique. Tel-Amara. No. 1+, 1965+. Magon, Sér. Sci. HI 67479

Maguinaria. Caracas. Maguinaria. See B–P–H 541/23. HI 56016

Magyar academiai értesítö. Budapest. Magyar Acad. Értes. See B–P–H 542/16. HI 56017

Magyar akademiai értesitö. A mathematikai és természettudományi osztályok közlönye. Pest [=Budapest, in part]. Magyar Akad. Értes. Math. Természettud. Oszt. Közl. See B–P–H 542/17. HI 56018

Magyar állami földtani intézet évi jelentese(i). Budapest. Magyar Állami Földt. Intéz. Évi Jel. See B–P–H 542/18. HI 56019

Magyar állami földtani intézet évkönyve. Budapest. Magyar Állami Földt. Intéz. Évk. See B–P–H 542/19. HI 56020

Magyar biológiai kutatóintézet évkönyve. Tihany, Hungary. Magyar Biol. Kutatóint. Évk. See B–P–H 542/21. HI 56022

Magyar biológiai kutatóintézet évi jelentése. Tihany, Hungary. Magyar Biol. Kutatóint. Évi Jel. See B–P–H 542/20. HI 56021

Magyar biológiai kutatóintézet munkái. Tihany, Hungary. Magyar Biol. Kutatóint. Munkái. See B–P–H 542/22. HI 56023

Magyar bor és gyümölcs. Budapest. Magyar Bor Gyüm. See B–P–H 542/23. HI 56024

Magyar botanikai lapok. Budapest. Magyar Bot. Lapok. See B–P–H 542/25. HI 56025

Magyar dohányujság. Budapest. Magyar Dohányujs. See B–P–H 543/1. HI 56026

Magyar erdész. Jolsva, Hungary [=Jelsava, Czechoslovakia]. Magyar Erdész (Jolsva). See B–P–H 543/2. HI 56027

Magyar erdész. Ungvar, Hungary [=Uzhgorod, Ukrainian S S R]. Magyar Erdész (Ungvar). See B–P–H 543/3. HI 56028

Magyar flóramüvek. Debrecen, Kolozsvar [=Cluj, Rumania], Pecs, & Veszprem, Hungary. Magyar Flóramüvek. See B–P–H 543/4. HI 56029

Magyar gazdák szemléje. Budapest. Magyar Gazd. Szemléje. See B–P–H 543/6. HI 56031

Magyar gazdasági növényvédelem. Grinad, & Poszony [=Bratislava, Czechoslovakia]. Magyar Gazd. Növényvéd. See B–P–H 543/5. HI 56030

Magyar gombászati lapok. Budapest. 1944-47. Magyar Gombász. Lapok. HI 67481

Magyar gyógyszerésztudományi társaság értesitöje. Magyar Gyógyszerésztud. Társ. Értes. See

B–P–H 543/7. HI 56032

Magyar gyümölcz. Budapest. Magyar Gyüm. See B–P–H 543/8. HI 56033

Magyar hirmondó. Bécs [=Vienna]. Magyar Hirmondó (Bécs). See B–P–H 543/9. HI 56034

Magyar hirmondó. Pozsony [=Bratislava, Czechoslovakia]. Magyar Hirmondó (Pozsony). See B–P–H 543/10. HI 56035

Magyar kertész. Budapest. Magyar Kert. (Budapest). See B–P–H 543/12. HI 56037

Magyar kertész. Nagyszöllös, Hungary [=Vinogradov, Ukrainian S S R]. Magyar Kert. (Nagyszöllös). See B–P–H 543/13. HI 56038

Magyar kertész. Pest [=Budapest, in part]. Magyar Kert. (Pest). See B–P–H 543/14. HI 56039

Magyar kertészet és faiskola. Budapest. Magyar Kert. Faisk. See B–P–H 543/11. HI 56036

Magyar kir[ályi] állami földtani intézet évkönyve. Budapest. Magyar Kir. Állami Földt. Intéz. Évk. See B–P–H 543/15. HI 56040

Magyar királyi állami földtani intézet évi jelentése(i). Budapest. Magyar Kir. Állami Földt. Intéz. Évi Jel. See B–P–H 543/16. HI 56041

M[agyar] kir[ályi] kertészeti akadémia közleményei. Budapest. Magyar Kir. Kert. Akad. Közlem. See B–P–H 543/17. HI 56042

M[agyar] kir[ályi] kertészeti és szölészeti föiskola közleményei. Budapest. Magyar Kir. Kert. Szölész. Föisk. Közlem. See B–P–H 543/18 HI 56043

M[agyar] kir[ályi] kertészeti tanintézet közleményei. Budapest. Magyar Kir. Kert. Tanintéz. Közlem. See B–P–H 544/1. HI 56044

Magyar királyi központi szölészeti kísérleti állomás és ampelologiai intézet évkönyve. Budapest. Magyar Kir. Közp. Szölész. Kísérl. Állomás Ampelol. Intéz. Évk. See B–P–H 544/2. HI 56045

M[agyar] kir[ályi] központi szölészeti kísérleti állomás és ampelologiai intézet közleményei. Budapest. Magyar Kir. Közp. Szölész. Kísérl. Állomás Ampelol. Intéz. Közlem. See B–P–H 544/3. HI 56046

Magyar könyvház. Magyar Könyvház. See B–P–H 544/4. HI 56047

Magyar méh. Budapest. Magyar Méh (Budapest). See B–P–H 544/5. HI 56048

Magyar méh. Buziasfürdö, Hungary [=Buzias, Rumania]. Magyar Méh (Buziasfürdö). See B–P–H 544/6. HI 56049

Magyar mezögazdaság. Budapest. Magyar Mezögazd. See B–P–H 544/7. HI 56050

Magyar muzeum. Pest & Kassa, Hungary [=Budapest, in part, & Kosice, Czechoslovakia]. Magyar Muz. See B–P–H 544/9. HI 56051

Magyar nemzeti múseum Albrecht kir. herceg biológiai állomásának kiadványa = Albertina. Belje, Yugoslavia. Albertina. See B–P–H 68/13.

Magyar nemzeti múzeum természetrajzi osztályainak folyóirata = Annales historico-naturales musei nationalis hungarici. Budapest. Ann. Hist.-Nat. Mus. Natl. Hung. See B–P–H 103/22.

Magyar nemzeti múzeum természettundományi múzeum évkönyve = Annales historico-naturales musei nationalis hungarici. Budapest. Ann. Hist.-Nat. Mus. Natl. Hung. See B–P–H 103/22.

Magyar növénytani lapok. Kolozsvar [=Cluj, Rumania]. Magyar Növénytani Lapok. See B–P–H 544/13. HI 56052

Magyar orvosok és természetvizsgálók nagy gyülésémek munkálatai. [Hungary.] Magyar Orv. Termész. Nagy Gyül. Munk. See B–P–H 544/14. HI 56180

Magyar orvosok és természetvizsgálók nagy gyülésének történeti vázlata és munkálatai. [Hungary.] Magyar Orv. Termész. Nagy Gyül. Tört. Vázl. Munk. See B–P–H 544/15. HI 56053

Magyar orvosok és természetvizsgálók vándorgyülésének történeti vázlata és munkálatai. [Hungary.] Magyar Orv. Termész. Vándorgyül. Tört. Vázl. Munk. See B–P–H 544/16. HI 56054

Magyar tájak növénytakarója = Die Vegetation ungarischer Landschaften. Budapest. Veg. Ung. Landschaften. See B–P–H 948/15.

Magyar tudomány. Budapest. Magyar Tud. See B–P–H 545/2. HI 56055

Magyar tudományegyejetenek biológiai intézeteinek évkönyve = Annales biologicae universitatum Hungariae. Budapest. Ann. Biol. Univ. Hungariae. See B–P–H 97/18.

Magyar tudományos akadémia agrártudományok osztálya monográfia sorozata. Budapest. Magyar Tud. Akad. Agrártud. Oszt. Monogr. Soroz. See B–P–H 545/3. HI 56056

Magyar tudományos akadémia biológiai és agrártudományi osztályanak közleményei. Budapest. Magyar Tud. Akad. Biol. Agrártud. Oszt. Közlem. See B–P–H 545/4. HI 56057

Magyar tudományos akadémia biológiai csoportjának közleményei. Budapest. Magyar Tud. Akad. Biol. Csoport. Közlem. See B–P–H 545/5. HI 56058

Magyar tudományos akadémia biológiai közleményei = Acta biologica academiae scientiarum hungaricae. Budapest.

Magyar tudományos akadémia biológiai osztályának

közleményei. Budapest. Magyar Tud. Akad. Biol. Oszt. Közlem. See B–P–H 545/6. HI 56059

Magyar tudományos akadémia biológiai tudományok osztályának közleményei. Budapest. Magyar Tud. Akad. Biol. Tud. Oszt. Közlem. See B–P–H 545/7. HI 56060

Magyar tudományos akadémia botanikai közleményei = Acta botanica academiae scientiarum hungaricae. Budapest.

Magyar tudományos akadémia értesitöje. Budapest. Magyar Tud. Akad. Értes. See B–P–H 545/8. HI 56061

Magyar tudományos akadémia évkönyvei. Buda [=Budapest, in part]. Magyar Tud. Akad. Évk. See B–P–H 545/9. HI 56062

Magyar tudományos akadémia mikrobiológiai közleményei = Acta microbiologica academiae scientiarum hungaricae. Budapest.

A magyar tudományos akadémia tihanyi biológiai kutatóintézetének évkönyve. Tihany. Vols. 20-?, 1950-77. Magyar Tud. Akad. Tihanyi Biol. Kutatóint. Evk. Preceded by: Magyar biológiai kutatóintézet evkönyve. HI 67482

Magyar tudós társaság évkönyvei. Buda [=Budapest, in part]. Magyar Tud. Társ. Évk. See B–P–H 545/12. HI 56063

Magyar ujság. Bécs [=Vienna]. Magyar Ujs. See B–P–H 545/14. HI 56064

Magyarhoni természetbarát. Nyitra, Hungary. Magyarhoni Természetbarát. See B–P–H 545/15. HI 56065

Magyarország bortermesztéset 's készitését tárgyazó folyóirás. Buda [=Budapest, in part]. Magyarorsz. Borterm. Készit. Tárgyazó Folyóir. See B–P–H 545/17. HI 56066

Magyarország kulturflórája. Budapest. Magyarorsz. Kulturfl. See B–P–H 545/19. HI 56068

Magyarország virágtalan növéneinek meghatározo kézikönyve. Budapest. Magyarorsz. Virágtalan Növén. Meghat. Kézikönyve. See B–P–H 545/20. HI 56069

Magyarországi Kárpát-egylet évkönyve. Kassa, Kesmark, & Iglo, Hungary [=Kosice, Kezmarok, & Spisska Nova Ves, Czechoslovakia]. Magyarorsz. Kárpát-Egyl. Évk. See B–P–H 545/18. HI 56067

M.A.H.A. magazine. Kuala Lumpur = M A H A magazine. Kuala Lumpur.

Ma'had al-Kuwayt li-al-akhat al-ilmiyyah. Kuwait City. 1979+. Ma'had al-Kuwayt li-al-akhat al-ilmiyyah HI 67483

Mahasagar; bulletin, national institute of oceanography, C S I R. New Delhi. Vol. 1+, 1968+. Mahasagar. HI 67484

Maharashtra Vidnyan Mandir patrika. Poona. 1966+. Maharashtra Vidnyan Mandir Patrika. HI 67485

Mahlaqa le-heqer 'asavim ba'im = Triennial report, division of weed research, institute of plant protection, Israel.

Maine agricultural experiment station. Abstracts of recent publications. Orono, ME. Maine Agric. Exp. Sta. Abstr. Recent Publ. See B–P–H 546/5. HI 56070

Maine agricultural experiment station. Annual report. Bangor, ME. 1885/86-1968? Maine Agric. Exp. Sta. Annual Rep. Superseded by: Report, life sciences and agricultural experiment station, university of Maine at Orono. HI 67486

Maine agricultural experiment station. Bulletin. Orono, ME. Nos. 1-26, 1885-86; ser. 2, nos. 1-684, 1889-1970. Maine Agric. Exp. Sta. Bull. Superseded by: Bulletin, Maine life sciences and agricultural experiment station. HI 67487

Maine agricultural experiment station. Bulletin, technical series. Orono, ME. Nos. 1-25, 1962-67. Maine Agric. Exp. Sta. Bull., Techn. Ser. Superseded by: Technical bulletin, life sciences agriculture experiment station, university of Maine at Orono. HI 67488

Maine agricultural experiment station. Mimeographed report. Orono, ME. Maine Agric. Exp. Sta. Mimeogr. Rep. See B–P–H 546/10. HI 56071

Maine applegrower. South Portland, ME. Maine Applegrower. See B–P–H 546/11. HI 56072

Maine farm research. Agricultural experiment station. Orono, ME. Maine Farm Res. See B–P–H 546/12. HI 56073

Maine forest facts. Bangor, ME. Maine Forest Facts. See B–P–H 546/14. HI 56075

Maine forester. Orono, ME. Maine Forest. See B–P–H 546/13. HI 56074

Maine naturalist; journal of the Knox academy of arts and sciences. Thomaston, Augusta, ME. Vols. 1-10(3), 1921-30. Maine Naturalist. HI 67489

Maine potato grower. Caribou, ME. Maine Potato Grower. See B–P–H 546/15. HI 56076

Maine potato yearbook. Fort Fairfield, ME. Maine Potato Yearb. See B–P–H 546/17. HI 56077

Maine state pomological society. Annual report. Augusta, ME. Maine State Pomol. Soc. Annual Rep. See B–P–H 546/19. HI 56078

Mainly about wildlife = Wildlife and the countryside. London.

Mainzer Anzeigen von gelehrten Sachen. Mainz. Mainzer Anz. Gel. Sachen. See B–P–H 546/20. HI 56079

Mainzer naturwissenschaftliches Archiv. Mainz. Vol. 5+, 1967+. Mainzer Naturwiss. Archiv. Preceded by: Zeitschrift der rheinischen naturforschenden Gesellschaft in Mainz. HI 67490

Mainzer naturwissenschaftliches Archiv. Beiheft. Mainz. Vol. 1+, 1978+. Mainzer Naturwiss. Archiv, Beiheft. HI 67491

Maize genetics cooperation news letter. Bloomington, IN, Ithaca, NY. 1929+. Maize Genet. Coop. Newslett. HI 67492

Maize quality protein abstracts. London. Vol. 1+, 1975+. Maize Qual. Protein Abstr. HI 67493

Majallah al-misriyah lil-mikrublyuluzhiya = Egyptian journal of microbiology. Cairo.

Majallat al-khalij al-Arabi lil-bu huth al-ilmiyah = Arab gulf journal of scientific research. Cairo.

Majallat al-mikrubiyul uzhiyalil-jumh uriyah al-Arabiyah at-Mutta hidah = United Arab Republic journal of microbiology. Cairo.

Majallat al-ulum al-saydal iyah lil-jumh ur iyah al-arab iyah al-mutta hidah = Journal of pharmaceutical sciences of the United Arab Republic. Cairo.

Majallat kulliyat al-'ulum = Bulletin of the college of science, university of Baghdad. Baghdad.

Majallat kulliyat al-ulum jami at al-Riyad if = Journal of the faculty of science, university of Riyadh. Riyadh.

Majallat kulliyat ulum al-Bihar = Journal of the faculty of marine science, King Abdulaziz university. Jeddah.

Makiling echo. Santa Cruz?, Philippines. Makiling Echo. See B–P–H 546/23. HI 56080

Maladies des plantes. St. Petersburg = Zhurnal "Bolĕzni Rastenii". St. Petersburg.

Malayan agricultural journal. Kuala Lumpur. Vols. 11-44, 1912-64. Malayan Agric. J. Preceded by: Agricultural bulletin of the Straits and Federated Malay States. Superseded by: Malaysian agricultural journal. 3-2527-1. HI 67494

Malayan agriculturist. Kuala Lumpur. Vols. 1-12, 1961-74. Malayan Agric. HI 67495

Malayan forest records. Singapore & Calcutta. Malayan Forest Rec. See B–P–H 547/10. HI 56081

Malayan forester. Kuala Lumpur. Vols. 1-35, 1931-72. Malayan Forester. Superseded by: Malaysian forester. 3-2527-1. HI 67496

Malayan garden plants. Singapore. Malayan Gard. Pl. See B–P–H 547/13. HI 56082

Malayan journal of tropical geography. Singapore. Malayan J. Trop. Geogr. See B–P–H 547/14. HI 56083

Malayan miscellanies. Bencoolen [=Benkulen, Indonesia]. Malayan Misc. See B–P–H 547/15. HI 56084

Malayan naturalist. [Supplement to: Malayan nature journal.] Kuala Lumpur. Vol. 33(2)+, 1979+. Malayan Naturalist. HI 67497

Malayan nature journal. Kuala Lumpur. Malayan Nat. J. See B–P–H 547/17. HI 56085

Malayan orchid review. Singapore Vol. 1+, 1931. Malayan Orchid Rev. 3-2527-2. HI 67498

Malayan science bulletin. Singapore & Calcutta. Malayan Sci. Bull. See B–P–H 547/20. HI 56086

Malayan scientist; journal of the persatuan sains university of Malaya. Kuala Lumpur. No. 1+, 1963/64+, 1964+. Malayan Sci. HI 67499

Malaysian agricultural journal. Kuala Lumpur. Vol. 45+, 1965+. Malaysian Agric. J. Preceded by: Malayan agricultural journal. HI 67500

Malaysian agricultural research. Kuala Lumpur. Vols. 1-5, 1972-76. Malaysian Agric. Res. Superseded by: Malaysian applied biology. HI 67501

Malaysian applied biology. Kuala Lumpur. Vol. 6+, 1977+. Malaysian Appl. Biol. Preceded by: Malaysian agricultural research. HI 67502

Malaysian forester. Kepong. Vol. 36+, 1973+. Malaysian Forester. Preceded by: Malayan forester. HI 67503

Malaysian journal of science. Kuala Lumpur. Vol. 1+, 1972+. Malaysian J. Sci. HI 67504

Malaysian journal of tropical geography. Kuala Lumpur. Vol. 1+, 1980+. Malaysian J. Tropl Geogr. Preceded by: Journal of tropical geography. HI 67505

Malaysian pineapple. Pekan Nevas. Vols. 1-2, 1971-72. Malaysian Pineapple. HI 67506

Malesia. [Edited by O. Beccari.] Genoa. Malesia. See B–P–H 547/21. HI 56087

Malomipar és terményforgalom. Budapest. Vol. ?-21, ?-1974. Malom. Terményforg. Superseded by: Gabonaipar. HI 67507

Malpighia; rassegna mensile di botanica. Messina, Italy. Malpighia. See B–P–H 547/22. HI 56088

Maltese naturalist. Sliema. Nos. 1-2(4), 1970-76. Maltese Naturalist. Incorporated: News and views, natural history society of Malta. Superseded by: Central mediterranean naturalist. HI 67508

Malus. Lisle, IL. Vol. 1(2)+, 1985+. Malus. Preceded by: Crab gab. HI 67509

Mammillaria. Prague. 1971+. Mammillaria. HI 67510

Mammillariae = Informationsbrief der zentralen Arbeitsgemeinschaft "Mammillaria". Potsdam.

Man chu kuo litsi Kung chu ling. Noji shikenjo kenkyu

jiho. Kunchuling. = Research bulletin of the agricultural experiment station, Kung-chu-ling. Kunchuling.

Management and nature = Amenagement et nature. Paris.

Manchester museum handbooks = Museum handbooks, Manchester museum. Manchester.

Manchester museum publications = Museum publications, Manchester museum. Manchester.

Manchester museum publications = Publications, Manchester museum. Manchester.

Manchester naturalists' field club annual report. Manchester, England. Manchester Naturalists' Field Club Annual Rep. See B–P–H 547/26. HI 56089

Manchu seibutsu gakkai kaiho = Transactions of the biological society of Manchoukuo. Mukden, Manchoukuo [China]. Trans. Biol. Soc. Manchoukuo. See B–P–H 881/7.

Mangroves et zone côtière; bulletin de liaison du groupe de travail, D G R S T. Petit-Bourg. Vol. 1+, 1977+. Mangroves Zone Côt. HI 67511

Manitoba nature. Winnipeg. Vols. 13(2)-14(4), 1972-82. Manitoba Nat. Preceded by: Zoolog [not entered]. HI 67512

Mannheimer Verein für Naturkunde. Jahresbericht. Mannheim. Mannheimer Verein Naturk. Jahresber. See B–P–H 547/32. HI 56090

Mannichfaltigkeiten. Berlin. Mannigfaltigkeiten. See B–P–H 548/3. HI 56092

Mannichfaltigkeiten aus dem Gebiete der Literatur, Kunst und Natur. Stuttgart. Mannichfaltigk. Lit. See B–P–H 548/1. HI 56091

Manshu hakubutsu doko-kai kaiho = Transactions of the natural history society of Manchuria. Mukden?, Manchoukuo [China]. Trans. Nat. Hist. Soc. Manchuria. See B–P–H 886/22.

Manshu igaku zasshi = Journal of oriental medicine. Dairen, China. J. Orient. Med. See B–P–H 475/11.

Manshu teikoku kokuritsu chuo hakubutsu kan runso = Bulletin of the central national museum of Manchoukuo. Hsinginjg, Manchoukuo [=Changchun, China]. Bull. Centr. Natl. Mus. Manchoukuo. See B–P–H 244/2.

Manurial times. [Hiryo jiho.] Tokyo. Manurial Times. See B–P–H 548/7. HI 56093

Mapas provinciales de suelos: información. Madrid. 1962+. Mapas Prov. Suelos Inform. HI 67513

Marathwada university journal of science. Aurangabad, Maharashtra. Vol. 1+, 1960+. Marathwada Univ. J. Sci. HI 67514

Marburger geographische Schriften. Marburg. No. 1+, 1949+. Marburg. Geogr. Schriften. HI 67515

Marburgische Beyträge zur Gelehrsamkeit nebst den Neuigkeiten der Universitäten Marburg und Rinteln, ... Marburg. Marburg. Beytr. Gelehrsamk. See B–P–H 548/14. HI 56098

Marcellia; rivista internazionale di cecidologia. Avellino, Italy. Marcellia. See B–P–H 548/16. HI 56099

Mardi report. Serdang. No. 1+, 1973+. Mardi Rep. HI 67516

Margaulx, Le. Perce, Québec. Vol. 1+, 1976+. Margaulx. HI 67517

Marine aquarist. Bayside, NY. Vols. 1-8(7), 1970-78. Mar. Aquar. HI 67518

Marine biological report; Union of South Africa. Cape Town. Mar. Biol. Rep. See B–P–H 548/8. HI 56094

Marine biology; international journal on life in oceans and coastal waters. Berlin, New York, etc. Vol. 1+, 1967+. Mar. Biol. HI 67519

Marine biology letters. Amsterdam. Vol. 1+, 1979+. Mar. Biol. Lett. HI 67520

Marine biotechnology abstracts. Bethesda, MD. Vol. 1(1-2), 1989. Mar. Biotechnol. Abstr. Superseded by: A F S A marine biotechnology abstracts. HI 75280

Marine briefs. Ocean Springs, MS. Vol. 1+, 1972+. Mar. Briefs. HI 67521

Marine ecology. [Forms section 1 of: Pubblicazioni della stazione zoologica di Napoli.] Berlin, Hamburg. Vol. 1+, 1980+. Mar. Ecol. HI 67522

Marine ecology: progress series. Halstenbek, Amelinghausen. Vol. 1+, 1979+. Mar. Ecol. Progr. Ser. HI 67523

Marine ecology research highlights. Narragensett, RI. 1973+. Mar. Ecol. Res. Highlights. HI 67524

Marine environmental research. Barking. Vol. 1+, 1978+. Mar. Environm. Res. HI 67525

Marine life; occasional papers. New York, NY. Mar. Life. See B–P–H 548/10. HI 56095

Marine parks journal. Tokyo. 1968+. Mar. Parks J. HI 67526

Marine pollution bulletin. London, Elmsford, NY. Nos. 1-18, 1968-69; n.s. vol. 1+, 1970+. Mar. Pollut. Bull. HI 67527

Marine pollution research titles. Plymouth. 1974+. Mar. Pollut. Res. Titles. HI 67528

Marine products month. [Shui ch'an yüeh k'an. (Fu kan).] Mar. Prod. Month. See B–P–H 548/11. HI 56096

Marine research in Indonesia = Penjelidikan laut di Indonesia. Jakarta, Indonesia. Penjel. Laut Indonesia. See B–P–H 700/21.

Marine resource bulletin. Gloucester Point, VA. Vol. 8+, 1976+. Mar. Resource Bull. Preceded by: Marine resource information bulletin. HI 67530

Marine resource information bulletin. Gloucester Point, VA. Vols. 1-7, 1971-76. Mar. Resource Inform. Bull. Superseded by: Marine resource bulletin. HI 67531

Marine resources: advisory series. Gloucester Point, VA. Vol. 1+, 1970+. Mar. Resources Advis. Ser HI 67532

Marine resources bulletin. Dauphin Island, AL. No. 1, 1963. Mar. Resources Bull. Superseded by: Alabama marine resources bulletin. HI 67533

Marine science communications. New York. Vols. 1-2, 1975-76. Mar. Sci. Commun. HI 67534

Marine sciences contents tables. Rome. Vol. 3+, 1967+. Mar. Sci. Tables. Preceded by: Current contents in marine science. HI 67535

Marine sciences newsletter. Djakarta. Nos. 1-3, 1959-60. Mar. Sci. Newslett. Superseded by: International marine science. HI 67536

Marine studies of San Pedro Bay. Los Angeles, CA. Vol. 1+, 1972+. Mar. Stud. San Pedro Bay. HI 67537

Maritimes. Kingston, RI. Vol. 1+, 1957+. Maritimes. HI 67538

Markaz buhut an-nahil wa-ăt-tumur = Technical bulletin, palms and dates research centre. Baghdad.

Markaz buhuth ulum al-hayah, Baghdad = Journal of biological sciences. Baghdad.

Market bulletin. Florida department of agriculture. Jacksonville, FL. Market Bull. See B–P–H 548/20. HI 56102

Märkische Heimat. Aus Natur und Geschichte der Bezirke Cottbus, Frankfurt (Oder), Potsdam. Potsdam. Märk. Heimat. See B–P–H 548/18. HI 56100

Märkische Naturschutz. Berlin. Märk. Naturschutz. See B–P–H 548/19. HI 56101

Marque, La. ?-1971-73-? Marque. HI 67539

Martonvásári növénytermelési kutató intézet évkönyve. Martonvasar, Hungary. Martonvásári Növényterm. Kutató Intéz. Évk. See B–P–H 548/23. HI 56103

Maryland; a journal of natural history. Baltimore, MD. 1944-47. Maryland. Preceded by: Bulletin, natural history society of Maryland. Superseded by: Maryland naturalist. 3-2547-1. HI 67540

Maryland agricultural college. Nature study bulletin. College Park, MD. Maryland Agric. Coll. Nat. Study Bull. See B–P–H 548/24. HI 56104

Maryland agricultural experiment station. Annual report. College Park, MD. Maryland Agric. Exp. Sta. Annual Rep. See B–P–H 548/25. HI 56105

Maryland agricultural experiment station. Bulletin. College Park, MD. Maryland Agric. Exp. Sta. Bull. See B–P–H 549/1. HI 56106

Maryland agricultural experiment station. Circular bulletin. College Park, MD. Maryland Agric. Exp. Sta. Circ. Bull. See B–P–H 549/2. HI 56107

Maryland agricultural experiment station. Miscellaneous publication. College Park, MD. Maryland Agric. Exp. Sta. Misc. Publ. See B–P–H 549/3. HI 56108

Maryland agricultural experiment station. Popular bulletin. College Park, MD. Maryland Agric. Exp. Sta. Popular Bull. See B–P–H 549/4. HI 56109

Maryland agricultural experiment station. Special bulletin. College park, MD. Maryland Agric. Exp. Sta. Special Bull. See B–P–H 549/5. HI 56110

Maryland agricultural experiment station. Special report. College Park, MD. Maryland Agric. Exp. Sta. Special Rep. See B–P–H 549/6. HI 56111

Maryland conservationist. Baltimore, MD. Vol. 1+, 1924+. Maryland Conservationist. 3-2545-3. HI 67541

Maryland fruit grower. College Park, Hancock, MD. Vol. 1+, 1931+. Maryland Fruit Grower. HI 67542

Maryland grapevine. Germantown, Burkittsville, MD. Vol. 1+, 1981+. Maryland Grapevine. HI 67543

Maryland naturalist. Baltimore, MD. 1948+. Maryland Naturalist. Preceded by: Maryland. 3-2547-1. HI 67544

Maryland research chronicle = University of Maryland graduate school chronicle. College Park, MD.

Maryland state board of agriculture bulletin. College Park, MD. Maryland State Board Agric. Bull. See B–P–H 549/7. HI 56112

Massachusetts agricultural experiment station. Annual report. Amherst, MA. Mass. Agric. Exp. Sta. Annual Rep. See B–P–H 549/11. HI 56113

Massachusetts agricultural experiment station. Biennial report. Amherst, MA. Mass. Agric. Exp. Sta. Bienn. Rep. See B–P–H 549/12. HI 56114

Massachusetts agricultural experiment station bulletin. Amherst, MA. Nos. 115-590, 1907-70. Mass. Agric. Exp. Sta. Bull. Preceded by: Hatch experiment station of Massachusetts agricultural college, bulletin. Superseded by: Research bulletin, university of Massachusetts agricultural experiment station. HI 67545

Massachusetts agricultural experiment station. Circular. Amherst, MA. Nos. 1-74, 1907-27. Mass. Agric. Exp. Sta. Circ. HI 56115

Massachusetts agricultural repository and journal. Boston, MA. Mass. Agric. Repos. J. See B–P–H 549/17. HI 56116

Massachusetts Audubon bulletin. Boston, MA. 1917-61. Mass. Audubon Bull. Superseded by: Massachusetts Audubon newsletter. HI 67546

Massachusetts Audubon newsletter. Lincoln, MA. Vols. 1-12, 1961-73. Mass. Audubon Newslett. Preceded by: Massachusetts Audubon bulletin. Superseded by: Newsletter, Massachusetts Audubon society. HI 67547

Massachusetts horticultural society. Address delivered on the occasion of its anniversary. Boston, MA. Mass. Hort. Soc. Address. See B–P–H 549/18. HI 56117

Massachusetts wildlife. Worcester, MA. Vol. 1+, 1950?+. Mass. Wildlife. HI 67548

Mata kuching bulletin. Paris. No. ?-2+, ?-1983+. Mata Kuching Bull. HI 67549

Matar′jaly pa ljasnoj das′ledčaj sprave Belaruskae S S R = Matar′yaly pa lyasnoi das′ledchai sprave Belaruskae S S R. Minsk.

Matar′jaly da vyvučen′nja flëry i faŭny Belarusi = Matar′yaly da vyvuchen′iya flëry i faŭny Belarusi. Minsk.

Matar′yaly pa lyasnoi das′ledchai sprave Belaruskae S S R. Minsk, Belorussian S S R. Vols. 1-6, 1927-30. Matar′y. Lyasn. Das′l. Sprave Belarusk. S.S.R. HI 67550

Matar′yaly da vyvuchen′iya flëry i fauny Belarusi. Minsk, Belorussian S S R. Vols. 1-7, 1927-32. Matar′y. Vyvuch. Flëry Fauny Belarusi. 3-2558-3. HI 67551

Matematikai és természettudományi értesitö. Budapest. Mat. Term. Értes. See B–P–H 549/22. HI 56118

Matematikai és természettudományi közlemények. Budapest. Mat. Term. Közlem. See B–P–H 550/2. HI 56119

Materia medica. [Honzo.] Tokyo. Mater. Med. See B–P–H 552/2. HI 56125

Materiae rudes plantarum = Trudy botanicheskogo instituta akademii nauk S S S R. Ser. 5, rastitel′noe syr′e. Moscow & Leningrad.

Materiae vegetabiles; acta culturae et praeparationis plantarum. The Hague. Mater. Veg. See B–P–H 552/20. HI 56128

Material reports, university museum, Tokyo. Tokyo. No. ?-2+, ?-1978+. Mater. Rep. Univ. Mus. Tokyo. HI 67552

Materialien für Erforschung der Flora und Fauna Weissrutheniens = Matar′yaly da vyvuchen′iya flëry i faŭny Belarusi. Minsk.

Materials for mycology and phytopathology = Materialy po mikologii i fitopatologii.

Materials and organisms = Materiel und organismen. Berlin.

Materialy k biobibliografii učenyh S S S R. Serija biologičeskih nauk; agrobiologija = Materialy k biobibliografii uchenykh S S S R. Seriya biologicheskikh nauk; agrobiologiya. Moscow & Leningrad.

Materialy k biobibliografii učenyh S S S R. Serija biologičeskih nauk; botanika = Materialy k biobibliografii uchenykh S S S R. Seriya biologicheskikh nauk; botanika. Moscow & Leningrad.

Materialy k biobibliografii učenyh S S S R. Serija biologičeskih nauk; fiziologija rastenij = Materialy k biobibliografii uchenykh S S S R. Seriya biologicheskikh nauk; fiziologiya rastenii. Moscow & Leningrad.

Materialy k biobibliografii učenyh S S S R. Serija biologičeskih nauk; paleontologija = Materialy k biobibliografii uchenykh S S S R. Seriya biologicheskikh nauk; paleontologiya. Moscow & Leningrad.

Materialy k biobibliografii uchenykh S S S R. Seriya biologicheskikh nauk; agrobiologiya. Moscow & Leningrad. 1 vol., 1953. Mater. Biobibliogr. Uchen. S.S.S.R., Ser. Biol. Nauk, Agrobiol. HI 67553

Materialy k biobibliografii uchenykh S S S R. Seriya biologicheskikh nauk; botanika. Moscow & Leningrad. Vol. 1+, 1946+. Mater. Biobibliogr. Uchen. S.S.S.R., Ser. Biol. Nauk, Bot. 1-112-3. HI 67554

Materialy k biobibliografii uchenykh S S S R. Seriya biologicheskikh nauk; fiziologiya rastenii. Moscow & Leningrad. Vol. 1+, 1948+. Mater. Biobibliogr. Uchen. S.S.S.R., Ser. Biol. Nauk, Fiziol. Rast. 1-112-3. HI 67555

Materialy k biobibliografii uchenykh S S S R. Seriya biologicheskikh nauk; paleontologiya. Moscow & Leningrad. Vol. 1, 1947. Mater. Biobibliogr. Uchen. S.S.S.R., Ser. Biol. Nauk, Paleontol. 1-112-3. HI 67556

Materialy po biogeografii S S S R. Moscow. Mater. Biogeogr. S.S.S.R. See B–P–H 550/12. HI 56120

Materialy k Bližajšemu Poznaniju Prozjabaemosti Rossijskoj Imperii = Beiträge zur Pflanzenkunde des Russischen Reiches. St. Petersburg. Beitr. Pflanzenk. Russ. Reiches. See B–P–H 172/6.

Materiały do fizjografii kraju. Cracow. Mater. Fizjogr. Kraju. See B–P–H 550/19. HI 56122

Materialy po flore i rastitel′nosti severnogo prikaskiya. Leningrad. Vols. 1-3?, 1964-68? Mater. Fl. Rastitel′n. Severn. Prikaskiya. HI 67557

Materialy po florističeskim i faunističeskim obsledovanijam Terskogo okruga = Materialy po floristicheskim i faunisticheskim obsledovaniyam Terskogo okruga. Pyatigorsk.

Materialy po floristicheskim i faunisticheskim obsledovaniyam Terskogo okruga. Pyatigorsk, Russian S F S R. 1 vol., 1928. Mater. Florist. Faunist. Obsl. Tersk. Okr. HI 67558

Materiały floristiczne i geobotaniczne = Fragmenta floristica et geobotanica. Cracow. Fragm. Florist. Geobot. See B–P–H 384/3.

Materiały do flory polski i krajów ośiennych = Planta polonica. Warsaw. Pl. Polon. See B–P–H 712/16.

Materialy po istorii fauny i flory Kazahstana = Materialy po istorii fauny i flory Kazakhstana. Alma-Ata.

Materialy po istorii fauny i flory Kazakhstana. Alma-Ata, Kazakh S S R. Vol. 1+, 1955+. Mater. Istorii Fauny Fl. Kazakhstana. HI 67559

Materialy po istorii fauny i rastitel'nosti S S S R. Moscow. Mater. Istorii Fl. Rastitel'n. S.S.S.R. See B–P–H 551/8. HI 56123

Materialy dlja izučenija estestvennyh proizvoditel'nyh sil Rossii = Materialy dlya izucheniya estestvennykh proizvoditel'nykh sil Rossii. Petrograd.

Materialy dlja izučenija estestvennyh proizvoditel'nyh sil S S S R, izdavaemye komissiej pri Rossijskoj [later Vsesojuznoj] akademii nauk = Materialy dlya izucheniya estestvennykh proizvoditel'nykh sil S S S R, izdavaemye komissiei pri Rossiiskoi [later Vsesoyuznoi] akademii nauk. Leningrad.

Materialy dlja izučenija fitoplanktona Volgi po nabljudenijam = Materialy dlya izucheniya fitoplanktona Volgi po nablyudeniyam. Saratov.

Materialy po izučeniju boleznej lesnyh porod Sibirskogo kraja = Materialy po izucheniyu boleznei lesnykh porod Sibirskogo kraya. Omsk.

Materialy po izučeniju Sibiri = Materialy po izucheniyu Sibiri. Tomsk.

Materialy k izučeniju žen'-šenja i limonnika = Materialy k izucheniyu zhen'-shenya i limonnika. Moscow & Leningrad.

Materialy po izucheniyu boleznei lesnykh porod Sibirskogo kraya. Omsk, Russian S F S R. 1929. Mater. Izuch. Boleznei Lesn. Porod Sibirsk. Kraya. HI 67560

Materialy dlya izucheniya estestvennykh proizvoditel'nykh sil Rossii. Petrograd. Vols. 1-51, 1917-24. Mater. Izuch. Estestv. Proizv. Sil Rossii. Superseded by: Materialy dlya izucheniya estestvennykh proizvoditel'nykh sil S S S R, izdavaemye komissiei pri Rossiiskoi [later: Vsesoyuznoi] akademii nauk. 1-118-2. HI 67561

Materialy dlya izucheniya estestvennykh proizvoditel'nykh sil S S S R, izdavaemye komissiei pri Rossiiskoi [later: Vsesoyuznoi] akademii nauk. Leningrad. Vols. 52-64, 1925-27. Mater. Izuch. Estestv. Proizv. Sil S.S.S.R. Preceded by: Materialy dlya izucheniya estestvennykh proizvoditel'nykh sil Rossii. Superseded by: Materialy. Komissiya po izucheniyu estestvennykh proizvoditel'nykh sil Soyuza akademii nauk S S S R. 1-118-2. HI 67562

Materialy dlya izucheniya fitoplanktona Volgi po nablyudeniyam. Saratov, Russian S F S R. 1902. Mater. Izuch. Fito-Plankt. Volgi Nablyud. Superseded by: Nablyudeniya nad Fitoplanktonum Volgi. HI 67563

Materialy po izucheniyu Sibiri. Tomsk, Russian S F S R. Vols. 1-4, 1930-33. Mater. Izuch. Sibiri. HI 67564

Materialy k izucheniyu zhen'-shenya i limonnika. Moscow & Leningrad. Vols. 1-2, 1952-55. Mater. Izuch. Zhen'-Shenya Limonnika. HI 67565

Materialy komissii ėkspedicionnyh issledovanij = Materialy komissii ékspeditsionnykh issledovanii. Leningrad.

Materialy komissii ékspeditsionnykh issledovanii. Leningrad. Vols. 1-30, 1928-30. Mater. Komiss. Ěksped. Issl. Preceded by: Materialy osobogo komiteta po issledovaniyu soyuznykh i avtonomnykh respublik. HI 67566

Materialy komissii po izučeniju Jakutskoj Avtonomnoj S S R = Materialy komissii po izucheniyu Yakutskoi Avtonomnoi S S R. Leningrad.

Materialy. Komissija po izučeniju estestvennyh proizvoditel'nyh sil Sojuza akademii nauk S S S R = Materialy. Komissiya po izucheniyu estestvennykh proizvoditel'nykh sil Soyuza akademii nauk S S S R. Leningrad.

Materialy. Komissiya po izucheniyu estestvennykh proizvoditel'nykh sil Soyuza akademii nauk S S S R. Leningrad. Vols. 65-81, 1928-30. Mater. Komiss. Izuch. Estestv. Proizv. Sil Soyuza Akad. Nauk S.S.S.R. Preceded by: Materialy dlya izucheniya estestvennykh proizvoditel'nykh sil S S S R, izdavaemye komissiei pri Rossiiskoi [later: Vsesoyuznoi] akademii nauk. Superseded by: Trudy Mongol'skoi komissii. 1-118-2. HI 67567

Materialy komissii po izucheniyu Yakutskoi Avtonomnoi S S R. Leningrad. Vols. 1-36, 1925-36. Mater. Komiss. Izuch. Yakutsk. Avton. S.S.R. 1-118-2. HI 67568

Materialy po lesnomu opytnomu delu Belorusskoj S S R = Matar'yaly pa lyasnoi das'ledchai sprave Belaruskae S S R. Minsk.

Materialy po mikologičeskomu obsledovaniju rossii. Petrograd = Materialy po mikologicheskomu obsledovaniyu rossii. Petrograd.

Materialy po matematicheskomu obespecheniyu E V M. Seriya ekomodel. Pushchino. Vol. ?-9+, ?-1984+.

Mater. Mat. Obespech. E. V. M. Ser. Ekomodel. HI 75112

Materialy po mikologicheskomu obsledovaniyu rossii. Petrograd. Vols. 1-5, 1914-22. Mater. Mikol. Obsl. Rossii. 3-2561-2. HI 67569

Materialy po mikologii i fitopatologii. Leningrad. Vols. 3-8, 1926-29. Mater. Mikol. Fitopatol. Preceded by: Materialy po mikologii i fitopatologii Rossii. Superseded by: Trudy po zashchite rastenii. Seriya 2. Fitopatologiya. 3-2391-1. HI 67570

Materialy po mikologii i fitopatologii Rossii. Petrograd. Mater. Mikol. Fitopatol. Rossii. See B–P–H 552/5. HI 56127

Materialy po ohoronu pryrody na Ukrajini = Materialy po okhoronu pryrody na Ukrayini. Kiev.

Materialy po okhoronu pryrody na Ukrayini. Kiev, Ukrainian S S R. Vols. 1-2, 1958-60. Mater. Okhor. Pryr. Ukrayini. Superseded by: Okhoronyaite ridnu pryrodu! HI 67571

Materialy osobogo komiteta po issledovaniju sojuznyh i avtonomnyh respublik = Materialy osobogo komiteta po issledovaniyu soyuznykh i avtonomnykh respublik. Leningrad.

Materialy osobogo komiteta po issledovaniyu soyuznykh i avtonomnykh respublik. Leningrad. Vols. 1-18 1926-28. Mater. Osob. Komiteta Issl. Soyuzn. Avton. Respubl. Superseded by: Materialy komissii ėkspeditsionnykh issledovanii. HI 67572

Materialy k poznaniju fauny i flory Rossii = Materialy k poznaniyu fauny i flory Rossii. Moscow.

Materialy k Poznaniju Fauny i Flory Rossijskoj Imperii. Otděl botaničeskij = Materialy k Poznaniyu Fauny i Flory Rossiiskoi Imperii. Otděl botanicheskii. Moscow.

Materialy k poznaniju fauny i flory S S S R, izdavaemye Moskovskim obščestvom ispytatelej prirody. Otdel botaničeskij = Materialy k poznaniyu fauny i flory S S S R, izdavaemye Moskovskim obshchestvom ispytatelei prirody. Otdel botanicheskii. Moscow.

Materialy k poznaniyu fauny i flory Rossii. Moscow. Vol. 8, 1918. Mater. Pozn. Fauny Fl. Rossii. Preceded by: Materialy k poznaniyu fauny i flory Rossiiskoi imperii. Otděl botanicheskii. Superseded by: Materialy k poznaniyu fauny i flory S S S R, izdavaemye Moskovskim obshchestvom ispytatelei prirody. Otdel botanicheskii. 3-2560-3. HI 67573

Materialy k poznaniyu fauny i flory Rossiiskoi imperii. Otděl botanicheskii. Moscow. Vols. 1-[7], 1890-1917. Mater. Pozn. Fauny Fl. Rossiisk. Imperii, Otd. Bot. Superseded by: Materialy k poznaniyu fauny i flory Rossii. 3-2560-3. HI 67574

Materialy k poznaniyu fauny i flory S S S R, izdavaemye Moskovskim obshchestvom ispytatelei prirody. Otdel botanicheskii. Moscow. Vol. 9 [also numbered n.s. vol. 1]+, 1940+. Mater. Pozn. Fauny Fl. S.S.S.R., Otd. Bot. Preceded by: Materialy k poznaniyu fauny i flory Rossii. 3-2560-3. HI 67575

Materialy ... sesji naukovej instytutu ochrony roslin. Poznan. Ca.1980?+. Mater. Sesyu. Nauk. Inst. Ochr. Rośl. HI 67576

Materialy Vsesojuznogo soveščanija botanikov i selekcionerov = Materialy Vsesoyuznogo soveshchaniya botanikov i selektsionerov. Moscow.

Materialy Vsesoyuznogo soveshchaniya botanikov i selektsionerov. Moscow. 1 vol., 1951. Mater. Vsesoyuzn. Soveshch. Bot. Selektsion. HI 67577

Materiały zakładu fitosocjologii stosowanej uniw. Warszawskiego. Warsaw = Phytocoenosis. Warsaw.

Materialy z′yizdu, ukrayins′koe botanichne tovarystvo. Kiev. Vol. ?-3+, ?-1964+. Mater. Z′yizdu Ukrayins′k. Bot. Tovar. HI 67578

Matériaux de la commission pour l′étude de la République Autonome Soviétique Socialiste Jakoute = Materialy komissii po izucheniyu Yakutskoi Avtonomnoi S S R. Leningrad.

Matériaux pour l′étude des ressources naturelles de la Russie = Materialy dlya izucheniya estestvennykh proizvoditel′nykh sil Rossii. Petrograd.

Matériaux pour la flore cryptogamique suisse = Beiträge zur Kryptogamenflora der Schweiz.

Matériaux pour le levé géobotanique de la Suisse. Zurich. Matér. Levé Géobot. Suisse. See B–P–H 552/1. HI 56124

Matériaux et organismes = Materiel und organismen. Berlin.

Materiel und organismen. Berlin. Vol. 1+, 1965+. Mater. & Organismen. HI 67579

Matériaux pour servir à l'étude de la flore et de la géographie botanique de l'Orient. Nancy. Matér. Etude Fl. Géogr. Bot. Orient. See B–P–H 550/17. HI 56121

Mathaf at-tarih at-taki'i al-iraqi = Bulletin of the Iraq natural history museum. Baghdad, and Iraq natural history museum's publications. Baghdad.

Mathematical biosciences. New York. Vol. 1+, 1967+ Math. Biosci. HI 67580

Mathematik und Naturwissenschaft in der neuen Schule. Berlin, Leipzig. Vol. 1, 1949. Math. Naturwiss. Neuen Schule. HI 67581

Mathematikai és természettudományi értesitö. Budapest. Math. Term. Értes. See B–P–H 553/10. HI 56130

Mathematikai és természettudományi közlemémyek. Pest [=Budapest, in part]. Math. Term. Közlem. See

B–P–H 553/11. HI 56131

Mathematische und naturwissenschaftliche Berichte aus Ungarn. Budapest, Berlin, & Leipzig. Math. Naturwiss. Ber. Ungarn. See B–P–H 553/5. HI 56129

Mathematischer und naturwissenschaftlicher Anzeiger der ungarischen Akademie der Wissenschaften = Matematikai és természettudományi értesitö, vols. 43-63. Budapest. Mat. Term. Értes. See B–P–H 549/22.

Matrodiana. Buenos Aires. 1970+. Matrodiana. HI 67582

Matsushima mycological memoirs. Kobe. No. 1+, 1980+. Matsushima Mycol. Mem. HI 67583

Matsuyama noka daigaku gakujutsu hokoku = Scientific reports of the Matsuyama Agricultural college. Matsuyama, Japan. Sci. Rep. Matsuyama Agric. Coll. See B–P–H 827/10.

Mauri ora. Christchurch, N.Z. Vol. 1+, 1973+. Mauri Ora. HI 67584

Mauritius institute bulletin. Port Louis. 1937+. Mauritius Inst. Bull. HI 67585

Maydica; rivista di tecnica e divulgazione maidicola ... dalla stazione sperimentale di maiscoltura di Bergamo. Bergamo, Italy. Maydica. See B–P–H 553/17. HI 56132

Mazama; a record of mountaineering in the pacific northwest. Portland, OR. Mazama. See B–P–H 553/18. HI 56133

Mazingira; world forum for environment and development. Oxford. No. 1+, 1977+. Mazingira. HI 67586

Mcenareta dacvis gankopilebis moambe. Tiflis, Georgian S S R. Mcenareta Dacvis Gankop. Moambe. See B–P–H 553/21. HI 56134

Mcenareta dacvis institutis šromebi = Trudy instituta zashchity rastenii, akademiya nauk Gruzinskoi S S R. Tiflis.

McGraw-Hill's biotechnology newswatch. New York. Vols. 1-7(22), 1981-87. McGraw-Hill's Biotechnol. Newswatch. Superseded by: Biotechnology newswatch. HI 74858

Mechanic's magazine. New York, NY. Mech. Mag. See B–P–H 553/25. HI 56135

Mecmuasi, Istanbul üniversitesi eczacilik fakültesi = Istanbul üniversitesi eczacilik fakültesi mecmuasi. Istanbul.

Medan ilmu pengetahuan madjalak filsafat, ilmu, ilmu sosial, budaja, pasti dan alam. Jakarta, Indonesia. Medan Ilmu Pengetahuan Madj. Filsafat. See B–P–H 556/1. HI 56170

Mededeelingen van de en de daaraan verbonden instituten. Wageningen. Meded. Rijks Landbouwhoogeschool. See B–P–H 561/7. HI 56229

Meddelanden af geografiska föreningen i Finland. Helsinki. Meddeland. Geogr. Fören. Finland. See B–P–H 556/25. HI 56175

Meddelanden från geografiska institutionen vid Stockholms universitet. Stockholm. ?-1979+. Meddel. Geogr. Inst. Stockholms Univ. HI 67587

Meddelanden från Göteborgs botaniska trädgård = Acta horti gothoburgensis. Göteborg.

Meddelanden från Gullåkers växtförädlingsanstalt. Hammenhög. Nos. 1-18, 1942-62. Meddel. Gullåkers Växtförädlingsanst. HI 67588

Meddelanden fran k. vetenskapsakdemiens Nobelinstitut. Stockholm. Meddeland. Vetenskapsakad. Nobelinst. See B–P–H 557/4. HI 56183

Meddelanden fran kongl. svenska vetenskaps-akademiens trädgard bergielund = Acta horti bergiani. Stockholm.

Meddelanden. Lantbruksakademiens trädgårdsavdelningen. Landskrona, Sweden. Meddeland. Lantbruksakad. Trädgårdsavd. See B–P–H 558/28. HI 56176

Meddelanden från Lunds botaniska museum. Lund. Meddeland. Lunds Bot. Mus. See B–P–H 556/29. HI 56177

Meddelanden från Lunds universitets limnologiska institution. Helsinki. Meddeland. Lunds Univ. Limnol. Inst. See B–P–H 556/30. HI 56178

Mededeelingen van het rijksboschbouwproefstation. Wageningen. Meded. Rijksboschbouwproefstat. See B–P–H 561/8. HI 56230

Mededeelingen van's rijks-herbarium. Leiden. Meded. Rijks-Herb. See B–P–H 561/9. HI 56231

Mededelingen van het rijskproefstation voor zaadcontrole. Wageningen. Meded. Rijksproefstat. Zaadcontrole. See B–P–H 561/10. HI 56232

Mededelingen van het rijkstuinbouwconsulentschap voor plantenziekten. Wageningen. Meded. Rijkstuinbouwconsulentschap Plantenziekten. See B–P–H 561/11. HI 56233

Mededeelingen over rubber. Buitenzorg, Dutch E. Indies [=Bogor, Indonesia]. Meded. Rubber. See B–P–H 561/12. HI 56234

Mededelingen van de schoolproeftuin van de rijkstuinbouwschool te Lisse. Lisse, Netherlands. Meded. Schoolproeftuin Rijkstuinbouwschool Lisse. See B–P–H 561/13. HI 56235

Meddelanden från skånes naturskyddsförening. Lund. Nos. 1-4, 1910-12. Meddel. Skånes Naturskyddsfören. Superseded by: Skånes naturskyddsförenings arsberättelse. HI 67589

Meddelanden af societas pro fauna et flora fennica.

Helsinki. Meddeland. Soc. Fauna Fl. Fenn. See B–P–H 556/31. HI 56179

Meddelanden, statens naturvårdsverk. Solna. 1977+. Meddel. Statens Naturvårdsverk. Preceded by: Publikationer, statens naturvårdsverk. HI 67590

Meddelanden från statens skogs-forskningsinstitut. Stockholm. Vol. 1+, 1947+. Meddeland. Statens Skogs-Forskningsinst. Preceded by: Meddelanden från statens skogsförsöksanstalt. HI 67591

Meddelanden från statens skogsförsöksanstalt. Stockholm. Vols. 1-34, 1903-46. Meddel. Statens Skogsförsöksanstalt. Superseded by: Meddelanden från statens skogsforskningsinstitut. HI 67592

Meddelanden. Statens växtskyddsanstalt. Stockholm. Meddeland. Statens Växtskyddsanst. See B–P–H 557/2. HI 56181

Mededelingen van de stichting Nederlands graancentrum. Wageningen. Meded. Stichting Ned. Graancentrum. See B–P–H 561/15. HI 56236

Mededelingen van de stichting voor plantenveredeling. Wageningen. Meded. Stichting Plantenveredel. See B–P–H 561/16. HI 56237

Mededelingen van de stichting proeftuinen tabak. Wageningen. Meded. Stichting Proeftuinen Tabak. See B–P–H 561/17. HI 56238

Mededelingen van de technische commissie voor duinonderzoek van het I.J.B.O.N. The Hague. Meded. Techn. Commiss. Duinonderz. I J B O N. See B–P–H 561/18. HI 56239

Mededelingen van het tuinbouwkundig laboratorium van Turkenburg's zaadhandel N.V. te Bodegraven. Bodegraven, Netherlands. Meded. Tuinbouwk. Lab. Turkenburg's Zaadhandel Bodegraven. See B–P–H 561/19. HI 56240

Mededelingen. Zeeuwse groentetelersvereniging. Goes, Netherlands. Meded. Zeeuwse Groentetelersver. See B–P–H 561/21. HI 56241

Meddelanden, stiftelsen för rasförädling av akogsträd = Tiedoituksia, metsäpuiden rodunjalostussaatio. Helsinki.

Meddelanden från stockholms högskolas botaniska institut. Stockholm. Meddeland. Stockholms Högskolas Bot. Inst. See B–P–H 557/3. HI 56182

Meddelanden från sveriges fröodlareforbund. Stockholm. 1st+, 1960+. Meddel. Sveriges Fröodlareforb. HI 67593

Meddelanden från växtbiologiska institutionen. Uppsala. ?-1982+. Meddel. Växtbiol. Inst. HI 67594

Meddelanden fran vetenskapsakdemiens Nobelinstitut. Stockholm. Meddeland. Vetenskapsakad. Nobelinst. See B–P–H 557/4. HI 56183

Meddelelser fra den biologiske station ved Drøbak. Drobak, Norway. Meddel. Biol. Stat. Drøbak. See B–P–H 556/12. HI 56172

Meddelelser, botanisk centralbibliotek, København. Copenhagen. Vol. 1+, 1964+. Meddel. Bot. Centralbibl. København. HI 67595

Meddelelser fra den botaniske forening i Kjøbenhavn. [Supplement to Botanisk tidsskrift.] Copenhagen. Vols. 1-2(10), 1882-91. Meddel. Bot. Foren. Kjøbenhavn. HI 67596

Meddelelser fra Carlsberg laboratoriet. Copenhagen. Vols. 1-20, 1876-1934. Medded. Carlsberg Lab. Superseded by: Comptes-rendus des travaux du Carlsberg laboratoriet. 2-929-1. HI 67597

Meddelelser fra C U B B I. Copenhagen. No. 1+, 1976+. Meddel. C. U. B. B. I. HI 67598

Meddelelser. Dansk botanisk forening. Copenhagen. Meddel. Dansk Bot. Foren. See B–P–H 556/14. HI 56173

Meddelelser. Dansk geologisk forening. Copenhagen. Meddel. Dansk Geol. Foren. See B–P–H 556/15. HI 56174

Meddelelser fra "Flora og Fauna". Copenhagen. 1889-91. Meddel. "Fl. & Fauna". Superseded by: Flora og fauna. HI 67599

Meddelelser fra foreningen til svampekundskabens fremme. Copenhagen. Vols. 1-4, 1912-30. Meddel. Forenin. Svampekundsk. Fremme. Superseded by: Friesia. 2-1602-3. HI 67600

Meddelelser om Groenland. Bioscience = Meddelelser om Grønland. Bioscience. Copenhagen.

Meddelelser om Grønland. Bioscience. Copenhagen. No. 1+, 1979+. Meddel. Grønland, Biosci. Preceded by: Meddelelser om Grønland, af kommissionen for ledelsen af de geologiske og geografiske undersøgelser i Grønland. HI 67601

Meddelelser om Grønland. Geoscience. Copenhagen. No. 1+, 1979+. Meddel. Grønland, Geosci. Preceded by: Meddelelser om Grønland, af kommissionen for ledelsen af de geologiske og geografiske undersøgelser i Grønland. HI 67602

Meddelelser om Grønland af kommissionen for ledelsen af de geologiske og geografiske undersøgelser i Grønland. Copenhagen. Vols. 1-206(5), 1879-1978. Meddel. Grønland. Superseded by: Meddelelser om Grønland. Bioscience and Meddelelser om Grønland. Geoscience. 3-2571-2. HI 67603

Meddelelser for Landmaen. Odense. Vols. 1-3, 1851-63. Meddel. Landmaen. HI 67604

Meddelelser, norges svalbard- og ishavs-undersøkelser. Oslo. 1926-47. Meddel. Norg. Svalbard Ishavs-Undersøk. Superseded by: Norsk polarinstitutt meddelelser. 4-3077-2. HI 67605

Meddelelser fra norsk farmaceutisk selskap. Oslo. Vols. 1-44, 1939-82. Meddel. Norsk Farm. Selskap. Superseded by: Norvegica pharmaceutica acta. HI 67606

Meddelelser fra norsk institutt for skogsforskning. Vol. 31+, 1974+. Meddel. Norsk. Inst. Skogsforskning. Preceded by: Meddelelser fra det skogforsøksvaesen and Meddelelser fra vestlandets forstlige forsøksstasjon. HI 67607

Meddelelser om norsk polarinstitutt = Norsk polarinstitutt meddelelser. Oslo.

Meddelelser fra det norske skogsforsøksvesen. Oslo, Vollebekk. Vols. 1(1)-30(126), 1928-74. Meddel. Norske. Skogsforsøksvesen. Preceded by: Meddelelser om skogsforsøksvaesen. Superseded by: Meddelelser fra norsk institutt for skogforskning. HI 67608

Meddelelser om skogsforsøksvaesen. Christiania. 1920-28. Meddel. Skogsforsøksvaesen. Superseded by: Meddelser fra det norske skogsforsøksvesen. HI 67609

Meddelelser fra statens forsøgsvirksohmed i plantekultur. Copenhagen, etc. Vols. 1-78(1303), 1898-1976. Meddel. Statens Forsøgsvirksohmed Plantekult. Superseded by: Meddelelser, statens planteavlsforsøg. HI 67610

Meddelelser, statens planteavlsforsøg. Lyngsby. Vol. 78(1304)+, 1976+. Meddel. Statens Planteavlsforsøg. Preceded by: Meddelelser, statens forsøgsvirksomhed i plantekultur. HI 67611

Meddelelser fra vestlandets forstlige forsøksstation. Bergen. 1917-78. Meddel. Vestl. Forstl. Forsøksstat. Incorporated in: Meddelelser fra norsk institutt for skogsforskning. HI 67612

Médecine moderne. Paris. Méd. Moderne. See B–P–H 555/2. HI 56157

Médecine traditionelle et pharmacopée. Paris. 1986+. Méd. Tradit. Pharmacopée. HI 67613

Mededelingen, see Mededeelingen.

Mededeelingen aangaande geneeskrachtige en aromatische gewassen, hun producten en handel. The Hague. Meded. Geneeskr. Aromat. Gewassen (The Hague). See B–P–H 558/14. HI 56199

Mededelingen aangaande geneeskrachtige en aromatische gewassen. Uitgave van de rijksvoorlichtingsdienst te Leeuwarden en de ned. vereniging voor geneeskruidentuinen. Leeuwarden, Netherlands. Meded. Geneeskr. Aromat. Gewassen (Leeuwarden). See B–P–H 558/15. HI 56200

Mededeelingen van de afdeeling handelsmuseum van het koloniaal instituut te Amsterdam. Amsterdam. Vols. 1-5, 1913-23; vols. 8-10, 1930-32; vols. 22-28, 1940-44. Meded. Afd. Handelsmus. Kolon. Inst. Preceded by: Bulletin, koloniaal museum. For vols. 6-7, 1925-29 and for vols. 11-21, 1932-40 see: Mededeelingen van de afdeeling handelsmuseum van de koninklijke vereeniging koloniaal instituut. Superseded by: Mededeelingen van de afdeeling handelsmuseum van de koninklijke vereeniging indisch instituut. 3-2311-2. HI 67614

Mededeelingen van de afdeeling handelsmuseum van de koninklijke vereeniging indisch instituut. Amsterdam. Vols. 29-31, 1945-4? Meded. Afd. Handelsmus. Kon. Ver. Indisch Inst. Preceded by: Mededeelingen van de afdeeling handelsmuseum van de koloniaal instituut te Amsterdam. Superseded by: Mededelingen van de afdeling tropische producten det koninklijke vereeniging indisch instituut. 3-2311-2. HI 67615

Mededeelingen van de afdeeling handelsmuseum van de koninklijke vereeniging koloniaal instituut. Amsterdam. Vols. 6-7, 1925-29; vols. 11-21, 1932-40. Meded. Afd. Handelsmus. Kon. Ver. Kolon. Inst. Preceded and superseded by (also for vols. 8-10 see): Mededeelingen van de afdeeling handelsmuseum van het koloniaal instituut. 3-2311-2. HI 67616

Mededeelingen van de afdeeling voor plantenziekten van het departement van landbouw, nijverheid en handel. Batavia, Dutch E. Indies [=Jakarta, Indonesia]. Meded. Afd. Plantenziekten Dept. Landb. See B–P–H 557/11. HI 56184

Mededelingen van de afdeling tropische producten van het koninklijk instituut voor de tropen. Amsterdam. Vols. 37-52, 1951-60. Meded. Afd. Trop. Prod. Kon. Inst. Tropen. Preceded by: Mededelingen van de afdeling tropische producten det koninklijke vereeniging indisch instituut. Superseded by: Communications of the department of agricultural research of the royal tropical institute. HI 67617

Mededelingen van de afdeling tropische producten det koninklijke vereeniging indisch instituut. Amsterdam. Vols. 32-36, 194?-50. Meded. Afd. Trop. Prod. Kon. Ver. Indisch Inst. Preceded by: Mededeelingen van de afdeeling handelsmuseum van de koninklijke vereeniging indisch instituut. Superseded by: Mededelingen van de afdeling tropische producten van het koninklijk instituut voor de tropen. 3-2311-2. HI 67618

Mededeelingen van de afdeeling voor de landbouwvoorlichtingsdienst. The Hague. Vols. 1-5, 1945-47. Meded. Afd. Landbouwvoorlichtingsdienst. Superseded by: Landbouw-documentatie. HI 67619

Mededeelingen van het agricultuur-chemisch laboratorium. Batavia, Dutch E. Indies [=Jakarta, Indonesia]. Meded. Agric.-Chem. Lab. See B–P–H 557/12. HI 56185

Mededelingen van het algemeen proefstation der A V R O

S. Algemeene ser. Batavia. Nos. 1-59, 1917-40. Meded. Alg. Proefstat. A. V. R. O. S., Alg. Ser. HI 67620

Mededelingen van het algemeen proefstation der A V R O S. Rubber ser. Batavia. Nos. 1-122, 1917-41. Meded. Alg. Proefstat. A. V. R. O. S., Rubber Ser. HI 67621

Mededeelingen van het algemeen proefstation op Java. Batavia, Dutch E. Indies [=Jakarta, Indonesia]. Meded. Alg. Proefstat. Java. See B–P–H 557/13. HI 56186

Mededelingen van het algemeen proefstation voor de landbouw. Buitenzorg [= Bogor]. Nos. 1-100, 1919-49. Meded. Alg. Proefstat. Landb., Buitenzorg. Superseded by: Contributions of the general agriculture research station. Bogor. HI 67622

Mededelingen van de Antwerpse mycologische kring. Antwerp. 1986+. Meded. Antwerpse Mycol. Kring. HI 67623

Mededelingen van het arboretum, landbouwhoogeschool te Wageningen. Wageningen. Vols. 1-3, 1935-37. Meded. Arbor. Landbouwhoogeschool Wageningen. HI 67624

Mededeelingen van het besoekisch proefstation. Besuki, Dutch E. Indies [Indonesia]. Meded. Besoekisch Proefstat. See B–P–H 557/16. HI 56187

Mededelingen van het biologisch station te Wijster. Wijster, Netherlands. Meded. Biol. Stat. Wijster. See B–P–H 557/17. HI 56188

Mededeelingen van het bosbouwproefstation. Buitenzorg [= Bogor]. Nos. 1-31, 1937-50. Meded. Bosbouwproefstat. Superseded by: Communications of the forest research institute. HI 67625

Mededeelingen van het botanisch instituut der rijksuniversiteit te Gent. Ghent. Meded. Bot. Inst. Rijksuniv. Gent. See B–P–H 557/19. HI 56189

Mededelingen van het botanisch laboratorium der rijksuniversiteit te Leiden. Leiden. Meded. Bot. Lab. Rijksuniv. Leiden. See B–P–H 557/20. HI 56190

Mededeelingen van het botanisch laboratorium der rijksuniversiteit te Utrecht. Utrecht. Meded. Bot. Lab. Rijksuniv. Utrecht. See B–P–H 557/21. HI 56191

Mededeelingen van het botanisch museum en herbarium van de rijks universiteit te Utrecht. Utrecht. Nos. 1-?, 1932-83? Meded. Bot. Mus. Herb. Rijks Univ. Utrecht. Superseded by: Miscellaneous publications of the university of Utrecht herbarium. 5-4347-2. HI 67626

Mededelingen van de botanische tuinen en het Belmonte arboretum der landbouwhogeschool te Wageningen. Wageningen. Vol. 1+, 1957+. Meded. Bot. Tuinen Belmonte Arbor. HI 67627

Mededelingen van het centraal instituut voor landbouwkundig onderzoek. Wageningen. Meded. Centr. Inst. Landbouwk. Onderz. See B–P–H 558/3. HI 56192

Mededeelingen van het centraal rubberstation. Batavia, Dutch E. Indies [=Jakarta, Indonesia]. Meded. Centr. Rubberstat. See B–P–H 558/4. HI 56193

Mededelingen, centrum voor onkruidonderzoek, rijksuniversiteit te Gent. Ghent. 1965+. Meded. Centrum Onkruidonderzoek Rijksuniv. Gent. HI 67628

Mededeelingen van den cultuurdienst, instituut voor plantenziekten en cultuures te Buitenzorg. Buitenzorg, Dutch E. Indies [=Bogor, Indonesia]. Meded. Cultuurdienst Inst. Plantenziekten Buitenzorg. See B–P–H 558/5. HI 56194

Mededeelingen uit den cultuurtuin. Buitenzorg, Dutch E. Indies [=Bogor, Indonesia]. Meded. Cultuurtuin. See B–P–H 558/6. HI 56195

Mededeelingen van het Deli proefstation. Medan?, Dutch E. Indies [Indonesia]. Meded. Deli Proefstat. See B–P–H 558/7. HI 56196

Mededeelingen uitgeven van het departement van landbouw in Nederlandsch-Indië. Batavia, Dutch E. Indies [=Jakarta, Indonesia]. Meded. Dept. Landb. Ned.-Indië. See B–P–H 558/11. HI 56197

Mededeeling. Departement van landbouw, nijverheid en handel in Suriname. Paramaribo, Dutch Guiana [Surinam]. Meded. Dept. Landb. Suriname. See B–P–H 558/12. HI 56198

Mededeelingen, directeur van de tuinbouw. The Hague. Vols. 1-31, 1937-68. Meded. Directeur Tuinb. Superseded by: Tuinbouw mededelingen. HI 67629

Mededelingen van de directie tuinbouw = Mededeelingen, directeur van de tuinbouw. The Hague.

Mededelingen, faculteit landbouwwetenschappen, rijksuniversiteit. Ghent. Vol. 35+, 1970+. Meded. Fac. Landbouwwetensch. Rijksuniv. Preceded by: Mededelingen, rijksfakulteit der landbouwwetenschappen, HI 67630

Mededeelingen uit het gebied van natuur, wetenschapen en kunst. Groningen. Meded. Natuur. See B–P–H 559/21. HI 56213

Mededeelingen, geologisch bureau voor het mijngebied te Heerlen. 1938-43. Meded. Geol. Bur. Mijngeb. Heerlen. Preceded by: Jaarverslag, geologisch bureau voor het Nederlandsche mijngebied te Heerlen. HI 67631

Mededelingen van de geologische stichting. Serie C, uitkomsten van het geol.-palaeontol. onderzoek van de ondergrond van Nederland. Leiden. Meded. Geol. Stichting, Ser. C. See B–P–H 558/16. HI 56201

Mededelingen van de hydrobiologische vereniging. Amsterdam. No. 17+, 1967+. Meded. Hydrobiol. Ver. Superseded by: Hydrobiological bulletin. HI 67632

Mededelingen van den inspecteur van den tuinbouw en het tuinbouwonderwijs = Mededeelingen, directeur van de tuinbouw. The Hague.

Mededelingen van het instituut voor bewaring en verwerking van tuinbouwproducten. Wageningen. 1956?-? Meded. Inst. Bewar. Verwerk. Tuinbouwprod. Superseded by: Mededelingen van het instituut voor plantenveredeling. HI 67633

Mededelingen. Instituut voor biologisch en scheikundig onderzoek van landbouwgewassen. Wageningen. Meded. Inst. Biol. Onderz. Landbouwgew. See B–P–H 558/17. HI 56202

Mededelingen van het instituut voor phytopathologie, laboratorium voor bloembollenonderzoek te Lisse. Lisse. Nos. ?-29-62-?, ?-1927-39-? Meded. Inst. Phytopathol. Lab. Bloembollenonderz. Lisse. Preceded by: Mededeelingen van het laboratorium voor bloembollenonderzoek, Lisse. Incorporated in: Mededelingen, laboratorium voor phytopathologie. HI 67634

Mededelingen van het instituut voor plantenveredeling. Wageningen. 19??+. Meded. Inst. Plantenveredel. Preceded by: Mededelingen van het instituut voor bewaring en verwerking van tuinbouwproducten. HI 67635

Mededeelingen van het instituut voor plantenziekten. Batavia, Dutch E. Indies [=Jakarta, Indonesia]. Meded. Inst. Plantenziekten. See B–P–H 558/18. HI 56203

Mededelingen van het instituut voor plantenziektenkundig onderzoek. Wageningen. No. 1+, 1950+. Meded. Inst. Plantenziektenk. Onderz. HI 67636

Mededelingen van het instituut voor de veredeling van tuinbouwgewassen. Wageningen. Meded. Inst. Veredel. Tuinbouwgew. See B–P–H 559/2. HI 56204

Mededelingen van het instituut voor rassenonderzoek van landbouwgewassen te Wageningen. Wageningen. Nos. 1-56?, 1949-69? Meded. Inst. Rassenonderz. Landbouwgew. Wageningen. Superseded by: Mededelingen van het rijksinstituut voor het rassenonderzoek van cultuurgewassen. HI 67637

Mededeelingen uit het instituut voor tropische geneeskunde = Acta leidensia, edita cura et sumptibus medicinae tropicae. Leiden. Acta Leidensia. See B–P–H 45/6.

Mededelingen uit de Java-suikerindustrie. Pasuruan. 1955-57. Meded. Java-Suikerindustr. Preceded by: Mededeelingen van het proefstation voor de Java-suikerindustrie. Landbouwkundige serie. Superseded by: Berita, balai penjelidikan perusahaan gula. HI 67639

Mededeelingen van het kina-proefstation. Batavia, Dutch E. Indies [=Jakarta, Indonesia]. Meded. Kina-Proefstat. See B–P–H 559/3. HI 56205

Mededelingen, Koffiebessenboeboek-fonds = Koffiebessenboeboek-fonds: mededelingen. Surabaya, Dutch E. Indies [Indonesia]. Koffiebessenboeboek-Fonds Meded. See B–P–H 515/16.

Mededelingen, koninklijk belgisch institut voor natuurwetenschappen = Bulletin de l'institut royal des sciences naturelles de Belgique. Brussels.

Mededelingen van de koninklijke academie voor wetenschappen, letteren en schone kunsten van België. Klasse der wetenschappen. Brussels. Vol. 34+, 1972+. Meded. Kon. Acad. Wetensch. Lett. Schone Kunsten België, Klasse Wetensch. Preceded by: Mededelingen van de k. vlaamse academie voor wetenschappen, letteren en schone kunsten van België. Klasse der wetenschappen. HI 67640

Mededeelingen van het koninklijk natuurhistorisch museum van België = Bulletin du musée royal d'histoire naturelle de Belgique. Brussels. Bull. Mus. Roy. Hist. Nat. Belgique. See B–P–H 264/24.

Mededelingen van de k[onklijke]. vlaamse academie voor wetenschappen, letteren en schone kunsten van België. Klasse der wetenschappen Ghent, Brussels. Vols. 1-33, 1939-71. Meded. Kon. Vlaamse Acad. Wetensch. België, Kl. Wetensch. Superseded by: Mededelingen van de koninklijke academie voor wetenschappen, letteren en schone kunsten van België. Klasse der wetenschappen. 5-4414-1. HI 67641

Mededeelingen van de laboratoria der gouvernement's kinaonderneming "Batavia". Batavia [=Jakarta, Indonesia]. Meded. Lab. Gouv. Kinaondern. "Batavia". See B–P–H 559/11. HI 56206

Mededelingen van het laboratorium voor bloembollenonderzoek. Lisse. Nos. 1-?, 1918-? Meded. Lab. Bloembollenonderz. Superseded by: Mededelingen van het instituut voor phytopathologie, laboratorium voor bloembollenonderzoek te Lisse. 3-2432-3. HI 67642

Mededeelingen van het laboratorium voor plantenphysiologie. Wageningen. Meded. Lab. Plantenphysiol. See B–P–H 559/12. HI 56207

Mededeelingen van het laboratorium voor plantenphysiologisch onderzoek. Wageningen. Meded. Lab. Plantenphysiol. Onderz. See B–P–H 559/13. HI 56208

Mededeelingen van het laboratorium voor plantenziekten. Batavia, Dutch E. Indies [=Jakarta, Indonesia]. Meded. Lab. Plantenziekten. See B–P–H 559/14. HI 56209

Mededeelingen van de landbouwhoogeschool en van de daaraan verbonden instituten. Wageningen. Vols. 14-27, 1918-20. Meded. Landbouwhoogeschool. Preceded by: Mededeelingen van de rijks hoogere land-, tuin- en boschbouwschool en van de daaraan verbonden instituten. Superseded by: Mededeelingen van de landbouwhoogeschool te Wageningen. 5-4437-2. HI 67643

Mededelingen van de landbouwhoogeschool te Wageningen = Mededeelingen van de rijks hoogere land-, tuin- en boschbouwschool en van de daaraan verbonden instituten. Wageningen.

Mededeelingen van de landbouwhoogeschool en de opzoekingsstation van de staat te Gent. Oudenaarde. Vols. 1-30, 1908-65. Meded. Landbouwhoogeschool Opzorkingsstat. Staat Gent. Superseded by: Mededelingen, faculteit landbouwwetenschappen, rijksuniversiteit. Ghent. 2-1724-1. HI 67644

Mededeelingen van de landbouwhoogeschool te Wageningen. Wageningen. Vol. 28-83, 192?-83. Meded. Landbouwhoogeschool. Preceded by: Mededeelingen van de landbouwhoogeschool en van de daaraan verbonden instituten. Superseded by: Agricultural university Wageningen papers. 5-4437-2. HI 67645

Mededeelingen. Landbouwproefstation Suriname. Paramaribo, Dutch Guiana [Surinam]. Meded. Landbouwproefstat. Suriname. See B–P–H 559/17. HI 56210

Mededeelingen van den landbouwvoorlichtingsdienst. [Dutch E. Indies] [=Indonesia]. Meded. Landbouwvoorlichtingsdienst. See B–P–H 558/18. HI 56211

Mededeelingen uit 's lands plantentuin. Batavia, Dutch E. Indies [=Jakarta, Indonesia]. Meded. Lands Plantentuin. See B–P–H 559/19. HI 56212

Mededeelingen van het natuurhistorisch genootschap in Limburg. Assen, Netherlands. Meded. Natuurhist. Genootsch. Limburg. See B–P–H 560/1. HI 56214

Mededelingen. Nederlands vlasinstituut Wageningen. Wageningen. Meded. Ned. Vlasinst. Wageningen. See B–P–H 560/5. HI 56217

Mededelingen. Nederlandse algemene keuringsdienst voor landbouwzaden en aardappelpootgoed. Wageningen. Meded. Ned. Alg. Keuringsdienst Landbouwz. See B–P–H 560/3. HI 56215

Mededelingen, nederlandsche botanische vereeniging. 1941. Meded. Ned. Bot. Ver. HI 67646

Mededeelingen nederlandsche mycologische vereeniging. Wageningen. Vols. 1-30, 1910-52. Meded. Ned. Mycol. Ver. HI 67647

Mededeelingen der nederlandsche pomologische vereeniging. Utrecht. Meded. Ned. Pomol. Ver. See B–P–H 560/4. HI 56216

Mededeelingen van het Nieuw Guinee comité en den Nieuw Guinee studiekring van het Molukken instituut. The Hague. Meded. Nieuw Guinee Comité Nieuw Guinee Studiekring Molukken Inst. See B–P–H 560/6. HI 56218

Mededeelingen uit het phytopathologisch laboratorium "Willie Commelin Scholten". Amsterdam. Meded. Phytopathol. Lab. "Willie Commelin Scholten". See B–P–H 560/8. HI 56219

Mededelingen, proefstation voor de akker- en weidebouw. Wageningen. Nos. 1-166, 1957-70. Meded. Proefstat. Akker-Weideb. Superseded by: Rapport, proefstation voor de rundveehouderij, Wageningen and Publikatie, proefstation voor de akkerbouw te Lelystad. HI 67648

Mededeelingen van het proefstation voor het boschwezen. Batavia, Dutch E. Indies [=Jakarta, Indonesia]. Meded. Proefstat. Boschw. See B–P–H 560/10. HI 56220

Mededelingen van het proefstation voor de fruitteelt in de volle grond. Wilhelminadorp. No. 1+, 1959+. Meded. Proefstat. Fruitteelt. Volle Grond. HI 67649

Mededelingen van het proefstation te Groenendaal. Groenendaal, Belgium. Meded. Proefstat. Groenendaal. See B–P–H 560/11. HI 56221

Mededeling, proefstation voor de groenteteelt in de volle grond in Nederland. Alkmaar. Nos. 1-68, 1955-77. Meded. Proefstat. Groentet. Volle Grond Ned. HI 67650

Mededeelingen van het proefstation voor de Java-suikerindustrie. Landbouwkundige serie. Surabaya. Nos. 1-?, 1907-34. Meded. Proefstat. Java-Suikerindustr., Landbouk. Ser. Preceded by: Mededeelingen van het proefstation oost-Java. Superseded by: Mededelingen uit de Java-suikerindustrie. 1-432-3. HI 67651

Mededeelingen van het proefstation Malang. Malang, Dutch E. Indies [Indonesia]. Meded. Proefstat. Malang. See B–P–H 560/13. HI 56222

Mededeelingen van het proefstation Midden-Java. Semarang & Klaten, Dutch E. Indies [Indonesia]. Meded. Proefstat. Midden-Java. See B–P–H 560/14. HI 56223

Mededeelingen van het proefstation Midden-Java te Salatiga. Salatiga. Nos. 1-44, 1911-27. Meded. Proefstat. Midden-Java Salatiga. HI 67652

Mededeelingen van het proefstation oost-Java. Surabaya. Nos. 1-50, 1887-93; n.s. nos. 1-50, 1893-98; ser. 3 nos. 1-50, 1898-1903; ser. 4 nos. 1-39, 1903-08. Meded. Proefstat. Oost-Java. Superseded by: Mededeelingen van het proefstation voor de Java suikerindustrie. HI 67653

Mededeelingen van het proefstation voor rijst. Batavia, Dutch E. Indies [=Jakarta, Indonesia]. Meded. Proefstat. Rijst. See B–P–H 560/16. HI 56224

Mededeelingen van het proefstation voor rubber. Buitenzorg, Dutch E. Indies [=Bogor, Indonesia]. Meded. Proefstat. Rubber. See B–P–H 560/17. HI 56225

Mededeelingen van het proefstation voor suikerriet in West-Java te Kagok-Tegal. Dresden. Meded. Proefstat. Suikerriet W.-Java Kagok-Tegal. See B–P–H 560/18. HI 56226

Mededeelingen van het proefstation voor thee. Buitenzorg, Dutch E. Indies [=Bogor, Indonesia]. Meded. Proefstat. Thee. See B–P–H 560/19. HI 56227

Mededeelingen. Proefstation vorstenlandsche tabak. Buitenzorg, Dutch E. Indies [=Bogor, Indonesia]. Meded. Proefstat. Vorstenl. Tabak. See B–P–H 561/1. HI 56228

Mededeelingen van de rijks hoogere land-, tuin- en boschbouwschool en van de daaraan verbonden instituten. Wageningen. Nos. 1-13, 1908-17. Meded. Rijks Hoogere Land- Boschbouwsch. Superseded by: Mededeelingen van de landbouwhoogeschool en van de daaraan verbonden instituten. 5-4437-2. HI 67654

Mededelingen, rijksfakulteit der landbouwwetenschappen. Ghent. Nos. 31-34, 1966-69. Meded. Rijksfak. Landbouwwetensch. Preceded by: Mededeelingen van de landbouwhogeschool en de opzoekingsstations van de staat de Gent. Superseded by: Mededelingen, faculteit landbouwwetenschappen, rijksuniversiteit. HI 67655

Mededeling, rijksinstitut voor onderzoek in de bos- en landschapsbouw "De Dorschkamp". Wageningen. No. 142+, 1975+. Meded. Rijksinst. Onderz. Bos-Landschapsbouw "De Dorschkamp". Preceded by: Mededeling, stichting bosbouwproefstation "De Dorschkamp" HI 67656.

Mededelingen van het rijksinstituut voor het rassenonderzoek van cultuurgewassen. Wageningen. No. 57+, 1969?+. Meded. Rijksinst. Rassenonderz. Cultuurgew. Preceded by: Mededelingen van het instituut voor rassenonderzoek van landbouwgewassen te Wageningen. HI 67657

Mededelingen, rijksstation voor landbouwtechniek. Merelbeke. 1964+. Meded. Rijksstat. Landbouwtechn. HI 67658

Mededeelingen van het rijksstation voor sierplantenteelt. Melle = Mededeling, station des plantes ornamentales, Melle.

Mededeelingen van het rubberproefstation "West-Java". Batavia. Vols. ?-[27]-42-?, ?-1920-23-? Meded. Rubberproefstat. "West-Java". HI 67660

Mededelingen, station des plantes ornamentales, Melle. Melle. No. 1+, 196?+. Med. Sta. Pl. Ornam., Melle. HI 67661

Mededelingen, station de recherches pour l'amélioration des plantes ornamentales, Melle = Mededeling, station des plantes ornamentales, Melle.

Mededelingen, stichting bosbouwproefstation "De Dorschkamp". Wageningen. Nos. ?-141, ?-1974. Meded. Stichting Bosbouwproefstat. "De Dorschkamp". Superseded by: Mededeling, rijksinstitut voor onderzoek in dee bos- en landschapsbouw "De Dorschkamp". HI 67662

Mededelingen van den tuinbouwvoorlichtingsdienst. The Hague. 19??-55? Meded. Tuinbouwvoorlicht. Superseded by: Tuinbouwvoorlichting. HI 67663

Mededelingenblad, hydrobiologische vereniging. Amsterdam. Vols. 1-6(3), 196?-72. Mededelingenblad Hydrobiol. Ver. Superseded by: Hydrobiological bulletin. HI 67664

Mededelingenblad, koninklijke vereniging voor natuur- en stedeschoon. Antwerp. Vols. ?-11-15, ?-1975-79. Mededelingenblad Kon. Ver. Nat. Stedesch. Superseded by: Natuur- en stedeschoon, koninklijke vereniging voor natuur- en stedeschoon. HI 67665

Mededelingenblad van de nederlandse jeugdbond voor natuurstudie = Amoeba. Amsterdam.

Median iris society newsletter = Newsletter, median iris society. Grafton, MA.

Medianite. Vol. 3+, 1962+. Medianite. Preceded by: Newsletter, median iris society. HI 67666

Medical age. Detroit, MI. Med. Age. See B–P–H 553/27. HI 56137

Medical and agricultural register. Boston, MA. Med. Agric. Reg. See B–P–H 553/28. HI 56138

Medical and biological illustration. London. Vols. 1-27, 1951-77. Superseded by: Journal of audiovisual media in medicine. HI 67667

Medical biology. Helsinki. Vol. 52+, 1974+. Med. Biol. Preceded by: Annales medicinae experimentalis et biologiae Fenniae. HI 67668

Medical clinics of North America. Philadelphia, PA, & London. Med. Clin. N. Amer. See B–P–H 554/7. HI 56143

Medical facts and observations. London. Med. Facts Observ. See B–P–H 554/18. HI 56152

Medical historical studies of medieval jewish medical works. Brooklyn, NY. Vol. 1+, 1961+. Med. Hist. Stud. Medieval Jew. Works. HI 67669

Medical history; a quarterly journal devoted to the history and bibliography of medicine and related sciences. London. Med. Hist. See B–P–H 554/19. HI 56153

Medical journal of Australia. Sydney. Med. J. Australia. See B–P–H 554/22. HI 56155

Medical microbiology and immunology. Berlin. Vol.

157+, 1971+. Med. Microbiol. Immunol. Preceded by: Zeitschrift für medizinische Mikrobiologie und Immunologie. HI 67670

Medical and physical journal. London. Med. Phys. J. See B–P–H 555/3. HI 56158

Medical press and circular. London and Dublin. Med. Press Circ. See B–P–H 555/6. HI 56159

Medical quarterly review. London. Med. Quart. Rev. See B–P–H 555/8. HI 56160

Medical record. New York, NY. Med. Rec. See B–P–H 555/9. HI 56161

Medical recorder of original papers and intelligence in medicine and surgery. Philadelphia, PA. Med. Rec. Original Pap. Intelligence Med. Surg. See B–P–H 555/10. HI 56162

Medical repository. New York, NY. Med. Repos. See B–P–H 555/12. HI 56163

Medical repository of original essays and intelligence relative to physic, surgery, chemistry, and natural history; ... New York, NY. Med. Repos. Original Essays Intelligence Phys. See B–P–H 555/15. HI 56164

Medical sentinel = Western journal of surgery, obstetrics and gynecology. Portland, OR. W. J. Surg. See B–P–H 970/4.

Medicina, cirugia, pharmacia. Rio de Janeiro. Med. Cirugia Pharm. See B–P–H 554/5. HI 56142

Medicina contemporânea. Lisbon. Med. Contemp. (Lisbon). See B–P–H 554/9. HI 56144

Medicina contemporánea. Madrid. Med. Contemp. (Madrid). See B–P–H 554/10. HI 56145

Medicina contemporanea. Naples. Med. Contemp. (Naples). See B–P–H 554/11. HI 56146

Medicina contemporánea. Reus, Spain. Med. Contemp. (Reus). See B–P–H 554/13. HI 56148

Medicina contemporânea. São Paulo. Med. Contemp. (São Paulo). See B–P–H 554/14. HI 56149

Medicina contemporanea. Turin. Med. Contemp. (Turin). See B–P–H 554/15. HI 56150

Medicina ibera; revista semanal de medicina y cirugía. Madrid. Med. Ibera. See B–P–H 554/20. HI 56154

Medicina moderna. Oporto, Portugal. Med. Moderna. See B–P–H 555/1. HI 56156

Medicine and biology. [Igaku to seibutsugaku.] [Japan]. Med. & Biol. See B–P–H 553/26. HI 56136

Médicine contemporaine. Paris. Med. Contemp. (Paris). See B–P–H 554/12. HI 56147

Médicine eclairée par les sciences physiques, ou journal des découvertes relatives aux différentes parties de l'art de guérir. Paris. Méd. Eclairée Sci. Phys. See B–P–H 554/17. HI 56151

Medicinal and aromatic plants abstracts; reporting current world literature. New Delhi. Vol. 1+, 1979+. Med. Aromat. Pl. Abstr. Preceded by: Bulletin of indian raw materials and their utilization. Series A, current literature on medicinal and aromatic plants. HI 67671

Medicinisch-chirurgische Monatshefte. Erlangen. Vols. 1-15, 1857-64. Med.-Chir. Monatsh. Preceded by: Neue medicinisch-chirurgische Zeitung. 3-2593-2. HI 67672

Medicinisch-chirurgische Zeitung. Salzburg. 1790-1839. [4 issues called "Band" published annually]. Med.-Chir. Zeitung. Superseded by: Neue medicinisch-chirurgische Zeitung. 4-2969-3. HI 50827

Medicinisch-chirurgische Zeitung. Ergänzungsband. Salzburg. Med.-Chir. Zeitung Ergänzungsband. See B–P–H 556/4. HI 56171

Medicinisch-practische Bibliothek; worin Nachrichten von den neuesten zur Ausübung der Heilkunde gehörigen Schriften und Vorfällen geliefert werden. Göttingen. Vols. 1-3, [1774] 1775-80. Med.-Pract. Biblioth. Superseded by: Medicinische Bibliothek. 3-2593-3. HI 67673

Medicinische Bibliothek. Göttingen. Vols. 1-3, 1783-95. Med. Biblioth. (Göttingen). Preceded by: Medicinisch-practische Bibliothek. 3-2594-1. HI 67674

Medicinische Zeitung. Herausgegeben von dem Verein für Heilkunde in Preussen. Berlin. Med. Zeitung. See B–P–H 555/24. HI 56169

Medizinischen Versuche, nebst Bemerkungen, welche von einer Gesellschaft in Edinburgh durchgesehen und herausgegeben werden. Altenburg. Med. Versuche Bemerk. Ges. Edinburgh . See B–P–H 555/19. HI 56165

Medicinisches Archiv von Wien und Oesterreich unter der Enns. Vienna. Med. Arch. Wien Oesterreich unter der Enns. See B–P–H 553/30. HI 56139

Medicinisches Journal. Göttingen. Vols. [1](1)-6(21), 1784-89. Med. J. (Göttingen). Superseded by: Medicinisches und physisches Journal. 4-3357-1. HI 53832

Medicinisches und physisches Journal. Göttingen. Vols. 6(22)-19(63), 1790-96. Med. Phys. J. (Göttingen). Preceded by: Medicinisches Journal. 4-3357-1. HI 53833

Medicinisches Wochenblatt für Aerzte, Wundärzte und Apotheker. Frankfurt a. M. Med. Wochenbl. Aerzte. See B–P–H 555/21. HI 56167

Medizinisches Wochenblatt oder fortgesetzte medizinische Annalen. Frankfurt a. M. Med. Wochenbl. Fortgesetzte Med. Ann. See B–P–H 555/

22. HI 56168

Medico veterinario. Turin. Med. Veterin. See B–P–H 555/20. HI 56166

Medio ambiente, instituto de ecologiá y evolución, universidad austral de Chile. Valdivia. Vol. 1+, 1975+. Medio Amb. Ecol. Evol. Univ. Austral Chile. HI 67675

Mediterranea; serie de estudios sobre biologia terrestre mediterranea. Revue des problemes agricoles mediterraneens. Alicante. No. 1+, 1976+. Mediterranea. Superseded by: Options méditerranéennes. HI 67676

Mediterranean naturalist; a monthly journal of natural science. Valletta, Malta. Medit. Naturalist. See B–P–H 562/1. HI 56242

Meditsinskaya promyshlennost S S S R. Moscow. 1947-66. Med. Promyshl. S.S.S.R. Superseded by: Khimiko farmatsevticheskii zhurnal. HI 67677

Medizinhistorisches Journal. Hildesheim. Vol. 1+, 1966+. Medizinhist. J. HI 67678

Medizinische und chirurgische Berlinische wöchentliche Nachrichten. Berlin. Med. Chir. Berlin. Wöchentl. Nachr. See B–P–H 554/4. HI 56141

Medizinische National-Zeitung für Deutschland, und die mit selbigem zunächst verbundenen Staaten. Altenburg. Vols. 1-2, 1798-99. Suppl. 1798. Med. Natl.-Zeitung Deutschl. Superseded by: Allgemeine medizinische Annalen. 1-144-1. HI 53393

Medlemsblad Lunds botanska föreningen. Lund. ?-1975+. Medlemsblad Lunds Bot. Fören. HI 67679

Meehans' monthly; a magazine of horticulture, botany and kindred subjects. Philadelphia, PA. Meehans' Monthly See B–P–H 562/7. HI 56243

Meer und Museum. Stralsund. Vol. 1+, 1980+. Meer & Mus. HI 67680

Meeresforschung. Hamburg. Vols. 25-30, 1976-85. Meeresforschung. Preceded by: Berichte, deutsche wissenschaftliche Kommission für Meeresforschung. Superseded by: Reports on marine research. HI 67681

Meereskunde. Berlin. Meereskunde. See B–P–H 562/8. HI 56244

Meereskundliche Beobachtungen und Ergebnisse, deutsches hydrographisches Institut. Hamburg. 1953+. Meereskundl. Beob. Ergebn. Deutsch. Hydogr. Inst. Preceded by: Meereskundliche Beobachtungen (und Ergebnisse) auf deutschen Feuerschiffen (der Nord und Ostsee). HI 67682

Meereskundliche Beobachtungen (und Ergebnisse) auf deutschen Feuerschiffen (der Nord und Ostsee). Hamburg. 1924-52. Meereskundl. Beob. Ergebn. Deutsch. Feuerschiffen N. Ostsee. Superseded by: Meereskundliche Beobachtungen und Ergebnisse, deutsches hydrographisches Institut. HI 67683

Méhészet. Budapest. Méhészet. See B–P–H 562/9. HI 56245

Mejores cosechas. Caracas. Mejores Cosechas. See B–P–H 562/11. HI 56246

Mekhanizmy peredvizheniya i orientatsii zhivotnykh, Kiev, 1968 = Bionika. Kiev.

Mélanges biologiques tirés du Bulletin (physico-mathématique) de l'académie impèriale des sciences de Saint-Pétersbourg. St. Petersburg. Mélanges Biol. Bull. Phys.-Math. Acad. Imp. Sci. Saint-Pétersbourg. See B–P–H 562/20. HI 56247

Mélanges botaniques, ou recueil d'observations, mémoires et notices sur la botanique. Bern. Mélanges Bot. See B–P–H 562/21. HI 56248

Mélanges de philosophie et de mathématique de la société royale de Turin. Turin. Mélanges Philos. Math. Soc. Roy. Turin. See B–P–H 562/22. HI 56249

Mélanges scientifiques; recueil de mémoires, discours, rapports, notices biographiques, etc., imprimés à diverses époques ou inédits. Limoges, France. Mélanges Sci. See B–P–H 562/23. HI 56250

Melbourne chemist & druggist. Melbourne. Melbourne Chem. Druggist. See B–P–H 562/26. HI 56251

Meldinger fra norges landbrukshøgskole. Vollebekk, Christiania. 1921+. Meld. Norg. Landbrukshøgskole. HI 67684

Meldinger fra statens plantepatologische institutt. Oslo. No. 1, 1945. Meld. Statens Plantepatol. Inst. Superseded by: Meldinger fra statens plantevern. HI 67685

Meldinger fra statens plantevern. Oslo. No. 2+, 1946+. Meld. Statens Plantevern. Preceded by: Meldinger fra statens plantepatologische institutt. HI 67686

Melhoramento; estudos da estação de melhoramento de plantas. Elvas. Vol. 1+, 1948+. Melhoramento. 3-2598-3. HI 67687

Melhoramento e producção de sementes de algumas hortaliças. Elvas, Portugal. Melhor. Prod. Sementes Hort. See B–P–H 562/29. HI 56252

Member newsletter, horticultural society of New York. New York. Nos. 1-?, 1969?-83. Memb. Newslett. Hort. Soc. New York. Superseded by: Newsletter, horticultural society of New York. HI 67688

Member's newsletter, flower essence society. Nevada City, CA. 1981+. Memb. Newslett. Fl. Essence Soc. HI 67689

Members' newsletter, Minnesota landscape arboretum. Chaska, MN. Vols. 1-7(1), 197?-81. Memb. Newslett. Minnesota Landscape Arbor. Superseded by: News,

Minnesota landscape arboretum. HI 75090

Mémoires de l'académie impériale de Metz. Metz. Mém. Acad. Imp. Metz. See B–P–H 563/16. HI 56255

Mémoires de l'académie impériale des sciences, belles-lettres et arts de Lyon. Section des lettres et arts = Mémoires de l'académie royale des sciences, belles-lettres et arts de Lyon. Section des lettres et arts. Lyons. Mém. Acad. Roy. Sci. Lyon, Sect. Lett. See B–P–H 565/13.

Mémoires de l'académie impériale des sciences, belles-lettres et arts de Lyon. Section des sciences = Mémoires de l'académie royale des sciences, belles-lettres et arts de Lyon. Section des sciences. Lyons. Mém. Acad. Roy. Sci. Lyon, Sect. Sci. See B–P–H 566/1.

Mémoires de l'académie impériale des sciences de Caen. Caen. 1852-71. Mém Acad. Imp. Sci. Caen. Preceded by: Mémoires de l'académie des sciences de Caen. Superseded by: Mémoires de l'académie nationale des sciences, Caen. HI 67690

Mémoires de l'académie impériale des sciences, littérature et beaux-arts de Turin. Sciences physiques et mathématiques. Turin. Mém. Acad. Imp. Sci. Turin, Sci. Phys. See B–P–H 564/2. HI 56260

Mémoires de l'académie impériale des sciences de St. Pétersbourg. Avec l'histoire de l'académie. St. Petersburg. Mém. Acad. Imp. Sci. St. Pétersbourg Hist. Acad. See B–P–H 563/26. HI 56257

Mémoires de l'académie impériale des sciences de Saint Pétersbourg, Classe des sciences physico-mathématiques = Zapiski Imperatorskoi Akademii Nauk po Fiziko-matematicheskomu Otděleniyu. St. Petersburg.

Mémoires de l'académie impériale des sciences de Saint Pétersbourg, Classe physico-mathématique = Zapiski Imperatorskoi Akademii Nauk po Fiziko-matematicheskomu Otděleniyu. St. Petersburg.

Mémoires de l'académie impériale des sciences de Saint Pétersbourg, Septième série. St. Petersburg. Vols. 1-42, 1859-97. Mém. Acad. Imp. Sci. Saint Pétersbourg, Sér. 7. Preceded by: Mémoires de l'académie impériale des sciences de Saint Pétersbourg. Sixième série. Sciences mathématiques, physiques et naturelles. Seconde partie: Sciences naturelles. Superseded by: Zapiski Imperatorskoi Akademii Nauk po Fiziko-matematicheskomu Otděleniyu. 1-113-1. HI 67691

Mémoires de l'académie impériale des sciences de St.-Pétersbourg. Sixième série. Sciences mathématiques, physiques et naturelles. St. Petersburg. Mém. Acad. Imp. Sci. St.-Pétersbourg, Sér. 6, Sci. Math. See B–P–H 564/1. HI 56259

Mémoires de l'académie impériale des sciences de Saint-Pétersbourg. Sixième série. Sciences mathématiques, physiques et naturelles. Seconde partie: Sciences naturelles. St. Petersburg. Mém. Acad. Imp. Sci. Saint-Pétersbourg, Sér. 6, Sci. Math., Seconde Pt. Sci. Nat. See B–P–H 563/24. HI 56256

Mémoires de l'académie impériale des sciences de Saint-Pétersbourg. Sixième série. Sciences naturelles = Mémoires de l'académie impériale des sciences de Saint-Pétersbourg. Sixième série. Sciences mathématiques, physiques et naturelles. Seconde partie: Sciences naturelles. St. Petersburg. Mém. Acad. Imp. Sci. Saint-Pétersbourg, Sér. 6, Sci. Math., Seconde Pt. Sci. Nat. See B–P–H 563/24.

Mémoires de l'académie international de géographie botanique. Le Mans. Vols. 1-23?, 1911-13? Mém. Acad. Int. Géogr. Bot. HI 67692

Mémoires de l'académie malgache. Tanavive, Madagascar [Malagasy Republic]. Mém. Acad. Malgache. See B–P–H 564/4. HI 56261

Mémoires de l'académie de médecine. Paris = Mémoires de l'académie royale de médecine. Paris, London, & Brussels. Mém. Acad. Roy. Méd. See B–P–H 564/21.

Mémoires de l'académie de Metz. Metz. Mém. Acad. Metz. See B–P–H 564/6. HI 56263

Mémoires de l'académie nationale de Metz. Lettres, sciences, arts, agriculture. Metz. Mem. Acad. Natl. Metz. See B–P–H 564/13. HI 56266

Mémoires de l'académie nationale des sciences, arts et belles-lettres, de Caen = Mémoires de l'académie royale des sciences, arts et belles-lettres, de Caen. Caen.

Mémoires de l'académie nationale des sciences, belles-lettres et arts de Lyon. Classe des sciences = Mémoires de l'académie royale des sciences, belles-lettres et arts de Lyon. Section des sciences. Lyons. Mém. Acad. Roy. Sci. Lyon, Sect. Sci. See B–P–H 566/1.

Mémoires de l'académie nationale des sciences, belles-lettres et arts de Lyon. Section des lettres et arts. = Mémoires de l'académie royale des sciences, belles-lettres et arts de Lyon. Section des lettres et arts. Lyons. Mém. Acad. Roy. Sci. Lyon, Sect. Lett. See B–P–H 565/13.

Mémoires de l'académie nationale des sciences de Caen. Caen. 1872+. Mém Acad. Nat. Sci. Caen. Preceded by: Mémoires de l'académie impériale des sciences de Caen. HI 67693

Mémoires de l'académie de Nîmes. Nîmes. Mém. Acad. Nîmes. See B–P–H 564/17. HI 56267

Mémoires de l'académie polonaise des sciences. Classe des sciences mathématiques et naturelles. Ser. B, sciences naturelles. Cracow. Nos. 1-18, 1928-52 [publication suspended 1940-48]. Mém. Acad. Polon.

Sci., Cl. Sci. Math. Nat., B. 4-3390-3. HI 67694

Mémoires de l'académie royale du Gard. Nîmes. Mém. Acad. Roy. Gard. See B–P–H 564/20. HI 56269

Mémoires de l'académie royale de médecine. Paris, London, & Brussels. Mém. Acad. Roy. Méd. See B–P–H 564/21. HI 56270

Mémoires de l'académie royale de Metz. Lettres, sciences, arts, agriculture. Metz. Mém Acad. Roy. Metz. See B–P–H 565/1. HI 56271

Mémoires de l'académie royale de Prusse. Concernant l'anatomie; la physiologie; la physique; l'histoire naturelle; la botanique; la mineralogie; ... Avignon. Mém. Acad. Roy. Prusse Anat. (Avignon). See B–P–H 565/2. HI 56272

Mémoires de l'académie royale de Prusse. Concernant l'anatomie; la physiologie; la physique; l'histoire naturelle; la botanique; la mineralogie, ... Paris. Mém. Acad. Roy. Prusse Anat. (Paris). See B–P–H 565/3. HI 56273

Mémoires de l'académie royale des sciences. Paris. Mém. Acad. Roy. Sci. (Paris). See B–P–H 565/4. HI 56274

Mémoires de l'académie royale des sciences. Turin. Mém. Acad. Roy. Sci. (Turin). See B–P–H 565/6. HI 56276

Mémoires de l'académie royale des sciences, arts et belles-lettres de Caen. Caen. 1820/23-47, 1823-47?; 2nd edn. 1823/24, 1825. Mém. Acad. Roy. Sci. Caen. Superseded by: Mémoires de l'académie des sciences de Caen. 1-32-2. HI 67695

Mémoires de l'académie royale des sciences, belles-lettres et arts de Lyon. Section des lettres et arts. Lyons. Mém. Acad. Roy. Sci. Lyon, Sect. Lett. See B–P–H 565/13. HI 56282

Mémoires de l'académie royale des sciences, belles-lettres et arts de Lyon. Section des sciences. Lyons. Mém. Acad. Roy. Sci. Lyon, Sect. Sci. See B–P–H 566/1. HI 56283

Mémoires de l'académie royale des sciences et Belles-lettres depuis l'avénement de Fréderic Guillaume II [later: III] au thrône. Avec l'histoire. Berlin. Mém. Acad. Roy. Sci. Hist. (Berlin). See B–P–H 565/11. HI 56280

Mémoires de l'académie royale des sciences de Caen = Mémoires de l'académie royale des sciences, arts et belles-lettres de Caen. Caen.

Mémoires de l'académie royale des sciences coloniales. Classe des sciences naturelles et médicales. Brussels. N.s. vols. 1-5, 1955-59. Mém. Acad. Roy. Sci. Colon., Sect. Sci. Nat. Preceded by: Mémoires de l'institut royal colonial belge; section des sciences naturelles et médicales (8vo). Superseded by: Mémoires de l'académie royale des sciences d'outre-mer. Classe des sciences naturelles et médicales. Collection in 8vo. 5-33-2. HI 67696

Mémoires de l'académie royale des sciences [Paris] contenant ..., avant son renouvellement en 1699. The Hague. Mém. Acad. Roy. Sci. (The Hague). See B–P–H 565/5. HI 56275

Mémoires de l'académie royale des sciences de l'institut de France. Paris. Mém. Acad. Roy. Sci. Inst. France. See B–P–H 565/12. HI 56281

Mémoires de l'académie royale des sciences, lettres et beaux arts de Belgique. Brussels. Mém. Acad. Roy. Sci. Beligique. See B–P–H 565/7. HI 56277

Mémoires de l'académie royale des sciences, lettres et beaux-arts de Belgique. Classe des sciences. [In octavo.] Brussels. Mém. Acad. Roy. Sci. Belgique, Cl. Sci. (8vo). See B–P–H 565/8. HI 56278

Mémoires de l'académie royale des sciences d'outre-mer. Classe des sciences naturelles et médicales. Collection in 8vo. Brussels. N.s. vol. 6+, 1960+. Mém. Acad. Roy. Sci. Outre-Mer, Cl. Sci. Nat. Méd., Collect. 8vo. Preceded by: Mémoires, académie royale des sciences coloniales. Classe des sciences naturelles et médicales. HI 67697

Mémoires de l'académie royale des sciences de Stockholm; concernant l'histoire naturelle, la physique, la médecine, l'anatomie, la chymie, l'oeconomie, les arts, etc. Paris. 1 vol., 1772. Mém. Acad. Roy. Sci. Stockholm. HI 67698

Mémoires de l'académie royale des sciences de Turin. Mém. Acad. Roy. Sci. Turin. See B–P–H 566/3. HI 56284

Mémoires de l'académie des sciences. Paris. Mém. Acad. Sci. (Paris). See B–P–H 566/7. HI 56285

Mémoires de l'académie des sciences, agriculture, commerce, belles-lettres, et arts du département de la Somme. Amiens, France. Mém. Acad. Sci. Dép. Somme. See B–P–H 566/15. HI 56289

Mémoires de l'académie des sciences, arts et belles-lettres de Dijon. Dijon. Mém. Acad. Sci. Dijon. See B–P–H 566/16. HI 56290

Mémoires de l'académie des sciences, arts et belles-lettres de Dijon. Partie des sciences. Dijon. Mém. Acad. Sci. Dijon, Pt. Sci. See B–P–H 566/17. HI 56291

Mémoires de l'académie des sciences et belles-lettres d'Angers. Angers. Mém. Acad. Sci. Angers. See B–P–H 566/10. HI 56287

Mémoires de l'académie des sciences, belles-lettres et arts d'Amiens. Amiens, France. Mém. Acad. Sci. Amiens. See B–P–H 566/9. HI 56286

Mémoires de l'académie des sciences, belles-lettres et arts d'Angers. Angers. Mém. Acad. Sci. Arts Angers. See

B–P–H 566/11. HI 56288

Mémoires de l'académie des sciences, belles-lettres et arts de Clermont-Ferrand. Clermont-Ferrand. N.s. vols. 1-29, 1859-87; 2. sér., vol. 1+, 1890+. Mém. Acad. Sci. Clermont-Ferrand. Preceded by: Annales scientifiques, littéraires et industrielles de l'Auvergne. HI 67699

Mémoires de l'académie des sciences, belles-lettres et arts de Lyon. Section des lettres et arts = Mémoires de l'académie royale des sciences, belles-lettres et arts de Lyon. Section des lettres et arts. Lyons. Mém. Acad. Roy. Sci. Lyon, Sect. Lett. See B–P–H 565/13.

Mémoires de l'académie des sciences, belles-lettres et arts de Lyon. Section des sciences = Mémoires de l'académie royale des sciences, belles-lettres et arts de Lyon. Section des sciences. Lyons. Mém. Acad. Roy. Sci. Lyon, Sect. Sci. See B–P–H 566/1.

Mémoires de l'académie des sciences de Caen. Caen. 1849-52. Mém. Acad. Sci. Caen. Preceded by: Mémoires de l'académie royale des sciences, arts et belles-lettres de Caen. Superseded by: Mémoires de l'académie impériale des sciences de Caen. HI 67700

Mémoires. Académie des sciences, inscriptions et belles-lettres de Toulouse. Toulouse. Mém. Acad. Sci. Toulouse. See B–P–H 567/6. HI 56292

Mémoires de l'académie des sciences de l'institut de France = Mémoires de l'académie royale des sciences de l'institut de France. Paris. Mém. Acad. Roy. Sci. Inst. France. See B–P–H 565/12.

Mémoires de l'académie des sciences, littérature et beaux-arts de Turin. Sciences physiques et mathématiques. Turin. Mém. Acad. Sci. Turin, Sci. Phys. See B–P–H 567/8. HI 56294

Mémoires de l'académie des sciences, Petrograd = Zapiski Akademii nauk.

Mémoires de l'académie des sciences de Russie = Zapiski Rossiiskoi akademii nauk. Petrograd.

Mémoires de l'académie des sciences de Turin. Turin. Mém. Acad. Sci. Turin. See B–P–H 567/7. HI 56293

Mémoires de l'académie des sciences de l'U S S R, Classe physique et mathématique = Zapiski Akademii nauk S S S R po fiziko-matematicheskomu otdeleniyu. Leningrad.

Mémoires de l'académie de Stanislas. Nancy. Mém. Acad. Stanislas. See B–P–H 567/10. HI 56295

Mémoires d'agriculture, d'économie rurale et domestique, pub. par la société d'agriculture du département de la Seine. Paris. Mém. Agric. Soc. Agric. Dép. Seine. See B–P–H 568/3. HI 56304

Mémoires d'agriculture, d'économie rurale et domestique, pub. par la société centrale d'agriculture de France. Paris. Mém. Agric. Soc. Centr. Agric. France. See B–P–H 568/4. HI 56305

Mémoires d'agriculture, d'économie rurale et domestique, publ. par la société impériale et centrale d'agriculture. Paris. Mém. Agric. Soc. Imp. Centr. Agric. See B–P–H 568/5. HI 56306

Mémoires d'agriculture, d'économie rurale et domestique, publ. par la société nationale et centrale d'agriculture. Paris. Mém. Agric. Soc. Natl. Centr. Agric. See B–P–H 568/6. HI 56307

Mémoires d'agriculture, d'économie rurale et domestique, pub. par la société royale d'agriculture de Paris. Paris. Mém. Agric. Soc. Roy. Agric. Paris. See B–P–H 568/7. HI 56308

Mémoires d'agriculture, d'économie rurale et domestique, pub. par la société royale et centrale d'agriculture. Paris. Mém. Agric. Soc. Roy. Centr. Agric. See B–P–H 568/8. HI 56309

Mémoires et analyse des travaux de la société d'agriculture, commerce, sciences et arts de la ville de Mende, chef-lieu du département de la Lozère. Mende, France. Mém. & Analyse Trav. Soc. Agric. Mende. See B–P–H 568/16. HI 56314

Mémoires de l'association des naturalistes de Nice et des Alpes-Maritimes. Nice? Mém. Assoc. Naturalistes Nice Alpes-Maritimes. See B–P–H 569/3. HI 56318

Mémoires de athénée oriental fondé en 1864. Paris. Mém. Athénée Orient. See B–P–H 569/4. HI 56319

Memoires, biological society of Nevada. Verdi, NV. Vol. 1+, 1962+. Mem. Biol. Soc. Nevada. HI 67701

Mémoires de biospéologie. Saint Girons, Moulis. N.s. no. 5+, 1978+. Mém. Biospéol. Preceded by: Série documents, laboratoire souterrain du C N R S. HI 67702

Mémoires de la classe des sciences mathématiques et physiques de l'institut de France = Mémoires de la classe des sciences mathématiques et physiques de l'institut national de France. Paris. Mém. Cl. Sci. Math. Inst. Natl. France. See B–P–H 570/3.

Mémoires de la classe des sciences mathématiques et physiques de l'institut national de France. Paris. Mém. Cl. Sci. Math. Inst. Natl. France. See B–P–H 570/3. HI 56329

Mémoires de la classe des sciences naturelles et techniques. Kiev, Ukrainian S S R = Trudy Pryrodnycho-technichnogo viddilu. Kiev.

Mémoires de la classe des sciences physiques et mathématiques. Kiev, Ukrainian S S R = Trudy fizychno-matematychnogo viddilu. Kiev.

Mémoires, commission historique du Cher = Mémoires de la société historique, litteraire et scientifique du Cher. Bourges, France. Mém. Soc. Hist. Cher. See B–P–H 583/20.

Mémoires et comptes rendus du congrès national de la

culture des plantes médicinales. Lons-le-Saulnier, Paris. Vols. 1-2, 1919-2? Mém. Compt. Rend. Congr. Natl. Cult. Pl. Méd. Superseded by: Compte rendu du congrès national de la culture des plantes médicinales. HI 67703

Mémoires et comptes rendus de la société d'émulation du Doubs. Besançon. Mém. Compt. Rend. Soc. Émul. Doubs. See B–P–H 570/20. HI 56332

Mémoires et comptes rendus de la société libre du Doubs = Mémoires et comptes rendus de la société d'émulation du Doubs. Besançon. Mém. Compt. Rend. Soc. Émul. Doubs. See B–P–H 570/20.

Mémoires concernant l'Asie orientale, Inde, Asie centrale, extrème-orient de l'académie des inscriptions et belles-lettres, Paris. Paris. Mém. Asie Orient. Acad. Inscript. Paris. See B–P–H 569/2. HI 56317

Mémoires concernant l'histoire naturelle de l'empire chinois. Shanghai. Mém. Hist. Nat. Empire Chin. See B–P–H 573/15. HI 56367

Mémoires concernant l'histoire, les sciences, les arts, les moeurs, les usages, &c. des Chinois: par les missionnaires de Pékin. Paris. Mém. Hist. Chin. See B–P–H 573/11. HI 56364

Mémoires du conservatoire de botanique et de l'institut de botanique systématique de l'université de Genève = Boissiera. Geneva. Boissiera. See B–P–H 201/20.

Mémoires couronnés par l'académie royale des sciences et belles-lettres de Bruxelles. [In quarto.] Brussels. Mém. Couronnés Acad. Roy. Sci. Bruxelles (4to). See B–P–H 571/1. HI 56335

Mémoires couronnés et autres mémoires. Académie royale des sciences, lettres et beaux-arts de Belgique. Brussels. Mém. Couronnés Autres Mém. Acad. Roy. Sci. Belgique. See B–P–H 571/2. HI 56336

Mémoires couronnés et mémoires des savants étrangers. Académie royale des sciences et belles-lettres de Bruxelles. [In quarto.] Brussels. Mém. Couronnés Mém. Savants Étrangers Acad. Roy. Sci. Bruxelles (4to). See B–P–H 571/3. HI 56337

Mémoires couronnés et mémoires des savants étrangers. Académie royale des sciences et belles-lettres de Bruxelles. [In octavo.] Brussels. Mém. Couronnés Mém. Savants Étrangers Acad. Roy. Sci. Bruxelles (8vo). See B–P–H 571/4. HI 56338

Mémoires et documents de l'académie des sciences, belles-lettres et arts, de Besançon. Besançon. Mém. Doc. Acad. Sci. Besançon. See B–P–H 571/12. HI 56342

Mémoires de la faculté des sciences de l'université de Lithuanie = Lietuvos universiteto matematikos gamtos fakulteto darbai. Kaunas.

Mémoires de la faculté des sciences de l'université de Vilnius = Vilnius universiteto matematikos-gamtos fakulteto darbai. Vilnius.

Mémoires de la faculté des sciences de l'université de Vytautas le Grand = Vytauto didžiojo universiteto matematikos-gamtos fakulteto darbai. Kaunas.

Mémoires de l'herbier Boissier suite au bulletin de l'herbier Boissier. Geneva. Mém. Herb. Boissier. See B–P–H 573/10. HI 56363

Mémoires d'histoire naturelle de la société Eduenne. Autun, France. Mém. Hist. Nat. Soc. Eduenne. See B–P–H 573/16. HI 56368

Mémoires pour l'histoire des sciences et des beaux arts. Trévoux, Lyons, Paris. Vols. 1-265, 1701-67. Mém. Hist. Sci. Beaux Arts. [For another edition, titled Mémoires de Trévoux, see Journal des sçavans.] Superseded by: Journal des beaux-arts et des sciences. 3-2600-2. HI 67704

Mémoires hors série. Société des naturalistes d'Oyonnax pour l'étude et la diffusion des sciences naturelles dans la région. No. 1+, 1955+. Mém. Hors Sér., Soc. Naturalistes Oyonnax Etude Diffusion Sci. Nat. Région HI 67705

Mémoires de l'institut agronomique de l'université catholique de Louvain. Louvain. Vols. 1-6, 1948-52. Mém. Inst. Agron. Univ. Catholique Louvain. 3-2480-3. HI 67706

Mémoires de l'institut agronomique à Voronèje = Zapiski voronezhskogo sel'sko-khozyaistvennogo instituta. Voronezh.

Mémoires de l'institut égyptien. Cairo. Mém. Inst. Égypt. See B–P–H 574/8. HI 56376

Mémoires de l'institut d'études centrafricaines. Paris, Brazzaville. Vols. 1-10?, 1948-63? Mém. Inst. Études Centrafr. 3-1996-1. HI 67707

Mémoires de l'institut fondamental d'Afrique noire. Dakar. No. 75+, 1966+. Mém. Inst. Fondam. Afrique Noire. Preceded by: Mémoires de l'institut français d'Afrique noire. HI 67708

Mémoires de l'institut français d'Afrique noire. Paris, Dakar. Nos. 1-74, 1939-65. Mém. Inst. Franç. Afrique Noire. Superseded by: Mémoires de l'institut fondamental d'Afrique noire. 3-1998-1. HI 67709

Mémoires de l'institut français d'Afrique noire. Centre de Cameroun. Serie sciences naturelles. Douala. Nos. 1-?, 1951-53. Mém. Inst. Franç. Afrique Noire. Superseded by: Mémoires de l'institut fondamental d'Afrique noire. HI 67710

Mémoires de l'institut géologique de l'université catholique de Louvain. Louvain. Mém. Inst. Géol. Univ. Catholique Louvain. See B–P–H 574/15. HI 56378

Mémoires de l'institut national genevois. Geneva. Mém.

Inst. Natl. Genevois. See B–P–H 574/20. HI 56381

Mémoires de l'institut national des sciences et arts. Sciences mathématiques et physiques. Paris. Mém. Inst. Natl. Sci., Sci. Math. See B–P–H 574/21. HI 56382

Mémoires, institut océanographique. Monaco-Ville. No. 1+, 1970+. Mém. Inst. Océanogr. Monaco. HI 67711

Mémoires de l'institut océanographique de Nhatrang. Nhatrang, Indochina [South Vietnam]. Mém. Inst. Océanogr. Nhatrang. See B–P–H 574/23. HI 56383

Mémoires de l'institut royal colonial belge; section des sciences naturelles et médicales. [In quarto.] Brussels. Mém. Inst. Roy. Colon. Belge, Sect. Sci. Nat. (4to). See B–P–H 575/4. HI 56385

Mémoires de l'institut royal colonial belge; section des sciences naturelles et médicales. [In octavo.] Brussels. Mém. Inst. Roy. Colon. Belge, Sect. Sci. Nat. (8vo). See B–P–H 575/5. HI 56386

Mémoires de l'institut royal des sciences naturelles de Belgique. Brussels. Mém. Inst. Roy. Sci. Nat. Belgique. See B–P–H 575/6. HI 56387

Mémoires de l'institut royal des sciences naturelles de Belgique. Deuxième série. Brussels. Nos. 30-186, 1949-71. Mém. Inst. Roy. Sci. Nat. Belgique, Sér. 2. Preceded by: Mémoires du musée royal d'histoire naturelle de Belgique. Deuxième série. Incorporated in: Mémoires de l'institut royal des sciences naturelles de Belgique. 1-807-1. HI 67712

Mémoires de l'institut des sciences, lettres et arts. Sciences mathématiques et physiques = Mémoires de l'institut national des sciences et arts. Sciences mathématiques et physiques, vol. 6 (1806). Paris. Mém. Inst. Natl. Sci., Sci. Math. See B–P–H 574/21.

Mémoires de l'institut scientifique de Madagascar. Série B, biologie végétale. Paris. Vols. 1-11, 1948-62. Mém. Inst. Sci. Madagascar, Sér. B, Biol. Vég. Superseded by: Cahiers O R S T O M [various series]. 5-4152-2. HI 67713

Mémoires de l'institut scientifique de Madagascar. Série F. Océanographie. Paris. Mém. Inst. Sci. Madagascar, Sér. F, Océanogr. See B–P–H 575/12. HI 56389

Mémoires, institut suisse de recherches forestières = Schweizerische Anstalt für das forstliche Versuchswesen. Mitteilungen. Zurich.

Mémoires de l'institute d'Égypte. Cairo. Mém. Inst. Égypte. See B–P–H 574/9. HI 56377

Mémoires du jardin botanique de Montréal. Montreal. Mém. Jard. Bot. Montréal. See B–P–H 574/17. HI 56392

Mémoires de laboratoire de biologie agricole. Institut Pasteur. Paris. Mém. Lab. Biol. Agric. Inst. Pasteur. See B–P–H 576/2. HI 56395

Mémoires et lettres de la société d'agriculture, sciences et arts du département de l'Aube. Troyes, France. Mém. Lett. Soc. Agric. Dép. Aube. See B–P–H 576/4. HI 56397

Mémoires sur la littérature du nord. Copenhagen & Geneva. Vols. 1-9, 1759-60. Mém. Litt. Nord. HI 67714

Mémoires de mathématique et de physique, présentés à l'académie royale des sciences, par divers sçavans ... Mém. Math. Phys. Acad. Roy. Sci. Divers Scavans. See B–P–H 576/18. HI 56407

Mémoires de mathématique et de physique, tiréz des régistres. Paris. Mém. Math. Phys. (Paris). See B–P–H 576/16. HI 56405

Mémoires de mathématique et de physique, tiréz des régistres de l'académie royale des sciences [Paris]. Amsterdam. Mém. Math. Phys. Acad. Roy. Sci. See B–P–H 576/17. HI 56406

Mémoires du musée royal d'histoire naturelle de Belgique. Brussels. Mém. Mus. Roy. Hist. Nat. Belgique. See B–P–H 577/24. HI 56413

Mémoires de musée royal d'histoire naturelle de Belgique. Deuzième série. Brussels. Mém. Mus. Roy. Hist. Nat Belgique, Sér. 2. See B–P–H 578/1. HI 56416

Mémoires du musée royal d'histoire naturelle de Belgique. Hors série. Brussels. Mém. Mus. Roy. Hist. Nat. Belgique, Hors Sér. See B–P–H 577/25. HI 56414

Mémoires du muséum d'histoire naturelle. Paris. Vols. 1-20, 1815-32. Mém. Mus. Hist. Nat. Preceded by: Annales du muséum national d'histoire naturelle. Superseded by: Nouvelles annales du muséum d'histoire naturelle, ... Paris. 4-3267-3. HI 56409

Mémoires du muséum national d'histoire naturelle. Paris. Mém. Mus. Natl. Hist. Nat. See B–P–H 577/19. HI 56411

Mémoires du muséum national d'histoire naturelle. Série B, botanique. Paris. N.s. vol. 1+, 1950+. Mém. Mus. Natl. Hist. Nat., B, Bot. Preceded by: Mémoires du muséum national d'histoire naturelle. HI 67715

Mémoires et observations recueilles par la société oeconomique de Berne. Bern. Mém. Observ. Soc. Oecon. Berne. See B–P–H 578/16. HI 56421

Mémoires, office de la recherche scientifique et technique outre-mer. Paris. 1961+. Mém. O. R. S. T. O. M. HI 67716

Mémoires de physique et de chimie de la société d'Arcueil. Paris. Mém. Phys Chim. Soc. Arcueil. See B–P–H 578/23. HI 56425

Mémoires présentés a l'académie impériale des sciences de St.-Pétersbourg par divers savans et lus dans ses assemblées. St. Petersburg. Mém. Acad. Imp. Sci. St.-Pétersbourg Divers Savans. See B–P–H 563/27. HI 56258

Mémoires présentés par divers savants à l'académie royale des sciences de l'institut royal de France et imprimés par son ordre. Sciences mathématiques et physiques. Paris. Mém. Divers Savants Acad. Roy. Sci. Inst. Roy. France, Sci. Math. See B–P–H 571/10. HI 56341

Mémoires présentés à l'institut des sciences, lettres et arts par divers savans, et lus dans ses assemblées; sciences mathématiques et physiques. Paris. Mém. Inst. Sci. Divers Savans, Sci. Math. See B–P–H 575/9. HI 56388

Mémoires présentés par divers savants à l'académie des sciences de l'institut de France et imprimés par son ordre. Sciences mathématiques et physiques = Mémoires présentés par divers savants à l'académie royale des sciences de l'institut royal de France et imprimés par son ordre. Sciences mathématiques et physiques. Paris. Mém. Divers Savants Acad. Roy. Sci. Inst. Roy. France, Sci. Math. See B–P–H 571/10.

Mémoires et publications de la société des sciences, des arts et des lettres du Hainaut. Mons, Belgium. Mém. & Publ. Soc. Sci. Hainaut. See B–P–H 579/4. HI 56427

Mémoires publiés par l'académie de Marseille. Marseilles. Mém. Acad. Marseille. See B–P–H 564/5. HI 56262

Mémoires et rapports sur les matières grasses. Marseilles. Vols. 1-3, 1922-28. Mém. Rapp. Matières Grasses. HI 67717

Mémoires et séance publique de l'académie des sciences, arts et belles-lettres. Dijon. Mém. & Séance Publique Acad. Sci. See B–P–H 581/6. HI 56453

Mémoires de la section botanique de la société des amis des sciences naturelles, d'anthropologie et d'ethnographie = Memuary Botanicheskogo otdeleniya Obshchestva lyubitelei estestvoznaniya, antropologii i etnografii. Moscow.

Mémoires de la section caucasienne de la société impériale russe de geographie = Zapiski Kavkazskago Otděla Imperatorskago Russkago Geograficeskago Obshchestva. Tiflis.

Mémoires de la section d'histoire des sciences et des techniques. Paris. Vol. 1+, 1984+. Mém. Sect. Hist. Sci. Techn., Paris. HI 67718

Mémoires de la section des sciences; académie des sciences et lettres de Montpellier. Montpellier. Mém. Sect. Sci. Acad. Sci. Montpellier. See B–P–H 581/12. HI 56456

Mémoires du service de la carte géologique d'Alsace et de Lorraine. Strasbourg. Vols. 1-34, 1927-71. Mém. Service Carte Géol. Alsace. Preceded by: Abhandlungen zur geologischen Specialkarte von Elsass-Lothringen. 5-4087-1. HI 67719

Mémoires, service de la carte de la végétation. Paris. No. 1+, 1969+. Mém. Serv. Carte Vég. HI 67720

Mémoires pour servir à l'histoire naturelle des animaux et des plantes. The Hague. Mém. Hist. Nat. Anim. Pl. See B–P–H 573/13. HI 56366

Mémoires de la société académique d'archéologie, sciences et arts du département de l'Oise. Beauvais, France. Mém. Soc. Acad. Archéol. Dép. Oise. See B–P–H 581/21. HI 56458

Mémoires de la société académique du département de l'Aube. Troyes, France. Mém. Soc. Acad. Dép. Aube. See B–P–H 581/23. HI 56459

Mémoires de la société académique de Maine et Loire. Angers. Mém. Soc Acad. Maine Loire. See B–P–H 581/24. HI 56460

Mémoires de la société académique de Savoie. Chambéry, France. Mém. Soc. Acad. Savoie. See B–P–H 581/26. HI 56461

Mémoires de la société académique des sciences, arts et belles-lettres de Falaise. Falaise & Paris. Mém. Soc. Acad. Sci. Falaise. See B–P–H 582/1. HI 56462

Mémoires de la société d'agriculture, sciences et arts d'Angers. Angers. Mém. Soc. Agric. Angers. See B–P–H 582/5. HI 56463

Mémoires de la société d'agriculture, des sciences et des arts, de l'arrondissement de Valenciennes. Valenciennes, France. Mém. Soc. Agric. Arrondissement Valenciennes. See B–P–H 582/6. HI 56464

Mémoires de la société d'agriculture, sciences, arts et belles-lettres du département de l'Aube = Mémoires de la société d'agriculture, sciences et arts du département de l'Aube. Troyes, France. Mém. Soc. Agric. Dép. Aube. See B–P–H 582/7.

Mémoires de la société d'agriculture, sciences et arts du département de l'Aube. Troyes, France. Mém. Soc. Agric. Dép. Aube. See B–P–H 582/7. HI 56465

Mémoires de la société d'agriculture, sciences, belles-lettres et arts d'Orléans. Orléans. Mém. Soc. Agric. Orléans. See B–P–H 582/10. HI 56466

Mémoires de la société des amateurs-naturalistes du nord de la Meuse. Montmédy. Vols. 1-17, 1889-1905. Mém. Soc. Amateurs-Naturalistes N. Meuse. Superseded by: Bulletin de la société des naturalistes et archéologiques du nord de la Meuse. 5-3954-3. HI 67721

Mémoires de la société des amis des sciences naturelles de la Nouvelle Russie = Zapiski Novorossiiskago

Obshchestva Estestvoispytatelei. Odessa.

Memoires de la société belge de géologie, de paléontologie. Brussels. Mem. Soc. Belge Géol. See B–P–H 582/14. HI 56467

Mémoires de la société de biogéographie. Paris. Mém. Soc. Biogéogr. See B–P–H 582/16. HI 56468

Mémoires de la société botanique de France. Paris. Vols. 1-7, 1905-17 [issued as part of: Bulletin de la société botanique de France]; 1949-73. Mém. Soc. Bot. France. 5-3933-3. HI 67722

Mémoires de la société botanique de Genève. Chambesy. Vol. 1+, 1979+. Mém. Soc. Bot. Genève. HI 67723

Mémoires pub. par la société centrale d'agriculture de France. Paris. Mém. Soc. Centr. Agric. France. See B–P–H 582/23. HI 56470

Mémoires de la société d'émulation d'Abbeville. Abbeville. Sér. 3, vols. 1-4, 1873-88; sér. 4, vols. 1-10, 1890-1927. Mém. Soc. Émul. Abbeville. Preceded by: Mémoires de la société impériale d'émulation d'Abbeville. 5-3938-3. HI 67724

Mémoires de la société d'émulation de Cambrai. (Agriculture, sciences et arts.) Séance publique. Cambrai. Vols. [1]+ [vols. 1-4 not numbered], 1820-33. Mém. Soc. Émul. Ville Cambrai. Preceded by: Société d'émulation Cambrai. Séance publique. 5-3938-3. HI 53441

Mémoires: Société d'émulation du Doubs. Besançon. Mém. Soc. Emul. Doubs. See B–P–H 583/11. HI 56474

Mémoire, société d'étude des sciences naturelles de Nîmes. Nîmes. No. ?-10+, ?-1975/76+. Mém. Soc. Étude Sci. Nat. Nîmes. HI 67725

Mémoires de la société des études de la région de l'Amour, la division Vladivostok de la section de l'Amour de la société impériale russe de géographie = Zapiski Obshchestva Izucheniya Amurskago Kraya Filial'nago Otdĕleniya Priamurskago Otdĕla Imperatorskago Russkago Geograficheskago Obshchestva. Vladivostok.

Mémoires de la société des études de la région de l'Amour, la division Vladivostok de la section de l'Amour de la société impériale russe de géographie = Zapiski Obshchestva Izucheniya Amurskago Kraya Vladivostokskago Otdĕleniya Priamurskago Otdĕla Imperatorskago Russkago Geograficheskago Obshchestva. Vladivostok.

Mémoire, société d'études scientifique d'Anjou. Angers. Vol. 1+, 1972+. Mém. Soc. Études Sci. Anjou. HI 67726

Mémoires de la société fribourgeoise des sciences naturelles. Bactériologie = Mémoires de la société fribourgeoise des sciences naturelles. Physiologie, hygiène, bactériologie. Fribourg.

Mémoires de la société fribourgeoise des sciences naturelles. Botanique. Fribourg. Vols. 1+, 1901/4+. Mém. Soc. Fribourg. Sci. Nat., Bot. 5-3960-3. HI 67727

Mémoires de la société fribourgeoise des sciences naturelles. Physiologie, hygiène, bactériologie. Fribourg. Vols. 1(1-5), 1908-23. Mém. Soc. Fribourg. Sci. Nat., Physiol. 5-3961-1. HI 67728

Mémoires. Société géologique de Belgique. Liége. Mém. Soc. Géol. Belgique. See B–P–H 583/17. HI 56475

Mémoires de la société géologique du Nord. Lille. Mém. Soc. Géol. Nord. See B–P–H 583/18. HI 56476

Mémoires de la société helvétique des sciences naturelles = Denkschriften der Schweizerischen Naturforschenden Gesellschaft. Zurich. Denkschr. Schweiz. Naturf. Ges. See B–P–H 342/2.

Mémoires de la société d'histoire naturelle de l'Afrique du nord. Algiers. Mém. Soc. Hist. Nat. Afrique N. See B–P–H 583/25. HI 56479

Mémoires de la société d'histoire naturelle de l'Afrique du nord. Hors série. Algiers. Mém. Soc. Hist. Nat. Afrique N., Hors Sér. See B–P–H 583/26. HI 56480

Mémoires de la société d'histoire naturelle d'Auvergne. Clermont-Ferrand. Nos. 1-6, 1940-58. Mém. Soc. Hist. Nat. Auvergne. 5-3942-2. HI 67729

Mémoires de la société d'histoire naturelle du département de la Moselle. Metz. Mém. Soc. Hist. Nat. Dép. Moselle. See B–P–H 584/3. HI 56481

Mémoires de la société d'histoire naturelle du Doubs. Besançon. 1899-1914. Mém. Soc. Hist. Nat. Doubs. Superseded by: Bulletin de la société d'histoire naturelle du Doubs. HI 67730

Mémoires de la société d'histoire naturelle de Paris. Paris. Mém. Soc. Hist. Nat. Paris. See B–P–H 584/5. HI 56482

Mémoires de la société d'histoire naturelle de Strasbourg. Paris, Strasbourg, & Brussels. Mém. Soc. Hist. Nat. Strasbourg. See B–P–H 584/6. HI 56483

Mémoires de la société historique, litteraire et scientifique du Cher. Bourges, France. Mém. Soc. Hist. Cher. See B–P–H 583/20. HI 56477

Mémoires de la société historique et scientifique des Deux-Sèvres. Niort, France. Mém. Soc. Hist. Deux-Sèvres. See B–P–H 583/21. HI 56478

Mémoires de la société impériale académique de Cherbourg. Cherbourg. Mém. Soc. Imp. Acad. Cherbourg. See B–P–H 584/9. HI 56485

Mémoires de la société impériale d'agriculture, sciences et arts d'Angers = Mémoires de la société nationale d'agriculture, sciences et arts d'Angers. Angers. Mém. Soc. Natl. Agric. Angers. See B–P–H 585/17.

Mémoires de la société impériale d'émulation d'Abbeville. Abbeville. 1852-68. Mém. Soc. Imp. Émul. Abbeville. Preceded by: Mémoires de la société royale d'émulation d'Abbeville. Superseded by: Mémoires de la société d'émulation d'Abbeville. 5-3938-3. HI 67731

Mémoires de la société impériale des naturalistes de Moscou. Moscow. Vols. 2-6, 1809-23; ed. 2. vol. 1, 1811; vols. 2-4, 1830; ed. 3. vol. 1, 1830. Mém. Soc. Imp. Naturalistes Moscou. Preceded by: Zapiski Obshchestva Ispytatelei Prirody, osnovannogo pri Imperatorskom Moskovskom Universitetě. Superseded by: Nouveaux mémoires de la société impériale des naturalistes de Moscou. 3-2770-2. HI 67732

Mémoires de la société impériale des naturalistes de Moscou vol. 7 = Nouveau mémoires de la société impériale des naturalistes de Moscou vol. 1.

Mémoires de la société impériale des sciences, de l'agriculture et des arts, de Lille. Lille, Paris. 1852, 1853; ser. 2, vols. 1-10, 1855-64; ser. 3, vols. 1-12, 1865-73. Mém. Soc. Imp. Sci. Agric. Arts Lille. Preceded by: Mémoires de la société nationale des sciences, de l'agriculture et des arts, de Lille. Superseded by: Mémoires de la société des sciences de l'agriculture et des arts de Lille. 5-3955-1. HI 67733

Mémoires de la société impériale des sciences naturelles de Cherbourg = Mémoires de la société des sciences naturelles de Cherbourg. Cherbourg. Mém. Soc. Sci. Nat. Cherbourg. See B–P–H 587/12.

Mémoires de la société des lettres, sciences et arts et d'agriculture de Metz. Metz. Mém. Soc. Lett. Metz. See B–P–H 584/19. HI 56488

Mémoires de la société des lettres, sciences et arts de Bar-le-Duc. Bar-le-Duc, France. Mém. Soc. Lett. Bar-le-Duc. See B–P–H 584/18. HI 56487

Mémoires de la société des lettres, sciences et arts "La Haute-Auvergne". Aurillae. 1937. Mém. Soc. Lettres Sci. Arts "La Haute-Auvergne". HI 67734

Mémoires de la société libre d'émulation de Liége. Liége. N.s. vols. 1-9, 1860-93. Mém. Soc. Libre Émul. Liége. Preceded by: Procès verbal de la séance publique par la société libre d'émulation de Liége, pour l'encouragement des lettres, des sciences et des arts [not entered]. 5-3964-3. HI 67735

Mémoires de la société libre d'émulation, de Rouen. Rouen. Mém. Soc. Libre Émul. Rouen. See B–P–H 584/20. HI 56489

Mémoires de la société linnéenne de Calvados. Caen. Mém. Soc. Linn. Calvados. See B–P–H 584/21. HI 56490

Mémoires de la société linnéenne du nord de la France. Amiens, France. Mém. Soc. Linn. N. France. See B–P–H 584/22. HI 56491

Mémoires de la société linnéenne de Normandie. Caen. 1827-28; ser. 2, vols. 1-26, 1829-1924. Mém. Soc. Linn. Normandie. Preceded by: Mémoires de la société linnéenne de Calvados. Superseded by: Mémoires de la société linnéenne de Normandie. Botanique. 5-3965-2. HI 67736

Mémoires de la société linnéenne de Normandie. Botanique. Caen. N.s. vol. 1+, 1925+. Mém. Soc. Linn. Normandie, Bot. Preceded by: Mémoires de la société linnéenne de Normandie. 5-3965-2. HI 67737

Mémoires de la société linnéenne de Paris, précédés de son histoire. Paris. Mém. Soc. Linn. Paris. See B–P–H 585/3. HI 56492

Mémoires de la société linnéenne de Provence. Marseilles. Mém. Soc. Linn. Provence. See B–P–H 585/4. HI 56493

Mémoires de la société littéraire de Grenoble. Grenoble. Mém. Soc. Litt. Grenoble. See B–P–H 585/5. HI 56494

Mémoires de la société médicale d'émulation séante à l'école de médecine, de Paris. Paris. Mém. Soc. Méd. Emul. Paris. See B–P–H 585/7. HI 56496

Mémoires de la société du muséum d'histoire naturelle de Strasbourg. Paris & Strasbourg. Mém. Soc. Mus. Hist. Nat. Strasbourg. See B–P–H 585/8. HI 56497

Mémoires de la société nationale académique de Cherbourg. Cherbourg. Mém. Soc. Natl. Acad. Cherbourg. See B–P–H 585/16. HI 56498

Mémoires pub. par la société nationale d'agriculture de France. Paris. Mém. Soc. Natl. Agric. France. See B–P–H 585/18. HI 56500

Mémoires de la société nationale d'agriculture, sciences et arts d'Angers. Angers. Mém. Soc. Natl. Agric. Angers. See B–P–H 585/17. HI 56499

Mémoires de la société nationale des sciences, de l'agriculture et des arts, de Lille. Lille. 1850-51, 1851-52. Mém. Soc. Natl. Sci. Agric. Arts Lille. Preceded by: Mémoires de la société royale des sciences, de l'agriculture et des arts, de Lille. Superseded by: Mémoires de la société impériale des sciences, de l'agriculture et des arts, de Lille. 5-3955-1. HI 67738

Mémoires de la société nationale des sciences naturelles de Cherbourg = Mémoires de la société des sciences naturelles de Cherbourg. Cherbourg. Mém. Soc. Sci. Nat. Cherbourg. See B–P–H 587/12.

Mémoires de la société des naturalistes d'Odessa = Zapiski Odesskogo obshchestva estestvoispytatelei. Odessa.

Mémoires de la société des naturalistes de Jaroslavl

(Jaroslaw) = Trudy Yaroslavskogo estestvenno-istoricheskogo i kraevedcheskogo obshchestva. Yaroslavl.

Mémoires de la société des naturalistes de Jaroslow = Trudy Yaroslavskogo estestvenno-istoricheskogo obshchestva. Yaroslavl.

Mémoires de la société des naturalistes de Kiew [or Kieff] = Zapiski Kievskago Obshchestva Estestvoispytatelei. Kiev.

Mémoires de la société des naturalistes de la Nouvelle Russie = Zapiski Novorossiiskago Obshchestva Estestvoispytatelei. Odessa.

Mémoires de la société des naturalistes de l'université impériale de Moscou = Zapiski Obshchestva Ispytatelei Prirody, osnovannogo pri Imperatorskom Moskovskom Universitetě. Moscow.

Mémoires de la société de physique et d'histoire naturelle de Genève. Geneva & Paris. Mém. Soc. Phys. Genève. See B–P–H 586/4. HI 56502

Mémoires de la société royale académique de Cherbourg. Cherbourg. Mém. Soc. Roy. Acad. Cherbourg. See B–P–H 586/7. HI 56503

Mémoires de la société royale académique de Savoie. Chambéry, France. Mém. Soc. Roy. Acad. Savoie. See B–P–H 586/8. HI 56504

Mémoires de la société royale d'agriculture, histoire naturelle et arts utiles de Lyon. Lyons. Mém. Soc. Roy. Agric. Lyon. See B–P–H 586/9. HI 56505

Mémoires de la société royale d'Arras, pour l'encouragement des sciences, des lettres et des arts. Arras, France. Mém. Soc. Roy. Arras. See B–P–H 586/10. HI 56506

Mémoires de la société royale de botanique de Belgique. Brussels. Vol. 1+, 1963+. Mém. Soc. Roy. Bot. Belg. HI 67739

Mémoires de la société royale d'émulation d'Abbeville. Abbeville. 1833-51. Mém. Soc. Roy. Émul. Abbeville. Preceded by: Analyse des travaux de la société d'émulation d'Abbeville. Superseded by: Mémoires de la société impériale d'émulation d'Abbeville. 5-3938-3. HI 67740

Mémoires de la société royale des sciences, de l'agriculture et des arts, à (de) Lille. Lille. 1827/28-49?, 1829-50. Mém. Soc. Roy. Sci. Lille. Preceded by: Recueil des travaux de la société des sciences de l'agriculture et des arts, de Lille. Superseded by: Mémoires de la société nationale des sciences, de l'agriculture et des arts, de Lille. 5-3955-1. HI 67741

Mémoires de la société royale des sciences, belles-lettres et arts d'Orléans. Orléans. Mém. Soc. Roy. Sci. Orléans. See B–P–H 586/17. HI 56509

Mémoires de la société royale des sciences, et belles-lettres de Nancy. Nancy. Mém. Soc. Roy. Sci. Nancy. See B–P–H 586/16. HI 56508

Mémoires de la société royale des sciences de Bohême. Classe des sciences = Věstník královské české společnosti nauk. Třida matematicko-přírodovědecké. Prague. Věstn. Král. Ceské Společn. Nauk, Tř. Mat.-Přír. See B–P–H 962/7.

Mémoires de la société royale des sciences, lettres et arts de Nancy = Mémoires de la société royale des sciences, et belles-lettres de Nancy. Nancy. Mém. Soc. Roy. Sci. Nancy. See B–P–H 586/16.

Mémoires de la société royale des sciences de Liège. Liège. Mém. Soc. Roy. Sci. Liège. See B–P–H 586/14. HI 56507

Mémoires de la société russe de géographie. Section de géographie générale = Zapiski russkogo geograficheskogo obshchestva po obshchei geografii., St. Petersburg.

Mémoires de la société des sciences de l'agriculture et des arts de Lille; et publications faites par ses soins. Paris, Lille. Ser. 3, vol. 13, 1874; ser. 4, vols. 1-12?, 1876-94; ser. 5, vol. 1-8, 1895-1925. Mém. Soc. Sci. Agric. Arts Lille. Preceded by: Mémoires de la société impériale des sciences, de l'agriculture et des arts, de Lille. 5-3955-1. HI 67742

Mémoires de la société des sciences, agriculture et arts, de Strasbourg. Strasbourg. Mém. Soc. Sci. Agric. Strasbourg. See B–P–H 587/5. HI 56510

Mémoires de la société des sciences, belles-lettres et arts d'Orléans. Orléans. Mém. Soc. Sci. Orléans. See B–P–H 587/21. HI 56519

Mémoires de la société des sciences de Bohême. Classe des sciences = Věstník královské české společnosti nauk. Třida matematicko-přírodovědecké. Prague. Věstn. Král. Ceské Společn. Nauk, Tř. Mat.-Přír. See B–P–H 962/7.

Mémoires de la société des sciences et lettres de Loir-et-Cher. Blois, France. Mém. Soc. Sci. Loir-et-Cher. See B–P–H 587/9. HI 56511

Mémoires de la société des sciences de Nancy. Nancy. Mém. Soc. Sci. Nancy. See B–P–H 587/10. HI 56512

Mémoires de la société des sciences naturelles de Cherbourg. Cherbourg. Mém. Soc. Sci. Nat. Cherbourg. See B–P–H 587/12. HI 56514

Mémoires de la société des sciences naturelles du Maroc. Rabat, Morocco. Meem. Soc. Sci. Nat. Maroc. See B–P–H 587/13. HI 56515

Mémoires de la société des sciences naturelles, des lettres et des beaux-arts de Cannes, et de l'arrondissement de Grasse. Cannes. Mém. Soc. Sci. Nat. Cannes. See B–P–H 587/11. HI 56513

Mémoires de la société des sciences naturelles et mathématiques de Cherbourg = Mémoires de la société des sciences naturelles de Cherbourg. Cherbourg. Mém. Soc. Sci. Nat. Cherbourg. See B–P–H 587/12.

Mémoires de la société des sciences naturelles et médicales de Seine-et-Oise. Versailles & Paris. Vols. [1]-20, 1835-1911; sér. 2, 1 vol. 1919. Mém. Soc. Sci. Nat. Seine-et-Oise. Superseded by: Bulletin de la société des sciences naturelles de Seine-et-Oise. 5-3957-1. HI 67743

Mémoires de la société des sciences naturelles de Neuchâtel. Neuchâtel. Mém. Soc. Sci. Nat. Neuchâtel. See B–P–H 587/16. HI 56517

Mémoires de la société des sciences naturelles et physiques du Maroc. Botanique. Rabat, Morocco. Mém. Soc. Sci. Nat. Maroc, Bot. See B–P–H 587/14. HI 56516

Mémoires de la société des sciences naturelles de Strasbourg. Paris & Strasbourg. Mém. Soc. Sci. Nat. Strasbourg. See B–P–H 587/18. HI 56518

Mémoire de la société des sciences naturelles de Tunisie. Tunis. Vols. 1-4, 1951-59. Mém. Soc. Sci. Nat. Tunisie. HI 67744

Mémoires de la société des sciences physiques de Lausanne. Lausanne. Mém. Soc. Sci. Phys. Lausanne. See B–P–H 587/22. HI 56520

Mémoires de la société des sciences physiques, de médecine et d'agriculture d'Orléans. Orléans. Mém. Soc. Sci. Phys. Orléans. See B–P–H 588/1. HI 56522

Mémoires de la société des sciences physiques et naturelles de Bordeaux. Bordeaux. Mém. Soc. Sci. Phys. Nat. Bordeaux. See B–P–H 587/23. HI 56521

Mémoires de la société vaudoise des sciences naturelles. Lausanne. Mém. Soc. Vaud. Sci. Nat. See B–P–H 588/3. HI 56524

Mémoires de la sous-section de Tchita = Zapiski Chitinskago Otdĕleniya Priamurskago Otdĕla Imperatorskago Russkago Geograficheskago Obshchestva. St. Petersburg.

Mémoires de la subdivision de Tchita = Zapiski Chitinskago Otdĕleniya Priamurskago Otdĕla Imperatorskago Russkago Geograficheskago Obshchestva. St. Petersburg.

Mémoires suisses de paléontologie = Schweizerische paläontologische Abhandlungen. Basel.

Mémoires et travaux de l'institut de Montpellier de l'école pratique des hautes études. Montpellier. 1973+. Mém. Trav. Inst. Montpellier École Prat. Hautes Études. HI 67745

Mémoires de Trévoux = Mémoires pour l'histoire des sciences et des beaux arts. Trévoux, Lyons, Paris.

Mémoires de l'université d'état à l'Extrême Orient. Géologie. Vladivostok, Russian S F S R = Trudy gosudarstvennogo dal'nevostochnogo universiteta. Ser. 11, geologiya. Vladivostok.

Mémoires de l'université d'état à l'Extrême Orient. Sciences forestières. Vladivostok, Russian S F S R = Trudy gosudarstvennogo dal'nevostochnogo universiteta. Ser. 4, lesnye nauk. Vladivostok.

Mémoires de l'université d'état à l'Extrême Orient. Ser. 5, Agronomie. Kiev, Ukrainian S S R = Trudy gosudarstvennogo dal'nevostochnogo universiteta. Ser. 5, sel'skoe khozyaistvo. Kiev.

Mémoires de l'université d'état à l'Extrême Orient. Ser. 8. Biologie. Vladivostok, Russian S F S R = Trudy gosudarstvennogo dal'nevostochnogo universiteta. Ser. 8, biologiya. Vladivostok.

Mémoires de l'université d'état à l'Extrême Orient. Ser. 13. Technique. Vladivostok, Russian S F S R = Trudy gosudarstvennogo dal'nevostochnogo universiteta. Ser. 13, tekhnika. Vladivostok.

Mémoires de l'université de Neuchâtel. [In quarto.] Neuchâtel. Mém. Univ. Neuchâtel (4to). See B–P–H 588/23. HI 56528

Memoirs, agricultural directorate, Mesopotamia = Memoirs, department of agriculture, Iraq. Baghdad.

Memoirs of agriculture and other oeconomical arts. London. Mem. Agric. Oecon. Arts. See B–P–H 568/2. HI 56303

Memoirs of the american academy of arts and science. Boston, MA. Mem. Amer. Acad. Arts. See B–P–H 568/12. HI 56310

Memoirs of the american association for the advancement of science. Salem, MA. Mem. Amer. Assoc. Advancem. Sci. See B–P–H 568/13. HI 56311

Memoirs of the american entomological society. Philadelphia, PA. Mem. Amer. Entomol. Soc. See B–P–H 568/14. HI 56312

Memoirs of the american philosophical society held at Philadelphia for promoting useful knowledge. Philadelphia, PA. Mem. Amer. Philos. Soc. See B–P–H 568/15. HI 56313

Memoirs of the asiatic society of Bengal. Calcutta. Mem. Asiat. Soc. Bengal. See B–P–H 569/1. HI 56316

Memoirs, association for plant protection of Kyushu. Chikugo City. No. 1+, 1963+. Mém Assoc. Pl. Protect. Kyushu. HI 67746

Memoirs of Beijing natural history museum. [Beijing.] No. 1+, 1979+. Mem. Beijing Nat. Hist. Mus. HI 67747

Memoirs of the Bernice P. Bishop museum of polynesian

ethnology and natural history = Memoirs of the Bernice Pauahi Bishop museum of polynesian ethnology and natural history. Honolulu, HI. Mem. Bernice Pauahi Bishop Mus. See B–P–H 569/7.

Memoirs of the Bernice Pauahi Bishop museum of polynesian ethnology and natural history. Honolulu, HI. Mem. Bernice Pauahi Bishop Mus. See B–P–H 569/7. HI 56320

Memoirs of the biological laboratory; Johns Hopkins university. Baltimore, MD. Mem. Biol. Lab. Johns Hopkins Univ. See B–P–H 569/9. HI 56321

Memoirs of the Boston society of natural history. Boston, MA. Mem. Boston Soc. Nat. Hist. See B–P–H 569/13. HI 56323

Memoirs of the botanical survey of South Africa. Pretoria. Vol. 1+, 1919+. Mem. Bot. Surv. S. Africa. HI 67748

Memoirs, Brooklyn botanic garden = Brooklyn botanic garden memoirs. Brooklyn, NY. Brooklyn Bot. Gard. Mem. See B–P–H 230/6.

Memoirs, caledonian horticultural society. Edinburgh. Vols. 1-5(1), 1814-32. Mem. Caledonian Hort. Soc. Superseded by: Memoirs, royal caledonian horticultural society. 2-871-3. HI 67749

Memoirs of the California academy of sciences. San Francisco, CA. Mem. Calif. Acad. Sci. See B–P–H 569/16. HI 56325

Memoirs of the Carnegie museum. Lancaster, PA. Mem. Carnegie Mus. See B–P–H 569/17. HI 56326

Memoirs of the college of agriculture, Ehime university. [Ehime daigaku nogakubu.] Matsuyama. 1968+. Mem. Coll. Agric. Ehime Univ. Preceded by: Memoirs of Ehime university. Section 6, agriculture. HI 67750

Memoirs of the college of science; Kyoto imperial university. Series B, biology. Kyoto. Vols. 1-33, 1924-67. Mem. Coll. Sci. Kyoto Imp. Univ., Ser. B, Biol. Superseded by: Memoirs of the faculty of science, Kyoto university. Series of biology. 3-2330-3. HI 67751

Memoirs of the college of agriculture; Kyoto university. [Kyoto daigaku nogakubu kiyo.] Kyoto, Japan. Mem. Coll. Agric. Kyoto Univ. See B–P–H 570/11. HI 56330

Memoirs of the college of agriculture; national Taiwan university. [Kuo li Taiwan ta hsüeh nung hsüeh yüan yen chiu pao k'an.] Taipei, Taiwan. Mem Coll. Agric. Natl. Taiwan Univ. See B–P–H 570/12. HI 56331

Memoirs of the Connecticut academy of arts and sciences. New Haven, CT. Mem. Connecticut Acad. Arts. See B–P–H 570/25. HI 56333

Memoirs, Cornell university agricultural experiment station = Cornell university agricultural experiment station memoir. Ithaca, NY. Cornell Univ. Agric. Exp. Sta. Mem. See B–P–H 332/8.

Memoirs of the cotton research station, Trinidad. Series A, genetics. London. Nos. 1-27, 1929-46. Mem. Cotton Res. Sta. Trinidad, A. HI 67752

Memoirs of the cotton research station, Trinidad. Series B, physiology. London. 1928-46. Mem. Cotton Res. Sta. Trinidad, B. HI 67753

Memoirs, department of agricultural science and applied biology, university of Cambridge. Cambridge. No. 42, 1970. Mem. Dept. Agric. Sci. Appl. Biol. Univ. Cambridge. Preceded by: Memoirs, school of agriculture, university of Cambridge. Superseded by: Memoirs, departmentof applied biology, university of Cambridge. HI 67754

Memoirs of the department of agriculture in India. Botanical series. Agricultural research institute. Pusa [=Samastipur], India. Mem. Dept. Agric. India, Bot. Ser. See B–P–H 571/7. HI 56339

Memoirs, department of agriculture, Iraq. Baghdad. Nos. 1-15, 1921-31. Mem. Dept. Agric. Iraq. HI 67755

Memoirs, department of agriculture, Mesopotamia = Memoirs, department of agriculture, Iraq. Baghdad.

Memoirs, department of applied biology, university of Cambridge. Cambridge. No. 43+, 1971+. Mem. Dept. Appl. Biol. Univ. Cambridge. Preceded by: Memoirs, department of agricultural science and applied biology, university of Cambridge. HI 67756

Memoirs from the department of botany of Columbia college. New York, NY. Mem. Dept. Bot. Columbia Coll. See B–P–H 571/8. HI 56340

Memoirs, ecological society of Australia. Canberra, A.C.T. Vol. 1+, 1973+. Mem. Ecol. Soc. Australia. HI 67757

Memoirs of Ehime university; section 2, natural science, series B. [Biology]. Matsuyama, Japan. Mem. Ehime Univ., Sect. 2, Nat. Sci., Ser. B. See B–P–H 571/17. HI 56344

Memoirs of Ehime university; section 6, agriculture. Matsuyama. Vols. 1-?, 1954-67. Mem. Ehime Univ., Sect. 6, Agric. Superseded by: Memoirs of the college of agriculture, Ehime university. 3-2566-2. HI 67758

Memoirs of the entomological society of Canada. Ottawa. Mem. Entomol. Soc. Canada. See B–P–H 571/18. HI 56345

Memoirs of the faculty of agriculture, Hokkaido university. [Hokkaido daigaku nogakubu kiyo.] Sapporo, Japan. Mem. Fac. Agric. Hokkaido Univ. See B–P–H 572/6. HI 56346

Memoirs of faculty of agriculture, Kagawa university. [Kagawa daigaku nogakubu kiyo.] Miki, Japan. Mem

Fac. Agric. Kagawa Univ. See B–P–H 572/7. HI 56347

Memoirs of the faculty of agriculture; Kagoshima university. [Kogoshima daigaku nogakubu kiyo.] Kagoshima, Japan. Mem Fac. Agric. Kagoshima Univ. See B–P–H 572/8. HI 56348

Memoirs of the faculty of agriculture of Kinki university. Osaka, Japan. Mem. Fac. Agric. Kinki Univ. See B–P–H 572/9. HI 56349

Memoirs of the faculty of agriculture; national Taiwan university. [Kuo li Tai wan ta hsüeh nung hsüeh yüan yen chiu pao kao.] Taipei, Taiwan. Mem. Fac. Agric. Natl. Taiwan Univ. See B–P–H 572/10. HI 56350

Memoirs, faculty of agriculture, Niigata university. [Niigata daigaku nogakubu kiyo.] Niigata. 1961+. Mem. Fac. Agric. Niigata Univ. HI 67759

Memoirs of the faculty of agriculture; Taihoku imperial university. [Taihoku teikoku daigaku nogakuku kiyo.] Taihoku [=Taipei, Taiwan]. Mem Fac. Agric. Taihoku Imp. Univ. See B–P–H 572/12. HI 56351

Memoirs of the faculty of agriculture; Tokyo university of education. Tokyo. Mem Fac Agric. Tokyo Univ. Educ. See B–P–H 572/13. HI 56352

Memoirs of the faculty of agriculture, university of Miyazaki. [Miyazaki daigaku nogakubu kiyo.] Miyazaki. 1955+. Mem. Fac. Agric. Univ. Miyazaki. HI 67760

Memoirs of the faculty of (liberal arts and) education, Akita university. Natural science. [Akita daigaku gakugeigakubu kenkyu kiyo.] Akita. 1950+. Mem. Fac. Liberal Arts Educ. Akita Univ., Nat. Sci. HI 67761

Memoirs, faculty of education, Fukui university. Series 5, applied science and agricultural science. [Fukui daigaku, kyoiku gakubu. 5bu, oyo kogaku.] Fukui. No. 5+, 1967+. Mem. Fac. Educ. Fukui Univ., Ser. 5, Appl. Sci. Agric. Sci. Preceded, in part, by: Memoirs, faculty of education, Fukui university. Series 5, applied sciences [not entered]. HI 67762

Memoirs of the faculty of education, Kumamoto university. Section 1, natural science. [Kumamoto daigaku kyoikugakubu kiyo, shizenkagaku.] Kumamoto. 1953+. Mem. Fac. Educ. Kumamoto Univ., Sect. 1, Nat. Sci. HI 67763

Memoirs of the faculty of education, Miyazaki university. Natural science. [Miyazaki daigaku kyoikugakubu kiyo, shizenkagaku.] Miyazaki. 1955+. Mem. Fac. Educ. Miyazaki Univ., Nat. Sci. HI 67764

Memoirs of the faculty of education, Niigata university. Natural science. [Niigata daigaku kyoikugakubu kiyo, shizenkagaku-hen.] Niigata. Vol. 1+, 1958+. Mem. Fac Educ. Niigata Univ., Nat. Sci. HI 67765

Memoirs of the faculty of education, Shiga university. Natural science. Otsu. No. 18+, 1968+. Mem. Fac. Educ. Shiga Univ., Nat. Sci. Preceded by: Memoirs of the faculty of liberal arts and education. Natural science. HI 67766

Memoirs of the faculty of education, Shimane university. Natural science. Matsue. Vol. 1+, 1967+. Mem. Fac. Educ. Shimane Univ., Nat. Sci. HI 67767

Memoirs of the faculty of fisheries; Hokkaido university. Hakodate, Japan. Mem. Fac. Fish. Hokkaido Univ. See B–P–H 572/14. HI 56353

Memoirs of the faculty of fisheries, Kagoshima university. [Kagoshima daigaku. Suisdan gakubu kiyo.] Kagoshima, Japan. Mem. Fac. Fish. Kagoshima Univ. See B–P–H 572/15. HI 56354

Memoirs of the faculty of integrated arts and sciences, Hiroshima university. Studies in environmental science. [Kiso kankyo kagaku kenkyu.] Hiroshima. Vol. 1+, 1975+. Mem. Fac. Integr. Arts Sci. Hiroshima Univ., Stud. Environm. Sci. HI 67768

Memoirs of the faculty of liberal arts and education, Shiga university. Part 2, natural science. [Shiga daigaku kyoiku gakubu kiyo, shizenkagaku.] Otsu. Nos. 3-15, 1954-65. Mem. Fac. Liberal Arts Shiga Univ., Pt. 2, Nat. Sci. Preceded by: Bulletin of the faculty of liberal arts and education, Shiga university. Part 2, natural science. Superseded by: Memoirs of the faculty of education, Shiga university. Natural science. HI 67769

Memoirs of the faculty of liberal arts, Fukui university. Ser. 2: Natural science. [Hukui daigaku gakugei gakubu kiyo. Shizen kagaku.] Fukui, Japan. Mem. Fac. Liberal Arts Fukui Univ., Ser. 2, Nat. Sci. See B–P–H 572/18. HI 56355

Memoirs of the faculty of literature and science, Shimane university. Natural sciences. [Shimane daigaku. Bunri gakubu.] Matsue. Nos. 1-11, 1968-77. Mem. Fac. Lit. Sci. Shimane Univ., Nat. Sci. Preceded by: Bulletin, Shimane university, natural science. Superseded by: Memoirs of the faculty of science, Shimane university. HI 67770

Memoirs of the faculty of science and agriculture; Taihoku imperial university. [Rino-gaku-bu kiyo.] Taihoku [=Taipei, Taiwan]. Mem Fac. Sci. Taihoku Imp. Univ. See B–P–H 572/25. HI 56356

Memoirs of the faculty of science, Kochi university. Series D, biology. Kochi. Vol. 1+, 1980+. Mem. Fac. Sci. Kochi Univ., D. HI 67771

Memoirs of the faculty of science, Kyoto university. Series of biology. Kyoto. N.s. vol. 1+, 1967+. Mem. Fac. Sci. Kyoto Univ., Ser. Biol. Preceded by: Memoirs of the college of science, Kyoto imperial university. Series B, biology. HI 67772

Memoirs of the faculty of science. Kyushu university.

Series E, biology. Fukuoka. Vols. 1-?, 1954-69. Mem. Fac. Sci. Kyushu Univ., Ser. E, Biol. HI 67773

Memoirs of the faculty of science, Shimane university. Matsue. No. 12+, 1978+. Mem. Fac. Sci. Shimane Univ. Preceded by: Memoirs of the faculty of literature and science, Shimane university. Natural sciences. HI 67774

Memoirs of the faculty of science; Taihoku imperial university; series 2, zoology, botany. Taihoku [=Taipei, Taiwan]. Mem. Fac. Sci. Taihoku Imp. Univ., Ser. 2, Zool. Bot. See B–P–H 573/1. HI 56357

Memoirs of gakugei faculty, Akita university. Natural science. [Akita daigaku gakugei gakubu kenkyu kiyo. Shizen kagaku.] Akita, Japan. Mem. Gakugei Fac. Akita Univ., Nat. Sci. See B–P–H 573/3. HI 56359

Memoirs of geography. [Ti li hsüeh t'zu liao.] Nanking. Mem. Geogr. See B–P–H 573/4. HI 56360

Memoirs of the geological society of America. Washington, DC. Mem. Geol. Soc. Amer. See B–P–H 573/7. HI 56361

Memoirs of the grassland research institute. Hurley. 1955. Mem. Grassland Res. Inst. HI 67775

Memoirs of the Gray herbarium of Harvard university. Cambridge, MA. Mem. Gray Herb. See B–P–H 573/8. HI 56362

Memoirs of the Hong Kong biological circle. Hong Kong. Nos. 1-3, 1953-55. Mem. Hong Kong Biol. Circ. Superseded by: Memoirs of the Hong Kong natural history society. HI 67776

Memoirs of the Hong Kong natural history society. Hong Kong. No. 4+, 1960+. Mem. Hong Kong Nat. Hist. Soc. Preceded by: Memoirs of the Hong Kong biological circle. HI 67777

Memoirs of the horticulture society of New York. New York, NY. Mem. Hort. Soc. New York. See B–P–H 573/19. HI 56369

Memoirs of the Hyogo prefectural university of agriculture. Sasayama. Vols. 1-23, 1950-68. Mem. Hyogo Pref. Univ. Agric. HI 67778

Memoirs of the Hyogo prefectural university of agriculture. Agronomical series. Sasayama. Nos. 1-9, 1950-68. Mem. Hyogo Pref. Univ. Agric., Agron. Ser. HI 67779

Memoirs of the Hyogo prefectural university of agriculture. Biological series. Sasayama. Nos. 1-6, 1962-64. Mem. Hyogo Pref. Univ. Agric., Biol. Ser. HI 67780

Memoirs of the Hyogo prefectural university of agriculture. Chemical series. Sasayama. Nos. 1-3, 1962-63. Mem. Hyogo Pref. Univ. Agric., Chem. Ser. HI 67781

Memoirs of the Hyogo prefectural university of agriculture. Phytopathological series. Sasayama. No. 1, 1961. Mem. Hyogo Pref. Univ. Agric., Phytopathol. Ser. HI 67782

Memoirs of the imperial college of tropical agriculture, Trinidad. Botanical series. Trinidad. Nos. 1-4, 1931-35. Mem. Imp. Coll. Trop. Agric. Trinidad, Bot. Ser. HI 67783

Memoirs of the imperial college of tropical agriculture, Trinidad. Economic series. Trinidad. No. ?-2+, ?-1953+. Mem. Imp. Coll. Trop. Agric. Trinidad, Econ. Ser. HI 67784

Memoirs of the imperial college of tropical agriculture, Trinidad. Mycological series. Trinidad. Nos. 1-8, 1930-44. Mem. Imp. Coll. Trop. Agric. Trinidad, Mycol. Ser. HI 67785

Memoirs of the imperial marine observatory. Kobe, Japan. Mem. Imp. Mar. Observ. See B–P–H 573/20. HI 56370

Memoirs of the indian botanical society. Bangalore. Vols. 1-4, 1958-63. Mem. Indian Bot. Soc. HI 67786

Memoirs of the indian museum. Calcutta. Mem. Indian Mus. See B–P–H 574/3. HI 56372

Memoir, Japan society of plant taxonomists. [Nihon shokubutsu bunrui gakki kaiho.] Tokyo. Mem. Japan Soc. Pl. Taxon. See B–P–H 575/16. HI 56391

Memoirs of the Kobe marine observatory. Kobe, Japan. Mem. Kobe Mar. Observ. See B–P–H 576/1. HI 56394

Memoirs of the Krasnojarsk section of the state russian geographical society = Zapiski Sredne-Sibirskogo otdela (byvshego Krasnoyarskogo) Gosudarstvennogo russkogo geograficheskogo obshchestva. Krasnoyarsk.

Memoirs of the literary and philosophical society of Manchester. Manchester, England. Mem. Lit. Soc. Manchester. See B–P–H 576/5. HI 56398

Memoir of the Malaysia geological survey, Borneo region. Kuching, Sarawak [Malaysia]. Mem. Malaysia Geol. Surv. Borneo Region. See B–P–H 576/7. HI 56400

Memoirs, marine biological association of India. Mandapam Camp. No. 1+, 1967+. Mem. Mar. Biol. Assoc. India. HI 67787

Memoirs of the medical society of London. London. Mem. Med. Soc. London. See B–P–H 577/1. HI 56408

Memoirs, Michigan agricultural experiment station = Michigan agricultural experiment station. Memoirs. East Lansing, MI. Michigan Agric. Exp. Sta. Mem. See B–P–H 592/13.

Memoirs of the middle-siberian section <formerly Krasnojarsk-Section> of the state russian geographical society = Zapiski Sredne-Sibirskogo otdela (byvshego Krasnoyarskogo) Gosudarstvennogo russkogo geograficheskogo obshchestva. Krasnoyarsk.

Memoirs of the Montreal botanical garden = Mémoires du jardin botanique de Montréal. Montreal. Mém. Jard. Bot. Montréal. See B–P–H 574/17.

Memoirs of the museum of Victoria = Memoirs of the national museum of Victoria. Melbourne, Vic.

Memoirs of the national academy of sciences. Washington, DC. Mem. Natl. Acad. Sci. See B–P–H 578/9. HI 56418

Memoirs of the national institute of polar research. Series E, biology and medical sciences. Tokyo. No. 32+, 1976+. Mem. Natl. Inst. Polar Res., E. Preceded by: J A R E scientific reports. Series E, biology. HI 67788

Memoirs of the national institute of polar research. Special issue. Tokyo. No. 1+, 1974+. Mem. Natl. Inst. Polar Res., Special Issue. HI 67789

Memoirs of the national institute of zoology and botany, academica sinica; botanical series. Mem. Natl. Inst. Zool. Acad. Sin., Bot. Ser. See B–P–H 578/10. HI 56419

Memoirs of the national museum. Melbourne, Vic. Vols. 1-14(1), 1906-44 [suspended 1915-26]. Mem. Natl. Mus. Superseded by: Memoirs of the national museum of Victoria. 5-4387-2. HI 67790

Memoirs of the national museum of Victoria. Melbourne, Vic. Vol. 14(2)+, 1946+. Mem. Natl. Mus. Victoria. Preceded by: Memoirs of the national museum, Melbourne. HI 67791

Memoirs of the national science museum, Tokyo. [Kokuritsu kagaku hakubutsukan senho.] Tokyo. Vol. 1+, 1968+. Mem. Natl. Sci. Mus. (Tokyo). HI 67792

Memoirs of the natural and cultural researches of the San-in region. Simane. No. ?-20+, ?-1980+. Mem. Nat. Cult. Res. San-in Region. HI 67793

Memoirs of the natural history foundation of Orange County. Newport Beach, CA. Vol. 1+, 1984+. Mem. Nat. Hist. Found. Orange County. HI 67794

Memoirs of natural sciences. [Museum of the Brooklyn institute of arts and sciences.] Brooklyn [=New York, in part], NY. Mem. Nat. Sci. See B–P–H 578/7. HI 56417

Memoirs of the New York botanical garden. New York, NY. Mem. New York Bot. Gard. See B–P–H 578/14. HI 56420

Memoirs, New York state museum = New York state museum memoirs. Albany, NY. New York State Mus. Mem. See B–P–H 656/27.

Memoirs of the New Zealand oceanographic institute. Wellington, N.Z. 1955+. Mem. New Zealand Oceanogr. Inst. HI 67795

Memoirs of Osaka gakugei university. B, natural science. Osaka. 1964-66. Mem. Osaka Gakugei Univ., B. Preceded by: Memoirs of the Osaka university of liberal arts and education. Superseded by: Memoirs of Osaka kyoiku university. III, natural science and applied science. HI 67796

Memoirs of Osaka kyoiku university. III, natural science & applied science. Osaka. 1967+. Mem. Osaka Kyoiku Univ., 3, Nat. Sci. Appl. Sci. Preceded by: Memoirs of Osaka gakugei university. B, natural science. HI 67797

Memoirs of Osaka university of the liberal arts and education. Series B, natural sciences. [Osaka gakugei daigaku kiyo. B. Shizen kahaku.] Osaka. Nos. 1-?, 1952-63. Mem. Osaka Univ. Liberal Arts, Ser. B, Nat. Sci. Superseded by: Memoirs of Osaka gakugei university. B, natural science. HI 67798

Memoirs of the Pacific tropical botanic garden. Lawai, HI. No. 1+, 1973+. Mem. Pacific Trop. Bot. Gard. HI 67799

Memoirs of the Peabody academy of sciences. Salem, MA. Mem. Peabody Acad. Sci. See B–P–H 578/19. HI 56422

Memoirs of the Peabody museum of Yale university. New Haven, CT. Mem. Peabody Mus. See B–P–H 578/20. HI 56423

Memoirs of the Philadelphia society for promoting agriculture. Philadelphia, PA. Mem. Philadelphia Soc. Promot. Agric. See B–P–H 578/21. HI 56424

Memoirs and proceedings of the Manchester literary and philosophical society. Manchester. Ser. 4, vols. 1-10 [also numbered vols. 31-40], 1888-96; vol. 41-120, 1897-1980. Mem. & Proc. Manchester Lit. Soc. Preceded by: Memoirs of the literary and philosophical society of Manchester and Proceedings of the Manchester literary and philosophical society. 3-2531-2. HI 67800

Memoirs published by the institute for plant protection = Posebna izdanja. Institut za zaštita bilja. Belgrade. Posebna Izd. Inst. Zašt. Bilja. See B–P–H 716/9.

Memoirs of the Queensland museum. Brisbane. Mem. Queensland Mus. See B–P–H 579/6. HI 56428

Memoirs, research division, department of agriculture, Uganda. Series 2, vegetation. Kampala. Nos. 1-6, 1959-60. Mem. Res. Div. Dept. Agric. Uganda., Ser. 2, Veg. HI 67801

Memoirs of the research institute for food science, Kyoto university. Kyoto. Vols. 1-?, 1951-72. Mem. Res. Inst. Food Sci. Kyoto Univ. HI 67802

Memoirs of the royal caledonian horticultural society. Edinburgh. Vol.1(1-2), 1905-08. Mem. Roy. Caledonian Hort. Soc. Preceded by: Memoirs, caledonian horticultural society. 4-3718-1. HI 67803

Memoirs of the royal society, being a new abridgement of the philosophical transactions from 1665-1735 inclusive. London. Mem. Roy. Soc. See B–P–H 580/8. HI 56445

Memoirs of the royal society of South Australia. Adelaide. Mem. Roy. Soc. South Australia. See B–P–H 580/9. HI 56446

Memoirs of the San Diego society of natural history. San Diego, CA. Mem. San Diego Soc. Nat. Hist. See B–P–H 580/15. HI 56449

Memoirs of the school of agriculture of Cambridge university. Cambridge. Nos. 1-41, 1929-69. Mem. School Agric. Cambridge Univ. Superseded by: Memoirs, depratment of agricultural science and applied biology, university of Cambridge. 2-893-3. HI 67804

Memoirs of science and the arts. Or, an abridgement of the transactions published by the principal learned and oeconomical societies established in Europe, Asia, and America. London. Mem. Sci. Arts. See B–P–H 580/19. HI 56450

Memoirs of the science department; university of Tokyo. Tokyo. Mem. Sci. Dept. Univ. Tokyo. See B–P–H 581/1. HI 56451

Memoirs of the science society of China. [Chung kuo k'o hsüeh shê yen chin ts'ung k'an.] Shanghai. Mem. Sci. Soc. China. See B–P–H 581/5. HI 56452

Memoirs, Sears foundation for marine research. Mem Sears Found. Mar. Res. See B–P–H 581/8. HI 56454

Memoirs of the Showa pharmaceutical college. Tokyo. Vol. 1+, 1963+. Mem. Showa Pharm. Coll. HI 67805

Memoirs of the south industrial science institute, Kagoshima university. [Nanpo-sango kagaku kenkyujo hokoku. Kagoshima daigaku.] Kagoshima, Japan. Mem. S. Industr. Sci. Inst. Kagoshima Univ. See B–P–H 580/13. HI 56448

Memoirs of the southern California academy of sciences. Los Angeles, CA. Mem. S. Calif. Acad. Sci. See B–P–H 580/10. HI 56447

Memoirs, swiss forest research institute = Schweizerische Anstalt für das forstliche Versuchswesen. Mitteilungen. Zurich.

Memoirs of the Tanaka citrus experiment station. [Tanaka kankitsu shiken-jo. Mino-mura, Fukuoka-ken.] Fukuoka?, Japan. Mem. Tanaka Citrus Exp. Sta. See B–P–H 588/7. HI 56525

Memoirs of the Tokyo university of agriculture. [Tokyo nogyo daigaku kiyo.] Tokyo. 1956+. Mem. Tokyo Univ. Agric. HI 67806

Memoirs of the Torrey botanical club. New York, NY. Mem. Torrey Bot. Club. See B–P–H 588/12. HI 56526

Memoirs of the Tottori agricultural college. [Tottori koto nogyo gakko gakujutsu hokoku.] Tottori, Japan. Mem. Tottori Agric. Coll. See B–P–H 588/15. HI 56527

Memoirs of the Wernerian natural history society. Edinburgh. Mem. Wern. Nat. Hist. Soc. See B–P–H 588/25. HI 56530

Memorabilia historiae naturalis = Iz istorii biologicheskii nauk. Moscow.

Memoranda, imperial bureau of plant genetics. Herbage plants. Aberystwyth. Nos. [1]-6, 1932-35. Mem. Imp. Bur. Pl. Genet., Herbage Pl. HI 67807

Memoranda, nature reserves investigation committee. London. 1943-45. Mem. Nat. Res. Invest. Committee. HI 67808

Memoranda societatis pro fauna et flora fennica. Helsinki. Memoranda Soc. Fauna Fl. Fenn. See B–P–H 588/31. HI 56531

Memoria, see Memorias

Memorial literario, instructivo y curioso de la corte de Madrid. Madrid. Vols. 1-53(15), 1784-1808 [publication suspended 1791-93, 1798-1800]. Mem. Lit. Instruct. Cur. Corte Madrid. 3-2601-2. HI 53984

Memórias da academia das ciências de Lisboa. Classe de ciências. Lisbon. Mem. Acad. Ci. Lisboa, Cl. Ci. See B–P–H 563/11. HI 56253

Memórias da academia das ciências de Lisboa. Classe de sciencias mathematicas, physicas e naturaes. Lisbon. Mem. Acad. Ci. Lisboa, Cl. Sci. See B–P–H 563/12. HI 56254

Memorias, academia nacional de ciencias "Antonia Alzate". Mexico City. Mem. Acad. Nac. Ci. "Antonia Alzate". See B–P–H 564/9. HI 56264

Memorias de la academia nacional de historia y geografia. Mexico City. Mem. Acad. Nac. Hist. See B–P–H 564/11. HI 56265

Memórias da academia real das sciencias de Lisboa. Lisbon. Mem. Acad. Real Sci. Lisboa. See B–P–H 564/18. HI 56268

Memórias da academia real das sciencias de Lisboa. Primeira classe de sciencias mathematicas, physicas e naturaes. Lisbon. N.s. vols. 1-7, 1854-1914. Mem. Acad. Real. Sci. Lisboa, 1 Cl. Sci. Math. Preceded by: Historia e memórias da academia r(eal) das sciencias de Lisboa. 1-17-3. HI 67809

Memórias da academia real das sciencias de Lisboa. Segunda classe de sciencias moraes, politicos e bellas lettras. Lisbon. N.s. vols. 1-?, 1854-57, 1863-98 [suspended 1858-62]. Mem. Acad. Real. Sci. Lisboa, 2 Cl. Sci. Moraes. Preceded by: Historia e memórias da academia r(eal) das sciencias de Lisboa. Superseded by: Historia e memórias da academia real das sciencias de Lisboa. Segunda classe de sciencias moraes, politicos e bellas lettras. 1-17-3. HI 67810

Memorias academicas de la real sociedad de medicina, y demàs ciencias de Sevilla. Seville. Vols. 1-11, 1766-1819. Mem. Acad. Real. Soc. Med. Sevilla. HI 53982

Memorias de agricultura y artes. Barcelona. Mem. Agric. Artes. See B–P–H 567/29. HI 56301

Memorias de agricultura premiadas pela academia real das sciencias de Lisboa. Lisbon. Mem. Agric. Acad. Real Sci. Lisboa. See B–P–H 567/28. HI 56300

Memorias anuales, instituto de biologia marina. Mar del Plata. 1960-72. Mem. Anuales Inst. Biol. Mar. Superseded by: Memorias, instituto nacional de investigacion y desarrollo pesquero. HI 67811

Memorias anuales del instituto de investigaciones agropecuarias. Santiago, Chile. No. 1+, 1964/65+. Mem. Anuales Inst. Invest. Agropecu. HI 67812

Memorias anuales, instituto del mar del Peru. Callao. 1970+, [1971]+. Mem. Anuales Inst. Mar Peru. HI 67813

Memorias anuales, junta de ciències naturals. Barcelona. 1919/20-1921/22. Mem. Anuales Junta Ci. Nat. Preceded by: Anuari, junta de ciències naturals de Barcelona. HI 67814

Memoria anuale, museo nacional de historia natural "Bernardino Rivadavia". Buenos Aires. Mem. Anual Mus. Nac. Hist. Nat. "Bernardino Rivadavia". See B–P–H 568/17. HI 56315

Memorias, asociación de investigación para la mejora de la alfalfa. Zaragoza. 1966+. Mem. Asoc. Invest. Mejora Alfalfa. HI 67815

Memorias de ciencias naturales y de industria nacional y estranjera. Lima. Mem. Ci. Nat. Industr. Nac. Estranjera. See B–P–H 569/19. HI 56327

Memórias. Classe de sciencias mathematicas, physicas e naturaes: Academia das ciencias de Lisboa. Lisbon. Mem. Cl. Sci. Math. Acad. Ci. Lisboa. See B–P–H 569/23. HI 56328

Memorias del consejo oceanográfico ibero-americano. Madrid. Mem. Cons. Oceanogr. Ibero-Amer. See B–P–H 570/26. HI 56334

Memorias, consejo superior de investigaciones cientificas. Barcelona. 1944+. Mem. Cons. Super. Invest. Ci. HI 67816

Mémorias economicas da academia real das sciencias de Lisboa, ... Lisbon. Mem. Econ. Acad. Real Sci. Lisboa. See B–P–H 571/14. HI 56343

Memorias de la facultad de ciencias. Universidad de la Habana. Serie ciencias biologicas. Havana. 1967+. Mem. Fac. Ci. Univ. Habana., Ser. Ci. Biol. HI 67817

Memorias de fomento. San José. 1912-13. Mem. Fomento. HI 67818

Memorias de historia natural, de quimica, de agricultura, artes, e medicina: lidas na academia real das ciencias de Lisboa. Lisbon. Mem. Hist. Nat. Acad. Real Ci. Lisboa. See B–P–H 573/12. HI 56365

Memórias do instituto de biociências universidade federal de Pernambuco. Recife. Vol. 1+, 1974+. Mem. Inst. Bioci. Univ. Fed. Pernambuco. HI 67819

Memorias, instituto de biología marina. Mar del Plata, Argentina. Mem. Inst. Biol. Mar. See B–P–H 574/5. HI 56374

Memórias do instituto biológico Ezequill Dias. Belo Horizonte, Brazil. Mem. Inst. Biol. Ezequill Dias. See B–P–H 574/4. HI 56373

Memórias do instituto Butantan. São Paulo. Mem. Inst. Butantan. See B–P–H 574/7. HI 56375

Memórias: instituto histórico, geográphico e ethnográphico de Brazil. Rio de Janeiro. Mem. Inst. Hist. Brazil. See B–P–H 574/16. HI 56379

Memórias do instituto de investigação agronômica de moçambique. Lourenço Marques. 1966+. Mem. Inst. Invest. Agron. Moçambique. HI 67820

Memórias do instituto de investigação científica de Moçambique. Lourenço Marques. Vols. 1-?, 1959-64. Mem. Inst. Invest. Ci. Moçambique. Superseded by: Memórias do instituto de investigação científica de Moçambique. Serie A, ciencias biologicas. HI 67821

Memórias do instituto de investigação científica de Moçambique. Serie A, ciencias biologicas. Lourenço Marques. 1965+. Mem. Inst. Ivest. Ci. Moçambique, A. Preceded by: Memórias do instituto de investigação científica de Moçambique. HI 67822

Memorias, instituto nacional para la conservacion de la naturaleza. Madrid. 1973+. Mem. Inst. Nac. Conserv. Naturaleza. HI 67823

Memorias, instituto nacional de investigacion y desarrollo pesquero. Mar del Plata. 1976/78+, 1978?+. Mem. Inst. Nac. Invest. Desarr. Pesq. Preceded by: Memorias anuales, instituto de biologia marina. HI 67824

Memórias do instituto Oswaldo Cruz. Rio de Janeiro. Mem. Inst. Oswaldo Cruz. See B–P–H 575/1. HI 56384

Memorias instructivas y curiosas sobre agricultura, comercio, industria, economia, chymica, botánica, historia natural, etc. Madrid. Mem. Instruct. Cur. Agric. See B–P–H 573/15. HI 56390

Memorias del jardin zoologico de La Plata. Rio de Janeiro. Vols. 1-7, 1927/28-36/37, 1924-38. Mem. Jard. Zool. La Plata. HI 67825

Memoria que la junta directiva de la sociedad nacional agraria. Lima. Mem. Junta Direct. Soc. Nac. Agrar. See B–P–H 575/18. HI 56393

Memorias, junta de investigações do ultramar. Segunda serie. Lisbon. 1958+. Mem. Junta Invest. Ultramar, 2 Ser. Preceded by: Memorias, junta de investigações do ultramar. Serie agronomia tropical and Serie botanica. HI 67826

Memorias, junta de investigações do ultramar. Serie botanica. Lisbon. Nos. 2-4, 1955-57. Mem. Junta Invest. Ultramar, Ser. Bot. Preceded by: Memorias, junta das missoes geograficas e de investigações colonais. Superseded by: Memorias de investigações do ultramar. Segunda serie. HI 67827

Memorias, junta de investigações do ultramar. Serie agronomia tropical. Lisbon. Nos. 1-2, 1956. Mem. Junta Invest. Ultramar, Ser. Agron. Trop. Superseded by: Memorias de investigações do ultramar. Segunda serie. HI 67828

Memorias, junta das missoes geograficas e de investigações colonais. Serie botanica. Lisbon. No. 1, 1943. Mem. Junta Missoes Geogr. Invest. Colon., Ser. Bot. Superseded by: Memorias, junta de investigações do ultramar. Serie botanica. HI 67829

Memórias de litteratura portugueza. Lisbon. Mem. Litt. Portug. See B–P–H 576/6. HI 56399

Memórias de mathematica e phisica da academia r. das sciencias de Lisboa. Lisbon. Mem. Math. Phis. Acad. Real Sci. Lisboa. See B–P–H 576/14. HI 56404

Memorias del museo de ciencias naturales de Barcelona, serie botanica. Barcelona. Vol. 1(2-3), 1924-25. Mem. Mus. Ci. Nat. Barcelona, Ser. Bot. Preceded by: Memòries del museu de ciències naturals de Barcelona. Sèrie botànica. 1-604-2. HI 67830

Memorias del museo de Entre Ríos. Paraná, Argentina. Mem. Mus. Entre Ríos. See B–P–H 577/11. HI 56415

Memorias del museo de historia natural "Javier Prado". Universidad de San Marcos. Lima. Mem. Mus. Hist. Nat. "Javier Prado". See B–P–H 577/13. HI 56410

Memorias del museo de Paraná. Paraná, Argentina. Mem. Mus. Paraná. See B–P–H 577/23. HI 56412

Memorias políticas y económicas sobre los frutos, comercio, fabricas y minas de España. Madrid. Mem. Polít. Econ. Frutos España. See B–P–H 578/24. HI 56426

Memorias de la real academia de ciencias y artes de Barcelona. Barcelona. Mem. Real Acad. Ci. Barcelona. See B–P–H 579/13. HI 56429

Memorias. Real academia de ciencias exactas, físicas y naturales de Madrid. Madrid. Mem. Real Acad. Ci. Exact. Madrid. See B–P–H 579/14. HI 56430

Memorias de la real academia médica de Madrid. Madrid. Mem. Real Acad. Méd. Madrid. See B–P–H 579/15. HI 56431

Memorias de la real sociedad económica de la Habana. Havana. Mem. Real Soc. Econ. Habana. See B–P–H 579/16. HI 56432

Memorias de la real sociedad económica de Madrid. Madrid. Mem. Real Soc. Econ. Madrid. See B–P–H 579/17. HI 56433

Memorias de la real sociedad económica mallorquina de amigos del país. Palma, Majorca. Mem. Real Soc. Econ. Mallorquina Amigos País. See B–P–H 579/18. HI 56434

Memorias de la real sociedad española de historia natural. Madrid. Mem. Real Soc. Esp. Hist. Nat. See B–P–H 579/19. HI 56435

Memorias de la real sociedad patriótica. Seville. Mem. Real Soc. Patriót. (Seville). See B–P–H 579/20. HI 56436

Memorias de la real sociedad patriótica de la Habana. Havana. Mem. Real Soc. Patriót. Habana. See B–P–H 579/21. HI 56437

Memoria de la secretaría general, consejo superior de investigaciones científicas. Madrid. Mem. Secr. Gen. Cons. Super. Invest. Ci. See B–P–H 581/9. HI 56455

Memórias de sociedade broteriana. Coimbra, Portugal. Mem. Soc. Brot. See B–P–H 582/21. HI 56469

Memoria de la sociedad de ciencias naturales La Salle. Caracas. Mem. Soc. Ci. Nat. La Salle. See B–P–H 583/1. HI 56471

Memorias, sociedad cientifica "Antonia Alzate" = Memorias, academia nacional de ciencias "Antonia Alzate". Mexico City. Mem. Acad. Nac. Ci. "Antonia Alzate". See B–P–H 564/9.

Memorias de la sociedad cubana de historia natural "Felipe Poey". Havana. Vols. 1-25, 1915-61 [suspended 1918-20; 1927-34]. Mem. Soc. Cub. Hist. Nat. "Felipe Poey". Superseded by: Poeyana. 5-3904-3. HI 67831

Memorias de la sociedad económica. Madrid. Mem. Soc. Econ. (Madrid). See B–P–H 583/8. HI 56472

Memorias de la sociedad económica de amigos del país. Caracas. Mem. Soc. Econ. Amigos País. See B–P–H 583/9. HI 56473

Memorias de la sociedad económica de la Habana. Havana. Nos. ?-40-63-?, 18??-1823?-25-? Mem. Soc. Econ. Habana. Preceded by?: Memorias de la real sociedad económica de la Habana. HI 52668

Memorias de la sociedad española de historia natural = Memorias de la real sociedad española de historia natural. Madrid. Mem. Real Soc. Esp. Hist. Nat. See B–P–H 579/19.

Memorias de la sociedad ibérica de ciencias naturales. Saragossa, Spain. Mem. Soc. Ibér. Ci. Nat. See B–P–H 584/7. HI 56484

Memorias de la sociedad patriótica de la Habana. Havana. Mem. Soc. Patriót. Habana. See B–P–H 586/1. HI 56501

Memoria de los trabajos. Instituto nacional de investigaciones científicas y museo de historia natural. Havana? Mem. Trab. Inst. Nac. Invest. Ci. Mus. Hist. Nat. See B–P–H 588/16. HI 56667

Memórias e trabalhos, centro de investigaçao cientifica Algodoeira. Lourenço Marques. Nos. 1-?, 1951-61. Mem. Trab. Centro Invest. Ci. Algodoeira. Preceded by: Trabalhos do centro de investigaçao cientifica Algodoeira. HI 67832

Memórias e trabalhos do instituto de investigação cientifíca de Angola. Luanda. Vol. 1+, 1960+. Mem. Trab. Inst. Invest. Ci. Angola. HI 67833

Mémoriaux de la société d'émulation d'Abbeville. Abbeville. 1816-17. Mém. Soc. Émul. Abbeville. Preceded by: Bulletin de la société d'émulation d'Abbeville. Superseded by: Analyse des travaux de la société d'émulation d'Abbeville. 5-3938-3. HI 67834

Memorie dell'accademia d'agricoltura, commercio ed arti di Verona. Verona. Mem. Accad. Agric. Verona. See B–P–H 567/17. HI 56296

Memorie dell'accademia imperiale delle scienze e belle arti di Genova. Genoa. Mem. Accad. Imp. Sci. Genova. See B–P–H 567/18. HI 56297

Memorie della accademia patavina di SS. LL. AA. Classe di scienze, matematiche e naturalisti = Atti e memorie, accademia patavina di scienze, lettere ed arti. Padua.

Memorie dell'accademia delle scienze, lettere et arti di Genova. Genoa. Mem. Accad. Sci. Genova. See B–P–H 567/21. HI 56298

Memorie della accademia di scienze, lettere ed arti di Padova. Padua. Mem. Accad. Sci. Padova. See B–P–H 567/24. HI 56299

Memorie di agricoltura, manifatture e commercio. Bologna. Mem. Agric. Manifatture. See B–P–H 568/1. HI 56302

Memorie dell'ateneo di salo. Seconda serie. ?-1988+. Mem. Ateneo Salo, 2 Ser. HI 67836

Memorie di biogeografica adriatica. Venice. Vols. 1-?, 1950-65. Mem. Biogeogr. Adriat. HI 67837

Memorie di biologia marina e di oceanografia, università di Messina. Messina, Italy. Mem. Biol. Mar. Oceanogr. Univ. Messina. See B–P–H 569/10. HI 56322

Memorie della comitato talassagrafico italiano = Memorie della reale comitato talassagrafico italiano. Venice. Mem. Reale Comitato Talassogr. Ital. See B–P–H 580/5.

Memorie sopra la fisica e istoria naturale di diversi valentuomini. Lucca, Italy. Mem. Fis. Istoria Nat. See B–P–H 573/2. HI 56358

Memorie, flora et vegetatio italica = Flora et vegetatio italica. Memoria. Milan.

Memorie fuori serie, museo civico di storia naturale di Verona. [Supplement to: Memorie del museo civico di storia naturale di Verona.] Verona. Vol. 1+, 1966+. Mem. Fuori Ser. Mus. Civico Storia Nat. Verona. HI 67838

Memorie dell'imperiale regio istituto del regno lombardo-veneto. Milan. Vols. 1-5, 1819-38. Mem. Imp. Regio Ist. Regno Lombardo. Preceded by: Memorie dell istituto nazionale italiano. Classe di fisica e matematica [not entered]. Superseded by: Memorie del reale istituto lombardo di scienze, lettere ed arti and by Memorie del reale istituto veneto di scienze, lettere ed arti. 3-2126-2. HI 67839

Memorie, istituto italiano di idrobiologia "Dott. Marco de Marchi". Novara. Vol. 1+, 1942+. Mem. Ist. Ital. Idrobiol. "Dott. Marco Marchi". HI 67840

Memorie dell'instituto ligure. Genoa. Mem. Inst. Ligure. See B–P–H 574/18. HI 56380

Memorie, instituto svizzero di ricerche forestali = Schweizerische Anstalt für das forstliche Versuchswesen. Mitteilungen. Zurich.

Memorie, istituto veneto di scienze, lettere ed arti. Classe de scienze matematiche e naturali. Venice. 1956+. Mem. Ist. Veneto Sci. Lett. Arti, Cl. Sci. Mat. Nat. Preceded by: Memorie del reale istituto veneto di scienze, lettere ed arti. HI 67841

Memorie di laboratorio di botanica; università, Bari. Bari, Italy. Mem. Lab. Bot. Univ. Bari. See B–P–H 576/3. HI 56396

Memorie di matematica e fisica della società italiana. Verona. Mem. Mat. Fis. Soc. Ital. See B–P–H 576/10. HI 56401

Memorie di matematica e di fisica della società italiana della scienze. Modena, Italy. Mem. Mat. Fis. Soc. Ital. Sci. See B–P–H 576/11. HI 56402

Memorie di matematica e di fisica della società italiana delle scienze residente in Modena. Parte contenente le memorie di fisica. Modena, Italy. Mem. Mat. Fis. Soc. Ital. Sci. Modena, Pt. Mem. Fis. See B–P–H 576/12. HI 56403

Memorie del museo civico di storia naturale di Verona. Verona. Vols. 1-20, 1897-1972. Mem. Mus. Civico Storia Nat. Verona. Partly superseded by: Memorie del museo civico di storia naturale di Verona. IIa serie, sezione sciènza della vita and Bollettino, museo civico di storia naturale di Verona. 5-4380-1. HI 67842

Memorie del museo civico di storia naturale di Verona. IIa serie, sezione sciènza della vita. Verona. No. 1+, 1977+. Mem. Mus. Civico Storia Nat. Verona, I2 Ser., Sez. Sci. Vita. Preceded by: Memorie del museo civico di storia naturale di Verona. HI 67843

Memorie del museo di storia naturale della Venezia tridentina. Trento. Vols. 1-15, 1931-65. Mem. Mus. Storia Nat. Venezia Tridentina. Superseded by: Memorie museo tridentina di science naturali. 5-4260-1. HI 67844

Memorie museo tridentina di science naturali. Trento. Vol. 16+, 1966+. Mem. Mus. Tridentia Sci. Nat. Preceded by: Memorie del museo di storia naturale della Venezia tridentina. HI 67845

Memorie della reale accademia di scienze, belle lettere ed arti. Mantua, Italy. Mem. Reale Accad. Sci. (Mantua). See B–P–H 579/22. HI 56438

Memorie della r[eale]. accademia delle scienze dell'istituto di Bologna. Bologna. Mem. Reale Accad. Sci. Ist. Bologna. See B–P–H 579/23. HI 56439

Memorie della reale accademia delle scienze dell'istituto di Bologna. Classe di scienze fisiche. Bologna. Ser. 6, vols. 5-10, 1907/08-1912/13; ser. 7, vols. 1-10, 1913/14-1922/23; ser. 8, vols. 1-10, 1923/24-1932/33; ser. 9, vols. 1-10, 1933/34-42/43; ser. 10, vols. 1-10, 1943/44-52/53. Mem. Reale Accad. Sci. Ist. Bologna, Cl. Sci. Fis. Preceded by: Memorie della reale accademia delle scienze dell'istituto di Bologna. Superseded by: Atti dell'accademia delle scienze dell'istituto di Bologna. Classe di scienze fisiche. Memorie. 1-40-1. HI 67846

Memorie della reale accademia di scienze, lettere e d'arti di Modena. Modena. Vols. 1-28, 1833-82; ser. 2, vols. 1-12, 1883-96; ser. 3, vols, 1-14, 1898-1922. Mem. Reale Accad. Sci. Modena. Superseded by: Atti e memorie, reale accademia de scienze, lettere ed arti, Modena. 1-41-3. HI 67847

Memorie della r[eale] accademia di scienze, lettere e arti degli Zelanti. Acireale. 1927-29. Mem. Reale Accad. Sci. Lett. Arti Zelanti. Preceded & superseded by: Rendiconti e memorie della r[eale] accademia di scienze, lettere ed arti degli Zelanti. HI 67848

Memorie della reale accademia delle scienze matematiche, scienze naturali, e scienze morali = Memorie delle reale accademia delle scienze di Napoli. Naples.

Memorie delle reale accademia delle scienze di Napoli. Naples. Vols. 1-2 [vol. 1 for the years 1852-54, vol. 2 for the years 1855-57]. 1856-57. Mem. Reale Accad. Sci. (Naples). Preceded by: Atti della reale accademia delle scienze, sezione della società reale borbonica. 1-39-3. HI 67849

Memorie della reale accademia delle scienze di Parigi spettanti alla storia naturale. Recate in italiana favella. Classe VI. [Botanica.] Venice. Mem. Reale Accad. Sci. Parigi, Cl. 6. See B–P–H 580/3. HI 56441

Memorie della reale accademia delle scienze di Torino. Turin. Mem. Reale Accad. Sci. Torino. See B–P–H 580/4. HI 56442

Memorie della reale comitato talassagrafico italiano. Venice. Mem. Reale Comitato Talassogr. Ital. See B–P–H 580/5. HI 56443

Memorie del reale istituto lombardo di scienze, lettere ed arti. Milan. Vols. 1-9, 1843-63 [vols. 7-9, 1859-63 also numbered ser. 2, vols. [1]-3]. Mem. Reale Ist. Lombardo Sci. Preceded by: Memorie dell'imperiale regio istituto del regno lombardo-veneto. Superseded in part by: Memorie del reale istituto lombardo di scienze e lettere. Serie 3, classe di scienze matematiche e naturali. 3-2127-3. HI 67850

Memorie del reale istituto lombardo di scienze e lettere. Serie 3, classe di scienze matematiche e naturali. Milan. Vol. 10 [also numbered vol. 1]+, 1867+. Mem. Reale Ist. Lombardo Sci., Ser. 3, Cl. Sci. Mat. Preceded by: Memorie del reale istituto lombardo di scienze, lettere ed arti. 3-2127-3. HI 67851

Memorie del reale istituto veneto di scienze, lettere ed arti. Venice. Vols. 1-30, 1843-1940. Mem. Reale Ist. Veneto Sc. Preceded by: Memorie dell'imperiale regio istituto del regno lombardo-veneto. Superseded by: Memorie, istituto veneto di scienze, lettere ed arti. Classe de scienze matematiche e naturali. 3-2130-3. HI 67852

Memorie della regia accademia di scienze, lettere ed arti in Modena = Memorie della reale accademia di scienze, lettere e d'arti di Modena. Modena.

Memorie e rendiconti della accademia di scienze, lettere e belle arti di Acireale. Classe di scienze. Acireale. Ser. 1, vol. 1+, 1943+. Mem. Rendiconti Accad. Sci. Acireale, Cl. Sci. Preceded by: Rendiconti e memorie della r[eale] accademia di scienze, lettere ed arti degli Zelanti. HI 67853

Memorie e rendiconti della società di naturalisti di Modena. Modena, Italy. Mem. Rendiconti Soc. Naturalisti Modena. See B–P–H 580/6. HI 56444

Memorie della società medica di Bologna. San Damiano, Italy. Mem. Soc. Med. Bologna. See B–P–H 585/6. HI 56495

Memorie della società dei naturalisti di Napoli. [Supplement to: Bollettino della società dei naturalisti di Napoli. See B–P–H 215/11.] Naples. Vol. 1+, 1970+. Mem. Soc. Naturalisti Napoli. HI 67854

Memorie valdarnesi. Pisa. Mem. Vald. See B–P–H 588/24. HI 56529

Memories de l'institució catalana d'historia natural. Barcelona. 1908; 1912-31. Mem. Inst. Catalana Hist. Nat. HI 67855

Memòries del museu de ciències naturals de Barcelona. Sèrie botànica. Barcelona. Vol. 1(1), 1922. Mem. Mus. Ci. Nat. Barcelona, Sèr. Bot. Superseded by: Memorias del museo de ciencias naturales de Barcelona, serie botanica. 1-604-2. HI 67856

Memoriile sectiunii ştiinţifice academia română. Bucharest. Mem Sect. Şti. Acad. Român ă. See B–P–H 581/14. HI 56457

Memphis medical monthly. Memphis, TN. Memphis Med. Monthly. See B–P–H 589/9. HI 56532

Memuary Botaničeskogo otdelenija Obščestva ljubitelej estestvoznanija, antropologii i etnografii = Memuary Botanicheskogo otdeleniya Obshchestva lyubitelei estestvoznaniya, antropologii i etnografii. Moscow.

Memuary Botanicheskogo otdeleniya Obshchestva lyubitelei estestvoznaniya, antropologii i etnografii. Moscow. 1 vol., 1925. Mem. Bot. Otd. Obshch. Lyubit. Estestv. 4-3131-1. HI 67857

Menara perkebunan. [Balai penjelikikan perkebunan besar di Indonesia.] Jakarta, Indonesia. Menara Perkebunan. See B–P–H 589/20. HI 56533

Mendel bulletin of the school of science of Villanova college. Villanova, PA. Mendel Bull. School Sci. Villanova Coll. See B–P–H 589/22. HI 56534

Mendel newsletter; archival resources for the history of genetics and allied sciences. Philadelphia, PA. No. 1+, 1968+. Mendel Newslett. HI 67858

Mensaje de café. [Instituto cubano de establización de café.] Havana. Mensaje Café. See B–P–H 589/24. HI 56535

Mensajero agrícola. Lima. Mensajero Agríc. See B–P–H 589/25. HI 56536

Mensajero forestal. Durango, Mexico. Mensajero Forest. See B–P–H 589/26. HI 56537

Mentzelia; journal of the northern Nevada native plant society. Reno, NV. No. 1+, 1975+. Mentzelia. HI 67859

Mer; bulletin de la société franco-japonaise d'océanographie. Tokyo. 1963+. Mer. HI 73806

Mercian; newsletter of the mercia group of the british iris society. ?-1977+. Mercian. HI 67860

Mercredi médical. Paris. Mercredi Méd. See B–P–H 589/28. HI 56538

Mercure de France, littéraire et politique. Paris. Mercure France Litt. Polit. See B–P–H 589/29. HI 56539

Mercurio chileno. Santiago. Mercurio Chileno. See B–P–H 589/31. HI 56540

Mercurio peruano de historia, literatura, y noticias públicas que da á luz la sociedad académica de amantes de Lima. Lima. Mercurio Peruano Hist. See B–P–H 589/32. HI 56541

Merentutkimuslaitoksen julkaisu. Helsinki. Nos. 1-243, 1920-78. Merentutkimuslait. Julk. Superseded by: Finnish marine research. HI 67861

Merkblätter, biologische Bundesanstalt für Land- und Forstwirtschaft. Berlin. 1953+. Merkbl. Biol. Bundesanst. Land- Forstw. Preceded by: Merkblätter, biologische Reichsanstalt für Land- und Forstwirtschaft. HI 67862

Merkblätter, biologische Reichsanstalt für Land- und Forstwirtschaft. Berlin. 1927-44. Merkbl. Biol. Reichsanst. Land- Forstw. Superseded by: Merkblätter, biologische Bundesanstalt für Land- und Forstwirtschaft. HI 67863

Merkur von Ungarn, oder Litterarzeitung für das Königreich Ungarn und dessen Kronländer. Budapest. Merkur Ungarn. See B–P–H 590/1. HI 56542

Merkwürdige Abhandlungen holländischer Ärzte. Leipzig. Merkwürd. Abh. Holl. Ärzte. See B–P–H 590/2. HI 56543

Merkwürdigkeiten der Natur, Kunst und des Menschlebens für allerlei Leser. Erfurt. Merkwürdigk. Natur. See B–P–H 590/3. HI 56544

Mesičnik kaktušarské. Prague. 1922+. Mesičnik Kaktuš. HI 67864

Mesemb study group bulletin. Belen, NM. 1986+. Mesemb Study Group Bull. HI 67865

Mesopotamia. Mosul. 1966-67. Mesopotamia. Superseded by: Mesopotamia agriculture. HI 67866

Mesopotamia agriculture. Mosul. 1968-69. Mesopotamia Agric. Preceded by: Mesopotamia. Superseded by: Mesopotamia journal of agriculture. HI 67867

Mesopotamia journal of agriculture. Mosul. 1970+. Mesopotamia J. Agric. Preceded by: Mesopotamia agriculture. HI 67868

Mesquita newsletter. Austin, TX. Ca.1983+. Mesquita Newslett. HI 67869

Messager des sciences et des arts. Ghent. Messager Sci. Arts Gand. See B–P–H 590/9. HI 56548

Messager des sciences et des arts de la Belgique. Ghent. Messager Sci. Arts Belgique. See B–P–H 590/8. HI 56547

Messager des sciences historiques de Belgique. Ghent.

Messager Sci. Hist. Belgique. See B–P–H 590/10. HI 56549

Messenger. Honolulu, HI = Ka 'elele. Honolulu, HI.

Messenger. New York. Vols. 1-8(3), 1924-31. Messenger. Preceded by: New York journal of pharmacy (1931-33). 4-3035-1. HI 67870

Meteor Forschungsergebnisse. Reihe D, Biologie. Berlin & Stuttgart. No. 1+, 1967+. Meteor Forschungsergebn., D. HI 67871

Meteorologické zprávy. Prague. Meteorol. Zprávy. See B–P–H 590/17. HI 56552

Meteorologisch-phänologische Beobachtungen aus der Fuldaer Gegend. Fulda. Meteorol.-Phänol. Beob. Fuldaer Gegend. See B–P–H 590/15. HI 56551

Meteorologische Ephemeriden. Munich. Meteorol. Ephem. See B–P–H 590/14. HI 56550

Meteorologische Rundschau. Berlin, Göttingen & Heidelberg. Vol. 1/2+, 1947/48+. Meteorol. Rundschau. 3-2621-2. HI 67872

Methods of biochemical analysis. New York, NY. Meth. Biochem. Analysis. See B–P–H 590/18. HI 56553

Methods in cell biology. New York & London. Vol. 6+, 1973+. Meth. Cell Biol. Preceded by: Methods in cell physiology. HI 67873

Methods in cell physiology. New York. Vols. 1-5, 1964-72. Meth. Cell Physiol. Superseded by: Methods in cell biology. HI 67874

Methods in membrane biology. New York. Vols. 1-10, 1974-79. Meth. Membr. Biol. Superseded by: Cell membranes, methods and reviews. HI 67875

Methods in microbiology. London, New York, etc. Vol. 1+, 1969+. Meth. Microbiol. HI 67876

Methods in molecular and cellular biology. New York. Vol. 1+, 1989+. Meth. Molec. Cell Biol. HI 74859

Methods in virology. New York & London. Vol. 1+, 1967+. Meth. Virol. HI 67877

Metodické prírucky experímentální botanicky. Prague. Vol. 1+, 1965+. Metod. Prír. Exp. Bot. HI 67878

Metrian; newsletter of the metropolitan tree improvement alliance. 19??-1979-81-? Metrian. HI 67879

Metrodiana. Buenos Aires. Vols. 1-6(2), 1970-77. Metrodiana. HI 67880

Metrodiana. Sonderheft. Buenos Aires. Vol. 1+, 1972+. Metrodiana Sonderh. HI 67881

Metroparks emerald necklace. Cleveland, OH. Vols. 24(6)-30, 1975-81. Metroparks Emerald Necklace. Preceded and superseded by: Emerald necklace. HI 67882

Metsanjalostussaatio. Helsinki. 1963+. Metsanjalostussaatio. Preceded by: Toiminta, metsäpuiden rodunjalostussaatio. HI 67883

Metsätieteellisen koelaitoken julkaisuja. Helsinki. 1919-? Metsätiet. Koelait. Julk. Superseded by: Metsätieteellisen tutkimuslaitoksen julkaisuja. HI 67884

Metsätieteellisen tutkimuslaitoksen julkaisuja; meddelanden från skogsforsknings-institutet i Finland. Helsinki. Vols. ?-15-40, ?-1930-52. Metsät. Tutkimuslait. Julk. Preceded by: Metsätieteellisen koelaitoken julkaisuja. Superseded by: Metsätutkimuslaitoksen julkausija. HI 67885

Metsätutkimuslaitoksen julkausija. Helsinki. Vols. 41-100?, 1953-82. Metsätutkimuslait. Julk. Preceded by: Metsätieteellisen tutkimuslaitoksen julkaisuja. Superseded by: Communicationes instituti forestalis fenniae. HI 67886

Metsandustikud uurimused. Tartu. Vol. 1+, 1957+. Metsand. Uurim. HI 67887

México agrícola. Mexico City. México Agríc. See B–P–H 590/19. HI 56554

México antiguo. Mexico City. Mexico Antiguo. See B–P–H 590/20. HI 56555

México forestal. Mexico City. México Forest. See B–P–H 590/21. HI 56556

México y sus bosques. Mexico City. México Sus Bosques. See B–P–H 590/23. HI 56557

Mezei gazda. Pest [=Budapest, in part]. Mezei Gazd. See B–P–H 591/1. HI 56558

Mezei gazdák barátja. Pest [=Budapest, in part] Mezei Gazd. Barátja. See B–P–H 591/2. HI 56559

Mezei gazdaságot tárgyazó jegyzések. Bécs [=Vienna]. Mezei Gazd. Tárgyazó Jegyzések. See B–P–H 591/3. HI 56560

Mezei naptár. Budapest. Mezei Naptár. See B–P–H 591/4. HI 56561

Mezögazdaság. Magyarovar [=Mosonmagyarovar, in part], Hungary. Mezögazdaság. See B–P–H 591/14. HI 56569

Mezögazdaság és kertészet. Magyarovar [=Mosonmagyarovar, in part], Hungary. Mezögazd. & Kert. See B–P–H 591/5. HI 56562

Mezögazdaságtudományi közlemények. Budapest. Mezögazdaságtud. Közlem. See B–P–H 591/16. HI 56570

Mezögazdasági és elelmezésügyi minisztérium, kísérletügyi. Közlemények. Budapest. Vol. ?-64/A+, ?-1973+. Mezögazd. Elelmezésügyi Miniszt. Kísérl. Közlem. HI 67888

Mezögazdasági kísérletügyi központ évkönyve. Budapest. Mezögazd. Kísérl. Közp. Évk. See B–P–H 591/6. HI 56563

Mezögazdasági közlöny. Budapest. Mezögazd. Közl. See B–P–H 591/7. HI 56564

Mezögazdasági kutatások. Budapest. Mezögazd. Kutatás. See B–P–H 591/9. HI 56565

Mezögazdasági növénynemesitési és növenytermesztési kutató intézet kózlemények. Budapest. Vols. 3(3)-6, 1967-70. Mezögazd. Növénynemes. Növenyterm. Kutató Intéz. Kózlem. Preceded by: Növénynemesítési és növénytermesztési kutató intézet közleményei Sopronhorpács. Superseded by: Répatermesztési kutató intézet, köslemények. HI 67889

Mezögazdasági szemle. Kolozsvar [=Cluj, Rumania]. Mezögazd. Szemle (Kolozsvar). See B–P–H 591/10. HI 56566

Mezögazdasági szemle. Magyarovar [=Mosonmagyarovar, in part] & Budapest. Mezögazd. Szemle (Magyarovar & Budapest) See B–P–H 591/11. HI 56567

Mezögazdasági tudományos közlemények. Budapest. Mezögazd. Tud. Közlem. See B–P–H 591/13. HI 56568

Mežsaimniecibas rakstu krajums. Riga. Vols. ?-9-15-?, ?-1931-37-? Mežsaimn. Rakstu Krajums. HI 67890

MGA Bulletin = M G A bulletin. Yaxley, Highfield.

MGCA Newsletter = M G C A newsletter.

Michelia; commentarium mycologicum fungos in primis italicos illustrans. Padua. Michelia. See B–P–H 592/4. HI 56571

Michigan academician; papers of the Michigan academy of science, arts and letters. Ann Arbor, MI. 1969+. Michigan Academician. Preceded by: Papers of the Michigan academy of science, arts and letters. HI 67891

Michigan agricultural experiment station. Annual report. East Lansing, MI. Michigan Agric. Exp. Sta. Annual Rep. See B–P–H 592/6. HI 56572

Michigan agricultural experiment station. Bulletin. East Lansing, MI. Michigan Agric. Exp. Sta. Bull. See B–P–H 592/8. HI 56573

Michigan agricultural experiment station. Circular. East Lansing, MI. Michigan Agric. Exp. Sta. Circ. See B–P–H 592/9. HI 56574

Michigan agricultural experiment station. Circular bulletin. East Lansing, MI. Michigan Agric. Exp. Sta. Circ. Bull. See B–P–H 592/10. HI 56575

Michigan agricultural experiment station. Elementary science bulletin. East Lansing, MI. Michigan Agric. Exp. Sta. Element. Sci. Bull. See B–P–H 592/11. HI 56576

Michigan agricultural experiment station. Folder. East Lansing, MI. Michigan Agric. Exp. Sta. Folder. See B–P–H 592/12. HI 56577

Michigan agricultural experiment station. Memoirs. East Lansing, MI. Michigan Agric. Exp. Sta. Mem. See B–P–H 592/13. HI 56578

Michigan agricultural experiment station. Press bulletin. East Lansing, MI. Michigan Agric. Exp. Sta. Press Bull. See B–P–H 592/14. HI 56579

Michigan agricultural experiment station. Quarterly bulletin. East Lansing, MI. Michigan Agric. Exp. Sta. Quart. Bull. See B–P–H 592/15. HI 56580

Michigan agricultural experiment station. Research bulletin. East Lansing, MI. Michigan Agric. Exp. Sta. Res. Bull. See B–P–H 592/16. HI 56581

Michigan agricultural experiment station. Research report. East Lansing, MI. Michigan Agric. Exp. Sta. Res. Rep. See B–P–H 592/17. HI 56582

Michigan agricultural experiment station. Special bulletin. East Lansing, MI. Michigan Agric. Exp. Sta. Special Bull. See B–P–H 592/18. HI 56583

Michigan agricultural experiment station. Technical bulletin. East Lansing, MI. Michigan Agric. Exp. Sta. Techn. Bull. See B–P–H 592/19. HI 56584

Michigan Audubon. Kalamazoo, MI. Vols. 26-31(6), 1978-83. Michigan Audubon. Preceded by: Michigan Audubon newsletter. Superseded by: Michigan Audubon news. HI 67892

Michigan Audubon news. Kalamazoo, MI. Vols. 1-6(3)?, 1956-58; vol. 32(1)+, 1984+. Michigan Audubon News. For 1958-83 see: Michigan Audubon newsletter and Michigan Audubon. HI 67893

Michigan Audubon newsletter. Kalamazoo, MI. Vols. 6(4)-25(5), 1958-77. Michigan Audubon Newslett. Preceded by: Michigan Audubon news. Superseded by: Michigan Audubon. HI 67894

Michigan beekeeper. Lansing, MI. Michigan Beekeeper. See B–P–H 592/20. HI 56585

Michigan botanist. Ann Arbor, MI. Michigan Bot. See B–P–H 592/21. HI 56586

Michigan conservation. Lansing, MI. Vols. 5-37, 1936-68. Michigan Conservation. Preceded by: Monthly bulletin, Michigan department of conservation. Superseded by: Michigan natural resources. 3-2642-2. HI 67895

Michigan geological and biological survey publication; biological series. East Lansing, MI. Michigan Geol. Biol. Surv. Publ., Biol. Ser. See B–P–H 592/22. HI 56587

Michigan horticulturist = American horticulturist. Detroit, MI. Amer. Hort. (Detroit). See B–P–H 80/14.

Michigan natural resources. Lansing, MI. Vols. 38-

46(1), 1969-77. Michigan Nat. Resources. Preceded by: Michigan conservation. Superseded by: Michigan natural resources magazine. HI 67896

Michigan natural resources magazine. Lansing, MI. Vol. 46(2)+, 1977+. Michigan Nat. Resources Mag. Preceded by: Michigan natural resources. HI 67897

Michigan science in action. East Lansing, MI. Nos. 1-46, 1968-81. Michigan Sci Action. Superseded by: Futures. HI 67898

Michurin seibutsugaku kankyu = Japanese journal of Michurin biology. Tokyo.

Micologia italiana; quadrimestrala dell'unione micologia italiana. Bologna. Vol. 1+, 1972+. Micol. Ital. HI 67899

Micologia e vegetazione mediterranea. Avezzano. Vol. 1+, 1986+. Micol. Veg. Medit. HI 67900

Microbia. Nancy. Vols. 1-5, 1975-79. Microbia. Preceded by: Bulletin, association des diplômes de microbiologie de la faculté de Nancy. HI 67901

Microbial ecology; an international journal. Berlin, New York, etc. Vol. 1+, 1974+. Microbial Ecol. HI 67902

Microbial genetics bulletin, department of genetics, Carnegie institute of Washington. Cold Spring Harbor, NY. Microbial Genet. Bull. See B–P–H 593/11. HI 56588

Microbiologia. Bucharest. Vol. 1+, 1970+. Microbiologia. HI 67903

Microbiologia española. Madrid. No. 1+, 1947+. Microbiol. Esp. HI 67904

Microbiologica. Bologna. Vol. 1+, 1978+. Microbiologica. HI 67905

Microbiological circular. Department of science and agriculture, Jamaica. Kingston, Jamaica. 1923-26. Microbiol. Circ. Dept. Sci. Agric. Jamaica. HI 67906

Microbiological reviews. Washington, DC. Vol. 42+, 1978+. Microbiol. Rev. Preceded by: Bacteriological reviews. HI 67907

Microbiological sciences. Oxford. Vol. 1+, 1984+. Microbiol. Sci. HI 67908

Microbiology. [Weishengwu tongbao.] Beijing. Vol. ?-8+, ?-1981+. Microbiology (Beijing). HI 67909

Microbiology. Moscow & Leningrad = Mikrobiologiya. Moscow & Leningrad.

Microbiology. [English translation of: Mikrobiologiya.] Washington, DC., New York. Vol. 26+, 1957+. Microbiology (U.S.S.R.). HI 67910

Microbiology. Washington, DC. 1974+, 1975+. Microbiology (Washington). HI 67911

Microbiology abstracts. Section A, industrial microbiology (later industrial and applied microbiology). Bethesda, MD. 1967+. Microbiol. Abstr., Sect. A, Industr. Microbiol. (Industr. Appl. Microbiol.) HI 67912

Microbiology abstracts: section B, general microbiology and bacteriology. London. Microbiol. Abstr., Sect. B, Gen. Microbiol. Bacteriol. See B–P–H 593/13. HI 56589

Microbiology abstracts. Section C, algology, mycology and protozoology. Bethesda, MD. Vol. 1+, 1972+. Microbiol. Abstr., Sect. C, Algol. Mycol. Protozool. HI 67913

Microbiology and immunology. Tokyo. Vol. 21(2)+, 1977+. Microbiol. Immunol. Preceded by: Japanese journal of microbiology. HI 67914

Microbiology letters, F E M S = F E M S microbiology letters. Amsterdam.

Microbios; a prestige international biomedical research journal of chemical and general microbiology. Cambridge. Vol. 1+, 1969+. Microbios. HI 67915

Microbios letters. Cambridge. Vol. 1+, 1976+. Microbios Lett. HI 67916

Microfilm abstracts. Ann Arbor, MI. Vols. 1-11, 1938-51. Microfilm Abstr. Superseded by: Dissertation abstracts. 2-1345-2. HI 67917

Micron; international quarterly journal of electron microscopy, electron probe micro-analysis and associated techniques. West Watford. Vols. 1-14(2), 1969-83. Micron. Superseded by: Micron and microscopica acta. HI 67918

Micron and microscopica acta. Oxford, Elmsford, NY. Vol. 14(3)+, 1983+. Micron Microscop. Acta. Preceded by: Micron and Microscopica acta. HI 67919

Micronesica; journal of the college of Guam. Agana, Guam. Micronesica. See B–P–H 593/17. HI 56591

Microorganisms and fermentation. [Misaengmul kwa parhyo.] Seoul. ?-1983+. Microorg. Ferment. HI 67920

Micropaleontology. New York, NY. Micropaleontology. See B–P–H 593/18. HI 56592

Microscopica acta; journal of microscopic equipment, methods and applications. Leipzig, Stuttgart. Vol. 71-87(3), 1971-83; Supplementa 1-6, 1977-83. Microscop. Acta. Preceded by: Zeitschrift für wissenschaftliche Mikroskopie und für mikroskopische Technik. Superseded by: Micron and microscopica acta. HI 67921

Microscope, Albany, NY. Microscope (Albany). See B–P–H 593/21. HI 56593

Microscope. Ann Arbor, MI. Microscope (Ann Arbor). See B–P–H 593/22. HI 56594

Microscope. Belfast. Microscope (Belfast). See B–P–H 593/23. HI 56595

Microscope. Louisville, KY. Microscope (Louisville). See B–P–H 593/25. HI 56596

Microscope. New Haven, CT. Microscope (New Haven). See B–P–H 593/26. HI 56597

Microscope and entomological monthly. London. Microscope Entomol. Monthly. See B–P–H 593/27. HI 56598

Microscopy; journal of the Quekett microscopical society. London. Vol. 30(7)+, 1966+. Microscopy. Preceded by: Journal of the Quekett microscopical club. HI 67922

Microskopion; l'actualité micrographique. Heerbrugg. No. 1+, 1964+. Microskopion. HI 67923

Middle-Thames naturalist. Slough, England. Middle-Thames Naturalist. See B–P–H 594/3. HI 56599

Midland druggist. Columbus, OH. Midl. Druggist. See B–P–H 594/5. HI 56600

Midland druggist and pharmaceutical review. Columbus, OH. Midl. Druggist & Pharm. Rev. See B–P–H 594/6. HI 56601

Midland florist and suburban horticulturist. London. Vols. 1-7, 1847-53. Midl. Florist Suburban Hort. HI 67924

Midland medical and surgical reporter, and topographical and statistical journal. Worcester, England. Midl. Med. Surg. Reporter & Topogr. Statist. J. See B–P–H 594/7. HI 56602

Midland naturalist. London. Vols. 1-16, 1878-93. Midl. Naturalist (London) 3-2651-3. HI 56603

Midland naturalist. Notre Dame, IL. Nos. 1-4, 1909. Midl. Naturalist (Notre Dame). Superseded by: American midland naturalist. HI 67925

Midwest cooperative conservation program newsletter = Newsletter, midwest cooperative conservation program.

Mid-west dahlia news. Madison, WI. Mid-W. Dahlia News (Madison). See B–P–H 594/12. HI 56604

Midwest dahlia news. Kansas City, KS. Midwest Dahlia News (Kansas City). See B–P–H 594/13. HI 56605

Mie daigaku nogaku-bu enshurin hokuku = Bulletin of the Mie university forests. Tsu, Japan. Bull. Mie Univ. Forests. See B–P–H 263/4.

Mie daigaku, nogakubu enshurin shiryo = Research data, Mie university forests. Tsu.

Mie-ken kankyo kagaku senta kenkyu hokoku = Report (Annual) of the environmental science institute of Mie prefecture. Yokkaichi.

Mie-ken kogai senta. Nempo = Report (Annual) of the environmental science institute of Mie prefecture. Yokkaichi.

Mie kenritsu daigaku nogakubu, gakujutsu hokoku = Bulletin of the faculty of agriculture; Mie university. Tsu, Japan. Bull. Fac. Agric. Mie Univ. See B–P–H 249/10.

Mijnwezen. The Hague. Mijnwezen. See B–P–H 594/16. HI 56606

Mikologiai közlemények. Budapest. Mikol. Közlem. See B–P–H 594/17. HI 56607

Mikologicheskie issledovaniya. Vol. 1+, 1970+. Mikol. Issl. HI 67926

Mikologiya i fitopatologiya. Leningrad. Vol. 1+, 1967+. Mikol. Fitopatol. HI 67927

Mikrobiologiakan žohovacu = Mikrobiologicheskii sbornik. Erivan.

Mikrobiologičeskij sbornik = Mikrobiologicheskii sbornik. Erivan.

Mikrobiologicheskii sbornik. Erivan [=Erevan], Armenian S S R. 1943-5? Mikrobiol. Sborn. Superseded by: Voprosy mikrobiologii. HI 67928

Mikrobiologicheskii zhurnal. Kiev. Vol. 40+, 1978+. Mikrobiol. Zhurn. Preceded by: Mikrobiologichnii zhurnal. HI 67929

Mikrobiologichnii zhurnal. Kiev. Vols. 1-39, 1934-77. Mikrobiol. Zhurn. Superseded by: Mikrobiologicheskii zhurnal. 3-2654-2. HI 67930

Mikrobiologičnii žurnal. Kiev = Mikrobiologichnii zhurnal. Kiev

Mikrobiologija. Belgrade. Vol. 1+, 1964+. Mikrobiologija. HI 67931

Mikrobiologija = Mikrobiologiya. Moscow & Leningrad.

Mikrobiologiya. Moscow & Leningrad. Vol. 1+, 1932+. Mikrobiologiya. 3-2654-2. HI 67932

Mikrobioloogilised uurimused. Tallinn. Vol. 1+, 1961+. Issled. Mikrobiol. HI 67933

Mikrochemica acta supplementum. Vienna. Mikrochem. Acta Suppl. See B–P–H 594/23. HI 56608

Mikroelementy v S S S R. Nauchnyi sovet po probleme "Biologicheskaya rol′mikroelementov v zhizni rastenii, zhivotnykh i cheloveka". Riga, Latvian S S R. Vol. 1+, 1961+. Mikroelem. S.S.S.R. HI 67934

Mikroelementy v S S S R. Naučnyj sovet po probleme "Biologičeskaja rol′ mikroelementov v žizni rastenij, životnyh i čeloveka" = Mikroelementy v S S S R. Nauchnyi sovet po probleme "Biologicheskaya rol′ mikroelementov v zhizni rastenii, zhivotnykh i cheloveka". Riga.

Mikroelementy v zhivotnovodstve i rastenievodstve. Frunze. Vol. 1+, 1962+. Mikroelem. Zhivotnov.

Rasteniev. HI 67935

Mikroevolyutsiya. Kazan. Vol. 1+, 1981+. Mikroevolyutsiya. HI 75113

Mikrographische Gesellschaft, Wien. Mitteilungen. Vienna. Mikrogr. Ges. Wien Mitt. See B–P–H 594/25. HI 56609

Mikrographische Gesellschaft, Wien. Mitteilungsblatt. Vienna. Mikrogr. Ges. Wien Mitteilungsbl. See B–P–H 595/1. HI 56610

Mikrokosmos. Stuttgart. Mikrokosmos. See B–P–H 595/2. HI 56611

Mikroskopie. Vienna. Mikroskopie. See B–P–H 595/4. HI 56613

Mikroskopie für Naturfreunde. Berlin. Mikroskop. Naturfr. See B–P–H 595/3. HI 56612

Milieu. Sainte-Foy, Quebec. No. 1+, 1972+. Milieu. HI 67936

Military medicine. Washington, DC. Military Med. See B–P–H 595/8. HI 56614

Military surgeon. Washington, DC. Military Surgeon. See B–P–H 595/9. HI 56615

Milotice nad Bečvou = Milotický hospodař. [Czechoslovakia]. Milotický Hospod. See B–P–H 595/10.

Milotický hospodař. [Czechoslovakia]. Milotický Hospod. See B–P–H 595/10. HI 56616

Milton Keynes natural history society journal. Milton Keynes. 1975+. Milton Keynes Nat. Hist. Soc. J. HI 67937

Milwaukee public museum publications in botany. Milwaukee, WI. Milwaukee Public Mus. Publ. Bot. See B–P–H 595/12. HI 56617

Mimeograph series, Alabama agricultural experiment station = Alabama agricultural experiment station. Progress report series. Auburn, AL.

Mimeograph series, department of forestry, Canada = Mimeographs, forest product laboratories, Canada. Ottawa.

Mimeographed circular, New Jersey agricultural college experiment station = New Jersey agricultural college experiment station. Mimeographed circular. New Brunswick, NJ. New Jersey Agric. Coll. Exp. Sta. Mimeogr. Circ. See B–P–H 653/17.

Mimeographed circular, Oklahoma agricultural experiment station = Oklahoma agricultural experiment station mimeographed circular. Stillwater, OK. Oklahoma Agric. Exp. Sta. Mimeogr. Circ. See B–P–H 687/5.

Mimeographed circular, Washington (state) agricultural experiment station = Washington (state) agricultural experiment station. Mimeographed circular. Pullman, WA. Wash. State Agric. Exp. Sta. Mimeogr. Circ. See B–P–H 971/15.

Mimeographed circular, Wyoming agricultural college agricultural experiment station = Wyoming agricultural college agricultural experiment station. Mimeographed circular. Laramie, WY. Wyoming Agric. Exp. Sta. Mimeogr. Circ. See B–P–H 980/27.

Mimeographed publications, commonwealth mycological institute = Imperial mycological institute. Mimeographed publications. Kew, England. Imp. Mycol. Inst. Mimeogr. Publ. See B–P–H 428/25.

Mimeographed publications, imperial mycological institute = Imperial mycological institute. Mimeographed publications. Kew, England. Imp. Mycol. Inst. Mimeogr. Publ. See B–P–H 428/25.

Mimeographed report, Maine agricultural experiment station = Maine agricultural experiment station. Mimeographed report. Orono, ME. Maine Agric. Exp. Sta. Mimeogr. Rep. See B–P–H 546/10.

Mimeographed series, Arkansas agricultural experiment station = University of Arkansas. The college of agriculture. Arkansas agricultural experiment station. Mimeographed series. Fayetteville, AR. Univ. Arkansas Coll. Agric. Arkansas Agric. Exp. Sta., Mimeogr. Ser. See B–P–H 934/3.

Mimeographed series, Clemson agricultural college, agronomy department = Clemson agricultural college, agronomy department. Mimeographed series. Clemson, SC. Clemson Agric. Coll. Agron. Dept., Mimeogr. Ser. See B–P–H 314/31.

Mimeographed series, Utah agricultural college, agricultural experiment station = Utah agricultural college, agricultural experiment station. Mimeographed series. Logan, UT. Utah Agric. Exp. Sta., Mimeogr. Ser. See B–P–H 945/32.

Mimeographs, dominion forest service, Canada = Mimeographs, forest product laboratories, Canada. Ottawa.

Mimeographs, forest product laboratories, Canada. Ottawa. 1943-63. Mimeogr. Forest Prod. Lab. Canada. Superseded by: Publications, department of forestry and rural development, Canada. HI 67938

Min-kuo i hsüeh tsa chih = General medical journal. Peking. Gen. Med. J. See B–P–H 396/3.

Mindenee gyüjtemény. [Hungary]. Mind. Gyüjt. See B–P–H 595/21. HI 56618

Mineralogisches Taschenbuch = Taschenbuch für die gesammte Mineralogie. Frankfurt a. M. Taschenb. Gesamamte Mineral. See B–P–H 866/20.

Minerva; or weekly literary, entertaining, and scientific journal containing a variety of original and select articles, ... [in new series "weekly" is omitted]. New York. Vols. 1-2, 1822-23 [1824]; n.s. vols. 1-3, 1824-25. Minerva (New-York, 1822-25). Superseded by: New York literary gazette. 3-2668-2. HI 53460

Minerva et maanedskrivt. Copenhagen. 1785-99. Minerva & Maanedskr. Superseded by: Maanedskriftet Minerva. 4-3123-3. HI 67939

Minimus. Prague. 1971+. Minimus. HI 67940

Mining and scientific press. San Francisco, CA. Mining Sci. Press. See B–P–H 595/24. HI 56619

Minnesota agricultural experiment station. Annual report. St. Anthony Park, MN. Minnesota Agric. Exp. Sta. Annual Rep. See B–P–H 596/6. HI 56620

Minnesota agricultural experiment station. Bulletin. St. Paul, MN. Minnesota Agric. Exp. Sta. Bull. See B–P–H 596/7. HI 56621

Minnesota agricultural experiment station. Miscellaneous report. St. Paul, MN. Minnesota Agric. Exp. Sta. Misc. Rep. See B–P–H 596/8. HI 56622

Minnesota agricultural experiment station. Technical bulletin. St. Paul, MN. Minnesota Agric. Exp. Sta. Techn. Bull. See B–P–H 596/9. HI 56623

Minnesota botanical studies. Minneapolis, MN. Minnesota Bot. Stud. See B–P–H 596/10, see also Report of the geological and natural history survey of Minnesota. Minneapolis, MN. HI 56624

Minnesota farm and home science. St. Paul, MN. Vols. 1-22, 1943-65? Minnesota Farm Home Sci. Superseded by: Minnesota science. 3-2680-3. HI 56625

Minnesota forestry notes. St. Paul, MN. Nos. 1-184, 1952-67? Minnesota Forest. Notes. Superseded by: Minnesota forestry research notes. HI 67941

Minnesota forestry research notes. St. Paul, MN. No. 185+, 1968+. Minnesota Forest. Res. Notes. Preceded by: Minnesota forestry notes. HI 67942

Minnesota horticultural magazine. St. Paul, MN. ?-1977+. Minnesota Hort. Mag. HI 67943

Minnesota horticulturist. Annual report of the Minnesota state horticultural society. Minnesota Hort. See B–P–H 596/14. HI 56626

Minnesota landscape arboretum news. Chanhassen, MN. 1981+. Minnesota Landscape Arbor. News. HI 67944

Minnesota naturalist. Minneapolis, MN. Vols. 1-8(2), 1950-57. Minnesota Naturalist. Superseded by: Naturalist (Minneapolis). HI 67945

Minnesota plant press; newsletter of the Minnesota native plant society. St. Paul, MN. Vol. 1(2)+, 1982+. Minnesota Pl. Press. Preceded by: Native plant news. St. Paul, MN. HI 67946

Minnesota plant studies. Minneapolis, MN. Minnesota Pl. Stud. See B–P–H 596/15. HI 56627

Minnesota science. St. Paul, MN. Vol. 23+, 1966+. Minnesota Sci. Preceded by: Minnesota farm and home science. HI 67947

Minnesota studies in the biological sciences. Minneapolis, MN. Minnesota Stud. Biol. Sci. See B–P–H 596/18. HI 56628

Minnesota studies in plant science. Minneapolis, MN. Minnesota Stud. Pl. Sci. See B–P–H 596/20. HI 56629

Minoufeia journal of agricultural research = Minufiya journal of agricultural research.

Minoufiya journal of agricultural research = Minufiya journal of agricultural research.

Minufiya journal of agricultural research. Shebin El-Kam. Vol. ?-5+, 19??-82+. Minufiya J. Agric. Res. HI 67948

Minulost přerovska. Prerov, Czechoslovakia. Minulost Přerovska. See B–P–H 596/21. HI 56630

Minutes of the annual meeting, Canada weed committee, western section = Minutes and proceedings, Canada weed committee, western section.

Minutes, Canada weed committee, eastern section. 1969+. Minutes Canada Weed Committee E. Sect. Preceded by: Minutes, national weed committee, Canada, eastern section. HI 67949

Minutes, national weed committee, Canada, eastern section. 1964-68. Minutes Natl. Weed Committee Canada E. Sect. Preceded by: Proceedings, national weed committee, Canada, eastern section. Superseded by: Minutes, Canada weed committee, eastern section. HI 67950

Minutes and proceedings, Canada weed committee, western section. Vol. 25+. 1971+. Minutes Proc. Canada Weed Committee W. Sect. Preceded by: Minutes and report of the research planning committee, Canada weed committee, western section. HI 67951

Minutes of the proceedings of the Dublin microscopical club. Dublin. Minutes Proc. Dublin Microscop. Club. See B–P–H 596/23. HI 56631

Minutes and report of the research planning committee, Canada weed committee, western section. Vols. 15-24. 1961-70. Minutes Rep. Res. Planning Committee Canada Weed Committe W. Sect. Preceded by: Proceedings, national weed committee, Canada, western section. Superseded by: Minutes and proceedings, Canada weed committee, western section. HI 67952

Mircen; journal of applied microbiology and biotechnology: A research journal for biotechnology in the developing world. Oxford. Vol. 1+, 1985+. Mircen. HI 67953

Misaengmul hakhoe chi = Korean journal of microbiology. Seoul.

Misaengmul kwa parhyo = Microorganisms and fermentation. Seoul.

Misaki marine biological station; contributions. Tokyo. Misaki Mar. Biol. Sta. Contr. See B–P–H 596/28. HI 56632

Miscelánea, academia nacional de ciencias. Córdoba. 1920+. Misc. Acad. Nac. Ci. HI 67954

Miscelánea, centro de pesquisa agropecuária de tropico umido. Belem. No. 1+, 1980+. Misc. Centro Pesq. Agropecu. Trop. Umido. HI 67955

Miscelánea, estacion experimental agropecuaria Balcarce. Balcarce. No. 1+, 1964+. Misc. Estac. Exp. Agropecu. Balcarce. HI 67956

Miscelánea, facultad de agronomia y zootecnica, universidad nacional de Tucuman. Tucuman. No. 1+, 1954+. Misc. Fac. Agron. Zootecn. Univ. Nac. Tucuman. HI 67957

Miscelánea, fundacion e instituto Miguel Lillo. Tucuman. No. ?-26-41, ?-1968-72. Misc. Fund. Inst. Miguel Lillo. Preceded by: Miscelánea, instituto Miguel Lillo. Superseded by: Miscelánea, fundacion Miguel Lillo. HI 67958

Miscelánea, fundacion Miguel Lillo. Tucuman. No. 42+, 1972+. Misc. Fund. Miguel Lillo. Preceded by: Miscelánea, fundacion e instituto Miguel Lillo. HI 67959

Miscelánea, instituto Miguel Lillo. Tucuman. Nos. 1-?, 1937-? Misc. Inst. Miguel Lillo. Superseded by: Miscelánea, fundacion e instituto Miguel Lillo. HI 67960

Miscelánea, instituto nacional de investigaciones agrarias. Puerto de Hierro, Madrid. ?-1976+. Misc. Inst. Nac. Invest. Agrar. HI 67961

Miscelánea instructiva, curiosa y agradable; o anales de literatura, ciencias y artes. Alcalá & Madrid. Vols. 1(1)-6(18), 1796-98. Misc. Instruct. HI 52797

Miscelanea, ministerio de cultura y educacion. Tucuman. ?-1975+. Misc. Minist. Cult. Educ. HI 67962

Miscelanea, museo argentino de ciencias naturales "Bernardino Rivadavia". Buenos Aires. Misc. Mus. Argent. Ci. Nat. "Bernardino Rivadavia". See B–P–H 598/2. HI 56645

Miscelaneas forestales. Buenos Aires. Misc. Forest. See B–P–H 597/13. HI 56640

Miscellanea of the american philosophical society, Philadelphia. Philadelphia, PA. Misc. Amer. Philos. Soc. See B–P–H 596/29. HI 56633

Miscellanea bryologica et lichenologica. [Sentai chii zappo.] Nichinan. Vols. 1-9, 1955-83. Misc. Bryol. Lichenol. HI 67963

Miscellanea congressus biologorum Jugoslaviae = Zbornik kongresa biology Jugoslavije. Zagreb, Yugoslavia. Zborn. Kongr. Biol. Jugoslavije. See B–P–H 996/18.

Miscellanea curiosa sive ephermeridum medico-physicarum germanicarum academiae caesarea-leopoldinae naturae-curiosorum. Leipzig & Frankfurt a. M. Misc. Cur. Ephem. Med.-Phys. German. Acad. Caes.-Leop. Nat.-Cur. See B–P–H 597/9. HI 56636

Miscellanea curiosa sive ephermeridum medico-physicarum germanicarum academiae imperialis leopoldinae naturae curiosorum. Nuremberg. Misc. Cur. Ephem. Med.-Phys. German. Acad. Imp. Leop. Nat. Cur. See B–P–H 597/10. HI 56637

Miscellanea curiosa sive ephermeridum medico-physicarum germanicarum academiae naturae curiosorum. Frankfurt & Leipzig Misc. Cur. Ephem. Med.-Phys. German. Acad. Nat. Cur. See B–P–H 597/11. HI 56638

Miscellanea curiosa medico-physica academiae naturae curiosorum sive ephemeridum medico-physicarum germanicarum curiosarum. Jena. Misc. Cur. Med.-Phys. Acad. Nat. Cur. See B–P–H 597/12. HI 56639

Miscellanea, det kongelige norske videnskabers selskab museet. Trondheim. Vols. 1-25, 1971-76. Misc. Knogel. Norske Vidensk. Selsk. Mus. Superseded by: Gunneria. HI 67964

Miscellanea lipsiensia nova ad incrementum scientiarum publicata. Leipzig. Misc. Lips. Nova. See B–P–H 597/16. HI 56643

Miscellanea mecklenburgica. Rostock. Misc. Mecklenburg. See B–P–H 598/1. HI 56644

Miscellanea philosophico-mathematica societatis privatae taurinensis. Turin. Misc. Philos.-Math. Soc. Privatae Taurinensis. See B–P–H 598/9. HI 56649

Miscellanea physico-medico-mathematica. Erfurt. Misc. Phys.-Med.-Math. See B–P–H 598/11. HI 56650

Miscellanea taurinensia = Mélanges de philosophie et de mathématique de la société royale de Turin. Turin. Mélanges Philos. Math. Soc. Roy. Turin. See B–P–H 562/22.

Miscellaneous bulletin, Adelaide botanic garden. Adelaide, S.A. No. 1+, 1967+. Misc. Bull. Adelaide Bot. Gard. HI 67965

Miscellaneous circular. United States department of agriculture. Washington, DC. Misc. Circ. U.S.D.A. See B–P–H 597/4. HI 56634

Miscellaneous contributions of the national herbarium. Ottawa. Misc. Contr. Natl. Herb. See B–P–H 597/6. HI 56635

Miscellaneous information. Tokyo university forests. [Enshurin. Tokyo daigaku nogakubu enshurin.] [Japan]. Misc. Inform. Tokyo Univ. Forests. See B–P–H 597/15. HI 56642

Miscellaneous paper, see Miscellaneous papers

Miscellaneous papers from the horticultural institute; Taihoku imperial university. [Taihoku teikoku daigaku rinogakubu engeigaku kiyoshitsu zappo.] Taihoku [=Taipei, Taiwan]. Misc. Pap. Hort. Inst. Taihoku Imp. Univ. See B–P–H 598/4. HI 56647

Miscellaneous papers, landbouwhoogeschool. Wageningen. Nos. 1-20?, 1968-81. Misc. Pap. Landbouwhoogeschool. HI 67966

Miscellaneous papers. Pineapple research institute of Hawaii. Honolulu, HI. Misc. Pap. Pineapple Res. Inst. Hawaii. See B–P–H 598/5. HI 56648

Miscellaneous papers, Texas memorial museum. Austin, TX. No. 1+, 1967+. Misc. Pap. Texas Mem. Mus. HI 67967

Miscellaneous publication, see Miscellaneous publications

Miscellaneous publications, agricultural experiment station, Storrs = Miscellaneous publications, Storrs agricultural experiment station. Storrs, CT.

Miscellaneous publications, agricultural experiment station, university of Minnesota. St. Paul, MN. No. 1+, 1980+. Misc. Publ. Agric. Exp. Sta. Univ Minnesota. HI 67968

Miscellaneous publications, Alaska forest research center. Juneau. 1960+. Misc. Publ. Alaska Forest Res. Center. HI 67969

Miscellaneous publications, American fuchsia society = American fuchsia society. Miscellaneous publications. San Francisco, CA. Amer Fuchsia Soc. Misc. Publ. See B–P–H 80/5.

Miscellaneous publications, Aomori apple experiment station. Aomori. 1956+. Misc. Publ. Aomori Apple Exp. Sta. HI 67970

Miscellaneous publications; Bernice Pauahi Bishop museum. Honolulu, HI. Misc. Publ. Bernice Pauahi Bishop Mus. See B–P–H 598/14. HI 56651

Miscellaneous publications, board of agriculture and fisheries. London. Nos. 1-?, 1905-19. Misc. Publ. Board Agric. Fish., London. Superseded by: Miscellaneous publications, ministry of agriculture and fisheries. London. HI 67971

Miscellaneous publication, bureau of plant industry, United States department of agriculture = U S department of agriculture. Bureau of plant industry. Miscellaneous publication. Washington, DC. U.S.D.A. Bur. Pl. Industr. Misc. Publ. See B–P–H 941/1.

Miscellaneous publications, commonwealth institute of biological control. Farnham Royal. Nos. 1-8, 1970-74. Misc. Publ. Commonw. Inst. Biol. Control. HI 67972

Miscellaneous publications, Commonwealth mycological institute = Commonwealth mycological institute. Miscellaneous publications. Kew, England. Commonw. Mycol. Inst. Misc. Publ. See B–P–H 320/18.

Miscellaneous publications, department of agriculture, Iraq. ?-1918-44-? Misc. Publ. Dept. Agric. Iraq. HI 67973

Miscellaneous publications, department of horticulture, Nottingham university school of agriculture. Nottingham. Nos. 1-14, 1953-57. Misc. Publ. Dept. Hort. Nottingham Univ. School Agric. HI 67974

Miscellaneous publications of the department of natural sciences, Los Angeles county museum. Los Angeles, CA. Misc. Publ. Dept. Nat. Sci. Los Angeles County Mus. See B–P–H 598/18. HI 56652

Miscellaneous publications of the entomological society. College Park, MD. Misc. Publ. Entomol. Soc. See B–P–H 598/19 HI 56668

Miscellaneous publications, forest research division, Canada. Ottawa. Nos. 1-5, 1950-5?, Misc. Publ. Forest Res. Div. Canada. Superseded by: Miscellaneous publications, forestry branch, division of forest research, Canada. HI 67975

Miscellaneous publications, forests commission, Victoria. Melbourne, Vic. No. 1+, 1939+. Misc. Publ. Forests Commiss. Victoria. HI 67976

Miscellaneous publications, forestry branch, division of forest research, Canada. Ottawa. Nos. 6-12, 1950-60. Misc. Publ. Forest. Branch Div. Forest Res. Canada. Preceded by: Miscellaneous publications, forest research division, Canada. Superseded by: Publications, department of forestry and rural development, Canada. HI 67977

Miscellaneous publications of the genetics society of Canada. Ottawa. No. 1+, 1971+. Misc. Publ. Genet. Soc. Canada. HI 67978

Miscellaneous publication. Horticultural research station. [Rinji Hokoku.] Morioka, Japan. Misc. Publ. Hort. Res. Sta. See B–P–H 598/20. HI 56653

Miscellaneous publications, institute of forestry. Teheran. 1951+. Misc. Publ. Inst. Forest., Teheran. HI 67980

Miscellaneous publications, intermountain forest and range experiment station, United States forest service. Ogden, UT. No. 1+, 1954+. Misc. Publ. Intermount. Forest Range Exp. Sta. U.S. Forest Serv. HI 67981

Miscellaneous publications, international association for angiosperm paleobotany = I A A P miscellaneous publications.

Miscellaneous publications, landbouwhogeschool. Wageningen. 1968+. Misc. Publ. Landbouwhogeschool. HI 67982

Miscellaneous publications, Maine agricultural experiment station. Orono, ME. Nos. 1-677, 19??-66. Misc. Publ. Maine Agric. Exp. Sta. HI 67983

Miscellaneous publication, Maryland agricultural experiment station = Maryland agricultural experiment station. Miscellaneous publication. College Park, MD. Maryland Agric. Exp. Sta. Misc. Publ. See B–P–H 549/3.

Miscellaneous publications, ministry of agriculture and fisheries. London. Nos. ?-70 [no. 68 never published], 1920-29. Misc. Publ. Minist. Agric. Fish., London. Preceded by: Miscellaneous publications, board of agriculture and fisheries. London. Superseded by: Bulletin of the ministry of agriculture and fisheries. London. HI 67984

Miscellaneous publication, Montana state college. Agricultural experiment station = Montana state college. Agricultural experiment station. Miscellaneous publication. Bozeman, MT. Montana Agric. Exp. Sta. Misc. Publ. See B–P–H 618/15.

Miscellaneous publication of the museum of natural history of the university of Kansas. Lawrence, KS. Misc. Publ. Mus. Nat. Hist. Univ. Kansas. See B–P–H 598/23. HI 56654

Miscellaneous publications, national agricultural research bureau, China. Nanking. Nos. 1-8-?, 1934-39. Misc. Publ. Natl. Agric. Res. Bur. China. HI 67985

Miscellaneous publications of the national institute of agricultural sciences. Series D, physiology and genetics. [Nogyo gijutsu kenkyujo. Shiryo D, seiri, iden.] Tokyo. No. 1+, 1974+. Misc. Publ. Natl. Inst. Agric. Sci., Tokyo, D. HI 67986

Miscellaneous publications, nature conservation, Orange Free State. Bloemfontein. No. 1+, [1971?]+. Misc. Publ. Nat. Conservation Orange Free State. HI 67987

Miscellaneous publication, Nebraska agricultural experiment station = Nebraska agricultural experiment station miscellaneous publication. Lincoln, NB. Nebraska Agric. Exp. Sta. Misc. Publ. See B–P–H 640/1.

Miscellaneous publications, northern Rocky Mountains forest and range experiment station. Missoula, MT. Nos. 1-6?, 1950-52. Misc. Publ. N. Rocky Mountains Forest Range Exp. Sta. HI 67988

Miscellaneous publication, Oklahoma agricultural experiment station = Oklahoma agricultural experiment station miscellaneous publication. Stillwater, OK. Oklahoma Agric. Exp. Sta. Misc. Publ. See B–P–H 687/6.

Miscellaneous publications, Queensland department of primary industry. Brisbane, Qld. No.1+, 1974+. Misc. Publ. Queensland Dept. Prim. Industr. HI 67989

Miscellaneous publications, Rhode Island agricultural experiment station. Kingston, RI. No. 1+, 1938+. Misc. Publ. Rhode Island Agric. Exp. Sta. HI 67990

Miscellaneous publications, Storrs agricultural experiment station. Storrs, CT. Nos. 1-2, 1957. Misc. Publ. Storrs Agric. Exp. Sta. HI 67991

Miscellaneous publications, tall timbers research station. Tallahassee, FL. No. 1+, 1961+. Misc. Publ. Tall Timbers Res. Sta. HI 67992

Miscellaneous publications, Texas agricultural experiment station. Austin, TX. No. 1+, 1946+. Misc. Publ. Texas Agric. Exp. Sta. HI 67993

Miscellaneous publications, Tohoku national agricultural experiment station. [Tohoku nogyo shikenjo, kenkyu shirjo.] Morioka. No. 1+, 1978+. Misc. Publ. Tohoku Natl. Agric. Exp. Sta. HI 67994

Miscellaneous publications, United States department of agriculture. Washington, DC. 1927+. Misc. Publ. U.S. Dept. Agric. Preceded by: United States department of agriculture. Miscellaneous circular. HI 67995

Miscellaneous publications of the university of Utrecht herbarium. Utrecht. Vol. 1+, 1983/86+, 1987+. Misc. Publ. Univ. Utrecht Herb. Preceded by: Mededelingen van het botanisch museum en herbarium van de rijksuniversiteit te Utrecht. HI 67996

Miscellaneous publication, Utah agricultural college, agricultural experiment station = Utah agricultural college, agricultural experiment station. Miscellaneous publication. Logan, UT. Utah Agric. Exp. Sta. Misc. Publ. See B–P–H 945/33.

Miscellaneous release, central states forest experiment station. Columbus, OH. No. 1+, 1948+. Misc. Release Centr. States Forest Exp. Sta. HI 67997

Miscellaneous report, agricultural experiment station, university of Arizona. Tucson, AZ. No. 1+, 1976+. Misc. Rep. Agric. Exp. Sta. Univ. Arizona. HI 67998

Miscellaneous report, applied forestry research institute = A F R I miscellaneous report. Syracuse, NY.

Miscellaneous reports, centro internacional de mejoramiento de maiz y trigo. Mexico, D.F. No. 1+, 1961+. Misc. Rep. Centro Int. Mejoram. Maiz Trigo. HI 67999

Miscellaneous reports of the Hiwa museum for natural history. Hiroshima. No. ?-8+, ?-1965+. Misc. Rep. Hiwa Mus. Nat. Hist. HI 68000

Miscellaneous reports of the lake states forest experiment

station. St. Paul, MN. No. 1+, 1947+. Misc. Rep. Lake States Forest Exp. Sta. HI 68001

Miscellaneous report, Minnesota agricultural experiment station = Minnesota agricultural experiment station. Miscellaneous report. St. Paul, MN. Minnesota Agric. Exp. Sta. Misc. Rep. See B–P–H 596/8.

Miscellaneous reports of the national science museum. [Kokuritsu kagaku hakubutsukan shuho.] Tokyo. Misc. Rep. Natl. Sci. Mus. See B–P–H 599/7. HI 56655

Miscellaneous reports of the phytopathological laboratory; faculty of science and agriculture, Taihoku imperial university. Taihoku [=Taipei, Taiwan]. Misc. Rep. Phytopathol. Lab. Fac. Sci. Taihoku Imp. Univ. See B–P–H 599/10. HI 56656

Miscellaneous reports of the research institute for natural resources. [Shigen kagaku kenkyujo iho.] Tokyo. Nos. 1-75, 1943-71. Misc. Rep. Res. Inst. Nat. Resources. 5-4230-3. HI 68002

Miscellaneous reports, Rodale research center. 1980+. Misc. Rep. Rodale Res. Center. HI 68003

Miscellaneous reports. Tokyo science museum. Tokyo. Misc. Rep. Tokyo Sci. Mus. See B–P–H 599/14. HI 56657

Miscellaneous series, Clemson agricultural college of South Carolina. Agricultural experiment station = Clemson agricultural college of South Carolina. Agricultural experiment station. Miscellaneous series. Clemson, SC. Clemson Agric. Exp. Sta., Misc. Ser. See B–P–H 315/2.

Miscellaneous series, dominion forest service, Canada. Ottawa. Nos. 1-6?, 1939-49. Misc. Ser. Domin. Forest Serv. Canada. HI 68004

Miscellaneous series, national museum of New Zealand. Wellington, N.Z. No. 1+, 1976+. Misc. Ser. Natl. Mus. New Zealand. HI 68005

Miscellaneous special report, Idaho agricultural experiment station = University of Idaho, college of agriculture, Idaho agricultural experiment station, miscellaneous special reports. Moscow, ID. Univ. Idaho Coll. Agric. Idaho Agric. Exp. Sta. Misc. Special Rep. See B–P–H 936/5.

Miscellaneous special report, U S department of agriculture = Department of agriculture. Miscellaneous special report. Washington, DC. Dept. Agric. Misc. Special Rep. See B–P–H 343/6.

Miscellaneous studies, centre for european agricultural studies, Wye college. Ashford, Kent. No. 1+, 1974+. Misc. Stud. Centre Eur. Agric. Stud. Wye Coll. HI 68006

Miscellen für Gartenfreunde, Botaniker und Gärtner. Leipzig. Misc. Gartenfr. See B–P–H 597/14. HI 56641

Miscellen aus der neuesten ausländischen Literatur. Jena. Misc. Neuesten Ausl. Lit. See B–P–H 598/3. HI 56646

Mises au point de biochimie pharmacologique. Paris. Vol. 1+, 1977+. Mises Point Biochim. Pharmacol. HI 68007

Mississippi agricultural college. Technical bulletin. Jackson, MS. Mississippi Agric. Coll. Techn. Bull. See B–P–H 599/22. HI 56658

Mississippi agricultural experiment station. Mississippi agricultural and mechanical college. Annual report. Jackson, MS. Mississippi Agric. Exp. Sta. Annual Rep. See B–P–H 599/23. HI 56659

Mississippi agricultural experiment station. Mississippi agricultural and mechanical college. Bulletin. Jackson, MS. Mississippi Agric. Exp. Sta. Bull. See B–P–H 599/24. HI 56660

Mississippi agricultural experiment station. Mississippi agricultural and mechanical college. Circular. Jackson, MS. Mississippi Agric. Exp. Sta. Circ. See B–P–H 599/25. HI 56661

Mississippi agricultural experiment station. Mississippi state college. Information sheet. Jackson, MS. Mississippi Agric. Exp. Sta. Inform. Sheet. See B–P–H 599/26. HI 56662

Mississippi department of agriculture, bulletin. Jackson, MS. Mississippi Dept. Agric. Bull. See B–P–H 599/27. HI 56663

Mississipi farm research. State College, MS. Vols. 1-35, 1938-72. Mississippi Farm Res. Superseded by: M A F E S research highlights. HI 68008

Mississippi medical monthly. Vicksburg, MS. Mississippi Med. Monthly. See B–P–H 600/2. HI 56664

Mississippi valley medical monthly. Memphis, TN. Mississippi Valley Med. Monthly. See B–P–H 600/4. HI 56665

Missouri agricultural college. Farm bulletin. Columbia, MO. Missouri Agric. Coll. Farm Bull. See B–P–H 600/13. HI 56666

Missouri agricultural experiment station. Annual report. Columbia, MO. 1896+ [not published 1904-09, from 1910+ published in Missouri agricultural experiment station. Bulletin]. Missouri Agric. Exp. Sta. Annual Rep. HI 68009

Missouri agricultural experiment station. Bulletin. Columbia, MO. Nos. 1-886, 1888-1970. Missouri Agric. Exp. Sta. Bull. HI 68010

Missouri agricultural experiment station. Circular. Columbia, MO. Nos. 36-360, 1898?-1951. Missouri

Agric. Exp. Sta. Circ. Preceded by: Missouri agricultural experiment station. Circular of information. HI 68011

Missouri agricultural experiment station. Circular of information. Columbia, MO. Nos. 1-35, 1896-98? Missouri Agric. Exp. Sta. Circ. Inform. Superseded by: Missouri agricultural experiment station. Circular. HI 68012

Missouri agricultural experiment station. Press bulletin. Columbia, MO. No. 1+, 1912+. Missouri Agric. Exp. Sta. Press Bull. HI 68013

Missouri agricultural experiment station. Research bulletin. Columbia, MO. No. 1+, 1910+. Missouri Agric. Exp. Sta. Res Bull. HI 68014

Missouri agricultural experiment station. Service circular. Columbia, MO. No. 1+, 1936+. Missouri Agric. Exp. Sta. Serv. Circ. HI 68015

Missouri agricultural experiment station. Special report. Columbia, MO. No. 1+, 1961+. Missouri Agric. Exp. Sta. Special Rep. HI 68016

Missouri botanical garden bulletin. St. Louis, MO. Missouri Bot. Gard. Bull. See B–P–H 601/2. HI 56669

Missouri conservationist. Columbia, MO. Vol. 1+, 1938+. Missouri Conservationist. HI 68017

Missouri prairie journal. Columbia, MO. 1981+. Missouri Prairie J. Preceded by: Missouri prairie news. HI 67835

Missouri prairie news. Columbia, MO. 1979-80. Missouri Prairie News. Superseded by: Missouri prairie journal. HI 68018

Missouri state agricultural college, bulletin. Columbia, MO. Missouri State Agric. Coll. Bull. See B–P–H 601/4. HI 56670

Missouri wildlife. Jefferson City, MO. Vol. 1+, 1939+. Missouri Wildlife. 3-2704-3. HI 68019

Missouriensis. St. Louis, MO. Vol. 1+, 1979+. Missouriensis. HI 68020

Missouri's environment. Jefferson City, MO. Vol. 1+, 1975+. Missouri Environm. HI 68021

Mitrephora. Sabah. Ca.1989+. Mitrephora. HI 68022

Mittheilungen der Aargauischen Naturforschenden Gesellschaft. Aarau, Switzerland. Mitth. Aargauischen Naturf. Ges. See B–P–H 611/9. HI 56782

Mitteilungen der Abteilung für Botanik am Landesmuseum "Joanneum" in Graz. Graz. Vol. 1(42)+, 1972+. Mitt. Abt. Bot. Landesmus. "Joanneum" Graz. Preceded by: Mitteilungen der Abteilung für Zoologie und Botanik am Landesmuseum Joanneum, Graz. HI 68023

Mitteilungen der Abteilung für Natur- und Wirtschaftskunde. Minsk = Zapiski Addzelu pryrody i gaspadarki. Minsk.

Mitteilungen der Abteilung für Zoologie und Botanik am Landesmuseum Joanneum, Graz. Graz. Nos. 7/8-41, 1958-71. Mitt. Abt. Zool. Bot. Landesmus. Joanneum Graz. Preceded by: Abteilung für Zoologie und Botanik am Landesmuseum Joanneum, Graz. Mitteilungsblatt. Superseded by: Mitteilungen der Abteilung für Botanik am Landesmuseum "Joanneum", Graz. HI 68024

Mitteilungen der Arbeitsgemeinschaft für Floristik in Schleswig-Holstein und Hamburg. Kiel. Vols. 1-21, 1950-72. Mitt. Arbeitsgem. Florist. Schleswig-Holstein. Superseded by: Mitteilungen der Arbeitsgemeinschaft Geobotanik in Schleswig-Holstein und Hamburg. 1-144-3. HI 68025

Mitteilungen der Arbeitsgemeinschaft zur floristischen Kartierung Bayerns. Munich. Nos. 1-10, 1971-80. Mitt. Arbeitsgem. Florist. Kart. Bayerns. HI 68026

Mitteilungen der Arbeitsgemeinschaft Geobotanik in Schleswig-Holstein und Hamburg. Kiel. No. 23+, 1973+. Mitt. Arbeitsgem. Geobot. Schleswig-Holstein & Hamburg. Preceded by: Mitteilungen der Arbeitsgemeinschaft für Floristik in Schleswig-Holstein und Hamburg. HI 68027

Mitteilungen der Arbeitsgemeinschaft internationaler phaenologischer Gärten = Arboreta phaenologica. Münden.

Mitteilungen des Arbeitskreises zur Beobachtung und zum Schutz heimischer Orchideen. Halle. 1965+. Mitt. Arbeitskreises Beob. Heimischer Orchideen. Superseded by?: Mitteilungsblatt Arbeitskreis heimische Orchideen Baden-Württemberg. HI 68028

Mittheilungen der Auslands-Hochschule. Berlin. Vols. 39-?, 1936-44? Mitth. Ausl.-Hochschule. Preceded by: Mittheilungen des Seminars für orientalische Sprachen an der Königlishen Universität. 1-645-2. HI 68029

Mitteilungen des Badischen Botanischen Vereins. Freiburg im Breisgau. Mitt. Bad. Bot. Vereins. See B–P–H 601/12. HI 56671

Mitteilungen des Badischen Botanischen Vereins = Mitteilungen des Badischen Landesvereins für Naturkunde. Freiburg im Breisgau. Mitt. Bad. Landesvereins Naturk. See B–P–H 601/13.

Mitteilungen des Badischen Landesvereins für Naturkunde. Freiburg im Breisgau. Mitt. Bad. Landesvereins Naturk. See B–P–H 601/13. HI 56672

Mitteilungen des Badischen Landesvereins für Naturkunde und Naturschutz. Freiburg im Breisgau. Mitt. Bad. Landesvereins Naturk. Naturschutz. See B–P–H 601/14. HI 56673

Mitteilungen des Badischen Landesvereins für

Naturkunde und Naturschutz e. V. in Freiburg im Breisgau. Freiburg im Breisgau. Mitt. Bad. Landesvereins Naturk. Naturschutz Freiburg. See B–P–H 601/15. HI 56674

Mitteilungen des baltisches Moorverein = Ezhegodnik baltiiskago obshchestva dlya pooshchreniya kul'tury bolot. Tartu.

Mitteilungen der Basler Botanischen Gesellschaft. Basel. Mitt. Basler Bot. Ges. See B–P–H 601/16. HI 56675

Mitteilungen der Bayerischen Botanischen Gesellschaft zur Erforschung der heimischen Flora a. V. Munich. Mitt. Bayer. Bot. Ges. See B–P–H 601/17. HI 56676

Mitteilungen der bayerischen Staatssammlung für Paläontologie und historische Geologie. Munich. Vol. 1+, 1961+. Mitt. Bayer. Staatssamml. Paläontol. HI 68031

Mitteilungen aus dem Bereich systematische Botanik und Pflanzengeographie der Sektion Biowissenschaften der Martin-Luther-Universität Halle-Wittenberg. Halle. Nos. 9-11, 1969-72. Mitt. Bereich Syst. Bot. Pflanzengeogr. Sekt. Biowiss. Martin-Luther-Univ. Halle-Wittenberg. Preceded by: Mitteilungen aus dem Institut für systematische Botanik und Pflanzengeographie der Martin-Luther-Universität Halle-Wittenberg. Superseded by: Mitteilungen aus dem Wissenschaftsbereich Geobotanik und botanischer Garten der Sektion Biowissenschaften der Martin-Luther-Universität Halle-Wittenberg. HI 68032

Mitteilungen, biogeographische Abteilung des geographischen Instituts der Universität des Saarlandes. Saarbrucken. Vol. 1+, 1971+. Mitt. Biogeogr. Abt. Geogr. Inst. Univ. Saarlandes. HI 68033

Mitteilungen der biologische Gesellschaft in der Deutschen Demokratischen Republik. Berlin. Vol. 1+, 1963+. Mitt. Biol. Ges. DDR. HI 68034

Mitteilungen aus der Biologischen Bundesanstalt für Land- und Forstwirtschaft Berlin-Dahlem. Berlin-Dahlem. Mitt. Biol. Bundesanst. Land- Forstw. Berlin-Dahlem. See B–P–H 601/19. HI 56677

Mitteilungen des biologischen Instituts am Goetheanum. Stuttgart. Mitt. Biol. Inst. Goetheanum. See B–P–H 601/20. HI 56678

Mitteilungen aus der Biologischen Reichsanstalt für Land- und Forstwirtschaft (Berlin-Dahlem). Berlin-Dahlem. Mitt. Biol. Reichsanst. Land- Forstw. Berlin-Dahlem. See B–P–H 601/21. HI 56679

Mitteilungen aus der Biologischen Zentralanstalt für Land- und Forstwirtschaft Berlin-Dahlem. Berlin-Dahlem. Mitt. Biol. Zentralanst. Land- Forstw. Berlin-Dahlem. See B–P–H 602/1. HI 56680

Mitteilungen der botanischen Arbeitsgemeinschaft am oberösterreichischen-Landesmuseum Linz. Linz. Vols. 1-6, 1969-74. Mitt. Bot. Arbeitsgem. Oberösterr.-Landesmus. Linz. Superseded by: Linzer biologische Beiträge. HI 68035

Mitteilungen aus dem Botanischen Garten und Museum Berlin-Dahlem. Berlin-Dahlem. Mitt. Bot. Gart. Berlin-Dahlem. See B–P–H 602/3. HI 56681

Mitteilungen aus dem Botanischen Garten und Museum der Universität Zürich = Mitteilungen aus dem Botanischen Museum der Universität Zürich. Zurich. Mitt. Bot. Mus. Univ. Zürich. See B–P–H 602/15.

Mitteilungen aus dem Botanischen Garten St. Gallen. St. Gallen, Switzerland. Mitt. Bot. Gart. St. Gallen. See B–P–H 602/7. HI 56682

Mitteilungen aus dem botanischen institut Graz. Graz. Vols. 1-2, 1886-88. Mitt. Bot. Inst. Graz. HI 68036

Mitteilungen aus der Botanischen Institut der Technischen Hochschule in Wien = Mitteilungen aus der Botanischen Laboratorium der Technischen Hochschule in Wien. Vienna. Mitt. Bot. Lab. TH Wien. See B–P–H 602/11.

Mitteilungen des Botanischen Instituts in Giessen. Giessen, Germany. Mitt. Bot. Inst. Giessen. See B–P–H 602/8. HI 56683

Mitteilungen des botanischen Instituts Sofia = Izvestiya na botanicheskaya institut. Sofia.

Mitteilungen aus der Botanischen Laboratorium der Technischen Hochschule in Wien. Vienna. Mitt. Bot. Lab. TH Wien. See B–P–H 602/11. HI 56684

Mitteilungen aus dem botanischen Museum in Hamburg. Hamburg. Mitt. Bot. Mus. Hamburg. See B–P–H 602/13. HI 56685

Mitteilungen aus dem Botanischen Museum der Universität Zürich. Zurich. Mitt. Bot. Mus. Univ. Zürich. See B–P–H 602/15. HI 56686

Mitteilungen aus den Botanischen Staatsinstituten in Hamburg. Hamburg. Mitt. Bot. Staatsinst. Hamburg. See B–P–H 602/19. HI 56687

Mitteilungen (aus) der Botanischen Staatssammlung München. Munich. Mitt. Bot. Staatssamml. München. See B–P–H 602/22. HI 56688

Mitteilungen des botanischen Vereins für den Kreis Freiburg und das Land Baden. Freiburg im Breisgau. Mitt. Bot. Vereins Kreis Freiburg. See B–P–H 602/23. HI 56689

Mitteilungen des brjansker Forstinstituts = Trudy brianskogo lesnogo instituta. Bryansk.

Mitteilungen der Bundesanstalt für Forst- und Holzwirtschaft. Reinbek. Vols. 11-34, 1950-54. Mitt. Bundesanst. Forst- Holzw. Preceded by: Mitteilungen des Zentralinstitutes für Forst- und Holzwirtschaft. Superseded by: Mitteilungen der

Bundesforschungsanstalt für Forst- und Holzwirtschaft, Reinbek bei Hamburg. 4-3561-1. HI 68037

Mitteilungen der Bundesforschungsanstalt für Forst- und Holzwirtschaft, Reinbek bei Hamburg. Reinbek. Vol. 35+, 1954+. Mitt. Bundesforschungsanst. Forst-Holzw. Reinbek. Preceded by: Mitteilungen des Bundesanstalt für Forst- und Holzwirtschaft. HI 68038

Mitteilungen des burgenländischen Heimat- und Naturschutzvereins. Eisenstadt, Austria. Mitt. Burgenl. Heimat-Naturschutzvereins. See B–P–H 602/24. HI 56690

Mitteilungen des burgenländischen Heimatschutzvereins. Eisenstadt, Austria. Mitt. Burgenl. Heimatschutzvereins. See B–P–H 602/25. HI 56691

Mittheilungen des Central-Instituts für Akklimatisation in Deutschland zu Berlin. Berlin. Mitth. Centr.-Inst. Akklim. Deutschl. Berlin. See B–P–H 611/10. HI 56783

Mitteilungen aus der Chirurgischen Klinik zu Tübingen = Beiträge zur klinischen Chirurgie. Tübingen. Beitr. Klin. Chir. See B–P–H 170/11.

Mittheilungen der Commission geologische Landes-Untersuchungen von Elsass-Lothringen. Strasbourg. Vols. 1-2, 1886-90. Mitt. Commiss. Geol. Landes-Untersuch. Elsass-Lothringen. Superseded by: Mittheilungen der geologischen Landesanstalt von Elsass-Lothringen. HI 68039

Mitteilungen der D L G = Mitteilungen der Deutschen Landwirtschaftsgesellschaft. Frankfurt am Main.

Mitteilungen der Deutschen Akademie der Naturforscher (Leopoldina) = Leopoldina. Amtliches Organ der Kaiserlichen Leopoldinisch-Carolinischen Detuschen Akademie der Naturforscher. Jena. Leopoldina. See B–P–H 529/1.

Mitteilungen des Deutschen Böhmerwaldbundes. Budweis, Bohemia [Czechoslovakia]. Mitt. Deutsch. Böhmerwaldbundes. See B–P–H 603/3. HI 56692

Mitteilungen der Deutschen Dendrologischen Gesellschaft. Berlin. Mitt. Deutsch. Dendrol. Ges. See B–P–H 603/4. HI 56693

Mitteilungen der deutschen Gesellschaft für Natur- und Völkerkunde Ostasiens. Yokohama. Vols. 1-36, 1873-1945. Mitt. Deutsch. Ges. Natur- Völkerk. Ostasiens. Superseded by: Mitteilungen der Gesellschaft für Natur- und Völkerkunde Ostasiens. 2-1717-3. HI 68040

Mitteilungen der deutschen Gesellschaft für Pilzkunde e. V. Seefeld. 1946. Mitt. Deutsch. Ges. Pilzk. HI 68041

Mitteilungen der Deutschen Landwirtschafts-Gesellschaft. Berlin. Vols. [1]-49(13), 1886-1934. Mitt. Deutsch. Landw.-Ges. Superseded by: Mitteilungen für die Landwirtschaft. 2-1307-2. HI 68042

Mitteilungen der Deutschen Landwirtschaftsgesellschaft. Frankfurt am Main. Vol. 66(40) [also numbered n.s. vol. 6(40)]-90(18), 1951-75. Mitt. Deutsch. Landwirtschaftsges. Preceded by: Neue Mitteilungen für die Landwirtschaft. Superseded by: D L G - Mitteilungen. 2-1307-2. HI 68043

Mitteilungen aus den deutschen Schutzgebieten. Deutsches Kolonialblatt. Wissenschaftliche Beihefte. Berlin. Mitt. Deutsch. Schutzgeb. See B–P–H 603/7. HI 56694

Mitteilungen aus den deutschen Schutzgebieten. Ergänzungshefte. Berlin. Mitt. Deutsch. Schutzgeb. Ergänzungsh. See B–P–H 603/8. HI 56695

Mitteilungen, eidgenossische Anstalt für das forstliche Versuchswesen. Birmensdorf. Vol. 50+, 1974+. Mitt. Eidgenoss. Anst. Forstl. Versuchswesen. Preceded by: Schweizerische Anstalt für das forstliche Versuchswesen, Mitteilungen. HI 68044

Mitteilungen der Fakultät für Garten- und Weinbau der Universität für Agrarwissenschaften. Budapest = Agrártudományi egyetem kert- és szölögazdaságtudományi karának közleményei. Budapest. Agrártud. Egyet. Kert- Szölögazdaságtud. Karának Közlem. See B–P–H 55/13.

Mittheilungen über Flora, Gesellschaft für Botanik und Gartenbau, in Dresden. Dresden & Leipzig. Mitth. Flora Ges. Bot. Dresden. See B–P–H 611/11. HI 56784

Mitteilungen der floristisch-soziologischen Arbeitsgemeinschaft in Niedersachsen. Hanover. Mitt. Florist.-Soziol. Arbeitsgem. Niedersachsen. See B–P–H 603/11. HI 56696

Mitteilungen zur floristischen Kartierung. Halle. Vol. 1+, 1975+. Mitt. Florist. Kart. HI 68045

Mitteilungen der floristisch-soziologischen Arbeitsgemeinschaft. Stolzenau/Weser. N.s. vols. 1-22, 1949-80. Mitt. Florist.-Soziol. Arbeitsgem. Preceded by: Mitteilungen der floristisch-soziologischen Arbeitsgemeinschaft in Niedersachsen. Superseded by: Tuezenia. HI 68046

Mitteilungen des Forschungsinstituts für Pflanzüchtung und Pflanzenbau in Sopronhorpács = Növénynemesítési és növénytermesztési kutató intézet közleményei Sopronhorpács. Budapest & Szombathely.

Mittheilungen von Forschungsreisenden und Gelehrten aus den Deutschen Schutzgebieten. Berlin. Mitth. Forschungsreisenden Gel. Deutsch. Schutzgeb. See B–P–H 611/12. HI 56785

Mittheilungen über das Forst- und Jagdwesen in Bayern. Munich. Vols. 2(4)-5(1), 1858-76. Mitt. Forst-

Jagdwesen Bayern. Preceded by: Forstliche Mittheilungen. Superseded by: Mitteilungen aus der Staatsforstverwaltung Bayerns. HI 68047

Mitteilungen der forstlichen Bundes-Versuchsanstalt Mariabrunn. Vienna. Mitt. Forstl. Bundes-Versuchsanst. Mariabrunn. See B–P–H 603/19. HI 56697

Mitteilungen der forstlichen Bundes-Versuchsanstalt Mariabrunn = Mitteilungen der forstlichen Bundes-Versuchsanstalt Mariabrunn. Vienna. Mitt. Forstl. Bundes-Versuchsanst. Mariabrunn. See B–P–H 603/19.

Mitteilungen der forstlichen Versuchanstalt Lettlands = Latvijas mežu petišanas stacijas raksti. Riga.

Mitteilungen aus dem forstlichen Versuchwesen. Moscow & Leningrad = Trudy po lesnomu opytnomu delu.

Mitteilungen aus dem forstlichen Versuchwesen. Omsk, Russian S F S R = Trudy po lesnomu opytnomu delu. Omsk, Russian S F S R. Trudy Lesn. Opytn. Delu (Omsk). See B–P–H 912/8.

Mitteilungen aus dem forstlichen Versuchswesen in der B S S R, belarussisches Institut für wissenschaftliche Forschung der Forstwirtschaft und Forstindustrie = Pratsy belaruskaga navukova-das'ledchaga instytutu liasnoi gaspadarki i liasnoi pramyslovas'chi. Trudy po lesnomu opytnomu delu B S S R. Minsk.

Mitteilungen aus dem Forstlichen Versuchswesen in der Belarussischen S S R = Matar'yaly pa lyasnoi das'ledchai sprave Belaruskae S S R. Minsk.

Mitteilungen aus dem forstlichen Versuchswesen Österreichs. Vienna. Mitt. Forstl. Versuchswesen Österreichs. See B–P–H 603/23. HI 56698

Mitteilungen aus dem forstlichen Versuchwesen in Ukraine = Trudy po lesnomu opytnomu delu Ukrainy. Kharkov, Ukrainian S S R. Trudy Lesn. Opytn. Delu Ukrainy. See B–P–H 912/11.

Mittheilungen des Forstvereins der österreichischen Alpenländer. Laibach [= Lubljana]. 1852-57. Mitth. Forstvereins Österr. Alpenländer. HI 68048

Mitteilungen aus Forstwirtschaft und Forstwissenschaft. Hanover. Vols. 1-14, 1930-43. Mitt. Forstw. Forstwiss. From 1944 incorporated: Forstarchiv. 3-2706-2. HI 68049

Mitteilungen der forstwissenschaftlichen Abteilung der Universität Tartu = Tartu ülikooli metsaosakonna toimetused. Tartu.

Mittheilungen aus dem Gebiete der Landwirthschaft. Leipzig. Mitth. Landw. See B–P–H 612/3. HI 56793

Mittheilungen aus dem Gebiete der Medizin, Chirurgie und Pharmacie. Altona [=Hamburg, in part]. Mitth. Med. See B–P–H 612/4. HI 56794

Mittheilungen aus dem Gebiete der theoretischen Erdkunde. Zurich. Mitth. Theor. Erdk. See B–P–H 613/1. HI 56805

Mitteilungen der geographisch-ethnographischen Gesellschaft in Zurich. Zurich. 1917-41. Mitt. Geogr.-Ethnogr. Ges. Zurich. Preceded by: Jahresbericht der geographisch-ethnographischen Gesellschaft in Zurich. Superseded by: Geographica helvetica. HI 68050

Mittheilungen der Geographischen Gesellschaft in Hamburg. Hamburg. Mitth. Geogr. Ges. Hamburg. See B–P–H 611/13. HI 56786

Mitteilungen der geographischen Gesellschaft Lübeck Lübeck. Vols. 1-12, 1882-89; n.s. vols. 48-?, 1958-? Mitt. Geogr. Ges. Lübeck. For n.s. vols. 1-40, 1890-1940 and vols. 44-47, 1953-57 see: Mitteilungen der geographischen Gesellschaft und des naturhistorischen Museums in Lübeck. For n.s. vols. 41-43, 1947-51 see: Forschungen der geographischen Gesellschaft und des naturhistorischen Museums in Lübeck. Superseded by: Bericht des Vereins Natur und Heimat und des naturhistorischen Museums zu Lübeck. 2-1696-2. HI 68051

Mitteilungen der Geographischen Gesellschaft in München. Munich. Mitt. Geogr. Ges. München. See B–P–H 604/5. HI 56699

Mitteilungen der Geographischen Gesellschaft und des Naturhistorischen Museums in Lübeck. Lübeck. Mitt. Geogr. Ges. Naturhist. Mus. Lübeck. See B–P–H 604/6. HI 56700

Mitteilungen der Geographischen Gesellschaft (für Thüringen) zu Jena. Jena. Mitt. Geogr. Ges. (Thüringen) Jena. See B–P–H 604/7. HI 56701

Mitteilungen der Geographischen Gesellschaft in Wien. Vienna. Mitt. Geogr. Ges. Wien. See B–P–H 604/8. HI 56702

Mittheilungen der geologischen Landesanstalt von Elsass-Lothringen. Strasbourg. Vols. 3-12(2), 1890-1920. Mitt. Geol. Landesanst. Elsass-Lothringen. Preceded by: Mittheilungen der Commission geologische Landes-Untersuchungen von Elsass-Lothringen. Superseded by: Bulletin du service de la carte géologique d'Alsace et de Lorraine. HI 68052

Mittheilungen aus dem Gesammtgebiete der Botanik. Leipzig. Mitth. Gesammtgeb. Bot. See B–P–H 611/15. HI 56787

Mitteilungen zur Geschichte der Medizin und Naturwissenschaften, herausgegeben von der Deutschen Gesellschaft für Geschichte der Medizin und der Naturwissenschaften. Hamburg & Leipzig. Mitt. Gesch. Med. Naturwiss. See B–P–H 604/13. HI 56705

Mitteilungen zur Geschichte der Medizin, der

Naturwissenschaften und der Technik = Mitteilungen zur Geschichte der Medizin und Naturwissenschaften, herausgegeben von der Deutschen Gesellschaft für Geschichte der Medizin und der Naturwissenschaften. Hamburg & Leipzig. Mitt. Gesch. Med. Naturwiss. See B–P–H 604/13.

Mitteilungen der Gesellschaft für Bibliothekswesen und Dokumentation des Landbaues. Stuttgart-Hohenheim. 1959+. Mitt. Ges. Bibliotheksw. Dokument. Landb. HI 68053

Mitteilungen der Gesellschaft deutscher Naturforscher und Arzte. Wuppertal. 1976+. Mitt. Ges. Deutsch. Naturf. Arzte. HI 68054

Mitteilungen der Gesellschaft für Gartenkultur. Zurich. No. 1+, 1983+. Mitt. Ges. Gartenkultur. HI 68055

Mitteilungen der Gesellschaft für Natur- und Völkerkunde Ostasiens. Hamburg. Vol. 37+, 1953+. Mitt. Ges. Natur- Völkerk. Ostasiens. Preceded by: Mitteilungen der deutschen Gesellschaft für Natur- und Völkerkunde Ostasiens. HI 68056

Mitteilungen der Gesellschaft für Salzburger Landeskunde. Salzburg. Mitt. Ges. Salzburger Landesk. See B–P–H 604/10. HI 56703

Mitteilungen der Gesellschaft Schweizer Kakteen-Freunde. Zurich. Mitt. Ges. Schweizer Kakteen-Freunde. See B–P–H 604/11. HI 56704

Mitteilungen der Gesellschaft für Voratsschutz E. V. Berlin. Vols. 1-20, 1925-44. Mitt. Ges. Voratsschutz. HI 68057

Mitteilungen aus den Grenzgebieten der Medizin und Chirurgie. Jena. Mitt. Grenzgeb. Med. Chir. See B–P–H 604/16. HI 56706

Mitteilungen. Herausgeber Ministerium für Land- und Forstwirtschaft als Zentrale Naturschutzverwaltung in Zusammenarbeit mit dem Institut für Landesforschung und Naturschutz der Deutschen Akademie der Landwirtschaftswissenschaften zu Berlin. Berlin. Mitt. Ministerium Land- Forstw. See B–P–H 606/5. HI 56718

Mitteilungen. Herausgeber Ministerium für Landwirtschaft, Erfassung und Forstwirtschaft als Zentrale Naturschutzverwaltung in Zusammenarbeit mit dem Institut für Landesforschung und Naturschutz der Deutschen Akademie der Landwirtschaftswissenschaften zu Berlin. Berlin. Mitt. Ministerium Landw. Erfassung Forstw. See B–P–H 606/6. HI 56719

Mitteilungen. Herausgeber Zentrale Naturschutzverwaltung in Zusammenarbeit mit dem Institut für Landesforschung und der Zentralen Kommission der Natur- und Heimatfreunde im Kulturbund. Berlin. Mitt. Zentr. Naturschutzverwalt. See B–P–H 611/4. HI 56780

Mitteilungen der Hermann-Göring-Akademie der Deutschen Forstwissenschaft. Frankfurt a. M. Mitt. Hermann-Göring-Akad. Deutsch. Forstwiss. See B–P–H 604/17. HI 56707

Mittheilungen aus Hohenheim. Stuttgart. Vols. 1-6, 1855-65; n.s. vol. 1, 1887. Mitth. Hohenheim. 3-2706-2. HI 68058

Mitteilungen der Höheren Bundeslehr- und Versuchsanstalten für Wein-, Obst- und Gartenbau, Wien-Klosterneuburg, und für Bienenkunde, Wien-Grinzing. Klosterneuburg, Austria. Mitt. Höheren Bundeslehr- Versuchsanst. Wein- Obst- Gartenbau Wien-Klosterneuburg. See B–P–H 604/18. HI 56708

Mitteilungen aus dem Institut für allgemeine Botanik in Hamburg. Hamburg. Vols. 1-10, 1916-39; vol. 15+, 1977+. Mitt. Inst. Allg. Bot. Hamburg. For vols. 11-14, 1957-73 see: Mitteilungen aus dem Staatsinstitut für allgemeine Botanik in Hamburg. 3-1788-1. HI 68059

Mitteilungen aus dem Institut für systematische Botanik und Pflanzengeographie der Martin-Luther-Universität Halle-Wittenberg. Halle. 1959/60-1967/68. Mitt. Inst. Syst. Bot. Martin-Luther-Univ. Halle-Wittenberg. Superseded by: Mitteilungen aus dem Bereich systematische Botanik und Pflanzengeographie der Sektion Biowissenschaften der Martin-Luther-Universität Halle-Wittenberg. HI 68060

Mitteilungen aus dem instituto Colombo-Aleman de investigaciones cientificas "Punta de Betin". Santa Marta. Nos. 1-8, 1967-76. Mitt. Inst. Colombo-Aleman Invest. Ci. "Punta de Betin". Superseded by: Anales del instituto de investigaciones marinas de Punta de Betin. HI 68061

Mitteilungen des Instituts für angewandte Mikrobiologie der Hochschule für Bodenkultur und der Versuchsanstalt für das Gärungsgewerbe. Vienna. Mitt. Inst. Angew. Mikrobiol. Hochschule Bodenk. Versuchsanst. Gärungsgewerbe. See B–P–H 605/1. HI 56709

Mitteilungen, internationale Vereinigung für theoretische und angewandte Limnologie. Stuttgart. 1953+. Mitt. Int. Vereinigung Theor. Angew. Limnol. HI 68062

Mitteilungen aus dem Jahrbuch der königlichen ungarischen geologischen Anstalt.] Budapest = A magyar kir[ályi] állami földtani intézet évkönyve. Budapest. Magyar Kir. Állami Földt. Intéz. Évk. See B–P–H 543/15.

Mitteilungen und Jahresbericht der naturhistorischen Gesellschaft Nürnberg. Nuremberg. Nos. 1-3, 1965/66-68. Mitt. Jahresber. Naturhist. Ges. Nürnberg. Preceded by: Jahresbericht der naturhistorischen Gesellschaft zu Nürnberg. Superseded by: Natur und Mensch. HI 68063

Mittheilungen aus Justus Perthes' Geographischer Anstalt über wichtige neue Erforschungen auf dem Gesammtgebiete der Geographie. Gotha. Mitth. Justus Perthes' Geogr. Anst. See B–P–H 611/16. HI 56788

Mittheilungen aus Justus Perthes' Geographischer Anstalt über wichtige neue Erforschungen auf dem Gesammtgebiete der Geographie. Ergänzungsband. Gotha. Mitth. Justus Perthes' Geogr. Anst. (Ergänzungsband). See B–P–H 611/17. HI 56789

Mittheilungen der K. K. Geographischen Gesellschaft in Wien. Vienna. Mitth. K. K. Geogr. Ges. Wien. See B–P–H 611/18. HI 56790

Mittheilungen der K. K. mährisch-schlesischen Gesellschaft zur Beförderung des Ackerbaues, der Natur- und Landeskunde in Brünn. Brünn [=Brno, Czechoslovakia]. Mitth. K. K. Mähr.-Schles. Ges. Beförd. Ackerbaues Brünn. See B–P–H 612/1. HI 56791

Mitteilungen des Kaukasischen Museum = Izvestiya kavkazskago muzeya. Tiflis.

Mitteilungen der Kayserlichen Freien Ökonomischen Gesellschaft zu St. Petersburg. Leipzig. Mitt. Kayserl. Freien Ökon. Ges. St. Petersburg. See B–P–H 605/5. HI 56710

Mitteilungen der Kommission für Heimatkunde der wissenschaftlichen Stefan Tisza Gesellschaft, Debrecen = A Debreceni Tisza István tudományos társaság honismertetö bizottságának kiadványai. Debrecen, Karcag & Budapest. Debreceni Tisza István Tud. Társ. Honism. Bizott. Kiadv. See B–P–H 339/13.

Mitteilungen der königl[ichen]. bayerischen Moorkulturanstalt. Stuttgart. Pts. 1-5. 1907-13. Mitt. Königl. Bayer. Moorkulturanst. 3-2784-2. HI 68064

Mitteilungen aus dem Königlich Mineralogisch-geologischen und Prähistorischen Museum zu Dresden. Dresden. Mitt. Königl. Mineral.-Geol. Mus. Dresden. See B–P–H 605/7. HI 56711

Mitteilungen aus den königl. naturwissenschaftlichen Instituten in Sofia = Izvestiya na tsarskite prirodonauchni instituti v Sofiya. Sofia.

Mitteilungen der k[öni]gl[ichen] ungarischen Gartenbau-Akademie. Busapest = A m[agyar] kir[ályi] kertészeti akadémia közleményei. Budapest. Magyar Kir. Kert. Akad. Közlem. See B–P–H 543/17.

Mitteilungen der k[öni]gl[ichen] ungarischen Gartenbau-Lehranstalt. Budapest = A m[agyar] kir[ályi] kertészeti tanintézet közleményei. Budapest. Magyar Kir. Kert. Tanintéz. Közlem. See B–P–H 544/1.

Mitteilungen der k[öni]gl[ichen] ungarischen Hochschule für Garten- und Weinbau. Budapest = M[agyar] kir[ályi] kertészeti és szölészeti föiskola közleményei. Budapest. Magyar Kir. Kert. Szölész. Föisk. Közlem. See B–P–H 543/18

Mittheilungen des krainisch-küstenländischen Forstvereins. Trieste, Austria [Italy]. Mitth. Krainisch-Küstenl. Forstvereins. See B–P–H 612/2. HI 56792

Mitteilungen aus den Laboratorien des Geologischen Dienstes Berlin. Berlin. Mitt. Lab. Geol. Dienstes Berlin. See B–P–H 605/12. HI 56712

Mitteilungen aus der Landesanstalt für Wasser-, Boden- und Lufthygiene = Mitteilungen aus der Landesanstalt für Wasserhygiene. Berlin. Mitt. Landesanst. Wasserhyg. See B–P–H 605/14.

Mitteilungen aus der Landesanstalt für Wasserhygiene. Berlin. Mitt. Landesanst. Wasserhyg. See B–P–H 605/14. HI 56713

Mitteilungen des Landesvereins "Sächsischer Heimatschutz" = Sächsischer Heimatschutz, Landesverein zur Pflege heimatlicher Natur, Kunst und Bauweise. Dresden. Sächs. Heimatschutz Mitt. Landesvereins Sächs. Heimatschutz. See B–P–H 809/20.

Mitteilungen zur Landschaftspflege. Dokumentation-Information. Bad Godesberg. N.s. vols. 1-10, 1961-70. Mitt. Landschaftspflege, Dokument.-Inform. Superseded by: Dokumentation für Umweltschutz und Landespflege. HI 68065

Mitteilungen für Landwirthe. Leipzig. Nos. 1-3, 1837. Mitth. Landwirthe. HI 68066

Mitteilungen für die Landwirtschaft. Frankfurt am Main, Berlin. Vols. 49(14)-59, 1934-44. Mitt. Landw. (Frankfurt). Preceded by: Mitteilungen der Deutschen LandwirtschaftsGesellschaft. Superseded by: Neue Mitteilungen für die Landwirtschaft. 2-1307-2. HI 68067

Mitteilungen der Landwirtschaftlichen Institute der K. Universität Breslau. Berlin. Mitt. Landw. Inst. Königl. Univ. Breslau. See B–P–H 605/15. HI 56714

Mitteilungen des landwirthschaftlichen Instituts der Universität Leipzig. Berlin. Mitt. Landw. Inst. Univ. Leipzig. See B–P–H 605/16. HI 56715

Mitteilungen der landwirtschaftlichen Versuchsstellen Ungarns. Budapest = Kísérletügyi kölemények. Budapest. Kísérl. Közlem. See B–P–H 514/5.

Mitteilungen des Leningrader Instituts für Wissenschaftliche Forschung auf dem Gebiet der Holzindustrie = Trudy i issledovaniya po lesnomu khozyaistvu i lesnoi promyshlennosti. Leningrad.

Mitteilungen aus dem Ludwig-Boltzmann Institut für Umweltwissenschaften und Naturschutz. Graz. Vol. 1+, 1965+. Mitt. Ludwig-Boltzmann Inst. Umweltwiss. Naturschutz. HI 68068

Mitteilungen der Märkischen Mikrobiologischen

Vereinigung (E.V. Berlin). Berlin. Mitt. Märk. Mikrobiol. Vereinigung. See B–P–H 605/22. HI 56716

Mitteilungen aus der Medicinischen Facultät der Kaiserlich-japanischen Universität. Tokyo. Mitt. Med. Fac. Kaiserl.-Jap. Univ. See B–P–H 606/1. HI 56717

Mitteilungen aus der medizinischen Fakultät der kaiserlichen Universität zu Tokyo = Mitteilungen aus der Medicinischen Facultät der Kaiserlich-japanischen Universität. Tokyo. Mitt. Med. Fac. Kaiserl.-Jap. Univ. See B–P–H 606/1.

Mitteilungen, Mikrographische Gesellschaft, Wien = Mikrographische Gesellschaft, Wien. Mitteilungen. Vienna. Mikrogr. Ges. Wien Mitt. See B–P–H 594/25.

Mitteilungen des Musealvereins für Krain. Laibach [=Ljubljana, Yugoslavia]. Mitt. Musealvereins Krain. See B–P–H 606/10. HI 56723

Mitteilungen aus dem Museum für Kulturgeschichte und dem naturwissenschaftlichen Arbeitskreis. Magdeburg. Mitt. Mus. Kulturgesch. See B–P–H 606/7. HI 56720

Mitteilungen aus dem Museum für Mineralogie, Geologie und Vorgeschichte zu Dresden. Dresden. Mitt. Mus. Mineral. Dresden. See B–P–H 606/8. HI 56721

Mitteilungen aus dem Museum für Naturkunde und Vorgeschichte und dem naturwissenschaftlichen Arbeitskreis. Halle. Mitt. Mus. Naturk. See B–P–H 606/9. HI 56722

Mitteilungen über Naturdenkmalpflege in der Provinz Grenzmark Posen-Westpreussen. Schneidemühl, Germany [=Pila, Poland]. Mitt. Naturdenkmalpflege Prov. Grenzmark Posen-Westpreussen. See B–P–H 606/12. HI 56724

Mitteilungen über Naturdenkmalpflege in der Provinz Westfalen. Münster. Mitt. Naturdenkmalpflege Prov. Westfalen. See B–P–H 606/13. HI 56725

Mittheilungen der Naturforschenden Gesellschaft Bern. Bern. Mitth. Naturf. Ges. Bern. See B–P–H 612/5. HI 56795

Mitteilungen der naturforschenden Gesellschaft in Freiburg, Schweiz = Mémoires de la société fribourgeoise des sciences naturelles. Botanique. Fribourg.

Mitteilungen der Naturforschenden Gesellschaft zu Halle a. d. Saale. Halle. Mitt. Naturf. Ges. Halle. See B–P–H 606/14. HI 56726

Mitteilungen der Naturforschenden Gesellschaft des Kantons Glarus. Glarus, Switzerland. Mitt. Naturf. Ges. Kantons Glarus. See B–P–H 606/15. HI 56727

Mitteilungen der Naturforschenden Gesellschaft in Luzern. Lucerne. Mitt. Naturf. Ges. Luzern. See B–P–H 606/16. HI 56728

Mitteilungen der Naturforschenden Gesellschaft Schaffhausen. Schaffhausen. Mitt. Naturf. Ges. Schaffhausen. See B–P–H 606/17. HI 56729

Mitteilungen der Naturforschenden Gesellschaft in Solothurn. Solothurn, Switzerland. Mitt. Naturf. Ges. Solothurn. See B–P–H 606/18. HI 56730

Mittheilungen der Naturforschenden Gesellschaft in Zürich. Zurich. Mitth. Naturf. Ges. Zürich. See B–P–H 612/6. HI 56796

Mittheilungen der Naturhistorischen Gesellschaft in Colmar. Colmar, Germany [France]. Mitth. Naturhist. Ges. Colmar. See B–P–H 612/7. HI 56797

Mitteilungen der Naturhistorischen Gesellschaft zu Nürnberg. Nuremberg. Mitt. Naturhist. Ges. Nürnberg. See B–P–H 607/5. HI 56731

Mitteilungen für Naturkunde und Naturschutz. Freiburg im Breisgau. Mitt. Naturk. Naturschutz. See B–P–H 607/6. HI 56732

Mitteilungen für Naturkunde und Vorgeschichte aus dem Museum für Kulturgeschichte in Magdeburg. Leipzig. Mitt. Naturk. Vorgesch. Mus. Kulturgesch. Magdeburg. See B–P–H 607/8. HI 56733

Mitteilungen für Naturkunde und Vorgeschichte aus dem Museum für Kulturgeschichte und dem naturwissenschaftlichen Arbeitskreis. Magdeburg. Mitt. Naturk. Vorgesch. Mus. Kulturgesch. Naturwiss. Arbeitskreis. See B–P–H 607/9. HI 56734

Mitteilungen der Naturwissenschaftlichen Arbeitsgemeinschaft am Haus der Natur in Salzburg. Botanische Arbeitsgruppe. Salzburg. Mitt. Naturwiss. Arbeitsgem. Haus Natur Salzburg, Bot. Arbeitsgr. See B–P–H 607/13. HI 56735

Mitteilungen des Naturwissenschaltlichen Arbeitskreises Kempten/Allgäu. Kempten. Vol. 2+, 1956+. Mitt. Naturwiss. Arbeitskreises Kempten/Allgäu. Preceded by: Mitteilungsblatt der biologischen Arbeitsgemeinschaft, Volksbildungskurse Kempten (Allgau). HI 68070

Mitteilungen aus der Naturwissenschaftlichen Gesellschaft "Isis" in Bautzen. Bautzen. Mitt. Naturwiss. Ges. "Isis" Bautzen. See B–P–H 607/14. HI 56736

Mitteilungen der Naturwissenschaftlichen Gesellschaft Thun. Thun, Switzerland. Mitt. Naturwiss. Ges. Thun. See B–P–H 607/15. HI 56737

Mitteilungen der Naturwissenschaftlichen Gesellschaft in Winterthur. Winterthur, Switzerland. Mitt. Naturwiss. Ges. Winterthur. See B–P–H 607/16. HI 56738

Mitteilungen des Naturwissenschaftlichen Museums der Stadt Aschaffenburg. Aschaffenburg. Mitt. Naturwiss. Mus. Stadt Aschaffenburg. See B–P–H 607/17.

HI 56739

Mittheilungen aus dem Naturwissenschaftlichen Vereine von Neu-Vorpommern und Rügen. (in Greifswald). Berlin. Mitth. Naturwiss. Vereine Neu-Vorpommern. See B–P–H 612/8. HI 56798

Mittheilungen aus dem Naturwissenschaftlichen Verein für Neu-Vorpommern und Rügen in Greifswald. Berlin. Mitth. Naturwiss. Verein Neu-Vorpommern Greifswald. See B–P–H 612/9. HI 56799

Mitteilungen des Naturwissenschaftlichen Vereines für Steiermark. Graz. Mitt. Naturwiss. Vereines Steiermark. See B–P–H 607/21. HI 56740

Mitteilungen des Naturwissenschaftlichen Verein(e)s in Aschaffenburg. Aschaffenburg. Mitt. Naturwiss. Vereins Aschaffenburg. See B–P–H 607/22. HI 56741

Mitteilungen des naturwissenschaftlichen Vereins zu Düsseldorf. Düsseldorf. Mitt. Naturwiss. Vereins Düsseldorf. See B–P–H 607/23. HI 56742

Mittheilungen des Naturwissenschaftlichen Vereins zu Freiberg in Sachsen. Freiberg, Germany. Mitth. Naturwiss. Vereins Freiberg. See B–P–H 612/11. HI 56800

Mittheilungen des Naturwissenschaftlichen Vereins zu Schneeberg. Schneeberg, Germany. Mitth. Naturwiss. Vereins Schneeberg. See B–P–H 612/12. HI 56801

Mitteilungen des Naturwissenschaftlichen Vereins in Troppau. Troppau [=Opava, Czechoslovakia]. Mitt. Naturwiss. Vereins Troppau. See B–P–H 608/1. HI 56743

Mitteilungen des Naturwissenschaftlichen Vereins der Universität Wien. Vienna. Mitt. Naturwiss. Vereins Univ. Wien. See B–P–H 608/2. HI 56744

Mittheilungen des Nordböhmischen Excursions-Clubs. Böhmisch-Leipa, Bohemia [Ceska Lipa, Czechoslovakia.] Mitth. Nordböhm. Excursions-Clubs. See B–P–H 612/13. HI 56802

Mitteilungen des Nordböhmischen Vereins für Heimatforschung und Wanderpflege. Böhmisch-Leipa, Bohemia [=Ceska Lipa, Czechoslovakia]. Mitt. Nordböhm. Vereins Heimatf. See B–P–H 608/5. HI 56745

Mitteilungen. Obst und Garten. Klosterneuburg, Austria. Mitt. Obst Gart. See B–P–H 608/6. HI 56746

Mitteilungen der ostalpin-dinarischen pflanzensoziologischen Arbeitsgemeinschaft. Trieste. ?-1968+. Mitt. Ostalpin-Dinarischen Pflanzensoziol. Arbeitsgem. HI 68072

Mittheilungen aus dem Osterlande. Altenburg. Vols. 1-19, 1837-69; ser. 2, vols. 1-22, 1880-1934 [suspended 1920-24]. Mitth. Osterl. Superseded by: Wissenschaftliche Beiträge aus dem Osterlande. 3-2706-1. HI 68073

Mitteilungen der Österreichischen Bodenkundlichen Gesellschaft. Vienna. Mitt. Österr. Bodenk. Ges. See B–P–H 608/7. HI 56747

Mitteilungen der Österreichischen Gartenbaugesellschaft. Vienna. Mitt. Österr. Gartenbauges. See B–P–H 608/8. HI 56748

Mitteilungen der Österreichischen Geographischen Gesellschaft = Mitteilungen der Geographischen Gesellschaft in Wien. Vienna. Mitt. Geogr. Ges. Wien. See B–P–H 604/8.

Mitteilungen der österreichischen Gesellschaft für Holzforschung. Vienna. Mitt. Österr. Ges. Holzf. See B–P–H 608/10. HI 56749

Mitteilungen der Österreichischen Mykologischen Gesellschaft. Vienna. Mitt. Österr. Mykol. Ges. See B–P–H 608/11. HI 56750

Mitteilungsblatt der Österreichischen Orchideengesellschaft. Vienna. Mitteilungsbl. Österr. Orchideenges. See B–P–H 611/6. HI 56781

Mitteilungen des Pfälzischen Vereins vür Naturkunde Pollichia. Dürkheim [=Bad Dürkheim], Germany. Mitt. Pfälz. Vereins Naturk. Pollichia. See B–P–H 608/12. HI 56751

Mitteilungen der Pflanzenschutzabteilung des Volkskomissariats für Landwirtschaft S S R Georgiens = Izvestiya Otdela zashchity rastenii. Tiflis.

Mittheilungen der Philomatischen Gesellschaft in Elsass-Lothringen. Strasbourg. Mitth. Philom. Ges. Elsass-Lothringen. See B–P–H 612/15. HI 56803

Mitteilungen der Phytopathologischen Versuchsstation der Universität Tartu = Tartu ülikooli taimehaiguste-katsejaama teated. Tartu, Estonia [Estonian S S R]. Tartu Ülik Taimeh.-Katsej. Teated. See B–P–H 866/12.

Mitteilungen der Pollichia, eines naturwissenschaftlichen Vereins der Rheinpfalz. Bad Dürkheim, Germany. Mitt. Pollichia Naturwiss. Vereins Rheinpfalz. See B–P–H 608/16. HI 56752

Mitteilungen der Pollichia, eines naturwissenschaftlichen Vereins der Rheinpfalz zu Bad Dürkheim. Bad Dürkheim, Germany. Mitt. Pollichia Naturwiss. Vereins Rheinpfalz Bad Dürkheim. See B–P–H 608/17. HI 56753

Mitteilungen der Pollichia des Pfälzischen Vereins für Naturkunde und Naturschutz. Bad Dürkheim, Germany. Mitt. Pollichia Pfälz. Vereins Naturk. See B–P–H 608/18. HI 56754

Mitteilungen Pommerschen Provinzialkomitees für Naturdenkmalpflege. Stettin [Poland]. Mitt. Pommerschen Provinzialkomitees Naturdenkmalpflege. See B–P–H 608/19. HI 56755

Mitteilungen der Provinzialstelle für Naturdenkmalpflege

Hannover. Hildesheim. Mitt. Provinzialstelle Naturdenkmalpflege Hannover. See B–P–H 609/1. HI 56756

Mitteilungen aus der Prüfungsanstalt für Wasserversorgung und Abwässerbeseitigung. Berlin. Mitt. Prüfungsanst. Wasserversorg. See B–P–H 609/2. HI 56757

Mitteilungen. Rebe und Wein. Klosterneuburg, Austria. Mitt. Rebe Wein. See B–P–H 609/3. HI 56758

Mitteilungen des Reichsamtes für Bodenforschung, Wien. Vienna. Mitt. Reichsamtes Bodenf. Wien. See B–P–H 609/4. HI 56759

Mitteilungen des Reichsinstitutes für Forst- und Holzwirtschaft. Reinbek. Vols. 1-7, 1947-48. Mitt. Reichsinst. Forst- Holzw. Superseded by: Mitteilungen des Zentralinstitutes für Forst- und Holzwirtschaft. 4-3559-2. HI 68074

Mitteilungen der Reichsstelle für Bodenforschung, Zweigstelle Wien. Neue Folge des Jahrbuches der Geologischen Bundesanstalt. Vienna. Mitt. Reichsstelle Bodenf. Zweigstelle Wien. See B–P–H 609/5. HI 56760

Mittheilungen des russischen Gartenbauvereins zu St. Petersburg. St. Petersburg. 1860. Mitt. Russ. Gartenbauver. St. Petersburg. HI 68075

Mitteilungen S W A wissenschaftliche Gesellschaft. Nuusbrief S W A wetenskaplike vereniging. Mitteilungsblatt S W A Wissenschaftliche Gesellschaft = Newsletter, S W A scientific society. Windhoek.

Mitteilungen des Saarpfälzischen Vereins für Naturkunde und Naturschutz Pollichia mit dem Sitz in Bad Dürkheim. Bad Dürkheim, Germany. Mitt. Saarpfälz. Vereins Naturk. Pollichia Bad Dürkheim. See B–P–H 609/6. HI 56761

Mitteilungen des Sächsisch-Thüringischen Vereins für Erdkunde zu Halle a. S. Halle. Mitt. Sächs.-Thüring. Vereins Erdk. Halle. See B–P–H 609/7. HI 56762

Mitteilungen der Sammelstelle für Schmarotzerbestimmung des V D E V. Aschaffenburg. 1930-50. Mitt. Sammelstelle Schmarotzerbestimm. V. D. E. V. Superseded by: Nachrichten der Sammelstelle für Schmarotzerbestimmung, naturwissenschaftliches Museum der Stadt Aschaffenburg. HI 68076

Mitteilungen, schweizerische Anstalt für das forstliche Versuchswesen = Schweizerische Anstalt für das forstliche Versuchswesen. Mitteilungen. Zurich.

Mitteilungen der Schweizerischen Anstalt für das forstliche Versuchswesen. Zurich. Mitt. Schweiz. Anst. Forstl. Versuchswesen. See B–P–H 609/8. HI 56763

Mittheilungen der Schweizerischen Centralanstalt für das forstliche Versuchswesen. Zurich. Mitth. Schweiz. Centralanst. Forstl. Versuchswesen. See B–P–H 612/16. HI 56804

Mitteilungen der Schweizerischen Kakteen-Gesellschaft. Zurich. Mitt. Schweiz. Kakteen-Ges. See B–P–H 609/10. HI 56764

Mitteilungen der Schweizerischen Zentralanstalt für das forstliche Versuchswesen = Mittheilungen der Schweizerischen Centralanstalt für das forstliche Versuchswesen. Zurich. Mitth. Schweiz. Centralanst. Forstl. Versuchswesen. See B–P–H 612/16.

Mitteilungen der Section für Naturkunde des Österreichischen Touristen-Clubs. Vienna. Mitt. Sect. Naturk. Österr. Touristen-Clubs. See B–P–H 609/12. HI 56765

Mitteilungen der Sektion Geobotanik und Phytotaxonomie der biologischen Gesellschaft der D D R. Berlin. ?-1975+. Mitt. Sekt. Geobot. Phytotax. Biol. Ges. D.D.R. HI 68077

Mitteilungen der Sektion spezielle Botanik der biologischen Gesellschaft in der D D R. Berlin. ?-1971+. Mitt. Sekt. Spez. Bot. Biol. Ges. D.D.R. HI 68078

Mittheilungen des Seminars für orientalische Sprachen an der Königlichen Universität. Berlin. Vols. 1-38, 1898-1935. Mitth. Seminars Orient. Sprachen Königl. Univ. Superseded by: Mittheilungen aus der Auslands-Hochschule. 1-645-2. HI 68079

Mitteilungen aus den Sitzungen, aturwissenschaftliche Gesellschaft "Isis" in Meissen = Naturwissenschaftliche Gesellschaft "Isis" in Meissen. Mitteilungen aus den Sitzungen. Meissen, Germany. Naturwiss. Ges. "Isis" Meissen Mitt. Sitzungen. See B–P–H 636/20.

Mitteilungen aus der Staatsforstverwaltung Bayerns. Munich. Vols. 1-15, 1894-1912. Mitt. Staatsforstverw. Bayerns. Preceded by: Mittheilungen über das Forst- und Jagdwesen in Bayern. HI 68080

Mitteilungen aus dem Staatsinstitut für allgemeine Botanik in Hamburg. Hamburg. Vols. 11-14, 1957-73. Mitt. Staatsinst. Allg. Bot. Hamburg. Preceded and superseded by: Mitteilungen aus dem Institut für allgemeine Botanik. HI 68081

Mitteilungen des Staatsinstituts für Wissenschaftliche Forschung auf dem Gebiete der Forstwirtschaft und Holzindustrie = Trudy i issledovaniya po lesnomu khozyaistvu i lesnoi promyshlennosti. Leningrad.

Mitteilungen aus dem Technisch-Mikroskopischen Laboratorium der Technischen Hochschule in Wien. Vienna. Mitt. Techn.-Mikroskop. Lab. TH Wien. See B–P–H 609/15. HI 56766

Mittheilungen der Thurgauischen Naturforschenden

Gesellschaft. Frauenfeld, Switzerland. Mitth. Thurgauischen Naturf. Ges. See B–P–H 613/2. HI 56806

Mittheilungen des Thurgauischen Naturforschenden Vereins über seine Thätigkeit. Frauenfeld, Switzerland. Mitth. Thurgauischen Naturf. Vereins Thätigk. See B–P–H 613/3. HI 56807

Mitteilungen der Thüringischen botanischen Gesellschaft. Weimar. Vols. 1-2, 1949-60. Mitt. Thüring. Bot. Ges. Superseded by: Häussknechtia. 5-4212-3. HI 68082

Mittheilungen der Thüringischen Botanischen Vereins. Weimar. Mitth. Thüring. Bot. Vereins. See B–P–H 613/4. HI 56808

Mitteilungen der Tomsker Abteilung der Russichen Botanischen Gesellschaft = Izvestiya Tomskogo otdeleniya Gosudarstvennogo russkogo botanicheskogo obshchestva. Tomsk.

Mittheilungen des ungarischen Forstvereines. Pressburg [=Bratislava, Czechoslovakia]. Mitth. Ung. Forstvereines. See B–P–H 613/5. HI 56809

Mitteilungen der ungarischen geographischen Gesellschaft = Földrajzi közlemények. Budapest. Földr. Közlem. See B–P–H 377/10.

Mittheilungen des ungarischen Landes-Forst-Vereines = Az országos erdészeti-egyesület közleményei. Pest [=Budapest, in part]. Orsz. Erdész.-Egyes. Közlem. See B–P–H 691/14.

Mitteilungen der Vereine: Landesverband der Gartenbauvereine Steiermarks, Fremdenverkehrsverein Graz, St. Veit und Umgebung, Arbeitsgemeinschaft der Förderer des Alpengartens, der Alpengarten-, Garten und Blumenfreunde = Alpengarten, Der. Graz. Alpengarten. See B–P–H 75/1.

Mitteilungen aus dem Vereine der Naturfreunde in Reichenberg. Reichenberg [=Liberec. Czechoslovakia]. Mitt. Vereine Naturfr. Reichenberg. See B–P–H 610/11. HI 56767

Mitteilungen des Vereines für Heimatkunde des Jeschken-Isergaues. Reichenberg [=Liberec. Czechoslovakia]. Mitt. Vereines Heimatk. Jeschken- Isergaues. See B–P–H 610/12. HI 56768

Mittheilungen des Vereins für Erdkunde zu Halle a. S. Halle. Mitth. Vereins Erdk. Halle. See B–P–H 613/7. HI 56810

Mitteilungen des Vereins für Forstliche Standortskunde und Forstpflanzenzüchtung. Stuttgart. Mitt. Vereins Forstl. Standortsk. See B–P–H 610/13. HI 56769

Mitteilungen des Vereins für Forstliche Standortskartierung. Stuttgart. Mitt. Vereins Forstl. Standortskart. See B–P–H 610/14. HI 56770

Mitteilungen des Vereins für Mathematik und Naturwissenschaften in Ulm a. D. Ulm. Mitt. Vereins Math. Ulm. See B–P–H 610/15. HI 56771

Mitteilungen des Vereins für Naturkunde und Naturschutz in der Westmark Pollichia. Kaiserslautern. Mitt. Vereins Naturk. Westmark Pollichia. See B–P–H 610/17. HI 56773

Mittheilungen des Vereins für Naturkunde in Reichenbach i. V. Reichenbach im Vogtland, Germany. Mitth. Vereins Naturk. Reichenbach. See B–P–H 613/8. HI 56811

Mitteilungen des Vereins für Naturkunde für Vegesack und Umgegend. Vegesack [=Bremen, in part]. Mitt. Vereins Naturk. Vegesack. See B–P–H 610/16. HI 56772

Mitteilungen des Vereins für Naturwissenschaft und Mathematik in Ulm a. D. (Donau) = Mitteilungen des Vereins für Mathematik und Naturwissenschaften in Ulm a. D. Ulm. Mitt. Vereins Math. Ulm. See B–P–H 610/15.

Mitteilungen des Vereins nördlich der Elbe zur Verbreitung naturwissenschaftlicher Kenntniss. Kiel. Mitt. Vereins Nördl. der Elbe Verbreit. Naturwiss. Kenntn. See B–P–H 610/19. HI 56774

Mittheilungen aus den Verhandlungen der Gesellschaft naturforschender Freunde zu Berlin. Berlin. Mitth. Verh. Ges. Naturf. Freunde Berlin. See B–P–H 613/9. HI 56812

Mittheilungen aus den Verhandlungen der Naturwissenschaftlichen Gesellschaft in Hamburg. Hamburg. Mitth. Verh. Naturwiss. Ges. Hamburg. See B–P–H 613/10. HI 56813

Mitteilungen der Versuchsanstalt für das Gärungsgewerbe und des Institutes für angewandte Mikrobiologie. Vienna. Mitt. Versuchsanst. Gärungsgewerbe Inst. Angew. Mikrobiol. See B–P–H 610/20. HI 56775

Mitteilungen der Versuchsanstalt für Pilzbau der Landwirtschaftskammer Rheinland (Krefeld-Grosshuttenhof). No. 1+, 1977+. Mitt. Versuchsanst. Pilzbau Landwirtschaftskammer Rheinland (Krefeld-Grosshuttenhof). HI 68083

Mitteilungen der Versuchsstation für das Gärungsgewerbe sowie des Institutes für angewandte Mikrobiologie. Vienna. Mitt. Versuchsstat. Gärungsgewerbe Inst. Angew. Mikrobiol. See B–P–H 610/21 HI 56776

Mitteilungen der Virusdokumentationsstelle im Hygiene-Institut der Universität Wien. Vienna. Mitt. Virusdokumentationsstelle Hyg.-Inst. Univ. Wien. See B–P–H 610/22. HI 56777

Mitteilungen der Vogtländischen Gesellschaft für Naturforschung. Plauen im Vogtland, Germany. Mitt. Vogtl. Ges. Naturf. See B–P–H 611/1. HI 56778

Mittheilungen des Vogtländischen Vereins für allgemeine und spezielle Naturkunde in Reichenbach i. V.

Reichenbach = Mittheilungen des Voigtländischen Vereins für allgemeine und specielle Naturkunde in Reichenbach. Reichenbach im Vogtland, Germany. Mitth. Voigtl. Vereins Allg. Naturk. Reichenbach. See B–P–H 613/12.

Mittheilungen des Voigtländischen Vereins für allgemeine und specielle Naturkunde in Reichenbach. Reichenbach im Vogtland, Germany. Mitth. Voigtl. Vereins Allg. Naturk. Reichenbach. See B–P–H 613/12. HI 56814

Mittheilungen zur Volks- und Heimatkunde des Schönhengster Landes. Mährisch Trübau, Moravia [=Moravska Trebova, Czechoslovakia]. Mitth. Volks-Heimatk. Schönhengster Landes. See B–P–H 613/13. HI 56815

Mittheilungen aus Wien. Vienna. Mitth. Wien. See B–P–H 613/14. HI 56816

Mitteilungen aus dem Wissenschaftsbereich Geobotanik und botanischer Garten der Sektion Biowissenschaften der Martin-Luther-Universität Halle-Wittenberg. Halle. No. 12+, 1974/75+. Mitt. Wissenschaftsbereich Geobot. Bot. Gart. Sekt. Biowiss. Martin-Luther-Univ. Halle-Wittenberg. Preceded by: Mitteilungen aus dem Bereich systematische Botanik und Pflanzengeographie der Sektion Biowissenschaften der Martin-Luther-Universität Halle-Wittenberg. HI 68084

Mitteilungen des Wissenschaftlichen Vereins Schneeberg. Schneeberg, Germany. Mitt. Wiss. Vereins Schneeberg. See B–P–H 611/2. HI 56779

Mitteilungen der Württembergischen Forstlichen Versuchsanstalt. Stuttgart. 1928-41; vols. 1-6, 1949-56. Mitt. Württemberg. Forstl. Versuchsanst. HI 68085

Mitteilungen der Zentralanstalt für Forst- und Holzwirtschaft. Reinbek. Vol. 10, 1949. Mitt. Zentralanst. Forst- Holzw. Preceded by: Mitteilungen des Zentralinstitutes für Forst- und Holzwirtschaft. Superseded by: Mitteilungen der Bundesanstalt für Forst- und Holzwirtschaft. HI 68086

Mitteilungen des Zentralinstitutes für Forst- und Holzwirtschaft. Reinbek. Vols. 8-9, 1949. Mitt. Zentralinst. Forst- Holzw. Preceded by: Mitteilungen des Reichsinstitutes für Forst- und Holzwirtschaft. Superseded by: Mitteilungen der Zentralanstalt für Forst- und Holzwirtschaft. HI 68087

Mitteilungsblatt, Arbeitskreis heimische Orchideen Baden-Württemberg; Beiträge zur Erhaltung und Erforschung heimische Orchideen. Stuttgart. Vol. 1+, 1969+. Mitt. Arebeitskreis Heimische Orchid. Baden-Württemberg. Preceded by?: Mitteilungen des Arbeitskreises zur Beobachtung und zum Schutz heimischer Orchideen. HI 68088

Mitteilungsblatt des Arbeitskreises für Mammillaria. Buchschlag. No. 1, 1972. Mitt. Arbeitskreises Mammillaria. Superseded by: Mitteilungsblatt des Arbeitskreises für Mammillarienfreunde e. V. HI 68089

Mitteilungsblatt des Arbeitskreises für Mammillarienfreunde e. V. Münster, Osnabrück. 1977+. Mitt. Arbeitskreises Mammillarienfr. Preceded by: Mitteilungsblat des Arbeitskreises Mammillaria. HI 68090

Mitteilungsblatt der biologischen Arbeitsgemeinschaft, Volksbildungskurse Kempten (Allgau). Kempten. Vol. 1, 1951. Mitt. Biol. Arbeitsgem. Volksbildungsk. Kempten (Allgau). Superseded by: Mitteilungen des naturwissenschaltlichen Arbeitskreises Kempten/Allgäu. HI 68091

Mitteilungsblatt, floristische Arbeitsgemeinschaft, naturwissenschaftlichen Verein für Steiermark. Graz. Nos. 5-26, 1967-74. Mitt. Florist. Arbeitsgem. Naturwiss. Ver. Steiermark. Preceded by: Hinweise an die Mitarbeiter. Superseded by: Notizen zur Flora der Steiermark. HI 68092

Mitteilungsblatt, Gesellschaft österreichischer Kakteenfreunde. Vienna. Vol. 1+, 1957+. Mitt. Ges. Österr. Kakteenfr. HI 68093

Mitteilungsblatt des Landesverbandes Gartenbau und Landwirtschaft Berlin. Berlin. Vol. ?-19+, ?-1968+. Mitt. Landesverb. Gart. Landw. Berlin. HI 68094

Mitteilungsblatt, Mikrographische Gesellschaft, Wien = Mikrographische Gesellschaft, Wien. Mitteilungsblatt. Vienna. Mikrogr. Ges. Wien Mitteilungsbl. See B–P–H 595/1.

Mitteilungsblatt der schweizerischen Akademie der Geisteswissenschaften und der schweizerischen naturforschenden Gesellschaft. Bern. 1985(2)+. 1985+. Mitt. Schweiz. Akad. Geisteswiss. Schweiz. Naturf. Ges. Preceded by: Mitteilungsblatt der schweizerischen Geisteswissenschaftlichen Gesellschaft und der schweizerischen naturforschenden Gesellschaft. HI 68095

Mitteilungsblatt der schweizerischen Geisteswissenschaftlichen Gesellschaft und der schweizerischen naturforschenden Gesellschaft. Bern. 1975-85. Mitt. Schweiz. Geisteswissenschaftl. Ges. Schweiz. Naturf. Ges. Superseded by: Mitteilungsblatt der schweizerischen Akademie der Geisteswissenschaften und der schweizerischen naturforschenden Gesellschaft. HI 68096

Mittheilungen, see Mitteilungen

Miyagi-ken nogyo tanki daigaku gakujutsu hokoku = Scientific reports of the Miyagi agricultural college. Sendai, Japan. Sci. Rep. Miyagi Agric. Coll. See B–P–H 827/12.

Miyagi-kenritsu nogyo shikenjo hokoku = Technical

bulletin of Miyagi prefectural agricultural experiment station. Sendai, Japan. Techn. Bull. Miyagi Prefect. Agric. Exp. Sta. See B–P–H 869/8.

Miyazaki daigaku kyoikugakubu kiyo, shizenkagaku = Memoirs of the faculty of education, Miyazaki university. Natural science. Miyazaki.

Miyazaki daigaku nogaku-bu kenkyo jiho = Bulletin of the faculty of agriculture; university of Miyazaki. Miyazaki, Japan. Bull. Fac. Agric. Univ. Miyazaki. See B–P–H 249/15.

Miyazaki daigaku nogakubu kiyo = Memoirs of the faculty of agriculture, university of Miyazaki.

Miyazaki daigaku ziho. Shizen kagaku = Bulletin of the Miyazaki university. Natural science. Miyazaki, Japan. Bull. Miyazaki Univ., Nat. Sci. See B–P–H 263/24.

Miyazaki-ken sogo nogyo shikenjo kenkyu hokoku = Bulletin of the Miyazaki agricultural experiment station. Miyazaki, Japan. Bull. Miyazaki Agric. Exp. Sta. See B–P–H 263/20.

Miyazaki koto norin gakko gakujutsu hokoku = Bulletin of the Miyazaki college of agriculture and forestry. Miyazaki, Japan. Bull. Miyazaki Coll. Agric. See B–P–H 263/22.

Mnemosyne. Mengelingen voor wetenschappen en fraaie letteren. Dordrecht, Netherlands. Mnemosyne. See B–P–H 614/9. HI 56817

MO; bulletin of botanical intelligence. St. Louis, MO. Vol. 1+, 1981+. MO. HI 68097

Modern beekeeping. Paducah, KY. Modern Beekeeping. See B–P–H 614/20. HI 56818

Modern cell biology. New York. Vol. 1+, 1983+. Modern Cell Biol. HI 68098

Modern gladiolus grower. Albany, NY. Modern Gladiolus Grower. See B–P–H 614/21. HI 56819

Modern medicine; a quarterly journal of modern therapeutics. Manila. Modern Med. (Manila). See B–P–H 614/22. HI 56820

Modern medicine; the newsmagazine of medicine. Minneapolis, MN. Modern Med. (Minneapolis). See B–P–H 614/23. HI 56821

Modern school store. Houston, TX. Modern School Store. See B–P–H 614/24. HI 56822

Modernization of plant protection = A növényvédelen korszerusitése. Budapest.

Moderno zooiatro. Bologna. Moderno Zooiatro. See B–P–H 614/25. HI 56823

Moenaretha sistematikisa da geographiis nakwewebi = Zametki po sistematike i geografii rastenii. Tiflis.

Möglinsche Annalen der Landwirthschaft. Berlin. Möglinsche Ann. Landw. See B–P–H 614/27. HI 56824

Moj mali svet. Ljubljana. Vol. ?-8+, ?-1976+. Moj Mali Svet. HI 68099

Mokuzai gakkaishi = Journal of the japanese wood research society. Tokyo. J. Jap. Wood Res. Soc. See B–P–H 471/3.

Mokuzai kenkyu. Kyoto daigaku mokuzai kenkyujo hokoku = Wood research; bulletin of the wood research institute. [Kyoto university.] Kyoto, Japan. Wood Res. See B–P–H 979/12.

Mokuzai kogyo = Wood industry. Minato-ku, Tokyo.

Molecular biology. [Translation of: Molekulyarnaya biologiya.] New York, Orlando, FL. 1967+. Molec. Biol. HI 68100

Molecular biology, biochemistry and biophysics. Berlin, New York, etc. Vol. 1+, 1967+. Molec. Biol. Biochem Biophys. HI 68101

Molecular biology and evolution. Chicago, IL. Vol. 1+, 1983+. Molec. Biol. Evol. HI 68102

Molecular biology reports; international journal for rapid communications in molecular biology. Dordrecht, The Hague. Vol. 1+, 1973+. Molec. Biol. Rep. HI 68103

Molecular and cellular biochemistry. Hague. Vol. 1+, 1973+. Molec. Cell. Biochem. Preceded by: Enzymologia. HI 68104

Molecular and cellular biology. Coimbra = Ciência biológica. Biologia molecular e celular. Coimbra.

Molecular and cellular biology. Washington, DC. Vol. 1+, 1981+. Molec. Cell. Biol. HI 68105

Molecular and general genetics. International journal. Berlin, Heidelberg, [etc.]. Vol. 99+, 1967+. Molec. Gen. Genet. Preceded by: Zeitschrift für Vererbungslehre. HI 68106

Molecular genetics, microbiology and virology. New York. 1986+. Molec. Genet. Microbiol. Virol. HI 74860

Molecular pharmacology. New York, NY. Molec. Pharmacol. See B–P–H 614/29. HI 56825

Molecular plant-microbe interactions. St. Paul, MN. Vol. 1+, 1988+. Molec. Pl.-Microbe Interact. HI 68107

Molekularbiologie, Biochemie und Biophysik = Molecular biology, biochemistry and biophysics. Berlin, New York, etc.

Molekuljarnaja biologija = Molekulyarnaya biologiya. Kiev.

Molekulyarnaya biologiya. Kiev. Vol. 1+, 1964+. Molek. Biol. (Kiev). HI 68108

Molekulyarnaya biologiya. Moscow. Vol. 1+, 1967+. Molek. Biol. (Moscow). HI 68108

Molekulyarnaya genetika i biofizika. Kiev. Vol. 1+,

1976+. Molek. Genet. Biofiz. HI 68109

Molina. Lima. Molina. See B–P–H 614/30. HI 56826

Molineria y panaderia; revista de las industrias de la harina. Barcelona. 1906+. Molineria Panaderia. 3-2723-2. HI 68111

Moliniana; universidad de Chile, instituto de botánico. Santiago. Moliniana. See B–P–H 614/31. HI 56827

Möller's Deutsche Gärtner-Zeitung. Erfurt. Möller's Deutsche Gärtn.-Zeitung. See B–P–H 614/33. HI 56828

Monaco-cactus; bulletin semestriel de la société des amis des cactées et plantes grasses de Monaco. Revue scientifique illustrée pour l'étude des cactées et plantes grasses. Monte Carlo. [Nos. 1-2], 1938-39. Monaco-Cactus. HI 68112

Monathliche Auszüge alt und neuer Gelehrten Sachen. Olmütz, Moravia [=Olomouc, Czechoslovakia]. Monathl. Auszüge Alt Neuer Gel. Sachen. See B–P–H 615/1. HI 56829

Monathliche Früchte einer gelehrten Gesellschaft in Ungern. Pest & Ofen [=Budapest]. Monathl. Früchte Gel. Ges. Ungern. See B–P–H 615/2. HI 56830

Monatliche Beiträge zur Naturkunde. Berlin. Monatl. Beitr. Naturk. See B–P–H 615/3. HI 56831

Monatliche Mittheilungen aus dem Gesammtgebiete der Naturwissenschaften. Frankfurt a. O. Monatl. Mitth. Gesammtgeb. Naturwiss. See B–P–H 615/4. HI 56832

Monatliche Mittheilungen des Naturwissenschaftlichen Vereins des Regierungsbezirk(e)s Frankfurt. Frankfurt a. O. Monatl. Mitth. Naturwiss. Vereins Regierungsbez. Frankfurt. See B–P–H 615/5. HI 56833

Monatsberichte der deutschen Akademie der Wissenschaften zu Berlin. Berlin. Vols. 1-13, 1959-71. Monatsber. Deutsch. Akad. Wiss. Berlin. HI 68113

Monatsberichte der Königlich Preussischen Akademie der Wissenschaften zu Berlin. Berlin. Monatsber. Königl. Preuss. Akad. Wiss. Berlin. See B–P–H 615/11. HI 56834

Monatsberichte über die Verhandlungen der Gesellschaft für Erdkunde zu Berlin. Berlin. Monatsber. Verh. Ges. Erdk. Berlin. See B–P–H 615/12. HI 56835

Monatsblatt der Königlich preussischen märkischen ökonomischen Gesellschaft zu Potsdam. Potsdam. Monatsbl. Königl. Preuss. Märk. Ökon. Ges. Potsdam. See B–P–H 615/13. HI 56836

Monatsblatt der Königlich preussischen märkischen ökonomischen Gesellschaften zu Potsdam und Frankfurth an der Oder. Potsdam. Monatsbl. Königl. Preuss. Märk. Ökon. Ges. Potsdam Frankfurth Oder. See B–P–H 615/14. HI 56837

Monats-Blatt des landwirthschaftlichen Vereins für den Oberdonau-Kreis im Königreiche Bayern. Augsburg. Monats-Blatt Landw. Vereins Oberdonau-Kreis Königr. Bayern. See B–P–H 615/17. HI 56840

Monatsblatt des Vereins für Landeskunde in Niederösterreich. Vienna. Monatsbl. Vereins Landesk. Niederösterreich. See B–P–H 615/15. HI 56838

Monats-Blätter des Vereins für Naturkunde und des ärztlichen Vereins zu Zwickau. Zwickau. Monats-Blätt. Vereins Naturk. Zwickau. See B–P–H 615/18. HI 56841

Monatsblätter des wissenschaftlichen Club in Wien. Vienna. Monatsbl. Wiss. Club Wien. See B–P–H 615/16. HI 56839

Monatshefte für Chemie und verwandte Teile andererer Wissenschaften. Vienna, Heidelberg. Vol. 1+, 1880+. Monatsh. Chem. Verwandte Teil And. Wiss. HI 68116

Monatshefte für den naturwissenschaftlichen Unterricht aller Schulgattungen und Natur und Schule. Leipzig. Vols. 7-17 [also numbered vols. 1-11], [1907] 1908-18. Monatsh. Naturwiss. Unterr. 4-2935-3. HI 68117

Monatshefte für Veterinärmedizin. Leipzig, Jena. 1946+. Monatsh. Veterinärmed. HI 68118

Monatschrift der Gesellschaft des Vaterländischen Museums in Böhmen. Prague. Monatschr. Ges. Vaterl. Mus. Böhmen. See B–P–H 616/1. HI 56842

Monatshefte für praktische Dermatologie. Leipzig. Monatsh. Prakt. Dermatol. See B–P–H 616/3. HI 56843

Monatsschrift der Deutschen Kakteen-Gesellschaft, e. V., Sitz Berlin. Berlin. Monatsschr. Deutsch Kakteen-Ges. See B–P–H 616/6. HI 56844

Monatsschrift für den elementaren naturwissenschaftlichen Unterricht. Hamburg. Monatsschr. Element. Naturwiss. Unterr. See B–P–H 616/7. HI 56845

Monatsschrift für das Forst- und Jagdwesen. Stuttgart. Monatsschr. Forst- Jagdwesen. See B–P–H 616/8. HI 56846

Monatsschrift des Gartenbau-Vereins in Darmstadt. Darmstadt, Germany. Monatsschr. Gartenbau-Vereins Darmstadt. See B–P–H 616/9. HI 56847

Monatsschrift für Kakteenkunde. Berlin. Monatsschr. Kakteenk. See B–P–H 616/10. HI 56848

Monatsschrift für Medicin, Augenheilkunde und Chirurgie. Dresden. Monatsschr. Med. See B–P–H 616/11. HI 56849

Monatsschrift für Pomologie und praktischen Obstbau.

Stuttgart. Monatsschr. Pomol. Prakt. Obstbau. See B–P–H 616/12. HI 56850

Monatsschrift der Reichsarbeitsgemeinschaft für Raumforschung = Raumforschung und Raumordnung. Heidelberg. Raumf. & Raumordn. See B–P–H 756/13.

Monatsschrift von und für Schlesien. Breslau [=Wroclaw, Poland]. Monatsschr. Schlesien. See B–P–H 616/14. HI 56851

Monatsschrift des Vereines zur Beförderung des Gartenbaues in den Königlich Preussischen Staaten für Gärtnerei und Pflanzenkunde. Berlin. Monatsschr. Vereines Beförd. Gartenbaues Königl. Preuss. Staaten. See B–P–H 616/17. HI 56853

Monatsschrift des Vereines zur Beförderung des Gartenbaues in den Königlich Preussischen Staaten und der Gesellschaft der Garten-freunde Berlins. Berlin. Monatsschr. Vereines Beförd. Gartenbaues Königl. Preuss. Staaten. Ges. Gartenfr. Berlins. See B–P–H 616/16. HI 56852

Mondes et cultures; comptes rendus trimestriels des séances de l'académie des sciences d'outre-mer. Paris. Vol. 38+, 1978+. Mondes & Cult. Preceded by: Comptes rendus trimestriels des séances de l'académie des sciences d'outre-mer. HI 68119

Monde des plantes; intermédiaire des botanistes; supplement aux bulletins de plusiers sociétés savantes. Le Mans. Vol. 1+, 1899+. Monde Pl. Preceded by: Monde des plantes, revue mensuelle de botanique. 1-32-1. HI 68120

Monde des plantes; revue mensuelle de botanique; organe de l'académie internationale de géographie botanique. Le Mans. Vols. 1-8, 1891-98. Monde Pl. Rev. Mens. Bot. Superseded by: Bulletin de l'académie internationale de géographie botanique and Monde des plantes; intermédiaire des botanistes. 1-32-1. HI 68121

Moniteur d'horticulture, arboriculture, viticulture, sciences, arts et industries horticoles; organe des amateurs de jardins. Paris. Vols. 1-38, 1877-1914. Monit. Hort. Arboric. Vitic. Sci. Arts Industr. Hort. HI 68122

Monitor de la farmacia y de la terapéutica; revista quincenal cientifico-profesional. Madrid. Vol. 1+, 1895+. Monit. Farm. Terap. HI 68123

Monitoring fonovogo zagryazneniya prirodnykh sred. Leningrad. Vol. 1+, 1982+. Monit. Fonov. Zagryazn. Prir. Sred. HI 75114

Monografias cientificas "Augusto Pi Suñer". Caracas. No. 1+, 1970+. Monogr. Ci. "Augusto Pi Suñer". HI 68124

Monografias cientificas, instituto pedagogico. Caracas = Monografias cientificas "Augusto Pi Suñer". Caracas.

Monografias, comision investigaciones cientificas de la provincia de Buenos Aires. La Plata. No. ?-10+, ?-1979+. Monogr. Comis. Invest. Ci. Prov. Buenos Aires. HI 68125

Monografías. Escuela nacional de agricultura. Colegio de postgraduados. Chapingo, Mexico. Monogr. Esc. Nac. Agric. See B–P–H 617/7. HI 56857

Monografias, estacion biologica de Doñana. Madrid. No. 1+, 1967+. Monogr. Estac. Biol. Doñana. HI 68126

Monografías de flora y vegetación bética. Granada. Vol. 1+, 1986+. Monogr. Fl. Veg. Bética. HI 68127

Monografias, instituto español de oceanografia. Madrid. No. 1+, 1986+. Monogr. Inst. Esp. Oceanogr. Preceded by: Trabajos del instituto español de oceanografía. HI 75281

Monografías del instituto de estudios geográficos. Tucumán, Argentina. Monogr. Inst. Estud. Geogr. See B–P–H 617/12. HI 56861

Monografias del instituto de estudios pirenaicos. Jaca. Nos. ?-104-?, ?-1975-86? Monogr. Inst. Estud. Piren. Superseded by: Monografías del instituto pirenaico de ecología. HI 68128

Monografias, instituto nacional para la conservacion de la naturaleza. Madrid. 1974+. Monogr. Inst. Nac. Conserv. Naturaleza. HI 68129

Monografías del instituto pirenaico de ecología. Jaca. Vol. 1+, 1987+. Monogr. Inst. Piren. Ecol. Preceded by: Publicaciones del centro pirenaico de biologia experimental and Monografías del instituto de estudios pirenaicos. HI 68130

Monografías. Laboratorio de biología marina. Universidad católica de Santo Tomás de Villanueva. Marianao, Cuba. Monogr. Lab. Biol. Mar. Univ. Católica Santo Tomás de Villanueva. See B–P–H 617/17. HI 56863

Monografias del museo de ciencias naturales. Vitoria-Gasteiz. 1988+. Monogr. Mus. Ci. Nat. HI 75283

Monografias tecnico-cientifica, universidad autonoma agraria "Antonio Narro". Saltillo. ?-1979+. Monogr. Tech.-Ci. Univ. Auton. Agrar. "Antonio Narro". HI 68131

Monografias, universidade rural de Pernambuco. Recife. Nos. 1-3?, 1954-56? Monogr. Univ. Rural Pernambuco. HI 68132

Monografie del centro internazionale di silvicultura = Silvae orbis. Schriftenreihe der Internationalen Forstzentrale. Berlin-Wannsee. Silvae Orbis. See B–P–H 836/15.

Monografie di episteme. Milan. Vol. 1+, 1968+. Monogr. Episteme. HI 68133

Monografie di genetica agraria. Pavia. Monogr. Genet. Agrar. See B–P–H 617/9. HI 56858

Monografie di "natura bresciana." Brescia. 1969+. Monogr. Nat. Bresciana. HI 75282

Monografie e rapporti coloniali. Rome. 1912-17; n.s. 1-2, 1922-24. Monogr. Rapp. Colon. HI 68134

Monografii. Izdavaemyja Komissiej po izučeniju estestvennyh proizvoditel'nyh sil Rossii pri Rossijskoj Adademii Nauk = Monografii. Izdavaemyya Komissiei po izucheniyu estestvennykh proizvoditel'nykh sil Rossii pri Rossiiskoi Adademii Nauk. Petrograd.

Monografii. Izdavaemyya Komissiei po izucheniyu estestvennykh proizvoditel'nykh sil Rossii pri Rossiiskoi Adademii Nauk. Petrograd. Vols. 1-10, 1920-28. Monogr. Komiss. Izuch. Estestv. Proizv. Sil Rossii Rossiisk. Akad. Nauk. 1-118-2. HI 68135

Monografii Volzhskoi biologicheskoi stantsii. Saratov, Russian S F S R. Vol. 3 [data on vol. 2 not ascertained], 1928. Monogr. Volzhsk. Biol. Stantsii. Preceded by: Monografii Volzhskoi biologicheskoi stantsii Saratovskogo obshchestva estestvoispytatelei. 5-3783-1. HI 68136

Monografii Volzhskoi biologicheskoi stantsii Saratovskogo obshchestva estestvoispytatelei. Saratov, Russian S F S R. Vol. 1, 1924. Monogr. Volzhsk. Biol. Stantsii Saratovsk. Obshch. Estestvoisp. Superseded by: Monografii Volzhskoi biologicheskoi stantsii. 5-3783-1. HI 68137

Monografii Volžskoj biologičeskoj stancii = Monografii Volzhskoi biologicheskoi stantsii. Saratov.

Monografii Volžskoj biologičeskoj stancii Saratovskogo obščestva estestvoispytatelej = Monografii Volzhskoi biologicheskoi stantsii Saratovskogo obshchestva estestvoispytatelei. Saratov.

Monograph, see Monographs

Monographiae biologicae. The Hague. Monogr. Biol. See B–P–H 617/3. HI 56855

Monographiae botanicae. Warsaw. Monogr. Bot. See B–P–H 617/6. HI 56856

Monographias del centro internacionale de silvicultura = Silvae orbis. Schriftenreihe der Internationalen Forstzentrale. Berlin-Wannsee. Silvae Orbis. See B–P–H 836/15.

Monographic series of the natural sciences bulletin, college of science; Sun Yatsen university. [Kuo li chung shan ta hsüeh tzu jan k'o hsüeh k'o.] Canton. Monogr. Ser. Nat. Sci. Bull. Coll. Sci. Sun Yatsen Univ. See B–P–H 617/20. HI 56866

Monographieen. Nederlandsche entomologische vereeniging. Amsterdam. Monogr. Ned. Entomol. Ver. See B–P–H 617/18. HI 56864

Monographien der Biologischen Wolga-Station = Monografii Volzhskoi biologicheskoi stantsii. Saratov.

Monographien der Biologischen Wolga-Station der Naturforscher-Gesellschaft zu Saratow = Monografii Volzhskoi biologicheskoi stantsii Saratovskogo obshchestva estestvoispytatelei. Saratov.

Monographies, académie serbe des sciences et des arts. Classe des sciences mathématiques et naturelles = Posebna izdanja, srpska akademija nauka i umetnosti. Odeljenje prirodno-matematičkih nauka. Belgrade.

Monographies du centre international de sylviculture = Silvae orbis. Schriftenreihe der Internationalen Forstzentrale. Berlin-Wannsee. Silvae Orbis. See B–P–H 836/15.

Monographs, academy of natural sciences of Philadelphia = Academy of natural sciences of Philadelphia monographs. Philadelphia, PA. Acad. Nat. Sci. Philadelphia Monogr. See B–P–H 36/21.

Monographs of the american society of plant physiologists. Lancaster, PA. Monogr. Amer. Soc. Pl. Physiol. See B–P–H 617/2. HI 56854

Monographs of the biological society of Iraq = Prospects of Iraq biology. [Baghdad.]

Monographs, forest science. Washington, DC = Forest science. Monograph. Washington, DC. Forest Sci. Monogr. See B–P–H 380/24.

Monographs on the improvement of the human plant of the Luther Burbank society. Santa Rosa, CA. Monogr. Improv. Human Pl. Luther Burbank Soc. See B–P–H 617/10. HI 56859

Monographs on the improvement of plant life of the Luther Burbank society. Santa Rosa, CA. Monogr. Improv. Pl. Life Luther Burbank Soc. See B–P–H 617/11. HI 56860

Monographs. Istanbul üniversitesi. Istanbul. Monogr. Istanbul Üniv. See B–P–H 617/14. HI 56862

Monographs, research division, Virginia polytechnic institute. Blacksburg, VA = Research division monograph, Virginia polytechnic institute. Blacksburg, VA.

Monograph series, american museum of natural history. New York, NY. Monogr. Ser., Amer. Mus. Nat. Hist. See B–P–H 617/19. HI 56865

Monograph series, australian institute of marine science. Townsville, Qld. 1976+. Monogr. Ser., Austral. Inst. Mar. Sci. HI 68138

Monographiae biologicae. The Hague. Monogr. Biol. See B–P–H 617/3. HI 56855

Monographiae biologicae canarienses. Las Palmas. Vols. 1-6, 1970-75. Monogr. Biol. Canar. HI 68139

Monographs, american phytopathological society. St.

Paul, MN. No. 1+, 1980+. Monogr. Amer. Phytopathol. Soc. HI 68140

Monographs, Birbal Sahni institute of palaeobotany. Lucknow. No. 1+. 1964+. Monogr. Birbal Sahni Inst. Palaebot. HI 68141

Monographs, british crop protection council. Croydon. No. 1+, 1970+. Monogr. Brit. Crop Protect. Council. HI 68142

Monographs, Grand Canyon natural history association. No. 1+, 1978+. Monogr. Grand Canyon Nat. Hist. Assoc. HI 68143

Monographs, Indiana academy of science. Indianapolis, IN. 1969+. Monogr. Indiana Acad. Sci. HI 68144

Monographs of the international forestry centre = Silvae orbis. Schriftenreihe der Internationalen Forstzentrale. Berlin-Wannsee. Silvae Orbis. See B–P–H 836/15.

Monographs of the malaysian branch of the royal asiatic society. Singapore. No. 1+, 1964+. Monogr. Malaysian Branch Roy Asiat. Soc. HI 68145

Monographs, museum of natural history, university of Kansas. Lawrence, KS. 1970+. Monogr. Mus. Nat. Hist. Univ. Kansas. HI 68146

Monographs, research branch, Canada department of agriculture. Ottawa. No. 1+, 1957+. Monogr. Res. Branch Canada Dept. Agric. HI 68147

Monographs, society for science and art of Montenegro. Section of natural sciences. Titograd = Posebna izdanja, drustvo za nauku i umjetnost Tsrne Gore. Odjeljenje prirodnikh nauka. Titograd.

Monographs in systematic botany from the Missouri botanical garden. St. Louis, MO. Vol. 1+, 1978+. Monogr. Syst. Bot. Missouri Bot Gard. HI 68148

Monographs of the university of Puerto Rico. Series B, physical and biological sciences. San Juan, PR. No. 1+, 1933+. Monogr. Univ. Puerto Rico, B. 4-3487-2. HI 68149

Monspeliensis Hippocrates. Revue de la société montpelliéraine d'histoire de la médecine. Montpellier. Monspel. Hippocrates. See B–P–H 618/7. HI 56867

Monsunia; Beiträge zur Kenntniss der Vegetation des Süd- und Ostasiatischen Monsungebietes. Leipzig. Vol. 1+, 1900. Monsunia. HI 68150

Montana college of agriculture and mechanic arts. Agricultural experiment station. Annual report. Bozeman, MT. Montana Agric. Exp. Sta. Annual Rep. See B–P–H 618/11. HI 56868

Montana college of agriculture and mechanic arts. Agricultural experiment station. Bulletin. Bozeman, MT. Montana Agric. Exp. Sta. Bull. See B–P–H 618/12. HI 56869

Montana college of agriculture and mechanic arts. Agricultural experiment station. Circular. Bozeman, MT. Montana Agric. Exp. Sta. Circ. See B–P–H 618/13. HI 56870

Montana college of agriculture and mechanic arts. Agricultural experiment station. Special circular. Bozeman, MT. Montana Agric. Exp. Sta. Special Circ. See B–P–H 618/17. HI 56874

Montana college of agriculture and mechanic arts. Science studies, botany. Bozeman, MT. Montana Coll. Agric. Sci. Stud., Bot. See B–P–H 619/2. HI 56876

Montana outdoors. Helena, MT. Vol. 1+, 1970+. Montana Outdoors. HI 68151

Montana state college. Agricultural experiment station. Department information circular. Bozeman, MT. Montana Agric. Exp. Sta. Dept. Inform. Circ. See B–P–H 618/14. HI 56871

Montana state college. Agricultural experiment station. Miscellaneous publication. Bozeman, MT. Montana Agric. Exp. Sta. Misc. Publ. See B–P–H 618/15. HI 56872

Montana state college. Agricultural experiment station. Research progress report. Bozeman, MT. Montana Agric. Exp. Sta. Res. Progr. Rep. See B–P–H 618/16. HI 56873

Montana state college. Agricultural experiment station. Special report. Bozeman, MT. Montana Agric. Exp. Sta. Special Rep. See B–P–H 618/18. HI 56875

Montana wild life. Helena, MT. Montana Wild Life. See B–P–H 619/11. HI 56877

Montanaro d'Italia, monti e boschi. Bologna. Vols. 28-32, 1977-81. Montan. Italia Monti Boschi. Preceded and superseded by: Montanaro d'Italia [not entered] and Monti e boschi. HI 68152

Montemar. Valparaiso. Vol. 11, 1961-64. Montemar. Preceded and superseded by: Revista de biologia marina. HI 68153

Montes. Madrid. Montes. See B–P–H 619/13. HI 56878

Month; a literary and critical journal. Sydney. Vols. 1-3, 1857-58; "Complete" edn., 1860. Month. 3-2743-3. HI 68154

Monthly agricultural bulletin, Palestine. Jerusalem. 1939-41. Monthly Agric. Bull. Palestine. Preceded by: Palestine gazette. Agricultural supplement. HI 68155

Monthly american journal of geology and natural science. Philadelphia, PA. Monthly Amer. J. Geol. Nat. Sci. See B–P–H 619/14. HI 56879

Monthly anthology and Boston review; containing sketches and reports of philosophy, religion, history, arts and manners. Boston, MA. Vols. 1-10, 1804-11. Monthly Anthol. HI 52510.

Monthly archives of the medical sciences. London. Monthly Arch. Med. Sci. See B–P–H 619/15. HI 56880

Monthly bulletin of agricultural intelligence and of plant diseases. Rome. Vols. 4-6, 1913-15. Monthly Bull. Agric. Intelligence Pl. Dis. Preceded by: Bulletin of the bureau of agricultural intelligence and of plant diseases. Superseded by: International review of the science and practice of agriculture. HI 68156

Monthly bulletin of agricultural science and practice. Rome. Vols. 20+, 1929+. Monthly Bull. Agric. Sci. Pract. Preceded by: International review of agriculture. HI 68157

Monthly bulletin, alpine garden club of British Columbia. Vancouver, B.C. Vols. ?-21-?, ?-1978-? Monthly Bull. Alpine Gard. Club British Columbia. Superseded by: Bulletin, alpine garden club of British Columbia. HI 68158

Monthly bulletin of the Bermuda department of agriculture (and fisheries). St. George's, Bermuda. Vol. 34+, 1964+. Monthly Bull. Bermuda Dept. Agric. (Fish). Preceded by: Agricultural bulletin, Bermuda department of agriculture. HI 68159

Monthly bulletin, California commission of horticulture. Sacramento, CA. Vols. 1-8(7), 1911-19. Monthly Bull. Calif. Commiss. Hort. Superseded by: Monthly bulletin, California department of agriculture. HI 68160

Monthly bulletin, California department of agriculture. Sacramento, CA. Vols. 8(8)-24(2), 1919-35. Monthly Bull. Calif. Dept. Agric. Preceded by: Monthly bulletin, California commission of horticulture. Superseded by: Bulletin of the California department of agriculture. HI 68161

Monthly bulletin, coffee board of Kenya. [Dates of publication not ascertained.] Monthly Bull. Coffee Board Kenya. HI 68162

Monthly bulletin, department of agriculture and fisheries, Bermuda = Monthly bulletin of the Bermuda department of agriculture (and fisheries). St. George's, Bermuda.

Monthly bulletin of the department of agriculture, state of California = Monthly bulletin, California department of agriculture. Sacramento, CA.

Monthly bulletin, Florida state plant board. Gainesville, FL. Vols. 12-15?, 1927-32. Monthly Bull. Florida State Pl. Board. Preceded by: Quarterly bulletin of the Florida state plant board. HI 68163

Monthly bulletin of the forest products research society. Madison, WI. Monthly Bull. Forest Prod. Res. Soc. See B–P–H 619/16. HI 56881

Monthly bulletin of the horticultural society of New York. New York. Vols. 1-21, 1930-50. Monthly Bull. Hort. Soc. New York. Superseded by: Gardening library news. 3-1887-3. HI 68164

Monthly bulletin, malayan orchid society. Nos. 12-10/67, 1957-67. Monthly Bull. Malayan Orchid Soc. Superseded by: Bulletin of the orchid society of south east Asia. HI 68165

Monthly bulletin, Michigan department of conservation. Lansing, MI. Vols. 1-4, 1931-? Monthly Bull. Michigan Dept. Conservation. Superseded by: Michigan conservation. 3-2642-2. HI 68166

Monthly bulletin, museums association. London. 1961-74. Monthly Bull. Mus. Assoc. Superseded by: Museums bulletin. HI 68167

Monthly bulletin, Ohio florists association = Ohio florists association, monthly bulletin. Columbus, OH. Ohio Florists Assoc. Monthly Bull. See B–P–H 685/15.

Monthly bulletin, Penrith and district natural history society. Penrith? 1951-53. Monthly Bull. Penrith Distr. Nat. HIst. Soc. Preceded by: News letter, Penrith and district natural history society. Superseded by: Field naturalist. HI 68168

Monthly critical gazette. London. Monthly Crit. Gaz. See B–P–H 619/18. HI 56882

Monthly current awareness service on pure and applied scientific articles. Series B, foreign edition. Ankara. Sample issue, March 1974; vol. 1, 1974. Monthly Curr. Aware. Serv. Pure Appl. Sci. Articles. Superseded by: Monthly current awareness service on scientific publications. Series B, foreign edition. HI 68169

Monthly current awareness service on scientific publications. Series B, foreign edition. Ankara. Vol. 2, 1975. Monthly Curr. Aware. Serv. Sci. Publ., B. Preceded by: Monthly current awareness service on pure and applied scientific articles. Series B, foreign edition. Superseded by: Current titles in turkish science. HI 68170

Monthly flora, or botanical magazine. New York. Vol. 1, 1846. Monthly Fl. Bot. Mag. 3-2746-1. HI 68171

Monthly information on nature protection in Poland = Chrońmy przyrode ojczysta. Cracow. Chrońmy Przyr. Ojczysta. See B–P–H 310/27.

Monthly microscopical journal: transactions of the royal microscopical society, and record of histological research at home and abroad. London. Monthly Microscop. J. See B–P–H 619/21. HI 56883

Monthly miscellany; or memoirs for the curious. London. Monthly Misc. See B–P–H 619/22. HI 56884

Monthly news, american association of botanical gardens

and arboretums. Vol. 1+, 1975+. Monthly News Amer. Assoc. Bot. Gard. Arbor. HI 68172

Monthly news letter, Louisiana society for horticultural research. Lafayette, LA. Vols. 1-8(2), 1960-67. Monthly News Lett. Louisiana Soc. Hort. Res. Superseded by: L S H R news letter. HI 68173

Monthly notes on cacti and other succulents. Worthing. Nos. 1-24, 1948-49. Monthly Notes Cacti Other Succ. Superseded by: Monthly notes on "the exotic collection" of cacti and other succulents. HI 68174

Monthly notes on "the collection" of cacti and other succulents = Monthly notes on "the exotic collection" of cacti and other succulents. Worthing.

Monthly notes on "the exotic collection" = Monthly notes on "the exotic collection" of cacti and other succulents. Worthing.

Monthly notes on "the exotic collection" of cacti and other succulents. Worthing. 1950+. Monthly Notes "Exot. Collect." Cact. Other Succ. Preceded by: Monthly notes on cacti and other succulents. HI 68175

Monthly notes, royal botanic garden, Edinburgh. Edinburgh. 1958-65. Monthly Notes Roy. Bot. Gard. Edinburgh Superseded by: Garden notes. Edinburgh. HI 68176

Monthly report of the department of agriculture. Washington, DC. Monthly Rep. Dept. Agric. See B–P–H 619/23. HI 56885

Monthly review. London. Monthly Rev. See B–P–H 619/24. HI 56886

Monthly review of dental surgery. London. Monthly Rev. Dental Surg. See B–P–H 619/25. HI 56887

Monthly summary of fungus and allied diseases. Harpenden. 1949+. Monthly Summ. Fungus Allied Dis. HI 68177

Monti e boschi; revista mensile di tecnica agraria. Milan, Bologna. Vols. 1-27, 1950-76; vol. 33+, 1982+. Monti & Boschi. For vols. 28-32 see: Montanaro d'Italia, monti e boschi. HI 68178

Monticello farmer and grape grower. Charlottesville, VA. Monticello Farmer Grape Grower. See B–P–H 620/2. HI 56888

Montpellier médical. Journal mensuel de médecine. Montpellier. Montpellier Méd. See B–P–H 620/3. HI 56889

Montreal botanic gardens, test garden results. Montreal. Montreal Bot. Gard. Test Gard. Results. See B–P–H 620/4. HI 56890

Montreal medical journal. Montreal. Montreal Med. J. See B–P–H 620/5. HI 56891

Monumenta nipponica. [Nihon bunka shiso.] Studies on japanese culture, past and present. Tokyo. Vol. 1+, 1938+. Monum. Nippon. 3-2756-2. HI 68179

Monumenta pharmaceutica. Amsterdam. Nos. 1-5, 1914-15. Monum. Pharm. 3-2756-2. HI 68180

Monumenta serica; journal of oriental studies of the Catholic university of Peking. Peiping [=Peking]. Monum. Ser. See B–P–H 620/6. HI 56892

Moorea; journal of the irish garden plant society. Dublin. Vol. 1+, 1982+. Moorea. HI 68181

Moosholzmännchen, heimatkundliches Beiblatt des lutterschen Stadtbüttels. Königslutter. Nos. 1-?, 1962-? Moosholzmänn. Heimatk. Beibl. Lutterschen Stadtbütt. HI 68182

Móra Ferenc múzeum évkönyve. Szeged, Hungary. Móra Ferenc Múz. Évk. See B–P–H 620/7. HI 56893

Morbi plantarum. Leningrad = Bolezni rastenii. Leningrad.

Morfologiya i anatomiya rastenii = Trudy botanicheskogo instituta imeni V. L. Komarova akademii nauk S S S R. Ser. 7, morfologiya i anatomiya rastenii. Moscow & Leningrad.

Morgagni. Naples. Morgagni. See B–P–H 620/12. HI 56894

Morgenblatt für gebildete Leser. Tubingen. Vols. ?-59, 1838?-65. Morgenbl. Gebildete Leser. Preceded by: Morgenblatt für gebildete Stände. 4-2761-1. HI 68183

Morgenblatt für gebildete Stände. Stuttgart, Tubingen. Vols. 1-?, 1807-37. Morgenbl. Gebildete Stände. Superseded by: Morgenblatt für gebildete Leser. 4-2761-1. HI 52481

Mori kinjin kenkyujo hokoku = Report of the Mori mycological institute. Kiryu.

Morioka kono dosokai gakujutsu iho = Bulletin of the scientific researches of the alumni association of the Morioka college of agriculture and forestry. Morioka, Japan. Bull. Sci. Res. Alumni Assoc. Morioka Coll. Agric. See B–P–H 272/12.

Morioka koto norin gakko gajukutsu hokoku = Bulletin of the imperial college of agriculture and forestry. Morioka, Japan. Bull. Imp. Coll. Agric. See B–P–H 254/17.

Morioka tabako shikenjo hokoku = Bulletin of the Morioka tobacco experiment station. Morioka, Japan. Bull. Morioka Tobacco Exp. Sta. See B–P–H 264/4.

Morisia. Sassari. Vol. 1+, 1967+. Morisia. HI 68184

Morphologische Arbeiten. Jena. Morphol. Arbeiten. See B–P–H 620/17. HI 56895

Morris arboretum annual report = Report (Annual), Morris arboretum. Philadelphia, PA.

Morris arboretum bulletin. Philadelphia, PA. Vols. 4-26, 1942-75. Morris Arbor. Bull. Preceded by:

Arboretum bulletin of the association [Morris arboretum]. Superseded by: Morris arboretum bulletin. Annual edition. 4-3294-3. HI 68185

Morris arboretum bulletin: Annual edition. Philadelphia, PA. Vol. 1+, 1976+. Morris Arbor., Annual Ed. Preceded by: Morris arboretum bulletin. HI 68186

Morris arboretum newsletter. Philadelphia, PA. Vol. 1+, 1972+. Morris Arbor. Newslett. HI 68187

Morton arboretum quarterly. Lisle, IL. Morton Arbor. Quart. See B–P–H 620/20. HI 56896

Moscosoa; contribuciones cientificas del jardin botanico nacional "Dr. Raphael M. Moscosa". Santo Domingo. Vol. 1+, 1976+. Moscosoa. HI 68188

Moscow university biological sciences bulletin. [Translation of: Vestnik moskovskogo universiteta. Seriya 6, biologiya, pochvovedenie.] New York. Vol. 29+, 1974+. Moscow Univ. Biol. Sci. Bull. HI 68189

Moskovskija Universitetskija Izvěstija = Moskovskiya Universitetskiya Izvěstiya. Moscow.

Moskovskiya Universitetskiya Izvěstiya. Moscow. 1865-72. Moskovsk. Univ. Izv. Superseded by: Protokoly Zasědanii Sověta Imperatorskago Moskovskago Universiteta. 3-2766-3. HI 68190

Moss Landing marine laboratories news. Moss Landing, CA. Vol. 1+, 1972+. Moss Landing Mar. Lab. News. HI 68191

Mountain echo. Manila. Vols. 1-2, 1919-20. Mountain Echo. Superseded by: Makiling echo. 3-2776-3. HI 68192

Mountain research. Chengdu. Vol. 1+, 1983+. Mountain Res. HI 68193

Mountain research and development. Boulder, CO. Vol. 1+, 1981+. Mountain Res. Developm. HI 68194

Mountaineer. Seattle, WA. Vol. 1+, 1907+. Mountaineer. 3-2777-1. HI 68195

Mouseion; bulletin de l'office international des musées. Paris. Nos. 1-58, 1927-46 [suspended 1941-44]. Mouseion. Superseded by: Museum. 3-2777-2. HI 68196

M.P.G. news. Presque Isle, ME = M P G news. Presque Isle, ME.

Mudpie; museum and university data program and information exchange. Washington, DC. 1967-72. Mudpie. HI 68197

Muelleria; an australian journal of botany. Melbourne. Muelleria. See B–P–H 621/2. HI 56897

Muhlenbergia; a journal of botany. Lancaster, PA & Los Gatos, CA. Vols. 1-9, 1900-15. Muhlenbergia. Preceded by: Contributions from the herbarium of Franklin and Marshall college. 3-2782-2. HI 68198

Mundo agrícola. São Paulo. Mundo Agríc. See B–P–H 621/8. HI 56898

Munibe; suplemento de ciencias naturales del Boletín de la real sociedad vascongada de los amigos del país. San Sebastián. Vol. 1+, 1949+. Munibe. 3-2784-2. HI 68199

Muntjac. Bedford. No. 31+, 1978+. Muntjac. Preceded by: Newsletter, Bedfordshire natural history society. HI 68200

Murhak yon'gu saengmurkak yonguhoe. Mulligwataehak Soul Taehakkyo = Bulletin of the biological institute, department of biology, college of liberal arts and sciences; Seoul (national) university. Seoul, Korea. Bull. Biol. Inst. Seoul Natl. Univ. See B–P–H 241/27.

Muscillanea. Belsele. No. 1+, 1981+. Muscillanea. HI 68201

Musealblatt. Zeitschrift für Geschichte, Kunst, Natur und Technologie Oesterreichs ob der Enns und Salzburgs. Linz. Musealblatt. See B–P–H 623/2. HI 56923

Musée botanique de Leide. Leiden. Mus. Bot. Leide. See B–P–H 621/14. HI 56899

Musée Heude. Notes de botanique chinoise. Shanghai. Mus. Heude Notes Bot. Chin. See B–P–H 621/21. HI 56904

Musée des sciences. Paris. Mus. Sci. (Paris). See B–P–H 622/13. HI 56917

Musées de Genève; revue mensuelle des musées et des collections de la ville de Genève. Geneva. Vols. 1-16, 1944-59; n.s. no. 1+, 1960+. Mus. Genève. HI 68202

Musées scientifiques. Dijon. Nos. 1-47, 1933-37. Mus. Sci. HI 68203

Musejní a vlastivědné práce. Prague. Mus. Vlastiv. Práce. See B–P–H 622/20. HI 56921

Musejní zprávy pražského kraje. Podebrady, Czechoslovakia. Mus. Zprávy Pražského Kraje. See B–P–H 623/1. HI 56922

Museo bresciano illustrato. Brescia, Italy. Mus. Bresciano Ill. See B–P–H 621/16. HI 56900

Museo canario, El. Las Palmas. Vol. [1]+, 1934+. Mus. Canario. 3-2791-3. HI 68204

Museo scientifico, letterario ed artistico; ovvero scelta raccolta di utili e svariate nozioni in fatto di scienze, lettere ed arti belle. Turin. Mus. Sci. (Turin). See B–P–H 622/14. HI 56918

Museum; a quarterly review. Paris. Vol. 1+, 1948+. Museum. Preceded by: Mouseion. 3-2792-3. HI 68205

Museum bulletin, see also Museums bulletin

Museum bulletin, department of lands and forests,

Ontario. Toronto. Nos. 1-?, 1962-65. Mus. Bull. Dept. Lands Forests Ontario. HI 68206

Museum bulletin, Santa Barbara museum of natural history. Santa Barbara, CA. No. 1+, 1976+. Mus. Bull. Santa Barbara Mus. Nat. Hist. HI 68207

Museum bulletin, South Carolina museum commission. Columbia, SC. 1976+. Mus. Bull. South Carolina Mus Commiss. HI 68208

Museum bulletin of the Staten Island association of arts and sciences. New Brighton, NY. Mus. Bull. Staten Island Assoc. Arts. See B–P–H 621/17. HI 56901

Museum bulletin of the Staten Island institute of arts and science. New Brighton, NY. Mus. Bull. Staten Island Inst. Arts. See B–P–H 621/18. HI 56902

Museum handbooks, Manchester museum. [Some are numbered as: Publications, Manchester museum.] Manchester. 1891-1933. Mus. Handb. Manchester Mus. HI 68209

Museum der Heilkunde. Zurich. Mus. Heilk. See B–P–H 621/20. HI 56903

Museum für Kulturgeschichte. Abhandlungen und Berichte für Naturkunde und Vorgeschichte. Magdeburg. Mus. Kulturgesch. Abh. Ber. Naturk. Vorgesch. See B–P–H 621/23. HI 56905

Museum für Kulturgeschichte Magdeburg. Abhandlungen und Berichte für Naturkunde und Vorgeschichte. Leipzig. Mus. Kulturgesch. Magdeburg Abh. Ber. Naturk. Vorgesch. See B–P–H 621/24. HI 56906

Museum and library bulletin of the Boston society of natural history. Boston, MA. Mus. Libr. Bull. Boston Soc. Nat. Hist. See B–P–H 621/25. HI 56907

Museum mego. Saint John. 1969+. Mus. Mego. HI 68210

Museum memoir, national museums and monuments of Rhodesia = Museum memoir, national museums of Rhodesia. Salisbury, Rhodesia.

Museum memoir, national museums of Rhodesia. Salisbury, Rhodesia. Vols. ?-4-9, ?-1971-79. Mus. Mem. Natl. Mus. Rhodesia. Superseded by: Museum memoir, national museums and monuments, Zimbabwe. HI 68211

Museum memoir, national museums and monuments, Zimbabwe. Salisbury, Rhodesia. Vol. 10+, 197?+. Mus. Mem. Natl. Mus. Monum. Zimbabwe. Preceded by: Museum memoir, national museums of Rhodesia. HI 68212

Museum memoir, trustees of the national museums and monuments of Rhodesia = Museum memoir, national museums of Rhodesia. Salisbury, Rhodesia.

Museum nationale hungaricum. Buda [=Budapest, in part]. Mus. Natl. Hung. See B–P–H 622/1. HI 56908

Museum der Naturgeschichte Helvetiens. Bern. Mus. Naturgesch. Helv. See B–P–H 622/4. HI 56910

Museum für Naturkunde und Heimatkunde zu Magdeburg. Abhandlungen und Berichte. Magdeburg. Mus. Natur- Heimatk. Magdeburg Abh. Ber. See B–P–H 622/2. HI 56909

Museum der neuesten und interessantesten Reisebeschreibungen für gebildete Leser. Vienna. Mus. Neuesten Interessantesten Reisebeschrieb. See B–P–H 622/6. HI 56911

Museum des Neuesten und Wissenswürdigsten aus dem Gebiete der Naturwissenschaft, der Künste, der Fabriken, der Manufakturen, der technischen Gewerbe, der Landwirthschaft, der Produkten-, Waaren- und Handelskunde, und der bürgerlichen Haushaltung; für gebildete Leser und Leserinnen aus allen Ständen. Berlin. Mus Neuesten Wissenswürd. Naturwiss. See B–P–H 622/7. HI 56912

Museum news. Washington, DC. Vol. 1+, 1924+. Mus. News. HI 68213

Museum noricum, oder Sammlung auserlesener kleiner Schriften, Abhandlungen und Nachrichten aus allen Theilen der Gelahrtheit. Altdorf, Germany. Mus. Noricum. See B–P–H 622/8. HI 56913

Museum notes. [Museum of northern Arizona.] Flagstaff, AZ. Mus. Notes N. Arizona. See B–P–H 622/9. HI 56914

Museum notes and views, Idaho state college. Pocatello, ID. Vol. 1(1-4), 1957-58. Tebiwa. Superseded by: Tebiwa. HI 68214

Museum publications, Manchester museum. Manchester. N.s. vol. 1+, 1941+. Mus. Publ. Manchester Mus. HI 68215

Museum publications, national museum of the Philippines. Manila. Vol. 1+, 1967+. Mus. Publ. Natl. Mus. Philippines. HI 68216

Museum rusticum et commerciale oder auserlesene Schriften, den Ackerbau, die Handlung, die Künste und Manufacturen betreffend. Leipzig. Mus. Rusticum Commerciale (Leipzig). See B–P–H 622/11. HI 56915

Museum rusticum et commerciale: or, select papers on agriculture, commerce, arts and manufactures. London. Mus. Rusticum Commerciale (London). See B–P–H 622/12. HI 56916

Museum of science newsletter. Boston, MA. ?-1985+. Mus. Sci. Newslett. HI 68217

Museum Senckenbergianum. Abhandlungen aus dem Gebiete der beschreibenden Naturgeschichte. Von Mitgliedern der Senckenbergischen naturforschenden Gesellschaft in Frankfurt am Main. Frankfurt a. M.

Mus. Senckenberg. See B–P–H 622/16. HI 56919

Museum für topographische Vaterlandskunde des Österreichischen Kaiserstaates. Vienna. Mus. Topogr. Vaterlandsk. Österr. Kaiserstaates. See B–P–H 622/19. HI 56920

Museums bulletin. London. Vol. 14+, 1974+. Museums Bull. Preceded by: Monthly bulletin, museums association. HI 68218

Museums journal. London. Vol. 1+, 1901+. Museums J. HI 68219

Museumsnytt. Oslo. 1951+. Museumsnytt. HI 68220

Mushroom; journal of wild mushrooming. Moscow, ID. Vol. 1+, 1983+. Mushroom. HI 68221

Mushroom culture. Pensacola, FL. Vol. 1+, 1984+. Mushroom Cult. HI 68222

Mushroom journal. London. No. 1+, 1973+. Mushroom J. Preceded by: M G A bulletin. HI 68223

Mushroom journal for the tropics. Shatin, Hong Kong. Vol. 7+, 1987+. Mushroom J. Trop. Preceded by: Mushroom newsletter for the tropics. HI 68224

Mushroom news. Worthing. Vols. 1-10, 1934-66. Mushroom News (Worthing). HI 68225

Mushroom news (Kennett Square). Kennett Square, PA. Vol. 1+, 1955+. Mushroom News (Kennett Square). HI 68226

Mushroom newsletter for the tropics. Bangkok, Shatin, Hong Hong. Vols. 1-6(4), 1981-86. Mushroom Newslett. Trop. Superseded by: Mushroom journal for the tropics. HI 68227

Mushroom science; proceedings, international congress on mushroom science [subtitle varies]. London, etc. 1950+. Mushroom Sci. HI 68228

Musk-ox. Saskatoon, Saskatchewan. No. 1+, 1967+ [suspended 1983-86]. Musk-ox. HI 68229

Mutagenez rastenii. Erevan. Vol. 1, 1971. Mutagen. Rast. Superseded by: Radiochuvstvitel′nost′ i mutabil′nost′ rastenii. HI 68230

Mutation breeding newsletter. Vienna. Vol. 1+, 1972+. Mutat. Breed. Newslett. HI 68231

Mutation research. Amsterdam. Mutat. Res. See B–P–H 623/5. HI 56924

Mutisia; acta botanica colombiana. Bogotá. Mutisia. See B–P–H 623/6. HI 56925

Múzeumi füzetek. Kolozsvar [= Cluj, Rumania]. Vols. 1-4, 1906-09; n.s. vols. 1-3, 1943-45. Múz. Füz. Preceded by: Ertesitö az erdélyi muzeum-egylet orvos-természet tudomanyi szakosztályából. 2-1469-3. HI 68233

Mycelium; newsletter, mycological society of Toronto. Toronto. Vol. ?-7(5)+, ?-1981+. Mycellium. HI 68234

Mycena news. San Francisco, CA. Vol. 35+, 1985+. Mycena News. HI 68235

Mycologia. Lancaster, PA. Mycologia. See B–P–H 623/22. HI 56931

Mycologia čechica = Česká mykologie. Prague. Česká Mykol. See B–P–H 304/19.

Mycologia helvetica. Berne. Vol. 1+, 1983+. Mycol. Helv. HI 68236

Mycologia memoirs. New York, Vaduz, Stuttgart. No. 1+, 1965+. Mycol. Mem. HI 68237

Mycological bulletin. [Ohio university.] Columbus, OH. Mycol. Bull. See B–P–H 623/10. HI 56926

Mycological leaflet. Morogoro, Tanganyika [Tanzania Republic]. Mycol. Leafl. See B–P–H 623/16. HI 56928

Mycological notes of the Lloyd library and museum. Cincinnati, OH. 1898-1925. Mycol. Notes Lloyd Libr. Mus. HI 68238

Mycological papers. Commonwealth mycological institute. Kew, England. Mycol. Pap. See B–P–H 623/17. HI 56929

Mycological research. Cambridge. Vol. 92+, 1989+. Mycol. Res. Preceded by: Transactions of the british mycological society. HI 68586

Mycological society of America, news letter. Ithaca, NY, Gainesville, FL. Vol. 1+, 1950+. Mycol. Soc. Amer. News Lett. Preceded, 1903-08, by: News letter of the mycological society of America. HI 68239

Mycological society of America. Year book. Ithaca, NY. Mycol. Soc. Amer. Year Book. See B–P–H 623/21. HI 56930

Mycologisches Centralblatt. Jena. Mycol. Centralbl. See B–P–H 623/13. HI 56927

Mycologist; bulletin of the british mycological society. Cambridge. Vol. 1+, 1987+. Mycologist. Preceded by: Bulletin of the British Mycological Society. HI 68240

Mycopathologia. The Hague. Vols. 1-4, 1938-49; vol. 55+, 1975+. Mycopathologia. For vols. 5-54, 1949-74 see: Mycopathologia et mycologia applicata. 3-2808-1. HI 68241

Mycopathologia et mycologia applicata. The Hague. Vols. 5-54, 1949-74; Supplementum 1959. Mycopathol. Mycol. Appl. Preceded and superseded by: Mycopathologia. 3-2808-1. HI 68242

Mycophile. Montreal. Vol. 1+, 1978+. Mycophile (Montreal). HI 68243

Mycophile. Portsmouth, OH. 1960-? Mycophile (Portsmouth). Superseded by: McIlvainea. HI 68244

Mycotaxon; international journal designed to expedite publication on research, on taxonomy, and on nomenclature of fungi and lichens. Ithaca, NY. Vol. 1+, 1974+. Mycotaxon. HI 68245

Mycovirus newsletter. Pittsburgh, PA. No. 1+, 1974+. Mycovirus Newslett. HI 68246

Myforest; quarterly journal of the Karnataka forest department. Bangalore. 1964+. Myforest. HI 68247

Mykologia. Prague. Mykologia. See B–P–H 624/9. HI 56938

Mykologický sborník. Časopis československých houbařů. Prague. 1935+. Mykol. Sborn. Preceded by: Časopis československých houbařů. HI 68248

Mykologický zpravodaj. Brno, Czechoslovakia. Mykol. Zprav. See B–P–H 624/8. HI 56937

Mykologische Berichte. Giessen, Germany. Mykol. Ber. See B–P–H 624/1. HI 56932

Mykologische Hefte, nebst einem allgemein-botanischen Anzeiger. Leipzig. Mykol. Hefte. See B–P–H 624/3. HI 56933

Mykologische mededelingen. Antwerp. No. 1+, 1968+. Mykol. Meded. HI 68249

Mykologische Schriftenreihe. Leipzig. 1963+. Mykol. Schriftenreihe. HI 68250

Mykologische Untersuchungen. Jena. Mykol. Untersuch. See B–P–H 624/7. HI 56936

Mykologisches Mitteilungsblatt. Halle. Mykol. Mitteilungsbl. See B–P–H 624/4. HI 56934

Mykosen; organ für experimentelle und klinische Mykologie. Berlin. Vol. 1+, 1957+. Mykosen. HI 68251

Mysore journal of agricultural sciences. Bangalore. Vol. 1+, 1967+. Mysore J. Agric. Sci. HI 68252

N A A S quarterly review. London. Nos. 1-97, 1948-71. N. A. A. S. Quart. Rev. Superseded by: A D A S quarterly review. HI 68253

N A - F B / P. Broomall, PA. No. 1+, 1980+. N. A. - F. B./P. HI 68254

N B G newsletter. Lucknow. Vols. 1-5, 1974-78. N. B. G. Newslett. Superseded by: N B R I newsletter. HI 68255

N B naturalist. St. John, N. B. Vol. 1+, 1970+. N. B. Naturalist. HI 68256

N B R I newsletter; quarterly house journal. Lucknow. Vol. 6+ 1979+. N. B. R. I. Newslett. Preceded by: N B G newsletter. HI 68257

N C T R H newsletter. Alexandria, VA, Gaithersburg, MD. Vols. 1-13, 1974-86? N. C. T. R. H. Newslett. Superseded by: A H T A newsletter. HI 68258

N E - I N F. Upper Darby, PA. No. ?-24+, ?-1976+. N E - I N F. HI 68671

N E R C news. Swindon. No. 1+, 1987+. N. E. R. C. News. HI 68259

N E R C news journal. London. No. 1+, 1970+. N. E. R. C. News J. HI 68260

N F T A news. Waimanolo, HI. Vol. 1+, 1985+. N. F. T. A. News. HI 68261

N F T R R = Nitrogen fixing tree research reports. Bangkok.

N & G C; journal of the horticultural trades' association = Nurseryman and garden centre. London.

N I B S bulletin of biological research. [Nihon seibutsu kagaku kenkyujo hokoku.] Tokyo. Vol. 1+, 1956+. N. I. B. S. Bull. Biol. Res. HI 68262

N J Audubon = New Jersey Audubon bulletin. Franklin Lakes, NJ.

N J M A news. Princeton, NJ. ?-1985+. N. J. M. A. News. HI 68263

N. Kec'xovelis sax. botanikis institutis šromebi. Tiflis. Vol. 30+, 1985+. N. Kec'xovelis Sax. Bot. Inst. Šromebi. Superseded by: Botanikis institutis šromebi. HI 75115

N L L announcement bulletin. Boston Spa. 1971-73. N. L. L. Announc. Bull. Superseded by: B L L announcement bulletin. HI 68264

N N N P S newsletter = Newsletter, northern Nevada native plant society. Reno, NV.

N P G S, national plant germplasm system, special report. Fort Collins, CO. No. 1+, 1982+. N. P. G. S. Natl. Pl. Germplasm Syst. Special Rep. HI 68265

N P S O newsletter. Ashland, OR. ?-1972-? N. P. S. O. Newslett. Superseded by: Native plant society of Oregon [series]. HI 68266

N R C research news. Ottawa. Vols. 1-22(1), 1948-69. N. R. C. Res. News. Superseded by: Science dimension. HI 53338

N R C P research bulletin = Research bulletin, national research council of the Philippines. Quezon City, Rizal.

N R D C newsletter. New York. 1971+. N. R. D. C. Newslett. HI 68267

N R D C newsline = N R D C newsletter. New York.

N S F bulletin = Bulletin, national science foundation. Washington, DC.

N T No. 1+, 1973+. N. T. HI 68268

N T U phytopathologist and entomologist. Taipei. No. 1+, 1971+. N. T. U. Phytopatholist Entomologist. HI 68269

N Z O I records. Wellington, N.Z. Vol. 1+, 1972+. N.

Z. O. I. Rec. HI 68270

Na okika o Hawaii. Honolulu, HI = Hawaii orchid journal. Honolulu, HI.

Näbatat institutunun äsärläri. Baku, Azerbaijan S S R = Trudy botanicheskogo instituta. Baku.

Nabljudenija nad Fitoplanktonum Volgi = Nablyudeniya nad Fitoplanktonum Volgi. Saratov.

Nablyudeniya nad Fitoplanktonum Volgi. Saratov, Russian S F S R. 1903. Nablyud. Fitoplankt. Volgi. Preceded by: Materialy dlya izucheniya fitoplanktona Volgi po nablyudeniyam. HI 68271

Nachricht von dem Fortgange der naturforschenden Gesellschaft zu Jena. Jena. Nachr. Fortgange Naturf. Ges. Jena. See B–P–H 625/34. HI 56948

Nachricht von dem Fortgange der Westphälischen naturforschenden Gesellschaft zu Brockhausen. Düsseldorf. Nachr. Fortgange Westphäl. Naturf. Ges. Brockhausen. See B–P–H 625/35. HI 56949

Nachrichten der Akademie der Wissenschaften in Göttingen. Mathematisch-Physikalische Klasse. Göttingen. Nachr. Akad. Wiss. Göttingen, Math.-Phys. Kl. See B–P–H 625/30. HI 56944

Nachrichten der Akademie der Wissenschaften in Göttingen. Mathematisch-Physikalische Klasse. Biologisch-Physiologisch-Chemische Abteilung. Göttingen. Nachr. Akad. Wiss. Göttingen, Math.-Phys. Kl., Biol.-Physiol.-Chem. Abt. See B–P–H 625/31. HI 56945

Nachrichten aus dem Blumen-Reiche. Eine Quartalschrift. Dessau & Leipzig. Nachr. Blumen-Reiche. See B–P–H 625/33. HI 56947

Nachrichten, Deutsche Gesellschaft für Natur- und Völkerkunde Ostasiens = Deutsche Gesellschaft für Natur- und Völkerkunde Ostasiens, Tokio. Nachrichten. Tokyo. Deutsche Ges. Natur- Völkerk. Ostasiens Tokio Nachr. See B–P–H 346/9.

Nachrichten zur floristischen Kartierung. Graz, Vienna. No. 1+, 1968+. Nachr. Florist. Kart. HI 68272

Nachrichten von gelehrten Sachen, herausgegeben von der Akademie zu Erfurt. Erfurt. Nachr. Gel. Sachen. See B–P–H 625/36. HI 56950

Nachrichten von der Georg-Augusts-Universität und der Königl. Gesellschaft der Wissenschaften zu Göttingen. Göttingen. Nachr. Georg-Augusts-Univ. Königl. Ges. Wiss. Göttingen. See B–P–H 625/37. HI 56951

Nachrichten der Gesellschaft für Natur- und Völkerkunde Ostasiens, Hamburg. Wiesbaden. Nachr. Ges. Natur-Völkerk. Ostasiens Hamburg. See B–P–H 626/1. HI 56952

Nachrichten von der Gesellschaft der Wissenschaften zu Göttingen. Geschäftliche Mitteilungen. Berlin. Nachr. Ges. Wiss. Göttingen Geschäftl. Mitt. See B–P–H 626/3. HI 56953

Nachrichten von der Gesellschaft der Wissenschaften zu Göttingen. Jahresbericht. Berlin. Nachr. Ges. Wiss. Göttingen Jahresber. See B–P–H 626/4. HI 56954

Nachrichten von der Gesellschaft der Wissenschaften zu Göttingen. Mathematisch-physikalische Klasse. Berlin. Nachr. Ges. Wiss. Göttingen, Math.-Phys. Kl. See B–P–H 626/5. HI 56955

Nachrichten von der Gesellschaft der Wissenschaften zu Göttingen. Mathematisch-Physikalische Klasse. Fachgruppe 6. Nachrichten aus der Biologie. Göttingen. Nachr. Ges. Wiss. Göttingen, Math.-Phys. Kl., Fachgr. 6, Nachr. Biol. See B–P–H 626/6. HI 56956

Nachrichten der Giessener Hochschulgesellschaft. Giessen, Germany. Nachr. Giessener Hochschulges. See B–P–H 626/7. HI 56957

Nachrichten für und über Kaiser Wilhelms-Land und den Bismarck-Archipel. Herausgegeben im Auftrage der Neu Guinea Compagnie zu Berlin. Berlin. Nachr. Kaiser Wilhelms-Land Bismarck-Archipel. See B–P–H 626/9. HI 56958

Nachrichten von der Königlichen Gesellschaft der Wissenschaften und von der Georg-Augusts-Universität. Göttingen. Nachr. Königl. Ges. Wiss. Georg-Augusts-Univ. See B–P–H 626/10. HI 56959

Nachrichten von der Königlichen Gesellschaft der Wissenschaften zu Göttingen. Geschäftliche Mitteilungen. Berlin. Nachr. Königl. Ges. Wiss. Göttingen Geschäftl. Mitt. See B–P–H 626/12. HI 56960

Nachrichten von der Königlichen Gesellschaft der Wissenschaften zu Göttingen. Mathematisch-Physikalische Klasse. Göttingen. Nachr. Königl. Ges. Wiss. Göttingen, Math.-Phys. Kl. See B–P–H 626/13. HI 56961

Nachrichten über die Lehranstalt für Obst- und Gartenbau zu Proskau = Bericht der Lehranstalt für Obst- und Gartenbau zu Proskau. Oppeln.

Nachrichten, den Naturforschenden Verein zu Riga betreffend. Riga [Latvian S S R]. Nachr. Naturf. Verein Riga. See B–P–H 626/14. HI 56962

Nachrichten des naturwissenschaftlichen Museum der Stadt Aschaffenburg. Aschaffenburg. Vol. 30+, 1951+. Nachr Naturwiss. Mus. Stadt. Preceded by: Nachrichten der Sammelstelle für Schmarotzerbestimmung, naturwissenschaftliches Museum der Stadt Aschaffenburg. HI 68273

Nachrichten über den Naturwissenschaftlichen Verein in Schleiz. Schleiz, Germany. Nachr. Naturwiss. Verein Schleiz. See B–P–H 626/15. HI 56963

Nachrichten der Sammelstelle für Schmarotzerbestimmung, naturwissenschaftliches Museum der Stadt Aschaffenburg. Aschaffenburg. Vol. 29, 1950. Nachr. Sammelstelle Schmarotzerbestimm. Naturwiss. Mus Stadt Aschaffenburg Preceded by: Mitteilungen der Sammelstelle für Schmarotzerbestimmung des V. D. E. V. Superseded by: Nachrichten des naturwissenschaftlichen Museum der Stadt Aschaffenburg. HI 68274

Nachrichten der steierischen Kakteenfreunde. Knittelfeld. Vols. 1-15, 1965-79. Nachr. Steier. Kakteenfr. HI 68275

Nachrichten von dem Zustande der Wissenschaften und Künste in den Königlich Dänischen Reichen und Ländern. Copenhagen & Leipzig. Nachr. Wiss. Künste Königl. Dän. Reichen Ländern. See B–P–H 626/16. HI 56964

Nachrichtenblätt der Deutschen Iris-Gesellschaft = Nachrichtenblätter der Deutschen Iris-Gesellschaft. Eldagsen, Germany. Nacharichtenbl. Deutsch. Iris-Ges. See B–P–H 627/1.

Nachrichtenblatt, Deutsche Iris-Gesellschaft e. V. = Deutsche Iris-Gesellschaft e. V. Nachrichtenblatt. Leonberg, Germany. Deutsche Iris-Ges. Nachrichtenbl. See B–P–H 346/16.

Nachrichtenblatt, Deutsche Iris- und Liliengesellschaft e. V. = Deutsche Iris- und Lilien-Gesellschaft e. V. Nachrichtenblatt. Leonberg, Germany. Deutsche Iris-Lilien-Ges. Nachrichtenbl. See B–P–H 346/14.

Nachrichtenblatt der Deutschen Gesellschaft für Geschichte der Medizin, Naturwissenschaft und Technik. Leipzig. Nachrichtenbl. Deutsch. Ges. Gesch. Med. See B–P–H 626/17. HI 56965

Nachrichtenblatt der Deutschen Kakteengesellschaft, e. V. Nuremberg. Nachrichtenbl. Deutsch. Kakteenges. See B–P–H 627/2. HI 56967

Nachrichtenblatt für den deutschen Pflanzenschutzdienst. Berlin, etc. Vols. 1-23(3), 1921-43; n.s. vols. 1-24, 1947-70. Nachrichtenbl. Deutsch. Pflanzenschutzdienst. For 1943-45 see: Reichspflanzenschutzblatt. Superseded by: Nachrichtenblatt für den Pflanzenschutzdienst in der D D R. 1-806-2. HI 68277

Nachrichtenblatt der deutschen pflanzenschutzdienstes. Stuttgart. No. 1+, 1950+. Nachrichtenbl. Deutsch. Pflanzenschutzdienstes. HI 68278

Nachrichtenblatt der Deutschen Vereinigung für Geschichte der Medizin, Naturwissenschaft und Technik. Leipzig. Nachrichtenbl. Deutsch. Vereinigung Gesch. Med. See B–P–H 627/3. HI 56968

Nachrichtenblatt für Naturschutz und Landschaftspflege. [Supplement to: Natur und Landschaft.] Bad Godesberg. Vols. 1-47, 1929-76. Nachrichtenbl. Naturschutz Landschaftsp. HI 68279

Nachrichten-Blatt der Oberbergischen Arbeitsgemeinschaft für naturwissenschaftliche Heimatforschung. Gummersbach, Germany. Nachr.-Blatt Oberberg. Arbeitsgem. Naturwiss. Heimatf. See B–P–H 625/32. HI 56946

Nachrichtenblatt für den Pflanzenschutz in der D D R. Berlin. Vol. 28+, 1974+. Nachrichtenbl. Pflanzenschutz D.D.R. Superseded by: Nachrichtenblatt für den Pflanzenschutzdienst in der D D R. HI 68280

Nachrichtenblatt für den Pflanzenschutzdienst in der D D R. Berlin. Vols. 25-27, 1971-73. Nachrichtenbl. Pflanzenschutzdienst D.D.R. Preceded by: Nachrichtenblatt für den deutschen Pflanzenschutzdienst. Superseded by: Nachrichtenblatt für den Pflanzenschutz in der D D R. HI 68281

Nachrichtenblätter der Deutschen Iris-Gesellschaft. Eldagsen, Germany. Nacharichtenbl. Deutsch. Iris-Ges. See B–P–H 627/1. HI 56966

Nachrichtenblätter, Deutsche Iris-Gesellschaft e. V. = Deutsche Iris-Gesellschaft e. V. Berlin-Dahlem. Nachrichtenblätter. Berlin-Dahlem. Deutsche Iris-Ges. Berlin-Dahlem Nachrichtenbl. See B–P–H 346/15.

Nagano-ken engei shikenjo hokoku = Report of the Nagano horticultural experiment station. Suzaka.

Nagano-ken engei shikenjo shiken seiseki = Research report of the Nagano horticultural experiment station. Susaka.

Nagano-ken nogyo shikenjo shuho = Proceedings of the Nagano prefectural agricultural experiment station. Nagano, Japan. Proc. Nagano Prefect. Agric. Exp. Sta. See B–P–H 732/16.

Nagano-ken shokubutsu kenkyukai-shi = Bulletin of the botanical society of Nagano. Nagano, Honshu.

Nagano kenritsu norin semmon gakko gakujutsu hokoku = Bulletin of the Nagano college of agriculture and forestry. Ina.

Nagao kenkyusho kinrui kenkyu hokoku = Nagaoa. Mycological journal of Nagao institute. [Nagao kenkyusho kinrui kenkyu hokoku.] Tokyo. Nagaoa. See B–P–H 627/7.

Nagaoa. Mycological journal of Nagao institute. [Nagao kenkyusho kinrui kenkyu hokoku.] Tokyo. Nagaoa. See B–P–H 627/7. HI 56969

Nagasaki daigaku kyoikugakubu shizenkagaku kenkyu hokoku = Science bulletin of the faculty of liberal arts and education; Nagasaki university.

Nagasaki daigaku kyoyobu kiyo, shizenkagaku = Bulletin of the faculty of liberal arts, Nagasaki university.

Natural science. Nagasaki.

Nagasaki daigaku suisangakubu kenkyu hokoku = Bulletin of the faculty of fisheries, Nagasaki university. Nagasaki, Japan. Bull. Fac. Fish. Nagasaki Univ. See B–P–H 249/20.

Nagasaki-ken sogo norin shikenjo. Kenkyu hokoku: nogyo bumon = Bulletin of the Nagasaki agricultural and forestry experiment station. Section of agriculture. Ishihaya.

Nagasaki-ken sogo norin shikenjo. Nogyo bumon. Kenkyu hokoku = Bulletin of the Nagasaki agricultural and forestry experiment station. Isahaya.

Nagoya daigaku nogakubu enshurin hokoku = Bulletin of the Nagoya university forests. Anjo.

Nagoya hishiryo kensajo jigyo hokoku hiryo no bu = Report (Annual) of the Nagoya fertilizer and feed inspection station, fertilizer section. Aichi.

Nagoya seibutsugakkai kiroku = Records of the biological association of Nagoya. Nagoya, Japan. Rec. Biol. Assoc. Nagoya. See B–P–H 757/1.

Nagoya-shiritsu daigaku kyoyobu kiyo, shizenkagaku-hen = Bulletin, department of general education, Nagoya city university. Natural science section. Nagoya.

Nagpur agricultural college magazine. Nagpur. Vol. ?-43+, ?-1970/71+. Nagpur Agric. Coll. Mag. HI 68283

Nakhlat al-tamr = Date palm journal. Baghdad.

Namib bulletin; bulletin of the desert ecological research unit, Gobabeb. Pretoria. No. 1+, 1976+. Namib Bull. HI 68284

Namib und Meer. Swakopmund. Vol. 1+, 1970+. Namib & Meer. HI 68285

Nan-ching chungshan chih wu yuan yen chu lun wei chi = Bulletin of the Nanjing botanical garden, Mem. Sun Yat Sen [Sun Yat-sen tomb and memorial]. Nanjing.

Nan-ching lin chan kung yeh hsueh yuan hsueh pao = Journal of Nanjing technological college of forest products. Nanching (Nanjing).

Nan-ching lin hsüeh yuan hsüeh pao = Journal of Nanjing institute of forestry. Nanching (Nanjing).

Nan Ching nung hsüeh yüan hsüeh pao = Acta instituti agriculturae nankinensis. Nanking. Acta Inst. Agric. Nankin. See B–P–H 44/15.

Nan-ching nung yeh ta hsueh = Journal of Nanjing agricultural university. Nanching (Nanjing).

Nan ching ta hsüeh pao = Acta universitatis nankinensis scientiarum naturalium. Nanking. Acta Univ. Nankin. Sic. Nat. See B–P–H 50/6.

Nan-ching yao hsueh yuan hsueh pao = Journal of Nanjing college of pharmacy. Nanching (Nanjing).

Nanjing daxue xuebao. Zaolei zhuankan. 1982+. Nanjing Daxue Xuebao, Zaolei Zhuankan. HI 68287

Nanjing daxue xuebao. Ziran kexue = Journal of Nanjing university. Natural sciences edition. Nanching (Nanjing).

Nanjing linchan gongye xueyuan xuebao = Journal of Nanjing technological college of forest products. Nanching (Nanjing).

Nanjing linxueyuan xuebao = Journal of Nanjing institute of forestry. Nanching (Nanjing).

Nanjing yaoxueyuan xuebao = Journal of Nanjing college of pharmacy. Nanching (Nanjing).

Nanjing zhongshan zhiwuyuan yanjiu lunwenji = Bulletin of the Nanjing botanical garden, Mem. Sun Yat Sen [Sun Yat-sen tomb and memorial]. Nanjing.

Nankai daxue xuebao. Ziran kexue ban = Acta scientiarum naturalium universitatis Nankaiensis. Nanking.

Nanki seibutsu = Journal of the Nanki biological society. Wakayama.

Nanking journal. [Chin ling hsüeh pao.] Nanking. Nanking J. See B–P–H 627/17. HI 56970

Nankyoku shiryo = Antarctic record. Tokyo.

Nanpo-sango kagaku kenkyujo hokoku. Kagoshima daigaku = Memoirs of the south industrial science institute, Kagoshima university. Kagoshima, Japan. Mem. S. Industr. Sci. Inst. Kagoshima Univ. See B–P–H 580/13.

Nanshi nanyo = Southern China and the south seas. S. China S. Seas. See B–P–H 808/17.

Napaea; revista de botânica. Porto Alegre. No. 1+, 1987+. Napaea. HI 68288

Napi közlöny. Pozsony [=Bratislava, Czechoslovakia]. Napi Közl. See B–P–H 627/21. HI 56971

Nara gakugei daigaku kiyo, shizenkagaku = Journal of Nara gakugei university. Natural science. Nara.

Nara joshi daigaku seibutsu gakkai-shi = Biological journal of Nara women's university. Nara.

Nara kyoiku daigaku kiyo shizen kagaku = Bulletin of Nara university of education. Series B, natural sciences. Nara.

Narodni šumar. Sarajevo. Vols. 1-30, 1959-76. Nar. Šumar. Superseded by: Šumarstvo i prerada drveta. HI 68289

Narragansett marine laboratory; collected reprints. Kingston, RI. Narragansett Mar. Lab. Collect. Repr. See B–P–H 627/25. HI 56972

Nas vrt. Ljubljana. Vols. ?-4-10, ?-1969-75. Nas Vrt. Incorporated in: Moj mali svet. HI 68290

Nasa poljoprivreda = Nasa poljoprivreda i sumarstvo. Titograd.

Nasa poljoprivreda i sumarstvo. Titograd. Vols. 1-9, 1955-63. Nasa Poljopr. Sumarstvo. Superseded by: Poljoprivreda i sumarstvo. HI 68291

Naše věda. Prague. Naše Věda. See B–P–H 627/26. HI 56973

Naše vlast. Prague. Naše Vlast. See B–P–H 627/27. HI 56974

Nashrat al'mahad al-qawmi al-l imi wa-al-fanni lil-uqyanus wa-al-sayd bi-Salambu = Bulletin de l'institut national scientifique et technique d'océanographie et de pêche de Salammbô. Tunis.

Nashville medical record. Nashville, TN. Nashville Med. Rec. See B–P–H 627/29. HI 56975

Nashville monthly record of medical and physical science. Nashville, TN. Nashville Monthly Rec. Med. Phys. Sci. See B–P–H 627/30. HI 56976

Nasrah kulliyat al-ulum = Bulletin of the faculty of science, Cairo university. Cairo.

Nasrah - markaz buhuth al-tarikh al-tabii = Publications, biological research centre, university of Baghdad. Baghdad.

Nasrat al-ma'had al-'ilmi, gami'at Muhammad al-hamis = Bulletin de l'institut scientifique, université Mohammed V. Rabat.

Nasturtium. Boston, MA. No. 1+, 1967+. Nasturtium. HI 68292

Natal agricultural journal = Natal agricultural journal and mining journal. Pietermaritzburg.

Natal agricultural journal and mining journal. Pietermaritzburg. Vols. 7-15, 1904-11. Natal Agric. J. & Mining J. Preceded by: Agricultural journal and mining record, Pietermaritzburg. Superseded by: Agricultural journal of the Union of South Africa. 4-2825-1. HI 68293

Natal colonial herbarium. Annual report. Durban, South Africa. Natal Colon. Herb. Annual Rep. See B–P–H 630/13. HI 57000

Natal sugar and cotton planter = African sugar and cotton planter. Durban, South Africa. African Sugar Cotton Pl. See B–P–H 54/1.

Natal wild life. Durban. Vol. ?-3+, ?-1962+. Natal Wild Life. HI 68294

National academy science letters. Allahabad. Vol. 1+, 1978+. Natl. Acad. Sci. Lett. HI 68295

National arboretum. Contribution. Washington, DC. Natl. Arbor. Contr. See B–P–H 630/19. HI 57001

National arboretum leaflet, forest service, U S department of agriculture. Washington, DC. 1954+. Natl. Arbor. Leafl. Forest Serv. U.S.D.A. HI 68296

National arborists' association newsletter. Wooster, OH. ?-1972+. Natl. Arbor. Assoc. Newslett. HI 68297

National botanic gardens. Bulletin. Lucknow, India. Natl. Bot. Gard. Bull. See B–P–H 630/20. HI 57002

National cactus and succulent journal. Leeds, Bradford. Vols. 1(3)-37, 1946-82. Natl. Cact. Succ. J. Preceded by: Yorkshire cactus journal. Superseded by: British cactus and succulent journal. 4-2850-1. HI 68298

National chrysanthemum society bulletin. Hackensack, NJ. Vol. 1+, 1945+. Natl. Chrysanthemum Soc. Bull. 4-2852-1. HI 68299

National farm and garden magazine. Portland, ME. Natl. Farm Gard. Mag. See B–P–H 630/23. HI 57003

National gardener. Pleasantville, NY, St. Louis, MO. 1948+. Natl. Gard. Preceded by: Bulletin of the national council of state garden clubs. 4-287-2. HI 68300

National gardening. Burlington, VT. Vol. 8(9)+, 1985+. Natl. Gardening. Preceded by: Gardens for all news. HI 68301

National gardening survey. Burlington, VT. 1979+. Natl. Gard. Surv. Preceded by: Gardening in America. HI 68302

National geographic = National geographic magazine. Washington, DC. Natl. Geogr. Mag. See B–P–H 630/25.

National geographic magazine. Washington, DC. Natl. Geogr. Mag. See B–P–H 630/25. HI 57004

National geographic research; a scientific journal. Washington, DC. Vol. 1+, 1985+. Natl. Geogr. Res. HI 68303

National geographic society research report = Research report, national geographic society. Washington, DC.

National gladiolus society newsletter. Vienna, VA. Natl. Gladiolus Soc. Newslett. See B–P–H 630/26. HI 57005

National horticultural magazine. Henning, MN. Natl. Hort. Mag. See B–P–H 630/27. HI 57006

National institute of oceanography; collected reprints. Wormley. 1953-73. Natl. Inst. Oceanogr. Collect. Repr. Superseded by: Collected reprints, institute of oceanographic sciences, Wormley. HI 68304

National library of Wales journal. Aberystwyth, Wales. Natl. Libr. Wales J. See B–P–H 631/3. HI 57007

National medical journal of China. Shanghai. Natl. Med. J. China. See B–P–H 631/4. HI 57008

National nurseryman. Rochester, NY. Natl. Nurseryman. See B–P–H 631/6. HI 57009

National nut news. Chicago, IL. Natl. Nut News. See B–P–H 631/7. HI 57010

National oleander society newsletter. Galveston, TX. ?-

1982+. Natl. Oleander Soc. Newslett. HI 68305

National parks; magazine of the national parks and conservation association. Washington, DC. Vol. 55+, 1981+. Natl. Parks. Preceded by: National parks and conservation magazine. HI 68306

National parks bulletin. Washington, DC. Vols. [5?]-15, 1923-41. Natl. Parks Bull. Preceded by: Bulletin of the national parks association. Superseded by: National parks magazine. 4-2896-1. HI 68307

National parks and conservation magazine; environmental journal. Washington DC. Vols. 44(4)-54, 1970-80. Natl. Parks Conservation Mag. Preceded by: National parks magazine. Superseded by: National parks. HI 68308

National parks magazine; official publication of the national parks association. Washington, DC. Vols. 16-44(3), 1942-70. Natl. Parks Mag. Preceded by: National parks bulletin. Superseded by: National parks and conservation magazine. 4-2896-1. HI 68309

National parks scientific series. Wellington, N.Z. 1977+. Natl. Parks Sci. Ser. HI 68310

National research council, annual report of the chairman of the division of biology and agriculture. Washington, DC. Natl. Res. Council Annual Rep. Chairm. Div. Biol. See B–P–H 631/8. HI 57011

National research council, division of biology and agriculture, cumulative report of the committee on effects of radiation upon living organisms. Washington, DC. Natl. Res. Council Div. Biol. Cumulative Rep. Committee Effects Radiat. See B–P–H 631/9. HI 57012

National research council, division of geology and geography, report of the committee on marine ecology as related to paleontology. Washington, DC. Natl. Res. Council Div. Geol. Rep. Committee Mar. Ecol. Related Paleontol. See B–P–H 631/10. HI 57013

National research council publications. Washington, DC. Natl. Res. Council Publ. See B–P–H 631/11. HI 57014

National review. Shanghai. Natl. Rev. See B–P–H 631/12. HI 57015

National science council monthly. [K'o hsüeh fa chan.] Taipei. Vol. 1+, 1973+. Natl. Sci. Council Monthly. HI 68311

National Tokai-kinki agricultural experiment station. Research progress report. Tsu, Japan. Natl. Tokai-Kinki Agric. Exp. Sta. Res. Progr. Rep. See B–P–H 631/15. HI 57016

National tree news. New York, NY. Natl. Tree News. See B–P–H 631/16. HI 57017

National trust. Magazine. London. Nos. 17-46, 1973-85. Natl. Trust. Preceded by: National trust news [not entered]. Superseded by: National trust magazine. HI 74544

National trust magazine. Magazine. London. No. 47+, 1986+. Natl. Trust Mag. Preceded by: National trust. HI 68312

National urban and community forestry forum. Washington, DC. Vols. ?-6(1), 1981-86?. Natl. Urban Community Forest Forum. Superseded by: National urban forest forum. HI 74861

National urban forest forum. Washington, DC. Vols. 6(2)-8(4), 1986-88. Natl. Urban Forest Forum. Preceded by: National urban and community forestry forum. Superseded by: Urban forest forum. HI 74862

National wildlife. Washington, DC. Vol. 1+, 1963+. Natl. Wildlife. HI 68313

National wildlife federation's conservation; national wildlife federation's weekly update of conservation legislation. Washington, DC. Vol. 1+, 1983+. Natl. Wildlife Fed. Conservation. Preceded by: Conservation report. Washington, DC. HI 68314

Nationalpark. Grafenau. No. ?-25+, ?-1979+. Nationalpark. HI 68315

Native orchid society of South Australia newsletter = Newsletter, native orchid society of South Australia. Everard Park, S.A.

Native plant news. St. Paul, MN. Vol. 1(1), 1982. Native Pl. News. Superseded by: Minnesota plant press. HI 75091

Native plant society of Oregon [series]. Ashland, OR. Vols. ?-13(12), ?-1976-80. Native Pl. Soc. Oregon. Preceded by: N P S O newsletter. Superseded by: Bulletin of the native plant society of Oregon. HI 68316

Native plant society of Texas news. Denton, TX. 1987+. Native Pl. Soc. Texas News. Preceded by: Texas native plant society news. HI 75092

Náttúrufraedingurinn. Alpýdlegt fraedsluritum náttúrufraedi. Reykjavik, Iceland. Náttúrufraedingurinn. See B–P–H 631/20. HI 57018

Natur. Halle. Natur (Halle). See B–P–H 631/22. HI 57019

Natur. Leipzig. Natur (Leipzig). See B–P–H 631/23. HI 57020

Natur. Schwäbisch Hall, Germany. Natur (Schwäbisch Hall). See B–P–H 631/24. HI 57021

Natur und Haus. Illustrierte Zeitschrift für alle Naturfreunde. Berlin. Natur & Haus. See B–P–H 632/2. HI 57023

Natur und Heimat. Aussig, Bohemia [=Usti nad Labem, Czechoslovakia]. Natur & Heimat (Aussig). See B–P–H 632/3. HI 57024

Natur und Heimat. Dresdner Verlag. Dresden. Natur & Heimat (Dresdner Verlag). See B–P–H 632/4. HI 57025

Natur und Heimat. Linz. Natur & Heimat (Linz). See B–P–H 632/5. HI 57026

Natur und Heimat. Münster. Natur & Heimat (Münster). See B–P–H 632/6. HI 57027

Natur und Heimat. Sachsenverlag. Dresden. Natur & Heimat (Sachsenverlag). See B–P–H 632/7. HI 57028

Natur, Kultur und Jagd; Beiträge zur Naturkunde Niedersachsens. Hanover. Vols. 18-26, 1965-73. Natur Kult. Jagd. Preceded and superseded by: Beitrage zur Naturkunde Niedersachsens. HI 68317

Natur- Kunstab = Gemeinnütziges Natur und Kunstmagazin oder Abhandlungen zur Beförderung der Naturkunde, der Künste, Manufacturen und Fabriken, 1764. Berlin. Gemeinnütz. Natur. Kunstmag. See B–P–H 395/26.

Natur- und Kunstkabinet: oder Sammlung nützlicher Nachrichten, zu Beförderung der Naturkunde, der Künste, und der Manufakturen. Jena. Natur- Kunstkab. See B–P–H 632/8. HI 57029

Natur und Land. Blätter für Naturkunde und Naturschutz. Vienna. Natur & Land. See B–P–H 632/9. HI 57030

Natur und Landschaft. Lüneburg, Germany. Natur & Landschaft. See B–P–H 632/10. HI 57031

Natur und Mensch; Jahresmitteilung der naturhistorischen Gesellschaft Nürnberg. Nuremberg. No. 4+, 1969+. Natur & Mensch (Nuremberg). Preceded by: Mitteilungen und Jahresbericht der naturhistorischen Gesellschaft Nürnberg. HI 68318

Natur und Mensch. Thayngen, Switzerland. Natur & Mensch (Thayngen). See B–P–H 632/11. HI 57032

Natur und Museum. Frankfurt a. M. Natur & Mus. See B–P–H 632/12. HI 57033

Natur og museum. Arhus. 1951+. Natur Mus., Arhus. HI 68319

Natur und Nahrung. Zentralblatt für Drogen und Wildfrüchte. Berlin. Natur & Nahrung. See B–P–H 632/13. HI 57034

Natur und Nationalparke; European bulletin. Stuttgart. Vols. ?-11(41), 1963-73; vol. ?(48)+, 1975+. Natur Nationalparke. Nos. (42)-(47) were published as Europäische Bulletin. In: Naturschutz- und naturparke, nos. 70-76. HI 68320

Natur und Naturschutz in Mecklenburg. Stralsund & Greifswald. Natur Naturschutz Mecklenburg. See B–P–H 632/14. HI 57035

Natur am Niederrhein. Krefeld, Germany. Natur am Niederrhein. See B–P–H 631/25. HI 57022

Natur und Recht. Hamburg, New York. 1979+. Natur & Recht. HI 68321

Natur und Schule. Leipzig. Vols. 1-6, 1902-07. Natur & Schule. Superseded by: Monatshefte für den naturwissenschaftlichen Unterricht aller Schulgattungen und Natur und Schule. 4-2935-3. HI 68322

Natur und Technik. Berlin. Natur & Techn. (Berlin). See B–P–H 632/15. HI 57036

Natur und Technik. Vienna. Natur & Techn. (Vienna). See B–P–H 632/16. HI 57037

Natur und Technik. Zurich. Natur & Techn. (Zurich). See B–P–H 632/17. HI 57038

Natur und Umwelt. Munich. Vol. 56+, 1976+. Natur & Umwelt. Preceded by: Blatter für Natur und Umwelt Schutz. HI 68323

Natur und Unterricht. Monatsschrift für den elementaren naturwissenschaftlichen Unterricht. Hamburg. Natur & Unterr. See B–P–H 632/19. HI 57039

Natur und Volk. Frankfurt a. M. Natur & Volk. See B–P–H 632/20. HI 57040

Natura. Amsterdam. Natura (Amsterdam). See B–P–H 632/21. HI 57041

Natura. Bucharest. Vols. 1-12, 1912-60. Natura (Bucharest). Superseded by: Natura. Seria biologie. 4-2926-1. HI 68324

Natura. Budapest. Natura (Budapest). See B–P–H 633/2. HI 57042

Natura. Buenos Aires. Natura (Buenos Aires). See B–P–H 633/3. HI 57043

Natura. Caracas. Natura (Caracas). See B–P–H 633/4. HI 57044

Natura. Ghent. Natura (Ghent). See B–P–H 633/5. HI 57045

Natura. Haarlem. Natura (Haarlem). See B–P–H 633/6. HI 57046

Natura. São Paulo. Natura (São Paulo). See B–P–H 633/7. HI 57047

Natura. Seria biologie. Revista societații de științe naturale și geografie din republica socialistă românia. Bucharest. Vols. 13-26, 1961-74. Natura, Ser. Biol. Preceded by: Natura. Bucharest. Superseded by: Revista ocrotirea mediului inconjurator natura, terra. Subtitlul natura. HI 68325

Natura; revista di scienze naturali. Milan. Vol. 1+, 1909+. Natura (Milan). HI 68326

Natura. Plovdiv. Vol. 1+, 1967+. Natura (Plovdiv). HI 68327

Natura alpina. Trento. 1974+. Nat. Alpina. HI 68328

Natura bresciana. Brescia. 1965+. Nat. Bresciana. HI 68329

Natura et cultura. [Shizen to bunka.] Kyoto. 1950-51. Nat. & Cult. HI 68330

Natura jutlandica. Aarhus. Vol. 1+, 1947+. Nat. Jutlandica. HI 68331

Natura-Limburg. Hasselt. No. 1+, 1955+. Natura-Limburg. HI 68332

Natura lusatica. Bautzen. Nat. Lusatica. See B–P–H 629/13. HI 56991

Natura e montagna; periodica trimestriale di divulgazione naturalistica. Bologna. No. 1+, 1954+. Nat. & Montagna. HI 68333

Natura mosana; vent être un trait d'union entre les sociétés de naturalistes des provinces Wallones. Liège. Vol. 1+, 1948+. Nat. Mosana. HI 68334

Natura – reis-en kampeergids. ?-1967. Nat. Reis Kampeergids. Superseded by: Natura, reizen en kampen. HI 68335

Natura, reizen en kampen. Hogwoud. 1968+. Nat. Reizen Kampen. Preceded by: Natura – reis-en kampergids. HI 68336

Natüra e l'uman = Natur und Mensch. Thayngen, Switzerland. Natur & Mensch. See B–P–H 632/11.

Natura e l'uomo = Natur und Mensch. Thayngen, Switzerland. Natur & Mensch. See B–P–H 632/11.

Naturae novitates. Bibliographie neuer Erscheinungen aller Länder auf dem Gebiete der Naturgeschichte und der exakten Wissenschaften. Berlin. Nat. Novit. See B–P–H 629/22. HI 56993

Natural and applied science bulletin; university of the Philippines. Manila. Nat. Appl. Sci. Bull. Univ. Philipp. See B–P–H 628/7. HI 56977

Natural areas journal; quarterly publication of the natural areas association. Rockford, IL. Vol. 2+, 1982+. Nat. Areas J. Preceded by: Journal of the natural areas association. HI 68337

Natural history. New York, NY. Nat. Hist. See B–P–H 628/21. HI 56981

Natural history book reviews. Richmond, Sy., Berkhamsted. Vol. 1+, 1976+. Nat. Hist. Book Rev. HI 68338

Natural history bulletin. Honolulu, Hawaii. Nat. Hist. Bull. See B–P–H 628/23. HI 56982

Natural history bulletin of the Grand Canyon natural history association. Grand Canyon, Arizona. Nat. Hist. Bull. Grand Canyon Nat. Hist. Assoc. See B–P–H 628/24. HI 56983

Natural history bulletin of the Siam society. Bangkok. Vol. 14(2)+, 1947+. Nat. Hist. Bull. Siam Soc. Preceded by: Natural history bulletin of the Thailand research society. HI 68339

Natural history bulletin of the Thailand research society. Bangkok. Vols. 13-14(1), 1942-44. Nat. Hist. Bull. Thailand Res. Soc. Preceded by: Journal of the Thailand research society. Natural history supplement. Superseded by: Natural history bulletin of the Siam society. 5-3868-1. HI 68340

Natural history division information, provincial museum and archives of Alberta. Edmonton. No. 1+, 1971+. Nat. Hist. Div. Inform. Prov. Mus. Arch. Alberta. HI 68341

Natural history guides of the Boston society of natural history. Boston, MA. Nat. Hist. Guides Boston Soc. Nat. Hist. See B–P–H 628/28. HI 56984

Natural history journal. York, England. Nat. Hist. J. See B–P–H 628/29. HI 56985

Natural history magazine. London. Nat. Hist. Mag. See B–P–H 628/30. HI 56986

Natural history miscellanea. Chicago, IL. Nat. Hist. Misc. See B–P–H 629/1. HI 56987

Natural history museum bulletin. San Diego, CA. Nos. 1-392, 1922-67. Nat. Hist. Mus. Bull. (San Diego). Superseded by: Environment southwest. 5-3771-2. HI 68342

Natural history museums' newsletter, I C O M. Wroclaw. No. 1+, 1978+. Nat. Hist. Mus. Newslett. I. C. O. M. HI 68343

Natural history notes. Kingston, Jamaica. Vols. 1-6, 1941-55. Nat. Hist. Notes. HI 68344

Natural history notes of the Worcester natural history society. Worcester, MA. Nat. Hist. Notes Worcester Nat. Hist. Soc. See B–P–H 629/4. HI 56988

Natural history papers. Ottawa. Vols. 1-45, 1958-69. Nat. Hist. Pap. HI 68345

Natural history report of Kanagawa. [Kanagawa shizenshi shiryo.] Yokohama. 1979+. Nat. Hist. Rep. Kanagawa. HI 68346

Natural history review; a quarterly journal of biological science. London & Dublin. Nat. Hist. Rev. See B–P–H 629/5. HI 56989

Natural history of Tosa. [Tosa no hakubutsu.] Kochi, Japan. Nat. Hist. Tosa. See B–P–H 629/9. HI 56990

Natural product reports; journal of current developments in bio-organic chemistry. London. Vol. 1+, 1984+. Nat. Prod. Rep. HI 68347

Natural resources. [Ziran ziyuan.] [Beijing]. 1981+. Nat. Resources. HI 68348

Natural rubber news. Washington, DC. Nat. Rubber News. See B–P–H 629/25. HI 56994

Natural science. London & New York, NY. Nat. Sci. See B–P–H 629/26. HI 56995

Natural science bulletin of the university of Amoy. [Hsia men ta hsüeh chih tzu jan k'o hsüeh tsung kan.] Amoy, China. Nat. Sci. Bull. Univ. Amoy. See B–P–H 629/27. HI 56996

Natural science and museum. [Shizen kagaku no hakubutsukan.] Tokyo. Nat. Sci. Mus. See B–P–H 629/28. HI 56997

Natural science report of the Ochanomizu university. [Ochanomizu joshi daigaku. Shizen Kagaku hokoku.] Tokyo. Nat. Sci. Rep. Ochanomizu Univ. See B–P–H 630/1. HI 56998

Natural sciences. Skodsborg. ?-1954+. Nat. Sci. HI 68349

Natural world; magazine of the royal society for nature conservation. Alford. No. 1+, 1981+. Nat. World. Preceded by: Conservation review. HI 68350

Naturaleza. Madrid. Naturaleza (Madrid). See B–P–H 633/13. HI 57048

Naturaleza, La. Madrid. Nos. 1-2, 1969. Naturaleza (Madrid 1969). HI 68351

Naturaleza. Mexico City. Naturaleza (Mexico City). See B–P–H 633/14. HI 57049

Naturaleza ecuatoriana. Guayaquil. Vol. 1+, 1966+. Naturaleza Ecuat. HI 68352

Naturalia. Lisbon. Naturalia. See B–P–H 633/16. HI 57050

Naturalia. São Paulo. Vol. 1+, 1975+. Naturalia (São Paulo). HI 68353

Naturalia hispanica. Madrid. Vol. 1+, 1974+. Naturalia Hispan. HI 68354

Naturalia monspeliensia. Série botanique. Montpellier. Vol. 8+, 19??+. Naturalia Monspel., Sér. Bot. Preceded by: Recueil des travaux des laboratoires de botanique, géologique et zoologie de la faculté des sciences de l'université de Montpellier. Série botanique. HI 68355

Naturalia monspeliensia. Hors-série. Montpellier. No. A1+, 1980+. Naturalia Monspel., Hors Sér. HI 68356

Naturalientausch. Prague. Naturalientausch. See B–P–H 633/19. HI 57051

Naturalist. Boston, MA. Naturalist (Boston). See B–P–H 633/20. HI 57052

Naturalist. Cascade. Vol. 3(10)+, 1981+. Naturalist (Cascade). Preceded by: Trinidad naturalist magazine. HI 68357

Naturalist. Durban. Nos. 1-6, 1910-11. Naturalist (Durban). HI 68358

Naturalist. London, Huddersfield, Hull, etc. Vols. 1-3, 1864/65-66/67, 1865-67; n.s. vols. 1-10, 1875-85; [ser. 3,] 1886+. Naturalist (Hull). 4-2928-3. HI 68359

Naturalist. London. Naturalist (London). See B–P–H 633/21. HI 57053

Naturalist. [Edited by Maund & Hall.] London. Naturalist (Maund & Hall). See B–P–H 633/22. HI 57054

Naturalist (Minneapolis). Minneapolis, MN. Vol. 8(3)+, 1957+. Naturalist (Minneapolis). Preceded by: Minnesota naturalist. HI 68360

Naturalist. Port Elizabeth. [Dates of publication not ascertained.] Naturalist (Port Elizabeth). Preceded by?: Eastern Cape naturalist. HI 68361

Naturalist. Trinidad & Tobago. [Dates of publication not ascertained.] Naturalist (Trinidad). HI 68362

Naturalist Man'chshurii. Harbin. 1936-37. Naturalist Man'chshurii. HI 68363

Naturalista Argentina; revista de historia natural. Buenos Aires. Vol. 1, nos. 1-11, 1875-78. Naturalista Argentina. HI 68364

Naturalista siciliano; giornale di scienze naturali. Palermo. Vols. 1-27, 1881-1930; ser. 4a, vol. 1+, 1978+. Naturalista Sicil. 4-2928-3. HI 68365

Naturaliste. Paris. Naturaliste. See B–P–H 633/24. HI 57055

Naturaliste amateur. Verviers. 1946-48. Naturaliste Amateur. Superseded by: Revue vervietoise d'histoire naturelle. HI 68366

Naturaliste canadien. Bulletin de recherches, observations et découvertes se rapportant à l'histoire naturelle du Canada. Quebec. Naturaliste Canad. See B–P–H 634/1. HI 57056

Naturaliste malgache. Tananarive. Vols. 1-13, 1949-62. Naturaliste Malgache. HI 68367

Naturalistes belges. Bulletin mensuel. Brussels. Naturalistes Belges. See B–P–H 634/3. HI 57057

Naturalistes orleanais. Orleans. 3rd ser., vol. ?+, 1971+. Naturalistes Orleanais. Preceded by: Bulletin de l'association des naturalistes orleanais et de la Loire Moyenne. HI 68368

Naturalists' directory. Salem, Boston, MA. 1877-1881, 1884, 1886-1890, 1894-1954. Naturalists' Directory. For the years 1882, 1885, 1892 & 1896 see: Scientists' international directory international. Superseded by: Naturalists' directory (international). HI 68369

Naturalists' directory (international). Boxford, MA. 1958-75. Naturalists' Directory (Int). Preceded by: Naturalist's directory. Superseded by: Naturalists' directory and almanac (international). HI 68370

Naturalists' directory and almanac (international). Baltimore, MD. 1979/80+, 1978+. Naturalists' Directory Alman. (Int.). Preceded by: Naturalists' directory (international). HI 68371

Naturalist's journal. London. Naturalist's J. (London 1767). See B–P–H 634/4. HI 57058

Naturalist's journal. London. Naturalist's J. (London 1893-99). See B–P–H 634/5. HI 57059

Naturalist's journal and miscellany. Edinburgh, London, & Dublin. Naturalist's J. Misc. See B–P–H 634/6. HI 57060

Naturalist's newsletter, Norfolk and Norwich naturalists' society. Norwich. 1978+. Naturalist's Newslett. Norfolk & Norwich Naturalists' Soc. HI 68372

Naturalist's pocket magazine; or, compleat cabinet of the curiosities and beauties of nature. London. Naturalist's Pocket Mag. See B–P–H 634/7. HI 57061

Naturalist's scrap book of the Liverpool district. Liverpool. Ser. 1, vol. 1- ser. 3, vol. 4, 1863-64. Naturalist's Scrap Book Liverpool Distr. HI 68373

Naturalists' universal directory = Naturalists' directory. Salem, Boston, MA.

Naturdenkmäler. Vorträge und Aufsätze. Berlin. Naturdenkmäler. See B–P–H 634/8. HI 57062

Naturdenkmalpflege und Naturschutz in Berlin und Brandenburg. Berlin-Lichterfelde. Naturdenkmalpflege Naturschutz Berlin Brandenburg. See B–P–H 634/9. HI 57063

Nature. [Daziran.] Beijing. No. 1+, 1980+. Nature (Beijing). HI 68377

Nature; a weekly illustrated journal of science. London. Vol. 1+, 1869/70+. Nature. For the years 1971-73 an extra series was added: Nature. New biology. 4-2929-3. HI 68374

Nature; revue des sciences et de leurs applications à l'art et à l'industrie [subtitle varies]. Paris. Vols. 1(1)-90(3332), 1873-62 [suspended 1942-44]. Nature (Paris). Superseded by: Science progrès, la nature. 4-2930-1. HI 68375

Nature. [Ta tzu jan.] Taipei. 1983+. Nature (Taipei). HI 68378

Nature. [Watakushitachi no shizen.] Tokyo. 1960+. Nature (Tokyo). HI 68379

Nature in art; newsletter of the society for wildlife art of the nations. Wallsworth Hall, Glos. Vol. ?-2+, ?-1988+. Nat. Art. HI 68380

Nature and art. London. 1866-67. Nat. & Art. HI 68381

Nature in Australia. Sydney. Nat. Australia. See B–P–H 628/9. HI 56978

Nature calédonienne; revue trimestrielle de l'association pour la sauvegarde de la nature neo-calédonienne. Nouméa. Vol. 1+, 1973+. Nat. Calédonienne. HI 68382

Nature in Cambridgeshire. Cambridge, England. Nat. Cambridgeshire. See B–P–H 628/13. HI 56979

Nature Canada. Ottawa. Vol. 1+, 1972+. Nat. Canada. Preceded by: Canadian Audubon. HI 68383

Nature center news. [Dates of publication not ascertained.] Nat. Center News. HI 68384

Nature conservancy news. Arlington, VA. Vol. 8(3)+, 1958+. Nat. Conservancy News (Arlington). Preceded by: Nature conservation news. HI 68385

Nature conservancy progress. London. 1964-68. Nat. Conservancy Progr. HI 68386

Nature conservation. Bratislava = Ochrana prirody. Bratislava.

Nature conservation. Prague = Ochrana přirody. Prague.

Nature conservation. [Chayon pojon.] Seoul. ?-1982+. Nat. Conservation (Seoul). HI 68387

Nature conservation news. Washington, DC. Vols. 1-8(2), 1951-58. Nat. Conservation News (Washington). Superseded by: Nature conservancy news. HI 68388

Nature conservation news-sheet. London, Banbury, Peterborough. ?-1975+. Nat. Conservation News-Sheet. HI 68389

Nature conservation in Warwickshire. Nos. 1-?, 1969-75. Nat. Conservation Warwickshire. Superseded by: Newsletter of the Warwickshire nature conservation trust. HI 68390

Nature in Devon. Exeter. No. 1+, 1980+. Nat. Devon. Preceded by: Journal of the Devon trust for nature conservation. HI 68391

Nature in east Africa; the bulletin of the east Africa natural history society. Nairobi. Nat. E. Africa. See B–P–H 628/15. HI 56980

Nature in focus; bulletin of the european information centre for nature conservation. Strasbourg. Nos. 1-19, 1968-74. Nat. Focus. Superseded by: Naturopa. HI 68392

Nature et l'homme = Natur und Mensch. Thayngen, Switzerland. Natur & Mensch. See B–P–H 632/11.

Nature of Illinois. Springfield, IL. Vol. 1+, 1986+. Nat. Illinois. HI 68393

Nature journal. [Ziran zazhi.] Beijing. Vol. 1+, 1978+. Nat. J. HI 68394

Nature in Lancashire. Birkenhead, Lancaster. Vols. 1-?, 1970-75? Nat. Lancashire. HI 68395

Nature and life in Southeast Asia. Kyoto. Vols. 1-7, 1961-76. Nat. Life S. E. Asia. HI 68396

Nature magazine. Washington, DC. Vols. 1-52, 1923-59. Nat. Mag. Preceded by: Nature study review. Incorporated with Natural history. 4-2930-3. HI 68397

Nature malaysiana. Kuala Lumpur. Vol. 1+, 1976+. Nat. Malaysiana. HI 68398

Nature. New biology. [Formed part of: Nature. See B–P–H 634/10.] London. Vols. 229-246, 1971-73. Nature, New Biol. HI 68376

Nature Niagara news. Niagara Falls, Ont. [1985]+. Nat. Niagara News. Preceded by: Bulletin, Niagara Falls nature club. HI 68399

Nature in north east Essex = Report and records of the Colchester and district natural history society and field club. Colchester.

Nature notes. Cape Town. Nos. ?-5-91-?, ?-1924-32-? Nat. Notes (Cape Town). HI 68400

Nature notes; Selborne society's magazine. London. Vols. 1-19, 1890-1908. Nat. Notes. Preceded by: Selborne magazine for lovers and students of living nature. Superseded by: Selborne magazine and nature notes. 5-3842-1. HI 68401

Nature notes from the Grand Canyon. Grand Canyon, AZ. Nat. Notes Grand Canyon. See B–P–H 629/21. HI 56992

Nature outlook. Worcester, MA. Vols. 1-10, 1930-40. Nat. Outlook. Superseded by: Quarterly of the Worcester natural history society. 5-4540-1. HI 68402

Nature and plants. [Shokubutsu to shizen.] Tokyo. Vol. ?-6(6)+, ?-1972+. Nat. & Pl. HI 68403

Nature et progrès; publication de l'asociation européenne d'agriculture et d'hygiène biologiques. Saint-Geneviève-des-Bois. 1964+. Nat. & Progr. HI 68404

Nature protection = Looduskaitse. Tallinn.

Nature and resources; bulletin of the international hydrological decade later International news about research on environmental resources and conservation of nature. Paris. Vol. 1+, 1965+. Nat. Resources. Preceded by: Arid zone. HI 68405

Nature et ressources = Nature and resources [French edition]. Paris.

Nature. Science, progrès = Nature. Paris.

Nature science report; annual review compiled by the staff of Nature. London. 1968-70. Nat. Sci. Rep. HI 68406

Nature study. Morehead, KY. 1947+. Nat. Study (Morehead). HI 68407

Nature study. [Osaka shizen kagaku kenkyukai.] Osaka. 1955+. Nat. Study (Osaka). HI 68408

Nature study bulletin. College Park, MD = Maryland agricultural college. Nature study bulletin. College Park, MD. Maryland Agric. Coll. Nat. Study Bull. See B–P–H 548/24.

Nature-study bulletin, New York state department of agriculture = New York state department of agriculture. Nature-study bulletin. Albany, NY. New York State Dept. Agric. Nat.-Study Bull. See B–P–H 656/19.

Nature-study leaflets, Cornell university = Cornell university nature-study leaflets. Ithaca, NY. Cornell Univ. Nat.-Study Leafl. See B–P–H 332/12.

Nature study review. Ithaca, NY. Vols. 1-19(9), 1905-23. Nat. Study Rev. Superseded by: Nature magazine. 4-2931-2. HI 68409

Nature vivante. Clermont-Ferrand. No. ?-10+, ?-1971+. Nat. Viv. HI 68410

Nature in Wales. Orielton, Wales. Nat. Wales. See B–P–H 630/10. HI 56999

Naturen; illustreret maanedsskrift for populaer naturvidenskab (later Populaervitenskapelig tidsskrift). Oslo, Bergen, Copenhagen. Vol. 1+, 1877+. Naturen. 4-2931-2. HI 68411

Naturen og mennesket. Copenhagen. Vols. 1-16, 1889-96. Nat. & Mennesk. 4-2931-2. HI 68412

Naturens verden. Copenhagen. Vol. 1+, 1917+. Nat. Verden. HI 68413

Naturescape. Wheaton, IL. Vol. 1+, 1980+. Naturescape. HI 68414

Naturschutz. Belgrade = Zaštita prirode. Belgrade. Zašt. Prir. See B–P–H 955/22.

Naturschutz und Landschaftspflege im Saarland. Veröffentlichungen der Landesstelle für Naturschutz und Landschaftspflege. Ottweiler, Germany. Naturschutz Landschaftspflege Saarland. See B–P–H 636/1.

Natureza em revista; publicação da fundação zoobotânica do Rio Grande do Sul. Porto Alegre. No. 1+, 1976+. Natureza Rev. HI 68415

Naturforschende Gesellschaft in Bamberg. Bericht. Bamberg, Germany. Naturf. Ges. Bamberg Ber. See B–P–H 634/16. HI 57064

Naturforscher. Berlin. Naturforscher (Berlin 1868-88). See B–P–H 634/19. HI 57065

Naturforscher. Berlin. Naturforscher (Berlin 1924-38). See B–P–H 634/20. HI 57066

Naturforscher. Görlitz. Naturforscher (Görlitz). See B–P–H 634/21. HI 57067

Naturforscher. Halle. Naturforscher (Halle). See B–P–H 634/22. HI 57068

Naturforscher. Leipzig. Naturforscher (Leipzig). See B–P–H 634/23 HI 57069

Naturfreund, oder Beiträge zur Schlesischen Naturgeschichte. Breslau [=Wroclaw, Poland]. Naturfreund (Breslau). See B–P–H 635/2. HI 57071

Naturfreund. Naumburg an der Saale & Halle. Naturfreund (Naumburg & Halle). See B–P–H 635/3. HI 57072

Naturfreund. Vienna. Naturfreund (Vienna). See B–P–H 635/4. HI 57073

Naturfreund. Witten, Germany. Naturfreund (Witten). See B–P–H 635/5. HI 57074

Naturfreund Ungarns. Neutra [=Nitra, Czechoslovakia]. Naturfr. Ungarns. See B–P–H 634/24. HI 57070

Naturhistorische Abhandlungen der Batavischen Gesellschaft der Wissenschaften zu Haarlem. Leipzig. Naturhist. Abh. Batav. Ges. Wiss. Haarlem. See B–P–H 635/8. HI 57075

Naturhistorische und chemisch-technische Notizen nach den neuesten Erfahrungen. Berlin. Naturhist. Chem.-Techn. Notizen. See B–P–H 635/9. HI 57076

Naturhistorische Gesellschaft Nürnberg e. V. Bericht. Nuremberg. Naturhist. Ges. Nürnberg Ber. See B–P–H 635/10. HI 57077

Naturhistorische Gesellschaft Nürnberg. Jahresbericht. Nuremberg. Naturhist. Ges. Nürnberg Jahresber. See B–P–H 635/11. HI 57078

Naturhistorisk tidsskrift. Copenhagen. Naturhist. Tidsskr. See B–P–H 635/12. HI 57079

Naturiaethur. Llanymddyfri. 1981+. Naturiaethur. HI 68416

Naturkundliche Jahresberichte des Museum Heineanum. Halberstadt. 1966-75. Naturk. Jahresber. Mus. Heineanum. Superseded by: Ornithologische Jahresberichte des Museum Heineanum (non botanical). HI 68417

Naturkundliche Mitteilungen aus Oberösterreich. Linz. Naturk. Mitt. Oberösterreich. See B–P–H 635/14. HI 57081

Naturkundliches Jahrbuch der Stadt Linz. Linz. Naturk. Jahrb. Stadt Linz. See B–P–H 635/13. HI 57080

Naturopa; bulletin of the european information centre for nature conservation. Strasbourg. No. 20+, 1970+. Naturopa. Preceded by: Nature in focus. HI 68418

Naturschutz aktuell. Bonn-Bad Godesburg. No. 1+, 1978+. Naturschutz Aktuell. HI 68419

Naturschutz und Landeskultur. Jahrbuch. Leipzig. Naturschutz & Landeskult. See B–P–H 635/17. HI 57082

Naturschutz und Landschaftspflege. Cologne. Naturschutz & Landschaftspflege. See B–P–H 635/18. HI 57197

Naturschutz und Landschaftspflege in Baden-Württemberg. Veröffentlichungen der Landesstelle für Naturschutz und Landschaftspflege Baden-Württemberg und der Württembergischen Bezirksstelle in Ludwigsburg. Ludwigsburg. Naturschutz Landschaftspflege Baden-Württemberg. See B–P–H 635/19. HI 57083

Naturschutz und Landschaftspflege im Saarland. Veröffentlichungen der Landesstelle für Naturschutz und Landschaftspflege. Ottweiler, Germany. Naturschutz Landschaftspflege Saarland. See B–P–H 636/1. HI 57084

Naturschutz und naturkundliche Heimatforschung in den Bezirken Halle und Magdeburg; Anleitungsmaterial für die Naturschutzmitarbeiter. Halle. Vol. 1+, 1964+. Naturschutz Naturk. Heimatf. Bez. Halle Magdeburg. HI 68420

Naturschutz- und Naturparke. Stuttgart. No. 37+, 1965+. Naturschutz- Naturparke. Preceded by: Naturschutzparke. HI 68421

Naturschutz in Nordhessen. Riethweg? Vol. 1+, 1975/76+, 1976+. Naturschutz Nordhessen. HI 68422

Naturschutzarbeit in Berlin und Brandenburg. Berlin. Naturschutzarbeit Berlin Brandenburg. See B–P–H 636/2. HI 57085

Naturschutzarbeit in Berlin und Brandenburg. Beiheft. Potsdam & Cottbus. Naturschutzarbeit Berlin Brandenburg Beih. See B–P–H 636/3. HI 57086

Naturschutzarbeit in Mecklenburg. Ludwigslust. Vol. 4+, 1961+. Naturschutzarbeit Mecklenburg. Preceded by: Naturschutzarbeit und naturkundliche Heimatforschung in den Bezirken Rostock-Schwerin-Neubrandenburg. HI 68423

Naturschutzarbeit und naturkundliche Heimatforschung in Sachsen. Dresden. Naturschutzarbeit Naturk. Heimatf. Sachsen. See B–P–H 636/5. HI 57087

Naturschutzarbeit und naturkundliche Heimatforschung in den Bezirken Rostock-Schwerin-Neubrandenburg. Greifswald. Vols. 1(1)-3(7), 1958-60. Naturschutzarbeit Naturk. Heimatf. Bez. Rostock. Superseded by: Naturschutzarbeit in Mecklenburg. HI 68424

Naturschutzparke. Stuttgart. Vols. 1-26, 1926-41; vols. 1947-65. Naturschutzparke. Superseded by: Naturschutz- und Naturparke. HI 68425

Naturschutzstelle Darmstadt, Institut zur Erforschung, Pflege u. Gestaltung d. Landschaft. Schriftenreihe. Darmstadt, Germany. Naturschutzstelle Darmstadt Inst. Erforsch. Landschaft Schriftenreihe. See B–P–H 636/7. HI 57088

Naturvetenskapliga forskningsrådets årsbok. Stockholm. 1976/77+, 1976+. Naturvetensk. Forskningsråd. Årsbok. Preceded by: Svensk naturvetenskap. HI 68426

Naturvidenskabeligt tidsskrift. Copenhagen. Vols. 1-2, 1877-79. Naturvidensk. Tidsskr. HI 68427

Naturwissenschaft und Landwirtschaft; Abhandlungen und Vorträge über Grundlagen und Probleme der Naturwissenschaft und Landwirtschaft. Freising-München. Vol. 1+, 1924+. Naturwiss. Landw. 4-2934-3. HI 68428

Naturwissenschaft und Medizin. Mannheim. Vol. 1+, 1964+. Naturwiss. Med. Preceded by: Therapie des Monats 1953-63 [not entered]. HI 68429

Naturwissenschaften. Berlin. Naturwissenschaften. See B–P–H 637/8. HI 57103

Naturwissenschaften im Unterricht. Teil Biologie. Cologne. Vol. ?-25+, ?-1977+. Naturwiss. Unterr., Biol. HI 68430

Naturwissenschaftliche Abhandlungen. Tübingen. Naturwiss. Abh. (Tübingen). See B–P–H 636/14. HI 57089

Naturwissenschaftliche Abhandlungen. Vienna. Naturwiss. Abh. (Vienna). See B–P–H 636/15. HI 57090

Naturwissenschaftliche Abhandlungen aus Dorpat. Berlin. Naturwiss. Abh. Dorpat. See B–P–H 636/16. HI 57091

Naturwissenschaftliche Abhandlungen herausgegeben von der württembergische Gesellschaft zur Förderung der Wissenschaften. Tübingen. No. 1+, 1922+. Naturwiss. Abh. Württemberg. Ges. Förd. Wiss. HI 68431

Naturwissenschaftliche Abhandlungen herausgegeben von der württembergische Gesellschaft der Wissenschaften = Naturwissenschaftliche Abhandlungen herausgegeben von der württembergische Gesellschaft zur Förderung der Wissenschaften. Tübingen.

Naturwissenschaftliche Bibliographie. Klein Machnow, Germany. Naturwiss. Bibliogr. See B–P–H 636/18. HI 57093

Naturwissenschaftliche Gesellschaft Bayreuth. Bericht. Naturwiss. Ges. Bayreuth Ber. See B–P–H 636/19. HI 57094

Naturwissenschaftliche Gesellschaft "Isis" in Meissen. Mitteilungen aus den Sitzungen. Meissen, Germany. Naturwiss. Ges. "Isis" Meissen Mitt. Sitzungen. See B–P–H 636/20. HI 57095

Naturwissenschaftliche Monatshefts für den biologischen, chemischen, geographischen Unterricht. Leipzig. Vols. 18-29 [also numbered vols. 1-12], 1919-1931/32. Naturwiss. Monstsh. Biol. Unterr. Preceded by: Monatshefte für den naturwissenschaftlichen Unterricht aller Schulgattungen und Natur und Schule. 4-2935-3. HI 68432

Naturwissenschaftliche Museumhefte = Múzeumi füzetek. Kolozsvar [= Cluj, Rumania].

Naturwissenschaftliche Nachrichten. Graz. Naturwiss. Nachr. See B–P–H 636/22. HI 57096

Naturwissenschaftliche Rundschau. Brunswick, Germany. Naturwiss. Rundschau (Brunswick). See B–P–H 636/24. HI 57097

Naturwissenschaftliche Rundschau. Stuttgart. Naturwiss. Rundschau (Stuttgart). See B–P–H 637/1. HI 57098

Naturwissenschaftliche Wochenschrift. Berlin. Naturwiss. Wochenschr. See B–P–H 637/3. HI 57100

Naturwissenschaftliche Zeitschrift für Forst und Landwirtschaft. Stuttgart. Naturwiss. Z. Forst- Landw. See B–P–H 637/5. HI 57101

Naturwissenschaftliche Zeitschrift Lotos. Prague. Naturwiss. Z. Lotos. See B–P–H 637/7. HI 57102

Naturwissenschaftlicher Anzeiger der allgemeinen Schweizerischen Gesellschaft für die gesammten Naturwissenschaften. Bern. Naturwiss. Anz. Allg. Schweiz. Ges. Gesammten Naturwiss. See B–P–H 636/17. HI 57092

Naturwissenschaftlicher Verein der Provinz Posen. Zeitschrift der botanischen Abteilung. Posen [=Poznan, Poland]. Naturwiss. Verein Prov. Posen Z. Bot. Abt. See B–P–H 637/2. HI 57099

Natuur. Utrecht. Natuur (Utrecht). See B–P–H 637/9. HI 57104

Natuur- en geneeskundig archief voor Nederlandsch-Indië. Batavia, Dutch E. Indies [=Jakarta, Indonesia]. Natuur- Geneesk. Arch. Ned.-Indië. See B–P–H 637/10. HI 57105

Natuur- en genees-kundige bibliotheek. The Hague. Natuur- Genees-Kundige Biblioth. See B–P–H 637/13. HI 57106

Natuur- en konstkabinet. Amsterdam. Natuur- Konstkab. See B–P–H 637/14. HI 57107

Natuur en landschap. Amsterdam. Natuur & Landschap. See B–P–H 637/15. HI 57108

Natuur en mensch. Amsterdam. Natuur & Mensch (Amsterdam). See B–P–H 637/16. HI 57109

Natuur en milieu. Amsterdam. Vols. ?-3-7, ?-1974-75; n.s. vol. 77/1+, 1977+. Natuur Milieu. HI 68433

Natuur en milieuzorg. Amsterdam. Vol. ?-1(3)+, ?-1973+. Natuur Milieuz. HI 68434

Natuur- en scheikundig archief. Rotterdam. Natuur- Scheik. Arch. See B–P–H 637/17. HI 57110

Natuur- en stedeschoon, bond beter leefmilieu. Brussels. 1970-76. Natuur- Stedesch. Bond Beter Leefmilieu. HI 68435

Natuur- en stedeschoon, koninklijke vereniging voor natuur- en stedeschoon. Antwerp. No. 1+, 1970+. Natuur- Stedesch. Kon. Ver. Natuur- Stedesch.

Preceded by: Mededelingenblad, koninklijke vereniging voor natuur- en stedeschoon. HI 68436

Natuurbehoud. Amsterdam. No. 1+, 1970+. Natuurbehoud. HI 68437

Natuurbeschouwer of verzameling van de nieuwste verhandelingen over de drie rijken de natuur. The Hague. Natuurbeschouwer. See B–P–H 637/18. HI 57111

Natuurhistorisch maandblad. Maastricht. Natuurhist. Maandbl. See B–P–H 637/20. HI 57112

Natuurkunde. Tijdschrift inhoudende physica, chemie, pharmacie, natuurlijke historie en literatuur. Arnhem, Netherlands. Natuurkunde. See B–P–H 638/7. HI 57120

Natuurkundig tijdschrift voor Nederlandsch-Indië. Batavia, Dutch E. Indies [=Jakarta, Indonesia]. Natuurk. Tijdschr. Ned.-Indië. See B–P–H 637/24. HI 57113

Natuurkundige verhandelingen van de bataafsche maatschappy der wetenschappen te Haarlem. Amsterdam. Natuurk. Verh. Bataafsche Maatsch. Wetensch. Haarlem. See B–P–H 638/1. HI 57114

Natuurkundige verhandelingen van de hollandsche maatschappij der wetenschappen te Haarlem. Amsterdam. Natuurk. Verh. Holl. Maatsch. Wetensch. Haarlem. See B–P–H 638/2. HI 57115

Natuurkundige verhandelingen van de koninglijke maatschappy der wetenschappen te Haarlem. Amsterdam. Natuurk. Verh. Kon. Maatsch. Wetensch. Haarlem. See B–P–H 638/3. HI 57116

Natuurkundige verhandelingen van de maatschappy der wetenschappen te Haarlem. Amsterdam. Natuurk. Verh. Maatsch. Wetensch. Haarlem. See B–P–H 638/4. HI 57117

Natuurkundige verhandelingen. Provinciaal utrechtsch genootschap van kunsten en wetenschappen. Utrecht. Natuurk. Verh. Prov. Utrechtsch Genootsch. Kunsten. See B–P–H 638/5. HI 57118

Natuurkundige voordrachten der maatschappij "Diligentia." The Hague. Natuurk. Voordrachten Maatsch. "Diligentia." See B–P–H 638/6. HI 57119

Natuurleven. Meppel & The Hague. Natuurleven. See B–P–H 638/8. HI 57121

Natuuronderzoeker. Orgaan van de vereeniging "De levende natuur" te Rotterdam. Rotterdam. Natuuronderzoeker. See B–P–H 638/9. HI 57122

Natuurvriend. Zutphen, Netherlands. Natuurvriend. See B–P–H 638/10. HI 57123

Natuurwetenschappilijk tijdschrift. Antwerp. Natuurw. Tijdschr. See B–P–H 638/12. HI 57124

Natuurwetenschappelijk tijdschrift voor Nederlandsch-Indië. Batavia, Dutch E. Indies [=Jakarta, Indonesia]. Natuurw. Tijdschr. Ned.-Indië. See B–P–H 638/14. HI 57125

Natuurwereld. Natuurwereld. See B–P–H 638/15. HI 57126

Nauchnaya Biblioteka. Otděl biologicheskii. St. Petersburg. 1907-11. Nauchn. Bibliot., Otd. Biol. HI 68438

Nauchni trudove, agronomicheski fakultet, vissh selskostopanski institut "Georgi Dimitrov." Sofia. Vols. 1-12, 1952-63. Nauchni Trudove Agron. Fak. Vissh Selskost. Inst. "George Dimitrov." Superseded by: Nauchni trudove, agronomicheski fakultet, vissh selskostopanski institut "Georgi Dimitrov." Seriya traini nasazhdeniya i rastitelna zashtita; Seriya rastenievudstvo. HI 68439

Nauchni trudove, agronomicheski fakultet, vissh selskostopanski institut "Georgi Dimitrov." Seriya gradinarstvo, lozarstvo i rastitelnazashchita. Sofia. Vol. 14+, 1965+. Nauchni Trudove Agron. Fak. Vissh Selskost. Inst. "George Dimitrov", Ser. Gradinarstvo, Loz. Rastitelnaz. Preceded by: Nauchni trudove, agronomicheski fakultet, vissh selskostopanski institut "Georgi Dimitrov." Seriya traini nasazhdeniya i rastitelna zashtita. HI 68440

Nauchni trudove, agronomicheski fakultet, vissh selskostopanski institut "Georgi Dimitrov." Seriya traini nasazhdeniya i rastitelna zashtita. Sofia. Vol. 13, 1964. Nauchni Trudove Agron. Fak. Vissh Selskost. Inst. "George Dimitrov", Ser. Traini Nasazhd. Rastiteln. Zashtita. Preceded by: Nauchni trudove, agronomicheski fakultet, vissh selskostopanski institut "Georgi Dimitrov." Superseded by: Nauchni trudove, agronomicheski fakultet, vissh selskostopanski institut "Georgi Dimitrov." Seriya gradinarstvo, lozarstvo i rastitelnazashchita. HI 68441

Nauchni trudove, agronomicheski fakultet, vissh selskostopanski institut "Georgi Dimitrov." Seriya rastenievudstvo. Sofia. Vols. 13-24, 1964-73. Nauchni Trudove Agron. Fak. Vissh Selskost. Inst. "George Dimitrov", Ser. Rasteniev. Preceded by: Nauchni trudove, agronomicheski fakultet, vissh selskostopanski institut "Georgi Dimitrov." HI 68442

Nauchni trudove, agronomicheski fakultet, vissh selskostopanski institut "Vasil Kolarov" = Nauchni trudove, vissh selskostopanski institut "Vasil Kolarov." Plovdiv.

Nauchni trudove, lesotekhnicheski fakultet, selskostopanska akademiya, Sofiya. Sofia. Vols. 1-2, 1952-53. Nauchni Trudove Lesotekhn. Fak. Selskost. Akad. Sofiya. Superseded by: Nauchi trudove, vissh lesotekhnicheski institut, Sofiya. HI 68444

Nauchni trudove, lozaro-gradinarski fakultet, vissh

selskostopanski institut "Vasil Kolarov" = Nauchni trudove, vissh selskostopanski institut "Vasil Kolarov." Plovdiv

Nauchni trudove, nauchno-izsledovatelski institut za zashtita na rasteniyata. Sofia. Vols. 1-3, 1958-60. Nauchni Trudove Nauchni-Izsl. Inst. Zasht. Rast. Superseded by: Izvestiya na tsentralniya nauchno-izsledovatelski institut za zashtita na rasteniyata. HI 68446

Nauchni trudove na Plovdivski universitet "Paisiĭ Khilendarski." Matematika, fizika, khimiya, biologiya. Plovdiv. Vol. 10+, 1972+. Nauchni Trudove Plovdivski Univ. "Paisiĭ Khilendarski," Mat. Fiz. Khim. Biol. Preceded by: Nauchni trudove, vissh pedagogicheski institut "Paisiĭ Khilendarski." HI 68447

Nauchi trudove, vissh lesotekhnicheski institut, Sofiya. Sofia. Vol. 3+, 1954+. Nauchi Trudove Vissh Lesotekhn. Inst. Sofiya. Preceded by: Nauchni trudove, lesotekhnicheski fakultet, selskostopanska akademiya, Sofiya. HI 68448

Nauchi trudove, vissh lesotekhnicheski institut, Sofiya. Seriya ozelenyavane. Sofia. Vol. 16+, 1968+. Nauchi Trudove Vissh Lesotekhn. Inst. Sofiya, Ser. Ozelen. Preceded, in part, by: Nauchi trudove, vissh lesotekhnicheski institut, Sofiya. HI 68449

Nauchni trudove, vissh pedagogicheski institut "Paisii Khilendarski." Matematika, fizika, khimiya, biologiya. Plovdiv. Vols. 1-9, 1964-71. Nauchni Trudove Pedagog. Inst. "Paisii Khilendarski," Mat. Fiz. Khim. Biol. Preceded by: Trudove, vissh pedagogicheskiii institut. Plovdiv. Superseded by: Nauchni trudove na plovdivski universitet "Paisiĭ Khilendarski." Matematika, fizika, khimiya, biologiya. HI 68450

Nauchni trudove, vissh selskostopanski institut "Vasil Kolarov." Plovdiv. Vol. 1+, 1952+. Nauchni Trudove Selskost. Inst. "Vasil Kolarov." HI 68451

Nauchno-tekhnicheskii byulleten' mironovskogo ordena Lenina nauchno-issledovatel'skogo instituta selektsii semenovodstva pshtenitsy = Selektsiya i semenovodstvo pshenitsy. Kiev.

Nauchno-tekhnicheskii byulleten', sibirskii nauchno-issledovatel'skii institut rastenievodstva i selektsii. Vol. ?-5/6+, ?-1978+. Nauchno-Tekhn. Byull. Sibirsk. Nauchno-Issl. Inst. Rasteniev. Selekts. HI 68452

Nauchno-tekhnicheskii byulleten' vsesoyuznogo ordena Lenina i ordena Druzhby Narodov nauchno-issledovatelstvo instituta rastenievodstva im. N. I. Vavilova. Leningrad. No. 127+, 198?+. Nauchno-Tekhn. Byull. Vsesoyuzn. Ordena Lenina Ordena Druzhby Narodov Nauchno-Issl. Inst. Rasteniev. N. I. Vavilova. Preceded by: Byulleten vsesoyuznogo instituta rastenievodstva. HI 68453

Nauchnoe Obozrěnie. Ezheneděl'nyi Nauchnyi Zhurnal. [Edited by V. V. Bitner.] St. Petersburg. 1911-12. Nauchn. Obozr. (Bitner). 4-2938-3. HI 68454

Nauchnye chteniya pamyati Mikhaila Grigor'evicha Popova. Irkutsk. Vol. 1/2+, 1956/57+. Nauchnye Chteniya Pamyati Mikhaila Grigor'evicha Popova. HI 68455

Nauchnye doklady vysshei shkoly. Biologicheskie nauki. Moscow. Vol. 1+, 1958+. Nauchnye Dokl. Vysshei Shkoly Biol. Nauki. HI 68456

Nauchnye osnovy okhrany prirody. Moscow. No. 1+, 1971+. Nauchnye Osnovy Okhr. Prir. HI 68457

Nauchnye soobshcheniya instituta biologii morya. Vladivostok Vols. 2-4, 1971-79. Nauchnye Soobshch. Inst. Biol. Morya. Preceded by: Referaty nauchnykh rabot instituta biologii morya. HI 68458

Nauchnye trudy aspirantov, odes'kyi sil's'kohospodars'kyi instytut. Rastenievodstvo. Vol. 1+, 1967+. Nauchnye Trudy Aspir. Odes'k. Sil'sk'ohosp. Inst. Rasteniev. HI 68459

Nauchnye trudy, gosudarstvennyi ordena trudovogo krasnogo znameni nikitskii botanicheskii sad. Moscow. Vol. 38, 1967. Nauchnye Trudy Gosud. Ordena Trudov. Krasn. Znam. Nikitsk. Bot. Sad. Preceded by: Sbornik nauchnykh trudov, gosudarstvennyi ordena trudovogo krasnogo znameni nikitskii botanicheskii sad. Superseded by: Trudy gosudarstvennogo nikitskogo botanicheskogo sada. HI 68460

Nauchnye trudy, instytut fiziolohiyi rastenii i ahrokhimiyi, akademiya nauk ukrayin'skoyi R S R. Kiev. Nos. 1-11, 1948-56. Nauchnye Trudy Inst. Fiziol. Rast. Ahrokhim. Akad. Nauk Ukrayin'sk. R.S.R. Superseded by: Nauchnye trudy, ukrainskii institut fiziologii rastenii. HI 68461

Nauchnye trudy kuibyshevskii gosudarstvennyi pedagogicheskii institut imeni V. V. Kuibysheva. Kuibyshev. Vol. 1+, 1970+. Nauchnye Trudy Kuibyshev. Gosud. Pedagog. Inst. Im. V. V. Kuibysheva. HI 68462

Nauchnye trudy kurskii gosudarstvennyi pedagogicheskii institut. Kursk. Vol. ?-6(=89?)+, ?-1972+ Nauchnye Trudy Kursk. Gosud. Pedagog. Inst. Preceded by: Uchenye zapiski, kurskii gosudarstvennyi pedagogicheskii institut. HI 68463

Nauchnye trudy, Moskovskii lesotekhnicheskii institut. Moscow. Vol. 1+, 1950+. Nauchnye Trudy Moskovsk. Lesotekhn. Inst. HI 75116

Nauchnye trudy nauchno-issledovatelskogo instituta rastenievodstvo v Prage-Ruzyne = Vědecké práce výzkumného ústavu rostlinné výrobý v Praze-Ruzynyibb. Prague.

Nauchnye trudy po ohrane prirody. Tartu. Vol. 1+,

1978+. Nauchn. Trudy Ohr. Prir. HI 75255

Nauchnye trudy, tambovskaya gosudarstvennaya oblastnaya sel'skokhozyaistvennaya opytnaya stantsiya. Vol. 1+, 1969+. Nauchnye Trudy Tambovsk. Gosud. Obl. Sel'skokhoz. Opytn. Stantsiya. HI 68464

Nauchnye trudy, Tashkentskii gosudarstvennyi universitet imeni V. I. Lenina. Tashkent. Nos. ?-160-484-?, ?-1960-75-? Nauchnye Trudy Tashkent. Gosud. Univ. Im. V. I. Lenina. Preceded by: Trudy sredne-Aziatskogo gosudarstvennogo universiteta imeni V. I. Lenina. Superseded by: Sbornik nauchnykh trudov, Tashkentskii gosudarstvennyi universitet. HI 68465

Nauchnye trudy, Tyumen universitet. Tyumen. [Dates of publication not ascertained.] Nauchnye Trudy Tyumen Univ. HI 68466

Nauchnye trudy, ukrainskii institut fiziologii rastenii. Kiev. Vol. 2-23?, 1959-62? Nauchnye Trudy Ukrainsk. Inst. Fiziol. Rast. Preceded by: Nauchnye trudy, institut fiziologii rastenii i agrokhimiki, akademiya nauk ukrainskoi S S R. HI 68467

Nauchnye trudy, ukrainskii nauchno-issledovatel'skii institut rastenievodstva selekt'sii i genetiki im V. Ya. Ure'eva. Vol. 13?+, 197?+. Nauchnye Trudy Ukrainsk. Nauchno-Issl. Inst. Rast. Selekt'sii Genet. V. Y. Ure'eva. Preceded by: Trudy instytutu henetyky i selekt'sii. HI 68468

Nauchnye trudy, vsesoyuznii nauchno-issledovatel'skii institut zernobovykh k krupianykh kul'tur. Moscow. Vol. 1+, 1976+. Nauchnye Trudy Vsesoyuzn. Nauchno-Issl. Inst. Zernob. Krup. Kul't. HI 68469

Nauchnye trudy, vsesoyuznii selektsionno-geneticheskii institut. Odessa. Vols. 1-?, 1949-7? Nauchnye Trudy Vsesoyuzn. Selektsion.-Genet. Inst. Superseded by: Sbornik naychnykh trudov, vsesoyuznii selektsionno-geneticheskii institut. HI 68470

Nauchnye trudy vysshikh uchebnykh zavedenii litovskoi S S R = Lietuvos T S R aukstuju mokyklu mokslo darbai. Biologija. Vilnius.

Naučnaja Biblioteka. Otdĕl biologičeskij = Nauchnaya Biblioteka. Otdĕl biologicheskii. St. Petersburg.

Naučni sbornik matice srpske. Serija prirodnih nauka. Novi Sad, Yugoslavia. Naučni Sborn. Matice Srpske, Ser. Prir. Nauka. See B–P–H 639/3. HI 57127

Naučni skupovi. Odelje prirodno-matematičkih nauka. Belgrade. Vol. 1+, 1974+. Naučni Skup. Od. Prir.-Mat. Nauka HI 68471

Naučnoe Obozrĕnie. Eženedĕl'nyj Naučnyj Žurnal. [Edited by V. V. Bitner] = Nauchnoe Obozrĕnie. Ezhenedĕl'nyi Nauchnyi Zhurnal. [Edited by V. V. Bitner.] St. Petersburg.

Naučnyj vestnik. Akademija Rumynskoj narodnoj respubliki = Buletin ştiintific. Academia republicii populare române [from 1954 ... romîne]. Bucharest. Bul. Sti. Acad. Republ. Populare Române. See B–P–H 234/1.

Nauka polska. Wroclaw. 1953+. Nauka Polska. HI 68472

Nauka dla wszystkich. Warsaw, Cracow. Vol. 1+, 1966+. Nauka Wszyst. HI 68473

Nauka i zhizn'. Moscow. Vol. 1+, 1934+. Nauka & Zhizn'. 4-2939-1. HI 68474

Nauka i žizn' = Nauka i zhizn'. Moscow.

Naukovi zapiski biologii. Cracow. Nauk. Zap. Biol. See B–P–H 639/5. HI 57128

Naukovi zapysky, biokhemichnyi instytut. Kyyiv. 1926-33. Nauk. Zap. Biokhem. Inst. Superseded by: Ukrayins'kyi biokhimichnyi zhurnal. HI 68475

Naukovi zapysky. Kyiivs'kyi derzhavnyi universytet imeny T. G. Shevchenka. Kiev, Ukrainian S S R. Vol. 1+, 1935+. Nauk. Zap. Kyiivs'k. Derzhavn. Univ. Shevchenka. 3-2289-3. HI 68476

Naukovi zapysky. Kyjivs'kyj deržavnyj universytet imeny T. G. Ševčenka = Naukovi zapysky. Kyiivs'kyi derzhavnyi universytet imeny T. G. Shevchenka. Kiev.

Naukovi zapysky. Odes'koi biologichnoi stantsii. Kiev, Ukrainian S S R. Vols. 1-3, 1959-61. Nauk. Zap. Odes'k. Biol. Stantsii. HI 68477

Naukovi zapysky. Odes'koji biologičnoji stanciji = Naukovi zapysky. Odes'koi biologichnoi stantsii. Kiev.

Naukovy zapysky. L'vivs'kyj deržavnyj universytet imeny Ivana Franka. Serija biologična = Uchenye zapiski. L'vovskii gosudarstvennyi universitet imeni Ivana Franko. Seriya biologicheskaya. Lvov.

Nautilus. Philadelphia, PA. Nautilus. See B–P–H 639/13. HI 57129

Naytenumero = Aika. Helsinki. Aika. See B–P–H 66/8.

Nazemnye i vodnye ekosistemy. Mezhvuzovskii sbornik. Gorky. Vol. 1+, 1977+. Nazem. Vodnye Mezhvuzov. Sborn. HI 68478

Nea agrotike zoe. Athens. Nea Agrotike Zoe. See B–P–H 639/15. HI 57130

Nea geoponika. Nauplia, Greece. Nea Geoponika. See B–P–H 639/16. HI 57131

Nebraska agricultural experiment station, annual report. Lincoln, NB. Nebraska Agric. Exp. Sta. Annual Rep. See B–P–H 639/24. HI 57132

Nebraska agricultural experiment station bulletin. Lincoln, NB. Nebraska Agric. Exp. Sta. Bull. See B–P–H 639/25. HI 57133

Nebraska agricultural experiment station circular. Lincoln, NB. Nebraska Agric. Exp. Sta. Circ. See

B–P–H 639/26. HI 57134

Nebraska agricultural experiment station horticultural progress report. Lincoln, NB. Nebraska Agric. Exp. Sta. Hort. Progr. Rep. See B–P–H 639/27. HI 57135

Nebraska agricultural experiment station miscellaneous publication. Lincoln, NB. Nebraska Agric. Exp. Sta. Misc. Publ. See B–P–H 640/1. HI 57136

Nebraska agricultural experiment station press bulletin. Lincoln, NB. Nebraska Agric. Exp. Sta. Press Bull. See B–P–H 640/2. HI 57137

Nebraska agricultural experiment station research bulletin. Lincoln, NB. Nebraska Agric. Exp. Sta. Res. Bull. See B–P–H 640/3. HI 57138

Nebraska agricultural experiment station wheat abstracts. Lincoln, NB. Nebraska Agric. Exp. Sta. Wheat Abstr. See B–P–H 640/4. HI 57139

Nebraska conservation bulletin. Lincoln, NE. No. 1+, 1928+. Nebraska Conservation Bull. Preceded by: Bulletin, conservation and survey division of university of Nebraska. 4-2946-1. HI 63428

Nebraska experiment station quarterly. Lincoln, NE. Vols. 1-11(2), 1952-64. Nebraska Exp. Sta. Quart. Superseded by: Quarterly, college of agriculture and home economics, university of Nebraska. HI 68479

Nebraska medical journal. Lincoln, NB. Nebraska Med. J. See B–P–H 640/6. HI 57140

Nederlandsch boschbouw-tijdschrift. Lo, Belgium. Ned. Boschbouw-Tijdschr. See B–P–H 640/9. HI 57143

Nederlandsch-Indië, oud en nieuw. Amsterdam. Ned.-Indië Oud Nieuw. See B–P–H 640/10. HI 57144

Nederlandsch-Indisch rubber- en thee-tijdschrift. Batavia, Dutch E. Indies [=Jakarta, Indonesia]. Ned.-Indisch Rubber-Thee-Tijdschr. See B–P–H 640/11. HI 57145

Nederlandsch-Indische geografische mededeelingen. Batavia, Dutch E. Indies [=Jakarta, Indonesia]. Ned.-Indische Geogr. Meded. See B–P–H 640/12. HI 57146

Nederlandsch koloniaal centraalblad. Leiden. Ned. Kolon. Centraalbl. See B–P–H 640/14. HI 57147

Nederlandsch kruidkundig archief. Verslagen en mededelingen der nederlandsche botanische vereeniging. Amsterdam. Ned. Kruidk. Arch. See B–P–H 640/17. HI 57148

Nederlandsch lancet. Amsterdam. Ned. Lancet. See B–P–H 640/19. HI 57149

Nederlandsch landbouw-weekblad. The Hague. Ned. Landb.-Weekbl. See B–P–H 640/20. HI 57150

Nederlandsch tijdschrift voor hygiene, microbiologie en serologie. Leiden. 1926-33. Ned. Tijdschr. Hyg. Microbiol. Serol. Preceded by: Tijdschrift voor vergelijkende geneeskunde, gezondheidsleer en parasitaire en infectieziekten. Superseded by: Antonie van Leeuwenhoek nederlandsch tijdschrift voor hygiëne, microbiologie en serologie. HI 68480

Nederlandsche aardappel. The Hague. Ned. Aardappel. See B–P–H 640/7. HI 57141

Nederlandsche boomkweker. Haarlem. Ned. Boomkweker. See B–P–H 640/8. HI 57142

Nederlandsche letter-courant. Leiden. Ned. Lett.-Courant. See B–P–H 640/21. HI 57151

Nederlandsche tuinbouwblad. Groningen. Ned. Tuinbouwbl. See B–P–H 640/23. HI 57153

Nederlandsche tuinbouwblad Sempervirens. Amsterdam. Ned. Tuinbouwbl. Sempervirens. See B–P–H 641/1. HI 57154

Neimenggu daxue xuebao ziran kexue = Acta scientiarum naturalium universitatis intramongolicae. Beijing.

Nematologica. Leiden. Nematologica. See B–P–H 641/6. HI 57155

Nemophila; meeting and field guide of the California botanical society. Berkeley, CA. Nemophila. See B–P–H 641/7. HI 57156

Nemouria; occasional papers, Delaware museum of natural history. Greenville, DE. No. 1+, 1970+. Nemouria. HI 68481

Nemzeti gazda. Bécs [=Vienna] & Pest [=Budapest, in part]. Nemz. Gazd. See B–P–H 641/8. HI 57157

Nép kertésze. Budapest, Korpina [=Krupina], & Selmec [=Banska Sharnica, Czechoslovakia]. Nép Kert. See B–P–H 641/9. HI 57158

Nepal agricultural abstracts. 1981+. Nepal Agric. Abstr. HI 68482

Nepalese journal of agriculture. Pulchok, Lalitpur. Vol. 1+, 1966+. Nepal. J. Agric. HI 68483

Nerine society bulletin. Clent. Nos. 1-4, 1966-70. Nerine Soc. Bull. HI 68484

Nerthus. Wochenschrift für Pflanzen- und Blumenfreunde. Altona [=Hamburg, in part]. Nerthus. See B–P–H 641/10. HI 57159

Ness series. Neston. Vol. 1+, 1986+. Ness Ser. HI 68485

Nestel's Rosengarten. Stuttgart. Nestel's Rosengart. See B–P–H 641/11. HI 57160

Nestlé research news. Lausanne, Vevey. 1971+. Nestlé Res. News. HI 68486

Netherlands journal of agricultural science. Wageningen. Netherlands J. Agric. Sci. See B–P–H 641/13. HI 57161

Netherlands journal of plant pathology; official organ of

the Netherlands phytopathological society. Wageningen. Vol. 69+, 1963+. Netherlands J. Pl. Pathol. Preceded by: Tijdschrift over plantenziekten. HI 68487

Netherlands journal of sea research. Helder, Netherlands. Netherlands J. Sea Res. See B–P–H 641/14. HI 57162

Nettai engei = Tropical horticulture. Taihoku [=Taipei, Taiwan]. Trop. Hort. See B–P–H 897/15.

Nettai nogaku kwaishi = Journal of the society of tropical agriculture. Taihoku [=Taipei, Taiwan]. J. Soc. Trop. Agric. See B–P–H 482/13.

Neu fortgesetzte Sammlung ökonomischer Schriften. Leipzig. Neu Fortgesetzte Samml. Ökon. Schriften. See B–P–H 641/19. HI 57163

Neubert's Deutsches Garten-Magazin. Stuttgart. Neubert's Deutsch. Gart.-Mag. See B–P–H 641/24. HI 57164

Neubert's Garten-Magazin. Munich. Neubert's Gart.-Mag. See B–P–H 641/25. HI 57165

Neue Abhandlungen und Nachrichten der Königlich Grossbritannischen Churfürstlich Braunschweig-Lüneburgischen Landwirthschafts-Gesellschaft. Celle & Hanover. Neue Abh. Nachr. Königl. Grossbritt. Churfürstl. Braunschweig-Lüneburg. Landw.-Ges. See B–P–H 641/27. HI 57166

Neue Abhandlungen aus der Naturlehre, Haushaltungskunst und Mechanik, der Königlich Schwedischen Akademie der Wissenschaften = Königl[lich]. schwedischen Akademie der Wissenschaften. Neue Abhandlungen aus der Naturlehre, Haushaltungskunst und Mechanik. Leipzig.

Neue Ackerbau-Zeitung der Ackerbau-Gesellschaft und der vier Bezirks-Comitien des Niederrheins. Strasbourg. Neue Ackerbau-Zeitung Ackerbau-Ges. Niederrheins See B–P–H 641/28. HI 57167

Neue allgemeine deutsche Bibliothek. Kiel. Neue Allg. Deutsche Biblioth. See B–P–H 642/1. HI 57168

Neue allgemeine deutsche Garten- und Blumenzeitung. Hamburg. Neue Allg. Deutsche Garten-Blumenzeitung. See B–P–H 642/2. HI 57169

Neue allgemeine geographische Ephemeriden. Weimar. Neue Allg. Geogr. Ephem. See B–P–H 642/4. HI 57170

Neue Alpina. Eine Schrift der Schweizerischen Naturgeschichte, Alpen- und Landwirthschaft gewiedmet. Winterthur, Switzerland. Neue Alpina. See B–P–H 642/6. HI 57171

Neue Analekten für Erd- und Himmels-Kunde. Munich. Neue Analekten Erd- Himmels-Kunde. See B–P–H 642/8. HI 57172

Neue Anmerkungen über alle Theile der Naturlehre, aus denen englischen Transactionen, denen Gedenkschriften der Akademie der Wissenschaften in Paris, und andern mehr zusammengezogen und gesamlet. Aus dem Französischen übersetzt. Copenhagen & Leipzig. Neue Anmerk. Naturl. See B–P–H 642/10. HI 57173

Neue Annalen der Blumisterei für Gartenbesitzer, Kunstgärtner, Samenhändler und Blumenfreunde. Nuremberg. Neue Ann. Blumisterei Gartenbesitz. See B–P–H 642/11. HI 57174

Neue Annalen der Botanick = Annalen der Botanick. ed. Usteri. Zurich. Ann. Bot. (Usteri). See B–P–H 98/12.

Neue Annalen der Botanik = Annalen der Botanick. ed. Usteri. Zurich. Ann. Bot. (Usteri). See B–P–H 98/12.

Neue Annalen der Mecklenburgischen Landwirthschafts-Gesellschaft. Rostock. Neue Ann. Mecklenburg. Landw.-Ges. See B–P–H 642/14. HI 57175

Neue Annalen aller Verhandlungen der vereinigten öconomisch-patriotischen Societät der Fürstenthümer Schweidnitz und Jauer. Jauer, Germany [=Jawor, Poland]. Neue Ann. Verh. Vereinigten Öcon.-Patriot. Soc. Fürstenth. Schweidnitz Jauer. See B–P–H 642/15. HI 57176

Neue Annalen der Wetterauischen Gesellschaft für die gesammte Naturkunde = Annalen der Wetterauischen Gesellschaft für die gesammte Naturkunde. Frankfurt a. M. Ann. Wetterauischen Ges. Gesammate Naturk. See B–P–H 118/12.

Neue Auszüge aus den besten ausländischen Wochen- und Monatschriften. Frankfurt am Main. Vols. 1-9/10, 1765-69. Neue Auszüge Besten Ausl. Wochen-Monatschr. HI 52910

Neue Bauwelt. Berlin. Vols. 1(37)-7(43<17>), 1946-52. Neue Bauwelt. Preceded and superseded by: Bauwelt. 1-611-3. HI 68488

Neue Beiträge zur Kenntniss von Afrika. Berlin. Neue Beitr. Kenntn. Afrika. See B–P–H 642/17. HI 57177

Neue Beiträge Zur Medizin und Chirurgie. Teplitz, Bohemia [=Teplice, Czechoslovakia]. Neue Beitr. Med. Chir. See B–P–H 642/18. HI 57178

Neue Beiträge zur Völker- und Länderkunde. Leipzig. Neue Beitr. Völker- Länderk. See B–P–H 642/19. HI 57179

Neue Berliner Beyträge zur Landwirthschafts-wissenschaft. Berlin. Neue Berliner Beytr. Landwirthschaftswiss. See B–P–H 642/22. HI 57181

Neue Berlinische Monatsschrift. Berlin. Neue Berlin. Monatsschr. See B–P–H 642/21. HI 57180

Neue Berlinische Zeitung von den merkwürdigsten Sachen aus dem Gebiete der Staaten, dem Reiche der Natur und der Wissenschaften. Berlin. Juli-Dezember,

1796. Neue Berlin. Zeitung Merkwürdigsten Sachen Geb. Staaten Reiche Natur Wiss. HI 68489

Neue Beyträge zur Botanik. Frankfurt a. M. Neue Beytr. Bot. Se: 643/1. HI 57182

Neue Bibliothek, oder Nachrichten und Urtheile von neuen Büchern. Frankfurt a. M. & Leipzig. Neue Biblioth. See B–P–H 643/2. HI 57183

Neue Bibliothek von seltenen und sehr seltenen Büchern und kleinen Schriften. Nuremberg. Neue Biblioth. Seltenen Sehr Seltenen Büchern Kleinen Schriften. See B–P–H 643/3. HI 57184

Neue Bibliothek der wichtigsten Reisebeschreibungen zur Erweiterung der Erd- und Völkerkunde. Weimar. Neue Biblioth. Wichtigsten Reisebeschreib. Erweit. Erd-Völkerk. See B–P–H 643/4. HI 57185

Neue Denkschriften der Allg. Schweizerischen Gesellschaft für die gesammten Naturwissenschaften. Neuchâtel. Neue Denkschr. Allg. Schweiz. Ges. Gesammten Naturwiss. See B–P–H 643/5. HI 57186

Neue Denkschriften der Physikalisch-Medicinischen Societät zu Erlangen, vol. 1 = Abhandlungen der Physikalisch-Medicinischen Societät zu Erlangen. Frankfurt a. M. Abh. Phys.-Med. Soc. Erlangen. See B–P–H 33/13.

Neue Denkschriften der Schweizerischen Naturforschenden Gesellschaft. Zurich. Neue Denkschr. Schweiz. Naturf. Ges. See B–P–H 643/8. HI 57187

Neue Entdeckungen und Beobachtungen aus der Physik, Naturgeschichte und Oekonomie. Frankfurt a. M. Neue Entdeck. Beob. Phys. See B–P–H 643/12. HI 57188

Neue Entdeckungen im ganzen Umfang der Pflanzenkunde. Leipzig. Neue Entdeck. Pflanzenk. See B–P–H 643/13. HI 57189

Neue-fortgesetzter Parnassus Boicus, oder Neu-eröffneter Musenberg, worauf verschiedene Denk- und Leswürdigkeiten aus der gelehrten Welt, zumalen aber aus den Landen zu Baiern, abgehandelt werden. Augsburg & Stadt am Hof. Neu-Fortgesetzter Parnassus Boicus. See B–P–H 651/5. HI 57304

Neue Hallische gelehrte Zeitungen. Halle. Neue Hallische Gel. Zeitungen. See B–P–H 643/14. HI 57190

Neue Hefte zur Morphologie. Weimar. Neue Hefte Morphol. See B–P–H 643/15. HI 57191

Neue historische Abhandlungen der baierischen Akademie der Wissenschaften. Munich. Neue Hist. Abh. Baier. Akad. Wiss. See B–P–H 643/16. HI 57192

Neue historische Abhandlungen der churfürstlichen baierischen Akademie der Wissenschaften. Munich. Neue Hist. Abh. Churfürstl. Baier. Akad. Wiss. See B–P–H 643/17. HI 57193

Neue Jahrbücher der Forstkunde. Mainz. Neue Jahrb. Forstk. See B–P–H 643/18. HI 57194

Neue Jahrbücher der Königlich Philosophisch-medicinischen Gesellschaft zu Würzburg. Abtheilung für Natur- und Heilkunde. Würzburg. Neue Jahrb. Königl. Philos.-Med. Ges. Würzburg. See B–P–H 643/19. HI 57195

Neue jenaische allgemeine Literatur-Zeitung. Leipzig. Vols. 1-7, 1842-48. Neue Jenaische Allg. Lit-Zeitung. Preceded by: Jenaische allgemeine Literatur-Zeitung. 4-2969-1. HI 68490

Neue Kielische gelehrte Zeitungen. Kiel. Neue Kiel. Gel. Zeitungen. See B–P–H 644/2. HI 57198

Neue landwirtschaftliche Zeitung. Stuttgart. Vols. 1-22, 1852-73. Neue Landw. Zeitung. Superseded by: Fühlings landwirtschaftliche Zeitung. 2-1651-3. HI 68491

Neue Lausizische Monatsschrift. Görlitz. Neue Lausiz. Monatsschr. See B–P–H 644/3. HI 57199

Neue Leipziger gelehrte Anzeigen oder Nachrichten von neuen Büchern und kleinen Schriften besonders der Chursächsischen Universitäten, Schulen und Lande. Leipzig. Neue Leipziger Gel. Anz. See B–P–H 644/6. HI 57200

Neue Leipziger gelehrte Zeitungen. Leipzig. Neue Leipziger Gel. Zeitungen. See B–P–H 644/7. HI 57201

Neue Leipziger Literaturzeitung. Leipzig. Neue Leipziger Literaturzeitung. See B–P–H 644/8. HI 57202

Neue Litteratur und Beyträge zur Kenntniss der Naturgeschichte; vorzüglich der Conchylien und Fossilien. Leipzig. Vols. 1-4, 1784-87. Neue Litt. Beytr. Kenntn. Naturgesch. Preceded by: Für die Litteratur und Kenntniss der Naturgeschichte. 4-2969-2. HI 68492

Neue Mannigfaltigkeiten. Berlin. Neue Mannigfaltigk. See B–P–H 644/10. HI 57203

Neue medicinisch-chirurgische Zeitung. Innsbruck & Leipzig. 1840-56 [4 issues called "Band" published annually. Vols. for 1843-50 also numbered n.s. vols. 1-8]. Neue Med.-Chir. Zeitung. Preceded by: Medicinisch-chirurgische Zeitung. Superseded by: Medicinisch-chirurgische Monatshefte. 4-2969-3. HI 51013

Neue medicinisch-chirurgische Zeitung. Ergänzungsband. Innsbruck & Leipzig. Neue Med.-Chir. Zeitung Ergänzungsband. See B–P–H 644/13. HI 57205

Neue medizinische Bibliothek. Göttingen. Neue Med. Biblioth. See B–P–H 644/11. HI 57204

Neue Mitteilungen für die Landwirtschaft. Frankfurt. Vols. 61-66(39) [also numbered n.s. vols. 1-6(39)], 1946-51. Neue Mitt. Landw. Preceded by: Mitteilungen für die Landwirtschaft. Frankfurt am Main. Superseded by: Mitteilungen der deutschen Landwirtschafts-Gesellschaft. 2-1307-2. HI 68493

Neue monatliche Beiträge zur Naturkunde. Schwerin. Neue Monatl. Beitr. Naturk. (Schwerin). See B–P–H 644/16. HI 57206

Neue monatliche Beyträge zur Naturkunde. Dessau. Neue Monatl. Beytr. Naturk. (Dessau). See B–P–H 644/17. HI 57207

Neue Museumskunde; Informationsorgan über die Arbeit der kulturgeschichtlichen und naturkundlichen Museen in der Deutschen Demokratischen Republik. Halle [Saale]. Vol. 1+, 1949+. Neue Museumsk. HI 68494

Neue Nachrichten von Abhandlungen der Grossbrit. Churfürstl.-Brandenburgischen Landwirthschafts-Gesellschaft zu Celle. Celle. Neue Nachr. Abh. Grossbrit. Churfürstl.-Brandenburg. Landw.-Ges. Celle. See B–P–H 644/18. HI 57208

Neue Nordische Beyträge zur physikalischen und geographischen Erd- und Völkerbeschreibung, Naturgeschichte und Oekonomie. St. Petersburg & Leipzig. Neue Nord. Beytr. Phys. Geogr. Erd-Völkerbeschreib. See B–P–H 644/21. HI 57209

Neue Notizen aus dem Gebiete der Natur- und Heilkunde. Weimar. Neue Not. Natur- Heilk. See B–P–H 645/1. HI 57210

Neue Nürnbergische Gelehrte Zeitung. Nuremberg. Neue Nürnberg. Gel. Zeitung. See B–P–H 645/4. HI 57211

Neue oberdeutsche allgemeine Litteraturzeitung. Munich. Neue Oberdeutsche Allg. Litteraturzeitung. See B–P–H 645/5. HI 57212

Neue Oeconomische Nachrichten. Leipzig. Neue Oecon. Nachr. See B–P–H 645/6. HI 57213

Neue philosophische Abhandlungen der baierischen Akademie der Wissenschaften. Munich. Neue Philos. Abh. Baier. Akad. Wiss. See B–P–H 645/7. HI 57214

Neue Physikalische Belustigungen. Prague. Neue Phys. Belust. See B–P–H 645/8. HI 57215

Neue Pommersche Provinzialblätter. Stettin [Poland]. Neue Pommersche Provinzialbl. See B–P–H 645/10. HI 57216

Neue Sammlung geographisch historisch-statistischer Schriften. Neue Samml. Geogr. Hist.-Statist. Schriften. See B–P–H 645/11. HI 57217

Neue Sammlung interessanter und zwekmässig abgefasster Reisebeschreibungen für die Jugend. Tübingen. Neue Samml. Interessanter Zwekmässig Abgefasster Reisebeschreib. Jugend. See B–P–H 645/12. HI 57218

Neue Sammlung merkwürdiger Reisebeschreibungen für die Jugend. Frankfurt a. M. & Leipzig. Neue Samml. Merkwürd. Reisebeschreib. Jugend. See B–P–H 645/13. HI 57219

Neue Sammlung der merkwürdigsten Reisegeschichten. Frankfurt a. M. Neue Samml. Merkwürdigsten Reisegesch. See B–P–H 645/14. HI 57220

Neue Sammlung physisch-ökonomischer Schriften herausgegeben von der ökonomischen Gesellschaft in Bern. Bern. Neue Samml. Phys.-Ökon. Schriften Ökon. Ges. Bern. See B–P–H 645/15. HI 57221

Neue Sammlung von Reisebeschreibungen. Hamburg. Neue Samml. Reisebeschreib. See B–P–H 645/16. HI 57222

Neue Sammlung der Reisen nach dem Orient. Königsberg [=Kaliningrad, Russian S F S R]. Neue Samml. Reisen Orient. See B–P–H 645/17. HI 57223

Neue Sammlung verschiedener Schriften der grössten Gelehrten in Schweden. Copenhagen. Neue Samml. Schriften Grössten Gel. Schweden. See B–P–H 645/18. HI 57224

Neue Sammlung, von Versuchen und Abhandlungen der naturforschenden Gesellschaft in Danzig. Danzig [=Gdansk, Poland]. Neue Samml. Versuchen Abh. Naturf. Ges. Danzig. See B–P–H 645/19. HI 57225

Neue Schelswig-Holstein-Lauenburgische Provinzialberichte. Kiel. Neue Schelswig-Holstein-Lauenburg Provinzialber. See B–P–H 645/21 HI 57227

Neue Schleswig-Holsteinische Provinzialberichte. Kiel. Neue Schleswig-Holst. Provinzialber. See B–P–H 645/20. HI 57226

Neue Schriften, Gesellschaft naturforschender Freunde zu Berlin = Der Gesellschaft naturforschender Freunde zu Berlin, Neue Schriften. Berlin. Ges. Naturf. Freunde Berlin Neue Schriften. See B–P–H 400/7.

Neue Schriften der Gesellschaft naturforschender Freunde Westphalens. Berlin. Neue Schriften Ges. Naturf. Freunde Westphalens. See B–P–H 646/3. HI 57228

Neue Schriften der kais. königl. patriotisch-ökonomischen Gesellschaft im Königreiche Böhmen. Prague. Neue Schriften K. K. Patriot.-Ökon. Ges. Königr. Böhmen. See B–P–H 646/4. HI 57229

Neue Schriften der naturforschenden Gesellschaft zu Halle. Halle. Neue Schriften Naturf. Ges. Halle. See B–P–H 646/5. HI 57230

Neue Versuche und Bemerkungen einer Gesellschaft zu Edinburg vorgelesen und von ihr herausgegeben Gelehrsamkeit. Altenburg. Vols. 1-2, 1756-57. Neue Versuche Bemerk. Arztneykunst. HI 68495

Neue Versuche nützlicher Sammlungen zu der Natur- und Kunst-Geschichte sonderlich von Ober-Sachsen. Schneeberg, Germany. Neue Versuche Nützl. Samml. Natur- Kunst-Gesch. Ober-Sachsen. See B–P–H 646/7. HI 57231

Neue wöchentliche Unterhaltungen grösstentheils über Gegenstände der Literatur und Kunst. Mitau. Vols. 1-2, 1808. Neue Wöchentl. Unterhalt. Gegenstände Lit. Kunst. HI 53963

Neue Würzburger Gelehrte Anzeigen. Würzburg. Neue Würzburger Gel. Anz. See B–P–H 646/9. HI 57232

Neue Zeitschrift des Ferdinandeums für Tirol und Vorarlberg. Innsbruck. Vols. 1-12, 1835-46. Neue Z. Ferdinandeums Tirol. Preceded by: Zeitschrift für Tirol und Vorarlberg Superseded by: Zeitschrift des Ferdinandeums für Tirol und Vorarlberg. 2-1553-2. HI 68496

Neue Zeitschrift für das Forst- und Jagdwesen. Bamberg. Vol. 1, 1823. Neue Z. Forst- & Jaagdwesen. Preceded by: Zeitschrift für das Forst- und Jagdwesen, für Cameral- u. Forstbeamte, Forst u. Jagdliebhaber. Superseded by: Zeitschrift für das Jagd- und Forstwesenn mit besonderer Rücksicht auf Baiern. 5-4598-2. HI 68497

Neue Zeitschrift für Geburtskunde. Berlin. Neue Z. Geburtsk. See B–P–H 646/11. HI 57233

Neue Zeitschrift für Natur- und Heilkunde. Dresden. Neue Z. Natur- Heilk. See B–P–H 646/12. HI 57234

Neue Zeitungen von gelehrten Sachen. Leipzig. Neue Zeitungen Gel. Sachen. See B–P–H 646/14. HI 57235

Neuer Almanach aller um Hamburg Liegenden Gärten. Hamburg. Neuer Alman. Aller um Hamburg Liegenden Gärt. See B–P–H 646/15. HI 57236

Neuer oder fortgesetzter allgemeiner literarischer Anzeiger. Leipzig. Neuer Fortgesetzter Allg. Lit. Anz. See B–P–H 646/17. HI 57237

Neuer und gar nützlicher Garten-Calender. Augsburg. Neuer Nützl. Gart.-Calend. See B–P–H 646/19. HI 57238

Neuere Abhandlungen der k. Böhmischen Gesellschaft der Wissenschaften. Prague. Neuere Abh. Königl. Böhm. Ges. Wiss. See B–P–H 646/20. HI 57239

Neuere Sammlung der merkwürdigsten Reisegeschichten. Frankfurt a. M. Neuere Samml. Merkwürd. Reisegesch. See B–P–H 646/21. HI 57240

Neueres Forstmagazin. Erste Abtheilung. Sammlung zerstreueter Forstschriften. Frankfurt a. M. Neueres Forstmag., Abth. 1. See B–P–H 646/22. HI 57241

Neueres Forstmagazin. Zweyte Abtheilung von neuen Aufsätzen, ... die Forstsachen und dahin einschlagende hülfreiche Wissenschaften betreffend auch von ältern, mittlern und neuern Büchern welche eigentlich das Forstwesen behandeln. Frankfurt a. M. Neueres Forstmag., Abth. 2. See B–P–H 646/23. HI 57242

Neues allgemeines Garten-Magazin oder gemeinnützige Beiträge für alle Theile des Teutschen Gartenwesens. Weimar. Neues Allg. Gart.-Mag. See B–P–H 647/1. HI 57243

Neues allgemeines Journal der Chemie. Berlin. Neues Allg. J. Chem. See B–P–H 647/2. HI 57244

Neues allgemeines Repertorium der neuesten in- und ausländischen Literatur. Leipzig. Neues Allg. Repert. Neuesten In- Ausl. Lit. See B–P–H 647/5. HI 57245

Neues Archiv für Geschichte, Staatenkunde, Literatur und Kunst. Vienna. Neues Arch. Gesch. See B–P–H 647/6. HI 57246

Neues Archiv für den Menschen und Bürger in allen Verhältnissen, ... Leipzig. Neues Arch. Menschen Bürger. See B–P–H 647/7. HI 57247

Neues Bergmännisches Journal. Freiberg, Germany. Neues Bergmänn. J. See B–P–H 647/9. HI 57248

Neues Berlinisches Jahrbuch für die Pharmacie = Berlinisches Jahrbuch für die Pharmacie und für die damit verbundenen Wissenschaften. Berlin. Berlin. Jahrb. Pharm. Verbundenen Wiss. See B–P–H 183/11.

Neues Berlinisches Magazin, oder gesammelte Schriften und Nachrichten für die Liebhaber der Arzneywissenschaft, Naturgeschichte und der angenehmen Wissenschaften überhaupt. Berlin. Neues Berlin. Mag. See B–P H 647/11. HI 57249

Neues Berlinisches Wocnenblatt zur Belehrung und Unterhaltung. Berlin. Neues Berlin. Wochenbl. Belehr. Unterhalt. See B–P–H 647/12. HI 57250

Neues Botanisches Garten-Journal. Eisenach, Germany. Neues Bot. Gart.-J. See B–P–H 647/13. HI 57251

Neues Botanisches Taschenbuch für die Anfänger dieser Wissenschaft und der Apothekerkunst. Nuremberg & Altdorf. Neues Bot. Taschenb. Anfänger Wiss. Apothekerkunst. See B–P–H 647/14. HI 57252

Neues Bremisches Magazin zur Ausbreitung der Wissenschaften, Künste und Tugend. Bremen. Neues Bremisches Mag. See B–P–H 647/17. HI 57253

Neues chemisches Archiv. Leipzig. Neues Chem. Arch. See B–P–H 647/18. HI 57254

Neues deutsches Magazin. Hamburg. Neues Deutsch. Mag. See B–P–H 647/20. HI 57255

Neues deutsches Museum. Leipzig. Neues Deutsch. Mus. See B–P–H 647/21. HI 57256

Neues Forst-Archiv zur Erweiterung der Forst- und Jagd-Wissenschaft und der Forrst- und Jagd-Literatur = Forst-Archiv zur Erweiterung der Forst- und Jagd-Wissenschaft und der Forrst- und Jagd-Literatur. Ulm.

Forst-Arch. Erweit. Forst- Jagd-Wiss. See B–P–H 382/13.

Neues fortgesetztes Westfälisches Magazin zur Geographie, Historie und Statistik. Lemgo, Germany. Neues Fortgesetztes Westfäl. Mag. Geogr. See 648/3. HI 57257

Neues gemeinnütziges Magazin für Freunde der nützlichen und schönen Wissenschaften und Künste. Hamburg. Neues Gemeinnütz. Mag. Freunde Wiss. Künste. See B–P–H 648/4. HI 57258

Neues Hamburgisches Archiv zur Verbreitung nützlicher und angenehmer Kenntnisse. Hamburg. Neues Hamburg. Arch. See B–P–H 648/5. HI 57259

Neues Hamburgisches Magazin, oder Fortsetzung gesammleter Schriften, aus der Naturforschung, der allgemeinen Stadt- und Land-Oekonomie, und den angenehmen Wissenschaften überhaupt. Hamburg & Leipzig. Neues Hamburg. Mag. See B–P–H 648/6. HI 57260

Neues Hannoverisches Magazin, worin kleine Abhandlungen, ... gesamlet und aufbewahrt sind. Hanover. Neues Hannover. Mag. See B–P–H 648/7. HI 57261

Neues Jahrbuch der Chemie und Physik, vols. 1-9 = Journal für Chemie und Physik, vols. 61-69, 1831-33. Nuremberg. J. Chem. Phys. (Nuremberg). See B–P–H 461/19.

Neues Jahrbuch für Geologie und Paläontologie. Abhandlungen. Stuttgart. Neues Jahrb. Geol. Paläontol., Abh. See B–P–H 648/14. HI 57267

Neues Jahrbuch für Geologie und Paläontologie. Monatshefte. Stuttgart. Neues Jahrb. Geol. Paläontol., Monatsh. See B–P–H 648/15. HI 57268

Neues Jahrbuch der Landwirthschaft. Breslau [=Wroclaw, Poland]. Neues Jahrb. Landw. See B–P–H 648/16. HI 57269

Neues Jahrbuch für Mineralogie, Geognosie, Geologie und Petrefaktenkunde. Stuttgart. Neues Jahrb. Mineral. Geognosie. See B–P–H 648/17. HI 57270

Neues Jahrbuch für Mineralogie, Geologie und Paläontologie. Stuttgart. Neues Jahrb. Mineral. Geol. See B–P–H 648/18. HI 57271

Neues Jahrbuch für Mineralogie, Geologie und Paläontologie. Abhandlungen. Abteilung B. Geologie und Paläontologie. Stuttgart. Neues Jahrb. Mineral. Geol., Abh., Abt. B, Geol. Paläontol. See B–P–H 648/19. HI 57272

Neues Jahrbuch für Mineralogie, Geologie und Paläontologie. Abteilung B. Geologie und Paläontologie. Referate. Stuttgart. Neues Jahrb. Mineral. Geol., Abt. B, Geol. Paläontol., Ref. See B–P–H 649/1. HI 57273

Neues Jahrbuch für Mineralogie, Geologie und Paläontologie. Beilageband. Stuttgart. Neues Jahrb. Mineral. Geol., Beilageband. See B–P–H 649/2. HI 57274

Neues Jahrbuch für Mineralogie, Geologie und Paläontologie. Beilageband. Abteilung B. Geologie und Paläontologie. Stuttgart. Neues Jahrb. Mineral. Geol., Beilageband, Abt. B, Geol. Paläontol. See B–P–H 649/3. HI 57275

Neues Jahrbuch für Mineralogie, Geologie und Paläontologie. Monatshefte. Abteilung B. Geologie, Paläontologie. Stuttgart. Neues Jahrb. Mineral. Geol., Monatsh., Abt. B, Geol. Paläontol. See B–P–H 649/4. HI 57276

Neues Jahrbuch für Mineralogie, Geologie und Paläontologie. Referate. Abteilung B. Geologie, Paläontologie. Stuttgart. Neues Jahrb. Mineral. Geol., Ref., Abt. B, Geol. Paläontol. See B–P–H 649/5. HI 57277

Neues Jahrbuch für Mineralogie, Geologie und Paläontologie. Referate. Teil 3, Historische und regionale Geologie, Paläontologie. Stuttgart. Neues Jahrb. Mineral. Geol., Ref., Teil 3, Hist. Regionale Geol. Paläontol. See B–P–H 649/6. HI 57278

Neues Jahrbuch der Pharmacie = Berlinisches Jahrbuch für die Pharmacie und für die damit verbundenen Wissenschaften. Berlin. Berlin. Jahrb. Pharm. Verbundenen Wiss. See B–P–H 183/11.

Neues Jahrbuch für Pharmacie und verwandte Fächer. Speyer, Germany. Neues Jahrb. Pharm. Verwandte Fächer. See B–P–H 649/8. HI 57279

Neues Journal für die Botanik. [Edited by Schrader.] Erfurt. Neues J. Bot. See B–P–H 648/8. HI 57262

Neues Journal der Erfindungen, Theorien und Widersprüche in der Natur- und Arzneiwissenschaft. Gotha. Neues J. Erfind. Natur- Arzneiwiss. See B–P–H 648/9. HI 57263

Neues Journal für Fabriken, Manufacturen, Handlung, Kunst und Mode. Leipzig. Neues J. Fabriken. See B–P–H 648/10. HI 57264

Neues Journal der Pharmacie für Aerzte, Apotheker und Chemiker. Leipzig. Neues J. Pharm. Aerzte. See B–P–H 648/11. HI 57265

Neues Journal der Physik. Halle. Neues J. Phys. See B–P–H 648/12. HI 57266

Neues Kielisches Litteraturjournal. Dessau & Leipzig. Neues Kiel. Litteraturj. See B–P–H 649/12. HI 57280

Neues Kunst- und Gewerbblatt. Munich. Neues Kunst-Gewerbbl. See B–P–H 649/13. HI 57281

Neues Lausitzisches Magazin. Görlitz. Neues Lausitz. Mag. See B–P–H 649/14. HI 57282

Neues Magazin für Aerzte. Leipzig. Neues Mag. Aerzte See B–P–H 649/15. HI 57283

Neues Magazin für die Botanik. [Edited by Römer.] Zurich. Neues Mag. Bot. See B–P–H 648/16. HI 57284

Neues Magazin für die gerichtliche Arzneikunde und medizinische Polizei. Stendal, Germany. Neues Mag. Gerichtl. Arzneik. Med. Polizei. See B–P–H 649/18. HI 57285

Neues Magazin aller neuen Erfindungen, Entdeckungen und Verbesserungen. Leipzig. Neues Mag. Neuen Erfind. See B–P–H 650/1. HI 57286

Neues und Nutzbares aus dem Gebiete der Haus- und Landwirthschaft. Weimar. Neues Nutzbares Geb. Haus- Landw. See B–P–H 650/2. HI 57287

Neues schwedisches Magazin kleiner Abhandlungen welche in die Natur- und Haushaltungskunde einschlagen. Nuremberg. Neues Schwed. Mag. See B–P–H 650/3. HI 57288

Neues Schweizerisches Museum. Zurich. Neues Schweiz. Mus. See B–P–H 650/5. HI 57289

Neues Taschenbuch für Natur-, Forst- und Jagdfreunde. Weimar. 1836-53, 1835-52. Neues Taschenb. Natur- Forst- Jagdfr. HI 68498

Neues vaterländisches Archiv, oder Beiträge zur allseitigen Kenntnis des Königreichs Hannover, wie es war und ist. Lüneburg, Germany. Neues Vaterl. Arch. See B–P–H 650/6. HI 57290

Neues westphälisches Magazin zur Geographie, Historie und Statistik. Bielefeld, Lemgo, Leipzig, & Bückeburg. Neues Westphäl. Mag. Geogr. See B–P–H 650/7. HI 57291

Neues Wittenbergisches Wochenblatt, eine Sammlung von Aufsätzen und Wahrnehmungen über die Witterungen, die Haushaltungskunde, das Gewerbe, die Naturkenntniss, Policey, und andere damit verknüpfte Wissenschaften. Wittenberg. Neues Wittenberg. Wochenbl. See B–P–H 650/8. HI 57292

Neues Wochenblatt des Landwirthschaftlichen Vereins in Baiern. Munich. Neues Wochenbl. Landw. Vereins Baiern. See B–P–H 650/9. HI 57293

Neueste Annalen der französischen Arzneykunde und Wundarzneykunst. Leipzig. Neueste Ann. Franz. Arzneyk. Wundarzneykunst. See B–P–H 650/12. HI 57294

Neueste aus der anmuthigen Gelehrsamkeit. Leipzig. 1751-62. Neueste Anmuth. Gelehrsamk. HI 52902

Neueste Beiträge zur Kunde der Insel Madagaskar. Weimar. Neueste Beitr. Kunde Insel Madagaskar. See B–P–H 650/13. HI 57295

Neueste Mannigfaltigkeiten. Berlin. Neueste Mannigfaltigk. See B–P–H 650/14. HI 57296

Neueste Nordische Beyträge zur physikalischen und geographischen Erd- und Völkerbeschreibung, Naturgeschichte und Oekonomie. St. Petersburg & Leipzig. Neueste Nord. Beytr. Phys. Geogr. Erd- Völkerbeschreib. See B–P–H 650/15. HI 57297

Neueste Sammlung von Abhandlungen und Beobachtungen. Bern. Neueste Samml. Abh. Beob. See B–P–H 650/16. HI 57298

Neueste Schriften der Naturforschenden Gesellschaft in Danzig. Danzig [=Gdansk, Poland]. Neueste Schriften Naturf. Ges. Danzig. See B–P–H 650/17 HI 57299.

Neuesten Entdeckungen in der Chemie. Danzig [=Gdansk, Poland]. Neuesten Entdeck. Chem. See B–P–H 650/18. HI 57300

Neuesten Entdeckungen in der Chemie, vol. 13 = Auswahl aller eigenthümlichen Abhandlungen und Beobachtungen aus den neuesten Entdeckungen in der Chemie. Leipzig. Auswahl Eigenthüml. Abh. Beob. Neuesten Entdeck. Chem. See B–P–H 165/13.

Neuesten Endeckungen französischer Gelehrten in den gemeinnützigen Wissenschaften und Künsten = Französische Annalen für die allgemeine Naturgeschichte, Physik, Chemie, Physiologie und ihre gemeinnützigen Anwendungen. Hamburg. Franz. Ann. Allg. Naturgesch. See B–P–H 384/24.

Neuestes chemisches Archiv. Weimar. Neuestes Chem. Arch. See B–P–H 651/2. HI 57301

Neuestes Garten-Jahrbuch, nach Le bon jardinier ... Weimar. Neuestes Gart.-Jahrb. See B–P–H 651/3. HI 57302

Neuestes Journal der Erfindungen, Theorien und Widersprüche in der gesammten Medizin. Gotha. Neuestes J. Erfind. Gesammten Med. See B–P–H 651/4. HI 57303

Neujahrsblatt. Herausgeber Naturforschende Gesellschaft Schaffhausen. Thayngen, Switzerland. Neujahrsbl. Naturf. Ges. Schaffhausen. See B–P–H 651/7. HI 57305

Neujahrsblatt der Naturforschenden Gesellschaft in Zürich. Zurich. Neujahrsbl. Naturf. Ges. Zürich. See B–P–H 651/8. HI 57306

Neujahrsgeschenck für Forst- und Jadgliebhaber. Marburg. 1795-99; ed. 2 for the years 1795-97. NeujahrsGeschenck Forst- Jagdliebh. Superseded by: Taschenbuch für Forst- und Jagdfreunde. 4-4155-2. HI 68499

Neuropsychologia; an international journal. Oxford, England. Neuropsychologia. See B–P–H 651/11. HI 57307

Neurospora newsletter. New York, NY. Neurospora Newslett. See B–P–H 651/12. HI 57308

Nevada agricultural experiment station. Annual report. Reno & Carson City, NV. Nevada Agric. Exp. Sta. Annual Rep. See B–P–H 651/15. HI 57309

Nevada agricultural experiment station, bulletin. Reno & Carson City, NV. Nos. 1-235, 1889-1963. Nevada Agric. Exp. Sta. Bull. Superseded by: Publications, Max C. Fleischmann college of agriculture. Series B. HI 68500

Nevada agricultural experiment station, circulars. Reno, NV. Nos. 1-45, 1952-61. Nevada Agric. Exp. Sta. Circ. Superseded by: Publications, Max C. Fleischmann college of agriculture. Series C. HI 68501

Nevada agricultural experiment station. Forage grasses circular. Reno, NV. Nevada Agric. Exp. Sta. Forage Grasses Circ. See B–P–H 651/17. HI 57310

Nevada agricultural experiment station. Irrigation bulletin. Reno, NV. Nevada Agric. Exp. Sta. Irrig. Bull. See B–P–H 651/18. HI 57311

Nevada agricultural experiment station, technical bulletin. Reno, NV. Nos. 1-232, 1899-1963. Superseded by: Publications, Max C. Fleischmann college of agriculture. Series T. HI 68502

New agricultural news. [Hsin nung t'ung hsün.] Nanking. New Agric. News. See B–P–H 651/20. HI 57312

New australian fruit grower. Brisbane. New Austral. Fruit Grower. See B–P–H 651/21. HI 57313

New biology. London & New York, NY. New Biol. See B–P–H 651/22. HI 57314

New biotech. Winnipeg. Vol. 1+, 1986+. New Biotech. HI 75247

New botanist; international quarterly journal of plant science research. New Delhi. Vol. 1+, 1974+. Int. Quart. J. Pl. Sci. Res. HI 68503

New Brunswick agricultural report = Report (Annual), department of agriculture, New Brunswick. Fredericton.

New Brunswick naturalist = N B naturalist. St. John, N. B.

New country life. Garden City, NY. New Country Life. See B–P–H 652/1. HI 57315

New development on agriculture. [Hatachi nogyo.] Tokyo. New Developm. Agric. See B–P–H 652/2. HI 57316

New ecologist. Wadebridge. Nos. 8-9(2), 1978-79. New Ecologist. Preceded and superseded by: Ecologist. HI 68504

New England conservationist. Canaan, ME. Vol. 1+, 1979+. New England Conservationist. HI 68505

New England environmental network news. Medford, MA. Vol. ?-4+, ?-1984+. New England Environm. Network News. HI 68506

New England farmer; containing essays, original and selected, relating to agriculture and domestic economy. Boston, MA. Vols. 1-5, 1822-27. New England Farmer (Boston, 1823-27). Superseded by: New England farmer, and horticultural journal. 4-2987-2. HI 52741

New England farmer; a monthly journal. Boston, MA. Vols. 1-16, 1848-64; n.s. vols. 1-5, 1867-71. New England Farmer (Boston, 1848-71). 4-2987-2. HI 68507

New England farmer, and gardener's journal; containing essays, original and selected, relating to agriculture and domestic economy. Boston, MA. Vols. 13-17(29) [also numbered n.s. vols. 4-8], 1835-39. New England Farmer Gard. J. Preceded by: New England farmer, and horticultural journal. Superseded by: New England farmer, and horticultural register. 4-2987-2. HI 52743

New England farmer, and horticultural journal; containing essays, original and selected, relating to agriculture and domestic economy. Boston, MA. Vols. 6-12 [vols. 10-12 also numbered n.s. vols. 1-3], 1828-34. New England Farmer Hort. J. Preceded by: New England farmer (Boston, 1823-27). Superseded by: New England farmer, and gardener's journal. 4-2987-2. HI 52742

New England farmer, and horticultural register; containing essays, original and selected, relating to agriculture and domestic economy. Boston, MA. Vols. 17(30)-24 [vols. 17-24 also numbered n.s. vols. 8-15], 1839-46. New England Farmer Hort. Reg. Preceded by: New England farmer, and gardener's journal. Superseded by: New England farmer (Boston, 1848-71). 4-2987-2. HI 53513

New England journal of medicine. Boston, MA. New England J. Med. See B–P–H 652/8. HI 57317

New England journal of medicine and surgery, and the collateral branches of science. Boston, MA. New England J. Med. Surg. See B–P–H 652/9. HI 57318

New-England medical review and journal. Boston, MA. New-England Med. Rev. J. See B–P–H 652/12. HI 57319

New England naturalist. Boston, MA. New England Naturalist. See B–P–H 652/13. HI 57320

New flora and silva. London. New Fl. & Silva. See B–P–H 652/15. HI 57321

New flowers. [Shin kaki.] Kyoto. 1954+. New Fl. HI 68508

New forests; international journal on the biology, biotechnology, and management of afforestation and reforestation. Boston, MA, Dordrecht & Hingham, MA. Vol. 1+, 1986+. New Forests. HI 68509

New Genessee farmer and gardener's journal. Rochester, NY. New Genessee Farmer Gard. J. See B–P–H 652/19. HI 57322

New Gold Coast farmer. Accra, Gold Coast [Ghana]. See B–P–H 652/20. HI 57323

New Guinea agricultural gazette. Port Moresby, New Guinea. New Guinea Agric. Gaz. See B–P–H 652/23. HI 57324

New Hampshire academy of science. Proceedings. Durham, NH. New Hampshire Acad. Sci. Proc. See B–P–H 652/25. HI 57325

New Hampshire agricultural experiment station. Annual report. Durham, NH. New Hampshire Agric. Exp. Sta. Annual Rep. See B–P–H 652/26. HI 57326

New Hampshire agricultural experiment station. Bulletin. Durham, NH. New Hampshire Agric. Exp. Sta. Bull. See B–P–H 653/1. HI 57327

New Hampshire agricultural experiment station. Circular. Durham, NH. New Hampshire Agric. Exp. Sta. Circ. See B–P–H 653/2. HI 57328

New Hampshire agricultural experiment station. Research mimeograph: botany. Durham, NH. Nos. 1-4?, 1955-6? New Hampshire Agric. Exp. Sta. Res. Mimeogr. Bot. Superseded by?: Research report, New Hampshire agricultural experiment station. HI 68511

New Hampshire agricultural experiment station. Science contributions. Durham, NH. New Hampshire Agric. Exp. Sta. Sci. Contr. See B–P–H 653/4. HI 57329

New Hampshire agricultural experiment station. Technical bulletin. Durham, NH. New Hampshire Agric. Exp. Sta. Techn. Bull. See B–P–H 653/5. HI 57330

New Hampshire plantsman. Durham, NH. New Hampshire Plantsman. See B–P–H 653/8. HI 57331

New Hampshire progress report. University of New Hampshire agricultural experiment station. Durham, NH. New Hampshire Progr. Rep. See B–P–H 653/9. HI 57332

New horizons from the horticultural research institute. Washington, DC. ?-1974+. New Horiz. Hort. Res. Inst. HI 68512

New information on horticulture. [Engei shinchishiki.] Kyoto. Vols. 1-?, 1946-73? New Inform. Hort. Superseded by: New information on horticulture. Flowers and New information on horticulture. Vegetables. HI 68513

New information on horticulture. Flowers. [Engei shinchishiki. Hana no go.] Kyoto. Vol. 1+, 1973+. New Inform. Hort., Fl. Preceded by: New information on horticulture. HI 68514

New information on horticulture. Vegetables. [Engei shinchishiki. Yasai go.] Kyoto. Vol. 1+, 1974+. New Inform. Hort., Veg. Preceded by: New information on horticulture. HI 68515

New Jersey agricultural college experiment station. Annual report. New Brunswick, NJ. New Jersey Agric. Coll. Exp. Sta. Annual Rep. See B–P–H 653/13. HI 57335

New Jersey agricultural college experiment station. Botany department report. New Brunswick, NJ. New Jersey Agric. Coll. Exp. Sta. Bot. Dept. Rep. See B–P–H 653/14. HI 57336

New Jersey agricultural college experiment station. Bulletin. New Brunswick, NJ. New Jersey Agric. Coll. Exp. Sta. Bull. See B–P–H 653/15. HI 57337

New Jersey agricultural college experiment station. Circular. New Brunswick, NJ. New Jersey Agric. Coll. Exp. Sta. Circ. See B–P–H 653/16. HI 57338

New Jersey agricultural college experiment station. Mimeographed circular. New Brunswick, NJ. New Jersey Agric. Coll. Exp. Sta. Mimeogr. Circ. See B–P–H 653/17. HI 57339

New Jersey agricultural college experiment station. Nursery disease notes. New Brunswick, NJ. New Jersey Agric. Coll. Exp. Sta. Nursery Dis. Notes. See B–P–H 653/18. HI 57340

New Jersey agriculture. New Brunswick, NJ. New Jersey Agric. See B–P–H 653/12. HI 57334

New Jersey Audubon bulletin. Franklin Lakes, NJ. No. 1+, 1913+. New Jersey Audubon Bull. 4-2999-2. HI 68516

New Jersey forester. Washington, DC. Vol. 1(1-3), 1895. New Jersey Forester. Superseded by: Forester. Washington. 1-223-2. HI 68517

New Jersey outdoors. Trenton, NJ. Vol. 1+, 1974+. New Jersey Outdoors. HI 68518

New journal of agriculture and forestry; college of agriculture, Chinling university. [Nung lin hsin pao.] [China]. New J. Agric. Forest. Coll. Agric. Chinling Univ. See B–P–H 653/11. HI 57333

New knowledge of forestry. [Ringyo shin chishiki.] Tokyo. 1953+. New Knowl. Forest. HI 74007

New medical and physical journal; or, annals of medicine, natural history, and chemistry. London. New Med. Phys. J. See B–P–H 654/3. HI 57341

New Mexico agricultural experiment station, annual report. State College of New Mexico. Las Cruces & Santa Fe, NM. New Mexico Agric. Exp. Sta. Annual Rep. See B–P–H 654/6. HI 57342

New Mexico agricultural experiment station, bulletin. Las Cruces & Santa Fe, NM. New Mexico Agric. Exp. Sta. Bull. See B–P–H 654/7. HI 57343

New Mexico agricultural experiment station, press bulletin. State College of New Mexico. Las Cruces & Santa Fe, NM. New Mexico Agric. Exp. Sta. Press Bull. See B–P–H 654/8. HI 57344

New Mexico agricultural experiment station, research report. New Mexico state college. University Park, NM. New Mexico Agric. Exp. Sta. Res. Rep. See B–P–H 654/9. HI 57345

New Mexico anthropologist. Albuquerque, NM. New Mexico Anthropol. See B–P–H 654/10. HI 57346

New monthly magazine, and universal register. London. New Monthly Mag. See B–P–H 654/13. HI 57422

New northwest. [Hsin hsi pei.] [China]. New NorthW. (China). See B–P–H 654/15. HI 57347

New northwest. Portland, OR. New NorthW. (Portland). See B–P–H 654/16. HI 57348

New northwest. Seattle, WA. New NorthW. (Seattle). See B–P–H 654/17. HI 57349

New Orleans journal of medicine. New Orleans, LA. New Orleans J. Med. See B–P–H 654/19. HI 57350

New Orleans medical journal. New Orleans, LA. New Orleans Med. J. See B–P–H 654/22. HI 57351

New Orleans medical and surgical journal. New Orleans, LA. New Orleans Med. Surg. J. See B–P–H 654/23. HI 57352

New phytologist; a british botanical journal. London. New Phytol. See B–P–H 655/3. HI 57353

New plants of the year. London. 1947-49. New Pl. Year. HI 68519

New remedies. New York, NY. New Remedies. See B–P–H 655/5. HI 57354

New scientist. London. New Sci. See B–P–H 655/6. HI 57355

New South Wales magazine. Sydney. New South Wales Mag. See B–P–H 655/8. HI 57356

New techniques in biophysics and cell biology. London. Vol. 1+, 1973+. New Techn. Biophys. Cell Biol. HI 68520

New York academy of sciences annals. New York, NY. New York Acad. Sci. Ann. See B–P–H 655/9. HI 57357

New York (state) agricultural experiment station. Annual report. Geneva, NY. New York Agric. Exp. Sta. Annual Rep. See B–P–H 655/10. HI 57358

New York (state) agricultural experiment station. Bulletin. Geneva, NY. Nos. 1-116, 1882-85; n.s. nos. 1-1032, 1885-1973. New York Agric. Exp. Sta. Bull. Superseded by: Search agriculture. HI 68521

New York (state) agricultural experiment station. Circular. Geneva, NY. New York Agric. Exp. Sta. Circ. See B–P–H 655/12. HI 57359

New York (state) agricultural experiment station, research circular series. Geneva, NY. Nos. 1-23, 1963-70. New York Agric. Exp. Sta. Res. Circ. Ser. Superseded by: New York's food and life sciences bulletin. HI 68522

New York (state) agricultural experiment station, seed research circular. Geneva, NY. Nos. 1-3, 1965-68. New York Agric. Exp. Sta. Seed Res. Circ. Superseded by: New York's food and life sciences bulletin. HI 68523

New York (state) agricultural experiment station. Technical bulletin. Geneva, NY. New York Agric. Exp. Sta. Techn. Bull. See B–P–H 655/14 HI 57360

New York botanical garden annual report = Report (Annual), New York botanical garden. Bronx, NY.

New York city notes on natural history. New York. No. 1+, 1977+. New York City Notes Nat. Hist. HI 68524

New York horticultural review. New York, NY. New York Hort. Rev. See B–P–H 655/22. HI 57361

New York lancet. New York, NY. New York Lancet. See B–P–H 655/26. HI 57362

New York literary gazette. New York. Vols. 1-2, 1825-26. New York Lit. Gaz. Preceded by: Minerva. New York. Superseded by: Literary gazette, and american athenaeum. 3-2438-2. HI 68525

New York medical journal. New York, NY. New York Med. J. See B–P–H 655/27. HI 57363

New-York medical and philosophical journal and review. New York, NY. New-York Med. Philos. J. Rev. See B–P–H 656/1. HI 57364

New-York medical and physical journal. New York, NY. New-York Med. Phys. J. See B–P–H 659/6. HI 57400

New York state botanist report. Albany, NY. New York State Bot. Rep. See B–P–H 656/14. HI 57366

New York state conservationist. Albany, NY. Vols. 1-14(4), 1946-60. New York State Conservationist. Superseded by: Conservationist. HI 68526

New York state department of agriculture. Nature-study bulletin. Albany, NY. New York State Dept. Agric. Nat.-Stud. Bull. See B–P–H 656/19. HI 57367

New York state environment. Albany, NY. Vol. 1+, 1971+. New York State Environm. HI 68527

New York state fruit growers' association. Proceedings. Fayetteville, NY. New York State Fruit Growers' Assoc. Proc. See B–P–H 656/20. HI 57368

New York state horticultural society proceedings. New York State Hort. Soc. Proc. See B–P–H 656/21. HI 57369

New York state journal of medicine. New York, NY. New York State J. Med. See B–P–H 656/22. HI 57370

New York state museum handbook. Albany, NY. New York State Mus. Handb. See B–P–H 656/25. HI 57371

New York state museum leaflet. New York. No. 21+, 1979+. New York State Mus. Leafl. Preceded by: Educational leaflet series, New York state museum. HI 68528

New York state museum memoirs. Albany, NY. New York State Mus. Mem. See B–P–H 656/27. HI 57372

New York's food and life sciences bulletin. Geneva, NY. Vol. 1+, 1970+. New York's Food Life Sci. Bull. Preceded by: New York (state) agricultural experiment station, research circular series and New York (state) agricultural experiment station, seed research circular. HI 68529

New York's food and life sciences quarterly. Ithaca, NY. Vol. 1+, 1968+. New York's Food Life Sci. Quart. Preceded by: Farm research. HI 68530

New York journal of pharmacy. New York. Vols. 1-3, 1852-54. New York J. Pharm. (1852-54). 4-3035-1. HI 68531

New York journal of pharmacy. New York. Vols. 8(4)-10, 1931-33. New York J. Pharm. (1931-33). Preceded by: Messenger. 4-3035-1. HI 68532

New Zealand agricultural science. Wellington, N.Z. New Zealand Agric. Sci. See B–P–H 657/7. HI 57373

New Zealand agriculturist. Fielding, N.Z. Vols. 1-21(2), 1948-76. New Zealand Agric. HI 68533

New Zealand antarctic research programme and report. Christchurch, N.Z. 1975/76+. New Zealand Antarc. Res. Programme Rep. HI 68534

New Zealand cactus and succulent journal. Auckland, N.Z. Vol. 6(6)+, 1953+. New Zealand Cact. Succ. J. Preceded by: Cactus and succulent journal. Auckland, N.Z. HI 68535

New Zealand camellia bulletin. Hamilton, Tauranga. Vol. 1+, 1958+. New Zealand Camellia Bull. Preceded by: Camellia bulletin, South Auckland camellia society. HI 68536

New Zealand department of scientific and industrial research information series. Wellington, N.Z. No. 1+, 1948+. New Zealand Dept. Sci. Industr. Res. Inform. Ser. HI 68537

New Zealand environment. No. ?-3+, ?-1973+; Supplement no. 1+, 1975+. New Zealand Environm. HI 68538

New Zealand forestry research notes. Wellington, N.Z. No. 1+, 1955+. New Zealand Forest. Res. Notes. Preceded by: Forest products research notes and Forest research notes. HI 68539

New Zealand gardener; national journal of horticulture. Wellington, N.Z. New Zealand Gard. See B–P–H 657/9. HI 57374

New Zealand geological survey, paleontological bulletin. Wellington, N.Z. New Zealand Geol. Surv. Paleontol. Bull. See B–P–H 657/10. HI 57375

New Zealand homes and gardening. Wellington, N.Z. Vol. 1+, 1928+. New Zealand Homes Gard. 4-3053-3. HI 68541

New Zealand iris society bulletin. Christchurch, N.Z., Upper Hutt. No. 1+, 1950+. New Zealand Iris Soc. Bull. HI 68542

New Zealand journal of agricultural research. Wellington, N.Z. New Zealand J. Agric. Res. See B–P–H 657/13. HI 57376

New Zealand journal of agriculture. Wellington, N.Z. Vol. 17+, 1918+. New Zealand J. Agric. Preceded by: Journal of agriculture, New Zealand department of agriculture. 4-3054-2. HI 68543

New Zealand journal of botany. Wellington, N.Z. New Zealand J. Bot. See B–P–H 657/14. HI 57377

New Zealand journal of ecology. Christchurch, N.Z. Vol. 1+, 1978+. New Zealand J. Ecol. Preceded by: Proceedings, New Zealand ecological society. HI 68544

New Zealand journal of experimental agriculture. Wellington, N.Z. Vol. 1+, 1973+. New Zealand J. Exp. Agric. HI 68545

New Jealand journal of forestry. Palmerston North, N.Z. New Zealand J. Forest. See B–P–H 657/15. HI 57378

New Zealand journal of forestry science. Rotorua. Vol. 1+, 1971+. New Zealand J. Forest. Sci. HI 68546

New Zealand journal of geology and geophysics. Wellington, N.Z. New Zealand J. Geol. Geophys. See B–P–H 657/16. HI 57379

New Zealand journal of marine and freshwater research. Wellington, N.Z. Vol. 1+, 1967+. New Zealand J. Mar. Freshwater Res. HI 68547

New Zealand journal of marine and freshwater research. Information series. Wellington, N.Z. ?-1972+. New Zealand J. Mar. Freshwater Res., Inform. Ser. HI 68548

New Zealand journal of science. Dunedin, N.Z. New Zealand J. Sci. (Dunedin). See B–P–H 657/18. HI 57381

New Zealand journal of science. Wellington, N.Z. New Zealand J. Sci. (Wellington). See B–P–H 657/17.

HI 57380

New Zealand journal of science and technology. Wellington, N.Z. New Zealand J. Sci. Technol. See B–P–H 657/19. HI 57382

New Zealand limnological society newsletter. No. ?-16+, ?-1981+. New Zealand Limnol. Soc. Newslett. HI 68549

New Zealand man and biosphere report. Canterbury, N.Z. No. 1+, 1979+. New Zealand Man Biosphere Rep. HI 68550

New Zealand marine sciences newsletter. Wellington, N.Z. 1961+. New Zealand Mar. Sci. Newslett. HI 68551

New Zealand oceanographic institute; collected reprints. Wellington, N.Z. New Zealand Oceanogr. Inst. Collect. Repr. See B–P–H 657/21. HI 57383

New Zealand orchid review. Wellington, N.Z. Vol. 1+, 1957+. New Zealand Orchid Rev. HI 68552

New Zealand plants and gardens. Wellington, N.Z. Vols. 1-7, 1955-68. New Zealand Pl. Gard. Superseded by: Journal of the royal New Zealand Institute of Horticulture. HI 68553

New Zealand rose annual. Wellington, N.Z. 1964+. New Zealand Rose Annual. Preceded by: Australian and New Zealand rose annual. HI 68554

New Zealand science abstracts. Wellington, N.Z. Vol. 1+, 1980+, 1982+. New Zealand Sci. Abstr. HI 68555

New Zealand science review. Wellington, N.Z. New Zealand Sci. Rev. See B–P–H 657/23. HI 57384

New Zealand tobacco growers journal. Motueka, N.Z. New Zealand Tobacco Growers J. See B–P–H 657/24. HI 57385

New Zealand veterinary journal. Wellington, N.Z. Vol. 1+, 1952+. New Zealand Veterin. J. HI 68556

New Zealand wheat review. Christchurch, N.Z. 1944/45+. New Zealand Wheat Rev. Preceded by: Report of the New Zealand wheat institute. HI 68557

Newark museum quarterly. Newark, NJ. 1949+. Newark Mus. Quart. HI 68558

News on agricultural research in Arkansas. Fayetteville, AR. Vols. 1-8(1), 1975-82. News Agric. Res. Arkansas. Superseded by: Research newsletter, agricultural experiment station, university of Arkansas. HI 75140

News bulletin, Alberta society of petroleum geologists = Bulletin of canadian petroleum geology. Calgary, Alberta. Bull. Canad. Petrol. Geol. See B–P–H 243/21.

News bulletin, aquatic plant management programme, Okanagan Lakes. Victoria, B.C. Vol. 1, 1977. News Bull. Aguatic Pl. Managem. Programme Okanagan Lakes. Superseded by: Information bulletin, aquatic plant management programme. HI 68559

News bulletin, british mycological society. Nos. 1-26, 1955-66. News Bull. Brit. Mycol. Soc. Superseded by: Bulletin of the british mycological society. HI 68560

News bulletin, Cheshire conservation trust ltd. Stockport. Vols. 1-4 1970-73. News Bull. Cheshire Conservation Trust. Superseded by: Grebe. HI 68561

News bulletin of the empire forest departments. Oxford. 1936-51. News Bull. Empire Forest Dept. HI 68562

News bulletin, Florida division of plant industry. Gainesville, FL. 1959+. News Bull. Florida Div. Pl. Industr. HI 68563

News bulletin of the Hampshire and Isle of Wight naturalists' trust ltd. Winchester. 1965-66. News Bull. Hampshire & Isle Wight Naturalists' Trust. HI 68564

News bulletin horticultural digest = Horticultural digest. Ames, IA. Hort. Digest. See B–P–H 421/5.

News bulletin, international association of wood anatomists = International association of wood anatomists, news bulletin. Zurich.

News bulletin, international seed testing association = I S T A news bulletin. Wageningen, Zürich.

News bulletin, middle-Thames natural history society. Maidenhead. No. ?-54+, ?-1981+. News Bull. Middle-Thames Nat. Hist. Soc. HI 68565

News bulletin of the national parks association = Bulletin of the national parks association. Washington, DC.

News bulletin, phycological society of America. Urbana, IL, Philadelphia, PA. Vols. 1-16, 1946-64. News Bull. Phycol. Soc. Amer. Superseded by: Journal of phycology. 4-3350-3. HI 68566

News, commonwealth agricultural bureaux = C A B news. Farnham Royal.

News in economic forestry = Noutati in economia forestiera. Brasov.

News, european biotechnology information project. London. Nos. 1-15?, 19??-87? News Eur. Biotechnol. Inform. Proj. Preceded by: Biotechnology information news. HI 75254

News, Florida citrus nurserymens association = Florida citrus nurserymens association news. Leesburg, FL. Florida Citrus Nurserymens Assoc. News. See B–P–H 375/9.

News, herb society of America = Newsletter, herb society of America. Boston, MA.

News, horticultural society of New York. New York, NY. Vols. 1(9)-7(1), 1951-57. News Hort. Soc. New York. Preceded by: Gardening library news.

Superseded by: Bulletin, horticultural society of New York. HI 68567

News, institute for environmental studies. Toronto. 1982+. News Inst. Environm. Stud. Preceded by: Research news, institute for environmental studies, Toronto. HI 68568

News letter, News-letter, see Newsletter

News of literature and fashion; or, journal of manners and society, the drama, the fine arts, literature, science, etc. ... London. News Lit. Fashion. See B–P–H 658/7. HI 57389

News, Minnesota landscape arboretum. Chaska, MN. Vol. 1+, 1981+. News Minnesota Landscape Arbor. Preceded by: Members newsletter, Minnesota landscape arboretum. HI 68572

News for naturalists. London. 1959-66. News Naturalists. Incorporated in: Report, council for nature. HI 68573

News and notes about Argentine scientific publications. Buenos Aires. No. 1+, 1968+. News Notes Argent. Sci. Publ. HI 68574

News report, national academy of sciences and national research council. Washington, DC. News Rep. Natl. Acad. Sci. See B–P–H 658/8. HI 57390

News review, Queensland succulent society. Chermside, Qld. Vol. 1+, 1965+. News Rev. Queensland Succ. Soc. HI 68576

News sheet, Illawarra cacti society. Illawarra, N.S.W. No. 1+, 1973+. News Sheet Illawarra Cact. Soc. HI 68577

News sheet, royal botanic garden, Edinburgh. Edinburgh. No. ?-36+, ?-1978+. News Sheet Roy. Bot. Gard. Edinburgh. HI 68578

News sheet, south Essex natural history society. Southend-on-Sea. 1936-38. News Sheet S. Essex Nat. Hist. Soc. Superseded by: Bulletin, south Essex natural history society. HI 68579

News, Somerset trust for nature conservation. Taunton. 1979+. News Somerset Trust Nat. Conservation. Preceded by: Newsletter, Somerset trust for nature conservation. HI 68580

News Taiwan forest experiment station. [Tai wan sheng lin yeh shih yen so tung hsün.] Taipei, Taiwan. News Taiwan Forest Exp. Sta. See B–P–H 658/9. HI 57391

News of vegetable and ornamental crops research station. [Nyusu yasai shikenjo.] No. 1+, 1973+. News Veg. Ornam. Crops Res. Sta. HI 68581

News and views, american fern society. East Orange, NJ. Nos. 1-12, 1971-73. News Views Amer. Fern Soc. Superseded by: Fiddlehead forum. HI 68582

News and views, american horticultural society; quarterly forum for gardeners. Alexandria, VA. Vol. 11+, 1968+. News & Views Amer. Hort. Soc. Preceded by: American horticultural society gardeners forum. HI 68583

News and views, national cactus and succulent society. Lincoln, Nos. ?-9-12, ?-1963-64. News Views Natl. Cact. Succ. Soc. HI 68584

News and views, natural history society of Malta. Sliema. 1970-72. News Views Nat. Hist. Soc. Malta. Incorporated in: Maltese naturalist. HI 68585

News, Wallop Valley field club = Newsletter, Wallop Valley field club. Over Wallop.

News, Yorkshire naturalists' trust = Newsletter, Yorkshire naturalists' trust ltd. York.

Newsletter, abundant life seed foundation. Port Townsend, WA. Vol. ?-7+, ?-1983+. Newslett. Abund. Life Seed Found. HI 68587

Newsletter, african committee for palynology. Bloemfontein. 1978+. Newslett. African Committee Palynol. HI 68588

Newsletter, agricultural research institute, Washington = A R I newsletter. Washington, DC.

Newsletter, Ahmadi natural history and field studies group. Ahmadi. No. 1+, 1970+. Newslett. Ahmadi Nat. Hist. Field Stud. Group. HI 68589

Newsletter, Alberta institute of agrostologists = A I A newsletter. Edmonton.

Newsletter, Alberta wilderness association. Calgary. 1968+. Newslett. Alberta Wildern. Assoc. HI 68590

Newsletter, aloe, cactus and succulent society of Rhodesia = Quarterly newsletter, aloe, cactus and succulent society of Rhodesia. Salisbury, Rhodesia.

Newsletter, aloe, cactus and succulent society of Zimbabwe = Quarterly newsletter, aloe, cactus and succulent society of Zimbabwe. Salisbury, Zimbabwe.

Newsletter, alpine garden society. Birdbrook, Essex. No. 1+, 1975+. Newslett. Alpine Gard. Soc. HI 68591

News letter, american association of botanical gardens and arboreta. Philadelphia, PA, Lisle, IL. Nos. 1-35, 1950-58; [n.s.] no. 1+, 1975+. News Lett. Amer. Assoc. Bot. Gard. For 1958-73 see: Quarterly newsletter, american association of botanical gardens and arboreta and Arboretum and botanical garden bulletin. HI 68569

Newsletter, american association of stratigraphic palynologists. College Station, TX. Vols. 1-12(3), 1969-79. Newslett. Amer. Assoc. Stratigr. Palynologists. Superseded by: A A S P newsletter. HI 68592

Newsletter, american bamboo society. Salona Beach,

CA. No. 1+, 1980+. Newslett. Amer. Bamboo Soc. HI 68593

Newsletter, american boxwood society. Boyce, VA. 1977+. Newslett. Amer. Boxwood Soc. HI 68594

Newsletter, american carnation society. Philadelphia, PA. ?-1972+. Newslett. Amer. Carnation Soc. HI 68595

Newsletter, american garden history society. Newark, OH. Vols. ?-7(3), 19??-83. Newslett Amer. Gard. Hist. Soc. Preceded by: Newsletter, american garden and landscape history society. HI 75094

Newsletter, american garden and landscape history society. Newark, OH. Vol. 7(4)+, 1983+. Newslett Amer. Gard. Landscape Hist. Soc. Preceded by: Newsletter, american garden history society. HI 68596

Newsletter, american hemerocallis society, region #3. 1966-67-? Newslett. Amer. Hemerocallis Soc., Region 3. HI 68597

Newsletter, american horticultural council = American horticultural council news letter. West Grove, PA.

Newsletter, american hosta society = American hosta society newsletter. Baldwin, NY.

Newsletter, american institute of biological sciences = A I B S newsletter. Washington, DC.

Newsletter, american iris society. 1980+. Newslett. Amer. Iris Soc. HI 68598

Newsletter, american littoral society. Highlands, NJ. 1969+. Newslett. Amer. Littoral Soc. HI 68599

Newsletter of the american magnolia society. Fraser, MI. Vols. 1-15, 1964-79. Newslett. Amer. Magnolia Soc. Superseded by: Magnolia. HI 68600

Newsletter, american nature study society. Elmsford, NY, etc. ?-1951+. Newslett. Amer. Nat. Stud. Soc. HI 68601

Newsletter, american plant life society. La Jolla, CA. ?-1984+. Newslett. Amer. Pl. Life Soc. HI 68602

Newsletter, American poinsettia society = American poinsettia society newsletter. Mission, TX Amer. Poinsettia Soc. Newslett. See B–P–H 84/18.

Newsletter, american society for horticultural science. St. Joseph, MI. ?-1956-72-? Newslett. Amer. Soc. Hort. Sci. Superseded by: A S H S newsletter. HI 68603

Newsletter, american society for horticultural science, tropical region. ?-1968-74-? Newslett. Amer. Soc. Hort. Sci., Trop. Region. HI 68604

Newsletter, american society for photobiology = A S P newsletter.

Newsletter, american society of plant physiologists = A S P P newsletter. Rockville, MD.

Newsletter on the application of nuclear methods in biology and agriculture. Wageningen. Nos. 1-8, 1973-77. Newslett. Applic. Nucl. Meth. Biol. Agric. HI 68605

Newsletter, applied naturalist guild. Strong, ME. 1963+. Newslett. Appl. Naturalist Guild. HI 68606

Newsletter, arabidopsis information service. Göttingen. No. 1+, 1964+. Newslett. Arabidopsis Inform. Serv. HI 68607

Newsletter of the arboretum of the Barnes foundation. Merion, PA. Nos. 1-22, 1968-79. Newslett. Arbor. Barnes Found. HI 68608

Newsletter, aril society international. Tujunga, CA., Albuquerque, NM. ?-1972+. Newslett. Aril Soc. Int. HI 68609

Newsletter of the Arizona native plant society. Tucson, AZ. Vols. 1-4, 1977-81. Newslett. Arizona Native Pl. Soc. Superseded by: Bulletin, Arizona native plant society. HI 75095

Newsletter, asian ecological society. Taichung. 1978+. Newslett. Asian Ecol. Soc. HI 68610

Newsletter of the asian Pacific weed science society. Kapaa, HI. Vol. 1+, 1969+. Newslett. Asian Pacific Weed Sci. Soc. HI 68611

Newsletter, aslib biological group. London. Nos. 1-8, 1967-78. Newslett. Aslib Biol. Group. Continued in: Aslib information. HI 68612

Newsletter, association of aquatic vascular plant biologists. Fairbanks, AK. Nos. ?-7, 19??-74. Newslett. Assoc. Aquatic Vasc. Pl. Biol. Superseded by: Newsletter, international association of aquatic vascular plant biologists. HI 75096

Newsletter, association of interpretive naturalists = A I N newsletter. East Lansing, MI.

Newsletter, association of official seed analysts. Fort Collins, CO, Reynoldsburg, OH. 1927+. Newslett. Assoc. Off. Seed Analysts. 1-533-2. HI 68613

Newsletter, association of Pacific systematists. Honolulu, HI. No. 1+, 1984+. Newslett. Assoc. Pacific Systematists. HI 68614

Newsletter, association of science-technology centers = A S T C newsletter. Washington, DC.

Newsletter, association of systematic collections = A S C newsletter. Lawrence, KS.

Newsletter, association for tropical biology inc. Washington, DC. Nos. 1-24/25, 1966-70. Newslett. Assoc. Trop. Biol. HI 68615

Newsletter, Auckland botanical society. Titirangi, Auckland. Vol. ?-37+, ?-1982+. Newslett. Auckland Bot. Soc. HI 68616

Newsletter, australian conservation foundaton. Canberra, A.C.T. No. 1+, 1967+. Newslett. Austral. Conservation Found. HI 68617

Newsletter, australian plant pathology society. Sydney, N.S.W. Vols. 1-6, 1972-76. Newslett. Austral. Pl. Pathol. Soc. Superseded by: Australasian plant pathology. HI 68618

Newsletter, australian plant society. Pasadena, CA = Australian plant society newsletter. Pasadena, CA.

Newsletter, australian society of limnology. Sydney, N.S.W. Vol. 1+, 1962+. Newslett. Austral. Soc. Limnol. HI 68619

Newsletter, australian society for microbiology. Melbourne, Vic. No. 1+, 1964+. Newslett. Austral. Soc. Microbiol. HI 68620

Newsletter, Australian systematic botany society. South Perth, W.A. No. 2+, 1974+. Newslett. Austral. Syst. Bot. Soc. Preceded by: A S B S newsletter. HI 68621

Newsletter B C G. St. Albans. Vol. 3+, 1981+. Newslett. B. C. G. Preceded by: B C G newsletter. HI 68622

Newsletter, B S I. Howrah. Vol. ?-5(2)+, ?-1979+. Newslett. B. S. I. HI 68623

Newsletter, Barnes foundation arboretum = Newsletter of the arboretum of the Barnes foundation. Merion, PA.

Newsletter, Bartlett arboretum. Stamford, CT. Vol. 1+, 1977+. Newslett. Bartlett Arbor. HI 68624

Newsletter, Bedfordshire natural history society. Bedford. Nos. 1-30, 1968-78. Newslett. Bedfordshire Nat. Hist. Soc. Superseded by: Muntjac. HI 68625

Newsletter, Bermuda biological station for research. St. George's West. 1971+. Newslett. Bermuda Biol. Sta. Res. HI 68626

Newsletter, biological council of Canada. Ottawa. 1972+. Newslett. Biol. Council Canada. HI 68627

Newsletter, Birmingham botanic gardens. Birmingham. 1972+. Newslett. Birmingham Bot. Gard. HI 68628

Newsletter, Blithewold gardens and arboretum. Bristol, RI. Vol. [1]+, 1980+. Newslett. Blithewold Gard. Arbor. HI 68629

Newsletter, botanical club of Wisconsin. Madison, WI. Vols. 1-10, 1969-78. Newslett. Bot. Club Wisconsin. Superseded by: Bulletin, botanical club of Wisconsin. HI 68630

Newsletter, botanical society of America. Ecological section. 1978+. Newslett. Bot. Soc. America, Ecol. Sect. HI 68631

Newsletter, botanical society of the British Isles. 1971 (July). Newslett. Bot. Soc. Brit. Isles. Superseded by: B S B I news. HI 68632

Newsletter of the botanical society of Edinburgh = B S E news. Edinburgh.

Newsletter, botanical society of Otago. Dunedin, N.Z. ?-1987+. Newslett. Bot. Soc. Otago. HI 68633

Newsletter, Brecknock county naturalists' trust. Brecon. Nos. 1-12, 1964-? Newslett. Brecknock County Naturalists' Trust. Superseded by: Breconshire naturalist. HI 68634

Newsletter, Brecon Beacons national park committee. Brecon. No. ?-32+, ?-1983+. Newslett. Brecon Beacons Natl. Park Committee. HI 68635

Newsletter, Bristol cactus society = Bristol cactus society newsletter. Bristol.

Newsletter, British cactus and succulent society. Oxford. March 1983-?, 1983-? Newslett. Brit. Cact. Succ. Soc. Continued in: British cactus and succulent journal. HI 68636

Newsletter, British Columbia nature council. Vancouver, B.C. Vols. 1-6(3), [1964]-68? Newslett. British Columbia Nat. Council. Superseded by: Newsletter, federation of British Columbia naturalists. HI 68637

Newsletter, british hosta and hemerocallis society. Boldre. No. 1+, 1982+. Newslett. Brit. Hosta Hemerocallis Soc. HI 68638

Newsletter, british phycological society. [London.] No. 1+, 1971+. Newslett. Brit. Phycol. Soc. HI 68639

Newsletter, british pteridological society. Loughton. No. 1+, 1963+. Newslett. Brit. Pteridol. Soc. HI 68640

Newsletter, Brookgreen gardens = Brookgreen gardens newsletter. Murrels Inlet, SC.

Newsletter, Brooklyn botanic garden = Brooklyn botanic garden newsletter. New York.

Newsletter, Buffalo botanical society. Buffalo, NY. Vols. 1-2(1), 1983-84. Newslett. Buffalo Bot. Soc. Superseded by: Newsletter, Niagara frontier botanical society. HI 68641

News letter of the bulb society. Los Angeles, CA. News Lett. Bulb Soc. See B–P–H 658/3. HI 57386

Newsletter, cactus and succulent society of America = C S S A newsletter. Laundale.

Newsletter, cactus and succulent society of Australia. Melbourne, Vic. Nos. 1-59, 1963-68. Newslett. Cact. Succ. Soc. Australia. HI 68642

Newsletter, cactus and succulent society. Bombay. Vol. 1+, 1965+. Newslett. Cact. Succ. Soc. HI 68643

Newsletter, cactus and succulent society of Hawaii. Honolulu, HI. No. 1-?, 1967-? Newslett. Cact. Succ. Soc. Hawaii. HI 68644

Newsletter, cactus and succulent society of Malta. Zejtun (Malta) 1967+. Newslett. Cact. Succ. Soc. Malta. HI 68645

Newsletter, cactus and succulent society of South

Australia. Norwood, S.A. 1967(May)+, 1967+. Newslett. Cact. Succ. Soc. S. Australia. HI 68646

Newsletter, California arboretum foundation. ?-1974+. Newslett. Calif. Arbor. Found. HI 68647

Newsletter, California grape and tree fruit league. San Francisco, CA. ?-1972+. Newslett. Calif. Grape Tree Fruit League. HI 68648

Newsletter, California rare fruit growers. Fullerton, CA. Vols. 3-18(2), 1971-86. Newslett. Calif. Rare Fruit Growers. Preceded by: Quarterly newsletter, California rare fruit growers. Superseded by: Fruit gardener. HI 75097

Newsletter, Callaway gardens = Callaway gardens newsletter. Pine Mountain, GA.

Newsletter, Camborne-Redruth natural history society. Camborne. No. 1+, 1984+. Newslett. Camborne-Redruth Nat. Hist. Soc. HI 68649

Newsletter, Cambridge university botanic garden association. Cambridge. ?-1973+. Newslett. Cambridge Univ. Bot. Gard. Assoc. HI 68650

Newsletter, Cambridgeshire and Isle of Ely naturalists trust. Brookside. 1-42, 1957-77. Newslett. Cambridgeshire Naturalists Trust. Superseded by: Cambient newsletter. HI 68651

Newsletter, campus arboretum association of Haverford. Vol. ?-2(2)+, ?-1975+. Newslett. Campus Arbor. Assoc. Haverford. HI 68652

Newsletter, canadian botanic garden = Canadian botanic garden newsletter. Edmonton, Alb.

Newsletter, canadian federation of biological societies. Saskatoon. No. 1+, 1982+. Newslett. Canad. Fed. Biol. Soc. Preceded by: Bulletin de la fédération canadienne. HI 68654

Newsletter, canadian iris society = Canadian iris society newsletter. Hannon, Willowdale, Ont.

Newsletter, canadian plant conservation programme = Canadian plant conservation programme newsletter. Edmonton, Alb.

Newsletter, canadian society of environmental biologists. Ottawa. ?-1985+. Newslett. Canad. Soc. Environm. Biol. HI 68656

Newsletter, canadian society for horticultural science. ?-1980+. Newslett. Canad. Soc. Hort. Sci. HI 68657

Newsletter, canadian society of microbiologists. London, Ont. Vol. 1+, 1952+. Newslett. Canad. Soc. Microbiologists. HI 68658

Newsletter, canadian society of plant physiologists. Edmonton. 1958+. Newslett. Canad. Soc. Pl. Physiologists. HI 68659

Newsletter, carnivorous plant society. 1984+. Newslett. Carniv. Pl. Soc. HI 68660

Newsletter, Cary arboretum = Cary arboretum of the New York botanical garden; newsletter published for friends of the arboretum by the public affairs department. Millbrook, NY.

Newsletter, Catherine Traill naturalists' club. Ste. Anne-de-Bellevue. No. ?-16+, ?-1975+. Newslett Catherine Traill Naturalists' Club. HI 68661

Newsletter, center for tropical agriculture, university of Florida. Gainesville, FL. Vol. 1+, 1967+. Newslett. Center Trop. Agric. Univ. Florida. HI 68662

Newsletter, centraalbureau voor schimmelcultures = C B S newsletter. Baarn.

Newsletter, chemical plant taxonomy = Chemical plant taxonomy newsletter. Cambridge, etc.

Newsletter, Clark garden of the Brooklyn botanic garden. Albertson, NY. ?-1983+. Newslett. Clark Gard. Brooklyn Bot. Gard. HI 68663

Newsletter, clover and special purpose legumes research. Madison, WI. Vols. 1-6?, 1967-73? Newslett. Clover Spec. Purp. Legumes Res. Superseded by: Progress report, clovers and special purpose legumes research. HI 75178

Newsletter, coast and wetlands society. Sydney, N.S.W. No. ?-3+, ?-1982+. Newslett. Coast Wetlands Soc. HI 68664

Newsletter, Colorado native plant society = Aquilegia; newsletter, Colorado native plant society. Fort Collins, CO.

Newsletter from the commission for scientific research in Greenland. Copenhagen. 1979+. Newslett. Commiss. Sci. Res. Greenland. HI 68666

Newsletter, committee on biological information = C O B I newsletter. London.

Newsletter, committee for the study of the scottish flora. 1972+. Newslett. Committee Stud. Scott. Fl. HI 68667

Newsletter, comparative investigations of tropical reef systems = C I T R E newsletter. [Washington, DC.]

Newsletter, conifer society of Australia. South Yarra, Vic. ?-1987+. Newslett. Conifer Soc. Australia. HI 68668

Newsletter of the Connecticut botanical society. Storrs, CT. Vol. 8+, 1980+. Newslett. Connecticut Bot. Soc. Preceded by: Connecticut botanical society newsletter. HI 68669

Newsletter, conservation council of New Brunswick = C C N B newsletter. Fredericton, N. B.

Newsletter, conservation council of Ontario. Toronto. Vol. 1(1-11), 1973-74. Newslett. Conservation Council Ontario. Superseded by: Conservation news. Toronto. HI 68670

Newsletter, conservation society. 1967+. Newslett. Conservation Soc. HI 68672

Newsletter, Cornwall garden society, group for the conservation of plants and gardens. Ca.1984+. Newslett. Cornwall Gard. Soc., Group Conservation Pl. Gard. HI 68673

Newsletter, Cornwall naturalists' trust. St. Ives. No. 1+, 1962+. Newslett. Cornwall Naturalists' Trust. HI 68674

Newsletter, Cornwall trust for nature conservation = Newsletter, Cornwall naturalists' trust. St. Ives.

Newsletter, council of biology editors. ?-1977. Newslett. Council Biol. Edit. Superseded by: C B E views. HI 68675

Newsletter, council on botanical and horticultural libraries. Beltsville MD, etc. No. 1+, 1970+. Newslett. Council Bot. Hort. Libr. HI 68676

Newsletter, Cox arboretum = Cox arboretum newsletter. Dayton, OH.

Newsletter, Cranbrook institute of science. Bloomfield Hills, MI. Vol. 1+, 1931+. Newslett. Cranbrook Inst. Sci. HI 68677

Newsletter, Creekmoor natural history society. Creekmoor. 1978+. Newslett. Creekmoor Nat. Hist. Soc. HI 68678

News letter, Cumbria naturalists' trust. Kirkby Stephen. Nos. 24-?, 1974-? News Lett. Cumbria Naturalists' Trust. Preceded by: News letter, Lake District naturalists' trust. Superseded by: Newsletter of the Cumbria trust for nature conservation. HI 68679

Newsletter of the Cumbria trust for nature conservation. Ambleside. N.s. no. 1+, 1981+. Newslett. Cumbria Trust Nat. Conservation. Preceded by: News letter, Cumbria naturalists' trust. HI 68680

Newsletter, the Dawes arboretum. Newark, OH. Vol. 1+, 1968+. Newslett. Dawes Arbor. HI 68681

Newsletter, Delaware museum of natural history. ?-1975+. Newslett. Delaware Mus Nat. Hist. HI 68682

Newsletter, Denver botanical garden. Denver, CO. Vol. ?-1(3)+, ?-1971+. Newslett. Denver Bot. Gard. HI 68683

Newsletter, Des Moines botanical center = Botanical center newsletter, Des Moines. Des Moines, IA.

Newsletter, Des Moines cactus and succulent society. Des Moines, IA. No. ?-4+, ?-1967+. Newslett. Des Moines Cact Succ. Soc. HI 68684

Newsletter, Devon trust for nature conservation. Exeter. 1977+. Newslett. Devon Trust Nat. Conservation. HI 68685

Newsletter, Dixon gallery and gardens. Memphis, TN. Vol. 1+, 1978+. Newslett. Dixon Gall. Gard. HI 68686

Newsletter, Dorset biological records centre. Dorchester. 1976-77. Newslett. Dorset Biol. Rec. Centre. Superseded by: Newsletter, Dorset environmental records centre. HI 68687

Newsletter, Dorset environmental records centre. Dorchester. 1977+. Newslett. Dorset Environm. Rec. Centre. Preceded by: Newsletter, Dorset biological records centre. HI 68688

Newsletter, Dorset natural history and archaeological society. Dorchester. No. 1+, 1973+. Newslett. Dorset Nat. Hist. Archaeol. Soc. HI 68689

Newsletter, Dorset naturalists' trust. Poole. No. ?-2+, ?-1962+. Newslett. Dorset Naturalists' Trust. HI 68690

Newsletter, East Africa natural history society = E A N H S newsletter. Nairobi.

Newsletter, East Grinstead and district natural science society. East Grinstead. Nos. 1-8, 1971-73. Newslett. E. Grinstead Distr. Nat. Sci. Soc. Superseded by: Newsletter, East Grinstead natural history society. HI 68691

Newsletter, East Grinstead natural history society. East Grinstead. No. 9+, 1973+. Newslett. E. Grinstead Nat. Hist. Soc. Preceded by: Newsletter, East Grinstead natural history society. HI 68692

Newsletter, ecology center. Berkeley, CA. 1969+. Newslett. Ecol. Center. HI 68693

Newsletter, Edinburgh natural history society. Edinburgh. Nos. [1]-?, 1965-71. Newslett. Edinburgh Nat. Hist. Soc. Superseded by: Journal, Edinburgh natural history society. HI 68694

Newsletter, elm research institute = Quarterly newsletter, elm research institute. Harrisville, NH.

Newsletter, empire state iris society. Oakfield, NY. Vol. 1+, 1953+. Newslett. Empire State Iris Soc. HI 68695

Newsletter, environmental mutagen society. Oak Ridge, TN. Nos. 1-6, 1969-72; supplement 1971-72. Newslett. Environm. Mutagen Soc. Continued in: Mutation research. HI 68696

Newsletter, environmental plant life services = Environmental plant life services newsletter. Versailles, Lexington, KY.

Newsletter, epiphyllum society of America. Monrovia, CA. ?-1972+. Newslett. Epiphyllum Soc. Amer. HI 68697

Newsletter, estación científica Charles Darwin = Carta informativa, estación científica Charles Darwin. Santa Cruz, Galapagos.

Newsletter, ethiopian natural history society. Addis Ababa. Nos. 1-22, 1966-68. Newslett. Ethiop. Nat. Hist. Soc. Continued as: Newsletter, ethiopian wildlife and natural history society. HI 68698

Newsletter, ethiopian wildlife and natural history society. Addis Ababa. No. 23+, 1968+. Newslett. Ethiop. Wildlife Nat. Hist. Soc. Preceded by: Newsletter, ethiopian natural history society. HI 68699

Newsletter, european-mediterranean division, I A B G. Durham, Berlin. No. 1+, [1983]+. Newslett. Eur.-Medit. Div. I. A. B. G. HI 68700

Newsletter, faba bean information service = F A B I S newsletter. Aleppo.

Newsletter, Fairchild tropical garden = Fairchild tropical garden newsletter. Cocnut Grove, FL. Fairchild Trop. Gard. Newslett. See B–P–H 366/23.

Newsletter, Fairfield experimental horticulture station. Kirkham. No. 14+, 1976+. Newslett. Fairfield Exp. Hort. Sta. Preceded by: Fairfield newsletter. HI 68701

Newsletter from the Falkland Islands Foundation. Godalming. No. 1+, 1983+. Newslett. Falkland Islands Found. HI 68702

Newsletter, federation of British Columbia naturalists. Victoria, B.C. Vols. 7-17, 1969-79. Newslett. Fed. Brit. Columbia Naturalists. Preceded by: Newsletter, British Columbia nature council. Superseded by: B C naturalist. HI 68703

Newsletter, federation of european societies of plant physiology = F E S P P newsletter. Copenhagen.

Newsletter, federation of Ontario naturalists. Don Mills, Ont. Vols. 1-15, 1960-74. Newslett. Fed. Ontario Naturalists. Incorporated in: Ontario naturalist. HI 68704

Newsletter, fish and wildlife reference service = Fish and wildlife reference service newsletter. Denver, CO.

Newsletter, flora of north America = Flora of north America newsletter. St. Louis, MO.

Newsletter, Florida ornamental growers' association. No. 1+, 1978+. Newslett. Florida Ornam. Growers Assoc. HI 68706

Newsletter, Florida turf grass association. Bradenton, FL. No. 1+, 1960+. Newslett. Florida Turf Grass Assoc. HI 68707

Newsletter, Folkestone natural history society. Folkestone. No. 1+, [1970]+. Newslett. Folkestone Nat. Hist. Soc. HI 68708

Newsletter, forest research institute and colleges. Dehra Dun. Vol. 9+, 1972+. Newslett. Forest Res. Inst. Coll. Preceded by: Quarterly newsletter, forest research institute and colleges, Dehra Dun. HI 68709

Newsletter, friends of the botanic garden, St. Andrews. St. Andrews. No. ?-15+, ?-1984+. Newslett. Friends Bot. Gard. St. Andrews. HI 68710

Newsletter, friends of the botanical garden, university of California. Berkeley, CA. Vols. 1-5, 1976?-80. Newslett. Friends Bot. Gard. Univ. Calif. Superseded by: Botanical garden quarterly, university of California. HI 68711

Newsletter, friends of the botanical garden, university of Georgia = Newsletter, friends of the university of Georgia botanical garden. Athens, GA.

Newsletter, friends of the Farlow. Cambridge, MA. Vol. 1+, 1982+. Newslett. Friends Farlow. HI 68712

Newsletter, friends of the Matthaei botanical gardens. Ann Arbor, MI. Vol. ?-10(2)+, ?-1983+. Newslett. Friends Matthaei Bot. Gard. HI 68713

Newsletter, friends of the U N C herbarium. Chapel Hill, NC. Vol. 1+, 1983+. Newslett. Friends U. N. C. Herb. HI 68715

Newsletter, friends of the university of Georgia botanical garden. Athens, GA. 1972+. Newslett. Friends Univ. Georgia Bot. Gard. HI 68716

Newsletter, garden centers of America = G C A newsletter. Washington, DC.

Newsletter, garden history society [1st series]. Reading. Extraordinary issue, 1971; nos. 13-17, 1971-72; [n.s.] no. 1+, 1981+. Newslett. Gard. Hist. Soc. Preceded by: Quarterly newsletter, garden history society. HI 68717

Newsletter, George Landis arboretum. Esperance, NY. Vol. 1+, 1982+. Newslett. George Landis Arbor. HI 68718

Newsletter, Glamorgan county naturalists' trust. [Swansea.] No. 1+, 1972+. Newslett. Glamorgan County Naturalists' Trust. Superseded by: Gnat. HI 68719

Newsletter, Gloucestershire trust for nature conservation. Gloucester. 1971+. Newslett. Gloucestershire Trust Nat. Conservation. HI 68720

Newsletter, greater New York orchid society. New York. 1976+. Newslett. Gr. New York Orchid Soc. HI 68721

Newsletter, guild of natural science illustrators = G N S I newsletter. Washington, DC.

Newsletter, Gwent trust for nature conservation. Newport, Gwent. 1974+. Newslett. Gwent Trust Nat. Conservation. Preceded by: Newsreel, Monmouthshire naturalists' trust ltd. HI 68722

Newsletter, Hampshire and Isle of Wight naturalists' trust ltd. Romsey. 1971+. Newslett. Hampshire Isle Wight Naturalists' Trust. HI 68723

Newsletter, hardy plant society. London. [Dates of publication not ascertained.] Newslett. Hardy Pl. Soc. HI 68724

Newsletter, Harrow natural history society. Harrow. ?-1975+. Newslett. Harrow Nat. Hist. Soc. HI 68725

Newsletter. Hawaiian botanical garden foundation. Honolulu, HI. Newslett. Hawaiian Bot. Gard. Found. See B–P–H 658/16. HI 57392

Newsletter of the hawaiian botanical society. Honolulu, HI. Newslett. Hawaiian Bot. Soc. See B–P–H 658/17. HI 57393

Newsletter, the haworthia society. Buckden, N. Yorks. Vol. 1+, 1987+. Newslett. Haworthia Soc. HI 68727

Newsletter of the hellenic society for the protection of nature. ?-1979+. Newslett. Hellen. Soc. Protect. Nat. HI 68728

Newsletter, herb society of America [title varies slightly]. Boston, MA. No. [1]+, 1935+. Newslett. Herb. Soc. Amer. HI 68729

Newsletter, Herefordshire and Radnorshire nature trust ltd. Hereford. 1963-75. Newslett. Herefordshire Radnorshire Nat. Trust. Superseded by: Flycatcher. HI 68730

Newsletter, Hertfordshire and Middlesex trust for nature conservation. St. Albans. No. ?-22+, ?-1971+. Newslett. Hertfordshire Middlesex Trust Nat. Conservation. HI 68731

News letter of himalayan botany. Tokyo. Vol. 1+, 1986+. News Lett. Himalayan Bot. HI 68732

Newsletter, holly society of America. College Park, MD. Nos. 1-3, 1952-53. Newslett. Holly Soc. Amer. Superseded by: Holly newsletter. HI 68733

Newsletter, horticultural society of New York. New York. 1983+. Newslett. Hort. Soc. New York. Preceded by: Member newsletter, horticultural society of New York. HI 68734

Newsletter, hydroponic society of America = Hydroponic society of America newsletter. Concord, CA.

Newsletter, Illinois state academy of science. Normal, IL. Vol. 1(1), 1977? Newslett. Illinois State Acad. Sci. Superseded by: Illinois science news. HI 67071

News letter of the Illinois state horticultural society. Springfield, IL. News Lett. Illinois State Hort. Soc. See B–P–H 658/4. HI 57387

Newsletter, indian society for nuclear technology in agriculture and biology = I S N A newsletter. New Delhi.

Newsletter, indigenous bulb growers' association of South Africa = I B S A newsletter. Durbanville.

Newsletter, institute of ecosystem studies = I E S newsletter. Millbrook, NY.

Newsletter, institute for systematic and evolutionary botany. St. Louis, MO. No. 1+, 1969+. Newslett. Inst. Syst. Evol. Bot. HI 68736

Newsletter, international aroid society, inc. Miami, FL. Vol. 1+, 1977+. Newslett. Int. Aroid Soc. HI 68737

Newsletter, international association for angiosperm paleobotany. Mexico City. Vols. 1-8(1), 1975-83. Newslett. Int. Assoc. Angiosp. Paleobot. HI 68738

Newsletter, international association of aquatic vascular plant biologists. Fairbanks, AK. No. 8+, 1975+ Newslett. Int. Assoc. Aquatic Vasc. Pl. Biol. Preceded by: Newsletter, association of aquatic vascular plant biologists. HI 68739

Newsletter, international association of botanic gardens, european-mediterranean division = Newsletter, european-mediterranean division, I A B G. Durham, Berlin.

Newsletter, international association of bryologists = I A B newsletter. Penicuik, etc.

Newsletter, international association for plant tissue culture. Calgary. No. 1+, 1971+. Newslett. Int. Assoc. Pl. Tissue Cult. HI 68740

Newsletter, international cactus and succulent society. San Angelo, TX. ?-1977+. Newslett. Int. Cact. Succ. Soc. HI 68742

Newsletter, international camellia society. U K and western european region. Worcester. No. 1+, 1976+. Newslett. Int. Camellia Soc. U.K. W. Eur. HI 68743

Newsletter, international center for arid and semi-arid land studies = I C A S A L S newsletter. Lubbock, TX.

Newsletter, international clematis society. No. 1+, 1984+. Newslett. Int. Clematis Soc. HI 68744

Newsletter, international commission for palynology. Tempe, AZ. Vols. 1-7(1), 1978-84. Newslett. Int. Commiss. Palynol. Superseded by: Palynos. HI 68745

Newsletter, international dendrology society. ?-1976+. Newslett. Int. Dendrol. Soc. HI 68746

Newsletter of the international geranium society. Gardena, CA. Newslett. Int. Geranium Soc. See B–P–H 658/18. HI 57394

Newsletter, international lilac society, inc. Medina, OH, Hamilton, Ont. Vol. 1+, 1971+. Newslett. Int. Lilac Soc. HI 68747

Newsletter, international nannoplankton association = I N A newsletter. Utrecht.

Newsletter, international organization for biological control of noxious animals and plants = I O B C newsletter. Zurich.

Newsletter, international organization of palaeobotany = I O P newsletter. London.

Newsletter, international organization of plant biosystematists. Reading, etc. No. 1+, 1984+. Newslett. Int. Organ. Pl. Biosyst. Preceded by: Plant biosystematics newsletter. HI 68748

Newsletter, international potato center = International potato center newsletter. Lima.

Newsletter. International rice commission. Bangkok. Newslett. Int. Rice Commiss. See B–P–H 658/19. HI 57395

Newsletter, international shade tree conference. ?-1979-? Newslett. Int. Shade Tree Conf. Superseded by: Newsletter, international society of arboriculture, Pen Del chapter. HI 68749

Newsletter, international society of arboriculture, Pen Del chapter. [Dates of publication not ascertained.] Newslett. Int. Soc. Arboric., Pen Del Chapt. Preceded by: Newsletter, international shade tree conference. HI 68750

Newsletter, international society for the history, philosophy and social studies of biology. Blacksburg, VA. No. 1+, 1989+. Newslett. Int. Soc. Hist. Philos. Social Stud. Biol. HI 68751

Newsletter, international society for oil palm breeders. Kuala Lumpur. Vol. ?-1(2)+, ?-1984+. Newslett. Int. Soc. Oil Palm Breed. HI 68752

Newsletter, international society of plant population biologists. Aiken, SC. No. 1+, 1975+. Newslett. Int. Soc. Pl. Populat. Biol. HI 68753

News-letter from the international union of biological sciences. Naples. Nos. 1-?, 1955-62. News-Lett. Int. Union Biol. Sci. Superseded by: I U B S news. HI 68754

Newsletter, international union of forestry research organizations. Asheville, NC. No. 1+, 1978+. Newslett. Int. Union Forest. Res. Organ. HI 68755

Newsletter, international union for the protection of new varieties of plants = U P O V newsletter. Geneva.

Newsletter, Inverness-shire survey 1970-74. Glasgow. ?-1973+. Newslett. Inverness-shire Surv. 1970-74. HI 68756

Newsletter, irish garden plant society. Glasnevin. No. 1+, 1981+. Newslett. Irish Gard. Pl. Soc. HI 68757

Newsletter, japanese society of plant physiologists. Ca.1989+. Newslett. Jap. Soc. Pl. Physiologists. HI 68758

Newsletter, Kansas wildflower society. Topeka, KS. Vol. 1+, 1979+. Newslett. Kansas Wildflower Soc. HI 68759

Newsletter, Kent naturalists' trust. East Malling. ?-1959-61-? Newslett. Kent Naturalists' Trust. HI 68760

Newsletter, Kerala forest research institute. Peechi. Nos. 1-4, 19??-78? Newslett. Kerala Forest Res. Inst. Superseded by: Evergreen. HI 68761

Newsletter, kiwifruit growers of California. Carmichael, CA. 1975+. Newslett. Kiwifruit Growers Calif. HI 68762

Newsletter, kokuritsu kagaku hakubutsukan, Japan = Newsletter, national science museum, Tokyo.

News letter, Lake District naturalists trust. Penrith, Windermere. Nos. 1-23, 1963-74. News Lett. Lake Distr. Naturalists Trust. Superseded by: News letter, Cumbria naturalists' trust. HI 68763

Newsletter, Leicestershire and Rutland trust for nature conservation. Leicester. 1973+. Newslett. Leicestershire Rutland Trust Nat. Conservation. HI 68764

Newsletter, lily committee, royal horticultural society. Place varies. No. ?-10+, ?-1983+. Newslett. Lily Committee Roy. Hort. Soc. HI 68765

Newsletter, limnological society of southern Africa. Grahamstown. Nos. 1-22, 1964-74. Newslett. Limnol. Soc. S. Africa. Superseded by: Journal, limnological society of southern Africa. HI 68766

Newsletter, Lincolnshire and south Humberside trust for nature conservation. Alford. Nos. ?-28-47, ?-1965-77. Newslett. Lincolnshire S. Humberside Trust Nat. Conservation. Preceded by: Newsletter, Lincolnshire trust for nature conservation. HI 68767

Newsletter, Lincolnshire trust for nature conservation. Nos. ?-28-47, ?-1965-77. Newslett. Lincolnshire Trust Nat. Conservation. Superseded by: Newsletter, Lincolnshire and south Humberside trust for nature conservation. HI 68768

Newsletter, Linnaean society of New York. New York. Vol. 1+, 1947+. Newslett. Linn. Soc. New York. HI 68769

Newsletter, London natural history society. London. 1980+. Newslett. London Nat. Hist. Soc. HI 68770

Newsletter, Louisiana native plant society. Shreveport, LA. Vol. 1+, 1983+. Newslett. Louisiana Native Pl. Soc. HI 68771

News letter, Louisiana society for horticultural research = L S H R news letter.

Newsletter, Lundy field society. Barnstaple. No. ?-15+, ?-1985+. Newslett. Lundy Field Soc. HI 68772

Newsletter, M A F F library services. London. No. 1+, 1981+. Newslett. M. A. F. F. Libr. Serv. HI 68773

Newsletter, Massachusetts Audubon society. Lincoln, MA. Vols. 13-19, 1973-80. Newslett. Mass. Audubon Soc. Preceded by: Massachusetts Audubon newsletter. Superseded by: Sanctuary. Bulletin of the Massachusetts Audubon society. HI 68774

Newsletter, median iris society. Grafton, MA. Vols. 1-2, 1960-61. Newslett. Median Iris Soc. Superseded by: Medianite. HI 68775

Newsletter, Memphis botanic garden. Memphis, TN. 1984?-85. Newslett. Memphis Bot. Gard. Superseded by: Garden appeal. HI 68776

Newsletter, midwest cooperative conservation program. Carbondale, IL. Nos. 1-11, 1984-85. Newslett. Midw. Coop. Conservation Progam. HI 68777

Newsletter, Missouri native plant society. Jefferson City, MO. Vol. 1(1-3), 1986. Newslett. Missouri Native Pl. Soc. Superseded by: Petal pusher. HI 68778

Newsletter, Morris arboretum of the university of Pennsylvania = Morris arboretum newsletter. Philadelphia, PA.

Newsletter, Moss Landing marine laboratories. Moss Landing, CA. 1968-69. Newslett. Moss Landing Mar. Lab. HI 68779

News letter of the mycological society of America. Columbus, OH. Vols. 1-6, 1903-08. News Lett. Mycol. Soc. Amer. Superseded (1950+) by: Mycological society of American news letter. HI 68571

Newsletter of the Natal cactus and succulent club. Durban. ?-1971+. Newslett. Natal Cact. Succ. Club. HI 68780

Newsletter of the Natal cactus and succulent society. Durban. 1964+. Newslett. Natal Cact. Succ. Soc. HI 68781

Newsletter, national cactus and succulent society. Bradford. Vol. 2+, 1947+. Newslett. Natl. Cact. Succ. Soc. Preceded by: Newsletter, Yorkshire cactus society. HI 68783

Newsletter, national council for the conservation of plants and gardens. Woking, Cerne Abbas. Vol. 1+, 1980+. Newslett. Natl. Council Conservation Pl. Gard. HI 68782

Newsletter, national council for therapy and rehabilitation through horticulture = N C T R H newsletter. Alexandria, VA, Gaithersburg, MD.

Newsletter, national gladiolus society = National gladiolus society newsletter. Vienna, VA. Natl. Gladiolus Soc. Newslett. See B–P–H 630/26.

Newsletter, national oleander society = National oleander society newsletter. Galveston, TX.

Newsletter, national resources defense council = N R D C newsletter. New York.

Newsletter, national science museum, Tokyo. Tokyo. No. 1+, 1973+. Newslett. Natl. Sci. Mus. Tokyo. HI 68785

Newsletter, native orchid society of South Australia. Everard Park, S.A. Vols. 1-2(2), 1977-78. Newslett. Native Orchid Soc. S. Australia. Superseded by: Journal, native orchid society of South Australia. HI 68786

Newsletter, native plant society of New Mexico. Santa Fe, NM. Ca.1983+. Newslett. Native Pl. Soc. New Mexico. HI 68787

Newsletter, native plant society of Oregon = N P S O newsletter. Ashland, OR.

Newsletter, natural history society of Prince Edward Island. Charlottetown, P.E.I. Nos. 1-17, 1974-76. Newslett. Nat. Hist. Soc. Prince Edward Island. Superseded by: Newsletter, Prince Edward Island natural history society. HI 68788

Newsletter - nature. Strasbourg. 1975+. Newslett. Nat. HI 68789

Newsletter of the nature printing society. Pittsburgh, PA. Vol. 1+, 1976+. Newslett. Nat. Print. Soc. HI 68790

Newsletter, Nepal nature conservation society. Kathmandu. ?-ca.1986+. Newslett. Nepal Nat. Conservation Soc. HI 68791

Newsletter, New England wild flower society. Framingham, MA. 1976+. Newslett. New England Wild Fl. Soc. HI 68792

Newsletter, New Jersey plant and flower growers' association. Vol. 1+, 1971+. Newslett. New Jersey Pl. Fl. Growers' Assoc. HI 68793

Newsletter, New York botanical garden. New York, NY. Vol. 1+, 1967+. Newslett. New York Bot. Gard. HI 68794

Newsletter, New Zealand native orchid group. Lower Hutt. No. 1+, 1982+. Newslett. New Zealand Native Orchid Group. HI 68795

Newsletter, New Zealand nature conservation council. Wellington, N.Z. No. 1+, 1972+. Newslett. New Zealand Nat. Conservation Council. HI 68796

Newsletter, New Zealand orchid society inc. Wellington, N.Z. 1968+. Newslett. New Zealand Orchid Soc. HI 68797

Newsletter, Niagara frontier botanical society. Buffalo, NY. Vols. 2-3, 1984-85. Newslett. Niagara Frontier Bot. Soc. Preceded by: Newsletter, Buffalo botanical society. Superseded by: Clintonia. HI 68798

Newsletter of the Norfolk botanical garden society. Norfolk, VA. Newslett. Norfolk Bot. Gard. Soc. See B–P–H 658/24. HI 57396

Newsletter, Norfolk naturalists' trust. Norwich. Nos. 1-9, 1974-78. Newslett. Norfolk Naturalists' Trust. Superseded by: Tern. HI 68799

Newsletter, North Carolina botanical garden = North Carolina botanical garden newsletter. Chapel Hill, NC.

Newsletter, North Carolina wildflower preservation society. Durham, NC. Vol. 1+, 1989+. Newslett. N. Carolina Wildflower Preserv. Soc. HI 68801

Newsletter of the north Wales naturalists' trust. Bangor. No. ?-32+, ?-1974+. Newslett. N. Wales Naturalists' Trust. HI 68802

Newsletter, Northamptonshire naturalists' trust = Newsletter, Northamptonshire trust for nature conservation. Northampton.

Newsletter, Northamptonshire trust for nature conservation. Northampton. No. 25+, 1981+. Newslett. Northamptonshire Trust Nat. Conservation. Preceded by: Bulletin, Northamptonshire trust for nature conservation. HI 68803

Newsletter, northern Nevada native plant society. Reno, NV. Vol. 1+, 1975+. Newslett. N. Nevada Native Pl. Soc. HI 68804

Newsletter, Northumberland and Durham naturalists' trust ltd. Newcastle-upon-Tyne. 1967-71. Newslett. Northumberland Durham Naturalists' Trust. Superseded by: Newsletter, Northumberland wildlife trust. HI 68805

Newsletter, Northumberland wildlife trust. Newcastle-upon-Tyne. 1971-72; Special supplement, 1971. Newslett. Northumberland Wildlife Trust. Preceded by: Newsletter, Northumberland and Durham naturalists' trust ltd. Superseded by: Roebuck. HI 68806

Newsletter, northwest herbarium consortium. Vancouver, B.C. No. 1+, 1976+. Newslett. NorthW. Herb. Consort. HI 68807

Newsletter, northwest ornamental horticultural society. Vols. 1-3, 19??-76. Newslett. NorthW. Ornam. Hort. Soc. Superseded by: Horticulture northwest. HI 68808

Newsletter, Nottinghamshire trust for nature conservation ltd. Nottingham. 1970+. Newslett. Nottinghamshire Trust Nat. Conservation. HI 68809

Newsletter, overseas development natural resources institute. London. No. 1+, 1987+. Newslett. Overseas Developm. Nat. Resources Inst. HI 68810

Newsletter, P E I natural history society = Newsletter, Prince Edward Island natural history society. Charlottetown, P.E.I.

Newsletter, palynological and palaeobotanical association of Australasia. Lower Hutt. No. 1+, 1980+. Newslett. Palynol. Palaeobot. Assoc. Australasia. HI 68811

Newsletter, pea growing research organisation. Thornhaugh. ?-1974? Newslett. Pea Growing Res. Organ. Superseded by: Newsletter, processors' and growers' research organisation. HI 68812

Newsletter, Pennsylvania bonsai society. Vol. 1+, 1971+. Newslett. Pennsylvania Bonsai Soc. HI 68813

News letter, Penrith and district natural history society. Penrith. 1948-50. News Lett. Penrith Distr. Nat. Hist. Soc. Superseded by: Monthly bulletin, Penrith and district natural history society. HI 68814

Newsletter, perennial plant association = Quarterly newsletter, perennial plant association. Kensington, CT.

Newsletter, phytochemical section, botanical society of America. Oxford, OH. Vols. 1-5(3), 1968?-72. Newslett. Phytochem. Sect. Bot. Soc. Amer. Superseded by: Phytochemical bulletin, botanical society of America. HI 68816

Newsletter, plant and flower growers' association, inc. Vol. 1+, 1971+. Newslett. Pl. Fl. Growers' Assoc. HI 68817

Newsletter, plant gene resources of Canada. No. ?-8+, ?-1980+. Newslett. Pl. Gene Resources Canada. HI 68818

Newsletter of the plant propagators society. New Brunswick, NJ. Newslett. Pl. Propag. Soc. See B–P–H 659/2. HI 57397

Newsletter, plant protection division, department of agriculture, Canada. Ottawa. Nos. 1-22. 1944-66. Newslett. Pl. Protect. Div. Dept. Agric. Canada. Superseded by: Plant quarantine news. HI 68819

News letter. Plant quarantine and control administration. United States department of agriculture. Washington, DC. News Lett. Pl. Quarant. Admin. U.S.D.A. See B–P–H 658/6. HI 57388

Newsletter of the plumeria society of America. Houston, TX. ?-1982+. Newslett. Plumeria Soc. Amer. HI 68820

Newsletter, porcupine [society]. South Shields. Vol. 1+, 1976+. Newslett. Porcupine Soc. HI 68821

Newsletter, Prince Edward Island natural history society. Charlottetown, P.E.I. Nos. 18-78, 1976-84. Newslett. Prince Edward Island Nat. Hist. Soc. Preceded by: Newsletter, natural history society of Prince Edward Island. Superseded by: Island naturalist. HI 68822

Newsletter, processors' and growers' research organisation. Thornhaugh. 1975-82. Newslett. Processors' Growers' Res. Organ. Preceded by: Newsletter, pea growing research organisation. Superseded by: Vegetable grower. HI 68823

Newsletter, provincial horticultural station, Brooks. Brooks, Alta. -1969? Newslett. Prov. Hort. Sta. Brooks. Superseded by: Crop protection newsletter. HI 68824

Newsletter, Queensland littoral society. Nos. 1-41, ?-1971. Newslett. Queensland Littoral Soc. Superseded by: Operculum. HI 68825

Newsletter, Rancho Santa Ana botanic garden. Claremont, CA. Vol. 1+, 1985+. Newslett. Rancho Santa Ana Bot. Gard. HI 68826

Newsletter, rare fruit council international. Miami, FL. 1955+. Newslett. Rare Fruit Council Int. HI 68827

Newsletter, regional committee for the south east asian programme, international board for plant genetic resources. Bogor. Vol. 1+, 1977+. Newslett. Regional Committee S. E. Asian Programme. HI 68828

Newsletter, rhododendron species foundation. ?-1977+. Newslett. Rhododendron Spec. Found. HI 68829

Newsletter, royal botanic gardens. Kew. 1967+. Newslett. Roy. Bot. Gard., Kew. HI 68830

Newsletter, royal horticultural society. London. No. 1+, 1986+. Newslett. Roy. Hort. Soc. HI 68831

Newsletter, royal New Zealand institute of horticulture. [Wellington, N.Z.] Vol. 1, nos. 1-2, 1972. Newslett. Roy. New Zealand Inst. Hort. Published during the suspension of the Journal of the royal New Zealand institute of horticulture. HI 68832

Newsletter, Ruislip and district natural history society. Ickenham. No. ?-51+, ?-1974+. Newslett. Ruislip Distr. Nat. Hist. Soc. HI 68833

Newsletter, S W A scientific society. Windhoek. No. 1+, 1960+. Newslett. S. W. A. Sci. Soc. HI 68834

Newsletter, San Gabriel Valley cactus and succulent society. Pasadena, CA. Vol. 1+, 1966+. Newslett. San Gabriel Valley Cact. Succ. Soc. HI 68835

Newsletter of the Sarasota succulent society. Sarasota, FL. ?-1973+. Newslett. Sarasota Succ. Soc. HI 68836

Newsletter, Saskatchewan natural history society. Saskatoon. Nos. 1-62, 1947-83. Newslett. Saskatchewan Nat. Hist. Soc. Superseded by: Blue jay news. HI 68837

Newsletter, scientific exploration society. [London.] Vol. ?-1(7)+, ?-1970+. Newslett. Sci. Explor. Soc. HI 68838

Newsletter, scottish orchid society. Edinburgh. Vol. 1+, 1982+. Newslett. Scott. Orchid Soc. HI 68839

Newsletter, the sedum society. Milton Keynes. ?-1988+. Newslett. Sedum Soc. HI 68840

Newsletter, Selbourne association. Selbourne. No. 1+, 1978+. Newslett. Selbourne Assoc. HI 68841

Newsletter, sempervivum fanciers' association. Randolph, MA. Vol. 1+, 1975+. Newslett. Sempervivum Fanciers' Assoc. HI 68842

Newsletter, societas internationalis plantarum demographia. Hickory Corners, MI. Ca.1977+. Newslett. Soc. Int. Pl. Demogr. HI 68843

Newsletter of the society of american bacteriologists. Baltimore, MD. Newslett. Soc. Amer. Bacteriol. See B–P–H 659/3. HI 57398

Newsletter, society for the bibliography of natural history. London. Nos. 1-17, 1979-83. Newslett. Soc. Bibliogr. Nat. Hist. Preceded by: British newsletter, society for the bibliography of natural history. Superseded by: Newsletter, society for the history of natural history. HI 68844

Newsletter, society for the history of natural history. London. Nos. 18-26, 1983-85; no. 31+, 1987+. Newslett. Soc. Hist. Nat. Hist. Preceded by: Newsletter, society for the bibliography of natural history. Nos. 27-30 contained in: Archives of natural history. HI 68845

Newsletter, society for industrial microbiology. Washington, DC. ?-1972+. Newslett. Soc. Industr. Microbiol. HI 68846

Newsletter of the society for Louisiana irises. Lafayette, LA. ?-1972+. Newslett. Soc. Louisiana Irises. HI 68847

Newsletter, Somerset trust for nature conservation ltd. Taunton. 1964-78. Newslett. Somerset Trust Nat. Conservation. Superseded by: News, Somerset trust for nature conservation. HI 68848

Newsletter of the south african aloe and succulent society. Pretoria. Vols. 1-6, 1976-81. Newslett. S African Aloe Succ. Soc. Superseded by: Kambroo. HI 68849

Newsletter, south-east asian rain forest research programme. London. No. 1+, 1985+. Newslett. S. E. Asian Rain Forest Res. Programme. HI 68850

Newsletter, south London botanical institute. London. 1982+. Newslett. S. London Bot. Inst. HI 68851

Newsletter, southern african weed science society. Sunnyside. ?-1983+. Newslett. S. African Weed Sci. Soc. HI 68852

Newsletter, species survival commission = Species. Gland.

Newsletter, Spelthorne natural history society. 1977+. Newslett. Spelthorne Nat. Hist. Soc. HI 68853

Newsletter, spuria iris society. Yorba Linda, CA, Purcell, OK. Vol. 1+, 1956?+. Newslett. Spuria Iris Soc. HI 68854

Newsletter, Staffordshire nature conservation trust. Stoke-on-Trent. No. ?-7+, ?-1973+. Newslett. Staffordshire Nat. Conservation Trust. HI 68855

Newsletter, succulent plant club. Southburgh. Nos. 1-2, 1977. Newslett. Succ. Pl. Club. Superseded by: Xerophyte. HI 68856

Newsletter, succulent plant institute. Coulsdon. 1962-77. Newslett. Succ. Pl. Inst. Superseded by: Newsletter, succulent plant trust. HI 68857

Newsletter, succulent plant trust. Coulsdon. 1978+. Newslett. Succ. Pl. Trust. Preceded by: Newsletter, succulent plant institute. HI 68858

Newsletter, Suffolk trust for nature conservation. Hundon. 1971-85. Newslett. Suffolk Trust Nat. Conservation. Superseded by: Magazine, Suffolk trust for nature conservation. HI 68859

Newsletter, Surrey flora committee. Esher. 1963+. Newslett. Surrey Fl. Committee. HI 68860

Newsletter, Surrey naturalists' trust. [Leatherhead], Dorking, etc. No. 1+, 1959+. Newslett. Surrey Naturalists' Trust. HI 68861

Newsletter, Surrey trust for nature conservation = Newsletter, Surrey naturalists' trust. [Leatherhead], Dorking, etc.

Newsletter, Sussex naturalists' trust. Henfield. Nos. ?-33-36, ?-1970-71. Newslett. Sussex Naturalists' Trust Superseded by: News letter, Sussex trust for nature conservation. HI 68862

Newsletter, Sussex trust for nature conservation. Henfield. No. 37+, 1971+. Newslett. Sussex Trust Nat. Conservation. Preceded by: Newsletter, Sussex naturalists' trust. HI 68863

Newsletter, tasmanian arboretum. Devonport, Tas. No. ?-2+, ?-1984+. Newslett. Tasmanian Arbor. HI 68864

Newsletter of the threatened plants committee, international union for conservation of nature and natural resources. Kew. Nos. 1-10, 1977-82. Newslett. Threat. Pl. Committee. Superseded by: Threatened plants newsletter. HI 68865

Newsletter of the Thunder Bay field naturalists' club. Thunder Bay. Vol. 1+, 1947+. Newslett. Thunder Bay Field Naturalists' Club. HI 68866

Newsletter, tissue culture for crops project, the international plant biotechnology network. Fort Collins, CO. No. 1+, 1982+. Newslett. Tissue Cult. Crops Proj. HI 74865

Newsletter, Toronto field naturalists' club. Toronto. Nos 1-332, 1938-70. Newslett. Toronto Field Naturalists' Club. Superseded by: Toronto field naturalist. 5-4240-3. HI 68867

Newsletter, transition zone horticultural institute. Flagstaff, AZ. Vol. 1+, 1984+. Newslett. Transit. Zone Hort. Inst. HI 68868

Newsletter, tropical development and research institute. London. 1983+. Newslett. Trop Developm. Res. Inst. Preceded by: Newsletter, tropical products institute. HI 68869

Newsletter, tropical products institute. London. No. 1+, 1974+. Newslett. Trop Prod. Inst. HI 68870

Newsletter U S department of agriculture library. Washington, DC. Newslett. U.S.D.A. Libr. See B–P–H 659/4. HI 57399

Newsletter, useful palms of tropical america. Brasilia. No. 1+, 1985+. Newslett. Useful Palms Trop. Amer. HI 68871

Newsletter of the Utah native plant society. Salt Lake City, UT. Vols. ?-4, 19??-81. Newslett. Utah Native Pl. Soc. Superseded by: Sego lily. HI 75102

Newsletter, Wallop Valley field club. Over Wallop. No. 1+, 1979+. Newslett. Wallop Valley Field Club. HI 68872

Newsletter, Warwick natural history society. Leamington Spa. No. ?-73+, ?-1974+. Newslett. Warwick Nat. Hist. Soc. HI 68873

Newsletter of the Warwickshire nature conservation trust. Warwick. Nos. 19-31, 1975-79. Newslett. Warwickshire Nat. Conservation Trust. Preceded by: Nature conservation in Warwickshire. Superseded by: Warnact newsletter. HI 68874

Newsletter, weed science society of America = W S S A newsletter. Champaign, IL.

Newsletter, West Virginia native plant society. Norton, WV. Vol. 1+, 1982+. Newslett. W. Virginia Native Pl. Soc. HI 68875

Newsletter, West Virginia university arboretum. Morgantown, WV. Vols. 4-5(2), 1954-55. Newslett. W. Virginia Univ. Arbor. Preceded by: Arboretum news, West Virginia university. Superseded by: Arboretum newsletter, West Virginia university. HI 68876

Newsletter, western society of naturalists. San Jose, CA. ?-1972+. Newslett. W. Soc. Naturalists. HI 68877

Newsletter, Wiltshire trust for nature conservation. Trowbridge. No. ?-2+, ?-1964+. Newslett. Wiltshire Trust Nat. Conservation. HI 68878

Newsletter, Worcestershire biological records centre. Worcester. No. 1+, (1979?)+. Newslett. Worcestershire Biol. Rec. Centre. HI 68879

News letter, Worcestershire naturalists' club. Malvern. 1965-79. News Lett. Worcestershire Naturalists' Club. Superseded by: Transactions of the Worcestershire naturalists' club. HI 68880

Newsletter, Worcestershire nature conservation trust. Bromsgrove. No. ?-10+, ?-1972+. Newslett. Worcestershire Nat. Conservation Trust. HI 68881

Newsletter, world wildlife fund Indonesia programme. Bogor. Vols. 1-2(3), 1977-78. Newslett. World Wildlife Fund Indonesia Programme. Superseded by: Conservation Indonesia. HI 68882

Newsletter, Yorkshire cactus society. Bradford. Vol. 1(1-4), 1945-46. Newslett. Yorkshire Cact. Soc. Superseded by: Newsletter, national cactus and succulent society. HI 68883

Newsletter, Yorkshire naturalists' trust ltd. York. 1971+. Newslett. Yorkshire Naturalists Trust. HI 68884

Newsletter, Yukon conservation society. Whitehorse, Yukon. 1979+. Newslett. Yukon Conservation Soc. HI 68885

Newspaper bulletin, Vermont state agricultural college, agricultural experiment station = Vermont state agricultural college, agricultural experiment station. Newspaper bulletin. Burlington, VT. Vermont Agric. Exp. Sta. Newspaper Bull. See B–P–H 957/2.

Newsreel, Monmouthshire naturalists' trust ltd. 1971-74. Newsr. Monmouthshire Naturalists' Trust. Superseded by: Newsletter, Gwent trust for nature conservation. HI 68886

Nexus; a quarterly publication by the university of Georgia botanical garden. Athens, GA. Vol. 1+, 1984+. Nexus. HI 68887

Nhatrang institut océanographique, contributions. Hue?, Indochina [South Vietnam]. Nhatrang Inst. Océanogr. Contr. See B–P–H 659/7. HI 57401

Nicko's fruit journal. Brisbane. Nicko's Fruit J. See B–P–H 659/9. HI 57402

Niederländische Kartoffel. The Hague. Vol. 1+, 1951+. Niederl. Kartoffel. HI 68888

Niederlausitzer floristische Mitteilungen. Guben. Vols. 1-4, 1965-68. Niederlausitzer Florist. Mitt. HI 68889

Niederrheinischer Anzeiger für Staats- und Landwirthschaftslehre, Natur- und Gewerbekunde. Bonn. Niederrhein. Anz. Staats- Landwirthschaftsl. See B–P–H 659/11. HI 57403

Niederrheinisches Taschenbuch für Liebhaber des Schönen und Guten. Düsseldorf. Niederrhein. Taschenb. Liebhaber Schönen Guten. See B–P–H 659/12. HI 57404

Niedersächsische Nachrichten von gelehrten neuen Sachen ... Hamburg. Niedersächs. Nachr. Gel. Neuen Sachen. See B–P–H 659/13. HI 57405

Niedersächsische neue Zeitungen von gelehrten Sachen. Hamburg. Niedersächs. Neue Zeitungen Gel. Sachen. See B–P–H 659/14. HI 57406

Nien-san; annals of the university of Cantho. Phong Dinh. Vol. 1, 1968. Nien-San. HI 68890

Nieuw algemeen magazijn voor wetenschappen, konst en smaak. Amsterdam. Nieuw Alg. Mag. Wetensch. See B–P–H 659/15. HI 57407

Nieuwe algemene konst- en letter-bode, ... Haarlem. Nieuwe Alg. Konst- Lett.-Bode. See B–P–H 659/16. HI 57408

Nieuwe algemeene vaderlandsche letteroefeningen. Amsterdam. Nieuwe Alg. Vaderl. Letteroefen. See B–P–H 659/17. HI 57409

Nieuwe genees-, natuur- en huishoud-kundige jaarboeken. Dordrecht, Netherlands. Nieuwe Genees- Natuur- Huishoud-Kundige Jaarb. See B–P–H 659/18. HI 57410

Nieuwe gids. Amsterdam. Nieuwe Gids (Amsterdam). See B–P–H 659/19. HI 57411

Nieuwe gids. Malang, Dutch E. Indies [Indonesia]. Nieuwe Gids (Malang). See B–P–H 659/20. HI 57412

Nieuwe gids bibliotheek; verzameling oorspronkelijke bijdragen op het gebied van letteren, kunst, wetenschap en wijsbegeerte. The Hague. Nieuwe Gids Biblioth. See B–P–H 660/1. HI 57413

Nieuwe natuur- en geneeskundige bibliotheek. Amsterdam. Nieuwe Natuur- Geneesk. Biblioth. See B–P–H 660/2. HI 57414

Nieuwe vaderlandsche letteroefeningen. Amsterdam. Nieuwe Vaderl. Letteroefen. See B–P–H 660/3. HI 57415

Nieuwe verhandelingen van de eerste klasse van het koninklijk nederlandsch instituut van wetenschappen te Amsterdam. Amsterdam. Nieuwe Verh. Eerste Kl. Kon. Ned. Inst. Wetensch. Amsterdam. See B–P–H 660/4. HI 57416

Nieuwe verhandelingen van het zeeuwsch genootschap der wetenschappen. Middelburg, Netherlands. Nieuwe Verh. Zeeuwsch Genootsch. Wetensch. See B–P–H 660/5. HI 57417

Nieuwe werken van het zeeuwsch genootschap der wetenschappen. Middelburg, Netherlands. Nieuwe Werken Zeeuwsch Genootsch. Wetensch. See B–P–H 660/7. HI 57418

Nieuwsbulletin, zeebiologisch museum. Hague. No. 1+, 1977+. Nieuwsbull. Zeebiol. Mus. HI 68891

Nigeria; quarterly magazine for everyone interested in the progress of the country. Lagos. Vols. 9-66, 1937-60 [publication suspended 1941-43, 1945]. Nigeria. Preceded by: Nigerian teacher. Superseded by: Nigerian magazine. 4-3065-1. HI 68892

Nigerian agricultural journal. Nsukka, Nigeria. Nigerian Agric. J. See B–P–H 660/9. HI 57419

Nigerian field. Stroud. Vol. 1+, 1931+. Nigerian Field. HI 68893

Nigerian forester. Ibadan. 1940-41. Nigerian Forester. Superseded by: Farm and forest. HI 68894

Nigerian geographical journal. Ibadan, Nigeria. Nigerian Geogr. J. See B–P–H 660/10. HI 57420

Nigerian grower and producer and west african farmer. Lagos, Nigeria. Nigerian Grower Producer. See B–P–H 660/11. HI 57421

Nigerian journal of forestry. Ibadan. Vol. 1+, 1971+.

Nigerian J. Forest. HI 68895

Nigerian journal of plant protection. Ibadan. Vol. 1+, 1975+. Nigerian J. Pl. Protect. HI 68896

Nigerian journal of science. Ibadan. Vols. 1-9, 1966-75. Nigerian J. Sci. Preceded by: Proceedings of the science association of Nigeria. HI 68897

Nigeria magazine. Lagos. No. 67+, 1960+. Nigeria. Preceded by: Nigeria. HI 68898

Nigerian teacher. Lagos. Vols. 1-8, 1934-36. Nigerian Teacher. Superseded by: Nigeria. 4-3065-1. HI 68899

Nihon ..., see also Nippon ...

Nihon bentosu kenkyukai shi = Bulletin of japanese association of benthology.

Nihon bunka shiso = Monumenta nipponica. Tokyo.

Nihon dojo-hiryo-gaku zasshi = Journal of the science of soil and manure. Tokyo. J. Sci. Soil Manure. See B–P–H 481/1.

Nihon gakujutsu kyokai hokoku = Proceedings of the japanese association for the advancement of science. Tokyo. Proc. Jap. Assoc. Advancem. Sci. See B–P–H 731/7.

Nihon gakujutsu kyokai hokoku = Report of the japanese association for the advancement of science. Tokyo. Rep. Jap. Assoc. Advancem. Sci. See B–P–H 767/4.

Nihon hiryo shinbun = Japan fertilizer news. Tokyo. Japan Fertilizer News. See B–P–H 499/22.

Nihon jozo kyokai zasshi = Journal of the society of brewing, Japan. Tokyo. J. Soc. Brew. Japan. See B–P–H 481/21.

Nihon kafun gakkai kaishi = Japanese journal of palynology. Shizuoka.

Nihon kaiyo gakkai shi = Journal of the oceanographical society of Japan. Tokyo.

Nihon kingakkai kaiho = Transactions of the mycological society of Japan. Tokyo. Trans. Mycol. Soc. Japan. See B–P–H 886/11.

Nihon nogaku bunken kiji sakuin = Japanese agricultural sciences index. Tokyo.

Nihon nogaku shinpo nenpo = Report (Annual) of the development of agriculture in Japan. Annual bibliography of japanese agriculture. Tokyo.

Nihon nogei kagakkai = Buletin of the agricultural chemical society of Japan. Tokyo.

Nihon nogei kagaku kaisi = Journal of the agricultural chemical society of Japan. Tokyo.

Nihon nogyo kenkyusho hokoku = Bulletin of the Nippon agricultural research institute. Tokyo. Bull. Nippon Agric. Res. Inst. See B–P–H 267/11.

Nihon purankuton gakkaiho = Bulletin of the plankton society of Japan. Hakodate.

Nihon ringyo = Japanese forestry. Tokyo.

Nihon saikingaku zasshi = Japanese journal of bacteriology. Tokyo.

Nihon sakumotsu gakkai kiji = Proceedings of the crop science society of Japan. Tokyo.

Nihon sanshigaku zasshi = Journal of sericultural science of Japan. Tokyo, Tsukubu.

Nihon seibutsu chiri gakkai kiji = Proceedings of the biogeographical society of Japan. Proc. Biogeogr. Soc. Japan. See B–P–H 726/24.

Nihon seibutsu chiri gakkaiho = Bulletin of the biogeographical society of Japan. Tokyo. Bull. Biogeogr. Soc. Japan. See B–P–H 241/19.

Nihon seibutsu kagaku kenkyujo hokoku = N I B S bulletin of biological research. Tokyo.

Nihon senbai kosha hatano tabako shikenjo hokoku = Bulletin of the Hatano tobacco experiment station. Kanagawa, Japan. Bull. Hatano Tobacco Exp. Sta. See B–P–H 252/22.

Nihon sentai-rui gakkai kaiho = Proceedings of the bryological society of Japan. Tokyo.

Nihon shida no kai kaiho = Journal, nippon fernist club. Tokyo.

Nihon shizen kagaku shuho = Recent progress of natural sciences in Japan. Tokyo.

Nihon shokubutsu bunrui gakkai kaiho = Proceedings of the Japan society of plant taxonomists. Tokyo.

Nihon shokubutsu bunrui gakki kaiho = Memoir, Japan society of plant taxonomists. [Nihon shokubutsu bunrui gakki kaiho]. Tokyo. Mem. Japan Soc. Pl. Taxon. See B–P–H 575/16.

Nihon shokubutsu seiri gakkai = Plant and cell physiology. Tokyo. Pl. Cell Physiol. See B–P–H 711/19.

Nihon shokuhin kogyo gakkai shi = Journal of food science and technology. Tokyo. J. Food Sci. Technol. (Tokyo). See B–P–H 466/6.

Nihon sochi gakkai shi = Japanese society of grassland science journal. Jap. Soc. Grassland Sci. J. See B–P–H 499/17.

Niigata agricultural science. [Niigata norin kenkyu.] Niigata. Vols. 1-28, 1951-76. Niigata Agric. Sci. Superseded by: Niigata daigaku, nogakubu kenkyu hokoku. HI 68900

Niigata daigaku kyoikugakubu kiyo, shizenkagaku-hen = Memoirs of the faculty of education, Niigata university. Natural science. Niigata.

Niigata daigaku nogakubu enshurin hokoku = Bulletin of

the Niigata university forests. Niigata.

Niigata daigaku nogakubu gakujutsu hokoku = Bulletin of the faculty of agriculture, Niigata university. Niigata, Japan. Bull. Fac. Agric. Niigata Univ. See B–P–H 249/11.

Niigata daigaku, nogakubu kenkyu hokoku. Niigata. No. 29+, 1977+. Niigata Daigaku Nogakubu Kenkyu Hokoku. Preceded by: Niigata agricultural science. HI 68901

Niigata daigaku nogakubu kiyo = Memoirs, faculty of agriculture, Niigata university. Niigata.

Niigata daigaku rigakubu kenkyu hokoku = Science reports of faculty of science, Niigata university. Series D, biology.

Niigata daigaku rigakubu kenkyu hokoku. D, seibutsugaku = Science reports of Niigata university. Series D, [biology]. Niigata.

Niigata daigaku rigakubu kenkyu hokoku. Dai-2-rui, seibutsugaku, chishitsugaku kobutsugaku = Journal of the faculty of science, Niigata university. Series II, biology, geology and mineralogy. Niigata.

Niigata-ken engei shikenjo kenkyu hokoku = Bulletin of the Niigata horticultural experiment station. Niitsu.

Niigata-ken engei shikenjo tokubetsu hokoku = Transactions of the Niigata horticultural experiment station. Niitsu.

Niigata-ken nogyo shikenjo kenkyu hokoku = Journal of the Niigata agricultural experiment station. Niigata, Japan. J . Niigata Agric. Exp. Sta. See B–P–H 475/3.

Niigata-ken ringyo shikenjo kenkyu hokoku = Bulletin of the Niigata prefectural forest experiment station. Murakami.

Niigata norin kenkyu = Niigata agricultural science. Niigata.

Nippon ..., see also Nihon ...

Nippon dobutsu-gaku iho = Annotationes zoologicae japonenses. Tokyo. Annot. Zool. Jap. See B–P–H 120/51.

Nippon engeikai zasshi = Journal of the japanese horticultural society. Tokyo. J. Jap. Hort. Soc. See B–P–H 470/21.

Nippon fernist club = Journal, nippon fernist club. Tokyo.

Nippon gakusiin kiji = Proceedings of the Japan academy. Tokyo.

Nippon ringakkai-shi = Transactions of the meeting of the Japanese forestry society, Meguro. Tokyo. Trans. Meeting Jap. Forest. Soc. See B–P–H 886/4.

Nippon ringaku kaishi = Journal of the japanese forestry society. Tokyo. J. Jap. Forest. Soc. See B–P–H 470/20.

Nippon sakumotsu gakkai kiji = Japanese journal of crop science. Tokyo.

Nippon seitai gakkaishi = Japanese journal of ecology. Sendai.

Nippon shokubutsu byori-gakkai ho = Annals of the phytopathological society of Japan. Tokyo. Ann. Phytopathol. Soc. Japan. See B–P–H 110/11.

Nippon shokubutsuen kyokwai kaiho = Report of the japanese botanical garden association. [Japan]. Rep. Jap. Bot. Gard. Assoc. See B–P–H 767/5.

Nippon shokubutsugaku shuho = Japanese journal of botany. Tokyo.

Nippon suisangakkaishi = Bulletin of the japanese society of scientific fisheries. Tokyo. Bull. Jap. Soc. Sci. Fish. See B–P–H 258/10.

Nippon suisangaku zasshi = Journal of fisheries. Hakodate, Japan. J. Fish. See B–P–H 465/26.

Nisseiken tayori = Journal, nippon institute for biological science. Tokyo.

Nitheroy; revista braziliense, sciencias, lettras e artes. Paris. Nitheroy. See B–P–H 661/1. HI 57423

Nitrogen fixing tree research reports. Bangkok. Vol. 1+, 1983+. Nitrogen Fixing Tree Res. Rep. HI 68903

Noesis; travaux du comité roumain d'histoire et de philosophie des sciences. Bucharest. No. 1+, 1972+. Noesis. HI 74866

Nogaku = Agricultural sciences. Tokyo. Agric. Sci. (Tokyo). See B–P–H 60/19.

Noyaku hyakuten = Agricultural chemicals. Tokyo. Agric. Chem. See B–P–H 56/17.

Nogaku kengyu = Agricultural research. Ohara institute for agricultural research. Kurashiki, Japan. Agric. Res. (Kurashiki). See B–P–H 60/6.

Nogaku kwai ho = Journal of the scientific agricultural society. [Nogaku kwai ho.] Tokyo. J. Sci. Agric. Soc. See B–P–H 480/4.

Nogaku shinpo nenpo. Tokyo. 1981+. Nogaku Shinpo Nenpo. Preceded by: Report (Annual) of the development of agriculture in Japan. Annual bibliography of japanese agriculture. HI 68904

Nogaku-bu enshurin hokoku = Bulletin of the Tokyo imperial university forests. Tokyo. Bull. Tokyo Imp. Univ. Forests. See B–P–H 284/2.

Nogakubu kiyo = Journal of the college of agriculture, imperial university of Tokyo. Tokyo.

Nogyo Aichi = Agriculture in Aichi. Aichi, Japan. Agric. Aichi. See B–P–H 56/5.

Nogyo-bu iho. Taiwan sotokufu chuo kenkyu-sho = Report of the department of agriculture, government research institute, Formosa. Taihoku [=Taipei, Taiwan]. Rep. Dept. Agric. Gov. Res. Inst. Formosa. See B–P–H 764/16.

Nogyo gijutsu kenkyujo hokoku, C = Bulletin of the national institute of agricultural sciences, Nishigahara. Series C, phytopathology and entomology. Tokyo.

Nogyo gijutsu kenkyujo hokoku, D = Bulletin of the national institute of agricultural sciences, Nishigahara. Series D, plant physiology, genetics and crops in general. Tokyo.

Nogyo gijutsu kenkyujo. Shiryo D, seiri, iden = Miscellaneous publications of the national institute of agricultural sciences. Series D, physiology and genetics. Tokyo.

Nogyo gijutsu kenkyusho hokoku, E. Engei = Bulletin of the national institute of agricultural sciences, Series E, Horticulture. Kanagawa, Japan. Bull. Natl. Inst. Agric. Sic., Ser. E, Hort. See B–P–H 265/25.

Nogyo gijutsu kyokai = Agricultural technology. Tokyo. Agric. Technol. See B–P–H 61/5.

Nogyo Gunma = Agriculture in Gunma. Maebashi, Japan. Agric. Gunma. See B–P–H 58/3.

Nogyo Hokkaido = Agriculture in Hokkaido. Sapporo, Japan. Agric. Hokkaido (Nogyo Hokkaido). See B–P–H 58/9.

Nogyo Kagawa = Agriculture in Kagawa. [Kagawa association for agricultural improvement.] Kagawa, Japan. Agric. Kagawa. See B–P–H 59/11.

Nogyo oyobi engei = Agriculture and horticulture. Tokyo. Agric. & Hort. See B–P–H 58/10.

Nogyo seibutsu shigen kenkyujo kenkyu hokoku = Bulletin of the national institute of agrobiological resources. Yatabe.

Nogyo sekai = Agricultural world. Tokyo. Agric. World. See B–P–H 61/12.

Nogyo shiken kenkyu nenju hokokusho = Year book of agricultural research. [Nogyo shiken kenkyu nenju hokokusho.] Tokyo. Year Book Agric. Res. See B–P–H 981/29.

Nogyo showa = Agriculture today. [Showa agricultural research association.] Tokyo. Agric. Today. See B–P–H 61/8.

Nogyo sosho = Agricultural series. [Ryukyu government, economic bureau.] Naha, Okinawa. Agric. Ser. See B–P–H 61/1.

Nogyo Yamaguchi = Agriculture in Yamaguchi. Yamaguchi, Japan. Agric. Yamaguchi. See B–P–H 61/13.

Noji geppo = Agricultural monthly. Sendai, Japan. Agric. Monthly. See B–P–H 59/19.

Noji shiken chosa shiryo = Agricultural experiment research report. Sapporo.

Noji shikenjo iho = Annals of the agricultural experiment station, government general of Chosen. Suigen [=Suwon], Korea. Ann. Agric. Exp. Sta. Gov. Gen. Chosen. See B–P–H 96/4.

Noji shikenjo iho = Journal of the imperial agricultural experiment station, Nishigahara, Tokyo. Tokyo. J. Imp. Agric. Exp. Sta. Nishigahara Tokyo. See B–P–H 469/24.

Noji shikenjo iho = Journal of the national agricultural experiment station, Nishigahara, Tokyo. Tokyo. J. Natl. Agric. Exp. Sta. Nishigahara Tokyo. See B–P–H 474/13.

Noji shikenjo kenkyu hokoku = Journal of the agricultural experiment station, government general Chosen. Suigen [=Suwon], Korea. J. Agric. Exp. Sta. Gov. Gen. Chosen. See B–P–H 455/8.

Noji shiken-jo tokubetsu hokoku = Special bulletin: Agricultural experiment station, government of Formosa. Taihoku [=Taipei, Taiwan]. Special Bull. Agric. Exp. Sta. Gov. Formosa. See B–P–H 850/8.

Noji shikenjo tokubetsu hokoku = Special report of the imperial agricultural experiment station. Tokyo. Special Rep. Imp. Agric. Exp. Sta. See B–P–H 851/18.

Noka daigaku gakujutsu hokoku = Bulletin of the college of agriculture, Tokyo imperial university. Tokyo.

Noma. Zaria. Vol. 1+, 1978+. Noma. Preceded by: Samaru agricultural newsletter. HI 68905

Nomenclatural forum. Kew. No. 1+, 1983+ ["Not effectively published: for private circulation."] Nomencl. Forum. HI 68906

Nonggwa taehak nonghak yongu, Soul taehakkyo. Seoul. Vol. 1+, 1976+. Nonggwa Taehak Nonghak Yongu Soul Taehakkyo. HI 68907

Nongso sihom yon'gu pogo = Research report of the office of rural development. (Japan).

Nord médical. Lille. N. Méd. See B–P–H 624/32. HI 56942

Nord-est agricole et horticole. Troyes. Vols. ?-2-3-?, ?-1877-78-? N.-Est Agric. Hort. HI 68908

Nord nature; buletin de la société fédérative pour l'étude et la protection de la nature dans le nord de la France. Villeneuve d'Arcq. No. 1+, 1974+. Nord Nat. HI 68909

Nordelbingen; Beiträge zur Heimatforschung in Schleswig-Holstein, Hamburg und Lübeck. Flensburg. Vol. 1+, 1923+. Nordelbingen. 4-3071-2. HI 68910

Nordic journal of botany. Copenhagen. Vol. 1+, 1980+. Nordic J. Bot. Preceded by: Botanisk tidsskrift, Friesia, Botanisk notiser, and Norwegian journal of botany. HI 68911

Nordische Beyträge zur physikalischen und geographischen Erd- und Völkerbeschreibung, Naturgeschichte und Oekonomie. St. Petersburg & Leipzig. Nord. Beytr. Phys. Geogr. Erd-Völkerbeschreib. See B–P–H 661/34. HI 57426

Nordische Beyträge zum Wachsthum der Naturkunde und der Wissenschaften, wie auch der nützlichen und schönen Künste überhaupt. Altona [=Hamburg, in part]. Nord. Beytr. Wachsth. Naturk. See B–P–H 661/36. HI 57427

Nordische Blätter für die Chemie. Halle. Nord. Blätt. Chem. See B–P–H 661/37. HI 57428

Nordisches Archiv für Natur- und Arzneywissenschaft. Copenhagen. Nord. Arch. Natur- Arzneywiss. See B–P–H 661/31. HI 57424

Nordisches Archiv für Naturkunde, Arzneywissenschaft und Chirurgie. Copenhagen. Nord. Arch. Naturk. See B–P–H 661/32. HI 57425

Nordisk farmaceutisk tidskrift. Copenhagen. Vols. 1-3, 1894-96. Nord. Farm. Tidskr. 4-3072-3. HI 68912

Nordisk have-tidende; organ for gaertnere. Aarhus. Nos. 1-38, 1883-84. Nord. Have-Tidende. HI 68913

Nordisk jordbruksforskning; organ for nordiske jordbrugsforskers forening. Copenhagen. Vol. 1+, 1919+. Nord. Jordbruksforskn. 4-3072-2. HI 68914

Nordisk jordbrugsforskning. Supplement. Stockholm. Nord. Jordbrugsforskn. Suppl. See B–P–H 656/3. HI 57365

Nordisk kaktus selskab meddelelser. 1978+. Nord. Kakt. Selsk. Meddel. HI 68915

Nordisk kaktus tidsskrift. Copenhagen, Odense. Vols. 1-2, 1921-23. Nord. Kakt. Tidsskr. HI 68916

Nordisk landvaesens og landhuusholdnings magasin; et maanedsskrivt. Copenhagen. Nord. Landvaes. Landhuushold. Mag. See B–P–H 661/39. HI 57429

Nordisk medicin. Helsinki. Vol. 1+, 1939+. Nord. Med. Preceded by: Hospitals-tidende (which merged with other journals to form this title). 4-3072-2. HI 68917

Nordisk tidsskrift for historie, litteratur og konst. Copenhagen. Nord. Tidsskr. Hist. See B–P–H 662/6. HI 57432

Nordisk trätidsskrift = N T

Nordiske magazin. Copenhagen. Nord. Mag. See B–P–H 662/1. HI 57430

Nordiskt medicinskt arkiv. Stockholm. Nord. Med. Ark. See B–P–H 662/3. HI 57431

Nordlyset; en folkelig tidende for naturvidenskab of kunstflid. Copenhagen. No. 1, 1848-49. Nordlyset. HI 68918

Nordmar. Helsinor. 1973+. Nordmar. HI 68919

Norfolk botanical garden society bulletin. Monthly newsletter of tidewater gardening. Norfolk, VA. 1940+. Norfolk Bot. Gard. Soc. Bull. HI 68920

Norfolk fair. Brundall. ?-1978+. Norfolk Fair. HI 68921

Norges Svalbard- og Ishavs-Undersøkelser skrifter = Skrifter om Svalbard og Nordishavet. Oslo.

Norin gijutsu senta, nyusu. Osaka. Nos. ?-58, ?-1973. Norin Gijutsu Senta Nyusu. Superseded by: Osaka-fu norin gijutsu senta nyusu. HI 68922

Norin-sho engei shikenjo kurume shijo rinji hokoku = Bulletin of the horticultural research station, Series D, Kurume, miscellaneous publication. Fukuoka, Japan. Bull. Hort. Res. Sta., Ser. D, Kurume, Misc. Publ. See B–P–H 254/4.

Norinsho Kansai rinboku Ikushujo nenpo = Report (Annual) of the Kansai forest tree breeding station, ministry of agriculture and forestry. Okayama.

Norinsho Kanto rinboku Ikushujo nenpo = Report (Annual) of Kanto forest tree breeding station. Mito.

Norin-sho tokai kinki nogyo shikenjo kenkyu sokuho = Research progress report, Tokai-kinki national agricultural experiment station. Mie, Japan. Res. Progr. Rep. Tokai-Kinki Natl. Agric. Exp. Sta. See B–P–H 775/18.

Norin-sho tokai kinki nogyo shikenjo tokubetsu hokoku = Special bulletin of the Tokai-kinki national agricultural experiment station. Mie, Japan. Special Bull. Tokai-Kinki Natl. Agric. Exp. Sta. See B–P–H 850/20.

Norin suisan shiken kenkyu nenpo. Ringyo-hen = Report (Annual) on experiments in agriculture, forestry and fisheries. Series forestry. Tokyo.

Norin suisansho nogyo seibutsu shigen kenkyujo = Report (Annual) of the national institute of agrobiological resources. Yatabe, Tsukuba.

Normandie médicale. Rouen. Normandie Méd. See B–P–H 662/12. HI 57433

Norois. Revue géographique de l'ouest et des pays l'antique nord. Poitiers, France. Norois. See B–P–H 662/16. HI 57434

Norrbottens natur. ?-1966+. Norrbottens Natur. HI 68923

Norrlinia. Helsinki. Vol. 1+, 1983+. Norrlinia. Preceded by: Helsingin yliopiston kasvimeseon monisteita. HI 68924

Norsk farmaceutisk tidsskrift. Oslo. Vol. 24+, 1916+. Norsk Farm. Tidsskr. Preceded by: Farmaceutisk tidende. Oslo. 4-3076-2. HI 68925

Norsk hagetidend. Oslo. Norsk Hagetid. See

B–P–H 662/17. HI 57435

Norsk havetidende; udgivet of selskabet havedyrkningens venner. Christiania [=Oslo, Norway]. Norsk Havetid. See B–P–H 662/18. HI 57436

Norsk landbruk. Oslo. Norsk Landbr. See B–P–H 662/19. HI 57437

Norsk magazin. Christiania [=Oslo, Norway]. Norsk Mag. See B–P–H 662/20. HI 57438

Norsk polarinstitutt årbok. Oslo. 1960+. Norsk Polarinst. Årbok. HI 68926

Norsk polarinstitutt meddelelser. Oslo. 1949+. Norsk Polarinst. Meddel. Preceded by: Meddelelser, norges svalbard- og ishavs-undersøkelser. 4-3077-2. HI 68927

Norsk polarinstitutt skrifter. Oslo. Norsk Polarinst. Skr. See B–P–H 662/21. HI 57439

Norsk tidsskrift for videnskab og litteratur. Christiania [=Oslo, Norway]. Norsk Tidsskr. Vidensk. Litt. See B–P–H 662/25. HI 57440

Norske skogsforsøksvesens virksomhet. ?-1971, ?-1972. Norsk Skogsforsøksvesens Virks. Superseded by: Arsberetning, norsk institutt for Skogsforskning. HI 68928

Norske videnskaps-akademi, matematisk-naturvidenskapelig klasse. Avhandlinger. Christiania. 1925+. Norske Vidensk.-Akad., Mat.-Naturvidensk. Kl., Avh. Preceded by: Forhandlinger i videnskabs-selskabet i Kristiania. 4-3079-3. HI 51707

North american newsletter, society for the bibliography of natural history. New York. Vols. 1-2(2), 1977-78. N. Amer. Newslett. Soc. Bibliogr. Nat. Hist. Continued in: Newsletter, society for the bibliography of natural history. HI 68929

North american pomona. Toronto, Portland, OR. Vols. 1-14(1), 1967-81. N.-Amer. Pomona. Superseded by: North american fruit explorers' quarterly. HI 68930

North American review. Boston, MA. Vols. 13-248 [vols. 13-30 also numbered as n.s. vols. 4-21], 1821-1940. N. Amer. Rev. Preceded by: North American review and miscellaneous journal. 4-3081-1. HI 52884

North American review and miscellaneous journal. Boston, MA. Vols. 1-12 [vols. 10=12 also numbered as n.s. vols. -3], 1815-21. N. Amer. Rev. Misc. J. Superseded by: North American review. 4-3081-1. HI 68931

North Borneo forest records. Sandakan. 1935-56. North Borneo Forest Rec. Superseded by: Sabah forest record. HI 68932

North british review. Edinburgh, London. Vols. 1(1)-53(106), 1844-71. N. Brit. Rev. 4-3082-2. HI 68933

North Carolina agricultural experiment station agronomy information circular. Raleigh, NC. North Carolina Agric. Exp. Sta. Agron. Inform. Circ. See B–P–H 663/4. HI 57441

North Carolina agricultural experiment station annual report. Raleigh, NC. North Carolina Agric. Exp. Sta. Annual Rep. See B–P–H 663/5. HI 57442

North Carolina agricultural experiment station bulletin. Raleigh, NC. North Carolina Agric. Exp. Sta. Bull. See B–P–H 663/6. HI 57443

North Carolina agricultural experiment station circular. Raleigh, NC. North Carolina Agric. Exp. Sta. Circ. See B–P–H 663/7. HI 57444

North Carolina agricultural experiment station, department of agronomy research report. Raleigh, NC. North Carolina Agric. Exp. Sta. Dept. Agron. Res. Rep. See B–P–H 663/8. HI 57445

North Carolina agricultural experiment station press bulletin. Raleigh, NC. North Carolina Agric. Exp. Sta. Press Bull. See B–P–H 663/9. HI 57446

North Carolina agricultural experiment station research and farming quarterly. Raleigh, NC. North Carolina Agric. Exp. Sta. Res. Farming Quart. See B–P–H 663/10. HI 57447

North Carolina agricultural experiment station research report. Raleigh, NC. North Carolina Agric. Exp. Sta. Res. Rep. See B–P–H 663/11. HI 57448

North Carolina agricultural experiment station special bulletin. Raleigh, NC. North Carolina Agric. Exp. Sta. Special Bull. See B–P–H 663/12. HI 57449

North Carolina agricultural experiment station special publication. Raleigh, NC. North Carolina Agric. Exp. Sta. Special Publ.. See B–P–H 663/13. HI 57450

North Carolina agricultural experiment station technical publication. Raleigh, NC. North Carolina Agric. Exp. Sta. Techn. Publ. See B–P–H 663/14. HI 57451

North Carolina agricultural experiment station tobacco abstracts. Raleigh, NC. North Carolina Agric. Exp. Sta. Tobacco Abstr. See B–P–H 663/15. HI 57452

North Carolina state horticultural society annual report. Raleigh, NC. North Carolina State Hort. Soc. Annual Rep. See B–P–H 664/5. HI 57453

North Carolina botanical garden newsletter. Chapel Hill, NC. 1971+. North Carolina Bot. Gard. Newslett. HI 68934

North Carolina grower. Charlotte, NC. 1967+. North Carolina Grower. HI 68935

North Carolina planter. Raleigh, NC. Vols. 1-4, 1858-61. North Carolina Pl. 4-3086-2. HI 68936

North China agriculture. [Hua pei nung yeh.] [China]. N. China Agric. See B–P–H 624/16. HI 56940

North China herald. Shanghai. N. China Herald. See B–P–H 624/17. HI 56941

North Dakota agricultural college, agricultural experiment station annual report. Fargo, ND. North Dakota Agric. Exp. Sta. Annual Rep. See B–P–H 664/8. HI 57454

North Dakota agricultural experiment station bimonthly bulletin. Fargo, ND. North Dakota Agric. Exp. Sta. Bimonthly Bull. See B–P–H 664/9. HI 57455

North Dakota agricultural experiment station bulletin. Fargo, ND. North Dakota Agric. Exp. Sta. Bull. See B–P–H 664/10. HI 57456

North Dakota agricultural experiment station circular. Fargo, ND. North Dakota Agric. Exp. Sta. Circ. See B–P–H 664/11. HI 57457

North Dakota agricultural experiment station, Dickinson sub-station annual report. Dickinson, ND. North Dakota Agric. Exp. Sta. Dickson Sub-Sta. Annual Rep. See B–P–H 664/12. HI 57458

North Dakota agricultural experiment station press bulletin. Fargo, ND. North Dakota Agric. Exp. Sta. Press Bull. See B–P–H 664/13. HI 57459

North Dakota agricultural experiment station research report. Fargo, ND. North Dakota Agric. Exp. Sta. Res. Rep. See B–P–H 664/14. HI 57460

North Dakota agricultural experiment station seed circular. Fargo, ND. North Dakota Agric. Exp. Sta. Seed Circ. See B–P–H 664/15. HI 57461

North Dakota agricultural experiment station special bulletin. Fargo, ND. North Dakota Agric. Exp. Sta. Special Bull. See B–P–H 664/16. HI 57462

North Dakota agricultural experiment station special seed bulletin. Fargo, ND. North Dakota Agric. Exp. Sta. Special Seed Bull. See B–P–H 664/17. HI 57463

North Dakota farm research. Bimonthly bulletin of the North Dakota agricultural experiment station. Fargo, ND. North Dakota Farm Res. See B–P–H 664/18. HI 57464

North Gloucestershire naturalists' society journal. Cheltenham. Vols. 1-25(1), 1957-74. N. Gloucestershire Naturalists' Soc. J. Preceded by: Cheltenham and district naturalists' society journal. Superseded by: Journal, Gloucestershire naturalists society.

North Queensland naturalist. Cairns, Qld. N. Queensland Naturalist. See B–P–H 625/1. HI 56943

North Staffordshire journal of field studies. Keele. 1961+. N. Staffordshire J. Field Stud. HI 68937

North western naturalist. Arbroath, Cheadle Hulme. Vols. 1-23, 1926-48; vols. 24-26 [i.e. n.s. vols. 1-3], 1953-55; 1973+. N. W. Naturalist. Preceded by: Lancashire and Cheshire naturalist. 4-3097-3. HI 68938

Northeast gulf science. [Dauphin Island, AL.] Vol. 1+, 1977+. NorthE. Gulf Sci. Preceded by: Journal of marine science. HI 68939

Northeast regional science review. Binghampton, NY. 1971?-83?. NorthE. Regional Sci. Rev. Superseded by: Regional science review. HI 68940

Northeastern environmental science. Troy, NY. 1982+. NorthE. Environm. Sci. HI 68941

Northeastern forest research notes, northeastern forest experiment station = Northeastern research notes, northeastern forest experiment station. Upper Darby, PA.

Northeastern research notes, northeastern forest experiment station. Upper Darby, PA. Nos. 1-21, 1950-53. NorthE. Res. Notes NorthE Forest Exp. Sta. Superseded by: Forest research notes, northeastern forest experiment station. HI 68942

Northern agriculture. [Hoppo nogyo.] Sapporo, Japan. N. Agric. See B–P–H 624/11. HI 56939

Northern forestry. [Hoppo ringyo.] = Journal of the northern forestry society. Sapporo.

Northern gardener. Manchester, Harrogate. Vol. 1+, 1947+. N. Gardener. HI 68943

Northern journal of applied forestry. Bethesda, MD. Vol. 1+, 1984+. N. J. Appl. Forest. HI 68944

Northern Territory botanical bulletin; botanical notes from the herbaria of the Northern Territory Alice Springs (NT) and Darwin (DNA). Alice Springs, N.T. No. 1+, 1976+. Northern Territory Bot. Bull. HI 68945

Northern Victoria fruitgrower. Shepperton, Vic. Vol. 1+, 1977+. N. Victoria Fruitgrower. HI 68946

Northwest agricultural bulletin; national northwest agricultural college. [Hsi pei nung pao.] Wukung, China. NorthW. Agric. Bull. See B–P–H 665/2. HI 57465

Northwest agricultural college journal. [Hsi pei nung hsüeh yuan hsüeh pao.] Sian, China. NorthW. Agric. Coll. J. See B–P–H 665/3. HI 57466

Northwest fuchsia facts. Portland, OR. Vol. 1+, 1948+. NorthW. Fuchsia Facts. HI 68947

Northwest medicine. Seattle, WA. NorthW. Med. See B–P–H 665/5. HI 57468

Northwest naturalist. Seattle, WA. NorthW. Naturalist. See B–P–H 665/7. HI 57469

Northwest orchid bulletin. Seattle, WA. NorthW. Orchid Bull. See B–P–H 665/8. HI 57470

Northwest rosarian. Vancouver, WA. NorthW. Rosarian. See B–P–H 665/9. HI 57471

Northwest science. Cheney, WA. NorthW. Sci. See B–P–H 665/10. HI 57472

Northwestern journal of agriculture and forestry. The northwestern college of agriculture. [Hsi pei nung lin. Kuo li hsi pei nung hsüeh yuan pi'en chi ch'u pan wei yuan hui Pien Hsing.] Wukung, China. NorthW. J. Agric. Forest. See B–P–H 665/4. HI 57467

Northwestern university studies in the biological sciences and medicine. Chicago, IL. Northwestern Univ. Stud. Biol. Sci. Med. See B–P–H 665/14. HI 57473

Norvegica pharmaceutica acta. Oslo. Vol. 45+, 1983+. Norveg. Pharm. Acta. Preceded by: Meddelelser fra norsk farmaceutisk selskap. HI 68948

Norwegian journal of botany. Oslo. Vols. 18-27, 1971-80. Norweg. J. Bot. Preceded by: Nytt magasin for botanikk. Superseded by: Nordic journal of botany. HI 68949

Nosan Kako Gijutsu kenkyukai Shi = Journal for utilization of agricultural products. Tokyo. J. Utiliz. Agric. Prod. See B–P–H 484/12.

Notarisia. Commentario ficologico generale. Venice. Notarisia. See B–P–H 666/23. HI 57484

Notas agrícolas. São Paulo. Notas Agríc. See B–P–H 666/24. HI 57485

Notas agronómicas. Palmira, Colombia. Notas Agron. See B–P–H 666/25. HI 57486

Notas do centro de biologia aquatica tropical. Lisbon. Nos. 47-?, 1968-73. Notas Centro Biol. Aquatica Trop. Preceded by: Notas mimeografadas do centro de biologia piscatoria. HI 68950

Notas y comunicaciones del instituto geológico y minero de España. Madrid. Notas Comun. Inst. Geol. España. See B–P–H 666/26. HI 57487

Notas didacticas del jardin botanico Fco. J. Clavijero. Xalapa. No. 1+, 1982+. Notas Didact. Jard. Bot. Fco. J. Clavijero. HI 75263

Notas de divulgación; instituto municipal de botanica. Buenos Aires. Notas Divulg. Inst. Munic. Bot. See B–P–H 666/27. HI 57488

Notas e estudos do instituto de biologia maritima. Lisbon. Vols. 1-42, 1952-75. Notas Estud. Inst. Biol. Marit. Superseded by: Boletim, instituto nacional de investigação das pescas [not entered]. HI 68951

Notas floristicas y ecologicas sobre la flora ibérica. Oviedo. Vol. 1+, 1978+. Notas Florist. Ecol. Fl. Ibér. HI 68952

Notas mimeografadas do centro de biologia piscatoria. Lisbon. Vols. 1-46, 1966-66. Notas Mimeogr. Centro Biol. Piscat. Superseded by: Notas do centro de biologia aquatica tropical. HI 68953

Notas del museo de La Plata. Buenos Aires. 198?+. Notas Mus. La Plata. Preceded by: Notas del museo de La Plata. Botânica. HI 68954

Notas del museo de La Plata. Botânica. Buenos Aires. Vols. 1-20, 1935-63. Notas Mus. La Plata, Bot. Preceded by: Notas preliminares del museo de La Plata. Superseded by: Notas del museo de La Plata. 3-2359-1. HI 68955

Notas preliminares e estudos. Divisão de geologia e mineralogia do Brasil. Notas Prelim. Estud. Div. Geol. Brasil. See B–P–H 667/7. HI 57489

Notas preliminares del museo de la Plata. Buenos Aires. Vols. 1-3, 1931-33. Notas Prelim. Mus. La Plata. Superseded by: Notas del museo de la Plata. Botânica. 3-2359-2. HI 68956

Notas y resumes, instituto español de oceanografia. Madrid. Ser. 2, nos. 1-140, 1924-47. Notas Resumes Inst. Esp. Oceanogr. Superseded by: Boletín del instituto español de oceanografia. HI 68957

Notas silvícolas, dirección de investigaciones forestales, Argentina. Buenos Aires. Notas Silvíc. See B–P–H 667/10. HI 57490

Notas tecnicas forestales = Notas tecnicas, instituto forestal. Santiago de Chile.

Notas técnicas. Instituto nacional de investigaciones forestales. Coyoacán, Mexico. Notas Técn. Inst. Nac. Invest. Forest. See B–P–H 667/12. HI 57491

Notas tecnico forestales. Santiago de Chile. No. 1+, 1965+. Notas Tecn. Forest. HI 68958

Notas tecnicas, instituto forestal. Santiago de Chile. No. 1+, 1963+. Notas Tecn. Inst. Forest. HI 68959

Notas tecnológicas forestales. Buenos Aires. Notas Tecnol. Forest. See B–P–H 667/13. HI 57492

Notationes biologicae. Bucharest. Vols. 1-?, 1933-48 [publication suspended 1936-45]. Notat. Biol. 4-3102-2. HI 68960

Note botanice; notulae botanicae horti agrobotanici Cluj-Napoca. [Issued with: Catalog de seminte (Index seminum).] Cluj. Vols. 1-7 1965-71. Note Bot. Superseded by: Notulae botanicae horti agrobotanici, Cluj-Napoca, institutum Agronomicum "Dr. Petru Groza". HI 68961

Note, institut d'élevage et de medécine vétérinaire des pays tropicaux. Cedex. ?-1980+. Note Inst. Élevage Med. Vétérin. Pags Trop. HI 68962

Note de l'institut océanographique de l'Indochine = Note du service océanographique des pêches de l'Indochine. Nhatrang, Indochina [South Vietnam]. Note Serv. Océanogr. Pêches Indochine. See B–P–H 667/21.

Note de l'institut océanographique de Nhatrang = Note du service océanographique des pêches de l'Indochine. Nhatrang, Indochina [South Vietnam]. Note Serv.

Océanogr. Pêches Indochine. See B–P–H 667/21.

Note dell' istituto italo-germanico di biologia marina di Rovigno d'Istria. Venice. Note Ist. Ital.-German. Biol. Mar. Rovigno d'Istria. See B–P–H 667/18. HI 57493

Note del laboratorio de biologia marina di Fano. Bologna. Note Lab. Biol. Mar. Fano. See B–P–H 667/20. HI 57494

Note de recherche, institut national de recherches forestières. Ariana. No. 1+, 1974+. Note Rech. Inst. Natl. Rech. Forest. HI 68963

Note du service océanographique des pêches de l'Indochine. Nhatrang, Indochina [South Vietnam]. Note Serv. Océanogr. Pêches Indochine. See B–P–H 667/21. HI 57495

Note technique, fonds de recherches forestières de l'université Laval. Laval. No. ?-2+, ?-1976+. Note Techn. Fonds Rech. Forest. Univ. Laval. HI 68964

Notes africaines; bulletin d'information et de correspondance de l'institut français d'Afrique noire. Dakar. No. 1+, 1939+. Notes Africaines. HI 68965

Notes from the agricultural research centre herbarium. Dokki. Vol. 1+, 1974+. Notes Agric. Res. Centre Herb. HI 68966

Notes de biogéographie. Dakar. 1986+. Notes Biogéogr. HI 75256

Notes biospéologiques. Paris. Nos. 7-13, 1947-58. Notes Biospéol. Nos. 1-6 were published as: Publications du muséum national d'histoire naturelle, Paris. Superseded by: Annales de spéléologie. HI 68967

Notes from the botanical school of Trinity college. Dublin. Notes Bot. School Trinity Coll. See B–P–H 667/26. HI 57496

Notes de botanique chinoise. Shanghai. 1931-50. Notes. Bot. Chin. HI 68968

Notes et documents, institut de cartographie de la végétation d'Algérie. Marseille. No. 1, 1964. Notes Doc. Inst. Cartogr. Vég. Algérie. HI 68969

Notes et documents d'océanographique, office de la recherche scientifique et technique outre-mer, centre de Papeete. Papeete. No. 1+, 1978+. Notes Doc. Océanogr. Off. Rech. Sci. Techn. Outre-Mer Centre Papeete. HI 68970

Notes et documents Voltaiques; bulletin trimestriels d'information scientifique. Ouagadougou. Vols. 1-4, 1967-70. Notes Doc. Volt. HI 68971

Notes. Herbarium. University of Oklahoma. Tulsa, OK. Notes Herb. Univ. Oklahoma. See B–P–H 668/1. HI 57490 HI 57497

Notes from the Jodrell laboratory. Kew. No. 1+, 1964+. Notes Jodrell Lab. HI 68972

Notes and memoirs, fisheries research directorate. Cairo. Nos. 1-20, 1933-37. Notes Mem. Fish. Res. Director. Superseded by: Notes and memoirs, hydrobiology and fisheries directorate. HI 68973

Notes and memoirs, Fouad I Institute (of) hydrobiology and fisheries = Notes and memoirs, hydrobiology and fisheries, Fouad I institute. Cairo.

Notes and memoirs, hydrobiological department, ministry of agriculture. Cairo. Nos. 39-67, 1958-61. Notes Mem. Hydrobiol. Dept. Minist. Agric. Preceded by: Notes and memoirs, hydrobiology and fisheries, Fouad I institute. Superseded by: Notes and memoirs, hydrobiological department, ministry of scientific research. Cairo. HI 68974

Notes and memoirs, hydrobiological department, ministry of scientific research. Cairo. Nos. 68-74, 1963-65. Notes Mem. Hydrobiol. Dept. Minist. Sci. Res. Preceded by: Notes and memoirs, hydrobiological department, ministry of agriculture. Cairo. Superseded by: Bulletin of the institute of oceanography and fisheries. Cairo. HI 68975

Notes and memoirs, hydrobiology and fisheries directorate. Cairo. Nos. 21-29, 1937. Notes & Mem. Hydrobiol. Fish. Directorate. Preceded by: Notes and memoirs, fisheries research directorate. Superseded by: Notes and memoirs, hydrobiology and fisheries, Fouad I institute. HI 68976

Notes and memoirs, hydrobiology and fisheries, Fouad I institute. Cairo. Vols. 31-38, 1937-40. Notes & Mem. Hydrobiol. Fish. Fouad I Inst. Preceded by: Notes and memoirs, hydrobiology and fisheries directorate. Superseded by: Notes and memoirs, hydrobiological department. HI 68977

Notes et mémoires, station océanographique de Salammbô. Tunis. No. 1, 1924. Notes Mém. Sta. Océanogr. Salammbô. Superseded by: Bulletin, station océanographique de Salammbô. HI 68978

Notes from the museum of northern Arizona. Flagstaff, AZ. Notes Mus. N. Arizona. See B–P–H 668/3. HI 57498

Notes, Perak museum. Taiping. Vols. 1-2, 1893-98. Notes Perak Mus. Superseded by: Journal of the Federated Malay States museums. 3-2242-3. HI 68979

Notes from the office of seed and plant introduction. Washington, DC. Notes Off. Seed Introd. See B–P–H 668/4. HI 57499

Notes and queries on China and Japan. Hong Kong. Notes Queries China Japan. See B–P–H 668/6. HI 57500

Notes and records of the royal society of London. London. Notes & Rec. Roy. Soc. London. See B–P–H 668/10. HI 57501

Notes from the royal botanic garden, Edinburgh. Edinburgh & Glasgow. Vols. 1-46, 1900-89. Notes Roy. Bot. Gard. Edinburgh. Superseded by: Edinburgh journal of botany. 2-1402-1. HI 57502

Notes from Strybing arboretum. San Francisco, CA. Vols. 1-6(1), 1963-68. Notes Strybing Arbor. Incorporated in: California horticultural journal. HI 68980

Notes techniques du centre d'écologie forestière. Gembloux. No. 1+, 1967?+. Notes Techn. Centre Écol. Forest. HI 68981

Notes and views, american fern society. [Dates of publication not ascertained.] Notes Views Amer. Fern Soc. Superseded by: Fiddlehead forum. HI 68982

Notes from the Waimea arboretum. Oahu, HI. Vol. 1+, 1974+. Notes Waimea Arbor. HI 68983

Nöthiger Beytrag zu den neuen Zeitungen von gelehrten Sachen; oder umständliche Auszüge aus denen gelehrten Monatschriften, welche in den neuen Zeitungen von gelehrten Sachen nicht Plass hatten. Leipzig. Vols. 1-8, 1734-43. Nöthiger Beytr. Neuen Zeitungen Gel. Sachen. HI 52695

Notice ou aperçu analitique des travaux les plus remarquables de l'académie royale du Gard = Notice des travaux de l'académie du Gard. Nîmes. Not. Trav. Acad. Gard. See B–P–H 666/21.

Notice des travaux de l'académie du Gard. Nîmes. Not. Trav. Acad. Gard. See B–P–H 666/21. HI 57482

Notice des travaux de la société des amateurs des sciences physiques et naturelles de Paris. Paris. Not. Trav. Soc. Amateurs Sci. Phys. Paris. See B–P–H 666/22. HI 57483

Notices et extraits des manuscrits de la bibliothèque du roi, ... Paris. Not. Extr. Manuscrits Biblioth. Roi. See B–P–H 665/20. HI 57476

Noticiario del instituto forestal. Santiago de Chile. No. 1+, 1962+. Not. Inst. Forest. HI 68984

Noticiário mensual, museo nacional de história natural. Santiago de Chile. No. 1+, 1956+. Not. Mens. Mus. Nac. Hist. Nat. HI 68985

Noticiário, palmeiras úteis da América tropical. Brasilia = Newsletter, useful palms of tropical america. Brasilia.

Noticiario tuberosas. Maracay. Vol. 1+, 1971+. Not. Tuber. HI 68986

Noticias agrícolas. [Servicio Shell para el agricultor.] Cagua, Venezuela. Not. Agríc. See B–P–H 665/16. HI 57474

Notícias agrostológicas Brasileiras. São Paulo = Correio agrostologico. São Paulo.

Noticias de Galapagos. Paris. Not. Galapagos. See B–P–H 665/21. HI 57477

Noticias ministerio de fomento. Caracas. Not. Minist. Fomento. See B–P–H 665/22. HI 57478

Noticias sobre las royas de los cereales de todos para todos = Robigo. Castelar.

Noticiero bibliotecario interamericano [O.E.A.]. Not. Bibliot. Interamer. See B–P–H 665/17. HI 57475

Noticiero botanico. Una hoja periódica dedicada a propagar información de interés y utilidad, y promover colaboración entre la comunidad botánica venezolana. Caracas. Vol. 1+, [1978]+. Not. Bot. HI 68988

Notifrut. Buenos Aires. Notifrut. See B–P–H 668/14. HI 57503

Notiser ur sällskapets pro fauna et flora fennica förhandlingar. Helsinki. Not. Sällsk. Fauna Fl. Fenn. Förh. See B–P–H 666/5. HI 57481

Notizblatt des Botanischen Gartens und Museums zu Berlin-Dahlem. Berlin-Dahlem. Notizbl. Bot. Gart. Berlin-Dahlem. See B–P–H 668/21. HI 57504

Notizblatt des Königlichen botanischen Gartens und Museums zu Berlin. Leipzig. Notizbl. Königl. Bot. Gart. Berlin. See B–P–H 668/25. HI 57505

Notizen des Deutsch-Italienischen Institutes für Meeresbiologie in Rovigno d'Istria = Note dell'istituto italo-germanico di biologia marina di Rovigno d'Istria. Venice. Note Ist. Ital.-German. Biol. Mar. Rovigno d'Istria. See B–P–H 667/18.

Notizen aus dem Gebiete der Natur- und Heilkunde. Erfurt. Not. Natur- Heilk. See B–P–H 665/28. HI 57472 HI 57479

Notizen aus dem Gebiete der practischen Pharmacie und deren Hülfswissenschaften. Krefeld, Germany. Not. Pract. Pharm. Hülfswiss. See B–P–H 666/1. HI 57480

Notiziario sulle malattie della piante. Pavia, Milan. No. 1+, 1949+. Not. Malatt. Piante. HI 68989

Notiziario periodico della vita dell'A N M S. Bologna. ?-1977+. Not. Period. Vita A. N. M. S. HI 68990

Notizen zur flora der Steiermark. Graz. No. 1+, 1974+. Not. Fl. Steiermark. Preceded by: Mitteilungsblatt der floristische Arbeitsgemeinschaft des naturwissenschaftlichen Verein für Steiermark. HI 68991

Notre vallée; bulletin du cercle des naturalistes corbeillois. [Corbeil?] Nos. ?-21-27, ?-1936-39. Notre Vallée. HI 68992

Notring-Almanach. Vienna. Notring-Almanach. See B–P–H 668/26. HI 57506

Notring-Jahrbuch. Vienna. Notring-Jahrbuch. See B–P–H 668/27. HI 57507

Nottingham naturalist. Nottingham. 1978+. Nottingham Naturalist. HI 68993

Notulae botanicae clujenses. Cluj = Note botanice. Cluj.

Notulae botanicae horti agrobotanici, Cluj-Napoca, institutum Agronomicum "Dr. Petru Groza". From 1976 to 1980 published with: Index seminum horti agrobotanici, Cluj-Napoca. Cluj. Vol. 8+, 1975/76+, 1976+. Notul. Bot. Horti Agrobot. Cluj-Napoca Inst., Agron. "Dr. Petru Groza". Preceded by: Note botanice. Cluj. HI 68994

Notulae naturae of the academy of natural sciences of Philadelphia. Philadelphia, PA. Notul. Nat. Acad. Nat. Sci. Philadelphia. See B–P–H 669/1. HI 57508

Notulae naturae, horti agrobotanicae Cluj-Napoca. Vol. ?-14+, ?-1984+. Notul. Nat. Horti Agrobot. Cluj-Napoca. HI 68995

Notulae systematicae ac geographicae instituti botanici tphilisiensis = Zametki po sistematike i geografii rastenii. Tiflis.

Notulae systematicae ex herbario horti botanici petropolitanae (rei publicae rossicae). Petrograd = Botanicheskiye materialy Gerbariya Glavnogo botanicheskogo sada R S F S R. Petrograd.

Notulae systematicae ex herbario horti botanici petropolitanae U S S R = Botanicheskiye materialy Gerbariya Glavnogo botanicheskogo sada S S S R. Leningrad.

Notulae systematicae ex herbario instituti botanici academiae scientiarum Kasachstanicae = Botanicheskiye materialy Gerbariya Instituta botaniki Akademii nauk Kazakhskoi S S R. Alma-Ata.

Notulae systematicae ex herbario instituti botanici nomine V. L. Komarovi academiae scientiarum U R S S = Botanicheskie materialy gerbariya botanicheskogo instituti imeni V. L. Komarova, akademii nauk S S S R. Leningrad.

Notulae systematicae, herbier du muséum national d'histoire naturelle. Phanérogamie = Notulae systematicae, herbier du muséum de Paris. Phanérogamie. Paris.

Notulae systematicae, herbier du muséum de Paris. Phanérogamie. Paris. Vols. 1-16, 1909-60. Notul. Syst. (Paris). Superseded by: Adansonia, n.s. 4-3268-1. HI 68996

Notulae systematicae ex instituto cryptogamico horti botanici petropolitani principalis U S S R = Botanicheskiye materialy Instituta sporovykh rastenii Glavnogo botanicheskogo sada R S F S R. Petrograd.

Notulae systematicae ex instituto cryptogamico horti botanici petropolitani reipublicae rossicae = Botanicheskiye materialy Instituta sporovykh rastenii Glavnogo botanicheskogo sada R S F S R. Petrograd.

Notulae systematicae e sectione cryptogamic instituti botanici academiae scientiarum U R S S = Botanicheskiye materialy Otdela sporvykh rastenii Botanicheskogo instituta Akademii nauk S S S R. Moscow & Leningrad.

Notulae systematicae e sectione cryptogamica = Botanicheskiye materialy Otdela sporovykh rastenii. Moscow & Leningrad.

Notulae systematicae e sectione cryptogamica instituti botanici nomine V. L. Komarovi academiae scientiarum U R S S = Botanicheskie materialy otdela sporovyh rastenii botanicheskogo instituta imeni V. L. Komarova akademii nauk S S S R. Moscow & Leningrad.

Notulen der algemeene vergadering van het proefstation Malang. Soerabaia. Vols. ?-12-31, ?-1918-31. Notul. Alg. Vergad. Proefstat. Malang. HI 68997

Noutati in economia forestiera. Brasov. Vols. 17-18, 1975-76. Nout. Econ. Forest. Preceded and superseded by: Buletinul, universitatii din Braşov. Seria B, economie forestiera. HI 68998

Nouveau bulletin des sciences, publié par la société philomatique de Paris. Paris. Nouv. Bull. Sci. Soc. Philom. Paris. See B–P–H 670/7. HI 57515

Nouveau bulletin de la société botanique de Lyon. Lyon. Vol. 1(1-2), 1913. Nouv. Bull. Soc. Bot. Lyon. HI 68999

Nouveau journal helvétique, ou annales littéraires et politiques de l'Europe et principalement de la Suisse. Neuchâtel. Nouv. J. Helv. See B–P–H 670/11. HI 57516

Nouveau journal des savans, dressé a Berlin. Berlin. Nouv. J. Savans Berlin. See B–P–H 670/12. HI 57517

Nouveau magazin d'histoire naturelle. Moscow. Nouv. Mag. Hist. Nat. See B–P–H 670/13. HI 57518

Nouveau Montpellier médical. Montpellier. Nouv. Montpellier Méd. See B–P–H 670/28. HI 57524

Nouveaux mémoires de l'académie de Dijon, pour la partie des sciences et arts. Dijon. Nouv. Mém. Acad. Dijon Pt. Sci. See B–P–H 670/14. HI 57519

Nouveaux mémoires de l'académie royale des sciences et belles-lettres de Bruxelles. Brussels. Nouv. Mém. Acad. Roy. Sci. Bruxelles. See B–P–H 670/15. HI 57520

Nouveaux mémoires de l'académie royale des sciences et belles-lettres. Avec l'histoire. Berlin. Nouv. Mém. Acad. Roy. Sci. Hist. (Berlin). See B–P–H 670/16. HI 57521

Nouveaux mémoires de la société helvétique des sciences naturelles = Neue Denkschriften der Allg. Schweizerischen Gesellschaft für die gesammten Naturwissenschaften. Neuchâtel. Neue Denkschr.

Allg. Schweiz. Ges. Gesammten Naturwiss. See B–P–H 643/5.

Nouveaux mémoires de la société helvétique des sciences naturelles = Neue Denkschriften der Schweizerischen Naturforschenden Gesellschaft. Zurich. Neue Denkschr. Schweiz. Naturf. Ges. See B–P–H 643/8.

Nouveaux mémoires de la société impériale des naturalistes de Moscou. Moscow. Vols. 1-18 [vols. 2-18(1) also numbered vols. 8-23(1)], 1829-1914. Nouv. Mém. Soc. Imp. Naturalistes Moscou. Preceded by: Mémoires de la société impériale des naturalistes de Moscou. Superseded by: Novye memuary Moskovskogo obshchestva ispytatelei prirody. 3-2770-2. HI 69000

Nouveaux mémoires de la société des naturalistes de Moscou = Novye memuary Moskovskogo obshchestva ispytatelei prirody. Moscow.

Nouveaux mémoires de la société des sciences, agriculture et arts du Bas-Rhin. Strasbourg. Nouv. Mém. Soc. Sci. Bas-Rhin. See B–P–H 670/26. HI 57522

Nouveaux mémoires de la société des sciences, agriculture et arts du département du Bas-Rhin. Strasbourg & Paris. Nouv. Mém. Soc. Sci. Dép. Bas-Rhin. See B–P–H 670/27. HI 57523

Nouvelle Bigarure, contenant ... The Hague. Nouv. Bigarure. See B–P–H 670/6. HI 57514

Nouvelles acquisitions, bibliothèque centrale du muséum national d'histoire naturelle. Paris. 1971+. Nouv. Acquis. Biblioth. Centrale Mus. Natl. Hist. Nat. Preceded by: Liste des acquisitions, bibliothèque centrale du muséum national d'histoire naturelle. HI 69001

Nouvelles annales du muséum d'histoire naturelle. Paris. Vols. 1-4, 1832-35. Nouv. Ann. Mus. Hist. Nat. Preceded by: Mémoires du muséum d'histoire naturelle. Paris. Superseded by: Archives du muséum d'histoire naturelle. Paris. 4-3268-1. HI 57509

Nouvelles annales des voyages, de la géographie et de l'histoire. Paris. Nouv. Ann. Voyages. See B–P–H 669/23. HI 57510

Nouvelles archives italiennes de biologie. Pisa. Nouv. Arch. Ital. Biol. See B–P–H 669/25. HI 57511

Nouvelles archives des missions scientifiques et littéraires. Paris. Nouv. Arch. Missions Sci. Litt. See B–P–H 669/28. HI 57512

Nouvelles archives du muséum d'histoire naturelle. Paris. Vols. 1-10, 1865-74; sér. 2, vols. 1-10, 1875-88; sér. 3, vols. 1-10, 1889-98; sér. 4, vols. 1-10, 1899-1908; sér. 5, vols. 1-6, 1909-14. Nouv. Arch. Mus. Hist. Nat. Preceded by: Archives du muséum d'histoire naturelle. Paris. Superseded by: Archives muséum national d'histoire naturelle. Paris. 4-3267-3. HI 57513

Nouvelles de l'E A W A G. Dübendorf. Vol. 1+, 1973+. Nouv. E. A. W. A. G. HI 69002

Nouvelles de l'I C O M = I C O M news. Paris.

Nouvelle revue encyclopédique. Paris. Vols. 1-5, 1846-47. Nouv. Rev. Encycl. Preceded by: Revue encyclopédique. 4-3108-3. HI 69003

Nouvelles, see Nouvelle.

Nouvelliste oeconomique et littéraire, ... The Hague. Nouvelliste Oecon. Litt. See B–P–H 671/1. HI 57525

Nouvelliste suisse, historique, politique, littéraire et amusant. Neuchâtel. Nouvelliste Suisse. See B–P–H 671/2. HI 57526

Nova acta academiae caesareae leopoldino-carolinae germanicae naturae curiosorum. Dresden. Nova Acta Acad. Caes. Leop.-Carol. German. Nat. Cur. See B–P–H 672/15. HI 57529

Nova acta academiae scientiarum imperialis petropolitanae. Praecedit historia ejusdem academiae. St. Petersburg. Nova Acta Acad. Sci. Imp. Petrop. Hist. Acad. See B–P–H 672/19. HI 57530

Nova acta eruditorum. Leipzig. Nova Acta Erud. See B–P–H 672/20. HI 57531

Nova acta eruditorum, quae Lipsiae publicantur, supplementa. Leipzig. Nova Acta Erud. Suppl. See B–P–H 672/21. HI 57532

Nova acta helvetica, physico-mathematico-botanico-medica. Basel. Nova Acta Helv. Phys.-Math. See B–P–H 672/23. HI 57533

Nova acta leopoldina. Halle. Nova Acta Leop. See B–P–H 672/24. HI 57534

Nova acta leopoldina. Supplementum. Supplement to: Nova acta Leopoldina. Leipzig. No. 1+, 1968+. Nova Acta Leop., Suppl. HI 69004

Nova acta physico-medica academiae caesareae leopoldino-carolinae naturae curiosorum. Nuremberg = Nova acta physico-medica academiae caesareae leopoldino-carolinae naturae curiosorum exhibentia ephemerides sive observationes historias et experimenta ... Nuremberg. Nova Acta Phys.-Med. Acad. Caes. Leop.-Carol. Nat. Cur. See B–P–H 673/5.

Nova acta physico-medica academiae caesareae leopoldino-carolinae naturae curiosorum exhibentia ephemerides sive observationes historias et experimenta ... Nuremberg. Nova Acta Phys.-Med. Acad. Caes. Leop.-Carol. Nat. Cur. See B–P–H 673/5. HI 57535

Nova acta regiae societatis scientiarum upsaliensis. Uppsala. Nova Acta Regiae Soc. Sci. Upsal. See B–P–H 673/7. HI 57536

Nova Guinea; a journal of botany, zoology, anthropology,

ethnography, geology and palaeontology of the papuan region [subtitle varies]. Leiden. Vols. 1-18 [vols. 10-11 not published], 1909-36; n.s. vols. 1-10, 1937-59. Nova Guinea. Superseded, in part, by: Nova Guinea. Botany. 4-3111-1. HI 69005

Nova Guinea. Botany; contributions to the anthropology, botany, geology and zoology of the papuan region. Leiden. Nos. 1-24, 1960-66. Nova Guinea, Bot. Preceded by: Nova Guinea. HI 69006

Nova Hedwigia. Zeitschrift für Kryptogamenkunde. Weinheim, Germany. Nova Hedwigia. See B–P–H 673/10. HI 57537

Nova literaria circuli franconici oder Fränkische Gelehrten-Histoire. Nuremberg. Nova Lit. Circuli Francon. See B–P–H 673/11. HI 57538

Nova literaria maris balthici et septentrionis. Lübeck. 1698-1708. Nova Lit. Maris Balt. Septentr. 4-3111-2. HI 69007

Nova Scotian institute of natural science proceedings and transactions. Halifax, N.S. Nova Scotian Inst. Nat. Sci. Proc. Trans. See B–P–H 673/16. HI 57539

Nova Thalassia. Venice. Nova Thalassia. See B–P–H 673/18. HI 57540

Novaya literatura po voprosam evolyutsii organicheskogo mira. Leningrad. Vols. 1-13, 1971-82. Novaya Lit. Vopr. Evol. Organich. Mira. Superseded by: Voprosy evolyutsii organicheskogo mira. HI 75117

Nové Obzory. Presov, Czechoslovakia. Nové Obzory. See B–P–H 673/19. HI 57541

Novedades, California avocado society. Saticoy, CA. 1978. Noved. Calif. Avocado Soc. HI 69008

Novedades científicas, contribuciones occasionales del museo de historia natural La Salle. Serie botánica. Caracas. Noved. Ci., Ser. Bot. See B–P–H 673/20. HI 57542

Novedades horticolas. Mexico City. Noved. Hort. See B–P–H 674/1. HI 57543

Novedades del museo de La Plata. La Plata. Vol. 1+, 1981+. Noved. Mus. La Plata. HI 69009

Növényegészégyügyi évkönyv. Budapest. Növényegész. Évk. See B–P–H 674/3. HI 57544

Növénygazdasági lapok. Budapest. Növénygazd. Lapok. See B–P–H 674/4. HI 57545

Növénynemesítési és növénytermesztési kutató intézet közleményei Sopronhorpács. Budapest & Szombathely. Vols. 1-3(2), 1961-65. Növénynemes. Növényterm. Kutató Intéz. Közlem. Sopronhorpács. Preceded by: Mezögazdasági növénynemesitési és növenytermesztési kutató intézet közlemények. HI 69010

Növenyrendszertani és novényföldrajzi tanszék. Eötvös Loránd tudományegyetem, Budapest. Budapest. Vol. 1+, 1971+. Növenyrendsz. Novényföldr. Tansz., Eötvös Loránd Tudományegyet. Budapest. HI 69011

Növénytani közlemények. Budapest. Növényt. Közlem. See B–P–H 674/6. HI 57546

Növénytermelés. Budapest. Növénytermelés. See B–P–H 674/7. HI 57540

Növényvédelem. Budapest. Növényvédelem. See B–P–H 674/12. HI 57552

Növényvédelem. [Edited by Lajos Kerekés.] Budapest. Növényvédelem (Kerekés). See B–P–H 674/13. HI 57553

Növényvédelem. [Edited by Gábor Ubrizsy, Lajos Zimmermann and András Kaczó.] Budapest. Növényvédelem (Ubrizsy). See B–P–H 674/14. HI 57554

Növényvédelem és kertészet. Budapest. Növényvéd. & Kert. See B–P–H 674/8. HI 57548

Növényvédelem idöszerü kerdései. Budapest. Növényvéd. Idöszerü Kérd. See B–P–H 674/9. HI 57549

A növényvédelen korszerusitése. Budapest. Vol. 1+, 1967+. Növényvéd. Korszerus. HI 69012

Növényvédelmi közlöny. Budapest. Növényvéd. Közl. See B–P–H 674/10. HI 57550

Növényvédelmi kutató intézet évkönyve. Budapest. Növényvéd. Kutató Intéz. Evk. See B–P–H 674/11. HI 57551

Növényvilág. Somorja, Hungary [=Samorin, Czechoslovakia]. Növényvilág. See B–P–H 674/15. HI 575505

Novi commentarii academiae scientiarum imperalis petropolitanae. St. Petersburg. Novi Comment. Acad. Sci. Imp. Petrop. See B–P–H 674/17. HI 57556

Novi commentarii academiae scientiarum instituti bononiensis. Bologna. Novi Comment. Acad. Sci. Inst. Bononiensis. See B–P–H 674/18. HI 57557

Novi commentarii societatis regiae scientiarum gottingensis. Göttingen & Gotha. Novi Comment. Soc. Regiae Sci. Gott. See B–P–H 674/20. HI 57558

Novices' gleanings in bee culture. Medina, OH. Novices' Gleanings Bee Cult. See B–P–H 675/1. HI 57559

Novitates botanicae cum delecta seminum, fructuum, sporarum plantarumque quas institutum botanicum et hortus botanicus universitatis carolinae pragensis = Novitates botanicae et delectus seminum horti botanici universitatis carolinae pragensis. Prague.

Novitates botanicae et delectus seminum horti botanici universitatis carolinae pragensis. Prague. 1958-70. Novit. Bot. Delect. Seminum Horti Bot. Univ. Carol. Prag. Superseded by: Novitates botanicae ex instituto et horto botanico universitatis carolinae pragensis. HI 69013

Novitates botanicae horti botanici universitatis carolinae pragensis = Novitates botanicae et delectus seminum horti botanici universitatis carolinae pragensis. Prague.

Novitates botanicae ex instituto et horto botanico universitatis carolinae pragensis. Prague. 1971-? Novit. Bot. Inst. Horto Bot. Univ. Carol. Prag. Preceded by: Novitates botanicae et delectus seminum horti botanici universitatis carolinae pragensis. HI 69014

Novitates botanicae ex universitate Carolina. Prague. Vol. 1+, 1982+. Novit. Bot. Univ. Carol. Preceded by: Novitates botanicae ex instituto et horto botanico universitatis carolinae pragensis. HI 69015

Novitates musei sarajevoensis. Sarajevo. Nos. 1-6-?, ?-1928-? Novit. Mus. Sarajev. HI 69016

Novitates systematicae plantarum non vascularium = Novosti sistematiki nizshikh rastenii. Moscow & Leningrad.

Novitates systematicae plantarum vascularium = Novosti sistematiki vysshikh rastenii. Moscow & Leningrad.

Novorum actorum academia caesareae leopoldinae-carolinae germanicae naturae curiosorum. Verhandlungen der Kaiserlichen Leopoldinisch-Carolinischen Deutschen Akademie der Naturforscher. Jena. Nov. Actorum Acad. Caes. Leop.-Carol. German. Nat. Cur. See B–P–H 671/15. HI 57527

Novorum actorum academiae caesareae leopoldinae-carolinae naturae curiosorum. Breslau [=Wroclaw, Poland] & Bonn. Nov. Actorum Acad. Caes. Leop.-Carol. Nat. Cur. See B–P–H 671/16. HI 57528

Novos annaes das sciencias, das artes, e das letras; por huma sociedade de portuguezes residentes em Parîs. Paris. Novos Ann. Sci. Soc. Portug. Parîs. See B–P–H 675/4. HI 57560

Novosti iz Bosansko-Khertsegovachakog muzeja = Novitates musei sarajevoensis. Sarajevo.

Novosti sistematiki nizshikh rastenii. Moscow & Leningrad. Vol. 1+, 1964+. Novosti Sist. Nizsh. Rast. Preceded by: Botanicheskie materialy otdela sporovykh rastenii botanicheskogo instituta imeni V. L. Komarova akademii nauk S S S R. HI 69017

Novosti sistematiki vysshikh i nizshikh rastenii. Kiev. 1974+. Novosti Sist. Vyssh. Nizsh. Rast. HI 69018

Novosti sistematiki vysshikh rastenii. Moscow & Leningrad. Vol. 1+, 1964+. Novosti Sist. Vyssh. Rast. Preceded by: Botanicheskie materialy gerbariya botanicheskogo instituti im. V. L. Komarova, akademii nauk S S S R. HI 69019

Novye knigi za rubezhom, kritiko-bibliograficheskii byulleten'. Seriya C, biologiya, meditsina, sel'skoe khozyaistvo. Moscow. 1957+. Novye Knigi Rubezh. Krit.-Bibliogr. Byull., Ser. C, Biol. Med. Sel'sk. Khoz. HI 69020

Novye memuary Moskovskogo obščestva ispytatelej prirody = Novye memuary Moskovskogo obshchestva ispytatelei prirody. Moscow.

Novye memuary Moskovskogo obshchestva ispytatelei prirody. Moscow. Vols. 23(2)-[25] [also numbered vols. 18(2)-20], 1925-40. Nov. Mem. Moskovsk. Obshch. Isp. Prir. Preceded by: Nouveau mémoires de la société impériale des naturalistes de Moscou. 3-2770-2. HI 69021

Novye vidy drevnikh rastenii i bespozvonochnykh S S S R. Moscow. Vol. 1+, 1960+. Novye Vidy Drevn. Rast. Bespozv. S.S.S.R. HI 75118

Novyi magazin estestvennoi istorii, fizikii, khimii i svedenii ekonomicheskikh. 1820-25. Novyi Mag. Estestv. Istorii. HI 52911

Novyj magazin estestvennoj istorii, fizikii, himii i svedenij ekonomičeskih = Novyi magazin estestvennoi istorii, fizikii, khimii i svedenii ekonomicheskikh.

Now. Bozeman, MT. Now. See B–P–H 675/6. HI 57561

Nuclear science abstracts. United States atomic energy commission. Oak Ridge, TN. Nucl. Sci. Abstr. See B–P–H 675/11. HI 57562

Nucleic acids abstracts. Bethesda, MD. 1971-79. Nucl. Acids Abstr. Superseded by: Biochemistry abstracts. Part 2, nucleic acids. HI 69022

Nucleic acids and molecular biology. Berlin, New York. Vol. 1+, 1987+. Nucl. Acids & Molec. Biol. HI 75223

Nucleic acids research. Arlington, VA. Vol. 1+, 1974+. Nucl. Acids Res. HI 69023

Nucleus. Boston, MA. Vol. 1+, 1924+. Nucleus (Boston). 4-3116-3. HI 69024

Nucleus; international journal of cytology and allied topics. Calcutta. Vol. 1+, 1958+. Nucleus (Calcutta). HI 69025

Nuestra agricultura. San José, Costa Rica. Nuestra Agric. See B–P–H 675/13. HI 57563

Nueva agricultura. Santiago de Chile. 1963+. Nueva Agric. Preceded by: Trigo y cereales. HI 69026

Nung chuan hsüeh pao = Bulletin of national Pingtung institute of agriculture. Pingtung.

Nung hsiao [hsüeh] tsa chih = Agricultura. Nanking.

Nung hsüeh ts'ung k'an = Agricultural research; college of agriculture, National Central university, Nanking. Nanking. Agric. Res. Nanking. See B–P–H 60/7.

Nung hsüeh yüeh k'an = Agricultural science. National

university of Peiping. Peiping [=Peking]. Agric. Sci. (Peiping). See B–P–H 60/18.

Nung lin chi k'an = Agricultural and forestry quarterly. Canton?, China. Agric. Forest. Quart. See B–P–H 57/14.

Nung lin hsin pao = New journal of agriculture and forestry; college of agriculture, Chinling university. [China]. New J. Agric. Forest. Coll. Agric. Chinling Univ. See B–P–H 653/11.

Nung lin hsüeh pao = Journal of agriculture and forestry. Taichung, Taiwan. J. Agric. Forest. (Taichung). See B–P–H 455/11.

Nung lin kung pao = Agricultural journal. [Department of Agriculture and forestry. Republic of China.] [China]. Agric. J. (China). See B–P–H 58/18.

Nung lin t'ung hsin = Agriculture and forestry news. Taipei, Taiwan. Agric. Forest. News. See B–P–H 57/13.

Nung shang pu chung yang nung shih shih yen ch'ang ti san ch'i ch'eng chi pao kao = Report of the central agricultural experiment station. Ministry of agriculture and commerce. Nanking. Rep. Centr. Agric. Exp. Sta. Minist. Agric. See B–P–H 763/18.

Nung sheng = Agricultural information. [College of agriculture; Sun Yatsen university.] Canton. Agric. Inform. See B–P–H 58/14.

Nung shih shuang yüeh k'an = Agricultural bimonthly. Canton. Agric. Bimonthly. See B–P–H 56/13.

Nung yeh hsüeh pao = Acta agriculturae sinica. Peking. Acta Agric. Sin. See B–P–H 39/18.

Nung yeh t'ui kuang t'ung hsin = Agricultural extension news. Chungking, China. Agric. Extens. News. See B–P–H 57/12.

Nung yeh t'ung hsin = Agricultural news. Nanking. Agric. News (Nanking). See B–P–H 59/22.

Nuova antologia; rivista di lettere, scienze ed arti. Florence & Rome. Nuova Antologia. See B–P–H 675/40. HI 57564

Nuova collezione di opuscoli e notizie di scienze, lettere ed arti. Fiesole, Italy. Nuova Collez. Opusc. Not. Sci. See B–P–H 675/41. HI 57565

Nuova collezione d'opuscoli scientifici. Bologna. Nuova Collez. Opusc. Sci. See B–P–H 676/1. HI 57566

Nuova Notarisia, rassegna consacrata allo studio della alghe. Modena & Padua. Nuova Notarisia. See B–P–H 676/2. HI 57567

Nuova scelta d'opuscoli interessanti sulle scienze e sulle arti. Milan. Nuova Scelta Opusc. Interessanti Sci. Arti. See B–P–H 676/3. HI 57568

Nuova veterinaria. Bologna. Nuova Veterin. See B–P–H 676/5. HI 57569

Nuovi annali d'igiene e microbiologia. Rome. 1950+. Nuovi Ann. Ig. Microbiol. Preceded by: Annali d'igiene sperimentale. HI 69027

Nuovi annali, istituto chimico-agrarico sperimentale. Gorizia. Vol. 1+, 1950+. Nuovi Ann. Ist. Chim.-Agrar. Sperim. HI 69028

Nuovi annali delle scienze naturali. Bologna. Nuovi Ann. Sci. Nat. See B–P–H 676/6. HI 57570

Nuovi commentarj di medicina e di chirurgia. Padua. Nuovi Comment. Med. Chir. See B–P–H 676/7. HI 57571

Nuovi saggi della cesareo-regia accademia di scienze, lettere ed arti di Padova. Padua. Nuovi Saggi Ces.-Regia Accad. Sci. Padova. See B–P–H 676/8. HI 57572

Nuovi saggi della imperiale regia accademia di scienze, lettere ed arti in Padova. Padua. Nuovi Saggi Imp. Regia Accad. Sci. Padova. See B–P–H 676/9. HI 57573

Nuovo Ercolani. Turin. Nuovo Ercolani. See B–P–H 676/10. HI 57574

Nuovo giornale di agricultura, industria e commercio; organo federale delle cooperative di consumo, roduzione e lavoro della provincia di Siena. Siena. Vol. 1-3-?, 1917-19-? Nuovo Giorn. Agric. Industr. Commercio. HI 69029

Nuovo giornale botanico italiano; e bollettino della società botanica italiano. Florence. Vols. 1-25, 1869-93; n.s. vols. 1-68, 1894-1961. Nuovo Giorn. Bot. Ital. Preceded & superseded by: Giornale botanico italiano. 4-3121-2. HI 69030

Nuovo giornale dei letterati. Pisa. Nuovo Giorn. Lett. See B–P–H 676/14. HI 57575

Nuovo giornale de' letterati d'Italia. Modena, Italy. Nuovo Giorn. Lett. Italia. See B–P–H 676/15. HI 57576

Nuovo giornale de' letterati. Scienze. Pisa. Nuovo Giorn. Lett., Sci. See B–P–H 676/16. HI 57577

Nuovo giornale ligustico di lettere, scienze ed arti. Genoa. Nuovo Giorn. Ligustico Lett. See B–P–H 676/17. HI 57578

Nuovo magazzino toscano. Florence. Nuovo Mag. Tosc. See B–P–H 676/18. HI 57579

Nürnbergische Gelehrte Zeitung. Nuremberg. Nürnberg. Gel. Zeitung. See B–P–H 676/22. HI 57580

Nürnbergisches Magazin zum Nutzen un Vergnügen. Nuremberg. Vol. 1, 1816. Nürnberg. Mag. Nutzen Vergnügen. HI 65239

Nursery disease notes, New Jersey agricultural college experiment station = New Jersey agricultural college experiment station. Nursery disease notes. New

Brunswick, NJ. New Jersey Agric. Coll. Exp. Sta. Nursery Dis. Notes. See B–P–H 653/18.

Nursery disease notes, Rutgers University agricultural experiment station = Rutgers University agricultural college experiment station. Nursery disease notes. New Brunswick, NJ. New Jersey Agric. Coll. Exp. Sta. Nursery Dis. Notes. See B–P–H 653/18.

Nursery manager. Fort Worth, TX. Vol. 1+, 1985+. Nursery Manager. Preceded by: S F and N. HI 65286

Nursery research journal. ?-1977+. Nursery Res. J. HI 69031

Nurseryman. Waterdown. Vols. 1, 1964. Nurseryman. Superseded by: Canadian nurseryman. HI 69032

Nurseryman and garden centre. London. Vol. 140+, 1965+. Nurseryman Gard. Centre. HI 69033

Nutrition. Hyderabad, India. Nutrition. See B–P–H 677/3. HI 57583

Nutrition abstracts and reviews. Aberdeen, Scotland. Nutr. Abstr. Rev. See B–P–H 677/1. HI 57581

Nutrition reviews. New York, NY. Nutr. Rev. See B–P–H 677/2. HI 57582

Nutshell. Knoxville, TN. Vol. 1+, 1948+. Nutshell. HI 69034

Nützliche und auserlesene Arbeiten der Gelehrten im Reich, d. i. in Francken, Schwaben, Ober-Rhein, Bayern, Österreich, Böhmen und angräntzenden Orten. Nuremberg. Nützl. Auserlesene Arbeiten Gel. Reich. See B–P–H 677/4. HI 57584

Nützliche Bemerkungen für Garten- und Blumenfreunde. Leipzig. Nützl. Bemerk. Garten- Blumenfr. See B–P–H 677/5. HI 57585

Nützliche Sammlungen. Hanover. Nützl. Samml. See B–P–H 677/6. HI 57586

Nützliches und unterhaltendes Berlinisches Wochenblatt für den gebildeten Bürger und denkenden Landmann. Berlin. Nützl. Unterhalt. Berlin. Wochenbl. Bürger Landmann. See B–P–H 677/7. HI 57587

Nuusbrief, limnologiese vereniging van suidelike Afrika = Newsletter, limnological society of southern Africa. Grahamstown.

Nuytsia; bulletin of the western australian herbarium. South Perth, W.A. Vol. 1+, 1970+. Nuytsia. HI 69035

Ny almeenyttig samler, ... Copenhagen. Vols. 1-4, 1824-26. Ny Almeennytt. Samler. Preceded by: Almeennytt samler. HI 69036

Ny journal uti hushållningen. Stockholm. Ny J. Hush. See B–P–H 677/12. HI 57588

Ny pharmaceutisk tidende. Copenhagen. Vols. 1-26, 1869-94. Ny Pharm. Tidende. Preceded by: Pharmaceutisk tidende. 4-3123-2. HI 69037

Nya handlingar, Kongl. vetenskaps academiens. Stockholm = Kongl. vetenskaps academiens nya handlingar. Stockholm. Kongl. Vetensk. Acad. Nya Handl. See B–P–H 517/14.

Nya handlingar af kongliga wettenskaps och witterhets samhället i Götheborg. Goteborg, Sweden. Nya Handl. Kongl. Wettensk. Samhället Götheborg. See B–P–H 677/14. HI 57589

Nya handlingar, svenska läkare-sällskapet = Svenska läkare-sällskapet. Nya handlingar. Stockholm. Svenska Läkare-Sällsk. Nya Handl. See B–P–H 861/15.

Nya lärda tidningar. Stockholm. Nya Lärda Tidn. See B–P–H 677/15. HI 57590

Nyala. Limbe, Zomba. 1975+. Nyala. HI 69038

Nyasaland agricultural quarterly journal. Blantyre. Vols. 1-10, 1941-51. Nyasaland Agric. Quart. J. Preceded by: Nyasaland tea association quarterly journal. HI 69039

Nyasaland farmer and forester. Blantyre, Nyasaland [Malawi]. Nyasaland Farmer Forester. See B–P–H 677/16. HI 57591

Nyasaland tea association quarterly journal. Cholo. Vols. 1-5(2), 1936-40. Nyasaland Tea Assoc. Quart. J. Superseded by: Nyasaland agricultural quarterly journal. HI 69040

Nye Hygaea. Copenhagen. Nye Hygaea. See B–P–H 677/17. HI 57592

Nye landoeconomiske tidender. Copenhagen. Nye Landoecon. Tidender. See B–P–H 677/18. HI 57593

Nye landoekonomiske tidender = Nye landoeconomiske tidender. Copenhagen. Nye Landoecon. Tidender. See B–P–H 677/18.

Nye oeconomiske annaler. Copenhagen. Nye Oecon. Ann. See B–P–H 677/19. HI 57594

Nye samling af det kongelige danske videnskabers selskabs skrifter. Copenhagen. Vols. 1-5, 1781-99. Nye Saml. Kongel. Danske Vidensk. Selsk.Skr. Preceded by: Skrifter, som udi det kongelige videnskabers selskab ere fremlagde, og nu til trykken befordrede. Superseded by: Kongelige danske videnskabernes selskabs skrifter. 2-1262-1. HI 69041

Nye samling af det kongelige norske videnskabers selskabs skrifter. Copenhagen. Nye Saml. Kongel. Norske Vidensk. Selsk. Skr. See B–P–H 677/21. HI 57595

Nye sundheds-tidende. Copenhagen. Nye Sundheds-Tidende. See B–P–H 677/22. HI 57596

Nye tidender om laerde og curieuse sager. Copenhagen. 1724-48. Nye Tidender Laerde Cur. Sager. Preceded by: Nye tidender om laerde sager. Superseded by: Kiøbenhavnske nye tidender om laerde og curieuse sager. 2-1260-1. HI 69042

Nye tidender om laerde sager. Copenhagen. 1720-23. Nye Tidender Laerde Sager. Superseded by: Nye tidender om laerde og curieuse sager. 2-1260-1. HI 69043

Nyeste kiøbenhavnske efterretninger. Laerde Sager. Copenhagen. 1783-90. Nyeste Kiøbenhavnske Efterretn. Laerde Sager. Preceded by: Kiøbenhavnske nye efterretninger. Laerde sager. Superseded by: Kjøbenhavnske laerde efterretninger. 2-1260-1. HI 69044

Nyeste samling af det kongelige norske videnskabers-selskabs skrifter. Copenhagen. Nyeste Saml. Kongel. Norske Vidensk.-Selsk. Skr. See B–P–H 677/23. HI 57597

Nymphaea; folia naturae bihariae. Oradea. Vol. 1+, 1973+. Nymphaea. HI 69045

Nyt bibliothek for laeger. Copenhagen. Nyt Biblioth. Laeger. See B–P–H 678/2. HI 57598

Nyt bibliothek for physik, medicin og oeconomie. Copenhagen. Nyt Biblioth. Phys. See B–P–H 678/3. HI 57599

Nyt magazin for naturvidenskaberne. Christiania [=Oslo, Norway]. Nyt Mag. Naturvidensk. See B–P–H 678/7. HI 57600

Nyt tidsskrift for fysik og kemi. Copenhagen. Vols. 1-3, 1896-98. Nyt Tidsskr. Fys. Kemi. Preceded by: Tidsskrift for physik og chemi samt disse videnskabers anvendelse. 4-3125-1. HI 69046

Nytt magasin for botanikk. Oslo. Vols. 1-17, 1952-70. Nytt. Mag. Bot. Preceded by: Nytt magazin for naturvidenskapene. Superseded by: Norwegian journal of botany. HI 69047

Nytt magazin for naturvidenskapene. Oslo. Vols. 75-88, 1935-51. Nytt Mag. naturvidensk. Preceded by: Nyt magazin for naturvidenskaberne. Superseded by: Nytt magasin for botanikk. 4-3125-1. HI 69048

O N R Tokyo scientific bulletin = Scientific bulletin, office of naval research, department of the navy, Tokyo.

O P T I M A leaflets. Geneva, Chambésy. No. 1+, 1975+. O. P. T. I. M. A. Leafl. HI 69049

O P T I M A meetings. Geneva. 1975+. O. P. T. I. M. A. Meetings. HI 69050

O P T I M A newsletter. Chambésy, Berlin, etc. No. 1+, 1975+. O. P. T. I. M. A. Newslett. HI 64485

O T S liana. Durham, NC = Liana. Durham, NC.

Oasis. Bimestrale di natura ecologia fotografia. Aosta. Vol. 1+, 1985+. Oasis. HI 69051

Obeche; journal of the tree club. Ibadan. ?-1981+. Obeche. HI 69052

Oberdeutsche allgemeine Litteraturzeitung. Salzburg, Mainz, & Vienna. Oberdeutsche Allg. LItteraturzeitung. See B–P–H 678/19. HI 57601

Oberdeutsche Beyträge zur Naturlehre und Oekonomie. Salzburg. Oberdeutsche Beytr. Naturl. Oekon. See B–P–H 678/21. HI 57602

Ober-Erzgebürgisches Journal oder Sammlung von allerhand in die hiesige Natur-Wissenschaft einschlagenden Abhandlungen. [Germany]. Ober-Erzgebürg. J. See B–P–H 678/22. HI 57603

Oberhessische naturwissenschaftliche Zeitschrift. Giessen. Vol. 38+, 1971+. Oberhess. Naturwiss. Z. Preceded by: Bericht der Oberhessischen Gesellschaft für Natur- und Heilkunde zu Giessen. Naturwissenschaftliche Abteilung. HI 69053

Oberlausitzscher Beytrag zur Gelahrtheit und deren Historie. Leipzig & Görlitz. Oberlausitzscher Beytr. Gelahrth. Hist. See B–P–H 678/25. HI 57604

Oberrheinische geologische Abhandlungen. Karlsruhe. Oberrhein. Geol. Abh. See B–P–H 678/26. HI 57605

Oberschlesische landwirthschaftliche Garten- und Blumen-Zeitung. Neisse [Poland]. Oberschles. Landw. Garten- Blumen-Zeitung. See B–P–H 678/27. HI 57606

Oberungerisches Weinbau-, Keller- und landwirtschaftliches Blatt = Felsö-magyarországi szölészeti, borászati és gazdasági lap. Kassa, Hungary [=Kosice, Czechoslovakia]. Felsö-Magyarorsz. Szölész. Borász. Gazd. Lap. See B–P–H 369/10.

Obeznik. Brno. 1944-46. Obeznik. HI 69054

Oběžnik, krouzku kaktusaru. Prague. Vols. 1-3, 1953-55. Oběžn. Krouzku Kakt. Superseded by: Věstnik, krouzku kaktusaru. HI 69055

Obituary notices of fellows of the royal society. London. Vols. 1-?, 1932-54. Obit. Not. Fellows Roy. Soc. Superseded by: Biographical memoirs of fellows of the royal society of London. HI 69056

Obmen opytom v sel'skom khozyaistve. Ser. ovoshchevod. Moscow. 1958-59. Obmen Opyt. Sel'sk., Ser. Ovoshchevod. Incorporated in: Kartofel' i ovoshchi. HI 69057

Obra del centenario del museo de La Plata. La Plata. Vol. 1+, 1977+. Obra Centen. Mus. La Plata. HI 69058

Observações gerais e contribuições ao estudo da flora e fitofisionomia do Brasil instituto de botânica. São Paulo. Observ. Gerais & Contr. Estudo Fl. Fitofision. Brasil Inst. Bot. See B–P–H 679/6. HI 57611

Observation. [Kuan ch'a.] Shanghai. Observation. See

B–P–H 679/9. HI 57613

Observationes botanicae et descriptiones plantarum novarum herbarii Van Heurckiani. Linz. Observ. Bot. Descript. Pl. Nov. Herb. Van Heurckiani. See B–P–H 679/2. HI 57607

Observationes botanicae et descriptiones plantarum novarum herbarii Van Heurckiani. Linz. Observ. Bot. Descript. Pl. Nov. Herb. Van Heurckiani. See B–P–H 679/2. HI 57608

Observationes curiosae naturae et artis. Hamburg. Observ. Cur. Nat. Artis. See B–P–H 679/3. HI 57609

Observationes et disputationes mycologicae. Paris, Sens. Vol. 1+, 1974+. Observ. Disp. Mycol. HI 69059

Observationes sobre la física, historia natural, y artes utiles. Observ. Fís. See B–P–H 679/4. HI 57610

Observations et mémoires sur la physique, sur l'histoire naturelle et sur les arts et métiers. Paris. Observ. Mém. Phys. See B–P–H 679/7. HI 57612

Observations sur la physique, sur l'histoire naturelle et sur les arts. Paris. Vols. [1]-2, 1772 [18 parts published, 2 of each forming a unit called "tome." Ed. 2 published as: Introduction au observations sur la physique, sur l'histoire naturelle et sur les arts, q.v.]; [sér. 2], vols. 2-43, 1773-93. Observ. Phys. For [sér. 2] vol. 1, 1773, see Observations et mémoires sur la physique, sur l'histoire naturelle et sur les arts et métiers. 3-2185-1. HI 51146

Observer. [Kuan ch'a.] Shanghai. Observer. See B–P–H 679/10. HI 57614

Observer of nature. Lawrence, KS. Observer Nat. See B–P–H 679/11. HI 57615

Obst und Garten (Karlsruhe); Monatsschrift für Obst-, Gemüse, Wein- und Gartenbau: Organ des Badischen Landesobstbauverbandes. Karlsruhe. 196?+. Obst & Gart. (Karlsruhe). Preceded by: Badische Obst- und Gartenbauer. HI 69060

Obst und Garten. Linz. Vols. 1-5, 1946-50. Obst & Gart. (Linz). Superseded by: Obst- und Gartenbau. Linz. 4-3132-2. HI 69061

Obst und Garten. Stuttgart. Vol. 22+, 1968+. Obst. & Gart. (Stuttgart). Preceded by: Obstbau. Ravensburg. HI 69062

Obst- und Gartenbau. Linz. Vol. 6+, 1951+. Obst-Gartenbau (Linz). Preceded by: Obst und Garten. Linz. HI 69063

Obst- und Gartenbau. Stuttgart. Vols. 1-3, [1946] 1947-49. Obst- Gartenbau (Stuttgart). Preceded and superseded by: Obstbau. Ravensburg. 4-3132-2. HI 69064

Obst- und Gemüsebau. Berlin. Obst- Gemüsebau. See B–P–H 679/14. HI 57616

Obst- und Weinbau. Graz. Obst- Weinbau. See B–P–H 679/16. HI 57617

Obstbau; Fachzeitschrift für Obstbau, Weinbau und Baumschule. [Ausgabe der Zeitschrift: Deutsche Gartenbau.] Berlin. Vols. 3-11(6), 1963-71. Obstbau (Berlin). Preceded by: Intensivobstbau. Superseded by: Gartenbau. HI 69065

Obstbau. Klosterneuburg, Austraia. Obstbau (Klosterneuburg). See B–P–H 679/17. HI 57618

Obstbau. Ravensburg. Vols. 1-62, 1881-1942; [n.s.] vols. 4-21, 1950-67 [publication suspended 1943-45]. Obstbau (Ravensburg). For [n.s.] vols. 1-3, 1947-49 see: Obst- Gartenbau (Stuttgart). Superseded by: Obst und Garten. Stuttgart. 4-3132-2. HI 69066

Obstbau. Scheibbs, Austria. Obstbau (Scheibbs). See B–P–H 679/18. HI 57619

Obstbau, Weinbau; Mitteilungen des Südtiroler Beratungsringes. Lana. Vol. 1+, 1964+. Obstbau Weinbau. HI 69067

Obstbaum-Freund. Frauendorf, Germany. Obstbaum-Freund. See B–P–H 679/19. HI 57620

Obstétrique. Paris. Obstétrique. See B–P–H 679/20. HI 57621

Obstgarten. Klosterneuburg, Austria. Obstgarten (Klosterneuburg, 1879-83). See B–P–H 679/21. HI 57622

Obstgarten. Klosterneuburg, Austria. Obstgarten (Klosterneuburg, 1893-1906). See B–P–H 680/1. HI 57623

Obstrundschau. Zug, Switzerland. Obstrundschau. See B–P–H 680/2. HI 57624

Obstzüchter. Zeitschrift für die Gesamt-Interessen des Obstbaues. Krems, Austria. Obstzüchter. See B–P–H 680/3. HI 57625

Obzor Botaničeskoj Dějatel'nosti v Rossii = Obzor Botanicheskoi Dĕyatel'nosti v Rossii. St. Petersburg.

Obzor Botanicheskoi Dĕyatel'nosti v Rossii. St. Petersburg. 1892-95. Obzor Bot. Dĕyatel'n. Rossii. 4-3133-2. HI 69068

Occasional bulletin, banana board research department. Kingston, Jamaica. No. 1+, 1958+. Occas. Bull. Banana Board Res. Dept. HI 69069

Occasional bulletin of the Barbados botanical station. Bridgetown, Barbados. Occas. Bull. Barbados Bot. Sta. See B–P–H 680/22. HI 57626

Occasional communications from the Utrecht university biohistorical institute = Incidentele mededelingen van het biohistorisch instituut der rijksuniversiteit te Utrecht. Utrecht.

Occasional notes of the Hongkong horticultural society. Hong Kong. Nos. 1-4, 1931-40. Occas. Notes

Hongkong Hort. Soc. 3-1882-1. HI 69070

Occasional notes, department of botany, king's college, Newcastle-upon-Tyne. Newcastle-upon-Tyne. Nos. 1-4, 1941-42. Occas. Notes Dept. Bot. King's Coll. Newcastle-Upon Tyne. HI 69071

Occasional, Nova Scotia museum. Halifax, N.S. 1973+. Occas Nova Scotia Mus. HI 69072

Occasional paper, see Occasional papers

Occasional papers, Allan Hancock foundation = Allan Hancock foundation. Publications. Occasional papers. Los Angeles, CA. Allan Hancock Found. Publ. Occas. Pap. See B–P–H 69/22.

Occasional papers, australian national parks and wildlife service. Canberra, A.C.T. Vol. 1+, 1978+. Occas. Pap. Austral. Natl. Parks Wildlife Serv. HI 69073

Occasional papers, Bell museum of natural history, university of Minnesota. Minneapolis, MN. Vol. 10+, 1971+. Occas. Pap. Bell Mus Nat. Hist. Univ Minnesota Preceded by: Occasional papers of the museum of natural history, university of Minnesota. HI 69074

Occasional papers of the Bernice P. Bishop museum = Occasional papers of the Bernice Pauahi Bishop museum of polynesian ethology and natural history. Honolulu, HI.

Occasional papers of the Bernice Pauahi Bishop museum of polynesian ethology and natural history. Honolulu, HI. Vols. 1-25, 1898/1902-84, 1899-84. Occas. Pap. Bernice Pauahi Bishop Mus. Superseded by: Bishop museum occasional papers. 1-651-1. HI 69075

Occasional papers; biological society of Nevada. Reno, NV. Occas. Pap. Biol. Soc. Nevada. See B–P–H 680/27. HI 57629

Occasional papers of the Boston scientific society. Boston, MA. Occas. Pap. Boston Sci. Soc. See B–P–H 680/28. HI 57630

Occasional papers of the Boston society of natural history. Boston, MA. Occas. Pap. Boston Soc. Nat. Hist. See B–P–H 680/29. HI 57631

Occasional papers of the British Columbia provincial museum. Victoria, B.C. 1939+. Occas. Pap. British Columbia Prov. Mus. HI 69076

Occasional papers of the C. C. Adams center for ecological studies. Kalamazoo, MI. Occas. Pap. Adams Center Ecol. Stud. See B–P–H 680/24. HI 57627

Occasional papers of the California academy of sciences. San Francisco, CA. Occas. Pap. Calif. Acad. Sci. See B–P–H 680/30. HI 57632

Occasional papers, Caradoc and Severn Valley field club. Shrewsbury. Vol. ?-2+, ?-1978+. Occas. Pap. Caradoc Severn Valley Field Club. HI 69077

Occasional papers, centre for north-west regional studies, university of Lancaster. Lancaster. No. 1+, 1976+. Occas. Pap. Centre N.-W. Regional Stud. Univ. Lancaster. HI 69078

Occasional papers, commission on ecology. Gland. No. 1+, 1982+. Occas. Pap. Commiss Ecol. HI 69079

Occasional papers, commonwealth forestry institute = C F I occasional papers. Oxford.

Occasional papers, Coryndon memorial museum. Nairobi. 1943-61. Occas. Pap. Coryndon Mem. Mus. Superseded by: Journal of the East Africa natural history society and Coryndon museum. HI 69080

Occasional papers, department of biology, college of Puget Sound. Tacoma, WA. Nos. 1-16, 1939-55. Occas. Pap. Dept. Biol. Coll. Puget Sound. Superseded by: Occasional papers, department of biology, university of Puget Sound. HI 69081

Occasional papers, department of biology, university of Papua New Guinea. Port Moresby. Nos. 1-7, 1971-75. Occas Pap. Dept. Biol. Univ. Papua New Guinea. HI 69082

Occasional papers, department of biology, university of Puget Sound. Tacoma, WA. Nos. 17-42, 1962-71; no. 52+, 1978+. Occas Pap. Dept. Biol. Univ. Puget Sound. Preceded by: Occasional papers, department of biology, college of Puget Sount. For nos. 43-51, 1972-78 see: Occasional papers, museum of natural history, university of Puget Sound. HI 69083

Occasional papers, department of geography, university college, London. London. Vol. ?-31+, ?-1976+. Occas. Pap. Dept. Geogr. Univ. Coll. London. HI 69084

Occasional papers by the division of systematic biology of Stanford University. Stanford, CA. Occas. Pap. Div. Syst. Biol. Stanford Univ. See B–P–H 680/31. HI 57633

Occasional papers, England field unit, nature conservancy council. Banbury. No. 1+, 1981+. Occas. Pap. England Field Unit Nat. Conservancy Council. HI 69085

Occasional paper. Fairchild tropical garden. Coconut Grove, FL. Occas. Pap. Fairchild Trop. Gard. See B–P–H 681/1. HI 57634

Occasional papers of the Farlow herbarium of cryptogamic botany. Cambridge, MA. Vols. 1-19, 1969-87. Occas. Pap. Farlow Herb. Cryptog. Bot. Superseded by: Harvard papers in botany. HI 69086

Occasional papers, garden history society. London. Nos. 1-2, 1969-70. Occas. Pap. Gard. Hist. Soc. HI 69087

Occasional papers of the Hongkong horticultural society. Hong Kong. Occas. Pap. Hongkong Hort. Soc. See

B–P–H 681/2. HI 57635

Occasional papers of the Hull scientific and field naturalists' club. Hull, England. Occas. Pap. Hull Sci. Club. See B–P–H 681/3. HI 57636

Occasional papers, Huron mountain wildlife foundation. Marquette, MI. No. 1+, 1971+. Occas. Pap. Huron Mount. Wildlife Found. HI 69088

Occasional papers of the marine laboratory; state university, Louisiana. Baton Rouge, LA. Occas. Pap. Mar. Lab. State Univ. Louisiana. See B–P–H 681/4. HI 57637

Occasional papers, Mauritius sugar industry research institute. Reduit. No. 1+, 1958+. Occas. Pap. Mauritius Sugar Industr. Res. Inst. HI 69089

Occasional papers, middle american research institute. New Orleans, LA. No. 1+, 1977+. Occas. Pap. Middle Amer. Res. Inst. HI 69090

Occasional papers, Milwaukee public museum. Natural history. Milwaukee, WI. No. 1+, 1968+. Occas. Pap. Milwaukee Public Mus. Nat. Hist. HI 69091

Occasional papers, Missouri academy of science. Columbia, MO. No. 3+, 1974+. Occas. Pap. Missouri Acad. Sci. Preceded by: Bulletin, Missouri academy of science. HI 69092

Occasional papers, museum of natural history, university of Kansas. Lawrence, KS. No. 1+, 1971+. Occas. Pap. Mus Nat. Hist. Univ. Kansas. HI 69093

Occasional papers of the museum of natural history, university of Minnesota. Minneapolis, MN. Vols. 1-9, 1916-63. Occas. Pap. Mus. Nat. Hist. Univ. Minnesota. Superseded by: Occasional papers, Bell museum of natural history, university of Minnesota. 3-2678-3. HI 69094

Occasional papers, museum of natural history, university of Puget Sound. Tacoma, WA. Nos. 43-51, 1972-78. Occas. Pap. Mus. Nat. Hist. Univ. Puget Sound. Preceded and superseded by: Occasional papers, department of biology, university of Puget Sound. HI 69095

Occasional papers, museum of southwestern biology. Albuquerque, NM. No. 1+, 1983+. Occas. Pap. Mus. SouthW. Biol. HI 69096

Occasional papers, the museum, Texas tech university. Lubbock, TX. No. 1+, 1972+. Occas. Pap. Mus Texas Tech Univ. HI 69097

Occasional papers from the museum of Victoria. Melbourne, Vic. No. 1+, 1984+. Occas. Pap. Mus. Victoria. Preceded by: Reports of the national museum of Victoria. HI 69098

Occasional papers, national botanic gardens, Glasnevin. Dublin. No. 1+, [1981]+. Occas. Pap. Natl. Bot. Gard. Glasnevin. HI 69099

Occasional papers of the national museum of Rhodesia. B. natural sciences. Bulawayo. Vols. 4(28B)-6(7), 1969-80. Occas. Pap. Natl. Mus. Rhodesia, B. Preceded by: Occasional papers of the national museum of Southern Rhodesia. B, natural sciences. Superseded by: Occasional papers of the national museums and monuments, Rhodesia. Series B, natural sciences. HI 69100

Occasional papers of the national museum of Southern Rhodesia. Bulawayo. Vols. 1-2, 1937-55. Occas. Pap. Natl. Mus. Southern Rhodesia. Superseded by: Occasional papers of the national museum of Southern Rhodesia. B, natural sciences. HI 69101

Occasional papers of the national museum of Southern Rhodesia. B, natural sciences. Bulawayo. Vols. 3-4(27B), 1956-64. Occas. Pap. Natl. Mus. Southern Rhodesia, B. Preceded by: Occasional papers of the national museum of Southern Rhodesia. Superseded by: Occasional papers of the national museum of Rhodesia. B, natural sciences. HI 69102

Occasional papers of the national museums and monuments, Rhodesia. Series B, natural sciences. Bulawayo. Vols. 6(8-10), 1981. Occas. Pap. Natl. Mus. Monu. Rhodesia, Ser. B. Preceded by: Occasional papers of the national museum of Rhodesia. Series B, natural sciences. Superseded by: Smithersia. HI 69103

Occasional papers of the natural history museum of Stanford university. Palo Alto, CA. Occas. Pap. Nat. Hist. Mus. Stanford Univ. See B–P–H 681/6. HI 57638

Occasional papers of the natural history society of New Brunswick. St. John, New Brunswick. Occas. Pap. Nat. Hist. Soc. New Brunswick. See B–P–H 681/7. HI 57639

Occasional papers of the natural history society of Wisconsin. Milwaukee, WI. Occas. Pap. Nat. Hist. Soc. Wisconsin. See B–P–H 681/8. HI 57640

Occasional papers, northern Nevada native plant society. Reno, NV. No. 1+, 1976+. Occas. Pap. N. Nevada Native Pl. Soc. HI 69104

Occasional papers from the Osaka museum of natural history. [Shizenshi kenkyu.] Osaka. Vol. 1+, 1968+. Occas. Pap. Osaka Mus. Nat. Hist. HI 69105

Occasional papers, Peterborough museum society = Peterborough museum society, occasional papers. Peterborough, England. Peterborough Mus. Soc. Occas. Pap. See B–P–H 702/5.

Occasional paper. Philippines bureau of forestry. Manila. Occas. Pap. Philipp. Bur. Forest. See B–P–H 681/9. HI 57641

Occasional paper, provincial museum of Alberta. Natural history. Edmonton. No. 1+, 1976+. Occas. Pap. Prov.

Mus. Alberta, Nat. Hist HI 69106

Occasional papers of the Rancho Santa Ana botanical garden. Claremont, CA. Occas. Pap. Rancho Santa Ana Bot. Gard. See B–P–H 681/10. HI 57642

Occasional papers, San Diego society of natural history. San Diego, CA. Occas. Pap. San Diego Soc. Nat. Hist. See B–P–H 681/11. HI 57643

Occasional papers, southern forest experiment station. New Orleans, LA. Nos. 1-194, 1921-62. Occas. Pap. S. Forest Exp. Sta. Superseded by: Research papers S O, United States forest service. HI 69107

Occasional papers, tropical science center. San Jose, Costa Rica. No. 1+, 1963+. Occas. Pap. Trop. Sci. Center. HI 69108

Occasional papers, Tulane university museum of natural history. Belle Chasse. No. 1+, 1977+. Occas. Pap. Tulane Univ. Mus. Nat. Hist. HI 69109

Occasional papers, university of Connecticut. Biological science series = University of Connecticut occasional papers. Biological science series. Storrs, CT.

Occasional papers, university of Hawaii. Honolulu, HI. Vols. 1-?, 1923-62. Occas. Pap. Univ. Hawaii. 3-1821-1. HI 69110

Occasional papers, university of Minnesota museum of natural history = Occasional papers of the museum of natural history, university of Minnesota. Minneapolis, MN.

Occasional papers, Warwickshire trust for nature conservation. Birmingham. No. 1+, 1974+. Occas. Pap. Warwickshire Trust Nat. Conservation. HI 69111

Occasional publication, arctic science and technology information system = A S T I S occasional publication. Calgary.

Occasional publications, australian conservation foundation. Melbourne, Vic. No. 1+, 1968+. Occas. Publ. Austral. Conservation Found. HI 69112

Occasional publications, australian national botanic gardens. Canberra, A.C.T. No. [1]+, 1980+. Occas. Publ. Austral. Natl. Bot. Gard. HI 69113

Occasional publications, freshwater biological associaton. Ambleside. No. 1+, 1976+. Occas. Publ. Freshwater Biol. Assoc. HI 69114

Occasional publications of the Hastings and St. Leonard's natural history society. Hastings. Vols. 1-5, 1907-11. Occas. Publ. Hastings St. Leonard's Nat. Hist. Soc. HI 69115

Occasional publications, institute of marine science, university of Alaska. Fairbanks, AK. Vol. 1+, 1971+. Occas. Publ. Inst. Mar. Sci. Univ. Alaska. HI 69116

Occasional publication, Narragansett marine laboratory, graduate school of oceanography, university of Rhode Island. Kingston, RI. Occas. Publ. Narragansett Mar. Lab. See B–P–H 681/22. HI 57644

Occasional publications, natural environment research council. Lowestoft. 1970+. Occas. Publ. Nat. Environm. Res. Council. HI 69117

Occasional publications, nigerian society for plant protection. Ibadan. No. 1+, 1975+. Occas. Publ. Nigerian Soc. Pl. Protect. HI 69118

Occasional publications, scottish horticultural research institute. Invergowrie. No. ?-2+, ?-1977+. Occas. Publ. Scott. Hort. Res. Inst. HI 69119

Occasional reports, Monks Wood experimental station. Abbots Ripton. No. 1+, 1973+. Occas. Rep. Monks Wood Exp. Sta. HI 69120

Occasional studies, Ruislip natural history society. Ruislip. No. 1+, 1977+. Occas. Stud. Ruislip Nat. Hist. Soc. HI 69121

Occasional symposium, British grassland society. Hurley. 1964+. Occas. Symp. Brit. Grassland Soc. HI 68714

Ocean science news. Washington, DC. 1958+. Ocean Sci. News. HI 69122

Oceanic abstracts. La Jolla, CA, Bethesda, MD. Vol. 1+, 1966+. Oceanic Abstr. HI 69123

Oceanic citation journal. La Jolla, CA. Vol. 5(6-8), 1968-71. Oceanic Citat. J. Preceded by: Oceanic index, citation journal. Continued in: Oceanic abstracts. HI 69124

Oceanic coordinate index. La Jolla, CA. Vols. 1-2, 1964-65. Oceanic Coord. Index. Citation section superseded by: Oceanic index citation journal. Superseded by: Oceanic index. HI 69125

Oceanic index. La Jolla, CA. Vols. 3-8, 1966-71. Oceanic Index. Continued in: Oceanic abstracts. HI 69126

Oceanic index, citation journal. La Jolla, CA. Vol. 5(1-5), 1968. Oceanic Index Citat. J. Preceded, in part, by: Oceanic index [citation section]. Superseded by: Oceanic citation journal. HI 69127

Oceanis; série de documets océanographiques. Paris. Vol. 1+, 1974+. Oceanis. HI 69128

Oceanographic abstracts = Deep-sea research and oceanographic abstracts. Oxford, etc.

Oceanographic abstracts and bibliography. Oxford, etc. Vols. 24-25, 1977-78. Oceanogr. Abstr. Bibliogr. Preceded by: Deep-sea research and oceanographic abstracts (oceanographic bibliography and abstracts sections). Superseded by: Deep sea research. Part B. Oceanographic literature review. HI 69129

Oceanographic index. Boston, MA. 1971/74+, 1976+. Oceanogr. Index. HI 69130

Oceanographica sinica. Taipei, Taiwan. Oceanogr. Sin. See B–P–H 681/26. HI 57647

Océanographie tropicale. Bondy. Vol. 17+, 1982+. Océanogr. Trop. Preceded by: Cahiers O R S T O M. Série océanographique. HI 69131

Oceanography and meteorology. [Nagasaki marine observatory.] Nagasaki, Japan. Oceanogr. & Meteorol. See B–P–H 681/23. HI 57645

Oceanography and marine biology. London. Oceanogr. Mar. Biol. See B–P–H 681/24. HI 57646

Oceanologia. Wroclaw. Vol. 1+, 1971+. Oceanologia. HI 69132

Oceanologica acta. Paris. Vol. 1+, 1978+. Oceanol. Acta. HI 69133

Oceanologia et limnologia sinica. [Hai yang yu hu zhao.] Peking (Beijing). Vol. 1+, 1957+. Oceanol. Limnol. Sin. HI 69134

Oceanology. [English translation of: Okeanologiya.] Washington, DC. Vols. 5-13, 1965-73. Oceanology. Preceded by: Soviet oceanography. HI 69135

Oceans magazine. La Jolla, San Francisco, CA. Vol. 1+, 1969+. Oceans Mag. HI 69136

Oceanus. Woods Hole, MA. Oceanus. See B–P–H 682/5. HI 57649

Očerki po izučeniju Jakutskogo kraja = Ocherki po izucheniyu Yakutskogo kraya. Irkutsk.

Ochanomizu joshi daigaku. Shizen Kagaku hokoku = Natural science report of the Ochanomizu university. Tokyo. Nat. Sci. Rep. Ochanomizu Univ. See B–P–H 630/1.

Ocherki po izucheniyu Yakutskogo kraya. Irkutsk, Russian S F S R. Vol. 1, 1927. Ocherki Izuch Yakutsk. Kraya. HI 69138

Ochrana prirody. Bratislava. Vol. 1+, 1980+. Ochr. Prir. (Bratislava) HI 69140

Ochrana přirody. Prague. Vols. 1-30, 1946-75. Ochr. Přir. (Prague). Continued in: Památky a příroda. 4-3135-3. HI 69139

Ochrana prírody a pamiatok. Bratislava. Ochr. Prír. Pamiatok. See B–P–H 682/9. HI 57650

Ochrona przyrody. Cracow, Warsaw. Vol. 1+, 1920+ [publication suspended 1937-48]. Ochr. Przyr. Superseded by?: Ochrona przyrody. [Series A]. 4-3135-3. HI 69141

Ochrona przyrody. [Series A]. Warsaw. Vol. 1+, 1967+. Ochr. Przyr., A. Preceded by?: Ochrona przyrody. HI 69142

Ocios de españoles emigrados, en Londres = Ocios de españoles emigrados. London. Ocios Esp. Emigrados. See B–P–H 682/16.

Ochrana rostlin. Prague. Ochr. Rostl. See B–P–H 682/12. HI 57651

Ocios de españoles emigrados. London. Ocios Esp. Emigrados. See B–P–H 682/16. HI 57652

Ocrotirea naturii. Bucharest. Vols. 1-18, 1955-74. Ocrot. Nat. Superseded by: Ocrotirea naturii si a mediului înconjurător. HI 69143

Ocrotirea naturii si a mediului înconjurător, natura, terra. Bucharest. Vol. 19+, 1975+. Ocrot. Nat. Mediului Inconj. Nat. Terra. Preceded by: Ocrotirea naturii. HI 69144

Oebalia. Taranto. Vol. 1+, 1975+. Oebalia. HI 69145

Oecologia. Berlin. Vol. 1+, 1968+. Oecologia. Preceded, in part, by: Zeitschrift für Morphologie und Ökologie der Tiere [not entered]. HI 69146

Oecologia aquatica. Barcelona. No. 1+, 1973+. Oecol. Aquatica. HI 69147

Oecologia plantarum. Paris. Vols. 1-14, 1966-79. Oecol. Pl. Superseded by: Acta oecologia. Oecologia plantarum. HI 69148

Oeconomisch-Physikalische Abhandlungen. Leipzig. Oecon.-Phys. Abh. See B–P–H 682/23. HI 57656

Oeconomische Nachrichten. Leipzig. Oecon. Nachr. See B–P–H 682/22. HI 57655

Oeconomiska tidningar. Uppsala. 1765. Oecon. Tidn. HI 69149

Oeconomiske annaler. Copenhagen. Oecon. Ann. See B–P–H 682/20. HI 57653

Oeconomiske correspondent, et blad til landboeres nytte. Copenhagen. Oecon. Corresp. See B–P–H 682/21. HI 57654

Oeconomiske, physiske og mechaniske afhandlinger, det kongelige svenske videnskabers academiens. Copenhagen = Det kongelige svenske videnskabers academiens oeconomiske, physiske og mechaniske afhandlinger. Copenhagen. Kongel. Svenske Vidensk. Acad. Oecon. Afh. See B–P–H 516/23.

Oefversigt af förhandlingar; finska vetenskaps-societeten. Helsinki. Oefvers. Förh. Finska Vetensk.-Soc. See B–P–H 682/25. HI 57657

Oeil; revue d'art. Paris. No. 1+, 1955+. Oeil. HI 74867

Oekonomisch-botanisches Garten-Journal. Eisenach, Germany. Oekon.-Bot. Gart.-J. See B–P–H 682/27. HI 57658

Oekonomische Neuigkeiten und Verhandlungen. Prague. Oekon. Neuigk. Verh. See B–P–H 682/29. HI 57659

Oekonomisch-physikalische Wochenschrift. Stuttgart. Oekon.-Phys. Wochenschr. See B–P–H 683/1. HI 57660

Oesterreich (etc.) see Österreich (etc.)

Of sea and shore. Port Gamble, WA. Vol. 1+, 1970+. Sea & Shore. HI 69150

Offene Briefe für Gartenbau, Land- und Forstwirtschaft. Prague. Offene Briefe Gartenbau. See B–P–H 684/4. HI 57671

Office of horticultural crops and diseases; semi-monthly news-letter. Bureau of plant industry. U. S. Department of agriculture. Washington, DC. Off. Hort. Crops Semi-Monthly News Lett. Seee: 683/29. HI 57669

Official bulletin, American gladiolus society = American gladiolus society official bulletin. Rochester, NY. Amer. Gladiolus Soc. Off. Bull. See B–P–H 80/10.

Official journal of the A P F O. Vol. 1+, 1976+. Off. J. A. P. F. O. HI 69151

Official journal, plant variety protection office. Hyattsville, Beltsville, MD. Vol. 1+, 1973+. Off. J. Pl. Var. Protect. Off. HI 69152

Official publication, conservation federation of Missouri = Missouri wildlife. Jefferson City, MO.

Official review, American gladiolus society = American gladiolus society official review. Goshen, IN. Amer. Gladiolus Soc. Off. Rev. See B–P–H 80/11.

Official year book of western Australia. New series. Perth, W.A. Vols. 1-5, 1957-65 1958-66? Off. Year Book W. Australia, New Ser. Superseded by: Western australian year book. HI 69153

Official yearbook of the scientific and learned societies of Great Britain and Ireland. London. 1931-39 [publication suspended 1940-50]. Off. Yearb. Sci. Learned. Soc. Great Britain & Ireland. Preceded by: Yearbook of the scientific and learned societies of Great Britain and Ireland. Superseded by: Scientific and learned societies of Great Britain. 5-3825-2. HI 74868

Officieele mededeelingen floralia. The Hague. Off. Meded. Fl. See B–P–H 684/1. HI 57670

Öfversigt af botaniska arbeten och upptäckter. Stockholm. 1821-25. Öfvers. Bot. Arbeten Upptäckter. Superseded by: Arsberättelse om framstegen uti botanik. HI 69154

Öfversigt af förhandlingar: Kongl. svenska vetenskaps-akademien. Stockholm. Öfvers. Förh. Kongl. Svenska Vetensk.-Akad. See B–P–H 684/9. HI 57672

Öfversigt af sällskapet hortikulturens vänners i Göteborg förhandlingar. Goteborg, Sweden. Öfvers. Sällsk. Hort. Vänners Göteborg Förh. See B–P–H 684/13. HI 57673

Ogasawara kenkyu = Ogasawara research. Tokyo.

Ogasawara research. [Ogasawara kenkyu.] Tokyo. 1978+. Ogasawara Res. HI 69155

Ogata jikken nojo jisshi gaiyo = Report of tobacco large experimental field. Tokyo.

Ogrodnictwo; organ stowarzyszenia inzynierów i techników ogrodnictwa i centrali spółdzielni ogrodniczych i pszczelarskich. Warsaw. Vol. 1+, 1964+. Ogrodnictwo. HI 69156

Ohara nogyo kenkyu-sho tokubetsu hokoku = Special report of the Ohara institute for agricultural research. Okayama, Japan. Special Rep. Ohara Inst. Agric. Res. See B–P–H 852/1.

Ohio agricultural experiment station annual report. Wooster, OH. Ohio Agric. Exp. Sta. Annual Rep. See B–P–H 684/20. HI 57674

Ohio agricultural experiment station bimonthly bulletin. Wooster, OH. Ohio Agric. Exp. Sta. Bimonthly Bull. See B–P–H 684/21. HI 57675

Ohio agricultural experiment station bulletin. Wooster, OH. 1883-1944. Ohio Agric. Exp. Sta. Bull. Superseded by: Ohio agricultural experiment station research bulletin. HI 69157

Ohio agricultural experiment station, bulletin of technical services. Wooster, OH. Ohio Agric. Exp. Sta. Bull. Techn. Serv. See B–P–H 684/23. HI 57676

Ohio agricultural experiment station forestry circular. Wooster, OH. Ohio Agric. Exp. Sta. Forest. Circ. See B–P–H 684/24. HI 57677

Ohio agricultural experiment station forestry publication. Wooster, OH. Ohio Agric. Exp. Sta. Forest. Publ. See B–P–H 684/25. HI 57678

Ohio agricultural experiment station, publications of the department of entomology and plant pathology. Wooster, OH. Ohio Agric. Exp. Sta. Publ. Dept. Entomol. See B–P–H 684/26. HI 57679

Ohio agricultural experiment station research bulletin. Wooster, OH. 1948-61. Ohio Agric. Exp. Sta. Res. Bull. Preceded by: Ohio agricultural experiment station bulletin. Superseded by: Research bulletin, Ohio agricultural research and development center. HI 69158

Ohio agricultural experiment station research circular. Wooster, OH. Nos. 1-138, 1948-65. Ohio Agric. Exp. Sta. Res. Circ. Superseded by: Research circular, Ohio agricultural research and development center. HI 69159

Ohio agricultural experiment station service bulletin. Wooster, OH. Ohio Agric. Exp. Sta. Service Bull. See B–P–H 685/2. HI 57680

Ohio agricultural experiment station special report series. Wooster, OH. Ohio Agric. Exp. Sta. Special Rep. Ser. See B–P–H 685/3. HI 57681

Ohio agricultural experiment station technical bulletin. Wooster, OH. Ohio Agric. Exp. Sta. Techn. Bull. See B–P–H 685/4. HI 57682

Ohio agricultural research development center research summary. Wooster, OH. Ohio Agric. Res. Developm. Center Res. Summary. See B–P–H 685/6. HI 57683

Ohio biological survey bulletin. Columbus, OH. Ohio Biol. Surv. Bull. See B–P–H 685/11. HI 57684

Ohio botanical garden society. [Bulletin.] Cincinnati, OH. Vol. 1(1-2), 1927-29. Ohio Bot. Gard. Soc., Bull. 4-3156-3. HI 69160

Ohio conservation bulletin; devoted to the conservation and restoration of Ohio's wildlife. Columbus, OH. Vols. 1-29(5), 1936-65. Ohio Conservation Bull. Incorporated in: Wonderful world of Ohio. 4-3157-2. HI 69161

Ohio farmer. Cleveland, OH. Ohio Farmer. See B–P–H 685/12. HI 57685

Ohio farmer and western horticulturist. Batavia, OH. Ohio Farmer & W. Hort. See B–P–H 685/13. HI 57686

Ohio florists association, bulletin. Columbus, OH. Ohio Florists Assoc. Bull. See B–P–H 685/14. HI 57687

Ohio florists association, monthly bulletin. Columbus, OH. Ohio Florists Assoc. Monthly Bull. See B–P–H 685/15. HI 57688

Ohio forester. Wooster, OH. Ohio Forester. See B–P–H 685/16. HI 57689

Ohio journal of science. Columbus, OH. Ohio J. Sci. See B–P–H 685/18. HI 57690

Ohio mycological bulletin. Columbus, OH. Ohio Mycol. Bull. See B–P–H 685/21. HI 57691

Ohio naturalist. Columbus, OH. Ohio Naturalist. See B–P–H 685/23. HI 57692

Ohio outdoorsman = Ohio conservation bulletin. Columbus, OH.

Ohio pomological society report. Wooster, OH. Ohio Pomol. Soc. Rep. See B–P–H 685/24. HI 57693

Ohio practical farmer = Ohio farmer. Cleveland, OH. Ohio Farmer. See B–P–H 685/12.

Ohio report on research and development. Wooster, OH. Ohio Rep. Res. Developm. See B–P–H 686/1. HI 57694

Ohio report on research and development in biology, agriculture and home economics. Wooster, OH. Ohio Rep. Res. Developm. Biol. See B–P–H 686/2. HI 57695

Ohio state medical journal. Columbus, OH. Ohio State Med. J. See B–P–H 686/6. HI 57696

Ohio state university biosciences colloquia = Annual biosciences colloquia, Ohio state university. Columbus, OH.

Ohio woodlands; conservation in action. Columbus, OH. Vols. 1-12(1), 1963-74; vol. 14(3)+, 1976+. Ohio Woodlands. Preceded by: Woodlands. HI 69162

Ohoronjajte ridnu pryrodu! = Okhoronyaite ridnu pryrodu! Kiev.

Ohrana prirody = Okhrana prirody. Moscow.

Oikos. Copenhagen. Oikos. See B–P–H 686/9. HI 57697

Oil and fat industries. Chicago, IL. Vols. 4-8, 1927-31. Oil Fat Industr. Preceded by: Journal of oil and fat industries. Superseded by: Oil and soap. 1-294-2. HI 69163

Oil palm news. London. No. 1+, 1966+. Oil Palm News. HI 69164

Oil seeds news. (South Africa). 1976+. Oil Seeds News. HI 69165

Oil and soap. Chicago, IL. Vols. 9-23, 1932-46. Oil & Soap. Preceded by: Oil and fat industries. Superseded by: Journal of the american oil chemists' society. 1-294-2. HI 69166

Oilseeds - situation and outlook. Canberra, A.C.T. 1977+. Oilseeds Situation Outlook. HI 69167

Oilseeds journal. Hyderabad. Vol. 3+, 1973+. Oilseeds J. Preceded by: Telhan patrika. HI 69168

Oinoue rinogaku kenkyusho. Kenkyu hokoku = Bulletin de l'institut Oinoue de recherches agronomiques et biologiques. Simo-Omi, Japan. Bull. Inst. Oinoue Rech. Agron. See B–P–H 256/10.

Oita daigaku gakugeigakubu kenkyu kiyo. Shizen kagaku = Research bulletin of the faculty of liberal arts; Oita university <Natural sciences>. Oita, Japan. Res. Bull. Fac. Liberal Arts Oita Univ., Nat. Sci. See B–P–H 774/5.

Oita daigaku kyoikugakubu kenkyu kiyo, shizenkagaku = Research bulletin of the faculty of education, Oita university. Natural science. Oita.

Oita-ken nogyo gijutsu senta kenkyu hokoku = Bulletin of the Oita prefectural agricultural research center. Usashi.

Oita-ken ringyo shikenjo kenkyu hokoku = Bulletin of the Oita prefectural forest experiment station.

Oji seishi k. k. rinkobu ikushu kenkyujo, kenkyu hokoku = Bulletin of the Oji institute for forest tree development.

Okayama daigaku nogakubu gakujutsu hokoku = Science report of the faculty of agriculture, Okayama university. Okayama, Japan. Sci. Rep. Fac. Agric. Okayama Univ. See B–P–H 826/23.

Okayama daigaku rigakubu seibutsu gaku kiyo = Biological journal of Okayama university. Okayama.

Okoyama daigaku Tamano rinkai jikkensho gyoseki =

Contributions from the Tamano marine laboratory, Okoyama university. Tamano.

Okayama-ken ringyo hokoku = Bulletin of the Okayama prefecture forest experiment station. Okayama.

Okayama-kenritsu nogyo shikenjo gyomu kotel = Bulletin of Okayama prefectural agricultural experiment station. Okayama, Japan. Bull. Okayama Prefect. Agric. Exp. Sta. See B–P–H 267/24.

Okayama-kenritsu nogyo shikenjo jiho = Journal of Okayama prefectural agricultural experiment station. Okayama, Japan. J. Okayama Prefect. Agric. Exp. Sta. See B–P–H 475/8.

Okayama kenritsu nogyo shikenjo. Kenkyu hokoku. Sanyo. No. 1+, 1976+. Okayama Kenritsu Nogyo Shikenjo Kenkyu Hokoku. HI 69169

Okayama kenritsu noji shikenjo rinji hokoku = Special bulletin. Okayama prefectural agricultural experiment station. Okayama, Japan. Special Bull. Okayama Prefect. Agric. Exp. Sta. See B–P–H 850/17.

Okayama tabako shikenjo hokoku = Bulletin of the Okayama tobacco experiment station. Okayama, Japan. Bull. Okayama Tobacco Exp. Sta. See B–P–H 267/25.

Okayama tobako shikenjo gyotei hokoku = Report of the Okayama tobacco experiment station. Okayama, Japan. Rep. Okayam Tobacco Exp. Sta. See B–P–H 769/5.

Okeanologija = Okeanologiya. Moscow.

Okeanologiya. Moscow. Vol. 1+, 1961+. Okeanologiya. HI 69170

Okhorona, vyvuchennia ta zbagachennia roslinnogo svitu. Kiev. Vols. 1-4, 1974-77. Okhor. Vyvuch. Zbagach. Rosl. Svitu. Superseded by: Okhrana, izuchenie i obogashchenie rastitel'nogo mira. HI 69171

Okhoronyaite ridnu pryrodu! Kiev, Ukrainian S S R. Vol. 3+, 1964+. Okhoronyaite Ridnu Pryr. Preceded by: Materialy po okhoronu pryrody na Ukrayini. HI 69172

Okhrana, izuchenie i obogashchenie rastitel'nogo mira. Kiev. Vol. 5+, 1978+. Okhr. Izuch. Obogashch. Rastitel'n. Mira. Preceded by: Okhorona, vyvuchennia ta zbagachennia roslinnogo svitu. HI 69173

Okhrana prirody. Bratislava = Ochrana prirody. Bratislava.

Okhrana prirody. Moscow. 1928-30. Okhr. Prir. Superseded by: Priroda i sotsialisticheskoe khozyaistvo. HI 69174

Okhrana prirody na dal'nem vostoke. Vladivostok. Vol. 1+, 1963+. Okhr. Prir. Dal'nem Vostoke. HI 69175

Okhrana prirody Gruzii. Tiflis. Vol. 1+, 1965+. Okhr. Prir. Gruzii. HI 69176

Okhrana prirody moldavii. Kishinev. Vols. 1-13, 1960-75. Okhr. Prir. Moldav. HI 69177

Okhrana prirody Sibiri i dal'nego vostoka. Novosibirsk. Vol. 1, 1962. Okhr. Prir. Sibiri Dal'nego Vostoka. HI 69178

Okhrana prirody tsentral'no-chernozemnoi polosy. Vol. ?-5+, ?-1964+. Okhr. Prir. Tsentr.-Chernoz. Polosy. HI 69179

Okhrana prirody na Urale. Sverdlovsk. Vol. 1+, 1960+. Okhr. Prir. Urale. HI 69180

Okhrana prirody i vosproizvodstvo estestvennykh Resursov. Chita. Vol. 1+, 1967+. Okhr. Prir. Vosproizv. Estestv. Resursov. HI 69181

Okhrana prirody i zapovednoe delo v S S S R. Byulleten'. Moscow. Vols. 1-7, 1956-69. Okhr. Prir. Zapov. S.S.S.R., Byull. HI 69182

Okinawa-ken nogyo shikenjo. Tokubetsu kenkyu hokoku. Naha. No. 1+, 1979+. Okinawa-ken Nogyo Shikenjo, Tokubetsu Kenkyu Hokoku. HI 69183

Okinawa kenritsu noji shikenjo = Bulletin. Okinawa prefectural agricultural experiment station. Naha?, Okinawa. Bull. Okinawa Prefect. Agric. Exp. Sta. See B–P–H 267/27.

Okinawa seibutsu gekhai = Biological magazine. Naha, Okinawa. Biol. Mag. See B–P–H 194/7.

Oklahoma agricultural experiment station. Abstracts of new publications. Stillwater, OK. Oklahoma Agric. Exp. Sta. Abstr. New Publ. See B–P–H 686/35. HI 57699

Oklahoma agricultural experiment station annual report. Stillwater, OK. Oklahoma Agric. Exp. Sta. Annual Rep. See B–P–H 686/36. HI 57700

Oklahoma agricultural experiment station biennial progress report of turf grass testing. Stillwater, OK. Oklahoma Agric. Exp. Sta. Bienn. Progr. Rep. Turf Grass Test. See B–P–H 686/37. HI 57701

Oklahoma agricultural experiment station bulletin. Stillwater, OK. Oklahoma Agric. Exp. Sta. Bull. See B–P–H 687/2. HI 57702

Oklahoma agricultural experiment station circular of information. Stillwater, OK. Oklahoma Agric. Exp. Sta. Circ. Inform. See B–P–H 687/3. HI 57703

Oklahoma agricultural experiment station forage crops leaflet. Stillwater, OK. Oklahoma Agric. Exp. Sta. Forage Crops Leafl.. See B–P–H 687/4. HI 57704

Oklahoma agricultural experiment station mimeographed circular. Stillwater, OK. Nos. 1-296, 1938-58. Oklahoma Agric. Exp. Sta. Mimeogr. Circ. Superseded by: Processed series, Oklahoma agricultural experiment station. HI 69184

Oklahoma agricultural experiment station miscellaneous publication. Stillwater, OK. Oklahoma Agric. Exp. Sta. Misc. Publ. See B–P–H 687/6. HI 57706

Oklahoma agricultural experiment station press bulletin. Stillwater, OK. Oklahoma Agric. Exp. Sta. Press Bull. See B–P–H 687/7. HI 57707

Oklahoma agricultural experiment station technical bulletin. Stillwater, OK. Oklahoma Agric. Exp. Sta. Techn. Bull. See B–P–H 687/9. HI 57708

Oklahoma agricultural and mechanical college. Botanical studies. Stillwater, OK. Oklahoma Agric. Coll. Bot. Stud. See B–P–H 686/34. HI 57698

Oklahoma cactus and succulent society magazine. Vol. 1(1)-?, 1945-? Oklahoma Cact. Succ. Soc. Mag. HI 69185

Oklahoma geology notes. Norman, OK. Oklahoma Geol. Notes. See B–P–H 687/10. HI 57709

Oklahoma state medical association. Journal. Muskogee, OK. Oklahoma State Med. Assoc. J. See B–P–H 687/11. HI 57710

Old colony naturalist. Plymouth, MA. Old Colony Naturalist. See B–P–H 687/19. HI 57711

Old Westbury gardens news. Vol. 1+, 1977+. Old Westbury Gard. News. HI 69186

Oldenburger Jahrbuch. Teil 2 (Naturkunde und Vorgeschichte). Oldenburg in Oldenburg, Germany. Oldenburger Jahrb., Teil 2. See B–P–H 687/21. HI 57713

Oldenburger Jahrbuch des Oldenburger Landesvereins für Geschichte, Natur- und Heimatkunde. Teil 2 (Vorgeschichte und Naturkunde). Oldenburg in Oldenburg, Germany. Oldenburger Jahrb. Oldenburger Landesvereins Gesch., Teil 2. See B–P–H 687/20. HI 57712

Oléagineux. Paris. Oléagineux. See B–P–H 687/22. HI 57714

Oleo. Madrid. Oleo. See B–P–H 688/1. HI 57715

Oliedade nuus = Oil seeds news. (South Africa).

Olivae. Madrid. No. 1+, 1984+. Olivae. HI 69776

Olive journal = Fig and olive journal. Los Angeles, CA. Fig Olive J. See B–P–H 371/6.

Olivicultura. Rome. Olivicultura. See B–P–H 688/3. HI 57716

Omi mustard seed. Aomi, Japan. Omi Mustard Seed. See B–P–H 688/4. HI 57717

Onderzoekverslag, instituut voor plantenziektenkundig onderzoek te Wageningen. Wageningen. Onderzoekverslag Inst. Plantenziektenk. Onderz. Wageningen. See B–P–H 688/5. HI 57718

Onira; botanical leaflets. Santiago, Chile. Ca.1989+. Onira. HI 69187

Online gardener. Watkinsville, GA. Vol. 1+, 1985+. Online Gard. HI 69188

Onsen kagaku = Berichte der japanischen Gesellschaft für Balneologie. Ber. Jap. Ges. Balneol. See B–P–H 175/15.

Onsen kagaku = Journal of the balneological society of Japan. J. Balneol. Soc. Japan. See B–P–H 459/1.

Onshitsu kenkyu = Magazine for greenhouse culture. Mag. Greenh. Cult. See B–P–H 539/18.

Onson kenkyu = Studies on thermal springs. [Japan]. Stud. Thermal Springs. See B–P–H 857/16.

Ontario agricultural college bulletin. Guelph, Ont. Ontario Agric. Coll. Bull. See B–P–H 688/9. HI 57719

Ontario conservation news. Toronto. Vol. 3(4)+, 1975+. Ontario Conservation News. Preceded by: Conservation news, conservation council of Ontario. HI 69189

Ontario field biologist. Willowdale, Ont. Ontario Field Biol. See B–P–H 688/10. HI 57720

Ontario natural science bulletin; journal of the Wellington field naturalists' club. Guelph, Ont. Ontario Nat. Sci. Bull. See B–P–H 688/11. HI 57721

Ontario naturalist. Don Mills, Ont. Vols. 1-19(4), 1969-79. Ontario Naturalist. Preceded by: Bulletin, federation of Ontario naturalists. Superseded by: Seasons. HI 69190

Ontogenez. Moscow. Vol. 1+, 1970+. Ontogenez. HI 69191

Öntözésügyi közlemények. Budapest. Öntözésügyi Közlem. See B–P–H 688/12. HI 57722

Onze stekelige vrienden. Bonheiden. 1987+. Onze Stekel. Vrienden. HI 69192

Onze tuinen. Geillustreerd weekblad. Haarlem. Onze Tuinen. See B–P–H 688/13. HI 57723

Open gates. Fontana, CA. 1966+. Open Gates. HI 69193

Opera academiae scientiarum et artium Slavorum meridionalium = Djela jugoslavenske akademije znanosti i umjetnosti. Zagreb, Croatia [Yugoslavia]. Djela Jugoslav. Akad. Znan. See B–P–H 349/4.

Opera botanica belgica. Meise. Vol. 1+, 1988+. Opera Bot. Belg. HI 69194

Opera botanica čechica. Prague. Vols. 1-6, 1939-48. Opera Bot. Čech. HI 69195

Opera botanica a societate botanice lundensi. Lund, Copenhagen. Vols. 1-59, 1953-80 [vols. 52-59, Oct. 1979 - Dec. 1980 issued as a supplement to Botaniska notiser]. Opera Bot. HI 69196

Opera botanica. Series B. Flora of Ecuador. Gothenberg. Vol. 1+, 1973+. Opera Bot., B. HI 69197

Opera corcontica. Vrchlabi, Czechoslovakia. Opera

Corcontica. See B–P–H 688/19. HI 57724

Opera. Institutum biologicum. Academia scientiarum et artium slovenica = Dela institut za biologijo. Slovenska akad. znanosti in umietnosti. Ljubljana, Yugoslavia. Dela Inst. Biol. Slov. Akad. Znan. See B–P–H 340/8.

Opera lilloana. Tucumán, Argentina. Opera Lilloana. See B–P–H 688/21. HI 57725

Operation report of Saga fruit tree experiment station. [Saga-ken kaju shikenjo gyomu nempo.] Saga. 1970+. Operat. Rep. Saga Fruit Tree Exp. Sta. HI 74465

Operculum; publication of the australian littoral society. Toowong, Qld. Vol. 1+, 19??+. Operculum. Preceded by: Newsletter, Queensland littoral society. HI 69198

Opetatud eesti seltsi toimetused. Tartu, Estonia [Estonian S S R]. Opet. Eesti Seltsi Toimet. See B–P–H 688/22. HI 57726

Ophelia. Helsingor, Denmark. Ophelia. See B–P–H 688/23. HI 57727

Ophthalmologische Klinik. Stuttgart. Ophthalmol. Klin. See B–P–H 688/26. HI 57728

Opora. Eine Zeitschrift zur Beförderung des Obstbaues in Deutschland. Zittau, Germany. Opora. See B–P–H 688/27. HI 57729

Optima. Johannesburg. Optima. See B–P–H 688/28. HI 57730

Options méditerranéennes; revue des problèmes agricole méditerranéennes. Paris. N.s. vol. 1+, 1970+. Options Médit. Preceded by: Mediterranea. HI 69199

Opuntia pad. Leicester. 1972+. Opuntia Pad. HI 69200

Opuscula botanica pharmaciae complutensis. Madrid. Vol. 1+, 1977+; 2nd ed., 1987+. Opusc. Bot. Pharm. Complut. HI 69201

Opuscula sparsa. Barcelona. Vol. 1+, 1970+. Opusc. Sparsa. HI 69202

Opuscules, institut des fruits et agrumes coloniaux. Paris. No. ?-2-11-?, ?-1952-? Opusc. Inst. Fruits Agrumes Colon. HI 69203

Opuscoli scientifici. Bologna. Opusc. Sci. See B–P–H 689/2. HI 57732

Opusculi scelti sulle scienze e sulle arti. Milan. Opusc. Scelti Sci. Arti. See B–P–H 689/1. HI 57731

Opyt kul′tury novykh syr′evykh rastenii. Tashkent. Vol. 1+, 1973+. Opyt Kul′t. Nov. Syr′ev. Rast. HI 69204

Orange Judd american agriculturist = American agriculturist. Ithaca, NY.

Orchadian. Sydney. Orchadian. See B–P–H 689/4. HI 57733

Orchard and garden. Little Silver, NJ. Orchard & Gard. See B–P–H 689/5. HI 57734

Orchardist of New Zealand. Wellington. Orchardist New Zealand. See B–P–H 689/6. HI 57735

Orchardist's companion. Philadelphia, PA. Orchardist's Companion. See B–P–H 689/7. HI 57736

Orchid advocate; official journal, cymbidium society of America. Santa Barbara, Arcadia, Carpentaria, CA. Vol. 1+, 1975+. Orchid Advocate. Preceded by: Cymbidium society news. HI 69205

Orchid album. London. Orchid Album. See B–P–H 689/8. HI 57737

Orchid biology; reviews and perspectives. Ithaca, NY, London. Vol. 1+, 1977+. Orchid Biol. HI 69206

Orchid digest. Berkeley, CA. Orchid Digest. See B–P–H 689/10. HI 57738

Orchid journal. Pasadena, Berkeley, CA, Coconut Grove, FL. Vols. 1-3(7), 1952-59. Orchid J. HI 69207

Orchid lore. Houston, TX. Vols. 1-4(3), 1947-50. Orchid Lore. HI 69208

Orchid monographs. Leiden. Vol. 1+, 1986+. Orchid Monogr. HI 69209

Orchid news; official publication of the orchid society of New South Wales. Vol. 1+, 1977+. Orchid News (N.S.W). HI 69210

Orchid news. Chandigarh. Vol. 1+, 1985+. Orchid News (India). HI 69211

Orchid research newsletter. Kew. No. 1+, 1983+. Orchid Res. Newslett. HI 69212

Orchid review. An illustrated monthly journal devoted to orchidology in all its branches. London. Orchid Rev. See B–P–H 689/14. HI 57739

Orchid weekly. Coconut Grove, FL. Orchid Weekly. See B–P–H 689/15. HI 57740

Orchid world; a monthly illustrated journal entirely devoted to orchidology. Haywards Heath, England. Orchid World. See B–P–H 689/16. HI 57741

Orchidata. New York, NY. Orchidata. See B–P–H 689/18. HI 57742

Orchidea. Budapest. Vol. 1+, 1981+. Orchidea (Budapest). HI 69213

Orchidea. Genoa. Vol. ?-25+, ?-1985+. Orchidea (Genoa). HI 69214

Orchidea. Niterói, Rio de Janeiro. Vols. 1-3(3), 1938-40. Orchidea (Niteroi). Superseded by: Orquídea. HI 69215

Orchidee. Bandoeng, Dutch E. Indies [=Bandung, Indonesia]. Orchidee (Bandoeng). See B–P–H 689/19. HI 57743

Orchidée; revue trimestrielle. Eaubonne. No. 1+, 1983+. Orchidée (France). HI 69216

Orchidee. Hamburg-Othmarschen & Hamburg. Orchidee (Hamburg). See B–P–H 689/20. HI 57744

Orchideën. Dordrecht, Netherlands. Orchideën. See B–P–H 689/21. HI 57745

Orchideen; Arbeitsmaterial für Fachgruppen und Interessengemeinschaften. Leipzig. Vols. 1-20, 1969-79. Orchideen. Superseded by: Orchideer. HI 69217

Orchideen beim Institut für Landesforschung und Naturschutz der deutschen Akademie der Landwirtschaftswissenschaften zu Berlin = Mitteilungen des Arbeitskreises zur Beobachtung und zum Schutz heimischer Orchideen. Halle.

Orchideer. [Vedbaek.] Vol. 1+, 1980+. Orchideer. Preceded by: Orchideen. HI 69218

Orchidologia zeylancia. Colombo, Ceylon. Orchidol. Zeylancia. See B–P–H 689/22. HI 57746

Orchidophile. Argenteuil, France. Vols. 1-15, 1881-95. Orchidophile (Argenteuil). 4-3190-3. HI 57747

Orchidophile; bulletin trimestriel, société française d'orchidophile. Asnières. No. 1+, 1970+. Orchidophile (Asnières). HI 69219

Orchidophile. Deuil-la-Barre. Vol. 1+, [1970]+. Orchidophile (Deuil-la-Barre). HI 69220

Orchids in New Zealand. New Plymouth. Vol. 1+, 1975+. Orchids New Zealand. HI 69221

Orchis. Monatsschrift der Deutschen Gesellschaft für Orchideenkunde. Berlin. Orchis. See B–P–H 689/24. HI 57748

Oréades. Belo Horizonte. Vol. 1+, 1970+. Oréades. HI 69222

Oregon agricultural experiment station biennial crop pest and horticulture report. Corvallis, OR. Oregon Agric. Exp. Sta. Bienn. Crop Pest Hort. Rep. See B–P–H 690/10. HI 57750

Oregon agricultural experiment station bulletin. Corvallis, OR. Oregon Agric. Exp. Sta. Bull. See B–P–H 690/11. HI 57751

Oregon agricultural experiment station circular. Corvallis, OR. Oregon Agric. Exp. Sta. Circ. See B–P–H 690/12. HI 57752

Oregon agricultural experiment station circular information. Corvallis, OR. Oregon Agric. Exp. Sta. Circ. Inform. See B–P–H 690/14. HI 57753

Oregon agricultural experiment station, director's biennial report. Corvallis, OR. 1890-1951. Oregon Agric. Exp. Sta., Director's Bienn. Rep. Superseded by: Report, Oregon agricultural experiment station. HI 69223

Oregon agricultural experiment station. Station bulletin. Corvallis, OR. Oregon Agric. Exp. Sta., Sta. Bull. See B–P–H 690/16. HI 57754

Oregon agricultural experiment station technical bulletin. Corvallis, OR. Oregon Agric. Exp. Sta. Techn. Bull. See B–P–H 690/18. HI 57755

Oregon agricultural experiment station. Station bulletin. Corvallis, OR. Oregon Agric. Exp. Sta., Sta. Bull. See B–P–H 690/16. HI 57754

Oregon agricultural experiment station technical bulletin. Corvallis, OR. Oregon Agric. Exp. Sta. Techn. Bull. See B–P–H 690/18. HI 57755

Oregon agriculture. Oregon agricultural college agricultural experiment station. Corvallis, OR. Oregon Agric. See B–P–H 690/9. HI 57749

Oregon orchid society bulletin. Portland, OR. Oregon Orchid Soc. Bull. See B–P–H 690/19. HI 57756

Oregon ornamental and nursery digest. Oregon agricultural college agricultural experiment station. Corvallie, OR. Oregon Ornam. Nursery Digest. See B–P–H 690/20. HI 57757

Oregon state monographs. Bacteriology. Corvallis, OR. Oregon State Monogr., Bacteriol. See B–P–H 690/21. HI 57758

Oregon state monographs. Studies in botany. Corvallis, OR. Oregon State Monogr., Stud. Bot. See B–P–H 690/22. HI 57759

Oregon state university, school of science, department of oceanography, collected reprints. Corvallis, OR. Oregon State Univ. School Sci. Dept. Oceanogr. Collect. Repr. See B–P–H 690/25. HI 57760

Oregon's agricultural progress. Oregon agricultural college agricultural experiment station. Corvallis, OR. Oregon's Agric. Progr. See B–P–H 690/26. HI 57761

Organic farmer. Emmaus, PA. Vols. 1-5, 1949-53. Org. Farmer. Superseded by: Organic gardening and farming. HI 69224

Organic gardening. Emmaus, PA. Vols. 1-21, 1942-53; vol. 25(7)+, 1978+. Org. Gard. From 1954-78 replaced by: Organic gardening and farming. HI 69225

Organic gardening and farming. Emmaus, PA. Vols. 1-25(6), 1954-78. Org. Gard. Farming. Preceded by: Organic gardening and Organic farmer. Superseded by: Organic gardening. HI 69226

Oriental economist; a monthly journal of practical finance and economy for Japan and Eastern Asia. Tokyo. Orient. Econ. See B–P–H 691/3. HI 57762

Oriental repertory. London. Orient. Repert. See B–P–H 691/5. HI 57763

Origins. Loma Linda, CA. 1974+. Origins. HI 69227

Origins of life. Dordrecht. Vols. 5-?, 1974-? Origins Life. Preceded by: Space life sciences. Superseded by: Origins of life and evolution of the biosphere.

HI 69228

Origins of life and evolution of the biosphere. Dordrecht & Hingham, MA. Vol. ?-19+, ?-1989+. Origins Life Evol. Biosphere. Preceded by: Origins of life. HI 69229

Orion. Murnay, Germany. Orion (Murnau). See B–P–H 691/6. HI 57764

Orion nature quarterly. New York. Vol. 1+, 1982+. Orion Nat. Quart. HI 69230

Ormancilik arastirma enstitusu dergisi. Ankara. 1954?+. Orman. Arastirma Enst. Derg. HI 69231

Ornamental horticulture. Farnham Royal. 1975+. Ornam. Hort. HI 69232

Ornamental horticulture abstracts. New York. [Dates of publication not ascertained.] Ornam. Hort. Abstr. HI 69233

Ornamentals newsletter. Ithaca, NY. Vols. [1]-17, 1957-73. Ornam. Newslett. Superseded by: Environmental enhancement. HI 69234

Ornamentals northwest. Corvallis, OR. Vol. 1+, 1975+. Ornam. NorthW. HI 69235

Orquidea; organo oficial de la sociedad mexicana "Amigos de las orquídeas". Mexico City. Vols. 1-?, 1943-?; n.s. vol. 1+, 1971+. Orquidea (Mexico City). HI 69236

Orquídea. Rio de Janeiro. Vol. 3(4)+, 1948+. Orquídea (Rio de Janeiro). Preceded by: Orchidea. HI 69237

Orquideologia; revista de la sociedad colombiana de orquideologia. Medellin. Vol. 3+, 1968+. Orquideologia. Preceded by: Revista de la sociedad columbiana de orquideologia. HI 69238

Orquidiota. Santo Domingo. Vol. 1(1), 1972. Orquidiota. Superseded by: Boletín de la sociedad dominicana de orquideología. HI 69239

Országos természettundományi múzeum évkönyve = Annales historico-naturales musei nationalis hungarici. Budapest. Ann. Hist.-Nat. Mus. Natl. Hung. See B–P–H 103/22.

Orsis; organismes i sistemes: Revista de botanica, zoologia i ecologia de la universitat autònoma de Barcelona. Barcelona. Vol. 1+, 1985+. Orsis. HI 69240

Országos erdészeti-egyesület évkönyve. Pest [=Budapest, in part]. Orsz. Erdész.-Egyes. Évk. See B–P–H 691/13. HI 57765

Országos erdészeti-egyesület közleményei. Pest [=Budapest, in part]. Orsz. Erdész.-Egyes. Közlem. See B–P–H 691/14. HI 57766

Országos fajtakiserletek. Vols 1-22, 1968-71. Orsz. Fajtakiserl. Superseded by: Fajtakiserletezes-fajtaminosites. HI 69241

Országos középiskolai tanáregyesület közlönye. Budapest. Orsz. Középisk. Tanáregyes. Közl. See B–P–H 691/15. HI 57767

Orvos-természettudományi értesitö. Kolozsvart. 1879-89. Orv.-Term. Értes. Preceded by: Ertesito a kolozsvari orvos-termeeszettudomanyi tarsulat. Superseded by: Ertesitö az erdélyi muzeum-egyesulet orvos-termeszettudomanyi szakosztalyabol. HI 69242

Orvosi hetilap. Budapest. Orv. Hetilap. See B–P–H 691/18. HI 57768

Orvosi tár. Pest [=Budapest, in part]. Orv. Tár. See B–P–H 691/19. HI 57769

Oryx. Hertford, London. 1950+. Oryx. Preceded by: Journal of the society for the preservation of the wild fauna of the empire [not entered]. HI 69243

Oryza; journal of the association of rice research workers. Cuttack. Vol. 1+, 1962/63+. Oryza. HI 69244

Osaka daigaku nanko, hokko rika hokoku = Science reports of South college, North college of Osaka university. Osaka, Japan. Sci. Rep. S. Coll. N. Coll. Osaka Univ. See B–P–H 828/7.

Osaka gakugei daigaku kiyo. B. Shizen kahaku = Memoirs of Osaka university of the liberal arts and education. Series B, natural sciences. Osaka.

Osaka plant protection. [Osaka shokubutsu boeki.] Osaka, Japan. Osaka Pl. Protect. See B–P–H 691/24. HI 57770

Osaka shiritsu daigaku. Rikogakubu = Journal of the institute of polytechnics, Osaka City university. Series D. Biology. Osaka, Japan. J. Inst. Polytechn. Osaka City Univ., Ser. D, Biol. See B–P–H 470/13.

Osaka-shiritsi shizenkagaku hakubutsukan = Report (Annual), Osaka municipal museum of natural history. Osaka.

Osaka-shiritsu shizenkagaku hakubutsukan kenkyu hokoku = Bulletin of the Osaka museum of natural history. Osaka.

Osaka shizen kagaku kenkyukai = Nature study. Osaka.

Osaka shokubutsu boeki = Osaka plant protection. Osaka, Japan. Osaka Pl. Protect. See B–P–H 691/24.

Osaka-fu norin gijutsu senta kenkyuh okoku = Bulletin of the Osaka agricultural center. Kibikino.

Osaka-fu norin gijutsu senta nyusu. Kibikino. No. 59+, 1973+. Osaka-fu Norin Gijutsu Senta Nyusu. Preceded by: Norin gijutsu senta nyusu. HI 69245

Oseanologi di Indonesia. Jakarta. No. 1+, 1974+. Oseanol. Indonesia. HI 69246

Osiris; studies on the history and philosophy of science, and on the history of learning and culture [new series subtitled: A research journal devoted to the history of science and its cultural influences.] Bruges,

Philadelphia, PA. Vols. 1-15, 1936-68; n.s. vol. 1+, 1985+. Osiris.

Osnabrücker naturwissenschaftliche Mitteilungen; Veröffentlichungen des naturwissenschaftlichen Vereins Osnabrück. Osnabrück. No. 1+, 1972+. Osnabrück. Naturwiss. Mitt. Preceded by: Veröffentlichungen des naturwissenschaftlicher Verein Osnabrück. HI 69247

Ospedale maggiore. Milan. Osped. Maggiore. See B–P–H 691/27. HI 57771

Osprey. St. John's. Vol. 1+, 1970+. Osprey. HI 69248

Ost-Friesische Mannigfaltigkeiten. Aurich, Germany. Ost-Fries. Mannigfaltigk. See B–P–H 693/6. HI 57788

Ost und West. Blätter für Kunst, Literatur und geselliges Leben. Prague. Ost & West. See B–P–H 691/28. HI 57772

Ostasiatische Rundschau. [Tung fang yu lun.] Berlin. Ostasiat. Rundschau. See B–P–H 692/6. HI 57773

Ostdeutscher Naturwart. Liegnitz [Poland]. Ostdeutsch. Naturwart. See B–P–H 692/7. HI 57774

Osterländische Blätter für Landes-, Natur- und Gewerbkunde, herausgegeben von den Secretären der naturforschenden Gesellschaft in Altenburg. Altenburg. Osterl. Blätt. Landes- Natur- Gewerbk. See B–P–H 692/8. HI 57775

Österreichisch-ungarischer Obstgarten. Vienna. Österr.-Ung. Obstgarten. See B–P–H 693/2. HI 57786

Österreichische Akademie der Wissenschaften. Handbuch der Österreichischen Wissenschaft. Vienna. Österr. Akad. Wiss. Handb. Öesterr. Wiss. See B–P–H 692/9. HI 57776

Österreichische Akademie der Wissenschaften. Jahrbuch der Österreichischen Wissenschaft. Vienna. Österr. Akad. Wiss. Jahrb. Österr. Wiss. See B–P–H 692/10. HI 57777

Österreichische Akademie der Wissenschaften. Mathematisch-naturwissenschaftliche Klasse. Anzeiger = Anzeiger, Österreichische Akademie der Wissenschaften. Mathematisch- naturwissenschaftliche Klasse.

Österreichische Akademie der Wissenschaften. Mathematisch-naturwissenschaftliche Klasse. Denkschriften. Vienna. Österr. Akad. Wiss., Math.-Naturwiss. Kl., Denkschr. See B–P–H 692/12. HI 57778

Österreichische Akademie der Wissenschaften. Mathematisch-naturwissenschaftliche Klasse. Sitzungsberichte. Abteilung 1. Biologie, Mineralogie, Erdkunde und verwandte Wissenschaften. Vienna. Österr. Akad. Wiss., Math.-Naturwiss. Kl., Sitzungsber,. Abt. 1, Biol. See B–P–H 692/13. HI 57779

Oesterreichische botanische Zeitschrift. Gemeinütziges Organ für Botanik. Vienna. Vols. 8-91, 1858-1942; vols. 94-122, 1947-73. Oesterr. Bot. Z. Preceded by: Oesterreichisches botanisches Wochenblatt. For vols. 92-93, 1943-44 see Wiener botanische Zeitschrift. Superseded by: Plant systematics and evolution. HI 69249

Österreichische Forst- und Jagd-Zeitung. Vienna. Österr. Forst- Jagd-Zeitung. See B–P–H 692/18. HI 57780

Oesterreichische Forstzeitung. Vienna. Oesterr. Forstzeitung. See B–P–H 683/18. HI 57663

Österreichische Garten-Zeitung. Organ der K. K. Gartenbau-Gesellschaft in Wien. Vienna. Österr. Gart.-Zeitung. See B–P–H 692/20. HI 57782

Österreichische gelehrte Anzeigen. Linz. Österr. Gel. Anz. (Linz). See B–P–H 692/21. HI 57783

Österreichische gelehrte Anzeigen. Vienna. Österr. Gel. Anz. (Vienna). See B–P–H 692/22. HI 57784

Oesterreichische Monatsschrift für Forstwesen. Vienna. Oesterr. Monatsschr. Forstwesen. See B–P–H 683/21. HI 57664

Oesterreichische Moorzeitschrift. Staab, Bohemia [=Stod, Czechoslovakia]. Oesterr. Moorz. See B–P–H 683/22. HI 57665

Österreichische Pflanzenschutzgesellschaft. Taschenbuch. Vienna. Österr. Pflanzenschutzges. Taschenb. See B–P–H 693/1. HI 57785

Oesterreichische Vierteljahresschrift für Forstwesen. Vienna. Oesterr. Vierteljahresschr. Forstwesen. See B–P–H 683/25. HI 57667

Oesterreichische Vierteljahresschrift für wissenschaftliche Veterinärkunde. Vienna. Oesterr. Vierteljahresschr. Wiss. Veterinärk. See B–P–H 683/24. HI 57666

Oesterreichische Zeitschrift für Geschichts- und Staatskunde. Vienna. Oesterr. Z. Geschichts- Staatsk. See B–P–H 683/26. HI 57668

Oesterreichische Zeitschrift für Pharmacie. Vienna. Vols. 1-16, 1847-62. Oesterr. Z. Pharm. Superseded by: Zeitschrift des allgemeinen oesterreichischen Apothekvereins. HI 69250

Österreichische Zeitschrift für Pilzkunde. Vienna. Österr. Z. Pilzk. See B–P–H 693/4. HI 57787

Österreichischer Gartenkalender. Vienna. Österr. Gartenkalend. See B–P–H 692/19. HI 57781

Oesterreichisches Archiv für Geschichte, Erdbeschreibung, Staatenkunde, Kunst und Literatur. Vienna. Oesterr. Arch. Gesch. See B–P–H 683/10. HI 57661

Oesterreichisches Botanisches Wochenblatt. Gemeinnütziges Organ für Botanik ... Vienna. Oesterr.

Bot. Wochenbl. See B–P–H 683/12. HI 57662

Osteuropa. Naturwissenschaft. Stuttgart. Vols. 1-12, 1957-68. Osteuropa, Naturwiss. Superseded by: Osteuropa. Naturwissenschaft und Technik. HI 69251

Osteuropa. Naturwissenschaft und Technik. Stuttgart. Vols. 13-15, 1969-71. Osteuropa Naturwiss. Techn. Preceded by: Osteuropa. Naturwissenschaft. HI 69252

Ostmärkischer Pflanzenschutzkalender. Vienna. Ostmärk. Pflanzenschutzkalender. See B–P–H 693/7. HI 57789

Otčet o dejatel'nosti Akademii nauk S S S R = Otchet o deyatel'nosti Akademii nauk S S S R. Moscow & Leningrad.

Otčet o Dějatel'nosti Tiflisskago Botaničeskago Sada = Otchet o Dĕyatel'nosti Tiflisskago Botanicheskago Sada. Tiflis.

Otčet o Dějatel'nosti Volžskoj Biologičeskoj Stancii = Otchet o Dĕyatel'nosti Volzhskoi Biologicheskoi Stantsii. Saratov.

Otčet o Prisuždenii Akademieju Nauk = Otchet o Prisuzhdenii Akademieyu Nauk. St. Petersburg.

Otčet o Prisuždenii Nagrad Grafa Uvarova = Otchet o Prisuzhdenii Akademieyu Nauk. St. Petersburg.

Otčet o rabotah Počvenno-botaničeskogo otrjada Kazakstanskoj èkspedicii Akademii nauk S S S R = Otchet o rabotakh Pochvenno-botanicheskogo otryada Kazakstanskoi èkspeditsii Akademii nauk S S S R. Leningrad.

Otčet o sostojanii i dejatel'nosti Glavnogo botaničeskogo sada R S F S R = Otchet o sostoyanii i deyatel'nosti Glavnogo botanicheskogo sada R S F S R. Petrograd.

Otčet Volžskoj Biologičeskoj Stancii Saratovskogo Obščestva Estestvoispytatelej i Ljubitelej Estestvoznanija = Otchet Volzhskoi Biologicheskoi Stantsii Saratovskogo Obshchestva Estestvoispytatelei i Lyubitelei Estestvoznaniya. Saratov.

Otčet o Zagraničnoj Komandirovke. Botaničeskij sad = Otchet o Zagranichnoi Komandirovke. Botanicheskii sad. Tiflis.

Otčety o Dějatel'nosti Saratovskago Obščestva Estestvoispytatelej = Otchety o Dĕyatel'nosti Saratovskago Obshchestva Estestvoispytatelei. Saratov.

Otchet o deyatel'nosti Akademii nauk S S S R. Moscow & Leningrad. 1923-34. Otchet Deyatel'n. Akad. Nauk S.S.S.R. 1-113-2. HI 69253

Otchet o deyatel'nosti astrakhanskoi stantsii zashchity rastenii ot vreditelei. Astrakhan. Vol. ?-14-?, ?-1923-24-? Otchet Deyatel'n. Astrakh. Stantsii Zashch. Rast. Vredit. HI 69254

Otchet o deyatel'nosti gerbariya tomskogo gosudarstvennogo universiteta. Tomsk. 1930-31. Otchet Deyatel'n. Gerb. Tomsk. Gosud. Univ. HI 69255

Otchet o Dĕyatel'nosti Tiflisskago Botanicheskago Sada. [Supplement to: Trudy Tiflisskogo Botanicheskogo Sada.] Tiflis, Georgian S S R. 1913. Otchet Dĕyatel'n. Tiflissk. Bot. Sada. HI 69256

Otchet o Dĕyatel'nosti Volzhskoi Biologicheskoi Stantsii. Saratov, Russian S F S R. 1903-19. Otchet Dĕyatel'n. Volzhsk. Biol. Stantsii. Preceded by: Otchet Volzhskoi Biologicheskoi Stantsii Saratovskogo Obshchestva Estestvoispytatelei i Lyubitelei Estestvoznaniya. HI 69257

Otchet o Prisuzhdenii Akademieyu Nauk. St. Petersburg. Vols. 1-56, 1857-1918. Otchet Prisuzhd. Akad. Nauk. Preceded by: Compte-rendu de l'académie impériale des sciences (de Saint Pétersbourg). 1-113-2. HI 69258

Otchet o rabotakh Pochvenno-botanicheskogo otryada Kazakhstanskoi èkspeditsii Akademii nauk S S S R. Leningrad. Vols. 1-7, 1928-29. Otchet Rabotakh Pochv.-Bot. Otryada Kazakhstansk. Eksped. Akad. Nauk S.S.S.R. HI 69259

Otchet o sostoyanii i deyatel'nosti Glavnogo botanicheskogo sada R S F S R. Petrograd. 1922-23. Otchet Sost. Deyatel'n. Glavn. Bot. Sada R.S.F.S.R. HI 69260

Otchet Volzhskoi Biologicheskoi Stantsii Saratovskogo Obshchestva Estestvoispytatelei i Lyubitelei Estestvoznaniya. Saratov, Russian S F S R. 1902. Otchet Volzhsk. Biol. Stantsii Saratovsk. Obshch. Estestvoisp. Superseded by: Otchet o Dĕyatel'nosti Volzhskoi Biologicheskoi Stantsii. HI 69261

Otchet o Zagranichnoi Komandirovke. Botanicheskii sad. Tiflis, Georgian S S R. 1912. Otchet Zagranichn. Komandir. HI 69262

Otchety o Dĕyatel'nosti Saratovskago Obshchestva Estestvoispytatelei. Saratov, Russian S F S R. 1901-14. Otchety Dĕyatel'n. Saratovsk. Obshch. Estestvoisp. HI 69263

Ottar. Tromso. No. 1+, 1954+. Ottar. HI 69264

Ottawa field-naturalists' club transactions. Ottawa. Nos. 1-7, 1879/80-87. Ottawa Field-Naturalists' Club Trans. Superseded by: Ottawa naturalist. HI 69265

Ottawa naturalist; transactions of the Ottawa field-naturalists' club. Ottawa. Vols. 1-32, 1887-1919. Ottawa Naturalist. Preceded by: Ottawa field-naturalists' club transactions. Superseded by: Canadian field naturalist. HI 69266

Our homes and gardens. London. Our Homes Gard. See B–P–H 694/4. HI 57790

Our public lands. Washington, DC. Vols. 1-31(3), 1951-

81. Our Public Lands. Superseded by: Your public lands. HI 75104

Outlook on agriculture. London. Outlook Agric. See B–P–H 694/5. HI 57791

Overseas newsletter from "the exotic collection." Worthing. 1967?+. Overseas Newslett. Exot. Collect. HI 69267

Overseas research publications. London. No. 1+, 1963+. Overseas Res. Publ. Preceded by: Colonial research publications. HI 69268

Oversigt over det kongelige danske videnskabernes selskabs forhandlinger og dets medlemmers arbeider. Copenhagen. Overs. Kongel. Danske Vidensk. Selsk. Forh. Medlemmers Arbeider. See B–P–H 694/6. HI 57792

Oversigt over videnskabs-selskabets møder. Christiania [=Oslo, Norway]. Overs. Vidensk.-Selsk. Møder. See B–P–H 694/7. HI 57793

Ovocnarsti a zelinarstvi. Prague. Vols. 1-14, 1953-66. Ovocnar. Zelin. Preceded by: Ovocnické rozhledy. Superseded by: Zahradnické listy. HI 69270

Ovocnické rozhledy. Prague. Vols. 1-43, 1910-52. Ovocn. Rozhl. Preceded by: Ceské listy zahradnické. Superseded by: Ovocnarsti a zelinarstvi. 4-322-3. HI 69271

Ovoščevodstvo = Ovoshchevodstvo. Moscow.

Ovoshchevodstvo. Moscow. 1939-40. Ovoshchevodstvo. Preceded by: Plodoovoshchnoe khozyaistvo. Superseded by: Sady i ogorody. 4-3222-3. HI 69272

Ovoshchnyie i bakhchevye kul'tury. Vol. 1+, 1970+. Ovoshchn. Bakhch. Kul't. HI 69273

Oxford botanical memoirs = Botanical memoirs. London. Bot. Mem. See B–P–H 223/8.

Oxford forestry memoirs. Oxford, England. Oxford Forest. Mem. See B–P–H 694/16. HI 57795

Oxford surveys in evolutionary biology. Oxford. Vol. 1+, 1984+. Oxford Surv. Evol. Biol. HI 69274

Oxford surveys of plant molecular and cell biology. Oxford. Vol. 1+, 1984+. Oxford Surv. Pl. Molec. Cell Biol. HI 69275

Oyo kinkagu = Journal of applied mycology; Hokkaido university. Sapporo, Japan. J. Appl. Mycol. Hokkaido Univ. See B–P–H 458/5.

P + A. Paris. Vol. 1+, 1984+. P. & A. HI 69276

P A A B S revista. Orlando, FL. Vol. 1+, 1972+. P. A. A. B. S. Rev. HI 69277

P A N E S A newsletter. Addis Ababa. No. 1+, 1985+. P. A. N. E. S. A. Newslett. HI 69278

P A N S. London. Vols. 20-25, 1974-79. P. A. N. S. Preceded by: Pest articles and news summaries. Section B, plant disease control and Section C, weed control. Superseded by: Tropical pest management. HI 69279

P A S C A L explore. E57, biologie marine. Paris. 1985+. P. A. S. C. A. L. Expl. E57, Biol. Mar. HI 69380

P A S C A L explore. E58, génétique. Paris. 1986+. P. A. S. C. A. L. Expl. E58, Génét. Preceded by: P A S C A L explore. E60, génétique. HI 69381

P A S C A L explore. E60, génétique. Paris. 1984-85. P. A. S. C. A. L. Expl. E60, Génét. Preceded by: Bulletin signalétique. 363. Génétique. Superseded by: P A S C A L explore. E58, génétique. HI 69382

P A S C A L explore. E61, microbiologie, [etc.]. Paris. Nos. 1-10-?, 1984-? P. A. S. C. A. L. Expl. E61, Microbiol. Preceded by: Bulletin signalétique. 340. Microbiologie, virologie, immunologie. HI 74540

P A S C A L folio. E61, microbiologie: bactériologie, virologie, mycologie, protozoaires pathogènes. Paris. 1985?+. P. A. S. C. A. L. Folio E61, Microbiol. HI 69383

P A S C A L folio. F52, biochimie; biophysique moléculaire; biologie moléculaire et cellulaire. Paris. 1984+. P. A. S. C. A. L. Folio F52 Biochim. Preceded by: Bulletin signalétique. 320. Biochimie, biophysique. HI 69384

P A S C A L folio. F55, biologie végétale. Paris. Nos. 4-10-?, 1984-? P. A. S. C. A. L. Folio F55, Biol. Vég. Preceded, in part, by: Bulletin signalétique. 370. Biologie et physiologie végétales (and later) Sylviculture. HI 69385

P A S C A L folio. F56, écologie animale et végétale (later écologie animale, végétale et microbienne, éthologie animale. Paris. Nos. 1-3-?, 1984-? P. A. S. C. A. L. Folio E56, Anim. Vég. Preceded by: Bulletin signalétique. 365. Zoologie des vertébrés, ecologie animale, physiologie appliquée humaine [not entered]. HI 69386

P A S C A L sigma. S2, sciences de la vie 1: biologie fondamentale et appliquée. Paris. 1985?+. P. A. S. C. A. L. Sigma, S2, Sci. Vie. HI 69387

P A S C A L thema. T210, industries agroalimentaires. Paris. 1985?+. P. A. S. C. A. L. Thema T210, Industr Agroaliment. HI 69388

P A S C A L thema. T215, biotechnologies. Paris. 1985?+. P. A. S. C. A. L. Thema, T215, Biotechnol. HI 69389

P A S C A L thema. T280, sciences agronomiques: production végétales. Paris. 1985?+. P. A. S. C. A. L. Thema, T280, Sci. Agron., Prod. Vég. HI 69390

P C E A. Boletín trimestral de experimentación

agropecuaria. Lima. Vol. 1+, 1952+. P. C. E. A. HI 69280

P F; boletín informativo de patrimonio forestal del estado. Madrid. No. 1+, 1963+. P. F. Bol. Inform. Patrimonio Forest. Estado. HI 69281

P G R bulletin = Plant growth regulator bulletin. St. Paul, MN, etc.

P G R C / E - I L C A germ plasm newsletter = Germ plasm newsletter. Addis Ababa.

P G R C newsletter. Ottawa. No. 1+, 1976+. P. G. R. C. Newslett. HI 69282

P H M - revue horticole = Pépiniéristes, horticulteurs, maraîchers. Paris.

P H S news. Philadelphia, PA. Vol. 1+, 1960+. P. H. S. News. HI 69283

P K V research journal. Akola. Vols. 1-3, 1972-75. P. K. V. Res. J. Superseded by: Journal of the Maharashtra agricultural universities, college of agriculture. HI 69284

P N H herbarium news. Manila. No. 1+, 1986+. P. N. H. Herb. News. HI 69285

P P = Památky a příroda. Prague.

P R C newsletter. No. 1+, 1973+. P. R. C. Newslett. HI 69286

P R O S E A newsletter. Ca.1989+. P. R. O. S. E. A. Newslett. HI 70102

P S L herbarium news = Plant science laboratories herbarium news, university of Reading. Reading.

P S Z N 1: Marine ecology = Marine ecology. Berlin, Hamburg.

Pacific coast archaeological society quarterly. Costa Mesa, CA. Vol. 1+, 1965+. Pacific Coast Archaeol. Soc. Quart. HI 69287

Pacific coast medicine. San Francisco, CA. Pacific Coast Med. See B–P–H 694/31. HI 57797

Pacific discovery. San Francisco, CA. Pacific Disc. See B–P–H 694/32. HI 57798

Pacific garden. Pasadena, CA. Pacific Gard. See B–P–H 694/33. HI 57799

Pacific horticulture; journal of the Pacific horticultural foundation. San Francisco, CA. Vol. 37+, 1976+. Pacific Hort. Preceded by: California horticultural journal. HI 69288

Pacific island studies and notes. Honolulu, HI. No. 1+, 1971+. Pacific Island Stud. Notes. HI 69289

Pacific medical record = Western journal of surgery, obstetrics and gynecology. Portland, OR. W. J. Surg. See B–P–H 970/4.

Pacific monthly. San Francisco, CA. Vols. 10(2)-11(11), 1863-64. Pacific Monthly. Preceded by: Hesperian. HI 69290

Pacific naturalist. Solway, Moss Landing, CA. Vols. 1-4(3), 1956-64. Pacific Naturalist. HI 69291

Pacific northwest. Seattle, WA. Vol. 14+, 1980+. Pacific NorthW. Preceded by: Pacific search. HI 69362

Pacific science; a quarterly devoted to the biological and physical sciences of the pacific region. Honolulu, HI. Pacific Sci. See B–P–H 695/2. HI 57800

Pacific science association. Information bulletin. Honolulu, HI. Pacific Sci. Assoc. Inform Bull. See B–P–H 695/3. HI 57801

Pacific science information. Honolulu, HI. Pacific Sci. Inform. See B–P–H 695/4. HI 57802

Pacific search; journal of natural science in the Pacific northwest. Seattle, WA. Vols. 1-13, 1966-80. Pacific Search. Superseded by: Pacific northwest. HI 69292

Pacific studies; journal devoted to the study of the Pacific — its islands and adjacent countries. Laie, HI. 1977+. Pacific Stud. HI 69293

Pacific viewpoint. Wellington, New Zealand. Pacific Viewpoint. See B–P–H 695/6. HI 57803

Paeonian. Hopkins, MN. ?-1982+. Paeonian. HI 69294

Pahove. Caldwell, OH. Vol. 1+, 1979+. Pahove. HI 69295

Pakistan cotton bulletin. Karachi. 1952-56. Pakistan Cotton Bull. Superseded by: Pakistan cottons. HI 69296

Pakistan cottons. Karachi. Vols. 1+, 1956+. Pakistan Cottons. Preceded by: Pakistan cotton bulletin. HI 69297

Pakistan journal of agricultural sciences. Vol. ?-13+, ?-1976+. Pakistan J. Agric. Sci. HI 69298

Pakistan journal of biological and agricultural sciences. Dacca. Vol. 1(1-12), 1957-69. Pakistan J. Biol. Agric. Sci. Superseded by: Bangladesh journal of biological and agricultural sciences. HI 69299

Pakistan journal of botany. Jamshoro, Karachi. Vol. 1+, 1969+. Pakistan J. Bot. HI 69300

Pakistan journal of forestry. Upper Topa, Pakistan. Pakistan J. Forest. See B–P–H 695/11. HI 57804

Pakistan journal of science. Lahore, Pakistan. Pakistan J. Sci. See B–P–H 695/12. HI 57805

Pakistan journal of scientific and industrial research. Karachi, Pakistan. Pakistan J. Sci. Industr. Res. See B–P–H 695/13. HI 57806

Pakistan journal of scientific research. Lahore. Vol. 1+, 1949+. Pakistan J. Sci. Res. Vols. 1-6 were included in: Pakistan journal of science. HI 69301

Pakistan science abstracts. Karachi. 1962+. Pakistan Sci. Abstr. Preceded by: Pakistan scientific literature: current bibliography. HI 69302

Pakistan scientific literature: current bibliography. Karachi. 1961. Pakistan Sci. Lit. Curr. Bibliogr. Superseded by: Pakistan science abstracts. HI 69303

Pakistan systematics; bulletin of the herbarium, Quaid-I-Azam university. Islamabad. Vol. 1+, 1977+. Pakistan Syst. HI 69304

Palacio, El. Santa Fe, NM. 1913+. Palacio. HI 69305

Palaeobiologica. Vienna. Palaeobiologica. See B–P–H 695/14. HI 57807

Palaebotanist. Lucknow, India. Palaeobotanist. See B–P–H 695/16. HI 57808

Palaeobotany in India. Progress report. Lucknow. 1939+. Palaeobot India Progr. Rep. HI 69306

Palaeoecology of Africa, and of the surrounding islands and Antarctica. [Vol. 1 is a reprint of: Palynology in Africa, nos. 1-8, 1950-63.] Cape Town, Amsterdam. Vol. 1+, 1966+. Palaeoecol. Africa. HI 69307

Palaeogeography, palaeoclimatology, palaeoecology. Amsterdam. Palaeogeogr. Palaeoclimatol. Palaeoecol. See B–P–H 695/17. HI 57809

Palaeontographica. Kassel. Palaeontographica. See B–P–H 695/21. HI 57811

Palaeontographica. Abteilung B. Paläophytologie. Stuttgart. Palaeontographica, Abt. B, Paläophytol. See B–P–H 695/22. HI 57812

Palaeontographica americana. Ithaca, NY. Palaeontogr. Amer. See B–P–H 695/20. HI 57810

Palaeontology. London. Palaeontology. See B–P–H 695/24. HI 57813

Palaios. Tulsa, OK. Vol. 1+, 1986+. Palaios. HI 69308

Palao tropical biological station studies. Tokyo. Palao Trop. Biol. Sta. Stud. See B–P–H 695/25. HI 57814

Paläontologische Abhandlungen. Berlin. Paläontol. Abh. See B–P–H 696/1. HI 57815

Paläontologische Abhandlungen. Abteilung B Paläobotanik. Berlin. Vols. 2-?, 1964-70. Paläontol. Abh., Abt. B, Paläobot. Preceded by: Paläontologische Abhandlungen. Superseded by: Zeitschrift für geologische Wissenschaften. HI 69309

Paläontologische Zeitschrift. Berlin. Paläontol. Z. See B–P–H 696/3. HI 57816

Paleo data banks; for the improvement of communication in palynology, paleobotany, and related sciences. Tucson, AZ. No. 1+, 1971+. Paleo Data Banks. HI 69310

Paleoantropologia y ciencias naturales = Kobie. Bilbao.

Paleobiologie continentale. Montpellier. Vol. 1+, 1970+. Paleobiol. Continentale. HI 69311

Paleobiology. Menlo Park, CA, etc. Vol. 1+, 1975+. Paleobiology. HI 69312

Paleobotânica latinoamericana. São Paulo. Vol. 1+, 1979+. Paleobot. Latinoamer. HI 69313

Paleobotanika. Moscow & Leningrad = Trudy botanicheskogo instituta imeni V. L. Komarova akademii nauk S S S R. Ser. 8, paleobotanika. Moscow & Leningrad.

Paleolimnology of Lake Biwa and the japanese pleistocene. Otsu. Vols. 1-?, 1971-76. Paleolimnol. Lake Biwa Jap. Pleistoc. HI 69314

Paleontológia mexicana. Mexico City. Paleontol. Mex. See B–P–H 696/8. HI 57817

Paleontological bulletin, New Zealand geological survey = New Zealand geological survey, paleontological bulletin. Wellington. New Zealand Geol. Surv. Paleontol. Bull. See B–P–H 657/10.

Paleontological journal. [Translation of Paleontologicheskii zhurnal.] Washington, DC. 1967+. Paleontol. J. HI 69315

Paleontologičeskij sbornik. Lvov = Paleontologicheskii sbornik. Lvov.

Paleontologičeskij sbornik. Moscow & Leningrad = Paleontologicheskii sbornik. Moscow & Leningrad.

Paleontologičeskij žurnal = Paleontologicheskii zhurnal. Moscow.

Paleontologicheskii sbornik. Lvov, Ukrainian S S R. Vol. 1+, 1961+. Paleontol. Sborn. (Lvov). HI 69316

Paleontologicheskii sbornik. Moscow & Leningrad. Vol. 1+, 1954+. Paleontol. Sborn. (Moscow & Leningrad). HI 69317

Paleontologicheskii zhurnal. Moscow. 1959+. Paleontol. Zhurn. HI 69318

Paleontologija. Moscow & Leningrad = Materialy k biobibliografii uchenykh S S S R. Seriya biologicheskikh nauk; paleontologiya. Moscow & Leningrad.

Paleontologiya, stratigrafiya, i litologiya. Sofia. 1975+. Paleontol. Stratigr. Litol. Preceded by: Izvestiya na geologicheskiya institut. Seriya paleontologiya. HI 69319

Palestine exploration quarterly. London. Vol. 1+, 1937+. Palestine Explor. Quart. 4-3246-2. HI 69320

Palestine gazette. Agricultural supplement. Jerusalem. Nos. 1-45, 1936-39. Palestine Gaz., Agric. Suppl. Superseded by: Monthly agricultural bulletin, Palestine. HI 69321

Palestine journal of botany. Jerusalem series. Jerusalem. Vols. 1-6, 1938/39-53. Palestine J. Bot., Jerusalem

Ser. Superseded by: Bulletin of the research council of Israel. Section D, botany. 4-3246-3. HI 69322

Palestine journal of botany. Rehovot series. Rehovot. Vols. 2-8(2), 1938-53. Palestine J. Bot., Rehovot Ser. Preceded by: Palestine journal of botany and horticultural science. Superseded by: Bulletin of the research council of Israel. Section D, botany. 4-3246-3. HI 69323

Palestine journal of botany and horticultural science. Rehovot series. Rehovot [Israel]. Palestine J. Bot. Hort. Sci., Rehovot Ser. See B–P–H 696/17. HI 57818

Palinologia; revista del instituto palinologico de León. León. Vol. 1+, 1979+; Num. extra. 1, 1978+. Palinologia. HI 69324

Palmengarten. Frankfurter Monatsschrift für Natur- und Gartenfreunde. Frankfurt a. M. Palmengarten. See B–P–H 696/21. HI 57819

Palmengarten-Mitteilungen = Der Palmengarten. Frankfurter Monatsschrift für Natur- und Gartenfreunde. Frankfurt a. M. Palmengarten. See B–P–H 696/21.

Palmeras utiles de América tropical, [noticiario]. Brasilia = Newsletter, useful palms of tropical america. Brasilia.

Palmetto. Winter Park, FL. Vol. 1+, 1981+. Palmetto. HI 69325

Palmiers utiles d'Amérique tropicale, [bulletin]. Brasilia = Newsletter, useful palms of tropical america. Brasilia.

Palms; newsletter of the palm society. Melbourne, FL. Vol. 1+, 1969+. Palms. HI 69326

Palynological bulletin. Lucknow. Vols. 1-6(1), 1965-70. Palynol. Bull. Incorporated in: Journal of palynology. HI 69327

Palynology. Austin, Dallas, TX. Vol. 1+, 1977+. Palynology. Preceded by: Proceedings, annual meeting, american association of stratigraphic palynologists. HI 69328

Palynology in Africa. Bloemfontein. Nos. 1-6, 1951-59. Palynol. Africa. HI 69329

Palynos; newsletter of the international federation of palynological societies. Tempe, AZ. Vol. 7(2)+, 1984+. Palynos. Preceded by: Newsletter, international commission for palynology. HI 69330

Památky a příroda. Prague. 1976+. Památky Přír. Preceded by: Ochrana přirody and Pamatkova pece [not entered]. HI 69331

Památky - příroda - život. Chomutov, Czechoslovakia. Památky Přír. Život. See B–P–H 696/25. HI 57820

Pamietnik akademji umiejetności w Krakowie. Wydział matematiczne-przyrodniczy. Cracow. Pamietn. Akad. Umiejetn. w Krakowie, Wydz. Mat.-Przyr. See B–P–H 696/26. HI 57821

Pamietnik fizyograficzny. Warsaw. Pamietn. Fizyogr. See B–P–H 696/27. HI 57822

Pamietnik pulawski; prace, instytut uprawy, nawczenia i gleboznawstwa. Warsaw. Vol. 1+, 1961+. Pamietn. Pulawski. HI 69332

Pamietnik zakladu genetycznego skoły glównej gospodarstwa wiejskiego. Warsaw. Vols. 1-2, 1921-24. Pamietn. Zakladu Genet. Skoły Glówn. Gosp. Wiejsk. HI 69333

Pamphlet, advisory council of science and industry Australia. Melbourne, Vic. 1918. Pam. Advis. Council Sci. Industr. Australia. Superseded by: Pamphlet, institute of science and industry. Melbourne. HI 69334

Pamphlet of the botanical museum, university of Helsinki = Helsingen yliopiston kasvimuseon monisteita. Helsinki.

Pamphlet, council for scientific and industrial research. Melbourne, Vic. Nos. 4-114, 1927-42. Pam. Council Sci. Industr. Res. Preceded by: Pamphlet, institute of science and industry. Melbourne. HI 69335

Pamphlet, institute of science and industry. Melbourne, Vic. 1919-23. Pam. Inst. Sci. Industr. Preceded by: Pamphlet, advisory council of science and industry Australia. Superseded by: Pamphlet, council for scientific and industrial research. Melbourne. HI 69336

Pamphlet, seed commissioner's branch, department of agriculture, Canada. Ottawa. ?-1922-35. Pam. Seed Commiss. Branch Dept. Agric. Canada. Preceded by: Bulletin, seed branch, department of agriculture, Canada. HI 69337

Pamphlet, tea research institute of Ceylon. Talawakelle. Nos. 1-3, 1955-58. Pam. Tea Res. Inst. Ceylon. HI 69338

Pamphlet, tea research institute of East Africa. Kericho, Nairobi. Nos. 1-24, 195?-? Pam. Tea Res. Inst. E. Africa. HI 69339

Pamphlet, Vermont state agricultural college, agricultural experiment station = Vermont state agricultural college, agricultural experiment station. Pamphlet. Burlington, VT. Vermont Agric. Exp. Sta. Pam. See B–P–H 957/3.

Pamphlet, Wau ecology institute. Wau. No. ?-2+, ?-1976+. Pam. Wau Ecol. Inst. HI 69340

Pan-american geologist. Des Moines, IA. Vols. 37-77, 1922-42. Pan-Amer. Geol. Preceded by: Economic geology (1905-21 only) [not entered]. For vols. 1-36 see American geologist. 4-3250-1. HI 69341

Pan pacific entomologist. San Francisco, CA. Pan Pacific Entomol. See B–P–H 697/3. HI 57823

Pandore. Paris. No. 1+, 1978+. Pandore. HI 69342

Pankia. Boletín informativo, J B L L. San Salvador. Vol. 7+, 1988+. Pankia. Preceded by: Boletín, jardín botánico La Laguna. HI 74869

Pannonia. Pecs, Hungary. Pannonia. See B–P–H 697/4. HI 57824

Panorama and news letter, national chrysanthemum society. London. No. 1+, 1967+. Pamor. News Lett. Chrysanthemum Soc. HI 69343

Pantnagar journal of research. Pantnagar. Vol. 1+, 1976+. Pantnagar J. Res. HI 69344

Papeis avulsos, herbario Hatschbach. Curitiba. No. ?-4, ?-1963. Pap. Avulsos Herb. Hatschbach. HI 69345

Papers of the agricultural history society. Washington, DC. Vols. 1-3, 1918-20. Pap. Agric. Hist. Soc. Superseded by: Agricultural history. 1-89-1. HI 69346

Papers on agriculture. Boston, MA = Massachusetts agricultural repository and journal. Boston, MA. Mass. Agric. Repos. J. See B–P–H 549/17.

Papers, bibliographical society of America. New York, Chicago, IL. Vol. 4+, 1909+, 1910+. Pap. Bibliogr. Soc. America. Preceded by: Proceedings and papers, bibliographical society of America. 1-667-2. HI 69347

Papers from the department of biology, university of Queensland. Brisbane, Qld. Vols. 1-2, 1937-50. Pap. Dept. Biol. Univ. Queensland. Superseded by: Papers from the department of botany, university of Queensland. HI 69348

Papers of the department of botany. McGill university. Montreal. Pap. Dept. Bot. McGill Univ. See B–P–H 697/10. HI 57825

Papers from the department of botany, university of Queensland. Brisbane, Qld. Vol. 3+, 1954+. Pap. Dept. Bot. Univ. Queensland. Preceded by: Papers from the department of biology, university of Queensland. HI 69349

Papers from the department of marine biology of the Carnegie institution of Washington. Washington, DC. Pap. Dept. Mar. Biol. Carnegie Inst. Wash. See B–P–H 697/11. HI 57826

Papers of the desert botanical laboratory. Carnegie institution of Washington. Washington, DC. Pap. Desert Bot. Lab. See B–P–H 697/12. HI 57827

Papers of the Eastbourne natural history society. Eastbourne, England. Pap. Eastbourne Nat. Hist. Soc. See B–P–H 697/13. HI 57828

Papers, laboratory of tree-ring research, university of Arizona. Tucson, AZ. Nos. 1-4, 1964-69. Pap. Lab. Tree-Ring Res. Univ. Arizona. HI 69350

Papers of the Michigan academy of sciences, arts and letters. New York. Vols. 1-53, 1921-67. Pap. Michigan Acad. Sci. Preceded by: Report (Annual) of the Michigan academy of science. Superseded by: Michigan academician. 3-2641-2. HI 69351

Papers on paleontology. Ann Arbor, MI. No. 1+, 1972+. Pap. Paleontol. HI 69352

Papers of the Peabody museum of american archaeology and ethnology. Cambridge, MA. Vol. 1+, 1888+. Pap. Peabody Mus. Amer. Archaeol. 3-1816-2. HI 69353

Papers on plant genetics. Cambridge, England. Pap. Pl. Genet. See B–P–H 697/23. HI 57829

Papers and proceedings of the Hampshire field club (and archaeological society). Southampton. Vols. 1-23, 1885-1968. Pap. Proc. Hampshire Field Club. Archaeol. Soc. Superseded by: Proceedings of the Hampshire field club. HI 69354

Papers and proceedings of the royal society of Tasmania. Hobart, Tas. Pap. & Proc. Roy. Soc. Tasmania. See B–P–H 697/27. HI 57830

Papers and proceedings of the royal society of Van Diemen's Land. Hobart, Tas. Pap. & Proc. Roy. Soc. Van Diemen's Land. See B–P–H 697/28. HI 57831

Papers and proceedings, technical meeting, international union for the conservation of nature and natural resources. Morges. 11th+, 1969+. Pap. Proc. Techn. Meetings Int. Union Conservation Nat. Nat. Res. Preceded by: Proceedings and papers, technical meeting, international union for conservation and natural resources. HI 69355

Papers and reports in oceanography. [Kobe marine observatory.] Kobe, Japan. Pap. Rep. Oceanogr. See B–P–H 698/1. HI 57832

Papers of the Robert S. Peabody foundation for archaeology. Andover, MA. Pap. Robert Peabody Found. Archaeol. See B–P–H 698/2. HI 57833

Papers from the Tortugas laboratory of the Carnegie institution of Washington. Washington, DC. Pap. Tortugas Lab. Carnegie Inst. Wash. See B–P–H 698/5. HI 57834

Papers, university of Kansas paleontological contributions = University of Kansas paleontological contributions. Papers. Lawrence, KS. Univ. Kansas Paleontol. Contr. Pap. See B–P–H 936/13.

Papers, university of Queensland, department of biology = University of Queensland, department of biology, papers. Brisbane. Univ. Queensland Dept. Biol. Pap. See B–P–H 937/26.

Papers from the Wellcome chemical research laboratories. London. Nos. ?-62-275-?, ?-1908-43-? Pap. Wellcome

Chem Res. Lab. HI 69356

Paphiopedilum world; international magazine of cypripedieae. San Marino, CA. Vol. 1+, 1971+. Paphiopedilum World. HI 69357

Pappelwirtschaft. Mitteilungen des Deutschen Pappelvereins. Bonn. Pappelwirtschaft. See B–P–H 698/16. HI 57835

Pappus; quarterly publication of the royal botanical gardens, Hamilton. Hamilton, Ont. Vol. 1+, 1981+. Pappus. HI 69358

Papua annual report. Melbourne, Vic. Papua Annual Rep. See B–P–H 698/18. HI 57836

Papua New Guinea = Papua and New Guinea agricultural journal. Port Moresby.

Papua and New Guinea agricultural gazette. Port Moresby, New Guinea. Papua New Guinea Agric. Gaz. See B–P–H 698/20. HI 57837

Papua and New Guinea agricultural journal. Port Moresby. Vol. 9+, 1955+. Papua New Guinea Agric. J. Preceded by: Papua and New Guinea agricultural gazette. HI 69359

Papua and New Guinea department of forests. Bulletin. Port Moresby, New Guinea. Papua New Guinea Dept. Forests Bull. See B–P–H 698/22. HI 57838

Papua New Guinea journal of agriculture, forestry and fisheries = Papua and New Guinea agricultural journal. Port Moresby.

Papyrus; a publication of the Rodef Shalom biblical botanical garden. Pittsburgh, PA. Vol. 1+, [1989]+. Papyrus. HI 50931

Paradisiaca. (Nigeria). Vol. 1+, 1976+. Paradisiaca. HI 69360

Paraguay agropecuario. Asunción. Paraguay Agropecu. See B–P–H 698/23. HI 57839

Parasitic diseases. New York. Vol. 1+, 1981+. Parasitic Dis. HI 69361

Parcs; revue internationale pour gestionnaires de parcs nationaux, de lieux historiques et autres lieux protégés. [French language edition of: Parks.] Washington, DC. Washington, D.C. Vols. 1-6(1), 1976-81. Parcs. HI 69363

Parcs nationaux; bulletin trimestriel de l'association Ardenne et Gaume. Brussels. Vol. ?-20(3)+, ?-1965+. Parcs Natx. HI 69364

Parfumerie moderne; revue scientifique et de défense professionelle, mensuelle illustrée. Paris. Vols. 1-54, 1908-40; 1946-57. Parfum. Moderne. Superseded by: Parfumerie moderne - bulletin technique de Gattefossé. HI 69365

Parfumerie moderne - bulletin technique de Gattefossé. Lyon. 1958+. Parfum. Moderne, Bull. Techn. Gattefossé. Preceded by: Parfumerie moderne. HI 69366

Park administration. Johannesburg. Vol. 1+, 1949+. Park Admin. (Johannesburg). HI 69367

Park administration. London. Vols. 28(3)-34(7), 1963-69. Park Admin. (London). Preceded by: Park administration, horticulture and recreation. Superseded by: Parks and recreation. HI 69368

Park administration, horticulture and recreation. London. Vols. 21(4)-28(2), 1956-63. Park Admin. Hort. Recreation. Preceded by: Journal of park administration, horticulture and recreation. Superseded by: Park administration. HI 69369

Park maintenance. Appleton, WI. Vol. 1+, 1948+. Park Maint. HI 69370

Park news; journal of the national and provincial parks association of Canada. Toronto. Vol. 1+, 1965+. Park News. HI 69371

Parks. Washington, DC. Vol. 1+, 1976+. Parks. HI 69372

Park's floral magazine; monthly journal of floriculture. La Park, Lancaster, PA. Vols. 1-61(8), 1871-1925. Park's Fl. Mag. 4-3271-3. HI 69373

Parks, golf courses and sports grounds. Staines. Vol. 1+, 1935+. Parks Golf Courses Sports Grounds. HI 69374

Parks magazine = Parks. Washington, DC.

Parks and recreation. London. Vol. 34(8)+, 1969+. Parks Recreation. Preceded by: Parks administration. HI 69375

Parks and sports grounds = Parks, golf courses and sports grounds. Staines.

Parma natura. Parma. Vol. 1+, 1985+. Parma Nat. HI 69376

Parnassus Boicus, oder Neueröffneter Musenberg, worauf verschiedene Denk- und Leswürdigkeiten aus der gelehrten Welt, zumalen aber aus den Landen zu Baiern, abgehandelt werden. Munich. Parnassus Boicus. See B–P–H 699/5. HI 57840

Parodiana; revista de la unidad botánica C E F A P R I N. Buenos Aires. Vol. 1+, 1981+. Parodiana. HI 69377

Parques y jardines; publicación del servicio municipal de parques y jardines de Barcelona. Barcelona. No. ?-2+, ?-1971+. Parques Jard. HI 69378

Parques nacionales. Santo Domingo. Vol. 1+, 1980+. Parques Nac. HI 69379

Parthenon, a weekly journal of english and foreign literature, the arts and sciences. London. Parthenon. See B–P–H 699/6. HI 57841

PASCAL, see P A S C A L

Pastos; revista de la sociedad española para el estudio de los pastos. Madrid. Vol. 1+, 1971+. Pastos. HI 69391

Pastos y forrajes; revista de la estacíon experimental de pastos y forrajes Indio Hatuey. Matanzas. Vol. 1+, 1978+. Pastos Forrajes. HI 69392

Pastos y forrajes. Paston, Colombia. Pastos & Forrajes. See B–P–H 699/7. HI 57842

Pasturos tropicales. Cali. Vol. ?-10+, ?-1988+. Pasturos Trop. HI 74539

Pathologia veterinaria. Basel. Pathol. Veterin. See B–P–H 699/9. HI 57843

Pathologica. Genoa. Pathologica. See B–P–H 699/10. HI 57844

Pathological herbarium notes, U S department of agriculture. Washington, DC. Nos. 1-6, 1920-23. Pathol. Herb. Notes, U.S.D.A. HI 69393

Pathology circular, division of plant industry, Florida department of agriculture. Gainesville, FL. Nos. 1-9, 1962-63. Pathol. Circ., Div. Pl. Industr., Fl. Dept. Agric. Superseded by: Plant pathology circular, division of plant industry, Florida department of agriculture. HI 69394

Pathology circular, Florida department of agriculture = Pathology circular, division of plant industry, Florida department of agriculture. Gainesville, FL.

Patologia comparata della tubercolosi. Milan. Patol. Comp. Tuberc. See B–P–H 699/11. HI 57845

Patrika; newsletter of the indian academy of sciences Bangalore. 1980+. Patrika. HI 74870

Patriotische Medicus. Hamburg. Patriot. Med. See B–P–H 699/13. HI 57847

Patriotisches Archiv der Herzogthümer Mecklenburg zur Aufbewahrung der Geschichte und Denkwürdigkeiten derselben. Rostock. Patriot. Arch. Herzogth. Mecklenburg. See B–P–H 699/12. HI 57846

Patriotisches Wochenblatt für Ungern. Pest [=Budapest, in part]. Patriot. Wochenbl. Ungern. See B–P–H 699/14. HI 57848

Paul Arendt's Monatsschrift für Kakteenkunde. Berlin-Friedenau. Paul Arendt's Monatsschr. Kakteenk. See B–P–H 699/15. HI 57849

Paxton's flower garden. London. Paxton's Fl. Gard. See B–P–H 699/20. HI 57850

Paxton's magazine of botany, and register of flowering plants. London. Paxton's Mag. Bot. See B–P–H 699/21. HI 57851

Paysage et aménagement. Paris = P + A. Paris.

Paysage Canada. Ottawa = Landscape Canada. Ottawa.

PCEA. Boletín trimestral de experimentación agropecuaria. Lima = P C E A. Boletín trimestral de experimentación agropecuaria. Lima.

Pchelovodstvo. Moscow. Vols. 1-40, 1909-63 [publication suspended 1941-45]; vol. 84+, 1964+ [volume numbering altered with vol. 40]. Pchelovodstvo. HI 69396

Peabody museum of natural history bulletin. New Haven, CT. Peabody Mus. Nat. Hist. Bull. See B–P–H 699/27. HI 57852

Peanut grower. Tifton, GA.. 1989+. Peanut Grower. HI 74871

Peanut promoter. Houston, TX. Peanut Promot. See B–P–H 699/28. HI 57853

Peanut research. Tifton, GA., Yoakum, TX. Vol. 1+, 1963+. Peanut Res. HI 69397

Peanut science. Suffolk, VA, Yoakum, TX. Vol. 1+, 1974+. Peanut Sci. HI 69398

Peat abstracts. Dublin. 1952+. Peat Abstr. HI 69399

Pecan growers' newsletter. Baton Rouge, LA. ?-1972+. Pecan Growers' Newslett. HI 69400

Pecan journal. Albany GA. Pecan J. See B–P–H 700/1. HI 57854

Pécsi pedagógiai föiskola évkönyv See B–P–H 700/2. HI 57855

Pectinifera. Prague. ?-1973+. Pectinifera. HI 69401

Pediatria archivio. Naples. Pediatria Arch. See B–P–H 700/3. HI 57856

Pedobiologia. Jena. Pedobiologia. See B–P–H 700/4. HI 57857

Pédologie. Ghent. Pédologie. See B–P–H 700/5. HI 57858

Pédologie. Moscow = Pochvovědění e. Moscow.

Pedology. Moscow = Pochvovědění e. Moscow.

Peelings. (Belgium.) 198?+. Peelings. HI 69402

Pei ta li k'o pao-kao = Science reports of the national university of Peking. Peiping [=Peking]. Sci. Rep. Natl. Univ. Peking. See B–P–H 827/18.

Pei-ching nung yüeh ta hsüeh pao = Science reports of Peking agricultural university. Peking. Sci. Rep. Peking Agric. Univ. See B–P–H 828/2.

Pei-ching shêng wu k'o hsüeh hui nien pao = Annual of the Peking biological science association. Peking. Annual Peking Biol. Sci. Assoc. See B–P–H 122/7.

Pei-ching ta hsüeh hsüeh p'ao, tzujan k'o hsüeh (chi kan) = Acta scientiarum naturalium universitatis pekinensis. Peking. Acta Sci. Nat. Univ. Pekin. See B–P–H 48/24.

Pei-ching tzu jan po wu kuan yen chiu pao kao = Memoirs of Beijing natural history museum. [Beijing.]

Pei-p'ing po wu tsa chih = Peking natural history bulletin. Peking. Peking Nat. Hist. Bull. See B–P–H 700/14.

Peking natural history bulletin. [Pei-p'ing po wu tsa chih.] Peking. Peking Nat. Hist. Bull. See B–P–H 700/14. HI 57859

Pelagos; bulletin de l'institut océanographiques d'Alger. Algiers. Pelagos. See B–P–H 700/15. HI 57860

Pelargonium news; official journal of the british pelargonium and geranium society. Welwyn Garden City, Beckenham. Vol. 16(2)+, 1967+. Pelargonium News. Preceded by: Bulletin, british pelargonium and geranium society. HI 69403

Peltophorum. Johannesburg. Vol. 1+, 1984+. Peltophorum. HI 69404

Pemberitaan, lembaga penelitian tanaman industri. Bogor. Nos. 15/16-38, 1973-81. Pemberit. Lemb. Penelitian Tanam. Industr. Superseded by: Pemberitaan penelitian tanaman industri. HI 69405

Pemberitaan penelitian tanaman industri. Bogor. Vol. 7(39)+, 1981+. Pemberit. Penelitian Tanam. Industr. Preceded by: Pemberitaan, lembaga penelitian tanaman industri. HI 69406

Pembina ilmu berkebun = Horticultura. Jakarta, Indonesia. Horticultura (Jakarta). See B–P–H 423/10.

Penelitian laut di Indonesia = Marine research in Indonesia. Djakarta.

Penerbitan, madjelis ilmu pengetahuan Indonesia. Jakarta. 1959. Penerbitan Madj. Pengetahuan Indonesia. HI 69407

Penggemar alam. Bogor. 1954-61. Pengg. Alam. Preceded by: Tropische natuur. HI 69408

Pengumuman balai besar penjelidikan kehutanan Indonesia. Bogor. Nos. 39-60, 1953-57. Pengum. Balai Besar Penjel. Kehut. Indonesia. Preceded by: Communications of the forest research institute (Indonesia). Superseded by: Pengumuman lembaga pusat penjelidikan kehutanan Indonesia. HI 69409

Pengumuman balai penjelidikan kehutanan = Mededeelingen van het bosbouwproefstation. Buitenzorg [= Bogor] and Communications of the forest research institute (Indonesia). Bogor.

Pengumuman isti mewa balai penjelidiken kehutanan = Bijzondere publicaties van het bosbouwproefstation. Buitenzorg, Dutch E. Indies [=Bogor, Indonesia]. Bijzondere Publ. Bosbouwproefstat. See B–P–H 191/24.

Pengumuman lembaga-lembaga penelitian kehutanan. Bogor. Nos. 75-?, 1961-72? Pengum. Lemb.-Lemb. Penelitian Kehut. Preceded by: Pengumuman lembaga pusat penjelidikan kehutanan. Superseded by: Pengumuman, lembaga penelitian hasil hutan. HI 69410

Pengumuman, lembaga penelitian hasil hutan. Bogor. No. 1+, 1973+. Pengum. Lemb. Penelitian Hasil Hutan. Preceded by: Pengumuman, lembaga penelitian kehutanan. HI 69411

Pengumuman, lembaga penelitian kehutanan = Pengumuman lembaga-lembaga penelitian kehutanan. Bogor.

Pengumuman lembaga pusat penjelidikan kehutanan. Bogor. Vols. 61-74, 1958-61. Pengum. Lemb. Pusat Penjel. Kehut. Preceded by: Pengumuman balai besar penjelidikan kehutanan Indonesia. Superseded by: Pengumuman lembaga-lembaga penelitian kehutanan. HI 69412

Penjelidikan laut di Indonesia. Jakarta, Indonesia. Penjel. Laut Indonesia. See B–P–H 700/21. HI 57861

Penn ar bed; bulletin trimestriel de la société pour l'étude de la protection de la nature en Bretagne (later Revue régionale de géographie, sciences naturelles, protection de la nature). Brest. N.s. vol. 1+, 1953+. Penn Ar Bed. HI 69413

Pennsylvania farmer and gardener. Philadelphia, PA. Pennsylvania Farmer Gard. See B–P–H 700/24. HI 57862

Pennsylvania flower growers bulletin. Kennett Square, PA. No. 1, 1950. Pennsylvania Fl. Growers Bull. HI 69414

Pennsylvania forests. Mechanicsburg, PA. Vol. 36(346)+, 1951+. Pennsylvania Forests. Preceded by: Forest leaves. 4-3299-1. HI 69415

Pennsylvania gardens. Harrisburg, PA. Vol. 1, nos. 1-4, 1937. Pennsylvania Gard. HI 69416

Pennsylvania horticultural society yearbook = Yearbook, Pennsylvania horticultural society. Philadelphia, PA.

Pennsylvania naturalist. Huntingdon, PA. Vol. 1+, 1978+. Pennsylvania Naturalist. HI 69417

Pennsylvania state college agricultural experiment station. Annual report. State College, PA. Pennsylvania State Coll. Agric. Exp. Sta. Annual Rep. See B–P–H 700/25. HI 57863

Pennsylvania state college agricultural experiment station. Bulletin. State College, PA. Pennsylvania State Coll. Agric. Exp. Sta. Bull. See B–P–H 700/27. HI 57865

Pennsylvania state college agricultural experiment station. Bulletin of information. State College, PA. Pennsylvania State Coll. Agric. Exp. Sta. Bull. Inform. See B–P–H 700/28. HI 57866

Pennsylvania state college agricultural experiment station circular. State College, PA. Pennsylvania State Coll. Agric. Exp. Sta. Circ. See B–P–H 701/1. HI 57867

Pennsylvania state college agricultural experiment station.

Progress report. State College, PA. Pennsylvania State Coll. Agric. Exp. Sta. Progr. Rep. See B–P–H 700/26. HI 57864

Pennsylvania state college, school of agriculture, department of botany. Contributions. State College, Pa. Pennsylvania State Coll. School Agric. Dept. Bot. Contr. See B–P–H 701/2. HI 57868

Pennsylvania state college studies. State College, PA. Pennsylvania State Coll. Stud. See B–P–H 701/3. HI 57869

Penny magazine. London. Penny Mag. (London). See B–P–H 701/4. HI 57870

Pensiero medico. Milan. Pensiero Med. See B–P–H 701/5. HI 57871

Peperomia gazette. St. Louis, MO. 1978+. Peperomia Gaz. HI 69418

Pépiniéristes, horticulteurs, maraîchers. Paris. Nos. 1-232, 1974-82. Pépiniéristes Hort. Maraîch. Preceded and superseded by: Revue horticole HI 69419

Peregrine; publication of the field section of the Isle of Man natural history and antiquarian society. Douglas, I.O.M. 1941+. Peregrine. HI 69420

Pereskia. Anvers. Vol. 1(1-14), 1946-47. Pereskia. HI 69421

Perez-arbelaezia. Bogota. Vol. 1+, 1985+. Perez-Arbelaezia. HI 69422

Perfumer and flavorist. Oak Park, Wheaton, IL. Vol. 1+, 1976+. Perfumer Flavorist. Preceded by: Cosmetics and perfumery. HI 69423

Perfumer and flavorist international = Perfumer and flavorist. Oak Park, Wheaton, IL.

Perfumers' journal and essential oil recorder. New York, NY. Perfumers' J. See B–P–H 701/7. HI 57872

Perfumery arts. New York, NY. Perfumery Arts. See B–P–H 701/8. HI 57873

Perfumery and essential oil record; incorporating Flavours. London. Vols. 1-60, 1910-69. Perfumery Essential Oil Rec. Superseded by: Flavour industry. 4-3308-3. HI 69424

Periodical of hydrology. Budapest = Hidrológiai közlöny. Budapest. Hidrol. Közl. See B–P–H 416/4.

Periodical of the society of nature investigators at the universtiy of Tartu = Eesti loodus. Tartu ülikooli juures oleva loodusuurijate seltsi teataja. Eesti loodus. Tartu, Estonia [Estonian S S R]. Eesti Loodus (Tartu). See B–P–H 355/3.

Periodichesko spisanie na bulgarskoto knizhovno druzhestvo. Sofia. Vols. 1-22, 1870-1910. Period. Spis. Bulg. Knizh. Druzh. Superseded by: Spisanie na bulgarskata akademiya na naukite. Kniga 2, klon prirodo-matematichen. 1-829-2. HI 69425

Periodicum biologorum. Zagreb. Vol. 72+, 1970+. Period. Biol. Preceded by: Biološki glasnik. HI 69426

Periodicum biologorum. Societas historico-naturalis croatica = Glasnik hrvatskog prirodoslovnoga društva. Zagreb, Croatia [Yugoslavia]. Glasn. Hrvatsk. Prir. Društva. See B–P–H 405/1.

Périodique d'information pour les associations françaises d'environment. Paris. Vol. ?-8+, ?-1975/78+. Périod. Inform. Assoc. Franç. Environm. HI 69427

Periplo; revista del instituto de la caza fotográfica y ciencias de la naturaleza. Madrid. Vol. ?-6(31)+, ?-1975?+. Periplo. Preceded by: Caza fotografica. HI 69428

Persoonia; a mycological journal. Leiden. Persoonia. See B–P–H 701/17. HI 57876

Perspectives in biology and medicine. Chicago, IL. Perspect. Biol. Med. See B–P–H 701/18. HI 57877

Perspectives in environmental botany. Lucknow. Vol. 1+, 1986+. Perspect. Environm. Bot. HI 69429

Pertanika. Serdang. Vol. 1+, 1978+. Pertanika. HI 69430

Perthshire courier. Notes on botany of Perthshire. Perth, Scotland. Perthshire Courier. See B–P–H 701/19. HI 57878

Perú agronómico. Lima. Perú Agron. See B–P–H 701/21. HI 57879

Pesquisas. Porto Alegre, Brazil. Pesquisas. See B–P–H 701/23. HI 57880

Pesquisas agropecuaria brasileira. Rio de Janeiro. Vols. 1-6, 1966-71; vol. 12+, 1977?+. Pesq. Agropecu. Brasil. Superseded by: Pesquisas agropecuaria brasileira. Serie agronomia. HI 69431

Pesquisas agropecuaria brasileira. Serie agronomia. Rio de Janeiro, Porto Alegre. Vols. 7-11, 1972-76. Pesq. Agropecu. Brasil., Ser. Agron. Preceded and superseded by: Pesquisas agropecuaria brasileira. HI 69432

Pesquisas agropecuaria pernambucana. Vol. ?-2+, ?-1978+. Pesq. Agropecu. Pernamb. HI 69434

Pesquisas. Botânica. Porto Alegre, Brazil. Pesquisas, Bot. See B–P–H 701/24. HI 57881

Pesquisas. Communications. Porto Alegre. Vols. 1-4, 1960-67. Pesq. Commun. HI 69435

Pest articles and news summaries. Section B, plant disease control. London. Vols. 11-19, 1965-68. Pest Articles News Summ., B, Pl. Dis. Control. Preceded by: Pesticides abstracts and news summary. Section B, fungicides. Superseded by: P A N S. HI 69436

Pest articles and news summaries. Section C, weed control. London. Vols. 11-14, 1974-79. Pest Articles

News Summ., Sec. C, Weed Control. Preceded by: Pesticides abstracts and news summary. Superseded by: P A N S. HI 69437

Pester medicinisch-chirurgische Presse. Budapest. Pester Med.-Chir. Presse. See B–P–H 702/1. HI 57882

Pesticide biochemistry and physiology. New York. Vol. 1+, 1971+. Pestic. Biochm. Physiol. HI 69440

Pesticide information. London. No. 1+, 1968+. Pestic. Inform. HI 69441

Pesticide review. Washington, DC. 1966+. Pestic. Rev. HI 69442

Pesticide science; agricultural science service research and development reports. London. 1979-82. Pestic. Sci., Agric. Sci. Serv. HI 69443

Pesticide science. London, Oxford. Vol. 1+, 1970+. Pestic. Sci., Agric. Sci. Serv. HI 69444

Pesticides. [Noyaku.] Tokyo. Pesticides. See B–P–H 702/2. HI 57883

Pesticides abstracts and news summary. Section B, fungicides (nematicides, application). London. Vols. 1-10, 1955-64. Pestic. Abstr. News Summ., Sec. B, Fungic. Superseded by: Pest articles and news summaries. Section B, plant disease control. HI 69438

Pesticides abstracts and news summary. Section C, herbicides, arboricides, defoliants. London. Vols. 1-10, 1955-64. Pestic. Abstr. News Summ., Sec. C, Herbic. Superseded by: Pest articles and news summaries. Section C, weed control. HI 69439

Pesticides documentation bulletin. National agricultural library. U.S.D.A. Washington, DC. Pesticides Doc. Bull. See B–P–H 702/3. HI 57884

Pest-o-gram: Insects, diseases, weeds. Brookings, SD. Vols. 1(1)-3(24), 1983-86. Pest-O-Gram. Superseded by: Field facts: Soils, insects, diseases, weeds, crops. HI 69445

Pestology. Vol. 1+, 1977+. Pestology. HI 69446

Petal pusher. Jefferson City, MO. Vol. 1(4)+, 1986+. Petal Pusher. Preceded by: Newsletter, Missouri native plant society. HI 68443

Petera Stučkas latvijas valsts universitätes botaniskä darzä raksti. Riga, Latvian S S R. Vol. 17+, 1961+. Petera Stučkas Latv. Valsts Univ. Bot. Darzä Raksti. Preceded by: Trudy botanicheskogo sada Latviiskogo gosudarstvennogo universiteta Petra Stuchki. HI 69447

Peterborough museum society, occasional papers. Peterborough, England. Peterborough Mus. Soc. Occas. Pap. See B–P–H 702/5. HI 57885

Petermanns geographische Mitteilungen. Gotha. Petermanns Geogr. Mitt. See B–P–H 702/9. HI 57886

Petermanns geographische Mitteilungen, Ergänzungsband. Gotha. Petermanns Geogr. Mitt. (Ergänzungsband). See B–P–H 702/11. HI 57887

Petermann's Mittheilungen aus Justus Perthes' Geographischer Anstalt. Gotha. Petermann's Mitth. Justus Perthes' Geogr. Anst. See B–P–H 702/13. HI 57888

Petermann's Mittheilungen aus Justus Perthes' Geographischer Anstalt. Ergänzungsband. Gotha. Petermann's Mitth. Justus Perthes' Geogr. Anst. (Ergänzungsband). See B–P–H 702/14. HI 57889

Petit apôtre de la botanique; bulletin trimestriel d'initiation à la botanique. Châlons-sur-Mer. Vol. ?-46+, 1940-51+. Petit Apôtre Bot. HI 69448

Petit jardin illustré; journal hebdomadaire de jardinage pratique. Paris. Vols. 1(1)-45(1576), 1893-1938. Petit Jard. Ill. HI 69449

Petrak's lists of fungi = Index of fungi. Kew.

Petrol. si gaze. Bucharest. Petrol. & Gaze. See B–P–H 702/17. HI 57890

Pewarta lembaga biologi nasional. Bogor. 1973+. Pewarta Lemb. Biol. Nasl. HI 69450

PF. Boletín informativo de patrimonio forestal del estado. Madrid = P F. Boletín informativo de patrimonio forestal del estado. Madrid.

Pfalzbaierisches Museum. Mannheim. Pfalzbaier. Mus. See B–P–H 703/2. HI 58125

Pfalzbayerische Beyträge zur Gelehrsamkeit. Mannheim. Pfalzbayer. Beytr. Gelehrsamk. See B–P–H 703/3. HI 57893

Pfälzer Heimat. Speyer, Germany. Pfälzer Heimat. See B–P–H 703/4. HI 57894

Pfälzische Gartenzeitung. Speyer, Germany. Pfälz. Gartenzeitung. See B–P–H 702/19. HI 57891

Pfälzisches Museum. Mannheim. Pfälz. Mus. See B–P–H 702/20. HI 57892

Pflanzenareale; Sammlung kartographischer Darstellungen von Verbreitungsbezirken der lebenden und fossilien Pflanzen-Familien, -Gattungen und -Arten. Jena. 1926-40. Pflanzenareale. HI 69451

Pflanzenarzt. Zeitschrift für Pflanzenschutz und Schädlingsbekämpfung. Vienna. Pflanzenarzt. See B–P–H 703/6. HI 57895

Pflanzenbau. Berlin. Pflanzenbau. See B–P–H 703/7. HI 57896

Pflanzenbau, Pflanzenschutz und Pflanzenzucht. Berlin. Pflanzenbau Pflanzenschutz Pflanzenzucht. See B–P–H 703/9. HI 57897

Pflanzenforschung. Jena. Pflanzenforschung. See B–P–H 703/10. HI 57898

Pflanzenphysiologische Untersuchungen. Zurich. Pflanzenphysiol. Untersuch. See B–P–H 703/11. HI 57899

Pflanzenschutzberichte. Vienna. Vols. 1+, 1947+. Pflanzenschutzberichte. 4-3321-2. HI 69452

Pflanzenschützer. Vienna. Pflanzenschützer. See B–P–H 703/14. HI 57901

Pflanzenschutzinformationen. Munich. Vol. 1+, 1961+. Pflanzenschutzinformationen. HI 69453

Pflanzenschutz-Kalender. Vienna. Pflanzenschutz-Kalender. See B–P–H 703/15. HI 57902

Pflanzenschutzkalender. Vienna. Pflanzenschutzkalender. See B–P–H 703/16. HI 57903

Pflanzenschutz-Kurier. Leverkusen. Vol. 13+, 1968+. Pflanzenschutz-Kurier. Preceded by: Bayer pflanzenschutz-Kurier. HI 69454

Pflanzenschutz-Nachrichten "Bayer". Leverkusen. Vol. 15+, 1962+. Pflanzenschutz-Nachr. Bayer. Preceded by: Höfchen-Briefe, Bayer Pflanzenschutz-Nachrichten. 3-1869-2. HI 69455

Pflanzenschutzpost. Vienna & Korneuburg. Pflanzenschutzpost. See B–P–H 703/18. HI 57905

Pflanzenschutz, Wissenschaft und Wirtschaft. Munich. Pflanzenschutz Wiss. Wirtsch. See B–P–H 703/12. HI 57900

Pflanzensoziologie. Jena. Pflanzensoziologie. See B–P–H 703/19. HI 57906

Pflanzer. Ratgeber für tropische Landwirtschaft. Tanga, German E. Africa [Tanzania Republic]. Pflanzer. See B–P–H 703/20. HI 57907

Phanerogamarum monographiae. Lehre, Vaduz, Stuttgart. Vol. 1+, 1969+. Phanerog. Monogr. HI 69457

Phanérogamie; notulae systematicae. Paris. 1909-90. Phanérogamie. Superseded by: Adansonia. HI 69458

Phänogenetische Variabilität = Fenogeneticheskaya izmenchivost′. Moscow.

Phänologische Mitteilungen. Darmstadt. Vols. 1-?, 1883-1940. Phänol. Mitt. Vols. for 1907-10 also issued as: Hessische landwirtschaftliche Zeitschrift, Beilage. Vols. for 1909-32 also issued as: Arbeiten, hessische Bauernkammer. HI 69459

Pharmaceutica acta Helvetiae. Zurich. Pharm. Acta Helv. See B–P–H 703/22. HI 57908

Pharmaceutical bulletin. Tokyo. Vols. 1-5, 1953-57. Pharm. Bull. Superseded by: Chemical and pharmaceutical bulletin. HI 69460

Pharmaceutical chemistry journal = Khimiko-farmatsevticheskii zhurnal. Moscow.

Pharmaceutical journal. London. Pharm. J. See B–P–H 704/4. HI 57911

Pharmaceutical journal and transactions. London. Pharm. J. Tans. See B–P–H 704/6. HI 57912

Pharmaceutical news. [Yo hsüeh t'ung pao.] [China]. Pharm. News. See B–P–H 704/13. HI 57914

Pharmaceutical review. Milwaukee, Wisconsin. Pharm. Rev. See B–P–H 704/15. HI 57916

Pharmaceutisch weekblad voor Nederland. Amsterdam, Hilversum. 1864+ [vols. 114-116, 1979-81 included issues titled "scientific edition" vols. 1-3]. Pharm. Weekbl. Ned. Superseded, in part, by: Pharmaceutisch weekblad voor Nederland. Scientific edition. HI 69461

Pharmaceutisch weekblad voor Nederland. Scientific edition. Utrecht. Vol. 4+, 1982+. Pharm. Weekbl. Ned., Sci. Ed. Previously included in: Pharmaceutisch weekblad voor Nederland. HI 69462

Pharmaceutische Monatsblätter. Schmalkalden, Germany. Pharm. Monatsbl. See B–P–H 704/12. HI 57913

Pharmaceutische Nachrichten. [Germany]. Vol. 1, 1827. Pharm. Nachr. HI 69463

Pharmaceutische Rundschau. Eine Monatsschrift für die wissenschaftlichen und gewerblichen Interessen der Pharmacie und verwandten Berufs- und Geschäftszweige in den Vereinigten Staaten. Berlin & New York, NY. Pharm. Rundschau (Berlin & New York). See B–P–H 704/16. HI 57917

Pharmaceutische Zeitung des Apotheker-Vereins im nördlichen Teutschland. Lemgo. Vols. 1-12?, 1827-38. Pharm. Zeitung Apotheker-Vereins Nördl. Teutschl. Previously included in: 1822-23(5/6) in Pharmaceutischer Monatsblätter. Schmalkalden, vols. 3-6, and 1823(7)-1824(3) in Archiv des Apothekervereins im nördlichen Teutschland. Schmalkalden, vols. 6-7. HI 53574

Pharmaceutisches Central-Blatt. Leipzig. Pharm. Central-Blatt. See B–P–H 703/23. HI 57909

Pharmaceutisches Correspondenzblatt für Süddeutschland nebst Anzeigeblatt. Erlangen. Pharm. Correspondenzbl. Süddeutschl. See B–P–H 704/2. HI 57910

Pharmaceutisches Intelligenzblatt = Repertorium für die Pharmacie. Nuremberg. Repert. Pharm. See B–P–H 772/10.

Pharmaceutisk tidende. Copenhagen. Vols. 1-7, 1861-68. Pharm. Tidende. Superseded by: Ny pharmaceutisk tidende. 4-3123-3. HI 69465

Pharmacia mediterranea. Bordeaux. ?-1971+. Pharm. Medit. HI 69466

Pharmacological reviews. Baltimore, MD. Pharmacol. Rev. See B–P–H 704/21. HI 57918

Pharmacy in history. Madison, WI. 1959+. Pharm. Hist. Preceded by: A I H P notes. HI 69467

Pharmakeftikon deltion = Pharmakeutikon deltion. Athens.

Pharmakeutikon deltion. Organ officiel de l'association panhellénique des pharmaciens. Athens. Vol. 1+, 1975+. Pharmak. Deltion. HI 69468

Pharmazeutische Praxis. Vienna & Leipzig. Pharm. Praxis. See B–P–H 704/14. HI 57915

Pharmazeutische Zeitschrift für Russland. St. Petersburg. Vols. 1-36, 1862-97. Pharm. Z. Russland. HI 69464

Pharmazeutische Zentralhalle für Deutschland. Dresden. Vols. 1-108, 1859-1969. Pharm. Zentralhalle Deutschl. Superseded by: Zentralblatt für Pharmazie, Pharmakotherapie und Laboratoriumsdiagnostik. HI 69469

Pharmazie. Berlin. Vol. 1+, 1946+. Pharmazie. HI 69470

Pharmazie. Beiheft. Ergänzungsband. Berlin. Pharmazie Beih. Ergänzungsband. See B–P–H 704/23. HI 57919

Philadelphia botanic sentinel and Thomsonian medical revolutionist. Philadelphia, PA. Philadelphia Bot. Sentinel & Thomsonian Med. Revolutionist. See B–P–H 705/4. HI 57920

Philadelphia florist and horticultural journal = Florist and horticultural journal. Philadelphia, PA. Florist Hort. J. See B–P–H 376/11.

Philadelphia journal of the medical and physical sciences. Philadelphia, PA. Philadelphia J. Med. Phys. Sci. See B–P–H 705/8. HI 57921

Philadelphia medical museum. Philadelphia, PA. Philadelphia Med. Mus. See B–P–H 705/9. HI 57922

Philadelphia medical and physical journal. Philadelphia, PA. Philadelphia Med. Phys. J. See B–P–H 705/10. HI 57923

Philadelphia Thomsonian sentinel and family journal of useful knowledge. Philadelphia, PA. Philadelphia Thompsonian Sentinel & Family J. Useful Knowl. See B–P–H 705/11. HI 57924

Philippia; Abhandlungen und Berichte aus dem Naturkundemuseum im Ottoneum zu Kassel. Kassel. Vol. 1+, 1970+. Philippia. HI 69471

Philippine abstracts. Manila. Philipp. Abstr. See B–P–H 705/24. HI 579255

Philippine agricultural review. Manila. Philipp. Agric. Rev. See B–P–H 705/27. HI 57927

Philippine agriculturist. Manila, Los Baños, Philippines. Philipp. Agric. See B–P–H 705/25. HI 57926

Philippine agriculturist and forester = Philippine agriculturist. Manila, Los Baños, Philippines. Philipp. Agric. See B–P–H 705/25.

Philippine biota. Vol. 1+, 1966+. Philipp. Biota. HI 69472

Philippine farms and gardens. Manila. Philipp. Farms Gard. See B–P–H 705/28. HI 57928

Philippine forest research journal = Sylvatrop. Laguna.

Philippine forests. Manila. Vol. 1+, 1967+. Philipp. Forests HI 69474

Philippine garden. Diliman. Vol. ?-2(2)+, ?-[1962]+. Philipp. Gard. HI 69475

Philippine geographical journal. Manila. 1953+. Philipp. Geogr. J. HI 69476

Philippine journal of agriculture. Manila. Vols. 1-28, 1930-65. Philipp. J. Agric. Preceded by: Philippine agricultural review. Superseded by: Philippine journal of plant industry. 4-3333-3. HI 69477

Philippine journal of coconut studies. Manila. Vol. 1+, 1976+. Philipp. J. Coconut Stud. HI 69478

Philippine journal of crop science. Laguna. Vol. 1+, 1976+. Philipp. J. Crop Sci. HI 69479

Philippine journal of forestry. Manila. Philipp. J. Forest. See B–P–H 706/1. HI 57929

Philippine journal of plant industry. Manila. Vol. 29+, 1965+. Philipp. J. Pl. Industr. Preceded by: Philippine journal of agriculture. HI 69480

Philippine journal of science. Manila. Vol. 1, 1906; vol. 14+, 1919+ [suspended 1941-47]. Philipp. J. Sci. Preceded by: Publications of the bureau of science government laboratories. For vols. 2-3, 1907-18 see: Philippine journal of science. Section C, botany. 4-3334-1. HI 69481

Philippine journal of science. Section C, botany. Manila. Vols. 2-13, 1907-18. Philipp. J. Sci., C. Preceded and superseded by: Philippine journal of science. 4-3334-1. HI 69482

Philippine journal of weed science. Manila. Vol. 5+, 1978+. Philipp. J. Weed Sci. Preceded by: Philippine weed science bulletin. HI 69483

Philippine orchid review. Manila. Philipp. Orchid Rev. See B–P–H 706/5. HI 57930

Philippine phytopathology. Vol. 1+, 1965+. Philipp. Phytopathol. HI 69484

Philippines rice and corn progress. Manila. Philipp. Rice Corn Progr. See B–P–H 706/6. HI 57931

Philippine sugar institute quarterly. Manila. Philipp. Sugar Inst. Quart. See B–P–H 706/7. HI 57932

Philippine tobacco review. Quezon, Philippines. Philipp. Tobacco Rev. See B–P–H 706/8. HI 57933

Philippine weed science bulletin. Laguna. Vols. 1-4,

1974-77. Philipp. Weed Sci. Bull. Superseded by: Philippine journal of weed science. HI 69485

Philobiblon. Vienna. Vols. 1-12, 1928-40. Philobiblon. 4-3337-3. HI 74872

Philosophia naturalis; Archiv für Naturphilosophie und die philosophischen Grenzgebiete der exakten Wissenschaften und Wissenschaftsgeschichte. Meisenheim. Vol. 1+, 1950+. Philos. Nat. HI 69486

Philosophical collections. [Edited by R. Hooke.] London. Philos. Collect. See B–P–H 706/22. HI 57934

Philosophical magazine; a journal of theoretical, experimental and applied physics. London. Philos. Mag. See B–P–H 706/23. HI 57935

Philosophical magazine, or annals of chemistry, mathematics, astronomy, natural history, and general science. London. Philos. Mag. Ann. Chem. See B–P–H 706/24. HI 57936

Philosophical magazine and journal. London. Philos. Mag. J. See B–P–H 706/25. HI 57937

Philosophical review. Boston, MA, Ithaca, NY. Vol. 1+, 1892+. Philos. Rev. 4-3341-1. HI 69487

Philosophical transactions: giving some account of the present undertakings, studies, and labours of the ingenious in many parts of the world. London. Philos. Trans. See B–P–H 706/29. HI 57939

Philosophical transactions abridged, with notes and biographical illustrations. [Edited by Hutton.] London. Philos. Trans. Abr. (Hutton). See B–P–H 707/1. HI 57940

Philosophical transactions (and collections) abridged and disposed under general heads. [Edited by Lowthrop (vols. 1-3), Jones (vols. 4-5), Reid & Gray (vol. 6), Eames & Martin (vols. 7-8) and Martyn (vols. 9-10). London. Philos. Trans. Abr. (Lowthrop). See B–P–H 707/2. HI 57941

Philosophical transactions, of the royal society of London = Philosophical transactions: giving some account of the present undertakings, studies, and labours of the ingenious in many parts of the world. London. Philos. Trans. See B–P–H 706/29.

Philosophical transactions of the royal society of London. Series B, containing papers of a biological character. London. Philos. Trans., Ser. B See B–P–H 707/5. HI 57942

Philosophy of science. Baltimore, MD. Philos. Sci. See B–P–H 706/26. HI 57938

Phipps folio; a publication of the Phipps [conservatory] friends. Pittsburgh, PA. Vol. 1+, 1988+. Phipps Folio. HI 69488

Photobiochemistry and photobiophysics. Amsterdam. Vol. 1+, 1979+. Photobiochem. Photobiophys. HI 69489

Photobiology bulletin. London. 1979+. Photobiol. Bull. HI 69490

Photochemical and photobiological reviews. New York. Vol. 1+, 1976+. Photochem. Photobiol. Rev. HI 69491

Photochemistry and photobiology. New York, NY. Photochem. & Photobiol. See B–P–H 707/9. HI 57943

Photogrammatic engineering and remote sensing. Falls Church, VA. Vol. 41+, 1975+. Photogramm. Engin. Remote Sensing. Preceded by: Photogrammatic engineering [not entered]. HI 75202

Photographic news. London. Vols. 1-53, 1858-1908. Photogr. News. 4-3348-1. HI 74873

Photographie und Forschung; Contaxphotographie in der Wissenschaft. Dresden, Stuttgart. Vol. 1+, 1935-44; 1952+. Photogr. Forsch. HI 69492

Photophile. Vol. ?-2-?, ?-1973-80-? Photophile. HI 69493

Photosynthesis bibliography. The Hague. Vol. 1+, 1966/70+, 1974+. Photosyn. Bibliogr. HI 69494

Photosynthesis research. The Hague. Vol. 1+, 1980+. Photosyn. Res. HI 69495

Photosynthetica; international journal for photosynthesis research. Prague. Vol. 1+, 1967+. Photosynthetica. HI 69496

Phycologia. Berkeley, CA. Phycologia. See B–P–H 707/15. HI 57947

Phycologia latino-americana. Vaduz. Vol. 1+, 1981+. Phycol. Latino-Amer. HI 69497

Phycological bulletin. Glasgow. Phycol. Bull. See B–P–H 707/11. HI 57944

Phycological newsletter. New York, NY. Phycol. Newslett. See B–P–H 707/13. HI 57945

Phycological studies. Austin, TX. Phycol. Stud. See B–P–H 707/14. HI 57946

Phykos. New Delhi. Phykos. See B–P–H 707/16. HI 57948

Physical and biological sciences, The. Washington, DC. 1961+. Phys. Biol. Sci. HI 69498

Physical techniques in biological research. New York. Vol. 1+, 1961+. Phys. Techn. Biol. Res. HI 69499

Physicalische Zeitung. Halle. Phys. Zeitung. See B–P–H 708/16. HI 57968

Physicalsk, oeconomisk og medico-chirurgisk bibliothek for Danmark og Norge. Copenhagen. Phys. Biblioth. Danmark Norge. See B–P–H 707/23. HI 57954

Physico-chemical biology. [Seibutsu butsuri kagaku.]

Chiba. Vol. 1+, 1951+. Phys.-Chem. Biol. HI 69500

Physikalisch-Oekonomische Auszüge aus den neuesten und besten Schriften, die zur Naturlehre, Haushaltungskunst, Policei, Kameral- auch andern damit verwandlen Wissenschaften gehören, ... Stuttgart. Phys.-Oekon. Auszüge Neuestern Besten Schriften Naturl. See B–P–H 708/7. HI 57959

Physikalisch-oekonomische Wochenschrift, welche als Realzeitung das nützlichste, zuverlässigte ... aus der Natur- und Haushaltungs-Wissenschaft enthält ... Stuttgart. Phys.-Oekon. Wochenschr. See B–P–H 708/8. HI 57960

Physikalisch-ökonomische Bibliothek worinn von den neuesten Büchern, welche die Naturgeschichte, ... betreffen, zuverlässige und vollständige Nachrichten ertheilet werden. [Edited by J. Beckmann.] Göttingen. Phys.-Ökon. Biblioth. See B–P–H 708/9. HI 57961

Physikalisch-ökonomische Monaths- und Quartalschrifft. Dresden & Leipzig. Phys.-Ökon. Monaths-Quartalschr. See B–P–H 708/10. HI 57962

Physikalisch-ökonomische Quartalschrifft. Dresden & Leipzig. Phys.-Ökon. Quartalschr. See B–P–H 708/12. HI 57964

Physikalisch-ökonomische Zeitung. Durch eine Gesellschaft Naturforscher und Oekonomen herausgegeben. Breslau [=Wroclow, Poland]. Phys.-Ökon. Zeitung. See B–P–H 708/13. HI 57965

Physikalische Abhandlungen der Königlichen Akademie der Wissenschaften zu Berlin = Abhandlungen der Königlichen Akademie der Wissenschaften in Berlin. Berlin. Abh. Königl. Akad. Wiss. Berlin. See B–P–H 30/2.

Physikalische Arbeiten der einträchtigen Freunde in Wien. Vienna. Phys. Arbeiten Eintr. Freunde Wien. See B–P–H 707/19. HI 57950

Physikalische Belustigungen. Berlin. Phys. Belust. See B–P–H 707/20. HI 57951

Physikalische Bibliothek. Göttingen. Phys. Biblioth. (Göttingen). See B–P–H 707/21. HI 57952

Physikalische Bibliothek. Rostock & Wismar. Phys. Biblioth. (Rostock & Wismar). See B–P–H 707/22. HI 57953

Physikalische Briefe. Stettin [Poland]. Phys. Briefe. See B–P–H 708/1. HI 57955

Physikalische, chemische, naturhistorische und mathematische Abhandlungen aus der neuen Sammlung der Schriften der Königlichen Dänischen Gesellschaft der Wissenschaften. Copenhagen. Phys. Abh. Königl. Dän. Ges. Wiss. See B–P–H 707/18. HI 57949

Physikalische und Medicinische Abhandlungen der Kayserlichen Akademie der Wissenschaften in Petersburg. Riga [Latvian S S R]. Phys. Med. Abh. Kayserl. Akad. Wiss. Petersburg. See B–P–H 708/3. HI 57956

Physikalische und Medicinische Abhandlungen der Königlichen Academie der WIssenschaften zu Berlin. Gotha. Phys. Med. Abh. Königl. Acad. Wiss. Berlin. See B–P–H 708/4. HI 57957

Physikalische und ökonomische Patriot oder Bemerkungen und Nachrichten aus der Naturhistorie, der allgemeinen Haushaltungskunst und der Handlungskunst. Hamburg. Phys. Ökon. Patriot. See B–P–H 708/11. HI 57963

Physikalische und philosophische Abhandlungen der Gesellschaft der Wissenschaften zu Manchester. Leipzig. Phys. Philos. Abh. Ges. Wiss. Manchester. See B–P–H 708/14. HI 57966

Physikalisches Taschenbuch für Freunde der Naturlehre und Künstler. Göttingen. Phys. Taschenb. Freunde Naturl. Künstler. See B–P–H 708/15. HI 57967

Physiographiska sällskapets årsberättelse. Lund. Physiogr. Sällsk. Årsberätt. See B–P–H 709/3. HI 57969

Physiographiska sällskapets tidskrift. Lund. Physiogr. Sällsk. Tidskr. See B–P–H 709/4. HI 57970

Physiographiska sälskapets handlingar. Stockholm. Physiogr. Sälsk. Handl. See B–P–H 709/6. HI 57971

Physiographiska sälskapets magazin. Lund. Physiogr. Sälsk. Mag. See B–P–H 709/7. HI 57972

Physiologia comparata et oecologia. The Hague. Physiol. Comp. Oecol. See B–P–H 709/9. HI 57974

Physiologia plantarum. Copenhagen. Physiol. Pl. (Copenhagen). See B–P–H 709/11. HI 58126

Physiologia plantarum supplementum. [Supplement to: Physiologia plantarum.] Copenhagen. Vols. 1-6, 1963-71. Physiol. Pl. Suppl. HI 69501

Physiological abstracts; issued by the physiological society (Great Britain and Ireland) with the co-operation of the american physiological society. London. Physiol. Abstr. See B–P–H 709/8. HI 57973

Physiological and molecular plant pathology. London & Orlando, FL. Vol. 28+, 1986+. Physiol. Molec. Pl. Pathol. Preceded by: Physiological plant pathology. HI 69502

Physiological plant pathology; an international journal of experimental plant pathology. London, New York. Vols. 1-27, 1971-85. Physiol. Pl. Pathol. Superseded by: Physiological and molecular plant pathology. HI 69503

Physiological reviews. Baltimore, MD. Physiol. Rev. See B–P–H 709/17. HI 57976

Physiologie végétale. Paris. Vols. 1-24, 1963-86?

Physiol. Vég. Superseded by: Plant physiology and biochemistry. Montrouge. HI 69504

Physiology and biochemistry of cultivated plants = Fiziologiya i biokhimiya kul'turnykh rastenii. Kiev.

Physiology and biochemistry of cultivated plants. [Translation of: Fiziologiya i biokhimiya kul'turnykh rastenii.] New York. Vols. 1-2, 1969-71. Physiol. Biochem Cult. Pl. HI 69505

Physiology and ecology. [Seiri-seitai.] Kyoto. Vol. 1+, 1947+. Physiol. & Ecol. Preceded by: Physiology and oecology. 5-3841-2. HI 69506

Physiology and oecology = Physiology and ecology. Kyoto.

Physiology of plants. [Shokubutsu seiri.] Kyoto, Japan. Physiol. Pl. (Kyoto). See B–P–H 709/12. HI 57975

Physis. Athens. Physis (Athens). See B–P–H 709/21. HI 57977

Physis. Barcelona. Physis (Barcelona). See B–P–H 709/22. HI 57978

Physis; revista de la sociedad argentina de ciencias naturales. Buenos Aires. Vols. 1-31(83), 1915-72. Physis (Buenos Aires). Preceded by: Boletin de la sociedad physis para el cultivo y difusion de las ciencias naturales en la Argentina. Superseded by: Physis. Sección A, Sección B, and Sección C. 4-3356-3. HI 69507

Physis. Florence. Physis (Florence). See B–P–H 709/24. HI 57979

Physis. Paris. Physis (Paris). See B–P–H 709/25. HI 57980

Physis. Stuttgart. Physis (Stuttgart). See B–P–H 709/27. HI 57981

Physis. Sección A, los océanos y sus organismos. Buenos Aires. Vols. 32(84)-35(91), 1973-76. Physis, A. Preceded by: Physis. Superseded by: Physis. Secciónes A, B y C. HI 69508

Physis. Sección B, los aguas continentales y sus organismos. Buenos Aires. Vols. 32(84)-35(91), 1973-76. Physis, B. Preceded by: Physis. Superseded by: Physis. Secciónes A, B y C. HI 69509

Physis. Sección C, los continentes y los organismos terrestres. Buenos Aires. Vols. 32(84)-35(91), 1973-76. Physis, C. Preceded by: Physis. Superseded by: Physis. Secciónes A, B y C. HI 69510

Physis. Secciónes A, B y C. Buenos Aires. Vol. 36(92)+, 1977+. Physis, A, B & C. Preceded by: Physis. Sección A, los océanos y sus organismos; Physis. Sección B, los aguas continentales y sus organismos; and Physis. Sección C, los continentes y los organismos terrestres. HI 69511

Physisch-medicinisches Journal. Leipzig. Phys.-Med. J. See B–P–H 708/6. HI 57958

Physische Abhandlungen, der Königlichen Akademie der Wissenschaften in Paris = Der Königlichen Akademie der Wissenschaften in Paris Physische Abhandlungen. Breslau [=Wroclaw, Poland]. Königl. Akad. Wiss. Paris Phys. Abh. See B–P–H 517/17.

Phyta. Peradeniya. Vol. 1+, 1976+. Phyta (Peradeniya). HI 69512

Phyta; journal of the society of plant taxonomists. Allahabad. Vol. 1+, 1978+. Phyta, J. Soc. Pl. Taxonomists. HI 69513

Phyta monograph. Allahabad. Vol. 1+, 1984+. Phyta Monogr. HI 69514

Phytiatrie--phytopharmacie. Revue française de médecine et de pharmacie des végétaux. Paris. Phytiatrie--Phytopharm. See B–P–H 710/3. HI 57982

Phytochemical bulletin, botanical society of America. Irvine, CA. Vol. 5(4)+, 1972+. Phytochem. Bull. Bot. Soc. Amer. Preceded by: Newsletter, phytochemical section, botanical society of America. HI 69515

Phytochemistry. Oxford, England. Phytochemistry. See B–P–H 710/5. HI 57984

Phytochemistry and phytobiology. Oxford, England. Phytochem. & Phytobiol. See B–P–H 710/4. HI 57983

Phytocoenologia; journal of the international society for plant geography and ecology. Lehre, Berlin, Stuttgart. Vol. 1+, 1973+. Phytocoenologia. HI 69516

Phytocoenosis; biuletyn fitosocjologiczny. Warsaw. Vol. 1, no. 2+, 1972+. Phytocoenosis. HI 69517

Phytographische Blätter. Göttingen. Phytogr. Blätt. See B–P–H 710/7. HI 57985

Phytologia. New York, NY. Phytologia. See B–P–H 710/12. HI 57986

Phytologia memoirs. Plainfield, NJ. Vol. 1+, 1980+. Phytologia Mem. HI 69518

Phytologist. London. Phytologist. See B–P–H 710/14. HI 57987

Phytology = Fitologiya. Sofia.

Phytoma; revue de phytomédecine appliquée. Paris. Phytoma. See B–P–H 710/16. HI 57988

Phytomorphology; an international journal of plant morphology. Delhi. Phytomorphology. See B–P–H 710/19. HI 57989

Phyton. Buenos Aires. Phyton (Buenos Aires). See B–P–H 710/21. HI 57990

Phyton. Horn, Austria. Phyton (Horn). See B–P–H 710/23. HI 57991

Phytoparasitica; Israel journal of plant protection sciences. Bet Dagan. Vol. 1+, 1973+; Supplement no.

1+, 1977+. Phytoparasitica. Preceded in Part by: Israel journal of agricultural research. HI 69519

Phytopathologia mediterranea. Bologna. Phytopathol. Medit. See B–P–H 710/32. HI 57993

Phytopathological classics. Ithaca, NY. Phytopathol. Class. See B–P–H 710/30. HI 57992

Phytopathological papers. Kew. No. 1+, 1956+. Phytopath. Pap. HI 69520

Phytopathologische Zeitschrift. Berlin. Vols. 1-114, 1929-65. Phytopathol. Z. Superseded by: Journal of phytopathology. 4-3357-2. HI 69521

Phytopathology. Ithaca, NY. Phytopathology. See B–P–H 711/2. HI 57994

Phytopathology news. St. Paul, MN. Vol. 1+, 1967+. Phytopath. News. HI 69522

Phytophthora newsletter. Madison, WI. No. 1+, 1973+. Phytophthora Newslett. HI 69523

Phytophylactica. Pretoria. Vol. 1+, 1969+. Phytophylactica. Preceded in part by: South african journal of agricultural science. HI 69524

Phytoprotection. Montreal. Vol. 44+, 1963+. Phytoprotection. Preceded by: Report (Annual) of the Quebec society for the protection of plants. HI 69525

Phytotherapy. Paris. Vol. 1+, 1982+. Phytotherapy. HI 69526

Phytotronic newsletter. Paris. No. 1+, 1971+. Phytotr. Newslett. HI 69527

Phytotronique. Paris. No. 1+, 1969+. Phytotronique. HI 69528

Piante grasse. Santo Spirito. Vol. 1+, 1981+. Piante Grasse. HI 69529

Pica. Calgary. Vol. 1+, 1979+. Pica. Preceded by: Calgary field naturalist. HI 69530

Picardie écologie. Poix. No. ?-2+, ?-1983+. Picardie Écol. HI 69531

Pilz. Berlin-Steglitz. Pilz. See B–P–H 711/5. HI 57996

Pilz- und Kräuterfreund. Nuremberg. Pilz-Kräuterfreund. See B–P–H 711/6. HI 57997

Pine plains. Albany, NY. Vol. 1+, 1977+. Pine Plains. HI 69532

Pineapple news. Honolulu, HI. Pineapple News. See B–P–H 711/7. HI 57998

Pineapple quarterly. Honolulu, HI. Pineapple Quart. See B–P–H 711/9. HI 57999

Ping tu hsüeh chi k'an = Acta virologica sinica. Beijing.

Pioneer pecan press. San Saba, TX. Pioneer Pecan Press. See B–P–H 711/12. HI 58000

Pionier; illustrierte Zeitschrift zur Unterhaltung und Belehrung für Jedermann. Dresden. Vol. 1, 1875. Pionier. HI 69533

Pipeline; international lilac society newsletter. Rumford, ME. Vol. 1+, 1974+. Pipeline. HI 69534

Pirineos; revista de la estación de estudios pirinaicos later Revista del instituto de estudios pirenaicos. Zaragoza, Jaca. Vol. 1+, 1952+. Pirineos. HI 69535

Pishchevaya promyshlennost' S S S R = Tabak. Moscow.

Pisum newsletter. Geneva, NY. Vol. 1+, 1969+. Pisum Newslett. HI 69536

Pitch pine naturalist. Central Islip, NY. No. 1+, 1972+. Pitch Pine Naturalist. HI 69537

Pittieria; publicación del herbario de la facultad de ciencias forestales de la universidad de los Andes. Merida. No. 1+, 1967+. Pittieria. HI 69538

Pittonia. Berkeley, CA. Pittonia. See B–P–H 711/14. HI 58001

Plankton jiho. Tokyo. Plankt. Jiho. See B–P–H 713/5. HI 58025

Plankton kontinental'nykh vodoemov. Leningrad. 1961/70+, 1970+. Plankt. Kontinental'n. Vodoemov. HI 75119

Plant bibliography. Lisle, IL. No. 1+, 1978+. Pl. Bibliogr. HI 69539

Plant biochemical journal. Kalyani, New Delhi. Vols. 1-7, 1974-80. Pl. Biochem. J. Superseded by: Plant physiology and biochemistry. HI 69540

Plant biochemistry news letter. New Delhi. No. 1+, 1973+. Pl. Biochem. News Lett. HI 69541

Plant biology. New York. Vol. 1+, 1986+. Pl. Biol. HI 71308

Plant biosystematics newsletter. Ann Arbor, MI. No. 1, 1969. Pl. Biochem. News Lett. Superseded by: Newsletter, international organization of plant biosystematists. HI 69542

Plant biotechnology. Sheffield. Vol. 1+, 1981+. Pl. Biotechnol. HI 69543

Plant breeding. Berlin, New York. Vol. 96+, 1986+. Pl. Breed. (New York). Preceded by: Zeitschrift für Pflanzenzüchtung. HI 69544

Plant breeding. [Chung-kuo k'o hsüeh yüan. Chih wu yüan kung tso wei yüan hui.] Peking. No. 1+, 1965+. Pl. Breed. (Peking). HI 58002

Plant breeding abstracts. Cambridge, England. Pl. Breed. Abstr. See B–P–H 711/17. HI 58003

Plant breeding reviews. Westport, CT, New York. Vol. 1+, 1983+. Pl. Breed. Rev. HI 69545

Plant cell, The. Rockville, MD. Vol. 1+, 1989+. Pl. Cell. HI 69546

Plant, cell and environment; cell physiology, whole plant physiology, community physiology. Oxford. Vol. 1+, 1978+. Pl. Cell Environm. HI 69547

Plant and cell physiology. [Nihon shokubutsu seiri gakkai.] Tokyo. Pl. Cell Physiol. See B–P–H 711/19. HI 58005

Plant cell reports. Heidelberg, New York, etc. Vol. 1+, 1981+. Pl. Cell Rep. HI 69548

Plant cell, tissue and organ culture; international journal on in vitro culture of higher plants. The Hague, Boston, MA. Vol. 1+, 1981+. Pl. Cell Tissue Organ Cult. HI 69549

Plant conservation bulletin. Howrah. No. 1+, 1981+. Pl. Conservation Bull. HI 69550

Plant disease; international journal of applied plant pathology. St. Paul, MN. Vol. 64+, 1980+. Pl. Dis. Preceded by: Plant disease reporter. HI 69551

Plant disease bulletin. Washington, DC. Vols. 1-6, 1917-22. Pl. Dis. Bull. Superseded by: Plant disease reporter. HI 69552

Plant disease knowledge. [Zhibing zhishi.] [China]. Pl. Dis. Knowl. See B–P–H 712/1. HI 58009

Plant disease reporter. Washington, DC, Beltsville, MD. Vols. 7-63, 1923-79. Pl. Dis. Reporter. Preceded by: Plant disease bulletin. Superseded by: Plant disease. HI 69553

Plant disease research. [Shokubutsu byogai kenkyu.] Kyoto. Vol. 1+, 1931+. Pl. Dis. Res. 2-1607-2. HI 54186

Plant diseases and pests in Denmark. Lyngby. Vol. 97+, 1980+. Pl. Dis. Pests Denmark. Preceded by: Plantesygdomme i Danmark. HI 69554

Plant ecology and phytogeography research series. [Chih wu shêng t'ai hsüeh yü ti chih wu hsüeh tzu liao t'sung k'an.] Peking. Pl. Ecol. Phytogeogr. Res. Ser. See B–P–H 712/5. HI 58010

Plant food review. Washington, DC. Vols. 1-15(1), 1955-69. Pl. Food Rev. HI 69555

Plant foods for human nutrition. Oxford, Dordercht. Vols. 1-2, 1968-72; vol. 37(3)+, 1987+. Pl. Foods Human Nutr. For 1973-87 see: Qualitas plantarum. HI 69556

Plant foods for man. Vols. 1-2, 1973-78. Pl. Foods Man. Superseded by: Journal of plant foods. HI 69557

Plant genetic resources newsletter. Rome. No. 25+, 1971+. Pl. Genet. Resources Newslett. Preceded by: Plant introduction newsletter. HI 69558

Plant growing. Sofia = Rastenievudni nauki. Sofia.

Plant growth regulation; international journal on natural and synthetic regulators. The Hague, Boston, MA. Vol. 1+, 1982+. HI 69559

Plant growth regulator abstracts. Farnham Royal. Vol. 1+, 1975+. Pl. Growth Regulator Abstr. HI 69560

Plant growth regulator bulletin. St. Paul, MN, etc. Vol. 1+, 1973+. Pl. Growth Regulator Bull. HI 69561

Plant health newsletter, european and mediterranean plant protection organisation = Publications, european and mediterranean plant protection organisation. Series B. Paris.

Plant immigrants, United States department of agriculture = U S department of agriculture. Plant immigrants. Washington, DC. U.S.D.A. Pl. Immigr. See B–P–H 944/11.

Plant information bulletin. [Supplement to: Morton arboretum quarterly.] Lisle. Il. No. 1+, 1974+. Pl. Inform. Bull. HI 69562

Plant introduction. Cairo. Nos. 15?, 1919-23? Pl. Introd. (Cairo). HI 75098

Plant introduction. Fiji. No. 1+, 1951+. Pl. Introd. HI 69563

Plant introduction newsletter. Rome. Nos. 1-24, 1957-70. Pl. Introd. Newslett. Superseded by: Plant genetic resources newsletter. HI 69564

Plant introduction review, division of plant industry, C S I R O. Canberra, A.C.T. Vols. 2-10(1), 1965-74. Pl. Introd. Rev. Div. Pl. Industr. C. S. I. R. O. Preceded by: Quarterly list of introductions, plant introduction section, C S I R O. Superseded by: Australian plant introduction review. HI 69565

Plant life. Stanford, CA. Pl. Life. See B–P–H 712/10. HI 58012

Plant lore. Geneseo, NY. Vols. 1-9(1), 1977-85. Pl. Lore. HI 69567

Plant magazine. [Zhiwuxue zazhi.] Beijing. Vol. 1-?, 1974-? Pl. Mag., Beijing. Superseded by: Plants. HI 69568

Plant material description. No. 1+, 19??+. Pl. Mater. Descr. HI 69569

Plant-microbe interactions; molecular and genetic perspectives. New York. Vol. 1+, 1984+. Pl.-Microbe Interact. HI 71808

Plant molecular biology; international journal on fundamental research and genetic engineering. The Hague, Boston, MA. Vol. 1+, 1982+. Pl. Molec. Biol. HI 69570

Plant molecular biology association newsletter = Plant molecular biology newsletter. Charlottesville, VA.

Plant molecular biology newsletter. Charlottesville, VA. Vol. 1-?, 1980-? Pl. Molec. Biol. Newslett. Superseded by: Plant molecular biology reporter. HI 69571

Plant molecular biology reporter; quarterly newsletter of

the international society for plant molecular biology New York. Vol. 1+, 1983+. Pl. Molec. Biol. Reporter. Preceded by: Plant molecular biology newsletter. HI 69572

Plant monograph reprints. Lehre. Vol. ?-7+, ?-1972+. Pl. Monogr. Repr. HI 69573

Plant and nature. Kanpur. Vol. 1+, 1983+. Pl. Nat. HI 69574

Plant news. Rome. No. 1+, 1963+. Pl. News. HI 69575

Plant patent. Washington, DC. No. 1+, 1931+. Pl. Patent. HI 72149

Plant pathologist's quarterly record. Harpenden. No. ?-3+, ?-1975+. Pl. Pathologist's Quart. Rec. HI 69576

Plant pathology. London. Pl. Pathol. See B–P–H 712/12. HI 58013

Plant pathology circular, division of plant industry, Florida department of agriculture. Gainesville, FL. No. 10+, 1963+. Pl. Pathol. Circ. Div. Pl. Industr., Florida Dep. Agric. Preceded by: Pathology circular, division of plant industry, Florida department of agriculture. HI 69577

Plant pathology circular, Florida department of agriculture = Plant pathology circular, division of plant industry, Florida department of agriculture. Gainesville, FL.

Plant pathology department pamphlet, South Dakota agricultural experiment station = South Dakota agricultural experiment station. Plant pathology department pamphlet. Brookings, SD. South Dakota Agric. Exp. Sta. Pl. Pathol. Dept. Pam. See B–P–H 848/3.

Plant pest news. Hyattsville, MD. Vol. 1+, 1981+. Pl. Pest News. HI 69578

Plant physiology. Lancaster, PA. Pl. Physiol. (Lancaster). See B–P–H 712/13. HI 58014

Plant physiology. ("Fiziologiya rasteny" in English translation). Washington, DC. Pl. Physiol. (Washington). See B–P–H 712/14. HI 58015

Plant physiology. Supplement. Rockville, MD. Vol. 30+, 1955+. Pl. Physiol., Suppl. 4-3370-2. HI 69579

Plant physiology and biochemistry; official journal of the french society of plant physiology and of the federation of european societies of plant physiology. Montrouge. Vol. 25+, 1987+. Pl. Physiol. Biochem. (Montrouge). Preceded by: Physiologie végétale. HI 69580

Plant physiology and biochemistry. New Delhi. Vol. 8+, 1981+. Pl. Physiol. Biochem. (New Delhi). Preceded by: Plant biochemical journal. HI 69581

Plant physiology communications. [Zhiwu shengli tongxun.] Shanghai. No. ?-95+, ?-1983+. Pl. Physiol. Commun. HI 69582

Plant portraits. [Offprints from: Journal of the Adelaide botanic gardens.] Adelaide, S.A. 1976+. Pl. Port. HI 69583

Plant press; plant conservation news from the C A B S. London. Vol. 38+, 1987+. Pl. Press (London). HI 69585

Plant press; field botany in Ontario. Mississauga. Vol. 1+, 1983+. Pl. Press (Mississauga). HI 69586

Plant press. Tucson, AZ. Vol. 5(2)+, 1981+. Pl. Press (Tucson). Preceded by: Bulletin, Arizona native plant society. HI 69584

Plant press. Washington, DC. No. 4+, 1973+. Pl. Press (Washington). Preceded by: Smithsonian botanical news. HI 69587

Plant production and protection papers, F A O = F A O plant production and protection papers. Rome.

Plant propagator. New Brunswick, NJ. Pl. Propag. See B–P–H 712/17. HI 58017

Plant protection. Belgrade = Zaštita bilja. Belgrade. Zašt. Bilja. See B–P–H 995/21.

Plant protection. [Translation of: Zaštita bilja.] Belgrade. Vol. 22+, 1973+. Pl. Protect. (Belgrade). HI 69588

Plant protection. Leningrad = Zashchita rastenii. Leningrad.

Plant protection. [Chih wupao hu Zhiwu baohu.] Peking. Pl. Protect. (Peking). See B–P–H 712/20. HI 58018

Plant protection. [Shokubutsu boeki.] Pl. Protect. (Tokyo). See B–P–H 712/21. HI 58019

Plant protection abstracts. Beersheva. Vol. 1+, 1965+; Supplement, 1967+. Pl. Protect. Abstr. HI 69590

Plant protection bulletin. Ankara = Bitki koruma bulteni. Ek yayin. Ankara. Bitki Koruma Bult. See B–P–H 197/18.

Plant protection bulletin. New Delhi. Vol. 1+, 1949+. Pl. Protect. Bull. (New Delhi). HI 69591

Plant protection bulletin. [Chih wu pao hu hsueh hui k'an.] Taipei. Vol. 1+, 1959+. Pl. Protect. Bull. (Taipei). HI 69592

Plant protection bulletin, F A O. Rome = F A O plant protection bulletin. Rome.

Plant protection courier = Pflanzenschutz-Kurier. Leverkusen.

Plant protection overseas review. London. Vols. 1-4, 1948-55. Pl. Protect. Overseas Rev. HI 69593

Plant protection quarterly. North Clayton?, Vic. Vol. 1+, 1985+. Pl. Protect. Quart. Preceded by: Australian weeds HI 69594

Plant quarantine and control administration report, United States department of agriculture = U S department of agriculture, plant quarantine and control administration report. Washington, DC. U.S.D.A. Pl. Quarant. Admin. Rep. See B–P–H 944/12.

Plant quarantine news. Ottawa. Vols. 1-3(2), 1967-70. Pl. Quarant. News. Preceded by: Newsletter, plant protection division, department of agriculture, Canada. HI 69595

Plant research. East Lansing, MI. Vols. 1-14, 1966-79. Pl. Res. Superseded by: Report (Annual) of the M S U - D O E plant research laboratory. HI 69596

Plant research and development. Tubingen. Vol. 1+, 1975+. Pl. Res. Developm. HI 72484

Plant science; international journal of experimental plant biology. Limerick, New York. Vol. 38+, 1985+. Pl. Sci. (Elsevier). Preceded by: Plant science letters. HI 69597

Plant science. Lucknow, Calcutta. Vol. 1+, 1969+. Pl. Sci. (Lucknow). HI 69598

Plant science. Sofia = Rastenievudni nauki. Sofia.

Plant science bulletin. Urbana, IL. Pl. Sci. Bull. See B–P–H 712/24. HI 58020

Plant science laboratories herbarium news, university of Reading. Reading. No. [1]+, 1983+. Pl. Sci. Lab. Herb. News Univ. Reading. HI 69599

Plant science letters; international journal of experimental plant biology. Amsterdam, Limerick. Vols. 1-37(3), 1973-85. Pl. Sci. Lett. Superseded by: Plant science. HI 69600

Plant science literature. U S department of agriculture library. Washington, DC. Pl. Sci. Lit. U.S.D.A. Libr. See B–P–H 712/27. HI 58021

Plant and soil; international journal of plant nutrition, plant chemistry, soil microbiology and soil-borne plant diseases. The Hague. Pl. & Soil. See B–P–H 713/2. HI 58022

Plant species biology; an international journal. Kyoto. Vol. 1+, 1986+. Pl. Spec. Biol. HI 69602

Plant systematics and evolution. Vienna, New York. Vol. 123+, 1974+; Supplementum 1977+. Pl. Syst. Evol. Preceded by: Oesterreichische botanische Zeitschrift. HI 69603

Plant systematics and evolution. Supplementum. Vienna, New York. 1977+. Pl. Syst. Evol., Suppl. HI 69604

Plant varieties journal. Canbera, A.C.T. Vol. 1+, 1988+. Pl. Var. J. HI 69605

Plant varieties and seeds. Oxford. Vol. 1+, 1988+. Pl. Var. Seeds. HI 72552

Plant varieties and seeds gazette. London. Pl. Var. Seeds Gaz. See B–P–H 713/3. HI 58023

Plant variety protection office official journal = Official journal, plant variety protection office. Hyattsville, Beltsville, MD.

Plant world. Binghamton, NY. Pl. World. See B–P–H 713/4. HI 58024

Planta. Archiv für wissenschaftliche Botanik. (Zeitschrift für wissenschaftliche Biologie. Abt. E.) Berlin. Planta. See B–P–H 713/15. HI 58026

Planta medica; Zeitschrift für Arzneipflanzenanwendung und Arzneipflanzenforschung [subtitle varies]. Stuttgart. Vols. 1-31, 1953-76; Supplement 1967-75. Pl. Med. (Stuttgart). Superseded by: Journal of medicinal plant research. HI 69606

Planta polonica. Warsaw. Pl. Polon. See B–P–H 712/16. HI 58016

Plantae cryptogamae. Moscow & Leningrad = Trudy botanicheskogo instituta akademii nauk S S S R. Ser. 2, sporovye rasteniya. Moscow & Leningrad.

Plantas cultivades en la República Argentina. Castelar, Argentina. Pl. Cult. Repúbl. Argent. See B–P–H 711/22. HI 58008

Plantas medicinales; revista. Havana. Vol. 1+, 1981+. Pl. Med. (Havana). HI 69607

Plantas novas cultivadas no jardim botanico do Rio de Janeiro. Rio de Janeiro. Vols. [1]-6, 1891-98. Pl. Novas Cult. Jard. Bot. Rio de Janeiro. Superseded by Contributions du jardin botanique de Rio de Janeiro. 4-3682-2. HI 69608

Plantation technical papers, Victoria forests commission. Melbourne, Vic. Nos. 1-5, 1955-58. Plantation Techn. Pap. Victoria. Superseded by: Forestry technical papers, forests commission, Victoria. HI 69609

Plante et l'homme. Paris. Vol. 1+, 1970+. Pl. Homme. HI 69610

Plante officinali. Rome. Plante Officinali. See B–P–H 711/4. HI 57995

Plantebeskyttelsesmidler. (Sweden). 1978+. Plantebeskyttelsesmidler. HI 69611

Planter. Kuala Lumpur. Planter. See B–P–H 713/21. HI 58027

Planters' bulletin, rubber research institut of Malaya. Kuala Lumpur. Pl. Bull. See B–P–H 711/18. HI 58004

Planter's chronicle. Bangalore, India. Pl. Chron. See B–P–H 711/20. HI 58006

Planters' gazette. London. 1882-97. Planter's Gaz. HI 69613

Plantes médicinales et phytothérapie. Angers. No. 1+, 1967+. Pl. Méd. Phytothérap. HI 69614

Plantes de montagne; bulletin de la société des amateurs de jardins alpins. Paris. Vol. 1+, [1952]+. Pl. Mont.

HI 69615

Plantesygdomme i Danmark. Copenhagen. Vols. 1-96, 1884-79. Plantesygd. Danmark. Superseded by: Plant diseases and pests in Denmark. HI 69616

Planteurs de betteraves. Paris. Nos. 1-104, 1931-40; n.s. nos. 1-83/84, 1945-52. Pl. Betteraves. Superseded by: Betteravier français. 4-3371-3. HI 69617

Plantfinder. Fort Lauderdale, FL. 1987+. Plantfinder. Preceded by: Florida plant finder. HI 69618

Planthunter. Eskilstuna. (Sweden). No. 1+, 1983+. Planthunter. HI 69619

Planting breeding reviews. Westport, CT. 1982+. Pl. Breed. Rev. HI 69620

Planto; bulletin of Darbhanga botanical society. Darbhanga. Vol. 1+, 1971+. Planto. HI 69621

Plants. [Zhiwu zazhi.] Peking (Beijing). 1977+. Plants. Preceded by: Plant magazine. HI 69622

Plants alive. [Seattle, WA.] Vol. 1+, 1972+. Pl. Alive. HI 69623

Plant's companion. [Shokubutsu no tomo.] Tokyo. Pl. Companion. See B–P–H 711/21. HI 58007

Plants and gardens. New York, NY. Pl. & Gard. See B–P–H 712/8. HI 58011

Plants and gardens news. New York. Vol. 1+, 1986+. Pl. Gard. News. Preceded by: Brooklyn botanic garden newsletter. HI 69624

Plants and the landscape. West Lafayette, IN. Vol. 1+, 1978+. Pl. Landscape. HI 69625

Plants today. Oxford. Vols. 1-2, 1988-89. Pl. Today. HI 69626

PlantSciences. Jamaica Plains, MA. Vol. [1]+, 1980+. PlantSciences. HI 69627

Plantsman. London. Vol. 1+, 1979+. Plantsman. HI 69628

Plasmid; an international journal devoted to extrachromosomal gene systems. San Diego, CA. Vol. 1+, 1977+. Plasmid. HI 73597

Plateau. Flagstaff, AZ. Plateau. See B–P–H 713/24. HI 58028

Plesse-Archiv. Bovenden. 1966+. Plesse-Archiv HI 69629

Plodoovoščnoe hozjajstvo = Plodoovoshchnoe khozyaistvo. Moscow.

Plodoovoshchnoe khozyaistvo. Moscow. 1932(7)-38. Plodoovoshchn. Khoz. Preceded by: Sotsialisticheskoe plodoovoshchnoe khozyaistvo. Superseded by: Ovoshchevodstvo and Sadovodstvo. 4-3374-2. HI 69630

Plodovodstvo. St. Petersburg. Plodovodstvo. See B–P–H 713/27. HI 58029

Plus trees in Korea. [Han'gok ui suhyongmok.] Suwon. No. 1+, 1961+. Plus Trees Korea. HI 69631

Pochvenno-biogeotsenologicheskie issledovaniya v priazov'e. Moscow. Vol. 1+, 1975+. Pochvenno-biogeotsenol. Issl. Priaz. HI 69632

Pochvovědĕnie. Moscow. Vol. 1+, 1899+ [suspended 1917-23; vols. 19-32, 1924-37 also numbered n.s. vols. 1-14]. Pochvovědĕnie. 4-3376-2. HI 69633

Pochvoznanie i agrochimiya. Sofia. Vol. 1+, 1966+. Pochvozn. Agrochim. HI 69634

Pochvy i produktivnost' rastitel'nykh soobschchestv. Vol. 1+, 1972+. Pochvy Produkt. Rastitel'n. Soobschch. HI 69635

Počvovědĕnie = Pochvovědĕnie. Moscow.

Póda a úroda. (Czechoslovakia). Vols. 18-22, 1970-75. Póda Úroda. Preceded and superseded by: Uroda. HI 69636

Pódyjí. Sborník musejních vlastivědných prací. Znojmo, Czechoslovakia. Pódyjí. See B–P–H 713/30. HI 58030

Poeyana. Series A and B. Havana. Vols. 1+, 1964+. Poeyana. Preceded by: Memorias de la sociedad cubana de historia natural "Felipe Poey." HI 69637

Pohl's Archiv der teutschen Landwirthschaft. Leipzig. Pohl's Arch. Teutsch. Landw. See B–P–H 714/2. HI 58031

Pojednání královské české společnosti nauk = Abhandlungen der königlichen Böhmischen Gesellschaft der Wissenschaften. Ser. 6. Prague. Abh Königl. Böhm. Ges. Wiss. See B–P–H 30/4.

Polabí, vlastivedný zpravodaj. Podebrady, Czechoslovakia. Polabí Vlastiv. Zprav. See B–P–H 714/4. HI 58032

Pol'ana. Casopis lesného a drevárského hospodarstva. Bratislava. Pol'ana. See B–P–H 714/5. HI 58033

Polar biology. Berlin, New York, etc. Vol. 1+, 1982+. Polar Biol. HI 69638

Polar geography. Silver Spring, MD. Vols. 1-3, 1977-79. Polar Geogr. Superseded by: Polar geography and geology. HI 69639

Polar geography and geology. Silver Spring, MD. Vol. 4+, 1980+. Polar Geogr. Geol. Preceded by: Polar geography. HI 69640

Polar record. Cambridge, England. Polar Rec. See B–P–H 714/6. HI 58034

Polar research. Oslo. Vol. 1+, 1982+. Polar Res. HI 69641

Polarflokken; medlemsblad for nord-norsk avdeling av norsk botanisk forening. Tromso. No. 1+, 1978+.

Polarflokken. Preceded by: Informasjon fra norsk botanikk forening, nord-norsk avdeling. HI 65770

Polarforschung; halbjahresschrift des Archivs für Polarforsching in Kiel. Holzminden, Kiel. Vol. 1+, 1931/43+. Polarforschung. 4-3381-2. HI 69642

Polarforschung. Beiheft. Kiel. Vol. 1+, 1956+. Polarforschung, Beih. 4-3381-2. HI 69643

Poleznye rasteniyata. Tiflis. 1965+. Polezn. Rast. HI 69644

Policlinico; sezione chirurgica. Rome. Policlinico, Sez. Chir. See B–P–H 714/7. HI 58035

Policlinico; sezione pratica. Rome. Policlinico, Sez. Prat. See B–P–H 714/9. HI 58036

Poligrafo. Milan. Poligrafo (Milan). See B–P–H 714/11. HI 58037

Poligrafo. Verona. Poligrafo (Verona). See B–P–H 714/12. HI 58038

Poligrafo de la Capitanata; giornale di scienze, lettere ed arti. Foggia, Italy. Poligrafo Capitanata. See B–P–H 714/13. HI 58039

Polish agricultural annual. Series A, plant production. Warsaw = Roczniki nauk rolniczych. Seria A, produkcja roslinna. Warsaw.

Polish agricultural annual. Series E, plant protection. Warsaw = Roczniki nauk rolniczych. Seria E, ochrona roslin. Warsaw.

Polish agricultural annual. Series F, land reclamation and grassland farming. Warsaw = Roczniki nauk rolniczych. Seria F, melioracji i uzytków zielonych. Warsaw.

Polish ecological bibliography. Warsaw. 1959+. Polish Ecol. Bibliogr. HI 69645

Polish ecological studies. Warsaw. Vol. 1+, 1975+. Polish Ecol. Stud. HI 69646

Polish journal of chemistry. Warsaw. Vol. 52+, 1978+. Polish J. Chem. Preceded by: Roczniki chemii. HI 69647

Polish journal of ecology = Ekologia polska. Warsaw.

Polish journal of genetics and plant breeding = Genetica polonica. Poznan.

Polish journal of theoretical and applied genetics = Genetica polonica. Poznan.

Polish polar research. Warsaw. Vol. 1+, 1980+. Polish Polar Res. HI 69648

Polish scientific periodicals; current contents with author directory. Warsaw. Nos. 1-126, 1963-75. Polish Sci. Period. HI 69649

Politecnica; revista de información técnica-cientifica. Quito. Vol. 1+, 1967+. Politecnica. HI 69650

Politics and the life sciences; journal of the association for politics and the life sciences. DeKalb, IL. Vol. 1+, 1982+. Polit. Life Sci. HI 69651

Politiske og physiske magazin. Copenhagen. Polit. Phys. Mag. See B–P–H 714/14. HI 58040

Poljodjelska znanstvena smotra. Zagreb. 1941-45. Poljod. Znanst. Smotra. Preceded by: Poljoprivredna nauchno smotra. Superseded by: Poljoprivredna znanstvena smotra. HI 69652

Poljoprivredna nauchno smotra. Zagreb. 1939-41. Poljopr. Nauchno Smotra. Superseded by: Poljodjelska znanstvena smotra. HI 69653

Poljoprivreda i sumarstvo. Titograd. Vols. 10+, 1964+. Poljopr. Sumarstvo. Preceded by: Nasa poljoprivreda i sumarstvo. HI 69654

Poljoprivredna znanstvena smotra. Zagreb. 1946+. Poljopr. Znanst. Smotra. Preceded by: Poljodjelska znanstvena smotra. HI 69655

Poljoprivredni pregled. Sarajeva. 1952+. Poljopr. Pregled. HI 69656

Pollen et spores. Paris. Pollen & Spores. See B–P–H 714/16. HI 58041

Pollen et spores. Supplement. Bibliographie. [Supplement to: Pollen et spores.] Vols. 7-?, 1965-73. Pollen Spores, Suppl., Bibliogr. Superseded by: Bibliographie. Palynologie. HI 69657

Pollustop; revue mensuelle contre la pollution. Paris. Vol. 1+, 1972+. Pollustop. HI 69658

Pollution research. New Delhi. Vol. 1+, 1982+. Pollut. Res. HI 69659

Polska akademja umiejetności. Rozprawy wydziału matematyczno-przyrodniczego. Dzial B. Nauki biologiczne. Cracow. Polska Akad. Umiejetn. Rozpr. Wydz. Mat.-Przyr., Dział B, Nauki Biol. See B–P–H 714/19. HI 58042

Polskie archivum weterynaryjne. Warsaw. Vol. 1+, 1951+. Polsk. Arch. Weteryn. HI 69661

Polskie badania polarne = Polish polar research. Warsaw.

Polytechnisches journal. Stuttgart. Polytechn. J. See B–P–H 714/20. HI 58043

Pome news. Portland, OR. Ca.1981+. Pome News. HI 69662

Pomikultura, vitikultura shi vineritul Moldovei = Sadovodstvo, vinogradarstvo i vinodelie Moldavii. Kishinev.

Pomme de terre française. Lille. Pomme Terre Franç. See B–P–H 714/23. HI 58044

Pommersche Bibliothek. Greifswald, Germany. Pommersche Biblioth. See B–P–H 715/1. HI 58045

Pommersche Provinzial-Blätter für Stadt und Land. Treptow an der Rega [=Trzebiatow, Poland]. Pommersche Prov.-Blätt. Stadt Land. See B–P–H 715/2. HI 58046

Pommersches Archiv der Wissenschaften und des Geschmaks. Stettin [Poland] & Anklam, Germany. Pommersches Arch. Wiss. Geschmaks. See B–P–H 715/3. HI 58047

Pomological magazine; or, figures and descriptions of the most important varieties of fruit cultivated in Great Britain. London. Pomol. Mag. See B–P–H 715/6. HI 58049

Pomologie française. Lyons. Vols. 1080(9), 1872-1954; n.s. vol. 1-19, 1959-77. Pomol. Franç. For 1960-58 see: Arboriculture frutière. 4-3394-1. HI 69663

Pomologische Blätter. Troja near Prague. Pomol. Blätt. See B–P–H 715/4. HI 58048

Pomologische Monatshefte. Stuttgart. Pomol. Monatsh. See B–P–H 715/7. HI 58050

Pomona; north american fruit explorers' quarterly. Bloomington, IL, Paducah, KY. Vol. 14(2)+, 1981+. Pomona (Paducah). Preceded by: North american pomona. HI 69664

Pomona; allgemeine deutsche Zeitschrift für den gesammten Obst- und Weinbau. Regensburg. Vols. 1-16, 1852-67 Pomona (Regensburg). 4-3394-2. HI 58051

Pomona college journal of economic botany, as applied to subtropical horticulture. Claremont, CA. Pomona Coll. J. Econ. Bot. See B–P–H 715/12. HI 58052

Pontificia academia scientiarum, acta. Rome. Pontif. Acad. Sci. Acta. See B–P–H 715/13. HI 58053

Poona agricultural college magazine. Poona. Vol. 1+, 1909+. 4-3394-2 HI 69665

Poplars = Popura. Tokyo.

Popular bulletin, Maryland agricultural experiment station = Maryland agricultural experiment station. Popular bulletin. College Park, MD. Maryland Agric. Exp. Sta. Popular Bull. See B–P–H 549/4.

Popular Bulletin, Washington (state) agricultural experiment station = Washington (state) agricultural experiment station. Popular bulletin. Pullman, WA. Wash. State Agric. Exp. Sta. Popular Bull. See B–P–H 971/16.

Popular gardening. Albany, NY. Vol. 1+, 1950+. Popular Gard. (Albany). HI 69666

Popular gardening. London. Vol. 22+, 1920+. Popular Gard. (London). Preceded by: Gardener. 4-3399-2. HI 69667

Popular information bulletin, Morton arboretum = Bulletin of popular information, Morton arboretum. Lisle, IL. Bull. Popular Inform. (Morton Arbor.). See B–P–H 269/12.

Popular science review. London. Popular Sci. Rev. See B–P–H 715/19. HI 58054

Popular science series, Illinois state museum. Springfield, IL. Vol. 1+, 1939+. Popular Sci. Ser. Illinois State Mus. 3-1933-2. HI 69669

Popular series, Chicago natural history museum. Botany. Chicago, IL. No. 26+, 1947+. Popular Ser. Chicago Nat. Hist. Mus., Bot. Preceded by: Leaflet, department of botany, Field museum of natural history. HI 69670

Populus = Topola. Belgrade.

Popura. Tokyo. [Dates of publication not ascertained.] Popura. HI 69671

Portefeuille des horticulteurs; journal pratique des jardins. Paris. Vols. 1-2, 1847-48. Portef. Hort. 4-3403-3. HI 69672

Portland roses. Portland, OR. Portland Roses. See B–P–H 715/22. HI 58055

Portland roses and flowers. Portland, OR. Portland Roses Fl. See B–P–H 715/23. HI 58056

Porto Rico agricultural experiment station bulletin. Mayagüez, Puerto Rico. Porto Rico Agric. Exp. Sta. Bull. See B–P–H 715/24. HI 58057

Porto Rico agricultural experiment station circular. Mayagüez, Puerto Rico. Porto Rico Agric. Exp. Sta. Circ. See B–P–H 715/25. HI 58058

Portugaliae acta biologica. Série A. Lisbon. Portugaliae Acta Biol., Sér. A. See B–P–H 716/4. HI 58059

Portugaliae acta biologica. Série B. Sistemática, ecologia, biogeografia e paleontologia. Lisbon. Portugaliae Acta Biol., Sér. B, Sist. See B–P–H 716/5. HI 58060

Posebna izdanja, biološki institut narodne republike Srbije. Belgrade. Vols. 1-?, 1954-63. Posebna Izd. Biol. Inst. nar. Republ. Srbije. HI 69673

Posebna izdanja, drustvo za nauku i umjetnost Tsrne Gore. Odjeljenje prirodnikh nauka. Titograd. Vol. 1+, 1975+. Posebna Izd. Drustvo Nauku Umjetnost Umjetn. Tsrne Gore, Odjeljenje Prir. Nauka. HI 69676

Posebna izdanja. Filoz. fak. na univerzitetot Skopje. Hidrobiološki zavod Ohrid. Ochrida. Vol. 1, 1957. Posebna Izd. Filoz. Fak. Univ. Skopje Hidrobiol. Zavod Ohrid. HI 69674

Posebna izdanja. Institut za zaštita bilja. Belgrade. Posebna Izd. Inst. Zašt. Bilja. See B–P–H 716/9. HI 58061

Posebna izdanja, naučno društvo NR Bosne i Hercegovine. Sajarevo. Vol. 1+, 1961+. Posebna Izd. Naučno Društvo NR Bosne Hercegovine. HI 69675

Posebna izdanja, odjeljenje prirodnikh nauka. Drustvo za nauku i umjetnost Tsrne Gore. Titograd = Posebna izdanja, drustvo za nauku i umjetnost Tsrne Gore. Odjeljenje prirodnikh nauka. Titograd.

Posebna izdanja, prirodnjacki musej u Beogradu. Belgrade. Vol. 1+, 1903+. Posebna Izd. Prir. Mus. Beogradu. HI 69677

Posebna izdanja, srpska akademija nauka i umetnosti. Odeljenje prirodno-matematičkih nauka. Belgrade. 1948-49; n.s. 1950+. Posebna Izd. Srpska Akad. Nauka Umetn., Odeljenje Prir.-Mat. Nauka. Preceded by: Posebna izdanija, srpska kralevska akademija nauka i umetnosti. Nauke prirodne i matematicke (spisi). HI 69678

Posebna izdanja, srpska kralevska akademija nauka i umetnosti. Nauke prirodne i matematicke (spisi). Belgrade. 1922-48. Posebna Izd. Srpska Kral. Akad. Nauka Umetn., Nauke Prir. Mat. (Spisi). Superseded by: Posebna izdanija, srpska akademija nauka i umetnosti. Odeljenje prirodno-matematičkih nauka.

Preceded by: Posebna izdanija, srpska kralevska akademija. Prirod. mat. spisi. HI 69679

Posebna izdanja, srpska kralevska akademija. Prirodnjački i matematički spisi. Belgrade. 1893-1912. Posebna Izd. Srpska Kral. Akad., Prir. Mat. Spisi. Superseded by: Posebna izdanija, srpska kralevska akademija nauka i umetnosti. Nauke prirodne i matematicke (spisi). HI 69680

Posebna izdanja, srpsko biološko društvo. Belgrade. 1955-58. Posebna Izd. Srpska Biol. Društvo. HI 69681

Posebni izdanija, hidrobiološki zavod na univerzitetot, Skopje = Posebna izdanja. Filoz. fak. na univerzitetot Skopje. Hidrobiološki zavod Ohrid. Ochrida.

Posebno izdanie. Musei macedonici scientiarum naturalium. Skoplje. No. 7+, 1976+. Posebno Izd., Mus Maced. Sci. Nat. Preceded by: Posebno izdanie. Prirodonaučen muzej Skopje. HI 69682

Posebno izdanie. Prirodonaučen muzej Skopje. Skoplje. Nos. 1-6, 1964-69. Posebno Izd. Prir. Muz. Skoplje. Superseded by: Posebno izdanie. Musei macedonici scientiarum naturalium. HI 69683

Postelsia; the year book of the Minnesota seaside station. St. Paul, MN. Postelsia. See B–P–H 716/10. HI 58062

Postepy biochemii. Warsaw. Postepy Biochem. See B–P–H 716/11. HI 58063

Postepy biologii komorki. Warsaw. Vol. 1+, 1974+. Postepy Biol. Komorki. HI 69684

Postepy microbiologii. Warsaw. Vol. 1+, 1962+. Postepy MIcrobiol. HI 69685

Postgraduate bulletin. Milwaukee, WI. Postgraduate Bull. See B–P–H 716/13. HI 58064

Postgraduate medical journal. London. Postgraduate Med. J. See B–P–H 716/14. HI 58065

Postharvest horticulture series. Davis, CA. No. 1+, 1986+. Postharv. Hort. Ser. HI 69686

Postilla. New Haven, CT. Postilla. See B–P–H 716/15. HI 58066

Poszonyi természettudományi és orvosi egyesület közleményei = Verhandlungen des Vereins für Natur- und Heilkunde zu Presburg. Pressburg [=Bratislava, Czechoslovakia]. Verh. Vereins Natur- Heilk. Presburg. See B–P–H 955/13.

Potamon; newsletter of the society for the study and conservation of nature. Valetta. No. 1+, 1979+. Potamon. HI 69687

Potash review. Bern. Potash Rev. See B–P–H 716/16. HI 58067

Potash and tropical agriculture. Amsterdam. Potash Trop. Agric. See B–P–H 716/17. HI 58068

Potasse. Mulhouse, France. Potasse. See B–P–H 716/18. HI 58069

Potato abstracts. Farnham Royal. Vol. 1+, 1976+. Potato Abstr. HI 69688

Potato journal. Christchurch, New Zealand. Potato J. See B–P–H 716/20. HI 58070

Potato news bulletin. Washington, DC. Potato News Bull. See B–P–H 716/21. HI 58071

Potato research. Wageningen. Vol. 13+, 1970+. Potato Res. Preceded by: European potato journal. HI 69689

Potchvovédénie. Moscow = Pochvovědĕnie. Moscow.

Pótfüzetek a Természettudományi közlönyhöz = Természettudományi közlöny. Budapest. Term. Közl. See B–P–H 871/20.

Potomac herb journal. Washington, DC. Potomac Herb J. See B–P–H 716/24. HI 58072

Pour nos jardins. Bulletin de la société d'horticulture et des jardins populaires de France. Lille. Pour Nos Jard. See B–P–H 716/25. HI 58073

Praca Navukovaga tovarystva pa vyvučen'nju Belarusi pry Belaruskaj dzjaržeaŭnaj akademii sel'skae haspadarki ŭ Gorkah = Pratsa navukovaga tovarystva pa vyvuchen'nyu Belarusi pry Belaruskai dzyarzheaŭnai akademii sel'skae khaspadarki ŭ Gorkakh. Gorki.

Praca Navukovaga tovarystva pa vyvučen'nju Belarusi pry Belaruskaj dzjaržeaŭnaj akademii sel'skae haspadarki z udzelam Gorackaga raennaga tovarystva krajaznaŭstva = Pratsa navukovaga tovarystva pa vyvuchen'nyu Belarusi pry Belaruskai dzyarzheaŭnai akademii sel'skae khaspadarki ŭ Gorkakh. Gorki.

Prace badawcze. Instytut badawczy leśnictwa. Warsaw. Prace Badawcze Inst. Badawczy Leśn. See B–P–H 717/18. HI 58074

Prace z biologii molekularnej. Vol. 1+, 1974+. Prace Biol. Molek. HI 69690

Prace botaniczne. Cracow = Zeszyty naukowe uniwersitetu jagiellónskiego; prace botaniczne. Cracow.

Prace botaniczne uniwersytet wrocławskie = Acta universitatis wratislaviensis. Prace botaniczne. Wroclaw.

Prace botaniczne, zeszyty naukowe uniwersytetu jagiellonskiego. Cracow = Zeszyty naukowe uniwersitetu jagiellónskiego; prace botaniczne. Cracow.

Práce brněnske základny československe akademie věd. Brno. Vols. 26-34, 1954-62. Práce Brněnske Zákl. Českoslov. Akad. Věd. Preceded by: Práce moravsko-slezské akademie věd přírodních. Superseded by: Přírodovědné práce ústavú československé akademie věd v Brné. HI 69691

Prace z dějin přírodních věd. Prague. No. 1+, 1969+. Prace Dějin Přir. Věd. HI 69692

Prace instytutu badawczego leśnictwa. Warsaw. Prace Inst. Badawczego Leśn. See B–P–H 717/22. HI 58075

Prace instytutu sadownictwa w Skierniewicach. Warsaw. Vol. 1+, 1955+. Prace Inst. Sadown. Skierniew. HI 69693

Prace instytutu sadownictwa w Skierniewicach. Seria D, monografie i rozprany. Warsaw. Nos. 1-5, 1973-76. Prace Inst. Sadown. Skierniew., D. HI 69694

Prace komisji biologiznej. Poznańskie towarzystwo przyjaciól nauk. Wydział matematiczno-przyrodniczy. Poznan, Poland. Prace Komis. Biol. See B–P–H 717/23. HI 58076

Prace komisji historii medycyny i nauk matematiczno-przyrodniczych. Ser. B. Nauki biologiczne. Cracow. Prace Komis. Hist. Med. Nauk Mat.-Przyr., Ser. B, Nauki Biol. See B–P–H 717/24. HI 58077

Prace komisji historii medycyny i nauk przyrodniczych-matematicznych. Cracow. Prace Komis. Hist. Med. Nauk Przyr.-Mat. See B–P–H 717/25. HI 58078

Prace komisji matematiczno-przyrodniczych. Poznańskie towarzystwo przyjaciól nauk, wydział matematyczno-przyrodniczych. Ser. B. Nauki biologiczne. Poznan, Poland. Prace Komis. Mat.-Przyr., Ser. B, Nauki Biol. See B–P–H 717/26. HI 58079

Prace komisji nauk roiniczych i leśnych. Poznan, Poland. Prace Komis. Nauk Roln. Leśn. See B–P–H 718/1. HI 58080

Práce krajského musea v Hradci Králové. Serie A, vedy přírodni. Hradec Králové. Vol. 1+. 1958+. Práce Krajsk. Mus. v Hradci Králové, A. HI 69695

Prace a materialy ustavu experimentalnej biologie ekologie, slovenskej akadémie vied = Acta geobiologica. Bratislava.

Prace monograficzne komisji fizjograficznej. Cracow. Prace Monogr. Komis. Fizjogr. See B–P–H 718/3. HI 58081

Prace monograficzne nad przyroda wielkopolskiego parku narodowego pod Poznaniem. Poznan, Poland. Prace Monogr. Przyr. Wielkopolsk. Parku Nar. Poznaniem. See B–P–H 718/4. HI 58082

Práce moravské přírodovědecké společnosti. Brno, Czechoslovakia. Práce Morav. Přír. Společn. See B–P–H 718/5. HI 58083

Práce moravsko-slezské akademie věd přirodních. Brno, Czechoslovakia. Práce Morav.-Slez. Akad. Věd Přír. See B–P–H 718/6. HI 58084

Prace muzeum przyrodniczego. Cracow. Prace Muz. Przyr. See B–P–H 718/7. HI 58085

Prace nauk matematyczno-przyrodniczych. Wydawnictwa łódzkiego towarzystwa naukowego. Wydział 3. Lodz, Poland. Prace Nauk Mat.-Przyr. See B–P–H 718/9. HI 58087

Prace naukowe, instytutu ochrony roslin. Poznan, Poland. Prace nauk. Inst. Ochr. Rosl. See B–P–H 718/8. HI 58086

Prace naukowe uniwersytetu slaşkiego w Katowicach = Acta biologica silesiana. Katowice.

Prăce z oboru botanicky. Brno. 1964/65-?, 1965-70. Prăce Oboru Bot. Superseded by: Prace z oboru botanicky a zoologie. HI 69697

Prăce z oboru botanicky a zoologie. Brno. 1976+. Prăce Oboru Bot. Zool. Preceded by: Prăce z oboru botanicky. HI 69698

Prăce 2. sekcie slovenskej akademie vied. Bratislava. Prăce 2. Sekc. Slov. Akad. Vied. See B–P–H 718/13. HI 58091

Prace towarzystwa przyjaciól nauk w Wilnie. Wydzial nauk matematycznych i przyrodniczych. Wilno. Nos. 1-12, 1923-38. Prace Towarz. Przyjac. Wilnie, Wydz. Nauk Mat. Przyr. HI 69699

Prace V Ú L H M = Prace výzkumného ústavu lesniho hospodářstvi a myslivosti. Zbraslav?

Prace výzkumného ústavu lesniho hospodářstvi a myslivosti. Zbraslav? Vol. 32+, 1966+. Prače Výzk. Lesn. Hospod. Mysliv. Preceded by: Práce výzkhumných ústavú lesnických. HI 69700

Práce výzkhumných ústavú lesnických. Prague. Vols. ?-31, 1952-65. Práce Výzk. Ustavú Lesn. Superseded by: Prace výzkumného ústavu lesniho hospodářstvi a myslivosti. HI 69701

Prace wrocławskiego towarzystwa naukowego. Wroclaw, Poland. Prace Wrocławsk. Towarz. Nauk. See B–P–H 718/11. HI 58089

Prace wydziału biologii i nauk o ziemi, uniwersytet imienia Adama Mickiewicza w Poznaniu. Seria biologia. Poznan. Nos. 1-3, 1961-67. Prace Wydz. Biol. Nauk Ziemi Uniw. Im. Adama Mickiewicza Poznaniu, Ser. Biol. Superseded by: Seria biologia. Poznan. HI 69702

Prace zakładu dendrologii i pomologii w Kórniku. Kornik, Poland. Prace Zakładu Dendrol. w Kórniku. See B–P–H 718/12. HI 58090

Prace zakładu systematyki roślin i ogrodu botanicznego. Wilno. Vols. ?-6-8(6)-?, ?-1931-34-? Prace Zakładu Syst. Rośl. Ogrodu Bot. HI 69703

Praci Biologo-gruntovogo fakul'tetu = Pratsy Biologo-gruntovogo fakul'tetu. Kiev.

Praci Central'nogo respublikans'kogo botaničnogo sadu = Pratsy tsentral'nogo respublikans'kogo botanichnogo sadu. Kiev.

Praci Naukova-doslidnogo instytuta biologiji. Kiev = Pratsy Naukova-doslidnogo instytuta biologii. Kiev.

Praci naukovo-doslidnogo instytutu biologiji. Kharkov, Ukrainian S S R = Trudy nauchno-issledovatel'skogo instituta biologii. Pratsy Naukovo-doslidnogo instytutu biologii. Kharkov.

Praci naukovo-doslidnogo instytutu biologiji i biologičnogo facul'tetu. Kharkov, Ukrainian S S R = Trudy nauchno-issledovatel'skogo instituta biologii. Pratsy Naukovo-doslidnogo instytutu biologii. Kharkov.

Praci Naukovo-doslidnogo zoologo-biologičnogo instytutu = Pratsy Naukovo-doslidnogo zoologo-biologichnogo instytutu. Kiev.

Praci Odes'kogo deržavnogo universytetu imeny I. I. Mečnikova = Pratsy Odes'kogo derzhavnogo universytetu imeny I. I. Mechnikova. Odessa.

Praci Zoobiologičnogo instytutu = Pratsy Zoobiologichnogo instytutu. Kiev.

Práctica médica. Madrid. Práct. Méd. See B–P–H 719/5. HI 58093

Practica otologica. Kyoto. Vol. ?-77(5)+, ?-1984+. Pract. Otol. HI 69704

Practical forestry. Lexington, TN. Vol. 1+, 1989+. Pract. Forest. HI 74874

Practical gardening. Market Harborough, Peterborough. Vol. 1+, 1960+. Pract. Gard. HI 69705

Practical horticulture. [Jissai engei.] [Japan]. Pract. Hort. See B–P–H 719/3. HI 58092

Practische und Kritische Mittheilungen aus dem Gebiete der Medizin, Chirurgie und Pharmacie = Mittheilungen aus dem Gebiete der Medizin, Chirurgie und Pharmacie. Altona [=Hamburg, in part]. Mitth. Med. See B–P–H 612/4.

Practisches Wochenblatt für Landwirthschaft, Gartenbau und Hauswirthschaft. Neubrandenburg. Pract. Wochenbl. Landw. See B–P–H 719/6. HI 58094

Practitioner. Baltimore, MD. Practitioner (Baltimore). See B–P–H 719/8. HI 58095

Practitioner. Lancaster, PA. Practitioner (Lancaster). See B–P–H 719/9. HI 58096

Practitioner. London. Practitioner (London). See B–P–H 719/10. HI 58097

Practitioner. New York, NY. Practitioner (New York). See B–P–H 719/11. HI 58098

Pracy Botaničnaga gabinetu Menskae céntral'nae das'ledčae balotnae stancyi = Pratsy Botanichnaga gabinetu Menskae tséntral'nae das'ledchae balotnae stanctsi. Minsk.

Pracy Gory-Goreckaga navukovaga tovarystva = Pratsy Gory-Goretskaga navukovaga tovarystva. Gory-Gorky.

Prag, Beiblätter zu Ost und West. Prague. Prag Beibl. Ost & West. See B–P–H 719/15. HI 58099

Prager gelehrte Nachrichten. Prague. Prager Gel. Nachr. See B–P–H 719/17. HI 58100

Prager medizinische Wochenschrift. Prague. Prager Med. Wochenschr. See B–P–H 719/18. HI 58101

Prairie garden. Winnipeg. Prairie Gard. See B–P–H 719/19. HI 58102

Prairie naturalist. Grand Forks, ND. Vol. 1+, 1968+. Prairie Naturalist. HI 69706

Prairie zephyr; newsletter of the Missouri prairie foundation. Columbia, MO. No. 1+, 1985+. Prairie Zephyr. HI 69707

Prakruti; Utkal university journal of science. Bhubaneswar. ?-1971+. Prakruti. HI 69708

Praktika tou hellenikou hydroviologikou institoutou. Athens. 19??-? Prakt. Hellen. Hydorviol. Inst. Superseded by: Helleniké okeanologia kai limnologia tou institoutou okeanographikon kai halieutikon ereunon. HI 69709

Praktische Blätter der bayerischen Landesanstalt für Pflanzenbau und Pflanzenschutz. Munich. Vols. 23-24, 1924-25. Prakt. Blätt. Bayer. Landesanst. Pflanzenbau. Preceded and superseded by: Praktische Blätter für Pflanzenbau und Pflanzenschutz. 5-4613-2. HI 69710

Praktische Blätter für Pflanzenbau und Pflanzenschutz. Stuttgart, Munich. Vols. 6-22, 1903-19; vols. 25-44, 1926-45; vols. 51-55, 1956-60 [suspended 1920-23 &

1946-49]. Prakt. Blätt. Pflanzenbau Pflanzenschutz. Preceded by: Praktische Blätter für Pflanzenschutz. For vols. 23-24 see Praktische Blätter der bayerischen Landesanstalt für Pflanzenbau und Pflanzenschutz. For vols. 45-50 see Zeitschrift für Pflanzenbau und Pflanzenschutz. 5-4613-3. HI 69711

Praktische Blätter für Pflanzenschutz. Stuttgart. Vols. 1-5, 1898-1902. Prakt. Blätt. Pflanzenschutz. Superseded by: Praktische Blätter für Pflanzenbau und Pflanzenschutz. 5-4613-6. HI 69712

Praktische Forstwirt für die Schweiz. Lenzburg, Aarau, Davos. Vols. 1-109, 1861-1973. Prakt. Forstw. Schweiz. Superseded by: Schweizer Forster. 5-3423-2. HI 69713

Praktische Mikroskopie. Berlin. Prakt. Mikroskop. See B–P–H 720/2. HI 58103

Praktische Obstzüchter. Klosterneuburg, Austria. Pract. Obstzüchter. See B–P–H 720/3. HI 58104

Praktische Ratgeber in Obst- und Gartenbau. Frankfurt a. O. Prakt. Ratgeber Obst- Gartenbau. See B–P–H 720/4. HI 58105

Pratsa navukovaga tovarystva pa vyvuchen'nyu Belarusi pry Belaruskai dzyarzheaünai akademii sel'skae khaspadarki ü Gorkakh. Gorki, Belorussian S S R. Vols. 1-4, 1926-27. Pratsa Navuk. Tovar. Vyvuch. Belarusi Belarusk. Dzyarzh. Akad. Sel'skae khaspad. Gorkah. Superseded by: Pratsy Gory-Goretskaga navukovaga tovarystva. HI 69714

Pratsy belaruskaga dzerzhnaunaga universytetu u Mensku. Minsk. Vols. 1-26, 1922-32. Pratsy Belarusk. Dzerzhn. Univ. Mensku. Preceded by: Trudy belorusskogo gosudarstvennogo universiteta v g. Minske. HI 69715

Pratsy belaruskaga navukova-das'ledchaga instytutu liasnoi gaspadarki i liasnoi pramyslovas'chi. Trudy po lesnomu opytnomu delu B S S R. Minsk. Vols. 1-39, 1927-31. Pratsy Belarusk. Navuk.-Das'l. Inst. Liasn. Gasp. Liasn. Pramysl., Trudy Lesn. Optyn. Delu B.S.S.R. HI 69716

Pratsy belaruskaga navukova-das'ledchaga instytutu sel'skae i lyasnoe gaspadarki im V. I. Lenina pri S N K-B S S R, menskaya tsentral'naya balotnaya stantsiya. Minsk. Vols. 1-24, 1928-30. Pratsy Belarusk. Navuk.-Das'l. Inst. Sel'sk. Lyasn. Gasp. V. I. Lenina S.N.K. B.S.S.R. Mensk. Tsentr. Balotn. Stantsiya. HI 69717

Pratsy Biologo-gruntovogo fakul'tetu. Kiev, Ukrainian S S R. Vols. 5-10, 1950-55. Pratsy Biol.-Gruntov. Fak. Preceded by: Biologichnyi zbirnyk. Superseded by: Zbirnyk Biologichnogo fakul'tetu. HI 69718

Pratsy Botanichnaga gabinetu Menskae tséntral'nae das'ledchae balotnae stantsyi. Minsk, Belorussian S S R. Vol. 1, 1930. Pratsy Bot. Gab. Menskae Tséntr. Das'l. Balotn. Stantsyi. HI 69719

Pratsy Gory-Goretskaga navukovaga tovarystva. Gory-Gorky, Belorussian S S R. Vols. 5-7, 1928-30. Pracy Gory-Goretsk. Navuk. Tovar. Preceded by: Pratsa navukovaga tovarystva pa vyvuchen'nyu Belarusi pry Belaruskai dzyarzheaünai akademii sel'skae khaspadarki ü Gorkakh. HI 69720

Pratsy Naukova-doslidnogo instytuta biologii. Kiev, Ukrainian S S R. Vols. 1-4, 1938-41. Pratsy Nauk.-Doslidn. Inst. Biol. HI 69721

Pratsy Naukovo-doslidnogo zoologo-biologichnogo instytutu. Kiev, Ukrainian S S R. Vol. 1, 1936; 3-10/11, 193?-41. Pratsy Nauk.-Doslidn. Zool.-Biol. Inst. Superseded by: Trudy nauchno-issledovatel'skogo instituta biologii. Pratsy Naukovo-doslidnogo instytutu biologii. For vol. 2 see: Pratsy Zoobiologichnogo instytutu. HI 69722

Pratsy Odes'kogo derzhavnogo universytetu imeny I. I. Mechnikova. Odessa, Ukrainian S S R. 1934-41; vols. 49-72, 1947-52; vol. 144+, 1955+ [suspended from 1942-46 and 1953-54; vol. nos. 73-143 not used; during second suspension, instead of numbered vols., 71 unnumbered monographs were published.] Pratsy Odes'k. Derzhavn. Univ. Mechnikova. HI 69723

Pratsy tsentral'nogo respublikans'kogo botanichnogo sadu. Kiev, Ukrainian S S R. Vol. 8+, 1962+. Pratsy Tsentr. Respubl. Bot. Sadu. Preceded by: Trudy botanicheskogo sada. Kiev. HI 69724

Pratsy Zoobiologichnogo instytutu. Kiev, Ukrainian S S R. Vol. 2, 1934. Pratsy Zoobiol. Inst. Preceded & superseded by: Pratsy Naukovo-doslidnogo zoologo-biologichnogo instytutu. HI 69725

Pražské hospodářské noviny. Prague. Pražské Hospod. Noviny. See B–P–H 720/5. HI 58106

Précis analytique de l'académie impériale des sciences, belles-lettres et arts de Rouen. Rouen. 1855/56-1869/70, 1856-70. Précis Analytique Acad. Imp. Sci. Rouen. Preceded and superseded by: Précis analytique de l'académie des sciences, belles-lettres et arts de Rouen. 1-30-1. HI 69726

Précis analytique de l'académie des sciences, belles-lettres et arts de Rouen. Rouen. 1848-54/55, 1848-55; 1870/71+, 1871+. Précis Analytique Acad. Sci. Rouen. Preceded by: Précis analytique des travaux de l'académie royale des sciences, des belles-lettres et des arts de Rouen. For 1855/56-1869/70 see: Précis analytique de l'académie impériale des sciences, belles-lettres et arts de Rouen. 1-30-1. HI 69727

Précis analytique des travaux de l'académie des sciences, belles-lettres et arts de Rouen, ..., précédé de l'histoire de l'académie. Rouen. Précis Analytique Trav. Acad. Sci. Rouen Hist. Acad. See B–P–H 720/7. HI 58107

Précis analytique des travaux de l'académie royale des sciences, des belles-lettres et des arts de Rouen. Rouen.

1816-47. Preceded by: Précis analytique des travaux de l'académie des sciences, des belles-lettres et des arts de Rouen. Superseded by: Précis analytique de l'académie des sciences, belles-lettres et arts de Rouen. 1-30-1. HI 53584

Précis analytique des travaux de l'académie des sciences, des belles-lettres et des arts de Rouen. Rouen. 1804-14, 1807-15. Précis Analytique Trav. Acad. Sci. Rouen. Superseded by: Précis analytique des travaux de l'académie royale des sciences, des belles-lettres et des arts de Rouen. 1-30-1. HI 69728

Précis analytique des travaux de la société académie des sciences, lettres et arts de Nancy. Nancy. Précis Analytique Trav. Soc. Acad. Sci. Nancy. See B–P–H 720/9. HI 58108

Précis analytique des travaux de la société académique des sciences, lettres, arts et agriculture de Nancy = Précis analytique des travaux de la société académie des sciences, lettres et arts de Nancy. Nancy. Précis Analytique Trav. Soc. Acad. Sci. Nancy. See B–P–H 720/9.

Précis analytique des travaux de la société des sciences, lettres, arts et agriculture de Nancy = Précis analytique des travaux de la société académie des sciences, lettres et arts de Nancy. Nancy. Précis Analytique Trav. Soc. Acad. Sci. Nancy. See B–P–H 720/9.

Précis analytique des travaux de la société des sciences de Nancy= Précis analytique des travaux de la société académie des sciences, lettres et arts de Nancy. Nancy. Précis Analytique Trav. Soc. Acad. Sci. Nancy. See B–P–H 720/9.

Précis des travaux de la société royale des sciences, lettres, arts et agriculture de Nancy. Nancy. Précis Trav. Soc. Roy. Sci. Nancy. See B–P–H 720/10. HI 58109

Précis des travaux de la société royale des sciences, lettres et arts de Nancy = Précis des travaux de la société royale des sciences, lettres, arts et agriculture de Nancy. [Title from 1819/23 (1825) ... lettres et arts de Nancy.] Nancy. Précis Trav. Soc. Roy. Sci. Nancy. See B–P–H 720/10.

Preisschriften und Abhandlungen der Kayserlichen freyen ökonomischen Gesellschaft zu St. Petersburg. Gotha & St. Petersburg. Preisschr. Abh. Kayserl. Freyen Ökon. Ges. St. Petersburg. See B–P–H 720/11. HI 58110

Preisschriften, gekrönt und herausgegeben von der Fürstlich Jablonowski'schen Gesellschaft zu Leipzig. Leipzig. Preisschr. Fürstl. Jablonowski'schen Ges. Leipzig. See B–P–H 720/12. HI 58111

Premiers travaux de la société libre d'agriculture et d'économie intérieure du département du Bas-Rhin, séante à Strasbourg. Strasbourg. Prem. Trav. Soc. Libre Agric. Dép. Bas-Rhin Strasbourg. See B–P–H 720/14. HI 58112

Prensa médica argentina. Buenos Aires. Prensa Méd. Argent. See B–P–H 720/15. HI 58113

President's address, papers and reports, west Kent natural history, microscopical and photographic society. 1892-1900. President's Address Pap. Rep. W. Kent Nat. Hist. Microscop. Photogr. Soc. Superseded by: Transactions of the west Kent natural history, microscopical and photographic society. HI 69729

President's letter, spuria iris society = Newsletter, spuria iris society. Yorba Linda, CA, Purcell, OK.

Preslia. Věstník (Časopis) československé botanické společnosti. Prague. Preslia. See B–P–H 720/16. HI 58114

Press bulletin, Colorado agricultural college, Colorado experiment station = Agricultural experiment station of the agricultural college of Colorado. Press bulletin. Fort Collins, CO. Agric. Exp. Sta. Agric. Coll. Colorado Press Bull. See B–P–H 57/4.

Press bulletin, Cornell university agricultural experiment station = Cornell university agricultural experiment station press bulletin. Ithaca, NY. Cornell Univ. Agric. Exp. Sta. Press Bull. See B–P–H 332/9.

Press bulletin, Florida experiment station = Florida experiment station press bulletin. Lake City, FL. Florida Exp. Sta. Press Bull. See B–P–H 375/12.

Press bulletin, Hawaii agricultural experiment station = Hawaii agricultural experiment station. Press bulletin. Honolulu, HI. Hawaii Agric. Exp. Sta. Press Bull. See B–P–H 413/1.

Press bulletin, Michigan agricultural experiment station = Michigan agricultural experiment station. Press bulletin. East Lansing, MI. Michigan Agric. Exp. Sta. Press Bull. See B–P–H 592/14.

Press bulletin, Missouri agricultural experiment station = Missouri agricultural experiment station. Press bulletin. Columbia, MO.

Press bulletin, Nebraska agricultural experiment station = Nebraska agricultural experiment station press bulletin. Lincoln, NB. Nebraska Agric. Exp. Sta. Press Bull. See B–P–H 640/2.

Press bulletin, New Mexico agricultural experiment station = New Mexico agricultural experiment station, press bulletin. State College of New Mexico. Las Cruces & Santa Fe, NM. New Mexico Agric. Exp. Sta. Press Bull. See B–P–H 654/8.

Press bulletin, North Carolina agricultural experiment station = North Carolina agricultural experiment station press bulletin. Raleigh, NC. North Carolina Agric. Exp. Sta. Press Bull. See B–P–H 663/9.

Press bulletin, North Dakota agricultural experiment station = North Dakota agricultural experiment station press bulletin. Fargo, ND. North Dakota Agric. Exp. Sta. Press Bull. See B–P–H 664/13.

Press bulletin, Oklahoma agricultural experiment station = Oklahoma agricultural experiment station press bulletin. Stillwater, OK. Oklahoma Agric. Exp. Sta. Press Bull. See B–P–H 687/7.

Press bulletin, Texas agricultural experiment station = Progress report of the Texas agricultural experiment station. College Station, TX.

Press bulletin, University of Florida agricultural experiment station = Florida experiment station press bulletin. Lake City, FL. Florida Exp. Sta. Press Bull. See B–P–H 375/12.

Press bulletin, Wyoming agricultural college agricultural experiment station = Wyoming agricultural college agricultural experiment station. Press bulletin. Laramie, WY. Wyoming Agric. Exp. Sta. Press Bull. See B–P–H 981/1.

Presse médicale. Paris. Presse Méd. (Paris). See B–P–H 721/16. HI 58115

Presse médicale belge. Brussels. Presse Méd. Belge. See B–P–H 721/17. HI 58116

Pressedient Wissenschaft; Informationen aus Lehre und Forschung an der freien Universität Berlin. Berlin. No. 1+, 1970+. Pressedient Wiss. HI 69730

Preussische Akademie der Wissenschaften zu Berlin. Vorträge und Schriften. Berlin. Preuss. Akad. Wiss. Berlin Vorträge Schriften. See B–P–H 721/19. HI 58117

Preussische Provinzialblätter. Königsberg [=Kaliningrad, Russian S F S R]. Preuss. Provinzialbl. See B–P–H 721/20. HI 58118

Preussische Sammler. Königsberg [=Kaliningrad, Russian S F S R]. Preuss. Sammler. See B–P–H 721/21. HI 58119

Preussische Sammler ... zur Känntnis der Naturgeschichte, ... Königsberg [=Kaliningrad, Russian S F S R]. Preuss. Sammler Känntn. Naturgesch. See B–P–H 721/22. HI 58120

Prikladnaya biokhimiya i mikrobiologiya. Moscow, Leningrad. Vol. 1+, 1965+. Prikl. BIokhem. Mikrobiol. HI 69731

Prilozi. Bitola. 1960+. Prilozi. HI 69732

Primroses; American primrose society quarterly. Tacoma, WA. Vol. 35(2)+, 1977+. Primroses. Preceded by: Quarterly of the american primrose society. HI 69733

Princeton college bulletin = Princeton university bulletin. Princeton, NJ.

Princeton university bulletin. Princeton, NJ. Vols. 1-15(4), 1889-1904. Princeton Univ. Bull. 4-3438-2. HI 69734

Principal's report, imperial college of tropical agriculture = Report, imperial college of tropical agriculture. St. Augustine, Trinidad.

Principes; journal of the (international) palm society. Miami, FL, Lawrence KS. Vol. 1+, 1956+. Principes. Preceded by: Bulletin, palm society. HI 69735

Příroda. Brünn [=Brno], Mährisch-Ostrau [=Ostrava], & Prague. Příroda (Brünn, Mährisch-Ostrau, & Prague). See B–P–H 722/14. HI 58135

Priroda. Ljubljana, Yugoslavia. Priroda (Ljubljana). See B–P–H 722/15. HI 58136

Priroda; populyarnyi estestvenno-istoricheskii zhurnal [Sub-title varies]. Moscow. Vol. 1+, 1912+. Priroda. 4-3442-1. HI 69736

Priroda. Sofia, Bulgaria. Priroda (Sofia). See B–P–H 722/19. HI 58137

Příroda jihovýchodní Moravy. Gottwaldov, Czechoslovakia. Přir. Jihových. Moravy. See B–P–H 722/4. HI 58129

Priroda i khoziaistvo severa. Murmansk. No. 1+, 1969+. Prir. Khoz. Severa. HI 69737

Priroda i ohota = Priroda i okhota. Moscow.

Priroda i okhota. Moscow. Vols. 6-40, 1878-1912. Prir. & Okhota. Preceded by: Priroda. Populyarnyi Estestvenno-istoricheskii Sbornik. [Edited by L. P. Sabaněev.] HI 69738

Priroda. Populjarnyj Estestvenno-istoričeskij Sbornik. [Edited by L. P. Sabaněev] = Priroda. Populyarnyi Estestvenno-istoricheskii Sbornik. [Edited by L. P. Sabaněev.] Moscow.

Priroda. Populyarnyi Estestvenno-istoricheskii Sbornik. [Edited by L. P. Sabaněev.] Moscow. Vols. 1-5, 1873-77. Priroda (Sabaněev). Superseded by: Priroda i okhota. HI 69739

Priroda i sel'skoe khozyaistvo zasushlivo-pustynnykh oblastei S S S R. Voronezh. Vols. 1-?, 1926-29. Prir. Sel'sk. Khoz. Zasushl.-Pustyn. Obl. S.S.S.R. HI 69740

Příroda a škola. Brünn & Mährisch-Ostrau, Moravia [=Brno & Ostrava, Czechoslovakia]. Přír. & Škola. See B–P–H 721/25. HI 58121

Priroda i socialističeskoe hozjajstvo = Priroda i sotsialisticheskoe khozyaistvo. Moscow.

Priroda i sotsialisticheskoe khozyaistvo. Moscow. Vols. 4-8, 1931-41. Prir. Sotsialist. Khoz. Preceded by: Okhrana prirody. HI 69741

Príroda tatranského národného parku. Turciansky Svaty Martin, Czechoslovakia. Prír. Tatransk. Nár. Parku. See B–P–H 722/11. HI 58134

Prirodonaucen muzej Skopje posebno izdanie. Skopje. Vol. 1+, 1964+. Prir. Muz. Skopje Posebno Izd. HI 69742

Prirodoslovna istraživanja Hrvatske i Slavonije. Zagreb, Croatia [Yugoslavia]. Prir. Istraž. Hrvatske Slavonije. See B–P–H 721/29. HI 58124

Prirodoslovna istraživanja kraljevine Jugoslavije. Zagreb, Yugoslavia. Prir. Istraž. Kral. Jugoslavije. See B–P–H 722/2. HI 58127

Prirodoslovna istraživanja nezavisne države Hrvatske. Zagreb, Yugoslavia. Prir. Istraž. Nezav. Države Hrvatske. See B–P–H 722/3. HI 58128

Prirodoslovne razprave. Ljubljana, Yugoslavia. Prir. Razpr. See B–P–H 722/16. HI 58131

Přírodovědecká edice. Opava, Czechoslovakia. Přír. Edice. See B–P–H 721/28. HI 58123

Přírodovědecký časopis jihočeský. Ceské Budějovice. Vol. 13, 1973. Přir. Čas. Jihoč. Preceded and superseded by: Sborník jihočeského muzea v českých budějovicích přírodni védy. HI 69743

Přírodovědecký časopis Slezský. Opava, Czechoslovakia. Přír. Cas. Slezský. See B–P–H 721/27. HI 58122

Přírodovědecký sborník ostravského kraje. Ostrava, Czechoslovaia. Přír. Sborn. Ostravsk. Kraje. See B–P–H 722/8. HI 58133

Prírodovedné práce slovenských múzei. Trnava, Czechoslovakia. Prír. Práce Slov. Múz. See B–P–H 722/5. HI 58130

Přírodovědné práce ústavú československé akademie věd v Brné. Brno. N.s. vol. 1+, 1967+. Přir. Práce Ústavú Českoslov. Akad. Věd Brné. Preceded by: Práce Brněské základny československé akademie věd. HI 69744

Prírodovedný sborník. Turciansky Svaty Martin, Czechoslavakia. Prír. Sborn. See B–P–H 722/7. HI 58132

Prírodovedny sborník slovenského múzea. Bratislava. Vols. 4-7, 1958-61. Prír. Sborn. Slov. Múz. Preceded by: Prírodovedné práce slovenských muzeí. Superseded by: Sborník slovenského národného múzea. Přírodné vědy. HI 69745

Prisma. Illustrierte Monatsschrift für Natur, Forschung und Technik. Zurich. Prisma. See B–P–H 723/3. HI 58138

Přitel zahrad; chovatel drobného zvířectva. Prague. Vol. ?-8-12-?, ?-1942- 48-? Přitel Zahrad. 4-3443-1. HI 69746

Prisuždenie Učreždennyh P. N. Demidovym Nagrad, (Obščij Otčet o Prisuždenii Demidovskih Nagrad) = Prisuzhdenie Uchrezhdennykh P. N. Demidovym Nagrad, (Obshchii Otchet o Prisuzhdenii Demidovskikh Nagrad). St. Petersburg.

Prisuzhdenie Uchrezhdennykh P. N. Demidovym Nagrad, (Obshchii Otchet o Prisuzhdenii Demidovskikh Nagrad). St. Petersburg. Vols. 1-34, 1832-65. Prisuzhd. Uchrezhd. Demidovym Nagrad. 1-113-3. HI 69747

Prize-essays and transactions of the Highland and agricultural society of Scotland. Edinburgh & London. Prize-Essays Trans. Highl. Agric. Soc. Scotland. See B–P–H 723/6. HI 58140

Prize essays and transactions of the Highland society of Scotland. Edinburgh. Prize Essays Trans. Highl. Soc. Scotland. See B–P–H 723/5. HI 58139

Pro natura. Organe de l'union internationale pour la protection de la nature. Organ of the international union for the protection of nature. Basel. Pro Natura (Basel). See B–P–H 723/7. HI 58141

Pro natura. Milan. Pro Natura (Milan). See B–P–H 723/8. HI 58142

Probe. Cape Town. Probe (Cape Town). See B–P–H 723/9. HI 58143

Probe. Philadelphia, PA. Probe (Philadelphia). See B–P–H 723/10. HI 58144

Probio; revue de l'association francophone des professeurs de biologie de Belgique. ?-1981+. Probio. HI 69748

Probleme agricole. Bucharest. Vols. 1-26(6), 1958-74. Probl. Agric. Superseded by: Producţia vegetală. Cereale şi plante tehnice. HI 69749

Probleme de agrofitotehnie teoretica şa aplicată. Fundulea. Vol. 1+, 1979+. Probl. Agrofitotehn. Teor. Apl. HI 69750

Probleme de genetica teoretica si aplicata. Bucharest. Vol. 1+, 1969+. Probl. Genet. Teor. Apl. HI 69751

Problèms biologiques. Collection de monographies publiées sous le patronage du comité technique des sciences naturelles des presses universitaires de France. Paris. Probl. Biol. See B–P–H 723/12. HI 58145

Problems of desert development. [Translation of: Problemy osvoeniya pustyn′.] New York. Vol. 1+, 1967+. Probl. Desert Developm. HI 69752

Problems of ecology and biocenology. Moscow & Leningrad = Voprosy ekologii i biotsenologii. Moscow & Leningrad.

Problems of evolution = Problemy evolyutsii. Novosibirsk.

Problems of the north. [Translation of: Problemy severa.] Ottawa. Vols. 1-14, 1958-70. Probl. North. HI 69753

Problems of pharmacognosy = Voprosy farmakognozii. Leningrad.

Problems of virology. [Translation of: Voprosy virusologii.] London, New York. Vols. 1-6, 1957-61. Probl. Virol. HI 69754

Problemy biohimii v Mičurinskoj biologii = Problemy biokhimii v Michurinskoi biologii. Moscow & Leningrad.

Problemy biokhimii v michurinskoi biologii. Moscow & Leningrad. Vol. 1, 1949. Probl. Biokhim. Michurinskoi Biol. 1-115-3. HI 69755

Problémy biológie krajiny = Quaestiones geobiologicae. Bratislava.

Problemy bioniki. Khar′kov. Vol. 1+, 1968+. Probl. Bioniki. HI 75120

Problemy botaniki. Moscow & Leningrad. Probl. Bot. See B–P–H 723/13. HI 58146

Problemy evolyutsii. Novosibirsk. Vols. 1-4, 1968-75. Probl. Evol. HI 69756

Problemy osvoeniya pustyn′. Ashkhabad. Vol. 1+, 1967+. Probl. Osvoeniya Pustyn′. HI 69757

Problemy palinologii. Kiev. Vol. 1+, 1971+. Probl. Palin. HI 69758

Problemy sovremennoi botaniki. Moscow, Leningrad. Vol. 1+, 1965+. Probl. Sovrem. Bot. HI 69759

Proceedings of the academy of natural sciences of Philadelphia. Philadelphia, PA. Proc. Acad. Nat. Sci. Philadelphia. See B–P–H 723/22. HI 58147

Proceedings, academy of sciences of the armenien S S R = Doklady akademiya nauk armyanskoi S S R. Erevan.

Proceedings, academy of sciences of the azerbaijan S S R = Doklady akademiya nauk azerbaidzhanskoi S S R. Baku.

Proceedings of the academy of sciences of the georgian S S R. Biological studies = Izvestiya akademii nauk gruzinskoi S S R. Seriya biologicheskaya. Tiflis.

Proceedings of the academy of sciences of the U S S R. Biochemistry section. [Translation of: Doklady akademii nauk S S S R.] New York, & Washington, DC. Vols. 112-117, 1957. Proc. Acad. Sci. U.S.S.R., Biochem. Sect. Superseded by: Doklady, biochemistry section. HI 69760

Proceedings of the academy of sciences of the united provinces of Agra and Oudh. Allahabad. Vols. 4-5, 1934-36. Proc. Acad. Sci. Unit. Prov. Agra Oudh. Preceded by: Bulletin of the academy of the united provinces of Agra and Oudh. Superseded by: Proceedings of the national academy of sciences of India. HI 69761

Proceedings of the acclimatization society of Victoria. Melbourne. Proc. Acclim. Soc. Victoria. See B–P–H 724/1. HI 58148

Proceedings of the Agassiz institute, of Sacramento, Cal. Sacramento, CA. Proc. Agassiz Inst. Sacramento. See B–P–H 724/2. HI 58149

Proceedings of the agricultural experiment station, Tel-Aviv = Yedeoth. Tel-Aviv.

Proceedings of the agricultural and horticultural society of India. Calcutta. 1837-41. Proc. Agric. Hort. Soc. India. Continued, as Monthly proceedings, in: Journal of the agricultural and horticultural society of India. HI 69762

Proceedings of the agricultural and horticultural society of St. Helena. Jamestown, St. Helena. Proc. Agric. Soc. St. Helena. See B–P–H 724/3. HI 58150

Proceedings of the agricultural society of Trinidad and Tobago. Port of Spain. Vols. 1-50, 1894-1950. Proc. Agric. Soc. Trinidad Tobago. Superseded by: Journal of the agricultural society of Trinidad and Tobago. 1-91-2. HI 69763

Proceedings of the agri-horticultural society of Madras. Madras. 1870-73; n.s. vols. 1-?, 1874-1919/20. Proc. Agri-Hort. Soc. Madras. Superseded by: Report and proceedings of the agri-horticultural society of Madras. HI 69764

Proceedings of the Alabama state horticultural society. Montgomery, AL. Proc. Alabama State Hort. Soc. See B–P–H 724/4. HI 58151

Proceedings of the Alaska science conference. Fairbanks, AK. Vol. 1+, 1950+. Proc. Alaska Sci. Conf. HI 69765

Proceedings of the All-Union scientific-research institute of plant protection. Moscow = Trudy. Vsesoyuznogo nauchno-issledovatel′skogo instituta zashchity rastenii. Moscow.

Proceedings of the Alloa society of natural history, science, and archaeology. Alloa, Scotland. Proc. Alloa Soc. Nat. Hist. See B–P–H 724/5. HI 58152

Proceedings of the american academy of arts and sciences. Boston, MA. Vols. 1-85, 1846-1958 [vol. 84, nos. 3-4 not published]. Proc. Amer. Acad. Arts. Superseded by: Daedalus. 2-1252-3. HI 69766

Proceedings, American antiquarian society = American antiquarian society. Proceedings. Worcester, MA. Amer. Antiq. Soc. Proc. See B–P–H 78/3.

Proceedings of the american association for the advancement of science. Philadelphia, PA, etc. 1st-73rd, 1848-1921. Proc. Amer. Assoc. Advancem. Sci. Preceded by: Abstracts of the proceedings of the association of american geologists and naturalists. Superseded by: Summarized proceedings of the meetings of the american association for the advancement of science. HI 69767

Proceedings, american association of cereal chemists. St. Paul, MN. 1915-39. Proc. Amer. Assoc. Cereal Chem.

Superseded by: Transactions, american association of cereal chemists. HI 69768

Proceedings, american association of feed microscopists. Mechanicsburg, PA. 1953+. Proc. Amer. Assoc. Feed Microscop. HI 69769

Proceedings of the american microscopical society. Lancaster, Pa. Proc. Amer. Microscop. Soc. See B–P–H 724/23. HI 58153

Proceedings, American breeders associaton = American breeders associaton proceedings. Washington, DC. Amer. Breed. Assoc. Proc. See B–P–H 78/23.

Proceedings, american peony society. Philadelphia, PA. Proc. Amer. Peony Soc. See B–P–H 724/24. HI 58154

Proceedings of the american pharmaceutical association. Baltimore, MD. Proc. Amer. Pharm. Assoc. See B–P–H 724/25. HI 58155

Proceedings of the american philosophical society, (held at Philadelphia, for promoting useful knowledge). Philadelphia, PA. Proc. Amer. Philos. Soc. See B–P–H 724/27. HI 58156

Proceedings, american phytopathological society. St. Paul, MN. Vols. 1-4, 1974-77. Proc. Amer. Phytopathol. Soc. Superseded, in part, by: Phytopathology. HI 69770

Proceedings of the american plant propagators association. Hatboro, PA. Proc. Amer. Pl. Propag. Assoc. See B–P–H 724/28. HI 58157

Proceedings of american pomological congress = Proceedings of the american pomological society. Madison, WI. Proc. Amer. Pomol. Soc. See B–P–H 724/31.

Proceedings of the american pomological society. Madison, WI. Proc. Amer. Pomol. Soc. See B–P–H 724/31. HI 58158

Proceedings of the american seed trade association. Washington, DC. Proc. Amer. Seed Trade Assoc. See B–P–H 725/1. HI 58159

Proceedings of the american society of agronomy. Washington, DC. Proc. Amer. Soc. Agron. See B–P–H 725/2. HI 58160

Proceedings of the american society of biological chemists = Journal of biological chemistry,1907. Baltimore, MD. J. Biol. Chem. See B–P–H 459/8.

Proceedings of the american society for horticultural science. Geneva, NY, College Park, MD. Vols. 1-93, 1903-69. Proc. Amer. Soc. Hort. Sci. Superseded by: Journal of the american society for horticultural science. 1-322-3. HI 69771

Proceedings of the american society of limnology and oceanography. St. Louis, MO. Proc. Amer. Soc. Limnol. See B–P–H 725/6. HI 58161

Proceedings of the american society of sugar beet technologists. Fort Collins, CO. Proc. Amer. Soc. Sugar Beet Technol. See B–P–H 725/7. HI 58162

Proceedings of the american soybean association. Chicago, IL. Proc. Amer. Soybean Assoc. See B–P–H 725/8. HI 58163

Proceedings, annual biology colloquium, Oregon state university. Corvallis, OR. 1943+. Proc. Annual Biol. Colloq. Oregon State Univ. Preceded by: Annual biology colloquium, Oregon state college. HI 69772

Proceedings, annual Colorado crop protection institute. [Fort Collins, CO.] Vol. 1+, [1972?]+. Proc. Annual Colorado Crop Protect. Inst. HI 69773

Proceedings of the annual conference, hydroponics society of America = Proceedings, hydroponics society of America. Concord, CA..

Proceedings of the annual convention of the american association of nurserymen. Proc. Annual Conv. Amer. Assoc. Nurserymen. See B–P–H 725/13. HI 58164

Proceedings of the annual exhibitions, Colorado agricultural society. Denver, CO. Proc. Annual Exhib. Colorado Agric. Soc. See B–P–H 725/14. HI 58165

Proceedings, annual forest vegetation management conference. 3rd+, 1981+ [earlier meetings proceedings not published]. Proc. Annual Forest Veg. Managem. Conf. HI 73785

Proceedings, annual forestry symposium, school of forestry, Louisiana university. Baton Rouge, LA. 1952+. Proc. Annual Forest. Symp. School Forest. Louisiana Univ. HI 69774

Proceedings at the annual meeting, American antiquarian society = American antiquarian society. Proceedings at the annual meeting. Worcester, MA. Amer. Antiq. Soc. Proc. Annual Meeting. See B–P–H 78/4.

Proceedings, annual meeting, american association of stratigraphic palynologists. [Included in: Geoscience and man.] Vols. 1-7, 1968-74, 1970-76. Proc. Annual Meeting Amer. Assoc. Stratigr. Palynologists. Superseded by: Palynology. HI 69775

Proceedings of annual meeting. American carnation society. Indianapolis, IN. Proc. Annual Meeting Amer. Carnation Soc. See B–P–H 725/15. HI 58166

Proceedings of annual meeting, american society of plant physiologists = Plant physiology. Supplement. Rockville, MD.

Proceedings of the annual meeting of the aquatic weed control society = Proceedings of the aquatic weed control society. Jackson, MI. Proc. Aquatic Weed Control. Soc. See B–P–H 726/2.

Proceedings of the annual meeting. Bristol institution.

Bristol, England. Proc. Annual Meeting Bristol. Inst. See B–P–H 725/16. HI 58167

Proceedings of the annual meeting, canadian society for horticultural science = Proceedings of the canadian society for horticultural science.

Proceedings of the annual meeting. Caribbean food crops society. Bridgetown, Barbados. Proc. Annual Meeting Caribbean Food Crops Soc. See B–P–H 725/17. HI 58168

Proceedings of the annual meeting, international shade tree conference = Proceedings of the annual meeting, national (later international) shade tree conference.

Proceedings of the annual meeting, national (later international) shade tree conference. 5th-37th [1st-4th not published], 1949-61. Proc. Annual Meeting Natl. (Int.) Shade tree Conf. Superseded by: Journal of arboriculture. 4-2912-2. HI 69777

Proceedings of annual meeting. Nebraska academy of sciences. Lincoln, NB. Proc. Annual Meeting Nebraska Acad. Sci. See B–P–H 725/19. HI 58169

Proceedings of the annual meeting of the northeastern weed science society = Proceedings of the northeastern weed science society. Ithaca, NY., Storrs, CT.

Proceedings of the annual meeting of the Ohio state horticultural society. Wooster, OH. Vol. 55+, 1923+. Proc. Annual Meeting Ohio State Hort. Soc. Preceded by: Report (Annual) of the Ohio state horticultural society. 4-3163-1. HI 58170

Proceedings of the annual meeting, society for the promotion of agricultural science. Syracuse, NY. Proc. Annual Meetings Soc. Promot. Agric. Sci. See B–P–H 725/21. HI 58171

Proceedings of the annual meeting, plant growth regulator working group = Proceedings of the plant growth regulator working group. Longmont, CO.

Proceedings of the annual meeting, southern weed conference = Proceedings, southern weed conference. Champaign, IL.

Proceedings of the annual meeting, southern weed science society = Proceedings, southern weed science society. Champaign, IL.

Proceedings, annual tall timbers fire ecology conference. Tallahassee, FL. 1962+. Proc. Annual Tall Timbers Fire Ecol. Conf. HI 69778

Proceedings of applied bacteriology. Reading, England. Proc. Appl. Bacteriol. See B–P–H 726/1. HI 58172

Proceedings of the aquatic weed control society. Jackson, MI. Proc. Aquatic Weed Control. Soc. See B–P–H 726/2. HI 58173

Proceedings of the Arkansas academy of science. Fayetteville, AR. Proc. Arkansas Acad. Sci. See B–P–H 726/5. HI 58174

Proceedings of the Arkansas garden institute. Fayetteville, AR. Proc. Arkansas Gard. Inst. See B–P–H 726/6. HI 58175

Proceedings of the Arkansas state horticultural society. Siloam Springs, AR. Proc. Arkansas State Hort. Soc. See B–P–H 726/7. HI 58176

Proceedings, asian-Pacific weed control interchange. 1st-?, 1967-? Proc. Asian-Pacific Weed Control Interchange. Superseded by: Proceedings of the conference, asian-Pacific weed science society. HI 69779

Proceedings of the asiatic society of Bengal. Calcutta. 1865-1904. Proc. Asiat. Soc. Bengal. For vols 1832-64 see: Journal of the asiatic society of Bengal. Superseded by: Journal and proceedings of the asiatic society of Bengal. 1-505-3. HI 69780

Proceedings, association of American geographers. Washington, DC. Vol. 1+, 1969+. Proc. Assoc. Amer. Geogr. HI 74875

Proceedings of the association of economic biologists. London. Proc. Assoc. Econ. Biol. See B–P–H 726/14. HI 58177

Proceedings of the association of economic biologists, Coimbatore. Coimbatore, India. Proc. Assoc. Econ. Biol. Coimbatore. See B–P–H 726/15. HI 58178

Proceedings of the association of military surgeons of the U.S. Washington, DC. Proc. Assoc. Military Surgeons U.S. See B–P–H 726/16. HI 58179

Proceedings of the association of official seed analysts of north America. New Brunswick, NJ. Vols. 1-65, 1908-75. Proc. Assoc. Off. Seed Analysts N. Amer. Superseded by: Journal of seed technology. 1-533-2. HI 69781

Proceeding of the association for plant protection of Hokuriku. [Hokuriku bykogaichu kenkyukai-ho.] Niijata, Japan. Proc. Assoc. Pl. Protect. Hokuriku. See B–P–H 726/18. HI 58180

Proceedings, association for plant protection of Kyushu. [Kyushu byogaichu kenkyukai kaiho.] Chikugo City. Vol. 1+, 1955+. Proc. Assoc. Pl. Protect. Kyushu. HI 69782

Proceedings of the association for promoting the discovery of the interior parts of Africa. London. Proc. Assoc. Promot. Disc. Interior Parts Africa. See B–P–H 726/19. HI 58181

Proceedings of the australian biochemical society. Clayton, Mitcham, Vic. Vol. 1+, 1968+. Proc. Austral. Biochem. Soc. HI 69783

Proceedings of the Barrow naturalists' field club. Barrow-in-Furness? 1946+. Proc. Barrow Naturalists' Field Club. Preceded by: Report (Annual) and proceedings of the Barrow naturalists' field club and literary and scientific association. HI 69784

Proceedings of the Bath natural history and antiquarian field club. Bath, England. Proc. Bath Nat. Hist. Field Club. See B–P–H 726/23. HI 58182

Proceedings of the Belfast natural history and philosophical society. Belfast. 1852-61, 1871-82. Proc. Belfast Nat. Hist. Philos. Soc. Superseded by: Report and proceedings of the Belfast natural history and philosophical society. HI 69785

Proceedings Bihar academy of agricultural sciences. Bangalore City. Vol. 1+, 1952+. Proc. Bihar Acad. Agric. Sci. HI 69786

Proceedings of the biogeographical society of Japan. [Nihon seibutsu chiri gakkai kiji.] Proc. Biogeogr. Soc. Japan. See B–P–H 726/24. HI 58183

Proceedings of the biological society of Washington. Washington, DC. Proc. Biol. Soc. Wash. See B–P–H 726/25. HI 58184

Proceedings of the Birmingham natural history and microscopical society. Birmingham. Vols. 1-2, 1869-70. Proc. Birmingham Nat. Hist. Soc. Superseded by: Report and transactions, Birmingham natural history and microscopical society. 1-711-2. HI 69787

Proceedings of the Birmingham natural history and philosophical society. Birmingham. Vols. 9-20(2), 1894-1963. Proc. Birmingham Nat. Hist. Philos. Soc. Preceded by: Proceedings of the Birmingham philosophical society. Superseded by: Proceedings of the Birmingham natural history society. HI 69788

Proceedings of the Birmingham natural history society. Birmingham. Vol. 20(3)+, 1964+. Proc. Birmingham Nat. Hist. Soc. Preceded by: Proceedings of the Birmingham natural history and philosophical society. HI 69789

Proceedings of the Birmingham philosophical society. Birmingham. 1876-93. Proc. Birmingham Philos. Soc. Superseded by: Proceedings of the Birmingham natural history and philosophical society. HI 69790

Proceedings of the board of agriculture of India. Calcutta. 1905-29. Proc. Board Agric. India. HI 69791

Proceedings of the Boston society of natural history. Boston, MA. Proc. Boston Soc. Nat. Hist. See B–P–H 727/5. HI 58185

Proceedings of the botanical institute, Kharkov A. Gorki university = Trudy nauchno-issledovatel'skogo instituta botaniki, Kharkivs'kogo derzhavnogo universiteti imeni O. M. Gor'kogo. Kharkov.

Proceedings of the botanical society of the British Isles = Botanical society of the British Isles. Proceedings. Arbroath.

Proceedings of the botanical society of Edinburgh. Edinburgh. Proc. Bot. Soc. Edinburgh. See B–P–H 727/9. HI 58186

Proceedings of the botanical society of London. Proc. Bot. Soc. London. See B–P–H 727/10. HI 58187

Proceedings of the Bournemouth natural science society. Bournemouth. Vols. 1+, 1908+. Proc. Bournemouth Nat. Sci. Soc. HI 69792

Proceedings of the Bristol naturalist's society. Bristol. England. Proc. Bristol Naturalist's Soc. See B–P–H 727/13. HI 58188

Proceedings, british commonwealth forestry conference. Various places. 6th-7th, 1952-57. Proc. Brit. Commonw. Forest. Conf. Preceded by: Proceedings, british empire forestry conference. Superseded by: Proceedings, commonwealth forestry conference. HI 69793

Proceedings, british empire forestry conference. Various places. 1st-5th, 1920-47. Proc. Brit. Empire Forest. Conf. Superseded by: Proceedings, british commonwealth forestry conference. HI 69794

Proceedings, british weed control conference. London. No. 1+, 1953+. Proc. Brit. Weed Control. Conf. HI 69795

Proceedings of the bryological society of Japan. [Nihon sentai-rui gakkai kaiho.] Tokyo. Vol. 1+, 1972+. Proc. Bryol. Soc. Japan. HI 69796

Proceedings of the California academy of natural sciences = Proceedings of the California academy of sciences. San Francisco, CA. Proc. Calif. Acad. Sci. See B–P–H 727/17.

Proceedings of the California academy of sciences. San Francisco, CA. Proc. Calif. Acad. Sci. See B–P–H 727/17. HI 58189

Proceedings of the California olive association annual technical conference. San Francisco, CA. Proc. Calif. Olive Assoc. Annual Techn. Conf. See B–P–H 727/19. HI 58190

Proceedings of the California weed conference. Sacramento, CA. Vol. 1+, 1949+. Proc. Calif. Weed Conf. 2-891-1. HI 69797

Proceedings of the Cambridge philosophical society. Cambridge. Vols. 1-25, 1843-1929. Proc. Cambridge Philos. Soc. Superseded by: Proceedings of the Cambridge philosophical society. Biological sciences. 2-895-2. HI 69798

Proceedings of the Cambridge philosophical society. Biological sciences. Cambridge. Vols. 1-2, 1923-25. Proc. Cambridge Philos. Soc., Biol. Sci. Preceded by: Proceedings of the Cambridge philosophical society. Superseded by: Biological reviews and biological proceedings of the Cambridge philosophical society. HI 69799

Proceedings, canadian federation of biological societies. London, Ont. 1958+. Proc. Canad. Fed. Biol. Soc. Superseded by: Programme and proceedings of the annual meeting, canadian federation of biological societies. HI 69800

Proceedings of the canadian institute. Toronto. 1879-1904. Proc. Canad. Inst. Preceded by: Canadian journal of science, literature and history. Superseded by: Transactions of the canadian institute. HI 69801

Proceedings, canadian phytopathological society. Winnipeg. Nos. 1-46, 1929-79. Proc. Canad. Phytopathol. Soc. 2-916-2. HI 69802

Proceedings of the canadian society for horticultural science. Vol. 5+, 1960+ [vols. 1-4 not published]. Proc. Canad. Soc. Hort. Sci. HI 69803

Proceedings. Canadian society of plant physiologists. Kingston, Ontario. Proc. Canad. Soc. Pl. Physiol. See B–P–H 727/25. HI 58191

Proceedings of the caribbean region, american society for horticultural science. Mexico City. Nos. 5-14 [nos. 1-4 not published], 1957-66. Proc. Caribbean Region Amer. Soc. Hort. Sci. Superseded by: Proceedings of the tropical region, american society for horticultural science. HI 69804

Proceedings, central states forest tree improvement conference. 1st+, 1959+. Proc. Centr. States Forest Tree Improv. Conf. Superseded by: Proceedings of the ... north central tree improvement conference. HI 75224

Proceedings of the Cheltenham natural science society. Cheltenham. N.s. vols. 1-5(1), 1906-30. Proc. Cheltenham Nat. Sci. Soc. HI 69805

Proceedings of the Chester society of natural science. Chester. 1874-1907. Proc. Chester Soc. Nat. Sci. 3-995-3. HI 69806

Proceedings of the Chester society of natural science, literature and art. Chester. Vols. 3-5, 1949-54. Proc. Chester Soc. Nat. Sci. Lit Art. Preceded by: Report and proceedings of the Chester society of natural science and literature and art. HI 69807

Proceedings of the chrysanthemum society of America. New York, NY. Proc. Chrysanthemum Soc. Amer. See B–P–H 727/28. HI 58192

Proceedings of the Cincinnati society of natural history. Cincinnati, OH. Proc. Cincinnati Soc. Nat. Hist. See B–P–H 728/1. HI 58193

Proceedings of the Cleveland academy of natural science. Cleveland, OH. Proc. Cleveland Acad. Nat. Sci. See 728/2. HI 58194

Proceedings of the Cleveland naturalists' field club. Middlesbrough. 1903-09. Proc. Cleveland Naturalists' Field Club. Preceded by: Record of proceedings, Cleveland naturalists' field club. HI 69808

Proceedings of the college field club and natural history society. Exeter. 1898-37. Proc. Coll. Field Club Nat. Hist. Soc. Superseded by: Proceedings, university college field club. HI 69809

Proceedings of the college of natural sciences. Section 4, life (biological) sciences. [Chayon taehak. Soul taehakkyo.] Seoul. Vols. 1-2(1), 1976-77; vol. 2(2)+, 1977+ [all sections united]. Proc. Coll. Nat. Sci., Sect. 4. HI 69810

Proceedings of the Colorado museum of natural history = Proceedings of the Denver museum of natural history. Denver, CO. Proc. Denver Mus. Nat. Hist. See B–P–H 728/20.

Proceedings of the Colorado scientific society. Denver, CO. Proc. Colorado Sci. Soc. See B–P–H 728/5. HI 58195

Proceedings of the Columbus horticultural society = Report (Annual) of the Columbus horticultural society. Columbus, OH.

Proceedings of the committee for the home geographical researches of the Count Stephan Tisza scientifical society of Debrecen. = A Debreceni Tisza István tudományos társaság honismertetö bizottságának kiadványai. Debrecen, Karcag & Budapest. Debreceni Tisza István Tud. Társ. Honism. Bizott. Kiadv. See B–P–H 339/13.

Proceedings, committee reports and recommendations, commonwealth forestry conference = Proceedings, commonwealth forestry conference.

Proceedings, commonwealth forestry conference. Various places. 8th-9th, 1962-68. Proc. Commonw. Forest. Conf. Preceded by: Proceedings, british commonwealth forestry conference. HI 69811

Proceedings of the conference, asian-Pacific weed science society. ?-6th+, ?-1977+. Proc. Conf. Asian-Pacific Weed Sci. Soc. Preceded by: Proceedings, asian-Pacific weed control interchange. HI 69812

Proceedings of the conference of fruit growers of the dominion of Canada. Ottawa. Proc. Conf. Fruit Growers Domin. Canada. See B–P–H 728/8. HI 58196

Proceedings, conference on Great Lakes research. Ann Arbor, MI. 10th-17th, 1967-74. Proc. Conf. Great Lakes Res. Preceded by: Publications, Great Lakes research division, university of Michigan, institute of science and technology. Superseded by: Abstracts, conference on Great Lakes research. HI 69813

Proceedings of the ... conference on southern forest tree improvement = Proceedings of the ... southern forest tree improvement conference. Macon, GA, New Orleans, LA, etc.

Proceedings of conference, tea research institute of East Africa. Nairobi. Vols. 1-7, 1954-60. Proc. Conf. Tea Res. Inst. East Africa. Superseded by: Report (Annual), tea research institute of East Africa. HI 69814

Proceedings of the Connecticut pomological society. New Haven, CT. Proc. Connecticut Pomol. Soc. See B–P–H 728/10. HI 58197

Proceedings of the convention of the american tung oil society. Poplarville, MS. Proc. Conv. Amer. Tung Oil Soc. See B–P–H 728/11. HI 58198

Proceedings of the Cork scientific and literary society. Cork, Ireland. Proc. Cork Sci. Soc. See B–P–H 728/12. HI 58199

Proceedings of the Cotteswold naturalists' field club. Gloucester, England. Proc. Coteswold Naturalists' Field Club. See B–P–H 728/14. HI 58200

Proceedings of the Coventry and district natural history and scientific society. Coventry. 1948+. Proc. Coventry Distr. Nat. Hist. Sci. Soc. Preceded by: Proceedings of the Coventry natural history and scientific society. HI 69815

Proceedings of the Coventry natural history society. Coventry. 1909-17; 1930-40. Proc. Coventry Nat. Hist. Soc. Superseded by: Proceedings of the Coventry and district natural history and scientific society. HI 69816

Proceedings of the crop science society of Japan. [Nihon sakumotsu gakkai kiji.] Tokyo. Vols. 1-?, 1929-77. Proc. Crop Sci. Soc. Japan. Superseded by: Japanese journal of crop science. 2-1233-2. HI 69817

Proceedings, crops and soils wing of the board of agriculture and animal husbandry in India. No. 1+, 1935+. Proc. Crops Soils Wing Board Agric. Anim. Husb. India. HI 69818

Proceedings of the Croydon natural history and scientific society. Croydon. 1925-65, 1970+. Proc. Croydon Nat. HIst. Sci. Soc. Preceded by: Proceedings and transactions of the Croydon natural history and scientific society. For 1966-69 see: Proceedings and transactions of the Croydon natural history and scientific society. HI 69819

Proceedings of the cuban national horticultural society = Report of the cuban national horticultural society. Havana. Rep. Cub. Natl. Hort. Soc. See B–P–H 764/9.

Proceedings of the Davenport academy of natural sciences. Davenport, IA. Proc. Davenport Acad. Nat. Sci. See B–P–H 728/18. HI 58201

Proceedings of Delaware county inst. of science. Media, PA. Vol. 1+, 1906+. Proc. Delaware County Inst Sci. HI 69820

Proceedings of the Denver museum of natural history. Denver, CO. Proc. Denver Mus. Nat. Hist. See B–P–H 728/20. HI 58202

Proceedings of the Dorset natural history and antiquarian field club. Dorchester. 1877+. Proc. Dorset Nat. Hist. Antiq. Field Club. Superseded by: Proceedings of the Dorset natural history and archaeological society. HI 69821

Proceedings of the Dorset natural history and archaeological society. Dorchester. 1928+. Proc. Dorset Nat. Hist. Archaeol. Soc. Preceded by: Proceedings of the Dorset natural history and antiquarian field club. HI 69822

Proceedings of the Dublin university biological association. Dublin. Proc. Dublin Univ. Biol. Assoc. See B–P–H 728/23. HI 58203

Proceedings of the Dublin university zoological and botanical association. Proc. Dublin Univ. Zool. Bot. Assoc. See B–P–H 728/24. HI 58204

Proceedings of the Dudley and Midland geological & scientific society and field club. Dudley?, England. Proc. Dudley Midl. Geol. Soc. See B–P–H 728/25. HI 58205

Proceedings of the Dyserth and district field club. Arbroath, Llanfairfechan. 1911+. Proc. Dyserth Distr. Field Club. HI 69823

Proceedings, east african weed control conference. Nairobi. ?-5th, ?-1974? Proc. E. African Weed Control Conf. Superseded by: Proceedings, east african weed science conference. HI 69824

Proceedings, east african weed science conference. Nairobi. 6th+, [1976?]+. Proc. E. African Weed Sci. Conf. Preceded by: Proceedings, east african weed control conference. HI 69825

Proceedings of the east of Scotland union of naturalists' societies. 1885-95. Proc. E. Scotland Union Naturalists' Soc. Preceded by: Reports of the east of Scotland union of naturalists' societies. HI 69826

Proceedings of the eastern New York horticultural society. Chatham, NY. Proc. E. New York Hort. Soc. See B–P–H 729/1. HI 58206

Proceedings, ecological society of Australia. Canberra, A.C.T. Vol. 1+, 1966+. Proc. Ecol. Soc. Australia. HI 69827

Proceedings of the egyptian academy of sciences. Cairo. Proc. Egypt. Acad. Sci. See B–P–H 729/2. HI 58207

Proceedings of the Elliott society of natural history of Charleston. Charleston, SC. Proc. Elliott Soc. Nat. Hist. Charleston. See B–P–H 729/3. HI 58208

Proceedings of the entomological society of Ontario. Toronto. Proc. Entomol. Soc. Ontario. See B–P–H 729/5. HI 58209

Proceedings of the entomological society of Philadelphia. Philadelphia, PA. Proc. Entomol. Soc. Philadelphia. See B–P–H 729/6. HI 58210

Proceedings of the entomological society of Washington. Washington, DC. Proc. Entomol. Soc. Wash. See B–P–H 729/7. HI 58211

Proceedings of the Essex institute. Salem, MA. Proc. Essex Inst. See B–P–H 729/9. HI 58212

Proceedings, european grassland federation. Wageningen. 1965+. Proc. Eur. Grassland Fed. HI 69828

Proceedings, european symposium on fruit tree virus diseases. Wädenswil, etc. Nos. 1-8, 1954-70. Proc. Eur. Symp. Fruit Tree Virus Dis. Superseded by: Proceedings, international symposium on fruit treee virus diseases. HI 69829

Proceedings of the finnish academy of science and letters = Sitzungsberichte der Finnischen Akademie der Wissenschaften. Helsinki.

Proceedings of the Florida academy of sciences. Gainesville, FL. Proc. Florida Acad. Sci. See B–P–H 729/13. HI 58213

Proceedings of the Florida fruit growers' association. [Florida, U.S.A.]. Proc. Florida Fruit Growers' Assoc. See B–P–H 729/14. HI 58214

Proceedings of the Florida state horticultural society. Tallahassee, FL. Proc. Florida State Hort. Soc. See B–P–H 729/15. HI 58215

Proceedings of the Florida turf association. Gainesville, FL. Proc. Florida Turf Assoc. See B–P–H 729/16. HI 58216

Proceedings of the Folkestone natural history and microscopical society. Folkestone. 1883-1901. Proc. Folkestone Nat. Hist. Microsop. Soc. HI 69830

Proceedings of the forage and grassland conference. Lexington, KY. 1978+. Proc. Forage Grassland Conf. HI 74188

Proceedings of the forest products research society national meeting. Madison, WI. Proc. Forest Prod. Res. Soc. Natl. Meeting See B–P–H 729/17. HI 58217

Proceedings, fruit growers' society of western New York = New York state horticultural society proceedings. New York State Hort. Soc. Proc. See B–P–H 656/21.

Proceedings of the general meeting, european grassland federation = Proceedings, european grassland federation. Wageningen.

Proceedings, genetics society of Canada. Ottawa. Vols. 1-3, 1956-58. Proc. Genet. Soc. Canada. Superseded by: Canadian journal of genetics and cytology. HI 69831

Proceedings of the geological association of Canada. Toronto. Proc. Geol. Assoc. Canada. See B–P–H 729/21. HI 58218

Proceedings of the geological society of London. London. Proc. Geol. Soc. London. See B–P–H 729/22. HI 58219

Proceedings of the Georgia state horticultural society. Athens, GA. Proc. Georgia State Hort. Soc. See B–P–H 729/23. HI 58220

Proceedings (of the annual congresses), grassland society of southern Africa. Pietermaritzburg, etc. Vol. 1+, 1966+. Proc. Grassland Soc. S. Africa. HI 69832

Proceedings of the Great Plains agricultural council. Laramie, WY. Proc. Great Plains Agric. Council. See B–P–H 730/1. HI 58221

Proceedings of the Hampshire field club and archaeological society. Southampton. Vols. 24+, 1967+. Proc. Hampshire Field Club Archaeol. Soc. Preceded by: Papers abd proceedings of the Hampshire field club. HI 69833

Proceedings of the hawaiian academy of science. Honolulu, HI. Proc. Hawaiian Acad. Sci. See B–P–H 730/3. HI 58222

Proceedings, Heacham and west Norfolk natural history society. Heacham. Vol. ?-2(2)+, ?-1968+. Proc. Heacham W Norfolk Nat. Hist. Soc. HI 69834

Proceedings of the holly society of America. Millbrook, NJ. Proc. Holly Soc. Amer. See B–P–H 730/4. HI 58223

Proceedings of the Holmesdale natural history club. Reigate. 1865-1919. Proc. Holmesdale Nat. Hist. Club. 3-1873-1. HI 69835

Proceedings of the horticultural and agricultural society of Jamaica. Kingston, Jamaica. Proc. Hort. Soc. Jamaica. See B–P–H 730/9. HI 58224

Proceedings of the horticultural society of London. London. Vols. 1-15, 1838-43. Proc. Hort. Soc. London. From 1844-55 incorporated in: Journal of the horticultural society of London; Suspended 1856-58. Superseded by: Proceedings of the royal horticultural society of London. HI 69836

Proceedings of the horticultural society of northern Illinois. Springfield, IL. Proc. Hort. Soc. N. Illinois. See B–P–H 730/10. HI 58225

Proceedings, hydroponics society of America. Concord, CA. ?-1982+. Proc. Hydrop. Soc. Amer. HI 69837

Proceedings of the imperial academy of Japan. [Teikoku gakushiin kiji.] Tokyo. Vols. 1-23, 1912-48 [suspended 1918-25]. Proc. Imp. Acad. Japan. Superseded by: Proceedings of the Japan academy. 3-2151-3. HI 69838

Proceedings of the indian academy of sciences. Bangalore. Vol. 1, 1935-35. Proc. Indian Acad. Sci. Superseded by: Proceedings of the indian academy of sciences. Section B, biological sciences. 3-1954-1. HI 69839

Proceedings, indian academy of sciences. Plant sciences. Bangalore. Vol. 89+, 1980+. Proc. Indian Acad. Sci., Pl. Sci. Preceded by: Proceedings, indian academy of sciences. Section B, biological sciences. HI 69840

Proceedings, indian academy of sciences. Section B, biological sciences. Bangalore. Vols. 2-88, 1935-1979. Proc. Indian Acad. Sci., B. Preceded by: Proceedings of the indian academy of sciences. Superseded by: Proceedings, indian academy of sciences. Plant sciences. HI 69841

Proceedings of the indian national science academy. Part B, biological sciences. New Delhi. Vol. 36+, 1970+. Proc. Indian Natl. Sci. Acad., B. Preceded by: Proceedings of the national institute of sciences of India. HI 69842

Proceedings of the indian science congress association. Calcutta. Proc. Indian Sci. Congr. Assoc. See B–P–H 730/18. HI 58226

Proceedings of the Indiana academy of science. Brookville, IN. Proc. Indiana Acad. Sci. See B–P–H 730/20. HI 58227

Proceedings, institut teknologi Bandung. Supplement. Bandung. Vol. 1+, 1971+. Proc. Inst. Teknol. Bandung, Suppl. HI 69843

Proceedings of the institute of sciences, Nihon university. Tokyo. No. ?-4+, ?-1968+. Proc. Inst. Sci. Nihon Univ. HI 69844

Proceedings, international association of theoretical and applied limnology = Internationale Vereinigung für theoretische und angewandte Limnologie. Verhandlungen. Stuttgart. Int. Vereinigung Theor. Limnol. Verh. See B–P–H 435/21.

Proceedings, international congress of ecology. 1st+, 1974+. Proc. Int. Congr. Ecol. HI 69845

Proceedings, international congress on photobiology. New York. Vol. 1+, 1954+. Proc. Int. Congr. Photobiol. HI 69846

Proceedings, international congress on photosynthesis research. 1st+, 1968+. Proc. Int. Congr. Photosyn. Res. HI 69847

Proceedings, international congress of plant tissue and cell culture. ?-3rd+, ?-1974+. Proc. Int. Congr. Pl. Tissue Cell Cult. HI 69848

Proceedings of the international lilac society inc. Hamilton, Ont. ?-1982+. Proc. Int. Lilac Soc. HI 69849

Proceedings, international photobiological congress = Proceedings, international congress on photobiology. New York.

Proceedings, international plant propagators' society. Various places. No. 1+, 1951+. Proc. Int. Pl. Propag. Soc. HI 69850

Proceedings, international seaweed symposium. London, etc. 1st+, 1952+. Proc. Int. Seaweed Symp. HI 69851

Proceedings of the international seed testing association. Washington, DC. Vols. 1-37, 1925-72. Proc. Int. Seed Testing Assoc. Superseded by: Seed science and technology. HI 69852

Proceedings, international shade tree conference = Proceedings of the annual meeting, national (later international) shade tree conference.

Proceedings of the international society of citriculture. Lake Alfred, FL. 1977?+. Proc. Int. Soc. Citricult. HI 69853

Proceedings, international specialists' workshop-conference on criteria and terminology in the classification of fungi imperfecti. 1st+, 1969+. Proc. Int. Specialists' Workshop-Conf. Criter. Terminol. Classific. Fungi Imperf. HI 69854

Proceedings, international symposium on aquatic weeds = Proceedings, symposium on aquatic weeds. Oldenburg, Oxford.

Proceedings, international symposium on biological control of weeds. 1st+, 1969+. Proc. Int. Symp. Biol. Control Weeds. HI 69856

Proceedings, international symposium on food microbiology. Lille, etc. 1st+, 1954+. Proc. Int. Symp. Food Microbiol. HI 69857

Proceedings, international symposium on fruit tree virus diseases. Wädenswil, etc. No. 9+, 1973+. Proc. Int. Symp. Fruit Tree Virus Dis. Preceded by: Proceedings, european symposium on fruit tree virus diseases. HI 69858

Proceedings, international symposium on microbiological standardization. Zagreb. Vols. 1-11, 1959-64. Proc. Int. Symp. Microbiol. Standardiz. Superseded: Symposia series in immunobiological standardization. HI 69859

Proceedings, international symposium on molecular biology. Baltimore, MD. 1st+, 1972+. Proc. Int. Symp. Molec. Biol. HI 69860

Proceedings, international symposium on plant pathology. 1st+, 1967+. Proc. Int. Symp. Pl. Pathol. HI 69861

Proceedings, international symposium on pollination = Meddelanden från sveriges fröodlareforbund. Stockholm.

Proceedings, international symposium on tropical root

crops. 1st-3rd, 1967-7? Proc. Int. Symp. Trop. Root Crops. Superseded by: Proceedings of the symposium, international society for tropical root crops. HI 69862

Proceedings of the international turfgrass research conference. Various places. 1st+, 1969+, [1970?]+. Proc. Int. Turfgrass Res. Conf. HI 69863

Proceedings, international union of biological sciences, general assemblies. Various places. 9th+, 1947+. Proc. Int. Union Biol. Sci. Gen. Assembl. Preceded by: Union internationale des sciences biologiques. Series A, générale. HI 69864

Proceedings, international wheat genetics symposium. 1st+, 1958+. Proc. Int. Wheat Genet. Symp. HI 69865

Proceedings of the Iowa academy of science. Des Moines, IA. Vols. 1-94, 1887/93-87. Proc. Iowa Acad. Sci. Superseded by: Journal of the Iowa academy of science. 3-2105-2. HI 58228

Proceedings of the Iowa state horticultural society. Des Moines. 1936+. Proc. Iowa State Hort. Soc. Preceded by: Report of the Iowa state horticultural society. 3-2109-3. HI 69866

Proceedings of the Isle of Man natural history and antiquarian society. Douglas. 1906+. Proc. Isle of Man Nat. Hist Antiq. Soc. Preceded by: Lioar manninagh, the journal of the Isle of Man natural history and antiquarian society. HI 69867

Proceedings of the Isle of Wight natural history and archaeological society. Newport, Isle of Wight. Proc. Isle of Wight Nat. Hist. Soc. See B–P–H 730/27. HI 58230

Proceedings of the Japan academy. [Nippon gakusiin kiji.] Tokyo. Vols. 24-53, 1948-77. Proc. Japan Acad. Preceded by: Proceedings of the imperial academy of Japan. Superseded by: Proceedings of the Japan academy. Series B, physical and biological sciences. 3-2151-3. HI 69868

Proceedings of the Japan academy. Series B, physical and biological sciences. Tokyo. Vol. 53+, 1977+. Proc. Japan Acad., B. Preceded by: Proceedings of the Japan academy . HI 69869

Proceedings of the Japan society of plant taxonomists. [Nihon shokubutsu bunrui gakkai kaiho.] Tokyo. Vol. 1+, 1967+. Proc. Japan Soc. Pl. Taxon. HI 69870

Proceedings of the japanese association for the advancement of science. [Nihon gakujutsu kyokai hokoku.] Tokyo. Proc. Jap. Assoc. Advancem. Sci. See B–P–H 731/7. HI 58231

Proceedings of the Kansai plant protection society. [Kansai byochugai kenkyukai-ho.] Mie, Japan. Proc. Kansai Pl. Protect. Soc. See B–P–H 731/9. HI 58232

Proceedings of the Kanto Tosan plant protection society. [Kanto Tosan byogaichu kenkyukai nenpo.] Urawa, Japan. Proc. Kanto Tosan Pl. Protect. Soc. See B–P–H 731/10. HI 58233

Proceedings of the Kharkov state university = Uchenye zapiski, khar'kovskii gosudarstvennyogo universiteta imeni A. M. Gor'kogo. Kharkov.

Proceedings, koninklijke nederlandse akademie van wetenschappen. Series B. Amsterdam. 1955+. Proc. Kon. Ned. Akad. Wetensch., B. Preceded by: Proceedings, section of sciences. Koninklijke nederlandse akademie van wetenschappen. HI 69871

Proceedings, koninklijke nederlandse akademie van wetenschappen. Series C, biological and medical sciences. Amsterdam. Vol. 54+, 1951+. Proc. Kon.Ned. Akad. Wetensch. C. Preceded by: Proceedings, section of sciences. Koninklijke nederlandse akademie van wetenschappen. HI 69872

Proceedings of the Kossino limnological station of the hydrometeorological service of the U S S R = Trudy Limnologicheskoi stantsii v Kosine. Moscow.

Proceedings, lake states forest tree improvement conference. 1st+, 1953+. Proc. Lake States Forest Tree Improv. Conf. Superseded by: Proceedings of the ... north central tree improvement conference. HI 75225

Proceedings of the Leeds philosophical and literary society. Scientific section. Leeds. Vol. 1+, 1925+. Proc. Leeds Philos. Lit. Soc. 3-2379-1. HI 69873

Proceedings of the Lenin academy of agricultural sciences of the U S S R. Moscow. Proc. Lenin Acad. Agric. Sci. U.S.S.R. See B–P–H 731/16. HI 58234

Proceedings of the Leningrad chemical-pharmaceutical institute = Trudy Leningradskogo khimiko-farmacevticheskogo instituta. Leningrad.

Proceedings of the linnaean society of New York. New York, NY. Proc. Linn. Soc. New York See B–P–H 731/22. HI 58237

Proceedings of the linnean society of London. London. Vols. 1-2, [1839-]1848-55; 1856-96; vols. 110-179, 1897/98-1968. Proc. Linn. Soc. Lond. Superseded by: Biological journal of the linnean society. 3-2429-3. HI 69874

Proceedings of the linnean society of New South Wales. Sydney. Proc. Linn. Soc. New South Wales. See B–P–H 731/21. HI 58236

Proceedings of the literary and philosophical society of Manchester. Manchester. 1857-77. Proc. Lit. Philos. Soc. Manchester. Superseded by: Proceedings of the Manchester literary and philosophical society. HI 69875

Proceedings of the Liverpool biological society = Proceedings and transactions of the Liverpool biological society. Liverpool. Proc. & Trans.

Liverpool Biol. Soc. See B–P–H 737/20.

Proceedings of the Liverpool botanical society. Liverpool. Proc. Liverpool Bot. Soc. See B–P–H 731/24. HI 58238

Proceedings of the Liverpool naturalists' field club. Liverpool. 1860-1942. Proc. Liverpool Naturalists' Field Club. Preceded by: Report of the Liverpool naturalists' field club. Superseded by: Proceedings and natural history notes for the area. Liverpool naturalists' field club. 3-2448-2. HI 69876

Proceedings of the Llandudno and district field club. Llanfairfechan. 1906-50. Proc. Llandudno Distr. Field Club. HI 69877

Proceedings of the Louisiana academy of sciences. Baton Rouge, LA. Proc. Louisiana Acad. Sci. See B–P–H 731/27. HI 58240

Proceedings of the Louisiana state agricultural society. Baton Rouge, LA. Proc. Louisiana State Agric. Soc. See B–P–H 730/28. HI 58241

Proceedings of the Louisiana state horticultural society. Baton Rouge, LA. Proc. Louisiana State Hort. Soc. See B–P–H 731/29. HI 58242

Proceedings of the lyceum of natural history in the city of New York. New York. Ser. 1-2, 1870-74. Proc. Lyceum Nat. Hist. City New York. HI 69878

Proceedings of the Manchester field club. Manchester. Vol. 1, 1899-1903. Proc. Manchester Field Club. HI 69879

Proceedings of the Manchester literary and philosophical society. Manchester. 1877-87. Proc. Manchester Lit. Soc. Preceded by: Proceedings of the literary and philosophical society of Manchester. Superseded by: Memoirs and proceedings of the Manchester literary and philosophical society. 3-2531-1. HI 69880

Proceedings of the Massachusetts horticultural society. [Published together with: Transactions of the Massachusetts horticultural society.] Boston, MA. 1847-51. Proc. Mass. Hort. Soc. 3-2552-1. HI 69881

Proceedings of the meeting of agricultural research workers in Kyushu. [Kyushu noji shiken kenkyu happyokai koen yoshi.] Tokyo. Proc. Meeting Agric. Res. Workers Kyushu. See B–P–H 732/4. HI 58244

Proceedings of the ... meeting of the Willi Hennig society = Advances in cladistics. New York.

Proceedings of the meteorological society. London. Proc. Meteorol. Soc. See B–P–H 732/5. HI 58245

Proceedings of the microscopical society of Canada. Toronto. Vol. 1+, 1974+. Proc. Microscop. Soc. Canada. HI 69882

Proceedings of the microscopical society of Victoria. Melbourne, Vic. Vol. 1+, 1912+. Proc. Microsc. Soc. Victoria. 3-2649-1. HI 69883

Proceedings of the Minnesota academy of science. St. Paul, MN. Vols. 1-31(2), 1892-1963. Proc. Minnesota Acad. Sci. Superseded by: Journal of the Minnesota academy of science. 3-2679-2. HI 69884

Proceedings, Missouri state horticultural society. Jefferson City, MO. Vols. 1-9, 1928/30-45/46, 1930-46. Proc. Missouri State Hort. Soc. Preceded by: Report (Annual), Missouri horticultural society. 3-2704-2. HI 69885

Proceedings of the Montana academy of sciences. Missoula, MT. Proc. Montana Acad. Sci. See B–P–H 732/10. HI 58246

Proceedings of the Nagano prefectural agricultural experiment station. [Nagano-ken nogyo shikenjo shuho.] Nagano, Japan. Proc. Nagano Prefect. Agric. Exp. Sta. See B–P–H 732/16. HI 58247

Proceedings of the Nairobi scientific and philosophical society. Nairobi. Proc. Nairobi Sci. Soc. See B–P–H 732/17. HI 58248

Proceedings of the national academy of sciences of India. Section B, biological sciences. Allahabad. Vol. 6+, 1936+. Proc. Natl. Acad. Sci. India, B. Preceded by: Proceedings of the academy of sciences of the united provinces of Agra and Oudh. HI 54982

Proceedings of the national academy of sciences of the United States of America. Washington, DC. Vol. 1(1)-1(3), 1863-94; vol. 1+, 1915+. Proc. Natl. Acad. Sci. U.S.A. 4-2827-3. HI 69886

Proceedings of the national institute of sciences of India. Calcutta. Vols. 1-20, 1935-54. Proc. Natl. Inst. Sci. India, B. Superseded by: Proceedings of the national institute of sciences of India. Part B, biological sciences. 4-2886-1. HI 69887

Proceedings of the national institute of sciences of India. Part B, biological sciences. Calcutta. Vols. 21-35, 1955-69. Proc. Natl. Inst. Sci. India, B. Preceded by: Proceedings of the national institute of sciences of India. Superseded by: Proceedings of the indian national science academy. Part B, biological sciences. HI 69888

Proceedings of the national science council, republic of China. [Kuo chia k'o hsüeh wei yüan hui.] Taipei. Vols. 1-4, 19??-80. Proc. Natl. Sci. Council Republ. China. Superseded by: Proceedings of the national science council, republic of China. Part A, applied science and Part b, basic science. HI 69889

Proceedings of the national science council, republic of China. Part A, applied science. Taipei. Vol. 5+, 1981+. Proc. Natl. Sci. Council Republ. China, A. Preceded by: Proceedings of the national science council, republic of China. HI 69890

Proceedings of the national science council, republic of China. Part B, basic science. Taipei. Vol. 5+, 1981+. Proc. Natl. Sci. Council Republ. China, B. Preceded by: Proceedings of the national science council, republic of China. HI 69891

Proceedings, national shade tree conference = Proceedings of the annual meeting, national (later international) shade tree conference.

Proceedings, national weed committee, Canada, eastern section. Vols. 1-15, 1948-61. Proc. Natl. Weed Committee Canada E. Sect. Superseded by: Minutes, national weed committee, Canada, eastern section. HI 69892

Proceedings, national weed committee, Canada, western section. Vols. 1-14, 1947-60. Proc. Natl. Weed Committee Canada W. Sect. Superseded by: Minutes and report of the research planning committee, Canada weed committee, western section. HI 69893

Proceedings, national weeds conference of South Africa. ?-2nd+, ?-1977+. Proc. Natl. Weeds Conf. South Africa. HI 69894

Proceedings and natural history notes for the area. Liverpool naturalists' field club. Liverpool. 1943-69. Proc. Nat. Hist. Notes Area Liverpool Naturalists' Field Club. Preceded by: Proceedings of the Liverpool naturalists' field club. HI 69895

Proceedings of the natural history society of Dublin. Dublin. Proc. Nat. Hist. Soc. Dublin. See B–P–H 732/20. HI 58249

Proceedings of the natural history society of Fukien christian university. Foochow, China. Proc. Nat. Hist Soc. Fukien Christian Univ. See B–P–H 732/22. HI 58250

Proceedings of the natural history society of Glasgow. Glasgow. Vols. 1-5, 1868-84. Proc. Nat. Hist. Soc. Glasgow. Superseded by: Proceedings and transactions of the natural history society of Glasgow. HI 66621

Proceedings of the natural history society of Maryland. Baltimore, MD. 1930+. Proc. Nat. Hist. Soc. Maryland. 4-2927-2. HI 69896

Proceedings of the natural history society of Montreal. Montreal. Vols. 1-53, 1828-81. Proc. Nat. Hist. Soc. Montreal. Continued in: Canadian record of science. 4-2927-2. HI 69897

Proceedings of the natural history society of Wisconsin. Milwaukee, WI. Proc. Nat. Hist. Soc. Wisconsin. See B–P–H 732/25. HI 58251

Proceedings of the natural science association of Staten Island. New York, NY. Proc. Nat. Sci. Assoc. Staten Island. See. 733/1. HI 58252

Proceedings of the Nebraska academy of sciences and affiliated societies. Lincoln, NE. 42nd+, 1932+ [Vols. 1, 18-26, 33-41 not published]. Proc. Nebraska Acad. Sci. Affil. Soc. Vols. 2-17, 27-32 published in the Publications of the Nebraska academy of sciences. HI 69898

Proceedings, New Hampshire academy of science = New Hampshire academy of science. Proceedings. Durham, NH. New Hampshire Acad. Sci. Proc. See B–P–H 652/25.

Proceedings of the New York state agricultural society. Albany, NY. Proc. New York State Agric. Soc. See B–P–H 733/14. HI 58256

Proceedings, New York state fruit growers' association = New York state fruit growers' association. Proceedings. Fayetteville, NY. New York State Fruit Growers' Assoc. Proc. See B–P–H 656/20.

Proceedings, New York state horticultural society = New York state horticultural society proceedings. New York State Hort. Soc. Proc. See B–P–H 656/21.

Proceedings, New Zealand ecological society. Wellington, N.Z. Nos. 1-24, 1952-77. Proc. New Zealand Ecol. Soc. Preceded by: Report of annual meeting, New Zealand ecological society. Superseded by: New Zealand journal of ecology. HI 69899

Proceedings of the New Zealand grasslands association. Wellington. Proc. New Zealand Grasslands Assoc. See B–P–H 733/17. HI 58258

Proceedings of the New Zealand weed conference. Wellington, N. Z. Vols. 1-16, 1948-63? Proc. New Zealand Weed Conf. Superseded by: Proceedings of the New Zealand weed and pest control conference. 4-3055-3. HI 69900

Proceedings of the New Zealand weed and pest control conference. Wellington, N. Z. Vol. 17+, 1964+. Proc. New Zealand Weed Pest Control Conf. Preceded by: Proceedings of the New Zealand weed conference. HI 69901

Proceedings of the Newport natural history society. Newport, Rhode Island. Proc. Newport Nat. Hist. Soc. See B–P–H 733/19. HI 58259

Proceedings, north american conference on mycorrhizae. 1st+, 1969+. Proc. N. Amer. Conf. Mycorrh. HI 69902

Proceedings of the ... north central tree improvement conference. Madison, WI. 1st+, 1979+. Proc. North Centr. Tree Improv. Conf. Preceded by: Proceedings, central states forest tree improvement conference and Proceedings, lake states forest tree improvement conference. HI 74950

Proceedings, north central weed control conference. Urbana, IL. 1st+, 1944+. Proc. N. Centr. Weed Control Conf. HI 69903

Proceedings of the North Dakota academy of science.

Grand Forks, ND. Proc. North Dakota Acad. Sci. See B–P–H 733/20. HI 58260

Proceedings, northeastern forest tree improvement conference. 1953?+. Proc. NorthE. Forest Tree Improv. Conf. HI 69904

Proceedings of the northeastern weed control conference. New Brunswick, NJ. 1st-24th, 1947-70. Proc. NorthE. Weed Control Conf. Superseded by: Proceedings of the northeastern weed science society. HI 69905

Proceedings of the northeastern weed science society. Ithaca, NY., Storrs, CT. 25th+, 1971+. Proc. NorthE. Weed Sci. Soc. Preceded by: Proceedings of the northeastern weed control conference. HI 69906

Proceedings of the northern nut growers' association. Ithaca, NY. Vol. 1+, 1911+. Proc. Northern Nut Growers' Assoc. 4-3093-1. HI 69907

Proceedings of the nova scotian institute of science. Halifax, Nova Scotia. Proc. Nova Scotian Inst. Sci. See B–P–H 733/22. HI 58261

Proceedings of the nut growers' society of Oregon and Washington. Tigard, OR. 42nd-62nd, 1958-77. Proc. Nut Growers' Soc. Oregon Washington. Proceedings of earlier conferences published in: Report (Annual), Oregon state horticultural society. Superseded by: Proceedings of the nut growers' society of Oregon, Washington and British Columbia. HI 69908

Proceedings of the nut growers' society of Oregon, Washington and British Columbia. Tigard, OR. 63rd+, 1978+. Proc. Nut Growers' Soc. Oregon Washington British Columbia. Preceded by: Proceedings of the nut growers' society of Oregon and Washington. HI 75130

Proceedings of the Ohio academy of science. Columbus, OH. Proc. Ohio Acad. Sci. See B–P–H 733/23. HI 58262

Proceedings of the Ohio state horticultural society. Columbus, OH. Proc. Ohio State Hort. Soc. See B–P–H 733/25. HI 58263

Proceedings of the Oklahoma academy of science. Norman, OK. Proc. Oklahoma Acad. Sci. See B–P–H 734/3. HI 58264

Proceedings of the Oregon academy of sciences. Corvallis, OR. Proc. Oregon Acad. Sci. See B–P–H 734/5. HI 58265

Proceedings of the of the Ottawa academy of natural sciences. Ottawa. Proc. Ottawa Acad. Nat. Sci. See B–P–H 734/6. HI 58266

Proceedings of the Pacific science congress. Various places. 1929+. Proc. Pacific Sci. Congr. Preceded by: Proceedings of the Pan-Pacific science congress. HI 69909

Proceedings of the Pakistan academy of sciences. Karachi. Vol. 1+, 1964+. Proc. Pakistan Acad. Sci. HI 69910

Proceedings of the Pakistan science conference. Lahore. 1949+. Proc. Pakistan Sci. Conf. HI 69911

Proceedings of the Pan-Pacific science congress. Various places. 1920-26. Proc. Pan-Pacific Sci. Congr. Superseded by: Proceedings of the Pacific science congress. HI 69912

Proceedings and papers, bibliographical society of America. New York. Vols. 1-3, 1904/05-08, 1906-09. Proc. Pap. Bibliogr. Soc. America. Superseded by: Papers, bibliographical society of America. 1-667-2. HI 69913

Proceedings and papers, international technical conference on the protection of nature. Brussels. [1st], 1949. Proc. Pap. Int. Techn. Conf. Protect. Nat. Superseded by: Proceedings and papers, technical meeting, international union for the protection of nature. HI 69914

Proceedings and papers, technical meeting, international union for conservation of nature and natural resources. Morges. 6th-10th, 1956-66. Proc. Pap. Techn. Meeting Int. Union Conservation Nat. Nat. Resources. Preceded by: Proceedings and papers, technical meeting, international union for the protection of nature. Superseded by: Papers and proceedings, technical meeting, international union for conservation of nature and natural resources. HI 69915

Proceedings and papers, technical meeting, international union for the protection of nature. Brussels. 2nd-5th, 1949-54. Proc. Pap. Techn. Meeting Int. Union Protect. Nat. Preceded by: Proceedings and papers, international technical conference on the protection of nature. Superseded by: Proceedings and papers, technical meeting, international union for conservation of nature and natural resources. HI 69916

Proceedings, Papua and New Guinea scientific society. Port Moresby, Konedobu. Vol. ?-21+, ?-1954-69+. Proc. Papua New Guinea Sci. Soc. HI 69917

Proceedings of the Pennsylvania academy of science. Harrisburg, PA. Proc. Pennsylvania Acad. Sci. See B–P–H 734/10. HI 58267

Proceedings of the Peoria academy of science. Peoria. 1968+. Proc. Peoria Acad. Sci. HI 69918

Proceedings of the Perthshire society of natural science. Perth, Scotland. Proc. Perthshire Soc. Nat. Sci. See B–P–H 734/14. HI 58268

Proceedings of the Philadelphia botanical club = Bartonia; a botanical annual. Philadelphia, PA. Bartonia. See B–P–H 168/4.

Proceedings of the pineapple technologists' society. Honolulu, HI. Proc. Pineapple Technol. Soc. See

B–P–H 734/17. HI 58269

Proceedings of the plant growth regulator society. Lake Alfred, FL, Longmont, CA. 8th+, 1981+. Proc. Pl. Growth Regulator Soc. Preceded by: Proceedings of the plant growth regulator working group. HI 69919

Proceedings of the plant growth regulator working group. Longmont, CO. 1st-7th?, 1977-80? Proc. Pl. Growth Regulator Working Group. Superseded by: Proceedings of the plant growth regulator society. HI 75131

Proceedings of the plant propagators' society. Cleveland, OH. Vols. 1-9, 1951-59. Proc. Pl. Propag. Soc. Superseded by: Combined proceedings, plant propogators society. HI 69920

Proceedings of the Plymouth athenaeum. Plymouth. 1961+. Proc. Plymouth Athenaeum. Preceded by: Report and transactions of the Plymouth institution and Devon and Cornwall natural history society. HI 69921

Proceedings of the Portland society of natural history. Portland, ME. Proc. Portland Soc. Nat. Hist. See B–P–H 734/18. HI 58270

Proceedings of the of the potato association of America. Washington, DC. Proc. Potato Assoc. Amer. See B–P–H 734/20. HI 58271

Proceedings of the prehistoric society. London. Proc. Prehist. Soc. See B–P–H 734/21. HI 58272

Proceedings and report of the ashmolean natural history society of Oxfordshire. Oxford. 1908-56, 1971+. Proc. & Rep. Ashmolean Nat. HIst. Soc. Oxfordshire. Preceded by: Report, ashmolean natural history society of Oxfordshire. 1-503-3. HI 66841

Proceedings and reports of the Belfast natural history and philosophical society. Belfast. 1920+. Proc. & Rep. Belfast Nat. Hist. Soc. Preceded by: Report and proceedings of the Belfast natural history and philosophical society. 1-630-3. HI 69922

Proceedings and reports of the Belfast naturalists' field club. Belfast. Ser. 2, vols. 8(5)-10, 1922-47. Proc. & Rep. Belfast Naturalists' Field Club. Preceded by: Report (Annual) and proceedings of the Belfast naturalists' field club. HI 69923

Proceedings of the Rhodesia scientific association. Bulawayo. Vols. 1-31, 1899/1900-1931/32, 1900-32. Proc. Rhodesia Sci. Assoc. Superseded by: Proceedings and transactions of the Rhodesia scientific association. 4-3675-3. HI 69924

Proceedings of the Rio Grande Valley horticultural society. Weslaco, TX. Proc. Rio Grande Valley Hort. Soc. See B–P–H 735/8. HI 58276

Proceedings of the Rochester academy of science. Rochester, NY. Proc. Rochester Acad. Sci. See B–P–H 735/9. HI 58277

Proceedings of the royal agricultural and horticultural society of South Australia. Adelaide, S.A. 1880/81-1939. Proc. Roy. Agric. Hort. Soc. South Australia. HI 69925

Proceedings of the royal asiatic society of Great Britain and Ireland. London. Proc. Roy. Asiat. Soc. Gr. Brit. See B–P–H 735/10. HI 58278

Proceedings of the royal canadian institute. Toronto. Proc. Roy. Canad. Inst. See B–P–H 735/11. HI 58279

Proceedings of the royal colonial institute. London. Vols. 1-40, 1869-1909. Proc. Roy. Colon. Inst. HI 69926

Proceedings of the royal entomological society of London. London. Proc. Roy. Entomol. Soc. London. See B–P–H 735/13. HI 58280

Proceedings of the royal geographical society of London. London. Vols. 1-22, 1855-78; n.s. vols. 1-14, 1879-92. Proc. Roy. Geogr. Soc. London. Superseded by: Geographical journal. London. HI 69927

Proceedings of the royal horticultural society of London. London. N.s. vols. [1]-5, 1859-65. Proc. Roy. Hort. Soc. London. Preceded by: Proceedings of the horticultural society of London.

From 1866 onwards incorporated in: Journal of the royal horticultural society. 4-3722-2. HI 69928

Proceedings of the royal institution of Great Britain. London. Proc. Roy. Inst. Gr. Brit. See B–P–H 735/15. HI 58282

Proceedings of the royal irish academy. Dublin. Vols. 1-10, [1837]1841-69; ser. 2, vols. 1-4, 1870-88; ser. 3, vols. 1-7, 1889-1901. Proc. Roy. Irish Acad. Superseded by: Proceedings of the royal irish academy. Section B, biological, geological and chemical sciences. 4-3724-1. HI 69929

Proceedings of the royal irish academy. Section B, biological, geological and chemical sciences. Dublin. Vol. 24+, 1902+. Proc. Roy. Irish Acad., B. Preceded by: Proceedings of the royal irish academy. 4-3724-1. HI 69930

Proceedings of the royal microscopical society. London. Proc. Roy. Microscop. Soc. See B–P–H 735/17. HI 58284

Proceedings of the royal philosophical society of Glasgow. Glasgow. Proc. Roy. Philos. Soc. Glasgow. See B–P–H 735/18. HI 58285

Proceedings of the royal physiographic society at Lund = Kungl. fysiografiska sällskapets i Lund förhandlingar. Lund. Kungl. Fysiogr. Sällsk. Lund Förh. See B–P–H 521/14.

Proceedings of the royal society of arts and sciences of Mauritius. Port Louis. Vol. 1+, 1949+. Proc. Roy.

Soc. Arts Mauritius. Preceded by: Transactions of the royal society of arts and sciences of Mauritius. 5-3972-2. HI 69931

Proceedings of the royal society of Canada. Montreal. Ser. 3, vol. 21+, 1927+. Proc. Roy. Soc. Canada. Preceded by: Proceedings and transactions of the royal society of Canada. 4-3727-1. HI 75184

Proceedings of the royal society of Edinburgh. Edinburgh. Vols. 1-60, 1832-1940. Proc. Roy. Soc. Edinburgh. Superseded by: Proceedings of the royal society of Edinburgh. Series B, biology. 4-3727-2. HI 69932

Proceedings of the royal society of Edinburgh. Series B, biology (later biological sciences). Edinburgh. Vol. 61+, 1941+. Proc. Roy. Soc. Edinburgh, B. Preceded by: Proceedings of the royal society of Edinburgh. 4-3727-2. HI 69933

Proceedings of the royal society of London. London. Proc. Roy. Soc. London. See B–P–H 735/24. HI 58286

Proceedings of the royal society of London. Series B. Biological sciences. London. Proc. Roy. Soc. London, Ser. B, Biol. Sci. See B–P–H 736/2. HI 58287

Proceedings of the royal society of medicine. London. Proc. Roy. Soc. Med. See B–P–H 736/3. HI 58289

Proceedings of the royal society of medicine. Section of the history of medicine. London. Vol. 6+, 1913+. Proc. Roy. Soc. Med., Sect. Hist. Med. HI 69934

Proceedings of the royal society of New Zealand. Wellington, N.Z. Proc. Roy. Soc. New Zealand. See B–P–H 736/6. HI 58290

Proceedings of the royal society of Queensland. Brisbane, Qld. Proc. Roy. Soc. Queensland. See B–P–H 736/8. HI 58291

Proceedings of the royal society of Victoria. Melbourne, Vic. Proc. Roy. Soc. Victoria. See B–P–H 736/13. HI 58292

Proceedings of the rubber research institute of malaysia planters' conference. Kuala Lumpur. ?-1974+. Proc. Rubber Res. Inst. Malaysia Planters' Conf. HI 75068

Proceedings of the Saudi biological society. 198?+. Proc. Saudi Biol. Soc. HI 69935

Proceedings of the science association of Nigeria. Ibadan. 1958-65. Proc. Sci. Assoc. Nigeria. Superseded by: Nigerian journal of science. HI 69936

Proceedings of the scientific association of Trinidad. Port of Spain, Trinidad. Proc. Sci. Assoc. Trinidad. See B–P–H 736/15. HI 58293

Proceedings, scientific society of Bosnia and Hercegovina, Yugoslavia = Posebna izdanja, naučno društvo NR Bosne i Hercegovine. Sajarevo.

Proceedings of the scientific society of London. London. Proc. Sci. Soc. London. See B–P–H 736/17. HI 58294

Proceedings of the section of sciences, koninklijke (Nederlandse) akademie van wetenschappen te Amsterdam. Amsterdam. Vols. 1-53, 1898-1954. Proc. Sect. Sci. Kon. Akad. Wetensch. Amsterdam. Preceded by: Verslagen van de gewone vergardering der Wis-en natuurkundige afdeeling der koninklijke akademie van wetenschappen te Amsterdam. Superseded by: Proceedings, koninklijke nederlandse akademie van wetenschappen. Series B & C. 1-108-2. HI 69937

Proceedings of the seminar for arabian studies. London. 1st+, 1971+. Proc. Seminar Arab. Stud. HI 69939

Proceedings series, international institute of tropical agriculture. Ibadan. Vol. 1+, 1978+. Proc. Ser. Int. Inst. Trop. Agric. HI 69938

Proceedings of the Sheffield naturalists' club. Sheffield. 1894-1915. Proc. Sheffield Naturalists' Club. HI 69940

Proceedings of the society of american florists and ornamental horticulturists. Boson, MA. Proc. Soc. Amer. Florists. See B–P–H 736/20. HI 58295

Proceedings of the society of american foresters. Washington, DC. Vols. 1-11, 1905-16; 1947+. Proc. Soc. Amer. Foresters. For 1916-46 incorporated in: Journal of forestry. Washington. 5-3982-1. HI 69941

Proceedings, society of biological chemists (India). Bangalore. Vol. 1+, 1936+. Proc. Soc. Biol. Chem. (India). HI 69942

Proceedings of the society for experimental biology and medicine. New York, NY. Proc. Soc. Exp. Biol. See B–P–H 736/22. HI 58297

Proceedings, society for general microbiology. Reading. Vols. 1-5, 1973-78. Proc. Soc. Gen. Microbiol. Superseded by: Quarterly, society for general microbiology. HI 69943

Proceedings of the society for the promotion of agricultural science. Syracuse, NY. Proc. Soc. Promot. Agric. Sci. See B–P–H 736/27. HI 58298

Proceedings of the society of soil and plant diagnosticians. Washington, DC. Proc. Soc. Soil Diagn. See B–P–H 736/28. HI 58299

Proceedings of the soil and crop science society of Florida. Hollywood, FL. Vol. 16+, 1956+. Proc. Soil. Crop Sci. Soc. Fla. Preceded by: Proceedings of the soil science society of Florida. 5-3992-2. HI 69944

Proceedings of the soil science society of America. Ann Arbor, MI. Vols. 1-39, 1936-75. Proc. Soil. Sci. Soc. Amer. Preceded by: Bulletin of the american soil survey association. Superseded by: Journal, soil science society of America. 5-3992-2. HI 69945

Proceedings of the soil science society of Florida. Hollywood, FL. Vols. 1-15, 1939-55. Proc. Soil Soc. Florida. Superseded by: Proceedings of the soil and crop science society of Florida. HI 69946

Proceedings of the Somersetshire archaeological and natural history society. Taunton. 1849+. Proc. Somersetshire Archaeol. Nat. Hist. Soc. HI 69947

Proceedings of the South Dakota academy of sciences. Vermillion, SD. Proc. South Dakota Acad. Sci. See B–P–H 737/3. HI 58301

Proceedings of the south London entomological and natural history society. London. 1897-1932. Proc. S. London Entomol. Nat. Hist. Soc. Preceded by: Abstracts of the proceedings of the south London entomological and natural history society. Superseded by: Transactions and proceedings of the south London entomological and natural history society. HI 69948

Proceedings of the south western naturalists' union. Bristol. 1929-31, 1968+. Proc. S. W. Naturalists' Union. HI 69949

Proceedings, southern african electron microscopy society. Pretoria. ?-1971+. Proc. S. African Electron Microscop. Soc. HI 69950

Proceedings of the ... southern forest tree improvement conference. Macon, GA, New orleans, LA, etc. ?-11th+, ?-1971+. Proc. S. Forest Tree Improv. Conf. HI 75226

Proceedings of the southern pasture and forage crop improvement conference. New Orleans, LA, etc. 30/31st+, 1974+. Proc. S. Pasture Forage Crop Improv. Conf. Preceded by: Report, southern pasture and forage crop improvement conference. HI 75227

Proceedings, southern weed conference. Champaign, IL. Vols. 1-21, 1948-68. Proc. S. Weed Conf. Superseded by: Proceedings, southern weed science society. HI 69951

Proceedings, southern weed science society. Champaign, IL. Vol. 22+, 1969+. Proc. S. Weed Sci. Soc. Preceded by: Proceedings, southern weed conference. HI 69952

Proceedings, specialized international symposium on yeasts. 1st+, 1971+. Proc. Special. Int. Symp. Yeasts. HI 69953

Proceedings of the staff meetings of the Mayo clinic. Rochester, MN. Proc. Staff Meetings Mayo Clin. See B–P–H 737/6. HI 58302

Proceedings of the Staten Island association of arts and sciences. New York, NY. Proc. Staten Island Assoc. Arts. See B–P–H 737/10. HI 58303

Proceedings of the Staten Island institute of arts and sciences. New York, NY. Proc. Staten Island Inst. Arts. See B–P–H 737/11. HI 58304

Proceedings of the sugar beet research association. [Tensai kenkyu kaiho.] (Japan). Vol. 15+, 1973+. Proc. Sugar Beet Res. Assoc. Preceded by: Bulletin of sugar beet research. Supplement. Tokyo. HI 69954

Proceedings of the sugar cane investigation committee, Trinidad. 1932-34. Proc. Sugar Can Invest. Committee Trinidad. Superseded by: Report of the sugar cane investigation committee, Trinidad. HI 69955

Proceedings of the Swansea scientific and field naturalists' society. Swansea. 1927-52. Proc. Swansea Sci. Field Naturalists' Soc. HI 69956

Proceedings, swedish weed conference. Uppsala. 1st+, 1960?+. Proc. Swed. Weed Conf. HI 69957

Proceedings of the symposium on antibiotics = Antibiotics annual. New York, NY. Antibiot. Annual. See B–P–H 130/10.

Proceedings, symposium on aquatic weeds. Oldenburg, Oxford. 1964+. Proc. Symp. Aquatic Weeds. HI 69855

Proceedings of the symposium, international society for tropical root crops. 4th+, 1976+. Proc. Symp. Int. Soc. Trop Root Crops. Preceded by: Proceedings, international symposium on tropical root crops HI 69958.

Proceedings of the symposium, phytochemical society of north America. New York, London, etc. Vol. 6+, 1966+. Proc. Symp. Phytochem. Soc. N. America. Preceded by: Proceedings of the symposium, plant phenolics group of north America. HI 69959

Proceedings of the symposium, plant phenolics group of north America. New York, London, etc. Vols. 1-5, 1960-65. Proc. Symp. Pl. Phenolics Group N. America. Superseded by: Proceedings of the symposium, phytochemical society of north America. HI 69960

Proceedings, tall timbers ecology and management conference. Tallahassee, FL. No. 16+, 1979+. Proc. Tall Timbers Ecol. Managem. Conf. HI 69961

Proceedings, tall timbers fire ecology conference = Proceedings, annual tall timbers fire ecology conference. Tallahassee, FL.

Proceedings of the tashiki-kwai, or natural history society. Tokyo. Vol. 1(1-2), 1888-90. Proc. Tashiki-kwai. HI 69962

Proceedings of the Teign naturalists' field club. Paignton. 1918+. Proc. Teign Naturalists' Field Club. Preceded by: Report and proceedings of the Teign naturalists' field club. HI 69963

Proceedings of the Tenasserim agri-horticultural society

of Moulmein. 1891-96-? Proc. Tenasserim Agri-Hort. Soc. Moulmein. HI 69964

Proceedings of the Thoreau museum of natural history. Concord, MA. Proc. Thoreau Mus. Nat. Hist. See B–P–H 737/14. HI 58305

Proceedings of the Tomsk state Kuybyshev-University = Trudy Tomskogo gosudarstvennogo universiteta imeni V. V. Kuibysheva. Tomsk.

Proceedings and transactions of the academy of science. Austin, TX = Transactions of the academy of science. Austin, TX. Trans. Texas Acad. Sci. See B–P–H 891/5.

Proceedings and transactions of the british entomological and natural history society. London. 1967/68+, 1968+. Proc. & Trans. Brit. Entomol. Nat. Hist. Soc. Preceded by: Proceedings and transactions, south London entomological and natural history society. HI 69965

Proceedings and transactions of the Croydon microscopical club = Proceedings and transactions of the Croydon natural history society. Croydon, England. Proc. & Trans. Croydon Nat. HIst. Soc. See B–P–H 737/19.

Proceedings and transactions of the Croydon microscopical and natural history club. Croydon. 1878-1901. Proc. & Trans. Croydon Nat. Hist. Soc. Preceded by: Report and abstracts of proceedings of the Croydon microscopical club. Superseded by: Proceedings and transactions of the Croydon natural history and scientific society. 2-1234-1. HI 69966

Proceedings and transactions of the Croydon natural history and scientific society. Croydon. 1901-25, 1966-69. Proc. & Trans. Croydon Nat. Hist. Sci. Soc. Preceded by: Proceedings and transactions of the Croydon microscopical and natural history club. For 1925-65 see: Proceedings of the Croydon natural history and scientific society. Superseded by: Proceedings of the Croydon natural history and scientific society. HI 69967

Proceedings and transactions of the Croydon natural history society. Croydon, England. Proc. & Trans. Croydon Nat. HIst. Soc. See B–P–H 737/19. HI 58306

Proceedings and transactions of the Liverpool biological society. Liverpool. Proc. & Trans. Liverpool Biol. Soc. See B–P–H 737/20. HI 58307

Proceedings and transactions of the natural history society of Glasgow. Glasgow. N.s. vols. 1-4, 1885-97. Proc. & Trans. Nat. Hist. Soc. Glasgow. Preceded by: Proceedings of the natural history society of Glasgow. Superseded by: Transactions of the natural history society of Glasgow. HI 69968

Proceedings and transactions, Nova Scotian institute of natural science = Nova Scotian institute of natural science proceedings and transactions. Halifax, Nova Scotia. Nova Scotian Inst. Nat. Sci. Proc. Trans. See B–P–H 673/16.

Proceedings and transactions of the nova scotian institute of science. Halifax, Nova Scotia. Proc. & Trans. Nova Scotian Inst. Sci. See B–P–H 737/22. HI 58308

Proceedings and transactions of the Rhodesia scientific association. Bulawayo. Vols. 32-55, 1933-74. Proc. & Trans. Rhodesia Sci. Assoc. Preceded by: Proceedings of the Rhodesia scientific association. Superseded by: Transactions of the Rhodesia scientific association. HI 69969

Proceedings and transactions of the royal society of Canada. Montreal. Vols. 1-12, 1882-94; ser. 2, vols. 1-12, 1895-1906; ser. 3, vols. 1-20, 1907-26. Proc. & Trans. Roy. Soc. Canada. Superseded by: Proceedings of the royal society of Canada and Transactions of the royal society of Canada. Section 5, biological sciences. 4-3727-1. HI 58309

Proceedings and transactions of the scottish microscopical society. 1889-95. Proc. & Trans. Scott. Microscop. Soc. HI 69970

Proceedings and transactions, south London entomological and natural history society. London. 1933-67. Proc. & Trans. S. London Entomol. Nat. Hist. Superseded by: Proceedings and transactions, british entomological and natural history society. HI 69971

Proceedings of the triennial conference, european association for potato research. Wageningen. 1st-?, 1960-?, 1961-? Proc. Trienn. Conf. Eur. Assoc. Potato Res. Superseded by: E A P R abstracts of conference papers. HI 69972

Proceedings of the tropical region, american society for horticultural science. Mexico City. No. 15+, 1967+. Proc. Trop. Region Amer. Soc. Hort. Sci. Preceded by: Proceedings of the caribbean region, american society for horticultural science. HI 75129

Proceedings of the United States national museum. Washington, DC. Vols. 1-125, 1878-1968. Proc. U. S. Natl. Mus. 5-4318-2. HI 69973

Proceedings, university college field club. Exeter. 1937-46. Proc. Univ. Coll. Field Club, Exeter. Superseded by: Proceedings of the college field club and natural history society. HI 69974

Proceedings of the university of Durham philosophical society. Durham. Vols. 1-?, 1896-1963. Proc. Univ. Durham Philos. Soc. Superseded by: Proceedings of the university of Newcastle-upon-Tyne philosophical society. 2-1374-3. HI 69975

Proceedings of the university of Newcastle-upon-Tyne philosophical society. Newcastle-upon-Tyne. 1964+.

Proc. Univ. Newcastle-upon-Tyne Philos. Soc. Preceded by: Proceedings of the university of Durham philosophical society. HI 69976

Proceedings of the Utah academy of sciences, arts and letters. Salt Lake City, UT. Vols. 3-53, 1929-76. Proc. Utah Acad. Sci. Preceded by: Transactions Utah academy of sciences, arts and letters. Superseded by: Encyclia. 5-4345-2. HI 69977

Proceedings of the Utah state horticultural society. Salt Lake City, UT. Proc. Utah State Hort. Soc. See B–P–H 738/4. HI 58311

Proceedings of the Victoria institute of Trinidad. Port of Spain. Parts 1-?, 1894-1904. Proc. Victoria Inst. Trinidad. Superseded by: Report, Victoria institute of Trinidad and Tobago. HI 69978

Proceedings of the viticultural science symposium and workshop. Tallahassee, FL. ?-3rd+, ?-1980+. Proc. Vitic. Sc. Symp. Workshop. HI 75143

Proceedings of the Warwickshire naturalists' field club. Warwick, England. Proc. Warwickshire Naturalists' Field Club. See B–P–H 738/9. HI 58312

Proceedings of the Washington academy of sciences. Washington, DC. Proc. Wash. Acad. Sci. See B–P–H 738/11. HI 58313

Proceedings of the Washington scientific association. Washington, DC. Proc. Wash. Sci. Assoc. See B–P–H 738/12. HI 58314

Proceedings of the Washington state horticultural association. Pullman, WA. Proc. Wash. State Hort. Assoc. See B–P–H 738/14. HI 58315

Proceedings, weed society of New South Wales. Sydney, N.S.W. Vol. 1+, 1967+. Proc. Weed Soc. New South Wales. HI 69980

Proceedings of the west indian agricultural conference. Kingston, Jamaica. 1899-1924. Proc. W. Indian Agric. Conf. Nos. 1-8 issued in: West indian bulletin. HI 69981

Proceedings of the west London scientific association and field club. London. 1975-77. Proc. W. London Sci. Assoc. Field Club. HI 69982

Proceedings of the West Virginia academy of science. Morgantown, WV. Proc. West Virginia Acad. Sci. See B–P–H 738/16. HI 58316

Proceedings of the western forest genetics association. Olympia, WA, etc. Vols. 11-15, 1965-68 [nos. 1-10 not published]. Proc. W. Forest Genet. Assoc. HI 69983

Proceedings, western forestry and conservation association. Portland, OR. 1948+. Proc. W. Forest. Conservation Assoc. HI 69984

Proceedings, western New York horticultural society = New York state horticultural society proceedings. New York State Hort. Soc. Proc. See B–P–H 656/21.

Proceedings, western society of weed science. Laramie, WY, Logan, UT. Vol. 22+, 1968+. Proc. W. Soc. Weed Sci. Preceded by: Proceedings, western weed control conference. HI 69985

Proceedings, western weed control conference. Salt Lake City, UT. Vols. 1-21, 1938-67. Proc. W. Weed Control Conf. Superseded by: Proceedings, western society of weed science. HI 69986

Proceedings of the Wisconsin natural history society. Milwaukee, WI. ?-1885-89. Proc. Wisconsin Nat. Hist. Soc. HI 69987

Proceedings of world mariculture society. Baton Rouge, LA. 1970-80. Proc. World Maric. Soc. Superseded by: Journal of mariculture society. HI 69988

Proceedings of the Yorkshire geological society. Leeds. Proc. Yorkshire Geol. Soc. See B–P–H 738/18. HI 58317

Proceedings of the zoological-biological institute for scientific research = Pratsy Naukovo-doslidnogo zoologo-biologichnogo instytutu. Kiev.

Procès-verbaux et mémoires de l'académie des sciences, belles-lettres et arts, Besançon. Besançon. Procès-Verbaux Mém. Acad. Sci. Besançon. See B–P–H 738/25. HI 58318

Procés-verbaux de l'académie des sciences de l'Ukraine = Visti Ukrayins'koi akademii nauk, and Visti Vseukrayins'koi akademii nauk. Kiev.

Procès verbaux, mémoires et documents de la société historique et scientifique des Deux-Sévres = Mémoires de la société historique et scientifique des Deux-Sévres. Niort, France. Mém. Soc. Hist. Deux-Sèvres. See B–P–H 583/21.

Procès-verbaux mensuels de la société dauphinoise d'ethnologie et d'anthropologie (later et d'archéologie). Grenoble. Vol. 1+, 1926+. Procès-Verbaux Mens. Soc. Dauphin. Ethnol. Anthropol. HI 69989

Procès verbaux et rapports, réunion technique, union internationale pour la conservation de la nature et de ses resources = Proceedings and papers, technical meeting, international union for conservation of nature and natural resources. Morges.

Procès-verbaux des séances générales de l'institut royal des Pays-Bas. Amsterdam. Procès-Verbaux Séances Gén. Inst. Roy. Pays-Bas. See B–P–H 738/27. HI 58319

Procès-verbaux des séances mensuelles de la société des sciences naturelles de Tunisie. Tunis. 1953-60. Procès-Verbaux Séances Mens. Soc. Sci. Nat. Tunisie. Previously and subsequently contained in: Bulletin de la société des sciences naturelles de Tunisie. HI 69990

Procès-verbaux des séances de la société des sciences

physiques et naturelles de Bordeaux. Paris & Bordeaux. Procès-Verbaux Séances Soc. Sci. Phys. Bordeaux. See B–P–H 739/1. HI 58320

Procès-verbaux de la société belge de géologie, de paléontologie et d'hydrologie. Brussels. Procès-Verbaux Soc. Belge Géol. See B–P–H 739/2. HI 58321

Procès-verbaux de la société dauphinoise d'études biologiques, Bio-Club. Grenoble. Vols. 1-328; 1922-39; n.s. vols. 1-12, 19??-?; n.s. vols. 1-35, 1954-61. Procès-Verbaux Soc. Dauphin. Études Biol. Bio-Club. HI 69991

Procès-verbaux de la société d'histoire naturelle de l'Ile Maurice. Port Louis. 1842-46. Procès-Verbaux Soc. Hist. Nat. l'Ile Maurice. Preceded by: Rapport annuel sur les travaux de la société d'histoire naturelle de l'Ile Maurice. HI 69992

Procès-verbaux de la société linnéenne de Bordeaux. Bordeaux. Vols. 1-101, 1876-1964. Procès-Verbaux Soc. Linn. Bordeaux. HI 69993

Process biochemistry. London, Rickmansworth. Vol. 1+, 1966+. Process Biochem. HI 75185

Processed series, Oklahoma agricultural experiment station. Stillwater, OK. Nos. 297-606, 1958-69. Processed Ser. Oklahoma Agric. Exp. Sta. Preceded by: Oklahoma agricultural experiment station mimeographed circular. Superseded by: Progress report, Oklahoma agricultural experiment station. HI 69994

Processi verbali della società toscana di scienze naturali in Pisa. Pisa. 1878-1948. Processi Verbali Soc. Tosc. Sci. Nat. Pisa. Continued in: Atti della società toscana di scienze naturali residente in Pisa. HI 69995

Processi verbali della società toscana di scienze naturali residente in Pisa = Atti della società toscana di scienze naturali residente in Pisa, processi verbali. Pisa.

Producers' review. [Royal agricultural society of Western Australia.] Perth, W.A. Producers' Rev. (Perth). See B–P–H 739/4. HI 58322

Producers' review. [Queensland cane growers' association.] Toowoomba, Qld. Producers' Rev. (Toowoomba). See B–P–H 739/5. HI 58323

Producţia vegetală. Cereale şei plante tehnice. Bucharest. Vol. 26(7)+, 1974+. Prod. Veg. Pl. Tehn. Preceded by: Probleme agricole. HI 69996

Producţia vegetală. Horticultura. Bucharest. Vol. 23(7)+, 1974+. Prod. Veg., Hort. Preceded by: Revista de horticultură şei viticultură. HI 69997

Professional gardener; official organ of the national association of gardeners. New York. Vol. 1+, 1949+. Profess. Gard. 4-3449-2. HI 69998

Professional geographer. Hamilton, NY. Profess. Geogr. See B–P–H 739/7. HI 58325

Professional horticulture. Oxford. Vol. 1+, 1987+. Profess. Hort. HI 69999

Professional papers, Boyce Thompson institute for plant research. Yonkers, NY. Vols. 1-2, nos. 1-14, 1925-55. Profess. Pap. Boyce Thompson Inst. Pl Res. HI 70000

Professional papers series, Florida department of natural resources marine laboratory. St. Petersburg, FL. Nos. 10-21, 1969-73. Profess. Pap. Ser. Florida Dept. Nat. Resources Mar. Lab. Preceded by: Professional papers series, Florida state board of conservation marine laboratory. Superseded by: Florida marine research publications. HI 70001

Professional papers series, Florida state board of conservation marine laboratory. St. Petersburg, FL. Nos. 1-9, 1960-67. Profess. Pap. Ser. Florida State Board Conservation Mar. Lab. Superseded by: Professional papers series, Florida department of natural resources marine laboratory. HI 70002

Professional papers, United States department of agriculture = United States department of agriculture bulletin. Washington, DC. U.S.D.A. Bull. (1915-23). See B–P–H 940/16.

Professional papers. United States geological survey. Washington, DC. Profess. Pap. U.S. Geol. Surv. See B–P–H 739/9. HI 58327

Program review, forest products laboratory. Vancouver. ?-1969/70, ?-1971. Program Rev. Forest Prod. Lab. Superseded by: Program review, western forest products laboratory. HI 70003

Program review, western forest products laboratory. Vancouver. 1971/72+, 1973?+. Program Rev. W. Forest Prod. Lab. Preceded by: Program review, forest products laboratory. Vancouver. HI 70004

Programme, Croydon natural history and scientific society. Croydon. No. 367+, 1980+. Programme Croydon Nat. Hist. Sci. Soc. Preceded by: Circular, Croydon natural history and scientific society. HI 70005

Programme and proceedings of the annual meeting, canadian federation of biological societies. Saskatoon. ?-1976+. Programme Proc. Annual Meeting Canad Fed. Biol. Soc. Preceded by: Proceedings, canadian federation of biological societies. HI 70006

Programme report, international genetic resources = International genetic resources programme report.

Programme of work, federal department of forest research, Nigeria. Ibadan. 1970/71+, 1970+. Programme Work Fed. Dept. Forest Res. Nigeria. HI 70007

Progrès agricole et viticole. Montpellier. Progr. Agric.

Vitic. See B–P–H 739/16. HI 58328

Progrès de la botanique. Jena = Progressus rei botanicae. Jena. Progr. Rei Bot. See B–P–H 740/1.

Progrès dans la chimie des substances organiques naturelles = Fortschritte der Chemie organischer Naturstoffe. Vienna, New York.

Progrès horticole. Bourg-en Bresse, France. Progr. Hort. See B–P–H 739/21. HI 58331

Progrès médical. Paris. Progr. Méd.. See B–P–H 739/22. HI 58332

Progress, A D A S. Winchester. ?-1981+. Progr. A. D. A. S. HI 70008

Progress, agricultural development and advisory service = Progress, A D A S. Winchester.

Progress in biochemical pharmacology. Basel. Vol. 1+, 1965+. Progr. Biochem. Pharmacol. HI 70009

Progress in the biological sciences in relation to dermatology. Cambridge. Vol. 1+, 1960+. Progr. Biol. Sci. Relat. Dermatol. HI 70010

Progress in biometeorology. Division C, progress in plant biometeorology. Amsterdam. Vol. 1+, 1963/74+, 197?+. Progr. Biometerol., Div. C, Progr. Pl. Biometerol. HI 70011

Progress in biophysics and biophysical chemistry. New York, NY. Progr. Biophys. Biophys. Chem. See B–P–H 739/17. HI 58329

Progress in biophysics and molecular biology = Progress in biophysics and biophysical chemistry. New York, NY. Progr. Biophys. Biophys. Chem. See B–P–H 739/17.

Progress in bioorganic chemistry. New York. Vol. 1+, 1971+. Progr. Bioorg. Chem. HI 70012

Progress in biotechnology. Amsterdam, New York. No. 1+, 1985+. Progr. Biotechnol. HI 75229

Progress in botany; morphology, physiology, genetics, taxonomy, geobotany. Berlin, etc. Vol. 36+, 1974+. Progr. Bot. Preceded by: Fortschritte der Botanik. HI 70013

Progress of botany. Jena = Progressus rei botanicae. Jena. Progr. Rei Bot. See B–P–H 740/1.

Progress in the chemistry of fats and other lipids. Oxford, Elmsford, NY. Vols. 1-16, 1952-78. Progr. Chem. Fats Other Lipids. Superseded by: Progress in lipid research. HI 70014

Progress in the chemistry of organic natural products = Fortschritte der Chemie organischer Naturstoffe. Vienna, New York.

Progress in ecology. New Delhi. Vol. 2+, 1977+. Progr. Ecol. Preceded by: Progress of plant ecology in India. HI 70015

Progress in industrial microbiology. Amsterdam. Vol. 1+, 1959+. Progr. Industr. Microbiol. HI 70016

Progress in lipid research. Oxford, Elmsford, NY. Vol. 17+, 1978+. Progr. Lipid Res. Preceded by: Progress in the chemistry of fats and other lipids. HI 70017

Progress in molecular and subcellular biology. Berlin, New York, etc. Vol. 1+, 1969+. Progr. Molec. Subcell. Biol. HI 70018

Progress, nature conservancy. London. 1964/68+, 1968+. Progr. Nat. Conservancy. HI 70019

Progress in nucleic acid research. New York, London, etc. Vols. 1-2, 1963. Progr. Nucl. Acid Res. Superseded by: Progress in nucleic acid research and molecular biology. HI 70020

Progress in nucleic acid research and molecular biology. New York, London, etc. Vol. 3+, 1964+. Progr. Nucl. Acid Res. Molec. Biol. Preceded by: Progress in nucleic acid research. HI 70021

Progress in oceanography. New York, NY. Progr. Oceanogr. See B–P–H 739/26. HI 58333

Progress in pesticide biochemistry. Chichester. Vols. 1-2, 19??-82?. Progr. Pestic. Biochem. Preceded by: Progress in pesticide biochemistry and toxicology. HI 75201

Progress in pesticide biochemistry and toxicology. Chichester. Vol. 3+, 1983+. Progr. Pestic. Biochem. Toxicol. Preceded by: Progress in pesticide biochemistry. HI 75200

Progress in pharmacology. Stuttgart, New York. Vol. 1+, 1975+. Progr. Pharmacol. HI 70022

Progress in phycological research. Amsterdam, New York. Vol. 1+, 1982+. Progr. Phycol. Res. HI 70023

Progress in phytochemistry. New York, London. Vol. 1+, 1968+. Progr. Phytochem. HI 70024

Progress in plant biometeorology = Progress in biometeorology. Division C, progress in plant biometeorology. Amsterdam.

Progress of plant ecology in India. New Delhi. Vol. 1, 1973. Progr. Pl. Ecol. India. Superseded by: Progress in ecology. HI 70025

Progress in plant research. New Delhi. Vols. 1-2, 1979. Progr. Pl. Res. HI 70026

Progress report, agricultural experiment station, College, AK = Alaska agricultural college and school of mines. Agricultural experiment station progress report. College, AK. Alaska Agric. Coll. School MInes Agric. Exp. Sta. Progr. Rep. See B–P–H 68/3.

Progress report on agricultural research and experimentation, faculty of agriculture, university of Manitoba. Winnipeg. 1954+. Progr. Rep. Agric. Res. Exp. Fac. Agric. Univ. Manitoba. HI 70027

Progress report on arboriculture in the North-West Frontier Province. Peshawar. 1904-19. Progr. Rep. Arboric. N.-W. Frontier Prov. HI 70028

Progress report on arboriculture in the Punjab. Lahore. 1876-1917? Progr. Rep. Arboric. Punjab. HI 70029

Progress report, asian vegetable research and development center = A V R D C progress report. Tainan.

Progress report, central potato research institute. Simla. 1949/56-1956-57, 1956-57. Progr. Rep. Centr. Potato Res. Inst. Superseded by: Scientific report of the central potato research institute. HI 70030

Progress report, central states forest experiment station. Columbus, OH. 1927-32. Progr. Rep. Centr. States Forest Exp. Sta. HI 70031

Progress report, cereal breeding laboratory. [From 1949-56 formed part of: Progress report, cereal crops division, department of agriculture, Canada.] Winnipeg, Ottawa. 1949-55/58, 194?-? Progr. Rep. Cereal Breed. Lab. Previously contained in: Progress report, cereal crops division, Canada. HI 70032

Progress report, cereal crops division, department of agriculture, Canada. Winnipeg. 1938/48-54/58, 194?-5? Progr. Rep. Cereal Crops Div. Dept. Agric. Canada. Preceded by: Progress report of the dominion cerealist, Canada. HI 70033

Progress report, clovers and special purpose legumes research. Madison, WI. Vol. 7?+, 1974+. Progr. Rep. Clovers Special Purp. Legumes Res. Preceded by: Newsletter, clover and special purpose legumes research. HI 75145

Progress report, cooperative forest genetics progress. Gainesville, FL. No. 1+, 1957+. Progr. Rep. Coop. Forest Genet. Progr. HI 70034

Progress report, cooperative forestry research unit, university of Maine at Orono. Orono, ME. ?-20+, ?-1982+. Progr. Rep. Coop. Forest. Res. Unit Univ Maine Orono. HI 70035

Progress report, Dehra Dun forest school. Dehra Dun. 1884-190? Progr. Rep. Dehra Dun Forest School. Superseded by: Progress report of the forest research institute, Dehra Dun. HI 70036

Progress report, division of botany, Canada. Ottawa. 1910-36/37, 1911-37? Progr. Rep. Div. Bot. Canada. HI 70037

Progress report, division of forage plants, experimental farms service, Canada. Ottawa. 1920/21-37/48, 1921-48? Progr. Rep. Div. Forage Pl. Exp. Farms Serv. Canada. Superseded by: Progress report, forage crops division, experimental farms service, Canada. HI 51741

Progress report, division of horticulture, experimental farms service, Canada. Ottawa. 1920/21-34/48, 1921?-50? Progr. Rep. Div. Hort. Exp. Farms Serv. Canada. Superseded by: Progress report, horticultural division, experimental farms service, Canada. HI 70038

Progress report of the dominion agricultural bacteriologist. Ottawa. 1923/24-37. 1925-39. Progr. Rep. Domin. Agric. Bacteriologist. Incorporated in: report of the science service, department of agriculture, Canada. HI 70039

Progress report of the dominion cerealist, Canada. Ottawa. 1930/33-34/37, 1933?-37? Progr. Rep. Domin. Cerealist Canada. Preceded by: Report of the dominion cerealist. Superseded by: Progress report, cereal crops division, canada. HI 70040

Progress report, dominion horticultural substation, McDonald's Corner. Ottawa. 1947/52-53/57, 1952?-57? Prog. Rep. Domin. Hort. Substa. McDonald's Corner. HI 70041

Progress report, elm research institute. [Supplement to: Elm bulletin.] Waldwick, NY. ?-1967+. Progr. Rep. Elm Res. Inst. HI 70042

Progress report, european plant protection organisation. Paris. 1951+. Progr. Rep. Eur. Pl. Protect. Organ. HI 70043

Progress report from experiment stations, cotton research corporation. London. 1965/66-69/70, 1967-70. Progr. Rep. Exp. Sta. Cotton Res. Corp. Preceded by: Progress report from experiment stations, empire cotton growing corporation. Superseded by: Cotton research report, Kenya; Cotton research report, Malawi; Cotton research report, Northern States, Nigeria; Cotton research report, Republic of the Sudan; Cotton research report, Swaziland; Cotton research report, Tanzania; Cotton research report, Uganda; and Cotton research report, Zambia. HI 70044

Progress report from experiment stations, empire cotton growing corporation. London. 1933-66. Progr. Rep. Exp. Sta. Empire Cotton Growing Corp. Preceded by: Report from experiment stations, empire cotton growing corporation. Superseded by: Progress report from experiment stations, cotton research corporation. HI 70045

Progress report, experimental horticulture stations, national agricultural advisory service. London. 1952-60. Progr. Rep. Exp. Hort. Sta. Natl. Agric. Advis. Serv. Incorporated in: Progress report, experimental husbandry farms and experimental horticulture stations, national agricultural advisory service. HI 70046

Progress report, experimental husbandry farms and experimental horticulture stations. London. 1960+. Progr. Rep. Exp. Husb. Farms Exp. Hort. Sta. Preceded by: Progress report, experimental horticulture stations, national agricultural advisory service.

HI 70047

Progress report, fibre division, experimental farms service, Canada. Ottawa. 1937/47, 1950. Progr. Rep. Fibre Div. Exp. Farms Serv. Canada. Preceded by: Report, division of economic fibre production, experimental farms service, Canada. HI 70048

Progress report, forage crops division, experimental farms service, Canada. Ottawa. 1949/53-1954/58, 1953?-58? Progr. Rep. Forage Crops Div. Exp. Farms Serv. Canada. Preceded by: Progress report, division of forage plants, experimental farms service, Canada. HI 70049

Progress report, forage crops laboratory. Saskatoon. 1955+. Progr. Rep. Forage Crops Lab. HI 70050

Progress report of forest administration in the Andamans. Calcutta. 1884/85+, 1885+. Progr. Rep. Forest Admin. Andamans. HI 70051

Progress report on forest administration in Assam. Shillong. 1874-1946. Progr. Rep. Forest Admin. Assam. HI 70052

Progress report on forest administration in Baluchistan. Calcutta. 1890-1934? Progr. Rep. Forest Admin. Baluchistan. HI 70053

Progress report, forest administration in Bengal. Calcutta. 1867+. Progr. Rep. Forest Admin. Bengal. HI 70054

Progress report on forest administration in Berar. Hyderabad. 1892-1902. Progr. Rep. Forest Admin. Berar. HI 70055

Progress report on forest administration in Burma = Report on forest administration in Burma. Rangoon.

Progress report on forest administration in Central Provinces. Nagpur? 1931-39. Progr. Rep. Forest Admin. Central Provinces. HI 70056

Progress report on forest administration in Coorg. Bangalore. 1893+. Prog. Rep. Forest Admin. Coorg. HI 70057

Progress report on forest administration in Madras. Madras? 1932-39. Prog. Rep. Forest Admin. Madras. Preceded by?: Administration report, forest department, Madras. HI 70058

Progress report on forest administration in North-West Frontier Province. Peshawar? 1902-24. Prog. Rep. Forest Admin. N.-W. Front. Prov. HI 70059

Progress report, forest administration in the Punjab. Lahore. 1882-1954. Progr. Rep. Forest Admin. Punjab. Superseded by: Progress report, forest administration, West Pakistan. HI 70060

Progress report, forest administration, West Pakistan. Peshawar. 1957+. Progr. Rep. Forest Admin. W. Pakistan. Preceded by: Progress report, forest administration in the Punjab. HI 70061

Progress report, forest conservator, Bihar. Patna. 1911-42. Progr. Rep. Forest Conserv. Bihar. HI 70062

Progress report, forest conservator, Orissa. Cuttack. 1939-41. Progr. Rep. Forest Conserv. Orissa. HI 70063

Progress report, forest department, Bombay. Bombay. 1849+. Progr. Rep. Forest Dep. Bombay. HI 70064

Progress report, forest department, United Provinces of Agra and Oudh. Allahabad. 1876-1927. Progr. Rep. Forest Dep. Unit. Prov. Agra Oudh. HI 70065

Progress report, forest products research, department of scientific and industrial research. London. 1928-29. Progr. Rep. Forest Prod. Res. Dep. Sci. Industr. Res. HI 70066

Progress report, forest products research laboratory. Princes Risborough. 1953+. Progr. Rep. Forest Prod. Res. Lab. HI 70067

Progress report of the forest research institute, Dehra Dun. Dehra Dun. 1906-30. Progr. Rep. Forest Res. Inst. Dehra Dun. Preceded by: Progress report, Dehra Dun forest school. Superseded by: Progress report of the Indian forest ranger college. HI 70068

Progress report, forest research laboratory, Oregon = Progress report, Oregon forest products laboratory. Corvallis, OR.

Progress report of forest research work in India and Burma = Forest research in India. Delhi.

Progress report of the forest surveys in India. Calcutta. 1882/83-1905/06, 1883-1906. Progr. Rep. Forest Surv. India. HI 70069

Progress report of the forest survey branch, India = Progress report of the forest surveys in India. Calcutta.

Progress report, forestry division, Trinidad. Port-of-Spain. 1966/72+, 1972?+. Progr. Rep. Forest. Div. Trinidad. HI 70070

Progress report, horticultural division, experimental farms service, Canada. Ottawa. 1949/53-54/58, 1955-60. Progr. Rep. Hort. Div. Exp. Farms Serv. Canada. Preceded by: Progress report, division of horticulture, experimental farms service, Canada. HI 70071

Progress report of the indian forest college, Dehra Dun. Delhi. 1940-55. Progr. Rep. Indian Forest Coll. Dehra Dun. Preceded by: Report of the indian forest college, Dehra Dun. Superseded by: Progress report of the northern forest rangers college, Dehra Dun. HI 70072

Progress report of the indian forest ranger college, Dehra Dun. Delhi. 1925-38. Progr. Rep. Indian Forest Ranger Coll. Dehra Dun. Preceded by?: Progress report of the forest research institute, Dehra Dun. Superseded by: Report of the indian forest college, Dehra Dun. HI 70073

Progress report, institute of plant industry, Indore. Indore. 1931+, 1932+. Progr. Rep. Inst. Pl. Industr. Indore. HI 70074

Progress report, institutes of the royal Netherlands academy of arts and sciences = Verhandelingen der koninklijke nederlandsche akademie van wetenschappen. Afdeeling natuurkunde; tweede sectie. Amsterdam.

Progress report, Kentucky agricultural experiment station. Lexington, KY. No. 1+, 195?+. Progr. Rep. Kentucky Agric. Exp. Sta. HI 70076

Progress report, land resources division, ministry of overseas development. Surbiton. 1971/74+. Progr. Rep. Land Resources Div. Minist. Overseas Developm. HI 70077

Progress report, Louisiana university and agricultural and mechanical college. Forestry department = Louisiana university and agricultural and mechanical college. Forestry department. Progress report. Baton Rouge, LA. Louisiana Univ. Agric. Coll. Forest. Dept. Progr. Rep. See B–P–H 536/10.

Progress report of the Madras forest college. Coimbatore. 1948-53. Progr. Rep. Madras Forest Coll. Superseded by: Progress report of the southern forest rangers college, Coimbatore. HI 70078

Progress report, marine life research program; Scripps institution of oceanography. La Jolla, CA. Progr. Rep. Mar. Life Res. Program Scripps Inst. Oceanogr. See B–P–H 740/7. HI 58335

Progress report, national botanic gardens, Lucknow. Lucknow. 1973/75+, 1975?+. Progr. Rep. Natl. Bot. Gard. Lucknow. HI 70079

Progress report of the northern forest rangers college, Dehra Dun. Dehra Dun. 1957-61. Progr. Rep. N. Forest Rangers Coll. Dehra Dun. Preceded by: Progress report of the indian forest college, Dehra Dun. HI 70080

Progress report, northern Rocky Mountain forest experiment station. Missoula, MT. 1945+. Progr. Rep. N. Rocky Mountain Forest Exp. Sta. Preceded by?: Report, northern Rocky Mountain forest experiment station. HI 70081

Progress report, Oklahoma agricultural experiment station. Stillwater, OK. Nos. 607-656, 1969-72. Progr. Rep. Oklahoma Agric. Exp. Sta. Preceded by: Processed series, Oklahoma agricultural experiment station. Superseded by: Research report, Oklahoma agricultural experiment station. HI 70082

Progress report, Oregon forest products laboratory. Corvallis, OR. Nos. 1-14, 1948-71. Progr. Rep. Oregon Forest Prod. Lab. HI 70083

Progress report, Pakistan forest college and research institute. Peshawar. 1960+. Progr. Rep. Pakistan Forest Coll. Res. Inst. HI 70084

Progress report, Pennsylvania state college agricultural experiment station = Pennsylvania state college agricultural experiment station. Progress report. State College, PA. Pennsylvania State Coll. Agric. Exp. Sta. Progr. Rep. See B–P–H 700/26.

Progress report, plant protection, United States department of agriculture. Hyattsville, MD. 1970. Progr. Rep. Pl. Protect. U.S.D.A. Preceded by: Cooperative plant pest control programs, United States department of agriculture. Superseded by: Progress report, plant protection and quarantine programs, United States department of agriculture. HI 70085

Progress report, plant protection and quarantine programs, U S department of agriculture. Hyattsville, MD. 1971+. Progr. Rep. Pl. Protect. Quarant. Programs U.S.D.A. Preceded by: Progress report, plant protection division, U S department of agriculture. HI 70086

Progress reports on research and development, home-grown cereals authority. London. 1967/68+, 1968+. Progr. Rep. Res. Developm. Home-Grown Cereals Author. HI 70087

Progress report, Saitama agricultural experiment station. Saitama. 1948+. Progr. Rep. Saitama Agric. Exp. Sta. HI 70088

Progress report of the scientific work, central potato research institute = Progress report, central potato research institute. Simla.

Progress report series, Alabama agricultural experiment station = Alabama agricultural experiment station. Progress report series. Auburn, AL.

Progress report, soil biotics division, soil bureau, D S I R, N Z. Wellington, N.Z. 1948+. Progr. Rep. Soil Biotics Div. Soil Bur. D.S.I.R., N.Z. HI 70089

Progress report of the southern forest rangers college, Coimbatore. Coimbatore. 1957+. Progr. Rep. S. Forest Rangers Coll. Coimbatore. Preceded by: Progress report of the Madras forest college. HI 70090

Progress report, Storrs agricultural experiment station. Storrs, CT. Nos. 1-57, 1954-64. Progr. Rep. Storrs Agric. Exp. Sta. HI 70091

Progress report of the Texas agricultural experiment station. College Station, TX. No. 1+, 1912+. Progr. Rep. Texas Agric. Exp. Sta. HI 70092

Progress report, tobacco division, experimental farms service, Canada. 1950-61. Progr. Rep. Tobacco Div. Exp. Farms Serv. Canada. Preceded by: Report, tobacco division, experimental farms, Canada. HI 70093

Progress report, tobacco research station, Mauritius.

1951+. Progr. Rep. Tobacco Res. Sta. Mauritius. HI 70094

Progress report, tropical forestry project, Florida State. Fort Myers, FL. [Dates of publication not ascertained.] Progr. Rep. Trop. Forest. Proj. Florida State. HI 70095

Progress report, university of Alaska agricultural experiment station = University of Alaska agricultural experiment station. Progress report. College, AK. Univ. Alaska Agric. Exp. Sta. Progr. Rep. See B–P–H 933/9.

Progress report, university of Guelph arboretum. Guelph. ?-1977+. Progr. Rep. Univ. Guelph Arbor. HI 70096

Progress in research, research branch, Canada department of agriculture. Ottawa. 1976+. Progr. Res. Res. Branch Canada Dep. Agric. HI 70097

Progress in theoretical biology. New York, London, etc. Vol. 1+, 1967+. Progr. Theor. Biol. HI 70098

Progress in veterinary microbiology and immunology. New York. Vol. 1+, 1985+. Progr. Veterin. Microbiol. Immunol. HI 70099

Progressi in biochemica. Vols. 1-10, 1964-75. Progr. Biochem. HI 70100

Progressive fish culturist. Washington, DC. Progr. Fish Cult. See B–P–H 739/20. HI 58330

Progressive horticulture. Lucknow. Vol. 1+, 1969/70+. Progr. Hort. HI 70101

Progresso; publicação scientifia e industrial offerecida as classes estudiosas e industriosas do Brasil. Paris. Progresso. See B–P–H 740/16. HI 58337

Progresso delle scienze, delle lettere e delle arti. Naples. Progr. Sci. See B–P–H 740/12. HI 58336

Progressus rei botanicae. Jena. Progr. Rei Bot. See B–P–H 740/1. HI 58334

Prometheus. Berlin. Prometheus. See B–P–H 740/19. HI 58338

Propagador das sciencias medicas. Rio de Janeiro. Propag. Sci. Med. See B–P–H 740/20. HI 58339

Propagatore ossia raccolta periodica delle cose appartenenti ai progressi dell'industria e specialmente di quelle riguardanti l'agricgltura, le arti e la medicina. Turin. Propagatore. See B–P–H 740/21. HI 58340

Prosea newsletter = P R O S E A newsletter.

Prospekt. Breslau. 1900-08. Prospekt. HI 70103

Prospects of Iraq biology. [Baghdad.] Vol. 1+, 1958+. Prosp. Iraq. Biol. HI 70104

Prosvetni glasnik. Belgrade. 1880-1928. Prosv. Glasn. HI 70105

Protection ecology; international journal devoted to the study and management of noxious organisms in plant and animal industries. Amsterdam. Vol. 1+, 1978+. Protect. Ecol. HI 70106

Protection de la nature. Basle = Schweizer Naturschutz. Basel.

Protection de la nature. Belgrade = Zaštita prirode. Belgrade. Zašt. Prir. See B–P–H 955/22.

Protection of nature. Cracow = Ochrona przyrody. Cracow, Warsaw.

Protection of nature. Ljubljana = Varstvo narove. Ljubljana.

Protection of nature. Warsaw = Ochrona przyrody. [Series A]. Warsaw.

Proteus. Erlangen. Proteus (Erlangen). See B–P–H 740/25. HI 58341

Proteus. Ljubljana. Vol. 1+, 1934+. Proteus (Ljubljana). HI 70107

Proteus. Stuttgart. Proteus (Stuttgart). See B–P–H 740/26. HI 58342

Proteus. Vienna. Proteus (Vienna). See B–P–H 740/27. HI 58343

Protistologia. Paris. Protistologia. See B–P–H 740/28. HI 58344

Protokoly Obščestva Ispytatelej Prirody pri Imperatorskom Har'kovskom Universitetě = Protokoly Obshchestva Ispytatelei Prirody pri Imperatorskom khar'kovskom Universitetě. Kharkov.

Protokoly Obščestva Estestvoispytatelej pri Imperatorskom Jur'evskom Universitetě = Protokoly Obshchestva Estestvoispytatelei pri Imperatorskom Yur'evskom Universitetě. Yurev.

Protokoly Obščestva Estestvoispytatelej i Vračej pri Imperatorskom Tomskom Universitetě = Protokoly Obshchestva Estestvoispytatelei i Vrachei pri Imperatorskom Tomskom Universitetě. Tomsk.

Protokoly Obshchestva Estestvoispytatelei pri Imperatorskom Yur'evskom Universitetě. Yurev [=Tartu], Estonian S S R. Vols. 12-23, 1901-16. Protok. Obshch. Estestvoisp. Imp. Yur'evsk. Univ. Preceded by: Sitzungsberichte der Naturforscher-Gesellschaft bei der Universität Jurjew. Superseded by: Tartu ülikooli juures oleva loodusuurijate seltsi aruanded. 3-2469-2. HI 70108

Protokoly Obshchestva Estestvoispytatelei i Vrachei pri Imperatorskom Tomskom Universitetě. Tomsk, Russian S F S R. 1898-1904; 1907-09; [for the year 1911], 1912 [1905, 1910 and 1911 no vols. published]. Protok. Obshch. Estestvoisp. Imp. Tomsk. Univ. Preceded by: Trudy Tomskogo Obshchestva Estestvoispytatelei. For 1906 see: Protokoly Zasědanii Tomskago Obshchestva Estestvoispytatelei i Vrachei. For the years 1908/10 see: Protokoly Zasědanii

Obshchestva Estestvoispytatelei i Vrachei pri Imperatorskom Tomskom Universitetě. Superseded by: Trudy Obshchestva Estestvoispytatelei i Vrachei pri Imperatorskom Tomskom Universitetě. HI 70109

Protokoly Obshchestva Ispytatelei Prirody pri Imperatorskom khar'kovskom Universitetě. Kharkov, Ukrainian S S R. 1912-15. Protok. Obshch. Isp. Prir. Imp. Khar'kovsk. Univ. HI 70110

Protokoly i Trudy Obščestva Estestvoispytatelej i Vračej Turkestanskago Kraja = Protokoly i Trudy Obshchestva Estestvoispytatelei i Vrachei Turkestanskago Kraya. Tashkent.

Protokoly i Trudy Obshchestva Estestvoispytatelei i Vrachei Turkestanskago Kraya. Tashkent, Uzbek S S R. Vols. 1-2, 1911-14. Protok. Trudy Obshch. Estestvoisp. Turkestansk. Kraya. HI 70111

Protokoly Zasědanii Imperatorskago Moskovskago Obshchestva Ispytatelei Prirody. Moscow. 1896-1916. Protok. Zasěd. Imp. Moskovsk. Obshch. Isp. Prir. HI 70112

Protokoly Zasědanii Kievskago Obshchestva Estestvoispytatelei. Kiev, Ukrainian S S R. 1901-18. Protok. Zasěd. Kievsk. Obshch. Estestvoisp. HI 70113

Protokoly Zasědanii Obshchestva Estestvoispytatelei pri Imperatorskom Kazanskom Universitetě. Kazan, Russian S F S R. Vols. 1-46, 1870-1915. Protok. Zasěd. Obshch. Estestvoisp. Imp. Kazansk. Univ. 3-2276-2. HI 70114

Protokoly Zasědanii Obshchestva Estestvoispytatelei i Vrachei pri Imperatorskom Tomskom Universitetě. Tomsk, Russian S F S R. 1908/10, 1912. Protok. Zasěd. Obshch. Estestvoisp. Imp. Tomsk. Univ. Preceded & superseded by: Protokoly Obshchestva Estestvoispytatelei i Vrachei pri Imperatorskom Tomskom Universitetě. HI 70115

Protokoly Zasědanii Rasporyaditel'nago Komiteta Krasnoyarskago Podotděla. Krasnoyarsk, Russian S F S R. Vol. [1(2)], 1901. Protok. Zasěd. Rasporyadit. Komiteta Krasnoyarsk. Podotd. Preceded by: Protokoly torzhestvennago publichnago Zasědanija Krasnojarskago Podotděla Vostochno-Sibirskago Otděla Imperatorskato Russkago Geograficheskago Obshchestva [not entered]. Superseded by: Izvestiya Krasnoyarskago Podotděla Vostochno-Sibirskago Otděla Imperatorskago Russkago Geograficheskago Obshchestva. 2-1693-1. HI 70116

Protokoly Zasědanii Sověta Imperatorskago Moskovskago Universiteta. Moscow. 1872-81. Protok. Zasěd. Sověta Imp. Moskovsk. Univ. Preceded by: Moskovskiya Universitetskiya Izvěstiya. Superseded by: Uchenya Zapiski Imperatorskago Moskovskago Universiteta. Otděl estestvenno-istoricheskii. HI 70117

Protokoly Zasědanii Tomskago Obshchestva Estestvoispytatelei i Vrachei. Tomsk, Russian S F S R. 1906. Protok. Zasěd. Tomsk. Obshch. Estestvoisp. Preceded & superseded by: Protokoly Obshchestva Estestvoispytatelei i Vrachei pri Imperatorskom Tomskom Universitetě. HI 70118

Protokoly Zasědanii i Trudy Obshchestva Estestvoispytatelei pri Imperatorskom Varshavskom Universitetě. Warsaw. Vols. 11-12, 1901-02. Protok. Zasěd. Trudy Obshch. Estestvoisp. Imp. Varshavsk. Univ. Preceded by: Trudy Obshchestva Estestvoispytatelei pri Imperatorskom Varshavskom Universitetě. Superseded by: Trudy i Protokoly Zasědanii Obshchestva Estestvoispytatelei pri Imperatorskom Varshovskom Universitetě. 5-4444-3. HI 70119

Protokoly Zasědanij Imperatorskago Moskovskago Obščestva Ispytatelej Prirody = Protokoly Zasědanii Imperatorskago Moskovskago Obshchestva Ispytatelei Prirody. Moscow.

Protokoly Zasědanij Kievskago Obščestva Estestvoispytatelej = Protokoly Zasědanii Kievskago Obshchestva Estestvoispytatelei. Kiev.

Protokoly Zasědanij Obščestva Estestvoispytatelej pri Imperatorskom Kazanskom Universitetě = Protokoly Zasědanii Obshchestva Estestvoispytatelei pri Imperatorskom Kazanskom Universitetě. Kazan.

Protokoly Zasědanij Obščestva Estestvoispytatelej i Vračej pri Imperatorskom Tomskom Universitetě = Protokoly Zasědanii Obshchestva Estestvoispytatelei i Vrachei pri Imperatorskom Tomskom Universitetě. Tomsk.

Protokoly Zasědanij Rasporjaditel'nago Komiteta Krasnojarskago Podotděla. Krasnoyarsk = Protokoly Zasědanii Rasporyaditel'nago Komiteta Krasnoyarskago Podotděla. Krasnoyarsk.

Protokoly Zasědanij Sověta Imperatorskago Moskovskago Universiteta = Protokoly Zasědanii Sověta Imperatorskago Moskovskago Universiteta. Moscow.

Protokoly Zasědanij Tomskago Obščestva Estestvoispytatelej i Vračej = Protokoly Zasědanii Tomskago Obshchestva Estestvoispytatelei i Vrachei. Tomsk.

Protokoly Zasědanij i Trudy Obščestva Estestvoispytatelej pri Imperatorskom Varšavskom Universitetě = Protokoly Zasědanii i Trudy Obshchestva Estestvoispytatelei pri Imperatorskom Varshavskom Universitetě. Warsaw.

Protoplasma. Internationale Zeitschrift für physikalische Chemie des Protoplasten. Leipzig. Protoplasma. See B–P–H 742/3. HI 58345

Protoplasma-Monographien. Berlin. Protoplasma-Monogr. See B–P–H 742/5. HI 58346

Protoplasmatologia. Vienna. Protoplasmatologia. See B–P–H 742/6. HI 58347

Provancheria; mémoires de l'herbier Louis-Marie, faculté d'agriculture de l'université Laval. Quebec. No. 1+, 1967+. Provancheria. HI 70120

Provence agricole et horticole. Toulon. Vol. 5/6, 1884/85. Prov. Agric. Hort. Preceded by: Provence agricole et horticole illustrée and Bulletin mensuel de la société d'agriculture, d'horticulture et d'acclimatation du Var à Toulon. HI 70121

Provence agricole et horticole illustrée; organe de l'agriculture et de l'horticulture méridionales. Toulon. Vols. 1-3, 1881-83. Prov. Agric. Hort. Ill. Preceded by: Bulletin mensuel de la société d'agriculture, d'horticulture et d'acclimatation du Var à Toulon. Superseded by: Provence agricole et horticole. HI 70122

Provinzialblätter oder Sammlungen zur Geschichte, Naturkunde, Moral und andern Wissenschaften, herausgegeben von der Oberlausizischen Gesellschaft der Wissenschaften. Leipzig & Dessau. Provinzialbl. Oberlausiz. Ges. Wiss. See B–P–H 742/8. HI 58348

Prüfenden Gesellschaft zu Halle fortgesetzte, zur Gelehrsamkeit gehörige, Bemühungen. Halle & Leipzig. Prüfenden Ges. Halle Fortgesetzte Bemüh. See B–P–H 742/9. HI 58349

Prüfenden Gesellschaft zu Halle herausgegebene Schriften. Halle. Prüfenden Ges. Halle Schriften. See B–P–H 742/10. HI 58350

Pryrodnycho-technichnyi viddil, Vseukrayins'ka akademiya nauk. Kiev, Ukrainian S S R. Vol. 8, 1932. Pryr.-Techn. Vidd. Vseukrayins'ka Akad. Nauk. Preceded & superseded by: Trudy Pryrodnycho-technichnogo viddilu. HI 70123

Pryrodnyčo-techničnyj viddil, Vseukrajins'ka akademija nauk = Pryrodnycho-technichnyi viddil, Vseukrayins'ka akademiya nauk. Kiev.

Przeglad geograficzny. Warsaw. Przeglad Geogr. See B–P–H 742/14. HI 58351

Przyroda polska. Warsaw. 1957+. Przyr. Polska. HI 70124

Przyroda polski zachodniej. Poznan, Poland. Przyr. Polski Zachodn. See B–P–H 742/15. HI 58352

Psyche; a journal of entomology. Boston & Cambridge, MA. Psyche. See B–P–H 742/16. HI 58353

Pteridologia. Washington, DC. No. 1+, 1979+. Pteridologia. HI 70125

Pteridologist. London. Vol. 1+, 1984+. Pteridologist. HI 70126

Pterocarpus; a philippine journal of forestry. Laguna. Vol. 1+, 1975+. Pterocarpus. HI 70127

Pubblicazioni del centro di sperimentazione agricola e forestale Rome. Vol. 1+, 1956+. Pubbl. Centro Sperim. Agric. HI 70128

Pubblicazioni, centro per lo studio della flora e della vegetazione italiana del consiglio nazionale delle ricerche. Florence. Nos. 1-125, 1948-57. Pubbl. Centro Stud. Fl. Veg. Ital. Cons. Naz. Ric. Superseded by: Pubblicazioni, fondazione Filippo Parlatore per lo studio della flora et della vegetazione italiana. HI 70129

Pubblicazioni del centro talassografico tirreno. Genoa. Pubbl. Centro Talassogr. Tirreno. See B–P–H 743/2. HI 58354

Pubblicazioni del civico museo di storia naturale di Ferrara. Ferrara, Italy. Pubbl. Civico Mus. Storia Nat. Ferrara. See B–P–H 743/4. HI 58355

Pubblicazioni, erbario tropicale di Firenze. Florence. No. 1+, 1967+. Pubbl. Erb. Trop. Firenze. Preceded by: Pubblicazioni dell'istituto botanico della università di Firenze e dell erbario coloniale. HI 70130

Pubblicazioni, fondazione Filippo Parlatore per lo studio della flora e della vegetazione italiano. Florence. Nos. 1-143?, 1958-73? Pubbl. Fondaz. Filippo Parlatore. Preceded by: Pubblicazioni, centro per lo studio della flora e delle vegetazione italiana del consiglio nazionale delle richerche. HI 70131

Pubblicazioni dell'istituto di biologia marina del Tirreno in San Bartolomeo. San Bartolomeo, Italy. Publ. Ist. Biol. Mar. Tirreno San Bartolomeo. See B–P–H 743/6. HI 58356

Pubblicazioni dell'istituto di botanica dell'università di Camerino. Camerino. Vol. ?-26+, ?-1965+. Pubbl. Ist. Bot. Univ. Camerino. HI 70132

Pubblicazioni dell'istituto di botanica dell'università di Catania. Catania. 1967+. Pubbl. Ist. Bot. Univ. Catania. HI 70133

Pubblicazioni, istituto di botanica, università degli studi di Trieste. Trieste. Nos. 1-?, 1962-? Pubbl. Ist. Bot. Univ. Stud. Trieste. Superseded by: Studia geobotanica. HI 70134

Pubblicazioni dell'istituto botanico "Hanbury" dell'università di Genova. Genoa. No. 1+, 1930+. Pubbl. Ist. Bot. "Hanbury" Univ. Genova. HI 70135

Pubblicazioni dell'istituto botanico della r[eale] università di Padova. Padua. Vols. 1-12, 1922-45. Pubbl. Ist. Bot. Reale Univ. Padova. Superseded by: Lavori di botanica. Istituto di botanica e di fisiologia vegetale dell'università di Padova. 4-3239-1. HI 70136

Pubblicazioni dell'istituto botanico della universita di Firenze. Florence. 1969+. Pubbl. Ist. Bot. Univ.

Firenze. Preceded by: Pubblicazioni dell'istituto botanico della università di Firenze e dell erbario coloniale. HI 70137

Pubblicazioni dell'istituto botanico della università di Firenze e dell erbario coloniale. Florence. Vols. 1(1)-4(115), 1931-52; n.s. 2, no. 1-n.s. 9, no. ?, 1951-69. Pubbl. Ist. Bot. Univ. Firenze Erb. Colon. Superseded by: Pubblicazioni dell'istituto botanico della universita di Firenze and Pubblicazioni, erbario tropicale di Firenze. HI 70138

Pubblicazioni dell'istituto ed orto botanico dell'università de Ferrara. Ferrara. Vol. ?-5+, ?-1968+. Pubbl. Ist. Orto Bot. Univ. Ferrara. HI 70139

Pubblicazioni dell'istituto sperimentale per la frutticoltura. Rome. Vol. 1, 1968. Pubbl. Ist. Sperim. Fruttic. Superseded by: Annali dell'istituto sperimentale per la frutticoltura. HI 70140

Pubblicazioni dell'istituto sperimentale per la selvicoltura. Arezzo. Nos. 14-18, 1965?-69. Pubbl. Ist. Sperim. Selvic. Preceded by: Pubblicazioni, stazione sperimentale di selvicoltura. Superseded by: Annali dell'istituto sperimentale per la selvicoltura. HI 70141

Pubblicazioni dell'istituto di zoologia, anatomia comparata e genetica e della stazione idrobiologica dell'università di Padova. Padua. Pubbl. Ist. Zool. Staz. Idrobiol. Univ. Padova. See B–P–H 743/8. HI 58357

Pubblicazioni mensili del reale osservatorio di fitopatologia. Turin. 19??-23? Pubbl. Mens. Reale Osserv. Fitopatol. Superseded by: Difesa delle piante contra le malattie ed i parassiti. HI 70142

Pubblicazioni, museo friulano di storia naturale. Udine. Vol. 1+, 1963+. Pubbl. Mus. Friul. Storia Nat. HI 70143

Pubblicazioni, stazione sperimentale di floricoltura "Orazio Raimondo". San Remo. Vols. ?-4-[90], ?-1935-64? Pubbl. Staz. Sperim. Floric. "Orazio Raimondo". HI 70144

Pubblicazioni, stazione sperimentale di selvicoltura. Arezzo. Nos. 1-13, 1932-65. Pubbl. Staz. Sperim. Selvic. Superseded by: Pubblicazioni dell'istituto sperimentale per la selvicoltura. HI 70145

Pubblicazioni della stazione zoologica di Napoli. Naples. Pubbl. Staz. Zool. Napoli. See B–P–H 743/9. HI 58358

Pubblicazioni della stazione zoologica di Napoli. Section 2, history and philosophy of the life sciences = History and philosophy of the life sciences. Naples, Florence.

Public garden; journal of the american association of botanical gardens and arboreta. Swarthmore, PA. Vol. 1+, 1986+. Public Gard. Preceded by: Bulletin, american association of botanical gardens and arboreta. HI 70146

Public health journal. Toronto. Public Health J. See B–P–H 748/12. HI 58420

Publicación, see Publicaciones.

Publicaciones de la academia Argentina de farmacia y bioquímica. [Buenos Aires]. Vol. 1+, 1974+. Publ. Acad. Argentina Farm. Bioquim. HI 70147

Publicaciones sobre biologia mediterranea, instituto español de estudios mediterraneos. Barcelona. Vols. 1-2, 1945-46. Publ. Biol. Medit. HI 70148

Publicaciones de biologia de la universidad de Navarra. Serie botánica. Pamplona. Vol. 1+, 1982+. Publ. Biol. Univ. Navarra, Ser. Bot. HI 70149

Publicaciones biologicas, instituto de investigaciones cientificas. Monterrey. Vol. 1+, 1973+. Publ. Biol. Inst. Invest. Ci. HI 70150

Publicaciones de la cátedra de historia de la medicina. Univerisdad de Buenos Aires. Buenos Aires. Publ. Cátedra Hist. Med. See B–P–H 744/14. HI 58373

Publicaciones, centro de experimentacion, servicio agropecuaria provincial, diputacion provincial de Tarragona. No. 1+, 1977+. Publ. Centro Exp. Serv. Agropecu. Prov. Diput. Tarragona. HI 70151

Publicaciones del centro pirenaico de biologia experimental. Jaca. Nos. [1]-?, 1964-86? Publ. Centro Piren. Biol. Exp. Superseded by: Monografías del instituto pirenaico de ecología. HI 70152

Publicaciones en ciencias agricolas, universidad de Chile. Santiago. No. 1+, 1967+. Publ. Ci. Agric. Univ. Chile. HI 70153

Publicaciones científicas, servicio oceanográfico y di pesca, ministerio de industrias y trabajo. Montevideo. Publ. Ci. Serv. Oceanogr. Minist. Industr. See B–P–H 744/17. HI 58375

Publicacion, diputacion provincial de Tarragona, servicio agropecuaria provincial, centro de experimentacion = Publicaciones, centro de experimentacion, servicio agropecuaria provincial, diputacion provincial de Tarragona.

Publicaciones, departamento forestal del Ecuador. Quito. Nos. 1-?, 1949-52. Publ. Dep. Forest. Ecuador. HI 70154

Publicaciones, dirección forestal parques y fauna. Montevideo? No. 1+, 1970+. Pub. Dirección Forest. Parques Fauna. HI 70155

Publicación. División de exploraciones e introducción de plantas. Instituto de fitotécnica. Buenos Aires. Publ. Div. Explor. Introd. Pl. See B–P–H 744/20. HI 58377

Publicaciones, escuela de agronomia, universidad de Buenos Aires. Buenos Aires. Nos. 1-3, 1952-54.

Publ. Esc. Agron. Univ. Buenos Aires. HI 70156

Publicaciones especiales, centro de ciencias del mar y limnologia, universidad nacional autonomia de Mexico. Mexico City, D.F. No. 1+, 1978+. Publ. Esp. Centro Ci. Mar Limnol. Univ. Nac. Auton. Mexico. HI 70157

Publicaciones especiales, instituto de biologia, universidad nacional autonoma de Mexico. Mexico City, D.F. 1969+. Publ. Esp. Inst. Biol. Univ. Nac. Auton. Mexico. HI 70158

Publicaciones especiales, instituto español de oceanografia. Madrid. No. 1+, 1988+. Publ. Espec. Inst. Esp. Oceanogr. HI 75248

Publicaciones especiales del instituto nacional de investigaciones forestales. Mexico, D.F. No. 1+, 1969+. Publ. Esp. Inst. Nac. Invest. Forest. HI 70159

Publicación especial. Laboratorio de biología marina. Universidad católica de Santo Tomás de Villanueva. Marianao, Cuba. Publ. Especial Lab. Biol. Mar. Univ. Católica Santo Tomás de Villanueva. See B–P–H 744/22. HI 58379

Publicaciones del establicimiento venezolana de ciencias naturales. Evencias, Caracas. 1945-47. Publ. Establ. Venez. Ci. Nat. HI 70160

Publicaciones, estacion experimental agropecuaria Manfredi. Manfredi. No. 1+, 1950+. Publ. Estac. Exp. Agropecu. Manfredi. HI 70161

Publicaciones de extensión cultural y didática, instituto nacional de investigación de las ciencias naturales y museo argentino de ciencias naturales "Bernardino Rivadavia". Buenos Aires. No. ?-7+, 1949-54, 1963+. Publ. Extens. Cult. Didát. Inst. Nac. Invest. Ci. Nat. Mus. Argent. Ci. Nat. Bernardino Rivadavia. Preceded by: Publicaciones de extension cultural y didactica, museo argentino de las ciencias "Bernardino Rivadavia". HI 70162

Publicaciones de extension cultural y didactica, museo argentino de ciencias naturales "Bernardino Rivadavia". Buenos Aires. 1947-48. Publ. Extens. Cult. Didact. Mus. Argent. Ci. Nat Bernardino Rivadavia. Superseded by: Publicaciones de extensión cultural y didática, instituto nacional de investigación de las ciencias naturales y museo argentino de ciencias naturales "Bernardino Rivadavia". HI 70163

Publicaciones "I N O C A R". Guayaquil. CM-B10-1-74+, 1974+. Publ. I. N. O. C. A. R. HI 70164

Publicaciones del instituto de biologia aplicada. Barcelona. Vols. 1-4, 1944-73. Publ. Inst. Biol. Aplicada. Superseded by: Treballs de l'institut botànic de Barcelona. 5-4038-1. HI 70165

Publicaciones del instituto botánico = Publicacions de l'institut botanic. Barcelona.

Publicaciones, instituto de ecologia. Mexico City. Vol. ?-3+, ?-1977+. Publ. Inst. Ecol. HI 70166

Publicaciones del instituto de fitotécnia. Castelar. No. 1+, 1947+. Publ. Inst. Fitotécn. HI 70167

Publicaciones, instituto de genetica, universidad de Buenos Aires. Buenos Aires. 1938-45. Publ. Inst. Genet. Univ. Buenos Aires. HI 70168

Publicaciones, instituto geografico "Augustin Codazzi", departamento agrologico. Bogota. Vol. 1+, 1965+. Publ. Inst. Geogr. Augustin Codazzi Dep. Agrol. HI 70169

Publicaciones, instituto geografico "Augustin Codazzi", direccion agriologica = Publicaciones, instituto geografico "Augustin Codazzi", departamento agrologico. Bogota.

Publicaciones del instituto de investigaciones geográficas, facultad de filosofia y letras, universidad de Buenos Aires. Serie A, Memorias originales y documentos. Buenos Aires. Vol. 1+, 1917+. Publ. Inst. Invest. Geogr. Fac. Filos. Letras Univ. Buenos Aires, A. HI 70170

Publicaciones, instituto de investigaçion de recursos naturales. Santiago de Chile. No. 1+, 1964+. Publ. Inst. Invest. Recurs. Nat. HI 70171

Publicaciones, instituto de investigaciones sobre recursos bióticos. Xalapa. Vol. 1(1-2), 1976. Publ. Inst. Invest. Recurs. Bióticos. Superseded by: Biótica. HI 70172

Publicaciones del instituto de la Patagonia. Série monografias. Punta Arenas. 1970+. Publ. Inst. Patagonia, Sér. Monogr. HI 70173

Publicaciones, inventario nacional forestal. Mexico City? No. 1+, 1967+. Publ. Invent. Nac. Forest. HI 70174

Publicaciones de la junta de ciències naturales de Barcelona. Musei barcinonensis scientiarum naturalium opera. Series biologico-oceanografia. Barcelona. Publ. Junta Ci. Nat. Barcelona, Ser. Biol.-Oceanogr. See B–P–H 746/11. HI 58393

Publicaciones de la junta de ciències naturales de Barcelona. Musei barcinonensis scientiarum naturalium opera. Series botanica. Barcelona. Publ. Junta Ci. Nat. Barcelona, Ser. Bot. See B–P–H 746/10. HI 58392

Publicaciones misceláneas, estacion experimental agropecuaria Pergamino. Pergamino. No. 1+, 1960+. Publ. Misc. Estac. Exp. Agropecu. Pergamino. HI 70175

Publicaciones misceláneas, estación experimental regional agropecuaria Anguil. La Pampa. No. ?-2+, ?-1980+. Publ. Misc. Estac. Exp. Regional Agropecu. Anguil. HI 70176

Publicaciones misceláneas, dirección de informaciones,

secretaría de agricultura y ganadería, Argentina. Buenos Aires. No. 1+, 1936+. Publ. Misc. Dirección Inform. Secr. Agric. Ganad. Argentina. HI 70177

Publicaciones misceláneas, museo botanico Cordoba. Cordoba. Nos. 1-11, 1954-58. Publ. Misc. Mus. Bot. Cordoba. HI 70178

Publicación de la misión de estudios de patología regional argentina. Jujuy, Argentina. Publ. Misión Estud. Patol. Regionale Argent. See B–P–H 747/1. HI 58400

Publicaciones del museo ecuatoriano de ciencias naturales. Quito. Vol. 1+, 1979+. Publ. Mus. Ecuat. Ci. Nat. HI 70179

Publicaciones del museo ecuatoriano de ciencias naturales. Serie miscelaneas. Quito. No. 1+, 1983+. Publ. Mus. Ecuat. Ci. Nat., Ser. Misc. HI 75257

Publicaciones del museo ecuatoriano de ciencias naturales. Serie monografia Quito. 1981/84+. Publ. Mus. Ecuat. Ci. Nat., Ser. Monogr. HI 75258

Publicaciones del museo ecuatoriano de ciencias naturales. Serie revista. Quito. No. 1+, 1979+. Publ. Mus. Ecuat. Ci. Nat., Ser. Revista. HI 75259

Publicaciones del museo de historia natural "Javier Prado". Serie B. Botánica. Lima. Publ. Mus. Hist. Nat. "Javier Prado", Ser. B, Bot. See B–P–H 747/3. HI 58401

Publicaciones de museo municipal de ciencias naturales y tradicional de Mar del Plata. Mar del Plata. Vols. 1-2(4), 1960-72. Publ. Mus. Munic. Ci. Nat. Tradic. Mar del Plata. Superseded by: Publicaciones del museo municipal de ciencias naturales "Lorenzo Scaglia". HI 70180

Publicaciones del museo municipal de ciencias naturales "Lorenzo Scaglia". Mar del Plata. Vol. 2(5)+, 1977+. Publ. Mus. Munic. Ci. Nat. Lorenzo Scaglia. Preceded by: Publicaciones de museo municipal de ciencias naturales y tradicional de Mar del Plata. HI 70181

Publicaciones del museo nacional de Buenos Aires. Biología general. Buenos Aires. Nos. 1-5, 1896-1911. Publ Mus. Nac. Buenos Aires, Biol. Gen. HI 70182

Publicaciones del museo nacional de Buenos Aires. Botánica. Buenos Aires. Nos. 1-87, 1895-1945. Publ Mus. Nac. Buenos Aires, Bot. HI 70183

Publicaciones del museo nacional de historia natural. Paleontología (paleobotánica). Buenos Aires. No. 1+, 1924+. Publ Mus. Nac. Hist. Nat. Buenos Aires, Paleontol. HI 70184

Publicaciones ocasionales del museo nacional de historia natural. Santiago de Chile. 1963+. Publ. Ocas. Mus. Nac. HIst. Nat. HI 70185

Publicaciones, parque biologico, reserva natural "Sierra de San Javier". Tucuman. No. ?-2+, ?-1976+. Publ. Parques Biol Reserva Nat. Sierra de San Javier. HI 70186

Publicaciones tecnicas, centro de investigaciones forestales. La Molina. No. 1+, 1970+. Publ. Tecn. Centro Invest. Forest. HI 70187

Publicación técnica. Esculea nacional de agricultura. Chapingo, Mexico. Publ. Técn. Esc. Nac. Agric. See B–P–H 748/2. HI 58413

Publicaciones técnicas, estación experimental agropecuaria Pergamino. Pergamino. No. 1+, 1939+. Publ. Técn. Estac. Exp. Agropecu. Pergamino. HI 70188

Publicaciones tecnicas, estacion experimental regional agropecuaria Balcarce. Balcarce. No. 1+, 1963+. Publ. Tecn. Estac. Exp. Regional Agropecu. Balcarce. HI 70189

Publicación técnica. Instituto de botánica. Buenos Aires. Publ. Técn. Inst. Bot. See B–P–H 748/3. HI 58414

Publicación técnica. Instituto de fitotécnia, Buenos Aires. Buenos Aires. Publ. Técn. Inst. Fitotécn. Buenos Aires. See B–P–H 748/4. HI 58415

Publicaciones tecnicas, instituto de patologia vegetal. Buenos Aires. No. 1+, 1958+. Publ. Tecn. Inst. Patol. Veg. Preceded by: Direccion de investigaciones, instituto de sanidad vegetal. Ser. A. HI 70190

Publicación. Universidad nacional. Facultad de ciencias exactas, físicas y naturales. Serie B. Científicas y técnicas. Buenos Aires. Publ. Univ. Nac. Fac. Ci. Exact., Ser. B, Ci. Técn. See B–P–H 748/5. HI 58416

Publicaciones de la universidad de Santo Domingo. Ciudad Trujillo [=Santo Domingo], Dominican Republic. Publ. Univ. Santo Domingo. See B–P–H 748/8. HI 58418

Publicacions de l'institut botanic. Barcelona. Vols. 1-4(1)-?, 1917-44. Publ. Inst. Bot. 1-604-1. HI 70191

Publicações avulsas, fundaçao zoobotanica do Rio Grande do Sul. Porto Alegre. Vol. 1+, 1976+. Publ. Avulsas Fund. Zoobot. Rio Grande do Sul. HI 70192

Publicações avulsas, Museu Paraense Emilio Goeldi. Belem. No. 1+, 1964+. Publ. Avulsas Mus. Paraense Emilio Goeldi. HI 70193

Publicações cientifícas, instituto de genética, escola superior de agricultura "Luiz de Queiroz". Piracicaba. No. 1+, 1960+. Publ. Ci. Inst. Genét. Esc. Super. Agric. Luiz de Queiroz. HI 70194

Publicações, commissão de combate as pragas da cana de açucar no estado de Pernambuco. Pernambuco. No. 1+, 1954+. Publ. Commiss. Combate Pragas Cana Açucar Estado Pernambuco. HI 70195

Publicações da commissão de estudo e debellação da praga caféeira. São Paulo. 1925-28. Publ. Commiss. Estudo Debell. Praga Café. HI 70196

Publicações, departamento de genética, instituto de biociências. Porto Alegre. No. 1+, 1971+. Publ. Dep. Genét. Inst. Bioci. HI 70197

Publicações, direcção géral dos serviços florestais e aquicolas. Lisbon. Vol. 1+, 1934+. Publ. Direcção Géral Serv. Florest. HI 70198

Publicações diversas, fomento geral de Angola. Lisbon. Nos. ?-3-10, ?-1923-28. Publ. Diversas Fomento Geral Angola. HI 70199

Publicações diversas, instituto botanico de universidade de Coimbra. Coimbra. Nos. 1-18, 1954-60. Publ. Diversas Inst. Bot. Univ. Coimbra. HI 70200

Publicações I P R N R. Porto Alegre. No. 1+, 1978+. Publ. I. P. R. N. R. HI 70201

Publicações de informação e divulgação, instituto de investigação cientifica de Moçambique. Lourenço Marques. No. 1+, 1966+. Publ. Inform. Divulg. Inst. Invest. Ci. Moçambique. HI 70202

Publicações do instituto de botânica "Dr. Gonçalo Sampaio". Oporto, Portugal. Publ. Inst. Bot. "Dr. Gonçalo Sampaio". See B–P–H 745/26. HI 58386

Publicações, instituto de micologia, universidade federal de Pernambuco. Recife. No. ?-506+, ?-1965+. Publ. Inst. Micol. Univ. Fed. Pernambuco. Preceded by: Publicações, instituto de micologia, universidade de Recife. HI 70203

Publicações, instituto de micologia, universidade de Recife. Recife. No. 1-463-?, 1956-65-? Publ. Univ. Recife Inst. Micol. Superseded by: Publicações, instituto de micologia, universidade federal de Pernambuco. HI 70204

Publicações, instituto nacional de pesquisas da Amazônia. Ser. Botânica. Manaus. Vols. 1-28, 1956-68. Publ. Inst. Nac. Pesq. Amazônia Bot. Superseded by: Boletim do instituto nacional de pesquisas da Amazônia. Serie, botânica and Boletim do instituto nacional de pesquisas da Amazônia. Florestais. HI 70205

Publicações, instituto nacional de pesquisas da Amazônia. Ser. Quimica. Manaus. Vols. 1-12, 1958-68. Publ. Inst. Nac. Pesq. Amazônia, Ser. Quim. Superseded by: Boletim do instituto nacional de pesquisas da Amazõnia. HI 70206

Publicações, instituto de pesquisas de recursos naturais renováveis = Publicações I P R N R. Porto Alegre.

Publicações, instituto de pesquisas agronomicas de Pernambuco. Pernambuco. Nos. 1-10-?, 1957-59-? Publ. Inst. Pesq. Agron. Pernambuco. HI 70207

Publicações ocacional, instituto de investigação agronómica de Angola. Nova Lisboa. Vol. 1+, 1969+. Publ. Ocac. Inst. Invest. Agron. Angola. HI 70208

Publicações, servicos de agricultura, Mozambique. Ser. A, cientifica tecnica. Lourenço Marques. No. 1+, 1956+. Publ. Serv. Agric. Mozambique, A. HI 70209

Publicações, servicos de agricultura, Mozambique. Ser. B, divulgação. Lourenço Marques. No. 1+, 1956+. Publ. Serv. Agric. Mozambique, B. HI 70210

Publicações, serviço de informação agricola. No. 1+, 1935+. Publ. Serv. Inform. Agric. HI 70211

Publicatiés voor cactusstudie = Blätter für Kakteenforschung. Volksdorf [=Hamburg, in part]. Blätt. Kakteenf. See B–P–H 199/14.

Publicatiés, hydrobiologische club. Amsterdam. 1932-41. Publ. Hydrobiol. Club. Superseded by: Publicatiés, hydrobiologische vereniging. HI 70212

Publicatiés, hydrobiologische vereniging. Amsterdam. 1952-60. Publ. Hydrobiol. Ver. Preceded by: Publicatiés, hydrobiologische club. HI 70213

Publicatiés van het natuur-historische genootschap in Limburg. Maastricht. Publ. Natuur-Hist. Genootsch. Limburg. See B–P–H 747/11. HI 58407

Publicatiés van het nederlandsch-indisch landbouw-syndicaat. Surabaya, Dutch E. Indies [Indonesia]. Publ. Ned.-Indisch Landb.-Synd. See B–P–H 747/13. HI 58409

Publicatiés, proefstation voor de akkerbouw en de groenteteelt in de vollegrond. Lelystad. No. 1+, 1977+. Publ. Proefstat. Akkerb. Groentet. Vollegrond. Preceded by: Publicatiés, proefstation voor de akkerbouw te Lelystad. HI 70214

Publicatiés, proefstation voor de akkerbouw te Lelystad. Lelystad. Nos. 1-23, 1972-76. Publ. Proefstat. Akkerb. Lelystad. Preceded in part by: Mededelingen, proefstation voor de akker- en weidebouw. Superseded by: Publicatiés, proefstation voor de akkerbouw en de groenteteelt in de vollegrond. HI 70215

Publicatiés van het proefstation voor de groenten- en fruitteelt onder glas te Naaldwijk. The Hague. No. 1+, 1954+. Publ. Proefstat. Groenten- Fruitteelt Glas Naaldwijk. HI 70216

Publicaţiile muzeului judatului Hunedoara. Hunedoara, Rumania. Publ. Muz. Judatului Hunedoara. See B–P–H 747/8. HI 58405

Publicationes academiae horti- et viticulturae = Kertészeti egyetem közleményei. Budapest.

Publicationes instituti botanici, universitatis jagellonicae cracoviensis. Cracow. 1931-56. Publ. Inst. Bot. Univ. Jagellon. Cracov. Superseded by: Publicationes instituti botanici, universitatis jagellonicae cracoviensis et academiae scientiarum poloniae. 3-2316-2. HI 70217

Publicationes instituti botanici, universitatis jagellonicae cracoviensis et academiae scientiarum poloniae. Cracow. 1957+. Publ. Inst. Bot. Univ.Jagellon. Cracov. Acad. Sci. Poloniae. Preceded by: Publicationes instituti botanici, universitatis jagellonicae cracoviensis. HI 70218

Publicationes membrorum instituti systematico-geobotanici, hortique botanici, universitatis L. Eotvos budapestinensis. Budapest. 1964/65-74/75. Publ. Memb. Inst. Syst.-Geobot. Hort. Bot. Univ. L. Eotvos Budapest. HI 70219

Publicationes, societas scientiarum islandica. Akureyri, Reykjavik = Rit vísindafjélags íslendinga. Akureyri, Reykjavik.

Publicationes universitatis horticulturae = Kertészeti és szölészeti föiskola évkönyve. Budapest.

Publication, see Publications

Publications of the Aberdeen natural history and antiquarian society. Aberdeen, Scotland. Publ. Aberdeen Nat. Hist. Soc. See B–P–H 743/11. HI 58359

Publication, advanced centre of palaeontology and himalayan geology = Publications, centre of advanced study in geology, Panjab university. Chandigarh.

Publications from the Akkeshi marine biological station. Sapporo, Japan. Publ. Akkeshi Mar. Biol. Sta. See B–P–H 743/13. HI 58360

Publications of the Allan Hancock Pacific expeditions. Publ. Allan Hancock Pacific Exped. See B–P–H 743/15. HI 58361

Publications from the Amakusa marine biological laboratory, Kyushu university. Kumamoto. No. 1+, 1966+. Publ. Amakusa Mar. Biol. Lab. Kyushu Univ. HI 70292

Publications in american archaeology and ethnology. Berkeley, CA. Publ. Amer. Archaeol. Ethnol. See B–P–H 743/17. HI 58362

Publications of the american association for the advancement of science. New York, NY. Publ. Amer. Assoc. Advancem. Sci. See B–P–H 743/18. HI 58363

Publications, American institute of biological sciences = American institute of biological sciences. Publication. Washington, DC. Amer. Inst. Biol. Sci. Publ. See B–P–H 80/21.

Publications. American university of Beirut. Faculty of agricultural sciences. Beirut. Publ. Amer. Univ. Beirut Fac. Agric. Sci. See B–P–H 743/19. HI 58364

Publications in anthropology. New Haven, CT. Publ. Anthropol. See B–P–H 743/20. HI 58365

Publications in anthropology of the university of New Mexico. Albuquerque, NM. Vol. 1+, 1945+. Publ. Anthropol. Univ. New Mexico. Preceded by: Bulletin of the university of New Mexico, anthropology series. 4-3007-1. HI 70220

Publications of the Arnold arboretum. Cambridge, MA. Publ. Arnold Arbor. See B–P–H 743/23. HI 58366

Publications of the association for the study of systematics in relation to general biology. London. 1942. Publ. Assoc. Stud. Syst. Relat. Gen. Biol. Superseded by: Publications, systematics association. HI 70221

Publications of the Atkins institution. Soledad, Cuba. Publ. Atkins Inst. See B–P–H 744/1. HI 58368

Publications, Barbados herbarium. Bridgetown. No. 1+, 1978+. Publ. Barbados Herb. HI 70222

Publications, Basrah natural history museum. Basrah. No. 1+, 1976+. Publ. Basrah Nat. HIst. Mus. HI 70223

Publications bimestrielle, institut belge pour l'amélioration de la betterave. Tirlemont. 1967-68. Publ. Bimestr. Inst. Belge Amélior. Betterave. Preceded by: Publications de vulgarisation de l'institut belge pour l'amélioration de la betterave and Publications techniques de l'institut belge pour l'amélioration de la betterave. Superseded by: Publications trimestrielle, institut belge pour l'amélioration de la betterave. HI 70224

Publications from the biological laboratory. [University of Toronto.] Toronto. Nos. 1-3, 1891-92. Publ. Biol. Lab., Toronto. Superseded by: University of Toronto studies, biological series. 5-4240-1. HI 70225

Publications in biological oceanography, national museum of natural sciences, Canada. Ottawa. Nos. 1-11, 1970-82. Publ. Biol. Oceanogr.Natl. Mus. Nat. Sci. Canada. Superseded by: Publications in natural sciences, national museum of natural sciences, Canada. HI 70226

Publications, biological research centre, university of Baghdad. Baghdad. No. 1+, 1967+. Publ. Biol. Res. Centre Univ. Baghdad. HI 70227

Publications in biological sciences. University of California at Los Angeles. Los Angeles, CA. Publ. Biol. Sci. Univ. Calif. Los Angeles. See B–P–H 744/3. HI 58369

Publications from the biological station, Espegrend. Bergen. 1951-61. Publ. Biol. Sta. Espegrend. HI 70228

Publications in biology and geology, Milwaukee public museum. Milwaukee, WI. No. 1+, 1973+. Publ. Biol. Geol. Milwaukee Public Mus. HI 70229

Publications in biology, King Abdulaziz university. Jeddah. 1978+. Publ. Biol. King Abdulaziz Univ.

HI 70230

Publications from the botanical institute, Kyoto imperial university. Kyoto. Nos. 1-51, 1924-41. Publ. Bot. Inst. Kyoto Imp. Univ. HI 70231

Publications of the botanical society of America. Ithaca, NY. Publ. Bot. Soc. Amer. See B–P–H 744/5. HI 58370

Publications en botanique, musée national des sciences naturelles, Canada = Publications in botany, national museum of natural sciences, Canada. Ottawa.

Publications in botany, national museum of natural sciences, Canada. Ottawa. Nos. 1-?, 1969-? Publ. Bot. (Ottawa). Superseded by: Publications in natural sciences, national museum of natural sciences, Canada. HI 70232

Publications in botany, Milwaukee public museum = Milwaukee public museum publications in botany. Milwaukee, WI. Milwaukee Public Mus. Publ. Bot. See B–P–H 595/12.

Publications. Bureau of plant industry. Taihoku [=Taipei, Taiwan]. Publ. Bur. Pl. Industr. See B–P–H 744/8. HI 58371

Publications of the bureau of science government laboratories. Manila. Vols. 1-36, 1902-05. Publ. Bur. Sci. Gov. Lab. Superseded by: Philippine journal of science. 4-333-3. HI 70233

Publications of the Bussey institution, Harvard university. Jamaica Plains, MA. ?-1980+. Publ. Bussey Inst. Harvard Univ. HI 70234

Publications from the Cairo university herbarium. Giza, Koenigstein/Taunus. Vols. 1-8, 1968-77. Publ. Cairo Univ. Herb. Superseded by: Taeckholmia. HI 70235

Publications, canadian forestry service = Publications, department of forestry and rural development, Canada. Ottawa.

Publications of the Carnegie institution of Washington. Washington, DC. Publ. Carnegie Inst. Wash. See B–P–H 744/12. HI 58372

Publications, centraalbureau voor schimmelcultures. Series A. Baarn. 1979+. Publ. Centraalbur. Schimmelcult., A. HI 70236

Publications, centraalbureau voor schimmelcultures. Series B. Baarn. 1979+. Publ. Centraalbur. Schemmelcult., B. HI 70237

Publications, centre of advanced study in geology, Panjab university. Chandigarh. No. 1+, 1966+. Publ. Centre Advanced Study Geol. Pubjab Univ. HI 70238

Publications du centre national pour l'exploitation des océans. Series, actes et colloques. Paris. Vol. 1+, 1974+. Publ. Centre Natl. Exploit. Océans, Ser. Actes Colloq. HI 70239

Publications, Chesapeake biological laboratory = Contributions, Chesapeake biological laboratory. Solomons Island, MD. Contr. Chesapeake Biol. Lab. See B–P–H 327/7.

Publications of the college of agriculture, Honan university. [Honan ta hsüeh nung hsüeh yüan yan k'an.] Kaifeng, China. Publ. Coll. Agric. Honan Univ. See B–P–H 744/18. HI 58376

Publications de la comité national français des recherches antartiques. Biologie. Paris. Vol. 1+, 1962+ [Biologie is a subseries of, and also numbered as part of, the whole series, i.e. 10+]. Publ. Comité Natl. Franç. Rech. Antart. Biol. HI 70240

Publications, Connecticut forest and park association. Hartford, CT. 1925+. Publ. Connecticut Forest Park Assoc. Preceded by: Publications of the Connecticut forestry association. HI 70241

Publications of the Connecticut forestry association. Hartford, CT. 1906-24. Publ. Connecticut Forest. Assoc. Preceded by: Annual bulletin, Connecticut forestry association. Superseded by: Publications, Connecticut forest and park association. HI 70242

Publications, conseil scientifique pour l'Afrique au sud du Sahara. Kikuyu, London. No. 1+, 1951+. Publ. Cons. Sci. Afrique Sud Sahara. HI 70243

Publications from the danish arctic station on Disko Islands, Greenland = Arbejder fra den danske arktische station pa Disko. Copenhagen.

Publications, department of agriculture, Canada. Ottawa. No. 475+, 1935+ [nos. 1-474 were not numbered]. Publ. Dep. Agric. Canada. HI 70244

Publications, department of agriculture, Malay states. General series = General series, department of agriculture, Straits Settlements and Federated Malay States. Kuala Lumpur.

Publications, department of agriculture, Malay states. Scientific series = Scientific series, department of agriculture, Straits Settlements and Federated Malay States. Kuala Lumpur.

Publications, department of biogeography and geomorphology, research school of Pacific studies, Australian national university. Canberra, A.C.T. Vol. BG/1+, 1969+. Publ. Dep. Biogeogr. Geomorph. Res. School Pacific Stud. Austral. Natl. Univ. HI 70247

Publications, department of botany, ministry of agriculture, Iran. Tehran. 1967. Publ. Dep. Bot. Minist. Agric. Iran. HI 70248

Publications from the department of botany, university of Helsinki = Helsingin yliopiston kasvitieteen laitoksen julkeisuja. Helsinki.

Publications from the department of botany, university of the Panjab. Lahore. 1929+. Publ. Dep. Bot. Univ.

Panjab. HI 70249

Publications from department of botany, university of Turku. Turku. No. 1+, 1956+. Publ. Dep. Bot. Univ. Turku. HI 70250

Publications of the department of entomology and plant pathology, Ohio agricultural experiment station = Ohio agricultural experiment station publications of the department of entomology and plant pathology. Wooster, OH. Ohio Agric. Exp. Sta. Publ. Dept. Entomol. See B–P–H 684/26.

Publications, department of forestry, Canada = Publications, department of forestry and rural development, Canada. Ottawa.

Publications, department of forestry and rural development, Canada. Ottawa. No. 1001+, 1963+. Publ. Dep. Forest. Rural Developm. Canada. Preceded by: Bulletin of the department of forestry, Canada; Mimeographs, forest product laboratories, Canada; Miscellaneous publications, forestry branch, division of forest research, Canada; and Technical notes, forest research division, forestry branch, Canada. HI 70251

Publications of the department herbarium, botany department, university of Dar es Salaam. Dar es Salaam. No. 1+, 1971+. Publ. Dep. Herb. Bot. Dep. Univ. Dar es Salaam. HI 70252

Publications de la direction générale des eaux et forêts de Portugal = Publicações, direcção géral dos serviços florestais e aquicolas. Lisbon.

Publications diverses du muséum national d'histoire naturelle. Paris. No. 24+, 1970+. Publ. Diverses Mus. Natl. His. Nat. Preceded by: Publications du muséum national d'histoire naturelle. HI 70253

Publications. Elisabethville. Université de l'état. Elizabethville [=Lubumbashi, Republic of the Congo]. Publ. Elisabethville Univ. État. See B–P–H 744/21. HI 58378

Publications, european and mediterranean plant protection organisation. Series A, reports of technical working parties and conferences. Paris. Nos. A14-A58, 1955-70. Publ. Eur. Medit. Pl. Protect. Organ., Ser. A, Rep. Techn. Working Parties Conf. Preceded by: Publications, european plant protection organisation. Series A. Superseded by: Bulletin O E P P. HI 70254

Publications, european and mediterranean plant protection organisation. Series B. Paris. No. 9+, 1954+. Publ. Eur. Medit. Pl. Protect. Organ., B. Preceded by: Publications, european plant protection organisation. Series B. Superseded by: Bulletin O E P P. HI 70255

Publications, european and mediterranean plant protection organisation. Series C, reports of working parties. Paris. No. 5+, 1955+. Publ. Eur. Medit. Pl. Protect. Organ., C. Preceded by: Publications, european plant protection organisation. Series C. Superseded by: Bulletin O E P P. HI 70256

Publications, european and mediterranean plant protection organisation. Series D. Paris. Nos. 1-18, 1955-71. Publ. Eur. Medit. Pl. Protect. Organ., D. Superseded by: Bulletin O E P P. HI 70257

Publications, european plant protection organisation. Series A. Paris. Nos. A1-A13, 1951-55. Publ. Eur. Pl. Protect. Organ., Ser. A Superseded by: Publications, european and mediterranean plant protection organisation. Series A, reports of technical working parties and conferences. HI 70258

Publications, european plant protection organisation. Series B. Paris. Nos. 1-8, 1949-53, 1950-54. Publ. Eur. Pl. Protect. Organ., B. Superseded by: Publications, european and mediterranean plant protection organisation. Series B. HI 70259

Publications, european plant protection organisation. Series C. Paris. Nos. 1-4, 1950-54. Publ. Eur. Pl. Protect. Organ., C. Superseded by: Publications, european and mediterranean plant protection organisation. Series C, reports of working parties. HI 70260

Publications de la faculté des sciences de l'université à Brno = Spisy přírodovědecké fakulty universit J. E. Purkinje v Brně. Brno.

Publications de la faculté des sciences de l'université Charles = Spisy vydávané přírodovědeckou fakultou Karlovy university. Prague. Spisy Přír. Fak. Karlovy Univ. See B–P–H 852/16.

Publications de la faculté des sciences de l'université Masaryk = Spisy vydávané přírodovědeckou fakultou Masarykovy university. Brünn [=Brno, Czechoslovakia]. Spisy Přír. Fak. Masarykovy Univ. See B–P–H 852/17.

Publications of the Far-Eastern state university. Ser. 4. Vladivostok, Russian S F S R = Trudy gosudarstvennogo dal′nevostochnogo universiteta. Ser. 4, lesnye nauk. Vladivostok.

Publications of the Far-Eastern state university. Ser. 5. Kiev, Ukrainian S S R = Trudy gosudarstvennogo dal′nevostochnogo universiteta. Ser. 5, sel′skoe khozyaistvo. Kiev.

Publications of the Far-Eastern state university. Ser. 8. Vladivostok, Russian S F S R = Trudy gosudarstvennogo dal′nevostochnogo universiteta. Ser. 8, biologiya. Vladivostok.

Publications of the Far-Eastern state university. Ser. 11. Vladivostok, Russian S F S R = Trudy gosudarstvennogo dal′nevostochnogo universiteta. Ser. 11, geologiya. Vladivostok.

Publications of the Far-Eastern state university. Ser. 13. Vladivostok, Russian S F S R = Trudy gosudarstvennogo dal'nevostochnogo universiteta. Ser. 13, tekhnika. Vladivostok.

Publications of the Field Columbian museum. Botanical series. Chicago, IL. Publ. Field Columbian Mus., Bot. Ser. See B–P–H 745/6. HI 58380

Publications of the Field columbian museum. Report series. Chicago, IL. Vols. 1-2, 1895-1905. Publ. Field Columb. Mus., Rep. Ser. Contnued as: Publications, Field museum of natural history. HI 70261

Publications of the Field museum of natural history. Botanical series. Chicago, IL. Publ. Field Mus. Nat. Hist., Bot. Ser. See B–P–H 745/10. HI 58381

Publications, Field museum of natural history. Report series. Chicago, IL. Vols. 3-13, 1906-42. Publ. Field Mus. Nat. Hist., Rep. Ser. Preceded by: Publications of the Field columbian museum. Report series. Superseded by: Report (Annual) of the Chicago natural history museum. HI 70262

Publications of forestry sciences. Sopron = Erdészettudományi közlemények. Sopron, Hungary. Erdészettud. Közlem. See B–P–H 360/11.

Publications of Fort Burgwin research center, Inc. Santa Fe, NM. Publ. Fort Burgwin Res. Center. See B–P–H 745/13. HI 58382

Publications, Ğami'at al-qahirah herbarium. No. 1+, 1970+. Publ. Ğami'at al-qahirah Herb. HI 70264

Publications, Great Lakes research division, university of Michigan, institute of science and technology. Ann Arbor, MI. 1st-9th, 1956-66. Publ. Great Lakes Res. Div. Univ. Michigan Inst. Sci. Techn. Superseded by: Proceedings, conference on Great Lakes research. HI 70265

Publications of the Hartley botanical laboratories. Liverpool. Publ. Hartley Bot. Lab. See B–P–H 745/ 16. HI 58383

Publications ... herbarium, department of biology, university of California, Santa Barbara. Santa Barbara, CA. No. 1+, 1981+. Publ. Herb. Dep. Biol. Univ. Calif. Santa Barbara. HI 70266

Publications, herbarium, egyptian university = Publications from the Cairo university herbarium. Giza, Koenigstein/Taunus.

Publications of the herbarium, southeastern Oklahoma state university. Durant, OK. No. 1+, 1977+. Publ. Herb. SouthE. Oklahoma State Univ. HI 70267

Publications from the herbarium, university of Uppsala. Uppsala. Nos. 1-19, 1978-86. Publ. Herb. Univ. Uppsala. Superseded by: Thunbergia. HI 70268

Publications hors-série des conservatoire et jardin botaniques de Genève. Geneva. Vol. 1+, 1977+. Publ. Hors-Ser. Conserv. Jard. Bot. Genève. HI 70269

Publications of the Hugo de Vries laboratory. [? Forms part of: Contributions from the botanical institute of the university of Amsterdam.] Amsterdam. Nos. ?-26-56-?, ?-19?? Publ. Hugo Vries Lab. HI 70270

Publications hydrologiques. Budapest = Hidrológiai közlöny. Budapest. Hidrol. Közl. See B–P–H 416/4.

Publications of the indian tea association; scientific department. Calcutta. Publ. Indian Tea Assoc., Sci. Dept. See B–P–H 745/19. HI 58384

Publications de l'institut belge pour l'amélioration de la betterave = Publications trimestrielle, institut belge pour l'amélioration de la betterave. Tirlemont.

Publications de l'institut de botanique systématique et de phytogéographie de l'université de Varsovie. Warsaw. Nos. 1-79, 1923-38. Publ. Inst. Bot. Syst. Phytogeogr. Univ. Varsovie. HI 70271

Publications de l'institut de botanique de l'université de Genève. Geneva. Publ. Inst. Bot. Univ. Genève. See B–P–H 745/27. HI 58387

Publications de l'institut national pour l'étude agronomique du Congo Belge. Hors série. Gembloux. 1938-72. Publ. Inst. Natl. Étude Agron. Congo Belge, Hors Sér. HI 70272

Publications de l'institut national pour l'étude agronomique de Congo Belge. Série technique. Brussels. Publ. Inst. Natl. Étude Agron. Congo Belge, Sér. Techn. See B–P–H 746/4. HI 58388

Publication de l'institut national pour l'étude agronomique de Congo Belge. Série scientifique. Brussels. Publ. Inst. Natl. Étude Agron. Congo Belge, Sér. Sci. See B–P–H 746/5 HI 58389

Publications de l'institut national de recherches scientifiques Butare. Butare. Nos. 1-16-?, 1966-? Publ. Inst. Natl. Rech. Sci. Butare. Superseded by: Publications de l'institut national de recherches scientifiques du Rwanda. HI 70273

Publications de l'institut national de recherches scientifiques du Rwanda. Butare. ?-1971+. Publ. Inst. Natl. Rech. Sci. Rwanda. Preceded by: Publications de l'institut national de recherches scientifiques Butare. HI 70274

Publications de l'institut de recherches hydrobiologiques de la faculté des sciences, université d'Istanbul = Istanbul üniversitesi fen fakültesi hidrobiologi arastirma enstitusü yayinlarindan. Ser. B. Istanbul.

Publications de l'institut royal grand-ducal de Luxembourg, section des sciences naturelles et mathématiques. Luxembourg. Publ. Inst. Roy. Grand-Ducal Luxembourg, Sect. Sci. Nat. See B–P–H 746/8. HI 58390

Publications of the institute of dendrology and pomology Kórnik = Prace zakładu dendrologii i pomologii w

Kórniku. Kornik, Poland. Prace Zakładu Dendrol. w Kórniku. See B–P–H 718/12.

Publications of the institute of marine science, university of Texas. Austin, TX. Vols. 1-11, 1945-66. Publ. Inst. Mar. Sci. Superseded by: Contributions in marine science. 5-4189-3. HI 70275

Publications, institute of mycology, university of Recife = Publicações, instituto de micologia, universidade de Recife. Recife.

Publications, institute of systematic botany, royal veterinary and agricultural college. Copenhagen. Nos. 1+, 1961+. Publ. Inst. Syst. Bot. Roy. Veterin. Agric. Coll. HI 70276

Publications of the Iwata institute of plant biochemistry. Tokyo. Publ. Iwata Inst. Pl. Biochem. See B–P–H 746/9. HI 58391

Publications de la laboratoire de botanique de l'université de Genève. Geneva. Publ. Lab. Bot. Univ. Genève. See B–P–H 746/12. HI 58394

Publication des laboratoires d'école normale supérieure. Paris. Publ. Lab. École Norm. Supér. See B–P–H 746/13. HI 58395

Publications, Manchester museum. Manchester = Museum publications, Manchester museum. Manchester.

Publication of the Maria Moors Cabot foundation for botanical research. Petersham, MA. Publ. Maria Moors Cabot Found. Bot. Res. See B–P–H 746/15. HI 58396

Publications of the marine biological station, Ghardaqa, Red Sea. Cairo. Vols. 1-?, 1939-67. Publ. Mar. Biol. Sta. Ghardaqa. Superseded by: Bulletin, institute of oceanography and fisheries, Cairo. 2-869-2. HI 70278

Publications of the Marsh botanical garden; Yale university. New Haven, CT. Publ. Marsh Bot. Gard. See B–P–H 746/16. HI 58374

Publications, Max C. Fleischmann college of agriculture. Series B [bulletin series]. Reno, NV. No. 1+, 1964+. Publ. Max C. Fleischmann Coll. Agric., B. Preceded by: Nevada agricultural experiment station, bulletin. HI 70279

Publications, Max C. Fleischmann college of agriculture. Series C [circular series]. Reno, NV. No. ?-21+, 19??+. Publ. Max C. Fleischmann Coll. Agric., C. Preceded by: Nevada agricultural experiment station, circulars. HI 70280

Publications, Max C. Fleischmann college of agriculture. Series T [technical series]. Reno, NV. No. 1+, 1964+. Publ. Max C. Fleischmann Coll. Agric., T. Preceded by: Nevada agricultural experiment station, technical bulletin. HI 70281

Publications of the maya society. Baltimore, MD. Publ. Maya Soc. See B–P–H 746/17. HI 58397

Publications of McGill university museums. Montreal. Publ. McGill Univ. Mus. See B–P–H 746/18. HI 58398

Publications of the McGill university. Series 2, botany. Montreal. Publ. McGill Univ., Ser. 2, Bot. See B–P–H 746/19. HI 58399

Publication, Michigan geological and biological survey; biological series = Michigan geological and biological survey publication; biological series. East Lansing, MI. Michigan Geol. Biol. Surv. Publ., Biol. Ser. See B–P–H 592/22.

Publication, ministère de l'agriculture et de la colonisation, Québec. Quebec. Nos. 1-401?, 1899?-1973. Publ. Minist. Agric. Colonis. Québec. HI 70282

Publication, ministère de l'agriculture, Québec = Publication, ministère de l'agriculture et de la colonisation, Québec. Quebec.

Publications du musée Hoangho Paiho de Tientsin. Tientsin, China. Publ. Mus. Hoangho Paiho Tientsin. See B–P–H 747/5. HI 58402

Publications du musée silésien à Katowice = Wydawnictwa muzeum ślaskiego w Katowicach. Dzial III. Katowice.

Publications of the museum, gulf coast research laboratory. Ocean Springs, MS. Vol. 1+, 1969+. Publ. Mus. Gulf Coast Res. Lab. HI 70283

Publications of the museum. Michigan state university. Biological series. East Lansing, MI. Publ. Mus. Michigan State Univ., Biol. Ser. See B–P–H 747/6. HI 58403

Publications du muséum national d'histoire naturelle. Paris. 1933-68. Publ. Mus. Natl. Hist. Nat. Superseded by: Publications diverses du muséum national d'histoire naturelle. HI 70284

Publications of the museum of natural history of the university of Kansas. Lawrence, KS. Publ. Mus. Nat. Hist. Univ. Kansas. See B–P–H 747/7. HI 58404

Publications of the Nantucket Maria Mitchell association. Nantucket, MA. Publ. Nantucket Maria Mitchell Assoc. See B–P–H 747/9. HI 58406

Publications, national research council. Washington, DC = National research council publications. Washington, DC. Natl. Res. Council Publ. See B–P–H 631/11.

Publications, national swedish nature conservancy office = Publikationer, statens naturvårdsverk. Stockholm.

Publications, natural environment research council. Series B. London. No. 1+, 1971+. Publ. Nat. Environm. Res. Council, B. HI 70285

Publications, natural environment research council. Series

C. London. No. 1+, 1970+. Publ. Nat. Environm. Res. Council, C. HI 70286

Publications, natural history research center. Baghdad. No. 30+, 1976+. Publ. Nat. Hist. Res. Center. Preceded by: Iraq natural history museum's publications. HI 70287

Publications in natural history, Texas academy of science. Non-technical series = Texas academy of science publications in natural history. Non-technical series. Austin, TX. Texas Acad. Sci. Publ. Nat. Hist., Non-Techn. Ser. See B–P–H 872/21.

Publications in natural sciences, national museum of natural sciences, Canada. Ottawa. No. 1+, 1983+. Publ. Nat. Sci. Natl. Mus. Nat. Sci. Canada. Preceded by: Publications in biological oceanography, national museum of natural sciences, Canada and Publications in botany, national museum of natural sciences, Canada. HI 70289

Publications of the Nebraska academy of science. Publ. Nebraska Acad. Sci. See B–P–H 747/12. HI 58408

Publications, north Queensland naturalists' club. Cairns, Qld. Nos. 1-7?, 1945-53. Publ. N. Queensland Naturalist's Club. HI 70291

Publications de l'O E P P. Série A = Publications, european and mediterranean plant protection organisation. Series A, reports of technical working parties and conferences. Paris.

Publications de l'O E P P. Série A = Publications, european plant protection organisation. Series A. Paris.

Publications de l'O E P P. Série B = Publications, european and mediterranean plant protection organisation. Series B. Paris.

Publications de l'O E P P. Série B = Publications, european plant protection organisation. Series B. Paris.

Publications de l'O E P P. Série C = Publications, european and mediterranean plant protection organisation. Series C, reports of working parties. Paris.

Publications de l'O E P P. Série C = Publications, european plant protection organisation. Series C. Paris.

Publications de l'O E P P. Série D = Publications, european and mediterranean plant protection organisation. Series D. Paris.

Publications d'océanographie biologique = Publications in biological oceanography, national museum of natural sciences, Canada. Ottawa.

Publications of the plant, pests and diseases research institute, botany department, Teheran. Tehran. ?-1980+. Publ. Pl. Pests Dis. Res. Inst. Bot. Dep. Teheran. HI 70293

Publications of the Puget Sound biological station of the university of Washington. Seattle, WA. Vols. 2-7, 1917-31. Publ. Puget Sound Biol. Sta. Preceded by: Puget Sound marine station publications. Superseded by: University of Washington publications in oceanography. 5-4450-1. HI 70294

Publications of the research institute for plant breeding and plant growing, Fertöd = A Fertödi növénynemesítési és növénytermesztési kutató intézet közleményei. Budapest. Fertödi Növénynemes. Növényterm. Kutató Intéz. Közlem. See B–P–H 369/16.

Publications, royal veterinary and agricultural college, institute of systematic botany = Publications, institute of systematic botany, royal veterinary and agricultural college. Copenhagen.

Publications, Saudi biological society. Riyad. No. ?-3+, ?-1979+. Publ. Saudi Biol. Soc. HI 70295

Publications des sciences forestières. Sopron = Erdészettudományi közlemények. Sopron, Hungary. Erdészettud. Közlem. See B–P–H 360/11.

Publications de sciences naturelles, musée national des sciences naturelles, Canada = Publications in natural sciences, national museum of natural sciences, Canada.

Publications scientifiques. Association internationale des botanistes. Jena. Publ. Sci. Assoc. Int. Bot. See B–P–H 747/16. HI 58410

Publications series, institute of botany, Ege university. Izmir. Vol. 1+, 1979+. Publ. Ser. Inst. Bot. Ege Univ. HI 70296

Publication series, Netherlands institute for sea research. [Supplement to Netherlands journal of sea research.] Texel. No. 1+, 1978+. Publ. Ser. Netherlands Inst. Sea Res. HI 70297

Publications du service des forêts de l'Algerie. Algiers. 1956+. Publ. Serv. Forêts Algerie. HI 70298

Publications of the Seto marine biological laboratory. Kyoto, Japan. Publ. Seto Mar. Biol. Lab. See B–P–H 747/19. HI 58411

Publications de la société botanique de Pologne = Acta societatis botanicorum Poloniae. Warsaw. Acta Soc. Bot. Poloniae. See B–P–H 49/5.

Publication de la société suisse d'histoire de la médecine et des sciences naturelles = Veröffentlichungen der schweizerischen Gesellschaft für Geschichte der Medizin und der Naturwissenschaften. Zurich.

Publications of the systematics association. London. Vols. 1-9, 1953-73. Publ. Syst. Assoc. Preceded by: Publications of the association for the study of systematics in relation to general biology. 5-4143-2. HI 70299

Publications techniques de l'institut belge pour l'amélioration de la betterave. Tirlemont. 1952-67.

Publ. Techn. Inst. Belge Amélior. Betterave. Preceded by: Publications trimestrielle, institut belge pour l'amélioration de la betterave. Superseded by: Publication bimestrielle, institut belge pour l'amélioration de la betterave. HI 70300

Publications trimestrielle, institut belge pour l'amélioration de la betterave. Tirlemont. 1932-51; 1968+. Publ. Trimestrielle Inst. Amélior. Betterave. From 1952-67 see: Publications techniques de l'institut belge pour l'amélioration de la betterave; Publications de vulgarisation de l'institut belge pour l'amélioration de la betterave and Publication bimestrielle, institut belge pour l'amélioration de la betterave. HI 70301

Publications, union internationale des sciences biologiques. Série A, générale = Proceedings, international union of biological sciences, general assemblies.

Publications, université libanaise. Sections des sciences naturelles. Beirut. Vol. 1+, [1968]+. Publ. Univ. Liban., Sect. Sci. Nat. HI 70302

Publications of the university of British Columbia. Biological sciences. Vancouver. No. 1+, 1945+. Publ. Univ. British Columbia, Biol. Sci. Preceded by: Reprints of the university of British Columbia. Biological sciences. 1-779-1. HI 70303

Publications of the university of Oklahoma biological survey. Norman, OK. Publ. Univ. Oklahoma Biol. Surv. See B–P–H 748/6. HI 58417

Publications de vulgarisation de l'institut belge pour l'amélioration de la betterave. Tirlemont. 1952-67. Publ. Vulg. Inst. Belge Amélior. Betterave. Superseded by: Publications bimestrielle, institut belge pour l'amélioration de la betterave. HI 70304

Publications of the Wagner free institute of science of Philadelphia. Philadelphia, PA. Publ. Wagner Free Inst. Sci. Philadelphia. See B–P–H 748/9. HI 58419

Publications, woodland ecology unit, C S I R O. Sydney, N.S.W. No. 1+, 1973+. Publ. Woodland Ecol. Unit C.S.I.R.O. HI 70305

Publicaţiunile asociatiei excursioniştilor români. Sectiunea ştiinţific. Bucharest. Publ. Asoc. Excurs. Români, Sect. Şti. See B–P–H 743/24. HI 58367

Publicaţiunile institutului botanic din Bucureşti. Bucharest. Publ. Inst. Bot. Bucureşti. See B–P–H 745/25. HI 58385

Publicaţiunile societăţii naturaliştilor din România. Bucharest. Publ. Soc. Nat. România. See B–P–H 747/21. HI 58412

Publicazioni idrologische. Budapest = Hidrológiai közlöny. Budapest. Hidrol. Közl. See B–P–H 416/4.

Publikace geobotanicke unie karpatske (sekce ceskoslovenska). Prague. Nos. 1-?, 1931-? Publ. Geobot. Karpat. (Sekce Ceskoslov.) HI 70306

Publikationer, statens naturvårdsverk. Stockholm. Nos. 1-?, 1969-76. Publ. Statens Naturvårdsverk. Superseded by: Meddelande, statens naturvårdsverk. HI 70307

PUDOC bulletin. [Centre for agricultural publications and documentation.] Wageningen. PUDOC Bull. See B–P–H 749/7. HI 58421

Puerto Rico agricultural experiment station. Annual report. Mayagüez, Puerto Rico. Puerto Rico Agric. Exp. Sta. Annual Rep. See B–P–H 749/10. HI 58422

Puerto Rico agricultural experiment station. Report. Mayagüez, Puerto Rico. Puerto Rico Agric. Exp. Sta. Rep. See B–P–H 749/13. HI 58423

Puerto Rico agricultural experiment station. Report on agricultural investigation in Puerto Rico. Mayagüez, Puerto Rico. Puerto Rico Agric. Exp. Sta. Rep. Agric. Invest. Puerto Rico. See B–P–H 749/14. HI 58424

Puerto Rico experiment station. United States department of agriculture. Circular. Mayagüez, Puerto Rico. Puerto Rico Agric. Exp. Sta. U.S.D.A. Circ. See B–P–H 749/16. HI 58425

Puerto Rico journal of public health and tropical medicine. New York, NY. Puerto Rico J. Public Health Trop. Med. See B–P–H 749/20. HI 58426

Puget Sound marine station publications. Seattle WA. Puget Sound Mar. Sta. Publ. See B–P–H 749/23. HI 58427

Punjab forest leaflets. Lahore. 1932-35. Punjab Forest Leafl. HI 70308

Punjab forest records. Lahore. 1936-52. Punjab Forest Rec. HI 70309

Punjab forestry notes. Lahore. 1936+. Punjab. Forest. Notes. HI 70310

Punjab fruit journal. Lyallpur, Pakistan. Punjab Fruit J. See B–P–H 749/24. HI 58428

Punjab horticultural journal. Patiala, India. Punjab Hort. J. See B–P–H 749/25. HI 58429

Punjab university botanical publication. Lahore?, Pakistan. Punjab Univ. Bot. Publ. See B–P–H 749/26. HI 58430

Punjab vegetable grower. Ludhiana. Vol. ?-11+, ?-1976/78+. Punjab Veg. Grower. HI 70311

Punjabrao krishi vidyapeeth research journal = P K V research journal. Akola.

Pure and applied chemistry. London. Pure Appl. Chem. See B–P–H 750/1. HI 58431

Pusan susan taehak yongu poko = Bulletin of Pusan fisheries college. Pusan, Korea. Bull. Pusan Fish. Coll. See B–P–H 269/23.

Pustyni S S S R i ikh osvoenie. Moscow & Leningrad. Pustyni S.S.S.R. Ikh Osvoenie. See B–P–H 750/4. HI 58432

Puutarha. Helsinki. Puutarha. See B–P–H 750/5. HI 58433

Pyramide. Innsbruck. Pyramide. See B–P–H 750/6. HI 58434

Pyrethrum post. London. Pyrethrum Post. See B–P–H 750/8. HI 58435

Pyrularia; journal of the botanical society of Westmoreland County, PA. Greensburg, PA. No. 1+, 1951+. Pyrularia. HI 70312

Quaderni della civica stazione idrobiologica di Milano. Milan. 1970+. Quad. Civica Staz. Idrobiol. Milano. HI 70313

Quaderni del frutticoltore. Ferrara, Italy. Quad. Fruttic. See B–P–H 750/27. HI 58436

Quaderni del giovane naturalista. Bra. No. ?-2+, ?-1978+. Quad. Giova. Naturalista. HI 70314

Quaderni; laboratorio crittogamico, istituto botanico della università di Pavia. Pavia. Vol. 1+, 1955+. Quad. Lab. Crittog. Ist. Bot. Univ. Pavia. HI 58439

Quaderni del museo di storia naturale di Livorno. Livorno. Vol. 1+, 1980+. Quad. Mus. Storia Nat. Livorno. HI 70315

Quaderni di sperimentazione. Lodi. No. 1+, 1970+. Quad. Sperim. HI 70316

Quaderni della sperimentazione frutticola italiana. No. ?-15+, ?-1964+. Quad. Sperim. Frutt. Ital. HI 70317

Quaderni di storia & critica della scienza. Pisa. Quad. Storia Crit. Sci. See B–P–H 750/28. HI 58437

Quaderni di storia naturale = Quaderni del museo di storia naturale di Livorno. Livorno.

Quaderni di storia della scienza. Rome. Quad. Storia Sci. See B–P–H 750/29. HI 58438

Quaderni di storia della scienze e della medicina. Ferrara. Vol. 1+, 1963+. Quad. Storia Sci. Med. HI 70318

Quaderno, see Quaderni.

Quaestiones geobiologicae. Bratislava. No. 1+, 1965+. Quaest. Geobiol. HI 70319

Qualitas plantarum et materiae vegetabiles. The Hague. Vols. 3/4-22, 1958-73. Qual. Pl. Mater. Veg. Preceded by: Materiae vegetabiles. Superseded by: Qualitas plantarum. Plant foods for human nutrition. HI 70320

Qualitas plantarum. Plant foods for human nutrition. The Hague. Vols. 23-37(2), 1973-87. Qual. Pl. Pl. Foods Human Nutr. Preceded by: Qualitas plantarum et materiae vegetabiles and Plant foods for human nutrition. Superseded by: Plant foods for human nutrition. HI 70321

Quarterly of the american primrose society. Portland, OR, Chenalis, WA. Vols. 1-35(1), 1943-77. Quart. Amer. Primrose Soc. Superseded by: Primroses. 1-307-1. HI 70322

Quarterly, Australian lilium society = Australian lilium society quarterly. Glenhuntly, Australia. Austral. Lilium soc. Quart. See B–P–H 164/10.

Quarterly bulletin of the Alpine garden society of Great Britain = Bulletin of the Alpine garden society of Great Britain. London. Bull. Alpine Gard. Soc. Gr. Brit. See B–P–H 238/1.

Quarterly bulletin of the american nature association. Washington, DC. Quart. Bull. Amer. Nat. Assoc. See B–P–H 751/1. HI 58440

Quarterly bulletin of the american rhododendron society. Portland, Tigard, OR. Vols. 2(2)-26, 1948-81. Quart. Bull. Amer. Rhododendron Soc. Preceded by: Bulletin of the american rhododendron society. Superseded by: Journal of the american rhododendron society. 1-314-2. HI 70323

Quarterly bulletin of chinese bibliography. [T'u shu chi k'an.] Shanghai. Quart. Bull. Chin. Bibliogr. See B–P–H 751/4. HI 58441

Quarterly, Colorado school of mines = Colorado school of mines quarterly. Golden, CO.

Quarterly bulletin of the faculty of science, Tehran university. Teheran. 1977+. Quart. Bull. Fac. Sci. Tehran Univ. HI 70324

Quarterly bulletin of the Florida state plant board. Gainesville, FL. Vols. 1-11(3), 1916-27. Quart. Bull. Florida State Pl. Board. Superseded by: Monthly bulletin, Florida state plant board. HI 70325

Quarterly bulletin of IAALD. [International association of agricultural librarians & documentalists.] The Hague. Quart. Bull. IAALD. See B–P–H 751/6. HI 58442

Quarterly bulletin of the international association of agricultural librarians and documentalists. Harpenden, etc. Vol. 1+, 1956+. Quart. Bull. Int. Assoc. Agaric. Librar. Documentalists. HI 70326

Quarterly bulletin, Michigan agricultural experiment station = Michigan agricultural experiment station. Quarterly bulletin. East Lansing, MI. Michigan Agric. Exp. Sta. Quart. Bull. See B–P–H 592/15.

Quarterly bulletin of the national chrysanthemum society. London. Quart. Bull. Natl. Chrysanthemum Soc. See B–P–H 751/8. HI 58443

Quarterly bulletin of the north american gladiolus council = Bulletin, north american gladiolus council. Bel Air, MD, etc.

Quarterly bulletin, north american lily society. Kennett

Square, PA, Geneva, NY, Waukesha, WI. Vol. ?-5+, ?-1951+. Quart. Bull. N. Amer. Lily Soc. HI 70327

Quarterly bulletin of science education. [K'o hsüeh chiao hsüeh chi k'an.] [China]. Quart. Bull. Sci. Educ. See B–P–H 751/9. HI 58444

Quarterly bulletin, south Pacific commission. Nouméa, Sydney, N.S.W. Vols. 1-9, 1951-59. Quart. Bull. S. Pacific Commiss. Superseded by: South Pacific bulletin. HI 70328

Quarterly bulletin, university of Hawaii = University of Hawaii, quarterly bulletin. Honolulu, HI. Univ. Hawaii Quart. Bull. See B–P–H 935/21.

Quarterly of the California historical society. San Francisco, CA. Vol. 1+, 1922+. Quart. Calif. Hist. Soc. 2-886-1. HI 70329

Quarterly, canadian gladiolus society. Guelph. 1934-37. Quart. Canad. Glad. Soc. Preceded by: Canadian gladiolus annual. Superseded by: Annual, canadian gladiolus society. 2-908-1. HI 70330

Quarterly of the Charleston museum. Charleston, SC. Quart. Charleston Mus. See B–P–H 751/11. HI 58445

Quarterly circular, Ceylon rubber research scheme. Peradeniya. 1924-50. Quart. Circ. Ceylon Rubber Res. Scheme. Superseded by: Quarterly circular, rubber. HI 70331

Quarterly circular, rubber research institute of Ceylon. Colombo. Vols. 27-34, 1951-58. Quart. Circ. Rubber Res. Inst. Ceylon. Preceded by: Quarterly circular, Ceylon rubber research scheme. Superseded by: Quarterly journal, rubber research institute of Ceylon. HI 70332

Quarterly, college of agriculture and home economics, university of Nebraska. Lincoln, NE. Vols. 11(3)-20(1), 1964-73. Quart. Coll. Agric. Home Econ. Univ. Nebraska. Preceded by: Nebraska experiment station quarterly. Superseded by: Farm, ranch and home quarterly. HI 70333

Quarterly information bulletin concerning the protection of nature = Kwartalny biuletyn informacyjny o ochronie przyrody. Cracow. Kwart. Biul. Inform. Ochr. Przyr. See B–P–H 522/15.

Quarterly journal of agriculture. Edinburgh & London. Quart. J. Agric. See B–P–H 751/13. HI 58446

Quarterly journal of the bank of Taiwan. [Tai wan yen hang chi k'an.] Taipei, Taiwan. Quart. J. Bank Taiwan. See B–P–H 751/14. HI 58447

Quarterly journal of the Calcutta medical and physical society. Calcutta. Quart. J. Calcutta Med. Soc. See B–P–H 751/15. HI 58448

Quarterly journal of chinese forestry. [Zhonghua a linxue iikan.] Taipei. Vol. 1+, 1967+. Quart. J. Chin. Forest. Preceded by: Tai-wan lin yeh chikan. HI 70334

Quarterly journal of crude drug research. Amsterdam. Vols. 1-19, 1961-81. Quart. J. Crude Drug Res. Superseded by: International journal of crude drug research. HI 70335

Quarterly journal, Devon trust for nature conservation. Exeter. Vols. ?-8, 1970-77. Quart J. Devon Trust Nat. Conservation. Preceded and superseded by: Journal of the Devon trust for nature conservation. HI 70336

Quarterly journal of the Florida academy of sciences. Gainesville, FL. Vols. 8-35, 1945-72. Quart. J. Florida Acad. Sci. Preceded by: Proceedings of the Florida academy of sciences. Superseded by: Florida scientist. 2-1578-2. HI 70337

Quarterly journal of forestry. London. Vol. 1+, 1907+. Quart. J. Forest. Preceded by: Transactions of the royal english arboricultural society. 4-3499-2. HI 70338

Quarterly journal, history of science and technology = Kwartalnik historii nauki i techniki. Warsaw.

Quarterly journal of the Illinois state agricultural society. Springfield, IL. Quart. J. Illinois State Agric. Soc. See B–P–H 751/20. HI 58449

Quarterly journal of the indian institute of science. Bangalore, India. Quart. J. Indian Inst. Sci. See B–P–H 751/21. HI 58450

Quarterly journal of the indian tea association; scientific department. Calcutta. Quart. J. Indian Tea Assoc., Sci. Dept. See B–P–H 751/22. HI 58451

Quarterly journal, institute of commercial research in the tropics. Liverpool. Vols. 1-3, 1906-08. Quart. J. Inst. Commercial Res. Tropics. HI 70339

Quarterly journal of international agriculture. Frankfurt am Main. Vol. 19+, 1980+. Quart. J. Int. Agric. Preceded by: Zeitschrift für ausländische Landwirtschaft. HI 70340

Quarterly journal of literature, science and the arts. London. Quart. J. Lit. Sci. Arts. See B–P–H 752/1. HI 58452

Quarterly journal of microscopical science. London. Vols. 1-8, 1852-60; n.s. vols. 1-106, 1861-1965. Quart. J. Microscop. Sci. Superseded by: Journal of cell science. 4-3500-1. HI 70341

Quarterly journal of the microscopical society of Victoria. Melbourne, Vic. Quart J. Microscop. Soc. Victoria. See B–P–H 752/4. HI 58453

Quarterly journal of pharmacy and allied sciences. London. Quart. J. Pharm. Allied Sci. See B–P–H 752/6. HI 58454

Quarterly journal of pharmacy and pharmacology. London. Vols. 1-21, 1928-48. Quart. J. Pharm.

Pharmacol. Preceded by: Yearbook of pharmacy. Superseded by: Journal of pharmacy and pharmacology. 5-4572-1. HI 70342

Quarterly journal, rubber research institute of Ceylon. Agalawatta. Vols. 35-49, 1959-72. Quart. J. Rubber Res. Inst. Ceylon. Preceded by: Quarterly circular, rubber research institute of Ceylon. Superseded by: Quarterly journal, rubber research institute of Sri Lanka. HI 70343

Quarterly journal, rubber research institute of Malaya. Kuala Lumpur. Vols. 1-2, 1929-31. Quart. J. Rubber Res. Inst. Malaya. Superseded by: Journal of the rubber research institute of Malaya. 3-2321-2. HI 70344

Quarterly journal, rubber research institute of Sri Lanka. Dartonfield, Agalawatta. Vols. 50-52, 1973-75. Quart. J. Rubber Res. Inst. Sri Lanka. Preceded by: Quarterly journal, rubber research institute of Ceylon. Superseded by: Journal, rubber research institute of Sri Lanka. HI 70345

Quarterly journal of science and the arts. London. Quart. J. Sci. Arts. See B–P–H 752/8. HI 58455

Quarterly journal of science, literature, and the arts. London. Quart. J. Sci. Lit. Arts. See B–P–H 752/9. HI 58456

Quarterly journal of science, Wu-han university. [Kuo li Wu-han ta hsüeh li k'o chi k'an.] Wuchang, China. Quart. J. Sci. Wu-han Univ. See B–P–H 752/10. HI 58457

Quarterly journal of the Taiwan museum. [Tai wan sheng li po wu kuan shi k'an.] Taipei. Vols. 1-35, 1948-82. Quart. J. Taiwan Mus. Superseded by: Journal of the Taiwan museum. 5-4150-2. HI 70346

Quarterly list of introductions, plant introduction section, C S I R O. Canberra, A.C.T. Nos. 1-76, 1945-64. Quart. List Introd. Pl. Introd. Sect. C. S. I. R. O. Superseded by: Plant introduction review, division of plant industry, C S I R O. HI 70347

Quarterly, Los Angeles county museum of natural history. Los Angeles, CA. Vols. 1-18(2), 1941-62. Quart. Los Angeles County Mus. Nat. Hist. Superseded by: Quarterly, Los Angeles county museum of natural history. Science and history. 3-2473-1. HI 70348

Quarterly, Los Angeles county museum of natural history. Science and history. Los Angeles, CA. Vols. 1-9(1), 1962-70. Quart. Los Angeles County Mus. Nat. Hist., Sci. Hist. Preceded by: Quarterly, Los Angeles county museum of natural history. Superseded by: Terra. HI 70349

Quarterly, Nebraska agricultural experiment station = Nebraska experiment station quarterly. Lincoln, NE.

Quarterly news journal of the Crosby arboretum. Hattiesburg, MS. 1983+. Quart. News J. Crosby Arbor. HI 75261

Quarterly newsletter, aloe, cactus and succulent society of Rhodesia. Salisbury, Rhodesia. Nos 1-44?, 1969-80? Quart. Newslett. Aloe Cact. Succ. Soc. Rhodesia. Superseded by: Quarterly newsletter, aloe, cactus and succulent society of Zimbabwe. HI 70350

Quarterly newsletter, aloe, cactus and succulent society of Zimbabwe. Salisbury, Zimbabwe. No. ?-47+, ?-1981+. Quart. Newslett. Aloe Cact. Succ. Soc. Zimbabwe. Preceded by: Quarterly newsletter, aloe, cactus and succulent society of Rhodesia. HI 70351

Quarterly newsletter, american association of botanical gardens and arboreta. Place varies. Nos. 36-68, 1958-66. Quart. Newslett. Amer. Assoc. Bot. Gard. Arbor. Preceded by: News letter, american association of botanical gardens and arboreta. Superseded by: Arboretum and botanical garden bulletin. HI 70352

Quarterly newsletter, american herb association. Rescue, CA. Vol. 1+, 1982+. Quart Newslett. Amer. Herb Assoc. HI 70353

Quarterly newsletter, american type culture collection. Rockville, MD. 1981+. Quart. Newslett. Amer. Type Cult. Collect. HI 70354

Quarterly newsletter, Asia and Pacific plant protection commission. Bangkok. Vol. 26+, 1983+ Quart. Newslett. Asia Pacific Pl. Protect. Commiss. Preceded by: Quarterly newsletter, plant protection committee for the south east Asia and Pacific region. HI 70355

Quarterly newsletter, California rare fruit growers. Fullerton, CA. Vols. 1-2, 1969-70. Quart. Newslett. Calif. Rare Fruit Growers. Superseded by: Newsletter, California rare fruit growers. HI 70356

Quarterly newsletter, elm research institute. Harrisville, NH. 1980+. Quart. Newslett. Elm Res. Inst. HI 70357

Quarterly newsletter, European aquaculture society. Bredene. No. ?-35/36+, ?-1985+. Quart. Newslett. Eur. Aquac. Soc. HI 70358

Quarterly newsletter, forest research institute and colleges. Dehra Dun. Vols. 1-8(4), 1964-71. Quart. Newslett. Forest Res. Inst. Coll. Superseded by: Newsletter of the forest research institute and college. Dehra Dun. HI 70359

Quarterly newsletter, garden history society. Hemel Hempstead, Berkhamsted. Nos. 1-12, 1966-70. Quart. Newslett. Gard. Hist. Soc. Superseded by: Newsletter, garden history society. HI 70360

Quarterly newsletter, international plant propogators' society. West Lafayette, IN. ?-1982+. Quart. Newslett. Int. Pl. Propogators' Soc. HI 70361

Quarterly newsletter, perennial plant association.

Kensington, CT. 1985+. Quart. Newslett. Perenn. Pl. Assoc. HI 68815

Quarterly newsletter, plant protection committee for the south east Asia and Pacific region. Bangkok. Vols. 8-25, 1965-82? Quart. Newslett. Pl. Protect. Committee S. E. Asia Pacific Region. Preceded by: Quarterly reports of the plant protection committee for the south east Asia and Pacific region. Superseded by: Quarterly newsletter, Asia and Pacific plant protection commission. HI 70362

Quarterly newsletter, Rye nature center. Rye, NY. ?-1984+. Quart. Newslett. Rye Nat. Center. HI 70363

Quarterly newsletter, tea research foundation of central Africa. Mulanje, Malawi. No. ?-14+, ?-1969+. Quart. Newslett. Tea Res. Found. Centr. Africa. HI 70364

Quarterly progress report, west african cacao (cocoa) research institute. Tafo. Nos. ?-34-63-?, 1944 -54-61-? Quart. Progr. Rep. W. African Cacao (Cocoa) Res. Inst. HI 70365

Quarterly progress report, west african institute for oil palm research. Benin City. No. 1, 1952. Quart. Progr. Rep. W. African Inst. Oil Palm Res. Superseded by: Quarterly report, nigerian institute for oil palm research. HI 70366

Quarterly record of the royal botanical society of London. London. Quart. Rec. Roy. Bot. Soc. London. See B–P–H 752/35. HI 58458

Quarterly report, Caribbean plant protection commission. Port of Spain. Vol. 1+, 1968+. Quart. Rep. Caribbean Pl. Protect. Commiss. HI 70367

Quarterly report, nigerian institute for oil palm research. Benin. Nos. ?-59?, ?-1966? Quart. Rep. Nigerian Inst. Oil Palm Res. Preceded by: Quarterly progress report, west african institute for oil palm research. HI 70368

Quarterly report of the plant protection committee for the south east Asia and Pacific region. Bangkok. Vols. 1-7, 19??-64? Quart. Rep. Pl. Protect. Committee S. E. Asia Pacific Region. Superseded by: Quarterly newsletter, plant protection committee for the south east Asia and Pacific region. HI 70369

Quarterly report, west african cacao (cocoa) research institute. Tafo. 1944+. Quart. Rep. W. African Cacao (Cocoa) Res. Inst. HI 70370

Quarterly report, scientific department, indian tea association. 1949-53. Quart. Rep. Sci. Dep. Indian Tea Assoc. HI 70371

Quarterly review, agricultural development and advisory service = A D A S quarterly review. London.

Quarterly review of biology. Baltimore, MD. Quart. Rev. Biol. See B–P–H 753/1. HI 58459

Quarterly review of biophysics. Cambridge. Vol. 1+, 1968+. Quart. Rev. Biophys. HI 70372

Quarterly review of the clove growers association. Zanzibar [Tanzania Republic]. Quart. Rev. Clove Growers Assoc. See B–P–H 753/2. HI 58460

Quarterly review, department of agriculture, Cyprus. Nicosia. 1945-55. Quart. Rev. Dep. Agric. Cyprus. Superseded by: Six-monthly review, department of agriculture, Cyprus. HI 70373

Quarterly of the San Bernardino county museum association. Redlands, Bloomington, CA. 1952+. Quart. Rev. San Bernardino County Mus. Assoc. HI 70374

Quarterly review of scientific publications of the polish academy of sciences, the Ossolineum and the polish scientific publishers. Series B. Biological sciences. Warsaw. Quart. Rev. Sci. Publ. Polish Acad. Sci., Ser. B, Biol. Sci. See B–P–H 753/3. HI 58461

Quarterly science bulletin of the national central university. [Kuo li chung yang ta hsüeh k'o hsüeh chi k'an.] Nanking. Quart. Sci. Bull. Natl. Centr. Univ. See B–P–H 753/5. HI 58462

Quarterly, society for general microbiology. Reading. Vol. 6+, 1978+. Quart. Soc. Gen. Microbiol. Preceded by: Proceedings, society for general microbiology. HI 70375

Quarterly summary and meteorological readings of the royal botanic society of London. London. Quart. Summary Meteorol. Readings Roy. Bot. Soc. London. See B–P–H 753/7. HI 58463

Quarterly supplement of the board of trade journal. London. Quart. Suppl. Board Trade J. See B–P–H 753/8. HI 58464

Quarterly transactions of the Barnsley naturalists' society. Barnsley. 1881-84. Quart. Trans. Barnsley Naturalists' Soc. Superseded by: Transactions of the Barnsley naturalists' Society. HI 70376

Quarterly of the Worcester natural history society. Worcester, MA. Vol. 1+, 1940+. Quart. Worcester Nat. Hist. Soc. Preceded by: Nature outlook. 5-4540-1. HI 70377

Quaternary international; journal of the international union for quaternary research. Oxford, New York. Vol. 1+, 1989+. Quatern. Int. HI 70378

Quaternary research; interdisciplinary journal. New York, London. Vol. 1+, 1970+. Quatern. Res. HI 70379

Quaternary science reviews; international review and research journal. Oxford, New York. Vol. 1+, 1982+. Quatern. Sci. Rev. HI 70380

Quaternary studies in Poland. Warsaw, Poznan. Vol. 1+, 1979+. Quatern. Stud. Poland. HI 70381

Quatember. Zeitschrift für naturwissenschaftliche, geschichtliche, philologische, literarische und gemischte Gegenstände. Mitau [=Jelgava, Latvian S S R]. Quatember. See B–P–H 753/11. HI 58465

Quatre-temps; bulletin de la société d'animation du jardin et de l'institut botaniques de Montréal. Montreal. Vol. 11+, 1987+. Quatre-Temps. Preceded by: Bulletin de la société d'animation du jardin et de l'institut botaniques. HI 70382

Québec horticole. Montreal. Quebec Hort. See B–P–H 753/12. HI 58466

Québec vert. Quebec. Vol. 1+, 1978+. Québec Vert. Preceded by: Fleuriste du Québec. HI 70383

Quebec science. Quebec. Vol. 8(3)+, 1970+. Quebec Sci. Preceded by: Jeune scientifique. HI 70384

Queens botanical garden news. Flushing, NY. 1980+. Queens Bot. Gard. News. HI 70385

Queensland agricultural journal. Brisbane, Qld. Queensland Agric. J. See B–P–H 753/21. HI 58467

Queensland botany bulletin. Brisbane, Qld. No. 1+, 1982+. Queensland Bot. Bull. Preceded by: Technical bulletin, botany branch, Queensland department of primary industries. HI 70386

Queensland department of agriculture and stock. Division of entomology and plant pathology. Bulletin. Brisbane, Qld. Queensland Dept. Agric. Div. Entomol. Bull. See B–P–H 753/23. HI 58468

Queensland department of agriculture and stock. Division of plant industury. Bulletin. Brisbane, Qld. Queensland Dept. Agric. Div. Pl. Industr. Bull. See B–P–H 753/24. HI 58469

Queensland government mining journal. Brisbane, Qld. Queensland Gov. Mining J. See B–P–H 753/25. HI 58470

Queensland journal of agricultural science. Brisbane, Qld. Vols. 1-21, 1944-64. Queensland J. Agric. Sci. Superseded by: Queensland journal of agricultural and animal sciences. 4-3506-3. HI 70387

Queensland journal of agricultural and animal sciences. Brisbane, Qld. Vol. 22+, 1965+. Queensland J. Agric. Anim. Sci. Preceded by: Queensland journal of agricultural science. HI 70388

Queensland naturalist. Brisbane, Qld. Queensland Naturalist. See B–P–H 754/4. HI 58471

Quellen und Studien zur Geschichte der Naturwissenschaften und der Medizin. Berlin. Quellen Stud. Gesch. Naturwiss. Med. See B–P–H 754/5. HI 58472

Quinzaine coloniale; organ de l'union coloniale française. Paris. Vols. 1(1)-18(450), 1897-14; vols. 32(511)-41(749), 1928-37 [publication suspended 1914-27]. Quinzaine Colon. For 1923-27 see: Bulletin de l'union coloniale française. Superseded by: Revue française d'outre mer. 4-3653-1. HI 70389

Quipu; revista latinoamericana de historia de las ciencias y la tecnologia. Mexico City. Vol. 1+, 1984+. Quipu (Mexico City). HI 70391

Quipu; museum of science and natural history quarterly journal [sub-title varies]. Tampa, FL. Vol. 1+, 1962+. Quipu (Tampa). HI 58473

R F C newsletter and yearbook = Newsletter, rare fruit council international. Miami, FL.

R H S garden club journal. Wisley. ?-1977+. R. H. S. Gard. Club J. HI 70392

R I C bulletin. Kepong. Vol. 1+, 1982+. R. I. C. Bull. HI 70393

R R I planters' bulletin = Planters' bulletin, rubber research institute of Malaysia. Kuala Lumpur.

R R I C bulletin. Agalawatta. N.s. vols. 1-7, 1966-72. R. R. I. C. Bull. Preceded by: Bulletin of the rubber research institute of Ceylon. Superseded by: R R I S L bulletin. HI 70394

R R I S L bulletin. Agalawatta. Vol. 8+, 1973+. R. R. I. S. L. Bull. Preceded by: R R I C bulletin. HI 70395

R S F newsletter. Federal Way, WA. 1977+. R. S. F. Newslett. HI 70396

R S R I abstracts. Philadelphia, PA. Vol. 1+, 1976+. R. S. R.I. Abstr. HI 70397

Raboty Kamskoi biologicheskoi stantsii. Perm, Russian S F S R. Vols. 1-2, 1936. Raboty Kamsk. Biol. Stantsii. HI 70398

Raboty Kamskoj biologičeskoj stancii = Raboty Kamskoi biologicheskoi stantsii. Perm.

Raboty murmanskoi biologicheskoi stantsii. Murmansk. 1925-29. Raboty Murmansk. Biol. Stantsii. Superseded by: Trudy murmanskoi biologicheskoi stantsii. HI 70399

Raboty Okskoi biologicheskoi stantsii v Gorode Murome. Murom, Russian S F S R. Vol. 1, 1921; vol. 2(2/3), 1922/23; vols. 3(2/3)-5, 1924-28. Raboty Oksk. Biol. Stantsii Gorode Murome. For vols. 2(1), 3(1) and 6 see: Raboty Okskoi biologicheskoi stantsii v Nizhnem Novogorode. 3-2789-3. HI 70400

Raboty Okskoi biologicheskoi stantsii v Nizhnem Novogorode. Murom, Russian S F S R. Vol. 2(1), 1922/23; vol. 3(1), 1924/25; vol. 6, 1931. Raboty Oksk. Biol. Stantsii Nizhnem Novgorode. For vols. 1, 2(2/3) and 3(2/3)-5 see: Raboty Okskoi biologicheskoi stantsii v Gorode Murome. 3-2789-3. HI 70401

Raboty Okskoj biologičeskoj stancii v Gorode Murome = Raboty Okskoi biologicheskoi stantsii v Gorode Murome. Murom.

Raboty Okskoj biologičeskoj stancii v Nižnem Novogorode = Raboty Okskoi biologicheskoi stantsii v

Nizhnem Novogorode. Murom.

Raboty Volzhsko-Kamskoi kraevoi promyslovoi biologicheskoi stantsii. Kazan, Russian S F S R. Vol. 1, 1931. Raboty Volzhsko-Kamsk. Kraev. Promysl. Biol. Stantsii. Superseded by: Raboty Volzhsko-Kamskoi zonal'noi okhotnich'e-promyslovoi biologicheskoi stantsii. HI 70403

Raboty Volzhsko-Kamskoi zonal'noi okhotnich'e-promyslovoi biologicheskoi stantsii. Kazan, Russian S F S R. Vols. 2-[5], 1932-36. Raboty Volzhsko-Kamsk. Zonal'n. Okhotn.-Promysl. Biol. Stantsii. Preceded by: Raboty Volzhsko-Kamskoi kraevoi promyslovoi biologicheskoi stantsii. HI 70404

Raboty Volzhskoi Biologicheskoi Stantsii. Saratov, Russian S F S R. Vol. 1, 1901; vols. 3-11, 1905-30. Raboty Volzhsk. Biol. Stantsii. Vol. 1, 1901 and 3(1), 1905 published as supplement in: Trudy Saratovskago Obshchestva Estestvoispytatelei i Lyubitelei Estestvozaniya. For vol. 2(1) see: Ezhegodnik Volzhskoi Biologicheskoi Stantsii Saratovskago Obshchestva Estestvoispytatelei i Lyubitelei Estestvoznaniya. For vol. 2(2) see: Lětniya raboty. HI 70402

Raboty Volžsko-Kamskoj kraevoj promyslovoj biologičeskoj stancii = Raboty Volzhsko-Kamskoi kraevoi promyslovoi biologicheskoi stantsii. Kazan.

Raboty Volžsko-Kamskoj zonal'noj ohotnič'e-promyslovoj biologičeskoj stancii = Raboty Volzhsko-Kamskoi zonal'noi okhotnich'e-promyslovoi biologicheskoi stantsii. Kazan.

Raboty Volžskoj Biologičeskoj Stancii = Raboty Volzhskoi Biologicheskoi Stantsii. Saratov.

Raccolta. Genoa. Raccolta. See B–P–H 755/4. HI 58477

Raccolta fisico-chimica italiana. Venice. Racc. Fis.-Chim. Ital. See B–P–H 754/18. HI 58474

Raccolta di memorie delle pubbliche accademia di agricoltura, arti, e commercio dello stato veneto. Venice. Racc. Mem. Pubbl. Accad. Agric. Stato Veneto. See B–P–H 754/19. HI 58475

Raccolta d'opuscoli scientifici e filologici. Venice. Racc. Opusc. Sci. Filol. See B–P–H 755/3. HI 58476

Rad fitopatoloskog zavoda u Sarajevu. Sarajevo. 1928-33-? Rad Fitopatol. Zavoda Sarajevu. Preceded by: Izvještaj o radu državnoga fitopatološka zavoda. HI 70405

Rad. Jugoslavenska akademije znanosti i umjetnost. Zagreb, Croatia [Yugoslavia]. Rad Jugoslav. Akad. Znan. See B–P–H 755/6. HI 58478

Rádcova práce v zahradě, na poli i na dvoře. Predmosti, Czechoslovakia. Rádcova Práce Zahradě. See B–P–H 755/11. HI 58479

Radiation botany; international journal devoted to plant radio-biology and closely related fields. London & New York. Vols. 1-15, 1961-75. Radiat. Bot. Superseded by: Environmental and experimental botany. HI 70406

Radiation and environmental biophysics. Berlin, New York, etc. Vol. 11+, 1974+. Radiat. Environm. Biophys. Preceded by: Biophysik. HI 70407

Radiation research. Official organ of the radiation research society. New York, NY. Radiat. Res. See B–P–H 755/13. HI 58480

Radiobiologiya. Moscow. Vol. 1+, 1961+. Radiobiologiya. HI 75121

Radiocarbon. New Haven, CT. Radiocarbon. See B–P–H 755/15. HI 58481

Radiochuvstvitel'nost' i mutabil'nost' rastenii. Erevan. Vol. ?-2+, ?-1974+. Radiochuvstv. Mutabil'. Rast. Preceded by: Mutagenez rastenii. HI 70408

Radovi poljoprivredno-šumarskog fakulteta univerziteta u Sarajevu. Sarajevo. Vols. 1-6, 1952-5? Rad. Poljopr.-Šumarsk. Fak. Univ. u Sarajevu. Superseded by: Radovi poljoprivredno-šumarskog fakulteta univerziteta u Sarajevu. Radovi A, poljoprivreda and Radovi B, šumarstvo]. HI 70409

Radovi poljoprivredno-šumarskog fakulteta univerziteta u Sarajevu. Radovi A, poljoprivreda. Sarajevo. Vols. 4-7, 195?-? Rad. Poljopr.-Šumersk. Fak. Univ. Sarajevu, A. Preceded by: Radovi poljoprivredno-šumarskog fakulteta univerziteta u Sarajevu. HI 70410

Radovi poljoprivredno-šumarskog fakulteta univerziteta u Sarajevu. Radovi B, šumarstvo. Sarajevo. Vols. 1-3, 1956-58. Rad. Poljopr.-Šumersk. Fak. Univ. Sarajevu, B. Preceded by: Radovi poljoprivredno-šumarskog fakulteta univerziteta u Sarajevu. Superseded by: Radovi šumarskog fakulteta i instituta za šumarstvo i drvnu industriju u Sarajevu. HI 70411

Radovi šumarskog fakulteta i instituta za šumarstvo i drvnu industriju u Sarajevu. Sarajevo. Vol. 4+, 1959+. Rad. Šumarsk. Fak. Inst. Šumarstvo Sarajevu. Preceded by: Radovi poljoprivredno-šumarskog fakulteta univerziteta u Sarajevu. Radovi B, šumarstvo. HI 70412

Raingan khonkwa wichai, mahawitthayalai Kasetsat = Research report, Kasetsart university. Bangkok.

Rajasthan journal of agricultural sciences. Jaipur. Vol. 1+, 1970+. Rajasthan J. Agric. Sci. HI 70413

Ran = Japan orchid society bulletin. Osaka.

Rancho mexicano. Mexico City. Rancho Mex. See B–P–H 755/17. HI 58482

Rancho Santa Ana botanic garden. Monographs. Botanical series. Anaheim, CA. Rancho Santa Ana Bot. Gard. Monogr., Bot. Ser. See B–P–H 755/19.

HI 58483

Rangelands. Denver, CO. Vol. 1+, 1979+. Rangelands. HI 75147

Rapport d'activité / C N R S. Laboratoire associé no. 218, taxonomie et écologie des flores tropicales. Paris. 1977/81+, 198?+. Rapp. Activité, C. N. R. S., Lab. Assoc. 218, Tax. Écol. Fl. Trop. HI 70414

Rapport d'activité, centre de recherches agronomiques de l'état-Gand. Merelbeke. 1976+, 1977+. Rapp. Activité Centre Rech. Agron. État-Gand. HI 70415

Rapport d'activité, centre suisse de recherches scientifiques en Côte-d'Ivoire. Adiopodoumé. 1979+, [1980]+. Rapp. Activité Centre Suisse Rech. Sci. Côte-d'Ivoire. HI 70416

Rapport d'activité, jardin botanique national de Belgique. Meise. 1976+, 1977+. Rapp. Activité Jard. Bot. Natl. Belgique. HI 70417

Rapport d'activité ... du laboratoire de techniques agricoles et horticoles et de la station phytosanitaire cantonale. Geneva. 1967+, 1968+. Rapp. Activité Lab. Techn. Agric. Hort. Stat. Phytosan. Canton. HI 70418

Rapport d'activité, office national des forêts. Paris. 1972+, 1973+. Rapp. Activité Off. Natl. Forêts. HI 70419

Rapport d'activité, station d'amélioration des plantes, Dijon. Dijon. 1964/67+, 196?+. Rapp. Activité Stat. Amélior. Pl. Dijon. HI 70420

Rapport d'activité, station d'amélioration des plantes, Gembloux = Rapport d'activité, station de recherches de l'état pour l'amélioration des plantes de grande culture, Gembloux. Gembloux.

Rapport d'activité, station d'amélioration des plantes maraîchères d'Avignon-Montfavet. Montfavet. 1969/70+, 197?+. Rapp. Activité Stat. Amélior. Pl. Maraîch. Avignon-Montfavet. HI 70421

Rapport d'activité, station centrale de génétique et d'amélioration des plantes, Versailles. Versailles. ?-1952+. Rapp. Activité Stat. Cent. Génét.Amélior. Pl. Versailles. HI 70422

Rapport d'activité, station fédéral de recherches agronomiques. Berne. 1966/68+, 1970+. Rapp. Activité Stat. Féd. Rech. Agron. Preceded by: Rapport d'activité, stations fédérales d'essais agricoles, Lausanne. HI 70423

Rapport d'activité, station de génétique et d'amélioration des plantes de Versailles = Rapport d'activité, station centrale de génétique et d'amélioration des plantes, Versailles. Versailles.

Rapport d'activité de la station de recherches de l'état pour l'amélioration des plantes fruitières et maraîchères = Rapport général de la station de recherches de l'état pour l'amélioration des plantes fruitières et maraîchères. Grand Manil.

Rapport d'activité, station de recherches de l'état pour l'amélioration des plantes de grande culture, Gembloux. Gembloux. 1964?+. Rapp. Activité Stat. Rech. État Amélior. Pl. Grande Cult. Gembloux. HI 70425

Rapport d'activité, stations fédérales d'essais agricoles, Lausanne. Berne. 1951-63/65, 1951?-66. Rapp. Activité Stat. Féd. Essais Agric. Lausanne. Preceded by: Rapport annuel, stations fédérale d'essais viticoles et arboricoles à Lausanne. Superseded by: Rapport d'activité, station fédérale de recherches agronomiques. HI 70424

Rapport annuaire et liste des membres; union internationale des institutes de recherches forestières = Jahresbericht und Mitgliederverzeichnis; internationaler Verband forstlicher Lehranstalten. Stockholm & Zurich.

Rapport annuel, association forêt-cellulose. Paris. 1972-75, 1973-76. Rapp. Annuel Assoc. Forêt-Cellulose. Superseded by: Annales de recherches sylvicoles. HI 70426

Rapport annuel, centre de recherches agronomiques de Boukoko. Boukoko. 1964/67+, 1967?+. Rapp. Annuel Centre Rech. Agron. Boukoko. HI 70427

Rapport annuel de la commission fédérale du parc national = Jahresbericht der Eidgenössischen Nationalparkkommission. Grosshöchstetten.

Rapport annuel, conseil consultatif des reserves écologiques. Quebec. Vol. 1+, 1975/76+, 1976+. Rapp. Annual Cons. Consultatif Reserves Écol. HI 70428

Rapport annuel, environment, Canada = Report (Annual), environment, Canada. Ottawa.

Rapport annuel d'experimentation agricole. Casablanca & Rabat. 1922/23-38/39, 1923-39. Rapp. Annuel Exp. Agric. HI 70429

Rapport annuel sur le fonctionnement du service de botanique (phanérogamie) du muséum d'histoire naturelle de Paris = Rapport annuel du service de botanique (phanérogamie) du muséum d'histoire naturelle de Paris.

Rapport annuel, institut français du café et du cacao. Bingerville. Vol. 1+, 1965/67+, 1967?+. Rapp. Annuel Inst. Franç. Café Cacao. HI 70430

Rapport annuel, institut français de recherches fruitières outre-mer. Paris. 1948+ [suspended 1950-56]. Rapp. Annuel Inst. Franç. Rech. Fruitières Outre-Mer. HI 70431

Rapport annuel de l'institut géologique de Hongrie. Budapest = A magyar állami földtani intézet évi jelentese(i). Budapest. Magyar Állami Földt. Intéz.

Évi Jel. See B–P–H 542/18.

Rapport annuel, institut national pour l'étude agronomique du Congo Belge. Brussels. 1934-59? Rapp. Annuel Inst. Natl. Étud Agron. Congo Belge. HI 70432

Rapport annuel, institut national de la recherche agronomique. Paris. 1949-52 Rapp. Annuel Inst. Natl. Rech. Agron. HI 70433

Rapport annuel, institut des recherches sur le caoutchouc au Cambodge. Vol. 1+, 1958/63+, 1963?+. Rapp. Annuel Inst. Rech. Caoutchouc Cambodge. HI 70434

Rapport annuel, institut pour la recherche scientifique en Afrique centrale. Brussels, Bukavu. Vol. 1+, 1948+. Rapp. Annuel Inst. Rech. Sci. Afrique Centrale. HI 70435

Rapport annuel, institut de recherche agronomiques tropicales et des cultures vivières. Saint-Denis, Réunion. 1946+. Rapp. Annuel Inst. Rech. Agron. Trop. Cult. Vivières. HI 70436

Rapport annuel, jardin botanique, Montréal = Report (Annual), Montreal botanic garden. Montreal.

Rapport annuel, ministère de l'agriculture et de la colonisation, Québec = Rapport, ministère de l'agriculture de la province de Québec. Quebec.

Rapport annuel, muséum national d'histoire naturelle de Paris. Paris. Vol. 1-4-?, 1909-13-? Rapp. Annuel Mus. Natl. Hist. Nat. Paris. HI 70437

Rapport annuel, organisation européenne et mediterranénne pour la protection des plantes = Report (Annual), european and mediterranean plant protection organisation. Paris.

Rapport annuel - S O U Q A R = Rapport annuel, section d'océanographie, université du Québec à Rimouski. Rimouski.

Rapport annuel, section d'océanographie, université du Québec à Rimouski. Rimouski. Vols. 1-4, 1973/74-77/78, 1974-78. Rapp. Annuel Sect. Océanogr. Univ. Québec Rimouski. Superseded by: Rapport bisannuel, département d'océanographie, université du Québec à Rimouski. HI 70438

Rapport annuel du service de botanique (phanérogamie) du muséum d'histoire naturelle de Paris. Paris. Vol. 1-4, 1909-13, 1910-14. Rapp. Annuel Serv. Bot. Mus. Hist. Nat. Paris. HI 70439

Rapport annuel, service des données sur le milieu marin = Report (Annual), marine environmental data service. Ottawa.

Rapport annuel, société Provancher d'histoire naturelle du Canada = Report of the Provancher society of natural history of Canada. Quebec.

Rapport annuel, station agronomique libano-française. Tell-Amara. 1954+. Rapp. Annuel Stat. Agron. Libano-Franç. HI 70440

Rapport annuel de la station de biologie marine Grande-Rivière. [Prior to 1961 published as part of: Contributions du département des pêcheries, Québec.] Quebec. 1953-68. Rapp. Annuel Stat. Biol. Mar. Grande-Rivière. Superseded by: Rapport, service de biologie, Québec. HI 70441

Rapport annuel, stations fédérale d'essais viticoles et arboricoles à Lausanne. Berne. 1898-1950. Rapp. Annuel Stat. Féd. Essais Vitic. Arbor. Lausanne. Superseded by: Rapport d'activité, stations fédérales d'essaie agricoles. HI 70442

Rapport annuel sur les travaux de la société d'histoire naturelle de l'Ile Maurice. Port-Louis. Vols. 1-13, 1830-42, 1831-43. Rapp. Annuel Trav. Soc. Hist. Nat. Ile Maurice. Superseded by: Procès-verbaux de la société d'histoire naturelle de l'Ile Maurice 5-3972-2. HI 70443

Rapporter och avhandlingar, institutionen för växtodling, sveriges lantbruksuniversitet. Uppsala. 1978+. Rapp. Avh. Inst. Växtodling Sveriges Lantbruksuniv. Preceded by: Rapporter och avhandlingar, landbrukshögskolan institutionen för växtodling. HI 70444

Rapporter och avhandlingar, landbrukshögskolan institutionen för växtodling. Uppsala. 1973-77. Rapp. Avh. Landbrukshögskolan Inst. Växtodling. Superseded by: Rapporter och avhandlingar, institutionen för växtodling, sveriges lantbruksuniversitet. HI 70445

Rapport biennial d'experimentation, centre de recherches agronomiques = Rapport annuel d'experimentation agricole. Casablanca & Rabat.

Rapport bisannuel, département d'océanographie, université du Québec à Rimouski. Rimouski. Vol. 1+, 1978/80+, [1980]+. Rapp. Bisannuel Dép. Océanogr. Univ. Québec Rimouski. Preceded by: Rapport annuel, section d'océanographie, université du Québec à Rimouski. HI 70446

Rapport van het bosbouwproefstation. Buitenzorg. Nos. 1-72-?, 1948-55-? Rapp. Bosbouwproefstat. Superseded by: Laporan dari lembaga pusat penjelidikan kehutanan. HI 70447

Rapport, botanisk museum universiteti Bergen. Bergen. ?-1979+. Rapp. Bot. Mus. Univ Bergen. HI 70448

Rapport, byggeteknisk utvalg, norges teknisk-naturvitenskapelige forsknings-råd. Oslo. Vols. 1-?, 1950-51. Rapp. Byggetekn. Utv. Norg. Tekn.-Naturvitensk. Forsknings-Råd. Superseded by: Rapport, norges byggforskningsinstitutt. HI 70449

Rapport sur le fonctionnement du laboratoire d'agriculture et de la station d'essais de semences et d'amélioration des plantes. Algiers. ?-1935-? Rapp. Fonct. Lab. Agric. Stat. Essais Semences Amélior. Pl. HI 70450

Rapport général, centre national de recherches herbàres et fourragères. Brussels. 1947+. Rapp. Gén. Centre Natl. Rech. Herb. Fourrag. HI 70451

Rapport général de la station de recherches de l'état pour l'amélioration des plantes fruitières et maraîchères. Grand Manil. 1948-67. Rapp. Gén. Stat. Rech. État Amélior. Pl. Fruitières Marîch. Superseded by: Rapport de la station des cultures fruitères et maraîchères Gembloux. HI 70452

Rapports généraux des travaux: société philomatique de Paris. Paris. Rapp. Gén. Trav. Soc. Philom. Paris. See B–P–H 755/26. HI 58484

Rapport d'information LAU - X. Sainte-Foy. No. 1+, 1973+. Rapp. Inform. LAU - X. HI 75160

Rapport de l'institut de biologie du gouvernement général de l'afrique occidentale française. Paris. 1920. Rapp. Inst Biol. Gouv. Gén. Afrique Occid. Franç. HI 70453

Rapports, institut féderal de recherches forestières = Bericht, eidgenössische Anstalt für das forstliche Versuchswesen. Birmensdorf.

Rapport, institut des parcs nationaux du Congo Belge. Brussels. 1935+. Rapp. Inst. Parcs Natx. Congo Belge. HI 70454

Rapport, institut Pasteur au Viêt-Nam = Institut Pasteur au Viêt-Nam. Rapport. Saigon. Inst. Pasteur Viêt-Nam Rapp. See B–P–H 434/6.

Rapport de l'institut des recherches agronomiques. Paris. 1922-32, 1923-33. Rapp. Inst. Rech. Agron. Superseded by: Recherches sur la fertilisation effectuée par les stations agronomiques. HI 70455

Rapport, institutionen för ekologi och miljövård, sveriges lantbruksuniversitet. Uppsala. Vol. 1+, 1978+. Rapp. Inst. Ekol. Miljövård Sveriges Lantbruksuniv. HI 70456

Rapport, institutionen for skoglig genetik och växtfysiologi, Sveriges lantbruksuniversitet. Umeå. No. 1+, 1979+. Rapp. Inst. Skoglig Genet. Växtfisiol. Sveriges Lantbruksuniv. HI 75159

Rapport, institutionen för skogsskötsel, sveriges lantbruksuniversitet. Umeå. No. 1+, 1979+. Rapp. Inst. Skogsskötsel Sveriges Lantbruksuniv. Preceded by: Rapport och uppsatser, institutionen för skogsforyngring, skogshögskolan, Rapport och uppsatser, institutionen för skogsproduction, skogshögskolan and Rapport och uppsatser, institutionen för skogsskötsel, skogshögskolan. HI 70457

Rapport, instutütt för marin biologi. Avd A. & C. Oslo. No. ?-2+, ?-1971+. Rapp. Inst. Mar. Biol., A & C. HI 70458

Rapport, kongelige norske videnskabers selskab, museet. Botanisk avdeling. Trondheim. 1974+. Rapp. Kongel. Norske Vidensk. Selsk. Mus., Bot. Avd. HI 70459

Rapport, ministère de l'agriculture de la province de Québec. Quebec. 195?+. Rapp. Minist. Agric. Prov. Québec. Preceded by: Report, department of agriculture, province of Quebec. HI 70460

Rapport, ministère des terres et forêts, Québec = Report of the department of lands and forests, province of Quebec. Quebec.

Rapport, musées d'histoire naturelle de Lausanne. Lausanne. 1887-? Rapp. Mus. Hist. Nat. Lausanne. Superseded by?: Rapports, musées et jardins botaniques, Lausanne. 3-2364-3. HI 70461

Rapports, musées et jardins botaniques. Lausanne. 1943-52, 1944-53. Rapp. Mus. Jard. Bot. Preceded by?: Rapport, musées d'histoire naturelle de Lausanne. HI 70462

Rapport, norges byggforskningsinstitutt. Oslo. Vols. ?-84 [64 & 76 not published], 1953-75. Rapp. Norg. Byggforskn. Inst. Preceded by: Rapport, byggeteknisk utvalg, norges teknisk-naturvitenskapelige forskningsråd. Superseded by: Arbeidsrapport, norges byggforskningsinstitutt. HI 70463

Rapport, norsk institutt for skogforskning = Årsberetning, norsk institutt for skogforskning. As.

Rapport, norsk institutt for tang- og tareforskning. Oslo. No. 1+, 1952+. Rapp. Norsk Inst. Tang- Tareforskning. HI 70464

Rapport particulier, centre de recherches du service de santé des armées = Rapport particulier, division de biologie générale et écologie, centre de recherches du service de santé des armées. Paris.

Rapport particulier, division de biologie générale et écologie, centre de recherches du service de santé des armées. Paris. No. ?-10+, ?-1967+. Rapp. Partic. Div. Biol. Gén. Écol. Centre Rech. Serv. Santé Armées. HI 70465

Rapport et procès-verbaux des réunions de la commission internationale pour l'exploration scientifique de la mer (Mediterranée). Paris, Monaco. Vol. 1+, 1926+. Rapp. Procès-Verbaux Réun. Commiss. Int. Explor. Sci. Mer. (Medit.) Preceded by: Bulletin de la commission internationale pour l'exploration de la mer Mediterranée. HI 70466

Rapport, proefstation voor de rundveehouderij. Wageningen. No. 1+, 1971+. Rapp. Proefstat. Rundveehoud. Preceded, in part, by: Mededeling, proefstation voor de akker- en weidebouw te Wageningen. HI 70467

Rapport de recherches, faculté d'agriculture, université

Laval. Quebec. No. 1+, 1965+. Rapp. Rech. Fac. Agric. Univ. Laval. HI 70468

Rapport scientifique, centre d'océanographie, institut français d'océanie = Rapports scientifiques et techniques, centre de Nouméa, office de la recherche scientifique et technique outre-mer. Océanographie. Nouméa.

Rapport scientifique. Institut française d'Océanie, section océanographie. Nouméa, New Caledonia. Rapp. Sci. Inst. Franç. Océanie, Sect. Océanogr. See B–P–H 755/27. HI 58485

Rapports scientifiques et techniques, centre de Nouméa, office de la recherche scientifique et technique outre-mer. Océanographie. Nouméa. Vol. 1+, 1978+. Rapp. Sci. Techn. Centre Nouméa Off. Rech. Sci. Techn. Outre-Mer, Océanogr. HI 70469

Rapport, section d'océanographie, université du Québec à Rimouski = Rapport annuel, section d'océanographie, université du Québec à Rimouski. Rimouski.

Rapport, service de biologie, Québec. Quebec. 1969. Rapp. Serv. Biol. Québec. Preceded by: Rapport, annuel de la station de biologie marine de Grande-Rivière. Superseded by: Rapport, direction générale des pêches maritimes [not entered]. HI 70470

Rapport i skogsekologi och skoglig marklära. Uppsala. Vol. 31+, 1978+. Rapp. Skogsekol. Skoglig Marklära. Preceded by: Rapport och uppsatser, institutionen för växtekologi och marklära, skogshögskolan. HI 70471

Rapport de la société d'agriculture, Le Caire. Cairo. ?-1932-47+. Rapp. Soc. Agric. Le Caire. HI 70472

Rapport, société d'encouragement de la culture des orges de brasserie et des houblons en France. Paris. 1942+. Rapp. Soc. Encour. Cult. Orges Brass. Houbl. France. HI 70473

Rapport, société de pomologie et de culture fruitière de la Province de Québec = Report of the pomological and fruit growing society of Quebec. Montreal.

Rapport de la société royale des arts et des sciences de Maurice. Port Louis. 1849-54/55, 1850-55. Rapp. Soc. Roy. Arts Maurice. Preceded by: Travaux de la société d'histoire naturelle de l'Ile Maurice. HI 70474

Rapport, statens skogsforskningsinstitut. Stockholm. 1957-62. Rapp. Statens Skogsforskningsinst. Superseded by: Rapport och uppsatser, institut för skogstechnik. HI 70475

Rapport, statens skogsforskningsinstitut. Avdelingen för skogsproduktion. Stockholm. Nos. 1-4, 1959-60. Rapp. Statens Skogsforskningsinst., Skogsprod. Superseded by: Rapport och uppsatser, institutionen för skogsproduction, skogshögskolan. HI 75132

Rapport, station agronomique de Guadeloupe. Point-à-Pitre. 1918-28. Rapp. Stat. Agron. Guadeloupe. HI 70476

Rapport, station biologique du St. Laurent à Trois Pistoles. Quebec. Vols. 1-8, 1931-49. Rapp. Stat. Biol. St. Laurent Trois Pistoles. HI 70477

Rapport, station centrale des cultures fruitières. Kinia. 1953-54. Rapp. Stat. Centrale Cult. Fruitières. HI 70478

Rapport de la station des cultures fruitières et maraîchères Gembloux. Grand Manil. 1969-72. Rapp. Stat. Cult. Fruitières Maraîch. Gembloux. Preceded by: Rapport général de la station de recherches de l'état pour l'amélioration des plantes fruitières et maraîchères. HI 70479

Rapport, station de génétique et de culture du mais. St. Martin-de-Hiux. 1932. Rapp. Stat. Gén. Cult. Mais. HI 70480

Rapport, station des plantes ornamentales, Melle. Melle. ?-1957/60+, ?-1960?+. Rapp. Stat. Pl. Ornam. Melle. HI 70481

Rapport, station de recherches pour l'amélioration des plantes ornamentales, Melle = Rapport, station des plantes ornamentales, Melle. Melle.

Rapport de la station de recherches des forêts de la Suède = Meddelanden från statens skogsförsöksanstalt. Stockholm.

Rapport technique, centre national d'experimentation agricole de Grignon. Paris. ?-1920-23. Rapp. Techn. Centre Natl. Exp. Agric. Grignon. Superseded by: Ferme extérieure de Grignon. HI 70482

Rapport sur les travaux de recherches effectues, service botanique et agronomique de Tunisie. Tunis. 1952-55. Rapp. Trav. Rech. Effect. Serv. Bot. Agron. Tunisie. Preceded by: Bulletin du service botanique et agronomique de Tunisie. HI 70483

Rapports de l'UNESCO sur les sciences de la mer = U N E S C O reports in marine science. Paris.

Rapport och uppsatser, institut för skogsbotanik. Stockholm. 1963+. Rapp. Uppsat. Inst. Skogsbot. HI 70484

Rapport och uppsatser, institutionen för skogsproduction, skogshögskolan. Uppsala. Nos. 5-45?, 1964-77? Rapp. Uppsat. Inst. Skogsprod. Superseded by: Rapport, institutionen för skogsskötsel, sveriges lantbruksuniversitet. HI 70485

Rapport och uppsatser, institut för skogstechnik. Stockholm. 1963+. Rapp. Uppsat. Inst. Skogstechn. Preceded by: Rapport, statens skogsforskningsinstitut. HI 70486

Rapport och uppsatser, institutionen för skogsföryngring, skogshögskolan. Stockholm. Nos. 1-103, 1963-78. Rapp. Uppsat, Inst. Skogsföryng. Skogshögskolan. Superseded by: Rapport, institutionen för skogsskötsel, sveriges lantbruksuniversitet. HI 70487

Rapport och uppsatser, institutionen för skogsskötsel, skogshögskolan. Stockholm. Nos. 1-11, 1973-78. Rapp. Uppsat. Inst. Skogsskötsel Skogshögskolan. Superseded by: Rapporter, institutionen för skogsskötsel, sveriges lantbruksuniversitet. HI 70488

Rapport och uppsatser, institutionen för växtekologi och marklära, skogshögskolan. Stockholm. Nos. 1-30, 1963-77. Rapp. Uppsat. Inst. Växtekol. Marklära Skogshögskolan. Superseded by: Rapport i skogsekologi och skoglig marklära. HI 70489

Rapporter, see Rapport

Rapports, see Rapport

Rare seeds newsletter. Silverton, OR. 1975+. Rare Seeds Newslett. HI 70490

Rasen, grünflächen, begrünungen. Bonn. Vol. 7+, 1976+. Rasen Grünfl. Begrün. Preceded by: Rasen, Turf, Gazon. HI 70491

Rasen, Turf, Gazon. Bonn. Vol. 1+, 1970+. Rasen Turf Gazon. Superseded by: Rasen, grünflächen, begrünungen. HI 70492

Rassegna geologica = Geologisches Zentralblatt. Anzeiger für Geologie, Petrographie, Palaeontologie und verwandte Wissenschaften. Leipzig. Geol. Zentralbl. See B–P–H 399/17.

Rassegna micologica ticinese. Chiasso. Vol. 1+, 1968+. Rassegna Micol. Ticin. HI 70493

Rassegna delle scienze biologiche. Florence. Rassegna Sci. Biol. See B–P–H 756/6. HI 58486

Rassenlijst voor groentengewassen. Wageningen. Rassenlijst Groentengew. See B–P–H 756/7. HI 58487

Rastenievodstvo. Budapest = Növénytermelés. Budapest. Növénytermelés. See B–P–H 674/7.

Rastenievudni nauki. Izvestiya na akademiyata na selskostopanskite nauki. Sofia. Vol. 1+, 1964+. Rasteniev. Nauki. HI 70494

Rastitel'noe syr'e = Trudy botanicheskogo instituta akademii nauk S S S R. Ser. 5, rastitel'noe syr'e. Moscow & Leningrad.

Rastitel'nost' Kazahstana. Materialy issledovanij rastitel'nosti Kazahstana = Rastitel'nost' Kazakhstana. Materialy issledovanii rastitel'nosti Kazakhstana. Moscow & Leningrad.

Rastitel'nost' Kazakhstana. Materialy issledovanii rastitel'nosti Kazakhstana. Moscow & Leningrad. Vols. 1-2, 19??-41. Rastitel'n. Kazakhstana. HI 70495

Rastitel'nost' Krainego Severa S S S R i eë osvoenie. Moscow & Leningrad. Vol. 1+, 1956+. Rastitel'n. Krainego Severa S.S.S.R. Eë Osvoenie. HI 70496

Rastitel'nost' Krajnego Severa S S S R i ee osvoenie = Rastitel'nost' Krainego Severa S S S R i eë osvoenie. Moscow & Leningrad.

Rastitel'nost' S S S R. Moscow & Leningrad. Rastitel'n. S.S.S.R. See B–P–H 756/12. HI 58488

Rastitel'nye belki. Kishinev. Vol. 9+, 1970+. Rastitel'n. Belki. Preceded by: Trudy po khimii prirodnykh soedinenii. HI 70497

Rastitel'nye resursy. Moscow, Leningrad. Vol. 1+, 1965+. Rastitel'n. Resursy. HI 70498

Rastitel'nye resursy Sibiri i dal'nego vostoka. Tekushchii ukazatel' literatury. Novosibirsk. 1974. Rastitel'n. Resursy Sibiri Dal'nego Vostoka, Tekushch. Ukaz. Lit. Superseded by: Rastitel'nyi mir Sibirii i dal'nego vostoka. Tekushchii ukazatel' literatury. HI 70499

Rastitel'nyi mir Sibirii i dal'nego vostoka. Tekushchii ukazatel' literatury. Novosibirsk. 1975+. Rastitel'n. Sibirii Dal'nego Vostoka, Tekushch. Ukaz. Lit. Preceded by: Rastitel'nye resursy Sibiri i dal'nego vostoka. Tekushchii ukazatel' literatury. HI 70500

Raumforschung und Raumordnung. Heidelberg. Raumf. & Raumordn. See B–P–H 756/13. HI 58489

Raymondiana. Lima. Vol. 1+, 1968 [1969]+. Raymondiana. HI 70501

Razprave matematično-prirodoslovnega razreda akademije znanosti in umetnosti v Ljubljani. Ljubljana. Vols. 1-2, 1940-42. Razpr. Mat.-Prir. Razr. Akad. Znan. v Ljubljani. Superseded by: Razprave, slovenska akademija znanosti in umetnosti. HI 70502

Razprave, slovenska akademija znanosti in umetnosti. Razrad za prirodoslovne in medicinske vede. Ljubljana. Vols. 1-5, 1951-59. Razpr. Slov. Akad. Znan. Umetn., Razr. Prir. Med. Vede Preceded by: Razprave matematično-prirodoslovnega razreda akademije znanosti in umetnosti v Ljubljani. Superseded by: Razprave, slovenska akademija znanosti in umetnosti. Razrad za prirodoslovne in medicinske vede. Oddelek za prirodoslovne vede. HI 70503

Razprave, slovenska akademija znanosti in umetnosti. Razrad za prirodoslovne in medicinske vede. Oddelek za prirodoslovne vede. Ljubljana. Vols. 6-17, 1961-74. Razpr. Slov. Akad. Znan. Umetn., Razr. Prir. Med. Vede, Oddel. Prir. Vede. Preceded by: Razprave, slovenska akademija znanosti in umetnosti. Razrad za prirodoslovne in medicinske vede. Superseded by: Razprave, slovenska akademija znanosti in umetnosti. Razrad za prirodoslovne vede. HI 70504

Razprave, slovenska akademija znanosti in umetnosti. Razrad za prirodoslovne vede. Ljubljana. Vol. 18+, 1975?+. Razpr. Slov. Akad. Znanosti Umetn., Razr. Prir. Vede. Preceded by: Razprave, slovenska akademija znanosti in umetnosti. Razrad za prirodoslovne in medicinske vede. Oddelek za prirodoslovne vede. HI 70505

Rea; bollettino di informazione del giardino sperimentale di acclimatazione. San Bernardino di Trana. Vol. ?-2+, ?-[1969]+. Rea. HI 70506

Reading naturalist. Reading, England. Reading Naturalist. See B–P–H 756/15. HI 58490

Readings on plant breeding = Vorträge für Pflanzenzuchtung. Bonn.

Real gardening. New Canaan, CT. Real Gard. See B–P–H 756/16. HI 58491

Reblooming iris recorder. Staten Island, NY. No. 1+, 1962+. Reblooming Iris Recorder. HI 70507

Reblooming iris reporter = Reblooming iris recorder. Staten Island, NY.

Recensent. Algemeen letterlievend maandschrift = De recensent, ook der recensenten. Amsterdam. Recensent. See B–P–H 757/29.

Recensent, ook der recensenten. Amsterdam. Recensent. See B–P–H 757/29. HI 58502

Recent advances in phytochemistry. Amsterdam, New York. Vol. 1+, 1966+. Recent Advances Phytochem. HI 70508

Recent progress in microbiology. Recent Progr. Microbiol. See B–P–H 757/30. HI 58503

Recent progress of natural sciences in Japan. [Nihon shizen kagaku shuho.] Tokyo. Vol. 1+, 1976+. Recent Progr. Nat. Sci. Japan. Preceded by: Japanese journal of botany and Record of oceanographical works in Japan. HI 70509

Recent publications in natural history. [From 1979-82 issued as a supplement to: Curator.] New York. 1979-82?; [n.s.] vol. 1+, 1983+. Recent Publ. Nat. Hist. HI 70510

Rechenschaftsberichte des Vereins für Pomologie und Gartenbau in Meiningen = Verhandlungen des Vereins für Pomologie und Gartenbau in Meiningen. Meiningen, Germany. Verh. Vereins Pomol. Meiningen. See B–P–H 955/17.

Recherche. Paris. Vol. 1+, 1970+. Recherche. Preceded by: Atomes [not entered]. Absorbed: Nucleus and Science progrès découverte. HI 70511

Recherche agronomique en Suisse = Schweizerische landwirtschaftliche Forschungen. Berne.

Recherches agronomiques. Quebec. Vol. 1+, 1957+. Rech. Agron. HI 70513

Recherches sur les bois; rapport annuel. Centre technique forestier tropical. Nogent-sur-Marne, France. Rech. Bois. See B–P–H 757/31. HI 58504

Recherches sur la fertilisation effectuée par les stations agronomiques. Paris. 1934-38. Rech. Fertilis. Effect. Stat. Agron. Preceded by: Rapport de l'institut des recherches agronomiques. Superseded by: Travaux effectués par les stations agronomiques, institut national de la recherche agronomique. HI 70514

Recherches d'hydrobiologie continentale. Paris. No. 1, 1969. Rech. Hydrobiol. Continentale. Superseded by: Annales d'hydrobiologie. HI 70515

Recherches marines = Cercetari marine. Constanta.

Reclamation and revegetation research. Amsterdam. Vol. 1+, 1982+. Reclam. Reveg. Res. HI 70516

Reclamation review. Elmsford, NY. Vol. 1+, 1978+. Reclam. Rev. HI 70517

Recombinant D N A. New York. 1981+. Recombinant DNA. HI 70518

Recombinant D N A technical bulletin. Washington, DC. 1977+. Recombinant DNA Techn. Bull. HI 70519

Record of agricultural research, ministry [later department] of agriculture for Northern Ireland. Belfast. Vol. 1+, 1963+. Rec. Agric. Res. Minist. Agric. Northern Ireland. HI 70521

Records of the agricultural research station. Jerusalem. Rec. Agric. Res. Sta. See B–P–H 756/18. HI 58492

Record of the Albany museum. Grahamstown. Vols. 1-4, 1903-35. Rec. Albany Mus. HI 70522

Records of the american society of naturalists. Boston, MA. Rec. Amer. Soc. Naturalists. See B–P–H 756/19. HI 58493

Records of the Auckland institute and museum. Auckland. Rec. Auckland Inst. Mus. See B–P–H 756/22. HI 58494

Record of the australian academy of science. Canberra, A.C.T. 1966-80. Rec. Austral. Acad. Sci. Superseded by: Historical records of australian science. HI 70523

Records of the australian museum. Sydney, N.S.W. Rec. Austral. Mus. See B–P–H 756/25. HI 58495

Record of bare facts, Caradoc and Severn Valley field club. Shrewsbury. 1894-1945. Rec. Bare Facts Caradoc Severn Valley Field Club. Preceded by: Caradoc record of bare facts. HI 70524

Records of the biological association of Nagoya. [Nagoya seibutsugakkai kiroku.] Nagoya, Japan. Rec. Biol. Assoc. Nagoya. See B–P–H 757/1. HI 58496

Records of the botanical survey of India. Calcutta. Rec. Bot. Surv. India. See B–P–H 757/4. HI 58497

Record, Brooklyn botanic garden = Brooklyn botanic garden record. Brooklyn, NY. Brooklyn Bot. Gard. Rec. See B–P–H 230/7.

Record of the Canterbury museum. Christchurch, N.Z. Vol. 1+, 1907+. Rec. Canterbury Mus. HI 70525

Record of the dominion museum. Wellington, N.Z. Vols. 1-8, 1942-75. Rec. Domin. Mus. Superseded by: Record, national museum of New Zealand. 5-4469-2. HI 70526

Records of general science. London. Rec. Gen. Sci. See B–P–H 757/11. HI 58498

Records of the genetics society of America. Rec. Genet. Soc. Amer. See B–P–H 757/13. HI 58499

Records of the indian museum. A journal of indian zoology. Calcutta. Rec. Indian Mus. See B–P–H 757/16. HI 58500

Records of the hungarian agricultural experiment stations, plant production = Mezögazdasági és elelmezésügyi minisztérium, kísérletügyi közlemények. Budapest.

Records of the hungarian agricultural research stations. Budapest = Kísérletügyi kölemények. Budapest. Kísérl. Közlem. See B–P–H 514/5.

Record of investigations, department of agriculture, Uganda. Entebbe. 1948+. Rec. Invest. Dep. Agric. Uganda. HI 70527

Record of investigations, department of science and agriculture, Barbados. Bridgetown. 1939+. Rec. Invest. Dep. Sci. Agric. Barbados. HI 70528

Record, national museum of New Zealand. Wellington, N.Z. Vol. 1+, 1975+. Rec. Natl. Mus. New Zealand. Preceded by: Record of the dominion museum. HI 70529

Record of oceanographical works in Japan. Tokyo. Vols. 1-10, 1928-38; n.s. vols. 1-13(1), 1953-75. Rec. Oceanogr. Work Japan. Superseded by: Recent progress of natural sciences in Japan. 4-3545-3. HI 70530

Record of proceedings, Cleveland naturalists' field club. Middlesbrough. 1889-1902. Rec. Proc. Cleveland Naturalists' Field Club. Superseded by: Proceedings of the Cleveland naturalists' field club. HI 70531

Record of the Queen Victoria museum. Launceston, Tas. Vols. 1-3, 1942-52; n.s. vol. 1+, 1952+. Rec. Queen Victoria Mus. HI 70532

Record of research of the east african agricultural and fisheries research council = Report of the east african agricultural and fisheries research council. Nairobi.

Record of research, east african agriculture and forestry research organisation. Annual report. Nairobi. 1954/55-76, 1956-76. Rec. Res. E. African Agric. Forest. Res. Organ., Annual Rep. Preceded by: Report, east african agriculture and forestry research organisation. Superseded by: Record of research, Kenya agricultural research institute. Annual report. HI 70533

Record of research, Kenya agricultural research institute. Annual report. Nairobi. 1977+. Rec. Res. Kenya Agric. Res. Inst., Annual Rep. Preceded by: Record of research, east african agriculture and forestry research organisation. Annual report. HI 70534

Record of researches in the faculty of agriculture, university of Tokyo. Tokyo. Nos. 1-13, 1952-63. Rec. Res. Fac. Agric. Univ. Tokyo. Preceded by: Journal of the college of agriculture, imperial university of Tokyo. HI 70535

Record of the royal society of London. London. No. 1+, 1897+. Rec. Roy. Soc. London. HI 70536

Record, scottish plant breeding station. Pentlandfield, Roslin. 1962-65. Rec. Scott. Pl. Breed. Sta. Preceded and superseded by: Report (Annual) of the scottish plant breeding station. HI 70537

Records of the south australian museum. Adelaide, S.A. Rec. S. Austral. Mus. See B–P–H 757/24. HI 58501

Records of the Tomsk state Kuybyshev university = Trudy Tomskogo gosudarstvennogo universiteta imeni V. V. Kuibysheva. Tomsk.

Records, see Record

Recreations in agriculture, natural-history, arts, and miscellaneous literature. London. Recreations Agric. See B–P–H 758/3. HI 58505

Recueil des actes de la séance publique de l'académie impériale des sciences de St.-Pétersbourg. St. Petersburg & Leipzig. Recueil Actes Séance Publique Acad. Imp. Sci. St.-Pétersbourg. See B–P–H 758/5. HI 58506

Recueil des actes de la séance publique de l'académie impériale des sciences de Saint-Pétersbourg. St. Petersburg & Leipzig. Recueil Actes Séance Publique Acad. Imp. Sci. Saint-Pétersbourg. See B–P–H 758/6. HI 58507

Recueil des actes de la séance solennelle de l'académie impériale des sciences de St.-Pétersbourg. St. Petersburg. Recueil Actes Séance Solennelle Acad. Imp. Sci. St.-Pétersbourg. See B–P–H 758/7. HI 58508

Recueil des actes des séances publiques de l'académie impériale des sciences de Saint-Pétersbourg ... St. Petersburg & Leipzig. Recueil Actes Séances Publiques Acad. Imp. Sci. Saint-Pétersbourg. See B–P–H 758/8 HI 58509.

Recueil biologique. Kiev = Biologichnii zbirnyk.

Recueil des comptes-rendus de la société botanique de Genève. [Supplement to Bulletin de la société botanique de Genève.] Geneva. 1901-08. Recueil Compt.-Rend. Soc. Bot. Genève. HI 70538

Recueil contenant les déliberations de la société royale d'agriculture de la généralité de Paris. Paris. Recueil Délib. Soc. Roy. Agric. Généralité Paris. See B–P–H 758/10. HI 58510

Recueil des éloges historiques lus dans les séances publiques de l'institut national des sciences et arts. Paris. Recueil Eloges Hist. Séances Publiques Inst. Natl. Sci. See B–P–H 758/11. HI 58511

Recueil des éloges historiques lus dans les séances publiques de l'institut royal de France = Recueil des éloges historiques lus dans les séances publiques de l'institut national des sciences et arts. Paris. Recueil Eloges Hist. Séances Publiques Inst. Natl. Sci. See B–P–H 758/11. HI 58513

Recueil général de médecine = Journal général de médecine, de chirurgie et de pharmacie, françaises et étrangères, vol. 58, 1816. Paris. J. Gén. Méd. Franç. Étrangères. See B–P–H 467/9.

Recueil général de médecine, de chirurgie et de pharmacie, françaises et étrangères = Journal général de médecine, de chirurgie et de pharmacie, françaises et étrangères, vol. 57, 1816. Paris. J. Gén. Méd. Franç. Étrangères. See B–P–H 467/9.

Recueil géobotanique. Kiev = Geobotanichnyi zbirnyk. Kiev.

Recueil de l'institut botanique. Brussels. Recueil Inst. Bot. See B–P–H 758/13. HI 58512

Recueil de l'institut botanique Léo Errera = Recueil de l'institut botanique. [Université libre de Bruxelles.] Brussels. Recueil Inst. Bot. See B–P–H 758/13.

Recueil des mémoires et actes de la société des sciences et arts du département du Mont-Tonnerre, séant à Mayence. Mainz. Recueil Mém. Actes Soc. Sci. Dép. Mont-Tonnerre. See B–P–H 758/16. HI 58514

Recueil de mémoires et autres pièces de prose et de vers, qui ont été lus dans les séances de la société des amis des sciences, des lettres, de l'agriculture et des arts, à Aix, département des Bouches-du-Rhône. Aix-en-Provence, France. Recueil Mém. Pièces Prose Vers. See B–P–H 759/2. HI 58518

Recueil de mémoires ou collection de pièces académiques. Dijon. Recueil Mém. Collect. Pièces Acad. See B–P–H 758/17. HI 58515

Recueil de mémoires et conférences sur les arts et les sciences. [Edited by J. Denis.] Paris. Recueil Mém. Conf. Arts Sci. See B–P–H 758/18. HI 58516

Recueil de mémoires de médecine, de chirurgie et de pharmacie militaires. Paris. Recueil Mém. Méd. See B–P–H 759/1. HI 58517

Recueil des mémoires les plus intéressans de chymie et d'histoire naturelle, contenus dans les actes de l'académie d'Upsal, et dans les mémoires de l'académie royale des sciences de Stockholm. Paris. Vols. 1-2, 1764. Recueil Mém. Chym. Hist. Nat. Acad. Upsal Acad. Roy. Sci. Stockholm. 4-3549-1. HI 70539

Recueil des mémoires et des travaux, société de botanique du Grand-Duché de Luxembourg. Luxemburg. Nos. 1-16, 1874-1905. Recueil Mém. Trav. Soc. Bot. Grand-Duché Luxembourg. Incorporated in: Bulletin mensuel de la société des naturalistes luxembourgeois. HI 70540

Recueil d'observations botaniques et de descriptiones de plantes nouvennes = Observationes botanicae et descriptiones plantarum novarum herbarii Van Heurckiani. Linz. Observ. Bot. Descript. Pl. Nov. Herb. Van Heurckiani. See B–P–H 679/2.

Recueil d'ophtalmologie. Paris. Recueil Ophtalmol. See B–P–H 759/4. HI 58519

Recueil périodique d'observations de médecine, de chirurgie et de pharmacie. Paris. Recueil Périod. Observ. Méd. See B–P–H 759/5. HI 58520

Recueil périodique de la société de médecine de Paris. Paris. Recueil Périod. Soc. Méd. Paris. See B–P–H 759/6. HI 58521

Recueil périodique de la société de santé de Paris. Paris. Recueil Périod. Soc. Santé Paris. See B–P–H 759/7. HI 58522

Recueil des pièces lues dans les séances publiques et particulières, de l'académie royale de Nismes. Nîmes. Recueil Pièces Séances Acad. Roy. Nismes. See 759/8. HI 58523

Recueil des prix remportés sur les questions proposées par l'académie de Bruxelles. Brussels. Recueil Prix Remportés Quest. Prop. Acad. Bruxelles. See B–P–H 759/9. HI 58524

Recueil des publications de la société havraise d'études diverses. Le Havre. Recueil Publ. Soc. Havraise Études Diverses. See B–P–H 759/10. HI 58525

Recueil des publications de la société nationale havraise d'études diverses. Le Havre. Recueil Publ. Soc. Natl. Havraise Études Diverses. See B–P–H 759/11. HI 58526

Recueil de la société d'agriculture, sciences, arts et belles-lettres du départment de l'Eure. Évreux, France. Recueil Soc. Agric. Dép. Eure. See B–P–H 759/12. HI 58527

Recueil de la société libre d'agriculture, sciences, arts et belles-lettres du départment de l'Eure = Recueil de la société d'agriculture, sciences, arts et belles-lettres du départment de l'Eure. Évreux, France. Recueil Soc. Agric. Dép. Eure. See B–P–H 759/12.

Recueil de la société des naturalistes et archéologues du nord de la Meuse. Montmédy. Vol. 25+, 1912+. Recueil Soc. Naturalistes N. Meuse. Preceded by: Bulletin de la société des naturalistes et archéologiques du nord de la Meuse. 5-3954-3. HI 70541

Recueil des travaux. Académie serbe des sciences. Institut d'écologie et de biogéographie = Zbornik radova.

Srpska akademija nauka. Institut za ekologiju i biogeografiju. Belgrade. Zborn. Rad. Srpska Akad. Nauka Inst. Ekol. See B–P–H 996/25.

Recueil des travaux botaniques néerlandais. Nijmegen, Netherlands. Recueil Trav. Bot. Néerl. See B–P–H 759/16. HI 58528

Recueil de travaux, centre de Nouméa, office de la recherche scientifique et technique outre-mer. Section océanographique. Nouméa. No. 1+, 1969+. Recueil Trav. Centre Nouméa Off. Rech. Sci. Techn Outre-Mer, Sect Océanogr. HI 70542

Recueil des travaux. Institut biologique. Belgrade = Zbornik radova. Biološki institut N.R. Srbije. Belgrade. Zborn. Rad. Biol. Inst. Nar. Republ. Srbije. See B–P–H 996/21.

Recueil des travaux de l'institut botanique. Annales de l'université de Montpellier et du Languedoc Méditerranéen-Roussillon. Supplément scientifique, série botanique. Montpellier. Vols. 1-4, 1944-46. Recueil Trav. Inst. Bot., Ann. Univ. Montpellier Languedoc Méditerranéen-Roussillon, Suppl. Sci., Sér. Bot. Superseded by: Recueil des travaux des laboratoires de botanique, géologie et zoologie de la faculté des sciences de l'université de Montpellier. Série botanique. HI 70543

Recueil des travaux du laboratoire de biologie végétale de la faculté des sciences de Bordeaux. Bordeaux. Vols. 1-5, 1960-65. Recueil Trav. Lab. Biol. Vég. Fac. Sci Bordeaux. Superseded by: Recueil des travaux du laboratoire de physiologie végétale de la faculté des sciences de Bordeaux. HI 70545

Recueil des travaux des laboratoires de biologie végétale, université de Dijon. Dijon. Vols. 1-?, 19??-? Recueil Trav. Lab. Biol. Vég. Univ. Dijon. HI 70546

Recueil des travaux du laboratoire de botanique de la faculté des sciences de Caen. Caen. 1948-65-? Recueil Trav. Lab. Bot. Fac. Sci. Caen. HI 70547

Recueil des travaux des laboratoires de botanique, géologie et zoologie de la faculté des sciences de l'université de Montpellier. Série botanique. Montpellier. Vols. 5-7, 1947-55. Recueil Trav. Lab. Bot. Fac. Sci. Univ. Montpellier, Sér. Bot. Preceded by: Recueil des travaux de l'institut botanique. Annales de l'université de Montpellier et du Languedoc Méditerranéen-Roussillon. Supplément scientifique, série botanique. Superseded by: Naturalia monspeliensia: série botanique. HI 70548

Recueil des travaux du laboratoire de physiologie végétale de la faculté des sciences de Bordeaux. Bordeaux. Vols. 6-9, 1966-72. Recueil Trav. Lab. Physiol. Vég. Fac. Sci. Bordeaux. Preceded by: Recueil des travaux du laboratoire de biologie végétale de la faculté des sciences de Bordeaux. Superseded by: Recueil des travaux du laboratoire de physiologie végétale et écophysiologie forestière de l'université de Bordeaux. HI 70549

Recueil des travaux du laboratoire de physiologie végétale et écophysiologie forestière de l'université de Bordeaux. Bordeaux. Vol. 10+, 1974+. Recueil Trav. Lab. Physiol. Vég. Écophys. Forest. Univ. Bordeaux. Preceded by: Recueil des travaux du laboratoire de physiologie végétale de la faculté des sciences de Bordeaux. HI 70550

Recueil des travaux des membres de la société des naturalistes Namur-Luxembourg. Namur. Vol. 1+, 1948+. Recueil Trav. Memb. Soc. Naturalistes Namur-Luxembourg. HI 70551

Recueil des travaux des professeurs de l'université d'état à Irkoutsk = Sbornik trudov professorov i prepodavatelei Gosudarstvennogo Irkutskogo universiteta. Irkutsk.

Recueil des travaux scientifiques de la société des naturalistes de Kiew = Zbirnyk naukovykh prats Kyiivs'kogo tovarystva pryrodnykiv. Kiev.

Recueil des travaux, société d'amateurs des sciences, de l'agriculture et des arts à Lille = Société d'amateurs des sciences, de l'agriculture et des arts à Lille. Recueil des travaux. Lille. Soc. Amateurs Sci. Lille Receuil Trav. See B–P–H 844/22.

Recueil des travaux de la société des sciences, de l'agriculture et des arts, de Lille. Lille. 1823-27. Recueil Trav. Soc. Sci. Lille. Superseded by: Mémoires de la société royale des sciences, de l'agriculture et des arts, à (de) Lille. 5-3955-1. HI 51237

Recueil, travaux de la station biologique du Lac d'Oredon. Toulouse. Nos. 1-2, 1962-63. Recueil Trav. Stat. Biol. Lac Oredon. Superseded by: Annales de limnologie. HI 70552

Recueil des travaux de la station marine d'Endoume. Marseilles. Vols. 1-48, 1949-69. Recueil Trav. Stat. Mar. Endoume. Superseded by: Téthys. HI 70553

Recueil de travaux de l'université d'état à Irkoutsk = Sbornik trudov gosudarstvennogo Irkutskogo universiteta.

Recueil de voyages et de mémoires: société de géographie, Paris. Paris. Recueil Voyages Mém. Soc. Géogr. Paris. See B–P–H 760/6. HI 58529

Recursos naturales. [Ministerio de recursos naturales.] Tegucigalpa, Honduras. Recursos Nat. See 760/7. HI 58530

Redia: giornale di entomologia. Florence. Redia. See B–P–H 760/8. HI 58531

Redai zuowu xuebao = Chinese journal of tropical crops. Tan-hsien.

Redogörelse för allmänna läroverken i Norrköping och Söderköpping. Norrköping. [Dates of publication not ascertained.] Redog. Allmänna Lärov. Norrköping Söderköpping. HI 70554

Redogörelse, arbetsutskott för undersökning af de finska insjöarnas vatten och plankton. Helsinki. 1914-18. Redog. Arbetsutsk. Undersök. Finska Vatt. Plankt. HI 70555

Redogörelse för verksamheten, Ultuna landtbruksinstitut. Uppsala. 1899-1936? Redog. Verksamh. Ultuna Landtbruksinst. HI 70556

Reef Point gardens bulletin. Bar Harbor, ME. Reef Point Gard. Bull. See B–P–H 760/9. HI 58532

Referativnyi byulleten′ bolgarskoi nauchnoi literatury. Sofia. Vols. 1-4, 1955-59. Ref. Byull. Bolg. Nauchn. Lit. Superseded by: Abstracts of bulgarian scientific literature: agriculture and forestry, veterinary medicine. HI 70557

Referativnyi byulleten′ bolgarskikh nauchnykh publikatsii = Referativnyi byulleten′ bolgarskoi nauchnoi literatury. Sofia.

Referativnyi zhurnal. Biofizika. Moscow. 1973-82. Ref. Zhurn., Biofiz. HI 70558

Referativnyi zhurnal. Biologicheskaya khimiya. Moscow. 1963+. Ref. Zhurn., Biol. Khim. Preceded by: Referativnyi zhurnal. Khimiya, biologicheskiya khimiya. HI 70559

Referativnyi zhurnal: Biologiya. Moscow. 1954+. Ref. Zhurn. Biol. HI 70560

Referativnyi zhurnal. Biologiya. Botanika. Moscow. 1968+. Ref. Zhurn., Biol. Bot. HI 70561

Referativnyi zhurnal: Biologiya. Obshchaya ekologiya, biotsenologiya, gidrobiologiya. Moscow. 1966+. Ref. Zhurn. Biol. Obshchaya Ekol. Biotsenol. Gidrobiol. HI 75122

Referativnyi zhurnal. Fitopatologiya. Moscow. 1978+. Ref. Zhurn., Fitopatol. Previously formed part of: Referativnyi zhurnal. Rastenievodstvo. HI 70562

Referativnyi zhurnal. Genetika i selektsiia vozdelyvaemykh rastenii. Moscow. 1978+. Ref. Zhurn., Genet. Selekts. Vozdelyvaem. Rast. HI 70563

Referativnyi zhurnal. Khimiya, biologicheskaya khimiya. Moscow. 1955-62. Ref Zhurn., Khim. Biol. Khim. Superseded by: Referativnyi zhurnal. Biologicheskaya khimiya. HI 70564

Referativnyi zhurnal. Lesovedenie i lesovodstvo. Moscow. Vol. 1+, 1963+. Ref. Zhurn., Lesoved. Lesov. HI 70565

Referativnyi zhurnal. Molekulyarnaya biologiya. Moscow. 1976-82. Ref. Zhurn., Molek. Biol. HI 70566

Referativnyi zhurnal. Okhrana prirody i vosproizvodstvo prirodnykh resursov. Moscow. Vol. 1+, 1975+. Ref. Zhurn., Okhr. Prir. Vosproizv. Prir. Resursov. Previously formed part of: referativnyi zhurnal. Geografiya [not entered]. HI 70567

Referativnyi zhurnal. 70. Radiatsionaya biologiya. Moscow. Vol. 1+, 1973+. Ref. Zhurn., 70. Radiats. Biol. HI 70568

Referativnyi zhurnal. Rastenievodstvo: biologicheskie osnovy. Moscow. Vol. 1+, 1963+. Ref. Zhurn., Rasteniev. Biol. Osnony. HI 70569

Referativnyi zhurnal. Tsvetovodstvo i dekorativnoe sadovodstvo. Moscow. Vol. 1+, 19??+. Ref. Zhurn., Tsvetovod. Dekorat. Sadov. HI 70570

Referativnyj žurnal: Biologija = Referativnyi zhurnal: Biologiya. Moscow.

Referaty nauchno-issledovatel′skikh rabot. Otdelenie biologicheskikh nauk. Moscow. 1942-47. Ref. Nauchno-Issl. Rabot, Otd. Biol. Nauk. HI 70571

Referaty nauchnykh rabot instituta biologiya morya. Vladivostok. Vol. 1, 1969. Ref. Nauchn. Rabot Inst. Biol. Morya. Superseded by: Nauchnye soobshcheniya instituta biologii morya. HI 70572

Referaty naučno-issledovatel′skih rabot. Otdelenie biologičeskih nauk = Referaty nauchno-issledovatel′skikh rabot. Otdelenie biologicheskikh nauk. Moscow.

Reflections. [Indoor Light Garden Society of America.] ?-1975-? Reflections. HI 70573

Reforma; revista de forestales y madereros colombianos. Medellín, Colombia. Reforma. See B–P–H 760/13. HI 58533

Refugium botanicum; or, figures and descriptions from living specimens of little known or new plants of botanical interest. London. Refug. Bot. See B–P–H 760/14. HI 58534

Regelia. Zeitschrift für Wissenschaft im Gartenbau. Klein Machnow, Germany. Regelia. See B–P–H 760/19. HI 58536

Regeneration. Emmaus, PA. Vol. 1+, 1985+. Regeneration. Preceded by: Cornucopia project newsletter. HI 70574

Regional science review. Binghampton, NY. Vol. 14+, 1984+. Regional Sci. Rev. Preceded by: Northeast regional science review. HI 70575

Registro trimestre ó colección de memorias de historia, lite ratura, ciencias y artes. Mexico City. Reg. Trimestre. See B–P–H 760/17. HI 58535

Règne végétal; revue mensuelle publ. par la société botanique du Limousin. Limoges. Vols. 1-3, 1890-92. Règne Vég. Superseded by: Revue scientifique du

Limousin. HI 70576

Regnum vegetabile; a series of handbooks for the use of plant taxonomists and plant geographers. Utrecht. Regnum Veg. See B–P–H 760/25. HI 58537

Regulatory horticulture. [Harrisburg, PA.] Vol. 1+, 1975+. Regulat. Hort. HI 70577

Reichspflanzenschutzblatt. Berlin. 1943-45. Reichspflanzenschutzblatt. Preceded and superseded by: Nachrichtenblatt für den deutschen Pflanzenschutzdienst. HI 70578

Reichtum und Not der Natur. Jahrbuch für Naturschutz und Landeskultur. Dresden. Reichtum Not Natur. See B–P–H 761/1. HI 58538

Reinwardtia. Bogor. Vol. 1+, 1950+. Reinwardtia. Preceded by: Bulletin du jardin botanique de Buitenzorg. 1-723-2. HI 70579

Relationes annuae instituti geologici publici hungarici. Budapest = A magyar állami földtani intézet évi jelentese(i). Budapest. Magyar Állami Földt. Intéz. Évi Jel. See B–P–H 542/18.

Relationes annuae instituti regii geologiae hungarici. Budapest = A magyar királyi állami földtani intézet évi jelentése(i). Budapest. Magyar Kir. Állami Földt. Intéz. Évi Jel. See B–P–H 543/16.

Relatório e annuário do instituto de cacáo de Bahia. Bahia. 1931-?, 1932-39? Relat. Annuário Inst. Cacáo Bahia. HI 70580

Relatório anual do departamento de botânica do estado São Paulo. São Paulo. 1939-41. Relat. Anual Dept. Bot. Estado São Paulo. Superseded by: Relatório anual do instituto de botânico. HI 70581

Relatório anual do instituto de botânico. São Paulo. 1942+. Relat. Anual Inst. Bot. Preceded by: Relatório anual do departamento de botânica do estado. São Paulo. 5-3779-3. HI 70582

Relatório anual, instituto de investigação agronomica de Moçambique. Lourenço Marques. 1972+. Relat. Anual Inst. Invest. Agron. Moçambique. HI 70583

Relatório. Aquario Vasco de Gama. Lisbon. Relat. Aquar. Vasco de Gama. See B–P–H 761/7. HI 58539

Relatório cientifico, departamento e instituto de genetica, escola superior de agricultura Luiz de Queiroz. São Paulo. 1970+. Relat. Ci. Dep. Inst. Genet. Esc. Super. Agric. Luiz Queiroz. HI 70584

Relatório, commissão das linhas telegráficas estratégicas de Matto Grosso ao Amazonas. Anexo. [Forms part of: Publicações, commissão das linhas telegráficas estratégicas de Matto Grosso ao Amazonas.] Rio de Janeiro. Nos. 1-13, 1910-22. Relat. Commiss. Linhas Telegr. Estratég. Matto Grosso Amazonas. HI 70585

Relatório, fundação zoobotânica do Rio Grande do Sul. Porto Alegre. 1974+. Relat. Fund. Zoobot. Rio Grande Sul. HI 70586

Relatório, instituto de cacäo de Bahia = Relatório e annuário do instituto de cacáo de Bahia. Bahia.

Relazione accademica dell'accademia degli zelanti di aci-reale di scienze, lettere ed arti. Palermo. Relaz. Accad. Accad. Zelanti Aci-Reale Sci. See B–P–H 761/9. HI 58540

Relazione tecnica, stazione sperimentale di floricoltura "Orazio Raimondo". San Remo. ?-1943-55-? Relaz. Tec. Staz. Sperim. Floric. Orazio Raimondo. HI 70587

Remote sensing of environment; an interdisciplinary journal. New York. Vol. 1+, 1969+. Remote Sensing Environm. HI 75186

Remote sensing yearbook. London. 1986+. Remote Sensing Yearb. HI 75187

Rendiconto delle adunanza e dei lavori dell'accademia delle scienze. Naples. N.s. vols. 1-5, 1852-56; ser. 3, (1859-61), 1860-61. Rendiconto Adunanza Lav. Accad. Sci. Preceded by: Rendiconto delle adunanza e dei lavori dell'accademia delle scienze, Sezione della società reale borbonica di Napoli. Superseded by: Rendiconto della reale accademia delle scienze fisiche e matematiche. HI 70588

Rendiconto delle adunanza e dei lavori dell'accademia delle scienze. Sezione della società reale borbonica di Napoli. Naples. Vols. 1-9, 1842-50. Rendiconto Accad. Sci. Soc. Borbon. Napoli. Superseded by: Rendiconto delle adunanza e dei lavori dell'accademia delle scienze. HI 70589

Rendiconti degli istututi scientifici della università di Camerino. Camerino, Italy. Rendiconti Ist. Sci. Univ. Camerino. See B–P–H 761/15. HI 58541

Rendiconti dell'istituto lombardo di scienze e lettere. Scienze biologiche e mediche. Milan. Vol. 92+, 1957+. Rendiconti Ist. Lombardo Sci., Sci. Biol. Preceded by: Rendiconti dell reale istituto lombardo di scienze e lettere. Classe di scienze, matematiche e naturali. HI 70590

Rendiconti e memorie della r[eale] accademia di scienze, lettere ed arti degli Zelanti. Memorie della classe di scienze. Acireale. Ser. 3, vol. 1, 1901/02, 1903; ser. 4, vol. 1, 1922-26, 1927. Rendiconti Mem. Reale Accad. Sci. Lett. Arti Zelanti, Mem. Cl. Sci. For 1927-29 see: Memorie della r[eale] accademia di scienze, lettere ed arti degli Zelanti. Superseded by: Memorie e rendiconti della accademia di scienze, lettere e belle arti di Acireale. HI 70591

Rendiconto della reale accademia delle scienze fisiche e matematiche. Naples. Vols. 1-25, 1862-86; ser. 2, vols. 1-8, 1887-94; ser. 3, vols. 1-36, 1895-1930; ser. 4, vol. 1+, 1931+. Rendiconto Reale Accad. Sci. Fis.

Preceded by: Rendiconto delle adunanza e dei lavori dell'accademia delle scienze. 1-40-3. HI 70592

Rendiconti dell reale istituto lombardo di scienze e lettere. Milan. Ser. 2, vols. 1-69, 1868-1936. Rendiconti Reale Ist. Lombardo Sci. Preceded and superseded in part by Rendiconti dell reale istituto di scienze e lettere. Classe di scienze, matematiche e naturali. 3-2127-3. HI 70593

Rendiconti dell reale istituto lombardo di scienze e lettere. Classe di scienze, matematiche e naturali. Milan. Vols. 1-4, 1864-67; ser. 3, vols. 1-21, 1937-56 [ser. 3, vols. 1-21 also numbered ser. 2, vols. 70-91]. Rendiconti Reale Ist. Lombardo Sci., Cl. Sci. Mat. Preceded by: Atti dell reale istituto lombardo di scienze, lettere ed arti. For ser. 2 see: Rendiconti dell reale istituto lombardo di scienze e lettere. Superseded in part by Rendiconti dell'istituto lombardo di scienze e lettere. Scienze biologiche e mediche. 3-2127-3. HI 70594

Rendiconti del seminario della facoltà di scienze della r. università di Cagliari. Cagliari. Vol. 1+, 1931+. Rendiconti Seminario Fac. Sci. Univ. Cagliari. HI 70595

Rendiconto delle sessioni ordinarie dell'accademia delle scienze dell' istituto di Bologna. Bologna. 1833-96; n.s. vols. 1-11, 1896-1907. Rendiconto Sess. Ordinarie Accad. Sci. Ist. Bologna. Superseded by: Rendiconto delle sessioni della r. accademia delle scienze dell'istituto di Bologna. Classe di scienze fisiche. 1-40-1. HI 51241

Rendiconto delle sessioni della reale accademia delle scienze dell'istituto di Bologna. Classe di scienze fisiche. Bologna. N.s. vols. 12-57, 1907-54. Rendiconto Sess. Reale Accad. Sci. Ist. Bologna, Cl. Sci. Fis. Preceded by: Rendiconto delle sessioni ordinarie dell'accademia delle scienze dell'istituto di Bologna. Superseded by: Atti dell'accademia delle scienze dell'istituto di Bologna. Classe di scienze fisiche. Rendiconti. 1-40-1. HI 70596

Rendiconto, see Rendiconti

Renewable resources journal. Bethesda, MD. Vol. 1(2)+, 1982+. Renew. Resources J. Preceded by: Resources evaluation journal. HI 74538

Renfermant divers mémoires publiées par la société royale d'agriculture de Paris. Paris. Renferm. Divers Mém. Soc. Roy. Agric. Paris. See B–P–H 761/18. HI 58542

Renner research report. Renner, TX. Vols. 1-11, [1968]-72. Renner Res. Rep. HI 70597

Repatermesztési kutató intezet közlemények. Budapest. Vol. 7+, 1972+. Repatermesztési Kutató Intez. Közlem. Preceded by: Mezögazdasági növénynemesitési kutató intézet, közlemények. HI 70598

Repertoria d'agricoltura e di scienze economiche ed industriali. Turin. Repert. Agric. Sci. Econ. Industr. See B–P–H 772/1. HI 58646

Repertorio di agricoltura pratica e di economica domestica. Turin. Repert. Agric. Prat. Econ. Domest. See B–P–H 771/30. HI 58645

Repertorio fisico-natural de la isla de Cuba. Havana. Vols. 1-2, 1865-68. Repert. Fis.-Nat. Isla Cuba. 4-3569-1. HI 70599

Repertorio médico-farmacéutico y de ciencias auxiliares. Havana. Repert. Méd.-Farm. Ci. Auxiliares. See B–P–H 772/6. HI 58650

Repertorium für Anatomie und Physiologie. Berlin. Repert. Anat. Physiol. See B–P–H 772/2. HI 58647

Repertorium annuum litteraturae botanicae periodicae. Haarlem. Repert. Annuum Litt. Bot. Period. See B–P–H 772/4. HI 58648

Repertorium der gesammten deutschen Literatur. Leipzig. Repert. Gesammten Deutsch. Lit. See B–P–H 772/5. HI 58649

Repertorium des Neuesten und Wissenswürdigsten aus der gesammten Naturkunde. Berlin. Repert. Neuesten Wissenswürd. Gesammten Naturk. See B–P–H 772/7. HI 58651

Repertorium novarum specierum regni vegetabilis = Repertorium specierum novarum regni vegetabilis. Berlin. Repert. Spec. Nov. Regni Veg. See B–P–H 772/20.

Repertorium der periodischen botanischen Literatur vom Beginn des Jahres 1864 an. Regensburg. Repert. Period. Bot. Lit. See B–P–H 772/9. HI 58652

Repertorium für die Pharmacie. Nuremberg. Repert. Pharm. See B–P–H 772/10. HI 58653

Répertoire de pharmacie, de chimie, de physique, d'hygiène publique, de la médecine légale et de thérapeutique; réimpression générale des ouvrages périodiques publiées en France sur ces sciences. Brussels. Répert. Pharm. Chim. See B–P–H 772/11. HI 58654

Repertorium plantarum succulentarum. [British section of the international organization for succulent plant study.] Leeds. Repert. Pl. Succ. See B–P–H 772/13. HI 58655

Repertorium specierum novarum regni vegetabilis. Berlin. Repert. Spec. Nov. Regni Veg. See B–P–H 772/20. HI 58656

Repertorium specierum novarum regni vegetabilis. Beihefte. Berlin. Repert. Spec. Nov. Regni Veg. Beih. See B–P–H 772/21. HI 58657

Repertorium specierum novarum regni vegetabilis. Sonderbeiheft A. Berlin-Dahlem. Repert. Spec. Nov. Regni Veg. Sonderbeih. A. See B–P–H 772/22.

HI 58658

Repertorium specierum novarum regni vegetabilis. Sonderbeiheft B. Berlin-Dahlem. Repert. Spec. Nov. Regni Veg. Sonderbeih. B. See B–P–H 773/1. HI 58659

Repertorium specierum novarum regni vegetabilis. Sonderbeiheft C. Berlin Dahlem. Repert. Spec. Nov. Regni Veg. Sonderbeih. C. See B–P–H 773/2. HI 58660

Repertorium specierum novarum regni vegetabilis. Sonderbeiheft D. Berlin-Dahlem. Repert. Spec. Nov. Regni Veg. Sonderbeih. D. See B–P–H 773/3. HI 58661

Repertorium specierum novarum regni vegetabilis. Sonderbeiheft E. Berlin-Dahlem. Repert. Spec. Nov. Regni Veg. Sonderbeih. E. See B–P–H 773/4. HI 58662

Report (Annual), A F R C institute of plant science research and the John Innes institute. Norwich. 1987+. Rep. (Annual) A. F. R. C. Inst. Pl. Sc. Res. & John Innes Inst. Preceded by: Report, John Innes institute. HI 75238

Report and abstract of proceedings of the Croydon microscopical club. Croydon. 1871-77. Rep. Abstr. Proc. Croydon Microscop. Club. Superseded by: Proceedings and transactions of the Croydon microscopical and natural history club. HI 70600

Report and abstracts of proceedings and papers of the Brighton and Sussex natural history and philosophical society = Report of the Brighton and Sussex natural history and philosophical society. Brighton.

Report (Annual), academy of natural sciences of Philadelphia. Philadelphia, PA. 1919/20-21/22, 1921-23. Rep. (Annual) Acad. Nat. Sci. Philadelphia. Superseded by: Yearbook of the academy of natural sciences, Philadelphia. HI 70601

Reports of the academy of sciences of the Ukrainian S S R = Dopovidi Akademiï nauk Ukrayins'koi R S R. Kiev.

Report (Annual) of the acclimatisation society of Victoria. Melbourne, Vic. Vols. 1-73, 1861-1936. Rep. (Annual) Acclim. Soc. Victoria. 4-3732-1. HI 70602

Report (Annual) and accounts, Birmingham botanical and horticultural society Ltd. Birmingham. 1977/78+. Rep. (Annual) Accounts Birmingham Bot. Hort. Soc. HI 70603

Report and accounts of the forestry commission = Report of the forestry commission. London.

Report (Annual) and accounts, Herbert Whitley trust and Paignton zoological and botanical gardens. Paignton. 1947+. Rep. (Annual) Accounts Herbert Whitley Trust Paignton Zool. Bot. Gard. HI 70604

Report and accounts of the national botanic gardens of South Africa = Report of the national botanic gardens of South Africa. Claremont.

Reports and accounts of the national institute of agricultural botany. Cambridge, England. Rep. Accounts Natl. Inst. Agric. Bot. See B–P–H 762/2. HI 58546

Report of activities, geological survey of Canada. Ottawa. 1972/73+, 1973?+. Rep. Activities Geol. Surv. Canada. HI 70605

Report on the activities of the international seed testing association. Copenhagen. 1928-36. Rep. Activities Int. Seed Test. Assoc. HI 70606

Report on activities, nature conservation, provincial administration, Orange Free State. Bloemfontein. 1970/71+, 1971+. Rep. Activities Nat. Conservation Prov. Admin. Orange Free State. HI 70607

Report of activities and special developments, plant products division, Canada. Ottawa. Nos. 1-?, 19??-65. Rep. Activities Special Developm. Pl. Prod. Div. Canada. Superseded by: Activities report, plant products division, Canada. HI 70608

Report on the administration of the department of agriculture, united provinces of Agra and Oudh = Report of the department of agriculture, united provinces of Agra and Oudh. Allahabad.

Report on administration, forest department, Sind. Karachi. 1948?+. Rep. Admin. Forest Dept., Sind. Preceded by: Administration report (Annual), forest department, Sind. HI 70609

Report of the advisory committee on forestry. London. 1954+. Rep. Advis. Committee Forest. HI 70610

Report (Annual), advisory council for scientific and industrial research, Canada. Ottawa. 1917/18-23/24, 1918?-24. Rep. (Annual) Advis. Council Sci. Industr. Res. Canada. Superseded by: Report (Annual) of the national research council of Canada. 4-2908-1. HI 70611

Report of the advisory department, Harper Adams agricultural college. Newport, Shropshire. Nos. 1+, 1923+. Rep. Advis. Dept. Harper Adams Agric. Coll. HI 70612

Report (Annual), agri-horticultural society of India. Calcutta. ?-1913-36-? Rep. (Annual) Agri-Hort. Soc. India. HI 70613

Report of the agri-horticultural society of Madras = Report and proceedings of the agri-horticultural society of Madras. Madras.

Report of the agricultural branch of the department of land records and agriculture, Bengal. Calcutta. 1901-04. Rep. Agric. Branch Dept. Land Rec. Agric. Bengal. Preceded by: Report of the department of land records and agriculture, Bengal. Superseded by: Report of the agricultural department, Bengal. HI 70614

Report (Annual) of the agricultural chemist, Burma. Rangoon. 1921/22-32/33, 1922-34. Rep. (Annual) Agric. Chem. Burma. Superseded by: Report (Annual) of the agricultural stations, the agricultural chemist, the agricultural engineer, the assistant entomologist and the economic botanist. HI 70615

Report of the agricultural chemist, department of agriculture, Mysore. Bangalore. 1900-10. Rep. Agric. Chem. Dept. Agric. Mysore. Superseded by: Report, agricultural department, Mysore. HI 70616

Report of the agricultural college of Sweden. Series A = Lantbrukshögskolans meddelanden. Serie A. Uppsala.

Report of the agricultural college of Sweden. Series B = Lantbrukshögskolans meddelanden. Serie B. Uppsala.

Report of the agricultural department, Antigua. St. John. 1913-64-? Rep. Agric. Dept. Antigua. Preceded by: Report on the botanic stations and experimental plots, Antiqua. HI 70617

Report of the agricultural department, Assam. Shillong. 1882-85, 1912+. Rep. Agric. Dept. Assam. For the years 1886-1905 see: Report of the department of land records and agriculture, Assam. For the years 1906-11 see: Report, agricultural department, eastern Bengal and Assam. HI 70618

Report of the agricultural department, Bengal. Calcutta. 1905-17/18. Rep. Agric. Dept. Bengal. Preceded by: Report of the agricultural branch of the department of land records and agriculture, Bengal. Superseded by: Report, department of agriculture, Bengal. HI 70619

Report of the agricultural department in Bihar and Orissa. Patna. 1911/12-1935/36, 1912-1936. Rep. Agric. Dept. Bihar Orissa. Superseded by: Report of the agricultural department, Bihar and Report of the agricultural department, Orissa. HI 70620

Report of the agricultural department, Bihar. Patna. 1935/36-1939/40. Rep. Agric. Dept. Bihar. Preceded by: Report of the agricultural department in Bihar and Orissa. HI 70621

Report of the agricultural department, British Virgin Islands. Tortola, Bridgetown. 1914-29, 1953-? Rep. Agric. Dept. British Virgin Islands. Preceded by: Report, Virgin Islands agricultural experiment station. HI 70622

Report on the agricultural department, Dominica = Report (Annual), agricultural and forestry department, Dominica.

Report, agricultural department, eastern Bengal and Assam. Shillong. 1906/07-10/11, 1907-11. Rep. Agric. Dept. E. Bengal Assam. Preceded by: Report of the department of land records and agriculture, Assam. Superseded by: Report of the agricultural department, Assam. HI 70623

Report (Annual) of the agricultural department, Gold Coast. Accra. 1908-53/54, 1908-54. Rep. (Annual) Agric. Dept. Gold Coast Superseded by: Report of the department of agriculture, Ghana. HI 70624

Report on the agricultural department, Grenada. [Bridgetown?], Barbados, [St. George]. 1911-?, 1913-63-? Rep. Agric. Dept. Grenada. Preceded by: Report on the botanic stations, experimental plots, agricultural instruction, and land settlement scheme, Grenada. HI 70625

Report of the agricultural department, Hong Kong. Hong Kong. 1947-50. Rep. Agric. Dept. Hong Kong. Preceded by: Report of the botanical and forestry department, Hong Kong. Superseded by: Annual departmental report, director of agriculture, fisheries and forestry, Hong Kong. HI 70626

Report on the agricultural department, Montserrat. Plymouth, Barbados. 1914-42, 1953+. Rep. Agric. Dept. Montserrat. Preceded by: Report on the botanic station and experiment plots, Montserrat. From 1943-52 incorporated in: Report, department of agriculture, Leeward Islands. HI 70627

Report (Annual), agricultural department, Mysore. Bangalore. 1910/11-35/36, 1911-36. Rep. (Annual) Agric. Dept. Mysor Preceded by: Report of the agricultural chemist, department of agriculture, Mysore. Superseded by: Report, department of agriculture, Mysore. HI 70628

Report (Annual) on the agricultural department, Nigeria. Lagos. 1921-52. Rep. (Annual) Agric. Dept. Nigeria. Preceded by: Report (Annual) on the agricultural department, Nigeria. Northern provinces and Report on the agricultural department, Nigeria. Southern provinces. Superseded, in part, by: Report (Annual), department of agricultural research, federation of Nigeria; Report (Annual) of the department of agriculture of the eastern region of Nigeria; Report (Annual) of the department of agriculture of the northern region of Nigeria and Report (Annual) of the department of agriculture of the western region of Nigeria. HI 70629

Report (Annual) on the agricultural department, Nigeria. Northern provinces. Kaduna. 1912-20. Rep. (Annual) Agric. Dept. Nigeria, N. Prov. Superseded by: Report (Annual) on the agricultural department, Nigeria. HI 70630

Report (Annual) on the agricultural department, Nigeria. Southern provinces. Lagos. 1911-21. Rep. (Annual) Agric. Dept. Nigeria, S. Prov. Superseded by: Report (Annual) on the agricultural department, Nigeria. HI 70631

Report of the agricultural department, Orissa. Cuttack. 1936/37-39/40. Rep. Agric. Dept. Orissa. Preceded by: Report of the agricultural department in Bihar and Orissa. HI 70632

Report on the agricultural department, St. Kitts & Nevis. Bridgetown, St. John's, Antiqua. 1914-43, 1953-54-? Rep. Agric. Dept. St. Kitts Nevis. Preceded by: Report on the botanic station, St. Kitts-Nevis. From 1944-52 incorporated in: Report, department of agriculture, Leeward Islands. HI 70633

Report, agricultural department, St. Lucia. Castries. 1911-45. Rep. Agric. Dept. St. Lucia. Preceded by: Reports on the botanic station, agricultural school, experiment station and experiment plots, St. Lucia. HI 70634

Report on the agricultural department, St. Vincent. Kingstown. 1911+. Rep. Agric. Dept. St. Vincent. Preceded by: Report on the botanic stations, St. Vincent. HI 70635

Report, agricultural department, Sierra Leone. Freetown. 1912-20. Rep. Agric. Dept. Sierra Leone. Superseded by: Report, department of lands and forests, Sierra Leone. HI 70636

Report of the agricultural department, colony of Singapore. Singapore. 1948+. Rep. Agric. Dept. Colony Singapore. HI 70637

Report on the agricultural department, Tortola. 1914/15-27/28. 1915-28. Rep. Agric. Dept. Tortola. Preceded by: Report on the botanic and experiment station, Tortola. HI 70638

Report, agricultural department, university college, Dublin = Report, faculty of general agriculture, university college, Dublin.

Report of the agricultural experiment station, agricultural and mechanical college = Bulletin, new series. Report of the agricultural experiment station, agricultural and mechanical college. Auburn AL. Bull. New Ser. Rep. Agric. Exp. Sta. Agric. Coll. See B–P–H 266/25.

Report of the agricultural experiment station, Baton Rouge. Baton Rouge, LA. Rep. Agric. Exp. Sta. (Baton Rouge). See B–P–H 762/4. HI 58547

Report on the agricultural experiment stations in the central circle, United Provinces of Agra and Oudh = Report on agricultural stations in the central circle, United Provinces of Agra and Oudh. Allahabad.

Report, agricultural experiment station, Christiansted, Virgin Islands of the United States = Report, Virgin Islands agricultural experiment station. Tortola, Washington, DC, St. Croix.

Report (Annual), agricultural experiment station, college of agriculture, university of California agricultural experiment station = University of California agricultural experiment station. Annual report. Berkeley, CA. Univ. Calif. Agric. Exp. Sta. Annual Rep. See B–P–H 934/7.

Report of the agricultural experiment station of the governor general of Taiwan. [Taiwan sotokufu nogyo shikenjo hokoku.] [Taiwan]. Rep. Agric. Exp. Sta. Gov. Gen. Taiwan. See B–P–H 762/5. HI 58548

Report of the agricultural experiment station of the university of California. Berkeley, CA. 1924/25-38/40, 1925-40. Rep. Agric. Exp. Sta. Univ. Calif. Preceded by: Report of the college of agriculture and the agricultural experiment station of the university of California. HI 70639

Report (Annual), agricultural experiment stations, university of Georgia = Report (Annual) of the experiment stations, college of agriculture, Georgia university Athens, GA.

Report (Annual) on the agricultural experimental stations in Assam. Shillong. 1911-12, 1912. Rep. (Annual) Agric. Exp. Sta. Assam. Preceded by: Report (Annual) of the agricultural stations in eastern Bengal and Assam. Superseded by: Report (Annual) on the agricultural experiments and demonstrations in Assam. HI 70640

Report (Annual) on the agricultural experiments and demonstrations in Assam. Shillong. 1911/13-17/18, 1913-18. Rep. (Annual) Agric. Exp. Demonst. Assam. Preceded by: Report (Annual) on the agricultural experimental stations in Assam. HI 70641

Report (Annual) of the agricultural extension division, ministry of agriculture, forests and wildlife, Tanzania. Dar-es-Salaam. Vol. 1+, 1963+. Rep. (Annual) Agric. Extens. Div. Minist. Agric. Forests Wildlife Tanzania. Preceded by: Report (Annual), department of agriculture, Tanganyika.

Report, agricultural faculty, university college, Dublin = Report, faculty of general agriculture, university college, Dublin.

Report (Annual), agricultural and forestry department, Dominica. Roseau. 1911-42. Rep. (Annual) Agric. Forest. Dept. Dominica. Preceded by: Report on the botanic station experiment plots and agricultural school, Dominica. Superseded by: Report of the department of agriculture, Dominica. HI 70643

Report (Annual), agricultural and home economics experiment station = Report of the Iowa agricultural experiment station. Ames, IA.

Report (Annual) of the agricultural and horticultural research station, university of Bristol. Long Ashton. 1913-61. Rep. (Annual) Agric. Hort. Res. Sta. Univ. Bristol. Preceded by: Report of the national fruit and cider institute, Long Ashton. Superseded by: Report (Annual), Long Ashton research station, university of

Bristol. HI 70644

Report on agricultural investigation in Puerto Rico, Puerto Rico agricultural experiment station = Puerto Rico agricultural experiment station. Report on agricultural investigation in Puerto Rico. Mayagüez, PR. Puerto Rico Agric. Exp. Sta. Rep. Agric. Invest. Puerto Rico. See B–P–H 749/14.

Report, agricultural research council. London. 1931/33+, 1933+. Rep. Agric. Res. Council. HI 70645

Report, agricultural research council radiobiological laboratory = Report, radiobiological laboratory, agricultural research council. Wantage.

Report, agricultural research division, university of Botswana, Lesotho and Swaziland. Malkerns. 1970/71-75/76. Rep. Agric. Res. Div. Univ. Botswana Lesotho Swaziland. HI 70646

Report, agricultural research institute of Amani = Report, east african agricultural research institute, Amani. Dar-es-Salaam.

Report of the agricultural research institute and college, Pusa. Calcutta. 1907/14-15/16, 1914-16. Rep. Agric. Res. Inst. Coll. Pusa. Superseded by: Scientific reports of the agricultural research institute, Pusa. HI 70647

Report, agricultural research institute of Northern Ireland. Hillsborough. 1927/28+, 1928+. Rep. Agric. Res. Inst. Northern Ireland. HI 70648

Report (Annual), agricultural research institute of Ontario. Toronto. 1962/63-71/72, 1963-72. Rep. (Annual) Agric. Res. Inst. Ontario. Superseded by: Research report, agricultural research institute of Ontario. HI 70649

Report of the agricultural research service, Anglo-Egyptian Sudan. Wad Medani. 1935-38 [suspended 1939-47]. Rep. Agric. Res. Serv. Anglo-Egyptian Sudan. Preceded by: Report of the Gezira agricultural research service, Anglo-Egyptian Sudan. Superseded by: Report of the research division, ministry of agriculture, Sudan. HI 70650

Report of the agricultural research service, Jordan. Amman. Nos. 1+, 1952+. Rep. Agric. Res. Serv. Jordan. HI 70651

Report on the agricultural resources and capabilities of Hawaii, office of experiment stations, United States department of agriculture = U S department of agriculture, office of experiment stations. Report on the agricultural resources and capabilities of Hawaii. Washington, DC. U.S.D.A. Off. Exp. Sta. Rep. Agric. Resources Capabil. Hawaii. See B–P–H 944/3.

Report of the agricultural society of Sizuokaken. [Sizuokaken nokai-ho.] Shizuoka, Japan. Rep. Agric. Soc. Sizuokaken. See B–P–H 762/11. HI 58549

Report on the agricultural stations of the Bundelkland circle, Jhansi. Allahabad. 1925-30. Rep. Agric. Sta. Bundelkland Circle Jhansi. HI 70652

Report on agricultural stations in the central circle, United Provinces of Agra and Oudh. Allahabad. 1917-1929/30, 1917-30. Rep. Agric. Sta. Centr. Circle Unit. Prov. Agra Oudh. Preceded by: Report on the Cawnpore agricultural station. HI 70653

Report on the agricultural stations in the Central provinces. Nagpur. 1905-14. Rep. Agric. Sta. Centr. Prov. Preceded by: Report on the experimental farms in the Central Provinces. Superseded by: Report on the experimental farm attached to the agricultural college, Nagpur. HI 70654

Report (Annual) of the agricultural stations, the agricultural chemist, the agricultural engineer, the assistant entomologist and the economic botanist [title varies], Burma. Rangoon. 1915/16-20/21, 1916-21. Rep. (Annual) Agric. Sta., Burma. Superseded by: Report of the agricultural stations, Burma; Report (Annual) of the agricultural chemist, Burma; and Report of the economic botanist, Burma. HI 70655

Report of the agricultural stations, Burma. Rangoon. 1933/34-39/40, 1934-40. Rep. Agric. Sta., Burma. HI 70656

Report (Annual) of the agricultural stations in eastern Bengal and Assam. Shillong. 1905-10/11, 1905-11. Rep. (Annual) Agric. Sta. E. Bengal Assam. Superseded by: Report (Annual) on the agricultural experimental stations in Assam. HI 70657

Report on the agricultural stations in the eastern circle, Central Provinces. Nagpur. 1914/15-24/25, 1915-26. Rep. Agric. Sta. E. Circle Centr. Prov. Incorporated in: Report (Annual) of the experimental farms of the southern and eastern circles, Central provinces. HI 70658

Report on the agricultural stations in the eastern circle, United Province of Agra and Oudh. Allahabad. 1916/17-30, 1917-30. Rep. Agric. Sta. E. Circle Unit. Prov. Agra Oudh. HI 70659

Report on the agricultural stations of the hill circle. Allahabad. 1927-30. Rep. Agric. Sta. Hill Circle. HI 70660

Report on the agricultural stations in the Madras Presidency. Madras. ?-1919/20-?, ?-1920-52. Rep. Agric. Sta. Madras Presidency. HI 70661

Report on the agricultural stations in the northeastern circle. Allahabad. 1920-30. Rep. Agaric. Sta. NorthE. Circle. HI 70662

Report on the agricultural stations in the northern circle, Central Provinces. Nagpur. 1914/15-24/25, 1915-25. Rep. Agric. Sta. N. Circle Centr. Prov. Superseded by: Report on the experimental farms in the Central

Provinces, northern and plateau circles. HI 70663

Report on the agricultural stations of the Rohilkhand circle. Shahjahanpur. 1923-30. Rep. Agric. Sta. Rohilkhand Circle. HI 70664

Report of the agricultural stations of the southern circle, Burma. Rangoon. 1914/15, 1915. Rep. Agric. Sta. S. Circle Burma. HI 70665

Report on the agricultural stations in the southern circle = Report on the experimental farms in the southern circle, Central provinces. Nagpur.

Report on the agricultural stations at Tarnab and Haripur. 1911-32? Rep. Agric. Sta. Tarnab Haripur. HI 70666

Report on the agricultural stations of the western circle, United Provinces. Allahabad. 1913/14-30, 1914-30. Rep. Agric. Sta. W. Circle Unit. Prov. HI 70667

Report on the agricultural stations of the western circle, Central Provinces. Nagpur. 1914-25. Rep. Agric. Sta. W. Circle Centr. Prov. HI 70668

Report on agricultural work in the botanical gardens, Georgetown. Georgetown, Guyana. 1890-91/92. Rep. Agric. Wk. Bot. Gard. Georgetown. HI 70669

Report of the agricultural work, imperial department of agriculture for the West Indies. London. 1899/1901-12. Rep. Agric. Wk. Imp. Dept. Agric. West Indies. Preceded by: Report of the results obtained on the experimental fields at Dodds reformatory, Barbados. HI 70670

Report of the agricultural work in the southern Shan States [title varies slightly]. Rangoon. 1892/93-1932/33, 1893-1933. Rep. Agric. Wk. S. Shan States. HI 70671

Report (Annual) on agriculture and crown lands, Seychelles. Victoria, Seychelles. 1911-22. Rep. (Annual) Agric. Crown Lands Seychelles. Superseded by: Report (Annual), department of agriculture, Seychelles Islands. HI 70672

Report of the agriculture department, Tonga = Report of the director of agriculture, Tonga. Nuku'alofa.

Report (Annual), agriculture division of the eastern region of Nigeria = Report (Annual) of the department of agriculture of the eastern region of Nigeria. Enugu.

Report, agriculture of Ethiopia. Addis Ababa. 1955+. Rep. Agric. Ethiopia. HI 70673

Report on agriculture, Fiji = Report, department of agriculture, Fiji. Suva.

Report on agriculture in Malaya. Singapore. 1940-47. Rep. Agric. Malaya. Preceded by: Report of the department of agriculture, Malaya. Superseded by: Report of the department of agriculture, Federation of Malaya. HI 70674

Report of the agronomy section of the research office of the sugar manufacturers' association of Jamaica. Mandeville. 1946+. Rep. Agron. Sect. Res. Off. Sugar Manufacturers* Assoc. Jamaica. HI 70675

Report of the agrostologist, United States department of agriculture = U S department of agriculture, division of agrostology. Report of the agrostologist. Washington, DC. U.S.D.A. Div. Agrostol. Rep. Agrostol. See B–P–H 941/20.

Report (Annual), Alabama agricultural experiment station of the agricultural and mechanical college = Alabama agricultural experiment station of the agricultural and mechanical college. Annual report. Auburn, AL. Alabama Agric. Exp. Sta. Agric. Coll. Annual Rep. See B–P–H 67/10.

Report (Annual), Alaska agricultural experiment station = U S department of agriculture. Annual report of the Alaska agricultural experiment station. Washington, DC. U.S.D.A. Annual Rep. Alaska Agric. Exp. Sta. See B–P–H 940/11.

Report (Annual), Alaska agricultural experiment stations = Alaska agricultural experiment stations, annual report. Washington, DC.

Report (Annual) of the Albany and Rensselaer horticultural society. Albany, NY. Vols. 1-3, 1848-50. Rep. (Annual) Albany Rensselaer Hort. Soc. 1-130-2. HI 70676

Report (Annual), Alberta special crops and horticultural research center. Brooks, Alta. 1987+. Rep. (Annual) Albany Rensselaer Hort. Soc. HI 70677

Report (Annual), Alberta tree nursery and horticulture centre. Edmonton, Alta. 1981+. Rep. (Annual) Alberta Tree Nursery Hort. Centre. HI 74537

Report, Alfred P. Sloan foundation. New York, NY. Rep. Alfred Sloan Found. See B–P–H 762/14. HI 58550

Report, Allegheny forest experiment station. Philadelphia, PA. 1929-43. Rep. Allegheny Forest Esp. Sta. Incorporated in: Report, northeastern forest experiment station, New Haven, Conn. HI 70678

Reports (Annual), american association of botanical gardens and arboretums. Pine Mountain, GA. ?-1965(Oct.)+. Rep. (Annual) Amer. Assoc. Bot. Gard. Arbor. HI 70679

Report (Annual), american breeders association. Washington, DC. Vols. 7-8, 1909-10. Rep. (Annual) Amer. Breed. Assoc. Preceded by: American breeders' association report. 1-228-1. HI 70680

Report (Annual), american museum of natural history. New York. No. 1+, 1869+, 1870+. Rep. (Annual) Amer. Mus. Nat. Hist. 1-290-2. HI 70681

Report (Annual), american type culture collection. Rockville, MD. 1975+, [1976?]+. Rep (Annual) Amer.

Type Cult. Collect. HI 70682

Report (Annual) ... [i.e. Annual report ...], N.B. "(Annual)" is disregarded in the filing order.

Report of the annual meeting, herb society of America = Report, herb society of America.

Report of annual meeting, New Zealand ecological society. Wellington, N.Z. 1953. Rep. Annual Meeting New Zealand Ecol. Soc. Preceded by: Report of ecological conference, New Zealand ecological society. Superseded by: Proceedings of the New Zealand ecological society. HI 70683

Report (Annual), annual meeting, Oregon horticultural society = Report (Annual), Oregon state horticultural society. Corvallis, Salem, Portland, OR.

Report of the annual meeting of the south african association for the advancement of science = Report of the (meeting of the) south african association for the advancement of science. Cape Town.

Report of the annual meeting of the Washington state horticultural association = Proceedings of the Washington state horticultural association. Pullman, WA. Proc. Wash. State Hort. Assoc. See B–P–H 738/14.

Report of the Antigua agricultural department = Report of the agricultural department, Antigua. St. John.

Report (Annual) of the Aomori apple experiment station. [Aomori-ken ringo shikenjo gyomu nenpo.] Aomori. ?-ca.1987+. Rep. (Annual) Aomori Apple Exp. Sta. HI 70684

Report (Annual), Arizona agricultural experiment station = University of Arizona. Arizona agricultural experiment station. Annual report. Tucson, AZ. Univ. Arizona Agric. Exp. Sta. Annual Rep. See B–P–H 933/13.

Report (Annual) of the Arkansas agricultural experiment station = Report (Annual) of the Arkansas agricultural experiment station at Arkansas industrial university. Fayetteville, AR.

Report (Annual) of the Arkansas agricultural experiment station at Arkansas industrial university. Fayetteville, AR. 1888+. Rep. (Annual) Arkansas Exp. Sta. Arkansas Industr. Univ. HI 70685

Report (Annual) of the Arkansas geological survey. Little Rock, AR. 1888+. Rep. (Annual) Arkansas Geol. Surv. HI 70686

Report (Annual) of the Arkansas state horticultural society. Fort Smith, AR. Vol. 1, 1893? Rep. (Annual) Arkansas State Hort. Soc. 1-481-3. HI 70687

Report of the Armagh natural history and philosophical society. Armagh, Northern Ireland. Rep. Armagh Nat. Hist. Soc. See B–P–H 762/16. HI 58551

Report, Arnold arboretum. Jamaica Plain, MA. 1984+. Rep. Arnold Arbor. HI 70688

Report, ashmolean natural history society of Oxfordshire. Oxford. 1901-07. Rep. Ashmolean Nat. Hist. Soc. Oxfordshire. Preceded by: Report, Oxfordshire natural history society. Superseded by: Proceedings and report, ashmolean natural history society of Oxfordshire. HI 70689

Report (Annual), asian vegetable research and development center. Tainan. 1972/73-74, 1973-75? Rep. (Annual) Asian Veg. Res. Developm. Center. Superseded by: A V R D C progress report. HI 70690

Report (Annual), asiatic society of Bengal. Calcutta. 1971+. Rep. (Annual) Asiat. Soc. Bengal. HI 70691

Reports of the association of american geologists and naturalists. Boston, MA. Rep. Assoc. Amer. Geol. See B–P–H 762/17. HI 58552

Report (Annual), association of official seed certifying agencies = Report (Annual), international crop improvement association. St. Paul, MN.

Report on the Atarra experimental station, district Banda. Allahabad. 1911-16. Rep. Atarra Exp. Sta. Distr. Banda. HI 70692

Report (Annual). Association internationale des botanistes, Jena. Jena. 1910/11-12/13. Rep. (Annual) Assoc. Int. Bot. Jena. 1-524-2. HI 70693

Report (Annual), Auckland acclimatisation society. Auckland. 1923-34. Rep. (Annual) Auckland Acclim. Soc. HI 70694

Report of the Auckland institute and museum. Auckland. Rep. Auckland Inst. Mus. See B–P–H 762/18. HI 58553

Report (Annual) of the australian conservation foundation. Eastwood, N.S.W. 1967/68+, 1968+. Rep. (Annual) Austral. Conservation Found. HI 70695

Report (Annual), australian national botanic gardens. Canberra, A.C.T. 1984/85+, 1985?+. Rep. (Annual) Austral. Natl. Bot. Gard. HI 70696

Report, australian national parks and wildlife service. Canberra, A.C.T. 1975/76+. Rep. Austral. Natl. Parks Wildlife Serv. HI 70697

Report (Annual), Ayuba agricultural research institute. Lyallpur. 1963/64+. Rep. (Annual) Ayuba Agric. Res. Inst. HI 70698

Report B C - X = Information report B C - X. Victoria, B.C.

Report and balance sheet of the national botanic gardens of South Africa = Report of the national botanic gardens of South Africa. Claremont.

Report (Annual), banana board research department,

Jamaica. Kingston, Jamaica. Vol. 1+, 1956/57+, 1959+. Rep. (Annual) Banana Board Res. Dept. Jamaica. HI 70699

Report (Annual), Bangladesh tea research institute. Srimangal. 1973+. Rep. (Annual) Bangladesh Tea Res. Inst. Preceded by: Report (Annual), Bangladesh tea research station. HI 70700

Report (Annual), Bangladesh tea research station. Srimangal. 19??-72? Rep. (Annual) Bangladesh Tea Res. Sta. Superseded by: Report (Annual), Bangladesh tea research institute. HI 70701

Report (Annual), bean improvement co-operative. Lincoln, NE, Geneva, NY. 1959+. Rep. (Annual) Bean Improv. Co-op. HI 70702

Report (Annual), Bedford institute of oceanography. Dartmouth, N.S. Nos. 1-5, 1962-66. Rep. (Annual) Bedford Inst. Oceanogr. Superseded by: Biennial review, Bedford institute of oceanography. HI 70703

Report of the Belfast natural history and philosophical society. Belfast. 1837-78-? Rep. Belfast Nat. Hist. Philos. Soc. Superseded by: Report and proceedings of the Belfast natural history and philosophical society. HI 70704

Report (Annual) of the Belfast naturalists' field club. Belfast. Nos. 1-10, 1863/64-72/73, 1863-73. Rep. (Annual) Belfast Naturalists' Field Club. Superseded by: Report (Annual) and proceedings of the Belfast naturalists' field club. HI 70705

Report, Benares agricultural station. Allahabad. 1912-16. Rep. Benares Agric. Sta. Incorporated in: Report on the agricultural stations in the eastern circle, United Provinces of Agra and Oudh. HI 70706

Report (Annual), Berkshire, Buckinghamshire & Oxfordshire naturalists' trust. Oxford. 1961+, [1962]+. Rep. (Annual) Berkshire Buckinghamshire Oxfordshire Naturalists' Trust. HI 70707

Report, Bermuda biological station. St. George's. 1975+. Rep. Bermuda Biol. Sta. HI 70708

Report (Annual), Bernice P[auahi] Bishop museum. Honolulu, HI. 1899+. Rep. (Annual) Bernice P. Bishop Mus. HI 70709

Report (Annual), Bickelhaupt arboretum. Clinton, IA. ?-1979/80+, ?-1980?+. Rep. (Annual) Bickelhaupt Arbor. HI 70710

Report (Annual) of the biological laboratory, Cold Spring Harbor. Cold Spring Harbor, NY. ?-1924-64. Rep. (Annual) Biol. Lab. Cold Spring Harbor. Superseded by: Report (Annual) of the Cold Spring Harbor laboratory of quantitative biology. HI 70711

Report (Annual) of the biological laboratory, Long Island biological association = Report (Annual) of the biological laboratory, Cold Spring Harbor. Cold Spring Harbor, NY.

Report (Annual), biological services division, department of agriculture, Western Australia. [Perth?], W.A. 1973/74+, 1974?+. Rep. (Annual) Biol. Serv. Div. Dept. Agric. W. Australia. HI 70712

Report (Annual) of biological works, faculty of science, Osaka university. Osaka. Vols. 1-19, 1953-72. Rep. Biol. Works. Ceased publication. HI 70713

Report, biology curators' group. No. 1+, 1980+. Rep. Biol. Curators' Group. HI 70714

Report (Annual), biology program, cooperative part studies unit, university of Washington. Seattle, WA. 1978+. Rep. (Annual) Biol. Program Coop. Part Stud. Unit Univ. Washington. HI 70715

Report, Birmingham botanical and horticultural society Ltd = Report (Annual) and accounts, Birmingham botanical and horticultural society Ltd. Birmingham.

Report (Annual) of the Birmingham natural history and microscopical society [title varies]. Birmingham. 1872-93. Rep. (Annual) Birmingham Nat. Hist. Microscop. Soc. 1-711-1. HI 70716

Report (Annual) of the Birmingham natural history and philosophical society. Birmingham. Vol. 1+, 1894+. Rep. (Annual) Birmingham Nat. Hist. Philos. Soc. 1-711-2. HI 70717

Report of the board of agriculture, Bermuda. Hamilton, Bermuda. 1908-24. Rep. Board Agric. Bermuda. Superseded by: Report (Annual), department of agriculture, Bermuda. HI 70718

Report of the board of agriculture, British Guiana. Georgetown. 1890-1904. Rep. Board Agric. British Guiana. Incorporated in: Report of the botanic gardens, British Guiana. HI 70719

Report of the board of agriculture and department of public gardens and plantations, Jamaica. Kingston, Jamaica. 1901-07. Rep. Board Agric. Dept. Public Gard. Pl. Jamaica. Preceded by: Report, public gardens and plantations, Jamaica. Superseded by: Report on the department of agriculture, Jamaica. HI 70720

Report, board of agriculture and forestry, Hawaii = Report, board of commissioners of agriculture and forestry, Hawaii. Honolulu, HI.

Report of the board of agriculture, Jamaica = Report of the board of agriculture and department of public gardens and plantations, Jamaica. Kingston, Jamaica.

Report (Annual), board of agriculture, New Brunswick. Fredericton. 1860-69. Rep. (Annual) Board Agric. New Brunswick. Superseded by: Report (Annual), department of agriculture, New Brunswick. HI 70721

Report (Annual) of the board of the botanic gardens of Adelaide and state herbarium. Adelaide, S.A. No.

124+, 1978/79+. Rep. (Annual) Board Bot. Gard. Adelaide State Herb. Preceded by: Report (Annual) of the board of governors of the botanic gardens (South Australia). HI 70722

Report, board of commissioners of agriculture and forestry, Hawaii. Honolulu, HI. 1900/01-56/58, 1901-59. Rep. Board Commiss. Agric. Forest. Hawaii. Superseded by: Report, department of agriculture, Hawaii. HI 70723

Report of the board of commissioners of agriculture and forestry, Hawaii. Division of plant inspection = Report of the division of plant inspection, board of agriculture and forestry, Hawaii. Honolulu, HI.

Report of the board of control of the state agricultural experiment station at Amherst = Massachusetts agricultural experiment station. Annual report. Amherst, MA. Mass. Agric. Exp. Sta. Annual Rep. See B–P–H 549/11.

Report (Annual) of the board of governors of the botanic garden (South Australia). Adelaide, S.A. 1948-77/78. Rep. (Annual) Board Gov. Bot. Gard. (S. Australia). Preceded by: Report on the progress and condition of the botanic garden and government plantations, Adelaide. Superseded by: Report (Annual) of the board of the botanic gardens of Adelaide and state herbarium. HI 70724

Report, board of greenkeeping research. Bingley. 1931-38. Rep. Board Greenkeeping Res. HI 70725

Report (Annual) of the board of regents, Smithsonian institution. Washington, DC. Vol. 1+, 1846+, 1847+. Rep. (Annual) Board Regents Smithsonian Inst. 5-3886-3. HI 70726

Report (Annual), board of scientific advice for India. Calcutta. 1902/03-22/23, 1903-23. Rep. (Annual) Board Sci. Advice India. HI 70727

Report by the board of trustees of the royal botanic gardens, Kew. Kew. 1984/87+, 1988+. Rep. Board Trustees Roy. Bot. Gard. Kew. Preceded by: Report (Annual), royal botanic gardens, Kew. HI 70729

Report (Annual) of the Bombay branch, royal asiatic society. Bombay. 1931-33. Rep. (Annual) Bombay Branch Roy. Asiat. Soc. HI 70730

Report of the Bose research institute. Calcutta. 1952+. Rep. Bose Res. Inst. HI 70731

Report on the botanic and experiment station, Tortola. 1913/14, 1914. Rep. Bot. Exp. Sta. Tortola. Preceded by: Report on the experiment station, Tortola. Superseded by: Report on the agricultural department, Tortola. HI 70732

Report on the botanic garden and agricultural school, Dominica = Report on the botanic station, experiment plots and agricultural school, Dominica. Roseau.

Report of the botanic garden, South Australia = Report (Annual) of the board of governors of the botanic garden (South Australia) and Report on the progress and condition of the botanic garden and government plantations. Adelaide, S.A.

Report (Annual) of the botanic garden syndicate, Cambridge. Cambridge. 1885+. Rep. (Annual) Bot. Gard. Syndic. Cambridge. HI 70733

Report of the botanic gardens, British Guiana. Georgetown, British Guiana. 1880-1919. Rep. Bot. Gard. British Guiana. Superseded by: Report of the department of science and agriculture, British Guiana. HI 70734

Report, botanic gardens conservation co-ordinating body. Kew. No. ?-14+, ?-1985+. Rep. Bot. Gard. Conservation Co-ord. Body. HI 70735

Report (Annual), botanic gardens department, colony of Singapore = Report (Annual), botanic gardens department, Singapore. Singapore.

Report (Annual), botanic gardens department, Singapore. Singapore. 1948-64. Rep. (Annual) Bot. Gard. Dept. Singapore. Preceded by: Report (Annual) of the director of gardens of the Straits Settlements. Superseded by: Report (Annual) of the botanic gardens, Singapore. HI 70736

Report on botanic gardens and domains. Sydney, N.S.W. 1848-58, 1870, 1878, 1897-1903. Rep. Bot. Gard. Domains. Superseded by: Report on botanic gardens and government domains of New South Wales. HI 70737

Report (Annual) of botanic garden, faculty of science, Kanazawa university. Kanazawa. Vol. 1+, 1968+. Rep. (Annual) Bot. Gard. Fac. Sci. Kanazawa. Univ. HI 70738

Report (Annual) on the botanic gardens and forest department, Singapore. Singapore. 1889-95. Rep. (Annual) Bot. Gard. Forest Dept. Singapore. Preceded by: Report (Annual) on the botanic gardens, Singapore and Report (Annual) on the forests departments, Singapore, Penang and Malacca. Superseded by: Report (Annual) on the botanic gardens, Singapore. HI 70739

Report on botanic gardens and government domains of New South Wales. Sydney, N.S.W. 1904-15. Rep. Bot. Gard. Gov. Domains New South Wales. Preceded by: Report on botanic gardens and domains. Superseded by: Report of director of the botanic gardens, government domains and centennial park. HI 70740

Report of the botanic gardens and government herbarium. Cape Town. 1856-1904-? Rep. Bot. Gard. Gov. Herb. HI 70741

Report (Annual) on the botanic gardens, Singapore.

Singapore. 1884-86, 1896-1902, 1965-70. Rep. Bot. Gard. Singapore. Preceded by: Report on the botanic and zoological gardens, Singapore. For 1889-95 see: Report (Annual) on the botanic gardens and forest department, Singapore. For 1905-14 see: Report (Annual) on the botanic gardens, Singapore and Penang. For 1915-39 see: Report (Annual) of the director of gardens of the Straits Settlements. For 1948-64 see: Report (Annual), botanic gardens department, Singapore. Superseded by: Singapore botanic gardens annual report. HI 70742

Report (Annual) on the botanic gardens, Singapore and Penang. Singapore. 1904-13, 1905-14. Rep. (Annual) Bot. Gard. Singapore Penang. Preceded by: Report (Annual) on the botanic gardens, Singapore. Superseded by: Report (Annual) of the director of gardens of the Straits Settlements. HI 70743

Report (Annual) of the botanic gardens, state of Singapore = Report (Annual) on the botanic gardens, Singapore. Singapore.

Report (Annual) of the botanic gardens, Straits Settlements = Report (Annual) of the director of gardens of the Straits Settlements. Singapore.

Report, botanic gardens and their work, British Guiana = Report of the botanic gardens, British Guiana. Georgetown, British Guiana.

Report on the botanic and zoological gardens, Singapore. Singapore. 1882-83. Rep. Bot. Zool. Gard. Singapore. Superseded by: Report (Annual) on the botanic gardens, Singapore. HI 70744

Report on the botanic station and agricultural education, St. Kitts-Nevis = Report on the botanic station, St. Kitts-Nevis. [Bridgetown?], Barbados.

Reports on the botanic station, agricultural school, experiment station and experiment plots, St. Lucia. Castries. 1901/02-10/11, 1902-11. Rep. Bot. Sta. Agric. School Exp. Sta. Exp. Plots St. Lucia. Preceded by: Report (Annual) on the botanical garden, St. Lucia. Superseded by: Report, agricultural department, St. Lucia. HI 70752

Report on the botanic station, agricultural school and land settlement scheme, St. Vincent = Report on the botanic stations, St. Vincent. [Bridgetown?], Barbados

Report on the botanic station, economic experiments and agricultural teaching, Antigua = Report on the botanic station and experimental plots, Antigua. [Bridgetown?], Barbados.

Report on the botanic stations, experimental plots, agricultural instruction, and land settlement scheme, Grenada. St. George. 1896-1911. Rep. Bot. Sta. Exp. Plots Agric. Instruction Land Settlem. Scheme Grenada. Superseded by: Report on the Agricultural Department, Grenada. HI 70750

Report on the botanic station, experiment plots and agricultural school, Dominica. Roseau. 1890/91-1910/11, 1891-1911. Rep. Bot. Sta. Exp. Plots Agric. School Dominica. Superseded by: Report, agricultural (and forestry) department, Dominica. HI 70747

Report on the botanic station and experimental plots, Antigua. [Bridgetown?], Barbados. 1890-1915, 1890-1916. Rep. Bot. Sta. Exp. Plots Antigua. Superseded by: Report of the agricultural department, Antigua. HI 70748

Report, botanic station and experimental plots, Montserrat. Barbados. 1900/01-13/14, 1900-14. Rep. Bot. Sta. Exp. Plots Montserrat. Superseded by: Report on the agricultural department, Montserrat. HI 70749

Report on the botanic station and experimental plots, St. Kitts-Nevis = Report on the botanic station, St. Kitts-Nevis. Bridgetown?], Barbados.

Report on the botanic station, Lagos. Lagos. 1894?-97? Rep. Bot. Sta. Lagos. HI 70745

Report on the botanic stations, Montserrat = Report, botanic station and experimental plots, Montserrat. Barbados.

Report on the botanic station, St. Kitts-Nevis. [Bridgetown?], Barbados. 1900-14. Rep. Bot. Sta. St. Kitts-Nevis Superseded by: Report on the agricultural department, St. Kitts & Nevis. HI 70751

Report on the botanic stations, St. Vincent. [Bridgetown?], Barbados. 1901-11. Rep. Bot. Sta. St. Vincent. Superseded by: Report on the agricultural department, St. Vincent. HI 70753

Report on the botanic station, Tobago. Tobago. 1901?-06?. Rep. Bot. Sta. Tobago. HI 70746

Report on the botanic stations, Virgin Islands = Report on the agricultural department, Tortola.

Report. Botanical club of Canada. Ottawa? Rep. Bot. Club Canada. See B–P–H 762/22. HI 58554

Report. Botanical exchange club. London. Rep. Bot. Exch. Club. See B–P–H 763/1. HI 58555

Report (Annual) on the botanical garden, St. Lucia. Castries. 1887-99. Rep. (Annual) Bot. Gard. St. Lucia. Superseded by: Reports on the botanic station, agricultural school, experiment station and experiment plots, St. Lucia. HI 70754

Report, botanical department, Trinidad. Trinidad. ?-1889-1908? Rep. Bot. Dept. Trinidad. HI 70755

Report of the botanical division, rubber research institute of Malaya. Kuala Lumpur. 1949-51. Rep. Bot. Div. Rubber Res. Inst. Malaya. Previously and subsequently incorporated in: Report, rubber research institute of Malaya. HI 70756

Report, botanical exchange club of the British Isles = Botanical exchange club of the British Isles. Report. Manchester, England. Bot. Exch. Club Brit. Isles Rep. See B–P–H 220/1.

Report, botanical exchange club and society of the British Isles = Botanical exchange club and society of the British Isles. Manchester, England. Bot. Exch. Club Soc. Brit. Isles See B–P–H 220/2.

Report of the botanical and forestry department, Hong Kong. Hong Kong. 1880-46. Rep. Bot. Dept. Hong Kong. Superseded by: Report of the agricultural department, Hong Kong and Report of the forestry department, Hong Kong. HI 70757

Report on the botanical gardens. Melbourne. 1851-59. Rep. Bot. Gard., Melbourne. Superseded by: Report of the government botanist and director of the botanic and zoologic garden, Melbourne. HI 70758

Reports from the botanical institute, university of Aarhus = Reports, botaniske institut, Aarhus universitet. Risskov.

Report on the botanical and natural history society of Oxford. Oxford, England. Rep. Bot. Soc. Oxford. See B–P–H 763/6. HI 58558

Report (First Second Third Annual) of the botanical office of the province of British Columbia. Victoria, B.C. Vol. 1(1-3), 1913-15, [1914-16]. Rep. (Annual) Bot. Off. Prov. British Columbia. HI 70759

Reports of the botanical record club. London. Rep. Bot. Rec. Club. See B–P–H 763/3. HI 58556

Report of the botanical society of the British Isles. Manchester, England. Rep. Bot. Soc. Brit. Isles. See B–P–H 763/4. HI 58557

Report (Annual), botanical society of South Africa. Kirstenbosch. 1932+. Rep. (Annual) Bot. Soc. South Africa. HI 70760

Report of the botanical survey of India. Calcutta. 1894+. Rep. Bot. Surv. India. Preceded by: Report of the director of the botanical survey of India. HI 70761

Reports, botaniske institut, Aarhus universitet. Risskov. No. 1+, 1976+. Rep. Bot. Inst., Aarhus Univ. HI 70762

Report of the botanist, U S department of agriculture = Department of agriculture. Report of the botanist. Washington, DC. Dept. Agric. Rep. Bot. See B–P–H 343/8.

Report (Annual) of the botany branch and Queensland herbarium. Brisbane, Qld. 1968-87. Rep. (Annual) Bot. Branch Queensland Herb. Preceded by: Report of botany section, government botanist, Queensland. Superseded by: Botany focus. HI 70763

Report, botany department, university college, Swansea. Swansea. 1957+. Rep. Bot. Dept. Univ. Coll. Swansea. HI 70764

Report, botany department, Victoria university of Wellington. Wellington, N.Z. No. ?-5+, ?-1976+. Rep. Bot. Dept. Victoria Univ. Wellington. HI 70765

Report (Annual), botany division, department of agriculture = Department of agriculture. Botany division. Annual report. Washington, DC. Dept. Agric. Bot. Div. Annual Rep. See B–P–H 342/16.

Report (Annual) of the botany section, government botanist, Queensland. [Brisbane, Qld.] 1966. Rep. (Annual) Bot. Sect. Gov. Bot. Queensland. Superseded by: Report (Annual) of botany branch and Queensland herbarium. HI 70766

Report, Botley fruit station. London. 1950-52. Rep. Botley Fruit Sta. HI 70767

Report of the Bournemouth natural science society = Report of the Bournemouth society of natural science. Bournemouth.

Report of the Bournemouth society of natural science. Bournemouth. 1904-14. Rep. Bournemouth Soc. Nat. Sci. HI 70768

Report (Annual), Bowman's Hill state wild flower preserve. Washington Crossing, PA. ?-1962-71+. Rep. (Annual) Bowman's Hill State Wild Fl. Preserve. HI 70769

Report (Annual), Boyce Thompson institute for plant research. Yonkers, NY. 1971+. Rep. (Annual) Boyce Thompson Inst. Pl. Res. HI 70770

Report, Brighton and Hove natural history and philosophical society. Brighton. 1938-40, 1949, 1953-54 [not issued for the intervening years]. Rep. Brighton Hove Nat. Hist. Philos. Soc. Preceded by: Abstracts of papers read before the Brighton and Hove natural history and philosophical society. HI 70771

Report of the Brighton and Sussex natural history and philosophical society. Brighton. 1855-87. Rep. Brighton Sussex Nat. Hist. Philos. Soc. Superseded by: Abstracts of papers read before the Brighton and Sussex natural history and philosophical society. HI 70772

Report (Annual), british antarctic survey. Cambridge. 1969/70+, 197?+. Rep. (Annual) Brit. Antarc. Surv. HI 70773

Reports of the british association for the advancement of science. London. Rep. Brit. Assoc. Advancem. Sci. See B–P–H 763/8. HI 58559

Report of the british bryological society. Cambridge. Vols. 1-4, 1923-45. Rep. Brit. Bryol. Soc. Preceded by: Report, moss exchange club. Superseded by: Transactions, british bryological society. 1-777-3. HI 70774

Report of the British Columbia department of agriculture.

Victoria, B.C. 1891+. Rep. British Columbia Dept. Agric. HI 70775

Report (Annual) of the British Columbia fruit-growers' association. Victoria, B.C. Vol. 1+, 1889+. Rep. British Columbia Fruit-Growers' Assoc. 1-779-3. HI 70776

Report of the British Columbia provincial museum of natural history and anthropology. Victoria, B.C. 1912-68. Rep. British Columbia Prov. Mus. Nat. Hist. Superseded by: Syesis. 1-779-1. HI 70777

Report (Annual), british cotton growing association. Manchester. Vols. 1-63, 1903-70. Rep. (Annual) Brit. Cotton Growing Assoc. HI 70778

Report on the British Museum (Natural History). London. 1963/65+, 1966+. Rep. Brit. Mus. (Nat. HIst.) HI 70779

Report (Annual) of British New Guinea. Melbourne, Vic. 1886-1906. Rep. (Annual) Brit. New Guinea. Superseded by: Reports (Annual) of Papua. HI 70780

Report on british palaeobotany and palynology. London. 1972/75+, 1976+. Rep. Brit. Paleobot. Palynol. HI 70781

Report (Annual), british society for the promotion of vegetable research. Warwick. Vol. 1+, 1951+. Rep. (Annual) Brit. Soc. Promot. Veg. Res. HI 70782

Report (Annual), British West Indies central sugar cane breeding station, Barbados. 1st-32nd, 1934-65. Rep. (Annual) British West Indies Centr. Sugar Cane Breed. Sta. Barbados. Superseded by: Report (Annual), West Indies central sugar cane breeding station. HI 70783

Report, British West Indies sugar research scheme. Trinidad. 1952+. Rep. British West Indies Sugar Res. Scheme. HI 70784

Report (Annual), Brooklyn botanic garden. New York. 1958/61+. Rep. (Annual) Brooklyn Bot. Gard. HI 70785

Report on bulb experiments, Kirton agricultural institute. Boston. Nos. 1-7, 1931-40. Rep. Bulb Exp. Kirton Agric. Inst. HI 70786

Report (Annual) and bulletin, museums of Malawi. Blantyre. 1980+. Rep. (Annual) Bull. Mus. Malawi. Preceded by: Report and bulletin, the Nyasaland museum. HI 70787

Report (Annual) and bulletin, the Nyasaland museum. Blantyre. 1960/61-?, 1961?-79? Rep. (Annual) Bull. Nyasaland Mus. Superseded by: Report and bulletin, museums of Malawi. HI 70788

Report on the Burdwan agricultural station. 1894-1913. Rep. Burdwan Agric. Sta. Incorporated in: Report of the agricultural department, Bengal. HI 70789

Report, bureau of agriculture, Philippine Islands. Manila. 1917-27. Rep. Bur. Agric. Philippine Islands. HI 70790

Report, bureau of entomology and plant quarantine, United States department of agriculture = U S department of agriculture. Bureau of entomology and plant quarantine. Report. Washington, DC. U.S.D.A. Bur. Entomol. Rep. See B–P–H 940/20.

Report (Annual), bureau of forestry, Pennsylvania. Harrisburg, PA. 1979+. Rep. (Annual) Bur. Forest. Pennsylvania. HI 70791

Report of the bureau of forestry, Philippine Islands. Report, Philippine forestry bureau = Report (Annual) of the director of forestry of the Philippine Islands. Manila.

Report of the bureau of government laboratories in the Philippine Islands. Manila. Nos. 1-4, 1901/02-04/05, 1902-06. Rep. Bur. Gov. Lab. Philippine Islands. Superseded by: Report of the bureau of science, Philippine Islands. HI 70792

Report, bureau of plant industry, Philippine Islands. Manila. 1930-39. Rep. Bur. Pl. Industr. Philippine Islands. HI 70793

Report (Annual) of the bureau of science, Philippine Islands. Manila. Nos. 5-37, 1905/06-38, 1905-38. Rep. (Annual) Bur. Sci. Philippine Islands. Preceded by: Report of the bureau of government laboratories in the Philippine Islands. HI 70794

Report (Annual), bureau of sugar experiment stations, Queensland. Brisbane, Qld. 1901+. Rep. (Annual) Bur. Sugar Exp. Sta. Queensland. HI 70795

Report of Burton-on-Trent natural history and archaeological society. Burton-on-Trent. 1884-92. Rep. Burton-on-Trent Nat. Hist. Archaeol. Soc. HI 70796

Report, Bury natural history society. Bury. 1868-71. Rep. Bury Nat. Hist. Soc. HI 70797

Report, Buxton field club. Buxton. 1969/71-?, 1971-77. Rep. Buxton Field Club. HI 70798

Report on cacao research. St. Augustine, Port of Spain. 1931+. Rep. Cacao Res. HI 70799

Report (Annual), calavo growers of California. Los Angeles, CA. Vol. 4+ 1927+. Rep. (Annual) Calavo Growers Calif. Preceded by: California calavo growers exchange. HI 70800

Report (Annual) of the Califor naturalist club. Charles City, IA. Vols. 1-6, 1916-19/20. Rep. (Annual) Califor Naturalist Club. 2-871-3. HI 70801

Report (Annual) of the California avocado association. Riverside, CA. 1915-25/26. Rep. (Annual) Calif. Avocado Assoc. 2-871-3. HI 70802

Report of the California college of agriculture and

agricultural experiment station = Report of the college of agriculture and the agricultural experiment station of the university of California. Sacramento, CA.

Report, California division of forestry. Sacramento, CA. ?-1945+. Rep. Calif. Div. Forest. Preceded by: Biennial report of the California state board of forestry. HI 70803

Report, California forest and range experiment station. Berkeley, CA. 1951-57? Rep. Calif. Forest Range Exp. Sta. Preceded by: Report, science - servant of agriculture. Superseded by: Report (Annual), Pacific southwest forest and range experiment station. HI 70804

Report (Annual), California institute of technology. Pasadena, CA. 1963+. Rep. (Annual) Calif. Inst. Technol. HI 70805

Report of the California state agricultural society. Sacramento, CA. Rep. Calif. State Agric. Soc. See B–P–H 763/15. HI 58560

Report, Camberley natural history society. Camberley. 1958+. Rep. Camberley Nat. Hist. Soc. HI 70806

Report of the canadian forestry association. Ottawa. Vols. 1-15, 1900-13. Rep. Canad. Forest. Assoc. Incorporated in: Canadian forestry journal. HI 70807

Report of the canadian forestry convention = Report of the canadian forestry association. Ottawa.

Report (Annual) of the canadian plant disease survey. Ottawa. Vols. 11-39, 1931-59. Rep. (Annual) Canad. Pl. Dis. Surv. Preceded by: Report (Annual) on the prevalence of plant diseases in the dominion of Canada. Superseded by: Canadian plant disease survey. HI 70808

Report of the canadian seed growers' association. Ottawa. Rep. Canad. Seed Growers' Assoc. See B–P–H 763/16. HI 58561

Report (Annual) of the Canebrake branch, agricultural experiment station. Montgomery, AL. 1888-1910 [suspended 1891-96]. Rep. (Annual) Canebrake Branch Agrc. Exp. Sta. HI 70809

Report of the Cape Cod cranberry growers' association. Wareham, MA. Rep. Cape Cod Cranberry Growers' Assoc. See B–P–H 763/17. HI 58562

Report (Annual) of the Cardiff naturalists' society = Report and transactions of the Cardiff naturalists' society. Cardiff, Wales. Rep. & Trans. Cardiff Naturalists' Soc.. See B–P–H 761/21.

Report, Carlisle natural history society. Carlisle. 1948-49. Rep. Carlisle Nat. Hist. Soc. HI 70810

Report (Annual) of the Carnegie museum. Pittsburgh, PA. Nos. 1-75, 1898-1972. Rep. (Annual) Carnegie Mus. Superseded by: Report (Annual) of the Carnegie museum of natural history. HI 70811

Report (Annual) of the Carnegie museum of natural history. Pittsburgh, PA. No. 76+, 1973+. Rep. (Annual) Carnegie Mus. Nat. Hist. Preceded by: Report (Annual) of the Carnegie museum. HI 70812

Report (Annual), cassava program. Cali. 1977+. Rep. (Annual) Cassava Program. HI 70813

Report of the Cawnpore agricultural station. Allahabad. 1878-1916. Rep. Cawnpore Agric. Sta. Incorporated in: Report on agricultural stations in the central circle, United Provinces of Agra and Oudh. HI 70814

Report (Annual), Cawthron institute. Nelson, N.Z. 1935-63/66, 1935-66. Rep. Cawthron Inst. HI 53283.

Report of the central agricultural experiment station. Ministry of agriculture and commerce. [Nung shang pu chung yang nung shih shih yen ch'ang ti san ch'i ch'eng chi pao kao.] Nanking. Rep. Centr. Agric. Exp. Sta. Minist. Agric. See B–P–H 763/18. HI 58563

Report (Annual), central cocoa research station, Tafo. Tafo. 1937/38-4?, 1937-42. Rep. (Annual) Centr. Cocoa Res. Sta. Tafo. Superseded by: Report (Annual), west african cacao research institute, Tafo. HI 70815

Report, central coconut research station, Kasaragod. Madras? 1957+. Rep. Centr. Coconut Res. Sta. Kasaragod. HI 70816

Report (Annual) of the central hygiene experiment station. Report of investigations. [Chung-yang wei-shong shih-yen so nein pao. Yen tiao ch'a pao kao.] Nanking? [Dates of publication not ascertained.] Rep. (Annual) Centr. Hyg. Exp. Sta. HI 70817

Report of the central research institute of the South Manchuria railway co[mpany]. [Chuo shiken-sho hokoku.] Rep. Centr. Res. Inst. S. Manchuria Railway Co. See B–P–H 763/19. HI 58564

Report (Annual), central rice research institute, Cuttack. New Delhi. 1946-62?, 1969-71. Rep. (Annual) Centr. Rice Res. Inst. Cuttack. For 1963-68 & 1972 see: Annual technical report, central rice research institute. Cuttack. HI 70818

Report (Annual), central sericultural research station. Berhampore. 1943+. Rep. (Annual) Centr. Seric. Res. Sta. HI 70819

Report, cereal crops division, Canada. Ottawa. Nos. 1-?, 188?-58? Rep. Cereal Crops Div. Canada. Superseded by: Report, genetics and plant breeding research institute. HI 70820

Report, cereal division, dominion experimental farms, Canada = Report of the dominion cerealist, Canada. Ottawa.

Report, Ceylon coconut research board = Report (Annual) of the coconut research board of the coconut research institute, Colombo. Colombo.

Report, Ceylon coconut research scheme = Report, coconut research scheme, Ceylon. Columbo.

Reports of the Ceylon marine biological laboratory. Colombo, Ceylon. Rep. Ceylon Mar. Biol. Lab. See B–P–H 763/21. HI 58565

Report (Annual) of the chairman of the division of biology and agriculture, national research council. Washington, DC = National research council, annual report of the chairman of the division of biology and agriculture. Washington, DC. Natl. Res. Council Annual Rep. Chairm. Div. Biol. See B–P–H 631/8.

Report of the chairman of the division of plant biology, Carnegie institution of Washington = Report (Annual) of the director of the department of plant biology, Carnegie institution of Washington. Washington, DC.

Report (Annual) of the Charles Darwin research station. Santa Cruz, Galapagos. 197?-80. Rep. (Annual) Charles Darwin Res. Sta. Superseded by: Informe anual estación científica Charles Darwin. HI 70821

Report of the Cheltenham and district naturalists' society. Cheltenham. 1951-56. Rep. Cheltenham Distr. Naturalists' Soc. Superseded by: Report, north Gloucestershire naturalists' society. HI 70822

Report (Annual) of the Chester society of natural science, literature and art. Chester. Nos. 1-27, 1871-98. Rep. (Annual) Chester Soc. Nat. Sci. Lit. Art. Superseded by: Report and proceedings, Chester society of natural science. HI 70823

Report of Chiba horticultural experiment station. [Chiba-ken danchi engei shikenjo kenkyu hokoku.] Tateyama. 1969+. Rep. Chiba Hort. Exp. Sta. HI 74958

Report of the chief of the bureau of plant industry, soils, and agricultural engineering; agricultural research administration, United States department of agriculture = U S department of agriculture. Report of the chief of the bureau of plant industry. Washington, DC. U.S.D.A. Rep. Chief Bur. Pl. Industr. See B–P–H 944/21.

Report of the chief of the bureau of plant industry, United States department of agriculture = U S department of agriculture. Report of the chief of the bureau of plant industry. Washington, DC. U.S.D.A. Rep. Chief Bur. Pl. Industr. See B–P–H 944/21.

Report of the chief of division of vegetable pathology, United States department of agriculture = U S department of agriculture. Division of vegetable pathology. Report of the chief of division of vegetable pathology. Washington, DC. U.S.D.A. Div. Veg. Pathol. Rep. Chief Div. Veg. Pathol. See B–P–H 942/4.

Report of the chief of division of vegetable physiology and pathology, United States department of agriculture = U S department of agriculture. Division of vegetable pathology. Report of the chief of division of vegetable physiology and pathology. Washington, DC. U.S.D.A. Div. Veg. Pathol. Rep. Chief Div. Veg. Physiol. See B–P–H 942/5.

Report of the chief of the forestry division, U S department of agriculture = Department of agriculture, forestry division. Report of the chief of the forestry division. Washington, DC. Dept. Agric. Forest. Div. Rep. Chief Forest. Div. See B–P–H 343/2.

Report of the chief of section of vegetable pathology, United States department of agriculture = United States department of agriculture. Division of vegetable pathology. Report of the chief of section of vegetable pathology. Washington, DC. U.S.D.A. Div. Veg. Pathol. Rep. Chief Sect. Veg. Pathol. See B–P–H 942/6.

Report (Annual) of the Chicago natural history museum. Chicago, IL. 1943/46-61/64, 1943-64. Rep. (Annual) Chicago Nat. HIst. Mus. Preceded by: Publications, Field museum of natural history. Report series. Superseded by: Report (Annual) of the Field museum of natural history. HI 70824

Report, Chichester and West Sussex natural history and microscopical society. Chichester. 1877-82. Rep. Chichester West Sussex Nat. Hist. Microscop. Soc. Superseded by: Transactions of the Chichester and West Sussex natural history and microscopical society. HI 70825

Report on chilean university life. Washington, DC. Nos. 1-18, 1979-84. Rep. Chil. Univ. LIfe. Superseded by: Chilean university life. HI 70826

Report of the city of London college science society. London. 1885-88. Rep. City London Coll. Sci. Soc. Superseded by: Journal of the city of London college science society. HI 70827

Report (Annual), Clemson agricultural college of South Carolina. Agricultural experiment station = Clemson agricultural college of South Carolina. Agricultural experiment station. Annual report. Clemson, SC. Clemson Agric. Exp. Sta. Annual Rep. See B–P–H 314/33.

Report, coastal ecology research station. Norwich. 1969/72+, 1973+. Rep. Coastal Ecol. Res. Sta. HI 70828

Report (Annual) of the cocoa research institute of Ghana. Tafo. 1962/63+, 1963+. Rep. (Annual) Cocoa Res. Inst. Ghana. Preceded by: Report (Annual) of the west african cacao research institute. HI 70829

Report on cocoa research. St. Augustine, Port of Spain = Report on cacao research. St. Augustine, Port of Spain.

Report (Annual) of the cocoa research institute of Nigeria. Ibadan. 1962/63+, 1963+. Rep. (Annual) Cocoa Res.

Inst. Nigeria. Preceded by: Report (Annual) of the west african cocoa research institute. HI 70830

Report, coconut industry board research department, Jamaica. No. 1+, 1959/61+, 1961+. Rep. Coconut Industr. Board Res. Dept. Jamaica. HI 70831

Report, coconut pests and diseases board, Fiji. 1954/55+, 1955+. Rep. Coconut Pests Dis. Board Fiji. HI 70832

Report (Annual) of the coconut research board of the coconut research institute, Colombo. [Forms part of: Ceylon coconut quarterly.] Colombo. 1951-55? Rep. (Annual) Coconut Res. Board Coconut Res. Inst. Colombo. Preceded by: Report, coconut research scheme, Ceylon. Superseded by: Report (Annual) of the coconut research institute of Ceylon. HI 70833

Report (Annual) of the coconut research institute of Ceylon. [Forms part of: Ceylon coconut quarterly.] Colombo. 27th-43rd, 1955-71. Rep. (Annual) Coconut Res. Inst. Ceylon. Preceded by: Report, coconut research board of the coconut research institute, Colombo. Superseded by: Report (Annual), coconut research institute of Sri Lanka. HI 70834

Report (Annual), coconut research institute of Sri Lanka. Colombo. 44th+, 1972+. Rep. (Annual) Coconut Res. Inst. Sri Lanka. Preceded by: Report (Annual), coconut research institute of Ceylon. HI 70835

Report, coconut research scheme, Ceylon. Columbo. 1932-50. Rep. Coconut Res. Scheme Ceylon. Superseded by: Report (Annual), coconut research board of the coconut research institute, Colombo. HI 70836

Report on coffee conference, department of agriculture, Kenya. Nairobi. 1927-37. Rep. Coffee Conf. Dept. Agric. Kenya. HI 70837

Report of the coffee planting experiment station, Sidapur. Madras. 1921-24. Rep. Coffee Pl. Exp. Sta. Sidapur. HI 70838

Report (Annual), coffee research and experimental station, Lyamungu = Report (Annual), coffee research station, Lyamungu. Dar-es-Salaam.

Report (Annual), coffee research foundation, Kenya. Vol. 1+, 1956/57+, 1957+. Rep. (Annual) Coffee Res. Found. Kenya. HI 70839

Report (Annual), coffee research station, Lyamungu. Dar-es-Salaam. 1934-61. Rep. (Annual) Coffee Res. Sta. Lyamungu. Superseded by: Research report, coffee research station, Lyamungu. HI 70840

Report, coffee research station, Ruiru coffee research services, Kenya. Ruiru. 1956+. Rep. Coffee Res. Sta. Ruiru Coffee Res. Serv. Kenya. HI 70841

Report, coffee scientific officer, department of agriculture, Mysore = Bulletin of the Mysore coffee experiment station. Bangalore.

Report (Annual) of the Cold Spring Harbor laboratory of quantitative biology. Cold Spring Harbor, NY. 1964+. Rep. (Annual) Cold Spring Harbor Lab. Quant. Biol. Preceded by: Report (Annual) of the biological laboratory, Cold Spring Harbor. HI 70842

Report of the college of agriculture and the agricultural experiment station of the university of California. Sacramento, CA. 1912/13-22/23, 1913-23. Rep. Coll. Agric. Agric. Exp. Sta. Univ. Calif. Preceded by: University of California agricultural experiment station. Reports. Superseded by: Report of the agricultural experiment station of the university of California. HI 70843

Report (Annual) of the college of agriculture and forestry, university of Nanking. Nanking. 191?-33/34, 191?-34. Rep. (Annual) Coll. Agric. Forest. Univ. Nanking. HI 70844

Report of the college of agriculture, Mauritius. Port Louis. 1923-38, 1946-52. Rep. Coll. Agric. Mauritius. HI 70845

Report (Annual) of the college of education, university of Iwate. [Iwate daigaku kyoikugakubu kenkyu nenpo.] Morioka. Vol. 26+, 1966+. Rep. (Annual) Coll. Educ. Univ. Iwate. Preceded by: Report of the gakugei faculty of the Iwate university. HI 70846

Report (Annual) of the college of liberal arts, university of Iwate = Report (Annual) of the gakugei faculty of the Iwate university. Morioka.

Report of the colonial botanical garden. Melbourne. Nos. ?-2-17- ?, ?-1874/75-85/86-?, ?-1875-86-? Rep. Colon. Bot. Gard. Preceded by?: Report (Annual) of the director, botanic gardens, Melbourne. HI 70847

Report (Annual) of the Colorado agricultural experiment station. Fort Collins, CO. 1888+. Rep. (Annual) Colorado Colorado Agric. Exp. Sta. HI 70848

Report, Colorado seed laboratory. Fort Collins, CO. 1918+. Rep. Colorado Seed Lab. HI 70849

Report (Annual) of the Columbus horticultural society. Columbus, OH. Vols. 1-27/28, 1886-1912/13. Rep. Annual Columbus Hort. Soc. 2-1124-3. HI 70850

Report of the commissioner of agriculture. Washington, DC. Rep. Commiss. Agric. See B–P–H 764/7. HI 58567

Report of the commissioner of crown lands, Ontario. Toronto. 1888-1904. Rep. Commiss. Crown Lands Ontario. Superseded by: Report (Annual), minister of lands, forests, and mines, Ontario. HI 70851

Report of the commissioner of patents, agriculture. Washington, DC. Rep. Commiss. Patents, Agric. See B–P–H 764/8. HI 58568

Report, committee of british palaeobotanists. London.

1959-62/63. Rep. Committee Brit. Palaeobotanists. HI 70852

Report of the committee in charge of the experiment station, hawaiian sugar planters' association = Report (Annual), hawaiian sugar planters' association experiment station. Honolulu, HI.

Report, committee for environmental conservation. London. Nos. 1-9, 1970-78. Rep. Committee Environm. Conservation. Superseded by: Report, council for environmental conservation. HI 70853

Report of the committee on horticultural research, canadian horticultural council. Ottawa. 1936+. Rep. Committee Hort. Res. Canad. Hort. Council. HI 70854

Report of the committee on marine ecology as related to paleontology, division of geology and geography, national research council. Washington, DC = National research council, division of geology and geography, report of the committee on marine ecology as related to paleontology. Washington, DC. Natl. Res. Council Div. Geol. Rep. Committee Mar. Ecol. Related Paleontol. See B–P–H 631/10.

Report of the committee of tropical bioclimatology. Leiden. 1960+. Rep. Committee Trop. Bioclimatol. HI 70855

Reports (Annual) of the committees and officers, american association of botanical gardens and arboretums = Reports (Annual), american association of botanical gardens and arboretums. Pine Mountain, GA.

Report, commonwealth agricultural bureaux. London, Farnham Royal. 1949+. Rep. Commonw. Agric. Bur. HI 70856

Report, commonwealth forestry association. Oxford. 1974+. Rep. Commonw. Forest. Assoc. HI 70857

Report (Annual), commonwealth forestry and timber bureau. Canberra, A.C.T., Melbourne, Vic. 1930-45. Rep. (Annual) Commonw. Forest. Timber Bur. Superseded by: Report (Annual), forestry and timber bureau, Australia. HI 70858

Report (Annual), commonwealth forestry institute, university of Oxford. Oxford. Vol. 38+, 1961/62+, 1963+. Rep. (Annual) Commonw. Forest. Inst. Univ. Oxford. Preceded by: Report (Annual), imperial forestry institute. HI 70859

Report on the commonwealth mycological conference. London, Kew. Vols. 4-6, 1948-60. Rep. Commonw. Mycol. Conf. Preceded by: Report on the imperial mycological conference. HI 70860

Report, commonwealth mycological institute. Kew = Year, commonwealth mycological institute (The). Kew.

Report of the conference of the botanical society of the British Isles. London. [1st-3rd], 1949-53. Rep. Conf. Bot. Soc. British Isles. Superseded by: Conference reports of the botanical society of the British Isles. 1-756-1. HI 70862

Report of the Connecticut agricultural experiment station. Hartford, CT. 1877+. Rep. Connecticut Agric. Exp. Sta. HI 70863

Report (Annual), conservation commission of the Northern Territory, Australia. Alice Springs, N.T. No. 1+, 1980+. Rep. (Annual) Conservation Commiss. Northern Territory Australia. HI 70864

Report (Annual), conservation foundation. New York, Washington, DC. 1950?+. Rep. (Annual) Conservation Found. HI 70865

Report of the conservators of forests, Cape of Good Hope. Cape Town. 1864-1905. Rep. Conserv. Forests Cape Good Hope. Superseded by: Report of the forest department, Cape of Good Hope. HI 70866

Report, Cornell university college of agriculture and experiment station. Albany, Ithaca, NY. 1888+. Rep. Cornell Univ. Coll. Agric. Exp. Sta. HI 70867

Report of the cotton breeding station, Coimbatore. Madras. 1920?+. Rep. Cotton Breed. Sta. Coimbatore. HI 70868

Report, cotton experiment station, Klongtan, Swankaloke. Bangkok. Vol. 1+, 1936/37+, 1937+. Rep. Cotton Exp. Sta. Klongtan Swankaloke. HI 70869

Report of the cotton experiment stations, Shanghai. Shanghai. [Dates of publication not ascertained.] Rep. Cotton Exp. Sta. Shanghai. HI 70870

Report of the cotton experiment station, Siam = Report, cotton experiment station, Klongtan, Swankaloke. Bangkok.

Report on cotton experimental work in Mesopotamia. Baghdad. 1918-19. Rep. Cotton Exp. Wk. Mesopotamia. HI 70871

Report, cotton research board, Egypt. Cairo. Vols. 1-7, 1920-28. Rep. Cotton Res. Board Egypt. HI 70872

Report, cotton research institute, Gatooma. Gatooma. 1973/74-75/76. Rep. Cotton Res. Inst. Gatooma. Preceded by: Report (Annual), Gatooma research station. Superseded by: Report (Annual), cotton research institute, Zimbabwe. HI 70873

Report, cotton research institute, Southern Rhodesia = Report, cotton research institute, Gatooma. Gatooma.

Report (Annual), cotton research institute, Zimbabwe. Kadoma. ?-1980/81+, ?-1981?+. Rep. (Annual) Cotton Res. Inst. Zimbabwe. Preceded by: Report, cotton research institute, Gatooma. HI 70874

Report (Annual) of the council and accounts, botanical

society of the British Isles. Horsham. 1976+. Rep. (Annual) Council Accounts Bot. Soc. British Isles. HI 70875

Report (Annual) of the council of the central and associated chambers of agriculture, London [title varies]. London. Vol. 1+, 1866+. Rep. (Annual) Council Centr. Assoc. Chambers Agric. 2-956-3. HI 70876

Report of the council, Cornwall naturalists' trust ltd. Penzance. 1969+. Rep. Council Cornwall Naturalists' Trust. HI 70877

Report, council for environmental conservation. London. No. 10+, 1979+. Rep. Council Environm. Conservation. Preceded by: Report, committee for environmental conservation. HI 70878

Report of the council, Essex naturalists' trust ltd. Felsted. 1970+. Rep. Council Essex Naturalists' Trust. HI 70879

Report, council for nature. London. 1965-69. Rep. Council Nat. HI 70880

Report, council for the promotion of field studies. London. 1946-54. Rep. Council Promot. Field Stud. Superseded by: Report, field studies council. HI 70881

Report (Annual) of the council of the royal institution of South Wales; with appendix of original papers on scientific subjects. Swansea. 1839. Rep. (Annual) Council Roy. Inst. S. Wales. HI 70882

Report (Annual) of the council of the Yorkshire philosophical society. York. 1823-75. Rep. (Annual) Council Yorkshire Philos. Soc. 5-4579-3. HI 70883

Report of the countryside commission. London. 1969+. Rep. Countryside Commiss. Preceded by: Report of the national parks commission. London. HI 70884

Report of the countryside commission for Scotland. Edinburgh. 1968+. Rep. Countryside Commiss. Scotland. HI 70885

Report on Cousin Island nature reserve and other islands. Oxford. 1968+. Rep. Cousin Island Nat. Reserve. HI 70886

Report (Annual) of the Cranbrook institute of science. Bloomfield Hills, MI. No. 1+, 1931+. Rep. (Annual) Cranbrook Inst. Sci. 2-1225-1. HI 70887

Report of the cuban national horticultural society. Havana. Rep. Cub. Natl. Hort. Soc. See B–P–H 764/9. HI 58569

Report of the curators, botanical exchange club of the British Isles. London. Rep. Curators Bot. Exch. Club Brit. Isles. See B–P–H 764/10. HI 58570

Report of the curators, London botanical exchange club. London. Rep. Curators London Bot. Exch. Club. See B–P–H 764/11. HI 58571

Report of the Cuttack agricultural experiment station. Calcutta. 1905-12. Rep. Cuttack Agric. Exp. Sta. Incorporated in: Report on experiment stations and scientific sections, department of agriculture, Bihar & Orissa. HI 70888

Report, Cyprus agricultural research institute. Nicosia. 1977+. Rep. Cyprus Agric. Res. Inst. HI 70889

Report (Annual), cytogenetics laboratory, department of botany, university of Calcutta = Research bulletin, cytogenetics laboratory, department of botany, university of Calcutta. Calcutta.

Report of the danish biological station to the ministry of agriculture and fisheries. Copenhagen. Rep. Danish Biol. Sta. Minist. Agric. See B–P–H 764/13. HI 58572

Report, danish forest experiment station = Forstlige forsøgvaesen i Danmark. Copenhagen.

Report (Annual), Darlington and Teesdale naturalists' field club. Darlington. 1969+, 1970+. Rep. (Annual) Darlington Teesdale Naturalists' Field Club. HI 70890

Report of the date growers' institute. Coachella, CA. Rep. Date Growers' Inst. See B–P–H 764/14. HI 58573

Report (Annual), the Dawes arboretum. Newark, OH. ?-1982+. Rep. (Annual) Dawes Arbor. HI 70891

Report, Deir Alla research station, agricultural research service, Jordan = Report of the agricultural research service, Jordan. Amman.

Report (Annual), Delaware college agricultural experiment station. Newark, DE. 1888-1921. Rep. (Annual) Delaware Coll. Agric. Exp. Sta. Superseded by: University of Delaware agricultural experiment station annual report. HI 70892

Report (Annual), Denver botanic gardens. Denver, CO. ?-1976+. Rep. (Annual) Denver Bot. Gard. HI 70893

Report (Annual), department of agricultural research, Federation of Nigeria. Lagos. 1952/53-59/60, 1953-60. Rep. (Annual) Dept. Agric. Res. Fed. Nigeria. Preceded by: Report (Annual) on the agricultural department, Nigeria. HI 70894

Report (Annual), department of agricultural research, Malawi. Zomba. 1970/71+. 1971+. Rep. (Annual) Dept. Agric. Res. Malawi. Preceded by: Report (Annual) of the department of agriculture, Malawi. HI 70895

Report (Annual) on the department of agricultural research, Nigeria = Report (Annual), department of agricultural research, Federation of Nigeria. Lagos.

Report (Annual), department of agricultural technical services, South Africa. Pretoria. 1958/59+, 1959+.

Rep. (Annual) Dept. Agric. Techn. Serv. South Africa. Preceded by: Report, department of agriculture, South Africa. HI 70896

Report (Annual) of the department of agriculture (for the fiscal year ended June 30 ...), United States of America. Washington, DC. 1894-1920. Rep. (Annual) Dept. Agric., U.S.A. (Included reports of the secretary and division chiefs.) HI 70897

Report of the department of agriculture, Aden Protectorate. 1946+. Rep. Dept. Agric. Aden Protectorate. HI 70898

Report of the department of agriculture, Alberta. Edmonton, Alta. 1905+. Rep. Dept. Agric. Alberta. HI 70899

Report of the department of agriculture, Antigua = Report of the agricultural department, Antigua. St. John.

Report, department of agriculture, Assam = Report of the agricultural department, Assam. Shillong.

Report of the department of agriculture, Barbados. Bridgetown. 1899-1928. Rep. Dept. Agric. Barbados. Preceded by: Report of the results obtained on the experimental fields at Dodds reformatory, Barbados. Superseded by: Report of the department of science and agriculture, Barbados. HI 70900

Report (Annual), department of agriculture, Basutoland. Bloemfontein, Maseru. 1935/36-63, 1936-63. Rep. (Annual) Dept. Agric. Basutoland. Superseded by: Report (Annual), ministry of agriculture, co-operatives and marketing, Lesotho. HI 70901

Report, department of agriculture, Bechuanaland. Vryburg, etc. 1939+. Rep. Dept. Agric. Bechuanaland. HI 70902

Report, department of agriculture, Bengal. Calcutta. 1918/19-43/44, 1919-44. Rep. Dept. Agric. Bengal. Preceded by: Report of the agricultural department, Bengal. HI 70903

Report (Annual), department of agriculture, Bermuda. Hamilton, Bermuda. 1925-68? Rep. (Annual) Dept. Agric. Bermuda. Preceded by: Report of the board of agriculture, Bermuda. Superseded by: Report, department of agriculture and fisheries, Bermuda. HI 70904

Report, department of agriculture, Bihar and Orissa = Report of the agricultural department in Bihar and Orissa. Patna.

Report, department of agriculture, Bombay. Poona. 1903-60. Rep. Dept. Agric. Bombay. Preceded by: Report of the department of land records and agriculture, Bombay. HI 70905

Report (Annual) of the department of agriculture, British East Africa. Nairobi. 1907/08-17/18, 1908-18 [report for 1918/19 not published]. Rep. (Annual) Dept. Agric. British East Africa. Superseded by: Report (Annual), department of agriculture, East Africa Protectorate. HI 70906

Report, department of agriculture, British Honduras. Belize. 1917-62. Rep. Dept. Agric. British Honduras. HI 70907

Report, department of agriculture, British North Borneo. Jesselton. 1920-61. Rep. Dept. Agric. British North Borneo. Superseded by: Report, department of agriculture, Sabah, Malaysia. HI 70908

Report, department of agriculture, British Solomon Islands. Honiara. 1954+. Rep. Dept. Agric. British Solomon Islands. HI 70909

Report, department of agriculture, British Virgin Islands = Report of the agricultural department, British Virgin Islands. Tortola, Bridgetown.

Report of the department of agriculture, Burma. Rangoon. 1886/87+, 1887+. Rep. Dept. Agric. Burma. HI 70910

Report, department of agriculture, Canada. (a) Agricultural bacteriology = Progress report of the dominion agricultural bacteriologist. Ottawa.

Report, department of agriculture, Canada. (b) Botany = Progress report, division of botany, Canada. Ottawa.

Report of the department of agriculture, Cape of Good Hope = Report of the department of agriculture, Cape Province. Cape Town.

Report of the department of agriculture, Cape Province. Cape Town. 1897-1910. Rep. Dept. Agric. Cape Province. Superseded by: Report, department of agriculture, South Africa. HI 70911

Report of the department of agriculture, Central Provinces and Berar. Nagpur. 1882-46? Rep. Dept. Agric. Central Provinces & Berar. Superseded by: Report on the working of the department of agriculture, Madhya Pradesh. HI 70912

Report of the department of agriculture, Ceylon. Colombo. 1911+. Rep. Dept. Agric. Ceylon. Preceded by: Report of the royal botanic gardens, Peradeniya. HI 70913

Report, department of agriculture and conservation, Hawaii = Report, department of agriculture, Hawaii. Honolulu, HI.

Report of the department of agriculture, Dominica. Roseau. 1943-50. Rep. Dept. Agric. Dominica. Preceded by: Report (Annual), agricultural and forestry department, Dominica. Superseded by: Report of the department of agriculture and forestry, Dominica. HI 70915

Report (Annual), department of agriculture, East Africa Protectorate. Nairobi. 1919/20-20/21, 1920-21. Rep.

(Annual) Dept. Agric. East Africa Protectorate. Preceded by: Report (Annual) of the department of agriculture, British East Africa. Superseded by: Report (Annual), department of agriculture, Kenya. HI 70916

Report (Annual) of the department of agriculture of the eastern region of Nigeria. Enugu. 1951/52+, 1952+. Rep. (Annual) Dept. Agric. E. Region Nigeria. Preceded by: Report (Annual) on the agricultural department, Nigeria. HI 70917

Report, department of agriculture, Falkland Islands. 1948-? Rep. Dept. Agric. Falkland Islands. HI 70918

Report, department of agriculture, Federated Malay States. Kuala Lumpur. 1907-34. Rep. Dept. Agric. Federated Malay States. Preceded by: Report, director of agriculture, Federated Malay States. Superseded by: Report of the department of agriculture, Malaya. HI 70919

Report of the department of agriculture, Federation of Malaya. Kuala Lumpur. 1948+. Rep. Dept. Agric. Federation Malaya. Preceded by: Report on agriculture in Malaya. HI 70920

Report, department of agriculture, Fiji. Suva. 1905-69. Rep. Dept. Agric. Fiji. HI 70921

Report of the department of agriculture and fisheries, Bermuda. Hamilton, Bermuda. 1969/73+ 1974+. Rep. Dept. Agric. Fish. Bermuda. Preceded by: Report (Annual), department of agriculture, Bermuda. HI 70922

Report, department of agriculture, forest and fisheries, Western Samoa. Apia. 1956+. Rep. Dept. Agric. Forest Fish. Western Samoa. HI 70923

Report of the department of agriculture and forestry, Dominica. Roseau. 1951-60. Rep. Dept. Agric. Forest. Dominica. Preceded by: Report of the department of agriculture, Dominica. HI 70924

Report of the department of agriculture and forestry, St. Helena. 1947-66. Rep. Dept. Agric. Forest. St. Helena. HI 70925

Report, department of agriculture and forests, New South Wales = Report, department of agriculture, New South Wales. Sydney, N.S.W.

Report, department of agriculture and forests, Palestine. Jerusalem. 1924-46. Rep. Dept. Agric. Forests Palestine. HI 70926

Report, department of agriculture and forests, Sudan. Khartoum. 1933/34-37/38, 1934-38. Rep. Dept. Agric. Forests Sudan. Partly superseded by: Report, forests department, Sudan and Report of the ministry of agriculture, Sudan. HI 70927

Report (Annual) of the department of agriculture, Gambia. London & Bathurst. 1923/24-48/49, 1924-49. Rep. (Annual) Dept. Agric. Gambia. Superseded by: Report (Annual) of the department of development and agriculture, Gambia. HI 70928

Report of the department of agriculture, Ghana. Accra. 1954/55+, 1956+. Rep. Dept. Agric. Ghana. Preceded by: Report of the agricultural department, Gold Coast. HI 70929

Report (Annual) of the department of agriculture, Gold Coast = Report (Annual) of the agricultural department, Gold Coast. Accra.

Report, department of agriculture, government of Bhopal. 1934-37. Rep. Dept. Agric. Gov. Bhopal. HI 70930

Report of the department of agriculture, government research institute, Formosa. [Nogyo-bu iho. Taiwan sotokufu chuo kenkyu-sho.] Taihoku [=Taipei, Taiwan]. Rep. Dept. Agric. Gov. Res. Inst. Formosa. See B–P–H 764/16. HI 58574

Report, department of agriculture, Grenada = Report on the agricultural department, Grenada.

Report, department of agriculture, Hawaii. Honolulu, HI. 1959/60+ 1960+. Rep. Dept. Agric. Hawaii. Preceded by: Report, board of commissioners of agriculture and forestry, Hawaii. HI 70914

Report, department of agriculture, Hong Kong = Report of the agricultural department, Hong Kong. Hong Kong.

Report, department of agriculture, Hyderabad. Hyderabad. 1936-41. Rep. Dept. Agric. Hyderabad. HI 70931

Report, department of agriculture, Iraq = Report, department of agriculture, Mesopotamia. Baghdad.

Report on the department of agriculture, Jamaica. Kingston, Jamaica. 1908-22. Rep. Dept. Agric. Jamaica. Preceded by: Report of the board of agriculture and department of public gardens and plantations, Jamaica. Superseded by: Report of the department of science and agriculture, Jamaica. HI 70932

Report (Annual), department of agriculture, Kenya. Nairobi. 1919/20-68, 1922-68. Rep. (Annual) Dept. Agric. Kenya. Preceded by: Report (Annual), department of agriculture, East Africa Protectorate. Superseded by: Report (Annual) of the research division, ministry of agriculture, Kenya. HI 70933

Report of the department of agriculture, Leeward Islands. Antigua. ?-1943-45. Rep. Dept. Agric. Leeward Islands. Preceded by: Report of the government laboratory and superintendent of agriculture of the Leeward Islands. Superseded by: Report of the director of agriculture, Leeward islands. HI 70934

Report, department of agriculture, Madras. Madras. 1871/72+, 1872+. Rep. Dept. Agric. Madras. HI 70935

Report, department of agriculture, Madhya Pradesh = Report on the working of the department of agriculture, Madhya Pradesh. Nagpur.

Report (Annual) of the department of agriculture, Malawi. Zomba. 1962/63-1968/69, 1963-69. Rep. (Annual) Dept. Agric. Malawi. Preceded by: Report (Annual), department of agriculture, Nyasaland. Superseded by: Report (Annual) of the department of agricultural research, Malawi. HI 70936

Report of the department of agriculture, Malaya. Singapore. 1935-39. Rep. Dept. Agric. Malaya. Preceded by: Report, department of agriculture, Federated Malay States. Superseded by: Report on agriculture in Malaya. HI 70937

Report of the department of agriculture, Malta. Valetta. 1911-50. Rep. Dept. Agric. Malta. HI 70938

Report (Annual) of the department of agriculture, Mauritius. Port Louis, Reduit. 1913-66. Rep. (Annual) Dept. Agric. Mauritius. Superseded by: Report of the ministry of agriculture and natural resources, Mauritius. HI 70939

Report, department of agriculture, Mesopotamia. Baghdad. 1919-21, 1920-22? Rep. Dept. Agric. Mesopotamia. HI 70940

Report, department of agriculture, Montserrat = Report on the agricultural department, Montserrat. Plymouth, Barbados.

Report of the department of agriculture, Mysore. Bangalore. 1937-40, 1949+. Rep. Dept. Agric. Mysore. Preceded by: Report, agricultural department, Mysore. HI 70941

Report, department of agriculture, Natal. Pietermaritzburg. 1895-? Rep. Dept. Agric. Natal. HI 70942

Report (Annual), department of agriculture, New Brunswick. Fredericton. 1870-1967. Rep. (Annual) Dept. Agric. New Brunswick. Preceded by: Report (Annual), board of agriculture, New Brunswick. Superseded by: Report (Annual), department of agriculture and rural development, New Brunswick. HI 70943

Report, department of agriculture, New Guinea. Melbourne, Vic. 1923-? Rep. Dept. Agric. New Guinea. Superseded by: Report, department of agriculture, stock and fisheries, territory of Papua and New Guinea. HI 70944

Report, department of agriculture, New South Wales. Sydney, N.S.W. 1890+. Rep. Dept. Agric. New South Wales. HI 70945

Report, department of agriculture, New Zealand. Wellington, N.Z. 1893+. Rep. Dept. Agric. New Zealand. HI 70946

Report (Annual), department of agriculture, Nigeria = Report (Annual) on the agricultural department, Nigeria. Lagos.

Report, department of agriculture, Nizam's government, Hyderabad = Report, department of agriculture, Hyderabad. Hyderabad.

Report, department of agriculture, North Borneo = Report, department of agriculture, British North Borneo. Jesselton.

Report (Annual), department of agriculture, northern provinces of Nigeria = Report (Annual) on the agricultural department, Nigeria. Northern provinces. Kaduna.

Report (Annual) of the department of agriculture of the northern region of Nigeria. Kaduna. 1955-68. Rep. (Annual) Dept. Agric. N. Region Nigeria. Preceded by: Report (Annual) on the agricultural department, Nigeria. HI 70947

Report, department of agriculture, Northern Rhodesia. Lusaka. 1926-58. Rep. Dept. Agric. Northern Rhodesia. Superseded by: Report, ministry of african agriculture, Northern Rhodesia. HI 70948

Report (Annual), department of agriculture, North-West Territories. Regina, Sask. 1898-1903. Rep. (Annual) Dept. Agric. North-West Territories. Superseded by: Report (Annual), department of agriculture, Saskatchewan. HI 70949

Report (Annual), department of agriculture, Nyasaland. Zomba. 1908/09-61/62, 1909-62. Rep. (Annual) Dept. Agric. Nyasaland. Superseded by: Report (Annual) of the department of agriculture, Malawi. HI 70950

Report (Annual) of the department of agriculture, Orange River Colony. Bloemfontein. Nos. 1-6, 1904-10. Rep. (Annual) Dept. Agric. Orange River Colony. HI 70951

Report, department of agriculture, Orissa = Report of the agricultural department, Orissa. Cuttack.

Report, department of agriculture, Palestine = Report, department of agriculture and forests, Palestine. Jerusalem.

Report of the department of agriculture, Philippine Islands = Report, bureau of agriculture, Philippine Islands. Manila.

Report, department of agriculture, province of Quebec. Quebec. 1868-19?? Rep. Dept. Agric. Prov. Quebec. Superseded by: Rapport, ministère de l'agriculture de la province de Québec. HI 70952

Report, department of agriculture, Punjab. Lahore. 1893-37/38, 1893-38. Rep. Dept. Agric. Punjab. HI 70953

Report (Annual) of the department of agriculture, Queensland. Brisbane, Qld. 1887/88-93/94, 1889-94. Rep. (Annual) Dept. Agric. Queensland. Superseded by: Report (Annual) of the department of agriculture and stock, Queensland. HI 70954

Report (Annual), department of agriculture and rural development, New Brunswick. Fredericton, N.B. 1968+. Rep. (Annual) Dept. Agric. Rural Developm. New Brunswick. Preceded by: Report (Annual), department of agriculture, New Brunswick. HI 70955

Report, department of agriculture, Sabah, Malaysia. Jesselton. 1962+. Rep. Dept. Agric. Sabah. Preceded by: Report, department of agriculture, British North Borneo. HI 70956

Report (Annual) of the department of agriculture, Sarawak. Kuching. 1945-72? Rep. (Annual) Dept. Agric. Sarawak. Superseded by: Report, research board of the department of agriculture, Sarawak. HI 70957

Report (Annual), department of agriculture, Saskatchewan. Regina, Sask. 1905-06. Rep. (Annual) Dept. Agric. Saskatchewan. Preceded by: Report (Annual), department of agriculture, North-West Territories. HI 70958

Report (Annual), department of agriculture, Seychelles Islands. Victoria. 1923+. Rep. (Annual) Dept. Agric. Seychelles Islands. Preceded by: Report (Annual) on agriculture and crown lands, Seychelles. HI 70959

Report of the department of agriculture, Sierra Leone. Freetown. 1929+. Rep. Dept. Agric. Sierra Leone. Preceded by: Report, department of lands and forests, Sierra Leone. HI 70960

Report, department of agriculture in Sind. Karachi, Bombay. 1930/31-40/41, 1931-41. Rep. Dept. Agric. Sind. HI 70961

Report, department of agriculture, Singapore = Report of the agricultural department, colony of Singapore. Singapore.

Report, department of agriculture, South Africa. Pretoria. 1910/11-57/58, 1911-58. Rep. Dept. Agric. South Africa. Superseded by: Report (Annual), department of agricultural technical services, South Africa. HI 70962

Report of the department of agriculture, South Australia. Adelaide, S.A. 1897+. Rep. Dept. Agric. South Australia. HI 70963

Report, department of agriculture, Southern Rhodesia = Report of the ministry of agriculture and lands, Southern Rhodesia. Salisbury, Rhodesia.

Report, department of agriculture, St. Christopher & Nevis = Report on the agricultural department, St. Kitts & Nevis. Bridgetown, St. John's, Antiqua.

Report, department of agriculture, St. Kitts & Nevis = Report on the agricultural department, St. Kitts & Nevis. Bridgetown, St. John's, Antiqua.

Report, department of agriculture, St. Lucia = Report, agricultural department, St. Lucia. Castries.

Report, department of agriculture, St. Vincent = Report on the agricultural department, St. Vincent. Kingstown.

Report (Annual), department of agriculture, southern provinces of Nigeria = Report (Annual) on the agricultural department, Nigeria. Southern provinces. Lagos.

Report, department of agriculture, stock and fisheries, territory of Papua and New Guinea. Melbourne, Vic. 1959-67/69, 1959-69. Rep. Dept. Agric. Stock Fish. Papua New Guinea. Preceded by: Report, department of agriculture, New Guinea. HI 70964

Report (Annual) of the department of agriculture and stock, Queensland. Brisbane, Qld. 1895?-1963. Rep. (Annual) Dept. Agric. Stock Queensland. Preceded by: Report (Annual) of the department of agriculture, Queensland. Superseded by: Report (Annual), department of primary industries, Queensland. HI 70965

Report, department of agriculture and stock, Tasmania = Report (Annual), department of agriculture, Tasmania. Hobart, Tas.

Report, department of agriculture, Sudan = Report, department of agriculture and forests, Sudan. Khartoum.

Report (Annual) of the department of agriculture, Swaziland. Mbabane. 1947?-66. Rep. (Annual) Dept. Agric. Swaziland. Superseded by: Report of the ministry of agriculture, Swaziland. HI 70966

Report (Annual), department of agriculture, Tanganyika. London & Dar-es-Salaam. 1922-62. Rep. (Annual) Dept. Agric. Tanganyika. Superseded by: Report (Annual) of the agricultural extension division, ministry of agriculture, forests and wildlife, Tanzania. HI 70967

Report (Annual), department of agriculture, Tasmania. Hobart, Tas. 1905/06+, 1906+. Rep. (Annual) Dept. Agric. Tasmania. HI 70968

Report, department of agriculture, territory of Papua and New Guinea = Report, department of agriculture, stock and fisheries, territory of Papua and New Guinea. Melbourne, Vic.

Report of the department of agriculture, Transvaal. Pretoria. 1902/03-08/09, 1903-09. Rep. Dept. Agric. Transvaal. Continued in: Report, department of agriculture, South Africa. HI 70969

Report of the department of agriculture, Trinidad and Tobago. Port of Spain. 1908/09-24/25, 1909-25. Rep. Dept. Agric. Trinidad Tobago. Superseded by: Administration report of director of agriculture, Trinidad andTobago. HI 70970

Report (Annual) of the department of agriculture,

Uganda. Entebbe. 1911-40, 1950-63. Rep. (Annual) Dept. Agric. Uganda. Superseded by: Record of investigations, department of agriculture, Uganda. HI 70971

Report, department of agriculture, Union of South Africa = Report, department of agriculture, South Africa. Pretoria.

Report of the department of agriculture, united provinces of Agra and Oudh. Allahabad. 1887-1951? Rep. Dept. Agric. Unit. Prov. Agra & Oudh. HI 70972

Report, department of agriculture, university institute of applied sciences = Rit landbúnaoardeildar, atvinnudeild háskólans. B-flokkur. Reykjavik.

Report of the department of agriculture, Victoria. Melbourne, Vic. 1874-1910. Rep. Dept. Agric. Victoria. HI 70973

Report, department of agriculture, Western Australia. Perth, W.A. 1896/97+, 1898+ [not published 1916-18, 1935, 1937, 1941-42]. Rep. Dept. Agric. Western Australia. HI 70974

Report (Annual) of the department of agriculture of the western region of Nigeria. Ibadan. 1951/52-58/59?, 1952-59? Rep. (Annual) Dept. Agric. W. Region Nigeria. Preceded by: Report (Annual) on the agricultural department, Nigeria. HI 70975

Report of the department of agriculture, Zanzibar. Zanzibar. 1898-1962, 1899-1963. Rep. Dept. Agric. Zanzibar. HI 70976

Report (Annual), department of arboreta and botanic gardens, Los Angeles county = Report (Annual), Los Angeles state and county arboretum. Arcadia, CA.

Report, department of botanical research, Carnegie institution of Washington = Report (Annual) of the director, department of botanical research, Carnegie institution of Washington. Washington, DC.

Report (Annual) of the department of development and agriculture, Gambia. Bathurst. 1949/50-51/52, 1950-52. Rep. (Annual) Dept. Developm. Agric. Gambia. Preceded by: Report (Annual) of the department of agriculture, Gambia. HI 70977

Report, department of developmental biology, australian national university research school of biological sciences. Canberra, A.C.T. 1969+. Rep. Dept. Developm. Biol. Austral. Natl. Univ. Res. School Biol. Sci. HI 70978

Report, department of the environment, Canada = Report (Annual), environment, Canada. Ottawa.

Report (Annual), department of the environment, government of India. New Delhi. 1982/83+, 1983?+. Rep. (Annual) Dept. Environm. Gov. India. HI 70979

Report (Annual), department of fisheries and forestry, Canada. Ottawa. 1968/69-70/71, 1969-71. Rep. (Annual) Dept. Fish. Forest. Canada. Preceded by: Report (Annual), department of forestry and rural development, Canada. Superseded by: Report (Annual), environment, Canada. HI 70980

Report, department of fisheries and wildlife, Western Australia. Perth, W.A. No. 15+, 1974+. Rep. Dept. Fish. Wildlife Western Australia. Preceded by: Report, department of fisheries and fauna [not entered]. HI 70981

Report, department of forest research, Nigeria = Report (Annual) of the federal department of forest research, Nigeria. Ibadan.

Report, department of forest yield research, forest research institute of Sweden = Rapport, statens skogsforskningsinstitut. Avdelingen för skogsproduktion. Stockholm.

Report, department of forestry, British North Borneo = Report, forest department, colony of North Borneo. Sandakan, Jesselton.

Report (Annual), department of forestry, California = Report, California division of forestry. Sacramento, CA.

Report (Annual), department of forestry, Canada. Ottawa. 1960/61-67/68. 1961-68. Rep. (Annual) Dept. Forest. Canada. Superseded by: Report (Annual), department of fisheries and forestry, Canada. HI 70982

Report, department of forestry, Fiji = Report, forest department, Fiji. Suva.

Report of the department of forestry, government research institute, Taihoku. [Taiwan sotokufu chuo kenkyu-sho, ringyo-bu hokoku.] Taihoku [=Taipei, Taiwan]. Rep. Dept. Forest. Gov. Res. Inst. Taihoku. See B–P–H 764/17. HI 58575

Report, department of forestry, Hong Kong = Report of the forestry department, Hong Kong. Hong Kong.

Report, department of forestry, New South Wales = Report of the forestry commission, New South Wales. Sydney, N.S.W.

Report, department of forestry, Nyasaland. Zomba. 1926+. Rep. Dept. Forest. Nyasaland. Preceded by: Report (Annual), department of agriculture, Nyasaland. HI 70983

Report (Annual) of the department of forestry, Queensland = Report (Annual), forestry branch, Queensland. Brisbane, Qld.

Report (Annual), department of forestry and rural development, Canada = Report (Annual), department of forestry, Canada. Ottawa.

Report (Annual), department of forestry, South Africa. Cape Town. 1912/13-77/78, 1913-79 [reports for 1913/14-15/16 and 1941/42-44/45 not published]. Rep.

(Annual) Dept. Forest. South Africa. Preceded by: Report of the forest department, Cape of Good Hope. Superseded by: Report of the director-general, water affairs, forestry and environmental conservation, South Africa. HI 66377

Report, department of forestry, St. Helena = Report of the department of agriculture and forestry, St. Helena.

Report, department of forests, Israel. Jerusalem. 1935/36+, 1936+. Rep. Dept. Forests Israel. Preceded by: Report of the forest service, Palestine. HI 70984

Report, department of forests, Palestine = Report of the forest service, Palestine. Jerusalem.

Report, department of forests, Papua and New Guinea = Report of the operations of the department of forests, Papua and New Guinea. Port Moresby.

Report (Annual), department of hop research, Wye college. Ashford, Kent. 1947-72. Rep. (Annual) Dept. Hop Res. Wye Coll. HI 70985

Report, department of horticulture, Mysore. Bangalore. 1930/31-38/39, 1931-39. Rep. Dept. Hort. Mysore. Preceded by: Report (Annual), government gardens department, Mysore. HI 70986

Report, department of land records and agriculture, Assam. Shillong. 1886-1905. Rep. Dept. Land Rec. Agric. Assam. Preceded by: Report of the agricultural department, Assam. Superseded by: Report, agricultural department, eastern Bengal and Assam. HI 70987

Report of the department of land records and agriculture, Bengal. Calcutta. 1883-1901. Rep. Dept. Land Rec. Agric. Bengal. Superseded by: Report of the agricultural branch of the department of land records and agriculture, Bengal. HI 70988

Report, department of land records and agriculture, Bombay. Bombay. 1883-1903. Rep. Dept. Land Rec. Agric. Bombay. Superseded by: Report, department of agriculture, Bombay. HI 70989

Report (Annual), department of lands and forests, Alberta. Edmonton. 1931-75. Rep. (Annual) Dept. Lands Forests Alberta. HI 70990

Report (Annual), department of lands and forests and mines, Ontario = Report (Annual), department of lands and forests, Ontario. Toronto.

Report, department of lands and forests, Nova Scotia. Halifax, N.S. 1912+. Rep. Dept. Lands Forests Nova Scotia. HI 70991

Report (Annual), department of lands and forests, Ontario. Toronto. 1921-72. Rep. (Annual) Dept. Lands Forests Ontario. Preceded by: Report (Annual), minister of lands, forests and mines, Ontario. Superseded by: Report (Annual), ministry of natural resources, Ontario. HI 70992

Report of the department of lands and forests, province of Quebec. Quebec. 1901+. Rep. Dept. Lands Forests Prov. Quebec. HI 70993

Report, department of lands and forests, Sierra Leone. Freetown. 1922-28. Rep. Dept. Lands Forests Sierra Leone. Preceded by: Report, agricultural department, Sierra Leone. Superseded by: Report of the department of agriculture, Sierra Leone and Report on forest administration, colony of Sierra Leone. HI 70994

Report, department of marine biology, university of Liverpool. Liverpool. No. 85+, 1972/73+. Rep. Dept. Mar. Biol. Univ. Liverpool. Preceded by: Report of the oceanic department of the university of Liverpool. HI 70995

Report, department of medicinal plants, ministry of forests, Nepal. Thapathali. 1973/74+. Rep. Dept. Med. Pl. Minist. Forests Nepal. HI 70996

Report, department of nature conservation, Union of South Africa. Cape Town. 1952+. Rep. Dept. Nat. Conservation South Africa. HI 70997

Report (Annual), department of oceanography, university of British Columbia. Vancouver, B.C. 1980+. Rep. (Annual) Dept. Oceanogr. Univ. British Columbia. HI 70998

Report (Annual), department of plant biology, Carnegie institution of Washington = Report (Annual) of the director of the department of plant biology, Carnegie institution of Washington. Washington, DC.

Report, department of plant pathology, Seale Hayne agricultural college. Newton Abbot. 1924-38. Rep. Dept. Pl. Pathol. Seale Hayne Agric. Coll. HI 70999

Report, department of plant pathology and section of biochemistry, Rothamsted experimental station. Harpenden. 1938-45. Rep. Dept. Pl. Pathol. Sect. Biochem, Rothamsted Exp Sta. HI 71000

Report, department of primary industries, botany branch and Queensland herbarium = Report (Annual) of the botany branch and Queensland herbarium. Brisbane, Qld.

Report (Annual), department of primary industries, Queensland. Brisbane, Qld. 1963/64+, 1964+. Rep. (Annual) Dept. Prim. Industr. Queensland. Preceded by: Report (Annual) of the department of agriculture and stock, Queensland. HI 71001

Report (Annual), department of public lands, sub-department of forestry = Report (Annual), forestry branch, Queensland. Brisbane, Qld.

Report of the department of state forests, Victoria. Melbourne, Vic. 1908-18. Rep. Dept. State Forests Victoria. Superseded by: Report (Annual), forests commission, Victoria. HI 71002

Report, department of science and agriculture, Barbados. Bridgetown. 1929-31, 1941-57. Rep. Dept. Sci. Agric. Barbados. Preceded by: Report, department of agriculture, Barbados. For the years 1932-40 see: Agricultural journal, department of science and agriculture, Barbados. HI 71003

Report of the department of science and agriculture, British Guiana. Georgetown. 1896-1927. Rep. Dept. Sci. Agric. British Guiana. Preceded by: Report of the botanic gardens, British Guiana. Superseded by: Administration report of the director of agriculture, British Guiana. HI 71004

Report of the department of science and agriculture, Jamaica. Kingston, Jamaica. 1923-56. Rep. Dept. Sci. Agric. Jamaica. Preceded by: Report on the department of agriculture, Jamaica. Superseded by: Report (Annual), ministry of agriculture and lands, Jamaica HI 71005.

Report, department of water affairs and the department of forestry and environmental conservation, South Africa = Report of the director-general, water affairs, forestry and environmental conservation, South Africa. Cape Town.

Report (Annual) of the development of agriculture in Japan. Annual bibliography of japanese agriculture. [Nihon nogaku shinpo nenpo.] Tokyo. 1954-80. Rep. (Annual) Developm. Agric. Japan, Annual Bibliogr. Superseded by: Nogaku shinpo nenpo. HI 71006

Report and development paper, forestry commission. London. No. 1+, 1951+. Rep. Developm. Pap. Forest. Commiss. HI 71007

Report, director of agriculture, Bermuda = Report (Annual), department of agriculture, Bermuda. Hamilton, Bermuda.

Report of the director of agriculture, British Guiana. Georgetown. 1950-65. Rep. Director Agric. British Guiana. Preceded by: Administration report of the director of agriculture, British Guiana. Superseded by: Report (Annual), ministry of agriculture and natural resources, Guyana. HI 71008

Report (Annual) of the director of agriculture, Cyprus. Nicosia. 1895+. Rep. (Annual) Director Agric. Cyprus. HI 71009

Report, director of agriculture, Federated Malay States. Kuala Lumpur. 1906. Rep. Director Agric. Federated Malay States. Superseded by: Report, department of agriculture, Federated Malay States. HI 71010

Report of the director of agriculture, Leeward Islands. Antigua. 1946-48. Rep. Director Agric. Leeward Islands. Preceded by: Report of the department of agriculture, Leeward Islands. HI 71011

Report of the director of agriculture, Tonga. Nuku'alofa. 1940+. Rep. Director Agric. Tonga. HI 71012

Report of the director of agriculture, Trinidad and Tobago = Report of the department of agriculture, Trinidad and Tobago. Port of Spain.

Report of director of the botanic gardens, government domains and centennial park (and Campbelltown state nursery of New South Wales). Sydney, N.S.W. 1915-23, 1916-24. Rep. Director Bot. Gard. Gov. Dom. Centen. Park New South Wales. Preceded by: Report on botanic gardens and government domains of New South Wales. Incorporated in: Report (Annual) of the department of agriculture, New South Wales. HI 71013

Report (Annual) of the director, botanic gardens. Melbourne, Vic. 1875-81? Rep. (Annual) Director Bot. Gard. Melbourne. Preceded by: Report of the government botanist and director of the botanic and zoologic garden. Superseded by?: Report of the colonial botanical garden, Melbourne. HI 71014

Report of the director of the botanical survey of India. Calcutta. 1893/94-1910/11, 1894-1911. Rep. Director Bot. Surv. India. Superseded by: Report of the botanical survey of India. HI 71015

Report (Annual) of the director, department of botanical research, Carnegie institution of Washington. [Reprinted from: Yearbook, Carnegie institution of Washington]. Washington, DC. 1906-22. Rep. (Annual) Director Dept. Bot. Res. Carnegie Inst. Wash. Superseded by: Report (Annual) of the director of the laboratory for plant physiology, Carnegie institution of Washington. HI 71016

Report (Annual) of the director, department of genetics, Carnegie institution of Washington. [Reprinted from: Yearbook, Carnegie institution of Washington]. Washington, DC. 1920-62. Rep. (Annual) Director Dept. Genet. Carnegie Inst. Wash. Superseded by: Report (Annual) of the director, genetics research unit, Carnegie institution of Washington. HI 75133

Report (Annual) of the director of the department of plant biology, Carnegie institution of Washington. [Reprinted from: Carnegie institution of Washington year book.] Washington, DC. 1928-83. Rep. (Annual) Director Dept. Pl. Biol. Carnegie Inst. Wash. Preceded by: Report (Annual) of the director of the laboratory for plant physiology, Carnegie institution of Washington. HI 71017

Report (Annual) of the director of the division of plant biology, Carnegie institution of Washington = Report (Annual) of the director of the department of plant biology, Carnegie institution of Washington. Washington, DC.

Report of the director, dominion experimental farms, Canada. Ottawa. 1920-49. Rep. Director Domin. Exp. Farms Canada. Preceded by: Report on the experimental farms, department of agriculture, Canada.

Superseded by: Report of the director, experimental farms service, Canada HI 71018.

Report of the director, experimental farms service, Canada. Ottawa. 1950+. Rep. Director Exp. Farms Serv. Canada. Preceded by: Report of the director, dominion experimental farms, Canada. HI 71019

Report of the director of forestry, Canada = Report of the forestry branch, department of the interior, Canada. Ottawa.

Report of the director of forestry, New Zealand forestry service = Report (Annual), New Zealand state forest service. Wellington, N.Z.

Report (Annual) of the director of forestry of the Philippine Islands. Manila. 1906/07-39, 1908-39. Rep. (Annual) Director Forest. Philippine Islands. HI 71120

Report (Annual) of the director of forests, Queensland = Report (Annual), forestry branch, Queensland. Brisbane, Qld.

Report (Annual) of the director of gardens of the Straits Settlements. Singapore. 1915-39. Rep. (Annual) Director Gard. Straits Settlements. Preceded by: Report (Annual) on the botanic gardens, Singapore and Penang. Superseded by: Report (Annual), botanic gardens department, Singapore. HI 71020

Report of the director-general, water affairs, forestry and environmental conservation, South Africa. Cape Town. 1979/80+, 1980+. Rep. Director-Gen. Water Affairs Forest. Environm. Conservation South Africa. Preceded by: Report (Annual), department of forestry, South Africa. HI 71021

Report (Annual) of the director, genetics research unit, Carnegie institution of Washington. Washington, DC. [Dates of publication not ascertained.] Rep. (Annual) Director Genet. Res. Unit Carnegie Inst. Wash. HI 71022

Report (Annual) of the director of the laboratory for plant physiology, Carnegie institution of Washington. [Reprinted from: Carnegie institution of Washington year book.] Washington, DC. 1923-27. Rep. (Annual) Director Lab. Pl. Physiol. Carnegie Inst. Wash. Preceded by: Report (Annual) of the director, department of botanical research, Carnegie institution of Washington. Superseded by: Report (Annual) of the director of the department of plant biology, Carnegie institution of Washington. HI 71023

Report of the director of research of the Philippine sugar association. Manila. 1932-39. Rep. Director Res. Philippine Sugar Assoc. Preceded by: Report, Philippine sugar association research bureau. HI 71024

Report of the director, science service, department of agriculture, Canada = Report of the science service, department of agriculture, Canada. Ottawa.

Report of the director, university arboretum, university of California = Report, university arboretum, university of California. Davis, CA.

Report of the director, West Virginia agricultural experiment station. Morgantown, WV. 1887+. Rep. Director West Virginia Agric. Exp. Sta. HI 71025

Report of the directorate of botanical and other public gardens, West Bengal. Alipore. 1955+. Rep. Director. Bot. Other Public Gard. West Bengal. HI 71026

Report (Annual) of the directors, Department of agriculture = Department of agriculture. Annual report of the directors. Washington, DC. Dept. Agric. Annual Rep. Directors See B–P–H 342/12.

Report (Annual), Ditton laboratory. 1959/60-67/68, 1960-68. Rep. (Annual) Ditton Lab. Incorporated with: Report (Annual), East Malling research station. HI 71027

Report, division of agriculture, Sierra Leone = Report of the department of agriculture, Sierra Leone. Freetown.

Report, division of bacteriology, experimental farms service, department of agriculture, Canada = Progress report of the dominion agricultural bacteriologist. Ottawa.

Report (Annual), division of economic botany, commonwealth scientific and industrial research organization. 1928/29-56/57?, 1929-57? Rep. (Annual) Div. Econ. Bot. Commonw. Sci. Industr. Res. Organ. Superseded by: Report (Annual), division of plant industry, commonwealth scientific and industrial research organization. HI 71028

Report, division of economic fibre production, experimental farms service, Canada. Ottawa. 1920/21-34/36, 1921-36. Rep. Div. Econ. Fibre Prod. Exp. Farms Serv. Canada. Superseded by: Progress report, fibre division, experimental farms service, Canada. HI 71029

Report (Annual), division of forest products, commonwealth scientific and industrial research organization. South Melbourne, Vic. 1928/29-68/69, 1930-69. Rep. (Annual) Div. Forest Prod. Commonw. Sci. Industr. Res. Organ. Superseded by: Research review, division of forest products, commonwealth scientific and industrial research organization. HI 71030

Report (Annual), division of forest research, commonwealth scientific and industrial research organisation. Canberra, A.C.T. 1975/76-78/79, 1976-80? Rep. (Annual) Div. Forest Res. Commonw. Sci. Industr. Res. Organ. Superseded by: Biennial report, division of forest research, C S I R O. HI 71031

Report, division of forest research, Zambia. Kitwe.

1968+. Rep. Div. Forest Res. Zambia. HI 71032

Report (Annual), division of plant industry, Colorado. Denver, CO. ?-1974/75+, ?-1975+. Rep. (Annual) Div. Pl. Industr. Colorado. HI 71033

Report (Annual), division of plant industry, commonwealth scientific and industrial research organization. Canberra, A.C.T. 1957/58+, 1958+. Rep. (Annual) Div. Pl. Industr. Commonw. Sci. Industr. Res. Organ. Preceded by: Report (Annual), division of economic botany, commonwealth scientific and industrial research organization. HI 71034

Report of the division of plant inspection, board of agriculture and forestry, Hawaii. Honolulu, HI. 1910-22-? Rep. Div. Pl. Inspect. Board Agric. Forest. Hawaii. HI 71035

Report (Annual), division of plant production, department of agriculture, Western Australia. Perth, W.A. ?-1977/78+, ?-1978+. Rep. (Annual) Div. Pl. Prod. Dept. Agric. Western Australia. Preceded by: Report (Annual), wheat and sheep division, department of agriculture, Western Australia. HI 71036

Report (Annual), division of tropical agronomy, commonwealth scientific and industrial research organization. Brisbane, Qld. 1973/74-74/75, 1974-[76]. Rep. (Annual) Div. Trop. Agron. Commonw. Sci. Industr. Res. Organ. Preceded by: Report (Annual), division of tropical pastures, commonwealth scientific and industrial research organization. Superseded by: Divisional report, division of tropical crops and pastures, commonwealth scientific and industrial research organisation. HI 71037

Report (Annual), division of tropical pastures, commonwealth scientific and industrial research organization. Brisbane, Qld. 1960/61-71/72, 1961-[72]. Rep. (Annual) Div. Trop. Pastures Commonw. Sci. Industr. Res. Organ. Superseded by: Report (Annual), division of tropical agronomy, commonwealth scientific and industrial research organization. HI 71038

Report of the dominion agrostologist, Canada. Ottawa. 1926-36. Rep. Domin. Agrostol. Canada. HI 71039

Report of the dominion cerealist, Canada. Ottawa. 1920/21-29, 1921-29? Rep. Domin. Cerealist Canada. Preceded by: Progress report, dominion cerealist, Canada. Superseded by: Progress report, cereal crops division, Canada. HI 71040

Report of the dominion grain research laboratory, Canada. Ottawa. 1913+. Rep. Domin. Grain Res. Lab. Canada. HI 71041

Report of the Dove marine laboratory of Kings College, Durham university. Cullercoats, England. Rep. Dove Mar. Lab. See B–P–H 764/19. HI 58576

Report, Dublin naturalists' field club. Dublin. 1898-1916. Rep. Dublin Naturalists' Field Club. HI 71042

Report, Duke university school of forestry. Durham, NC. 1930+. Rep. Duke Univ. School Forest. HI 71043

Report of the Dumraon agricultural experiment station. Calcutta. 1894-1912. Rep. Dumraon Agric Exp. Sta. Incorporated in: Report of the agricultural department, Bengal. HI 71044

Report, Dundee naturalists' society. Dundee. Nos. ?-4-12, ?-1876-85. Rep. Dundee Naturalists' Soc. HI 71045

Report of the Ealing microscopical and natural history society. London. Nos. ?-3-17, ?-1880-93. Rep. Ealing Microscop. Nat. Hist. Soc. Superseded by: Report of the Ealing natural science and microscopical society. HI 71046

Report of the Ealing natural science and microscopical society. London. Nos. 18-27, 1894-1904. Rep. Ealing Nat. Sci. Microscop. Soc. Preceded by: Report of the Ealing microscopical and natural history society. Superseded by: Report of the Ealing scientific and microscopical society. HI 71047

Report of the Ealing scientific and microscopical society. London. 1904-35. Rep. Ealing Sci. Microscop. Soc. Preceded by: Report of the Ealing natural science and microscopical society. HI 71048

Report, east african agricultural research institute, Amani. Dar-es-Salaam. 1928/29-47, 1929-47. Rep. E. African Agric. Res. Inst. Amani. Superseded by: Report, east african agriculture and forestry research organisation. HI 71049

Report of the east african agricultural and fisheries research council. Nairobi. 1954+. Rep. E. African Agric. Fish. Res. Council. HI 71050

Report, east african agriculture and forestry research organisation. Nairobi. 1948-56. Rep. E. African Agric. Forest. Res. Organ. Preceded by: Report, east african agricultural research institute, Amani. Superseded by: Record of research, east african agriculture and forestry research organisation. HI 71051

Report (Annual), east anglian institute of agriculture = Report and papers, east anglian institute of agriculture. Chelmsford.

Report, east Kent natural history society. Canterbury. 1858-95. Rep. E. Kent Nat. Hist. Soc. Superseded by: Report and transactions, east Kent scientific and natural history society. HI 71052

Report (Annual) of the East Malling research station. East Malling. Vol. 1+, 1913+. Rep. (Annual) East Malling Res. Sta. 2-1380-3. HI 71053

Reports of the east of Scotland union of naturalists' societies. 1884. Rep. E. Scotland Union Naturalists'

Soc. Superseded by: Proceedings of the east of Scotland union of naturalists' societies. HI 71054

Report of ecological conference, New Zealand ecological society. [Formed part of: New Zealand science review.] Wellington, N.Z. 1952. Rep. Ecol. Conf. New Zealand Ecol. Soc. Superseded by: Report of annual meeting, New Zealand ecological society. HI 71055

Report of the economic botanist, Burma. Rangoon. 1922/23-32/33, 1923-33. Rep. Econ. Bot. Burma. Preceded by: Report (Annual) of the agricultural stations, the agricultural chemist, the agricultural engineer, the assistant entomologist and the economic botanist. Continued in: Report, department of agriculture, Burma. HI 71056

Report on economic mycology. Wye. 1906-13. Rep. Econ. Mycol. HI 71057

Report (Annual), economic poisons. Agricultural experiment station, university of Arizona, special bulletin. Tucson, AZ. 1945+. Rep. (Annual) Econ. Poisons Agric. Exp. Sta. Univ. Arizona Special Bull. HI 71058

Report (Annual), Efford experimental horticulture station. Lymington. 1964-78. Rep. (Annual) Efford Exp. Hort. Sta. Preceded by: Station report, Efford experimental horticulture station. Superseded by: Annual review, Efford experimental horticulture station. HI 71059

Report, empire forestry association. London. 1923+. Rep. Empire Forest. Assoc. HI 71060

Report (Annual) of the entomological society of Ontario. Guelph, Ont. Vols. 1-89, 1869/70-1958. Rep. (Annual) Entomol. Soc. Ontario. Superseded by: Proceedings of the entomological society of Ontario. 2-1461-2. HI 71061

Report (Annual), environment, Canada. Ottawa. 1971/72-75/76, 1972-76. Rep. (Annual) Environm. Canada. Preceded in part by: Report (Annual), department of fisheries & forestry, Canada. Superseded by: Report, fisheries and environment, Canada. HI 71062

Report (Annual) of the environment science institute of Mie prefecture = Report (Annual) of the environmental science institute of Mie prefecture. Yokkaichi.

Report (Annual) of the environmental science institute of Mie prefecture. [Mie-ken kogai senta. Nempo.] Yokkaichi. No. 1+, 1973+. Rep. (Annual) Environm. Sci. Inst. Mie Pref. HI 71063

Report of the Erith and Belvedere natural history and scientific society. Nos. ?-5-6, ?-1882-84. Rep. Erith Belvedere Nat. Hist. Sci. Soc. HI 71064

Report (Annual), Essex naturalists' trust. 1970+. Rep. (Annual) Essex Naturalists' Trust. HI 71065

Report (Annual), estación científica Charles Darwin = Report (Annual) of the Charles Darwin research station. Santa Cruz, Galapagos.

Report (Annual), european and mediterranean plant protection organisation. [Forms part of: Publications, european and mediterranean plant protection organisation. Series C.] Paris. 1951+. Rep. (Annual) Eur. Medit. Pl. Protect. Organ. HI 71066

Report on european paleobotany. Stockholm. 1939-49. Rep. Eur. Paleobot. HI 71067

Report from the executive director, wilderness society = Report, wilderness society. Washington, DC.

Report (Annual), experiment station, Georgia. Experiment, GA. Vols. 1-62, 1890-50. Rep. (Annual) Exp. Sta. Georgia. Superseded by: Report (Annual) of the experiment stations, college of agriculture, Georgia university HI 71068

Report, experiment station of the south african sugar association. Durban. 1952+. Rep. Exp. Sta. S. African Sugar Assoc. HI 71069

Report on the experiment station, Tortola, Virgin Islands. 1902-12/13, 1903-13. Rep. Exp. Sta. Tortola Virgin Islands. Superseded by: Report on the botanic and experiment station, Tortola. HI 71070

Report (Annual) of the experiment stations, college of agriculture, Georgia university Athens, GA. 1950/51-68, 1951-68. Rep. (Annual) Exp. Sta. Coll. Agric. Georgia Univ. Preceded by: Report (Annual), experiment station, Georgia. Superseded by: Biennial report, Georgia agricultural experiment station. HI 71071

Reports from experiment stations, empire cotton growing corporation. London. 1923/25-33, 1925-33. Rep. Exp. Sta. Empire Cotton Growing Corp. Superseded by: Progress report from experiment stations, empire cotton growing corporation. HI 71072

Report on experiment stations and scientific sections, department of agriculture, Bihar and Orissa. Patna. 1912/13-35/36, 1913-36 [reports for 1919/20 & 1926/27 not published]. Rep. Exp. Sta. Sci. Sect. Dept. Agric. Bihar & Orissa. HI 71073

Report, experimental farms in Bihar and Orissa = Report on experiment stations and scientific sections, department of agriculture, Bihar and Orissa. Patna.

Report of experimental farms in the Central Provinces, southern and eastern circles. Nagpur. 1925/26-30/31, 1926-31. Rep. Exp. Farms Central Provinces S. & E. Circles. Preceded by: Report on the experimental farms in the southern circle, Central provinces. HI 71074

Report on the experimental farms in the central provinces. Nagpur. 1903/04-04/05, 1904-05. Rep. Exp. Farms Central Provinces. Preceded by: Report of the Nagpur experimental farm. Superseded by: Report of the agricultural stations in the central provinces. HI 71075

Report (Annual) of experimental farms of the Central Provinces, northern and plateau circles. Nagpur. 1925/26-30/31, 1926-31. Rep. (Annual) Exp. Farms Central Provinces N. & Plateau Circles. HI 71076

Report on the experimental farms, department of agriculture, Canada. Ottawa. 1887-191?, 1888-1919. Rep. Exp. Farms Dept. Agric. Canada. Superseded by: Report of the director, dominion experimental farms, Canada. HI 71077

Report (Annual) of experimental farms of the northern and plateau circles, Central Provinces = Report (Annual) of experimental farms of the Central Provinces, northern and plateau circles. Nagpur.

Report on the experimental farms in the southern circle, Central provinces. Nagpur. 1914/15-24/25, 1915-25. Rep. Exp. Farms S. Circle Central Provinces. Superseded by: Report of experime HI 71078

Report (Annual) of the experimental farms of the southern and eastern circles, Central provinces (Tharsa & Raipur) = Report of experimental farms in the Central Provinces, southern and eastern circles. Nagpur.

Report, experimental and research station, nursery and market garden industries development society ltd. Cheshunt. Vols. 1-40, 1915-54. Rep. Exp. Res. Sta. Nursery Market Gard. Industr. Developm. Soc. Superseded by: Report of the glasshouse crops research institute. HI 71079

Report of experimental work of the agricultural stations in the Bombay Presidency. Bombay. 1894-1920. Rep. Exp. Work Agric. Sta. Bombay Presidency. HI 71080

Report on experimental work on cotton, department of agriculture, Queensland. Brisbane, Qld. 1924-30. Rep. Exp. Work Cotton Dept. Agric. Queensland. HI 71081

Report on experimental work by the economic botanist and his staff, Bombay. Poona. 1913-16. Rep. Exp. Work Econ. Bot. Staff Bombay. HI 71082

Report of the experimental work of the Ganeshkind botanical garden. Bombay. 1913-16. Rep. Exp. Work Ganeshkind Bot. Gard. Superseded by: Short report of the experimental work of the Ganeshkind botanical garden. HI 71083

Report, experimental work on hops, Rosemaund experimental husbandry farm. Preston Wynne. 1959-72. Rep. Exp. Work Hops Rosemaund Exp. Husb. Farm. Superseded by: Hop review, Rosemaund experimental husbandry farm. HI 71084

Report on the experimental work of the sugar cane experiment station, Jamaica. Kingston, Jamaica. 1905+. Rep. Exp. Work Sugar Cane Exp. Sta. Jamaica. HI 71085

Report on the experimental work of the sugar experiment station, Jamaica = Report on the experimental work of the sugar cane experiment station, Jamaica. Kingston, Jamaica.

Report (Annual) on experiments in agriculture, forestry and fisheries. Series forestry. [Norin suisan shiken kenkyu nenpo. Ringyo-hen.] Tokyo. 1964+. Rep. (Annual) Exp. Agric. Forest. Fish., Ser. Forest. HI 75142

Report of experiments. Government botanical garden, Nikita, Yalta, Crimea = Kratkii obzor nauchno-opytnykh rabot Gosudarstvennogo Nikitskogo opytnogo botanicheskogo sada. Yalta.

Report on experiments, Kirton agricultural institute. Boston. 1922-40. Rep. Exp. Kirton Agric. Inst. HI 71086

Report (Annual), faculty of agriculture, imperial college of tropical agriculture = Report, imperial college of tropical agriculture. St. Augustine, Trinidad.

Report of the faculty of agriculture; Shizuoka university. [Shizuoka daigaku nogakubu kenkyu hokoku.] Iwata, Japan. Rep. Fac. Agric. Shizuoka Univ. See B–P–H 764/23. HI 58577

Report of the faculty of agriculture, university of the West Indies. St. Augustine, Trinidad. 1966+. Rep. Fac. Agric. Univ. West Indies. Preceded by: Report, imperial college of tropical agriculture. HI 71087

Report (Annual) of the faculty of education, university of Iwate = Report (Annual) of the college of education, university of Iwate. Morioka.

Report of the faculty of fisheries; prefectural university of Mie. Otanimachi, Japan. Rep. Fac. Fish. Pref. Univ. Mie. See B–P–H 764/24. HI 58578

Report, faculty of general agriculture, university college, Dublin. Dublin. 1950?-68? Rep. Fac. Gen. Agric. Univ. Coll. Dublin. Superseded by: Research report, faculty of general agriculture, university college, Dublin. HI 71088

Reports of the faculty of science, Kagoshima university. Earth sciences and biology. [Kagoshima daigaku rigakubu kiyo. Chigaku, seibutsugaku.] Kagoshima. No. 1+, 1968+. Rep. Fac. Sci. Kagoshima Univ, Earth Sci. Biol. Preceded by: Science reports of the Kagoshima university. HI 71089

Reports of the faculty of science, Shizuoka university. [Shizuoka daigaku rigakubu.] Shizuoka. Vol. 1+, 1965+. Rep. Fac. Sci. Shizuoka Univ. Preceded by: Report, liberal arts faculty, Shizuoka university. Series B, natural science. HI 71090

Report, Fairfield experimental horticulture station. Preston. 1st+ 1959+. Rep. Fairfield Exp. Hort. Sta.

HI 71091

Report (Annual) of the Fan memorial institute of biology. Peiping. Vols. 1-9, 1928/29-37, 1928-38. Rep. (Annual) Fan Mem. Inst. Biol. HI 71092

Report and farm guide, Rosemaund experimental husbandry farm = Report, Rosemaund experimental husbandry farm. Preston Wynne.

Reports on the fauna and flora of Wisconsin. Stevens Point, WI. No. 1+, 1969+. Rep. Fauna Fl. Wisconsin. HI 71093

Report (Annual) of the federal department of forest research, Nigeria. Ibadan. 1956+. Rep. (Annual) Fed. Dept. Forest Res. Nigeria. Preceded by: Report on forest research and the forest school, Nigeria. HI 71094

Report on federal forest administration in the Federation of Malaya = Report of the forest department, Federation of Malaya. Kuala Lumpur.

Report (Annual) on fermentation processes. New York. Vol. 1+, 1977+. Rep. (Annual) Ferment. Processes. HI 71095

Report of the fermentation research institute. [Kogyo gijutsu-in hakko kenkyusho kenkyu hokoku.] Chiba. 1943+. Rep. Ferment. Res. Inst. HI 71096

Report, fermentation research institute, Osaka = Report (Annual) of the institute for fermentation research, Osaka. Osaka, Nenpo.

Report (Annual), field experiments on sugar-cane in Trinidad. Port of Spain. 1937-44. Rep. (Annual) Field Exp. Sugar-Cane Trinidad. Preceded by: Report, sugar agronomy division, department of agriculture, Trinidad and Tobago. Incorporated in: Report of the sugar cane investigation committee. HI 71097

Report (Annual) of the Field museum of natural history. Chicago, IL. 1965-78/79, 1966?-79? Rep. (Annual) Field Mus. Nat. Hist. Preceded by: Report (Annual), Chicago natural history museum. Superseded by: Biennial report, Field museum of natural history. HI 71098

Report (Annual), field naturalists' club of Victoria. Melbourne, Vic. Nos. ?-4-12, ?-1884/5-91/92. Rep. (Annual) Field Naturalists' Club Victoria. Incorporated in: Victorian naturalist. 2-1557-2. HI 71099

Report, field studies council. London. 1952+. Rep. Field Stud. Council. Preceded by: Report, council for the promotion of field studies. HI 71100

Report on field trials and observations, north of Scotland college of agriculture. Aberdeen. 1962/63-64/65, 1963-65. Rep. Field Trials Observ. N. Scotland Coll. Agric. Preceded by: Investigations, research and field trials, north of Scotland college of agriculture. Superseded by: Research investigations and field trials, north of Scotland college of agriculture. HI 71101

Report (Annual), fisheries and environment, Canada. Ottawa. 1976/77+. Rep. (Annual) Fish. Environm. Canada. Preceded by: Report (Annual), environment, Canada, HI 71102

Report, Florida agricultural experiment station. Gainesville, FL. 1908-67. Rep. Florida Agric. Exp. Sta. HI 71103

Report (Annual) of the Florida geological survey. Tallahassee, FL. Vol. 1+, 1956+. Rep. (Annual) Florida Geol. Surv. HI 71104

Report (Annual) of the Florida state geological survey. Tallahassee, FL. 1907-30. Rep. (Annual) Florida State Geol. Surv. HI 71105

Report of the Folkestone natural history society. Folkestone. 1870-71, 1950-58. Rep. Folkestone Nat. Hist. Soc. HI 71106

Report (Annual) of forage research in the northeast United States = Report (Annual) of the United States regional pasture research laboratory. State College, PA.

Report (Annual), foreign students' college of Chiba university. Yoyoicho, Chibashi. ?-1972+. Rep. (Annual) Foreign Stud. Coll. Chiba Univ. HI 71107

Report on forest administration, Ajmere-Merwara. Mount Abu, N. Bombay. 1876-1941. Rep. Forest Admin. Ajmere-Merwara. HI 71108

Report of the forest administration in the Andamans = Progress report of forest administration in the Andamans. Calcutta.

Report (Annual) of the forest administration, Bombay = Administration report (Annual), forest department, Bombay.

Report on forest administration in Burma. Rangoon. 1861-1939/40, 1861-40. Rep. Forest Admin. Burma. HI 71109

Report on forest administration, colony of Sierra Leone. Freetown. 1913+. Rep. Forest Admin. Colony Sierra Leone. HI 71110

Report of the forest administration in Cyprus. Nicosia. 1921/30+, 1925+. Rep. Forest Admin. Cyprus. HI 71111

Report on forest administration, department of agriculture, forests and fisheries, West Bengal. Alipore. 1947+. Rep. Forest Admin. Dept. Agric. Forests Fish. West Bengal. HI 71112

Report (Annual) on the forest administration of the eastern region of Nigeria. Enugu. 1953+, 1954+. Rep. (Annual) Forest Admin. E. Region Nigeria. Preceded by: Report (Annual), forestry department, Nigeria. HI 71113

Report on forest administration in the Malay union =

Report of the forest department, Malayan Union. Kuala Lumpur.

Report (Annual) on the forest administration of Nigeria = Report, forestry department, Nigeria. Lagos.

Report of the forest administration of the northern region of Nigeria. Kaduna. 1951+. Rep. Forest Admin. N. Region Nigeria. Preceded by: Report (Annual), forest department, Nigeria. HI 71114

Report on forest administration in Orissa. Angul, Cuttack. 1936-41. Rep. Forest Admin. Orissa. HI 71115

Report on forest administration, forest department, Poona. Poona. 1849-70, 1936-59? Rep. Forest Admin. Forest Dept. Poona. Superseded by: Administration report (Annual), forest department, Maharashtra. HI 71116

Report on forest administration in the utilization circle, Burma. Rangoon. 1925-40, 1944-47 [suspended 1941-44]. Rep. Forest Admin. Utiliz. Circle Burma. HI 71117

Report (Annual) on the forest administration of the western region of Nigeria. Lagos. 1953+. Rep. (Annual) Forest Admin. W. Region Nigeria. Preceded by: Report (Annual), forest department, Nigeria. HI 71118

Report (Annual) of the forest board, South Australia = Report, woods and forests department, South Australia. Adelaide, S.A.

Report of the forest branch of the department of lands, British Columbia. Victoria, B.C. 1912-? Rep. Forest Branch Dept. Lands British Columbia. Superseded by?: Report of the ministry of forests, British Columbia. HI 71119

Report, forest department, Andhya Pradesh. 1958+. Rep. Forest Dept. Andhya Pradesh. HI 71121

Report, forest department, Bechuanaland. Pharing. 1940+. Rep. Forest Dept. Bechuanaland. HI 71122

Report, forest department, Belize. Belize. 1967-71. Rep. Forest Dept. Belize. Preceded by: Report (Annual) of the forest department, British Honduras. HI 71123

Report, forest department, British Guiana = Report, forestry department, British Guiana. Georgetown, British Guiana.

Report (Annual) of the forest department, British Honduras. Belize. 1923?-66 [1950-51 not published]. Rep. (Annual) Forest Dept. British Honduras. Superseded by: Report, forest department, Belize. HI 71124

Report, forest department, British Solomon Islands = Report, forestry department, British Solomon Islands. Honiara.

Report of the forest department, Cape of Good Hope. Cape Town. 1905-09. Rep. Forest Dept. Cape Good Hope. Preceded by: Report of the conservators of forests, Cape of Good Hope. Superseded by: Report (Annual), department of forestry, South Africa. HI 71125

Report, forest department, colony of North Borneo. Sandakan, Jesselton. 1915-61. Rep. Forest Dept. Colony North Borneo. Superseded by: Report (Annual), forest department, Sabah, Malaysia. HI 71126

Report of the forest department, colony of Singapore. Singapore. 1949/51-1951/52, 1951-52. Rep. Forest Dept. Colony Singapore. HI 71127

Report (Annual), forest department, Cyprus = Report of the forest administration in Cyprus. Nicosia.

Report of the forest department, Federated Malay States. Kuala Lumpur. 1904-45. Rep. Forest Dept. Federated Malay States. Superseded by: Report of the forest department, Malayan Union. HI 71128

Report of the forest department, Federation of Malaya. Kuala Lumpur. 1948+. Rep. Forest Dept. Malaya. Preceded by: Report of the forest department, Malayan Union. Superseded by: Report of the forest department, Malaysia. HI 71129

Report, forest department, Fiji. Suva. 1938+. Rep. Forest Dept. Fiji. HI 71130

Report, forest department, Gambia = Report of the forestry adviser, Gambia. Bathurst.

Report (Annual), forest department, Jamaica. Kingston, Jamaica. 1938+. Rep. (Annual) Forest Dept. Jamaica. HI 71131

Report, forest department, Johore. 1927-39. Rep. Forest Dept. Johore. Incorporated in: Report of the forest department, Federated Malay States. HI 71132

Report (Annual), forest department, Kenya. Nairobi. 1921-29, 1944+. Rep. (Annual) Forest Dept. Kenya. Preceded by: Report, forestry department, East Africa Protectorate. HI 71133

Report, forest department, Madhya Pradesh. Part 1, silviculture. Nagpur. 1928+. Rep. Forest Dept. Madhya Pradesh, Part 1 Silvic. HI 71134

Report, forest department, Madhya Pradesh. Part 2, forest utilization. Nagpur. 1939+. Rep. Forest Dept. Madhya Pradesh, Part 2 Forest Utiliz. HI 71135

Report of the forest department, Malayan Union. Kuala Lumpur. 1946-48. Rep. Forest Dept. Malayan Union. Preceded by: Report of the forest department, Federated Malay States. Superseded by: Report of the forest department, Federation of Malaya. HI 71136

Report of the forest department, Malaysia. (Sarawak). 1967+. Rep. Forest Dept. Malaysia. Preceded by: Report of the forest department, Federation of Malaya.

HI 71137

Report, forest department, Mauritius = Report of the forests and gardens department, Mauritius. Port Louis, Curepipe.

Report (Annual), forest department, ministry of rural development, Zambia. Part 2, research activities. Kitwe. 1968+, 1969+. Rep. (Annual) Forest Dept. Minist. Rural Developm., Part 2 Res. Activities. HI 71138

Report, forest department, Northern Rhodesia. Mazabuka. 1946-? Rep. Forest Dept. Northern Rhodesia. Previously contained in: Report, department of agriculture, Northern Rhodesia. HI 71139

Report, forest department, Nyasaland. Zomba. 1926+. Rep. Forest Dept. Nyasaland. Previously contained in: Report, department of agriculture, Nyasaland. HI 71140

Report, forest department, Orissa = Report on forest administration in Orissa. Angul, Cuttack.

Report, forest department, Rajasthan. Jaipur. 1956+. Rep. Forest Dept. Rajasthan. HI 71141

Report (Annual), forest department, Sabah, Malaysia. Jesselton. 1962-? Rep. (Annual) Forest Dept. Sabah Malaysia. Preceded by: Report, forest department, colony of North Borneo. HI 71142

Report of the forest department, Sarawak. Kuching. 1922-62. Rep. Forest Dept. Sarawak. Superseded by: Report of the forest research officer, forest department, Sarawak. HI 71143

Report, forest department, Sierra Leone = Report on forest administration, colony of Sierra Leone. Freetown.

Report, forest department, Sind = Report on administration, forest department, Sind. Karachi.

Report of the forest department, Singapore = Report of the forest department, colony of Singapore. Singapore.

Report, forest department, state of Sabah = Report (Annual), forest department, Sabah, Malaysia. Jesselton.

Report of the forest department, Straits Settlements. Kuala Lumpur. 1905-34. Rep. Forest Dept. Straits Settlements. Incorporated in: Report of the forest department, Federated Malay States. HI 71144

Report, forest department, Tanganyika. Dar-es-Salaam. 1921+. Rep. Forest Dept. Tanganyika. HI 71145

Report, forest department, Trinidad and Tobago. Port of Spain. 1906/07+, 1907+. Rep. Forest Dept. Trinidad Tobago. HI 71146

Report (Annual) of the forest department, Uganda. Entebbe. 1906+. Rep. (Annual) Forest Dept. Uganda. HI 71147

Report of the forest department, Union of South Africa = Report (Annual), department of forestry, South Africa. Cape Town.

Report, forest department, West Pakistan. Peshawar. 1955+. Rep. Forest Dept. West Pakistan. HI 71149

Report, forest division, Bechuanaland = Report, forest department, Bechuanaland. Pharing.

Reports in forest ecology and forest soils = Rapport i skogsekologi och skoglig marklära. Uppsala.

Report of the forest experiment station, bureau of plant industry, government of Formosa. [Ringyo shiken-jo hokoku, Shokusan-kyoku.] Taihoku [=Taipei, Taiwan]. Rep. Forest. Exp. Sta. Bur. Pl. Industr. Gov. Formosa. See B–P–H 765/9. HI 58581

Report, forest products research board. London. 1927/28-38, 1928-38. Rep. Forest Prod. Res. Board. Superseded by: Forest products research. HI 71150

Report of the forest products research institute. [Ringyo shidojo kenkyu hokoku.] Asahigawa, Japan. No. ?-3+, ?-1952+. Rep. Forest Prod. Res. Inst. (Asahigawa). HI 58582

Report (Annual), forest products research institute. Kumasi. 1965+. Rep. Forest Prod. Res. Inst. (Kumasi). HI 71151

Report of the forest research, forestry education and extension in Cyprus = Report of the forest research, Cyprus. Nicosia.

Report of the forest research institute. Buitenzorg, Dutch E. Indies [=Bogor, Indonesia] Rep. Forest Res. Inst. See B–P–H 765/11. HI 58583

Report (Annual), forest research laboratory, Oregon state university. Corvallis, OR. 1959+. Rep. (Annual) Forest Res. Lab. Oregon State Univ. Preceded by?: Report, Oregon forest products laboratory. HI 71152

Report of the forest research branch, ministry of natural resources, Malawi. Zomba. 1967/68+. Rep. Forest Res. Branch Minist. Nat. Resources Malawi. HI 71153

Report (Annual), forest research council of British Columbia. No. 1+, 1982+. Rep. (Annual) Forest Res. Council British Columbia. HI 71154

Report of the forest research, Cyprus. Nicosia. 1963+. Rep. Forest Res. Cyprus. HI 71155

Report on forest research, forest research branch, Canada. Ottawa. 1956/57+, 1957+. Rep. Forest Res. Forest Res. Branch Canada. HI 71156

Report on forest research, forest research division, Canada = Report on forest research, forest research branch, Canada. Ottawa.

Report on forest research and the forest school, Nigeria. Lagos. 1954-55. Rep. Forest Res. Forest School

Nigeria. Preceded by: Report, forestry department, Nigeria. Superseded by: Report, department of forest research, Nigeria. HI 71157

Report on forest research, forestry commission. London. 1948/49+, 1949+. Rep. Forest Res. Forest. Commiss. HI 71158

Report of the forest research institute, Federated Malay States. Kepong. 1934-39. Rep. Forest Res. Inst. Federated Malay States. Incorporated in: Report of the forest department, Federated Malay States HI 71159.

Report, forest research institute of Sweden, department of forest yield research = Rapport, statens skogsforskningsinstitut. Avdelingen för skogsproduktion. Stockholm.

Report of the forest research institute, Wellington. Wellington, N.Z. 1954+. Rep. Forest Res. Inst. Wellington. Preceded by: Forest research notes, Wellington. HI 71160

Report of the forest research institute of west Norway = Meddelelser fra vestlandets forstlige forsøksstation. Bergen.

Report of the forest research officer, forest department, Sarawak. Kuching. 1965+. Rep. Forest Res. Off. Forest Dept. Sarawak. Preceded by: Report of the forest department, Sarawak. HI 71161

Report of the forest service, British Columbia. Victoria, B.C. 1945-77. Rep. Forest Serv. British Columbia. Preceded by: Report of the forest branch of the department of lands, British Columbia. Superseded by: Report of the ministry of forests, British Columbia. HI 75099

Report, forest service, New Zealand = Report (Annual), New Zealand state forest service. Wellington, N.Z.

Report of the forest service, Palestine. Jerusalem. 1927-36. Rep. Forest Serv. Palestine. Preceded by: Report, department of agriculture and forests, Palestine. Superseded by: Report, department of forests, Israel. HI 71163

Report (Annual) of the forest service, Queensland = Report (Annual), forestry branch, Queensland. Brisbane, Qld.

Report of the forest service, United States department of agriculture. Washington, DC. 1923+. Rep. Forest Serv. U.S.D.A. Preceded by: Department of agriculture, division of forestry, report of the forester. HI 71164

Report, forest trust, British Honduras = Report (Annual) of the forest department, British Honduras. Belize.

Report of the forester, United States department of agriculture = Department of agriculture, division of forestry, report of forester. Washington, DC.

Report of the forestry adviser, Gambia. Bathurst. 1950+. Rep. Forest. Adviser Gambia. HI 71165

Report of the forestry branch, department of the interior, Canada. Ottawa. 1899-36. Rep. Forest. Branch Dept. Interior Canada. Superseded by: Report of the lands, parks and forests branch, department of mines, Canada. HI 71166

Report forestry branch, department of natural resources, Saskatchewan. Regina, Sask. 1955+. Rep. Forest. Branch Dept. Nat. Resources Saskatchewan. HI 71167

Report (Annual), forestry branch, Queensland. Brisbane, Qld. 1906+. Rep. (Annual) Forest. Branch Queensland. HI 71168

Report of the forestry commission, Great Britain. London. No. 1+, 1919/20+, 1920+ [None published for the years 1938-44]. Rep. Forest. Commiss. Great Britain. HI 71169

Report of the forestry commission, New South Wales. Sydney, N.S.W. 1916+. Rep. Forest. Commiss. New South Wales. Preceded by: Report of the forestry department, New South Wales. HI 71170

Report, forestry commission, Southern Rhodesia. Salisbury, Rhodesia. 1936-? Rep. Forest. Commiss. Southern Rhodesia. HI 71171

Report, forestry commission, Tasmania. Hobart, Tas. 1946/47+, 1947+. Rep. Forest. Commiss. Tasmania. Preceded by: Report, forestry department, Tasmania. HI 71172

Report, forestry department, British Guiana. Georgetown, British Guiana. 1925-66. Rep. Forest. Dept. British Guiana. Superseded by: Report, forestry department, Guyana. HI 71173

Report, forestry department, British North Borneo = Report, forest department, colony of North Borneo. Sandakan, Jesselton.

Report, forestry department, British Solomon Islands. Honiara. 1952+. Rep. Forest. Dept. British Solomon Islands. HI 71174

Report of the forestry experiment station. [Ringyo shiken iho.] Tokyo. Rep. Forest. Exp. Sta. See B–P–H 765/8. HI 58580

Report (Annual), forestry department, East Africa Protectorate. Nairobi. 1909/10-20/21?, 1910-21. Rep. (Annual) Forest. Dept. East Africa Protectorate. Superseded by: Report (Annual), forest department, Kenya. HI 71175

Report (Annual), forestry department, Federation of Nigeria. Lagos. 1951-? Rep. (Annual) Forest. Dept. Federation Nigeria. Preceded by: Report, forestry department, Nigeria. Superseded by: Report (Annual) of the federal department of forest research, Nigeria. HI 71176

Report, forestry department, Ghana. Accra. 1957-58.

Rep. Forest. Dept. Ghana. Preceded by: Report, forestry department, Gold Coast. Superseded by: Report, forestry division, ministry of agriculture, Ghana. HI 71177

Report, forestry department, Gold Coast. Accra. 1919/20-55/56, 1920-57. Rep. Forest. Dept. Gold Coast. Superseded by: Report, forestry department, Ghana. HI 71178

Report, forestry department, Guyana. Georgetown, Guyana. 1967. Rep. Forest. Dept. Guyana. Preceded by: Report, forestry department, British Guiana. HI 71179

Report of the forestry department, Hong Kong. Hong Kong. 1946-50. Rep. Forest. Dept. Hong Kong. Preceded by: Report of the botanical and forestry department, Hong Kong. Superseded by: Annual departmental report, director of agriculture and fisheries, and forestry, Hong Kong. HI 71180

Report of the forestry department, New South Wales. Sydney, N.S.W. 1882-1916. Rep. Forest. Dept. New South Wales. Superseded by: Report of the forestry commission, New South Wales. HI 71181

Report, forestry department, Nigeria. Lagos. 1911-53. Rep. Forest. Dept. Nigeria. Superseded by: Report (Annual), forestry department, Federation of Nigeria; Report on forest research and the forest school, Nigeria; Report on the forest administration of the eastern region of Nigeria; Report on the forest administration of the northern region of Nigeria and Report on the forest administration of the western region of Nigeria. HI 71182

Report, forestry department, Nyasaland = Report, forest department, Nyasaland. Zomba.

Report, forestry department, Tasmania. Hobart, Tas. 1920/21-45/46, 1921-46. Rep. Forest. Dept. Tasmania. Superseded by: Report, forestry commission, Tasmania. HI 71183

Report (Annual), forestry department, Uganda = Report (Annual) of the forest department, Uganda. Entebbe.

Report, forestry division, department of lands, Irish Free State (later Eire). Dublin. 1933/38+, 1938?+. Rep. Forest. Div. Dept. Lands Irish Free State. HI 71184

Report (Annual), forestry division, eastern region, Nigeria = Report (Annual) on the forest administration of the eastern region of Nigeria. Enugu.

Report, forestry division, Trinidad = Report, forest department, Trinidad and Tobago. Port of Spain.

Report (Annual), forestry and timber bureau, Australia. 1946-74/75, 1946-76. Rep. (Annual) Forest. Timber Bur. Australia. Preceded by: Report (Annual), commonwealth forestry and timber bureau. Superseded by: Forest resources. HI 71185

Report on forestry, U S department of agriculture = Department of agriculture. Report on forestry. Washington, DC. Dept. Agric. Rep. Forest. See B–P–H 343/11.

Report, forests administration, Sudan = Report, forests department, Sudan. Khartoum.

Report, forests commission, Victoria. Melbourne, Vic. 1919+. Rep. Forests Commiss. Victoria. Preceded by: Report of the department of state forests, Victoria. HI 71186

Report, forests department, Sudan. Khartoum. 1948+. Rep. Forests Dept. Sudan. Preceded by: Report, department of agriculture and forests, Sudan. Superseded by: Report, forests administration, Sudan. HI 71187

Report of the forests department, Western Australia. Perth, W.A. 1895+. Rep. Forests Dept. Western Australia. HI 71188

Report (Annual) on the forests departments, Singapore, Penang and Malacca. Singapore. 1888, 1889. Rep. (Annual) Forests Dept. Singapore Penang Malacca. Superseded by: Report (Annual), botanic gardens and forest department, Straits Settlements. HI 71189

Report on the forests of Ethiopia. Addis Ababa. 1955+. Rep. Forests Ethiopia. HI 71190

Report of the forests and gardens department, Mauritius. Port Louis, Curepipe. 1901+. Rep. Forests Gard. Dept. Mauritius. HI 71191

Report, Franz Theodore Stone institute of hydrobiology. Columbus, OH. 1950-54. Rep. Franz Theodore Stone Inst. Hydrobiol. Preceded by: Report, Franz Theodore Stone laboratory. HI 71192

Report, Franz Theodore Stone laboratory. Columbus, OH. 1926-50. Rep. Franz Theodore Stone Lab. Superseded by: Report, Franz Theodore Stone institute of hydrobiology. HI 71193

Report (Annual), freshwater biological association = Report (Annual), freshwater biological association of the British Empire. London, Ambleside.

Report (Annual), freshwater biological association of the British Empire. London, Ambleside. Vol. 1+, 1932+. Rep. (Annual) Freshwater Biol. Assoc. Brit. Empire. 2-1642-1. HI 71194

Report (Annual) of the fruit growers' association of Nova Scotia. Kentville, Halifax, N.S. No. 28+, 1892+. Rep. (Annual) Fruit Growers Assoc. Nova Scotia. Preceded by: Transactions and reports, fruit growers' association of Nova Scotia. HI 71195

Report (Annual) of the fruit growers' association of Ontario. Toronto. Vols. 1-87, 1869-1946 [suspended 1931-38]. Rep. (Annual) Fruit Growers' Assoc. Ontario. Incorporated in: Grower. 2-1650-3.

HI 71196

Report of the fruit research station, Kodur. Anantharajpet. 1935/36-51/52, 1937-52. Rep. Fruit Res. Sta. Kodur. HI 71197

Report, fruit research station, Saharanpur. 1950/53, 1953. Rep. Fruit Res. Sta. Saharanpur. HI 71198

Reports of the Fuji bamboo gardens. [Fuji take-rui shokubutsuen hokoku.] Shizuoka, Japan. Rep. Fuji Bamboo Gard. See B–P–H 765/15. HI 58584

Report (Annual) of the gakugei faculty of the Iwate university. Morioka. Vols. 1-25, 1941-65. Rep. (Annual) Gakugei Fac. Iwate Univ. Superseded by: Report of the college of education, university of Iwate. HI 71199

Report (Annual), garden club of America. New York. 1979/80+, 1980?+. Rep. (Annual) Gard Club. Amer. HI 71200

Report (Annual), Gatooma research station. Gatooma. 1967-73. Rep. (Annual) Gatooma Res. Sta. Superseded by?: Report, cotton research institute, Gatooma. HI 71201

Report of general assemblies, international union of biological sciences = Union internationale des sciences biologiques. Series A: générale.

Report from the general experimental farms, department of agriculture, Tanganyika. Dar-es-Salaam. 1936-38? Rep. Gen. Exp. Farms Dept. Agric. Tanganyika. HI 71202

Report, genetics and plant breeding research institute. Ottawa. Nos. 1-?, 1960-64. Rep. Genet. Pl. Breed. Res. Inst. Preceded by: Report, cereal crops division, Canada. Superseded by: Report, Ottawa research station. HI 71203

Report of the geological and natural history survey of Minnesota. Minneapolis, MN. 1872-1912 [Vols. 2, 4, & 6 entitled Minnesota botanical studies vols. 1-3]. Rep. Geol. Nat. Hist. Surv. Minnesota. HI 71204

Report (Annual) of the geological survey of Louisiana. New Orleans, LA. Vols. 1-3, 1870-72. Rep. (Annual) Geol. Surv. Louisiana. HI 71205

Report (Annual), geological survey of New York. Albany, NY. No. 1+, 1882+, 1884+. Rep. (Annual) Geol. Surv. New York. HI 71206

Report (Annual) of the Georgia agricultural experiment station. Athens, GA. 1975+. Rep. (Annual) Georgia Agric. Exp. Sta. Preceded by: Biennial report, Georgia agricultural experiment station. HI 71207

Report of the Gezira agricultural research service, Anglo-Egyptian Sudan. Wad Medani. 1932-35. Rep. Gezira Agric. Res. Serv. Anglo-Egyptian Sudan. Superseded by: Report of the agricultural research service, Anglo-Egyptian, Sudan. HI 71208

Report (Annual) of the Gezira research station and substations. Khartoum. 1964/65+, 1965?+. Rep. (Annual) Gezira Res. Sta. Substa. HI 71209

Report of the glasshouse crops research institute. Littlehampton. 1957+. Rep. Glasshouse Crops Res. Inst. Preceded by: Report, experimental and research station, nursery and market garden industries development society ltd. HI 71210

Report, glasshouse investigational unit for Scotland. Auchincruive. 1977+. Rep. Glasshouse Invest. Unit Scotland. HI 71211

Report, Gloucestershire trust for nature conservation. No. ?-5+, ?-1965+. Rep. Gloucestershire Trust Nat. Conservation. HI 71212

Report of the governing body and the principal's report, imperial college of tropical agriculture = Report, imperial college of tropical agriculture. St. Augustine, Trinidad.

Report. Government agricultural research institute, Taiwan, Nippon. [Taiwan sotokufu nogyo shikenjo hokoku.] Taihoku [=Taipei, Taiwan]. Rep. Gov. Agric. Res. Inst. Taiwan. See B–P–H 765/16. HI 58585

Reports of the government botanical garden Nikita = Kratkii obzor nauchno-opytnykh rabot Gosudarstvennogo Nikitskogo opytnogo sada. Yalta.

Report (Annual) of the government botanic gardens and parks, Ootacamund. Madras. 1876-1941? Rep. (Annual) Gov. Bot. Gard. Parks Ootacamund. HI 71213

Report of the government botanical gardens, Saharanpur. Allahabad. 1878-1921. Rep. Gov. Bot. Gard. Saharanpur. Incorporated in: Report on the working and administration of the government gardens, Allahabad. HI 71214

Report of the government botanist and curator, Cape Town. Cape Town. 1897-1903. Rep. Gov. Bot. Curator Cape Town. HI 71215

Report of the government botanist and director of the botanic and zoologic garden. Melbourne, Vic. 1860/61-74, 1861-74? Rep. Gov. Bot. Director Bot. Zool. Gard. Preceded by: Report on the botanical gardens, Melbourne. Superseded by: Report (Annual) of the director, botanic gardens, Melbourne. HI 71216

Report of the government cinchona plantation and factory in Bengal. Calcutta. 1882-1939. Rep. Gov. Cinchona Plantation Fact. Bengal. HI 71217

Report (Annual) of the government forest experiment station. [Ringyo shiken syuho.] Tokyo. Rep. (Annual) Gov. Forest Exp. Sta. See B–P–H 765/18. HI 58586

Report (Annual), government gardens department, Mysore. Bangalore. 1897/98-1929/30, 1898-1930.

Rep. (Annual) Gov. Gard. Dept. Mysore. Superseded by: Report, department of horticulture, Mysore. HI 71218

Report on the government horticultural gardens, Lucknow. Allahabad. 1879-1921. Rep. Gov. Hort. Gard. Lucknow. Superseded by: Report on the working and administration of the government gardens, Allahabad. HI 71219

Report of the government laboratory and superintendent of agriculture of the Leeward Islands. [Bridgetown], Barbados. 1911-14-? Rep. Gov. Lab. Superintendent Agric. Leeward Islands. Superseded by: Report of the department of agriculture, Leeward Islands. HI 71220

Report of the government research institute, Formosa. [Taiwan sotoku-fu kenkyu-sho hokoku.] Taihoku [=Taipei, Taiwan]. Rep. Gov. Res. Isnt. Formosa. See B–P–H 765/19. HI 58587

Report of the government sugar experiment station. Taihoku [=Taipei]. Nos. 1-8, 1934-40. Rep. Gov. Sugar Exp. Sta. Superseded by: Report of the Taiwan sugar experiment station. HI 71221

Report, grain research laboratory. Winnipeg. 1920, 1927+. Rep. Grain Res. Lab. HI 71222

Report of the grassland improvement station. Hurley. 1948-55. Rep. Grassland Improv. Sta. Superseded by: Report of the grassland research institute. HI 71223

Report (Annual), grassland research institute. Hurley. 1955+. Rep. (Annual) Grassland Res. Inst. Preceded by: Report of the grassland improvement station and Experiments in progress, grassland research institute. HI 71224

Report of the grassland research station, Kitale. Kitale. 1955+. Rep. Grassland Res. Sta. Kitale. HI 71225

Report, grasslands agricultural research station, Marandellas. Marandellas. 1952-73. Rep. Grasslands Agric. Res. Sta. Marandellas. HI 71226

Report (Annual), Great Lakes institute. Toronto. 1967-71, 1968-71? Rep. (Annual) Great Lakes Inst. HI 71227

Report, Guam agricultural experiment station, office of experiment stations, United States department of agriculture = U S department of agriculture, office of experiment stations; the Guam agricultural experiment station, report. Washington, DC. U.S.D.A. Off. Exp. Sta. Guam Agric. Exp. Sta. Rep. See B–P–H 943/28.

Report, Gwent trust for nature conservation. Newport, Gwent. 1973/74+. Rep. Gwent Trust Nat. Conservation. Preceded by: Report, Monmouthshire naturalists' trust ltd. HI 71228

Report, Hamilton natural history society. Hamilton. No. 1+, 1969+. Rep. Hamilton Nat. Hist. Soc. HI 71229

Report (Annual), Hampshire and Isle of Wight naturalists' trust ltd. Portsmouth. 1960/61+, 1961?+. Rep. (Annual) Hampshire Isle Wight Naturalists' Trust. HI 71230

Report, Hampstead naturalists' club. London. Nos. ?-3-4, ?-1882-84. Rep. Hampstead Naturalists' Club. HI 71231

Report, Harrogate and district naturalists society. Harrogate. 1960+. Rep. Harrogate Distr. Naturalists Soc. HI 71232

Report, Harvard forest. Petersham, MA. 1966/67+. Rep. Harvard Forest. HI 71233

Report of the Haslemere microscope and natural history society. 1897-1905. Rep. Haslemere Microscope Nat. Hist. Soc. Preceded by: Record of lectures and addresses, Haslemere microscope and natural history society. Superseded by: Report of the Haslemere natural history society. HI 71234

Report of the Haslemere natural history society. 1907+. Rep. Haslemere Nat. Hist. Soc. Preceded by: Report of the Haslemere microscope and natural history society. HI 71235

Report, Hastings natural history society. Hastings. 1933-48. Rep. Hastings Nat. Hist. Soc. Preceded by: Report of the Hastings and St. Leonard's natural history society. HI 71236

Report of the Hastings and St. Leonard's natural history society. Hastings. 1899-1933. Rep. Hastings St. Leonard's Nat. Hist. Soc. Superseded by: Report, Hastings natural history society. HI 71237

Report (Annual) of the Hastings university school naturalists' field club. Hastings. Vols. 1-2, 1874-76. Rep. (Annual), Hastings Univ. School. Naturalists' Field Club. HI 71238

Report (Annual) , Hawaii agricultural experiment station = Hawaii agricultural experiment station. Annual report. Honolulu, HI. Hawaii Agric. Exp. Sta. Annual Rep. See B–P–H 412/26.

Report (Annual) of the hawaiian agricultural experiment station, office of experiment stations, United States department of agriculture = U S department of agriculture; office of experiment stations. Annual report of the hawaiian agricultural experiment station. Washington, DC. U.S.D.A. Off. Exp. Sta. Annual Rep. Hawaiian Agric. Exp. Sta. See B–P–H 943/22.

Report (Annual), hawaiian sugar planters' association experiment station. Honolulu, HI. 1895+. Rep. (Annual) Hawaiian Sugar Pl. Assoc. Exp. Sta. HI 71239

Report, herb society of America. Boston, MA. 1st+, 19??+. Rep. Herb Soc. Amer. HI 71240

Report, herbarium jutlandicum: A A U. Aarhus. No. 1+, 1978+. Rep. Herb. Jutlandicum A. A. U. HI 71241

Report (Annual), herbarium of the university of California. Santa Barbara, CA. 1977/78?+, 1978?+. Rep. (Annual) Herb. Univ. Calif. HI 71242

Report (Annual), herbarium of the university of Malaya "KLU". Kuala Lumpur. 1979+. Rep. (Annual) Herb. Univ. Malaya, KLU. HI 71243

Report, Herbert Whitley trust = Report (Annual) and accounts, Herbert Whitley trust and Paignton zoological and botanical gardens. Paignton.

Report, herbicide committee, Canada, eastern section = Report, herbicide and weed classification committee, Canada, eastern section.

Report, herbicide and weed classification committee, Canada, eastern section. 1949-56. Rep. Herbic. Weed Classific. Committee Canada, E. Sect. Superseded by: Weed control recommendations for eastern Canada. HI 71244

Report, herbicide and weed classification committee, Canada, western section. 1947-55. Rep. Herbic. Weed Classific. Committee, Canada, W. Sect. Superseded by: Weed control recommendations for western Canada. HI 71245

Report (Annual), Herefordshire & Radnorshire nature trust ltd. Hereford. 1970+, [1970]+. Rep. (Annual) Herefordshire & Radnorshire Nat. Trust. HI 71246

Report (Annual), Hertfordshire and Middlesex trust for nature conservation ltd. St. Albans. 1970+, 1971+. Rep. (Annual) Hertfordshire Middlesex Trust Nat. Conservation. HI 71247

Report, Hillingdon natural history society. West Drayton. 1971/73+. Rep. Hillingdon Nat. Hist. Soc. Preceded by: Journal, Hillingdon natural history society. HI 71248

Report of historico-geographical studies of Taiwan. [Wên hsien chuan k'an.] Taipei, Taiwan. Rep. Hist.-Geogr. Stud. Taiwan. See B–P–H 766/1. HI 58588

Report (Annual) of the Hokkaido branch, forestry and forest products research institute = Report of the Hokkaido branch, government forest experiment station. Sapporo.

Report of the Hokkaido branch, government forest experiment station. Sapporo. 1954+. Rep. Hokkaido Branch Gov. Forest Exp. Sta. Preceded by?: Report of the Hokkaido branch of the imperial forestry and estates bureau. HI 71249

Report of the Hokkaido branch of the imperial forestry and estates bureau. Sapporo. [Dates of publication not ascertained.] Rep. Hokkaido Branch Imp. Forest. Estates Bur. Superseded by?: Report of the Hokkaido branch, government forest experiment station. HI 71250

Report, Hokkaido national agricultural experiment station. [Hokkaido nogyo shikenjo hokoku.] Sapporo, Japan. Rep. Hokkaido Natl. Agric. Exp. Sta. See B–P–H 766/6. HI 58589

Report of the Hokkaido prefectural agricultural experiment station. [Hokkaidoritsu nogyo shikenjo hokoku.] Sapporo, Japan. Rep. Hokkaido Prefect. Agric. Exp. Sta. See B–P–H 766/7. HI 58590

Report (Annual), honorary advisory council for scientific and industrial research, Canada = Report (Annual), advisory council for scientific and industrial research, Canada. Ottawa.

Report on the honourable company's botanic gardens. Calcutta royal botanic garden. Calcutta. Rep. Hon. Company's Bot. Gard. Calcutta. See B–P–H 766/9. HI 58591

Report on hops, Rosemaund experimental husbandry farm = Report, experimental work on hops, Rosemaund experimental husbandry farm. Preston Wynne.

Report, horticultural experiment station of Ontario = Report (Annual) of the horticultural experiment station and products laboratory, vineland station, Ontario. Toronto.

Report (Annual) of the horticultural experiment station and products laboratory, Vineland station, Ontario. Toronto. 1906+. Rep. (Annual) Hort. Exp. Sta. Prod. Lab. Vineland Sta. Ontario. HI 71251

Report of the horticultural societies of Ontario. Toronto. Rep. Hort. Soc. Ontario. See B–P–H 766/11. HI 58592

Report (Annual) of the horticultural society of New York. New York. 1900/01-04/05. Rep. (Annual) Hort. Soc. New York. Superseded by: Year book of the horticultural society of New York. 3-1887-3. HI 71252

Report (Annual) of the hungarian geological institute. Budapest = A magyar állami földtani intézet évi jelentese(i). Budapest. Magyar Állami Földt. Intéz. Évi Jel. See B–P–H 542/18.

Report (Annual), Huntingdonshire fauna & flora society. Huntingdon. 1949+. Rep. (Annual) Huntingdonshire Fauna Fl. Soc. HI 71253

Report of the hydrobiological research unit, university college of Khartoum. Khartoum. 1st+, 1953+. Rep. Hydrobiol. Res. Unit Univ. Coll. Khartoum. HI 71254

Report (Annual) of hydrographical observations, fishery research station, Fusan. Fusan. Vols. 1-8, 1926-? Rep. (Annual) Hydrogr. Observ. 2-1657-2. HI 71255

Report of the Ibaraki prefectural forest experiment station = Bulletin of the Ibaraki prefectural forest experiment station. Mito.

Report (Annual), Idaho agricultural experiment station =

University of Idaho, college of agriculture, Idaho agricultural experiment station annual report. Moscow, ID. Univ. Idaho Coll. Agric. Idaho Agric. Exp. Sta. Annual Rep. See B–P–H 936/3.

Report (Annual) of the Illinois agricultural experiment station. Champaign, IL. 1889+. Rep. (Annual) Illinois Agric. Exp. Sta. HI 71256

Report (Annual), Illinois natural history survey. Springfield, IL. ?-1949/50+, ?-1950+. Rep. (Annual) Illinois Nat. Hist. Surv. HI 71257

Report, Illinois nature preserves commission. Rockford, IL. 1977/78+, 1978+. Rep. Illinois Nat. Preserves Commiss. HI 71258

Report (Annual), Illinois state museum of natural history. Springfield, IL. 1909+. Rep. (Annual) Illinois State Mus. Nat. Hist. Preceded by: Biennial report of the Illinois state museum of natural history. HI 71259

Report, imperial agricultural bureau. London. 1929-48. Rep. Imp. Agric. Bur. Superseded by: Report, commonwealth agricultural bureaux. HI 71260

Report of imperial bureau of fisheries. Scientific investigation. Tokyo. Rep. Imp. Bur. Fish. Sci. Invest. See B–P–H 766/15. HI 58593

Report, imperial college of tropical agriculture. St. Augustine, Trinidad. 1922-59. Rep. Imp. Coll. Trop. Agric. Superseded by: Report of the faculty of agriculture, university of the West Indies. HI 71261

Report (Annual), imperial council of agricultural research. Calcutta. 1929-46. Rep. (Annual) Imp. Council Agric. Res. Superseded by: Report (Annual), indian council of agricultural research. HI 71262

Report of the imperial department of agriculture, India. Calcutta. 1904-07. Rep. Imp. Dept. Agric. India. Superseded by: Report on the progress of agriculture in India. HI 71263

Report (Annual), imperial forestry institute, university of Oxford. Oxford. Vols. 1-37, 1924/25-60/61, 1925-62. Rep. (Annual) Imp. Forest. Inst. Univ. Oxford. Superseded by: Report (Annual), commonwealth forestry institute, university of Oxford. HI 71264

Report on the imperial mycological conference. London. Vols. 1-3, 1924-? Rep. Imp. Mycol. Conf. Superseded by: Report on the commonwealth mycological conference. HI 71265

Report (Annual) of the indian agricultural research institute. Delhi. Rep. Indian Agric. Res. Inst. See B–P–H 766/16. HI 58594

Report of the indian botanic garden and the gardens in Calcutta, (parks and gardens in Cooch Behar) and the Lloyd botanic garden, Darjeeling. Alipore. 1950-55. Rep. Indian Bot. Gard. Gard. Calcutta Lloyd Bot. Gard. Preceded by: Report of the royal botanic gardens, Calcutta, etc. Superseded by: Report of the directorate of botanical and other public gardens, West Bengal. HI 71267

Report (Annual), indian council of agricultural research. New Delhi, Calcutta. 1945/46+, 1946+. Rep. (Annual) Indian Council Agric. Res. Preceded by: Report (Annual), imperial council of agricultural research. HI 71268

Report, indian ecological society. Bombay. 1941+. Rep. Indian Ecol. Soc. HI 71269

Report of the indian forest college, Dehra Dun. Delhi. 1940. Rep. Indian Forest Coll. Dehra Dun. Preceded by: Progress report of the indian forest ranger college, Dehra Dun. Superseded by: Progress report of the indian forest college, Dehra Dun. HI 71270

Report (Annual), indian institute of sugarcane research. Lucknow. Vol. 1+, 1954/55+, 1955+. Rep. (Annual) Indian Inst. Sugarcane Res. HI 71271

Report (Annual) of the Indian museum, natural history section. Calcutta. 1880/81-?, 1881?-1933. Rep. (Annual) Indian Mus. Nat. Hist. Sect. HI 71272

Report (Annual) of the Indiana agricultural experiment station. Lafayette, IN. 1887+. Rep. (Annual) Indiana Agric. Exp. Sta. HI 71273

Report (Annual) of the Indiana corn growers' association. Lafayette, IN. Vol. 1+, 1900+. Rep. (Annual) Indiana Corn Growers' Assoc. 3-1966-1. HI 71274

Report, institute of agricultural research, Addis Ababa. Addis Ababa. 1966/68+, 1968+. Rep. Inst. Agric. Res. Addis Ababa. HI 71275

Report, institute for agricultural research and special services, Ahmadu Bello university. Zaria. 1962/63+. Rep. Inst. Agric. Res. Special Serv. Ahmadu Bello Univ. HI 71276

Report, institute of agricultural research and training, university of Ife. Ife. 1973+. Rep. Inst. Agric. Res. Training Univ. Ife. HI 71277

Report of the institute of agricultural research, Tohoku university. [Science reports of the research institutes, Tohoku university. Series D, agriculture.] Sendai. Vol. 1+, 1950+. Rep. Inst. Agric. Res. Tohoku Univ. HI 71278

Report (Annual), institute of agriculture, Anand. Anand. No. 1+, 1942+. Rep. (Annual) Inst. Agric. Anand. HI 71279

Report, institute of agriculture and natural history. Tel-Aviv. 1921-26. Rep. Inst. Agric. Nat. Hist., Tel Aviv. HI 71280

Report, institute of applied microbiology, university of Tokyo. Tokyo. 1961+. Rep. Inst. Appl. Microbiol. Univ. Tokyo. HI 71281

Report, institute of arable crops research. Harpenden. 1988+. Rep. Inst. Arable Crops Res. HI 75084

Report (Annual), institute of biological resources (Australia). Dickson, A. C. T. 1979/80+, 1980?+. Rep. (Annual) Inst. Biol. Resources (Australia). HI 71282

Report (Annual), institute of botany, academia sinica. [Zhongyang yanjiuyuan.] Nankang, Taipei. 1944/47+, 1947+. Rep. (Annual) Inst. Bot. Acad. Sin.

Report of the institute for breeding research, Tokyo university of agriculture. Tokyo. ?-1975+. Rep. Inst. Breed. Res. Tokyo Univ. Agric. HI 71283

Report (Annual) of the institute for fermentation research, Osaka. Osaka, Nenpo. 1961/62+, 1963+. Rep. (Annual) Inst. Ferment. Res. Osaka. HI 71284

Report of the institute of fishery biology of ministry of economic affairs & national Taiwan university. Taipei, Taiwan. Rep. Inst. Fish. Biol. Minist. Econ. Affairs Natl. Taiwan Univ. See B–P–H 766/18. HI 58595

Report (Annual) of the institute of food microbiology, Chiba university. [Chiba daigaku fuhai kenkyusho hokoku.] Chiba. 1948+. Rep. (Annual) Inst. Food Microbiol. Chiba Univ. HI 71285

Report, institute of freshwater research, Drottningholm. Stockholm. Rep. Inst. Freshwater Res. Drottningholm. See B–P–H 766/19. HI 58596

Report (Annual), institute of genetics, academia sinica. [Zhongguo kexueyuan yichuan yanjiusuo jikan.] 1979+. Rep. (Annual) Inst. Genet. Acad. Sin. HI 71286

Report, institute for marine environmental research. Plymouth. 1971/73+, 1973+. Rep. Inst. Mar. Environm. Res. HI 71287

Report, institute of marine research. Series biology. Goteborg, Sweden. Rep. Inst. Mar. Res., Ser. Biol. See B–P–H 766/20. HI 58597

Report, institute of marine science, university of Alaska. Fairbanks, AK. 1978+. Rep. Inst. Mar. Sci. Univ. Alaska. HI 71288

Report (Annual), institute of microbiology, Rutgers university. New Brunswick, NJ. Vol. 1+, 1954+. Rep. (Annual) Inst. Microbiol. Rutgers Univ. HI 71289

Report, institute of ocean sciences, Patricia Bay. Sidney, B.C. 1979+. Rep. Inst. Ocean Sci. Patricia Bay. HI 71290

Report (Annual), institute of oceanographic sciences. Wormley. 1973/74+, 1976?+. Rep. (Annual) Inst. Oceanogr. Sci. Preceded by: Report (Annual), national institute of oceanography. HI 71291

Report (Annual), institute of phytopathological research = Jaarverslag, instituut voor plantenziektenkundig onderzoek. Wageningen.

Report, institute of polar studies, Ohio state university research foundation. Columbus, OH. 1962+. Rep. Inst. Polar Stud. Ohio State Univ. Res. Found. HI 71292

Report of the institute of scientific research, Manchoukuo. Hsinking, Manchoukuo. Vols. 1-4, 1936-40. Rep. Inst. Sci. Res. Manchoukuo. 3-2013-3. HI 71293

Report, institute of seaweed research. Edinburgh. 1951-68. Rep. Inst. Seaweed Res. Preceded by: Report, scottish seaweed research association. HI 71294

Report (Annual), institute of terrestrial ecology. London. 1974+, 1975+. Rep. (Annual) Inst. Terrestr. Ecol. Preceded by: Report, Merlewood research station and Report, Monks Wood experimental station. HI 71295

Report, institute of tropical forestry. Washington, DC. 1940+. Rep. Inst. Trop. Forest. Reports for 1950-63 included in: Caribbean forester. HI 71296

Report, institute for tropical agriculture, university of Puerto Rico. Mayagüez. 1942/43-45/46, 1943-46. Rep. Inst. Trop. Agric. Univ. Puerto Rico. HI 71297

Report (Annual), institute for virus research, Kyoto university. [Kyoto daigaku uiurusu kenkyusho nenkan liyo.] Kyoto. Vol. 1+, 1958+. Rep. (Annual) Inst. Virus Res. Kyoto Univ. HI 71298

Report (Annual), instituut voor bewaring en verwerking van tuinbouwprodukten. [An English translation of: Mededelingen van het instituut voor bewaring en verwerking van tuinbouwproducten.] Wageningen. 1956-65. Rep. (Annual) Inst. Bewar. Verwerk. Tuinbouwprod. Superseded by: Report (Annual), Sprenger instituut. HI 71299

Report (Annual), instituut voor plantenziektenkundig onderzoek. Wageningen. 1972-75. Rep. (Annual) Inst. Plantenziektenk. Onderz. Preceded by: Jaarverslag, instituut voor plantenziektenkundig onderzoek. Superseded by: Report, research institute for plant protection. HI 71300

Report, inter-american institute of agricultural sciences. Washington, DC, Turrialba. 1943+. Rep. Inter-Amer. Inst. Agric. Sci. HI 71301

Report (Annual), intermountain forest and range experiment station. Ogden, UT. 1938+. Rep. (Annual) Intermount. Forest Range Exp. Sta. HI 71302

Report (Annual), international board for plant genetic resources. Rome. 1977+, 1978+. Rep. (Annual) Int. Board Pl. Genet. Resources. HI 71303

Report (Annual), international center for agricultural research in the dry areas = I C A R D A annual report. Aleppo.

Report, international council for research in agroforestry. Vol. 1+, 1978/79+, 1979?+. Rep. Int. Council Res.

Agroforest. HI 71304

Report (Annual), international crop improvement association. St. Paul, MN. 1919+. Rep. (Annual) Int. Crop Improv. Assoc. HI 71305

Report (Annual), international hydrographic bureau. Monaco. Vol. 1+, 1921+. Rep. (Annual) Int. Hydrogr. Bur. HI 71306

Report, international institute of tropical agriculture. Ibadan. 1971+. Rep. Int. Inst. Trop. Agric. HI 71307

Report, international maize and wheat improvement center. Mexico City. 1966/67+, 1967+. Rep. Int. Maize Wheat Improv. Center HI 71309

Report (Annual), international potato center. Lima. 1975+. Rep. (Annual) Int. Potato Center. HI 71310

Report (Annual), international rice research institute. Manila. 1961/62+, 1962+. Rep. (Annual) Int. Rice Res. Inst. HI 71311

Report (Annual), international union for the conservation of nature and natural resources. Morges. 1961-69, 1962-[70]. Rep. (Annual) Int. Union Conservation Nature Nat. Resources. Superseded by: I U C N yearbook. HI 71312

Report, international union of forest research organizations = Jahresbericht und Mitgliederverzeichnis; internationaler Verband forstlicher Lehranstalten. Stockholm & Zurich.

Report of investigations, bureau of mines. Washington, DC. No. 2000+, 1919+. HI 71313

Report of investigations. Minnesota geological survey. Minneapolis, MN. Rep. Invest. Minnesota Geol. Surv. See B–P–H 766/23. HI 58598

Report on investigations and research, north of Scotland college of agriculture. Aberdeen. 1963-65. Rep. Invest. Res. N. Scotland Coll. Agric. Preceded by: Investigations, research and field trials, north of Scotland college of agriculture. Superseded by: Research investigations and field trials, north of Scotland college of agriculture. HI 71314

Report of investigations of the South Australia geological survey. Adelaide. Rep. Invest. South Australia Geol. Surv. See B–P–H 767/1. HI 58599

Report of investigations, United States bureau of mines = Report of investigations, bureau of mines. Washington, DC.

Report of the Iowa agricultural experiment station. Ames, IA. 1914+. HI 71315

Report, Iowa corn research institute. Ames, IA. 1936+. Rep. Iowa Corn Res. Inst. HI 71316

Report of the Iowa state agricultural society. Ames, IA. Rep. Iowa State Agric. Soc. See B–P–H 767/2. HI 58600

Report of the Iowa state horticultural society. Des Moines, IA. Vols. 26-70, 1891-1936. Rep. Iowa State Hort. Soc. Preceded by: Transactions, Iowa state horticultural society. Superseded by: Proceedings, Iowa state horticultural society. HI 71317

Report of the Iraq natural history museum. Baghdad. 1950/67. Rep. Iraq Nat. Hist. Mus. Superseded by: Report of the natural history research center, Baghdad. HI 71318

Report, Isle of Thanet field club. Ramsgate. 1948-51. Rep. Isle Thanet Field Club. HI 71319

Report (Annual), Israel oceanographic and limnological research. Haifa. Vols. 1-2, 1971/72-73/75, 1972-75. Rep. (Annual) Israel Oceanogr. Limnol. Res. HI 71320

Report (Annual), Jamaica agricultural society. Kingston, Jamaica. 1913?-62. Rep. (Annual) Jamaica Agric. Soc. Incorporated in: Farmer. Kingston, Jamaica. HI 71321

Report, Jamaica research department, banana board = Report (Annual), banana board research department, Jamaica. Kingston, Jamaica.

Report of the japanese association for the advancement of science. [Nihon gakujutsu kyokai hokoku.] Tokyo. Rep. Jap. Assoc. Advancem. Sci. See B–P–H 767/4. HI 58601

Report of the japanese botanical garden association. [Nippon shokubutsuen kyokwai kaiho.] [Japan]. Rep. Jap. Bot. Gard. Assoc. See B–P–H 767/5. HI 58602

Report of the japanese national committee for I B P. Tokyo. [Dates of publication not ascertained.] Rep. Jap. Natl. Committee I. B. P. Preceded by: Interim report of the japanese national committee for I B P HI 71322

Report (Annual), John Innes horticultural institution. London, Bayfordbury. 1910-59. Rep. (Annual) John Innes Hort. Inst. Superseded by: Report (Annual), John Innes institute. HI 71323

Report (Annual), John Innes institute. Norwich. 51st-76th?, 1960-86? Rep. (Annual) John Innes Inst. Preceded by: Report (Annual), John Innes horticultural institution. Superseded by: Report (Annual), A F R C institute of plant science research and the John Innes institute. HI 71324

Report (Annual), jute agricultural research institute. Calcutta. 1941/42+, 1942+. Rep. (Annual) Jute Agric. Res. Inst. HI 71325

Report of the Kagoshima prefectural forest experiment station. [Kagoshima-ken ringyo shikenjo gyomu hokoku.] Gamo. [Dates of publication not ascertained.] Rep. Kagoshima Pref. Forest Exp. Sta. HI 71326

Report of the Kagoshima tobacco experiment station. [Kagoshima tabako shikenjo gyotei hokoku.] Kagoshima, Japan. Rep. Kagoshima Tobacco Exp. Sta. See B–P–H 767/7. HI 58603

Report of the Kanagawa-ken agricultural experiment station. [Kanagawa-kenritsu noji shiken-jo moji shiken seiseki.] Yokohama, Japan. Rep. Kanagawa-Ken Agric. Exp. Sta. See B–P–H 767/9. HI 58604

Report (Annual) of the Kansai branch, government forest experiment station. [Ringyo shikenjo Kansai shijo nenpo.] Kyoto. 1959+. Rep. (Annual) Kansai Branch Gov. Forest Exp. Sta. HI 75144

Report (Annual) of the Kansai forest tree breeding station, ministry of agriculture and forestry. [Kansai rinboku Ikushujo nenpo.] Okayama. 1973+. Rep. (Annual) Kansai Forest Tree Breed. Sta. HI 75143

Report (Annual), Kansas agricultural experiment station = Kansas agricultural experiment station. Annual report. Manhattan, KS.

Report (Annual) of Kanto forest tree breeding station. [Kanto rinboku Ikushujo nenpo.] Mito. 1960+. Rep. (Annual) Kanto Forest Tree Breed Sta. HI 75145

Report (Annual), Kasetsart university research activities. Bangkok = Kasetsart university research activities. Bangkok.

Report of the keihaushin branch of the cactus and succulent society of Japan. Keihaushin. Nos. 1-10, 1950-53. Rep. Keihaushin Branch Cact. Succ. Soc. Japan. HI 71328

Report (Annual) of the Kenana research station. Khartoum. 1964/65+, 1965?+. Rep. (Annual) Kenana Res. Sta. HI 71329

Report (Annual), Kent trust for nature conservation. Maidstone. 1969+, [1970]+. Rep. (Annual) Kent Trust Nat. Conservation. HI 71330

Report (Annual) of the Kentucky agricultural experiment station. Lexington, KY. 1888-1948. Rep. (Annual) Kentucky Agric. Exp. Sta. Superseded by: Results of research, agricultural experiment station, university of Kentucky. HI 71331

Reports from the Kevo subarctic research station. [Forms part of: Annales universitatis turkuensis. Sarja A, 2, Biologica-geographica.] Turku. Vol. 1+, 1964+. Rep. Kevo Subarctic Res. Sta. HI 71332

Report of the Kihara institute for biological research. [Sei-ken jiho.] Yokohama, Japan. Rep. Kihara Isnt. Biol. Res. See B–P–H 767/11. HI 58605

Report (Annual), king's park and botanic garden. Perth, W.A. 1962/63+ 1963?+. Rep. (Annual) King's Park Bot. Gard. HI 71333

Report (Annual), Kirton experimental horticulture Station. Boston. 1st-15th, 1961/64-78, 1964-78. Rep. (Annual) Kirton Exp. Hort. Sta. Superseded by: Annual review, Kirton experimental horticulture station. HI 71334

Reports of the Kochi university; natural science (biology, geology, agriculture). [Kochi daigaku kenkyu hokokui. Shizen kagaku.] Kochi, Japan. Rep. Kochi Univ., Nat. Sci. See B–P–H 767/12. HI 58606

Report of the Kumaun government gardens. Allahabad. 1909-22. Rep. Kumaun Gov. Gard. Superseded by: Report on the working and administration of the government gardens, Allahabad. HI 71335

Report, Kwangsi agricultural experiment station. Linchow. 1938+. Rep. Kwangsi Agric. Exp. Sta. HI 71336

Report of the Kyoto university forest. [Kyoto daigaku nogakubu enshurin iho.] Kyoto, Japan. Rep. Kyoto Univ. Forest. See B–P–H 767/14. HI 58607

Report (Annual) of the Kyushu branch, government forest experiment station. [Ringyo shikenjo Kyushu shijo nenpo.] Kumamoto. 1959+. Rep. (Annual) Kyushu Branch Gov. Forest Exp. Sta. HI 75146

Report of the Kyushu university forests. [Kyushu daigaku nogakubu enshurin shuho.] Fukuoka. No. 1+, 1953+. Rep. Kyushu Univ Forests. HI 58608

Report (Annual) of the L M B committee. Liverpool. Nos. 16-33, 1902-19. Rep. (Annual) L. M. B. Committee. Preceded by: Report (Annual) of the Liverpool marine biological committee. Superseded by: Report (Annual) of the oceanography department of the university of Liverpool. 3-2447-1. HI 71337

Report of the laboratory of biology. Taipei = Report of the laboratory of hydrobiology. Taipei, Taiwan. Rep. Lab. Hydrobiol. See B–P–H 767/18.

Report, laboratory of biology, Taiwan fisheries research institute = Taiwan fisheries research institute, laboratory of biology. Report. [Tai wan sheng sui ch'an shih yen so. Sui ch'an shêng wu hsi yen chiu pao kao.] Taipei, Taiwan. Taiwan Fish. Res. Inst. Lab. Biol. Rep. See B–P–H 865/1.

Report (Annual) of the laboratory of experimental algology and department of applied algology. Třeboň. 1966+, 1967+. Rep. (Annual) Lab. Exp. Algol. HI 71338

Report of the laboratory of hydrobiology. [Taiwan fisheries research institute.] [Shui chan shêng wu hsi yen chiu pao kao Tai wan shen shui chan shih yen shoin hsin.] Taipei, Taiwan. Rep. Lab. Hydrobiol. See B–P–H 767/18. HI 58609

Report (Annual) of the Laguna marine laboratory. Laguna Beach, CA. 1912. Rep. (Annual) Laguna Mar. Lab. HI 71339

Reports of the Lake Sevan station = Trudy Sevanskoi

ozernoi stantsii. Erivan.

Report (Annual), lake states forest experiment station. St. Paul, MN. 1929-63. Rep. (Annual) Lake States Forest Exp. Sta. HI 71340

Report, Lancashire naturalists' trust. 1978+. Rep. Lancashire Naturalists' Trust. HI 71341

Report of the latvian forest research station = Latvijas mežu petišanas stacijas raksti. Riga.

Report (Annual), laws and transactions, botanical society of Edinburgh. Edinburgh. Vol. 1, 1836/37. Rep. (Annual) Laws trans. Bot. Soc. Edinburgh. Superseded by: Report (Annual) and proceedings of the botanical society of Edinburgh. 1-755-3. HI 71342

Report, Lee Valley experimental horticulture station. Hoddesdon. 1958/59-?, 1959?-? Rep. Lee Valley Exp. Hort. Sta. Superseded by: Annual review, Lee Valley experimental horticulture station. HI 71343

Report (Annual), Leicestershire and Rutland trust for nature conservation ltd. Leicester. 1972/73+, 1973?+. Rep. (Annual) Leicestershire Rutland Trust Nat. Conservation. HI 71344

Report, liberal arts faculty, Shizuoka university. Series B, natural science. Shizuoka. Nos. 1-?, 1950-65. Rep. Liberal Arts Fac. Shizuoka Univ., B. Superseded by: Report of the faculty of science, Shizuoka university. HI 71345

Report, liberal arts and science faculty, Shizuoka university = Report, liberal arts faculty, Shizuoka university. Shizuoka.

Report of the lichen exchange club of the British isles. Leicester. 1908-11. Rep. Lichen Exch. Club British Isles. HI 71346

Report, life sciences and agricultural experiment station, university of Maine at Orono. Orono, ME. 1974+, 1975+. Rep. Life Sci. Agric. Exp. Sta. Univ. Maine Orono. Preceded by: Maine agricultural experiment station. Annual report. HI 71347

Reports of the limnological station of Lake Sevan = Trudy Sevanskoi ozernoi stantsii. Erivan.

Report of the Lincolnshire naturalists' trust ltd. Alford. 1949+. Rep. Lincolnshire Naturalists' Trust. HI 71348

Report (Annual) and list of members; international union of forest research organizations = Jahresbericht und Mitgliederverzeichnis; internationaler Verband forstlicher Lehranstalten. Stockholm & Zurich.

Report (Annual) of the Liverpool marine biological station on Puffin Island. Liverpool. Nos. 1-5, 1887-91, 1888-92. Rep. (Annual) Liverpool Mar. Biol. Sta. Puffin Island. Superseded by: Report (Annual) of the Liverpool marine biology committee. 3-2447-1. HI 71349

Report (Annual) of the Liverpool marine biology committee and their biological station at Port Erin. Liverpool. Nos. 6-15, 1893-1902. Rep. (Annual) Liverpool Mar. Biol. Committee Biol. Sta. Port Erin. Preceded by: Report (Annual) of the Liverpool marine biological station on Puffin Island. Superseded by: Report (Annual) of the L. M. B. committee. 3-2447-1. HI 71350

Report (Annual) of the Liverpool microscopical society. Liverpool. Vols. 1-52, 1919/20-? Rep. (Annual) Liverpool Microscop. Soc. 3-2448-1. HI 71351

Report of the Liverpool natural history society. Liverpool. Rep. Liverpool Nat. Hist. Soc. See B–P–H 768/2. HI 58610

Report of the Liverpool naturalists' field club. Liverpool. 1860-68. Rep. Liverpool Naturalists' Field Club. Superseded by: Proceedings of the Liverpool naturalists' field club. HI 71352

Report of the Llandudno and district field club. Llandudno. Nos. ?-3-6-?, ?-1908-12. Rep. Llandudno Distr. Field Club. HI 71353

Report. London botanical exchange club. London. Rep. London Bot. Exch. Club. See B–P–H 768/3. HI 58611

Report, London natural history society. London. 1914. Rep. London Nat. Hist. Soc. Superseded by: Transactions of the London natural history society. HI 71354

Report (Annual), Long Ashton agricultural and horticultural research statio = Report (Annual) of the agricultural and horticultural research station, university of Bristol. Long Ashton.

Report (Annual), Long Ashton research station, university of Bristol. Bristol. 1962+. Rep. (Annual) Long Ashton Res. Sta. Univ. Bristol. Preceded by: Report (Annual) of the agricultural and horticultural research station, university of Bristol. HI 71355

Report (Annual), Los Angeles state and county arboretum. [Reports for 1953/54-1956/57 form vols. 2-5 of: Lasca miscellanea.] Arcadia, CA. 1953/54-58/59, 1954-59. Rep. (Annual) Los Angeles State County Arbor. Preceded by: Lasca miscellanea. Superseded by: Biennial report, Los Angeles state and county arboretum. HI 71356

Report, Loughborough naturalists' club. Loughborough. 1966+. Rep. Loughborough Naturalists' Club. HI 71357

Report (Annual), Loughgall horticultural centre. Belfast. 1968+. Rep. (Annual) Loughgall Hort. Centre. HI 71358

Report (Annual), Louisiana agricultural experiment station. North Louisiana experiment station =

Louisiana agricultural experiment station. North Louisiana experiment station, annual report. Calhoun, LA. Louisiana Agric. Exp. Sta. N. Louisiana Exp. Sta. Annual Rep. See B–P–H 536/2.

Report (Annual), Louisiana agricultural experiment station. Rice experiment station = Louisiana agricultural experiment station. Rice experiment station. Annual report. Crowley, LA. Louisiana Agric. Exp. Sta. Rice Exp. Sta. Annual Rep. See B–P–H 536/4.

Report (Annual), Louisiana state seminary of learning and military academy = Report (Annual) of the board of supervisors of the Louisiana state seminary of learning and military academy.

Report, Louth antiquarian and naturalists' society. Louth. 1895-1939. Rep. Louth Antiq. Naturalists' Soc. HI 71359

Report, Lowestoft and north Suffolk field naturalists' club. Lowestoft? 1946+. Rep. Lowestoft N. Suffolk Field Naturalists' Club. HI 71360

Report (Annual), Luddington experimental horticulture station. Stratford-on-Avon. 1950+. Rep. (Annual) Luddington Exp. Hort. Sta. HI 71361

Report (Annual), Lundy field society. Exeter. 1947+. Rep. (Annual) Lundy Field Soc. HI 71362

Report (Annual) of the Lu-Shan arboretum and botanical garden of the Fan memorial institute of the Kiangsi provincial agricultural institute. Lu-Shan. Vols 1-4?, 1935-37? Rep. (Annual) Lu-Shan Arbor. Bot. Gard. HI 71363

Report on the Lyallpur agricultural station. Lyallpur. 1901-10. Rep. Lyallpur Agric. Sta. HI 71364

Report (Annual) of the M S U - D O E plant research laboratory. East Lansing, MI. Vol. 15+, 1980+. Rep. (Annual) M. S. U.-D. O. E. Pl. Res. Lab. Preceded by: Plant research. HI 71365

Report (Annual), Macaulay land use research institute. Aberdeen. 1987+. Rep. (Annual) Macaulay Land Use Res. Inst. Preceded by: Report (Annual), Macaulay institute for soil research [not entered]. HI 75239

Report, Maidenhead naturalists' field club and Thames valley antiquarian society. Maidenhead. 1865-1946. Rep. Maidenhead Naturalists' Field Club Thames Valley Antiq. Soc. HI 71366

Report (Annual), Maine agricultural experiment station = Maine agricultural experiment station. Annual report. Bangor, ME.

Report (Annual), Maine state pomological society = Maine state pomological society. Annual report. Augusta, ME. Maine State Pomol. Soc. Annual Rep. See B–P–H 546/19.

Report of the Malton field naturalists' and scientific society. 1884-87. Rep. Malton Field Naturalists' Sci. Soc. HI 71367

Report, Malvern field club. 1931-37. Rep. Malvern Field Club. HI 71368

Report (Annual), Manchester microscopical society. Manchester. 1883-1900. Rep. (Annual) Manchester Microscop. Soc. Preceded by: Transactions and annual report of the Manchester microscopical society. HI 71369

Report (Annual), Manchester naturalists' field club = Manchester naturalists' field club annual report. Manchester, England. Manchester Naturalists' Field Club Annual Rep. See B–P–H 547/26.

Report of the Manchester scientific students' association. Manchester. 1862-76. Rep. Manchester Sci. Stud. Assoc. Superseded by: Report and proceedings of the Manchester scientific students' association. HI 71370

Report, mango research scheme, Uttar Pradesh. Basti. 1958+. Rep. Mango Res. Scheme Uttar Pradesh. HI 71371

Report, Manx nature conservation trust. Kirk Michael. 1977+. Rep. Manx Nat. Conservation Trust. HI 71372

Report (Annual) of the marine biological association of China. Nanking. Vols. 1-3, 1932-34. Rep. (Annual) Mar. Biol. Assoc. China. 3-2538-3. HI 71373

Report, marine biological association of the west of Scotland. Glasgow. 1900-13. Rep. Mar. Biol. Assoc. W. Scotland. Preceded by: Report, Scottish marine biological association. Superseded by: Report, Millport marine biological association. HI 71374

Report(s), marine biological laboratory. Woods Hole, MA. Rep. Mar. Biol. Lab. See B–P–H 768/5. HI 58612

Report (Annual) of the marine biological station at Port Erin. Liverpool. Nos. 47-84, 1934-71. Rep. (Annual) Mar. Biol. Sta. Port Erin. Preceded by: Report of the oceanography department of the university of Liverpool. Superseded by: Report of the department of marine biology, university of Liverpool. 3-2447-1. HI 71375

Report (Annual), marine environmental data service. Ottawa. 1976+. Rep. (Annual) Mar. Environm. Data Serv. HI 71376

Report, marine laboratories, department of biological sciences, university of Delaware. Newark, DE. Rep. Mar. Lab. Dept. Biol. Sci. Univ. Delaware. See B–P–H 768/6. HI 58613

Reports on marine research. Berlin, New York. Vol. 31+, 1986+. Rep. Mar. Res. Preceded by: Meeresforschung. HI 71377

Report (Annual), Maryland agricultural experiment

station = Maryland agricultural experiment station. Annual report. College Park, MD. Maryland Agric. Exp. Sta. Annual Rep. See B–P–H 548/25.

Report of the Maryland agricultural society. Baltimore, MD. 1916+. Rep. Maryland Agric. Soc. Preceded, in part, by: Report of the Maryland state horticultural society. HI 71378

Report of the Maryland state horticultural society. Baltimore, MD. Vols. 1-17?, 1899-1915. Rep. Maryland State Hort. Soc. Incorporated in: Report of the Maryland agricultural society. HI 71379

Report (Annual), Massachusetts agricultural experiment station = Massachusetts agricultural experiment station. Annual report. Amherst, MA. Mass. Agric. Exp. Sta. Annual Rep. See B–P–H 549/11.

Report (Annual) of the Massachusetts horticultural society. Boston, MA. Vol. 1, 1920-22. Rep. (Annual) Mass. Hort. Soc. Preceded by: Pt. 2 of Transactions of the Massachusetts horticultural society. Superseded by: Yearbook of the Massachusetts horticultural society with the annual report. 3-2552-2. HI 71380

Report of the Mauritius institute. Port Louis. 1901-13, 1934+. Rep. Mauritius Inst. HI 71381

Report (Annual), Mauritius sugar industry research institute. Port Louis, Reduit. No. 1+, 1954+. Rep. (Annual) Mauritius Sugar Industr. Res. Inst. Preceded by: Report, sugar cane research station, Mauritius. HI 71382

Reports of meetings of the australasian association for the advancement of science. Sydney, N.S.W. Rep. Meetings Australas. Assoc. Advancem. Sci. See B–P–H 768/8. HI 58614

Report (Annual) of the Melbourne botanic gardens = Report of the colonial botanical garden. Melbourne, Vic.

Report (Annual) of the Melbourne botanic and domain gardens = Report of the colonial botanical garden. Melbourne, Vic.

Report, Merlewood research station. Grange-over-Sands. 1967/69+, 1970+. Rep. Merlewood Res. Sta. Superseded by: Report (Annual), Institute of terrestrial ecology. HI 71383

Report (Annual) of the Michigan academy of science, (arts, and letters). Lansing, MI. Vols. 1-22, 1894/99-1920, 1900-21. Rep. (Annual) Michigan Acad. Sci. Superseded by: Papers of the Michigan academy of science, arts and letters. 3-2641-3. HI 71384

Report (Annual), Michigan agricultural experiment station = Michigan agricultural experiment station. Annual report. East Lansing, MI. Michigan Agric. Exp. Sta. Annual Rep. See B–P–H 592/6.

Report (Annual) of the Michigan state horticultural society. East Lansing, MI. Vol. 11?+, 1881+. Rep. (Annual) Michigan State Hort. Soc. Preceded by: Report (Annual), state pomological society of Michigan. 3-2643-3 HI 71385

Report of the microscopist, microscopy division, U S department of agriculture = Department of agriculture. Microscopy division. Report of the microscopist. Washington, DC. Dept. Agric. Microscop. Div. Rep. Microscop. See B–P–H 343/5.

Report, Millport marine biological association. Glasgow. 1896-99. Rep. Millport Mar. Biol. Assoc. Superseded by: Report, marine biological association of the west of Scotland. HI 71386

Report (Annual), minister of lands, forests, and mines, Ontario. Toronto. 1905-20. Rep. (Annual) Minist. Lands Forests Mines Ontario. Preceded by: Report of the commissioner of crown lands, Ontario. Superseded by: Report (Annual), department of lands and forests (and mines), Ontario. HI 71387

Report, ministry of african agriculture, Northern Rhodesia. Lusaka. 1959-63. Rep. Minist. African Agric. Northern Rhodesia. Preceded by: Report, department of agriculture, Northern Rhodesia. Superseded by: Report, ministry of agriculture, Zambia. HI 71388

Report (Annual), ministry of agriculture, co-operatives and marketing, Lesotho. Maseru. Vol. [1]+, 1966+. Rep. (Annual) Minist. Agric. Co-op. Marketing Lesotho. Preceded by: Report (Annual), department of agriculture, Basutoland. HI 71389

Report (Annual), ministry of agriculture, forests and wildlife. Dar-es-Salaam. Vol. 1+, 1963+. Rep. (Annual) Minist. Agric. Forests WIldlife. Preceded by: Report (Annual), ministry of agriculture, Tanganyika. HI 71390

Report (Annual), ministry of agriculture and lands, Jamaica. Kingston, Jamaica. 1957+. Rep. (Annual) Minist. Agric. Lands Jamaica. Preceded by: Report of the department of science and agriculture, Jamaica. HI 71391

Report of the ministry of agriculture and lands, Southern Rhodesia. Salisbury, Rhodesia. 1900-54. Rep. Minist. Agric. Lands Southern Rhodesia. Superseded by: Report, ministry of agriculture, Rhodesia & Nyasaland. HI 71392

Report (Annual), ministry of agriculture and natural resources, Guyana. Georgetown. 1966+. Rep. (Annual) Minist. Agric. Nat. Resources Guyana Preceded by: Report of the director of agriculture, British Guiana. HI 71393

Report of the ministry of agriculture and natural resources, Mauritius. Port Louis. 1969+. Rep. Minist. Agric. Nat. Resources Mauritius. Preceded by: Report of the department of agriculture, Mauritius. HI 71394

Report (Annual), ministry of agriculture and natural resources, Sierra Leone. Freetown. 1930-65. Rep. (Annual) Minist. Agric. Nat. Resources Sierra Leone. Preceded by: Report, department of agriculture, Sierra leone. HI 71395

Report (Annual), ministry of agriculture and natural resources of the western region of Nigeria = Report (Annual) of the department of agriculture of the western region of Nigeria. Ibadan.

Report, ministry of agriculture, Rhodesia = Report of the secretary, ministry of agriculture, Rhodesia. Zomba.

Report, ministry of agriculture, Rhodesia & Nyasaland. Zomba. 1953/54-62/63, 1954-63. Rep. Minist. Agric. Rhodesia Nyasaland. Preceded by: Report of the ministry of agriculture and lands, Southern Rhodesia and Report (Annual), department of agriculture, Nyasaland. Superseded by: Report of the secretary, ministry of agriculture, Rhodesia and Report (Annual) of the department of agriculture, Malawi. HI 71396

Report (Annual) of the ministry of agriculture, Sudan. Khartoum. 1938-54. Rep. (Annual) Minist. Agric. Sudan. Preceded by: Report, department of agriculture and forests, Sudan. HI 71397

Report (Annual) of the ministry of agriculture, Swaziland. Mbabane. 1967-75. Rep. (Annual) Minist. Agric. Swaziland. Preceded by: Report of the department of agriculture, Mbabane. HI 71398

Report (Annual), ministry of agriculture, Tanganyika. Dar-es-Salaam. 1922-62. Rep. (Annual) Minist. Agric., Tanganyika. Superseded by: Report (Annual), ministry of agriculture, forests and wildlife. HI 71399

Report, ministry of agriculture, Zambia. Lusaka. 1964+. Rep. Minist. Agric. Zambia. Preceded by: Report, ministry of african agriculture, Northern Rhodesia. HI 71400

Report of the ministry of forests, British Columbia. Victoria, B.C. 1978+. Rep. Minist. Forests British Columbia. Preceded by: Report (Annual) of the forest service, British Columbia. HI 71401

Report (Annual), ministry of lands, natural resources and tourism, republic of Zambia. Lusaka. 1974+. Rep. (Annual) Minist. Lands Nat. Resources Tourism Republ. Zambia. Preceded by: Report (Annual), HI 71402

Report (Annual), ministry of natural resources, Ontario. [Toronto]. 1972/73+, 1973+. Rep. (Annual) Minist. Nat. Resources Ontario. Preceded by: Report (Annual), department of lands and forests, Ontario. HI 71403

Report (Annual), Minnesota agricultural experiment station = Minnesota agricultural experiment station. Annual report. St. Anthony Park, MN. Minnesota Agric. Exp. Sta. Annual Rep. See B–P–H 596/6.

Report (Annual) of the Minnesota state horticultural society. St. Paul & Minneapolis, MN. 1884; vols. 13-26, 1885-98. Rep. (Annual) Minnesota State Hort. Soc. Preceded by: Transactions of the Minnesota state horticultural society. Superseded by: Trees, fruits and flowers of Minnesota. 3-2684-2. HI 71404

Report (Annual) of the Minnesota state horticultural society = Minnesota horticulturist. Annual report of the Minnesota state horticultural society. Minnesota Hort. See B–P–H 596/14.

Report (Annual), Mississippi agricultural experiment station. Mississippi agricultural and mechanical college = Mississippi agricultural experiment station. Mississippi agricultural and mechanical college. Annual report. Jackson, MS. Mississippi Agric. Exp. Sta. Annual Rep. See B–P–H 599/23.

Report (Annual), Mississippi agricultural experiment station. Mississippi state college = Mississippi agricultural experiment station. Mississippi agricultural and mechanical college. Annual report. Jackson, MS. Mississippi Agric. Exp. Sta. Annual Rep. See B–P–H 599/23.

Report (Annual), Missouri agricultural experiment station = Missouri agricultural experiment station. Annual report. Columbia, MO.

Report (Annual) of the Missouri botanical garden. St. Louis, MO. Vols. 1-23, 1889-1912; 1987+. Rep. (Annual) Missouri Bot. Gard. 5-3759-1. HI 71405

Report (Annual), Missouri horticultural society. St. Louis, MO. Nos. 1-55, 1859-1912. Rep. (Annual) Missouri Hort. Soc. Superseded by: Proceedings, Missouri state horticultural society. 3-2704-1. HI 71406

Report (Annual), Mitsubishi-kasei institute of life sciences. Tokyo. 1971/72+, 1973+. Rep. (Annual) Mitsubishi-Kasei Inst. Life Sci. HI 71407

Report, Monks Wood experimental station. Abbots Ripton. 1960/65+, 1966+. Rep. Monks Wood Exp. Sta. Superseded by: Report (Annual), institute of terrestrial ecology. HI 71408

Report, Monmouthshire naturalists' trust ltd. Newport, Gwent. 1970-73. Rep. Monmouthshire Naturalists' Trust. Superseded by: Report, Gwent trust for nature conservation. HI 71409

Report of the monsoon committee. [Kisetsuhu chosakai shi.] [Japan]. Rep. Monsoon Committee. See B–P–H 768/15. HI 58615

Report (Annual), Montana college of agriculture and mechanic arts. Agricultural experiment station = Montana college of agriculture and mechanic arts.

Agricultural experiment station. Annual report. Bozeman, MT. Montana Agric. Exp. Sta. Annual Rep. See B–P–H 618/11.

Report (Annual), Montreal botanic garden. Montreal. ?-1964+. Rep. (Annual) Montreal Bot. Gard. HI 71410

Report (Annual) of the Montreal horticultural society and fruit growers association of the province of Quebec. Montreal. Vols. 1-17, 1875-95. Rep. (Annual) Montreal Hort. Soc. 3-2752-2. HI 71411

Report of the Montrose natural history and antiquarian society. Montrose, Scotland. Rep. Montrose Nat. Hist. Soc. See B–P–H 768/16. HI 58616

Report of the Mori mycological institute. [Mori kinjin kenkyujo hokoku.] Kiryu. [Dates of publication not ascertained.] Rep. Mori Mycol. Inst. HI 71412

Report (Annual), Morris arboretum. Philadelphia, PA. 1979+. Rep. (Annual) Morris Arbor. HI 71413

Report, moss exchange club. London. Nos. 1-27, 1896-1922. Rep. Moss Exch. Club. Superseded by: Report, british bryological society. HI 71414

Report (Annual) of the museum of applied arts and sciences = Report (Annual) of the trustees of the museum of applied arts and sciences. Sydney, N.S.W.

Report, museum of natural history, university of Wisconsin. Stevens Point, WI. 1974/75+. Rep. Mus. Nat. Hist. Univ. Wisconsin. HI 71415

Report (Annual), museum, Texas tech university. Lubbock, TX. 1970+, 1972+. Rep. (Annual) Mus. Texas Tech Univ. HI 71416

Report, mushroom research association. 1950-54. Rep. Mushr. Res. Assoc. HI 71417

Report, mushroom research station. Yaxley. 1949-55. Rep. Mushr. Res. Sta. HI 71418

Report of the mycologist, Burma. Rangoon. 1921/22-32/33, 1922-33. Rep. Mycol., Burma. HI 71419

Report of the mycologist, division of vegetable pathology, United States department of agriculture = U S department of agriculture. Division of vegetable pathology. Report of the mycologist. Washington, DC. U.S.D.A. Div. Veg. Pathol. Rep. Mycol. See B–P–H 942/7.

Report of the Nagano horticultural experiment station. [Nagano-ken engei shikenjo hokoku.] Suzaka. No. ?-2+, ?-1960+. Rep. Nagano Hort. Exp. Sta. HI 75157

Report (Annual) of the Nagoya fertilizer and feed inspection station, fertilizer section. [Nagoya hishiryo kensajo jigyo hokoku hiryo no bu.] Aichi. Vol. 1+, 1948+. Rep. (Annual) Nagoya Fertilizer Inspect. Sta. Fertilizer Sect. HI 71420

Report of the Nagpur experimental farm. Nagpur. 1878/79-1902/03, 1879-1903. Rep. Nagpur Exp. Farm. Superseded by: Report on the experimental farms in the central provinces HI 71421.

Report on the Natal botanic gardens and colonial herbarium. Durban, South Africa. Rep. Natal Bot. Gard. Colon. Herb. See B–P–H 768/23. HI 58618

Report (Annual), Natal colonial herbarium = Natal colonial herbarium. Annual report. Durban, South Africa. Natal Colon. Herb. Annual Rep. See B–P–H 630/13.

Report (Annual) of the national agricultural library. Beltsville, MD. 1988+, 1989+. Rep. (Annual) Natl. Agric. Libr. HI 71422

Report of the national agricultural research bureau of the ministry of industry, China. 1934-38. Rep. Natl. Agric. Res. Bur. Minist. Industr. China. HI 71423

Report (Annual), national arboretum. Washington, DC = Report (Annual), United States national arboretum. Washington, DC.

Report (Annual), national botanic gardens, Lucknow. Lucknow. 1966-77. Rep. (Annual) Natl. Bot. Gard. Lucknow. Superseded by: Report (Annual), national botanical research institute, Lucknow. HI 71425

Report of the national botanic gardens of South Africa. Claremont. 1913+. Rep. Natl. Bot. Gard. South Africa. HI 71426

Report (Annual), national botanical research institute, Lucknow. Lucknow. 1978+. Rep. (Annual) Natl. Bot. Res. Inst. Lucknow. Preceded by: Report (Annual), national botanic gardens, Lucknow. HI 71427

Report of the national fruit and cider institute, Long Ashton. Bristol. Nos. 1-10, 1903/04-12/13, 1904-13. Rep. Natl. Fruit Cider Inst. Long Ashton. Superseded by: Report of the agricultural and horticultural research station, university of Bristol. HI 71428

Report (Annual), national fruit trials, Brogdale. Faversham. 1st-7th, 1972-79. Rep. (Annual) Natl. Fruit Trials Brogdale. Superseded by: Annual review, national fruit trials, Brigdale. HI 71429

Report (Annual) of the national grassland research institute. [Sochi shikenjo nenpo.] Tochigi. No. 1+, 1970+. Rep. (Annual) Natl. Grassland Res. Inst. HI 75147

Report (Annual) of the national institute of agrobiological resources. [Norin suisansho nogyo seibutsu shigen kenkyujo]. Yatabe, Tsukuba. 1985+. Rep. (Annual) Natl. Inst. Agrobiol. Resources. HI 75100

Report (Annual) of the national institute of genetics. [Kokuritsu idengaku kenkyusho, nenpo.] Mishima. Vol. 1+, 1949+. Rep. (Annual) Natl. Inst. Genet. HI 71430

Report (Annual), national institute of oceanography. Cambridge. 1949-50, 1965-68. Rep. (Annual) Natl.

Inst. Oceanogr. From 1950-65 continued as: Report (Annual) of the national oceanographic council. Superseded by: Report (Annual), institute of oceanographic sciences. HI 71431

Report (Annual), national institute of oceanography, India. New Delhi. 1965/66+, 1966+. Rep. (Annual) Natl. Inst. Oceanogr. India. HI 71432

Report (Annual), national museum of Botswana. Gaberones. 1967+, [1968]+. Rep. (Annual) Natl. Mus. Botswana. HI 71433

Report (Annual), national museum. Manila. 1966/67+, 1967+. Rep. (Annual) Natl. Mus., Manila. HI 71434

Report (Annual), national museum of Tanzania. Dar-es-Salaam. 1969/70+, [1970/]+. Rep. (Annual) Natl. Mus. Tanzania. HI 71435

Reports of the national museum of Victoria. Melbourne, Vic. No. 1, 1982. Rep. Natl. Mus. Victoria. Superseded by: Occasional papers from the museum of Victoria. HI 71436

Report (Annual), national museum of Wales. Cardiff. 1907+. Rep. (Annual) Natl. Mus. Wales. HI 71437

Report (Annual) of the national oceanographic council. Cambridge. 1950-65. Rep. (Annual) Natl. Oceanogr. Council. Preceded and superseded by: Report (Annual) of the national institute of oceanography. HI 71438

Report (Annual), national park service, United States department of the interior. Washington, DC. 1916+. Rep. (Annual) Natl. Park Serv. U.S. Dept. Interior. HI 71439

Report of the national parks commission. London. 1931-68. Rep. Natl. Parks Commiss. Superseded by: Report of the countryside commission. HI 71440

Report of the national parks committee = Report of the national parks commission. London.

Report (Annual) of the national research council of Canada. Ottawa. 1925/26+, 1926+. Rep. (Annual) Natl. Res. Council Canada. Preceded by: Report (Annual), advisory council for scientific and industrial research. 4-2908-1. HI 71441

Report (Annual), national science foundation. Washington, DC. 1950+. Rep. (Annual) Natl. Sci. Found. HI 71442

Report, national sugar institute, Kanpur. Calcutta. 1955+. Rep. Natl. Sugar Inst. Kanpur Preceded by: Report, indian institute of sugar technology. HI 71443

Report (Annual), national vegetable research station. Wellesbourne. 1950-60. Rep. (Annual) Natl. Veg. Res. Sta., Wellesbourne. HI 71444

Report, natural environment research council. London. 1965+. Rep. Nat. Environm. Res. Council. HI 71445

Report (Annual) of the natural history club of Philadelphia. Philadelphia, PA. Vols. 1-7, 1868-74. Rep. (Annual) Nat. Hist. Club Philadelphia. 4-2927-1. HI 71446

Report of the natural history museum of Stanford university. Palo Alto, CA. 1937-48. Rep. Nat. Hist. Mus. Stanford Univ. HI 71447

Report, natural history museum, university of Khartoum. Khartoum. 1956-66. Rep. Nat. HIst. Mus. Univ. Khartoum. HI 71448

Report of the natural history research center, Baghdad. Baghdad. 1973/74+. Rep. Nat. Hist. Res. Center Baghdad. Preceded by: Report of the Iraq natural history museum. HI 71449

Report of the natural history section, Prince of Wales museum of western India. Bombay. 1923-25. Rep. Nat. Hist. Sect. Prince of Wales Mus W. India. Superseded by: Report, Prince of Wales museum of western India. HI 71450

Report, natural history section, Wiltshire archaeological and natural history society. Devizes. 1947-68. Rep. Nat. lHist. Sect. Wiltshire Archaeol. Nat. Hist. Soc. Preceded by: Wiltshire bird (and plant) notes. Superseded by: Report, Wiltshire archaeological and natural history society. HI 71451

Report (Annual) of the natural history society of New Brunswick. St. John. 1863-80. Rep. (Annual) Nat. Hist. Soc. New Brunswick. 4-2927-2. HI 71452

Report of the natural history society of Maryland = Proceedings of the natural history society of Maryland. Baltimore, MD.

Report of the natural history society of Northumberland, Durham and Newcastle-upon-Tyne. Newcastle-upon-Tyne. 1830-51; 1921?-74. Rep. Nat. Hist. Soc. Northumberland. Superseded by: Report, natural history society of Northumbria. 4-2927-3. HI 71453

Report, natural history society of Northumbria. Newcastle-upon-Tyne. 1975+. Rep. Nat. Hist. Soc. Northumbria. Preceded by: Report of the natural history society of Northumberland, Durham and Newcastle-upon-Tyne. HI 71454

Report (Annual), natural history survey division, Illinois = Report (Annual), Illinois natural history survey. Springfield, IL.

Report, natural monuments investigation, plants. Tokyo. Nos. ?-10-18-?, ?-1930-37-? Rep. Nat. Monum. Invest. Pl. HI 71455

Report, natural science and archaeological society, Littlehampton = Report of proceedings, natural science and archaeological society, Littlehampton. Littlehampton.

Report (Annual), nature conservancy, United States.

[Forms part of: Nature conservancy news.] Arlington, VA. ?-1982+. Rep. (Annual) Nat. Conservancy, U.S.A. HI 71456

Report, nature conservancy. London. 1953-64. Rep. Nat. Conservancy, Great Britain. Superseded by: Report, nature conservancy council. HI 71457

Report, nature conservancy council, Great Britain. London. 1st+, 1973/75+, 1975+. Rep. Nat. Conservancy Council, Great Britain. Preceded by: Report, nature conservancy. HI 71458

Report, nature conservancy research in Scotland. Edinburgh. 1968/70+, 1970+. Rep. Nat. Conservancy Res. Scotland. HI 71459

Report (Annual) of the nature conservation branch, Transvaal provincial administration. Pretoria. 1965-66. Rep. (Annual) Nat. Conservation Branch Transvaal. Superseded by: Report, Transvaal nature conservation division. HI 71460

Report, nature conservation, provincial administration of the Orange Free State. Bloemfontein. 1971+. Rep. Nat. Conservation Orange Free State. HI 71461

Report of the nature reserves committee, department of the environment, Northern Ireland. Belfast. ?-13th+, ?-1978/79+. Rep. Nat. Res. Committee Northern Ireland. HI 71462

Report (Annual), Nebraska agricultural experiment station = Nebraska agricultural experiment station, annual report. Lincoln, NB. Nebraska Agric. Exp. Sta. Annual Rep. See B–P–H 639/24.

Report, Netherlands foundation for the advancement of tropical research. The Hague. ?-1964/65+, ?-1966+. Rep. Netherlands Found. Advancem. Trop. Res. HI 71463

Report (Annual), Nevada agricultural experiment station = Nevada agricultural experiment station. Annual report. Reno & Carson City, NV. Nevada Agric. Exp. Sta. Annual Rep. See B–P–H 651/15.

Report of the New Cross microscopical and natural history society. London. Nos. ?-9-12-?, ?-1881-84-? Rep. New Cross Microscop. Nat. Hist. Soc. HI 71464

Report (Annual), New England museum of natural history. Boston, MA. 1936/37-37/38, 1936-38. Rep. (Annual) New England Mus. Nat. Hist. HI 71465

Report (Annual), New Hampshire agricultural experiment station = New Hampshire agricultural experiment station. Annual report. Durham, NH. New Hampshire Agric. Exp. Sta. Annual Rep. See B–P–H 652/26.

Report (Annual), New Jersey agricultural college experiment station = New Jersey agricultural college experiment station. Annual report. New Brunswick, NJ. New Jersey Agric. Coll. Exp. Sta. Annual Rep. See B–P–H 653/13.

Report (Annual), New Jersey conservation foundation. Morristown, NJ. 196?+. Rep. (Annual) New Jersey Conservation Found. HI 71466

Report (Annual), New Mexico agricultural experiment station = New Mexico agricultural experiment station, annual report. State College of New Mexico. Las Cruces & Santa Fe, NM. New Mexico Agric. Exp. Sta. Annual Rep. See B–P–H 654/6.

Report, New South Wales department of agriculture = Report, department of agriculture, New South Wales. Sydney, N.S.W.

Report (Annual), New York botanical garden. Bronx, NY. 1966/67+, 1967+. Rep. (Annual) New York Bot. Gard. HI 71467

Report (Annual), New York (state) agricultural experiment station = New York (state) agricultural experiment station. Annual report. Geneva, NY. New York Agric. Exp. Sta. Annual Rep. See B–P–H 655/10.

Report (Annual) of the New York state agricultural society = Proceedings of the New York state agricultural society. Albany, NY. Proc. New York State Agric. Soc. See B–P–H 733/14.

Report, New York state botanist = New York state botanist report. Albany, NY. New York State Bot. Rep. See B–P–H 656/14.

Report of the New York State college of agriculture at Cornell university agricultural experiment station = Report, Cornell university college of agriculture and experiment station. Albany, Ithaca, NY.

Report, New York state college of forestry. Syracuse, NY. 1941-44. Rep. New York State Coll. Forest. HI 71469

Report (Annual) on the New York state museum of natural history = Report (Annual) of the regents of the university of the state of New York on the New York state museum. Albany, NY.

Report (Annual) of the New York state museum of natural history by the regents of the university of the state of New York. Albany, NY. Vols. 24-37, 1870-83, 1872-84; vol. 40, 1886, 1887. Rep. (Annual) New York State Mus. Nat. Hist. Preceded by: Report (Annual) of the regents of the university of the state of New York on the condition of the state cabinet of natural history. For vols. 38-39 see: Report (Annual) of the trustees of the state museum of natural history. Superseded by: Report (Annual) of the trustees of the state museum of natural history. 4-3045-1. HI 71470

Report, New Zealand ecological society = Report of annual meeting, New Zealand ecological society and Report of ecological conference, New Zealand ecological society. Wellington, N.Z.

Report on the New Zealand forest service = Report (Annual), New Zealand state forest service. Wellington, N.Z.

Report, New Zealand institute of horticulture. Wellington, N.Z. 1923-28. Rep. New Zealand Inst. Hort. Superseded by: Journal of the New Zealand institute of horticulture. HI 71471

Report (Annual), New Zealand state forest service. Wellington, N.Z. 1922+. Rep. (Annual) New Zealand State Forest Serv. Preceded by: State forestry report, New Zealand. HI 71472

Report of the New Zealand wheat research institute. Wellington, N.Z. 1930-40. Rep. New Zealand Wheat Res. Inst. Superseded by: New Zealand wheat review. HI 71473

Report, nigerian institute for oil palm research. Benin City. Vol. 1+, 1964+. Rep. Nigerian Inst. Oil Palm Res. Preceded by: Report, west african institute for oil palm research. HI 71474

Report, Norfolk naturalists trust. 1966+. Rep. Norfolk Naturalists Trust. HI 71475

Report (Annual), North Carolina agricultural experiment station = North Carolina agricultural experiment station annual report. Raleigh, NC. North Carolina Agric. Exp. Sta. Annual Rep. See B–P–H 663/5.

Report (Annual), North Carolina state horticultural society = North Carolina state horticultural society annual report. Raleigh, NC. North Carolina State Hort. Soc. Annual Rep. See B–P–H 664/5.

Report (Annual), North Dakota agricultural college, agricultural experiment station = North Dakota agricultural college, agricultural experiment station annual report. Fargo, ND. North Dakota Agric. Exp. Sta. Annual Rep. See B–P–H 664/8.

Report (Annual), North Dakota agricultural experiment station, Dickinson sub-station = North Dakota agricultural experiment station, Dickinson sub-station annual report. Dickinson, ND. North Dakota Agric. Exp. Sta. Dickson Sub-Sta. Annual Rep. See B–P–H 664/12.

Report, north Gloucestershire naturalists' society. Cheltenham. 1957-63/67. Rep. N. Gloucestershire Naturalists' Soc. Preceded by: Report, Cheltenham and district naturalists' society. HI 71476

Report, north London natural history society. London. 1904-13. Rep. N. London Nat. Hist. Soc. HI 71477

Report of the north of Scotland horticultural and arboricultural association. Aberdeen. 1882-1915. Rep. N. Scotland Hort. Arboric. Assoc. HI 71478

Report, north Staffordshire naturalists' field club and archaeological society. 1883-86. Rep. N. Staffordshire Naturalists' Field Club Archaeol. Soc. Superseded by: Report and transactions, north Staffordshire (naturalists') field club. HI 71479

Report (Annual), north Wales naturalists' trust ltd. Bangor. No. ?-5+, ?-1967/68+, ?-1968+. Rep. (Annual) N. Wales Naturalists' Trust. HI 71480

Report, north west wheat research institute. Narrabri, N.S.W. 1967+. Rep. N. W. Wheat Res. Inst. HI 71481

Report (Annual), Northamptonshire naturalists' trust ltd. Northampton. Nos. 1-6?, 1971-76. Rep. (Annual) Northamptonshire Naturalists' Trust. Superseded by: Chronicle, Northamptonshire naturalists' trust.. HI 71482

Report, northeastern forest experiment station. New Haven, CT., Upper Darby, Broomall, PA. 1927+. Rep. NorthE. Forest Exp. Sta. HI 71483

Report (Annual) of the northern academy of arts and sciences. Hanover, NH. Vols. 1-2, 1842-43. Rep. (Annual) N. Acad. Arts. 4-3091-3. HI 71484

Report of the northern forest rangers college, Dehra Dun = Progress report of the northern forest rangers college, Dehra Dun. Dehra Dun.

Report (Annual) of the northern nut growers' association. Ames, IA, St. Catherines, Ont. 33rd-42nd, 1942-51. Rep. (Annual) N. Nut Growers' Assoc. Preceded by: Report of proceedings, northern nut growers' association. HI 71485

Report (Annual), northern Rocky Mountain forest (and range) experiment station. Missoula, MT. 1929-53. Rep. (Annual) N. Rocky Mountain Forest (Range) Exp. Sta. Superseded by?: Progress report, Northern Rocky Mountain Forest (& Range) Experiment Station. HI 71486

Report, norwegian institute of seaweed research = Rapport, norsk institutt for tang- og tareforskning. Oslo.

Report and notes, Montgomeryshire field society. Welshpool. 1978+. Rep. Notes Montgomeryshire Field Soc. HI 71487

Report (Annual) of the Noto marine laboratory of the faculty of science, university of Kanazawa. Kanazawa. Vol. 1+, 1961+. Rep. (Annual) Noto Mar. Lab. HI 71488

Report (Annual), Nottinghamshire trust for nature conservation ltd. Nottingham. 1969+. Rep. (Annual) Nottinghamshire Trust Nat. Conservation. HI 71489

Report (Annual) of the Nova Scotia fruit growers' association = Report (Annual) of the fruit growers' association of Nova Scotia. Kentville, Halifax, N.S.

Report (Annual), Nova Scotia research foundation. Halifax, N.S. Vol. 1+, 1947+. Rep. (Annual) Nova Scotia Res. Found. 4-311-3. HI 71490

Report, oceanographic institute, Florida state university. Talahassee, FL. 1956+. Rep. Oceanogr. Inst. Florida State Univ. HI 71491

Report (Annual) of the oceanography department of the university of Liverpool. Liverpool. Nos. 34-46, 1920-33. Rep. (Annual) Oceanogr. Dept. Univ. Liverpool. Preceded by: Report (Annual) of the L. M. B. committee. Superseded by: Report (Annual) of the marine biological station at Port Erin. 3-2447-1. HI 71492

Report of the office of ministry of railroads. [Tetsudo daijin kanbo kendyu-jo. Gyomu kenkyu shiryo.] [Japan]. Rep. Off. Minist. Railroads. See B–P–H 768/29. HI 58619

Report, office of the secretary, United States department of agriculture = U S department of agriculture, office of the secretary. Report. Washington, DC. U.S.D.A. Off. Secr. Rep. See B–P–H 944/7.

Report (Annual) of the official seed testing station for England and Wales. Cambridge. 4th+, 1920/21+, 1921+. Rep. (Annual) Off. Seed Test. Sta. England Wales. 1st-3rd published in: Journal of the board of agriculture. Incorporated in: Journal of the national institute of agricultural botany. HI 71493

Report of the Ohara institute for agricultural biology = Berichte des Ohara Instituts für landwirtschaftliche Biologie.

Report, Ohio agricultural experiment station = Ohio agricultural experiment station annual report. Wooster, OH. Ohio Agric. Exp. Sta. Annual Rep. See B–P–H 684/20.

Report, Ohio pomological society = Ohio pomological society report. Wooster, OH. Ohio Pomol. Soc. Rep. See B–P–H 685/24.

Report (Annual) of the Ohio state forestry bureau. Columbus, OH. 1885-90. Rep. (Annual) Ohio State Forest. Bur. HI 71494

Report (Annual) of the Ohio state horticultural society. Wooster, OH. Vols. 1-54, 1867-1922. Rep. (Annual) Ohio State Hort. Soc. Superseded by: Proceedings of the annual meeting of the Ohio state horticultural society. 4-3163-1. HI 71495

Report (Annual), oil palm research station, Nigeria. Benin City. 1939-47. Rep. (Annual) Oil Palm Res. Sta. Superseded by: Report, west african institute for oil palm research. HI 71496

Report of the Okayama tobacco experiment station. [Okayama tobako shikenjo gyotei hokoku.] Okayama, Japan. Rep. Okayam Tobacco Exp. Sta. See B–P–H 769/5. HI 58620

Report (Annual), Oklahoma agricultural experiment station = Oklahoma agricultural experiment station annual report. Stillwater, OK. Oklahoma Agric. Exp. Sta. Annual Rep. See B–P–H 686/36.

Report of the Ontario agricultural college and experimental farm. Toronto. 1875+. Rep. Ontario Agric. Coll. Exp. Farm. HI 71497

Report, Ontario agricultural and experimental union. Toronto. 1879+. Rep. Ontario Agric. Exp. Union. HI 71498

Report, Ontario horticultural experiment station = Report (Annual) of the horticultural experiment station and products laboratory, vineland station, Ontario. Toronto.

Report on the operation of the department of agriculture, Madras = Report, department of agriculture, Madras. Madras.

Report on the operations of the department of agriculture, Bengal = Report, department of agriculture, Bengal. Calcutta.

Report of the operations of the department of forests, Papua and New Guinea. Port Moresby. 1957+. Rep. Operat. Dept. Forests Papua New Guinea. HI 71499

Report on the operations of the tobacco board, Mauritius = Report, tobacco board, Mauritius. Port Louis.

Report, Oregon agricultural experiment station. Corvallis, OR. 1952+. Rep. Oregon Agric. Exp. Sta. Preceded by: Oregon agricultural experiment station director's biennial report. HI 71500

Report, Oregon forest research center = Report (Annual), forest research laboratory, Oregon state university. Corvallis, OR.

Report (Annual), Oregon state horticultural society. Corvallis, Salem, Portland, OR. 1909+. Rep. (Annual) Oregon State Hort. Soc. HI 71501

Report (Annual) of the Orkney natural history society. Edinburgh. 1839. Rep. (Annual) Orkney Nat. Hist. Soc. HI 71502

Report (Annual), Osaka municipal museum of natural history. Osaka. 1964+, [1965+]. Rep. (Annual) Osaka Munic. Mus. Nat. Hist. HI 71503

Report, Ottawa research station. Ottawa. 1965+. Rep. Ottawa Res. Sta. Preceded by: Report, genetics and plant breeding research institute, Ottawa. HI 71504

Report (Annual), Oxford university exploration club. Oxford. Nos. 1-10, 1928-38. Rep. (Annual) Oxford Univ. Explor. Club. Superseded by: Bulletin, Oxford university exploration club. HI 71505

Report, Oxfordshire natural history society and field club. Oxford. 1900. Rep. Oxfordshire Nat. Hist. Soc. Field Club. Superseded by: Report, ashmolean natural history society of Oxfordshire. HI 71506

Report of the P F R A tree nursery. Indian Head, Sask. 1973+. Rep. P. F. R. A. Tree Nursery. Preceded by:

Summary report of the P F R A tree nursery. HI 71507

Report (Annual), Pacific northwest forest experiment station and range experiment station. Portland, OR. 1930+. Rep. (Annual) Pacific NorthW. Forest Exp. Sta. Range Exp Sta. HI 71508

Report (Annual), Pacific southwest forest and range experiment station. Berkeley, CA. 1958-? Rep. (Annual) Pacific SouthW. Forest Range Exp. Sta. Preceded by: Report, California forest and range experiment station. Superseded by: Research progress, Pacific southwest forest and range experiment station. HI 71509

Report, Pakistan forest college and research institute. Abbotabad. 1950+. Rep. Pakistan Forest Coll. Res. Inst. HI 71510

Reports (Annual) of Papua. Melbourne, Vic. 1907+. Rep. (Annual) Papua. Preceded by: Reports (Annual), British New Guinea. HI 71511

Report and papers, east anglian institute of agriculture. Chelmsford. 1912/13-38/39, 1912-39. Rep. Pap. E. Anglian Inst. Agric. HI 71512

Report of the Partabargh agricultural station of the United Provinces of Agra and Oudh. Allahabad. 1908/09-16, 1909-18. Rep. Partabargh Agric. Sta. Unit. Prov. Agra Oudh. Incorporated in: Report of the agricultural stations in eastern Bengal and Assam. HI 71513

Report, pathological division of the rubber research institute of Malaya. Kuala Lumpur. 1949-51. Rep. Pathol. Div. Rubber Res. Inst. Malaya. Previously and subsequently incorporated in: Report, rubber research institute of Malaya. HI 71514

Report (Annual), pea growing research organisation. Yaxley. 1956-72, 1957-73. Rep. (Annual) Pea Growing Res. Organ. Superseded by: Report (Annual), processors' and growers' research organization. HI 71515

Report (Annual), peanut collaborative research support program. Experiment, GA. 1982+. Rep. (Annual) Peanut Collab. Res. Support Program. HI 71516

Report of the Pennsylvania agricultural state college. Harrisburg, PA. Rep. Pennsylvania Agric. State Coll. See B–P–H 769/8. HI 58621

Report (Annual), Pennsylvania bureau of forestry = Report (Annual), bureau of forestry, Pennsylvania. Harrisburg, PA.

Report (Annual), Pennsylvania state college agricultural experiment station = Pennsylvania state college agricultural experiment station. Annual report. State College, PA. Pennsylvania State Coll. Agric. Exp. Sta. Annual Rep. See B–P–H 700/25.

Report, Penzance natural history and antiquarian society. Penzance. 1893-99. Rep. Penzance Nat. Hist. Antiq. Soc. HI 71517

Report, Perthshire natural history museum. Perth. 1902-13. Rep. Perthshire Nat. Hist. Mus. HI 71518

Report of the Peterborough natural history, scientific and archaeological society. Peterborough. 1882-1936. Rep. Peterborough Nat. Hist. Sci. Archaeol. Soc. HI 71519

Report, Philippine forestry bureau = Report (Annual) of the director of forestry of the Philippine Islands. Manila.

Report, Philippine sugar association research bureau. Manila. 1930-35. Rep. Philippine Sugar Assoc. Res. Bur. HI 71520

Report (Annual), physiologic race survey (cereal pathogens). Cambridge. 1967-76, 1968?-77. Rep. (Annual) Physiol. Race Surv. Cereal Pathog. Superseded by: Report (Annual), United Kingdom cereal pathogen virulence survey. HI 71521

Report (Annual), plant breeding institute. Cambridge. 1958/59+, 1959+. Rep. (Annual) Pl. Breed. Inst. HI 71522

Report of the ... plant breeding research forum. Des Moines, IA. 1982?+. Rep. Pl. Breed. Res. Forum. HI 71523

Report, plant breeding section, ministry of agriculture, Egypt. Cairo. 1925-31. Rep. Pl. Breed. Sect. Minist. Agric. Egypt HI 71524

Report of the plant breeding station of university college, Aberystwyth. Aberystwyth. 1950/56-?, 1956-? Rep. Pl. Breed. Sta. Univ. Coll. Aberystwyth. Superseded by: Report (Annual) of the welsh plant breeding station. HI 71525

Report (Annual) of the plant disease survey, Canada = Report (Annual) of the canadian plant disease survey. Ottawa.

Report, plant introduction gardens, Canal Zone. Panama. 1924-40. Rep. Pl. Introd. Gard. Canal Zone. HI 71526

Report (Annual) of the plant pathologist, Brunei. Jalan Tutong? ?-ca.1986+. Rep. (Annual) Pl. Pathologist Brunei. HI 71527

Report of the plant pathologist, department of agriculture, Bermuda. Hamilton, Bermuda. 1945-46. Rep. Pl. Pathologist Dept. Agric. Bermuda. Incorporated in: Report, department of agriculture, Burmuda. HI 71528

Report on the plant pathology laboratory. Harpenden. 1930-42. Rep. Pl. Pathol. Lab. HI 71529

Report (Annual), plant protection division, agricultural research service, United States department of agriculture. Hyattsville, MD. 1969+, 1970+. Rep. (Annual) Pl. Protect. Div. Agric. Res. Serv. U.S.D.A.

HI 71530

Report (Annual), plant protection research institute, Zimbabwe. Salisbury, Zimbabwe. ?-1977/78+, ?-1978+. Rep. (Annual) Pl. Protect. Res. Inst. Zimbabwe HI 71531

Report, plant research division, department of agriculture, Western Australia. Perth, W.A. 1970/71+, 1971+. Rep. Pl. Res. Div. Dept. Agric. Western Australia. HI 71532

Report (Annual), plant research laboratory, Victorian plant institute. Mildura, Vic. 1st-2nd, 1974-75. Rep. (Annual) Pl. Res. Lab. Victorian Pl. Inst. Superseded by: Biannual report, Mildura horticultural research station. HI 71533

Reports on polar research = Berichte zur Polarforschung. Bremerhaven.

Report, pollution research and the research councils. London. 1st+, 1971+. Rep. Pollut. Res. Res. Councils. HI 69660

Report of the pomological and fruit growing society of Quebec. Montreal. 1894-60. Rep. Pomol. Fruit Growing Soc. Quebec. HI 71534

Report (Annual) of the pomological station at Coonoor (and Burliar and Kallar fruit stations). Madras. 1920/21+, 1921+ [vols. for 1926/27-40/41 include Burliar & Kallar stations]. Rep. (Annual) Pomol. Sta. Coonoor. HI 71535

Report of the pomologist, United States department of agriculture = U S department of agriculture. Report of the pomologist. Washington, DC. U.S.D.A. Rep. Pomol. See B–P–H 945/3.

Report (Annual), Port Elizabeth museum. Port Elizabeth. 1967+, 1968+. Rep. (Annual) Port Elizabeth Mus. HI 71536

Report, Porto Rico sugar producers' association experiment station. 1911-12. Rep. Porto Rico Sugar Producers' Assoc. Exp. Sta. HI 71537

Report (Annual) on the prevalence of plant diseases in the dominion of Canada. Ottawa. Vols. 5-10, 1921-30. Rep. (Annual) Preval. Pl. Dis. Canada. Preceded by: Survey of the prevalence of common plant diseases in the Dominion of Canada. Superseded by: Report (Annual) of the canadian plant disease survey. HI 71538

Report, prickly pear experimental station, Queensland. Brisbane, Qld. 1912/13-15/16, 1913-16. Rep. Prickly Pear Exp. Sta. Queensland. HI 71539

Report of the Prince of Wales museum of western India. Bombay. 1927-66/69. Rep. Prince of Wales Mus W. India. Preceded by: Report of the natural history section, Prince of Wales museum of western India. HI 71540

Report and proceedings of the agri-horticultural society of Madras. Madras. 1920/21-47. Rep. Proc. Agri-Hort. Soc. Madras. Preceded by: Proceedings of the agri-horticultural society of Madras. HI 71541

Report of proceedings of the american association of nurserymen = Proceedings of the annual convention of the american association of nurserymen. Proc. Annual Conv. Amer. Assoc. Nurserymen. See B–P–H 725/13.

Report (Annual) and proceedings of the Barrow naturalists' field club and literary and scientific association. Barrow-in-Furness. Vols. 1-20, 1876/77-1910/12; n.s. vols. 1-?, 1928/29-? Rep. (Annual) Proc. Barrow Naturalists' Field Club. Superseded by: Proceedings of the Barrow naturalists' field club. 1-606-3. HI 71542

Report and proceedings of the Belfast natural history and philosophical society. Belfast. 1882-1920. Rep. & Proc. Belfast Nat. Hist. Soc. Preceded by: Report of the Belfast natural history and philosophical society and Proceedings of the Belfast natural history and philosophical society. Superseded by: Proceedings and reports of the Belfast natural history and philosophical society. 1-630-3. HI 71543

Report (Annual) and proceedings of the Belfast naturalists' field club. Belfast. 1873-1922. Rep. (Annual) & Proc. Belfast Naturalists' Field Club. Preceded by: Report (Annual) of the Belfast naturalists' field club. Superseded by: Proceedings and report of the Belfast naturalists' field club. 1-630-3. HI 71544

Report (Annual) and proceedings of the botanical society of Edinburgh. Edinburgh. Vols. 2-8, 1837-44. Rep. (Annual) & Proc. Bot. Soc. Edinburgh. Preceded by: Report (Annual), laws and transactions, botanical society of Edinburgh. 1-755-3. HI 71545

Report and proceedings of the Chester society of natural science, literature and art. Chester. Nos. 28-50, 1898-1947. Rep. Proc. Chester Soc. Nat. Sci. Lit. Art. Preceded by: Report of the Chester society of natural science, literature and art. Superseded by: Proceedings of the Chester society of natural science, literature and art. HI 71546

Report of proceedings, Conference on maple products = Conference on maple products. Report of proceedings. Philadelphia, PA. Conf. Maple Prod. Rep. Proc. See B–P–H 324/14.

Report and proceedings, Liverpool science students' association. Liverpool. Nos. ?-9-26, ?-1889-1907. Rep. Proc. Liverpool Sci. Stud. Assoc. HI 71547

Report and proceedings of the Manchester field naturalists' and archaeologists society. Manchester. 1860-1914, 1932. Rep. Proc. Manchester Field Naturalists' Archaeologists Soc. HI 71548

Report and proceedings of the Manchester scientific students' association. Manchester. 1878-87. Rep. & Proc. Manchester Sci. Stud. Assoc. Preceded by: Report of the Manchester scientific students' association. 3-2532-1. HI 71549

Report and proceedings of the meeting, agri-horticultural society of Madras = Report and proceedings of the agri-horticultural society of Madras. Madras.

Report of proceedings, natural science and archaeological society, Littlehampton. Littlehampton. 1928-45. Rep. Proc. Nat. Sci. Archaeol. Soc. Littlehampton. Preceded by: Report of proceedings, nature and archaeological circle, Littlehampton. HI 71550

Report of proceedings, nature and archaeological circle, Littlehampton. Littlehampton. 1924-27. Rep. Proc. Nat. Archaeol. Circle Littlehampton. Superseded by: Report of proceedings, natural science and archaeological society, Littlehampton. HI 71551

Report of proceedings, northern nut growers' association. Ithaca, NY. 1st-32nd, 1911-41. Rep. Proc. N. Nut Growers Assoc. Superseded by: Report (Annual), northern nut growers' association. 4-3093-1. HI 71552

Report (Annual) and proceedings, Nova Scotia fruit growers' association = Report (Annual) of the fruit growers' association of Nova Scotia. Kentville, Halifax, N.S.

Report (Annual), proceedings, Oregon state horticultural society = Report (Annual), Oregon state horticultural society. Corvallis, Salem, Portland, OR.

Report (Annual) and proceedings, Papua and New Guinea scientific society = Proceedings, Papua and New Guinea scientific society. Port Moresby, Konedobu.

Reports of the proceedings of the pathological society of London = Transactions of the pathological society of London. London. Trans. Pathol. Soc. London. See B–P–H 887/21.

Report (Annual), Puerto Rico agricultural experiment station = Puerto Rico agricultural experiment station. Annual report. Mayagüez, PR. Puerto Rico Agric. Exp. Sta. Annual Rep. See B–P–H 749/10.

Report of proceedings, Queensland cane growers' association. Brisbane, Qld. Vol. 1+, 1927?+. Rep. Proc. Queensland Cane Growers Assoc. HI 71553

Report and proceedings, Reading literary and scientific society. Reading. 1880-1926. Rep. Proc. Reading Lit. Sci. Soc. HI 71554

Report and proceedings, south western naturalists' union. Bristol. 1924-27. Rep. Proc. S. W. Naturalists' Union. Superseded by: Proceedings of the south western naturalists' union. HI 71555

Report of the proceedings of the Teign naturalists' field club. Exeter. 1858-1917. Rep. Proc. Teign Naturalists' Field Club. Superseded by: Proceedings of the Teign naturalists' field club. 5-4171-1. HI 71556

Report of the proceedings of the western canadian society for horticulture. Regina, Sask. Rep. Proc. W. Canad. Soc. Hort. See B–P–H 769/21. HI 58622

Report and proceedings, western nut growers' association. Gresham, OR. 1919-25. Rep. Proc. W. Nut Growers' Assoc. Superseded by: Report, Oregon state horticultural society. HI 71557

Report (Annual), processors' and growers' research organization. Peterborough. 1973+, 1975+. Rep. (Annual) Processors Growers' Res. Organ. Preceded by: Report (Annual), pea growing research organisation. HI 71558

Report on the progress of agriculture in India. Calcutta. 1904/05-18/19, 1907-19. Rep. Progr. Agric. India. Preceded by: Report of the imperial department of agriculture, India. Superseded by: Review of agricultural operations in India. HI 71559

Report on the progress and condition of the botanic garden and government plantations. Adelaide, S.A. 1871?-89?, 1919-30? Rep. Progr. Condition Bot. Gard. Gov. Plantations. Superseded by: Report (Annual) of the board of governors of the botanic gardens (South Australia), Adelaide. HI 71560

Report on the progress and condition of the royal botanic gardens at Kew. London. 1862-82. Rep. Progr. Condition Roy. Bot. Gard. Kew. Preceded by: Report of the royal botanic garden at Kew. 3-2284-1. HI 58623

Report (Annual) of progress in forage crops research. Stillwater, OK. 1951-56. Rep. (Annual) Progr. Forage Crops Res. HI 71561

Report of the progress of rust research, dominion rust research laboratory. Winnipeg. 1925-30. Rep. Progr. Rust Res. Domin. Rust Res. Lab. Continued in: Progress report, division of botany, Canada. HI 71562

Report of progress, tobacco research board of Rhodesia = Report of progress, tobacco research board of Southern Rhodesia. Salisbury, Rhodesia.

Report of progress, tobacco research board of Rhodesia and southern Nyasaland = Report of progress, tobacco research board of Southern Rhodesia. Salisbury, Rhodesia.

Report of progress, tobacco research board of Southern Rhodesia. Salisbury, Rhodesia. 1951/53+, 1953+. Rep. Progr. Tobacco Res. Board Southern Rhodesia. Preceded by: Report of progress, tobacco research board of Southern Rhodesia. HI 71563

Report of progress, university of Alaska agricultural

experiment station = University of Alaska agricultural experiment station. Report of progress. College, AK. Univ. Alaska Agric. Exp. Sta. Rep. Progr. See B–P–H 933/10.

Report of the Provancher society of natural history of Canada. Quebec. 1918-50/51, 1918-51. Rep. Provancher Soc. Nat. Hist. Canada. HI 71564

Report of the provincial museum of natural history and anthropology. Victoria, B.C. 1912-69. Rep. Prov. Mus. Nat. Hist. Incorporated in: Report, department of recreation and conservation, British Columbia. 1-779-1. HI 71565

Report, public gardens and plantations. Kingston, Jamaica. 1880-1900. Rep. Public Gard. Plantations. Superseded by: Report of the board of agriculture and department of public gardens and plantations, Jamaica. HI 71566

Report of the public museum of the city of Milwaukee. Milwaukee, WI. Vols. 1-29 & 37 [vols. 30-36 never published], 1883-1911, 1919. Rep. Public Mus. City Milwaukee. Superseded by: Year book of the public museum of the city of Milwaukee. 3-2663-2. HI 71567

Report, Puerto Rico agricultural experiment station = Puerto Rico agricultural experiment station. Report. Mayagüez, PR. Puerto Rico Agric. Exp. Sta. Rep. See B–P–H 749/13.

Report of the Puerto Rico experiment station. U.S. department of agriculture. Office of experiment stations. Mayagüez, PR. Rep. Puerto Rico Exp. Sta. U.S.D.A. Off. Exp. Sta. See B–P–H 769/28. HI 58624

Report of the pyrethrum board of Kenya. Nakuru. 1946+, 1947+. Rep. Pyrethrum Board Kenya. HI 71568

Report (Annual) of the Quebec society for the protection of plants. Montreal. Vols. 1-43, 1908/09-62. Rep. (Annual) Quebec Soc. Protect. Pl. Superseded by: Phytoprotection. 4-3505-2. HI 71569

Report, Queensland acclimatisation society. Brisbane, Qld. 1866-1951. Rep. Queensland Acclim. Soc. HI 71570

Report, Queensland cane growers' association = Report of proceedings, Queensland cane growers' association. Brisbane, Qld.

Report of the Queensland geological survey. Brisbane, Qld. Rep. Queensland Geol. Surv. See B–P–H 770/1. HI 58625

Report of the Quekett microscopical club. London. 1st-16th, 1866-81. Rep. Quekett Microscop. Club. HI 71571

Report, radiobiological laboratory, agricultural research council. Wantage. Nos. 1-20, 1959-70. Rep. Radiobiol. Lab. Agric. Res. Council. HI 71572

Report (Annual) of the Rancho Santa Ana botanic garden. Claremont, CA. ?-1960-74+. Rep. (Annual) Rancho Santa Ana Bot. Gard. HI 71573

Report, rangelands research union, commonwealth scientific and industrial research organization. Canberra, A.C.T. 1968/69+, 1970+. Rep. Rangelands Res. Union Commonw. Sci. Industr. Res. Organ. HI 71574

Report of the Ray society. London. 1855+. Rep. Ray Soc. HI 71575

Report of the Reading natural history society. Reading. 1903-04. Rep. Reading Nat. Hist. Soc. HI 71576

Reports received from experiment stations, empire cotton growing corporation = Reports from experiment stations, empire cotton growing corporation. London.

Report on recent collections, studies in the vegetation of the state, Nebraska university botanical survey = Botanical survey of Nebraska. Lincoln, NE.

Report and records of the Colchester and district natural history society and field club. Colchester. 1964/65+. Rep. Rec. Colchester Distr. Nat. Hist. Soc. Field Club. HI 71577

Report and reference book, mid-Somerset naturalist society. Bridgewater. 1949/51-65. Rep. Refer. Book Mid-Somerset Naturalist Soc. HI 71578

Report (Annual) of the regents of the university of the state of New York on the condition of the state cabinet of natural history, [etc.]. Albany, NY. Vols. 1-23, 1847-69, 1848-73. Rep. (Annual) Regents Univ. State New York State Cab. Nat. Hist. Superseded by: Report (Annual) on the New York state museum of natural history by the regents of the university of the state of New York. HI 71579

Report (Annual) of the regents of the university of the state of New York on the New York state museum. Albany, NY. Vols. 43-72(3) [vol. 72(1) not published], 1889-1918, 1890-1918. Rep. (Annual) Regents Univ. State New York New York State Mus. Preceded by: Report (Annual) of the trustees of the state museum of natural history. 4-3045-1. HI 71580

Report (Annual) of the regional research laboratory, Jammu. Jammu. 1977?+. Rep. (Annual) Regional Res. Lab. Jammu. HI 71581

Report of research activites, division of technical services, department of forestry, Queensland. Brisbane, Qld. No. 1+, 1977+. Rep. Res. Activites Div. Techn. Serv. Dept. Forest. Queensland. HI 71582

Report, research and advisory department, south-eastern agricultural college, Wye. Wye. 1920/21-29/30, 1921-30. Rep. Res. Advis. Dept. S.-E. Agric. Coll. Wye.

HI 71583

Report of the research appraisal and research planning committee, Canada weed committee, eastern section. 14th+, 1969+, 1970+. Rep. Res. Appraisal Res. Planning Committee Canada Weed Committee E. Sect. Preceded by: Reports of the research appraisal and research planning committee, national weed committee, Canada, eastern section. HI 71584

Reports of the research appraisal and research planning committee, Canada weed committee, western section. Regina, Sask. 1971+. Rep. Res. Appraisal Res. Planning Committee Canada Weed Committee W. Sect. Preceded by: Report of the research appraisal committee for western Canada, national weed committee, western section. HI 71585

Reports of the research appraisal and research planning committee, national weed committee, Canada, eastern section. 8th-13th, 1964-69. Rep. Res. Appraisal Res. Planning Committee Natl. Weed Committee Canada E. Sect. Preceded by: Weed control recommendations for eastern Canada. Superseded by: Reports of the research appraisal and research planning committee, Canada weed committee, eastern section. HI 71586

Report of the research appraisal committee for western Canada, national weed committee, western section. 1964-68. Rep. Res. Appraisal Committee W. Canada Natl. Weed Committee W. Sect. Preceded by: Weed control recommendations for western Canada. Superseded by: Reports of the research appraisal committee and research planning committee, Canada weed committee, western section. HI 71587

Report (Annual), research branch of the department of agriculture, Sarawak. Kuching. 1977+. Rep. (Annual) Res. Branch Dept. Agric. Sarawak. Preceded by: Report (Annual) of the department of agriculture, Sarawak. HI 71588

Report, research bureau, Philippine sugar association = Report, Philippine sugar association research bureau. Manila.

Report on research carried out under the commonwealth (later university) research grant, university of Western Australia = Report on research, university of Western Australia. Perth, W.A.

Report of research committee, north central weed control conference = Research report, north central weed control conference. Champaign, IL.

Report on research, department of agriculture, Canada. Ottawa. 1956-57. Rep. Res. Dept. Agric. Canada. Preceded by: Report of the science service, department of agriculture, Canada. HI 71589

Report, research department, coconut industry board, Jamaica = Report, coconut industry board research department, Jamaica.

Report (Annual), research department, indian coffee board. Bangalore. No. 1+, 1947+. Rep. (Annual) Res. Dept. Indian Coffee Board. HI 71590

Report (Annual), research department, sugar manufacturers' association of Jamaica. Mandeville. 194?-72, 1948-73. Rep. (Annual) Res. Dept. Sugar Manufacturers' Assoc. Jamaica. Superseded by: Report (Annual) of the sugar industry research institute and sugar research department. HI 71591

Report (Annual), research and development department, banana board, Jamaica = Report (Annual), banana board research department, Jamaica. Kingston, Jamaica.

Report, research and development, ministry of agriculture, fisheries and food. London. 1976+, 1978+. Rep. Res. Developm. Minist. Agric. Fish. Food. HI 71592

Report of the research division, ministry of agriculture, Sudan. Khartoum. 1951+. Rep. Res. Div. Minist. Agric. Sudan. Preceded by: Report of the agricultural research service, Anglo-Egyptian Sudan. HI 71593

Report of research and experimental work of the department of agriculture, Hyderabad-Deccan. 1932-38. Rep. Res. Exp. Work Dept Agric. Hyderabad-Deccan. HI 71594

Report (Annual) of research, Govind Ballabh Pant university of agriculture and technology. Pantnagar. 1968/69+, 1969?+. Rep. (Annual) Res. Govind Ballabh Pant Univ. Agric. Technol. HI 71595

Report, research institute on agriculture, Tokyo. Tokyo. 1946/48+, 1948+. Rep. Res. Inst. Agric. Tokyo. HI 71596

Report of the research institute of brewing. [Jozo shikenjo hokoku.] Tokyo. Rep. Res. Inst. Brew. See B–P–H 770/3. HI 58626

Report of the research institute of marine algae of the faculty of physical sciences, Hokkaido imperial university. [Rigaku-bu Kaiso kenkyujo hokoku.] Sapporo, Japan. Rep. Res. Inst. Mar. Algae Fac. Phys. Sci. Hokkaido Imp. Univ. See B–P–H 770/5. HI 58627

Report, research institute for plant protection. Wageningen. 1977+. Rep. Res. Inst. Pl. Protect. Preceded by: Report (Annual), instituut voor plantenziektenkundig onderzoek. HI 71597

Report on research and investigations, office of rural development = Research report of the office of rural development. [Nongso sihom yon'gu pogo.] Suwon.

Report (Annual) of research, Moss Landing marine laboratories. [Moss Landing, CA.] No. 1, 1970. Rep. (Annual) Res. Moss Landing Mar. Lab. HI 71598

Report of research for planning a green belt in Tokyo.

[Tokyo ryokuchi keikaku chosa iho.] Tokyo. Rep. Res. Planning Green Belt Tokyo. See B–P–H 770/7. HI 58628

Report (Annual), research projects, institute for horticultural plant breeding = Jaarverslag, projecten van onderzoek, instituut voor de veredeling van tuinbouwgewassen. Wageningen.

Report (Annual), research school of biological sciences, Australian national university. Canberra, A.C.T. 1969+, 1970+. Rep. (Annual) Res. School Biol. Sci. Austral. Natl. Univ. HI 71599

Report (Annual) on research and technical work of the department (ministry) of agriculture for Northern Ireland. Belfast. 1963+. Rep. (Annual) Res. Techn. Work Dept. (Minist.) Agric. Northern Ireland. Preceded by: Research and experimental record, ministry of agriculture, Northern Ireland. HI 71600

Report (Annual) on the research of the unit of comparative plant ecology, department of botany, university of Sheffield = Report, unit of comparative plant ecology (N E R C). Sheffield.

Report on research, university of Western Australia. Perth, W.A. 1950-71. Rep. Res. Univ. Western Australia. Superseded by: Research report, university of Western Australia. HI 71601

Report on research work, British West Indies sugar association. Port of Spain. 1943-51? Rep. Res. Work British West Indies Sugar Assoc. HI 71602

Report on research work on sugarcane agriculture, British West Indies sugar association. = Report on research work, British West Indies sugar association. Port of Spain.

Report of the results obtained on the experimental fields at Dodds reformatory, Barbados. Bridgetown. 1891-99. Rep. Results Obtained Exp. Fields Dodds Reformatory Barbados. Superseded by: Report of the agricultural work, imperial department of agriculture for the West Indies. HI 71603

Report of the Rhode Island agricultural experiment station. Kingston, RI. 1888+. Rep. Rhode Island Agric. Exp. Sta. HI 71604

Report, rice research institute, Cuttack = Report (Annual), central rice research institute, Cuttack. New Delhi.

Report, rice research institute, Rokupr. Rokupr. ?-1971/72+. ?-1973+. Rep. Rice Res. Inst. Rokupr. Preceded by: Report (Annual), west african rice research station, Rokupr and Experimental report, Rokpur rice research station. HI 71605

Report of the rice research officer, Burma. Rangoon. 1932/33-33/34, 1933-34. Rep. Rice Res. Officer Burma. Incorporated in: Report, department of agriculture, Burma. HI 71606

Report, rice research station, Rokupr = Report (Annual), west african rice research station, Rokupr. Freetown.

Report (Annual), Rocky Mountains forest and range experiment station. Fort Collins, CO. 1938-68. Rep. (Annual) Rocky Mountains Forest Range Exp. Sta. Superseded by: Forestry research highlights, Rocky Mountains forest and range experiment station. HI 71607

Report, Rosemaund experimental husbandry farm. Preston Wynne. 1950/60-73, 1960?-73. Rep. Rosemaund Exp. Husb. Farm. Superseded by: Review, Rosemaund experimental husbandry farm. HI 71608

Report (Annual), Rosewarne experimental horticulture station. Camborne. 1st-24th, 1952/55-79, 1952-79. Rep. (Annual) Rosewarne Exp. Hort. Sta. Superseded by: Annual review, Rosewarne experimental horticulture station. HI 71609

Report, Rothamsted experimental station. St. Albans, Harpenden. 1906+; [From 1968+ published in two parts.] Rep. Rothamsted Exp. Sta. HI 71610

Report (Annual), royal agri-horticultural society of India = Report (Annual), agri-horticultural society of India. Calcutta.

Report (Annual) of the royal botanic garden, Calcutta. Calcutta. 1882-1949. Rep. (Annual) Roy. Bot. Gard. Calcutta. Superseded by: Report (Annual) of the indian botanic garden and the gardens in Calcutta, (parks and gardens in Cooch Behar) and the Lloyd botanic garden. HI 71611

Report (Annual) of the royal botanic garden, Calcutta, and of the Lloyd botanic garden, Darjeeling = Report (Annual) of the royal botanic garden, Calcutta.

Report, royal botanic garden, Edinburgh. Edinburgh. 1870+. Rep. Roy. Bot. Gard. Edinburgh. HI 71612

Report (Annual) of the royal botanic garden and other gardens in Calcutta, and of the Lloyd botanic garden, Darjeeling = Report (Annual) of the royal botanic garden, Calcutta.

Report of the royal botanic garden at Kew. London. 1844-61. Rep. Roy. Bot. Gard. Kew. Superseded by: Report on the progress and condition of the royal botanic gardens at Kew. 3-2284-1. HI 58629

Report (Annual), royal botanic gardens, Kew. Richmond, Sy. 1977-83? Rep. (Annual) Roy. Bot. Gard. Kew. Superseded by: Report by the board of trustees of the royal botanic gardens, Kew. HI 71613

Report (Annual), royal botanic gardens and national herbarium. Sydney, N.S.W. 1968+, 1970+. Rep. (Annual) Roy. Bot. Gard. Natl. Herb. HI 71614

Report of the royal botanic gardens, Peradeniya.

Colombo. 1883-1910/11. Rep. Roy. Bot. Gard. Peradeniya. Incorporated in: Report of the department of agriculture, Ceylon. HI 71615

Report (Annual) on the royal botanic gardens, Trinidad. Port of Spain. 1887-1907, 1888?-1908? Rep. (Annual) Roy. Bot. Gard. Trinidad. HI 74545

Report (Annual), royal botanical gardens, Ontario. Hamilton, Ont. ?-ca.1986+. Rep. (Annual) Roy Bot. Gard. Ontario. HI 71616

Report, royal Cornwall polytechnic society. Falmouth. 1833+. Rep. (Annual) Roy. Cornwall Polytech. Soc. 4-3729-1. HI 71617

Report, royal horticultural society. London. 1895-1933? Rep. Roy. Hort. Soc. Incorporated in: Journal, royal horticultural society. HI 71618

Report, royal Jersey agricultural and horticultural society. Jersey, C.I. 1912+. Rep. Roy. Jersey Agric. Hort. Soc. HI 71619

Report, royal microscopical society. London. 1976+. Rep. Roy. Microscop. Soc. HI 71620

Report (Annual), royal Ontario museum. Toronto. 1956+. Rep. (Annual) Roy. Ontario Mus. Preceded by: Report, royal Ontario museum of zoology and palaeontology [not entered]. HI 71621

Report (Annual), royal society for nature conservation. Lincoln. 1981/82+, 1982+. Rep. (Annual) Roy Soc. Nat. Conservation. Preceded by: Report (Annual), society for the promotion of nature conservation. HI 71622

Report (Annual) of the royal society of South Australia. Adelaide, S.A. Vols. 1-13, 1854-65/66? Rep. (Annual) Roy. Soc. South Australia. HI 71623

Report of the royal society of Tasmania. Hobart, Tas. Rep. Roy. Soc. Tasmania. See B–P–H 770/10. HI 58630

Report of the royal society of Van Diemen's Land. Hobart, Tas. Rep. Roy. Soc. Van Diemen's Land. See B–P–H 770/11. HI 58631

Report (Annual) of the royal zoological and acclimatisation society of Victoria = Report (Annual) of the acclimatisation society of Victoria. Melbourne, Vic.

Report, rubber growers' association. London. 1909+. Rep. Rubber Growers' Assoc. HI 71624

Report of the rubber planting experiment station, Mooply. Madras. 1921-24. Rep. Rubber Pl. Exp. Sta. Mooply. HI 71625

Report of the rubber planting experiment station, Tenmalai. Madras. 1921-24. Rep. Rubber Pl. Exp. Sta. Tenmalai. HI 71626

Report, rubber research board (scheme), Ceylon. Colombo. 1923-55. Rep. Rubber Res. Board (Scheme) Ceylon. Superseded by: Report, rubber research institute of Ceylon. HI 71627

Report, rubber research institute of Ceylon. Colombo, Agalawatta. 1956-59. Rep. Rubber Res. Inst. Ceylon. Preceded by: Report, rubber research board (scheme), Ceylon. Superseded by: Annual review, rubber research institute of Ceylon. HI 71628

Report, rubber research institute of Malaya (later Malaysia). Kuala Lumpur. 1926/28+. 1928+. Rep. Rubber Res. Inst. Malaya (later Malaysia). HI 71629

Report of the Rugby school natural history society. Rugby, England. Rep. Rugby School Nat. Hist. Soc. See B–P–H 770/13. HI 58632

Report (Annual), Rutgers University agricultural experiment station = New Jersey agricultural college experiment station. Annual report. New Brunswick, NJ. New Jersey Agric. Coll. Exp. Sta. Annual Rep. See B–P–H 653/13.

Report of the Rutland archaeological and natural history society. Oakham. 1902+. Rep. Rutland Archaeol. Nat. Hist. Soc. HI 71630

Report (Annual) of the Sado marine biological station. [Sado rinkai jikkenjo. Kenkyu nempo.] Niigata. No. 1+, 1971+. Rep. (Annual) Sado Mar. Biol. Sta. HI 74546

Report, Saghalien central experiment station, department of forestry. Konuma. 1930/31-31/32, 1931-33. Rep. Saghalien Centr. Exp. Sta. Dept. Forest. HI 71631

Report of the Saikyo university forest. [Saikyo daigaku nogakubu enshurin shuho.] Kyoto, Japan. Rep. Saikyo Univ. Forest. See B–P–H 770/16. HI 58633

Report, Saito Ho-on Kai = Report (Annual) of the work of Saito Ho-on Kai. Sendai.

Report (Annual) of the Santa Barbara botanic garden. Santa Barbara, CA. Vol. 1-?, 1950-? Rep. (Annual) Santa Barbara Bot. Gard. Superseded by: Biennial report, Santa Barbara botanic garden. HI 71632

Report, Sarawak department of agriculture = Report (Annual) of the department of agriculture, Sarawak. Kuching.

Report, Sarawak museum. Kuching. 1900-12, 1936-39. Rep. Sarawak Mus. HI 71633

Report of Sarobetsu agricultural experiment farm. [Sarobetsu jikken nojo chosa hokokusho.] Tokyo. Rep. Sarobetsu Agric. Exp. Farm. See B–P–H 770/18. HI 58634

Report of the Saskatchewan department of agriculture. Regina, Sask. Vol. 1+, 1905+. Rep. Saskatchewan Dept. Agric. HI 71634

Report, saudi arabian natural history society. Jeddah.

Vol. 1(1), 1971. Rep. Saudi Arabian Nat. Hist. Soc. Superseded by: Journal, saudi arabian natural history society. HI 71635

Report (Annual), Savanna forestry research station. Samaru. ?-1973+. Rep. (Annual) Savanna Forest. Res. Sta. HI 71636

Report, science – servant of agriculture. 1938-40. Rep. Sci. Servant Agric. Superseded by: Report, California forest and range experiment station. HI 71637

Report (Annual), science arm, agricultural development and advisory service. London. 1972, 1974. Rep. (Annual) Sci. Arm Agric. Developm. Advis. Serv. Superseded by: Report (Annual), science service, agricultural development and advisory service. HI 71638

Report (Annual), science museum of Minnesota. St, Paul, MN. 1972/73+, 1973+. Rep. (Annual) Sci. Mus. Minnesota. HI 71639

Report (Annual), science service, agricultural development and advisory service. London. 1975+, 1976+. Rep. (Annual) Sci. Serv. Agric. Developm. Advis. Serv. Preceded by: Report (Annual), science arm, agricultural development and advisory service. HI 71640

Report of the science service, department of agriculture, Canada. Ottawa. 1943-57. Rep. Sci. Serv. Dept. Agric. Canada. Superseded by: Report on research, department of agriculture, Canada. HI 71641

Report on scientific activities, institute of ecology, polish academy of sciences. Warsaw. 1963+. Rep. Sci. Activities Inst. Ecol. Polish Acad. Sci. HI 71642

Report, scientific department, indian tea association. Calcutta. 1930+. Rep. Sci. Dept. Indian Tea Assoc. HI 71643

Report of the scientific and education departments, american museum of natural history. New York. 1977/78+, [1978]+. Rep. Sci. Educ. Dept. Amer. Mus. Nat. Hist. Preceded by: Research report and bibliography, american museum of natural history. HI 71644

Report of the scientific investigation group on Hainan island from the Taihoku imperial university. [Taihoku teikoku daigaku. Kainanto gakujitsu chosa hokoku.] Taihoku [=Taipei, Taiwan]. Rep. Sci. Invest. Group Hainan Island Taihoku Imp. Univ. See B–P–H 770/19. HI 58635

Report (Annual) of scientific works from the faculty of science, Osaka university. Osaka. Vols. 1-2, 1952-63. Rep. (Annual) Sci. Works Fac. Sci. Osaka Univ. HI 71645

Report (Annual), scottish crop research institute. Mylnefield. 1st+, 1981+. Rep. (Annual) Scott. Crop Res. Inst. Preceded by: Report (Annual), scottish horticultural research institute. HI 71646

Report, scottish field studies association. Glasgow. 1956-77. Rep. Scott. Field Stud. Assoc. Superseded by: Scottish field studies. HI 71647

Report (Annual), scottish horticultural research institute. Dundee. Nos. 1-?, 1954-80? Rep. (Annual) Scott. Hort. Res. Inst. Superseded by: Report (Annual), scottish crop research institute. HI 71648

Report (Annual), scottish marine biological association. Glasgow. 1914+. Rep. (Annual) Scott. Mar. Biol. Assoc. Preceded by: Report marine biological association of the west of Scotland. HI 71649

Report (Annual) of the scottish plant breeding station. Edinburgh. 1951+. Rep. (Annual) Scott. Pl. Breed. Sta. Preceded by: Report (Annual), scottish society for research in plant breeding. From 1962-65 entitled: Record of the scottish plant breeding station. 5-3832-1. HI 71650

Report, scottish seaweed research association. Edinburgh. 1945-50. Rep. Scott. Seaweed Res. Assoc. Superseded by: Report, institute of seaweed research. HI 71651

Report (Annual), scottish society for research in plant breeding. Edinburgh. 1922-50. Rep. (Annual) Scott. Soc. Res. Pl. Breed. Superseded by: Report (Annual) of the scottish plant breeding station. HI 71652

Report, Scripps institution for biological research. Berkeley, CA. 1923-26. Rep. Scripps Inst. Biol. Res. Superseded by: Report, Scripps institution of oceanography. HI 71653

Report, Scripps institution of oceanography. Berkeley, CA. 1924+. Rep. Scripps Inst. Oceanogr. Preceded by: Scripps Institution for Biological Research. HI 71654

Report, Seale Hayne agricultural college, department of plant pathology = Report, department of plant pathology, Seale Hayne agricultural college. Newton Abbot.

Report of the secretary of agriculture. Washington, DC. Rep. Secr. Agric. See B–P–H 770/23. HI 58636

Report of the secretary, ministry of agriculture, Rhodesia. Zomba. 1963+, 1964+. Rep. Secr. Minist. Agric. Rhodesia. Preceded by: Report, ministry of agriculture, Rhodesia & Nyasaland. HI 71655

Report (Annual), seed division, U. S. department of agriculture = Department of agriculture. Seed division. Annual report. Washington, DC. Dept. Agric. Seed Div. Annual Rep. See B–P–H 343/15.

Report (Annual) of the seed testing laboratory. Tokyo. Vol. 1+, 1951+. Rep. (Annual) Seed Test. Lab. HI 71656

Report of the seed testing and plant registration station for

Scotland. Edinburgh. 1925/27+, 1929+. Rep. Seed Test. Pl. Registr. Sta. Scotland. HI 71657

Report, Sheffield botanical and horticultural society. Sheffield. Nos. ?-50-53, ?-1894-97. Rep. Sheffield Bot. Hort. Soc. HI 71658

Report. Sheffield literary and philosophical society. Sheffield. Rep. Sheffield Lit. Soc. See B–P–H 770/24. HI 58637

Report, Sheffield naturalists' club. Sheffield. 1914-15. Rep. Sheffield Naturalists' Club. HI 71659

Report of the Shizuoka citrus experiment station. [Shizuoka-ken kankitsu shikenjo hokoku.] Shizuoka, Japan. Rep. Shizuoka Citrus Exp. Sta. See B–P–H 771/1. HI 58638

Report (Annual) of the Shizuoka prefecture forest experiment station. [Shizuoka-ken ringyo shikenjo gyomu seiseki hokokusho.] Hamakita. 1957+. Rep. (Annual) Shizuoka Pref. Forest Exp. Sta. HI 75148

Report of Shizuoka prefecture tea research station. [Shizuoka-ken chagyo shikenjo kenkyu hokoku.] Shizuoka, Japan. Rep. Shizuoka Prefect. Tea Res. Sta. See B–P–H 771/2. HI 58639

Report (Annual) of the Shropshire and north Wales natural history and antiquarian society. Shrewsbury. Nos. 1-40, 1835-76. Rep. (Annual) Shropshire Nat. Hist. Soc. HI 71660

Report of the Sibpur experimental farm. Calcutta. 1894+. Rep. Sibpur Exp. Farm. HI 71661

Report, Sidcup natural history society. Sidcup. 1950-63. Rep. Sidcup Nat. HIst. Soc. HI 71662

Report, sisal experimental station, Tanganyika = Report, Tanganyika sisal growers' association. Tanga.

Report, sisal research station, Mlingano, Tanganyika. Mlingano. 1936-58. Rep. Sisal Res. Sta. Mlingano Tanganyika. Incorporated in: Report, Tanganyika sisal growers' association. HI 71663

Report of the Slough natural history society. Slough. 1947. Rep. Slough Nat. Hist. Soc. Superseded by: Middle Thames naturalist. HI 71664

Report (Annual), Smithsonian institution = Report (Annual) of the board of regents, Smithsonian institution. Washington, DC.

Report (Annual), society of plant protection of north Japan. Omagari. 1950+. Rep. (Annual) Soc. Pl. Protect. N. Japan HI 71665

Report, society for the promotion of nature conservation. Nettleham. 1975/76-80/81, 1976-81. Rep. Soc. Promot. Nat. Conservation. Preceded by: Handbook (and annual report) of the society for the promotion of nature reserves. Superseded by: Report (Annual), royal society for nature conservation. HI 71666

Report (Annual) of the society for the protection of native plants. [Boston, MA.] ?-197?+. Rep. (Annual) Soc. Protect. Nat. Pl. HI 71667

Report, soil survey of Great Britain. London. No. 1+, 1950+. Rep. Soil Surv. Britain. HI 71668

Report (Annual), Somerset trust for nature conservation. Taunton. No. ?-7+, ?-1971+. Rep. (Annual) Somerset Trust Nat. Conservation. HI 71669

Report, Sorby natural history society. Sheffield. 1942-50. Rep. Sorby Nat. Hist. Soc. HI 71670

Report of the (meeting of the) south african association for the advancement of science. Cape Town. Vols. 1-6, 1903-08. Rep. S. African Assoc. Advancem. Sci. Incorporated in: South african journal of science. HI 71671

Report (Annual) of the south african institution. Cape Town. 1830. Rep. (Annual) S. African Inst. HI 71672

Report of the south african museum. Cape Town. 1855+. Rep. South African Mus. HI 71673

Report of the south african sugar association experiment station = Report, experiment station of the south african sugar association. Durban.

Report (Annual) of the South Carolina agricultural experiment station. Clemson College, SC. Vol. 1+, 1888+. Rep. (Annual) South Carolina Agric. Exp. Sta. HI 71674

Report (Annual), South Dakota agricultural experiment station = South Dakota agricultural experiment station. Annual report. Brookings, SD. South Dakota Agric. Exp. Sta. Annual Rep. See B–P–H 847/25.

Report (Annual) of the South Dakota state horticultural society. Aberdeen, SD. Vol. 1+, 1903/04+. Rep. (Annual) South Dakota State Hort. Soc. 5-4008-2. HI 71675

Report of the south-eastern agricultural college, Wye. London & Wye. 1895-1939. Rep. S.-E. Agric. Coll. Wye. HI 71676

Report of the south London entomological and natural history society. London. 1884. Rep. S. London Entomol. Nat. Hist. Soc. Superseded by: Abstracts of the proceedings of the south London entomological and natural history society. HI 71677

Report of the south London microscopical and natural history club. London. 1872-92. Rep. S. London Microscop. Nat. Hist. Club. HI 71678

Report (Annual) of the south west african scientific society. Windhoek. 1973+, 1974+. Rep. (Annual) S. W. African Sci. Soc. HI 71679

Report (Annual) of the southern forest experiment station. New Orleans, LA. Vols. 1+, 1929-63. Rep. (Annual)

S. Forest Exp. Sta. HI 71680

Report, southern pasture and forage crop improvement conference. New Orleans, LA, etc. ?-29th, ?-197? Rep. S. Pasture Forage Crop Improv. Conf. Superseded by: Proceedings of the southern pasture and forage crop improvement conference. HI 75228

Report, Southport scientific society. Southport. 1933-46. Rep. Southport Sci. Soc. Preceded by: Report of the Southport society of natural history. HI 71681

Report of the Southport society of natural science. Southport. 1890-1934. Rep. Southport Soc. Nat. Sci. Superseded by: Report, Southport scientific society. HI 71682

Report, southwestern forest and range experiment station. Tucson, AZ. 1930-52. Rep. SouthW. Forest Range Exp. Sta. HI 71683

Report (Annual), Sprenger instituut. Wageningen. 1966+. Rep. (Annual) Sprenger Inst. Preceded by: Report (Annual), instituut voor bewaring en verwerking van tuinbouwprodukten. HI 71684

Reports of St. Bartholomew's hospital. London. Rep. St. Bartholomew's Hosp. See B–P–H 771/6. HI 58640

Report (Annual), Staffordshire nature conservation trust ltd. Stoke-on-Trent. 1970+. Rep. (Annual) Staffordshire Nat. Conservation Trust. HI 71685

Report on state afforestation in New Zealand. Wellington, N.Z. ?-1912-15? Rep. State Afforest. New Zealand. Preceded by: Report on state nurseries and plantations, New Zealand. Superseded by: State forestry report. HI 71686

Report (Annual), state agricultural college agricultural experiment station, Colorado = State agricultural college agricultural experiment station annual report. Fort Collins, CO. State Agric. Coll. Agric. Exp. Sta. Annual Rep. See B–P–H 855/2.

Report (Annual), state agricultural experiment station [Title varies]. Auburn, AL. 1884-94. Rep. (Annual) State Agric. Exp. Sta. Superseded by: Alabama agricultural experiment station of the agricultural and mechanical college, annual report. HI 71687

Report of the state biological survey of Kansas. Lawrence, KS. 1964/65+. Rep. State Biol. Surv. Kansas. HI 71688

Report of the state botanist, New York state museum. [Published in: Report (Annual) of the regents of the university of the state of New York on the condition of the state cabinet of natural history, [etc.], Report (Annual) New York state museum of natural history by the regents of the university of the state of New York, Report (Annual) of the trustees of the state museum of natural history, New York, Report (Annual) of the regents of the university of the state of New York on the New York state museum, and New York state museum bulletin, q.v.] Albany, NY. 1868-1924. Rep. State Bot. New York State Mus. HI 71689

Report, state forest department, Victoria = Report of the department of state forests, Victoria. Melbourne, Vic.

Report (Annual), on state forests, New Zealand. Nos. 1-?, 1896/97-1910?, 189?-191? Rep. (Annual) State Forests New Zealand. Superseded by: Report on state nurseries and plantations, New Zealand. HI 71690

Report (Annual) of the state geologist of New Jersey. Trenton, NJ. 1867-72. Rep. (Annual) State Geol. New Jersey. HI 71691

Report (Annual) of the state geologist, New York = Report (Annual), geological survey of New York. Albany, NY.

Report (Annual), state natural history survey, Illinois = Report (Annual), Illinois natural history survey. Springfield, IL.

Report on state nurseries and plantations in New Zealand. Wellington, N.Z. 1911-? Rep. State Nurseries Plantations New Zealand. Preceded by: Report (Annual) on state forests, New Zealand. Superseded by: Report on state afforestation in New Zealand. HI 71692

Report (Annual), state pomological society of Michigan. East Lansing, MI. Vols. 1-?, 1871-80. Rep. (Annual) State Pomol. Soc. Michigan. Superseded by: Report (Annual) of the Michigan state horticultural society. 3-2643-3. HI 71693

Report, station agronomique, Mauritius. Port Louis. 1897-1913. Rep. Sta. Agron. Mauritius. Incorporated in: Report (Annual) of the department of agriculture, Mauritius. HI 71694

Report, Stockbridge House experimental horticulture station. Cawood. 1949/59-78?, 1959?-78. Rep. Stockbridge House Exp. Hort. Sta. Superseded by: Annual review, Stockbridge House experimental horticulture station. HI 71695

Report (Annual) of the Storrs agricultural experiment station. Storrs, CT. 1888-1927 Rep. (Annual) Storrs Agric. Exp. Sta. Incorporated in: Storrs agricultural experiment station bulletin. HI 71696

Report (Annual), sub-department of quaternary research, university of Cambridge. Cambridge. ?-9th+, ?-1957/58+, ?-1958+. Rep. (Annual) Sub-Dept. Quatern. Res. Univ. Cambridge. HI 71697

Report, Suffolk trust for nature conservation. Ipswich. 1969/70+. Rep. Suffolk Trust Nat. Conservation. HI 71698

Report of the sugar agronomist, department of agriculture, Trinidad and Tobago = Report, sugar agronomy division, department of agriculture, Trinidad and

Tobago. Port of Spain.

Report, sugar agronomy division, department of agriculture, Trinidad and Tobago. Port of Spain. 1948-55? Rep. Sugar Agron. Div. Dept. Agric. Trinidad & Tobago. Preceded by: Report (Annual), field experiments on sugar-cane in Trinidad. Superseded by: Report (Annual), sugar cane section, department of agriculture, Trinidad and Tobago. HI 71699

Report on the sugar beet experiments, Irish Free State. Dublin. 1925-32. Rep. Sugar Beet Exp. Irish Free State. HI 71700

Report of sugar-cane experiments, department of agriculture, Barbados. Bridgetown. 1900-24/26, 1900-26. Rep. Sugar-Cane Exp. Dept. Agric. Barbados. Incorporated in: Report, department of science and agriculture, Barbados. HI 71701

Report on the sugar cane experiments, Ste Madeleine sugar co. Trinidad. 1924+. Rep. Sugar Cane Exp. Ste Madeleine Sugar Co. HI 71702

Report of the sugar-cane investigation committee, Trinidad. Port of Spain. 1935+. Rep. Sugar-Cane Invest. Committee, Trinidad. Preceded by: Proceedings of the sugar-cane investigation committee, Trinidad. HI 71703

Report of the sugar-cane research scheme, Padegaon. New Delhi. 1932+. Rep. Sugar-Cane Res. Scheme Padegaon. HI 71704

Report (Annual), sugar cane section, department of agriculture, Trinidad and Tobago. Port of Spain. 1957-59. Rep. (Annual) Sugar Cane Sect. Dept. Agric. Trinidad & Tobago. Preceded by: Report, sugar agronomy division, department of agriculture, Trinidad and Tobago. Superseded by: Annual technical report, sugar cane agronomy section, department of agriculture, Trinidad and Tobago. HI 71705

Report, sugar-cane research station, Mauritius. Port Louis, Reduit. Nos. 1-23, 1930-52. Rep. Sugar-Cane Res. Sta. Mauritius. Superseded by: Report, Mauritius sugar industry research institute. HI 71706

Report (Annual) of the sugar industry research institute and sugar research department. [Mandeville?] (Jamaica). 1973+, 1974?+. Rep. (Annual) Sugar Industr. Res. Inst. Sugar Res. Dept. Preceded by: Report (Annual), research department, sugar manufacturers' association of Jamaica. HI 71707

Report, sugar industry research, Mauritius = Report (Annual), Mauritius sugar industry research institute. Port Louis, Reduit.

Report, sugarcane breeding institute, Coimbatore. New Delhi. 1950/51+, 1951+. Rep. Sugarcane Breed. Inst. Coimbatore. HI 71708

Report of the superintendent of the bureau of government laboratories in the Philippine Islands = Report of the bureau of government laboratories in the Philippine Islands. Manila.

Report of the superintendent of forestry, Canada = Report of the forestry branch, department of the interior, Canada. Ottawa.

Report, superintendent of gardens, Hong Kong. Hong Kong. 1950+. Rep. Superintendent Gard. Hong Kong. HI 71709

Report (Annual), Sussex naturalists' trust ltd. Henfield. Nos.1-?, 1961-70. Rep. (Annual) Sussex Naturalists' Trust. Superseded by: Report (Annual), Sussex trust for nature conservation. HI 71710

Report (Annual), Sussex trust for nature conservation. Henfield. 1971+. Rep. (Annual) Sussex Trust Nat. Conservation. Preceded by: Report (Annual), Sussex naturalists' trust ltd. HI 71711

Report, Sutton Bridge experiment station, potato marketing board. London. No. 1+, 1967+. Rep. Sutton Bridge Exp. Sta. Potato Marketing Board. HI 71712

Reports (Annual) of the Taihoku botanic garden. [Rino-gaku-bu fuzoku shokubutsuen nenpo.] Taihoku. Vols. 1-3, 1931-33. Rep. (Annual) Taihoku Bot. Gard. 5-4149-3. HI 71713

Report of the Taisetsuzan institute of science. [Hokkaido kyoiku daigaku Taisetsuzan kagaku kenkyusho hokoku.] Asahigawa. 1962+. Rep. Taisetsuzan Inst. Sci. HI 71714

Report (Annual) of the Taiwan sugar experiment station. [Taiwan tangye shiyansuo yanjiu huibao.] Taipei. Vols. 1-59, 1946-73. Rep. Taiwan Sugar Exp. Sta. Preceded by: Report of the government sugar experiment station, Taihoku. Superseded by: Report (Annual) of the Taiwan sugar research institute. HI 71715

Report (Annual) of the Taiwan sugar research institute. [Taiwan tangye yanjiusuo yanjiu huibao.] Taipei. 1972/73+. 1974+. Rep. (Annual) Taiwan Sugar Res. Inst. Preceded by: Report (Annual) of the Taiwan sugar experiment station. HI 71716

Report, Tanganyika sisal growers' association. Tanga. 1959/60-67/68, 1960-68. Rep. Tanganyika Sisal Growers' Assoc. Preceded by: Report, sisal research station, Mlingano, Tanganyika. HI 71717

Report (Annual), Tanzania coffee board. Dar-es-Salaam. ?-1966-67? Rep. (Annual) Tanzania Coffee Board. HI 71718

Report, tasmanian museum and art gallery. Hobart, Tas. 1972/73+, 1973+. Rep. Tasmanian Mus. Art Gall. Preceded by: Report, tasmanian museum and botanical gardens. HI 71719

Report, tasmanian museum and botanical gardens. Hobart, Tas. 1886-? Rep. Tasmanian Mus. Bot. Gard. Superseded by: Report, tasmanian museum and art gallery. HI 71720

Report, Tatura horticultural research station. Tatura, Vic. No. 1+, 1964+. Rep. Tatura Hort Res. Sta. HI 71721

Report on tea culture in Eastern Bengal and Assam. Shillong. 1873+. Rep. Tea Cult. Eastern Bengal Assam. HI 71722

Report of the tea institute, Taiwan. Taipei. 1952+. Rep. Tea Inst. Taiwan. HI 71723

Report (Annual), tea research foundation of central Africa. ?-1972/73+, ?-1973+. Rep. (Annual) Tea Res. Found. Centr. Africa. HI 71724

Report (Annual), tea research institute of Ceylon. Kandy, Talawakelle. 1st-45th, 1926-71. Rep. (Annual) Tea Res. Inst. Ceylon. Superseded by: Technical report for the year, tea research institute of Sri Lanka. HI 71725

Report (Annual), tea research institute of East Africa. Kericho, Nairobi. 1950+. Rep. (Annual) Tea Res. Inst. East Africa. Preceded by: Proceedings of conference, tea research institute of East Africa. HI 71726

Report, tea research station, Nyasaland. Zomba. 1952+. Rep. Tea Res. Sta. Nyasaland. HI 71727

Report (Annual), tea scientific department, united planters' association of southern India = Annual administration report, tea section, united planters' association of southern India. Madras, Coimbatore.

Report, technological, industrial and sanitary museum, Sydney = Report of the technological museum, Sydney. Sydney, N.S.W.

Report of the technological museum, Sydney. Sydney, N.S.W. 1906-45. Rep. Technol. Mus. Sydney. Supersed by: Report (Annual) of the trustees of the museum of technology and applied science. HI 71728

Report (Annual) of the Tennessee agricultural experiment station. Knoxville, TN. 1888+. Rep. (Annual) Tennessee Agric. Exp. Sta. HI 71729

Report, Texas agricultural experiment station. College Station, TX. 1888+. Rep. Texas Agric. Exp. Sta. HI 71730

Report, Texas forest service = Report, Texas state forest service. College Station, TX.

Report, Texas state forest service. College Station, TX. 1940-44. Rep. Texas State Forest Serv. HI 71731

Report. Thirsk natural history society. Thirsk, England. Rep. Thirsk Nat. Hist. Soc. See B–P–H 771/9. HI 58641

Reports of the Timiryazev agricultural academy = Izvestiya timiryazevskoi sel'skokhozyaistvennoi akademii. Moscow.

Report, tissue culture association = T C A report. Lake Placid, NY.

Report, tobacco board, Mauritius. Port Louis. 1932+, 1933+. Rep. Tobacco Board Mauritius. HI 71732

Report, tobacco division, experimental farms, Canada. Ottawa. 1913-36. Rep. Tobacco Div. Exp. Farms Canada. Previously contained in: Tobacco bulletin, tobacco division, Canada. Superseded by: Progress report, tobacco division, experimental farms service, Canada. HI 71733

Report of the tobacco institute of Puerto Rico. San Juan, PR. 1936/37-41/43, 1937-43. Rep. Tobacco Inst. Puerto Rico. HI 71734

Report of tobacco large experimental field. [Ogata jikken nojo jisshi gaiyo.] Tokyo. ?-1970+. Rep. Tobacco Large Exp. Field. HI 75159

Report on the Tobago botanic station = Report on the botanic station, Tobago.

Report, Tocklai experimental station of the indian tea association. Calcutta. 1935+. Rep. Tocklai Exp. Sta. Indian Tea Assoc. HI 71735

Report (Annual) of the Tohoku branch, government forest experiment station. [Ringyo shikenjo Tohoku shijo nenpo.] Morioka. 1960+. Rep. (Annual) Tohoku Branch Forest Exp. Sta. HI 75149

Report of the tomato genetics cooperative. West Lafayette, IN. No. 1+, 1951+. Rep. Tomato Genet. Coop. HI 71736

Reports of the Tottori mycological institute. [Kinjin kenkyusho kenkyu hokoku.] Tottori, Japan. Rep. Tottori Mycol. Inst. See B–P–H 771/10. HI 58642

Report and transactions, Birmingham natural history and microscopical society. Birmingham. 1872-86. Rep. Trans. Birmingham Nat. Hist. Microscop. Soc. Preceded by: Proceedings of the Birmingham natural history and microscopical society. HI 71737

Report and transactions of the Cardiff naturalists' society. Cardiff, Wales. Rep. & Trans. Cardiff Naturalists' Soc.. See B–P–H 761/21. HI 58543

Report and transactions of the Devonshire association for the advancement of science, literature and art. Plymouth, England. Rep. & Trans. Devonshire Assoc. Advancem. Sci. See B–P–H 761/22. HI 58544

Reports and transactions of the East Kent scientific and natural history society. Canterbury. Rep. & Trans. E. Kent Sci. Soc. See B–P–H 761/23. HI 58545

Report and transactions of the Glasgow society of field naturalists. Glasgow. 1872-78. Rep. Trans. Glasgow Soc. Field Naturalists. HI 71738

Report and transactions, Guernsey society of natural science (and local research). Guernsey, C.I. 1882-

1921. Rep. Trans. Guernsey Soc. Nat. Sci. Superseded by: Report and transactions, société guernésiaise. HI 71739

Report and transactions of the Manchester microscopical society. Manchester. 1901-30. Rep. Trans. Manchester Microscop. Soc. Preceded by: Report (Annual), Manchester microscopical society and Transactions and annual report of the Manchester microscopical society. HI 71740

Report of the transactions of the Massachusetts horticultural society = Transactions of the Massachusetts horticultural society. Boston, MA.

Report and transactions, north Staffordshire field club. Stafford. 1887-1915. Rep. Trans. N. Staffordshire Field Club. Preceded by: Report, north Staffordshire naturalists' field club and archaeological society. Superseded by: Transactions and annual report, north Staffordshire field club. HI 71741

Report and transactions, north Staffordshire naturalists' field club = Report and transactions, north Staffordshire field club.

Report and transactions, Nottingham naturalists' society. Nottingham. 1877-1918. Rep. Trans. Nottingham Naturalists' Soc. HI 71742

Report and transactions of the Penzance natural history society and antiquarian society = Transactions of the Penzance natural history society and antiquarian society. Plymouth.

Report and transactions of the Plymouth institution and Devon and Cornwall natural history society. Plymouth. 1855-1961. Rep. Trans. Plymouth Inst. Devon Cornwall Nat. Hist. Soc. Superseded by: Proceedings of the Plymouth athenaeum. HI 71743

Report and transactions, société guernésiaise. Guernsey, C.I. 1922+. Rep. Trans. Soc. Guernésiaise. Preceded by: Report and transactions, Guernsey society of natural science (and local research). HI 71744

Report and transactions of the south eastern union of scientific societies. Canterbury? 1898-99. Rep. Trans. S. E. Union Sci. Soc. Preceded by: Transactions of the south-eastern union of scientific societies. Superseded by: South-eastern naturalist. HI 71745

Report (Annual) and transactions Worcester (Mass.) agricultural society. Worcester, MA. 1843-1901. Rep. (Annual) & Trans. Worcester Agric. Soc. 5-4539-2. HI 71746

Report (Annual) and transactions Worcester county agricultural society = Report (Annual) and transactions Worcester (Mass.) agricultural society. Worcester, MA.

Report (Annual), Transvaal nature conservation division. Pretoria. 1966+. Rep. (Annual) Transvaal Nat. Conservation Div. Preceded by: Report (Annual) of the nature conservation branch, Transvaal provincial administration. HI 71747

Report (Annual) of the Trelawny tobacco research station. Salisbury, Rhodesia. 1940-49. Rep. (Annual) Trelawny Tobacco Res. Sta. Superseded by: Results of experiments, tobacco research board of Southern Rhodesia. HI 71748

Report on the trial of new varieties of hops. East Malling. 1917-49. Rep. Trial New Var. Hops. HI 71749

Report of the tropical development and research instiutute. London. 1983/84+, 1984+. Rep. Trop. Developm. Res. Inst. Preceded by: Report, tropical products institute. HI 71750

Report, tropical products institute. London. 1959-8?, 1960-8? Rep. Trop Prod. Inst. Superseded by: Report of the tropical development and research institute. HI 71751

Report (Annual) of the trustees of the museum of applied arts and sciences. Sydney, N.S.W. 1950+. Rep. (Annual) Trustees Mus. Appl. Arts Sci. Preceded by: Report (Annual) of the trustees of the museum of technology and applied science. HI 71752

Report (Annual) of the trustees of the museum of technology and applied science. Sydney, N.S.W. 1946-49. Rep. (Annual) Trustees Mus. Techn. Appl. Sci. Preceded by: Report of the technological museum, Sydney. Superseded by: Report (Annual) of the trustees of the museum of applied arts and sciences. HI 71753

Report (Annual) of the trustees of the public museum and art gallery of Papua and New Guinea. Port Moresby. 1966+, 1967+. Rep. (Annual) Trustees Public Mus. Art Gall. Papua New Guinea. HI 71754

Report (Annual) of the trustees of the state museum of natural history, New York. Albany, NY. Vols. 38-39, 1884-85, 1886-87; vols. 41-42, 1887-88, 1888-89. Rep. (Annual) Trustees State Mus. Nat. Hist., New York. Preceded by: Report (Annual) of the New York state museum of natural history by the regents of the university of the state of New York. For vol. 40 see: Report (Annual) of the New York state museum of natural history by the regents of the university of the state of New York. Superseded by: Report (Annual) of the regents of the university of the state of New York on the New York state museum. 4-3045-1. HI 71755

Report, Tunbridge Wells natural history and philosophical society. Tunbridge Wells. 1891-1933. Rep. Tunbridge Wells Nat. Hist. Philos. Soc. HI 71756

Report of the uniform alfalfa nurseries. Washington, DC, Beltsville, MD. 1937-54. Rep. Uniform Alfalfa Nurseries. HI 71757

Report, unit of comparative plant ecology (N E R C).

Sheffield. 1975+. Rep. Unit Comp. Pl. Ecol. HI 71758

Report (Annual), United Kingdom cereal pathogen virulence survey. Cambridge. 1978+, 1979+. Rep. (Annual) United Kingdom Cereal Pathog. Virul. Surv. Preceded by: Report (Annual), physiologic race survey (cereal pathogens). HI 71759

Report, United States department of agriculture = U S department of agriculture. Report. Washington, DC. U.S.D.A. Rep. See B–P–H 944/18.

Report, United States forest service = Report of the forest service, United States department of agriculture. Washington, DC.

Report (Annual), United States national arboretum. Washington, DC. ?-1973/74+, ?-1974+. Rep. (Annual) U.S. Natl. Arbor. HI 71760

Report (Annual) of the United States national museum [Smithsonian institution]. Washington, DC. 1881-1964. Rep. (Annual) U.S. Natl. Mus. 5-3886-3. HI 71761

Report (Annual) of the United States regional pasture research laboratory. State College, PA. No. 1+, 1937+. Rep. (Annual) U.S. Regional Pasture Res. Lab. HI 71762

Report (Annual), university of agricultural sciences. Hebbal, Bangalore. Vol. 1+, 1964/65+, 1965?+. Rep. (Annual) Univ. Agric. Sci. HI 71763

Report, university arboretum, university of California. Davis, CA. 1975/76+, 1976?+. Rep. Univ. Arbor. Univ. Calif. HI 71764

Report (Annual), university of California agricultural experiment station = University of California agricultural experiment station. Annual report. Berkeley, CA. Univ. Calif. Agric. Exp. Sta. Annual Rep. See B–P–H 934/7.

Report (Annual), university of Delaware agricultural experiment station = University of Delaware agricultural experiment station. Annual report. Newark, DE. Univ. Delaware Agric. Exp. Sta. Annual Rep. See B–P–H 935/11.

Report (Annual) of the university of Kansas museum of natural history. Lawrence, KS. 1968+. Rep. (Annual) Univ. Kansas Mus. Nat. Hist. HI 71765

Report, university marine biological station, Millport. 1970/71+. Rep. Univ. Mar. Biol. Sta. Millport. HI 71766

Research (Annual), university of Western Australia = Research report, university of Western Australia. Perth, W.A.

Reports of the Usa marine biological institute, Kochi university. [Kochi daigaku. Kaiyo seibutsu kyoiku kenkyu senta.] Kochi. No. 1+, 1979+. Rep. Usa Mar. Biol. Inst. Kochi Univ. Preceded by: Reports of the Usa marine biological station, Kochi university. HI 71767

Reports of the Usa marine biological station, Kochi university. Kochi. Vols. 1-?, 1954-77. Rep. Usa Mar. Biol. Sta. Kochi Univ. Superseded by: Reports of the Usa marine biological institute, Kochi university. HI 71768

Report of the Utsunomiya tobacco experiment station. [Utsunomiya tabako shikenjo gyotei hokoku.] Tochigi, Japan. Rep. Utsunomiya Tobacco Exp. Sta. See B–P–H 771/22. HI 58643

Report (Annual) of vegetables and flowers research works in Japan. [Sosai kaki shiken kenkyu nenpo.] Tokyo. Vol. 1+, 1954+. Rep. (Annual) Veg. Fl. Res. Works Japan. HI 71769

Report (Annual), Vermont state agricultural college, agricultural experiment station = Vermont state agricultural college, agricultural experiment station. Annual report. Burlington, VT. Vermont Agric. Exp. Sta. Annual Rep. See B–P–H 956/21.

Report, Victoria conservation trust. Melbourne, Vic. 1973/74+, 1974+. Rep. Victoria Conservation Trust HI 71770

Report, Victoria institute of Trinidad and Tobago. Port of Spain. 1905-08. Rep. Victoria Inst. Trinidad & Tobago. Preceded by: Proceedings of the Victoria institute of Trinidad. HI 71771

Report, Vineland horticultural experiment station = Report (Annual) of the horticultural experiment station and products laboratory, vineland station, Ontario. Toronto.

Report, Virgin Islands agricultural experiment station. Tortola, Washington, DC, St. Croix. 1902-14, 1919-32, 1974+. Rep. Virgin Islands Agric. Exp. Sta. HI 71772

Report of the Virgin Islands agricultural research and extension programs, officew of experiment stations, United States department of agriculture = U S department of agriculture, office of experiment stations. Report of the Virgin Islands agricultural research and extension programs. Washington, DC. U.S.D.A. Off. Exp. Sta. Rep. Virgin Islands Agric. Res. Extens. Programs. See B–P–H 944/4.

Report of the Virginia agricultural experiment station. Blacksburg, VA. 1889/90-1946/47, 1890-1947. Rep. Virginia Agric. Exp. Sta. Superseded by: Agricultural research report, agricultural experiment station, Blacksburg. HI 71773

Report of the Virginia state horticultural society. Winchester, VA. 1898+. Rep. Virginia State Hort. Soc. HI 71774

Report of the Waite agricultural research institute, university of Adelaide. Adelaide, S.A. 1925/32-65? 1932-65? Rep. Waite Agric. Res. Inst. Univ. Adelaide. Superseded by: Biennial report of the Waite agricultural research institute, university of Adelaide. HI 71775

Report, Walthamstow natural history and microscopical society. London. Nos. 1-3, 1882-83. Rep. Walthamstow Nat. Hist. Microscop. Soc. HI 71776

Report (Annual) of the Warwick natural history society. Warwick. 1955+. Rep. (Annual) Warwick Nat. Hist. Soc. HI 71777

Report (Annual) of the Warwickshire natural history and archaeological society. Warwick. Vols. 1-49, 1836/37-84/85. Annual Rep. Warwickshire Nat. Hist. Soc. 5-4445-3. HI 71778

Report of the Warwickshire nature conservation trust. Warwick. 1976+. Rep. Warwickshire Nat. Conservation Trust. HI 71779

Report (Annual), Washington (state) agricultural experiment station = Washington (state) agricultural experiment station. Annual report. Pullman, WA. Wash. State Agric. Exp. Sta. Annual Rep. See B–P–H 971/11.

Report (Annual) of the Watson botanical exchange club. York, Cambridge. Vols. 1-4(5), 1884-1934. Rep. (Annual) Watson Bot. Exch. Club. HI 71780

Report, wattle research institute, university of Natal. Pietermaritzburg. 1948+. Rep. Wattle Res. Inst. Univ. Natal. HI 71781

Report, Wau ecology institute. 1971/73+. Rep. Wau Ecol. Inst. HI 71782

Report of the Wellcome research laboratories at the Gordon memorial college. Khartoum [Republic of the Sudan]. Rep. Wellcome Res. Lab. See B–P–H 771/25. HI 58644

Report of the Wellcome tropical research laboratories at the Gordon memorial college = Report of the Wellcome research laboratories at the Gordon memorial college. Khartoum [Republic of the Sudan]. Rep. Wellcome Res. Lab. See B–P–H 771/25.

Report (Annual) of the welsh plant breeding station. Aberystwyth. 1950/56+, 1956?+. Rep. (Annual) Welsh Pl. Breed. Sta. Preceded by: Report of the plant breeding station of university college, Aberystwyth. HI 71783

Report (Annual), west african cacao research institute. Tafo. 1944/45-61/62, 1945-64. Rep. (Annual) W. African Cacao Res. Inst. Preceded by: Report (Annual) of the central cocoa research station, Tafo. Superseded by: Report (Annual) of the cocoa research institute of Nigeria. HI 71784

Report (Annual), west african cocoa research institute = Report (Annual), west african cacao research institute. Tafo.

Report, west african institute for oil palm research. Benin City. Vols. 1-12, 1952-64. Rep. W. African Inst. Oil Palm Res. Preceded by: Report (Annual) of the oil palm research station. Superseded by: Report, nigerian institute for oil palm research. HI 71785

Report, west african maize rust research unit. Ibadan. 1953+. Rep. W. African Maize Rust Res. Unit. HI 71786

Report (Annual), west african rice research station, Rokupr. Freetown. 1953-64/65, 1954-66. Rep. (Annual) W. African Rice Res. Sta. Rokupr. Superseded by: Report, rice research institute, Rokupr. HI 71787

Report, west Dumbartonshire naturalist. Helensburgh. 1974+. Rep. W. Dumbartonshire Naturalist. HI 71788

Report (Annual), West Indies central sugar cane breeding station. 33rd+, 1966+. Rep. (Annual) West Indies Central Sugar Cane Breed. Sta. Preceded by: Report (Annual), British West Indies central sugar cane breeding station, Barbados. HI 71789

Report, West Pakistan forest department = Report, forest department, West Pakistan.

Report, west Wales field society. Tenby. 1945-56. Rep. W. Wales Field Soc. HI 71791

Report (Annual), west Wales naturalists' trust ltd. Haverfordwest. 1968+, [1969]+. Rep. (Annual) W. Wales Naturalists' Trust. HI 71792

Report (Annual) of the western australian herbarium. South Perth, W.A. 1976+. Rep. (Annual) W. Austral. Herb. HI 71793

Report, western nut growers' association = Report and proceedings, western nut growers' association. Gresham, OR.

Report, wheat research institute, New Zealand = Report of the New Zealand wheat research institute. Wellington, N.Z.

Report (Annual), wheat research unit, C S I R O, Australia. Melbourne, Vic. 1960/61+, 1961+. Rep. (Annual) Wheat Res. Unit C. S. I. R. O. Australia. HI 71794

Report (Annual), wheat and sheep division, department of agriculture, Western Australia. Perth, W.A. 1964-? Rep. (Annual) Wheat Sheep Div. Dept. Agric Western Australia. Superseded by: Report, division of plant production, department of agriculture, Western Australia. HI 71795

Report (Annual) of the Whitby literary and philosophical society. Whitby. Vols. 1-54, 1823-76. Rep. (Annual) Whitby Lit. Soc. HI 71796

Report, wilderness society. Washington, DC. Vols. 1-5, 1964-69. Rep. Wildern. Soc. Superseded by: Wilderness report. HI 71797

Report, Wiltshire archaeological and natural history society. Devizes. 1977+. Rep. Wiltshire Archaeol. Nat. Hist. Soc. Preceded by: Report, natural history section, Wiltshire archaeological and natural history society. HI 71798

Report (Annual), Wiltshire trust for nature conservation. Trowbridge. No. ?-2+, ?-1964+. Rep. (Annual) Wiltshire Trust Nat. Conservation. HI 71799

Report, winter congress, international institute for sugar beet research = Compte rendu, congrès d'hiver, institut international de recherches betteravières. Brussels.

Report (Annual), Wisconsin (state) agricultural experiment station = Wisconsin (state) agricultural experiment station. Annual report. Madison, WI. Wisconsin Agric. Exp. Sta. Annual Rep. See B–P–H 975/29.

Report of the Woburn agricultural experiment station. London. 1877-1938. Rep. Woburn Agric. Exp. Sta. Incorporated in: Report, Rothamsted experimental station. HI 71800

Report, Woburn experimental farm = Report of the Woburn agricultural experiment station. London.

Report of the Woburn experimental fruit farm. London. Nos. 1-18, 1897-1921. Rep. Woburn Exp. Fruit Farm. Incorporated in: Journal of the royal agricultural society of England. HI 71801

Report, Woburn fruit farm = Report of the Woburn experimental fruit farm. London.

Report, woods and forests department, South Australia. Adelaide, S.A. 1879+. Rep. Woods Forests Dept. South Australia. HI 71802

Report of the woods and forests department, Western Australia = Report of the forests department, Western Australia. Perth, W.A.

Report (Annual) of the Worcester natural history society. Worcester. 1898-1900. Rep. (Annual) Worcester Nat. Hist. Soc. 5-4540-1. HI 71803

Report, Worcestershire nature conservation trust. Birmingham. 1977/78+. Rep. Worcestershire Nat. Conservation Trust. HI 71804

Reports on the work of the agricultural stations, department of agriculture, Madras = Report on the agricultural stations in the Madras Presidency. Madras.

Report (Annual) of the work of Saito Ho-on Kai. Sendai. Vol. 1+, 1923/24+. Rep. (Annual) Work Saito Ho-On Kai. 5-3763-1. HI 71805

Report on the working and administration of the government gardens. Allahabad. 1922+. Rep. Working Admin. Gov. Gard., Allahabad. Preceded by: Report of the government botanical gardens, Saharanpur, Report on the government horticultural gardens, Lucknow and Report of the Kumaun government gardens. HI 71806

Report on the working of the department of agriculture, Madhya Pradesh. Nagpur. 1947+. Rep. Working Dept. Agric. Madhya Pradesh. Preceded by: Report of the department of agriculture, Central Provinces and Berar. HI 71807

Report (Annual), world wildlife fund – U. S. Washington, DC. 1981?+. Rep. (Annual) World Wildlife Fund, U.S. HI 71809

Report (Annual), wrapper and hookah tobacco research station. Duikata. 1963/64+. Rep. (Annual) Wrapper Hookah Tobacco Res. Sta. HI 71810

Report (Annual), Wyoming agricultural college agricultural experiment station = Wyoming agricultural college agricultural experiment station. Annual report. Laramie, WY. Wyoming Agric. Exp. Sta. Annual Rep. See B–P–H 980/24.

Report, Yale Peabody museum of natural history. New Haven, CT. 1868+. Rep. Yale Peabody Mus. Nat. Hist. HI 71811

Report, Yorkshire natural science association. York. 1918-28. Rep. Yorkshire Nat. Sci. Assoc. HI 71812

Report (Annual), Yorkshire naturalists' trust ltd. York. 1970+, [1971]+. Rep. (Annual) Yorkshire Naturalists' Trust. HI 71813

Report (Annual), Yorkshire naturalists' union. Leeds & Halifax. 1887-1910, 19??+ [from 1911-? contained in Naturalist (Hull).], 1972+. Rep. (Annual) Yorkshire Naturalists' Union. HI 71814

Report (Annual), Zamorano; escuela agricola panamericana. Tegucigalpa, Honduras. ?-1984+. Rep. (Annual) Zamorano Esc. Agric. Panamer. HI 71815

Report (Annual) of the zoological and acclimatisation society of Victoria = Report (Annual) of the acclimatisation society of Victoria. Melbourne, Vic.

Reporte de investigación del instituto de botánica, academia de ciencias de Cuba. Havana. Vol. 1+, 1982+. Rep. Invest. Inst. Bot. Acad. Ci. Cuba. HI 71816

Reporte de investigación del instituto de investigaciones fundamentales en agricultura tropical. Havana. No. 1+, 1982+. Rep. Invest. Inst. Invest. Fundam. Agric. Trop. HI 71817

Reporte de investigación del instituto de oceanologia. Havana. Vol. 1+, 1982+. Rep. Invest. Inst. Oceanol. HI 71818

Reporter, international rice research institute (IRRI) = I R R I reporter.

Reportes misceláneos, centro internacional de mejoramiento de maiz y trigo = Miscellaneous reports, centro internacional de mejoramiento de maiz y trigo. Mexico, D.F.

Reports, see Report

Reprints of the university of British Columbia. Biological sciences. Vancouver, B.C. Nos. 1-57, 1939-43. Repr. Univ. British Columbia, Biol. Sci. Superseded by: Publications of the university of British Columbia. Biological sciences. 1-779-2. HI 71819

Research. A journal of science and its applications. London. Research. See B–P–H 776/24. HI 58694

Research accomplished, southern forest experiment station. New Orleans, LA. 1976+. Res. Accomp. S. Forest Exp. Sta. HI 71820

Research abstract report of the agricultural department of St. Lucia. Castries. 1929-36. Res. Abstr. Rep. Agric. Dept. St. Lucia. HI 71821

Research annual report, university of Alaska. Fairbanks, AK. 1976/77+, 1977?+. Res. Annual Rep. Univ. Alaska. HI 71822

Research applied in industry. London. Res. Appl. Industr. See B–P–H 773/14. HI 58664

Research branch papers, forestry commission. London. Nos. 1-30?, 1951-65. Res. Branch Pap. Forest. Commiss. Superseded by: Research and development papers, forestry commission. HI 71823

Research branch report, biosystematics research institute = Research report, biosystematics research institute. Ottawa.

Research branch report, department of agriculture, Canada = Report on research, department of agriculture, Canada. Ottawa.

Research briefs, international institute of tropical agriculture = I I T A research briefs. Ibadan.

Research in british universities, polytechnics and colleges. London. 1979-? Res. Brit. Univ. Superseded by: Current research in Britain. Biological sciences. HI 71824

Research bulletin of the agricultural experiment station, Kung-chu-ling. [Man chu kuo litsi Kung chu ling. Noji shikenjo kenkyu jiho.] Nos. 25-31, 1938-40. Res. Bull. Agric. Exp. Sta. Kung-chu-ling Manchoukuo. Preceded by: Research bulletin of the agricultural experiment station of the South Manchuria Railway Co. HI 71825

Research bulletin of the agricultural experiment station of the South Manchuria Railway Co. Kunchuling. Nos. 1-24, 1930-38. Res. Bull. Agric. Exp. Sta. South Manchuria Railway Co. Superseded by: Research bulletin of the agricultural experiment station, Kung-chu-ling. HI 71826

Research bulletin agricultural experiment station, university of Idaho. Moscow, ID. Res. Bull. Agric. Exp. Sta. Univ. Idaho. See B–P–H 773/19. HI 58666

Research bulletin of the Aichi-ken agricultural research center. Series A, field crops. [Aichi-ken engei shikenjo kenkyu hokoku.] Aichi. No. 1+, 1969+. Res. Bull. Aichi-ken Agric. Res. Center, A. Preceded by: Bulletin of Aichi horticultural experiment station. HI 71827

Research bulletin of the Aichi-ken agricultural research center. Series B, horticulture. [Aichi-ken engei shikenjo kenkyu hokoku.] Aichi. No. 1+, 1969+. Res. Bull. Aichi-ken Agric. Res. Center, B. Preceded by: Bulletin of Aichi horticultural experiment station. HI 71828

Research bulletin, college of agriculture, Alberta university. Edmonton, Alta. 1924-36. Res. Bull. Coll. Agric. Alberta Univ. HI 71829

Research bulletins of the college experiment forests, college of agriculture, Hokkaido imperial university. [Hokkaido teikoku daigaku nogakubu enshurin kenkyu hokoku.] Sapporo, Japan. Res. Bull. Coll. Exp. Forests Coll. Agric. Hokkaido Imp. Univ. See B–P–H 773/24. HI 58667

Research bulletins of the college experiment forests, Hokkaido university. [Hokkaido teikoku daigaku nogakubu. Enshurin kenkyu hokoku.] Sapporo, Japan. Res. Bull. Coll. Exp. Forests Hokkaido Univ. See B–P–H 774/1. HI 58668

Research bulletin, Colorado greenhouse growers' association. Denver, CO. Nos. 355-?, 1980-89. Res. Bull. Colorado Greenh. Growers' Assoc. Preceded by: Bulletin, Colorado flower growers' association. HI 71842

Research bulletin, cooperative forestry research unit, university of Maine at Orono. Orono, ME. No. 1+, 1979+. Res. Bull. Coop. Forest. Res. Unit Univ. Maine Orono. HI 75139

Research bulletin, cytogenetics laboratory, department of botany, university of Calcutta. Calcutta. Vol. 1+, 1966+. Res. Bull. Cytogen. Lab. Dept. Bot. Univ. Calcutta. HI 71830

Research bulletin, department of agriculture, stock and fisheries, Papua New Guinea. Port Moresby. Nos. 1-16, 1963-75. Res. Bull. Dept. Agric. Stock Fish., Papua New Guinea. HI 71831

Research bulletin, department of agriculture, stock and fisheries, Papua-New Guinea. Crop production series. Port Moresby. No. 1+, 1969+. Res. Bull. Dept. Agric. Stock Fish. Papua New Guinea, Crop Prod. Ser. HI 71832

Research bulletin, department of forests, Papua New Guinea. Port Moresby. Vol. 1+, 197?+. Res. Bull.

Dept. Forests Papua New Guinea. HI 71833

Research bulletin, division of forest research, forest department, Zambia. Ndola. 1960+. Res. Bull. Div. Forest Res. Forest Dept. Zambia. HI 71834

Research bulletin, division of life sciences research, Louisiana tech university. Ruston, LA. 1978+. Res. Bull. Div. Life Sci. Res. Louisiana Tech Univ. Preceded by: Research bulletin, school of agriculture and forestry, polytechnic institute, Ruston. HI 71835

Research bulletin, faculty of agriculture, Ain Shams university. Cairo. ?-1983+. Res. Bull. Fac. Agric. Ain Shams Univ. HI 71836

Research bulletin, faculty of agriculture, Alberta university = Research bulletin, college of agriculture, Alberta university. Edmonton.

Research bulletin of the faculty of education, Oita university. Natural science. [Oita daigaku kyoikugakubu kenkyu kiyo, shizenkagaku.] Oita. 1952. Res. Bull. Fac. Educ. Oita Univ., Nat. Sci. HI 71837

Research bulletin of the faculty of liberal arts; Oita university (Natural sciences). [Oita daigaku gakugeigakubu kenkyu kiyo. Shizen kagaku.] Oita, Japan. Res. Bull. Fac. Liberal Arts Oita Univ., Nat. Sci. See B–P–H 774/5 HI 58669.

Research bulletin, forest department, Kenya = Research bulletin, forestry department, Kenya. Nairobi.

Research bulletin of the forest experiment station. [Ryuku seifu, Keizai-kyoku ringyo shikenjo.] [Ryukyu government, economics department, forest experiment station.] Naha?, Okinawa. Res. Bull. Forest Exp. Sta. See B–P–H 774/7. HI 58670

Research bulletin, forest research laboratory. Corvallis, OR. Vol. 11+, 1958+. Res. Bull. Forest Res. Lab., Corvallis. Preceded by: Research bulletin, Oregon state board of forestry. HI 71838

Research bulletin, forestry department, Kenya. Nairobi. 1936+. Res. Bull. Forest. Dept., Kenya. HI 71839

Research bulletin, Fukien agricultural and forestry experiment station. Fukien. ?-1942+. Res. Bull. Fukien Agric. Forest. Exp. Sta. HI 71840

Research bulletin, Georgia agricultural experiment stations. Athens, GA. No. 1+, 1967+. Res. Bull. Georgia Agric Exp. Sta. HI 71841

Research bulletin of the Gifu agricultural experiment station. [Gifu-ken nogyo shikenjo hokoku.] Gifu, Japan. Res. Bull. Gifu Agric. Exp. Sta. See B–P–H 774/9. HI 58671

Research bulletin of the Gifu university; faculty of agriculture. Gifu, Japan. Res. Bull. Gifu Univ. Fac. Agric. See B–P–H 774/12. HI 58672

Research bulletin, Hawaii agricultural experiment station. Honolulu, HI. No. 143+, 1971+. Res. Bull. Hawaii Agric. Exp. Sta. Preceded by: Bulletin, Hawaii agricultural experiment station. HI 61703

Research bulletin of the Hokkaido national agricultural research station. [Hokkaido nogyo shikenjo kenkyu hokoku.] Sapporo. No. 62+, 1952+. Res. Bull. Hokkaido Natl. Agric. Res. Sta. Preceded by: Bulletin of the Hokkaido agricultural experiment station. HI 71843

Research bulletin, horticultural research station, Okitsu = Research bulletin, imperial horticultural research station, Okitsu. Okitsu.

Research bulletin, imperial horticultural research station, Okitsu. Okitsu. Nos. 1-19, 192?-44. Res. Bull. Imp. Hort. Res. Sta. Okitsu. HI 71844

Research bulletin of the institute of zoology and botany: Fukien academy. [Yuan chiu huei pao. Fu chian seng yuan chiu yuan.] Foochow, China. Res. Bull. Inst. Zool. Bot. Fukien Acad. See B–P–H 774/15. HI 58673

Research bulletin, international crops research institute for the semi-arid tropics. Hyderabad. No. ?-3+, ?-1980+. Res. Bull. Int. Crops Res. Inst Semi-Arid Trop. HI 71845

Research bulletin international maize and wheat improvement center. Chapingo. Nos. 1+, 1965+. Res. Bull. Int. Maize Wheat Improv. Center. HI 71846

Research bulletin, Iowa agricultural experiment station = Iowa agricultural experiment station. Research bulletin. Ames, IA. Iowa Agric. Exp. Sta. Res. Bull. See B–P–H 436/35.

Research bulletin of the Kentucky agricultural experiment station = Bulletin of the Kentucky agricultural experiment station. Lexington, KY.

Research bulletin, Michigan agricultural experiment station = Michigan agricultural experiment station. Research bulletin. East Lansing, MI. Michigan Agric. Exp. Sta. Res. Bull. See B–P–H 592/16.

Research bulletin, Missouri agricultural experiment station = Missouri agricultural experiment station. Research bulletin. Columbia, MO.

Research bulletin, national research council of the Philippines. Quezon City, Rizal. Vol. 36+, 1981+. Res. Bull. Natl. Res. Council Philippines. Preceded by: Bulletin of the national research council of the Philippines. HI 62088

Research bulletin, Nebraska agricultural experiment station = Nebraska agricultural experiment station research bulletin. Lincoln, NB. Nebraska Agric. Exp. Sta. Res. Bull. See B–P–H 640/3.

Research bulletin of Obihiro university. Series 1.

Obihiro. Vol. ?-9+, ?-1974+. Res. Bull. Obihiro Univ., Ser. 1. Preceded by: Research bulletin, Obihiro zootechnical university [not entered]. HI 71847

Research bulletin, Ohio agricultural experiment station = Ohio agricultural experiment station research bulletin. Wooster, OH.

Research bulletin, Ohio agricultural research and development center. Wooster, OH. 1965?+. Res. Bull. Ohio Agric. Res. Developm. Center. Preceded by: Ohio agricultural experiment station research bulletin. HI 71848

Research bulletin, Oregon agricultural experiment station. Corvallis, OR. Nos. 1-3, 1913-14. Res. Bull. Oregon Agric. Exp. Sta. Incorporated in: Oregon agricultural experiment station, station bulletin. HI 71849

Research bulletin, Oregon state board of forestry. Corvallis, OR. Vols. 1-10, 1949-56. Res. Bull. Oregon State Board Forest. Superseded by: Research bulletin, forest research laboratory. Corvallis. HI 71850

Research bulletin of the Panjab university. Science (section). Chandigarh. Vols. 1-9, 1950-58; n.s. vol. 10+, 1959+. Res. Bull. Panjab Univ. Sci. HI 71851

Research bulletin, Pennsylvania department of forests and waters. Harrisburg, PA. Nos. 1-4, 1930. Res. Bull. Pennsylvania Dept. Forests Waters. HI 71852

Research bulletin of plant protection service, Japan. [Shokubutsu boekisho chosa kenkyu hokoku.] Kanagawa, Japan. Res. Bull. Pl. Protect. Serv. Japan. See B–P–H 774/23. HI 58674

Research bulletin Porto Rico university agricultural experiment station. Rio Piedras, PR. Nos. 1-5, 1941-45. Res. Bull. Porto Rico Univ. Agric. Exp. Sta. HI 71853

Research bulletin, Puerto Rico university agricultural experiment station = Research bulletin Porto Rico university agricultural experiment station. Rio Piedras, PR.

Research bulletin of the Punjab university. New series: Science. Chandigarh, India. Res. Bull. Punjab Univ., New Ser., Sci. See B–P–H 774/24. HI 58675

Research bulletin, Saitama agricultural experiment station. Saitama. No. 1+, 1951+. Res. Bull. Saitama Agric. Exp. Sta. HI 62075

Research bulletin. Saito Ho-on Kai museum. [Saito Ho-on-kai gakujutsu kenkyu hokoku.] Sendai, Japan. Res. Bull. Saito Ho-On Kai Mus. See B–P–H 775/1. HI 58676

Research bulletin, school of agriculture and forestry, polytechnic institute, Ruston. Ruston, LA. 1967-? Res. Bull. School Agric. Forest. Polytechn. Inst. Ruston. Superseded by: Research bulletin, division of life sciences research, Louisiana tech university. HI 71854

Research bulletin of the Seoul national university forests = Bulletin, Seoul national university forests. Seoul.

Research bulletin, sisal experimental station, Ngomeni. Ngomeni? 1949+. Res. Bull. Sisal Exp. Sta. Ngomeni. HI 71855

Research bulletin, state university of Oklahoma. Norman, OK. Res. Bull. State Univ. Oklahoma See B–P–H 775/2. HI 58677

Research bulletin, university of Massachusetts agricultural experiment station. Amherst, MA. No. 591+, 1971+. Res. Bull. Univ. Massachusetts Agric. Exp. Sta. Preceded by: Massachusetts agricultural experiment station bulletin. HI 71856

Research bulletin, west of Scotland college of agriculture. Glasgow. No. 1+, 1928+. Res. Bull. W. Scotland Coll. Agric. HI 71857

Research bulletin, Wisconsin (state) agricultural experiment station = Wisconsin (state) agricultural experiment station. Research bulletin. Madison, WI. Wisconsin Agric. Exp. Sta. Res. Bull. See B–P–H 976/3.

Research circular, forestry commission. London. 1934-36. Res. Circ. Forest. Commiss. HI 71858

Research circular, Ohio agricultural research and development center. Wooster, OH. 1965?+. Res. Circ. Ohio Agric. Res. Devlopm. Center. Preceded by: Ohio agricultural experiment station research circular. HI 71859

Research circular, Pennsylvania department of forests and waters. Harrisburg, PA. Nos. 1-3, 1930. Res. Circ. Pennsylvania Dept. Forests Waters. HI 65219

Research communications, institute for fermentation, Osaka. [Hakko kenkyusho hokoku.] Osaka. No. ?-7+, ?-1975+. Res. Commun. Inst. Ferment. Osaka. Preceded by: Report, institute for fermentation research, Osaka. HI 71860

Research data, Mie university forests. [Mie daigaku, nogakubu enshurin shiryo.] Tsu. No. 1+, 1968+. Res. Data Mie Univ. Forests. HI 71861

Research and development in agriculture. Harlow. Vol. 1+, 1984+. Res. Developm. Agric. HI 75161

Research and development papers, forestry commission. London, Edinburgh. No. 31?+, 1966+. Res. Developm. Pap. Forest. Commiss. Preceded by: Research branch papers, forestry commission. HI 71862

Research and development publication, west of Scotland agricultural college. Glasgow. No. 1+, 1977+. Res. Developm. Publ. W Scotland Agric. Coll. HI 71863

Research division bulletin, Virginia polytechnic institute. Blacksburg, VA. Vol. 1+, 1967+. Res. Div. Bull. Virginia Polytechn. Inst. HI 71864

Research division monograph, Virginia polytechnic institute. Blacksburg, VA. No. ?-2+, ?-1970+. Res. Div. Monogr. Virginia Polytechn. Inst. HI 73752

Research and experimental record, ministry of agriculture, Northern Ireland. Belfast. Vols. 1-12, 1951-62. Res. Exp. Rec. Minist. Agric. Northern Ireland. Preceded by: Journal of the ministry of agriculture for Northern Ireland. Superseded by: Report (Annual) on research and technical work of the department (ministry) of agriculture for Northern Ireland. HI 71865

Research for farmers. Ottawa. Vols. 1-11(1), 1956-66. Res. Farmers. Superseded by: Canada agriculture. HI 71866

Research and farming; progress report. Raleigh, NC. Vols. 1-?, 1942-80? Res. & Farming. Superseded by: Research perspectives. Raleigh, NC. 4-3573-2. HI 58663

Research and farming quarterly, North Carolina agricultural experiment station = North Carolina agricultural experiment station research and farming quarterly. Raleigh, NC. North Carolina Agric. Exp. Sta. Res. Farming Quart. See B–P–H 663/10.

Research highlights, international center for agricultural research in dry areas = I C A R D A research highlights. Aleppo.

Research information digest. Asheville, NC. ?-1975? Res Inform. Digest. Superseded by: Forest research in the southeast. HI 71867

Research information note, forestry commission research and development division. Farnham. ?-1978+. Res. Inform. Note Forest. Commiss Res. Developm. Div. HI 71868

Research investigations and field trials, north of Scotland college of agriculture. Aberdeen. 1965/66+, 1966+. Res. Invest. Field Trials N. Scotland Coll. Agric. Preceded by: Report on investigations and research, north of Scotland college of agriculture and Report on field trials and observations, north of Scotland college of agriculture. HI 71869

Research journal, agricultural experiment station, university of Wyoming. Laramie, WY. Vol. 1+, 1966+. Res. J. Agric. Exp. Sta. Univ. Wyoming. HI 71870

Research journal of the Hindi science academy. Allahabad, India. Res. J. Hindi Sci. Acad. See B–P–H 775/7. HI 58678

Research journal, Jawaharlal Nehru Krishi Vishwa Vidylaya = J N K V V research journal. Jabalpur.

Research journal of Mahatma Phule agricultural university. Poona. Vols. 1-6, 1970-75. Res. J. Mahatma Phule Agric. Univ. Superseded by: Journal of Maharashtra agricultural universities. HI 71871

Research laboratory notes, western pine association. Portland, OR. 1932-41. Res. Lab. Notes W. Pine Assoc. HI 71872

Research leaflet, forest research institute, New Zealand. Rotorua. 1963+. Res. Leafl. Res. Inst. New Zealand. HI 71873

Research leaflet, New Zealand forest research institute. Rotorua, New Zealand. Res. Leafl. New Zealand Forest Res. Inst. See B–P–H 775/8. HI 58679

Research leaflet, Oregon state college school of forestry. Corvallis, OR. 1941+. Res. Leafl. Oregon State Coll. School Forest. HI 71874

Research letter, horticultural research institute, Washington. Washington, DC. ?-1982+. Res. Lett. Hort. Res. Inst. Washington. HI 71875

Research in Melanesia. Port Moresby. Vol. 1+, 1975+. Res. Melanesia. HI 71876

Research memoirs, cotton research corporation. Nos. 64-92, 1967-74. Res Mem. Cotton Res. Corp. Preceded by: Empire cotton growing corporation, research memoirs. HI 71877

Research in microbiology. Amsterdam, New York. Vol. 140+, 1989+. Res. Microbiol. HI 71878

Research in molecular biology. Wiesbaden, Mainz. Vol. 1+, 1973+. Res. Molec. Biol. HI 71879

Research monographs, Texas agricultural experiment station. College Station, TX. Vol. 1+, 1973+. Res. Monogr. Texas Agric. Exp. Sta. HI 71880

Research news, Arkansas agricultural experiment station. Fayetteville, AR. Vol. 15(5)+, 1989+. Res. News Arkansas Agric. Exp. Sta. Preceded by: Research newsletter, agricultural experiment station, university of Arkansas. HI 72035

Research news, department of forestry and rural development, Canada. Ottawa. Vols. 1-16(2), 1958-73. Res. News Dept. Forest. Rural Developm. Canada. HI 71881

Research news, institute for environmental studies. Toronto. Vol. ?-7, ?-1981. Res. News Inst. Environm. Stud. Superseded by: News, institute for environmental studies, Toronto. HI 71882

Research news, southeastern forest experiment station. Asheville, NC. Nos. 1-14, 1948-51. Res. News SouthE. Forest Exp. Sta. HI 71883

Research news, university of Michigan. Ann Arbor, MI. Vol. 9+, 196?+. Res. News Univ. Michigan. Preceded by: Engineering news [not entered]. HI 71884

Research newsletter, agricultural experiment station,

university of Arkansas. Fayetteville, AR. Vols. 8(2)-15(4), 1982-89. Res. Newslett. Agric. Exp. Sta. Univ. Arkansas. Preceded by: News on agricultural research in Arkansas. Superseded by: Research news, Arkansas agricultural experiment station. HI 75141

Research newsletter, Arkansas agricultural experiment station = Research newsletter, agricultural experiment station, university of Arkansas. Fayetteville, AR.

Research newsletter, forest research laboratory. Sault Ste. Marie = Research newsletter, Great Lakes forest research centre. Sault Ste. Marie.

Research newsletter, Great Lakes forest research centre. Sault Ste. Marie. Vols. 1-2(3), 1971-72. Res. Newslett. Great Lakes Forest Res. Centre. Superseded by: Forestry research newsletter, Great Lakes forest research centre. HI 71885

Research note, see Research notes

Research notes. [Agricultural research science, U.S. department of agriculture and the agricultural experiment station, university of Puerto Rico.] Río Piedras, PR. Res. Notes. See B–P–H 775/11. HI 58680

Research notes, British Columbia forest service. Victoria, B.C. No. 1+, 1937+. Res. Notes British Columbia Forest Serv. HI 71886

Research notes C S, United States forest service. Columbus, OH Nos. 1-45, 1963-65. Res. Notes C. S., U.S. Forest Serv. Preceded by: Station note, central states forest experiment station. Superseded by: Research notes NC, United States forest service. HI 71887

Research notes, California forest and range experiment station = Forest research notes, California forest and range experiment station. Berkeley, CA.

Research notes, canadian forestry service. Ottawa. Vol. 1+, 1981+. Res. Notes Canad. Forest. Serv. Preceded by: Bi-monthly research notes, forestry service, Canada. HI 71888

Research notes, Cansapscal forest research station. Quebec. 1954+. Res. Notes Cansapscal Forest Res. Sta. HI 71889

Research notes, central states forest experiment station = Research notes C S, United States forest service. Columbus, OH.

Research notes, Colorado school of forestry and range management. Fort Collins, CO. 1955+. Res. Notes Colorado School Forest. Range Managem. HI 71890

Research notes, cooperative forestry research unit, university of Maine at Orono. Orono, ME. No. ?-8+, ?-1982?+. Res. Notes Coop. Forest. Res. Unit Univ. Maine Orono. HI 71891

Research notes, department of forest yield research, royal college of forestry = Rapport och uppsatser, institutionen för skogsproduction, skogshögskolan. Uppsala.

Research notes, department of forestry, Nanking national central university. Dendrological series. Nanking. 1947+. Res. Notes Dept. Forest. Nanking Natl. Centr. Univ., Dendrol. Ser. HI 71892

Research notes, department of reforestation, royal college of forestry = Rapport och uppsatser, institutionen för skogsföryngring, skogshögskolan. Stockholm.

Research notes, department of silviculture, royal college of forestry = Rapport och uppsatser, institutionen för skogsskötsel, skogshögskolan. Stockholm.

Research notes, division of forest management, forestry commission, New South Wales. Sydney, N.S.W. 1958+. Res. Notes Div. Forest Managem. Forest. Commiss. New South Wales. HI 71893

Research notes, division of forest management, New South Wales = Research notes, forestry commission of New South Wales. Sydney, N.S.W.

Research notes, division of forest research, forest department, Zambia. Kitwe. No. 1+, 1968+. Res. Notes Div. Forest Res. Forest Dept. Zambia. HI 71894

Research notes F P L, United States forest service. Madison, WI. No. 01+, 1963+. Res. Notes F. P. L., U.S. Forest Serv. HI 71895

Research notes, faculty of forestry, university of British Columbia. Vancouver, B.C. 1950+. Res. Notes Fac. Forest. Univ. British Columbia. HI 71896

Research notes, Ford forestry center, Michigan technological university. L'Anse, MI. 1966+. Res. Notes Ford Forest. Center Michigan Technol. Univ. HI 71897

Research notes, forest experiment station, Oregon state college. Corvallis, OR. 1955+. Res. Notes Forest Exp. Sta. Oregon State Coll. HI 71898

Research notes, forest product laboratories, Canada. Ottawa. Nos. 1-?, 1924-? Res. Notes Forest Prod. Lab. Canada. Superseded by: Silvicultural research note, forestry branch, Canada. HI 71899

Research notes, forest products laboratory. Madison, WI = Research notes F P L, United States forest service. Madison, WI.

Research notes, forest research division, bureau of forestry, Philippines. Manila. 1964+. Res. Notes Forest Res. Div. Bur. Forest. Philippines. HI 71900

Research notes, forest research division, department of agriculture and natural resources, Philippines = Research notes, forest research division, bureau of forestry, Philippines. Manila.

Research notes, forest research division, Oregon state college. Corvallis, OR. 1959+. Res. Notes Forest Res. Div. Oregon State Coll. HI 71901

Research notes, forest service, British Columbia. Victoria, B.C. No. 1+, 1937+. Res. Notes Forest Serv. British Columbia. HI 71902

Research notes, forest, wildlife and range experiment station, university of Idaho. Moscow, ID. Nos. 1-21, 1951-64. Res. Notes Forest Wildlife Range Exp. Sta. Univ. Idaho. Superseded by: Station note, forest, wildlife and range experiment station, university of Idaho and Station paper, forest, wildlife and range experiment station, university of Idaho. HI 71903

Research notes, forestry commission of New South Wales. Sydney, N.S.W. No. 1+, 1958+. Res. Notes Forest. Commiss. New South Wales. HI 71904

Research notes, forestry institute; national central university, Nanking. Dendrological series. [Kuo li chung yang ta hsüeh shen lin hsüeh yen chiu so yen chiu pao-kao. Shu mu hsüeh.] Nanking. Res. Notes Forest. Inst. Natl. Centr. Univ. Nanking, Dendrol. Ser. See B–P–H 775/14. HI 58681

Research notes, forestry, wildlife and range experiment station, university of Idaho = Research notes, forest, wildlife and range experiment station, university of Idaho. Moscow, ID.

Research notes I N T, United States forest service. Ogden, UT. No. 1+, 1963+. Res. Notes I. N. T., U.S. Forest Serv. Preceded by: Research notes, intermountain forest and range experiment station. HI 71905

Research notes I T F, United States forest service. Rio Piedras, PR. No. 1+, 1964+. Res. Notes I. T. F., U.S. Forest Serv. Preceded by: Tropical forest note, institute of tropical forestry. Rio Piedras, PR. HI 60207

Research notes, institute of tropical forestry. Rio Piedras, PR = Research notes I T F, United States forest service. Rio Piedras, PR.

Research notes, institute of tropical forestry, United States forest service. Rio Piedras, PR = Research notes I T F, United States forest service. Río Piedras, PR.

Research notes, intermountain forest and range experiment station. Ogden, UT. Nos. 1-104, 1951-62. Res. Notes Intermount. Forest Range Exp. Sta. Superseded by: Research notes I N T, United States forest service. HI 71907

Research notes L S, United States forest service. St. Paul, MN. Nos. 1-74, 1963-65. Res. Note L. S. U.S. Forest Serv. Preceded by: Technical notes, lake states forest experiment station. Superseded by: Research note N C, United States forest service. HI 71908

Research notes, lake states forest experiment station = Research notes L S, United States forest service. St. Paul, MN.

Research notes N C, United States forest service. St. Paul, MN. No. 1+, 1966+. Res. Notes N. C., U.S. Forest Serv. Preceded by: Research notes C S, United States forest service and Research notes L S, United States forest service. HI 71909

Research notes N E, United States forest service. Upper Darby, PA. No. 1+, 1963+. Res. Notes N. E., U.S. Forest Serv. Preceded by: Forest research notes, northeastern forest experiment station. HI 71910

Research notes N O R, United States forest service. Juneau, AK. No. 1+, 1963+. Res. Notes N. O. R., U.S. Forest Serv. Preceded by: Technical notes, northern forest experiment station. HI 71911

Research notes, north central forest experiment station = Research notes N C, United States forest service. St. Paul, MN.

Research notes, northeastern forest experiment station = Research notes N E, United States forest service. Upper Darby, PA.

Research notes, northern forest experiment station = Research notes N O R, United States forest service. Juneau, AK.

Research notes, northern Rocky Mountain forest experiment station. Missoula, MT. Nos. 1-137?, 1939-54. Res. Notes N. Rocky Mountain Forest Exp. Sta. Preceded by: Applied forestry notes, northern Rocky Mountain experiment station. HI 71912

Research notes, Oregon state board of forestry. Salem, OR. 1948+. Res. Notes Oregon State Board Forest. HI 71913

Research notes P N W, United States forest service. Portland, OR. No. 1+, 1963+. Res. Notes P. N. W., U.S. Forest Serv. Preceded by: Research notes, Pacific northwest forest and range experiment station. HI 71914

Research notes P S W, United States forest service. Berkeley, CA. Nos. 1-19, 1963-66. Res. Notes P. S. W., U.S. Forest Serv. Preceded by: Research notes, Pacific southwest forest and range experiment station. HI 71915

Research notes, Pacific northwest forest and range experiment station. Portland, OR. Nos. 46-327, 1948-62. Res. Notes Pacific NorthW. Forest Range Exp. Sta. Preceded by: Forest research notes, Pacific northwest forest and range experiment station. Superseded by: Research notes P N W, United States forest service. HI 71916

Research notes, Pacific southwest forest and range experiment station. Berkeley, CA. Nos. 148-211,

1959?-62. Res. Notes Pacific SouthW. Forest Range Exp. Sta. Preceded by: Forest research notes, California forest and range experiment station. Superseded by: Research notes P S W, United States forest service. HI 71917

Research notes, Queensland forest service. Brisbane, Qld. 1954+. Res. Notes Queensland Forest Serv. HI 71918

Research notes R M, United states forest service. Fort Collins, CO. No. 1+, 1963+. Res. Notes R. M., U.S. Forest Serv. Preceded by: Research note, Rocky Mountain forest and range experiment station. HI 71919

Research notes, Rocky Mountain forest and range experiment station. Fort Collins, CO. Nos. 1-84, 1947-62. Res. Notes Rocky Mountain Forest Range Exp. Sta. Superseded by: Research notes R M, United states forest service. HI 71920

Research notes S E, United States forest service. Asheville, NC. No. 1+, 1963+. Res. Notes S. E., U.S. Forest Serv. Preceded by: Research notes, southeastern forest experiment station. HI 71921

Research note S O, United States forest service. New Orleans, LA. No. 1+, 1963+. Res. Notes S. O., U.S. Forest Serv. Preceded by: Southern forestry notes, southern forest experiment station. HI 71922

Research notes, school of forestry, Florida university. Gainesville, FL. 1953+. Res. Notes School Forest. Florida Univ. HI 71923

Research notes, southeastern forest experiment station. Asheville, NC. Nos. 1-183, 1952-62. Res. Notes SouthE. Forest Exp. Sta. Superseded by: Research notes S E, United States forest service. HI 71924

Research notes, southern forest experiment station = Research note S O, United States forest service. New Orleans, LA.

Research notes, southwestern forest experiment station. Tucson, AZ. Nos. 1-123?, 1936-53. Res. Notes SouthW. Forest Exp. Sta. HI 71925

Research notes, Texas forest service. College Station, TX. 1952-68. Res. Notes Texas Forest Serv. Superseded by: Circular, Texas forest service. HI 71926

Research notes, U B C forest club. Vancouver, BC. 1950+. Res. Notes U. B. C. Forest Club. HI 71927

Research notes W O, United States forest service. Washington, DC. No. ?-4+, ?-1964+. Res. Notes W. O., U.S. Forest Serv. HI 71928

Research notes, western australian herbarium. South Perth, W.A. Nos. 1-12, 1978-86. Res. Notes W. Austral. Herb. Superseded by: Kingia. HI 71929

Research notes, western pine association = Research laboratory notes, western pine association. Portland, OR.

Research pamphlet, division of forest research, forest department, Zambia. Kitwe. No. ?-15+, ?-1968+. Res. Pam. Div. Forest Res. Forest Dept. Zambia. HI 71930

Research pamphlet, faculty of agriculture, university of Aleppo. Aleppo. No. 1+, 1971+. Res. Pam. Fac. Agric. Univ. Aleppo. HI 71931

Research pamphlet, forest research institute, Federation of Malaya. Kepong. No. 1+, 1953+. Res. Pam. Forest Res. Inst. Federation of Malaya. HI 71932

Research pamphlet, forest research institute, West Malaysia = Research pamphlet, forest research institute, Federation of Malaya. Kepong.

Research paper, see Research papers

Research papers C S, United States forest service. Columbus, OH. Nos. 1-21, 1963-65. Res. Pap. C. S., U.S. Forest Serv. Preceded by: Technical paper, central states forest experiment station. Superseded by: Research paper N C, United States forest service. HI 71933

Research papers, central states forest experiment station = Research papers C S, United States forest service. Columbus, OH.

Research paper, department of forestry, Queensland. Brisbane, Qld. 1971+. Res. Pap. Dept. Forest. Queensland. HI 71934

Research papers, division of agricultural biochemistry. University of Minnesota. St. Paul, MN. Res. Pap. Div. Agric. Biochem. Univ. Minnesota. See B–P–H 775/15. HI 58682

Research papers F P L, United States forest service. Madison, WI. No. 1+, 1963+. Res. Pap. F. P. L., U.S. Forest Serv. HI 71935

Research papers, faculty of forestry, university of British Columbia. Vancouver, B.C. No. 1+, 1952+. Res. Pap. Fac. Forest. Univ. British Columbia. HI 71936

Research papers, federal department of forest research, Nigeria. Forest series. Ibadan. No. 1-33?, 1971-76? Res. Pap. Fed. Dept. Forest Res. Nigeria Forest Ser. Superseded by: Research papers, forestry research institute of Nigeria. Forestry series. HI 71937

Research papers, federal department of forest research, Nigeria. Savanna series. Ibadan. No. ?-26+, ?-1973+. Res. Pap. Fed. Dept. Forest Res. Nigeria, Savanna Ser. HI 71938

Research papers, forest products laboratory, United States forest service = Research papers F P L, United States forest service. Madison, WI.

Research papers, forest research laboratory. Corvallis,

OR. 1965+. Res. Pap. Forest Res. Lab. HI 71939

Research papers, forestry commission of New South Wales. Beecroft, N.S.W. No. 1+, 1982+. Res. Pap. Forest. Commiss. New South Wales. HI 71940

Research papers, forestry research institute of Nigeria. Forestry series. Ibadan. No. 34+, 1977+. Res. Pap. Forest. Res. Inst. Nigeria, Forest. Ser. Preceded by: Research papers, federal department of forest research, Nigeria. Forest series. HI 71941

Research papers, forests department, Western Australia. Perth, W.A. No. 1+, 1971+. Res. Pap. Forests Dept. Western Australia. HI 71942

Research papers I N T, United States forest service. Ogden, UT. No. 1+, 1963+. Res. Pap. I. N. T., U.S. Forest Service. Preceded by: Research papers, intermountain forest and range experiment station, United States forest service. HI 71943

Research papers I T F, United States forest service. Rio Piedras, PR. No. 1+, 1964+. Res. Pap. I. T. F., U.S. Forest Serv. HI 71944

Research papers, institute of tropical forestry = Research papers I T F, United States forest service.

Research papers, intermountain forest and range experiment station. Ogden, UT. Nos. 1-71?, 1943-62. Res. Pap. Intermount. Forest Range Exp. Sta. Superseded by: Research paper I N T, United States forest service. HI 71945

Research papers L S, United States forest service. St. Paul, MN. Nos. 1-21, 1963-65. Res. Pap. L. S., U.S. Forest Serv. Preceded by: Station papers, lake states forest experiment station. Superseded by: Research paper N C, United States forest service. HI 71946

Research papers, lake states forest experiment station = Research papers L S, United States forest service. St. Paul, MN.

Research papers N C, United States forest service. St. Paul, MN. No. 1+, 1966+. Res. Pap. N. C., U.S. Forest Serv. Preceded by: Research papers C S, United States forest service and Research papers L S, United States forest service. HI 71947

Research papers N E, United States forest service. Upper Darby, PA. No. 1+, 1963+. Res. Pap. N. E., U.S. Forest Serv. Preceded by: Station notes, northeastern forest experiment station and Station papers, north eastern forest experiment station. HI 71948

Research papers N O R, United States forest service. Juneau, AK. Nos. 1-3, 1964-67. Res. Pap. N. O. R., U.S. Forest Service. HI 71949

Research papers, north central forest experimental station = Research papers N C, United States forest service. St. Paul, MN.

Research papers, north eastern forest experiment station = Research papers N E, United States forest service. Upper Darby, PA.

Research papers, northern forest experiment station = Research papers N O R, United States forest service. Juneau, AK.

Research papers, Oregon agricultural experiment station, forest research division = Research papers, forest research laboratory. Corvallis, OR.

Research papers P N W, United States forest service. Portland, OR. No. 1+, 1963+. Res. Pap. P. N. W., U.S. Forest Serv. Preceded by: Research papers, Pacific northwest forest and range experiment station. HI 71950

Research papers P S W, United States forest service. Berkeley, CA. No. 1+, 1963+. Res. Pap. P. S. W., U.S. Forest Serv. Preceded by: Technical papers, Pacific southwest forest and range experiment station. HI 71951

Research papers, Pacific northwest forest and range experiment station. Portland, OR. Nos. 46-221, 1948-62. Res. Pap. Pacific NorthW. Forest Range Exp. Sta. Preceded by: Forest research notes, Pacific northwest forest and range experiment station. Superseded by: Research papers P N W, United States forest service. HI 71952

Research papers, Pacific southwest forest and range experiment station = Research papers P S W, United States forest service. Berkeley, CA.

Research papers, Pennsylvania state forest school. State College, PA. 1940+. Res. Pap. Pennsylvania State Forest School. HI 71953

Research papers R M, United States forest service. Fort Collins, CO. No. 1+, 1963+. Res. Pap. R. M., U.S. Forest Serv. Preceded by: Station papers, Rocky Mountain forest and range experiment station. HI 71954

Research papers, Rocky Mountain forest and range experiment station = Research papers R M, United States forest service. Fort Collins, CO.

Research papers S E, United States forest service. Asheville, NC. No. 1+, 1963+. Res. Pap. S. E., U.S. Forest Serv. Preceded by: Station papers, southeastern forest experiment station. HI 71955

Research papers S O, United States forest service. New Orleans, LA. No. 1+, 1963+. Res. Pap. S. O., U.S. Forest Serv. Preceded by: Occasional papers, southern forest experiment station. HI 71956

Research paper series, international rice research institute = I R R I research paper series. Manila.

Research papers, southeastern forest experiment station = Research papers S E, United States forest service. Asheville, NC.

Research papers, southern forest experiment station = Research papers S O, United States forest service. New Orleans, LA.

Research papers, southern forest experiment station = Research papers W O, United States forest service. Washington, DC.

Research papers W O, United States forest service. Washington, DC. No. 1+, 1964+. Res. Pap. W. O., U.S. Forest Serv. HI 71957

Research perspectives. Raleigh, NC. Vol. 1+, 1981+. Res. Perspect. Preceded by: Research and farming; progress report. Raleigh, NC. HI 71958

Research on plant diseases = Plant disease research. Kyoto.

Research proceedings of the marine laboratory. Athens = Thalassina epistemonika phylla. Athens. Thalassina Epistemonika Phylla. See B–P–H 873/11.

Research and progress; two-monthly review of german science. [English edition of: Forschungen und Fortschritte.] Berlin. Vols. 1-5?, 1935-39. Res. Progr. HI 71959

Research progress, Pacific southwest forest and range experiment station. Berkeley, CA. 19??+. Res. Progr. Pacific SouthW. Forest Range Exp. Sta. Preceded by: Report, Pacific southwest forest and range experiment station. HI 71960

Research progress report, agricultural experiment station, university of Illinois at Urbana-Champaign. Urbana, IL. 1888+. Res. Progr. Rep. Agric. Exp. Sta. Univ. Illinois Urbana-Champaign. HI 71961

Research progress report, australian institute of marine science. Townsville, Qld. 1975/76+. Res. Progr. Rep. Austral. Inst. Mar. Sci. HI 71962

Research progress report, Indiana experiment station. Lafayette, IN. Nos. 1-417, 1962-73. Res. Progr. Rep. Indiana Exp. Sta. HI 71963

Research progress report, Montana state college. Agricultural experiment station = Montana state college. Agricultural experiment station. Research progress report. Bozeman, MT. Montana Agric. Exp. Sta. Res. Progr. Rep. See B–P–H 618/16.

Research progress report, national Tokai-kinki agricultural experiment station = National Tokai-kinki agricultural experiment station. Research progress report. Tsu, Japan. Natl. Tokai-Kinki Agric. Exp. Sta. Res. Progr. Rep. See B–P–H 631/15.

Research progress report, Tokai-kinki national agricultural experiment station. [Norin-sho tokai kinki nogyo shikenjo kenkyu sokuho.] Mie, Japan. Res. Progr. Rep. Tokai-Kinki Natl. Agric. Exp. Sta. See B–P–H 775/18. HI 58684

Research progress report, western society of weed science. Reno, NV, Logan, UT. 1952+. Res. Progr. Rep. W. Soc. Weed Sci. HI 71964

Research progress report, western weed control conference = Research progress report, western society of weed science. Reno, NV, Logan, UT.

Research publications, Kansas agricultural experiment station. Manhattan, KS. No. 162+, 1970+. Res. Publ. Kansas Agric. Exp. Sta. Preceded by: Kansas state agricultural college, agricultural experiment station technical bulletin. HI 71965

Research publications, Texas technological college. Lubbock, TX. No. 3+, 1942+. Res. Publ. Texas Technol. Coll. Preceded by: Bulletin, Texas technological college. Scientific series. HI 71966

Research publications; university of Hawaii. Honolulu, HI. Res. Publ. Univ. Hawaii. See B–P–H 775/20. HI 58685

Research publications of the university of Minnesota; studies in the biological sciences. Minneapolis, MN. Res. Publ. Univ. Minnesota, Stud. Biol. Sci. See B–P–H 775/22. HI 58686

Research record, Malawi forest research institute. Dedza. Nos. ?-31-56-?, ?-1969-72-? Res. Rec. Malawi Forest Res. Inst. HI 71967

Research report, agricultural experiment station, Burlington. Burlington, VT. No. ?-105+, 1968-79+. Res. Rep. Agric. Exp. Sta. Burlington. Preceded by: Miscellaneous publications series, agricultural experiment station, Burlington. HI 71968

Research report, agricultural experiment station, Georgia = Research report, college of agriculture experiment stations, university of Georgia. Athens, GA.

Research report, agricultural experiment station, Storrs = Research report, Storrs agricultural experiment station. Storrs, CT.

Research report, agricultural experiment station, university of Florida. Gainesville, FL. No. 1+, 1956+. Res. Rep. Agric. Exp. Sta. Univ. Florida. HI 58687

Research report, agricultural experiment station, Utah. Logan, UT. 1972+. Res. Rep. Agric. Exp. Sta. Utah. HI 71969

Research report, agricultural experiment stations, Georgia = Research report, college of agriculture experiment stations, university of Georgia. Athens, GA.

Research report of the agricultural research institute of Ontario. Toronto. 1972/73+, 1973+. Res. Rep. Agric. Res. Inst. Ontario. Preceded by: Report (Annual), agricultural research institute of Ontario. HI 71970

Research report of the Anhwei university college of agriculture. [Sheng li An huei ta shhsüeh nung hsüeh yüan yen chiu pao-kao.] [China]. Res. Rep. Anhwei Univ. Coll. Agric. See B–P–H 776/2. HI 58688

Research report, applied forestry research institute = A F R I research report. Syracuse, NY.

Research report, Balsgård fruit breeding institute. Fjälkestad. 1942?+. Res. Rep. Balsgård Fruit Breed. Inst. HI 71971

Research report and bibliography, american museum of natural history. New York. 1974/75-76/77, 1975?-77? Res. Rep. Bibliogr. Amer. Mus. Nat. Hist. Superseded by: Report of the scientific and education departments, american museum of natural history. HI 71972

Research report, biological and chemical research institute, department of agriculture, New South Wales. Sydney, N.S.W. 1976+. Res. Rep. Biol. Chem. Res. Inst. Dept. Agric. New South Wales. HI 71973

Research report, biosphere reserves = Biosphere reserves research report.

Research report, biosystematics research institute. Ottawa. 1971/75+, 1975+. Res. Rep. Biosyst. Res. Inst. Preceded by: Research report of the plant research institute, Ottawa. HI 71974

Research reports, biotechnical faculty, University Edvard Kardelj of Ljubljana = Zbornik biotehniske fakultete univerze Edvarda Kardelja v Ljubljani. Kmetijstvo.

Research report, Canada weed committee, eastern section. Ottawa. 14th-22nd, 1969-77. Res. Rep. Canada Weed Committee E. Sect. Preceded by: Research report, national weed committee, Canada, eastern section. Superseded by: Research report, expert committee on weeds, Canada, eastern section. HI 71975

Research report, Canada weed committee, western section. 16th-24th, 1968-77. Res. Rep. Canada Weed Committee W. Sect. Preceded by: Research report, national weed committee, Canada, western section. Superseded by: Research report, expert committee on weeds, western Canada. HI 71976

Research report, chemistry and biology research institute, department of agriculture, Canada = Research report, microbiology research institute, department of agriculture, Canada. Ottawa.

Research report, coffee research station, Lyamungu. Arusha. 1960+. Res. Rep. Coffee Res. Sta. Lyamungu. Preceded by: Report (Annual), coffee research station, Lyamungu. HI 71977

Research report, college of agriculture experiment stations, university of Georgia. Athens, GA. No. 1+. 1967+. Res. Rep. Coll. Agric. Exp. Sta. Univ. Georgia. HI 71978

Research report of cultural properties of Kagoshima prefecture. [Kagoshima-ken bunkazai chosa hokoku.] Kagoshima, Japan. Res. Rep. Cult. Propert. Kagoshima Prefect. See B–P–H 776/4. HI 58689

Research report, department of agronomy and horticulture, university of Sydney. Sydney, N.S.W. No. 1+, 1972/73+, 1974+. Res. Rep. Dept. Agron. Hort. Univ. Sydney. HI 71979

Research report, department of horticulture, university of Nottingham. Sutton Bonington. 1954. Res. Rep. Dept. Hort. Univ. Nottingham. HI 71980

Research report, division of forest research, department of lands and forests, Ontario. Toronto. Nos. 1-90, 1945-69. Res. Rep. Div. Forest Res. Dept. Lands Forests Ontario. Superseded by: Research report, fish and wildlife research branch, ministry of natural resources, Ontario. HI 71981

Research report, expert committee on weeds, Canada, eastern section. Ottawa. 23rd+, 1978+. Res. Rep. Expert Committee Weeds Canada E. Sect. Preceded by: Research report, Canada weed committee, eastern section. HI 71982

Research report, expert committee on weeds, western Canada. Ottawa. 25th+, 1978+. Res. Rep. Expert Committee Weeds W. Canada. Preceded by: Research report, Canada weed committee, western section. HI 71983

Research report, faculty of general agriculture, university college, Dublin. Dublin. 1971+. Res. Rep. Fac. Gen. Agric. Univ. Coll. Dublin. Preceded by: Report, faculty of general agriculture, university college, Dublin. HI 71984

Research report, fish and wildlife research branch, ministry of natural resources, Ontario. Toronto. No. 92+, 1972+. Res. Rep. Fish Wildlife Res. Branch Minist. Nat. Resources Ontario. Preceded by: Research report, division of forest research, department of lands and forests, Ontario. HI 71985

Research report of foreign wood. 1972+. Res. Rep. Foreign Wood. HI 71986

Research report of the forest experiment station. [Imop sihom yon'gu pogo.] Seoul. Nos. 10-14, 1965-68. Res. Rep. Forest Exp. Sta. Preceded by: Bulletin of the forest experiment station, Seoul. Superseded by: Research report of the forest research institute, Seoul. HI 71987

Research report of the forest research institute, Seoul. [Imop sihomjang yongu pogo.] Seoul. No. 15+, 1968+. Res. Rep. Forest Res. Inst. Seoul. Preceded by: Research report of the forest experiment station, Seoul. HI 71988

Research report, genetics and plant breeding research institute. Ottawa. Nos. 1-?, 1960-64? Res. Rep. Genet. Pl. Breed. Res. Inst. Superseded by: Research report, Ottawa research station. HI 71989

Research report, Hebrew university of Jerusalem. Science, agriculture. Jerusalem. 1963/64+. Res. Rep.

Hebrew Univ. Jerusalem, Sci. Agric. HI 71990

Research report, horticulture, agricultural institute, Dublin. Dublin. 1971+, 1972+. Res. Rep. Hort. Agric. Inst. Dublin. Preceded by: Research report, horticulture and forestry division, agricultural institute. Dublin. HI 71991

Research report, horticulture and forestry division, agricultural institute, Dublin. Dublin. 1959/60-70, 1961-71. Res. Rep. Hort. Forest. Div. Agric. Inst. Foras Talúntais. Superseded by: Research report, horticulture, agricultural institute. Dublin. HI 71992

Research report; Hyogo agricultural college. [Hyogo nogyo tanki daigaku kenkyu shuroku.] Kobe, Japan. Res. Rep. Hyogo Agric. Coll. See B–P–H 776/5. HI 58690

Research report, imperial college of science and technology. London. 1971/74+, [1975?]+. Res. Rep. Imp. Coll. Sci. Technol. Preceded by: Research report, royal college of science, London. HI 71993

Research report of the institute of forest genetics. [Imop sihomjang yon'gu pogi.] Suwon. No. 1+, 1959+. Res. Rep. Inst. Forest Genet. HI 71994

Research report, Kanagawa prefectural museum. Natural history. Yokohama. No. 1+, 1970+. Res. Rep. Kanagawa Pref. Mus., Nat. Hist. HI 71995

Research report, Kasetsart university. Bangkok. 1971/72+, 1973?+. Res. Rep. Kasetsart Univ. Preceded by: Kasetsart university research activities. HI 71996

Research reports of the Kochi university. [Kochi daigaku gakujutsu kenkyu hokoku.] Kochi. Vols. 1(1)-8(26), 1952-59. Res. Rep. Kochi Univ. Superseded by: Research reports of the Kochi university. 2, natural science. HI 71997

Research reports of the Kochi university. 2, natural science. [Kochi daigaku gakujutsu kenkyu hokoku. Shizen kagaku.] Kochi. Vol. 9+, 1960+. Res. Rep. Kochi Univ., 2, Nat. Sci. Preceded by: Research reports of the Kochi university. HI 71998

Research report, Michigan agricultural experiment station = Michigan agricultural experiment station. Research report. East Lansing, MI. Michigan Agric. Exp. Sta. Res. Rep. See B–P–H 592/17.

Research report, microbiology research institute, department of agriculture, Canada. Ottawa. 1959/62+, 1962+. Res. Rep. Microbiol. Res. Inst. Dept. Agric. Canada. HI 71999

Research report, Mississipi agricultural and forestry experiment station. State College, MS. Vol. 1+, 1975+. Res. Rep. Mississipi Agric. Forest. Exp. Sta. HI 72000

Research report, molecular and cellular biology unit, commonwealth scientific and industrial research organisation. Sydney, N.S.W. 1976+. Res. Rep. Molec. Cell. Biol. Unit Commonw. Sci. Indust. Res. Organ. HI 72001

Research report of the Nagano horticultural experiment station. [Nagano-ken engei shikenjo shiken seiseki.] Susaka. Vol. ?-49+, ?-[1975?]+. Res. Rep. Nagano Hort. Exp. Sta. HI 72002

Research report, national geographic society. Washington, DC. 1963+. Res. Rep. Natl. Geogr. Soc. HI 72003

Research report, national weed committee, Canada, eastern section. 1st-13th, 1957-68. Res. Rep. Natl. Weed Committee Canada E. Sect. Superseded by: Research report, Canada weed committee, eastern section. HI 72004

Research report, national weed committee, Canada, western section. 1st-13th, 1954-67. Res. Rep. Natl. Weed Committee Canada W. Sect. Superseded by: Research report, Canada weed committee, western section. HI 72005

Research report, New Hampshire agricultural experiment station. Durham, NH. No. 1+, 1961+. Res. Rep. New Hampshire Agric. Exp. Sta. HI 72006

Research report, New Mexico agricultural experiment station = New Mexico agricultural experiment station, research report. New Mexico state college. University Park, NM. New Mexico Agric. Exp. Sta. Res. Rep. See B–P–H 654/9.

Research report, North Carolina agricultural experiment station = North Carolina agricultural experiment station research report. Raleigh, NC. North Carolina Agric. Exp. Sta. Res. Rep. See B–P–H 663/11.

Research report, North Carolina agricultural experiment station, department of agronomy = North Carolina agricultural experiment station, department of agronomyresearch report. Raleigh, NC. North Carolina Agric. Exp. Sta. Dept. Agron. Res. Rep. See B–P–H 663/8.

Research report, north central weed control conference. Champaign, IL. 1st+, 1944+. Res. Rep. N. Centr. Weed Control Conf. For 1944-47 included in: Proceedings, north central weed control conference. HI 72007

Research report, North Dakota agricultural experiment station = North Dakota agricultural experiment station research report. Fargo, ND. North Dakota Agric. Exp. Sta. Res. Rep. See B–P–H 664/14.

Research report of the office of rural development. [Nongso sihom yon'gu pogo.] Suwon. Vol. 1+, 1958+. Res. Rep. Off. Rural Developm. HI 72008

Research report, Oklahoma agricultural experiment station. Stillwater, OK. No. p657+, 1972+. Res. Rep.

Oklahoma Agric. Exp. Sta. Preceded by: Progress report, Oklahoma agricultural experiment station. HI 72009

Research report, Ottawa research station. Ottawa. 1961/66+ 1965?+. Res. Rep. Ottawa Res. Sta. Preceded by: Research report, genetics and plant breeding research institute, Ottawa. HI 72010

Research report, Papua New Guinea university of technology. Lae. ?-1981+. Res. Rep. Papua New Guinea Univ. Technol. HI 72011

Research report, plant pathology laboratory, department of agriculture, Canada. Edmonton, Alta. 1956/61, 1961? Res. Rep. Pl. Pathol. Lab. Dept. Agric. Canada. HI 72012

Research report of the plant research institute. Ottawa. 1959/63-68/70. 1964-72. Res. Rep. Pl. Res. Inst. Superseded by: Research report, biosystematics research institute, Ottawa. HI 72013

Research report, plant sciences and crop husbandry division, agricultural institute. Dublin. Vol. 1+, 1959/60+, 1961+. Res. rep. Pl. Sci. Div. Agric. Inst. Dublin. HI 72014

Research report, research division, Virginia polytechnic institute. Blacksburg, VA. No. 116+, 1967+. Res. Rep. Res. Div. Virginia Polytechn. Inst. Preceded by: Research report, Virginia agricultural experiment station. HI 72015

Research report, royal college of science. London. 1950/52-7?, 1952-7? Res. Rep. Roy. Coll. Sci. Superseded by: Research report, imperial college of science and technology. HI 72016

Research report, school of forest resources and conservation, university of Florida. Gainesville, FL. No. 20+, 19??+. Res. Rep. School Forest Resources Conservation Univ. Florida. Preceded by: Research report, school of forestry, Florida university. HI 72017

Research report, school of forestry, Florida university. Gainesville, FL. Nos. 1-19, 1952-? Res. Rep. School Forest. Florida Univ. Superseded by: Research report, school of forest resources and conservation, university of Florida. HI 72018

Research report series, Alabama agricultural experiment station = Alabama agricultural experiment station. Research report series. Auburn, AL.

Research report of the Shizuoka prefecture forestry experiment station. [Shizuoka-ken ringyo shikenjo kenkyu hokoku.] Hamakita. ?-1975+. Res. Rep. Shizuoka Pref. Forest. Exp. Sta. HI 75160

Research report, Smithsonian institution. Washington, DC. No. 1+, 1972+. Res. Rep. Smithsonian Inst. HI 72019

Research report, south african avocado growers' association. Izaneen. Vol. ?-2+, ?-1978+. Res. Rep. S. African Avocado Growers' Assoc. HI 72020

Research report, southern weed conference. 1962-69. Res. Rep. S. Weed Conf. Superseded by: Research report, southern weed science society. HI 72021

Research report, southern weed science society. Raleigh, NC, Auburn, AL. 1970+. Res. Rep. S. Weed Sci. Soc. Preceded by: Research report, southern weed conference. HI 72022

Research report, southwestern forest and range experiment station. Tucson, AZ. Nos. 1-12?, 1940-53. Res. Rep. SouthW. Forest Range Exp. Sta. HI 72023

Research report, Storrs agricultural experiment station. Storrs, CT. No. 1+, 1964+. Res. Rep. Storrs Agric. Exp. Sta. HI 72024

Research report of tea. [Chagyo kenkyu hokoku.] Tokyo, Shizuoka. No. 1+, 1953+. Res. Rep. Tea. HI 75158

Research report, timber development association. Biology. London. 1959+. Res. Rep. Timber Developm. Assoc., Biol. HI 72025

Research report of Tottori prefecture forestry experiment station. [Ringyo shikenjo shiken kenkyu hokoku, Tottori.] Tottori. 1958+. Res. Rep. Tottori Pref. Forest. Exp. Sta. HI 75150

Research report, university of Melbourne. Botany. Melbourne, Vic. 1974+. Res. Rep. Univ. Melbourne, Bot. HI 72026

Research report, university of Western Australia. Perth, W.A. 1972+. Res. Rep. Univ. Western Australia. Preceded by: Report on research, university of Western Australia. HI 72027

Research report, Virginia agricultural experiment station. Blacksburg, VA. Nos. 1-115, 1956-66. Res. Rep. Virginia Agric. Exp. Sta. Superseded by: Research report, research division, Virginia polytechnic institute. HI 72028

Research report, western pine association. Portland, OR. 1960+. Res. Rep. W. Pine Assoc. HI 72029

Research report, Wisconsin (state) agricultural experiment station = Wisconsin (state) agricultural experiment station, research report. Madison, WI. Wisconsin Agric. Exp. Sta. Res. Rep. See B–P–H 976/4.

Research reports digest, nature conservancy council. Banbury. 1979+. Res. Rep. Digest Nat. Conservancy Council. HI 72030

Research in review. Amherst, MA. Res. Rev. See B–P–H 776/13. HI 58691

Research review, agricultural research council = A R C research review. London.

Research review, commonwealth bureau of horticulture and plantation crops. East Malling. No. 1+, 19??+.

Res. Rev. Commonw. Bur. Hort. Plantation Crops. HI 72031

Research review, division of forest products, commonwealth scientific and industrial research organisation. 1969/70- 74/75? Res. Rev. Div. Forest Prod. Commonw. Sci. Industr. Res. Organ. Preceded by: Report (Annual), division of forest products, commonwealth scientific and industrial research organisation. HI 72032

Research review. East Malling commonwealth bureau of horticulture and plantation crops. East Malling, England. Res. Rev. East Malling Commonw. Bur. Hort. See B–P–H 776/14. HI 58692

Research review of Kyungpook national university. Taegu, Korea. Vol. ?-39+, ?-1985+. Res. Rev. Kyungpook Natl. Univ. HI 72033

Research review of the south african council for scientific and industrial research. Pretoria. Vols. 1-2, 1951-52. Res. Rev. S. African Council Sci. Industr. Res. Superseded by: C S I R research review. HI 72034

Research in review, university of Massachusetts agricultural experiment station = University of Massachusetts agricultural experiment station. Research in review. Amherst, MA. Univ. Mass. Agric. Exp. Sta. Res. Rev. See B–P–H 936/29.

Research series, Hawaii institute of tropical agriculture and human resources. Honolulu, HI. 1982?+. Res. Ser. Hawaii Inst. Trop. Agric. Human Resources. HI 72036

Research series, museum of systematic biology, university of California. Irvine, CA. Vol. 1+, 1968+. Res. Ser. Mus. Syst. Biol. Univ. Calif. HI 72037

Research studies of the state college of Washington. Pullman, WA. Vols. 1-?, 1929-? Res. Stud. State Coll. Wash. Superseded by: Research studies, Washington state university. 5-4446-1. HI 72038

Research studies, Washington state university. Pullman, WA. 19??+. Res. Stud. Washington State Univ. Preceded by: Research studies of the state college of Washington. HI 72039

Research studies, West Pakistan agricultural university = West Pakistan agricultural university. Research studies. Lyallpur, Pakistan. W. Pakistan Agric. Univ. Res. Stud. See B–P–H 970/10.

Research of tobacco leaf. [Hatabako kenkyu.] Tokyo. Res. Tobacco Leaf. See B–P–H 776/23. HI 58693

Research in virology. Amsterdam, New York. Vol. 140+, 1989+. Res. Virol. HI 72040

Researches on essential oils of the australian flora. Sydney, N.S.W. Vols. 1-3, 1948-53. Res. Essential Oils Austral. Fl. HI 72041

Researches on population ecology. Kyoto, Japan. Res. Populat. Ecol. See B–P–H 775/16. HI 58683

Reseñas cientificas de la sociedad española de historia natural. Madrid. Vols. 6?-10?, 1931-36. Reseñas Ci. Soc. Esp. Hist. Nat. Preceded by: Conferencias y reseñas cientificas de la real sociedad española de historia natural. HI 72042

Resource bulletin C S, United States forest service. Columbus, OH. No. 1+, 1963+. Resource Bull. C. S., U.S. Forest Serv. HI 72043

Resource bulletin F P L, United States forest service. Madison, WI. No. 4+, 1978+. Resource Bull. F. P. L., U.S. Forest Serv. Preceded by: Resource report, forest products laboratory, United States forest service. HI 75162

Resource bulletin I N T, United States forest service. Ogden, UT. No. 1+, 1963+. Resource Bull. I. N. T., U.S. Forest Serv. Preceded by: Forest survey release, intermountain forest and range experiment station. HI 72044

Resource bulletin N C, United States forest service. St. Paul, MN. No. 1+, 1966+. Resource Bull. N. C., U.S. Forest Serv. HI 72045

Resource bulletin N E, United States forest service. Upper Darby, PA. No. 1+, 1963+. Resource Bull. N. E., U.S. Forest Serv. HI 72046

Resource bulletin N O R, United States forest service. Juneau, AK. No. 1+, 1963+. Resource Bull. N. O. R., U.S. Forest Serv. HI 72047

Resource bulletin P N W, United States forest service. Portland, OR. No. 1+, 1963+. Resource Bull. P. N. W., U.S. Forest Serv. Preceded by: Forest survey report, Pacific northwest forest and range experiment station. HI 72048

Resource bulletin P S W, United States forest service. Berkeley, CA. No. 1+, 1965+. Resource Bull. P. S. W., U.S. Forest Serv. HI 72049

Resource bulletin R M, United States forest service. Fort Collins, CO. No. 1+, 1980+. Resource Bull. R. M., U.S. Forest Serv. HI 75163

Resource bulletin S E, United States forest service. Asheville, NC. No. 1+, 1963+. Resource Bull. S. E., U.S. Forest Serv. HI 72050

Resource bulletin S O, United States forest service. New Orleans, LA. No. 1+, 1963+. Resource Bull. S. O., U.S. Forest Serv. Preceded by: Forest survey release, southern forest experiment station. HI 72051

Resource and environmental review; current awareness journal. Woodbridge, CT. Vol. 1+, 1985+. Resource Environm. Rev. HI 72052

Resource report, forest products laboratory, United States forest service. Madison, WI. Nos. 1-3?, 1977-78? Resource Rep. Forest Prod. Lab., U.S. Forest Serv.

Superseded by: Resource bulletin F P L, United States forest service. HI 75230

Resources. Washington, DC. 1959+. Resources. HI 72053

Resources evaluation journal. Bethesda, MD. Vol. 1(1), 1982. Resources Eval. J. Superseded by: Renewable resources journal. HI 72054

Resultationes investigationis rerum naturalium montium Bakony = A Bakony természettudományi kutatásának eredménye kutatásának eredményei. Veszprem, Hungary. Bakony Term. Kutatás. Eredm. See B–P–H 167/9.

Résultats des recherches scientifiques entreprises au parc national suisse = Ergebnisse der wissenschaftlichen Untersuchung des schweizerischen Nationalparkes. Aarau, Switzerland. Ergebn. Wiss. Untersuch. Schweiz. Nationalparkes. See B–P–H 361/22.

Results of experiments, tobacco research board of Southern Rhodesia. Salisbury, Rhodesia. 1949/51, 1951. Results Exp. Tobacco Res. Board Southern Rhodesia. Preceded by: Report (Annual) of the Trelawny tobacco research station. Superseded by: Report of progress, tobacco research board of Southern Rhodesia. HI 72055

Results, international bean rust nursery. 1975/76+, 1976+. Results Int. Bean Rust Nursery. HI 72056

Results of research, agricultural experiment station, university of Kentucky. Lexington, KY. 1950+. Results Res. Agric. Exp. Sta. Univ. Kentucky. Preceded by: Report of the Kentucky agricultural experiment station. HI 72057

Resumé analytique des travaux scientifiques et techniques publiées en Egypte = Abstracts of scientific and technical papers published in Egypt. Dokki-Cairo.

Resumé analytique des travaux scientifiques et techniques publiées en R A U = Abstracts of scientific and technical papers published in U A R [United Arab Republic]. Dokki-Cairo.

Résumé analytique des travaux de la société nationale havraise d'études diverses. Le Havre. Résumé Analytique Trav. Soc. Natl. Havraise Études Diverses. See B–P–H 776/28. HI 58695

Résumé des mémoires, envoyés au concours ouvert par la société vaudoise des sciences naturelles. Lausanne. Résumé Mém. Soc. Vaud. Sci. Nat. See B–P–H 777/1. HI 58696

Résumés des communications, fédération canadienne des sociétés de biologie = Proceedings, canadian federation of biological societies and Programme and proceedings of the annual meeting, canadian federation of biological societies. Saskatoon.

Resumés des communications de la société de microscopie du Canada = Proceedings of the microscopical society of Canada. Toronto.

Resumptio genetica. The Hague. Resumptio Genet. See B–P–H 777/2. HI 58697

Resursy biosfery. Leningrad. Vol. 1+, 1975+. Resursy Biosfery. HI 72058

Reunião biológica portuguesa. Lisbon. Reunião Biol. Portug. See B–P–H 777/4. HI 58698

Réunion rapport, institut de recherche agronomiques tropicales et des cultures vivières = Rapport annuel, institut de recherche agronomiques tropicales et des cultures vivières. Saint-Denis, Réunion.

Reunión de sociedad argentina de pathologia regional del norte. Buenos Aires. Reunión Soc. Argent. Pathol. Regional N. See B–P–H 777/5. HI 58699

Review of the academy of natural sciences of Philadelphia. Philadelphia, PA. 1932+, 1933+. Rev. Acad. Nat. Sci. Philadelphia. Preceded by: Year book, academy of natural sciences of Philadelphia. HI 72060

Review of agricultural operations in India. Calcutta. 1919/20-1923/24, 1921-33. Rev. Agric. Operat. India. Preceded by: Report on the progress of agriculture in India. HI 72061

Review of agriculture and forestry. Ljubljana = Zbornik za kmetijstvo in gozdarstvo. Ljubljana.

Review of applied mycology. Kew. Vols. 1-48, 1922-69. Rev. Appl. Mycol. Superseded by: Review of plant pathology. 4-3583-3. HI 72062

Review of applied mycology. Supplement. Kew. Nos. 1-15, 1940-47. Rev. Appl. Mycol. Suppl. Superseded by: Index of fungi. HI 72063

Review of the faculty of agriculture Ege university = Ziraat fakultesi dergisi, ege universitesi. Bornova.

Review of history and geography; national Chekiang university. [Shih ti tsa chi. Kuo li Che kiang ta hsüeh.] Hangchow, China. Rev. Hist. Geogr. Natl. Chekiang Univ. See B–P–H 781/14. HI 58739

Review of medical and veterinary mycology. Kew. Vol. 1+, 1962+. Rev. Med. Veterin. Mycol. Preceded by: Annotated bibliography of medical mycology. HI 72064

Review of palaeobotany and palynology; an international journal. Amsterdam. Vol. 1+, 1967+. Rev. Palaeobot. Palynol. HI 72065

Review of plant pathology; consisting of abstracts and reviews of current literature on plant pathology. Kew. Vol. 49+, 1970+. Rev. Pl. Pathol. Preceded by: Review of applied mycology. HI 72066

Review of plant protection research. Nishigahara, Tokyo. Vol. 1-?, 1968-81. Rev. Pl. Protect. Res. HI 72067

Review of the polish academy of sciences. Warsaw.

Rev. Polish Acad. Sci. See B–P–H 784/10. HI 58771

Review, Rosemaund experimental husbandry farm. Preston Wynne. 1974+. Rev. Rosemaund Exp. Husb. Farm. Superseded by: Report, Rosemaund experimental husbandry farm. HI 72068

Review series, commonwealth bureau of pastures and field crops. Farnham Royal. Vol. 1+, 1968+. Rev. Ser. Commonw. Bur. Pastures. Field Crops. HI 72069

Review, society for japanese irises. Terre Haute, IN. Vol. 1+, 1964+. Rev. Soc. Jap. Irises. HI 72070

Review of tropical agriculture = Revista de agricultura tropical. Mexico City. Revista Agric. Trop. (Mexico City). See B–P–H 788/7.

Review of tropical plant pathology. New Delhi. Vol. 1+, 1984+. Rev. Trop. Pl. Pathol. HI 72071

Review, tussock grasslands and mountain lands institute. Canterbury, N.Z. No. 1+, 1961+. Rev. Tussock Grasslands Mount. Lands Inst. HI 72073

Review of works in progress, department of agriculture, Cyprus. Nicosia. 1956+. Rev. Works Progr. Dept. Agric. Cyprus. HI 72072

Review of world's forestry. Neudamm & Berlin = Zeitschrift für Weltforstwirtschaft. Neudamm & Berlin. Z. Weltforstw. See B–P–H 988/15.

Reviews in aquatic science. Boca Raton, FL. Vol. 1+, 1989+. Rev. Aquatic Sci. HI 72074

Reviews in biochemical toxicology. New York. Vol. 1+, 1979+. Rev. Biochem. Toxicol. HI 75188

Reviews of weed science. Champaign, IL. Vol. 1+, 19??+. Rev. Weed Sci. HI 72075

Revista de la academia de ciencias exactes, físicas y naturales de Madrid = Revista de la real academia de ciencias exactes, físicas y naturales de Madrid. Madrid. Revista Real Acad. Ci. Madrid. See B–P–H 795/9.

Revista de academia de ciencias exactas, físico-químicas y naturales de Zaragoza. Saragossa, Spain. Revista Acad. Ci. Exact. Zaragoza. See B–P–H 786/18. HI 58790

Revista de la academia colombiana de ciencias exactas, físicas y naturales. Bogota. Vol. 1+, 1936+. Revista Acad. Colomb. Ci. Exact. Superseded by: Revista de la sociedad colombiana de ciencias naturales. 1-16-3. HI 72076

Revista de la academia nacional de ciencias de Bolivia. La Paz. Vol. 1+, 1978+. Revista Acad. Nac. Ci. Bolivia. HI 72077

Revista agrícola. Bogotá. Vols. 1-88, 1915-23. Revista Agríc. (Bogotá) (1915-23). 4-3589-2. HI 72078

Revista agrícola. Bogotá. Vols. 14+, 1967+. Revista Agríc. (Bogotá) (1953+). Preceded by: Revista Esso agrícola. HI 72079

Revista agrícola. Chicago, IL. Revista Agríc. (Chicago). See B–P–H 786/26. HI 58791

Revista agrícola. Cúcuta, Colombia. Revista Agríc. (Cúcuta). See B–P–H 786/28. HI 58793

Revista agrícola. Guatemala City, Guatemala. Revista Agríc. (Guatemala City). See B–P–H 786/29. HI 58794

Revista agrícola. Havana. Vols. 1-15, 1879-95. Revista Agric. (Havana 1879-95). 4-3595-1. HI 72080

Revista agrícola. Lisbon. Revista Agríc. (Lisbon). See B–P–H 786/31. HI 58795

Revista agrícola. Mexico City. Revista Agríc. (Mexico City). See B–P–H 786/32. HI 58796

Revista agrícola. São Paulo. Revista Agríc. (São Paulo, 1895-1907). See B–P–H 787/6. HI 58802

Revista agrícola de Caldas. Manizales, Colombia. Revista Agríc. Caldas. See B–P–H 787/8. HI 58804

Revista agrícola de Filipinas. Manila. Revista Agríc. Filip. See B–P–H 787/15. HI 58809

Revista agrícola y ganadería. Cali, Colombia. Revista Agríc. Ganad. (Cali). See B–P–H 787/19. HI 58813

Revista agrícola; periodico quincenal. Bucaramanga, Colombia. Revista Agríc. Period. Quincenal. See B–P–H 788/3. HI 58818

Revista agrícola salvadoreña. San Salvador, Salvador. Revista Agríc. Salvadoreña. See B–P–H 788/5. HI 58820

Revista agrícola del Tolima. Ibagué, Colombia. Revista Agríc. Tolima. See B–P–H 788/6. HI 58821

Revista agrícola y veterinaria. Buenos Aires. Revista Agríc. Veterin. (Buenos Aires). See B–P–H 788/9. HI 58824

Revista agrícolo-veterinaria. Olinda, Brazil. Revista Agríc.-Veterin. (Olinda). See B–P–H 788/10. HI 58825

Revista de agricultura. Cochabama, Bolivia. Revista Agric. (Cochabama). See B–P–H 786/27. HI 58792

Revista de agricultura. Havana. Vol. 14(22)+, 1934+. Revista Agric. (Havana 1934+). Preceded by: Revista de agricultura, comércio y trabajo. Havana. 4-3595-1. HI 72081

Revista de agricultura. Havana. Vols. 1-9, 1967-76. Revista Agric. (Havana 1967-76). Superseded by: Ciencias de la agricultura. HI 72082

Revista de agricultura. Piracicaba, Brazil. Revista Agaric. (Piracicaba). See B–P–H 787/1. HI 58797

Revista de agricultura. Recife, Brazil. Revista Agric. (Recife). See B–P–H 787/2. HI 58798

Revista de agricultura. San José, Costa Rica. Revista Agric. (San José). See B–P–H 787/3. HI 58799

Revista de agricultura. Santiago. Revista Agric. (Santiago). See B–P–H 787/4. HI 58800

Revista de agricultura. Santo Domingo, Dominican Republic. Revista Agric. (Santo Domingo). See B–P–H 787/5. HI 58801

Revista de agricultura. São Paulo. Revista Agric. (São Paulo, 1920). See B–P–H 787/7. HI 58803

Revista de agricultura y comércio. Havana. Revista Agric. Comércio (Havana). See B–P–H 787/9. HI 58805

Revista de agricultura y comércio. Panama City, Panama. Revista Agric. Comércio (Panama). See B–P–H 787/10. HI 58806

Revista de agricultura y comércio. Santo Domingo, Dominican Republic = Revista de agricultura. Santo Domingo, Dominican Republic. Revista Agric. (Santo Domingo). See B–P–H 787/5.

Revista de agricultura, comercio i trabajo. Havana. Vols. 1-14(21), 1918-34. Revista Agric. Comércio Trab. Superseded by: Revista de agricultura. Havana 1934+. 4-3595-1. HI 72083

Revista de agricultura y cría. Maracaibo, Venezuela. Revista Agric. Cría. See B–P–H 787/13. HI 58807

Revista de agricultura cubana. Havana. Revista Agric. Cub. See B–P–H 787/14. HI 58808

Revista de agricultura y ganadería. Asunción. Vols. 1-2, 1937-38. Revista Agric. Ganad. (Asunción 1937-38). 4-3595-2. HI 58811

Revista de agricultura y ganadería. Asunción. Nos. 1-19, 1944-46. Revista Agric. Ganad. (Asunción 1944-46). 4-3595-2. HI 58810

Revista de agricultura y ganadería. Buenos Aires. Revista Agric. Ganad. (Buenos Aires). See B–P–H 787/18. HI 58812

Revista de agricultura y ganadería. Havana. Revista Agric. Ganad. (Havana). See B–P–H 787/20. HI 58814

Revista de agricultura y ganadería. La Paz, Bolivia. Revista Agric. Ganad. (La Paz). See B–P–H 787/21. HI 58815

Revista de agricultura ganadería nicaraguense. Managua, Nicaragua. Revista Agric. Ganad. Nicarag. See B–P–H 788/1. HI 58816

Revista de agricultura e industrias. Bucaramanga, Colombia. Revista Agric. Industr. See B–P–H 788/2. HI 58817

Revista de agricultura de Puerto Rico. San Juan, PR. Revista Agric. Puerto Rico. See B–P–H 788/4. HI 58819

Revista de agricultura tropical. Mexico City. Revista Agric. Trop. (Mexico City). See B–P–H 788/7. HI 58822

Revista de agricultura tropical. San Salvador, Salvador. Revista Agric. Trop. (San Salvador). See B–P–H 788/8. HI 58823

Revista de agronomía. Lima. Revista Agron. (Lima). See B–P–H 788/11. HI 58826

Revista de agronomía. Puerto Bertoni, Paraguay. Revista Agron. (Puerto Bertoni). See B–P–H 788/14. HI 58829

Revista de agronomia y boletín de la escuela nacional de agricultura de la Asunción del Paraguay = Revista de agronomía y de ciencias aplicadas. Asunción.

Revista de agronomía y de ciencias aplicadas. Asunción. Vols. 1-5, 1897-1913. Revista Agron. Ci. Aplicadas. HI 72084

Revista de agronómíca. Lisbon. Revista Agron. (Lisbon). See B–P–H 788/12. HI 58827

Revista de agronómica. Porto Alegre, Brazil. Revista Agron. (Porto Alegre). See B–P–H 788/13. HI 58828

Revista de agronómica. del noroeste argentino. Tucumán, Argentina. Revista Agron. Noroeste Argent. See B–P–H 788/16. HI 58830

Revista agropecuaria. Bucaramanga, Colombia. Revista Agropecu. See B–P–H 788/17. HI 58831

Revista agropecuaria. San José, Costa Rica. No. 1+, 1970+. Revista Agropecu. (San José). HI 72085

Revista agropecuaria. Santa Clara, Cuba. No. 1+, 1970?+. Revista Agropecu. (Santa Clara). HI 72086

Revista de agroquímica y tecnología de alimentos. Valencia, Spain. Revista Agroquím. Tecnol. Aliment. See B–P–H 788/18. HI 58832

Revista argentina de agronomía. Buenos Aires. Revista Argent. Agron. See B–P–H 788/19. HI 58833

Revista argentina de botánica. La Plata, Argentina. Revista Argent. Bot. See B–P–H 788/20. HI 58834

Revista argentina de microbiologia. Buenos Aires. Vol. 11+, 1979+. Revista Argent. Microbiol. Preceded by: Revista de la asociacion argentina de microbiologia. HI 72087

Revista de la asociacion argentina de microbiologia. Buenos Aires. Vols. ?-10, 19??-79. Revista Asoc. Argent. Microbiol. Superseded by: Revista argentina de microbiologia. HI 72088

Revista de la asociación bioquímica argentina. Buenos Aires. Revista Asoc. Bioquím. Argent. See B–P–H 789/1. HI 58835

Revista de la asociación escuela ingeniería agronómica. Quito, Ecudor. Revista Soc. Esc. Ing. Agron. See

B–P–H 789/2. HI 58836

Revista de la asociación médica argentina. Buenos Aires. Revista Asoc. Méd. Argent. See B–P–H 789/3. HI 58837

Revista azucarera. Buenos Aires. Revista Azucarera (Buenos Aires). See B–P–H 789/4. HI 58838

Revista de bibliografía chilena. Santiago de Chile. 1927-29. Revista Bibliogr. Chilena. Preceded by: Revista de bibliografía chilena y extranjera. HI 72089

Revista de bibliografía chilena y extranjera. Santiago de Chile. Vols. 1-6, 1913-18. Revista Bibliogr. Chilena Extran. Superseded by: Revista de bibliografía chilena. HI 72090

Revista de biologia. Lisbon. Vol. 1+, 1956+. Revista Biol. (Lisbon). HI 72091

Revista de biologia. Lourenço Marques, Mozambique. Vol. 1+, 1956+. Revista Biol. (Lourenço Marques). HI 58839

Revista de biologia e hygiene. São Paulo. Revista Biol Hyg. See B–P–H 789/6. HI 58840

Revista de biología marina. Valparaiso. Vols. 1-10, 1948-60; vol. 12+, 1965+. Revista Biol. Mar. For vol. 11, 1961, see Montemar. 4-3596-2. HI 72092

Revista de biología tropical. San José, Costa Rica. Revista Biol. Trop. See B–P–H 789/8. HI 58841

Revista de biologico del Uruguay. Montevideo. Vol. 1+, 1973+. Revista Biol. Uruguay. HI 72093

Revista de la bolsa de cereales. Corrientes, Argentina. Revista Bolsa Cereales. See B–P–H 789/9. HI 58842

Revista brasileira. Rio de Janeiro. Vols. 1-10, 1879-81; n.s. vols. 1-19, 1895-99. Revista Brasil. (1879-81). 4-3591-2. HI 72094

Revista brasileira de armazenamento. Viçosa. No. 1+, 1976+. Revista Brasil. Armazen. HI 72095

Revista brasileira de biologia. Rio de Janeiro. Revista Brasil. Biol. See B–P–H 789/11. HI 58843

Revista brasileira de botánica; publicação oficial da sociedade botánica do Brasil, regional de São Paulo. São Paulo. Vol. 1+, 1978+. Revista Brasil. Bot. HI 72096

Revista brasileira de genética. Ribeirão Preto. Vol. 1+, 1978+. Revista Brasil. Genét. HI 72097

Revista brasileira de geografía. Rio de Janeiro. Revista Brasil. Geogr. See B–P–H 789/12. HI 58844

Revista brasileira de orquideas. São Paulo. Vols. 1-3, 1948-51. Revista Brasil. Orquideas. HI 72098

Revista brasileira de oto-rinolaringologia. São Paulo. Revista Brasil. Oto-Rinolaringol. See B–P–H 789/13. HI 58845

Revista brasileira de pesquisas médicas e biológicas. São Paulo. Vols. 1-13, 1968-80. Revista Brasil. Pesq. Méd. Biol. Superseded by: Brazilian journal of medical and biological research. HI 75211

Revista brasileira e portuguesa de biologia em geral = Revista de biologia. Lisbon.

Revista brasileira de sementes. Brasilia. Vol. 1+, 1979+. Revista Brasil. Sementes. HI 72099

Revista brazileira; jornal de sciencias, lettras e artes. Rio de Janeiro. Vols. 1-4, 1857-61. Revista Brazil. Preceded by: Guanabara. 4-3591-2. HI 72100

Revista del café. Ponce, PR. Revista Café. See B–P–H 789/14. HI 58846

Revista do café portugues. Lisbon. Revista Café Portug. See B–P–H 789/15. HI 58847

Revista Cafetalera. Guatemala City, Guatemala. Revista Cafetalera. See B–P–H 789/16. HI 58848

Revista cafetera de Colombia. Bogotá. Revista Cafetera Colombia. See B–P–H 789/17. HI 58849

Revista de la cámara de agricultura. San José, Costa Rica. Revista Cámara Agric. See B–P–H 789/18. HI 58850

Revista de la cámara de agricultura de la primera zona. Quito. 1937-47. Revista Cámara Agric. Primera Zona. Preceded by: Boletín de la cámara de agricultura de la primera zona. Superseded by: Agricultor (Quito). 1-85-3. HI 72101

Revista Catalana de geografia; publicacio trimestral de la societat catalana. Barcelona. Vol. 1+, 1978+. Revista Catalana Geogr. HI 72102

Revista del centro de estudiantes de agronomía y veterinaría. Buenos Aires. Vols. 1-23, 1908-30. Revista Centro Estud. Agron. Superseded by: Centro de estudiantes de agronomía. 1-93-1. HI 72103

Revista del centro de estudiantes del doctorado en ciencias naturales. Buenos Aires. Revista Centro Estud. Doct. Ci. Nat. See B–P–H 789/19. HI 58851

Revista del centro de estudiantes de la facultad de ciencias medicas, farmacia y ramos menores. Tucuman. Vols. ?-16-19-?, ?-1936-39-? Revista Centro Estud. Fac. Ci. Med. Farm. Ramos Menores. HI 72104

Revista del centro estudiantes de farmacia. Cordoba. Vols. 1+, 1925+. Revista Centro Estud. Farm. HI 72105

Revista do centro de estudos de Cabo Verde. Serie de ciencias biologicas. Praia. Vol. 1+, 1972+. Revista Centro Estudos Cabo Verde, Ser. Ci. Biol. HI 72106

Revista del centro nacional de agricultura. San José, Costa Rica. Revista Centro Nac. Agric. See B–P–H 789/20. HI 58852

Revista do centro de sciencias, letras e artes de Campinas.

Campinas, Brazil. Revista Centro Sci. Campinas. See B–P–H 789/21. HI 58853

Revista Ceres; publicação de ensinamentos teóricos e práticos sôbre agricultura, veterinária, indústrias rurais. Viçosa, Brazil. Revista Ceres. See B–P–H 790/1. HI 58854

Revista chapingo. Chapingo. Año 7(33/34)+, 1982+ Revista Chapingo. Preceded by: Chapingo. HI 72107

Revista chilena de historia y geografía. Santiago. Revista Chilena Hist. Geogr. See B–P–H 790/4. HI 58855

Revista chilena de historia natural. Valparaiso. Revista Chilena Hist. Nat. See B–P–H 790/5. HI 58856

Revista de chimica pura e applicada. Oporto, Portugal. Revista Chim. Pura Appl. See B–P–H 790/6. HI 58857

Revista de ciencias. Lima. Revista Ci. (Lima). See B–P–H 790/9. HI 58860

Revista de ciências agronómicas. Lourenço Marques. Vols. 6-7, 1973-74. Revista Ci. Agron. (Lourenço Marques). Superseded by: Revista de ciências agronómicas. Lourenço Marques. Série A and Série B. HI 72108

Revista de ciências agronómicas. Lourenço Marques. Série A. Lourenço Marques. Vols. 1-5, 1968-72. Revista Ci. Agron. (Lourenço Marques), A. Superseded by: Revista de ciências agronómicas. Lourenço Marques. HI 72109

Revista de ciências agronómicas. Lourenço Marques. Série B. Lourenço Marques. Vols. 1-5, 1968-72. Revista Ci. Agron. (Lourenço Marques), B Superseded by: Revista de ciências agronómicas. Lourenço Marques. HI 72110

Revista de ciências biológicas. Faculdade de ciências, universidade de Loucenço Marques. Serie A. Lourenço Marques. Vols. 1-7, 1968-75. Revista Ci. Biol., Fac. Ci. Univ. Loucenço Marques, A. Preceded by: Revista dos estudos gerais universitarios de Moçambique. Serie 2, ciencias biologicas e agronomicas. HI 72111

Revista de ciências biológicas. Faculdade de ciências, universidade de Loucenço Marques. Serie B. Lourenço Marques. Vols. 1-3, 1970-75. Revista Ci. Biol., Fac. Ci. Univ. Loucenço Marques, B. HI 72112

Revista de ciencias biologicas, universidade do Pará. Belém. Vol. 1+, 1963+. Revista Ci. Biol. HI 72113

Revista de ciencias i letras. Santiago. Revista Ci. Letras. See B–P–H 790/11. HI 58862

Revista de ciencias, literatura y artes. Seville. Revista Ci. Lit. See B–P–H 790/12. HI 58863

Revista di ciencias naturales de Madrid. Madrid. Revista Ci. Nat. Madrid. See B–P–H 790/15. HI 58866

Revista científica. Buenos Aires. Revista Ci. (Buenos Aires). See B–P–H 790/7. HI 58858

Revista científica. Caracas. Revista Ci. (Caracas). See B–P–H 790/8. HI 58859

Revista científica de investigaciones del museo de historia natural de San Rafael. San Rafael, Argentina. Revista Ci. Invest. Mus. Hist. Nat. San Rafael. See B–P–H 790/10. HI 58861

Revista científica y literaria de México. Mexico City. Revista Ci. Lit. México. See B–P–H 790/13. HI 58864

Revista cientifica mensuel. de la universidad central de Venezuela. Caracas. Vols. 1-?, 1887-91. Revista Ci. Mens. Univ. Centr. Venezuela. HI 72114

Revista científica mexicana. Mexico City. Revista Ci. Mex. See B–P–H 790/14. HI 58865

Revista do circulo paulista de orquidofilas. São Paulo. Revista Circ. Paul. Orquidofilas. See B–P–H 790/16. HI 58867

Revista clínica de São Paulo. São Paulo. Revista Clín. São Paulo. See B–P–H 790/17. HI 58868

Revista (de) clinica de São Paulo. São Paulo. Revista Clin. São Paulo. See B–P–H 790/18. HI 58869

Revista del consejo oceanográfico ibero-americana. Madrid. Revista Cons. Oceanogr. Ibero-Amer. See B–P–H 790/19. HI 58870

Revista cubana de ciencia agrícola. Havana. Vols. 1-6, 1967-72. Revista Cub. Ci. Agric. Preceded by: Cuban journal of agricultural science. HI 72115

Revista de difusión, instituto antártico chileno. Santiago. No. 9+, 1976+. Revista Difusión Inst. Antárt. Chileno. Preceded by: Boletin de difusión, instituto antártico chileno. HI 72116

Revista española de biología. Madrid. Revista Esp. Biol. See B–P–H 790/20. HI 58871

Revista española de fisiología. Barcelona. Revista Esp. Fisiol. See B–P–H 790/21. HI 58872

Revista Esso agrícola. Bogotá. Vols. 1-13, 1951-66? Revista Esso Agríc. Superseded by: Revista agrícola. Bogota (1953+). HI 72117

Revista dos estudos gerais universitários de Moçambique. Série 2, ciências biológicas e agronómicas. Lourenço Marques. Vols. 1-4, 1964-67. Revista Estudos Gerais Univ. Moçambique, Sér. 2. Superseded by: Revista de ciências biológicas. Faculdade de ciências, universidade de Loucenço Marques. Serie A. HI 72118

Revista da faculdade de agronomia da universidade federal do Rio Grande do Sul. Porto Alegre. Vol. 1+, 1975+. Revista Fac. Agron. Univ. Fed. Rio Grande do Sul. Preceded by: Revista da faculdade de agronomia e veterinaria da universidade (federal) do Rio Grande do Sul. HI 72119

Revista da faculdade de agronomia e veterinaria da universidade (federal) do Rio Grande do Sul. Porto Alegre. Vols. 1-10, 1954-71. Revista Fac. Agron. Univ. Rio Grande do Sul. Superseded by: Revista da faculdade de agronomia da universidade federal do Rio Grande do Sul. HI 72120

Revista. Faculdade ciências. Universidad de Lisboa. Lisbon. Revista Fac. Ci. Univ. Lisboa. See B–P–H 791/11. HI 58879

Revista de la faculdade de ciências, universidade de Lisboa. Sér. 2, C, ciências naturais. Lisbon. Revista Fac. Ci. Univ. Lisboa, Sér. 2, C, Ci. Nat. See B–P–H 791/12. HI 58880

Revista, faculdade de farmácia e bioquímica, universidade do São Paulo. São Paulo. Vols. 1-7(2), 1963-69. Revista Fac. Farm. Bioquim. Univ. São Paulo. Superseded by: Revista de farmacía e bioquímica, universidade de São Paulo. HI 72121

Revista da faculdade de filosofia, ciencias e letras "Manoel de Nobrega". Vol. 1, 1948. Revista Fac. Filos. Ci. Letras Manoel Nobrega. HI 72122

Revista de la facultad de agricultura. Universidad central de Venezuela. Maracay, Venezuela. Revista Fac. Agric. Univ. Centr. Venezuela. See B–P–H 791/1. HI 58873

Revista de la facultad de agronomía. [Universidad central de Venezuela.] Maracay, Venezuela. Revista Fac. Agron. (Maracay). See B–P–H 791/2. HI 58875

Revista de la facultad de agronomía; universidad nacional de La Plata. La Plata, Argentina. Revista Fac. Agron. Univ. Nac. La Plata. See B–P–H 791/5. HI 58876

Revista de la facultad de agronomía, universidad de la república, Uruguay. Montevideo. Nos. 1-12, 1907-13; ser. 2, nos. 1-3, 1918; ser. 3, no. 1+, 1928+. Revista Fac. Agron., Univ. Repúbl. Uruguay. 3-2742-3. HI 72123

Revista de la facultad de agronomía, universidad del Zulia. Maracaibo. Vol. 1+, 1968+. Revista Fac. Agron. Univ. Zulia. HI 72124

Revista de la facultad de agronomía y veterinaria. Buenos Aires. Revista Fac. Agron. Veterin. See B–P–H 791/7. HI 58877

Revista de la facultad de ciencias agrarias, universidad nacional de Cuyo. Mendoza, Argentina. Revista Fac. Ci. Agrar. Univ. Nac. Cuyo. See B–P–H 791/9. HI 58878

Revista de la facultad de ciencias biológicas, universidad nacional de Trujillo. Trujillo. Vol. 1+, 1964+. Revista Fac. Ci. Biol. HI 72125

Revista de la facultad de ciencias exactas, físicas y naturales, universidad nacional de Cordoba. Córdoba. 1948+. Revista Fac. Ci. Exact. Preceded by: Boletín de la facultad de ciencias exactas, físicas y naturales. 2-1200-1. HI 72126

Revista de la facultad de ciencias naturales de Salta, universidad nacional de Tucumán. Salta. Vol. 1+, 1959+. Revista Fac. Ci. Nat. Salta. HI 72127

Revista de la facultad de ciencias químicas. La Plata = Revista. Facultad de química y farmacía. La Plata, Argentina. Revista Fac. Quím. See B–P–H 791/16.

Revista de la facultad de ciencias, universidad de Oviedo. Oviedo. N.s. vol. 1+, 1960+. Revista Fac. Ci. Univ. Oviedo. HI 72129

Revista de la facultad de farmácia, universidad de los Andes. Merida. No. 1+, 1958+. Revista Fac. Farm. Univ. Andes. HI 72130

Revista. Facultad de farmacía. Universidad central. Caracas. Revista Fac. Farm. Univ. Centr. See B–P–H 791/13. HI 58881

Revista. Facultad nacional de agronomía. Medellín universidad de Antioquia. Medellín, Colombia. Revista Fac. Nac. Agron. Medellín Univ. Antioquia. See B–P–H 791/14. HI 58882

Revista. Facultad nacional de agronomía. Universidad, Bogotá. Bogotá. Revista Fac. Nac. Agron. Univ. Bogotá. See B–P–H 791/15. HI 58883

Revista. Facultad de química y farmacía. La Plata, Argentina. Revista Fac. Quím. See B–P–H 791/16. HI 58884

Revista farmacéutica. Buenos Aires. Revista Farm. See B–P–H 791/17. HI 58885

Revista de farmacia e bioquimica da Amazonia. Pará. Vols. 1-2, 1968-69. Revista Farm. Bioquim. Amazonia. HI 72131

Revista de farmácia e bioquímica, universidade de São Paulo. São Paulo. Vol. 8+, 1970+. Revista Farm Bioquim. Univ. São Paulo. Preceded by: Revista, faculdade de farmácia e bioquímica, universidade do São Paulo. HI 72132

Revista fitosanitaria mundial. London. Revista Fitosan. Mundial. See B–P–H 791/19. HI 58886

Revista de flora medicinal. Rio de Janeiro. Revista Fl. Med. See B–P–H 791/20. HI 58887

Revista de fomento. [Venezuela. Ministerio del fomento.] Caracas. Revista Fomento. See B–P–H 791/22. HI 58888

Revista forestal argentina. [Argentina]. Revista Forest. Argent. See B–P–H 791/23. HI 58889

Revista forestal Baracoa. Havana. Vol. ?-10+, ?-1980+. Revista Forest. Baracoa. HI 72133

Revista forestal chilena. [Chile]. Revista Forest. Chilena. See B–P–H 792/1. HI 58890

Revista forestal mexicana. Mexico City. Revista Forest. Mex. See B–P–H 792/2. HI 58891

Revista forestal del Perú. Lima. Vol. 1+, 1967+. Revista Forest. Perú. HI 72134

Revista forestal latinoamericana. Merida. No. 01/81+, 1981+. Revista Forest. Latinoamer. Preceded by: Boletin, instituto forestal latino-americano de investigación y capitación. HI 72135

Revista forestal venezolana. Mérida, Venezuela. Revista Forest. Venez. See B–P–H 792/3. HI 58892

Revista geográfica. Barranquilla, Colombia. Revista Geogr. (Barranquilla). See B–P–H 792/4. HI 58893

Revista geográfica. Mérida, Venezuela. Revista Geogr. (Mérida). See B–P–H 792/5. HI 58894

Revista geográfica. Mexico City. Revista Geogr. (Mexico City). See B–P–H 792/6. HI 58895

Revista de guimãraes. Oporto, Portugal. Revista Guimãraes. See B–P–H 792/7. HI 58896

Revistă horticolă. Bucharest. Revistă Hort. (Bucharest). See B–P–H 792/8. HI 58897

Revista horticola; el cultivador de plantes y flores. Buenos Aires. Revista Hort. (Buenos Aires). See B–P–H 792/9. HI 58898

Revistă de horticultura şi viticultură. Bucharest. Vols. 16-23(6), 1967-74. Revistă Hort. Vitic. Preceded by: Grădina, via şi livada. Superseded by: Producţia vegetală. Horticultura. HI 72136

Revista I B P T. Curitiba, Brazil. Revista I. B. P. T. See B–P–H 792/10. HI 58899

Revista ibérica de micología. Barcelona. Vol. 1+, 1984+. Revista Ibér. Micol. HI 72137

Revista de las Indias. Bogotá. Revista Indias. See B–P–H 792/11. HI 58900

Revista industrial y agrícola de Tucumán. San Miguel. Vol. 1+, 1910+. Revista Industr. Agric. Tucumán. HI 72138

Revista del instituto agrícola catalán de San Isidro. Barcelona. Revista Inst. Agríc. Catalán San Isidro. See B–P–H 792/12. HI 58901

Revista do instituto de antibióticos. Recife, Brazil. Revista Inst. Antibiót. See B–P–H 792/13. HI 58902

Revista del instituto bacteriológico de Chile. Santiago. Revista Inst. Bacteriol. Chile. See B–P–H 792/14. HI 58903

Revista del instituto bacteriológico del departamento nacional di higiene. Buenos Aires. Revista Inst. Bacteriol. Dept. Nac. Hig. See B–P–H 792/15. HI 58904

Revista del instituto bacteriológico "Dr. Carlos G. Malbrán". Buenos Aires. Revista Inst. Bacteriol. "Dr. Carlos G. Malbrán". See B–P–H 792/16. HI 58905

Revista do instituto de biologia da universidade federal do Paraná = Acta biológica paranaense. Curitiba.

Revista do instituto de café de estado de São Paulo. São Paulo. Vols. 6-16, 1931-41. Revista Inst. Café Estado São Paulo. Preceded by: Boletim do instituto de café do estado de São Paulo. 3-2022-2. HI 72139

Revista, instituto colombiano agropecuario. Bogotá. 1966+. Revista Inst. Colomb. Agropecu. HI 72140

Revista del instituto de defensa del café. San José, Costa Rica. Revista Inst. Defensa Café. See B–P–H 792/17. HI 58906

Revista del instituto Malbrán. Buenos Aires. Revista Inst. Malbrán. See B–P–H 792/19. HI 58907

Revista, instituto municipal de botánica. Buenos Aires. Vols. 1-3, 1961-69. Revista Inst. Munic. Bot. HI 72141

Revista. Instituto nacional del café. Caracas. Revista Inst. Nac. Café. See B–P–H 793/1. HI 58908

Revista del instituto nacional de investigación de las ciencias naturales anexo al museo argentino de ciencias naturales "Bernardino Rivadavia". Ciencias botánicas. Buenos Aires. Vols. 1-2(4), 1948-51. Revista Inst. Nac. Invest. Ci. Nat., Ci. Bot. Preceded by: Anales del museo argentino de ciencias naturales "Bernardino Rivadavia." Superseded by: Revista del museo argentino de ciencias naturales "Bernardino Rivadavia" e instituto nacional de investigación de las ciencias naturales. Botánica. 1-818-1. HI 72142

Revista del instituto nacional de investigación de las ciencias naturales anexo al museo argentino de ciencias naturales "Bernardino Rivadavia". Ciencias geológicas. Buenos Aires. Revista Inst. Nac. Invest. Ci. Nat., Ci. Geol. See B–P–H 793/2. HI 58909

Revista del instituto nacional de investigación de las ciencias naturales anexo al museo argentino de ciencias naturales "Bernardino Rivadavia". Ecologia. Buenos Aires. Revista Inst. Nac. Invest. Ci. Nat., Ecol. See B–P–H 793/4. HI 58910

Revista del instituto nacional de investigación de las ciencias naturales anexo al museo argentino de ciencias naturales "Bernardino Rivadavia". Hidrobiologia. Buenos Aires. Vols. 1-?, 1963-? Revista Inst. Nac. Invest. Ci. Nat., Hidrobiol. Superseded by: Revista del museo argentino de ciencias naturales "Bernardino Rivadavia" e instituto nacional de investigación de las ciencias naturales. Hidrobiologia. HI 72143

Revista del instituto de salubridad y enfermedades tropicales. Mexico City. Revista Inst. Salubr. Enferm. Trop. See B–P–H 793/7. HI 58912

Revista interamericana de bibliografía. Washington, DC. Revista Interamer. Bibliogr. See B–P–H 793/8.

HI 58913

Revista interamericana de ciencias agricolas. Turrialba. Vol. ?-25+, ?-1975+. Revista Interamer. Ci. Agric. HI 72144

Revista internacional de fitosociologia, ecologia y fitogeografia. The Hague = Vegetatio; acta geobotanica. The Hague. Vegetatio. See B–P–H 948/18.

Revista internationale pentru agricultura. Sofia. No. 1+, 1958+. Revista Int. Pentru Agric. HI 72145

Revista de investigación en salud publica. Mexico City. Revista Invest. Salud Publica. See B–P–H 793/16. HI 58920

Revista de investigaciones agrícolas. Buenos Aires. Revista Invest. Agríc. See B–P–H 793/10. HI 58914

Revista de investigaciones agropecuarias. Ser. 2: Biología y producción vegetal. Buenos Aires. Revista Invest. Agropecu., Ser. 2, Biol. Prod. Veg. See B–P–H 793/11. HI 58915

Revista de investigaciones agropecuarias. Serie 3: Clima y suelo. Buenos Aires. Revista Invest. Agropecu., Ser. 3, Clima Suelo. See B–P–H 793/12. HI 58916

Revista de investigaciones agropecurias. Ser. 5: Patología vegetal. Buenos Aires. Revista Invest. Agropecu., Ser. 5, Patol. Veg. See B–P–H 793/13. HI 58917

Revista de investigaciones forestales. Buenos Aires. Revista Invest. Forest. See B–P–H 793/14. HI 58918

Revista de investigaciones ganaderas. Buenos Aires. Revista Invest. Ganad. See B–P–H 793/15. HI 58919

Revista de investigaciones marinas. Havana. Vol. 1+, 1980+. Revista Invest. Mar. Preceded by: Ciencias. Serie 8, investigaciones marinas. HI 72146

Revista del jardín botánico y museo de historia natural del Paraguay. Asunción. Vols. 1-3, 1921-33. Revista Jard. Bot. Mus. Hist. Nat. Paraguay. 1-544-2. HI 72147

Revista del jardin botanico nacional, universidad de la Habana. Havana. Vol. 1+, 1980+. Revista Jard. Bot. Nac. Univ. Habana. HI 72148

Revista latinoamericana de historia de las ciencias y la tecnología. Mexico, D.F. Vol. 1+, 1984+. Revista Latinoamer. Hist Ci. Technol. HI 72150

Revista latinoamericana de microbiología. Mexico City. Revista Latinoamer. Microbiol. See B–P–H 793/18. HI 58921

Revista litteraria; periodico de litteratura, philosophia, viagem, e ciencias e bellas artes. Oporto, Portugal. Revista Litt. See B–P–H 793/19. HI 58922

Revista médica de Chile. Santiago. Vol. 1+, 1872+. Revista Méd. Chile. 4-3616-2. HI 72152

Revista médica gallega. Santiago. Revista Méd. Gallega. See B–P–H 794/2. HI 58924

Revista médica latino-americana. Buenos Aires. Revista Méd. Latino-Amer. See B–P–H 794/3. HI 58925

Revista médica militar. Mexico City. Revista Méd. Militar (Mexico City). See B–P–H 794/4. HI 58926

Revista médica del Uruguay. Montevideo. Revista Méd. Uruguay. See B–P–H 794/8. HI 58930

Revista médica de Valparaiso. Valparaiso. Vol. 1+, 1948+. Revista Méd. Valparaiso. 4-3616-3. HI 72153

Revista medicalǎ. Turgu-Mures. Vol. 1+, 1955+. Revista Med. (Turgu-Mures). HI 72151

Revista de medicina y cirugía de la Habana. Havana. Revista Med. Cirugía Habana. See B–P–H 794/1. HI 58923

Revista de medicina militar. Oporto, Portugal. Revista Med. Militar (Oporto). See B–P–H 794/5. HI 58927

Revista de medicina tropical. Bogotá. Revista Med. Trop. (Bogotá). See B–P–H 794/6. HI 58928

Revista de medicina tropical. Havana. Revista Med. Trop. (Havana). See B–P–H 794/7. HI 58929

Revista médico-quirúrgica de patologie femenina. Buenos Aires. Revista Méd.-Quir. Patol. Femenina. See B–P–H 794/9. HI 58931

Revista mexicana. Mexico. Revista Mex. See B–P–H 794/10. HI 58932

Revista mexicana = Revista científica y literaria de México. Mexico City. Revista Ci. Lit. México. See B–P–H 790/13.

Revista mexicana de alergología. Monterrey, Mexico. Revista Mex. Alergol. See B–P–H 794/11. HI 58933

Revista mexicana de biología. Mexico City. Revista Mex. Biol. See B–P–H 794/12. HI 58934

Revista mexicana de estudios antropológicos. Mexico City. Revista Mex. Estud. Antropol. See B–P–H 794/13. HI 58935

Revista mexicana de estudios historicas. Mexico City. Revista Mex. Estud. Hist. See B–P–H 794/14. HI 58936

Revista mexicana de micologia. Mexico City. Vol. 1+, 1985+. Revista Mex. Micol. Preceded by: Boletín de la sociedad mexicana de micologia. HI 72154

Revista de microbiologica; orgao oficial da sociedade brasileira de microbiologia. São Paulo. Vol. 1+, 1971+. Revista Microbiol. HI 72155

Revista del ministerio de agricultura. [Colombia]. Bogotá. Revista Minist. Agric. See B–P–H 794/15. HI 58937

Revista del museo argentino de ciencias naturales "Bernardino Rivadavia" e instituto nacional de investigación de las ciencias naturales. Botánica. Buenos Aires. Vol. 2(5)+, 1961+. Revista Mus. Argent. Ci. Nat., Bernardino Rivadavia Inst. Nac. Invest. Ci. Nat., Bot. Preceded by: Revista del instituto nacional de investigación de las ciencias naturales anexo al museo argentino de ciencias naturales "Bernardino Rivadavia". Ciencias botánicas. HI 72156

Revista del museo argentino de ciencias naturales "Bernardino Rivadavia" e instituto nacional de investigación de las ciencias naturales. Hidrobiologia. Buenos Aires. Vol. ?-2+, ?-1966+. Revista Mus. Argent. Ci. Nat., Bernardino Rivadavia Inst. Nac. Invest. Ci. Nat., Hidrobiol. Preceded by: Revista del instituto nacional de investigación de las ciencias naturales anexo al museo argentino de ciencias naturales "Bernardino Rivadavia". Hidrobiologia. HI 72157

Revista del museo Canario. Las Palmas. ?-1966+. Revista Mus. Canario. HI 72158

Revista del museo civico di scienze naturali E. Caffi. Bergamo. 1979+. Revista Mus Civico Sci. Nat. E. Caffi. HI 72159

Revista del museo de la ciudad Eva Peron. Sección botánica = Revista de museo de La Plata. Sección botánica. La Plata.

Revista del museo de la ciudad Eva Peron. Sección paleontología = Revista de museo de La Plata. Sección paleontología. La Plata.

Revista del museo de historia natural de Mendoza. Mendoza. Vol. 1+, 1947+. Revista Mus. Hist. Nat. Mendoza. HI 72160

Revista de museo de La Plata. Sección botánica. La Plata. N.s. vol. 1+, 1936+. Revista Mus. La Plata, Secc. Bot. 3-2359-3. HI 72161

Revista de museo de La Plata. Sección paleontología. La Plata. N.s. vol. 1+, 1936+. Revista Mus. La Plata, Secc. Paleontol. HI 72162

Revista del museo nacional. Lima. Vol. 1+, 1932+. Revista Mus. Nac., Lima. 3-2424-1. HI 58938

Revista do museu nacional. Rio de Janeiro. Vols. 1(1)-2(5), 1944-45. Revista Mus. Nac., Rio de Janeiro. HI 72163

Revista del museo provincial de ciencias naturales de Córdoba. Córdoba [Argentina.] Vol. ?-3+, ?-1938+. Revista Mus. Prov. Ci. Nat. Córdoba. HI 72164

Revista do museu paulista, universidade de São Paulo. São Paulo. Vols. 1-26, 1895-1942; n.s. vol. 1+, 1947+. Revista Mus. Paul. Univ. São Paulo. 5-3778-1. HI 72165

Revistă muzeelov. (Romania). Vol. ?-2(3)+, ?-1965+. Revistă Muzeelov. HI 72166

Revista nacional de agricultura. Bogotá. Vol. 1+, 1906+. Revista Nac. Agric. Preceded by: Agricoltor. Bogotá (1880-1901). 4-3620-1. HI 72167

Revista ocrotirea mediului inconjurator natura, terra. Subtitlul natura. Bucharest. Vol. 28+, 1977+. Revista Ocrot. Mediului Inconj. Nat. Terra, Nat. Preceded by: Natura. Serie biologie. HI 72168

Revista oto-laringologica de São Paulo. São Paulo. Revista Oto-Laringol. São Paulo. See B–P–H 794/22. HI 58939

Revistă pădurilor. Bucharest. Revistă Pădur. See B–P–H 795/1. HI 58940

Revista, pan-american association of biochemical societies = P A A B S revista. Orlando, FL.

Revista paulista de medicina. São Paulo. Revista Paul. Med. See B–P–H 795/2. HI 58941

Revista peruana de entomología. Lima. Revista Peruana Entomol. See B–P–H 795/3. HI 58942

Revista plantas medicinales. Havana. Vol. ?-2+, ?-1982+. Revista Pl. Med. HI 72169

Revista de la potasa. Bern. Revista Potasa. See B–P–H 795/4. HI 58943

Revista de los progresos de las ciencias exactas, físicas y naturales. Madrid. Revista Progr. Ci. Exact. See B–P–H 795/6. HI 58944

Revista de química e farmácia. Rio de Janeiro. Revista Quím. Farm. (Rio de Janeiro). See B–P–H 795/8. HI 58946

Revista químico farmacéutica. Maracaibo, Venezuela. Revista Quím. Farm. (Maracaibo). See B–P–H 795/7. HI 58945

Revista de la real academia de ciencias exactes, físicas y naturales de Madrid. Madrid. Revista Real Acad. Ci. Madrid. See B–P–H 795/9. HI 58947

Revista semillas. Tunja, Colombia. Revista Semillas. See B–P–H 795/10. HI 58948

Revista. Setor de ciencias agrarias, universidade do Parana. Curitiba. Vol. ?-2+, ?-1980+. Revista Setor Ci. Agrar. Univ. Parana. HI 72170

Revista, sociedad boliviana de historia natural. Cochabamba. Vol. 1+, 1974+. Revista Soc. Boliv. Hist. Nat. HI 72171

Revista de la sociedad científica del Paraguay. Asunción. Revista Soc. Ci. Paraguqy. See B–P–H 795/13. HI 58951

Revista de la sociedad colombiana de ciencias naturales. Bogota. Nos. 100-110, 1929-31. Revista Soc. Colomb. Ci. Nat. Preceded by: Boletín de la sociedad colombiana de ciencias naturales. Superseded by: Revista de la academia colombiana de ciencias exactas, físicas y naturales. HI 72172

Revista de la sociedad colombiana de orquideologia. Medellin. Nos. 1-2, 1966-67. Revista Soc. Colomb. Orquideol. Superseded by: Orquideologia. HI 72173

Revista de la sociedad cubana de botánica; organo oficial del jardin botanico de la universidad de la Habana. Havana. Vols. 1-17, 1944-60. Revista Soc. Cub. Bot. 5-3904-2. HI 72174

Revista de la sociedad médica argentina = Revista de la asociación médica argentina. Buenos Aires. Revista Asoc. Méd. Argent. See B–P–H 789/3.

Revista de la sociedad mexicana de historia natural. Mexico City. Revista Soc. Mex. Hist. Nat. See B–P–H 795/16. HI 58952

Revista de la sociedad química de México. Mexico City. Revista Soc. Quím. México. See B–P–H 795/17. HI 58953

Revista de la sociedad uruguaya de entomología. Montevideo. Revista Soc. Uruguaya Entomol. See B–P–H 795/18. HI 58954

Revista da sociedade dos agronomos e veterinarios do Pará. Belém, Brazil. Revista Soc. Agron. Pará. See B–P–H 795/11. HI 58949

Revista da sociedade brasileira de agronomia. Rio de Janeiro. Revista Soc. Brasil. Agron. See B–P–H 795/12. HI 58950

Revista da sociedade scientifica de São Paulo. São Paulo. Vols. 1-7, 1905-13. Revista Soc. Sci. São. Paulo. HI 72175

Revista de stiinta şi tehnica horti-viticola = Revista de horticultura şi viticultură. Bucharest.

Revistă ştinţifica V. Adamachi. Jassy, Rumania. Revistă Sti. Adamachi. See B–P–H 795/19. HI 58955

Revista sudamericana de botánica. Montevideo. Revista Sudamer. Bot. See B–P–H 795/21. HI 58956

Revista sud-americana de endocrinologia, inmunologia y quimioterapía. Buenos Aires. Revista Sud-Amer. Endocrinol. See B–P–H 795/22. HI 58957

Revista sudamericana de morfologia. Buenos Aires. Revista Sudamer. Morfol. See B–P–H 795/23. HI 58958

Revista svizzera di biotecnologia; organo d'informazione della commissione svizzera di coordinamento per la biotecnologia = Swiss biotech; Schweizerische zeitschrift für biotechnologie; Informationsorgan des schweizerische koordinationsausschusses für biotechnologie. Küsnacht.

Revista terapeutica. Rio de Janeiro. Revista Terap. See B–P–H 796/1. HI 58959

Revista theobroma. Itabuna. Vol. 1+, 1971+. Revista Theobroma. HI 72176

Revista trimensal de historia e géografia; jornal; instituto histórico, geográfico e ethnográfico de Brazil. Rio de Janeiro. Revista Trimensal Hist. Géogr. See B–P–H 796/2. HI 58960

Revista U N E L L E Z de ciencia y tecnologia. Barinas. Vol. 1+, 1983+. Revista U. N. E. L. L. E. Z. Ci. Techn. HI 72177

Revista de la universidad de Caldas. Manizales, Colombia. Revista Univ. Caldas. See B–P–H 796/6. HI 58963

Revista de la universidad de Costa Rica. San José, Costa Rica. Revista Univ. Costa Rica. See B–P–H 796/7. HI 58964

Revista da universidade de Coimbra. Coimbra. Vol. 1+, 1912+. Revista Univ. Coimbra. 2-1084-2. HI 75284

Revista universitaria. Cuzco, Peru. Revista Univ. (Cuzco). See B–P–H 796/3. HI 58961

Revista universitaria. Universidad catolica de Chile. Santiago. Vol. 1+, 1915+. Revista Univ. (Santiago). 5-3778-3. HI 72178

Revista universitatii "Al. I. Cuza" si a institului politehnic din Iasi. Jassy. Vol. 1+, 1954+. Revista Univ. "Al. I. Cuza" Inst. Politehn. Iasi. HI 72179

Revistă universitatii C. I. Parhon Bucureşti. Seria ştiinţelor naturi. Bucharest. Revistă Univ. C. I. Parhon Bucurešti, Ser. Šti. Nat. See B–P–H 796/5. HI 58962

Revista uruguaya de dermatologia y sifilografia. Montevideo. Revista Uruguaya Dermatol. Sifilogr. See B–P–H 795/11. HI 58965

Revista uruguaya de geografía. Montevideo. Revista Uruguaya Geogr. See B–P–H 796/12. HI 58966

Revista vinícola y de agricultura. Saragossa, Spain. Revista Viníc. Agric. See B–P–H 796/13. HI 58967

Revue de l'académie polonaise des sciences. Warsaw. Rev. Acad. Polon. Sci. See B–P–H 777/12. HI 58700

Revue acadienne. Montreal. Rev. Acadienne. See B–P–H 777/13. HI 58701

Revue africaine de la protection des végétaux = African journal of plant protection. Yaounde.

Revue agricole. Nouméa, New Caledonia. Rev. Agric. (Noumea). See B–P–H 777/18. HI 58702

Revue agricole. Pointe-à-Pitre, Guadeloupe. Rev. Agric. (Pointe-à-Pitre). See B–P–H 777/19. HI 58703

Revue agricole de l'Afrique du Nord. Algiers. Rev. Agric. Afrique N. See B–P–H 777/20. HI 58704

Revue agricole d'Haiti. Port-au-Prince, Haiti. Rev. Agric. Haiti. See B–P–H 777/22. HI 58705

Revue agricole illustrée. Bordeaux. Nos. 1-18, 1904-? Rev. Agric. Ill. Superseded by: Revue agricole, viticole, horticole, illustrée et organe de la féderation des sociétés horticoles et viticoles du sud-ouest. HI 72180

Revue agricole de l'Ile Maurice. Port Louis. Vols. 1-33, 1922-54. Rev. Agric. Ile Maurice. Superseded by: Revue agricole et sucrière de l'Ile Maurice. 4-3625-2. HI 72181

Revue agricole, industrielle, historique et artistique. Valenciennes, France. Rev. Agric. Industr. See B–P–H 777/25. HI 58706

Revue agricole et sucrière de l'Ile Maurice. Réduit. Vol. 34+, 1955+. Rev. Agric. Sucr. Ile Maurice. Preceded by: Revue agricole de l'Ile Maurice. HI 72182

Revue agricole, viticole, horticole, illustrée et organe de la féderation des sociétés horticoles et viticoles du sud-ouest. Bordeaux. Nos. 20-30, 19??-05. Rev. Agric. Vitic. Hort. Ill. Organe Féd. Soc. Hort. Vitic. Sud-Ouest. Preceded by: Revue horticole illustrée. HI 72183

Revue de l'agriculture. Brussels. Vol. 1+, 1948+. Rev. Agric. (Brussels). 4-3636-1. HI 72184

Revue agrologique et botanique du Kivu. Brussels. Rev. Agrol. Bot. Kivu. See B–P–H 777/26. HI 58707

Revue agronomique. Louvain. Vols. 1-5, 1892-96. Rev. Agron. Superseded by: Revue générale agronomique. 1-86-1. HI 72185

Revue agronomique canadienne = Scientific agriculture. Ottawa.

Revue algérienne. Algiers. Rev. Algér. See B–P–H 778/1. HI 58708

Revue algologique. Paris. Vols. 1-13, 1924-42; n.s. vol. 1-?, 1954-79 [suspended 1942-53]. Rev. Algol. Superseded by: Cryptogamie - algologie. 4-3625-3. HI 72186

Revue d'Alsace. [Edited by Börsch.] Strasbourg. Rev. Alsace (Börsch). See B–P–H 778/4. HI 58709

Revue analytique des publications, département des recherches forestières, C N R F. Seichamps. 1964/75+, 1976+. Rev. Analytique Publ. Dép. Rech. Forest. C. N. R. F. HI 72187

Revue of applied biology. Boston, MA. Rev. Appl. Biol. See B–P–H 778/5. HI 58710

Revue arboricole = Obstrundschau. Zug, Switzerland. Obstrundschau. See B–P–H 680/2.

Revue des archives italiennes de biologie. Pisa. Rev. Arch. Ital. Biol. See B–P–H 778/8. HI 58711

Revue de l'association nationale des agronomes haitiens. Port-au-Prince, Haiti. Rev. Assoc. Natl. Agron. Haitiens. See B–P–H 778/14. HI 58712

Revue belge de géographie. Brussels. Vol. 1+, 1962+. Rev. Belge Géogr. Preceded by: Bulletin de la société royale belge de géographie. HI 72188

Revue bibliographique pour servir de complément aux Annales des sciences naturelles. Paris. Rev. BIbliogr. Complément Ann. Sci. Nat. See B–P–H 778/16. HI 58713

Revue bimestrielle de recherches, service des forêts, Canada. Ottawa. Vols. 28-36, 1972-80. Rev. Bimestr. Rech. Serv. Forêts Canada. Superseded by: Revue de recherches du service canadien des forêts. HI 72189

Revue de biologie. Bucharest. Vol. 2, 1954; [n.s.] vols. 1-8, 1956-63. Rev. Biol. (Bucharest). Preceded by: Science dans la république populaire roumaine. Superseded by: Revue roumaine de biologie. Série de botanique. HI 72190

Revue biologique du nord de la France. Lille. Rev. Biol. N. France. See B–P–H 778/20. HI 58714

Revue de botanique appliquée et d'agriculture coloniale. Paris. Vols. 1-9, 1921-29. Rev. Bot. Appl. Agric. Colon. Preceded by: Journal d'agriculture tropicale. Superseded by: Revue internationale de botanique appliquée et d'agriculture tropicale. 4-3658-2. HI 72191

Revue de botanique appliquée et d'agriculture tropicale. Paris. Vols. 10-25, 1929-45. Rev. Bot. Appl. Agric. Trop. Preceded by: Revue de botanique appliquée et d'agriculture coloniale. Superseded by: Revue international de botanique appliquée et d'agriculture tropicale. 4-3658-2. HI 72192

Revue de botanique, bulletin mensuel. [Société française de botanique.] Toulouse. Rev. Bot. Bull. Mens. See B–P–H 778/26. HI 58715

Revue botanique; recueil mensuel. [Edited by P. Duchartre.] Paris. Rev. Bot. Recueil Mens. See B–P–H 778/27. HI 58716

Revue de botanique systématique et de géographie botanique. Paris. Rev. Bot. Syst. Géogr. Bot. See B–P–H 778/29. HI 58717

Revue bretonne de botanique pure et appliquée. Rennes. Vols. 1-29, 1906-37. Rev. Bretonne Bot. Pure Appl. 4-3628-1. HI 72193

Revue bryologique. Caen. Vols. 1-58, 1874-1931 [vols. 42, 1915; 46, 1919, & 54, 1927 never published. Vol. 55-58, 1928-31 also designated n.s. vols. 1-4]. Rev. Bryol. Superseded by: Revue bryologique et lichénologique. HI 72194

Revue bryologique et lichénologique. Caen. N.s. vols. 5-48, [1932]-[79]. Rev. Bryol. Lichénol. Preceded by: Revue bryologique. Superseded by: Cryptogamie - bryologie et lichénologie. 4-3628-2. HI 72195

Revue de la C S T A = C S T A review. Ottawa.

Revue canadienne de biologie. Montreal. Vols. 1-40, 1942-81. Rev. Canad. Biol. Superseded by: Revue canadienne de biologie experimentelle. 4-3628-2. HI 72196

Revue canadienne de biologie experimentelle. Montreal. Vols. 41-42, 1982-83. Rev. Canad. Biol. Exp. Preceded by: Revue canadienne de biologie. Superseded by: Experimental biology. HI 72197

Revue canadienne de géographie. Montreal. Vols. 1-17, 1947-63. Rev. Canad. Géogr. Preceded by: Bulletin des sociétés de géographie de Québec et Montréal. Superseded by: Revue de géographie de Montréal. 4-3628-2. HI 72198

Revue canadienne de phytologie = Canadian journal of plant pathology. Ottawa.

Revue chrysanthémiste. Paris. Rev. Chrysanthémiste. See B–P–H 779/14. HI 58718

Revue coloniale. Paris. Rev. Colon. See B–P–H 779/20. HI 58719

Revue des cours scientifiques de la France et de l'étranger. Paris. Rev. Cours Sic. France Étranger. See B–P–H 779/21. HI 58720

Revue des cultures coloniales. Paris. Vols. 1(1)-15(153), 1897-1904. Rev. Cultures Colon. Incorporated in: Quinzaine coloniale. 4-3644-2. HI 72199

Revue de cytologie et de biologie végétales. Paris. Vols. 11-40, 1945-77. Rev. Cytol. Biol. Vég. Preceded by: Revue de cytologie et de cytophysiologie végétales. Superseded by: Revue de cytologie et de biologie végétales - le botaniste. 4-3634-1. HI 72200

Revue de cytologie et de biologie végétales - le botaniste. Paris. Vol. 1+, 1978+. Rev. Cytol. Biol Vég. Botaniste. Preceded by: Revue de cytologie et de biologie végétales and (Le) Botaniste. HI 72201

Revue de cytologie et de cytophysiologie végetales. Paris. Rev. Cytol. Cytophysiol. Vég. See B–P–H 779/25. HI 58721

Revue Desjardins. Quebec. Rev. Desjardins. See B–P–H 779/33. HI 58722

Revue des eaux et forêts. Paris. Vols. 1-86, 1862-1948. Rev. Eaux Forêts. Superseded by: Revue forestière française. 4-3645-2. HI 72202

Revue d'écologie; la terre et la vie. Paris. Vol. 36+, 1982+. Rev. Écol. Preceded by: Revue d'écologie appliquée à la protection de la nature. HI 74519

Revue d'écologie. Supplément. La terre et la vie. Paris. Vol. 3+, 1986+. Rev. Écol., Suppl. Preceded by: Terre et la vie. Supplément. HI 74876

Revue d'écologie appliquée à la protection de la nature. Paris. Vols. 34-35, 1980-81. Rev. Écol. Appl. Protect. Nat. Preceded by: Terre et la vie. Superseded by: Revue d'écologie. HI 74877

Revue d'écologie et de biologie du sol. Paris. Rev. Écol. Biol. Sol. See B–P–H 780/5. HI 58723

Revue économique forestière universelle. Neudamm & Berlin = Zeitschrift für Weltforstwirtschaft. Neudamm & Berlin. Z. Weltforstw. See B–P–H 988/15.

Revue de l'élevage et des productions animales françaises; bétail & bassecour. Paris. Rev. Élevage Prod. Anim. Franç. See B–P–H 780/8. HI 58724

Revue encyclopédique, ou analyse raisonné des productions les plus remarquables dans la littérature, les sciences et les arts. Paris. Vols. 1-61, 1819-35. Rev. Encycl. Superseded by: Nouvelle revue encyclopédique. 4-3652-1. HI 51257

Revue entomologique. Strasbourg. Rev. Entomol. See B–P–H 780/11. HI 58725

Revue d'études corses. Ajaccio = Revue d'études historiques, littéraires et scientifiques corses. Ajaccio.

Revue d'études historiques, littéraires et scientifiques corses. Ajaccio. Vol. 1(1-4), 1961. Rev. Études Hist. Litt. Sci. Corses. Preceded by: Etudes corses. Superseded by: Corse historique, archéologique, littéraire, scientifique. HI 72203

Revue européenne d'océanologie = Oceanologica acta. Paris.

Revue de la faculté des sciences forestières de l'université d'Istanbul. Série A = Istanbul üniversitesi orman fakültesi dergisi. Ser. A. Istanbul.

Revue de la faculté des sciences de l'université d'Istanbul = Istanbul üniversitesi fen fakültesi mecmuasi. Istanbul.

Revue de la faculté des sciences de l'université d'Istanbul. Série B, sciences naturelles = Istanbul üniversitesi fen fakültesi mecmuasi. Seri B, tabii ilimler. Istanbul.

Revue de la fédération française des sociétés des sciences naturelles. Paris. 1962-73. Rev. Féd. Franç. Soc. Sci. Nat. Preceded by: Bulletin trimestriel, fédération française des sociétés des sciences naturelles. Superseded by: Bulletin, société versaillaise de sciences naturelles. HI 72204

Revue des fermentations et des industries alimentaires. Brussels. Vols. 1+, 1946+. Rev. Ferment. Industr. Aliment. Preceded by: Annales de zymologie. 4-3646-2. HI 72205

Revue de la fondation océanographique Ricard. Six-Fours-les-Plages. No. ?-4+, ?-1980+. Rev. Fond. Océanogr. Ricard. HI 72206

Revue forestière. Zagreb = Šumarski list. Zagreb.

Revue forestière croate = Hrvatski šumarski list. Zagreb.

Revue forestière de France. Paris. Rev. Forest. France. See B–P–H 780/26. HI 58726

Revue forestière française. Nancy. Vol. 1+, 1949+. Rev.

Forest. Franç. Preceded by: Revue des eaux et forêts. HI 72207

Revue forestière Slovène = Gozdarski vestnik. Ljubljana.

Revue der Fortschritte der Naturwissenschaften in theoretischer und praktischer Beziehung. Cologne & Leipzig. Rev. Fortschr. Naturwiss. See B–P–H 780/27. HI 58727

Revue française. Paris. Nos. 1-16, 1828-30; n.s. vols. 1-12, 1837-39. Rev. Franç. (Paris, 1828-30, 1837-39). 4-3652-3. HI 52563

Revue française de l'agriculture. Paris. Vol. 1+, 1963+. Rev. Franç. Agric. HI 72208

Revue française d'apiculture. Paris. Vol. 1+, 1946+. Rev. Franç. Apic. HI 72209

Revue française de l'étranger et des colonies. Exploration et gazette geógraphique. Paris. Rev. Franç. Étranger Colon. See B–P–H 780/28. HI 58728

Revue française d'outre mer. Paris. Vols. 42(750)-43(768), 1938-39. Rev. Franç. Outre Mer. Preceded by: Quinzaine coloniale. 4-3653-1. HI 72210

Revue française des sciences de l'eau. Paris. Vol. 1+, 1982+. Rev. Franç. Sci. Eau. HI 72211

Revue générale agronomique. Louvain. Vols. 6-32(1), 1897-1928 [suspended 1914-20]. Rev. Gén. Agron. Preceded by: Revue agronomique. Superseded by: Agricultura. Louvain. 1-86-1. HI 72212

Revue générale de botanique. Paris. Rev. Gén. Bot. See B–P–H 780/31. HI 58729

Revue générale du caoutchouc. Paris. Rev. Gén. Caoutchouc. See B–P–H 780/32. HI 58730

Revue générale d'histologie. Paris. Rev. Gén. Histol. See B–P–H 780/34. HI 58731

Revue générale de médecine vétérinaire. Toulouse. Rev. Gén. Med. Vétérin. See B–P–H 781/1. HI 58732

Revue générale d'ophtalmologie. Paris. Rev. Gén. Ophtalmol. See B–P–H 781/2 HI 58733

Revue générale des sciences pures et appliquées. Paris. Rev. Gén. Sci. Pures Appl. See B–P–H 781/4. HI 58734

Revue de géographie alpine. Grenoble. Vol. 8+, 1920+. Rev. Géogr. Alpine. Preceded by: Recueil des travaux de l'institut de géographie alpine. 4-3635-2. HI 72213

Revue de géographie de Montréal. Montreal. Vols. 18-30, 1964-77. Rev. Géogr. Montréal. Preceded by: Revue canadienne de géographie. Superseded by: Géographie physique et quaternaire. HI 72214

Revue de géographie physique et de géologie dynamique. Paris. Rev. Géogr. Phys. Géol. Dynam. See B–P–H 781/8. HI 58737

Revue géographique. Paris. Rev. Géogr. See B–P–H 781/5. HI 58735

Revue géographique internationale. Paris. Rev. Géogr. Int. See B–P–H 781/6. HI 58736

Revue géologique = Geologisches Zentralblatt. Anzeiger für Geologie, Petrographie, Palaeontologie und verwandte Wissenschaften. Leipzig. Geol. Zentralbl. See B–P–H 399/17.

Revue hebdomadaire de chimie scientifique et industrielle. Paris. Rev. Hebd. Chim. Sci. Industr. See B–P–H 781/12. HI 58738

Revue d'histoire de la médecine hébraique. Paris. Rev. Hist. Méd. Hébraique. See B–P–H 781/15. HI 58740

Revue d'histoire de la pharmacie. Bulletin de la société de la pharmacie. Paris. Rev. Hist. Pharm. See B–P–H 781/16. HI 58741

Revue d'histoire des sciences; et de leurs applications. Paris. Vol. 1+, 1947+. Rev. Hist. Sci. Applic. HI 72215

Revue historique. Paris. Vol. 1+, 1876+. Rev. Hist. HI 72216

Revue horticole. Marseilles = Revue horticole des Bouches-du-Rhône. Marseilles. Rev. Hort. Bouches-du-Rhône. See B–P–H 781/24.

Revue horticole; journal d'horticulture pratique. Paris. Vols. 1(1)-146(2327), 1829-1974; [n.s.] no. 233+, 1983+. Rev. Hort. For 1974-82 see: Pépiniéristes, horticulteurs, maraîchers. 4-3657-2. HI 72217

Revue horticole de l'Algérie, Tunisie, Maroc. Algiers. 1897+. Rev. Hort. Algérie Tunisie Maroc. From 1930-47 see: Revue d'horticulture et agriculture de l'Afrique du Nord. 4-3657-3. HI 74878

Revue horticole des Bouches-du-Rhône. Marseilles. Rev. Hort. Bouches-du-Rhône. See B–P–H 781/24. HI 58744

Revue horticole suisse. Geneva. Vols. 1-46, 1928-73. Rev. Hort. Suisse. HI 72218

Revue horticole de la Suisse romande. Geneva. Rev. Hort. Suisse Romande. See B–P–H 782/1. HI 58745

Revue d'horticulture et agriculture de l'Afrique du Nord. Algiers. 1930-47. Rev. Hort. Afrique N. Preceded and superseded by: Revue horticole de l'Algérie, Tunisie, Maroc. 4-3632-3. HI 58742

Revue de l'horticulture belge et étrangère. Ghent. Rev. Hort. Belge Etrangère. See B–P–H 781/22. HI 58743

Revue hydrologique. Budapest = Hidrológiai közlöny. Budapest. Hidrol. Közl. See B–P–H 416/4.

Revue d'hydrobiologie tropicale. Bondy, Paris. Vol. 14+, 1981+. Rev. Hydrobiol. Trop. Preceded by: Cahiers O R S T O M. Série hydrobiologie. HI 72219

Revue d'hydrologie. Aarau = Zeitschrift für Hydrologie. Aarau, Switzerland. Z. Hydrol. (Aarau). See B–P–H 985/17.

Revue industrielle. Paris. Rev. Industr. See B–P–H 782/6. HI 58746

Revue d'information agricole. Leopoldville [=Kinshasa, Republic of the Congo]. Rev. Inform. Agric. See B–P–H 782/7. HI 58747

Revue de l'institut français du pétrole et annales des combustibles liquides. Paris. Rev. Inst. Franç. Pétrole & Ann. Combust. Liquides. See B–P–H 782/11. HI 58748

Revue de l'institut Pasteur de Lyon. Lyons. Vol. 1+, 1967+. Rev. Inst. Pasteur Lyon. HI 72220

Revue international des sciences biologiques = Revue international des sciences. Paris. Rev. Int. Sci. See B–P–H 782/17.

Revue internationale d'agriculture. Rome. 1927-46. Rev. Int. Agric. Preceded by: Revue internationale de renseignements agricoles. HI 72221

Revue internationale du bois. Paris. Rev. Int. Bois. See B–P–H 782/13. HI 58749

Revue internationale de botanique appliquée et d'agriculture tropicale. Paris. Vols. 26-33, 1946-53. Rev. Int. Bot. Appl. Agric. Trop. Preceded by: Revue de botanique appliquée et d'agriculture tropicale. Superseded by: Journal d'agriculture tropicale et de botanique appliquée. 4-3658-2. HI 72222

Revue internationale d'océanographie et médicale. Nice. Rev. Int. Océanogr. Méd. See B–P–H 782/15. HI 58750

Revue internationale de phytosociologie, ecologie et phytogéographie. The Hague = Vegetatio; acta geobotanica. The Hague. Vegetatio. See B–P–H 948/18.

Revue internationale des products tropicaux et du matériel tropica. Paris. Rev. Int. Prod. Trop. Matér. Trop. See B–P–H 782/16. HI 58751

Revue internationale de renseignements agricoles. Rome. Vols. 7-13, 1916-22; n.s. vols. 1-4, 1923-26. Rev. Int. Renseign. Agric. Preceded by: Bulletin mensuel des renseignements agricoles et des maladies des plantes. Superseded by: Revue internationale d'agriculture. HI 72223

Revue internationale des sciences. Paris. Rev. Int. Sci. See B–P–H 782/17. HI 58752

Revue internationale des tobaco. Paris. Rev. Int. Tobaco. See B–P–H 782/19. HI 58753

Revue des jardins. Quebec. Rev. Jard. See B–P–H 782/24. HI 58754

Revue des jardins et des champs; journal mensuel d'horticulture et d'agriculture. Paris. Vols. 1-11, 1860-70. Rev. Jard. Champs. HI 72224

Revue de Madagascar; organe du comité de Madagascar. Paris. Vols. 1-13, 1899-1911. Rev. Madagascar. HI 72225

Revue médicale. Louvain. Rev. Méd. (Louvain). See B–P–H 782/28. HI 58755

Revue médicale. Paris. Rev. Méd. (Paris). See B–P–H 782/29. HI 58756

Revue médicale de l'Est. Paris & Nancy. Rev. Méd. Est. See B–P–H 782/32. HI 58757

Revue médicale, française et étrangère. Paris. Rev. Méd. Franç. Étrangère. See B–P–H 783/3. HI 58759

Revue médicale, française et étrangère, et Journal de clinique de l'Hôtel-Dieu et de la Charité de Paris. Paris. Rev. Méd. Franç. Étrangère & J. Clin. Hôtel-Dieu Charité Paris. See B–P–H 783/2. HI 58758

Revue médicale, historique et philosphique. Paris. Rev. Méd. Hist. Philos. See B–P–H 783/5. HI 58760

Revue médicale de Nancy. Paris & Nancy. Rev. Méd. Nancy. See B–P–H 783/8. HI 58761

Revue médicale de la Suisse romande. Lausanne & Geneva. Rev. Méd. Suisse Romande. See B–P–H 783/11. HI 58762

Revue mensuelle illustrée de botanique cryptogamique et d'anatomie végétale = Brébissonia. Paris. Brébissonia. See B–P–H 227/32.

Revue mensuelle d'information et de techniques apicoles. Rhisnes. 1967+. Rev. Mens. Inform Techn. Apicoles. HI 72226

Revue de micropaléontologie. Paris. Rev. Micropaléontol. See B–P–H 783/18. HI 58763

Revue de mycologie. Paris. N.s. vols. 1-43, 1936-79. Rev. Mycol. (Paris). Preceded by: Annales de cryptogamie exotique. Superseded by: Cryptogamie - mycologie. 4-3640-3. HI 72227

Revue de mycologie; supplément colonial. Paris. Rev. Mycol. Suppl. Colon. See B–P–H 783/28. HI 58765

Revue mycologique; recueil trimestriel illustré, consacré à l'étude des champignons et des lichens. Toulouse. Rev. Mycol. (Toulouse). See B–P–H 783/27. HI 58764

Revue d'Oka, agronomie, médecine, vétérinaire de l'institut agricole. Oka, Quebec. Rev. Oka Agron. Inst. Agric. See B–P–H 783/30. HI 58766

Revue de l'Orient, de l'Algérie et des colonies. Paris. Rev. Orient. See B–P–H 784/1. HI 58767

Revue de pathologie végétale et d'entomologie agricole de France. Paris. Rev. Pathol. Vég. Entomol. Agric. France. See B–P–H 784/5. HI 58768

Revue philosophique, littéraire et politique. Paris. Rev. Philos. See B–P–H 784/7. HI 58769

Revue physiologique. Rev. Physiol. See B–P–H 784/8. HI 58770

Revue des questions scientifiques. Louvain. Rev. Quest. Sci. See B–P–H 784/14. HI 58772

Revue de la recherche agronomique marocaine = El Awamia. Rabat, Morocco. Awamia. See B–P–H 166/9.

Revue de recherches du service canadien des forêts. Ottawa. Vol. 1+, 1981+. Rev. Rech. Serv. Canad. Forêts. Preceded by: Revue bimestrielle de recherches, service des forêts, Canada. HI 72228

Revue romande d'agriculture, de viticulture et d'arboriculture. Lausanne. Vols. 1-17, 1945-61. Rev. Romande Agric. Superseded by: Agriculture romande. 4-3667-1. HI 72229

Revue roumaine de biochimie. Bucharest. Vol. 1+, 1964+. Rev. Roumaine Biochim. HI 72230

Revue roumaine de biologie. Bucharest. Vol. 19-20, 1974-75. Rev. Roumaine Biol. Preceded by: Revue roumaine de biologie. Série de botanique. Superseded by: Revue roumaine de biologie. Série de biologie végétale. HI 72231

Revue roumaine de biologie. Série de biologie végétale. Bucharest. Vol. 21+, 1976+. Rev. Roumaine Biol., Sér. Biol. Vég. Preceded by: Revue roumaine de biologie. Série de botanique. HI 72232

Revue roumaine de biologie. Série de botanique. Bucharest. Vols. 9-18, 1964-73. Rev. Roumaine Biol., Sér. Bot. Preceded by: Revue de biologie. Superseded by: Revue roumaine de biologie. Série de biologie végétale. HI 72233

Revue roumaine d'embryologie et de cytologie. Série de cytologie. Jassy. 1964-73. Rev. Roumaine Embryol. Cytol., Sér. Cytol. Superseded by: Revue roumaine de morphologie et d'embryologie. HI 72234

Revue roumaine d'inframicrobiologie. Académie de la République Populaire Roumaine. Bucharest. Rev. Roumaine Inframicrobiol. See B–P–H 784/15. HI 58773

Revue roumaine de morphologie et d'embryologie. Bucharest. Vol. 19, 1974. Rev. Roumaine Morphol. Embryol. Preceded by: Revue roumaine d'embryologie et de cytologie. Série de cytologie. Superseded by: Revue roumaine de morphologie et de physiologie. HI 72235

Revue roumaine de morphologie, d'embryologie et physiologie. Série morphologie et embryologie. Bucharest. Vol. 21+, 1975+. Rev. Roumaine Morphol. Embryol. Physiol., Sér., Morphol. Embryol. Preceded by: Revue roumaine de morphologie et de physiologie. HI 72236

Revue roumaine de morphologie et de physiologie. Bucharest. Vol. 20, 1974. Rev. Roumaine Morphol. Physiol. Preceded by: Revue roumaine de morphologie et d'embryologie. Superseded by: Revue roumaine de morphologie, d'embryologie et physiologie. Série morphologie et embryologie. HI 72237

Revue sanitaire de la province Bordeaux. Bordeaux. Rev. Sanit. Prov. Bordeaux. See B–P–H 784/16. HI 58774

Revue savoisienne; journal publié par l'association florimontane d'Annecy, histoire - sciences - arts - industrie - littérature. Annecy. Vol. [1]+, 1860+. Rev. Savois. HI 72238

Revue des sciences naturelles. Montpellier, Paris. Vols. 1-7, 1872-78; 2nd ser., vols. 1-2, 1879-81; 3rd ser., vols. 1-4, 1881-85 [2nd & 3rd series also numbered as vols. gén. 8-13]. Rev. Sci. Nat. Montpellier. HI 72239

Revue des sciences naturelles de l'ouest. Paris. Rev. Sci. Nat. Ouest. See B–P–H 785/6. HI 59063

Revue des sciences naturelles. St. Petersburg = Věstnik Estestvoznaniya. St. Petersburg.

Revue des sciences naturelles, société impériale des naturalistes de Moscou. Moscow. = Věstnik Estestvennykh Nauk, izdavaemiya Moskovskim Obshchestvom Ispytatelei Prirody. Moscow.

Revue des sciences naturelles appliquées. Paris. Vols. 36-42, 1889-95. Rev. Sci. Nat. Appl. Preceded by: Bulletin de la société zoologique d'acclimatation [not entered]. Superseded by: Bulletin de la société nationale d'acclimatation de France. HI 72240

Revue des sciences naturelles d'Auvergne. Clermont Ferrand. Vols. 1+, 1935+. Rev. Sci. Nat. Auvergne. Preceded by: Bulletin de la société d'histoire naturelle d'Auvergne. 4-3648-3. HI 72241

Revue scientifique. Paris. Rev. Sci. See B–P–H 784/18. HI 58775

Revue scientifique du Bourbonnais et du centre de la France. Moulins, France. Rev. Sci. Bourbonnais Centr. France. See B–P–H 784/22. HI 58776

Revue scientifique hebdomadaire de Montpellier. [Edited by Béchamp et al.] Paris. Rev. Sci. Hebd. Montpellier. See B–P–H 784/26. HI 58777

Revue scientifique et industrielle. [Edited by Breton.] Paris. Rev. Sci. Industr. (Breton). See B–P–H 784/27. HI 58778

Revue scientifique et industrielle des faits les plus utiles et les plus curieux observés dans la médecine, ... [Edited by Quesneville.] Paris. Rev. Sci. Industr. (Quesneville). See B–P–H 784/28. HI 58779

Revue scientifique italienne. Paris. Rev. Sci. Ital. See B–P–H 785/1. HI 58780

Revue scientifique du Limousin. Limoges. Vols. 1(1)-38(387), 1893/94-1936/38, 1896-1938. Rev. Sci. Limousin. Preceded by: Règne végétal. Revue mensuelle publ. par la société botanique du Limousin. 5-3940-3. HI 72242

Revue de séricicultrure comparée. Paris. Rev. Séricic. Comp. See B–P–H 785/9. HI 58781

Revue de la société montpelliéraine d'histoire de la médecine = Monspeliensis Hippocrates. Revue de la société montpelliéraine d'histoire de la médecine. Montpellier. Monspel. Hippocrates. See B–P–H 618/7.

Revue des sociétés savantes des départements = Revue des société savantes de la France et de l'étranger. Paris. Rev. Soc. Savantes France Etranger. See B–P–H 785/15.

Revue des société savantes de la France et de l'étranger. Paris. Rev. Soc. Savantes France Etranger. See B–P–H 785/15. HI 58782

Revue des sociétés savantes de Haute Normandie. Rouen. Rev. Soc. Savantes Haute Normandie. See B–P–H 785/16. HI 58783

Revue suisse d'agriculture. Lausanne. Vol. 1-?, 1969-? Rev. Suisse Agric. Preceded by: Agriculture romande. Superseded by: Revue suisse de viticulture, arboriculture et horticulture. HI 72243

Revue suisse de biotechnologie; organe d'information du comité de coordination de la biotechnmologie en Suisse = Swiss biotech; Schweizerische zeitschrift für biotechnologie; Informationsorgan des schweizerische koordinationsausschusses für biotechnologie. Küsnacht.

Revue suisse d'hydrologie = Schweizerische Zeitschrift für Hydrologie. Basel.

Revue suisse de viticulture et d'arboriculture. Lausanne. Vol. 1-3, 1969-71. Revue Suisse Vitic. Arboric. Preceded by: Agriculture romande. Superseded by: Revue suisse de viticulture, arboriculture et horticulture. HI 72244

Revue suisse de viticulture, arboriculture et horticulture. Nyon. Vol. 4+, 1972+. Rev. Suisse Vitic. Arboric. Hort. Preceded by: Revue suisse de viticulture et arboriculture. HI 72380

Revue des tabacs. Paris. Rev. Tabacs. See B–P–H 785/23. HI 58784

Revue des tabacs helléniques. Kavalla, Greece. Rev. Tabacs Hellén. See B–P–H 785/24. HI 58785

Revue technique des industries du cuir; scientifique et industrielle. Paris. Rev. Techn. Industr. Cuir. See B–P–H 785/25. HI 58786

Revue des travaux de l'institut de géographie alpine. Grenoble. Vols. 1-7, 1913-19. Rev. Trav. Inst. Géogr. Alpine. Superseded by: Revue de géographie alpine. 4-3635-2. HI 70544

Revue des travaux scientifiques. Paris. Rev. Trav. Sci. See B–P–H 785/27. HI 58787

Revue des travaux scientifiques tchécoslovaques. Section deuxième. Prague. Rev. Trav. Sci. Tchécoslov., Sect. 2. See B–P–H 785/28. HI 58788

Revue trimestrielle consacrée à la protection des plantes en Pologne = Choroby i szkodniki roślin. Warsaw.

Revue trimestrielle des recherches des matières premières = Quarterly journal of crude drug research. Amsterdam.

Revue trimestrielle des sociétés des amis des arbres et du reboisement a Nice et d'histoire naturelle de Haute-Provence a Digne. Digne. 1966-67. Rev. Trimestrielle Soc. Amis Arbres Rebois. Nice Hist. Nat. Haute-Provence Digne. Preceded by: Bulletin trimestriel de la société des amis des arbres et du reboisement et société d'histoire naturelle de Haute-Provence. Superseded by: Revue trimestrielle de la société d'histoire naturelle de Haute-Provence. HI 72245

Revue trimestrielle de la société d'histoire naturelle de Haute-Provence. Digne. 1968-70. Rev. Trimestrielle Soc. Hist. Nat. Haute-Provence. Preceded by: Revue trimestrielle des sociétés des amis des arbres et du reboisement a Nice et d'histoire naturelle de Haute-Provence a Digne. HI 72246

Revue universelle d'économie forestière, de la science du sol et communications coloniales forestières = Zeitschrift für Weltforstwirtschaft waldwirtschaftliche und bodenkundliche Grossraumforschung zugleich Kolonialforstliche Mitteilungen. Neudamm [=Debno, Poland] & Berlin. Z. Weltforstw. Waldw. Bodenk. Grossraumf. & Kolonialforstl. Mitt. See B–P–H 988/16.

Revue de l'université officielle de Bujumbura. Bujumbura. Vol. 1+, 1967+. Rev. Univ. Off. Bujumbura. HI 72247

Revue de l'université d'Ottawa. Ottawa. Vol. 1+, 1931+. Rev. Univ. Ottawa. 4-3215-1. HI 72248

Revue Valdôtaine d'histoire naturelle. Aoste. Vol. 25+, 1972+. Rev. Valdôtaine Hist. Nat. Preceded by: Bulletin, société de la flore Valdôtaine. HI 72249

Revue verviétoise d'histoire naturelle. Verviers. 1949+. Rev. Verviét. Hist. Nat. Preceded by: Naturaliste amateur. HI 72250

Revue de viticulture. Paris. Rev. Vitic. See B–P–H 786/10. HI 58789

Revue de zoologie et de botanique africaines. Brussels & Ostend. Vols. 18-87, 1928-73. Rev. Zool. Bot.

Africaines. Preceded by: Revue zoologique africaine [Not entered.] Superseded by: Revue de zoologie africaine [Not entered.] 4-3643-1. HI 72251

Revuo de biostatisko. [Seibutsu tokeigaku zasshi.] Tokyo. Vols. 1-10, 1952-68. Rev. Biostat. HI 72252

Rezervy rastenievodstva. Vol. 1(13)+, 1979+. Rezervy Rasteniev. HI 72253

Rheinische Beiträge zur Gelehrsamkeit. Mannheim. Rhein. Beitr. Gelehrsamk. See B–P–H 798/35. HI 58968

Rheinische Heimatpflege. Düsseldorf. Rhein. Heimatpflege. See B–P–H 798/36. HI 58969

Rheinisches Jahrbuch für Gartenkunde und Botanik. Bonn. Rhein. Jahrb. Gartenk. Bot. See B–P–H 798/37. HI 58970

Rheinisches Magazin zur Erweiterung der Naturkunde. Giessen, Germany. Rhein. Mag. Erweit. Naturk. See B–P–H 798/38. HI 58971

Rheinische Mannigfaltigkeiten. Mainz. Rhein. Mannigfaltigk. See B–P–H 798/39. HI 58972

Rheinische Provinzialblätter. Cologne. Rhein. Provinzialbl. See B–P–H 798/40. HI 58973

Rheinische Zeitschrift für Landwirthschaft. Mainz. Rhein Z. Landw. See B–P–H 798/41. HI 58974

Rheinländische Gartenzeitung = Rheinländische (landwirthschaftliche) Gartenzeitung. Neuwied, Germany. Rheinl. Gartenzeitung. See B–P–H 798/42.

Rheinländische landwirthschaftliche Gartenzeitung. Neuwied, Germany. Rheinl. Gartenzeitung. See B–P–H 798/42. HI 58975

Rhizome reporter. Miami, FL. ?-1982+. Rhizome Reporter. HI 72254

Rhode Island agriculture. Kingston, RI. Rhode Island Agric. See B–P–H 798/46. HI 58976

Rhodesia agricultural journal; official journal of the ministry of agriculture. Salisbury, Rhodesia. Vols. 1-76(3), 1903-79. Rhodesia Agric. J. Superseded by: Zimbabwe Rhodesia agricultural journal. 4-3675-3. HI 72255

Rhodesia agricultural journal. Technical bulletin. No. ?-7+, ?-1969+. Rhodesia Agric. J., Techn. Bull. HI 72256

Rhodesia bulletin of forest research. Salisbury, Rhodesia. No. 1+, 1968+. Rhodesia Bull. Forest. Res. HI 72257

Rhodesia research index. Salisbury, Rhodesia. 1971-78, 1972-79. Rhodesia Res. Index. Superseded by: Zimbabwe research index. HI 72258

Rhodesia science news. Salisbury, Rhodesia. Vols. 1-13(4), 1967-79. Rhodesia Sci. News. Superseded by: Zimbabwe Rhodesia science news. HI 72259

Rhodesia, Zambia and Malawi journal of agricultural research. Salisbury, Rhodesia. Vols. 3(2)-5(3), 1965?-67. Rhodesia Zambia Malawi J. Agric. Res. Preceded and superseded by: Rhodesian journal of agricultural research. HI 72260

Rhodesian journal of agricultural research. Salisbury, Rhodesia. Vols. 1-3(1), 1963-65; vols. 5(4)-17, 1967-69. Rhodesia J. Agric. Res. From 1965-67 titled: Rhodesia, Zambia and Malawi journal of agricultural research. Superseded by: Zimbabwe journal of agricultural research. HI 72261

Rhodesian tobacco journal. Salisbury, Rhodesia. Rhodesian Tobacco J. See B–P–H 799/5. HI 58977

Rhododendron. Melbourne, Vic. 1962+. Rhododendron (Melbourne). Preceded by: Journal of the australian rhododendron society. HI 72262

Rhododendron and azalea news. Portland, Tigard, OR. Vol. 1+, 1980+. Rhododendron Azalea News. HI 72263

Rhododendron and camellia year book. London. Vols. 9-26, 1954-71. Rhododendron Camellia Year Book. Preceded by: Rhododendron year book. Superseded by: Rhododendrons, with magnolias and camellias. HI 72264

Rhododendron und Immergrüne Laubgehölze Jahrbuch. Berlin, Bremen. 1938-42, 1955+. Rhododendron Immergrüne Laubgeh. From 1952-54 entitled: Jahrbuch der Rhododendron-Gesellschaft. HI 72265

Rhododendron notes and records. Federal Way, WA. Vol. 1+, 1984+. Rhododendron Notes Rec. HI 72266

Rhododendron society notes. Ipswich, England. Rhododendron Soc. Notes. See B–P–H 799/14. HI 58978

Rhododendron species foundation newsletter = R S F newsletter.

Rhododendron year book. London. 1932-?, 1946-53. Rhododendron Year Book. Superseded by: Rhododendron and camellia year book. HI 72267

Rhododendrons, with magnolias and camellias. London. 1972+. Rhododendron Magnolias Camellias. Preceded by: Rhododendron and camellia year book. HI 72268

Rhodora; journal of the New England botanical club. Lancaster, PA. Rhodora. See B–P–H 799/17. HI 58979

Rhönwacht; Zeitschrift des Rhönklubs. Fulda. ?-1969+. Rhönwacht. HI 72269

Rice abstracts. Farnham Royal. Vol. 1+, 1977+. Rice Abstr. HI 72270

Rice journal. New Orleans, LA. Rice J. See B–P–H 800/1. HI 58987

Rice report, department of agriculture, Solomon Islands. 1968-71. Rice Rep. Dept. Agric. Solomon Islands. HI 72271

Rice review. Georgetown, Guyana. Vol. 1+, 1959+. Rice Rev. HI 72272

Rice, sugar and coffee journal. New Orleans, LA. Rice Sugar Coffee J. See B–P–H 800/2. HI 58988

Rice and sugar journal. New Orleans, LA. Rice Sugar J. See B–P–H 800/3. HI 58989

Rice and wheat. Mitsubishi agricultural journal. [Ine to mugi.] Tokyo. Rice & Wheat. See B–P–H 799/24. HI 58986

Ricerca e documentazioni tessile. Milan. Ric. Doc. Tessile. See B–P–H 799/18. HI 58980

Ricerca scientifica, La. Rome. Vol. 18+, 1948+. Ric. Sci. Preceded by: Ricerca scientifica e recostruzione. 4-3676-3. HI 72273

Ricerca scientifica ed il progresso tecnico. Rome. Vols. 12-14, 1941-43. Ric. Sci. Progr. Tecn. Preceded by: Ricerca scientifica ed il progresso tecnico nell'economia nazionale. Superseded by: Ricerca scientifica e recostruzione. Memorie. 4-3676-3. HI 72274

Ricerca scientifica ed il progresso tecnico nell'economia nazionale. Rome. Vols. 3-11, 1932-40. Ric. Sci. Progr. Tecn. Econ. Naz. Preceded by: Bollettino d'informazioni, consiglio nazionale delle richerche. Superseded by: Ricerca scientifica ed il progresso tecnico. 4-3676-3. HI 72275

Ricerca scientifica. Quaderni. Rome. Ric. Sci., Quad. See B–P–H 799/20. HI 58982

Ricerca scientifica e recostruzione. Memorie. Rome. Vols. 15-17, 1945-47. Ric. Sci. Recostruz. Preceded by: Ricerca scientifica ed il progresso tecnico. Superseded by: Ricerca scientifica, La. 4-3676-3. HI 72276

Ricerca scientifica. Serie 2a. Parte I: Rivista. Rome. Ric. Sci., Ser. 2a, Pt. 1, Rivista. See B–P–H 799/21. HI 58983

Ricerca scientifica. Serie 2a. Parte II: Rendiconti, sezione B: Biologica. Rome. Ric. Sci., Ser. 2a, Pt. 2, Rendiconti, Sez. B, Biol. See B–P–H 799/22. HI 58984

Ricerca scientifica. Serie 2a. Supplemento. Rome. Ric. Sic., Ser. 2a, Suppl. See B–P–H 799/23. HI 58985

Ricerche e lavori. Istituto botanico, università di Pisa. Pisa. Ric. Lav. Ist. Bot. Univ. Pisa. See B–P–H 799/19. HI 58981

Ricerche e lavori del r. museo ed orto botanico di Firenze. Florence. 1896-97. Ric. Lav. Mus Orto Bot. Firenze. HI 72277

Ricerche sulla riserva naturale di Torricchio. Camerino. Vol. 1+, 1976+. Ric. Riserva Nat. Torricchio. HI 72278

Rickia. São Paulo. Rickia. See B–P–H 800/4. HI 58990

Rickia. Supplemento. São Paulo. Rickia Suppl. See B–P–H 800/5. HI 58991

Riesengebirge im Wort und Bild. Marschendorf, Bohemia [=Marsov, Czechoslovakia]. Riesengebirge im Wort & Bild. See B–P–H 800/6. HI 58992

Riforma medica. Naples. Riforma Med. See B–P–H 800/8. HI 58993

Rigaku = Science. Tokyo. Science (Tokyo/Rigaku). See B–P–H 829/24.

Rigaku-bu Kaiso kenkyujo hokoku = Report of the research institute of marine algae of the faculty of physical sciences, Hokkaido imperial university. Sapporo, Japan. Rep. Res. Inst. Mar. Algae Fac. Phys. Sci. Hokkaido Imp. Univ. See B–P–H 770/5.

Rigaku-bu kiyo = Journal of the faculty of science of the Hokkaido imperial university. Series 5, botany. Sapporo.

Rigaku-bu kiyo = Journal of the faculty of science, university of Tokyo. Section III. Botany. Tokyo. J. Fac. Sci. Univ. Tokyo, Sect. 3, Bot. See B–P–H 465/18.

Rigakukai = Science world. Tokyo?

Rika kyoiku = Science education. Tokyo. Sci. Educ. (Tokyo). See B–P–H 824/8.

Rikkyo daigaku kenkyu hokoku, shizenkagaku = St. Paul's review of science. Tokyo.

Rikusui gakkai shi = Japanese journal of limnology. Tokyo. Jap. J. Limnol. See B–P–H 498/23.

Rimba Indonesia; penerbitan populer. Bogor. Vols. 1-?, 1952-59. Rimba Indonesia. Superseded by: Suara rimbawan. HI 72279

Rinboku no ikushu = Forest tree breeding. Tokyo.

Ringaku kiho = Sylvia. Taihoku [=Taipei, Taiwan]. Sylvia. See B–P–H 862/19.

Ringakukai zasshi = Journal of the society of forestry. [Japan]. J. Soc. Forest. See B–P–H 482/1.

Ringyo gijitsu = Forest technique. Tokyo. Forest Techn. See B–P–H 381/1.

Ringyo shidojo kenkyu hokoku = Report of the forest products research institute. Asahigawa, Japan. Rep. Forest Prod. Res. Inst. See B–P–H 765/10.

Ringyo shiken hokoku = Bulletin of the imperial forest experiment station. Tokyo.

Ringyo shiken iho = Report of the forestry experiment station. Tokyo. Rep. Forest. Exp. Sta. See

B–P–H 765/8.

Ringyo shiken syuho = Report (Annual) of the government forest experiment station. Tokyo. Rep. (Annual) Gov. Forest Exp. Sta. See B–P–H 765/18.

Ringyo shikenjo Hokkaido shijo nempo = Report (Annual) of the Hokkaido branch, government forest experiment station. Sapporo.

Ringyo shikenjo hokoku, Shokusan-kyoku = Report of the forest experiment station, bureau of plant industry, government of Formosa. Taihoku [=Taipei, Taiwan]. Rep. Forest. Exp. Sta. Bur. Pl. Industr. Gov. Formosa. See B–P–H 765/9.

Ringyo shikenjo Kansai shijo nenpo = Report (Annual) of the Kansai branch, government forest experiment station. Kyoto.

Ringyo shikenjo kenkyu hokoku = Bulletin of the government forest experiment station. Tokyo. Bull. Gov. Forest Exp. Sta. See B–P–H 252/5.

Ringyo shikenjo Kyushu shijo nenpo = Report (Annual) of the Kyushu branch, government forest experiment station. Kumamoto.

Ringyo shikenjo shiken kenkyu hokoku, Tottori = Research report of Tottori prefecture forestry experiment station. Tottori.

Ringyo shikenjo Tohoku shijo nenpo = Report (Annual) of the Tohoku branch, government forest experiment station. Morioka.

Ringyo shin chishiki = New knowledge of forestry. Tokyo.

Rinji Hokoku = Miscellaneous publication. Horticultural research station. Morioka, Japan. Misc. Publ. Hort. Res. Sta. See B–P–H 598/20.

Rinkobu ikushujo kenkyu hokoku. Ibaraki = Bulletin of the forest tree breeding institute. Ibaraki.

Rino-gaku-bu fuzoku shokubutsuen nenpo = Reports (Annual) of the Taihoku botanic garden. Taihoku.

Rino-gaku-bu kiyo = Memoirs of the faculty of science and agriculture; Taihoku imperial university. Taihoku [=Taipei, Taiwan]. Mem Fac. Sci. Taihoku Imp. Univ. See B–P–H 572/25.

Rinyu = Friends of the forest. Tokyo. Friends Forest. See B–P–H 385/10.

Risicoltura. Vercelli, Italy. Risicoltura. See B–P–H 800/31. HI 58994

Riso, Il; rassegna trimestrale di studi. Milan. Vol. 1+, 1952+. Riso. HI 72280

Rit landbúnaoardeildar, atvinnudeild háskólans. B-flokkur. Reykjavik. Nos. 1-19, 1943-63. Rit Landbúnaoard. Atvinnud. Háskólans, B. HI 72281

Rit vísindafjélags íslendinga. Akureyri, Reykjavik. Vols. 1-4, 1923-29. Rit Visindafj. Islend. Superseded by: Vísindafjélags íslendinga. HI 72282

Rivièra scientifique; revue mensuelle des sciences physiques et naturelles. Nice. Vol. 1+, 1914+. Rivièra Sci. Preceded by: Bulletin des naturalistes des Alpes-Maritimes. 4-3684-2. HI 72283

Rivista di agricoltura subtropicale e tropicale. Florence. Vol. 1+, 1907+. Rivista Agric. Subtrop. Trop. Preceded by: Agricoltura coloniale. 4-3686-2. HI 72284

Rivista di agronomia. Bologna. Vol. 1+, 1967+. Rivista Agron. HI 72285

Rivista di agrumicoltura. Stazione sperimentale di agrumicolture e frutticoltura. Acireale, Italy. Rivista Agrumic. See B–P–H 801/10. HI 58995

Rivista alpina italiana; periodico mensile del club alpino italiano. Turin. Vols. 1-3, 1882-84. Rivista Alpina Ital. Superseded by: Rivista mensile, club alpino italiano. 1-151-2. HI 72286

Rivista degli archivi italiani di biologie. Pisa. Rivista Arch. Ital. Biol. See B–P–H 801/11. HI 58996

Rivista di biologia. Rome Rivista Biol. See B–P–H 801/12. HI 58997

Rivista di biologia coloniale. Rome. Rivista Biol. Colon. See B–P–H 801/13. HI 58998

Rivista di biologia generale. Turin. Rivista Biol. Gen. See B–P–H 801/14. HI 58999

Rivista de biología del Uruguay. Montevideo. Vol. 1+, 1973+. Rivista Biol. Uruguay. HI 72287

Rivista del club alpino italiano. Turin. Vols. 26-57(4), 1908-38. Rivista Club Alpino Ital. Preceded by: Rivista mensile, club alpino italiano. Superseded by: Revista mensile del centro alpinistico italiano. 1-151-2. HI 72288

Rivista di fisica, matematica e scienze naturali. Pavia. Rivista Fis. See B–P–H 801/16. HI 59000

Rivista forestale italiana. Florence. Vols. 1-?, 1939-43. Rivista Forest. Ital. Preceded by: Alpe. HI 72289

Rivista di frutticoltura. Ravenna. Vols. 1-16, 1937-54. Rivista Fruttic. Superseded by: Frutticoltura. 4-3687-3. HI 72290

Rivista de frutticoltura e di ortofloricultura. Bologna. Vol. 45+, 1983+. Rivista Fruttic. Ortofloric. Preceded by: Frutticoltura. HI 72291

Rivista di idrobiologia. Perugia. 1960+. Rivista Idrobiol. HI 72292

Rivista idrologica. Budapest = Hidrológiai közlöny. Budapest. Hidrol. Közl. See B–P–H 416/4.

Rivista internazionale di agricoltura. Milan. Rivista Int. Agric. See B–P–H 801/19. HI 59001

Rivista italiana essenze, profumi, piante officinali, aromi, saponi, cosmetici, aerosol. Milan. 1938+. Rivista Ital. Essenze Profumi. HI 72293

Rivista italiana di paleontologia e stratigrafia. Milan. Rivista Ital. Paleontol. Stratigr. See B–P–H 801/20. HI 59002

Rivista italiana di scienze, lettere ed arti. Milan. Rivista Ital. Sci. Lett. See B–P–H 801/21. HI 59003

Rivista italiana di scienze naturale e loro applicazioni. Naples. Vols. 1-2, 1885-86. Rivista Ital. Sci. Nat. Applic. HI 72294

Rivista italiana de scienze naturali. Siena. Rivista Ital. Sci. Nat. See B–P–H 801/22. HI 59004

Rivista italiana delle sostanze grasse. Milan. 1967+. Rivista Ital. Sostanze Grasse. HI 72295

Rivista italiana di stomatologia. Rome. Rivista Ital. Stomatol. (Rome). See B–P–H 801/23. HI 59005

Rivista italiana di stomatologie. Venice. Rivista Ital. Stomatol. (Venice). See B–P–H 801/24. HI 59006

Rivista ligure. Genoa. Rivista Ligure. See B–P–H 801/25. HI 59007

Rivista ligure di scienze, lettere ed arti. Genoa. Rivista Ligure Sci. See B–P–H 801/26. HI 59008

Rivista mensile del centro alpinistico italiano. Turin. Vol. 57(5-12), 1938. Rivista Mens. Centro Alpinist. Ital. Preceded by: Rivista del club alpino italiano. Superseded by: Alpi. 1-151-2. HI 72296

Rivista mensile, club alpino italiano. Turin. Vols. 4-26, 1885-1907. Rivista Mens. Club Alpino Ital. Preceded by: Rivista alpina italiana. Superseded by: Rivista del club alpino italiano. 1-151-2. HI 72297

Rivista di micologia; bollettino dell'associazione micologica Bresadola. Trento. Vol. 30+, 1987+. Rivista Micol. Preceded by: Bollettino del gruppo micologico G. Bresadola. HI 72298

Rivista del museo civico di scienze naturali "E. Caffi". Bergamo. Vol. 1+, 1979+. Rivista Mus. Civico Sci. Nat. E. Caffi. HI 72299

Rivista della ortoflorofrutticultura italiana. Florence. Anno 73-93[-112] [anno numbering stopped with 93,1967], vols. 33-70, 1948-86. Rivista Ortoflorofruttic. Ital. Preceded by: Rivista della r[eale] società toscana di orticultura. Superseded by: Advances in horticultural science. 4-3686-1. HI 59009

Rivista di parassitologia. Rome. Rivista Parassitol. See B–P–H 802/1. HI 59010

Rivista di patologia vegetale. Padua. Vols. 1-10, 1892-1902; ser. 2, vol. 1+, 1905+. Riv. Patol. Veg. 4-3688-3. HI 72300

Rivista pellagrologica italiana. Udine, Italy. Rivista Pellagrol. Ital. See B–P–H 802/2. HI 59011

Rivista periodico della r. scuola di viticoltura d'enologia di Conegliano. Conegliano. Vols. 1-4, 1895-98. Rivista Period. Reale Scuola Vitic. Enol. Conegliano. Preceded by: Annali della r. scuola di viticoltura e di enologia in Conegliano. Superseded by: Rivista periodico quindicinale organo della r. scuola di viticoltura d'enologia e del comizio agrario di Conegliano. HI 72301

Rivista periodico quindicinal organo della r. scuola di viticoltura d'enologia e del comizio agrario di Conegliano. Conegliano. Vols. 5-17, 1899-1912. Rivista Period. Quindic. Organo Reale Scuola Vitic. Enol. Comiz. Agrar. Conegliano. Preceded by: Rivista periodico della r. scuola di viticoltura d'enologia di Conegliano. Superseded by: Rivista periodico quindicinale di viticoltura, enologia ed agraria. HI 72302

Rivista periodico quindicinale di viticoltura, enologia ed agraria. Conegliano. Vols. 18-23, 1912-17. Rivista Period. Quindic. Vitic. Enol. Agrar. Preceded by: Rivista periodico quindicinal organo della r. scuola di viticoltura d'enologia e del comizio agrario di Conegliano. HI 72303

Rivista di radiologia e fisica medica. Bologna. Rivista Radiol. Fis. Med. See B–P–H 802/3. HI 59012

Rivista della r[eale] società toscana di orticultura. Florence. Anno. 64-72, vols. 24-32, 1939-47. Rivista Reale Soc Tosc. Ortic. Preceded by: Bollettino della r[eale] società toscana d'orticultura. Superseded by: Rivista ortoflorofrutticoltura italiana. HI 72304

Rivista sanitaria siciliana. Palermo & Milan. Rivista Sanit. Sicil. See B–P–H 802/4. HI 59013

Rivista de scienza. Bologna. Rivista Sci. See B–P–H 802/5. HI 59014

Rivista di scienze biologiche. Turin. Rivista Sci. Biol. See B–P–H 802/6. HI 59015

Rivista scientifica. R. accademia dei fisiocritici. Siena. Rivista Sci. Reale Accad. Fisiocrit. See B–P–H 802/7. HI 59016

Rivista della società toscana di orticultura. Florence. Rivista Soc. Tosc. Ortic. See B–P–H 802/8. HI 59017

Rivon verhandelingen. Zeist, Netherlands. Rivon Verh. See B–P–H 802/9. HI 59018

Riz et riziculture. Paris. Riz & Rizic. See B–P–H 802/10. HI 59019

Riz et riziculture et cultures vivières tropicales. [Supplement a l'agronomie tropicale. See B–P–H 64/12.] Nogent-sur-Marne, Paris. N.s. vols. 1-7, 1955-61. Riz Rizic. Cult. Vivières Trop. Preceded by: Riz et riziculture. Incorporated in: Agronomie tropicale. 4-

3693-2. HI 72305

Robigo; noticias sobre las royas de los cereales de todos para todos. Castelar. No. 0, 1959; nos. 1-19, 1956-67. Robigo. HI 72306

Ročenka vlastivědné společnosti jihočeské v Ceských Budějovicích. Budweis, Czechoslovakia. Ročenka Vlastiv. Společn. Jihočeské v Ceských Budějovicích. See B–P–H 802/13. HI 59020

Rock garden; journal of the scottish rock garden club. Aberdeen. Vol. 18(3)+, 1983+. Rock Gard. Preceded by: Journal of the scottish rock garden club. HI 72307

Rockefeller institute review. New York, NY. Rockefeller Inst. Rev. See B–P–H 802/15. HI 59021

Rockefeller university review = Rockefeller institute review. New York, NY. Rockefeller Inst. Rev. See B–P–H 802/15.

Rocznik akademii rolniczej w Poznaniu. Poznan. Vol. 54+, 1972+. Roczn. Akad. Roln. Poznaniu. Preceded by: Roczniki wyzszej szkoly rolniczej w poznaniu. Wydzial ogrodniczy. HI 72308

Rocznik fenologiczny. Warsaw. Roczn. Fenol. See B–P–H 802/20. HI 59022

Rocznik polskiego towarzystwa dendrologicznego. Lwów. Vols. 1-5, 1926-35. Roczn. Polsk. Towarz. Dendrol. Superseded by: Rocznik sekcji dendrologicznej polskiego towarzystwa botanicznego. HI 72309

Rocznik sekcji dendrologicznej polskiego towarzystwa botanicznego. Warsaw. Vol. 1+, 1945+. Roczn. Sekc. Dendrol. Polsk. Towarz. Bot. Preceded by: Rocnik polskiego towarzystwa dendrologicznego. HI 72310

Rocznik towarzystwa przyjaciól nauk poznańskiego. Poznan. Vols. 1-11, 1860-81; vols. 15-49, 1887-1922. Roczn. Towarz. Przyjac. Nauk Poznańsk. For vols. 12-14 see: Sprawozdanie z cynności towarzystwa przyjaciól nauk poznańskiego. 4-3416-2. HI 72311

Rocznik towarzystwa przyjaciól nauk poznańskiego. Sekcya lekarska = Rocznik towarzystwa przyjaciól nauk poznańskiego. Poznan.

Roczniki chemii. Warsaw. Vols. 1-51, 1921-77 [publication suspended 1940-45]. Roczn. Chem. Superseded by: Polish journal of chemistry. 4-3700-1. HI 72312

Roczniki gleboznawcze. Warsaw Vol. 1+, 1950+. Roczn. Glebozn. HI 72313

Roczniki nauk lesnych. Warsaw. Roczn. Nauk Lesn. See B–P–H 802/22. HI 59023

Roczniki nauk rolniczych. Warsaw. Roczn. Nauk Roln. See B–P–H 802/23. HI 59024

Roczniki nauk rolniczych. Seria A, produkcja roslinna. Warsaw. Vol. 96+, 1969+. Roczn. Roln., Ser. A, Prod. Rosl. Preceded by: Roczniki nauk rolniczych. Seria A, roslinna. HI 72314

Roczniki nauk rolniczych. Seria A, roslinna. Warsaw. Vols. 66-95?, 1952-68? Roczn. Nauk Roln., Ser. A, Rosl. Preceded by: Roczniki nauk rolniczych. Superseded by: Roczniki nauk rolniczych. Seria A, produkcja roslinna. HI 72315

Roczniki nauk rolniczych. Seria E, ochrona roslin. Warsaw. Vol. 1+, 1970+. Roczn. Roln., Ser. E, Ochr. Rosl. HI 72316

Roczniki nauk rolniczych. Seria F, melioracji i uzytków zielonych. Warsaw. Vol. 71+, 1955+. Roczn. Roln., Ser. F, Melior. Uzytk. Zielon. HI 72317

Roczniki nauk rolniczych i lesnych. Warsaw. Roczn. Nauk Roln. Lesn. See B–P–H 802/24. HI 59025

Roczniki poznańskiego towarzystwa przyjaciół nauk. Poznan. Vol. 50, 1928. Roczn. Poznańsk. Towarz. Przyjac. Nauk. Preceded by: Rocznik towarzystwa przyjaciól nauk poznańskiego. HI 72318

Roczniki towarzystwa naukowego krakowskiego. Cracow. Roczn. Towarz. Nauk. Krakowsk. See B–P–H 803/2. HI 59026

Roczniki towarzystwa przyjaciela nauk. Warsaw. Roczn. Towarz. Przyjac. Nauk. See B–P–H 803/3. HI 59027

Roczniki wyzszej szkoly rolniczej w poznaniu. Wydzial ogrodniczy. Poznan. Vols. 1-53?, 1957-71? Roczn. Szkoly Roln. Pozn., Wydz. Ogrodn. Superseded by: Rocznik akademii rolniczej w Poznaniu. HI 72319

Rodale's environment action bulletin = Environmental action. Washington, DC.

Rodriguésia; revista do instituto de biologia vegetal, jardim botânico e estação biologica do Itatiaya. Rio de Janeiro. Rodriguésia. See B–P–H 803/8. HI 59028

Roebuck; journal of the Northumberland wildlife trust. Newcastle-upon-Tyne. 1973+. Roebuck. Preceded by: Newsletter, Northumberland wildlife trust. HI 72320

Roessleria. Porto Alegre. Vol. 1+, 1977+. Roessleria. HI 72321

Roezliana. Brno. ?-1982+. Roezliana. HI 72322

Romagna agricola e zootecnica. Ravenna, Italy. Romagna Agric. Zootecn. See B–P–H 803/17. HI 59030

Romania viticola. Bucharest. Romania Vitic. See B–P–H 803/18. HI 59031

Römisch Kaiserlichen Akademie der Naturforscher auserlesene Medicinisch- Chirurgisch- Anatomisch- Chymisch- und Botanische Abhandlungen. Nuremberg. Röm. Kaiserl. Akad. Naturf. Auserlesene Med.-Chir. Abh. See B–P–H 803/13. HI 59029

Rosarium; eerste nederlandsche maandschrift gewijd aan rozencultuur. Wageningen, Utrecht. Nos. 1-51, 1891-

1941. Rosarium. HI 72323

Rosarian's year-book. Edinburgh & London. Rosarian's Year-Book. See B–P–H 803/20. HI 59032

Rose; quarterly journal for all rose lovers. London. Vols. 1-18(1), 1952-69. Rose. HI 72324

Rose annual. Croydon, England. Rose Annual. See B–P–H 803/22. HI 59033

Rose bulletin of the canadian rose society. Toronto. Vols. 1-19, 1946-74. Rose Bull. Canad. Rose Soc. Superseded by: Canadian rosarian. 4-3709-3. HI 72325

Rose bulletin of the royal national rose society. St. Albans. No. 1+, 1970+. Rose Bull. Roy. Natl. Rose Soc. HI 72326

Rose bush. Cleveland, OH. Rose Bush. See B–P–H 803/24. HI 59034

Rose journal. Fishkill-on-Hudson, NY. Rose J. See B–P–H 803/25. HI 59035

Rose, the shamrock, and the thistle. Edinburgh. Rose Shamrock Thistle. See B–P–H 803/26. HI 59036

Rose technic. Terre Haute, IN. Rose Techn. See B–P–H 803/27. HI 59037

Rosemaund on record = Report, Rosemaund experimental husbandry farm. Preston Wynne.

Rosenbogen, Der; Mitteilungen des Vereins deutscher Rosenfreunde e. V. Zweibrücken. Vol. [1]+, 1964+. Rosenbogen. HI 72327

Rosenjahrbuch. Berlin. Rosenjahrbuch. See B–P–H 804/2. HI 59038

Rosenzeitung. Frankfurt a. M. Rosenzeitung. See B–P–H 804/4. HI 59039

Roses, Les; bulletin de la société française des rosiéristes. Lyon. Vols. ?-17-61, 1898-19??; sér. 2, vol. 1-12, 19??-? Roses. Superseded by: Amis des roses, Les. HI 72328

Rostlinna vyroba. Prague. Rostol. Vyroba. See B–P–H 804/5. HI 59040

Rostnövények. Kompolt. 1967-71? Rostnövények. HI 72329

Rotem; bulletin of the Israel plant information center. Tel Aviv. Vol. 1+, 1981+. Rotem. HI 72330

Rotunda; bulletin of the royal Ontario museum. Toronto. Vol. 1+, 1968+. Rotunda. HI 72331

Rovartani lapok. Budapest. Rovart. Lapok. See B–P–H 804/8. HI 59041

Royal botanic gardens newsletter = Newsletter, royal botanic gardens. Kew.

Royal horticultural society, lily year book. London = Lily year-book. London.

Royal horticultural society, rhododendron year book. London = Rhododendron year book. London.

Rozpravy české akademie císaře Františka Josefa pro vědy, slovesnost a umení. Třida 2. Vědy mathematické, přírodní. Prague. Rozpr. České Akad. Císaře Františka Josefa Vědy, Tř. 2, Vědy Math. Přír. See B–P–H 805/3. HI 59045

Rozpravy české akademie věd a umění. Třida 2. Vědy mathematické, přírodní. Prague. Rozpr. České Akad. Věd, Tř. 2, Vědy Mat. Přír. See B–P–H 805/4. HI 59046

Rozpravy československé akademie věd a umění. Prague. Rozpr. Českoslov. Akad. Věd. See B–P–H 805/6. HI 59047

Rozprava královské české společnosti nauk, třida mat.-prírodovědecké. Prague. Rozpr. Král. České Společn. Nauk, Tř. Mat.-Přír. See B–P–H 805/8. HI 59048

Rozpravy královské české společnosti nauk = Abhandlungen der königlichen Böhmischen Gesellschaft der Wissenschaften. Ser. 7. Prague. Abh Königl. Böhm. Ges. Wiss. See B–P–H 30/4.

Rozprawy akademii umiejetności. Wydział matematyczno-przyrodniczy. Cracow. Rozpr. Akad. Umiejetn., Wydz. Mat.-Przyr. See B–P–H 804/46. HI 59044

Rozprawy, akademiya rolnicza w szczecinie. Sacarzin. No. 32+, 1973+. Rozpr. Akad. Roln. Szczecinie. Preceded by: Rozprawy, wyzsza szkoka rolnicza w Szczecinie. HI 72332

Rozprawy habilitacyjne. Cracow. Vol. 1+, 1977+. Rozpr. Habilit. HI 72333

Rozprawy i sprawozdania. Instytut badawczy lasów państwowych. Warsaw. Rozpr. Spraw. Inst. Badawczy Lasów Państw. See B–P–H 805/11. HI 59049

Rozprawy i sprawozdania. Instytut badawczy leśnictwa. Warsaw. Rozpr. Spraw. Inst. Badawczy Leśn. See B–P–H 805/12. HI 59050

Rozprawy i sprawozdania z posiedzeń wydziału matematyczno-przyrodniczego akademii umiejetności. Cracow. Rozpr. Spraw. Posiedzeń Wydz. Mat.-Przyr. Akad. Umiejetn. See B–P–H 805/13. HI 59051

Rozprawy i sprawozdania. Zakład doświadczalny lasów państwowych w Warszawie. Warsaw. Rozpr. Spraw. Zakład Doświadcz. Lasów Państw. w Warszawie. See B–P–H 805/15. HI 59052

Rosprawy wydziału matematyczno-przyrodniczego akademii umiejetności. Dział B. Nauki biolgiczne. Cracow. Rozpr. Wydz. Mat.-Przyr. Akad. Umiejetn., Dział B, Nauki Biol. See B–P–H 805/17. HI 59053

Rosprawy wydziału matematyczno-przyrodniczego polskiej akademji umiejetności. Dział A/B. Nauki matematyczno-fizyezne oraz biolgiczne. Cracow.

Rozpr. Wydz. Mat.-Przyr. Polsk. Akad. Umiejetn., Dział A/B, Nauki Mat.-Fiz. Biol. See B–P–H 806/1. HI 59054

Rozprawy, wyzsza szkoka rolnicza w Szczecinie. Sacarzin. Nos. 1-31, 19??-72? Rozpr. Wyzsza Szkoka Roln. Szczecinie. Superseded by: Rozprawy, akademiya rolnicza w szczecinie. HI 72334

Rózsa újság. Temesvar [=Timisoara, Rumania]. Rózsa Újs. See B–P–H 806/3. HI 59055

Rubber developments. London. Rubber Developm. See B–P–H 806/9. HI 59056

Rufaca. Puigcerdá. ?-1981+. Rufaca. HI 72335

Ruizia; monografías del jardín botánico. Madrid. Vol. 1+, 1984+. Ruizia. HI 72336

Rumanian scientific abstracts. Natural sciences. Bucharest. Rumanian Sci. Abstr., Nat. Sci. See B–P–H 806/13. HI 59057

Rumbos. Universidad de Antioquia. Medellín, Colombia. Rumbos. See B–P–H 806/14. HI 59058

Rural environmental conservation. Washington, DC. ?-1974? Rural Environm. Conservation. Superseded by: Agricultural conservation, program accomplishments. HI 72337

Rural India; journal of the Madras forest panachayats. Madras. Rural India. See B–P–H 806/15. HI 59059

Rural research. Dickson, A.C.T. No. 78+, 1978+. Rural Res. Preceded by: Rural research in C S I R O. HI 72338

Rural research in C S I R O. Melbourne, Vic. Nos. 1-77, 1952-72. Rural Res. C. S. I. R. O. Superseded by: Rural research. HI 72339

Russische Hydrobiologische Zeitschrift. Sarativ = Russkii gidrobiologicheskii zhurnal. Saratov.

Russische Sammlung für Naturwissenschaft und Heilkunst. Riga [Latvian S S R] & Leipzig. Russ. Samml. Naturwiss. Heilkunst. See B–P–H 806/21. HI 59061

Russisches Jahrbuch für die Chemie und Pharmacie = Russisches Jahrbuch der Pharmacie. Riga, [Latvian S S R]. Russ. Jahrb. Pharm. See B–P–H 806/20.

Russisches Jahrbuch der Pharmacie. Riga, [Latvian S S R]. Russ. Jahrb. Pharm. See B–P–H 806/20. HI 59060

Russkaja Bibliografija po Estestvoznaniju i Matematikě, sostavlennaja sostojaščim pri Imperatorskoj (Rossijeskoj) Akademii Nauk Sankt-Peterburgskim Bjuro Meždunoradnoj Bibliografii = Russkaya Bibliografiya po Estestvoznaniyu i Matematikě, sostavlennaya sostoyashchim pri Imperatorskoi (Rossiieskoi) Akademii Nauk Sankt-Peterburgskim Byuro Mezhdunoradnoi Bibliografii. St. Petersburg.

Russkie Subtropiki. Batum [Georgian S S R]. Russk. Subtrop. See B–P–H 807/4. HI 59062

Russkii arkhiv protistologii. Moscow. Vols. 3-8, 1925-29. Russk. Arh. Protistol. Preceded by: Arkhiv Russkogo protistologicheskogo obshchestva. 4-3743-2. HI 72340

Russkaya Bibliografiya po Estestvoznaniyu i Matematikě, sostavlennaya sostoyashchim pri Imperatorskoi (Rossiieskoi) Akademii Nauk Sankt-Peterburgskim Byuro Mezhdunoradnoi Bibliografii. St. Petersburg. Vols. 1-9, 1904-17. Russk. Bibliogr. Estestv. Mat. 4-3742-3. HI 72341

Russkii botanicheskii zhurnal. St. Petersburg. 1908-15. Russk. Bot. Zhurn. 4-3743-2. HI 72342

Russkii gidrobiologicheskii zhurnal. Saratov, Russian S F S R. Vols. 1-8, 1921-29. Russk. Gidrobiol. Zhurn. Superseded by: Gidrobiologicheskii zhurnal S S S R. 2-1724-3. HI 72343

Russkie Subtropiki. Batum, Georgian S S R. 1912-17. Russk. Subtrop. Preceded by: Batumskii Sel'skii Khozyain. 4-3742-2. HI 72344

Russkij arhiv protistologii = Russkii arkhiv protistologii. Moscow.

Russkij botaničeskij žurnal = Russkii botanicheskii zhurnal. St. Petersburg.

Russkij gidrobiologičeskij žurnal = Russkii gidrobiologicheskii zhurnal. Saratov.

Ryukyu daigaku bunrui gakubu kiyo = Bulletin of arts & science division, Ryukyu university. Naha, Okinawa. Bull. Arts Sci. Div. Ryukyu Univ. See B–P–H 240/14.

Ryukyu daigaku kenkyu fukyubu = University of Ryukyus, extension service. Naha?, Okinawa. Univ. Ryukyus Extens. Serv. See B–P–H 937/28.

Ryukyu daigaku nogakubu gakujutsu hokoku = Science bulletin of the college of agriculture, university of the Ryukyus. Naha, Okinawa.

Ryukyu daigaku nogakubu gakujutsu hokoku = Science bulletin of the faculty of agriculture; university of the Ryukyus. Naha, Okinawa. Sci. Bull. Fac. Agric. Univ. Ryukyus. See B–P–H 823/18.

Ryukyu daigaku nokasei kogakubu gakujutsu hokoku = Science bulletin of the agriculture and home economics division, university of the Ryukyus. Naha, Okinawa.

Ryukyu ringyo shikenjo shuho = Collections of the Ryukyu forestry experiment station. Naha?, Okinawa. Collect. Ryukyu Forest. Exp. Sta. See B–P–H 317/14.

Ryukyu seifu, Keizai-kyoku ringyo shikenjo = Research bulletin of the forest experiment station. Naha?, Okinawa. Res. Bull. Forest Exp. Sta. See

B–P–H 774/7.

S A digest/opinion = South african digest. Pretoria.

S A B R A O journal. Taipei. Vol. ?-15+, ?-1983+. S. A. B. R. A. O. J. HI 75164

S A G; a magazine of outdoor interests. Cape Town. Vols. 26(4)-30, 1937-40. "S.A.G." Preceded by: South african country life. See B–P–H 807/25. 5-3749-2. HI 72345

S A R H boletin tecnico = Boletín tecnico, secretaria de agricultura y ganaderia, subsecretaria forestal y de la fauna. Mexico City.

S B N H north american newsheet. New York. No. 1, 1982. S. B. N. H. N. Amer. Newsheet. HI 72346

S C A R bulletin. [Reprinted from: Polar record.] Cambridge. 1959+. S. C. A. R. Bull. HI 72347

S C R A L. Tucumán?, Argentina. Vol. ?-12+, 19??-66+. Scral. HI 72348

S D A T. [Costa Rica]. 1954-57. SDAT. HI 72349

S E P A S A T newsletter; survey of economic plants for arid and semi-arid tropics. Kew. No. 1+, 1983+. S. E. P. A. S. A. T. Newslett. HI 72350

S F and N. Fort Worth, TX. Vol. 97, 1984. S. F. & N. Preceded by: Southern florist and nurseryman. Superseded by: Nursery manager. HI 72351

S I A T S A bulletin. La Lima. (Honduras.) ?-1978+. S. I. A. T. S. A. Bull. HI 72352

S L A - tidskriften; organ för skogs- och lantarbetsgivareföreningen. Stockholm. 1970+, 1971+. S. L. A. Tidskr. Preceded by: Skogs- och lantarbetsqivareföreningens tidskrift. HI 72353

S N A nursery research journal. Nashville, TN. Vol. 1+, 1974+. S. N. A. Nursery Res. J. HI 72354

S O N G news. Niagara-on-the-Lake. No. 1+, 1972+. S. O. N. G. News. HI 72355

S P C quarterly bulletin = Quarterly bulletin, south Pacific commission and South Pacific bulletin. Nouméa, Sydney, NSW.

S Ş Ç I elmler aqademijas' Azerbajçan filial' = Azerbaidzhanskii filial Akademii nauk S S S R. Baku.

S S S R ylymlar akademijasynyn Turkmenistan filialynyn harbarlary = Izvestiya Turkmenskogo filiala Akademii nauk S S S R. Ashkhabad.

S W A News = SWANEWS. Norman, OK.

S U K meddelande. No. ?-55+, ?-1980+. S. U. K. Meddel. HI 72356

S W A news, see SWANEWS

S W A wetenskaplige vereniging. Windhoek = Journal of the South West Africa scientific society. Windhoek, South-West Africa. J. South West Africa Sci. Soc. See B–P–H 482/22.

S W A Wissenschaftliche Gesellschaft. Windhoek = Journal of the South West Africa scientific society. Windhoek, South-West Africa. J. South West Africa Sci. Soc. See B–P–H 482/22.

Saatgut-Wirtschaft. Stuttgart. Saatgut-Wirtschaft. See B–P–H 809/12. HI 59087

Sabah forest record. Sandakan. 1964+. Sabah Forest Rec. Preceded by: North Borneo forest records. HI 72357

Saboten. Hyogo = Cacti. Hyogo.

Saboten. Tokyo = Cactus Tokyo. Tokyo.

Saboten nihon = Succulentarum japonica. Tokyo. Succ. Jap. See B–P–H 858/15.

Sabouraudia. Edinburgh. Vols. 1-23, 1961-85. Sabouraudia. Superseded by: Journal of medical and veterinary mycology. HI 59088

Saccardoa: monographiae mycologicae. Pavia. Saccardoa. See B–P–H 809/16. HI 59089

Saccharum; boletim técnico do serviço técnico-agronómico. Rio de Janeiro. Saccharum. See B–P–H 809/17. HI 59090

Sächsischer Heimatschutz, Landesverein zur Pflege heimatlicher Natur, Kunst und Bauweise. Dresden. Sächs. Heimatschutz Mitt. Landesvereins Sächs. Heimatschutz. See B–P–H 809/20. HI 59091

Sad i igorod. Moscow. 1885-1910; [1918-23 not published]; 1924-31(2/3); 1946-59. Sad & Ogorod. Preceded by: Zhurnal Sadovodstva, izdavaemyi Rossiiskim Obshchestvom Lyubitelei Sadovodstva v Moskvě. For 1911-17 see: Zhurnal Sad i ogorod. For 1931(4)-1932(6) see: Sotsialisticheskoe plodoovoshchnoe khozyaistvo. For 1932(7)-38 see: Plodoovoshchnoe khozyaistvo. For 1939-40 see: Ovoshchevodstvo and Sadovodstvo. For 1941 see: Sady i ogorody. Superseded by: Sadovodstvo. Moscow. 5-3752-3. HI 72358

Sado rinkai jikkenjo. Kenkyu nempo = Report (Annual) of the Sado marine biological station. Niigata.

Sado rinkai jikkenjo. Tokubetsu hokoku. Niigata. Series 1+, 1978+. Sado Rinkai Jikkenjo Tokubetsu Hokoku. HI 72359

Sadovodstvo. Kiev. Vol. 1+, 1964+. Sadovodstvo. HI 72360

Sadovodstvo; ezhemesyachnyi nauchno-proizvodstvennyi zhurnal Ministerstva sel'skogo khozyaistva S S S R. Moscow. 1939-40; [1942-45 not published]; 1960+. Sadovodstvo. Preceded by Plodoovoščn. Hoz. For 1941 see: Sady i ogorody. For 1946-59 see: Sad i ogorod. 5-3753-1. HI 72361

Sadovodstvo promyshlennogo tipa. Vol. 1+, 1977+.

Sadov. Promyshl. Tipa. HI 72362

Sadovodstvo, vinogradstvo i ovoshchevodstvo na gruboskeletnykh pochvakh. Vol. ?-14+, ?-1978+. Sadov. Vinograd. Ovoshchevod. Gruboskel. Pochvakh. HI 72363

Sadovodstvo, vinogradarstvo i vinodelie Moldavii. Kishinev. Vol. 1+, 1946+. Sadov. Vinograd. Vinod. Moldavii. HI 72364

Sady i ogorody. Moscow. 1941. Sady & Ogorody. Preceded by: Ovoshchevodstvo. Superseded by: Sad i ogorod. 5-3753-1. HI 72365

Saengmul hak'kae-po = Korean journal of biology. Seoul, Korea. Korean J. Biol. See B–P–H 518/9.

"S.A.G."; a magazine of outdoor interests. Cape Town = S A G; a magazine of outdoor interests. Cape Town.

Saga daigaku nogakubu iho = Agricultural bulletin of the Saga university. Saga.

Saga-ken kaju shikenjo gyomu nempo = Operation report of Saga fruit tree experiment station. Saga.

Saga-ken nogyo shikenjo kenkyu hokoku = Bulletin of the Saga agricultural experiment station. Saga, Japan. Bull. Saga Agric. Exp. Sta. See B–P–H 270/27.

Saggi e dissertazioni accademiche. Accademia etrusca dell'antichissima città di Cortona. Rome. Saggi Diss. Accad. Accad. Etrusca Cortona. See B–P–H 810/5. HI 59092

Saggi scientifici e letterarj dell'accademia di Padova. Padua. Saggi Sci. Lett. Accad. Padova. See B–P–H 810/6. HI 59093

Saggi della società letteraria ravennate. Cesena, Italy. Saggi Soc. Lett. Ravennate. See B–P–H 810/7. HI 59094

Sagtevrugteboer = Deciduous fruit grower. Bellville.

Saguaroland bulletin. Phoenix, AZ. Saguaroland Bull. See B–P–H 810/8. HI 59095

Saibo kagaku shimpojumu = Symposia for cellular chemistry.

Saibo seibutsugaku shimpojum = Symposia for cell biology. Okayama.

Saikyo daigaku gakujutsu hokoku = Scientific reports of the Saikyo university. Natural science and living science (or Mathematics and natural science). Kyoto.

Saikyo daigaku gakujutsu hokoku. Nogaku = Scientific reports of the Saikyo university. Agriculture. Kyoto.

Saikyo daigaku nogakubu enshurin shuho = Report of the Saikyo university forest. Kyoto, Japan. Rep. Saikyo Univ. Forest. See B–P–H 770/16.

Sains malaysiana; Malaysian journal of natural sciences. Vol. 1+, 1972+. Sains Malaysiana. HI 72366

Saint, see also St.

Saintpaulian. Golden, CO. Saintpaulian. See B–P–H 810/11. HI 59096

Saishu to shiiku = Collecting and breeding. Tokyo. Collect. & Breed. See B–P–H 317/7.

Saiski der Odessar Naturforschergesellschaft = Zapiski Odesskogo obshchestva estestvoispytatelei. Odessa.

Saitama daigaku kiyo. Shizenkagaku-hen = Jaurnal of Saitama university, Natural science. Saitama.

Saitama-ken engei shikenjo kenkyu hokoku = Bulletin of the Saitama horticultural experiment station. Kuki.

Saitama-ken engei shikenjo nenpo = Annals of the Saitama horticultural experiment station. Kuki.

Saitama-kenritsu shizenshi hakubutsukan kenkyu hokoku = Bulletin of the Saitama museum of natural history. Saitama.

Saito Ho-on-kai gakujutsu kenkyu hokoku = Research bulletin. Saito Ho-on Kai museum. Sendai, Japan. Res. Bull. Saito Ho-On Kai Mus. See B–P–H 775/1.

Sakharnaya svekla. Moscow. 1956-87. Sakharn. Svekla. Superseded by: Sakharnaya svekla - proizvodstvo i pererabotka. HI 72369

Sakharnaya svekla - proizvodstvo i pererabotka. Moscow. 1988+. Sakharn. Svekla Proizvodstvo Perarabot. Preceded by: Sakharnaya svekla and Sakharnaya promyshlennosti [not entered]. HI 72369

Sakartvelos S S R mecnierebata akademiis macne. Biologiis seria = Izvestiya akademii nauk gruzinskoi S S R. Seriya biologicheskaya. Tiflis.

Sakharthwelos S S R mecnierebatha akademiis moambe. [For Russian edition see: Soobščenija Gruzinskogo filiala Akademii nauk S S S R.] Tiflis, Georgian S S R. Vol. 2+, 1941+. Sakharthw. S.S.R. Mecniereb. Akad. Moambe. Preceded by: Soobshcheniya Gruzinskogo filiala Akademii nauk S S S R. HI 72370

Sakharthwelos S S R mecnierebatha akademiis Sakharthwelos philialis moambe = Soobshcheniya Gruzinskogo filiala Akademii nauk S S S R. Tiflis.

Sakyu kenkyu = Sand-dune research. Tattori, Japan. Sand-Dune Res. See B–P–H 812/9.

Salzburgische gelehrte Unterhaltungen. Salzburg. Salzburg. Gel. Unterhalt. See B–P–H 810/17. HI 59097

Samaru agricultural newsletter. Samaru. Vols. 1-19, 1959-77. Samaru Agric. Newslett. Superseded by: Noma. HI 72371

Samaru journal of agriculture research. Samaru. Vol. 1+, 1981+. Samaru J. Agric. Res. HI 72372

Samaru miscellaneous papers. Samaru. No. 1+, 1963+. Samaru Misc. Pap. HI 72373

Samaru research bulletin. Zaria, Nigeria. Samaru Res.

Bull. See B–P–H 810/18. HI 59098

Samling af minde-taler, holdne i det kongelige norske videnskabers-selskab over adskillige af dets afdode medlemmer. Copenhagen. Saml. Minde-Taler Kongel. Norske Vidensk.-Selsk. See B–P–H 810/22. HI 59100

Samling af rön och uptäkter, gjorde i senare tider, uti physik, medecin, chirurgie, natural-historia, chemie, hushållning ... Goteborg, Sweden. Saml. Rön Uptäckter Phys. See B–P–H 810/23. HI 59101

Samlingar i blandade ämnen för läkarevetenskapen och naturforskningen. Stockholm. Saml. Blandade Ämnen Läkarevetensk. Naturf. See B–P–H 810/21. HI 59099

Samlinger for venner af naturen og haugekonsten. [Edited by N. F. Lassen.] Copenhagen. Saml. Venner Nat. Haugek. See B–P–H 810/25. HI 59102

Sammelschrift der Mathematisch-naturwissenschaftlich-ärztlichen Sektion der Ševčenko-Gesellschaft der Wissenschaften in Lemberg = Zbirnyk Sektsii Matematychno-pryrodopysno-likars'koi Naukovogo Tovarystva imeny Shevchenka. Lvov.

Sammlung von Abhandlungen ökonomischen und technologischen Inhalts. Halle. Samml. Abh. Ökon. Technol. Inhalts. See B–P–H 811/1. HI 59103

Sammlung auserlesener Abhandlungen zum Gebrauche praktischer Aerzte. Leipzig. Samml. Auserlesener Abh. Gebrauche Prakt. Aerzte. See B–P–H 811/2. HI 59104

Sammlung auserlesener Abhandlungen über die interessantesten Gegenstände der Chemie. Leipzig. Samml. Auserlesener Abh. Interessantesten Gegenstände Chem. See B–P–H 811/3. HI 59105

Sammlung auserlesener Schriften von Staats- und landwirthschaftlichen Inhalte. Bern. Samml. Auserlesener Schriften Staats- Landw. Inhalt. See B–P–H 811/4. HI 59106

Sammlung auserlesener Wahrnehmungen aus der Arzney-Wissenschaft, der Wund-Arzney und der Apotheker-Kunst. Hamburg. Samml. Auserlesener Wahrnehm. Arzney-Wiss. See B–P–H 811/5. HI 59107

Sammlung der besten und neuesten Reisebeschreibungen. Berlin. Samml. Besten Neuesten Reisebeschreib. See B–P–H 811/6. HI 59108

Sammlung der Deutschen Abhandlungen welche in der Königlichen Akademie der Wissenschaften zu Berlin vorgelesen wurden. Berlin. Samml. Deutsch. Abh. Königl. Akad. Wiss. Berlin. See B–P–H 811/7. HI 59109

Sammlung forstwirtschaftlicher Schriften = Mežsaimniecibas rakstu krajums. Riga.

Sammlung interessanter und durchgängig zwekmässig abgefasster Reisebeschreibungen für die Jugend. Reutlingen, Germany. Samml. Interessanter Zwekmässig Abgefasster Reisebeschreib. Jugend. See B–P–H 811/8. HI 59110

Sammlung kleiner Ausführungen aus verschieden Wissenschaften. Hanover. Samml. Kleiner Ausführ. Verschiedenen Wiss. See B–P–H 811/9. HI 59111

Sammlung kurzer Reisebeschreibungen, und anderer zur Erweiterung der Länder- und Menschenkenntniss dienender Nachrichten. [Edited by Johann Bernoulli.] Berlin. Samml. Kurzer Reisebeschreib. See B–P–H 811/10. HI 59112

Sammlung der merkwürdigsten Reisen in den Orient. Jena. Samml. Merkwürdigsten Reisen Orient. See B–P–H 811/11. HI 59113

Sammlung naturwissenschaftlicher Vorträge. Berlin. Vols. ?-3, 188?-90. Samml. Naturwiss. Vorträge. Superseded by: Abhandlungen und Vorträge aus dem Gesammtgebiete der Naturwissenschaften. HI 72374

Sammlung neuer und nützlicher Abhandlungen und Versuche aus der Ökonomie, Mechanik und Naturlehre. Nuremberg. Samml. Neuer Nützl. Abh. Versuche Ökon. See B–P–H 811/15. HI 59116

Sammlung Physikalischer Aufsätze, besonders die Böhmische Naturgeschichte betreffend. Dresden. Samml. Phys. Aufsätze Böhm. Naturgesch. See B–P–H 811/17. HI 59118

Sammlung physikalisch-oekonomischer Aufsätze. Zur Aufnahme der Naturkunde und deren damit verwandten Wissenschaften in Böhmen. Prague. Samml. Phys.-Oekon. Aufsätze. See B–P–H 812/4. HI 59120

Sammlung zwangloser Abhandlungen aus dem Gebiete der Dermatologie, der Syphilidologie und der Krankheiten des Urogenitalapparates. Halle. Samml. Zwangloser Abh. Dermatol. See B–P–H 812/6. HI 59121

Sammlungen von Natur- und Medicin-, wie auch hierzu gehörigen Kunst- und Litteratur-Geschichten so sich in Schlesien und anderen Ländern begeben haben. Breslau [=Wroclaw, Poland]. Samml. Natur- Litt.-Gesch. Schlesien. See B–P–H 811/12. HI 59114

Sammlungen aus der Naturgeschichte, Oekonomie-Polizey-Kameral- und Finanzwissenschaft. [Edited by I. K. H. Börner.] Dresden. Samml. Naturgesch. See B–P–H 811/14. HI 59115

Sammlungen nützlicher und angenehmer Gegenstände aus allen Theilen der Natur-Geschichte, Arzneywissenschaft und Haushaltungskunst. Leipzig. Samml. Nützl. Angenehmer Gegenstände Natur-Gesch. See B–P–H 811/16. HI 59117

Sammlungen zur Physik und Naturgeschichte von einigen Liebhabern dieser Wissenschaften. Leipzig. Samml. Phys. Naturgesch. See B–P–H 812/1. HI 59119

Sampote; journal of the Nara cactus club. Nos. 1-2, 1955-56. Sampote. HI 72375

Sanctuary; bulletin of the Massachusetts Audubon society. Lincoln, MA. Vol. 20+, 1980+. Sanctuary. Preceded by: Newsletter, Massachusetts Audubon society. HI 72376

Sand-dune research. [Sakyu kenkyu.] Tattori, Japan. Sand-Dune Res. See B–P–H 812/9. HI 59122

Sangaku = Journal of the japanese alpine club. J. Jap. Alpine Club. See B–P–H 470/18.

Sangyo shiryo = Industrial contributions. [South Manchuria railway company. Bureau of agriculture.] [China]. Industr. Contr. See B–P–H 432/5.

Sanidad vegetal. San Cristóbel. No. 1+, 1971+. Sanid. Veg. HI 72377

Sankt-Petersburger medicinische Wochenschrift. St. Petersburg. St.-Petersburger Med. Wochenschr. See B–P–H 854/4. HI 59554

Sankt-Petersburger medicinische Zeitschrift. St. Petersburg. St.-Petersburger Med. Z. See B–P–H 854/5. HI 59555

Sanrin = Forestry. Tokyo.

Sanshi nenkan = Year book of sericulture. Tokyo. Yearb. Seric. See B–P–H 983/4.

Sanshi shikenjo hokoku = Bulletin of the sericultural experiment station. Tokyo. Bull. Seric. Exp. Sta. See B–P–H 273/7.

Santa Cruz naturalist. Scotts Valley, CA. 1975+. Santa Cruz Naturalist. HI 72378

São Paulo médico. São Paulo. São Paulo Méd. See B–P–H 812/16. HI 59123

Sapporo bulletin of the botanic garden. Sapporo. Nos. 1-2, 1963-69? Sapporo Bull. Bot. Gard. HI 72471

Sapporo hakubutsu gakkai kaiho = Transactions of the Sapporo natural history society. Sapporo, Japan. Trans. Sapporo Nat. Hist. Soc. See B–P–H 890/8.

Sapporo noringakkai-ho = Journal of the Sapporo society of agriculture and forestry. Sapporo, Japan. J. Sapporo Soc. Agric. See B–P–H 479/13.

Sarawak museum journal. Kuching, Sarawak [Malaysia]. Sarawak Mus. J. See B–P–H 812/22. HI 59124

Sargentia; continuation of the contributions from the Arnold arboretum of Harvard university. Jamaica Plain, MA. Sargentia. See B–P–H 812/24. HI 59125

Sargetia. Series scientia naturae [Deva.] No. ?-8+, ?-1971+. Sargetia, Ser. Sci. Nat. HI 72472

Sarobetsu jikken nojo chosa hokokusho = Report of Sarobetsu agricultural experiment farm. Tokyo. Rep. Sarobetsu Agric. Exp. Farm. See B–P–H 770/18.

Sarracenia. Montreal. Sarracenia. See B–P–H 812/26. HI 59126

Sarsia. Bergen. Sarsia. See B–P–H 812/27. HI 59127

Saturday evening post. Philadelphia, PA. Saturday Eve. Post. See B–P–H 812/29. HI 59128

Saugar university journal. Saugar. 1951-56. Saugar Univ. J. Superseded by: Journal of the university of Saugar. Part 2, section B, natural sciences. HI 62519

Saussurea; travaux de la société botanique de Genève. Geneva. Vol. 1+, 1970+. Saussurea. Preceded by: Travaux de la société botanique de Genève. HI 72473

Sauteria. Ca.1989+. Sauteria. HI 72474

Sauvage; nouvel observateur - écologie. Supplement to: Nouvel observateur. Paris. Vol. 1+, 1973+. Sauvage. HI 72475

Savanna research series. Montreal. 1964+. Savanna Res. Ser. HI 72476

Savaria. Szombathely, Hungary. Savaria. See B–P–H 812/30. HI 59129

Savoie littéraire et scientifique; revue trimestrielle. Chambéry. Vols. 1-18, 1906-24. Savoie Litt. & Sci. HI 72477

Savon luonto. Kuopio. 1974+. Savon Luonto. Preceded by: Savonia. HI 72478

Savonia. Kuopio. 1972. Savonia. Preceded by: Kuopion luonnon ystavain yhdistyksen julkaisuja. Superseded by: Savon luonto. HI 72479

Savrema jugoslovenska poljoprivredny bibliografija. Belgrade. Vol. 1+, 1947+. Savrema Jugoslav. Poljopr. Bibliogr. HI 72480

Saxiflora. Summit, NJ. Saxiflora. See B–P–H 813/1. HI 59130

Sayaña; revista boliviana de agricultura. La Paz, Bolivia. Sayaña. See B–P–H 813/2. HI 59131

Sborník biologických a geologických věd pedagogických fakulty. Budweis, Czechoslovakia. Sborn. Biol. Geol. Věd Pedagog. Fak. See B–P–H 813/24. HI 59132

Sbornik botanicheskikh rabot, belorusskoe otdelenie, vsesoyuznogo botanicheskogo obshchestva. Minsk. Vols. 1-4, 1959-62. Sborn. Bot. Rabot Belorussk. Otdelenie Vsesoyuzn. Bot. Obshch. Superseded by: Botanika, issledovaniya belorusskoe otdelenie, vsesoyuznogo botanicheskogo obshchestvo. HI 72481

Sborník české akademie technické. Prague. Sborn. Ceské Akad. Techn. See B–P–H 813/28. HI 59133

Sborník české společnosti zeměvědné. Prague. Sborn. Ceské Společn. Zeměvědné. See B–P–H 813/29. HI 59134

Sborník československé akademie zemědelské. Prague Sborn. Českoslov. Akad. Zeměd. See B–P–H 813/30. HI 59135

Sborník československé akademie zemědělských věd. Prague. Sborn. Českoslov. Akad. Zeměd. Věd. See B–P–H 813/31. HI 59136

Sborník československé akademie zemědělských věd. [Ser.] Lesnictví. Prague. Sborn. Českoslov. Akad. Zeměd. Věd, Lesn. See B–P–H 813/32. HI 59137

Sborník československé akademie zemědělských věd. [Ser.] Rostlinna vyroba. Prague. Sborn. Českoslov. Akad. Zeměd. Věd, Rostl. Vyroba. See B–P–H 813/33. HI 59138

Sborník československé společnosti zeměpisné. Prague. Sborn. Českoslov. Společn. Zeměp. See B–P–H 813/34. HI 59139

Sborník pro dějinyk přírodních věd a techniky. Prague. Sborn. Dějiny Přír. Věd Techn. See B–P–H 814/2. HI 59140

Sbornik, institut botaniki, akademiya nauk gruzinskoi S S R. Tiflis. 1965+. Sborn. Inst. Bot. Akad. Nauk Gruzinsk. S.S.R. HI 72482

Sborník jihočeského muzea v českých budějovicích, přírodni védy. Ceské Budějovice. Vols. 4-12, 1964-72; vol. 14+, 1974+; Supplementum vol. 1+, 1974+. Sborn. Jihočesk. Muz. Českých. Budějovicích Přir. Vedy. Preceded by: Sbornik krajskeho vlastivedneho musea v ceskych budejovicich, přírodni védy. For 1974 see: Přírodovědecký časopis jihočeský. HI 72483

Sbornik po karantinu rastenii. Moscow. Vols. 1/2-6, 1957-58. Sborn. Karant Rast. Superseded by: Sbornik rabot po voprosam karantina rastenii. HI 72485

Sbornik po karantinu rastenij = Sbornik po karantinu rastenii. Moscow.

Sborník klubu přírodovědeckého v Brně. Brünn [=Brno, Czechloslovakia] Sborn. Klubu Přír. v Brně. See B–P–H 814/5. HI 59141

Sborník klubu přírodovědeckého v Praze. Prague. Sborn. Klubu Přír. v Praze. See B–P–H 814/6. HI 59142

Sborník krajského múzea v Trnave. Trnava, Czechoslovakia. Sborn. Krajsk. Múz. v Trnave. See B–P–H 814/8. HI 59143

Sbornik krajskeho vlastivedneho musea v ceskych budejovicich, přírodni védy. Ceské Budějovice. Vols. 1-3, 1958-61. Sborn. Krajsk. Vlastiv.Mus. Ceskych Budejovicích Přir. Védy. Superseded by: Sborník jihočeského muzea v českých budějovicích, přírodni védy. HI 72486

Sbornik, krymskoe obshchestvo estestvoispytatelei i lyubitelei prirody. Simferopol. 1914-18. Sborn. Krymskoe Obshch. Estestvoisp. Lyubit. Prir. HI 72487

Sborník lesnické fakulty vysoké školy zemědělské v Praze. Prague. Sborn. Lesn. Fak. Vysoké Školy Zeměd. v Praze. See 814/9. HI 59144

Sborník Masarykovy akademie práce. Prague. Sborn. Masarykovy Akad. Práce. See B–P–H 814/11. HI 59145

Sborník matice moravské. Brno, Czechoslovakia. Sborn. Matice Morav. See B–P–H 814/12. HI 59146

Sborník muzeálinej slovenskej společnosti. Turocszentmarton, Hungary [=Turciansky Svaty Martin, Czechoslovakia]. Sborn. Muz. Slov. Společn. See B–P–H 814/13. HI 59147

Sborník národního muzea v Praze. Řada B: Prírodni vědy (Přírodovědný). Prague. 1938+. Sborn. Nár. Mus. v Praze, Řada B, Přír. Vědy. 4-3421-1. HI 72488

Sbornik nauchnykh rabot, nauchno-issledsovatelskii institut sadovodstva, Michurinsk. Michurinsk. Vols. 1-11, 1956-65. Sborn. Nauchn. Nauchno-Issl. Inst. Sadov. Michurinsk. Superseded by: Sbornik nauchnykh rabot, vsesoyuznii nauchno-issledsovatelskii institut sadovodstva im. I. V. Michurina. HI 72489

Sbornik nauchnykh rabot, tsentral'nii botanicheskii sad. Minsk. Vols. 1-2, 1960-61. Sborn. Nauchn. Rabot Tsentr. Bot. Sad., Minsk. HI 72490

Sbornik nauchnykh rabot, vsesoyuznii nauchno-issledsovatelskii institut sadovodstva im. I. V. Michurina. Michurinsk. Vols. 12-25?, 1967-77? Sborn. Nauchn. Rabot Vsesoyuzn. Nauchno-Issl. Inst. Sadov. I. V. Michurina. Preceded by: Sbornik nauchnykh rabot, nauchno-issledsovatelskii institut sadovodstva, Michurinsk. Superseded by: Sbornik nauchnykh trudov, vsesoyuznii nauchno-issledovatelskii institut sadovodstva im. I. V. Michurina. HI 72491

Sbornik nauchnykh soobshchenii, vsesoyuznoe botanicheskoe obshchestvo. Dagestanskoe otdelenie. Vol. ?-3+, ?-1972+. Sborn. Nauchn. Soobshch. Vsesoyuzn. Bot. Obshch., Dagestanskoe Otdelenie. HI 72492

Sbornik nauchnykh statei. Vilnius = Straipsniu rinkinys. Vilnius, Lithuanian S S R. Straipsniu Rinkinys. See B–P–H 856/9.

Sbornik nauchnykh trudov, Blagoveshchenskii sel'skokhozyaistvennyi institut. Blagoveshchensk. Vol. 1+, 1973+. Sborn. Nauch. Trudov Blagoveshchenskii Sel'skokhoz. Inst. HI 72493

Sbornik nauchnykh trudov, gosudarstvennyi ordena trudovogo krasnogo znameni nikitski botanicheskii sad. Yalta? Vol. 37, 1964. Sborn. Nauch. Trudov Gosud. Ordena Trud. Krasn. Znam. Nikitsk. Bot. Sad. Preceded by: Trudy, gosudarstvennyi nikitskii botanicheskii sad. Superseded by: Nauchnye trudy, gosudarstvennyi ordena trudovogo krasnogo znameni nikitskii botanicheskii sad. HI 72494

Sbornik nauchnykh trudov. Institut biologii, Akademiya nauk Belorusskoi S S R. Minsk, Belorussian S S R.

Vols. 1-3, 1950-52. Sborn. Nauchn. Trudov Inst. Biol. Akad. Nauk Belorussk. S.S.R. HI 72495

Sbornik nauchnykh trudov instituta, Leningradskii farmatsevticheskii institut. Leningrad. Vol. 1, 1947. Sborn. Nauchn. Trudov Inst. Leningradsk. Superseded by: Sbornik nauchnykh trudov, Leningradskii khimikofarmatsevticheskii institut. HI 72496

Sbornik nauchnykh trudov, karantinnye i drugie opasnye vrediteli i bolezni rastenii. Vol. ?-2+, ?-1975+. Sborn. Nauchn. Trudov Karant. Drugie Opasnye Vredit. Bolezni Rast. HI 72497

Sbornik nauchnykh trudov, Leningradskii khimiko-farmacevticheskii institut. Leningrad. Vols. 2-3, 1957. Sborn. Nauchn. Trudov Leningradsk. Khim.-Farm. Inst. Preceded by: Sbornik nauchnykh trudov instituta, Leningradskii farmacevticheskii institut. HI 72498

Sbornik nauchnykh trudov, mironovskii nauchno-issledovatel'skii institut selektsii i semenovodstva pshenitsy. Belaya Tserkov'. Vol. ?-2+, ?-1978+. Sborn. Nauchn. Trudov Mironovsk. Naucho-Issl. Inst. Selekts. Semenov. Pshenitsy. HI 72499

Sbornik nauchnykh trudov, permskaya gosudarstvennaya sel'skokhozyaistvennaya opytnaya stantsiya. Vol. ?-5+, ?-1977+. Sborn. Nauchn. Trudov Permsk. Gosud. Selskokhoz. Opytn. Stantsiya. HI 72500

Sbornik nauchnykh trudov po prikladnoi botanike, genetike i selektsii. Moscow & Leningrad. Ca.1989+. Sborn. Nauchn. Trudov Prikl. Bot. Genet. Selekts. Preceded by: Trudy po prikladnoi botanike, genetike i selektsii. HI 72501

Sbornik nauchnykh trudov, sibirskii nauchno-issledovatel'skii institut kormov. Vol. 1+, 1974+. Sborn. Nauchn. Trudov Sibirsk. Nauchno-Issl. Inst. Kormov. HI 72502

Sbornik nauchnykh trudov, Tashkentskii gosudarstvennyi universitet. Tashkent. No. 494+, 1975+. Sborn. Nauchn. Trudov Tashkent. Gosud. Univ. Preceded by: Nauchnye trudy, Tashkentskii gosudarstvennyi universitet im V. I. Lenina. HI 72503

Sbornik nauchnykh trudov, voprosy rastenievodstva i zemledeliya. Vol. ?-5+, ?-1977+. Sborn. Nauchn. Trudov Vopr. Rasteniev. Zemled. HI 72504

Sbornik nauchnykh trudov, vsesoyuznii nauchno-issledovatelskii institut sadovodstva im. I. V. Michurina. Michurinsk. Vol. 26+, 1978+. Sborn. Nauchn. Trudov Vsesoyuzn. Nauchno-Issl. Inst. Sadov. I. V. Michurina. Preceded by: Sbornik nauchnykh rabot, vsesoyuznii nauchno-issledsovatelskii institut sadovodstva im. I. V. Michurina. HI 72505

Sbornik naychnykh trudov, vsesoyuznii selektsionno-geneticheskii institut. Odessa. Vol. ?-11+, ?-1974+. Sborn. Naychn. Trudov Vsesoyuzn. Selektsion.-Genet. Inst. Preceded by: Nauchnye trudy, vsesoyuznii selektsionno-geneticheskii institut. HI 72506

Sbornik naučnyh rabot Kievskogo obščestva estestvoispytatelej = Zbirnyk naukovyh prac' Kyjivs'kogo tovarystva pryrodnykiv. Kiev, Ukrainian S S R. Zbirn. Nauk. Prac' Kyjivs'k. Tovar. Pryr. See B–P–H 996/1.

Sbornik naučnyh trudov. Institut biologii, Akademija nauk Belorusskoj S S R = Sbornik nauchnykh trudov. Institut biologii, Akademiya nauk Belorusskoi S S R. Minsk.

Sborník pedagogického inštitútu v Banskej Bystrici. Banska Bystrica, Czechoslovakia. Sborn. Pedagog. Inšt. v Banskej Bystrici. See B–P–H 814/19. HI 59148

Sborník pedagogického inštitútu v Brandýse. Brandysek, Czechoslovakia. Sborn. Pedagog. Inst. v Brandýse. See B–P–H 814/20. HI 59149

Sborník pedagogického inštitútu v Jihlavě. Jihlava, Czechoslovakia. Sborn. Pedagog. Inst. v Jihlavě See B–P–H 814/21. HI 59150

Sborník pedagogického inštitútu v Košiciach. Kosice, Czechoslovakia. Sborn. Pedagog. Inšt. v Košiciach. See B–P–H 814/22. HI 59151

Sborník pedagogického inštitútu v Martine. Turciansky Svaty Martin, Czechoslovakia. Sborn. Pedagog, Inšt. v Martine See B–P–H 815/1. HI 59152

Sborník pedagogického inštitútu v Nitre. Nitra. Czechoslovakia. Sborn. Pedagog. Inšt. v Nitre. See B–P–H 815/2. HI 59153

Sborník pedagogického inštitútu v Olomouci. Ser. B. Olomouc, Czechoslovakia. Sborn. Pedagog. Inst. v Olomouci, Ser. B. See B–P–H 815/3. HI 59154

Sborník pedagogického inštitútu v Plzni. Pilsen. Czechoslovakia. Sborn. Pedagog. Inst. v Plzni. See B–P–H 815/4. HI 59155

Sborník pedagogického inštitútu v Prešove. Presov, Czechoslovakia. Sborn. Pedagog. Inšt. v Prešove. See B–P–H 815/5. HI 59156

Sborník pol'nohosp. múzea. Nitra, Czechoslovakia = Agrikultúra. Agrikultúra. See B–P–H 63/5.

Sborník prác lesníckeho a drevárskeho múzea ve Zvolene. Svolen, Czechoslovakia. Sborn. Prác Lesn. Múz. ve Svolene. See B–P–H 815/6. HI 59157

Sborník prác ochrony prírody v západoslovenskem kraji. Bratislava. Sborn. Prác Ochr. Prír. Západoslov. Kraji. See B–P–H 815/7. HI 59158

Sborník prác prírodovedeckej fakulty slovenskej university v Bratislave. Bratislava. Sborn. Prác Prír. Fak. Slov. Univ. v Bratislave. See B–P–H 815/8. HI 59159

Sborník prác o tatranskom národnom parku = Zbornik

prac o Tatranskom narodnom parku. Martin.

Sbornik praci pedagogicke fakulty v Ostrave. Prirodni vedy. 1972+. Sborn. Praci Pedagog. Fak. Ostrave, Prir. Vedy. HI 72507

Sborník prací pedagogického inštitútu v Ostravě. Ostrava, Czechoslovakia. Sborn. Prací Pedagog. Inst. v Ostravě. See B–P–H 815/10. HI 59160

Sborník přírodovědecké společnosti v Moravské Ostravě. Ostrava, Czechoslovakia. Sborn. Přír. Společn. v Moravské Ostravě. See B–P–H 815/14. HI 59164

Sborník prírodovedeckého klubu v Košiciach. Kosice, Czechoslovakia. Sborn. Prír. Klubu v Košiciach. See B–P–H 815/12. HI 59162

Sbornik prirodovedeckeho klubu v Trebici. Třebič. 1936-54. Sborn. Prir. Klubu Trebici. HI 72508

Sborník přírodovědecký. Prague. Sborn. Přír. (Prague). See B–P–H 815/11. HI 59161

Sborník prírodovedného odboru slovenského vlastivedného múzea v Bratislave. Bratislava. Sborn. Prír. Odb. Slov. Vlastiv. Múz. v Bratislave. See B–P–H 815/13. HI 59163

Sbornik rabot po biohimii kul'turnyh rastenij = Sbornik rabot po biokhimii kul'turnykh rastenii. Leningrad.

Sbornik rabot po biokhimii kul'turnykh rastenii. Leningrad. Vol. 5, 1936. Sborn. Rabot Biokhim. Kul't. Rast. Preceded by: Institut rastenievodstva. HI 72509

Sbornik rabot, institut biologii morya. Vladivostok. No. 1+, 1974+. Sborn. Rabot Inst. Biol. Morya. HI 72510

Sbornik rabot po voprosam karantina rastenii. Moscow. Vol. 7+, 1961+. Sborn. Rabot Vopr. Karant. Rast. Preceded by: Sbornik po karantinu rastenii. HI 72511

Sbornik rabot po voprosam karantina rastenij = Sbornik rabot po voprosam karantina rastenii. Moscow.

Sborník severočeskeho musea. Přírodní vědy. Liberec, Czechoslovakia. Sborn. Severočesk. Mus., Přír. Vědy. See B–P–H 815/19. HI 59165

Sborník slovenského národného múzea. Přírodné vědy. Bratislava. Vols. 8-13(1), 1962-67. Sborn. Slov. Nár. Múz., Přír. Vědy. Preceded by: Prírodovedný sborník slovenského múzea. Superseded by: Zbornik slovenského narodného muzea. Přírodné vědy. HI 72512

Sborník s[tudejniho a] l[idovychovného] ú[stavu] k[raje] o[lomouckého]. Oddil přír. vědy. Olomouc, Czechoslovakia. Sborn. Stud. Lidovychovného Ústavu Kraje Olomouc., Odd. Přír. Vědy. See B–P–H 815/21. HI 59166

Sbornik trudov aspirantov i molodykh nauchnykh sotrudnikov, vsesoyuzny institut rastenievodstva. Leningrad. Vol. 1+, 1958+. Sborn. Trudov Aspir. Molod. Nauchn. Sotrudn. Vsesoyuzn. Inst. Rasteniev. HI 72513

Sbornik trudov gosudarstvennogo Irkutskogo universiteta. Irkutsk, Russian S F S R. Vols. 7-18, 1924-31. Sborn. Trudov. Gosud. Irkutsk. Univ. Preceded by: Sbornik trudov professorov i prepodavatelei Gosudarstvennogo Irkutskogo universiteta. 3-2116-3. Superseded by: Trudy Vostochno-sibirskogo gosudarstvennogo universiteta. HI 72514

Sbornik trudov, institut zashchity rastenii. Erevan. Vol. 1+, 1970+. Sborn. Trudov Inst. Zashch. Rast. HI 72515

Sbornik trudov Moldavskoi stantsii Vsesoyuznogo instituta zashchity rastenii. Kishinev, Moldavian S S R. Vol. [1], 1954+. Sborn. Trudov Moldavsk. Stantsii Vsesoyuzn. Inst. Zashch. Rast. HI 72516

Sbornik trudov Moldavskoj stancii Vsesojuznogo instituta zaščity rastenij = Sbornik trudov Moldavskoi stantsii Vsesoyuznogo instituta zashchity rastenii. Kishinev.

Sbornik trudov molodykh nauchnykh rabotnikov, institut botaniki, akademiya nauk gruzinskoi S S R. Tiflis. Vol. ?-7+, ?-1976+. Sborn. Trudov Molod. Nauchn. Rabotn. Inst. Bot. Akad. Nauk Gruzinsk. S.S.R. HI 72517

Sbornik trudov molodykh uchenykh, nauchno-issledovatel'skii institut sadovodstva, vinogradarstva i vinodelya. Tiflis. No. ?-2+, ?-1972+. Sborn. Trudov Molod. Uchen. Nauchno-Issl. Inst. Sadov. Vinograd. Vinod. HI 72518

Sbornik trudov professorov i prepodavatelei Gosudarstvennogo Irkutskogo universiteta. Irkutsk, Russian S F S R. Vols. 1-6, 1921-23. Sborn. Trudov Profess. Prepodav. Gosud. Irkutsk. Univ. Superseded by: Sbornik trudov gosudarstvennogo Irkutskogo universiteta. 3-2116-3. HI 72519

Sbornik trudov professorov i prepodavatelej Gosudarstvennogo Irkutskogo universiteta = Sbornik trudov professorov i prepodavatelei Gosudarstvennogo Irkutskogo universiteta. Irkutsk.

Sbornik trudov volgogradskaya opytnaya stantsiya vir. Volgograd. Vols. ?-3-6, ?-1963-69. Sborn. Trudov Volgograd. Opytn. Stantsiya Vir. HI 72520

Sbornik trudov Vsesojuznogo instituta zaščity rastenij = Sbornik trudov Vsesoyuznogo instituta zashchity rastenii. Moscow & Leningrad.

Sbornik trudov Vsesoyuznogo instituta zashchity rastenii. Moscow & Leningrad. Vol. 1, 1948. Sborn. Trudov Vsesoyuzn. Inst. Zashch. Rast. Superseded by: Trudy Vsesoyuznogo instituta zashchity rastenii. 3-2391-1. HI 72521

Sbornik trudov po zaščite rastenij Vostočnoj Sibiri = Sbornik trudov po zashchite rastenii Vostochnoi Sibiri.

Irkutsk.

Sbornik trudov po zashchite rastenii Vostochnoi Sibiri. Irkutsk. Russian S F S R. Vol. 5, 1937. Sborn. Trudov Zashch. Rast. Vost. Sibiri. Preceded by: Trudy po zashchite rastenii Vostochnoi Sibiri. HI 72522

Sbornik trudy selodykh uchenykh, kirgizskii nauchno-issledovatel'skii institut zemledeliya. Vol. ?-2+, ?-1969+. Sborn. Trudy Selod. Uchen. Kirgizsk. Nauchno-Issl. Inst. Zemled. HI 72523

Sborník Ú V T I. Genetika a šlechtění. Prague. Vols. 1(38)-11(48), 1965-75. Sborn. Ú. V. T. I., Genet. Šlecht. Preceded by: Sbornik československé akademie zemědělskych věd. Superseded by: Sbornik Ú V T I Z. Genetika a šlechtěni. HI 72524

Sbornik Ú V T I. Ochrana rostlin. Prague. Vol. 1 [= ročn. 38]+, 1965+. Sborn. Ú. V. T. I., Ochr. Rostl. HI 72525

Sbornik Ú V T I. Zahradnictvi. Prague. Vol. ?-2(5)+, ?-1975+. Sborn. Ú. V. T. I., Zahradn. HI 72526

Sbornik Ú V T I Z. Genetika a šlechtěni. Prague. Vol. 12(49)+, 1976+. Sborn. Ú. V. T. I. Z., Genet. Šlecht. Preceded by: Sborník Ú V T I. Genetika a šlechtění. HI 72527

Sbornik ústav védeckotechnických informací = Sbornik Ú V T I. Prague.

Sbornik ústav védeckotechnických informací. Genetika a šlechtění = Sborník Ú V T I. Genetika a šlechtění. Prague.

Sbornik ústav védeckotechnických informací pro zemědělství. Genetika a šlechtěni = Sbornik Ú V T I Z. Genetika a šlechtěni. Prague.

Sbornik, vedeckeho lesnikeho ustavu skoly zemedelske v Praze. Prague. Vol. ?-17+, ?-1974+. Sborn. Ved. Lesn. Ustavu Skoly Zemed. Praze. HI 72528

Sborník vědeckých prác, Zvolen. Bratislava. Vol. 6+, 1964+. Sborn. Věd. Prác Zvolen. Preceded by: Sborník vědeckých prác vysokej školy lesníckej a drevárskej vo Zvolene. HI 72529

Sborník vědeckých prác vysokej školy lesníckej a drevárskej vo Zvolene. Zvolen. Vols. 1-5, 1960-63. Sborn. Věd. Prac. Vysokej. Školy Lesn. Drev. Zvolene. Superseded by: Sbornik vedeckych prac, Bratislava. HI 72530

Sborník vědeckých prací fakulty lesnické. Prague. Sborn. Věd. Prací Fak. Lesn. See B–P–H 816/7. HI 59167

Sborník vlastivědného musea v Olomouci. Olomouc. Czechoslovakia. Sborn. Vlastiv. Mus. v Olomouce. See B–P–H 816/8. HI 59168

Sborník vlastivědného musea v Prostějově. Prostejov, Czechoslovakia. Sborn. Vlastiv. Mus. v Prostějově. See B–P–H 816/9. HI 59169

Sborník vlastivědného ústavu v Olomouci. Olomouc, Czechoslovakia. Sborn. Vlastiv. Ústavu v Olomouci. See B–P–H 816/11. HI 59171

Sborník vlastivédných prací z Podblanícka. Benesov, Czechoslovakia. Sborn. Vlastiv. Prací z Podblanícka. See B–P–H 816/10. HI 59170

Sbornik, voprosy regulyatsii fotosinteza. Vol. ?-3+, ?-1973+. Sborn.Vopr. Regulyat. Fotosin. HI 72531

Sbornik Vsesojuznogo instituta zaščity rastenij = Sbornik Vsesoyuznogo instituta zashchity rastenii. Leningrad.

Sbornik Vsesoyuznogo instituta zashchity rastenii. Leningrad. Vols. 1-8, 1932-34. Sborn. Vsesoyuzn. Inst. Zashch. Rast. Preceded & superseded by: Zashchita rastenii. 3-2391-1. HI 72532

Sborník východoslovenského múzea v Košiciach. Séria A, prirodné vedy. Kosice. Vols. 1-7, 1960-66. Sborn. Východoslov. Múz. v Košiciach, Sér. A. Superseded by: Zborník východoslovenského múzea v Košiciach. Séria A, geologické vedy and Séria B, zoologia, botanika. HI 72533

Sborník vysoké školy pedagogické v Olomouci. Přírodní vědy. Prague. Sborn. Vysoké Školy Pedagog. v Olomouci, Přír. Vědy. See B–P–H 816/15. HI 59172

Sborník vysokej školy pol'nohospodárskej v Nitre. Seria A. Bratislava. Vols. 1-12, 1958-65. Sborn. Vysokej Školy Pol'nohosp. v Nitre, A. Superseded by: Acta fytotechnica. HI 72534

Sborník vysoké školy pedagogické v Praze. Přírodní vědy. Prague. Sborn. Vysoké Školy Pedagog. v Praze, Přír. Vědy See B–P–H 816/16. HI 59173

Sborník vysoké školy zemědělské v Brně. Brno, Czechoslovakia. Sborn. Vysoké Školy Zeměd. v Brně. See B–P–H 817/1. HI 59175

Sborník vysoké školy zemědělské a lesnickě. Spisy fakulty agronomické a zootechnické. Spisy fakultry lesnické. Brno, Czechoslovaika. Sborn. Vysoké Školy Zeměd. Lesn. See B–P–H 816/17. HI 59174

Sborník vysoké školy zemědělské v Praze. Prague. Sborn. Vysoké Školy Zeměd. v Praze. See B–P–H 817/2. HI 59176

Sborník vyššej pedagogické školy v Plzni. Pilsen, Czechoslovakia. Sborn. Vyššej Pedagog. Školy v Plzni. See B–P–H 817/4. HI 59177

Sbornik vyzkumnych praci z odborne celulosy a papiru. Bratislava. Vol. 1+, 1956+. Sborn. Vysk. Praci Odb. Celulosy Pap. HI 72535

Sbornik zapadoceske museum v Plzni. Zapadoceske muzeum, prirodoveda. Plzen. Ca.1989+. Sborn. Zapadoceske Mus. Plzni, Zapadoceske Muz. Prir. HI 72536

Scandinavian journal of forest research. Stockholm. Vol. 1+, 1986+. Scand. J. Forest Res. HI 72537

Scanning electron microscopy; international journal of scanning electron microscopy, related techniques, and applications. Chicago, IL. 1978+. Scan. Electron Microscop. Preceded by: Proceedings, scanning electron microscope symposium. HI 72538

Scelta di memorie agrarie inedite o estratte dalle opere de' più valenti agronomi italiani ed esteri. Bologna. Scelta Mem. Agrar. See B–P–H 817/13. HI 59178

Scelta di opuscoli interessanti tradotti da varie lingue. Milan. Scelta Opusc. Interessanti (Milan). See B–P–H 817/15. HI 59179

Scelta di opuscoli interessanti tradotti di varie lingue. Turin. Scelta Opusc. Interessanti (Turin). See B–P–H 817/16. HI 59180

Schedae ad floram ibericam selectam. Barcelona. Cent. 1, 1934. Sched. Fl. Iber. Selectam. HI 72539

Schedae ad herbarium florae reipublicae sowjeticae Ucrainicae = Spisok roslin gerbari flori U S R R. Kiev.

Schedae ad herbarium florae rossicae a museo botanico academiae imperialis scientiarum petropolitanae editum = Spisok Rastenii Gerbariya Russkoi Flory izdavaemago Botanicheskim Muzeem Imperatorskoi Akademii Nauk. St. Petersburg.

Schedae ad herbarium florae rossicae a museo botanico academiae scientiarum petropolitanae editum = Spisok rastenii Gerbariya Russkoi flory izdavaemogo Botanicheskim muzeem Rossiiskoi akademii nauk. Petrograd.

Schedae ad herbarium florae U R S S ab instituto botanico academiae scientiarum URSS editum = Spisok rastenii Gerbariya flory S S S R izdavaemogo Botanicheskim institutom Vsesoyuznogo akademii nauk. Leningrad.

Schedulae orchidianae. Boston, MA. Nos. 1-10, 1922-30. Schedul. Orchid. HI 72540

Schlechteriana. Lahnau. ?-1987+. Schlechteriana. HI 72541

Schlern. Illustrierte Monatsschrift für Heimat- und Volkskunde. Bolzano, Italy. Schlern. See B–P–H 817/21. HI 59181

Schlesische Garten- und Blumen-Zeitung. Breslau [=Wroclaw, Poland]. Schles. Garten- Blumen-Zeitung. See B–P–H 817/22. HI 59182

Schlesische Landwirthschaftliche Monatsschrift. Breslau [=Wroclaw, Poland]. Schles. Landw. Monatsschr. See B–P–H 817/23. HI 59183

Schlesische Landwirthschaftliche Zeitschrift. Breslau [=Wroclaw, Poland]. Schles. Landw. Z. See B–P–H 817/24. HI 59184

Schlesische Provinzialblätter. Breslau [=Wroclaw, Poland]. Schles. Provinzialbl. See B–P–H 818/3. HI 59186

Schlesischer Lehrer-Verein für Naturkunde. Jahres-Bericht. Görlitz. Schles. Lehrer-Verein Naturk. Jahres-Ber. See B–P–H 818/1. HI 59185

Schleswig-Holsteinisch-Lauenburgische Provinzialberichte. Kiel. Schleswig-Holst.-Lauenburg. Provinzialber. See B–P–H 818/11. HI 59194

Schleswig-Holstein-Lauenburgischer Gewerbefreund. Schleswig. Schleswig-Holstein-Lauenburg. Gewerbefr. See B–P–H 818/10. HI 59193

Schleswig-Holsteinische Anzeigen. Schleswig-Holst. Anz. See B–P–H 818/5. HI 59188

Schleswig-Holsteinische Blätter. Schleswig. Schleswig-Holst. Blätt. See B–P–H 818/6. HI 59189

Schleswig-Holsteinische Provinzialberichte. Altona [=Hamburg, in part]. Schleswig-Holst. Provinzialber. See B–P–H 818/9. HI 59192

Schleswig-Holsteinischer Acker- und Garten-Allmanach. Altona [=Hamburg, in part]. Schleswig-Holst. Acker-Gart.-Allman. See B–P–H 818/4. HI 59187

Schleswig-Holsteinischer historischer Almanach. Altona [=Hamburg, in part]. Schleswig-Holst. Hist. Alman. See B–P–H 818/7. HI 59190

Schleswig-Holsteinisches Magazin oder Sammlung vermischter Schriften zur Aufnahme der Wissenschaften und Künste. Glückstadt, Germany. Schleswig-Holst. Mag. See B–P–H 818/8. HI 59191

School nature league bulletin. New York. Nos. 1-?, 1930-? School Nat. League Bull. Superseded by: Audubon nature bulletin. 1-555-3. HI 72542

Schriften der Albertus-Universität. Naturwissenschaftliche Reihe. Königsberg [=Kaliningrad, R S F S R]. Schriften Albertus-Univ., Naturwiss. Reihe. See B–P–H 818/36. HI 59195

Schriften des arbeitskreises für naturwissenschaftliche Heimatforschung in Wedel und Umgeburg e. V. Wedel. Vol. 1+, 1965+. Schriften Arbeitskeises Naturwiss. Heimatf. Wedel Umgeburg. HI 72543

Schriften der Berlinishchen Gesellschaft naturforschender Freunde. Berlin. Schriften Berlin. Ges. Naturf. Freunde. See B–P–H 819/1. HI 59196

Schriften des Botanischen Gartens der Universität. Riga = Acta horti botanici universitatis. Riga, Latvian S S R.

Schriften des Bremer naturwissenschaftlichen Vereins Reihe G = Bremer Beiträge zur Naturwissenschaft. Bremen. Bremer Beitr. Naturwiss. See B–P–H 227/34.

Schriften der Bremer Wissenschaftlichen Gesellschaft Reihe B = Abhandlungen herausgegeben vom naturwissenschaftlichen Vereine zu Bremen, vol. 13+.

Bremen. Abh. Naturwiss. Vereine Bremen. See B–P–H 32/21.

Schriften der Bremer wissenschaftlichen Gesellschaft Reihe G = Bremer Beiträge zur Naturwissenschaft. Bremen. Bremer Beitr. Naturwiss. See B–P–H 227/34.

Schriften der Dänischen Naturforschenden Gesellschaft. Copenhagen. Schriften Dän. Naturf. Ges. See B–P–H 819/6. HI 59197

Schriften des Deutschen Pappelvereins und Lignikultur e. V. Bonn. Schriften Deutsch. Pappelvereins. See B–P–H 819/7. HI 59198

Schriften der Duisburgischen gelehrten Gesellschaft. Duisburg, Germany. Schriften Duisburg, Germany. Schriften Duisburg. Gel. Ges. See B–P–H 819/8. HI 59199

Schriften der Freien Vereinigung von Freunden der Mikroskopie. Berlin. Schriften Freien Vereinigung Freunden Mikroskop. See B–P–H 819/9. HI 59200

Schriften aus dem ganzen Gebiete der Botanik. St. Petersburg. Vols. 1-2, 1853. Schriften Ganzen Geb. Bot. HI 72544

Schriften des geographischen Instituts der Universität Kiel. Kiel. 1932-42, 1947+. Schriften Geogr. Inst. Univ. Kiel. HI 72545

Schriften der Gesellschaft zur Beförderung der gesammten Naturwissenschaften zu Marburg. Marburg. Schriften Ges. Beförd. Gesammten Naturwiss. Marburg. See B–P–H 819/10. HI 59201

Schriften der Gesellschaft naturforschender Freunde zu Berlin. Berlin. Schriften Ges. Naturf. Freunde Berlin. See B–P–H 819/11. HI 59202

Schriften, herausgegeben von der Naturforscher-Gesellschaft bei der Universität Dorpat. Dorpat [=Tartu, Estonian S S R]. Vols. 1-9, 1884-96. Schriften Naturf.-Ges. Univ. Dorpat. Superseded by: Trudy Obshchestva Estestvoispytatelei pri Imperatorskom Yur′evskom Universitetě. 5-4154-2. HI 72546

Schriften, herausgegeben von der Naturforscher-Gesellschaft bei der Universität Jurjeff. Dorpat = Schriften, herausgegeben von der Naturforscher-Gesellschaft bei der Universität Dorpat. Dorpat.

Schriften, herausgegeben von der Naturforscher-Gesellschaft bei der Universität Jurjeff (or Jurjew) = Trudy Obshchestva Estestvoispytatelei pri Imperatorskom Yur′evskom Universitetě. Moscow.

Schriften, herausgegeben von der Naturforscher-Gesellschaft bei der Universität Jurjew <Dorpat> = Schriften, herausgegeben von der Naturforscher-Gesellschaft bei der Universität Dorpat. Dorpat.

Schriften herausgegeben von der Naturforscher-Gesellschaft bei der Universität Tartu = Tartu ülikooli juures oleva loodusuurijate seltsi kirjatööd.

Schriften der Königlich Sächsischen Weinbau-Gesellschaft. Grimma, Germany. Schriften Königl. Sächs. Weinbau-Ges. See B–P–H 819/13. HI 59204

Schriften der Königlichen Physikalisch-Ökonomischen Gesellschaft zu Königsberg [=Kaliningrad, R S F S R]. Schriften Königl. Phys.-Ökon. Ges. Königsberg. See B–P–H 819/12. HI 59203

Schriften der Königsberger Gelehrten Gesellschaft. Naturwissenschaftliche Klasse. Königsberg [=Kaliningrad, R S F S R]. Schriften Königsberger Gel. Ges., Naturwiss. Kl. See B–P–H 819/14. HI 59205

Schriften der kurfürstlichen deutschen Gesellschaft in Mannheim. Mannheim. Schriften Kurfürstl. Deutsch. Ges. Mannheim. See B–P–H 819/15. HI 59206

Schriften der Leipziger ökonomischen Societät. Friedrichstadt [=Dresden, in part]. Schriften Leipziger Ökon. Soc. See B–P–H 819/16. HI 59207

Schriften der Naturforschenden Gesellschaft in Danzig. Danzig [=Gdansk, Poland]. Schriften Naturf. Ges. Danzig. See B–P–H 819/17. HI 59208

Schriften der Naturforschenden Gesellschaft zu Jena. Dresden. Schriften Naturf. Ges. Jena. See B–P–H 819/18. HI 59209

Schriften der naturforschenden Gesellschaft zu Kopenhagen. Copenhagen. Schriften Naturf. Ges. Kopenhagen. See B–P–H 819/19. HI 59210

Schriften der Naturforschenden Gesellschaft zu Leipzig. Acta societatis naturae scrutatorum lipsiensis. Leipzig. Schriften Naturf. Ges. Leipzig. See B–P–H 819/20. HI 59211

Schriften des Naturwissenschaftlichen Vereins des Harzes in Wernigerode. Wernigerode, Germany. Schriften Naturwiss. Vereins Harzes Wernigerode. See B–P–H 820/8. HI 59212

Schriften des Naturwissenschaftlichen Vereins zu Passau. Munich. Schriften Naturwiss. Vereins Passau. See B–P–H 820/9. HI 59213

Schriften des Naturwissenschaftlichen Vereins für Schleswig-Holstein. Kiel. Schriften Naturwiss. Vereins Schleswig-Holstein. See B–P–H 820/10. HI 59214

Schriften der Physikalisch-Ökonomischen Gesellschaft zu Königsberg = Schriften der Königlichen Physikalisch-Ökonomischen Gesellschaft zu Königsberg. Königsberg [=Kaliningrad, R S F S R]. Schriften Königl. Phys.-Ökon. Ges. Königsberg. See B–P–H 819/12.

Schriften der physikalischen Klasse der Königlich Dänischen Gesellschaft der Wissenschaften in Kopenhagen. Copenhagen & Leipzig. Schriften Phys.

Kl. Königl. Dän. Ges. Wiss. Kopenhagen. See B–P–H 820/11. HI 59215

Schriften der Regensburgischen Botanischen Gesellschaft = Geschichte der Regensburgischen Botanischen Gesellschaft nebst einigen Aufsätzen, Reden und Abhandlungen. Regensburg. Gesch. Regensburg. Bot. Ges. See B–P–H 400/13.

Schriften für Süsswasser- und Meereskunde. Büsum, Germany. Schriften Süsswasser- Meeresk. See B–P–H 820/14. HI 59216

Schriften der Thüringischen Landesarbeitsgemeinschaft für Heilpflanzenkunde und Heilpflanzenbeschaffung in Weimar und Mitteilungen der Thüringischen Botanischen Gesellschaft (ehemals Thüringischer Botanischer Verein). Beiheft. Weimar. Schriften Thüring. Landesarbeitsgem. Heilpflanzenk. Weimar & Mitt. Thüring. Bot. Ges. Beih. See B–P–H 820/15. HI 59217

Schriften der Thüringischen Landesarbeitsgemeinschaft für Heilpflanzenkunde und Heilpflanzenbeschaffung Weimar. Weimar. Schriften Thüring. Landesarbeitsgem. Heilpflanzenk. Weimar. See B–P–H 820/16. HI 59218

Schriften des Vereins für Geschichte und Naturgeschichte der Baar und der angrenzenden Landestheile in Donaueschingen. Karlsruhe. Schriften Vereins Gesch. Baar Donaueschingen. See B–P–H 820/17. HI 59219

Schriften des Vereins für Naturkunde an der Unterweser. Wesermünde [=Bremerhaven]. Schriften Vereins Naturk. Unterweser. See B–P–H 820/18. HI 59220

Schriften des Vereins zur Verbreitung Naturwissenschaftlicher Kenntnisse in Wien. Vienna. Schriften Vereins Verbreit. Naturwiss. Kenntn. Wien. See B–P–H 821/1 HI 59221

Schriften und Verhandlungen der Ökonomischen Gesellschaft im Königreich Sachsen. Dresden. Schriften Verh. Ökon. Ges. Königr. Sachsen. See B–P–H 821/2. HI 59222

Schriften des Würtembergischen naturhistorischen Reisevereins, enthaltend Reisebeschreibungen und Mittheilungen aus der Natur- und Völkerkunde. Stuttgart. Schriften Württemberg. Naturhist. Reisevereins. See B–P–H 821/3. HI 59223

Schriftenreihe der forstlichen Abteilung der Albert-Ludwigs-Universität, Freiburg im Bresgau. Freiburg im Bresgau. Vol. 1+, 1962+. Schriftenreihe Forstl. Abt. Albert-Ludwigs-Univ. Freiburg. HI 72547

Schriftenreihe der forstlichen Bundesversuchsanstalt Mariabrunn in Wien. Vienna & Munich. Schriftenreihe Forstl. Bundesversuchsanst. Mariabrunn Wien. See B–P–H 821/4. HI 59224

Schriftenreihe der Hermann-Göring-Akademie der deutschen Forstwissenschaft. Berlin. Schriftenreihe Hermann-Göring-Akad. Deutsch. Forstwiss. See B–P–H 821/5. HI 59225

Schriftenreihe, Institut für Naturschutz Darmstadt = Institut für Naturschutz Darmstadt. Schriftenreihe. Darmstadt, Germany. Inst. Naturschutz Darmstadt Schriftenreihe. See B–P–H 433/24.

Schriftenreihe der Internationalen Forstzentrale = Silvae orbis. Schriftenreihe der Internationalen Forstzentrale. Berlin-Wannsee. Silvae Orbis. See B–P–H 836/15.

Schriftenreihe des Landesamtes für Naturschutz und Landschaftspflege Schleswig-Holstein. Kiel. Vol. 1+, 1977+. Schriftenreihe Landesamtes Naturschutz Landschaftspflege Schleswig-Holstein. HI 72548

Schriftenreihe der Landesanstalt für Oekologie, Landschaftsentwicklung und Forstplanung Nordrhein-Westfalen. Recklinghausen. 1975?+. Schriftenreihe Landesanst. Oekol. Landschaftsentw. Forstplan. Nordrhein-Westfalen. HI 72549

Schriftenreihe für Landschaftspflege und Naturschutz. Bonn-Bad Godesberg. Vol. 1+, 1966+. Schriftenreihe Landschaftspflege Naturschutz. HI 72550

Schriftenreihe der landwirtschaftlichen Fakultät der Christian-Albrechts-Universität Kiel. Kiel. Vol. 1+, 1949+. Schriftenreihe Landw. Fak. Christian-Albrechts-Univ. Kiel. HI 72551

Schriftenreihe der landwirtschaftlichen Fakultät der Universität Kiel = Schriftenreihe der landwirtschaftlichen Fakultät der Christian-Albrechts-Universität Kiel. Kiel.

Schriftenreihe der Naturschutzstelle Darmstadt. Darmstadt, Germany. Schriftenreihe Naturschutzstelle Darmstadt. See B–P–H 821/6. HI 59226

Schriftenreihe der Naturschutzstelle Darmstadt-Stadt. Darmstadt, Germany. Schriftenreihe Naturschutzstelle Darmstadt-Stadt. See B–P–H 821/7. HI 59227

Schriftenreihe der Österreichischen Gesellschaft für Holzforschung. Vienna. Schriftenreihe Österr. Ges. Holzf. See B–P–H HI 59228

Schriftenreihe Umweltschutz. Berne = Cahiers de l'environnement. Berne.

Schriftenreihe für Vegetationskunde. Bundesanstalt für Vegetationskunde, Naturschutz und Landschaftspflege. Bad Godesberg. Schriftenreihe Vegetationsk. See B–P–H 821/9. HI 59229

Schrifttum der Bodenkultur. Vienna. Schrifttum Bodenkult. See B–P–H 821/10. HI 59230

Schwäbisches Magazin von gelehrten Sachen. Stuttgart. Schwäb. Mag. Gel. Sachen. See B–P–H 821/13. HI 59232

Schwedische Annalen der Medicin und Naturgeschichte.

Berlin & Stralsund. Schwed. Ann. Med. Naturgesch. See B–P–H 821/14. HI 59233

Schwedisches Magazin oder gesammelte Schriften der grössten Gelehrten in Schweden, für die Liebhaber der Arzneywissenschaft, der Naturgeschichte, Chemie und Oekonomie. Copenhagen. Schwed. Mag. Schriften Gel. Schweden. See B–P–H 821/16. HI 59234

Schwedisches Magazin, oder Schriften aus der Naturforschung, Stadt- und Landwirthschaft. Copenhagen. Schwed. Mag. Schriften Naturf. See B–P–H 821/17. HI 59235

Schwedisches oeconomisches Wochenblatt. [Translation of: Oekonomiska tidningar.] Greifswald. Vols. 1-4, 1765-66. Schwed. Oecon. Wochenbl. HI 72553

Schweitzerischen Gesellschaft in Bern Sammlungen von Landwirthschaftlichen Dingen. Zurich. Schweitz. Ges. Bern Samml. Landw. Dingen. See B–P–H 821/18. HI 59236

Schweiz, Der. Pflanzen-Freund; ein praktischer Ratgeber für jeden, welcher Blumen-, Gemüse- und Obstzucht zu seinem Vergnügen oder zu seinem Nutzen treibt. Rüti. 1902-03. Schweiz. HI 72554

Schweizer Forster. Aarau. Vol. 110+, 1974+. Schweizer Forster. Preceded by: Praktische Forstwirt für die Schweiz. HI 72555

Schweizer Garten. Zurich. 1937-56; 1964(1-10). Schweizer Gart. For 1957-63 see: Schweizer Garten und Wohnkultur. HI 72556

Schweizer Garten und Wohnkultur. Zurich. 1957-63. Schweizer Gart. Wohnkultur. Preceded and superseded by: Schweizer Garten. HI 72557

Schweizer Naturschutz. Basel. Vol. 1+, 1935+. Schweizer Naturschutz. Preceded by: Schweizerische Blätter für Naturschutz. 5-3808-3. HI 72558

Schweizerische Anstalt für das forstliche Versuchswesen. Mitteilungen. Zurich. Vols. 32-49, 1956-73. Schweiz. Anst. Forstl. Versuchswesen Mitt. Preceded by: Mitteilungen der schweizerischen Anstalt für das forstliche Versuchswesen. Superseded by: Mitteilungen eidgenossische Anstalt für das forstliche Versuchswesen. HI 72559

Schweizerische Bauern-Zeitung. Brugg. Vol. 1+, 1900+. Schweiz. Bauern-Zeitung. 5-3809-2. HI 72560

Schweizerische Beiträge zur Dendrologie. Horgen, Switzerland. Schweiz. Beitr. Dendrol. See B–P–H 822/3. HI 59237

Schweizerische Blätter für Naturschutz. Basel. Vols. 1-8, 1926-33. Schweiz. Blätt. Naturschutz. Superseded by: Schweizer Naturschutz. HI 72561

Schweizerische Chemikerzeitung = Chimia. Zurich. Chimia. See B–P–H 307/39.

Schweizerische Gartenbau; ein praktischer Führer für Gärtner, Garten- und Blumen- Freunde. Zurich. Vols. ?-9-11-?, ?-1896-98-? Schweiz. Gartenbau. HI 72562

Schweizerische Gärtnerzeitung. Zurich. Vols. 1-78, 1898-1975. Schweiz. Gärtnerzeitung. Superseded by: Gärtner-Meister. 5-3810-1. HI 72563

Schweizerische landwirtschaftliche Forschungen. Berne. Vol. 1+, 1962+. Schweiz. Landw. Forsch. HI 72564

Schweizerische landwirthschaftliche Zeitschrift "Die Grüne". Aarau, Switzerland. Schweiz. Landw. Z. "Die Grüne". See B–P–H 822/7. HI 59239

Schweizerische paläontologische Abhandlungen. Basel. Vol. 63+, 1940+. Schweiz. Paläontol. Abh. Preceded by: Abhandlungen der schweizerischen paläontologischen Gesellschaft. 5-3812-2. HI 72565

Schweizerische Zeitschrift für Forstwesen. Bern. Schweiz. Z. Forstwesen. See B–P–H 822/8. HI 59240

Schweizerische Zeitschrift für Gartenbau. Zurich. Schwiez. Z. Gartenbau. See B–P–H 822/9. HI 59241

Schweizerische Zeitschrift für Hydrologie. Basel. Vols. 11-50?, 1948-88? Schweiz. Z. Hydrol. Preceded by: Zeitschrift für Hydrologie. Superseded by: Aquatic sciences. 5-3813-2. HI 72566

Schweizerische Zeitschrift für Land- und Gartenbau. Zurich. Schweiz. Z. Land- Gartenbau. See B–P–H 822/11. HI 59242

Schweizerische Zeitschrift für Natur- und Heilkunde. Zurich & Heilbronn. Schweiz. Z. Natur- Heilk. See B–P–H 822/12. HI 59243

Schweizerische Zeitschrift für Pilzkunde. Bern-Bümpliz. Schweiz. Z. Pilzk. See B–P–H 822/13. HI 59244

Schweizerisches Forst-Journal. Bern. Schweiz. Forst-J. See B–P–H 822/5. HI 59238

Schweizerisches Gartenbaublatt. Solothurn. Vol. 1+, 1952+. Schweiz. Gartenbaubl. HI 72567

SchweizerJugend forscht = Science appelle les jeunes. Winterthur.

Science; An illustrated journal (later A weekly journal devoted to the advancement of science). [American Association for the Advancement of Science.] Cambridge, MA, New York, Washington, DC. Vols. 1-22, 1883-94; n.s. vols. 1-113, 1895-1951; ser. 2, vols. 114+, 1951+. Science. 5-3816-1. HI 72568

Science. Moscow. Science (Moscow). See B–P–H 829/20. HI 59326

Science. New York. Vols. 1(1)-3(82), 1880-82. Science (New York). 5-3816-3. HI 72569

Science. [Chayon.] Seoul. No. 1+, 1973+. Science (Seoul). HI 72570

Science. [K'o hsüeh.] Shanghai. Science (Shanghai). See B–P–H 829/22. HI 59327

Science. [Kagaku.] Tokyo. Science (Tokyo/Kagaku). See B–P–H 829/23. HI 59328

Science. [Rigaku.] Tokyo. Science (Tokyo/Rigaku). See B–P–H 829/24. HI 59329

Science 80 [i.e. year, e.g. 81, 82 etc.] Washington, DC. Vol. 1+, 1979+. Sci. 80. [etc.] HI 72571

Science abstracts of China: Biological sciences. Peking. Sci. Abstr. China, Biol. Sci. See B–P–H 822/21. HI 59249

Science - Afrique; information bulletin of the C C T A/C S A secretariat. London, Bukavu. Nos. 1-23, 195?-61. Sci. Afrique. HI 72572

Science in Alaska = Proceedings of the Alaska science conference. Fairbanks, AK.

Science appelle les jeunes. Winterthur. Vols. ?-5-11(2), ?-1972-78. Sci. Appelle Jeunes. HI 72573

Science of biology journal. Stillwater, OK. Vol. 1+, 1975+. Sci Biol. J. HI 72574

Science booklist of the american association for the advancement of science. New York, NY. Sci. Booklist Amer. Assoc. Advancem. Sci. See B–P–H 823/9. HI 59254

Science bulletin. [K'o hsüeh hui pao.] Taipei, Taiwan. Sci. Bull. (Taipei). See B–P–H 823/10. HI 59255

Science bulletin of the agriculture and home economics division, university of the Ryukyus. Naha, Okinawa. Nos. 2-?, 1955-? Sci. Bull. Agric. Div. Univ. Ryukyus. Preceded by: Science bulletin of the faculty of agriculture, university of the Ryukyus. Superseded by: Science bulletin of the college of agriculture, university of the Ryukyus. HI 72575

Science bulletin, Brigham Young university, biological series = Brigham Young university. Science bulletin. Biological series. Provo, UT. Brigham Young Univ. Sci. Bull., Biol. Ser. See B–P–H 228/10.

Science bulletin, Brigham Young university, geological series = Brigham Young university. Science bulletin. Geological series. Provo, UT. Brigham Young Univ. Sci. Bull., Geol. Ser. See B–P–H 228/11.

Science bulletin of the college of agriculture, university of the Ryukyus. Naha, Okinawa. No. ?-19+, 1967?-72+. Sci. Bull. Coll. Agric. Univ. Ryukyus. Preceded by: Science bulletin of the agriculture and home economics division, university of the Ryukyus. HI 72576

Science bulletin, department of agriculture and fisheries, South Africa. Pretoria. Nos. ?-27-97-?, ?-1923-31-? Sci. Bull. Dept. Agric. Fish. South Africa. Preceded by: Scientific bulletin, department of agricultural technical services, South Africa. HI 72579

Science bulletin, department of agriculture and forestry, South Africa = Science bulletin, department of agriculture, South Africa. Pretoria.

Science bulletin, department of agriculture, New South Wales. Sydney, N.S.W. No. 1+, 1911+. Sci. Bull. Dept. Agric. New South Wales. HI 72577

Science bulletin, department of agriculture, South Africa. Pretoria. Nos. 1-379-?, 1911-58. Sci. Bull. Dept. Agric. S. Africa. Superseded by: Scientific bulletin, department of agricultural technical services, South Africa. HI 72578

Science bulletin, department of agriculture, Thailand. Bangkok. 1966+. Sci. Bull. Dept. Agric. Thailand. HI 72580

Science bulletin, department of agriculture, Union of South Africa = Science bulletin, department of agriculture, South Africa. Pretoria.

Science bulletin of the faculty of agriculture, Kyushu university. [Kyusu daigaku nogakubu. Gakugei zasshi.] Fukuoka, Japan. Sci. Bull. Fac. Agric. Kyushu Univ. See B–P–H 823/17. HI 59256

Science bulletin of the faculty of agriculture; university of the Ryukyus. [Ryukyu daigaku nogakubu gakujutsu hokoku.] Naha, Okinawa. Sci. Bull. Fac. Agric. Univ. Ryukyus. See B–P–H 823/18. HI 59257

Science bulletin of the faculty of liberal arts and education; Nagasaki university. [Nagasaki daigaku kyoikugakubu shizenkagaku kenkyu hokoku.] No. 6+, 1957+. Sci. Bull. Fac. Liberal Arts Nagasaki Univ. Preceded by: Science reports of the faculty of arts and literature, Nagasaki university. HI 72581

Science bulletin, natural history museum of Los Angeles County. Los Angeles, CA. Nos. 13-30, 1972-78. Sci. Bull. Nat. Hist. Mus. Los Angeles County. Preceded by: Bulletin of the Los Angeles County museum (of natural history). Science. Incorporated in: Contributions in science, natural history museum of Los Angeles county. HI 72582

Science bulletin, Pahlavi university. Vol. 1+, 1966+. Sci. Bull. Pahlavi Univ. HI 72583

Science bulletin, university of Kansas = University of Kansas science bulletin. Univ. Kansas Sci. Bull. See B–P–H 936/14.

Science in China. Series B: Chemistry, life science and earth sciences. Beijing, Oxford, Elmsford, NY. Vol. 32+, 1989+. Sci. China, B. HI 74536

Science contributions, New Hampshire agricultural experiment station = New Hampshire agricultural experiment station. Science contributions. Durham, NH. New Hampshire Agric. Exp. Sta. Sci. Contr. See B–P–H 653/4.

Science and culture. Calcutta. Sci. & Cult. See

B–P–H 822/18. HI 59246

Science citation index. Philadelphia, PA. Sci. Citation Index. See B–P–H 823/24. HI 59258

Science counselor; quarterly journal for teachers of science in the catholic high schools. Pittsburgh, PA. Sci. Counselor. See B–P–H 824/3. HI 59260

Science digest from Israel. Jerusalem. Sci. Digest Israel. See B–P–H 824/5. HI 59261

Science dimension. Ottawa. Vol. 1+, 1969+. Sci. Dimension. Preceded by: N R C research news. Ottawa. HI 72584

Science education. [K'o hsüeh chiao yü.] [Taiwan]. Sci. Educ. (Taiwan). See B–P–H 824/7. HI 59262

Science education. [Rika kyoiku.] Tokyo. Sci. Educ. (Tokyo). See B–P–H 824/8. HI 59263

Science for the farmer. University Park, PA. Sci. for Farmer. See B–P–H 824/12. HI 59266

Science of food and agriculture. Ames, IA. Vol. 1+, 1983+. Sci. Food Agric. HI 72585

Science-gossip; an illustrated monthly record of nature, country lore and applied science. London. Sci.-Gossip. See B–P–H 830/30. HI 59335

Science and government report. Washington, DC. Vol. 1+, 1971+. Sci. Gov. Rep. HI 72586

Science in Iceland. Reykjavik. Vols. 1-2, 1968-70. Sci. Iceland. HI 72587

Science journal. [College of science; Sun Yatsen university.] [Tzu jan k'o hsüeh.] Canton. Sci. J. (Canton). See B–P–H 824/16. HI 59269

Science journal. London. Vols. 1-7(1), 1966-71. Sci. J. (London). Preceded by: Discovery. Incorporated in: New scientist and science journal. HI 72588

Science journal. Mysore. ?-1982+. Sci. J. (Mysore). HI 72589

Science letters, national academy of sciences, India. Allahabad = National academy science letters. Allahabad.

Science library. Bibliographical series. London. Sci. Libr., Bibliogr. Ser. See B–P–H 824/23. HI 59272

Science on the march. Buffalo, NY. Vols. 38(5)-55, 1958-76. Sci. March. Preceded by: Hobbies. Superseded by: Collections. HI 72591

Science monograph, agricultural experiment station, university of Wyoming. Laramie, WY. No. 1+, 1966+. Sci. Monogr. Agric. Exp. Sta. Univ. Wyoming. HI 72592

Science monthly bulletin. [World science association.] [K'o hsüeh shih pao.] Peiping [=Peking]. Sci. Monthly Bull. See B–P–H 825/7. HI 59276

Science museum news. No. ?-27+, ?-1989+. Sci. Mus. News. HI 72593

Science et nature. Paris. Sci. & Nat. See B–P–H 822/19. HI 59247

Science et nature par la photographie et par l'image; revue de la société des amis du Muséum. Paris. Nos. 1-121, 1954-74. Sci. Nat. Photogr. Image. HI 72594

Science in Nevada ranching. Reno, NV. Sci. Nevada Ranching. See B–P–H 825/8. HI 59277

Science in New Guinea. Port Moresby. Vol. 1+, 1972+. Sci. New Guinea. HI 72595

Science news. Karachi, Pakistan. Sci. News (Karachi). See B–P–H 825/10. HI 59278

Science news. London. Sci. News (London). See B–P–H 825/11. HI 59279

Science news. [K'o hsüeh t'ung pao.] Peking. Sci. News (Peking). See B–P–H 825/12. HI 59280

Science news. Salem, MA. & New York, NY. Sci. News (Salem & New York). See B–P–H 825/13. HI 59281

Science news bulletin. Washington, DC. Sci. News Bull. See B–P–H 825/14. HI 59282

Science news-letter. Washington, DC. Sci. News-Lett. See B–P–H 825/16. HI 59283

Science now. London. ?-1983+. Sci. Now. HI 72597

Science papers, Haslemere microscope and natural history society. Guildford. 1903-04. Sci. Pap. Haslemere Microscope Nat. Hist. Soc. Continued as: Science papers, Haslemere natural history society. HI 72598

Science papers, Haslemere natural history society. Guildford? Nos. 4-11, 1909-34. Sci. Pap. Haslemere Nat. Hist. Soc. Preceded by: Science papers, Haslemere microscope and natural history society. HI 72599

Science and practice of mushroom growing. Santa Cruz, CA. 1969+. Sci. Pract. Mushr. Growing. HI 72600

Science progrès, la nature. Paris. Nos. 3333-3413, 1963-69. Sci. Progr. Nat. Superseded by: Science progrès, découverte. HI 72601

Science progrès découverte. Paris. Nos. 3414-3450, 1969-72. Sci. Progr. Découv. Preceded by: Science progrès, la nature. Incorporated in: Recherche. HI 72602

Science progress; a quarterly review of current scientific investigation. London. Sci. Progr. (London, 1894-1900). See B–P–H 826/5. HI 59289

Science progress; a quarterly journal of scientific thought, work and affairs. London. Sci. Progr. (London, 1906+). See B–P–H 826/6. HI 59290

Science in progress. New Haven, CT. Sci. Progr. (New Haven). See B–P–H 826/7. HI 59291

Science progress in the twentieth century = Science

progress; a quarterly journal of scientific thought, work and affairs. London. Sci. Progr. (London, 1906+). See B–P–H 826/6.

Scientific publication of the freshwater biological association of the british empire. Ambleside. Sci. Publ. Freshwater Biol. Assoc. Brit. Empire. See B–P–H 826/9. HI 59292

Science quarterly. [National central university.] [K'o hsüeh chi k'an.] Chungking, China. Sci. Quart. See B–P–H 826/11. HI 59293

Science quarterly of the national university of Peking. [Tzu-jan k'o-hsüeh chi-k'an.] Peiping [=Peking]. Sci. Quart. Natl. Univ. Peking. See B–P–H 826/14. HI 59294

Science record. [Academia sinica.] [K'o hsüeh chi lu.] Chungking, China. Sci. Rec. (Chungking). See B–P–H 826/15. HI 59295

Science record. [Otago university science students' association.] Dunedin, New Zealand. Sci. Rec. (Dunedin). See B–P–H 826/16. HI 59296

Science reports, college of general education, Osaka university. Osaka. Vol. 1+, 1952+. Sci. Rep. Coll. Gen. Educ. Osaka Univ. HI 72603

Science reports of the faculty of agriculture, Kobe university. [Kobe daigaku nogakubu kenkyu hokoku.] Kobe. Vol. 9+, 1971+. Sci. Rep. Fac. Agric. Kobe Univ. Preceded by: Science reports of the Hyogo university of agriculture. HI 72604

Science report of the faculty of agriculture, Okayama university. [Okayama daigaku nogakubu gakujutsu hokoku.] Okayama, Japan. Sci. Rep. Fac. Agric. Okayama Univ. See B–P–H 826/23. HI 59299

Science reports of the faculty of arts and literature, Nagasaki university. Nagasaki. 1951-56. Sci. Rep. Fac. Arts Lit. Nagasaki Univ. Superseded by: Science bulletin of the faculty of liberal arts and education, Nagasaki university. HI 72605

Science report of the faculty of arts and science; Fukushima university. [Fukushima daigaku gakugei gakubu rika hokoku.] Fukushima, Japan. Sci. Rep. Fac. Arts Fukushima Univ. See B–P–H 826/25. HI 59300

Science reports of the faculty of education, Gunma university. Maebashi. Vol. 15+, 1966+. Sci. Rep. Fac. Educ. Gunma Univ. Preceded by: Science reports of the Gunma university, natural science series. HI 72606

Science report of the faculty of liberal arts and education; Gifu university. Natural science. [Gifu daigaku gakugei by kenkyu hokoku. Shizen kagaku.] Gifu, Japan. Sci. Rep. Fac. LIberal Arts Gifu Univ., Nat. Sci. See B–P–H 827/2. HI 59301

Science reports of faculty of science, Niigata university. Series D, biology. [Niigata daigaku rigakubu kenkyu hokoku.] Niigata. No. ?-22+, ?-1985+. Sci. Rep. Fac. Sci. Niigata Univ., D. Preceded by: Science reports of Niigata university. Series D, [biology]. HI 72607

Science reports of the Gunma university, natural science series. [Gunma daigaku kiyo, shizen kagaku hen.] Maebashi. Vols. ?-3-14, ?-1953-66. Sci. Rep. Gunma Univ., Nat. Sci. Ser. Preceded by: Science reports of the faculty of education, Gunma university. HI 72608

Science reports of the Hyogo university of agriculture (and faculty of agriculture, Kobe university). Sasayama. Vol. 8, 1967-68. Sci. Rep. Hyogo Univ. Agric. Preceded by: Science reports of the Hyogo university of agriculture. Series, agriculture and horticulture; Science reports of the Hyogo university of agriculture. Series, natural science; and Science reports of the Hyogo university of agriculture. Series, plant protection. Superseded by: Science reports of the faculty of agriculture, Kobe university. HI 72609

Science reports of the Hyogo university of agriculture. Series, agriculture. Sasayama. Vols. 1-6, 1953-64. Sci. Rep. Hyogo Univ. Agric., Ser. Agric. Superseded by: Science reports of the Hyogo university of agriculture. Series, agriculture and horticulture. HI 72610

Science reports of the Hyogo university of agriculture. Series, agriculture and horticulture. Sasayama. Vol. 7(1), 1965. Sci. Rep. Hyogo Univ. Agric., Ser. Agric. Hort. Preceded by: Science reports of the Hyogo university of agriculture. Series, agriculture. Superseded by: Science reports of the Hyogo university of agriculture (and faculty of agriculture, Kobe university). HI 72611

Science reports of the Hyogo university of agriculture. Series, natural science. Sasayama. Vols. 1-7, 1953-66. Sci. Rep. Hyogo Univ. Agric., Ser. Nat. Sci. Superseded by: Science reports of the Hyogo university of agriculture (and faculty of agriculture, Kobe university). HI 72612

Science reports of the Hyogo university of agriculture. Series, plant protection. Sasayama. Vol. 7, 1965-66. Sci. Rep. Hyogo Univ. Agric., Ser. Pl. Protect. Superseded by: Science reports of the Hyogo university of agriculture (and faculty of agriculture, Kobe university). HI 72613

Science reports of the Kagoshima university. [Kagoshima daigaku rika hokoku.] Kagoshima. Nos. 1-16, 1952-67. Sci. Rep. Kagoshima Univ. Superseded by: Report of the faculty of science, Kagoshima university. HI 72614

Science reports of the Kanazawa university; biology. [Kanazawa daigaku rika hokoku. Seibutsugaku.] Kanazawa, Japan. Sci. Rep. Kanazawa Univ., Biol.

See B–P–H 827/9. HI 59302

Science reports of the National central university. Nanking. Sci. Rep. Natl. Centr. Univ. See B–P–H 827/15. HI 59305

Science reports of the national Taiwan university. = Acta botanica taiwanica. Taipei, Taiwan. Acta Bot. Taiwan. See B–P–H 41/24.

Science reports of national Tsing Hua university. Series B. Biological and psychological sciences. [Kuo li ch'ing-hua ta hsüeh li k'o pao kao.] Peiping [=Peking]. Sci. Rep. Natl. Tsing Hua Univ., Ser. B, Biol. Sci. See B–P–H 827/17. HI 59306

Science reports of the national university of Peking. [Pei ta li k'o pao-kao.] Peiping [=Peking]. Sci. Rep. Natl. Univ. Peking. See B–P–H 827/18. HI 59307

Science reports of Niigata university. Series D, [biology]. [Niigata daigaku rigakubu kenkyu hokoku. D, seibutsugaku.] Niigata. Vol. 1+, 1964+. Sci. Rep. Niigata Univ., Ser. D. Preceded by: Journal of the faculty of science, Niigata university. Series II, biology, geology and mineralogy. Superseded by: Science reports of the faculty of science, Niigata university. Series D, biology. HI 72615

Science reports of Peking agricultural university. [Pei ching nung yüeh ta hsüeh pao.] Peking. Sci. Rep. Peking Agric. Univ. See B–P–H 828/2. HI 59308

Science reports of the research institute; Tohoku university. Ser. D Agriculture. Sendai, Japan. Sci. Rep. Res. Inst. Tohoku Univ., Ser. D, Agric. See B–P–H 828/3. HI 59309

Science reports of the Saitama university. Series B, biology and earth sciences. Urawa. Vols. 1-?, 1952-70. Sci. Rep. Saitama Univ., Ser. B, Biol. HI 72616

Science reports of South college, North college of Osaka university. [Osaka daigaku nanko, hokko rika hokoku.] Osaka, Japan. Sci. Rep. S. Coll. N. Coll. Osaka Univ. See B–P–H 828/7. HI 59310

Science reports of the Takao museum of natural history. Takaoka. ?-1976+. Sci. Rep. Takao Mus. Nat. Hist. HI 72617

Science reports of the Tohoku imperial university. Ser. 4, Biology. [Tohoku teikoku-daigaku rikwa Hokoku.] Sendai, Japan. Sci. Rep. Tohoku Imp. Univ., Ser. 4, Biol. See B–P–H 828/15. HI 59313

Science reports of the Tokyo bunrika daigaku. Section B, Tokyo university of literature and science. Tokyo. Vols. 1-?, 1932-54. Sci. rep. Tokyo Bunrika Daigaku, Sect. B. Superseded by: Science reports of the Tokyo kyoiku daigaku. Section B, zoology and botany. 5-4232-2. HI 72618

Science reports of the Tokyo kyoiku daigaku. Section B, zoology and botany. Tokyo. 1955-77. Sci. Rep. Tokyo Kyoiku Daigaku., B. Preceded by: Science reports of the Tokyo bunrika daigaku. Section B, Tokyo university of literature and science. HI 72619

Science reports of the Yamaguchi university. [Yamaguchi daigaku rika hokoku.] Yamaguchi. Vol. 10+, 1959+. Sci. Rep. Yamaguchi Univ. Preceded by: Yamaguchi journal of science. HI 72620

Science reports of the Yokohama national university. Section II. Biological and geological sciences. [Yokohama kokuritsu daigaku rika kiyo.] Kamakura, Japan. Sci. Rep. Yokohama Natl. Univ., Sect. 2, Biol. Sci. See B–P–H 828/22. HI 59314

Science report of the Yokosuka city museum. [Yokosukashi hakubutsu kan kenkyu hokoku.] Yokosuka, Japan. Sci. Rep. Yokosuka City Mus. See B–P–H 829/1. HI 59315

Science reporter. New Delhi. Sci. Reporter. See B–P–H 829/3. HI 59316

Science dans la république populaire roumaine. Bucharest. Vol. 1, 1953 (1954). Sci. Républ. Populaire Roumaine. Superseded, in part, by: Revue de biologie. HI 72621

Science of the sea. [Kaiyo no kagaku.] Tokyo. Sci. Sea. See B–P–H 829/9. HI 59320

Science series of the bulletin of the institute of Jamaica. Kingston, Jamaica. Sci. Ser. Bull. Inst. Jamaica. See B–P–H 829/10. HI 59321

Science serves your farm and home. Morgantown, WV. 1950-58. Sci. Serves Your Farm Home. Superseded by: West Virginia agriculture and forestry. HI 72622

Science du sol; bulletin, association française pour l'étude du sol. Versailles. 1963-72. Sci. Sol. Incorporated in: Bulletin, association française pour l'étude du sol. HI 72623

Science studies, botany, Montana college of agriculture and mechanic arts = Montana college of agriculture and mechanic arts. Science studies, botany. Bozeman, MT. Montana Coll. Agric. Sci. Stud., Bot. See B–P–H 619/2.

Science studies, Montana agricultural college. Bozeman. 1904-05. Sci. Stud. Montana Agric. Coll. HI 72624

Science teacher. Dobbs Ferry, NY, Orange, NJ. Vol. 1(1-7), 1898. Sci. Teacher. Superseded by: Science teacher and journal of nature study. 5-3822-2. HI 72625

Science teacher and journal of nature study. Orange, NJ. Vols. 1(8)-2(6), 1898-99. Sci. Teacher J. Nat. Study. Preceded by: Science teacher. 5-3822-2. HI 72059

Science & technology. [Zhongguo kexue zengkan.] No. 1+, 1978+. Sci. Technol. HI 72626

Science and technology in China. [Chung kuo k'o hsüeh

yü chien she.] Nanking. Sci. Technol. China. See B–P–H 829/15. HI 59323

Science and technology in Japan. Tokyo. Sci. Technol. Japan. See B–P–H 829/16. HI 59324

Science of the total environment. Amsterdam. Vol. 1+, 1972+. Sci. Total Environm. HI 72627

Science et (la) vie; magazine des sciences et de leurs applications à la vie moderne. Paris. Sci. & Vie. See B–P–H 822/20. HI 59248

Science world. [Rigakukai.] Tokyo? 1903-40. Sci. World. HI 72628

Sciences. New York, NY. Sciences (New York). See B–P–H 830/4. HI 59330

Sciences. Paris. Sciences (Paris). See B–P–H 830/5. HI 59331

Sciences et avenir. Paris. Sci. & Avenir. See B–P–H 822/16. HI 59245

Sciences et l'enseignement des sciences. Paris. Sci. Enseignem. Sci. See B–P–H 824/11. HI 59265

Sciences pharmaceutiques et biologiques de Lorraine. Nancy. Vol. 1+, 1973+. Sci. Pharm. Biol. Lorraine. Preceded by: Bulletin, société de pharmacie de Nancy. HI 72630

Sciencia medica. Rio de Janeiro. Sci. Med. See B–P–H 824/24. HI 59273

Scientia. Bologna. Scientia (Bologna). See B–P–H 830/25. HI 59332

Scientia. [K'o hsüeh t'ung pao.] Peking (Beijing). Vols. 1-17, 1950-66 [suspended 1967-72]. Scientia (China). Superseded by: Kexue Tongbao. HI 72631

Scientia agricultura sinica. [Zhongguo nongye kexue.] Beijing. 1977+. Sci. Agric. Sin. Preceded by: Chinese agricultural science. HI 72632

Scientia agriculturae bohemoslovaca. Prague. Vol. 1+, 1969+. Sci. Agric. Bohemoslov. HI 72633

Scientia canadensis; journal of the history of canadian science, technology and medicine. Thornhill, Ont. Vol. 8(1)+, 1984+. Sci. Canadensis. Preceded by: H S T C bulletin. HI 75275

Scientia genetica. Turin. Sci. Genet. See B–P–H 824/13. HI 59267

Scientia Guaianae; a series on natural sciences of the Guayana region. Caracas. Vol. 1+, 1969+. Sci. Guaianae. HI 72634

Scientia horticulturae; international journal sponsored by the international society for horticultural sciences. Amsterdam. Vol. 1+, 1973+. Sci. Hort. HI 72635

Scientia scandinavica. Stockholm. Sci. Scand. See B–P–H 829/8. HI 59319

Scientia silvae. Peking = Scientia silvae sinicae. [Linye kexue.] Peking (Beijing).

Scientia silvae sinicae. [Linye kexue.] Peking (Beijing). Vols. 1-11(3), 1955-66; vol. 15+, 1979+ [suspended 1966-75]. Sci. Silvae Sin. For 1976-78 see: Chinese forestry science. HI 72636

Scientia sinica. [European edition.] [Zhongguo kexue.] Beijing. Vols. 3(3)-24, 1954-81. Sci. Sin. Preceded by: Acta scientia sinica. Superseded by: Scientia sinica. Series B, Chem. biol. ag. med. earth sci. HI 72637

Scientia sinica. Series B, Chem. biol. ag. med. earth sci. [European edition; also published in a Chinese edition.] [Zhongguo kexue. B (huaxue, shengwuxue, nongxue, yixue, dixue).] Beijing. Vol. 25+, 1982+. Sci. Sin., B. Preceded by: Scientia sinica. [European edition]. HI 72638

Scientiarum historia. Antwerp. Sci. Hist. See B–P–H 824/14. HI 59268

Scientific activities, institute of field and garden crops. Bet Dagan, Israel. 1974+. Sci. Activities Inst. Field Gard Crops. HI 72639

Scientific agriculture. Ottawa. Vols. 1-32, 1921-52. Sci. Agric. (Ottawa). Superseded by: Canadian journal of agricultural science. 2-910-3. HI 72640

Scientific agriculture. [K'o hsüeh nung yeh.] Taipei, Taiwan. Sci. Agric. (Taiwan). See B–P–H 823/2. HI 59250

Scientific american. New York, NY. Sci. Amer. See B–P–H 823/3. HI 59251

Scientific american monthly. New York, NY. Sci. Amer. Monthly. See B–P–H 823/4. HI 59252

Scientific american supplement. New York, NY. Sci. Amer. Suppl. See B–P–H 823/5. HI 59253

Scientific annals of the faculty of physics and mathematics, aristotelian university of Thessaloniki. Salonika. Vol. ?-15+, ?-1975+. Sci. Ann. Fac. Phys. Math. Aristotelian Univ. Thessaloniki. HI 72641

Scientific bulletin, department of agricultural technical services, South Africa. Pretoria. No. 364-?, 1964-? Sci. Bull. Dept. Agric. Techn. Serv. South Africa. Preceded by: Science bulletin, department of agriculture, South Africa. Superseded by: Science bulletin, department of agriculture and fisheries, South Africa. HI 72642

Scientific bulletin, office of naval research, department of the navy, Tokyo. Arlington, VA. Vol. 1+, 1976+. Sci. Bull. Off. Naval Res. Dept. Navy Tokyo. HI 72643

Scientific contributions. Tropical plant research foundation. Washington, DC. Sci. Contr. Trop. Pl. Res. Found. See B–P–H 824/2. HI 59259

Scientific elements of nature conservation = Nauchnye

osnovy okhrany prirody. Moscow.

Scientific enquirer; a monthly medium for the supply of information on all scientific subjects. London. Sci. Enquirer. See B–P–H 824/10. HI 59264

Scientific horticulture; journal of the horticultural education association. Wye. Vol. 3+, 1935+. Sci. Hort. Preceded by: H E A year book. 3-1887-1. HI 72644

Scientific journal of the royal college of science. London. Sci. J. Roy. Coll. Sci. See B–P–H 824/18. HI 59270

Scientific knowledge. [Kagaku chishiki.] Tokyo. Sci. Knowl. See B–P–H 824/21. HI 59271

Scientific and learned societies of Great Britain; a handbook compiled from official sources. London. 57th-61st, 1951-64. Sci. Learned. Soc. Great Britain & Ireland. Preceded by: Official yearbook of the scientific and learned societies of Great Britain and Ireland. 5-3825-2. HI 74879

Scientific memoirs by officers of the medical and sanitary department of the government of India. Calcutta. Sci. Mem. Off. Med. Dept. Gov. India. See B–P–H 825/2. HI 59275

Scientific memoirs, selected from the transactions of foreign academies of science and learned societies, and from foreign journals. London. Sci. Mem. See B–P–H 824/25. HI 59274

Scientific memoirs of the university of Perm = Uchenye zapiski. Permskii gosudarstvennyi universitet imeni A. M. Gor'kogo. Perm.

Scientific monographs, council of agricultural research, India. Calcutta. 1931+. Sci. Monogr. Council Agric. Res. India. HI 72645

Scientific monographs, Pakistan association for the advancement of science. Lahore. No. 1+, 1952+. Sci. Monogr. Pakistan Assoc. Advancem. Sci. HI 72646

Scientific monthly. Washington, DC. Vols. 1-85, 1915-57. Sci. Monthly. Incorporated in: Science. Cambridge. 5-3826-1. HI 72647

Scientific museums = Musées scientifiques. Dijon.

Scientific papers of the applied sections of the Tiflis botanical garden = Zapiski Nauchno-prikladnykh Otdělov Tiflisskago Botanicheskago Sada. Tiflis.

Scientific papers of the college of arts and sciences, university of Tokyo. Tokyo. Vol. 33+, 1983+. Sci. Pap. Coll. Arts Sci. Univ. Tokyo. Preceded by: Scientific papers of the college of general education, university of Tokyo. HI 72648

Scientific papers of the college of general education, university of Tokyo. Tokyo. Vols. 1-32, 1951-82. Sci. Pap. Coll. Gen. Educ. Univ. Tokyo. Superseded by: Scientific papers of the college of arts and sciences, university of Tokyo. HI 72649

Scientific papers of the Hokkaido fisheries scientific institution. [Hokkaido sui san shikenjo hokoku.] Yoichi, Japan. Sci. Pap. Hokkaido Fish. Sci. Inst. See B–P–H 825/19. HI 59284

Scientific papers, Illinois state museum. Springfield, Ill. Vol. 1+, 1940+. Sci. Pap. Illinois State Mus. 3-1933-2. HI 72650

Scientific papers of the institute of algological research; faculty of science of the Hokkaido (imperial) university. [Hokkaido daigaku rigakubu kaiso kenkyusho obun hokoku.] Sapporo, Japan. Sci. Pap. Inst. Algol. Res. Fac. Sci. Hokkaido Imp. Univ. See B–P–H 825/21. HI 59285

Scientific papers of the Namib desert research station. Pretoria. 1961-69. Sci. Pap. Namib Desert Res. Sta. Superseded by: Madoqua. Series 2. HI 72651

Scientific papers of the Tokyo institute of physical and chemical research. Tokyo. Sci. Pap. Tokyo Inst. Phys. Res. See B–P–H 825/22. HI 59286

Scientific pest control. Bulletin of insect control. [Bochukagaku.] Kyoto, Japan. Sci. Pest Control. See B–P–H 825/23. HI 59287

Scientific press. San Francisco, CA = Mining and scientific press. San Francisco, CA. Mining Sci. Press. See B–P–H 595/24.

Scientific proceedings of the royal Dublin society. Dublin. Sci. Proc. Roy. Dublin Soc. See B–P–H 825/26. HI 59288

Scientific proceedings of the royal Dublin society. Series B. Dublin. Vols. 1-3, 1960/66-1971/76. Sci. Proc. Roy. Dublin Soc. Preceded by: Scientific proceedings of the royal Dublin society. Superseded by: Journal of life sciences, royal Dublin society. HI 72652

Scientific reports of the agricultural research institute, Pusa. Calcutta. 1916/17-28/29, 1917-29. Sci. Rep. Agric. Res. Inst. Pusa. Preceded by: Report of the agricultural research institute and college, Pusa. Superseded by: Scientific reports of the imperial institute of agricultural research, Pusa. HI 72654

Scientific reports, british antarctic survey. London. No. 36+, 1962+. Sci. Rep. Brit. Antarc. Surv. Preceded by: Scientific reports, Falkland Island Dependencies survey. HI 72655

Scientific report of the central potato research institute. Simla. 1957/58+, 1958?+. Sci. Rep. Centr. Potato Res. Inst. Preceded by: Progress report, central potato research institute. HI 72656

Scientific reports of the Ehime agricultural college. [Ehime kenritsu norin senmon gakko gakujutsu hokoku.] Matsuyama, Japan. Sci. Rep. Ehime Agric. Coll. See B–P–H 826/19. HI 59297

Scientific reports of the faculty of agriculture, Ibaraki university. [Ibaraki daigaku nogakubu gakujutsu hokoku.] Ibaraki. Vol. 1+, 1953+. Sci. Rep. Fac. Agric. Ibaraki Univ. HI 72657

Scientific reports of the faculty of agriculture, Naniwa university. Sakai, Japan. Sci. Rep. Fac. Agric. Naniwa Univ. See B–P–H 826/22. HI 59298

Scientific reports of the faculty of science, Ege university = Ege üniversitesi fen fakultesi ilmi raporlar serisi. Bornova, Izmir.

Scientific reports, Falkland Island Dependencies survey. London. Nos. 1-35, 1953-62. Sci. Rep. Falkland Island Dependencies Surv. Superseded by: Scientific reports, british antarctic survey. HI 72658

Scientific reports of the imperial institute of agricultural research, Pusa. Calcutta. 1930-35. Sci. Rep. Imp. Inst. Agric. Res. Pusa. Preceded by: Scientific reports of the agricultural research institute, Pusa. Superseded by: Scientific reports of the indian agricultural research institute. HI 72659

Scientific reports of the indian agricultural research institute. New Delhi. 1936-37. Sci. Rep. Indian Agric. Res. Inst. Preceded by: Scientific reports of the imperial institute of agricultural research, Pusa. 3-1954-3. HI 72660

Scientific reports, Japan Antarctic research expedition = J A R E scientific reports. Series E, biology. Tokyo.

Scientific reports, Japan artarctic research expedition. Series E, biology. Tokyo = J A R E scientific reports. Series E, biology. Tokyo.

Scientific reports of the Kyoto prefectural university. Agriculture. [Kyoto furitsu daigaku. Gakujutsu hokoku. Nogaku.] Kyoto. 1959+. Sci. Rep. Kyoto Pref. Univ., Agric. Preceded by: Scientific reports of the Saikyo university. Agriculture. HI 72661

Scientific reports of the Kyoto prefectural university. Natural science and living science. [Kyoto furitsu daigaku. Gakujutsu hokoku. Nogaku.] Kyoto. Vols. 3+, 1959+. Sci. Rep. Kyoto Pref. Univ., Nat. Sci. Liv. Sci. Preceded by: Scientific reports of the Saikyo university. Natural science and living science. HI 72662

Scientific reports of the Matsuyama Agricultural college. [Matsuyama noka daigaku gakujutsu hokoku.] Matsuyama, Japan. Sci. Rep. Matsuyama Agric. Coll. See B–P–H 827/10. HI 59303

Scientific reports of the Miyagi agricultural college. [Miyagi-ken nogyo tanki daigaku gakujutsu hokoku.] Sendai, Japan. Sci. Rep. Miyagi Agric. Coll. See B–P–H 827/12. HI 59304

Scientific reports of the Saikyo university. Agriculture. [Saikyo daigaku gakujutsu hokoku. Nogaku.] Kyoto. 1951-58. Sci. Rep. Saikyo Univ., Agric. Superseded by: Scientific reports of the Kyoto prefectural university. Agriculture. HI 72663

Scientific reports of the Saikyo university. Natural science and living science (or Mathematics and natural science). [Saikyo daigaku gakujutsu hokoku.] Kyoto. Vols. 1-2, 1952-54; 1957-58. Sci. Rep. Saikyo Univ., Nat. Sci. Liv. Sci. For. 1955-56 series title altered to: Mathematics and natural science. Superseded by: Scientific reports of the Kyoto prefectural university. Natural science and living science. HI 72664

Scientific reports of Shiga agricultural experiment station. [Shiga-ken nogyo shikenjo kenkyu hokoku.] Otsu, Japan. Sci. Rep. Shiga Agric. Exp. Sta. See B–P–H 828/12. HI 59311

Scientific reports of the Shiga fisheries experiment station. Special report. [Shigaken suisan shikenjo tokubetsu hokoku.] Otsu?, Japan. Sci. Rep. Shiga Fish. Exp. Sta. Special Rep. See B–P–H 828/14. HI 59312

Scientific research abstracts in Republic of China. Medicine & agriculture. Taipei. ?-1974/76?+. Sci. Res. Abstr. Republic of China, Med. Agric. HI 72653

Scientific research in british universities and colleges. London. 1951/52-1974/75, 1952-75. Sci. Res. Brit. Univ. Coll. Superseded by: Research in british universities, polytechnics and colleges. HI 72665

Scientific researches, Bangladesh council of scientific and industrial research laboratories. Dacca. Vol. 7(2-4), 1970-[72]. Sci. Res. Bangladesh Council Sci. Industr. Res. Lab. Preceded by: Scientific researches, east research laboratories, Pakistan council of scientific and industrial research. Superseded by: Bangladesh journal of scientific and industrial research. HI 72666

Scientific researches, east research laboratories, Pakistan council of scientific and industrial research. Dacca. Vols. 1-7(1), 1964-70. Sci. Res. E. Res. Lab. Pakistan Council Sci. Industr. Res. Superseded by: Scientific researches, Bangladesh council of scientific and industrial research laboratories. HI 72667

Scientific results of marine biological research = Hvalrådets skrifter. Oslo. Hvalrådets Skr. See B–P–H 424/31.

Scientific reviews on arid zone research. Jodhpur. Vol. 1+, 1982+. Sci. Rev. Arid Zone Res. HI 72668

Scientific roll and magazine of systematized notes. London. Sci. Roll Mag. Syst. Notes. See B–P–H 829/6. HI 59317

Scientific roll and magazine of systematized notes. Botanical section. Bacteria. London. Sci. Roll Mag. Syst. Notes, Bot. Sect., Bact. See B–P–H 829/7. HI 59318

Scientific series, department of agriculture, Federation of Malaya. Kuala Lumpur. 1949-52. Sci. Ser. Dept. Agric. Federation of Malaya. Preceded by: Scientific series, department of agriculture, Straits Settlements and Federated Malay States. HI 72669

Scientific series, department of agriculture, Straits Settlements and Federated Malay States. Kuala Lumpur. 1930-39 [suspended 1940-48]. Sci. Ser. Dept. Agric. Straits Settlements & Federated Malay States. Superseded by: Scientific series, department of agriculture, Federation of Malaya. HI 72670

Scientific series, national parks authority. Wellington, N.Z. No. 1+, 1976+. Sci Ser. Natl. Parks Author. HI 72671

Scientific studies, central research institute for plant production, Prague = Vědecké práce výzkumného ústavu rostlinné výrobý v Praze-Ruzynyibb. Prague.

Scientific survey of Porto Rico and the Virgin Islands. New York. 1919-70. Sci. Surv. Porto Rico & Virgin Islands. HI 72672

Scientific and technical proceedings = Index to scientific and technical proceedings. Philadelphia, PA.

Scientific works, institute of plant protection = Nauchni trudove, nauchno-izsledovatelski institut za zashtita na rasteniyata. Sofia.

Scientific world. Chungking & Nanking. [K'o hsüeh shih chieh. Chung hua tsu jao k'o hsüeh shê.] Sci. World (Chunking & Nanking). See B–P–H 829/17. HI 59325

Scientific world. Tokyo = Science world. Tokyo?

Scientist. Philadelphia, PA. Vol. 1+, 1986+. Scientist (Philadelphia). HI 72673

Scientists' international directory. Boston, MA. 1882, 1885, 1892 & 1896. Scientists' Int. Directory. Preceded and superseded by: Naturalists' directory. HI 72674

Scienzia e gioventù = Science appelle les jeunes. Winterthur.

Ščoričnyk Ukrajins'kogo botaničnogo tovarystva = Shchorichnyk Ukrayins'kogo botanichnogo tovarystva. Kiev.

Scotlands gardens. Edinburgh. 1931+. Scotlands Gard. HI 72675

Scots farmer; or, select essays on agriculture, adapted to the soil and climate of Scotland. Edinburgh. Scots Farmer. See B–P–H 830/35. HI 59336

Scots magazine; monthly miscellany of scottish life and letters. Dundee. Vol. 1+, 1924+. Scots Mag. 5-3829-1. HI 72676

Scottish agriculture. Edinburgh. Scott. Agric. See B–P–H 830/36. HI 59337

Scottish botanical review. Edinburgh. Scott. Bot. Rev. See B–P–H 830/37. HI 59338

Scottish field studies. Glasgow. 1979+. Scott. Field Stud. Preceded by: Report, scottish field studies association. HI 72677

Scottish forestry. Edinburgh. Vol. 1+, 1947+. Scott. Forest. Preceded by: Scottish forestry journal. 5-3830-1. HI 72678

Scottish forestry journal. Edinburgh. Vols. 41-60, 1927-46. Scott. Forest. J. Preceded by: Transactions of the royal scottish arboricultural society. Superseded by: Scottish forestry. HI 72679

Scottish gardener; magazine of horticulture and floriculture. Edinburgh. Vols. 1-15, 1852-66. Scott. Gard. Superseded by: Gardener. 5-3830-2. HI 72680

Scottish geographical magazine. Edinburgh. Vol. 1+, 1885+. Scott. Geogr. Mag. HI 72681

Scottish journal of agriculture. Edinburgh. Scott. J. Agric. See B–P–H 831/1. HI 59339

Scottish marine biological association, collected reprints. Edinburgh, Millport. 1948-74. Scott. Mar. Biol. Assoc. Collect. Repr. HI 72682

Scottish medical and surgical journal. Edinburgh. Scott. Med. Surg. J. See B–P–H 831/3. HI 59340

Scottish naturalist. Edinburgh. Scott. Naturalist (Edinburgh). See B–P–H 831/6. HI 59341

Scottish naturalist. Perth, Scotland. Scott. Naturalist (Perth). See B–P–H 831/7. HI 59342

Scral. Tucumán?, Argentina = S C R A L. Tucumán?, Argentina.

Scrinia florae selectae. St.-Quentin, France. Scrinia Fl. Select. See B–P–H 831/12. HI 59343

Scripps insitutue of oceanography, contributions. La Jolla, CA. Scripps Inst. Oceanogr. Contr. See B–P–H 831/13. HI 59344

Scripta academica hierosolymitana. Scientific report. Tel-Aviv. Vol. 1, 1938. Scripta Acad. Hierosolymitana, Sci. Rep. 5-3834-3. HI 72683

Scripta botanica. Tartu = Botaanilised uurimused. Tartu.

Scripta botanica horti universitatis imperialis petrogradensis = Botanicheskiya zapiski. St. Petersburg.

Scripta botanica horti universitatis imperialis petropolitane = Botanicheskiya zapiski. St. Petersburg.

Scripta botanica musei transsilvanici. Kolozsvar [=Cluj, Rumania] & Debrecen. Scripta Bot. Mus. Transsilv. See B–P–H 831/22. HI 59347

Scripta facultatis scientarum naturalium universitatis J. E. Purkynianae brunensis. Biologia. Brno. Vols. 1-9, 1971-79. Scripta Fac. Sci. Nat. Univ. J. E.

Purkynianae Brun., Biol. Preceded by: Spisy přírodovědecke fakulty university J. E. Purkyne v Brne. Superseded by: Scripta facultatis scientarum naturalium universitatis J. E. Purkynianae brunensis [without series]. HI 72684

Scripta facultatis scientarum naturalium universitatis J. E. Purkynianae brunensis. Brno. Vol. 10+, 1980+. Scripta Fac. Sci. Nat. Univ. J. E. Purkynianae Brun. Preceded by: Scripta facultatis scientarum naturalium universitatis J. E. Purkynianae brunensis. Biologia. HI 72685

Scripta geobotanica. Göttingen. Vol. 1+, 1970+. Scripta Geobot. HI 72686

Scripta horti botanici Tallinnensis = Tallinna botaanikaaia uurimused. Tallinn.

Scripta horti botanici universitatis Vilnensis = Vilnius universiteto botanikos sodo rašti. Vilnius.

Scripta horti botanici universitatis Vytauti Magni. Kaunas, Lithuania [Lithuanian S S R]. Scripta Horti Bot. Univ. Vytauti Magni. See B–P–H 831/23. HI 59348

Scripta instituti et horti botaniki universitatis Vilniensis = Vilnius universiteto botanikos instituto ir sodo raštai. Vilnius.

Scripta mycologica. Tartu. 1970+. Scripta Mycol. HI 72687

Scriptorum a societate hafniensi bonis artibus promovendis dedita danice editorum. Copenhagen. Script. Soc. Hafn. Bonis Artibus Promov. See B–P–H 831/17. HI 59345

Scrophulariaceae research newsletter. Vol. 1+, 1985+. Scrophulariaceae Res. Newslett. HI 72688

SDAT. [Costa Rica] = S D A T. [Costa Rica].

Sea frontiers. Miami, FL. Vol. 1+, 1955+. Sea Frontiers. HI 72689

Sea grant abstracts. Narragansett, RI. Vol. 1+, 1986+. Sea Grant Abstr. HI 72690

Sea grant quarterly. Honolulu, HI. 1979+. Sea Grant Quart. HI 72691

Séance publique de la société d'émulation de Rouen = Séance publique de la société libre d'émulation de Rouen. Rouen. Séance Publique Soc. Libre Emul. Rouen. See B–P–H 831/27.

Séance publique de la société libre d'émulation de Rouen. Rouen. Séance Publique Soc. Libre Emul. Rouen. See B–P–H 831/27. HI 59349

Séance publique de la société d'émulation de la ville de Cambrai. Cambrai. 1808-09, 1817. Soc. Émul. Cambrai Séance Publique. Superseded by: Société d'émulation Cambrai. Séance publique. HI 72692

Séance publique de la société linnéenne de Normandie. Caen & Paris. Séance Publique Soc. Linn. Normandie. See B–P–H 831/28. HI 59350

Séances publiques de l'académie des sciences, belles-lettres et arts de Besançon. Besançon. Séances Publiques Acad. Sci. Besançon. See B–P–H 832/1. HI 59351

Séances publiques de la société des amateurs des sciences, de l'agriculture et des arts de Lille. Lille. Séances Publiques Soc. Amateurs Sci. Lille. See B–P–H 832/2. HI 59352

Search; science, technology and society, journal of the australian and New Zealand association for the advancement of science. Sydney, N.S.W. Vol. 1+, 1970+. Search (Sydney). Preceded by: Australian journal of science. HI 72693

Search. Agriculture. Ithaca, NY. Vol. 1+, 1970+. Search (Ithaca), Agric. HI 72694

Seasons; nature and outdoors magazine. Don Mills, Ont. Vol. 20+, 1980+. Seasons. Preceded by: Ontario naturalist. HI 72695

Seaweed research and utilisation. Madurai. Vol. 1+, 1971+. Seaweed Res. Utilis. HI 72696

Secrest arboretum notes. Wooster, OH. Vol. 1+, 1972+. Secrest Arbor. Notes. HI 72697

Sections romandes du club alpin suisse. Geneva. Sect. Romandes Club Alpin Suisse. See B–P–H 832/4. HI 59353

Seed abstracts. Farnham Royal. Vol. 1+, 1978+. Seed Abstr. HI 72698

Seed bulletin, agricultural experiment station, university of California = University of California, college of agriculture, agricultural experiment station. Seed bulletin. Berkeley, CA. Univ. Calif. Coll. Agric. Exp. Sta. Seed Bull. See B–P–H 934/15.

Seed circular, North Dakota agricultural experiment station = North Dakota agricultural experiment station seed circular. Fargo, ND. North Dakota Agric. Exp. Sta. Seed Circ. See B–P–H 664/15.

Seed notes. Cambridge. Nos. 1-84, 1942-75. Seed Notes. HI 72699

Seed and nursery trader of Australia and New Zealand. Melbourne. Vols. 25(4)-79(5), 1927-81. Seed Nursery Trader Australia & New Zealand. Preceded by: Australasian international nurseryman, seedsman, and florist. Superseded by: Australian horticulture. 5-3840-1. HI 72700

Seed pathology news. Copenhagen. No. 1+, 1971+. Seed Pathol. News. HI 72701

Seed pod. Jacksonville, FL. Seed Pod. See B–P–H 832/11. HI 59354

Seed potato. London. Seed Potato. See B–P–H 832/12.

HI 59355

Seed producer's review. Melbourne, Vic. Vol. 1(1-3), 1968. Seed Producer's Rev. Superseded by: Australian seed producers' review. HI 72703

Seed research. New Delhi. Vol. 1+, 1973+. Seed Res. HI 72704

Seed savers exchange. Princeton, MO. ?-1985+. Seed Savers Exch. Preceded by: True seed exchange. HI 72705

Seed science and technology; proceedings of the international seed testing association. Zurich, Wageningen. Vol. 1+, 1973+. Seed Sci. Techn. Preceded by: Proceedings, international seed testing association. HI 72706

Seed scoop. Ottawa. Seed Scoop. See B–P–H 832/13. HI 59356

Seed technologist news. Louisville, KY, Lincoln, NE. ?-1958+. Seed Technol. News. 5-3840-1. HI 72707

Seed trade news. Chicago, IL. Vol. 1+, 1923+. Seed Trade News. 5-3840-1. HI 72708

Seed world. Chicago, IL. Seed World. See B–P–H 832/15. HI 59358

Seedhead news. Tucson, AZ. No. 1+, 1983+. Seedhead News. HI 72709

Seedling canes and manurial experiments, Barbados = Report of sugar-cane experiments, department of agriculture, Barbados. Bridgetown.

Seedlings and horticulture. [Shubyo to engei.] Urawa, Japan. Seedlings & Hort. See B–P–H 832/16. HI 59359

Sego lily. Salt Lake City, UT. Vol. 5+, 1982+. Sego Lily. Preceded by: Newsletter of the Utah native plant society. HI 72710

Seibutsu. Sapporo, Japan. Seibutsu. See B–P–H 832/19. HI 59360

Seibutsu butsuri = Biophysics. Kyoto.

Seibutsu butsuri kagaku = Physico-chemical biology. Chiba.

Seibutsu gakkaishi. The biological society of Hiroshima university. Hiroshima. Seibutsu gakkaishi. See B–P–H 832/21. HI 59361

Seibutsu gakushi kenkyu = Japanese journal of the history of biology. Tokyo.

Seibutsu kagaku = Biological science. Tokyo. Biol. Sci. (Tokyo). See B–P–H 195/7.

Seibutsu kagaku nyusu = Biological sciences news. Japan.

Seibutsu kankyo chosetsu = Environment control in biology. Tokyo.

Seibutsu kenkyu = Life study. Fukui.

Seibutsu tokeigaku zasshi = Revuo de biostatisko. Tokyo.

Seibutsukai = Biosphaera. [Japanese biological association.] Tokyo. Biosphaera. See B–P–H 197/6.

Seikagaku = Journal of japanese biochemical society. Tokyo.

Seiken jiho = Report of the Kihara institute for biological research. Yokohama, Japan. Rep. Kihara Inst. Biol. Res. See B–P–H 767/11.

Seiri-seitai = Physiology and ecology. Kyoto.

Seitaigaku kenkyu = Ecological review. [Mt. Hakkoda botanical laboratory.] Sendai, Japan. Ecol. Rev. See B–P–H 353/14.

Séiva. Viçosa. Año 1+, 1944+. Séiva. 5-3842-1. HI 72711

Selborne magazine. London. Vols. 70-72, 1958-60; vols. 79-?, 1967-74. Selborne Mag. Preceded by: Selborne magazine and nature notes. Vols. 73-78 included in: Birds and country. HI 72712

Selborne magazine for lovers and students of living nature. London. 1888-89. Selborne Mag. Lovers Stud. Liv. Nat. Superseded by: Nature notes. HI 72713

Selborne magazine and nature notes. London. Vols. 20-29, 1909-25. Selborne Mag. Nat. Notes. Preceded by: Nature notes. Superseded by: Selborne magazine. HI 72714

Selbyana; journal of the Marie Selby botanical gardens. Sarasota, FL. Vol. 1+, 1975+. Selbyana. HI 72715

Selecciones avicolas. Lima. Selecc. Avic. See B–P–H 832/27. HI 59363

Seleções agrícolas. Rio de Janeiro. Seleçôes Agríc. See B–P–H 832/26. HI 59362

Select papers on agriculture, commerce, arts and manufactures = Museum rusticum et commerciale: or, select papers on agriculture, commerce, arts and manufactures. London. Mus. Rusticum Commerciale (London). See B–P–H 622/12.

Selected bibliography on algae. Halifax, Nova Scotia. Nos. 1-14, 1953-73. Select. Bibliogr. Algae. HI 72716

Selected symposia series, A A A S. Boulder, CO. Vol. 1+, 1978+ [vols. 2, 3 & 5 published in 1978]. Select. Symp. Ser. A. A. A. S. HI 72717

Selekcija i semenovodstvo = Selektsiya i semenovodstvo. Moscow.

Selektsiya i semenovodstvo. Kiev. Vol. 1+, 1964+. Selekts. Semenov. HI 72718

Selektsiya i semenovodstvo. Moscow. 1935(1)-40;

1945-53; 1956(3)+ [suspended 1941-44 & 1954-February 1956; only 4 nos. published in 1935, thereafter 12 nos. annually.] Selekts. & Semenov. Preceded by: Semenovodstvo. 5-3843-3. HI 72719

Selektsiya i semenovodstvo pshenitsy. Kiev. Vol. 1+, 1970+. Selekts. Semenov. Pshenitsy. HI 72720

Sellowia; anais botânicos do herbário "Barbosa Rodrigues". Hajaí, Brazil. Sellowia. See B–P–H 833/2. HI 59364

Selmeczbányai gyógyászati és természettudományi egyesület évkönyve. Selmec, Hungary [=Banska Stiavnica, Czechoslovakia]. Selmeczbányai Gyógy. Term. Egyes. Évk. See B–P–H 833/4. HI 59365

Sel′skokhozyaistvennaya biologiya. Moscow. Vol. 1+, 1966+. Sel′skokhoz. Biol. Preceded by: Agrobiologiya. HI 72721

Sel′skokhozyaistvennaya literatura S S S R. Sistematicheskii ukazatel′. Moscow. Vol. 1+, 1948+. Sel′skokhoz. Lit. S.S.R., Sist. Ukaz. HI 72722

Semaine horticole. Brussels. Semaine Hort. See B–P–H 833/7. HI 59366

Semana médica. Buenos Aires. Semana Méd. See B–P–H 833/8. HI 59367

Semana vitivinícola. Valencia, Spain. Semana Vitiviníc. See B–P–H 833/10. HI 59368

Semanario de agricultura y artes. Madrid. Semanario Agric. Artes. See B–P–H 833/11. HI 59369

Semenovodstvo. Moscow. Semenovodstvo. See B–P–H 833/12. HI 59370

Sementi elette. Bologna, Milan. Vol. 1+, 1955+. Sementi Elette. HI 72723

Semi-tropic California. Los Angeles, CA. Vols. 3-5, 1879-82. Semi-Trop. Calif. Preceded by: Southern California horticulturist. HI 72724

Seminar series, society for experimental biology. New York. Vol. 1+, 1976+. Seminar Ser. Soc. Exp. Biol. HI 72725

Seminarium. Leipzig. Seminarium. See B–P–H 833/13. HI 59371

Seminarium transmarinum. Hamburg. Vols. 1-2(8), 1954-55. Seminarium Transmarinum. Superseded by: F L K Berichte. HI 72726

Sempervirens. Amsterdam. Sempervirens. See B–P–H 833/14. HI 59372

Sempervivum fanciers' association newsletter = Newsletter, sempervivum fanciers' association. Randolph, MA.

Sempervivum society journal = Journal of the sempervivum society. Bishops Stortford, Burgess Hill.

Sempervivum year book. Burgess Hill. 1975+. Sempervivum Year Book. HI 72727

Sen lin = Journal of forestry. Tsinan, China. J. Forest. (Tsinan). See B–P–H 466/7.

Senckenbergiana. Frankfurt a. M. Senckenbergiana. See B–P–H 833/21. HI 59375

Senckenbergiana biologica. Frankfurt a. M. Senckenberg. Biol. See B–P–H 833/19. HI 59373

Senckenbergiana lethaea. Frankfurt a. M. Senckenberg. Leth. See B–P–H 833/20. HI 59374

Senckenbergiana maritima; Zeitschrift für Meeresgeologie und Meeresbiologie. Frankfurt am Main. Vol. 1+, 1969+ [vol. 1 also numbered vol. 50 in continuation of vol. numbering of its predecessor]. Senckenberg. Marit. Preceded by: Senckenbergiana. HI 72729

Senshokutai = Kromosomo. Tokyo. Kromosomo. See B–P–H 519/21.

Sentai chii zappo = Miscellanea bryologica et lichenologica. Nichinan.

Seoul national university faculty papers. Biology and agriculture series. Seoul. 1971+. Seoul Natl. Univ. Fac. Pap., Biol Agric. Ser. Preceded by: Seoul university journal. Biology and agriculture series. HI 72730

Seoul taehak-kyo nonmun-jip soengnong-kae = Seoul university journal. Biology and agriculture series. Seoul.

Seoul taehak-kyo nonmun-jip. Chayon kwahak = Universitas seoulensis. Collectio theseon. Scientia naturalis. Chayon kwahak.] Seoul, Korea. Univ. Seoul. Collect. Theseon, Sci. Nat. See B–P–H 938/4.

Seoul university journal. Biology and agriculture series. [Seoul taehak-kyo nonmun-jip soengnong-kae.] Seoul. 1958-71? Seoul Univ. J., Biol. Ser. Preceded by: Universitas seoulensis. Collectio theseon. Scientia naturalis. Superseded by: Seoul national university faculty papers. Biology and agriculture series. HI 72731

Separate Abhandlungen des Vereins für Naturkunde an der Unterweser = Separate Schriften des Vereins für Naturkunde an der Unterweser. Leipzig. Separate Schriften Vereins Naturk. Unterweser. See B–P–H 833/27.

Separate Schriften des Vereins für Naturkunde an der Unterweser. Leipzig. Separate Schriften Vereins Naturk. Unterweser. See B–P–H 833/27. HI 59376

Sera; Panorama Aquariophile actuel, international, moderne. Heinsberg. 1984+. Sera. HI 72732

Seri bibliografi, lembaga perpustakaan biologi dan pertanian, bibliotheca bogoriense. Bogor. Vols. 1-?, 1967-79. Ser. Bibliogr. Lemb. Perpust. Biol. Pert.

Bibliot. Bogor. Superseded by: Seri bibliografi, pusat perpustakaan biologi dan pertinian. HI 72733

Seri bibliografi, pusat perpustakaan biologi dan pertinian. Bogor. 1980+. Ser. Bibliogr. Pusat Perpust. Biol. Pert. Preceded by: Seri bibliografi, lembaga perpustakaan biologi dan pertanian, bibliotheca bogoriense. HI 72734

Seria biologia. Poznan. No. 4+, 1974+. Ser. Biol., Poznan. Preceded by: Prace wydziału biologii i nauk o ziemi, uniwersytet imienia Adama Mickiewicza w Poznaniu. Seria biologia. HI 72735

Sericicultura. Bucharest. Sericicultura. See B–P–H 834/7. HI 59380

Série B, botanica. Estacion agronomica de moca. Santo Domingo Nos. 1-17. 1925-30. Sér. B, Bot., Estac. Agron. Moca. HI 72736

Série biologica, instituto de biologia, academia de ciencias de Cuba. Havana. No. 1+, 1967+. Sér. Biol. Inst. Biol. Acad. Ci. Cuba. HI 72737

Série botanica applicada, estación agro-forestal del pontifico de Calima-Buenaventura. Cali. Vol. ?-1(2)-?, ?-[1948]-? Sér. Bot. Appl. Est. Agro-Forest. Pontif. Calima-Buenaventura. HI 72738

Série botanica e fisiologia vegetal. Instituto de pesquisas e expérimentaçao agropecuárias do norte. Belém. Vol. 1(1-2), 1970-72. Sér. Bot. Fisiol. Veg., Inst. Pesq. Exp. Agropecu. N. HI 72739

Serie botanica de museu de ciències naturales. Barcelona. Ser. Bot. Mus. Ci. Nat. See B–P–H 834/3. HI 59377

Série botánica, museo nacional de Costa Rica. San José. Vol. 1(1-4), 1937-40. Sér. Bot. Mus. Nac. Costa Rica. 5-3773-1. HI 72740

Serie ciencias naturales facultad de ciencias exactas fisicas y naturales. Cordoba. 1954-69. Ser. Ci. Nat. Fac. Ci. Exact. Fis. Nat. HI 72741

Serie científica; museo nacional. Mexico City. Ser. Ci. Mus. Nac. See B–P–H 834/4. HI 59378

Série cientifico, instituto antartico chileno. Santiago. Vol. 1+, 1970+. Sér. Ci. Inst. Antart. Chileno. HI 72742

Série científica, instituto de investigação agronómica de Angola. Luanda. Nos. 1-35, 1968-74. Sér. Ci. Inst. Invest. Agron. Angola. HI 72743

Serie conferences et documents, laboratoire d'océanographie biologique, université de Rennes = Conférences et documents, laboratoire d'océanographie biologique, université de Rennes. Rennes.

Serie conferences, laboratoire d'océanographie biologique, université de Rennes = Conférences, laboratoire d'océanographie biologique, université de Rennes. Rennes.

Serie conservación de la naturaleza, fundación Miguel Lillo. Tucuman. No. 1+, 1979+. Ser. Conserv. Naturaleza Fund. Miguel Lillo. HI 72746

Série culturas de Amazônia. Instituto de pesquisas e expérimentação agropecuárias do norte. Belém. Vols. 1-2(1), 1970-71. Sér. Cult. Amazônia, Inst. Pesq. Exp. Agropecu. N. Superseded by: Boletim técnico do instituto de pesquisas e expérimentação agropecuárias do norte. HI 72747

Serie divulgação, fundação brasileira para a conservação da natureza. Rio de Janeiro. No. ?-10+, ?-1979+. Ser. Divulg. Fund. Brasil. Conserv. Nat. HI 72748

Serie divulgação, projeto de desenvolvimento e pesquisa florestal. Brasilia. Ca.1980+. Ser. Divulg. Proj. Desenvolv. Pesq. Florest. HI 72749

Série documentaire, conservatoire et jardin botaniques de la ville de Genève. Geneva. Vol. 1+, 1980+. Sér. Doc. Conserv. Jard. Bot. Genève. HI 72750

Série documents, laboratoire souterrain du C N R S. Saint Girons, Moulis. N.s. nos. 1-4, 1970-75. Sér. Doc. Lab. Souterr. C. N. R. S. Superseded by: Mémoires de biospéologie. HI 72751

Série estudos sôbre forrageiras na Amazônia. Instituto de pesquisas e expérimentaçao agropecuárias do norte. Belém. Vols. 1-2(1), 1970-71. Sér. Estudos Forrag. Amazônia, Inst Pesq. Exp. Agropecu. N. HI 72752

Série: Fertilidade de solo. Belém. Vol. 1+, 1971+. Sér. Fertil. Solo. HI 72753

Série fitogeográfica, ministerio de agricultura y ganaderia. Buenos Aires. Nos. 1-9?, 1951-68? Sér. Fitogeogr. Minist. Agric. Ganad. HI 72754

Série fitotecnia. Instituto de pesquisas e expérimentaçao agropecuárias do norte. Belém. Vols. 1-2, 1970-71. Sér. Fitotecn. Inst. Pesq. Exp. Agropecu. N. HI 72755

Série florestal, academia de ciencias de Cuba, departamento de ecologia forestal. Havana. ?-1976+. Sér. Florest. Acad. Ci. Cuba Dept. Ecol. Forest. HI 72756

Serie informatica, instituto forestal. Santiago. No. 1+, 1981+. Ser. Inform. Inst. Forest. HI 72757

Série oceanologica, instituto de oceanologia. Havana. No. 1+, 1968+. Sér. Oceanol. Comis. Nac. Ci. Republ. Cuba. HI 72758

Serie de pubblicazioni sulla protezione dell'ambiente. Berne = Cahiers de l'environnement. Berne.

Série quimica de solos, instituto de pesquisas e expérimentaçao agropecuárias do norte. Belém. Vols. 1-5(1), 1970-7? Sér. Quim. Solos Inst. Pesq. Exp. Agropecu. N. Superseded by: Boletim técnico do instituto de pesquisas e expérimentaçao agropecuárias do norte. HI 72759

Série solos da Amazônia, instituto de pesquisas e expérimentaçao agropecuárias do norte. Belém. Vols. 1-3, 1967-71. Sér. Solos Amazônia Inst. Pesq. Exp. Agropecu. N. HI 72760

Série systématique, institut supérieur de biologie de la Guyane Française. São Cristóvão. Nos. ?-3-11, ?-1971-72. Sér. Syst. Inst. Supér. Biol. Guyane Franç. HI 72761

Série tecnica F U P E F. Curitiba. No. ?-2+, ?-1980+. Sér. Tecn. F. U. P. E. F. HI 72762

Série tecnica y didactica, facultad de ciencias naturales y museo de La Plata. La Plata. 1949-70. Sér. Tecn. Didact. Fac. Ci. Nat. Mus. Plata. HI 72763

Série tecnica, instituto de investigação agronomica de Angola. Luanda. Nos. 1-44, 1968-74. Sér. Tecn. Inst. Invest. Agron. Angola. HI 72764

Serie tecnica, projeto de desenvolvimento e pesquisa florestal. Brasilia. Ca.1980+. Ser. Tecn. Proj. Desenvolv. Pesq. Florest. HI 72765

Série técnico do instituto de pesquisas e expérimentaçao agropecuárias do norte. Belém. Nos. 1-56, 19??-73. Sér. Técn. Inst. Pesq. Exp. Agropecu. N. Preceded by: Boletim técnico do instituto de pesquisas e expérimentaçao agropecuárias do norte. HI 72766

Série técnologia do instituto de pesquisas e expérimentaçao agropecuárias do norte. Belém. Vol. 1, 1970. Sér. Técn. Inst. Pesq. Exp. Agropecu. N. Superseded by: Boletim técnico do instituto de pesquisas agropecuarias do norte (IPEAN). HI 72767

Série transformacion de la naturaleza. Havana. No. ?-5+, ?-1968+. Sér. Transform. Naturaleza. HI 72768

Serie universitaria, fundación Juan March. Ciencias agrarias. Madrid. ?-1981+. Ser. Univ. Fund. Juan March, Ci. Agrar. HI 72769

Series of monographs on general physiology. Normandy, MO. Ser. Monogr. Gen. Physiol. See B–P–H 834/6. HI 59379

Series on mycology. New York. Vol. 1+, 1979+. Ser. Mycol. HI 72770

Seriya broshyur, botanicheskii kabinet i botanicheskii sad imperatorskago nikitskago sada. Yalta. Nos. ?-2-7-?, ?-1916-17-? Ser. Brosh. Bot. Kab. Bot. Sad Imp. Nikitsk. Sada. HI 72771

Service bulletin, Ohio agricultural experiment station = Ohio agricultural experiment station service bulletin. Wooster, OH. Ohio Agric. Exp. Sta. Service Bull. See B–P–H 685/2.

Service circular, Missouri agricultural experiment station = Missouri agricultural experiment station. Service circular. Columbia, MO.

Servicio forestal. Madrid. Serv. Forest. See B–P–H 834/11. HI 59381

Sessioni pubbliche dell'ateneo veneto. Venice. Sess. Pubbliche Ateneo Veneto. See B–P–H 834/13. HI 59382

Settimana medica. Palermo & Milan. Settimana Med. See B–P–H 834/14. HI 59383

Seu ch'uan ta hsüeh hsüeh pao, tzu jân k'o hsüeh = Acta scientiarum naturalium universitatis szechuanensis. Chengtu, China. Acta Sci. Nat. Univ. Szechuan. See B–P–H 48/25.

Severnaja Mongolija = Severnaya Mongoliya. Leningrad.

Severnaya Mongoliya. Leningrad. Vols. 1-3, 1926-28. Severn. Mongoliya. 5-3854-1. HI 72772

Severní Morava, vlastivědný sborník. Zabreh, Czechoslovakia. Severní Morava, Vlastiv. Sborn. See B–P–H 834/19. HI 59384

Severočeskou přírődou; sborník severočeské pobočky československé botaniké společnosti. Litoměřice. 1969+. Severoc'haeskou Přír. HI 72773

Sexual plant reproduction. Berlin, New York. Vol. 1+, 1988+. Sexual Pl. Reprod. HI 72774

S''ězd Russkikh Estestvoispytatelei i Vrachei. St. Petersburg. Vol. 8, 1890. S''ězd Russk. Estestvoisp. Vrachei. HI 72775

S''ězd Russkih Estestvoispytatelej i Vračej = S''ězd Russkikh Estestvoispytatelei i Vrachei. St. Petersburg.

Sha-ling-wei = Bulletin of the central cotton improvement institute. Nanking. Bull. Centr. Cotton Improv. Inst. See B–P–H 243/26.

Shaboten = Journal of the cactus and succulent society of Japan.

Shaboten. [Shaboten sha.] Nagoya City. Nos. 1-69, 1934-40. Shaboten (Nagoya). Preceded by: Shaboten sha. Superseded by: Acta succulentologica. HI 72776

Shaboten. Zushi, Kanagawa. No. 1+, 1955+. Shaboten (Zushi). HI 72777

Shaboten gumma = Cactus and succulent journal of the gumma cactus club. Gumma.

Shaboten no kenkyu = Study of cacti. Osaka.

Shaboten newsletter. ?-ca.1957+. Shaboten Newslett. HI 72778

Shaboten sha = Journal of the cactus and succulent club of Tokyo. Tokyo.

Shaboten sha = Shaboten. Nagoya City.

Shade tree. Vol. ?-28(11)+, ?-1955+. Shade Tree. HI 72779

Shade tree digest. ?-1959-79. Shade Tree Digest. HI 72780

Shamba; journal of agriculture for Zanzibar. Nos. 1-24, 1897-1903. Shamba. HI 72781

Shan-hsi nung yeh = Shensi agriculture. Sian, China. Shensi Agric. See B–P–H 835/4.

Shan hsi sze fan hsüeh yuan hsüeh pao = Journal of Shansi Normal College. Taiyüan, China. J. Shansi Norm. Coll. See B–P–H 481/12.

Shan-hsi ta hsüeh hsüeh pao. Tzu jan k'o hsüeh pan = Shanxi university journal. Natural science edition.

Shan tung ta hsüeh pao = Shantung university journal. Tsinan, China. Shantung Univ. J. See B–P–H 834/28.

Shandong haiyangxue yuan xuebao = Journal of Shandong college of oceanology. Beijing.

Shang Ch'ang ho chi hua sha lien hui k'an = China cotton journal. Shanghai. China Cotton J. See B–P–H 309/4.

Shanghai educational review. [Chung-hua chiao yü chieh.] Shanghai. Shanghai Educ. Rev. See B–P–H 834/25. HI 59385

Shanghai shizen kwagaku kenkyu-sho iho = Bulletin of the Shanghai science institute. Shanghai. Bull. Shanghai Sci. Inst. See B–P–H 273/16.

Shangtai xuebao = Acta ecologica sinica. Shanghai.

Shantung university journal. [Shan tung ta hsüeh pao.] Tsinan, China. Shantung Univ. J. See B–P–H 834/28. HI 59386

Shanxi daxue xeubao. Ziran kexue ban = Shanxi university journal. Natural science edition.

Shanxi guoshu. No. ?-11+, ?-1983+. Shanxi Guoshu. HI 72782

Shanxi shida xuebao. Ziran kexue ban. No. ?-6+, ?-1979+. Shanxi Shida Xuebao. Ziran Kexue Ban. HI 72783

Shanxi university journal. Natural science edition. [Shanxi daxue xeubao. Ziran kexue ban.] No. 1+, 1955+. Shanxi Univ. J., Nat. Sci. Ed. HI 72784

Sharon cactus guide. Sharon, PA. Sharon Cact. Guide. See B–P–H 834/29. HI 59387

Shchorichnyk ukrayinsʹkogo botanichnogo tovarystva. Kiev. Nos. [1]-4, 1959-6? Shchor. Ukrayinsʹk. Bot. Tovar. Superseded by: Dosyahnennya botanichnoyi nauky na Ukrayini. HI 72785

Shelk. Tashkent. ?-1982+. Shelk. HI 72786

Sheng li An huei ta shhsüeh nung hsüeh yüan yen chiu pao-kao = Research report of the Anhwei university college of agriculture. [China]. Res. Rep. Anhwei Univ. Coll. Agric. See B–P–H 776/2.

Sheng li hsüeh pao = Acta physiologica sinica. Peiping [=Peking]. Acta Physiol. Sin. See B–P–H 47/19.

Sheng t'ai hsueh tsa chih = Journal of ecology. Beijing.

Shêng wu hsüeh t'ung pao = Biological news. Peking. Biol. News. See B–P–H 194/16.

Sheng wu hsueh tung pao = Shengwuxue tongbao. Peking (Beijing).

Sheng wu hua yu sheng wu li hsüeh pao = Acta biochemica et biophysica sinica. Peking. Acta Biochem. Biophys. Sin. See B–P–H 40/8.

Shengtaixue zazhi = Journal of ecology. Beijing.

Shengwuxue tongbao. Peking = Biological news. [Shêng wu hsüeh t'ung pao.] Peking. Biol. News. See B–P–H 194/16.

Shensi agriculture. [Shan-hsi nung yeh.] Sian, China. Shensi Agric. See B–P–H 835/4. HI 59388

Shibakusa kenkyu = Journal of the Japan turf grass research association. Tokyo.

Shiga daigaku gakugeibu kenkyu ranshu = Bulletin of the faculty of liberal arts and education, Shiga university. Part 2, natural science. Hikone, Japan. Bull. Fac. Liberal Arts Shiga Univ., Pt. 2, Nat. Sci. See B–P–H 250/3.

Shiga daigaku kyoiku gakubu kiyo, shizenkagaku = Memoirs of the faculty of liberal arts and education, Shiga university. Part 2, natural science. Otsu.

Shiga kogen seibutsu kenkyusho kemkyu gyoseki = Bulletin of the institute of biology in Shiga Heights. Nagano.

Shiga shizen kyoiku kenkyu shisetsu kenkyu gyoshi = Bulletin of institute of natural education in Shiga Heights. Nagano.

Shiga-ken nogyo shikenjo kenkyu hokoku = Scientific reports of Shiga agricultural experiment station. Otsu, Japan. Sci. Rep. Shiga Agric. Exp. Sta. See B–P–H 828/12.

Shigaken suisan shikenjo tokubetsu hokoku = Scientific reports of the Shiga fisheries experiment station. Special report. Otsu?, Japan. Sci. Rep. Shiga Fish. Exp. Sta. Special Rep. See B–P–H 828/14.

Shigen kagaku kenkyujo iho = Miscellaneous reports of the research institute for natural resources. Tokyo.

Shigen kagaku kenkyusho obun hokoku = Journal of the Sigenkagaku Kenkyusyo. Tokyo. J. Sigenkagaku Kenkyusyo. See B–P–H 481/16.

Shih ta hsüeh pao = Bulletin of Taiwan normal university. Taipei, Taiwan. Bull. Taiwan Normal Univ. See B–P–H 283/16.

Shih ta sheng wu hsüeh pao = Biological bulletin of national Taiwan normal university. Taipei.

Shih ti tsa chi. Kuo li Che kiang ta hsüeh = Review of history and geography; national Chekiang university. Hangchow, China. Rev. Hist. Geogr. Natl. Chekiang Univ. See B–P–H 781/14.

Shih yeh pu chung yang nung yeh shih yen so k'an mu lu = Special publications of the national agricultural research bureau. Nanking. Special Publ. Natl. Agric. Res. Bur. See B–P–H 851/13.

Shih yen sheng wu hsueh pao = Chinese journal of experimental biology. Peking (Beijing).

Shiitake news. Lanesboro, MN. Vol. 1+, 1984+. Shiitake News. HI 72788

Shikoku nogyo shikenjo hokoku = Bulletin of the Shikoku agricultural experiment station. Kagawa, Japan. Bull. Shikoku Agric. Exp. Sta. See B–P–H 273/17.

Shimane daigaku. Bunri gakubu = Memoirs of the faculty of literature and science, Shimane university. Natural sciences. Matsue.

Shimane daigaku nogakubu enshurin hokoku = Bulletin of the Shimane university forests. Matsue.

Shimane daigaku ronshu. Shizen kagaku = Bulletin of the Shimane university. Natural science. Matsue.

Shimane-kenritsu noji shikenjo kenkyu hokoku = Bulletin of the Shimane agricultural experiment station. Shimane, Japan. Bull. Shimane Agric. Exp. Sta. See B–P–H 273/20.

Shimane noka daigaku kenkyu hokoku = Bulletin of the Shimane agricultural college. Matsue, Japan. Bull. Shimane Agric. Coll. See B–P–H 273/19.

Shin kaki = New flowers. Kyoto.

Shinano hakubutsu-gaku zasshi = Journal of the Shinano natural science society. Shinano, Japan. J. Shinano Nat. Sci. Soc. See B–P–H 481/14.

Sh'ing nien k'o hsüeh = Youth and science. Chungking, China. Youth & Sci. See B–P–H 983/27.

Shinkin to shinkinsho = Japanese journal of medical mycology. Tokyo.

Shinrin kosoku = Forest aerial survey. Tokyo.

Shinrin rekuryeshon kenkyu = Forest recreation research. Tokyo.

Shinrin ritchi = Forest and environment. Tokyo.

Shinshu daigaku nogaku-bu enshurin hokoku = Bulletin of the Shinshu university forest. Uyeda, Japan. Bull. Shinshu Univ. Forest. See B–P–H 273/22.

Shinshu daigaku nogakubu gakujitsu hokoku = Bulletin of the faculty of agriculture, Shinshu university. Ina, Japan. Bull. Fac. Agric. Shinshu Univ. See B–P–H 249/13.

Shinshu daigaku nogakubu kiyo = Journal of the faculty of agriculture, Shinshu university. Ina.

Shinshu horticulture. Nagano, Japan. Shinshu Hort. See B–P–H 835/21. HI 59389

Shiyan shengwu xuebao = Acta biologiae experimentalis sinica. Peking (Beijing).

Shizen to bunka = Natura et cultura. Kyoto.

Shizen kagaku no hakubutsukan = Natural science and museum. Tokyo. Nat. Sci. Mus. See B–P–H 629/28.

Shizenshi kenkyu = Occasional papers from the Osaka museum of natural history. Osaka.

Shizuoka daigaku kyoikugakubu kenkyu hokoku. Shizenkagaku-hen = Bulletin, education faculty, Shizuoka university. Natural science series. Shizuoka.

Shizuoka daigaku nogakubu kenkyu hokoku = Report of the faculty of agriculture; Shizuoka university. Iwata, Japan. Rep. Fac. Agric. Shizuoka Univ. See B–P–H 764/23.

Shizuoka daigaku rigakubu = Reports of the faculty of science, Shizuoka university. Shizuoka.

Shizuoka-ken chagyo shikenjo kenkyu hokoku = Bulletin of the Shizuoka prefecture tea experiment station.

Shizuoka-ken kankitsu shikenjo hokoku = Report of the Shizuoka citrus experiment station. Shizuoka, Japan. Rep. Shizuoka Citrus Exp. Sta. See B–P–H 771/1.

Shizuoka-ken kankitsu shikenjo kenkyu hokoku = Bulletin of Shizuoka prefectural citrus experiment station. Shizuoka, Japan. Bull. Shizuoka Prefect. Citrus Exp. Sta. See B–P–H 274/2.

Shizuoka-ken kankitsu shikenjo tokubetsu hokoku = Special bulletin of Shizuoka prefectural citrus experiment station. Shizuoka.

Shizuoka-ken nogyo shikenjo kenkyu hokoku = Bulletin of Shizuoka prefecture agricultural experiment station. Shizuoka, Japan. Bull. Shizuoka Prefect Agric. Exp. Sta. See B–P–H 274/1.

Shizuoka-ken ringyo shikenjo gyomu seiseki hokokusho = Report (Annual) of the Shizuoka prefecture forest experiment station. Hamakita.

Shizuoka-ken ringyo shikenjo kenkyu hokoku = Research report of the Shizuoka prefecture forestry experiment station. Hamakita.

Shokubutsu boeki = Plant protection. Pl. Protect. (Tokyo). See B–P–H 712/21.

Shokubutsu boekisho chosa kenkyu hokoku = Research bulletin of plant protection service, Japan. Kanagawa, Japan. Res. Bull. Pl. Protect. Serv. Japan. See B–P–H 774/23.

Shokubutsu bunrui chiri = Acta phytotaxonomica et geobotanica. Kyoto, Japan. Acta Phytotax. Geobot. See B–P–H 48/5.

Shokubutsu byogai kenkyu = Plant disease research. Kyoto.

Shokubutsu chiri bunrui kenkyu = Journal of phytogeography and taxonomy. Ishikawa.

Shokubutsu-gaku zasshi = Botanical magazine. Tokyo. Bot. Mag. (Tokyo). See B–P–H 221/17.

Shokubutsu-kai = Botanical world. [Edited by J. Sujimoto.] [Japan]. Bot. World. See B–P–H 225/16.

Shokubutsu kenkyu zasshi = Journal of japanese botany. Tokyo. J. Jap. Bot. See B–P–H 470/19.

Shokubutsu kwagaku zasshi = Acta phytochimica. Tokyo. Acta Phytochim. See B–P–H 47/20.

Shokubutsu no kagaku chosetsu = Chemical regulation in plants. [The society of chemical regulation of plants]. Tokyo. Chem. Regulat. Pl. See B–P–H 307/2.

Shokubutsu no tomo = Plant's companion. Tokyo. Pl. Companion. See B–P–H 711/21.

Shokobutsu seiri = Physiology of plants. Kyoto, Japan. Physiol. Pl. (Kyoto). See B–P–H 709/12.

Shokubutsu seitai-gaku kaiho = Bulletin of the society of plant ecology. Sendai.

Shokubutsu shumi = Amatores herbarii. Kobe, Japan. Amatores Herb. See B–P–H 77/18.

Shokubutsu to shizen = Nature and plants.

Shokubutsu zasshi; [Botanical magazine.] Peking (Beijing). No. ?-2+, ?-1978+. Shok. Zasshi. HI 72789

Short guide to experiments, Stockbridge house experimental horticulture station = Guide to experiments, Stockbridge house experimental horticulture station. Cawood.

Short report of the experimental work of the Ganeshkind botanical garden. Bombay. 1916-20. Short Rep. Exp. Wk. Ganeshkind Bot. Gard. Preceded by: Report of the experimental work of the Ganeshkind botanical garden. HI 72790

Shropshire wildlife. Ca.1989+. Shropshire Wildlife. HI 72791

Shubyo kensa nenpo = Report (Annual) of the seed testing laboratory. Tokyo.

Shui chan shêng wu hsi yen chiu pao kao Tai wan shen shui chan shih yen shoin hsin = Report of the laboratory of hydrobiology. Taipei, Taiwan. Rep. Lab. Hydrobiol. See B–P–H 767/18.

Shui ch'an yüeh k'an. (Fu kan) = Marine products month. Mar. Prod. Month. See B–P–H 548/11.

Shui sheng sheng wu hsüeh chi k'an = Acta hydrobiologica sinica. Peking. Acta Hydrobiol. Sin. (Peking). See B–P–H 44/13.

Si-ka-wei, shanghai = Bulletin de l'université l'Aurore. Shanghai. Bull. Univ. Aurore. See B–P–H 285/27.

Sichuan daxue xuebao = Journal of Sichuan university. Natural science edition. Chengtu.

Sida; contributions to botany. Dallas, TX. Sida. See B–P–H 835/48. HI 59390

Siebenbürgisches Archiv. Cologne & Graz. Siebenbürg. Arch. See B–P–H 836/1. HI 59391

Siebenbürgische Quartalschrift. Hermannstadt [=Sibiu, Rumania]. Siebenbürg. Quartalschr. See B–P–H 836/2. HI 59392

Sieboldia. Leiden. Sieboldia (Leiden). See B–P–H 836/4. HI 59394

Sieboldia; acta biologia. Fukuoka, Japan. Sieboldia (Fukuoka). See B–P–H 836/3. HI 59393

Sienilehti. Helsinki. Vol. 25+, 1973+. Sienilehti. Preceded by: Sienitietoja. HI 72792

Sienitietoja - svampnytt. Helsinki. Vols. ?-24, ?-1972? Sienitietoja Svampnytt. Superseded by: Sienilehti. HI 72793

Sierra; the sierra club bulletin. San Francisco, CA. Vol. 62(8)+, 1977+. Sierra. Preceded by: Sierra club bulletin. HI 72794

Sierra club bulletin. San Francisco, CA. Vols. 1-62(7), 1893-1977. Sierra Club Bull. Superseded by: Sierra. 5-3870-1. HI 59395

Sigen kagaku kenkyusyo hokoku = Bulletin of the research institut for natural resources. Tokyo. Bull. Res. Inst. Nat. Resources. See B–P–H 270/4.

Sigma Xi quarterly. Champaign, IL. Sigma Xi Quart. See B–P–H 836/8. HI 59396

Signa. Hannon, Ont. No. 1+, 1968+. Signa. HI 72795

Sikmul hak'kae-chi = Korean journal of botany. Seoul, Korea. Korean J. Bot. See B–P–H 518/10.

Sikmul poho = Journal of plant protection. Suwon.

Silva. Darmstadt. Germany. Silva . See B–P–H 836/12. HI 59397

Silva fennica. Helsinki. Silva Fenn. See B–P–H 836/13. HI 59398

Silva revue. Deutsche Ausgabe. Zurich. Vol. ?-5-9+, ?-19??-55+. Silva Rev., Deutsche Ausgabe. HI 72796

Silva mediterranea. Florence. 1924-35. Silva Medit. HI 72797

Silvae genetica. Zeitschrift für Forstgenetik und Forstpflanzenzüchtung ... Frankfurt a. M. Silvae Genet. See B–P–H 836/14. HI 59399

Silvae orbis. Berlin-Wannsee. Silvae Orbis. See B–P–H 836/15. HI 59400

Silvaecultura tropica et subtropica. Prague. 1969+. Silvaec. Trop. Subtrop. HI 72798

Silvery leaves. [Gin'yo.] Hakodate. 1948+. Silvery Leaves. HI 75151

Silvical leaflet, bureau of forestry, department of agriculture and natural resources, Philippines Islands.

Manila. 1958+. Silvical Leafl. Bur. Forest. Dept. Agric. Nat. Resources Philippines Islands. HI 72799

Silvical leaflets, forest service, United States department of Agriculture. Washington, DC. 1907-09. Silvical Leafl. Forest Serv. U.S.D.A. HI 72800

Silvical series, Pacific northwest forest and range experiment station. Portland, OR. 1957+. Silvical Ser. Pacific NorthW. Forest Range Exp. Sta. HI 72801

Silvicultura. Vol. 1+, 1975+. Silviculatura (Brazil). HI 72802

Silvicultura. Maldonado, Uruguay. Vol. 1+, 1951+. Silvicultura (Uruguay). HI 59402

Silvicultura em São Paulo. São Paulo. Silvic. São Paulo. See B–P–H 836/16. HI 59401

Silvicultural leaflet, dominion forest service, Canada = Silvicultural leaflet, forest research branch, Canada. Ottawa.

Silvicultural leaflet, forest research branch, Canada. Ottawa. Vols. 1-100, 1941-54. Silvic. Leafl. Forest Res. Branch, Canada. Superseded by: Technical note, forest research branch (division), Canada. HI 72803

Silvicultural notes, forest research institute, Whakarewarewa. Whakarewarewa. 1958+. Silvic. Notes Forest Res. Inst. Whakarewarewa. HI 72804

Silvicultural research note, dominion forest service, Canada = Silvicultural research note, forestry branch, Canada. Ottawa.

Silvicultural research note, forestry branch, Canada. Ottawa. Nos. 1-105, 1924-54. Silvic. Res. Note Forest. Branch, Canada. HI 72805

Simiente. Santiago. Simiente. See B–P–H 836/18. HI 59403

Sinar pasoh. Kepong. No. ?-2+, ?-1980+. Sinar Pasoh. HI 72806

Sind university research journal. Science series. Publication of the faculty of sciences, university of Sind, Jamshoro campus. Hyderabad. Vol. 1+, 1965+. Sind Univ. Res. J., Sci. Ser. HI 72807

Sinensia. Nanking. Sinensia. See B–P–H 836/19. HI 59404

Sinet. Addis Ababa. Vol. 1+, 1978+. Sinet. HI 72808

Singapore botanic gardens annual report. Singapore. 1971+. Singapore Bot. Gard. Annual Rep. Preceded by: Report (Annual) of the botanic gardens, Singapore. HI 72809

Singapore journal of tropical geography. Singapore. Vol. 1+, 1980+. Singapore J. Trop. Geogr. Preceded by: Journal of tropical geography. HI 72810

Singapore naturalist. Singapore. Singapore Naturalist. See B–P–H 836/22. HI 59405

Singmul chojik paeyang hakhoe chi = Korean journal of plant tissue culture. Seoul.

Sinugasa san-ho = Bulletin des soies Kinugasa. Bull. Soies Kinugasa. See B–P–H 282/23.

Sisal mexicano. Mérida, Mexico. Sisal Mex. See B–P–H 836/23. HI 59406

Sisal review. London. Vols. 1-6(3), 1938-40. Sisal Rev. HI 72811

Sisal de Yucatán. Henequeneros de Yucatán. Mérida, Mexico. Sisal Yucatán. See B–P–H 836/24. HI 59407

Sistematičeskie zametki po materialam Gerbarii imeni P. N. Krylova pri Tomskom gosudarstvennom universitete imeni V. V. Kujbyševa = Sistematicheskie zametki po materialam Gerbarii imeni P. N. Krylova pri Tomskom gosudarstvennom universitete imeni V. V. Kuybysheva. Tomsk.

Sistematicheskie zametki po materialam Gerbarii imeni P. N. Krylova pri Tomskom gosudarstvennom universitete imeni V. V. Kuybysheva. Tomsk, Russian S F S R. 1927-39 (-1948?); pt. 73/74+, 1949+. Sist. Zametki Mater. Gerb. Krylova Tomsk. Gosud. Univ. Kuybysheva. HI 72814

Sistematicheskii ukazatel′ statei v inostrannykh zhurnalakh: Biofizika. Biokhimiya. Fiziogiya. Mikrobiologiya. (Genetika). Moscow. Vols. 1-71, 1952-59. Sist. Ukaz. Statei Inostr. Zhurn., Bikofiz. Preceded by: Sistematicheskii ukazatel′ statei v inostrannykh zhurnalakh: Biologicheskie nauki. HI 72812

Sistematicheskii ukazatel′ statei v inostrannykh zhurnalakh: Biologicheskie nauki. Moscow. Vols. 1-3, 1950-51. Sist. Ukaz. Statei Inostr. Zhurn., Biol. Nauki. Superseded by: Sistematicheskii ukazatel′ statei v inostrannykh zhurnalakh: Biofizika. Biokhimiya. Fiziogiya. Mikrobiologiya. (Genetika). HI 72813

Sistematičeskij ukazatel′ statej v inostrannyh žurnalah: Biofizika. Biohimija. Fiziogija. Mikrobiologija. (Genetika) = Sistematicheskii ukazatel′ statei v inostrannykh zhurnalakh: Biofizika. Biokhimiya. Fiziogiya. Mikrobiologiya. (Genetika). Moscow.

Sistematičeskij ukazatel′ statej v inostrannyh žurnalah: Biologičeskie nauki = Sistematicheskii ukazatel′ statei v inostrannykh zhurnalakh: Biologicheskie nauki. Moscow.

Sistematicheskie zametki po materialam Gerbariya imeni Tomskogo universiteta = Sistematicheskie zametki po materialam Gerbarii imeni P. N. Krylova pri Tomskom gosudarstvennom universitete imeni V. V. Kuybysheva.

Situación del coco. Rome. Situación Coco. See

B–P–H 837/3. HI 59408

Situaţia daunatorilor animali ai plantelor cultivate. Bucharest. Situaţia Daunatorilor Anim. Pl. Cult. See B–P–H 837/4. HI 59409

Sitzungsberichte und Abhandlungen, königlich Sächsische Gesellschaft für Botanik und Gartenbau "Flora" zu Dresden = Königlich Sächsische Gesellschaft für Botanik und Gartenbau "Flora" zu Dresden. Sitzungsberichte und Abhandlungen. Dresden. Königl. Sächs. Ges. Bot. "Flora" Dresden Sitzungsber. Abh. See B–P–H 518/1.

Sitzungsberichte und Abhandlungen der Naturforschenden Gesellschaft zu Rostock. Rostock. Sitzungsber. Abh. Naturf. Ges. Rostock. See B–P–H 838/18. HI 59410

Sitzungsberichte und Abhandlungen der naturwissenschaftlichen Gesellschaft Isis zu Bautzen. Bautzen. Sitzungsber. Abh. Naturwiss. Ges. Isis Bautzen. See B–P–H 839/1. HI 59411

Sitzungsberichte und Abhandlungen der Naturwissenschaftlichen Gesellschaft Isis in Dresden. Dresden. Sitzungsber. Abh. Naturwiss. Ges. Isis Dresden. See B–P–H 839/2. HI 59412

Sitzungsberichte, Akademie der Wissenschaften in Wien = Akademie der Wissenschaften in Wien. Sitzungsberichte. Mathematisch-naturwissenschaftliche Klasse. Abteilung 1. Vienna. Akad. Wiss. Wien Sitzungsber., Math.-Naturwiss, Kl., Abt. 1. See B–P–H 66/26.

Sitzungsberichte der bayerischen Akademie der Wissenschaften. Mathematisch-naturwissenschaftliche Klasse. Munich. 1955+. Sitzungsber. Bayer. Akad. Wiss., Math.-Naturwiss. Kl. Preceded by: Sitzungsberichte der mathematisch-naturwissenschaftlichen (Abteilung) Klasse der bayerischen Akademie der Wissenschaften zu München. 1-105-2. HI 72816

Sitzungsberichte der Bernischen Botanischen Gesellschaft. Bern. Sitzungsber. Bern. Bot. Ges. See B–P–H 839/5. HI 59413

Sitzungsberichte, botanische Gesellschaft zu Stockholm. [Forms part of: Botanisches centralblatt.] Cassel. Vols. 1-3?, 1883-85?, 1884-86. Sitzungsber. Bot. Ges. Stockholm. HI 72817

Sitzungsberichte, botanischer Verein in München. [Forms part of: Flora. and Botanisches centralblatt.] Cassel. 1881-95?, 1882?-96? Sitzungsber. Bot. Verein München. HI 72818

Sitzungsberichte der Dermatologischen Vereinigung zu Berlin. Vienna. Sitzungsber. Dermatol. Vereinigung Berlin. See B–P–H 839/6. HI 59414

Sitzungsberichte der deutschen Akademie der Landwirtschaftswissenschaften zu Berlin. Leipzig. Vol. 1+, 1952+. Sitzungsber. Deutschen Akad. landwirtschaftswiss. Berlin. HI 72819

Sitzungsberichte der Deutschen Akademie der Wissenschaften zu Berlin. Klasse für Chemie, Geologie und Biologie. Berlin. Sitzungsber. Deutsch. Akad. Wiss. Berlin, Kl. Chem. See B–P–H 839/7. HI 59415

Sitzungsberichte der Deutschen Akademie der Wissenschaften zu Berlin. Klasse für landwirtschaftliche Wissenschaften. Berlin. Sitzungsber. Deutsch. Akad. Wiss. Berlin, Kl. Landw. Wiss. See B–P–H 839/8. HI 59416

Sitzungsberichte der Deutschen Akademie der Wissenschaften zu Berlin. Klasse für Mathematik und allgemeine Naturwissenschaften. Berlin. Sitzungsber. Deutsch. Akad. Wiss. Berlin, Kl. Math. See B–P–H 839/9. HI 59417

Sitzungsberichte der Deutschen Akademie der Wissenschaften zu Berlin. Mathematisch-naturwissenschaftliche Klass. Berlin. Sitzungsber. Deutsch. Akad. Wiss. Berlin, Math,-Naturwiss. Kl. See B–P–H 839/10. HI 59418

Sitzungsberichte des deutschen naturwissenschaftlich-medicinischen Vereins für Böhmen "Lotos" in Prag. Prague. Sitzungsber. Deutsch. Naturwiss.-Med. Vereins Böhmen "Lotos" Prag. See B–P–H 839/11. HI 59419

Sitzungsberichte der Finnischen Akademie der Wissenschaften. Helsinki. 1908-76. Sitzungsber. Finn. Akad. Wiss. Superseded by: Vuoskirja, suomalainen tiedeakatemia. For edition in Finnish see: Esitelmiä ja pöytäkirjoja. 5-4116-3. HI 72820

Sitzungsberichte der Gelehrten Estnischen Gesellschaft. Dorpat [=Tartu, Estonian S S R]. Sitzungsber. Gel. Estn. Ges. See B–P–H 839/13. HI 59420

Sitzungsberichte der Gelehrten Estnischen Gesellschaft zu Dorpat. Dorpat [=Tartu, Estonian S S R]. Sitzungsber. Gel. Estn. Ges. Dorpat. See B–P–H 839/14. HI 59421

Sitzungsberichte der Gesellschaft zur Beförderung der gesammten Naturwissenschaften in Marburg; [From 1908 "gesammten" became "gesamtem"]. Marburg. 1866-1926; vols. 61-87, 1927-66. Sitzungsber. Ges. Beförd. Gesammten Naturwiss. Marburg. Superseded by: Sitzungsberichte der wissenschaftlichen Gesellschaft zu Marburg. 2-1720-2. HI 72821

Sitzungsberichte der Gesellschaft für Natur- und Heilkude zu Dresden. Dresden. Sitzungsber. Ges. Natur- Heilk. Dresden. See B–P–H 839/16. HI 59422

Sitzungsberichte der Gesellschaft Naturforschender Freunde zu Berlin. Berlin. Sitzungsber. Ges. Naturf. Freunde Berlin. See B–P–H 840/1. HI 59423

Sitzungsberichte der Heidelberger Akademie der Wissenschaften. Jahresheft. Heidelberg. Sitzungsber. Heidelberger Akad. Wiss. Jahresh. See B–P–H 840/3. HI 59424

Sitzungsberichte der Heidelberger Akademie der Wissenschaften. Mathematisch-naturwissenschaftliche Klasse. Berlin & Leipzig. Sitzungsber. Heidelberger Akad. Wiss., Math.-Naturwiss. Kl. See B–P–H 840/4. HI 59425

Sitzungsberichte der Heidelberger Akademie der Wissenschaften. Mathematisch-naturwissenschaftliche Klasse. Abteilung B. Berlin & Leipzig. Sitzungsber. Heidelberger Akad. Wiss., Math.-Naturwiss. Kl., Abt. B. See B–P–H 840/5. HI 59426

Sitzungsberichte der Heidelberger Akademie der Wissenschaften. Stiftung Heinrich Lanz. Jahresheft. Heidelberg. Sitzungsber. Heidelberger Akad. Wiss. Stiftung Heinrich Lanz Jahresh. See B–P–H 840/6. HI 59427

Sitzungsberichte der Heidelberger Akademie der Wissenschaften. Stiftung Heinrich Lanz. Mathematisch-naturwissenschaftliche Klasse. Heidelberg. Sitzungsber. Heidelberger Akad. Wiss. Stiftung Heinrich Lanz, Math.-Naturwiss. Kl. See B–P–H 840/7. HI 59428

Sitzungsberichte der Jenaischen Gesellschaft für Medicin und Naturwissenschaft. Leipzig. Sitzungsber. Jenaischen Ges. Med. See B–P–H 840/8. HI 59429

Sitzungsberichte der Kaiserlichen Akademie der Wissenschaften. Mathematisch-naturwissenschaftliche Classe. Vienna. Sitzungsber. Kaiserl. Akad. Wiss., Math.-Naturiwss. Cl. See B–P–H 840/10. HI 59430

Sitzungsberichte der Kaiserlichen Akademie der Wissenschaften. Mathematisch-naturwissenschaftliche Classe. Abteilung 1. Vienna. Sitzungsber. Kaiserl. Akad. WIss., Math.-Naturiwss. Cl., Abt. 1. See B–P–H 840/11. HI 59431

Sitzungsberichte der königlich böhmischen Gesellschaft der Wissenschaften in Prag. Prague. Sitzungsber. Königl. Böhm. Ges. Wiss. Prag. See B–P–H 840/12. HI 59432

Sitzungsberichte der Königlich Preussischen Akademie der Wissenschaften zu Berlin. Berlin. Sitzungsber. Königl. Preuss. Akad. Wiss. Berlin. See B–P–H 841/1. HI 59434

Sitzungsberichte der königl. bayerischen Akademie der Wissenschaften zu München. Munich. 1860-70. Sitzungsber. Königl. Bayer. Akad. Wiss. München. Superseded by: Sitzungsberichte der mathematisch-physikalischen Classe der königlichen bayerischen Akademie der Wissenschaften. 1-105-2. HI 72822

Sitzungsberichte der königlichen böhmischen Gesellschaft der Wissenschaften in. Mathematisch-naturwissenschaftliche Classe. Prague. Sitzungsber. Königl. Böhm. Ges. Wiss. Prag., Math.-Naturwiss. Cl. See B–P–H 840/13. HI 59433

Sitzungsberichte der mathematisch-naturwissenschaftlichen Abteilung der bayerischen Akademie der Wissenschaften zu München. Munich. 1924-54. Sitzungsber. Math.-Naturwiss. Abt. Bayer. Akad. Wiss. München. Preceded by: Sitzungsberichte der mathematisch-physikalischen Klasse der bayerischen Akademie der Wissenschaften. Superseded by: Sitzungsberichte der mathematisch-naturwissenschaftlichen Klasse der Bayerischen Akademie der Wissenschaften zu München. 1-106-1. HI 72823

Sitzungsberichte der Mathematisch-Naturwissenschaftlich-Ärztlichen Sektion. Lvov, Galicia [Ukrainian S S R]. Sitzungsber. Math.-Naturwiss.-Ärztl. Sekt. See B–P–H 841/3. HI 59436

Sitzungsberichte der mathematisch-naturwissenschaftlichen Klasse der bayerischen Akademie der Wissenschaften zu München. Munich. 1945/46-54. Sitzungsber. Math.-Naturwiss. Kl. Bayer. Akad. Wiss. München. Preceded by: Sitzungsberichte der mathematisch-naturwissenschaftlichen Abteilung der bayerischen Akademie der Wissenschaften zu München. Superseded by: Sitzungsberichte der bayerischen Akademie der Wissenschaften. Mathematisch-naturwissenschaftliche Klasse 1-106-1. HI 59437

Sitzungsberichte der mathematisch-physikalischen Classe [later Klasse] der königlichen bayerischen Akademie der Wissenschaften. Munich. Vols. 1-38 [vols after 1908 unnumbered], 1871-1917. Sitzungsber. Math.-Phys. Cl. Königl. Bayer. Akad. Wiss. München. Preceded by: Sitzungsberichte der königl. bayerischen Akademie der Wissenschaften zu München. Superseded by: Sitzungsberichte der mathematisch-physikalischen Klasse der Bayerischen Akademie der Wissenschaften München. 1-105-2. HI 72824

Sitzungsberichte der mathematisch-physikalischen Klasse der bayerischen Akademie der Wissenschaften München. Munich. 1918-23. Sitzungsber. Math.-Phys. Kl. Bayer. Akad. Wiss. München. Preceded by: Sitzungsberichte der mathematisch-physikalischen Classe der königlichen bayerischen Akademie der Wissenschaften. Superseded by: Sitzungsberichte der mathematisch-naturwissenschaftlichen Abteilung der bayerischen Akademie der Wissenschaften zu München. 1-106-1. HI 59439

Sitzungsberichte der Medicinisch-naturwissenschaftliche Section des Siebenbürgischen Museum-Vereins = Ertesitö az erdélyi muzeum-egylet orvos-természet tudomanyi szakosztályából. Kolozsvar [= Cluj, Rumania].

Sitzungsberichte der Naturforschenden Gesellschaft zu

Leipzig. Leipzig. Sitzungsber. Naturf. Ges. Leipzig. See B–P–H 841/8. HI 59441

Sitzungsberichte der Naturforschenden Gesellschaft zu Rostock. Rostock. Sitzungsber. Naturf. Ges. Rostock. See B–P–H 841/9. HI 59442

Sitzungsberichte der Naturforscher-Gesellschaft zu Dorpat. Dorpat [=Tartu, Estonian S S R]. Sitzungsber. Naturf.-Ges. Dorpat. See B–P–H 841/7. HI 59440

Sitzungsberichte der Naturforscher-Gesellschaft bei der Universität Dorpat. Dorpat [=Tartu, Estonian S S R]. Sitzungsber. Naturf.-Ges. Univ. Dorpat. See B–P–H 841/10. HI 59443

Sitzungsberichte der Naturforscher-Gesellschaft bei der Universität Jurjeff (Dorpat). Yur'ev [=Tartu, Estonian S S R]. Sitzungsber. Naturf.-Ges. Univ. Jurjeff. See B–P–H 841/11. HI 59444

Sitzungsberichte der Naturforscher-Gesellschaft bei der Universität Jurjew <Dorpat>. Yur'ev [=Tartu], Estonian S S R. Vol. 11, 1896. Sitzungsber. Naturf.-Ges. Univ. Jurjew. Preceded by: Sitzungsberichte der Naturforscher-Gesellschaft bei der Universität Jurjeff <Dorpat>. See B–P–H 841/11. Superseded by: Protokoly Obshchestva Estestvoispytatelei pri Imperatorskom Yur'evskom Universitetě. 3-2469-2. HI 72825

Sitzungsberichte, herausgegeben vom Naturhistorischen Verein der preussischen Rheinlande und Westfalens. Bonn. Sitzungsber. Naturhist. Vereins Preuss. Rheinl. See B–P–H 842/1. HI 59445

Sitzungs-Berichte der Naturwissenschaftlichen Gesellschaft Isis zu Dresden. Dresden. Sitzungs-Ber. Naturwiss. Ges. Isis Dresden. See B–P–H 842/3. HI 59446

Sitzungsberichte der Niederrheinischen Gesellschaft für Natur- und Heilkunde zu Bonn. Bonn. Sitzungsber. Niederrhein. Ges. Natur- Heilk. Bonn. See B–P–H 842/4. HI 59447

Sitzungsberichte, österreichische Akademie der Wissenschaften. Mathematisch-naturwissenschaftliche Klasse = Österreichische Akademie der Wissenschaften. Mathematisch-naturwissenschaftliche Klasse. Sitzungsberichte. Abteilung 1. Biologie, Mineralogie, Erdkunde und verwandte Wissenschaften. Vienna. Österr. Akad. Wiss., Math.-Naturwiss. Kl., Sitzungsber,. Abt. 1, Biol. See B–P–H 692/13.

Sitzungsberichte der Physikalisch-Medicinischen Gesellschaft zu Würburg. Würzburg. Sitzungsber. Phys.-Med. Ges. Würzburg. See B–P–H 842/5. HI 59448

Sitzungsberichte der Physikalisch-Medicinischen Societät zu Erlangen. Erlangen. Sitzungsber. Phys.-Med. Soc. Erlangen. See B–P–H 842/6. HI 59449

Sitzungsberichte der Physikalisch-Medizinischen Sozietät zu Erlangen = Sitzungsberichte der Physikalisch-Medicinischen Societät zu Erlangen. Erlangen. Sitzungsber. Phys.-Med. Soc. Erlangen. See B–P–H 842/6.

Sitzungsberichte der Preussischen Akademie der Wissenschaften. Berlin. Sitzungsber. Preuss. Akad. Wiss. See B–P–H 842/7. HI 59450

Sitzungsberichte der Preussischen Akademie der Wissenschaften. Physikalisch-mathematische Klasse. Berlin. Sitzungsber. Preuss. Akad. Wiss., Phys.-Math. Kl. See B–P–H 842/8. HI 59451

Sitzungsberichte der Sächsischen Akademie der Wissenschaften zu Leipzig. Mathematisch-naturwissenschaftliche Klasse. Berlin. Sitzungsber. Sächs. Akad. Wiss. Leipzig, Math.-Naturwiss. Kl. See B–P–H 842/9. HI 59452

Sitzungsberichte der wissenschaftlichen Gesellschaft zu Marburg. Marburg. 1967. Sitzungsber. Wiss. Ges. Marburg. Preceded by: Sitzungsberichte der Gesellschaft zur Beförderung der gesammten Naturwissenschaften zur Marburg. HI 72826

Six-monthly review, department of agriculture, Cyprus. Nicosia. 1955-56. Six-Monthly Rev. Dept. Agric. Cyprus. Preceded by: Quarterly review, department of agriculture, Cyprus. Incorporated in: Review of works in progress, department of agriculture, Cyprus. HI 72827

Sizuokaken nokai-ho = Report of the agricultural society of Sizuokaken. Shizuoka, Japan. Rep. Agric. Soc. Sizuokaken. See B–P–H 762/11.

Skalky skalničky. Prague. ?-1977+. Skalky Skalničky. HI 72828

Skalničky. Prague. 1971+. Skalničky. HI 72829

Skandinavisches Archiv für Physiologie. Leipzig. Skand. Arch. Physiol. See B–P–H 842/12. HI 59453

Skandinavisk folkemagazin; et billed-blad for alle staender. Copenhagen. Vols. 1-?, 1851-76. Skand. Folkemag. HI 72830

Skandinavisk literaturselskab; skrifter. Copenhagen. Skand. Literaturselsk. Skr. See B–P–H 842/13. HI 59454

Skandinavisk museum. Ved et selskab. Copenhagen. Skand. Mus. See B–P–H 842/14. HI 59455

Skandia. Uppsala. Skandia. See B–P–H 842/16. HI 59456

Skånes natur; Skånes naturskyddsförenings arsskrift. Malmö. No. 9+, 1921+. Skånes Natur. Preceded by: Skånes naturskyddsförenings årsberättelse. HI 72831

Skånes naturskyddsförenings årsberättelse. Malmö. Nos. 5-8, 1913-15. Skånes Naturskyddssören. Årsberätt.

Preceded by: Meddelanden från Skånes naturskyddsförening. Superseded by: Skánnes natur. HI 72832

Skånska correspondenten. Skånska Corresp. See B–P–H 842/18. HI 59457

Skaraborgsnatur. ?-1966+. Skaraborgsnatur. HI 72833

Skenectada. Albany, NY. Vol. 1+, 1979+. Skenectada. HI 72834

Skogen. Stockholm. Skogen. See B–P–H 842/19. HI 59458

Skogliga rön; ur meddelanden fran statens skogs-Försöksanstalt. Stockholm. Vols. 1-?, 1921-40-? Skogliga Rön. HI 72835

Skogsägaren. Stockholm, Jönköping. Vols. 1-54, 1930-78. Skogsägaren. Incorporated in: Skogen. HI 72836

Skogsáret. ?-1973, ?-1974. Skogsáret. Superseded by: Swedish forestry. HI 72837

Skogs- och lantarbetsgivareföreningens tidskrift. Stockholm. 1964-70. Skogs- Lantarbetsgivareför. Tidskr. Preceded by: Svenska lantarbetsgivare-föreningens tidskrift. Superseded by: S L A - tidskriften. HI 72838

Skrifter, det. k. norske videnskabers selskabs. Trondheim = Det. k. norske videnskabers selskabs skrifter. Trondheim. Kongel. Norske Vidensk. Selsk. Skr. (Trondheim). See B–P–H 516/21.

Skrifter, som udi det kiøbenhavnske selskab af laerdoms og videnskabers elskere ere fremlagte og oplaeste. Copenhagen. Skr. Kiøbenhavnske Selsk. Laerd. Elsk. See B–P–H 842/20. HI 59459

Skrifter, det kongelige danske landhuusholdings-selskabs. Copenhagen = Det kongelige danske landhuusholdings-selskabs skrifter. Copenhagen. Kongel. Danske Landhuushold.-Selsk. Skr. See B–P–H 516/12.

Skrifter, det kongelige norske videnskabers selskabs. Copenhagen = Det kongelige norske videnskabers selskabs skrifter. Copenhagen. Kongel. Norske Vidensk. Selsk. Skr. (Copenhagen). See B–P–H 516/20.

Skrifter, som udi det kongelige videnskabers selskab ere fremlagde, og nu til trykken befordrede. Copenhagen. Vols. 11-12, 1777-79. Skr. Kongel. Vidensk. Selsk. Preceded by: Skrifter, som udi det kiøbenshavnske selskab af laerdoms og videnskabers elskere ere fremlagte og oplaeste. Superseded by: Nye samling af det kongelige danske videnskabers selskabs skrifter. HI 72839

Skrifter af naturhistorie-selskabet. Copenhagen. Skr. Naturhist.-Selsk. See B–P–H 843/3. HI 59460

Skrifter i naturskyddsärenden. Stockholm. Skr. Naturskyddsärenden. See B–P–H 843/4. HI 59461

Skrifter i det 19de aarhundrede, det kongelige norske videnskabersselskabs. Copenhagen = Det kongelige norske videnskabersselskabs skrifter i det 19de aarhundrede. Copenhagen. Kongel. Norske Videnskabersselsk. Skr. 19de Aarhundr. See B–P–H 516/22.

Skrifter om norsk polarinstitutt. Oslo. 1948+. Skr. Norsk Polarinst. Preceded by: Skrifter om Svalbard og Ishavet. HI 72840

Skrifter utgitt av det norske videnskaps-akademi i Oslo. Matematisk-naturvidenskapelig klasse. Oslo. Skr. Norske Vidensk.-Akad. Oslo, Mat.-Naturvidensk. Kl. See B–P–H 843/7. HI 59462

Skrifter, skandinavisk literaturselskab = Skandinavisk literaturselskab; skrifter. Copenhagen. Skand. Literaturselsk. Skr. See B–P–H 842/13.

Skrifter om Svalbard og Ishavet = Skrifter om Svalbard og Nordishavet. Oslo.

Skrifter om Svalbard og Nordishavet. Oslo. Nos. 1-93, 1922-50. Skr. Svalbard Nordishavet. Superseded by: Skrifter om norsk polarinstitutt. HI 72841

Skrifter udgivne af videnskabs-selskabet i Christiana. Mathematisk-naturvidenskabelig klasse. Christiania [=Oslo, Norway]. Skr. Vidensk.-Selsk. Christiana, Math,-Naturvidensk. Kl. See B–P–H 843/13. HI 59463

Skýrska, rannssóknastofnunin nedri As. Hveragerdi. Vol. 1+, 1969+. Skýrska Rannssóknastofn. Nedri As. HI 72842

Slezsky sborník. Opava, Czechoslovakia. Slez. Sborn. See B–P–H 843/21. HI 59464

Sloanea; occasional papers of the natural history division of the institute of Jamaica. Kingston, Jamaica. No. 1+, 1977+. Sloanea. HI 72843

Slovenian journal of forestry = Gozdarski vestnik. Ljubljana.

Slovenské lesné a drevárske hospodárstvo. Bratislava. Slov. Lesné Drev. Hospod. See B–P–H 843/23. HI 59465

Slovenské liečivé rastliny. Bratislava. Slov. Lieč. Rastl. See B–P–H 843/24. HI 59466

Slowenische Forstzeitschrift = Gozdarski vestnik. Ljubljana.

Słupskie prace matematyczno-przyrodnicze. Slupsk. Vol. 1+, 1980+. 1982+. Słupskie Prace Mat.-Przyr. HI 72844

Small fruit grower. Rochester, NY. Small Fruit Grower. See B–P–H 843/25. HI 59467

Smithersia. Bulawayo. No. 1+, 1983+. Smithersia. Preceded by: Occasional papers of the national museums and monuments. Series B, natural sciences. HI 72845

Smithsonian. Washington, DC. Vol. 1+, 1970+. Smithsonian. HI 72846

Smithsonian botanical news. Washington, DC. Nos. 1-3, 1973. Smithsonian Bot. News. Superseded by: Plant press. Washington. HI 72847

Smithsonian contributions to botany. Washington, DC. Vol. 1+, 1969+. Smithsonian Contr. Bot. Preceded by: Contributions from the United States national herbarium. HI 72848

Smithsonian contributions to knowledge. Washington, DC. Smithsonian Contr. Knowl. See B–P–H 844/7. HI 59468

Smithsonian contributions to the marine sciences. Washington, DC. Vol. 1+, 1977+. Smithsonian Contr. Mar. Sci. HI 72849

Smithsonian contributions to the paleobiology. Washington, DC. Vol. 1+, 1969+. Smithsonian Contr. Paleobiol. HI 72850

Smithsonian miscellaneous collections. Washington, DC. Vols. 1-153, 1862-1969. Smithsonian Misc. Collect. Superseded by: Smithsonian contributions to botany, Smithsonian contributions to the marine sciences. 5-3888-1. HI 72851

Smithsonian year. Washington, DC. Smithsonian Year. See B–P–H 844/9. HI 59469

Sobranie trudov o tatranskom natsional′nom parke = Zbornik prac o Tatranskom narodnom parku. Martin.

Sobrevivencia. Porto Alegre. No. 1+, 1973+. Sobrevivencia. HI 72852

Sochi gakkaishi. Tochigi. 1955+. Sochi Gakkaishi. HI 72853

Sochi shikenjo kenkyu hokoku = Bulletin, national grassland research institute. Nishinasuno.

Sochi shikenjo nenpo = Report (Annual) of the national grassland research institute. Tochigi.

Sochineniya i perevody k pol′ze i uveseleniyu sluzhashchiya. St. Petersburg. 1758-62. Sochin. Perev. Pol′ze Uvesel. Sluzha. HI 52833

Socialističeskoe plodoovoščnoe hozjajstvo = Sotsialisticheskoe plodoovoshchnoe khozyaistvo. Moscow.

Societatis medicae havniensis collectanea. Copenhagen. Soc. Med. Havn. Collect. See B–P–H 845/17. HI 59477

Société agricole, scientifique et littéraire des Pyrénées-orientales. Soc. Agric. Pyrénées-Orientales. See B–P–H 844/20. HI 59470

Société d'amateurs des sciences, de l'agriculture et des arts à Lille. Recueil des travaux. Lille. Soc. Amateurs Sci. Lille Receuil Trav. See B–P–H 844/22. HI 59471

Sociedad botánica de México. Boletín. Mexico City. Soc. Bot. México Bol. See B–P–H 844/29. HI 59472

Sociedad científica y literaria de Campeche. Campeche, Mexico. Soc. Ci. Lit. Campeche. See B–P–H 844/33. HI 59473

Société des amis des sciences naturelles de Rouen = Bulletin de la société des amis des sciences naturelles de Rouen. Rouen. Bull. Soc. Amis Sci. Nat. Rouen. See B–P–H 274/18.

Société d'émulation Cambrai. Séance publique. Cambrai. 1817-18. Soc. Émul. Cambrai Séance Publique. Preceded by: Société d'émulation de la ville de Cambrai. Séance publique. Superseded by: Mémoires de la société d'émulation de Cambrai. (Agriculture, sciences et arts.) Séance publique. HI 72855

Société d'émulation du département du Jura. Lons-le-Saunier, France. Soc. Émul. Dép. Jura. See B–P–H 844/39. HI 59474

Société d'émulation de la ville de Cambrai. Séance publique = Séance publique de la société d'émulation de la ville de Cambrai.

Société helvétique pour l'echange des plantes. Neuchatel. Vols. 1-16, 1871-85. Soc. Helv. Echange Pl. HI 72856

Société d'histoire naturelle du Creusot bulletin = Bulletin de la société d'histoire naturelle du Creusot. Le Cruesot.

Société des lettres, sciences et arts de Metz. Metz. Soc. Lett. Metz. See B–P–H 845/11. HI 59475

Société linnéenne du nord de la France. Compte-rendu de la session. Amiens, France. Soc. Linn. N. France Compt.-Rend. Session. See B–P–H 845/14. HI 59476

Société mycologique de l'ouest. Le Mans. Vols. 1-8, 1954-61. Soc. Mycol. Ouest. HI 72857

Société des sciences, agriculture et arts de la Basse-Alsace (Gesellschaft zur Beförderung der Wissenschaften, des Ackerbaues und der Künste im Unter-Elsass). Bulletin mensuel. Strasbourg. Soc. Sci. Basse-Alsace Bull. Mens. See B–P–H 845/30. HI 59478

Société des sciences, agriculture et arts de la Basse-Alsace. Bulletin de la société et de la station agronomique. Strasbourg. Soc. Sci. Basse-Alsace Bull. Soc. Stat. Agron. See B–P–H 845/31. HI 59479

Société des sciences, agriculture et arts de la Basse-

Alsace. Bulletin trimestriel. Strasbourg. Soc. Sci. Basse-Alsace Bull. Trimestriel. See B–P–H 845/32. HI 59480

Société des sciences, agriculture et arts de la Basse-Alsace. Bulletin trimestriel de la société et de la station agronomique. Strasbourg. Soc. Sci. Basse-Alsace Bull. Trimestriel Soc. Stat. Agron. See B–P–H 845/33. HI 59481

Société des sciences, agriculture et arts du Bas-Rhin. Bulletin trimestriel de la société et de la station agronomique. Strasbourg. Soc. Sci. Dép. Bas-Rhin Bull. Trimestriel Soc. Stat. Agron. See B–P–H 846/1. HI 59482

Société des sciences naturelles. Grand-duché de Luxembourg. Luxembourg. Soc. Sci. Nat. Grand-Duché Luxembourg. See B–P–H 846/2. HI 59483

Society for applied bacteriology symposium series = Symposia series, society for applied bacteriology. Orlando, FL.

Society for applied bacteriology technical series = Technical series, society for applied bacteriology. London, New York, Orlando, FL.

Society for economic botany newsletter. Vol. ?-3+, ?-1989+. Soc. Econ. Bot. Newslett. HI 72858

Society for general microbiology quarterly = Quarterly, society for general microbiology. Reading.

Society news, Brunei nature society. Brunei. Vol. ?-2(5)+, ?-1979+. Soc. News Brunei Nat. Soc. HI 72859

Sočinenija i perevody k pol'zě i uveseleniju služaščija. St. Petersburg = Sochineniya i perevody k pol'ze i uveseleniyu sluzhashchiya. St. Petersburg.

Södermanlands-nerikes nation majhälsning till landsmännen. ?-1966+. Södermanlands-Nerikes Nation Majhälsn. Landsmänn. HI 72860

Sogo kenkyu shiryokan = Bulletin, university museum, university of Tokyo. Tokyo.

Soil biology. Paris = Biologie du sol. Paris.

Soil biology and biochemistry. Oxford, Elmsford, NY., etc. Vol. 1+, 1969+. Soil Biol. Biochem. HI 72861

Soil bulletin. [T'u jang chuan pao.] Peiping [=Peking]. Soil Bull. See B–P–H 846/8. HI 59484

Soil conservation. Washington, DC. Soil Conservation. See B–P–H 846/9. HI 59485

Soil and health. Wilmthorpe, London. Vols. 1-2, 1946-48. Soil & Health. HI 72862

Soil and plant = Zemljište i biljka. Belgrade.

Soil and plant analyst. Athens, GA. 1970+. Soil Pl. Analyst. HI 72863

Soil and plant food. Tokyo. Soil Pl. Food. See B–P–H 846/13. HI 59486

Soil research report. Nanjing. No. 1+, 1980+. Soil Res. Rep. HI 72864

Soil science. New Brunswick, NJ. Soil Sci. See B–P–H 846/15. HI 59487

Soil science annual. Warsaw = Roczniki gleboznawcze. Warsaw.

Soil science and agrochemistry = Pochvoznanie i agrochimiya. Sofia.

Soil science annual. Warsaw = Roczniki gleboznawcze. Warsaw.

Soil science and plant nutrition. Tokyo. Soil Sci. Pl. Nutr. See B–P–H 846/16. HI 59488

Soil and tillage research. Amsterdam. Vol. 1+, 1980+. Soil Tillage Res. HI 72865

Soil and water. Wellington, N.Z. Soil & Water. See B–P–H 846/18. HI 59489

Soilless culture. Wageningen. Vol. 1+, 1985+. Soilless Culture. HI 72866

Soils quarterly. Nanking. Soils Quart. See B–P–H 846/21. HI 59490

Solanaceae newsletter. Birmingham, St. Louis, MO. No. 1+, 1974+. Solanaceae Newslett. HI 72867

Sols africains. Paris. Sols Africains. See B–P–H 846/22. HI 59491

Somali journal of range science. Mogadishu. Vol. 1+, 1986+. Somali J. Range Sci. HI 72868

Somersetshire archaeology and natural history = Proceedings of the Somersetshire archaeological and natural history society. Taunton.

Sommerfeltia. Oslo. Vol. 1+, 1985+. Sommerfeltia. HI 72869

Sonderbände des naturwissenschaftlichen Vereins in Hamburg. Hamburg. Vol. 1+, 1976+. Sonderb. Naturwiss. Vereins Hamburg. HI 72870

Sonderhefte, naturwissenschaftlichen vereins in Hamburg. Hamburg. ?-1979+. Sonderh. Naturwiss. Vereins Hamburg. HI 72871

Song news = S O N G news. Niagara-on-the-Lake.

Sonorensis. Tucson, AZ. Vol. 1+, 1978+. Sonorensis. HI 72872

Soobščenija Akademii nauk Gruzinskoj S S R = Soobshcheniya Akademii nauk Gruzinskoi S S R. Tiflis.

Soobščenija Bjuro po Častnomu Rastenievodstvu Učenago Komiteta Glavnago Upravlenija Zemleustrojstva i Zemledělija = Soobshcheniya Byuro po Chastnomu Rastenievodstvu Uchenago Komiteta Glavnago Upravleniya Zemleustroistva i Zemledělija.

St. Petersburg.

Soobščenija Bjuro po Častnomu Rastenievodstvu Učenago Komieteta Ministerstva Zemledělija = Soobshcheniya Byuro po Chastnomu Rastenievodstvu Uchenago Komieteta Ministerstva Zemleděliya. Petrograd.

Soobščenija Dal'nevostočnogo filiala imeni V. L. Komarova Akademii nauk S S S R = Soobshcheniya Dal'nevostochnogo filiala imeni V. L. Komarova Akademii nauk S S S R. Vladivostok.

Soobščenija Gruzinskogo filiala Akademii nauk S S S R = Soobshcheniya Gruzinskogo filiala Akademii nauk S S S R. Tiflis.

Soobščenija Otdela rastenievodstva Sel'skohozjajstvennogo učenogo komiteta = Soobshcheniya Otdela rastenievodstva Sel'skokhozyaistvennogo uchenogo komiteta. Petrograd.

Soobščenija Sahalinskogo filiala Akademii nauk S S S R = Soobshcheniya Sakhalinskogo filiala Akademii nauk S S S R. Yuzhno-Sakhalinsk.

Soobščenija. Sahalinskij kompleksnyj naučno-issledovatel'skij institut = Soobshcheniya Sakhalinskogo kompleksnogo nauchno-issledovatel'skogo instituta. Yuzhno-Sakhalinsk.

Soobščenija Sahalinskogo kompleksnogo naučno-issledovatel'skogo instituta = Soobshcheniya Sakhalinskogo kompleksnogo nauchno-issledovatel'skogo instituta. Yuzhno-Sakhalinsk.

Soobshcheniya Dal'nevostochnogo filiala Akademii nauk S S S R = Soobshcheniya Dal'nevostochnogo filiala imeni V. L. Komarova Akademii nauk S S S R.

Soobshcheniya Akademii nauk Gruzinskoi S S R. Tiflis, Georgian S S R. Vol. 2+, 1941+. Soobshch. Akad. Nauk Gruzinsk. S.S.R. Preceded by: Soobshcheniya Gruzinskogo filiala Akademii nauk S S S R. 1-110-1. HI 72873

Soobshcheniya Byuro po Chastnomu Rastenievodstvu Uchenago Komiteta Glavnago Upravleniya Zemleustroistva i Zemleděliya. St. Petersburg. 1914-15. Soobshch. Byuro Chastn. Rasteniev. Uchen. Komiteta Glavn. Upravl. Zemleustr. Superseded by: Soobshcheniya Byuro po Chastnomu Rastenievodstvu Uchenago Komieteta Ministerstva Zemleděliya. HI 72874

Soobshcheniya Byuro po Chastnomu Rastenievodstvu Uchenago Komieteta Ministerstva Zemleděliya. Petrograd. 1916. Soobshch. Byuro Chastn. Rasteniev. Uchen. Komiteta Minist. Zemled. Preceded by: Soobshcheniya Byuro po Chastnomu Rastenievodstvu Uchenago Komiteta Glavnago Upravleniya Zemleustroistva i Zemleděliya. Superseded by: Soobshcheniya Otdela rastenievodstva Sel'skokhozyaistvennogo uchenogo komiteta. HI 72875

Soobshcheniya Dal'nevostochnogo filiala imeni V. L. Komarova Akademii nauk S S S R. Vladivostok, Russian S F S R. Vol. 1+, 1950+. Soobshch. Dal'nevost. Fil. Komarova Akad. Nauk S.S.S.R. HI 72876

Soobshcheniya Gruzinskogo filiala Akademii nauk S S S R. Tiflis, Georgian S S R. Vol. 1, 1940. Soobshch. Gruzinsk. Fil. Akad. Nauk S.S.S.R. Superseded by: Soobshcheniya Akademii nauk Gruzinskoi S S R and Sakharthwelos S S R mecnierebatha akademiis moambe. 1-110-1. HI 72877

Soobshcheniya instituta lesa. Moscow. Vols. 1-13, 1953-59. Soobshch. Inst. Lesa. HI 72878

Soobshcheniya laboratorii lesovedeniya. Moscow. Vol. 1, 1959. Soobshch. Lab. Lesoved. HI 72879

Soobshcheniya Otdela rastenievodstva Sel'skokhozyaistvennogo uchenogo komiteta. Petrograd. Vols. 1-3, 1918-22. Soobshch. Otd. Rasteniev. Sel'skokhoz. Uchen. Komiteta. Preceded by: Soobshcheniya Byuro po Chastnomu Rastenievodstvu Uchenago Komieteta Ministerstva Zemleděliya. HI 72880

Soobshcheniya Sakhalinskogo filiala Akademii nauk S S S R. Yuzhno-Sakhalinsk, Russian S F S R. Vols. 1-3, 1954-56. Soobshch. Sakhalinsk. Fil. Akad. Nauk S.S.S.R. Superseded by: Soobshcheniya Sakhalinskogo kompleksnogo nauchno-issledovatel'skogo instituta. HI 72881

Soobshcheniya Sakhalinskogo kompleksnogo nauchno-issledovatel'skogo instituta. Yuzhno-Sakhalinsk, Russian S F S R. Vols. 4-8, 1956-59. Soobshch. Sakhalinsk. Kompl. Nauchno-Issl. Inst. Preceded by: Soobshcheniya Sakhalinskogo filiala Akademii nauk S S S R. Superseded by: Trudy. Sakhalinskii kompleksnyi nauchno-issledovatel'skii institut. HI 72882

Soproni szemle. Sopron, Hungary. Soproni Szemle. See B–P–H 847/12. HI 59492

Sorbifolia. Helsinki. Vol. 13+, 1982+. Sorbifolia. Preceded by: Dendrologian seuran tiedotuksia. HI 72883

Sorby record; journal of the Sorby natural history society of Sheffield. Sheffield. Vol. 1+, 1958+. Sorby Rec. HI 72884

Sorghum and millets abstracts. Farnham Royal. Vol. 1+, 1976+. Sorg. Millets Abstr. HI 72885

Sorghum newsletter. Fort Hays, KS, Tucson, AZ. Vol. 1+, 1958+. Sorg. Newslett. HI 72886

Sörmländska Handlingar. Nyköping. 1936+. Sörmländska Handl. HI 72887

Sorter af landbrugsplanter. [Denmark.] 1975+. Sorter Landbrugspl. HI 72889

Sorui = Bulletin of the japanese society of phycology. Sapporo.

Sorui = Japanese journal of phycology. Tokyo.

Sosai = Vegetables. Okayama, Japan. Vegetables. See B–P–H 948/17.

Sosai kaki shiken kenkyu nenpo = Report (Annual) of vegetables and flowers research works in Japan. Tokyo.

Sotsialisticheskoe plodoovoshchnoe khozyaistvo. Moscow. 1931(4)-1932(6). Sotsialist. Plodoovoshchn. Khoz. Preceded by: Sad i ogorod. Superseded by: Plodoovoshchnoe khozyaistvo. HI 72890

Soul taehakkyo nongkwa taehak yonsumnim yongu pogo = Bulletin, Seoul national university forests. Seoul.

South african biological society pamphlet. Pretoria. Nos. 1-21, 1931-59. S. African Biol. Soc. Pam. Preceded by: South african journal of natural history. Superseded by: Journal of the south african biological society. 5-4000-1. HI 72891

South african citrus journal. Nos. 1-450, 1925-71. S. African Citrus J. Superseded by: Citrus and sub-tropical fruit journal. HI 72892

South african country life. Cape Town. S. African Country Life. See B–P–H 807/25. HI 59064

South african digest. Pretoria. Vol. 9+, 1962+. S. African Digest. Preceded by: Fortnightly digest of south african affairs, comment and opinion. HI 72893

South african forestry journal; journal of the south african forestry association. Pretoria. No. 41+, 1962+. S. African Forest. J. Preceded by: Journal of the south african forestry association. HI 72894

South african forum botanicum. Kirstenbosch, Grahamstown. Vols. 1-7, 1962-69. S. African Forum Bot. Superseded by: Forum botanicum. HI 72895

South african gardening. Cape Town. S. African Gard. See B–P–H 807/26. HI 59065

South african horticultural journal. Johannesburg. Vols. 1-3(1), 1938-40. S. African Hort. J. HI 72896

South african journal of agricultural science. Pretoria. Vols. 1-11, 1958-68. S. African J. Agric. Sci. Superseded by: Agroplantae, and Phytophylactica and Agroanimalia and Agrochemophysica [not entered]. HI 72897

South african journal of antarctic research. Pretoria. No. 1+, 1971+. S. African J. Antarc. Res. HI 72898

South african journal of botany. Pretoria. Vols. 1-3, 1982-84; vol. 51+, 1985+ [absorbed Journal of of south african botany and continued its volume numbering]. S. African J. Bot. HI 72899

South african journal of marine science. Cape Town. No. 1+, 1983+. S. African J. Mar. Sci. HI 72900

South african journal of natural history. Pretoria. Vols. 1-6, 1918-30. S. African J. Nat. Hist. Superseded by: South african biological society pamphlet. 5-4001-3. HI 72901

South african journal of plant and soil. Pretoria. Vol. 1+, 1984+. S. African J. Pl. Soil. HI 72902

South african journal of science. Johannesburg. S. African J. Sci. See B–P–H 807/28. HI 59066

South african medical journal. Cape Town. S. African Med. J. See B–P–H 807/30. HI 59067

South african medical record. Cape Town. S. African Med. Rec. See B–P–H 807/31. HI 59068

South african orchid journal. Johannesburg. Vol. 1+, 1970+. S. African Orchid J. HI 72903

South african plant variety journal. Pretoria. No. 1+, 1979+. S. African Pl. Var. J. HI 72904

South african quarterly journal. Cape Town. S. African Quart. J. See B–P–H 807/32. HI 59069

South african report of science. Johannesburg. S. African Rep. Sci. See B–P–H 807/33. HI 59070

South african science. Johannesburg. Vols. 1-2, 1947-48. S. African Sci. Incorporated in: South african journal of science. HI 72905

South african sugar journal. Durban, South Africa. S. African Sugar J. See B–P–H 808/2. HI 59071

South african sugar year book. Durban. 1930+. S. African Sugar Year Book. HI 72906

South australian naturalist. Adelaide, S.A. S. Austral. Naturalist. See B–P–H 808/5. HI 59072

South China agricultural science. [Hua nan nung yüeh k'o hsüeh.] Canton. S. China Agric. Sci. See B–P–H 808/16. HI 59077

South Dakota agricultural experiment station. Agronomy department pamphlet. Brookings, SD. South Dakota Agric. Exp. Sta. Agron. Dept. Pam. See B–P–H 847/24. HI 59493

South Dakota agricultural experiment station. Annual report. Brookings, SD. South Dakota Agric. Exp. Sta. Annual Rep. See B–P–H 847/25. HI 59494

South Dakota agricultural experiment station. Bulletin. Brookings, SD. South Dakota Agric. Exp. Sta. Bull. See B–P–H 848/1. HI 59495

South Dakota agricultural experiment station. Circular. Brookings, SD. South Dakota Agric. Exp. Sta. Circ. See B–P–H 848/2. HI 59496

South Dakota agricultural experiment station. Plant pathology department pamphlet. Brookings, SD.

South Dakota Agric. Exp. Sta. Pl. Pathol. Dept. Pam. See B–P–H 848/3. HI 59497

South Dakota agricultural experiment station. Technical bulletin. Brookings, SD. South Dakota Agric. Exp. Sta. Techn. Bull. See B–P–H 848/4. HI 59498

South eastern naturalist; journal of the associated natural history societies of the south east of England. Canterbury. Vols. 1-2(2), 1890-99. S. E. Naturalist. 5-4010-2. HI 72907

South eastern naturalist; transactions of the south-eastern union of scientific societies. London. 1900-27. S. E. Naturalist. Preceded by: Report and transactions of the south eastern union of scientific societies. Superseded by: South-eastern naturalist and antiquary. HI 72908

South-eastern naturalist and antiquary. London. 1928+. S.-E. Naturalist Antiq. Preceded by: South eastern naturalist. London. HI 72909

South Essex naturalist. Westcliff-on-Sea. Nos. 1+, 1951+. S. Essex Naturalist. HI 72910

South indian horticulture. Coimbatore. Vol. 1+, 1953+. S. Indian Hort. HI 72911

South Pacific bulletin. Nouméa, Sydney, N.S.W. Vol. 10+, 1960+. S. Pacific Bull. Preceded by: Quarterly bulletin, south Pacific commission. HI 72912

Southeast asian studies. [Tonan ajia kenkyu.] Kyoto. Vol. 1+, 1963+. SouthE. Asian Stud. HI 72913

Southern beekeeper. Hapeville, GA. S. Beekeeper. See B–P–H 808/9. HI 59073

Southern California crops. Los Angeles, CA. S. Calif. Crops. See B–P–H 808/10. HI 59074

Southern California horticulturist. Los Angeles, CA. Vols. 1-2, 1877-79. S. Calif. Hort. Superseded by: Semi-tropical California. 5-3847-3. HI 72914

Southern California practitioner. Los Angeles, CA. S. Calif. Practitioner. See B–P–H 808/11. HI 59075

Southern California turf grass culture. Los Angeles, CA. S. Calif. Turf Grass Cult. See B–P–H 808/12. HI 59076

Southern China and the south seas. [Nanshi nanyo.] S. China S. Seas. See B–P–H 808/17. HI 59078

Southern florist = Southern florist and nurseryman. Fort Worth, TX.

Southern florist and nurseryman. Fort Worth, TX. Vols. 1-96?, 1915-83 [suspended 1918-20]. S. Florist Nurseryman. Superseded by: S F and N. 5-4016-2. HI 72915

Southern forestry notes, southern forest experiment station. New Orleans, LA. Nos. 1-142, 1933?-62. S. Forest. Notes S. Forest Exp. Sta. Superseded by: Research note S O, United States forest service. HI 72916

Southern garden. New Orleans, LA. S. Gard. (New Orleans). See B–P–H 808/26. HI 59079

Southern garden. Raleigh, NC. S. Gard. (Raleigh). See B–P–H 808/27. HI 59080

Southern home and garden. Raleigh, NC. S. Home Gard. See B–P–H 808/28. HI 59081

Southern journal of applied forestry. Washington, D.C. Vol. 1+, 1977+. S. J. Appl. Forest. HI 72917

Southern journal of the medical and physical sciences. Nashville & Knoxville, TN. S. J. Med. Phys. Sci. See B–P–H 809/1. HI 59082

Southern life, home and garden magazine. Raleigh, NC. S. Life Home Gard. Mag. See B–P–H 809/2. HI 59083

Southern medical journal. La Grange, NC. S. Med. J. (La Grange). See B–P–H 809/4. HI 59084

Southern medical journal. Nashville, TN. S. Med. J. (Nashville). See B–P–H 809/5. HI 59085

Southern science record. Melbourne, Vic. Vols. 1-3, 1880-83; n.s. vols. 1-2(1), 1885-86. S. Sci. Rec. 5-4024-3. HI 72918

Southern science record and magazine of natural history = Southern science record. Melbourne, Vic.

Southern spine; newsletter of the associated southern clubs of the cactus and succulent society of New Zealand. Oamaru. Vol. 1+, 1959+. S. Spine. HI 72919

Southwest agricultural science. [Hsi nan nung yeh k'o hsüeh.] Chungking, China. SouthW. Agric. Sci. See B–P–H 848/15. HI 59499

Southwest orchid review. Vol. ?-4+, ?-1963+. SouthW. Orchid Rev. HI 72920

Southwest science bulletin. Los Angeles, CA. SouthW. Sci. Bull. See B–P–H 848/23. HI 59505

Southwestern crop and stock. Lubbock, TX. SouthW. Crop Stock. See B–P–H 848/16. HI 59500

Southwestern journal; a magazine of science, literature and miscellany. Jefferson college and Washington lyceum. Natchez, MS. SouthW. J. See B–P–H 848/17. HI 59501

Southwestern journal of anthropology; university of New Mexico. Albuquerque, NM. SouthW. J. Anthropol. See B–P–H 848/18. HI 59502

Southwestern Louisiana journal. Lafayette, LA. Vols. 1-7(4), 1957-67. SouthW. Louisiana J. HI 72921

Southwestern medicine. Las Cruces, NM. SouthW. Med. See B–P–H 848/20. HI 59503

Southwestern naturalist. Dallas, TX. SouthW. Naturalist. See B–P–H 848/22. HI 59504

Sovetskaja botanika = Sovetskaya botanika. Moscow &

Leningrad.

Sovetskaya botanika. Moscow & Leningrad. 1933-47. Sovetsk. Bot. Preceded by: Izvestiya Botanicheskogo sada Akademii nauk S S S R. 5-4031-2. HI 72922

Sovětske zemědělství. Prague. Vols. 1-?, 1951-56. Sovětsk. Zeměd. Superseded by: Zemědělství v zahraniči. HI 72923

Sovetskie subtropiki. Sukhumi, Georgian S S R. Sovetsk. Subtrop. (Sukhumi). See B–P–H 848/29. HI 59506

Soviet agricultural biology. Part 1, plant biology. [Translation of: Sel'skokhozyaistvennaya biologiya.] New York. Vol. 1+, 1986+. Soviet Agric. Biol., Part 1, Pl. Biol. HI 72924

Soviet agricultural sciences. [Translation of: Doklady vsesoyuznoi ordena Lenina akademii sel'skokhozyaistvennykh nauk im V. I. Lenina.] New York. 1976+. Soviet Agric. Sci. HI 72925

Soviet biotechnology. [Translation of: Biotekhnologiya.] New York. 1986+. Soviet Biotechnol. HI 72926

Soviet forest sciences. [Translation of: Lesovedenie.] New York. ?-1987+. Soviet Forest Sci. HI 72927

Soviet genetics. [Translation of: Genetika.] New York. Vol. 2+, 1966+. Soviet Genet. Preceded by: Genetics. HI 72928

Soviet geography: review and translation. New York, NY. Soviet Geogr. See B–P–H 849/2. HI 59507

Soviet journal of bioorganic chemistry. New York. Vol. 1+, 1975+. Soviet J. Bioorg. Chem. HI 75199

Soviet journal of developmental biology. New York. Vol. 1+, 1970+. Soviet J. Developm. Biol. HI 75165

Soviet journal of ecology. [Translation of: Ekologiya.] New York. Vol. 2+, 1972+. Soviet J. Ecol. Preceded by: Ecology. HI 72929

Soviet journal of marine biology. [Translation of: Biologiya moriya.] New York. Vol. 1+, 1975+. Soviet J. Mar. Biol. HI 72930

Soviet oceanography; transactions of the marine hydrophysical institute, academy of sciences of the Ukrainian S S R and Oceanology sections, Doklady of the academy of sciences of the U S S R. Washington, DC. 1964. Soviet Oceanogr. Superseded by: Oceanology. HI 72931

Soviet plant industry record = Vestnik sotsialisticheskogo rastenievodstva. Moscow.

Soviet plant physiology. Washington, DC. Soviet Pl. Physiol. See B–P–H 849/3. HI 59508

Soviet progress in virology. [Translation of: Voprosy virusologii.] New York. No. 1+, 1980+. Soviet Progr. Virol. HI 72932

Soviet science review. Guildford. Vols. 1-3, 1970-72. Soviet Sci. Rev. HI 72933

Soviet soil science. Washington, DC. Soviet Soil Sci. See B–P–H 849/4. HI 59509

Soviet scientific reviews. Section D, biology reviews: Physico-chemical aspects. Chur, New York. Vol. 1+, 1980+. Soviet Sci. Rev., D. HI 72934

Soviet subtropics = Sovetskie subtropiki. Sukhumi, Georgian S S R. Sovetsk. Subtrop. (Sukhumi). See B–P–H 848/29.

Sovietskaia botanika = Sovetskaya botanika. Moscow & Leningrad.

Sowjetwissenschaft. Berlin. Sowjetwissenschaft. See B–P–H 849/9. HI 59512

Sowjetwissenschaft. Naturwissenschaftliche Abteilung. Berlin. Sowjetwiss., Naturwiss. Abt. See B–P–H 849/7. HI 59510

Sowjetwissenschaft. Naturwissenschaftliche Beiträge. Berlin. Sowjetwiss., Naturwiss. Beitr. See B–P–H 849/8. HI 59511

Soyabean abstracts. Farnham Royal. Vol. 1+, 1978+. Soyabean Abstr. HI 72935

Soybean digest. Hudson, IA. Soybean Digest. See B–P–H 849/11. HI 59513

Soybean genetics newsletter. Ames, IA. Vol. 1+, 1974+. Soybean Genet Newslett. HI 72936

Soybean news. Urbana, IL. 1949+. Soybean News. HI 72937

Soybean rust newsletter. Tainan. Vol. 1+, 1977+. Soybean Rust Newslett. HI 72938

Sozialistische Forstwirtschaft. Berlin. Vol. 12+, 1962+. Sozialist. Forstw. Preceded by: Forst und Jagd. HI 72939

Space life sciences; international journal of space biology and medicine. Dordrecht. Vols. 1-4, 1968-73. Space Life Sci. Superseded by: Origins of life. HI 72940

Span; Shell public health and agricultural news. London. Span. See B–P–H 849/12. HI 59514

Spasmodic monthly. Barstow, CA. Nos. 1-7, 1965-71. Spasmodic Monthly. HI 72941

Spawn. Richmond, CA. ?-1984+. Spawn. HI 72942

Specchio delle scienze o giornale enciclopedico di Sicilia ... Palermo. Specchio Sci. See B–P–H 850/7. HI 59515

Special bulletin: Agricultural experiment station, government of Formosa. [Noji shiken-jo tokubetsu hokoku.] Taihoku [=Taipei, Taiwan]. Special Bull. Agric. Exp. Sta. Gov. Formosa. See B–P–H 850/8. HI 59516

Special bulletin of the Aichi-ken agricultural research

center. [Aichi-ken nogyo sogo shikenjo tokubetsu hokoku.] Nagakute. 1979+. Special Bull. Aichi-ken Agric. Res. Center. HI 72943

Special bulletin, college of agriculture, Utsunomiya university. [Utsunomiya daigaku nogakubu gakujutsu hokoku tokushu.] Utsunomiya, Japan. Special Bull. Coll. Agric. Utsunomiya Univ. See B–P–H 850/10. HI 59517

Special bulletin of the forest products laboratory. Kaiting, China. Special Bull. Forest Prod. Lab. See B–P–H 850/11. HI 59518

Special bulletin, horticultural station, Tokai Kinki agricultural experiment staion. [Tokai kinki nogyo shikenjo engei-bu tokubetsu hokoku.] Shizuoka, Japan. Special Bull. Hort. Sta. Tokai Kinki Agric. Exp. Sta. See B–P–H 850/12. HI 59519

Special bulletin of the Ishikawa-ken agricultural experiment station. [Ishikawa-ken nogyo shikenjo, tokubetsu kenkyu hokoku.] Nonoichi. No. 1+, 1975+. Special Bull. Ishikawa-ken Agric. Exp. Sta. HI 72944

Special bulletin of the Kanagawa prefectural horticultural experiment station. [Kanagawa-ken engei shikenjo.] Ninomiya-machi. Vol. 1+, 1978+. Special Bull. Kanagawa Pref. Hort. Exp. Sta. HI 72945

Special bulletin, Maryland agricultural experiment station = Maryland agricultural experiment station. Special bulletin. College park, MD. Maryland Agric. Exp. Sta. Special Bull. See B–P–H 549/5.

Special bulletin, Michigan agricultural experiment station = Michigan agricultural experiment station. Special bulletin. East Lansing, MI. Michigan Agric. Exp. Sta. Special Bull. See B–P–H 592/18.

Special bulletin, middle east biological scheme. Cairo. Nos. ?-7-14-?, ?-1943-46-? Special Bull. Middle E. Biol. Scheme. HI 72946

Special bulletin, North Carolina agricultural experiment station = North Carolina agricultural experiment station special bulletin. Raleigh, NC. North Carolina Agric. Exp. Sta. Special Bull. See B–P–H 663/12.

Special bulletin, North Dakota agricultural experiment station = North Dakota agricultural experiment station special bulletin. Fargo, ND. North Dakota Agric. Exp. Sta. Special Bull. See B–P–H 664/16.

Special bulletin. Okayama prefectural agricultural experiment station. [Okayama kenritsu noji shikenjo rinji hokoku.] Okayama, Japan. Special Bull. Okayama Prefect. Agric. Exp. Sta. See B–P–H 850/17. HI 59520

Special bulletin of the royal botanical gardens. Hamilton, Ont. Special Bull. Roy. Bot. Gard. See B–P–H 850/18. HI 72947

Special bulletin of Shizuoka prefectural citrus experiment station. [Shizuoka-ken kankitsu shikenjo tokubetsu hokoku.] Shizuoka. No. 1+, 1967+. Special Bull. Shizuoka Pref. Citrus Exp. Sta. HI 75152

Special bulletin. South China botanical institute. Academia sinica. Series C. [Chung kuo k'o hsüeh yuan. Hua nan chih we yen chiu so, ping chung chuan k'an, ti i hao.] [China]. Special Bull. S. China Bot. Inst., Ser. C. See B–P–H 850/19. HI 59522

Special bulletin of the Tokai-kinki national agricultural experiment station. [Norin-sho tokai kinki nogyo shikenjo tokubetsu hokoku.] Mie, Japan. Special Bull. Tokai-Kinki Natl. Agric. Exp. Sta. See B–P–H 850/20. HI 59523

Special bulletin of the Tottori agricultural experiment station. [Tottori-ken nogyo shikenjo. Tokubetsu kenkyu hokoku.] Tottori. No. 1+, 1971+. Special Bull. Tottori Agric. Exp. Sta. HI 72948

Special bulletin of the Utsunomiya tobacco experiment station. [Utsunomiya tabako shikenjo tokubetsu hokoku.] Oyama. No. 1+, 1979+. Special Bull. Utsunomiya Tobacco Exp. Sta. HI 72949

Special bulletin, Wisconsin (state) agricultural experiment station = Wisconsin (state) agricultural experiment station. Special bulletin. Madison, WI. Wisconsin Agric. Exp. Sta. Special Bull. See B–P–H 976/5.

Special bulletin of the Yamaguchi agricultural experiment station. [Yamaguchi-ken nogyo shikenjo tokubetsu kenkyu hokoku.] Yamaguchi, Japan. Special Bull. Yamaguchi Agric. Exp. Sta. See B–P–H 850/23. HI 59524

Special circular, Clemson agricultural college of South Carolina. Agricultural experiment station = Clemson agricultural college of South Carolina. Agricultural experiment station. Special circular. Clemson, SC. Clemson Agric. Exp. Sta. Special Circ. See B–P–H 315/3.

Special circular, experimental farms service, Canada. Ottawa. Nos. 1-20, 1917-18. Special Circ. Exp. Farms Serv. Canada. HI 72950

Special circular, Montana college of agriculture and mechanic arts. Agricultural experiment station = Montana college of agriculture and mechanic arts. Agricultural experiment station. Special circular. Bozeman, MT. Montana Agric. Exp. Sta. Special Circ. See B–P–H 618/17.

Special issue. Sylviculture and education. [Taiwan forest experiment station.] [Lin yeh t'ui kuang tsuan k'an.] Taipei, Taiwan. Special Issue Sylvic. Educ. See B–P–H 851/2. HI 59525

Special number of the educational miscellany of Hunan. [Hu-nan chiao-yü tsa-chih.] Special Number Educ. Misc. Hunan. See B–P–H 851/3. HI 59526

Special papers of the geological society of America. Washington, DC. Special Pap. Geol. Soc. Amer. See B–P–H 851/4. HI 59527

Special papers, Ohio state academy of sciences. Columbus, OH. Special Pap. Ohio State Acad. Sci. See B–P–H 851/5. HI 59528

Special publications, academy of natural sciences of Philadelphia. Philadelphia, PA. No. 1+, 1922+. Special Publ. Acad. Nat. Sci. Philadelphia. 1-37-1. HI 72951

Special publications, agronomy society of New Zealand. Christchurch, N.Z. No. 1+, 1982+. Special Publ. Agron. Soc. New Zealand. HI 72952

Special publications, american geographical society. New York. No. 1+, 1915+. Special Publ. Amer. Geogr. Soc. HI 72953

Special publications; american society of limnology and oceanography. Ann Arbor, MI. Special Publ. Amer. Soc. Limnol. See B–P–H 851/6. HI 59529

Special publication, australian conservation foundation. [Canberra, A.C.T.]. No. 1+, [1968]+. Special Publ. Austral. Conservation Found. HI 72954

Special publications, australian national parks and wildlife service. Canberra, A.C.T. Vol. 1+, 1979+. Special Publ. Austral. Natl. Parks Wildlife Serv. HI 72955

Special publications; Bernice Pauahi Bishop museum. Honolulu, HI. Special Publ. Bernice Pauahi Bishop Mus. See B–P–H 851/7. HI 59530

Special publications in biology and geology. Milwaukee, WI. No. 1+, 1974+. Special Publ. Biol. Geol. HI 72956

Special publications, Birbal Sahni institute of palaeobotany. Lucknow. Nos. 1-5, 1974. Special Publ. Birbal Sahni Inst. Palaeobot. HI 72957

Special publications, botanical society of Bengal. Calcutta. 1950+. Special Publ. Bot. Soc. Bengal. HI 72958

Special publications of the British Columbia provincial museum of natural history and anthropology. Vancouver. Special Publ. British Columbia Prov. Mus. Nat. Hist. See B–P–H 851/8. HI 59531

Special publications, british ecological society. Oxford. No. 1+, 1982+. Special Publ. Brit. Ecol. Soc. HI 72959

Special publication, british pteridological society. London. No. 1+, 1984+. Special Publ. Brit. Pteridol. Soc. HI 72960

Special publications, California native plant society. Berkeley, CA. No. 1+, 1981+. Special Publ. Calif. Native Pl. Soc. HI 72961

Special publications, Carnegie museum of natural history. Pittsburgh, PA. No. 1+, 1975+. Special Publ. Carnegie Mus. Nat. Hist. HI 72962

Special publications of the Chicago academy of sciences. Chicago, IL. Special Publ. Chicago Acad. Sci. See B–P–H 851/9. HI 59532

Special publications, college of agriculture, national Taiwan university. Taipei. No. 1+, 1955+. Special Publ. Coll. Agric. Natl. Taiwan Univ. HI 72963

Special publications, college of agriculture, university of Illinois at Urbana-Champaign. Urbana, IL. No. 1+, 1960+. Special Publ. Coll. Agric. Univ. Illinois Urbana-Champaign. HI 72964

Special publication, commonwealth bureau of soils. Farnham Royal. No. 1+, 1975+. Special Publ. Commonw. Bur. Soils. HI 72965

Special publications, Illinois natural history survey. Carbondale, IL. No. 1+, 1976+. Special Publ. Illinois Nat. Hist. Surv. HI 72966

Special publications of the Indo-Pacific fisheries council. Bangkok. Special Publ. Indo-Pacific Fish. Council. See B–P–H 851/10. HI 59533

Special publication, institute of oceanographic and fisheries research = Eidike ekdose, institouton okeanographikon kai halieutikon ereunon. Athens.

Special publications, institute of oceanography, national Taiwan university. Taipei. 1972+. Special Publ. Inst. Oceanogr. Natl. Taiwan Univ. HI 72967

Special publication of the laboratory of palaeobotany and palynology. Amsterdam. No. 1+, 1979+. Special Publ. Lab. Palaeobot. Palynol. HI 72968

Special publications, limnological society of America. St. Louis, MO. Special Publ. Limnol. Soc. Amer. See B–P–H 851/11. HI 59534

Special publications. Lingnan natural history survey and museum; Lingnan university. Canton. Special Publ. Lingnan Nat. Hist. Surv. Mus. See B–P–H 851/12. HI 59535

Special publication of the macedonian museum of natural history = Posebno izdanie. Musei macedonici scientiarum naturalium. Skoplje.

Special publications, museum of natural history, university of Oregon. Eugene, OR. 1975+. Special Publ. Mus. Nat. Hist. Univ. Oregon. HI 72969

Special publications, museum, Texas tech university. Lubbock, TX. No. 1+, 1972+. Special Publ. Mus. Texas Tech Univ. HI 72970

Special publications of the national agricultural research bureau. [Shih yeh pu chung yang nung yeh shih yen so k'an mu lu.] Nanking. Special Publ. Natl. Agric. Res. Bur. See B–P–H 851/13. HI 59536

Special publications, national museum of natural science, Taiwan. Taichung. No. 1+, 1986+. Special Publ. Natl. Mus. Nat. Sci. Taiwan. HI 72971

Special publication, North Carolina agricultural experiment station = North Carolina agricultural experiment station special publication. Raleigh, NC. North Carolina Agric. Exp. Sta. Special Publ. See B–P–H 663/13.

Special publications, Osaka municipal museum of natural history. Osaka. Vol. 1+, 1969+. Special Publ. Osaka Munic. Mus. Nat. Hist. HI 72972

Special publications, Ottawa field naturalists' club. Ottawa. No. 1+, 1979+. Special Publ. Ottawa Field Naturalists' Club. HI 72973

Special publications, Pymatuning laboratory of ecology = Special publications, Pymatuning laboratory of field biology. Pittsburgh, PA.

Special publications, Pymatuning laboratory of field biology. Pittsburgh, PA. No. 1+, 1956+. Special Publ. Pymatuning Lab. Field Biol. HI 72974

Special publications series of the british ecological society = Special publications, british ecological society.

Special publications series, Fiji museum. Suva. No. 1+, 1970+. Special Publ. Ser. Fiji Mus. HI 72976

Special publications from the Seto marine biological laboratory. [Nos. 1-17, 1959-62 comprised: Biological results of the japanese Antarctic research expedition]. Sirahama. No. 1+, 1959+. Special Publ. Seto Mar. Biol. Lab. HI 72977

Special publications, society for the bibliography of natural history. London. No. 1+, 1981+. Special Publ. Soc. Bibliogr. Nat. Hist. HI 72978

Special publications, South West Africa scientific society. Windhoek. 1964+. Special Publ. S. W. Africa Sci. Soc. HI 72979

Special publication, university of Kansas museum of natural history. Lawrence, KS. No. 1+, 1976+. Special Publ. Univ. Kansas Mus. Nat. Hist. HI 72980

Special publication, Utrecht micropalaeontological bulletins. Utrecht. No. 1+, 1974+. Special Publ. Utrecht Micropalaeontol. Bull. HI 72981

Special report, agricultural experiment station, New York state. Geneva, N.Y. No. 1+, 1970+. Special Rep. Agric. Exp. Sta. New York State. HI 72982

Special report, Arizona agricultural experiment station = University of Arizona. Arizona agricultural experiment station. Special report. Tucson, AZ. Univ. Arizona Agric. Exp. Sta. Special Rep. See B–P–H 933/17.

Special report of the forest experiment station. Office of industry, government general of Taiwan. [Taiwan sotokufu shokusan kyoku ringyo shikenjo tokubetsu hokoku.] Taipei, Taiwan. Special Rep. Forest Exp. Sta. See B–P–H 851/17. HI 59537

Special report, forest products research. London. Nos. 1-6, 1927-45. Special Rep. Forest Prod. Res. HI 72983

Special report, Formosa agricultural experiment station. Taihoku. 1910-17. Special Rep. Formosa Agric. Exp. Sta. HI 72984

Special report, Idaho agricultural experiment station = University of Idaho, college of agriculture, Idaho agricultural experiment station, special reports. Moscow, ID. Univ. Idaho Coll. Agric. Idaho Agric. Exp. Sta. Special Rep. See B–P–H 936/6.

Special report of the imperial agricultural experiment station. [Noji shikenjo tokubetsu hokoku.] Tokyo. Special Rep. Imp. Agric. Exp. Sta. See B–P–H 851/18. HI 59538

Special report, Iowa agricultural experiment station = Iowa agricultural experiment station. Special report. Ames, IA. Iowa Agric. Exp. Sta. Special Rep. See B–P–H 436/36.

Special report, Maryland agricultural experiment station = Maryland agricultural experiment station. Special report. College Park, MD. Maryland Agric. Exp. Sta. Special Rep. See B–P–H 549/6.

Special report, Missouri agricultural experiment station = Missouri agricultural experiment station. Special report. Columbia, MO.

Special report, Montana state college. Agricultural experiment station = Montana state college. Agricultural experiment station. Special report. Bozeman, MT. Montana Agric. Exp. Sta. Special Rep. See B–P–H 618/18.

Special report, New York agricultural experiment station. Geneva, NY. No. 1+, 1970+. Special Rep. New York Agric. Exp. Sta. HI 72985

Special report of the Ohara institute for agricultural research. [Ohara nogyo kenkyu-sho tokubetsu hokoku.] Okayama, Japan. Special Rep. Ohara Inst. Agric. Res. See B–P–H 852/1. HI 59539

Special report, U S department of agriculture = Department of agriculture. Special report. Washington, DC. Dept. Agric. Special Rep. See B–P–H 343/16.

Special report, Utah agricultural college, agricultural experiment station = Utah agricultural college, agricultural experiment station. Special report. Logan, UT. Utah Agric. Exp. Sta. Special Rep. See B–P–H 945/34.

Special report series, Ohio agricultural experiment station = Ohio agricultural experiment station special report series. Wooster, OH. Ohio Agric. Exp. Sta. Special Rep. Ser. See B–P–H 685/3.

Special research bulletin of the Aichi-ken agricultural research center = Special bulletin of the Aichi-ken agricultural research center. Nagakute.

Special research bulletin. North China agricultural science research institute. [Hua pei nung yeh k'o hsüeh yen chiu so. Yen chu tsuan k'an.] Special Res. Bull. N. China Agric. Sci. Res. inst. See B–P–H 852/5. HI 59540

Special scientific report, Florida department of natural resources marine laboratory. St. Petersburg, FL. Nos. 23-40, 1969-73. Special Sci. Rep. Florida Dept. Nat. Resources Mar. Lab. Preceded by: Special scientific report, Florida state board of conservation marine laboratory. Superseded by: Florida marine research publications. HI 72986

Special scientific report, Florida state board of conservation, marine laboratory. St. Petersburg, FL. Nos. 1-22, 1959-69. Special Sci. Rep. Florida State Board Conservation Mar. Lab. Superseded by: Special scientific report, Florida department of natural resources, marine laboratory. HI 72987

Special scientific report, Virginia institute of marine science, Gloucester Point. Gloucester Point, VA. 1948+. Special Sci. Rep. Virginia Inst. Mar. Sci. Gloucester Point. HI 72988

Special seed bulletin, North Dakota agricultural experiment station = North Dakota agricultural experiment station special seed bulletin. Fargo, ND. North Dakota Agric. Exp. Sta. Special Seed Bull. See B–P–H 664/17.

Special series, Texas research foundation. Renner, TX. Nos. 1-5, 1951-68. Special Ser. Texas Res. Found. HI 72989

Special soils publication. [Tu jan t'e k'an.] Chungking, China. Special Soils Publ. See B–P–H 852/8. HI 59541

Special symposia, american society of limnology and oceanography. Lawrence, KS. Vol. 1+, 1971+, 1972+. Special Symp. Amer. Soc. Limnol. Oceanogr. HI 72990

Special volume, british bryological society. Cardiff. Vol. 1+, 1985+. Special Vol. Brit. Bryol. Soc. HI 72991

Speciella skrifta, sveriges lantbruksuniversitet. Uppsala. Vol. 1+, 1979+. Special Skrifta Sveriges Lantbruksuniv. HI 65439

Species; newsletter of the I U C N species survival commission. Gland. 198?+. Species. HI 65441

Species lupinorum. Saratoga, CA. Nos. 1-44, 1938-53. Spec. Lupinorum. HI 72992

Spectra; the international journal of computer applications in museums. East Winthrop, ME, Syracuse, NY. Vol. 1+, 1974+. Spectra (Syracuse). HI 75232

Spectrum. Berlin. Vol. 10+, 1979+. Spectrum. Preceded by: Spektrum. HI 72993

Speculations in science and technology. Lausanne, Kew. Vol. 1+, 1977+. Specul. Sci. Technol. HI 72994

Spektrum. Berlin. Vols. 1-9, 1970-78. Spektrum. Superseded by: Spectrum (Berlin). HI 72995

Spektrum der wissenschaften. [German language edition of: Scientific american.] Heidelberg. 1978+. Spektrum Wiss. HI 72996

Spices bulletin. Ernakulam, India. Spices Bull. See B–P–H 852/9. HI 59542

Spinal column; bulletin of the Detroit cactus and succulent society. Detroit, MI. 1946+ [suspended 1951-55]. Spinal Column (Detroit). HI 72997

Spinal column. Englewood, NJ. Vols. 1-7, 1961-68. Spinal Column (Englewood). Superseded by: Succulent journal. HI 72998

Spine; journal of the cactus and succulent society of Australia. Rosanna, Vic. Vol. 1, 1948; vol. 2+, 1953. Spine. For 1949-52 see: Bulletin of the cactus and succulent society of Australia. 5-4047-2. HI 72999

Spine spiel; Colorado's cactus journal and succulent bulletin. Denver, CO. Vols. 1-5, 1948-52. Spine Spiel. HI 73000

Spines and glochids. San Fernando, CA. Vols. 1-2, 1934-36. Spines & Glochids. HI 73001

Spinette. Melbourne, Vic. Vol. 1+, 1968+. Spinette. HI 73002

Spiny tips. Bedford, IN. Vols. 1-2, 1940-42. Spiny Tips. HI 73003

Spisanie na bulgarskata akademiya na naukite. Sofia. Nos. 1-71, 1911-50. Spis. Bulg. Akad. Nauk. Preceded by: Periodichesko spisanie na bulgarskoto knizhovno druzhestvo. Superseded by: Spisanie na bulgarskata akademiya na naukite. Kniga 2, klon prirodo-matematichen. 1-829-2. HI 73004

Spisanie na bulgarskata akademiya na naukite. Kniga 2, klon prirodo-matematichen. Sofia. Vol. [1]-[4], 5+, 1953+. Spis. Bulg. Akad. Nauk., Kniga 2, Prir.-Mat. Preceded by: Spisanie na bulgarskata akademiya na naukite. HI 73005

Spisok rastenii Gerbariya flory S S S R izdavaemogo Botanicheskim institutom Vsesoyuznogo akademii nauk. Leningrad. Vols. 9-10, 1932-36. Spisok Rast. Gerb. Fl. S.S.S.R. Bot. Inst. Vsesoyuzn. Akad. Nauk. Preceded by: Spisok rastenii Gerbariya Russkoi flory izdavaemogo Botanicheskim muzeem Rossiiskoi akademii nauk. 1-114-2. HI 73006

Spisok rastenii Gerbariya Russkoi flory izdavaemago Botanicheskim Muzeem Imperatorskoi Akademii Nauk. St. Petersburg. Vols. 1-7, 1898-1911. Spisok Rast.

Gerb. Russk. Fl. Bot. Muz. Imp. Akad. Nauk. Superseded by: Spisok rastenii Gerbariya Russkoi flory izdavaemogo Botanicheskim muzeem Rossiiskoi akademii nauk. 1-114-2. HI 73007

Spisok rastenii Gerbariya Russkoi flory izdavaemogo Botanicheskim muzeem Rossiiskoi akademii nauk. Petrograd. Vol. 8, 1922. Spisok Rast. Gerb. Russk. Fl. Bot. Muz. Rossiisk. Akad. Nauk. Preceded by: Spisok Rastenii Gerbariya Russkoi Flory izdavaemago Botanicheskim Muzeem Imperatorskoi Akademii Nauk. Superseded by: Spisok rastenii Gerbariya flory S S S R izdavaemogo Botanicheskim institutom Vsesoyuznogo akademii nauk. 1-114-2. HI 73008

Spisok rastenij Gerbarija flory S S S R izdavaemogo Botaničeskim institutom Vsesojuznogo akademii nauk = Spisok rastenii Gerbariya flory S S S R izdavaemogo Botanicheskim institutom Vsesoyuznogo akademii nauk. Leningrad.

Spisok Rastenij Gerbarija Russkoj Flory izdavaemago Botaničeskim Muzeem Imperatorskoj Akademii Nauk = Spisok Rastenii Gerbariya Russkoi Flory izdavaemago Botanicheskim Muzeem Imperatorskoi Akademii Nauk. St. Petersburg.

Spisok roslin gerbari flori U S R R. Kiev. 1934+. Spisok Rosl. Gerb. Fl. U.S.R.R. HI 73009

Spisy přírodovědecké fakulty universit J. E. Purkinje v Brně. Brno. Nos. 400-418, 1959-70. Spisy Přír. Fak. Univ. J. E. Purkinje Brně. Preceded by: Spisy vydávané přírodovědeckou fakultou Masarykovy university. Superseded by: Scripta facultatis scientiarum naturalium universitatis Purkynianae Brunensis. HI 73010

Spisy vydávané přírodovědeckou fakultou Karlovy university. Prague. Spisy Přír. Fak. Karlovy Univ. See B–P–H 852/16. HI 59543

Spisy vydávané přírodovědeckou fakultou Masarykovy university. Brünn [=Brno, Czechoslovakia]. Spisy Přír. Fak. Masarykovy Univ. See B–P–H 852/17. HI 59544

Spolia zeylanica. Colombo, Ceylon. Spolia Zeylanica. See B–P–H 853/5. HI 59545

Spore research. London. 1971+. Spore Res. HI 73011

Spolkový časopis pro lesnictví, myslivost a přírodovědu = Vereinsschrift für die Forst-, Jagd- und Naturkunde. Prague. Vereinsschr. Forst- Naturk. See B–P–H 949/3.

Spores; proceedings of the international spore conference. Ann Arbor, MI, Washington, DC. 1st+, 1956+. Spores. HI 73012

Sporovye rasteniya = Trudy botanicheskogo instituta akademii nauk S S S R. Ser. 2, sporovye rasteniya. Moscow & Leningrad.

Sports turf bulletin. Bingley. No. 1+, 1951+. Sports Turf Bull. HI 73013

Sports turf research institute journal. Bingley, England. Sports Turf Res. Inst. J. See B–P–H 853/7. HI 59546

Spotlight on community gardening. Shelburne, VT. Vol. 1(1-2), 1978. Spotlight Community Gard. Superseded by: Gardens for all news. HI 75101

Spravočnyj Listok Biologa. Yur'ev [=Tartu, Estonian S S R]. Sprav. Listok Biol. See B–P–H 853/8. HI 59547

Sprawozdanie z cynności towarzystwa przyjaciól nauk poznańskiego. Poznan, Poland. Spraw. Cynn. Towarz. Przyjac. Nauk Poznańsk. See B–P–H 853/9. HI 59548

Sprawozdanie komisji fizjograficznej. Cracow. Spraw. Komis Fizjogr. See B–P–H 853/13. HI 59549

Sprawozdanie komisji fizjograficznej c.k. towarzystwa naukowego krakowskiego. Cracow. Spraw. Komis Fizjogr. C.K. Towarz. Nauk. Krakowsk. See B–P–H 853/14. HI 59550

Sprawozdanie poznańskiego towarzystwa przyjaciól nauk. Poznan, Poland. Spraw. Poznańsk. Towarz. Przyjac. Nauk. See B–P–H 853/17. HI 59551

Springfield museum of natural history bulletin. Springfield, MA. Springfield Mus. Nat. Hist. Bull. See B–P–H 853/20. HI 59552

Spuria iris society newsletter = Newsletter, spuria iris society. Yorba Linda, CA, Purcell, OK.

Sri-Lanka forester; journal of the Sri-Lanka forest department. Colombo. N.s. vol. 10(3/4)+, 1972+. Sri-Lanka Forester. Preceded by: Ceylon forester. HI 73014

Ssu chuan chih sheng lin. = Szechuan forests. [China]. Szechuan Forests. See B–P–H 863/10.

Ssu-chuan ta hsueh hsueh pao. Tzu jan ko hsueh = Journal of Sichuan university. Natural science edition. Chengtu.

St., see also Saint

St. Louis medical and surgical journal. St. Louis, MO. St. Louis Med. Surg. J. See B–P–H 853/26. HI 59553

St. Paul's review of arts and sciences. Natural science. Tokyo. Vol. 1(1-10), 1956-61. St. Paul's Rev. Arts Sci., Nat. Sci. Superseded by: St. Paul's review of science. HI 72367

St. Paul's review of science. [Rikkyo daigaku kenkyu hokoku, shizenkagaku.] Tokyo. Vol. 1(11)+, 1962+. St. Paul's Rev. Sci. Preceded by: St. Paul's review of arts and sciences. Natural science. HI 72368

Staats-Anzeigen. Göttingen. Staats-Anzeigen. See B–P–H 854/15. HI 59558

Staatsinstitut für angewandte Botanik, Hamburg. Jahresbericht. Hamburg. Staatsinst. Angew. Bot. Hamburg Jahresber. See B–P–H 854/16. HI 59559

Stachelpost; Mitteilungsblätter für Kakteenfreunde. Stuttgart, Mainz/Klosterhein. Vols. 1-10, 1964-74. Stachelpost. HI 73015

Stadler genetics symposium. Columbia, MO. 1971+. Stadler Genet. Symp. HI 75166

Stahlia; miscellaneous papers of the museum of biology. Río Piedras. Nos. 1-?, 1961-67. Stahlia. HI 73016

Stain technology; a journal for microtechnic and histochemistry. Geneva, NY. Stain Technol. See B–P–H 854/18. HI 59560

Stancija zaščity sel'skohozjajstvennyh rastenij ot vreditelej Voronežskogo gubernskogo zemel'nogo upravlenija = Stantsiya zashchity sel'skokhozyaistvennykh rastenii ot vreditelei Voronezhskogo gubernskogo zemel'nogo upravleniya. Voronezh.

Stanford studies in geology. Stanford, CA. Stanford Stud. Geol. See B–P–H 854/20. HI 59561

Stanford university publications. Biological sciences. Palo Alto, CA. Stanford Univ. Publ., Biol. Sci. See B–P–H 854/23. HI 59562

Stantsiya zashchity sel'skokhozyaistvennykh rastenii ot vreditelei Voronezhskogo gubernskogo zemel'nogo upravleniya. Voronezh, Russian S F S R. Vols. 4-5, 1925. Stantsiya Zashch. Sel'skokhoz. Rast. Vredit. Voronezhsk. Gub. Zemel'n. Upravl. Preceded by: Izvestiya Voronezhskoi stantsii po bor'be s vreditelyalmi rastenii. Superseded by: Byulleteni Voronezhskoi stantsii zashchity rastenii. HI 73017

Stapfia; Publikation der botanischen Arbeitsgemeinschaft am O. Ö. Landesmuseum, Linz. Linz. Vol. 1+, 1977+. Stapfia. HI 73018

Starea fitosanitară in Republica Populară Romînă. Bucharest. Starea Fitosan. Republ. Populară Romînă. See B–P–H 855/1. HI 59563

Starch. Stuttgart, etc. = Stärke. Stuttgart, Weinheim/Bergst.

Stärke; Fachzeitschrift für Erforschung, Herstellung und Verwendung von Stärke und Stärkeerzeugnissen. Stuttgart, Weinheim/Bergst. Vol. 1+, 1949+. Stärke. HI 73019

Starunia; studia ad poloniae diluvium cognoscendum pertinentia. Cracow. Nos. 1-30, 1934-53 [publication suspended 1940-44]. Starunia. HI 73020

State agricultural college agricultural experiment station annual report. Fort Collins, CO. State Agric. Coll. Agric. Exp. Sta. Annual Rep. See B–P–H 855/2. HI 59564

State agricultural college agricultural experiment station bulletin. Fort Collins, CO. State Agric. Coll. Agric. Exp. Sta. Bull. See B–P–H 855/3. HI 59565

State agricultural college of Colorado. Bulletin. Fort Collins, CO. State Agric. Coll. Colorado Bull. See B–P–H 855/4. HI 59566

State agricultural college experiment station bulletin. Fort Collins, CO. State Agric. Coll. Exp. Sta. Bull. See B–P–H 855/5. HI 59567

State cabinet of natural history. Albany, NY. State Cab. Nat. Hist. See B–P–H 855/6. HI 59568

State farms bulletin, Wyoming agricultural college agricultural experiment station = Wyoming agricultural college agricultural experiment station. State farms bulletin. Laramie, WY. Wyoming Agric. Exp. Sta. State Farms Bull. See B–P–H 981/2.

State forestry report, New Zealand. Wellington, N.Z. 1916-22. State Forest. Rep. New Zealand. Preceded by: Report on state afforestation in New Zealand. Superseded by: Report (Annual), New Zealand state forest service. HI 73021

State university studies in natural history, Iowa = State university of Iowa studies in natural history. Iowa City, IA.

State of the world; worldwatch institute report on progress toward a sustainable society. New York. 1984+. State World. HI 73022

State university of Iowa studies in natural history. Iowa City, IA. Vols. 19?-?, 1948-71. State Univ. Iowa Stud. Nat. Hist. Preceded by: Studies in natural history, Iowa university. HI 73023

Statens naturvetenskapliga forskningsrads årsbok. Bromma. Vols. 1-11, 1946-54. Statens Naturventensk. Forskningsrads Årsbok. Superseded by: Svensk naturvetenskap. HI 73024

Statens plantevern flygeskrift. Oslo. Vol. 39+, 1948+. Statens Plantevern Flygeskr. Preceded by: Flygeskrift, statens plantepatologiske institut. HI 73025

Station bulletin, Nebraska agricultural experiment station = Nebraska agricultural experiment station bulletin. Lincoln, NB. Nebraska Agric. Exp. Sta. Bull. See B–P–H 639/25.

Station Bulletin, New Hampshire agricultural experiment station = New Hampshire agricultural experiment station. Bulletin. Durham, NH. New Hampshire Agric. Exp. Sta. Bull. See B–P–H 653/1.

Station bulletin, Oregon agricultural experiment station = Oregon agricultural experiment station. Station bulletin. Corvallis, OR. Oregon Agric. Exp. Sta., Sta. Bull. See B–P–H 690/16.

Station circular: Forestry facts, agricultural experiment station, Washington state college. Pullman, WA.

1953+. Sta. Circ. Forest. Facts Agric. Exp. Sta. Washington State Coll. HI 73026

Station circular, Nebraska agricultural experiment station = Nebraska agricultural experiment station circular. Lincoln, NB. Nebraska Agric. Exp. Sta. Circ. See B–P–H 639/26.

Station circular, Oregon agricultural experiment station = Oregon agricultural experiment station circular. Corvallis, OR. Oregon Agric. Exp. Sta. Circ. See B–P–H 690/12.

Station circular, Washington (state) agricultural experiment station = Washington (state) agricultural experiment station. Station circular. Pullman, WA. Wash. State Agric. Exp. Sta. Sta. Circ. See B–P–H 971/17.

Station guide, Cleppa Park horticulture demonstration station. Newport. 1953-65. Sta. Guide Cleppa Park Hort. Demonst. Sta. HI 73027

Station guide, Efford experimental horticulture station. Lymington. 1954?-5? Sta. Guide Efford Exp. Hort. Sta. Superseded by: Experimental programme, Efford experimental horticulture station. HI 73028

Station guide, Lee Valley experimental horticulture station. Hoddesdon. 1961+. Sta. Guide Lee Valley Exp. Hort. Sta. HI 73029

Station guide, Stockbridge house experimental horticulture station. Cawood. 195?-66. Sta. Guide Stockbride House Exp. Hort. Sta. Superseded by: Guide to experiments, Stockbridge house experimental horticulture station. HI 73030

Station notes, central states forest experiment station. Columbus, OH. Nos. 1-157?, 1933-62. Sta. Notes Centr. States Forest Exp. Sta. Superseded by: Research notes C S, United States forest service. HI 73031

Station notes, forest, wildlife and range experiment station, university of Idaho. Moscow, ID. No. 1+, 1965+. Sta. Notes Forest Wildlife Range Exp. Sta. Univ. Idaho Preceded by: Research notes, forest, wildlife and range experiment station, university of Idaho. HI 73032

Station notes, northeastern forest experiment station. Philadelphia, PA. Nos. 1-171?, 1947-62. Sta. Notes NorthE. Forest Exp. Sta. Superseded by: Research paper NE, United States forest service. HI 73033

Station paper, Alaska forest research center = Station papers, forest research center. Juneau, AK.

Station papers, forest research center. Juneau, AK. Nos. 1-13, 1953-60. Sta. Pap. Forest Res. Center, Juneau. HI 73034

Station papers, forest, wildlife and range experiment station, university of Idaho. Moscow, ID. No. 1+, 1966+. Sta. Pap. Forest Wildlife Range Exp. Sta. Univ. Idaho. Preceded by: Research notes, forest, wildlife and range experiment station, university of Idaho. HI 73035

Station papers, lake states forest experiment station. St. Paul, MN. Nos. 1-106?, 1945-62. Sta. Pap. Lake States Forest Exp. Sta. Superseded by: Research papers L S, United States forest service. HI 73036

Station papers, Northeastern forest experiment station. Upper Darby, PA. Nos. 1-171?, 1941-62. Sta. Pap. NorthE. Forest Exp. Sta. Superseded by: Research papers N E, United States forest service. HI 73037

Station papers, northern Rocky Mountain forest and range experiment station. Missoula, MT. Nos. 1-35?, 1939-53. Sta. Pap. N. Rocky Mountain Forest Range Exp. Sta. HI 73038

Station papers, Rocky Mountain forest and range experiment station. Fort Collins, CO. Nos. 1-73, 1949-62. Sta. Pap. Rocky Mountain Forest Range Exp. Sta. Superseded by: Research papers R M, United States forest service. HI 73039

Station papers, southeastern forest experiment station. Asheville, NC. Nos. 1-156, 1949-62. Sta. Pap. SouthE. Forest Exp. Sta. Preceded by: Technical notes, southeastern forest experiment station. Superseded by: Research papers S E, United States forest service. HI 73040

Station guide and programme of experiments, Efford experimental horticulture station. Lymington. 1960. Sta. Guide Programme Exp. Efford Exp. Hort. Sta. Preceded by: Experimental programme, Efford experimental horticulture station. Superseded by: Guide to experiments, Efford experimental horticulture station. HI 73041

Station progress notes. University agricultural experiment station. Honolulu, HI. Sta. Progr. Notes Univ. (Hawaii) Agric. Exp. Sta. See B–P–H 854/13. HI 59556

Station report, Efford experimental horticulture station. Lymington. 1959-63. Sta. Rep. Efford Exp. Hort. Sta. Superseded by: Report (Annual), Efford experimental horticulture station. HI 73042

Station report. Horticultural research station. Tatura, Australia. Sta. Rep. Hort. Res. Sta. See B–P–H 854/14. HI 59557

Staudengarten. Leonberg bei Stuttgart. 1973(2)+, 1973+. Staudengarten. Preceded by: Iris und Lilien. HI 73043

Stazioni sperimentali agrari italiane. Turin. Staz. Sperim. Agrar. Ital. See B–P–H 855/12. HI 59569

Steiermärkische Zeitschrift. Graz. Steiermärk. Z. See B–P–H 855/13. HI 59570

Steirischer Naturschutzbrief. Graz. Steir. Naturschutzbrief. See B–P–H 855/14. HI 59571

Stem cells; international journal of cellular differentiation and proliferation. Basel. 1981+. Stem Cells. HI 73044

Stencil bulletin, Wisconsin (state) agricultural experiment station = Wisconsin (state) agricultural experiment station. Stencil bulletin. Madison, WI. Wisconsin Agric. Exp. Sta. Stencil Bull. See B–P–H 976/6.

Stěpař. Rakonitz, Bohemia [=Rakovnik, Czechoslovakia]. Stěpař. See B–P–H 855/16. HI 59572

Sterbeeckia; orgaan van de Antwerpse mycologische kring. Antwerp. Jaarg 1-?, 1961-69. Sterbeeckia. HI 73045

Steroids; an international journal. San Francisco, CA. Steroids. See B–P–H 855/17. HI 59573

Stesicoro; opera periodica. Catania, Italy. Stesicoro. See B–P–H 855/18. HI 59574

Stettiner entomologische Zeitung = Entomologische Zeitung. Stettin [Poland]. Entomol. Zeitung. See B–P–H 358/35.

Stikstof; dutch nitrogenous fertilizer review. The Hague. Stikstof. See B–P–H 855/22. HI 59575

Stockholm contributions in geology = Acta universitatis stockholmiensis. Stockholm. Acta Univ. Stockholm. See B–P–H 50/8.

Stockholm studies in English. Stockholm. Vol. 1+, 1937+. Stockholm Stud. English. HI 73046

Stockholmisches Magazin, darinnen kleine schwedische Schriften, welche die Geschichte, Staatsklugheit und Naturforschung betreffen, ... mitgetheilet werden. Stockholm. Stockholm. Mag. See B–P–H 855/26. HI 59576

Stockholms historiska bibliothek. Stockholm. Stockholms Hist. Biblioth. See B–P–H 855/27. HI 59577

Stockholms lärda tidningar. Stockholm. Stockholms Lärda Tidn. See B–P–H 856/1. HI 59578

Stockholms magazin. Stockholm, Uppsala, & Turku, Finland. Stockholms Mag. See B–P–H 856/2. HI 59579

Stomatologia. Athens. Stomatologia (Athens). See B–P–H 856/3. HI 59580

Stomatologia. Milan. Stomatologia (Milan). See B–P–H 856/4. HI 59581

Stomatologia. Rome. Stomatologia (Rome). See B–P–H 856/5. HI 59582

Storrs agricultural experiment station bulletin. Mansfield, CT. Storrs Agric. Exp. Sta. Bull. See B–P–H 856/6. HI 59583

Storrs school agricultural experiment station bulletin. Mansfield, CT. Storrs School Agric. Exp. Sta. Bull. See B–P–H 856/8. HI 59584

Straipsniu rinkinys. Vilnius, Lithuanian S S R. Straipsniu Rinkinys. See B–P–H 856/9. HI 59585

Stralsundisches Magazin, oder Sammlungen auserlesener Neuigkeiten, zur Aufnahme der Naturlehre, Arzneywissenschaft und Haushaltungskust. Berlin & Stralsund. Stralsund. Mag. See B–P–H 856/10. HI 59586

Strawberry culturist. Salisbury, MD. Strawberry Cult. See B–P–H 856/13. HI 59587

Strawberry culturist and small fruit grower. Salisbury, MD. Strawberry Cult. Small Fruit Grower. See B–P–H 856/14. HI 59588

Stromatolite newsletter. Canberra, A.C.T. No. 1+, 1972+. Stromatolite Newslett. HI 73047

Struktura i biosintez belkov. Pushchino. Vol. 1+, 1987+. Strukt. Biosintez Belkov. HI 75123

Struktura i ul'trastruktura plodov. Kishinev. Vol. ?-2, ?-1968. Strukt. Ul'trastrukt. Plodov. HI 73048

Studia geograficzne uniwersytet wrocławskie = Acta universitatis wratislaviensis. Studia geograficzne. Wroclaw.

Studiensamlung der Staatsuniversität in Irkutsk = Sbornik trudov gosudarstvennogo Irkutskogo universiteta.

Studies, biological laboratory of Owens college (Manchester university) = Biological laboratory of Owens college (Manchester university), studies. Manchester, England. Biol. Lab. Owens Coll. Stud. See B–P–H 194/3.

Studies of the east siberian state university = Trudy Vostochno-sibirskogo gosudarstvennogo universiteta. Moscow.

Studies on the flora of Curaçāo and other Caribbean islands = Uitgaven van de natuurwetenschappelijke studiekring voor Suriname en de Nederlandse Antillen. Natuurhistorische reeks. Utrecht.

Studies of the Irkutsk state university = Sbornik trudov gosudarstvennogo Irkutskogo universiteta.

Studies in science. Madison, WI. = Bulletin of the university of Wisconsin. Madison, WI. Bull. Univ. Wisconsin Stud. Sci. See B–P–H 286/21.

Studies in the vegetation of the state, Nebraska university botanical survey. Report on recent collections = Botanical survey of Nebraska. Lincoln, NE.

Studium Generale. Zeitschrift für die Einheit der Wissenschaften im Zusammenhang ihrer Begriffsbildungen und Forschungsmethoden. Berlin,

Göttingen, & Heidelberg. Stud. Gen. See B–P–H 857/7. HI 59599

Studi sasseressi. Sezione 2. Sassari, Sardinia. Stud. Sassaressi, Sez. 2. See B–P–H 857/12. HI 59601

Studi sassaresi. Sez. 3. Annali della facolta di agraria dell'universita di Sassari. Sassari. Vol. 1+, 1953+. Stud. Sassar., Sez. 3. HI 73049

Studi trentini; rivista trimestrale della societa per gli studi trentini. Trento. Vols. 1-12, 1920-30. Stud. Trent. Superseded by: Studi trentini di scienze naturali. HI 73050

Studi trentini di scienze naturali. Trento. Vols. 12-41, 1931-64. Stud. Trent. Sci. Nat. Preceded by: Studi trentini. Superseded by: Studi trentini di scienze naturali. HI 73051

Studi trentini di scienze naturali. Acta biologica. Trento. Vol. 54+, 1977+. Stud. Trent. Sci. Nat., Acta Biol. Preceded by: Studi trentini di scienze naturali. Sez. B, biologica. HI 73052

Studi trentini di scienze naturali. Sez. B, biologica. Trento. Vols. 42-53, 1965-76. Studi. Trent. Sci. Nat., B. Preceded by: Studi trentini di scienze naturali. Superseded by: Studi trentini di scienze naturali. Acta biologica. HI 73053

Studia biologica hungarica. Budapest. Stud. Biol. Hung. See B–P–H 856/15. HI 59589

Studia botanica. Salamanca. [Vol. 1]+, 1982+. Stud. Bot. HI 73054

Studia botanica čechica. Prague. Stud. Bot. Čech. See B–P–H 856/16. HI 59590

Studia botanica čechoslovaca. Prague. Stud. Bot. Čechoslov. See B–P–H 856/18. HI 59591

Studia botanica hungarica. Budapest. Vol. 8+, 1973+. Stud. Bot. Hung. Preceded by: Fragmenta botanica (Hungary). HI 73055

Studia botanica, universidad de Salamanca. Salamanca. Vol. [1]+, 1982+. Stud. Bot. Univ. Salamanca. HI 73056

Studia ethnographica upsaliensia. Uppsala. 1950+. Stud. Ethnogr. Upsal. HI 73057

Studia forestalia suecica. Stockholm. Stud. Forest. Suec. See B–P–H 857/6. HI 59598

Studia geobotanica. Trieste. Vol. 1+, 1980+. Stud. Geobot. Preceded by: Pubblicazioni, istituto di botanica, università degli studi di Trieste. HI 73058

Studia geomorphologica carpatho-balcanica. Cracow. Vol. 1+, 1967+. Stud. Geomoph. Carpatho-Balcan. HI 73059

Studia instituti physiologiae plantarum universitatis caroline. Prague. 1945+. Stud. Inst. Physiol. Pl. Univ. Caroline. Preceded by: Studies from the plant physiological laboratory of Charles university. HI 73060

Studia i materialy z dziejów nauki Polskiej. Seria B, historia nauk biologicznych i medycznych. Warsaw. Vol. 1+, 1957+. Stud. Mater. Dziejów Nauki Polsk., B. HI 73061

Studia i materialy oceanologiczne. Warsaw, [etc.] No. 1+, 1972+. Stud. Mater. Oceanol. HI 73062

Studia microbiologica. Pretoria. Vols. 1-2, 1972-73. Stud. Microbiol. HI 73063

Studia naturae, Cracow. Seria A, wydawnictwa naukowe. Cracow. No. 1+, 1967+. Stud. Nat., Cracow, A. HI 73064

Studia naturae, Cracow. Seria B, wydawnictwa popularnonaukowe. Cracow. No. 1+, 19??+. Stud. Nat., Cracow, B. HI 73065

Studia oecologica. Salamanca. Vol. 1+, 1980+. Stud. Oecol. HI 73066

Studia osrodka dokumentacki fizjogracznej. Cracow. 1972+. Stud. Osrod. Dokument. Fizjogr. HI 73067

Studia societatis scientiarum torunensis. Sectio D. Botanica. Torun, Poland. Stud. Soc. Sci. Torun., Sect. D, Bot. See B–P–H 857/14. HI 59602

Studia universitatis Babeş-Bolyai. Series 2: Biologia. Cluj, Rumania. Stud. Univ. Babeş-Bolyai, Ser. 2, Biol. See B–P–H 857/20. HI 59606

Studia universitatum Victor Babeş et Bolyai. Series biologia. Cluj, Rumania. Stud. Univ. Victor Babeş Bolyai, Ser. Biol. See B–P–H 857/21. HI 59607

Studie Československa akademie ved. Prague. 1971+. Stud. Českoslov. Akad. Ved. HI 73068

Studies in the biological sciences, university of Minnesota. Minneapolis, MN. Nos. 1-6, 1918-27. Stud. Biol. Sci. Univ. Minnesota. HI 73069

Studies in forest environment. Canberra, A.C.T. No. 1+, 1976+. Stud. Forest Environm. HI 73070

Studies in the history of biology. Baltimore, MD. Vols. 1-7, 1977-84. Stud. Hist. Biol. HI 73071

Studies in history of medicine. New Delhi. Vol. 1+, 1977+. Stud. Hist. Med. HI 73072

Studies in history and philosophy of science. London. Vol. 1+, 1970+. Stud. Hist. Philos. Sci. HI 73073

Studies from the laboratories. Philadelphia general hospital. Philadelphia, PA. Stud. Lab. Philadelphia Gen. Hosp. See B–P–H 857/9. HI 59600

Studies in mycology. Baarn. No. 1+, 1972+. Stud. Mycol. HI 73074

Studies in natural history, Iowa university. Iowa City, IA. Vols. 8-18, 1918-48, 1919-48. Stud. Nat. Hist. Iowa Univ. Preceded by: Bulletin of the laboratories of natural history, Iowa state university. Superseded by: State university of Iowa studies in natural history. 3-2102-2. HI 73075

Studies in natural sciences, eastern New Mexico university. Portales, NM. 1971+. Stud. Nat. Sci. E. New Mexico Univ. HI 73076

Studies from the plant physiological laboratory of Charles university. Prague. Vols. 1-5, 1923-35. Stud. Pl. Physiol. Lab. Charles Univ. Superseded by: Studia instituti physiologiae plantarum universitatis caroline. HI 73077

Studies in natural sciences, natural sciences research institute, eastern New Mexico university. Portales, NM. Vol. 1+, 1971+. Stud. Nat. Sci. Nat. Sci. Res. Inst. E. New Mexico Univ. HI 73078

Studies in plant ecology. Uppsala, Stockholm. Vol. 16+, 1985+. Stud. Pl. Ecol. Preceded by: Växtekologiska studier. HI 73079

Studies on thermal springs. [Onson kenkyu.] [Japan]. Stud. Thermal Springs. See B–P–H 857/16. HI 59604

Studies in tropical oceanography. Coral Gables, FL. Stud. Trop. Oceanogr. See B–P–H 857/17. HI 59605

Studies from the Tokugawa institute. Tokyo. Vols. 1-5(3), 1924-40. Stud. Tokugawa Inst. HI 73080

Studii şi cercetări de biochimie. Bucharest. Stud. Ceret. Biochim. See B–P–H 856/21. HI 59592

Studii şi cercetări de biologie. Bucharest. Vols. 7-13, 1956-61; vols. 26-27, 1974-75. Stud. Cercet. Biol. (Bucharest). Preceded by: Buletin stiintific, academia republicii populare romåne. For vols. 14-15, 1962-63 see: Studii şi cercetări de biologie. Seria biologie vegetală. Superseded by: Studii şi cercetări de biologie. Seria biologie vegetală. HI 73082

Studii şi cercetări de biologie. Academia republicii populare romine. Filiala Cluj. Cluj, Rumania. Stud. Cercet. Biol. (Cluj). See B–P–H 856/23. HI 59593

Studii şi cercetări de biologie. Seria biologie vegetală. Bucharest. Vols. 14-15, 1962-63; vol. 28+, 1976+. Stud. Cercet. Biol. (Bucharest), Ser. Biol. Veg. Preceded by: Studii şi cercetări de biologie. For vols. 16-25, 1964-73 see: Studii şi cercetări de biologie. Seria botanică. For vols. 26-27, 1974-75 see: Studii şi cercetări de biologie (Bucureşti). HI 73083

Studii şi cercetări de biologie. Seria botanică. Bucharest. Vols. 7-25, 1956-73. Stud. Cercet. Biol. (Bucharest), Ser. Bot. Preceded by: Buletin stiintific, academia republicii populare romåne. Incorporated in: Studii şi cercetări de biologie (Bucureşti). HI 73084

Studii şi cercetări de biologie, geographie, biologie, muzeologie. Piatra Neamt. Vol. 1, 1970. Stud. Cercet. Biol. Geogr. Biol. Muzeol. Superseded by: Studii şi cercetări de geologie, geographie, biologie. Seria botanica-zoologie. HI 73085

Studii şi cercetări de geologie, geographie, biologie. Seria botanica-zoologie. Piatra Neamt. Vol. 2, 1974. Stud. Cercet. Geol. Geogr. Biol., Ser. Bot. Zool. Preceded by: Studii şi cercetări de biologie, geographie, biologie, muzeologie. Superseded by: Anuarul muzeului de ştiinte naturale. Seria botanica-zoologie. HI 73086

Studii şi cercetări de inframicrobiologie, microbiologie şi parazitologie. Bucharest. Vols. 1-22, 1950-71. Stud. Cercet. Inframicrobiol. Microbiol. Parazitol. Superseded by: Studii şi cercetări di virusologie. HI 73087

Studii şi cercetări, institutul de cercetări forestiere. Bucharest. Vol. 23A+, 1963+. Stud. Cercet. Inst. Cercet. Forest. Preceded by: Analele institutului de cercetări silvice. HI 73088

Studii şi cercetări, institutul de cercetări silvice. Seria 1. Bucharest. Vols. ?-12-15, ?-1951-54. Stud. Cercet. Inst. Cercet. Silvice, Ser. 1. Superseded by: Analele institutului de cercetări silvice. HI 73089

Studii şi cercetări ştiinţifice. Cluj, Rumania. Stud. Cercet. Şti. (Cluj). See B–P–H 857/1. HI 59594

Studii si cercetări ştiinţifice. Jassy, Rumania. Stud. Cercet. Şti. (Jassy). See B–P–H 857/2. HI 59595

Studii si cercetări ştiinţifice. Seria 2. Ştiinţe biologicae, agricole şi medicale. Cluj, Rumania. Stud. Cercet. Şti. Ser. 2, Şti. Biol. See B–P–H 857/3 HI 59596

Studii şi cercetări silvicultură. Bucharest. ?-1971+. Stud. Cercet. Silvic. HI 73090

Studii şi cercetări di virusologie. Bucharest. Vol. 23+, 1972+. Stud. Cercet. Virusol. Preceded by: Studii şi cercetări de inframicrobiologie, microbiologie şi parazitologie. HI 73091

Studia citrologica. [Kankitsu kenkyu.] Fukuoka, Japan. Stud. Citrol. See B–P–H 857/4. HI 59597

Studii şi comunicări muzeul Brukenthal. Sibiu. Vol. ?-15+, ?-1970+. Stud. Comun. Muz. Brukenthal. HI 73092

Studii şi comunicări, muzeul judetean Suceava. Ştiinţe naturale. Suceava. Vols. 1-3, 1970-76? Stud. Comun. Muz. Judet. Suceava, Şti. Nat. Superseded by: Anuarul muzeului judetean Suceava. Fascicola stiintele naturii. HI 73093

Studii şi comunicări, muzeul de ştiintele naturii. Bacău. Vols. 1-3, 1968-70. Stud. Comun. Muz. Şti. Nat. Superseded by: Studii şi comunicări, muzeul de ştiintele naturii. Botanică. HI 73094

Studii şi comunicări, muzeul de ştiintele naturii. Biologie vegetală. Bacău. Vol. 6+, 1973+. Stud. Comun. Muz. Şti. Nat., Biol. Veg. Preceded by: Studii şi comunicări, muzeul de ştiintele naturii. Botanică.

HI 73095

Studii şi comunicări, muzeul de ştiintele naturii. Botanică. Bacău. Vols. 4-5, 1971-72. Stud. Comun. Muz. Şti. Nat., Bot. Preceded by: Studii si comunicari, muzeul de ştiintele naturii, Bacău. Superseded by: Studii şi comunicări, muzeul de ştiintele naturii. Biologie vegetală. HI 73096

Studii şi comunicări de ocrotirea naturii. Suceava. Vol. 1+, 1970+. Stud. Comun. Ocrot. Nat. HI 73097

Studio. London. Vol. 1+, 1893+. Studio. 5-4104-1. HI 74880

Study of cacti. Journal of the amateur cactus group of Japan. Yamatosi, Kanagawa. Vol. 1+, 1955+. Stud. Cacti. HI 73098

Study of cacti. [Shaboten no kenkyu.] Osaka. No. ?-4(6)+, ?-1933+. Stud. Cacti (Osaka). HI 73099

Study of tea. [Chagyo gijutsu kenkyu.] Shizuoka, Japan. Stud. Tea. See B–P–H 857/15. HI 59603

Stultifera navis. Basel. Vols. 1-14, 1944-57. Stultifera Navis. Superseded by: Librarium. 5-4105-1. HI 73100

Stuttgarter Beiträge zur Naturkunde aus dem staatlichen Museum für Naturkunde in Stuttgart. Stuttgart. Vols. 1-253, 1957-72. Stuttgarter Beitr. Naturk. Superseded by: Stuttgarter Beiträge zur Naturkunde. Seria A, Biologie; Stuttgarter Beiträge zur Naturkunde. Serie B, Geologie und Paläontologie and Stuttgarter Beiträge zur Naturkunde. Serie C, Allgemeinverstandliche Aufsatze. HI 73101

Stuttgarter Beiträge zur Naturkunde. Seria A, Biologie. Stuttgart. No. 254+, 1973+. Stuttgarter Beitr. Naturk., A. Preceded by: Stuttgarter Beiträge zur Naturkunde aus dem staatlichen Museum für Naturkunde in Stuttgart. HI 73102

Stuttgarter Beiträge zur Naturkunde. Serie B, Geologie und Paläontologie. Stuttgart. Vol. 1+, 1972+. Stuttgarter Beitr. Naturk., B. Preceded by: Stuttgarter Beiträge zur Naturkunde aus dem staatlichen Museum für Naturkunde in Stuttgart. HI 73103

Stuttgarter Beiträge zur Naturkunde. Serie C, Allgemeinverstandliche Aufsatze. Stuttgart. Vol. 1+, 1974+. Stuttgarter Beitr. Naturk., C. Preceded by: Stuttgarter Beiträge zur Naturkunde aus dem staatlichen Museum für Naturkunde in Stuttgart. HI 73104

Suara rimbawan. Bogor. 1959+. Suara Rimbawan. Preceded by: Rimba Indonesia. HI 73105

Subalpino; giornale de scienze, lettere ed arti. Turin. Subalpino. See B–P–H 858/10. HI 59608

Sub-cellular biochemistry. New York. Vol. 1+, 1971+. Sub-Cell. Biochem. HI 73106

Subtropicheskie kul'tury. Makharadze. Vol. 1+, 1960+. Subtrop. Kul't. HI 73107

Subtropics = Subtropiki. Sukhumi, Georgian S S R. Subtropiki. See B–P–H 858/13.

Subtropiki. Sukhumi, Georgian S S R. Subtropiki. See B–P–H 858/13. HI 59609

Suburban California = Pacific garden. Pasadena, CA. Pacific Gard. See B–P–H 694/33.

Succulent. Tokyo. No. 1+, 1980+. Succulent (Tokyo). HI 73108

Succulent journal. Englewood, NJ. Vol. 8+, 1969+. Succ. J. Preceded by: Spinal column. (Englewood). HI 73109

Succulent plant club newsletter = Newsletter, succulent plant club. Southburgh.

Succulent plant trust newsletter = Newsletter, succulent plant trust. Coulsdon.

Succulent scene. Burgess Hill. No. 1+, 1981+. Succ. Scene. HI 73110

Succulenta; orgaan van de nederl. vereeniging van vetplantenverzamelaars. Maanblad van de Nederlands-Belgische vereniging van liefhebbers van cactussen en andere vetplanten. Huizum, etc. Vols. 1-25(4), 1919-43; [n.s.] 1947+ [from 1964 onwards numbered vol. 43+]. Succulenta (Netherlands). 5-4107-2. HI 73111

Succulenta. Ludhiana. Vols. 1-4, 1975-78. Succulenta (India). HI 73112

Succulentarum bibliographia. Oakland, CA. Vols. 1-4, 1971-74. Succ. Bibliogr. HI 73113

Succulentarum japonica. [Saboten nihon.] Tokyo. Succ. Jap. See B–P–H 858/15. HI 59610

Succulentenkunde; maanbladgewijd aan de studie, het verzamelen en kweeken van succulente planten. Voorburg. 1 vol., 1931. Succulentenkunde. 5-4107-2. HI 73114

Succulentenland; tijdschrift van de afdeeling Ijsselstreek van succulenta. Zutphen, Lochum. Vol. 1(1-6)?, 1976? Succulentenland. HI 73115

Succulentes. Monte-Carlo. Ser. 1, nos. 1-4, 1977-78; ser. 2, nos.1-4, 1979-80; ser. 3, nos. 1-4; [n.s.] no. 1+, 1983+. Succulentes. HI 73116

Sudan agricultural journal. Khartoum. Vols. 1+, 1965+. Sudan Agric. J. HI 73117

Sudan notes and records; incorporating proceedings of the philosophical society of the Sudan. Khartoum. Vols. 1+, 1918+. Sudan Notes Rec. HI 73118

Sudan silva. Khartoum. Vol. 1+, 1949+. Sudan Silva. HI 73119

Süddeutscher Erwerbsgärtner. Stuttgart. Vols. 1-21, 1947-67. Süddeutsch. Erwerbsgärtn. Superseded by: Erwerbsgärtner. HI 73120

Sudhoffs Archiv; Vierteljahrschrift für Geschichte der Medizin und der Naturwissenschaften der Pharmazie und der Mathematik. Wiesbaden. Vol. 50+, 1966+. Sudhoffs Arch. Preceded by: Sudhoffs Archiv für Geschichte der Medizin und der Naturwissenschaften. HI 73121

Sudhoffs Archiv. Beiheft. Leipzig. Vol. 5+, 1966+. Sudhoffs Arch., Beih. Preceded by: Sudhoffs Archiv für Geschichte der Medizin und der Naturwissenschaften. Beiheft. HI 73122

Sudhoffs Archiv für Geschichte der Medizin. Leipzig. Sudhoffs Arch. Gesch. Med. See B–P–H 858/21. HI 59611

Sudhoffs Archiv für Geschichte der Medizin und der Naturwissenschaften. Leipzig. Vols. 27-49, 1934-64 [suspended 1943-52]. Sudhoffs Arch. Gesch. Med. Naturwiss. Preceded by: Sudhoffs Archiv für Geschichte der Medizin. Superseded by: Sudhoffs Archiv. 5-4108-2. HI 73123

Sudhoffs Archiv für Geschichte der Medizin und der Naturwissenschaften. Beiheft. Leipzig. Vols. 1-4, 1961-64. Sudhoffs Arch. Gesch. Med. Naturwiss., Beih. Superseded by: Sudhoffs Archiv. Beiheft. HI 73124

Sudwestdeutsche Bienenzeitung = Hessische Biene. Giessen.

Suffolk natural history; transactions of the Suffolk naturalists' society. Bury St. Edmunds. 1969+. Suffolk Nat. Hist. Preceded by: Transactions of the Suffolk naturalists' society. HI 73125

Sugar beet journal. Saginaw, MI. Vols. 1+, 1935+. Sugar Beet J. HI 73126

Sugar beet review and british beet grower. London. Sugar Beet Rev. See B–P–H 858/27. HI 59614

Sugar bulletin. New Orleans, LA. Sugar Bull. See B–P–H 858/28. HI 59615

Sugar cane; monthly magazine devoted to the interests of the sugar cane industry. Manchester. Vols. 1-30, 1869-98. Sugar Cane. Superseded by: International sugar journal. HI 73127

Sugar central and planters' news. Manila. Sugar Centr. Pl. News. See B–P–H 859/2. HI 59616

Sugar journal. New Orleans, LA. Sugar J. See B–P–H 859/3. HI 59617

Sugar journal congress and exhibition number = South african sugar journal, 1923-24. Durban, South Africa. S. African Sugar J. See B–P–H 808/2.

Sugar maple research news. Burlington, VT. 1979+. Sugar Maple Res. News. HI 73128

Sugar news. Manila. Sugar News. See B–P–H 859/5. HI 59618

Sugarcane documentation news. Coimbatore. Vol. 1+, 1978+. Sugarcane Doc. News. HI 73129

Sugarcane farmers' bulletin. Manila. Sugarcane Farmers' Bull. See B–P–H 859/6. HI 59619

Sugarcane herald. New Delhi. Sugarcane Herald. See B–P–H 859/7. HI 59620

Suginami-machi = Bulletin of the imperial sericultural experiment station, Japan. Tokyo. Bull. Imp. Seric. Exp. Sta. Japan See B–P–H 255/3.

Suid-afrikaanse bosboutydskryf = South african forestry journal. Pretoria.

Suid-Afrikaanse orgideejoernaal = South african orchid journal. Johannesburg.

Suid-afrikaanse tijdskrif vir wetenskap = South african journal of science. Johannesburg. S. African J. Sci. See B–P–H 807/28.

Suid-afrikanse tydskrif vir landbouwetenskap = South african journal of agricultural science. Pretoria.

Suid-Afrikaanse tydskrif vir plantkunde = South african journal of botany. Pretoria.

Suid-Afrikaanse tydskrif vir seewetenskap = South african journal of marine science. Cape Town.

Suisan koshujo kenkyu-hokoku = Journal of the Shimonoseki college of fisheries. Shimonoseki, Japan. J. Shimonoseki Coll. Fish. See B–P–H 481/13.

Suisan koshujo shiken hokoku = Journal of the imperial fisheries institute. Tokyo. J. Imp. Fish. Inst. See B–P–H 470/1.

Sukkulentenkunde; Jahrbücher der Schweizerischen Kakteen-Gesellschaft. Zurich. Vols. 1-8, 1947-63. Sukkulentenkunde. 5-4112-1. HI 73130

Sukkulenti. Zejtun. No. 1, 1963. Sukkulenti. Superseded by: Kakti u sukkulenti ohra. HI 73131

Sultania; botanical sciences bulletin. Peshawar. Vol. 1+, 1975+. Sultania. HI 73132

Šumarski list. Zagreb. Vols. 1-64, 1877-1940; vol. 69+, 1945+. Šumarski List. For vols. 65-68, 1941-44 see: Hrvatski šumarski list. HI 73133

Sumarski pregled. Skopje. Vol. 1+, 1953+. Sumarski Pregled. HI 73134

Šumarstvo. Belgrade. 1948+. Šumarstvo. HI 73135

Sumarstvo i prerada drveta. Sarajevo. Vol. 31+, 1977+. Sumarstvo Prerada Drveta. Preceded by: Narodni šumar. HI 73136

Summa brasiliensis biologiae. Rio de Janeiro. Vol. 1(1-17), 1945-48. Summa Brasil. Biol. HI 73137

Summaries of scientific progress. Biology series, microbiology. Vol. 1+, 1974+. Summ. Sci. Progr., Biol. Ser. Microbiol. HI 73138

Summarized proceedings of the american association for the advancement of science. New York, NY. Summarized Proc. Amer. Assoc. Advancem. Sci. See B–P–H 859/16. HI 59621

Summarized proceedings of the meetings of the american association for the advancement of science. Philadelphia, PA. 74th-85th, 1921-29. Summarized Proc. Meetings Amer. Assoc. Advancem. Sci. Preceded by: Proceedings of the american association for the advancement of science. HI 73139

Summary of the results of the cultivation of seedlings and other canes, Barbados = Report of sugar-cane experiments, department of agriculture, Barbados. Bridgetown.

Summary report of the P F R A tree nursery. Indian Head, Sask. 1971-72. Summ. Rep. P. F. R. A. Tree Nursery. Superseded by: Report of the P F R A tree nursery. HI 73140

Summary report of researches, national research institute of agriculture, Tokyo = Report, research institute on agriculture, Tokyo. Tokyo.

Sunflower newsletter. Zevenaar. ?-1980?+. Sunflower Newslett. HI 73141

Sundhed og underholdning. Copenhagen. Sundhed & Underhold. See B–P–H 859/19. HI 59622

Sundheds-journal. Et medicinsk-diaetisk og underholdende ugeblad. Copenhagen. Sundheds-J. See B–P–H 859/21. HI 59624

Sundhedsblade. Copenhagen. Sundhedsblade. See B–P–H 859/20. HI 59623

Sundhedstidende. Et medicinsk ugeskrift af blandet indhold. Copenhagen. Sundhedstidende. See B–P–H 859/22. HI 59625

Sunshine state agricultural research report. Gainesville, FL. Vols. 1-?, 1956-82. Sunshine State Agric. Res. Rep. Superseded by: Florida agricultural research. HI 73142

Sun Yatsen university science and technique correspondence. Biological edition. Canton? ?-1979+. Sun Yatsen Univ. Sci. Techn. Corresp., Biol. Ed. HI 73143

Sunyatsenia. Journal of the botanical institute; college of agriculture, Sun Yatsen university. Canton. Sunyatsenia. See B–P–H 859/25. HI 59626

Suo. Helsinki. Suo. See B–P–H 859/26. HI 59627

Suomalainen lääküriseura Duodecim = Acta societas medicorum fennica Duodecim. Helsinki. Acta Soc. Med. Fenn. Duodecim. See B–P–H 49/13.

Suomalaisen elain-ja kasvitieteellisen seuran vanamon julkaisuja. Helsinki. Vols. 1-15, 1923-31. Suom. Elain-ja Kasvit. Seuran Van. Julk. Superseded by: Suomalaisen elain-ja kasvitieteellisen seuran vanamon kasvitieteellisiä julkaisuja. HI 73144

Suomalaisen elain-ja kasvitieteellisen seuran vanamon kasvitieteellisiä julkaisuja. Helsinki. Vols. 1-35, 1931-64. Suom. Elain-ja Kasvit. Seuran Van. Kasvit. Julk. Preceded by: Suomalaisen elain-ja kasvitieteellisen seuran vanamon julkaisuja. Superseded by: Annales botanici fennici. HI 73145

Suomalaisen elain-ja kasvitieteellisen seuran vanamon tiedonannot. Helsinki. Vols. 1-18, 1951-64. Suom. Elain-ja Kasvit. Seuran Van. Tiedon. Preceded by: Suomalaisen elain-ja kasvitieteellisen seuran vanamon tiedonannot ja pöytäkirjat. Superseded by: Annales botanici fennici. 5-4116-2. HI 73146

Suomalaisen elain-ja kasvitieteellisen seuran vanamon tiedonannot ja pöytäkirjat. Helsinki. 1946-50. Suom. Elain-ja Kasvit. Seuran Van. Tiedon. Pöytäkirjat. Superseded by: Suomalaisen elain-ja kasvitieteellisen seuran vanamon tiedonannot. 5-4116-2. HI 73147

Suomalaisen tiedeakatemian toimtuksia = Annales academiae scientiarum fennicae. Ser. A. Helsinki.

Suomen kemistilehtl = Acta chemica fennica. Helsinki. Acta Chem. Fenn. See B–P–H 42/1.

Suomen luonto. Helsinki. Vol. 1+, 1941+. Suom. Luonto. HI 73148

Suomi. Helsinki. Suomi. See B–P–H 859/29. HI 59628

Suplemento bibliográfico de Turrialba. [Instituto interamericano de ciencias agrícolas.] Turrialba, Costa Rica. Supl. Bibliogr. Turrialba. See B–P–H 860/1. HI 59629

Suplemento de ciencias del boletin del instituto de estudios asturianos. [Supplement to: Boletin del instituto de estudios asturianos.] Oviedo. No. 18+, 1973+. Supl. Ci. Bol. Inst. Estud. Asturianos. Preceded by: Boletín del instituto de estudios asturianos. Suplemento de ciencias. HI 73149

Supplément à la bibliothèque universelle de Genève; Archives des sciences physiques et naturelles. Geneva & Paris. Vols. 1-6, 1846-47. Suppl. Biblioth. Universelle Genève. Preceded by: Bibliothèque universelle de Genève. Geneva & Paris. Superseded by: Bibliothèque universelle de Genève. Archives des sciences physiques et naturelles. 1-698-1. HI 74541

Supplement to the bulletin of the tasmanian field naturalists' club = Tasmanian naturalist. Hobart.

Supplement to the rhododendron association year book. London. Suppl. Rhododendron Assoc. Year Book. See B–P–H 860/6. HI 59633

Supplementary papers of the royal geographical society. London. Suppl. Papers Roy. Geogr. Soc. See B–P–H 860/5. HI 59632

Supplementband der Allgemeinen medizinischen Annalen des ersten Jahrzendes des neunzehnten Jahrhunderts. Altenburg = Allgemeine medizinische Annalen des neunzehnten Jahrhunderts. Altenburg.

Supplemento memorias de instituto Oswaldo Cruz. Rio de Janeiro. Suppl. Mem. Inst. Oswaldo Cruz. See B–P–H 860/4. HI 59631

Supplementum curieuser und nutzbarer Anmerkungen von Natur- und Kunstgeschichte. Leipzig. Suppl. Cur. Nutzbarer Anmerk. Natur- Kunstgesch. See B–P–H 860/2. HI 59630

Surco. [Escuela de agricultura del estado de Tamaulipas.] [Taumalipas, Mexico]. Surco. See B–P–H 860/8. HI 59634

Surgical clinic. Chicago, IL. Surg. Clin. See B–P–H 860/9. HI 59685

Surgical clinic of Chicago. Philadelphia, PA. Sur. Clin. Chicago. See B–P–H 860/10. HI 59635

Surgical clinics of North America. Philadelphia, PA. Surg. Clin. N. Amer. See B–P–H 860/11. HI 59636

Surgery, gynecology and obstetrics; an international magazine. Chicago, IL. Surg. Gynecol. Obstet. See B–P–H 860/13. HI 59637

Surinaamse landbouw. Paramaribo, Dutch Guiana [Surinam]. Surinaamse Landb. See B–P–H 860/14. HI 59638

Surrey archaeological collections. London, Guildford. Vol. 1+, 1854+. Surrey Archaeol. Collect. HI 73150

Surrey naturalist. Hindhead, Croydon. 1965-73. Surrey Naturalist. HI 73151

Surrey naturalist annual report. Dorking. 1966/67+, 1968+. Surrey Naturalist Annual Rep. HI 73152

Surtsey research progress report. Reykjavik. No. 1+, 1965+. Surtsey Res. Progr. Rep. HI 73153

Survey of biological progress. New York, NY. Surv. Biol. Progr. See B–P–H 860/16. HI 59639

Survey bulletin, Great Lakes forest research centre. Sault Ste. Marie. 1972+. Surv. Bull. Great Lakes Forest Res. Centre. HI 73154

Survey bulletin, Wisconsin geology and natural history. Madison, WI. 1914? Surv. Bull. Wisconsin Geol. Nat. Hist. HI 73155

Survey notes. Cairo = Cairo scientific journal. Cairo. Cairo Sci. J. See B–P–H 292/20.

Survey of potato research in the U. K. London. 1970+. Surv. Potato Res. U.K. HI 73156

Survey of the prevalence of common plant diseases in the dominion of Canada. Annual report. Ottawa. Vols. 1-4, 1920-23. Surv. Preval. Common Pl. Dis. Canada, Annual Rep. Superseded by: Report (Annual) on the prevalence of plant diseases in the dominion of Canada. HI 73157

Survey of the prevalence of plant diseases in the dominion of Canada. Annual report = Survey of the prevalence of common plant diseases in the dominion of Canada. Annual report. Ottawa.

Survey of sources newsletter. Philadelphia, PA. Vol. 1+, 1975+. Surv. Sources Newslett. HI 73158

Survival international news. London. No. 1+, 1983+. Survival Int. News. Preceded by: Survival international review. HI 73159

Survival international review. London. Vols. 1-?, 1976-82? Survival Int. Rev. Superseded by: Survival international news. HI 75103

Suzugamine joshi tandai kenkyu shuho, shizenkagaku = Bulletin of the Suzugamine women's college. Natural science. Hiroshima.

Svampe. Ballerup. 1980+. Svampe. HI 73160

Svea. Uppsala. Svea. See B–P–H 860/30. HI 59640

Svensk botanisk tidskrift utgifven af svenska botaniska föreningen. Stockholm. Svensk Bot. Tidskr. See B–P–H 861/1. HI 59641

Svensk farmaceutisk tidskrift. Stockholm. Svensk Farm. Tidskr. See B–P–H 861/2. HI 59642

Svensk frötidning. Örebro. Vol. 1+, 1932+. Svensk Frötidn. Preceded by: Sveriges fröodlareförbunds årsskrift. 5-4127-2. HI 73161

Svensk literatur-tidning. Stockholm. Svensk Lit.-Tidn. See B–P–H 861/4. HI 59643

Svensk naturvetenskap. Stockholm. Vols. 11-28, 1957-75. Svensk Naturvetensk. Preceded by: Statens naturvetenskapliga forskningsrads arsbok. Superseded by: Naturvetenskapliga forskningsrådets årsbok. HI 73162

Svensk papperstidning. Stockholm. Svensk Papperstidn. See B–P–H 861/5. HI 59644

Svensk tidskrift. Stockholm. Svensk Tidskr. (Stockholm). See B–P–H 861/6. HI 59645

Svensk tidskrift. Uppsala. Svensk Tidskr. (Uppsala). See B–P–H 861/7. HI 59646

Svenska archivum. [Edited by Giörwell.] Stockholm. Svenska Arch. (Giörwell). See B–P–H 861/11. HI 59647

Svenska archivum. Stockholm. Svenska Arch. (Stockholm). See B–P–H 861/12. HI 59648

Svenska läkare-sällskapet. Årsberättelse. Stockholm. Svenska Läkare-Sällsk. Årsberätt. See B–P–H 861/13. HI 59649

Svenska läkare-sällskapet. Handlingar. Stockholm. Svenska Läkare-Sällsk. Handl. See B–P–H 861/14.

HI 59650

Svenska läkare-sällskapet. Nya handlingar. Stockholm. Svenska Läkare-Sällsk. Nya Handl. See B–P–H 861/15. HI 59651

Svenska lantarbetsgivareföreningens tidskrift. Stockholm. Vols. ?-30-?, ?-1940-64. Svenska Lantarbetsgivareför. Tidskr. Superseded by: Skogs- och lantarbetsgivareföreningens tidskrift. HI 73163

Svenska Linné-sällskapets årsskrift. Uppsala. Svenska Linné-Sällsk. Årsskr. See B–P–H 861/16. HI 59652

Svenska literatur-föreningen; tidning. Uppsala. Svenska Lit.-Fören. Tidn. See B–P–H 861/18. HI 59653

Svenska mosskulturföreningens tidskrift. Jonkoping, Sweden. Svenska Mosskulturfören. Tidskr. See B–P–H 861/19. HI 59654

Svenska parnassen. Stockholm. Svenska Parnassen. See B–P–H 861/20. HI 59655

Svenska trädgårds-föreningens tidskrift. Lund. Svenska Trädg.-Fören. Tidskr. See B–P–H 861/21. HI 59656

Svenska trägårds-föreningens års-skrift. Stockholm. Svenska Träg.-Fören. Års-Skr. See B–P–H 861/22. HI 59657

Svenska växtsociologiska sällskapets handlingar. Uppsala. Svenska Växtsociol. Sällsk. Handl. See B–P–H 861/23. HI 59658

Svenska vetenskapsakademiens årsbok. Stockholm. 1903-68 [1969-72 not published]. Svenska Vetenskapsakad. Årsbok. Superseded by: Documenta kungl. vetenskaps-akademien. 5-4131-2. HI 73164

Sveriges fröodlareförbunds årsskrift. Uppsala. ?-1921-29. Sveriges Fröodlareforb. Årsskr. Superseded by: Svensk frötidning. HI 73165

Sveriges natur; svenska naturskyddsföreningens tidskrift. Stockholm. Sveriges Natur. See B–P–H 862/3. HI 59659

Sveriges pomologiska förenings årsskrift. Stockholm. Sveriges Pomol. Fören. Årsskr. See B–P–H 862/4. HI 59660

Sveriges skogsvårdsförbunds tidskrift. Stockholm, Djursholm. Vol. 64+, 1966+. Sveriges Skogsvårdsförb. Tidskr. HI 73166

Sveriges utsadesforenings tidskrift. Malmö, Svaloev. Vol. 1+, 1891+. Sveriges Utsadesforen. Tidskr. HI 73167

Svet vedy. Bratislava. Svet Vedy. See B–P–H 862/5. HI 59661

Svetovodstvo. Moscow. Vol. ?-4+, ?-1973+. Svetovodstvo. HI 73168

Swadaya = Swadhyaya. Djakarta.

Swadhyaya; journal of the oriental institute. Djakarta. Vol. 1+, 1964+. Swadhyaya. HI 73169

SWANEWS. Norman, OK. 1960-74. SWANEWS. HI 73170

Swedish forestry. Stockholm. [1975]+. Swed. Forest. Preceded by: Skogsácret. HI 73171

Swedish journal of agricultural research. Stockholm. Vol. 1+, 1971+. Swed. J. Agric. Res. Preceded by: Lantbrukshögskolans annaler. HI 73172

Sweet pea annual. Burnley, London. 1905+. Sweet Pea Annual. HI 73173

Swenska wetenskaps academiens handlingar. Stockholm. Swenska Wetensk. Acad. Handl. See B–P–H 862/8. HI 59662

Świat kaktusów. Warsaw. 1935-39. Śwait Kakt. (Warsaw [1935-39]). HI 73174

Świat kaktusów. Warsaw. 1966-79? Świat Kakt. (Warsaw [1966+]). HI 73175

Świat kaktusów. Katowice. 1975+. Świat Kakt. (Katowice). HI 73176

Świecie kaktusów = Świat kaktusów. Warsaw.

Swiss biotech; Schweizerische zeitschrift für biotechnologie; informationsorgan des schweizerische koordinationsausschusses für biotechnologie. Küsnacht. Vol. 1+, 1983+. Swiss Biotech. HI 73177

Swiss cross, monthy magazine of the Agassiz association. New York, NY. Swiss Cross. See B–P–H 862/9. HI 59663

Swiss journal of hydrology = Schweizerische Zeitschrift für Hydrologie. Basel.

Sydney magazine of science and art. Sydney. Sydney Mag. Sci. Art. See B–P–H 862/12. HI 59664

Sydowia. Annales mycologici editi in notitiam scientiae mycologicae universalis. Horn, Austria. Sydowia. See B–P–H 862/13. HI 59665

Syesis. Victoria, B.C. Vol. 1+, 1968+. Syesis. Preceded by: Report of the British Columbia provincial museum of natural history and anthropology. HI 73178

Sylloge plantarum novarum itemque minus cognitarum. Regensburg. Syll. Pl. Nov. See B–P–H 862/15. HI 59666

Syllogeus. Ottawa. Vol. 1+, 1972+. Syllogeus. HI 73179

Sylva africana. Nairobi. ?-1979+. Sylva Africana. HI 73180

Sylva of Formosa. [Taiwan no sanrin.] Sylva Formosa. See B–P–H 862/18. HI 59667

Sylvan; ein Jahrbuch für Forstmänner, Jäger, und Jagdfreunde. Marburg, Heidelberg & Cassel. Vols. [1]-9, 1813-23; n.s. vols. 1-4, 1823-28. Sylvan. Preceded by: Taschenbuch für Forst- und Jagdfreunde.

Superseded by: Taschenbuch zur belehrung und Unterhaltung für Wald- und Jagdfreunde. 5-4155-2. HI 73181

Sylvatrop; Philippine forest research journal. Laguna. Vol. 1+, 1976+. Sylvatrop. HI 73182

Sylvia. [Ringaku kiho.] Taihoku [=Taipei, Taiwan]. Sylvia. See B–P–H 862/19. HI 59668

Sylvicole tunisienne. Tunis. 1939?+. Sylvicole Tunis. Preceded by: Bulletin, société internationale des amis des arbres de Tunisie. HI 73183

Sylvicultural researches from the Tokai district forest bureau. Tokai, Malaya [Malaysia]. Sylvic. Res. Tokai Distr. Forest Bur. See B–P–H 862/20. HI 59669

Sylwan. Lvov, Galicia [Ukrainian S S R]. Sylwan (Lvov). See B–P–H 682/23.. HI 59670

Sylwan. Warsaw. Sylwan (Warsaw). See B–P–H 862/ 24. HI 59671

Symbiosis. Philadelphia, PA. Vol. 1+, 1985+. Symbiosis. HI 73184

Symbolae botanicae upsalienses; arbeten från botaniska institutionen i Uppsala. Uppsala. Symb. Bot. Upsal. See B–P–H 862/25. HI 59672

Symposia of the biochemical society, London. Cambridge, England. Symp. Biochem. Soc. London. See B–P–H 862/27. HI 59673

Symposia biologica hungarica. Budapest. 1958+. Symp. Biol. Hung. HI 73185

Symposia, british ecological society. Oxford. Vol. 1+, 1960+. Symp. Brit. Ecol. Soc. HI 73186

Symposia, british society for developmental biology. Cambridge. 1st+, 1972+, 1973+. Symp. Brit. Soc. Developmental Biol. HI 73187

Symposia for cell biology. [Saibo seibutsugaku shimpojum.] Okayama. No. 20+, 1969+. Symp. Cell Biol. Preceded by: Symposia for cellular chemistry. HI 73188

Symposia for cellular chemistry. [Saibo kagaku shimpojumu.] Nos. 1-9, 1953-68. Symp. Cell. Chem. Superseded by: Symposia for cell biology. HI 73189

Symposia genetica et biologica italica. Pavia. Vol. 1+, 1951+. Symp. Genet. Biol. Ital. HI 73190

Symposia, institute for terrestrial ecology. Cambridge. No. 1+, 1981+. Symp. Inst. Terrestr. Ecol. HI 73191

Symposia, international society for cell biology. New York, London, etc. 1962-70. Symp. Int. Soc. Cell Biol. HI 73192

Symposia series, British mycological society. Cambridge. No. 1+, 1977+. Symp. Ser. Brit. Mycol. Soc. HI 75167

Symposia series in immunobiological standardization. Basel. Vols. 1-22, 1965-73. Symp. Ser. Immunbiol. Standardiz. Preceded: Proceedings, international symposium on microbiological standardization. Superseded by: Developments in biological standardization. HI 73193

Symposia series, phytochemical society of Europe = Annual proceedings of the phytochemical society of Europe. Oxford, London.

Symposia series, society for applied bacteriology. Orlando, FL. No. 1+, 1971+. Symp. Ser. Soc. Appl. Bacteriol. HI 73194

Symposia of the society for experimental biology. Cambridge. Vol. 1+, 1947+. Symp. Soc. Exp. Biol. 5-3976-3. HI 73195

Symposia of the society for general microbiology. Cambridge. Vols. 1+, 1949+. Symp. Soc. Gen. Microbiol. HI 73196

Symposium, see Symposia

Synthese. Utrecht & Bussum, Netherlands. Synthese. See B–P–H 863/1. HI 59674

Synthesis, japanese national committee, international biological programme = J I B P synthesis. Tokyo.

Syobubutu - iho. [Annals of the botanic gardens, Buitenzorg. Vol. hors ser.] Batavia. Vol. 1(1), 2603 (=1943). Syobubutu-Iho. HI 73197

Syokubutsu oyobi dobutsu = Botany and zoology; theoretical and applied. Tokyo. Bot. & Zool. See B–P–H 226/6.

Syoyakugaku zasshi = Japanese journal of pharmacognosy. Chiyoda.

Systema ascomycetum. Umeå. Vol. 1+, 1982+. Syst. Ascomycetum. HI 73198

Systematic and applied microbiology. Stuttgart & New York. Vol. 4+, 1983+. Syst. Appl. Microbiol. Preceded by: Zentralblatt für Bakteriologie, Mikrobiologie und Hygiene. 1 Abteilung. Originale. Reihe C, allgemeine, angewandte und ökologische Mikrobiologie. HI 73199

Systematic botany; quarterly journal of the american society of plant taxonomists. Tallahassee, FL. Vol. 1+, 1976+. Syst. Bot. Preceded, in part, by: Brittonia. HI 73200

Systematic botany monographs; monographic series of the american society of plant taxonomists. Ann Arbor, MI. No. 1+, 1980+. Syst. Bot. Monogr. HI 73201

Systematic lists illustrative of the flora, fauna, palaeontology, and archaeology of the north of Ireland. Belfast. Syst. Lists Ill. Fl. N. Ireland. See B–P–H 863/5. HI 59675

Systematic zoology. Washington, DC. Syst. Zool. See B–P–H 863/6. HI 59676

Szata roslinna polski. Warsaw. 1977+. Szata Rosl. Polsk. HI 73203

Sze ta hsüeh k'an = Journal of school of education. Peiping [=Peking]. J. School Educ. See B–P–H 480/2.

Szechuan forests. [Ssu chuan chih sheng lin.] [China]. Szechuan Forests. See B–P–H 863/10. HI 59677

Szegedi pedagógiai föiskola évkönyve. Szeged, Hungary. Szegedi Pedagóg. Föisk. Évk. See B–P–H 863/11. HI 59678

Szegedi tanárképsö föiskola tudományos közleményei. Szeged, Hungary. Szegedi Tanárképzö Föisk. Tud. Közlem. See B–P–H 863/12. HI 59679

Szegedi tudományegyetem biológiai intézeteinek évkönyve = Annales biologicae universitatis szegediensis. Szeged, Hungary. Ann. Biol. Univ. Szeged. See B–P–H 97/19.

Szemle. A kisérletügyi közlemények melléklete. Budapest. Szemle. See B–P–H 863/14. HI 59680

Szölöszeti, borászati és gazdasági lap. Kassa, Hungary [=Kosice, Czechoslovakia]. Szölösz. Borász. Gazd. Lap. See B–P–H 863/17. HI 59682

Szölöszeti és borászati közlemények. Pest [=Budapest, in part]. Szölösz. Borász. Közlem. See B–P–H 863/18. HI 59683

Szölöszeti és borászati lap. Kassa, Hungary [=Kosice, Czechoslovakia], Budapest & Kecskemet. Szölösz. Borász. Lap. See B–P–H 863/19: HI 59684

Szölészeti kutató intézet évkönyve. Budapest. Szölész. Kutató Intéz. Évk. See B–P–H 863/16. HI 59681

Szölö- és gyümölcstermesztés. Budapest. Vols. 1-8, 1965-73. Szölö- Gyümölcsterm. Superseded by: Gyümölcstermesztés. HI 73204

T C A manual. Gaithersburg, MD. 1975+. T. C. A. Manual. Superseded by: Journal of tissue culture methods. HI 73205

T C A newsletter. Lake Placid, NY. Vols. 1-11, 1968-77. T. C. A. Newslett. Preceded by: T C A report. HI 73206

T C A report. Lake Placid, NY. Vol. 12+, 1978+. T. C. A. Rep. Preceded by: T C A newsletter. HI 73207

T D W G newsletter. Chambésy. No. 1+, 1988+. T. D. W. G. Newslett. HI 73208

T F news; official publication of the Texas forest service. College Station, TX. Vols. ?-68(2), 1970-89. T. F. News. Preceded by: Texas forest news. Superseded by: Texas trees. HI 73209

T I P S journal = International permaculture solutions journal. Orange, MA.

T N S times; herbarium news from national science museum, Tokyo. Tokyo. No. 1+, 1984+. T. N. S. Times. HI 73210

T P C newsletter. Kew = Newsletter of the threatened plants committee, international union for conservation of nature and natural resources. Kew.

T R A F F I C (international) bulletin [i.e. Trade Records Analysis of Flora and Fauna In Commerce]. London. Vol. ?-2+, ?-1980+. T. R. A. F. F. I. C. (Int.) Bull. HI 73211

T R A F F I C (USA) newsletter [i.e. Trade Records Analysis of Flora and Fauna In Commerce]. Washington, DC. Vol. 1+, 1979+. T. R. A. F. F. I. C. (USA) Newslett. HI 73212

T V I S news. Tainan. Vol. 1+, 1985+. T. V. I. S. News. HI 73213

Ta lu k'o hsüeh yuan hui pao = Bulletin of the institute of scientific reserach, Manchoukuo. Hsinking, Manchoukuo [=Changchun, China]. Bull. Inst. Sci. Res. Manchoukuo. See B–P–H 257/4.

Ta tzu jan = Nature. Taipei.

Tabaco. Maracay, Venezuela. Tabaco. See B–P–H 863/23. HI 59686

Tabacco; bollettino [istituto scientifico sperimentale per i tabacchi]. Rome. Vols. 1(1)-77(743), 1897-1973. Tabacco. 5-4147-2. HI 73214

Tabachnaya promyshlennost′ S S S R. Moscow. 1930-36. Tabachn. Promyshl. S.S.S.R. Superseded by: Tabak. Moscow. HI 73215

Tabacologia; Zeitschrift der internationalen tabakwissenschaftlichen Gesellschaft. Cologne. 1951+. Tabacologia. Preceded by: Chronica nicotiana. 5-4146-3. HI 73216

Tabak. Berlin. 1937-40. Tabak (Berlin). Preceded by: Chronica nicotiana. HI 73217

Tabak. Moscow. 1937-40; 1950+. Tabak (Moscow). Preceded by: Tabachnaya promyshlennost′ S S S R. HI 73218

Tabakpflanzer Österreichs. Linz. Vol. 1+, 1950+. Tabakpfl. Österr. HI 73219

Tabulae biologicae. Berlin. Tabulae Biol. See B–P–H 864/1. HI 59687

Tachibana = Citrus. Hiroshima. Citrus (Hiroshima). See B–P–H 314/22.

Taeckholmia. Koenigstein. No. 9+, 1978+. Taeckholmia. Preceded by: Publications from Cairo university herbarium. HI 73220

Taeckholmia additional series. Contains: Flora of Egypt. Koenigstein, Cairo. No. 1+, 1980+. Taeckholmia Addit. Ser. HI 73221

Taehan misaengmul hakhoe chi = Journal of the korean

society for microbiology. Seoul.

Taehan pairosu hakhoe chi = Korean journal of virology. Seoul.

Taehan saenghwahakhoe = Korean journal of biochemistry. Seoul.

Tagblatt der Versammlung der Gesellschaft Deutscher Naturforscher und Aerzte. Erlangen. Tagbl. Versamml. Ges. Deutsch. Naturf. See B–P–H 864/3. HI 59688

Tageblatt für die Versammlung deutscher Naturforscher und Aerzte. Pyrmont [=Bad Pyrmont, Germany]. Tagebl. Versamml. Deutsch. Naturf. Aerzte. See B–P–H 864/4. HI 59689

Tageblatt bei der Versammlung der Naturforscher und Aerzte Deutschlands. Jena. Tagebl. Versamml. Naturf. Aerzte Deutschl. See B–P–H 864/5. HI 59690

Tägliches Notizenblatt für die Theilnehmer an der Versammlung der Naturforscher und Aerzte zu Bonn. Bonn. Tägl. Notizenbl. Theilnehmer Versamml. Naturf. Aerzte Bonn. See B–P–H 864/6. HI 59691

Tagsberichte über die Fortschritte der Natur- und Heilkunde. Abtheilung für Botanik. Weimar. Tagsber. Fortschr. Natur- Heilk., Abth. Bot. See B–P–H 864/7. HI 59692

Tagungsbericht der Gesellschaft für Ökologie. The Hague. Vol. 1, 1973? Tagungsber. Ges. Ökol. Superseded by: Verhandlungen der Gesellschaft für Ökologie. HI 73222

Tagungsbericht, Ludwig-Boltzmann-Instituts für Umweltwissenschaften und Naturschutz in der Universität Graz. Graz. No. 1+, 1976+. Tagungsber. Ludwig-Boltzmann-Inst. Umweltwiss. Naturschutz Univ. Graz. HI 73223

Tai-lin. [Taiwan province agriculture and forestry bureau.] [Taiwan]. Tai-Lin. See B–P–H 864/24. HI 59693

Tai t'ang t'ung hsun = Taiwan sugar news. Taipei, Taiwan. Taiwan Sugar News. See B–P–H 865/27.

Tai wan, see Taiwan.

Taihoku noringakkaiho = Journal of the Taihoku society of agriculture and forestry. Taihoku [=Taipei].

Taihoku teikoku daigaku fuzoku norin senmon-bu Gajutsu hokoku = Bulletin of the school of agriculture and forestry, Taihoku imperial university. Taihoku [=Taipei, Taiwan]. Bull. School Agric. Taihoku Imp. Univ. See B–P–H 271/13.

Taihoku teikoku daigaku nogakuku kiyo = Memoirs of the faculty of agriculture; Taihoku imperial university. Taihoku [=Taipei, Taiwan]. Mem Fac. Agric. Taihoku Imp. Univ. See B–P–H 572/12.

Taihoku teikoku daigaku rinogakubu engeigaku kiyoshitsu zappo = Miscellaneous papers from the horticultural institute; Taihoku imperial university. Taihoku [=Taipei, Taiwan]. Misc. Pap. Hort. Inst. Taihoku Imp. Univ. See B–P–H 598/4.

Taiwan agriculture and forest monthly. [Tai wan nung lin yüeh k'an.] Taipei, Taiwan. Taiwan Agric. Forest Monthly. See B–P–H 864/27. HI 59694

Taihoku teikoku daigaku. Kainanto gakujitsu chosa hokoku = Report of the scientific investigation group on Hainan island from the Taihoku imperial university. Taihoku [=Taipei, Taiwan]. Rep. Sci. Invest. Group Hainan Island Taihoku Imp. Univ. See B–P–H 770/19.

Taihoku teikoku daigaku rinogakubu engeigaku kyoshitsu kiyo = Contributions from the horticultural institute of Taihoku imperial university. Taihoku [=Taipei, Taiwan]. Contr. Hort. Inst. Taihoku Imp. Univ. See B–P–H 328/9.

Taihoku teikoku daigaku rinogakubu engeigaku kyoshitsu ronhyo = Horticultural institute, Taihoku imperial university. Essay. Taihoku [=Taipei, Taiwan]. Hort. Inst. Taihoku Imp. Univ. Essay. See B–P–H 421/8.

Tairiku kagakuin kenkyu hokoku = Report of the institute of scientific research, Manchoukuo. Hsinking, Manchoukuo.

Taiwan agriculture quarterly. Taipei?, Taiwan. Taiwan Agric. Quart. See B–P–H 864/28. HI 59695

Taiwan fisheries research institute, laboratory of biology. Report. [Tai wan sheng sui ch'an shih yen so. Sui ch'an shêng wu hsi yen chiu pao kao.] Taipei, Taiwan. Taiwan Fish. Res. Inst. Lab. Biol. Rep. See B–P–H 865/1. HI 59696

Taiwan forestry. [Taiwan lin yeh.] Taipei, Taiwan. Taiwan Forest. See B–P–H 865/2. HI 59697

Taiwan forestry journal. [Taiwan linye.] Taipei. Vol. 1+, 1974+. Taiwan Forest. J. Preceded by: Taiwan mucai gongye. HI 73224

Taiwan forests. [Tai-wan sen lin.] Taipei. 1955-64. Taiwan Forests. Superseded by: Tai-wan lin yeh chikan. HI 73225

Taiwan k'o hsüeh = Formosan science. Taipei, Taiwan. Formosan Sci. See B–P–H 381/25.

Taiwan lin yeh = Taiwan forestry. Taipei, Taiwan. Taiwan Forest. See B–P–H 865/2.

Taiwan lin yeh = Taiwan timberman. Taipei, Taiwan. Taiwan Timberman. See B–P–H 865/30.

Taiwan linye = Taiwan forestry journal. Taipei.

Tai-wan lin yeh chikan. Taipei. Vols. 1-3, 1964-67. Taiwan Lin Yeh Chikan. Preceded by: Taiwan forests. Superseded by: Quarterly journal of chinese forestry. HI 73226

Taiwan mucai gongye. Taipei? Nos. ?-30, ?-1974. Taiwan Mucai Gongye. Superseded by: Taiwan forestry journal. HI 73227

Taiwan mushrooms. [Tai-wan yang ku.] Taipei. Vol. 1+, 1977+. Taiwan Mushr. HI 73228

Taiwan no sanrin = Sylva of Formosa. Sylva Formosa. See B–P–H 862/18.

Taiwan nojiho = Formosan agricultural review. Taihoku [=Taipei, Taiwan]. Formosan Agric. Rev. See B–P–H 381/24.

Tai wan nung lin yüeh k'an = Taiwan agriculture and forest monthly. Taipei, Taiwan. Taiwan Agric. Forest Monthly. See B–P–H 864/27.

Taiwan research bulletin. [Tai wan yen chiu ts'ung k'an.] Taipei, Taiwan. Taiwan Res. Bull. See B–P–H 865/10. HI 59698

Taiwan ringyo shikenjo hokoku = Bulletin of forest experiment station. Government of Taiwan. Taihoku [=Taipei, Taiwan]. Bull. Forest Exp. Sta. Gov. Taiwan. See B–P–H 251/1.

Taiwan sanrinkai kaiho = Journal of formosan forestry. Taihoku [=Taipei, Taiwan]. J. Formosan Forest. See B–P–H 466/12.

Tai-wan sen lin = Taiwan forests. Taipei.

Tai-wan sen lin yueh k'an = Taiwan forests. Taipei.

Taiwan shêng li po wu k'uan k'o hsüeh nien k'an = Annual of the Taiwan provincial museum. Taipei, Taiwan. Annual Taiwan Prov. Mus. See B–P–H 129/22.

Ta'i-wan sheng li po wu kuan pan nien k'an = Journal of the Taiwan museum. Taipei.

Tai wan sheng li po wu kuan shi k'an = Quarterly journal of the Taiwan museum. Taipei.

Taiwan shêng lin yeh shih yen so pao kao = Bulletin of the Taiwan forest research institute. Taipei, Taiwan. Bull. Taiwan Forest Res. Inst. See B–P–H 283/15.

Tai wan sheng lin yeh shih yen so tung hsün = News Taiwan forest experiment station. Taipei, Taiwan. News Taiwan Forest Exp. Sta. See B–P–H 658/9.

Tai wan sheng sui ch'an shih yen so. Sui ch'an shêng wu hsi yen chiu pao kao = Taiwan fisheries research institute, laboratory of biology. Report. Taipei, Taiwan. Taiwan Fish. Res. Inst. Lab. Biol. Rep. See B–P–H 865/1.

Taiwan sotokufu chuo kendyu jo, nogyo-bu iho = Bulletin of the department of agriculture, government research institute, Formosa. Taihoku [=Taipei, Taiwan]. Bull. Dept. Agric. Gov. Res. Inst. Formosa. See B–P–H 247/29.

Taiwan sotokufu chuo kenkyu-sho, ringyo-bu hokoku = Report of the department of forestry, government research institute, Taihoku. Taihoku [=Taipei, Taiwan]. Rep. Dept. Forest. Gov. Res. Inst. Taihoku. See B–P–H 764/17.

Taiwan sotokufu chuo kendyu-sho, ringyobu iho = Bulletin. Department of forestry, government research institute, Taihoku, Taiwan Taihoku [=Taipei, Taiwan]. Bull. Dept. Forest. Gov. Res. Inst. Taihoku. See B–P–H 248/4.

Taiwan sotoku-fu kenkyu-sho hokoku = Report of the government research institute, Formosa. Taihoku [=Taipei, Taiwan]. Rep. Gov. Res. Isnt. Formosa. See B–P–H 765/19.

Taiwan sotokufu nogyo shikenjo hokoku = Report of the agricultural experiment station of the governor general of Taiwan. [Taiwan]. Rep. Agric. Exp. Sta. Gov. Gen. Taiwan. See B–P–H 762/5.

Taiwan sotukufu nogyobu, shikenjo iho = Bulletin of the government agricultural research institute of Formosa. Taihoku.

Taiwan sotokufu nogyo shikenjo hokoku = Report. Government agricultural research institute, Taiwan, Nippon. Taihoku [=Taipaei, Taiwan]. Rep. Gov. Agric. Res. Inst. Taiwan. See B–P–H 765/16.

Taiwan sotokufu shokusan kyoku ringyo shikenjo tokubetsu hokoku = Special report of the forest experiment station. Office of industry, government general of Taiwan. Taipei, Taiwan. Special Rep. Forest Exp. Sta. See B–P–H 851/17.

Taiwan sotokufu togyo shikenjo hokoku = Report of the government sugar experiment station. Taihoku [=Taipei].

Taiwan sugar. Taipei, Taiwan. Taiwan Sugar. See B–P–H 865/24. HI 59699

Taiwan sugar journal quarterly. [T'ai wan t'ang yeh chi k'an.] Taipei, Taiwan. Taiwan Sugar J. Quart. See B–P–H 865/25. HI 59700

Taiwan sugar news. [Tai t'ang t'ung hsun.] Taipei, Taiwan. Taiwan Sugar News. See B–P–H 865/27. HI 59701

Taiwan suisan zasshi = Formosa journal of aquatic products. Taihoku [=Taipei, Taiwan]. Formosa J. Aquatic Prod. See B–P–H 381/21.

T'ai wan t'ang yeh chi k'an = Taiwan sugar journal quarterly. Taipei, Taiwan. Taiwan Sugar J. Quart. See B–P–H 865/25.

Taiwan tangye shiyansuo yanjiu huibao = Report (Annual) of the Taiwan sugar experiment station. Taipei.

Taiwan tangye yanjiusuo yanjiu huibao = Report (Annual) of the Taiwan sugar research institute. Taipei.

Taiwan t'ê ch'an ts'ung k'an. = Bulletin of special

products in Taiwan. See B–P–H 283/1.

Taiwan timberman. [Taiwan lin yeh.] Taipei, Taiwan. Taiwan Timberman. See B–P–H 865/30. HI 59702

Tai-wan yang ku = Taiwan mushrooms. Taipei.

Tai wan yen chiu ts'ung k'an = Taiwan research bulletin. Taipei, Taiwan. Taiwan Res. Bull. See B–P–H 865/10.

Tai wan yen hang chi k'an = Quarterly journal of the bank of Taiwan. Taipei, Taiwan. Quart. J. Bank Taiwan. See B–P–H 751/14.

Taiwania. Taipei, Taiwan. Taiwania. See B–P–H 685/31. HI 59703

Tajekoztatoja; Pester vasas muvelodesi haz kaktuszkedvelo szakkorenek. Budapest. Vol. 1+, 1964+. Tajekoztatoja. HI 73229

Takara kai kaiho = Alumni report of the Taihoku imperial university agronomy graduates. Taihoku [=Taipei, Taiwan]. Alumni Rep. Taihoku Imp. Univ. Agron. Graduates. See B–P–H 75/14.

Tal ... hållet för kongl. svenska vetenskaps academien. Stockholm = Tal ... hållet för kongl. vetenskaps academien. Stockholm.

Tal ... hållet för kongl. vetenskaps academien. Stockholm. 1739-92? [several items were re-published in 2nd or 3rd editions]. Tal Hållet Kongl. Vetensk. Acad. HI 51352

Tal ... hållet för svenska vetenskaps academien. Stockholm = Tal ... hållet för kongl. vetenskaps academien. Stockholm.

Tal ... hållet för swenska wetenskaps academien. Stockholm = Tal ... hållet för kongl. vetenskaps academien. Stockholm.

Tal ... hållet för wetenskaps academien. Stockholm = Tal ... hållet för kongl. vetenskaps academien. Stockholm.

Tal ... hållet vid praesidii ... för kongl. svenska vetenskaps academien. Stockholm = Tal ... hållet för kongl. vetenskaps academien. Stockholm.

Tall timbers report. Tallahassee, FL. 1978+. Tall Timbers Rep. HI 73230

Tallinna botaanikaaia uurimused. Tallinn. Vols. 1-3, 1962-69. Tallinna Bot. Uurim. HI 73231

Tamagawa daigaku nogakabu kenkyu hokoku = Bulletin of faculty of the agriculture, Tamagawa university. Machida.

Tanaka kankitsu shiken-jo. Mino-mura, Fukuoka-ken = Memoirs of the Tanaka citrus experiment station. Fukuoka?, Japan. Mem. Tanaka Citrus Exp. Sta. See B–P–H 588/7.

Tanap. Sborník prác o tatranskom národnom parku. Tatranska Lomnica, Czechoslovakia. Tanap. See B–P–H 866/4. HI 59704

Tane; journal of the Auckland university college field club ["college" later omitted]. Auckland. Vol. 1+, 1948+. Tane. HI 73232

Tanganyika coffee news. Moshi, Tanganyika [Tanzania Republic]. Tanganyika Coffee News. See B–P–H 866/5. HI 59705

Tanganyika notes and records. Dar-es-Salaam. Nos. 1-64, 1936-65. Tanganyika Notes Rec. Superseded by: Tanzania notes and records. 5-4152-3. HI 73233

Tany malagasy = Terre malgache. Tananarive.

Tanzania notes and records; journal of the Tanzania society. Dar-es-Salaam. No. 65+, 1966+. Tanzania Notes Rec. Preceded by: Tanganyika notes and records. HI 73234

Tanzania silviculture technical note. Lushoto. ?-1981+. Tanzania Silvic. Techn Note. Preceded by: Technical notes, silviculture section, forest division, ministry of lands, forests and wildlife, Tanganyika. HI 73235

Tâp chi sinh vât hoc. Hanoi. Vol. ?-1(2)+, ?-1979+. Tâp Chi Sinh Vât Hoc. HI 73236

Tâp chi sinh vât-dia hoc. Hanoi. Vols. 1-16(1), 1960-78. Tâp Chi Sinh Vât-Dia Hoc. Superseded by: Tâp san sinh vât-dia hoc. HI 73237

Tâp san sinh vât-dia hoc. Hanoi. Vol. 16(2)+, 1978+. Tâp San Sinh Vât-Dia Hoc. Preceded by: Tâp chi sinh vât-dia hoc. HI 73238

Täppan. Skånska trädgårds-föreningen. Malmo, Sweden. Täppan. See B–P–H 866/7. HI 59706

Tarim bakanliği; ziraí mücadele ve ziraí karantina genel müdürlüğü. Arastirma subesi. Vols. ?-7, ?-1973. Tarim Bakanliği. Superseded by: Tarim ve hayavancilik bakanliği. HI 73239

Tarim ve hayvancilik bakanliği; ziraí mücadele ve ziraí karantina genel müdürlüğü. Arastirma. Vol. 8+, 1974+. Tarim Hayvanc. Bakanliği. Preceded by: Tarim bakanliği. HI 73240

Társalkodó. [Edited by? Helmaczy.] Társalkodó (Helmaczy). See B–P–H 866/8. HI 59707

Tartu riikliku ülikooli toimetised = Uchenye zapiski Tartuskogo gosudarstvennogo universiteta. Tartu.

Tartu ülikooli juures oleva loodusuurijate seltsi kirjatööd. Tartu, Estonia [Estonian S S R]. Vol. 24, 1925. Tartu Ülik. Juures Oleva Loodusuur. Seltsi Kirjatööd. Preceded by: Trudy Obshchestva Estestvoispytatelei pri Imperatorskom Yur'evskom Universitetě. HI 73241

Tartu ülikooli juures oleva loodusuurijate seltsi aruanded. Yur'ev [=Tartu], Estonian S S R. Vols. 24-47, 19??-43. Tartu Ülik. Juures Oleva Loodusuur. Seltsi Aruanded.

Preceded by: Protokoly Obshchestva Estestvoispytatelei pri Imperatorskom Yur′evskom Universitetě. 3-2469-2. HI 73242

Tartu ülikooli metsaosakonna toimetused. Tartu. Vols. 1-28, 1924-37. Tartu Ülik Metsaosak. Toimet. HI 73243

Tartu ülikooli taimehaiguste-katsejaama teated. Phytopathological experiment station of the university of Tartu. Tartu, Estonia [Estonian S S R]. Tartu Ülik Taimeh.-Katsej. Teated. See B–P–H 866/12. HI 59708

Tartu ülikooli taimehaiguste-katsejaama tööd. Tartu, Estonia [Estonian S S R]. Tartu Ülik. Taimeh.-Katsej. Tööd. See B–P–H 866/13. HI 59709

Taschen-Bibliothek der wichtigsten und interessantesten See- und Land-Reisen. Nuremberg. Taschen-Biblioth. Wichtigsten Interessantesten See- Land-Reisen. See B–P–H 867/7. HI 59721

Taschenbuch zur Belehrung und Unterhaltung für Wald- und Jagdfreunde. Marburg. 1831. Taschenb. Belehr. Unterhalt. Wald- Jagdfr. Preceded by: Sylvan. 5-4155-2. HI 73244

Taschenbuch für Forst- und Jagdfreunde. Marburg. 1800-12. Taschenb. Forst- Jagdfr. Preceded by: Neujahrsgeschenck für Forst- und Jagdliebhaber. Superseded by: Sylvan. 5-4155-2. HI 73245

Taschenbuch für Garten-Freunde. [Edited by Becker.] Leipzig. Taschenb. Gart.-Freunde (Becker). See B–P–H 866/14. HI 59710

Taschenbuch für Gartenbesitzer und für Blumenfreunde. Leipzig. Taschenb. Gartenbesitz. Blumenfr. See B–P–H 866/15. HI 59711

Taschenbuch für Gartenbesitzer und Blumenliebhaber. Leipzig. Taschenb. Gartenbesitz. Blumenliebh. See B–P–H 866/16. HI 59712

Taschenbuch für Gartenfreunde. Kiel & Leipzig. Taschenb. Gartenfr. (Kiel & Leipzig). See B–P–H 866/17. HI 59713

Taschenbuch für Gartenfreunde und für Blumenliebhaber. Leipzig. Taschenb. Gartenfr. Blumenliebh. See B–P–H 866/18. HI 59714

Taschenbuch für die gesammte Mineralogie. Frankfurt a. M. Taschenb. Gesamamte Mineral. See B–P–H 866/20. HI 59715

Taschenbuch für Küchen-, Garten-, Blumen- und Landwirthschaftsfreunde. Halle. Taschenb. Küchen-Landwirthschaftsfr. See B–P–H 867/1. HI 59716

Taschenbuch für Natur- und Gartenfreunde. Tübingen. Taschenb. Natur- Gartenfr. See B–P–H 867/2. HI 59717

Taschenbuch, österreichische Pflanzenschutzgesellschaft = Österreichische Pflanzenschutzgesellschaft. Taschenbuch. Vienna. Österr. Pflanzenschutzges. Taschenb. See B–P–H 693/1.

Taschenbuch für Pomologen, Gärtner und Gartenfreunde. Ravensburg. Vols. 1-10, 1860-70. Taschenb. Pomol. Superseded by: Jahrbuch für Pomologen, Gärtner und Gartenfreunde. 3-2142-2. HI 73246

Taschenbuch der Reisen, oder unterhaltende Darstellung der Entdeckungen des 18. Jahrunderts. Leipzig. Taschenb. Reisen. See B–P–H 867/4. HI 59718

Taschenbuch zur Verbreitung geographischer Kenntnisse. Prague. Taschenb. Verbreit. Geogr. Kenntn. See B–P–H 867/5. HI 59719

Taschenbuch der Wunder und Seltenheiten in der Natur, Kunst und im Menschenleben. Leipzig. Taschenb. Wunder Seltenh. Natur. See B–P–H 867/6. HI 59720

Taschenkalender für Natur- und Gartenfreunde. Tübingen. Taschenkalend. Natur- Gartenfr. See B–P–H 867/8. HI 59722

Tasks for vegetation science. The Hague, Boston, MA. Vol. 1+, 1981+. Tasks Veg. Sci. HI 73247

Tasmanian fruitgrower and farmer. Franklin, Tas. Vol. 1+, 1915+. Tasmanian Fruitgrower Farmer. 5-4157-1. HI 73248

Tasmanian journal of agriculture. Hobart, Tas. Vols. 1-49(3), 1929-78. Tasmanian J. Agric. Superseded by: Journal of agriculture (Tasmania). 5-4157-1. HI 73249

Tasmanian journal of natural science, agriculture, statistics, etc. Hobart, Tasmania. Tasmanian J. Nat. Sci. See B–P–H 867/15. HI 59723

Tasmanian naturalist. [Supplement to: Bulletin of the tasmanian field naturalists' club, 1965+.] Hobart. 1907-11; 1924-28; 1946-55; 1965+. Tasmanian Naturalist. HI 73250

Tätigkeitsbericht der Bundesanstalt für Pflanzenschutz, Wien. Vienna. Tätigkeitsber. Bundesanst. Pflanzenschutz Wien. See B–P–H 867/17. HI 59724

Tätigkeitsbericht der Naturforschenden Gesellschaft Baselland = Thätigkeitsbericht der Naturforschenden Gesellschaft Baselland. Liestal, Switzerland. Thätigkeitsber. Naturf. Ges. Baselland. See B–P–H 873/14.

Tätigkeitsbericht der Naturwissenschaftlichen Vereines in Aussig. Aussig, Bohemia [=Usti nad Labem, Czechoslovakia]. Tätigkeitsber. Naturwiss. Vereines Aussig. See B–P–H 867/19. HI 59725

Taxodium. Vol. 1+, 1943+. Taxodium. HI 73251

Taxometrics; newsletter dealing with mathematical and statistical aspects of classification. Milan, London. Nos. 1-12, 1962-69. Taxometrics. HI 73252

Taxon; official news bulletin of the international society for plant taxonomy. Utrecht. Taxon. See B–P–H 867/21. HI 59726

Taxonomic index. Cambridge, MA. Taxon. Index. See B–P–H 867/22. HI 59727

Te Kwa Ngahere = New Jealand journal of forestry. Palmerston North, New Zealand. New Zealand J. Forest. See B–P–H 657/15.

Tea. Nairobi. Vols. 1-12(3), 1959-73; [n.s.] vol. 1+, 1980+. Tea (Nairobi). For 1974-75 see: Tea in East Africa. HI 73253

Tea of China. [Zhongguo chaye.] Hangzhou. Vol. ?-3+, ?-1981+. Tea China. HI 73254

Tea in East Africa. Nairobi. Vols. 12(4)-15(1)?, 1973-75? Tea East Africa. Preceded and superseded by: Tea (Nairobi). HI 73255

Tea journal of Bangladesh. Srimangal. Vol. ?-13+, ?-1977+. Tea J. Bangladesh. Preceded by: Tea journal of Pakistan. HI 73256

Tea journal of Pakistan. Dacca. Vols. 1-6, 1963?-68. Tea J. Pakistan. Superseded by: Tea journal of Bangladesh. HI 73257

Tea phytologist. Cambridge. 1934+. Tea Phytol. HI 73258

Tea quarterly. Nuwara Eliya, Ceylon. Tea Quart. See B–P–H 867/25. HI 59728

Tea research journal, Kyoto. [Kyoto-fu chagyo kenkyusho gyomu hokokusho.] Kyoto, Japan. Tea Res. J. Kyoto. See B–P–H 868/2. HI 59729

Tea research journal = Research report of tea. [Chagyo kenkyu hokoku.] Tokyo, Shizuoka.

Tea surveys. Calcutta. ?-1954/56+, ?-1956?+. Tea Surv. HI 73259

Teacher's leaflets on nature-study. [Cornell university. New York state college of agriculture.] Ithaca, NY. Teacher's Leafl. Nat.-Study. See B–P–H 868/3. HI 59730

Tebiwa. Miscellaneous papers of the Idaho state university museum of natural history. Pocatello, ID. Vols. 2-18, 1959-76; [n.s.] vol. 1+, 1976+. Tebiwa. Preceded by: Museum notes and views, Idaho state college. HI 73260

Tech notes: Life sciences. Springfield, VA. ?-1983-86. Tech Notes, Life Sci. HI 75244

Techne; news from the North Carolina biotechnology center. Research Triangle Park, NC. 1986+. Techne. HI 75245

Technical bulletin, agricultural experiment station, Washington state institute of agricultural sciences. Pullman, WA. 1950-74. Techn. Bull. Agric. Exp. Sta. Washington State Inst. Agric. Sci. Superseded by: Technical bulletin, college of agriculture research center, Washington state university. HI 73261

Technical bulletin, Arizona agricultural experiment station = University of Arizona. Arizona agricultural experiment station. Technical bulletin. Tucson, AZ. Univ. Arizona Agric. Exp. Sta. Techn. Bull. See B–P–H 933/18.

Technical bulletin, asian vegetable research and development center. Tainan. No. 1+, 1975+. Techn. Bull. Asian Veg. Res. Developm. Center. HI 73262

Technical bulletin, botany branch, Queensland department of primary industries. Brisbane, Qld. Nos. 1-6, 1977-80. Techn. Bull. Bot. Branch Queensland Dept. Prim. Industr. Superseded by: Queensland botany bulletin. HI 73263

Technical bulletin, botany division, ministry of agriculture, Iraq. Abu Ghraib. ?-1971+. Techn. Bull. Bot. Div. Minist. Agric. Iraq. HI 73264

Technical bulletin. Bureau of entomology and phytopathology. Hangchow, China. Techn. Bull. Bur. Entomol. See B–P–H 868/20. HI 59731

Technical bulletin, centro internacional de agricultura tropical. Cali. No. 1+, 1971+. Techn. Bull. Centro Int. Agric. Trop. HI 73265

Technical bulletin, cocoa research institute. Tafo. No. 9+, 1975+. Techn. Bull. Cocoa Res. Inst. Preceded by: Technical bulletin, west african cacao (later cocoa) research institute. HI 73266

Technical bulletin, college of agriculture. Bangalore. No. 1+, 1952+. Techn. Bull. Coll. Agric., Bangalore. HI 73267

Technical bulletin, college of agriculture research center, Washington state university. Pullman, WA. No. 81+, 1975+. Techn. Bull. Coll. Agric. Res. Center Washington State Univ. Preceded by: Technical bulletin, agricultural experiment station, Washington state institute of agricultural sciences. HI 73268

Technical bulletin. Colorado agricultural college. Colorado experiment station. Fort Collins, CO. Techn. Bull. Colorado Agric. Coll. Colorado Exp. Sta. See B–P–H 868/21. HI 59732

Technical bulletin, commonwealth institute of biological control. Ottawa, Farnham Royal. Nos. 1-18, 1961-77. Techn. Bull. Commonw. Inst. Biol. Control. HI 73269

Technical bulletin, conservation commission of the Northern Territory. Alice Springs, N.T. ?-1981+. Techn. Bull. Conservation Commiss. Northern Territory. HI 73270

Technical bulletin, Cyprus agricultural research institute. Nicosia. No. 1+, 1966+. Techn. Bull. Cyprus Agric. Res. Inst. HI 73271

Technical bulletin, danish government institute of seed pathology for developing countries. Hellerup. Vol. 1+, 1986+. Techn. Bull. Danish Gov. Inst. Seed Pathol. Developing Countries. HI 73272

Technical bulletin of the department of agriculture. Bangkok. Techn. Bull. Dept. Agric. See B–P–H 868/22. HI 59733

Technical bulletin, department of agriculture, Canada. Ottawa. Nos. 1-77 (no. 72 not published], 1935-50. Techn. Bull. Dept. Agric. Canada Superseded by: Publication, department of agriculture, Canada. HI 73273

Technical bulletin, department of agriculture and commerce, Philippine Islands. Manila. Nos. 1-57, 1934-38. Techn. Bull. Dept. Agric. Philipp. HI 73274

Technical bulletin, department of agriculture, Thailand. Bangkok. No. 1+, 1936+. Techn. Bull. Dept. Agric. Thailand. HI 73275

Technical bulletin, department of agriculture, Western Australia. Perth, W.A. No. 1+, 1969+. Techn. Bull. Dept. Agric. Western Australia. HI 73276

Technical bulletin, department of conservation, Illinois. Springfield, IL. No. 1+, 1958+. Techn. Bull. Dept. Conservation Illinois. HI 73277

Technical bulletin of the faculty of agriculture, Kagawa university. Mikita, Japan. Techn. Bull. Fac. Agric. Kagawa Univ. See B–P–H 868/24. HI 59735

Technical bulletin of the faculty of horticulture; Chiba university. [Chiba daigaku engeigakubu gakujutsu hokoku.] Matsudo, Japan. Techn. Bull. Fac. Hort. Chiba Univ. See B–P–H 868/25. HI 59736

Technical bulletin. Forestry experiment station, national Taiwan university. [Yen chiu pau, Kuo li Tai wan ta hsüeh shih yan lin ch'ang] Taipei, Taiwan. Techn. Bull. Forest. Exp. Sta. Natl. Taiwan Univ. See B–P–H 868/27. HI 59737

Technical bulletin of the forests products laboratory; national bureau of industrial research. [Ching chi pu chung yang kung yeh shih yen so mu to'ai shih yen kuan chuan pao.] Kaiting, China. Techn. Bull. Forests Prod. Lab. See B–P–H 868/28. HI 59738

Technical bulletin, Hawaii agricultural experiment station, university of Hawaii = Hawaii agricultural experiment station, university of Hawaii, technical bulletin. Honolulu, HI. Hawaii Agric. Exp. Sta. Univ. Hawaii Techn. Bull. See B–P–H 413/5.

Technical bulletin, international institute of tropical agriculture. Ibadan. No. 1+, 1975+. Techn. Bull. Int. Inst. Trop. Agric. HI 73278

Technical bulletin, international rice research institute. Los Baños, Philippines = International rice research institute; technical bulletin. Los Baños, Philippines. Int. Rice Res. Inst. Techn. Bull. See B–P–H 435/16.

Technical bulletin of the Kagawa-ken agricultural college. Hirai, Japan. Techn. Bull. Kagawa-Ken Agric. Coll. See B–P–H 869/3. HI 59739

Technical bulletin, Kansas state agricultural college, agricultural experiment station = Kansas state agricultural college, agricultural experiment station technical bulletin. Manhattan, KS.

Technical bulletin of the Lafayette natural history museum. Lafayette, LA. No. 1+, 1969+. Techn. Bull. Lafayette Nat. Hist. Mus. HI 73279

Technical bulletin, life sciences agriculture experiment station, university of Maine at Orono. Orono. No. 26+, 1967+. Techn. Bull. Life Sci. Agric. Exp. Sta. Univ. Maine Orono. Preceded by?: Maine agricultural experiment station. Bulletin, technical series. HI 73280

Technical bulletin, Maine agricultural experiment station = Technical bulletin, life sciences agriculture experiment station, university of Maine at Orono. Orono.

Technical bulletin, Michigan agricultural experiment station = Michigan agricultural experiment station. Technical bulletin. East Lansing, MI. Michigan Agric. Exp. Sta. Techn. Bull. See B–P–H 592/19.

Technical bulletin, ministry of agriculture and agrarian reform, United Arab Republic. Cairo. No. 1+, 1967+. Techn. Bull. Minist. Agric. Agrar. Reform United Arab Republic HI 73281

Technical bulletin, ministry of agriculture, Iraq. Baghdad. Nos. 1-20, 1960-65. Techn. Bull. Minist. Agric. Iraq. Superseded by: Bulletin, ministry of agriculture, Iraq. HI 73282

Technical bulletin of Miyagi prefectural agricultural experiment station. [Miyagi-kenritsu nogyo shikenjo hokoku.] Sendai, Japan. Techn. Bull. Miyagi Prefect. Agric. Exp. Sta. See B–P–H 869/8. HI 59740

Technical bulletin, Montana college of agriculture and mechanic arts = Montana college of agriculture and mechanic arts. Agricultural experiment station. Bulletin. Bozeman, MT. Montana Agric. Exp. Sta. Bull. See B–P–H 618/12.

Technical bulletin of the national forestry research bureau. [Yen chiu chuan k'an.] Nanking. Techn. Bull. Natl. Forest. Res. Bur. See B–P–H 869/10. HI 59741

Technical bulletin, New Hampshire agricultural experiment station = New Hampshire agricultural experiment station. Technical bulletin. Durham, NH. New Hampshire Agric. Exp. Sta. Techn. Bull. See B–P–H 653/5.

Technical bulletin, New York (state) agricultural experiment station = New York (state) agricultural experiment station. Technical bulletin. Geneva, NY. New York Agric. Exp. Sta. Techn. Bull. See B–P–H 655/14

Technical bulletin, North Carolina agricultural experiment station = North Carolina agricultural experiment station technical publication. Raleigh, NC. North Carolina Agric. Exp. Sta. Techn. Publ. See B–P–H 663/14.

Technical bulletin, Ohio agricultural experiment station = Ohio agricultural experiment station technical bulletin. Wooster, OH. Ohio Agric. Exp. Sta. Techn. Bull. See B–P–H 685/4.

Technical bulletin, Oklahoma agricultural experiment station = Oklahoma agricultural experiment station technical bulletin. Stillwater, OK. Oklahoma Agric. Exp. Sta. Techn. Bull. See B–P–H 687/9.

Technical bulletin, Oregon agricultural experiment station = Oregon agricultural experiment station technical bulletin. Corvallis, OR. Oregon Agric. Exp. Sta. Techn. Bull. See B–P–H 690/18.

Technical bulletin, palms and dates research centre. Baghdad. No. 1/75+, 1975+. Techn. Bull. Palms Dates Res. Centre. HI 73283

Technical bulletin, Qubba botanic garden. Cairo. No. 1+, 1977+. Techn. Bull. Qubba Bot. Gard. HI 73284

Technical bulletin. Rhodesia agricultural journal. [Supplement to: Rhodesia agricultural journal.] Causeway, Rhodesia. Vols. 1+, 1963+. Techn. Bull. Rhodesia Agric. J. HI 73285

Technical bulletin, royal botanic gardens. Hamilton, Ont. No. [1], 2+, (1957), 1963+ [No. <1>, 1957 lacks any serial title]. Techn. Bull. Roy. Bot. Gard. HI 73286

Technical bulletin, South Dakota agricultural experiment station = South Dakota agricultural experiment station. Technical bulletin. Brookings, SD. South Dakota Agric. Exp. Sta. Techn. Bull. See B–P–H 848/4.

Technical bulletin, T A R C. Tokyo. Nos. 1-11, 1971-78. Techn. Bull. T. A. R. C. Superseded by: Technical bulletin, tropical agriculture research center. HI 73287

Technical bulletin, Texas agricultural experiment station = Bulletin, Texas agricultural experiment station. College Station, TX.

Technical bulletin, tropical agriculture research center. Tokyo. No. 12+, 1979+. Tech. Bull. Trop. Agric. Res. Center Preceded by: Technical bulletin, T A R C. HI 73288

Technical bulletin, United States department of agriculture. Washington, DC. Techn. Bull. U.S.D.A. See B–P–H 869/23. HI 59742

Technical bulletin of the university of British Columbia botanical garden. Vancouver, B.C. No. 1+, 1972+. Techn. Bull. Univ. British Columbia Bot. Gard. HI 73289

Technical bulletin, Virginia agricultural experiment station. Blacksburg. 1915+. Techn. Bull. Virginia Agric. Exp. Sta. HI 73290

Technical bulletin, Welsh plant breeding station. Aberystwyth. Nos. 1-3, 1967-69. Techn. Bull. Welsh Pl. Breed. Sta. HI 73291

Technical bulletin, west african cacao (later cocoa) research institute. Tafo. Nos. 1-8, 1954-61. Techn. Bull. W. African Cacao Res. Inst. Superseded by: Technical bulletin, cocoa research institute. HI 73292

Technical communications, commonwealth bureau of horticulture and plantation crops. Farnham Royal. 1930+. Techn. Commun. Commonw. Bur. Hort. Plantation Crops. HI 73293

Technical communications, commonwealth institute of biological control. Farnham Royal. No. 1+, 1960+. Techn. Commun. Commonw. Inst. Biol. Control. HI 73294

Technical communications, department of agricultural technical services, South Africa. Pretoria. No. 1+, 1960+. Techn. Commun. Dept. Agric. Techn. Serv. South Africa HI 73295

Technical communications of I S H S = Acta horticulturae. The Hague. Acta Hort. See B–P–H 43/14.

Technical communications of international society for horticultural science = Acta horticulturae. The Hague. Acta Hort. See B–P–H 43/14.

Technical communication of the national botanic gardens. Lucknow. No. [1]+, 1968+. Techn. Commun. Natl. Bot. Gard. HI 73296

Technical document, plant protection committee for the south east Asia and Pacific region. Bangkok. No. ?-127+, ?-1980+. Techn. Doc. Pl. Protect. Committee S. E. Asia Pacific Region. HI 73297

Technical handbook, Rhodesia agricultural journal. [Supplement to: Rhodesia agricultural journal.] Salisbury, Rhodesia. Vol. 1+, 1978+. Techn. Handb. Rhodesia Agric. J. HI 73298

Technical leaflet, national institute of agricultural botany. Cambridge. No. 1+, 1977+. Techn. Leafl. Natl. Inst. Agric. Bot. HI 73299

Technical leaflet, weed research organisation. Oxford. No. 1+, 1978+ [no. 1 issued, May 1979]. Techn. Leafl. Weed Res. Organ. HI 73300

Technical monograph, Texas agricultural experiment station = Texas agricultural experiment station. Technical monograph. College Station, TX. Texas Agric. Exp. Sta. Techn. Monogr. See B–P–H 872/22.

Technical news, institute of radiation breeding. [Tekunikaru nyusu hoshasen ikushujo.] Ohmiya. No. 1+, 1969+. Techn. News Inst. Radiat. Breed. HI 73301

Technical newsletter, forest products research institute, Ghana. Kumasi. Vol. ?-4+, ?-1970+. Techn. Newslett. Forest Prod. Res. Inst. Ghana. HI 73302

Technical notes of the Alaska forest research center. Juneau, AK. Nos. 1-50, 1949-60. Techn. Notes Alaska Forest Res. Center. Superseded by: Technical notes, northern forest experiment station. HI 73303

Technical notes, australian forestry and timber bureau. Canberra, A.C.T. ?-1976+. Techn. Notes Austral. Forest. Timber Bur. HI 73304

Technical notes, department of forestry, Queensland. Brisbane, Qld. No. 1+, 1978+. Techn. Notes Dept. Forest. Queensland. HI 73305

Technical notes, dominion forest service, Ottawa = Technical notes, forest research division, forestry branch, Canada. Ottawa.

Technical notes, federal department of forest research, Nigeria. Ibadan. No. 1+, 1960?+. Techn. Notes Fed. Dept. Forest Res. Nigeria. HI 73306

Technical notes, forest products research institute, Ghana. Kumasi. Vol. 1+, 1969+. Techn. Notes Forest Prod. Res. Inst. Ghana. HI 73307

Technical notes, forest products research laboratory. Princes Risborough. 1966-71. Techn. Notes Forest Prod. Res. Lab., Princes Risborough. Superseded by: Technical notes, Princes Risborough laboratory. HI 73308

Technical notes, forest research division, forestry branch, Canada. Ottawa. Nos. 1-123, 1955-62. Techn. Notes Forest Res. Div. Forest. Branch Canada. Preceded by: Silvicultural leaflet, dominion forest service, Canada. Superseded by: Publications, department of forestry and rural development, Canada. HI 73309

Technical notes, forestry research branch, Canada = Technical notes, forest research division, forestry branch, Canada. Ottawa.

Technical notes, lake states forest experiment station. St. Paul, MN. Nos. 1-629, 1935-62. Techn. Notes Lake States Forest Exp. Sta. Superseded by: Research notes L S, United States forest service. HI 73310

Technical note, Laval university forest research foundation = Note technique, fonds de recherches forestières de l'université Laval. Laval.

Technical notes, northern forest experiment station. Juneau, AK. Nos. 51-54, 1961-62. Techn. Notes N. Forest Exp. Sta. Preceded by: Technical notes, Alaska forest research center. Superseded by: Research notes N O R, United States forest service. HI 73311

Technical notes, Princes Risborough laboratory. Princes Risborough. 1972+. Techn. Notes Princes Risborough Lab. Preceded by: Technical notes, forest products research laboratory, Princes Risborough. HI 73312

Technical notes, silviculture section, forest division, ministry of lands, forests and wildlife, Tanganyika. Lushoto. Nos. 1-?, 1956-? Techn. Notes Silvic. Sect. Forest Div. Tanganyika. Superseded by: Tanzania silviculture technical note. HI 73313

Technical notes, south coast glasshouse and mushroom advisory unit. Chichester. No. ?-97+, ?-1985+. Techn. Notes S. Coast Glasshouse Mushr. Advis. Unit. HI 73314

Technical notes, southeastern forest experiment station. Asheville, NC. Nos. 1-72, 1934-49. Techn. Notes SouthE. Forest Exp. Sta. Superseded by: Station papers, southeastern forest experiment station. HI 73315

Technical papers, agricultural experiment station, Puerto Rico = Technical papers, Puerto Rico agricultural experiment station. Rio Piedras, PR.

Technical papers, agricultural experiment station, university of California = University of California, college of agriculture, agricultural experiment station. Technical papers. Berkeley, CA. Univ. Calif. Coll. Agric. Exp. Sta. Techn. Pap. See B–P–H 934/16.

Technical papers, agricultural research institute, Cyprus. Nicosia. No. 1+, 1972+. Techn. Pap. Agric. Res. Inst. Cyprus. HI 73316

Technical papers, arctic institute of north America. Montreal. No. 1+, 1956+. Techn. Pap. Arctic Inst. N. Amer. HI 73317

Technical papers, central states forest experiment station. Columbus, OH. Nos. 1-192, 1943-62. Techn. Pap. Centr. States Forest Exp. Sta. Superseded by: Research papers C S, United States forest service. HI 73318

Technical papers, department of forestry, Queensland. Brisbane, Qld. No. 1+, 1974+. Techn. Pap. Dept. Forest. Queensland. HI 73319

Technical papers, division of horticultural research, C S I R O. No. 1+, 1970+. Techn. Pap. Div. Hort. Res. C. S. I. R. O. HI 73320

Technical papers. Division of plant industry, C S I R O. Melbourne, Vic. Techn. Pap. Div. Pl. Industr. C.S.I.R.O. See B–P–H 869/26. HI 59743

Technical papers, division of tropical agronomy, C S I R O. Melbourne, Vic. No. 15+, 1974+. Techn. Pap. Div. Trop. Agron. C. S. I. R. O. Preceded by: Technical papers, division of tropical crops and pastures, C S I R O. HI 73321

Technical papers, division of tropical crops and pastures,

C S I R O. Melbourne, Vic. Nos. ?-14, ?-1974. Techn. Pap. Div. Trop. Crops Pastures C. S. I. R. O. Preceded by: Technical papers, division of tropical pastures, C S I R O. Superseded by: Technical papers, division of tropical agronomy, C S I R O. HI 73322

Technical papers, division of tropical pastures, C S I R O. Nos. 1-?, 1961-? [suspended between no. 3, 1964 and no., 4, 1970]. Techn. Pap. Div. Trop. Pastures C. S. I. R. O. Superseded by: Technical papers, division of tropical crops and pastures, C S I R O. HI 73323

Technical papers, forest product research, Department of Scientific & Industrial Research. London. Nos. 1-2, 1926. Techn. Pap. Forest Prod. Res. D. S. I. R. HI 73324

Technical papers, forest research institute, New Zealand. Wellington, N.Z. No. 1+, 1954+. Techn. Pap. Forest Res. Inst. New Zealand. HI 73325

Technical paper, herbarium australiense = Herbarium australiense technical paper.

Technical papers in hydrology. Paris. No. 1+, 1969+ [no. 1 issued in 1970, nos. 3 & 4 issued in 1969]. Techn. Pap. Hydrol. HI 75134

Technical paper, international rice research institute. Los Baños, Philippines = International rice research institute. Technical paper. Los Baños, Philippines. Int. Rice Res. Inst. Techn. Pap. See B–P–H 435/17.

Technical papers in marine science, UNESCO = U N E S C O technical papers in marine science. New York, NY.

Technical papers, Pacific southwest forest and range experiment station. Berkeley, CA. Nos. 1-75, 1953-62. Techn. Pap. Pacific SouthW. Forest Range Exp. Sta. Superseded by: Research papers P S W, United States forest service. HI 73326

Technical papers, Puerto Rico agricultural experiment station. Rio Piedras, PR. No. 1+, 1946+. Techn. Pap. Puerto Rico Agric. Exp. Sta. HI 73327

Technical progress report, horticulture and forestry division, agricultural institute, Dublin = Research report, horticulture and forestry division, agricultural institute, Dublin. Dublin.

Technical publication, college of environmental science and forestry, state university of New York. Syracuse, NY. Nos. 96-97, [1973]-75. Techn. Pap. Coll. Environm. Sci. Forest. State Univ. New York. Preceded by: Technical publications, New York state university college of forestry. HI 73328

Technical publication, college of forestry at Syracuse, state university of New York = Technical publications, New York state university college of forestry. Syracuse, NY.

Technical publications, Moss Landing marine laboratories. [Moss Landing, CA.] No. 71-1+, 1971+. Techn. Publ. Moss Landing Mar. Lab. HI 73329

Technical publications, New York state university college of forestry. Syracuse, NY. Vols. 1-95, 1914-69. Techn. Publ. New York State Univ. Coll. Forest. Superseded by: Technical publications, college of environmental science and forestry, state university of New York. HI 73330

Technical publication, North Carolina agricultural experiment station = North Carolina agricultural experiment station technical publication. Raleigh, NC. North Carolina Agric. Exp. Sta. Techn. Publ. See B–P–H 663/14.

Technical publications R8-TP. Atlanta, GA. 1983+. Techn. Publ. R8-TP. HI 73331

Technical publications, society for the promotion of nature reserves. Alford. 1969+. Techn. Publ. Soc. Promot. Nat. Reserves. HI 73332

Technical publications of the state biological survey of Kansas. Lawrence, KS. No. 1-4+, ?-1977+. Techn. Publ. State Biol. Surv. Kansas. HI 73333

Technical publications, Thorne ecological institute. Boulder, CO. No. 10+, 1974+. Techn. Publ. Thorne Ecol. Inst. Preceded by: Bulletin, Thorne ecological institute. HI 73334

Technical report, Bell museum of natural history. Minneapolis, MN. Nos. 1-15, 1963-68. Techn. Rep. Bell Mus. Nat. Hist. HI 73335

Technical report, central rice research institute = Annual technical report, central rice research institute. New Delhi.

Technical report; Chesapeake Bay institute. [Johns Hopkins univeristy.] Baltimore, MD. Techn. Rep. Chesapeake Bay Inst. See B–P–H 869/31. HI 59744

Technical report, college of forestry, wildlife and range sciences, university of Idaho. Moscow, ID. No. 1+, 1975+. Techn. Rep. Coll. Forest. Wildlife Range Sci. HI 73336

Technical report, conservation commission of the Northern Territory. Alice Springs, N.T. No. ?-2+, ?-1982+. Techn. Rep. Conservation Commiss. Northern Territory. HI 62137

Technical report, cooperative national park resources studies unit. Honolulu, HI. No. 1+, Ca.198?+. Techn. Rep. Coop. Natl. Park Resources Stud. Unit. HI 73337

Technical report, department of biological sciences, university of southern California. Los Angeles, CA. No. 1+, 1970+. Techn. Rep. Dept. Biol. Sci. Univ. S. Calif. HI 73338

Technical report, grassland research institute. Hurley. 1965+. Techn. Rep. Grassland Res. Inst. HI 73339

Technical report, Hawaii institute of marine biology. Kaneohe, HI. 1964+. Techn. Rep. Hawaii Inst. Mar. Biol. HI 73340

Technical report, institute of marine science, university of Alaska. Fairbanks, AK. No. 1+, 1971+. Techn. Rep. Inst. Mar. Sci. Univ. Alaska. HI 73341

Technical report, international institute of tropical agriculture. Ibadan. No. 1+, 1976+. Techn. Rep. Int. Inst. Trop. Agric. HI 73342

Technical reports, island ecosystems integrated research program. Honolulu, HI. No. ?-3+, ?-1972+. Techn. Rep. Island Ecosyst. Integr. Res. Program. HI 73343

Technical report, natural science research center, university of the Philippines. Quezon City. No. ?-2+, ?-1973+. Techn. Rep. Nat. Sci. Res. Center Univ. Philipp. HI 73344

Technical report series, international atomic energy agency = International atomic energy agency. Technical report series. Vienna. Int. Atomic Energy Agency, Techn. Rep. Ser. See B–P–H 434/19.

Technical report. Weed research organization. Kidlington, England. Techn. Rep. Weed Res. Organ. See B–P–H 870/2. HI 59745

Technical report for the year, tea research institute of Sri Lanka. St. Coombs, Talawakele. 1973+, 1974+. Techn. Rep. Year Tea Res. Inst. Sri Lanka. Preceded by: Report (Annual), tea research institute of Ceylon. HI 73345

Technical series, board of conservation, state of Florida = Technical series; Florida state board of conservation marine laboratory (later department of natural resources). Coral gables, FL.

Technical series of the bulletin of the Scripps institution of oceanography. La Jolla, CA. Techn. Ser. Bull. Scipps Inst. Oceanogr. See B–P–H 870/3. HI 59746

Technical series; Florida state board of conservation marine laboratory (later department of natural resources). Coral gables, FL. Vols. 1-69, 1951-72. Techn. Ser. Florida State Board Conservation Mar. Lab. Superseded by: Florida marine research publications. HI 73346

Technical series, Max C. Fleischmann college of agriculture = Publications, Max C. Fleischmann college of agriculture. Series T [technical series]. Reno, NV.

Technical series, society for applied bacteriology. London, New York, Orlando, FL. No. 1+, 1969+. Techn. Ser. Soc. Appl. Bacteriol. HI 73347

Technical translations, office of technical services, United States department of commerce. Washington, DC. Vols. 1-18, 1959-1967. Techn. Transl. Off. Techn. Serv. U.S. Dept. Commerce. Preceded by: Translations monthly. Superseded by: Translations register index. HI 73348

Techniques in pure and applied microbiology. New York, NY. Vol. 1+, 1969+. Techn. Pure Appl. Microbiol. HI 73349

Technisch-wirtschaftliche Zeitung für die mitteleuropäische Landwirtschaft. Vienna. Techn.-Wirtschaftl. Zeitung Mitteleurop. Landw. See B–P–H 870/5. HI 59747

Technologia; annali universali di tecnologia, di agricoltura, di economia rurale e domestica, di arti e di mestiere. Milan. Technologia. See B–P–H 870/6. HI 59748

Tecnica agricola. Catania. Vol. 1+, 1949+. Tecn. Agric. 5-4170-1. HI 73351

Tectona. Buitenzorg, Dutch E. Indies [=Bogor, Indonesia]. Tectona. See B–P–H 870/7. HI 59749

Tegniese mededeling, department van landbou, Suid Afrika = Technical communications, department of agricultural technical services, South Africa. Pretoria.

Teikoku daigaku ika kiyo = Mitteilungen aus der Medicinischen Facultät der Kaiserlich-japanischen Universität. Tokyo. Mitt. Med. Fac. Kaiserl.-Jap. Univ. See B–P–H 606/1.

Teikoku gakushiin kiji = Proceedings of the imperial academy of Japan. Tokyo.

Teion kagaku = Low temperature science. Sapporo.

Teion kagaku kenkyujo obun hokoku = Contributions from the institute of low-temperature science. Series B, biological science. Sapporo.

Teishitsu rinyakyoku Hokkaido ringyo shikenjo hokoku = Report of the Hokkaido branch of the imperial forestry and estates bureau. Sapporo.

Tekunikaru nyusu hoshasen ikushujo = Technical news, institute of radiation breeding.

Telegen abstracts. New York. Vol. ?-7+, ?-1989+. Telegen Abstr. HI 73352

Telegen reporter; worldwide biotechnology and genetic engineering intelligence network. New York. Vol. 1+, 1982+. Telegen Reporter. HI 73353

Tełekagir: Bnakan gitouthjounner = Izvestiya: Estestvennye nauki. Erivan.

Tełekagir: Biologiakan gitouthjounner = Izvestiya akademiya nauk armyanskoi S S R. Biologicheskie nauki. Erevan.

Tełkagir: Biologiakan ew gjoułatntesakan gitouthjounner = Izvestiya akademiya nauk armyanskoi S S R. Biologicheskie i sel′skokhozyaistvennie nauki. Erevan.

Tełekagir. S S R M gitouthjounneri akademiaji Hajkakan filial = Izvestiya. Armyanskii filial Akademii nauk S S

S R. Erivan.

Tełekagir S S R M gitouthjounneri akademiaji Hajkakan filiali: Bnakan gitouthjounner = Izvestiya Armyanskogo filiala Akademii nauk S S S R: Estestvennye nauki. Erivan.

Telhan patrika. Hyderabad. Vols. 1-2, 1969-70. Telhan Patrika. Superseded by: Oilseeds journal. HI 73354

Telma. Hannover. Vol. 1+, 1971+. Telma. HI 73355

Telopea; contributions from the national herbarium of New South Wales. Sydney, N.S.W. Vol. 1+, 1975+. Telopea. Preceded by: Contributions from the New South Wales national herbarium. HI 73356

Tematicheskii sbornik, otdel fiziologii i biofiziki rastenii, akademiya nauk tadzhikskoi S S R. Dushanbe. No. 1-?, 1962-67. Temat. Sborn. Otd. Fiziol. Biofiz. Akad. Nauk Tadzhiksk. S.S.R. HI 73357

Tempo; giornale italiano di medicina, chirurgia e scienze affini. Florence. Tempo. See B–P–H 870/16. HI 59750

Temporary descriptive entries for new varieties of cereals (and flax) = Fiches descriptives provisoires des nouvelles variétés de cereales proposées à l'inscription au catalogue (Liste de commercialisation en France). Versailles.

Temporary descriptive entries for new varieties of forage plants = Fiches descriptives provisoires des nouvelles variétés de plantes fourragères proposées à l'inscription au catalogue (Liste de commercialisation en France). Versailles.

Tenerife agrícola. Santa Cruz de Tenerife, Canary Islands. Tenerife Agríc. See B–P–H 870/17. HI 59751

Tenger. Budapest. Tenger. See B–P–H 870/18. HI 59752

Tennessee horticulture. Knoxville, TN. Tennessee Hort. See B–P–H 870/21. HI 59753

Tensai kenkyu hokoku = Bulletin of sugar beet research. Tokyo.

Tensai kenkyu kaiho = Proceedings of the sugar beet research association. (Japan).

Természet. Buda [=Budapest, in part]. Természet (Buda). See B–P–H 871/6. HI 59757

Természet. Budapest. Természet (Budapest). See B–P–H 871/7. HI 59758

Természet. Pest [=Budapest, in part]. Természet (Pest). See B–P–H 871/8. HI 59759

Természet. gazdasági és mesterségi esméretek tára. Pest [=Budapest, in part]. Term. Gazd. Mest. Esm. Tára. See B–P–H 870/26. HI 59756

Természet és társadalom. Budapest. Term. & Társad. See B–P–H 870/23. HI 59754

Természet és technika. Budapest. Term. & Techn. See B–P–H 870/24. HI 59755

Természetbarát. Budapest. Természetbarát (Budapest). See B–P–H 871/11. HI 59760

Természetbarát. Kolozsvar [=Cluj, Rumania]. Természetbarát (Kolozsvar). See B–P–H 871/12. HI 59761

Természetjárás. Budapest. Természetjárás. See B–P–H 871/13. HI 59762

Természetrajzi füzetek. Budapest. Természetrajzi Füz. See B–P–H 871/16. HI 59763

Természettudomány. Budapest. Természettudomány. See B–P–H 871/24. HI 59769

Természettudományi dokumentáció. Budapest. Term. Dokument. See B–P–H 871/18. HI 59764

Természettudományi füzetek. Temesvar [=Timisoara, Rumania] & Budapest. Term. Füz. See B–P–H 871/19. HI 59765

Természettudományi közlöny. Budapest. Term. Közl. See B–P–H 871/20. HI 59766

Természettudományi pálya munkak. Pest [=Budapest, in part]. Term. Pálya Munk. See B–P–H 871/22. HI 59767

Természettudományok tanítása. Budapest. Term. Tanít. See B–P–H 871/23. HI 59768

Tern. Norwich. No. 1+, 1978+. Tern. Preceded by: Newsletter, Norfolk naturalists' trust. HI 73358

Terra; museum alliance quarterly. Los Angeles, CA. Vol. 9(2)+, 1970+. Terra. Preceded by: Quarterly, Los Angeles county museum of natural history. Science and technology. HI 73359

Terra cimbra. ?-1987+. Terra Cimbra. HI 73360

Terrarium topics. Norwalk, CT. ?-1977+. Terrar. Topics. HI 73361

Terre, air, mer, la géographie = Géographie. Paris. Géographie. See B–P–H 398/9.

Terre malgache. Tananarive. No. 1+, 1966+. Terre Malgache. HI 73362

Terre marocaine. Rabat, Morocco. Terre Maroc. See B–P–H 872/1. HI 59770

Terre sauvage. Levallois-Perret. 1986+. Terre Sauv. HI 73363

Terre sauvage. Hors série Levallois-Perret. 1988+. Terre Sauv., Hors Sér. HI 75285

Terre vaudoise; journal agricole. Lausanne. 1909+. Terre Vaud. Preceded by: Chronique agricole du Canton de Vaud. HI 73364

Terre et la vie; revue d'histoire naturelle. Paris. Vols. 1-

9, 1931-37; sér. 2, vols. 8-33, 1947-79. Terre & Vie. Série 2 preceded by: Bulletin de la société nationale d'acclimatation de France. Superseded by: Revue d'écologie appliquée à la protection de la nature. Paris. 5-4183-3. HI 59771

Terre et la vie. Supplément. Paris. Vols. 1-2?, 19??-? Terre & Vie, Suppl. Superseded by: Revue d'écologie. Supplément. HI 74881

Terre vive; bulletin de la société d'histoire naturelle et de préhistoire de Mâcon. Macon. No. 1+, 198?+. Terre Vive. HI 73365

Terrestrische ökologie. Sonderheft. Halle (Salle). Vol. 1+, 1981+. Terrestr. Ökol., Sonderh. HI 73366

Testing report. Forestry experiment station, Tyukyu government. [Tyukyu seifu keizai kyoku ringyo shikenjo kenkyu hokoku.] Naha?, Okinawa. Test. Rep. Forest. Exp. Sta. Ryukyu Gov. See B–P–H 872/4. HI 59772

Tests of agrochemicals and cultivars. London. No. 1+, 1980+. Tests Agrochemicals Cultivars. HI 73367

Téthys. Marseille. Vol. 1+, 1969+. Téthys. Preceded by: Recueil des travaux de la station marine d'Endoume. HI 73368

Tetrahedron. London & New York, NY. Tetrahedron. See B–P–H 872/5. HI 59773

Tetrahedron letters. New York, NY. Tetrahedron Lett. See B–P–H 872/6. HI 59774

Tetsudo daijin kanbo kendyu-jo. Gyomu kenkyu shiryo = Report of the office of ministry of railroads. [Japan]. Rep. Off. Minist. Railroads. See B–P–H 768/29.

Teutsche Obstgärtner. Weimar. Teutsche Obstgärtn. See B–P–H 872/9. HI 59775

Teutsche Zeitschrift für die gesammte Thierheilkunde. Giessen, Germany. Teutsche Z. Gesammte Thierheilk. See B–P–H 872/10. HI 59776

Tev'a va-arets. Tel Aviv, Israel. Tev'a Va-Arets. See B–P–H 872/11. HI 59777

Texas A & M agriculturist. College Station, TX. Texas A. & M. Agric. See B–P–H 872/19. HI 59778

Texas A & M scientific review. College Station, TX. Texas A. & M. Sci. Rev. See B–P–H 872/20. HI 59779

Texas academy of science publications in natural history. Non-technical series. Austin, TX. Texas Acad. Sci. Publ. Nat. Hist., Non-Techn. Ser. See B–P–H 872/21. HI 59780

Texas agricultural experiment station. Technical monograph. College Station, TX. Texas Agric. Exp. Sta. Techn. Monogr. See B–P–H 872/22. HI 59781

Texas agricultural progress. College Station, TX. Texas Agric. Progr. See B–P–H 872/23. HI 59782

Texas forest news. College Station, TX. ?-1926-69 [suspended May 1933-Dec. 1935]. Texas Forest News. Superseded by: T F news. 5-4192-1. HI 73369

Texas forestry paper. Nacogdoches, TX. No. 1+, 1970+. Texas Forest. Pap. HI 73370

Texas horticulturist. College Station, TX. Vol. 1+, 1974+. Texas Hort. HI 73371

Texas journal of science. San Marcos, TX. Texas J. Sci. See B–P–H 872/24. HI 59783

Texas native plant society news. Denton, TX. Vols. 1-5(3), 1983?-87. Texas Native Pl. Soc. News. Superseded by: Native plant society of Texas news. HI 73372

Texas natural history. Dallas, TX. Vol. 1+, 1985+. Texas Nat. Hist. HI 73373

Texas reports on biology and medicine. Galveston, TX. Texas Rep. Biol. Med. See B–P–H 872/26. HI 59784

Texas trees. College Station, TX. Vol. 68(3)+, 1989+. Texas Trees. Preceded by: T F news. HI 74882

Texas university publications of the institute of marine science. Austin, TX. Texas Univ. Publ. Inst. Mar. Sci. See B–P–H 873/1. HI 59785

Texas wildflower newsletter. Kerrville, TX. Vol. 1+, 19??+ [Dates of publication not ascertained.] Texas Wildflower Newslett. HI 73374

Textile research journal. Lancaster, PA. Textile Res. J. See B–P–H 873/3. HI 59786

Teysmannia; magazijn van horticultuur en landbouw der tropen. Batavia, Dutch E. Indies [=Jakarta, Indonesia]. Teysmannia. See B–P–H 873/5. HI 59787

Tezisy dokladov, vsesoyuznoe soveshchanie po voprosam izucheniya i osvoeniya flory i rastitel'nosti vysokogovii. Vol. ?-5+, ?-1971+. Tezisy Dokl. Vsesoyuzn. Soveshch. Vopro. Izuch. Osvoeniya Fl. Rastitel'n. Vysokogov. HI 73375

Thai abstracts. Series A, science and technology. Bangkok. No. 1+, 1974+. Thai Abstr., A. HI 73376

Thai forest bulletin. Botany. Bangkok. Vol. 1+, 1954+. Thai Forest Bull., Bot. HI 73377

Thai journal of agricultural science. Bang Khen, Bangkok. Vol. 1+, 1968+. Thai J. Agric. Sci. HI 73378

Thai national scientific papers. Biological series. Bangkok. No. 1+, 1971+. Thai Natl. Sci. Pap., Biol. Ser. HI 73379

Thailand illustrated. Bangkok. Vol. 1+, 1975+. Thailand Ill. HI 73380

Thalassas; review of marine sciences. Revista de ciencias del mar. Santiago. Vol. 1+, 1983+. Thalassas.

HI 73381

Thalassia. Venice & Jena. Thalassia. See B–P–H 873/7. HI 59788

Thalassia jonica. Istituto sperimentale talassografico di Taranto. Taranto, Italy. Thalassia Jonica. See B–P–H 873/8. HI 59789

Thalassia jugoslavica. Zagreb, Yugoslavia. Thalassia Jugoslav. See B–P–H 873/10. HI 59790

Thalassia salentina. Porto Cesareo. No. 1+, 1966+. Thalassia Salentina. HI 73382

Thalassina epistemonika phylla. Athens. Thalassina Epistemonika Phylla. See B–P–H 873/11. HI 59791

Thalassina phylla. Athens. Thalassina Phylla. See B–P–H 873/12. HI 59792

Thalassographica. Athens. Vol. 1+, 1976+. Thalassographica. Preceded by: Hellenikė okeanologia kai limnologia tou institoutou okeanographikon kai halieutikon ereunon. HI 73383

Tharandter forstliches Jahrbuch. Berlin. Tharandter Forstl. Jahrb. See B–P–H 873/13. HI 59793

Thätigkeitsbericht der Naturforschenden Gesellschaft Baselland. Liestal, Switzerland. Thätigkeitsber. Naturf. Ges. Baselland. See B–P–H 873/14. HI 59794

Thbilisi botanikur bagis mecnierul-bamogenebithi ganqophilebatha nacerebi = Zapiski Nauchno-prikladnykh Otdělov Tiflisskago Botanicheskago Sada. Tiflis.

Thbilisis botanikuri bagis moambe = Vestnik Tbilisskogo botanicheskogo sada. Tiflis.

Thbilisis botanikuri bagis šromebi = Trudy Tbilisskogo botanicheskogo sada. Tiflis.

Thbilisis botanikuri institutis šromebi = Trudy Tbilisskogo botanicheskogo instituta. Tiflis.

Thee. Korte aanteekeningen van het algemeen proefstation voor thee. [Netherlands]. Thee. See B–P–H 873/23. HI 59795

Thema; recherche scientifique dans les hautes écoles et universités suisse. Berne. No. 1+, 1986+. Thema. HI 73384

Theoretical and applied genetics; international journal of breeding research and cell genetics. Berlin, New York, etc. Vol. 38+, 1968+. Theor. Appl. Genet. Preceded by: Züchter. HI 73385

Theoretical population biology; an international journal. New York. Vol. 1+, 1970+. Theor. Populat. Biol. HI 73386

Therapeutic gazette. Detroit, MI. Therap. Gaz. See B–P–H 873/24. HI 59796

Therapia; revista de medicina e cirurgia. São Paulo. Therapia. See B–P–H 873/25. HI 59797

Thesis abstracts. Hissar. Vol. 1+, 1975+. Thesis Abstr. HI 73388

Thicket; newsletter of the Long Island pine barren society. [Smithtown, NY.] Vol. 1+, 1982+. Thicket. HI 73389

This week in the arboretum, botany and pathology division, Canada. Vols. 1-11, 19??-56. This Week Arbor. Bot. Pathol. Div. Canada. Superseded by: Arboretum notes, botany and plant pathology division, Canada. HI 73390

Thomsonian recorder, or impartial advocate of botanic medicine, ... Columbus, OH. Thomsonian Rec. See B–P–H 874/1. HI 59798

Threatened plants committee newsletter = Newsletter of the threatened plants committee, international union for conservation of nature and natural resources. Kew.

Threatened plants newsletter. Kew. No. 11+, 1983+. Threat. Pl. Newslett. Preceded by: Newsletter of the threatened plants committee, international union for conservation of nature and natural resources. HI 73391

Three year report, royal botanic gardens, Kew = Report by the board of trustees of the royal botanic gardens, Kew. Kew.

Thunbergia. Uppsala. Vol. 1+, 1986+. Thunbergia. Preceded by: Publications from the herbarium, university of Uppsala. HI 73392

Thüringens Merkwürdigkeiten aus dem Gebiete der Natur, der Kunst, des Menschenlebens. Arnstadt, Germany. Thüringens Merkwürdigk. Natur. See B–P–H 874/2. HI 59799

Ti hsüeh tsa-chih = Geographical journal. Peking. Geogr. J. (Peking). See B–P–H 397/20.

Ti-li = Geography. Nanking. Geography (Nanking). See B–P–H 398/11.

Ti li hsüeh pao = Acta geographica sinica. Peiping [=Peking]. Acta Geogr. Sin. See B–P–H 43/5.

Ti li hsüeh t'zu liao.] [Academia sinica = Memoirs of geography. Nanking. Mem. Geogr. See B–P–H 573/4.

Ti-li tsa-chih = Geographical review. Nanking. Geogr. Rev. (Nanking). See B–P–H 397/25.

Ti ts'eng hsüeh tsa chih = Acta stratigraphica sinica. Nanking.

Ticorquidea. San José. Vol. 1+, 1970+. Ticorquidea. HI 73393

Tidning för trädgårdsodlare. Stockholm. Tidn. Trädgårdsodlare. See B–P–H 874/5. HI 59800

Tidning för trädgårdsskötsel och allmän wextkultur. Lund. Tidn. Trädgårdsskötsel Allmän Wextkult. See

B–P–H 874/6. HI 59801

Tidskrift för jägare och naturforskare. Stockholm. Tidskr. Jägare Naturf. See B–P–H 874/11. HI 59802

Tidskrift för läkare och pharmaceuter. Stockholm. Tidskr. Läkare Pharm. See B–P–H 874/13. HI 59803

Tidskrift för landtmän. Lund. Tidskr. Landtm. See B–P–H 874/14. HI 59804

Tidskrift för landtmänna- och kommunalekonomien. Uppsala. Tidskr. Landtmänna- Kommun.-Ekon. See B–P–H 874/15. HI 59805

Tidskrift för lantmän och andelsfolk. Helsinki. Tidskr. Lantm. Andelsfolk. See B–P–H 874/16. HI 59806

Tidskrift. Skogs- och lantbruksakademien. Stockholm. Tidskr. Skogs- Lantbruksakad. See B–P–H 874/17. HI 59807

Tidskrift för skogshushållning. Uppsala. Tidskr. Skogshush. See B–P–H 874/18. HI 59808

Tidskrift för vetenskapelig och praktisk mykologi = Karstenia. Sienitieteellinen ja sienitalondellinen aikakauskirja. Helsinki. Karstenia. See B–P–H 511/ 18.

Tidskrift för veterinär-medicin och husdjursskötsel. Stockholm. Tidskr. Veterin.-Med. Husdjursskötsel. See B–P–H 874/20. HI 59809

Tidsskrift for den danske bondestand, indeholdende bidrag til dens oplysning, nytte og fornøielse. Odense. 1861/62-1864/65, 1861-65. Tidsskr. Danske Bondest. HI 73394

Tidsskrift for frugttraedyrkere. Copenhagen. Nos. 1-3, 1865-66. Tidsskr. Frugttraeddyrk. HI 73395

Tidsskrift for Havevaesen. Copenhagen. Vols. 1-14, 1866-80? Tidsskr. Havevaes. 5-4215-2. HI 73396

Tidsskrift för historisk botanik. Copenhagen. Vol. 1, 1918-21. Tidsskr. Hist. Bot. HI 73397

Tidsskrift for landbrukets planteavl. Copenhagen. 1895-1913. Tidsskr. Landbr. Planteavl. Superseded by: Tidsskrift for planteavl. HI 73398

Tidsskrift for landoekonomie. Copenhagen. Tidsskr. Landoekon. See B–P–H 874/22. HI 59810

Tidsskrift for landøkonomi = Tidsskrift for landoekonomie. Copenhagen. Tidsskr. Landoekon. See B–P–H 874/22.

Tidsskrift for litteratur og kritik. Copenhagen. Tidsskr. Litt. Krit. See B–P–H 874/24. HI 59811

Tidsskrift for naturvidenskaberne. Copenhagen. Tiddskr. Naturvidensk. See B–P–H 874/26. HI 59812

Tidsskrift for physik og chemi samt disse videnskabers anvendelse. Copenhagen. Vols. 1-18, 1862-79; ser. 2, vols. 1-12, 1800-91; ser. 3, vols. 1-3, 1892-94 [also numbered vols. 1-33]. Tidsskr. Phys. Chem. Superseded by: Nyt tidsskrift for fysik og kemi. 5-4215-3. HI 73399

Tidsskrift for planteavl. Copenhagen, Lyngby. 1914+. Tidsskr. Planteavl. Preceded by: Tidsskrift for landbrukets planteavl. HI 73400

Tidsskrift for populaere fremstillinger af naturvidenskaben. Copenhagen. Tidsskr. Populaere Fremstill. Naturvidensk. See B–P–H 875/3. HI 59813

Tidsskrift för skogbruk. Christiania [=Oslo, Norway]. Tidsskr. Skogbr. See B–P–H 875/4. HI 59814

Tidsskrift for skovbrug. Copenhagen. Vols. 1-12, 1875-91. Tidsskr. Skovbr. 5-4216-1. HI 73401

Tidsskrift for Skovvaesen. Copenhagen. Vols. 1-30, 1889-1918. Tidsskr. Skovvaes. 5-4216-1. HI 73402

Tiedoituksia, metsäpuiden rodunjalostussaatio. Helsinki. Nos. 1-3, 1955-57. Tiedoit. Metsäp. Rodunjalost. Superseded by: Toiminta, metsäpuiden rodunjalostussaatio. HI 73403

Tierärztliche Rundschau. Berlin. Tierärztl. Rundschau. See B–P–H 875/7. HI 59815

Tierra; revista ... de agricultura y ganadería. Mexico City. Tierra. See B–P–H 875/8. HI 59816

Tigerpaper. Bangkok. Vol. 1+, 1974+. Tigerpaper. HI 73404

Tijdschrift, door het antwerpsch kruidkundig genootschap uitgegeven, over land- en tuinbouwkunde en verdere natuurwetenschappen. Antwerp. Tijdschr. Antwerpsch Kruidk. Genootsch. Land- Tuinbouwk. See B–P–H 875/12. HI 59817

Tijdschrift over boomteeltkunde, bloementeelt en moeshovenierderij. Ghent. Tijdschr. Boomteeltk. See B–P–H 875/13. HI 59818

Tijdschrift voor economische geographie = Tijdschrift voor economische en sociale geografie. The Hague & Rotterdam. Tijdschr. Econ. Sociale Geogr. See B–P–H 875/14.

Tijdschrift voor economische en sociale geografie. The Hague & Rotterdam. Tijdschr. Econ. Sociale Geogr. See B–P–H 875/14. HI 59819

Tijdschrift voor internationaal succulentenliefhebbers. Antwerp. Vols. 1-2, 1967-68. Tidschr. Int. Succ.-Liefhebb. Preceded by: Cactusweelde. Superseded by: Cactus. (Antwerp). HI 73405

Tijdschrift van het koninklijk nederlandsch aardrijkskundig genootschap. Amsterdam. Tijdschr. Kon. Ned. Aardrijksk. Genootsch. See B–P–H 875/16. HI 59820

Tijdschrift voor kunsten en wetenschappen van het departement der Zuiderzee. Amsterdam. Tijdschr. Kunsten Dept. Zuiderzee. See B–P–H 875/17.

HI 59821

Tijdschrift voor land-, tuin-, en boschbouwcultures in Nederlandsch Oost Indië. Semarang, Dutch E. Indies [Indonesia]. Tijdschr. Land- Tuin- Boschbouwcult. Ned. Oost Indië. See B–P–H 875/18. HI 59822

Tijdschrift voor land- en tuinbouw en boschkultuur. Semarang, Dutch E. Indies [Indonesia]. Tijdschr. Land-Tuinb. Boschkult. See B–P–H 875/19. HI 59823

Tijdschrift voor landbouwkunde. Groningen. Tijdschr. Landbouwk. See B–P–H 875/20. HI 59824

Tijdschrift van den leeraarskring tot volmaaktere boomteeltkunde in Belgié. Ghent. Tijdschr. Leeraarskring Volmaaktere Boomteeltk. België . See B–P–H 875/21. HI 59825

Tijdschrift voor natuurkundige wetenschappen en kunsten. Amsterdam. Tijdschr. Natuurk. Wetensch. Kunsten. See B–P–H 876/2. HI 59826

Tijdschrift voor natuurlijke geschiedenis en physiologie. Amsterdam. Tijdschr. Natuurl. Gesch. Physiol. See B–P–H 876/4. HI 59827

Tijdschrift voor Nederlandsch-Indië. Batavia, Dutch E. Indies [=Jakarta, Indonesia]. Tijdschr. Ned.-Indië. See B–P–H 876/7. HI 59829

Tijdschrift van de nederlandsche heidemaatschappij. Arnhem. Netherlands. Tijdschr. Ned. Heidemaatsch. See B–P–H 876/5. HI 59828

Tijdschrift der nederlandsche maatschappij ter bevordering van nijverheid. Leiden. Vols. 1-4, 1897-1900. Tijdschr. Ned. Maatsch. Bevord. Nijverh. HI 73406

Tijdschrift "Nieuw-Guinea". The Hague. Tijdschr. "Nieuw-Guinea". See B–P–H 876/8. HI 59830

Tijdschrift voor nijverheid en landbouw in Nederlandsch-Indië. Batavia, Dutch E. Indies [=Jakarta, Indonesia]. Tijdschr. Nijverh. Landb. Ned.-Indië. See B–P–H 876/9. HI 59831

Tijdschrift voor nijverheid in Nederlandsch-Indië. Batavia, Dutch E. Indies [=Jakarta, Indonesia]. Tijdschr. Nijverh. Ned.-Indië. See B–P–H 876/10. HI 59832

Tijdschrift orgaan der maatschappij afdeeling, koloniaal museum. Haarlem. 1889-91. Tijdschr. Orgaan Maatsch. Afd. Kolon. Mus. HI 73407

Tijdschrift over plantenziekten. Amsterdam. Vols. 1-68, 1895-1962. Tijdschr. Plantenziekten. Superseded by: Netherlands journal of plant pathology. 5-4218-3. HI 73408

Tijdskrif van die suid-afrikaanse biologiese vereniging = Journal of the south african biological society. Pretoria. J. S. African Biol. Soc. See B–P–H 479/5.

Tijdschrift voor tuinbouw. Assen, Netherlands. Tijdschr. Tuinb. (Assen). See B–P–H 876/12. HI 59833

Tijdschrift voor tuinbouw. Groningen. Tijdschr. Tuinb. (Groningen). See B–P–H 876/13. HI 59834

Tijdschrift voor vergelijkende geneeskunde, gezondheidsleer en parasitaire en infectieziekten. Leiden. 1915-25. Tijdschr. Vergelijk. Geneesk. Gezondheid. Parasit. Infectiez. Superseded by: Nederlandsch tijdschrift voor hygiene, microbiologie en serologie. HI 73409

Tijdschrift voor de wis- en natuurkundige wetenschappen, uitgegeven door de eerste klasse van het kon. ned. instituut van wetenschappen, letteren en schoone kunsten. Amsterdam. Tijdschr. Wis- Natuurk. Wetensch. Eerste Kl. Kon. Ned. Inst. Wetensch. See B–P–H 876/17. HI 59835

Tijdskrif van die suid-afrikaanse bosbouwvereniging. Pretoria. Tijdsckr. Suid-Afrikaanse Bosbouwver. See B–P–H 876/19. HI 59836

Tilth; biological agriculture in the northwest. Arlington, WA. Vols. ?-8(3), ?-1982; vol. 10(1-2), 1984. Tilth. For 1983-84 see: Tilth newsletter. HI 73410

Tilth newsletter. Olympia, WA. Vols. 8(4)-9, 1983-84. Tilth Newslett. Preceded and superseded by: Tilth. HI 73411

Tilton's journal of horticulture. Boston, MA. 1870-71. Tilton's J. Hort. Preceded by: American journal of horticulture and florist's companion. 5-4220-3. HI 73412

Timberlab papers. Princes Risborough. Nos. 3-6, 1969-?; no. 8+, 19??+. Timberlab Pap. Preceded by, and no. 7 entitled: F P R L papers. HI 73413

Timbers of Ethiopia. Addis Ababa, Ethiopia. Timbers Ethiopia. See B–P–H 876/25. HI 59962

Timehri. Georgetown, Demarara. Vols. 1-5, 1882-86; n.s. vols. 1-[15?], 1887-1902; n.s. vols. 1-2, 1911-12. Timehri. HI 73414

Timely hints, Arizona agricultural experiment station = University of Arizona. Arizona agricultural experiment station. Timely hints. Tucson, AZ. Univ. Arizona Agric. Exp. Sta. Timely Hints. See B–P–H 933/19.

Timiryazevskie chteniya. Moscow. Nos. 1+, 1940+. Timiryazev. Chteniya. HI 73415

Tiómarit um íslenzka grasafraedi = Acta botanica islandica. Reykjavik.

Tiroler Heimatblätter. Innsbruck. Tiroler Heimatbl. See B–P–H 877/1. HI 59837

Tiroler landwirthschaftliche Blätter. San Michele all'Adige, [Italy]. Tiroler Landw. Blätt. See B–P–H 877/2. HI 59838

Tiscia. Dissertationes biologicae. Szeged, Hungary. Tiscia. See B–P–H 877/4. HI 59839

Tisia. Debrecen, Hungary. Tisia. See B–P–H 877/5. HI 59840

Tissue culture association newsletter = T C A newsletter. Lake Placid, NY.

Tissue culture bibliography. Bethesda, MD. Tissue Cult. Bibliogr. See B–P–H 877/6. HI 59841

Tissue culture for crops. Fort Collins, CO. No. ?-2+, ?-1983+. Tissue Cult. Crops. HI 73416

Titles of dissertations approved for the Ph.D, M.Sc., and M.Litt. degrees in the university of Cambridge. Cambridge. 1957+. Titles Diss., Univ. Cambridge. Preceded by: Abstracts of dissertations approved for the Ph.D, M.Sc., and M.Litt. degrees in the university of Cambridge. HI 73417

Tiszántuli gazdák. Debrecen, Hungary. Tiszántuli Gazd. See B–P–H 877/8. HI 59842

Tiszántúli öntözésügyi közlemények. Debrecen, Hungary. Tiszántúli Öntözésügyi Közlem. See B–P–H 877/9. HI 59843

Toa shokubutsu zusetsu = Iconographia plantarum Asiae orientalis. Tokyo.

Tobacco. New York, NY. Tobacco. See B–P–H 877/12. HI 59844

Tobaco. Maracay, Venezuela. Tobaco. See B–P–H 877/17. HI 59845

Tobacco abstracts, North Carolina agricultural experiment station = North Carolina agricultural experiment station tobacco abstracts. Raleigh, NC. North Carolina Agric. Exp. Sta. Tobacco Abstr. See B–P–H 663/15.

Tobacco bulletin, tobacco division, Canada. Ottawa. Nos. A1-A14, 1906-12. Tobacco Bull Tobacco Div. Canada. HI 73418

Tobacco quarterly. London. Vol. 1+, 1979+. Tobacco Quart. HI 73419

Tobacco research. Rajahmundry. Vol. 1+, 1975+. Tobacco Res. HI 73420

Tochigi-ken nogyo shikenjo kenkyu hokoku = Bulletin of the Tochigi agricultural experiement station. Tochigi, Japan. Bull. Tochigi Agric. Exp. Sta. See B–P–H 283/23.

Toegepaste plantwetenskap = Applied plant science. Sunnyside, Botswana.

Toho gakuho = Journal of oriental studies. Kyoto, Japan. J. Orient. Stud. See B–P–H 475/12.

Tohoku agricultural research. [Tohoku nogyo kenkyu.] Morioka, Japan. Tohoku Agric. Res. See B–P–H 877/21. HI 59847

Tohoku agriculture. [Tohoku nogyo.] Miyagi, Japan. Tohoku Agric. See B–P–H 877/20. HI 59846

Tohoku daigaku nogaku kenkyujo iho = Bulletin of the institute of agricultural research; Tohoku university. Sendai, Japan. Bull. Inst. Agric. Res. Tohoku Univ. See B–P–H 255/17.

Tohoku journal of agricultural research. Sendai, Japan. Tohoku J. Agric. Res. See B–P–H 877/23. HI 59848

Tohoku nogyo = Tohoku agriculture. Miyagi, Japan. Tohoku Agric. See B–P–H 877/20.

Tohoku nogyo kenkyu = Tohoku agricultural research. Morioka, Japan. Tohoku Agric. Res. See B–P–H 877/21.

Tohoku nogyo shikenjo kenkyu hokoku = Bulletin of the Tohoku national agricultural experiment station. Morioka, Japan. Bull. Tohoku Natl. Agric. Exp. Sta. See B–P–H 283/24.

Tohoku nogyo shikenjo, kenkyu shirjo = Miscellaneous publications, Tohoku national agricultural experiment station. Moriaka.

Tohoku teikoku-daigaku rikwa hokoku = Science reports of the Tiohoku imperial university. Ser. 4, Biology. Sendai, Japan. Sci. Rep. Tohoku Imp. Univ., Ser. 4, Biol. See B–P–H 828/15.

Toiminta, metsäpuiden rodunjalostussaatio. Helsinki. 1959-62. Toiminta Metsäp. Rodunjalost. Preceded by: Tiedoituksia, metsäpuiden rodunjalostussaatio. Superseded by: Metsanjalostussaatio. HI 73421

Tokai kinki nogy kenkyu = Journal of the agricultural researches in Tokai-kinki region. Mie, Japan. J. Agric. Res. Tokai-Kinki Region. See B–P–H 456/6.

Tokai kinki nogyo shikenjo engei-bu tokubetsu hokoku = Special bulletin, horticultural station, Tokai Kinki agricultural experiment staion. Shizuoka, Japan. Special Bull. Hort. Sta. Tokai Kinki Agric. Exp. Sta. See B–P–H 850/12.

Tokai kinki nogyo shikenjo kenkyu hokoku = Bulletin of the Tokai-kinki national agricultural experiment station. Tsu.

Tokai kinki nogyo shikenjo kenkyu hokoku, engei-bu = Horticultural station, Tokai Kinki agricultural experiment station, bulletin. Shizuoka, Japan. Hort. Sta. Tokai Kinki Agric. Exp. Sta. Bull. See B–P–H 422/20.

Tokushima daigaku gakubeibu kiyo. Shizen kagaku = Journal of gakugei, Tokushima university. Natural science. Tokushima, Japan. J. Gakugei Tokushima Univ., Nat. Sci. See B–P–H 466/19.

Tokushima-ken nogyo daigako tokubetsu kenkyu hokoku. Myosai. Vol. 1+, 1976+. Tokushima-ken Daigako Tokubetsu Kenkyu Hokoku. HI 73422

Tokyo daigaku, nogakubu. No. ?-2+, ?-1968+. Tokyo Daigaku Nogakubu. HI 73423

Tokyo daigaku nogakubu enshurin hokoku = Bulletin of

the Tokyo university forest. Tokyo. Bull. Tokyo Univ. Forest See B–P–H 284/6.

Tokyo daigaku ritchi shizen kagaku kenkyujo hokoku = Bulletin of the physiographical science research institute. Tokyo.

Tokyo daigaku, sogo kenkyu shiryokan hyohon shiryo hokoku. Tokyo. No. ?-2+, ?-1978+. Tokyo Daigaku Sogo Kenkyu Shiryokan Hyohon Shiryo Hokoku. HI 73424

Tokyo hakubutsukan iho = Tokyo museum news. Tokyo. Tokyo Mus. News. See B–P–H 878/2.

Tokyo museum news. [Tokyo hakubutsukan iho.] Tokyo. Tokyo Mus. News. See B–P–H 878/2. HI 59849

Tokyo nogyo daigaku kiyo = Memoirs of the Tokyo university of agriculture. Tokyo.

Tokyo nogyo daigaku kiyo = Journal of the Tokyo agricultural college. Tokyo. J. Tokyo Agric. Coll. See B–P–H 483/22.

Tokyo nogyo daigaku. Nogaku shuho, tokubetsugo = Journal of agricultural science, Tokyo nogyo daigaku. Supplement. Tokyo.

Tokyo nogyo daigaku nogyo shuho = Journal of agricultural science. Tokyo. J. Agric. Sci. (Tokyo). See B–P–H 456/9.

Tokyo noko daigaku nogakubu enshurin hokoku = Bulletin of the experiment forests, Tokyo university of agriculture and technology. Tokyo.

Tokyo noko daigaku nogakubu gakujutsu hokoku = Bulletin of the faculty of agriculture, Tokyo university of agriculture and technology. Tokyo.

Tokyo ryokuchi keikaku chosa iho = Report of research for planning a green belt in Tokyo. Tokyo. Rep. Res. Planning Green Belt Tokyo. See B–P–H 770/7.

Tokyo-to nogyo shikenjo kenkyu hokoku = Bulletin of the Tokyo-to agricultural experiment station. Tokyo. Bull. Tokyo-To Agric. Exp. Sta. See B–P–H 284/8.

Tomato genetics cooperative report. Davis, CA. 1951+. Tomato Genet. Coop. Rep. HI 73425

Tommasi. Naples. Tommasi. See B–P–H 878/8. HI 59850

Tonan ajia kenkyu = Southeast asian studies. Kyoto.

Top fruit times. Tunbridge Wells. No. 1+, 1980+. Top Fruit Times. HI 73426

Topics in geobiology. New York. Vol. 1+, 1980+. Topics Geobiol. HI 75168

Topics in mycobiology. Guelph. No. 1+, 1977+. Topics Mycobiol. HI 73427

Topics in photosynthesis. Amsterdam, etc. Vol. 1+, 1976+. Topics Photosyn. HI 73428

Topics in plant physiology. London, Boston, MA. Vol. 1+, 1988+. Topics Pl. Physiol. HI 73429

Topographisk journal for Norge. Christiania [=Oslo, Norway]. Topogr. J. Norge. See B–P–H 878/9. HI 59851

Topola. Belgrade. 1956+. Topola. HI 73430

Torfyanaya promyshlennost. Moscow. 1941+. Torfyanaya Promyshl. Preceded by: Za torfyanuyu undustriyu. HI 73431

Torfyanoe delo. Moscow. 1924-35. Torfyanoe Delo. Superseded by: Za torfyanuyu industriyu. HI 73432

Toronto field naturalist. Toronto. No. 333+, 1980+. Toronto Field Naturalist. Preceded by: Newsletter, Toronto field naturalists' club. HI 73433

Torreia; museo Poey, universidad de la Habana. Havana. Torreia. See B–P–H 878/11. HI 59852

Torreya; a monthly journal of botanical notes and news. New York, NY. Torreya. See B–P–H 878/13. HI 59853

Tosa no hakubutsu = Natural history of Tosa. Kochi, Japan. Nat. Hist. Tosa. See B–P–H 629/9.

Tottori daigaku gakugeibu kenkyu hokoku = Liberal arts journal, Tottori university. Natural science. Tottori.

Tottori daigaku kyoikugakubu kenkyu hokoku, shizenkagaku = Journal of the faculty of education, Tottori university. Natural science. Tottori.

Tottori daigaku nogaku-bu Fusoku enshurin hokoku = Bulletin of the Tottori university forests. Tottori, Japan. Bull. Tottori Univ. Forests. See B–P–H 284/21.

Tottori daigaku nogakubu kenkyu hokoku = Bulletin of the faculty of agriculture, Tottori university. Tottori.

Tottori daigaku nogakubu kiyo = Journal of the faculty of agriculture, Tottori university. Tottori.

Tottori koto nogyo gakko gakujutsu hokoku = Memoirs of the Tottori agricultural college. Tottori, Japan. Mem. Tottori Agric. Coll. See B–P–H 588/15.

Tottori-ken kaju shikenjo kenkyu hokoku = Bulletin of the Tottori fruit tree experiment station. Tottori, Japan. Bull. Tottori Fruit Tree Exp. Sta. See B–P–H 284/19.

Tottori-ken nogyo shikenjo kenkyu hokoku = Bulletin of the Tottori agricultural experiment station. Tottori, Japan. Bull. Tottori Agric. Exp. Sta. See B–P–H 284/18.

Tottori-ken nogyo shikenjo. Tokubetsu kenkyu hokoku = Special bulletin of the Tottori agricultural experiment station. Tottori.

T'oung pao; ou, archives concernant l'historie, les langues, la géographie et l'ethnographie de l'Asie orientale. Leiden. T'oung Pao. See B–P–H 878/23. HI 59854

Tour du monde; nouveau journal des voyages. Paris. Tour du Monde. See B–P–H 878/24. HI 59855

Tout votre jardin, encyclopédie pratique permanente. Paris. Nos. 1-94, 1969-72. Tout Votre Jard. Encycl. Prat. Perman. HI 73434

Toxicological and environmental chemistry. London, New York. Vol. 3(3/4)+, 1981+. Toxicol. Environm. Chem. Preceded by: Toxicological and environmental chemistry reviews. HI 75233

Toxicological and environmental chemistry reviews. London, New York. Vols. 1-3(2), 1972-80. Toxicol. Environm. Chem. Rev. Superseded by: Toxicological and environmental chemistry. HI 75190

Toxicology; international journal concerned with the effects of chemicals on living systems. Amsterdam. Vol. 1+, 1973+. Toxicology. HI 73435

Toxicology abstracts. Bethesda, MD. 1978+. Toxicol. Abstr. HI 73436

Toxicology and applied pharmacology. New York, NY. Toxicol. Appl. Pharmacol. See B–P–H 878/26. HI 59856

Toxicon. London. Toxicon. See B–P–H 878/27. HI 59857

Toyama daigaku kyoy obu kiyo = Journal of the college of liberal arts, Toyama university. Toyama

Toyo gakugei zasshi = Journal of the oriental arts and sciences. Tokyo. J. Orient. Arts Sci. See B–P–H 475/10.

Trabajos de la catedra botanica, escuela tecnica superior de ingenieros de montes. Madrid. 1974+. Trab. Catedra Bot. Esc. Tecn. Super. Ing. Montes. HI 73437

Trabajos compostelanos de biologia. Santiago de Compostela, Spain. No. 1+, 1971+. Trab. Compostelanos Biol. HI 73438

Trabajos del departamento de botanica y fisiologià vegetal, universidad de Madrid. Madrid. Vols. 1-6, 1968-73. Trab. Dept. Bot. Fisiol. Veg. Madrid. Superseded by: Trabajos del departamento de botanica, universidad complutense de Madrid. HI 73439

Trabajos del departamento de botanica, universidad complutense de Madrid. Madrid. Vol. 7+, 1975+. Trab. Dept. Bot. Univ. Complut. Madrid. Preceded by: Trabajos del departamento de botanica y fisiologia vegetal, universidad de Madrid. HI 73440

Trabajos del departamento de botánica, universidad de Granada. Granada. Vol. 1+, 1972+. Trab. Dept. Bot. Univ. Granada. HI 73441

Trabajos del departamento de botanica, universidad de Oviedo. Oviedo. Vol. 1+, 1977+. Trab. Dept. Bot. Univ. Oviedo. HI 73442

Trabajos del departamento de botanica, universidad de Salamanca. Salamanca. Vol. [1]+, 1976+. Trab. Dept. Bot. Univ. Salamanca. HI 73443

Trabajos, estación agrícola experimental de Léon. Léon. Vol. 1+, 1964+. Trab. Estac. Agríc. Exp. Léon. HI 73444

Trabajos del instituto de botánica y farmacología. Buenos Aires. Nos. 29-60, 1913-39; ser. 2, no. 1+, 1939+. Trab. Inst. Bot. Farmacol. Preceded by: Trabajos del instituto de farmacología. 1-823-1. HI 73445

Trabajos del instituto Cajal de investigaciones biológicas. Madrid. Trab. Inst. Cajal Invest. Biol. See B–P–H 879/12. HI 59863

Trabajos del instituto de ciencias naturales "José de Acosta". Serie biológica. Madrid. Trab. Inst. Ci. Nat. "José de Acosta", Ser. Biol. See B–P–H 878/36. HI 59859

Trabajos del instituto español de oceanografía. Madrid. Vols. 1-?, 1929-? Trab. Inst. Esp. Oceanogr. Superseded by: Monografias, instituto español de oceanografia. 5-4039-1. HI 59860

Trabajos del instituto de farmacología. Buenos Aires. Nos. 25-28, 1910-12. Trab. Inst. Farmacol. Preceded by: Trabajos del museo farmacología. Superseded by: Trabajos del instituto de botánica y farmacología. 1-823-1. HI 73446

Trabajos del jardin botánico. Santiago de Compostela. No. 1-9, 1950-63. Trab. Jard. Bot., Santiago de Compostela. HI 73447

Trabajos del laboratorio de investigaciones biológicas de la universidad de Madrid. Madrid. Trab. Lab. Invest. Biol. Univ. Madrid. See B–P–H 879/3. HI 59861

Trabajos y monografias, cátedra de botánica. El Ejido, Malaga. Vol. 1+, 1980+. Trab. Monogr. Cátedra Bot. HI 73448

Trabajos del museo botánico, universidad nacional de Córdoba. Córdoba. Vols. 1+, 1947+. Trab. Mus. Bot. Córdoba. HI 73449

Trabajos del museo de ciencias naturales. Madrid. Nos. 1, 1912. Trab. Mus. Ci. Nat. Superseded by: Trabajos del museo de ciencias naturales. Serie botánica. HI 73450

Trabajos del museo de ciencias naturales de Barcelona. Barcelona. Trab. Mus. Ci. Nat. Barcelona. See B–P–H 879/4. HI 59862

Trabajos del museo de ciencias naturales. Serie botánica. Madrid. Nos. 2-27, 1913-34. Trab. Mus. Ci. Nat., Ser. Bot. Superseded by: Trabajos del museo nacional de ciencias naturales y jardin botanico. Serie botánica. HI 73451

Trabajos del museo comercial de Venezuela. Caracas. Vols. 1-8, 1927-31. Trab. Mus. Comercial Venezuela.

HI 73452

Trabajos del museo de farmacología de la facultad de ciencias médicas de Buenos Aires. Buenos Aires. Nos. 1-24, 1903-09. Trab. Mus. Farmacol. Superseded by: Trabajos del instituto de farmacología. 1-823-1. HI 73453

Trabajos del museo nacional de ciencias naturales y jardin botanico. Serie botánica. Madrid. Nos. 28-33, 1935-36. Trab. Mus. Nac. Ci. Nat., Ser. Bot. Preceded by: Trabajos del museo de ciencias naturales. Serie botánica. HI 73454

Trabalhos do centro de botanica da junta de investigaçoes do ultramar. Lisbon. Vols. 1-33, 1960-70. Trab. Centro Bot. Junta Invest. Ultramar. HI 73455

Trabalhos do centro de investigaçao cientifica algodoeira. Lourenço Marques. 1949-51. Trab. Centro. Invest. Ci. Algodoeira. Superseded by: Memorias y trabalhos, centro de investigaçao cientifica Algodoeira. HI 73456

Trabalhos do instituto de biologia marítima e oceanográfia, universidade do Recife. Recife. Vols. 1-2, 1959-60. Trab. Inst. Biol. Mar. Oceanogr. Univ. Recife. Superseded by: Trabalhos do instituto oceanográfico da universidade do Recife. HI 73457

Trabalhos instituto de botanico Dr. Goncalo Sampaio. Porto. 1943+. Trab. Inst. Bot. Dr. Goncalo Sampaio. HI 73458

Trabalhos. Instituto botânico. Universidade de Lisboa. Lisbon. Trab. Inst. Bot. Univ. Lisboa. See B–P–H 878/35. HI 59858

Trabalhos do instituto de investigação cientifica de Moçambique. Lourenço Marques. 1961-73. Trab. Inst. Invest. Ci. Moçambique. HI 73459

Trabalhos do instituto oceanografico, universidade federal de Pernambuco. Recife. Vol. 7/8, 1967. Trab. Inst. Oceanogr. Univ. Fed. Pernambuco. Preceded by: Trabalhos do instituto oceanográfico da universidade do Recife. Superseded by: Trabalhos oceanograficos, laboratorio de ciencias do mar, universidade federal de Pernambuco. HI 73460

Trabalhos do instituto oceanografico da universidade do Recife. Recife. Vols. 3-6, 1963-66. Trab. Inst. Oceanogr. Univ. Recife. Preceded by: Trabalhos do instituto de biologia maritima e oceanografia, universidade do Recife. Superseded by: Trabalhos do instituto oceanografico, universidade federal de Pernambuco. HI 73461

Trabalhos oceanograficos, laboratorio de ciencias do mar, universidade federal de Pernambuco. Recife. Vol. 9/11+, 1970+. Trab. Oceanogr. Lab. Ci. Mar Univ. Fed. Pernambuco. Preceded by: Trabalhos do instituto oceanografico, universidade federal de Pernambuco. HI 73462

Trädgardstidningen. Stockholm. Trädgardstidningen. See B–P–H 879/13. HI 59864

Traffic (USA) = T R A F F I C (USA) newsletter. Washington, DC.

Traffic bulletin international = T R A F F I C (international) bulletin. London.

Trail and landscape; publication concerned with natural history and conservation. Ottawa. Vol. 1+, 1967+. Trail & Landscape. HI 73463

Trail and timberline. Denver, CO. Vol. 1+, 1918+. Trail & Timberline. HI 73464

Transactions of the Aberdeen philosphical society. Aberdeen, Scotland. Trans. Aberdeen Philos. Soc. See B–P–H 879/14. HI 59865

Transactions of the Aberdeen working men's natural history and scientific society. Aberdeen. 1901-06. Trans. Aberdeen Working Man's Nat. Hist. Sci. Soc. HI 73465

Transactions of the academy of science. Austin, TX. Trans. Texas Acad. Sci. See B–P–H 891/5. HI 59997

Transactions of the academy of sciences of the Estonian S S R = Izvestiya Akademii nauk Ėstonskoi S S R. Tallinn.

Transactions of the academy of science of St. Louis. St. Louis, MO. Trans. Acad. Sci. St. Louis. See B–P–H 879/19. HI 59866

Transactions of the agricultural and horticultural society of India. Calcutta. Trans. Agric. Soc. India. See B–P–H 879/22. HI 59867

Transactions of the agricultural and horticultural society of Jamaica. Kingston, Jamaica. Trans. Agric. Soc. Jamaica. See B–P–H 879/23. HI 59868

Transactions of the agricultural society, university college of Wales. Aberystwyth. 1900-07. Trans. Agric. Soc. Univ. Coll. Wales. Superseded by: Journal of the agricultural department of the university college of Wales. HI 73466

Transactions of the Albany institute. Albany, NY. Trans. Albany Inst. See B–P–H 879/24. HI 59869

Transactions of the american academy of otolaryngic allergy. Miami, FL. Vol. 22+, 1981+. Trans. Amer. Acad. Otolaryngic Allergy. Preceded by: Transactions of the american society of ophthalmologic and otolaryngologic allergy. HI 64028

Transactions; american agricultural association. Trans. Amer. Agric. Assoc. See B–P–H 879/30. HI 59870

Transactions, american association of cereal chemists. St. Paul, MN. Vols. 1-13, 1942-55. Trans. Amer. Assoc. Cereal Chem. Preceded by: Proceedings, american association of cereal chemists. Superseded by: Cereal science today. HI 73467

Transactions of the american crystallographic association. Pittsburgh, PA. Trans. Amer. Crystallogr. Assoc. See B–P–H 880/1. HI 59871

Transactions of the american entomological society. Philadelphia, PA. Trans. Amer. Entomol. Soc. See B–P–H 880/2. HI 59872

Transactions of the american horticultural society. Indianapolis, IN. Trans. Amer. Hort. Soc. See B–P–H 880/3. HI 59873

Transactions of the american medical association. Chicago, IL. Trans. Amer. Med. Assoc. See B–P–H 880/4. HI 59874

Transactions of the american microscopical society. Buffalo, NY. Trans. Amer. Microscop. Soc. See B–P–H 880/7. HI 59875

Transactions of the american philosophical society held at Philadelphia for promoting useful knowledge. Philadelphia, PA. Trans. Amer. Philos. Soc. See B–P–H 880/9. HI 59876

Transactions of the american society of agricultural engineers. St. Joseph, MI. Trans. Amer. Soc. Agric. Engin. See B–P–H 880/10. HI 59877

Transactions of the american society of landscape architects. Brookline, MA. Trans. Amer. Soc. Landscape Architects. See B–P–H 880/11. HI 59878

Transactions of the american society of ophthalmologic and otolaryngologic allergy. Indianapolis, IN. Vols. 1-21, 1960-81. Trans. Amer. Soc. Ophthalmol. Otolaryngol. Allergy. Superseded by: Transactions of the american academy of otolaryngic allergy. HI 73468

Transactions of the Amesbury and Salisbury agricultural and horticultural society. Easton, MA. Trans. Amesbury Salisbury Agric. Soc. See B–P–H 880/12. HI 59879

Transactions of the Anglesey antiquarian society and field club. Llangefni, Wales. Trans. Anglesey Antiq. Soc. See B–P–H 880/19. HI 59880

Transactions of the annual exhibitions, Colorado agricultural society. Denver, CO. Trans. Annual Exhib. Colorado Agric. Soc. See B–P–H 880/20. HI 59881

Transactions of the annual meeting of the Gulf-coast association of geological societies. New Orleans, LA. Trans. Annual Meeting Gulf-Coast Assoc. Geol. Soc. See B–P–H 880/21. HI 59882

Transactions of the annual meeting of the Indiana horticultural society. Indianapolis, IN. Trans. Annual Meeting Indiana Hort. Soc. See B–P–H 880/22. HI 59883

Transactions and annual report of the Manchester microscopical society. Manchester. 1884-1900. Trans. & Annual Rep. Manchester Microscop. Soc. Preceded by: Report, Manchester microscopical society. Superseded by: Report and transactions of the Manchester microscopical society. HI 73469

Transactions and annual report, north Staffordshire field club. [Newcastle-under-Lyme.] 1915-60. Trans. & Annual Rep. N. Staffordshire Field Club. Preceded by: Report and transactions, north Staffordshire (naturalists') field club. HI 73470

Transactions and annual report of the Nottingham naturalists' society. Nottingham, England. Trans. & Annual Rep. Nottingham Naturalists' Soc. See B–P–H 880/23. HI 59884

Transactions and archaeological record of the Cardiganshire antiquarian society = Transactions of the Cardiganshire antiquarian society. Aberystwyth, Wales. Trans. Cardiganshire Antiq. Soc. See B–P–H 882/7.

Transactions of the arctic institute. Leningrad = Trudy arkticheskogo instituta. Leningrad.

Transactions of the arctic institute of the chief administration of the northern sea route = Trudy arkticheskogo nauchno-issledovatel′skogo instituta glavnogo upravleniya severnogo morskogo puti pri sovete S S S R. Moscow, Leningrad.

Transactions of the Ashmolean society. Oxford, England. Trans. Ashmolean Soc. See B–P–H 881/1. HI 59885

Transactions of the asiatic society of Japan. Tokyo. Trans. Asiat. Soc. Japan. See B–P–H 881/2. HI 59886

Transactions of the association of american physicians. Philadlphia, PA. Trans. Assoc. Amer. Physicians. See B–P–H 881/4. HI 59887

Transactions of the association of military surgeons of the U.S. Washington, DC. Trans. Assoc. Military Surgeons U.S. See B–P–H 881/5. HI 59889

Transactions of the Barnsley naturalists' society. Barnsley. Vol. 5, 1885-86. Trans. Barnsley Naturalists' Soc. Preceded by: Quarterly transactions of the Barnsley naturaliats' society. HI 73471

Transactions, biochemical society. Colchester. Vol. 1+, 1973+. Trans. Biochem Soc. HI 73472

Transactions of the biogeographical society of Japan = Biogeographica. Tokyo. Biogeographica. See B–P–H 192/28.

Transactions of the biological society of Manchoukuo. [Manchu seibutsu gakkai kaiho.] Mukden, Manchoukuo [China]. Trans. Biol. Soc. Manchoukuo. See B–P–H 881/7. HI 59890

Transactions of the Birmingham and Midland institute scientific society. Birmingham, England. Trans. Birmingham Midl. Inst. Sci. Soc. See B–P–H 881/8.

HI 59891

Transactions of the Bishop's Stortford and district natural history society. Bishop's Stortford. 1950-53. Trans. Bishop's Stortford Distr. Nat. Hist. Soc. HI 73473

Transactions of the Bombay geographical society. Bombay. Trans. Bombay Geogr. Soc. See B–P–H 881/9. HI 59892

Transactions of the Bose research institute, Calcutta. Calcutta. Trans. Bose Res. Inst. Calcutta. See B–P–H 881/10. HI 59893

Transactions of the botanical society of Edinburgh. Edinburgh. Vols. 1-11, 1844-73; vols. 16-18, 1886-91; vol. 41+, 1970+. Trans. Bot. Soc. Edinburgh. For vols. 12-15, 1876-84 and vols. 19-40, 1893-1970 see: Transactions and proceedings of the botanical society Edinburgh. 1-755-3. HI 59894

Transactions of the Brecknock society = Brycheiniog. Brecknock, Wales. Brycheiniog. See B–P–H 230/28.

Transactions of the british bryological society. Cambridge. Vols. 1-6, 1947-71. Trans. Brit. Bryol. Soc. Preceded by: Report of the british bryological society. Superseded by: Journal of bryology. 1-777-3. HI 73474

Transactions of the british and foreign institute. London. Trans. Brit. Foreign Inst. See B–P–H 881/16. HI 59895

Transactions of the british mycological society. Worcester. Vols. 1-91, 1896-1988. Trans. Brit. Mycol. Soc. Superseded by: Mycological research. 1-790-1. HI 59896

Transactions of the British pharmaceutical conference = Yearbook of pharmacy. London.

Transactions of the Buchan club. Peterhead. 1887+. Trans. Buchan Club. HI 73475

Transactions of the Buchan field club = Transactions of the Buchan club. Peterhead.

Transactions of the Burnley literary and scientific club (later society). Burnley. Vols. 1-36, 1874-1919. Trans. Burnley Lit. Sci. Club. HI 73476

Transactions of the Burton-on-Trent natural history and archaeological society. London. 1889-1933. Trans. Burton-on-Trent Nat. Hist. Archaeol. Soc. HI 73477

Transactions of the Buteshire natural history society. Rothesay. Vol. 1+, 1907+. Trans. Buteshire Nat. Hist. Soc. HI 73478

Transactions of the Caernarvonshire historical society. Caernarvon, Wales. Trans. Caernarvonshire Hist. Soc. See B–P–H 881/20. HI 59897

Transactions of the Calcutta medical society. Calcutta. Trans. Calcutta Med. Soc. See B–P–H 881/21. HI 59898

Transactions of the California society of natural history. San Diego, CA. Trans. Calif. Soc. Nat. Hist. See B–P–H 881/22. HI 59899

Transactions of the California state agricultural society. Sacramento, CA. Trans Calif. State Agric. Soc. See B–P–H 882/1. HI 59900

Transactions of the Cambridge philosophical society. Cambridge, England. Trans. Cambridge Philos. Soc. See B–P–H 882/4. HI 59901

Transactions of the Caradoc and Severn Valley field club. Shrewsbury. Vols. 1-?, 1893-1945. Trans. Caradoc Severn Valley Field Club. HI 73479

Transactions of the Cardiff naturalists' society. Cardiff, Wales. Trans. Cardiff Naturalists' Soc. See B–P–H 882/6. HI 59902

Transactions of the Cardiganshire antiquarian society. Aberystwyth, Wales. Trans. Cardiganshire Antiq. Soc. See B–P–H 882/7. HI 59903

Transactions of the Carlisle natural history society. Carlisle. 1909+. Trans Carlisle Nat. Hist. Soc. HI 73480

Transactions of the Carmarthenshire antiquarian society and field club. Carmarthen, Wales. Trans. Carmarthenshire Antiq. Soc. See B–P–H 882/9. HI 59904

Transactions of the Chicago pathological society. Chicago, IL. Trans. Chicago Pathol. Soc. See B–P–H 882/11. HI 59905

Transactions of the Chichester and West Sussex natural history and microscopical society. Chichester. 1882-89. Trans. Chichester West Sussex Nat. Hist. Microscop. Soc. Preceded by: Report, Chichester and West Sussex natural history and microscopical society. HI 73481

Transactions of the China branch of the royal asiatic society. Hong Kong. Trans. China Branch Roy. Asiat. Soc. See B–P–H 882/16. HI 59908

Transactions of the chinese association for the advancement of science. Shanghai. Trans. Chin. Assoc. Advancem. Sci. (Shanghai). See B–P–H 882/14. HI 59906

Transactions of the chinese association for the advancement of science. Taipei, Taiwan. Trans. Chin. Assoc. Advancem. Sci. (Taipei). See B–P–H 882/15. HI 59907

Transactions of the Chosen natural history society. [Chosen hakubutsu gakkai kaiho.] [Korea]. Trans. Chose Nat. Hist. Soc. See B–P–H 882/17. HI 59909

Transactions of the city of London entomological and natural history society. London. 1890-1913. Trans. City London Entomol. Nat. Hist. Soc. Superseded by: Transactions of the London natural history society.

HI 73482

Transactions of the Connecticut academy of arts and sciences. New Haven, CT. Trans. Connecticut Acad. Arts. See B–P–H 882/19. HI 59910

Transactions of the county of Middlesex natural history and science society. [London.] 1886/87-90/91, 1887-92. Trans. County MIddlesex Nat. Hist. Sci. Soc. HI 73483

Transactions of the Cumberland association for the advancement of literature and science. Keswick, Carlisle. Vols. 1-17, 1876-93. Trans. Cumberland Assoc. Advancem. Lit. Sci. HI 73484

Transactions of the Cumberland and Westmorland association for the advancement of literature and science = Transactions of the Cumberland association for the advancement of literature and science. Keswick, Carlisle.

Transactions of the Derby natural history society. Derby. 1960+. Trans. Derby Nat. Hist. Soc. HI 73485

Transactions of the east Kent natural history society. Canterbury. N.s. vols. 1-4, 1885-89. Trans. E. Kent Nat. Hist. Soc. HI 73486

Transactions of the East Lothian antiquarian and field naturalists' society. Edinburgh. 1924+. Trans. East Lothian Antiq. Field Nat. Soc. HI 73487

Transactions of the Eastbourne natural history society. Eastbourne. N.s. vols. 1-4, 1881-1912. Trans. Eastbourne Nat. Hist. Soc. Preceded by: Papers of the Eastbourne natural history society. Superseded by: Transactions and journal of the Eastbourne natural history, photographic, literary (and archaeological) society. HI 73488

Transactions of the Eastbourne natural history scientific and literary society = Transactions of the Eastbourne natural history society. Eastbourne.

Transactions of the Edinburgh field naturalists' and microscopical society. Edinburgh. Trans. Edinburgh Field Naturalists' Soc. See B–P–H 882/24. HI 59911

Transactions of the Edinburgh naturalists' field club = Transactions of the Edinburgh field naturalists' and microscopical society. Edinburgh. Trans. Edinburgh Field Naturalists' Soc. See B–P–H 882/24.

Transactions of the english arboricultural society. Newcastle. Vols. 1-4, 1884-189? Trans. Engl. Arboric. Soc. Superseded by: Transactions of the royal english arboricultural society. HI 73489

Transactions of the entomological society of London. London. Trans. Entomol. Soc. London. See B–P–H 883/2. HI 59912

Transactions of the Epping Forest and county of Essex naturalists' field club. Buckhurst Hill. Vols. 1-2, 1880-82. Trans. Epping Forest County Essex Naturalists' Field Club. Superseded by: Transactions of the Essex field club. HI 73490

Transactions of the Essex field club. Buckhurst Hill. Vols. 3-4, 1882-87. Trans. Essex Field Club. Preceded by: Transactions of the Epping Forest and county of Essex naturalists' field club. Superseded by: Essex naturalist. HI 73491

Transactions of the ethnological society of London. London. Trans. Ethnol. Soc. London. See B–P–H 883/4. HI 59913

Transactions of faculty of horticulture, Chiba university. [Chiba daigaku engei gakabu tokubetsu hokoku.] Matsudo. Nos. 1-12, 1967-75. Trans. Fac. Hort. Chiba Univ. HI 75153

Transactions of the Folkestone natural history society. Folkestone. 1949-55? Trans. Folkestone Nat. Hist. Soc. HI 73492

Transactions of the geological society. London. Trans. Geol. Soc. (London). See B–P–H 883/5. HI 59914

Transactions of the geological society of Glasgow. Glasgow. Trans. Geol. Soc. Glasgow. See B–P–H 883/7. HI 59915

Transactions of the geological society of Pennsylvania. Philadelphia & Pittsburgh, PA. Trans. Geol. Soc. Pennsylvania. See B–P–H 883/8. HI 59916

Transactions of the geological society of South Africa. Johannesburg. Trans. Geol. Soc. South Africa. See B–P–H 883/9. HI 59917

Transactions of the Georgia naturalists club = Bulletin of the Georgia society of naturalists. Atlanta, GA. Bull. Georgia Soc. Naturalists. See B–P–H 252/1.

Transactions of the Georgia state agricultural society. Macon, GA. Trans. Georgia State Agric. Soc. See B–P–H 883/11. HI 59918

Transactions of the government for exploring meadows institute named after W. R. Williams = Trudy gosudarstvennogo lugovogo instituta imeni professora V. R. Vil′yamsa. [Lobnya.]

Transactions of the Hertfordshire natural history society and field club. London. Vol. 1+, 1879+. Trans. Hertfordshire Nat. Hist. Soc. Preceded by: Transactions of the Watford natural history society and Hertfordshire field club. 3-1845-2. HI 73493

Transactions of the Himeji natural history association. Himeji. Vol. ?-7+, ?-1981+. Trans. Himeji Nat. HIst. Assoc. HI 73494

Transactions of the Hingham agricultural and horticultural society. Hingham, MA. Trans. Hingham Agric. Soc. See B–P–H 883/16. HI 59919

Transactions of the honourable society of Cymmrodorion. London. Trans. Hon. Soc. Cymmrodorion. See

B–P–H 883/17. HI 59920

Transactions, of the horticultural society of London. London. Trans. Hort. Soc. London. See B–P–H 883/19. HI 59921

Transactions of the Hull scientific and field naturalists' club. Hull, England. Trans. Hull Sci. Club. See B–P–H 883/20. HI 59922

Transactions of the Illinois academy of science = Transactions of the Illinois state academy of science. Springfield, IL. Trans. Illinois State Acad. Sci. See B–P–H 884/2.

Transactions of the Illinois natural history society. Springfield, IL. Trans. Illinois Nat. Hist. Soc. See B–P–H 884/1. HI 59923

Transactions of the Illinois state academy of science. Springfield, IL. Trans. Illinois State Acad. Sci. See B–P–H 884/2. HI 59924

Transactions of the Illinois state agricultural society. Springfield, IL. Trans. Illinois State Agric. Soc. See B–P–H 884/3. HI 59925

Transactions of the Illinois state horticultural society = Transactions of the Illinois state agricultural society. Springfield, IL. Trans. Illinois State Agric. Soc. See B–P–H 884/3.

Transactions of the Illinois state horticultural society. Springfield, IL. Trans. Illinois State Hort. Soc. See B–P–H 884/4. HI 59926

Transactions of indian society of desert technology and university centre of desert studies. Jodhpur. Vol. 1+, 1976+. Trans. Indian Soc. Desert Technol. Univ. Centre Desert Stud. HI 73495

Transactions of the Indiana horticultural society. Indianapolis, IN. Trans. Indiana Hort. Soc. See B–P–H 884/5. HI 59927

Transactions of the institute of Jamaica. Kingston, Jamaica. 1884-85. Trans. Inst. Jamaica. HI 73496

Transactions of the institute of geography, Moscow = Trudy instituta geografii, akademiya nauk S S S R. Moscow.

Transactions of the institute for the history of science and technology. Moscow & Leningrad = Trudy instituta istorii nauki i tekhniki. Seriya 1, arkhiv istorii nauki i tekhniki. Moscow & Leningrad.

Transactions of the institute of oceanology. Moscow = Trudy Instituta okeanologii. Moscow. Trudy Inst.Okeanol. See B–P–H 909/10.

Transactions of the institute for scientific exploration of the north = Trudy nauchno-issledovatel'skogo instituta po izucheniyu severa. Leningrad.

Transactions of the Inverness scientific society and field club. Inverness. 1875-1925. Trans. Inverness Sci. Soc. Field Club. HI 73497

Transactions, Iowa state horticultural society. Des Moines, IA. Vols. 1-25. 1867-90. Trans. Iowa State Hort. Soc. Superseded by: Report of the Iowa state horticultural society. HI 73498

Transactions of the Isle of Man natural history and antiquarian society. Douglas. 1879-84. Trans. Isle of Man Nat. Hist. Antiq. Soc. HI 73499

Transactions of the Jamaica society of arts. Kingston, Jamaica. Vols. 1, 1854-55? Trans. Jamaica Soc. Arts. Superseded by: Transactions of the royal society of arts and agriculture of Jamaica. HI 73500

Transactions and journal of the Eastbourne natural history and archaeological society. Eastbourne. Vol. 12(2-4), 1939-46. Trans. & J. Eastbourne Nat. Hist. Archaeol. Soc. Preceded by: Transactions and journal of the Eastbourne natural history, photographic, literary and archaeological society. Superseded by: Journal and transactions of the Eastbourne natural history and archaeological society. HI 73501

Transactions and journal of the Eastbourne natural history, photographic, literary (and archaeological) society. Eastbourne. Vols. 5-12(1), 1927-38. Trans. & J. Eastbourne Nat. Hist. Photogr. Lit. Soc. Preceded by: Transactions of the Eastbourne natural history, (scientific and literary) society. Superseded by: Transactions of the eastbourne natural history and archaeological society. HI 73502

Transactions and journal of proceedings of the Dumfriesshire and Galloway natural history and antiquarian society. Dumfries, Scotland. Trans. & J. Proc. Dumfriesshire Nat. Hist. Soc. See B–P–H 884/8. HI 59928

Transactions of the Kansas academy of science. Topeka, KS. Trans. Kansas Acad. Sci. See B–P–H 884/110. HI 59929

Transactions of the Kent field club. Maidstone, England. Trans. Kent Field Club. See B–P–H 884/11. HI 59930

Transactions of the Kentucky academy of science. Lexington, KY. Trans. Kentucky Acad. Sci. See B–P–H 884/13. HI 59931

Transactions of the Kentucky state horticultural society. Frankfort, KY. Trans. Kentucky State Hort. Soc. See B–P–H 884/14. HI 59932

Transactions of the Korea branch of the royal asiatic society of Great Britain and Ireland. Seoul, Korea. Trans. Korea Branch Roy. Asiat. Soc. Gr. Brit. See B–P–H 884/16. HI 59933

Transactions of the Leeds naturalists' club and scientific association. Leeds. Vols. 1-2, 1886-92. Trans. Leeds Naturalists' Club Sci. Assoc. HI 73503

Transactions of the Leicester literary and philosophical society. Leicester, England. Trans. Leicester Lit. Soc. See B–P–H 884/21. HI 59934

Transactions of the Lincolnshire naturalists union. Louth, England. Trans. Lincolnshire Naturalists Union. See B–P–H 884/24. HI 59935

Transactions of the linnean society of London. London. Trans. Linn. Soc. London. See B–P–H 885/8. HI 59936

Transactions of the linnean society of London. Botany. London. Trans. Linn. Soc. London, Bot. See B–P–H 885/9. HI 59937

Transactions of the linnean society of New York. New York, NY. Trans. Linn. Soc. New York. See B–P–H 885/10. HI 59938

Transactions of the literary society of Bombay. London. Trans. Lit. Soc. Bombay. See B–P–H 885/11. HI 59939

Transactions of the literary and philosophical society of New-York. New York, NY. Trans. Lit. Soc. New-York. See B–P–H 885/12. HI 59940

Transactions of the literary ahd historical society of Quebec. Quebec. Trans. Lit. Soc. Quebec. See B–P–H 885/13. HI 59941

Transactions of the Liverpool botanical society. Liverpool. Vol. 1, 1909. Trans. Liverpool Bot. Soc. HI 73504

Transactions of the Liverpool nautical society. Liverpool. Trans. LIverpool Naut. Soc. See B–P–H 885/14. HI 59942

Transactions of the London natural history society. London. 1914-20. Trans. London Nat. Hist. Soc. Preceded by: Transactions of the city of London entomological and natural history society. Superseded by: London naturalist. HI 73505

Transactions of the Maidstone and Mid-Kent natural history and philosophical society. Maidstone, England. Trans. Maidstone Mid-Kent Nat. Hist. Soc. See B–P–H 885/16. HI 59943

Transactions of the Malvern field club. Worcester, England. Trans. Malvern Field Club. See B–P–H 885/17. HI 59944

Transactions of the Maryland academy of science and literature. Baltimore, MD. Trans. Maryland Acad. Sci. See B–P–H 885/18. HI 59945

Transactions of the Massachusetts horticultural society. Boston, MA. 1837-1919. Trans. Mass. Hort. Soc. Preceded by: Anniversary of the Massachusetts horticultural society. Vol. for 1847/51 published with: Proceedings of the Massachusetts horticultural society. Superseded by: Bulletin of the Massachusetts horticultural society and Report (Annual) of the Massachusetts horticultural society. 3-2552-2. HI 73506

Transactions of the medical and physical society of Bombay. Bombay. Trans. Med. Soc. Bombay. See B–P–H 885/25. HI 59948

Transactions of the medical and physical society of Calcutta. Calcutta. Trans. Med. Soc. Calcutta. See B–P–H 886/1. HI 59949

Transactions of the medical society of London. London. Trans. Med. Soc. London. See B–P–H 886/2. HI 59950

Transactions of the medical society of New Jersey. [New Jersey]. Trans. Med. Soc. New Jersey. See B–P–H 886/3. HI 59951

Transactions médicales: journal de médecine pratique. Paris. Trans. Méd. See B–P–H 885/21. HI 59946

Transactions of the medico-chirurgical society of Edinburgh. Edinburgh. Trans. Med.-Chir. Soc. Edinburgh. See B–P–H 885/22. HI 59947

Transactions of the meeting of the Japanese forestry society, Meguro. [Nippon ringakkai-shi.] Tokyo. Trans. Meeting Jap. Forest. Soc. See B–P–H 886/4. HI 59952

Transactions of the meteorological society of London. London. Trans. Meteorol. Soc. London. See B–P–H 886/5. HI 59953

Transactions of the Michigan state agricultural society. Lansing, MI. Trans. Michigan State Agric. Soc. See B–P–H 886/6. HI 59954

Transactions of the Midland scientific association. Birmingham, England. Trans. Midl. Sci. Assoc. See B–P–H 886/8. HI 59955

Transactions of the Minnesota state horticultural society. St. Paul & Minneapolis, MN. Trans. Minnesota State Hort. Soc. See B–P–H 886/9. HI 59956

Transactions of the Missouri academy of science. St. Louis, Columbia, MO. Vol. 1+, 1967+. Trans. Missouri Acad. Sci. HI 73507

Transactions of the mycological society of India = Kavaka. Madras.

Transactions of the mycological society of Japan. [Nihon kingakkai kaiho.] Tokyo. Trans. Mycol. Soc. Japan. See B–P–H 886/11. HI 59957

Transactions of the national chrysanthemum society. London. 1911-13, 1918-19. Trans. Natl. Chrysanthemum Soc. HI 73508

Transactions of the national institute of sciences of India. New Delhi. Tans. Natl. Inst. Sci. India. See B–P–H 887/5. HI 59964

Transactions of the natural history and antiquarian society of Penzance. Penzance. Vols. 1-3(1), 1845-65. Trans.

Nat. Hist. Antiq. Soc. Penzance. Superseded by: Transactions of the Penzance natural history and antiquarian society. HI 73509

Transactions of the natural history society of Aberdeen. Aberdeen. 1878-85. Trans. Nat. Hist. Soc. Aberdeen. HI 73510

Transactions of the natural history society of Glasgow. Glasgow. N.s. vols. 5-8, 1896/97-1907/08, 1897-1911. Trans. Nat. Hist. Soc. Glasgow. Preceded by: Proceedings and transactions of the natural history society of Glasgow. Superseded by: Glasgow naturalist. HI 73511

Transactions of the natural history society of Hartford. Harford, CT. Trans. Nat. Hist. Soc. Hartford. See B–P–H 886/19. HI 59959

Transactions of the natural history society, Kagoshima imperial college of agriculture and forestry. [Kagoshima koto norin-gakko hakubutsu gakkai kaiho.] Kagoshima, Japan. Trans. Nat. Hist. Soc. Kagoshima Imp. Coll. Agric. See B–P–H 886/21. HI 59960

Transactions of the natural history society of Manchuria. [Manshu hakubutsu doko-kai kaiho.] Mukden?, Manchoukuo [China]. Trans. Nat. Hist. Soc. Manchuria. See B–P–H 886/22. HI 59961

Transactions of the natural history society of Northumberland, Durham, and Newcastle-upon-Tyne. Newcastle-upon-Tyne. Vols. 1-2, 1831-38; 1904-73. Trans. Nat. Hist. Soc. Northumberland. Superseded by: Transactions of the natural history society of Northumbria. HI 73512

Transactions of the natural history society of Northumbria. Newcastle-upon-Tyne. Vol. 42(2)+, 1974+. Trans. Nat. Hist. Soc. Northumbria. Preceded by: Transactions of the natural history society of Northumberland, Durham & Newcastle-upon-Tyne. HI 73513

Transactions of the natural history society of Taiwan. Taihoku [=Taipei, Taiwan]. Trans. Nat. Hist. Soc. Taiwan. See B–P–H 887/1. HI 59963

Transactions of the natural-technic class. Kiev, Ukrainian S S R = Trudy Pryrodnycho-technichnogo viddilu. Kiev.

Transactions, Nebraska academy of sciences. Lincoln, NE. Vol. 1+, 1972+. Trans. Nebraska Acad. Sci. HI 73514

Transactions of the New York academy of sciences. New York. Vols. 1-16, 1881-97; ser. 2, vols. 1-37, 1938-74. Trans. New York Acad. Sci. Transactions for 1897-1938 incorporated in: Annals of the New York academy of sciences. HI 73515

Transactions of the New York state agricultural society = Proceedings of the New York state agricultural society. Albany, NY. Proc. New York State Agric. Soc. See B–P–H 733/14.

Transactions of the Newbury district field club. Newbury. 1870-1911, 1930+. Trans. Newbury Distr. Field Club. HI 73516

Transactions of the Niigata horticultural experiment station. [Niigata-ken engei shikenjo tokubetsu hokoku.] Niitsu. Vol. 1+, 1969+. Trans. Niigata Hort. Exp. Sta. HI 75154

Transactions of the Norfolk and Norwich naturalists' society. Norwich, England. Trans. Norfolk Norwich Naturalists' Soc. See B–P–H 887/14. HI 59965

Transactions of the North Carolina state agricultural society. Raleigh, NC. 1857. Trans. North Carolina State Agric. Soc. 4-3086-2. HI 73517

Transactions of the northern naturalists' union. Newcastle-upon-Tyne. 1931-53. Trans. N. Naturalists' Union. HI 73518

Transactions of the northern scientific and economic expedition = Trudy severnoi nauchno-promyslovoi ekspeditsii V S N Kh. Petrograd.

Transactions of the Nova Scotia literary and scientific society. Halifax, Nova Scotia. Trans. Nova Scotia Lit. Soc. See B–P–H 887/15. HI 59966

Transactions of the ophthalmological society of the United Kingdom. London. Trans. Ophthalmol. Soc. U.K. See B–P–H 887/19. HI 59967

Transactions, Ottawa field-naturalists' club = Ottawa field-naturalists' club transactions. Ottawa.

Transactions of the Ottawa natural history society. Ottawa. Nos. [1]-3, [1867]-68; no.[1] reprinted 1968. Trans. Ottawa Nat. Hist. Soc. HI 73519

Transactions of the Paisley naturalists' society. Paisley. Vols. 1-6, 1912-54/60. Trans. Paisley Naturalists' Soc. HI 73520

Transactions, Papua and New Guinea scientific society. Port Moresby. Vols. 1-?, 1960-69. Trans. Papua New Guinea Sci. Soc. Superseded by: Proceedings, Papua and New Guinea scientific society. HI 73521

Transactions of the pathological society of London. London. Trans. Pathol. Soc. London. See B–P–H 887/21. HI 59968

Transactions of the Penzance natural history society and antiquarian society. Plymouth. N.s. vols. 1-4(2), 1880-94. Trans. Penzance Nat. Hist. Soc. Antiq. Soc. Preceded by: Transactions of the natural history and antiquarian society of Penzance. HI 73522

Transactions of the Perthshire society of natural science. Perth. Vol. 13+, 1979/80+. Trans. Perthshire Soc. Nat. Sci. Preceded by: Transactions and proceedings of the Perthshire society of natural science. HI 73523

Transactions of the philosophical and literary society of Leeds. Leeds. Trans. Philos. Soc. Leeds. See B–P–H 887/26. HI 59969

Transactions of the philosophical society of New South Wales. Sydney. 1862-66. Trans. Philos. Soc. New South Wales. Superseded by: Transactions of the royal society of New South Wales. HI 73524

Transactions of the philosophical society of Victoria. Melbourne, Vic. Vol. 1, 1855. Trans. Philos. Soc. Victoria. Preceded by: Transactions and proceedings of the Victorian institute for the advancement of science. Superseded by: Transactions and proceedings of the philosophical institute of Victoria. 4-3341-2. HI 73525

Transactions philosophiques de la société royale de Londres. [Translated by Brémond.] Paris. Trans. Philos. Soc. Roy. Londres (Brémond). See B–P–H 887/28. HI 59970

Transactions philosophiques de la société royale de Londres. [Translated by Demours.] Paris. Trans. Philos. Soc. Roy. Londres (Demours). See B–P–H 887/29. HI 59971

Transactions of the Plinian society. Edinburgh. Trans. Plinian Soc. See B–P–H 888/1. HI 59972

Transactions of the Plumstead and district natural history society. Plumstead. Nos. 1-?, 1927-32. Trans. Plumstead Distr. Nat. Hist. Soc. HI 73526

Transactions of the Plymouth and district field club. Plymouth. Nos. 1-5, 1913-17. Trans. Plymouth Distr. Field Club HI 73527

Transactions and proceedings of the botanical society Edinburgh. Edinburgh. Vols. 12-15, 1876-84; vols. 19-40, 1893-1970. Trans. & Proc. Bot. Soc. Edinburgh. For vols. 1-11, 1844-73, vols. 16-18, 1886-91, and vol. 41+ see: Transactions of the botanical society of Edinburgh. 1-755-3. HI 59973

Transactions and proceedings of the botanical society of Pennsylvania. Philadelphia, PA. Trans. & Proc. Bot. Soc. Pennsylvania. See B–P–H 888/6. HI 59974

Transactions and proceedings of the California association of nurserymen. Los Angeles, CA. Trans. & Proc. Calif. Assoc. Nurserymen. See B–P–H 888/8. HI 59975

Transactions and proceedings of the Japan society. London. Trans. & Proc. Japan Soc. See B–P–H 888/9. HI 59976

Transactions and proceedings, Liverpool botanical society = Transactions of the Liverpool botanical society. Liverpool.

Transactions and proceedings of Natal scientific society. Durban. Vols. ?-2(1-4)-?, ?-1911-12-? Trans. & Proc. Natal Sci. Soc. HI 73528

Transactions and proceedings of the New Zealand institute. Wellington, N.Z. Vols. 1-63, 1868-1934. Trans. & Proc. New Zealand Inst. Superseded by: Transactions and proceedings of the royal society of New Zealand. 4-3730-2. HI 73529

Transactions and proceedings of the palaeontological society of Japan. Tokyo. Trans. & Proc. Palaeontol. Soc. Jap. See B–P–H 888/15. HI 59977

Transactions and proceedings of the Perthshire society of natural science. Perth. Vols. 1-12, 1886-1965/68. Trans. & Proc. Perthshire Soc. Nat. Sci. Preceded by: Proceedings of the Perthshire society of natural science. Superseded by: Transactions of the Perthshire society of natural science. HI 73530

Transactions and proceedings of the philosophical institute of Victoria. Melbourne, Vic. Vols. 1-4, 1857-60. Trans. & Proc. Philos. Inst. Victoria. Preceded by: Transactions of the philosophical society of Victoria. Superseded by: Transactions of the royal society of Victoria. 4-3731-2. HI 73531

Transactions, proceedings and report, philosophical society of Adelaide. Adelaide, S.A. Vols. 1-2, 1877-79. Trans. & Proc. Rep. Philos. Soc. Adelaide. Superseded by: Transactions and proceedings and report of the royal society of South Australia. HI 73532

Transactions, proceedings and report, royal society of South Australia. Adelaide, S.A. Vols. 3-35, 1880-1911. Trans. & Proc. Rep. Roy. Soc. South Australia. Preceded by: Transactions, proceedings and report, philosophical society of Adelaide. Superseded by: Transactions and proceedings of the royal society of South Australia. 4-3730-3. HI 73533

Transactions and proceedings of the royal society of New South Wales. Sydney, N.S.W. Vol. 9, 1875. Trans. & Proc. Roy. Soc. New S. Wales. Preceded by: Transactions of the royal society of New South Wales. Superseded by: Journal and proceedings of the royal society of New South Wales. 4-3730-1. HI 73534

Transactions and proceedings of the royal society of New Zealand. Wellington, N.Z. Vols. 64-79, 1934-52. Trans. & Proc. Roy. Soc. New Zealand. Preceded by: Transactions and proceedings of the New Zealand institute. Superseded by: Transactions of the royal society of New Zealand. 4-3730-2. HI 73535

Transactions and proceedings of the royal society of South Australia. Adelaide, S.A. Vols. 36-61, 1912-37. Trans. & Proc. Roy. Soc. South Australia. Preceded by: Transactions, proceedings and report, royal society of South Australia. Superseded by: Transactions of the royal society of South Australia. 4-3730-3. HI 73536

Transactions and proceedings of the royal society of Victoria. Melbourne, Vic. Vols. 6-24, 1865-87. Trans.

& Proc. Roy. Soc. Victoria. Preceded and superseded by: Transactions of the royal society of Victoria. Also superseded by: Proceedings of the royal society of Victoria. 4-3731-1. HI 73537

Transactions and proceedings of the south London entomological and natural history society. London. 1932-33. Trans. & Proc. S. London. Entomol. Nat. Hist. Soc. Preceded by: Proceedings of the south London entomological and natural history society. Superseded by: Proceedings and transactions of the south London entomological and natural history society. HI 73538

Transactions and proceedings, Torquay natural history society. Torquay. 1922-38. Trans. & Proc. Torquay Nat. Hist. Soc. Preceded by: Journal of the Torquay natural history society. HI 73539

Transactions and proceedings of the victorian institute for the advancement of science. Melbourne, Vic. 1854-55. Trans. & Proc. Victorian Inst. Advancem. Sci. Superseded by: Transactions of the philosophical society of Victoria. 5-4390-2. HI 73540

Transactions of the provincial medical and surgical association. London. Trans. Prov. Med. Assoc. See B–P–H 889/1. HI 59978

Transactions of the Radnorshire society. Presteign, Llandrindod Wells. ?-1931+. Trans. Radnorshire Soc. HI 73541

Transactions and reports, fruit growers' association of Nova Scotia. Kentville, Halifax, N.S. Nos. 1, 19-26, 1874, 1883-90 [not published for 1875-82 & 1891]. Trans. Rep. Fruit Growers' Assoc. Nova Scotia. Superseded by: Report (Annual) of the fruit growers' association of Nova Scotia. HI 73542

Transactions of the Rhodesia scientific association. Salisbury, Rhodesia. Vols. 56-59, 1974-80. Trans. Rhodesia Sci. Assoc. Preceded by: Proceedings and transactions of the Rhodesia scientific association. Superseded by: Transactions of the Zimbabwe scientific association. HI 73543

Transactions of the Rochdale literary and scientific society. Rochdale. 1878-1949. Trans. Rochdale Lit. Sci. Soc. HI 73544

Transactions of the royal asiatic society of Great Britain and Ireland. London. Trans. Roy. Asiat. Soc. Gr. Brit. See B–P–H 889/13. HI 59979

Transactions, royal caledonian horticultural society. Edinburgh. Vols. 1-?, 1921-37. Trans. Roy. Caledonian Hort. Soc. Superseded by: Journal of the royal caledonian horticultural society. 4-3718-1. HI 73545

Transactions of the royal canadian institute. Toronto. Vols. 1+, 1889+. Trans. Roy. Canad. Inst. Preceded by: Proceedings of the canadian institute. 4-3718-1. HI 73546

Transactions of the royal Dublin society. Dublin. Trans. Roy. Dublin Soc. See B–P–H 889/16. HI 59980

Transactions of the royal english arboricultural society. Newcastle. Vols. 5-7(1), 189?-1907. Trans. Roy. Engl. Arboric. Soc. Preceded by: Transactions of the english arboricultural society. Incorporated in: Quarterly journal of forestry. HI 73547

Transactions of the royal english forestry society. London. Trans. Roy. Engl. Forest. Soc. See B–P–H 889/17. HI 59981

Transactions of the royal geological society of Cornwall. Penzance, England. Trans. Roy. Geol. Soc. Cornwall. See B–P–H 889/18. HI 59982

Transactions of the royal irish academy. Dublin. Trans. Roy. Irish Acad. See B–P–H 889/20. HI 59983

Transactions of the royal medico-botanical society of London. London. Trans. Roy. Med.-Bot. Soc. London. See B–P–H 889/21. HI 59984

Transactions of the royal microscopical society of London. London. Trans. Roy. Microscop. Soc. London. See B–P–H 889/22. HI 59985

Transactions of the royal scottish arboricultural society. Edinburgh. Vols. 1-40, 1858-1926. Trans. Roy Scott. Arboric. Soc. Superseded by: Scottish forestry journal. HI 73548

Transactions of the royal society of arts and agriculture of Jamaica = Transactions of the royal society of arts of Jamaica. Kingston, Jamaica.

Transactions of the royal society of arts of Jamaica. Kingston, Jamaica. Vols. 2-4(2), 1856-61; [n.s.] vols 1(1-4), 186?-68. Trans. Roy. Soc. Arts Jamaica. Preceded by: Transactions of the Jamaica society of arts. HI 73549

Transactions of the royal society of arts and sciences of Mauritius. Port Louis, Mauritius. Vols. 1-2, 1848-52; n.s. vols. 1-20, 1860-89; ser. C [= ser. 3], nos. 1-15, 1929-49. Trans. Roy. Soc. Arts Mauritius. Superseded by: Proceedings of the royal society of arts and sciences of Mauritius. 5-3972-2. HI 73550

Transactions of the royal society of Canada. Ottawa. Ser. 4, vol. 1+, 1963+. Trans. Roy. Soc. Canada, Ser. 4. Preceded by: Transactions of the royal society of Canada. Section 3, science. Part 2, biological sciences. 4-3727-2. HI 75192

Transactions of the royal society of Canada. Section 3, science. Part 2, biological sciences. Ottawa. Ser. 3, vol. 56, 1962. Trans. Roy. Soc. Canada, Sect. 3, Part 2, Biol. Sci. Preceded by: Transactions of the royal society of Canada. Section 5, biological sciences. Superseded by: Transactions of the royal society of Canada. Series 4. 4-3727-2. HI 75193

Transactions of the royal society of Canada. Section 5, biological sciences. Ottawa. Ser. 3, vols. 21-55, 1927-61. Trans. Roy. Soc. Canada, Sect. 5, Biol. Sci. Preceded by: Proceedings and transactions, royal society of Canada. Superseded by: Transactions of the royal society of Canada. Section 3, science. Part 2, biological sciences. 4-3727-2. HI 75194

Transactions of the royal society of Edinburgh. Edinburgh. Trans. Roy. Soc. Edinburgh. See B–P–H 889/28. HI 59986

Transactions of the royal society of New South Wales. Sydney, N.S.W. Vols. 1-8, 1867-74. Trans. Roy. Soc. New South Wales. Preceded by: Transactions of the philosophical society of New South Wales. Superseded by: Transactions and proceedings of the royal society of New South Wales. 4-3730-1. HI 73551

Transactions of the royal society of New Zealand. Dunedin, N.Z. Vols. 80-88, 1952-61. Trans. Roy. Soc. New Zealand. Preceded by: Transactions and proceedings of the royal society of New Zealand. Superseded by: Transactions of the royal society of New Zealand. Botany and Transactions of the royal society of New Zealand. General. HI 73552

Transactions of the royal society of New Zealand. Biological sciences. Dunedin, N.Z. Vols. 11-12, 1968-70. Trans. Roy. Soc. New Zealand, Biol. Sci. Preceded by: Transactions of the royal society of New Zealand. Botany. Superseded by: Journal of the royal society of New Zealand. HI 73553

Transactions of the royal society of New Zealand. Botany. Wellington, N.Z. Vols. 1-3, 1961-68. Trans. Roy. Soc. New Zealand, Bot. Preceded by: Transactions of the royal society of New Zealand. Superseded by: Transactions of the royal society of New Zealand. Biological sciences. HI 73554

Transactions of the royal society of New Zealand. General. Dunedin, N.Z. Vols. 1-2, 1962-71. Trans. Roy. Soc. New Zealand, Gen. Preceded by: Transactions of the royal society of New Zealand. Superseded by: Journal, royal society of New Zealand. HI 73555

Transactions of the royal society of South Africa. Cape Town. Vol. 1+, 1908+, 1909+. Trans. Roy. Soc. South Africa. Preceded by: Transactions of the south african philosophical society. HI 73556

Transactions of the royal society of South Australia. Adelaide, S.A. Vol. 62+, 1938+. Trans. Roy. Soc. South Australia. Preceded by: Transactions and proceedings of the royal society of South Australia. 4-3730-3. HI 73557

Transactions of the royal society of tropical medicine and hygiene. London. Vol. 14+, 1920+. Trans. Roy. Soc. Trop. Med. Hyg. Preceded by: Transactions of the society of tropical medicine and hygiene. 4-3731-1. HI 64659

Transactions of the royal society of Victoria. Melbourne, Vic. Vol. 5, 1860[-61?]; [n.s.] vols. 1-6, 1888-1914. Trans. Roy. Soc. Victoria. Preceded by: Transactions and proceedings of the philosophical institute of Victoria. For 1861-87 see: Transactions and proceedings of the royal society of Victoria. 4-3731-2. HI 73558

Transactions of the Salisbury field club. Salisbury. Vol. 1, 1890-93. Trans. Salisbury Field Club. HI 73559

Transactions of the San Diego society of natural history. San Diego, CA. Trans. San Diego Soc. Nat. Hist. See B–P–H 890/7. HI 59987

Transactions of the Sapporo natural history society. [Sapporo hakubutsu gakkai kaiho.] Sapporo, Japan. Trans. Sapporo Nat. Hist. Soc. See B–P–H 890/8. HI 59988

Transactions of the science society of China. Nanking. Trans. Sci. Soc. China. See B–P–H 890/12. HI 59990

Transactions of the scientific association. Richmond, IN. Trans. Sci. Assoc. See B–P–H 890/11. HI 59989

Transactions of the scientific society of Turkestan = Trudy Turkestanskogo nauchnogo obshchestva. Tashkent.

Transactions of the scientific society of Turkestan at Middle Asiatic university = Trudy Turkestanskogo nauchnogo obshchestva pri Sredne-Aziatskom gosudarstvennom u universitete. Tashkent.

Transactions of the scottish natural history society. Edinburgh. Vols. 1-2(1), 1898-1903. Trans. Scott. Nat. Hist. Soc. HI 73560

Transactions of the Severn valley naturalists' field club. Wellington, N. Z. Trans. Severn Valley Naturalists' Field Club. See B–P–H 890/18. HI 59991

Transactions of the Shropshire archaeological and natural history society. Shrewsbury. 1877-1935. Trans. Shropshire Archaeol. Nat. Hist. Soc. Superseded by: Transactions of the Shropshire archaeological society [not entered]. HI 73561

Transactions of the siberian institute of agriculture and forestry = Trudy sibirskogo instituta sel'skogo khozyaistva i lesovodstva. Omsk.

Transactions of the society, instituted at London, for the encouragement of arts, manufactures, and commerce. London. Trans. Soc. London Encour. Arts. See B–P–H 890/20. HI 59992

Transactions of the society, instituted in the state of New-York, for the promotion of agriculture, arts, and manufactures. New York, NY. Trans. Soc. State New-York Promot. Agric. See B–P–H 890/24. HI 59995

Transactions of the society for the promotion of agriculture, arts and manufactures, instituted in the state of New-York. Albany, NY. Trans. Soc. Promot. Agric. State New-York. See B–P–H 890/22. HI 59993

Transactions of the society for the promotion of useful arts. Albany, NY. Trans. Soc. Promot. Useful Arts. See B–P–H 890/23. HI 59994

Transactions of the society of tropical medicine and hygiene. London. Vols. 1-13, 1907-20? Trans. Soc. Trop. Med. Hyg. Superseded by: Transactions of the royal society of tropical medicine and hygiene. 4-3731-1. HI 75198

Transactions of the south african philosophical society. Cape Town. Vols. 1-18, 1878-1908. Trans. S. African Philos. Soc. Superseded by: Transactions of the royal society of South Africa. 4-3730-3. HI 73562

Transactions of the south-eastern union of scientific societies. Tunbridge Wells. 1897. Trans. S.-E. Union Sci. Soc. Superseded by: Report and transactions of the south-eastern union of scientific societies. HI 73563

Transactions of the state V. M. Molotov university of Rostov-on-Don. Works of the biological faculty = Uchenye zapiski Rostovskogo-na-Donu gosudarstvennogo universiteta im. V. M. Molotova. Trudy biologicheskogo fakul'teta. Rostov.

Transactions of the Stirling field club. Stirling. 1878-82. Trans. Stirling Field Club. Superseded by: Transactions of the Stirling natural history and archaeological society. HI 73564

Transactions of the Stirling natural history and archaeological society. Stirling. 1882-1932. Trans. Stirling Nat. Hist. Archaeol. Soc. Preceded by: Transactions of the Stirling field club. HI 73565

Transactions and studies of the college of physicians of Philadelphia. Philadelphia, PA. Vols. 1-3, 1841-50; n.s. vols. 1-4, 1850-74; ser. 3 vols. 1-54, 1875-1932; ser. 4 vol. 1+, 1933+. Trans. Stud. Coll. Physicians Philadelphia. 2-1101-2. HI 73566

Transactions of the Suffolk naturalists' society. Norwich. Vols. 1-?, 1930-69. Trans. Suffolk Naturalists' Soc. Superseded by: Suffolk natural history. HI 73567

Transactions of the Tennessee academy of science. Nashville, TN. Trans. Tennessee Acad. Sci. See B–P–H 891/3. HI 59996

Transactions of Tomsk state university = Izvestiya Tomskogo gosudarstvennogo universiteta. Tomsk.

Transactions of the Tottori society of agricultural science. Tottori. Vols. 1-21, 1927-69. Trans. Tottori Soc. Agric. Sci. Superseded by: Bulletin of the faculty of agriculture, Tottori university. 5-4242-1. HI 73568

Transactions of the Tyneside naturalists' field club. Newcastle-upon-Tyne. Trans. Tyneside Naturalists' Field Club. See B–P–H 891/11. HI 59998

Transactions of the Utah academy of sciences, arts and letters. Salt Lake City, UT. Trans. Utah Acad. Sci. See B–P–H 891/12. HI 59999

Transactions of the Wagner free institute of science of Philadelphia. Philadelphia, PA. Trans. Wagner Free Inst. Sci. Philadelphia. See B–P–H 891/14. HI 60000

Transactions of the Watford natural history society and Hertfordshire field club. Watford. Vols. 1-2, 1875-79. Trans. Watford Nat. Hist. Soc. Superseded by: Transactions of the Hertfordshire natural history society and field club. 5-4459-2. HI 73569

Transactions, Weardale naturalists' field club. Bishop Auckland. Vol. 1(1-2), 1900-04. Trans. Weardale Naturalists' FIeld Club. HI 73570

Transactions of the west Kent natural history, microscopical and photographic society. London. 1900-11. Trans. W. Kent Nat. Hist. Microscop. Photgr. Soc. Preceded by: President's address, papers and reports, west Kent natural history, microscopical and photographic society. HI 73571

Transactions of the Wisconsin academy of sciences, arts and letters. Madison, WI. Trans. Wisconsin Acad. Sci. See B–P–H 891/18. HI 60001

Transactions of the Wisconsin state agricultural society. Madison, WI. Vol. 1+, 1852+. Trans. Wisconsin State Agric. Soc. HI 73572

Transactions of the Woolhope naturalists' field club. Hereford, England. Trans Woolhope Naturalists' Field Club. See B–P–H 891/25. HI 60002

Transactions of the Worcestershire naturalists' club. Worcester. 1847-1966; n.s. vol. 1+, 1979/80+. Trans. Worcestershire Naturalists' Club. For 1966-79 see: Newsletter, Worcestershire naturalists' club. HI 73573

Transactions of the Yorkshire naturalists' union. Leeds. 1877-1946. Trans. Yorkshire Naturalists' Union. HI 73574

Transactions, Zimbabwe scientific association. [Salisbury, Zimbabwe.] Vol. 60+, 1980+. Trans. Zimbabwe Sci. Assoc. Preceded by: Transactions, Rhodesia scientific association. HI 73575

Transdex; bibliograpy and index to the United States joint publications research service translations. New York. Vols. 1-12, 1962-74. Transdex. HI 73576

Transsilvania. Periodische Zeitschrift für Landeskunde. Hermannstadt [=Sibiu, Rumania]. Transsilvania. See B–P–H 892/1. HI 60003

Translations monthly. Chicago, IL. Vols. 1-4, 1955-58. Transl. Monthly. Superseded by: Technical translations, office of technical services, United States department of commerce. HI 73577

Translations register-index. Chicago, IL. Vol. 1+, 1967+. Transl. Reg.-Index. Preceded by: Technical translations, office of technical services, United States department of commerce. HI 73578

Translations, United States forest service. Division of forest management research = United States forest service. Division of silvics. Translations. Washington, DC. U.S. Forest Serv. Div. Silvics Transl. See B–P–H 939/30.

Translations, United States forest service. Division of silvics = United States forest service. Division of silvics. Translations. Washington, DC. U.S. Forest Serv. Div. Silvics Transl. See B–P–H 939/30.

Transvaal agricultural journal. Pretoria. Transvaal Agric. J. See B–P–H 892/3. HI 60004

Transvaal gardener; journal of the Transvaal horticultural society. Johannesburg. Vol. 32?+, 1983?+. Transvaal Gard. Preceded by: Journal of the Transvaal horticultural society. HI 73579

Transvaal museum bulletin. Pretoria. No. 10+, 1971+. Transvaal Mus. Bull. Preceded by: Bulletin of the Transvaal museum. HI 73580

Transvaal university college bulletin. Faculty of agriculture. Pretoria. Transvaal Univ. Coll. Bull. Fac. Agric. See B–P–H 892/4. HI 60005

Transylvania journal of medicine and the associate sciences. Lexington, KY. Transylvania J. Med. Assoc. Sci. See B–P–H 892/6. HI 60006

Travail de l'institut de botanique de l'université de Montpellier et de la station zoologique de Cette. Serie mixte. [Forms part of: Travaux de l'institut de zoologie de l'université de Montpellier et de la station zoologique de Cette.] Cette. Nos. ?-2-5, ?-1905-16. Trav. Inst. Bot. Univ. Montpellier Stat. Zool. Cette, Ser. Mixte. HI 73581

Travaux de l'association des instituts scientifiques du Caucase du Nord = Trudy Severo-Kavkazskoi assotsiatsii nauchno-issledovatel'skikh institutov. Krasnodar.

Travaux, association internationale de limnologie théorique et appliquée = Internationale Vereinigung für theoretische und angewandte Limnologie. Verhandlungen. Stuttgart. Int. Vereinigung Theor. Limnol. Verh. See B–P–H 435/21.

Travaux biologiques de l'institut J. B. Carnoy. Trav. Biol. Inst. Carnoy See B–P–H 892/8. HI 60007

Travaux du centre d'études phytosociologiques et ecologiques. Montpellier. 1958-60. Trav. Centre Études Phytosoc. Ecol. HI 73582

Travaux du centre de recherches et d'études océanographiques. Paris. Trav. Centr. Rech. Etudes Océanogr. See B–P–H 892/12. HI 60008

Travaux des chercheurs de la station de Lamto. N'douci. No. 1+, 1983+. Trav. Cherch. Stat. Lamto. HI 73583

Travaux de la commission pour l'étude de la république autonome Iakoute = Trudy komissii po izucheniyu Yakutskoi avtonomnof S S R. Leningrad.

Travaux de la commission des sciences mathématiques et naturelles. Société scientifique de Poznań = Prace komisji matematiczno-przyrodniczych. Poznańskie towarzystwo przyjaciól nauk, wydział matematyczno-przyrodniczych. Ser. B. Nauki biologiczne. Poznan, Poland. Prace Komis. Mat.-Przyr., Ser. B, Nauki Biol. See B–P–H 717/26.

Travaux et documents, O R S T O M. Paris. No. 1+, 1969+. Trav. Doc. O. R. S. T. O. M. HI 73584

Travaux effectués par les stations agronomiques, institut national de la recherche agronomique. Paris. 1939-49. Trav. Effect. Stat. Agron. Inst. Natl. Rech. Agron. Preceded by: Recherches sur la fertilisation effectué par les stations agronomiques. Superseded by: Annales de l'institut national de la recherche agronomique. Sér. A, annales agronomiques. HI 73585

Travaux de la faculté des sciences, université de Rennes, Série océanographique biologique. Rennes. Nos. 1-4, 1968-70. Trav. Fac. Sci. Univ. Rennes, Sér. Océanogr. Biol. Superseded by: Travaux du laboratoire de biologie halieutique, université de Rennes. Série biologie halieutique. HI 73586

Travaux sur la géologie de Bulgarie. Serie paléontologie = Trudove varkhu geologiyata na Bulgariya. Seriya paleontologiya. Sofia.

Travaux, institut de bactériologie. Louvain. Trav. Inst. Bactériol. See B–P–H 892/18. HI 60009

Travaux de l'institut biologique de Péterhof = Trudy Petergofskogo biologicheskogo instituta. Moscow.

Travaux de l'institut botanique Léo Errera. Brussels. Trav. Inst. Bot. Léo Errera. See B–P–H 892/20. HI 60010

Travaux de l'institut botanique de Tibilissi = Trudy Tbilisskogo botanicheskogo instituta. Tiflis.

Travaux de l'institut botanique, université de Kharkov = Trudy institutu botaniky, Kharkivs'kii derzhavnii universitet. Kharkov.

Travaux de l'institut de botanique de l'université de Lausanne. Lausanne. Trav. Inst. Bot. Univ. Lausanne. See B–P–H 892/22. HI 60011

Travaux de l'institut botanique de l'université de Neuchâtel. Neuchâtel. Trav. Inst. Bot. Univ. Neuchâtel. See B–P–H 892/23. HI 60012

Travaux de l'institut de botanique de l'université de

Stockholm. = Meddelanden från stockholms högskolas botaniska institut. Stockholm. Meddeland. Stockholms Högskolas Bot. Inst. See B–P–H 557/3.

Travaux de l'institut français d'études andines. Paris & Lima. Trav. Inst. Franç. Études Andines. See B–P–H 892/25. HI 60013

Travaux de l'institut de l'histoire de la science et de la technique. Moscow & Leningrad = Trudy instituta istorii nauki i tekhniki. Seriya 1, arkhiv istorii nauki i tekhniki. Moscow & Leningrad.

Travaux de l'institut océanographique de l'Indochine; mémoire. Nhatrang, Indochina [South Vietnam]. Trav. Inst. Océanogr. Indochine Mém. See B–P–H 892/27. HI 60014

Travaux de l'institut des recherches biologiques de Molotov = Trudy biologicheskogo nauchno-issledovatel'skogo instituta pri Molotovskom gosudarstvennom universitete imeni M. Gor'kogo. Molotov.

Travaux de l'institut des recherches biologiques de Perm = Trudy biologicheskogo nauchno-issledovatel'skogo instituta pri Permskom gosudarstvennom universitete. Perm.

Travaux de l'institut des recherches biologiques de Perm = Trudy biologicheskogo nauchno-issledovatel'skogo instituta Tomskogo gosudarstvennom universitete. Tomsk.

Travaux de l'institut des recherches biologiques de Perm = Trudy biologicheskogo nauchno-issledovatel'skogo instituta pri Permskom gosudarstvennom universitete imeni M. Gor'kogo. Perm.

Travaux de l'institut des recherches biologiques de Perm = Trudy Permskogo biologicheskogo nauchno-issledovatel'skogo instituta. Perm.

Travaux de l'institut des recherches biologiques et de la station biologique à l'université de Perm = Trudy biologicheskogo nauchno-issledovatel'skogo instituta i biologicheskoi stantsii pri Permskom gosudarstvennom universitete. Perm.

Travaux de l'institut des recherches scientifiques à l'université d'état Voronèje. Voronezh, Russian S F S R = Trudy Nauchno-issledovatel'skogo instituta pri Voronezhskom gosudarstvennom universitete. Voronezh.

Travaux de l'institut des sciences naturelles de Peterhof = Trudy Petergofskogo estestvenno-nauchnogo instituta. Leningrad.

Travaux de l'institut scientifique Chérifien. Série botanique. Tanger. Nos. 4, 6, 1952-53; [n.s.] nos. 1-29, 1954-64. Trav. Inst. Sci. Chérifien, Sér. Bot. Superseded by: Travaux de l'institut scientifique Chérifien. Série botanique et biologique végétale. HI 73587

Travaux de l'institut scientifique Chérifien. Série botanique et biologique végétale. Tanger, Rabat. Nos. 30-33, 1967-8? Trav. Inst. Sci. Chérifien, Sér. Bot. Biol. Vég. Preceded by: Travaux de l'institut scientifique Chérifien. Série botanique. Superseded by: Travaux de l'institut scientifique, université mohammed V. Série botanique. HI 73588

Travaux de l'institut scientifique Chérifien. Série générale. Tangier. Trav. Inst. Sci. Chérifien, Sér. Gén. See B–P–H 893/8. HI 60015

Travaux de l'institut scientifique chérifien et de la faculté des sciences de Rabat. Série botanique et biologie végétale = Travaux de l'institut scientifique Chérifien. Série botanique et biologique végétale. Tanger & Rabat.

Travaux de l'institut scientifique, université mohammed V. Série botanique. Rabat. No. 34+, 1984 [i.e. 1985]+. Trav. Inst. Sci. Univ. Mohammed V, Sér. Bot. Preceded by: Travaux de l'institut scientifique Chérifien. Série botanique et biologique végétale. HI 73589

Travaux du jardin botanique du Tibilissi = Trudy Tbilisskogo botanicheskogo sada. Tiflis.

Travaux du jardin botanique de Tiflis = Trudy Tiflisskogo Botanicheskogo Sada. Tiflis.

Travaux du laboratoire Arago. Banyuls-sur-Mer, France. Trav. Lab. Arago. See B–P–H 893/12. HI 60016

Travaux du laboratoire de biogeokhimie près l'académie des sciences de l'URSS = Trudy biogeokhimicheskoi laboratorii, akademiya nauk S S S R. Moscow & Leningrad.

Travaux du laboratoire de biologie halieutique, université de Rennes. Série biologie halieutique. Rennes. No. 5+, 1971+. Trav. Lab. Biol. Halieut. Univ. Rennes, Sér. Biol. Halieut. Preceded by: Travaux de la faculté des sciences, université de Rennes. Série océanographique biologique. HI 73590

Travaux du laboratoire de botanique générale et appliquée de l'université d'Alger. Alger. 1950-51, 1952. Trav. Lab. Bot. Gén. Appl. Univ. Alger. HI 73591

Travaux du laboratoire de botanique systematique et de phytogéographique de l'université libre de Bruxelles. Brussels. 1953+. Trav. Lab. Bot. Syst. Univ. Libre Bruxelles. HI 73592

Travaux du laboratoire forestier de Toulouse. Toulouse. Tome 1, Vols. 1-8, 1928-70. Trav. Lab. Forest. Toulouse. Superseded by: Gaussenia. 4-4242-3. HI 73593

Travaux de laboratoire de géologie de la faculté des sciences de l'université de Grenoble. Grenoble. Trav. Lab. Géol. Fac. Sci. Univ. Grenoble. See B–P–H 893/

17. HI 60017

Travaux. Laboratoire de géologie, faculté des sciences, université de Lyon. Lyons. Trav. Lab. Géol. Fac. Sci. Univ. Lyon. See B–P–H 893/18. HI 60018

Travaux du laboratoire et de l'institut botanique. Bordeaux. 1956. Trav. Lab. Inst. Bot. HI 73594

Travaux du laboratoire de "La Jaysinia" à Samoëns (Haute-Savoie). Paris. Vols. [1]-4, 1957-72. Trav. Lab. "La Jaysinia" Samoëns. HI 73595

Travaux de laboratoire de matière médicale de l'école supérieure de pharmacie de Paris. Paris. Trav. Lab. Matière Méd. Ecole Supér. Pharm. Paris. See B–P–H 893/22. HI 60019

Travaux du laboratoire de microbiologie de la faculté de pharmacie de Nancy. Nancy. Vols. 1-9, [1928]-[56]. Trav. Lab. Microbiol. Fac. Pharm. Nancy. 4-2818-2. HI 73596

Travaux du laboratoire de recherches biologiques de l'université de Madrid. Madrid. Trav. Lab. Rech. Biol. Univ. Madrid. See B–P–H 893/23. HI 60020

Travaux & mémoires des facultés des Lille. Lille. Trav. & Mém. Fac. Lille. See B–P–H 893/24. HI 60021

Travaux du musée botanique de l'académie impériale des sciences de Pétrograd = Trudy botanicheskogo muzeya imperatorskoi akademii nauk. St. Petersburg.

Travaux du musée botanique de l'académie impériale des sciences de Saint Pétersbourg = Trudy botanicheskogo muzeya imperatorskoi akademii nauk. St. Petersburg.

Travaux du musée botanique de l'académie des sciences de Russie = Trudy botanicheskogo muzeya rossiiskoi akademii nauk. Petrograd.

Travaux du musée botanique, Kiev = Zbirnyk prats Botanichnogo muzeyu. Kiev.

Travaux du musée botanique. Leningrad = Trudy botanicheskogo muzeya. Leningrad.

Travaux, musée geologique et minéralogique Pierre le Grand = Trudy geologicheskago i mineralogicheskago muzeya imeni imperatora Petra Velikago, imperatorskoi akademii nauk. St. Petersburg.

Travaux, musée geologique Pierre le Grand = Trudy geologicheskago muzeya imeni imperatora Petra Velikago, imperatorskoi akademii nauk. St. Petersburg.

Travaux du museum d'histoire naturelle "Gr. Antipa". Bucharest. Trav. Mus. Hist. Nat. "Gr. Antipa". See B–P–H 894/5. HI 60022

Travaux des naturalistes de la vallée du Loing. Moret-sur-Loing. 1927-58. Trav. Naturalistes Vallée Loing. HI 73598

Travaux et notices de l'académie d'agriculture de France. Paris. Trav. & Notices Acad. Agric. France. See B–P–H 894/8. HI 60023

Travaux scientifiques, école normale supérieure ... Plovdiv = Nauchni trudove, vissh pedagogicheski institut "Paisii Khilendarski". Matematika, fizika, khimiya, biologiya. Plovdiv.

Travaux scientifiques. Institut de recherches hortiviticoles = Lucrări ştiinţifice din cursul anului. Institutul de cercetări horti-viticole. Bucharest. Lucr. Şti. Curs. Anului Inst. Cercet. Horti-Vitic. See B–P–H 536/17.

Travaux scientifiques du musée d'histoire naturelle de Luxembourg. Luxemburg. Vol. 1+, 1981+. Trav. Sci. Mus Hist. Nat. Luxembourg. HI 73599

Travaux scientifiques du parc national des Ecrins. Gap. Vol. 1+, 1981+. Trav. Sci. Parc Natl. Ecrins. HI 73600

Travaux scientifiques du parc national de Port-Cros. Hyères. Vol. 1+, 1975+. Trav. Sci. Parc. Natl. Port-Cros. HI 73601

Travaux scientifiques du Parc national de la Vanoise. Chambéry. Vol. 1+, 1970+. Trav. Sci. Parc Natl. Vanoise. HI 73602

Travaux scientifiques, parc naturel régional de Corse. Ajaccio. Vol. ?-1(3/4)+, ?-198?+. Trav. Sci. Parc Nat. Rég. Corse. HI 73603

Travaux scientifiques de la société royale des sciences de Bohême = Rozprava královské české společnosti nauk, třida mat.-prírodovědecké. Prague. Rozpr. Král. České Společn. Nauk, Tř. Mat.-Přír. See B–P–H 805/8.

Travaux scientifiques de l'université de l'Ubekistan = Trudy Uzbekskogo gosudarstvennogo universiteta. Samarkand.

Travaux de la section géologique du cabinet de Sa Majesté (Ministère de la maison de l'empereur). St. Petersburg = Trudy geologicheskoi chasti kabineta ego imperatorskago velichestva. St. Petersburg.

Travaux de la section de mycologie et de phytopathologie de la société botanique de Russie = Trudy Sektsii po mikologii i fitopatologii Russkogo botanicheskogo obshchestva. Petrograd.

Travaux de la section de pédologie de la société des sciences naturelles et physiques du Maroc. Rabat. Vols. 1-14, 1950-59. Trav. Sect. Pédol. Soc. Sci. Nat. Phys. Maroc. HI 73604

Travaux de la section scientifique et technique, institut français de Pondichéry. Pondichéry. Vols. 1+, 1957+. Trav. Sect. Sci. Techn. Inst. Franç. Pondichéry. HI 73605

Travaux de la section scientifique et technique, institut français de Pondichéry. Extraits hors série. Pondichéry. Vol. 1+, 1961+. Trav. Sect. Sci. Techn. Inst. Franç. Pondichéry, Extr. Hors Sér. HI 73606

Travaux, service de la conservation de la nature, Belgique. Brussels. No. 8+, 1977+. Trav. Serv.

Conservation Nat., Belgique. Preceded by: Travaux, service des réserves naturelles domaniales et de la conservation de la nature. HI 73607

Travaux de service océanographique des pêches de l'Indochine. Nhatrang, Indochina [South Vietnam]. Trav. Serv. Océanogr. Pêches Indochine. See B–P–H 894/18. HI 60025

Travaux, service des réserves naturelles domaniales et de la conservation de la nature. Brussels. Nos. 1-7, 1966-74. Trav. Serv. Réserves Nat. Doman. Conservation Nat. Superseded by: Travaux, service de la conservation de la nature. Belgium. HI 73608

Travaux de la société allemande d'agriculture. Berlin = Arbeiten der Deutschen Landwirtschafts-Gesellschaft. Prenzlau, Germany. Arbeiten Deutsch. Landw.-Ges. See B–P–H 135/4.

Travaux de la société botanique de Genève. Chateleine, Geneva. Vols. 1/2-9, 1954-68. Trav. Soc. Bot. Genève. Preceded by: Bulletin de la société botanique de Genève. Superseded by: Saussurea. HI 73609

Travaux de la société bulgare des sciences naturelles. Sofia, Bulgaria. Trav. Soc. Bulg. Sci. Nat. See B–P–H 894/21. HI 60026

Travaux de la société d'émulation du département du Jura. Lons-le-Saunier, France. Trav. Soc. Émul. Dép. Jura. See B–P–H 894/23. HI 60027

Travaux de la société d'histoire naturelle de l'Ile Maurice. Port Louis, Mauritius. 1842-46. Trav. Soc. Hist. Nat. Ile Maurice. Preceded by: Rapport annuel sur les travaux de la société d'histoire naturelle de Ile Maurice. Superseded by: Rapport de la société royale des arts et des sciences de Maurice. HI 73610

Travaux de la société impériale des naturalistes de Pétrograd. Section de botanique = Trudy imperatorskago Petrogradskago obshchestva estestvoispytatelei. Vypusk 3, otdělenie botaniki. Petrograd.

Travaux de la société impériale des naturalistes de Saint-Pétersbourg. Section de botanique = Trudy imperatorskago S.-Peterburgskago obshchestva estestvoispytatelei. Vypusk 3, otdělenie botaniki. St. Petersburg.

Travaux de la société impériale des naturalistes de Saint-Pétersbourg. Section de botanique = Trudy imperatorskago S.-Peterburgskago obshchestva estestvoispytatelei. Vypusk 4, otdělenie botaniki. St. Petersburg.

Travaux de la société des naturalistes et des amateurs des sciences naturelles de Bessarabie. Kishinev [Moldavian S S R]. Trav. Soc. Naturalistes Bessarabie. See B–P–H 895/1. HI 60028

Travaux de la société des naturalistes de Léningrad. Section de botanique = Trudy Leningradskogo obshchestva estestvoispytatelei. Vypusk. 3: Otdelenie botaniki. Leningrad.

Travaux de la société des naturalistes de Pétrograd. Section de botanique = Trudy Petrogradskogo obshchestva estestvoispytatelei. Vypusk 3... Otdelenie botaniki. Petrograd.

Travaux de la société des naturalistes de Saint-Pétersbourg. Section de botanique = Trudy imperatorskago S.-Peterburgskago obshchestva estestvoispytatelei. Vypusk 2, otdělenie botaniki. St. Petersburg.

Travaux de la société des naturalistes de Saint-Pétersbourg. Section de botanique = Trudy imperatorskago S.-Peterburgskago obshchestva estestvoispytatelei. Vypusk 3, otdělenie botaniki. St. Petersburg.

Travaux de la société des naturalistes de Saint-Pétersbourg. Section de botanique = Trudy imperatorskago S.-Peterburgskago obshchestva estestvoispytatelei. Vypusk 4, otdělenie botaniki. St. Petersburg.

Travaux de la société des naturalistes de Saint-Pétersbourg: Section de botanique = Trudy S.-Peterburgskago Obshchestva Estestvoispytatelei: Otdělenie botaniki. St. Petersburg.

Travaux de la société des naturalistes à Saratow = Trudy Saratovskago Obshchestva Estestvoispytatelei i Lyubitelei Estestvozaniya. Saratov.

Travaux de la société des naturalistes à l'université impériale de Kharkow = Trudy Obshchestva Ispytatelei Prirody pri Imperatorskom khar'kovskom Universitetě. Kharkov.

Travaux de la société des sciences et des lettres de Wilno. Classe des sciences mathématiques et naturelles = Prace towarzystwa przyjaciól nauk w Wilnie. Wydzial nauk matematycznych i przyrodniczych. Wilno.

Travaux de la société des sciences et des lettres de Wrocław = Prace wrocławskiego towarzystwa naukowego. Wroclaw, Poland. Prace Wrocławsk. Towarz. Nauk. See B–P–H 718/11.

Travaux de la sous-section Troitzkossawsk-Kiakhta, section du pays d'Amour de la société impériale russe de géographie = Trudy Troitskosavsko-Kyakhtinskago Otděleniya Priamurskago Otděla Imperatorskago Russkago Geograficheskago Obshchestva. Moscow.

Travaux de la sous-section Troitzkossawsk-Kiakhta, section du pays d'Amour de la société impériale russe de géographie = Trudy Troitskosavsko-Kyakhtinskago Otděleniya Priamurskago Otděla Russkago Geograficheskago Obshchestva. Irkutsk.

Travaux de la station de biologie maritime de Lisbonne.

Lisbon. Trav. Stat. Biol. Marit. Lisbonne. See B–P–H 895/13. HI 60029

Travaux de la station biologique du Dnièpre = Zbirnyk prats Dniprovs'koi biologichnoi stantsii. Kiev.

Travaux de la station biologique du Dnièpre = Zbirnyk prats Dnipryans'koi biologichnoi stantsii. Kiev.

Travaux de la station biologique de Roscoff. Roscoff, France. Trav. Stat. Biol. Roscoff. See B–P–H 895/14. HI 60030

Travaux de la station biologique (de) Sebastopol = Trudy, sevastopol'skoi biologicheskoi stantsii (imeni A. O. Kovalevskogo). Moscow, Leningrad.

Travaux de la station biologique à Volga = Raboty Volzhskoi Biologicheskoi Stantsii. Saratov.

Travaux de la station d'hydrobiologie lacustre de l'I N R A à Thonon. Thonon. Vol. ?-99+, ?-1971+. Trav. Stat. Hydrobiol. Lacustre I. N. R. A. Thonon. HI 73611

Travaux de la station hydrobiologique. Kiev, Ukrainian S S R = Trudy gidrobiologichnoi stantsii. Kiev.

Travaux de la station du Lac Sevane = Trudy Sevanskoi ozernoi stantsii. Erivan.

Travaux de la station limnologique du lac Baikal = Trudy Baikal'skoi limnologicheskoi stantsii. Moscow & Leningrad.

Travaux de la station de recherches des eaux et forêts. Série D, hydrobiologie. Groenendaal-Hoeilaart. 1943+. Trav. Stat. Rech. Eaux Forêts, D. HI 73612

Travaux de l'université d'état du Turkestan = Trudy Turkestanskogo gosudarstvennogo universiteta. Tashkent, Uzbek S S R. Trudy Turkestansk. Gosud. Univ. See B–P–H 922/4.

Travaux de l'université de Turkestan = Trudy Turkestanskogo gosudarstvennogo universiteta. Tashkent, Uzbek S S R. Trudy Turkestansk. Gosud. Univ. See B–P–H 922/4.

Travaux des vacances de la station biologique du Wolga. Saratov = Lĕtniya raboty. Saratov.

Traveller. London. ?-1981+. Traveller. HI 73613

Treatises concerning the Tatra national park = Zbornik prac o Tatranskom narodnom parku. Martin.

Treballs de l'institut botànic de Barcelona. Barcelona. No. 5+, 1979+. Treb. Inst. Bot. Barcelona. Preceded by: Publicaciones del instituto de biologia aplicada. Barcelona. HI 73614

Treballs de l'institució catalana d'historia natural. Barcelona. No. 1+, 1915-23, 1976+. Treb. Inst. Catalana Hist. Nat. 3-1994-2. HI 73615

Treballs de la societat de biología de Barcelona. Barcelona. 1913-34. Treb. Soc. Biol. Barcelona. HI 73616

Tree farmer. Washington, DC. 1989+. Tree Farmer. HI 74534

Tree genetics newsletter. 1948-54. Tree Genet. Newslett. HI 73617

Tree improvement bulletin. Kitwe. 1976+. Tree Improv. Bull. HI 73618

Tree lover; A quarterly magazine. London. 1932 (Oct.)-45 (Spring). Tree Lover (London). 5-4259-1. HI 73619

Tree lover. Sydney. Tree Lover (Sydney). See B–P–H 895/32. HI 60031

Tree news. London. 1981+. Tree News. HI 73620

Tree pamphlet, forestry branch, Canada. Ottawa. Nos. 1-14, 1923-30. Tree Pam. Forest. Branch Canada. HI 73621

Tree physiology. Victoria, B.C. Vol. 1+, 1986+. Tree Physiol. HI 75169

Tree planter. Cabazon, CA. 1978+. Tree Planter. HI 73622

Tree planters' notes. Washington, DC. Tree Pl. Notes. See B–P–H 895/33. HI 60032

Tree-ring bulletin. Tucson, Flagstaff, AZ. Vols. 1+, 1934+. Tree-Ring Bull. 5-4259-1. HI 73623

Tree topics. Montgomery, AL. Vol. 1+, 1980+. Tree Topics. HI 73624

Tree trimmer. Los Angeles, CA. 1987+. Tree Trimmer. HI 73625

Trees; journal of the men of the trees. Crawley, Southampton. Vols. 1-19, 1936-56; vol. 24(4)+, 1961+. Trees (Crawley). For vols. 20-24(3), 1956-60 see: Trees and life. HI 73626

Trees; journal of american arboriculture. Santa Monica, CA. Vols. ?-35(2), 1940-76. Trees (Santa Monica). Preceded by: Trees magazine. 5-4259-2. HI 73627

Trees, fruits and flowers of Minnesota. St. Paul & Minneapolis, MN. Trees Minnesota. See B–P–H 896/4. HI 60033

Trees and life. Southampton. Vols. 20-24(3), 1956-60. Trees & Life. Preceded and superseded by: Trees. HI 73628

Trees magazine. Santa Monica, CA. 1938-40. Trees Mag. Preceded by: Western trees, parks and forests. Superseded by: Trees. Santa Monica, CA. 5-4259-2. HI 73629

Trees in South Africa. Johannesburg. Vol. 1+, 1949+. Trees South Africa. 5-4259-2. HI 73630

Trees: structure and function. New York. Vol. 1+, 1987+. Trees Struct. Funct. HI 73631

Trees and Victoria's resources. Springvale, Vic. Vol. 23+, 1981+. Trees & Victoria's Resources. Preceded by: Victoria's resources. HI 73632

Trencsénvármegyei múzeum-egyesület értesitöje. Trencsén, Hungary [=Trencin, Czechoslovakia]. Trencsénvárm. Mús.-Egyes. Értes. See B–P–H 896/6. HI 60034

Trencsénvármegyei természettudományi egylet évkönyve. Trencsén, Hungary [=Trencin, Czechoslovakia]. Trencsénvárm. Term. Egyl. Évk. See B–P–H 896/7. HI 60035

Trends in biochemical sciences. Amsterdam, New York. Vol. 1+, 1976+. Trends Biochem. Sci. HI 73633

Trends in biochemical sciences. Reference edition. Amsterdam. Vol. 1+, 1976+. Trends Biochem Sci., Ref. Ed. HI 73634

Trends in biotechnology. New York. Vol. 1+, 1983+. Trends Biotechnol. HI 73635

Trends in ecology and evolution. Amsterdam, Cambridge. Vol. 1+, 1986+. Trends Ecol. Evol. HI 73636

Trends in food science and technology. Cambridge. Vol. 1+, 1990+. Trends Food Sci. Technol. HI 75249

Trends in genetics. (Personal edition). Amsterdam. Vol. 1+, 1985+. Trends Genet. (Personal Ed.) HI 75170

Trends in genetics. (Reference edition). Amsterdam. Vol. 1+, 1985+. Trends Genet. (Refer. Ed.) HI 75171

Trends in pharmacological sciences. Amsterdam. Vol. 1+, 1979+. Trends Pharmacol. Sci. HI 73637

Treubia; recueil de travaux zoologiques, hydrobiologiques, et océanographiques. Buitenzorg, Dutch E. Indies [=Bogor, Indonesia]. Treubia. See B–P–H 896/8. HI 60036

Trianea; acta cientifica y tecnologia Inderena. Bogota. No. 1+, 1988+. Trianea. HI 73639

Tribuna farmacéutica. Curitiba, Brazil. Tribuna Farm. See B–P–H 896/10. HI 60037

Tribune horticole. Brussels. Tribune Hort. See B–P–H 896/11. HI 60038

Tribune médicale. Revue mensuelle de médecine, de chirurgie et des sciences biologiques. Paris. Tribune Méd. (Paris). See B–P–H 896/12. HI 60039

Triennial report, botany division, department of scientific and industrial research, New Zealand. Christchurch, N.Z. 1957/59+, 1960?+. Trienn. Rep. Bot. Div. Dept. Sci. Industr. Res. New Zealand. HI 73640

Triennial report, Cawthron institute = Report (Annual), Cawthron institute. Nelson, N.Z.

Triennial report, division of weed research, institute of plant protection, Israel. 1971/74. Trienn. Rep. Div. Weed Res. Inst. Pl. Protect. Israel. HI 73641

Triennial report, Israel oceanographic and limnological research. Haifa. 1976/78+. Trienn. Rep. Israel Oecanogr. Limnol. Res. HI 73642

Triennial review of research at the marine laboratory, Aberdeen. Edinburgh. 1973/75+, 1975+. Trienn. Rev. Res. Mar. Lab. Aberdeen. HI 73643

Trigo. Santiago de Chile. 1954-55. Trigo. Superseded by: Chile triguero. HI 73644

Trigo y cereales. Santiago de Chile. 1957-62. Trigo & Cereales. Preceded by: Chile triguero. Superseded by: Nueva agricultura. HI 73645

Trigo e soja. ?-1981+. Trigo & Soja. HI 73646

Trillia; proceedings of the botanical society of western Pennsylvania. Pittsburgh, PA. Trillia. See B–P–H 896/13. HI 60040

Trimen's journal of botany = Journal of botany, british and foreign. London.

Trinidad horticultural magazine. Port of Spain, Trinidad. Trinidad Hort. Mag. See B–P–H 896/16. HI 60041

Trinidad naturalist magazine. Port of Spain, Woodbrook, Cascade. Vols. 1-3(9), 1975-81. Trinidad Naturalist Mag. Superseded by: Naturalist. Cascade. HI 73647

Tri-ology [entomology, nematology, pathology] technical report, division of plant industry, department of agriculture, Florida. [Gainesville, FL.] Vol. 1+, 1962+. Tri-ology Techn. Rep. Div. Pl. Indust., Florida. HI 73638

Triticale abstracts. Farnham Royal. Vols. 1-?, 1975-83? Triticale Abstr. Superseded by: Wheat, barley and triticale abstracts. HI 73649

Tromsø museums aarshefter. Tromso, Norway. Tromsø Mus. Aarsh. See B–P–H 896/20. HI 60042

Tromsø museum årsberetning. Tromsø. 1872/1922-1931-35, 1873-1935; 1952-78. Tromsø Mus. Årsberetn. Superseded by: Tromura. Fellesserie. HI 73650

Tromsø museums skrifter. Tromso, Norway. Tromsø Mus. Skr. See B–P–H 896/24. HI 60043

Tromura. Fellesserie. Tromsø. No. 1+, 1979+. Tromura, Felless. Preceded by: Tromsø museum årsberetning. HI 73651

Tromura. Naturvitenskap. Tromsø. No. 1+, 1978+. Tromura, Naturvitensk. HI 73652

Trondheimske selskabs skrifter. Copenhagen. Trondh. Selsk. Skr. See B–P–H 896/25. HI 60044

Tropenlandwirt; Zeitschrift für die landwirtschaft in den Tropen und Subtropen. Witzenhausen. ?-1980+. Tropenlandwirt. HI 73653

Tropenpflanzer. Berlin. Tropenpflanzer. See

B–P–H 897/30. HI 60056

Tropenpflanzer. Beiheft. Berlin. Tropenpflanzer Beih. See B–P–H 898/1. HI 60057

Tropgarden news. Trivandrum. Vol. 1+, 1986+. Tropgarden News. HI 73654

Tropic oceanology. [Je tai hai yang] Beijing. Vol. 1+, 1982+. Trop. Oceanol. HI 73655

Tropical abstracts. Amsterdam. Vols. 8-29, 1953-74. Trop. Abstr. Preceded by: Documentatieblad van de afdeling tropische production, k. instituut voor de tropen. Superseded by: Abstracts on tropical agriculture. HI 73656

Tropical agriculture; journal of the imperial college of tropical agriculture. St. Augustine, Trinidad. Trop. Agric. (Trinidad). See B–P–H 897/8. HI 60046

Tropical agriculturist. Journal of the Ceylon agricultural society. Peradeniya, Ceylon. Trop. Agric. (Ceylon). See B–P–H 897/7. HI 60045

Tropical bryology. Duisburg. Vol. 1+, 1989+. Trop. Bryol. HI 73657

Tropical crops research bulletin. [Je tai tso wu yan chiu t'ung hsun.] Canton. Trop. Crops Res. Bull. See B–P–H 897/10. HI 60047

Tropical diseases bulletin. London. Vol. 1+, 1912+. Trop. Diseases Bull. Preceded by: Bulletin, sleeping sickness bureau and by Kala azar bulletin [not entered]. 5-4266-1. HI 73658

Tropical ecology; official publication of the international society for tropical ecology. Allahabad, Varanasi. Vol. 2+, 1961+. Trop. Ecol. Preceded by: Bulletin of the international society for tropical ecology. HI 73659

Tropical forest note, institute of tropical forestry. Rio Piedras, PR. Nos. 1-14, 1959-62. Trop. Forest Note Inst. Trop. Forest. Superseded by: Research notes I T F, United States forest service. HI 73660

Tropical forestry papers, commonwealth forestry institute. Oxford. No. 7+, 1975+. Trop. Forest. Pap. Commonw. Forest. Inst. Preceded by: Fast-growing timber trees of the lowland tropics. HI 73661

Tropical gardening. Miami, FL. Trop. Gard. See B–P–H 897/13. HI 60048

Tropical grain legume bulletin. Ibadan. No. 1+, 1975+. Trop. Grain Legume Bull. HI 73662

Tropical grasslands. St. Lucia, Qld. Vol. 1+, 1967+. Trop. Grasslands. HI 73663

Tropical homes and gardening. Miami, FL. Trop. Homes Gard. See B–P–H 897/14. HI 60049

Tropical horticulture. [Nettai engei.] Taihoku [=Taipei, Taiwan]. Trop. Hort. See B–P–H 897/15. HI 60050

Tropical life. London. Trop. Life. See B–P–H 897/16. HI 60051

Tropical oil seeds abstracts. Farnham Royal. Vol. 1+, 1976+. Trop. Oil Seed Abstr. HI 73664

Tropical pest bulletin. London. No. 1+, 1972+. Trop. Pest Bull. HI 73665

Tropical pest management. London. Vol. 26+, 1980+. Trop. Pest Managem. Preceded by: P A N S. Pest articles and news summaries. Section B: Plant disease control. HI 73666

Tropical plant research foundation bulletin. Washington, DC. Trop. Pl. Res. Found. Bull. See B–P–H 897/20. HI 60053

Tropical plant science research. New Delhi. Vol. 1+, 1983+. Trop. Pl. Sci. Res. HI 73667

Tropical root and tuber crops newsletter. Nayaquez. (Puerto Rico.) No. 1+, 1968+. Trop. Root Tuber Crops Newslett. HI 73668

Tropical science; quarterly journal of the tropical products institute. London. Vol. 1+, 1959+. Trop. Sci. Preceded by: Colonial plant and animal products. HI 73669

Tropical storage abstracts. Farnham Royal. No. 1+, 1973+. Trop. Storage Abstr. HI 73670

Tropical and sub-tropical America. New York, NY. Trop. Sub-Trop. Amer. See B–P–H 897/25. HI 60054

Tropical and subtropical forest ecosystem. Guangzhou. Vol. 1+, 1982+. Trop. Subtrop. Forest Ecosyst. HI 73671

Tropical woods; a technical journal devoted to the furtherance of knowledge of tropical woods and forests and to the promotion of forestry in the tropics. New Haven, CT. Trop. Woods. See B–P–H 897/26. HI 60055

Tropics magazine. Coconut Grove, FL. Trop. Mag. See B–P–H 897/17. HI 60052

Tropicus. Washington, DC. 1988+. Tropicus. HI 73673

Tropinet. [Published as a supplement to: Biotropica.] Durham, NC. Vol. [1]+, 1989+. Tropinet. HI 75246

Tropische natuur; orgaan van de Nederlandsch-Indische natuur-historische vereeniging. Buitenzorg [= Bogor], Weltevreden. Vols. 1-3?, 1912-53. Trop. Natuur. Superseded by: Penggemar alam. 5-4266-3. HI 73674

Tropische und subtropische Pflanzenwelt. [Forms part of: Akademie der Wissenschaften und der Literatur, Mainz. Abhandlungen der mathematisch-naturwissenschaftlichen Klasse.] Wiesbaden. Vol. 1+, 1973+. Trop. Subtrop. Pflanzenwelt. HI 73675

Trudove na balgarskoto prirodizpatelno druzhestvo = Travaux de la société bulgare des sciences naturelles.

Sofia, Bulgaria. Trav. Soc. Bulg. Sci. Nat. See B–P–H 894/21.

Trudove na mikrobiologičeskaja institut = Trudove na mikrobiologicheskaya institut. Sofia.

Trudove na mikrobiologicheskaya institut. Sofia, Bulgaria. Vol. 1+, 1950+. Trudove Mikrobiol. Inst. HI 73676

Trudove, vissh pedagogicheskii institut. Matematika, fizika, chimiya, biologiya. Plovdiv. 1963. Trudove Vissh Pedagog. Inst., Mat. Fiz. Chim. Biol. Superseded by: Nauchnie trudove, vissh pedagogicheskii institut "Paisii Khilevdarski." Matematika, fizika, khimiya, biologiya. HI 73677

Trudove varkhu geologiyata na Bulgariya. Seriya paleontologiya. Sofia. Vols. 1-8, 1959-66. Trudove Varkhu Geol. Bulgariya, Ser. Paleontol. HI 73678

Trudy akademii nauk litovskoi S S R. Ser. biologicheskii nauk = Lietuvos T S R mokslu akademijos biologijos instituto darbai. Vilnius.

Trudy akademii nauk litovskoi S S R. V s. = Lietuvos T S R mokslu akademijos darbai. C s. Vilnius.

Trudy akademija nauk tadžikskoi S S R. Stalinabad = Trudy, akademiya nauk tadzhikskoi S S R. Stalinabad.

Trudy, akademiya nauk tadzhikskoi S S R. Stalinabad [= Dushanbe]. Vols. 1-120, 1951-60. Trudy Akad. Nauk Tadzhiksk. S.S.R. HI 73679

Trudy Alma-Atinskogo botaničeskogo sada = Trudy Alma-Atinskogo botanicheskogo sada. Alma-Ata.

Trudy Alma-Atinskogo botanicheskogo sada. Alma-Ata, Kazakh S S R. Vols. 2-7, 1954-63. Trudy Alma-Atinsk. Bot. Sada. Preceded by: Trudy respublikanskogo botanicheskogo sada. Alma-Ata. Superseded by: Trudy botanicheskikh sadov. Alma-Ata. HI 73680

Trudy, altaiskii nauchno-issledovatel'ski institut zemledeliya i selektsii sel'skokhozyaistvennykh kul'tur. Vol. ?-5+, ?-1979+. Trudy Altaisk. Nauchno-Issl. Inst. Zemled. Selekts. Sel'skokhoz. Kul't. HI 73681

Trudy arhiva = Trudy arkhiva. Moscow & Leningrad.

Trudy arhiva Akademii nauk S S S R = Trudy arkhiva Akademii nauk S S S R. Moscow & Leningrad.

Trudy arkhiva. Moscow & Leningrad. Vols. 1-5, 1933-46. Trudy Arkh. Superseded by: Trudy arkhiva Akademii nauk S S S R. 1-113-3. HI 73682

Trudy arkhiva Akademii nauk S S S R. Moscow & Leningrad. Vol. 6+, 1946+. Trudy Arh. Akad. Nauk S.S.S.R. Preceded by: Trudy arkhiva. 1-113-3. HI 73683

Trudy arkticheskogo i antarkticheskogo nauchno-issledovatel'skogo instituta glavnogo upravleniya severnogo morskogo puti ministerstva morskogo flota S S S R. Leningrad. Vols. 218, 233+, 1960+. Trudy Arktich. Antarktich. Nauchno-Issl. Inst. Glavn. Upravl. Severn. Morsk. Minist. Morsk. Flota S.S.S.R. Preceded by: Trudy arkticheskogo nauchno-issledovatel'skogo instituta glavnogo upravleniya severnogo morskogo puti ministerstva morskogo flota S S S R. HI 73684

Trudy arkticheskogo instituta. Leningrad. Vols. 1-28, 30-42, 44-52, 54-59, 61-80, 104-105, 107, 116, 122, 1931-38. Trudy Arktich. Inst., Leningrad. Preceded by: Trudy po izucheniyu severa. For vols. 81-83, 86-88, 91-101, 103, 106, 108-109 see: Trudy, vsesoyuznyi arkticheskii institut. Superseded by: Trudy arkticheskogo nauchno-issledovatel'skogo instituta glavnogo upravleniya severnogo morskogo puti pri sovete S S S R. HI 73685

Trudy arkticheskogo nauchno-issledovatel'skogo instituta glavnogo upravleniya severnogo morskogo puti ministerstva morskogo flota S S S R. Moscow, Leningrad. Vols. 203-217, 219-221, 1956-58. Trudy Arktich. Nauchno-Issl. Inst. Glavn. Upravl. Severn. Morsk. Minist. Morsk. Flota S.S.S.R. Preceded by: Trudy arkticheskogo nauchno-issledovatel'skogo instituta glavnogo upravleniya severnogo morskogo puti pri sovete S S S R. Superseded by: Trudy arkticheskogo i antarkticheskogo nauchno-issledovatel'skogo instituta glavnogo upravleniya severnogo morskogo puti ministerstva morskogo flota S S S R. HI 73686

Trudy arkticheskogo nauchno-issledovatel'skogo instituta glavnogo upravleniya severnogo morskogo puti pri sovete S S S R. Moscow, Leningrad. Vols. 29, 43, 53, 60, 84-85, 89-90, 102, 110-115, 117-121, 123-202, 1935-49. Trudy Arktich. Nauchno-Issl. Inst. Glavn. Upravl. Severn. Morsk. Sovete S.S.S.R. Preceded by: Trudy arkticheskogo instituta. Superseded by: Trudy arkticheskogo nauchno-issledovatel'skogo instituta glavnogo upravleniya severnogo morskogo puti ministerstva morskogo flota S S S R. HI 73687

Trudy armyanskogo filiala akademii nauk S S S R. Seriya biologicheskaya. Erevan. Vol. 1+, 1936?+. Trudy Armyansk. Fil. Akad. Nauk S.S.S.R., Ser. Biol. HI 73688

Trudy Azerbaidzhanskogo filiala, akademiya nauk S S S R. Baku, Azerbaijan S S R. Vols. 12-13, 1935; vols. 15-18, 1935-36; vols. 20-63, 1936-39 [vols. 15 and 16 published 1936, vols. 17 and 18 1935]. Trudy Azerbaidzhansk. Fil., Akad. Nauk S.S.S.R. Preceded by: Trudy Azerbaidzhanskogo otdeleniya. For vol. 14 see: Trudy Azerbaidzhanskogo otdeleniya. For vol. 19 see: Azerbaidzhanskii filial Akademii nauk S S S R. 1-109-3. HI 73689

Trudy Azerbaidzhanskogo otdeleniya. Baku, Azerbaijan S S R. Vol. 11, 1935; vol. 14, 1935. Trudy

Azerbaidzhansk. Otd. Preceded by: Trudy Azerbaidzhanskogo otdeleniya Zakavkazskogo filiala Akademii nauk S S S R. Superseded by: Trudy Azerbaidzhanskogo filiala, akademiya nauk S S S R. For vols. 12-13 see: Trudy Azerbaidzhanskogo filiala, akademiya nauk S S S R. 1-109-3. HI 73690

Trudy Azerbaidzhanskogo otdeleniya Zakavkazskogo filiala Akademii nauk S S S R. Baku, Azerbaijan S S R. Vols. 1-10, 1933-36. Trudy Azerbaidzhansk. Otd. Zakavkazsk. Fil. Akad. Nauk S.S.S.R. Superseded by: Trudy Azerbaidzhanskogo otdeleniya. 1-109-3. HI 73691

Trudy Azerbaidzhanskogo stantsii vsesoyuznogo instituta zashchity rastenii. Baku, Azerbaijan S S R. Vol. 1+, 1960+. Trudy Azerbaidzhansk. Stantsii Vsesoyuzn. Inst. Zashch. Rast. HI 73692

Trudy Azerbajdžanskogo filiala = Trudy Azerbaidzhanskogo filiala, akademiya nauk S S S R. Baku.

Trudy Azerbajdžanskogo otdelenija = Trudy Azerbaidzhanskogo otdeleniya. Baku.

Trudy Azerbajdžanskogo otdelenija Zakavkazskogo filiala Akademii nauk S S S R = Trudy Azerbaidzhanskogo otdeleniya Zakavkazskogo filiala Akademii nauk S S S R. Baku.

Trudy Azerbajdžanskogo stancii vsesojuznogo instituta zaščity rastenij = Trudy Azerbaidzhanskogo stantsii vsesoyuznogo instituta zashchity rastenii. Baku.

Trudy Azovo-Černomorskogo kraevogo biologičeskogo obščestva = Trudy Azovo-Chernomorskogo kraevogo biologicheskogo obshchestva. Rostov.

Trudy Azovo-Chernomorskogo kraevogo biologicheskogo obshchestva. Rostov, Russian S F S R. Vols. 1-2, 1935-37. Trudy Azovo-Chernomorsk. Kraev. Biol. Obshch. Superseded by: Trudy Rostovskogo oblastnogo biologicheskogo obshchestva. HI 73693

Trudy Baikal'skoi limnologicheskoi stantsii. Moscow & Leningrad. Vols. 1-20, 1931-61. Trudy Baikal'sk. Limnol. Stantsii. Preceded by: Trudy komissii po izucheniyu ozera Baikala. Superseded by: Trudy limnologicheskogo instituta. Moscow. 2-1651-2. HI 73694

Trudy Bajkal'skoj limnologičeskoj stancii. Moscow & Leningrad = Trudy Baikal'skoi limnologicheskoi stantsii. Moscow & Leningrad.

Trudy belaruskaga sel'ska-haspadarchaga instytuta. Gorki. Vols. 1-13, 1935-48. Trudy Belarusk. Sel'ska-hasp. Inst. HI 73695

Trudy Belomorskoi biologicheskoi stantsii Moskovskogo gosudarstvennogo universiteta. Voronezh, Russian S F S R. Vol. 1+, 1962+. Trudy Belomorsk. Biol. Stantsii Moskovsk. Gosud. Univ. HI 73696

Trudy Belomorskoj biologičeskoj stancii Moskovskogo gosudarstvennogo universiteta = Trudy Belomorskoi biologicheskoi stantsii Moskovskogo gosudarstvennogo universiteta. Voronezh.

Trudy belorusskogo gosudarstvennogo universiteta v g. Minske. Minsk. Vol. 1, 1922. Trudy Belorussk. Gosud. Univ. Minske. Superseded by: Pratsy belaruskaga dzerzhnaunaga universitetu u Mensku. HI 73697

Trudy belorusskogo nauchno-issledovatel'skogo instituta sel'skogo i lesnogo khozyaistva imeni V. I. Lenina pri S N K B S S R = Pratsy belaruskaga navukova-das'ledchaga instytutu liasnoi gaspadarki i liasnoi pramyslovas'chi. Trudy po lesnomu opytnomu delu B S S R. Minsk.

Trudy belorusskogo sel'skokhozyaistvennogo instituta = Trudy belaruskaga sel'ska-haspadarchaga instytuta. Gorki.

Trudy Bessarabskago Obščestva Estestvoispytatelej i Ljubitelej Estestvoznanija. Kishinev = Travaux de la société des naturalistes et des amateurs des sciences naturelles de Bessarabie. Kishinev [Moldavian S S R]. Trav. Soc. Naturalistes Bessarabie. See B–P–H 895/1.

Trudy biogeokhimicheskoi laboratorii, akademiya nauk S S S R. Moscow, Leningrad. Vols. 1+, 1930+. Trudy Biogeokhim. Lab. Akad. Nauk S.S.S.R. HI 73698

Trudy biologičeskogo fakul'teta po genetike i zoologii. Kharkov, Ukrainian S S R = Trudy biologicheskogo fakul'teta po genetike i zoologi. Kharkov, Ukrainian S S R. Kharkov.

Trudy biologičeskogo fakul'teta Tomskogo gosudarstvennogo universiteta = Trudy biologicheskogo fakul'teta Tomskogo gosudarstvennogo universiteta. Tomsk.

Trudy biologičeskogo instituta Frunze = Trudy biologicheskogo instituta Frunze. Frunze.

Trudy biologičeskogo instituta. Novosibirsk, Russian S F S R = Trudy biologicheskogo instituta. Novosibirsk, Russian S F S R.

Trudy biologičeskogo naučno-issledovatel'skogo instituta i biologičeskoj stancii pri Permskom gosudarstvennom universitete = Trudy biologicheskogo nauchno-issledovatel'skogo instituta i biologicheskoi stantsii pri Permskom gosudarstvennom universitete. Perm.

Trudy biologičeskogo naučno-issledovatel'skogo instituta pri Molotovskom gosudarstvennom universitete imeni M. Gor'kogo = Trudy biologicheskogo nauchno-issledovatel'skogo instituta pri Molotovskom gosudarstvennom universitete imeni M. Gor'kogo. Molotov.

Trudy biologičeskogo naučno-issledovatel'skogo instituta

pri Permskom gosudarstvennom universitete = Trudy biologicheskogo nauchno-issledovatel'skogo instituta pri Permskom gosudarstvennom universitete. Perm.

Trudy biologičeskogo naučno-issledovatel'skogo instituta pri Permskom gosudarstvennom universitete imeni M. Gor'kogo = Trudy biologicheskogo nauchno-issledovatel'skogo instituta pri Permskom gosudarstvennom universitete imeni M. Gor'kogo. Perm.

Trudy biologičeskogo naučno-issledovatel'skogo instituta Tomskogo gosudarstvennom universitete = Trudy biologicheskogo nauchno-issledovatel'skogo instituta Tomskogo gosudarstvennom universitete. Tomsk.

Trudy Biologičeskoj Laboratorii. Otděl estestvenno-istoričeskij. Moscow = Uchenye Zapiski. Moskovskago Gorodskago Narodnago Universiteta imeni A. L. Sanyavskago. Moscow.

Trudy biologičeskoj stancii "Borok" imeni N. A. Morozova = Trudy biologicheskoi stantsii "Borok" imeni N. A. Morozova. Moscow & Leningrad.

Trudy biologicheskogo fakul'teta po genetike i zoologii. Kharkov, Ukrainian S S R. Vol. 36, 1963. Trudy Biol. Fak. Genet. Preceded & superseded by: Trudy nauchno-issledovatel'skogo instituta biologii. Pratsi Naukovo-doslidnogo instytutu biologii. HI 73699

Trudy biologicheskogo fakul'teta Tomskogo gosudarstvennogo universiteta. Tomsk, Russian S F S R. Vol. 1 [also numbered vol. 83], 1930. Trudy Biol. Fak. Tomsk. Gosud. Univ. Preceded & superseded by: Izvestiya Tomskogo gosudarstvennogo universiteta. 5-4234-3. HI 73700

Trudy biologicheskogo instituta. Novosibirsk, Russian S F S R. Vol. 1+, 1956+. Trudy Biol. Inst. HI 73701

Trudy biologicheskogo instituta Frunze. Frunze, Kirghiz S S R. Vols. 1-4, 1947-51. Trudy Biol. Inst. Frunze. 1-111-1. HI 73702

Trudy biologicheskogo instituta, Sibirskoe otdelenie, akademiya nauk S S S R = Trudy biologicheskogo instituta. Novosibirsk, Russian S F S R.

Trudy biologicheskogo instituta, Tomskii gosudarstvennyi universitet im V. V. Kuibysheva = Trudy biologicheskogo nauchno-issledovatel'skogo instituta Tomskogo gosudarstvennom universitete.

Trudy biologicheskogo nauchno-issledovatel'skogo instituta i biologicheskoi stantsii pri Permskom gosudarstvennom universitete. Perm, Russian S F S R. Vols. 1-3(2), 1927-31. Trudy Biol. Nauchno-Issl. Inst. Biol. Stantsii Permsk. Gosud. Univ. Superseded by: Trudy Permskogo biologicheskogo nauchno-issledovatel'skogo instituta. 4-3310-2. HI 73703

Trudy biologicheskogo nauchno-issledovatel'skogo instituta pri Molotovskom gosudarstvennom universitete imeni M. Gor'kogo. Molotov [=Perm], Russian S F S R. Vol. 9, 1940. Trudy Biol. Nauchno-Issl. Inst. Molotovsk. Gosud. Univ. Gor'kogo. Preceded by: Trudy biologicheskogo nauchno-issledovatel'skogo instituta pri Permskom gosudarstvennom universitete imeni M. Gor'kogo. Superseded by: Trudy estestvenno-nauchnogo (byvshego biologicheskogo) instituta pri Molotovskom gosudarstvennom universitete imeni A. M. Gor'kogo. 4-3310-2. HI 73704

Trudy biologicheskogo nauchno-issledovatel'skogo instituta pri Permskom gosudarstvennom universitete. Perm, Russian S F S R. Vols. 6-7(2), 1934-35. Trudy Biol. Nauchno-Issl. Inst. Permsk. Gosud. Univ. Preceded by: Trudy Permskogo biologicheskogo nauchno-issledovatel'skogo instituta. Superseded by: Trudy biologicheskogo nauchno-issledovatel'skogo instituta pri Permskom gosudarstvennom universitete imeni M. Gor'kogo. 4-3310-2. HI 73705

Trudy biologicheskogo nauchno-issledovatel'skogo instituta pri Permskom gosudarstvennom universitete imeni M. Gor'kogo. Perm, Russian S F S R. Vols. 7(3/4)-8, 1937-39. Trudy Biol. Nauchno-Issl. Inst. Permsk. Gosud. Univ. Gor'kogo. Preceded by: Trudy biologicheskogo nauchno-issledovatel'skogo instituta pri Permskom gosudarstvennom universitete. Superseded by: Trudy biologicheskogo nauchno-issledovatel'skogo instituta pri Molotovskom gosudarstvennom universitete imeni M. Gor'kogo. 4-3310-2. HI 73706

Trudy biologicheskogo nauchno-issledovatel'skogo instituta Tomskogo gosudarstvennom universitete. Tomsk, Russian S F S R. Vols. 1-7, 1935-40. Trudy Biol. Nauchno-Issl. Inst. Tomsk. Gosud. Univ. 5-4234-3. HI 73707

Trudy biologicheskoi stantsii "Borok" imeni N. A. Morozova. Moscow & Leningrad. Vols. 1-3, 1950-58. Trudy Biol. Stantsii "Borok" Morozova. Superseded by: Trudy instituta biologii vodokhranilishch. HI 73708

Trudy biologo-pochvennogo instituta AN S S S R. Vladivostok. No. 1+, 1969+. Trudy Biol.-Pochv. Inst. A.N. S.S.S.R. HI 73709

Trudy bjuro po mikologii i fitopatologii učenago komiteta = Trudy byuro po mikologii i fitopatologii uchenago komiteta. St. Petersburg.

Trudy bjuro po mikologii i fitopatologii učenago komiteta glavnago upravlenija zemleustrojstva i zemledělija = Trudy byuro po mikologii i fitopatologii uchenago komiteta glavnago upravleniya zemleustroistva i zemleděliya. St. Petersburg.

Trudy bjuro po prikladnoj botanike. St. Petersburg = Trudy byuro po prikladnoi botanike. St. Petersburg.

Trudy Borodinskoi biologicheskoi stantsii. Leningrad. Vol. [10], 1839. Trudy Borodinskoi Biol. Stantsii. Preceded by: Trudy Borodinskoi biologicheskoi stantsii v Karelii. 3-2394-3. HI 73710

Trudy Borodinskoi biologicheskoi stantsii v Karelii. Leningrad. Vols. 6(2)-9, 1933-36. Trudy Borodinskoi Biol. Stantsii Karelii. Preceded by: Trudy Borodinskoi presnovodnoi biologicheskoi stantsii v Karelii. Superseded by: Trudy Borodinskoi biologicheskoi stantsii. 3-2394-3. HI 73711

Trudy Borodinskoi biologicheskoi stantsii Petrogradskago obshchestva estestvoispytatelei. Petrograd. Vol. 4, 1917. Trudy Borodinskoi Biol. Stantsii Petrogradsk. Obshch. Estestvoisp. Preceded by: Trudy Presnovodnoi Biologicheskoi Stantsii Imperatorskago S.-Peterburgskago Obshchestva Estestvoispytatelei. Superseded by: Trudy Borodinskoi presnovodnoi biologicheskoi stantsii v Karelii. 3-2394-3. HI 73712

Trudy Borodinskoi presnovodnoi biologicheskoi stantsii v Karelii. Leningrad. Vols. 5-6(1), 1927-32. Trudy Borodinskoi Presnovodn. Biol. Stantsii Karelii. Preceded by: Trudy Borodinskoi biologicheskoi stantsii Petrogradskago obshchestva estestvoispytatelei. Superseded by: Trudy Borodinskoi biologicheskoi stantsii v Karelii. 3-2394-3. HI 73713

Trudy Borodinskoj biologičeskoj stancii = Trudy Borodinskoi biologicheskoi stantsii. Leningrad.

Trudy Borodinskoj biologičeskoj stancii v Karelii = Trudy Borodinskoi biologicheskoi stantsii v Karelii. Leningrad.

Trudy Borodinskoj biologičeskoj stancii Petrogradskago obščestva estestvoispytatelej = Trudy Borodinskoi biologicheskoi stantsii Petrogradskago obshchestva estestvoispytatelei. Petrograd.

Trudy Borodinskoj presnovodnoj biologičeskoj stancii v Karelii = Trudy Borodinskoi presnovodnoi biologicheskoi stantsii v Karelii. Leningrad.

Trudy botaničeskih sadov. Alma-Ata, Kazakh S S R = Trudy botanicheskikh sadov. Alma-Ata.

Trudy botaničeskii institut imeni V. L. Komarova, akademija nauk S S S R. Moscow & Leningrad = Trudy botanicheskogo instituta akademii nauk S S S R. Moscow & Leningrad.

Trudy botaničeskij institut imeni V. L. Komarova, akademija nauk S S S R. Moscow & Leningrad = Trudy botanicheskogo instituta imeni V. L. Komarova akademii nauk S S S R. Moscow & Leningrad.

Trudy botaničeskogo instituta, akademii nauk Armjanskoj S S R = Trudy botanicheskogo instituta, akademii nauk Armyanskoi S S R. Erivan.

Trudy botaničeskogo instituta akademii nauk S S S R. Moscow & Leningrad = Trudy botanicheskogo instituta akademii nauk S S S R. Moscow & Leningrad.

Trudy botaničeskogo instituta, akademija nauk tadžikskoi S S R. Dushanbe = Trudy, botanicheskogo instituta, akademiya nauk tadzhikskoi S S R. Dushanbe.

Trudy botaničeskogo instituta im. akademika V. L. Komarova. Baku, Azerbaijan S S R = Trudy botanicheskogo instituta imeni akademika V. L. Komarova. Baku.

Trudy botaničeskogo instituta Armjanskogo filiala akademii nauk S S R = Trudy botanicheskogo instituta Armyanskogo filiala akademii nauk S S R. Erivan.

Trudy botaničeskogo instituta, Baku = Trudy botanicheskogo instituta. Baku.

Trudy botaničeskogo instituta Moscow & Leningrad = Trudy botanicheskogo instituta. Moscow & Leningrad.

Trudy botaničeskogo instituta, Tiflis = Trudy botanicheskogo instituta. Tiflis.

Trudy botaničeskogo instituta im. V. L. Komarova akademij nauk S S S R. Moscow & Leningrad = Trudy botanicheskogo instituta imeni V. L. Komarova akademii nauk S S S R. Moscow & Leningrad

Trudy Botaničeskogo kabineta Minskoj central'noj opytnoj bolotnoj stancii = Pratsy Botanichnaga gabinetu Menskae tséntral'nae das'ledchae balotnae stantsyi. Minsk.

Trudy botaničeskogo muzeja imperatorskoj akademii nauk. St. Petersburg = Trudy botanicheskogo muzeya imperatorskoi akademii nauk. St. Petersburg.

Trudy botaničeskogo muzeja, Leningrad = Trudy botanicheskogo muzeya. Leningrad.

Trudy botaničeskogo muzeja rossijskoj akademii nauk. Petrograd = Trudy botanicheskogo muzeya rossiiskoi akademii nauk. Petrograd.

Trudy botaničeskogo sada Akademii nauk S S S R = Trudy botanicheskogo sada Akademii nauk S S S R. Moscow & Leningrad.

Trudy botaničeskogo Sada Imperatorskago Jur'evskago Universiteta = Trudy botanicheskogo Sada Imperatorskago Yur'evskago Universiteta. Yur'ev.

Trudy Botaničeskogo sada, Kiev = Trudy Botanicheskogo sada. Kiev.

Trudy botaničeskogo sada Latvijskogo gosudarstvennogo universiteta Petra Stučki = Trudy botanicheskogo sada Latviiskogo gosudarstvennogo universiteta Petra Stuchki. Riga.

Trudy botaničeskogo sada, moskovskij ordena Lenina gosudarstvennij universitet imeni M. V. Lomonosova. Moscow = Trudy botanicheskogo sada, moskovskii ordena Lenina gosudarstvennii universitet imeni M. V. Lomonosova. Moscow.

Trudy botaničeskogo sada Moskovskogo

gosudarstvennogo universiteta = Trudy botanicheskogo sada Moskovskogo gosudarstvennogo universiteta. Moscow.

Trudy botaničeskogo sada, Novosibirsk = Trudy botanicheskogo sada. Novosibirsk.

Trudy botaničeskogo sada Sredne-Aziatskogo gosudarstvennogo universiteta = Trudy botanicheskogo sada Sredne-Aziatskogo gosudarstvennogo universiteta. Tashkent.

Trudy botaničeskogo sada, Tashkent = Trudy botanicheskogo sada. Tashkent.

Trudy botaničeskoj opytnoj stancii B. A. Kellera = Trudy botanicheskoi opytnoi stantsii B. A. Kellera. Voronezh.

Trudy botanicheskii institut imeni V. L. Komarova, akademiya nauk S S S R. Ser 1, flora i sistematika vysshikh rastenii = Trudy botanicheskogo instituta akademii nauk S S S R. Ser. 1, flora i sistematika vysshikh rastenii. Moscow & Leningrad.

Trudy botanicheskii institut imeni V. L. Komarova, akademiya nauk S S S R. Ser 2, sporovye rasteniya = Trudy botanicheskogo instituta akademii nauk S S S R. Ser. 2, sporovye rasteniya. Moscow & Leningrad.

Trudy botanicheskii institut imeni V. L. Komarova, akademiya nauk S S S R. Ser 3, geobotanika = Trudy botanicheskogo instituta akademii nauk S S S R. Ser. 3, geobotanika. Moscow & Leningrad.

Trudy botanicheskii institut imeni V. L. Komarova, akademiya nauk S S S R. Ser 4, eksperimental'naya botanika = Trudy botanicheskogo instituta akademii nauk S S S R. Ser. 4, eksperimental'naya botanika. Moscow & Leningrad.

Trudy botanicheskii institut imeni V. L. Komarova, akademiya nauk S S S R. Ser 5, rastitel'noe syr'e = Trudy botanicheskogo instituta akademii nauk S S S R. Ser. 5, rastitel'noe syr'e. Moscow & Leningrad.

Trudy botanicheskii institut imeni V. L. Komarova, akademiya nauk S S S R. Ser 6, introdukciya rastenii i zelenoe stroitel'stvo = Trudy botanicheskogo instituta imeni V. L. Komarova akademii nauk S S S R. Ser. 6, introduktsiya rastenii i zelenoe stroitel'stvo. Moscow & Leningrad.

Trudy botanicheskii institut imeni V. L. Komarova, akademiya nauk S S S R. Ser 7, morfologiya i anatomiya rastenii = Trudy botanicheskogo instituta imeni V. L. Komarova akademii nauk S S S R. Ser. 7, morfologiya i anatomiya rastenii. Moscow & Leningrad.

Trudy botanicheskii institut imeni V. L. Komarova, akademiya nauk S S S R. Ser 8, paleobotanika = Trudy botanicheskogo instituta imeni V. L. Komarova akademii nauk S S S R. Ser. 8, paleobotanika. Moscow & Leningrad.

Trudy botanicheskikh sadov. Alma-Ata, Kazakh S S R. Vol. 8+, 1964+. Trudy Bot. Sadov. Preceded by: Trudy Alma-Atinskogo botanicheskofo sada. HI 73714

Trudy botanicheskogo instituta. Baku, Azerbaijan S S R. Vols. 1-4, 1936-38; vols. 6-14, 1939-49. Trudy Bot. Inst. (Baku). For vol. 5 see: Trudy khimicheskogo i botanicheskogo institutov. Superseded by: Trudy botanicheskogo instituta imeni akademika V. L. Komarova. 1-109-3. HI 73715

Trudy botanicheskogo instituta. Tiflis, Georgian S S R. Vol. 14, 1952. Trudy Bot. Inst. (Tiflis). Preceded & superseded by: Trudy Tbilisskogo botanicheskogo instituta. HI 73716

Trudy botanicheskogo instituta, akademii nauk Armyanskoi S S R. Erivan [=Erevan], Armenian S S R. Vol. 4+, 1946+. Trudy Bot. Inst. Akad. Nauk Armyansk. S.S.R. Preceded by: Trudy botanicheskogo instituta Armyanskogo filiala akademii nauk S S R. HI 73717

Trudy botanicheskogo instituta akademii nauk S S S R. Ser. 1, flora i sistematika vysshikh rastenii. Moscow & Leningrad. Vols. 1-13, 1933-64. Trudy Bot. Inst. Akad. Nauk S.S.S.R, Ser. 1, Fl. Sist. Vyssh. Rast. Preceded by: Trudy botanicheskogo muzeya. Leningrad. 1-114-2. HI 73718

Trudy botanicheskogo instituta akademii nauk S S S R. Ser. 2, sporovye rasteniya. Moscow & Leningrad. Vols. 1-12, 1933-59. Trudy Bot. Inst. Akad. Nauk S.S.S.R, Ser. 2, Sporov. Rast. Preceded by: Trudy botanicheskogo muzeya. Leningrad. 1-114-2. HI 73719

Trudy botanicheskogo instituta akademii nauk S S S R. Ser. 3, geobotanika. Moscow & Leningrad. Vols. 1-18, 1934-70. Trudy Bot. Inst. Akad. Nauk S.S.S.R, Ser. 3, Geobot. Preceded by: Trudy botanicheskogo muzeya. Leningrad. 1-114-2. HI 73720

Trudy botanicheskogo instituta akademii nauk S S S R. Ser. 4, eksperimental'naya botanika. Moscow & Leningrad. Vols. 1-20, 1934-70. Trudy Bot. Inst. Akad. Nauk S.S.S.R, Ser. 4, Eksper. Bot. Preceded by: Trudy botanicheskogo muzeya. Leningrad. 1-114-2. HI 73721

Trudy botanicheskogo instituta akademii nauk S S S R. Ser. 5, rastitel'noe syr'e. Moscow & Leningrad. Vols. 1-16, 1938-72. Trudy Bot. Inst. Akad. Nauk S.S.S.R, Ser. 5, Rastitel'n. Syr'e. Preceded by: Trudy botanicheskogo muzeya. Leningrad. 1-114-2. HI 73722

Trudy, botanicheskogo instituta, akademiya nauk tadzhikskoi S S R. Dushanbe. Vol. 18, 1962 [vols. [1]-[17], 1952-59 not published as separate volumes but included in Trudy, akademiya nauk tadzhikskoi S S R.

Trudy Bot. Inst. Akad. Nauk Tadzhiksk. S.S.R. Superseded by: Trudy pamirskoi biologicheskoi stantsii. Botanicheskogo instituta, akademiya nauk tadzikskoi S S R. HI 73723

Trudy botanicheskogo instituta Armyanskogo filiala akademii nauk S S R. Erivan [=Erevan], Armenian S S R. Vols. 1-3, 1941. Trudy Bot. Inst. Armyansk. Fil. Akad. Nauk S.S.S.R. Superseded by: Trudy botanicheskogo instituta, akademii nauk Armyanskoi S S R. HI 73724

Trudy botanicheskogo instituta imeni akademika V. L. Komarova. Baku, Azerbaijan S S R. Vol. 15, 1945. Trudy Bot. Inst. Komarova. Preceded by: Trudy botanicheskogo instituta. Baku. Superseded by: Trudy instituta botaniki. Baku. 1-109-3. HI 73725

Trudy botanicheskogo instituta, Azerbaidzhanskii filial, Akademiya nauk S S S R. Baku = Trudy botanicheskogo instituta. Baku.

Trudy botanicheskogo instituta, Dushanbe = Trudy, botanicheskogo instituta, akademiya nauk tadzhikskoi S S R. Dushanbe.

Trudy botanicheskogo instituta im. V. L. Komarova akademii nauk S S S R. Ser. 1, flora i sistematika vysshikh rastenii = Trudy botanicheskogo instituta akademii nauk S S S R. Ser. 1, flora i sistematika vysshikh rastenii. Moscow & Leningrad.

Trudy botanicheskogo instituta im. V. L. Komarova akademii nauk S S S R. Ser. 2, sporovye rasteniya = Trudy botanicheskogo instituta akademii nauk S S S R. Ser. 2, sporovye rasteniya. Moscow & Leningrad.

Trudy botanicheskogo instituta im. V. L. Komarova akademii nauk S S S R. Ser. 3, geobotanika = Trudy botanicheskogo instituta akademii nauk S S S R. Ser. 3, geobotanika. Moscow & Leningrad.

Trudy botanicheskogo instituta im. V. L. Komarova akademii nauk S S S R. Ser. 4, eksperimental'naya botanika = Trudy botanicheskogo instituta akademii nauk S S S R. Ser. 4, eksperimental'naya botanika. Moscow & Leningrad.

Trudy botanicheskogo instituta im. V. L. Komarova akademii nauk S S S R. Ser. 5, rastitel'noe syr'e = Trudy botanicheskogo instituta akademii nauk S S S R. Ser. 5, rastitel'noe syr'e. Moscow & Leningrad.

Trudy botanicheskogo instituta imeni V. L. Komarova akademii nauk S S S R. Ser. 6, introduktsiya rastenii i zelenoe stroitel'stvo. Moscow & Leningrad. Vols. 1-10, 1950-70. Trudy Bot. Inst. Komarova Akad. Nauk S.S.S.R, Ser. 6, Introd. Rast. HI 73726

Trudy botanicheskogo instituta imeni V. L. Komarova akademii nauk S S S R. Ser. 7, morfologiya i anatomiya rastenii. Moscow & Leningrad. Vols. 1-5?, 1950-70. Trudy Bot. Inst. Komarova Akad. Nauk S.S.S.R, Ser. 7, Morfol. Anat. Rast. HI 73727

Trudy botanicheskogo instituta imeni V. L. Komarova akademii nauk S S S R. Ser. 8, paleobotanika. Moscow & Leningrad. Vols. 1-6, 1956-67. Trudy Bot. Inst. Komarova Akad. Nauk S.S.S.R, Ser. 8, Paleobot. HI 73728

Trudy botanicheskogo muzeya. Leningrad. Vols. 19-25, 1928-32. Trudy Bot. Muz. Preceded by: Trudy botanicheskogo muzeya rossiiskoi akademii nauk. Superseded by: Trudy botanicheskogo instituta akademii nauk S S S R. Ser. 1, flora i sistematika vysshikh rastenii; Trudy botanicheskogo instituta akademii nauk S S S R. Ser. 2, sporovye rasteniya; Trudy botanicheskogo instituta akademii nauk S S S R. Ser. 3, geobotanika; Trudy botanicheskogo instituta akademii nauk S S S R. Ser. 4, eksperimental'naya botanika; and Trudy botanicheskogo instituta akademii nauk S S S R. Ser. 5, rastitel'noe syr'e. 1-114-3. HI 73729

Trudy botanicheskogo muzeya, Akademiya nauk S S S R. Leningrad = Trudy botanicheskogo muzeya. Leningrad.

Trudy botanicheskogo muzeya imperatorskoi akademii nauk. St. Petersburg. Vols. 1-16, 1902-16. Trudy Bot. Muz. Imp. Akad. Nauk. Superseded by: Trudy botanicheskogo muzeya rossiiskoi akademii nauk. 1-114-3. HI 73730

Trudy botanicheskogo muzeya rossiiskoi akademii nauk. Petrograd. Vols. 17-18, 1918-20. Trudy Bot. Muz. Rossiisk. Akad. Nauk. Preceded by: Trudy botanicheskogo muzeya imperatorskoi akademii nauk. Superseded by: Trudy botanicheskogo muzeya. 1-114-3. HI 73731

Trudy Botanicheskogo sada. Kiev, Ukrainian S S R. Vols. 2-7, 1953-60. Trudy Bot. Sada (Kiev). Preceded by: Trudy botanichnogo sadu akademii nauk Ukrayins'koi R S R. Superseded by: Pratsy tsentral'nogo respublikans'kogo botanichnogo sadu. HI 73733

Trudy botanicheskogo sada. Novosibirsk, Russian S F S R. Vols. 1-2, 1956-57. Trudy Bot. Sada (Novosibirsk). Superseded by: Trudy tsentral'nogo sibirskogo botanicheskogo sada. HI 73734

Trudy botanicheskogo sada. Tashkent, Uzbek S S R. Vols. 1-5, 1949-56. Trudy Bot. Sada (Tashkent). 5-4156-3. HI 73735

Trudy botanicheskogo sada Akademii nauk S S S R. Moscow & Leningrad. Vol. 42(2), 1931; vols. 43(2)-44, 1931. Trudy Bot. Sada Akad. Nauk S.S.S.R. Preceded & superseded by: Trudy glavnago botanicheskago sada. For vol. 43(1), 1930 see: Trudy glavnago botanicheskago sada. 3-2389-3. HI 73736

Trudy botanicheskogo sada, akademiya nauk Ukrayins'koi R S R = Trudy botanichnogo sadu akademii nauk Ukrayins'koi R S R Kiev.

Trudy botanicheskogo sada imeni prof. B. M. Kozo-Polyanskogo. Voronezh. Vol. ?-2+, ?-1963+. Trudy Bot. Sada B. M. Kozo-Polyanskogo. HI 73737

Trudy Botanicheskogo sada Imperatorskago Yur'evskago universiteta. Yur'ev [=Tartu], Estonian S S R. Vols. 1-14, 1900-14. Trudy Bot. Sada Imp. Yur'evsk. Univ. Superseded by: Vestnik Russkoi flory. 5-4154-2. HI 73738

Trudy botanicheskogo sada Latviiskogo gosudarstvennogo universiteta Petra Stuchki. Riga, Latvian S S R. Vols. 15-16, 1954-58. Trudy Bot. Sada Latviisk. Gosud. Univ. Petra Stuchki. Preceded by: Acta horti botanici universitatis. Riga. Superseded by: Petera Stučkas latvijas valsts universitätes botaniskä darzä raksti. HI 73739

Trudy botanicheskogo sada, moskovskii ordena Lenina gosudarstvennii universitet imeni M. V. Lomonosova. Moscow. Vols. 4-?, 194?-75. Trudy Bot. Sada Moskovsk. Ordena Lenina Gosud. Univ. Lomonosova. Preceded by: Trudy botanicheskogo sada Moskovskogo gosudarstvennogo universiteta. HI 73740

Trudy botanicheskogo sada Moskovskogo gosudarstvennogo universiteta. Moscow. Vols. 1-3, 1937-40. Trudy Bot. Sada Moskovsk. Gosud. Univ. Superseded by: Trudy botanicheskogo sada, moskovskii ordena Lenina gosudarstvennii universitet imeni M. V. Lomonosova. 3-2767-1. HI 73741

Trudy botanicheskogo sada Sredne-Aziatskogo gosudarstvennogo universiteta. Tashkent, Uzbek S S R. Vols. [1]-[2], 1928-29; vol. 3, 1929; vols. [4]-[7], 1930; vol. 8, 1931 Trudy Bot. Sada Sredne-Aziatsk. Gosud. Univ. 5-4156-3. HI 73742

Trudy botanicheskogo sada. Zapadnosibirskii filial, akademiya nauk S S S R = Trudy botanicheskogo sada. Novosibirsk, Russian S F S R.

Trudy botanicheskogo sadu akademii imeny O. V. Fomina = Trudy botanichnogo sadu imeny akad. O. V. Fomina. Kiev.

Trudy botanicheskoi laboratorii imperatorskago Varshavskago universitet. Varshava. Vol. 1, 1875. Trudy Bot. Lab. Imp. Varshavsk. Univ. HI 73743

Trudy botanicheskoi opytnoi stantsii B. A. Kellera. Voronezh, Russian S F S R. Vols. 1-4, 1929-41. Trudy Bot. Opytn. Stantsii Kellera. HI 73732

Trudy botanichnogo sadu imeny akad. O. V. Fomina. Kiev, Ukrainian S S R. Vols. 19-20, 1948-49. Trudy Bot. Sadu Fomina. Preceded by: Izvestiya Kievskogo botanicheskogo sada imeni akad. A. V. Fomina. Superseded by: Trudy botanichnogo sadu akademii nauk Ukrayins'koi R S R. HI 73745

Trudy botanichnogo sadu akademii nauk Ukrayins'koi R S R. Kiev, Ukrainian S S R. Vol. 1, 1949. Trudy Bot. Sadu Akad. Nauk Ukrayins'k R.S.R. Preceded by: Trudy botanichnogo sadu imeny akad. O. V. Fomina. Superseded by: Trudy Botanicheskogo sada. Kiev. 1-121-3. HI 73744

Trudy botaničnogo sadu im. akad. O. V. Fomina = Trudy botanichnogo sadu imeny akad. O. V. Fomina. Kiev.

Trudy botaničnogo sadu akademii nauk Ukrajins'koji R S R = Trudy botanichnogo sadu akademii nauk Ukrayins'koi R S R Kiev.

Trudy brianskogo lesnogo instituta. Bryansk. Vols. 1-8, 1936-57. Trudy Briansk. Lesn. Inst. HI 73746

Trudy brianskogo lesokhozyaistvennogo instituta = Trudy brianskogo lesnogo instituta. Bryansk.

Trudy byuro po evgenike. Leningrad. 1922-29. Trudy Byuro Evgen. Superseded by: Trudy byuro po genetike. HI 73747

Trudy byuro po genetike. Leningrad. 1930-31. Trudy Byuro Genet. Preceded by: Trudy byuro po evgenike. Superseded by: Trudy laboratorii po genetike. HI 73748

Trudy byuro po mikologii i fitopatologii uchenago komiteta. St. Petersburg. Pts. 6-7, 1909-10; pt. 12, 1916. Trucy Byuro Mikol. Uchen. Komiteta. For pts. 1-5 and 8-11 see: Trudy byuro po mikologii i fitopatologii uchenago komiteta glavnago upravleniya zemleustroistva i zemlediliya. HI 73749

Trudy byuro po mikologii i fitopatologii uchenago komiteta glavnago upravleniya zemleustroistva i zemleděliya. St. Petersburg. Pts. 1-5, 1908-09; pts. 8-11, 1911-?. Trudy Byuro Mikol. Uchen. Komiteta Glavn. Upravl. Zemleustr. For pts. 6-7 and pt. 12 see: Trudy byuro po mikologii i fitopatologii uchenago komiteta. HI 73750

Trudy Byuro po prikladnoi botanike. St. Petersburg. Vols. 1-10, 1908-17; Prilož. vols. 1-16, 1908-16. Trudy Byuro Prikl. Bot. Superseded by: Trudy po prikladnoi botanike i selektsii. 5-4265-3. HI 73751

Trudy central'nogo sibirskogo botaničeskogo sada = Trudy tsentral'nogo sibirskogo botanicheskogo sada. Novosibirsk.

Trudy Černomorskogo gosudarstvennogo zapovednika = Trudy Chernomorskogo gosudarstvennogo zapovednika. Kiev.

Trudy Chernomorskogo gosudarstvennogo zapovednika. Kiev. Ukrainian S S R. Vol. 1, 1950. Trudy Chernomorsk. Gosud. Zapov. HI 73753

Trudy Dal'nevostochnoi bazy imeni akademika V. L. Komarova. Moscow & Leningrad. Vol. 1, 1947. Trudy Dal'nevost. Bazy Komarova. Preceded by: Trudy Dal'nevostochnogo filiala Akademii nauk S S S

R. Seriya botanicheskaya. Superseded by: Trudy Dal'nevostochnogo filiala imeni V. L. Komarova. Seriya botanicheskaya. 1-115-1. HI 73754

Trudy Dal'nevostochnogo filiala Akademii nauk S S S R. Seriya botanicheskaya. Moscow & Leningrad. Vols. 1-2, 1935-37. Trudy Dal'nevost. Fil. Akad. Nauk S.S.S.R., Ser. Bot. Superseded by: Trudy Dal'nevostochnoi bazy imeni akademika V. L. Komarova. 1-115-1. HI 73755

Trudy Dal'nevostochnogo filiala imeni V. L. Komarova. Seriya botanicheskaya. Moscow & Leningrad. Vols. 2-3 [also numbered vols. 4-5], 1956. Trudy Dal'nevost. Fil. Komarova, Ser. Bot. Preceded by: Trudy Dal'nevostochnoi bazy imeni akademika V. L. Komarova. Superseded by: Trudy Dal'nevostochnyi filial imeni V. L. Komarova, Sibirskoe otdelenie, akademiya nauk S S S R. Seriya botanicheskaya. HI 73756

Trudy Dal'nevostochnogo filiala imeni V. L. Komarova, Akademiya nauk S S S R. Seriya botanicheskaya = Trudy dal'nevostochnogo filiala imeni V. L. Komarova. Seriya botanicheskaya. Moscow & Leningrad.

Trudy dal'nevostochnogo gosudarstvennogo universiteta. Vladivostok, Russian S F S R = Trudy gosudarstvennogo dal'nevostochnogo universiteta. Vladivostok, Russian S F S R

Trudy dal'nevostochnogo pedagogicheskogo instituta. Ser. 5, biologiya. Vladivostok, Russian S F S R. Vols. 1-2 [also numbered vols. 3-4], 1931-32. Trudy Dal'nevost. Pedagog. Inst., Ser. 5, Biol. Preceded by: Trudy gosudarstvennogo dal'nevostochnogo universiteta. Ser. 8, biologiya. 5-4414-3. HI 73758

Trudy dal'nevostochnyi filiala imeni V. L. Komarova, Sibirskoe otdelenie, akademiya nauk S S S R. Seriya botanicheskaya. Moscow. & Leningrad. Vols. 4 [also numbered vol. 6]+, 1958. Trudy Dal'nevost. Fil. Komarova Sibirsk. Otd. Akad. Nauk S.S.S.R., Ser. Bot. Preceded by: Trudy dal'nevostochnogo filiala imeni V. L. Komarova. Seriya botanicheskaya. HI 73757

Trudy dal'nevostočnogo filiala Akademii nauk S S S R. Serija botaničeskaja = Trudy dal'nevostočnogo filiala Akademii nauk S S S R. Seriya botanicheskaya. Moscow & Leningrad.

Trudy dal'nevostočnogo filiala im. V. L. Komarova. Serija botaničeskaja = Trudy dal'nevostochnogo filiala imeni V. L. Komarova. Seriya botanicheskaya. Moscow & Leningrad.

Trudy dal'nevostočnogo pedagogičeskogo instituta. Ser. 5, biologija = Trudy dal'nevostochnogo pedagogicheskogo instituta. Ser. 5, biologiya. Vladivostok.

Trudy dal'nevostočnoj bazy imeni akademika V. L. Komarova = Trudy dal'nevostochnoi bazy imeni akademika V. L. Komarova. Moscow & Leningrad.

Trudy dal'nevostočnyj filial imeni V. L. Komarova, Sibirskoe otdelenie, akademija nauk S S S R. Serija botaničeskaja = Trudy dal'nevostochnyi filial imeni V. L. Komarova, Sibirskoe otdelenie, akademiya nauk S S S R. Seriya botanicheskaya. Moscow. & Leningrad.

Trudy Dneprovskoi biologicheskoi stantsii. Kiev, Ukrainian S S R. Vols. 1-2, 1914-15. Trudy Dneprovsk. Biol. Stantsii. HI 73759

Trudy Dneprovskoj biologičeskoj stancii = Trudy Dneprovskoi biologicheskoi stantsii. Kiev.

Trudy Dnyprovs'ka biolohichna stantsiya. Kyyiv, Ukrainian, S S R = Trudy Dneprovskoi biologicheskoi stantsii.

Trudy estestvenno-istoričeskago muzeja Tavričeskago gubernskago zemstva = Trudy estestvenno-istoricheskago muzeya Tavricheskago gubernskago zemstva. Simferopol.

Trudy estestvenno-istoričeskogo otdelenija central'nogo muzeja Tavridy = Trudy estestvenno-istoricheskogo otdeleniya tsentral'nogo muzeya Tavridy. Simferopol.

Trudy estestvenno-istoricheskago muzeya Tavricheskago gubernskago zemstva. Simferopol, Ukrainian S S R. Vols. 1-4, 1913-16. Trudy Estestv.-Istorich. Muz. Tavrichesk. Gub. Zemstva. Superseded by: Trudy estestvenno-istoricheskogo otdeleniya tsentral'nogo muzeya Tavridy. HI 74061

Trudy estestvenno-istoricheskogo otdeleniya tsentral'nogo muzeya Tavridy. Simferopol, Ukrainian S S R. Vol. 1 [also numbered vol. 5], 1927. Trudy Estestv.-Istorich Otd. Tsentr. Muz. Tavridy. Preceded by: Trudy estestvenno-istoricheskago muzeya Tavricheskago guvernskago zemstva. HI 74062

Trudy estestvenno-nauchnogo (byvshego biologicheskogo) instituta pri Molotovskom gosudarstvennom universitete imeni A. M. Gor'kogo. Molotov [=Perm], Russian S F S R. Vol. 10, 1948-53. Trudy Estestv.-Nauchn. Inst. Molotovsk. Gosud. Univ. Gor'kogo. Preceded by: Trudy biologicheskogo nauchno-issledovatel'skogo instituta pri Molotovskom gosudarstvennom universitete imeni M. Gor'kogo. 4-3310-2. HI 74063

Trudy estestvenno-naučnogo (byvšego biologičeskogo) instituta pri Molotovskom gosudarstvennom universitete imeni A. M. Gor'kogo = Trudy estestvenno-nauchnogo (byvshego biologicheskogo) instituta pri Molotovskom gosudarstvennom universitete imeni A. M. Gor'kogo. Molotov.

Trudy fitopatologičeskoj stancii pri Moskovskom sel'skohozjajstvennom institutě = Trudy fitopatologicheskoi stantsii pri Moskovskom sel'skokhozyaistvennom institutě. Moscow.

Trudy fitopatologicheskoi stantsii pri Moskovskom sel'skokhozyaistvennom institutě. Moscow. Vol. 1, 1916. Trudy Fitopatol. Stantsii Moskovsk. Sel'skokhoz. Inst. HI 73760

Trudy fizychno-matematychnogo viddilu. Kiev, Ukrainian S S R. Vols. 1-15, 1923-30. Trudy Fiz.-Mat. Vidd. Superseded by: Trudy Pryrodnycho-technichnogo viddilu. 1-121-3. HI 73761

Trudy fizyčno-matematyčnogo viddilu. Kiev, Ukrainian S S R = Trudy fizychno-matematychnogo viddilu. Kiev.

Trudy po geobotaničeskomu obsledovaniju pastbišč S S R Azerbajdžana. Serija A, zimnie pastbišč. Baku = Trudy po geobotanicheskomu obsledovaniyu pastbishch S S R Azerbaidzhana. Seriya A, zimnie pastbishch. Baku.

Trudy po geobotaničeskomu obsledovaniju pastbišč S S R Azerbajdžana. Serija B, letnie pastbišča. Baku = Trudy po geobotanicheskomu obsledovaniyu pastbishch S S R Azerbaidzhana. Seriya B, letnie pastbishcha. Baku.

Trudy po geobotanicheskomu obsledovaniyu pastbishch S S R Azerbaidzhana. Seriya A, zimnie pastbishch. Baku, Azerbaijan S S R. Vols. 1-6, 1929-30. Trudy Geobot. Obsl. Pastb. S.S.R. Azerbaidzhna, Ser. A, Zimn. Pastb. 5-4267-3. HI 73762

Trudy po geobotanicheskomu obsledovaniyu pastbishch S S R Azerbaidzhana. Seriya B, letnie pastbishcha. Baku. Vols. 1-4-?, 1929-33. Trudy Geobot. Pastb. S.S.R. Azerbaidzhana, Ser. B, Letn. Pastb. 5-4267-3. HI 73763

Trudy geografičeskago otdělenija. [Imperatorskoe obščestvo ljubitelej estestvoznanija, antropologii i etnografii] = Trudy geograficheskago otděleniya. [Imperatorskoe obshchestvo lyubitelei estestvoznaniya, antropologii i etnografii.] Moscow.

Trudy geograficheskago otděleniya. [Imperatorskoe obshchestvo lyubitelei estestvoznaniya, antropologii i etnografii.] Moscow. Vols. 1-2, 1894-1910. Trudy Geogr. Otd., Moscow. HI 73764

Trudy geologičeskoj časti kabineta ego imperatorskago veličestva. St. Petersburg = Trudy geologicheskoi chasti kabineta ego imperatorskago velichestva. St. Petersburg.

Trudy geologicheskogo instituta, akademiya nauk S S S R. Leningrad, Moscow. Vols. 1-9, 1932-39; n.s. vol. 1+, 1956+. Trudy Geol. Inst. Akad. Nauk S.S.S.R. Preceded by: Trudy geologicheskogo muzeya, akademiya nauk S S S R. For 1939-56 incorporated in: Trudy instituta geologicheskikh nauk akademiya nauk S S S R. HI 73765

Trudy geologicheskago i mineralogicheskago muzeya imeni imperatora Petra Velikago, imperatorskoi akademii nauk. St. Petersburg. Vols. 1-5, 1915-26. Trudy Geol. Mineral. Muz. Imp. Petra Velikago Imp. Akad Nauk. Preceded by: Trudy geologicheskago muzeya imeni imperatora Petra Velikago, imperatorskoi akademii nauk. Superseded by: Trudy geologicheskogo muzeya, akademiya nauk S S S R. HI 73766

Trudy geologicheskago muzeya imeni imperatora Petra Velikago, imperatorskoi akademii nauk. St. Petersburg. Vols. 1-8, 1907-14. Trudy Geol. Muz. Imp. Petra Velikago Imp. Akad. Nauk. Superseded by: Trudy geologicheskago i mineralogicheskago muzeya imeni imperatora Petra Velikago, imperatorskoi akademii nauk. HI 73767

Trudy geologicheskogo muzeya, akademiya nauk S S S R. Leningrad. Vols. 1-8, 1926-31. Trudy Geol. Muz. Akad. Nauk S.S.S.R. Preceded by: Trudy geologicheskago i mineralogicheskago muzeya imeni imperatora Petra Velikago, imperatorskoi akademii nauk. Superseded by: Trudy geologicheskogo instituta, akademiya nauk S S S R. HI 73768

Trudy geologicheskoi chasti kabineta ego imperatorskago velichestva. St. Petersburg. Vols. 1-8, 1895-1915. Trudy Geol. Chasti Kab. Ego Imp. Velich. HI 73769

Trudy gidrobiologičeskoj stancii na glubokom ozere = Trudy gidrobiologicheskoi stantsii na glubokom ozere. Moscow.

Trudy gidrobiologičnoji stanciji. Kiev, Ukrainian S S R = Trudy gidrobiologichnoi stantsii. Kiev.

Trudy gidrobiologicheskoi stantsii na glubokom ozere. Moscow. 1900-23/30. Trudy Gidrobiol. Stantsii Glubokom Ozere. HI 73770

Trudy gidrobiologichnoi stantsii. Kiev, Ukrainian S S R. Vols. 7-19, 1934-40. Trudy Gidrobiol. Stantsii. Preceded by: Zbirnyk prats Dnipryans'koi biologichnoi stantsii. Superseded by: Trudy institutu gidrobiologii. Kiev. 1-122-2. HI 73771

Trudy glavnago botaničeskago sada. Petrograd, Moscow = Trudy glavnago botanicheskago sada. Petrograd, Moscow.

Trudy glavnago botanicheskago sada. Petrograd, Moscow. Vols. 33-42(1), 1915-29; vol. 43(1), 1930; [n.s.] vols. 1-9, 1949-63 [suspended from 1932-48]. Trudy Glavn. Bot. Sada. Preceded by: Trudy imperatorskago S.-Peterburgskago botanicheskago sada. For vols. 42(2) and 43(2)-44 see: Trudy botanicheskogo sada akademii nauk S S S R. 3-2389-3. HI 73772

Trudy glavnogo botanicheskogo sada = Trudy glavnago botanicheskago sada. Petrograd, Moscow.

Trudy Gornotaezhnoi stantsii. Khabarovsk, Russian S F S R. Vol. 1, 1936. Trudy Gornotaezhn. Stantsii. Superseded by: Trudy Gornotaezhnoi stantsii dal'nevostochnogo filiala akademii nauk S S S R. 1-

115-1. HI 73773

Trudy Gornotaezhnoi stantsii dal'nevostochnogo filiala akademii nauk S S S R. Vladivostok, Russian S F S R. Vols. 2-3, 1938-39. Trudy Gornotaezhn. Stantsii Dal'nevost. Fil. Akad. Nauk S.S.R. Preceded by: Trudy Gornotaezhnoi stantsii. Superseded by: Trudy Gornotaezhnoi stantsii imeni akademika Komarova. 1-115-1. HI 73774

Trudy Gornotaezhnoi stantsii imeni akademika Komarova. Vladivostok, Russian S F S R. Vols. 4-5, 194?-46. Trudy Gornotaezhn. Stantsii Komarova. Preceded by: Trudy Gornotaezhnoi stantsii dal'nevostochnogo filiala akademii nauk S S S R. 1-115-1. HI 73775

Trudy Gornotaežnoj stancii = Trudy Gornotaezhnoi stantsii. Khabarovsk.

Trudy Gornotaežnoj stancii im. akademika Komarova = Trudy Gornotaezhnoi stantsii imeni akademika Komarova. Vladivostok.

Trudy Gornotaežnoj stancii dal'nevostočnogo filiala akademii nauk S S S R = Trudy Gornotaezhnoi stantsii dal'nevostochnogo filiala akademii nauk S S S R. Vladivostok.

Trudy gosudarstvennogo dal'nevostochnogo universiteta. Ser. 4, lesnye nauki. Vladivostok, Russian S F S R. Vols. 1-8, 1926-29. Trudy Gosud. Dal'nevost. Univ., Ser. 4, Lesn. Nauki. 5-4414-3. HI 73776

Trudy gosudarstvennogo dal'nevostochnogo universiteta. Ser. 5, sel'skoe khozyaistvo. Kiev, Ukrainian S S R. Vols. 1-9, 1929-30. Trudy Gosud. Dal'nevost. Univ., Ser. 5, Sel'sk. Khoz. 5-4414-3. HI 73777

Trudy gosudarstvennogo dal'nevostochnogo universiteta. Ser. 8, biologiya. Vladivostok, Russian S F S R. Vols. 1-2, 1926-27. Trudy Gosud. Dal'nevost. Univ., Ser. 8, Biol. Superseded by: Trudy dal'nevostochnogo pedagogicheskogo instituta. Ser. 5, biologiya. 5-4415-1. HI 73778

Trudy gosudarstvennogo dal'nevostochnogo universiteta. Ser. 11, geologiya. Vladivostok, Russian S F S R. Vols. 1-4, 1926-27. Trudy Gosud. Dal'nevost. Univ., Ser. 11, Geol. 5-4415-1. HI 73779

Trudy gosudarstvennogo dal'nevostochnogo universiteta. Ser. 13, tekhnika. Vladivostok, Russian S F S R. Vols. 1-15, 1926-30. Trudy Gosud. Dal'nevost. Univ., Ser. 13, Tekhn. 5-4415-1. HI 73780

Trudy gosudarstvennogo dal'nevostočnogo universiteta. Ser. 4, lesnye nauki = Trudy gosudarstvennogo dal'nevostochnogo universiteta. Ser. 4, lesnye nauk. Vladivostok.

Trudy gosudarstvennogo dal'nevostočnogo universiteta. Ser. 5, sel'skoe hozjajstvo = Trudy gosudarstvennogo dal'nevostochnogo universiteta. Ser. 5, sel'skoe khozyaistvo. Kiev.

Trudy gosudarstvennogo dal'nevostočnogo universiteta. Ser. 8, biologija = Trudy gosudarstvennogo dal'nevostochnogo universiteta. Ser. 8, biologiya. Vladivostok.

Trudy gosudarstvennogo dal'nevostočnogo universiteta. Ser. 11, geologija = Trudy gosudarstvennogo dal'nevostochnogo universiteta. Ser. 11, geologiya. Vladivostok.

Trudy gosudarstvennogo dal'nevostočnogo universiteta. Ser. 13, tehnika = Trudy gosudarstvennogo dal'nevostochnogo universiteta. Ser. 13, tekhnika. Vladivostok.

Trudy gosudarstvennogo lugovogo instituta imeni professora V. R. Vil'yamsa. [Lobnya.] Vols. 1-5-?, 1927-28-? Trudy Gosud. Lugov. V. R. Vil'yamsa. HI 73781

Trudy gosudarstvennogo Nikitskogo botaničeskogo sada. Yalta = Trudy gosudarstvennogo Nikitskogo botanicheskogo sada. Yalta.

Trudy Gosudarstvennogo Nikitskogo botaničeskogo sada imeni V. M. Molotova = Trudy gosudarstvennogo Nikitskogo botanicheskogo sada. Yalta.

Trudy gosudarstvennogo Nikitskogo botanicheskogo sada. Yalta, Ukrainian S S R. Vols. 14(2)-15, 1930?; 18-25, 1934-53. Trudy Gosud. Nikitsk. Bot. Sada. Preceded by: Zapiski Gosudarstvennogo Nikitskogo opytnogo botanicheskogo sada. Superseded by: Trudy gosudarstvennyi Nikitskii botanicheskii sad. 3-1912-2. HI 73782

Trudy gosudarstvennogo okeanograficheskogo instituta. Moscow, Leningrad. 1931-34; vols. 1(13)+, 1947+. Trudy Gosud. Okeanogr. Inst. Preceded by: Trudy morskogo nauchnogo instituta. HI 73783

Trudy gosudarstvennyi Nikitskii botanicheskii sad. Yalta, Ukrainian S S R. Vol. 26+, 1956+. Trudy Gosud. Nikitsk. Bot. Sad. Preceded by: Trudy gosudarstvennogo Nikitskogo botanicheskogo sada. HI 73784

Trudy gosudarstvennyj Nikitskij botaničeskij sad = Trudy gosudarstvennyi Nikitskii botanicheskii sad. Yalta.

Trudy hidrobiolohichnoyi stantsiyi. Kyyiv, Ukrainian S S R = Trudy gidrobiologichnoi stantsii. Kiev.

Trudy himičeskogo i botaničeskogo institutov. Baku, Azerbaijan S S R = Trudy khimicheskogo i botanicheskogo institutov. Baku.

Trudy imperatorskago botaničeskago sada Petra Velikago. St. Petersburg = Trudy imperatorskago botanicheskago sada Petra Velikago. St. Petersburg.

Trudy imperatorskago botanicheskago sada Petra Velikago. St. Petersburg. Vol. 30(2), 1913; vol. 31(3),

1915. Trudy Imp. Bot. Sada Petra Velikago. Preceded by: Trudy imperatorskago S.-Peterburgskago botanicheskago sada. For vols. 31(1)-31(2) and 32 see: Trudy imperatorskago S.-Peterburgskago botanicheskago sada. 3-2389-3. HI 73786

Trudy imperatorskago Petrogradskago obščestva estestvoispytatelej. Vypusk 3, otdělenie botaniki = Trudy imperatorskago Petrogradskago obshchestva estestvoispytatelei. Vypusk 3, otdělenie botaniki. Petrograd.

Trudy imperatorskago Petrogradskago obshchestva estestvoispytatelei. Vypusk 3, otdělenie botaniki. Petrograd. Vol. 46, 1916. Trudy Imp. Petrogradsk. Obshch. Estestvoisp., Vyp. 3, Otd. Bot. Preceded by: Trudy imperatorskago S.-Peterburgskago obshchestva estestvoispytatelei. Vypusk 3, otdělenie botaniki. Superseded by: Trudy Petrogradskogo obshchestva estestvoispytatelei. Vypusk 3... Otdelenie botaniki. 3-2394-3. HI 73787

Trudy imperatorskago S.-Peterburgskago botaničeskago sada = Trudy imperatorskago S.-Peterburgskago botanicheskago sada. St. Petersburg.

Trudy imperatorskago S.-Peterburgskago botanicheskago sada. St. Petersburg. Vols. 1-30(1), 1871-1909; vol. 31(1)-31(2), 1912-13; vol. 32, 1912. Trudy Imp. S.-Peterburgsk. Bot. Sada. For vols. 30(2), 1913 and 31(3), 1915 see: Trudy imperatorskago botanicheskago sada Petra Velikago. Superseded by: Trudy glavnago botanicheskago sada. Petrograd. 3-2389-3. HI 73788

Trudy imperatorskago S.-Peterburgskago obščestva estestvoispytatelej. Vypusk 2, otdělenie botaniki = Trudy imperatorskago S.-Peterburgskago obshchestva estestvoispytatelei. Vypusk 2, otdělenie botaniki. St. Petersburg.

Trudy imperatorskago S.-Peterburgskago obščestva estestvoispytatelej. Vypusk 3, otdělenie botaniki = Trudy imperatorskago S.-Peterburgskago obshchestva estestvoispytatelei. Vypusk 3, otdělenie botaniki. St. Petersburg.

Trudy imperatorskago S.-Peterburgskago obščestva estestvoispytatelej. Vypusk 4, otdělenie botaniki = Trudy imperatorskago S.-Peterburgskago obshchestva estestvoispytatelei. Vypusk 4, otdělenie botaniki. St. Petersburg.

Trudy imperatorskago S.-Peterburgskago obshchestva estestvoispytatelei. Vypusk 2, otdělenie botaniki. St. Petersburg. Vol. 27, 1897. Trudy Imp. S.-Peterburgsk. Obshch. Estestvoisp., Vyp. 2, Otd. Bot. Preceded & superseded by: Trudy imperatorskago S.-Peterburgskago obshchestva estestvoispytatelei. Vypusk 3, otdělenie botaniki. 3-2394-3. HI 73789ě

Trudy imperatorskago S.-Peterburgskago obshchestva estestvoispytatelei. Vypusk 3, otdělenie botaniki. St. Petersburg. Vol. 26, 1896; vols. 28-34, 1898-1908; [vols. 38-39 not published] vols. 41-44/45, 1910-13/14. Trudy Imp. S.-Peterburgsk. Obshch. Estestvoisp., Vyp. 3, Otd. Bot. For vol. 27 see: Trudy imperatorskago S.-Peterburgskago obshchestva estestvoispytatelei. Vypusk 2, otdělenie botaniki. For vols. 35-37 see: Botanicheskii zhurnal. St. Petersburg. For vol. 40 see: Trudy imperatorskago S.-Peterburgskago obshchestva estestvoispytatelei. Vypusk 4, otdělenie botaniki. Preceded by: Trudy S.-Peterburgskago Obshchestva Estestvoispytatelei: Otdělenie botaniki. Superseded by: Trudy imperatorskago Petrogradskago obshchestva estestvoispytatelei. Vypusk 3, otdělenie botaniki. 3-2394-3. HI 73790

Trudy imperatorskago S.-Peterburgskago obshchestva estestvoispytatelei. Vypusk 4, otdělenie botaniki. St. Petersburg. Vol. 40 [vols. 38-39 not published], 1909. Trudy Imp. S.-Peterburgsk. Obshch. Estestvoisp., Vyp. 4, Otd. Bot. Preceded by: Botanicheskii zhurnal. St. Petersburg. Superseded by: Trudy imperatorskago S.-Peterburgskago obshchestva estestvoispytatelei. Vypusk 3, otdělenie botaniki. 3-2394-3. HI 73791

Trudy. Institut biologii, akademija nauk Latvijskoj S S R = Trudy. Institut biologii, akademiya nauk Latviiskoi S S R. Riga.

Trudy institut biologii. Akademija nauk Litovskoj S S R = Trudy institut biologii. Akademiya nauk Litovskoi S S R. Vilnius.

Trudy. Institut biologii, akademiya nauk Latviiskoi S S R. Riga, Latvian S S R. Vols. 2-13, 1955-60; vols. 16-21, 1960-61 [vols. [1] (1954) & [23] (1963) published as monographs; 2 vols. bearing vol. no. 8 published in 1959; vol. 11 not published]. Trudy Inst. Biol. Akad. Nauk Latviisk. S.S.R. For vols. 14 and 22 see: Trudy instituta. Institut biologii, akademiya nauk Latviiskoi S S R. For vol. 15 see: Trudy institut biologii i Botanicheskii sad. Akademii nauk Latviiskoi S S R. HI 73792

Trudy institut biologii. Akademiya nauk Litovskoi S S R. Vilnius, Lithuanian S S R. Vols. 1-3, 1951-58. Trudy Inst. Biol. Akad. Nauk Litovsk. S.S.R. Superseded by: Trudy Laboratorii fiziologii životnyh Instituta biologii Akademii nauk Litovskoj S S R, vol. 4, 1954 [not entered]. HI 73793

Trudy Instituta biologii. Akademija nauk Litovskoj S S R = Trudy institut biologii. Akademiya nauk Litovskoi S S R. Vilnius.

Trudy instituta biologii akademija nauk S S S R, jakutskij filial. Moscow & Leningrad = Trudy instituta biologii akademiya nauk S S S R, yakutskii filial. Moscow & Leningrad.

Trudy instituta biologii. Akademija nauk Turkmenskoj S S R = Trudy instituta biologii. Akademiya nauk Turkmenskoi S S R. Ashkhabad.

Trudy instituta biologii akademiya nauk S S S R, sibirskogo otdeleniya = Trudy instituta biologii akademiya nauk S S S R, yakutskii filial. Moscow & Leningrad.

Trudy instituta biologii, akademija nauk S S S R, Ural'skij filial. Moscow, Leningrad, Sverdlovsk = Trudy instituta biologii, akademiya nauk S S S R, Ural'skii filial. Moscow, Leningrad, Sverdlovsk

Trudy instituta biologii, akademiya nauk S S S R, Ural'skii filial. Moscow, Leningrad, Sverdlovsk. Vols. 1-51, 56, 1947-66. Trudy Inst. Biol. Akad. Nauk S.S.S.R. Ural'sk. Fil. Superseded by: Trudy instituta ekologii rastenii i zhivotnykh, Ural'skii filial, akademiya nauk S S S R. 1-120-3. HI 73794

Trudy instituta biologii, akademiya nauk Turkmenskoi S S R. Ashkhabad, Turkmen S S R. Vols. 1-3, 1954-55. Trudy Inst. Biol. Akad. Nauk Turkmensk. S.S.R. Superseded by: Trudy instituta botaniki, akademiya nauk Turkmenskoi S S R. Ashkhabad. HI 73795

Trudy instituta biologii akademiya nauk S S S R, yakutskii filial. Moscow & Leningrad. Vols. 1-?, 1955-62. Trudy Inst. Biol. Akad. Nauk S.S.S.R. Yakutsk Fil. HI 73796

Trudy institut biologii i Botaničeskij sad Akademii nauk Latvijskoj S S R = Trudy institut biologii i Botanicheskii sad. Akademii nauk Latviiskoi S S R. Riga.

Trudy institut biologii i Botanicheskii sad. Akademii nauk Latviiskoi S S R. Riga, Latvian S S R. Vol. 15, 1960. Trudy Inst. Biol. Bot. Sad Akad. Nauk Latviisk. S.S.R. Preceded by: Trudy instituta. Institut biologii. Akademiya nauk Latviiskoi S S R. Superseded by: Trudy. Institut biologii. Akademiya nauk Latviiskoi S S R. HI 73797

Trudy instituta biologii vnutrennikh vod, akademii nauk S S S R. Moscow. Vol. 6(9)+, 1963+. Trudy Inst. Biol. Vnutrenn. Vod Akad. Nauk S.S.S.R. Preceded by: Trudy instituta biologii vodokhranilishch. HI 73798

Trudy Instituta Biologii Vodohranilišč = Trudy instituta biologii vodokhranilishch.

Trudy instituta biologii vodokhranilishch. Moscow & Leningrad. Vols. 1-5 [also numbered 4-8], 1959-63. Trudy Inst. Biol. Vodokhr. Preceded by: Trudy biologicheskoi stantsii "Borok" imeni N. A. Morozova. Superseded by: Trudy instituta biologii vnutrennikh vod, akademii nauk S S S R. HI 73799

Trudy instituta botaniki. Alma-Ata, Kazakh S S R. Trudy Inst. Bot. (Alma-Ata). See B–P–H 908/5. HI 60058

Trudy instituta botaniki. Tashkent, Uzbek S S R. Trudy Inst. Bot. (Tashkent). See B–P–H 908/8. HI 60061

Trudy instituta botaniki, akademiya nauk Azerbaidzhanskoi S S R. Baku, Azerbaijan S S R. Vol. 15+ [vol. 15 used twice, in 1945 for Trudy botanicheskogo instituta imeni akademika V. L. Komarova, and in 1950 for this title] 1950+. Trudy Inst. Bot. (Baku). Preceded by: Trudy botanicheskogo instituta imeni akademika V. L. Komarova. HI 73800

Trudy instituta botaniki, akademiya nauk gruzinskoi S S R. Seriya flora i sistematika. Tiflis. ?-1966+. Trudy Inst. Bot. Akad. Nauk Gruzinsk. S.S.R., Ser. Fl. Sist. HI 73801

Trudy instituta botaniki, akademiya nauk gruzinskoi S S R. Seriya geobotanika. Tiflis. ?-1968+. Trudy Inst. Bot. Akad. Nauk Gruz. S.S.R., Ser. Geobot. HI 73802

Trudy instituta botaniki, akademiya nauk gruzinskoi S S R. Seriya kul'turnaya rasteniya. Tiflis. ?-1968+. Trudy Inst. Bot. Akad. Nauk Gruz. S.S.R., Ser. Kul't. Rast. HI 73803

Trudy instituta botaniki, akademiya nauk Kirgizskoi S S R. Frunze. Nos. 1-3, 1955-[58]. Trudy Inst. Bot. Akad. Nauk Kirgizsk. S.S.R. Preceded by: Trudy instituta botaniki i rastenievodstva, akademiya nauk Kirgizskoi S S R. HI 73804

Trudy instituta botaniki, akademiya nauk Tadzhikskoi S S R = Trudy, akademiya nauk tadzhikskoi S S R. Stalinabad [= Dushanbe].

Trudy instituta botaniki, akademiya nauk Turkmenskoi S S R. Ashkhabad, Turkmen S S R. Vols. 4-7, 1958-62. Trudy Inst. Bot. (Ashkhabad). Preceded by: Trudy instituta biologii, akademiya nauk Turkmenskoi S S R. HI 73805

Trudy instituta botaniki, akademiya nauk uzbekskoi S S R. Tashkent, Uzbek S S R = Trudy Inst. Bot. (Tashkent). See B–P–H 908/8.

Trudy instituta botaniki i rastenievodstva, akademiya nauk Kirgizskoi S S R. Frunze. Vol. 1, 1954. Trudy Inst. Bot. Rasteniev. Superseded by: Trudy instituta botaniki, akademiya nauk Kirgizskoi S S R. HI 73807

Trudy instituta ekologii rastenii i zhivotnykh, akademiya nauk S S S R, Ural'skii filial. Sverdlovsk. Vols. 53-54, 57+, 1967+. Trudy Inst. Ekol. Rast. Preceded by: Trudy instituta biologii, akademiya nauk S S S R, Ural'skii filial. HI 73808

Trudy instituta fiziologii rastenii imeni K. A. Timiryazeva. Moscow & Leningrad. Vols. 1(2)-10, 1937-55. Trudy Inst. Fiziol. Rast. Timiryazeva. Preceded by: Trudy laboratorii fiziologii i biokhimii rastenii. HI 73809

Trudy instituta fiziologii rastenij imeni K. A. Timirjazeva = Trudy instituta fiziologii rastenii imeni K. A. Timiryazeva. Moscow & Leningrad.

Trudy instituta genetiki, akademiya nauk S S S R.

Moscow, Leningrad. Nos. 10+, 1935+. Trudy Inst. Genet. Akad. Nauk S.S.S.R. Preceded by: Trudy laboratorii po genetike. HI 73810

Trudy instituta genetiki i selektsii Azerbaidzhanskoi S S R. Baku. Vol. 1+, 1959+. Trudy Inst. Genet. Selekts. Azerbaidzansk. S.S.R. HI 73811

Trudy instituta geografii, akademiya nauk S S S R. Moscow. Vol. 25+, 1937+. Trudy Inst. Geogr. Preceded by: Trudy instituta fizicheskoi geografii [not entered]. HI 73812

Trudy instituta geologičeskih nauk. Serija stratigrafii i paleontologii = Trudy instituta geologicheskikh nauk. Seriya stratigrafii i paleontologii. Kiev.

Trudy instituta geologicheskikh nauk, akademiya nauk S S S R. Moscow. Vols. 1-165, 1939-56. Trudy Inst. Geol. Nauk Akad. Nauk S.S.S.R. Preceded and superseded by: Trudy geologicheskogo instituta, akademiya nauk S S S R. HI 73813

Trudy instituta geologicheskikh nauk. Seriya stratigrafii i paleontologii. Kiev, Ukrainian S S R. Vol. 1+, 1950+. Trudy Inst. Geol. Nauk, Ser. Stratigr. HI 73814

Trudy Instituta gidrobiologii. Kiev, Ukrainian S S R = Trudy institutu gidrobiologii. Kiev.

Trudy institutu gidrobiologii. Kiev, Ukrainian S S R. Vol. 20+, 194?+. Trudy Inst. Gidrobiol. Preceded by: Trudy gidrobiologichnoi stantsii. Kiev. 1-122-2. HI 73815

Trudy institutu gidrobiologiji. Kiev, Ukrainian S S R = Trudy institutu gidrobiologii. Kiev.

Trudy instituta imeni Pastera. Leningrad. Vol. 43+, 1975+. Trudy Inst. Pastera. Preceded by: Trudy leningradskogo nauchnoissledovatel'skogo instituta epidemiologii i mikrobiologii imeni Pastera. HI 73816

Trudy instituta. Institut biologii, akademija nauk Latvijskoj S S R = Trudy instituta. Institut biologii, akademiya nauk Latviiskoi S S R. Riga.

Trudy instituta. Institut biologii, akademiya nauk Latviiskoi S S R. Riga, Latvian S S R. Vol. 14, 1960; vol. 22, 1962 (zoological). Trudy Inst. Inst. Biol. Akad. Nauk Latviisk. S.S.R. Preceded by: Trudy. Institut biologii. Akademiya nauk Latviiskoi S S R. For vol. 15, 1960 see: Trudy institut biologii i Botanicheskii sad. Akademii nauk Latviiskoi S S R. For vols. 2-13, 1955-60 and vols. 16-21, 1960-61 see: Trudy. Institut biologii. Akademiya nauk Latviiskoi S S R. HI 73817

Trudy instituta. Institut mikrobiologii, akademija nauk Latvijskoj S S R = Trudy instituta. Institut mikrobiologii, akademiya nauk Latviiskoi S S R. Riga.

Trudy instituta. Institut mikrobiologii, akademiya nauk Latviiskoi S S R. Riga, Latvian S S R. Vol. 4, 1956; vol. 6+, 1958+. Trudy Inst. Mikrobiol. Akad. Nauk Latviisk. S.S.R. For vols. 1-3 (1952-55) and vol. 5 (1956) see: Trudy instituta mikrobiologii. Riga. HI 73818

Trudy instituta istorii estestvoznanija. Moscow & Leningrad = Trudy instituta istorii estestvoznaniya. Moscow & Leningrad.

Trudy instituta istorii estestvoznanija i tehniki. Moscow & Leningrad = Trudy instituta istorii estestvoznaniya i tekhniki. Moscow & Leningrad.

Trudy instituta istorii estestvoznaniya. Moscow & Leningrad. Vols. 1-5, 1947-53. Trudy Inst. Istorii Estestv. Superseded by: Trudy instituta istorii estestvoznaniya i tekhniki. 1-116-3. HI 73819

Trudy instituta istorii estestvoznaniya i tekhniki. Moscow & Leningrad. Vol. 1+, 1954+. Trudy Inst. Istorii Estestv. Tekhn. Preceded by: Trudy instituta istorii estestvoznaniya. HI 73820

Trudy instituta istorii nauki i tehniki. Serija 1, arhiv istorii nauki i tehniki = Trudy instituta istorii nauki i tekhniki. Seriya 1, arkhiv istorii nauki i tekhniki. Moscow & Leningrad.

Trudy instituta istorii nauki i tekhniki. Seriya 1, arkhiv istorii nauki i tekhniki. Moscow & Leningrad. Vols. 1-9, 1933-36. Trudy Inst. Istorii Nauki, Ser. 1, Arkh. Istorii Nauki Tekhn. 1-116-3. HI 73821

Trudy instituta po izučeniju lesa = Trudy instituta po izucheniyu lesa. Leningrad.

Trudy instituta po izucheniyu lesa. Leningrad. 1 vol., 1933. Trudy Inst. Izuch. Lesa. HI 73822

Trudy instituta po izucheniyu severa. Moscow. Nos. 31-49, 1926-31. Trudy Inst. Izuch. Severa. Preceded by: Trudy nauchno-issledovatel'skogo instituta po izucheniyu severa. Superseded by: Trudy arkticheskogo instituta. HI 73823

Trudy instituta lesa, akademiya nauk S S S R. Moscow, Leningrad. Vols. 1-49 [vol. 20 not published], 1947-59. Trudy Inst. Lesa. Superseded by: Trudy instituta lesa i drevesiny, akademiya nauk S S S R. 1-116-3. HI 73824

Trudy instituta lesa i drevesiny, akademiya nauk S S S R. Moscow, Leningrad. Vols. 51-66, 1962-63. Trudy Inst. Lesa Drevesiny Akad. Nauk S.S.S.R. Preceded by: Trudy instituta lesa, akademiya nauk S S S R. HI 73825

Trudy instituta mikrobiologii. Moscow. Trudy Inst. Mikrobiol. (Moscow). See B–P–H 909/7. HI 60062

Trudy instituta mikrobiologii. Riga, Latvian S S R. Vols. 1-3, 1952-55; vol. 5, 1956. Trudy Inst. Mikrobiol. (Riga). For vol. 4 (1956) and vol. 6+ (1958+) see: Trudy instituta. Institut mikrobiologii, Akademiya nauk Latviiskoi S S R. HI 73826

Trudy instituta mikrobiologii i virusologii, akademiya

nauk Kazakhskoi S S R. Alma-Ata. Vol. 1+, 1956+. Trudy Inst. Mikrobiol. Virusol. HI 73827

Trudy instituta novogo lubyanogo ser'ya. Moscow. Vols. 1-9(1), 1931-34. Trudy Inst. Nov. Lubyan. Ser. HI 73828

Trudy instituta okeanologii. Moscow. Trudy Inst. Okeanol. See B–P–H 909/10. HI 60064

Trudy instituta pochvovedeniya, akademiya nauk Kazakhskoi S S R. Alma-Ata. Vol. 1+, 1952+. Trudy Inst. Pochvov. Akad. Nauk Kazakhsk. S.S.R. HI 73829

Trudy instituta pochvovedeniya i geobotaniki Sredne-Aziatskogo gosudarstvennogo universiteta. Kazakhstanskaya seriya. Tashkent, Uzbek S S R. Vol. 1+, 1929+. Trudy Inst. Pochvov. Sredne-Aziatsk. Gosud. Univ., Kazakhstansk. Ser. 1-116-3. HI 73830

Trudy instituta pochvovedeniya i geobotaniki Sredne-Aziatskogo gosudarstvennogo universiteta. Turkmenistanskaya seriya. Tashkent, Uzbek S S R. Vol. 1/2, 1930. Trudy Inst. Pochvov. Sredne-Aziatsk. Gosud. Univ., Turkmenistansk. Ser. 5-4156-3. HI 73831

Trudy instituta počvovedenija i geobotaniki Sredne-Aziatskogo gosudarstvennogo universiteta. Kazahstanskaja serija = Trudy instituta pochvovedeniya i geobotaniki Sredne-Aziatskogo gosudarstvennogo universiteta. Kazakhstanskaya seriya. Tashkent.

Trudy instituta počvovedenija i geobotaniki Sredne-Aziatskogo gosudarstvennogo universiteta. Turkmenistanskaja serija = Trudy instituta pochvovedeniya i geobotaniki Sredne-Aziatskogo gosudarstvennogo universiteta. Turkmenistanskaya seriya. Tashkent.

Trudy instituta sel'skokhozyaistvennoi mikrobiologii. Moscow, Leningrad. Vols. 4(2), 1930; vol. 10, 1938. Trudy Inst. Sel'skokhoz. Mikrobiol. Preceded by: Trudy otdela sel'sko-khozyaistvennoi mikrobiologii, gosudarstvennyi institut opytnoi agronomii. For intervening vols. see: Trudy vsesoyuznogo instituta sel'skokhozyaistvennoi mikrobiologii. Superseded by: Trudy vsesoyuznogo nauchno-issledovatel'skogo instituta sel'skokhozyaistvennoi mikrobiologii. HI 73832

Trudy instituta zashchity rastenii, akademiya nauk Gruzinskoi S S R. Tiflis. Vols. 4-23?, 1947-72? Trudy Inst. Zashch. Rast. Preceded by: Izvestiya Otdela zashchity rastenii and Mcenareta dacvis gankopilebis moambe. Superseded by: Trudy nauchno-issledovatel'skogo instituta zashchity rastenii, ministerstvo sel'skogo khozyaistva Gruzinskoi S S R. HI 73833

Trudy instituta zaščity rastenij, akademija nauk Gruzinskoj S S R. Tiflis = Trudy instituta zashchity rastenii, akademiya nauk Gruzinskoi S S R. Tiflis.

Trudy institutu botaniky, Kharkivs'kii derzhavnii universitet. Kharkov. Vols. 1-2, 1936-38. Trudy Inst. Bot. Kharkivs'k. Derzhavn. Univ. Superseded by: Trudy nauchno-issledovatel'skogo instituta botaniki, Khar'kov. HI 73834

Trudy Instytutu geologičnyh nauk. Kiev = Trudy instituta geologicheskikh nauk. Kiev

Trudy instytutu henetyky i selekt'sii. Kyyiv. Vols. ?-5-12, ?-196?-70. Trudy Inst Henet. Selekt's. Superseded by: Nauchnye trudy, ukrainskii nauchno-issledovatel'skii institut rastenievodstva selekt'sii i genetiki im V. Ya. Ure'eva. HI 73835

Trudy introdukcionnogo pitomnika subtropičeskih kul'tur = Trudy introduktsionnogo pitomnika subtropicheskikh kul'tur. Sukhumi.

Trudy introduktsionnogo pitomnika subtropicheskikh kul'tur. Sukhumi, Georgian S S R. Vols. 1-10, 1937-38. Trudy Introd. Pitomn. Subtrop. Kul't. 5-4112-1. HI 73836

Trudy Irkutskogo gosudarstvennogo universiteta imeni A. A. Ždanova. Serija biologičeskaja = Trudy Irkutskogo gosudarstvennogo universiteta imeni A. A. Zhdanova. Seriya biologicheskaya. Irkutsk.

Trudy Irkutskogo gosudarstvennogo universiteta imeni A. A. Zhdanova. Seriya biologicheskaya. Irkutsk, Russian S F S R. Vols. 3-4, 1948-49; vol. 1/2, 1953; 1954. Trudy Irkutsk. Gosud. Univ. Zhdanova, Ser. Biol. Preceded by: Trudy Vostochno-sibirskogo gosudarstvennogo universiteta. 3-2116-3. HI 73837

Trudy Irkutskogo obščestva estestvoispytatelej = Trudy Irkutskogo obshchestva estestvoispytatelei. Irkutsk.

Trudy Irkutskogo obshchestva estestvoispytatelei. Irkutsk, Russian S F S R. 1 vol., 1923. Trudy Irkutsk. Obshch. Estestvoisp. 3-2117-1. HI 73838

Trudy i issledovanija po lesnomu hozjajstvu i lesnoj promyšlennosti = Trudy i issledovaniya po lesnomu khozyaistvu i lesnoi promyshlennosti. Leningrad.

Trudy i issledovaniya po lesnomu khozyaistvu i lesnoi promyshlennosti. Leningrad. Vols. 9-20, 1931. Trudy Issl. Lesn. Khoz. Lesn. Promyshl. Preceded by: Trudy po lesnomu opytnomu delu. Moscow & Leningrad. 3-2391-3. HI 73839

Trudy Jaroslavskogo estestvenno-istoričeskogo i kraevedčeskogo obščestva = Trudy Yaroslavskogo estestvenno-istoricheskogo i kraevedcheskogo obshchestva. Yaroslavl.

Trudy Jaroslavskogo estestvenno-istoričeskogo obščestva = Trudy Yaroslavskogo estestvenno-istoricheskogo obshchestva. Yaroslavl.

Trudy, kabardino-balkarskaya opytnaya stantsiya

sadovodstva. Nal'chik. Vol. 1+, 1977+. Trudy Kabardino-Balkarsk. Opytn. Stantsiya Sadov. HI 73840

Trudy Kandalakshskogo gosudarstvennogo zapovednika. Vologda. Vol. 1+, 1958+. Trudy Kandalaksk. Gosud. Zapov. HI 73841

Trudy Karadagskoi biologicheskoi stantsii. Kiev, Ukrainian S S R. Vol. [2], 19??; vol. 5, 1939; vol. 7+, 1949+. Trudy Karadagsk. Biol. Stantsii. Preceded by: Trudy Karadagskoi nauchnoi stantsii. For vols. 3-4 see: Trudy Karadagskoi biologicheskoi stantsii Moskovskogo obshchestva ispytatelei prirody. For vol. 6 see: Trudy Karadahs'koyi biolohichnoyi stantsiyi. 3-2272-2. HI 73842

Trudy Karadagskoi biologicheskoi stantsii Moskovskogo obshchestva ispytatelei prirody. Kiev, Ukrainian S S R. Vols. 3-4, 1930-31. Trudy Karadagsk. Biol. Stantsii Moskovsk. Obshch. Isp. Prir. Preceded & superseded by: Trudy Karadagskoi biologicheskoi stantsii. 3-2272-2. HI 73843

Trudy Karadagskoi nauchnoi stantsii. Kiev, Ukrainian S S R. Vol. 1, 1917. Trudy Karadagsk. Nauchn. Stantsii. Superseded by: Trudy Karadagskoi biologicheskoi stantsii. 3-2272-2. HI 73845

Trudy Karadagskoj biologičeskoj stancii = Trudy Karadagskoi biologicheskoi stantsii. Kiev.

Trudy Karadagskoj biologičeskoj stancii Moskovskogo obščestva ispytatelej prirody = Trudy Karadagskoi biologicheskoi stantsii Moskovskogo obshchestva ispytatelei prirody. Kiev.

Trudy Karadagskoj naučnoj stancii = Trudy Karadagskoi nauchnoi stantsii. Kiev.

Trudy Karadags'koji biologičeskoji stanciji = Trudy Karadahs'koyi biolohichnoyi stantsiyi. Kiev.

Trudy Karadahs'koyi biolohichnoyi stantsiyi. Kiev, Ukrainian S S R. Vol. 6, 1940. Trudy Karadags'k. Biol. Stantsiyi. Preceded & superseded by: Trudy Karadagskoi biologicheskoi stantsii. 3-2272-2. HI 73844

Trudy Karadahs'koyi nauchnoyi stantsiyi imeni T. I. Vyazems'koho = Trudy Karadagskoi nauchnoi stantsii. Kiev.

Trudy Karagandinskogo botaničeskogo sada = Trudy Karagandinskogo botanicheskogo sada. Alma-Ata.

Trudy Karagandinskogo botanicheskogo sada. Alma-Ata, Kazakh S S R. 1 vol., 1960. Trudy Karagandinsk. Bot. Sada. HI 73846

Trudy Karelo-Finskogo filiala akademii nauk S S S R. Petrozavodsk, Russian S F S R. Trudy Karelo-Finsk. Fil. Akad. Nauk S.S.S.R. See B–P–H 910/12. HI 60065

Trudy Karel'skogo filiala akademii nauk S S S R. Petrozavodsk, Russian S F S R. Trudy Karel'sk. Fil. Akad. Nauk S.S.S.R. See B–P–H 910/13. HI 60066

Trudy Kazahstanskoj bazy akademii nauk S S S R = Trudy Kazakhstanskoi bazy akademii nauk S S S R. Moscow & Leningrad.

Trudy Kazakhskogo nauchno-issledovatel'skogo instituta lesnogo khozyaistva. Alma-Ata. Vol. 1+, 1959+. Trudy Kazakhsk. Nauchno-Issl. Inst. Lesn. Khoz. HI 73847

Trudy, Kazakhskogo nauchno-issledovatel'skogo instituta zaschity rastenii. Alma-Ata. Vol. 9+, 1965+. Trudy Kazakhsk. Nauchno-Issl. Inst. Zashch. Rast. Preceded by: Trudy nauchno-issledovatel'skogo instituta zashchity rastenii, Kazakhskaya akademiya sel'skokhozyaistvennykh nauk. HI 73848

Trudy, kazakhskii nauchno-issledovatel'skii institut zemledeliya. Vols. ?-7-10, ?-19?? Trudy Kazakhsk. Nauchno-Issl. Inst. Zemled. HI 73849

Trudy Kazakhstanskoi bazy akademii nauk S S S R. Moscow & Leningrad. Vol. 1, 1934. Trudy Kazakhstansk. Bazy Akad. Nauk S.S.S.R. Superseded by: Trudy. Kazakhstanskii filial, akademii nauk S S S R. 1-118-1. HI 73850

Trudy. Kazakhstanskii filial, akademii nauk S S S R. Moscow & Leningrad. Vols. 2-20, 1937-41. Trudy Kazakhstansk. Fil. Akad. Nauk S.S.S.R. Preceded by: Trudy Kazakhstanskoi bazy akademii nauk S S S R. 1-118-1. HI 73851

Trudy. Kazahstanskij filial, akademii nauk S S S R = Trudy. Kazakhstanskii filial, akademii nauk S S S R. Moscow & Leningrad.

Trudy khimicheskogo i botanicheskogo institutov. Baku, Azerbaijan S S R. Vol. 5, 1938. Trudy Khim. Bot. Inst. Preceded & superseded by: Trudy botanicheskogo instituta. Baku. 1-109-3. HI 73852

Trudy po khimii prirodnykh soedinenii. Kishinev. Vols. ?-2-8, ?-1959-69. Trudy Khimii Prir. Soedin. Superseded by: Rastitel'nye belki. HI 73853

Trudy Kirgizskogo gosudarstvennogo pedagogičeskogo instituta imeni M. V. Frunze = Trudy Kirgizskogo gosudarstvennogo pedagogicheskogo instituta imeni M. V. Frunze. Frunze.

Trudy Kirgizskogo gosudarstvennogo pedagogicheskogo instituta imeni M. V. Frunze. Frunze, Kirghiz S S R. Vols. 1-2, 1947-48. Trudy Kirgizsk. Gosud. Pedagog. Inst. Frunze. Superseded by: Uchenye zapiski Biologo-pochvennogo fakul'teta. 2-1651-2. HI 73854

Trudy komi filiala akademii nauk S S S R. Syktyvkar. Vol. 1+, 1953+. Trudy Komi Fil. Akad. Nauk S.S.S.R. HI 73855

Trudy komissii po izučeniju četvertičnogo perioda = Trudy komissii po izucheniyu chetvertichnogo perioda.

Moscow & Leningrad.

Trudy komissii po izučeniju Jakutskoj avtonomnof S S R = Trudy komissii po izucheniyu Yakutskoi avtonomnof S S R.

Trudy komissii po izučeniju ozera Bajkala = Trudy komissii po izucheniyu ozera Baikala. Petrograd.

Trudy komissii po izucheniyu chetvertichnogo perioda. Moscow & Leningrad. Vol. 1+ [vol. 11 not published], 1932+. Trudy Komiss. Izuch. Chetvert. Perioda. 1-118-1. HI 73856

Trudy komissii po izucheniyu Yakutskoi avtonomnof S S R. Leningrad. Vols. 1-16, 1926-30. Trudy Komiss. Izuch. Yakutsk. Avton. S.S.R. 1-118-2. HI 73857

Trudy komissii po izucheniyu ozera Baikala. Petrograd. Vols. 1-3, 1918-30. Trudy Komiss. Izuch. Ozera Baikala. Superseded by: Trudy Baikal'skoi limnologicheskoi stantsii. 1-118-2. HI 73858

Trudy komissii po ohrane prirody = Trudy komissii po okhrane prirody. Sverdlovsk.

Trudy komissii po okhrane prirody. Sverdlovsk, Russian S F S R. Vol. 2, 1963. Trudy Komiss. Okhr. Prir. HI 73859

Trudy komissii po sanitarno-biologičeskomu obsledovaniju reka S.-Donca i ego pritikov (Lopani i Udy) = Trudy komissii po sanitarno-biologicheskomu obsledovaniyu reka S.-Donca i ego pritikov (Lopani i Udy). Kharkov.

Trudy komissii po sanitarno-biologicheskomu obsledovaniyu reka S.-Donca i ego pritikov (Lopani i Udy). Kharkov, Ukrainian S S R. Vols. 1-2, 1926. Trudy Komiss. Sanit.-Biol. Obsl. Reka S.-Donca. HI 73860

Trudy Kosinskoi biologicheskoi stantsii Moskovskogo obshchestva ispytatelei prirody. Moscow. Vols. 1-11, 1924-30. Trudy Kosinsk. Biol. Stantsii Moskovsk. Obshch.Isp. Prir. Superseded by: Trudy Limnologicheskoi stantsii v Kosine. 3-2314-3. HI 73861

Trudy Kosinskoj biologičeskoj stancii Moskovskogo obščestva ispytatelej prirody = Trudy Kosinskoi biologicheskoi stantsii Moskovskogo obshchestva ispytatelei prirody. Moscow.

Trudy Kostromskago nauchnago obshchestva po izucheniyu mĕstnago kraya. Kostroma, Russian S F S R. Vols. 1-43, 1914-43. Trudy Kostromsk. Nauchn. Obshch. Izuch. Mĕstn. Kraya. 3-2315-2. HI 73862

Trudy Kostromskago naučnago obščestva po izučeniju mĕstnago kraja = Trudy Kostromskago nauchnago obshchestva po izucheniyu mĕstnago kraya. Kostroma.

Trudy Kostromskogo naučnogo obščestva po izučeniju Mestnogo kraja = Trudy Kostromskago nauchnago obshchestva po izucheniyu mĕstnago kraya. Kostroma.

Trudy Krymskogo nauchno-issledovatel'skogo instituta. Simferopol'. 1926-27. Trudy Krymsk. Nauchno-Issl. Inst. HI 73863

Trudy, krymskaya opytno-selektsionnnnaya stantsiya. Moscow. Vol. 5+, 1970+. Trudy Krymsk. Opytn.-Selekts. Stantsiya. Preceded by: Trudy plodoovoshchnoi opytno-selektsionnoi stantsii. HI 73864

Trudy laboratorii evoljucionnoj ekologii rastenij = Trudy laboratorii evolyutsionnoi ekologii rastenii. Moscow.

Trudy laboratorii evoljucionnoj i ekologicheskoj fiziologij imeni B. A. Kellera. Moscow = Trudy laboratorii evolyutsionnoi i ekologicheskoi fiziologii imeni B. A. Kellera. Moscow.

Trudy laboratorii evolyutsionnoi i ekologicheskoi fiziologii imeni B. A. Kellera. Moscow. Vols. 2-6, 1950-67? Trudy Lab. Evol. Ekol. Fiziol. B. A. Kellera. Preceded by: Trudy laboratorii evolyutsionnoi ekologii rastenii. HI 73865

Trudy laboratorii evolyutsionnoi ekologii rastenii. Moscow. Vol. 1, 1940. Trudy Lab. Evol. Ekol. Rast. Superseded by: Trudy laboratorii evolyutsionnoi i ekologicheskoi fiziologii imeni B. A. Kellera. HI 73866

Trudy laboratorii fiziologii i biohimii rastenij = Trudy laboratorii fiziologii i biokhimii rastenii. Moscow & Leningrad.

Trudy laboratorii fiziologii i biokhimii rastenii. Moscow & Leningrad. Vol. 1(1), 1934. Trudy Lab. Fiziol. Rast. Superseded by: Trudy instituta fiziologii rastenii imeni K. A. Timiryazeva. 1-116-1. HI 73867

Trudy laboratorii po genetike. Leningrad, Moscow. Vol. 9, 1932. Trudy Lab. Genet. Preceded by: Trudy byuro po genetike. Superseded by: Trudy instituta genetiki, akademiya nauk S S R. HI 73868

Trudy laboratorii lesovedeniya, akademiya nauk S S S R. Moscow. Vol. 1-?, 1960-62. Trudy Lab. Lesoved. Akad. Nauk S.S.S.R. HI 73869

Trudy po lekarstvennym i aromatičeskim rastenijam = Trudy po lekarstvennym i aromaticheskim rasteniyam. Moscow.

Trudy po lekarstvennym i aromaticheskim rasteniyam. Moscow. Vol. 1, 1932. Trudy Lekarstv. Aromat. Rast. Superseded by: Trudy po lekarstvennym i lekarstvenno-tekhnicheskim rasteniyam. 5-4267-3. HI 73870

Trudy po lekarstvennym i lekarstvenno-tehničeskim rastenijam = Trudy po lekarstvennym i lekarstvenno-tekhnicheskim rasteniyam. Moscow.

Trudy po lekarstvennym i lekarstvenno-tekhnicheskim rasteniyam. Moscow. Vols. 2-4, 1933-35. Trudy Lekarstv. Lekarstv.-Tekhn. Rast. Preceded by: Trudy po lekarstvennym i aromaticheskim rasteniyam. Superseded by: Vsesoyuznyi nauchno-issledovatel'skii institut lekarstvennykh rastenii. 5-4267-3. HI 73871

Trudy, leningradskii institut epidemiologii i bakteriologii imeni Pastera. Leningrad. Vols. 1-7, 1935-40. Trudy Leningradsk. Inst. Epidem. Bakteriol. Pastera. Superseded by: Trudy leningradskogo nauchno-issledovatel'skogo instituta epidemiologii i mikrobiologii imeni Pastera. HI 73872

Trudy Leningradskogo khimiko-farmacevticheskogo instituta. [From vol. 12+, 1961+ some vols. also titled: Voprosy farmakognozoo vol. 1+]. Leningrad. Vol. 4+, 1958+. Trudy Leningradsk. Khim.-Farm. Inst. Preceded by: Sbornik nauchnykh trudov, Leningradskii khimiko- farmacevticheskii institut. HI 73873

Trudy leningradskogo nauchno-issledovatel'skogo instituta epidemiologii i mikrobiologii imeni Pastera. Leningrad. Vols. 8-42, 194?-74? Trudy Leningradsk. Nauchno-Issl. Inst. Epidem. Mikrobiol. Pastera. Preceded by: Trudy, leningradskii institut epidemiologii i bakteriologii imeni Pastera. Superseded by: Trudy instituta imeni Pastera. HI 73874

Trudy Leningradskogo obščestva estestvoispytatelej = Trudy Leningradskogo obshchestva estestvoispytatelei. Leningrad.

Trudy Leningradskogo obščestva estestvoispytatelej. Vypusk. 3: Otdelenie botaniki = Trudy Leningradskogo obshchestva estestvoispytatelei. Vypusk. 3: Otdelenie botaniki. Leningrad.

Trudy Leningradskogo obshchestva estestvoispytatelei. Leningrad. Vol. 61+, 1932+. Trudy Leningradsk. Obshch. Estestvoisp. Preceded by: Trudy Leningradskogo obshchestva estestvoispytatelei. Vypusk. 3: Otdelenie botaniki. 3-2394-3. HI 73876

Trudy Leningradskogo obshchestva estestvoispytatelei. Vypusk. 3: Otdelenie botaniki. Leningrad. Vols. 54-60, 1924-30. Trudy Leningradsk. Obshch. Estestvoisp., Vyp. 3, Otd. Bot. Preceded by: Trudy Petrogradskogo obshchestva estestvoispytatelei. Vypusk 3... Otdelenie botaniki. Superseded by: Trudy Leningradskogo obshchestva estestvoispytatelei. HI 73875

Trudy Leningradskoi ordena Lenina lesotekhnicheskoi akademii imeni S. M. Kirova. Leningrad. Vols. 72-75, 1955+. Trudy Leningradsk. Ordena Lenina Lesotekhn. Akad. S. M. Kirova. Preceded by: Trudy lesotekhnicheskoi akademii imeni S. M. Kirova. HI 73877

Trudy po lesnomu khozyaistva Sibiri. Novosibirsk. Vols. ?-8, ?-1964. Trudy Lesn. Khoz. Sibiri. Superseded by: Trudy po lesnomy khozyaistvu zapadnoi Sibiri. HI 73878

Trudy po lesnomu khozyaistvu zapadnoi Sibiri. Novosibirsk. Vol. 9+, 1971+. Trudy Lesn. Khoz. Zapadnoi Sibiri. Preceded by: Trudy po lesnomu khozyaistva Sibiri. HI 73879

Trudy po lesnomu opytnomu delu. Moscow & Leningrad. Vols. 65-70 [also numbered vols. 1-6], 1925-29 [data on vols. 71-74 not ascertained]; vols. 75-76 [also numbered vols. 1-2], 1929; vols. 1-8, 1930. Trudy Lesn. Opytn. Delu (Moscow & Leningrad). Preceded by: Trudy po lěsnomu opytnomu dělu v Rossii. Superseded by: Trudy i issledovaniya po lesnomu khozyaistvu i lesnoi promyshlennosti. 3-2391-3. HI 73880

Trudy po lesnomu opytnomu delu. Omsk, Russian S F S R. Trudy Lesn. Opytn. Delu (Omsk). See B–P–H 912/8. HI 60067

Trudy po lěsnomu opytnomu dělu v Rossii. St. Petersburg. Vols. 1-64, 1907-24. Trudy Opytn. Lěsnich. Preceded by: Trudy Opytnykh Lěsnichestv. Superseded by: Trudy po lesnomu opytnomu delu. Moscow & Leningrad. HI 73881

Trudy po lesnomu opytnomu delu Ukrainy. Kharkov, Ukrainian S S R. Trudy Lesn. Opytn. Delu Ukrainy. See B–P–H 912/11. HI 60068

Trudy po lesnomu opytnomu delu. Ural'skoi oblasti. Moscow & Sverdlovsk, Russian S F S R. 1 vol., 1930. Trudy Lesn. Opytn. Delu Ural'sk. Obl. HI 73882

Trudy po lesnomu opytnomu delu. Ural'skoj oblasti = Trudy po lesnomu opytnomu delu. Ural'skoi oblasti. Moscow & Sverdlovsk.

Trudy po lesnomu opytnomu delu zasek. Moscow. 1939. Trudy Lesn. Opytn. Delu Zasek. HI 73883

Trudy lesotekhnicheskoi akademii. Leningrad. Vols. 39-41, 1932-33. Trudy Lesotekhn. Akad. Preceded and superseded by: Izvestiya lesotekhnicheskoi akademii. HI 73884

Trudy lesotekhnicheskoi akademii imeni S. M. Kirova. Moscow, Leningrad. Vols. 45-71, 1936-53. Trudy Lesotekhn. Akad. S. M. Kirova. Preceded by: Izvestiya lesotekhnicheskoi akademii. Superseded by: Trudy Leningradskoi ordena Lenina lesotekhnicheskoi akademii imeni S. M. Kirova. HI 73885

Trudy limnologicheskogo instituta. Moscow, Leningrad. 1962. Trudy Limnol. Inst. Preceded by: Trudy Baikal'skoi limnologicheskoi stantsii. Leningrad. HI 73886

Trudy Limnologicheskoi stantsii v Kosine. Moscow. Vols. 12-22, 1931-39. Trudy Limnol. Stantsii Kosine. Preceded by: Trudy Kosinskoi biologicheskoi stantsii Moskovskogo obshchestva ispytatelei prirody. 3-2314-3. HI 73887

Trudy Limnologičeskoj stancii v Kosine = Trudy Limnologicheskoi stantsii v Kosine. Moscow.

Trudy Minskoi bolotnoi stantsii. Minsk. Vol. ?-8(7)-?, ?-1925-? Trudy Minsk. Bolotn. Stantsii. HI 73888

Trudy molodykh uchenykh i aspirantov po selektsii i semenovodstvu ovoshchnykh kul'tur, vsesoyuznii nauchno-issledovatel'skii institut selektsii i semenovodstvo voshchnykh kul'tur. Vol. ?-4+, ?-1971+. Trudy Molod. Uchen. Aspir. Selekts. Semenov. Ovoshch. Kul't. HI 73889

Trudy Mongol'skoi komissii. Moscow & Leningrad. Vol. 1+, 1928+. Trudy Mongol'sk. Komiss. Preceded by: Materialy. Komissiya po izucheniyu estestvennykh proizvoditel'nykh sil Soyuza akademii nauk S S S R. 1-119-2. HI 73890

Trudy Mongol'skoj komissii = Trudy Mongol'skoi komissii. Moscow & Leningrad.

Trudy mordovskogo gosudarstvennogo zapovednika im P. G. Smidovicha. Saransk. Vol. 1+, 1960+. Trudy Mordovsk. Gosud. Zapov. P. G. Smidovicha. HI 73891

Trudy morskogo nauchnogo instituta. Moscow, Leningrad. Vols. 3-4, 1927-30. Trudy Morsk. Nauchn. Inst. Preceded by: Trudy plovuchego morskogo nauchnogo instituta. Superseded by: Trudy gosudarstvennogo okeanograficheskogo instituta. HI 73892

Trudy Moskovskogo obščestva ispytatelej prirody = Trudy Moskovskogo obshchestva ispytatelei prirody. Moscow.

Trudy Moskovskogo obshchestva ispytatelei prirody. Moscow. Vol. 1+, 1951+. Trudy Moskovsk. Obshch. Isp. Prir. HI 73893

Trudy Murmanskogo morskogo biologičeskogo instituta = Trudy Murmanskogo morskogo biologicheskogo instituta. Moscow & Leningrad.

Trudy Murmanskogo morskogo biologicheskogo instituta. Moscow & Leningrad. Vol. 5 [also numbered vol. 1]+, 1960+. Trudy Murmansk. Morsk. Biol. Inst. Preceded by: Trudy Murmanskoi biologicheskoi stantsii. HI 73894

Trudy Murmanskoi biologicheskoi stantsii. Moscow, Leningrad. Vols. 1-4, 1948-58. Trudy Murmansk. Biol. Stantsii. Preceded by: Raboty Murmanskoi biologicheskoi stantsii. Superseded by: Trudy Murmanskogo morskogo biologicheskogo instituta. HI 73895

Trudy Murmanskoj biologičeskoi stancii. Moscow, Leningrad = Trudy Murmanskoi biologicheskoi stantsii. Moscow, Leningrad.

Trudy nauchno-issledovatel'skogo instituta biologii. Pratsy Naukovo-doslidnogo instytutu biologii. Kharkov, Ukrainian S S R. Vols. 12-35, 1947-63; vol. 37+, 1963+. Trudy Nauchno-Issl. Inst. Biol. Preceded by: Pratsy Naukovo-doslidnogo zoologo-biologichnogo instytutu. For vol. 36 see: Trudy biologicheskogo fakul'teta po genetike i zoologii. HI 73896

Trudy nauchno-issledovatel'skogo instituta botaniki, Kharkivs'kogo derzhavnogo universiteti imeni O. M. Gor'kogo. Kharkov. Vols. 3-4, 1938-41. Trudy Nauchno-Issl. Inst. Bot. Preceded by: Trudy institutu botaniky, Kharkivs'kii derzhavnii universitet. 3-2286-1. HI 73897

Trudy nauchno-issledovatel'skogo instituta po izucheniyu severa. Leningrad. Vols. 25-30, 1921-26. Trudy Nauchno-Issl. Inst. Izuch. Severa. Preceded by: Trudy severnoi nauchno-promyslovoi ekspeditsii V S N Kh. Superseded by: Trudy instituta po izucheniyu severa. HI 73898

Trudy Nauchno-issledovatel'skogo instituta pri Voronezhskom gosudarstvennom universitete. Voronezh, Russian S F S R. Vols. 1-4, 1927-30. Trudy Nauchno-Issl. Inst. Voronezhsk. Gosud. Univ. 5-4429-3. HI 73899

Trudy nauchno-issledovatel'skogo instituta zashchity rastenii, Kazakhskaya akademiya sel'skokhozyaistvennykh nauk. Alma-Ata. Vols. 4-8, 1958-64. Trudy Nauchno-Issl. Inst. Zashch. Rast. Kazakhsk. Akad. Sel'skokhoz. Nauk Preceded by: Trudy respublikanskoi stantsii zashchity rastenii. Superseded by: Trudy, Kazakhskogo nauchno-issledovatel'skogo instituta zaschity rastenii. HI 73900

Trudy nauchno-issledovatel'skogo instituta zashchity rastenii, ministerstvo sel'skogo khozyaistva gruzinskoi S S R. Tiflis. Vols. 14-15, 1973. Trudy Nauchno-Issl. Inst. Zashch. Rast. Minist. Sel'sk. Khoz. Gruzinsk. S.S.R. Preceded by: Trudy instituta zashchity rastenii, akademiya nauk gruzinskoi S S R. HI 73901

Trudy Nauchno-issledovatel'skogo torfyanogo instituta (Instorfa). Moscow & Leningrad. Vols. 1-19, 1929-30. Trudy Nauchno-Issl. Torfyan. Inst. HI 73902

Trudy naučno-issledovatel'skogo instituta biologii i biologičeskogo fakul'teta. Kharkov, Ukrainian S S R = Trudy nauchno-issledovatel'skogo instituta biologii. Pratsy Naukovo-doslidnogo instytutu biologii. Kharkov.

Trudy naučno-issledovatel'skogo instituta biologii. Praci Naukovo-doslidnogo instytutu biologiji = Trudy nauchno-issledovatel'skogo instituta biologii. Pratsy Naukovo-doslidnogo instytutu biologii. Kharkov.

Trudy naučno-issledovatel'skogo instituta botaniki, Kharkivs'kogo derzhavnogo universiteti imeni O. M. Gor'kogo. Kharkov = Trudy nauchno-issledovatel'skogo instituta botaniki, Kharkivs'kogo derzhavnogo universiteti imeni O. M. Gor'kogo. Kharkov.

Trudy Naučno-issledovatel'skogo instituta pri Voronežskom gosudarstvennom universitete = Trudy Nauchno-issledovatel'skogo instituta pri Voronezhskom gosudarstvennom universitete. Voronezh.

Trudy Naučno-issledovatel'skogo torfjanogo instituta (Instorfa) = Trudy Nauchno-issledovatel'skogo torfyanogo instituta (Instorfa). Moscow & Leningrad..

Trudy Naučno-issledovatel'skogo zoologo-biologičeskogo instituta = Pratsy Naukovo-doslidnogo zoologo-biologichnogo instytutu. Kiev.

Trudy Obščestva Estestvoispytatelej pri Imperatorskom Jur'evskom Universitetě = Trudy Obshchestva Estestvoispytatelei pri Imperatorskom Yur'evskom Universitetě. Moscow.

Trudy Obščestva Estestvoispytatelej pri Imperatorskom Kazanskom Universitetě = Trudy Obshchestva Estestvoispytatelei pri Imperatorskom Kazanskom Universitetě. Kazan.

Trudy Obščestva Estestvoispytatelej pri Imperatorskom Varšavskom Universitetě = Trudy Obshchestva Estestvoispytatelei pri Imperatorskom Varshavskom Universitetě. Warsaw.

Trudy Obščestva estestvoispytatelej pri Kazanskom gosudarstvennom universitete = Trudy Obshchestva estestvoispytatelei pri Kazanskom gosudarstvennom universitete. Kazan.

Trudy Obščestva estestvoispytatelej pri Kazanskom gosudarstvennom universitete imeni V. I. Ul'janova-Lenina = Trudy Obshchestva estestvoispytatelei pri Kazanskom gosudarstvennom universitete imeni V. I. Ul'yanova-Lenina. Kazan.

Trudy Obščestva Estestvoispytatelej i Vračej pri Imperatorskom Tomskom Universitetě = Trudy Obshchestva Estestvoispytatelei i Vrachei pri Imperatorskom Tomskom Universitetě. Tomsk.

Trudy Obščestva Ispytatelej Prirody pri Imperatorskom Har'kovskom Universitetě = Trudy Obshchestva Ispytatelei Prirody pri Imperatorskom khar'kovskom Universitetě. Kharkov.

Trudy Obščestva izučenija Orenburgskogo kraja = Trudy Obshchestva izucheniya Orenburgskogo kraya. Orenburg.

Trudy Obshchestva Estestvoispytatelei pri Imperatorskom Yur'evskom Universitetě. Moscow. Vols. 10-23, 1902-16. Trudy Obshch. Estestvoisp. Imp. Yur'evsk. Univ. Preceded by: Schriften, herausgegeben von der Naturforscher-Gesellschaft bei der Universität Dorpat. Superseded by: Tartu ülikooli juures oleva loodusuurijate seltsi kirjatööd. 5-4154-2. HI 73903

Trudy Obshchestva Estestvoispytatelei pri Imperatorskom Kazanskom Universitetě. Kazan, Russian S F S R. Vols. 1-49(1) [vol. 48(5) published in 1917], 1871-1916. Trudy Obshch. Estestvoisp. Imp. Kazansk. Univ. Superseded by: Trudy Obshchestva estestvoispytatelei pri Kazanskom gosudarstvennom universitete. 3-2276-3. HI 73904

Trudy Obshchestva Estestvoispytatelei i Vrachei pri Imperatorskom Tomskom Universitetě. Tomsk, Russian S F S R. 1913-15. Trudy Obshch. Estestvoisp. Imp. Tomsk. Univ. Preceded by: Protokoly Obshchestva Estestvoispytatelei i Vrachei pri Imperatorskom Tomskom Universitetě. HI 73905

Trudy Obshchestva Estestvoispytatelei pri Imperatorskom Varshavskom Universitetě. Warsaw. Vols. 1-10, 1889-1900; vol. 16/19, 1910. Trudy Obshch. Estestvoisp. Imp. Varshavsk. Univ. For vols. 11-12 see: Protokoly Zasědanii i Trudy Obshchestva Estestvoispytatelei pri Imperatorskom Varshavskom Universitetě. For vols. 13-15 see: Trudy i Protokoly Zasědanii Obshchestva Estestvoispytatelei pri Imperatorskom Varshovskom Universitetě. 5-4444-3. HI 73906

Trudy Obshchestva estestvoispytatelei pri Kazanskom gosudarstvennom universitete. Kazan, Russian S F S R. Vols. 49(3)-57 [vol. 49(2) not published], 1818-47. Trudy Obshch. Estestvoisp. Kazansk. Gosud. Univ. Preceded by: Trudy Obshchestva Estestvoispytatelei pri Imperatorskom Kazanskom Universitetě. Superseded by: Trudy Obshchestva estestvoispytatelei pri Kazanskom gosudarstvennom universitete imeni V. I. Ul'yanova-Lenina. 3-2276-3. HI 73907

Trudy Obshchestva estestvoispytatelei pri Kazanskom gosudarstvennom universitete imeni V. I. Ul'yanova-Lenina. Kazan, Russian S F S R. Vol. 58+, 1951+. Trudy Obshch. Estestvoisp. Kazansk. Gosud. Univ. Ul'yanova-Lenina. Preceded by: Trudy Obshchestva estestvoispytatelei pri Kazanskom gosudarstvennom universitete. HI 73908

Trudy Obshchestva Ispytatelei Prirody pri Imperatorskom khar'kovskom Universitetě. Kharkov, Ukrainian S S R. Vols. 1-49, 1870-1918. Trudy Obshch. Isp. Prir. Imp. Khar'kovsk. Univ. 4-3130-3. HI 73909

Trudy Obshchestva izucheniya Orenburgskogo kraya. Orenburg [=Chkalov], Russian S F S R. Vol. 27 [also numbered vol. 2], 1930. Trudy Obshch. Izuch. Orenburgsk. Kraya. Preceded by: Trudy Orenburgskogo otdela Gosudarstvennogo geograficheskogo obshchestva. 2-1692-2. HI 73910

Trudy Odesskogo gosudarstvennogo universiteta imeni I. I. Mečnikova = Pratsy Odes'kogo derzhavnogo universytetu imeny I. I. Mechnikova. Odessa.

Trudy omskogo instituta sel'skogo khozyaistva imeni S. M. Kirova. Omsk. Vol. 1(1), 1935. Trudy Omsk. Inst. Sel'sk. Khoz. S. M. Kirova. Preceded by: Trudy sibirskogo instituta sel'skogo khozyaistva i lesovodstva. HI 73911

Trudy Opytnyh Lěsničestv = Trudy Opytnykh Lěsnichestv. St. Petersburg.

Trudy Opytnyh Stancij pri Moskovskom Sel'skohozjajstvennom Institutě: Selekcionnaja Stancija = Trudy Opytnykh Stantsii pri Moskovskom Sel'skokhozyaistvennom Institutě: Selektsionnaya Stantsiya. Moscow.

Trudy Opytnykh Lěsnichestv. St. Petersburg. 1900-01; vols. 1-4, 1902-16. Trudy Opytn. Lěsnich. Superseded by: Trudy po lěsnomu opytnomu dělu v Rossii HI 73912

Trudy Opytnykh Stantsii pri Moskovskom Sel'skokhozyaistvennom Institutě: Selektsionnaya Stantsiya. Moscow. Vols. 2-6, 1914-16. Trudy Opytn. Stantsii Moskovsk. Sel'skokhoz. Inst. Selektsion. Stantsiya. Preceded by: Trudy Selektsionnoi Stantsii pri Moskovskom Sel'skokhozyaistvennom Institutě. 3-2766-1. HI 73913

Trudy Orenburgskogo otdela Gosudarstvennogo geografičeskogo obščestva = Trudy Orenburgskogo otdela Gosudarstvennogo geograficheskogo obshchestva. Orenburg.

Trudy Orenburgskogo otdela Gosudarstvennogo geograficheskogo obshchestva. Orenburg [=Chkalov], Russian S F S R. Vol. 26 [also numbered vol. 1], 1928. Trudy Orenburgsk. Otd. Gosud. Geogr. Obshch. Preceded by: Izvestiya Orenburgskago Otděla Imperatorskago Russkago Geograficheskago Obshchestva. Superseded by: Trudy Obshchestva izucheniya Orenburgskogo kraya. 2-1692-2 HI 73914

Trudy, otdel fiziologii i biofiziki rastenii, akademiya nauk Tadzhikskoi S S R. Dushanbe. 1962+. Trudy Otd. Fiziol. Biofiz. Rast. Akad. Nauk Tadzhiksk. S.S.R. HI 73915

Trudy Otdela rastitle'nyh resursov. Makhachkala, Russian S F S R = Trudy Otdela rastitle'nykh resursov. Makhachkala, Russian S F S R. Makhachkala.

Trudy Otdela rastitle'nykh resursov. Makhachkala, Russian S F S R. Vol. 2, 1960. Trudy Otd. Rastitel'n. Resursov. Preceded by: Trudy Sektorov botaniki i pochvovedeniya. HI 73916

Trudy otdela sel'sko-khozyaistvennoi mikrobiologii, gosudarstvennyi institut opytnoi agronomii. Leningrad. Vols. 1-4(1), 1926-29. Trudy Otd. Sel'skokhoz. Mikrobiol. Gosud. Inst. Opytn. Agron. Superseded by: Trudy instituta sel'skokhozyaistvennoi mikrobiologii. HI 73917

Trudy pamirskoi biologicheskoi stantsii. Botanicheskogo instituta, akademiya nauk tadzikskoi S S R. Dushanbe. Vol. 19+, 1962+. Trudy Pamirsk. Biol. Stantsii. Preceded by: Trudy, botanicheskogo instituta, akademiya nauk tadzhikskoi S S R. HI 73918

Trudy Permskogo biologičeskogo naučno-issledovatel'skogo instituta = Trudy Permskogo biologicheskogo nauchno-issledovatel'skogo instituta. Perm.

Trudy Permskogo biologicheskogo nauchno-issledovatel'skogo instituta. Perm, Russsian S F S R. Vols. 3(3)-5, 1930-33. Trudy Permsk. Biol. Nauchno-Issl. Inst. Preceded by: Trudy biologicheskogo nauchno-issledovatel'skogo instituta i biologicheskoi stantsii pri Permskom gosudarstvennom universitete. Superseded by: Trudy biologicheskogo nauchno-issledovatel'skogo instituta pri Permskom gosudarstvennom universitete. 4-3310.2. HI 73919

Trudy, pervomaiskaya sveklovichnaya opytno-selektsionnaya stantsiya. Krasnodar. Vol. 1(5)+, 1964+. Trudy Pervvomaisk. Sveklov. Opytn.-Selekts. Stantsiya. Preceded by: Trudy pervomaiskoi zonal'noi opytno-selektsionnoi. HI 73920

Trudy pervomaiskoi zonal'noi opytno-selektsionnoi. Krasnodar. Vols. ?-4, 19??-? Trudy Pervomaisk. Zonal'n. Opytn.-Selekts. Superseded by: Trudy, pervomaiskaya sveklovichnaya opytno-selektsionnaya stantsiya. HI 73921

Trudy Petergofskogo biologičeskogo instituta = Trudy Petergofskogo biologicheskogo instituta. Moscow.

Trudy Petergofskogo biologičeskogo instituta Leningradskogo gosudarstvennogo universiteta = Trudy Petergofskogo biologicheskogo instituta Leningradskogo gosudarstvennogo universiteta. Leningrad.

Trudy Petergofskogo biologicheskogo instituta. Moscow. Vols. 9-13/14, 1932-35; vols. 16-17, 1938-39. Trudy Petergofsk. Biol. Inst. Preceded by: Trudy Petergofskogo estestvenno-nauchnogo instituta. For vol. 15 see: Trudy Petergofskogo biologicheskogo instituta Leningradskogo gosudarstvennogo universiteta. 4-3318-3. HI 73922

Trudy Petergofskogo biologicheskogo instituta Leningradskogo gosudarstvennogo universiteta. Leningrad. Vol. 15, 1935. Trudy Petergofsk. Biol. Inst. Leningradsk. Gosud. Univ. Preceded & superseded by: Trudy Petergofskogo biologicheskogo instituta. 4-3318-3. HI 73923

Trudy Petergofskogo estestvenno-nauchnogo instituta. Leningrad. Vols. 1-8, 1925-32. Trudy Petergofsk. Estestv.-Nauchn. Inst. Superseded by: Trudy Petergofskogo biologicheskogo instituta. 4-3318-3. HI 73924

Trudy Petergofskogo estestvenno-naučnogo instituta = Trudy Petergofskogo estestvenno-nauchnogo instituta. Leningrad.

Trudy Petrogradskogo obščestva estestvoispytatelej. Vypusk 3... Otdelenie botaniki = Trudy Petrogradskogo obshchestva estestvoispytatelei.

Vypusk 3... Otdelenie botaniki. Petrograd.

Trudy Petrogradskogo obshchestva estestvoispytatelei. Vypusk 3... Otdelenie botaniki. Petrograd. Vol. 47/53, 1924. Trudy Petrogradsk. Obshch. Estestvoisp., Vyp. 3, Otd. Bot. Preceded by: Trudy imperatorskago Petrogradskago obshchestva estestvoispytatelei. Vypusk 3, otdělenie botaniki. Superseded by: Trudy Leningradskogo obshchestva estestvoispytatelei. Vypusk. 3: Otdelenie botaniki. 3-2394-3. HI 73925

Trudy, petrovskaya selektsionno-opytnaya stantsiya. Vol. ?-4+, ?-1971+. Trudy Petrovsk. Selekts.-Opytn. Stantsiya. HI 73926

Trudy plodoovoshchnoi opytno-selektsionnoi stantsii. Moscow. Vols. 1-4, 1956-6? Trudy Plodoovoshch. Opytn.-Selekts. Stantsii Superseded by: Trudy, krymskaya opytno-selektsionnnnaya stantsiya. HI 73927

Trudy plovuchego Morskogo nauchnogo instituta. Moscow. Vols. 1-2, 1923-27. Trudy Plovuch. Morsk. Nauchn. Inst. Superseded by: Trudy morskogo nauchnogo instituta. HI 73928

Trudy Pochvenno-botanicheskikh Ekspeditsii po Izslědovaniyu Kolonizatsionnykh Raionov Aziatskoi Rossii. Chast′ 2. Botanicheskiya izslědovaniya. St. Petersburg. Vols. 1-12, 19??-12. [vol. 8 published in 1913]. Trudy Pochv.-Bot. Eksped. Izsl. Kolon. Raionov Aziatsk. Rossii, Chast′ 2, Bot. Izsl. HI 73929

Trudy Počvenno-botaničeskih Ekspedicii po Izslědovaniju Kolonizacionnyh Rajonov Aziatskoj Rossii. Cast′ 2. Botaničeskija izslědovanija = Trudy Pochvenno-botanicheskikh Ekspeditsii po Izslědovaniyu Kolonizatsionnykh Raionov Aziatskoi Rossii. Chast′ 2. Botanicheskiya izslědovaniya. St. Petersburg.

Trudy Poljarnoj komissii = Trudy Polyarnoi komissii. Moscow & Leningrad.

Trudy Polyarnoi komissii. Moscow & Leningrad. Vols. 1-31 [vol. 23 not published], 1930-37. Trudy Polyarn. Komiss. 1-120-1. HI 73930

Trudy Presnovodnoi Biologicheskoi Stantsii Imperatorskago S.-Peterburgskago Obshchestva Estestvoispytatelei. St. Petersburg. Vols. 1-3, 1901-12. Trudy Presnovodn. Biol. Stantsii Imp. S.-Peterburgsk. Obshcha. Estestvoisp. Superseded by: Trudy Borodinskoi biologicheskoi stantsii Petrogradskago obshchestva estestvoispytatelei. 3-2394-3. HI 73931

Trudy Presnovodnoj Biologičeskoj Stancii Imperatorskago S.-Peterburgskago Obščestva Estestvoispytatelej = Trudy Presnovodnoi Biologicheskoi Stantsii Imperatorskago S.-Peterburgskago Obshchestva Estestvoispytatelei. St. Petersburg.

Trudy po prikladnoi botanike, genetike i selektsii. Moscow & Leningrad. Vols. 17(3)-27. 1927-31; vol. 28(3)-?, 1950-?; Prilož. vols. 30-84, 192?-37 [published in series and later suspended, from 1932-49; vol. 28(1)-28(2) and Prilož. vols. 57, 59 & 82 not published. Trudy Prikl. Bot. Preceded by: Trudy po prikladnoi botanike, genetike i selektsii. Seriya A. Sotsialisticheskoe rastenievodstva. Superseded by: Sbornik nauchnykh trudov po prikladnoi botanike, genetike i selektsii. 5-4267-3. HI 73932

Trudy po prikladnoi botanike i selektsii. Petrograd. Vols. 11(5/6)-17(2) [vol. 11(1-4) not published], 1918-27. Trudy Prikl. Bot. Selekts. Preceded by: Trudy byuro po prikladnoi botanikě. Superseded by: Trudy po prikladnoi botanike, genetike i selektsii. 5-4267-3. HI 73933

Trudy po prikladnoi botanike, genetike i selektsii. Seriya A. Sotsialisticheskoe rastenievodstva. Moscow & Leningrad. Vols. 1-21, 1932-37. Trudy Prikl. Bot., Ser. A, Sotsialist. Rasteniev. Preceded & superseded by: Trudy po prikladnoi botanike, genetike i selektsii. 5-4267-3. HI 73934

Trudy po prikladnoi botanike, genetike i selektsii. Seriya 1. Sistematika, ekologiya i geografii rastenii. Moscow & Leningrad. Vol. 1, 1933. Trudy Prikl. Bot., Ser. 1, Sist. Rast. Preceded by: Trudy po prikladnoi botanike, genetike i selektsii. Superseded by: Trudy po prikladnoi botanike, genetike i selektsii. Seriya 1. Sistematika, ekologiya i geografii rastenii i obshchie voprosy rastenievodstva. 5-4268-1. HI 73935

Trudy po prikladnoi botanike, genetike i selektsii. Seriya 1. Sistematika, ekologiya i geografii rastenii i obshchie voprosy rastenievodstva. Moscow & Leningrad. Vol. 2, 1937. Trudy Prikl. Bot., Ser. 1, Sist. Rast. Obshchie Vopr. Rasteniev. Preceded by: Trudy po prikladnoi botanike, genetike i selektsii. Seriya 1. Sistematika, ekologiya i geografii rastenii. Superseded by: Trudy po prikladnoi botanike, genetike i selektsii. 5-4268-1. HI 73936

Trudy po prikladnoi botanike, genetike i selektsii. Seriya 2. Genetika selektsiya i tsitologiya rastenii. Moscow & Leningrad. Vols. 1-11, 1932-37. Trudy Prikl. Bot., Ser. 2, Genet. Rast. Preceded & superseded by: Trudy po prikladnoi botanike, genetike i selektsii. 5-4268-1. HI 73937

Trudy po prikladnoi botanike, genetike i selektsii. Seriya 3. Fiziologiya, biokhimiya i anatomiya rastenii. Moscow & Leningrad. Vols. 1-15, 1933?-36. Trudy Prikl. Bot., Ser. 3, Fiziol. Rast. Preceded & superseded by: Trudy po prikladnoi botanike, genetike i selektsii. 5-4268-1. HI 73938

Trudy po prikladnoi botanike, genetike i selektsii. Seriya 4. Semenovedenie i semennoi kontrol′. Moscow & Leningrad. Vols. 1-2, 1936-37. Trudy Prikl. Bot., Ser. 4, Semenov. Preceded & superseded by: Trudy po prikladnoi botanike, genetike i selektsii. 5-4268-1. HI 73939

Trudy po prikladnoi botanike, genetike i selektsii. Seriya 5. Zernovye kul'tury. Moscow & Leningrad. Vols. 1-2, 1932-34. Trudy Prikl. Bot., Ser. 5, Zernov. Kul't. Preceded by: Trudy po prikladnoi botanike, genetike i selektsii. Superseded by: Trudy po prikladnoi botanike, genetike i selektsii. Seriya 5a. Pshenica. 5-4268-1. HI 73940

Trudy po prikladnoi botanike, genetike i selektsii. Seriya 5a. Pshenica. Moscow & Leningrad. Vols. 1-3, 1935-37. Trudy Prikl. Bot., Ser. 5a, Pshenica. Preceded by: Trudy po prikladnoi botanike, genetike i selektsii. Seriya 5. Zernovye kul'tury. Superseded by: Trudy po prikladnoi botanike, genetike i selektsii. 5-4268-1. HI 73941

Trudy po prikladnoi botanike, genetike i selektsii. Seriya 6. Ovoshchnye kul'tury. Moscow & Leningrad. 1 vol., 1936. Trudy Prikl. Bot., Ser. 6, Ovoshchn. Kul't. Preceded & superseded by: Trudy po prikladnoi botanike, genetike i selektsii. 5-4268-1. HI 73942

Trudy po prikladnoi botanike, genetike i selektsii. Seriya 7. Kormovye kul'tury. Moscow & Leningrad. 1 vol., 1934. Trudy Prikl. Bot., Ser. 7, Kormov. Kul't. Preceded & superseded by: Trudy po prikladnoi botanike, genetike i selektsii. 5-4268-1. HI 73943

Trudy po prikladnoi botanike, genetike i selektsii. Seriya 8. Plodovye i yagodnye kul'tury. Moscow & Leningrad. Vols. 1-6, 1932-36. Trudy Prikl. Bot., Ser. 8, Plodovye Yagodnye. Kul't. Preceded & superseded by: Trudy po prikladnoi botanike, genetike i selektsii. 5-4268-1. HI 73944

Trudy po prikladnoi botanike, genetike i selektsii. Seriya 9. Tekhnicheskie kul'tury. Moscow & Leningrad. Vols. 1-2, 1932. Trudy Prikl. Bot., Ser. 9, Tekhn. Kul't. Preceded & superseded by: Trudy po prikladnoi botanike, genetike i selektsii. 5-4268-2. HI 73945

Trudy po prikladnoi botanike, genetike i selektsii. Seriya 10. Dendrologiya i dekorativnoe sadovodstvo. Moscow & Leningrad. Vols. 1-2, 1933-35. Trudy Prikl. Bot., Ser. 10, Dendrol. Preceded & superseded by: Trudy po prikladnoi botanike, genetike i selektsii. 5-4268-2. HI 73946

Trudy po prikladnoi botanike, genetike i selektsii. Seriya 11. Novye kul'tury i voprosy introduktsii. Moscow & Leningrad. Vols. 1-2, 1936. Trudy Prikl. Bot., Ser. 11, Nov. Kul't. Preceded & superseded by: Trudy po prikladnoi botanike, genetike i selektsii. 5-4268-2. HI 73947

Trudy po prikladnoi botanike, genetike i selektsii. Seriya 13. Referaty i bibliografiya. Moscow & Leningrad. Vols. 1-9, 1933-36. Trudy Prikl. Bot., Ser. 13, Ref. Preceded & superseded by: Trudy po prikladnoi botanike, genetike i selektsii. 5-4268-2. HI 73948

Trudy po prikladnoi botanike, genetike i selektsii. Seriya 14. Osvoenie pustyn'. Moscow & Leningrad. Vols. 1-6, 1933-37. Trudy Prikl. Bot., Ser. 14, Osvoenie Pustyn'. Preceded & superseded by: Trudy po prikladnoi botanike, genetike i selektsii. HI 73949

Trudy po prikladnoi botanike, genetike i selektsii. Seriya 15. Problemy severnogo rastenievodstva. Moscow & Leningrad. Vols. 1-4, 1932-33. Trudy Prikl. Bot., Ser. 15, Probl. Severn. Rasteniev. Preceded by: Trudy po prikladnoi botanike, genetike i selektsii. Superseded by: Trudy po prikladnoi botanike, genetike i selektsii. Seriya 15. Severnoe (pripolyarnoe) zemledelie. 5-4268-2. HI 73950

Trudy po prikladnoi botanike, genetike i selektsii. Seriya 15. Severnoe (pripolyarnoe) zemledelie. Moscow & Leningrad. Vol. 5, 1936. Trudy Prikl. Bot., Ser. 15, Severn. (Pripolyarn.) Zemled. Preceded by: Trudy po prikladnoi botanike, genetike i selektsii. Seriya 15. Problemy severnogo rastenievodstva. Superseded by: Trudy po prikladnoi botanike, genetike i selektsii. 5-4268-2. HI 73951

Trudy po prikladnoj botanike, genetike i selekcii = Trudy po prikladnoi botanike, genetike i selektsii. Moscow & Leningrad.

Trudy po prikladnoj botanike i selekcii = Trudy po prikladnoi botanike i selektsii. Petrograd.

Trudy po prikladnoj botanike, genetike i selekcii. Serija 1. Sistematika, ekologija i geografii rastenij = Trudy po prikladnoi botanike, genetike i selektsii. Seriya 1. Sistematika, ekologiya i geografii rastenii. Moscow & Leningrad.

Trudy po prikladnoj botanike, genetike i selekcii. Serija 1. Sistematika, ekologija i geografii rastenij i obščie voprosy rastenievodstva = Trudy po prikladnoi botanike, genetike i selektsii. Seriya 1. Sistematika, ekologiya i geografii rastenii i obshchie voprosy rastenievodstva. Moscow & Leningrad.

Trudy po prikladnoj botanike, genetike i selekcii. Serija 2. Genetika selekcija i citologija rastenij = Trudy po prikladnoi botanike, genetike i selektsii. Seriya 2. Genetika selektsiya i tsitologiya rastenii. Moscow & Leningrad.

Trudy po prikladnoj botanike, genetike i selekcii. Serija 3. Fiziologija, biohimija i anatomija rastenij = Trudy po prikladnoi botanike, genetike i selektsii. Seriya 3. Fiziologiya, biokhimiya i anatomiya rastenii. Moscow & Leningrad.

Trudy po prikladnoj botanike, genetike i selekcii. Serija 4. Semenovedenie i semennoj kontrol' = Trudy po prikladnoi botanike, genetike i selektsii. Seriya 4. Semenovedenie i semennoi kontrol'. Moscow & Leningrad.

Trudy po prikladnoj botanike, genetike i selekcii. Serija 5. Zernovye kul'tury = Trudy po prikladnoi botanike, genetike i selektsii. Seriya 5. Zernovye kul'tury. Moscow & Leningrad.

Trudy po prikladnoj botanike, genetike i selekcii. Serija 5a. Pšenica = Trudy po prikladnoi botanike, genetike i selektsii. Seriya 5a. Pshenica. Moscow & Leningrad.

Trudy po prikladnoj botanike, genetike i selekcii. Serija 6. Ovoščnye kul'tury = Trudy po prikladnoi botanike, genetike i selektsii. Seriya 6. Ovoshchnye kul'tury. Moscow & Leningrad.

Trudy po prikladnoj botanike, genetike i selekcii. Serija 7. Kormovye kul'tury = Trudy po prikladnoi botanike, genetike i selektsii. Seriya 7. Kormovye kul'tury. Moscow & Leningrad.

Trudy po prikladnoj botanike, genetike i selekcii. Serija 8. Plodovye i jagodnye kul'tury = Trudy po prikladnoi botanike, genetike i selektsii. Seriya 8. Plodovye i yagodnye kul'tury. Moscow & Leningrad.

Trudy po prikladnoj botanike, genetike i selekcii. Serija 9. Tehničeskie kul'tury = Trudy po prikladnoi botanike, genetike i selektsii. Seriya 9. Tekhnicheskie kul'tury. Moscow & Leningrad.

Trudy po prikladnoj botanike, genetike i selekcii. Serija 10. Dendrologija i dekorativnoe sadovodstvo = Trudy po prikladnoi botanike, genetike i selektsii. Seriya 10. Dendrologiya i dekorativnoe sadovodstvo. Moscow & Leningrad.

Trudy po prikladnoj botanike, genetike i selekcii. Serija 11. Novye kul'tury i voprosy introdukcii = Trudy po prikladnoi botanike, genetike i selektsii. Seriya 11. Novye kul'tury i voprosy introduktsii. Moscow & Leningrad.

Trudy po prikladnoj botanike, genetike i selekcii. Serija 13. Referaty i bibliografija = Trudy po prikladnoi botanike, genetike i selektsii. Seriya 13. Referaty i bibliografiya. Moscow & Leningrad.

Trudy po prikladnoj botanike, genetike i selekcii. Serija 14. Osvoenie pustyn' = Trudy po prikladnoi botanike, genetike i selektsii. Seriya 14. Osvoenie pustyn'. Moscow & Leningrad.

Trudy po prikladnoj botanike, genetike i selekcii. Serija 15. Problemy rastenievodstva Krajnego Severa = Trudy po prikladnoi botanike, genetike i selektsii. Seriya 15. Problemy severnogo rastenievodstva. Moscow & Leningrad.

Trudy po prikladnoj botanike, genetike i selekcii. Serija 15. Problemy severnogo rastenievodstva = Trudy po prikladnoi botanike, genetike i selektsii. Seriya 15. Problemy severnogo rastenievodstva. Moscow & Leningrad.

Trudy po prikladnoj botanike, genetike i selekcii. Serija 15. Severnoe (pripoljarnoe) zemledelie = Trudy po prikladnoi botanike, genetike i selektsii. Seriya 15. Severnoe (pripolyarnoe) zemledelie. Moscow & Leningrad.

Trudy po prikladnoj botanike, genetike i selekcii. Serija A. Socialističeskoe rastenievodstva = Trudy po prikladnoi botanike, genetike i selektsii. Seriya A. Sotsialisticheskoe rastenievodstva. Moscow & Leningrad.

Trudy i Protokoly Zasědanii Obshchestva Estestvoispytatelei pri Imperatorskom Varshovskom Universitetě. Warsaw. Vols. 13-15, 1905. Trudy Protok. Zasěd. Obshch. Estestvoisp. Imp. Varshavsk. Univ. Preceded by: Protokoly Zasědanii i Trudy Obshchestva Estestvoispytatelei pri Imperatorskom Varshavskom Universitetě. Superseded by: Trudy Obshchestva Estestvoispytatelei pri Imperatorskom Varshavskom Universitetě. 5-444-3. HI 73952

Trudy i Protokoly Zasědanij Obščestva Estestvoispytatelej pri Imperatorskom Varšovskom Universitetě = Trudy i Protokoly Zasědanii Obshchestva Estestvoispytatelei pri Imperatorskom Varshovskom Universitetě. Warsaw.

Trudy Pryrodnycho-technichnogo viddilu. Kiev, Ukrainian S S R. Vols. 1-7, 1930-31; vols. 9-14, 1931. Trudy Pryr.-Techn. Vidd. Preceded by: Trudy fizychno-matematychnogo viddilu. For vol. 8, 1932, see Pryrodnycho-technichnyi viddil, Vseukrayins'ka akademiya nauk. 1-123-2. HI 73953

Trudy Pryrodnyčo-techničnogo viddilu = Trudy Pryrodnycho-technichnogo viddilu. Kiev.

Trudy, respublikanskaya kormovaya opytnaya stantsiya. Frunze. Vol. 1+, 1970+. Trudy Respubl. Kormov. Opytn. Stantsiya. HI 73954

Trudy Respublikanskogo botaničeskogo sada = Trudy Respublikanskogo botanicheskogo sada. Alma-Ata.

Trudy Respublikanskogo botanicheskogo sada. Alma-Ata, Kazakh S S R. Vol. 1, 1948. Trudy Respubl. Bot. Sada. Superseded by: Trudy Alma-Atinskogo botanicheskofo sada. 1-111-1. HI 73955

Trudy respublikanskoi stantsii zashchity rastenii. Alma-Ata, Kazakh S S R. Vols. 1-3, 1953-56. Superseded by: Trudy nauchno-issledovatel'skogo instituta zashchity rastenii, Kazakhskaya akademiya sel'skokhozyaistvennykh nauk. HI 73956

Trudy Respublikanskoj stancii zaščity rastenij. Alma-Ata, Kazakh S S R = Trudy Respublikanskoi stantsii zashchity rastenii. Alma-Ata.

Trudy Rostovskogo oblastnogo biologičeskogo obščestva = Trudy Rostovskogo oblastnogo biologicheskogo obshchestva. Rostov.

Trudy Rostovskogo oblastnogo biologicheskogo

obshchestva. Rostov, Russian S F S R. Vols. 3-4, 193?-40. Trudy Rostovsk. Obl. Biol. Obshch. Preceded by: Trudy Azovo-Chernomorskogo kraevogo biologicheskogo obshchestva. HI 73957

Trudy Rostovskogo otdelenija Vsesojuznogo botaničeskogo obščestva = Trudy Rostovskogo otdeleniya Vsesoyuznogo botanicheskogo obshchestva. Rostov.

Trudy Rostovskogo otdeleniya Vsesoyuznogo botanicheskogo obshchestva. Rostov, Russian S F S R. 1 vol., 1960. Trudy Rostovsk. Otd. Vsesoyuzn. Bot. Obshch. HI 73958

Trudy Russkago Obščestva Akklimatizacii Životnyh i Rastenij = Trudy Russkago Obshchestva Akklimatizatsii Zhivotnykh i Rastenii. Moscow.

Trudy Russkago Obshchestva Akklimatizatsii Zhivotnykh i Rastenii. Moscow. Vols. 1-6, 1895-99. Trudy Russk. Obshch. Akklim. Zhivotn. Superseded by: Věstnik Russkago Obshchestva Akklimatizatsii Zhivotnykh i Rastenii. 4-3745-2. HI 73959

Trudy sakhalinskaya oblastnaya stantsiya zashchity rastenii. Vol. 1+, 1970+. Trudy Sakhalinsk. Obl. Stantsiya Zashch. Rast. HI 73960

Trudy S.-Peterburgskago Obščestva Estestvoispytatelej = Trudy S.-Peterburgskago Obshchestva Estestvoispytatelei. St. Petersburg.

Trudy S.-Peterburgskago Obščestva Estestvoispytatelej: Otdělenie botaniki = Trudy S.-Peterburgskago Obshchestva Estestvoispytatelei: Otdělenie botaniki. St. Petersburg.

Trudy S.-Peterburgskago Obshchestva Estestvoispytatelei. St. Petersburg. Vols. 1-18, 1870-87. Trudy S.-Peterburgsk. Obshch. Estestvoisp. Superseded by: Trudy S.-Peterburgskago Obshchestva Estestvoispytatelei: Otdělenie botaniki. 3-2394-3. HI 73961

Trudy S.-Peterburgskago Obshchestva Estestvoispytatelei: Otdělenie botaniki. St. Petersburg. Vols. 19-25, 1888-95. Trudy S.-Peterburgsk. Obshch. Estestvoisp., Otd. Bot. Preceded by: Trudy S.-Peterburgskago Obshchestva Estestvoispytatelei. Superseded by: Trudy imperatorskago S.-Peterburgskago obshchestva estestvoispytatelei. Vypusk 3, otdělenie botaniki. 3-2394-3. HI 73962

Trudy. Sahalinskij kompleksnyj naučno-issledovatel'skij institut = Trudy. Sakhalinskii kompleksnyi nauchno-issledovatel'skii institut. Yuzhno-Sakhalinsk

Trudy. Sakhalinskii kompleksnyi nauchno-issledovatel'skii institut. Yuzhno-Sakhalinsk, Russian S F S R. Vol. 9+, 1960+. Trudy Sakhalinsk. Kompl. Nauchno-Issl. Inst. Preceded by: Soobshcheniya Sakhalinskogo kompleksnogo nauchno-issledovatel'skogo instituta. HI 73963

Trudy Sankt-Peterburgskago Obshchestva Estestvoispytatelei = Trudy S.-Peterburgskago Obshchestva Estestvoispytatelei.

Trudy Saratovskago Obščestva Estestvoispytatelej i Ljubitelej Estestvozanija = Trudy Saratovskago Obshchestva Estestvoispytatelei i Lyubitelei Estestvozaniya. Saratov.

Trudy Saratovskago Obshchestva Estestvoispytatelei i Lyubitelei Estestvozaniya. Saratov, Russian S F S R. Vols. 1-12, 1901-29. Trudy Saratovsk. Obshch. Estestvoisp. 5-3783-2. HI 73964

Trudy Saratovskogo obščestva estestvoispytatelej i ljubitelej estestvoznanija = Trudy Saratovskago Obshchestva Estestvoispytatelei i Lyubitelei Estestvozaniya. Saratov.

Trudy Sekcii po mikologii i fitopatologii Russkogo botaničeskogo obščestva = Trudy Sektsii po mikologii i fitopatologii Russkogo botanicheskogo obshchestva. Petrograd.

Trudy sektora agrobotaniki, akademiya nauk Kazakhskoi S S R. Alma-Ata. Vols. 1-8, 1953-60. Trudy Sektora Agrobot. HI 73965

Trudy Sektorov botaniki i pochvovedeniya. Makhachkala, Russian S F S R. Vol. 1, 1950. Trudy Sektorov Bot. Pochvov. Superseded by: Trudy Otdela rastitle'nykh resursov. HI 73967

Trudy Sektorov botaniki i počvovedenija = Trudy Sektorov botaniki i pochvovedeniya. Makhachkala.

Trudy Sektsii po mikologii i fitopatologii Russkogo botanicheskogo obshchestva. Petrograd. 1 vol., 1923. Trudy Sekts. Mikol. Russk. Bot. Obshch. 4-3744-3. HI 73966

Trudy Selekcionnoj Stancii pri Moskovskom Sel'skohozjajstvennom Institutě = Trudy Selektsionnoi Stantsii pri Moskovskom Sel'skokhozyaistvennom Institutě. Moscow.

Trudy po selektsii ovoshchnykh kul'tur. Moscow. Vols. 8-14, 1978-81. Trudy Selekts. Ovoshchn. Kul't. Preceded by: Trudo po semenovodstvu ovoshchnykh kul'tur. Superseded by: Trudy po selektsii i semenovodstvu ovoshchnykh kul'tur. HI 73968

Trudy po selektsii i semenovodstvu ovoshchnykh kul'tur. Moscow. Vol. 15+, 1979+. Trudy Selekts. Semenov. Ovoshchn. Kul't. Preceded by: Trudy po selektsii ovoshchnykh kul'tur. HI 73969

Trudy Selektsionnoi Stantsii pri Moskovskom Sel'skokhozyaistvennom Institutě. Moscow. Vol. 1, 1913. Trudy Selektsion. Stantsii Moskovsk. Sel'skokhoz. Inst. Superseded by: Trudy Opytnykh Stantsii pri Moskovskom Sel'skokhozyaistvennom Institutě: Selektsionnaya Stantsiya. 3-2766-1. HI 73970

Trudy selsko-gospodarskoi botaniki. Kharkov. Vol. 1(1-3), 1926-27. Trudy Selsko-Gosp. Bot. HI 73971

Trudy po semenovodstvu ovoshchnykh kul'tur. Moscow. Vols. ?-3-7, ?-1975-78. Trudy Semenov. Ovoshch. Kul't. Superseded by: Trudy po selektsii ovoshchnykh kul'tur. HI 73972

Trudy Sevanskoi gidrobiologicheskoi stantsii. Tiflis, Georgian S S R. Vol. 5+, 1938+. Trudy Sevansk. Gidrobiol. Stantsii. Preceded by: Trudy Sevanskoi ozernoi stantsii. 1-109-3. HI 73973

Trudy Sevanskoi ozernoi stantsii. Erivan [=Erevan], Armenian S S R. Vols. 1-4, 1927-32. Trudy Sevansk. Ozernoi Stantsii. Superseded by: Trudy Sevanskoi gidrobiologicheskoi stantsii. 1-109-3. HI 73974

Trudy Sevanskoj gidrobiologičeskoj stancii = Trudy Sevanskoi gidrobiologicheskoi stantsii. Tiflis.

Trudy Sevanskoj ozernoj stancii = Trudy Sevanskoi ozernoi stantsii. Erivan.

Trudy, sevastopol'skoi biologicheskoi stantsii (imeni A. O. Kovalevskogo). Moscow, Leningrad. Vols. 1-17, 1929-64. Trudy Sevastopol'sk. Biol. Stantsii. Superseded by: Biologiya morya. 1-120-2. HI 73975

Trudy sevastopol'skoj biologičeskoj stancii (imeni A. O. Kovalevskogo). Moscow, Leningrad = Trudy, sevastopol'skoi biologicheskoi stantsii (imeni A. O. Kovalevskogo). Moscow, Leningrad.

Trudy severnoi nauchno-promyslovoi ekspeditsii V S N Kh. Petrograd. Nos. 1-24, 1920-24. Trudy Severn. Nauchno-Promysl. Eksped. V. S. N. Kh. Superseded by: Trudy nauchno-issledovatel'skogo instituta po izucheniyu severa. HI 73976

Trudy Severo-Kavkazskogo instituta zaščity rastenij = Trudy Severo-Kavkazskogo instituta zashchity rastenii. Rostov.

Trudy Severo-Kavkazskoj associacii naučno-issledovatel'skih institutov = Trudy Severo-Kavkazskoi assotsiatsii nauchno-issledovatel'skikh institutov. Krasnodar.

Trudy Severo-Kavkazskoi assotsiatsii nauchno-issledovatel'skikh institutov. Krasnodar, Russian S F S R. Vols. 1-90, 1926-30. Trudy Severo-Kavkazsk. Assoc. Nauchno-Issl. Inst. 5-3854-1. HI 73977

Trudy Severo-Kavkazskogo instituta zashchity rastenii. Rostov, Russian S F S R. Vol. 8 [also numbered vol. 1], 1932/33. Trudy Severo-Kavkazsk. Inst. Zashch. Rast. Preceded by: Izvestiya Severo-Kavkazskoi kraevoi stantsii zashchity rastenii. Superseded by: Izvestiya Rostovskoi stantsii zashchity rastenii. 4-3711-1. HI 73978

Trudy severo-vostochnogo kompleksnogo nauchno-issledovatel'skogo instituta. Magadan. Vol. 1+, 1962+. Trudy Severo-Vost. Kompl. Nauchno-Issl. Inst. HI 73979

Trudy S''ězda Russkih Estestoispytatelej = Trudy S''ězda Russkikh Estestoispytatelei. Kazan.

Trudy S''ězda Russkikh Estestoispytatelei. Kazan, Russian S F S R. Vol. 4, 1875. Trudy S''ězda Russk. Estestvoisp. HI 73980

Trudy sibirskogo instituta sel'skogo khozyaistva i lesovodstva. Omsk. Vols. 7-13, 1927-34. Trudy Sibirsk. Inst. Sel'sk. Khoz. Lesov. Preceded by: Trudy sibirskoi sel'skokhozyaistvennoi akademii. Superseded by: Trudy omskogo instituta sel'skogo khozyaistva imeni S. M. Kirova. HI 73981

Trudy sibirskoi sel'skokhozyaistvennoi akademii. Omsk. Vols. 1-6, 1922-26. Trudy Sibirsk. Sel'skokhoz. Akad. Superseded by: Trudy sibirskogo instituta sel'skogo khozyaistva i lesovodstva. HI 73982

Trudy Sihote-Alinskogo gosudarstvennogo zapovednika = Trudy Sikhote-Alinskogo gosudarstvennogo zapovednika. Moscow.

Trudy Sikhoté-Alinskogo gosudarstvennogo zapovednika. Moscow. Vols. 1-2, 1938. Trudy Sikhote-Alinsk. Gosud. Zapov. HI 73983

Trudy sil's'ko-gospodars'koi botaniky. Kharkov. Ukrainian S S R. Vols. 1-2, 1926-29. Trudy Sil's'ko-Gosp. Bot. HI 73984

Trudy sil's'ko-gospodars'koji botaniky = Trudy sil's'ko-gospodars'koi botaniky. Kharkov.

Trudy sil's'ko-hospodars'koyi botaniky = Trudy sil's'ko-gospodars'koi botaniky. Kharkov.

Trudy Smolenskogo obščestva estestvoispytatelej i vračej pri Smolenskom gosudarstvennom universitete = Trudy Smolenskogo obshchestva estestvoispytatelei i vrachei pri Smolenskom gosudarstvennom universitete. Smolensk.

Trudy Smolenskogo obshchestva estestvoispytatelei i vrachei pri Smolenskom gosudarstvennom universitete. Smolensk, Russian S F S R. Vols. 1-4, 1926-30. Trudy Smolensk. Obshch. Estestvoisp. Smolensk. Gosud. Univ. 5-3889-3. HI 73985

Trudy soveta po izučeniju prirodnyh resursov. Serija dal'ne-vostočnaja = Trudy soveta po izucheniyu prirodnykh resursov. Seriya dal'ne-vostochnaya. Moscow & Leningrad.

Trudy soveta po izučeniju prirodnyh resursov. Serija jakutskaja = Trudy soveta po izucheniyu prirodnykh resursov. Seriya yakutskaya. Moscow & Leningrad.

Trudy soveta po izučeniju prirodnyh resursov. Serija sibirskaja = Trudy soveta po izucheniyu prirodnykh resursov. Seriya sibirskaya. Moscow & Leningrad.

Trudy soveta po izučeniju prirodnyh resursov. Serija turkmenskaja = Trudy soveta po izucheniyu prirodnykh resursov. Seriya turkmenskaya. Moscow & Leningrad.

Trudy soveta po izučeniju proizvoditel'nyh sil. Serija dal'nevostočnaja = Trudy soveta po izucheniyu prirodnykh resursov. Seriya dal'ne-vostochnaya. Moscow & Leningrad.

Trudy soveta po izučeniju proizvoditel'nyh sil. Serija jakutskaja = Trudy soveta po izucheniyu prirodnykh resursov. Seriya yakutskaya. Moscow & Leningrad.

Trudy soveta po izučeniju proizvoditel'nyh sil. Serija sibirskaja = Trudy soveta po izucheniyu prirodnykh resursov. Seriya sibirskaya. Moscow & Leningrad.

Trudy soveta po izučeniju proizvoditel'nyh sil. Serija turkmenskaja = Trudy soveta po izucheniyu prirodnykh resursov. Seriya turkmenskaya. Moscow & Leningrad.

Trudy soveta po izucheniyu prirodnykh resursov. Seriya dal'ne-vostochnaya. Moscow & Leningrad. Vols. 1-3, 1933-34. Trudy Soveta Izuch. Prir. Resursov, Ser. Dal'ne-Vost. 1-120-2. HI 73986

Trudy soveta po izucheniyu prirodnykh resursov. Seriya Yakutskaya. Moscow & Leningrad. Vols. 1-25, 1931-35. Trudy Soveta Izuch. Prir. Resursov, Ser. Yakutsk. 1-120-2. HI 73987

Trudy soveta po izucheniyu prirodnykh resursov. Seriya sibirskaya. Moscow & Leningrad. Vols. 1-24, 1932-37. Trudy Soveta Izuch. Prir. Resursov, Ser. Sibirsk. 1-120-2. HI 73988

Trudy soveta po izucheniyu prirodnykh resursov. Seriya turkmenskaya. Moscow & Leningrad. Vols. 1-9, 1932-34. Trudy Soveta Izuch. Prir. Resursov, Ser. Turkmensk. 1-120-2. HI 73989

Trudy Sredne-Aziatskogo gosudarstvennogo universiteta. Seriya 8b. Botanika. Moscow. Vols. 1-35, 1928-37. Trudy Sredne-Aziatsk. Gosud. Univ., Ser. 8b, Bot. Superseded by: Trudy sredne-Aziatskogo gosudarstvennogo universiteta imeni V. I. Lenina. 5-4156-2. HI 73990

Trudy sredne-Aziatskogo gosudarstvennogo universiteta imeni V. I. Lenina. Tashkent. N.s. vols. 1-112-?, 1945-57-? Trudy Sredne-Aziatsk. Gosud. Univ. Lenina. Preceded by: Trudy Sredne-Aziatskogo gosudarstvennogo universiteta. Seriya 8b. Botanika and Trudy Sredne-Aziatskogo gosudarstvennogo universiteta. Seriya 8c, ekologiya. Superseded by: Nauchnye trudy Tashkentskii gosudarstvennyi universitet imeni V. I. Lenina. HI 73991

Trudy sredne-Aziatskogo gosudarstvennogo universiteta. Seriya 8c, ekologiya. Moscow. Vols. 1-2, 1933-39. Trudy Sredne-Aziatsk. Gosud. Univ., Ser. 8C. Superseded by: Trudy sredne-Aziatskogo gosudarstvennogo universiteta imeni V. I. Lenina. HI 73992

Trudy sredne-Aziatskoi nauchno-issledovatel'skoi opytnoi stantsii efirno-maslichnykh rastenii O M P K. Tashkent. Vols. 1-6, 1932-35. Trudy Sredne-Aziatsk. Nauchno-Issl. Opytn. Stantsii Efirno-Masl. Rast. O. M. P. K. HI 73993

Trudy Sredneaziatskoi opytnoi stantsii Vsesoyuznogo instituta rastenievodstva. Tashkent, Uzbek S S R. 1 vol., 1963. Trudy Sredneaziatsk. Opytn. Stantsii Vsesoyuzn. Inst. Rasteniev. HI 73994

Trudy Sredneaziatskoj opytnoj stancii Vsesojuznogo instituta rastenievodstva = Trudy Sredneaziatskoi opytnoi stantsii Vsesoyuznogo instituta rastenievodstva. Tashkent.

Trudy Stavropol'skago Obščestva dlja Izučenija Severo-Kavkazskago Kraja v estestvnno-istoričeskom, geografičeskom i antropologičeskom Otnošenijah = Trudy Stavropol'skago Obshchestva dlya Izucheniya Severo-Kavkazskago Kraya v estestvnno-istoricheskom, geograficheskom i antropologicheskom Otnosheniyakh. St. Petersburg.

Trudy Stavropol'skago Obshchestva dlya Izucheniya Severo-Kavkazskago Kraya v estestvenno-istoricheskom, geograficheskom i antropologicheskom Otnosheniyakh. St. Petersburg. Vols. 1-3, 1911-14. Trudy Stavropol'sk. Obshch. Izuch. Severo-Kavkazsk. Kraya. HI 73995

Trudy Studenčeskih Kružkov Fiziko-matematičeskago Fakul'teta Imperatorskago S.-Peterburgskago Universiteta = Trudy Studencheskikh Kruzhkov Fiziko-matematicheskago Fakul'teta Imperatorskago S.-Peterburgskago Universiteta. St. Petersburg.

Trudy Studencheskikh Kruzhkov Fiziko-matematicheskago Fakul'teta Imperatorskago S.-Peterburgskago Universiteta. St. Petersburg. 1 vol., 1909-13. Trudy Stud. Kruzhk. Fiz.-Mat. Fak. Imp. S.-Peterburgsk. Univ. HI 73996

Trudy Suhumskogo botaničeskogo sada = Trudy Sukhumskogo botanicheskogo sada. Sukhumi.

Trudy, sukhumskaya opytnaya stantsiya efirnomaslichnykh kul'tur. Sukhumi. Vol. ?-3+, ?-1969+. Trudy Sukhumsk. Opytn. Stantsiya Efirnomasl. Kul't. HI 73997

Trudy, sukhumskaya opytnaya stantsiya subtropicheskikh kul'tur. Sukhumi. Vol. 1+, 1967+. Trudy Sukhumsk. Opytn. Stantsiya Subtrop. Kul't. HI 73998

Trudy Sukhumskogo botanicheskogo sada. Sukhumi, Georgian S S R. Vol. 6+ [vols. [1]-[4] (1948-50) published as monographs; data for vol. 5 not available], 1951+. Trudy Sukhumsk. Bot. Sada. HI 73999

Trudy, sverdlovskii sel'skokhyaistvennii institut. Sverdlovsk. Vol. 1+, 1957+. Trudy Sverdlovsk. Sel'skokh. Inst. HI 74000

Trudy Tadzhikistanskoi bazy. Moscow & Leningrad. Vols. 8-11, 193?-40. Trudy Tadzhikistansk. Bazy. Preceded by: Trudy Tadzhikskoi bazy. Superseded by: Trudy Tadzhikskoi filial Akademii nauk S S S R. 1-120-3. HI 74001

Trudy Tadzhikistanskoi filiala. Moscow & Leningrad. Vols. 14-16, 1935-45. Trudy Tadzhikistansk. Fil. Preceded & superseded by: Trudy Tadzhikskoi filial Akademii nauk S S S R. 1-120-3. HI 74002

Trudy tadzhikskogo botanicheskogo sada = Acta horti botanici tadshikistanici. Moscow. Acta Horti Bot. Tadshik. See B–P–H 43/25.

Trudy Tadzhikskoi bazy. Moscow & Leningrad. Vols. 1-7, 1935-38. Trudy Tadzhiksk. Bazy. Superseded by: Trudy Tadzhikistanskoi bazy. 1-120-3. HI 74003

Trudy Tadzhikskoi filial Akademii nauk S S S R. Moscow & Leningrad. Vol. 13, 1951; vols. 17-25, 194?-49 [data for vol. 12 not available]. Trudy Tadzhiksk. Fil. Akad. Nauk S.S.S.R. Preceded by: Trudy Tadzhikistanskoi bazy. Moscow & Leningrad. For vols. 14-16 see: Trudy Tadzhikistanskoi filiala. 1-120-3. HI 74004

Trudy Tadžikistanskoj bazy = Trudy Tadzhikistanskoi bazy. Moscow & Leningrad.

Trudy Tadžikistanskoj filiala = Trudy Tadzhikistanskoi filiala. Moscow & Leningrad.

Trudy Tadžikskoj bazy = Trudy Tadzhikskoi bazy. Moscow & Leningrad.

Trudy Tadžikskoj filial Akademii nauk S S S R = Trudy Tadzhikskoi filial Akademii nauk S S S R. Moscow & Leningrad.

Trudy tallinskogo botanicheskogo sada = Tallinna botaanikaaia uurimused. Tallinn.

Trudy Tbilisskogo botaničeskogo instituta = Trudy Tbilisskogo botanicheskogo instituta. Tiflis.

Trudy Tbilisskogo botaničeskogo sada = Trudy Tbilisskogo botanicheskogo sada. Tiflis.

Trudy Tbilisskogo botanicheskogo instituta. Tiflis, Georgian S S R. Vols. 2-13, 1938-49; vol. 15+, 1953+. Trudy Tbilissk. Bot. Inst. Preceded by: Trudy Tiflisskogo botanicheskogo instituta. For vol. 14 see: Trudy botanicheskogo instituta. Tiflis. 1-110-3. HI 74005

Trudy Tbilisskogo botanicheskogo sada. Tiflis, Georgian S S R. Vols. 27-38 [also numbered ser. 2, vols. 7-10, and ser. 2, vols. 6-13], 1938-49 [ser. 2, vols. nos. 6-10 used twice]. Trudy Tbilissk. Bot. Sada. Preceded by: Trudy Tiflisskogo Botanicheskogo Sada. HI 74006

Trudy Tiflisskogo botaničeskogo instituta = Trudy Tiflisskogo botanicheskogo instituta. Tiflis.

Trudy Tiflisskogo Botaničeskogo Sada = Trudy Tiflisskogo Botanicheskogo Sada. Tiflis.

Trudy Tiflisskogo botanicheskogo instituta. Tiflis, Georgian S S R. Vol. 1, 1933/34. Trudy Tiflissk. Bot. Inst. Superseded by: Trudy Tbilisskogo botanicheskogo instituta. 1-110-3. HI 74008

Trudy Tiflisskogo Botanicheskogo Sada. Tiflis, Georgian S S R. Vols. 1-20, 1895-1917; vols. 21-26 [also numbered ser. 2, vols. 1-6], 1920-33/34. Trudy Tiflissk. Bot. Sada. Superseded by: Trudy Tbilisskogo botanicheskogo sada. 5-4217-2. HI 74009

Trudy Tomskogo gosudarstvennogo universiteta. Tomsk, R S F S R. Vols. 85-86, 1932-34. Trudy Tomsk. Gosud. Univ. Preceded by: Izvestiya Tomskogo gosudarstvennogo universiteta. Superseded by: Trudy Tomskogo gosudarstvennogo universiteta imeni V. V. Kuibysheva. 5-4234-3. HI 74010

Trudy Tomskogo gosudarstvennogo universiteta imeni V. V. Kujbyševa = Trudy Tomskogo gosudarstvennogo universiteta imeni V. V. Kuibysheva. Tomsk.

Trudy Tomskogo gosudarstvennogo universiteta imeni V. V. Kujbyševa. i Tomskogo gosudarstvennogo pedagogičeskogo instituta = Trudy Tomskogo gosudarstvennogo universiteta imeni V. V. Kuibysheva. i Tomskogo gosudarstvennogo pedagogicheskogo instituta. Tomsk.

Trudy Tomskogo gosudarstvennogo universiteta imeni V. V. Kuibysheva. Tomsk, R S F S R. Vols. 87-97, 1935-4?; vol. 99+, 194?+. Trudy Tomsk. Gosud. Univ. Kuibysheva. Preceded by: Trudy Tomskogo gosudarstvennogo universiteta. For vol. 98, 1947 see: Trudy Tomskogo gosudarstvennogo universiteta imeni V. V. Kuibysheva. i Tomskogo gosudarstvennogo pedagogicheskogo instituta. 5-4234-3. HI 74011

Trudy Tomskogo gosudarstvennogo universiteta imeni V. V. Kuibysheva. i Tomskogo gosudarstvennogo pedagogicheskogo instituta. Tomsk, R S F S R. Vol. 98, 1947. Trudy Tomsk. Gosud. Univ. Kuibysheva Tomsk. Gosud. Pedagog. Inst. Preceded & superseded by: Trudy Tomskogo gosudarstvennogo universiteta imeni V. V. Kuibysheva. 5-4234-3. HI 74012

Trudy Tomskogo Obsččestva Estestvoispytatelej = Trudy Tomskogo Obshchestva Estestvoispytatelei. Tomsk.

Trudy Tomskogo Obshchestva Estestvoispytatelei. Tomsk, R S F S R. 1890-94. Trudy Tomsk. Obshch. Estestvoisp. Superseded by: Protokoly Obshchestva Estestvoispytatelei i Vrachei pri Imperatorskom Tomskom Universitetě. 5-4235-1. HI 74013

Trudy Troickosavsko-Kjahtinskago Otdělenija Priamurskago Otděla Imperatorskago Russkago Geografičeskago Obsččestva = Trudy Troitskosavsko-Kyakhtinskago Otděleniya Priamurskago Otděla Imperatorskago Russkago Geograficheskago Obshchestva. Moscow.

Trudy Troickosavsko-Kjahtinskago Otdělenija Priamurskago Otděla Russkago Geografičeskago Obščestva = Trudy Troitskosavsko-Kyakhtinskago Otdělenjya Priamurskago Otděla Russkago Geograficheskago Obshchestva. Irkutsk.

Trudy Troitskosavsko-Kyakhtinskago Otděleniya Priamurskago Otděla Imperatorskago Russkago Geograficheskago Obshchestva. Moscow. Vols. 1-15, 1898-1914. Trudy Troitskos.-Kyakhtinsk. Otd. Priamursk. Otd. Imp. Russk. Geogr. Obshch. Superseded by: Trudy Troitskosavsko-Kyakhtinskago Otděleniya Priamurskago Otděla Russkago Geograficheskago Obshchestva. 2-1692-3. HI 74014

Trudy Troitskosavsko-Kyakhtinskago Otděleniya Priamurskago Otděla Russkago Geograficheskago Obshchestva. Irkutsk, R S F S R. Vols. 16-17, 19?-22. Trudy Troitskos.-Kyakhtinsk. Otd. Priamursk. Otd. Russk. Geogr. Obshch. Preceded by: Trudy Troitskosavsko-Kyakhtinskago Otděleniya Priamurskago Otděla Imperatorskago Russkago Geograficheskago Obshchestva. HI 74015

Trudy, tsentral'naya geneticheskaya laboratoriya imeni I. V. Michurina. Vol. ?-7+, ?-1961+. Trudy Tsentr. Genet. Lab. I. V. Michurina. HI 74016

Trudy tsentral'nogo sibirskogo botanicheskogo sada. Novosibirsk, Russian S F S R. Vol. 3, 1960+. Trudy Tsentr. Sibirsk. Bot. Sada. Preceded by: Trudy botanicheskogo sada. Novosibirsk. HI 74017

Trudy Turkestanskogo gosudarstvennogo universiteta. Tashkent, Uzbek S S R. Trudy Turkestansk. Gosud. Univ. See B–P–H 922/4. HI 60069

Trudy Turkestanskogo nauchnogo obshchestva. Tashkent, Uzbek S S R. Vol. 1, 1923. Trudy Turkestansk. Nauchn. Obshch. Superseded by: Trudy Turkestanskogo nauchnogo obshchestva pri Sredne-Aziatskom gosudarstvennom universitete. 5-4280-1. HI 74018

Trudy Turkestanskogo nauchnogo obshchestva pri Sredne-Aziatskom gosudarstvennom universitete. Tashkent, Uzbek S S R. Vol. 2, 1925. Trudy Turkestansk. Nauchn. Obshcha. Sredne-Aziatsk. Gosud. Univ. Preceded by: Trudy Turkestanskogo nauchnogo obshchestva. 5-4280-1. HI 74019

Trudy Turkestanskogo naučnogo obščestva = Trudy Turkestanskogo nauchnogo obshchestva. Tashkent.

Trudy Turkestanskogo naučnogo obščestva pri Sredne-Aziatskom gosudarstvennom u universitete = Trudy Turkestanskogo nauchnogo obshchestva pri Sredne-Aziatskom gosudarstvennom u universitete. Tashkent.

Trudy Turkmenskogo botaničeskogo sada = Trudy Turkmenskogo botanicheskogo sada. Ashkhabad.

Trudy Turkmenskogo botanicheskogo sada. Ashkhabad, Turkmen S S R. Vols. 2+, 1956+. Trudy Turkmensk. Bot. Sada. Preceded by: Trudy Turkmenskogo gosudarstvennogo botanicheskogo sada. HI 74020

Trudy Turkmenskogo gosudarstvennogo botaničeskogo sada = Trudy Turkmenskogo gosudarstvennogo botanicheskogo sada. Ashkhabad.

Trudy Turkmenskogo gosudarstvennogo botanicheskogo sada. Ashkhabad, Turkmen S S R. Vol. 1, 1941. Trudy Turkmensk. Gosud. Bot. Sada. Superseded by: Trudy Turkmenskogo botanicheskogo sada. 1-503-3. HI 74021

Trudy Tuvinskoi kompleksnoi ekspeditsii. Moscow. Vols. 1-4, 19??-56 [vols. 3 published in 1957]. Trudy Tuvinsk. Kompl. Eksped. HI 74022

Trudy Tuvinskoj kompleksnoj ekspedicii = Trudy Tuvinskoi kompleksnoi ekspeditsii. Moscow.

Trudy Ukrajins'kogo instytutu prykladnoji botaniky = Trudy Ukrayins'kogo instytutu prykladnoi botaniky. Kharkov.

Trudy Ukrayins'kogo instytutu prykladnoi botaniky. Kharkov, Ukrainian S S R. 1 vol., 1930. Trudy Ukrayins'k. Inst. Prykl. Bot. 5-4292-3. HI 74023

Trudy Uzbekistanskogo filiala akademii nauk S S S R. Seriya 11, botanika. Tashkent. [Dates of publication not ascertained.] Trudy Uzbekistansk. Fil. Akad. Nauk S.S.S.R., Ser. 11, Bot. HI 74024

Trudy Uzbekistanskogo geografičeskogo obščestva = Trudy Uzbekistanskogo geograficheskogo obshchestva. Tashkent.

Trudy Uzbekistanskogo geograficheskogo obshchestva. Tashkent, Uzbek S S R. Vols. 21-22 [also numbered vols. 1-2], 1937-48. Trudy Uzbekistansk. Geogr. Obshch. Preceded by: Izvestiya Sredne-Aziatskogo geograficheskogo obshchestva. 5-4054-2. HI 74025

Trudy, uzbekskii nauchno issledovatel'skii institut risa. Vol. ?-7+, ?-1975+. Trudy Uzbeksk. Nauchn. Issl. Inst. Risa. HI 74026

Trudy, uzbekskii nauchno issledovatel'skii institut zerna. Vol. ?-15+, ?-1978+. Trudy Uzbeksk. Nauchn. Issl. Inst. Zerna. HI 74027

Trudy Uzbekskogo gosudarstvennogo universiteta. Samarkand. Vols. 1-28-?, 1935-41-? Trudy Uzbeksk. Gosud. Univ. Superseded by: Trudy Uzbekiskogo gosudarstvennogo universiteta imena Alishera Navoi. Novaya seriya, biologiya. HI 74028

Trudy Uzbekskogo gosudarstvennogo universiteta imena Alishera Navoi. Novaya seriya, biologiya. Samarkand. Vol. ?-92+, ?-1958+. Trudy Uzbeksk. Gosud. Univ. Alishera Navoi. N.S., Biol. Preceded by: Trudy Uzbekiskogo gosudarstvennogo universiteta. HI 74029

Trudy vladimirskago obshchestva lyubitelei estestvoznaniya. Vladimir'. Vols. 1-4, 1907-14. Trudy

Vladim. Obshch. Lyubit. Estestv. HI 74030

Trudy, volzhsko-Kamskii gosudarstvennyi zapovednik. Kazan'. Vol. 1+, 1968+. Trudy Volzhsko-Kamskii Gosud. Zapov. HI 74031

Trudy Voronezhskogo gosudarstvennogo universiteta. Leningrad. Vols. 1-10, 1925-39; vol. 13+, 1945+. Trudy Voronezhsk. Gosud. Univ. Preceded by: Uchenya Zapiski Imperatorskago Yur'evskago Universiteta. For vols. 11-12 see: Trudy Voronezhskogo gosudarstvennogo universiteta. Botanicheskii otdel. 5-4429-3. HI 74032

Trudy Voronezhskogo gosudarstvennogo universiteta. Botanicheskii otdel. Leningrad. Vol. 11 [vol. 12 not published], 1939. Trudy Voronezhsk. Gosud. Univ., Bot. Otd. Preceded & superseded by: Trudy Voronezhskogo gosudarstvennogo universiteta. 5-4429-3. HI 74033

Trudy Voronezhskoi Stantsii po Bor'bě s Vreditelyami Rastenii. Voronezh, R S F S R. Vol. 1, 1915. Trudy Voronezhsk. Stantsii Bor'bě Vredit. Rast. Superseded by: Izvestiya Voronezhskoi stantsii po bor'be s vreditelyalmi rastenii. HI 74034

Trudy voronezhskogo gosudarstvennogo zapovednika. Moscow. Vol. 1+, 1938+. Trudy Voronezhsk. Gosud. Zapov. HI 74035

Trudy Voronezhskoi stantsii zashchity rastenii. Voronezh, R S F S R. Vols. 12-13, 19??; vol. 15+, 1960+. Trudy Voronezhsk. Stantsii Zashch. Rast. Preceded by: Byulleteni Voronezhskoi stantsii zashchity rastenii. For vol. 14 see: Itogi Voronezhskaya stantsii zashchity rastenii. HI 74036

Trudy Voronežskogo gosudarstvennogo universiteta = Trudy Voronezhskogo gosudarstvennogo universiteta. Leningrad.

Trudy Voronežskogo gosudarstvennogo universiteta. Botaničeskij otdel = Trudy Voronezhskogo gosudarstvennogo universiteta. Botanicheskii otdel. Leningrad.

Trudy Voronežskoj Stancii po Bor'bě s Vrediteljami Rastenij = Trudy Voronezhskoi Stantsii po Bor'bě s Vreditelyami Rastenii. Voronezh.

Trudy Voronežskoj stancii zaščity rastenij = Trudy Voronezhskoi stantsii zashchity rastenii. Voronezh.

Trudy Vostochno-sibirskogo gosudarstvennogo universiteta. Moscow. Vols. 1-2, 1932-34. Trudy Vostochnosibirsk. Gosud. Univ. Preceded by: Sbornik trudov gosudarstvennogo Irkutskogo universiteta. Superseded by: Trudy Irkutskogo gosudarstvennogo universiteta imeni A. A. Zhdanova. Seriya biologicheskaya. 3-2116-3. HI 74037

Trudy Vostochno-Sibirskogo filiala. Moscow. Vol. 1+, 1954+. Trudy Vost.-Sibirsk. Fil. HI 74038

Trudy Vostochno-Sibirskogo Otděla Imperatorskago Russkago Geograficheskago Obshchestva. Irkutsk, R S F S R. Vols. 1-8 [vol. 6 (1911) erroneously numbered vol. 5], 1897-1924. 2-1693-1. Trudy Vost.-Sibirsk. Otd. Imp. Russk. Geogr. Obshch. 2-1693-1. HI 74039

Trudy Vostočno-Sibirskogo filiala = Trudy Vostochno-Sibirskogo filiala. Moscow.

Trudy Vostočno-sibirskogo gosudarstvennogo universiteta = Trudy Vostochno-sibirskogo gosudarstvennogo universiteta. Moscow.

Trudy Vostočno-Sibirskogo Otděla Imperatorskago Russkago Geografičeskago Obščestva = Trudy Vostochno-Sibirskogo Otděla Imperatorskago Russkago Geograficheskago Obshchestva. Irkutsk.

Trudy, vserossiiskogo nauchno-issledovatel'skogo instituta zashchita rastenii. Voronezh. Vol. 1+, 1971+. Trudy Vserosssiisk. Nauchn.-Issl. Inst. Zashch. Rast. HI 74040

Trudy Vsesojuznogo gidrobiologičeskogo obščestva = Trudy Vsesoyuznogo gidrobiologicheskogo obshchestva. Moscow.

Trudy Vsesojuznogo instituta zaščity rastenij = Trudy Vsesoyuznogo instituta zashchity rastenii. Moscow & Leningrad.

Trudy. Vsesojuznogo naučno-issledovatel'skogo instituta lekarstvennyh rastenij = Trudy. Vsesoyuznogo nauchno-issledovatel'skogo instituta lekarstvennykh rastenii. Moscow.

Trudy. Vsesojuznogo naučno-issledovatel'skogo instituta zaščity rastenij = Trudy. Vsesoyuznogo nauchno-issledovatel'skogo instituta zashchity rastenii. Moscow.

Trudy. Vsesojuznogo s''ezda po ohrane prirody v S S S R = Trudy. Vsesoyuznogo s''ezda po okhrane prirody v S S S R. Moscow.

Trudy Vsesojuznoj akademii sel'sko-hozjajstvennyh nauk imeni V. I. Lenina. Serija 1 = Trudy Vsesoyuznoi akademii sel'sko-khozyaistvennykh nauk imeni V. I. Lenina. Seriya 1. Moscow & Leningrad.

Trudy Vsesojuznoj akademii sel'sko-hozjajstvennyh nauk imeni V. I. Lenina. Serija 4. Selekcija rastenij = Trudy Vsesoyuznoi akademii sel'sko-khozyaistvennykh nauk imeni V. I. Lenina. Seriya 4. Selektsiya rastenii. Moscow & Leningrad.

Trudy Vsesojuznoj akademii sel'sko-hozjajstvennyh nauk imeni V. I. Lenina. Serija 6. Tehničeskie kul'tury = Trudy Vsesoyuznoi akademii sel'sko-khozyaistvennykh nauk imeni V. I. Lenina. Seriya 6. Tekhnicheskie kul'tury. Moscow & Leningrad.

Trudy Vsesojuznoj akademii sel'sko-hozjajstvennyh nauk imeni V. I. Lenina. Serija 7. Kormovye kul'tury = Trudy Vsesoyuznoi akademii sel'sko-khozyaistvennykh nauk imeni V. I. Lenina. Seriya 7. Kormovye kul'tury. Moscow & Leningrad.

Trudy Vsesojuznoj akademii sel'sko-hozjajstvennyh nauk imeni V. I. Lenina. Serija 9. Bor'ba s sornjakami = Trudy Vsesoyuznoi akademii sel'sko-khozyaistvennykh nauk imeni V. I. Lenina. Seriya 9. Bor'ba s sornyakami. Moscow & Leningrad.

Trudy Vsesojuznoj akademii sel'sko-hozjajstvennyh nauk imeni V. I. Lenina. Serija 10 = Trudy Vsesoyuznoi akademii sel'sko-khozyaistvennykh nauk imeni V. I. Lenina. Seriya 10. Moscow & Leningrad.

Trudy Vsesojuznoj akademii sel'sko-hozjajstvennyh nauk imeni V. I. Lenina. Serija 17. Bor'ba s vrediteljami sel'skohozjajstvennyh rastenij [later: Sel'skohozjajstvennaja mikrobiologija] = Trudy Vsesoyuznoi akademii sel'sko-khozyaistvennykh nauk imeni V. I. Lenina. Seriya 17. Bor'ba s vreditelyami sel'skokhozyaistvennykh rastenii [later: Sel'skokhozyaistvennaya mikrobiologiya.] Moscow & Leningrad.

Trudy. Vsesojuznyh soveščanij po ohrane prirody = Trudy. Vsesoyuznykh soveshchanii po okhrane prirody. Vilnius.

Trudy. Vsesojuznyj naučno-issledovatel'skij institut lekarstvennyh i aromatičeskih rastenij (V I L A R) = Trudy. Vsesoyuznyi nauchno-issledovatel'skii institut lekarstvennykh i aromaticheskikh rastenii (V I L A R). Moscow.

Trudy, vsesoyuznyi arkticheskii institut. Leningrad. Vols. 81-83, 86-88, 91-101, 103, 106, 108-109, 1937-1938? Trudy Vsesoyuzn. Arktich. Inst. Preceded and superseded by: Trudy arkticheskogo instituta. Leningrad. HI 74041

Trudy. Vsesoyuznyi nauchno-issledovatel'skii institut lekarstvennykh i aromaticheskikh rastenii (V I L A R). Moscow. Vol. 12+, 1959+. Trudy Vsesoyuzn. Nauchno-Issl. Inst. Lekarstv. Aromat. Rast. Preceded by: Trudy. Vsesoyuznogo nauchno-issledovatel'skogo instituta lekarstvennykh rastenii. HI 74042

Trudy, vsesoyuznii nauchno-issledovatel'skii institut mikrobiologicheskikh sredstv zashchity rastenii i bakterial'nykh preparatov. Vol. ?-2+, ?-1973+. Trudy Vsesoyuzn. Inst. Nauchn.Issl. Inst. Mikrobiol. Sredstv. Zashch. Rast. Bakt. HI 74043

Trudy, vsesoyuznii nauchno-issledovatel'skii institut risa. Vols. 1-3, 1971-73. Trudy Vsesoyuzn. Nauchn. Issl. Inst. Risa. HI 74044

Trudy Vsesoyuznogo gidrobiologicheskogo obshchestva. Moscow. Vol. 1+, 1949+. Trudy Vsesoyuzn. Gidrobiol. Obshch. 5-4434-2. HI 74045

Trudy vsesoyuznogo instituta sel'skokhozyaistvennoi mikrobiologii. Leningrad, Moscow. Vols. 4(3)-6, 7(2)-8(1), 1931-36. Trudy Vsesoyuzn. Inst. Sel'skokhoz. Mikrobiol. Preceded and superseded by: Trudy instituta sel'skokhozyaistvennoi mikrobiologii. HI 74046

Trudy Vsesoyuznogo instituta zashchity rastenii. Moscow & Leningrad. Vols. 2-11, 1949-58; vols. 13-16, 1958-60. Trudy Vsesoyuzn. Inst. Zashch. Rast. Preceded by: Sbornik trudov Vsesoyuznogo instituta zashchity rastenii. For vol. 12 see: Trudy. Vsesoyuznogo nauchno-issledovatel'skogo instituta zashchity rastenii. Superseded by: Trudy. Vsesoyuznogo nauchno-issledovatel'skogo instituta zashchity rastenii. 3-2391-1. HI 74047

Trudy. Vsesoyuznogo nauchno-issledovatel'skogo instituta lekarstvennykh rastenii. Moscow. Vols. 8-11, 194?-59. Trudy Vsesoyuzn. Nauchno-Issl. Inst. Lekarstv. Rast. Preceded by: Vsesoyuznyi nauchno-issledovatel'skii institut lekarstvennykh rastenii. Superseded by: Trudy. Vsesoyuznyi nauchno-issledovatel'skii institut lekarstvennykh i aromaticheskikh rastenii (V I L A R). 3-2391-1. HI 74048

Trudy vsesoyuznogo nauchno-issledovatel'skogo instituta tekhnologii torfa = Trudy Nauchno-issledovatel'skogo torfyanogo instituta (Instorfa). Moscow & Leningrad.

Trudy. Vsesoyuznogo nauchno-issledovatel'skogo instituta zashchity rastenii. Moscow. Vol. 12, 1958; vol. 17+, 1963. Trudy Vsesoyuzn. Nauchno-Issl. Inst. Zashch. Rast. For vols. 2-11 and vols. 13-16 see: Trudy Vsesoyuznogo instituta zashchity rastenii. HI 74049

Trudy. Vsesoyuznogo s''ezda po okhrane prirody v S S S R. Moscow. 1 vol., 1935. Trudy Vsesoyuzn. S''ezda Okhr. Prir. S.S.S.R. HI 74050

Trudy Vsesoyuznoi akademii sel'sko-khozyaistvennykh nauk imeni V. I. Lenina. Seriya 1. Moscow & Leningrad. Vols. 2-31 [data on vol. 1 not available], 1936-37. Trudy Vsesoyuzn. Akad. Sel'sko-Khoz. Nauk Lenina, Ser. 1. HI 74051

Trudy Vsesoyuznoi akademii sel'sko-khozyaistvennykh nauk imeni V. I. Lenina. Seriya 4. Selektsiya rastenii. Moscow & Leningrad. Vols. 1-22, 1935. Trudy Vsesoyuzn. Akad. Sel'sko-Khoz. Nauk Lenina, Ser. 4, Selekts. Rast. HI 74052

Trudy Vsesoyuznoi akademii sel'sko-khozyaistvennykh nauk imeni V. I. Lenina. Seriya 6. Tekhnicheskie kul'tury. Moscow & Leningrad. Vols. 5-10 [data on earlier vols. not available], 1936? Trudy Vsesoyuzn. Akad. Sel'sko-Khoz. Nauk Lenina, Ser. 6, Tekhn. Kul't. HI 74053

Trudy Vsesoyuznoi akademii sel'sko-khozyaistvennykh nauk imeni V. I. Lenina. Seriya 7. Kormovye kul'tury. Moscow & Leningrad. Vols. 3-9 [data on earlier vols. not available], 1935-36. Trudy Vsesoyuzn. Akad.

Sel'sko-Khoz. Nauk Lenina, Ser. 7, Kormov. Kul't. HI 74054

Trudy Vsesoyuznoi akademii sel'sko-khozyaistvennykh nauk imeni V. I. Lenina. Seriya 9. Bor'ba s sornyakami. Moscow & Leningrad. Vols. 1-5, 1935-36. Trudy Vsesoyuzn. Akad. Sel'sko-Khoz. Nauk Lenina, Ser. 9, Bor'ba Sornyak. HI 74055

Trudy Vsesoyuznoi akademii sel'sko-khozyaistvennykh nauk imeni V. I. Lenina. Seriya 10. Moscow & Leningrad. Vol. 5 [data on earlier vols. not available], 1936. Trudy Vsesoyuzn. Akad. Sel'sko-Khoz. Nauk Lenina, Ser. 10. HI 74056

Trudy Vsesoyuznoi akademii sel'sko-khozyaistvennykh nauk imeni V. I. Lenina. Seriya 17. Bor'ba s vreditelyami sel'skokhozyaistvennykh rastenii [later: Sel'skokhozyaistvennaya mikrobiologiya.] Moscow & Leningrad. Vols. 3-20 [data on earlier vols. not available], 1936. Trudy Vsesoyuzn. Akad. Sel'sko-Khoz. Nauk Lenina, Ser. 17, Bor'ba Vredit. HI 74057

Trudy. Vsesoyuznykh soveshchanii po okhrane prirody. Vilnius, Lithuanian S S R. Vol. 2, 1960 [data on vol. 1 not available]. Trudy Vsesoyuzn. Soveshch. Okhr. Prir. HI 74058

Trudy Yaroslavskogo estestvenno-istoricheskogo i kraevedcheskogo obshchestva. Yaroslavl, Russian S F S R. Vols. 4-6, 1925-30. Trudy Yaroslavsk. Estestv.-Istorich. Kraev. Obshch. Preceded by: Trudy Yaroslavskogo estestvenno-istoricheskogo obshchestva. 3-1912-2. HI 74059

Trudy Yaroslavskogo estestvenno-istoricheskogo obshchestva. Yaroslavl, Russian S F S R. Vols. 1-3, 1902-22. Trudy Yaroslavsk. Estestv.-Istorich. Obshch. Superseded by: Trudy Yaroslavskogo estestvenno-istoricheskogo i kraevedcheskogo obshchestva. 3-1912-3. HI 74060

Trudy Zapadno-Sibirskoi kraevoi stantsii zashchity rastenii. Tomsk, R S F S R. Vols. 10-11 [also numbered vols. 2-3], 1937. Trudy Zapadno-Sibirsk. Kraev. Stantsii Zashch. Rast. Preceded by: Izvestiya Zapdno-Sibirskoi kraevoi stantsii zashchity rastenii. Superseded by: Trudy Zapadnosibirskoi stantsii zashchity rastenii. HI 74064

Trudy Zapadno-sibirskoi stantsii zashchity rastenii. Tomsk, R S F S R. Vol. 12 [also numbered vol. 4], 1938. Trudy Zapadnosibirsk. Stantsii Zashch. Rast. Preceded by: Trudy Zapadno-Sibirskoi kraevoi stantsii zashchity rastenii. HI 74065

Trudy Zapadno-Sibirskoj kraevoj stancii zaščity rastenij = Trudy Zapadno-Sibirskoi kraevoi stantsii zashchity rastenii. Tomsk.

Trudy Zapadno-sibirskoj stancii zaščity rastenij = Trudy Zapadno-sibirskoi stantsii zashchity rastenii. Tomsk.

Trudy po zaščite rastenij. Serija 2. Fitopatologija = Trudy po zashchite rastenii. Seriya 2. Fitopatologiya. Leningrad.

Trudy po zaščite rastenij Sibiri = Trudy po zashchite rastenii Sibiri. Tomsk.

Trudy po zaščite rastenij Vostočnoj Sibiri = Trudy po zashchite rastenii Vostochnoi Sibiri. Moscow & Irkutsk.

Trudy po zashchite rastenii. Seriya 2. Fitopatologiya. Leningrad. Vols. 1-8, 1932-35. Trudy Zashch. Rast., Ser. 2, Fitopatol. Preceded by: Materialy po mikologii i fitopatologii. 3-2391-1. HI 74066

Trudy po zashchite rastenii Sibiri. Tomsk, R S F S R. Vol. 8 [also numbered vol. 1], 1931. Trudy Zashch. Rast. Sibiri. Preceded by: Izvestiya Sibirskoi kraevoi stantsii zashchity rastenii ot vreditelei. Superseded by: Izvestiya Zapdno-Sibirskoi kraevoi stantsii zashchity rastenii. HI 74067

Trudy po zashchite rastenii Vostochnoi Sibiri. Moscow & Irkutsk, R S F S R. Vols [3]-4 [also numbered vols. [1]-2], 1933-35. Trudy Zashch. Rast. Vost. Sibiri. Preceded by: Izvestiya Irkutskoi stantsii zashchity rastenii ot vreditelei. Superseded by: Sbornik trudov po zashchite rastenii Vostochnoi Sibiri. HI 74068

Trudy Zoobiologičeskogo instituta. Kiev = Pratsy Zoobiologichnogo instytutu. Kiev.

Trutnovsko. Sborník vlastivědných prací okresního musea v Trutnově. Trutnov, Czechoslovakia. Trutnovsko. See B–P–H 925/8. HI 60070

Ts'ing cheng cheng wou tiao tch'a so houei pao = Bulletin of the Fan memorial institute of biology. Peiping [=Peking].

Tsing Hua journal. [Ch'ing-hua hsüeh pao.] Peking. Tsing Hua J. See B–P–H 925/9. HI 60071

Tsing Hua weekly. [Ch'ing-hua chou k'an.] Peking. Tsing Hua Weekly. See B–P–H 925/12. HI 60072

Tsinghai agriculture and forestry. [Ching-hai nung ling. Quingbai nonglin.] Sining, China. Tsinghai Agric. Forest. See B–P–H 925/13. HI 60073

Tsitologiya. Moscow & Leningrad. Vol. 1+, 1959+. Tsitologiya. HI 74069

Tsitologiya i genetika. Kiev. Vol. 1+, 1967+. Tsitol. Genet. HI 74070

Tsitologiya i selektsiya kul'turnykh rastenii. Vol. ?-2+, ?-1964+. Tsitol. Selekts. Kul't. Rast. HI 74071

Tso wu hsüeh pao. Zuown zuebao = Crop science. Peking. Crop Sci. (Peking). See B–P–H 335/11.

Tsukuba jikken shokubutsuen kenkyu hokoku = Annals of the Tsukuba botanical garden. Ibaraki.

Tsvetovodstvo. Vol. ?-4+, ?-1961+. Tsvetovodstvo. HI 74072

"Tsvetenie" vody; voprosy fiziologii, biokhimii, toksikologii i ispol′zovaniya sinezelenykh vodroslei. Kiev. Vol. ?-2+, ?-1969+. Tsvetenie Vody. HI 74073

Tu jan t'e k'an = Special soils publicaiton. [National geological survey of China.] Chungking, China. Special Soils Publ. See B–P–H 852/8.

T'u jang chuan pao = Soil bulletin. [Geological survey of China.] Peiping [=Peking]. Soil Bull. See B–P–H 846/8.

T'u jang hsüeh hsüeh p'ao = Acta pedologica sinica. Peking. Acad Pedol. Sin. See B–P–H 47/7.

T'u shu chi k'an = Quarterly bulletin of chinese bibliography. Shanghai. Quart. Bull. Chin. Bibliogr. See B–P–H 751/4.

Tuatara; journal of the biological society; Victoria university college. Wellington. Tuatara. See B–P–H 925/20. HI 60074

Tübinger Blätter für Naturwissenschaften und Arzneykunde. Tübingen. Tübinger Blätt. Naturwiss. Arzneyk. See B–P–H 925/23. HI 60075

Tübingische Berichte von gelehrten Sachen. Tubingen. Tübingische Ber. Gel. Sachen. See B–P–H 925/25. HI 60076

Tübingische gelehrten Anzeigen. Tubingen. Tübingische Gel. Anz. See B–P–H 925/26. HI 60077

Tudomänyos gyüjtemény. Pest [=Budapest, in part]. Tud. Gyüjt. See B–P–H 925/27 HI 60078.

Tudománytár. Buda [=Budapest, in part]. Tudománytár. See B–P–H 925/29. HI 60079

Tudománytár. Értekezerek. Buda [=Budapest, in part]. Tudománytár Értek. See B–P–H 926/1. HI 60080

Tudománytár. Litteratura. Buda [=Budapest, in part]. Tudománytár Litt. See B–P–H 926/2. HI 60081

Tuexenia. Göttingen. N.s. vol. 1+, 1981+. Tuexenia. Preceded by: Mitteilungen der floristisch-soziologischen Arbeitsgemeinschaft. HI 74074

Tufts college studies. Scientific series. Boston, MA. Tufts Coll. Stud., Sci. Ser. See B–P–H 926/4. HI 60082

Tuinbode. Ghent. Tuinbode. See B–P–H 926/9. HI 60085

Tuinbouw. Amsterdam. Tuinbouw (Amsterdam). See B–P–H 926/10. HI 60086

Tuinbouw. The Hague. Tuinbouw (The Hague). See B–P–H 926/11. HI 60087

Tuinbouw-flora van Nederland en zijne overzeesche bezittingen. Leiden. Tuinb.-Fl. Ned. See B–P–H 926/7. HI 60083

Tuinbouw-illustratie. Tijdschrift voor tuinbouw en plantkunde. Haarlem. Tuinb.-Ill. See B–P–H 926/8. HI 60084

Tuinbouw mededelingen. The Hague. Vol. 32, 1969. Tuinb. Med. Preceded by: Mededeelingen, directeur van de tuinbouw. Superseded by: Bedrijfsontwikkeling: Editie tuinbouw. HI 74075

Tuinbouwberichten. Groningen. Tuinbouwberichten (Groningen). See B–P–H 926/12. HI 60088

Tuinbouwberichten. Louvain. Vols. 1-39, 1947-75. Tuinbouwberichten (Louvain). 5-4276-2. HI 74076

Tuinbouwblad De nederlandsche boomkwerker. Haarlem. Tuinbouwbl. Ned. Boomkweker. See B–P–H 926/14. HI 60089

Tuinbouwgids. Grammont, Belgium. Tuinbouwgids. See B–P–H 926/16. HI 60090

Tuinbouwtechniek. Wageningen. Tuinbouwtechniek. See B–P–H 926/17. HI 60091

Tuinbouwvoorlichting. The Hague. Nos. 1+, 1956+. Tuinbouwvoorlichting. Preceded by: Mededelingen van den tuinbouwvoor lichtingsdienst. HI 74077

Tuinderij / vollegrand. [Supplement to: Tuinderij.] Vol. 1+, 1979+. Tuinderij Vollegrand. HI 74078

Tukar-menukar. Gofuku. Vol. 1+, 1982+. Tukar-menukar. HI 74079

Tulane studies in zoology and botany. New Orleans, LA. Vol. 15+, 1969+. Tulane Stud. Zool. Bot. Preceded by: Tulane studies in zoology [not entered]. HI 74080

Tulip tidings. New York, NY. Tulip Tidings. See B–P–H 926/18. HI 60092

Tulsa geological society digest. Tulsa, OK. Tulsa Geol. Soc. Digest. See B–P–H 926/20. HI 60093

Tung chih wu chi pu = Research bulletin of the institute of zoology and botany: Fukien academy. Foochow, China. Res. Bull. Inst. Zool. Bot. Fukien Acad. See B–P–H 774/15.

Tung fang yu lun = Ostasiatische Rundschau. Berlin. Ostasiat. Rundschau. See B–P–H 692/6.

Tung fong hsüeh tsa chih = Far eastern medical journal. Peking. Far E. Med. J. See B–P–H 367/4.

Tung-hai ta hsüeh = Biological bulletin; department of biology, college of science, Tunghai university. Taichung, Taiwan. Biol. Bull. Dept. Biol. Coll. Sci. Tunghai Univ. See B–P–H 193/9.

Tung pei lin hsüeh yuan chih wu piao pen shih hui ken = Bulletin of the herbarium of the north-eastern forestry academy. Harbin.

Tung-pei lin hsueh yuan hsueh pao = Journal of north-eastern forestry institute. Harbin.

Tung-wu hsüeh-pao = Journal of Soochow university. College of arts and sciences. Soochow, China. J. Soochow Univ. Coll. Arts. See B–P–H 482/21.

Turf-grass times. Elm Grove, WI. Vols. 1-16(1), 1965-80. Turf-grass Times. Superseded by: Landscape and turf industry. HI 74081

Turista közlöny. Budapest. Turist. Közl. See B–P–H 927/1. HI 60096

Turisták lapja. Budapest. Turist. Lapja. See B–P–H 927/2. HI 60097

Turistaság és alpinismus. Budapest. Turist. & Alpinism. See B–P–H 926/29. HI 60094

Turistik, Alpinismus, WIntersport. Kezmarok, Czechoslovakia. Turist. Alpinism. Wintersp. See B–P–H 926/30. HI 60095

Türk biologi dergisi. Istanbul. Vols. 8-19, 1958-69. Türk Biol. Derg. Preceded by: Biologi. Superseded by: Türk biyoloji dergisi. HI 74082

Türk biologi derneği'nin yayin organi = Türk biologi dergisi. Istanbul.

Türk biyoloji dergisi. Istanbul. Vols. 20-23, 1970-73. Türk Biyol. Derg. Preceded by: Türk biologi dergisi. Superseded by: Biyologi dergisi (Biologi derneği'nim yayim organi). HI 74083

Turkish journal of biology = Türk biologi dergisi. Istanbul.

Turkish journal of plant protection = Turkiye bitki koruma dergisi. Izmir.

Turkish journal of plant science. Izmir = Bitki. Izmir.

Turkiye bitki koruma dergisi. Izmir. 1977+. Turk. Bitki Derg. HI 74084

Turkiye'deki fitopatolojik yayinlar listesi. No. 1+, 1975+. Turk. Fitopatol. Yayinl. List. HI 74085

Turkmenistan S S R ylymlar akademijasynyn khabarlary. Biologik ylmlaryn serijasy = Izvestiya akademiya nauk turkmenskoi S S R. Seriya biologicheskikh nauk. Ashkhabad.

Turkmenistan S S Rnin ylymlar akademijasynyn khabarlary = Izvestiya akademiya nauk turkmenskoi S S R. Ashkhabad.

Turrialba; revista interamericana de ciencias agrícolas. Turrialba, Costa Rica. Turrialba. See B–P–H 927/5. HI 60098

Turtox news. Chicago, IL. Turtox News. See B–P–H 927/6. HI 60099

Turun suomalainen Yliopiston Julkaisuja = Annales universitatis fennicae åboensis. Series A, physico-mathematica-biologica. Turku.

Turun yliopiston julkaisuja = Annales universitatis turkuensis. Ser. A, biologica-geographica. Turku.

Two and a bud. Jorhat. Vol. 1+, 1954+. Two & Bud. HI 74086

Tydskrif vir natuurwetenskappe. Pretoria. Vol. 1+, 1961+. Tydskr. Natuurwetensk. Preceded by: Tydskrif vir wetenskap en kuns. HI 74087

Tydskrif van die suid-afrikaanse bosbouvereeniging = Journal of the south african forestry association. Pretoria.

Tydskrif vir wetenskap en kuns. Bloemfontein & Pretoria. Vols. 1-17, 1923-39; n.s. deel 1-20, 1940-60. Tydskr. Wetensk. Kuns. Superseded by: Tydskrif vir natuurwetenskappe. 5-4276-2. HI 74088

Tyukyu seifu keizai kyoku ringyo shikenjo kenkyu hokoku = Testing report. Forestry experiment station, Tyukyu government. Naha?, Okinawa. Test. Rep. Forest. Exp. Sta. Ryukyu Gov. See B–P–H 872/4.

Tzu jan k'o hsüeh = Science journal. [College of science; Sun Yatsen university.] Canton. Sci. J. (Canton). See B–P–H 824/16.

Tzu-jan k'o-hsüeh chi-k'an = Science quarterly of the national university of Peking. Peiping [=Peking]. Sci. Quart. Natl. Univ. Peking. See B–P–H 826/14.

U A S research review bulletin. Bangalore. No. 1+, 1977+. U. A. S. Res. Rev. Bull. HI 74089

U A S miscellaneous series. Bangalore. No. 1+, 1965+. U. A. S. Misc. Ser. HI 74090

U B I wendingen. Circulaire = Circulaire, biohistorisch instituut der rijksuniversiteit te Utrecht. Utrecht.

U C L A symposia on molecular and cellular biology. New York. Vol. 24+ [= n.s. vol. 1+], 1982+. U. C. L. A. Symp. Molec. Cell. Biol. Preceded by: I C N - U C L A symposia on molecular and cellular biology. HI 75234

U L science magazine. Monrovia. Vol. 1+, 1972+. U. L. Sci. Mag. HI 74091

U-M botany times. Ann Arbor, MI. No. 1+, 1990+. U-M Bot. Times. HI 56935

U N E P news. Nairobi. Vols. 1-2?, 1974-75?; 1985+. U. N. E. P. News. For 1976-82 see: Uniterra. HI 74092

U N E S C O reports in marine science. Paris. Vol. 1+, 1977+. UNESCO Rep. Mar. Sci. HI 74093

U N E S C O technical papers in marine science. New York, NY. Vol. 1+, 1965+. UNESCO Techn. Pap. Mar. Sci. HI 74094

U P A S I scientific department advisory leaflet = Advisory leaflet, U P A S I scientific department. Madras.

U P research digest. Quezon City. Vols. 1-6(2), 1962-67. U. P. Res. Digest. HI 74095

U P O V newsletter. Geneva. No. 1+, 1975+. U. P. O. V. Newslett. HI 74096

U R T. Copenhagen. 1/77+, 1977+. U. R. T. Risskov -

Botanisk Institut. HI 74097

U S D A Bureau of plant industry. Miscellaneous publication = U S department of agriculture. Bureau of plant industry. Miscellaneous publication. Washington, DC. U.S.D.A. Bur. Pl. Industr. Misc. Publ. See B–P–H 941/1.

U S D A. United States department of agriculture. Washington, DC. USDA. See B–P–H 940/6. HI 60209

U S department of agriculture. Annual report of the Alaska agricultural experiment station. Washington, DC. U.S.D.A. Annual Rep. Alaska Agric. Exp. Sta. See B–P–H 940/11. HI 60210

U S department of agriculture. Bulletin. Washington, DC. U.S.D.A. Bull. (1895-1901). See B–P–H 940/15. HI 60211

U S department of agriculture bulletin. Washington, DC. U.S.D.A. Bull. (1915-23). See B–P–H 940/16. HI 60212

U S department of agriculture. Bulletin of foreign plant introductions. Washington, DC. U.S.D.A. Bull. Foreign Pl. Introd. See B–P–H 940/19. HI 60213

U S department of agriculture. Bureau of entomology and plant quarantine. Report. Washington, DC. U.S.D.A. Bur. Entomol. Rep. See B–P–H 940/20. HI 60214

U S department of agriculture, bureau of forestry. Bulletin. Washington, DC. Nos. 32-63, 1902-05. U.S.D.A. Bur. Forest. Bull. Preceded by: U S department of agriculture, division of forestry bulletin. Superseded by: Bulletin, United States forest service. HI 74098

U S department of agriculture. Bureau of plant industry. Bulletin. Washington, DC. U.S.D.A. Bur. Pl. Industr. Bull. See B–P–H 940/24. HI 60215

U S department of agriculture. Bureau of plant industry. Circular. Washington, DC. U.S.D.A. Bur. Pl. Industr. Circ. See B–P–H 940/25. HI 60216

U S department of agriculture. Bureau of plant industry. Inventory of seeds and plants imported by the office of foreign seed and plant introduction. Washington, DC. U.S.D.A. Bur. Pl. Industr. Invent. Seeds. See B–P–H 940/26. HI 60217

U S department of agriculture. Bureau of plant industry. Miscellaneous publication. Washington, DC. U.S.D.A. Bur. Pl. Industr. Misc. Publ. See B–P–H 941/1. HI 60218

U S department of agriculture. Circular. Washingon, DC. U.S.D.A. Circ. See B–P–H 941/6. HI 60219

U S department of agriculture. Department bulletin. Washington, DC. U.S.D.A. Dept. Bull. See B–P–H 941/17. HI 60221

U S department of agriculture. Department circular. Washington, DC. U.S.D.A. Dept. Circ. See B–P–H 941/18. HI 60222

U S department of agriculture. Departmental circular. Washington, DC. U.S.D.A. Departmental Circ. See B–P–H 941/16. HI 60220

U S department of agriculture, division of agrostology. Report of the agrostologist. Washington, DC. U.S.D.A. Div. Agrostol. Rep. Agrostol. See B–P–H 941/20. HI 60223

U S department of agriculture. Division of botany. Bulletin. Washington, DC. U.S.D.A. Div. Bot. Bull. See B–P–H 941/21. HI 60224

U S department of agriculture. Division of botany. Circular. Washington, DC. U.S.D.A. Div. Bot. Circ. See B–P–H 941/22. HI 60225

U S department of agriculture. Division of botany. Inventory. Washington, DC. U.S.D.A. Div. Bot. Invent. See B–P–H 941/23. HI 60226

U S department of agriculture, division of forestry bulletin. Washington, DC. U.S.D.A. Div. Forest. Bull. See B–P–H 941/24. HI 60227

U S department of agriculture. Division of pomology. Bulletin. Washington, DC. U.S.D.A. Div. Pomol. Bull. See B–P–H 941/27. HI 60228

U S department of agriculture. Division of pomology. Circular. Washington, DC. U.S.D.A. Div. Pomol. Circ. See B–P–H 942/1. HI 60229

U S department of agriculture. Division of vegetable pathology. Bulletin. Washington, DC. U.S.D.A. Div. Veg. Pathol. Bull. See B–P–H 942/2. HI 60230

U S department of agriculture. Division of vegetable pathology. Circular. Washington, DC. U.S.D.A. Div. Veg. Pathol. Circ. See B–P–H 942/3. HI 60231

U S department of agriculture. Division of vegetable pathology. Report of the chief of division of vegetable pathology. Washington, DC. U.S.D.A. Div. Veg. Pathol. Rep. Chief Div. Veg. Pathol. See B–P–H 942/4. HI 60232

U S department of agriculture. Division of vegetable pathology. Report of the chief of division of vegetable physiology and pathology. Washington, DC. U.S.D.A. Div. Veg. Pathol. Rep. Chief Div. Veg. Physiol. See B–P–H 942/5. HI 60233

U S department of agriculture. Division of vegetable pathology. Report of the chief of section of vegetable pathology. Washington, DC. U.S.D.A. Div. Veg. Pathol. Rep. Chief Sect. Veg. Pathol. See B–P–H 942/6. HI 60234

U S department of agriculture. Division of vegetable pathology. Report of the mycologist. Washington, DC. U.S.D.A. Div. Veg. Pathol. Rep. Mycol. See

B–P–H 942/7. HI 60236

U S department of agriculture. Farmers bulletin. Washington, DC. U.S.D.A. Farmers Bull. See B–P–H 942/13. HI 60237

U S department of agriculture, forestry division bulletin. Washington, DC. U.S.D.A. Forest. Div. Bull. See B–P–H 942/16. HI 60238

U S department of agriculture. Inventory. Washington, DC. U.S.D.A. Invent. See B–P–H 943/1. HI 60239

U S department of agriculture. Library list. Washington, DC. U.S.D.A. Libr. List. See B–P–H 943/11. HI 60240

U S department of agriculture. Miscellaneous circular. Washington, DC. U.S.D.A. Misc. Circ. See B–P–H 943/16. HI 60241

U S department of agriculture; office of experiment stations. Annual report of the hawaiian agricultural experiment station. Washington, DC. U.S.D.A. Off. Exp. Sta. Annual Rep. Hawaiian Agric. Exp. Sta. See B–P–H 943/22. HI 60242

U S department of agriculture. Office of experiment stations. Experiment station work. Washington, DC. U.S.D.A. Off. Exp. Sta. Exp. Sta. Work. See B–P–H 943/24. HI 60243

U S department of agriculture, office of experiment stations; the Guam agricultural experiment station, bulletin. Washington, DC. U.S.D.A. Off. Exp. Sta. Guam Agric. Exp. Sta. Bull. See B–P–H 943/25. HI 60244

U S department of agriculture, office of experiment stations; the Guam agricultural experiment station, circular. Washington, DC. U.S.D.A. Off. Exp. Sta. Guam Agric. Exp. Sta. Circ. See B–P–H 943/26. HI 60245

U S department of agriculture, office of experiment stations; the Guam agricultural experiment station, extension circular. Washington, DC. U.S.D.A. Off. Exp. Sta. Guam Agric. Exp. Sta. Extens. Circ. See B–P–H 943/27. HI 60246

U S department of agriculture, office of experiment stations; the Guam agricultural experiment station, report. Washington, DC. U.S.D.A. Off. Exp. Sta. Guam Agric. Exp. Sta. Rep. See B–P–H 943/28. HI 60247

U S department of agriculture, office of experiment stations; the Guam agricultural experiment station and its work. Washington, DC. U.S.D.A. Off. Exp. Sta. Guam Agric. Exp. Sta. Work. See B–P–H 944/1. HI 60248

U S department of agriculture, office of experiment stations. Report on the agricultural resources and capabilities of Hawaii. Washington, DC. U.S.D.A. Off. Exp. Sta. Rep. Agric. Resources Capabil. Hawaii. See B–P–H 944/3. HI 60249

U S department of agriculture, office of experiment stations. Report of the Virgin Islands agricultural research and extension programs. Washington, DC. U.S.D.A. Off. Exp. Sta. Rep. Virgin Islands Agric. Res. Extens. Programs. See B–P–H 944/4. HI 60250

U S department of agriculture. Office of the secretary. Circular. Washington, DC. U.S.D.A. Off. Secr. Circ. See B–P–H 944/6. HI 60251

U S department of agriculture, office of the secretary. Report. Washington, DC. U.S.D.A. Off. Secr. Rep. See B–P–H 944/7. HI 60252

U S department of agriculture. Plant immigrants. Washington, DC. U.S.D.A. Pl. Immigr. See B–P–H 944/11. HI 60253

U S department of agriculture, plant quarantine and control administration report. Washington, DC. U.S.D.A. Pl. Quarant. Admin. Rep. See B–P–H 944/12. HI 60254

U S department of agriculture. Report. Washington, DC. U.S.D.A. Rep. See B–P–H 944/18. HI 60255

U S department of agriculture. Report of the chief of the bureau of plant industry. Washington, DC. U.S.D.A. Rep. Chief Bur. Pl. Industr. See B–P–H 944/21. HI 60256

U S department of agriculture. Report of the pomologist. Washington, DC. U.S.D.A. Rep. Pomol. See B–P–H 945/3. HI 60257

U S department of agriculture, section of seed and plant introduction. Circular. Washington, DC. Nos. ?+, 19?? U.S.D.A. Sect. Seed Introd. Invent. HI 74099

U S department of agriculture. Section of seed and plant introduction. Inventory. Washington, DC. U.S.D.A. Sect. Seed Introd. Invent. See B–P–H 945/5. HI 60258

U S forest service research notes; institute of tropical forestry. Rio Piedras, PR = Research notes I T F, United States forest service. Rio Piedras, PR.

U S / I B P desert biome monograph. Logan, UT. No. 1+, [1976?]+. U.S. / I. B. P. Desert Biome Monogr. HI 74100

U S / I B P synthesis series. Stroudsberg, PA. 1976+. U.S. / I. B. P. Synth. Ser. HI 74101

U S S R and east Europe scientific abstracts. Biomedical sciences. Arlington, VA. Nos. 1-31, 1973-75. U.S.S.R. E. Eur. Sci. Abstr., Biomed. Sci. Preceded by: U S S R scientific abstracts. Biology and medicine. Superseded by: U S S R and eastern Europe scientific abstracts. Biomedical and behavioral sciences. HI 74102

U S S R and eastern Europe scientific abstracts. Biomedical and behavioral sciences. Arlington, VA. No. 1+, 1973+. U.S.S.R. E. Eur. Sci. Abstr., Biomed. Behav. Sci. Preceded by: U S S R and east Europe scientific abstracts. Biomedical sciences. HI 74103

U S S R report. Life sciences. Agrotechnology and food resources. Arlington, VA. No. 1+, 1980+. U.S.S.R. Rep., Life Sci., Agrotechnol. Food Resources. Preceded (in part) by: U S S R report. Biomedical and behavioral sciences [not entered]. HI 74104

U S S R scientific abstracts. Biology and medicine. Arlington, VA. Nos. 1-125, 1964-73. U.S.S.R. Sci. Abstr., Biol. Med. Superseded by: U S S R and east Europe scientific abstracts. Biomedical sciences. HI 74105

U S S R scientific abstracts: Bio-medical science = U S S R scientific abstracts. Biology and medicine. Arlington, VA.

Ueber das Bestehen und Wirken der Naturforschenden Gesellschaft zu Bamberg. Bamberg, Germany. Ueber Bestehen Wirken Naturf. Ges. Bamberg. See B–P–H 930/15. HI 60102

Ueber das Bestehen und Wirken des Naturforschenden Vereins zu Bamberg. Bamberg, Germany. Ueber Bestehen Wirken Naturf. Vereins Bamberg. See B–P–H 930/16. HI 60103

Übersetzungen und deutsche Abhandlungen welche bey der churfürstlich mainzischen Akademie der Wissenschaften nach und nach übergeben worden. Erfurt. Übersetz. Deutsche Abh. Churf. Mainz. Akad. Wiss. See B–P–H 927/16. HI 60101

Uebersicht der Arbeiten und Veränderungen der Schlesischen Gesellschaft für vaterländische Cultur. Breslau [=Wroclaw, Poland]. Uebers. Arbeiten Veränd. Schles. Ges. Vaterl. Cult. See B–P–H 931/2. HI 60104

Übersicht der Leistungen auf dem Gebiete der Botanik in Russland. St. Petersburg. Übers. Leist. Bot. Russland. See B–P–H 927/15. HI 60100

Uebersicht der neuesten Fortschritte, Entdeckungen, Meinungen und Gründe in den spekulativen und positiven Wissenschaften. Erfurt. Uebers. Neuesten Fortschr. Wiss. See B–P–H 931/4. HI 60105

Uebersicht der neuesten pomologischen Literatur. Frankfurt a. M. Uebers. Neuesten Pomol. Lit. See B–P–H 931/5. HI 60106

Uebersicht der Verhandlungen der Naturhistorischen Gesellschaft in Solothurn. Solothurn, Switzerland. Uebers. Verh. Naturhist. Ges. Solothurn. See B–P–H 931/6. HI 60107

Uebersicht der Verhandlungen der St. Gallischen naturwissenschaftlichen Gesellschaft. St. Gallen, Switzerland. Ubers. Verh. St. Gallischen Naturwiss. Ges. See B–P–H 931/7. HI 60108

Učeni zapysky harkivs'kij (ordena trudovogo červonogo prapora) deržavnii universytet imeny O. M. Gor'kogo. Kharkov = Uchenye zapiski, khar'kovskii gosudarstvennogo universiteta imeni A. M. Gor'kogo. Kharkov.

Učeni zapysky harkivs'kogo deržavnogo universytetu imeny O. M. Gor'kogo. Kharkov = Uchenye zapiski, khar'kovskii gosudarstvennogo universiteta imeni A. M. Gor'kogo. Kharkov.

Učenyja Zapiski Imperatorskago Jur'evskago Universiteta = Uchenyya Zapiski Imperatorskago Yur'evskago Universiteta. Yur'ev.

Učenyja Zapiski Imperatorskago Kazanskago Universiteta = Uchenyya Zapiski, izdavaemyya Imperatorskim Kazanskim Universitetom. Dazan.

Učenyja Zapiski Imperatorskago Kazanskago Universiteta po Otděleniju Fiziko-matematičeskih i Medicinskih Nauk = Uchenyya Zapiski Imperatorskago Kazanskago Universiteta po Otděleniyu Fiziko-matematicheskikh i Meditsinskikh Nauk. Kazan.

Učenyja Zapiski Imperatorskago Moskovskago Universiteta = Uchenyya Zapiski Imperatorskago Moskovskago Universiteta. Moscow.

Učenyja Zapiski Imperatorskago Moskovskago Universiteta. Otděl estestvenno-istoričeskij = Uchenyya Zapiski Imperatorskago Moskovskago Universiteta. Otděl estestvenno-istoricheskii. Moscow.

Učenyja Zapiski, izdavaemyja Imperatorskim Kazanskim Universitetom = Uchenyya Zapiski, izdavaemyya Imperatorskim Kazanskim Universitetom. Dazan.

Učenyja Zapiski Kazanskago Universiteta = Uchenyya Zapiski Kazanskago Universiteta. Kazan.

Učenye zapiski Biologičeskogo fakul'teta. Frunze, Kirghiz S S R = Uchenye zapiski Biologicheskogo fakul'teta. Frunze, Kirghiz S S R. Frunze.

Učenye zapiski Biologo-počvennogo fakul'teta. Frunze, Kirghiz S S R = Uchenye zapiski Biologo-pochvennogo fakul'teta. Frunze, Kirghiz S S R. Frunze.

Učenye zapiski Fakul'teta estestvoznanija = Uchenye zapiski Fakul'teta estestvoznaniya. Leningrad.

Učenye zapiski Fakul'teta estestvoznanija. Leningradskij gosudarstvennyj pedagogičeskij institut = Uchenye zapiski Fakul'tet estestvoznaniya. Leningradskii gosudarstvennyi pedagogicheskii institut. Leningrad.

Učenye zapiski Gosudarstvennogo Saratovskogo imeni N. G. Černyševskogo universiteta = Uchenye zapiski. Saratovskogo gosudarstvennogo imeni N. G. Chernyshevskogo universiteta. Saratov.

Učenye zapiski Gosudarstvennogo Saratovskogo universiteta imeni N. G. Černyševskogo = Uchenye zapiski. Saratovskogo gosudarstvennogo imeni N. G. Chernyshevskogo universiteta. Saratov.

Učenye zapiski, har'kovskii gosudarstvennogo universiteta imeni A. M. Gor'kogo. Kharkov = Uchenye zapiski, khar'kovskii gosudarstvennogo universiteta imeni A. M. Gor'kogo. Kharkov.

Učenye zapiski, har'kovskij (ordena trudovogo krasnogo znameni) gosudarstvennyj universitet imeni A. M. Gor'kogo. Kharkov = Uchenye zapiski, khar'kovskii gosudarstvennogo universiteta imeni A. M. Gor'kogo. Kharkov.

Učenye zapiski Karelo-Finskogo gosudarstvennogo universiteta = Uchenye zapiski Karelo-Finskogo gosudarstvennogo universiteta. Petrozavodsk.

Učenye zapiski Karelo-Finskogo gosudarstvennogo universiteta. Vypusk 3. Biologičeskie nauki = Uchenye zapiski Karelo-Finskogo gosudarstvennogo universiteta. Vypusk 3. Biologicheskie nauki. Petrozavodsk.

Učenye zapiski. Kazanskij gosudarstvennyj universitet imeni V. I. Ul'janova-Lenina = Uchenye zapiski. Kazanskii gosudarstvennyi universitet imeni V. I. Ul'yanova-Lenina. Kazan.

Učenye zapiski. Kazanskij gosudarstvennyj universitet imeni V. I. Ul'janova-Lenina. Botanika = Uchenye zapiski. Kazanskii gosudarstvennyi universitet imeni V. I. Ul'yanova-Lenina. Botanika. Kazan.

Učenye zapiski Kazanskogo gosudarstvennogo universiteta imeni V. I. Ul'janova-Lenina = Uchenye zapiski. Kazanskii gosudarstvennyi universitet imeni V. I. Ul'yanova-Lenina. Kazan.

Učenye zapiski Kazanskogo gosudarstvennogo universiteta imeni V. I. Ul'janova-Lenina. Botanika = Uchenye zapiski. Kazanskii gosudarstvennyi universitet imeni V. I. Ul'yanova-Lenina. Botanika. Kazan.

Učenye Zapiski: Krasnojarskij gosudarstvennyj pedagogičeskij instituta = Uchenye Zapiski: Krasnoyarskii gosudarstvennyi pedagogicheskii instituta. Krasnoyarsk.

Učenye zapiski. Leningradskij gosudarstvennyj pedagogičeskij instituta imeni A. I. Gercena = Uchenye zapiski. Leningradskii gosudarstvennyi pedagogicheskii instituta imeni A. I. Gertsena. Leningrad.

Učenye zapiski. Leningradskij gosudarstvennyj universitet imeni A. S. Bubnova. Serija biologičeskaja = Uchenye zapiski. Leningradskii gosudarstvennyi universitet imeni A. S. Bubnova. Seriya biologicheskikh nauk. Leningrad.

Učenye zapiski. Leningradskij gosudarstvennyj universitet imeni A. S. Bubnova. Serija biologičeskih nauk = Uchenye zapiski. Leningradskii gosudarstvennyi universitet imeni A. S. Bubnova. Seriya biologicheskikh nauk. Leningrad.

Učenye zapiski. Leningradskogo (ordena Lenina) gosudarstvennogo universiteta (imeni A. A. Ždanova). Serija biologičeskih nauk = Uchenye zapiski. Leningradskogo (ordena Lenina) gosudarstvennogo universiteta (imeni A. A. Zhdanova). Seriya biologicheskikh nauk. Leningrad.

Učenye zapiski. L'vovskij gosudarstvennyj universitet imeni Ivana Franko. Serija biologičeskaja = Uchenye zapiski. L'vovskii gosudarstvennyi universitet imeni Ivana Franko. Seriya biologicheskaya. Lvov.

Učenye zapiski. Molotovskij gosudarstvennyj universitet imeni A. M. Gor'kogo = Uchenye zapiski. Molotovskii gosudarstvennyi universitet imeni A. M. Gor'kogo. Molotov.

Učenye Zapiski. Moskovskago Gorodskago Narodnago Universiteta imeni A. L. Sanjavskago = Uchenye Zapiski. Moskovskago Gorodskago Narodnago Universiteta imeni A. L. Sanyavskago. Moscow.

Učenye zapiski. Moskovskij gosudarstvennyj universitet = Uchenye zapiski. Moskovskii gosudarstvennyi universitet. Moscow & Leningrad.

Učenye zapiski. Moskovskij ordena Lenina gosudarstvennyj universitet imeni M. V. Lomonosova = Uchenye zapiski. Moskovskii gosudarstvennyi universitet. Moscow & Leningrad.

Učenye zapiski. Permskij gosudarstvennyj universitet imeni A. M. Gor'kogo = Uchenye zapiski. Permskii gosudarstvennyi universitet imeni A. M. Gor'kogo. Perm.

Učenye zapiski. Petrozavodskogo gosudarstvennogo universiteta. Vypusk 3. Biologičeskie nauki = Uchenye zapiski. Petrozavodskogo gosudarstvennogo universiteta. Vypusk 3. Biologicheskie nauki. Petrozavodsk.

Učenye zapiski. Petrozavodskogo gosudarstvennogo universiteta. Vypusk 3. Biologičeskie i sel'skohozjajstvennye nauki = Uchenye zapiski. Petrozavodskogo gosudarstvennogo universiteta. Vypusk 3. Biologicheskie i sel'skokhozyaistvennye nauki. Petrozavodsk.

Učenye zapiski. Saratovskij gosudarstvennyj universitet imeni N. G. Černyševskogo = Uchenye zapiski. Saratovskogo gosudarstvennogo universiteta imeni N. G. Chernyshevskogo. Saratov.

Učenye zapiski. Saratovskogo gosudarstvennogo imeni N. G. Černyševskogo universiteta = Uchenye zapiski. Saratovskogo gosudarstvennogo imeni N. G. Chernyshevskogo universiteta. Saratov.

Učenye zapiski. Saratovskogo gosudarstvennogo

universiteta imeni N. G. Černyševskogo = Uchenye zapiski. Saratovskogo gosudarstvennogo universiteta imeni N. G. Chernyshevskogo. Saratov.

Učenye zapiski. Saratovskogo gosudarstvennogo universiteta imeni N. G. Černyševskogo. Ser. Biologičeskaja = Uchenye zapiski. Saratovskogo gosudarstvennogo universiteta imeni N. G. Chernyshevskogo. Ser. Biologicheskaya. Saratov.

Učenye zapiski Tartuskogo gosudarstvennogo universiteta = Uchenye zapiski Tartuskogo gosudarstvennogo universiteta. Tartu.

Ucheni zapysky kharkivs'kii (ordena trudovogo chervonogo prapora) derzhavnii universytet imeny O. M. Gor'kogo = Uchenye zapiski, khar'kovskii gosudarstvennogo universiteta imeni A. M. Gor'kogo. Kharkov.

Ucheni zapysky kharkivs'kogo derzhavnogo universytetu imeny O. M. Gor'kogo = Uchenye zapiski, khar'kovskii gosudarstvennogo universiteta imeni A. M. Gor'kogo. Kharkov.

Uchenya Zapiski Imperatorskago Kazanskago Universiteta po Otděleniyu Fiziko-matematicheskikh i Meditsinskikh Nauk. Kazan, R S F S R. 1862-64. Uchen. Zap. Imp. Kazansk. Univ. Otd. Fiz.-Mat. Nauk. Preceded by: Uchenya Zapiski Kazanskago Universiteta. Superseded by: Izvestiya i Ucheniya Zapiski Imperatorskago Kazanskago Universiteta. 3-2276-2. HI 74111

Uchenya Zapiski Imperatorskago Moskovskago Universiteta. Moscow. 1833-35. Uchen. Zap. Imp. Moskovsk. Univ. 3-2766-3. HI 51399

Uchenya Zapiski Imperatorskago Moskovskago Universiteta. Otděl estestvenno-istoricheskii. Moscow. Vols. 1-43, 1880-1917. Uchen. Zap. Imp. Moskovsk. Univ., Otd. Estestv.-Istorich. Preceded by: Protokoly Zasědanii Sověta Imperatorskago Moskovskago Universiteta. Superseded by: Uchenye zapiski. Moskovskii gosudarstvennyi universitet. 3-2768-1. HI 74112

Uchenya Zapiski Imperatorskago Yur'evskago Universiteta. Yur'ev [=Tartu], Estonian S S R. Vols. 1-25, 1893-1917. Uchen. Zap. Imp. Yur'evsk. Univ. Superseded by: Trudy Voronezhskogo gosudarstvennogo universiteta (vols. 1-12, 1925-41, only); afterwards superseded by: Uchenye zapiski Tartuskogo gosudarstvennogo universiteta. 5-4157-1. HI 74113

Uchenya Zapiski Kazanskago Universiteta. Dazan, R S F S R. 1834-61; vols. 51-84(2), 1884-1917. Uchen. Zap. Imp. Kazansk. Univ. For 1862-64 see: Uchenya Zapiski Imperatorskago Kazanskago Universiteta po Otděleniyu Fiziko-matematicheskikh i Meditsinskikh Nauk. For vols. 32-50 see: Izvestiya i Ucheniya Zapiski Imperatorskago Kazanskago Universiteta. Superseded by: Uchenya Zapiski Kazanskago Universiteta. 3-2276-2. HI 74118

Uchenya Zapiski Kazanskago Universiteta. Kazan, R S F S R. Vol. 84(3-12), 1917. Uchen. Zap. Kazansk. Univ. Preceded by: Uchenya Zapiski Kazanskago Universiteta. Superseded by: Uchenye zapiski. Kazanskii gosudarstvennyi universitet imeni V. I. Ul'yanova-Lenina. 3-2276-2. HI 74121

Uchenye zapiski Biologicheskogo fakul'teta. Frunze, Kirghiz S S R. Vol. 18+, 1957+. Uchen. Zap. Biol. Fak. Preceded by: Uchenye zapiski Biologo-pochvennogo fakul'teta. HI 74106

Uchenye zapiski biologicheskogo nauchno-issledovatel'skogo instituta pri Rostovskom na Donu. Rostov. Vols. 1-5, 1938-46. Uchen. Zap. Biol. Nauchn.-Issl. Inst. Rostovsk. Donu. HI 74107

Uchenye zapiski Biologo-pochvennogo fakul'teta. Frunze, Kirghiz S S R. Vols. 3-7, 1952-55. Uchen. Zap. Biol.-Pochv. Fak. Preceded by: Trudy Kirgizskogo gosudarstvennogo pedagogicheskogo instituta imeni M. V. Frunze. Superseded by: Uchenye zapiski Biologicheskogo fakul'teta. HI 74108

Uchenye zapiski Fakul'teta estestvoznaniya. Leningrad. Vol. [1], 1939. Uchen. Zap. Fak. Estestv. Superseded by: Uchenye zapiski Fakul'tet estestvoznaniya. Leningradskii gosudarstvennyi pedagogicheskii institut. HI 74109

Uchenye zapiski Fakul'teta estestvoznaniya. Leningradskii gosudarstvennyi pedagogicheskii institut. Leningrad. Vols. 2-6, 1949-56. Uchen. Zap. Fak. Estestv. Leningradsk. Gosud. Pedagog. Inst. Preceded by: Uchenye zapiski Fakul'teta estestvoznaniya. HI 74110

Uchenye zapiski Kabardino-Balkarskogo gosudarstvennogo universiteta. Nal'chik. Vol. 1+, 1957+. Uchen. Zap. Kabardino-Balkarsk. Gosud. Univ. HI 74114

Uchenye zapiski Karelo-Finskogo gosudarstvennogo universiteta. Petrozavodsk, R S F S R. Vol. 1, 1947. Uchen. Zap. Karelo-Finsk. Gosud. Univ. Superseded by: Uchenye zapiski Karelo-Finskogo gosudarstvennogo universiteta. Vypusk 3. Biologicheskie nauki. 4-3320-1. HI 74115

Uchenye zapiski Karelo-Finskogo gosudarstvennogo universiteta. Vypusk 3. Biologicheskie nauki. Petrozavodsk, R S F S R. Vols. 2-6, 1948-54. Uchen. Zap. Karelo-Finsk. Gosud. Univ., Vyp. 3, Biol. Nauki. Preceded by: Uchenye zapiski Karelo-Finskogo gosudarstvennogo universiteta. Superseded by: Uchenye zapiski. Petrozavodskogo gosudarstvennogo universiteta. Vypusk 3. Biologicheskie nauki. 4-3320-1. HI 74116

Uchenye zapiski kazakhskogo gosudarstvennogo universiteta im S. M. Kirova. Alma-Ata. 1938-61. Uchen. Zap. Kazakhsk. Gosud. Univ. S. M. Kirova. HI 74117

Uchenye zapiski, Kazanskii gosudarstvennyi universitet imeni V. I. Ul'yanova-Lenina. Kazan, R S F S R. Vol. 85+, 1925+. Uchen. Zap. Kazansk. Gosud. Univ. Ul'yanova-Lenina. Preceded by: Uchenya Zapiski Kazanskago Universiteta. 3-2276-2. HI 74119

Uchenye zapiski. Kazanskii gosudarstvennyi universitet imeni V. I. Ul'yanova-Lenina. Botanika. Kazan, R S F S R. Vols. 1-7, 1933-48. Uchen. Zap. Kazansk. Gosud. Univ. Ul'yanova-Lenina, Bot. HI 74120

Uchenye zapiski, khar'kovskii gosudarstvennogo universiteta imeni A. M. Gor'kogo. Kharkov. Vols. 1-?, 1935-63? Uchen. Zap. Khar'kovsk. Gosud. Univ. Gor'kogo. Superseded by: Vestnik Khar'kovskogo universiteta. 3-2285-3. HI 74122

Uchenye zapiski, khar'kovskii (ordena trudovogo krasnogo znameni) gosudarstvennogo universiteta imeni A. M. Gor'kogo = Uchenye zapiski, khar'kovskii gosudarstvennogo universiteta imeni A. M. Gor'kogo. Kharkov.

Uchenye Zapiski: Krasnoyarskii gosudarstvennyi pedagogicheskii instituta. Krasnoyarsk, R S F S R. Vol. 1+, 1952+. Uchen. Zap. Krasnoyarsk. Gosud. Pedagog. Inst. HI 74123

Uchenye zapiski, kurskii gosudarstvennyi pedagogicheskii institut. Kursk. Vols. 1-71, 1941-70. Uchen. Zap. Kursk. Gosud. Pedagog. Inst. Superseded by: Nauchnye trudy kurskii gosudarstvennyi pedagogicheskii institut. HI 74124

Uchenye zapiski. Leningradskii gosudarstvennyi pedagogicheskii instituta imeni A. I. Gertsena. Leningrad. Vols. 1-96, 1936-54; vol. 100+, 1955+ [vols. 97-99 not published]. Uchen. Zap. Leningradsk. Gosud. Pedagog. Inst. Gertsena. 3-2391-3. HI 74125

Uchenye zapiski. Leningradskii gosudarstvennyi universitet imeni A. S. Bubnova. Seriya biologicheskikh nauk. Leningrad. Vols. 1(1)-3(5) (= vols. 1-5], 1935-37. Uchen. Zap. Leningradsk. Gosud. Univ. Bubnova, Ser. Biol. Nauk. Superseded by: Uchenye zapiski. Leningradskogo (ordena Lenina) gosudarstvennogo universiteta (imeni A. A. Zhdanova). Seriya biologicheskikh nauk. 3-2393-1. HI 74126

Uchenye zapiski. Leningradskogo (ordena Lenina) gosudarstvennogo universiteta (imeni A. A. Zhdanova). Seriya biologicheskikh nauk. Leningrad. Vol. 6+, 1938+. Uchen. Zap. Leningradsk. Gosud. Univ., Ser. Biol. Nauk. Preceded by: Uchenye zapiski. Leningradskii gosudarstvennyi universitet imeni A. S. Bubnova. Seriya biologicheskikh nauk. 3-2393-1. HI 74127

Uchenye zapiski. L'vovskii gosudarstvennyi universitet imeni Ivana Franko. Seriya biologicheskaya. Lvov, Ukrainian S S R. Vols. 1-7, 1946-54. Uchen. Zap. L'vovsk. Gosud. Univ. Ivana Franko, Ser. Biol. HI 74128

Uchenye zapiski. Molotovskii gosudarstvennyi universitet imeni A. M. Gor'kogo. Molotov [=Perm], R S F S R. Vols. 4-10, 1940-57; vol. 11(2)-11(4), 1957 Uchen. Zap. Molotovsk. Gosud. Univ. Gor'kogo. For vol. 11(1) see: Uchenye zapiski. Permskii gosudarstvennyi universitet imeni A. M. Gor'kogo. Preceded & superseded by: Uchenye zapiski. Permskii gosudarstvennyi universitet imeni A. M. Gor'kogo. 4-3310-2. HI 74129

Uchenye Zapiski. Moskovskago Gorodskago Narodnago Universiteta imeni A. L. Sanyavskago. Moscow. Vols. 1-2, 1915-16. Uchen. Zap. Moskovsk. Gorodsk. Nar. Univ. Shanyavskago. HI 74130

Uchenye zapiski. Moskovskii gosudarstvennyi universitet. Moscow & Leningrad. Vol. 1+, 1933+. Uchen. Zap. Moskovsk. Gosud. Univ. Preceded by: Uchenya Zapiski Imperatorskago Moskovskago Universiteta. Otděl estestvenno-istoricheskii. 3-2766-3. HI 74131

Uchenye zapiski. Permskii gosudarstvennyi universitet imeni A. M. Gor'kogo. Perm, R S F S R. Vols. 1-3, 1935-39; vol. 11(1), 1959; vol. 12+, 1958+. Uchen. Zap. Permsk. Gosud. Univ. Gor'kogo. For vols. 4-10 and vol. 11(2-4) see: Uchenye zapiski. Molotovskii gosudarstvennyi universitet imeni A. M. Gor'kogo. 4-3310-2. HI 74132

Uchenye zapiski. Petrozavodskogo gosudarstvennogo universiteta. Vypusk 3. Biologicheskie nauki. Petrozavodsk, R S F S R. Vols. 7-8, 1956-57. Uchen. Zap. Petrozavodsk. Gosud. Univ., Vyp. 3, Biol. Nauki. Preceded by: Uchenye zapiski Karelo-Finskogo gosudarstvennogo universiteta. Vypusk 3. Biologicheskie nauki. Superseded by: Uchenye zapiski. Petrozavodskogo gosudarstvennogo universiteta. Vypusk 3. Biologicheskie i sel'skokhozyaistvennye nauki. HI 74133

Uchenye zapiski. Petrozavodskogo gosudarstvennogo universiteta. Vypusk 3. Biologicheskie i sel'skokhozyaistvennye nauki. Petrozavodsk, R S F S R. Vol. 9, 1958. Uchen. Zap. Petrozavodsk. Gosud. Univ., Vyp. 3, Biol. Sel'skokhoz. Nauki. Preceded by: Uchenye zapiski. Petrozavodskogo gosudarstvennogo universiteta. Vypusk 3. Biologicheskie nauki. HI 74134

Uchenye zapiski pyatigorskogo farmatsevticheskogo instituta. Stavropol. Vol. 1+, 1952+. Uchen. Zap. Pyatigorsk. Farm. Inst. HI 74135

Uchenye zapiski Rostovskogo-na-Donu gosudarstvennogo universiteta im. V. M. Molotova. Trudy biologicheskogo fakul'teta. Rostov. Vol. ?-1(5)-

?, ?-1946-? Uchen. Zap. Rostovsk. Donu Gosud. Univ. V. M. Molotova, Trudy Biol. Fak. HI 74136

Uchenye zapiski. Saratovskogo gosudarstvennogo imeni N. G. Chernyshevskogo universiteta. Saratov, R S F S R. Vols. 10-17 [also numbered vols. 1-8], 1923-30; vols. 9-13, 1931-35. Uchen. Zap. Saratovsk. Gosud. Chernyshevskogo Univ. Superseded by: Uchenye zapiski. Saratovskogo gosudarstvennogo universiteta imeni N. G. Chernyshevskogo. Ser. Biologicheskaya. 5-3783-1. HI 74137

Uchenye zapiski. Saratovskogo gosudarstvennogo universiteta imeni N. G. Chernyshevskogo. Saratov, R S F S R. Vol. 15+, 1940+. Uchen. Zap. Saratovsk. Gosud. Univ. Chernyshevskogo. Preceded by: Uchenye zapiski. Saratovskogo gosudarstvennogo universiteta imeni N. G. Chernyshevskogo. Ser. Biologicheskaya.; see also Uchenye zapiski. Saratovskogo gosudarstvennogo imeni N. G. Chernyshevskogo universiteta. 5-3783-1. HI 74138

Uchenye zapiski. Saratovskogo gosudarstvennogo universiteta imeni N. G. Chernyshevskogo. Ser. Biologicheskaya. Saratov, R S F S R. Vols. 1-2, 1937-39. Uchen. Zap. Saratovsk. Gosud. Univ. Chernyshevskogo, Ser. Biol. This series is part of vol. 14 (also numbered vol. 1) of: Uchenye zapiski. Saratovskogo gosudarstvennogo universiteta imeni N. G. Chernyshevskogo. Preceded by: Uchenye zapiski. Saratovskogo gosudarstvennogo imeni N. G. Chernyshevskogo universiteta. Superseded by: Uchenye zapiski. Saratovskogo gosudarstvennogo universiteta imeni N. G. Chernyshevskogo. HI 74139

Uchenye zapiski Tartuskogo gosudarstvennogo universiteta. Tartu, Estonian S S R. Vol. 37(5), 1941; 1946-50; vol. 35+, 1954+. Uchen. Zap. Tartusk. Gosud. Univ. Preceded by: Uchenya Zapiski Imperatorskago Yur'evskago Universiteta and by Trudy Voronežsk. Gosud. Univ. (vols. 1-12, 1925-41 only); numbering of vol. 37(1941) is computed from the addition of the volume numbers of the two preceding journals; from 1946-50, 34 unnumbered vols. were published. HI 74140

Uchenye zapiski, vladimirskii pedagogicheskii institut imeni P. I. Lebedeva-Polyanskogo. Seriya botanika. Vladimir. Vol. 1+, 1968+. Uchen. Zap. Vladim. Pedagog. Inst. P. I. Lebedeva-Polyanskogo, Ser. Bot. Vladimir - Vladimirskii Pedagogicheskii Institut imeni P. I. Lebedeva-Polyanska. HI 74141

Uganda journal. Kampala. 1934-42, 1946+. Uganda J. For 1943-45, see Bulletin of the Uganda society. 3-2285-3. HI 74142

Ugeskrift for agronomer. Copenhagen. 1967-71. Ugeskr. Agronomer. Preceded by: Ugeskrift for landmaend. Superseded by: Ugeskrift for agronomer og hortonomer. HI 74143

Ugeskrift for agronomer og hortonomer. Copenhagen. Vols. 1(117)-4(120), 1972-75. Ugeskr. Agronomer Hortonomer. Preceded by: Ugeskrift för agronomer and Horticultura. Superseded by: Ugeskrift for agronomer, hortonomer, forstkandidater og licentiater. HI 74144

Ugeskrift for agronomer, hortonomer, forstkandidater og licentiater. Copenhagen. Vols. 121-?, 1976-79. Ugeskr. Agronomer Hortonomer Forstkand. Licent. Preceded by: Ugeskrift for agronomer og hortonomer. Superseded by: Ugeskrift for jordbrug. HI 74145

Ugeskrift for jordbrug. Copenhagen. 1980+. Ugeskr. Jordbrug. Preceded by: Ugeskrift for agronomer, hortonomer, forstkandidater og licentiater. HI 74146

Ugeskrift for laeger. Copenhagen. Ugeskr. Laeger. See B–P–H 931/9. HI 60109

Ugeskrift for Landbostanden. Odense. ?-1868-? Ugeskr. Landbost. HI 74147

Ugeskrift for landmaend. Copenhagen. 1855-1902 [vols. for 1855-96 also numbered in 7 ser. of 10 vols. each; vols. for1897-1902 also numbered ser. 8, vols. 1-6]; 1919-66. Ugeskr. Landmaend. For 1903-15 superseded by: Foreningen af danskelandbrugskandidater. Medlemsblad [not entered]; for 1915-18 superseded by: Kobenhavns amt landboforening. Medlemsblad [not entered]. 5-4290-3. HI 74148

Uirusu = Virus. Tokyo.

Uitgaven. Natuurwetenschappelijke studiekring voor Suriname en Curaçao. The Hague. Uitgaven Natuurw. Studiekring Suriname Curaçao. See B–P–H 931/12. HI 60110

Uitgaven. Natuurwetenschappelijke studiekring voor Suriname en de Nederlandse Antillen. The Hague. Uitgaven Natuurw. Studiekring Suriname Ned. Antillen. See B–P–H 931/13. HI 60111

Uitgaven van de natuurwetenschappelijke studiekring voor Suriname en de Nederlandse Antillen. Natuurhistorische reeks. [Forms: Uitgaven van de natuurwetenschappelijke studiekring voor Suriname en de Nederlandse Antillen, no. 97+.] Utrecht. No. 1+, 1979+. Uitgaven Natuurwetensch. Studiekring Suriname Ned. Antillen, Natuurhist. Reeks. Preceded by: Uitgaven van de "Naturwetenschappelijke werkgroep Nederlandse Antillen." HI 74149

Uitgaven van de "Naturwetenschappelijke werkgroep Nederlandse Antillen." Willenstad. Nos. 1-22, 1951-76. Uitgaven Naturw. Werkgroep Ned. Antillen. Superseded by: Uitgaven van de natuurwetenschappelijke studiekring voor Suriname en de Nederlandse Antillen. Natuurhistorische reeks. HI 74150

Uitgezogte verhandelingen uit de nieuwste werken van de

societeiten der wetenschappen in Europa en van andere mannen. Amsterdam. Uitgez. Verh. Nieuwste Werken Soc. Wetensch. Eur. See B–P–H 931/15. HI 60112

Új magyar föld. Budapest. Új Magyar Föld. See B–P–H 931/17. HI 60113

Új magyar múzeum. Pest [=Budapest, in part]. Új Magyar Múz. See B–P–H 931/18. HI 60114

Ukrainian biochemistry (Soviet progress in biochemistry). [Translation of: Ukrayins'kyi biokhimichnyi zhurnal.] New York. Vol. ?-59+, ?-1987+. Ukrain. Biochem. HI 74151

Ukrainian botanical review = Ukrayins'kyi botanichnyi zhurnal. Kiev.

Ukrainian journal of biochemistry. [Translation of: Ukrayins'kyi biokhimichnyi zhurnal.] Jerusalem. Vol. 39(1-6), 1967. Ukrain. J. Biochem. HI 74152

Ukrainskii biokhimicheskii zhurnal = Ukrayins'kyi biokhimichnyi zhurnal. Kyyiv.

Ukrajins'kyj botaničnyj žurnal = Ukrayins'kyi botanichnyi zhurnal. Kiev.

Ukrayins'kyi biokhimichnyi zhurnal. Kyyiv. Vol. 7+, 1934-37; 1946+. Ukrayins'k. Biokhim. Zhurn. Preceded by: Naukovi zapysky, biokhemichnyi instytut. Kyyiv. For 1938-41 see: Biokhimichnyi zhurnal. HI 74153

Ukrayins'kyi botanichnyi zhurnal. Kiev, Ukrainian S S R. Vols. 1-5, 1921-29; [n.s.] vol. 13+, 1956+. Ukrayins'k. Bot. Zhurn. For [n.s.] vols. 1-12 see: Botanichnyi zhurnal. Kiev. 1-121-3. HI 74154

Ulmus; Helsingin yliopiston kasvitieteellinen puutarha. Helsinki. Vol. 1+, 1985+. Ulmus. HI 74155

Ultramicroscopy. Amsterdam. Vol. 1+, 1975+. Ultramicroscopy. HI 74156

Umbelliferae newsletter. Kew. No. 1+, 1970 [1971]+. Umbelliferae Newslett. HI 74157

Umi = Sea. Tokyo.

Umozritel'nyja Izslědovanija Imperatorskoj Sanktperterburgskoj Akademii Nauk = Umozritel'nya Izslědovaniya Imperatorskoi Sanktperterburgskoi Akademii Nauk. St. Petersburg.

Umozritel'nya Izslědovaniya Imperatorskoi Sanktpeterburgskoi Akademii Nauk. St. Petersburg. Vols. 1-5, 1808-19. Umozr. Izsl. Imp. Sanktpeterburgsk. Akad. Nauk. 1-113-3. HI 74158

Umschau. Übersicht über Fortschritte und Bewegungen auf dem Gesamtgebiet der Wissenschaft. Frankfurt a. M. Umschau. See B–P–H 932/8. HI 60115

Umschau in Wissenschaft und Technik. Frankfurt a. M. Umschau Wiss. Techn. See B–P–H 932/9. HI 60116

Umtali museum society newsletter. Nos. 1-10-?, 1968-72-? Umtali Mus. Soc. Newslett. HI 74159

Unasylva. Washington, DC. Unasylva. See B–P–H 932/12. HI 60117

Under glass. Irvington, NY. Vol. 1+, 1947+. Under Glass. HI 74160

Under glass for the commercial grower. Irvington, NY. Vol. 1, 1948-56. Under Glass Commercial Grower. 5-4295-2. HI 74161

Underwater information bulletin. Guildford. Vol. 6+, 1974+. Underwater Inform. Bull. Preceded by: Underwater journal and information bulletin. HI 74162

Underwater journal and information bulletin. Guildford. Vols. 3-5, 1971-73. Underwater J. Inform. Bull. Preceded by: Underwater science and technology information bulletin and Underwater science and technology journal. Superseded by: Underwater information bulletin. HI 74163

Underwater naturalist; bulletin of the american littoral society. Highlands, NJ. Vol. 1+, 1962+. Underwater Naturalist. HI 74164

Underwater science and technology information bulletin. Guildford. 1970. Underwater Sci. Technol. Inform. Bull. Superseded by: Underwater journal and information bulletin. HI 74165

Underwater science and technology journal. Guildford. 1970. Underwater Sci. Techn. J. Superseded by: Underwater journal and information bulletin. HI 74166

Underwater world. New Malden. 1966-67. Underwater World. HI 74167

UNESCO reports in marine science = U N E S C O reports in marine science. Paris.

UNESCO technical papers in marine science. New York, NY = U N E S C O technical papers in marine science. New York, NY.

Ungarische Agrar-Rundschau. Budapest. Ung. Agrar-Rundschau. See B–P–H 932/14. HI 60118

Ungarische botanische Blätter = Magyar botanikai lapok. Budapest. Magyar Bot. Lapok. See B–P–H 542/25.

Ungarische medicinisch-chirurgische Presse. Budapest. Ung. Med.-Chir. Presse. See B–P–H 932/18. HI 60119

Ungarische Revue. Budapest. Ung. Rev. See B–P–H 932/19. HI 60120

Ungarische Rosenzeitung. Temeschburg [=Timisoara, Rumania]. Ung. Rosenzeitung. See B–P–H 932/20. HI 60121

Ungarische Staats- und gelehrte Nachrichten. Ofen [=Budapest, in part]. Ung. Staats- Gel. Nachr. See B–P–H 932/21. HI 60122

Ungrisches Magazin, oder Beyträge zur ungarischen Geschichte, Geographie, Naturwissenschaft und der dahin einschlagenden Litteratur. Pressburg [=Bratislava, Czechoslovakia]. Ungrisches Mag. See B–P–H 932/23. HI 60123

Ungzami bulletin. Loughborough. No. 1+, 1976+. Ungzami Bull. HI 74168

Unicornis. ?-1982+. Unicornis. HI 74169

Union of Burma journal of life sciences. Rangoon. Vol. 1+, 1968+. Union Burma J. Life Sci. HI 74170

Union internationale des sciences biologiques. Series A, générale. [Place varies.] 1st-8th, 1919-35. Union Int. Sci. Biol. Sér. A, Gén. Superseded by: Proceedings, international union of biological sciences, general assemblies. HI 74171

Union internationale des sciences biologiques. Série B: Colloques. Paris. Union Int. Sci. Biol., Sér. B, Colloques. See B–P–H 932/26. HI 60124

Union internationale des sciences biologiques. Série C: Publications diverses. Paris. Union Int. Sci. Biol., Sér. C, Publ. Diverses. See B–P–H 932/27. HI 60125

Unión médica. Caracas. Unión Méd. (Caracas). See B–P–H 933/1. HI 60126

Unión médica. Castellón, Spain. Unión Méd. (Castellón). See B–P–H 933/2. HI 60127

Unión médica. Lérida, Spain. Unión Méd. (Lérida). See B–P–H 933/3. HI 60129

Unión médica. San Salvador, Salvador. Unión Méd. (San Salvador). See B–P–H 933/4. HI 60130

Unión médica. Santiago. Unión Méd. (Santiago). See B–P–H 933/5. HI 60131

Unitab; revue de l'union internationale des planteurs et producteurs de tabac. Paris. Vol. 1+, 1954+. Unitab. HI 74172

United Arab Republic journal of botany. Cairo. Vols. 3-12, 1960-71. U.A.R. J. Bot. Preceded & superseded by: Egyptian journal of botany. HI 74173

United Arab Republic journal of microbiology. Cairo. Vols. 5-6, 1970-71. U.A.R. J. Microbiol. Preceded by: Journal of microbiology of the United Arab Republic. Superseded by: Egyptian journal of microbiology. HI 74174

United Arab Republic journal of pharmaceutical sciences. Cairo. Vols. 11-12, 1970-71. U.A.R. J. Pharm. Sci. Preceded by: Journal of the pharmaceutical sciences of the United Arab Republic. Superseded by: Egyptian journal of pharmaceutical sciences. HI 74175

United avocado growers bulletin. La Habra, CA. United Avocado Growers Bull. See B–P–H 933/6.HI 60133

United florists news. Toronto. United Florists News. See B–P–H 933/7. HI 60134

United gardeners' and land-stewards' journal. London. Vols. 1(1)-3(26), 1845-47. United Gard. Land-Stewards' J. Superseded by: Gardeners' and farmers' journal. 5-4309-2. HI 74176

United States, see also U S

United States department of agriculture ..., see U S department of agriculture

United States department of agriculture, forest service, see United States forest service.

United States forest service. America's woods (series). Washington, DC. 1927+. U.S. Forest Serv., Amer. Woods. HI 74177

United States forest service. Division of silvics. Translations. Washington, DC. U.S. Forest Serv. Div. Silvics Transl. See B–P–H 939/30. HI 60206

United States forest service research notes; institute of tropical forestry. Rio Piedras, PR = Research notes I T F, United States forest service. Rio Piedras, PR.

United States naval medical bulletin. Washington, DC. U.S. Naval Med. Bull. See B–P–H 940/3. HI 60208

Uniterra. Nairobi. Vols. 1-7, 1976-82. Uniterra. Preceded & superseded by: U N E P news. HI 74178

Univers du vivant. Paris. Vol. 1+, 1985+. Univers Viv. HI 74179

Universe alive = Univers du vivant. Paris.

Universitas carolina biologica. Prague. Univ. Carol.. Biol. See B–P–H 935/3. HI 60154

Universitas seoulensis. Collectio theseon. Scientia naturalis. [Seoul taehak-kyo nonmun-jip. Chayon kwahak.] Seoul, Korea. Univ. Seoul. Collect. Theseon, Sci. Nat. See B–P–H 938/4. HI 60181

Universitäten-Almanach für das Jahr 1811. Neustrelitz = Jahrbuch der Universitäten Deutschlands. Neustrelitz.

Universitätes botaniskä darzä raksti. Riga = Acta horti botanici universitatis. Riga, Latvian S S R.

Université de Besançon. Institut de botanique, Prof. A. Mangin. Lyon. Vols. 1-7, 1899-1900. Univ. Besançon, Inst. Bot. Prof. A. Mangin. HI 74180

Universitet i Bergen årbok. Naturvidenskapelig rekke. Bergen. 1940-60. Univ. Bergen Årbok, Naturvidensk. Rekke. Preceded by: Bergens museums årbok. Superseded by: Aarbok for universitet i Bergen. Matematisk-naturvitenskapelig serie. HI 74181

Universitet i Bergen årsberetning = Årsberetning, universitet i Bergen, botanisk museum. Bergen.

Universitet i Bergen skrifter. Bergen. 1948-60. Univ. Bergen Skr. Preceded by: Bergens museums skrifter. HI 74182

Universitetskija Izvěstija. Kiev [Ukrainian S S R]. Univ.

Izv. See B–P–H 936/10. HI 60168

University of Alaska agricultural experiment station. Progress report. College, AK. Univ. Alaska Agric. Exp. Sta. Progr. Rep. See B–P–H 933/9. HI 60135

University of Alaska agricultural experiment station. Report of progress. College, AK. Univ. Alaska Agric. Exp. Sta. Rep. Progr. See B–P–H 933/10. HI 60136

University of Arizona. Arizona agricultural experiment station. Annual report. Tucson, AZ. Univ. Arizona Agric. Exp. Sta. Annual Rep. See B–P–H 933/13. HI 60137

University of Arizona. Arizona agricultural experiment station. Bulletin. Tucson, AZ. Univ. Arizona Agric. Exp. Sta. Bull. See B–P–H 933/14. HI 60138

University of Arizona. Arizona agricultural experiment station. Circular. Tucson, AZ. Univ. Arizona Agric. Exp. Sta. Circ. See B–P–H 933/15. HI 60139

University of Arizona. Arizona agricultural experiment station. Special report. Tucson, AZ. Univ. Arizona Agric. Exp. Sta. Special Rep. See B–P–H 933/17. HI 60140

University of Arizona. Arizona agricultural experiment station. Technical bulletin. Tucson, AZ. Univ. Arizona Agric. Exp. Sta. Techn. Bull. See B–P–H 933/18. HI 60141

University of Arizona. Arizona agricultural experiment station. Timely hints. Tucson, AZ. Univ. Arizona Agric. Exp. Sta. Timely Hints. See B–P–H 933/19. HI 60142

University of Arkansas. The college of agriculture. Arkansas agricultural experiment station. Bulletin. Fayetteville, AR. Univ. Arkansas Coll. Agric. Arkansas Agric. Exp. Sta. Bull. See B–P–H 934/1. HI 60143

University of Arkansas. The college of agriculture. Arkansas agricultural experiment station. Circular. Fayetteville, AR. Univ. Arkansas Coll. Agric. Arkansas Agric. Exp. Sta. Circ. See B–P–H 934/2. HI 60144

University of Arkansas. The college of agriculture. Arkansas agricultural experiment station. Mimeographed series. Fayetteville, AR. Univ. Arkansas Coll. Agric. Arkansas Agric. Exp. Sta., Mimeogr. Ser. See B–P–H 934/3. HI 60145

University of Auckland gazette. Auckland. Vol. 1+, 1959+. Univ. Auckland Gaz. HI 74183

University of Baghdad, natural history research publications = Publications, natural history research center. Baghdad.

University of British Columbia publications. Biological sciences = Publications of the university of British Columbia. Biological sciences. Vancouver.

University of California agricultural experiment station. Annual report. Berkeley, CA. Univ. Calif. Agric. Exp. Sta. Annual Rep. See B–P–H 934/7. HI 60146

University of California agricultural experiment station, bulletin. Berkeley, CA. Nos. 1-870, 1883-1975. Univ. Calif. Agric. Exp. Sta. Bull. Superseded by: Bulletin, division of agricultural sciences, university of California. HI 74184

University of California agricultural experiment station. Reports. Berkeley, CA. 1883-1904 [suspended 1905-12]. Univ. Calif. Agric. Exp. Sta. Rep. Superseded by: Report of the college of agriculture and the agricultural experiment station of the university of California. HI 74185

University of California botanical garden newsletter = Newsletter, friends of the botanical garden, university of California.

University of California, college of agriculture, agricultural experiment station, bulletin = University of California agricultural experiment station, bulletin. Berkeley, CA.

University of California, college of agriculture, agricultural experiment station. Circular. Berkeley, CA. Univ. Calif. Coll. Agric. Exp. Sta. Circ. See B–P–H 934/14. HI 60147

University of California, college of agriculture, agricultural experiment station. Seed bulletin. Berkeley, CA. Univ. Calif. Coll. Agric. Exp. Sta. Seed Bull. See B–P–H 934/15. HI 60148

University of California, college of agriculture, agricultural experiment station. Technical papers. Berkeley, CA. Univ. Calif. Coll. Agric. Exp. Sta. Techn. Pap. See B–P–H 934/16. HI 60149

University of California, occasional papers. Berkeley, CA. Univ. Calif. Occas. Pap. See B–P–H 934/17. HI 60150

University of California publications in agricultural sciences. Berkeley, CA. Univ. Calif. Publ. Agric. Sci. See B–P–H 934/20. HI 60151

University of California publications in botany. Berkeley, CA. Univ. Calif. Publ. Bot. See B–P–H 934/21. HI 60152

University of California publications in entomology. Berkeley, CA. Vol. 1+, 1906+. Univ. Calif. Publ. Entomol. HI 74186

University of California publications in geography. Berkeley, CA. 1913+. Univ. Calif. Publ. Geogr. HI 74187

University of California publications in geological sciences. Berkeley, CA. Univ. Calif. Publ. Geol. Sci. See B–P–H 934/22. HI 60153

University of Colorado studies. Boulder, CO. Vols. 1-29,

1902/03-57. Univ. Colorado Stud. HI 74189

University of Colorado studies. Series in bibliography. Boulder, CO. No. 1-31-?, 1950-70-?. Univ. Colorado Stud., Ser. Bibliogr. HI 74190

University of Colorado studies. Series in biology. Boulder, CO. Nos. 1-31, 1950-70. Univ. Colorado Stud. Ser. Biol. Preceded by: University of Colorado studies. Series D, physical and biological sciences. HI 74191

University of Colorado studies. Series D, physical and biological sciences. Boulder, CO. Vols. 1-2, 1940-47. Univ. Colorado Stud., Ser. D, Phys. Sci. Preceded by: University of Colorado studies. Superseded by: University of Colorado studies. Series in biology. 2-1109-3. HI 74192

University of Connecticut occasional papers. Biological science series. Storrs, CT. Vol. 1+, 1966+. Univ. Connecticut Occas. Pap., Biol. Sci. Ser. HI 74193

University of Delaware agricultural experiment station. Annual report. Newark, DE. Univ. Delaware Agric. Exp. Sta. Annual Rep. See B–P–H 935/11. HI 60155

University of Delaware agricultural experiment station bulletin. Newark, DE. Univ. Delaware Agric. Exp. Sta. Bull. See B–P–H 935/12. HI 60156

University of Florida agricultural experiment station bulletin. Gainesville, FL. Univ. Florida Agric. Exp. Sta. Bull. See B–P–H 935/14. HI 60157

University of Hawaii agricultural experiment station. Circular. Honolulu, HI. Univ. Hawaii Agric. Exp. Sta. Circ. See B–P–H 935/16. HI 60158

University of Hawaii bulletin. Honolulu, HI. Univ. Hawaii Bull. (Honolulu, 1917-19). See B–P–H 935/17. HI 60159

University of Hawaii bulletin. Honolulu, HI. Univ. Hawaii Bull. (Honolulu, 1927+). See B–P–H 935/18. HI 60160

University of Hawaii, quarterly bulletin. Honolulu, HI. Univ. Hawaii Quart. Bull. See B–P–H 935/21. HI 60161

University of Ibadan botanical studies. Ibadan, Nigeria. Univ. Idaban Bot. Stud. See B–P–H 935/24. HI 60162

University of Idaho, college of agriculture, Idaho agricultural experiment station annual report. Moscow, ID. Univ. Idaho Coll. Agric. Idaho Agric. Exp. Sta. Annual Rep. See B–P–H 936/3. HI 60163

University of Idaho, college of agriculture, Idaho agricultural experiment station bulletin. Moscow, ID. Univ. Idaho Coll. Agric. Idaho Agric. Exp. Sta. Bull. See B–P–H 936/4. HI 60164

University of Idaho, college of agriculture, Idaho agricultural experiment station, miscellaneous special reports. Moscow, ID. Univ. Idaho Coll. Agric. Idaho Agric. Exp. Sta. Misc. Special Rep. See B–P–H 936/5. HI 60165

University of Idaho, college of agriculture, Idaho agricultural experiment station, special reports. Moscow, ID. Univ. Idaho Coll. Agric. Idaho Agric. Exp. Sta. Special Rep. See B–P–H 936/6. HI 60166

University of Illinois agricultural experiment station bulletin. Champaign, IL. Univ. Illinois Agric. Exp. Sta. Bull. See B–P–H 936/7. HI 60167

University of Illinois bulletin. Urbana, IL. Vol. 1+, 1909+. Univ. Illinois Bull. HI 74194

University of Iowa studies in natural history = Studies in natural history, Iowa university. Iowa City, IA.

University of Kansas paleontological contributions. Papers. Lawrence, KS. Univ. Kansas Paleontol. Contr. Pap. See B–P–H 936/13. HI 60169

University of Kansas science bulletin. Univ. Kansas Sci. Bull. See B–P–H 936/14. HI 60170

University of London bulletin. London. ?-1975+. Univ. London Bull. HI 74195

University of Maryland graduate school chronicle. College Park, MD. Vol. 8+, 1974?+ [vols. 7(4), 8(4), 16(4) & 17(2-4) not published]. Univ. Maryland Graduate School Chron. Preceded by: Graduate school chronicle, university of Maryland. HI 74196

University of Massachusetts agricultural experiment station. Progress report. Amherst, MA. Univ. Mass. Agric. Exp. Sta. Progr. Rep. See B–P–H 936/28. HI 60171

University of Massachusetts agricultural experiment station. Research in review. Amherst, MA. Univ. Mass. Agric. Exp. Sta. Res. Rev. See B–P–H 936/29. HI 60172

University of Michigan studies. Scientific series. Ann Arbor, MI. Univ. Michigan Stud., Sci. Ser. See B–P–H 937/2. HI 60173

University of Missouri studies; science series. Columbia, MO. Univ. Missouri Stud., Sci. Ser. See B–P–H 937/9. HI 60174

University of Nanking magazine. [Chin ling kuang.] Nanking. Univ. Nanking Mag. See B–P–H 937/11. HI 60175

University of New Mexico. Bulletin. Anthropology series = Bulletin of the university of New Mexico, anthropology series. Albuquerque, NM.

University of New Mexico. Bulletin. Archaeology series = Bulletin of the university of New Mexico, anthropology series. Albuquerque, NM.

University of New Mexico. Bulletin. Biological series.

Albuquerque, NM. Vols. 1-3, 1901-41. Univ. New Mexico Bull., Biol. Ser. Superseded by: University of New Mexico publications in biology. HI 74197

University of New Mexico. Bulletin. Monograph series. Albuquerque, NM. Univ. New Mexico Bull., Monogr. Ser. See B–P–H 937/18. HI 60176

University of New Mexico publications in biology. Albuquerque, NM. Vol. 1+, 1946+. Univ. New Mexico Publ. Biol. Preceded by: University of New Mexico. Bulletin. Biological series. HI 74198

University of Oklahoma bulletin = Proceedings of the Oklahoma academy of science. Norman, OK. Proc. Oklahoma Acad. Sci. See B–P–H 734/3.

University of Oregon publications. Plant biology series. Eugene, OR. Vol. 1, no. 1, 1929. Univ. Oregon Publ., Pl. Biol. Ser. Superseded by: Oregon state monographs. Studies in botany. HI 74199

University of Pretoria, faculty of agriculture. Bulletin. Pretoria. Univ. Pretoria Fac. Agric. Bull. See B–P–H 937/24. HI 60177

University of Queensland, department of biology, papers. Brisbane. Univ. Queensland Dept. Biol. Pap. See B–P–H 937/26. HI 60178

University of Queensland papers, department of biology = Papers from the department of biology, university of Queensland. Brisbane, Qld.

University of Queensland papers, department of botany = Papers from the department of botany, university of Queensland. Brisbane, Qld.

University of Rajasthan studies. Biological sciences. Jaipur. 1958-62/63. Univ. Rajasthan Stud., Biol. Sci. Preceded by: University of Rajputana studies. Biological sciences. HI 74200

University of Rajputana studies. Biological sciences. Jaipur. 1951-55. Univ. Rajputana Stud., Biol. Sci. Superseded by: University of Rasjasthan studies. Biological sciences. HI 74201

University of Rochester library bulletin. Rochester, NY. Vol. 1+, 1945+. Univ. Rochester Libr. Bull. HI 74883

University of Ryukyus, extension service. [Ryukyu daigaku kenkyu fukyubu.] Naha?, Okinawa. Univ. Ryukyus Extens. Serv. See B–P–H 937/28. HI 60179

University science journal, university of Dar es Salaam. Dar es Salaam. Vol. 1+, 1975+. Univ. Sci. J. Univ. Dar es Salaam. HI 74202

University of south Florida botanical garden newsletter = Botanical garden newsletter, university of south Florida botanical garden. Tampa, FL.

University of southern California biological science series. Los Angeles, CA. Univ. S. Calif. Biol. Sci. Ser. See B–P–H 937/29. HI 60180

University studies, university of Karachi. Karachi. Vol. 1+, 1964+. Univ. Stud. (Karachi). HI 74203

University of Tennessee arboretum society bulletin. Oak Ridge, TN. Vol. 1+, 1965+. Univ. Tennessee Arbor. Soc. Bull. HI 74204

University of Toronto studies. Biological series. Toronto. Nos. 1-62, 1898-1956. Univ. Toronto Stud., Biol. Ser. Preceded by: Publications from the biological laboratory. [University of Toronto]. 5-4238-1. HI 74205

University of Utah biological series. Salt Lake City, UT. Univ. Utah Biol. Ser. See B–P–H 938/5. HI 60182

University of Washington arboretum bulletin = Arboretum bulletin, university of Washington. Seattle, WA.

University of Washington publications in biology. Seattle, WA. Univ. Wash. Publ. Biol. See B–P–H 938/9. HI 60183

University of Washington publications in botany. Seattle, WA. Univ. Wash. Publ. Bot. See B–P–H 938/10. HI 60184

University of Washington publications in oceanography. Seattle. 1932-60. Univ. Washington Publ. Oceanogr. Preceded by: Publications of the Puget Sound biological station of the university of Washington. HI 74206

University of Waterloo biology series. Waterloo, Ont. No. 1+, 1971+. Univ. Waterloo Biol. Ser. HI 74207

Universum. Welt, Wissen, Fortschritt. Vienna. Universum (Vienna). See B–P–H 938/14. HI 60187

University of Wyoming publications. Laramie, WY. Univ. Wyoming Publ. See B–P–H 938/12. HI 60185

University of Wyoming publications in science. Botany. Laramie, WY. Univ. Wyoming Publ. Sci., Bot. See B–P–H 938/13. HI 60186

Unser Egerland. Eger, Bohemia [=Cheb, Czechoslovakia]. Unser Egerland. See B–P–H 938/15. HI 60188

Unser Ostland. Heimatkundliche Arbeiten, herausgegeben vom Preussischen Botanischen Verein zu Königsberg. Königsberg [=Kaliningrad, R S F S R]. Unser Ostl. See B–P–H 938/16. HI 60189

Unser Pommerland. [Germany.] Vol. ?-15-?, ?-1930-? Unser Pommerland. HI 74208

Unsere Heimat. Vienna. Unsere Heimat (Vienna). See B–P–H 938/17. HI 60190

Unsere Heimat. Zwickau. Unsere Heimat (Zwickau). See B–P–H 938/18. HI 60191

Unsere Welt. Illustrierte Monatsschrift zur Förderung der Naturerkenntnis. Bonn. Unsere Welt. See B–P–H 938/19. HI 60192

Unterhaltendes Magazin zur Verbreitung der Natur- und Weltkenntniss. Vol. ?-2-?, ?-1807-? Unterhalt. Mag. Verbreit. Natur- Weltkenntn. HI 74209

Unterhaltungen aus dem Gebiethe der Naturwissenschaften. Brünn [=Brno, Czechoslovakia]. Unterhalt. Naturwiss. See B–P–H 938/21. HI 60193

Untersuchungen zur angewandten Bodenbiologie. Göttingen. Vol. 1+, 1954+. Untersuch. Angew. Bodenbiol. HI 74210

Untersuchungen aus dem Botanischen Institut zu Tübingen. Leipzig. Untersuch. Bot. Inst. Tübingen. See B–P–H 938/22. HI 60194

Untersuchungen aus dem botanischen Laboratorium der Universität Göttingen. Berlin. Untersuch. Bot. Lab. Univ. Göttingen. See B–P–H 938/23. HI 60195

Untersuchungen aus dem Forstbotanischen Institut zu München. Berlin. Untersuch. Forstbot. Inst. München. See B–P–H 938/24. HI 60196

Untersuchungen über die Natur des Menschen, der Thiere und der Pflanzen. Heidelberg. Untersuch. Natur Menschen. See B–P–H 939/1. HI 60197

Uomo e la natura; bollettino del comitato di Vicenza del movimento Italiano protezione della natura. ?-1967+. Uomo & Nat. HI 74211

Uppsala newsletter history of science. Uppsala. Vol. 1+, 1984+. Uppsala Newslett. Hist. Sci. HI 74212

Uppsala universitets årsskrift. Uppsala. Uppsala Univ. Årsskr. See B–P–H 939/3. HI 60199

Uránia. Budapest. Uránia (Budapest). See B–P–H 939/6. HI 60200

Urania. Jena. Urania (Jena). See B–P–H 939/7. HI 60201

Urania. Vienna. Urania (Vienna). See B–P–H 939/8. HI 60202

Urban forest forum. Washington, DC. Vols. 8(5)-10(2), 1988-90. Urban Forest Forum. Preceded by: National urban forest forum. Superseded by: Urban forests. HI 74533

Urban forests. Washington, DC. Vols. 10(3)+, 1990+. Urban Forests. Preceded by: Urban forest forum. HI 74884

Urda, et norsk antiqvarisk-historisk tidsskrift. Bergen. Urda. See B–P–H 939/10. HI 60203

Uredineana; recueil d'études systématiques et biologiques sur les urédinées du globe. Paris. Uredineana. See B–P–H 939/11. HI 60204

Úroda. Vols. 16-17, 1968-69; vol. 24+, 1976+. Úroda. Preceded by: Za vysokou úrodu. For vols. 18-22, 1970-75 see: Póda a úroda. HI 74213

Urologic and cutaneous review. St. Louis, MO. Urol. Cutan. Rev. See B–P–H 939/14. HI 60205

Useful palms of tropical america, newsletter. Brasilia = Newsletter, useful palms of tropical america. Brasilia.

Usimliklarni dastlab eka boshlash va iqlimlashtirish = Introduktsiya i akklimatizatsiya rastenii. Tashkent.

Uspehi mikrobiologii = Uspekhi mikrobiologii. Moscow.

Uspehi sovremennoj biologii = Uspekhi sovremennoi biologii. Moscow.

Uspekhi biologicheskoi khimii. Moscow. Vol. 1+, 1950+. Uspekhi Biol. Khim. HI 74214

Uspekhi mikrobiologii. Moscow. No. 1+, 1964+. Uspekhi Mikrobiol. HI 74215

Uspekhi sovremennoi biologii. Moscow. Vol. 1+, 1932+. Uspekhi Sovrem. Biol. 5-4344-2. HI 74216

Uspekhi sovremennoi genetiki. Moscow. Vol. 1+, 1967+. Uspekhi Sovrem. Genet. HI 75124

Utafiti; occasional papers of the national museums of Kenya. Nairobi. ?-1988+. Utafiti. HI 74217

Utah agricultural college, agricultural experiment station. Biennial report. Logan, UT. Utah Agric. Exp. Sta. Bienn. Rep. See B–P–H 945/28. HI 60259

Utah agricultural college, agricultural experiment station. Bulletin. Logan, UT. Utah Agric. Exp. Sta. Bull. See B–P–H 945/29. HI 60260

Utah agricultural college, agricultural experiment station. Circular. Logan, UT. Utah Agric. Exp. Sta. Circ. See B–P–H 945/30. HI 60261

Utah agricultural college, agricultural experiment station. Leaflet. Logan, UT. Utah Agric. Exp. Sta. Leafl. See B–P–H 945/31. HI 60262

Utah agricultural college, agricultural experiment station. Mimeographed series. Logan, UT. Utah Agric. Exp. Sta., Mimeogr. Ser. See B–P–H 945/32. HI 60263

Utah agricultural college, agricultural experiment station. Miscellaneous publication. Logan, UT. Utah Agric. Exp. Sta. Misc. Publ. See B–P–H 945/33. HI 60264

Utah agricultural college, agricultural experiment station. Special report. Logan, UT. Utah Agric. Exp. Sta. Special Rep. See B–P–H 945/34. HI 60265

Utah farm and home science. Logan, UT. Vols. 23-27, 1962-66. Utah Farm Home Sci. Preceded by: Farm and home science. Superseded by: Utah science. HI 74218

Utah science. Logan, UT. Vol. 28+, 1967+. Utah Sci. Preceded by: Utah farm and home science. HI 74219

Utrecht micropaleontological bulletins. Utrecht. Vol. 1+,

1969+. Utrecht Micropaleontol. Bull. HI 74220

Utrecht micropaleontological bulletins. Special publication. Utrecht. Vol. 1+, 1974+. Utrecht Micropaleontol. Bull., Special Publ. HI 74221

Utsunomiya daigaku nogakubu enshurin hokoku = Bulletin of the Utsunomiya university forests. Tochigi.

Utsunomiya daigaku nogakubu gakujutsu hokoku = Bulletin of the college of agriculture, Utsunomiya university. Utsunomiya.

Utsunomiya daigaku nogakubu gakujutsu hokoku tokushu = Special bulletin, college of agriculture, Utsunomiya university. Utsunomiya, Japan. Special Bull. Coll. Agric. Utsunomiya Univ. See B–P–H 850/10.

Utsunomiya koto norin gakko gakujutsu hokoku = Bulletin of the Utsunomiya agricultural college. Utsunomiya, Japan. Bull. Utsunomiya Agric. Coll. See B–P–H 287/8.

Utsunomiya tabako shikenjo hokoku = Bulletin of Utsunomiya tobacco experiment station. Tochigi, Japan. Bull. Utsunomiya Tobacco Exp. Sta. See B–P–H 287/11.

Utsunomiya tabako shikenjo tokubetsu hokoku = Special bulletin of the Utsunomiya tobacco experiment station. Oyama.

Utvalda; allmänt nyttiga och nyare merendels rön och samlingar i medicin, pharmacie, chemie, naturkunnighet, landhushållning, handel och slögder jämte utdrag af nöjsamare ämmen i naturalhistorie, verldsoch resebeskrifningen. Stockholm. Utvalda. See B–P–H 946/10. HI 60266

Uusi jakso = Virittäjä. Helsinki. Virittäjä. See B–P–H 966/6.

Uzbek biological journal = Uzbekskii biologicheskii zhurnal. Tashkent.

Ŭzbekiston filialining ahboroti. Tashkent = Izvestiya Uzbekistanskogo filiala Akademii nauk S S S R. Tashkent.

Ŭzbekiston S S R fanlar akademijasining ahboroti = Izvestiya Akademii nauk Uzbekskoi S S R. Tashkent.

Ŭzbekiston S S R fanlar akademijasining ahboroti = Izvestiya Akademii nauk Uzbekskoi S S R. Seriya biologicheskaya. Tashkent.

Ŭzbekiston S S R fanlar akademijasining bjulleteni = Byulleten′ Akademii nauk Uzbekskoi S S R. Tashkent.

Ŭzbekiston S S R fanlar akademijasining dokladlari = Doklady Akademii nauk Uzbekskoi S S R. Tashkent.

Uzbekskii biologicheskii zhurnal. Tashkent, Uzbek S S R. 1958+. Uzbeksk. Biol. Zhurn. HI 74222

Uzbekskij biologičeskij žurnal = Uzbekskii biologicheskii zhurnal. Tashkent.

V K W Mitteilungen. Stuttgart. Vols. 1-8, 1962-69. V. K. W. Mitt. Incorporated in: Stachelpost. HI 74223

V N K nieuws. Coöp. bond van nederl. coöp. kruidteleis verenigingen. Laren, Netherlands. 1952. V.N.K. Nieuws. HI 74224

V pomošč′ nabljudateljam prirody = V pomoshch′ nablyudatelyam prirody. Tartu.

V pomoshch′ nablyudatelyam prirody. Tartu, Estonian S S R. Vol. 1+, 1951+. V Pomoshch′ Nablyud. Prir. HI 74225

Vaderlandsche letteroefeningen. Amsterdam. Vaderl. Letteroefen. See B–P–H 946/22. HI 60268

Vaderlandsche letteroefeningen of tijdschrift voor kunsten en wetenschappen. Amsterdam. Vaderl. Letteroefen. Tijdschr. Kunsten. See B–P–H 946/23. HI 60269

Vaderlandsche magazijn voor wetenschap, konst en smaak. Amsterdam. Vaderl. Mag. Wetensch. See B–P–H 946/24. HI 60270

Vakblad voor biologen. Amsterdam. Vakbl. Biol. See B–P–H 946/25. HI 60271

Vakblad voor bloembollenteelt en handel. Hillegom, Netherlands. Vakbl. Bloembollenteelt Handel. See B–P–H 946/26. HI 60272

Vakblad voor de bloemisterij. Wageningen. ?-ca.1988+. Vakbl. Bloemist. HI 74226

Valvoja. Helsinki. Valvoja. See B–P–H 946/28. HI 60273

Valvoja-Aika = Valvoja. Helsinki. Valvoja. See B–P–H 946/28.

Van vigyan. Dehra Dun. Vol. ?-5+, ?-1967+. Van Vigyan. HI 74227

Van zee tot land; rijksdienst voor de Ijsselmeerpoldens. The Hague. Nos. 1-54, 1951-74. Van Zee Tot Land. HI 74228

Vanasarn, The. Bangkok. Vol. 1?-24(2)+, ?-1966+. Vanasarn. HI 74229

Våre nyttevekster. Heggedal, Norway. Våre Nyttevekster. See B–P–H 946/32. HI 60274

Vargasia; boletin de la sociedad de ciencias fisicas y naturales de Carácas. Carácas. Nos. 1-7, 1868-70. Vargasia. HI 74230

Variedades de ciencias, literatura y artes. Madrid. Varied. Ci. See B–P–H 946/33. HI 60275

Variedades; ó mensagero de Londres: Periódico trimestre. London. Variedades (London). See B–P–H 947/4. HI 60276

Variétés scientifiques recueilles par la société des sciences naturelles et physiques du Maroc. Rabat. 1921-60. Var. Sci. Soc. Sci. Nat. Phys. Maroc. HI 74231

Varsity; annual magazine of the university of Malaya, Kuala Lumpur. Kuala Lumpur. ?-1962+. Varsity. HI 74232

Varstvo narove. Ljubljana. Vol. 1+, 1962+. Varstvo Narove. HI 74233

Vasculum. London. Vasculum. See B–P–H 947/6. HI 60277

Vasi szemle. Szombathely, Hungary. Vasi Szemle. See B–P–H 947/7. HI 60278

Vasvármegyei múzeum természetrajzi osztályának évi jelentése. Szombathely, Hungary. Vasvárm. Múz. Term. Oszt. Évi Jel. See B–P–H 947/9. HI 60279

Vasvármegye és Szombathely város kultúregyesülete és a vasvármegyei múzeum évokönyve. Szombathely, Hungary. Vasvárm. Szombathely Város Kultúregyes. Vasvárm. Múz. Évk. See B–P–H 947/10. HI 60280

Vaterländische Blätter für den Oesterreichischen Kaiserstaat. Vienna. Vaterl. Blätt. Oesterr. Kaiserstaat. See B–P–H 947/15. HI 60284

Vaterländisches Archiv, oder Beiträge zur allseitigen Kenntniss des Königreichs Hannover, wie es war und ist. Celle. Vaterl. Arch. See B–P–H 947/11. HI 60281

Vaterländisches Archiv für das Herzogthum Lauenburg. Ratzeburg, Germany. Vaterl. Arch. Herzogth. Lauenburg. See B–P–H 947/12. HI 60282

Vaterländisches Archiv für Wissenschaft, Kunst, Industrie und Agricultur, oder Preussische Provinzialblätter. Königsberg [=Kaliningrad, R S F S R]. Vaterl. Arch. Wiss. See B–P–H 947/13. HI 60283

Vaterländisches Magazin für diejenigen Landwirthe, Forstmänner, ... welche über ihren Beruf nachdenken und die neuesten ihnen nützlichen Fortschritte ... kennen lernen wollen ... Prague. Vaterl. Mag. Landwirthe Forstmänner. See B–P–H 947/16. HI 60285

Växt-nährings-nytt; gödsel- och kalkindustriernas samarbetsdelegation efter sområd med statens nordbruksforsok. Stockholm. 1945-70; Summaries of articles 1960-70. Växt-Nährings-Nytt. HI 74235

Växtekologiska studier. Uppsala. Nos. 1-15, 1972-84? Växtekol. Stud. Superseded by: Studies in plant ecology. HI 74234

Växtodling. Uppsala. Váxtodling. See B–P–H 947/18. HI 60286

Växtskyddsnotiser. Statens växtskyddsanstalt. Stockholm. Växtskyddsnotiser. See B–P–H 947/19. HI 60287

Växtskyddsrapporter: Jordbruk. Uppsala. Vol. 1+, 1977+. Växtskyddsrapp. Jordbruk. HI 74236

Včlař. Prague. Včelař. See B–P–H 947/20. HI 60288

Včlár a ovocinár. Turocszentmarton, Hungary [=Turciansky Svaty Martin, Czechoslovakia]. Včelár & Ovocinár. See B–P–H 948/1. HI 60289

Včlařské listy. Prague. Včelařské Listy. See B–P–H 948/2. HI 60290

Včelařstvi. Prague. Vol. 1+, 1948+. Včelařstvi. HI 74237

Vecko-skrift för läkare och naturforskare. Stockholm. Vecko-Skr. Läkare Naturf. See B–P–H 948/3. HI 60291

Věda česká. Prague. Věda Česká. See B–P–H 948/8. HI 60294

Věda přírodni. Prague. Věda Přír. See B–P–H 948/9. HI 60295

Vědecké práce výzkumného ústavu bramborářského ČSAZV v Havlíčkově Broděe. Prague. Věd. Práce Výzk. Ústavu Bramboarářsk. ČSAZV v Havlíčkově Brodě. See B–P–H 948/6. HI 60292

Vědecké práce výzkumného ústavu kraivářského. Pohorelice, Czechoslovakia. Věd. Práce Výzk. Ústavu Kraivářsk. See B–P–H 948/7. HI 60293

Vědecké práce výzkumného ústavu okrasného zahradnictvi v průhonich. Prague. Vol. 1+, 1960+. Věd. Práce Výzk. Ústavu Okrasn. Zahradn. Průnhon. HI 74238

Vědecké práce výzkumného ústavu rostlinné výrobý v Praze-Ruzynyibb. Prague. Vol. 1+, 1955+. Věd. Práce Výzk. Ústavu Rostl. Výrobý Praze-Ruzyn. HI 74239

Vedetta agricola, La. Siena. 1910-19. Vedetta Agric. HI 74240

Vegetable crops bibliographies; a comprehensive series. West Covina, CA. Veg. Crops Bibliogr. See B–P–H 948/10. HI 60296

Vegetable grower. Thornhaugh. 1982(Summer)+. Veg. Grower. Preceded by: Newsletter, processors' and growers' research organisation. HI 74241

Vegetable improvement newsletter. Ithaca, NY. No.1+, 1959+. Veg. Improv. Newslett. HI 74242

Vegetable outlook and situation. Washington, DC. No. 219+, 1981+. Veg. Outlook Situation. Preceded by: Vegetable situation. HI 74243

Vegetable science. New Delhi. Vol. 1+, 1974+. Veg. Sci. HI 74244

Vegetable situation. Washington, DC. Vols. TVS6-TVS218, 1937-80. Veg. Situation. Preceded by: Fruit and vegetable situation. Superseded by: Vegetable outlook and situation. HI 74245

Vegetables. [Sosai.] Okayama, Japan. Vegetables. See B–P–H 948/17. HI 60299

Vegetables for the hot, humid tropics. New Orleans, LA.

Vol. 1+, 1978+. Veg. Hot Humid Trop. HI 74246

Vegetables newsletter. Fredericton. 1967+. Veg. Newslett. HI 74247

Vegetace C S S R. Reihe A. Prague. Vol. 1+, 1965+. Veg. C.S.S.R., A. HI 74248

Vegetační problémy při budování vodních děl. Prague. Veg. Probl. Budov. Vodních Dél. See B–P–H 948/14. HI 60297

Vegetatio; acta geobotanica. The Hague. Vegetatio. See B–P–H 948/18. HI 60300

Vegetatio U R S S = Rastitel′nost′ S S S R. Moscow & Leningrad. Rastitel′n. S.S.S.R. See B–P–H 756/12.

Vegetation of the Far North of U S S R and its utilization. Moscow & Leningrad = Rastitel′nost′ Krainego Severa S S S R i eë osvoenie. Moscow & Leningrad.

Vegetation ungarischer Landschaften. Budapest. Veg. Ung. Landschaften. See B–P–H 948/15. HI 60298

Vegetationsbilder. Jena. Vols. 1-26, 1903-44. Vegetationsbilder. HI 74249

Veld. Kirstenbosch. Vol. 1(1), 1970. Veld. Superseded by: Veld & flora (1971-74). HI 74250

Veld & flora (1971-74). Kirstenbosch Vols. 1(2)-4(34), 1971-74. Veld Fl. (1971-74). Preceded by: Veld. Superseded by: Veld & flora (1975+). HI 74251

Veld & flora (1975+); journal of the botanical society of South Africa. Kirstenbosch. Vol. 61+, 1975+. Veld Fl. (1975+). Preceded by: Journal, botanical society of South Africa. HI 74252

Vellosia; contribuições do museu botanico do Amazonas. Rio de Janeiro. Vols. 1-4, 1885-88; ed. 2, vols. 1-4, 1891-92. Vellosia. 5-4358-1. HI 74253

Vellozia; publicação do centro de pesquisas florestais e conservação do natureza. Rio de Janeiro. Vellozia. See B–P–H 948/21. HI 60301

Vendetta agricola. Siena, Italy. Vendetta Agric.. See B–P–H 948/22. HI 60302

Verein Deutscher Rosenfreunde e. V. in der Deutschen Gesellschaft für Gartenkultur e. V. Rosenjahrbuch. Berlin. Verin Deutsch Rosenfr. Rosenjahrb. See B–P–H 948/26. HI 60303

Verein für Naturkunde Mannheim. Jahres-Bericht. Mannheim. Verein Naturk. Mannheim Jahres-Ber. See B–P–H 949/1. HI 60304

Vereinigte Frauendorfer Blätter. Passau, Germany. Vereinigte Frauendorfer Blätt. See B–P–H 949/2. HI 60305

Vereins zum Schutze der Alpenpflanzen und -Tiere e. v. Munich. ?-1978+. Vereins Schutze Alpenpfl. Tiere. HI 74254

Vereinsschrift für die Forst-, Jagd- und Naturkunde. Prague. Vereinsschr. Forst- Naturk. See B–P–H 949/3. HI 60306

Vereniging voor het onderwijs in de biologie. Brussels. ?-1982+. Ver. Onderwijs Biol. HI 74255

Verhandelingen van het bataviaasch genootschap van kunsten en wetenschappen. Batavia, Dutch E. Indies [=Jakarta, Indonesia]. Verh. Batav. Genootsch. Kunsten. See B–P–H 949/13. HI 60310

Verhandelingen der eerste klasse van het hollandsch instituut van wetenschappen, letterkunde en schoone kunsten te Amsterdam. Amsterdam. Verh. Eerste Kl. Holl. Inst. Wetensch. Amsterdam. See B–P–H 950/12. HI 60321

Verhandelingen der eerste klasse van het koninklijk nederlandsch instituut van wetenschappen, letterkunde en schoone kunsten te Amsterdam. Amsterdam. Verh. Eerste Kl. Kon. Ned. Inst. Wetensch. Amsterdam. See B–P–H 950/13. HI 60322

Verhandelingen ... hollandsche maatschappye der weetenschappen, te Haarlem = Verhandelingen uitgegeeven door de hollandse maatschappy der weetenschappen, te Haarlem. Haarlem. Verh. Holl. Maatsch. Weetensch. Haarlem. See B–P–H 951/12.

Verhandelingen ... hollandse maatschappy der weetenschappen, te Haarlem. Haarlem. Verh. Holl. Maatsch. Weetensch. Haarlem. See B–P–H 951/12. HI 60333

Verhandelingen van het koninklijk belgisch instituut voor natuurwetenschappen = Mémoires de l'institut royal des sciences naturelles de Belgique. Brussels. Mém. Inst. Roy. Sci. Nat. Belgique. See B–P–H 575/6.

Verhandelingen van het koninklijk belgisch instituut voor natuurwetenschappen. Tweede reeks = Mémoires de l'institut royal des sciences naturelles de Belgique. Deuxième série. Brussels.

Verhandelingen. Koninklijk belgisch koloniaal instituut; afdeeling der natuur- en geneeskundige wetenschappen. [In octavo.] = Mémoires de l'institut royal colonial belge; section des sciences naturelles et médicales. Brussels. Mém. Inst. Roy. Colon. Belge, Sect. Sci. Nat. (8vo). See B–P–H 575/5.

Verhandelingen. Koninklijk belgisch koloniaal instituut; afdeeling der natuur- en geneeskundige wetenschappen. [In quarto.] = Mémoires de l'institut royal colonial belge; section des sciences naturelles et médicales. Brussels. Mém. Inst. Roy. Colon. Belge, Sect. Sci. Nat. (4to). See B–P–H 575/4.

Verhandelingen van het koninklijk museum van natuurlijke historie van Belgie = Mémoires du musée royal d'histoire naturelle de Belgique. Brussels. Mém. Mus. Roy. Hist. Nat. Belgique. See B–P–H 577/24.

Verhandelingen van het koninklijk natuurhistorisch

museum van Belgie. Buiten reeks = Mémoires du musée royal d'histoire naturelle de Belgique. Hors série. Brussels. Mém. Mus. Roy. Hist. Nat. Belgique, Hors Sér. See B–P–H 577/25.

Verhandelingen von het koninklijk natuurhistorisch museum van Belgie. Tweede reeks = Mémoires de musée royal d'histoire naturelle de Belgique. Deuzième série. Brussels. Mém. Mus. Roy. Hist. Nat Belgique, Sér. 2. See B–P–H 578/1.

Verhandelingen. Koninklijke academie voor koloniale wetenschappen. Klasse der natuur- en geneeskundige wetenschappen = Mémoires de l'académie royale des sciences coloniales. Classe des sciences naturelles et médicales. Brussels.

Verhandelingen der koninklijke nederlandsche akademie van wetenschappen. Afdeeling natuurkunde. Amsterdam. Vols. 1-29, 1854-92. Verh. Kon. Ned. Akad. Wetensch., Afd. Natuurk. Superseded by: Verhandelingen der koninklijke nederlandsche akademie van wetenschappen. Afdeeling natuurkunde; tweede sectie. 1-108-2. HI 74256

Verhandelingen der koninklijke nederlandsche akademie van wetenschappen. Afdeeling natuurkunde; tweede sectie Amsterdam. Vol. 1+, 1892+. Verh. Kon. Ned. Akad. Wetensch., Afd. Natuurk., Tweede Sect. Preceded by: Verhandelingen der koninklijke nederlandsche akademie van wetenschappen. Afdeeling natuurkunde. 1-108-2. HI 74257

Verhandelingen van de koninklijke vlaamsche academie voor wetenschappen, letteren en schone kunsten van België. Klasse der wetenschappen. Antwerp. Vol. 1+, 1939+. Verh. Kon. Vlaamse Acad. Wetensch. België, Kl. Wetensch. 5-4414-1. HI 74258

Verhandelingen ... maatschappij ter bevordering van den landbouw te Amsterdam. Amsterdam. Ver. Maatsch. Bevord. Landb. Amsterdam. See B–P–H 952/18. HI 60338

Verhandelingen van de natuur- en geneeskundige correspondentie-societeit in de vereenigde Nederlanden. The Hague. Verh. Natuur- Geneesk. Corresp.-Soc. Ver. Nederl. See B–P–H 954/2. HI 60354

Verhandelingen der natuurkundige vereeniging in Nederlandsch-Indië. = Acta societatis regiae scientiarum indo-neerlandicae. Batavia. Acta Soc. Regiae Sci. Indo-Neerl. See B–P–H 49/20.

Verhandelingen van het provinciaal utrechtsch genootschap van kunsten en wetenschappen. Utrecht. Verh. Prov. Utrechtsch Genootsch. Kunsten. See B–P–H 954/8. HI 60357

Verhandelingen, rijksinstituut voor natuurbeheer. Wageningen. No. 1+, 1970+. Verh. Rijksinst. Natuurbeheer. HI 74259

Verhandelingen ... Teyler's tweede genootschap. Haarlem. Verh. Teyler's Tweede Genootsch. See B–P–H 955/1. HI 60362

Verhandelingen uitgegeeven door de hollandsche maatschappye der weetenschappen, te Haarlem = Verhandelingen uitgegeeven door de hollandse maatschappy der weetenschappen, te Haarlem. Haarlem. Verh. Holl. Maatsch. Weetensch. Haarlem. See B–P–H 951/12.

Verhandelingen uitgegeeven door de hollandse maatschappy der weetenschappen, te Haarlem. Haarlem. Verh. Holl. Maatsch. Weetensch. Haarlem. See B–P–H 951/12. HI 60333

Verhandelingen uitgegeven door de maatschappij ter bevordering van den landbouw te Amsterdam. Amsterdam. Ver. Maatsch. Bevord. Landb. Amsterdam. See B–P–H 952/18. HI 60338

Verhandelingen, uitgegeeven door Teyler's tweede genootschap. Haarlem. Verh. Teyler's Tweede Genootsch. See B–P–H 955/1. HI 60362

Verhandelingen uitgegeven door het zeeuwsch genootschap der wetenschappen te Vlissingen. Middelburg, Netherlands. Verh. Zeeuwsch Genootsch. Wetensch. Vlissingen. See B–P–H 956/3. HI 60375

Verhandlungen der allgemeinen schweizerischen Gesellschaft für die gesammten Naturwissenschaften. Solothurn, Switzerland. Verh. Allg. Schweiz. Ges. Gesammten Naturwiss. See B–P–H 949/5. HI 60307

Verhandlungen und Arbeiten der vereinigten ökonomisch-patriotischen Societät der Fürstenthümer Schweidnitz und Jauer. Jauer, Germany [=Jawor, Poland]. Verh. Arbeiten Vereinigten Ökon.-Patriot. Soc. Fürstenth. Schweidnitz Jauer. See B–P–H 949/6. HI 60308

Verhandlungen und Aufsätze der K. K. Landwirthschafts-Gesellschaft in Steyermarck. Graz. Verh. Aufsätze K. K. Landw.-Ges. Steyermarck. See B–P–H 949/7. HI 60309

Verhandlungen des Berliner botanischen Vereins. Berlin. Vol. 1+, 1982+. Verh. Berliner Bot. Vereins. Preceded by: Verhandlungen des botanischen Vereins der Provinz Brandenburg. HI 74260

Verhandlungen der Berliner Dermatologischen Gesellschaft. Berlin. Verh. Berliner Dermatol. Ges. See B–P–H 949/15. HI 60311

Verhandlungen der Berliner medicinischen Gesellschaft. Berlin. Verh. Berliner Med. Ges. See B–P–H 949/16. HI 60312

Verhandlungen des botanischen Vereins der Provinz Brandenburg. Berlin. Vols. 1-115, 1859-1980. Verh. Bot. Vereins Prov. Brandenburg. Superseded by: Verhandlungen des Berliner botanischen Vereins.

HI 74261

Verhandlungen des botanischen Vereins für die Provinz Brandenburg (und die angrenzenden Länder) = Verhandlungen des botanischen Vereins der Provinz Brandenburg. Berlin.

Verhandlungen des Congresses für innere Medicin [or Medizin]. Wiesbaden. Verh. Congr. Innere Med. See B–P–H 949/23. HI 60313

Verhandlungen der Dermatologischen Vereinigung zu Berlin. Leipzig. Verh. Dermatol. Vereinigung Berlin. See B–P–H 950/1. HI 60314

Verhandlungen des deutschen Geographentages. Berlin. Verh. Deutsch. Geographentages. See B–P–H 950/3. HI 60316

Verhandlungen des Deutschen Gesellschaft für innere Medizin. Munich. Verh. Deutsch. Ge. Innere Med. See B–P–H 950/5. HI 60317

Verhandlungen des deutschen Kongresses für innere Medizin. Wiesbaden. Verh. Deutsch. Kongr. Innere Med. See B–P–H 950/6. HI 60318

Verhandlungen des Deutschen Pathologischen Gesellschaft. Berlin. Verh. Deutsch. Pathol. Ges. See B–P–H 950/8. HI 60320

Verhandlungen des deutschen wissenschaftlichen Verein zu Santiago de Chile. Valparaiso. Vols. 1-7, 1885/88-1910/13, 1887-1913?; n.s. vol. 1+, 1931+. Verh. Deutsch. Wiss. Verein Santiago de Chile. HI 74262

Verhandlungen deutscher Beauftragter für Naturschutz und Landschaftpflege. Bonn. Verh. Deutsch. Beauftragter Naturschutz. See B–P–H 950/2. HI 60315

Verhandlungen deutscher Landes- und Bezirkes-beauftragter für Naturschutz und Landschaftspflege. Egestorf, Germany. Verh. Deutsch. Landes-Bezirksbeauftragter Naturschutz. See B–P–H 950/7. HI 60319

Verhandlungen des Gartenbau-Vereins zu Erfurt. Erfurt. Verh. Gartenbau-Vereins Erfurt. See B–P–H 950/15. HI 60323

Verhandlungen des Gartenbau-Vereins für das Königreich Hannover. Hanover. Verh. Gartenbau-Vereins Königr. Hannover. See B–P–H 950/16. HI 60324

Verhandlungen der Gelehrten Esthnischen Gesellschaft zu Dorpat. Dorpat [=Tartu, Estonian S S R]. Verh. Gel. Esthn. Ges. Dorpat. See B–P–H 950/17. HI 60325

Verhandlungen der Gesellschaft zur Beförderung der Naturkunde und Industrie Schlesiens. Breslau [=Wroclaw, Poland]. Verh. Ges. Beförd. Naturk. Schlesiens. See B–P–H 950/18. HI 60326

Verhandlungen der Gesellschaft Deutscher Naturforscher und Ärzte. Leipzig, etc. Verh. Ges. Deutsch. Naturf. See B–P–H 951/2. HI 60327

Verhandlungen der Gesellschaft für Erdkunde zu Berlin. Berlin. Verh. Ges. Erdk. Berlin. See B–P–H 951/4. HI 60328

Verhandlungen der Gesellschaft von Freunden der Naturwissenschaften in Gera und des naturwissenschaftlichen Kränzchens in Schleiz. Gera, Germany. Verh. Ges. Freunden Naturwiss. Gera. See B–P–H 951/6. HI 60329

Verhandlungen der Gesellschaft naturforschender Freunde zu Berlin. Berlin. Verh. Ges. Naturf. Freunde Berlin. See B–P–H 951/8. HI 60330

Verhandlungen der Gesellschaft für Ökologie. The Hague. Vol. 2+, 1973+, 1974+. Verh. Ges. Ökol. Preceded by: Tagungsbericht der Gesellschaft für Ökologie. HI 74263

Verhandlungen der Gesellschaft des vaterländischen Museums in Böhmen. Prague. Verh. Ges. Vaterl. Mus. Böhmen. See B–P–H 951/9. HI 60331

Verhandlungen des Grossherzoglich Badischen Landwirthschaftlichen Vereins zu Ettlingen und Karlsruhe. Pforzheim. Verh. Grossherzogl. Bad. Landw. Vereins Ettlingen. See B–P–H 951/10. HI 60332

Verhandlungen des heil- und naturwissehschaftlichen Vereins zu Bratislava (Pressburg). Pressburg [=Bratislava, Czechoslovakia].

Verhandlungen der internationalen Vereinigung für theoretische und angewandte Limnologie = Internationale Vereinigung für theoretische und angewandte Limnologie. Verhandlungen. Stuttgart. Int. Vereinigung Theor. Limnol. Verh. See B–P–H 435/21.

Verhandlungen der Kaiserlich-Königlichen Gartenbaugesellschaft in Wien. Vienna. Verh. K. K. Gartenbauges. Wien. See B–P–H 951/19. HI 60334

Verhandlungen der Kaiserlich-Königlichen Landwirthschaftsgesellschaft in Wien. Vienna. Verh. K. K. Landwirthschaftsges. Wien. See B–P–H 951/20. HI 60335

Verhandlungen der kaiserlich-königlichen zoologisch-botanischen Gesellschaft in Wien. Vienna. Verh. K. K. Zool.-Bot. Ges. Wien. See B–P–H 952/1. HI 60336

Verhandlungen der Kaiserlichen Leopoldinisch-Carolinischen Academie der Naturforscher. Bonn = Nova acta physico-medica academiae caesareae leopoldino-carolinae naturae curiosorum exhibentia ephemerides sive observationes historias et experimenta ... Nuremberg. Nova Acta Phys.-Med. Acad. Caes. Leop.-Carol. Nat. Cur. See B–P–H 673/5.

Verhandlungen der Kaiserlichen Leopoldinisch-

Carolinischen Akademie der Naturforscher. Wroclaw & Bonn = Novorum actorum academiae caesareae leopoldinae-carolinae naturae curiosorum. Breslau [=Wroclaw, Poland] & Bonn. Nov. Actorum Acad. Caes. Leop.-Carol. Nat. Cur. See B–P–H 671/16.

Verhandlungen der Kaiserlich Leopoldinisch-Carolinischen Deutschen Akademie der Naturforscher. Dresden = Nova acta academiae caesareae leopoldino-carolinae germanicae naturae curiosorum. Dresden. Nova Acta Acad. Caes. Leop.-Carol. German. Nat. Cur. See B–P–H 672/15.

Verhandlungen der Kaiserlichen Leopoldino-Carolinischen deutschen Akademie der Naturforscher. Jena = Novorum actorum academia caesareae leopoldinae-carolinae germanicae naturae curiosorum. Verhandlungen der Kaiserlichen Leopoldinisch-Carolinischen Deutschen Akademie der Naturforscher. Jena. Nov. Actorum Acad. Caes. Leop.-Carol. German. Nat. Cur. See B–P–H 671/15.

Verhandlungen des Kongresses für innere Medizin. Wiesbaden. Verh. Kongr. Innere Med. See B–P–H 952/15. HI 60337

Verhandlungen der Leopoldinisch-Carolinischen Academie der Naturforscher. Erlangen = Nova acta physico-medica academiae caesareae leopoldino-carolinae naturae curiosorum exhibentia ephemerides sive observationes historias et experimenta ... Nuremberg. Nova Acta Phys.-Med. Acad. Caes. Leop.-Carol. Nat. Cur. See B–P–H 673/5.

Verhandlungen der Mecklenburgischen Naturforschenden Gesellschaft. Rostock. Verh. Mecklenburg. Naturf. Ges. See B–P–H 952/19. HI 60339

Verhandlungen der Meklenburgischen Naturforschenden Gesellschaft = Verhandlungen der Mecklenburgischen Naturforschenden Gesellschaft. Rostock. Verh. Mecklenburg. Naturf. Ges. See B–P–H 952/19.

Verhandlungen und Mitteilungendes siebenbürgischen Vereins für Naturwissenschaften zu Hermannstadt = Verhandlungen und Mittheilungen des siebenbürgischen Vereins für Naturwissenschaften zu Hermannstadt. Hermannstadt [=Sibiu, Rumania]. Verh. Mitth. Siebenbürg. Vereins Naturwiss. Hermannstadt. See B–P–H 952/23.

Verhandlungen und Mittheilungen der K. K. Patriotisch-ökonomischen Gesellschaft im Königreiche Böhmen. Prague. Verh. Mitth. K. K. Patriot.-Ökon. Ges. Königr. Böhmen. See B–P–H 952/21. HI 60340

Verhandlungen und Mittheilungen des siebenbürgischen Vereins für Naturwissenschaften zu Hermannstadt. Hermannstadt [=Sibiu, Rumania]. Verh. Mitth. Siebenbürg. Vereins Naturwiss. Hermannstadt. See B–P–H 952/23. HI 60341

Verhandlungen der Naturforschenden Gesellschaft in Basel. Basel. Verh. Naturf. Ges. Basel. See B–P–H 953/1. HI 60342

Verhandlungen des Naturforschenden Vereins in Brünn. Brünn [=Brno, Czechoslovakia]. Verh. Naturf. Vereins Brünn. See B–P–H 953/4. HI 60343

Verhandlungen des Naturhistorisch-Medicinischen Vereins zu Heidelberg. Heidelberg. Verh. Naturhist.-Med. Vereins Heidelberg. See B–P–H 953/8. HI 60344

Verhandlungen des Naturhistorischen Vereines für Anhalt in Dessau. Dessau. Verh. Naturhist. Vereins Anhalt Dessau. See B–P–H 953/12. HI 60348

Verhandlungen des Naturhistorischen Vereines für das Grossherzogthum Hessen und Umgebung. Darmstadt, Germany. Verh. Naturhist. Vereins Grossherzogth. Hessen. See B–P–H 953/13. HI 60349

Verhandlungen des Naturhistorischen Vereines der preussischen Rheinlande. Bonn. Verh. Naturhist. Vereines Preuss. Rheinl. See B–P–H 953/9. HI 60345

Verhandlungen des Naturhistorischen Vereines der preussischen Rheinlande und Westphalens. Bonn. Verh. Naturhist. Vereines Preuss. Rheinl. Westphalens. See B–P–H 953/11. HI 60347

Verhandlungen des Naturhistorischen Vereines der preussischen Rheinlande, Westfalens und des Reg.-Bezirkes Osnabrück. Bonn. Verh. Naturhist. Vereines Preuss. Rheinl. Westfalens Reg.-Bez. Osnabrück. See B–P–H 953/10. HI 60346

Verhandlungen des Naturhistorischen Vereins der preussischen Rheinlande und Westfalens = Verhandlungen des Naturhistorischen Vereines der preussischen Rheinlande und Westphalens. Bonn. Verh. Naturhist. Vereines Preuss. Rheinl. Westphalens. See B–P–H 953/11.

Verhandlungen der naturwissenschaftlichen Gesellschaft zu Aachen. Aachen. Verh. Naturwiss. Ges. Aachen. See B–P–H 953/17. HI 60350

Verhandlungen des Naturwissenschaftlichen Vereins in Carlsruhe. Karlsruhe. Verh. Naturwiss. Vereins Carlsruhe. See B–P–H 953/18. HI 60351

Verhandlungen des Naturwissenschaftlichen Vereins in Karlsruhe = Verhandlungen des Naturwissenschaftlichen Vereins in Carlsruhe. Karlsruhe. Verh. Naturwiss. Vereins Carlsruhe. See B–P–H 953/18.

Verhandlungen des naturwissenschaftlichen Vereins in Hamburg. Hamburg. Ser. 3, vols. 1-29, 1894-1921; ser. 4, vols. 1-5, 1922-35; n.s. vol. 23+, 1979+. Verh. Naturwiss. Vereins Hamburg. Preceded by: Verhandlungen des Naturwissenschaftlichen Vereins von Hamburg-Altona. For 1937-78 see: Abhandlungen und Verhandlungen des naturwissenschaftlichen

Vereins in Hamburg. 4-2936-3. HI 60352

Verhandlungen des Naturwissenschaftlichen Vereins von Hamburg-Altona. Hamburg. Verh. Naturwiss. Vereins Hamburg-Altona. See B–P–H 953/20. HI 60353

Verhandlungen der Physikalisch-Medicinischen Gesellschaft in Würzburg. Erlangen. Verh. Phys.-Med. Ges. Würzburg. See B–P–H 954/6. HI 60355

Verhandlungen der Physicalisch-Medicinischen Societät zu Erlangen. Erlangen. Verh. Phys.-Med. Soc. Erlangen. See B–P–H 954/7. HI 60356

Verhandlungen und Schriften der Hamburgischen Gesellschaft zur Beförderung der Künste und nützlichen Gewerbe. Hamburg. Verh. Schriften Hamburg. Ges. Beförd. Künste. See B–P–H 954/9. HI 60358

Verhandlungen und Schriften der ökonomischen Section der Schlesischen Gesellschaft für vaterländische Cultur. Breslau [=Wroclaw, Poland]. Verh. Schriften Ökon. Sect. Schles. Ges. Vaterl. Cult. See B–P–H 954/10. HI 60359

Verhandlungen der schweizerischen Gesellschaft für die gesammten Naturwissenschaften. Aarau, Switzerland. Verh. Schweiz. Ges. Gesammten Naturwiss. See B–P–H 954/11. HI 60360

Verhandlungen der schweizerischen naturforschenden Gesellschaft. Basel. Vols. 23-24, 1838-39; vols. 26-27, 1842-43; vol. 29, 1845; vols. 31-43, 1847-60; vols. 46-140, 1863-1960; Wissenschaftlicher Teil, vols. 141-157, 1961-1977. Verh. Schweiz. Naturf. Ges. For vols. 25, 28 and 30 see: Actes de la société helvétique des sciences naturelles; for vol. 44 see: Atti della società elvetica delle scienze naturali; for vol. 45 see: Actes de la société suisse des sciences naturelles. Preceded by: Actes de la société helvétique des sciences naturelles. Superseded by: Jahrbuch der schweizerischen naturforschenden Gesellschaft. Wissenschaftlicher Teil. 5-3812-1. HI 74265

Verhandlungen und Sitzungsberichte der Physikalisch-Medizinischen Gesellschaft zu Würzburg. Würzburg. Verh. Sitzungsber. Phys.-Med. Ges. Würzburg. See B–P–H 954/15. HI 60361

Verhandlungen des Vereins zur Beförderung des Garten- und Feldbaues. Berlin. Verh. Vereins Beförd. Garten-Feldbaues. See B–P–H 955/8. HI 60363

Verhandlungen des Vereins zur Beförderung des Gartenbaues in den Königlich Preussischen Staaten. Berlin. Verh. Vereins Beförd. Gartenbaues Königl. Preuss. Staaten. See B–P–H 955/9. HI 60364

Verhandlungen des Vereins zur Beförderung des Gewerbefleisses. Berlin. Verh. Vereins Beförd. Gewerbefl. See B–P–H 955/10. HI 60365

Verhandlungen des Vereins zur Beförderung des Gewerbefleisses in Preussen. Berlin. Verh. Vereins Beförd. Gewerbefl. Preussen. See B–P–H 955/11. HI 60366

Verhandlungen des Vereins zur Beförderung der Landwirthschaft zu Königsberg. Königsberg [=Kaliningrad, R S F S R]. Verh. Vereins Beförd. Landw. Königsberg. See B–P–H 955/12. HI 60367

Verhandlungen des Vereins für Natur- und Heilkunde zu Presburg. Pressburg [=Bratislava, Czechoslovakia]. Verh. Vereins Natur- Heilk. Presburg. See B–P–H 955/13. HI 60368

Verhandlungen des Vereins für Naturkunde zu Presburg. Pressburg [=Bratislava, Czechoslovakia]. Verh. Vereins Naturk. Presburg. See B–P–H 955/14. HI 60369

Verhandlungen des Vereins für naturwissenschaftliche Heimatforschung zu Hamburg. Hamburg. Verh. Vereins Naturwiss. Heimatf. Hamburg. See B–P–H 955/15. HI 60370

Verhandlungen des Vereins für naturwissenschaftliche Unterhaltung zu Hamburg. Hamburg. Verh. Vereins Naturwiss. Unterhalt. Hamburg. See B–P–H 955/16. HI 60371

Verhandlungen des Vereins für Pomologie und Gartenbau in Meiningen. Meiningen, Germany. Verh. Vereins Pomol. Meiningen. See B–P–H 955/17. HI 60372

Verhandlungen der Wandergesellschaft sächsischer Landwirthe und Naturforscher. Dresden & Leipzig. Verh. Wanderges. Sächs. Landwirthe. See B–P–H 955/18. HI 60373

Verhandlungen und wissenschaftliche Abhandlungen des deutschen Geographentages. Breslau [=Wroclaw, Poland]. Verh. Wiss. Abh. Deutsch. Geographentages. See B–P–H 956/1. HI 60374

Verhandlungen des wissenschaftlichen Torf-Forschungsinstitutes. Moscow & Leningrad = Trudy Nauchno-issledovatel′skogo torfyanogo instituta (Instorfa). Moscow & Leningrad.

Verhandlungen der zoologisch-botanischen Gesellschaft in Österreich. Vienna. Vol. 116+, 1978+. Verh. Zool.-Bot. Ges. Österreich. Preceded by: Verhandlungen der zoologisch-botanischen Gesellschaft in Wien. HI 74266

Verhandlungen der zoologisch-botanischen Gesellschaft in Wien. Vienna. Vols. 68-115, 1918-76 [publication suspended 1942-50]. Verh. Zool.-Bot. Ges. Wien. Preceded by: Verhaldlungen der kaiserlich-königlichen zoologisch-botanischen Gesellschaft in Wien. Superseded by: Verhandlungen der zoologisch-botanischen Gesellschaft in Österreich. 5-4641-3. HI 74267

Verhandlungen der zoologisch-botanischen Vereins in

Wien. Vienna. Verh. Zool.-Bot. Vereins Wien. See B–P–H 956/7. HI 60377

Verhandlungsblätter der Gesellschaft für vaterländische Cultur im Kanton Aargau. Aarau, Switzerland. Verhandlungsbl. Ges. Vaterl. Cult. Kanton Aargau. See B–P–H 956/14. HI 60378

Verkündiger; oder Wochenschrift zur Belehrung, Unterhaltung und Bekanntmachung für alle Stände. Nuremberg. Verkündiger. See B–P–H 956/16. HI 60379

Verlag van die direkteur-generaal, waterwese, bosbou en omgewingsbewaring = Report of the director-general, water affairs, forestry and environmental conservation, South Africa. Cape Town.

Vermischte Aufsätze und Dialogen zum Nutzen und Vernügen. Leipzig. 1790. Vermischte Aufsätze Dialogen Nutzen Vernügen. HI 68069

Vermischte Beyträge zur physikalischen Erdbeschreibung. Brandenburg. Vermischte Beytr. Phys. Erdbeschreib. See B–P–H 956/17. HI 60380

Vermischte Landwirthschaftliche Schriften aus den Annalen der Niedersächsischen Landwirthschaft. Hanover. Vermischte Landw. Schriften Ann. Niedersächs. Landw. See B–P–H 956/18. HI 60381

Vermischte Schriften der Ackerbaugesellschaft in Tyrol. Innsbruck. Vermischte Schriften Ackerbauges. Tyrol. See B–P–H 956/19. HI 60382

Vermischte Schriften aus der Naturwissenschaft, Chymie und Arzneygelahrtheit. Frankfurt a. O. Vermischte Schriften Naturwiss. See B–P–H 956/20. HI 60383

Vermont botanical club bulletin = Bulletin of the Vermont botanical club. Burlington, VT.

Vermont farm and home science. University of Vermont agricultural experiment station. Burlington, VT. Vermont Farm Home Sci. See B–P–H 957/4. HI 60389

Vermont natural history. Woodstock, VT. 1973+. Vermont Nat. Hist. HI 74268

Vermont state agricultural college, agricultural experiment station. Annual report. Burlington, VT. Vermont Agric. Exp. Sta. Annual Rep. See B–P–H 956/21. HI 60384

Vermont state agricultural college, agricultural experiment station. Bulletin. Burlington, VT. Vermont Agric. Exp. Sta. Bull. See B–P–H 956/22. HI 60385

Vermont state agricultural college, agricultural experiment station. Circular. Burlington, VT. Vermont Agric. Exp. Sta. Circ. See B–P–H 956/23. HI 60386

Vermont state agricultural college, agricultural experiment station, miscellaneous publications. Burlington, VT. Nos. 1-?, 1951-? Vermont Agric. Exp. Sta. Misc. Publ. Superseded by: Research report, Vermont agricultural experiment station. HI 74269

Vermont state agricultural college, agricultural experiment station. Newspaper bulletin. Burlington, VT. Vermont Agric. Exp. Sta. Newspaper Bull.. See B–P–H 957/2. HI 60387

Vermont state agricultural college, agricultural experiment station. Pamphlet. Burlington, VT. Vermont Agric. Exp. Sta. Pam. See B–P–H 957/3. HI 60388

Veröffentlichung, see Veröffentlichungen

Veröffentlichungen der Alexander Kohut memorial foundation. Vienna. Vols. ?-2-4, ?-1924-26. Veröff. Alexander Kohut Mem. Found. HI 74270

Veröffentlichungen der Arbeitsgemeinschaft Donauforschung der societas internationalis limnologiae. [Archiv für Hydrobiologie, Supplement.] Stuttgart. Vol. 1+, 1964+. Veröff. Arbeitsgem. Donauforsch. Soc. Int. Limnol. HI 74271

Veröffentlichungen der Arbeitsgemeinschaft für Forschung des Landes Nordrhein-Westfalen. Cologne & Opladen. Veröff. Arbeitsgem. Forsch. Landes Nordrhein-Westfalen. See B–P–H 957/7. HI 60390

Veröffentlichungen der Arbeitsgemeinschaft für Forschung des Landes Nordrhein-Westfalen. Abt. Naturwissenschaften. Cologne & Opladen. Pts. 8-203, 1953-69 [pts. 8-14 published in 1952]. Veröff. Arbeitsgem. Forsch. Landes Nordrhein-Westfalen, Abt. Naturwiss. Preceded by: Veröffentlichungen der Arbeitsgemeinschaft für Forschung des Landes Nordrhein-Westfalen. Superseded by: Vorträge, rheinisch-westfalische Akademie der Wissenschaften. HI 74272

Veröffentlichungen des Bezirksheimatmuseums Potsdam. Potsdam. Vol. 1+, 1962+. Veröff. Bezirksheimatmus. Potsdam. HI 74273

Veröffentlichungen der Bundesanstalt für alpine Landwirtschaft in Admont. Vienna. Veröff. Bundesanst. Alpine Landw. Admont. See B–P–H 957/ 9. HI 60391

Veröffentlichungen des Deutsch-Dominikanischen Tropenforschungsinstituts, Hamburg. Jena. Veröff. Deutsch-Dominikan. Tropenforschungsinst. Hamburg. See B–P–H 957/10. HI 60392

Veröffentlichungen der Deutschen Gesellschaft für Natur- und Völkerkunde Ostasiens = Mitteilungen der deutschen Gesellschaft für Natur- und Völkerkunde Ostasiens. Yokohama.

Veröffentlichungen aus dem Deutschen Kolonial- und Überseemuseum in Bremen. Bremen & Berlin. Veröff. Deutsch. Kolonial- Überseemus. Bremen. See

B–P–H 957/12. HI 60393

Veröffentlichungen der Fachgruppe andere Sukkulenten = F a S. Grafenhausen-Balzhausen.

Veröffentlichungen des Forschungsinstitutes für Hydrobiologie der naturwissenschaftlichen Fakultät Universität Istanbul = Istanbul üniversitesi fen fakültesi hidrobiologi arastirma enstitusü yayinlarindan. Ser. B. Istanbul.

Veröffentlichungen. Forstliche Bundesversuchsanstalt, Institut für Standort. Vienna. Veröff. Forstl. Bundesversuchsanst. Inst. Standort. See B–P–H 957/13. HI 60394

Veröffentlichungen der Forstlichen Bundesversuchsanstalt Mariabrunn in Schönbrunn, Abteilung für Standortserkundung und -kartierung. Vienna. Veröff. Forstl. Bundesversuchsanst. Mariabrunn Schönbrunn Abt. Standortserkund. See B–P–H 957/14. HI 60395

Veröffentlichungen des Geobotanischen Institutes der Eidg. Techn. Hochschule, Stiftung Rübel, in Zürich. Bern. Veröff. Geobot. Inst. ETH Stiftung Rübel Zürich. See B–P–H 957/16. HI 60396

Veröffentlichungen des Geobotanischen Institutes Rübel in Zürich. Zurich. Veröff. Geobot. Inst. Rübel Zürich. See B–P–H 957/17. HI 60397

Veröffentlichungen des Geographischen Instituts der Albertus-Universität zu Königsberg. Hamburg. Veröff. Geogr. Inst. Albertus-Univ. Königsberg. See B–P–H 958/2. HI 60398

Veröffentlichungen des Geographischen Instituts der Albertus-Universität zu Königsberg. Reihe Geographie. Königsberg [=Kaliningrad, R S F S R]. Veröff. Geogr. Inst. Albertus-Univ. Königsbergk, Reihe Geogr. See B–P–H 958/3. HI 60399

Veröffentlichung der Haupt-Piltzstelle am botanischen museum der universität in Berlin-Dahlem. Berlin. No. 1, 1941. Veröff. Haupt-Piltzst. Bot. Mus. Univ. Berlin-Dahlem. HI 74274

Veröffentlichungen aus dem Haus der Natur in Salzburg. Salzburg. Veröff. Haus Natur Salzburg. See B–P–H 958/5. HI 60400

Veröffentlichungen des Heimatmuseums Pritzwalk. Pritzwalk. Vol. 1+, 1957+. Veröff. Heimatmus. Pritzwalk. HI 74275

Veröffentlichungen des Instituts für Meeresforschung in Bremerhaven. Bremen. Vol. 1+, 1952+. Veröff. Inst. Meeresf. Bremerhaven. HI 74276

Veröffentlichungen des Instituts für Meeresforschung in Bremerhaven. Sonderband. Bremen. Vols. [1]-3, 1963-68. Veröff. Inst. Meeresf. Bremerhaven, Sonderb. Superseded by: Veröffentlichungen des Instituts für Meeresforschung in Bremerhaven. Supplement. HI 74277

Veröffentlichungen des Instituts für Meeresforschung in Bremerhaven. Supplement. Bremen. Vol. 4+, 1974+. Veröff. Inst. Meeresf. Bremerhaven, Suppl. Preceded by: Veröffentlichungen des Instituts für Meeresforschung in Bremerhaven. Sonderband. HI 74278

Veröffentlichungen, Joachim Jungius-Gesellschaft der Wissenschaften Hamburg. Göttingen. No. ?-44+, ?-1981+. Veröff. Joachim Jungius-Ges. Wiss. Hamburg. HI 74279

Veröffentlichungen der Landesstelle für Naturschutz und Landschaftspflege. Ottweiler = Naturschutz und Landschaftspflege im Saarland. Ottweiler, Germany. Naturschutz Landschaftspflege Saarland. See B–P–H 636/1.

Veröffentlichungen der Landesstelle für Naturschutz und Landschaftspflege Baden-Württemberg. Ludwigsburg. Vols. 29-42, 1961-75? Veröff. Landesstelle Naturschuz Baden-Württemberg. Preceded by: Naturschutz und Landschaftspflege in Baden-Württemberg. Veröffentlichungen der Landesstelle für Naturschutz und Landschaftspflege Baden-Württemberg und der Württembergischen Bezirksstelle in Ludwigsburg. Superseded by: Veröffentlichungen für naturschutz und landschaftspflege in Baden-Württemberg. HI 60401

Veröffentlichungen der Landesstelle für Naturschutz und Landschaftspflege Baden-Württemberg und der württembergischen Bezirksstellen in Ludwigsburg und Tübingen. Ludwigsburg. Veröff. Landesstelle Naturschuz Baden-Württemberg Württemberg. Bezirksstellen Ludwigsburg Tübingen. See B–P–H 958/7. HI 60402

Veröffentlichungen der Landesstelle für Naturschutz und Landschaftspflege Baden-Württemberg und der württembergischen Bezirksstellen in Stuttgart und Tübingen. Ludwigsburg. Veröff. Landesstelle Naturschuz Baden-Württemberg Württemberg. Bezirksstellen Stuttgart Tübingen. See B–P–H 958/8. HI 60403

Veröffentlichungen der Museen der Stadt Gera. Naturwissenschaftliche Reihe. Gera. Vol. 1+, 1973+. Veröff. Mus. Stradt Gera, Naturwiss. Reihe. HI 74280

Veröffentlichungen des Museum Ferdinandeum. Innsbruck. Veröff. Mus. Ferdinandeum. See B–P–H 958/9. HI 60404

Veröffentlichungen aus dem Museum für Natur-, Völker- und Handelskunde Bremen. Reihe A. Naturwissenschaften. Bremen. Veröff. Mus. Natur-Handelsk. Bremen, Reihe A, Naturwiss. See B–P–H 958/10. HI 60405

Veröffentlichungen des Museums für Naturkunde, Karl-Marx-Stadt. Chemnitz [=Karl-Marx-Stadt]. Vol. 1+, 1961+. Veröff. Mus. Naturk. Karl-Marx-Stadt. HI 74281

Veröffentlichungen der Naturforschenden Gesellschaft in Emden. Emden. Veröff. Naturf. Ges. Emden. See B–P–H 958/12. HI 60406

Veröffentlichungen aus dem Naturhistorischen Museum. Vienna. Veröff. Naturhist. Mus. See B–P–H 958/13. HI 60407

Veröffentlichungen aus dem Naturkunde-Museum Bielefeld. Bielefeld. Vol. 1+, 1986+. Veröff. Naturk-Mus. Bielefeld. HI 74282

Veröffentlichungen des naturkundemuseums Erfurt. Erfurt. Vol. 1+, 1982+. Veröff. Naturkundemus. Erfurt. HI 75286

Veröffentlichungen für naturschutz und landschaftspflege in Baden-Württemberg. Ludwigsburg, later Karlsruhe. Vol. 43+, 1975+, 1976+. Veröff. Naturschutz Landschaftspflege Baden-Württemberg. Preceded by: Veröffentlichungen der landesstelle für naturschutz und landschaftspflege Baden-Württemberg. HI 74283

Veröffentlichungen des Naturwissenschaftlichen Vereins zu Osnabrück. Osnabrück. Veröff. Naturwiss. Vereins Osnabrück. See B–P–H 958/14. HI 60408

Veröffentlichungen des naturwissenschaftlichen Vereins zu Osnabruck. Jahresberichte. Osnabrück. Vols. 27-33, 1955-70. Veröff. Naturwiss. Vereins Osnabrück Jahresber. Preceded by: Veröffentlichungen des naturwissenschaftlichen Vereins zu Osnabrück. Superseded by: Osnabrücker naturwissenschaftliche Mitteilungen. HI 74284

Veröffentlichungen des Niedersächsischen Landesverwaltungsamtes -- Naturschutz und Landschaftspflege. Hanover. Veröff. Niedersächs. Landesverwaltungsamtes Naturschutz & Landschaftspflege. See B–P–H 959/1. HI 60409

Veröffentlichungen der Österreichischen Mykologischen Gesellschaft. Horn, Austria. Veröff. Österr. Mykol. Ges. See B–P–H 959/2. HI 60410

Veröffentlichungen der Reichsarbeitsgemeinschaft für Heilpflanzenkunde und Heilpflanzenbeschaffung. Berlin. Veröff. Reichsarbeitsgem. Heilpflanzenk. See B–P–H 959/3. HI 60411

Veröffentlichungen der schweizerischen Gesellschaft für Geschichte der Medizin und der Naturwissenschaften. Zurich. Vol. 1+, 1922+. Veröff. Schweiz. Ges. Gesch. Med. Naturwiss. HI 74285

Veröffentlichungen der Staatlichen Botanischen Anstalten Krakau. Cracow. Veröff. Staatl. Bot. Anst. Krakau. See B–P–H 959/4. HI 60412

Veröffentlichungen der Staatlichen Stelle für Naturschutz beim Württ. Landesamt für Denkmalpflege. Stuttgart. Veröff. Staatl. Stelle Naturschutz Württemberg. Landesamt Denkmalpflege. See B–P–H 959/5. HI 60413

Veröffentlichungen der Staatlichen Universität des Fernen Ostens. Ser. 4. Vladivostok, Russian S F S R = Trudy gosudarstvennogo dal'nevostochnogo universiteta. Ser. 4, lesnye nauk. Vladivostok.

Veröffentlichungen der Staatlichen Universität des Fernen Ostens. Ser. 5. Kiev, Ukrainian S S R = Trudy gosudarstvennogo dal'nevostochnogo universiteta. Ser. 5, sel'skoe khozyaistvo. Kiev.

Veröffentlichungen der Staatlichen Universität des Fernen Ostens. Ser. 8. Vladivostok, Russian S F S R = Trudy gosudarstvennogo dal'nevostochnogo universiteta. Ser. 8, biologiya. Vladivostok.

Veröffentlichungen der Staatlichen Universität des Fernen Ostens. Ser. 11. Vladivostok, Russian S F S R = Trudy gosudarstvennogo dal'nevostochnogo universiteta. Ser. 11, geologiya. Vladivostok.

Veröffentlichungen der Staatlichen Universität des Fernen Ostens. Ser. 13. Vladivostok, Russian S F S R = Trudy gosudarstvennogo dal'nevostochnogo universiteta. Ser. 13, tekhnika. Vladivostok.

Veröffentlichungen des städtischen Museums zur Geschichte von Natur und Gesellschaft der Stadt Halberstadt. Halberstadt. Vols. 1-6, 1955-61. Veröff. Städt. Mus. Gesch. Natur Ges. Stadt Halberstadt. Superseded by: Veröffentlichungen des städtischen Museums Halberstadt. HI 74286

Veröffentlichungen des städtischen Museums Halberstadt. Halberstadt. Vol. 7+, 1964+. Veröff. Städ. Mus. Halberstadt. Preceded by: Veröffentlichungen des städtischen Museums zur Geschichte von Natur und Gesellschaft der Stadt Halberstadt. HI 74287

Veröffentlichungen des Tiroler Landesmuseum Ferdinandeum. Innsbruck. Veröff. Tiroler Landesmus. Ferdinandeum. See B–P–H 959/7. HI 60414

Veröffentlichungen aus dem Überseemuseum Bremen. Reihe A. Naturwissenschaften. Bremen. Veröff. Überseemus. Bremen, Reihe A, Naturwiss. See B–P–H 959/8. HI 60415

Veröffentlichungen aus dem Übersee-Museum Bremen. Reihe C. Bremen. 1977+. Veröff. Übersee-Mus. Bremen, C. Preceded by: Veröffentlichungen aus dem Museum für Natur- Volker- und Handelskunde in Bremen. HI 74288

Veröffentlichungen der Wissenschaftlichen Vereinigung in Südwestafrika. Windhoek, South-West Africa. Veröff. Wiss. Vereinigung Südwestafrika. See B–P–H 959/9. HI 60416

Veröffentlichungen der Württembergischen Landesstelle für Naturschutz. Stuttgart. Veröff. Württemberg. Landesstelle Naturschutz. See B–P–H 959/10. HI 60417

Veröffentlichungen der Württembergischen Landesstelle

für Naturschutz und Landschaftspflege. Stuttgart. Veröff. Württemberg. Landesstelle Naturschutz Landschaftspflege. See B–P–H 959/11. HI 60418

Veröffentlichungen der Württ. Landesstellen für Naturschutz und Landschaftspflege in Ludwigsburg und Tübingen. Ludwigsburg. Veröff. Württemberg. Landesstelle Naturschutz Ludwigsburg Tübingen. See B–P–H 959/12. HI 60419

Verrigtings, elektronmikroskopie vereniging van Suidelike Afrika = Proceedings, southern african electron microscopy society. Pretoria.

Versammlung deutscher Naturforscher und Aerzte zu Regensburg. Regensburg. Versamml. Deutsch. Naturf. Aerzte Regensburg. See B–P–H 960/1. HI 60420

Versammlung ungarischer Ärzte und Naturforscher = Magyar orvosok és természetvizsgálók nagy gyülésémek munkálatai. [Hungary.] Magyar Orv. Termész. Nagy Gyül. Munk. See B–P–H 544/14.

Versammlung ungarischer Ärzte und Naturforscher = A magyar orvosok és természetvizsgálók nagy gyülésének történeti vázlata és munkálatai. [Hungary.] Magyar Orv. Termész. Nagygyül. Tört. Vázl. Munk. See B–P–H 544/15.

Versammlung ungarischer Ärzte und Naturforscher = A magyar orvosok és természetvizsgálók vándorgyülésének történeti vázlata és munkálatai. [Hungary.] Magyar Orv. Termész. Vándorgyül. Tört. Vázl. Munk. See B–P–H 544/16.

Verslag, see Verslagen

Verslagen omtrent de te Buitenzorg gevestigde technische afdeelingen van het departement van landbouw. Batavia. 1905, 1906. Verslagen Buitenzorg Gevestigde Techn. Afd. Dept. Landb. Preceded by: Verslagen omtrent den staat van 's lands plantentuin te Buitenzorg. Superseded by: Jaarboek van het departement van landbouw in Nederlandsch-Indië. HI 74289

Verslagen, caraïbisch marien-biologisch instituut. Curaçao. 1967/69, 1969. Verslagen Caraïb. Marien-Biol. Inst. Superseded by: Jaarverslag, caraïbisch marien-biologisch instituut. HI 74290

Verslag van het comité inzake bestudering en bestrijding van de iepenzichte. Wageningen. Verslag Comité Bestud. Iepenzichte. See B–P–H 960/6. HI 60421

Verslagen, departement landbouwproefstation in Suriname. Paramaribo. 1907-31/32. Verslagen Dept. Lanbouwproefstat. Suriname. HI 74291

Verslagen van de eerste (tweede - twaalfde) openbare vergadering der eerste klasse, van het koninklijk-nederlandsch instituut van wetenschappen, letterkunde en schoone kunsten. Amsterdam. 1817-39. Verslagen Eerste [etc.] Openb. Vergad. Eerste Kl. Kon.-Ned. Inst. Wetensch. Letterk. Schoone Kunsten. Preceded by: Verslag van de werkzaamheden der eerste klasse van het koninklijke instituut van wetenschappen, letterkunde en schoone kunsten. HI 74299

Verslagen van de gewone vergadering des wis-en natuurkundige afdeeling der koninklijke akademie van wetenschappen te Amsterdam. Amsterdam. Vols. 5-6, 1897-98. Verslagen Gewone Vergad. Wis-en Natuurk. Afd. Kon. Akad. Wetensch. Amsterdam. Preceded by: Verslagen der Zittingen van de wis-en natuurkundige afdeeling der koninklijke akademie van wetenschappen. Superseded by: Proceedings of the section of sciences, koninklijke (Nederlandse) akademie van wetenschappen te Amsterdam. HI 74292

Verslagen omtrent de gouvernements kinaonderneming te tjinjiroean. Bandoeng. 1915-27. Verslagen Gouv. Kinaondern. Tjinjiroean. HI 74293

Verslagen van de hoogleraar-directeur, rijksherbarium. Leiden = Verslagen van het rijksherbarium. Leiden.

Verslagen, koloniaal museum. Haarlem. 1890/91. Verslag Kolon. Mus. Superseded by: Bulletin, koloniaal museum. 3-1781-3. HI 74294

Verslagen van landbouwkundige onderzoekingen van rijkslandbouwproefstations in Nederland. The Hague. Verslagen Landbouwk. Onderz. Rijkslandbouwproefstat. Ned. See B–P–H 960/12. HI 60426

Verslagen en mededeelingen van de afdeeling natuurkunde; koninklijke akademie van wetenschappen. Amsterdam. Verslagen Meded. Afd. Natuurk. Kon. Akad. Wetensch. See B–P–H 960/13. HI 60427

Verslagen en mededeelingen van de k[onklijke]. vlaamse academie voor wetenschappen, letteren en schone kunsten van België. Klasse der wetenschappen = Mededelingen van de k[onklijke]. vlaamse academie voor wetenschappen, letteren en schone kunsten van België. Klasse der wetenschappen. Ghent, Brussels.

Verslagen en mededelingen, commissie voor het botanische onderzoek van de Zuiderzee en omgeving. Bilthoven. Nos. 1-40, 1929-39. Verslagen Meded. Commiss. Bot. Onderz. Zuiderzee Omgev. HI 74295

Verslagen en mededelingen van de koninklijke nederlandse botanische vereniging. Jaarboek. Amsterdam. 1959+. Verslagen Meded. Kon. Ned. Bot. Ver., Jaarb. HI 74296

Verslagen, nederlands instituut voor onderzoek der Zee. Texel. 1976/1+, 1976+. Verslagen Ned. Inst. Onderz. Zee. HI 74297

Verslagen, nederlandsch-indische vereeniging tot natuurbescherming. Buitenzorg. 1912/13-3?/34?, 1914-34? Verslagen Ned.-Indische Ver. Natuurbesch. HI 74298

Verslagen van het proefstation voor de Java-Suikerindustrie. Soerbaia. 1907-56. Verslagen Proefstat. Java-Suikerindus. Superseded by: Laporan dari balai penjelidikan perusahaan gula. HI 74300

Verslagen van het proefstation voor suikerriet in West-Java "Kagok." Tegal, Semarang. 1894-1906. Verslagen Proefstat. Suikerriet W.-Java "Kagok". HI 74301

Verslag omtrent s'rijks museum voor natuurlijke historie. Leiden. Verslag Rijks Mus. Natuurl. Hist. See B–P–H 960/8. HI 60422

Verslagen van het rijksherbarium. Leiden. 1933+. Verslagen Rijksherb. HI 74302

Verslagen omtrent den staat van 's lands plantentuin te Buitenzorg. Buitenzorg, Dutch E. Indies [=Bogor, Indonesia]. 1869-1905. Verslag Staat Lands Plantentuin Buitenzorg. Superseded by: Verslag omtrent de te Buitenzorg gevestigde technische afdeelingen van het departement van landbouw. 1-723-3. HI 74303

Verslag van de vereeniging "De proeftuin". Boskoop, Netherlands. Verslag Ver. "Proeftuin". See B–P–H 960/10. HI 60424

Verslagen van de werkzaamheden der eerste klasse van het koninklijke Instituut van wetenschappen, letterkunde en schoone kunsten. Amsterdam. 1809-16. Verslagen Werkzaamh. Eerste Kl. Kon. Inst. Wetensch. Letterk. Schoone Kunsten. Superseded by: Verslagen van de eerste (tweede - twaalfde) openbare vergadering der eerste klasse, van het koninklijk-nederlandsch instituut van wetenschappen, letterkunde en schoone kunsten. HI 74304

Verslag van de werkzaamheden van het natuur- en scheikundig genootschap te Groningen. Groningen. Verslag Werkzaamh. Natuur- Scheik. Genootsch. Groningen. SeeL 960/11. HI 60425

Verslagen van de werkzaamheden, rijksinstituut voor veldbiologisch onderzoek ten behoven van het natuurbehoud. Bilthoven. 1959-61. Verslagen Werkzaamh. Rijksinst. Veldiol. Onderz. Behoven Natuurbehoud. Superseded by: Jaarverslag, rijks instituut voor veldbiologisch onderzoek ten behoven van het natuurbehoud. HI 74305

Verslagen der zittingen van de wis-en natuurkundige afdeeling. Amsterdam. Vols. 1-4, 1892-96. Verslagen Zittingen Wis- Natuurk. Afd. Superseded by: Verslagen van de gewone vergadering der wis-en natuurkundige afdeeling koninklijke akademie van wetenschappen te Amsterdam. 1-108-3. HI 74306

Versuche und Abhandlungen der Naturforschenden Gesellschaft in Dantzig. Danzig [=Gdansk, Poland]. Versuche Abh. Natuf. Ges. Dantzig. See B–P–H 960/15. HI 60428

Versuche und Versuchsergebnisse im Gartenbau. 1964-65. Versuche Versuchsergebn. Gartenbau. Superseded by: Versuchsergebnisse im Gartenbau. HI 74307

Versuchsergebnisse der Bundesanstalt für alpine Landwirtschaft. Admont, Gumpenstein. Vols. 1-28, 1949-55. Versuchsergebn. Bundesanst. Alpine Landw. Superseded by: Versuchsergebnisse der Bundesversuchsanstalt für alpenländische Landwirtschaft Gumperstein. HI 74308

Versuchsergebnisse der Bundesversuchsanstalt für alpenländische Landwirtschaft Gumperstein. Gumperstein. Vol. 29+, 1955+. Versuchsergebn. Bundesversuchsanst. Alpenländ. Landw. Gumperstein. Preceded by: Versuchsergebnisse der Bundesanstalt für Alpenländische Landwirtschaft. HI 74309

Versuchsergebnisse im Gartenbau. 1966-72. Versuchsergebn. Gartenbau. Preceded by: Versuche und Versuchsergebnisse im Gartenbau. HI 74310

Verzeichnis der Vorlesungen an den koniglichen Akademie zu Braunsberg. Braunsberg. 1912-22. Verzeichnis Vorles. Konigl. Akad. Braunsberg. Preceded by: Verzeichnis der Vorlesungen am koniglichen Lyceum hosianum zu Braunsberg. HI 74311

Verzeichnis der Vorlesungen am koniglichen Lyceum hosianum zu Braunsberg. Braunsberg. 1906-09. Verzeichnis Vorles. Konigl. Lyceum Hosianum Braunsberg. Superseded by: Verzeichnis der Vorlesungen an der koniglichen Akademie zu Braunsberg. HI 74312

Vesci Akademii navuk Belaruskaj S S R = Vestsi Akademii navuk Belaruskai S S R. Minsk.

Vesci Akademii navuk Belaruskaj S S R. Seryja bijalagičnyh navuk = Vestsi Akademii navuk Belaruskai S S R. Seryya biyalagichnykh navuk. Minsk.

Vesci Akademii navuk Belaruskaj S S R. Seryja bijalagičnyh i sel′skagaspadarčyh navuk = Vestsi Akademii navuk Belaruskai S S R. Seryya biyalagichnykh i sel′skagaspadarchykh navuk. Minsk.

Vesmír. Prague. Vesmír. See B–P–H 961/2. HI 60429

Vestnik Akademii nauk Kazahskoj S S R = Vestnik Akademii nauk Kazakhskoi S S R. Alma-Ata.

Vestnik Akademii nauk Kazakhskoi S S R. Alma-Ata, Kazakh S S R. 1939-43; 1946(6)-1946(12); vol. 4+, 1947+. Vestn. Akad. Nauk Kazakhsk. S.S.R. For 1944-1946(5) see: Vestnik Kazakhskogo filiala Akademii nauk S S S R. 1-111-1. HI 74313

Vestnik, akademiya nauk Kirgizskoi S S R. Frunze. 1961. Vestn. Akad. Nauk Kirgizsk. S.S.R. HI 74314

Vestnik Akademii nauk S S S R. Moscow & Leningrad. Vestn. Akad. Nauk S.S.S.R. See B–P–H 961/20.

HI 60430

Vestnik Akademii nauk Ukrainskoj S S R = Visnyk Akademii nauk Ukrayins'koi R S R. Kiev.

Věstnik Bakteriologo-agronomičeskoj Stancii imeni Vladimira Karloviča Ferrejn = Věstnik Bakteriologo-agronomicheskoi Stantsii imeni Vladimira Karlovicha Ferrein. Moscow.

Věstnik Bakteriologo-agronomicheskoi Stantsii imeni Vladimira Karlovicha Ferrein. Moscow. Vols. 13-22, 1907-16. Věstn. Bakteriol.-Agron. Stantsii Vladimira Karlovicha Ferrein. Preceded by: Věstnik Russkago Obshchestva Akklimatizatsii Zhivotnykh i Rastenii. 3-2763-1. HI 74315

Vestnik, belorusskogo gosudarstvennogo universitet im V. I. Lenina. Seriya II, khimiya, biologiya, geologiya, geografiya. Minsk. Vol. 1+, 1972+. Vestn. Belorussk. Gosud. Univ. V. I. Lenina, Ser. 2. HI 74316

Vestnik botanicheskogo obshchestva gruzinskoi S S R. Tiflis. Vol. 1, 1962. Vestn. Bot. Obshch. Gruzinsk. S.S.R. Superseded by: Vestnik gruzinskogo botanicheskogo obshchestva. HI 74317

Věstnik české akademie Cisare Frantiska Josefa pro vedy, slovesnost a umeni v Praze. Prague. 1891-1920. Věstn. České Akad. Cisare Frantiska Josefa Vedy Slovesn. Umeni Praze. Superseded by: Věstnik české akademie ved a umeni v Praze. HI 74318

Věstnik české akademie ved a umeni v Praze. Prague. 1926-52. Věstn. České Akad. Ved Umeni Praze. Preceded by: Věstnik české akademie Cisare Frantiska Josefa pro vedy, slovesnost a umeni v Praze. Superseded by: Věstnik československe akademie ved. HI 74319

Věstník české společnosti nauk. Třida matematicko-přírodovědecké = Věstník královské české společnosti nauk. Třida matematicko-přírodovědecké. Prague. Věstn. Král. Ceské Společn. Nauk, Tř. Mat.-Přír. See B–P–H 962/7.

Věstnik československe akademie ved. Prague. 1953+. Věstn. Českoslov. Akad. Ved. Preceded by: Věstnik české akademie ved a umeni v Praze. HI 74320

Věstnik československé akademie zemědelské. Prague. 1926-51. Věstn. Českoslov. Akad. Zeměd. Superseded by: Věstnik československé akademie zemědělskych věd. HI 74321

Věstnik československé akademie zemědělskych věd. Prague. Vol. 1+, 1954+. Věstn. Českoslov. Akad. Zeměd. Věd. Preceded by: Věstnik československé akademie zemědelské. HI 74322

Věstnik československeho zemědělskeho museum v Praze. Prague. 1928-39. Věstn. Českoslov. Zeměd. Mus. Praze. HI 74323

Vestnik Dal'nevostochnogo filiala Akademii nauk S S S R. Vladivostok, R S F S R. Vols. 3/4-33, 1932-39. Vestn. Dal'nevost. Fil. Akad. Nauk S.S.S.R. Preceded by: Vestnik Dal'nevostochnogo otdeleniya Akademii nauk S S S R. 1-115-1. HI 74324

Vestnik Dal'nevostochnogo otdeleniya Akademii nauk S S S R. Vladivostok, R S F S R. Vol. 1/2, 1932. Vestn. Dal'nevost. Otd. Akad. Nauk S.S.S.R. Superseded by: Vestnik Dal'nevostochnogo filiala Akademii nauk S S S R. 1-115-1. HI 74325

Vestnik Dal'nevostočnogo filiala Akademii nauk S S S R = Vestnik Dal'nevostochnogo filiala Akademii nauk S S S R. Vladivostok.

Vestnik Dal'nevostočnogo otdelenija Akademii nauk S S S R = Vestnik Dal'nevostochnogo otdeleniya Akademii nauk S S S R. Vladivostok.

Věstnik Donskago Otděla Imperatorskago Rossiiskago Obshchestva Sadovodstva. Novocherkassk, R S F S R. 1901-16. Věstn. Donsk. Otd. Imp. Rossiisk. Obshch. Sadov. HI 74326

Věstnik Donskago Otděla Imperatorskago Rossijskago Obščestva Sadovodstva = Věstnik Donskago Otděla Imperatorskago Rossiiskago Obshchestva Sadovodstva. Novocherkassk.

Věstnik Estestvennyh Nauk, izdavaemyja Moskovskim Obščestvom Ispytatelej Prirody = Věstnik Estestvennykh Nauk, izdavaemyya Moskovskim Obshchestvom Ispytatelei Prirody. Moscow.

Věstnik Estestvoznznija = Věstnik Estestvoznzniya. St. Petersburg.

Věstnik Estestvoznaniya. St. Petersburg. 1890-93. Věstn. Estestv. 5-4382-3. HI 74327

Věstnik Estestvennykh Nauk, izdavaemiya Moskovskim Obshchestvom Ispytatelei Prirody. Moscow. 1854-65. Věstn. Estestv. Nauk Moskovsk. Obshch. Isp. Prir. 5-4382-3. HI 74328

Vestnik gibridizacii = Vestnik gibridizatsii. Moscow.

Vestnik gibridizatsii. Moscow. 1941. Vestn. Gibridiz. HI 74329

Vestnik gosudarstvennogo Moskovskogo universiteta = Vestnik Moskovskogo universiteta. Moscow.

Vestnik, gosudarstvennogo muzeya Gruzii. Tiflis. Vol. 15+, 1951+. Vestn. Gosud. Muz. Gruzii. Preceded by: Byulleten' gosudarstvennogo muzeya Gruzii. HI 74330

Vestnik gruzinskogo botanicheskogo obshchestva. Tiflis. Vol. 2+, 196?+. Vestn. Gruzinsk. Bot. Obshch. Preceded by: Vestnik botanicheskogo obshchestva gruzinskoi S S R. HI 74331

Věstnik Imperatorskago Rossiiskago Obshchestva Sadovodstva. St. Petersburg. 1870-81; 1894-1906. Věstn. Imp. Rossiisk. Obshch. Sadov. Preceded by:

Věstnik Rossiiskogo Obshchestva Sadovodstva v S.-Peterburgě. For 1882-93 see: Věstnik Sadovodstva, Plodovodstva i Ogorodnichestva. Superseded by: Věstnik Sadovodstva, Plodovodstva i Ogorodnichestva. 4-3710-3. HI 74332

Věstnik Imperatorskago Rossijskago Obščestva Sadovodstva = Věstnik Imperatorskago Rossiiskago Obshchestva Sadovodstva. St. Petersburg.

Věstnik Imperatorskago Russkago Geografičeskago Obščestva = Věstnik Imperatorskago Russkago Geograficheskago Obshchestva. St. Petersburg.

Věstnik Imperatorskago Russkago Geograficheskago Obshchestva. St. Petersburg. 1851-60. Věstn. Imp. Russk. Geogr. Obshch. Preceded by: Geograficheskiya Izvěstiya, vydavaemyya ot Russkago Geograficheskago Obshchestva. Superseded by: Zapiski Imperatorskago Russkago Geograficheskago Obshchestva. 2-1692-2. HI 74333

Vestnik instituta experimental'noi agronomii Gruzii. Tiflis. Vols. ?-2-7-?, ?-1929-30-? Vestn. Inst. Exp. Agron. Gruzii. HI 74334

Vestnik Kazahskogo filiala Akademii nauk S S S R = Vestnik Kazakhskogo filiala Akademii nauk S S S R. Alma-Ata.

Vestnik Kazakhskogo filiala Akademii nauk S S S R. Alma-Ata, Kazakh S S R. 1944-46(5). Vestn. Kazakhsk. Fil. Akad. Nauk S.S.S.R. Preceded & superseded by: Vestnik Akademii nauk Kazakhskoi S S R. 1-111-1. HI 74335

Vestnik Khar'kovskogo universiteta. Kharkov. Vols. 12?-68, 1964-70; Vol. 108+, 1974+. Vestn. Khar'kovsk. Univ. Preceded by: Uchenye zapiski, khar'kovskii gosudarstvennogo universiteta imeni A. M. Gor'kogo. For 1971-73 see: Visnyk Kharkivs'kogo universitetu. HI 74336

Věstník klubu českých farmaceutu v Praze. Prague. Věstn. Klubu Ceských Farm. v Praze. See B–P–H 962/5. HI 60431

Věstnik, klubu kaktusářů Praha. Prague. 1970+. Věstn. Klubu Kakt. Praha. HI 74337

Věstník klubu přírodovědeckého v Prostějově. Prossnitz, Moravia [=Prostejov, Czechoslovakia]. Věstn. Klubu Přír. v Prostějově. See B–P–H 962/6. HI 60432

Věstník královské české společnosti nauk. Třida matematicko-přírodovědecké = Sitzungsberichte der königlichen böhmischen Gesellschaft der Wissenschaften in. Mathematisch-naturwissenschaftliche Classe. Prague. Sitzungsber. Königl. Böhm. Ges. Wiss. Prag., Math.-Naturwiss. Cl. See B–P–H 840/13.

Věstník královské české společnosti nauk. Třida matematicko-přírodovědecké. Prague. Věstn. Král. Ceské Společn. Nauk, Tř. Mat.-Přír. See B–P–H 962/7. HI 60433

Věstnik, krouzku kaktusářů. Prague. Vols. 4-12, 1956-64. Věstn. Krouzku Kakt. Preceded by: Oběžník krouzku kaktusářů. Superseded by: Kaktusy krouzku kaktusaru. HI 74338

Vestník lekársko-prírodovedeckého spolku v Bratislave = Verhandlungen des heil- und naturwissehschaftlichen Vereins zu Bratislava <Pressburg>. Pressburg [=Bratislava, Czechoslovakia]. Verh. Heil- Naturwiss. Vereins Bratislava. See B–P–H 951/11.

Vestnik Leningradskogo gosudarstvennogo universiteta. Leningrad = Vestnik Leningradskogo universiteta. Leningrad.

Vestnik leningradskogo gosudarstvennogo universiteta. Geologiya, geografiya = Vestnik Leningradskogo universiteta. Geologiya, geografiya. Leningrad.

Vestnik Leningradskogo universiteta. Leningrad. Vols. 1-8, 1946-53. Vestn. Leningradsk. Univ. Superseded by: Vestnik Leningradskogo universiteta. Seriya biologii, geografii i geologii. HI 60434

Vestnik Leningradskogo universiteta. Geologiya, geografiya. Leningrad. 1956+. Vestn. Leningradsk. Univ. Geol. Geogr. Preceded by: Vestnik leningradskogo universiteta. Seriya biologii, geografii i geologii. HI 74339

Vestnik Leningradskogo universiteta. Seriya biologii. Leningrad. 1956+. Vestn. Leningradsk. Univ., Ser. Biol. Preceded by: Vestnik Leningradskogo universiteta. Seriya biologii, geografii i geologii. HI 74340

Vestnik Leningradskogo universiteta. Seriya biologii, geografii i geologii. Leningrad. 1954-55. Vestn. Leningradsk. Univ., Ser. Biol. Geogr. Preceded by: Vestnik Leningradskogo universiteta. Superseded by: Vestnik Leningradskogo universiteta. Seriya biologii. HI 74341

Věstník Masarykovy akademie práce. Prague. Věstn. Masarykovy Akad. Práce. See B–P–H 962/12. HI 60435

Věstník matice opavske. Troppau [=Opava, Czechoslovakia]. Věstn. Matice Opavske. See B–P–H 962/13. HI 60436

Věstnik moskovskogo universiteta. Serija 6, biologija, počvovedenie. Moscow = Vestnik moskovskogo universiteta. Seriya 6, biologiya, pochvovedenie. Moscow.

Vestnik Moskovskogo universiteta. Seriya biologii, pochvovedeniya, geologii, geografii. Moscow. 1957-59. Vestn. Moskovsk. Univ., Ser. Biol. Preceded by: Vestnik Moskovskogo universiteta. Seriya fiziko-matematicheskikh i estestvennykh nauk. Superseded by: Vestnik moskovskogo universiteta. Seriya 6, biologiya, pochvovedenie. HI 74342

Vestnik Moskovskogo universiteta. Seriya fiziko-matematicheskikh i estestvennykh nauk. Moscow. 1946-56. Vestn. Moskovsk. Univ., Ser. Fiz.-Mat. Nauk. Superseded by: Vestnik Moskovskogo universiteta. Seriya biologii, pochvovedeniya, geologii, geografii. HI 74343

Vestnik moskovskogo universiteta. Seriya 6, biologiya, pochvovedenie. Moscow. Vols. 15-?, 1960-76. Vestn. Moskovsk. Univ., Ser. 6, Biol. Preceded by: Vestnik moskovskogo universiteta. Seriya biologii, pochvovedeniya, geologii, geografii. Superseded by: Vestnik moskovskogo universiteta. Seriya 16, biologiya and Vestnik moskovskogo universiteta. Seriya 17, pochvovedenie. HI 74344

Vestnik moskovskogo universiteta. Seriya 16, biologiya. Moscow. 1977+. Vestn. Moskovsk. Univ., Ser. 16, Biol. Preceded by: Vestnik moskovskogo universiteta. Seriya 6, biologiya, pochvovedenie. HI 74345

Vestnik moskovskogo universiteta. Seriya 17, pochvovedenie. Moscow. 1977+. Vestn. Moskovsk. Univ., Ser. 17, Pochvov. Preceded by: Vestnik moskovskogo universiteta. Seriya 6, biologiya, pochvovedenie. HI 74346

Vestnik respublikanskogo instituta za ochrany prirody i estestvennonauchnnogo muzeya v Titograde = Glasnik, republičkog zavoda za zaštitu prirode i prirodnjäcke zbirke u Titogradu. Titograd.

Věstnik Rossiiskogo Obshchestva Sadovodstva v S.-Peterburgě. St. Petersburg. 1860-69. Věstn. Rossiisk. Obshch. Sadov. S.-Peterburgě. Superseded by: Věstnik Imperatorskago Rossiiskago Obshchestva Sadovodstva. 4-3710-3. HI 74347

Věstnik Rossijskago Obščestva Sadovodstva v S.-Peterburgě = Věstnik Rossiiskogo Obshchestva Sadovodstva v S.-Peterburgě. St. Petersburg.

Věstnik Russkago Obščestva Akklimatizacii Životnyh i Rastenij = Věstnik Russkago Obshchestva Akklimatizatsii Zhivotnykh i Rastenii. Moscow.

Věstnik Russkago Obshchestva Akklimatizatsii Zhivotnykh i Rastenii. Moscow. Vols. 7-12, 1900-05. Věstn. Russk. Obshch. Akklim. Zhivotn. Preceded by: Trudy Russkago Obshchestva Akklimatizatsii Zhivotnykh i Rastenii. Superseded by: Věstnik Bakteriologo-agronomicheskoi Stantsii imeni Vladimira Karlovicha Ferrein. HI 74348

Vestnik Russkoi flory. Yur'ev [= Tartu] Vols. 1-3, 1915-17. Věstn. Russk. Fl. Preceded by: Trudy botanicheskogo Sada Imperatorskago Yur'evskago Universiteta. HI 74349

Vestnik Russkoj flory. Yur'ev = Vestnik Russkoi flory. Yur'ev.

Věstnik Sadovodstva, Plodovodstva i Ogorodničestva = Věstnik Sadovodstva, Plodovodstva i Ogorodnichestva. St. Petersburg.

Věstnik sadovodstva, plodovodstva i ogorodnichestva. St. Petersburg. 1882-93; 1907-17. Věstn. Sadov. Preceded by: Věstnik Imperatorskago Rossiiskago Obshchestva Sadovodstva. For 1894-1906 see: Věstnik Imperatorskago Rossiiskago Obshchestva Sadovodstva. 4-3710-3. HI 74350

Vestnik sel'skokhozyaistvennoi nauki; ezhemesyachnyi nauchnyi zhurnal ministerstva sel'skogo khozyaistva S S S R. Moscow. Vol. 1+, 1940+ [From 1940-55 appeared in several series]. Vestn. Sel'skokhoz. Nauki. HI 74351

Vestnik sotsialisticheskogo rastenievodstva. Moscow. Vols. 1+, 1940+. Vestn. Sotsialist. Rasteniev. HI 74352

Věstnik státní péče o ochranu přirody = Ochrana přirody. Prague.

Věstnik státniho geologického ústavu Československé republiky. Prague. Nos. 1-25, 1925-50. Věstn. Státn. Geol. Ústavu Českoslo. Republ. Superseded by: Věstnik ustredniho ústavu geologického. HI 74353

Vestnik Tbilisskogo botaničeskogo sada = Vestnik Tbilisskogo botanicheskogo sada. Tiflis.

Vestnik Tbilisskogo botanicheskogo sada. Tiflis, Georgian S S R. Vol. 57+, 1948+. Vestn. Tbilissk. Bot. Sada. Preceded by: Věstnik Tiflisskago Botanicheskago Sada. 5-4217-2. HI 74354

Věstnik Tiflisskago Botaničeskago Sada = Věstnik Tiflisskago Botanicheskago Sada. Tiflis.

Věstnik Tiflisskago botanicheskago sada. Tiflis, Georgian S S R. Vols. 1-51, 1905-21; n.s. vols. 1-5, 1925-31. Věstn. Tiflissk. Bot. Sada. Superseded by: Vestnik Tbilisskogo botanicheskogo sada. 5-4217-2. HI 74355

Věstnik ustredniho ústavu geologického. Prague. No. 26+, 1951+. Věstn. Ustredn. Geol. Preceded by: Věstnik státniho geologického ústavu Československé republiky. HI 74356

Věstník vinařského spolku okolí Mělníka. Melnik & Dolni Berkovice, Bohemia [Czechoslovakia]. Věstn. Vinařsk. Spolku Okolí Mělníka. See B–P–H 963/7. HI 60437

Vestnik zaščity rastenij = Vestnik zashchity rastenii. Moscow & Leningrad.

Vestnik zashchity rastenii. Moscow & Leningrad. Vol. 20 [also numbered vol. 1], 1939; 1940-41. Vestn. Zashch. Rast. Preceded by: Zashchita rastenii. 5-4384-3. HI 74357

Vestsi Akademii navuk Belaruskai S S R. Minsk, Belorussian S S R. 1948(3)-55 [1948(1)-1948(2) not published.] Vestsi Akad. Navuk Belarusk. S.S.R. Superseded by: Vestsi Akademii navuk Belaruskai S S R. Seryya biyalagichnykh navuk. 1-124-1. HI 74358

Vestsi Akademii navuk Belaruskai S S R. Seryya biyalagichnykh navuk. Minsk, Belorussian S S R. 1956-57. Vestsi Akad. Navuk Belarusk. S.S.R., Ser. Biyal. Navuk. Preceded by: Vestsi Akademii navuk Belaruskai S S R. Superseded by: Vestsi Akademii navuk Belaruskai S S R. Seryya biyalagichnykh i sel'skagaspadarchykh navuk. HI 74359

Vestsi Akademii navuk Belaruskai S S R. Seryya biyalagichnykh i sel'skagaspadarchykh navuk. Minsk, Belorussian S S R. 1957+, Vestsi Akad. Navuk Belarusk. S.S.R, Ser. Biyal. Sel'skagasp. Navuk. Preceded by: Vestsi Akademii navuk Belaruskai S S R. Seryya biyalagichnykh navuk. HI 74360

Vetenskap för alla. Stockholm. Vetensk. för Alla. See B–P–H 963/16. HI 60438

Vetenskapliga meddelanden af geografiska föreningen i Finland = Meddelanden af geografiska föreningen i Finland. Helsinki. Meddeland. Geogr. Fören. Finland. See B–P–H 556/25.

Veterinarian. Oxford, England. Veterinarian. See B–P–H 963/22. HI 60440

Veterinary record. London. Veterin. Rec. See B–P–H 963/21. HI 60439

Viata agricola. Bucharest. Viata Agric. See B–P–H 963/23. HI 60441

Vick's family magazine = Vick's monthly magazine. Rochester, NY. Vick's Monthly Mag. See B–P–H 963/26.

Vick's magazine = Vick's monthly magazine. Rochester, NY. Vick's Monthly Mag. See B–P–H 963/26.

Vick's monthly magazine. Rochester, NY. Vick's Monthly Mag. See B–P–H 963/26. HI 60506

Victoria Naturalist. Melbourne, Vic. Victoria Naturalist. See B–P–H 964/1. HI 60442

Victorian historical magazine. Melbourne, Vic. Vol. 1+, 1911+. Victorian Hist. Mag. 5-4390-2. HI 74361

Victorian naturalist; journal and magazine of the field naturalists' club of Victoria. Melbourne, Springvale, Vic. Vol. 1+, 1884+. Victorian Naturalist. 5-4390-2. HI 74362

Victorian rose news. Melbourne, Vic. Vol. 1+, 1964+. Victorian Rose News. HI 74363

Victoria's resources. Springvale. 1959-80. Victoria's Resources. Superseded by: Trees and Victoria's resources. HI 74364

Vida agrícola; revista mensual. Organo de los intereses agrícolas y ganaderos del Perú. Lima. Vida Agríc. See B–P–H 964/9. HI 60443

Vida silvestre; revista del instituto nacional para la conservacion de la naturaleza. Madrid. No. 1+, 1972+. Vida Silvest. HI 74365

Videnskabelige meddelelser fra dansk naturhistorisk forening i Kjøbenhavn. Copenhagen. Vidensk. Meddel. Dansk Naturhist. Foren. Kjøbenhavn. See B–P–H 964/12. HI 60444

Vidya. B - sciences; journal of the Gujarat university. Ahmedabad. Vol. 1+, 1956+. Vidya, B. Sci. HI 74366

Vie marine; annales de la fondation océanographique Ricard. Six-Fours-les-Plages. Vol. 1+, 1979+. Vie Mar. HI 74367

Vie marine hors série; annales de la fondation océanographique Ricard. Vol. 1+, 1980+. Vie Mar. Hors Sér. HI 74368

Vie et milieu. Bulletin du laboratoire Arago, université de Paris. Série A. Biologie marine. Banyuls-sur-Mer, France. Vie & Milieu, Sér. A, Biol. Mar. See B–P–H 964/23. HI 60445

Vie et milieu. Bulletin du laboratoire Arago, université de Paris. Série C. Biologie terrestre. Banyuls-sur-Mer, France. Vie & Milieu, Sér. C, Biol. Terrestre. See B–P–H 964/24. HI 60446

Vie des sciences. Paris = Compte rendu de l'académie des sciences, Paris. Série générale, vie des sciences. Paris.

Vieraea; folia scientiarum biologicarum canariensium. La Laguna, Santa Cruz de Ténérife. Vol. 1+, 1970+. Vieraea. HI 55885

Vierring-Reihe. Brixlegg & Innsbruck. Vierring-Reihe. See B–P–H 964/25. HI 60447

Vierteljahres-Revue der Fortschritte der Naturwissenschaften in theoretischer und praktischer Beziehung = Revue der Fortschritte der Naturwissenschaften in theoretischer und praktischer Beziehung. Cologne & Leipzig. Rev. Fortschr. Naturwiss. See B–P–H 780/27.

Vierteljahresschrift für Dermatologie und Syphilis. Vienna, Leipzig, Berlin. Vols. 6-20, 1874-88. Vierteljahresschr. Dermatol. Syph. Preceded and superseded by: Archiv für Dermatologie und Syphilis. 1-445-2. HI 74369

Vierteljärliche Zeitschrift für Drogenforschung = Quarterly journal of crude drug research. Amsterdam.

Vierteljahrschrift für Forst-, Jagd- und Naturkunde. Prague. Vierteljahrschr. Forst- Naturk. See B–P–H 965/1. HI 60448

Vierteljahrsschrift für Geschichte und Landeskunde Vorarlbergs. Bregenz, Austria. Vierteljahrsschr. Gesch. Landesk. Vorarlbergs. See B–P–H 965/2. HI 60449

Vierteljahrsschrift der Naturforschenden Gesellschaft in Zürich. Zurich. Vierteljahrsschr. Naturf. Ges. Zürich. See B–P–H 965/4. HI 60450

Viewpoint series, australian conservation foundation. Canberra, A.C.T. Vol. 1+, 1967+. Viewpoint Ser. Austral. Conservation Found. HI 74370

Viewpoints in biology. London. Viewpoints Biol. See B–P–H 965/14. HI 60451

Vigne américaine et la viticulture en Europe. Paris & Mâcon. Vigne Amér. Vitic. Eur. See B–P–H 965/16. HI 60452

Vignevini. Bologna. Vol. 1+, 1974+. Vignevini. HI 74371

Vigyan pragati. New Delhi. Vigyan Pragati. See B–P–H 965/17. HI 60453

Vilnius universiteto botanikos instituto ir sodo raštai. Vilnius. Vol. 2/VIII+, 1942+. Vilnius Univ. Bot. Inst. Sodo Raštai. Preceded by: Vilnius universiteto botanikos sodo raštai. HI 65654

Vilnius universiteto botanikos sodo rašti. Vilnius. Vol. ?-1/VII, ?-1941. Vilinius Univ. Bot. Sodo Rašti. Superseded by: Vilnius universiteto botanikos instituto ir sodo raštai. HI 74372

Vilnius universiteto matematikos-gamtos fakulteto darbai. Vilnius. Vol. ?-1/XIV-? ?-1941-? Vilinus Univ. Mat.-Gamtos Fak. Darb. HI 74373

Vilnius valstybinis V. Kapsuko varo universitetas mokslo darbai. Biologiya, geografiya, geologiya. Vilnius. Vol. ?-36(7)+, ?-1960+. Vilnius Valst. V. Kapsuko Varo Univ. Mokslo Darb., Biol., Geogr. Geol. HI 74374

Vil'nyusskii gosudarstvennyi universitet imeni V. Kapsukasa uchenye zapiski = Vilnius valstybinis V. Kapsuko varo universitetas mokslo darbai. Biologiya, geografiya, geologiya. Vilnius.

Vinea et vino Portugaliae documenta. Ser. I. Viticultura. Lisbon. Vinea Vino Portug. Doc., Ser. 1, Vitic. See B–P–H 965/19. HI 60454

Vinea et vino Portugaliae documenta. Ser. II, enologia. Lisbon. Vinea Vino Portug. Doc., Ser. 2, Enol. See B–P–H 965/20. HI 60455

Vineyard. Melbourne, Vic. Vineyard. See B–P–H 965/21. HI 60456

Vinos, viñas y frutas. Buenos Aires. Vinos Viñas Frutas. See B–P–H 965/22. HI 60457

Viola. Stockholm. Viola. See B–P–H 965/23. HI 60458

Virágkedvelök lapja. Budapest. Virágkedv. Lapja. See B–P–H 965/24. HI 60459

Virágos Budapest--Virágos magyarország. A kertészeti lapok melléklete. Budapest. Virágos Budapest--Virágos Magyarorsz. See B–P–H 965/25. HI 60460

Virchow's Archiv für pathologische Anatomie und Physiologie und klinische Medizin. Berlin. Virchow's Arch. Pathol. Anat. See B–P–H 965/30. HI 60461

Vireya vine. Federal Way, WA. No. 1+, 1982+. Vireya Vine. HI 74375

Virgin Islands agricultural experiment station. Bulletin. St. Croix Island, Virgin Islands. Virgin Islands Agric. Exp. Sta. Bull. See B–P–H 966/2. HI 60462

Virginia forests. Richmond, VA. Vol. 1+, 1946+. Virginia Forests. 5-4405-3. HI 74376

Virginia forests magazine = Virginia forests. Richmond, VA.

Virginia fruit. Staunton, VA. ?-ca.1988+. Virginia Fruit. HI 74377

Virginia journal of science. Charlottesville, VA. Virginia J. Sci. See B–P–H 966/4. HI 60463

Virittäjä. Helsinki. Virittäjä. See B–P–H 966/6. HI 60464

Virology. Baltimore, MD. Virology. See B–P–H 966/7. HI 60465

Virology abstracts. Cambridge. Vol. 1+, 1967+. Virol. Abstr. HI 74378

Virology monographs. New York, Berlin, etc. Vol. 1+, 1968+. Virol. Monogr. Preceded by: Handbuch der Virusforschung. HI 74379

Virology reviews. [Soviet medical reviews: Section E.] New York, Chur. Vol. 1+, 1987+. Virol. Rev. HI 74380

Virus. [Uirusu.] Tokyo. Vol. 1+, 1951+. Virus. HI 74381

Virus genes. Dordrecht & Hingham, MA. Vol. 1+, 1987+. Virus Genes. HI 74382

Virus research; international journal of molecular and cellular virology. Amsterdam. Vol. 1+, 1984+. Virus Res. HI 74383

Virusforschung in Einzeldarstellungen = Virology monographs. New York, Berlin, etc.

Virusnye bolezni plodovo-yagadnykh kul'tur i Vinograda v Moldavii. Vols. 1-2, 1973. Virusn. Bol. Plodovo-Yagadnykh Kul't. Vinograda Moldav. HI 74384

Vi'sgálódó magyar gazda. Bécs [=Vienna]. Vi'sgálódó Magyar Gazd. See B–P–H 966/8. HI 60466

Vísindafjélags íslendinga. Reykjavik. Vol. 5+, 1930+. Visindafj. Islend. Preceded by: Rit vísindafjélags íslendinga. HI 74385

Visnik kyyivs'kogo universitety. Kiev. 1958+. Visn. Kyyivs'k. Univ. HI 74386

Visnyk Akademiji nauk Ukrajins'koji R S R = Visnyk Akademii nauk Ukrayins'koi R S R. Kiev.

Visnyk Akademii nauk Ukrayins'koi R S R. Kiev, Ukrainian S S R. Vol. 20+, 1947+. Visn. Akad. Nauk Ukrayins'k. R.S.R. Preceded by: Visti Akademii nauk Ukrayins'koi R S R. 1-121-2. HI 74387

Visnyk botanichnoho sadu, akademiya nauk Ukrayins'koyi R S R. Kyyiv. 1959-62. Visn. Bot. Sadu Akad. Nauk Ukrayins'k. R.S R. HI 74388

Visnyk Kharkivskogo universitetu. Kharkiv. Nos. 69-102, 1971-73. Visn. Kharkivsk. Univ. Preceded and superseded by: Vestnik Khar'kovskogo universiteta. HI 74389

Visnyk Kyyivs'kogo botaničnogo sadu imeny akad. O. V. Fomina = Izvestiya Kievskogo botanicheskogo sada imeni akad. A. V. Fomina. Kiev.

Visnyk Kyyivsk'kogo botanichnogo sadu = Izvestiya Kievskogo botanicheskogo sada. Kiev.

Visnyk Kyyivs'koho botanichnoho sadu. Kiev. Vols. 1-17, 1923-34. Visn. Kyyivs'koho Bot. Sadu. HI 74390

Visnyk kyyivs'koho universytetu. Seriya biolohiyi = Visnik kyyivs'kogo universitety. Kiev.

Visnyk prykladnoi botaniky. Kharkov, Ukrainian S S R. 1 vol., 1930. Visn. Prykl. Bot. 1-121-2. HI 74391

Visnyk prykladnoji botaniky = Visnyk prykladnoi botaniky. Kharkov.

Vistas in botany. London & New York, NY. Vistas Bot. See B–P–H 966/15. HI 60467

Vistas in plant sciences. Hissar. Vol. 1+, 1972+. Vistas Pl. Sci. HI 74392

Visti Akademii nauk Ukrayins'koi R S R. Kiev, Ukrainian S S R. Vols. 9(3)-19, 1936-46. Visti Akad. Nauk Ukrayins'k. R.S.R. Preceded by: Visti Ukrayins'koi akademii nauk. Superseded by: Visnyk Akademii nauk Ukrayins'koi R S R. 1-121-2. HI 74393

Visti Akademiji nauk Ukrajins'koji R S R = Visti Akademii nauk Ukrayins'koi R S R. Kiev.

Visti Ukrayins'koi akademii nauk. Kiev, Ukrainian S S R. Vols. 8-9(1/2), 1935-36. Visti Ukrayins'k. Akad. Nauk. Preceded by: Visti Vseukrayins'koi akademii nauk. Superseded by: Visti Akademii nauk Ukrayins'koi R S R. 1-121-2. HI 74394

Visti Vseukrajins'koji akademiji nauk = Visti Vseukrayins'koi akademii nauk. Kiev.

Visti Vseukrayins'koi akademii nauk. Kiev, Ukrainian S S R. Vols. 1-7, 1929-34. Visti Vseukrayins'k. Akad. Nauk. Superseded by: Visti Ukrayins'koi akademii nauk. 1-121-2. HI 74395

Vita marina. Den Haag. 1964+. Vita Mar. HI 74396

Viticulture - arboriculture. Paris = Revue de viticulture. Paris. Rev. Vitic. See B–P–H 786/10.

Vitis; the Glasnevin grapevine: journal of the botanic gardens educational society. Dublin. Vol. 1+, 1982+. Vitis (Dublin). HI 74397

Viticultura chilena. Santiago. Vitic. Chilena. See B–P–H 966/23. HI 60469

Viticulture algérienne. Algiers. Vitic. Algér. See B–P–H 966/22. HI 60468

Viticulture française. Paris. Vitic. Franç. See B–P–H 966/24. HI 60470

Vitis. Berichte über Rebenforschung. Geilweilerhof, Germany. Vitis. See B–P–H 966/25. HI 60471

Vitis munsoniana; report of the T. V. Munson memorial vineyard. Denison, TX. Vol. ?-4(2)+, ?-19??+. Vitis Munsoniana. HI 74399

Vlaamsche genees- en heelkundige bladen. Amsterdam & Ghent. Nos. 1-9/11, 1902-05. Vlaamsche Genees-Heelk. Bladen. 5-4414-2. HI 74400

Vlaamsche gids; algemeen maandschrift. Antwerp. Vol. 1+, 1905+. Vlaamsche Gids. 5-4414-2. HI 74401

Vlaamsche kunstbode; maandelijksch tijdschrift voor kunsten, letteren en wetenschappen. Antwerp. Vols. 1-34, 1871-1904. Vlaamsche Kunstbode. 5-4414-2. HI 74402

Vlaamsche wacht; veertiendaagsch tijdschrift voor nederelandsche letteren, kunst, wetenschap en bibliographie. Ghent. Nos. 1-8, 1878-86. Vlaamsche Wacht. 5-4414-2. HI 74403

Vlastivěda ostravského kraje. Ostrava, Czechoslovakia. Vlastiv. Ostravsk. Kraje. See B–P–H 967/3. HI 60474

Vlastivedný časopis. Bratislava. Vlastiv. Cas. (Bratislava). See B–P–H 967/1. HI 60472

Vlastivedný obzor. Trencin, Czechoslovakia. Vlastiv. Obzor. See B–P–H 967/2. HI 60473

Vlastivědný sborník Českobraodka. Prague. Vlastiv. Sborn. Českobrodska. See B–P–H 967/4. HI 60475

Vlastivědný sborník v Košiciach. Kosice, Czechoslovakia. Vlastiv. Sborn. v Košiciach. See B–P–H 967/6. HI 60477

Vlastivědný sborník Považia. Banska Bystrica, Czechoslovakia. Vlastiv. Sborn. Považia. See B–P–H 967/5. HI 60476

Vlastivědný sborník východní Čechy. Havlickuv Brod, Czechoslovakia. Vlastiv. Sborn. Východní Čechy. See B–P–H 967/7. HI 60478

Vlastivědný sborník Vysočiny. Oddil ved přírodních. Jihlava, Czechoslovakia. Vlastiv. Sborn. Vysočiny, Odd. Věd Přír. See B–P–H 967/8. HI 60479

Vlastivedný sborník žílnského kraje . Turciansky Svaty Martin, Czechoslovakia. Vlastiv. Sborn. Žílinsk.

Kraje. See B–P–H 967/9. HI 60480

Vlastivedný zpravodaj podjavorinského múzea. Nove Mesto nad Vahom, Czechoslovakia. Vlastiv. Zprav. Podjavorinsk. Múz. See B–P–H 967/10. HI 60481

Vlastivěné zprávy okrasu Nový Bor. Novy Bor, Czechoslovakia. Vlastiv. Zprávy Okrasu Nový Bor. See B–P–H 967/11. HI 60482

Vlastivěné zprávy z Adamova a okolí. Brno, Czechoslovakia. Vlastiv. Zprávy z Adamova Okolí. See B–P–H 967/12. HI 60483

Vlugschriften van de directie van den landbouw. The Hague. Vol. 1+, 1948+. Vlugschr. Direct. Landb. 3-2349-3. HI 74404

Vlugschriften van het instituut voor phytopathologie. Wageningen. Vlugschr. Inst. Phytopathol. See B–P–H 967/13. HI 60484

Vlugschriften van den phytopathologischen dienst. Wageningen. Vlugschr. Phytopathol. Dienst. See B–P–H 967/14. HI 60485

Vlugschriften van de plantenkundige dienst. Wageningen. Vlugschr. Plantenk. Dienst. See B–P–H 967/15. HI 60486

Vlugschriften, deli proefstation te Medan. Medan. Nos. 1-69, 1920-41. Vlugschr. Deli Proefstat. Medan. HI 74405

V.N.K. nieuws. Coöp. bond van nederl. coöp. kruidteleis verenigingen. Laren, Netherlands = V N K nieuws. Coöp. bond van nederl. coöp. kruidteleis verenigingen. Laren, Netherlands.

Vodní hospodářství. Prague. Vodní Hospod. See B–P–H 967/18. HI 60487

Vodnii zhurnal. Budapest = Hidrológiai közlöny. Budapest. Hidrol. Közl. See B–P–H 416/4.

Vodorosli i griby zapadnoi Sibiri. Novosibirsk. Vols. 1-2, 1964-65. Vodorosli Griby Zapadnoi Sibiri. Superseded by: Vodorosli i griby sibiri i dal'nego vostoka. HI 74406

Vodorosli i griby sibiri i dal'nego vostoka. Novosibirsk. Vol. 1(3)+, 1970+. Vodorosli Griby Sibiri Dal'nego Vostoka. Preceded by: Vodorosli i griby zapadnoi Sibiri. HI 74407

Vole. London. Vol. 1+, 1977+. Vole. HI 74408

Volk und Heimat. Eisenstadt, Austria. Volk & Heimat. See B–P–H 967/22. HI 60488

Voprosy biofiziki. Trudy. Vol. ?-4+, ?-1967+. Vopr. Biofiz., Trudy. HI 74409

Voprosy biokhimii. Erevan. Vol. 1+, 1960+. Vopr. Biokhim. HI 74410

Voprosy biologii. Erevan. ?-1983+. Vopr. Biol. HI 74411

Voprosy biologii rastenii. Cheliabinsk. Vol. ?-2+, ?-1970+. Vopr. Biol. Rast. HI 74412

Voprosy biologii, turkmenskii gosudarstvennyi universitet. Ashkhabad. 1973-74. Vopr. Biol. Turkmensk. Gosud. Univ. Superseded by: Biologiya zhivotnykh i rastenii Turkmenistana. HI 74413

Voprosy biologii zhivotnykh i rastenii Turkmenistana = Voprosy biologii, turkmenskii gosudarstvennyi universitet. Ashkhabad.

Voprosy botaniki. Moscow & Leningrad. Vopr. Bot. See B–P–H 968/1. HI 60489

Voprosy botaniki. Vilnius = Botanikos klausimai. Vilnius, Lithuanian S S R. Bot. Klaus. See B–P–H 221/1.

Voprosy botaniki yugo-vostoka. Saratov. Vol. 1+, 1975+. Vopr. Bot. Yugo-Vostoka. HI 74414

Voprosy ekologii. Kyyiv. 1957-62. Vopr. Ekol. HI 74415

Voprosy ekologii i biocenologii = Voprosy ekologii i biotsenologii. Moscow & Leningrad.

Voprosy ekologii i biotsenologii. Moscow & Leningrad. Vols. [1]-7, 1934-39. Vopr. Ekol. Biotsenol. Preceded by: Zhurnal ekologii i biotsenologii. 5-4428-2. HI 74416

Voprosy ekologii i okhrany prirody. Leningrad. No. 1+, 1981+. Vopr. Ekol. Okhrany Prir. HI 74417

Voprosy evolyutsii organicheskogo mira. Leningrad. Vol. 14+, 1983+. Vopr. Evol. Organich. Mira. Preceded by: Novaya literatura po voprosam evolyutsii organicheskogo mira. HI 75125

Voprosy farmakognozii. [Forms part of: Trudy, leningradskii khimiko-farmatsevticheskii institut.] Leningrad. Vol. ?-5+, ?-1968+. Vopr. Farmakogn. HI 74418

Voprosy farmakognozoo = Trudy Leningradskogo khimiko-farmacevticheskogo instituta. Leningrad.

Voprosy fiziologii i biokhimii kul'turnykh rastenii. Kishinev. Vol. 1+, 1962+. Vopr. Fiziol. Biokhim. Kul't. Rast. Preceded by: Biokhimii kul'turnykh rastenii Moldavii. HI 74419

Voprosy fiziologii rastenii i mikrobiologii. Minsk, Belorussian S S R. Vol. 2, 1961. Vopr. Fiziol. Rast. Mikrobiol. Vol. [1], 1959 published as monograph. HI 74420

Voprosy fiziologii rastenij i mikrobiologii = Voprosy fiziologii rastenii i mikrobiologii. Minsk.

Voprosy geografii. Moscow. Vopr. Geogr. See B–P–H 968/5. HI 60490

Voprosy geografii Dal'nego Vostoka. Khabarovsk, R S F S R. Vol. 1+, 1949+. Vopr. Geogr. Dal'nego Vostoka. Preceded by: Zapiski Priamurskago Otděla

Imperatorskago Russkago Geograficheskago Obshchestva. HI 74421

Voprosy geografii Sibiri. Tomsk, R S F S R. Vopr. Geogr. Sibiri. See B–P–H 968/8. HI 60492

Voprosy introduktsii rastenii i zelenogo stroitel'stva. Tiflis. Vols. 1(70)-13(82), 1965-82. Vopr. Introd. Rast. Zelen. Stroit. Superseded by: Introduktsiya rastenii i zelenoe stroitel'stvo. HI 74422

Voprosy istorii estestvoznanija i tehniki = Voprosy istorii estestvoznaniya i tekhniki. Moscow.

Voprosy istorii estestvoznaniya i tekhniki. Moscow. Vol. 1+, 1956+. Vopr. Istorii Estestv. Tekhn. HI 74423

Voprosy mikrobiologii. Erevan. Vol. 1+, 1961+. Vopr. Mikrobiol. Preceded by: Mikrobiologicheskii sbornik. HI 75126

Voprosy mikropaleontologii. Moscow. Vopr. Mikropaleontol. See B–P–H 968/10. HI 60493

Voprosy pochvozashchitnogo zemledeliya, nauchno-tekhnicheskii byulleten'. Tselinograd. Vol. 1+, 1974+. Vopr. Istorii Estestv. Tekhn. HI 74424

Voprosy virusologii. Moscow. Vol. 1+, 1956+. Vopr. Virusol. HI 74425

Vorallberger Museums-Verein. Rechenschaftsbericht des Ausschusses. Bregenz. Vols. 1-13, 1858-72. Vorallberger Mus.-Verein Rechenschaftsber. Ausschusses. HI 74426

Vorlesungen der Churpfälzischen physicalisch-ökonomischen Gesellschaft. Mannheim. Vorles. Churpfälz. Phys.-Okon. Ges. See B–P–H 968/13. HI 60494

Voronezhskaya stantsiya zashchity rastenii. Voronezh, R S F S R. Vol. 8, 1927. Voronezhsk. Stantsiya Zashch. Rast. Preceded & superseded by: Byulleteni Voronezhskoi stantsii zashchity rastenii. HI 74427

Voronežskaja stancija zaščity rastenij = Voronezhskaya stantsiya zashchity rastenii. Voronezh.

Vort havebrug. [Edited by P. Hansen]. Copenhagen. Vols. 1-?, 1897-99. Vort Havebrug (Hansen). 5-4430-1. HI 74428

Vort havebrug. [Edited by N. P. Jensen]. Skjern. Vols. 1-3, 1914-19. Vort Havebrug (Jensen). HI 74429

Vort havebrug. [Edited by J. Uldall]; medlemsblad for foreningen for udførsel af havesager. Copenhagen. Vols. [1]-6, 1885-90. Vort Havebrug (Uldall). HI 74430

Vort Landbrug. Copenhagen. Vols. 1-55, 1882-1936. Vort Landbr. Superseded by: Dansk Landbrug. 2-1260-1. HI 67458

Vorträge aus dem Gebiete der Naturwissenschaften und der Ökonomie gehalten vor einem Kreise gebildeter Zuhörer in der physikalisch-ökonomischen Gesellschaft zu Königsberg. Königsberg [=Kaliningrad, R S F S R]. Vorträge Naturwiss. Ökon. See B–P–H 968/17. HI 60495

Vorträge aus dem Gesamtgebiet der Botanik herausgegeben von der deutschen botanischen Gesellschaft. Berlin. Vols. 1-4, 1914-19; n.s. vols. 1-?, 1962-70. Vortr. Gesamtgeb. Bot. 2-1296-3. HI 74431

Vorträge für Pflanzenzuchtung. Bonn. Vol. 1+, 1982+. Vortr. Pflanzenzucht. HI 74432

Vorträge, rheinisch-Westfälische Akademie der Wissenschaften. Natur-, Ingenieuru und Wirtschaftswissenschaften. Opladen. No. 204+, 1970+. Votr. Rhein.-Westfäl. Akad. Wiss. Natur- Ing. Wirtschaftswiss. Preceded by: Veröffentlichungen der Arbeitsgemeinschaft für Forschung des Landes Nordrhein-Westfalen. HI 74433

Vorträge und Schriften = Preussische Akademie der Wissenschaften zu Berlin. Vorträge und Schriften. Berlin. Preuss. Akad. Wiss. Berlin Vorträge Schriften. See B–P–H 721/19.

Vorträge und Schriften, Deutsche Akademie der Wissenschaften zu Berlin = Deutsche Akademie der Wissenschaften zu Berlin. Vorträge und Schriften. Berlin. Deutsche Akad. Wiss. Berlin Vorträge Schriften. See B–P–H 345/11.

Vorzüglichsten Vorlesungen, welche in der Königlichen Schwedischen Akademie der Wissenschaften zu Stockholm gehalten worden sind. Leipzig. Vorzügl. Vorles. Königl. Schwed. Akad. Wiss. Stockholm. See B–P–H 968/19. HI 60496

Vragen van den dag. Amsterdam. Vragen Dag. See B–P–H 968/20. HI 60497

Vsesojuznyj naučno-issledovatel'skij institut lekarstvennyh rastenij = Vsesoyuznyi nauchno-issledovatel'skii institut lekarstvennykh rastenii. Moscow.

Vsesoyuznyi nauchno-issledovatel'skii institut lekarstvennykh rastenii. Moscow. Vols. 5-7, 1936-42. Vsesoyuzn. Nauchno-Issl. Inst. Lekarstv. Rast. Preceded by: Trudy po lekarstvennym i lekarstvenno-tekhnicheskim rasteniyam. Superseded by: Trudy. Vsesoyuznogo nauchno-issledovatel'skogo instituta lekarstvennykh rastenii. 5-4267-3. HI 67529

Vsesoyuznoe obshchestvo fiziologov rastenii. Moscow. Vol. 1+, 1988+. Vsesoyuzn. Obshch. Fiziol. Rast. HI 75127

Vulgarisation de l'horticulture. Vulg. Hort. See B–P–H 969/4. HI 60498

Vulgarisation, renseignements agricoles. Algiers. Vols. 1-2, 1961?-? Vulg. Renseign. Agric. HI 71327

Vuosikirja, lapin tutkimusseura. Rovaniemi? 1960+.

Vuosik. Lapin Tutkimusseura. HI 74434

Vuosikirja, suomalainen tiedeakatemia. Helsinki. 1977+. Vuosik. Suom. Tiedakat. Preceded by: Sitzungsberichte der Finnischen Akademie der Wissenschaften. HI 74435

Východočeský botanický zpravodaj. Pardubice, Czechoslovakia. Východočeský Bot. Zprav. See B–P–H 969/5. HI 60499

Výroční zpráva klubu přírodovědeckého v Praze. Prague. Výr. Zpráva Klubu Přír. v Praze. See B–P–H 969/7. HI 60500

Výroční zpráva klubu přírodovědeckého v Prostějově. Prossnitz, Moravia [=Prostejov, Czechoslovakia]. Výr. Zpráva Klubu Přír. v Prostějově. See B–P–H 969/8. HI 60501

Výroční zpráva komise pro přírodovědecké prozkoumání Moravy. Brünn [=Brno, Czechoslovakia]. Výr. Zpráva Komis. Přír. Prozk. Moravy. See B–P–H 969/9. HI 60502

Výroční zpráva komise na přírodovědecký výzkum Moravy a Slezska. Brünn [=Brno, Czechoslovakia]. Výr. Zpráva Komis. Přír. Výzk. Moravy Slezska. See B–P–H 969/10. HI 60503

Výrochni zpráva královské České společnosti nauk. Prague. 1918-39. Výr. Zpráva Král. České Společn. Nauk. Preceded by: Jahresbericht der koniglich Bohmischen Gesellschaft der Naturwissenschaften. HI 74438

Výroční zpráva moravské přírodovědecké společnosti. Brno, Czechoslovakia. Výr. Zpráva Morav. Přír. Společn. See B–P–H 969/11. HI 60504

Výroční zpráva moravsko-slezské akademie věd přírodních. Brno, Czechoslovakia. Výr. Zpráva Morav.-Slez. Akad. Věd Přír. See B–P–H 969/12. HI 60505

Vygie. [Supplement to: Veld & flora.] Kirstenbosch. Vol. 1+, 1980+. Vygie. HI 74436

Vykopni fauna i flora Ukrainy. Kiev. Vol. 1+, 1973+. Vykopni Fauna Fl. Ukrainy. HI 74437

Vykopni fauna i flora Ukroyiny = Vykopni fauna i flora Ukrainy. Kiev.

Výskumný ústav lesného hodspodárstva = Acta instituti forestalis zvolenensis. Bratislava.

Vytauto didziojo universiteto botanikos sodo rastai = Scripta horti botanici universitatis Vytauti Magni. Kaunas, Lithuania [Lithuanian S S R]. Scripta Horti Bot. Univ. Vytauti Magni. See B–P–H 831/23.

Vytauto didžiojo universiteto matematikos-gamtos fakulteto darbai. Kaunas. Vols. 5-13?, 1931-39 [issued in series: Biologijas skyrius/section de biologie; botanikos skyrius/section de botanique; fizikos bei chemijos skyrius/section de physique et de chemie; geografijos skyrius/section de la géographie; geologijos skyrius/section de géologie]. Vytauto Didžiojo Univ. Mat.-Gamtos Fak. Darbai. Preceded by: Lietuvos universiteto matematikos-gamtos fakulteto darbai. 3-2275-2. HI 59042

W A R D A technical newsletter. Monrovia. Vol. 1+, 1979+. W. A. R. D. A. Techn. Newslett. HI 74439

W N P S newsletter = Douglasia. Seattle, WA.

W S S A newsletter. Champaign, IL. 1973+. W. S. S. A. Newslett. HI 68071

Wachendorffia; chronica horti botanici Rheno-Trajectini. Utrecht. Vol. 1+, 1972+. Wachendorffia. HI 74440

Wahlenbergia; scripta botanica Umensia. Umea. Vol. 1+, 1975+. Wahlenbergia. HI 74441

Wald; Zeitschrift für Forstwirtschaft Holzwirtschaft. Berlin. Vols. 1-3, 1951-53. Wald. Preceded by: Forstwirtschaft, Holzwirtschaft. Superseded by: Forst und Jagd. HI 74442

Wald und Holz. Solothurn, Switzerland. Wald & Holz. See B–P–H 970/18. HI 60527

Waldeckische gemeinnützige Zeitschrift. Arolsen, Germany. Waldeck. Gemeinnütz. Z. See B–P–H 970/19. HI 60528

Waldhygiene. Würzburg. Vol. 1+, 1954+. Waldhygiene. HI 74443

Waldschutzbrief. Mitteilungen des Steiermärkischen Waldschutzverbandes. Graz. Waldschutzbrief. See B–P–H 970/20. HI 60529

Walia. Addis Ababa. No. 1+, 1969+, 1970+. Walia. HI 74444

Walla Walla college publications in the department of biological sciences and the biological station. College Place, WA. Walla Walla Coll. Publ. Dept. Biol. Sci. See B–P–H 970/21. HI 60530

Wallaceana; an ecology newsletter for south east Asia (later A global newsletter for tropical ecology). Kuala Lumpur, Utrecht. Vol. 1+, 1973+. Wallaceana. HI 59333

Wallerstein laboratories communications. New York, NY. Wallerstein Lab. Commun. See B–P–H 971/1. HI 60531

Wandelaar. Laren, Netherlands. Wandelaar. See B–P–H 971/2. HI 60532

Wanderer im Riesengebirge. Hirschberg, Germany. Wanderer Riesengebirge. See B–P–H 971/3. HI 60533

Warasan khanawitthayasat mahawitthayalai Chiangmai = Journal of the science faculty of Chiangmai university. [Chiangmai.]

Warnact newsletter. Warwick. No. 32+, 1979+. Warnact Newslett. Preceded by: Newsletter of the Warwickshire nature conservation trust. HI 59334

Warrah. Port Stanley, Falkland Is. No. 1+, 1981+. Warrah. HI 74445

Waseda biology. [Waseda seibutsu.] Tokyo. 1953+. Waseda Biol. HI 75155

Waseda daigaku rikogaku kenkyujo hokoku = Bulletin of the science and engineering research laboratory, Waseda university. Tokyo.

Waseda seibutsu = Waseda biology. Tokyo.

Washington (state) agricultural experiment station. Annual report. Pullman, WA. Wash. State Agric. Exp. Sta. Annual Rep. See B–P–H 971/11. HI 60535

Washington (state) agricultural experiment station. Bulletin, special series. Pullman, WA. Wash. State Agric. Exp. Sta. Bull., Special Ser. See B–P–H 971/13. HI 60536

Washington (state) agricultural experiment station. Circular. Pullman, WA. Wash. State Agric. Exp. Sta. Circ. See B–P–H 971/14. HI 60537

Washington (state) agricultural experiment station. Mimeographed circular. Pullman, WA. Wash. State Agric. Exp. Sta. Mimeogr. Circ. See B–P–H 971/15. HI 60538

Washington (state) agricultural experiment station. Popular bulletin. Pullman, WA. Wash. State Agric. Exp. Sta. Popular Bull. See B–P–H 971/16. HI 60539

Washington (state) agricultural experiment station. Station circular. Pullman, WA. Wash. State Agric. Exp. Sta. Sta. Circ. See B–P–H 971/17. HI 60540

Washington daffodil society yearbook. Washington, DC. Wash. Daffodil Soc. Yearb. See B–P–H 971/8. HI 60534

Washington Park arboretum bulletin. Seattle, WA. Vol. 49(2)+, 1986+. Washington Park Arbor. Bull. Preceded by: Arboretum bulletin, university of Washington. HI 74446

Washington (state) agricultural experiment station (later Stations). Bulletin. Pullman, WA. Nos. 1-792, 1892-74. Wash. State Agric. Exp. Sta. Bull. Superseded by: Bulletin, college of agriculture research center, Washington state university. HI 55852

Wasmann club collector. San Francisco, CA. Wasmann Club Collect. See B–P–H 971/18. HI 60541

Wasmann collector. San Francisco, CA. Wasmann Collect. See B–P–H 971/19. HI 60542

Wasmann journal of biology. San Francisco, CA. Wasmann J. Biol. See B–P–H 971/20. HI 60543

Wasserwirtschaft. Stuttgart. Wasserwirtschaft. See B–P–H 971/23. HI 60544

Waste materials biodeterioration research titles. Birmingham. Vol. 1+, 1974+. Waste Mater. Biodeterior. Res. Titles. Superseded by: International biodeterioration. HI 55853

Watakushitachi no shizen = Nature. Tokyo.

Watch over Essex; journal of the Essex naturalists' trust. 1982+. Watch Over Essex. Preceded by: Bulletin, Essex naturalist's trust. HI 74447

Water-in-plants bibliography. The Hague. Vol. 1+, 1975+, 1977+. Water in Pl. Bibliogr. HI 74448

Water farming journal; North America's aquaculture newspaper. Metairie, LA. ?-1986(July)+. Water Farming J. HI 74449

Water garden journal; official publication of the water lily society. Buckeystown, MD. Vol. 1(4)+, 1985+. Water Gard. J. Preceded by: Water lily journal. HI 74450

Water, land and life; publication of western Pennsylvania conservancy. Pittsburgh, PA. Vols. 1-13(2), 1959-71. Water Land Life. Superseded by: Conserve. HI 74451

Water lily journal. Buckeystown, MD. Vol. 1(1-3), 1985. Water Lily J. Superseded by: Water garden journal. HI 74452

Water and sewage works. Chicago, IL. Water Sewage Works. See B–P–H 972/1. HI 60545

Watersheds. Toronto. Watersheds. See B–P–H 972/2. HI 60546

Watsonia; journal of the botanical society of the British Isles (later Journal and proceedings of the botanical society of the British Isles). Arbroath & London. Vol. 1+, 1949/50+. Watsonia. Preceded by: Botanical society and exchange club of the British Isles. From vol. 8+ incorporated: Botanical society of the British Isles proceedings. 5-4459-2. HI 74453

Webbia; raccolta di scritti botanici. Florence. Webbia. See B–P–H 972/4. HI 60547

Wecko-skrift för läkare och naturforskare. Stockholm. Wecko-Skr. Läkare Naturf. See B–P–H 972/5. HI 60548

Weed abstracts. Farnham Royal, England. Weed Abstr. See B–P–H 972/6. HI 60549

Weed control recommendations for eastern Canada. 1957-63. Weed Control Recommend. E. Canada. Preceded by: Report, herbicide and weed classification committee, Canada, eastern section. Superseded by: Report of the research appraisal and research planning committee, national weed committee, Canada, eastern section. HI 74454

Weed control recommendations, national weed committee, Canada, eastern section = Weed control recommendations for eastern Canada.

Weed control recommendations, national weed committee, Canada, western section = Weed control recommendations for western Canada.

Weed control recommendations for western Canada. 1956-63. Weed Control Recommend. W. Canada. Preceded by: Report, herbicide and weed classification committee, western section, Canada. Superseded by: Report of the research appraisal and research planning committee, national weed committee, Canada, western section. HI 74455

Weed research. Oxford. Vol. 1+, 1961+. Weed Res. HI 74456

Weed research. [Zasso kenkyu.] Tokyo. Weed Res. See B–P–H 972/7. HI 60550

Weed science. Geneva, NY, Champaign, IL. Vol. 16+, 1968+. Weed Sci. Preceded by: Weeds. HI 74457

Weed technology; a journal of the weed science society of America. Champaign, IL. Vol. 1+, 1987+. Weed Technol. HI 75172

Weeds. New York. Vols. 1-15, 1951-67. Weeds. Superseded by: Weed science. HI 74458

Weeds in Indonesia; communication of the weed science society of Indonesia. Bogor. Vol. ?-2(2)+, ?-1971+. Weeds Indonesia. HI 60769

Weeds today. Minneapolis, MN. Vol. 1+, 1970+. Weeds Today. HI 60770

Weeds, trees, and turf. Cleveland, OH. Weeds Trees Turf. See B–P–H 972/10. HI 60551

Weeds and weed control = Proceedings, swedish weed conference. Uppsala.

Weekblad von de koninklijke m[aatschapp]i j voor tuinbouw en plantkunde. The Hague. Weekbl. Kon. Maatsch. Tuinb. Plantk. See B–P–H 972/11. HI 60552

Weekly bulletin. California state board of health. Sacramento, CA. Weekly Bull. Calif. State Board Health. See B–P–H 972/13. HI 60553

Weekly-bulletin, Paiforce naturalists' club. Nos. ?-30-48-?, ?-1945-46-? Weekly-Bull. Paiforce Naturalists' Club. HI 60771

Weekly bulletin of the St. Louis medical society. St. Louis, MO. Weekly Bull. St. Louis Med. Soc. See B–P–H 972/14. HI 60554

Weekly florists' review = Florists' review; a weekly journal for florists, seedsmen and nurserymen. Chicago, IL. Florists' Rev. See B–P–H 376/21.

Weekly memorials for the ingenious; or, an account of books lately set forth in several languages, with other accounts relating to arts and sciences. [Published by R. Chiswell]. London. Nos. 1-29, [1681-82] 1683. Weekly Memorials Ingenious (Chiswell). 5-4463-2. HI 74460

Weekly memorials for the ingenious. [Published by Faithorne, Henry & John Kersey]. London. Nos. 1-50, 1682-83. Weekly Memorials Ingenious (Faithorne & Kersey). 5-4463-3. HI 74461

Weekly miscellany for the improvement of husbandry, trade, arts, and science. London. Weekly Misc. Improv. Husb. See B–P–H 972/16. HI 60555

Weekly news letter to the crop correspondents. United States department of agriculture. Washington, DC. Weekly News Lett. Crop Corresp. U.S.D.A. See B–P–H 972/17. HI 60556

Weekly news letter. United States department of agriculture. Washington, DC. Weekly News Lett. U.S.D.A. See B–P–H 972/18. HI 60557

Weekly observations of the royal Dublin society. Dublin. Weekly Observ. Roy. Dublin Soc. See B–P–H 972/19. HI 60558

Weekly reports on insects, diseases and crop development. Ithaca, NY. Vols. ?-63(15), 19??-82. Weekly Rep. Insects Dis. Crop Developm. Superseded by: Weekly report on pests and crop development. HI 74462

Weekly report on pests and crop development. Ithaca, NY. Vol. 63(16)+, 1982+. Weekly Rep. Pests Crop Developm. Preceded by: Weekly reports on insects, diseases and crop development. HI 74463

Weekly weather and crop bulletin. 1924+. Weekly Weath. Crop Bull. HI 74464

Wei shêng wu hsüeh p'ao = Acta microbiologia sinica. Peking. Acta Microbiol. Sin. See B–P–H 46/10.

Wei sheng wu hsueh tung pao = Microbiology. Beijing.

Wei ti gu sheng wu xue-bao = Acta micropalaeontologica sinica. Beijing.

Wein und Rebe. Mainz. Wein & Rebe. See B–P–H 972/21. HI 60559

Weishengwu tongbao = Microbiology. Beijing.

Weleda Nachrichten. Arlesheim. Vol. ?-100-112+, ?-[1963]-[67]+. Weleda Nachr. HI 74466

Welsh bulletin, botanical society of the British Isles. Aberystwyth. No. 26+, 1977+. Welsh Bull. Bot. Soc. British Isles. Preceded by: Welsh regional bulletin, botanical society of the British Isles. HI 74467

Welsh journal of agriculture. Cardiff, Wales. Welsh J. Agric. See B–P–H 972/22. HI 60560

Welsh regional bulletin, botanical society of the British Isles. Aberystwyth. Nos. 1-25, 1964-76. Welsh Regional Bull. Bot. Soc. British Isles. Superseded by: Welsh bulletin, botanical society of the British Isles. HI 74468

Wên hsien chuan k'an = Report of historico-geographical studies of Taiwan. Taipei, Taiwan. Rep. Hist.-Geogr. Stud. Taiwan. See B–P–H 766/1.

Wentia. Amsterdam. Wentia. See B–P–H 973/2. HI 60561

Werdenda. Beiträge zur Pflanzenkunde. Bingen, Germany & Washington, DC. Werdenda. See B–P–H 973/3. HI 60562

Werken, dienst domaniale natuurreservaten en natuurbescherming = Travaux, service des réserves naturelles domaniales et de la conservation de la nature. Brussels.

Werken, dienst natuurbescherming = Travaux, service de la conservation de la nature. Belgium. Brussels.

Werken van het genootschap tot bevordering der natuurgenees- en heelkunde te Amsterdam. Amsterdam. Werken Genootsch. Bevord. Naturr- Heelk. Amsterdam. See B–P–H 973/4. HI 60563

Wesley naturalist; journal of the Wesley scientific society. London. Wesley Naturalist. See B–P–H 973/5. HI 60564

West african journal of biological and applied chemistry. Ibadan, NIgeria. W. African J. BIol. Appl. Chem. See B–P–H 969/15. HI 60507

West african journal of biological chemistry. Ibadan, NIgeria. W. African J. BIol. Chem. See B–P–H 969/16. HI 60508

West american scientist. San Diego, CA. W. Amer. Sci. See B–P–H 969/17. HI 60509

West australian gardener. Perth, Balcatta, W.A. Vols. 1-25, 1932-58; [n.s.] vol. 1+, 1969+. W. Austral. Gard. HI 70277

West australian naturalist Perth, W.A. 1939. W. Austral. Naturalist. Superseded by: Western australian naturalist. HI 74469

West of England journal of science and literature. Bristol. Bath, Exeter, Cardiff, & London. W. England J. Sci. Lit. See B–P–H 969/19. HI 60511

West indian bulletin; the journal of the imperial department of agriculture for the West Indies. Bridgetown, Barbados. W. Indian Bull. See B–P–H 970/1. HI 60514

West indian review; magazine of the Caribbean. Dorking, Kingston, Jamaica. Vols. 1-6(10), 1934-40; n.s. vols. 1-6, 1944-49; vol. 9(2)+, 1972+. W. Indian Rev. For 1963-72 see: Jamaica and west indian review. 5-4476-2. HI 74470

West Indië. Vol. ?-4-?, ?-1919-? West Indië. HI 74471

West London medical journal. London. W. London Med. J. See B–P–H 970/5. HI 60518

West Pakistan agricultural university. Research studies. Lyallpur, Pakistan. W. Pakistan Agric. Univ. Res. Stud. See B–P–H 970/10. HI 60522

West Pakistan journal of agricultural research. Lahore, Pakistan. W. Pakistan J. Agric. Res. See B–P–H 970/11. HI 60523

West Virginia agriculture and forestry. Morgantown, WV. Vol. 1+, 1968+. West Virginia Agric. Forest. Preceded by: Science serves your farm and home. HI 74472

West Virginia forestry notes. Morgantown, WV. No. 1+, 19??+. West Virginia Forest. Notes. HI 74473

West Virginia geological survey. Educational series. Morgantown, WV. West Virginia Geol. Surv., Educ. Ser. See B–P–H 973/13. HI 60565

West Virginia medical journal. Charlestown, WV. West Virginia Med. J. See B–P–H 973/14. HI 60566

Westdeutscher Naturwart. Mitteilungsblatt für floristisch-vegetationskundliche und faunistische Erforschung Westdeutschlands. Backnang, Germany. Westdeutsch. Naturwart. See B–P–H 973/15. HI 60567

Western australian herbarium research notes = Research notes, western australian herbarium. South Perth, W.A.

Western australian naturalist. Perth, Nedlands, W.A. Vol. 1+, 1947+. W. Austral. Naturalist. Preceded by: West australian naturalist. 5-4483-1. HI 74474

Western australian year book. Perth, W.A. Vols. 6+, 1967+, 1968?+. W. Austral. Year Book. Preceded by: Official year book of western Australia. New series. HI 74475

Western blister rust news letter. U.S. Department of agriculture. Washington, DC. W. Blister Rust News Lett. U.S.D.A. See B–P–H 969/18. HI 60510

Western canadian weed control conference = Proceedings, national weed committee, Canada, western section.

Western farmer. Cincinnati, OH = Western farmer and gardener, devoted to agriculture, horticulture, and rural economy. Cincinnati, OH. W. Farmer Gard. See B–P–H 969/21.

Western farmer. Madison, WI. Nos. 1-10(14), 1881-91. W. Farmer (Madison). Superseded by: Wisconsin farmer. 5-4523-1. HI 67659

Western farmer and gardener, devoted to agriculture, horticulture, and rural economy. Cincinnati, OH. W. Farmer Gard. See B–P–H 969/21. HI 60512

Western forestry = Western journal of applied forestry. Bethesda, MD.

Western fruit grower. San Francisco, CA, Willoughby Ave, OH. 1950+. W. Fruit Grower. Preceded by: California fruit and grape grower. 5-4486-3. HI 74476

Western horticultural review. Cincinnati, OH. W. Hort.

Rev. See B–P–H 969/22. HI 60513

Western Illinois university bulletin. Macomb, IL. ?-1972+. W. Illinois Univ. Bull. HI 74477

Western journal of applied forestry. Bethesda, MD. Vol. 1+, 1986+. W. J. Appl. Forest. HI 74478

Western journal of the medical and physical sciences. Cincinnati, OH. W. J. Med. Phys. Sci. See B–P–H 970/2. HI 60515

Western journal of medicine and surgery. Louisville, KY. W. J. Med. Surg. See B–P–H 970/3. HI 60516

Western journal of surgery, obstetrics and gynecology. Portland, OR. W. J. Surg. See B–P–H 970/4. HI 60517

Western medical gazette. Cincinnati, OH. W. Med. Gaz. See B–P–H 970/6. HI 60519

Western medical and physical journal, original and eclectic. Cincinnati, OH. W. Med. Phys. J. See B–P–H 970/7. HI 60520

Western monthly magazine, and literary journal. Cincinnati, OH. W. Monthly Mag. See B–P–H 970/8. HI 60521

Western naturalist; journal of scottish natural history. Paisley. Vol. 1+, 1972+. W. Naturalist. HI 74479

Western quarterly reporter of medical, surgical and natural science. Cincinnati, OH. W. Quart. Reporter Med. Sci. See B–P–H 970/12. HI 60524

Western review and miscellaneous magazine. Lexington, KY. W. Rev. & Misc. Mag. See B–P–H 970/13. HI 60525

Western trees, parks and forests. Santa Monica, CA. 1937-38. W. Trees. Superseded by: Trees magazine. 5-4259-2. HI 56126

Western wood products research notes. Portland, OR. W. Wood Prod. Res. Notes. See B–P–H 970/16. HI 60526

Westfälischer Anzeiger, oder Vaterländisches Archiv zur Beförderung und Verbreitung des Guten und Nützlichen. Dorsten, Germany. Westfäl. Anz. See B–P–H 973/21. HI 60568

Westfälische Pilzbriefe. Recklinghausen, Germany. Westfäl. Pilzbriefe. See B–P–H 973/22. HI 60569

Westphälisches Magazin zur Geographie, Historie und Statistik. Dessau, Leipzig, & Minden. Westphäl. Mag. Geogr. See B–P–H 973/24. HI 60570

Westphälische Provinzial-Blätter. Verhandlungen der Westphälischen Gesellschaft für vaterländische Cultur. Minden. Westphäl. Prov.-Blätt. See B–P–H 973/25. HI 60571

Wetenschappelijk tijdschrift van Antwerpen = Natuurwetenschappilijk tijdschrift. Antwerp. Natuurw. Tijdschr. See B–P–H 638/12.

Wetenschappelijke mededelingen van de koninklijke nederlandse natuurhistorische vereniging. Hoogwoud, Netherlands. Wetensch. Meded. Kon. Ned. Natuurhist. Ver. See B–P–H 973/26. HI 60572

Wetenskaplike bydraes van die Potchefstrooms universiteit vir C. H. O. Reeks B, natuurwetenskappe. Potchefstroom. No. 1+, 1968+. Wetenskapl. Bydr. Potchefstrooms Univ. C.H.O., Reeks B, Natuurwetensk. HI 74480

Wetenskaplike publikasies van die Namib voestynnavorsingstasie = Madoqua. Series 2, scientific papers of the Namib desert research station. Windhoek.

Wetlands; journal of the coast and wetlands society Sydney, N.S.W. Vol. 1+, 1981+. Wetlands (Sydney). HI 74481

Wetlands; journal of the society of the wetlands scientists. Wilmington, NC. Vol. 1+, 1981+. Wetlands (Wilmington). HI 74482

Wetter und Leben. Vienna. Wetter & Leben. See B–P–H 973/28. HI 60573

What's new in crops and soils. Madison, WI. What's New Crops Soils. See B–P–H 974/1. HI 60574

What's new in forest research. Rotorua. Vol. 1+, 1973+. What's New Forest Res. HI 74483

What's new in plant physiology. Gaithersburg, MD. 1970+. What's New Pl. Physiol. HI 74484

Wheat abstracts, Nebraska agricultural experiment station = Nebraska agricultural experiment station wheat abstracts. Lincoln, NB. Nebraska Agric. Exp. Sta. Wheat Abstr. See B–P–H 640/4.

Wheat, barley and triticale abstracts. Farnham Royal? Vol. 1+, 1984+. Wheat Barley Triticale Abstr. Preceded by: Triticale abstracts. HI 74485

Wheat information service, biological laboratory, Kyoto university = Wheat information service, Kihara institute for biological research. Kyoto, Mishima.

Wheat information service, Kihara institute for biological research. Kyoto, Mishima. No. 1+, 1954+. Wheat Inform. Serv. HI 74486

Wheat information service, laboratory of genetics, biological institute, Kyoto university = Wheat information service, Kihara institute for biological research. Kyoto, Mishima.

Wheat research. Melbourne, Vic. Vol. 1+, 1975+. Wheat Res. HI 74487

Whistle punk. Victoria, BC. 1984+. Whistle Punk. HI 74488

Whitwell Wood N[atural] H[istory] G]roup]. Carburton. 1979+. Whitwell Wood Nat. Hist. Group. HI 74489

Whole earth review. Sausalito, CA. No. 44+, 1985+. Whole Earth Rev. Preceded by: CoEvolution quarterly and Whole earth software review [not entered]. HI 74490

Who's who in floriculture. Washignton, DC. Who's Who Floric. See B–P–H 974/6. HI 60576

Wiadomości botaniczne. Cracow. Vol. 1+, 1957+. Wiadom. Bot. HI 74491

Wiadomości ekologiczne. Warsaw. Vol. 16+, 1970+. Wiadom. Ekol. Preceded by: Ekologia polska. Seria B, referate dyshusje. HI 74492

Wiadomości farmaceutyczne. Warsaw. Wiadom. Farm. See B–P–H 974/9. HI 60577

Wiadomości zielarskie. Warsaw. 1974+. Wiadom. Zielarsk. Preceded by: Zielarski biuletyn informacyjny. HI 74493

Wiener allgemeine Forst- und Jagd-Zeitung. Vienna. Wiener Allg. Forst- Jagd-Zeitung. See B–P–H 974/27. HI 60578

Wiener Allgemeine Literaturzeitung. Vienna. Wiener Allg. Literaturzeitung. See B–P–H 974/28. HI 60579

Wiener botanische Zeitschrift. Vienna. Wiener Bot. Z. See B–P–H 974/29. HI 60580

Wiener Illustrirte Garten-Zeitung. Organ der K. K. Gartenbau-Gesellschaft in Wien. Vienna. Wiener Ill. Gart.-Zeitung. See B–P–H 974/31. HI 60581

Wiener Journal für's gesammte Pflanzenreich. Bunzlau, Germany [=Boleslawiec. Poland]. Wiener J. Gesammte Pflanzenr. See B–P–H 974/32. HI 60582

Wiener landwirthschaftliche Zeitung. Vienna. Wiener Landw. Zeitung. See B–P–H 974/33. HI 60583

Wiener landwirtschaftliche Zeitung = Wiener landwirthschaftliche Zeitung. Vienna. Wiener Landw. Zeitung. See B–P–H 974/33.

Wiener medizinische Monatschrift. Vienna. Wiener Med. Monatschr. See B–P–H 974/34. HI 60584

Wiener Moden-Zeitung und Zeitschrift für Kunst, schöne Litteratur und Theater. Vienna. Wiener Moden-Zeitung & Z. Kunst. See B–P–H 975/1. HI 60585

Wiener Obst- und Garten-Zeitung. Vienna. Wiener Obst-Gart.-Zeitung. See B–P–H 975/3. HI 60586

Wiener Urania. Vienna. Wiener Urania. See B–P–H 975/4. HI 60587

Wiener Zeitschrift für Kunst, Litteratur, Theater und Mode. Vienna. Wiener Z. Kunst. See B–P–H 975/5. HI 60588

Wild flower. Cincinnati, OH. Vols. 1-40, 1924-64. Wild Fl. 5-4508-2. HI 74494

Wild flower magazine. Tunbridge Wells. Nos. ?-198-346-?, ?-1930-66-? Wild Fl. Mag. HI 74495

Wild flower notes and news, New England wild flower society. Framingham, MA. 1922+. Wild Fl. Notes News New England Wild Fl. Soc. HI 74496

Wild life observer. London. 1957-66. Wild Life Observ. Incorporated in: Wildlife and the countryside. HI 74497

Wild plants. [Yaso shumi.] Tokyo. Wild Pl. See B–P–H 975/8. HI 60589

Wilderness. Washington, DC. Vol. 46(158)+, 1982+. Wilderness. Preceded by: Living wilderness. HI 74498

Wilderness report. Washington, DC. Vol. 6+, 1970+. Wildern. Rep. Preceded by: Report, wilderness society HI 74499

Wildflower; newsletter of the national wildflower research center. Austin, TX. Vol. 1+, 1984+. Wildflower (Austin). HI 74500

Wildflower; North America's magazine of wild flora. Unionville, Ont. Vol. 1+, 1985+. Wildflower (Unionville). HI 74501

Wildland news. Toronto. Vol. 1+, 1968+. Wildland News. HI 74502

Wildlife. Brisbane. Vol. 1(1), 1963. Wildlife. Superseded by: Wildlife in Australia. HI 74503

Wildlife art news; the national magazine of wildlife art. Elk River, MN. Vol. 1+, 1982+. Wildlife Art News. HI 74885

Wildlife in Australia. Brisbane, Qld. Vol. 1(2)+, 1963+. Wildlife Australia. Preceded by: Wildlife. Brisbane. HI 74504

Wildlife in Bahrain. Manama. 1977+. Wildlife Bahrain. HI 74505

Wildlife and the countryside. London. 1966-70. Wildlife Countryside. HI 74506

Wildlife in North Carolina. Raleigh, NC. Vol. 10+, 1946+. Wildlife North Carolina. Preceded by: North Carolina wildlife conservation [not entered]. HI 74886

Wildlife research bulletin, Western Australia. Perth, W.A. No. 3+, 1975+. Wildlife Res. Bull. Western Australia. Preceded by: Fauna bulletin, fisheries department, Western Australia [not entered]. HI 74507

Wilhelm Roux' Archiv für Entwicklungsmechanik der Organismen. Leipzig, etc. Vols. 105-176. 1925-75. Wilhelm Roux' Arch. Entwicklungsmech. Organismen. Preceded by: Archiv für mikroskopische Anatomie und Entwicklungsmechanik. Superseded by: Wilhelm Roux' archives of developmental biology. 5-4509-2. HI 74508

Wilhelm Roux' archives of developmental biology. Berlin, etc. Vol. 177+, 1975+. Wilhelm Roux' Arch. Developmental Biol. Preceded by: Wilhelm Roux'

Archiv für Entwicklungsmechanik der Organismen. HI 74509

Willdenowia. Berlin-Dahlem. Willdenowia. See B–P–H 975/9. HI 60590

Willdenowia. Beihefte. Berlin-Dahlem. Vols. 1-11, 1963-77. Willdenowia, Beih. Superseded by: Englera. HI 69473

William L. Hutcheson memorial forest bulletin. New Brunswick, NJ. 1957+. William L. Hutcheson Mem. Forest Bull. HI 60423

Wiltshire archaeological and natural history magazine. Devizes. Vols. 1-64, 1854-1969. Wiltshire Archaeol. Nat. Hist. Mag. Superseded by: Wiltshire archaeological and natural history magazine. Part A, natural history. HI 72596

Wiltshire archaeological and natural history magazine. Part A, natural history. Devizes. Vols. 65-69, 1970-74. Wiltshire Archaeol. Nat. Hist. Mag., A. Preceded by: Wiltshire archaeological and natural history magazine. Superseded by: Wiltshire archaeological magazine [not entered] and Wiltshire natural history magazine. HI 72629

Wiltshire natural history magazine. Devizes. Vol. 70+, 1975+. Wiltshire Nat. Hist. Mag. Preceded by: Wiltshire archaeological and natural history magazine. Part A, natural history. HI 65753

Windahlia; Göteborgs svampklubbs årsskkrift. Oslo. Vol. 12/13+, 1982/83+, 1983+. Windahlia. Preceded by: Årsskrift, Göteborgs svampklubb. HI 59086

Wine and fruit grower. Little Silver, NJ. Wine Fruit Grower. See B–P–H 975/16. HI 60591

Winged bean flyer. Laguna. Vol. 1+, 1977+. Winged Bean Flyer. HI 72441

Winnipeg flower garden. Winnipeg. Winnipeg Fl. Gard. See B–P–H 975/18. HI 60592

Winter congress report, international institute for sugar beet research = Compte rendu, congrès d'hiver, institut international de recherches betteravières. Brussels.

Winter yearbook, seed savers' exchange. Decorah, IA. 1984?+. Winter Yearb. Seed Savers' Exch. HI 72702

Winterthur portfolio; A journal of American material culture. Winterthur, DE. Vol. 1+, 1964+. Winterthur Portfol. HI 74887

Winzer; Fachblatt des Österreichische Weinhaus. Vol. 1+, 1945+. Winzer. HI 59357

Wirtembergisches Repertorium der Litteratur. Eine Vierteljahr-Schrift. Heilbronn & Ulm. Wirtemberg. Repert. Litt. See B–P–H 975/19. HI 60593

Wirtschaftseigene Futter; Erzeugung - Konservierung - Verwetung. Frankfurt am Main. 1955+. Wirtschaftseig. Futter. HI 72728

Wirzburger gelehrte Anzeigen. Würzburg. Wirzburger Gel. Anz. See B–P–H 975/20. HI 60594

Wis- en natuurkundige verhandelingen van het genootschap. Amsterdam. Wis- Natuurk. Verh. Genootsch. See B–P–H 975/26. HI 60595

Wis- en natuurkundige verhandelingen van de kon. akademie van wetenschappen te Amsterdam. Amsterdam. Wis- Natuurk. Verh. Kon. Akad. Wetensch. Amsterdam. See B–P–H 975/27. HI 60596

Wisconsin (state) agricultural experiment station. Annual report. Madison, WI. Wisconsin Agric. Exp. Sta. Annual Rep. See B–P–H 975/29. HI 60597

Wisconsin (state) agricultural experiment station. Bulletin. Madison, WI. Wisconsin Agric. Exp. Sta. Bull. See B–P–H 976/1. HI 60598

Wisconsin (state) agricultural experiment station. Circular of information. Madison, WI. Wisconsin Agric. Exp. Sta. Circ. Inform. See B–P–H 976/2. HI 60599

Wisconsin (state) agricultural experiment station. Research bulletin. Madison, WI. Wisconsin Agric. Exp. Sta. Res. Bull. See B–P–H 976/3. HI 60600

Wisconsin (state) agricultural experiment station. Research report. Madison, WI. Wisconsin Agric. Exp. Sta. Res. Rep. See B–P–H 976/4. HI 60601

Wisconsin (state) agricultural experiment station. Special bulletin. Madison, WI. Wisconsin Agric. Exp. Sta. Special Bull. See B–P–H 976/5. HI 60602

Wisconsin (state) agricultural experiment station. Stencil bulletin. Madison, WI. Wisconsin Agric. Exp. Sta. Stencil Bull. See B–P–H 976/6. HI 60603

Wisconsin agriculturist. Racine, WI. Vols. 1-53(21), 1877-1919. Wisconsin Agric. Superseded by: Wisconsin agriculturist and farmer. 5-4521-1. HI 59435

Wisconsin agriculturist and farmer. Racine, WI. Vol. 58(22)+, 1929+. Wisconsin Agric. Farmer. Preceded by: Wisconsin farmer and Wisconsin agriculturist. HI 59438

Wisconsin farmer. Madison, WI. Vols. 10(14)-58(21), 1891-1929. Wisconsin Farmer. Preceded by: Western farmer. Superseded by: Wisconsin agriculturist and farmer. 5-4523-1. HI 59521

Wisconsin horticulture. Madison, WI. Wisconsin Hort. See B–P–H 976/7. HI 60604

Wisconsin medical journal. Milwaukee, WI. Wisconsin Med. J. See B–P–H 976/9. HI 60605

Wissen und Wissenschaft. [Hsüeh yi tsa chih.]. Shanghai. Wissen & Wiss. See B–P–H 978/9. HI 60632

Wissenschaft. Braunschweig. Vol. 1+,1904+. Wissenschaft. HI 73202

Wissenschaftliche Abhandlungen, Deutsche Akademie

der Landwirtschaftswissenschaften zu Berlin = Deutsche Akademie der Landwirtschaftswissenschaften zu Berlin. Wissenschaftliche Abhandlungen. Berlin. Deutsche Akad. Landwirtschaftswiss. Berlin Wiss. Abh. See B–P–H 345/9.

Wissenschaftliche Abhandlungen der Staatsuniversität in Irkutsk = Sbornik trudov gosudarstvennogo Irkutskogo universiteta.

Wissenschaftliche Annalen. Zur Verbreitung neuer Forschungsergebnisse herausgegeben von der Deutschen Akademie der Wissenschaften zu Berlin. Berlin. Wiss. Ann. See B–P–H 976/12. HI 60606

Wissenschaftliche Arbeiten aus dem Burgenland. Eisenstadt, Austria. Wiss. Arbeiten Burgenland. See B–P–H 976/14. HI 60607

Wissenschaftliche Arbeiten, Fakultät für Ackerbau, landwirtschaftliche Hochschule "Wassil Kolarov" = Nauchni trudove, vissh selskostopanski institut "Vasil Kolarov." Plovdiv.

Wissenschaftliche Arbeiten, Fakultät für Wein- und Gartenbau, landwirtschaftliche Hochschule "Wassil Kolarov" = Nauchni trudove, vissh selskostopanski institut "Vasil Kolarov." Plovdiv.

Wissenschaftliche Arbeiten, landwirtschaftliche Hochschule "Wassil Kolarov" = Nauchni trudove, vissh selskostopanski institut "Vasil Kolarov". Plovdiv.

Wissenschaftliche Beihefte, Mitteilungen aus den deutschen Schutzgebieten. Deutsches Kolonialblatt = Mitteilungen aus den deutschen Schutzgebieten. Deutsches Kolonialblatt. Wissenschaftliche Beihefte. Berlin. Mitt. Deutsch. Schutzgeb. See B–P–H 603/7.

Wissenschaftliche Beiträge, Martin-Luther-Universität Halle-Wittenberg. Halle. Vol. 1+, 1971+. Wiss. Beitr. Martin-Luther-Univ. Halle-Wittenberg. HI 73648

Wissenschaftliche Beiträge, Martin-Luther-Universität Halle-Wittenberg. P, biowissenschaftliche Beiträge. Halle. Vol. 1+, 1973+. Wiss. Beitr. Martin-Luther-Univ. Halle-Wittenberg, P, Biowiss. Beitr. HI 59734

Wissenschaftliche Beiträge aus dem Osterlande. Altenburg. Vols. 1?-23/24-?, 1934-41-? Wiss. Beitr. Osterl. Preceded by: Mitteilungen aus dem Osterlande. Superseded by: Abhandlungen und Berichte des naturkundlichen Museums "Mauritianum". 3-2706-1. HI 59958

Wissenschaftliche Berichte der Biologischen Fakultät der Tomsker Staats-Universität = Trudy biologicheskogo fakul'teta Tomskogo gosudarstvennogo universiteta. Tomsk.

Wissenschaftliche Berichte der Moskauer Staats-Universität = Uchenye zapiski. Moskovskii gosudarstvennyi universitet. Moscow & Leningrad.

Wissenschaftliche Berichte der Tomsker Staats-Universität = Trudy Tomskogo gosudarstvennogo universiteta.

Wissenschaftliche Meeresuntersuchungen. Kiel. Wiss. Meeresuntersuch. See B–P–H 976/20. HI 60609

Wissenschaftliche Meeresuntersuchungen. Abteilung Helgoland. Kiel. Wiss. Meeresuntersuch., Abt. Helgoland. See B–P–H 976/21. HI 60610

Wissenschaftliche Meeresuntersuchungen. Abteilung Kiel. Kiel & Leipzig. Wiss. Meeresuntersuch., Abt. Kiel. See B–P–H 976/22. HI 60611

Wissenschaftliche Mitteilungen der agrarwissenschaftlichen Hochschule = Debreceni agrártudományi föiskola tudományos közleményei. Debrecen, Hungary. Debreceni Agrártud. Föisk. Tud. Közlem. See B–P–H 339/9.

Wissenschaftliche Mitteilungen aus Bosnien und der Herzegovina. Vienna. 1893-1912. Wiss. Mitt. Bosnien & Herzegovina. Superseded by: Wissenschaftliche Mitteilungen des Bosnisch-Herzegowinischen Landesmuseums. HI 73672

Wissenschaftliche Mitteilungen des Bosnisch-Herzegowinischen Landesmuseums. Heft C, Naturwissenschaft. Sarajevo. 1971+. Wiss. Mitt. Bosnisch-Herzegowinischen Landesmus., C. Preceded by: Wissenschaftliche Mitteilungen aus Bosnien und Herzegovina. HI 60059

Wissenschaftliche Mittheilungen der Physicalisch-Medicinischen Societät zu Erlangen. Erlangen. Wiss. Mitth. Phys.-Med. Soc. Erlangen. See B–P–H 977/3. HI 60613

Wissenschaftliche Mitteilungen der Universität für Forst- und Holzwirtschaft = Az erdészeti és faipari egyetem tudományos közleményei. Sopron, Hungary. Erdész. Faip. Egyet. Tud. Közlem. See B–P–H 359/24.

Wissenschaftliche Mitteilungen des Vereins für Natur- und Heimatkunde in Köln. Cologne. Wiss. Mitt. Vereins Natur- Heimatk. Köln. See B–P–H 977/2. HI 60612

Wissenschaftliche Veröffentlichungen des Deutschen Instituts für Länderkunde. Leipzig. Wiss. Veröff. Deutsch. Inst. Länderk. See B–P–H 977/4. HI 60614

Wissenschaftliche Veröffentlichungen der Deutschen Museums für Länderkunde. Leipzig. Wiss. Veröff. Deutsch. Mus. Länderk. See B–P–H 977/5. HI 60615

Wissenschaftliche Veröffentlichungen der Gesellschaft für Erdkunde zu Leipzig = Wissenschaftliche Veröffentlichungen des Vereins für Erdkunde zu Leipzig. Leipzig. Wiss. Veröff. Vereins Erdk. Leipzig. See B–P–H 977/7.

Wissenschaftliche Veröffentlichungen des Vereins für Erdkunde zu Leipzig. Leipzig. Wiss. Veröff. Vereins Erdk. Leipzig. See B–P–H 977/7. HI 60616

Wissenschaftliche Zeitschrift der Ernst-Moritz-Arndt-Universität Greifswald. Mathematisch-naturwissenschaftliche Reihe. Greifswald, Germany. Wiss. Z. Ernst-Moritz-Arndt-Univ. Greifswald, Math.-Naturwiss. Reihe. See B–P–H 977/8. HI 60617

Wissenschaftliche Zeitschrift der Friedrich-Schiller-Universität Jena/Thüringen. Jena. Wiss. Z. Friedrich-Schiller-Univ. Jena. See B–P–H 977/9. HI 60618

Wissenschaftliche Zeitschrift der Friedrich-Schiller-Universität Jena/Thüringen. Mathematisch-naturwissenschaftliche Reihe. Jena. Wiss. Z. Friedrich-Schiller-Univ. Jena, Math.-Naturwiss. Reihe. See B–P–H 977/10. HI 60619

Wissenschaftliche Zeitschrift der Humboldt-Universität zu Berlin. Mathematisch-naturwissenschaftliche Reihe. Berlin. Wiss. Z. Humboldt-Univ. Berlin, Math.-Naturwiss. Reihe. See B–P–H 977/11. HI 60620

Wissenschaftliche Zeitschrift der Karl-Marx-Universität Leipzig. Leipzig. Wiss. Z. Karl-Marx-Univ. Leipzig. See B–P–H 977/12. HI 60621

Wissenschaftliche Zeitschrift der Karl-Marx-Universität Leipzig. Mathematisch-naturwissenschaftliche Reihe. Leipzig. Wiss. Z. Karl-Marx-Univ. Leipzig, Math.-Naturwiss. Reihe. See B–P–H 977/13. HI 60622

Wissenschaftliche Zeitschrift der Martin-Luther-Universität Halle-Wittenberg. Halle. Wiss. Z. Martin-Luther-Univ. Halle-Wittenberg. See B–P–H 977/15. HI 60623

Wissenschaftliche Zeitschrift der Martin-Luther-Universität Halle-Wittenberg. Mathematisch-naturwissenschaftliche Reihe. Halle. Wiss. Z. Martin-Luther-Univ. Halle-Wittenberg, Math.-Naturwiss. Reihe. See B–P–H 977/16. HI 60624

Wissenschaftliche Zeitschrift der Pädagogischen Hochschule Potsdam. Mathematisch-naturwissenschaftliche Reihe. Potsdam. Wiss. Z. Pädagog. Hochschule Potsdam, Math.-Naturwiss. Reihe. See B–P–H 977/17. HI 60625

Wissenschaftliche Zeitschrift der Technischen Hochschule Dresden. Dresden. Wiss. Z. TH Dresden. See B–P–H 978/1. HI 60626

Wissenschaftliche Zeitschrift der Universität Greifswald. Greifswald, Germany. Wiss. Z. Univ. Greifswald. See B–P–H 978/2. HI 60627

Wissenschaftliche Zeitschrift der Universität Greifswald. Mathematisch-naturwissenschaftliche Reihe. Greifswald, Germany. Wiss. Z. Univ. Greifswald, Math.-Naturwiss. Reihe. See B–P–H 978/3. HI 60628

Wissenschaftliche Zeitschrift der Universität Leipzig. Leipzig. Wiss. Z. Univ. Leipzig. See B–P–H 978/4. HI 60629

Wissenschaftliche Zeitschrift der Universität Rostock. Rostock. Wiss. Z. Univ. Rostock. See B–P–H 978/5. HI 60630

Wissenschaftliche Zeitschrift der Universität Rostock. Reihe Mathematik/Naturwisenschaften. Rostock. Wiss. Z. Univ. Rostock, Reihe Math. See B–P–H 978/6. HI 60631

Wittenbergisches Wochenblatt zum Aufnehmen der Naturkunde und des Ökonomischen Gewerbes. Wittenberg. Wittenberg. Wochenbl. Naturk. Ökon. Gewerbes. See B–P–H 978/14. HI 60633

Wochenblatt für die gesammte Heilkunde. Berlin. Wochenbl. Gesammte Heilk. See B–P–H 978/21. HI 60634

Wochenblatt für Land-, Forst- und Hauswirthschaft. Prague. Wochenbl. Land- Forst- Hausw. See B–P–H 978/22 HI 60635

Wochenblatt für Land- und Forstwirthschaft. Stuttgart. Vols. 1-29, 1849-77. Wochenbl. Land- Forstw. Preceded by: Wochenblatt für Land- und Hauswirthschaft, Gewerbe und Handel. Superseded by: Württembergisches Wochenblatt für Landwirtschaft. 5-4529-3. HI 72590

Wochenblatt für Land- und Hauswirthschaft, Gewerbe und Handel. Stuttgart & Tübingen. Vols. 1-15, 1834-48. Wochenbl. Land- Hausw. Superseded by: Wochenblatt für Land- und Forstwirthschaft. 5-4529-3. HI 68735

Wochenblatt der Landesbauernschaft Württemberg. Stuttgart. Vols. 101(8)-111, 1934-44. Wochenbl. Landesbauernschaft Württemberg. Preceded and superseded by: Württembergisches Wochenblatt für Landwirtschaft. 5-4557-2. HI 68286

Wochenblatt des landwirthschaftlichen Vereins in Baiern. Munich. Wochenbl. Landw. Vereins Baiern. See B–P–H 978/25. HI 60636

Wochenblatt der Viehzucht, Thierarzneykunde, Reitkunst und des Thierhandels. Nuremberg. Wochenbl. Viehzucht. See B–P–H 979/1. HI 60637

Wochenschrift für Brauerei. Berlin. Wochenschr. Brauerei. See B–P–H 979/2. HI 60638

Wochenschrift für Gärtnerei und Pflanzenkunde. Berlin. Wochenschr. Gärtnerei Pflanzenk. See B–P–H 979/5. HI 60639

Wochenschrift des Vereines zur Beförderung des Gartenbaues in den Königlich Preussischen Staaten für Gärtnerei und Pflanzenkunde. Berlin. Wochenschr. Vereines Beförd. Gartenbaues Königl. Preuss. Staaten. See B–P–H 979/8. HI 60640

Wöchentliche Anzeigen zum Vortheil der Liebhaber der Wissenschaften und Künste. Zurich. Wöchentl. Anz. Vortheil Liebhaber WIss. Künste. See B–P–H 979/9.

HI 60641

Wöchentliche gelehrte Neuigkeiten. Tübingen. Wöchentl. Gel. Neuigk. See B–P–H 979/10. HI 60642

Wöchentliche Nachrichten von gelehrten Sachen. Regensburg. Wöchentl. Nachr. Gel. Sachen. See B–P–H 979/11. HI 60643

Wonderful world of Ohio. Columbus, OH. Vol. 29(6)+, 1965+. Wonderful World Ohio. Preceded by: Ohio conservation bulletin. HI 68784

Wonderful world of Ohio magazine = Wonderful world of Ohio. Columbus, OH.

Wood; Forestry, marketing, application. London. Vol. 1+, 1936+. Wood. 5-4536-1. HI 68800

Wood duck. Hamilton, Ont. Vol. 1+, 1947+. Wood Duck. HI 60060

Wood and fiber. Madison, WI. Vols. 1-14, 1969-82. Wood Fiber. Preceded by: Wood science. Superseded by: Wood and fiber science. HI 60063

Wood and fiber science; Journal of the society of wood science and technology. Madison, WI. Vol. 15+, 1983+. Wood Fiber Sci. Preceded by: Wood and fiber. HI 74264

Wood industry. [Mokuzai kogyo] Minato-ku, Tokyo. ?-1976+. Wood Industry. HI 60376

Wood research; bulletin of the wood research institute. [Kyoto university.] [Mokuzai kenkyu. Kyoto daigaku mokuzai kenkyujo hokoku.] Kyoto, Japan. Wood Res. See B–P–H 979/12. HI 60644

Wood science. Madison, WI. Vols. 1-15(2), 1968-82. Wood Sci. Superseded by: Wood and fiber. HI 74547

Wood science and technology; Journal for wood and pulp of the international academy of wood sciences. New York, Heidelberg, Berlin. Vol. 1+, 1967+. Wood Sci. Techn. HI 74548

Wood thrush. Washington, DC. Wood Thrush. See B–P–H 979/14. HI 60645

Wood and wood products. Chicago, IL. Wood & Wood Prod. See B–P–H 979/15. HI 60646

Woodcock; Report and journal of the Wyre Forest society. Bewdley. Vol. 1+, 1979+. Woodcock. HI 74549

Woodlands. Columbus, OH. Vols. 12(2)-14(2), 1974-76. Woodlands. Preceded and superseded by: Ohio woodlands. HI 74550

Woodlands forum. Woodlands, TX. Vol. 1+, 1984+. Woodlands Forum. HI 74551

Woods and forests. London. Vols. 1-2(68), 1884-85. Woods Forests. Incorporated in: Garden. HI 74552

Woods Hole oceanographic institution; collected reprints. Lancaster, PA. Woods Hole Oceanogr. Inst. Collect. Repr. See B–P–H 979/16. HI 60647

Woodward's record of horticulture. New York. Nos. 1-2, 1866, 1867-68. Woodward's Rec. Hort. HI 74553

Work of the experiment station and laboratories of the hawaiian sugar planters' association = Report (Annual), hawaiian sugar planters' association experiment station. Honolulu, HI.

Work of the Hawaiian experiment station = Report (Annual), hawaiian sugar planters' association experiment station. Honolulu, HI.

Works of the agricultural faculty of the university of Sarajevo = Radovi poljoprivredno-šumarskog fakulteta univerziteta u Sarajevu. Radovi B, šumarstvo. Sarajevo.

Workshop summaries, american society of plant physiologists. Rockville, MD. Vol. 1+, 1982+. Workshop Summ. Amer. Soc. Pl. Physiologists. HI 74554

World agriculture. Washington, DC. World Agric. See B–P–H 979/25. HI 60648

World coffee and tea. New York, NY. World Coffee Tea. See B–P–H 980/1. HI 60649

World conservation strategy in action. [Supplement to: I U C N bulletin.] Gland. 1983+. World Conservation Strategy Action. Previously contained in: I U C N bulletin. HI 74555

World crops. London. Vols. 1-28, 1949-76. World Crops. Superseded by: World crops and livestock. 5-4545-3. HI 74556

World crops and livestock. London. Vol. 29+, 1977+. World Crops & Livestock. Preceded by: World crops. HI 74557

World ecology 2000; Ecosystems and environmental management. Washington, DC. 1970+. World Ecol. 2000. HI 74558

World of enzyme. [Koso no sekai.] Otsu, Japan. World of Enzyme. See B–P–H 980/4. HI 60651

World farming; the magazines of modern agriculture. Kansas City, MO. World Farming. See B–P–H 980/3. HI 60650

World index of scientific translations. Delft. Vols. 1-5, 1967-71. World Index Sci. Transl. Superseded by: World index of scientific translations and list of translations notified to E T C. HI 74559

World index of scientific translations and list of translations notified to E T C. Delft. Vols. 6-11, 1972-77. World Index Sci. Transl. List Transl. Not. E. T. C. Preceded by: World index of scientific translations. Superseded by: World transindex. HI 74560

World index of scientific translations and list of translations notified to the international translations centre = World index of scientific translations and list of translations notified to E T C. Delft.

World maize facts and trends. Mexico, D.F. Vol. 1+, 1981+. World Maize Facts Trends. HI 74561

World meetings: Outside U S A and Canada. New Hartford, NY. Vol. 1+, 1968+. World Meetings, Outs. U.S.A & Canada. HI 74562

World meetings: United States and Canada. New Hartford, NY. Vol. 1+, 1968+. World Meetings, U.S. & Canada. HI 74563

World pollen flora. Stockholm. Nos. 1-4, 1970. World Pollen Fl. Superseded by: World pollen and spore flora. HI 59043

World pollen and spore flora. Stockholm. 1973+. World Pollen Spore Fl. Preceded by: World pollen flora. HI 74564

World review of pest control. London. World Rev. Pest Control. See B–P–H 980/5. HI 60652

World transindex. Delft. Vol. 1+, 1978+. World Transindex. Preceded by: World index of scientific translations. HI 74565

World wheat facts and trends. Mexico, D.F. Vol. 1+, 1981+. World Wheat Facts Trends. HI 59322

World wildlife news. London. No. 1+, 1962+. World Wildlife News. HI 70246

World wood. San Francisco, CA. World Wood. See B–P–H 980/6. HI 60653

World's work. New York, NY. World's Work. See B–P–H 980/7. HI 60654

Worthing naturalist. Worthing. 1978/79+. Worthing Naturalist. HI 74566

Wrightia; a botanical journal. Dallas, TX. Wrightia. See B–P–H 980/9. HI 60655

Wuhan botanical research. Wuhan, Hubei. Vol. 1(1-2), 1983-84. Wuhan Bot. Res. Superseded by: Journal of Wuhan botanical research. HI 74567

Wu-han chih wu hsüeh yen chiu = Journal of Wuhan botanical research. Wuhan, Hubei.

Wuhan daxue xuebao. Ziran kexue ban = Journal of Wuhan university. Natural science edition. Wuhan.

Wu-han ta hsüeh hsüeh pao. Tzu jan k'o hsüeh pan = Journal of Wuhan university. Natural science edition. Wuhan.

Württembergisches Wochenblatt für Landwirtschaft. Stuttgart. 1878-1934; vol. 112+, 1945+ [vol. nos. 51-94 not used]. Württemberg. Wochenbl. Landw. Preceded by: Wochenblatt für Land- und Forstwirthschaft. For vols. 101(8)-111, 1934-44 see: Wochenblatt der Landesbauernschaft Württemberg. 5-4557-2. HI 74568

Würzburger Naturwissehschaftliche Zeitschrift. Würzburg. Würzburger Naturwiss. Z. See B–P–H 980/18. HI 60656

Würzburger wöchentliche Anzeigen von gelehrten Sachen ... und andern gemeinnützigen Gegenständen. Würzburg. Würzburger Wöchentl. Anz. Gel. Sachen. See B–P–H 980/19. HI 60657

Wuyi science journal. Vol. 1+, 1981+. Wuyi Sci. J. HI 74569

Wydawnictwa muzeum ślaskiego w Katowicach. Dzial III. Katowice. Vol. ?-3(1-8)-?, ?-1930-35-? Wydaw. Muz. Śląsk. Katowicach. Dzial III. HI 74570

Wydawnictwa okregowego komitetu ochrony przyrody na wielkopolske i Pomorze w Poznaniu. Poznan. 1930-37. Wydaw. Okreg. Komitetu Ochr. Przyr. Wielkopolske Pomorze Poznaniu. HI 74571

Wydawnictwa popularnonaukowe. Crakow. Nos. 1-24, 1952-67. Wydaw. Popularnonauk. HI 74572

Wynboer. Stellenbosch. Vol. 1+, 1931+. Wynboer. HI 74573

Wyoming agricultural college agricultural experiment station. Annual report. Laramie, WY. Wyoming Agric. Exp. Sta. Annual Rep. See B–P–H 980/24. HI 60658

Wyoming agricultural college agricultural experiment station. Bulletin. Laramie, WY. Wyoming Agric. Exp. Sta. Bull. See B–P–H 980/25. HI 60659

Wyoming agricultural college agricultural experiment station. Circular. Laramie, WY. Wyoming Agric. Exp. Sta. Circ. See B–P–H 980/26. HI 60660

Wyoming agricultural college agricultural experiment station. Mimeographed circular. Laramie, WY. Wyoming Agric. Exp. Sta. Mimeogr. Circ. See B–P–H 980/27. HI 60661

Wyoming agricultural college agricultural experiment station. Press bulletin. Laramie, WY. Wyoming Agric. Exp. Sta. Press Bull. See B–P–H 981/1. HI 60662

Wyoming agricultural college agricultural experiment station. State farms bulletin. Laramie, WY. Wyoming Agric. Exp. Sta. State Farms Bull. See B–P–H 981/2. HI 60663

Wyoming range management. University of Wyoming agricultural experiment station. Laramie, WY. Wyoming Range Managem. See B–P–H 981/3. HI 60664

Xenobiotica. London. Vol. 1+, 1971+. Xenobiotica. HI 75191

Xerophyte. Thetford. Vol. 1+, 1978+. Preceded by: Newsletter, succulent plant club. HI 74574

Xia men da xue xue bao = Acta scientiarum naturalium universitatis amoiensis. Amoy?, China. Acta Sci. Nat Univ. Amoiensis. See B–P–H 48/23.

Yale journal of biology and medicine. Hew Haven, CT. Yale J. Biol. Med. See B–P–H 981/5. HI 60665

Yale Sheffield monthly. New Haven, CT. Yale Sheffield Monthly. See B–P–H 981/8. HI 60666

Yale studies in the history of science and medicine. New Haven, CT. Vol. 1+, 1965+. Yale Stud. Hist. Sci. Med. HI 71266

Yale university library gazette. New Haven, CT. Vol. 1+, 1926+. Yale Univ. Libr. Gaz. 5-4566-3. HI 58617

Yalova bahce kulturleri arastirma ve egitim merkezi dergisi. Istanbul. Vols. 1-7, 1968-77. Bahce. Superseded by: Bahce. HI 58665

Yakagaku zasshi = Journal of the pharmaceutical society of Japan. Tokyo. J. Pharm. Soc. Japan. See B–P–H 476/11.

Yalin Keji = Asian forestry science and technology.

Yamagata daigaku kiyo = Bulletin of the Yamagata university. Tsuruoka, Japan. Bull. Yamagata Univ. See B–P–H 287/30.

Yamagata daigaku kiyo nogaku = Bulletin of the faculty of agriculture; Yamagata university. Shimonoseki, Japan. Bull. Fac. Agric. Yamagata Univ. See B–P–H 249/16.

Yamagata kenritsu hakabutsukan. Kenkyu hokoku = Bulletin of Yamagata prefectural museum. Yamagata.

Yamagata norin gakkaiho = Journal of the Yamagata agriculture and forestry society. Tsuruoka, Japan. J. Yamagata Agric. Soc. See B–P–H 485/8.

Yamaguchi daigaku nogakubu gakujutsu hokoku = Bulletin of the faculty of agriculture, Yamaguchi university. Shimonoseki.

Yamaguchi daigaku rika hokoku = Science reports of the Yamaguchi university. Yamaguchi.

Yamaguchi journal of science. Yamaguchi. Vols. 1-9, 1950-58. Yamaguchi J. Sci. Superseded by: Science reports of the Yamaguchi University. HI 74575

Yamaguchi-ken nogyo shikenjo kenkyu hokoku = Bulletin of the Yamaguchi agricultural experiment station. Yamaguchi, Japan. Bull. Yuamaguchi Agric. Exp. Sta. See B–P–H 287/31.

Yamaguchi-ken nogyo shikenjo tokubetsu kenkyu hokoku = Special bulletin of the Yamaguchi agricultural experiment station. Yamaguchi, Japan. Special Bull. Yamaguchi Agric. Exp. Sta. See B–P–H 850/23.

Yamanashi-ken kaju shikenjo kenkyu hokoku = Bulletin of the Yamanashi fruit tree experiment station. Yamanashi.

Yamanashi-ken nogyo shikenjo hokoku = Bulletin of the Yamanashi prefectural agricultural experiment station. Kofu, Japan. Bull. Yamanashi Prefect. Agric. Exp. Sta. See B–P–H 288/1.

Yamanashi-ken ringyo shikenjo hokoku = Bulletin of the Yamanashi prefectural forest experiment station. Fujiyoshida.

Yao hsüeh hsüeh pao = Acta pharmaceutica sinica. Peking. Acta Pharm. Sin. See B–P–H 47/11.

Yao hsüeh t'ung pao = Chinese pharmaceutical bulletin. Peking (Beijing).

Yaoxue tongbao = Chinese pharmaceutical bulletin. Peking (Beijing).

Yarovizatsiya; Zhurnal po biologii i razvitiya rastenii. Moscow & Odessa. 1935-41. Yarovizatsiya. Preceded by: Byulleten′ yarovizatsii. Superseded by: Agrobiologiya. 3-1912-3. HI 74576

Yasai go = [New information on horticulture. Vegetables.] Kyoto.

Yasai shikenjo hokoku. A = Bulletin of the vegetable and ornamental crops research station, Tsu. Series A. Tsu.

Yasai shikenjo hokoku. B, Morioka = Bulletin of the vegetable and ornamental crops research station. Series B. Morioka.

Yasai shikenjo hokoku. C, Kurume = Bulletin of the vegetable and ornamental crops research station. Series C, (Kurume). Kurume.

Yasai shikenjo nyusu = News of vegetable and ornamental crops research station.

Yaso shumi = Wild plants. Tokyo. Wild Pl. See B–P–H 975/8.

Year book, Year-book, see Yearbook

Year book, academia scientiarum fennica = Vuosikirja, suomalainen tiedeakatemia. Helsinki.

Yearbook of the academy of natural sciences of Philadelphia. Philadelphia, PA. 1923-31, 1924-32. Yearb. Acad. Nat. Sci. Philadelphia. Preceded by: Report (Annual), academy of natural sciences of Philadelphia. Superseded by: Review of the academy of natural sciences of Philadelphia. HI 74577

Year book of agricultural research. [Nogyo shiken kenkyu nenju hokokusho.] Tokyo. Year Book Agric. Res. See B–P–H 981/29. HI 60667

Yearbook of agriculture, United States department of agriculture. Washington, DC. 1926+. Yearb. Agric. U.S.D.A. Preceded by: Agricultural yearbook, United States department of agriculture. HI 74578

Year book of the alpine garden society. London. Year Book Alpine Gard. Soc. See B–P–H 981/30. HI 60668

Year book of the american amaryllis society. Orlando, FL. Vols. 1-2, 1934-35. Year Book Amer. Amaryllis Soc. Superseded by: Herbertia. 3-1843-1. HI 74579

Year book of the american orchid society. Cambridge, MA. 1932-81. Year Book Amer. Orchid Soc. HI 74580

Year book of the american philosophical society. Philadelphia, PA. Yearb. Amer. Philos. Soc. See B–P–H 982/12. HI 60675

Year book of the american rock garden society. New York, NY. Year Book Amer. Rock Gard. Soc. See B–P–H 981/32. HI 60669

Year book of the american rock garden society. Bulletin = Bulletin of the american rock garden society. Plainfield, NJ. Bull. Amer. Rock Gard. Soc. See B–P–H 239/6.

Yearbook of the asiatic society of Bengal. Calcutta. 1951+. Yearb. Asiat. Soc. Bengal. Preceded by: Yearbook of the royal asiatic society of Bengal. HI 71906

Year book of the bibliograpical society, Chicago. Chicago, IL. Year Book Bibliogr. Soc. Chicago. See B–P–H 982/2. HI 60670

Yearbook of the botanical society of America. New York. 1938/39+, 1939+. Yearb. Bot. Soc. Amer. HI 58911

Year book of the botanical society of the British Isles. Arbroath, Scotland. Year Book Bot. Soc. Brit. Isles. See B–P–H 982/3. HI 60671

Yearbook, British delphinium society = British delphinium society's yearbook. London. Brit. Delphinium Soc. Yearb. See B–P–H 228/20.

Yearbook, british pelargonium and geranium society. London. No. 10+, 1965+. Yearb. Brit. Pelargonium Geranium Soc. Preceded by: Geranium year book. HI 70390

Year book of the bureau of entomology, Hangchow. [Chekiang shěng k'un ch'ung chü nien k'an.] Hangchow, China. Yearb. Bur. Entomol. Hangchow. See B–P–H 982/16. HI 60676

Yearbook C R F G = California rare fruit growers' yearbook. Bonsall, Fullerton, CA.

Yearbook of the cactus and succulent society of America. [Supplemental volume of: Cactus and succulent journal.] Los Angeles, CA. 1966+. Yearb. Cact. Succ. Soc. Amer. HI 72128

Yearbook and calendar, Essex field club. Buckhurst Hill. 1905/06-12. Yearb. Calend. Essex Field Club. HI 50791

Year book of the California avocado association. Riverside, CA. Yearb. Calif. Avocado Assoc. See B–P–H 982/17. HI 60677

Yearbook, California macdamia society = California macdamia society. Yearbook. Carlsbad, CA. Calif. Macadamia Soc. Yearb. See B–P–H 294/6.

Yearbook, California rare fruit growers = California rare fruit growers' yearbook. Bonsall, Fullerton, CA.

Year book. Canadian rose society. Toronto. Year Book Canad. Rose Soc. See B–P–H 982/4. HI 60672

Yearbook, Carnegie institution of Washington. Washington, DC. 1902+. Yearb. Carnegie Inst. Wash. HI 74581

Yearbook of the Ceylon agricultural society. Colombo. 1914-20. Yearb. Ceylon Agric. Soc. Superseded by: Yearbook of the department of agriculture, Ceylon. HI 74582

Year book of Connecticut botanical society, inc. New Haven, CT. Yearb. Connecticut Bot. Soc. See B–P–H 982/18. HI 60678

Yearbook, Delphinium society = Delphinium society yearbook. London. Delphinium Soc. Yearb. See B–P–H 340/16.

Yearbook of the department of agriculture, Ceylon. Colombo. 1923-27. Yearb. Dept. Agric. Ceylon. Preceded by: Yearbook of the Ceylon agricultural society. HI 71790

Yearbook of the department of agriculture, Madras. Madras. 1917-29. Yearb. Dept. Agric. Madras. HI 71468

Yearbook of the faculty of agriculture and forestry, university of Belgrade = Godišnjak poljoprivredno-šumarskog fakulteta, universitet u Beogradu. Belgrade.

Yearbook, faculty of agriculture, university of Ankara. Ankara. 1958-73. Yearb. Fac. Agric. Univ. Ankara. HI 74583

Yearbook of the faculty of agriculture, university of Ege. Bornova. Vol. 1+, 1970+. Yearb. Fac. Agric. Univ. Ege. HI 74584

Yearbook of forest products. Rome. Vol. 1+, 1947+. Yearb. Forest Prod. HI 74585

Yearbook of forest products statistics. Rome. 1946-65. Yearb. Forest Prod. Statist. HI 68030

Yearbook, genetical society. London. 1946+. Yearb. Genet. Soc. HI 74586

Year book of the Hawaii orchid society. Hilo, HI. Yearb. Hawaii Orchid Soc. See B–P–H 982/19. HI 60679

Year book of the heather society. Southwark [=London, in part]. Year Book Heather Soc. See B–P–H 982/5. HI 60673

Year book, horticultural education association = H E A year book. Wye.

Year book of the horticultural society of New York. New York. 1924/25+. Year Book Hort. Soc. New York. Preceded by: Report (Annual) of the horticultural society of New York. 3-1887-3. HI 74587

Yearbook of the hungarian geological institute. Budapest = A magyar állami földtani intézet évkönyve. Budapest. Magyar Állami Földt. Intéz. Évk. See B–P–H 542/19.

Yearbook of the hungarian institute of forest science = As erdészeti tudományos intézet évkönyve. Budapest. Erdész. Tud. Intéz. Évk. See B–P–H 360/5.

Yearbook of the hungarian research institute for plant protection. Budapest = Növényvédelmi kutató intézet évkönyve. Budapest. Növényvéd. Kutató Intéz. Evk. See B–P–H 674/11.

Yearbook, indian national science academy. New Delhi. 1970+. Yearb. Indian Natl. Sci. Acad. Preceded by: Yearbook, national institute of sciences of India. HI 74588

Year book, international council of scientific unions. ?-1986+. Year Book Int. Council Sci. Unions. HI 71148

Yearbook, international dendrology society = International dendrology society year book. London.

Yearbook, international society of plant morphologists. Delhi. 1962+. Yearb. Int. Soc. Pl. Morphol. HI 69612

Yearbook, international society for tropical ecology. Varanasi. 1967+. Yearb. Int. Soc. Trop. Ecol. HI 74589

Year book, iris society = Iris society year book. Liverpool. Iris Soc. Year Book. See B–P–H 437/24.

Year book and journal, royal caledonian horticultural society. Edinburgh. 1968-78, 1969-79. Year Book J. Roy. Caledonian Hort. Soc. Preceded by: Journal of the royal caledonian horticultural society. HI 74590

Yearbook of the Massachusetts horticultural society with the annual report. Boston, MA. 1924-61. Yearb. Mass. Hort. Soc. Preceded by: Report (Annual) of the Massachusetts horticultural society and Bulletin of the Massachusetts horticultural society. 3-2552-2. HI 74591

Year book, mycological society of America = Mycological society of America. Year book. Ithaca, NY. Mycol. Soc. Amer. Year Book. See B–P–H 623/21.

Yearbook, national auricula and primula society, northern section. Norden. 1957+. Yearb. Natl. Auricula Primula Soc. N. Sect. Preceded by: Yearbook, national auricula society, northern section. HI 74592

Yearbook, national auricula and primula society, southern section. London. 1957+. Yearb. Natl. Auricula Primula Soc. S. Sect. HI 74593

Yearbook, national auricula society, northern section. Norden. 1946?-56. Yearb. Natl. Auricula Soc. N. Sect. Superseded by: Yearbook, national auricula and primula society, northern section. HI 74594

Yearbook of the national chrysanthemum society. London. 1901-66 [publication suspended 1940-44, 1947-48]. Yearb. Natl. Chrysanthemum Soc. Superseded by: Chrysanthemum year book. HI 74595

Yearbook, national institute of sciences of India. New Delhi. 1960-69. Yearb. Natl. Inst. Sci. India. Superseded by: Yearbook, indian national science academy. HI 74596

Yearbook, New York microscopical society. New York. 1952+. Yearb. New York Microscop. Soc. HI 74597

Yearbook, north american lily society. Indianapolis, IN. Vol. 1+, 1947/48+, 1948?+. Yearb. N. Amer. Lily Soc. HI 74598

Yearbook of the official phytosanitary service. Budapest = Növényegészégyügyi évkönyv. Budapest. Növényegész. Évk. See B–P–H 674/3.

Year book of the Ontario rose society. Streetsville, Ontario. Yearb. Ontario Rose Soc. See B–P–H 982/5. HI 60680

Yearbook of the pelargonium and geranium society. Carshalton Beeches. ?-1971+. Yearb. Pelargonium Geranium Soc. HI 74599

Yearbook, Pennsylvania horticultural society. Philadelphia, PA. ?-1976+. Yearb. Pennsylvania Hort. Soc. HI 74600

Yearbook of pharmacy; A practical summary of researches in pharmacy, materia medica, and pharmaceutical chemistry. London. 1864-1926/7. Yearb. Pharm. From 1881 onwards includes: Transactions of the British pharmaceutical conference. Superseded by: Quarterly journal of pharmacy and pharmacology. 5-4572-1. HI 74601

Yearbook of plant protection research. Baghdad. 1974/75+. Yearb. Pl. Protect. Res. HI 74602

Year book of the public museum of the city of Milwaukee. Milwaukee, WI. Vols. 1-10, 1921-30. Year Book Public Mus. City Milwaukee. Preceded by: Report of the public museum of the city of Milwaukee. 3-2663-3. HI 74603

Yearbook of the research institute for ampelology = Szölészeti kutató intézet évkönyve. Budapest. Szölész. Kutató Intéz. Évk. See B–P–H 863/16.

Year book of the rhododendron association. London. Year Book Rhododendron Assoc. See B–P–H 982/9. HI 60674

Yearbook of the royal asiatic society of Bengal. Calcutta.

1935-50. Yearb. Roy. Asiat. Soc. Bengal. Superseded by: Yearbook of the asiatic society of Bengal. HI 74604

Yearbook, royal horticultural society of Ireland. ?-1965+. Yearb. Roy. Hort. Soc. Ireland. HI 74605

Year book of the royal society of Edinburgh. Edinburgh. Yearb. Roy. Soc. Edinburgh. See B–P–H 983/3. HI 60681

Yearbook of the royal society of London. London. No. 1+, 1896/97+, 1897+. Yearb. Roy. Soc. London. HI 74606

Yearbook of the scientific and learned societies of Great Britain and Ireland. London. 1884-1930. Yearb. Sci. Learned. Soc. Great Britain & Ireland. Superseded by: Official yearbook of the scientific and learned societies of Great Britain and Ireland. 5-3825-2. HI 74888

Yearbook, scottish rock garden club. Edinburgh. 1950+. Yearb. Scott. Rock Gard. Club. HI 74607

Yearbook of the sempervivum society. Burgess Hill. 1974+. Yearb. Sempervivum Soc. HI 74608

Year book of sericulture. [Sanshi nenkan.] Tokyo. Yearb. Seric. See B–P–H 983/4. HI 60682

Yearbook of the United States department of agriculture. Washington, DC. 1894-1922. Yearb. U.S. Dept. Agric. Superseded by: Agricultural yearbook, United States department of agriculture. HI 67399

Year book of the vegetable growers' association of New South Wales. Sydney. Yearb. Veg. Growers' Assoc. New South Wales. See B–P–H 983/6. HI 60683

Yearbook, Washington daffodil society = Washington daffodil society yearbook. Washington, DC. Wash. Daffodil Soc. Yearb. See B–P–H 971/8.

Yearbook, west australian nutgrowing society. Subiaco, W.A. Vol. 1+, 1975+. Yearb. W. Austral. Nutgrow. Soc. HI 70512

Yearbook, Worcestershire nature conservation trust. Worcester. 1972. Yearb. Worcestershire Nat. Conservation Trust. Superseded by: Journal of the Worcestershire nature conservation trust. HI 59346

Year, commonwealth mycological institute (The). [Issued with: Review of plant pathology.] Kew. 1971/72-75/76. 1972-76. Year Commonw. Mycol. Inst. Superseded by: Annual review, commonwealth mycological institute. HI 70861

Yeast. Chichester. Vol. 1+, 1985+. Yeast (Chichester). HI 72745

Yeast. New York. Vol. 1+, 1985+. Yeast (New York). HI 58235

Yeast; Dried yeast and their derivatives. St. Louis, MO, Davis, CA. Vol. 1+, 1950+. Yeast (St. Louis). HI 58229

Yedeoth; Proceedings of the agricultural experiment station, Tel-Aviv. Tel-Aviv. 1926-39. Yedeoth. HI 58239

Yellow dragon. Hong Kong. Yellow Dragon. See B–P–H 983/9. HI 60684

Yellowstone nature notes. Yellowstone Park, WY. Yellowstone Nat. Notes. See B–P–H 983/10. HI 60685

Yen chiu chuan k'an = Technical bulletin of the national forestry research bureau. Nanking. Techn. Bull. Natl. Forest. Res. Bur. See B–P–H 869/10.

Yen chiu pao kao, kuo li Taiwan ta hsüeh shih yen lin = Bulletin of the experimental forest of national Taiwan university. Taipei.

Yen chiu pau, Kuo li Tai wan ta hsüeh shih yan lin ch'ang = Technical bulletin. Published by the forestry experiment station, national Taiwan university. Taipei, Taiwan. Techn. Bull. Forest. Exp. Sta. Natl. Taiwan Univ. See B–P–H 868/27.

Yen ta shêng wu pu ts'ung k'an = Bulletin of the department of biology, Yenching university. Peiping [=Peking]. Bull. Dept. Biol. Yenching Univ. See B–P–H 248/2.

Yetmag. London. 1977+. Yetmag. Preceded by: Bulletin, youmg explorers' trust and Newsletter, young explorers' trust. HI 58243

Yichuan = Hereditas. Beijing.

Yichuan xuebao = Acta genetica sinica. Peking (Beijing).

Yichuan yu yuzhong. Beijing. Vols. 1-3, 1976-78. Yichuan Yu Yuzhong. Superseded by: Hereditas. Beijing. HI 72744

Yillik bülteni, kavak ve hizli gelişen yabanci tur orman agaclari araştirma enstitüsü. Izmir. No. 8+, 1973+. Yillik Bült. Kavak Hizli Gelişen Yabanci Orman Agacl. Araşt. Enst. Preceded by: Yillik bülteni. Kavakçeilik araştirma enstitüsü. HI 58088

Yillik bülteni, kavakçilik araştirma enstitüsü. Izmir. Vols. 1-7, 1966-72. Yillik Bült. Kavakç. Araşt. Enst. Superseded by: Yillik bülteni, kavak ve hizli gelişen yabanci tur orman agaclari araştirma enstitüsü. HI 72787

Ymer. Svenska sällskapet för antropologi och geografi. Stockholm. Ymer. See B–P–H 983/14. HI 60686

Yo hsüeh t'ung pao = Pharmaceutical news. [China]. Pharm. News. See B–P–H 704/13.

Yokohama kokuritsu daigaku. Kankyo kagaku kenkyu senta = Bulletin, institute of environmental science and technology, Yokohama national university. Kiyo.

Yokohama kokuritsu daigaku rika kiyo = Science reports of the Yokohama national university. Section II.

Biological and geological sciences. Kamakura, Japan. Sci. Rep. Yokohama Natl. Univ., Sect. 2, Biol. Sci. See B–P–H 828/22.

Yokosuka-shi hakubutsukan. Shiryoshu. Yokosuka. No. 1+, 1978+. Yokosuka-shi Hakubutsukan, Shiryoshu. HI 58253

Yokohama shiritsu daigaku kiyo = Journal of the Yokohama municipal university. Series C, natural sciences. Yokohama, Japan. J. Yokohama Munic. Univ., Ser. C, Nat. Sci. See B–P–H 485/9.

Yokohama-shiritsu daigaku ronso. Shizenkagaku keiritsu = Bulletin, Yokohama municipal (later city) university society. Natural science. Yokohama.

Yokosukashi hakubutsu kan kenkyu hokoku = Science report of the Yokosuka city museum. Yokosuka, Japan. Sci. Rep. Yokosuka City Mus. See B–P–H 829/1.

Yongu pogo. Kwanak sumokwon. Suwin. No. 1+, 1976+. Yongu Pogo, Kwanak Sumokwon. HI 58254

Yonsumnim yongu pogo = Bulletin, Seoul national university forests. Seoul.

Yorkshire cactus journal. Bradford, England. Yorkshire Cact. J. See B–P–H 983/22. HI 60687

Yorkshire naturalists' recorder. Manchester, England. Yorkshire Naturalists' Rec. See B–P–H 983/25. HI 60688

Yoshoku kenkyujo kenkyu hokoku = Bulletin of national research institute of aquaculture. Mie.

Your environment. London. Vols. 1-4(2), 1970-73. Your Environm. HI 58255

Your garden and home. Toronto. Your Gard. Home. See B–P–H 983/26. HI 60689

Your public lands. Washington, DC. Vol. 31(4)+, 1981+. Your Public Lands. Preceded by: Our public lands. HI 58257

Youth and science. [Sh'ing nien k'o hsüeh.] Chungking, China. Youth & Sci. See B–P–H 983/27. HI 60690

Yu shan chih pao = Yushania. Tunghai.

Yuan chiu huei pao. Fu chian seng yuan chiu yuan = Research bulletin of the institute of zoology and botany: Fukien academy. Foochow, China. Res. Bull. Inst. Zool. Bot. Fukien Acad. See B–P–H 774/15.

Yüan-i = Hortus. Nanking. Hortus. See B–P–H 423/19.

Yuan-i chuan k'an = Horticultural bulletin. College of agriculture, Sun Yatsen university. Canton. Hort. Bull. Coll. Agric. Sun Yatsen Univ. See B–P–H 421/3.

Yuan i hsüeh pao = Journal of horticulture. Peking. J. Hort. (Peking). See B–P–H 469/4.

Yuanyi xuebao = Acta horticulturae sinica. Peking (Beijing).

Yüksek ziraat enstitüsü, calismalari. Ankara. Nos. 1-153, 1934-48. Yüks. Zir. Enst. Calism. HI 69589

Yüksek ziraat enstitüsü dergisi. Ankara. 1939+. Yüks. Zir. Enst. Derg. HI 71162

Yun nan, see Yunnan

Yun-nan chih wu yen chiu = Acta botanica Yunnanica. Kunming.

Yunnan industrial news. [Yun nan shih yeh tung hsin.] Kunming, China. Yunnan Industr. News. See B–P–H 983/34. HI 60691

Yunnan nung lin chih wu yen chiu so t'sung k'an = Bulletin of the Yunnan botanical research institute. Kunming, China. Bull. Yunnan Bot. Res. Inst. See B–P–H 288/2.

Yun nan shih yeh tung hsin = Yunnan industrial news. Kunming, China. Yunnan Industr. News. See B–P–H 983/34.

Yun nan ta hsüeh hsüeh pao. Lui = Journal of the Yunnan university. Series A. Kunming, China. J. Yunnan Univ., Ser. A. See B–P–H 485/10.

Yunnan zhiwu yanjiu = Acta botanica Yunnanica. Kunming.

Yunnania. Kunming?, China. Yunnania. See B–P–H 984/1. HI 60692

Yushania. Tunghai. Vol. 1+, 1984+. Yushania. HI 58273

Z A G Wasserpflanzen A M. Weimar. 1983+. Z. A. G. Wasserpflanzen A. M Preceded by: Informationen Z A G Wasserpflanzen. HI 65507

Z B = Zierpflanzenbau mit internationaler Gartenbautechnik.

Z lesu a luhu. Saar, Moravia [=Zdar, Czechoslovakia]. Z Lesu Luhu. See B–P–H 986/16. HI 60733

Za torfyanuyu industriyu. Moscow. 1935-40. Za Torfyan. Industr. Preceded by: Torfyanoe delo. Superseded by: Torfyanaya promyshlennost. HI 57196

Za vysokou úrodu. Roc. 11, vols. 12-15, 1963-67. Za Vysokou Úrodu. Superseded by: Úroda. HI 58274

Zaadbelangen. The Hague. Zaadbelangen. See B–P–H 989/6. HI 60766

Zabaikal'skii otdel Russkogo geograficheskogo obshchestva i Kraevoi muzei imeni A. K. Kuznetsova v gorode Chite. Chita, R S F S R. 1924. Zabaikal'sk. Otd. Russk. Geogr. Obshch. Kraev. Muz. Kuznetsova Gorode Chite. Preceded by: Zapiski Chitinskago Otděleniya Priamurskago Otděla Imperatorskago Russkago Geograficheskago Obshchestva. HI 68575

Zabajkal'skij otdel Russkogo geografičeskogo obščestva i Kraevoj muzej imeni A. K. Kuznecova v gorode Čite =

Zabaikal'skii otdel Russkogo geograficheskogo obshchestva i Kraevoi muzei imeni A. K. Kuznetsova v gorode Chite. Chita.

Zahrada domácí a s'hakolní. Chrudim, Bohemia [Czechoslovakia]. Zahrada Domácí Školní. See B–P–H 989/9. HI 60767

Zahradnické listy. Prague. Vols. ?-68, 1967-75. Zahradn. Listy. Preceded by: Ovocnarsti a zelinarstvi. Superseded by: Záhradnictvo. 5-4590-1. HI 58275

Záhradníctvo. Bratislava. Vol. 7+, 1976+. Záhradníctvo. Preceded by: Záhradník and Zahradnické listy. HI 58281

Zahradník. Prague? Vols. 1(44)-9(52), 1972-75. Zahradník. Preceded by: Ovocnarstvi a zelinarstvi. Superseded by: Záhradníctvo. HI 58283

Zambia journal of science and technology. Lusaka. Vol. 1+, 1976+. Zambia J. Sci. Techn. HI 58296

Zambia museum papers. London. No. 1+, 1967+. Zambia Mus. Pap. HI 58300

Zambia science abstracts; Key to Zambia's science and technology literature. Lusaka. 1973+, [1977]+. Zambia Sci. Abstr. HI 58310

Zametki po sistematike i geografii rastenii. Tiflis. Vol. 1+, 1938+. Zametki Sist. Geogr. Rast. HI 69979

Zanco; Scientific journal of Salahaddin university. Erbil, Iraq. Vol. 1+, 1988+. Zanco. HI 74835

Zandera; Mitteilungen aus der Bücherei des deutschen Gartenbaues e. V. Berlin. Vol. 1+, 1982+. Zandera. HI 58324

Zanku. Erbil = Zanco. Erbil.

Západočeské vlastivěda. Pilsen, Czechoslovakia. Západočeské Vlastiv. See B–P–H 995/4. HI 60772

Zapiski Addzelu pryrody i gaspadarki. Minsk, Belorussian S S R. Vol. 1, 1928. Zap. Addz. Pryr. Gasp. Superseded by: Zapiski Addzelu pryrody i narodnai gaspadarki. 1-124-1. HI 58326

Zapiski Addzelu pryrody i narodnai gaspadarki. Minsk, Belorussian S S R. Vols. 2-4, 1929-30. Zap. Addz. Pryr. Nar. Gasp. Preceded by: Zapiski Addzelu pryrody i gaspadarki. 1-124-1. HI 51743

Zapiski Addzelu pryrody i narodnaj gaspadarki = Zapiski Addzelu pryrody i narodnai gaspadarki. Minsk.

Zapiski Akademii nauk. Petrograd. Vol. 35(1), 1917. Zap. Akad. Nauk. Preceded by: Zapiski Imperatorskoi Akademii Nauk po Fiziko-matematicheskomu Otdĕleniyu. Superseded by: Zapiski Rossiiskoi akademii nauk. 1-114-3. HI 72854

Zapiski Akademii nauk S S S R po fiziko-matematičeskomu otdeleniju = Zapiski Akademii nauk S S S R po fiziko-matematicheskomu otdeleniyu. Leningrad.

Zapiski Akademii nauk S S S R po fiziko-matematicheskomu otdeleniyu. Leningrad. Vol. 21(7)-21(8), 1929-30; vol. 37(2)-37(3), 1926-30. Zap. Akad. Nauk S.S.S.R. Fiz.-Mat. Otd. Preceded by: Zapiski Imperatorskoi Akademii Nauk po Fiziko-matematicheskomu Otdĕleniyu. For vols. 22-34 see: Zapiski Imperatorskoi Akademii Nauk po Fiziko-matematicheskomu Otdĕleniyu. For vol. 35(1) see: Zapiski Akademii nauk. For vols. 35(2)-37(1) see: Zapiski Rossiiskoi akademii nauk. 1-114-3. HI 72975

Zapiski Akademii nauk Soyuza S S R po Otdeleniyu fiziko-matematicheskikh nauk = Zapiski Akademii nauk S S S R po fiziko-matematicheskomu otdeleniyu.

Zapiski belaruskai déyarzhaynai akadémii sel'skai haspadarki imya kastrychnikavai revalyutsyi = Zapiski belorusskoi gosudarstvennoi akademii sel'skogo khozyaistva im. oktyabr'skoi revolyutsii. Gorki.

Zapiski belorusskoi gosudarstvennoi akademii sel'skogo khozyaistva im. oktyabr'skoi revolyutsii. Gorki. Vols. ?-4-?, 1926-27-? Zap. Belorussk. Gosud. Akad. Sel'sk. Khoz. Im. Okt. Revolyutsii Preceded by: Zapiski belorusskogo gosudarstvennogo instituta sel'skogo i lesnogo khozyaistva v pamyat' Oktyarbr'skoi revolyutsii and Zapiski harétskaha instytutu sel'skai haspadarki. HI 55129

Zapiski belorusskogo gosudarstvennogo instituta sel'skogo i lesnogo khozyaistva v pamyat' Oktyarbr'skoi revolyutsii. Minsk. 1923-25. Zap. Belorussk. Gosud. Inst. Sel'sk. Lesn. Khoz. Pamyat' Okt. Revolyutsii. Superseded by: Zapiski belorusskoi gosudarstvennoi akademii sel'skogo khozyaistva im. oktyabr'skoi revolyutsii. HI 67480

Zapiski Biologičeskoj stancii Obščestva ljubitelej estestvoznanija, antropologii i etnografii v Bolševe Moskovskoj gubernii = Zapiski Biologicheskoi stantsii Obshchestva lyubitelei estestvoznaniya, antropologii i etnografii v Bolsheve Moskovskoi gubernii. Moscow.

Zapiski Biologicheskoi stantsii Obshchestva lyubitelei estestvoznaniya, antropologii i etnografii v Bolsheve Moskovskoi gubernii. Moscow. Vols. 1-4, 1925-30. Zap. Biol. Stantsii Obshch. Lyubit. Estestv. Bolsheve Moskovsk. Gub. Superseded by: Zapiski Bolshevskoi biologicheskoi stantsii. 4-3131-1. HI 68653

Zapiski Bolshevskoi biologicheskoi stantsii. Bolshevo, R S F S R. Vols. 5-11, 193?-39. Zap. Bolshevsk. Biol. Stantsii. Preceded by: Zapiski Biologicheskoi stantsii Obshchestva lyubitelei estestvoznaniya, antropologii i etnografii v Bolsheve Moskovskoi gubernii. 4-3131-1. HI 74609

Zapiski Bolševskoj biologičeskoj stancii = Zapiski Bolshevskoi biologicheskoi stantsii. Bolshevo.

Zapiski Chitinskago Otdĕleniya Priamurskago Otdĕla Imperatorskago Russkago Geograficheskago

Obshchestva. St. Petersburg. 1896-1911. Zap. Chitinsk. Otd. Priamursk. Otd. Imp. Russk. Geogr. Obshch. Superseded by: Zabaikal'skii otdel Russkogo geograficheskogo obshchestva i Kraevoi muzei imeni A. K. Kuznetsova v gorode Chite. 2-1692-3. HI 74610

Zapiski Čitinskago Otdělenija Priamurskago Otděla Imperatorskago Russkago Geografičeskago Obščestva = Zapiski Chitinskago Otděleniya Priamurskago Otděla Imperatorskago Russkago Geograficheskago Obshchestva. St. Petersburg.

Zapiski Central'no-Kavkazskogo otdelenija Vsesojuznogo botaničeskgo obščestva = Zapiski tsentral'no-Kavkazskogo otdeleniya Vsesoyuznogo botanicheskgo obshchestva. Ordzhonikidze.

Zapiski Dal'nevostochnaya kraevaya planovaya komissiya. Primorskoi filial Gosudarstvennogo geograficheskogo obshchestva. Vladivostok, R S F S R. Vol. 23 [also numbered vol. 6], 1936. Zap. Dal'nevost. Kraevl Planov. Komiss Primorsk. Fil. Gosud. Geogr. Obshch. Preceded by: Zapiski Vladivostokskogo otdela Gosudarstvennogo russkogo geograficheskogo obshchestva izucheniya Amurskogo kraya. HI 74611

Zapiski Dal'nevostočnaja kraevaja planovaja komissija. Primorskoj filial Gosudarstvennogo geografičeskogo obščestva = Zapiski Dal'nevostochnaya kraevaya planovaya komissiya. Primorskoi filial Gosudarstvennogo geograficheskogo obshchestva. e. Vladivostok.

Zapiski geograficheskogo obshchestva S S S R. Novaya seriya. Moscow. 1948+. Zap. Geogr. Obshch. S.S.S.R., N.S. Preceded by: Zapiski russkogo geograficheskogo obshchestva po obshchi geografii. HI 74612

Zapiski Gosudarstvennogo Nikitskogo botaničeskogo sada = Zapiski Gosudarstvennogo Nikitskogo botanicheskogo sada. Yalta.

Zapiski Gosudarstvennogo Nikitskogo botanicheskogo sada. Yalta, Ukrainian S S R. Vol. 8, 1925. Zap. Gosud. Nikitsk. Bot. Sada. Preceded by: Zapiski Imperatorskago Nikitskago Sada. Superseded by: Zapiski Gosudarstvennogo Nikitskogo opytnogo botanicheskogo sada. 3-1912-2. HI 68570

Zapiski Gosudarstvennogo Nikitskogo opytnogo botaničeskogo sada = Zapiski Gosudarstvennogo Nikitskogo opytnogo botanicheskogo sada. Yalta.

Zapiski Gosudarstvennogo Nikitskogo opytnogo botanicheskogo sada. Yalta, Ukrainian S S R. Vols. 9-14[(1)], 1926-30; vols. 16-17, 1931. Zap. Gosud. Nikitsk. Opytn. Bot. Sada. Preceded by: Zapiski Gosudarstvennogo Nikitskogo botanicheskogo sada. For vols. 14(2)-15 see: Trudy gosudarstvennogo Nikitskogo botanicheskogo sada. 3-1912-2. HI 74802

Zapiski harétskaha instytutu sel'skai haspadarki. Gorki. 1924-25. Zap. Haréts. Inst. Sel'sk. Haspad. Superseded by: Zapiski belorusskoi gosudarstvennoi akademii sel'skogo khozyaistva im. oktyabr'skoi revolyutsii. HI 74613

Zapiski Imperatorskago Nikitskago Sada. Yalta, Ukrainian S S R. Vols. 1-7, 1908-16. Zap. Imp. Nikitsk. Sada. Superseded by: Zapiski Gosudarstvennogo Nikitskogo botanicheskogo sada. 3-1912-2. HI 74614

Zapiski Imperatorskago Novorossiiskago Universiteta. Odessa, Ukrainian S S R. Vols. 1-113, 1868-1914. Zap. Imp. Novorossiisk. Univ. 4-3136-3. HI 74615

Zapiski Imperatorskago Novorossiiskago Universiteta. Fiziko-matematicheskago Fakul'teta. Odessa, Ukrainian S S R. Vols. 1-12, 1910-19. Zap. Imp. Novorossiisk. Univ., Fiz.-Mat. Fak. 4-3136-3. HI 74616

Zapiski Imperatorskago Novorossijskago Universiteta = Zapiski Imperatorskago Novorossiiskago Universiteta. Odessa.

Zapiski Imperatorskago Novorossijskago Universiteta. Fiziko-matematičeskago Fakul'teta = Zapiski Imperatorskago Novorossiiskago Universiteta. Fiziko-matematicheskago Fakul'teta. Odessa.

Zapiski Imperatorskago Russkago Geografičeskago Obščestva = Zapiski Imperatorskago Russkago Geograficheskago Obshchestva. St. Petersburg.

Zapiski Imperatorskago Russkago Geograficheskago Obshchestva. St. Petersburg. 1861-64. Zap. Imp. Russk. Geogr. Obshch. Preceded by: Věstnik Imperatorskago Russskago Geograficheskago Obshchestva. Superseded by: Izvestiya Imperatorskogo Russkogo Geograficheskogo Obshchestva. 2-1692-2. HI 74617

Zapiski Imperatorskoi Akademii Nauk. St. Petersburg. Vols. 1-76, 1862-94. Zap. Imp. Akad. Nauk. Superseded by: Zapiski Imperatorskoi Akademii Nauk po Fiziko-matematicheskomu Otděleniyu. 1-113-3. HI 74618

Zapiski Imperatorskoi Akademii nauk po Fiziko-matematicheskomu otděleniyu. St. Petersburg. Ser. 8, vols. 1-21(6) 1894-1914; vols. 22-34, 1907-16. Zap. Imp. Akad. Nauk Fiz.-Mat. Otd. Preceded by: Zapiski Imperatorskoi Akademii Nauk and Mémoires de l'académie impériale des sciiences de Saint Pétersbourg, Septième série. For vol. 21(7)-21(8) see: Zapiski Imperatorskoi Akademii Nauk po Fiziko-matematicheskomu Otděleniyu. Superseded by: Zapiski Akademii nauk. 1-114-3. HI 74619

Zapiski Imperatorskoj Akademii Nauk. St. Petersburg = Zapiski Imperatorskoi Akademii Nauk. St. Petersburg.

Zapiski Imperatorskoj Akademii Nauk po Fiziko-matematičeskomu Otdĕleniju = Zapiski Imperatorskoi Akademii Nauk po Fiziko-matematicheskomu Otdĕleniyu. St. Petersburg.

Zapiski Južno-Ussurijskogo otdela Gosudarstvennogo russkogo geografičeskogo obščestva = Zapiski Yuzhno-Ussuriiskogo otdela Gosudarstvennogo russkogo geograficheskogo obshchestva. Vladivostok.

Zapiski Južno-Ussurijskogo otdelenija Priamurskogo otdela Russkogo geografičeskogo obščestva = Zapiski Yuzhno-Ussuriiskogo otdeleniya Priamurskogo otdela Russkogo geograficheskogo obshchestva. Vladivostok.

Zapiski Kavkazskago Otdĕla Imperatorskago Russkago Geograficeskago Obščestva = Zapiski Kavkazskago Otdĕla Imperatorskago Russkago Geograficeskago Obshchestva. Tiflis.

Zapiski Kavkazskago Otdĕla Imperatorskago Russkago Geograficheskago Obshchestva. Tiflis, Georgian S S R. Vols. 1-30 [vol. 29(2) published in 1915, vol. 29(3/4) in 1916], 1852-1914. Zap. Kavkazsk. Otd. Imp. Russk. Geogr. Obshch. 1-1692-2. HI 68987

Zapiski Kievskago Obščestva Estestvoispytatelej = Zapiski Kievskago Obshchestva Estestvoispytatelei. Kiev.

Zapiski Kievskago obshchestva estestvoispytatelei. Kiev, Ukrainian S S R. Vols. 1-26(1), 1870-1917. Zap. Kievsk. Obshch. Estestvoisp. Superseded by: Zapysky Kyyivs'kogo tovarystva pryrodoznavtsiv. 3-2290-1. HI 69269

Zapiski Komiteta Akklimatizacija, učreždennago pri Imperatorskom Moskovskom Obščestvĕ Sel'skago Hozjajstva = Zapiski Komiteta Akklimatizatsiya, uchrezhdennago pri Imperatorskom Moskovskom Obshchestvĕ Sel'skago Khozyaistva. Moscow.

Zapiski Komiteta Akklimatizatsiya, uchrezhdennago pri Imperatorskom Moskovskom Obshchestvĕ Sel'skago Khozyaistva. Moscow. 1858-59. Zap. Komiteta Akklim. Superseded by: Akklimatizatsiya. Ezhemĕsyachnoe izdanie Komiteta Akklimatizatsiya. HI 69433

Zapiski Krasnojarskago Podotdĕla Vostočno-Sibirskago Otdĕla Imperatorskago Russkago Geografičeskago Obščestva. Po Meteorologii = Zapiski Krasnoyarskago Pod'otdĕla Vostochno-Sibirskago Otdĕla Imperatorskago Russkago Geograficheskago Obshchestva. Po Meteorologii. Irkutsk.

Zapiski Krasnojarskogo otdela Gosudarstvennogo russkogo geografičeskogo obščestva = Zapiski Sredne-Sibirskogo otdela (byvshego Krasnoyarskogo) Gosudarstvennogo russkogo geograficheskogo obshchestva. Krasnoyarsk.

Zapiski Krasnoyarskago Podotdĕla Vostochno-Sibirskago Otdĕla Imperatorskago Russkago Geograficheskago Obshchestva. Po Meteorologii. Irkutsk, R S F S R. Vols. 1-3, 1902-06. Zap. Krasnoyarsk. Podotd. Vost.-Sibirsk. Otd. Imp. Russk. Geogr. Obshch., Meteorol. 2-1692-2. HI 69566

Zapiski Krymskago Obščestva Estestvoispytatelej i Ljubitelej Prirody = Zapiski Krymskago Obshchestva Estestvoispytatelei i Lyubitelei Prirody. Simferopol.

Zapiski Krymskago Obshchestva Estestvoispytatelei i Lyubitelei Prirody. Simferopol, Ukrainian S S R. Vols. 1-12, 1912-30. Zap. Krymsk. Obshch. Estestvoisp. 3-2321-1. HI 69601

Zapiski Laboratorii po semenovedeniju pri Glavnom botaničeskom sade R S F S R = Zapiski Laboratorii po semenovedeniyu pri Glavnom botanicheskom sade R S F S R. Petrograd.

Zapiski Laboratorii po semenovedeniyu pri Glavnom botanicheskom sade R S F S R. Petrograd. Vol. 4(4), 1921. Zap. Lab. Semenov. Glavn. Bot. Sade R.S.F.S.R. Preceded by: Zapiski Stantsii dlya ispytaniya semyan pri Glavnom botanicheskom sade R S F S R. Superseded by: Zapiski po semenovedeniyu. 3-2389-3. HI 70520

Zapiski Nauchno-prikladnykh otdĕlov Tiflisskago botanicheskago sada. Tiflis, Georgian S S R. Vols. 1-7, 1919-30. Zap. Nauchno-Prikl. Otd. Tiflissk. Bot. Sada. 5-4217-2. HI 70728

Zapiski Naučno-prikladnyh Otdĕlov Tiflisskago Botaničeskago Sada = Zapiski Nauchno-prikladnykh Otdĕlov Tiflisskago Botanicheskago Sada. Tiflis.

Zapiski Novorossiiskago Obshchestva Estestvoispytatelei. Odessa, Ukrainian S S R. Vols. 1-42, 1872-1918. Zap. Novorossiisk. Obshch. Estestvoisp. Superseded by: Zapiski Odesskogo obshchestva estestvoispytatelei. 4-3137-1. HI 68110

Zapiski Novorossijskago Obščestva Estestvoispytatelej. Odessa.

Zapiski Novorossijskago Universiteta. Fiziko-matematičeskago Fakul'teta = Zapiski Imperatorskago Novorossiiskago Universiteta. Fiziko-matematicheskago Fakul'teta. Odessa.

Zapiski Obščestva Ispytatelej Prirody, osnovannogo pri Imperatorskom Moskovskom Universitetĕ = Zapiski Obshchestva Ispytatelei Prirody, osnovannogo pri Imperatorskom Moskovskom Universitetĕ. Moscow.

Zapiski Obščestva Izučenija Amurskago Kraja Filial'nago Otdĕlenija Priamurskago Otdĕla Imperatorskago Russkago Geografičeskago Obščestva = Zapiski Obshchestva Izučeniya Amurskago Kraya Filial'nago Otdĕleniya Priamurskago Otdĕla Imperatorskago Russkago Geograficheskago Obshchestva. Vladivostok.

Zapiski Obščestva Izučenija Amurskago Kraja

sostojaščago pod Avgustejšim Pokrovitel'skom Ego Imperatorskago Vysočajšago Veličestva Knazja Aleksandra Mihajloviča Obščestva Izučenija Amurskago Kraja = Zapiski Obshchestva Izucheniya Amurskago Kraya Filial'nago Otděleniya Priamurskago Otděla Imperatorskago Russkago Geograficheskago Obshchestva. Vladivostok.

Zapiski Obščestva Izučenija Amurskago Kraja Vladivostokskago Otdělenija Priamurskago Otděla Imperatorskago Russkago Geografičeskago Obščestva = Zapiski Obshchestva Izucheniya Amurskago Kraya Vladivostokskago Otděleniya Priamurskago Otděla Imperatorskago Russkago Geograficheskago Obshchestva. Vladivostok.

Zapiski Obščestva izučenija Amurskogo kraja Vladivostokskogo otdělenija Priamurskogo otdela Russkogo geografičeskogo obščestva = Zapiski Obshchestva izucheniya Amurskogo kraya Vladivostokskogo otděleniya Priamurskogo otdela Russkogo geograficheskogo obshchestva. Vladivostok.

Zapiski Obščestva Podol'skih Estestvoispytatelej i Ljubitelej Prirody = Zapiski Obshchestva Podol'skikh Estestvoispytatelei i Lyubitelei Prirody. Kamenets-Podolsk.

Zapiski Obshchestva ispytatelei prirody, osnovannogo pri imperatorskom Moskovskom universitetě. Moscow. Vol. 1, 1806. Zap. Obshch. Isp. Prir. Imp. Moskovsk. Univ. Superseded by: Mémoires de la société impériale des naturalistes de Moscou. 3-2770-2. HI 72379

Zapiski Obshchestva Izucheniya Amurskago Kraya Filial'nago Otděleniya Priamurskago Otděla Imperatorskago Russkago Geograficheskago Obshchestva. Vladivostok, R S F S R. Vols. 1-10, 1888-1907. Zap. Obshch. Izuch. Amursk. Kraya Fil. Otd. Priamursk. Otd. Imp. Russk. Geogr. Obshch. Superseded by: Zapiski Obshchestva Izucheniya Amurskago Kraya Vladivostokskago Otděleniya Priamurskago Otděla Imperatorskago Russkago Geograficheskago Obshchestva. 2-1692-3. HI 68726

Zapiski Obshchestva Izucheniya Amurskago Kraya Vladivostokskago Otděleniya Priamurskago Otděla Imperatorskago Russkago Geograficheskago Obshchestva. Vladivostok, R S F S R. Vols. 11-15, 1907-16. Zap. Obshch. Izuch. Amursk. Kraya Vladivostoksk. Otd. Priamursk. Otd. Imp. Russk. Geogr. Obshch. Preceded by: Zapiski Obshchestva Izucheniya Amurskago Kraya Filial'nago Otděleniya Priamurskago Otděla Imperatorskago Russkago Geograficheskago Obshchestva. Superseded by: Zapiski Obshchestva izucheniya Amurskogo kraya Vladivostokskogo otděleniya Priamurskogo otdela Russkogo geograficheskogo obshchestva. 2-1692-2. HI 65793

Zapiski Obshchestva izucheniya Amurskogo kraya Vladivostokskogo otděleniya Priamurskogo otdela Russkogo geograficheskogo obshchestva. Vladivostok, R S F S R. Vols. 16-17, 1917-22. Zap. Obshch. Izuch. Amursk. Kraya Vladivostoksk. Otd. Priamursk. Otd. Russk. Geogr. Obshch. Preceded by: Zapiski Obshchestva Izucheniya Amurskago Kraya Vladivostokskago Otděleniya Priamurskago Otděla Imperatorskago Russkago Geograficheskago Obshchestva. Superseded by: Zapiski Vladivostokskogo otdela Gosudarstvennogo russkogo geograficheskogo obshchestva izucheniya Amurskogo kraya. 2-1692-3. HI 67309

Zapiski Obshchestva Podol'skikh Estestvoispytatelei i Lyubitelei Prirody. Kamenets-Podolsk [=Kamenets-Podolski], Ukrainian S S R. 1 vol., 1912. Zap. Obshch. Podol'sk. Estestvoisp. HI 72888

Zapiski Odesskogo obščestva estestvoispytatelej = Zapiski Odesskogo obshchestva estestvoispytatelei. Odessa.

Zapiski Odesskogo obshchestva estestvoispytatelei. Odessa, Ukrainian S S R. Vols. 43-45, 1927-29. Zap. Odessk. Obshch. Estestvoisp. Preceded by: Zapiski Novorossiiskago Obshchestva Estestvoispytatelei. 4-3137-1. HI 74620

Zapiski Priamurskago Otděla Imperatorskago Russkago Geografičeskago Obščestva = Zapiski Priamurskago Otděla Imperatorskago Russkago Geograficheskago Obshchestva. Khabarovsk.

Zapiski Priamurskago Otděla Imperatorskago Russkago Geograficheskago Obshchestva. Khabarovsk, R S F S R. Vols. 1-10, 1894-1914. Zap. Priamursk. Otd. Imp. Russk. Geogr. Obshch. Superseded by: Vopr. Geogr. Dal'nego Vostoka. 2-1692-3. HI 74621

Zapysky Pryrodnychco-tekhnichnogo viddulu. Kiev, Ukrainian S S R. Vols. 1-3, 1931. Zap. Pryr.-Tekhn. Vidd. 1-123-2. HI 74623

Zapiski Rossiiskoi akademii nauk. Petrograd. Vols. 35(2)-37(1), 19??-25. Zap. Rossiisk. Akad. Nauk. Preceded by: Zapiski akademii nauk. Petrograd. Superseded by: Zapiski Akademii nauk S S S R po fiziko-matematicheskomu otdeleniyu. 1-114-3. HI 74624

Zapiski Rossiiskoi akademii nauk po fiziko-matematicheskomu otdeleniyu = Zapiski Rossiiskoi akademii nauk. Petrograd.

Zapiski Rossijskoj akademii nauk = Zapiski Rossiiskoi akademii nauk. Petrograd.

Zapiski russkogo geografičeskogo obščestva po obščei geografii. St. Petersburg = Zapiski russkogo geograficheskogo obshchestva po obshchei geografii. St. Petersburg.

Zapiski russkogo geograficheskogo obshchestva po

obshchei geografii. St. Petersburg. [Dates of publication not ascertained.] Zap. Russk. Geogr. Obshch. Obshchei Geogr. Superseded by: Zapiski geograficheskogo obshchestva S S S R. Novaya seriya. 2-1692-2. HI 74625

Zapiski po semenovedeniju = Zapiski po semenovedeniyu. Leningrad.

Zapiski po semenovedeniyu. Leningrad. Vols. 4(5)-8, 1923-31. Zap. Semenov. Preceded by: Zapiski Laboratorii po semenovedeniyu pri Glavnom botanicheskom sade R S F S R. 3-2389-3. HI 74626

Zapiski Semipalatinskogo otdela Gosudarstvennogo russkogo geografičeskogo obščestva = Zapiski Semipalatinskogo otdela Gosudarstvennogo russkogo geograficheskogo obshchestva. Semipalatinsk.

Zapiski Semipalatinskogo otdela Gosudarstvennogo russkogo geograficheskogo obshchestva. Semipalatinsk, Kazakh S S R. Vols. 16-17. 1927-28. Zap. Semipalatinsk. Otd. Gosud. Russk. Geogr. Obshch. Preceded by: Zapiski Semipalatinskogo otdela Russkogo geograficheskogo obshchestva. Superseded by: Zapiski Semipalatinskogo otdela Obshchestva izucheniya Kazakhstana. 4-3130-3. HI 74627

Zapiski Semipalatinskogo otdela Obščestva izučenija Kazahstana = Zapiski Semipalatinskogo otdela Obshchestva izucheniya Kazakhstana. Semipalatinsk.

Zapiski Semipalatinskogo otdela Obshchestva izucheniya Kazakhstana. Semipalatinsk, Kazakh S S R. Vols. 18-20, 1929-31. Zap. Semipalatinsk. Otd. Obshch. Izuch. Kazakhstana. Preceded by: Zapiski Semipalatinskogo otdela Gosudarstvennogo russkogo geograficheskogo obshchestva. 4-3130-3. HI 74628

Zapiski Semipalatinskogo otdela Russkogo geografičeskogo obščestva = Zapiski Semipalatinskogo otdela Russkogo geograficheskogo obshchestva. Semipalatinsk.

Zapiski Semipalatinskogo otdela Russkogo geograficheskogo obshchestva. Semipalatinsk, Kazakh S S R. Vol. 15, [1925]1931. Zap. Semipalatinsk. Otd. Russk. Geogr. Obshch. Preceded by: Zapiski Semipalatinskogo podotděla Zapadno-Sibirskogo otdela Russkogo geograficheskogo obshchestva. Superseded by: Zapiski Semipalatinskogo otdela Gosudarstvennogo russkogo geograficheskogo obshchestva. 4-3130-3. HI 74629

Zapiski Semipalatinskago Podotděla Zapadno-Sibirskago Otděla Imperatorskago Russkago Geografičeskago Obščestva = Zapiski Semipalatinskago Pod'otděla Zapadno-Sibirskago Otděla Imperatorskago Russkago Geograficheskago Obshchestva. Semipalatinsk.

Zapiski Semipalatinskago Pod'otděla Zapadno-Sibirskago Otděla Imperatorskago Russkago Geograficheskago Obshchestva. Semipalatinsk, Kazakh S S R. Vols. 1-9/10, 1903-15. Zap. Semipalatinsk. Pod'otd. Zapadno-Sibirsk. Otd. Imp. Russk. Geogr. Obshch. Superseded by: Zapiski Semipalatinskogo pod'otděla Zapadno-Sibirskogo otdela Russkogo geograficheskogo obshchestva. 4-3130-3. HI 68510

Zapiski Semipalatinskogo podotděla Zapadno-Sibirskogo otdela Russkogo geografičeskogo obščestva = Zapiski Semipalatinskogo podotděla Zapadno-Sibirskogo otdela Russkogo geograficheskogo obshchestva. Semipalatinsk.

Zapiski Semipalatinskogo podotděla Zapadno-Sibirskogo otdela Russkogo geograficheskogo obshchestva. Semipalatinsk, Kazakh S S R. Vols. 11-14, 1917-23. Zap. Semipalatinsk. Podotd. Zapadno-Sibirsk. Otd. Russk. Geogr. Obshch. Preceded by: Zapiski Semipalatinskago Podotděla Zapadno-Sibirskago Otděla Imperatorskago Russkago Geograficheskago Obshchestva. Superseded by: Zapiski Semipalatinskogo otdela Russkogo geograficheskogo obshchestva. 4-3130-3. HI 74630

Zapiski Sěvero-Zapadnago Otděla Imperatorskago Russkago Geografičeskago Obščestva = Zapiski Sěvero-Zapadnago Otděla Imperatorskago Russkago Geograficheskago Obshchestva. Vilnius.

Zapiski Sěvero-Zapadnago Otděla Imperatorskago Russkago Geograficheskago Obshchestva. Vilnius, Lithuanian S S R. Vols. 1-4, 1910-14. Zap. Sěvero-Zapadn. Otd. Imp. Russk. Geogr. Obshch. 2-1692-3. HI 74631

Zapiski Sibirskago Otděla Imperatorskago Russkago Geografičeskago Obščestva = Zapiski Sibirskago Otděla Imperatorskago Russkago Geograficheskago Obshchestva. St. Petersburg.

Zapiski Sibirskago Otděla Imperatorskago Russkago Geograficheskago Obshchestva. St. Petersburg. Vols. 1-11, 1856-74. Zap. Sibirsk. Otd. Imp. Russk. Geogr. Obshch. Superseded by: Zapiski Vostochno-Sibirskago Otděla Imperatorskago Russkago Geograficheskago Obshchestva. 2-1693-1. HI 74632

Zapiski Simbirskago Oblastnago Estestvenno-istoričeskago Muzeja = Zapiski Simbirskago Oblastnago Estestvenno-istoricheskago Muzeya. St. Petersburg.

Zapiski Simbirskago Oblastnago Estestvenno-istoricheskago Muzeya. St. Petersburg. Vols. 1-2, 1913-15. Zap. Simbirsk. Obl. Estestv.-Istorich. Muz. HI 74633

Zapiski Sredne-Sibirskogo otdela (byvšego Krasnojarskogo) Gosudarstvennogo russkogo geografičeskogo obščestva = Zapiski Sredne-Sibirskogo otdela (byvshego Krasnoyarskogo) Gosudarstvennogo russkogo geograficheskogo obshchestva. Krasnoyarsk.

Zapiski Sredne-Sibirskogo otdela (byvshego Krasnoyarskogo) Gosudarstvennogo russkogo geograficheskogo obshchestva. Krasnoyarsk, R S F S R. Ser. 2, vol. 1, 1927 [Data for ser. 1 not ascertained]. Zap. Sredne-Sibirsk. Otd. Gosud. Russk. Geogr. Obshch. 5-4054-2. HI 74634

Zapiski Stancii dlja ispytanija semjan pri Glavnom botaničeskom sade R S F S R = Zapiski Stantsii dlya ispytaniya semyan pri Glavnom botanicheskom sade R S F S R. Petrograd.

Zapiski Stancii dlja Ispytanija Sěmjan pri Imperatorskom Botaničeskom Sadě = Zapiski Stantsii dlya Ispytaniya Sěmyan pri Imperatorskom Botanicheskom Sadě. St. Petersburg.

Zapiski Stancii dlja Ispytanija Sěmjan pri Imperatorskom Botaničeskom Sadě Petra Velikago = Zapiski Stantsii dlya Ispytaniya Sěmyan pri Imperatorskom Botanicheskom Sadě Petra Velikago. St. Petersburg.

Zapiski Stantsii dlya ispytaniya semyan pri Glavnom botanicheskom sade R S F S R. Petrograd. Vol. 4(1)-4(3), 1918-19. Zap. Stantsii Ispytan. Semyan Glavn. Bot. Sade R.S.F.S.R. Preceded by: Zapiski Stantsii dlya Ispytaniya Sěmyan pri Imperatorskom Botanicheskom Sadě Petra Velikago. Superseded by: Zapiski Laboratorii po semenovedeniyu pri Glavnom botanicheskom sade R S F S R. 3-2389-3. HI 74635

Zapiski Stantsii dlya Ispytaniya Sěmyan pri Imperatorskom Botanicheskom Sadě. St. Petersburg. Vol. 1, 1912-13. Zap. Stantsii Ispytan. Semyan Imp. Bot. Sadě. Superseded by: Zapiski Stantsii dlya Ispytaniya Sěmyan pri Imperatorskom Botanicheskom Sadě Petra Velikago. 3-2389-3. HI 74636

Zapiski Stantsii dlya Ispytaniya Sěmyan pri Imperatorskom Botanicheskom Sadě Petra Velikago. St. Petersburg. Vols. 2-3, 1914-17. Zap. Stantsii Ispytan. Sěmyan Imp. Bot. Sadě Petra Velikago. Preceded by: Zapiski Stantsii dlya Ispytaniya Sěmyan pri Imperatorskom Botanicheskom Sadě. Superseded by: Zapiski Stantsii dlya ispytaniya semyan pri Glavnom botanicheskom sade R S F S R. 3-2389-3. HI 74637

Zapiski Sverdlovskogo otdelenija Vsesojuznogo botaničeskogo obščestva = Zapiski Sverdlovskogo otdeleniya Vsesoyuznogo botanicheskogo obshchestva. Sverdlovsk.

Zapiski Sverdlovskogo otdeleniya Vsesoyuznogo botanicheskogo obshchestva. Sverdlovsk, R S F S R. Vol. 1+, 1960+. Zap. Sverdlovsk. Otd. Vsesoyuzn. Bot. Obshch. HI 74638

Zapiski tsentral'no-Kavkazskogo otdeleniya Vsesoyuznogo botanicheskgo obshchestva. Ordzhonikidze, R S F S R. Vol. 1+, 1963+. Zap. Tsentr.-Kavkazsk. Otd. Vsesoyuzn. Bot. Obshch. HI 74639

Zapiski Ural'skago Obščestva Ljubitelej Estestvozanija = Zapiski Ural'skago Obshchestva Lyubitelei Estestvozaniya. Ekaterinburg.

Zapiski Ural'skago obščestva ljubitelej estestvozanija v gorode Sverdlovske = Zapiski Ural'skago obshchestva lyubitelei estestvozaniya v gorode Sverdlovske. Sverdlovsk.

Zapiski Ural'skago Obshchestva Lyubitelei Estestvozaniya. Ekaterinburg [Sverdlovsk], R S F S R. Vols. 1-38, 1874-1922. Zap. Ural'sk. Obshch. Lyubit. Estestv. Superseded by: Zapiski Ural'skago obshchestva lyubitelei estestvozaniya v gorode Sverdlovske. 5-4342-1. HI 74640

Zapiski Ural'skago obshchestva lyubitelei estestvozaniya v gorode Sverdlovske. Sverdlovsk, R S F S R. Vols. 39-40, 1924-27. Zap. Ural'sk. Obshch. Lyubit. Estestv. Gorode Sverdlovske. Preceded by: Zapiski Ural'skago Obshchestva Lyubitelei Estestvozaniya. 5-4342-1. HI 74641

Zapiski Vladivostokskogo otdela Gosudarstvennogo russkogo geografičeskogo obščestva izučenija Amurskogo kraja = Zapiski Vladivostokskogo otdela Gosudarstvennogo russkogo geograficheskogo obshchestva izucheniya Amurskogo kraya. Vladivostok.

Zapiski Vladivostokskogo otdela Gosudarstvennogo russkogo geograficheskogo obshchestva izucheniya Amurskogo kraya. Vladivostok, R S F S R. Vols. 18-22 [also numbered vols. 1-5], 1928-30. Zap. Vladivostoksk. Otd. Gosud. Russk. Geogr. Obshch. Izuch. Amursk. Kraya. Preceded by: Zapiski Obshchestva izucheniya Amurskogo kraya Vladivostokskogo otděleniya Priamurskogo otdela Russkogo geograficheskogo obshchestva. Superseded by: Zapiski Dal'nevostochnaya kraevaya planovaya komissiya. Primorskoi filial Gosudarstvennogo geograficheskogo obshchestva. 2-1692-3. HI 74642

Zapiski voronezhskogo sel'sko-khozyaistvennogo instituta. Voronezh. Vol. ?-6-?, ?-1926-? Zap. Voronezhsk. Sel'sk.-Khoz. Inst. HI 74643

Zapiski Vostochno-Sibirskago Otděla Imperatorskago Russkago Geograficheskago Obshchestva. Irkutsk, R S F S R. Vol. 12, 1886; n.s. vols. 1-3, 1889-96. Zap. Vost.-Sibirsk. Otd. Imp. Russk. Geogr. Obshch. Preceded by: Zapiski Sibirskago Otděla Imperatorskago Russkago Geograficheskago Obshchestva. 2-1693-1. HI 74644

Zapiski Vostočno-Sibirskago Otděla Imperatorskago Russkago Geografičeskago Obščestva = Zapiski Vostochno-Sibirskago Otděla Imperatorskago Russkago Geograficheskago Obshchestva. Irkutsk.

Zapiski, vsesoyuznoe botanicheskoe obshchestvo, sverdlovsko otdelenie. Sverdlovsk. Vol. 1+, 1960+. Zap. Vsesoyuzn. Obshch. Sverdlovsk. Otd. HI 74645

Zapiski Yuzhno-Ussuriiskogo otdela Gosudarstvennogo russkogo geograficheskogo obshchestva. Vladivostok, R S F S R. Vols. 2-3, 192?-29. Zap. Yuzhno-Ussuriisk. Otd. Gosud. Russk. Geogr. Obshch. Preceded by: Zapiski Yuzhno-Ussuriiskogo otdeleniya Priamurskogo otdela Russkogo geograficheskogo obshchestva. 2-1692-2. HI 74646

Zapiski Yuzhno-Ussuriiskogo otdeleniya Priamurskogo otdela Russkogo geograficheskogo obshchestva. Vladivostok, R S F S R. Vol. 1, 1922. Zap. Yuzhno-Ussuriisk. Otd. Priamursk. Otd. Russk. Geogr. Obshch. Superseded by: Zapiski Yuzhno-Ussuriiskogo otdela Gosudarstvennogo russkogo geograficheskogo obshchestva. 2-1692-2. HI 74647

Zapiski Zapadno-Sibirskago otděla Gosudarstvennogo russkogo geografičeskogo obščestva = Zapiski Zapadno-Sibirskago otděla Gosudarstvennogo russkogo geograficheskogo obshchestva. Omsk.

Zapiski Zapadno-Sibirskago otděla Gosudarstvennogo russkogo geograficheskogo obshchestva. Omsk, R S F S R. Vol. 39, 1927. Zap. Zapadno-Sibirsk. Otd. Gosud. Russk. Geogr. Obshch. Preceded by: Zapiski Zapadno-Sibirskogo Otděla Imperatorskago Russkago Geograficheskago Obshchestva. 2-1693-1. HI 74648

Zapiski Zapadno-Sibirskogo Otděla Imperatorskago Russkago Geografičeskago Obščestva = Zapiski Zapadno-Sibirskogo Otděla Imperatorskago Russkago Geograficheskago Obshchestva. Moscow.

Zapiski Zapadno-Sibirskogo Otděla Imperatorskago Russkago Geograficheskago Obshchestva. Moscow. Vols. 1-38, 1879?-1916. Zap. Zapadno-Sibirsk. Otd. Imp. Russk. Geogr. Obshch. Superseded by: Zapiski Zapadno-Sibirskago otděla Gosudarstvennogo russkogo geograficheskogo obshchestva. 2-1693-1. HI 74649

Zapysky Kyjivs'kogo tovarystva pryrodoznavciv = Zapysky Kyyivs'kogo tovarystva pryrodoznavtsiv. Kiev.

Zapysky Kyyivs'kogo tovarystva pryrodoznavtsiv. Kiev, Ukrainian S S R. Vols. 26(2)-27, 1919-29. Zap. Kyyivs'k. Tovar. Pryr. Preceded by: Zapiski Kievskago Obshchestva Estestvoispytatelei. 3-2290-1. HI 74650

Zapysky odes'kogo tovarystva pryrodnavciv = Zapiski Odesskogo obshchestva estestvoispytatelei. Odessa.

Zapysky odes'kogo tovarystva pryrodoslidnykiv = Zapiski Odesskogo obshchestva estestvoispytatelei. Odessa.

Zapysky Pryrodnyčco-tehničnogo viddulu = Zapysky Pryrodnychco-tekhnichnogo viddulu. Kiev.

Zara'at i Pakistan. Karachi. 1961+. Zara'at Pakistan. HI 74651

Zaschita lesa. Moscow. Ca.1930s. Zashch. Lesa. Superseded by: Lesnoe khozyaistvo. HI 74652

Zashchita lesa: Mezhvuzovskii sbornik nauchnykh trudov. Leningrad. Vol. 1+, 1975+. Zashch. Lesa Mezhvuzov. Sborn. Nauchn. Trudov. Superseded by: Lesnoe khozyaistvo. HI 74653

Zashchita rastenii. Leningrad. Vols. 8-[9], 1931-32; n.s. vols. 1-19, 1935-39. Zashch. Rast. (Leningrad). Preceded by: Zashchita rastenii ot vreditelei. For 1932-34 see: Sbornik Vsesoyuznogo instituta zashchity rastenii. Superseded by: Vestnik zashchity rastenii. 5-4591-2. HI 74654

Zashchita rastenii. Moscow. Vol. 11+, 1966+. Zashch. Rast. (Moscow). Preceded by: Zashchita rastenii ot vredetelei i boleznei. HI 74655

Zashchita rastenii; sbornik nauchnykh rabot. Minsk. Vol. 1+, 1976+. Zashch. Rast. Sborn. Nauchn. Rabot. HI 74656

Zashchita rastenii ot vreditelei. Leningrad. Vols. 1-7, 1924-30. Zashch. Rast. Vredit. Superseded by: Zashchita rastenii. Leningrad. 5-4591-2. HI 74657

Zashchita rastenii ot vredetelei i boleznei. Moscow. Vols. 1-10, 1956-65. Zasch. Rast. Vredit. Boleznei. Superseded by: Zashchita rastenii. Moscow. HI 74658

Zaščita rastenij. Leningrad = Zashchita rastenii. Leningrad.

Zaščita rastenij ot vreditelej = Zashchita rastenii ot vreditelei. Leningrad.

Zasshi kifi sakuin, kagaku gijutsu-hen = Japanese periodicals index, science and technology. Tokyo.

Zasso kenkyu = Weed research. Tokyo. Weed Res. See B–P–H 972/7.

Zaštita bilja. Belgrade. Zašt. Bilja. See B–P–H 995/21. HI 60773

Zaštita prirode. Belgrade. Zašt. Prir. See B–P–H 955/22. HI 60774

Zavod za ratarstvo N R H. Zagreb. No. 1+, 1948+. Zavod Za Rat. N. R. H. HI 74659

Zbirnyk Biologichnogo fakul'tetu. Kiev, Ukrainian S S R. Vol. 11+, 1954. Zbirn. Biol. Fak. Preceded by: Pratsy Biologo-gruntovogo fakul'tetu. HI 74660

Zbirnyk Biologičnogo fakul'tetu = Zbirnyk Biologichnogo fakul'tetu. Kiev.

Zbirnyk Matematychno-pryrodopysno-likars'koi Sektsii Naukovogo Tovarystva imeny Shevchenka. Lvov, Galicia, Ukrainian S S R. Vols. 3-32, 1898-1939. Zbirn. Mt.-Pryr.-Lakars'k Sekts. Nauk. Tovar. Shevchenka. Preceded by: Zbirnyk Sektsii

Matematychno-pryrodopysno-likars'koi Naukovogo Tovarystva imeny Shevchenka. HI 60768

Zbirnyk Matematyčno-pryrodopysno-likars'koji Sekciji Naukovogo Tovarystva imeny Ševc:haenka = Zbirnyk Matematychno-pryrodopysno-likars'koi Sektsii Naukovogo Tovarystva imeny Shevchenka. Lvov.

Zbirnyk naukovyh prac' Kyjivs'kogo tovarystva pryrodnykiv = Zbirnyk naukovykh prats Kyiivs'kogo tovarystva pryrodnykiv. Kiev.

Zbirnyk naukovykh prats Kyiivs'kogo tovarystva pryrodnykiv. Kiev, Ukrainian S S R. 1924. Zbirn. Nauk. Prats Kyiivs'k. Tovar. Pryr. 3-2290-1. HI 68276

Zbirnyk prac' Botaničnogo muzeju = Zbirnyk prats Botanichnogo muzeyu. Kiev.

Zbirnyk prac' Dniprjans'koji biologičnoji stanciji = Zbirnyk prats Dnipryans'koi biologichnoi stantsii. Kiev.

Zbirnyk prac' Dniprovs'koji biologičnoji stanciji = Zbirnyk prats Dniprovs'koi biologichnoi stantsii. Kiev.

Zbirnyk prats Botanichnogo muzeyu. Kiev, Ukrainian S S R. 1 vol., 1929. Zbirn. Prats Bot. Muz. HI 60575

Zbirnyk prats Dnipryans'koi biologichnoi stantsii. Kiev, Ukrainian S S R. Vol. 6, 1931. Zbirn. Prats Dnipryans'k. Biol. Stantsii. Preceded by: Zbirnyk prats Dniprovs'koi biologichnoi stantsii. Superseded by: Trudy gidrobiologichnoi stantsii. Kiev. 1-122-2. HI 74459

Zbirnyk prats' Dniprovs'koi biologichnoi stantsii. Kiev, Ukrainian S S R. Vols. 1-5, 1926-29. Zbirn. Prats' Dniprovs'k. Biol. Stantsii. Superseded by: Zbirnyk prats' Dnipryans'koi biologichnoi stantsii. 1-122-2. HI 60491

Zbirnyk Sekciji Matematyčno-pryrodopysno-likars'koji Naukovogo Tovarystva imeny Ševčenka = Zbirnyk Sektsii Matematychno-pryrodopysno-likars'koi Naukovogo Tovarystva imeny Shevchenka. Lvov.

Zbirnyk Sektsii Matematychno-pryrodopysno-likars'koi Naukovogo Tovarystva imeny Shevchenka. Lvov, Galicia, Ukrainian S S R. Vols. 1-2, 1897. Zbirn. Sekts. Mat.-Pryr.-Likars'k. Nauk. Tovar. Shevchenka. Superseded by: Zbirnyk Matematychno-pryrodopysno-likars'koi Sektsii Naukovogo Tovarystva imeny Shevchenka. HI 74398

Zbornik biotehniske fakultete univerze Edvarda Kardelja v Ljubljani. Kmetijstvo. Ljubljana. Vol. 34+, 1979+. Zborn. Bioteh. Fak. Univ. Edvarda Kardelja Ljubljani. Preceded by: Zbornik biotehniske fakultete univerze v Ljubljani. HI 74661

Zbornik biotehniske fakultete univerze v Ljubljani. Ljubljana. Vols. 8-33?, 1961-7? Zborn. Bioteh. Fak. Univ. Ljubljani. Preceded by: Zbornik za kmetijstvo in gozdarstvo. Superseded by: Zbornik biotehniske fakultete univerze Edvarda Kardelja v Ljubljani. Kmetijstvo. HI 75105

Zbornik gozdarstva in lesarstva. Ljubljana. Vol. 11+, 1973+. Zborn. Gozd. Lesarstva. Preceded by: Zbornik instituta za gozdo in lesno gospodarstvo Slovanije. HI 74662

Zbornik instituta za gozdno in lesno gospodarstvo Slovanije. Ljubljana. 19??-72? Zborn. Inst. Gozdno Lesno Gosp. Slovanije. Preceded by: Izvestja, gozdarski institut slovenije. Superseded by: Zbornik gozdarstva in lesarstva. HI 74663

Zbornik za kmetijstvo in gozdarstvo. Ljubljana. Vols. 1-7, 1953-60. Zborn. Kmet. Gozd. Superseded by: Zbornik biotehniske fakultete univerze v Ljubljani. HI 75106

Zbornik kongresa biology Jugoslavije. Zagreb, Yugoslavia. Zborn. Kongr. Biol. Jugoslavije. See B–P–H 996/18. HI 60775

Zbornik prác arboréta Mlyňany. [Nos. 1-6 formed part of: Biologické práce, slovenskej akademie vied.] Slepčany. No. 1+, 1958+. Zborn. Prác Arbor. Mlyňany. HI 74664

Zbornik prac Biolëgičnaga instytutu = Zbornik prats Biolëgichnaga instytutu. Minsk.

Zbornik prac Biolëgičnyj instytut = Zbornik prats Biolëgichnaga instytutu. Minsk.

Zbornik prac Instytutu biolëgii = Zbornik prats Instytutu biolëgii. Minsk.

Zbornik prac o Tatranskom narodnom parku. Martin. Vol. 1+, 1957+. Zborn. Prac Tatransk. Nar. Parku. HI 74667

Zbornik prats Biolëgichnaga instytutu. Minsk, Belorussian S S R. Pts. 1-3, 1932-33; vol. 3, 1933 [vol. nos. 1-2 not used]. Zborn. Prats Biol. Inst. Superseded by: Zbornik prats Instytutu biolëgii. HI 74665

Zbornik prats Instytutu biolëgii. Minsk, Belorussian S S R. Vols. 4-7, 193?-38. Zborn. Prats Inst. Biol. Preceded by: Zbornik prats Biolëgichnaga instytutu. HI 74666

Zbornik rabot, belaruski sel'ska-haspadarchy instytut. Gory-Gorki. Vols. 1-2, 1934. Zborn. Rabot Belarusk. Sel'skahasp. Inst. HI 74668

Zbornik radova. Biološki institut N.R. Srbije. Belgrade. Zborn. Rad. Biol. Inst. Nar. Republ. Srbije. See B–P–H 996/21. HI 60776

Zbornik radova, institut za farmakognoziju farmaceutskog fakulteta u Beogradu. Lekovite sirovine. Belgrade. Vols. 1-2, 1951-? Zborn. Rad. Inst. Farmakogn. Farm. Fak. Beogradu, Lekov. Sirovine. Superseded by: Zbornik radova, institut za ispitivanje lekovitpog bilja nr srbije u Beogradu. Lekovite sirovine. HI 74669

Zbornik radova, institut za ispitivanje lekovitpog bilja nr srbije u Beogradu. Lekovite sirovine. Belgrade. Vols. 3-5-?, 195?-60-? Zborn. Rad. Inst. Isp. Lekov. Bilja Srbije Beogradu, Lekov. Sironine. Preceded by: Zbornik radova, institut za farmakognoziju farmaceutskog fakulteta u Beogradu. Lekovite sirovine. HI 74670

Zbornik radova, institut za oceanografiju i ribarstvo. Split. Vol. 1+, 1968/69+, [1969]+. Zborn. Rad. Inst. Oceanogr. Ribarst. HI 74671

Zbornik radova poljoprivrednog fakulteta, universitet u Beogradu. Belgrade. Vol. 5+, 1953+. Zborn. Rad. Poljopr. Fak. Univ. Beogradu. Preceded by: Godišjnak poljoprivredno-šumarskog fakulteta, universitet u Beogradu. HI 74672

Zbornik radova prirodno-matematichkog fakulteta, universiteta novom sadu. Serija zabiologiju. Novi Sad. Vol. 11+, 1981+. Zborn. Rad. Prir.-Mat. Fak. Univ. Novom Sadu, Ser. Zabiol. HI 74673

Zbornik radova. Srpska akademija nauka. Belgrade. Zborn. Rad. Srpska Akad. Nauka. See B–P–H 996/24. HI 60777

Zbornik radova. Srpska akademija nauka. Institut za ekologiju i biogegrafiju. Belgrade. Zborn. Rad. Srpska Akad. Nauka Inst. Ekol. See B–P–H 996/25. HI 60778

Zbornik radova, zavod za ratarstvo. Sarajevo. Vol. 1-?, 1961-? Zborn. Rad. Zavod Ratarstvo. HI 74674

Zbornik slovenského narodného muzea. Přírodné vědy. Bratislava. Vol. 13(1)+, 1968+. Zborn. Slov. Nar. Muz., Přir. Vědy. Preceded by: Sbornik slovenskeho narodneho muzea. Prirodne vedy. HI 74675

Zbornik vědeckých prác lesníckej fakulty vysokej školy lesníckej a drevárskej vo Zvolene = Sborník vědeckých prác, Zvolen. Bratislava.

Zbornik východoslovenského muzea v Košiciach. Prírodné vedy. Košice. Vol. 20+, 1979+. Zborn. Vychodoslov. Muz. Košiciach, Prir. Vedy. Preceded by: Zbornik východoslovenského muzea v Košiciach. Séria AB, prírodné vedy. HI 74676

Zborník východoslovenskeho múzea v Košiciach. Séria A, geologické vedy. Košice. Vols. 8-10, 1967-69. Zborn. Vychodoslov. Muz. Košiciach, A. Preceded by: Sborník východoslovenského múzea v Košiciach. Continued in: Zborník východoslovenského múzea v Košiciach. Séria AB, prírodné vedy. HI 74677

Zbornik východoslovenského muzea v Košiciach. Séria AB, prírodné vedy. Košice. Vols. 11/14-19, 1973-79. Zborn. Vychodoslov. Muz. Košiciach, AB. Preceded by: Zborník východoslovenského múzea v Košiciach. Séria A, geologické vedy and Séria B, zoologia, botanika. by: Zbornik východoslovenského muzea v Košiciach. Prírodné vedy. HI 74678

Zborník východoslovenského múzea v Košiciach. Séria B, zoologia, botanika. Košice. Vols. 7-11/12, 1966-72. Zborn. Vychodoslov. Muz. Košiciach, B. Preceded by: Sborník východoslovenského múzea v Košiciach. Séria A, prirodné vedy. Continued in: Zbornik východoslovenského muzea v Košiciach. Séria AB, prírodné vedy. HI 74679

Zbornik, zemjodelskli institut, Skopje. Skopje. Vol. 1+, 1966?+. Zborn. Zemjod. Inst. Skopje. Preceded by: Zbornik, zemjodelsko-ispitatelen institut, Skopje. HI 74680

Zbornik, zemjodelsko-ispitatelen institut, Skopje. Skopje. Vols. 1-?, 1952-65? Zborn. Zemjod.-Isp. Inst. Skopje. Superseded by: Zbornik, zemjodelskli institut, Skopje. HI 74681

Zdravotnická pracovnice. Prague. 1951+. Zdrav. Pracovn. HI 74682

Zeeuws fruittelersblad. Vols. ?-20-28(10), ?-1964-72. Zeeuws Fruittelersblad. HI 74683

Zeiss-Mitteilungen über Fortschritte der technischen Optik. Stuttgart. Vol. 1+, 1957+. Zeiss-Mitt. Fortschr. Techn. Optik. HI 74684

Zeitblatt für Gewerbetreibende und Freunde der Gewerbe. Berlin. Zeitbl. Gewerbetreibende Freunde Gewerbe. See B–P–H 997/10. HI 60779

Zeitgemässe Obstbaufragen. Vienna. Zeitgemässe Obstbaufragen. See B–P–H 997/11. HI 60780

Zeitschrift für Acclimatisation. Berlin. Z. Acclim. See B–P–H 984/2. HI 60693

Zeitschrift für Acker- und Pflanzenbau. Berlin. Vols. 92?-155, 1949-1985. Z. Acker- Pflanzenbau. Preceded by: Journal für Landwirtschaft. Superseded by: Journal of agronomy and crop science. 5-4593-3. HI 74685

Zeitschrift für Agrargeschichte und Agrdaroziologie. Frankfurt a. M. Z. Agrargesch. Agrarsoziol. See B–P–H 984/4. HI 60694

Zeitschrift für Akklimatisation. Berlin. Z. Akklim. See B–P–H 984/5. HI 60695

Zeitschrift für allgemeine Erdkunde. Berlin. Z. Allg. Erdk. See B–P–H 984/6. HI 60696

Zeitschrift für die allgemeine Geographie. Breslau [=Wroclaw, Poland]. Z. Allg. Geogr. See B–P–H 984/7. HI 60697

Zeitschrift für allgemeine Mikrobiologie. Berlin. Vols. 21(5)-24, 1981-84. Z. Allg. Mikrobiol. Preceded by: Zeitschrift für allgemeine Mikrobiologie, Morphologie, Physiologie und Ökologie der Mikroorganismen. Superseded by: Journal of basic microbiology.

HI 74686

Zeitschrift für allgemeine Mikrobiologie, Morphologie, Physiologie, Genetik und Ökologie der Mikroorganismen = Zeitschrift für allgemeine Mikrobiologie, Morphologie, Physiologie und Ökologie der Mikroorganismen. Berlin.

Zeitschrift für allgemeine Mikrobiologie, Morphologie, Physiologie und Ökologie der Mikroorganismen. Berlin. Vols. 1-21(4), 1960-81. Z. Allg. Mikrobiol. Morphol. Physiol. Ökol. Mikroorgan. Superseded by: Zeitschrift für allgemeine Mikrobiologie. HI 74687

Zeitschrift des allgemeinen oesterreichischen Apothekervereins. Vienna. Vols. 17-?, 1863-1921. Z. Allg. Oesterr. Apothekervereins. Preceded by: Oesterreichische Zeitschrift für Pharmacie. 1-147-1. HI 74688

Zeitschrift für Angewandte Botanik. Kharkov, Ukrainian S S R = Visnyk prykladnoi botaniky. Kharkov.

Zeitschrift für angewandte Entomologie. Berlin. Z. Angew. Entomol. See B–P–H 984/10. HI 60698

Zeitschrift für angewandte Mikroskopie. Berlin. Z. Angew. Mikroskop. See B–P–H 984/12. HI 60699

Zeitschrift für angewandte Mikroskopie und klinische Chemie = Zeitschrift für angewandte Mikroskopie. Berlin. Z. Angew. Mikroskop. See B–P–H 984/12.

Zeitschrift für den Ausbau der Entwicklungslehre. Stuttgart. Z. Ausbau Entwicklungsl. See B–P–H 984/13. HI 60700

Zeitschrift für ausländische Landwirtschaft. Frankfurt am Main. Vols. 1-18, 1962-79. Z. Ausl. Landw. Superseded by: Quarterly journal of international agriculture. HI 74689

Zeitschrift für Baiern und die angränzenden Länder. Munich. Z. Baiern Angränzenden Länder. See B–P–H 984/14. HI 60701

Zeitschrift der Basler Botanischen Gesellschaft = Bauhinia. Basel. Bauhinia. See B–P–H 168/9.

Zeitschrift für bildende Gartenkunst. Organ des Vereins Deutscher Gartenkünstler. Berlin. Z. Bildende Gartenkunst. See B–P–H 984/15. HI 60702

Zeitschrift für biologie. Bucharest = Revue de biologie. Bucharest.

Zeitschrift für Biologie. Munich & Berlin. Vols. 1-116(6), 1865-44, 1949-71. Z. Biol. 5-4597-1. HI 74690

Zeitschrift für biologische Technik und Methodik. Strasbourg. Z. Biol. Techn. Meth. See B–P–H 984/18. HI 60703

Zeitschrift für Botanik. Jena. Z. Bot. See B–P–H 984/19. HI 60704

Zeitschrift der botanischen Abteilung, naturwissenschaftlicher Verein der Provinz Posen = Naturwissenschaftlicher Verein der Provinz Posen. Zeitschrift der botanischen Abteilung. Posen [=Poznan, Poland]. Naturwiss. Verein Prov. Posen Z. Bot. Abt. See B–P–H 637/2.

Zeitschrift für Chemie und Industrie der Kolloide. Dresden. Z. Chem. Industr. Kolloide. See B–P–H 984/20. HI 60705

Zeitschrift der Deutschen Mykologischen Gesellschaft. Vienna. Z. Deutsch. Mykol. Ges. See B–P–H 984/21. HI 60706

Zeitschrift des Ferdinandeums für Tirol und Vorarlberg. Innsbruck. Vols. 1-60, 1853-1920. Z. Ferdinandeums Tirol. Preceded by: Neue Zeitschrift des Ferdinandeums für Tirol und Vorarlberg. 2-1553-2. HI 74691

Zeitschrift für Fischerei und deren Hilfswissenschaften. Berlin. Z. Fischerei Hilfswiss. See B–P–H 985/1. HI 60707

Zeitschrift für Forst- und Jagdwesen. Berlin. Z. Forst-Jagdwesen. See B–P–H 985/2. HI 60708

Zeitschrift für das Forst- und Jagdwesen in Baiern = Zeitschrift für das Forst- und Jagdwesen, für Cameral- u. Forstbeamte, Forst u. Jagdliebhaber. Munich.

Zeitschrift für das Forst- und Jagdwesen, für Cameral- u. Forstbeamte, Forst u. Jagdliebhaber. Munich. Vols. 1-5, 1813-18. Z. Forst Jagdwesen Cameral Forstbeamte. Superseded by: Neue Zeitschrift für das Forst- und Jagdwesen. HI 74692

Zeitschrift für Forstgenetik und Forstpflanzenzüchtung. Frankfurt a. M. Z. Forstgenet. Forstpflanzenzücht. See B–P–H 985/3. HI 607009

Zeitschrift für die Forstwissenschaft. Copenhagen. Z. Forstwiss. See B–P–H 985/4. HI 60710

Zeitschrift für Garten- und Obstbau. Vienna. Z. Garten-Obstbau. See B–P–H 985/5. HI 60711

Zeitschrift für Gartenbau und Gartenkunst. Neue Folge des Jahrbuches für Gartenkunde und Botanik. Organ des Vereins Deutscher Gartenkünstler. Neudamm [=Debno, Poland]. Z. Gartenbau Gartenkunst. See B–P–H 985/6. HI 60712

Zeitschrift des Gartenbau-Vereins für das Königreich Hannover. Hanover. Z. Gartenbau-Vereins Königr. Hannover. See B–P–H 985/7. HI 60713

Zeitschrift für Gärtner, Botaniker und Blumenfreunde, oder repertorium botanicae exoticae systematicae, sistens diagnoses generum et specierum novarum. Jena. Vols. 1-5, 1840-50. Z. Gärtn. Bot. 5-4605-2. HI 74693

Zeitschrift für Gärtner und Gartenfreunde. Vienna. Z. Gärtn. Gartenfr. See B–P–H 985/9. HI 60714

Zeitschrift für Gärungsphysiologie, allgemeine, landwirtschaftliche und technische Mykologie. Berlin. Z. Gärungsphysiol. See B–P–H 985/10. HI 60715

Zeitschrift für geologische Wissenschaften. Berlin. 1973+. Z. Geol. Wiss. Preceded by: Paläontologische Abhandlungen. Abteilung B, Paläobotanik. HI 74694

Zeitschrift für die gesammte Botanik = Bonplandia. Hanover. Bonplandia. See B–P–H 217/2.

Zeitschrift für die gesammte Thierheilkunde und Viehzucht. Giessen, Germany. Z. Gesammte Thierheilk. Viehzucht. See B–P–H 985/12. HI 60717

Zeitschrift für die gesammten Naturwissenschaften. Halle. Z. Gesammten Naturwiss. (Halle). See B–P–H 985/13. HI 60718

Zeitschrift für das gesamte Forstwesen. Berlin. Z. Gesamte Forstwesen. See B–P–H 985/14. HI 60719

Zeitschrift für gesamte Getreidewesen; technische und wissenschaftliche Monatshefte für Landwirtschaft, Müllerei und Bäckerei. Rostock. 1909-44. Z. Gesamte Getreidew. HI 74695

Zeitschrift für die gesamte Naturwissenschaft einschliesslich Naturphilosophie und Geschichte der Naturwissenschaft und Medizin. Brunswick, Germany. Z. Gesamte Naturwiss. (Brunswick). See B–P–H 985/15. HI 60720

Zeitschrift für Geschichte der Naturwissenschaften, Technik und Medizin. Berlin. Z. Gesch. Naturwiss. See B–P–H 985/16. HI 60721

Zeitschrift für Gesellschaft für Erdkunde zu Berlin. Berlin. Z. Ges. Erdk. Berlin. See B–P–H 985/11. HI 60716

Zeitschrift für Hydrologie. Aarau, Switzerland. Z. Hydrol. (Aarau). See B–P–H 985/17. HI 60722

Zeitschrift für Hydrologie. Budapest = Hidrológiai közlöny. Budapest. Hidrol. Közl. See B–P–H 416/4.

Zeitschrift für Hygiene. Leipzig. Z. Hyg. See B–P–H 985/19. HI 60723

Zeitschrift für Hygiene und Infektionskrankheiten. Berlin, Leipzig. Vols. 11-147, 1892-1961. Z. Hyg. Infektionskrankh. Preceded by: Zeitschrift für Hygiene. Superseded by: Zeitschrift für Hygiene und Infektionskrankheiten, medizinische Mikrobiologie, Immunologie und Virologie. 5-4606-2. HI 74696

Zeitschrift für Hygiene und Infektionskrankheiten, medizinische Mikrobiologie, Immunologie und Virologie. Berlin. Vols. 148-151, 1961-65. Z. Hyg. Infektionskrankh. Med. Mikrobiol. Immunol. Virol Preceded by: Zeitschrift für Hygiene und Infektionskrankheiten. Superseded by: Zeitschrift für medizinische Mikrobiologie und Immunologie. HI 74697

Zeitschrift für induktive Abstammungs- und Vererbungslehre. Berlin. Z. Indukt. Abstammungs-Vererbungsl. See B–P–H 986/4. HI 60724

Zeitschrift für Infektionskrankheiten, parasitäre Krankheiten und Hygiene der Haustiere. Berlin. Z. Infektionskrankh. See B–P–H 986/5. HI 60725

Zeitschrift für das Jagd- und Forstwesen mit besonderer Rücksicht auf Baiern. Bamberg. N.s. vols. 2-4, 1824-26; n.s. vols. 1-11, 1826-40; n.s. vols. 1-7, 1841-47. Z. Jagd Forstwesen Rücks. Baiern. Preceded by: Neue Zeitschrift für das Forst- und Jagdwesen. 5-4598-2. HI 74698

Zeitschrift der japanischen Gartenbau-Gesellschaft = Journal of the japanese horticultural society. Tokyo. J. Jap. Hort. Soc. See B–P–H 470/21.

Zeitschrift der K. K. Landwirthschafts-Gesellschaft für Tirol. und Vorarlberg. Innsbruck. Z. K. K. Landw.-Ges. Tirol. See B–P–H 986/7. HI 60726

Zeitschrift für klinische Medizin. Berlin. Z. Klin. Med. (Berlin). See B–P–H 986/8. HI 60727

Zeitschrift für Krebsforschung. Berlin. Z. Krebsf. See B–P–H 986/9. HI 60728

Zeitschrift des Landes-Obstbaumzucht-Vereines für das Königreich Böhmen = Časopis zemského spolku štěpařkého pro království České. Prague. Čas. Zemsk. Spolku Štěp. Král. České. See B–P–H 301/14.

Zeitschrift der landwirtschaftlichen Sektion der königl. ungarischen naturwissenschaftlichen Gesellschaft = Mezögazdaságtudományi közlemények. Budapest. Mezögazdaságtud. Közlem. See B–P–H 591/16.

Zeitschrift für die landwirthschaftlichen Vereine des Grossherzogthums Hessen. Darmstadt, Germany. Z. Landw. Vereine Grossherzoght. Hessen. See B–P–H 986/12. HI 60729

Zeitschrift des Landwirthschaftlichen Vereins in Bayern. Munich. Z. Landw. Vereins Bayern. See B–P–H 986/13. HI 60730

Zeitschrift des landwirthschaftlichen Vereins für Rheinpreussen. Bonn. Z. Landw. Vereins Rehinpreussen. See B–P–H 986/14. HI 60731

Zeitschrift für landwirthschaftliches Versuchs- und Untersuchungswesen. Berlin. Z. Landw. Versuchs-UIntersuchungswesen. See B–P–H 986/15. HI 60732

Zeitschrift des mährischen Landesmuseums. Brünn [=Brno, Czechoslovakia]. Z. Mähr. Landesmus. See B–P–H 986/17. HI 60734

Zeitschrift für medizinische Mikrobiologie und Immunologie. Berlin, New York. Vols. 152-156, 1966-71. Z. Med. Mikrobiol. Immunol. Preceded by: Zeitschrift für Hygiene und Infektionskrankheiten, medizinische Mikrobiologie, Immunologie und Virologie. Superseded by: Medical microbiology and immunology. HI 74699

Zeitschrift für Mineralogie = Taschenbuch für die gesammte Mineralogie. Frankfurt a. M. Taschenb. Gesamamte Mineral. See B–P–H 866/20.

Zeitschrift der Moorversuchsstation zu Minsk = Bolotovedenie; Vestnik minskoi bolotnoi opytnoi stantsii. Minsk.

Zeitschrift für Mykologie. Tubingen, Karlsruhe. Vol. 44+, 1978+. Z. Mykol. Preceded by: Zeitschrift für Pilzkunde. HI 74700

Zeitschrift für Mykologie und Pilzwirtschaft = Karstenia. Sienitieteellinen ja sienitalondellinen aikakauskirja. Helsinki. Karstenia. See B–P–H 511/18.

Zeitschrift für Natur- und Heilkunde in Ungarn. Pest [=Budapest, in part] & Oedenburg [=Sopron]. Z. Natur- Heilk. Ungarn. See B–P–H 986/21. HI 60735

Zeitschrift für Naturforschung. Wiesbaden. Vol. 1, 1946. Z. Naturf. Superseded by: Zeitschrift für Naturforschung. Teil A, Astrophysik, Physik, physikalische Chemie and Teil B, Chemie, Biochemie, Biophysik, Biologie. 5-4612-1. HI 74701

Zeitschrift für Naturforschung. Teil B, Chemie, Biochemie, Biophysik, Biologie. Wiesbaden. Vols. 2B-27B, 1947-73. Z. Naturf., B. Preceded by: Zeitschrift für Naturforschung. Superseded by: Zeitschrift für Naturforschung. Teil C, Biochemie, Biophysik, Biologie, Virologie. 5-4612-1. HI 74702

Zeitschrift für Naturforschung. Teil C, Biochemie, Biophysik, Biologie, Virologie. Tübingen. Vol. 28+, 1973+. Z. Naturf., C. Preceded, in part, by: Zeitschrift für Naturforschung. Teil B, Chemie, Biochemie, Biophysik, Biologie. HI 74703

Zeitschrift für Naturwissenschaften. Berlin. Z. Naturwiss. See B–P–H 986/25. HI 60736

Zeitschrift der naturwissenschaftlichen Abteilung (des naturwissenschaftlichen Vereins), Deutsche Gesellschaft für Kunst und Wissenschaft in Posen = Deutsche Gesellschaft für Kunst und Wissenschaft in Posen. Zeitschrift der naturwissenschaftlichen Abteilung (des naturwissenschaftlichen Vereins). Posen [=Poznan, Poland]. Deutsche Ges. Kunst Posen Z. Naturwiss. Abt. See B–P–H 346/8.

Zeitschrift des Niederrheinischen Landwirthschaftlichen Vereins. Bonn. Z. Niederrhein. Landw. Vereins. See B–P–H 987/1. HI 60737

Zeitschrift für die Ophthalmologie. Dresden. Z. Ophthalmol. See B–P–H 987/3. HI 60738

Zeitschrift für Parasitenkunde. (Zeitschrift für wissenschaftliche Biologie. Abt. F.) Berlin. Z. Parasitenk. (Berlin). See B–P–H 987/4. HI 60739

Zeitschrift für Parasitenkunde. Jena. Z. Parasitenk. (Jena). See B–P–H 987/5. HI 60740

Zeitschrift für Pflanzenbau und Pflanzenschutz. Munich. Z. Pflanzenbau Pflanzenschutz. See B–P–H 987/6. HI 60741

Zeitschrift für Pflanzenernährung und Bodenkunde. Leipzig & Berlin. Vol. 116+, 1967+. Z. Pflanzenernähr. Bodenk. Preceded by: Zeitschrift für Pflanzenernährung, Dügung und Bodenkunde. HI 74704

Zeitschrift für Pflanzenernährung und Düngung. A. Wissenschaftlicher Teil. Leipzig & Berlin. Z. Pflanzenernähr. Düngung, A, Wiss. Teil. See B–P–H 987/8. HI 60742

Zeitschrift für Pflanzenernährung, Dügung und Bodenkunde. Leipzig & Berlin. Vols. 37-45, 1935-36; vol. 82-144 [also numbered n.s. vol. 37-99], vols. 100-115, 1946-66. Z. Pflanzenernähr. Dügung Bodenk. Preceded by: Zeitschrift für Pflanzenernährung, Dügung und Bodenkunde. A, wissenschaftlicher Teil. For vols. 6-36 see: Bodenkunde und Pflanzenernährung. Superseded by: Zeitschrift für Pflanzenernährung und Bodenkunde. 5-4613-3. HI 74705

Zeitschrift für Pflanzenernährung, Düngung und Bodenkunde. A. Wissenschaftlicher Teil. Leipzig & Berlin. Z. Pflanzenernähr. Düngung Bodenk., A, Wiss. Teil. See B–P–H 987/10. HI 60743

Zeitschrift für Pflanzenkrankheiten. Stuttgart. Z. Pflanzenkrankh. See B–P–H 987/13. HI 60744

Zeitschrift für Pflanzenkrankheiten und Gallenkunde. Stuttgart. Z. Pflanzenkrankh. Gallenk. See B–P–H 987/14. HI 60745

Zeitschrift für Pflanzenkrankheiten (Pflanzenpathologie) und Pflanzenschutz. Stuttgart. Z. Pflanzenkrankh. Pflanzenschutz. See B–P–H 987/15. HI 60746

Zeitschrift für Pflanzenkrankheiten und Pflanzenschutz = Zeitschrift für Pflanzenkrankheiten (Pflanzenpathologie) und Pflanzenschutz. Stuttgart. Z. Pflanzenkrankh. Pflanzenschutz. See B–P–H 987/15.

Zeitschrift für Pflanzenkrankheiten und Pflanzenschutz. Sonderhefte. [Forms part of: Zeitschrift für Pflanzenkrankheiten und Pflanzenschutz.] Stuttgart. Vols. 1-6, 1964-72. Z. Pflanzenkrankh. Pflanzenschutz, Sonderh. HI 74706

Zeitschrift für Pflanzenpathologie und Pflanzenschutz. Sonderhefte = Zeitschrift für Pflanzenkrankheitenund Pflanzenschutz. Sonderhefte. Stuttgart.

Zeitschrift für Pflanzenphysiologie. Stuttgart. Vols. 53-114, 1965-84. Z. Pflanzenphysiol. Preceded by: Zeitschrift für Botanik. Superseded by: Journal of plant physiology. HI 74707

Zeitschrift für Pflanzenzüchtung. Berlin. Vols. 1-14, 1913-29; vols. 23-95, 1941-85. Z. Pflanzenzücht.

Superseded by: Plant breeding. For vols. 15-22, 1930-38 see: Zeitschrift für Züchtung, Reihe A, Pflanzenzüchtung. 5-4614-1. HI 74708

Zeitschrift für Physiologie. Heidelberg = Untersuchungen über die Natur des Menschen, der Thiere und der Pflanzen. Heidelberg. Untersuch. Natur Menschen. See B–P–H 939/1.

Zeitschrift für Pilzfreunde; populäre Mitteilungen über essbare und schädliche Pilze. Dresden. Vols. 1-2, 1883-85. Z. Pilzfr. HI 74709

Zeitschrift für Pilzkunde. Heilbronn, Karlsruhe. N.s. vols. 1-43, 1922-77. Z. Pilzk. Preceded by: Pilz-Kräuterfreund. Superseded by: Zeitschrift für Mykologie. 5-4615-3. HI 60747

Zeitschrift für die politisch-statistische Geographie. Breslau [=Wroclaw, Poland]. Z. Polit.-Statist. Geogr. See B–P–H 987/21. HI 60748

Zeitschrift für die Provinz Hanau. Hanau. Z. Prov. Hanau. See B–P–H 988/1. HI 60749

Zeitschrift der Rheinischen Naturforschenden Gesellschaft. Mainz. Vols. 1-4, 1961-66. Z. Rhein. Naturf. Ges. Mainz. Superseded by: Mainzer naturwissenschaftliches Archiv. HI 74710

Zeitschrift des Rheinpreussischen Landwirthschaftlichen Vereins. Bonn. Z. Rheinpreuss. Landw. Vereins. See B–P–H 988/3. HI 60750

Zeitschrift der Sektion für Botanik, Deutsche Gesellschaft für Kunst und Wissenschaft in Posen. Naturwissenschaftliche Abteilung = Deutsche Gesellschaft für Kunst und Wissenschaft in Posen. Naturwissenschaftliche Abteilung (Naturwiss. Verein). Zeitschrift der Sektion für Botanik. Posen [=Poznan, Poland]. Deutsche Ges. Kunst Posen Naturwiss. Abt. Z. Sekt. Bot. See B–P–H 346/7.

Zeitschrift für die Staatsarzneikunde. Erlangen. Z. Staatsarzneik. See B–P–H 988/4. HI 60751

Zeitschrift für Sukkulentenkunde. Berlin. Z. Sukkulentenk. See B–P–H 988/5. HI 60752

Zeitschrift für technische Biologie. Berlin. Z. Techn. Biol. See B–P–H 988/6. HI 60753

Zeitschrift für Tirol und Vorarlberg. Innsbruck. Vols. 1-8, 1825-34. Z. Tirol Vorarlberg. Superseded by: Neue Zeitschrift des Ferdinandeums für Tirol und Vorarlberg. 2-1553-2. HI 74711

Zeitschrift für tropische Landwirtschaft. Der Tropenpflanzer. Berlin. Z. Trop. Landw. Tropenpflanzer. See B–P–H 988/7. HI 60754

Zeitschrift von und für Ungern zur Beförderung der vaterländischen Geschichte, Erdkunde und Literatur. Pest [=Budapest, in part]. Z. Ungern. See B–P–H 988/8. HI 60755

Zeitschrift für Vegetationstechnik im Landschafts- und Sportstättenbau. Berlin, Hanover. Vol. 1+, 1978+. Z. Vegetationstechn. Landschafts- Sportstättenbau. HI 74712

Zeitschrift des Vereins der deutschen Zucker-Industrie. Allgemeiner Teil. Berlin. Vols. 48-84, 1898-1934. Z. Vereins Deutsch. Zucker-Industr., Allg. Teil. Preceded by: Zeitschrift des Vereins für die Rübenzuckerindustrie des deutschen Reiches. Superseded by: Zeitschrift für Wirtschaftsgruppe Zuckerindustrie. HI 74713

Zeitschrift des Vereins für Naturbeobachter und Sammler. Vienna. Z. Vereins Naturbeob. See B–P–H 988/9. HI 60756

Zeitschrift des Vereins für die Rübenzuckerindustrie des deutschen Reiches. Berlin. Vols. 23-47, 1873-97. Z. Vereins Rübenzuckerindustr. Deutsch. Reiches. Preceded by: Zeitschrift des Vereins für die Rübenzuckerindustrie im Zollverein. Superseded by: Zeitschrift des Vereins der deutschen Zucker-Industrie. Allgemeiner Teil. HI 74714

Zeitschrift des Vereins für die Rübenzuckerindustrie im Zollverein. Berlin. Vols. 1-22, 1851-72. Z. Vereins Rübenzuckerindustr. Zollv. Superseded by: Zeitschrift des Vereins für die Rübenzuckerindustrie des deutschen Reiches. HI 74715

Zeitschrift für Vererbungslehre. Berlin. Vols. 89-98, 1958-66. Z. Vererbungsl. Preceded by: Zeitschrift für induktive Abstammungs- und Vererbungslehre. Superseded by: Molecular and general genetics. 5-4620-1. HI 74716

Zeitschrift für Veterinärkunde. Berlin. Z. Veterinärk. See B–P–H 988/12. HI 60757

Zeitschrift für Weinbau und Weinbereitung in Ungarn und Siebenbürgen, für Weinbergbesitzer, Winzer, Landwirthe und Weinhändler. Pest [=Budapest, in part]. Z. Weinbereitung Ungarn Siebenbürgen. See B–P–H 988/14. HI 60759

Zeitschrift für Weinbau und Weinbereitung in Ungarn, für Weinbergbesitzer, Winzer und Landwirthe. Buda & Pest [=Budapest]. Z. Weinbereitung Ungarn. See B–P–H 988/13. HI 60758

Zeitschrift für Weltforstwirtschaft. Neudamm & Berlin. Z. Weltforstw. See B–P–H 988/15. HI 60760

Zeitschrift für Weltforstwirtschaft waldwirtschaftliche und bodenkundliche Grossraumforschung zugleich Kolonialforstliche Mitteilungen. Neudamm [=Debno, Poland] & Berlin. Z. Weltforstw. Waldw. Bodenk. Grossraumf. & Kolonialforstl. Mitt. See B–P–H 988/16. HI 60761

Zeitschrift für Wirtschaftsgruppe Zuckerindustrie. Berlin. Vols. 85-94, 1935-44. Z. Wirtschaftsgr. Zuckerindustr. Preceded by: Zeitschrift des Vereins der deutschen

Zucker-Industrie. Allgemeiner Teil. HI 74717

Zeitschrift für wissenschaftliche Biologie. Abteilung D, vols. 105-130 = Wilhelm Roux' Archiv für Entwicklungsmechanik der Organismen. Leipzig, etc.

Zeitschrift für wissenschaftliche Botanik. Zurich. Z. Wiss. Bot. See B–P–H 989/1. HI 60762

Zeitschrift für wissenschaftliche Mikroskopie und für mikroskopische Technik. Brunswick, Germany. Z. Wiss. Mikroskop. Mikroskop. Techn. See B–P–H 989/3. HI 60763

Zeitschrift für Zellforschung und mikroskopische Anatomie. Berlin. Vols. 1-147, 1925-74. Z. Zellf. Mikroskop. Anat. Superseded by: Cell and tissue research. HI 74718

Zeitschrift für Zellforschung und mikroskopische Anatomie. Abteilung B. Chromosoma. Berlin. Z. Zellf. Mikroskop. Anat., Abt. B, Chromosoma See B–P–H 989/4. HI 60764

Zeitschrift für zoologische Systematik und Evolutionsforschung. Frankfurt am Main. Vol. 1+, 1963+. Z. Zool. Syst. Evolutionsforsch. HI 74719

Zeitschrift für Züchtung. Reihe A: Pflanzenzüchtung. Berlin. Z. Zücht., Reihe A, Pflanzenzücht. See B–P–H 989/5. HI 60765

Zeitung für Geognosie, Geologie und innere Naturgeschichte der Erden. Weimar. Vols. 1-11, 1826-31. [Also numbered "Jahrgung" 1826-31, 2 vols. each forming a Jahrgung (3 in 1828)]. Zeitung Geognosie. 5-4623-3. HI 53825

Želmenija; illustrated journal of general botany. Boston, MA. Želmenija. See B–P–H 997/52. HI 60781

Zemědělský archiv. Prague. Zeměděl. Arch. See B–P–H 997/54. HI 60782

Zemědělství v zahraniči. Prague. 1957+. Zeměděl. Zahran. Preceded by: Sovětske zemědělství. HI 74720

Zeměpisný sborník. Prague. Zeměp. Sborn. See B–P–H 997/55. HI 60783

Žemés ükio akademijos metraštis. Kaunas. Vols. 1-13-?, 1924-[1935]-[40]-? Žemés Ükio Akad. Metraštis. HI 74721

Žemés ukis; žemés ükio mokslo žurnalas. Kaunas. Vols. ?-2-?, ?-[1928]-? Žemés Ukis. HI 74722

Zemledĕl'českaja Gazeta. St. Petersburg = Zemledĕl'cheskaya Gazeta, izdavaemaya po Vysochaisemu Povelĕniyu. St. Petersburg.

Zemledĕl'českaja Gazeta, izdavaemaja po Vysočajsemu Povelĕniju = Zemledĕl'cheskaya Gazeta, izdavaemaya po Vysochaisemu Povelĕniyu. St. Petersburg.

Zemledĕl'cheskaya Gazeta, izdavaemaya po Vysochaisemu Povelĕniyu. St. Petersburg. 1834-1905; 1913-16. Zemled. Gaz. 5-4625-1. HI 74723

Zemlevĕdĕnie. Periodičeskoe izdanie Geografičeskago Otdĕlenija Imperatorskago Obščestva Ljubitelej Estestvozanija, Antropologii i Etnografii = Zemlevĕdĕnie. Periodicheskoe izdanie Geograficheskago Otdĕleniya Imperatorskago Obshchestva Lyubitelei Estestvozaniya, Antropologii i Etnografii. Moscow.

Zemlevĕdĕnie. Periodicheskoe izdanie Geograficheskago Otdĕleniya Imperatorskago Obshchestva Lyubitelei Estestvozaniya, Antropologii i Etnografii. Moscow. Vols. 1-24, 1894-1917. Zemlevĕdĕnie. Superseded by: Zemlevedenie. 5-4625-1. HI 74724

Zemlevedenie. Moscow. Zemlevedenie. See B–P–H 998/4. HI 60784

Zemljište i biljka. Belgrade. Vol. 1+, 1952+. Zemljište i Biljka. HI 74725

Zentralblatt für allgemeine Pathologie und pathologische Anatomie = Centralblatt für allgemeine Pathologie und pathologische Anatomie. Jena. Centralbl. Allg. Pathol. Pathol. Anat. See B–P–H 302/27.

Zentralblatt für allgemeine Pathologie und pathologische Anatomie. Ergänzungsheft = Verhandlungen des Deutschen Pathologischen Gesellschaft. Berlin. Verh. Deutsch. Pathol. Ges. See B–P–H 950/8.

Zentralblatt für Bakteriologie. 1 Abteilung. Originale. Reihe A, medizinische Mikrobiologie, Infektionskrankheiten und Parasitologie. Stuttgart & New York. Vol. 246, 1980. Zentralbl. Bakteriol., 1 Abt., Originale, Reihe A. Preceded by: Zentralblatt für Bakteriologie, Parasitenkunde, Infektionskrankheiten und Hygiene. 1 Abteilung, medizinisch-hygienische Bakteriologie, Virusforschung und tierische Parasitologie. Originale. Reihe A, medizinische Mikrobiologie und Parasitologie. Superseded by: Zentralblatt für Bakteriologie, Mikrobiologie und Hygiene. 1 Abteilung. Originale. Medizinische Mikrobiologie, Infektionskrankheiten und Parasitologie. HI 74726

Zentralblatt für Bakteriologie. 1 Abteilung, Originale. Reihe B, Hygiene, Krankenhaushygiene, Betriebshygiene, präventive Medizin. Stuttgart & New York. Vol. 170, 1980. Zentralbl. Bakteriol., 1 Abt., Originale, Reihe B. Preceded by: Zentralblatt für Bakteriologie, Parasitenkunde, Infektionskrankheiten und Hygiene. 1 Abteilung, Originale. Reihe B, Hygiene, Krankenhaushygiene, Betriebshygiene, präventive Medizin. Superseded by: Zentralblatt für Bakteriologie, Mikrobiologie und Hygiene. Abteilung Originale. B, Hygiene. HI 74727

Zentralblatt für Bakteriologie, Mikrobiologie und Hygiene. Abteilung Originale. B, Hygiene. Stuttgart & New York. Vols. 171-181, 1980-85. Zentralbl.

Bakteriol. Mikrobiol. Hyg., Abt. Originale, B. Preceded by: Zentralblatt für Bakteriologie. 1 Abteilung, Originale. Reihe B, Hygiene, Krankenhaushygiene, Betriebshygiene, präventive Medizin. Superseded by: International journal of microbiology and hygiene. Series B, environmental hygiene, hospital hygiene, industrial hygiene, preventive medicine. HI 74728

Zentralblatt für Bakteriologie, Mikrobiologie und Hygiene. 1 Abteilung. Originale. Medizinische Mikrobiologie, Infektionskrankheiten und Parasitologie. Stuttgart & New York. Vol. 247+, 1980+. Zentralbl. Bakteriol. Mikrobiol. Hyg., 1 Abt., Originale, Med. Mikrobiol. Preceded by: Zentralblatt für Bakteriologie. 1 Abteilung. Originale. Reihe A, medizinische Mikrobiologie, Infektionskrankheiten und Parasitologie. HI 74729

Zentralblatt für Bakteriologie, Mikrobiologie und Hygiene. 1 Abteilung. Originale. Medizinische Mikrobiologie, Infektionskrankheiten, Virologie und Parasitologie = Zentralblatt für Bakteriologie, Mikrobiologie und Hygiene. 1 Abteilung. Originale. Medizinische Mikrobiologie, Infektionskrankheiten und Parasitologie. Stuttgart & New York.

Zentralblatt für Bakteriologie, Mikrobiologie und Hygiene. 1 Abteilung. Originale. Reihe C, allgemeine, angewandte und ökologische Mikrobiologie. Stuttgart & New York. Vols. 2-3, 1981-82. Zentralbl. Bakteriol. Mikrobiol. Hyg., 1 Abt., Originale, Reihe C. Preceded by: Zentralblatt für Bakteriologie, Parasitenkunde, Infektionskrankheiten und Hygiene. 1 Abteilung, medizinisch-hygienische Bakteriologie, Virusforschung und tierische Parasitologie. Originale. Reihe C, allgemeine, angewandte und ökologische Mikrobiologie. Superseded by: Systematic and applied microbiology. HI 74730

Zentralblatt für Bakteriologie, Mikrobiologie und Hygiene. 1 Abteilung. Referate. Medizinische Mikrobiologie, Parasitologie, Hygiene, präventive Medizin. Stuttgart. Vols. 271-286, 1981-84. Zentralbl. Bakteriol. Mikrobiol. Hyg., 1 Abt., Referate. Preceded by: Centralblatt für Bakteriologie, Parasitenkunde und Infektionskrankheiten. 1 Abteilung, medizinisch-hygienische Bakteriologie, Virusforschung und tierische Parasitologie. Referate. Superseded by: International journal of microbiology and hygiene. Abstracts, medical microbiology, virology, parasitology, hygiene, preventive medicine. HI 74731

Zentralblatt für Bakteriologie, Parasitenkunde und Infektionskrankheiten = Centralblatt für Bakteriologie, Parasitenkunde und Infektionskrankheiten. Jena.

Zentralblatt für Bakteriologie, Parasitenkunde und Infektionskrankheiten. Zweite Abteilung. Jena. Zentralbl. Bakteriol., 2. Abt. See B–P–H 998/12. HI 60785

Zentralblatt für Bakteriologie, Parasitenkunde, Infektionskrankheiten und Hygiene. 1 Abteilung, medizinisch-hygienische Bakteriologie, Virusforschung und tierische Parasitologie. Originale. Jena. Vols. 152-217, 1947-71. Zentralbl. Bakteriol. Parasitenk. Infektionskrankh. Hyg., 1 Abt., Originale. Preceded by: Centralblatt für Bakteriologie, Parasitenkunde und Infektionskrankheiten. 1 Abteilung, medizinisch-hygienische Bakteriologie, Virusforschung und tierische Parasitologie. Originale. Superseded by: Zentralblatt für Bakteriologie, Parasitenkunde, Infektionskrankheiten und Hygiene. 1 Abteilung, medizinisch-hygienische Bakteriologie, Virusforschung und tierische Parasitologie. Originale. Reihe A, medizinische Mikrobiologie und Parasitologie i Zentralblatt für Bakteriologie, Parasitenkunde, Infektionskrankheiten und Hygiene. 1 Abteilung, medizinisch-hygienische Bakteriologie, Virusforschung und tierische Parasitologie. Originale. Reihe B, Hygiene, präventive Medizin and Zentralblatt für Bakteriologie, Parasitenkunde, Infektionskrankheiten und Hygiene. Abteilung 1, medizinisch-hygienische Bakteriologie, Virusforschung und tierische Parasitologie. Originale. Reihe C, allgemeine, angewandte und ökologische Mikrobiologie. 5-4626-3. HI 74732

Zentralblatt für Bakteriologie, Parasitenkunde, Infektionskrankheiten und Hygiene. 1 Abteilung, medizinisch-hygienische Bakteriologie, Virusforschung und tierische Parasitologie. Originale. Reihe A, medizinische Mikrobiologie und Parasitologie. Jena, Stuttgart. Vols. 217-245, 1971-79. Zentralbl. Bakteriol. Parasitenk. Infektionskrankh. Hyg., 1 Abt., Originale, Reihe A, Med. Mikrobiol. Parasitol. Preceded by: Zentralblatt für Bakteriologie, Parasitenkunde, Infektionskrankheiten und Hygiene. 1 Abteilung, medizinisch-hygienische Bakteriologie, Virusforschung und tierische Parasitologie. Originale. Superseded by: Zentralblatt für Bakteriologie. 1 Abteilung. Originale. Reihe A, medizinische Mikrobiologie, Infektionskrankheiten und Parasitologie. HI 74733

Zentralblatt für Bakteriologie, Parasitenkunde, Infektionskrankheiten und Hygiene. 1 Abteilung, Originale. Reihe B, Hygiene, Betriebshygiene, präventive Medizin. Stuttgart & New York. Vols. 162-167(3), 1976-78. Zentralbl. Bakteriol. Parasitenk. Infektionskrankh. Hyg., 1 Abt., Originale, Reihe B, Hyg. Betriehshyg. Präv. Med. Preceded by: Zentralblatt für Bakteriologie, Parasitenkunde, Infektionskrankheiten und Hygiene. 1 Abteilung, Originale. Reihe B, Hygiene, präventive Medizin. Superseded by: Zentralblatt für Bakteriologie, Parasitenkunde, Infektionskrankheiten und Hygiene. 1

Abteilung, Originale. Reihe B, Hygiene, Krankenhaushygiene, Betriebshygiene, präventive Medizin. HI 62545

Zentralblatt für Bakteriologie, Parasitenkunde, Infektionskrankheiten und Hygiene. 1 Abteilung, Originale. Reihe B, Hygiene, Krankenhaushygiene, Betriebshygiene, präventive Medizin. Stuttgart & New York. Vols. 167(4)-169(6), 1978-79. Zentralbl. Bakteriol. Parasitenk. Infektionskrankh. Hyg., 1 Abt., Originale, Reihe B, Hyg. Krankenhaushyg. Betriehshyg. Präv. Med. Preceded by: Zentralblatt für Bakteriologie, Parasitenkunde, Infektionskrankheiten und Hygiene. 1 Abteilung, Originale. Reihe B, Hygiene, Betriebshygiene, präventive Medizin. Superseded by: Zentralblatt für Bakteriologie. 1 Abteilung, Originale. Reihe B, Hygiene, Krankenhaushygiene, Betriebshygiene, präventive Medizin. HI 74734

Zentralblatt für Bakteriologie, Parasitenkunde, Infektionskrankheiten und Hygiene. 1 Abteilung, Originale. Reihe B, Hygiene, präventive Medizin. Stuttgart & New York. Vols. 155-161, 1971-76. Zentralbl. Bakteriol. Parasitenk. Infektionskrankh. Hyg., 1 Abt., Originale, Reihe B, Hyg. Präv. Med. Preceded by: Archiv für Hygiene und Bakteriologie and Zentralblatt für Bakteriologie, Parasitenkunde, Infektionskrankheiten und Hygiene. Abteilung 1, medizinisch-hygienische Bakteriologie, Virusforschung und tierische Parasitologie. Originale. Superseded by: Zentralblatt für Bakteriologie, Parasitenkunde, Infektionskrankheiten und Hygiene. 1 Abteilung, Originale. Reihe B, Hygiene. HI 74735

Zentralblatt für Bakteriologie, Parasitenkunde, Infektionskrankheiten und Hygiene. 1 Abteilung, medizinisch-hygienische Bakteriologie, Virusforschung und tierische Parasitologie. Originale. Reihe C, allgemeine, angewandte und ökologische Mikrobiologie. Stuttgart & New York. Vol. 1(1-4), 1980. Zentralbl. Bakteriol. Parasitenk. Infektionskrankh. Hyg., 1 Abt., Originale, Reihe C, Allg. Angew. Ökol. Mikrobiol. Preceded by: Zentralblatt für Bakteriologie, Parasitenkunde, Infektionskrankheiten und Hygiene. 1 Abteilung, medizinisch-hygienische Bakteriologie, Virusforschung und tierische Parasitologie. Originale. Superseded by: Zentralblatt für Bakteriologie, Mikrobiologie und Hygiene. 1 Abteilung. Originale. Reihe C, allgemeine, angewandte und ökologische Mikrobiologie. HI 74736

Zentralblatt für Bakteriologie, Parasitenkunde, Infektionskrankheiten und Hygiene. 2 naturwissenschaftliche Abteilung, allgemeine, landwirtschaftliche und technische Mikrobiologie. Jena. Vols. 107-136, 1952-81. Zentralbl. Bakteriol. Parasitenk. Infektionskrankh. Hyg., 2 Abt., Allg. Landwirtschaftliche Techn. Mikrobiol. Preceded by: Zentralblatt für Bakteriologie, Parasitenkunde und Infektionskrankheiten. 2 Abteilung. Superseded by: Zentralblatt für Mikrobiologie. HI 74737

Zentralblatt für Bakteriologie, Parasitenkunde, Infektionskrankheiten und Hygiene. 2 naturwissenschaftliche Abteilung, Mikrobiologie der Landwirtschaft, der Technologie und des Umweltschutzes = Zentralblatt für Bakteriologie, Parasitenkunde, Infektionskrankheiten und Hygiene. 2 naturwissenschaftliche Abteilung, allgemeine, landwirtschaftliche und technische Mikrobiologie. Jena.

Zentralblatt für Chirurgie = Centralblatt für Chirurgie. Leipzig. Centralbl. Chir. See B–P–H 303/10.

Zentralblatt für den deutschen Erwerbsgartenbau. Vols. ?-59/60, ?-1969. Zentralbl. Deutsch. Erwerbsgartenbau. Superseded by: Deutsche Gartenbauwirtschaft. HI 74738

Zentralblatt für Geologie und Paläontologie. Teil 2. Historische Geologie und Paläontologie. Stuttgart. Zentralbl. Geol. Paläontol., Teil 2, Hist. Geol. Paläontol. See B–P–H 998/16. HI 60786

Zentralblatt für die gesamte Forst- und Holzwirtschaft. Vienna. Zentralbl. Gesamte Forst- Holzw. See B–P–H 998/17. HI 60787

Zentralblatt für die gesamte Forstwesen. Vienna. Zentralbl. Gesamte Forstwesen. See B–P–H 998/18. HI 60788

Zentralblatt für die gesamte Gynäkologie und Geburtshilfe. Berlin. Zentralbl. Gesamte Gyn. Geburtsh. See B–P–H 998/19. HI 60789

Zentralblatt für Haut- und Geschlechtskrankheiten sowie deren Grenzgebiete. Berlin. Zentralbl. Haut-Geschlechtskrankh. Grenzgeb. See B–P–H 998/20. HI 60790

Zentralblatt für Landwirthschaft. Brünn [=Brno, Czechoslovakia]. Zentralbl. Landw. See B–P–H 998/21. HI 60791

Zentralblatt für Mikrobiologie. Jena. Vol. 137+, 1982+. Zentralbl. Mikrobiol. Preceded by: Zentralblatt für Bakteriologie, Parasitenkunde, Infektionskrankheiten und Hygiene. 2 naturwissenschaftliche Abteilung, allgemeine, landwirtschaftliche und technische Mikrobiologie. HI 74739

Zentralblatt für Mineralogie, Geologie und Paläontologie. Teil 3. Historische und regionale Geologie, Paläontologie. Stuttgart. Zentralbl. Mineral, Teil 3, Hist. Regionale Geol. Paläontol. See B–P–H 998/22. HI 60792

Zentralblatt für Mineralogie, Geologie und Paläontologie. Teil 4. Paläontologie. Stuttgart. Zentralbl. Mineral, Teil 4, Hist. Paläontol. See B–P–H 999/1. HI 60793

Zentralblatt für Pharmazie, Pharmakotherapie und Laboratoriumsdiagnostik. Berlin. Vol. 109+, 1970+. Zentralbl. Pharm. Pharmakother. Lab.Diagn. Preceded by: Pharmazeutische Zentralhalle für Deutschland. HI 74740

Zentralblatt für Sammlung und Veröffentlichung von Einzeldiagnosen neuer Pflanzen = Repertorium specierum novarum regni vegetabilis. Berlin. Repert. Spec. Nov. Regni Veg. See B–P–H 772/20.

Zentralblatt für Wasserbau und Wasserwirthschaft. Berlin. Zentralbl. Wasserbau Wasserw. See B–P–H 999/2. HI 60794

Zeszyty naukowe akademii rolniczej w Szczecinie. Rolnictwo. Szczecin. Vol. 38+, 1972+. Zesz. Nauk. Akad. Roln. Szczecinie, Roln. Preceded by: Zeszyty naukowe wyzszej szkoły rolniczej w Szczecinie. HI 74741

Zeszyty naukowe akademii rolniczo-technicznej w Olsztynie. Rolnictwo. Olsztyn. Vols. 1-40?, 1973-84? Zesz. Nauk. Akad. Roln.-Techn. Olsztynie, Roln. Superseded by: Acta academiae agriculturae ac technicae olstenensis. Agricultura. HI 74742

Zeszyty naukowe akademii rolniczej w Warszawie. Ogrodictwo. Warsaw. Vol. 8+, 1974+. Zesz. Nauk. Akad. Roln. Warszawie, Ogrod. Preceded by: Zeszyty naukowe skoły głownej gospodarstwa wiejskiego w Warszawie. Ser. ogrodnictwo. HI 74743

Zeszyty naukowe skoły głownej gospodarstwa wiejskiego w Warszawie. Ser. ogrodnictwo. Warsaw. Vols. 1-7, 1957-72. Zesz. Nauk. Skoły Głown. Gosp. Wiejsk. Warszawie, Ser. Ogrodn. Superseded by: Zeszyty naukowe akademii rolniczej w Warszawie. Ogrodnictwo. HI 74744

Zeszyty naukowe uniwersitet Gdanski wydzialu biologii i nauk o ziemi. Geografia. Gdansk. Vol. 1+, 1970(1971)+. Zesz. Nauk. Uniw. Gdanski Wydz. Biol. Nauk Ziemi, Geogr. HI 74745

Zeszyty naukowe uniwersytet Gdanski, wydziału biologii i nauk o ziemi. Oceanografia. Gdansk. Vol. 1+, 1973+. Zesz. Nauk. Uniw. Gdanski Wydz. Biol. Nauk Ziemi, Oceanogr. HI 74746

Zeszyty naukowe universytetu im. Adama Mickiewicza, biologia. Poznan, Poland. Zesz. Nauk. Univ. Adama Mickiewicza, Biol. See B–P–H 999/5. HI 60795

Zeszyty naukowe uniwersitetu jagiellónskiego; prace botaniczne. Cracow. Vol. 1+, 1973+. Zesz. Nauk. Uniw. Jagiellon. Prace Bot. HI 74747

Zeszyty naukowe uniwersytetu łodziego. Seria 2, nauki matematyczno przyrodnicze. Łódź. Vols. 1-63, 1955-75. Zesz. Nauk. Uniw. Łodziego, Ser. 2. Superseded by: Acta universitatis lodziensis ... Seria 2, nauki matematyczno-przyrodnicze. HI 74748

Zeszyty naukowe uniwersytetu mikołaja Kopernika w Toruniu. Nauki matematyczno-przyrodnicze. Biologia. Torun. Nos. 1-30, 1956-72. Zesz. Nauk. Uniw. Mikołaja Kopernika Toruniu, Nauk Mat.-Przyr. Biol. Superseded by: Acta universitatis Nicolai Copernici. Biologia. HI 74749

Zeszyty naukowe wydzialu biologii i nauk o ziemi. Oceanografia. Gdansk = Zeszyty naukowe uniwersytet Gdanski, wydziału biologii i nauk o ziemi. Oceanografia. Gdansk.

Zeszyty naukowe wyzszej szkoły rolniczej w Szczecinie. Szczecin. Vols. 1-37, 1958-72? Zesz. Nauk. Wyszszej Szkoły Roln. Szczecinie. Superseded by: Zeszyty naukowe akademii rolniczej w Szczecinie. Rolnictwo. HI 74751

Zeszyty przyrodnicze. Opole. Vol. 3+, 1963+. Zesz. Przyr. Preceded by: Kwartalnik opolski. HI 57152

Zhejiang linxueyuan xuebao = Journal of Zhejiang forestry college. Che-chiang Lin-an.

Zhenjun xuebao = Acta mycologica sinica. Beijing.

Zhibing zhishi = Plant disease knowledge. [China]. Pl. Dis. Knowl. See B–P–H 712/1.

Zhiwi yanjiu = Bulletin of botanical research. Harbin.

Zhiwu baohu = Plant protection. Beijing. Pl. Protect. (Peking). See B–P–H 712/20.

Zhiwu baohu xuebao = Acta phytophylacica sinica. Beijing.

Zhiwu bingli xuebao = Acta phytopathologica sinica. Peking (Beijing).

Zhiwu fenlei xuebao = Acta phytotaxonomica sinica. Additamentum. Peking (Beijing).

Zhiwu shengli tongxun = Plant physiology communications. Shanghai.

Zhiwu shengli xuebao = Acta phytophysiologia sinica. Shanghai.

Zhiwu shengtaixue yu dizhixue xuebao = Acta phytoecologia et geobotanica sinica. Beijing.

Zhiwu xuebao = Acta botanica sinica. Peking (Beijing).

Zhiwu zazhi = Plant magazine. Peking (Beijing). and Plants. Peking (Beijing).

Zhiwuxue zazhi = Plant magazine. Peking (Beijing). and Plants. Peking (Beijing).

Zhiwuzazhi = Plant magazine. Peking (Beijing). and Plants. Peking (Beijing).

Zhong cao yao = Chinese traditional and herbal drugs. Shao-yang.

Zhongcaoyao tongxun = Chung tsao yao t'ung hsün. Shao-yang?

Zhongguo caoyuan = Grassland of China. Beijing.

Zhongguo chaye = Tea of China. Hangzhou.

Zhongguo huahui penjing = Chung-kuo hua hui p'en ching. Beijing.

Zhongguo kexue = Scientia sinica. [European edition]. Beijing.

Zhongguo kexue. B (huaxue, shengwuxue, nongxue, yixue, dixue) = Scientia sinica. Series B, Chem. biol. ag. med. earth sci. [European edition; also published in a Chinese edition]. Beijing.

Zhongguo kexue jishu daxue xuebao = Journal of China university of science and technology. Hofei.

Zhongguo kekue shiliao = China historical materials of science and technology. Peking (Beijing).

Zhongguo kexue zengkan = Science & technology.

Zhongguo kexueyuan huanan zhiwu yanjiusuo jikan = Acta botanica austro sinica. Guangzhou.

Zhongguo kexueyuan linye turang yanjiusuo jikan = Bulletin of the institute of forestry and pedology, academia sinica. Shenyang.

Zhongguo kexueyuan yichuan yanjiusuo jikan = Report (Annual), institute of genetics, academia sinica.

Zhongguo linye = Chinese forestry. Peking (Beijing).

Zhongguo linye kexue = Chinese forestry science. Peking (Beijing).

Zhongguo mazuo = Fiber crops in China.

Zhongguo mianhua = China cottons. Honan.

Zhongguo nongye kexue = Chinese agricultural science. Peking (Beijing).

Zhongguo nongye kexue = Scientia agricultura sinica. Beijing.

Zhongguo shengwu huaxue hui = Journal of the chinese biochemical society. Taipei.

Zhongguo yaoli xuebao = Acta pharmacologica sinica. Shanghai.

Zhonghua a linxue iikan = Quarterly journal of chinese forestry. Taipei.

Zhonghua minguo wei shengwu ji mianyixue zazhi = Chinese journal of microbiology and immunology. Taipei.

Zhongshan daxue xuebao. Ziran kexue ban = Acta scientiarum naturalium universitatis sunyatseni. Guangzhou.

Zhongwen baozhang zazhi keji lunwen suoyin = Index to science and technical articles in chinese periodicals and newspapers.

Zhongyang yanjiuyuan = Report (Annual), institute of botany, academia sinica. Nankang, Taipei.

Zhongyao tongbao = Bulletin of chinese materia medica. Peking (Beijing).

Zhongyao tongbao = Chung yao t'ung pao. Peking (Beijing).

Zhongyaocai keji. No. ?-26+, ?-1983+. Zhongyaocai Keji. HI 74752

Zhurnal Bio-botanichnogo tsyklu Vseukraïns'koi akademii nauk. Kiev, Ukrainian S S R. Vols. 1-2 [also numbered vols. 1-8], 1931-33. Zhurn. Bio-Bot. Tsyklu Vseukraïns'k. Akad. Nauk. Superseded by: Zhurnal Instytuto botaniky Vseukraïns'koi akademii nauk. 1-122-1. HI 74753

Zhurnal Bio-botanichnogo tsyklu Vseukrayins'koi akademii nauk = Zhurnal Bio-botanichnogo tsyklu Vseukraïns'koi akademii nauk. Kiev.

Zhurnal "Bolĕzni Rastenii". St. Petersburg. Vols. 1-12, 1907-22. Zhurn. "Bolĕzni Rast." Preceded by: Listok dlya Bor'by s Bolĕznyamy i Povrezhdeniyamy Kul'turnykh i Dikorastushchikh Poleznykh Rastenii. Superseded by: Bolezni rastenii. 1-731-2. HI 74754

Zhurnal ekologii i biotsenologii. Moscow & Leningrad. 1931. Zhurn. Ekol. Biotsenol. Superseded by: Voprosy ekologii i biotsenologii. HI 74755

Zhurnal epidemiologii i mikrobiologii. Moscow. 1932-34. Zhurn. Epidem. Mikrobiol. Superseded by: Zhurnal mikrobiologii, epidemiologii i immunobiologii. HI 74756

Zhurnal evolutsionnoi biokhimii i fiziologii. Moscow. Vol. 1+, 1965+. Zhurn. Evol. Biokhim. Fiziol. HI 74757

Zhurnal Instytuto botaniky Vseukraïns'koi akademii nauk. Kiev, Ukrainian S S R. Vols. 9-31 [also numbered vols. [1]-23], 1934-39. Zhurn. Inst. Bot. Vseukraïns'k. Akad. Nauk. Preceded by: Zhurnal Bio-botanichnogo tsyklu Vseukraïns'koi akademii nauk. Superseded by: Botanichnyi zhurnal. Kiev. 1-122-1. HI 74758

Zhurnal Instytuto botaniky Vseukrayins'koi akademii nauk = Zhurnal Instytuto botaniky Vseukraïns'koi akademii nauk. Kiev.

Zhurnal Mikrobiologii. Petrograd. Vols. 1-4, 1914-19. Zhurn. Mikrobiol. 5-4635-3. HI 74759

Zhurnal mikrobiologii, epidemiologii i immunobiologii. Moscow. 1935+. Zhurn. Mikrobiol. Epidem. Immunobiol. Preceded by: Zhurnal epidemiologii i mikrobiologii and Zhurnal mikrobiologii i immunobiologii. HI 74760

Zhurnal mikrobiologii i immunobiologii. Moscow. 1930-34. Zhurn. Mikrobiol. Immunobiol. Preceded by: Zhurnal mikrobiologii, patologii i infektsionnykh boleznei. Superseded by: Zhurnal mikrobiologii, epidemiologii i immunobiologii. HI 74761

Zhurnal mikrobiologii, patologii i infektsionnykh boleznei. Moscow. 1924-29. Zhurn. Mikrobiol. Patol. Infekts. Bolezni. Superseded by: Zhurnal mikrobiologii i immunobiologii. HI 74762

Zhurnal Moskovskogo otdeleniya Russkogo botanicheskogo obshchestva. Moscow & Petrograd. 1 vol., 1922. Zhurn. Moskovsk. Otd. Russk. Bot. Obshch. 4-3744-3. HI 74763

Zhurnal novocherkasskago otdeleniya russkago botanicheskago obshchestva. Novocherkassk. Vol. 1, 1919. Zhurn. Novocherkassk. Otd. Russk. Bot. Obshch. HI 74764

Zhurnal obshchei biologii. Moscow. Vol. 1+, 1940+. Zhurn. Obshchei Biol. 5-4635-3. HI 74767

Zhurnal Obshchestva Lyubitelei Komnatnykh Rastenii i Akvariumov. St. Petersburg. 1902-08. Zhurn. Obshch. Lyubit. Komnatn. Rast. Preceded by: Zhurnaly Obshchestva Lyubitelei Komnatnykh Rastenii i Akvariumov v S.-Peterburgě. HI 74765

Zhurnaly Obshchestva Lyubitelei Komnatnykh Rastenii i Akvariumov v S.-Peterburgě. St. Petersburg. 1901. Zhurn. Obshch. Lyubit. Komnatn. Rast. S.-Peterburgě. Preceded by: Zhurnaly Sobranii Obshchestva Lyubitelei Komnatnoi Kul'tury Rastenii i Akvariumov v S.-Peterburgě. Superseded by: Zhurnal Obshchestva Lyubitelei Komnatnykh Rastenii i Akvariumov. HI 74766

Zhurnal rezinovoi promyshlennosti. Moscow. Vols. 1-10, 1927-36. Zhurn. Rezinovoi Promyshl. Superseded by: Kauchuk i rezina. 3-2275-1. HI 74768

Zhurnal Rossiiskago Obshchestva Lyubitelei Sadovodstva v Moskvě. Moscow. 1864-66. Zhurn. Rossiisk. Obshch. Lyubit. Sadov. Moskvě. Preceded & superseded by: Zhurnal Sadovodstva, izdavaemyi Rossiiskim Obshchestvom Lyubitelei Sadovodstva v Moskvě. 5-4636-2. HI 74769

Zhurnal Rossiiskago Sadovodstva. St. Petersburg. 1838-40. Zhurn. Rossiisk. Sadov. HI 74770

Zhurnal Russkogo botanicheskogo obshchestva. Moscow. Vols. 14-16, 1930-31. Zhurn. Russk. Bot. Obshch. Preceded by: Zhurnal Russkogo botanicheskogo obshchestva pri Akademii nauk S S S R. Superseded by: Botanicheskii zhurnal S S S R. 1-756-1. HI 74771

Zhurnal Russkago Botanicheskago Obshchestva pri Akademii Nauk. Petrograd. Vols. 2-9, 1917-24. Zhurn. Russk. Bot. Obshch. Akad. Nauk. Preceded by: Zhurnal Russkago Botanicheskago Obshchestva pri Akademii Nauk. Superseded by: Zhurnal Russkogo botanicheskogo obshchestva pri Akademii nauk S S S R. 1-756-1. HI 74772

Zhurnal Russkogo botanicheskogo obshchestva pri Akademii nauk S S S R. Moscow & Leningrad. Vols. 10-13, 1925-28. Zhurn. Russk. Bot. Obshch. Akad. Nauk S.S.S.R. Preceded by: Zhurnal Russkago Botanicheskago Obshchestva pri Akademii Nauk. Superseded by: Zhurnal Russkogo botanicheskogo obshchestva. 1-756-1. HI 74773

Zhurnal Russkago Botanicheskago Obshchestva pri Akademii Nauk. Petrograd. Vol. 1, 1916-17. Zhurn. Russk. Bot. Obshch. Imp. Akad. Nauk. Superseded by: Zhurnal Russkago Botanicheskago Obshchestva pri Akademii Nauk. 1-756-1. HI 74774

Zhurnal Russkago Fiziko-khimicheskago Obshchestva pri Imperatorskom Petrogradskom Universitetě. Chast' khimicheskaya. Petrograd. Vols. 46(5)-48, 1914-17. Zhurn. Russk. Fiz.-Khim. Obshch. Imp. Petrogradsk. Univ., Chast' Khim. Preceded by: Zhurnal Russkago Fiziko-khimicheskago Obshchestva pri Imperatorskom S.-Peterburgskom Universitetě. Chast' khimicheskaya. Superseded by: Zhurnal Russkago Fiziko-khimicheskago Obshchestva pri Petrogradskom Universitetě. Chast' khimicheskaya. 4-3744-3. HI 74775

Zhurnal Russkago Fiziko-khimicheskago Obshchestva pri Imperatorskom S.-Peterburgskom Universitetě. St. Petersburg. Vols. 11-38, 1879-1906. Zhurn. Russk. Fiz.-khim. Obshch. Imp. S.-Peterburgsk. Univ. Preceded by: Zhurnal Russkago khimicheskago i Fizicheskago Obshchestva pri Imperatorskom S.-Peterburgskom Universitetě. Superseded by: Zhurnal Russkago Fiziko-khimicheskago Obshchestva pri Imperatorskom S.-Peterburgskom Universitetě. Chast' khimicheskaya. 4-3744-3. HI 74776

Zhurnal Russkago Fiziko-khimicheskago Obshchestva pri Imperatorskom S.-Peterburgskom Universitetě. Chast' khimicheskaya. St. Petersburg. Vols. 39-46(4), 1907-14. Zhurn. Russk. Fiz.-Khim. Obshch. Imp. S.-Peterburgsk. Univ., Chast' Khim. Preceded by: Zhurnal Russkago Fiziko-khimicheskago Obshchestva pri Imperatorskom S.-Peterburgskom Universitetě. Superseded by: Zhurnal Russkago Fiziko-khimicheskago Obshchestva pri Imperatorskom Petrogradskom Universitetě. Chast' khimicheskaya. 4-3744-3. HI 74777

Zhurnal Russkago Fiziko-khimicheskago Obshchestva pri Leningradskom Universitete. Moscow & Leningrad. Vols. 56-62, 1924-30. Zhurn. Russk. Fiz.-Khim. Obshch. Leningradsk. Univ. Preceded by: Zhurnal Russkago Fiziko-khimicheskago Obshchestva pri Petrogradskom Universitetě. Chast' khimicheskaya. 4-3744-3. HI 74778

Zhurnal Russkago Fiziko-khimicheskago Obshchestva pri Petrogradskom Universitetě. Chast' khimicheskaya. Petrograd. Vols. 49-55, 1917-23. Zhurn. Russk. Fiz.-Khim. Obshch. Petrogradsk. Univ., Chast' Khim. Preceded by: Zhurnal Russkago Fiziko-khimicheskago Obshchestva pri Imperatorskom Petrogradskom Universitetě. Chast' khimicheskaya. Superseded by: Zhurnal Russkago Fiziko-khimicheskago Obshchestva pri Leningradskom Universitete. 4-3744-3. HI 74779

Zhurnal Russkago khimicheskago i Fizicheskago Obshchestva pri Imperatorskom S.-Peterburgskom Universitetě. St. Petersburg. Vols. 5-10, 1873-78. Zhurn. Russk. Khim. Fiz. Obshch. Imp. S.-Peterburgsk. Univ. Preceded by: Zhurnal Russkago khimicheskago Obshchestva. Superseded by: Zhurnal Russkago Fiziko-khimicheskago Obshchestva pri Imperatorskom S.-Peterburgskom Universitetě. 4-3744-3. HI 74780

Zhurnal Russkago khimicheskago Obshchestva. St. Petersburg. Vols. 1-4, 1869-72. Zhurn. Russk. Khim. Obshch. Superseded by: Zhurnal Russkago khimicheskago i Fizicheskago Obshchestva pri Imperatorskom S.-Peterburgskom Universitetě. 4-3744-3. HI 74781

Zhurnal sad i ogorod. Moscow. 1911-16. Zhurn. Sad Ogorod. Preceded & superseded by: Sad i ogorod. 5-3752-3. HI 74782

Zhurnal Sadovodstva, izdavaemyi Rossiiskim Obshchestvom Lyubitelei Sadovodstva v Moskvě. Moscow. 1838-59; n.s. 1861-63; 1874-76 [suspended 1867-73]. Zhurn. Sadov. For 1864-66 see: Zhurnal Rossiiskago Obshchestva Lyubitelei Sadovodstva v Moskvě. Superseded by: Sad i Ogorod. 5-4637-2. HI 74783

Zhurnaly Sobranii Obshchestva Lyubitelei Komnatnoi Kul'tury Rastenii i Akvariumov v S.-Peterburgě. St. Petersburg. 1893-1900. Zhurn. Sobr. Obshch. Lyubit. Komnatn. Kul't. Rast. S.-Peterburgě. Superseded by: Zhurnaly Obshchestva Lyubitelei Komnatnykh Rastenii i Akvariumov v S.-Peterburgě. HI 74784

Zhurnaly zasedanii soveta imperatorskago S.-Peterburgskago universiteta. St. Petersburg. Vols. ?-58-62-? ?-1903-07-? Zhurn. Zased. Soveta Imp. S.-Peterburgsk. Univ. HI 74785

Zhurnaly zasedanii soveta petrovskoi sel'skokhozyaistvennoi akademii. Moscow. 1894. Zhurn. Zased. Soveta Petrovsk. Sel'skokhoz. Akad. Preceded by: Izvestiya petrovskoi sel'skokhoz-yaistvennoi akademii (1889-93). Superseded by: Izvestiya moskovskogo sel'skokhozyaistvennago instituta. 3-2766-1. HI 74786

Zhuzi yanjiu huikan = Journal of bamboo research. Hangzhou.

Zielarski biuletyn informacyjny. Warsaw. Nos. ?-124-183, ?-1969-73. Zielarski Biul. Inform. Superseded by: Wiadomosci zielarskie. HI 74787

Ziemleviedieniie. Moscow = Zemlevedenie. Moscow. Zemlevedenie. See B–P–H 998/4.

Zierpflanzenbau mit internationaler Gartenbautechnik. Vol. ?-15+, ?-1975+. Zierpflanzenbau Int. Gartenbautechnik. HI 74788

Zig-Zag forestière. Bucharest. Zig-Zag Forest. See B–P–H 999/16. HI 60796

Zimbabwe agricultural journal. Salisbury. Vol. 77(3)+, 1980+. Zimbabwe Agric. J. Preceded by: Zimbabwe Rhodesia agricultural journal. HI 74789

Zimbabwe journal of agricultural research. Salisbury. Vol. 18+, 1980+. Zimbabwe J. Agric. Res. Preceded by: Rhodesian journal of agricultural research. HI 74790

Zimbabwe research index. Salisbury. 1979+, 1980+. Zimbabwe Res. Index. Preceded by: Rhodesia research index. HI 74791

Zimbabwe Rhodesia agricultural journal. Salisbury. Vols. 76(4)-77(2), 1979-80. Zimbabwe Rhodesia Agric. J. Preceded by: Rhodesia agricultural journal. Superseded by: Zimbabwe agricultural journal. HI 74792

Zimbabwe Rhodesia science news. Salisbury. Vols. 13((5/6)-14(3), 1979-80. Zimbabwe Rhodesia Sci. News. Preceded by: Rhodesia science news. Superseded by: Zimbabwe science news. HI 74793

Zimbabwe science news; journal of the Zimbabwe scientific association. Salisbury. Vol. 14(4/5)+, 1980+. Zimbabwe Sci. News. Preceded by: Zimbabwe Rhodesia science news. HI 74794

Zion-Bryce museum bulletin. Springdale, UT. Zion-Bryce Mus. Bull. See B–P–H 999/17. HI 60797

Ziraat fakultesi dergisi, Ege universitesi. Bornova. Vol. ?-12+, ?-1975+. Zir. Fak. Derg. Ege Univ. HI 74795

Zirai mücedale araştirma yilliği. Ankara. 1970+. Zir. Müc. Araşt. Yilliği. HI 74796

Ziran zazhi = Nature journal. Beijing.

Ziran ziyuan = Natural resources. [Beijing].

Živa. Prague. Živa. See B–P–H 999/18. HI 60798

Živena. Prague. Živena (Prague). See B–P–H 999/19. HI 60799

Živena. Turocszentmarton, Hungary [=Turciansky Svaty Martin, Czechoslovakia]. Živena (Turocszentmarton). See B–P–H 999/20. HI 60800

Zoe; a biological journal. San Francisco, CA. Zoe. See B–P–H 999/21. HI 60801

Zongguo nongye huaxue hui = Journal of the chinese agricultural chemical society. Taipei.

Zooleo; bulletin de la société botanique et de zoologie congolaise. Léopoldville. Vols. 1-3(1), 1938-40; ser. 3, nos. 1-39-?, 1949-60. Zooleo. For 1940-42 see: Bulletin de la société de botanique et de zoologie congolaise. HI 74797

Zoological and acclimatisation society of Victoria. Melbourne. Zool. Acclim. Soc. Victoria. See

B–P–H 999/22. HI 60802

Zoologicheskii zhurnal. Moscow. Vol. 11+, 1932+. Zool. Zhurn. Preceded by: Russkii zoologicheskii zhurnal [not entered]. HI 56097

Zoon. Uppsala. Vols. 1-6, 1973-76. Zoon. HI 74798

Zootecnia. Nova Odessa. Vol. ?-22(3)+, ?-1984+. Zootecnia. Preceded by: Boletim de zootecnia [not entered]. HI 74799

Zorin jikken hokokusho. Nos. ?-2, ?-1969. Zorin Jikken Hokokusho. Incorporated in: Aomori eirinkyoku gijutsu kaihatsu hokokusho. HI 74800

Zpráva o činnosti Masarykovy akademie práce. Prague. Zpráva Činnosti Masarykovy Akad. Práce. See B–P–H 1000/3. HI 60809

Zpráva o činnosti Masarykovy akademie práce k oslavě. 10, výročí republiky. Prague. Zpráva Činnosti Masarykovy Akad. Práce Oslavě, 10. Výročí Republ. See B–P–H 1000/4. HI 60810

Zpravodaj jihočeských botaniku. Budweis, Czechoslovakia. Sprav. Jihočesk. Bot. See B–P–H 999/24. HI 60803

Zpravodaj sekce mexikofilu. Prague. 1971+. Zprav. Sekce Mexikofilu. HI 74801

Zpravodaj severočeského musea v Liberci. Liberec, Czechoslovakia. Zprav. Severočesk. Mus. v Liberci See B–P–H 999/26. HI 60805

Zpravodaj ústředí zemědělských výzkumníku. Prague. Zprav. Ústředí Zeměd. Výzk. See B–P–H 999/28. HI 60806

Zpravodaj vlastivedného krúžku pri vlastivednom múzeu ve Zvolene. Svolen, Czechoslovakia. Sprav. Vlastiv. Krúžku Vlastiv. Múz. ve Zvolene. See B–P–H 1000/1. HI 60807

Zpravodaj západočeské pobočky československé botanické společnosti v Plzni. Pilsen, Czechoslovakia. Sprav. Západočeské Pobočky Českoslov. Bot. Společn. v Plzni. See B–P–H 1000/2. HI 60808

Zprávy botanické zahrady československé akademie věd v Pruhonichích. Pruhonice, Czechoslovakia. Zprávy Bot. Zahrady Českoslov. Akad. Věd v Pruhonichích. See B–P–H 1000/5. HI 60811

Zprávy botanické zahrady v Plzni. Pilsen, Czechoslovakia. Zprávy Bot. Zahrady v Plzni. See B–P–H 1000/6. HI 60812

Zprávy československé botanické společnosti při československé akademii věd. Prague. Zprávy Českoslov. Bot. Společn. Českoslov. Akad. Věd. See B–P–H 1000/7. HI 60813

Zprávy, československe kaktusarske spolecnosti. Prague. 1946-47. Zprávy Českoslov. Kaktusar. Společn. Preceded and superseded by: Kaktusarske listy. HI 68232

Zprávy československé společnosti pro dejiny věd. a techniky. Prague. Zprávy Českoslov. Společn. Dejiny Věd. See B–P–H 1000/8. HI 60814

Zprávy dendrologické sekce československé botanické společnosti. Prague. Správy Dendrol. Sekce Českoslov. Bot. Společn. See B–P–H 1000/9 HI 60815

Zprávy geografického ústavu Č S A V. Opava. Vol. 1+, 1959?+. Zprávy Geogr. Ústavu Českoslov. Akad. Véd. Preceded by: Zprávy slezského ústavu Č S A V v Opave. Prirodni vedi. HI 56590

Zprávy komise pro dejiny přírodních, lékařských a technických věd. Prague. Zprávy Komis. Dejiny Přír. Věd. See B–P–H 1000/11. HI 60816

Zprávy krajského arboreta. Novy Dvur near Opava, Czechoslovakia. Zprávy Krajsk. Arbor. See B–P–H 1000/12. HI 60817

Zprávy krajského vlastivědného musea v Českých Budějovicích. Budweis, Czechoslovakia. Zprávy Krajsk. Vlastiv. Mus. v Českých Budějovicích. See B–P–H 1000/13. HI 60818

Zprávy krkonošského národního parku. Vrchlabi, Czechoslovakia. Zprávy Krkonšsk. Nár. Parku. See B–P–H 1000/14. HI 60819

Zprávy lesnického výzkumu. Zbraslav, Czechoslovakia. Zprávy Lesn. Výzk. See B–P–H 1000/15. HI 60820

Zprávy moravského zemského výzkumného ústavu zemědělského v Brně. Brünn [=Brno, Czechoslovakia]. Zprávy Morav. Zemsk. Výzk. Ústavu Zeměd. v Brně. See B–P–H 1000/16. HI 60821

Zprávy museí jihočeského kraje. Budweis, Czechoslovakia. Zprávy Mus. Jihočesk. Kraje. See B–P–H 1000/17. HI 60822

Zpravy muzei zapodočeskeho kraje. Priroda. Plzen. Vol. ?-2+, ?-1964+. Zprávy Muz. Zapodočeskeho Kraje, Prir. HI 56545

Zprávy oblastního musa jihovýchodní Moravy. Gottwaldov, Czechoslovakia. Zprávy Obl. Mus. Jihových. Moravy. See B–P–H 1000/18. HI 60823

Zprávy okresního musea v Trutnově. Trutnov, Czechoslovakia. Zprávy Okresn. Mus. v Trutnově. See B–P–H 1001/1. HI 60824

Zprávy okresního musea ve Valašském Meziríčí. Valasske Meziríči, Czechoslovakia. Zprávy Okresn. Mus. ve Valašském Meziríčí. See B–P–H 1001/2. HI 60825

Zprávy okresního musea ve Vsetíně. Vestin, Czechoslovakia. Zprávy Okresn. Mus. ve Vsetině. See B–P–H 1001/3. HI 608206

Zprávy okresního musea ve Vyškově. Vyskov,

Czechoslovakia. Zprávy Okresn. Mus. ve Vyškově. See B–P–H 1001/4. HI 60827

Zprávy památkové péče. Prague. Zprávy Památk. Péče. See B–P–H 1001/5. HI 60828

Zprávy slezského ústavu Č S A V v Opave. Prirodni vedi. Opava. Nos. ?-124B-129B, ?-1963. Zprávy Slez. Ústavu Č.S.A.V. Opave, Prir. Vedi. Superseded by: Zprávy geografického ústavu Č S A V. HI 56546

Zprávy vlastivědného musea v Prostějově. Prostejov, Czechoslovakia. Zprávy Vlastiv. Mus. v Prostějově. See B–P–H 1001/6. HI 60829

Zprávy vlastivědného ústavu v Olomouci. Olomouc, Czechoslovakia. Zprávy Vlastiv. Ústavu v Olomouci. See B–P–H 1001/7. HI 60830

Zprávy východočeského musea. Pardubice, Czechoslovakia. Zprávy Východočesk. Mus. See B–P–H 1001/8. HI 60831

Zprávy výzkumného ústavu lesného hospodárstva. Banska Stiavnica, Czechoslovakia. Zprávy Výzk. Ústavu Lesn. Hospod. See B–P–H 1001/9. HI 60832

Zprávy západoslovenského múzea v Trnave. Trnava, Czechoslovakia. Zprávy Západoslov. Múz. v Trnave. See B–P–H 1001/10. HI 60833

Zprávy o zasedání královské české společnosti nauk v Praze = Sitzungsberichte der königlich böhmischen Gesellschaft der Wissenschaften in Prag. Prague. Sitzungsber. Königl. Böhm. Ges. Wiss. Prag. See B–P–H 840/12.

Zprávy o zasedání královské české společnosti nauk v Praze. Třida mathematicko-přírodovědeckě = Sitzungsberichte der königlichen böhmischen Gesellschaft der Wissenschaften in. Mathematisch-naturwissenschaftliche Classe. Prague. Sitzungsber. Königl. Böhm. Ges. Wiss. Prag., Math.-Naturwiss. Cl. See B–P–H 840/13.

Zprávy zemské hospodářské výzkumné stanice pro pěstováni rostlin. Brünn [=Brno, Czechoslovakia]. Zprávy Zemské Hospod. Výzk. Stanice Pěstováni Rostl. See B–P–H 1001/14. HI 60835

Zprávy zemského výzkumného ústavu hospodařského pro pěstováni rostlin v Brne. Brünn [=Brno, Czechoslovakia]. Zprávy Zemsk. Výzk. Ústavu Hospod. Pěstováni Rostl. v Brně. See B–P–H 1001/13. HI 60834

Züchter; Zeitschrift für theoretische und angewandte Genetik [subtitle varies]. Berlin. Vols. 1-37, 1929-67. Züchter. Superseded by: Theoretical and applied genetics. 5-4644-1. HI 56440

Zugabe zu den Göttingischen Anzeigen von gelehrten Sachen. Göttingen. Zugabe Gött. Anz. Gel. Sachen. See B–P–H 1001/22. HI 60836

Zuowu xuebao = Acta agronomica sinica. Beijing.

Žurnal Bio-botaničnogo cyklu Vseukrajins'koji akademiji nauk = Zhurnal Bio-botanichnogo tsyklu Vseukraïns'koi akademii nauk. Kiev.

Žurnal "Bolězni Rastenij" = Zhurnal "Bolězni Rastenii". St. Petersburg.

Žurnal ekologii i biocenologii = Zhurnal ekologii i biotsenologii. Moscow & Leningrad.

Žurnal Instytuto botaniky Ukrajins'koji akademiji nauk = Zhurnal Instytuto botaniky Vseukraïns'koi akademii nauk. Kiev.

Žurnal Instytuto botaniky Vseukrajins'koji akademiji nauk = Zhurnal Instytuto botaniky Vseukraïns'koi akademii nauk. Kiev.

Žurnal Instytutu botaniky Akademiji nauk Ukrajins'koji R S R = Zhurnal Instytuto botaniky Vseukraïns'koi akademii nauk. Kiev.

Žurnal Mikrobiologii = Zhurnal Mikrobiologii. Petrograd.

Žurnal Moskovskogo otdelenija Russkogo botaničeskogo obščestva = Zhurnal Moskovskogo otdeleniya Russkogo botanicheskogo obshchestva. Moscow & Petrograd.

Žurnal obščej biologii = Zhurnal obshchei biologii. Moscow.

Žurnal Obščestva Ljubitelej Komnatnyh Rastenij i Akvariumov = Zhurnal Obshchestva Lyubitelei Komnatnykh Rastenii i Akvariumov. St. Petersburg.

Žurnal Rossijskago Obščestva Ljubitelej Sadovodstva v Moskvě = Zhurnal Rossiiskago Obshchestva Lyubitelei Sadovodstva v Moskvě. Moscow.

Žurnal Rossijskago Sadovodstva = Zhurnal Rossiiskago Sadovodstva. St. Petersburg.

Žurnal Russkago Botaničeskago Obščestva pri Akademii Nauk = Zhurnal Russkago Botanicheskago Obshchestva pri Akademii Nauk. Petrograd.

Žurnal Russkago Botaničeskago Obščestva pri Akademii Nauk = Zhurnal Russkago Botanicheskago Obshchestva pri Akademii Nauk. Petrograd.

Žurnal Russkago Fiziko-himičeskago Obščestva pri Imperatorskom Petrogradskom Universitetě. Čast' himičeskaja = Zhurnal Russkago Fiziko-khimicheskago Obshchestva pri Imperatorskom Petrogradskom Universitetě. Chast' khimicheskaya. Petrograd.

Žurnal Russkago Fiziko-himičeskago Obščestva pri Imperatorskom S.-Peterburgskom Universitetě = Zhurnal Russkago Fiziko-khimicheskago Obshchestva pri Imperatorskom S.-Peterburgskom Universitetě. St. Petersburg.

Žurnal Russkago Fiziko-himičeskago Obščestva pri Imperatorskom S.-Peterburgskom Universitetě. Čast' himičeskaja = Zhurnal Russkago Fiziko-khimicheskago Obshchestva pri Imperatorskom S.-Peterburgskom Universitetě. Chast' khimicheskaya. St. Petersburg.

Žurnal Russkago Fiziko-himičeskago Obščestva pri Leningradskom Universitete = Zhurnal Russkago Fiziko-khimicheskago Obshchestva pri Leningradskom Universitete. Moscow & Leningrad.

Žurnal Russkago Fiziko-himičeskago Obščestva pri Petrogradskom Universitetě. Čast' himičeskaja = Zhurnal Russkago Fiziko-khimicheskago Obshchestva pri Petrogradskom Universitetě. Chast' khimicheskaya. Petrograd.

Žurnal Russkago Himičeskago i Fizičeskago Obščestva pri Imperatorskom S.-Peterburgskom Universitetě = Zhurnal Russkago khimicheskago i Fizicheskago Obshchestva pri Imperatorskom S.-Peterburgskom Universitetě. St. Petersburg.

Žurnal Russkago Himičeskago Obščestva = Zhurnal Russkago khimicheskago Obshchestva. St. Petersburg.

Žurnal Russkogo botaničeskogo obščestva = Zhurnal Russkogo botanicheskogo obshchestva. Moscow.

Žurnal Russkogo botaničeskogo obščestva pri Akademii nauk = Zhurnal Russkago Botanicheskago Obshchestva pri Akademii Nauk. Petrograd.

Žurnal Russkogo botaničeskogo obščestva pri Akademii nauk S S S R = Zhurnal Russkogo botanicheskogo obshchestva pri Akademii nauk S S S R. Moscow & Leningrad.

Žurnal Russkogo fiziko-himičeskogo obščestva pri Petrogradskom universitete. Čast' himičeskaja = Zhurnal Russkago Fiziko-khimicheskago Obshchestva pri Petrogradskom Universitetě. Chast' khimicheskaya. Petrograd.

Žurnal Sad i Ogorod = Zhurnal Sad i Ogorod. Moscow.

Žurnal Sadovodstva. Moscow = Zhurnal Sadovodstva, izdavaemyi Rossiiskim Obshchestvom Lyubitelei Sadovodstva v Moskvě. Moscow.

Žurnal Sadovodstva, izdavaemyj Rossijskim Obščestvom Ljubitelej Sadovodstva = Zhurnal Sadovodstva, izdavaemyi Rossiiskim Obshchestvom Lyubitelei Sadovodstva v Moskvě. Moscow.

Žurnal Sadovodstva, izdavaemyj Rossijskim Obščestvom Ljubitelej Sadovodstva v Moskvě = Zhurnal Sadovodstva, izdavaemyi Rossiiskim Obshchestvom Lyubitelei Sadovodstva v Moskvě. Moscow.

Žurnaly Obščestva Ljubitelej Komnatnyh Rastenij i Akvariumov v S.-Peterburgě = Zhurnaly Obshchestva Lyubitelei Komnatnykh Rastenii i Akvariumov v S.-Peterburgě. St. Petersburg.

Žurnaly Sobranij Obščestva Ljubitelej Komnatnoj Kul'tury Rastenij i Akvariumov v S.-Peterburgě = Zhurnaly Sobranii Obshchestva Lyubitelei Komnatnoi Kul'tury Rastenii i Akvariumov v S.-Peterburgě. St. Petersburg.

Zustand der Wissenschaften und Künste in Schwaben. Augsburg. Zustand Wiss. Künste Schwaben. See B–P–H 1003/11. HI 60837

Zuverlässige Nachrichten von dem gegenwärtigen Zustande, Veränderungen und Wachsthum der Wissenschaften. Leipzig. Zuverlässige Nachr. Gegenwärt. Zustande WIss. Leipzig. See B–P–H 1003/12. HI 60838

Zygon; journal of religion and science. Chicago, IL. Vol. 1+, 1966+. Zygon. HI 56523

Zyma-journal; bulletin thérapeutique. (Therapeutische Zeitschrift). Nyon. Vol. 1+, 1936+. Zyma-J. HI 56486

Zymologica e chemica dei colloidi. Bologna. Vols. 1-2(1), 1926-27. Zymologica Chem. Colloid. Superseded by: Zymologica e chemica dei colloidi e degli zuccheri. 5-4648-3. HI 56324

Zymologica e chemica dei colloidi e degli zuccheri. Bologna. Vols. 2(2)-5, 1927-30. Zymologica Chem. Colloid. Zucch. Preceded by: Zymologica e chemica dei colloidi. Superseded by: Giornale di biologia applicata alla industria chimica. 5-4648-3. HI 56371

Zymotechnisk tidende; annoncetidende for Norge, Sverige, Finlend og Denmark. No. 1+, 1885+. Zymotechn. Tidende. HI 56279

Zymotechnisk tidsskrift. Copenhagen. 1885-1922? Zymotechn. Tidsskr. HI 56140

Appendix A

Words, Phrases and Abbreviations

Listed here are words and phrases that occur in the titles of periodicals cited in *B-P-H* and *B-P-H/S*, together with the corresponding forms used in the recommended title abbreviations. Hyphens are ignored in the alphabetical arrangement. Geographical nouns are rarely abbreviated (see *B-P-H* Introduction, page 14, for exceptions) and (apart from those exceptions) are omitted from this list. However, geographical adjectives are abbreviated when appropriate, and are included here.

Aalstersche — Aalstersche
Aalwyn — Aalwyn
Aanmoediging — Aanm.
Aanteekeningen — Aanteek.
Aarbog — Aarbog
Aarbok — Aarbok
Aardappel(dagen) — Aardappel(dagen)
Aardappelstudiecentrum — Aardappelstudiecentrum
Aardrijkskundig — Aardrijksk.
Aargauischen — Aargauischen
Aarhundrede — Aarhundr.
Aarsberetning — Aarsberetn.
Aarshefter — Aarsh.
Aastaraamat — Aastar.
Abbildungen — Abbild.
Abeille(s) — Abeille(s)
Aberrant — Aberr.
Abgefasster — Abgefasster
Abhandlungen — Abh.
Abiologica — Abiol.
Abkhazskoi — Abkhaz.
Åboensis — Åbo.
Abrégé, Abreviadas, Abridged — Abr.
Absorption — Absorpt.
Abstract(a,s) — Abstr.
Abstracts of bioanalytic technology — A. B. T.
Abteilung — Abt.
Abtheilung — Abth.
Abwässer — Abwässer
Academia [etc.], Academiai, Academiái, Académiai, Academic(e,i) — Acad.
Academician — Academician
Academie, Académie, Academiei, Academien(s), Académique, Academy — Acad.
Acadian — Acadian
Acadienne — Acadienne
Accademia, Accademiche — Accad.
Access — Access
Accession — Accession
Accessorie(s) — Accessorie(s)
Acción — Acción
Acclimatation, Acclimatisation, Acclimatization — Acclim.
Accomplished, Accomplishment — Accomp.
Accounts — Accounts
Achievements — Achievem.
Acid(s) — Acid(s)
Aci-Reale — Aci-Reale
Ackerbau(es) — Ackerbau(es)
Ackerbaugesellschaft — Ackerbauges.
Ackerbaus — Ackerbaus
Aclimatación — Aclim.
Acqua — Acqua
Acquisitions — Acquis.
Act — Act
Acta(s) — Acta(s)
Actes — Actes
Action — Action
Actividades — Actividades
Activité — Activité
Activities — Activities
Actorum — Actorum
Actualidades, Actualité(es) — Actual.
Acuatica — Acuatica
Açucar — Açucar
Açucareio — Açucareio
Additamentum, Additional, Additive — Addit.
Address — Address
Addzelu — Addz.
Administration, Administrative(s) — Admin.
Admiralty — Admiralty
Adriatica — Adriat.
Adumbratio — Adumbr.
Adunanza — Adunanza
Adunanze — Adunanze
Adundant — Abund.
Advanced — Advanced
Advancement — Advancem.
Advances — Advances
Advancing — Advancing
Advertiser — Advertiser
Advice — Advice
Adviser — Adviser
Advisory — Advis.
Advocate — Advocate
Aerial — Aerial
Aereo — Aereo
Aeronautics — Aeron.
Aerzte — Aerzte
Aethiopicae — Aethiop.
Afdeeling(en), Afdeling, Afdelning — Afd.
Affairs — Affairs
Affiliate — Affil.
Affini — Affini
Afforestation — Afforest.
Afhandlingar, Afhandlinger — Afh.
Africaines — Africaines
Africains — Africains
African — African
Africana — Africana
Africanarum — African.
Africano — Africano
Africans — Africans
Afrika — Afrika
Afrique — Afrique
Agaclari — Agacl.
Age — Age
Agency — Agency
Agenda — Agenda
Agent(s) — Agent(s)
Agerdyrknings-tidende — Agerdyrkn.-Tidende
Agrargeschichte — Agrargesch.
Agraria(n,s), Agrario(s) — Agrar.
Agrárirodalmi — Agrárirod.
Agrarisch — Agrar.
Agrarsoziologie — Agrarsoziol.
Agrártörténeti — Agrártört.
Agrártudományi — Agrártud.
Agrarwissenschaftlichen — Agrarwiss.
Agrégée — Agrég.
Agri — Agri
Agricell — Agricell
Agrichemical — Agrichem.
Agricola, Agrícola(s), Agricole(s), Agricolo, Agricoltura, Agriculteurs, Agricultores, Agricultura [etc.], Agricultural — Agric.
Agricultural and veterinary chemicals — A. V. C.
Agriculturchemie — Agriculturchem.
Agriculture, Agriculturist — Agric.
Agroalimentaires — Agroaliment.
Agrobiologija — Agrobiol.
Agrobotanica, Agrobotánica — Agrobot.
Agrochemicals — Agrochemicals
Agrochemie — Agrochem.
Agrochimie, Agrochimija — Agrochim.
Agroecologia — Agroecol.
Agrofitotechnie — Agrofitotechn.
Agroforestry — Agroforest.
Agrokémia — Agrokém.
Agrologique — Agrol.
Agronomer — Agronomer
Agronomia, Agronomía, Agronómiai, Agronomic, Agronómica(s), Agronomice, Agronomicheskoi,

Agronomické, Agronómico, Agronomicum, Agronomie, Agronomii, Agronomique(s), Agronomy, Agronoomia — Agron.
Agropecuaria(s), Agropecuario — Agropecu.
Agroquímica — Agroquím
Agros — Agros
Agrostologist, Agrostology — Agrostol.
Agrotechnology — Agrotechnol.
Agrotecnia — Agrotecn.
Agrotekhnika — Agrotekhn.
Agrotike — Agrotike
Agrumes — Agrumes
Agrumicoltura — Agrumic.
Ahboroti — Ahbor.
Ährenlese — Ährenlese
Ahrokhimiyi — Ahrokhim.
Aikakuskrija — Aikak.
Ailments — Ailments
Aizsardzibas — Aizsard.
Ajgu — Ajgu
Akadeemia, Akadémia, Akademia(e), Akademiai, Akademiaji, Akademičeskija, Akademicheskiya, Akademie, Akademieju, Akademien, Akademieyu, Akademii(s), Akademiï, Akaděmii, Akademija(i,s), Akademijasining, Akademijasnyn, Akademijasynyn, Akademije, Akademiji, Akademijos, Akademika, Akademisch, Akademiska, Akademiya, Akademj(a,i) — Akad.
Akkerbouw — Akkerb.
Akklimatisation, Akklimatizacii [etc.], Akklimatizatsii [etc.] — Akklim
Aklimatation, Aklimatyzacji — Aklim.
Aktuelle — Aktuelle
Akvaristické, Akvarium, Akvárium — Akvar., Akvár.
Alam — Alam
Albertina — Albertina
Album — Album
Alergologia — Alergol.
Alerta — Alerta
Alfalfa — Alfalfa
Alföldi — Alföldi
Algae [etc.] — Algae
Algemeen, Algemene — Alg.
Algen — Algen
Algérienne — Algér.
Algodón — Algodón
Algodonera — Algodonera
Algological, Algologie, Algologique — Algol.
Aliment(s) — Aliment(s)
Alimentaires, Alimentation, Alimento(s) — Aliment.
Alimurgiche — Alimurgiche
Aliquando — Aliquando
Alive — Alive
Alla — Alla
Államas — Államas
Állami — Állami
Államilag — Államilag
Állapotáról — Állapot
Állattani — Állatt.
Allen — Allen
Aller — Aller
Allergnädigst — Allergnäd.
Allergologica — Allergol.
Allergy — Allergy
Allerneueste — Allerneueste
Allgemeine(n,r,s) — Allg.
Allied — Allied
Allmän — Allmän
Allmanach — Allman.
Allmänna — Allmänna
Allmogen — Allmogen
Állomás — Állomás
Alma-Atinskogo — Alma-Atinsk.
Almanach, Almanachja, Almanack, Almanaque — Alman.
Almeenyttig — Almeennytt.
Almindelig(e) — Almind.
Almond — Almond
Alpenländische — Alpenländ.
Alpenpflanzen — Alpenpfl.
Alpenrosen — Alpenrosen
Alpentiere — Alpentiere
Alpina [etc.] — Alpina
Alpine — Alpine
Alpinismus — Alpinism.
Alpinistico — Alpinist.
Alpino — Alpino
Alsacienne — Alsac.
Alsófehérmegyei — Alsófehérm.
Alt — Alt
Altaiskii — Altaisk.
Altajskij — Altajsk.
Általynos — Általynos
Altenburgischen — Altenburg.
Ältere — Ältere
Altmärkisches — Altmärk.
Alumni — Alumni
Amaranth — Amaranth
Amaryllis — Amaryllis
Amateur(s) — Amateur(s)
Amateurs-Naturalistes — Amateurs-Naturalistes
Amatores — Amatores
Amazonica — Amazon.
Ambas — Ambas
Ambiente — Amb.
Ameliorarea, Amélioration — Amelior., Amélior.
Amenagement — Amenagem.
America(s), América, Américaine, American(a) — Amer., Amér.
American Association for the Advancement of Science — A. A. A. S.
American institute of biological sciences — A. I. B. S.
American society of professional biologists — A. S. P. B.
Americanische, Amérique — Amer., Amér.
Ami(s) — Ami(s)
Amico — Amico
Amigo(s) — Amigo(s)
Amino Acid — Amino Acid
Ämnen — Ämnen
Amoenitates — Amoen.
Amoiensis — Amoiensis
Ampelographische — Ampelogr.
Ampelologiai, Ampélologique — Ampelol., Ampélol.
Amtlicher — Amtl.
Amurskago, Amurskogo — Amursk.
Amusante — Amusante
An — An
Anais — Anais
Anale(s) — Anale(s)
Analecta — Analecta
Analecten — Analecten
Analekten — Analekten
Analele — Analele
Anali — Anali
Analile — Analile
Analisis — Analisis
Analitika — Analitika
Analyse — Analyse
Analysis — Analysis
Analyst(s) — Analyst(s)
Analytical — Analytical
Analytique — Analytique
Anatomical, Anatomie, Anatomija, Anatomique, Anatomische, Anatomist(s), Anatomy — Anat.
Ancient — Ancient
Andelsfolk — Andelsfolk
Andere — And.
Andersonian — Andersonian
Andina — Andina
Andines — Andines
Anexos — Anexos
Anfänger — Anfänger
Angenehmer — Angenehmer
Angewandte — Angew.
Angiosperm — Angiosp.
Anglais(e) — Anglais(e)
Anglian — Anglian
Angolana, Angolensis — Angol.
Angränzenden — Angränzenden
Anhaltische — Anhalt.
Animadversiones — Animadv.
Animal(e,es,i,s) — Anim.
Animation — Animat.
Animaux — Anim.
Anmerkungen — Anmerk.
Annaes, Annale(n,r,s), Annali, Annals — Ann.
Année — Année
Anniversary — Anniv.
Annonces — Annonces
Annotated, Annotationes — Annot.

Announcement — Announc.
Annuaire — Annuaire
Annual(es) — Annual(es)
Annuario — Annuario
Annuel — Annuel
Annuelles — Annuelles
Annus — Ann.
Annuum — Annuum
Anstalt [etc.] — Anst.
Antarctic(a) — Antarc.
Antarkticheskogo — Antarktich.
Antartico, Antartigues — Antart.
Ante — Ante
Anthropologica(l), Anthropologie, Anthropology — Anthropol.
Antibacterial — Antibact.
Antibiotics — Antibiot.
Antica — Antica
Antifungal — Antifungal
Antiguo — Antiguo
Antimicrobic — Antimicrob.
Antioqueña — Antioqueña
Antiquarian, Antiquaries, Antiquary — Antiq.
Antique — Antique
Antiquities — Antiquities
Antiquity — Antiquity
Antiseptic — Antiseptic
Antiviral — Antiviral
Antologia — Antologia
Antropologi, Antropologia, Antropología, Antropológicas, Antropológicos — Antropol.
Antwerpsch — Antwerpsch
Anual(es) — Anual(es)
Anuar — Anuar
Anuario, Anuário — Anuario
Anuarul — Anuarul
Anului — Anului
Anzeige(n,r) — Anz.
Apícola — Apícola
Apicole(s) — Apicole(s)
Apicultural, Apiculture — Apic.
Aplicada(s) — Aplicada(s)
Aplicatá — Apl.
Aplicazioni — Applic.
Apotheker — Apotheker
Apothekerkunde — Apothekerk.
Apothekerkunst — Apothekerkunst
Apothekervereins — Apothekervereins
Apôtre — Apôtre
Apparatus, Appareil — Appar.
Appeal — Appeal
Appelle — Appelle
Appendices, Appendix(es) — Append.
Apple — Apple
Applegrower — Applegrower
Applicada, Applicata — Appl.
Application(s) — Applic.
Applied, Appliquée(s) — Appl.
Appraisal — Appraisal
Approved — Approv.
Apuntes — Apuntes
Aqademijasb — Aqad.
Aqua(e) — Aqua(e)
Aquaculture — Aquac.
Aquariculture — Aquaric.
Aquarien, Aquarist, Aquarium — Aquar.
Aquaterra — Aquaterra
Aquatic(a) — Aquatic(a)
Aque — Aque
Aquicolas, Aquicultura — Aquic.
Arab(ian) — Arab
Arable — Arable
Aragonesa — Aragonesa
Araştirma — Araşt.
Arbeider — Arbeider
Arbeidsrapport — Arbeidsrapp.
Arbeit(en) — Arbeit(en)
Arbeitsgemeinschaft — Arbeitsgem.
Arbeitsgruppe — Arbeitsgr.
Arbeitskreis(es) — Arbeitskreis(es)
Arbeitsmaterial — Arbeitsmater.
Arbejder — Arbejder
Arbeten — Arbeten
Arbetsutskatt — Arbetsutsk.
Årbok — Årbok
Arbol — Arbol
Arbor(ea) — Arbor(ea)
Arboreta, Arboretum(s), Arboricales — Arbor.
Arboricoltura, Arboriculture — Arboric.
Arborist's — Arborist's
Arbre(s) — Arbre(s)
Arcadico — Arcadico
Arcana — Arcana
Archaeological — Archaeol.
Archaeologist(s) — Archaeologist(s)
Archaeology — Archaeol.
Archeion — Archeion
Archéologie, Archéologique — Archéol.
Archipelago — Archipel.
Architect(s) — Architect(s)
Architecture — Architect.
Archiv(es), Archivio — Arch.
Archivist — Archivist
Archivni, Archivo, Archivum, Archiwum — Arch.
Arctic(a) — Arctic(a)
Area — Area
Arealy — Arealy
Argentina, Argentino — Argent.
Argumenta — Argum.
Arhiiv, Arhiv(a) — Arh.
Aria — Aria
Arid — Arid
Aridas — Aridas
Aridus — Aridus
Arkhiv(a) — Arkh.
Arkif, Arkiv — Ark.
Arkticheskogo — Arktich.
Arktische — Arktische
Arm — Arm
Armazenamento — Armazen.
Armchair — Armchair
Armées — Armées
Armenian — Armen.
Armjanskij [etc.] — Armjansk.
Armoricain — Armoricain
Armyanskii [etc.] — Armyansk.
Armyanskoi S S R — Armyanskoi S.S.R.
Aromatic, Aromatičeskih [etc.], Aromaticheskikh [etc.]. — Aromat.
Around — Around
Arquivos — Arq.
Arrondissement — Arrondissement
Arroz — Arroz
Arrozeira — Arroz.
Ars — Arš
Årsberättelse(r) — Årsberätt.
Årsberetning — Årsberetn.
Årsbok — Årsbok
Arsenal — Arsenal
Årshefter — Årsh.
Årsmelding — Årsmeld.
Årsskrift, Års-Skrift — Årsskr., Års-Skr.
Art(s) — Art(s)
Arte(s) — Arte(s)
Arti(s) — Arti(s)
Artibus — Art.
Article(s) — Article(s)
Artillera — Artillera
Artist, Artistica, Artistico — Artist
Artium — Art.
Aruanded — Aruanded
Arxivs — Arxivs
Arzneikunde, Arzneikundige — Arzneik.
Arzneikunst — Arzneikunst
Arzneiwissenschaft — Arzneiwiss.
Arzneygelahrtheit — Arzneygelahrth.
Arzneykunde, Arzneykundige — Arzneyk.
Arzneykunst — Arzneykunst
Arzneywissenschaft — Arzneywiss.
Ärzte — Ärzte
Ärztliche — Ärztl.
Äsärlari — Äsärl.
Asean — Asean
Asian — Asian
Asiatic(a), Asiatique — Asiat.
Asociación — Asoc.
Asociacion interamericana de bibliotecarios y documentalistas agricolas — A. I. B. D. A.
Asociatiei — Asoc.
Aspas — Aspas
Aspects — Aspects
Aspiranti, Aspirantov — Aspir.
Assemblies — Assembl.
Assessment — Assessm.
Associação, Associacii, Associate(d) — Assoc.
Association(s), Associazione, Associé — Assoc.
Association canadienne-française pour l'avancement des sciences — A. C. F.

A. S.
Association of Southeastern Biologists — A. S. B.
Association of special libraries and information bureaux. — Aslib
Assotsiatsii — Assots.
Assoziation — Assoz.
Astrakhanskaya — Astrakh.
Astrobotaniki — Astrobot.
Astronomique, Astronomy — Astron.
Asturianos — Asturianos
Astúrica — Astúrica
Asuntos — Asuntos
Atas — Atas
Ateneo — Ateneo
Athenaeum — Athenaeum
Athénée — Athénée
Atheniensis — Athen.
Atlantic(a) — Atlantic(a)
Atlántico — Atlántico
Atlantide — Atlantide
Atlas(es) — Atlas(es)
Atoll — Atoll
Atomic — Atomic
Atti — Atti
Atualidades — Atual.
Atvinnudeild — Atvinnud.
Audiovisual — Audiovis.
Aufsätze — Aufsätze
Augenheilkunde — Augenheilk.
Augstskolas — Augstsk.
Augu — Augu
Aujourd'hui — Aujourd'hui
Aukstuju — Aukst.
Aus — Aus
Ausbau — Ausbau
Ausbildung — Ausbild.
Ausbreitung — Ausbreit.
Auserlesene(r) — Auserlesene(r)
Ausführungen — Ausführ.
Ausgabe — Ausgabe
Ausgrabungen — Ausgrabungen
Auskunftsblatt — Auskunftsbl.
Ausland(es), Ausländischen — Ausl.
Auslandskunde — Auslandsk.
Ausschusses — Ausschusses
Austral — Austral
Australasian — Australas.
Australian, Australiense — Austral.
Austro- — Austro-
Auswahl — Auswahl
Auszug — Auszug
Auszüge(n) — Auszüge(n)
Author — Author
Authority — Author.
Autonome, Autonomia, Autónomo — Auton., Autón.
Autori — Aut.
Autres — Autres
Auxiliares — Auxiliares
Avancement — Avancem.
Avdelning(en) — Avd.
Avenir — Avenir
Avhandlinger — Avh.
Avicolas, Avicole, Avicoltura, Aviculture — Avic.
Avocado — Avocado
Avtonomnoi S S R — Avton. S.S.R.
Avtonomnoj [etc.] — Avton.
Avulsa(s) — Avulsa(s)
Avulsos — Avulsos
Avvenire — Avvenire
Award — Award
Awareness — Aware.
Ayri — Ayri
Ayurveda — Ayurveda
Azerbaidzhanskii [etc.] — Azerbaidzhansk.
Azerbaidzhanskoi S S R — Azerbaidzhnsk. S.S.R.
Azerbajdžana — Azerbajdžana
Azerbajdžanskij [etc.] — Azerbajdžansk.
Aziatskoj — Aziatsk.
Azoricae — Azor.
Azovo-Černomorskogo — Azovo-Černomorsk.
Azovo-Chernomorskogo — Azovo-Chernomorsk.
Azucarera — Azucarera
Babylonian — Babylonian
Back — Back
Bacteria(l) — Bact.
Bacteriological, Bacteriológico, Bactériologie — Bacteriol.
Bacteriologist — Bacteriologist
Bacteriology — Bacteriol.
Bactériophage(s) — Bactérioph.
Badania — Badan.
Badawcze(go) — Badawcze(go)
Badawczy — Badawczy
Badger — Badger
Badischen — Bad.
Baer — Baer
Bags — Bags
Baierische(n) — Baier.
Bajkal'skoj — Bajkal'sk
Bakanlĭgi — Bakanlĭgi
Bakhchevye — Bakhch.
Bakinskogo — Bakinsk.
Bakterial'nykh — Bakt.
Bakteriologie — Bakteriol.
Bakteriologo-agronomičeskoj — Bakteriol.-Agron
Balai — Balai
Balatonicum — Balaton.
Balcanica — Balcan.
Bălgarsko(to) — Bălg.
Balneologie — Balneol.
Balotnaya — Balot.
Balotne — Balotn.
Balthici, Baltiiskogo, Baltische — Balt.
Bamboo — Bamboo
Banana — Banana
Banano — Banano
Banater — Banater
Bank(o,s) — Bank(o,s)
Barátja — Barátja
Barcinonensis — Barcinon.
Bare — Bare
Bark — Bark
Barki — Barki
Barley — Barley
Bascongada — Bascongada
Basic — Basic
Basin — Basin
Basket — Basket
Baster — Basler
Bataafsche — Bataafsche
Bataviaasch, Batavischen — Batav.
Batteriologia — Batteriol.
Batumskii [etc.], Batumskij [etc.] — Batumsk.
Bauern-Zeitung — Bauern-Zeitung
Baumschule — Baumschule
Bavarica — Bavar.
Bay — Bay
Bayerische(n,r,s) — Bayer.
Bayrische — Bayr.
Bazy — Bazy
Bean — Bean
Beauftragter — Beauftragter
Beautiful — Beautiful
Beaux — Beaux
Bed — Bed
Bedeutender — Bedeut.
Bee(s) — Bee(s)
Beekeeper(s) — Beekeeper(s)
Beekeeping — Beekeeping
Beet — Beet
Beförderung — Beförd.
Begrünungen — Begrün.
Behavioral — Behav.
Behouve — Behouve
Behoven — Behoven
Beiblatt [etc.] — Beibl.
Beiden — Beiden
Beiheft [etc.] — Beih.
Beilage — Beil.
Beilageband — Beilageband
Beiträge — Beitr.
Bekanntmachung — Bekanntm.
Békés — Békés
Bekiendtgiörelse, Bekiendtgiørelse — Bekiendtg.
Belaruskai S S R — Belarusk. S.S.R.
Belaruskai [etc.], Belaruskaj [etc.] — Belarusk.
Belarussischen — Belaruss.
Belehrende — Belehrende
Belehrung — Belehr.
Belge(s) — Belge(s)
Belgica, Belgique(s) [adj.], Belgische — Belg.
Belle arti — Belle Arti
Belles-lettres — Belles-Lettres
Belomorskoi, Belomorskoj — Belomorsk.

Belorusskoi S S R — Belorussk. S.S.R.
Belorusskoi [etc.], Belorusskoj [etc.] — Belorussk.
Belt — Belt
Belustigung — Belust.
Bemerkungen — Bemerk.
Bemühungen — Bemüh.
Benthology — Benthol.
Beobachtung(en) — Beob.
Berättelse — Berätt.
Bereich — Bereich
Beretning — Beretn.
Bergbau — Bergbau
Bergcultures — Bergcultures
Bergiani — Berg.
Bergmännisches — Bergmänn.
Bergstreken — Bergstreken
Berichte — Ber.
Berita — Berita
Berliner — Berliner
Berlinisches — Berlin.
Bernerisches — Berner.
Bernische(s) — Bern.
Berolinensia [etc.] — Berol.
Berry — Berry
Besar — Besar
Beschäftigungen — Beschäft
Beschrijvende, Beschrijving — Beschr.
Besoekisch — Besoekisch
Besorgte — Besorgte
Bespozvonochnykh — Bespozv.
Bessarabskago, Bessarabskogo — Bessarabsk.
Besseres — Besseres
Bestehen — Bestehen
Besten — Besten
Bestimmung — Bestimm.
Bestrijding — Bestrijd.
Bestudering — Bestud.
Beter — Beter
Bética — Bética
Better — Better
Betterave(s) — Betterave(s)
Betteravier(es) — Betterav.
Betuwsche — Betuwsche
Bevordering — Bevord.
Bewaring — Bewar.
Beyträge — Beytr.
Bezirk(en,es,s) — Bez.
Bezirks-Naturkundemuseums — Bez.-Naturkundemus.
Bezirksbeauftragter — Bezirksbeauftragter
Bezirksheimatmuseum — Bezirksheimatmus.
Bezirksstellen — Bezirksstellen
Bibelgesellschaft(en) — Bibelges.
Biblical — Biblical
Bibliografi(a), Bibliografía, Bibliograficheskie, Bibliografico, Bibliográfico, Bibliografie, Bibliografiya, Bibliographical, Bibliographique, Bibliography — Bibliogr.
Biblioteca — Bibliot.
Bibliotecologia — Bibliotecol.
Bibliotek(a,i) — Bibliot.
Bibliotheca, Bibliothek(a,et) — Biblioth.
Bibliothekswesen — Bibliotheksw.
Bibliothèque — Biblioth.
Bidrag — Bidrag
Biedribas — Biedribas
Bienengesellschaft — Bienenges
Bienenkunde — Bienenk.
Biennale, Biennial — Bienn.
Bifidobacteria — Bifidobact.
Bifidus — Bifidus
Bihang — Bih.
Bijalagičnyh — Bijal.
Bijblad — Bijbl.
Bijdragen — Bijdr.
Bijzondere — Bijzondere
Bild(ende) — Bild(ende)
Bildungs-Anstalt — Bildungs-Anst.
Bilja — Bilja
Biljeske — Biljeske
Biljka — Biljka
Biljna — Biljna
Bilten — Bilten
Bimestriel(le) — Bimestr.
Bimonthly, Bi-Monthly — Bimonthly, Bi-Monthly
Bioactive — Bioactive
Biobibliografii — Biobibliogr.
Bio-Botaničnogo — Bio-Bat.
Bio-botanique — Bio-Bot.
Biocenologii, Biocenology — Biocenol.
Biochemica(l), Biochemie, Biochemii, Biochemische, Biochemistry — Biochem.
Biochimica, Biochimie — Biochim.
Biociências — Bioci.
Bioclimatology — Bioclimatol.
Biocommerce — Biocommerce
Biocommunication — Biocommun.
Biocosme — Biocosme
Biodeterioration — Biodeterior.
Bioelectronics — Bioelectronics
Bioenergetics — Bioenerget.
Bio-energy — Bio-energy
Bioengineering — Bioengin.
Bioethics — Bioethics
Biofeedback — Biofeedb.
Biofizika — Biofiz.
Biogeochemical — Biogeochem.
Biogeográfica, Biogeográfico, Biogeografii, Biogeographical, Biogeographie, Biogéographie — Biogeogr., Biogéogr.
Biogeokhimicheskoi — Biogeokhim.
Biogeotsenologicheskii — Biogeotsenol.
Biographical — Biogr.
Biohimii — Biohim.
Biohistorisch — Biohist.
Biokhimicheskie [etc.], Biokhimichnyi, Biokhimii, Biokhimiya — Biokhim.
Bioklimatische — Bioklimat.
Biolëgichnaga — Biol.
Biologa, Biologe(n), Biologi, Biologia, Biológia, Biología, Biologiai, Biologic(a,ae,al,as), Biologic(o,um), Biologičeskij [etc.], Biologiche, Biologicheskii [etc.], Biologichnii [etc.], Biologické, Biologický, Biologických, Biológico(s) — Biol.
Biologico-Oceanografia — Biol.-Oceanogr.
Biologiczne, Biologie, Biologii, Biologija, Biologijo, Biologique(s), Biologische(e,en,es), Biologiske, Biologist(s) — Biol.
Biologo-Geografičeskogo — Biol.-Geogr.
Biologo-Gruntovogo — Biol.-Gruntov.
Biologo-Počvennji [etc.] — Biol.Počv.
Biologo-pochvennii [etc.] — Biol.-Pochv.
Biologorum, Biology, Biologyznych, Biolohichnoyi, Biološk(e,i,ih,o,og) — Biol.
Biome — Biome
Biomedica(l), Bio-medical — Biomed., Bio-Med.
Biomembranes — Biomembr.
Biometeorology — Biometeorol.
Biometrical, Biométrie, Biometrische — Biometr., Biométr.
Biomolecular — Biomolec.
Bioniki — Bioniki
Bionomics — Bionom.
Bio-nouvelles — Bio-Nouv.
Biontologie — Biontol.
Bioorganic, Bioorganicheskii — Bioorg.
Biophysica(l), Biophysics, Biophysik, Biophysique — Biophys.
Bioprocess — Bioprocess
Bioprocessing — Bioprocess.
Bioquímica — Bioquim.
Bioresearch — Biores.
Bioscience(s) — Biosci.
Biosfery — Biosfery
Biosintez — Biosintez
Biosource(s) — Biosource(s)
Biospéologie, Biospéologiques — Biospéol.
Biosphere — Biosphere
Biostatisko — Biostat.
Biosystematics, Biosystematist — Biosyst.
Biota — Biota
Biotech — Biotech
Biotechnological, Biotechnologie(s), Biotechnology — Biotechnol.
Biotehniske — Bioteh.
Biotekhnologiya — Biotekhnol.
Bioteknisk — Biotekn.
Biotheoretica — Biotheor.
Bioticos — Bioticos

Biotics — Biotics
Biotrop — Biotrop
Biotsenologii — Biotsenol.
Biowissenschaften, Biowissenschaftliche — Biowiss.
Bisannuel — Bisannuel
Bitki — Bitki
Bitkileni — Bitkil.
Biuletyn — Biul.
Biyalagichnykh — Biyal.
Biyoloji — Biyol.
Bizottságának — Bizott.
Bjuleten — Bjul.
Bjulleten′ [etc.] — Bjull.
Bjuro — Bjuro
Blad — Blad
Blandade — Blandade
Blatt [etc.] — Blatt
Blister — Blister
Bližajšemu — Bližajš.
Bloembollencultuur — Bloembollencult.
Bloembollenhandelaren — Bloembollenhandelaren
Bloembollenonderzoek — Bloembollenonderz.
Bloembollentalt — Bloembollentalt
Bloembollenteelt — Bloembollenteelt
Bloemhof — Bloemhof
Bloemisterij — Bloemist.
Blühende — Blüh.
Blumen — Blumen
Blumenbau — Blumenbau
Blumenfreunde — Blumenfr.
Blumenkunde — Blumenk.
Blumenliebhaber — Blumenliebh.
Blumen-Reiche — Blumen-Reiche
Blumenzeitung — Blumenzeitung
Blumisterei — Blumisterei
Blüten — Blüten
Bnakan — Bnakan
Board — Board
Bochumer — Bochum.
Bodendenkmalpflege — Bodendenkmalpflege
Bodenforschung — Bodenf.
Bodenfruchtbarkeit — Bodenfruchtb.
Bodenkultur — Bodenkult.
Bodenkunde — Bodenk.
Body — Body
Bogoriense(s), Bogoriensis — Bogor.
Bohemica — Bohem.
Bohemoslovaca — Bohemoslov.
Böhmerwaldbundes — Böhmerwaldbundes
Böhmische(n) — Böhm.
Boicus — Boicus
Bois — Bois
Bolěznyakh, Bolěznyamy — Bolězny.
Boletim, Boletín, Boletinos — Bol.
Boleznej — Boleznej
Bolězni, Bolěznjah, Bolěznjamy — Bolězni
Bolgarskoi — Bolg.
Boliviana, Boliviano — Boliv.
Bollettino — Boll.
Bolotnoi, Bolotnoj — Bolotn.
Bolševskoj — Bolševsk.
Bolshevskoi — Bolshevsk.
Bolyaiana — Bolyaiana
Bon — Bon
Bond — Bond
Bondestand — Bondest.
Bonis — Bonis
Bonner — Bonner
Bononiensi — Bononiensi
Bononiensis — Bononiensis
Bonsai — Bonsai
Book(s) — Book(s)
Booklet — Booklet
Booklist — Booklist
Booklore — Booklore
Bookman — Bookman
Boomkweekerij — Boomkweekerij
Boomkweker(ij) — Boomkweker(ij)
Boomsoorten — Boomsoorten
Boomteeltkunde — Boomteeltk.
Boor — Boor
Bor — Bor
Borászati — Borász.
Bor′ba — Bor′ba
Bor′bě — Bor′bě
Borbonica — Borbon.
Bor′by — Bor′by
Border — Border
Borealis — Boreal.
Borodinskoi — Borodinskoi
Boron — Boron
Börsenhalle — Börsenhalle
Bortermésztéset — Borterm.
Bosbouwproefstation — Bosbouwproefstat.
Bosbouwvereninging — Bosbouwver.
Boschbouwcultures — Boschbouwcult.
Boschbouwkundig — Boschbouwk.
Boschbouwproefstation — Boschbouwproefstat.
Boschbouwschool — Boschbouwsch.
Boschbouw-Tijdschrift — Boschbouw-Tijdschr.
Boschi — Boschi
Boschkultuur — Boschkult.
Boschwezen — Boschw.
Bosque(s) — Bosque(s)
Boswezen — Bosw.
Botaanilised — Bot.
Botani, Botanic, Botanica(e,l), Botánica(s), Botânica, Botanice, Botaničeskij [etc.], Botanicheskiii [etc.], Botanichnyi [etc.], Botanici, Botanické, Botaničkog, Botaničnyj [etc.], Botanico(rum) — Bot.
Botanico-géologique — Bot.-Géol.
Botanico-medical — Bot.-Med.
Botanicum, Botanicus, Botanicze, Botanicznego, Botanika(i), Botanike(r), Botaniki, Botanikk, Botanikis, Botanikos, Botanikov, Botanique(s), Botanisch(e,en,er,es) — Bot.
Botanisch-phaenologische — Bot.-Phaenol.
Botanisch-Zoologischen — Bot.-Zool.
Botanisk(a), Botanist(e), Botany — Bot.
Bouriate — Bour.
Boxwood — Boxwood
Brackishwater — Brackishwater
Bramborářského — Bramborářsk.
Branch — Branch
Brandenburgischen — Brandenburg.
Brasileira, Brasilerio, Brasiliensis — Brasil.
Brasserie — Brass.
Braunschweigischen [etc.] — Brunschweig.
Brazilian — Brazil.
Breeders, Breeding — Breed.
Bremer — Bremer
Bremisches — Bremisches
Bresciano — Bresciano
Breton(ne) — Breton(ne)
Breviora — Brevia
Brewer — Brewer
Brewing — Brew.
Brianskogo — Briansk.
Briefe — Briefe
Briefs — Briefs
Britain, Britannique, British — Brit.
Brněnske — Brněnske
Bromatologia, Bromatologico — Bromatol.
Bromeliad — Bromeliad
Bromeliana — Bromeliana
Broom — Broom
Broshyar — Brosh.
Broteriana — Brot.
Brunensis — Brun.
Bryologica(l), Bryologici, Bryologique, Bryologische, Bryologist, Bryology — Bryol.
Bryophytorum — Bryophyt.
Buch — Buch
Büchern — Büchern
Buckskin — Buckskin
Bucurestiensis — Bucurest.
Budapestinensis — Budapest.
Budissina — Budissina
Budování — Budov.
Bugd — Bugd
Builder — Builder
Bulb — Bulb
Buletin, Buletinul — Bul.
Bulgare, Bulgarian, Bulgarska(ta), Bulgarskoto — Bulg.
Bulletin(s), Bullettino — Bull.
Bulteni, Bulteno — Bult.
Bundesanstalt — Bundesanst.
Bundesforschungsanstalt —

Bundesforschungsanst.
Bundesversuchsanstalt — Bundesversuchsanst.
Bündig — Bündig
Bureau(x) — Bur.
Burgenländische — Burgenl.
Bürger — Bürger
Bürgerlichen — Bürgerl.
Busabanakan — Busaban.
Bush — Bush
Business — Business
Butlletí — Butl.
Bydraes — Bydr.
Byggeteknisk — Byggetekn.
Byggforskningsinstitutt — Byggforskn.Inst.
Byulleten′ [etc.] — Byull.
Byuro — Byuro
Byvšego — Byvš.
Cabinet — Cab.
Cacao — Cacao
Cacau — Cacau
Cactaceas — Cact.
Cacticulture — Cacticult.
Cactus(sen) — Cact.
Cadernos — Cad.
Cadrele — Cadr.
Caesareae [etc.] — Caes.
Caesareo-leopoldinae — Caes.-Leop.
Café, Caféeira — Café
Cafetalera — Cafetalera
Cafetera — Cafetera
Cahiers — Cah.
Calavo — Calavo
Caledonian — Caledonian
Caledonienne — Caledonienne
Calendar, Calendrier — Calend.
California — Calif.
Calismalari — Calism.
Cámara — Cámara
Camellia(s) — Camellia(s)
Camellian — Camellian
Cameralwissenschaft — Cameralwiss.
Camerounaises — Cameroun.
Cana — Cana
Canadenses — Canadenses
Canadensis — Canadensis
Canadian, Canadien(ne) — Canad.
Canariensis — Canar.
Cancer — Cancer
Cane — Cane
Canegrower — Canegrower
Canopus — Canopus
Canton — Cant.
Cantonale — Canton.
Caoutchouc — Caoutchouc
Capabilities — Capabil.
Capacitacion — Capac.
Caraïbisch — Caraïb.
Carbohydrate(s) — Carbohyd.
Cardiovascular — Cariovasc.
Care — Care
Cargo — Cargo
Carnation — Carnation
Carnet — Carnet
Carnivorous — Carniv.
Carolinae, Carolinische — Carol.
Carpatho — Carpatho
Carskite — Carsk.
Carsologica — Carsol.
Carta — Carta
Carte(s) — Carte(s)
Cartilla — Cartilla
Cartographie — Cartogr.
Caryosystematique — Caryosyst.
Casas — Casas
Cashew — Cashew
Casket — Casket
Časopis — Čas.
Čast′ — Čast′
Castellana — Castellana
Časti — Časti
Castnomu — Castn.
Catalana — Catalana
Catalans — Catalans
Catalog, Catalógo, Catalogue — Cat.
Catalysis — Catal.
Catalyst — Catalyst
Catedra — Catedra
Catholic — Catholic
Catholique — Catholique
Catolica — Catolica
Caucasian, Caucasien(ne) — Caucas.
Causeries — Causeries
Cearense — Cear.
Čechica — Čech.
Čechoslovaca, Čechoslovakisch, Čechoslovenicae — Čechoslov.
Cecidologia — Cecidol.
Celebration, Celébration — Celebr., Celébr.
Cell(s) — Cell(s)
Cellulaire, Cellular — Cell.
Cellulosa — Cellulosa
Cellulose — Cellulose
Čelovek — Čelovek
Celular — Cel.
Celulosy — Celulosy
Cenni — Cenni
Centenario, Centennial — Centen.
Center — Center
Centraalblad — Centraalbl.
Centraalbureau — Centraalbur.
Centrafricaines — Centrafr.
Central — Centr.
Centralanstalt(en) — Centralanst.
Centralbibliotek — Centralbibl.
Centralblatt — Centralbl.
Centrale — Centr.
Central-Moor-Versuchsstation — Centr.-Moor-Versuchsstat.
Central′nae — Centr.
Centralnija — Centr.
Central′no-Kavkazskogo — Centr.-Kavkazsk.
Central′noj [etc.] — Centr.
Centre — Centre
Centre de co-opération pour les recherches scientifiques relatif au tabac — C. O. R. E. S. T. A.
Centre technique interprofessionnel des oléagineux metropolitains — C. E. T. I. O. M.
Centro — Centro
Centrum — Centrum
Cercetari, Cercetarui — Cercet.
Cercle — Cercle
Cereal(e,s) — Cereal(s)
Cerealicoltura — Cerealicol.
Cerealist — Cerealist
Černomorskogo — Černomorsk.
Černonogo — Červon.
Cerrado — Cerrado
Cesareo — Ces.
Česka — Česka
České(ho) — České(ho)
Českobrodska — Českobrodska
Československa, Československé(ho), Československý(ch) — Českoslov.
Český(ch) — Český(ch)
Cetvertičnogo — Cetvert.
Chácaras — Chácaras
Chaire — Chaire
Chairman — Chairm.
Châlonnais — Châlonnais
Chamber(s) — Chamber(s)
Chambre — Chambre
Champignoncultur — Champignoncultur
Champignons — Champignons
Champs — Champs
Change — Change
Chapingo — Chapingo
Chapters — Chapt.
Characteristic — Charact.
Characteristik — Charac.
Chast′ — Chast′
Chastnomu — Chastn.
Chauds — Chauds
Chelovek(a) — Chelovek(a)
Chemi(e), Chemical(s), Chemische(s), Chemist(ry,s) — Chem.
Chemotherapy — Chemotherapy
Chercheurs — Cherch.
Chérifien — Chérifien
Chernomorskogo — Chernomorsk.
Chernozemoi — Chernoz.
Chernyshevskogo — Chernyshevskogo
Chetvertichnogo — Chetvert.
Chief — Chief
Chikuhoana — Chikuho.
Children — Children
Chilean — Chil.
Chilena — Chilena
Chileno — Chileno
Chimia, Chimica, Chimico — Chim.
Chimico-agrarico — Chim.-Agrar.

Chimie, Chimique(s) — Chim.
Chinese, Chinesische, Chinois(e,es) — Chin.
Chirurgia, Chirurgica(e), Chirurgici, Chirurgie, Chirurgischen, Chirurgisk — Chir.
Chitinskago — Chitinsk.
Chmelařské, Chmelařský — Chmel.
Choix — Choix
Choroby — Choroby
Chorologica — Chorol.
Christian — Christian
Chromatographic, Chromatography — Chromatogr.
Chromosoma — Chromosoma
Chromosome — Chromosome
Chronica, Chronicle, Chronika, Chronique — Chron.
Chrońmy — Chrońmy
Chronobiology — Chronobiol.
Chrudimsky — Chrud.
Chrysanthème — Chrysanthème
Chrysanthemiste — Chrysanthemiste
Chrysanthemum — Chrysanthemum
Chteniya — Chteniya
Churfürstlich — Churfürstl.
Churlande — Churl.
Churpfälzischen — Churpfälz.
Chursächsischen — Chursächs.
Chymisch(e) — Chym.
Ciba — Ciba
Cider — Cider
Ciel — Ciel
Ciencia(s), Ciência, Cienciès, Científica(s), Cientifice, Cientifico(s), Científico — Ci.
Cimento — Cimento
Cinchona — Cinchona
Činnosti — Činnosti
Circadian — Circad.
Circle — Circle
Circolare, Circulaire, Circular — Circ.
Circuli — Circuli
Circulo — Circulo
Cirugia, Cirugía — Cirugia
Cirujia — Cirujia
Cirurgia — Cirurgia
Císaře — Císaře
Cistula — Cistula
Citation — Citat.
Cities — Cities
Čitinskago — Čitinsk.
Citrícola — Citríc.
Citriculture — Citricult.
Citrograph — Citrogr.
Citrologica — Citrol.
Citrus — Citrus
City — City
Civica — Civica
Civico — Civico
Civil(i) — Civil(i)
Cladistics — Cladist.
Class(e) — Cl.
Classification — Classific.
Classique, Classischen — Class.
Claudiopolitanae — Claudiopol.
Climatologia, Climatological, Climatology — Climatol.
Clinic(a,al), Cliniques — Clin.
Clippings — Clippings
Clock(s) — Clock(s)
Cloning — Cloning
Clove — Clove
Clover — Clover
Club(s) — Club(s)
Coast(al) — Coast(al)
Coco(a,nut) — Coco(a,nut)
Coffee — Coffee
Cognitarum — Cogn.
Cold — Cold
Colecção, Colecçion, Colecçionistas — Colecç.
Colegio — Colegio
Coleopterists' — Coleopterists'
Coletanea — Colet.
Collaborative — Collab.
Collana — Collana
Collateral — Collat.
Collecção — Collecç.
Collectanea, Collected, Collecting, Collectio, Collection(s), Collector — Collect.
College — Coll.
Collegit — Colleg.
Collezione — Collez.
Colloid — Colloid
Colloques, Colloquis, Colloquium — Colloq.
Colombiana, Colombiano — Colomb.
Colonais, Cólonia, Colonial, Coloniale(s), Coloniaux, Colonies — Colon.
Colonisation — Colonis.
Colonización — Coloniz.
Colony — Colony
Coloratae — Color.
Coloured — Coloured
Coltivazione — Coltiv.
Columbian(a) — Columb.
Column — Column
Combate — Combate
Combined — Comb.
Combustibles — Combust.
Comenianae — Comen.
Comerciales — Comerciales
Comércio — Comércio
Comice — Comice
Comisión — Comis.
Comissão — Comiss.
Comitato — Comitato
Comité — Comité
Comizio — Comiz.
Commemoration — Commem.
Commentaries, Commentarii, Commentario, Commentarj — Comment.
Commentationes — Commentat.
Comments — Comments
Commerce — Commerce
Commercial(e) — Commercial
Commercio — Commercio
Commercium — Commercium
Commissão, Commissie — Commiss.
Commission on Undergraduate Education in Biological Sciences — C. U. E. B. S.
Commission(er) — Commiss.
Committee — Committee
Common — Common
Commonwealth — Commonw.
Communication(es,s), Communicavit — Commun.
Communique — Communique
Community — Community
Comoara — Comoara
Compact — Compact
Compañía — Co.
Companion — Companion
Company — Co.
Comparata, Comparative — Comp.
Compendium — Compend.
Complément(aire) — Complém.
Complet — Compl.
Complutense [etc.] — Complut.
Compost — Compost
Compostelanos — Compostelanos
Comprehensive — Compreh.
Compte(s) Rendu(s) — Compt. Rend.
Computer — Computer
Comunicaciones, Comunicações — Comun.
Comunicado — Comunicado
Comunicări(le) — Comun.
Condition — Condition
Conference — Conf.
Congolaise — Congol.
Congrès, Congress(o) — Congr.
Conifer — Conifer
Connaissance(s) — Connaissance(s)
Conoscenza — Conosc.
Conseil, Consejo, Conselho — Cons.
Conservação, Conservacion — Conserv.
Conservancy — Conservancy
Conservation(ist) — Conservation(ist)
Conservatoire, Conservator(ium) — Conserv.
Consiglio — Cons.
Consortium — Consort.
Consultant — Consultant
Consultatif — Consultatif
Consultatio — Consultatio
Consulting — Consulting
Consumer — Consumer
Contacter — Contacter
Contaminant(s), Contamination — Contam.
Contemporaine, Contemporánea, Contemporânea, Contemporary —

Contemp.
Contentes — Contentes
Contents — Contents
Continental(e,es) — Continental(e,es)
Continuation, Continuazione, Continued — Cont.
Contribuciones, Contribuições, Contributi, Contribuțiiuni, Contribution(es,s) — Contr.
Contributor — Contributor
Contribuzioni — Contr.
Control — Control
Convegno, Convention — Conv.
Conversationsblatt — Conversationsbl.
Cooperation, Cooperative, Co-operative, Cooperator — Coop., Co-op.
Coordinate, Coordinating, Co-ordinating — Coord., Co-ord.
Coral — Coral
Corn — Corn
Corny — Corny
Corporación, Corporated, Corporation — Corp.
Corps — Corps
Correio — Correio
Correspondence, Correspondent(en,es,ie) — Corresp.
Correspondentieblad — Correspondentiebl.
Correspondentiesocieteit — Correspondentiesoc.
Correspondenzblatt — Correspondenzbl.
Corses — Corses
Cortica — Cort.
Cosmetic(i,s) — Cosmet.
Cosmochimica — Cosmochim.
Costarricense — Costarric.
Côtière — Côt.
Coton — Coton
Cotonnière — Cotonn.
Cottage — Cottage
Cotton — Cotton
Council(s) — Council(s)
Counselor — Counselor
Counties — Counties
Countries — Countries
Country — Country
Country-side — Country-side
Countryman — Countryman
Countryside — Countryside
County — County
Coup-d'oeil — Coup-d'oeil
Courant — Courant
Courier — Courier
Couronnes — Couronnes
Courrier — Courrier
Cours — Cours
Course(s) — Course(s)
Cracoviensia — Cracov.
Craft — Craft
Cranberries — Cranberries
Cranberry — Cranberry
Cream — Cream
Creation, Création — Creation, Création
Cría — Cría
Criação — Criação
Criador — Criador
Criteria — Criter.
Critica(e,l), Criticism, Critische — Crit.
Crittogamica, Crittogamico, Crittogamologica — Crittog.
Croatica — Croat.
Croisières — Crois.
Crónica — Crón.
Cronichetta — Cronich.
Crop(s) — Crop(s)
Croquis — Croq.
Crown — Crown
Crude — Crude
Cruiser — Cruiser
Cryptogam(ae,ic,ica,ici,ico,ie) — Cryptog.
Crystallographic — Crystallogr.
Csoportjának — Csoport.
Čtenija — Čtenija
Cuadernos — Cuad.
Cuban(a), Cubano — Cub.
Cukrovarnické — Cukrovarn.
Cultivadas — Cult.
Cultivador(es) — Cultivador(es)
Cultivar — Cultivar
Cultivari, Cultivate(d) — Cult.
Cultivateur — Cultivateur
Cultivation — Cultivation
Cultivator('s) — Cultivator('s)
Cultur(al,e,es,ist), Cultura(s) — Cult.
Cultuur — Cultuur
Cultuurgewassen — Cultuurgew.
Cultuurtuin — Cultuurtuin
Cumulative — Cumulative
Curator(s) — Curator(s)
Cure — Cure
Curiosa(s), Curiosorum, Curiosum — Cur.
Current — Curr.
Currents — Currents
Cursul — Curs.
Cutaneous — Cutan.
Cybele — Cybele
Cycle(s) — Cycle(s)
Cyklu — Cyklu
Cynnosci — Cynn.
Cytobiologica — Cytobiol.
Cytochemica(e) — Cytochem.
Cytochemistry — Cytochem.
Cytogenetics — Cytogen.
Cytogénétique — Cytogén.
Cytologie, Cytology — Cytol.
Cytophysiologie — Cytophysiol.
Cytotaxonomic(al) — Cytotax.
Czechoslovakian — Czech.
Dabas — Dabas
Dacvis — Dacvis
Daffodil — Daffodil
Dag — Dag
Dagestana — Dagestana
Dagestanskogo — Dagestansk.
Dahlia — Dahlia
Dahlie(n) — Dahlie(n)
Daigako — Daigako
Daigaku — Daigaku
Dairy — Dairy
Dal'ne-Vostočnaja — Dal'ne-Vost.
Dal'nego — Dal'nego
Dal'nevostočnyj [etc.], Dal'nevostochnoi [etc.] — Dal'nevost
Dal'nem — Dal'nem
Danica — Danica
Dänische(n,s) — Dän.
Danish — Danish
Dansk(e) — Dansk(e)
Darbai — Darb.
Dari — Dari
Darstellung — Darstellung
Darza — Darza
Das'ledčae, Das'ledčaj, Das'ledchae, Das'ledchai, Das'ledchaga — Das'l.
Dasonomica — Dasonomica
Data — Data
Date — Date
Daunatorilor — Daunatorilor
Dauphinoise — Dauphin.
Daylily — Daylily
Dějin — Dějiny
Debellação — Debell.
Debreceniensis — Debrecen.
Debricina — Debricina
Decade, Décade — Decade, Décade
Deciduous — Decid.
Découverte(s) — Découv.
Dedicada, Dedication — Dedic.
Deductive — Deduct.
Deep-Sea — Deep-Sea
Defensa — Defensa
Défense — Défense
Defesa — Defesa
Deforestation — Deforest.
Deistvie — Deistvie
Dějatel'nosti — Dějatel'n.
Dejini — Dejini
Dějiny — Dějiny
Dekadenblatt — Dekadenbl.
Dekorativnoe — Dekorat.
Dél — Dél
Dela — Dela
Delectus — Delect.
Deli — Deli
Déliberations — Délib.
Delo — Delo
Delphinium(s) — Delphinium(s)
Deltion — Deltion
Delu, Dělu — Delu,Dělu
Demographia — Demogr.
Demonstrations — Demonst.
Dendrologi(i), Dendrologian, Dendrologica(l), Dendrologické,

Dendrologický, Dendrologicznego, Dendrologiczny, Dendrologija, Dendrologische, Dendrologisk(a) — Dendrol.
Dendrologist — Dendrologist
Dendrologiya, Dendrology — Dendrol.
Denkende — Denkende
Denkmalpflege — Denkmalpflege
Denkschriften — Denkschr.
Denkwürdigkeiten — Denkwürdigk.
Departamento, Département, Départementale, Departementet — Dep., Dép.
Department(al,s) — Dept.
Dépendences — Dépend
Derev'ev — Derev.
Dergisi — Derg.
Derivado, Derivato — Deriv.
Dérivé — Dérivé
Dermatologica, Dermatologie, Dermatologische(n), Dermatology — Dermatol.
Deržavnyj [etc.] — Deržavn.
Derzhavnyi — Derzhavn.
Desarrollo — Desarr.
Descripsión, Description(s), Descriptives — Descr.
Desenvolvimento — Desenvolv.
Desert(s) — Desert(s)
Design — Design
Deutsch(e) — Deutsch(e)
Deutsch-Dominikanischen — Deutsch-Dominikan.
Deutsche Demokratische Republik — D.D.R.
Deutscher [etc.] — Deutsch.
Deutschfreiburger — Deutschfreiburger
Deutschland — Deutschl.
Deutschösterreichische — Deutschösterr.
Developing — Developing
Development — Developm.
Developmental — Developmental
Developpement — Developpem.
Dhauner — Dhauner
Dia — Dia
Diagnoses, Diagnostica, Diagnostik — Diagn.
Dialogen — Dialogen
Diatomologica — Diatomol.
Dictionnaire — Dict.
Didactic(a,as,e) — Didact.
Didatica, Didatico — Didat.
Didžiojo — Didžiojo
Dienst(e,es) — Dienst(e,es)
Difesa — Difesa
Differentiation — Different.
Diffusion — Diffusion
Difusión — Difusión
Digest(s) — Digest(s)
Digestive — Digestive
Dikorastuščih, Dikorastushchikh — Dikorast.
Dimension — Dimension
Dinantais — Dinantais
Dingen — Dingen
Dipartmento — Dip.
Diputacion — Diput.
Dirección — Dirección
Directeur — Directeur
Direction, Directiva — Direct.
Director(s,s') — Director(s,s')
Directorate — Director.
Directory — Directory
Discours — Discours
Discoveries, Discovery — Disc.
Disease (s) — Dis.
Disputation — Disp.
Dissertation(s), Dissertazioni — Diss.
Distillerie — Distill.
Distretti, Distretto — Distr.
Distribution(es,s), Distributorum, Distribuzione — Distrib.
District(us) — Distr.
Divers — Divers
Diversas — Diversas
Diverses — Diverses
Divisão — Divisão
Division(al) — Div.
Divulgação, Divulgaças, Divulgación — Divulg.
Djela — Djela
Dneprovskoi, Dneprovskoj — Dneprovsk.
Dnevnik — Dnevn.
Dniprjans'koji — Dniprjans'k.
Dniprovs'koji — Dniprovs'k
Dnipryans'koi — Dnipryans'k.
Docencia — Docen.
Doctorado, Doctoral — Doct.
Document(a,os,s), Documentaire — Doc.
Documentalist — Documentalist
Documentatie — Doc.
Documentatieblad — Documentatieblad
Documentation(s) — Doc.
Dodatek — Dodatek
Dohányujság — Dohányujs.
Dokladlari, Dokladov, Doklady — Dokl.
Dokumentacija, Dokumentácio, Dokumentacki, Dokumentation — Dokument.
Domácí — Domácí
Domaine(s) — Dom.
Domani — Domani
Domaniales — Doman.
Domestic(a) — Domest.
Dominicana — Dominic.
Dominikanischen — Dominikan.
Dominion — Domin.
Domova — Domova
Donskago — Donsk.
Dopovidi — Dopov.
Dorpatensis — Dorpatensis
Dorpater — Dorpater
Dortmundisches — Dortmund.
Dosjahnennia — Dosjahn.
Dosvidnoji — Dosvidn.
Doświadczalny — Doświadcz.
Down — Down
Dragon — Dragon
Dresdnisches — Dresdnisches
Drevárske, Drevarského, Drevarski — Drev.
Drevesiny — Drevesiny
Drevnikh — Drevn.
Driemandelijke — Driemaand.
Drivesny — Drivesny
Drontheimischen — Drontheim.
Drug — Drug
Druggist(s) — Druggist(s)
Drugic — Drugic
Društva — Društva
Družestv, Družestva, Družestvo — Druž.
Druzhestvo — Druzh.
Drveta — Drveta
Drvna — Drvna
Dryland — Dryland
Države — Države
Državnog(a) — Državn.
Due — Due
Duinonderzoek — Duinonderz.
Duisburgischen — Duisburg.
Dunántúli — Dunántúli
Dune — Dune
Düngung — Düngung
Düngungsversuche — Düngungsversuche
Durchforschung — Durchforsch.
Dyers — Dyers
Dynamics, Dynamique — Dynam.
Dzerzhnaunaga — Dzerzhn.
Dział — Dział
Dziejów — Dziejów
Dziennik — Dziennik
Dzjarzaŭnaj — Dzjarz.
Dzyarzheaŭnai — Dzyarzh.
Earth — Earth
East(ern) — E.
Eau(x) — Eau(x)
Éburnéennes — Éburn.
Échange — Échange
Écho — Écho
Éclairée — Éclairée
Éclaireur — Éclaireur
Eclectic, Eclettica — Ecl.
Eclogae — Eclog.
École — École
Ecologae, Ecologia, Ecological, Écologie, Ecologique — Ecol., Écol.
Ecologist — Ecologist
Ecology — Ecol.
Economia, Economía, Economic(s), Económica, Economiche, Económico, Économie, Économique(s), Economist, Economista, Economy — Econ.
Ecophysiologie — Ecophys.
Ecosystem(s) — Ecosyst.
Ecotoxicology — Ecotoxicol.
Ecotrópica — Ecotróp.

Ecuatoria(no), Ecuatoriales — Ecuat.
Eczacilik — Eczac.
Edafología — Edafol.
Edice — Edice
Editie, Editions, Édition(s) — Ed., Éd.
Editor(s) — Edit.
Editrice — Editr.
Educacion, Education(al) — Educ.
Ee — Ee
Eerste — Eerste
Eesti — Eesti
Effects — Effects
Effectues — Effect.
Effemeridi — Effem.
Efirno-maslichnykh — Efirno-masl.
Ego — Ego
Egri — Egri
Egyesület(e,enek) — Egyes.
Egyetem(ek) — Egyet.
Egylet(i) — Egyl.
Egyptian, Égyptien(ne) — Egypt.
Eidgenössische — Eidgenöss.
Eidike — Eidike
Eigenthümlichen — Eigenthüml.
Einschliesslich — Einschl.
Einträchtigen — Eintr.
Einwohner — Einwohner
Eizeitalter — Eizeitalter
Ekdose — Ekd.
Ekologia, Ekologiczne, Ěkologii, Ekologiju, Ekologiya — Ekol., Ěkol.
Ekomodel — Ekomodel
Ěkskursii — Ěkskurs.
Ěkspedicii, Ěkspedicionnyh, Ekspeditsii, Ěkspeditsii, Ěkspeditsionnykh — Ěksped.
Eksperimental'naja, Eksperimentalno — Eksper.
Elain-ja — Elain-ja
Electoralis — Elect.
Electron — Electron
Elelmezésugyi — Elelmezésugyi
Element — Elem.
Élémentaire(s) — Élément.
Elementaren, Elementary — Element.
Élet — Élet
Élevage — Élevage
Eliömaailmaa — Eliöm.
Ellenikes — Ellen.
Éloges — Éloges
Elskere — Elsk.
Elvetica — Elvet.
Ember — Ember
Embetsmän — Embetsmän
Embryologiae, Embryology — Embryol.
Emerald — Emerald
Emigrados — Emigrados
Emirates — Emirates
Emission — Emiss.
Emlékbeszédek — Emlékbeszédek
Empereur — Emp.
Empire — Empire
Emporium — Empor.
Empresa — Empresa
Émulation — Émul.
Enciclopedico — Encicl.
Encouragement — Encour.
Encyclopädisches, Encyclopaedisches, Encyclopedia, Encyclopédie, Encyclopédique(s) — Encycl.
Encyklopädischer — Encykl.
Endangered — Endang.
Endocrinologia — Endocrinol.
Energy — Energy
Enfants — Enfants
Enfermedade(s), Enfermidade(s) — Enferm.
Engineering, Engineers — Engin.
Engländer, Englische, English — Engl.
Enhancement — Enhancem.
Ennemi — Ennemi
Enologia, Enology — Enol.
Enquirer — Enquirer
Ensayo — Ensayo
Enseñanza — Enseñ.
Ensiegnement — Enseignem.
Enstitüsü — Enst.
Entdeckungen — Entdeck.
Entertaining — Entertaining
Enthusiast — Enthus.
Entomologia, Entomológica, Entomologica(l), Entomologie, Entomologique(s), Entomologisk, Entomologist(s,s'), Entomologo, Entomology — Entomol.
Entretion — Entret.
Entries — Entries
Entwicklungsgeschichte — Entwicklungsgescg.
Entwicklungslehre — Entwicklungsl.
Environment(al) — Environm.
Enzyme(n) — Enzyme(n)
Enzymology — Enzymol.
Ephemeriden, Ephemerides — Ephem.
Epidemiologii — Epidem.
Épiphyties — Épiphyt.
Epistemonika — Epistemonika
Equatorial — Equator.
Erbario, Erbarului — Erb.
Erbländern — Erbländern
Erboristeria — Erborist.
Erdbeschreibung — Erdbeschreib.
Erde — Erde
Érdekeink — Érdek.
Erdélyi — Erdélyi
Erdész, Erdészet(i) — Erdész
Erdészettudomanyi — Erdészettud.
Erdkunde — Erdk.
Erdögazdasag — Erdögazd.
Erdömérnoki — Erdömérn
Erdwissenschaft(liche) — Erdwiss.
Eredmenyei — Eredm.
Ereunon — Ereunon
Erfahrungen — Erfahr.
Erfassung — Erfassung
Erfelijkheid — Erfelijkheid
Erfindungen — Erfind.
Erfolgreicher — Erfolgr.
Erforschung — Erforsch.
Erfurtische — Erfurt.
Ergänzungsband — Ergänzungsband
Ergänzungshefte — Ergänzungsh.
Ergebnisse — Ergebn.
Ergötzlichkeiten — Ergötzlichk.
Erlanger — Erlanger
Erlangische — Erlangische
Ernährung — Ernähr.
Erneuerte — Erneuerte
Eröffnungsrede — Eröffnungsrede
Erste(n) — Erste(n)
Értekezések — Értek.
Értesitó, Értesitö, Értesitöje — Értes.
Erudita, Eruditorum — Erud.
Erweiterung — Erweit.
Erwerbsgartenbau — Erwerbsgartenbau
Erwerbsgärtner — Erwerbsgärtn.
Erzherzogthum — Erzherzogh.
Escola, Escuela — Esc.
Esercitazioni — Esercit.
Eserleri — Eserleri
Esitelmiä — Esitelmiä
Eşlere — Eşlere
Esméretek — Esm.
Español(a), Españoles — Esp.
Especial(es) — Espec.
Esperienzi — Esper.
Espinas — Espinas
Espirit — Espirit
Essais — Essais
Essay(s) — Essay(s)
Essence — Essence
Essential — Essential
Essenze — Essenze
Est — Est
Esta — Esta
Establicimiento — Establ.
Estação, Estaçión — Estaç.
Estadística(s), Estadístico — Estadíst.
Estado — Estado
Estates — Estates
Estestvenno- — Estestv.-
Estestvenno-Istoričeskij [etc.] — Estestv.-Istorič.
Estestvenno-Naučnogo — Estestv.-Naučn.
Estestvennye [etc.] — Estestv.
Estestvoispytatelei, Estestvoispytatelej — Estestvoisp.
Estestvoznanija, Estestvoznzniya — Estestv.
Esthnischen — Esthn.
Estonian — Estonian
Estonica [etc.] — Estonica(e)
Estonicarum — Estonicarum
Ěstonskoi S S R — Ěstonsk. S.S.R.
Ěstonskoj — Ěstonsk.

Estranjera — Estranjera
Estratégicas — Estratég.
Estuarine — Estuarine
Estudiantes, Estudios, Estudis — Estud.
Estudo(s) — Estudo(s)
Esuela — Esc.
Etaireias — Etair.
État — État
Ethiopian — Ethiop.
Ethnobiology — Ethnobiol.
Ethnobotanique — Ethnobot.
Ethnographica, Ethnographisches — Ethnogr.
Ethnological, Ethnologie — Ethnol.
Ethnozoologie — Ethnozool.
Etnôgrafía — Etnôgr.
Etnologia — Etnol
Ĕtomologicheskogo — Ĕtomol.
Étranger(s) — Étranger(s)
Étrangère(s) — Étrangère(s)
Etrusca — Etrusca
Étude(s) — Étude(s)
Etwelche — Etwelche
Eucalypt — Eucalypt
Europa, Europäische, Europe(an), Européenne(s) — Eur.
Evaluación, Evaluation — Eval.
Evangelischen — Evang.
Evening — Eve.
Events — Events
Evgenike — Evgen.
Évi — Évi
Évkönye, Évkönyve(i) — Évk.
Ĕvoljucionnoj — Ĕvol.
Evolucion — Evol.
Evolution(ary), Evolutionis — Evol.
Evolutionsforschung — Evolutionsforsch.
Evolyutsia [etc.], Evolyutsionnoi — Evol.
Exactas, Exactes — Exact.
Excellence — Excellence
Excerpta — Excerpta
Exchange — Exch.
Excursion(en,s), Excursioniştilor — Excurs.
Exhibitions — Exhib.
Exkursionen — Exkurs.
Exotic, Exotique — Exot.
Expedition — Exped.
Experiment station — Exp. Sta.
Experiment(al,ale,alis), Experiment(s), Experimentação, Experimentación, Experimentaţie, Experimentation, Experimentell(en), Expérimentelle — Exp.
Expert — Expert
Exploitation — Exploit.
Exploration — Explor.
Explore — Expl.
Explorer — Explor.
Exsiccata, Exsiccatorum — Exsicc.
Extension — Extens.
Extérieure — Extér.
Extracts, Extraits — Extr.
Extranjera — Extran.
Extraordinaire — Extraoid.
Ežegodnik, Ezhegodnik — Ežeg., Ezheg.
Fabrik(en) — Fabrik(en)
Fächer — Fächer
Fachgruppe — Fachgr.
Facoltà — Fac.
Fact(s) — Fact(s)
Factory — Fact.
Faculdade, Facultad(e), Facultas, Facultatii, Facultătii, Facultatis, Facultătis, Facultato, Faculté, Faculteit, Faculty — Fac.
Faellesudvalget — Faellesudv.
Fagne — Fagne
Faipar(i) — Faip.
Fair — Fair
Faiskola — Faisk.
Fajtakiserletek — Fajtakiserl.
Fakulta, Fakultät, Fakultesi, Fakultet, Fakul'tet, Fakulteta, Fakul'teta, Fakultete, Fakul'tete, Fakulteto, Fakultetu, Fakultou, Fakulty — Fak.
Falusi — Falusi
Fama — Fama
Familienkunde — Familienk.
Family — Family
Fan — Fan
Fancier(s) — Fancier(s)
Fanerogâmicos — Fanerog.
Far — Far
Farm(s) — Farm(s)
Farmaceutica, Farmacéutica, Farmaceutícký, Farmacéutico, Farmacêutico, Farmaceutisk, Farmaceutů, Farmaceutyczne, Farmaci, Farmacia, Farmácia — Farm.
Farmacognosia — Farmacogn.
Farmacología, Farmacologiche, Farmacoterapia — Farmacol.
Farmakognozii — Farmakogn.
Farmakologi(ia) — Farmakol.
Farmatsevticheskii — Farm.
Farmer('s,s,s') — Farmer('s,s,s')
Farming — Farming
Farmosi — Farm.
Fascicola, Fasciculus — Fasc.
Faserforschung — Faserf.
Fat — Fat
Fauna — Fauna
Faunističeskim, Faunisticheskim, Faunistisch — Faunist.
Fauny, Faŭny — Fauny, Faŭny
Feather — Feather
Federación, Federal, Fédéral, Federated, Federation, Fédération — Fed., Féd.
Feld(bau) — Feld(bau)
Felett — Felett
Felleserie — Felless.
Fellows — Fellows
Felsö-Magyarországi — Felsö-Magyarorsz.
Felsöoktatási — Felsöokt.
Fen — Fen
Fenniae — Fenniae
Fennica(e) — Fenn.
Fenogenetičeskaja, Fenogeneticheskaya — Fenogenet.
Fenologiczny — Fenol.
Fermentări, Fermentation — Ferment.
Fern — Fern
Fernist — Fernist
Fertilidade — Fertil.
Fertilisation — Fertilis.
Fertility — Fertil.
Fertilizers — Fertilizers
Fertödi — Fertödi
Feuerschiffen — Feuerschiffen
Feuille(s) — Feuille(s)
Fiber(s) — Fiber(s)
Fibre(s) — Fibre(s)
Ficha — Ficha
Fiche(s) — Fiche(s)
Field — Field
Filial, Filial' [etc.], Filialining, Filial'n'n, Filialov, Filialynyn — Fil.
Filicum — Filicum
Filipiano — Filip.
Filogici — Filol.
Filosofia, Filosófica — Filos.
Finder — Finder
Finnischen, Finnish — Finn.
Finnländische — Finnl.
Finska — Finska
Finskoi — Finsk.
Fiorito — Fior.
Fire — Fire
Fischereywesen — Fischereywesen
Fish — Fish
Fisheries, Fishery — Fish.
Fisica(s), Física, Fisiche, Físico, Fisikie — Fis., Fís.
Fisiocritici — Fisiocrit.
Fisiologia — Fisiol.
Fito-planktona — Fito-Plankt.
Fitofarmacologia — Fitofarmacol.
Fitofisionomia — Fitofision.
Fitogeografica, Fitogeografii — Fitogeogr.
Fitopatologia, Fitopatologičeskoj, Fitopatologicheskoi, Fitopatologii, Fitopatologija, Fitopatologiya, Fitopatolojik, Fitopatološka, Fitopatoloskog — Fitopatol.
Fitoplanktonom — Fitoplankt.
Fitopublicaciones — Fitopubl.
Fitosanitară, Fitosanitaria — Fitosan.
Fitosenozakh — Fitosenozakh
Fitosociologii, Fitosociologia — Fitosociol.
Fitosotsiologii — Fitosotsiol.
Fitossanitário — Fitossan.
Fitotécnia — Fitotéc.

Fitotécnica, Fitotecnistas — Fitotécn., Fitotecn.
Fizičeskago, Fizicheskago, Fizika — Fiz.
Fiziko-Himičeskago — Fiz.-Him.
Fiziko-khimicheskago — Fiz.-Khim.
Fiziko-matematičeskih [etc.], Fiziko-matematicheskikh [etc.] — Fiz.-Mat.
Fiziologii, Fiziologija, Fiziologiya — Fiziol.
Fizjogracznej, Fizjograficzenej, Fizjografii — Fizjogr.
Fizychno-matematychnogo, Fizycno-matematyčnogo — Fiz.-Mat.
Fizyczne — Fiz.
Fizyograficznej, Fizyograficzny — Fizyogr.
Flash — Flash
Flavorist — Flavorist
Flavour — Flav.
Flëry — Flëry
Fleur(s) — Fleur(s)
Fleuriste — Fleur.
Flora(e,l,m) — Fl.
Flóramüvek — Flóramüvek
Flore, Floreale, Flôres — Fl.
Florestais, Florestal — Florest.
Floricoltori, Floricultura(l), Floriculture — Floric.
Florimontane — Florimont.
Florist('s,s), Florističeskim, Floristicheskim, Floristik, Floristilized, Floristique, Floristisch(e,en) — Florist(s)
Floristŭ — Floristŭ
Flory — Fl.
Flota — Flota
Flower(s), Flowering — Fl.
Flugblatt — Flugbl.
Fluszstation — Fluszstat.
Flyer — Flyer
Flygblad — Flygblad
Flygeskrift — Flygeskr.
Focus — Focus
Föiskola — Föisk.
Földmüvelési — Földmuv.
Földrajzi — Földr.
Földtani — Földt.
Folia — Folia
Foliage — Foliage
Folio — Folio
Folizoski — Filoz.
Folleto(s) — Folleto(s)
Fölött — Fölött
Folyóira(t,s) — Folyóir.
Fomento — Fomento
Fonctionnement — Fonct.
Fondamental — Fondam.
Fondation — Fond.
Fondazione — Fondaz.
Fonds — Fonds
Fonovogo — Fonov.
Fontes — Fontes
Food(s) — Food(s)
Forage — Forage
Foras — Foras
Forbildung — Forbild.
Förderung — Förd.
Foreign — Foreign
Föreldrar — Föreldrar
Forening, Förening(en,ens) — Foren., Fören.
Forensic — Forensic
Forest(s) — Forest(s)
Forestal(e,es,ia,is) — Forest.
Forester(s,'s) — Forester(s,'s)
Forestier, Forestieră, Forestiere, Forestières, Forestry — Forest.
Forêt(s) — Forêt(s)
Förhandlingar, Forhandlinger, Förhandlinger — Förh., Forh.
Forma — Forma
Formatio(n) — Format.
Formosan(orum) — Formosan
Forrageiras — Forrag.
Forrajes — Forrajes
Forschung(en) — Forsch.
Forschungsanstalt — Forschungsanst.
Forschungsberichte — Forschungsber.
Forschungsergebnisse — Forschungsergebn.
Forschungshefte — Forschungsh.
Forschungsinstitut(e) — Forschungsinst.
Forschungsreisenden — Forschungsreisenden
Forskning — Forskning
Forskningsinstituttet — Forskningsinst.
Forskningsråd(s) — Forskningsråd(s)
Forskningsrådets — Forskningsråd.
Forsøgsvirksohmed — Forsøgsvirksohmed
Forsøgvirksomhed — Forsøgvirksomh.
Forsøksstation — Forsøksstat.
Försöksväsendet — Försöksväs.
Forstbotanischen — Forstbot.
Forster — Forster
Forstgenetik — Forstgenet.
Forstinstitutes — Forstinst.
Forstkandidater — Forstkand.
Forstkultur — Forstkult.
Forstkunde — Forstk.
Forstlich(e,en,er), Forstliga — Forstl.
Forstmagazin — Forstmag.
Forstmann [etc.]. — Forstmann [etc.].
Forstpflanzen — Forstpflanzen
Forstpflanzenzüchtung — Forstpflanzenzücht.
Forstplanung — Forstplan.
Forstsamen — Forstsamen
Forstverein(s,es) — Forstverein(s,es)
Forstwesen — Forstwesen
Forstwirt, Forstwirtschaft(lichen) — Forstw.
Forstwissenschaft, Forstwissenschaftliche [etc.]. — Forstwiss.
Forstzeitschrift — Forstz.
Fortgange — Fortgange
Fortgesetzte(r,s) — Fortgesetzte(r,s)
Fortschritte — Fortschr.
Fortsetzung — Fortsetz.
Forum — Forum
Fossilium — Foss.
Fotografica — Fotogr.
Fotosinteza — Fotosin.
Foundation — Found.
Fourragères — Fourrag.
Fragment(a,e) — Fragm.
Fragrance — Fragr.
Framsteg — Framsteg
Français(e,es), Francesas — Franç., Franc.
Francisceum — Francis.
Franckfurtische — Franckfurt.
Frankfurter — Frankfurter
Fränksiche(s) — Fränk.
Französische(n) — Franz.
Fraternity — Fraternity
Freiburger — Freiburger
Fremme — Fremme
Fremstillinger — Fremstill.
Freshwater — Freshwater
Freunde(n) — Freunde(n)
Freye(n) — Freye(n)
Freywillige — Freywillige
Fribourgeoise — Fribourg.
Friend(s) — Friend(s)
Frijol — Frijol
Friulano — Friul.
Friutteelt — Friutteelt
Friuttelersblad — Friuttelersblad
Frøanl — Frøanl
Frontier(s) — Frontier(s)
Fröodlareforbund(s) — Fröodlareforb.
Früchte — Früchte
Fructus — Fructus
Fruechte — Fruechte
Frugtavleren — Frugtavl.
Frugttraedyrkere — Frugttraedyrk.
Fruit(s) — Fruit(s)
Fruitgewesen — Fruitgew.
Fruitgrower — Fruitgrower
Fruitière(s) — Fruitière(s)
Fruitist — Fruitist
Frutas — Frutas
Fruticultura — Frutic.
Frutteti, Frutticola — Frutt.
Frutticoltura — Fruttic.
Führer — Führer
Funciare — Func.
Functio — Functio
Function(al) — Funct.
Fund — Fund
Fundação, Fundacion — Fund.
Fundamenta(les) — Fundam.
Funde — Funde
Fundulea — Fundul.
Fungi — Fungi

Fungicide(s) — Fungic.
Fungorum — Fungorum
Fungus — Fungus
Fuori — Fuori
Furstenthum(s), Fürstenthümer — Furstenth.
Fürstlich — Fürstl.
Further — Further
Futter — Futter
Füzetei, Füzetek — Füz.
Fysik — Fys.
Fysiografiska — Fysiogr.
Gabinetu — Gab.
Gabonica — Gabon.
Gabrauche — Gebrauche
Gaceta — Gac.
Gallenkunde — Gallenk.
Gallery — Gall.
Gallica — Gallica
Gamete — Gamete
Gamtos — Gamtos
Ganadería — Ganad.
Gandavensis — Gand.
Gankopielebis — Gankop.
Ganzen — Ganzen
Garden(s), Gardener(s), Gardening — Gard.
Gärten, Garten(s) — Gärt., Gart.
Gartenbau(es) — Gartenbau(es)
Gartenbaublatt — Gartenbaubl.
Gartenbaugesellschaft — Gartenbauges.
Gartenbautechnik — Gartenbautechnik
Gartenbauvereins — Gartenbauver.
Gartenbauwirtschaft — Gartenbauwirt.
Gartenbauwissenschaft — Gartenbauwiss.
Gartenbeisitzer — Gartenbeisitz.
Gartenbibliothek — Gartenbiblioth.
Gartenborse — Gartenborse
Gartenfreunde — Gartenfr.
Gartenkunde — Gartenk.
Gartenkunst — Gartenkunst
Gartenkuntur — Gartenkultur
Gartenwelt — Gartenwelt
Gartenzeitschrift — Gartenz.
Gartenzeitung — Gartenzeitung
Gartner, Gärtner — Gartn., Gärtn
Gärtnerbörse — Gärtnerb.
Gärtnerei — Gärtnerei
Gärtnerey — Gärtnerey
Gartneri, Gärtnerisch — Gartn., Gärtn.
Gärtnerlehranstalt — Gärtnerlehranst.
Gärtnerzeitung — Gärtnerzeitung
Gärungsgewerbe — Gärungsgewerbe
Gärungsphysiologie — Gärungsphysiol.
Gaspadarki — Gasp.
Gaumais — Gaumais
Gavnlig — Gavnlig
Gazda — Gazda
Gazdák, Gazdaságban, Gazdasági, Gazdaságok, Gazdaszáti — Gazd.
Gazeta, Gazette — Gaz.
Gazon — Gazon
Gazzetta — Gazz.
Gebiete — Geb.
Gebrauche — Gebrauche
Geburtshilfe — Geburtsh.
Geburtskunde — Gebursk.
Gedanensis — Gedan.
Gegend — Gegend
Gegenstande — Gegenstande
Gegenwart — Gegenwart
Gegenwärtigen — Gegenwärt.
Geillustreerd — Geill.
Geisteswissenschaften — Geisteswiss.
Geisteswissenschaftlichen — Geisteswissenschaftl.
Gelahrtheit — Gelahrth.
Gelehrsamkeit(s) — Gelehrsamk.
Gelehrte(n,r) — Gel.
Gelişen — Gelişen
Gemeinnützige(r,s) — Gemeinnütz.
Gemeinsame — Gemeinsame
Gemuese — Gemuese
Genavensium — Genav.
Gene(s) — Gene(s)
Geneeskrachtige — Geneeskr.
Geneeskundig(e) — Geneesk.
General(e,es), Géneral(e,es), Generalis — Gen., Gén.
Généralité — Généralité
Genetic(s), Genetica, Genética, Geneticheskii, Genetika, Genetike, Génétique, Genetycznego — Genet., Genét.
Genèvois(e) — Genèvois(e)
Genitik(i) — Genet.
Genito-urinary — Genito-Urin.
Genius — Genius
Genootschap(s) — Genootsch.
Genossenschaft — Genossensch.
Gentes — Gentes
Gentleman's — Gent.
Geobiologicae, Geobiológico, Geobiology — Geobiol.
Geobotanica, Geobotaničeskomu, Geobotanicheskoe, Geobotanicheskomu, Geobotanichnyi, Geobotanici, Geobotaničnyj, Geobotanik(a), Géobotanique, Geobotanischen, Geobotany — Geobot.
Geochemistry — Geochem.
Geochimica — Geochim.
Geociencias — Geoci.
Geofisica — Geofis.
Geofizika — Geofiz.
Geognosie — Geognosie
Geografía(s), Geográfica, Geografičeskija [etc.], Geograficheskiya [etc.], Geografico, Geograficzny, Geografii, Geografinis, Geografisk — Geogr.
Geographentag(es) — Geographentag(es)
Geographer, Geographica(l), Geographicum, Geographie, Géographie, Géographique(s), Geographische(n,s), Geography — Geogr., Géogr.
Geologičeskoj, Geologica(l), Geologicheskii [etc.], Geologichnyi, Geologickych, Geologicos, Geologie, Geologija, Geologique, Geologische [etc.], Geologiska, Geologist(s), Geologiya, Geology, Geoloogilised — Geol.
Geomicrobiology — Geomicrobiol.
Geomorphological, Geomorphology — Geomorph.
Geophysic(al,s) — Geophys.
Geoponika — Geoponika
Geoponon — Geoponon
Geórgikon(s) — Geórgikon(s)
Georgofili — Georgof.
Geoscience — Geosci.
Gerais — Gerais
Geral — Geral
Geranium(s) — Geranium(s)
Gerbari(i), Gerbarija, Gerbariya — Gerb.
Gerichtliche(n,r) — Gerichtl.
Germ — Germ
Germanica, Germanischen — German.
Germplasm — Germplasm
Gesammte(n) — Gesammte(n)
Gesammtgebeite — Gesammtgeb.
Gesamte [etc.] — Gesamte(n)
Gesamtgebiet — Gesamtgeb.
Geschäftliche — Geschäftl.
Geschäftsbericht — Geschäftsber.
Geschehenen — Geschehenen
Geschichte, Geschiedenis — Gesch.
Geschlechtskrankheiten — Geschlechtskrankh.
Geschmacks — Geschmacks
Geschmaks — Geschmaks
Geselliges — Geselliges
Gesellschaft — Ges.
Gesneriad — Gesneriad
Gestifteten — Gestifteten
Gestis — Gestis
Gesunde — Gesunde
Getah — Getah
Getreidewesen — Getreidew.
Gewächskunde — Gewächsk.
Gewassen — Gewassen
Gewässer — Gesässer
Gewebe — Gewebe
Gewerbblatt — Gewerbbl.
Gewerbe(s) — Gewerbe(s)
Gewerbefleiss(es) — Gewerbefl.
Gewerbefreund — Gewerbefer.
Gewerbkunde — Gewerbk.
Gewerbskunde — Gewerbsk.
Gewerfbetreibende — Gewerbetreibende
Gewone — Gewone
Gewürz-Pflanzen — Gewürz-Pflanzen
Gezondheidsleer — Gezondheid.
Ghanaian — Ghanaian
Giardinaggio, Giardino — Giard.

Gibridizacii — Gibridiz.
Gidrobiologičeskij [etc.], Gidrobiologičnoji, Gidrobiologicheskii [etc.], Gidrobiologichnoi, Gidrobiologii, Gidrobiologiji — Gidrobiol.
Giessener — Giessener
Gimtasai — Gimt.
Ginecologia — Ginecol.
Giornale — Giorn.
Giovane — Giova.
Glad — Glad
Gladiolen — Gladiolen
Gladiolus — Gladiolus
Glas — Glas
Glasnik — Glasn.
Glasshouse — Glasshouse
Glastechnische — Glastechn.
Glavnago, Glavnom — Glavn.
Gleaner — Gleaner
Gleanings — Gleanings
Gleboznawcze — Gledozn.
Global — Global
Globe — Globe
Glównej — Główn.
Godičnyj — Godičn.
Godišen — Godišen
Godišnik, Godišnjak, Godišnji — God.
Golf — Golf
Gombászati — Gombász.
Good — Good
Gordoskago — Gorodsk.
Gornoe — Gornoe
Gornotaežnoj — Gornotaežn.
Gornotaezhnoi — Gornotaezhn.
Gorode — Gorode
Gorskostopanska — Gorskost.
Gory-Goreckaga — Gory-Goreck.
Gory-Goretskaga — Gory-Goretsk.
Gospodars'koi, Gospodarstvo, Gospodarstwa — Gosp.
Gossip — Gossip
Gosudarstvennii [etc.], Gosudarstvennyj — Gosud.
Gothaische(s) — Gothaische(s)
Götheborgska — Götheborgska
Gothoburgensi(a,s) — Gothob.
Gottingensis — Gott.
Göttinger — Göttinger
Göttingsche — Gött.
Gourd — Gourd
Gouvernement(s) — Gouv.
Government, Governor — Gov.
Gozdarski — Gozd.
Gozdno — Gozdno
Graancentrum — Graancentrum
Graanonderzoekers — Graanonderz.
Grădina — Grăd.
Gradinarska — Grad.
Gradinarstvo — Gradinarstvo
Grădinii — Grăd.
Graduate(s) — Graduate(s)
Graeca — Graeca
Grain(s) — Grain(s)
Grana — Grana
Grand-Ducal — Grand-Ducal
Grand-Duché — Grand-Duché
Grande — Grande
Grant — Grant
Grape — Grape
Grapevine — Grapevine
Graphia — Graphia
Grasas — Grasas
Grass(e,es) — Grass(e,es)
Grassland(s) — Grassland(s)
Great Britain — Gr. Brit.
Greater — Gr.
Green — Green
Greenhouse — Greenh.
Greenkeeping — Greenkeeping
Greifswaldische(s) — Greifswald.
Grenzgebiete — Grenzgeb.
Grenzmärkischen — Grenzmärk.
Gribnye — Gribn.
Griby — Griby
Groentegewassen — Groentegew.
Groenten — Groenten
Groentengewassen — Groentengew.
Groenteteelt — Groentet.
Groentetelersvereniging — Groentetelersver.
Grond — Grond
Groninganae — Groning.
Grossbritannischen — Grossbrit.
Grossbrittannischen — Grossbritt.
Grossherzoglich — Grossherzogl.
Grossherzogthum(s) — Grossherzogth.
Grossramforschung — Grossraumf.
Grössten — Grössten
Ground(s) — Ground(s)
Group(e) — Group(e)
Grower('s,s,s') — Grower('s,s,s')
Growing — Growing
Growth — Growth
Gruboskeletnykh — Gruboskel.
Grundlagen — Grundl.
Gründung — Gründung
Grunflächen — Grunfl.
Gruppo — Gruppo
Gruzii — Gruzii
Gruzinskoi S S R, Gruzinskoj S S R — Gruzinsk. S.S.R.
Gruzinskoi [etc.], Gruzinskoj [etc.] — Gruzinsk.
Gubernii, Gubernskago, Gubernskogo — Gub.
Guernésiaise — Guernésiaise
Guide(s) — Guide(s)
Guild — Guild
Gula — Gula
Gulf — Gulf
Gummi — Gummi
Guten — Guten
Gutta-Percha — Gutta-Percha
Gyarapodási — Gyarap.
Gynaecology, Gynäkologie, Gynecology — Gyn.
Gyógyszeres — Gyógy.
Gyógyszerésztudományi — Gyógyszerésztud.
Gyüjtemény — Gyüjt.
Gyülésémek, Gyülésének — Gyül.
Gyümölcs — Gyüm.
Gyümölcsészeti — Gyümölcsesz.
Gyümölcskultura — Gyümölcskult.
Gyümölcstermesztés — Gyümölcsterm.
Gyümölez — Gyüm.
Habarovskogo — Habarovsk.
Habilitacyjne — Habilit.
Hafniensi(a,s) — Hafn.
Hage — Hage
Hagetidende — Hagetid.
Haitiens — Haitiens
Háj — Háj
Haladó — Haladó
Half — Half
Half-Hard — Half-Hard
Half-Yearly — Half-Yearly
Halieutikon, Halieutique — Halieut.
Hallérienne — Hallér.
Hallische(n) — Hallische(n)
Hållne — Hållne
Hamburger — Hamburger
Hamburgische(n,r,s) — Hamburg.
Hanauisches — Hanauisches
Handbook, Handbuch — Handb.
Handel — Handel
Handelingen — Handel.
Handels- — Handels-
Handelsblatt — Handelsbl.
Handelskunde — Handelsk.
Handelsmuseum — Handelsmus.
Handlingar, Handlinger — Handl.
Hannoverische, Hannoversche — Hannover.
Hanseatisches — Hanseat.
Happenings — Happen.
Hardy — Hardy
Harétskaha — Haréts.
Harkivs'kogo — Harkivs'k.
Har'kovskij [etc.] — Har'kovsk.
Harvest(s) — Harvest(s)
Hasil — Hasil
Háskolans — Háskolans
Haspadarki — Haspad.
Hassiaca — Hassiaca
Hasznos — Hasznos
Haugekonsten — Haugek.
Haugetidende — Haugetid.
Haugevaesen — Haugevaes.
Hauniense, Hauniensi(s) — Haun.
Hauptstelle — Hauptstelle
Hauptversammlung — Hauptversamml.
Haus — Haus
Haushaltungskunde — Haushaltungsk.
Hauswirthschaft — Hausw.

Haute(s) — Haute(s)
Have(s) — Have(s)
Havebrugs-Tidende — Havebr.-Tidende
Havetidende — Havetid.
Havevaesen — Havevaes.
Havniensis — Havn.
Havraise — Havraise
Hawaiian — Hawaiian
Hayat — Hayat
Hayvancilik — Hayvanc.
Hazai — Hazai
Health — Health
Heather — Heather
Hebdomadaire(s) — Hebd.
Hébraique — Hébraique
Hedendaagsche — Hedend.
Hedeselskab — Hedeselsk.
Heelkunde — Heelk.
Hefte — Hefte
Heidelberger — Heidelberger
Heidelbergische — Heidelberg.
Heidemaatschappij — Heidemaatsch.
Heilkunde — Heilk.
Heilkunst — Heilkunst
Heilpflanze — Heilpflanze
Heilpflanzenkunde — Heilpflanzenk.
Heilwissenschaft — Heilwiss.
Heimat — Heimat
Heimatblätter — Heimatbl.
Heimatforschung — Heimatf.
Heimatkunde — Heimatk.
Heimatmuseums — Heimatmus.
Heimatpflege — Heimatpflege
Heimatschutz — Heimatschutz
Heimatschutzvereins — Heimatschutzvereins
Heimische(n) — Heimische(n)
Helgoländer — Helgoländer
Hellencia, Hellenic, Hellenike, Hellenikou, Helléniques — Hellen., Hellén.
Helminthological, Helminthology — Helminthol.
Helveti(ae,ca,cae), Helvetiens, Helvétique, Helvetische(s) — Helv.
Hemerocallis — Hemerocallis
Herald — Herald
Herb — Herb
Herba(age,l) — Herba(age,l)
Herbari(es,o,um) — Herb.
Herbarsbeilage — Herbarsbeil.
Herbaryumundaki — Herbaryum.
Herbicide(s) — Herbic.
Herbier — Herb.
Herbologica — Herbol.
Heredity — Heredity
Herzogthum(s) — Herzogth.
Hessische — Hess.
Hetilap — Hetilap
Hidrobiologia, Hidrobiološki — Hidrobiol.
Hidrográficos — Hidrogr.
Hidrologia — Hidrol.
Highland — Highl.
Highlights — Highlights
Higiene — Hig.
Hilfswissenschaften — Hilfswiss.
Hill — Hill
Himalayan — Himalayan
Himičeskaja [etc.] — Him.
Himmel — Himmel
Himmelskunde — Himmelsk.
Hindi — Hindi
Hindu — Hindu
Hints — Hints
Hirmondó — Hirmondó
Hispanica(us) — Hispan.
Hispanoamerico — Hispanoamer.
Histochemica, Histochemistry — Histochem.
Histoire — Hist.
Histologie — Histol.
Histori(a,ae,am,i), História, Historica(l,s), Historico, Historique, Historische, Historisk(a), History — Hist.
Hiver — Hiver
Hizli — Hizli
Hjemmet — Hjemmet
Hlasy — Hlasy
Hobbyist — Hobbyist
Hochschule — Hochschule
Hochschulgesellschaft — Hochschulges.
Hodowii — Hodowii
Hodowla — Hodowla
Hofmuseums — Hofmus.
Högskola — Högskola
Höheren — Höheren
Hoja — Hoja
Hoje — Hoje
Holarctic — Holarc.
Holländische [etc.], Hollandsche — Holl.
Holly — Holly
Holmensis — Holm.
Holsteinische — Holst.
Holz — Holz
Holzforschung — Holzf.
Holzindustrie — Holzindustr.
Holzverwertung — Holzverwert.
Holzwirt — Holzwirt
Holzwirtschaft(lichen) — Holzw.
Hombre — Hombre
Home(s) — Home(s)
Home-grown — Home-grown
Homemaker — Homemaker
Homeopathy — Homeopathy
Homme — Homme
Hongrois(e) — Hongr.
Honi — Honi
Honismertetö — Honism.
Honorable, Honourable — Hon.
Hoogere — Hoogere
Hookah — Hookah
Hoosier — Hoosier
Hop(s) — Hop(s)
Hôpital, Hôpitaux — Hôp.
Horizons — Horiz.
Hors — Hors
Horti — Horti
Horticolă, Horticole — Hort.
Hortícolo-agrícola — Hort.-Agric.
Horticulteur(s), Horticultur(a,ae), Horticultural(ia), Horticulture, Horticulturist, Hortikultura — Hort.
Hortiviticole(s), Horti-viticole — Hortivitic., Horti-Vitic.
Horto — Horto
Hortonomer — Hortonomer
Hortorum — Hort.
Hortulus — Hortulus
Hortus — Hortus
Hosianum — Hosianum
Hospital — Hosp.
Hospodař, Hospodářské(ho), Hospodářský, Hospodářství, Hospodárstva, Hospodárstvo — Hospod.
Hosta — Hosta
Hot — Hot
Houbarů — Houb.
Houblons — Houblons
Houillère — Houillère
House — House
Household — Household
Hozjain — Hozjain
Hozjajstva, Hozjajstvo, Hozjatstvu — Hoz.
Hradecký — Hradecký
Hristianskogo — Hristiansk.
Hrvatskih, Hrvatskog(a) — Hrvatsk.
Huis — Huis
Huishoudkundige — Huishoudk.
Hülfswissenschaft — Hülfswiss.
Humaine — Humaine
Human — Human
Humanities — Humanit.
Humboldtiana — Humboldt.
Humid — Humid
Hungarica(e), Hungarici, Hungarisches — Hung.
Hunyadmergyei — Hunyadmegyei
Husbandry — Husb.
Husdyrbrugsforsoeg — Husdyrbrugsfors.
Hushållningen — Hush.
Hushåll-Sällskapets — Hush.-Sällsk.
Hushållstidningen — Hushållstidn.
Hutan — Hutan
Hüttenwesen — Hüttenwesen
Huudrobioloogilised — Huudrobiol.
Huusmaend — Huusm.
Hvalrådet — Hvalrådet
Hybrid — Hybrid
Hydrobiologica(l), Hydrobiologie, Hydrobiologique, Hydrobiologische, Hydrobiologji, Hydrobiology — Hydrobiol.
Hydrographie — Hydrogr.

Hydrographisch-biologische — Hydrogr.-Biol.
Hydrographisches — Hydrogr.
Hydrology — Hydrol.
Hydrometeorological, Hydrometeorologischen — Hydrometeorol.
Hydroponic(s) — Hydrop.
Hydroscience — Hydrosci.
Hydroviologikou — Hydroviol.
Hygiene — Hyg.
Hygiene-Institut — Hyg.-Inst.
Hygienic — Hyg.
Iberica(m), Ibérica — Iber., Ibér.
Ice — Ice
Icones — Icon.
Iconografia, Iconographia — Iconogr.
Idöszeru — Idöszeru
Idrobiologia, Idrobiologica — Idrobiol.
Idrografici — Idrogr.
Iepenzichte — Iepenzichte
Ifjuság — Ifjuság
Igieneh — Ig.
Ih — Ih
Ile — Ile
Illini — Illini
Illustrate(d), Illustratie, Illustration, Illustrator, Illustré(e), Illustrierte, Illustrirter — Ill.
Ilmi — Ilmi
Ilustrato — Ill.
Image — Image
Imbunatatiri — Imbunat.
Imeni — Im.
Immediate — Immed.
Immergrüne — Immergrüne
Immigrant, Immigration — Immigr.
Immunität — Immunität
Immunitatsforschung — Immunitatsforsch.
Immunity — Immun.
Immunobiological — Immunobiol.
Immunologia, Immunologie, Immunology — Immunol.
Impact — Impact
Imperatorskoi [etc.], Imperatorskoj [etc.] — Imp.
Imperfecti — Imperf.
Imperial, Imperiale, Impériale, Imperialis — Imp.
Imperii — Imperii
Importanti — Import.
Imprimerie — Imprimerie
Improvement — Improv.
Impulsora — Impuls.
Inalte — Inalte
Incidental(e) — Incidental(e)
Incompatibility — Incompat.
Inconjurator — Inconj.
Incoraggiamento — Incoragg.
Incrementum — Incrementum
Indeks — Indeks
Independent — Indep.
Index(ed) — Index(ed)
Indian — Indian
Indica — Indica
Indice — Indice
Indigo — Indigo
Indisch(e) — Indisch(e)
Indonesian — Indones.
Indoor — Indoor
Induktive — Indukt.
Industri, Industria(s), Industrial(e,i), Industrie(lles,s) — Industr.
Industrieplanter — Industriepl.
Industriya, Industriyu, Industry — Industr.
Ineditos, Inéditos — Ined., Inéd.
Infectieziekten — Infectiez.
Infection, Infectious — Infect.
Infektionskrankheiten — Infektionskrankh.
Infektsionnykh — Infekts.
Inferential — Inferential
Inférieure — Infér.
Influences — Influences
Infoletter — Infolett.
Informa — Informa
Informação, Informacii, Información(es), Informacyjny, Informasi, Informasjon, Informatica, Informaţii, Information(al,en,es,s) — Inform.
Informationsbrief — Informationsbrief
Informativa, Informative, Informativo, Informatore, Informatsionnii, Informazioni — Inform.
Informe — Informe
Inframicrobiologie — Inframicrobiol.
Ingeniería, Ingenieros, Ingénieurs — Ing.
Inhalt(s) — Inhalt(s)
Inheemsche — Inheemsche
Initiations — Init.
Injury — Injury
Inländische(n,s) — Inl.
Innere — Innere
Inostrannyh, Inostrannykh — Inostr.
Inscriptions — Inscript.
Insect(s) — Insect(s)
Insectivorous — Insectiv.
Insel — Insel
Insight — Insight
Inspección — Inspecc.
Inspectie, Inspection — Inspect.
Institoutou, Institució, Institución, Institut(a,s), Institute(n,s), Instituti, Institution(en,s), Institutis, Institutit, Instituto(r), Institutt, Inštitútu, Institutul(ui), Institutum, Instituut — Inst.
Instruction, Instructivas — Instruct.
Instrument(e) — Instrum.
Instruzione — Instruz
Instytut(o,om,u) — Inst.
Intecol — Intecol
Integrated, Integrierten — Integr.
Intelligence(r) — Intelligence(r)
Intelligenzblatt [etc.] — Intelligenzbl.
Intensivobstbau — Intensivobstbau
Interaction(s) — Interact.
Interafrican — Interafr.
Interamerican(a,o) — Interamer.
Interchange — Interchange
Interdisciplinary — Interdiscipl.
Interdisziplinäre — Interdiszipl.
Interese — Interese
Interessantesten — Interessante(r).
Interessanti — Interessanti
Interesting — Interesting
Intérêts — Intérêts
Interface — Interface
Interim — Interim
Interior — Interior
Intermountain — Intermount.
Internacional — Int.
Internal — Intern.
International(e,er,es), Internazionale — Int.
International Union for Conservation of Nature and Natural Resources — I. U. C. N.
Interpretation — Interpret.
Intertropiques — Intertrop.
Intézet(enek), Intézöbizottságának — Intéz.
Intramongolica — Intramongol.
Introduction(s), Introdukcija [etc.], Introduktsii [etc.], Introduktsionnogo — Introd.
Inventario(s), Inventions, Inventory — Invent.
Invertebrate — Invert.
Investigação, Investigacion(es), Investigações, Investigandis, Investigandum, Investigation(al,is,s), Investigative — Invest.
Investigator — Investigator
Ipar — Ipar
Iranica — Iran.
Iregszemcse — Iregszemcse
Irish — Irish
Irkutskoi [etc.], Irkutskoj [etc.] — Irkutsk.
Irodalmi — Irod.
Irregular — Irreg.
Irrigation — Irrig.
Island(s) — Island(s)
Islandica — Islandica
Isle(s) — Isle(s)
Islendinga — Islend.
Ismertetö — Ismert.
Isotopes — Isotopes
Ispitatelei, Ispitivanje — Isp.
Ispytanija, Ispytaniya — Ispytan.
Ispytatelei, Ispytatelej — Isp.
Issledovanii [etc.], Issledovanija [etc.], Issledovatel'skoi [etc.], Issledovatel'skoj [etc.] — Issl.

Issue — Issue
Istituti, Istituto — Ist.
Istoria — Istoria
Istoricheskii [etc.] — Istorich.
Istorii — Istorii
Istraživanja — Istraž.
Iszlenzkar — Iszlen.
Italian(a,o), Italica, Italienischen — Ital.
Italo-Argentina — Italo-Argent.
Item — Item
Itogi — Itogi
Izdanie, Izdanija, Izdanja — Izd.
Izlézü — Izlézü
Izmenčivost' — Izmenčiv.
Izmenchivost' — Izmenchiv.
Izslědovanija [etc.], Izslědovaniya [etc.] — Izsl.
Izučenija, Izučeniju — Izuč.
Izuchenie — Izuch.
Izucheniya [etc.] — Izuch.
Izvešča, Izveštaj, Izvestija, Izvěstija, Izvestiya, Izvestja — Izv.
Izvješča — Izvj.
Jaarblad — Jaarbl.
Jaarboek(en) — Jaarb.
Jaarverslag — Jaarversl.
Jablonovianae — Jablonov.
Jägare — Jägare
Jagd — Jagd
Jagd-Archiv — Jagd-Arch.
Jagdbibliothek — Jagdbiblioth.
Jagdkunde — Jagdk.
Jagdwesen — Jagdwesen
Jagdwissenschaft, Jagd-Wissenschaft — Jagdwiss., Jagd-Wiss.
Jagdzeitung, Jagd-Zeitung — Jagdzeitung, Jagd-Zeitung
Jagellonicae — Jagellon.
Jagiellonskiego — Jagiellon.
Jagodyne — Jagodnye
Jahrbuch [etc.] — Jahrb.
Jahres — Jahres
Jahresbericht(e) — Jahresber.
Jahrescheft(e) — Jahresh.
Jahresgabe — Jahresg.
Jahreskatalog — Jahreskat.
Jahresschrift — Jahresschr.
Jahresverhandlungen — Jahresverh.
Jahresversammlung — Jahresversamml.
Jährhunderts — Jährh.
Jährliche — Jährl.
Jahrs — Jahrs
Jakutskij [etc.] — Jakutsk.
Janakari — Janak.
Japanese, Japanische(n), Japonaise, Japonennes, Japonenses, Japonicae — Jap.
Jardim, Jardin(s), Jardinage, Jardinier(s) — Jard.
Jaroslavskogo — Jaroslavsk.
Jarovizacii — Jarov.
Jedermann — Jedermann
Jegyzéke — Jegyzéke
Jegyzések — Jegyzések
Jelentés(e) — Jel.
Jenaische(n) — Jenaische(n)
Jenesis — Jenesis
Jersiaise — Jersiaise
Jeune(s) — Jeune(s)
Jewish — Jew.
Jiho — Jiho
Jihočeského, Jihočeských — Jihočesk.
Jihovýchodní — Jihových.
Jilinensis — Jilin.
Joint — Joint
Jonica — Jonica
Jord — Jord
Jordbrug — Jordbrug
Jordbrugsforskning — Jordbrugsforskn.
Jordbruk — Jordbruk
Jordbruksområdet — Jordbruksomr.
Jornal — Jorn.
Josephinae — Joseph.
Journal(en), Journaux — J.
Jours — Jours
Judatului — Judatului
Judetean — Judet.
Jugend — Jugend
Jugoslavenske, Jugoslavica — Jugoslav.
Jukutskago — Jakutsk.
Julkaisu(ja) — Julk.
Junge — Junge
Junta(s) — Junta(s)
Jur'evskom [etc.] — Jur'evks.
Jurassienne — Jurass.
Jurnal — J.
Jute — Jute
Juures — Juures
Južno-Russkago — Južno-Russk.
Južno-Ussurijskago — Južno-Ussurijsk.
Kabardino-Balkarskogo — Kabardino-Balkarsk.
Kabinet(a,s) — Kab.
Kafedra — Kafedra
Kaiserlich(e,en) — Kaiserl.
Kaiserlich-Königlichen — K. K.
Kaiserstaat(es) — Kaiserstaat(es)
Kakteen — Kakteen
Kakteenforschung — Kakteenf.
Kakteenfreund(e) — Kakteenfr.
Kakteengesellschaft — Kakteenges.
Kakteenkunde — Kakteenk.
Kakti — Kakti
Kaktiletter — Kaktilett.
Kaktos — Kaktos
Kaktus(ar) — Kakt.
Kaktusářské — Kaktusář.
Kaktusaru, Kaktusow — Kakt.
Kaktusy — Kaktusy
Kaktusz — Kaktusz
Kaktuszgyujto — Kaktuszg.
Kalender — Kalend.
Kampen — Kampen
Kamskoi — Kamskoi
Kännedom — Kännedom
Känntnis — Känntn.
Kanton(s) — Kanton(s)
Kantonal-Gesellschaft — Kantonal-Ges.
Kar — Kar
Karadags'koji — Karadags'k.
Karadagskoi, Karadagskoj — Karadagsk.
Karadahs'koyi — Karadahs'k.
Karagandinskogo — Karagandinsk.
Karának — Kar.
Karantina, Karantinu, Karantinye — Karant.
Karel'skogo — Karel'sk
Karelo-Finskogo, Karelo-Finskoj — Karelo-Finsk.
Karlovarská — Karlovarska
Kärntner — Kärntner
Kárpát-Egylet — Kárpát-Egyl.
Karpatske — Karpat.
Kartenwerk — Kartenw.
Kartierung — Kart.
Kartofel' — Kartofel'
Kartografirovanie — Kartogr.
Kasachstanicae — Kasachst.
Kaspiiskogo — Kaspiisk.
Kasvimuseon — Kasvismus.
Kasvitieteellisiä, Kasvitieteen, Kasvitietellisen — Kasvit.
Katsejaama — Katsej.
Kaučuk — Kaučuk
Kaukasusländern — Kaukasusländern
Kautschuk — Kautschuk
Kavak — Kavak
Kavakçilik — Kavakç.
Kavkazskoi [etc.] — Kavkazsk.
Kayserlichen — Kayserl.
Kazahskoj [etc.] — Kazahsk.
Kazahstanskij [etc.] — Kazahstansk.
Kazakhskoi S S R — Kazakhsk. S.S.R.
Kazakhskoi [etc.] — Kazakhsk.
Kazakhstana — Kazakhstana
Kazakhstanskoi [etc.] — Kazakhstansk.
Kazakstanskoj [etc.] — Kazakstansk.
Kazanskii [etc.], Kazanskij [etc.] — Kazansk.
Kehutanan — Kehut.
Kemi — Kemi
Kemiska — Kem.
Kémlö — Kémlö
Kenkyusyo — Kenkyusyo
Kennis — Kennis
Kenntnis(s,se) — Kenntn.
Kérdései — Kérd.
Kereskedésben, Kereskedó — Keresk.
Kert — Kert
Kertész(e,et,eti) — Kert.
Kertészgazdasági — Kertészgazd.
Készítését — Készit
Keszthelyi — Keszthelyi
Keuringsdienst — Keuringsdienst
Kewensis — Kew.
Key — Key

Kézikönyve — Kézikönyve
Khabarovskogo — Khabarovsk.
Khar'kovskom [etc.] — Khar'kovsk.
Kharkivs'kii — Kharkivs'k.
Khaspadarki — Khaspad.
Khedivial, Khédiviale — Khediv., Khédiv.
Khimicheskaya, Khimii, Khimiya — Khim.
Khorologiya — Khorol.
Khozyain — Khozyain
Khozyaistva [etc.] — Khoz.
Khristianskogo — Khristiansk.
Kiadványai — Kiadv.
Kieler — Kieler
Kielische(s) — Kiel.
Kievskago, Kievskogo — Kievsk.
Kinaonderneming, Kina-Onderneming — Kinaondern., Kina-Ondern.
Kina-Profstation — Kina-Proefstat.
Kinetics — Kinet.
Kiøbenhavnske — Kiøbenhavnske
Kiralyi, Királyi — Kir.
Kirgizskii [etc.], Kirgizskoj — Kirgizsk.
Kirgizskoi S S R, Kirgizskoj S S R — Kirgizsk. S.S.R
Kirjatööd — Kirjatööd
Kísérletek, Kísérleti, Kísérletügyi — Kísérl.
Kivonatai — Kivonatai
Klasse — Kl.
Klausimai — Klaus.
Kleine(n,r,s) — Kleine(n,r,s)
Kleingärtner — Kleingärtn.
Kleinhandel — Kleinhandel
Kleintierhof — Kleintierhof
Kleintierzucht — Kleintierzucht
Klinische — Klin.
Kluba — Kluba
Kluborgan — Kluborgan
Klubu — Klubu
Kniga [etc.] — Kniga
Knizhovno — Knizh.
Knizni — Knizni
Know — Know
Knowledge — Knowl.
Koelaitoken — Koelait.
Koffiebessenboeboek-Fonds — Koffiebessenboeboek-Fonds
Koffiecultuur — Koffiecult.
Koisikavenses — Koisikav.
Kol'skogo — Kol'sk.
Kolektivu — Kolektivu
Kollegium — Koll.
Kolloid(e) — Kolloid(e)
Kolloidnyi — Kolloid.
Kölnisches — Köln
Koloniaal — Kolon.
Kolonialblatt — Kolonialbl.
Kolonialdeutsche — Kolonialdeutsche
Koloniale — Kolon.
Kolonialforschung — Kolonialf.
Kolonialforstliche — Kolonialforstl.
Kolonialinstituts — Kolonialinst.
Kolonie — Kolon.
Kolonizacionnyh — Kolonizac.
Kolonizatsionnykh — Kolonizat.
Koloszvári — Koloszvári
Komandirovke — Komandir.
Komise, Komisji — Komis.
Komissiei, Komissiej, Komissii, Komissija, Komissiya — Komiss.
Komitees — Komitees
Komiteta — Komiteta
Komitete — Komitete
Komitetu — Komitetu
Kommission — Kommiss.
Kommunal-Ekonomien — Kommun.-Ekon.
Komnatnoi [etc.], Komnatnoj [etc.] — Komnatn.
Komorki — Komorki
Kompleksnoi [etc.], Kompleksnoj [etc.] — Kompl.
Konchyliologie — Konchyliol.
Kongelige — Kongel.
Kongliga — Kongl.
Kongres(a), Kongresses — Kongr.
Königlich [etc.] — Königl.
Königlich-Baierischen — Königl.-Baier.
Königlich-Bestätigten — Königl.-Bestätigten.
Königlich-Ospreussisch-Mohrungschen — Königl.-Ospreuss-Mohrungschen
Koniglig — Kon.
Königreich — Königr.
Königsberger — Königsberger
Koninklijk(e) — Kon.
Konst — Konst
Konstkabinet — Konstkab.
Kontinental'nykh — Kontinental'n.
Kontrol — Koltrol
Konyhakertészeti — Konyhakert.
Könyvház — Könyvhaz
Korean — Korean
Köréböl — Köréb.
Kormo — Kormo
Kormovaya, Kormovyne — Kormov.
Kormovyrobnyststvo — Kormovyrobn.
Kórnickie — Kórnickie
Korrespondent — Korresp.
Korrespondenzblatt — Korrespondenzbl.
Korszerusitase — Korszerus.
Korte — Korte
Koruma — Koruma
Kosinskoi, Kosinskoj — Kosinsk.
Kostromskago, Kostromskogo — Kostromsk.
Kött — Kött
Középiskolai — Középisk.
Közleményei, Közlemények — Közlem.
Közlöny(e), Közlönyhoz — Közl.
Központ(i) — Közp.
Kraevaja, Kraevedčeskogo, Kraevedcheskogo, Kraevedenie, Kraevoi — Kraev.
Krainego — Krainego
Krainisch-Küstenländischen — Krainisch-Küstenl.
Krainisches — Krain.
Kraivárského Kraivářského — Kraivársk.
Kraj — Kraj
Kraja — Kraja
Kraje — Kraje
Kraji — Kraji
Krajnego — Krajnego
Krajského — Krajsk.
Kraju — Kraju
Krajums — Krajums
Krakowskiego — Krakowsk.
Kralevska, Kraljevine, Kraljevska, Královské, Království — Kral., Král.
Kranke — Kranke
Kras — Kras
Krása — Krása
Krasnogo — Krasn.
Krasnojarskij [etc.] — Krasnojarsk.
Krasnoyarskii [etc.] — Krasnoyarsk
Kraštas — Krašt.
Kratkij — Kratk.
Kravevoj — Kraev.
Kraya — Kraya
Krebsforschung — Krebsf.
Kreis — Kreis
Kreismuseums — Kreismus.
Kring — Kring
Kritik, Kritische, Kritiske — Krit.
Krkonošského — Krkonošk.
Krouzku — Krouzku
Krugom — Krugom
Kruidkundig — Kruidk.
Krupainykh — Krup.
Kruzhkov — Kruzhkov
Kružkov — Kružk.
Krúžku — Krúžku
Krymskago, Krymskogo — Krymsk.
Kryptogamenflora — Kryptogamenfl.
Kryptogamische — Kryptog.
Küchenfreunde — Küchenfr.
Kuhrpfälzischen — Kuhrpfälz.
Kul'turnaya — Kul't.
Külföldi — Külf.
Külgazdságban — Külgazd.
Kul'tur, Culture — Kul't., Kult.
Kultúregyesülete — Kultúregyes.
Kulturflórája — Kulturfl.
Kulturgeschichte — Kulturgesch.
Kul'turnyh, Kul'turnykh — Kul't.
Kulturpionier — Kulturpionier
Kul'tury — Kul't.
Kunde — Kunde
Kundskab — Kundskab
Kungliga — Kungl.
Kuns — Kuns
Kunst [etc.] — Kunst
Künste — Künste

Kunstfreunde — Kunstfr.
Kunstgärtner — Kunstgärtn.
Kunstgeschichte — Kunstgesch.
Kunsthandels — Kunsthandels
Kunstkabinet — Kunstkab.
Künstler — Künstler
Kunstmagazin — Kunstmag.
Kurfürstlich(en) — Kurfürstl.
Kurier — Kurier
Kurländische — Kurl.
Kurpfälzisce(n) — Kurpfälz.
Kurskii — Kursk.
Kurz(e,er) — Kurz(e,er)
Kurzmitteilungen — Kurzmitt.
Kustarnikov — Kustarnik.
Kutaró — Kutaró
Kutatásának, Kutatások — Kutatás.
Kutató — Kutató
Kutatóintézet(enek) — Kutatóint.
Kuybyshev-University — Kuybyshev-Univ.
Kwartalnik, Kwartalny — Kwart.
Kyiivs'kyi [etc.] — Kyiivs'k.
Kyjivs'kyj [etc.] — Kyjivs'k
L'vovskii — L'vovsk.
Laboratoire, Laboratoria, Laboratories, Laboratoriet, Laboratorii, Laboratorio, Laboratorium, Laboratory, Labores — Lab.
Lacustre — Lacustre
Ladies' — Ladies'
Lady's — Lady's
Laeger — Laeger
Laegevidenskaben — Laegevidensk.
Laerde — Laerde
Laerdom(s) — Laerd.
Laesning — Laesn.
Laitoksen — Laitok.
Läkare(n) — Läkare(n)
Läkaresällskapet — Läkaresällsk.
Läkarevetenskapen — Läkarevetensk.
Lake(s) — Lake(s)
Län — Län
Lancet — Lancet
Lancisiana — Lancisiana
Land(s) — Land(s)
Landbaues — Landb.
Landbostanden — Landbost.
Landbotinde — Landbotid.
Landbouw — Landb.
Landbouwblad — Landbouwbl.
Landbouwetenskap — Landbouwetensk.
Landbouw-Genootschap — Landb.-Genootsch.
Landbouwgewassen — Landbouwgew.
Landbouwhogeschool — Landbouwhogeschool
Landbouwhoogeschool — Landbouwhoogeschool
Landbouwjournaal — Landbouwj.
Landbouwkunde, Landbouwkundig — Landbouwk.
Landbouwproefstation — Landbouwproefstat.
Landbouw-Syndicate — Landb.-Synd.
Landbouwtechniek — Landbouwtechn.
Landbouwvoorlichtingsdienst — Landbouwvoorlichtingsdienst
Landbouwweekblad — Landbouwweekbl.
Landbouw-Weekblad — Landb.-Weekbl.
Landbouwwetenschappen — Landbouwwetensch.
Landbouwzaden — Landbouwz.
Landbrug — Landbr.
Landbrugsplanter — Landbrugspl.
Landbrukets, Landbruks — Landbr.
Landbrukshøgskolans — Landbrukshøgskolans
Landbrukshøgskole — Landbrukshøgskole
Landbruksuniversitet — Landbruksuniv.
Landbunada — Landbun.
Landbunaoardeildar — Landbunaoard.
Lande [etc.] — Lande
Länderkunde — Länderk.
Landesamt(es) — Landesamt(es)
Landesanstalt — Landesanst.
Landesarbeitsgemeinschaft — Landesarbeitsgem.
Landesaufnahme — Landesaufn.
Landescultur — Landescult.
Landesdurchforschung — Landesdurchf.
Landeskultur — Landesskult.
Landeskunde — Landesk.
Landesmuseums, Landes-Museum — Landesmus., Landes-Mus.
Landesmuseumsvereins — Landesmuseumsvereins
Landespflege — Landespflege
Landesstelle(n) — Landesstelle(n)
Landesverbandes — Landesverb.
Landesvereins — Landesvereins
Landesverwaltungsamtes — Landesverwaltungsamtes
Landhusholding(s) — Landhushold.
Landhuusholding(s) — Landhuushold.
Landhuusholdings-Selskabs — Landhuushold.-Selsk.
Landinrichting — Landinricht.
Landmaen(d) — Landmaen(d)
Landmands-Bog — Landmands-Bog
Landmann — Landmann
Landoeconomisk(e) — Landoecon.
Landoekonomi — Landoekon.
Landreisen — Landreisen
Landscape — Landscape
Landschaft(en) — Landschaft(en)
Landschaftsentwicklung — Landschaftsentw.
Landschaftsforschung — Landschaftsf.
Landschaftspflege — Landschaftspflege
Landschap — Landschap
Landschapsbouw — Landschapsbouw
Landsmännen — Landsmänn.
Landtbruks — Landtbr.
Landtbruks-Academiens — Landtbr.-Acad.
Landtbruksinstitut — Landtbruksinst.
Landtmann [etc.] — Landtm.
Landwirte — Landwirte
Landwirth(e) — Landwirth(e)
Landwirthschaft [etc.] — Landw.
Landwirthschaftgesellschaft — Landwirthschaftsges.
Landwirthschaftliche [etc.] — Landwirthschaftliche(s)
Landwirthschaftskalender — Landwirthschaftskalend.
Landwirthschaftslehre — Landwirthschaftsl.
Landwirthschaftswissenschaft — Landwirthschafswiss.
Landwirtschaft, Landwirtschaftliche [etc.] — Landw.
Landwirtschafts-Gesellschaft — Landw.-Ges.
Landwirtschafts-Recht — Landw.-Recht
Landwirtschaftsfreunde — Landwirtschaftsfr.
Landwirtschaftskammer — Landwirtschaftskammer
Landwirtschaftswissenschaften — Landwirtschaftswiss.
Lantarbetsgivareföreningens — Lantarbetsgivareför.
Lantbruksakademiens — Lantbruksakad.
Lantbrukshögskolans — Lantbrukshögskolans
Lantbrukstidskrift — Lantbrukstidskr.
Lantbruksuniversitet — Lantbruksuniv.
Lantmän — Lantmän
Lanvaesens — Landvaes.
Lap — Lap
Lapin — Lapin
Lapja — Lapja
Lapok — Lapok
Laporan — Laporan
Lapponicus — Lappon.
Lärda — Lärda
Large — Large
Laringologia — Laringol.
Larngologie — Laryngol.
Läroverken — Lärov.
Laryngology — Laryngol.
Larynx — Larynx
Läsning — Läsn.
Lasów — Lasów
Latin(a,e) — Latin(a,e)
Latinoamericana — Latinoamer.
Latviensis — Latv.
Latviiskogo — Latviisk.
Latviiskoi S S R — Latviisk. S.S.R.
Latvijas — Latv.
Latvijskoj [etc.] — Lajvijsk.
Laubgehölze — Laubgeh.
Lauenburgische(r) — Lauenburg.

Lauksaimniecibas — Lauksaimn.
Laurentianae — Laurent.
Lausitzer — Lausitzer
Lausitzische(s) — Lausitz.
Lausizische — Lausiz.
Laut — Laut
Lavori — Lav.
Lavoura — Lavoura
Law — Law
Lawn — Lawn
Laws — Laws
Lĕsnichestv — Lĕsnich.
Lĕtniya — Lĕtn.
Leaf — Leaf
Leaflet(s) — Leafl.
League — League
Learned — Learned
Leaves — Leaves
Leben(den,diges) — Leben(den,diges)
Lectionum — Lect.
Lecture — Lecture
Leczniczych — Leczn.
Ledger — Ledger
Leefmilieu — Leefmilieu
Leeraaskring — Leeraaskring
Legno — Legno
Legume(s), Légume(s) — Legume(s), Légume(s)
Legumicultura — Legumic.
Lehranstalt — Lehranst.
Lehre — Lehre
Lehrerklubs — Lehrerklubs
Lehrerseminars — Lehrersem.
Lehrer-Verein — Lehrer-Verein
Lehrerverein(es) — Lehreverein(es)
Lehrmeister — Lehrmeister
Lei — Lei
Leidensia — Leidensia
Leidse — Leidse
Leipziger — Leipziger
Leistungen — Leist.
Lekar — Lekar
Lékařskych — Lékař.
Lékárnické — Lékárnické
Lékárnictva — Lékárnictva
Lekarstvenno-Tehniceskim — Lekarstv.-Tehn.
Lekarstvennyh, Lekarstvennykh, Lekarstvennym — Lekarstv.
Lékařův — Lékařův
Lekovite, Lekovitpog — Lekov.
Lembaga — Lemb.
Lemon — Lemon
Lengua — Lengua
Leningrader — Leningrader
Leningradskii [etc.], Leningradskij [etc.] — Leningradsk.
Lentil — Lentil
Leodiensis — Leod.
Leopoldenses, Leopoldina, Leopoldinisch, Leopoldino — Leop.
Leopoldino-Carolinae — Leop.-Carol.
Leopoldino-Franciscanae — Leop.-Francisc.
Leopolitanae — Leop.
Lepidoptera, Lepidopterist — Lepid.
Lesa — Lesa
Lesarstva — Lesarstva
Leśna, Lesné(ho) — Leśn., Lesn.
Lĕsničestv — Lĕsnič.
Lesnická, Lesnické(ho), Lesnický(ch), Lesnictví, Lesnictwa, Lesniho, Lesnikeho — Lesn.
Lesno — Lesno
Lesnoi [etc.], Lesnoj [etc.], Lĕsnoj, Lesnye, Lesnyh — Lesn., Lĕsn.
Lesotekhnicheskii [etc.] — Lesotekhn.
Lesovedenie [etc.] — Lesoved.
Lesovodstva, Lesovodstvennie, Lesovodstvo — Lesov.
Lesswürdigkeiten — Lesswürdigk.
Lesů — Lesů
Lethaea — Leth.
Letnie — Letn.
Lĕtnija — Lĕtn.
Leto — Leto
Letopis — Let.
Letras — Letras
Letter(s), Letterari(o), Letterarie — Lett.
Letterario-Scientifico — Lett.-Sci.
Letterati, Letteratura — Lett.
Letterbode, Letter-Bode — Letterbode, Lett.-Bode
Letter-Courant — Lett.-Courant
Lettere(n) — Lett.
Letterkundig — Letterk.
Letteroefeningen — Letteroefen.
Lettres — Lett.
Lettuce — Lettuce
Letzbeuger — Letzeb.
Levé — Levé
Levende — Levende
Liaison — Liais.
Liasnoi — Liasn.
Libanaise — Liban.
Liberal — Liberal
Liberärgeschichte — Literärgesch.
Libico — Libico
Libraire — Libr.
Librarian — Librar.
Librarie(s), Library — Libr.
Libre — Libre
Licentiater — Licent.
Lichen — Lichen
Lichenologica, Lichénologie, Lichénologique, Lichenology — Lichenol., Lichénol.
Lidovychovného — Lidovychovneho
Liebhaber — Liebhaber
Liečivé — Lieč.
Liefhebbers — Liefhebb.
Liefländischen — Liefl.
Liegenden — Liegenden
Life — Life
Light — Light
Ligure — Ligure
Ligustico — Ligustico
Likars'koi — Likars'k.
Lilac — Lilac
Lilien — Lilien
Liliengesellschaft, Lilien-Gesellschaft — Lilienges., Lilien-Ges.
Lilium(s) — Lilium(s)
Lilloana — Lilloana
Lily — Lily
Lime — Lime
Limnologia, Limnologica(l), Limnologičeskoj, Limnologicheskoi, Limnologie(se), Limnologique, Limnologischen, Limnologiska, Limnology — Limnol.
Limonnika — Limonnika
Limousin — Limousin
Lincei — Lincei
Linhas — Linhas
Linnaean(a) — Linn.
Linné-Sallskapets — Linné-Sallsk.
Linnean, Linnéenne, Linnéska — Linn.
Lipid — Lipid
Lipsiensia, Lipsiensis — Lips.
Liquides — Liquides
List(s) — List(s)
Listesi — List.
Listok — Listok
Listy — Listy
Literaria, Literarie, Literarii, Literario, Literarische(r), Literarum, Literary, Literatur(a) — Lit.
Literaturberichte — Literaturber.
Literatur-Blatt [etc.] — Lit.-Blätt.
Literaturblatt [etc.] — Literaturbl.
Literature — Lit.
Literatur-Föreningen — Lit.Fören.
Literaturinformation — Lit.Inform.
Literaturschau — Lit.-Schau
Literaturselskab — Literaturselsk.
Literatur-Tidning — Lit.-Tidn.
Literatury — Lit.
Literaturzeitung — Literaturzeitung
Literatur-Zeitung — Lit.-Zeitung
Lithosphere — Lithosphere
Litologiya — Lit.
Litovskoi S S R, Litovskoj S S R — Litovsk. S.S.R.
Litovskoi, Litovskoj — Litovsk.
Littéraire(s), Litteraria, Littéraries, Litterarische, Litterarium, Litterarum, Litteratur(en) — Litt.
Litteraturblatt [etc.] — Literaturbl.
Litteraturgeschichte — Litteraturgesch.
Litteraturjournal — Litteraturj.
Litteraturzeitung — Litteraturzeitung
Littoral — Littoral
Lituanica — Lituanica
Livada — Livada
Livestock — Livestock

Living — Liv.
Livländische — Livl.
Livres — Livres
Livros — Livros
Ljasnoj — Ljasn.
Ljubitelej — Ljubit.
Ljudej — Ljudej
Locator — Locator
Lodziensis, Lodzkie — Lodz.
Logis — Logis
Loisir — Loisir
Londinensis — Lond.
London — London
Long — Long
Loodus — Loodus
Looduskaitse — Loodusk.
Loodusteaduse — Loodustead.
Loodusteaduste — Loodustead.
Loodusuurijate — Loodusuur.
Lore — Lore
Los Angeles State and County Arboretum — Lasca
Louisiana State University — L. S. U.
Lov(a) — Lov(a)
Lovaniensis — Lovan.
Lover — Lover
Lowland — Lowl.
Lozaro, Lozarska, Lozarstvo — Loz.
Lubyavogo — Lubyav.
Lucrări(le) — Lucr.
Lufthygiene — Lufthyg.
Lugovogo — Lugov.
Luhů — Luhů
Lumberman — Lumberman
Luminary — Luminary
Lundensis — Lund.
Lüneburgischen — Lüneburg.
Luonnon — Luonnon
Luonto — Luonto
Lusatica — Lusatica
Lusitana — Lusit.
Lustgården — Lustgården
Luxembourgeois — Luxemb.
Luxus — Luxus
L'vis'kyj — L'vivs'k.
L'vovskij — L'vovsk.
Lyasnoi — Lyasn.
Lycée — Lycée
Lyceum(s) — Lyceum(s)
Lyonnais — Lyon.
Lysine — Lysine
Lyubitel', Lyubitelei — Lyubit.
Lyudei — Lyudei
Maandblad — Maandbl.
Maandelijksche — Maandel.
Maandschrift — Maandschur.
Maanedsoversigt — Maanedsovers.
Maanedsskrift — Maanedsskr.
Maataloustieteellinen — Maataloust.
Maatschappij, Maatschappy — Maatsch.
Macaronésica — Macaronés.
Macedonici — Maced.
Macromolecules — Macromolec.
Maderas — Maderas
Madjalah, Madjalak, Madjelis — Madj.
Magasb — Magasb
Magasin, Magazin(e), Magazzino — Mag.
Magni — Magni
Magyar — Magyar
Magyarhoni — Magyarhoni
Magyarország — Magyarorsz.
Mährisch(e,en) — Mähr.
Mahrisch-Schlesischen — Mahr.-Schles.
Maine Potato Growers — M. P. G.
Maintenance — Maint.
Mainzer — Mainzer
Mainzischen — Mainz.
Maison — Maison
Maíz — Maíz
Maize — Maize
Maizuru — Maizuru
Majhälsning — Majhälsn.
Major — Major
Malacitana — Malac.
Maladies — Malad.
Malattie — Malatt.
Malayan — Malayan
Malayan Agri-Horticultural Association — M. A. H. A.
Malayensium — Malayensium
Malaysian(a) — Malaysian(a)
Maler — Maler
Malesiana — Males.
Malgache — Malgache
Mali — Mali
Mallorquina — Mallorquina
Malomipar — Malom.
Mammillaria — Mammillaria
Mammillarienfreunde — Mammillarienfr.
Man — Man
Management — Managem.
Manager — Manager
Mangrove — Mangrove
Manifatture — Manifatture
Manipulation — Manipul.
Mannheimer — Mannheimer
Mannichfaltigkeiten — Mannichfaltigk
Mannigfaltigkeiten — Mannigfaltigk.
Manual — Manual
Manufacturer — Manufacturer
Manufactures — Manufactures
Manure — Manure
Manuscripts — Manuscripts
Manuscrits — Manuscrits
Maple — Maple
Mar — Mar
Maraîchers — Maraîch.
Marburger, Marburgische — Marburg.
March — March
Marichères — Marich.
Mariculture — Maric.
Marina(s), Marine, Marinha, Marinka — Mar.
Maritima, Maritime(s) — Marit.
Market(ing) — Market(ing)
Märkisch(e,en) — Märk.
Marklära — Marklära
Märkmed — Märkmed
Marocaine — Maroc.
Marquerite — Marquerite
Marsh — Marsh
Martonvásári — Martonvasari
Masarykovy — Masarykovy
Massachusetts — Mass.
Matar'jaly — Matar'j.
Matar'yaly — Matar'y.
Mate, Maté — Mate, Maté
Matematičeskih, Matematički(h), Matematično — Mat.
Matematično-Prirodoslovnega — Mat.-Prir.
Matematiche(n), Matematicheskikh, Matematicheskomu, Matematick(é,o), Matematika, Matematike, Matematiki, Matematikos — Mat.
Matematisk-Naturvidenskapelig — Mat.-Naturvidensk.
Matematisk-Naturvitenskapelig — Mat.-Naturvitensk.
Matematisko-Přírodovědecké — Mat.-Přír.
Matematyčno-Pryodopysno-Likars'koji — Mat.-Pryr.-Likars'k
Matematyczno-Fizyczne — Mat.-Fiz.
Matematyczno-Przyrodnicz(ego,y,ych) — Mat.-Przyr.
Matematycznych, Matematyzno — Mat.
Materia(ae), Materiały, Materialam, Materialien, Materials, Materialy, Matériaux, Materiel — Mater., Matér.
Mathematical, Mathemáticas, Mathematické, Mathematics, Mathematik(ai), Mathematiques, Mathematisch(e) — Math.
Mathematisch-Naturwissenschaftlich-Ärztlichen — Math.-Naturwiss.-Ärztl.
Mathematisch-Physikalische, Mathematisch-Psysische — Math.-Phys.
Mathematische-Naturwissenschaftliche — Math.-Naturwiss.
Mathematisk-Naturvidenskabelig — Math.-Naturvidensk.
Mathmaticarum — Math.
Matice — Matice
Matière — Matière
Mauritensis — Maurit.
Mayenne — Mayenne
Mayenne-Sciences — Mayenne-Sci.
Mayor — Mayor
Mcenareta — Mcenareta
Mechanic(al,'s) — Mech.
Mechanism — Mechanism
Mecklenburgischen — Mecklenburg.
Mecmuasi — Mecm.

Mecnierebatha — Mecniereb.
Medan — Medan
Meddeland(e,en), Meddelelser — Meddel.
Médecale(e), Médecine — Méd.
Mededeelingen — Meded.
Mededelingenblad — Mededelingenblad
Media — Media
Medic(a,al,al,ale), Médicale(s), Médicas, Medicin(a,ais,al,ales,alium), Medicine(s), Medicinische, Medicinskih, Medico, Médico(s) — Med., Méd.
Medico-Botanical — Med.-Bot.
Médico-Chirurgicale, Medico-Chirurigici — Méd.-Chir., Med.-Chir.
Medico-Farmaceutica, Médico-Farmacéutico — Med.-Farm., Méd.-Farm.
Médico-Nacional — Méd.-Nac.
Médico-Physique — Méd.-Phys.
Médico-Quirúrgica — Méd.-Quir.
Medicorum, Medicus — Med.
Medio — Medio
Mediterranea(n), Méditerranéenne(s), Mediterraneos — Medit., Médit.
Meditsina — Med.
Mediului — Mediului
Medizin — Med.
Medizinhistorisches — Medizinhist.
Medizinisch(e,en,er) — Med.
Medlemmers — Medlemmers
Medlemsblad — Medlemsblad
Medycyny — Med.
Meeresforschungen — Meeresf.
Meereskunde — Meeresk.
Meereskundliche — Meereskundl.
Meeresuntersuchungen — Meeresuntersuch.
Meeting(s) — Meeting(s)
Meghatározo — Meghat.
Megyei — Megyei
Méh — Méh
Meister — Maister
Mejora — Mejora
Mejoramiento — Mejoram.
Mejores — Mejores
Mélanges — Mélanges
Meldinger — Meld.
Melhoramento — Melhor.
Melioracji, Melioratsii — Melior.
Melitense — Melit.
Member — Memb.
Membrane(s) — Membr.
Membrany — Membrany
Membres, Membrorum — Memb.
Memoir(s), Mémoire(s), Memoranda, Mémori(al,aux), Memori(s), Memoria(l), Memórias, Memorie, Memoriile, Memuary — Mem., Mém.
Men('s) — Men('s)
Menara — Menara
Mennesket — Mennesk.
Menores — Menores
Mens Garden Club of America — M. G. C. A.
Mensajero — Mensajero
Mensal — Mensal
Mensch(en) — Mensch(en)
Menschheit — Menschh.
Mensile, Mensili — Mens.
Menskae — Menskae
Menskaya — Mensk.
Mensual, Mensuel(s), Mensuelle — Mens.
Mer — Mer
Mercredi — Mercredi
Mercure — Mercure
Mercurio — Mercurio
Mercuur — Mercuur
Merentutkimuslaitoksen — Merentutkimuslait.
Merkantilischer — Merkantil.
Merkblätter — Merkbl.
Merkur — Merkur
Merkwürdige(n) — Merkwürd.
Merkwürdigkeiten — Merkwürdigk.
Merkwürdigsten — Merkwürdigsten
Mesičnik — Mesičnik
Mésogéen — Mésogéen
Message(r) — Message(r)
Messenger — Messenger
Mesterségi — Mest.
Metallurgischen — Metallurg.
Meteorological, Meteorologické, Meteorologii, Météorologique, Meteorologische(n), Meteorology — Meteorol.
Methodik, Methods — Meth.
Metiers — Metiers
Metodické — Metod.
Metraštis — Metraštis
Metroparks — Metroparks
Metropolitan — Metrop.
Metsanduse, Metsanduslikud — Metsand.
Metsaosakonna — Metsaosak.
Metsäpuiden — Metsäp.
Metsätieteellisen — Metsätiet.
Metsätutkimuslaitoksen — Metsätutkimuslait.
Mexicana, Méxicana, Mexicano — Mex., Méx
Mexikofilu — Mexikofilu
Mezei — Mezei
Mezhvuzovskii — Mezhvuzov.
Meziriči — Meziriči
Mezögazdaság(i) — Mezögazd.
Mezögazdaságtudományi — Mezögazdaságtud.
Mežsaimniecibas — Mežsaimn.
Mežu — Mežu
Michurinskoi — Michurinskoi
Micologia — Micol.
Microbe — Microbe
Microbial — Microbial
Microbiologia, Microbiologica(l), Microbiologie, Microbiologique — Microbiol.
Microbiologist — Microbiologist
Microbiologiya, Microbiology — Microbiol.
Microfilm — Microfilm
Microflora — Microfl.
Micrographie — Microgr.
Micrology — Microl.
Micron — Micron
Microorganisms — Microorg.
Micropalaeontological, Micropalaeontologist, Micropalaeontology — Micropalaeontol.
Micropaleontologia, Micropaléontologie — Micropaleontol., Micropaléontol.
Microscope — Microscope
Microscopica(l), Microscopie, Microscopist, Microscopy — Microscop.
Mičurinskoj — Mičurinskoj
Mid-America — Mid-Amer.
Middle — Middle
Midland — Midl.
Midwest, Mid-West — Midw., Mid-W.
Migliore — Migl.
Mijngebied — Mijngeb.
Mijnwesen — Mijnw.
Mikologiae, Mikologičeskje [etc.], Mikologicheskie [etc.], Mikologii [etc.] — Mikol.
Mikrobiológiai, Mikrobiologiakan, Mikrobiologičeskij [etc.], Mikrobiologičnyj, Mikrobiologicheskii [etc.], Mikrobiologie, Mikrobiologii, Mikrobiologija, Mikrobiologikes — Mikrobiol.
Mikrochemica — Mikrochem.
Mikroelementry — Mikroelem.
Mikrographische — Mikrogr.
Mikrokosmos — Mikrokosmos
Mikrologischen — Mikrol.
Mikroorganismen — Mikroorgan.
Mikropaleontologii — Mikropalentol.
Mikroskop — Mikroskop
Mikroskopie, Mikroskopiker, Mikroskopische — Mikroskop.
Milanese — Milan.
Milieu — Milieu
Milieuzorg — Milieuz.
Militaires — Militaires
Militar(y) — Militar(y)
Miljövård — Miljövård
Millet(s) — Millet(s)
Milotický — Milot.
Mimeografades, Mimeograph(ed) — Mimeogr.
Mindenee — Mind.
Mineiros — Mineir.
Mineralogicheskago — Mineral.

Mineralogie, Minéralogie — Mineral., Minéral.
Mineralogisk-Geologiska — Mineral.-Geol.
Mineralogy — Mineral.
Mineria — Mineria
Minero — Minero
Mines — Mines
Mining — Mining
Minister(e), Ministére, Ministeria, Ministerio, Ministerium, Ministerstva, Ministry — Minist.
Minisztérium — Miniszt.
Minösitett — Minös.
Minulost — Minulost
Minutes — Minutes
Mira — Mira
Mironovskii — Mironovsk.
Mirror — Mirror
Miscelânea(s), Miscellanea, Miscellaneous, Miscellanies, Miscellany, Miscellen — Misc.
Mise(s) — Mise(s)
Mision, Misión — Mision, Misión
Miskolciensis — Miskolc.
Missionary — Missionary
Missions — Missions
Missoes — Missoes
Mitglieder — Mitgl.
Mitgliederverzeichnis — Mitgliederverz.
Mitteilungen — Mitt.
Mitteilungsblatt [etc.] — Mitteilungsbl.
Mitteilungsheft — Mitteilungsh.
Mittel — Mittel
Mitteldeutsche — Mitteldeutsche
Mitteldeutschland — Mitteldeutschl.
Mitteleuropäische — Mitteleurop.
Mittheilungen — Mitth.
Mixte — Mixte
Mjarkmed — Mjarkmed
Moambe — Moambe
Mode(n) — Mode(n)
Modelling — Modelling
Modenese — Modenese
Moden-Zeitung — Moden-Zeitung
Møder — Møder
Modern(e) — Modern(e)
Moderno — Moderno
Moete — Moete
Möglinsche — Möglinsche
Moguntinae — Mogunt.
Moj — Moj
Mokslo — Mokslo
Mokslu — Mokslu
Mokyklu — Mokyklu
Moldavii — Moldav.
Moldavskoi, Moldavskoj — Moldavsk.
Moleculaire, Molecular, Molecules — Molec.
Molekularnej [etc], Molekularnii [etc.] — Molek.
Molineria — Molineria
Moliniana — Moliniana
Molodyh, Molodykh — Molod.
Molotovskii [etc.], Molotovskij [etc.] — Molotovsk.
Monathliche — Monathl.
Monatliche — Monatl.
Monats — Monats
Monatsberichte — Monatsber.
Monatsblatt [etc.] — Monatsbl.
Monats-Blatt [etc.] — Monats-Blätt
Monatschrift — Monatschr.
Monatshefte — Monatsh.
Monatsschrift — Monatsschr.
Monde(s) — Monde(s)
Mondstuk — Mondstuk
Mongol'skoi, Mongol'skoj — Mongol'sk.
Monisteita — Monist.
Moniteur, Monitor(ing) — Monit.
Monografias, Monografie, Monografii, Monograph(s), Monographiae, Monographic, Monographien — Monogr.
Monsoon — Monsoon
Monspeliensia, Monspeliensis — Monspel.
Montagna — Montagna
Montagne, Montana [Italian] — Mont.
Montanaro — Montan.
Montes — Montes
Month(ly) — Month(ly)
Monti — Monti
Monumenta, Monuments — Monum.
Moorkulturstation — Moorkulturstat.
Moorkunde — Moork.
Moorzeitschrift — Moorz.
Moraes — Moraes
Morale — Morale
Moravska, Moravské(ho) — Morav.
Moravsko-Slezské — Morav.-Slez.
Moravský — Morav.
Morbi(d) — Morbi(d)
Mordovskogo — Mordovsk.
Morfologia, Morfologija, Morfologiya — Morfol.
Morphogenesis — Morphogen.
Morphologiae, Morphologie, Morphology — Morphol.
Morskogo — Morsk.
Morya — Morya
Mosana — Mosana
Moskauer — Moskauer
Moskovskii [etc.], Moskovskij [etc.] — Moskovsk.
Mosquensem — Mosq.
Mosskulturföreningen — Mosskulturfören.
Möte — Möte
Mount — Mount
Mountain — Mount.
Mountaineer — Mountaineer
Movement(o) — Movem.
Movimento — Movim.
Mücedale — Müc.
Müködéséröl, Muküdése — Mük.
Mulatságok — Mul.
Mundi — Mundi
Municipal — Munic.
Munkái — Munkái
Munkak, Munkalatai — Munk.
Murithienne — Murith.
Murmanskoi [etc.], Murmanskoj [etc.] — Murmansk.
Musea — Mus.
Musealblatt — Musealblatt
Musealverein(s) — Musealverein(s)
Musée, Museen, Museet(s), Musei, Museí, Museja, Musejní, Musejního, Museo(rum), Museu, Museum(s), Muséum — Mus.
Museumshefte — Museumsh.
Museumskunde — Museumsk.
Museums-Vereins — Mus.-Verein
Mushroom Growers' Association — M. G. A.
Mushroom(s) — Mushr.
Mustard — Mustard
Müszaki — Müsz.
Mutabil'nost — Mutabil.
Mutagen(es,ez) — Mutagen
Mutation — Mutat.
Mutuel — Mutuel
Müveszetek — Müv.
Múzea, Muzeálnej, Muzeem, Muzeia, Muzej, Múzej, Muzeja, Muzeju — Múz., Muz.
Muzeologie — Muzepl.
Múzeu, Muzeul(ui), Muzeum, Múzeum — Múz.
Muzeumának, Múzeumi, Muzeya, Muzeyu — Muz.
Múzeum-Egylet — Múz.-Egyl.
Múzeum-Egysület — Múz.-Egyes
Mycena — Mycena
Mycologia, Mycologica(e,l), Mycologie, Mycologique, Mycologist, Mycologue, Mycology — Mycol.
Mycopathologia — Mycopathol.
Mycorrhizae — Mycorrh.
Mycovirus — Mycovirus
Mykologia, Mykologický, Mykologie, Mykologische(n,s) — Mykol.
Mykosen — Mykosen
Myslivosti — Mysliv.
Nabljudenija, Nabljudateljam, Nabljudenijam — Nabljud.
Nablyudeniya, Nablyudeniyam, Nablyudatelyam — Nablyud.
Nachrichten — Nachr.
Nachrichten-Blatt — Nachr.-Blatt
Nachrichtenblatt [etc.] — Nachrichtenbl.
Nacional(es) — Nac.
Nagrad — Nagrad
Nagy — Nagy
Nagygülese — Nagyg.

Nagygyülésének — Nagygyül.
Näheren — Näheren
Nahrung — Nahrung
Nankaiensis — Nankai.
Nankinensis — Nankin.
Nantaise — Nantaise
Napi — Napi
Napolitana — Napol.
Naptar — Naptár
Naravoslovnoga — Nar.
Näringarne — Näring.
Narodnago, Narodnai, Narodnaji, Narodne, Národného, Národního, Narodnyi, Národom, Narodowego — Nar., Nár.
Nas(a) — Nas(a)
Nasazhdeniya — Nasazhd.
Naše — Naše
Našeho — Našeho
Nasiennictwo — Nasienn.
Nasional — Nasl.
Nassauischen — Nassauischen
Nation — Nation
National(e,is) — Natl.
National-Museums — Natl.-Mus.
Nationalpark(es) — Nationalpark(es)
Nationaux — Natx.
Native — Native
Nátturufraedingurinn — Nátturufraedingurinn
Natur — Natur
Natura(e) — Nat.
Naturae-Curiosum — Nat.-Cur.
Naturais, Natural — Nat.
Natural history — Nat. Hist.
Natural-Technic — Nat.-Techn.
Naturale(s) — Nat.
Naturaleza — Naturaleza
Naturali — Nat.
Naturalia — Naturalia
Naturaliensammlung — Naturaliensamml.
Naturalientausch — Naturalientausch
Naturalis — Nat.
Naturaliştilor — Naturaliştilor
Naturalist(a,e,es,i,s,'s,s') — Naturalist(a,e,es,i,s,'s,s')
Naturalium — Nat.
Naturbeobachter — Naturbeob.
Naturdenkmäler — Naturdenkmäler
Naturdenkmalpflege — Naturdenkmalpflege
Nature — Nat.
Nature-Study — Nat.-Stud.
Naturelle(s), Naturen, Natureza — Nat.
Naturforschenden, Naturforscher — Naturf.
Naturforscher-Gesellschaft — Naturf.-Ges.
Naturforscher-Vereins — Naturf.-Vereins
Naturforschergesellschaft — Naturforscherges.
Naturforscherverein — Naturforscherverein
Naturforskare, Naturforskeres — Naturf.
Naturforskermøde — Naturforskermøde
Naturfreund(e) — Naturfr.
Naturgeschichte, Natur-Geschichte — Naturgesch., Natur-Gesch.
Naturhistorica — Naturhist.
Naturhistorie — Naturhist.
Naturhistorie-Selskabet — Naturhist.-Selsk.
Naturhistorisch-Medicinischen — Naturhist.-Med.
Naturhistorische [etc.], Natur-Historische [etc.] — Naturhist., Natur-Hist.
Naturii — Nat.
Naturkunde — Naturk.
Naturkundemuseums — Naturkundemus.
Naturkundlichen — Naturk.
Naturkundskab — Naturkundsk.
Naturlehre, Natürlichen — Naturl., Natürl.
Naturparke — Naturparke
Naturpflege — Naturpflege
Naturschurzbrief — Naturschurzbrief
Naturschutz — Naturschutz
Naturschutzarbeit — Naturschutzarbeit
Naturschutzblatt [etc.] — Naturschutzbl.
Naturschutzgebiete — Naturschutzgeb.
Naturschutzkommission — Naturschutzkommiss.
Naturschutzparke — Naturschutzparke
Naturschutzstelle — Naturschutzstelle
Naturschutztag — Naturschutztag
Naturschutzvereins — Naturschutzvereins
Naturschutzverwaltung — Naturschutzverwalt.
Naturskyddsärende(n) — Naturskyddsärende(n)
Naturskyddsförenings — Naturskyddsfören.
Naturstoffe — Naturst.
Naturvårdsverk — Naturvårdsverk
Naturvetenskap(liga), Naturvidenskab, Naturvidenskabeligt, Naturvidenskaben, Naturvidenskaberne(n,s), Naturvidenskapens, Naturvitenskap(elig) — Naturvetensk.
Naturwart — Naturwart
Natur-Wetenskaperne — Natur-Wetensk.
Naturwissenschaft(en) — Naturwiss.
Naturwissenschaftlich-Medicinischen — Naturwiss.-Med.
Naturwissenschaftliche(n,r,s) — Naturwiss.
Natuur — Natuur
Natuurbeheer — Natuurbeheer
Natuurbehoud — Natuurbehoud
Natuurbescherming — Natuurbesch.
Natuurbeschouwer — Natuurbeschouwer
Natuurhistorisch — Natuurhist.
Natuurkunde, Natuurkundige — Natuurk.
Natuurleven — Natuurleven
Natuurlijke — Natuurl.
Natuurmonumenten — Natuurmonum.
Natuuronderzoeker — Natuuronderzoeker
Natuurvriend — Natuurvriend
Natuurwereld — Natuurwereld
Natuurwetenschappen, Natuurwetenschappilijk(e) — Natuurwetensch.
Natuurwetenskappe — Naturwetensk.
Natyres, Natyrore — Nat.
Nauchni — Nauchni
Nauchno — Nauchno
Nauchno-issledovatel'skii [etc.] — Nauchno-Issl.
Nauchnyi [etc.] — Nauchn.
Naučnaja [etc.] — Naučn.
Naučni — Naučni
Naučno-Issledovatel'skij [etc.] — Naučno-Issl.
Naučno-Opytnyh — Naučno-Opytn.
Naučno-Prikladnyh — Naučno-Prikl.
Naučno-Promyslovoj — Naučno-Promysl.
Naučno-Tehničeskoj — Naučno-Tehn.
Nauk(a) — Nauk(a)
Nauke — Nauke
Nauki — Nauki
Naukite — Nauk.
Naukova-Doslidnogo — Nauk.-Doslidn.
Naukovi, Naukovogo, Naukovy(h), Naukowe(go), Naukowy — Nauk.
Nauku — Nauku
Nautical — Naut.
Naval — Naval
Navuk(ova) — Navuk.
Navy — Navy
Nazemnye — Nazem.
Nazionale — Naz.
Nea — Nea
Necklace — Necklace
Nederland(en) — Ned.
Nederlands-Belgische — Ned.-Belg.
Nederlandsch(e) — Ned.
Nederlandsch-Indië — Ned.-Indië
Nederlandsch-Indisch(e) — Ned.-Indische(e)
Nederlandse — Ned.
Needle — Needle
Néerlandaise, Néerlande, Neerlandica(e) — Néerl., Neerl.
Neglected — Negl.
Neitronov — Neitron.
Nemesítés — Nemes.
Nemzeti — Nemz.
Neotropica(l) — Neotrop.
Nép — Nép
Népalais, Nepalese — Népal., Nepal.
Nervose, Nervous — Nerv.
Neskorsie — Neskorsie
Network — Network
Neu [etc.] — Neu [etc.].

Neuchâteloise — Neuchâtel.
Neuerrichteten — Neuerrichteten
Neuerung(en) — Neuerung(en)
Neueste(n,s) — Neueste(n,s)
Neu-Fortgesetzter — Neu-Fortgesetzter
Neuigkeiten — Neuigk.
Neujahrsblatt — Neujahrsbl.
Neunzehnten — Neunzehnten
Neurologia, Neurology — Neurol.
Neuropsychología — Neuropsychología
Neurospora — Neurospora
New series — N.S.
New South Wales — New South Wales
New(s) — New(s)
Newsbriefs — Newsbr.
Newsheet, News-sheet — Newsheet, News-Sheet
Newsletter, News-Letter — Newslett., News-Lett.
Newspaper — Newspaper
Newsreel — Newsr.
Newswatch — Newswatch
Nezavisne — Nezav.
Nicaraguense — Nicarag.
Nickel — Nickel
Nicotiana — Nicotiana
Niederösterreichischen — Niederösterr.
Niederrheinische(n,r,s) — Niederrhein.
Niedersächsische(n) — Niedersächs.
Nieuwsblad — Nieuwsblad
Nieuwsbulletin — Nieuwsbull.
Nieuwste — Nieuw(e,s)
Nigerian — Nigerian
Nijverheid — Nijverh.
Nikitskii [etc.], Nikitskij [etc.] — Nikitsk.
Nipponica — Nippon.
Nisipurilor — Nisipur.
Nitrogen — Nitrogen
Nizhnem — Nizhnem
Nizših — Nizš.
Nizshikh — Nizsh.
Noire — Noire
Nomenclature(al) — Nomencl.
Nominatae, Nominati — Nomin.
Non-Technical — Non-Techn.
Noorboltens — Noorboltens
Noord, Nord — N.
Nordböhmischen, Nordboehmischen — Nordböhm., Nordboehm.
Nordeste — Nordeste
Nordische(s) — Nord.
Nordishavet — Nordishavet
Nordisk(e) — Nord.
Nordkaukasischen — Nordkaukas.
Nördlichen — Nördl.
Nordoberfränkischen — Nordoberfränk.
Norges — Norg.
Noricum — Noricum
Normal(e) — Norm.
Noroeste — Noroeste
Nororossijskago — Novorossijsk.
Norrbottens — Norrbottens
Norsk(e) — Norsk(e)
Norte, North — N.
Northeast(ern) — NorthE.
Northern — N.
Northwest — NorthW.
Norvegica — Norveg.
Norwegisch(en) — Norweg.
Nos — Nos
Notas — Notas
Notationes — Notat.
Note(s) — Note(s)
Notice(r,s), Noticias, Notified — Not.
Notizblatt — Notizbl.
Notizen — Not.
Notizenblatt [etc.] — Notizenbl.
Notiziario — Notiz.
Notulae — Notul.
Nouă — Nouă
Noutati — Nout.
Nouveau(x), Nouvelle(s), Nouvelliste — Nouv.
Nova — Nova
Nova serie — N.S.
Novae — Novae
Novam — Novam
Novarum — Nov.
Novas — Novas
Novaya — Novaya
Novaya seriya — N.S.
Nové — Nové
Novedades — Noved.
Novel — Novel
Növéneubek — Növén.
Növényegeszegühi — Növényegész.
Növényfajták — Növényfajt.
Novenyfoldrajzi — Novenyfoldr.
Növénygazdasági — Növénygazd.
Növénynemesitési — Növénynemes.
Növenyrendszertani — Növenyrendsz.
Növénytani — Növényt.
Növénytermeles(i), Növénytermeszies, Növénytermesztés(i) — Növényterm.
Növényvédelem, Növényvédelen — Növényvéd.
Növényvilág — Növényvilág
Novi — Novi
Novia — Novia
Novices' — Novices'
Noviny — Noviny
Novis — Novis
Novissimae — Noviss.
Novitates — Novit.
Novočerkasskago — Novočerkassk.
Novogo — Nov.
Novorossiiskago — Novorossiisk.
Novorum — Nov.
Novosti — Novosti
Nový — Nový
Novye — Novye
Novykh — Nov.
Now — Now
Nuclear, Nucleic — Nucl.
Nucleo — Nucleo
Nueva Epoca — N.E.
Nueva — Nueva
Number(s) — Number(s)
Nuova — Nuova
Nuovi — Nuovi
Nuovo — Nuovo
Nürnbergische(n) — Nürnberg.
Nurseries — Nurseries
Nursery — Nursery
Nurseryman — Nurseryman
Nurserymen(s) — Nurserymen(s)
Nut — Nut
Nutgrowing — Nutgrow.
Nutrición, Nutrition, Nutrizione — Nutr.
Nutzbar(er,es) — Nutzbar(er,es)
Nutzen — Nutzen
Nützliche(n,r) — Nützl.
Nuusbrief — Nuusbrief
Ny — Ny
Nya — Nya
Nye(ste) — Nye(ste)
Nyt(t) — Nyt(t)
Nyttevekster — Nyttevek.
Oberbergisch(en) — Oberberg.
Ober-Erzgebürgisches — Ober-Erzgebürg.
Oberhessische — Oberhess.
Oberlausitzischen — Oberlausitz.
Oberlauszisch(en) — Oberlausiz.
Oberösterreichische(n,r) — Oberösterr.
Oberrheingebiet — Oberrheingeb.
Oberrheinische(n) — Oberrhein.
Oberschlesische — Oberschles.
Oberungarisches — Obereung.
Obespecheniyu — Obespech.
Oběžnik — Oběžn.
Obituary — Obit.
Objets — Objets
Oblast(i), Oblastniho, Oblastnoi [etc], Oblastnoj [etc.] — Obl.
Obmen — Obmen
Obogashchenie — Obogashch.
Oboru — Oboru
Obozrěnie — Obozr.
Obra(s) — Obra(s)
Obščee — Obščee
Obščej — Obščej
Obščestva, Obščestvom — Obšč.
Obščie — Obščie
Observation(es,s) — Observ.
Observatoire — Observatoire
Observatory — Observatory
Observer — Observ.
Obshchaya — Obshchaya
Obshchie — Obshchie
Obshchestva, Obshchestvo(m) — Obshch.
Obsledovanijam — Obsl.
Obsledovaniju, Obsledovaniyam, Obsledovaniyu — Obsl.
Obst — Obst
Obstbau — Obstbau

Obstbaufragen — Obstbaufragen
Obstbaum-Freund — Obstbaum-Freund
Obstétricale — Obstétr.
Obstetrics — Obstet.
Obstgarten — Obstgarten
Obstgärtner — Obstgärtn.
Obstkunde — Obstk.
Obstrundschau — Obstrundschau
Obstsort — Obstsort
Obstzüchter — Obstzüchter
Obtained — Obtained
Obzor — Obzor
Obzoy — Obzoy
Ocacional — Ocac.
Ocasionales — Ocas.
Occasional — Occas.
Occidental(e,is) — Occid.
Ocean(s) — Ocean(s)
Oceanic — Oceanic
Océaniennes — Océanien.
Océanistes — Océanistes
Oceanografía, Oceanográfica(s), Oceanografiyu — Oceanogr.
Océanographes — Océanographes
Oceanographica(a,l), Océanographie, Océanographique(s), Oceanography — Oceanogr., Océanogr.
Oceanologia, Oceanologica(l), Oceanologiczne, Oceanology — Oceanol.
Očerki — Očerki
Ocherki — Ocherki
Ochrana, Ochronie, Ochrony — Ochr.
Ocios — Ocios
Öconomisch-Patriotischen — Öcon.-Patriot
Öconomischen, Öconomisk — Öcon.
Ocrotirea — Ocrot.
Oculistica — Oculist.
Odboru — Odb.
Oddel — Oddel
Oddelek — Oddel.
Oddíl — Odd.
Odelje — Od.
Odes'koi [etc.], Odes'koji [etc.] — Odes'k.
Odessaer — Odessaer
Odesskogo — Odessk.
Odjeljenje — Odjeljenje
Odlare — Odlare
Odontoiatria, Odontologia — Odont.
Oecologia, Oecologica, Oecology — Oecol.
Oeconomica, Oeconomie, Oeconomique — Oecon.
Oeconomisch-Physikalische — Oecon.-Phys.
Oeconomisches, Oeconomiske — Oecon.
Oefversigt — Oefvers.
Oekologie — Oekol.
Oekonomie, Oekonomisch(e,en) — Oekon.
Oekonomisch-Botanisches — Oekon.-Bot.
Oekonomisch-Physikalische — Oekon.-Phys.
Oelandicus — Oeland.
Oesterreichisch(en,s) — Oesterr.
Öffentliche — Öffentl.
Office — Off.
Office de la recherche scientifique et technique d'Outre-Mer — O. R. S. T. O. M.
Officers, Official, Officieele, Officielle, Officinale, Officinali — Off.
Oficial, Oficina — Of.
Oftalmologica — Oftalmol.
Öfversigt — Öfvers.
Ogorody — Ogorod
Ogrodnicze, Ogrodniczo, Ogrodniczy — Ogrodn.
Ogrodu — Ogrodu
Ohoronjajte — Ohoronjajte
Ohoronu — Ohor.
Ohotnič'e-Promyslovoj — Ohotn.-Promysl.
Ohra — Ohra
Ohrana, Ohrane — Ohr.
Ohrenheilkunde — Ohrenheilk.
Oil — Oil
Oilseed(s) — Oilseed(s)
Oitu — Oitu
Ojczysta — Ojczysta
Oka-Station — Oka-Stat.
Okeanograficheskogo, Okeanografii, Okeanographikon — Okeanogr.
Okeanologii — Okeanol.
Okhorona, Okhoronu — Okhor.
Okhoronyaite — Okhoronyaite
Okhota — Okhota
Okhotnich'e — Okhotn.
Okhrana, Okhrane, Okhrany — Okhr.
Okolí — Okolí
Ökologie — Ökol.
Ökonomische(n) — Ökon.
Okrasného — Okrasn.
Okrasu — Okrasu
Okregowego — Okreg.
Okresniho — Okresn.
Okruga — Okr.
Okskoi, Okskoj — Oksk.
Oktyabr'skoi — Okt.
Old — Old
Oleicultura — Oleic.
Oleo — Oleo
Oleva — Oleva
Olim — Olim
Olimpica — Olimpica
Olive — Olive
Olivicultura — Olivicultura
Olomoucensis, Olomouckeho — Olomouc.
Olonetskoi — Olonetsk.
Olstenensis — Olsten.
Omeiesium — Omeiesium
Omeinensium — Omeinensium
Omgeving — Omgev.
Onderneming — Ondern.
Onderwijs — Onderwijs
Onderzoek — Onderz.
Onderzoekingstations — Onderzoekingstat.
Onderzoekverslag — Onderzoekverslag
Online — Online
Öntozésügyi — Öntozésügyi
Onze — Onze
Opasnye — Opasnye
Opavske — Opavske
Open — Open
Openbare — Openb.
Opera — Opera
Operation(s) — Operat.
Opetatud — Opet.
Ophtalmologie — Ophtalmol.
Ophthalmologic, Ophthalmologie, Ophthalmology — Ophthalmol.
Opinion — Opin.
Opolski — Opolski
Optical — Opt.
Options — Options
Opuscoli, Opuscula, Opuscules — Opusc.
Opyleniya — Opyl.
Opyt — Opyt
Opyten — Opyt.
Opytnoi [etc.], Opytnoj [etc.] — Opytn.
Opytnologo — Opytn.
Oral — Oral
Orbis — Orbis
Orchard(ist) — Orchard(ist)
Orchid(s) — Orchid(s)
Orchideen — Orchid.
Orchideen-Gesellschaft — Orchideen-Ges.
Orchideengesellschaft — Orchideenges.
Orchidées — Orchidées
Orchidianae — Orchid.
Orchidist — Orchidist
Orchidologia, Orchidology — Orchidol.
Ordena — Ordena
Ordinarie — Ordinarie
Ordinary — Ordinary
Oreille — Oreille
Orenburgskago, Orenburgskogo — Orenburgsk.
Orgaan — Orgaan
Organ(e,i,ic,s) — Organ(e,i,ic,s)
Organicheskaya — Organich.
Organisatie, Organisation — Organ.
Organischer — Organischer
Organism(os,s) — Organism(os,s)
Organismen — Organismen
Organizacionnye, Organization, Organizatsionnye — Organ.
Orges — Orges
Oriental(is,ische,s) — Orient.
Oriente — Oriente

Orients — Orients
Orientvereins — Orientvereins
Origin(s) — Origin(s)
Original(e,es) — Original(e,es)
Orlovskii — Orlovsk.
Orman — Orman
Ormancilik — Orman.
Ornamentais — Ornam.
Ornamental(s) — Ornam.
Orquideas — Orquideas
Orquideologia — Orquideol.
Orquidofilas — Orquidofilas
Országos — Orsz.
Orthopädie — Orthopäd
Orthopedic — Orthop.
Orticola, Orticoltura, Orticulture — Ortic.
Orto — Orto
Ortofloricoltura — Ortofloric.
Ortoflorofrutticoltura — Ortoflorofruttic
Orvos — Orv.
Orvosi, Orvosok — Orv.
Orvos-Termĕszettudományi — Orv.-Természettud.
Oscensis — Oscensis
Oseanologi — Oseanol.
Osnabrücker — Osnabrück.
Osnovy — Osnovy
Osobogo — Osob.
Ospedale, Ospedali — Osped.
Osrodka — Osrod.
Osservatore, Osservatorio — Osserv.
Ostasiatische — Ostasiat.
Ostbau — Ostbau
Ostdeutscher — Ostdeutsch.
Osterlande, Osterländische — Osterl.
Österreichisch-Ungarischer — Österr.-Ung.
Österreichische(n,r) — Österr.
Ostetricia — Ostet.
Ostfriesische, Ost-Friesische — Ostfries., Ost-Fries.
Ostmärkischer — Ostmärk.
Ostpreussisch — Ostpreuss.
Ostpreussisch-Mohrungschen — Ostpreuss.-Mohrungschen
Ostravského — Ostrav.
Ostseeprovinzen — Ostseeprov.
Ost-Sibirischen — Ost-Sibir.
Osvoenie — Osvoenie
Osvoeniya — Osvoeniya
Osztálya, Osztályának, Osztályok — Oszt.
Otčet(y) — Otčet(y)
Otchet(y) — Otchet(y)
Otděl(a) — Otd.
Otdělenie — Otdělenie
Otdělenija — Otdělenija
Otděleniju — Otděleniju
Otdělov — Otdělov
Other — Other
Otolaryngologic, Oto-Laringologica — Otolaryngol., Oto-Laringol.
Otologica, Otology — Otol.
Otorinolaringoliatrica, Oto-Rinolaringologia — Otorinolaringol., Oto-Rinolaringol.
Otrjada — Otrjada
Otryada — Otryada
Ottalmologia — Ottalmol.
Ouest — Ouest
Ouluensis — Oulu.
Ouralienne — Oural.
Outdoors — Outdoors
Outlook — Outlook
Outre-Mer — Outre-Mer
Outside — Outs.
Ouvrage(s) — Ouvrage(s)
Ouvrant — Ouvrant
Over — Over
Overseas — Overseas
Oversigt — Overs.
Ovocinár — Ovocinár
Ovocnarstvi — Ovocnar.
Ovocnické, Ovocnického, Ovocnictvi, Ovocnictvi — Ovocn.
Ovošči — Ovošči
Ovoščnye — Ovoščn.
Ovoshchevod — Ovoshchevod
Ovoshchnye — Ovoshchn.
Ovoshtarstvo — Ovoshtarstvo
Ovoštarstvo — Ovoštarstvo
Ozelenyavane — Ozelen.
Ozera — Ozera
Ozere — Ozere
Ozernoj — Ozernoj
Pachter — Pachter
Pacific(a,um) — Pacific(a,um)
Pacifique — Pacifique
Pädagogisch(en) — Pädagog.
Pădurilor — Pădur.
Paedagogicae — Paedagog.
País — País
Paisaje — Paisaje
Palackianae — Palack.
Palaebiologica — Palaebiol.
Palaeobotanica, Palaeobotanika — Palaeobot.
Palaeobotanist(s) — Palaeobotanist(s)
Palaeobotany — Palaeobot.
Palaeoclimatology — Palaeoclimatol.
Palaeoecology — Palaeoecol.
Palaeogeography — Palaeogeogr.
Palaeontographica — Palaeontogr.
Palaeontologica(l) — Palaeontol.
Paläobotanik — Paläobot.
Paläontologie, Paläontologische(r,s) — Paläontol.
Paläophytologie — Paläophytol.
Palatina — Palat.
Paleo — Paleo
Paleobiology — Paleobiol.
Paleobotanika, Paleobotany — Paleobot.
Paleolimnology — Paleolimnol.
Paleontologičeskij, Paleontologica, Paleontologicheskii, Paleontologija, Paleontologiya, Paleontology — Paleontol.
Palermitana — Palermitana
Palinologica, Palinologii, Palinologos — Palin.
Palm(a) — Palm(a)
Palmen — Palmen
Palmengarten — Palmengart.
Pálya — Pálya
Pályamunkak — Pályamunkak
Palynologia, Palynologica(l) — Palynol.
Palynologist(s) — Palynologist(s)
Palynology — Palynol.
Památkové — Památk.
Památky — Památky
Pamiatok — Pamiatok
Pamiętnik — Pamiętn.
Pamirskoi — Pamirsk.
Pamphlet(s) — Pam.
Pamyati — Pamyati
Pan — Pan
Panaderia — Panaderia
Panamericana — Panamer.
Pannonius — Pannon.
Panorama — Panor.
Panstwowy, Państwowych — Panstw., Państw.
Papeis, Paper(s), Papiru — Pap.
Pappelvereins — Pappelvereins
Pappelwirtschaft — Pappelw.
Papperstidning — Papperstidn.
Papuan — Papuan
Paraense — Paraense
Paraguayos — Parag.
Paranaense — Paran.
Parasitaire — Parasit.
Parasitenkunde — Parasitenk.
Parasitic — Parasitic
Parasitologiá, Parasitologie, Parasitology — Parasitol.
Parassiti — Parassiti
Parassitologia — Parassitol.
Parazitologie — Parazitol.
Parc(s) — Parc(s)
Pardubicensis — Pardubic.
Parfumerie — Parfum.
Park('s,s) — Park('s,s)
Parku — Parku
Parkvård — Parkvård
Parmense — Parmense
Parques — Parques
Pars — Pars
Part(s) — Part(s)
Particulier — Partic.
Partie(s) — Partie(s)
Pasoh — Pasoh
Pastbisč(a), Pastbishch — Pastb.
Pastos — Pastos
Pasture(s) — Pasture(s)
Pasturos — Pasturos
Patata — Patata

Patavina — Patavina
Patent(s) — Patent(s)
Pathobiology — Pathobiol.
Pathogen(es), Pathogenic — Pathog.
Pathologica(l), Pathologie, Pathologique — Pathol.
Pathologist — Pathologist
Pathology — Pathol.
Patologia, Patologie, Patologii — Patol.
Patrika — Patrika
Patrimonio — Patrim.
Patriotica, Patriótica — Patriot., Patriót.
Patriotisch-Ökonomischen — Patriot.-Ökon.
Patriotische(n,s) — Patriot.
Paturages — Patur.
Paulistas — Paul.
Pays(age) — Pays(age)
Peach — Peach
Peanut — Peanut
Pear — Pear
Peat — Peat
Pecan — Pecan
Péče — Péče
Pêches — Pêches
Pecuaria — Pecu.
Pedagógiai, Pedagogical, Pedagogičeskij [etc.], Pedagogicheskii [etc.], Pedagogické(ho), Pedagogických, Pédagogique — Pedagóg., Pedagog., Pédagog.
Pediatria — Pediatria
Pedobiologia — Pedobiol.
Pedologica, Pedologie, Pédologie — Pedol., Pédol.
Pekinensis — Pekin.
Pelagos — Pelagos
Pellagrologica — Pellagrol.
Pelotas — Pelotas
Pemberitaan — Pemberit.
Penelitian — Penelitian
Penerbitan — Penerbitan
Pengembangan — Pengemb.
Pengetahuan — Penget.
Penggemar — Pengg.
Pengumuman — Pengum
Penicillin — Penicillin
Peninsula — Penins.
Penjelidikan — Penjel.
Penn — Penn
Penny — Penny
Penstemon — Penstemon
Pentru — Pentru
Penyelidikan — Penyel.
Peony — Peony
Peptide — Peptide
Perargonium — Pelargonium
Perarabotka — Perarabot.
Perennial — Perenn.
Perfume — Perfume
Perfumer(s,y) — Perfumer(s,y)
Perioda — Perioda
Periodic(al,als,um), Periodichesko, Periódico, Périodique, Periodischen — Period., Periód., Périod.
Peritos — Peritos
Perkebunan — Perkebunan
Permaculture — Permacult.
Permanent(e) — Perman.
Permskii [etc.], Permskij [etc.] — Permsk.
Pernambucana — Pernamb.
Perpustakaan — Perpust.
Personal — Personal
Perspectives — Perspect.
Pertanian — Pert.
Peruanas — Peruanas
Peruano — Peruano
Perusahaan — Perus.
Pervestigandae — Pervestig.
Pervogo — Perv.
Pervomaiskaya — Pervomaisk.
Pesca — Pesca
Pesciatina — Pesciat.
Pesqueria, Pesquero, Pesquisa(s) — Pesq.
Pest(s) — Pest(s)
Pesticide(s) — Pestic.
Petergofskogo — Petergofsk.
Petišanas — Petiš.
Petit — Petit
Petitos — Petitos
Petrogradskoi [etc.] — Petrogradsk.
Petrol — Petrol
Pétrole — Pétrole
Petroleum — Petrol.
Petropolitanae, Petropolitani — Petrop.
Petrovskoj — Petrovsk.
Petrozavodskogo — Petrozavodsk.
Pewarta — Pewarta
Pfalzbaierisches — Pfalzbaier.
Pfalzbayerische — Pfalzbayer.
Pfälzische(n,s) — Pfälz.
Pflanze(n) — Pflanze(n)
Pflanzenanatomie — Pflanzenanat.
Pflanzenarzt — Pflanzenarzt
Pflanzenbau — Pflanzenbau
Pflanzenernährung — Pflanzenernähr.
Pflanzenforschung — Pflanzenf.
Pflanzengeographie, Pflanzengeographischen — Pflanzengeogr.
Pflanzenkrankheiten — Pflanzenkrankh.
Pflanzenkunde — Pflanzenk.
Pflanzenliebhaber — Pflanzenliebh.
Pflanzenphysiologie — Pflanzenphysiol.
Pflanzenproduktion — Pflanzenprod.
Pflanzenreich(e,es) — Pflanzenr.
Pflanzenschutz(es) — Pflanzenschutz(es)
Pflanzenschutzabteilung — Pflanzenschutzabt.
Pflanzenschutzämter — Pflanzenschutzämter
Pflanzenschutzbestimmungen — Pflanzenschutzbestimm.
Pflanzenschutzdienst(es) — Pflanzenschutzdienst(es)
Pflanzenschützer — Pflanzenschützer
Pflanzenschutzgesellschaft — Pflanzenschutzges.
Pflanzenschutz-Literatur — Pflanzenschutz-Lit.
Pflanzenschutznachrichten — Pflanzenschutznachr.
Pflanzenschutzpost — Pflanzenschutzpost
Pflanzenschutzstelle — Pflanzenschutzstelle
Pflanzenschutz-Vereins — Pflanzenschutz-Vereins
Pflanzensoziologie, Pflanzensoziologischen — Pflanzensoziol.
Pflanzenwelt — Pflanzenwelt
Pflanzenzelle — Pflanzenzelle
Pflanzenzüchtung — Pflanzenzücht.
Pflanzliche — Pflanzl.
Pflege — Pflege
Phaenologische(n) — Phaenol.
Phanerogamarum, Phanérogamie — Phanerog., Phanérog.
Phänogenetische — Phänogenet.
Phänologie — Phänol.
Pharmaceutic, Pharmaceutica(l), Pharmaceutiques, Pharmaceutisch(es), Pharmaceutisk, Pharmaci(a,ae,e) — Pharm.
Pharmacien — Pharmacien
Pharmacist — Pharmacist
Pharmacobiodynamics — Pharmacobiodyn.
Pharmacognosie, Pharmacognostische, Pharmacognosy — Pharmacogn.
Pharmacologica(l), Pharmacologie, Pharmacologiques, Pharmacology — Pharmacol.
Pharmacy — Pharm.
Pharmakeutikon — Pharmak.
Pharmakologie — Pharmakol.
Pharmakotherapie — Pharmakother.
Pharmazeutische — Pharm.
Phase — Phase
Phenolics — Phenolics
Philippinae, Philippine(s) — Philipp.
Philologique — Philol.
Philomatique — Philom.
Philosophia, Philosophical — Philos.
Philosophico-Mathematica — Philos.-Math.
Philosophiques — Philos.
Philosophisch-Medicinischen — Philos.-Med.
Philosophy — Philos.
Phisikalisch — Phis.
Photobiochemistry — Photobiochem.
Photobiological, Photobiology — Photobiol.
Photobiophysics — Photobiophys.
Photochemical, Photochemistry —

Photochem.
Photogrammatic — Photogramm.
Photographic, Photographie, Photography — Photogr.
Photosynthesis — Photosyn.
Phycologia, Phycologica(l), Phycologique, Phycologium, Phycology — Phycol.
Phylla — Phylla
Physica(l) — Phys.
Physicalisch-Medicinischen — Phys.-Med.
Physicalisch-Öconomomischen — Phys.-Öcon.
Physicaliske, Physicalske — Phys.
Physician(s) — Physician(s)
Physico — Phys.
Physicochemic(al), Physicochemie — Physicochem.
Physico-Chimique — Phys.-Chim.
Physico-Économique — Phys.-Écon.
Physico-Mathématique, Physico-Mathematique(s) — Phys.-Math.
Physico-Medicae — Phys.-Med.
Physico-Medico-Mathematica — Phys.-Med.-Math.
Physics, Physik — Phys.
Physikalisch-Chemische — Phys.-Chem.
Physikalische(n), Physiker — Phys.
Physiographica, Physiographiska — Physiogr.
Physiolog(ia,ie,y), Physiological, Physiologique, Physiologischen — Physiol.
Physiologist(s) — Physiologist(s)
Physique(s), Physischen — Phys.
Phytiatrie — Phytiat.
Phytobiology — Phytobiol.
Phytochemical, Phytochemistry — Phytochem.
Phytochimica, Phytochimique — Phytochim.
Phytoecologia — Phytoecol.
Phytogénétique — Phytogénét.
Phytogeographica, Phytogéographie, Phytogeography — Phytogeogr., Phytogéogr.
Phytographische, Phytography — Phytogr.
Phytologist — Phytol.
Phytomedica — Phytomed.
Phytomorphogenese — Phytomorphogenese
Phytomorphology — Phytomorphol.
Phytopalaeontologische — Phytopalaeontol.
Phytopathologic(a,al), Phytopathologie, Phytopathologique — Phytopathol.
Phytopathologist — Phytopathologist
Phytopathology — Phytopathol.
Phytopharmacie — Phytopharm.
Phytophylacica — Phytophyl.
Phytophysiologia — Phytpphysiol.
Phytoplankton — Phytoplankt.
Phytoprotection — Phytoprotection
Phytosanitaire, Phytosanitary — Phytosan.
Phytosociologica, Phytosociologique(s), Phytosociology — Phytosoc.
Phytotaxonomica, Phytotaxonomie — Phytotax.
Phytotechnica(l) — Phytotechn.
Phytotherapeutic(a,s), Phytotheraphy, Phytotherapie — Phototherap.
Phytotronic — Phytotr.
Piacere — Piacere
Piante — Piante
Pièces — Pièces
Piemontese — Piemont.
Piltzstelle — Piltzst.
Pilz(en) — Pilz(en)
Pilzbau — Pilzbau
Pilzbriefe — Pilzbriefe
Pilzfreunde — Pilzfr.
Pilzklub — Pilzklub
Pilzkunde — Pilzk.
Pine — Pine
Pineapple — Pineapple
Pioneer — Pioneer
Pirenaico(s) — Piren.
Pisano — Pisano
Piscatoria — Piscat.
Piscicoltura — Piscicolt.
Pistoles — Pistoles
Pitaniya — Pitan.
Pitesti — Pitesti
Pitomnika — Pitomn.
Plagas — Plagas
Plain(s) — Plain(s)
Planetary — Planet.
Planktology — Planktol.
Plankton, Planktonique — Plankt.
Planktonkunde — Planktonk.
Planktonology — Planktonol.
Planning — Planning
Planovaja, Planovaya — Planov.
Plant(a,as,er,es,s), Plantarum — Pl.
Plantation(s) — Plantation(s)
Planteavl — Planteavl
Planteavlsudvalg — Planteavlsudv.
Plantekultur — Plantekult.
Plantelor — Pl.
Plantenkundige — Plantenk.
Plantenphysiologie — Plantenphysiol.
Plantenrijk — Plantenrijk
Plantentuin — Plantentuin
Plantenveredeling — Plantenveredel.
Plantenziekten — Plantenziekten
Plantenziektenkundig — Plantenziektenk.
Plantepatologiske — Plantepatol.
Planter(s,'s,s') — Planter(s,'s,s')
Plantesygdomme — Plantesygd.
Plantetilsyn — Plantetilsyn
Plantevern — Plantevern
Planting — Pl.
Plantkunde — Plantk.
Plantsman — Plantsman
Plasm — Plasm
Plasmologie — Plasmol.
Plateau — Plateau
Pleistocene — Pleistoc.
Plénière — Plén.
Plinian — Plinian
Plodoovoščnoe — Plodoovoščn
Plodoovoshchnoe — Plodoovoshchn.
Plodov(ye) — Plodov(ye)
Plot(s) — Plot(s)
Plovuchego — Plovuch.
Pobočky — Pobočky
Pochvakh — Pochvakh
Pochvenno-, Pochvennogo — Pochv.-, Pochv.
Pochvovedeniya — Pochvov.
Pochvozashchitnogo — Pochvozashch.
Pochvy — Pochvy
Pocket — Pocket
Počumavských — Počumavsk.
Počvenno-Botaničeskih, Počvenno-Botaničeskogo — Počv.-Bot.
Počvovedenija — Počvov.
Počvoznanie — Počvozn.
Póda — Póda
Podblanícka — Podblanícka
Podjavorinskeho — Podjavorin.
Podol´skih, Podol´skikh — Podol´sk
Podotděla, Podotdela — Podotd.
Pódyzi — Pódyzi
Poeyana — Poeyana
Poinsettia — Poinsettia
Point — Point
Poisons — Poisons
Pokrovitel´skom — Pokrov.
Pokrytosemmennyki — Pokryt.
Pokuse — Pokuse
Pol´nohospodárskej — Pol´nohosp.
Polabí — Polabí
Polar — Polar
Polarforschung — Polarforsch.
Polarinstituut — Polarinst.
Poleznyh, Poleznykh — Polezn.
Polgári — Polg.
Policlinico — Policlin.
Policy — Policy
Poligrafo — Poligrafo
Polish — Polish
Politechnic — Politechn.
Politehnic — Politehn.
Politekhnicheskogo — Politekhn.
Politicas, Políticas, Politicos, Politics, Politik, Politiques(s) — Polit., Polít.
Politisch-Statistische — Polit.-Statist.
Politischen — Polit.
Polizei — Polizei
Polizey — Polizey
Poljarnoj — Poljarn.
Poljodjelska — Poljod.

Poljoprivredna, Poljoprivredne, Poljoprivredni, Poljoprivredno — Poljopr.
Poljoprivredno-Šumarskog — Poljopr.-Šumarsk.
Poljoprivredny — Poljopr.
Pollen — Pollen
Pollution — Pollut.
Polonaise — Polon.
Poloniae — Poloniae
Polonica — Polon.
Polosy — Polosy
Polska — Polska
Polski(e,ego,ej) — Polsk.
Polyarnoi, Polyarnym — Polyarn.
Polynesian — Polynes.
Polytechnia, Polytechnic, Polytechnique, Polytechnische(n,s) — Polytechn.
Polytecnica — Polytecn.
Pome — Pome
Pomicultura — Pomic.
Pomme — Pomme
Pomme de terre — Pomme de terre
Pommersche(n,s) — Pommersche(n,s)
Pomological, Pomologické, Pomologie, Pomologii, Pomologischen, Pomologist — Pomol.
Pomona — Pomona
Pomone — Pomone
Pomorze — Pomorze
Pomošč — Pomošč
Pomoshch′ — Pomoshch′
Pondkeeper — Pondkeeper
Pontaniana — Pontan.
Pontificia, Pontifico — Pontif.
Pooshchreniya — Pooshchr.
Poppy — Poppy
Populaere — Populaere
Populaire — Populaire
Popular(a,ă,e) — Popular(a,ă,e)
Popularnonaukowe — Popularnonauk.
Population — Populat.
Populyatsii — Populyat.
Poriadki — Poriad.
Porod — Porod
Portatif — Portatif
Portefeuille — Portef.
Portfolio — Portfol.
Portraits — Port.
Portugaises, Portugesea, Portugueza — Portug.
Posebna — Posebna
Posebno — Posebno
Posiedzeń — Posiedzeń
Post — Post
Postepy — Postepy
Postharvest — Postharv.
Potagère — Potag.
Potato — Potato
Potosina — Potos.
Pour — Pour
Považia — Považia
Povreždenijah, Povreždenijamy — Povrežden.
Povrezhdeniyakh, Povrezhdeniyamy — Povrezhden.
Pöytäkirjat — Pöytäkirjat
Pöytäkirjoja — Pöytäkirj.
Poznaniju, Poznaniyu — Pozn.
Poznańskiego — Poznańsk.
Přírodovědné(ho), Přírodovědny, Prírodovedný, Prírodvedeckej, Prirody, Přírody, Prírucky — Prir., Prír., Přír.
Prac, Prác, Prac′ — Prac, Prác, Prac′
Prace, Práce, Pracé — Prace, Práce, Pracé
Praci, Prací — Praci, Prací
Pracovnice — Pracovn.
Practica, Práctica, Practical, Practice, Practici, Practicien, Practischer — Pract., Práct.
Practitioner — Practitioner
Pracy — Pracy
Praga(s,ti) — Praga(s)
Pragensis — Prag.
Prager — Prager
Prairie — Prairie
Praktika, Praktische(n,r) — Prakt.
Pramyslovas′chi — Pramysl.
Prapora — Prapora
Pratica — Prat.
Praticien — Praticien
Pratique(s) — Prat.
Prats(a,y) — Prats(a,y)
Präventive — Präv.
Praximetrie — Praximetr.
Praxis — Praxis
Praze — Praze
Pražske — Pražske
Pražského — Pražského
Preceded — Preceded
Precios — Precios
Précis — Précis
Prefectural, Prefecture — Pref.
Pregled — Pregled
Prehistoric, Préhistorique — Prehist.
Preise — Preise
Preisschriften — Preisschr.
Preliminares, Preliminary — Prelim.
Premi — Premi
Première — Prem.
Premios — Premios
Prensa — Prensa
Preparation — Prepar.
Prepodavatelei, Prepodavatelej — Prepodav.
Prerada — Prerada
Přerovska — Přerovska
Preservation — Preserv.
Preserve(s) — Preserve(s)
Preserving — Preserving
Presidencia — Presid.
President('s) — President('s)
Presnovodnoi, Presnovodnoj — Presnovodn.
Presove — Presove
Press(e) — Press(e)
Pressedienst — Pressedienst
Pressoir — Pressoir
Preussische(n) — Preuss.
Prevalence — Preval.
Preventive — Prev.
Preview — Preview
Priamurskago, Priamurskogo — Priamursk.
Priazov′e — Priaz.
Pribavlenie — Pribavl.
Pribrezhya — Pribr.
Prichernomov′ya — Prichern.
Prikladnoi [etc.], Prikladnoj [etc.] — Prikl.
Prilozenie — Priloz.
Prima — Prima
Primary — Prim.
Primera — Primera
Primeurs — Primeurs
Primorskoj — Primorsk.
Primrose — Primrose
Principal(is) — Princ.
Printing — Print.
Pripoljarnoe — Pripoljarn.
Priroda, Príroda, Příroda, Prirode, Prirodne, Přírodní(ch), Prirodnih, Prirodnik(h), Prirodnjačkog, Prirodnjacki, Prirodno-, Přírodny, Prirodnyh, Prirodnykh, Prirodo, Prirodoizpatelno, Prirodonaučkog, Prirodonaučni, Prirodonaučniot, Prirodonauchni, Prirodoslovna [etc.], Přírodovědecké(ho), Přírodovědéckou, Přírodovědecký — Prir., Prír., Přír., Prir.-
Prisma — Prisma
Prisuždenii [etc.] — Prisužd.
Prisuzhdenii [etc.] — Prisuzhd.
Přítel — Přítel
Privatae — Privatae
Privatgesellschaft — Privatges.
Privilegeride, Privilegierte — Privileg.
Prix — Prix
Prize — Prize
Prize-Essays — Prize-Essays
Probe — Probe
Problem(e,s,y) — Probl.
Proceedings — Proc.
Procès — Procès
Procès-Verbaux — Procès-Verbaux
Process(ed,es,i) — Process(ed,es)
Processor(s) — Processor(s)
Produçao, Producçao, Producciones — Prod.
Produce — Produce
Producer(s,s') — Producer(s,s')
Product(en,s), Producţia, Production(s), Produkcja, Produktion — Prod.
Produktivnost′ — Produkt.
Produtos — Prod.

Proefstation — Proefestat.
Proeftuin(en) — Proeftuin(en)
Professeurs, Professional(e), Professoral, Professorov — Profess.
Profumi — Profumi
Program(a,s) — Program(a,s)
Programma — Programma
Programme — Programme
Progrés, Progresos, Progress(i,ive,o,us) — Progr.
Proizvodstvo — Proizvodstvo
Proizvoditel'nye [etc.]. — Proizv.
Project(en,o) — Proj.
Promise — Promise
Promoción — Promoc.
Promoting, Promotion — Promot.
Promovendis — Promov.
Promyšlennosti — Promyšl.
Promyshlennogo, Promyshlennost(i) — Promyshl.
Promyslovoi, Promyslovoj — Promysl.
Propagador, Propagation, Propagatore, Propagators — Propag.
Properties — Propert.
Property — Property
Proposals, Proposés — Prop.
Prose — Prose
Prospects — Prosp.
Prosvetni — Prosv.
Protecção, Proteccion — Protecc.
Protectia, Protection, Protectrice — Protect.
Protein — Protein
Protistenkunde — Protistenk.
Protistologia, Protistologičeskogo, Protistologicheskogo, Protistologie, Protistologii — Protistol.
Protocollen — Protoc.
Protokollen, Protokoly — Protok.
Protozoology — Protozool.
Provençal, Provence, Province(s), Provincia(al,l), Provinz(ial) — Prov.
Provinzialberichte — Provinzialber.
Provinzialblatt [etc.] — Provinzialbl.
Provinzialkomitees — Provinzialkomitees
Provinzialstelle — Provinzialstelle
Provisoires — Provis.
Proyecto(s) — Proy.
Prozkoumání — Prozk.
Prüfende(n) — Prüfende(n)
Prüfung — Prüfung
Prüfungsanstalt — Prüfungsanst.
Průhonich — Průhonich
Pruhoniciana — Pruhon.
Prykladnoi, Prykladnoji — Prykl.
Pryrodnavciv, Pryrodnu — Pryr.
Pryrodnyčo-Tehničnyj [etc.] — Pryr.-Tehn.
Pryrodnychno-technichnogo — Pryr.-Techn.
Pryrodnykiv, Pryrodoslidnykiv, Pryrodoznavciv, Pryrodoznavtsiv, Pryrodu, Pryrody — Pryr.
Przeglad — Przeglad
Przemyslu — Przemyslu
Przyjaciól — Przyjac.
Przyroda, Przyrode, Przyrodnice, Przyrodniczego, Przyrodniczych, Przyrody — Przyr.
Pšenica — Pšenica
Pshenica — Pshenica
Pshenitsy — Pshenitsy
Psychedelic — Psychedelic
Psychiatry — Psychiat.
Psychoactive — Psychoact.
Psychology — Psychol.
Psychopathology — Psychopathol.
Pteridologica(l), Ptéridologiques — Pteridol., Ptéridol.
Pubblicazione, Pubbliche — Pubbl.
Public — Public
Pública — Pública
Publicaciones, Publicações, Publicati(es), Publication(es,s), Publicatiunile — Publ.
Público — Público
Publikace, Publikationer — Publ.
Publique(s) — Publique(s)
Published — Published
Publishing — Publishing
Pulawski — Pulawski
Pulmonary — Pulmon.
Punk — Punk
Pura — Pura
Pure(s) — Pure(s)
Purpose — Purp.
Pusat — Pusat
Pustyn' — Pustyn'
Pustyni — Pustyni
Pustynnykh — Pustyn.
Puutarhan — Puut.
Pyatigorskogo — Pyatigorsk.
Pyramide — Pyramide
Pyrénéenne — Pyrén.
Pyrethrum — Pyrethrum
Quaderni — Quad.
Quaestiones — Quaest.
Qualitas, Quality — Qual.
Quantitative — Quant.
Quarantine — Quarant.
Quartalschrift — Quartalschr.
Quarterly — Quart.
Quaternaire — Quatern.
Quaternaria — Quaternaria
Quaternary — Quatern.
Québecoise — Québec.
Quellen — Quellen
Queries — Queries
Questions — Quest.
Quimica, Química — Quim., Quím.
Quimioterapia — Quimioter.
Quincenal — Quincenal
Quindicinal — Quindic.
Quintais — Quintais
Quirúrgica — Quir.
R S F S R — R.S.F.S.R
Rabot(ah) — Rabot(ah)
Rabotnikov — Rabotn.
Raboty — Raboty
Raccolta — Racc.
Race — Race
Rad(a) — Rad(a)
Rádcova — Rádcova
Řadi — Řadi
Radiation — Radiat.
Radiatsonnaya — Radiats.
Radio — Radio
Radiobiological — Radiobiol.
Radiobotanica — Radiobot.
Radiocarbon — Radiocarbon
Radiochuvstvitel'nost' — Radiochuvstv.
Radiologica — Radiol.
Radium — Radium
Radova, Radovi — Rad.
Radu — Radu
Railroad(s) — Railroad(s)
Railway(s) — Railway(s)
Rain — Rain
Raionov — Raionov
Rajonirovanie — Rajonirov.
Rajonov — Rajonov
Raksti — Raksti
Ramos — Ramos
Ramsoknir — Ramsok.
Ranching — Ranching
Rancho — Rancho
Range — Range
Rangeland(s) — Rangeland
Ranger — Ranger
Rannssóknastofnunin — Rannssóknastofn.
Raporlar — Raporlar
Rapport(s) — Rapp.
Rare — Rare
Rariores — Rar.
Rasen — Rasen
Rasporjaditel'nago — Rasporjadit.
Rasporyaditel'nago — Rasporyadit.
Rassegna — Rassegna
Rassenkreuze — Rassenkreuze
Rassenlijst — Rassenlijst
Rassenonderzoek — Rassenonderz.
Raštai — Raštai
Rasteni, Rastenii [etc.] — Rast.
Rastenievadni, Rastenievodstva, Rastenievodstvu, Rastenievudstvo — Rasteniev.
Rasteniyata — Rast.
Rašti — Rašti
Rastitel'nazashchita — Rastitel'naz.
Rastitel'nie [etc.], Rastitel'nost', Rastitel'nyh, Rastitle'nykh — Rastitel'n.
Rastliny — Rastl.
Ratarstvo — Rat.
Ratgeber — Ratgeber

Ratisbonensi(a,s) — Ratisb.
Raumforschung — Raumf.
Raumordnung — Raumordn.
Razpravama, Razprave — Razpr.
Razrad, Razred(a) — Razr.
Reading(s) — Reading(s)
Real(e) — Real(e)
Realgymnasium — Realgymn.
Realschule — Realschule
Realzeitung — Realzeitung
Reauthorization — Reauthoriz.
Rebe — Rebe
Reblooming — Reblooming
Reboisement — Rebois.
Rebus — Reb.
Received — Receiv.
Recensent — Recensent
Recent — Recent
Recentiores — Recent.
Rechenschaftsbericht — Rechenschaftsber.
Recherche(s) — Rech.
Recht — Recht
Reclamation — Reclam.
Recommendations — Recommend.
Reconstructs — Reconstructs
Record(er,s) — Rec.
Recostruzione — Recostruz.
Recreation(s) — Recreation(s)
Recueil — Recueil
Reçues — Reçues
Recursos — Recurs.
Redogörelse — Redog.
Reef — Reef
Reeks — Reeks
Referate, Referativnyi, Referativnyj, Referaty — Ref.
Reference — Refer.
Reflorestadores — Reflorestad.
Reforestation — Reforest.
Reform(er) — Reform(er)
Refugium — Refug.
Regensburgischen — Regensburg.
Regents — Regents
Régészeti — Régész.
Regia(e) — Regia(e)
Regierungs-Bezirkes — Reg.-Bez.
Regierungsbezirk(es) — Regierungsbez.
Régies — Régies
Reginaehradecensis — Reginaehrad.
Regio — Regio
Region(s), Région — Region(s), Région
Regional — Regional
Régionale — Régionale
Register, Registi — Reg.
Registration — Registr.
Registro — Reg.
Regni — Regni
Regno — Regno
Regnum — Regnum
Regulated, Regulation — Regulat.
Regulator(y) — Regulator(y)
Regulyatsii — Regulyat.
Rei — Rei
Reiche(n,s) — Reiche(n,s)
Reichsamts — Reichsamts
Reichsanstalt — Reichsanst.
Reichsarbeitsgemeinschaft — Reichsarbeitsgem.
Reichsinstitutes — Reichsinst.
Reichsstelle — Reichsstelle
Reichtum — Reichtum
Reihe — Reihe
Reipublicae — Reipubl.
Reisebeschreibungen — Reisebeschreib.
Reisegeschichten — Reisegesch.
Reisen — Reisen
Reisevereins — Reisevereins
Reizen — Reizen
Reka — Reka
Rekke — Rekke
Relación, Relaction — Relac.
Related — Relat.
Relating — Relating
Relation, Relatório — Relat.
Relazione — Relaz.
Release — Release
Religion — Relig.
Rem — Rem
Remote - Remote
Remporté(s) — Remporté(s)
Renaissance — Renaiss.
Rendiconti — Rendiconti
Rendiconto — Rendiconto
Rendsertani — Rendsert.
Rendus — Rend.
Renewable — Renew.
Renfermant — Renferm.
Renovables — Renov.
Renseignement(s) — Renseign.
Repair — Repair
Repertario, Repertatio, Répertoire, Repertorio, Repertorium — Repert., Répert.
Replication — Replic.
Report(e,s) — Rep.
Reporter — Reporter
Repository — Repos.
Reprint(s) — Repr.
Reproduction, Reproductive — Reprod.
Reptilian — Reptil.
Republic, Republica, República, Republicii, Republičkog, Republike, République(s) — Republ., Repúbl., Républ.
Rerum — Rerum
Research(es) — Res.
Reseñas — Reseñas
Reserve, Réserve(s) — Reserve, Réserve(s)
Reservoir — Reservoir
Resistance — Resist.
Resource(s) — Resource(s)
Respiratory — Resp.
Respublik, Respublikans'kogo, Respublikanskoj — Respubl.
Ressource(s) — Ressource(s)
Resultationes, Résultats — Result., Résult.
Results — Results
Resume(s) — Resume(s)
Resumos — Resumos
Resumptio — Resumptio
Resursov — Resursov
Resursy — Resursy
Réunion — Réun.
Revegetation — Reveg.
Review(s) — Rev.
Revista, Revistă — Revista, Revistă
Revolutionist — Revolutionist
Revolyutsii — Revolyutsii
Revue — Rev.
Rezervy — Rezervy
Rezina — Rezina
Rezinovoj — Rezinov.
Rhedonica — Rhedon.
Rheinische(n) — Rhein
Rheinland(e), Rheinländische — Rheinl.
Rheinpressichen — Rheinpreuss.
Rheinuniversität — Rheinuniv.
Rheno-Trajectinae — Rheno-Traject.
Rhizome — Rhizome
Rhodesian — Rhodesian
Rhododendron — Rhododendron
Rhythm(s) — Rhythm(s)
Ribarstvo — Ribarst.
Rice — Rice
Ricerca, Ricerche — Ric.
Ridnu — Ridnu
Riforma — Riforma
Rigischen — Rigischen
Riikliku — Riikl.
Rijetkosti — Rijet.
Rijks — Rijks
Rijks-Herbarium — Rijks-Herb.
Rijksboschbouwproefstation — Rijksboschbouwproefstat.
Rijksfakulteit — Rijksfak.
Rijksherbarium — Rijksherb.
Rijksinstituut — Rijksinst.
Rijkslandbouwproefstations — Rijkslandbouwproefstat.
Rijksproefstation — Rijksproefstat.
Rijksstation — Rijksstat.
Rijkstuinbouwconsulentschap — Rijkstuinbouwconsulentschap
Rijksuniversiteit — Rijksuniv.
Rijst — Rijst
Rimba — Rimba
Rimbawan — Rimbawan
Ring — Ring
Rinkinys — Rinkinys
Rio-Grandenses — Rio-Grandenses
Risa — Risa
Riserva — Riserva
Risicoltura — Risic.

Riso — Riso
Rit — Rit
Riunione — Riunione
River — River
Rivière — Rivière
Rivista — Riv.
Rivon — Rivon
Riz — Riz
Riziculture — Rizic.
Ročenka — Ročenka
Rochelaise — Rochel.
Rock — Rock
Rocznik(i) — Roczn.
Rodunjalostussaatio — Rodunjalost.
Roentgenology — Roentgenol.
Roestbericht — Roestber.
Rohonczi — Rohonczi
Roi — Roi
Roll — Roll
Rolnické, Rolnictwo, Rolnicza, Rolniczej, Rolniczy(ch) — Roln.
Romana, Română — Romana, Română
Romand(e,s) — Romand(e,s.)
Romåne — Romåne
Români — Români
Romină — Romină
Romîne — Romîne
Römisch — Röm.
Römisch-Germanischen — Röm.-German
Rön — Rön
Root — Root
Rosarian(s) — Rosarian(s)
Rose(s) — Rose(s)
Rosenfreunde — Rosenfr.
Rosengarten — Rosengart.
Rosenjahrbuch — Rosenjahrb.
Roślin, Roslinna, Roslinnogo, Roslinnych — Rośl., Rosl.
Rossicae — Ross.
Rossiiskoi — Rossiisk.
Rossijskago, Rossijskoj — Rossijsk.
Rostlin(na,né) — Rostl.
Rostockschen — Rostockschen
Rostovskoi [etc.], Rostovskoj [etc.] — Rostovsk.
Rotondo — Rotondo
Roumain(e,es) — Roumain(e,es)
Rovartani — Rovart.
Roveretana — Roveretana
Rövid — Rövid
Royal(e) — Roy.
Rozhledy — Rozhl.
Rozprava, Rozpravy, Rozprawy — Rozpr.
Rözsa — Rözsa
Rubber — Rubber
Rubbercultuur — Rubbercult.
Rubberproefstation — Rubberproefstat.
Rubberstation — Rubberstat.
Rubezhom — Rubezh.
Rücksicht — Rücks.
Rudes — Rudes
Rumanian — Rumanian
Rundbriefe — Rundbr.
Rundschau — Rundschau
Rundveehouderij — Rundveehoud.
Rural(e) — Rural(e)
Russe, Russian, Russische(n,s) — Russ.
Russkih, Russkii [etc.], Russkij [etc.], Russkikh — Russk.
Rust(ica) — Rust(ica)
Rusticarum — Rustic.
Rusticum — Rusticum
Ruzynyibb. — Ruzyn.
Rybactwa — Rybactwa
Rybinskago, Rybinskogo — Rybinsk.
Rybnogo — Rybn.
S S S R — S.S.S.R
S.-Peterburgskoi [etc.], S.-Peterburgskoj [etc.] — S.-Peterburgsk.
Saarpfälzischen — Saarpfälz.
Sabarienses — Sabar.
Sabouradia — Sabouradia
Sachen — Sachen
Sächsische(n.r) — Sächs.
Sächsisch-Thuringischen — Sächs.-Thuring.
Sad(a) — Sad(a)
Sade, Sadě — Sade, Sadě
Sadoi — Sadoi
Sadov, Sadovodstva, Sadovodstvo — Sadov
Sadownicta — Sadown.
Sadu — Sadu
Sady — Sady
Saecularia — Saecularia
Safety — Safety
Sager — Sager
Saggi — Saggi
Saguaroland — Saguaroland
Sahalinskij, Sahalinskogo — Sahalinsk.
Sains — Sains
Saint — St.
St. Gallischen — St. Gallischen
Sakhalinskii, Sakhalinskogo — Sakhalinsk.
Sakharthwelos — Sakharthw.
Sakhernaya — Sakhern.
Salle — Salle
Sallskapets — Sallsk.
Salo — Salo
Sälskapet(s) — Sälsk.
Salubridad — Salubr.
Salutaire — Salut.
Salvadoreña — Salvadoreña
Salzburgische — Salzburg.
Samenhändler — Samenhändl.
Samenprüfung — Samenprüfung
Sämereien — Sämereien
Samfundets — Samfund
Samhället(s) — Samhället(s)
Samler — Samler
Samling(er) — Saml.
Sammelschrift(en) — Sammelschr.
Sammelstelle — Sammelstelle
Sammler — Sammler
Sammlung(en) — Samml.
Sämmtlichen — Sämmtl.
Sand — Sand
Sanidad — Sanid.
Sanitarium — Sanit.
Sanitarno-Biologičeskomu — Sanit.-Biol.
Sanitatis — Sanit.
Sanktpeterburgskoj — Sanktpeterburgsk.
Sante — Sante
Sapropelevogo, Saprop̄élites — Sapropel., Saprop̄él.
Saratovskij [etc.] — Sartovsk.
Saraviensis — Sarav.
Sassaresi — Sassar.
Saturday — Sat.
Sauvage — Sauv.
Sauvegarde — Sauveg.
Savanna — Savanna
Savans — Savans
Savant(es,s) — Savant(es,s)
Savariensia — Savar.
Saver(s') — Saver(s')
Savoisienne — Savois.
Savremena — Savremena
Sayajirao — Sayajirao
Sbornik, Sborník — Sborn.
Scandinav(ia,ian,ica) — Scand.
Scanning — Scan.
Sçavans — Sçavans
Scelta — Scelta
Scelti — Scelti
Scene — Scene
Schädlingskunde — Schädlingsk.
Schadlungsbekämpfungsmittel — Schadlungsbekämpf.
Schedae — Sched.
Schedulae — Schedul.
Scheikundig — Scheik
Scheme — Scheme
Schimmelcultures — Schimmelcult.
Schlesien(s) — Schlesien(s)
Schlesische(n) — Schles.
Schleswig-Holsteinische(r,s) — Schleswig-Holst.
Schmarotzerbestimmung — Schmarotzerbestimm.
Schönen — Schönen
Schönheit — Schönh.
Schönhengster — Schönhengster
School — School
Schoolproeftuin — Schoolproeftuin
Schoone — Schoone
Schriften — Schriften
Schriftenreihe — Schriftenreihe
Schrifttum(s) — Schrifttum
Schultz(e) — Schultz(e)
Schutze — Schutze
Schutzgebieten — Schutzgeb.
Schwabens — Schwabens
Schwäbisches — Schwäb.
Schwarzen — Schwarzen

Schwedische(n,s) — Schwed.
Schweizer — Schweizer
Schweizerische(n) — Schweiz.
Schweizerlandes — Schweizerl.
Schwyz — Schwyz
Science(s) — Sci.
Science-Gossip — Sci.-Gossip
Sciencias, Scientia(e), Scientiarum, Scientific(a,al,o), Scientifiche, Scientifici, Scientifique(s), Scientist(s), Scienze — Sci.
Scoop — Scoop
Sčoričnyk — Sčor.
Scotia(n) — Scotia(n)
Scots — Scots
Scottish — Scott.
Scrap book — Scrap Book
Scrinia — Scrinia
Scripta — Scripta
Scriptorum — Script.
Scrutatorum — Scrut.
Scuola — Scuola
Sea(s) — Sea(s)
Séance(s) — Séance(s)
Search — Search
Seasons — Seasons
Seaweed — Seaweed
Seçao — Seçao
Secção, Secció(n) — Secç.
Seconda — Seconda
Seconde — Seconde
Secretaría(t), Secretary — Secr.
Sectio, Section(e,is,s), Sectiunea, Secţiunii — Sect.
Sediment — Sediment
Sedimentary — Sediment.
Sedute — Sedute
Seed(s) — Seed(s)
Seedsman — Seedsman
Seeländischen — Seel.
Seereisen — Seereisen
Seeria — Seer.
Segunda — Segunda
Sehr — Sehr
Seidenbau — Seidenbau
Sekai — Sekai
Sekce — Sekce
Sekcii, Sekcije, Sekciji — Sekc.
Sekcji — Sekcji
Sektion — Sekt.
Sektora — Sektora
Sektorov — Sektorov
Sel'skagaspadarchykh — Sel'skagasp.
Sel'skii — Sel'skii
Sel'skokhozyaistvennogo, Sel'skokhozyaistvennom, Sel'skokhozyaistvennye, Sel'skokhozyaistvennykh, Sel'sko-khozyaistvennykh — Sel'skokhoz., Sel'sko-Khoz.
Selecciones — Selecc.
Seleções — Seleções
Selecta(m) — Selecta(m)
Selection — Select.
Selects — Selects
Selekcii, Selekcija — Selekc.
Selekcionerov, Selekcionnaja, Selekcionnoj — Selekcion.
Selektsii — Selekts.
Selektsionerov, Selektsionnaya, Selektsionno — Selektsion.
Selektsiya — Selekts.
Selmeczbányai — Selmeczbányai
Selodykh — Selod.
Selskab(et), Sel'skae — Selsk.
Sel'skagaspadarchykh, Sel'skagaspadarčyh — Sel'skagasp.
Sel'skii, Sel'skij — Sel'sk.
Sel'skoe [etc.]. — Sel'sk.
Sel'sko-gospodarskoi — Sel'sk-Gosp.
Sel'skohozjajstvennaja, Sel'skohozjajstvennogo, Sel'skohozjajstvennoj, Sel'skohozjajstvennom, Sel'skohozjajstvennye, Sel'skohozjajstvennyh — Sel'skohoz.
Sel'sko-Hozjajstvennyh — Sel'sko-Hoz.
Sel'skokhozyaistvennogo, Sel'skokhozyaistvennom, Sel'skokhozyaistvennykh, Sel'skokhozyaistvennyye — Sel'skokhoz.
Sel'sko-khozyaistvennykh — Sel'sko-Khoz.
Sel'skom(u) — Sel'sk.
Selskostopanskite — Selskost.
Selststudiums — Selbststud.
Seltenen — Seltenen
Seltenheiten — Seltenh.
Seltsi — Seltsi
Selva — Selva
Selvicoltura — Selvic.
Semaine — Semaine
Semana(l) — Semana(l)
Semanario — Semanario
Semences — Semences
Semenovedenie, Semenovedeniju, Semenovedeniyu, Semenovodstva, Semenovodstvo — Semenov.
Sementes — Sementes
Sementi — Sementi
Semi-Monthly — Semi-Monthly
Semiarid, Semi-arid — Semiarid, Semi-arid
Semina — Semina
Seminar(io,ium,s) — Seminar(io,ium,s)
Seminum — Seminum
Semipalatinskago, Semipalatinskogo — Semipalatinsk.
Sĕmjan — Sĕmjan
Sempervirens — Sempervirens
Semyan — Semyan
Senckenbergian(a,um), Senckenbergischen — Senckenberg.
Sénégalaises — Sénégal.
Sense — Sense
Sensing — Sensing
Sentinel — Sentinel
Seoulensis — Seoul.
Separate — Separate
Separation — Separation
Septentrionis — Septentr.
Serias, Serica — Ser.
Sériciculture — Séricic.
Sericulture — Seric.
Série, Series, Serija, Serisi, Seriya — Sér., Ser.
Serological, Serologie — Serol.
Serres — Serres
Servant — Servant
Serves — Serves
Service(s), Servicio, Serviço — Serv.
Seryja — Ser.
Session(i) — Sess.
Setor — Setor
Settlement(s) — Settlem.
Seucana — Suec.
Seuran — Seuran
Sevanskoj — Sevansk.
Sevastopol'skoj — Sevastopol'sk.
Ševčenko-Gesellschaft — Ševčenko-Ges.
Severa — Severa
Severnaja, Severnaya, Severnega — Severn.
Severní — Severní
Severnyi [etc.], Severnyj [etc.] — Severn.
Severočeské(ho) — Severočesk.
Severočeskou — Severočeskou
Severo-Kavkazshoi [etc.], Severo-Kavkazskoj [etc.] — Severo-Kavkazsk.
Sĕvero-Zapadnago — Sĕvero-Zapadn.
Sewage — Sewage
Sexual — Sexual
Sĕzd(a) — Sĕzd(a)
Sezda — Sezda
Sezione — Sez.
Shade — Shade
Shamrock — Shamrock
Shchorichnyk — Shchor.
Sheep — Sheep
Sheet — Sheet
Shelf — Shelf
Shkencat — Shkencat
Shkencave — Shkencave
Shkoly — Shkoly
Shokubutsu — Shok.
Shore — Shore
Short — Short
Shrub — Shrub
Shtetëror — Shtetëror
Siberian — Siber.
Sibirskoi [etc.], Sibirskoj [etc.] — Sibirsk.
Siciliano — Sicil.
Sidda — Sidda
Siebenbürgische(n,s) — Siebenbürg.
Sienitietoja — Sienitietoja

Sieroterapico — Sieroterap.
Sifilografia — Sifilogr.
Sifilologia — Sifilol.
Sigma — Sigma
Signalétique — Signal.
Sihote-Alinskogo — Sihote-Alinsk.
Sil — Sil
Sil's'ko-Gospodars'koji — Sil's'ko-Gosp.
Sil'sk'ohospodars'kyi — Sil'sk'ohosp.
Silesiacae — Siles.
Silk — Silk
Silva — Silva
Silvaecultura — Silvaec.
Silvestre — Silvest.
Silvical — Silvical
Silvice — Silvice
Silvícola, Silvicolas — Silvíc., Silvic.
Silvics — Silvics
Silvicultur(a,e), Silvicultural — Silvic.
Sily — Sily
Simbirskago — Simbirsk.
Sinar — Sinar
Sindical, Sindicato — Sind.
Sinensia — Sin.
Singapore — Singapore
Sinic(a,ae,orum) — Sin.
Sirovine — Sirovine
Sisal — Sisal
Sistematic(a,ae), Sistematičeskij [etc.], Sistematicheskii [etc.], Sistematika, Sistematiki — Sist.
Sitten — Sitten
Sittenlehre — Sittenl.
Situación — Situación
Situaţia — Situaţia
Situation — Situation
Sitzungen — Sitzungen
Sitzungsberichte — Sitzungsber.
Sitzungs-Berichte — Sitzungs-Ber.
Sitzungs-Protokollen — Sitzungs-Protok.
Skalky — Skalky
Skalničky — Skalničky
Skandinavische(r,s), Skandinavisk(a,e,) — Skand.
Skånska — Skånska
Skierniewicach — Skierniew.
Skilling — Skilling
Skog(en) — Skog(en)
Skogavdelingen — Skogavd.
Skogbruk — Skogbr.
Skogdirektoratet — Skogdirektorat.
Skoglig(a) — Skoglig(a)
Skogsbotanik — Skogsbot.
Skogsekologi — Skogsekol.
Skogs-Forskiningsinstitut — Skogs-Forskiningsinst.
Skogsforskning — Skogsforskning
Skogsforsøksvesen — Skogsforsøksvesen
Skogsföryngring — Skogsföryng.
Skogshögskolan — Skogshögskolan
Skogshushållning — Skogshush.
Skogsproduction — Skogsprod.
Skogsskötsel — Skogsskötsel
Skogstechnik — Skogstechn.
Skogsvårdsförbunds — Skogsvårdsförb.
Skogsvårdsföreningen — Skogsvårdsforen.
Škola — Škola
Školní — Školní
Školy, Szkoły — Školy, Szkoły
Skopskog — Skopsk.
Skötsel — Skötsel
Skovbrug — Skovbr.
Skovforenings — Skovforen.
Skovvaesen — Skovvaes.
Skrifta — Skrifta
Skrifter, Skrivter — Skr.
Skupovi — Skup.
Skyrius — Skyr.
Skýrska — Skýrska
Slaskiego — Slaskiego
Śląsku — Śląsku
Slaves — Slaves
Slavonije — Slavonije
Šlechtěni — Šlechtěni
Slezské(ho) — Slez.
Slezsky — Slezsky
Slope — Slope
Slovenica, Slovenicorum, Slovenska, Slovenské(ho), Slovenskej, Slovenskéslovenskej, Slovenských — Slov.
Slovesnost — Slovesn.
Słupskie — Słupskie
Small — Small
Smallholder — Smallholder
Smaskrift — Smaskr.
Smolenskogo, Smolenskom — Smolensk.
Smotra — Smotra
Snapdragon — Snapdragon
Sobranii, Sobranij — Sobr.
Social(e) — Social(e)
Socialiste — Socialiste
Socialistic, Socialističeskoe, Socialistische — Socialist.
Sociedad Dasonomica de la America Tropical — S. D. A. T.
Sociedad(e), Societa(s), Societá, Società, Societät, Societăţii, Societatis, Société(s), Society — Soc.
Sociologica — Sociol.
Sodo — Sodo
Soedenenii — Soedin.
Sofijskija — Sofijsk
Soil(s) — Soil(s)
Soilless — Soilless
Sojuza — Sojuza
Sojuznyh — Sojuzn.
Sol(s) — Sol(s)
Solenne(lle) — Solenne(lle)
Solenni — Solenni
Solid — Solid
Solo(s) — Solo(s)
Solutions — Solut.
Sommaire — Sommaire
Sonderband(e) — Sonderb.
Sonderbeiheft — Sonderbeih.
Sonderheft(e) — Sonderh.
Songaricus — Song.
Sonnblick-Vereins — Sonnblick-Vereins
Soobščenija — Soobšč.
Soobshcheniya, Soobshchestva — Soobshch.
Sopronhorpacs — Soproni
Sorghum — Sorg.
Sornjakami — Sornjak.
Sornyakami — Sornyak.
Sorozat — Soroz.
Sörter — Sörter
Sostanze — Sostanze
Sostojanii, Sostoyanii — Sost.
Sotrudnikov — Sotrudn.
Sotsialisticheskoe, Sotsialisticheskogo — Sotsialist.
Sound — Sound
Source(s) — Source(s)
Souterrain — Souterr.
South — S.
South East — S.E.
Southeast(ern) — SouthE.
Southern — S.
Southwest(ern) — SouthW.
Southwestern Association of Naturalists [News] — S. W. A. N. E. W. S.
Soveščanij, Soveščanija — Sovešč.
Soveshchanii [etc.]. — Soveshch.
Sověta — Sověta
Sovete — Sovete
Sovetskaja, Sovetskaya, Sovetskie — Sovetsk.
Soviet — Soviet
Soviétique — Soviét.
Sovremennoi, Sovremennoj — Sovrem.
Soyabean — Soyabean
Soybean — Soybean
Soyuza — Soyuza
Soyuznykh — Soyuzn.
Soziologischen — Soziol.
Space — Space
Spade — Spade
Spasmodic — Spasmodic
Specchio — Specchio
Special — Special
Spéciale — Spéciale
Specialist(s,s') — Specialist(s,s')
Specialized — Special.
Spéciaux, Specierum, Species — Spéc.
Spectantia — Spectantia
Spectroscopy — Spectrosc.
Speculation — Specul.
Spektrum — Spektrum
Spekulativen — Spekulativen
Spéléologie, Speleology — Spéléol., Speleol.
Sperimentale, Sperimentazione — Sperim.

Spezialkarte — Spezialkarte
Spezielle — Spez.
Spezifischen — Spezif.
Spices — Spices
Spiel — Spiel
Spinal — Spinal
Spine(s) — Spine
Spiny — Spiny
Spisanie — Spis.
Spisok — Spisok
Spisy — Spisy
Společnost(i) — Společn.
Spolken, Spolkový — Spolk.
Spolku — Spolku
Spomenika — Spomen.
Spore(s) — Spore(s)
Sporovye, Sporovyh, Sporovykh — Sporov.
Sports — Sports
Sportsman — Sportsman
Spotlight — Spotlight
Sprachen — Sprachen
Sprave — Sprave
Spravočnyj — Sprav.
Spravy — Spravy
Sprawozdania, Sprawozdanie — Spraw.
Spring(s) — Spring(s)
Sred — Sred
Sredne- — Sredne-
Sredneaziatskoi, Sredneaziatskoj, Sredne-Aziatskogo, Sredne-Aziatskom — Sredneaziatsk., Sredne-Aziatsk.
Sredne-Sibirskogo — Sredne-Sibirsk.
Srodowiska — Srodow.
Šromebi — Šromebi
Srpska — Srpska
Srpske — Srpske
Staat(en,s) — Staat(en,s)
Staatenkunde — Staatenk.
Staatlich(en) — Staatl.
Staats-Anzeigen — Staats-Anzeigen
Staats-Universitat — Staats-Univ.
Staatsanstalt — Staatsanst.
Staatsarzneikunde — Staatsarzneik.
Staatsforstverwaltung — Staatsforstverw.
Staatsinstitute(n,s) — Staatsinst.
Staatskunde — Staatsk.
Staatssmmlung — Staatssamml.
Staatsuniversität — Staatsuniv.
Staatswissenchaft — Staatswiss.
Stacijas — Sta.
Stad(t) — Stad(t)
Stadtischen — Stadt.
Staff — Staff
Stain — Stain
Stancii — Stancii
Stancij(a,i) — Stancij(a,i)
Stancyi — Stancyi
Standardization — Standardiz.
Standorskunde — Standortsk.
Standort — Standort
Standortserkundung — Standortserkund.
Standortskartierung — Standortrskart.
Stanice — Stanice
Stantsii — Stantsii
Stantsiya — Stantsiya
Stantsiyi — Stantsiyi
Stantsyi — Stantsyi
Starea — Starea
State(s) — State(s)
Statej — Statej
Statement — Statem.
Statens — Statens
Station(en) [non-English] — Stat.
Station(s) [English] — Sta.
Statistical, Statistics, Statistik, Statistiques, Statistische — Statist.
Statiunea — Stat.
Statniho — Statn.
Stato — Stato
Stattsarzneykunde — Staatsarzneyk.
Stauden — Stauden
Staudenkunde — Staudenk.
Stavropol'skago — Stavropol'sk
Stawów — Stawów
Stazione — Staz.
Stedeschoon — Stedesch.
Steierischen — Steier.
Steiermärkische — Steiermärk.
Steinreiches — Steinreiches
Steirischer — Steir.
Stekelige — Stekel.
Stelle — Stelle
Stellen-Anzeiger — Stellen-Anz.
Stem — Stem
Stencil — Stencil
Step' — Step'
Štěpařského — Štěp.
Steroids — Steroids
Steyermärkische — Steyermärk.
Stichting — Stichting
Stiftung — Stiftung
Ştiinţe(lor), Ştiinţific(a,e) — Şti.
Stock — Stock
Stockholmisches — Stockholm.
Stomatologia — Stomatol.
Storage — Storage
Store — Store
Stored — Stored
Storia — Storia
Storici — Storici
Strahlen — Strahlen
Straipsnių — Straipsnių
Stralsundisches — Stralsund.
Straniera — Straniera
Strasbourgeoise — Strasbourg
Strategy — Strategy
Stratigrafia, Stratigraphic — Stratigr.
Strawberry — Strawberry
Stream — Stream
Strigonienses — Strigon.
Stroitel'stva — Stroit.
Stručne — Stručne
Structural — Struct.
Structure — Struct.
Struktura — Strukt.
Studenčeskih, Studencheskikh, Student(s), Studenţilor, Studi(i), Studia — Stud.
Studiekring — Studiekring
Studien — Stud.
Studiensammlung — Studiensamml.
Studier, Studies, Studimeve, Studio — Stud.
Study — Study
Stuttgarter — Stuttgarter
Suara — Suara
Sub-cellular — Sub-cell.
Subalpino — Subalpino
Subarctic — Subarctic
Subcellular — Subcell.
Subdivision — Subdiv.
Subjects — Subj.
Submicroscopic — Submicrosc.
Sub-Station — Sub-Sta.
Subtropen, Subtropica, Sub-Tropical, Subtropičeskih [etc.], Subtropicheskie [etc.], Subtropiki — Subtrop., Sub-Trop.
Suburban — Suburban
Succulent(arum,en,s) — Succ.
Succulentologica — Succulentol.
Sucriène — Sucr.
Suculentas, Suculentos — Suc.
Sudamericana, Sud-America — Sudamer., Sud-Amer.
Süddeutscher — Süddeutsch.
Süddeutschland — Süddeutschl.
Sud-Ouest — Sud-Ouest
Südwestdeutschland — Südwestdeutschl.
Sueciae — Sueciae
Suelos — Suelos
Sugar — Sugar
Sugarcane — Sugarcane
Sughero — Sughero
Suhumskogo — Suhumsk.
Suid-Afrikaanse — Suid-Afrikaanse
Suidelike — Suidel.
Suikerindustrie — Suikerindustr.
Suikerriet — Suikerriet
Suisse — Suisse
Sukhumskoi [etc.] — Sukhumsk.
Sukkulent(en,i,y) — Sukk.
Sukkulentenkunde — Sukkulentenk.
Sukkulentenpflege — Sukkulentenpflege
Sul — Sul
Šumar — Šumar
Sumarski — Sumarski
Šumarskog — Šumarsk.
Šumarstva — Šumarstva
Šumarstvo — Šumarstvo
Summa — Summa
Summaries — Summ.
Summarized — Summarized
Summary — Summ.
Sumske — Šumske
Sun — Sun

Sundhed — Sundhed
Sundhed-Tidende — Sundhed-Tidende
Sundheds-journal — Sundheds-J.
Sunshine — Sunshine
Sunyatseni — Sunyatseni
Suolo — Suolo
Suomalainen, Suomalaisen, Suomen — Suom.
Supérieur(e) — Supér.
Superintendencia — Superintend.
Superintendent — Superintendent
Superior — Super.
Suplemento — Supl.
Supplement, Supplementa(ry), Supplementum — Suppl.
Support — Support
Supramolecular — Supramolec.
Surgery, Surgical — Surg.
Surowcow — Surow.
Survey(s) — Surv.
Survival — Survival
Sus — Sus
Süsswasser — Süsswasser
Svampekundskabens — Svampekundsk.
Svampnytt — Svampnytt
Svaty — Svaty
Svěděnii — Svěd.
Svecicorum — Svecicorum
Svědenij — Svěd.
Sveedsche — Sveedsche
Svekla — Svekla
Sveklovichnaya — Sveklov.
Svensk(a,e) — Svensk(a)
Sverdlovskogo — Sverdlovsk.
Sveriges — Sveriges
Svet — Svet
Sveučilišta — Sveučilišta
Svitu — Svitu
Swedish — Swed.
Sweedsche — Sweedsche
Swenska — Swenska
Swiat — Swiat
Swiss — Swiss
Sylloge — Syll.
Sylva — Sylva
Sylvestris — Sylvest.
Sylvicole — Sylvicole
Sylvicultur(al,e) — Sylvic.
Symbolae — Symb.
Symposia, Symposium — Symp.
Syndicale, Syndicate — Syndic.
Synopsis — Syn.
Syntaxonomica — Syntax.
Synthése(s), Synthesis — Synth.
Syokubutsu — Syok.
Syphiligraph(ia,ie) — Syphiligr.
Syphilis — Syph.
Syr'e — Syr'e
Syr'evykh — Syr'ev.
Systema, Systematic(a,ae,o,s), Systematik, Systématique, Systematische(n) — Syst.
Systematist(s) — Systematist(s)
Systematized, Systematyki — Syst.
Systems — Systems
Szakirodalmi — Szakirod.
Szakkor — Szakkor
Szakosztályabál — Szakosztályabál
Szata — Szata
Szechuanensis — Szechuanensis
Szedgediensis — Szeged.
Szemle — Szemle
Szemléje — Szemléje
Szkodniki — Szkodn.
Szkoka — Szkoka
Szkole — Szkole
Szkoly — Szkoly
Szölészet(i) — Szölész.
Szölö — Szölö
Szölögazdaságtudományi — Szölögazdaságtud.
Szölöszéti — Szölösz.
Szombathely — Szombathely
Szorgalom(ban) — Szorg.
Szövetseg(e) — Szöv.
Tabacchi — Tabacchi
Tabacco — Tabacco
Tabachnaya — Tabachn.
Tabacs — Tabacs
Tabak — Tabak
Tabakpflanzer — Tabakpfl.
Table(s) — Table(s)
Tabulae — Tabulae
Tadshikistanici — Tadshik.
Tadzhikistanskoi — Tadzhikistansk.
Tadzhikskoi S S R — Tadzhiksk. S.S.R.
Tadžikistanskogo, Tadžikistanskoj — Tadžikistansk.
Tadžikskij [etc.] — Tadžiksk.
Tagblatt — Tagbl.
Tageblatt — Tagebl.
Tagebuch — Tageb.
Tagesbericht(e,en) — Tagesber.
Tagjai — Tagjai
Tagjairól — Tagjairól
Tägliches — Tägl.
Tagsberichte — Tagsber.
Tagungsbericht — Tagungsber.
Tahunan — Tahunan
Taimehaiguste-Katsejaama — Taimeh.-Katsej.
Taimeharguste — Taimeh.
Taiwanica — Taiwan.
Tájékoztató — Tájékozt.
Talajtan — Talajtan
Talassografico — Tallassogr.
Talinna — Tallinna
Talks — Talks
Tall — Tall
Talúntais — Talúntais
Tanaman — Tanam.
Tanáregyesület — Tanáregyesulet
Tanárképsö — Tanárképsö
Tanarkepzo — Tanarkepzo
Tanintéz(et) — Tanint.
Tanítása — Tanít.
Tanszek — Tansz.
Tára — Tára
Tareforskning — Tareforskning
Targyazó — Targyazó
Tarim — Tarim
Társadalom — Társad.
Társalkodó — Társalkodó
Társaság — Társ.
Társodalom — Társod.
Társulat — Társ.
Tartuensis — Tartuensis
Tartuskogo — Tartusk.
Taschen-Bibliotheka — Taschen-Biblioth.
Taschenbuch — Taschenb.
Taschenkalender — Taschenkalend.
Tashkentskii — Tashkent.
Task(s) — Task(s)
Tasmanian — Tasmanian
Taste — Taste
Tätigkeit — Tätigk.
Tätigkeitsbericht — Tätigkeitsber.
Tatranského, Tatranskom — Tatransk.
Tauriensia, Taurinensis — Taur.
Tauschanstalt — Tauschanst.
Tauschevereins — Tauschvereins
Tavola — Tavola
Tavričeskago — Tavričesk.
Tavricheskago — Tavrichesk.
Taxa — Taxa
Taxon — Taxon
Taxonomi(e,que), Taxonomic(a,o) — Taxon.
Taxonomist — Taxonomist
Taxonomy — Taxon.
Tbilisskogo — Tbilissk.
Tchécoslovaques — Tchécoslov.
Tea — Tea
Teacher(s) — Teacher(s)
Teaching — Teaching
Teated — Teated
Technic, Technica(l), Technicarum, Techniche — Techn.
Technichnyi — Technichnyi
Technicka, Technicke, Technické, Technickych, Technik(i), Techniky, Technique(s) — Techn.
Technische(n) — Techn.
Technischen Hochschule — T. H.
Technisch-Mikroskopischen — Techn.-Mikroskop.
Technisch-Wirtschaftliche — Techn.-Wirtschaftl.
Technological, Technologicas, Technologie — Technol.
Technologie-Korrespondent — Technol.-Korresp.
Technologiques, Technologischen, Technology — Technol.
Técnica(s), Tecnico(s), Técnico — Técn., Tecn.

Tecnología, Tecnológicas — Tecnol.
Tehnice, Tehničeskie, Tehnika, Tehniki — Tehn.
Teil — Teil
Tekhnicheskie [etc.], Tekhnika — Tekhn.
Teknisk — Tekn.
Teknologi — Teknol.
Tekushchii — Tekushch.
Telegráficas, Telégrafo — Telegr., Telégr.
Tełekagir — Tełekagir
Tematica, Tematicheskii — Temat.
Temperate, Temperature — Temp.
Teoretica, Teórica — Teor.
Terapeutica — Terap.
Tergestinus — Tergest.
Terkultura — Terk.
Termenyforgalom — Termenyforg.
Természet(barát,i), Természetjárás, Természetrajzi, Természettudomány(i) — Term.
Természettudományok, Természet-Tudományok — Természettud., Természet-Tud.
Természetvédelmi — Természetvédel.
Természetvizsgálók — Term.
Terminology — Terminol.
Terra — Terra
Terrarien — Terrar.
Terrarien-Zeitschrift — Terrar.-Z.
Terrarium, Terrárium — Terrar., Terrár.
Terre — Terre
Terrestres, Terrestrial, Terrestrische — Terrestr.
Terskogo — Tersk.
Tertiary — Tert.
Tessile — Tessile
Test(s) — Test(s)
Testing — Test.
Tetrahedron — Tetrahedr.
Teutsche [etc.] — Teutsche
Teutschland(s) — Teutschl.
Tev'a — Tev'a
Texana — Texana
Textile(s) — Textile(s)
Tezisy — Tezisy
Thalassia — Thalassia
Thalassina — Thalassina
Tharandt(er) — Tharandt(er)
Thätigkeit — Thätigk.
Thätigkeitsbericht — Thätigkeitsber.
Thé — Thé
Theatern — Theatern
Thee — Thee
Theecultuur — Theecult.
TheeTijdschrift, Thee-Tijdschrift — Thee-Tijdscher.
Theilnehmer — Theilnehmer
Thema — Thema
Theodoro — Theod.
Theodoro-Palatinae — Theod.-Palat.
Theoretical, Theoretische(n) — Theor.
Theory — Theory
Therapeutics, Thérapeutique — Therap.
Thérapie — Thérapie
Thermal — Thermal
Theseon — Theseon
Theses — Theses
Thesis — Thesis
Thier(e) — Thier(e)
Thiergeschichte — Thiergesch.
Thierheilkunde — Thierheilk.
Thiermedizin — Thiermed.
Thistle — Thistle
Thoracic — Thorac.
Threatened — Threat.
Thrush — Thrush
Thumb — Thumb
Thun — Thun
Thurgauischen — Thurgauischen
Thüringische(n) — Thüring.
Ticinese — Ticin.
Tidende(r) — Tidende(r)
Tidings — Tidings
Tidning(ar) — Tidn.
Tidskrift(en) — Tidskr.
Tidsskrift — Tidsskr.
Tiedeakatemia — Tiedeakat.
Tiedeseura — Tiedes.
Tiedoituksia — Tiedoit.
Tiedonannot — Tiedon.
Tierärztliche — Tierärztl.
Tiergartenvereins — Tiergartenvereins
Tiflisskago, Tiflisskogo — Tiflissk.
Tihookeanskij [etc.] — Tihookeansk.
Tijdschrift — Tijdschr.
Tijdskrif — Tijdskr.
Tikhookeanskoi [etc.] — Tikhookeansk.
Tillage — Tillage
Timber(s) — Timber(s)
Timberman — Timberman
Timely — Timely
Times — Times
Timiryazevskie [etc.] — Timiryazev.
Tipa — Tipa
Tipografica — Tipogr.
Tips — Tips
Tireeno — Tireeno
Tissue — Tissue
Tiszántúli — Tiszántúli
Title(s) — Title(s)
Tobacco — Tobacco
Tobaco — Tobaco
Today — Today
Todomany(i) — Tud.
Toilet — Toilet
Toimenta — Toimenta
Toimetised, Toimetused — Toimet
Tomato — Tomato
Tomskensis — Tomsk.
Tomsker — Tomsker
Tomskoe [etc.] — Tomsk.
Tööd — Tööd
Top — Top
Topiara — Topiara
Topic(s) — Topic(s)
Topographical, Topographisk — Topogr.
Torfjanogo, Torfyanaya [etc.] — Torfjan.
Torinese — Torin.
Történelmi, Történeti — Tört.
Torunensis — Torun.
Toscana, Toscano — Tosc.
Total — Total
Totem — Totem
T'oung — T'oung
Tour(ism) — Tour(ism)
Touristen-Clubs — Touristen-Clubs
Tovarystva — Tovar.
Towarzystw(a,o) — Towarz.
Town — Town
Toxicological, Toxicologie, Toxicology — Toxicol.
Trabajo(s), Trabalhos — Trab.
Trace — Trace
Trade(s) — Trade(s)
Trader — Trader
Trädgårds — Trädg.
Trädgårds-Föreningens — Trädg.-Fören.
Trädgårds-Odlare — Trädg.-Odlare
Trädgårdsavdelningen — Trädgårdsavd.
Trädgårdsodlare — Trädgårdsodlare
Trädgårdsskötsel — Trädgårdsskotsel
Trädgårdstidningen — Trädgårdstidningen
Tradicional — Tradic.
Traditional, Traditionelle — Tradit.
Traducción — Trad.
Traini — Traini
Training — Training
Trajectinae — Traject
Transactions — Trans.
Transalpina — Transalpina
Transformacion — Transform.
Transindex — Transindex
Transition — Transit.
Translation(s) — Transl.
Transmarinum — Transmarinum
Transportation — Transport.
Transsilvanici — Transsilv.
Travail, Travaux — Trav.
Travel(s) — Travel(s)
Treasures — Treas.
Treatment — Treatm.
Trebajos, Treballs — Treb.
Tree(s) — Tree(s)
Tree-Ring — Tree-Ring
Trencsénvármegyei — Trencsénvárm.
Trend(s) — Trend(s)
Trentini — Trent.
Trial(s) — Trial(s)
Tribuna — Tribuna
Tribune — Tribune
Třida — Tř.
Tridentina — Tridentina
Triennale — Triennale
Triennial — Trienn.
Trigo — Trigo

Triguera — Trig.
Trimensal — Trimensal
Trimestre — Trimestre
Trimestriel(s) — Trimestriel(s)
Trimestrielle(s) — Trimestrielle(s)
Trimmer — Trimmer
Trip — Trip
Troickosavsko-Kjahtinskago — Troickos.-Kjahtinsk.
Trois — Trois
Troitskosavsko-Kyakhtinskago — Troitskos.-Kyakhtinsk.
Trondhiemske — Trondh.
Trondhjemska — Trondhj.
Tropen — Tropen
Tropenforschungsinstituts — Tropenforschungsinst.
Tropenhygiene — Tropenhyg.
Tropenlandwirt — Tropenlandwirt
Tropenmedizinische — Tropenmed.
Tropenpflanzer — Tropenpflanzer
Tropic(a,s), Tropical(e,es), Tropicanarum, Trópico — Trop., Tróp.
Tropics — Tropics
Tropiques, Tropische — Trop.
Truck — Truck
Trudov(e) — Trudov(e)
Trudovogo — Trudov.
Trudy — Trudy
Trust — Trust
Trustees — Trustees
Tsarskite — Tsarsk.
Tschechischen — Tschech.
Tsentr — Tsentr
Tséntral'nae, Tsentral'nii [etc.], Tsentral'no- — Tséntr., Tsentr., Tsentr.-
Tsetovodstvo — Tsetovod.
Tsing(hai) — Tsing(hai)
Tsitologiya — Tsitol.
Tsyklu — Tsyklu
Tuatara — Tuatara
Tuber — Tuber
Tuberculosis — Tuberc.
Tuberkulose — Tuberkulose
Tuberosas — Tuber.
Tübinger — Tübinger
Tübingische — Tübingische
Tudomány — Tud.
Tudományegyetem(i) — Tudományegyet.
Tudományos, Tudománytár, Tudós(itások) — Tud.
Tuinbouw — Tuinb.
Tuinbouw-Flora — Tuinb.-Fl.
Tuinbouw-Illustratie — Tuinb.-Ill.
Tuinbouwblad — Tuinbouwbl.
Tuinbouwgewassen — Tuinbouwgew.
Tuinbouwgids — Tuinbouwgids
Tuinbouwkundig — Tuinbouwk.
Tuinbouwproducten — Tuinbouwprod.
Tuinbouwvoorlichtingsdienst — Tuinbouwvoorlicht.
Tuinderij — Tuinderij
Tuinen — Tuinen
Tulip — Tulip
Tung — Tung
Tunisienne — Tunis.
Turf — Turf
Turfgrass — Turfgrass
Turista, Turisták, Turistaság, Turistik, Turistů — Turist.
Turkestanskago, Turkestanskogo — Turkestansk.
Turkish, Turkiye — Turk.
Turkmenistana — Turkmen.
Turkmenistanskaja, Turkmenistanskaya — Turkmenistansk
Turkmenskoi S S R — Turkmensk. S.S.R.
Turkmenskoi [etc.], Turkmenskoj [etc.] — Turkmensk.
Turkuensis — Turku.
Turtox — Turtox
Tussock — Tussock
Tutkija — Tutkija
Tutkimuksia — Tutkimuksia
Tutkimuslaitoksen — Tutkimuslait.
Tutkimusseura — Tutkimusseura
Tutunului — Tutunului
Tuvinskoi, Tuvinskoj — Tuvinsk.
Tweede — Tweede
Tydskrif — Tydskr.
Type — Type
Typographique — Typogr.
Tyrnaviensis — Tyrnav.
Tzélozó — Tzél.
Uananimi — Uananimi
Überseemuseum — Überseemus.
Übersetzungen — Übersetz.
Übersicht — Übers.
Učenie [etc.], Učenyi [etc.] — Učen.
Uchenie [etc.] — Uchenago
Uchenyi [etc.] — Uchen.
Uchrezhdennykh — Uchrezhd.
Učreždennyh — Učrežd.
Ueber — Ueber
Uebersicht — Uebers.
Ufficiale — Uff.
Ugeskrift — Ugeskr.
Uitgaven — Uitgaven
Uitgezogte — Uitgez.
Uj — Uj
Ujság — Ujs.
Ukazatel' — Ukaz.
Ukrainian, Ukrainischen — Ukrain.
Ukrains'koi R S R — Ukrains'k. R.S.R.
Ukrains'koi [etc.], Ukrains'koj [etc.] — Ukrains'k.
Ukrayini — Ukrayini
Ukrayins'koi R S R — Ukrayins'k. R.S.R.
Ukrayins'kyi [etc.] — Ukrayins'k.
Ülikooli — Ülik.
Ultramar — Ultramar
Ultrastructure — Ultrastruct.
Ul'trastruktura — Ul'trastrukt.
Umeni — Umeni
Umetnosti — Umetn.
Umido — Umido
Umiejętności — Umiejętn.
Umietnosci — Umietn.
Umijetnost(i) — Umijetn.
Umjetnost(i) — Umjetn.
Umozritel'nyja, Umozritel'nyya — Umozr.
Umschau — Umschau
Umwelt — Umwelt
Umweltforschung — Umweltforsch.
Umweltwissenschaften — Unweltwiss.
Undeholdende — Undeh.
Under — Under
Underholding, Underholdning — Underhold.
Undersøgelse — Undersøg.
Undersøkelser, Undersökning — Undersøk., Undersök.
Underwater — Underwater
Ungarische(en,s) — Ung.
Ungrisches — Ungrisches
Unidade — Unid.
Uniform — Uniform
Union internationale pour la conservation de la nature et ses resources — U. I. C. N.
Union list of serials — U. L. S.
Union of Soviet Socialist Republics — U.S.S.R.
Union, Unión — Union, Unión
Unit — Unit
United Arab Republic — U.A.R.
United Kingdom — U.K.
United Nations Educational, Scientific and Cultural Organization — UNESCO
United States — U.S.
United States Department of Agriculture — U.S.D.A.
United States of America — U.S.A.
United — Unit.
Univers — Univers
Universali — Universali
Universel(le) — Universel(le)
Universidad(e), Università, Universitaria, Universitarios, Universität, Universitatea, Universitatem, Universitatii, Universitatis, Universitatum, Université, Üniversitesi, Universitet(a,at,e,s), Universitetskija, Universitetskiya, University(s') — Univ., Üniv.
Universum — Universum
Universytet(u), Univerze — Univ.
Uniwersíteit — Uniw.
Unserer — Uns.
Unterhaltendes, Unterhaltung(en) — Unterhalt.
Unterhandlungen — Unterhandl.
Unterredungen — Unterred.
Unterricht — Unterr.

Untersuchung(en) — Untersuch.
Untersuchungswesen — Untersuchungswesen
Untuk — Untuk
Uomo — Uomo
Update — Update
Upfostrings — Upfostr.
Uppsätser — Uppsät.
Upptäckter — Upptäckter
Upravlenija, Upravleniya — Upravl.
Upsalienses, Upsaliensia, Upsaliensis — Upsal.
Uptäckter — Uptäckter
Ural'skoi [etc.], Ural'skoj [etc.] — Ural'sk.
Urban(o) — Urb.
Úroda — Úroda
Úroda — Úrodu
Urologia, Urology — Urol.
Uruguaya — Uruguaya
Urwelt — Urwelt
Use — Use
Useful — Useful
Useulles — Usuelles
Usloviya — Uslov.
Uspehi — Uspehi
Uspekhi — Uspekhi
Ussuriiskogo — Ussuriisk.
Ústavu, Ústavů — Ústavu, Ústavů
Ustoichivost' — Ustoichivost'
Ustredí — Ustredí
Ustredniho — Ustredn.
Utflygter — Utflygter
Utilisation — Utilis.
Utilité, Utilium — Util.
Utilization — Utiliz.
Utilizzazione — Util.
Utsädesföreningens — Utsädesfören
Utvalg — Utv.
Uurimused, Uurimuzed — Uurim.
Uzbekistana — Uzbekist.
Uzbekistanskoi [etc.], Uzbekistanskoj [etc.] — Uzbekistansk.
Uzbekskoi S S R, Uzbekskoj S S R — Uzbeksk. S.S.R.
Uzbekskoi [etc.], Uzbekskoj [etc.] — Uzbeksk.
Uzytkow — Uzytk.
Va-Arets — Va-Arets
Vacances — Vacances
Vaderlandsche — Vaderl.
Vagyis — Vagyis
Vakblad — Vakbl.
Valaissanne — Valais.
Valdarnesi — Vald.
Valdôtaine — Valdôtaine
Vallée — Vallée
Valley — Valley
Vallon — Vallon
Valsts — Valsts
Valstybinis — Valst.
Vanamon — Van.
Vandorgyülésének — Vandorgyül.
Vänners — Vänners
Vanof — Vanof
Våre — Våre
Variabilität — Variabil.
Variation — Variat.
Variedades — Varied.
Variétés, Varieties, Variety — Var.
Varkhu — Varkhu
Város — Vár.
Varšavskom — Varšavsk.
Varshavskom — Varshavsk.
Vasculaires, Vascular(es) — Vasc.
Vasi — Vasi
Vasvármegye — Vasvárm.
Vaterland, Vaterländische(n,s) — Vaterl.
Vaterlandskunde — Vaterlandsk.
Vatten — Vatten
Vaudoise — Vaud.
Växtekologi(ska) — Växtekol.
Växtfisiologi — Växtfisiol.
Vaxtförädlingsanstalt — Vaxtförädlingsanst.
Växtodling — Växtodling
Växtskyddsanstalt — Växtskyddsanst.
Vaxtskyddsmedel — Vaxtskyddsmedel
Växtskyddsrapporter — Växtskydsrapp.
Växtsociologiska — Växtsciol.
Vázlata — Vázl.
Včelař, Včelár — Včelař, Včelár
Včelařske — Včelařske
Veckoskrift — Veckoskr.
Vector — Vector
Věd(a) — Věd(a)
Vědecké(ho), Vědeckých — Ved., Věd.
Vedrørende — Vedrør.
Vedy, Vědy — Vedy, Vědy
Vegesack — Vegesack
Vegetabile, Vegetabilis, Vegetable(s), Vegetace, Vegetacio, Vegetační, Vegetal(ă), Végétal(e,es), Vegetale, Vegetatic, Vegetatio, Vegetation, Végétation — Veg., Vég.
Vegetationskunde — Vegetationsk.
Vegetationstechnik — Vegetationstechn.
Végétaux, Vegetazione — Vég., Veg.
Veldbiologisch — Veldbiol.
Veličestva — Velič.
Velichestva — Velich.
Velikago — Velikago
Vendetta — Vendetta
Veneer — Veneer
Veneral — Veneral
Venereology, Venereologica — Venereol.
Veneto — Veneto
Veneziana — Veneziana
Venezolana, Venezuelica — Venez.
Venner — Venner
Veränderungen — Veränd.
Verbali — Verbali
Verband(s) — Verbands
Verbaux — Verbaux
Verbesserungen — Verbess.
Verbonden — Verbonden
Verbreitung — Verbreit.
Verbundenen — Verbundenen
Verden — Verden
Verdische — Verdische
Veredeling — Veredel.
Vereenigde, Vereeniging — Ver.
Verein(e,es,s) — Verein(e,es,s)
Vereinigten — Vereinigten
Vereinigung — Vereinigung
Vereinsschrift — Vereinsschr.
Verenigde, Vereniging — Ver.
Vererbungsforschung — Verebungsf.
Vererbungslehre — Verebungsl.
Vergadering — Vergad.
Vergelijkende — Vergelijk.
Vergers — Vergers
Vergleichende — Vergleichende
Vergnügen — Vergnügen
Verhandelingen, Verhandlungen — Verh.
Verhandlungsblätter — Verhandlungsbl.
Verksamheten — Verksamh.
Verkündiger — Verkündiger
Verlag — Verlag
Vermischte(n) — Vermischte(n)
Vernügen — Vernügen
Veröffentlichungen — Veröff.
Verrichtungen — Verricht.
Verrslagen — Verslagen
Versaillaise — Versaill.
Versammlung — Versamml.
Verschiedenen — Verschiedenen
Verslag(en) — Verslag(en)
Versuche(n) — Versuche(n)
Versuchsanstalt — Versuchsanst.
Versuchsstation(en) — Versuchsstat.
Versuchswesen — Versuchswesen
Vert — Vert
Vertreter — Vertreter
Vervietoise — Verviet.
Verwandte(n,r) — Verwandte(n,r)
Verwanten — Verw.
Verwerking — Verwerk.
Verzameling — Verzamel.
Verzeichnis — Verzeichnis
Vesci — Vesci
Vestlandets — Vestl.
Vestnik, Vestník, Věstnik, Věstník — Vestn., Věstn.
Vestsi — Vestsi
Vestures — Vestures
Vetenskapernas, Vetenskaperne, Vetenskaps — Vetensk.
Vetenskaps-Akademien — Vetensk.-Akad.
Vetenskaps-Societeten — Vetensk.-Soc.
Vetenskapsakademien(s) — Vetenskapsakad.
Vétérinaire(s), Veterinaria — Vétérin., Veterin.
Veterinärkunde — Veterinärk.

Veterinärmedizin — Veterinärmed.
Veterinary — Veterin.
Vetömagvizsgáló — Vetöm.
Vetplanten — Vetpl.
Vetplantvereniging — Vetplantvereniging
Viata — Viata
Vida — Vida
Viddil(u) — Vidd.
Videnskabernes, Videnskabers — Vidensk.
Videnskabersselskabs — Videnskabersselsk.
Videnskabs-Selskabet — Vidensk.-Selsk.
Videnskaps — Vidensk.
Videnskaps-Akademi — Vidensk.-Akad.
Vidnyan — Vidnyan
Vidy — Vidy
Vie — Vie
Vied — Vied
Vichzucht — Vichzucht
Viertelijahrsschrift — Vierteljahrsschr.
Viewpoint(s) — Viewpoint(s)
Views — Views
Vigne(ron) — Vigne(ron)
Vigyan — Vigyan
Vilag — Vilag
Vin — Vin
Vinař — Vinař
Vinařského — Vinařsk.
Vinas, Viñas — Vinas, Viñas
Vine(a) — Vine(a)
Vineyard — Vineyard
Vinho — Vinho
Vinícola, Vinicolo, Vinicultor — Viníc.
Vinificatie — Vinif.
Vino(s) — Vino(s)
Vinodelie, Vinodelya — Vinod.
Vinogradarstva — Vinograd.
Viola — Viola
Violet — Violet
Vir — Vir
Virágkedvelök — Virágkedv.
Virágos — Virágos
Virágtalan — Virágtalan
Viral — Viral
Virgiliana — Virgil.
Virksohmed, Virksomhet — Virks.
Virological, Virologie, Virology — Virol.
Virttäjä — Virttäjä
Virulence — Virul.
Virus — Virus
Virusdokumentationsstelle — Virusdokumentationsstelle
Virusforschung — Virusforsch.
Virusnye — Virusn.
Virusologii — Virusol.
Virusov — Virusov
Vi'sgálódó — Vi'sgálódó
Visindafjelags — Visindafj.
Visnyk(h) — Visn.
Vissh — Vissh
Vistas — Vistas
Visti — Visti
Visual — Visual
Vita — Vita
Viticole(s), Viticoltura, Viticultura(e), Viticulture — Vitic.
Vitivinícola, Vitiviniculture — Vitiviníc., Vitivinic.
Vitterhets- — Vitterh.-
Vivanti, Vivants — Viv.
Vive — Vive
Vivières — Vivières
Vlaamsch — Vlaamsch
Vlaamse — Vlammse
Vladimirskago — Vladim.
Vladivostoksakago — Vladivostoksk.
Vlasinstituut — Vlasinst.
Vlast — Vlast
Vlasteneckého, Vlasteneckeho — Vlasten.
Vlastivěda, Vlastivědné(ho), Vlastivednom, Vlastivedný, Vlastivědný — Vlastiv.
Vlugschriften — Vlugschr.
Vnutrennykh — Vnutrenn.
Vocabulaire — Vocab.
Vodních — Vodní
Vodnye — Vodnye
Vodobmena — Vodobmena
Vodoemov — Vodoemov
Vodohranilišč — Vodohr.
Vodokhranilishch — Vodokhr.
Vodorosli — Vodorosli
Vogéso-Rhénane — Vogéso-Rhénane
Vogtländischen — Vogtl.
Voigtländischen — Voigtl.
Voiny — Voiny
Voisins — Voisins
Volga-Kama — Volga-Kama
Volk — Volk
Völker — Völker
Völkerbeschreibung — Völkerbeschreib.
Völkerkunde — Völkerk.
Volks — Volks
Volksbildungskurse — Volksbildungsk.
Volkskunde — Volksk.
Vollc — Vollc
Vollegrand — Vollegrand
Volmaaktere — Volmaaktere
Voltaiques — Volt.
Volume — Vol.
Volzhsko- — Volzhsko-
Volžsko-Kamskoj — Volžsko-Kamsk.
Volžskoj — Volžsk.
Voordrachten — Voordrachten
Voorstellen — Voorstellen
Voprosy [etc.]. — Vopr.
Voratsschutz — Voratsschutz
Vorderen — Vorderen
Vorgeschichte — Vorgesch.
Vorlesungen — Vorles.
Vormals — Vormals
Voronegiensis — Voroneg.
Voronezhskoi [etc.] — Voronezhsk.
Voronežskoj [etc.] — Voronežsk.
Vorstenlandsche — Vorstenl.
Vort — Vort
Vortheil — Vortheil
Vorträge — Vorträge
Vorwelt — Vorwelt
Vorzüglichsten — Vorzügl.
Vosproizvodstvo — Vosproizv.
Vostochno-, Vostochnoi [etc.] — Vost.-, Vost.
Vostočnoj [etc.] — Vost.
Vostočnosibirskoi [etc.], Vostočno-Sibirskoj [etc.] — Vost.Sibirsk., Vost.-Sibirsk.
Vostoka — Vostoka
Vostoke — Vostoke
Voyages — Voyages
Vozdelyvaemykh — Vozdelyvaem.
Vračej — Vračej
Vrachei — Vrachei
Vragen — Vragen
Vratislaviensis — Vratislav.
Vreditelei [etc.], Vreditelej [etc.] — Vredit.
Vrienden — Vrienden
Vrt — Vrt
Vserossiiskogo — Vserossiisk.
Vserossijskogo — Vserossijsk.
Vsesojuznyj [etc.] — Vsesojuzn.
Vsesoyuznyi [etc.] — Vsesoyuzn.
Vsetine — Vsetine
Vseukrajins'ka — Vseukrajins'ka
Vseukrajins'koji — Vseukrajins'k.
Vseukrayins'ka — Vseukrayins'ka
Vseukrayins'koi — Vseukrayins'k.
Vulgarisation — Vulg.
Vuosikirja — Vuosik.
Východní — Východní
Východočeského — Východočesk.
Východočeský — Východočeský
Vychodoslovenského — Vychodoslov.
Vykopni — Vykopni
Vypusk — Vyp.
Vyroba — Vyroba
Vyroci — Vyroci
Výrocní — Výr.
Vyškové — Vyškové
Vysočajšago — Vysočajš.
Vysociny — Vysociny
Vysoka — Vysoka
Vysoké — Vysoké
Vysokej — Vysokej
Vysokogovii — Vysokogov.
Vysokou — Vysokou
Vysshii [etc.]. — Vysshie
Vyššij [etc.] — Vyššej
Vytauti — Vytauti
Vytauto — Vytauto
Vyvučěn'nja [etc.] — Vyvuč.
Vyvuchen'iya [etc.], Vyvuchennia — Vyvuch.

Výzkoum, Výzkum, Výzkumné, Výzkumníků, Výzkumnu, Výzkumnych — Výzk.
Vzaimodeistviya — Vzaimod.
Wachsthum — Wachsth.
Wahrnehmungen — Wahrnehm.
Wald(e,en) — Wald(e,en)
Waldeckische — Waldeck.
Waldschutzbrief — Waldschutzbrief
Waldwirtschaftlich(e) — Waldw.
Walk — Walk
Walnut — Walnut
Wandelaar — Wandelaar
Wanderer — Wanderer
Wandergesellschaft — Wanderges.
Wanderpflege — Wanderpflege
Warm — Warm
Warszawskie — Warsz.
Warszawskiego — Warszawsk.
Warzywniczy — Warzywn.
Washington [city] — Wash.
Wasser — Wasser
Wasserbau — Wasserbau
Wasserhygiene — Wasserhyg.
Wasserpflanzen — Wasserpflanzen
Wasserversorgung — Wasserversorg.
Wasserwirtschaft — Wasserw.
Waste(s) — Waste(s)
Watch — Watch
Water(s) — Water(s)
Weather — Weath.
Wecko-Skrift — Wecko-Skr.
Weed(s) — Weed(s)
Weekblad — Weekbl.
Weekly — Weekly
Weetenschappen — Weetensch.
Weidebouw — Weideb.
Weidingsgenootskap — Weidingsgenootsk.
Weinbau(s) — Weinbau(s)
Weinbaugesellschaft — Weinbauges.
Weinbau-Gesellschaft — Weinbau-Ges.
Weinbau-Zeitung — Weinbau-Zeitung
Weinbereitung — Weinbereitung
Weinberg — Weinberg
Welsh — Welsh
Welt — Welt
Weltforstwirtschaft — Weltforstw.
Weltkenntniss — Weltkenntn.
Wereld — Wereld
Wereldnieuws — Wereldnieuws
Werken — Werken
Werkgroep — Werkgroep
Werksamheten — Werksamh.
Werkzaamheden — Werkzaamh.
Wernerian — Wern.
West — W.
Westdeutscher — Westdeutsch.
Western — W.
Westfalen(s) — Westfalen(s)
Westfälische(n,s) — Westfäl.
Westmark — Westmark
Westphälische(n,s) — Westphäl.
Westpreussischen — Westpreuss.
Wetenschappelijke, Wetenschappen — Wetensch.
Wetenskaplike — Wetenskapl.
Wetenskaps — Wetensk.
Weterynaryjne — Weteryn.
Wetland — Wetland
Wettenskaps — Wettensk.
Wetter(auer) — Wetter(auer)
Wetterauischen — Wetterauischen
Wextkultur — Wextkult.
What('s) — What('s)
Wheat — Wheat
Whistle — Whistle
Who('s) — Who('s)
Whole — Whole
Wiadomości — Wiadom.
Wichtigsten — Wichtigsten
Wiejskiego — Wiejsk.
Wielkopolske — Wielkopolske
Wielkopolskiego — Wielkopolsk.
Wiener — Wiener
Wild — Wild
Wilderness — Wildern.
Wildland — Wildland
Wildlife — Wildlife
Wilénski — Wilénski
Wine — Wine
Wing(ed) — Wing(ed)
Winter — Winter
Wintersport — Wintersp.
Winzer — Winzer
Wirken — Wirken
Wirksamkeit — Wirksamk.
Wirtembergisches — Wirtemberg.
Wirthschaftlichen — Wirthschaftl.
Wirthschaftsberater — Wirtschaftsberater
Wirtschaft — Wirtsch.
Wirtschaftseigene — Wirtschaftseig.
Wirtschaftsgruppe — Wirtschaftsgr.
Wirtschaftskunde — Wirtschaftsk.
Wirtschaftswissenschaften — Wirtschaftswiss.
Wiskundige — Wisk.
Wissen — Wissen
Wissenschaft(en), Wissenschaftliche(n) — Wiss.
Wissenschaftsbereich — Wissenschaftsbereich
Wissenschaftsberichte — Wissenschaftsber.
Wissenswürdigsten — Wissenswürd.
Wittenbergisches — Wittenberg.
Wochenblatt — Wochenbl.
Wochenschrift(en) — Wochenschr.
Wöchentliche(r) — Wöchentl.
Wohnkultur — Wohnkultur
Women's — Women's
Wonderful — Wonderful
Wood(s) — Wood(s)
Woodland(s) — Woodland(s)
Work(er,s) — Work(er,s)
Working — Working
Workshop — Workshop
World('s) — World('s)
Wort — Wort
Wrapper — Wrapper
Wreath — Wreath
Writer('s) — Writer('s)
Writings — Writings
Wrocławskie(go) — Wrocławsk.
Wszystikich — Wszyst.
Wundarzneykunst — Wundarzneykunst
Wundärzte — Wundärzte
Wunder — Wunder
Württembergische(n) — Württemberg.
Wydawnictwa — Wydaw.
Wydział(u) — Wydz.
Wyzsza — Wyzsza
Yabanci — Yabanci
Yagodnye — Yagodnye
Yagodnykh — Yagodnykh
Yagodnyye — Yagodnyye
Yakutskaya [etc.] — Yakutsk.
Yaroslavskago, Yaroslavskogo — Yaroslavsk.
Yarovizatsii — Yarov.
Yayinlari — Yayinl.
Year — Year
Year Book, Year-book — Year Book, Year-book
Yearbook — Yearb.
Yearly — Yearly
Yeast(s) — Yeast(s)
Yellow — Yellow
Yestyestvenno-, Yestyestvennye — Yestyestv.-, Yestyestv.
Yestyestvoispytatelei — Yestyestvoisp.
Yestyestvoznzniya — Yestyestv.
Yield — Yield
Yilliği — Yilliği
Yillik — Yillik
Yliopiston — Yliop.
Ylymlar — Ylymlar
Yoga — Yoga
Young — Young
Your — Your
Youth — Youth
Yüksek — Yüks.
Yunnanica — Yunnan.
Yur'evskago, Yur'evskom — Yur'evsk.
Yuzhno- — Yuzhno-
Za — Za
Zaadbelangen — Zaadbelangen
Zaadcontrole — Zaadcontrole
Zaadhandel — Zaadhandel
Zabaikal'skii, Zabajkal'skij — Zabaikal'sk.
Zabiologiju — Zabiol.
Zachodniej — Zachodn.
Zagraničnoj — Zagraničn.
Zagranichnoi — Zagranichn.
Zagryazneniya — Zagryazn.

Zagrebiensis — Zagreb.
Zahrad(a) — Zahrad(a)
Zahradě — Zahradě
Zahradnické — Zahradn.
Zahradnictvi — Zahrad.
Zahradník(ů) — Zahradn.
Zahrady — Zahrady
Zahraniči — Zahraniči
Zakavkazskogo — Zakavkazsk.
Zakład — Zakład
Základny — Zákl.
Zakładu — Zakładu
Zametki — Zametki
Zapadno- — Zapadno-
Zapadnosibirskoi [etc.], Zapadno-Sibirskoi [etc.], Zapadnosibirskoj [etc.], Zapadno-Sibirskoj [etc.] — Zapadnosibirsk., Zapadno-Sibirsk.
Západoceské Západočeské — Západočeské
Západoslovenského, Západoslovenskem — Západoslov.
Zapiskam, Zapiski — Zap.
Zapovednika, Zapovednoe — Zapov.
Zapysky — Zap.
Zara'at — Zara'at
Zaščita, Zaščite, Zaščity — Zašč.
Zasědanii — Zasěd.
Zasedání, Zasedanij, Zasědanij — Zased., Zasěd.
Zashchite, Zashchity — Zashch.
Zashtita — Zashtita
Zasshi — Zasshi
Zaštita — Zaštita
Zaštite — Zaštite
Zaštitu — Zaštitu
Zasushlivo — Zasushl.
Zavod(a) — Zavod(a)
Zbagachennia — Zbagach.
Zbirnyk — Zbirn.
Zbornik — Zborn.
Zdravotnicka — Zdrav.
Zee — Zee
Zeebiologisch — Zeebiol.
Zeeuws(e) — Zeeuws(e)
Zeeuwsch — Zeeuwsch
Zeit — Zeit
Zeitblatt — Zeitbl.
Zeitgemässe — Zeitgemässe
Zeitschrift(en) — Z.
Zeitung(en) — Zeitung(en)
Zelanti — Zelanti
Zelenoe [etc.] — Zelen.
Zelinarstvi — Zelin.
Zelle — Zelle
Zellforschung — Zellf.
Zemalijskog, Zemaljskih — Zemaljsk.
Zemedelska, Zemědělské(ho), Zemědělsky(ch) — Zemed., Zeměd.
Zemědělstvi — Zemědělstvi
Zemel'nogo — Zemel'n.
Zemeleustrojstva — Zemleutstr.
Zeměpisne, Zeměpisny — Zeměp.
Žemes — Žemes
Zemevedné — Zemevedné
Zemjodelsko- — Zemjod.-
Zemjodelsko-šumarski — Zemjod.-Šumarski
Zemleděl'cheskaya, Zemleděliya, Zemleděl'českaja, Zemleděl'českoj, Zemledelie, Zemledelija, Zemlědělija — Zemled.
Zemleustroistva — Zemleustr.
Zemlevedenie, Zemlevěděnie — Zemlevedenie, Zemlevěděnie
Zemlje — Zemlje
Zemlješte — Zemlješte
Zemské — Zemské
Zemského — Zemsk.
Zemstva — Zemstva
Žen'-Šenja — Žen'-Šenja
Zentralanstalt — Zentralanst.
Zentralblatt — Zentralbl.
Zentrale(n) — Zentr.
Zentralhalle — Zentralhalle
Zentralinstitutes — Zentralinst.
Zentralmuseums — Zentralmus.
Zephyr — Zephyr
Zerna — Zerna
Zernobovykh — Zernob.
Zernovye — Zernov.
Zeszyty — Zesz.
Zeylancia — Zeylancia
Zhen'-shenya — Zhen'-shenya
Zhivotnovodstve — Zhivotnov.
Zhivotnykh — Zhivot.
Zhizn' — Zhizn'
Zhurnal(akh,y) — Zhurn.
Zielarskie(go) — Zielarsk.
Zielonych — Zielon.
Ziemi — Ziemi
Ziergehölze — Ziergeh.
Zierpflanzenbau — Zierpfl.
Zig-Zag — Zig-Zag
Žilenského — Žilinsk.
Zimmerpflanzen — Zimmerpfl.
Zimmerpflanzen-Zeitschift — Zimmerpfl.-Z.
Zimnie — Zimn.
Ziraat, Zirai — Zir.
Zittingen — Zittingen
Živa — Živa
Živena — Živena
Život — Život
Život — Životn.
Žizn' — Žizn'
Znameni — Znam.
Znanosti — Znan.
Znanstvena — Znanst.
Zöglingen — Zögl.
Žohovacu — Žohov.
Zöldségtermesztesi — Zöldségterm.
Zollverein — Zollv.
Zona(s) — Zona(s)
Zonal — Zonal
Zonal'noi, Zonal'noj — Zonal'n.
Zone — Zone
Zoobiologichnogo, Zoobiologicnogo — Zoobiol.
Zoobotanica, Zoobotanico — Zoobot.
Zooiatro — Zooiatro
Zoologia, Zoologica(e,l) — Zool.
Zoologicae-Botanicae — Zool.-Bot.
Zoologicheskii, Zoologico, Zoologie, Zoologii, Zoologique, Zoologisch — Zool.
Zoologisch-Botanischen — Zool.-Bot.
Zoologo-Biologicnogo — Zool.-Biol.
Zoology — Zool.
Zootechnie — Zootechn.
Zootecnica — Zootecn.
Zpráva — Zpráva
Zpravodaj(ský) — Zprav.
Zprávy — Zprávy
Zuccheri — Zucch.
Zücherisch(e,en) — Zücherisch(e,en)
Züchter — Züchter
Züchtung — Zücht.
Zucker — Zucker
Zuerkennung — Zuerkenn.
Zugabe — Zugabe
Zuiderzeeonderzoek — Zuiderzeeonderz.
Zürcherische(n) — Zürcherische(n)
Žurnal(ah,y) — Žurn.
Zusammenkunft — Zusammenk.
Zustand(e,es) — Zustand(e,es)
Zuverlässige — Zuverlässige
Zvolenensis — Zvolen.
Zweige — Zweige
Zweigstelle — Zweigstelle
Zwekmässig — Zwekmässig
Zwolle — Zwolle
Zwoten — Zwoten
Zwy(ten) — Zwy(ten)
Zymologie — Zymol.
Zymotechnisk. — Zymotechn.
Zynia — Zynia

Appendix B

Acronyms and Initialisms

In recent years, acronyms and initialisms have been included as parts of periodical titles with increasing frequency. Many are likely to be unfamiliar to the general reader, with the unfortunate result that their use obscures the descriptive function that titles usually have. The following list provides expansions of many of these cryptic contractions found in titles that are cited in B-P-H and B-P-H/S.

A A A S — American Association for the Advancement of Science
A A B G A — American Association of Botanical Gardens & Arboreta
A A M — American Association of Museums
A A S P — American Association of Stratigraphic Palynologists
A C F A S — Association Canadienne-Française pour l'Avancement des Sciences
A C I A R — Australian Centre for International Agricultural Research
A D A S — Agricultural Development & Advisory Service
A E T F A T — Association pour l'Étude Taxonomique de la Flore d'Afrique Tropicale
A F A — Association Françaises des Aquariophiles
A F R I — Applied Forestry Research Institute
A G G S — American Gloxinia & Gesneriad Society
A G P — Arabinogalactan protein
A G R E P — Agricultural research projects
A G R I C O L A — Agricultural Online Access
A H C — American Horticultural Council
A H S — American Horticultural Society
A H T A — American Horticultural Therapy Association
A I A — Alberta Institute of Agrostologists
A I B D A — Associación Interamericana de Bibliotecarios y Documentalistas Agrícolas
A I B S — American Institute of Biological Sciences
A I C — Agricultural Institute of Canada
A I C E — American Institute of Crop Ecology
A I H P — American Institute of the History of Pharmacy
A I N — Association of Interpretive Naturalists
A M A — American Medical Association
A M I — American Mushroom Institute
A N M S — Associazione Nazionale dei Musei Scientifici
A P P S — Australian Plant Pathology Society
A R C — Agricultural Research Council
A R E R S — Association Regionale pour l'Étude et la Recherche Scientifiques
A R I — Agricultural Research Institute
A R T J B — Associations des Responsables et Techniciens de Jardins Botaniques
A S B — Association of the Southeastern Biologists
A S B P — Association of Systematic Biologists of the Phillipines
A S B S — Australian Systematic Botany Society
A S C — Association of Systematics Collections
A S H S — American Society for Horticultural Science
A S L I B — Association of Special Libraries & Information Bureaux
A S M — American Society for Microbiology
A S P — American Society for Photobiology
A S P B — American Society of Professional Biologists
A S P P — American Society of Plant Physiologists
A S P T — American Society of Plant Taxonomists
A S T C — Association of Science-Technology Centers
A S T I S — Arctic Science & Technology Information System
A V R D C — Asian Vegetable Research & Development Center
A V R O S — Algemeene Vereeniging van Rubberplanters ter Oostkust van Sumatra
A W A — Alberta Wilderness Association
B A S I C — Biological Abstracts' Subjects In Context
B C G — Biology Curators' Group
B I O — Bedford Institute of Oceanography
B I O S I S — BioSciences Information Service
B L L — British Library Lending
B L L D — British Library, Lending Division
B R A I S — Brackish Water Aquaculture Information System
B S B I — Botanical Society of the British Isles
B S C P — Biological Sciences Communication Project
B S C S — Biological Sciences Curriculum Study
B S E — Botanical Society of Edinburgh
B S I — Botanical Survey of India
C A — Chemical Abstracts
C A B — Commonwealth Agricultural Bureaux
C A B I — Commonwealth Agricultural Bureaux International
C A B S — Conservation Association of Botanical Societies
C A N O P — Canadian Association of Nature & Outdoor Photographers
C A T I E — Centro Agronómico Tropical de Investigación y Enseñanza
C B A — Canadian Botanical Association
C B E — Council of Biology Editors
C B H L — Council on Botanical & Horticultural Libraries
C B S — Centraalbureau voor Schimmelcultures
C B S — Connecticut Botanical Society
Č Č H — Časopis Československých Houbařů
C C N B — Conservation Council of New Brunswick
C E F A P R I N — Consejo Nacional de Investigaciones Cientificas y Tecnicas, Centro de Estudios Farmacológicos y de Principios Naturales
C E T I O M — Centre Technique Interprofessionnel des Oléagineux Metropolitains
C F B S — Canadian Federation of Biological Societies
C F I — Commonwealth Forestry Institute
C F R U — Cooperative Forestry Research Unit
C I K A R D — Center for Indigenous Knowledge for Agriculture & Rural Development
C I M M Y T — Centro Internacional de Mejoramiento de Maiz y Trigo
C I P — Centro Internacional de la Papa
C I T R E — Comparative Investigations of Tropical Reef Ecosystems
C M I — Commonwealth Mycological Institute

C N F R A — Comité National Français des Recherches Antartiques
C N R S — Centre National de la Recherche Scientifique
C O B I — Committee on Biological Information
C O D E C A P — Commissão Executiva de Defesa Fitossanitaria da Lavoura Canavieira de Pernambuco
C O N A — Comité Oceanografico Nacional
C O R E S T A — Centre de Co-opération pour les Recherches Scientifiques rélatives au Tabac
C P — Cornell Plantations
C P S — Canadian Phytopathological Society
C P T — Central Plains Turfgrass
C R C — Chemical Rubber Company
C R F G — California Rare Fruit Growers
C S C — Commonwealth Science Council
C S I R — Council for Scientific & Industrial Research
C S I R O — Commonwealth Scientific & Industrial Research Organization
C S S A — Cactus & Succulent Society of America
C S T A — Canadian Society of Technical Agriculturists
C U B B I — Copenhagen University, Balearic Botanic Investigation
C U E B S — Commission on Undergraduate Education in the Biological Sciences
C V F — Corporación Venezolana de Fomento
D G R S T — Delegation Générale à la Recherche Scientifique et Technique
D I A — División de Investigaciones Agropecuarias
D L G — Deutsche Landwirtschaftliche-Gesellschaft
D N T — Devon Naturalists' Trust
D S I R — Department of Scientific & Industrial Research
E A N H S — East Africa Natural History Society
E A P R — European Association for Potato Research
E A W A G — Eidgenössische Anstalt für Wasserversorgung, Abwasserreinigung und Gewasserschutz
E I S — Environmental Impact Statements
E M B O — European Molecular Biological Organization
E P A — Environmental Protection Agency
E P P O — European Plant Protection Organisation
E S A — Epiphyllum Society of America
E S N — European science notes
E S S A — Environmental Science Services Administration
E T C — European Translations Centre
E T H — Eidgenössische Technische Hochschule
F A B I S — Faba Bean Information Service
F A C E N A. — Facultad de Ciencias Exactas y Naturales y Agrimensura
F A O — Food & Agriculture Organisation
F A S — Fachgruppe Andere Sukkulenten
F A S E B — Federation of American Societies for Experimental Biology
F B C N — Fundacao Brasileira para a Conservacao da Natureza
F B C N — Fundação Brasileira para a Conservação da Natureza
F C A P — Facultado de Ciências Agrárias do Pará
F E B S — Federation of European Biochemical Societies
F E E M A — Fundação Estadual de Engenharia do Meio Ambiente
F E M S — Federation of European Microbiological Societies
F E S P P — Federation of European Societies of Plant Physiology
F L K — Forschungs-Läbor für Kaffee und Kakao
F N G A — Florida Nurserymen & Growers' Association
F O N A — Friends of National Arboretum
F P L — Forest Products Laboratory
F P R D I — Forest Products Research & Development Institute
F P R L — Forest Products Research Laboratory
F R I — Forest Research Institute
F U P E F — Fundação de Pesquisas Florestais
G C A — Garden Centers of America
G C V A — Garden Club of Virginia
G F G — German Forestry Group
G N S I — Guild of Natural Science Illustrators
H B L — Hunt Botanical Library
H E A — Horticultural Education Association
I A A P — International Association for Angiosperm Paleobotany
I A B — International Association of Bryologists
I A B G — International Association of Botanic Gardens
I A P — Indian Association of Palynostratigraphers
I A P — International Association of Pteridologists
I A T E M — Instituto Agrotécnico Económico de Misiones
I A W A — International Association of Wood Anatomists
I B P — International Biological Programme
I B P G R — International Board for Plant Genetic Resources
I B P T — Instituto de Biología e Pesquisas Tecnológicas
I B S A — Indigenous Bulb Growers' Association of South Africa
I C A R — Indian Council for Agricultural Research
I C A R D A — International Center for Agricultural Research in the Dry Areas
I C A S A L S — International Center for Arid & Semi-Arid Land Studies
I C B — Instituto Central de Biociências
I C E S — International Council for the Exploration of the Sea
I C O M — International Council of Museums
I C P — International Commission for Palynology
I E S — Institute of Ecosystem Studies
I F C C — Institut Français du Café et du Cacao
I I C A — Instituto Interamericano de Ciencias Agrícolas
I I R B — Institut International de Recherches Betteravières
I I T A — International Institute of Tropical Agriculture
I M A — International Mycological Association
I N A — International Nannoplankton Association
I N I A — Instituto Nacional de Investigaciones Agrarias
I N I R E B — Instituto Nacional de Investigaciones sobre Recursos Bióticos
I N P A — Instituto Nacional de Pesquisas da Amazônia
I N P A B O — Instituto Paranaense de Botanica
I N R A — Institut National de la Recherche Agronomique
I N T A — Instituto Nacional Tecnología Agropecuaria
I O B C — International Organization for Biological Control of Noxious Animals & Plants
I O P — International Organization of Palaeobotany
I O P B — International organization of plant biosystematists
I O S — International Organization for the Study of Succulent Plants
I P A — Indian Potato Association
I P E A N — Instituto de Pesquisas Agropecuárias do Norte
I P E F — Instituto de Pesquisas e

Estudos Florestais
I P P C — International Plant Protection Center
I P R — Institute of Parks & Recreation
I P R N R — Instituto de Pesquisas de Recursos Naturais Renováveis
I R A T — Institut de Recherches Agronomiques Tropicales
I R R I — International Rice Research Institute
I S H A — Institute for the Study of Science in Human Affairs
I S H S — International Society for Horticultural Science
I S I — Institute for Scientific Information
I S N A — Indian Society for Nuclear Techniques in Agriculture
I S N A R — International Service for National Agricultural Research
I S T A — International Seed Testing Association
I S T F — International Society of Tropical Foresters
I T E — Institute for Terrestrial Ecology
I T F — Institute of Tropical Forestry
I U B S — International Union of Biological Sciences
I U C N — International Union for Conservation of Nature & Natural Resources
I W E — Inland Water Ecosystems
J A R E — Japan Antarctic Research Expedition
J I B P — International Biological Programme, Japanese National Committee
J N K V V — Jawaharlal Nehru Krishi Vishwa Vidylaya
J S P P — Japanese Society of Plant Physiologists
K F R I — Kerala Forest Research Institute
L A I F S — Los Angeles International Fern Society
L A S C A — Los Angeles State & County Arboretum
L S H R — Louisiana Society for Horticultural Research
L S S A — Limnological Society of Southern Africa & South Africa
L S U — Louisiana State University
M A F E S — Mississipi Agricultural & Forestry Experiment Station
M A F F — Ministry of Agriculture, Fisheries & Food
M A H A — Malayan Agri-Horticultural Association
M A S C A — Museum Applied Science Center for Archeology
M A V I S — Maize Virus Information Service
M C C P — Midwest Cooperative Conservation Program
M G A — Mushroom Growers' Association
M G C A — Men's Gardens Clubs of America
M N H N — Museo Nacional de História Natural
M N H N — Museum National d'Histoire Naturelle
M P G — Maine Potato Growers, Inc
N A A S — National Agricultural Advisory Service
N A L — National Agricultural Library
N B G — National Botanic Gardens
N B R I — National Botanical Research Institute
N C T R H — National Council for Therapy & Rehabilitation through Horticulture
N E R C — Natural Environment Research Council
N F T A — Nitrogen Fixing Tree Association
N & G C — Nurseryman & Garden Centre
N I B S — Nippon Institute for Biological Science
N J M A — New Jersey Mycological Association
N L L — National Lending Library
N N N P S — Northern Nevada Native Plant Society
N P G S — National Plant Germplasm System
N P S O — Native Plant Society of Oregon
N R C — National Research Council
N R C P — National Research Council of the Philippines
N R D C — Natural Resources Defense Council
N S F — National Science Foundation
N T U — National Taiwan University
N Z O I — New Zealand Oceanographic Institute
O E P P — Organisation Européenne et Mediterranéenne pour la Protection des Plantes
O N R — Office of Naval Research
O P T I M A — Organisation for the Phyto-Taxonomic Investigation of the Mediterranean Area
O R S T O M — Office de la Recherche Scientifique et Technique d'Outre-Mer
O T S — Organization for Tropical Studies
P A A B S — Pan-American Association of Biochemical Societies
P A N E S A — Pasture Network for Eastern Southern Africa
P A S C A L — Programme Appliqué à la Sélection et à la Compilation Automatique de la Littérature
P C E A — Programa Cooperativo de Experimentación Agropecuaria
P F — Patrimonio Forestal
P F R A — Prairie Farm Rehabilitation Administration
P G R — Plant Growth Regulator
P G R C — Plant Gene Resources of Canada
P H S — Pennsylvania Horticultural Society
P K V — Punjabrao Krishi Vidyapeeth
P N H — Philippine National Herbarium
P R C — Plant Records Center
P R O S E A — Plant Resources of Southeast Asia
P S L — Plant Science Laboratories
P S Z N — Pubblicazioni della stazione zoologica di Napoli
P U C R G S — Pontifíca Universidade Católica do Rio Grande do Sul
R F C — Rare Fruit Council
R H S — Royal Horticultural Society
R I C — Rattan Information Centre
R O S T L A C — Regional Office for Science & Technology for Latin America & the Caribbean
R P C — Ressources Phytogénétiques du Canada
R R I — Rubber Research Institute
R R I C — Rubber Research Institute of Ceylon
R R I S L — Rubber Research Institute of Sri Lanka
R S F — Rhododendron Species Foundation
R S R I — Regional Science Research Institute
S A B R A O — Society for the Advancement of Breeding Researches in Asia & Oceania
S A J I B — Société d'Animation du Jardin et de l'Institut Botaniques
S A R H — Secretaria de Agricultura y Recursos Hidraulicos
S B N H — Society for the Bibliography of Natural History
S C A R — Special Committee for Antarctic Research
S C R A L — Sociedad Cooperativa Rural Argentina Ltda
S D A T — Sociedad Dasonomica de la America Tropical
S I A T S A — Servicios de Investigaciones Agrícolas Tropicales, S A
S N A — Southern Nurserymen's Associetion
S O N G — Society of Ontario Nut Growers
S O U Q A R — Section d'Océanographie, Université du

Québec À Rimouski
S P A — Sumatra Planters' Association
S P C — South Pacific Commission
S R O P — Section Régionale Ouest Paléarctique
S U D E N E — Superintendência do Desenvolvimento do Nordeste
S U K — Statens Utsädeskontroll
T A R C — Tropical Agriculture Research Center
T C A — Tissue Culture Association
T D W G — Taxonomic Databases Working Group
T P C — Threatened Plants Committee
T R A F F I C — Trade Records Analysis of Flora & Fauna In Commerce
T V I S — Tropical Vegetable Information Service
U A R — United Arab Republic
U A S — University of Agricultural Sciences
U B C — University of British Columbia
U B I — Utrecht - Biohistorisch Instituut
U I C N — Union Internationale pour la Conservation de la Nature et de ses Ressources
U I C N — Union International pour la Conservation de la Nature et de ses Ressources
U L — University of Liberia
U N C — University of North Carolina
U N E L L E Z — Universidad Nacional Experimental de los Llanos Occidentales "Ezequiel Zamora."
U N E P — United Nations Environment Programme
U N E S C O — United Nations Educational, Scientific, & Cultural Organization
U P — University of the Philippines
U P A S I — United Planters' Association of Southern India
U P O V — Union International pour la Protection des Obtentions Végétales
U S D A — United States Department of Agriculture
U V T I Z — Ústav VédeckoTechnických Informací pro Zemědělství
V K W — Vereinigung der Kakteenfreunde Wurttembergs
W A R D A — West African Rice Development Association
W N P S — Washington Native Plant Society
W S S A — Weed Science Society of America
Z A G — Zentral Arbeitsgruppe